Living Figure 16.16: Phosphofru...

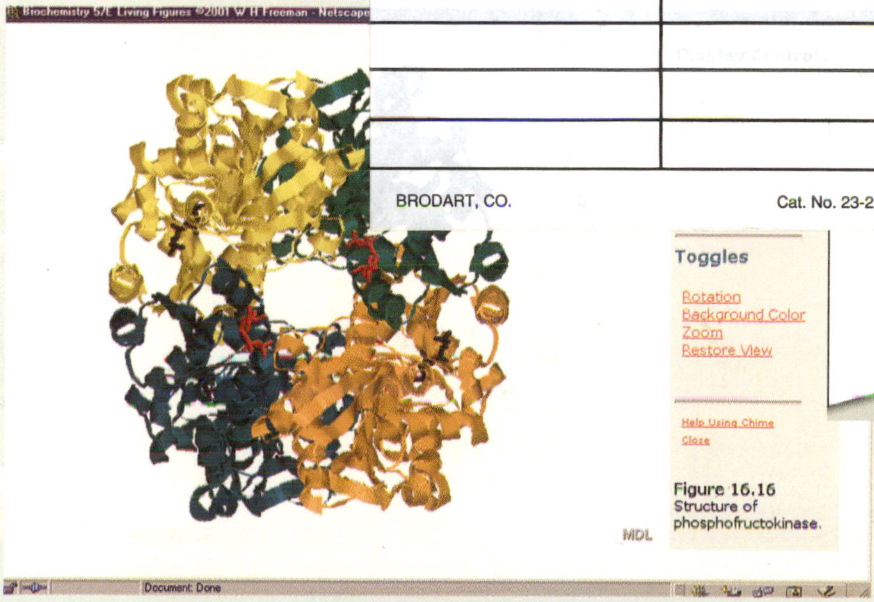

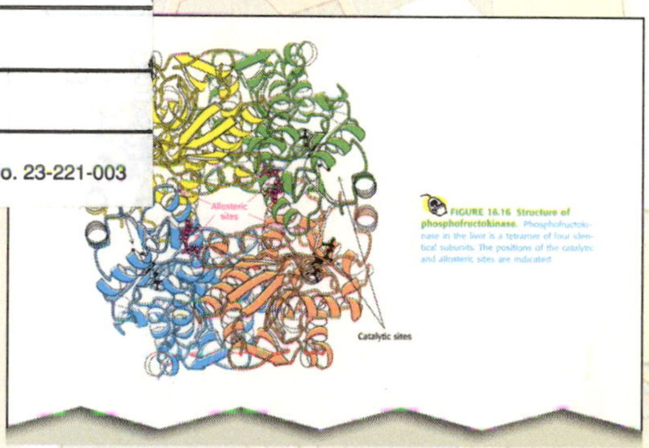

Figure 16.16 Structure of phosphofructokinase.

Every textbook figure of a protein structure can also be viewed on-line in interactive...

Available at www.whfreeman.com/biochem5, the official *Biochemistry*, Fifth Edition, Web s...

Additional On-Line Resources

- **QUIZZES:** Designed to reinforce key terms and concepts from the text, the chapter-specific online quizzes help students to prepare for tests and exams. Each question provides corrective feedback and text references for further study. With the power of Question Mark's Perception™, instructors can access their students' scores to monitor progress through a password-protected Web site.

- **WEB LINKS:** Annotated links to outside Web sites guide students to useful public sources of data and information related to topics covered in the text.

- **FLASHCARDS:** Enable students to quickly test their recall of key terms from the text.

nature publishing group **npg**   W. H. Freeman and Company is proud to present a weekly collection of abstracts and a monthly featured article from the Nature family of journals, including *Nature,* its associated monthly titles, and the recently launched *Nature Review Journals.*

Conceptual Insights (Chapter 32) Signaling Pathways

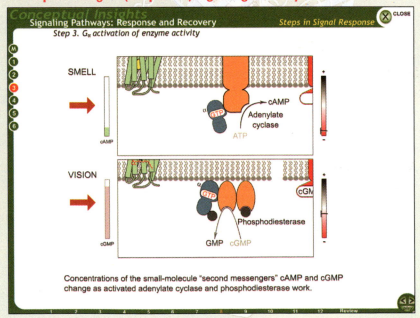

Concentrations of the small-molecule "second messengers" cAMP and cGMP change as activated adenylate cyclase and phosphodiesterase work.

Available at **www.whfreeman.com/biochem5**, the official *Biochemistry,* Fifth Edition, Web site.

BIOCHEMISTRY

• FIFTH EDITION •

Jeremy M. Berg
Johns Hopkins University School of Medicine

John L. Tymoczko
Carleton College

Lubert Stryer
Stanford University

Web content by
Neil D. Clarke
Johns Hopkins University School of Medicine

W. H. Freeman and Company
New York

PUBLISHER: **Michelle Julet**

DEVELOPMENT EDITOR: **Susan Moran**

NEW MEDIA DEVELOPMENT EDITOR: **Sonia Divittorio**

NEW MEDIA AND SUPPLEMENTS EDITOR: **Mark Santee**

MEDIA DEVELOPERS: **CADRE design; molvisions.com—
3D molecular visualization**

MARKETING MANAGER: **Carol Coffey**

PROJECT EDITOR: **Georgia Lee Hadler**

MANUSCRIPT EDITOR: **Patricia Zimmerman**

COVER AND TEXT DESIGN: **Victoria Tomaselli**

COVER ILLUSTRATION: **Tomo Narashima**

ILLUSTRATION COORDINATOR: **Cecilia Varas**

ILLUSTRATIONS: **Jeremy Berg with Network Graphics**

PHOTO EDITOR: **Vikii Wong**

PHOTO RESEARCHER: **Dena Betz**

PRODUCTION COORDINATOR: **Julia DeRosa**

COMPOSITION: **TechBooks**

MANUFACTURING: **RR Donnelley & Sons Company**

Library of Congress Cataloguing-in-Publication Data

Berg, Jeremy Mark
    Biochemistry / Jeremy Berg, John Tymoczko, Lubert Stryer.—5th ed.
      p. cm.
    Fourth ed. by Lubert Stryer.
    Includes bibliographical references and index.
    ISBN 0-7167-4954-8 (CH1–31 only)—ISBN 0-7167-3051-0 (CH1–34)
    ISBN 0-7167-4955-6 (CH32–34 only)—ISBN 0-7167-4684-0 (CH1–34,
                                                    International edition)
    1. Biochemistry. I. Tymoczko, John L. II. Stryer, Lubert. III. Title.

QP514.2 .S66 2001
   572—dc21

                                           2001051259

Printed in the United States of America

First printing 2001

W. H. Freeman and Company
41 Madison Avenue, New York, New York 10010
Houndmills, Basingstoke RG21 6XS, England

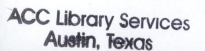

TO OUR TEACHERS AND OUR STUDENTS

About the authors

JEREMY M. BERG has been Professor and Director (Department Chairperson) of Biophysics and Biophysical Chemistry at Johns Hopkins University School of Medicine since 1990. He received his B.S. and M.S. degrees in Chemistry from Stanford (where he learned X-ray crystallography with Keith Hodgson and Lubert Stryer) and his Ph.D. in chemistry from Harvard with Richard Holm. He then completed a post-doctoral fellowship with Carl Pabo. He is recipient of the American Chemical Society Award in Pure Chemistry (1994), the Eli Lilly Award for Fundamental Research in Biological Chemistry (1995), the Maryland Outstanding Young Scientist of the Year (1995), and the Harrison Howe Award (1997). While at Johns Hopkins, he has received the W. Barry Wood Teaching Award (selected by medical students as), the Graduate Student Teaching Award and the Professor's Teaching Award for the Preclinical Sciences. He is co-author, with Stephen Lippard, of the text *Principles of Bioinorganic Chemistry.*

JOHN L. TYMOCZKO is the Towsley Professor of Biology at Carleton College, where he has taught since 1976. He currently teaches Biochemistry, Biochemistry Laboratory, Oncogenes and the Molecular Biology of Cancer, and, Exercise Biochemistry and co-teaches an introductory course, Bioenergetics and Genetics. Tymoczko received his B.A. from the University of Chicago in 1970 and his Ph.D. in Biochemistry from the University of Chicago with Shutsung Liao at the Ben May Institute for Cancer Research. He followed that with a post-doctoral position with Hewson Swift of the Department of Biology at the University of Chicago. Tymoczko's research has focused on steroid receptors, ribonucleoprotein particles, and proteolytic processing enzymes.

LUBERT STRYER is currently Winzer Professor in the School of Medicine and Professor of Neurobiology at Stanford University, where he has been on the faculty since 1975. He received his M.D. from Harvard Medical School. Professor Stryer has received many awards for his research, including the Eli Lilly Award for Fundamental Research in Biological Chemistry and the Distinguished Inventors Award of the Intellectual Property Owners' Association. He was elected to the National Academy of Sciences in 1984. He was formerly the President and Scientific Director of the .Affymax Research Institute. He is a founder and a member of the Scientific Advisory Board of Senomyx, a company that is using biochemical knowledge to develop new and improved flavor and fragrance molecules for use in consumer products. This publication of the first edition of his text *Biochemistry* in 1975 transformed biochemistry textbooks.

Preface

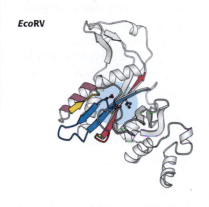

EcoRV

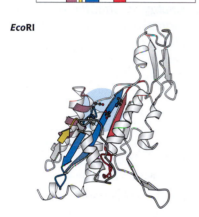

EcoRI

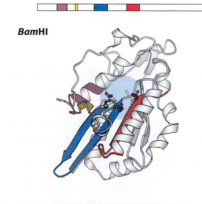

BamHI

FIGURE 9.44 A conserved structural core in type II restriction enzymes.

For more than 25 years, and through four editions, Stryer's *Biochemistry* has laid out this beautiful subject in an exceptionally appealing and lucid manner. The engaging writing style and attractive design have made the text a pleasure for our students to read and study throughout our years of teaching. Thus, we were delighted to be given the opportunity to participate in the revision of this book. The task has been exciting and somewhat daunting, doubly so because of the dramatic changes that are transforming the field of biochemistry as we move into the twenty-first century. Biochemistry is rapidly progressing from a science performed almost entirely at the laboratory bench to one that may be explored through computers. The recent ability to determine entire genomic sequences has provided the data needed to accomplish massive comparisons of derived protein sequences, whose results may be used to formulate and test hypotheses about biochemical function. The power of these new methods is explained by the impact of evolution; many molecules and biochemical pathways have been generated by duplicating and modifying existing ones. Our challenge in writing the fifth edition of this text has been to introduce this philosophical shift in biochemistry while maintaining the clear and inviting style that has been a distinguishing feature of the preceding four editions.

A New Molecular Evolutionary Perspective

How should these evolution-based insights affect the teaching of biochemistry? Often macromolecules with a common evolutionary origin play diverse biological roles yet have many structural and mechanistic features in common. An example is a protein family containing macromolecules that are crucial to moving muscle, to transmitting the information that adrenaline is present in the bloodstream, and to driving the formation of chains of amino acids. The key features of such a protein family, presented to the student once in detail, become a model that the student can apply each time that he or she encounters a new member of the family. The student is then able to focus on how these features, observed in a new context, have been adapted to support other biochemical processes. Throughout the text, a stylized tree icon 🌳 is positioned at the start of discussions focused primarily on protein homologies and evolutionary origins.

TWO NEW CHAPTERS. To enable students to grasp the power of these insights, two completely new chapters have been added. The first, "Biochemical Evolution" (Chapter 2), is a brief tour from the origin of life to the development of multicellular organisms. On one level, this chapter provides an introduction to biochemical molecules and pathways and their cellular context. On another level, it attempts to deepen student understanding by examining how these molecules and pathways arose in response to key biological challenges. In addition, the evolutionary perspective of Chapter 2 makes some apparently peculiar aspects of biochemistry more sensible to the student. For example, the presence of ribonucleotide fragments in biochemical cofactors can be accounted for by the likely occurrence of an early world based largely on RNA. The second new chapter, "Exploring Evolution" (Chapter 7), develops the conceptual basis for the comparison of protein and nucleic acid sequences. This chapter parallels the chapters on "Exploring Proteins" (Chapter 4) and "Exploring Genes" (Chapter 6), which have thoughtfully examined experimental techniques in earlier editions. Its goal is to enable students to use the vast information available in sequence and structural databases in a critical and effective manner.

ORGANIZATION OF THE TEXT. The evolutionary approach influences the organization of the text, which is divided into four major parts. As it did in the preceding edition, Part I introduces the language of biochemistry and the structures of the most important classes of biological molecules. The remaining three parts correspond to three major evolutionary challenges—namely, the interconversion of different forms of energy, molecular reproduction, and the adaptation of cells and organisms to changing environments. This arrangement parallels the evolutionary path outlined in Chapter 2 and naturally flows from the simple to the more complex.

PART I, THE MOLECULAR DESIGN OF LIFE, introduces the most important classes of biological macromolecules, including proteins, nucleic acids, carbohydrates, and lipids, and presents the basic concepts of catalysis and enzyme action. Here are two examples of how an evolutionary perspective has shaped the material in these chapters:

- **Chapter 9,** on catalytic strategies, examines four classes of enzymes that have evolved to meet specific challenges: promoting a fundamentally slow chemical reaction, maximizing the absolute rate of a reaction, catalyzing a reaction at one site but not at many alternative sites, and preventing a deleterious side reaction. In each case, the text considers the role of evolution in fine-tuning the key property.

- **Chapter 13,** on membrane channels and pumps, includes the first detailed three-dimensional structures of an ion channel and

an ion pump. Because most other important channels and pumps are evolutionarily related to these proteins, these two structures provide powerful frameworks for examining the molecular basis of the action of these classes of molecules, so important for the functioning of the nervous and other systems.

PART II, TRANSDUCING AND STORING ENERGY, examines pathways for the interconversion of different forms of energy. Chapter 15, on signal transduction, looks at how DNA fragments encoding relatively simple protein modules, rather than entire proteins, have been mixed and matched in the course of evolution to generate the wiring that defines signal-transduction pathways. The bulk of Part II discusses pathways for the generation of ATP and other energy-storing molecules. These pathways have been organized into groups that share common enzymes. The component reactions can be examined once and their use in different biological contexts illustrated while these reactions are fresh in the students' minds.

- **Chapter 16** covers both glycolysis and gluconeogenesis. These pathways are, in some ways, the reverse of each other, and a core of enzymes common to both pathways catalyze many of the steps in the center of the pathways. Covering the pathways together makes it easy to illustrate how free energy enters to drive the overall process either in the direction of glucose degradation or in the direction of glucose synthesis.

- **Chapter 17,** on the citric acid cycle, ties together through evolutionary insights the pyruvate dehydrogenase complex, which feeds molecules into the citric acid cycle, and the α-ketoglutarate dehydrogenase complex, which catalyzes one of the key steps in the cycle itself.

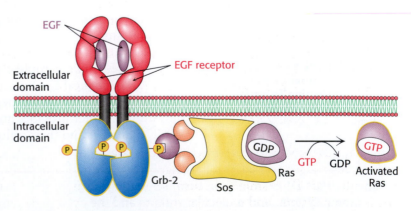

FIGURE 15.34 EGF signaling pathway.

Oxidative phosphorylation, in Chapter 18, is immediately followed in Chapter 19 by the light reactions of photosynthesis to emphasize the many common chemical features of these pathways.

The discussion of the light reactions of photosynthesis in Chapter 19 leads naturally into a discussion of the dark reactions—that is, the components of the Calvin cycle—in Chapter 20. This pathway is naturally linked to the pentose phosphate pathway, also covered in Chapter 20, because in both pathways common enzymes interconvert three-, four-, five-, six-, and seven-carbon sugars.

PART III, SYNTHESIZING THE MOLECULES OF LIFE, focuses on the synthesis of biological macromolecules and their components.

Chapter 24, on the biosynthesis of amino acids, is linked to the preceding chapter on amino acid degradation by a family of enzymes that transfer amino groups to and from the carbon frameworks of amino acids.

Chapter 25 covers the biosynthesis of nucleotides, including the role of amino acids as biosynthetic precursors. A key evolutionary insight emphasized here is that many of the enzymes in these pathways are members of the same family and catalyze analogous chemical reactions. The focus on enzymes and reactions common to these biosynthetic pathways allows the student to understand the logic of the pathways, rather than having to memorize a set of seemingly unrelated reactions.

Chapters 27, 28, and 29 cover DNA replication, recombination, and repair, RNA synthesis and splicing, and protein synthesis. Evolutionary connections between prokaryotic systems and eukaryotic systems reveal how the basic biochemical processes have been adapted to function in more-complex biological systems. The recently elucidated structure of the ribosome gives the student a glimpse into a possible early RNA world, in which nucleic acids, rather than proteins, played almost all the major roles in catalyzing important pathways.

PART IV, RESPONDING TO ENVIRONMENTAL CHANGES, looks at how cells sense and adapt to changes in their environments. Part IV examines, in turn, sensory systems, the immune system, and molecular motors and the cytoskeleton. These chapters illustrate how signaling and response processes, introduced earlier in the text, are integrated in multicellular organisms to generate powerful biochemical systems for detecting and responding to environmental changes. Again, the adaptation of proteins to new roles is key to these discussions.

Integrated Chemical Concepts

We have attempted to integrate chemical concepts throughout the text. They include the mechanistic basis for the action of selected enzymes, the thermodynamic basis for the folding and assembly of proteins and other macromolecules, and the structures and chemical reactivity of the common cofactors. These fundamental topics underlie our understanding of all biological processes. Our goal is not to provide an encyclopedic examination of enzyme reaction mechanisms. Instead, we have selected for examination at a more detailed chemical level specific topics that will enable students to understand how the chemical features help meet the biological needs.

Chemical insight often depends on a clear understanding of the structures of biochemical molecules. We have taken considerable care in preparing stereochemically accurate depictions of these molecules where appropriate.

Aspirin (acetylsalicyclic acid)

These structures should make it easier for the student to develop an intuitive feel for the shapes of molecules and comprehension of how these shapes affect reactivity.

Newly Updated to Include Recent Discoveries

Given the breathtaking pace of modern biochemistry, it is not surprising that there have been major developments since the publication of the fourth edition. Foremost among them is the sequencing of the human genome and the genomes of many simpler organisms. The text's evolutionary framework allows us to naturally incorporate information from these historic efforts. The determination of the three-dimensional structures of proteins and macromolecular assemblies also has been occurring at an astounding pace.

As noted earlier, the discussion of excitable membranes in Chapter 13 incorporates the detailed structures of an ion channel (the prokaryotic potassium channel) and an ion pump (the sacroplasmic reticulum calcium ATPase).

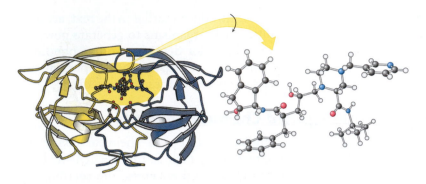

FIGURE 9.21   HIV protease-Crixivan complex.

• Great excitement has been generated in the signal transduction field by the first determination of the structure of a seven-transmembrane-helix receptor—the visual system protein rhodopsin—discussed in Chapters 15 and 32.

• The ability to describe the processes of oxidative phosphorylation in Chapter 18 has been greatly aided by the determination of the structures for two large membrane protein complexes: cytochrome c oxidase and cytochrome bc_1.

• Recent discoveries regarding the three-dimensional structure of ATP synthase are covered in Chapter 18, including the remarkable fact that parts of the enzyme rotate in the course of catalysis. The determination of the structure of the ribosome transforms the discussion of protein synthesis in Chapter 29.

• The elucidation of the structure of the nucleosome core particle—a large protein–DNA complex—facilitates the description in Chapter 31 of key processes in eukaryotic gene regulation.

Finally, each of the three chapters in Part IV is based on recent structural conquests.

• The student's ability to grasp key concepts in sensory systems (Chapter 32) is aided by the structures of rhodopsin and the aforementioned ion channel.

• Chapter 33, on the immune system, now includes the more recently determined structure of the T-cell receptor and its complexes.

• The determination of the structures of the molecular motor proteins myosin and kinesin first revealed the evolutionary connections on which Chapter 34, on molecular motors, is based.

New and Improved Illustrations

The relation of structure and function has always been a dominant theme of *Biochemistry*. This relation becomes even clearer to students using the fifth edition through the extensive use of molecular models. These models are superior to those in the fourth edition in several ways.

• All have been designed and rendered by one of us (JMB), with the use of MOLSCRIPT, to emphasize the most important structural features.

• We have chosen ribbon diagrams as the most effective, clearest method of conveying molecular structure. All molecular diagrams are rendered in a consistent style. Thus students are able to compare structures easily and to develop familiarity and facility in interpreting the models.

• Many new molecular models have been added, serving as sources of structural insight into additional molecules and in some cases affording multiple views of the same molecule.

In addition to the molecular models, the fifth edition includes more diagrams providing an overview of pathways and processes and setting processes in their biological context.

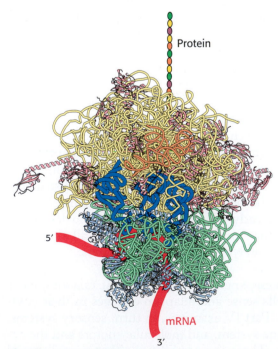

The ribosome (Chapter 29 opener).

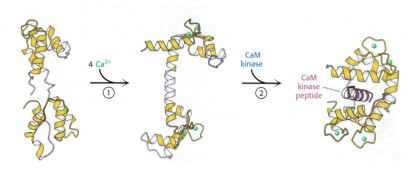

FIGURE 15.23   Calmodulin binds to amphipathic alpha-helices.

New Pedagogical Features

The fifth edition supplies additional tools to assist the student in learning the subject matter.

ICONS. Icons are used to highlight three categories of material, making these topics easier to locate for the interested student or teacher.

 A *caduceus* signals the beginning of a clinical application.

 A *stylized tree icon* marks sections or paragraphs that primarily or exclusively explore evolutionary aspects of biochemistry.

 A *mouse and finger* point to references to animations on the text's Web site (www.whfreeman.com/biochem5) for those students who wish to reinforce their understanding of concepts by using the electronic media.

MORE PROBLEMS. The number of problems has increased by 50%. Four new categories of problem have been created to develop specific skills.

    Mechanism problems ask students to suggest or elaborate a chemical mechanism.

    Data interpretation problems ask questions about a set of data provided in tabulated or graphic form. These exercises give students a sense of how scientific conclusions are reached.

    Chapter integration problems require students to use information from multiple chapters to reach a solution. These problems reinforce student awareness of the interconnectedness of the different aspects of biochemistry.

    Media problems encourage and assist students in taking advantage of the animations and tutorials provided on our Web site.

NEW CHAPTER OUTLINE AND KEY TERMS. An outline at the beginning of each chapter giving major headings serves as a framework for the student to use in organizing the information in the chapter. The major headings appear again in the chapter's summary, helping to organize the information for easier review. A set of key terms also helps students focus on the important concepts and review them.

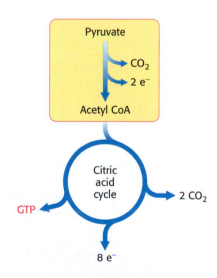

FIGURE 17.4   Glycolysis and the citric acid cycle are linked.

Tools and Techniques

The fifth edition of *Biochemistry* offers three chapters that present the tools and techniques of biochemistry: "Exploring Proteins" (Chapter 4), "Exploring Genes" (Chapter 6), and "Exploring Evolution" (Chapter 7). Additional experimental techniques are presented throughout the text, as appropriate.

Exploring Proteins (Chapter 4)

Protein purification   Section 4.1
Differential centrifugation   Section 4.1.2
Salting out   Section 4.1.3
Dialysis   Section 4.1.3
Gel-filtration chromatography   Section 4.1.3
Ion-exchange chromatography   Section 4.1.3
Affinity chromatography   Section 4.1.3
High-pressure liquid chromatography   Section 4.1.3
Gel electrophoresis   Section 4.1.4
Isoelectric focusing   Section 4.1.4
Two-dimensional electrophoresis   Section 4.1.4
Qualitative and quantitative evaluation of protein
    purification   Section 4.1.5
Ultracentrifugation   Section 4.1.6
Mass spectrometry (MALDI-TOF)   Section 4.1.7
Peptide mass fingerprinting   Section 4.1.7
Edman degradation   Section 4.2
Protein sequencing   Section 4.2
Production of poplyclonal antibodies   Section 4.3.1
Production of monoclonal antibodies   Section 4.3.2
Enzyme-linked immunosorbet assay (ELISA)
    Section 4.3.3
Western blotting   Section 4.3.4
Fluorescence microscopy   Section 4.3.5
Green fluorescent protein as a marker   Section 4.3.5
Immunoelectron microscopy   Section 4.3.5
Automated solid-phase peptide synthesis   Section 4.4
Nuclear magnetic resonance spectroscopy   Section 4.5.1
NOESY spectroscopy   Section 4.5.1
X-ray crystallography   Section 4.5.2

Exploring Proteins (other chapters)

Basis of fluorescence in green fluorescent protein
    Section 3.6.5
Time-resolved crystallography   Section 8.3.2
Using fluorescence spectroscopy to analyze enzyme–
    substrate interactions   Section 8.3.2
Using irreversible inhibitors to map the active site
    Section 8.5.2
Using transition state analogs to study enzyme active
    sites   Section 8.5.3
Enzyme studies with catalytic antibodies   Section 8.5.4

Exploring Genes (Chapter 6)

Restriction-enzyme analysis   Sections 6.1.1 and 6.1.2
Southern and Northern blotting techniques   Section 6.1.2
Sanger dideoxy method of DNA sequencing
    Section 6.1.3
Solid-phase analysis of nucleic acids   Section 6.1.4
Polymerase chain reaction (PCR)   Section 6.1.5

Recombinant DNA technology   Sections 6.2-6.4
    DNA cloning in bacteria   Sections 6.2.2 and 6.2.3
    Chromosome walking   Secction 6.2.4
    Cloning of eukaryotic genes in bacteria   Section 6.3.1
    Examining expression levels (gene chips)   Section 6.3.2
    Introducing genes into eukaryotes   Section 6.3.3
    Transgenic animals   Section 6.3.4
    Gene disruption   Section 6.3.5
    Tumor-inducing plasmids   Section 6.3.6
    Site-specific mutagenesis   Section 6.4

Exploring Genes (other chapters)

Density-gradient equilibrium sedimentation
    Section 5.2.2
Footprinting technique for isolating and characterizing
    promoter sites   Section 28.1.1
Chromatin immunoprecipitation (ChIP)   Section 31.2.3

Exploring Evolution (Chapter 7)

Sequence-comparison methods   Section 7.2
    Sequence-alignment methods   Section 7.2
        Estimating the statistical significance of alignments (by
            shuffling)   Section 7.2.1
    Substitution matrices   Section 7.2.2
Sequence templates   Section 7.3.2
Self-diagonal plots for finding repeated motifs   Section 7.3.3
Mapping secondary structures through RNA sequence
    comparisons   Section 7.3.5
Construction of evolutionary trees   Section 7.4
Combinatorial chemistry   Section 7.5.2

Other Techniques

Sequencing of carbohydrates by using MALDI-TOF
    mass spectroscopy   Section 11.3.7
Use of liposomes to investigate membrane permeability
    Section 12.4.1
Use of hydropathy plots to locate transmembrane
    helices   Section 12.5.4
Fluorescence recovery after photobleaching (FRAP)
    for measuring lateral diffusion in membranes
    Section 12.6
Patch-clamp technique for measuring channel activity
    Section 13.5.1
Measurement of redox potential   Section 18.2.1
Functional magnetic resonance imaging (fMRI)
    Section 32.1.3

Animated Techniques: Animated explanations of experimental techniques used for exploring genes and proteins are available at
www.whfreeman.com/biochem5.

Clinical Applications

 This icon signals the start of a clinical application in the text. Additional, briefer clinical correlations appear in the text as appropriate.

Molecular Evolution

 This icon signals the start of many discussions that highlight protein commonalities or other molecular evolutionary insights that provide a framework to help students organize information.

Supplements Supporting *Biochemistry*, Fifth Edition

The fifth edition of *Biochemistry* offers a wide selection of high-quality supplements to assist students and instructors.

For the Instructor

Print and Computerized Test Banks NEW

Marilee Benore Parsons, University of Michigan-Dearborn Print Test Bank 0-7167-4384-1; Computerized Test Bank CD-ROM (Windows/Macintosh hybrid) 0-7167-4386-8

The test bank offers more than 1700 questions posed in multiple choice, matching, and short-answer formats. The electronic version of the test bank allows instructors to easily edit and rearrange the questions or add their own material.

Instructor's Resource CD-ROM NEW

© W. H. Freeman and Company and Sumanas, Inc. 0-7167-4385-X

The Instructor's Resource CD-ROM contains all the illustrations from the text. An easy-to-use presentation manager application, Presentation Manager Pro, is provided. Each image is stored in a variety of formats and resolutions, from simple jpg and gif files to preformatted PowerPoint® slides, for instructors using other presentation programs.

Overhead Transparencies

0-7167-4422-8

Full-color illustrations from the text, optimized for classroom projection, in one volume.

For the Student

Student Companion

Richard I. Gumport, College of Medicine at Urbana-Champaign, University of Illinois; Frank H. Deis, Rutgers University; and Nancy Counts Gerber, San Fransisco State University. Expanded solutions to text problems provided by Roger E. Koeppe II, University of Arkansas 0-7167-4383-3

More than just a study guide, the *Student Companion* is an essential learning resource designed to meet the needs of students at all levels. Each chapter starts with

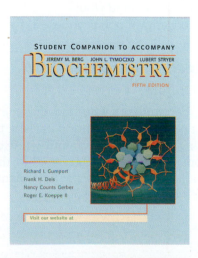

STUDENT COMPANION TO ACCOMPANY

JEREMY M. BERG   JOHN L. TYMOCZKO   LUBERT STRYER

BIOCHEMISTRY

FIFTH EDITION

Richard I. Gumport
Frank H. Deis
Nancy Counts Gerber
Roger E. Koeppe II

Visit our website at

a summarized abstract of the related textbook chapter. A comprehensive list of learning objectives allows students to quickly review the key concepts. A self-test feature allows students to quickly refresh their understanding, and a set of additional problems requires students to apply their knowledge of biochemistry. The complete solution to every problem in the text is provided to help students better comprehend the core ideas.

Individual chapters of the Student Companion can be purchased and downloaded from www.whfreeman.com/biochem5.

Clinical Companion NEW

Kirstie Saltsman, Ph.D., and Gordon Tomaselli, M.D., Johns Hopkins University School of Medicine 0-7167-4738-3

Designed for students and instructors interested in clinical applications, the *Clinical Companion* is a rich compendium of medical case studies and clinical discussions. It contains numerous problems and references to the textbook. Such topics as glaucoma, cystic fibrosis, Tay-Sachs disease, and autoimmune diseases are covered from a biochemical perspective.

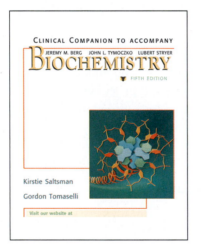

Lecture Notebook NEW

0-7167-4682-4

For students who find that they are too busy writing notes to pay attention in class, the *Lecture Notebook* brings together a black-and-white collection of illustrations from the text, arranged in the order of their appearance in the textbook, with plenty of room alongside for students to take notes.

Experimental Biochemistry, Third Edition

Robert L. Switzer, University of Illinois, and Liam F. Garrity, Pierce Chemical Corporation 0-7167-3300-5

The new edition of *Experimental Biochemistry* has been completely revised and updated to make it a perfect fit for today's laboratory course in biochemistry. It provides comprehensive coverage of important techniques used in contemporary biochemical research and gives students the background theory that they need to understand the experiments. Thoroughly classroom tested, the experiments incorporate the full range of biochemical materials in an attempt to simulate work in a research laboratory. In addition, a comprehensive appendix provides detailed procedures for preparation of reagents and materials, as well as helpful suggestions for the instructor.

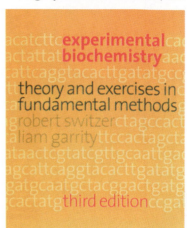

ALSO AVAILABLE THROUGH THE W. H. FREEMAN CUSTOM PUBLISHING PROGRAM, *Experimental Biochemistry* is designed to meet all your biochemistry laboratory needs. Visit http://custompub.pam to learn more about creating your own laboratory manual.

Student Media Resources

This icon links materials from the book to our Web site. See the inside front cover for a complete description of the resources available at www.whfreeman.com/biochem5.

nature publishing group **npg**   W. H. Freeman and Company is proud to present a weekly collection of abstracts and a monthly featured article from the *Nature* family of journals, including *Nature*, its associated monthly titles, and the recently launched *Nature Review Journals*. Please visit www.whfreeman.com/biochem5.

Acknowledgments

There is an old adage that says that you never really learn a subject until you teach it. We now know that you learn a subject even better when you write about it. Preparing the fifth edition of *Biochemistry* has provided us with a wonderful opportunity to unite our love of biochemistry and teaching and to share our enthusiasm with students throughout the world. Nonetheless, the project has also been a daunting one because so many interesting discoveries have been made since the publication of the fourth edition. The question constantly confronted us: What biochemical knowledge is most worth having? Answering this question required attempting to master as much of the new material as possible and then deciding what to include and, even harder, what to exclude.

However, we did not start from scratch. We feel both fortunate and intimidated to be writing the fifth edition of Stryer's *Biochemistry*. Fortunate, because we had, as our starting point, the best biochemistry book ever produced. Intimidated, because we had, as our starting point, the best biochemistry book ever produced, with the challenge of improving it. To the extent that we have succeeded, we have done so because of the help of many people.

Thanks go first and foremost to our students at Johns Hopkins University and Carleton College. Not a word was written or an illustration constructed without the knowledge that bright, engaged students would immediately detect vagueness and ambiguity. One of us (JMB) especially thanks the members of the Berg lab who have cheerfully tolerated years of neglect and requests to review drafts of illustrations when they would rather have been discussing their research. Particular thanks go to Dr. Barbara Amann and Kathleen Kolish who helped preserve some order in the midst of chaos. We also thank our colleagues at Johns Hopkins University and Carleton College who supported, advised, instructed, and simply bore with us during this arduous task. One of us (JLT) was graciously awarded a grant from Carleton College to relieve him of some of his academic tasks so that he could focus more fully on the book.

We are also grateful to our colleagues throughout the world who served as reviewers for the new edition. Their thoughtful comments, suggestions, and encouragement have been of immense help to us in maintaining the excellence of the preceding editions. These reviewers are:

Mark Alper
University of California at Berkeley

Mario Amzel
Johns Hopkins University

Paul Azari
Colorado State University

Ruma Banerjee
University of Nebraska

Michael Barbush
Baker University

Douglas Barrick
Johns Hopkins University

Loran L. Bieber
Michigan State University

Margaret Brosnan
University of Newfoundland

Lukas K. Buehler
University of California at San Diego

C. Allen Bush
University of Maryland, Baltimore County

Tom Cech
Howard Hughes Medical Institute

Oscar P. Chilson
Washington University

Steven Clarke
University of California at Los Angeles

Philip Cole
Johns Hopkins University School of Medicine

Paul A. Craig
Rochester Institute of Technology

David L. Daleke
Indiana University

David Deamer
University of California at Santa Cruz

Frank H. Deis
Rutgers University

Eric S. Eberhardt
Vassar College

Duane C. Eichler
University of San Francisco School of Medicine

Stephen H. Ellis
Auburn University

Nuran Ercal
University of Missouri at Rolla

Gregg B. Fields
Florida Atlantic University

Gregory Gatto
Johns Hopkins University

Nancy Counts Gerber
San Francisco State University

Claiborne Glover III
University of Georgia

E. M. Gregory
Virginia Polytechnic Institute and State University

Mark Griep
University of Nebraska at Lincoln

Hebe M. Guardiola-Diaz
Trinity College

James R. Heitz
Mississippi State University

Neville R. Kallenbach
New York University

Harold Kasinsky
University of British Columbia

Dan Kirschner
Boston College

G. Barrie Kitto
University of Texas at Austin

James F. Koerner
University of Minnesota

John Koontz
University of Tennessee

Gary R. Kunkel
Texas A&M University

David O. Lambeth
University of North Dakota

Timothy Logan
Florida State University

Douglas D. McAbee
California State University at Long Beach

William R. Marcotte, Jr.
Clemson University

Alan Mellors
University of Guelph

Dudley G. Moon
Albany College of Pharmacy

Kelley W. Moremen
University of Georgia

Scott Napper
University of Saskatchewan

Jeremy Nathans
Johns Hopkins University School of Medicine

James W. Phillips
University of Health Sciences

Terry Platt
University of Rochester Medical Center

Gary J. Quigley
Hunter College, City University of New York

Carl Rhodes
Howard Hughes Medical Institute

Gale Rhodes
University of Southern Maine

Mark Richter
University of Kansas

Anthony S. Serianni
University of Notre Dame

Ann E. Shinnar
Barnard College

Jessup M. Shively
Clemson University

Roger D. Sloboda
Dartmouth University

Carolyn M. Teschke
University of Connecticut

Dean R. Tolan
Boston University

Gordon Tollin
University of Arizona

Jeffrey M. Voigt
Albany College of Pharmacy

M. Gerard Waters
Princeton University

Linette M. Watkins
Southwest Texas State University

Gabriele Wienhausen
University of California at San Diego

James D. Willett
George Mason University

Gail R. Willsky
State University of New York at Buffalo

Dennis Winge
University of Utah

Charles F. Yocum
University of Michigan

Working with our colleagues at W. H Freeman and Company has been a wonderful experience. We would especially like to acknowledge the efforts of the following people. Our developmental editor, Susan Moran, has contributed immensely to the success of this project. During this process, Susan became a committed biochemistry student. Her understandings of how the subject matter, text and illustrations, would be perceived by students and her commitment to excellence were a true inspiration. Our project editor, Georgia Lee Hadler, managed the flow of the entire project—from manuscript to final product—sometimes with a velvet glove and other times more forcefully, but always effectively. The careful manuscript editor, Patricia Zimmerman, enhanced the text's literary consistency and clarity. Designers Vicki Tomaselli and Patricia McDermond produced a design and layout that are organizationally clear and esthetically pleasing. The tireless search of our photo researchers, Vikii Wong and Dena Betz, for the best possible photographs has contributed effectively to the clarity and appeal of the text. Cecilia Varas, the illustration coordinator, ably oversaw the rendering of hundreds of new illustrations, and Julia DeRosa, the production manager, astutely handled all the difficulties of scheduling, composition, and manufacturing.

Neil Clarke of Johns Hopkins University, Sonia DiVittorio, and Mark Santee piloted the media projects associated with the book. Neil's skills as a teacher and knowledge of the power and pitfalls of computers, Sonia's editing and coordination skills and her stylistic sense, and Mark's management of an ever-changing project have made the Web site a powerful supplement to the text and a lot of fun to explore. We would also like to acknowledge the media developers who transformed scripts into the animations you find on our Web site. For the Structural Insights modules we thank Nick McLeod, Koreen Wykes, Dr. Roy Tasker, Robert Bleeker, and David Hegarty, all at CADRE design. For the three-dimensional molecular visualizations in the Structural Insights modules we thank Timothy Driscoll (molvisions.com—3D molecular visualization). Daniel J. Davis of the University of Arkansas at Fayetteville prepared the online quizzes.

Publisher Michelle Julet was our cheerleader, taskmaster, comforter, and cajoler. She kept us going when we were tired, frustrated, and discouraged. Along with Michelle, marketing mavens John Britch and Carol Coffey introduced us to the business of publishing. We also thank the sales people at W. H. Freeman and Company for their excellent suggestions and view of the market, especially Vice President of Sales Marie Schappert, David Kennedy, Chris Spavins, Julie Hirshman, Cindi WeissGoldner, Kimberly Manzi, Connaught Colbert, Michele Merlo, Sandy Manly, and Mike Krotine. We thank Elizabeth Widdicombe, President of W. H. Freeman and Company, for never losing faith in us.

Finally, the project would not have been possible without the unfailing support of our families—especially our wives, Wendie Berg and Alison Unger. Their patience, encouragement, and enthusiasm have made this endeavor possible. We also thank our children, Alex, Corey, and Monica Berg and Janina and Nicholas Tymoczko, for their forbearance and good humor and for constantly providing us a perspective on what is truly important in life.

Brief Contents

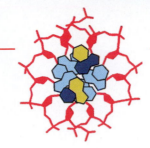

Contents

PART I THE MOLECULAR DESIGN OF LIFE

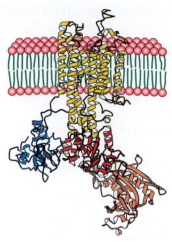

PART II TRANSDUCING AND STORING ENERGY

PART III SYNTHESIZING THE MOLECULES OF LIFE

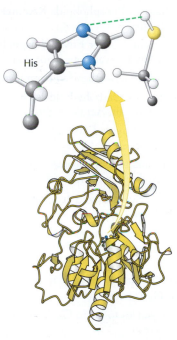

His

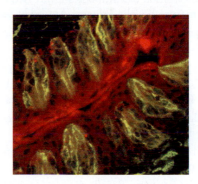

PART IV RESPONDING TO ENVIRONMENTAL CHANGES

Chapters 32, 33, and 34 are available on the Web at www.whfreeman.com/biochem5 or may be obtained separately in printed form using ISBN 0-7167-4955-6

CHAPTER 32 Sensory Systems

PART I

The Molecular Design of Life

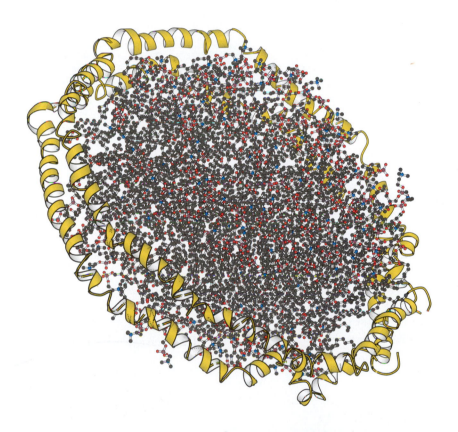

Part of a lipoprotein particle. A model of the structure of apolipoprotein A-I (yellow), shown surrounding sheets of lipids. The apolipoprotein is the major protein component of high-density lipoprotein particles in the blood. These particles are effective lipid transporters because the protein component provides an interface between the hydrophobic lipid chains and the aqueous environment of the bloodstream. [Based on coordinates provided by Stephen Harvey.]

CHAPTER 1

Prelude: Biochemistry and the Genomic Revolution

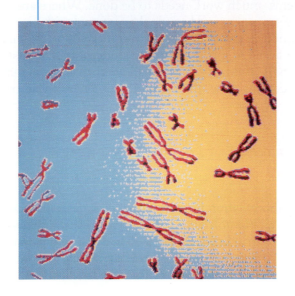

← q31.2

Disease and the genome. Studies of the human genome are revealing disease origins and other biochemical mysteries. Human chromosomes, left, contain the DNA molecules that constitute the human genome. The staining pattern serves to identify specific regions of a chromosome. On the right is a diagram of human chromosome 7, with band q31.2 indicated by an arrow. A gene in this region encodes a protein that, when malfunctioning, causes cystic fibrosis. [(Left) Alfred Pasieka/Peter Arnold.]

GACTTCACTTCTAATGATGATTATGGGAGAACTGGAGCCT TCAGAGGGTAAAAATTAAGCACAGTGGAAGAATTTCATTC TGTTCTCAGTTTTCCTGGATTATGCCTGGCACCATTAAAG AAAATATCTTTGGTGTTTCCTATGATGAATATAGATACAG AAGCGTCATCAAAGCATGCCAACTAGAAGAG.... This string of letters A, C, G, and T is a part of a DNA sequence. Since the biochemical techniques for DNA sequencing were first developed more than three decades ago, the genomes of dozens of organisms have been sequenced, and many more such sequences will be forthcoming. The information contained in these DNA sequences promises to shed light on many fascinating and important questions. What genes in *Vibrio cholera*, the bacterium that causes cholera, for example, distinguish it from its more benign relatives? How is the development of complex organisms controlled? What are the evolutionary relationships between organisms?

Sequencing studies have led us to a tremendous landmark in the history of biology and, indeed, humanity. *A nearly complete sequence of the entire human genome* has been determined. The string of As, Cs, Gs, and Ts with which we began this book is a tiny part of the human genome sequence, which is more than 3 billion letters long. If we included the entire sequence, our opening sentence would fill more than 500,000 pages.

The implications of this knowledge cannot be overestimated. By using this blueprint for much of what it means to be human, scientists can begin the identification and characterization of sequences that foretell the appearance of specific diseases and particular physical attributes. One consequence will be the development of better means of diagnosing and treating diseases. Ultimately, physicians will be able to devise plans for preventing or managing heart disease or cancer that take account of individual variations. Although the sequencing of the human genome is an enormous step toward a complete understanding of living systems, much work needs to be done. Where are the functional genes within the sequence, and how do they interact with one another? How is the information in genes converted into the functional characteristics of an organism? Some of our goals in the study of biochemistry are to learn the concepts, tools, and facts that will allow us to address these questions. It is indeed an exciting time, the beginning of a new era in biochemistry.

1.1 DNA ILLUSTRATES THE RELATION BETWEEN FORM AND FUNCTION

The structure of DNA, an abbreviation for *deoxyribonucleic acid*, illustrates a basic principle common to all biomolecules: the intimate relation between structure and function. The remarkable properties of this chemical substance allow it to function as a very efficient and robust vehicle for storing information. We begin with an examination of the covalent structure of DNA and its extension into three dimensions.

1.1.1 DNA Is Constructed from Four Building Blocks

DNA is a *linear polymer* made up of four different monomers. It has a fixed backbone from which protrude variable substituents (Figure 1.1). The backbone is built of repeating sugar-phosphate units. The sugars are molecules of *deoxyribose* from which DNA receives its name. Joined to each deoxyribose is one of four possible bases: adenine (A), cytosine (C), guanine (G), and thymine (T).

Adenine (A)	Cytosine (C)	Guanine (G)	Thymine (T)

All four bases are planar but differ significantly in other respects. Thus, the monomers of DNA consist of a sugar-phosphate unit, with one of four bases attached to the sugar. *These bases can be arranged in any order along a strand of DNA.* The order of these bases is what is displayed in the sequence that begins this chapter. For example, the first base in the sequence shown is G

FIGURE 1.1 Covalent structure of DNA. Each unit of the polymeric structure is composed of a sugar (deoxyribose), a phosphate, and a variable base that protrudes from the sugar-phosphate backbone.

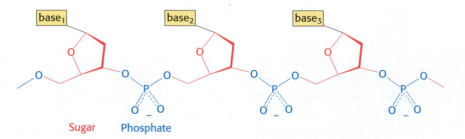

Sugar   Phosphate

(guanine), the second is A (adenine), and so on. *The sequence of bases along a DNA strand constitutes the genetic information*—the instructions for assembling proteins, which themselves orchestrate the synthesis of a host of other biomolecules that form cells and ultimately organisms.

1.1.2 Two Single Strands of DNA Combine to Form a Double Helix

Most DNA molecules consist of not one but two strands (Figure 1.2). How are these strands positioned with respect to one another? In 1953, James Watson and Francis Crick deduced the arrangement of these strands and proposed a three-dimensional structure for DNA molecules. This structure is a *double helix* composed of two intertwined strands arranged such that the sugar-phosphate backbone lies on the outside and the bases on the inside. The key to this structure is that the bases form *specific base pairs* (bp) held together by *hydrogen bonds* (Section 1.3.1): adenine pairs with thymine (A–T) and guanine pairs with cytosine (G–C), as shown in Figure 1.3. Hydrogen bonds are much weaker than covalent bonds such as the carbon–carbon or carbon–nitrogen bonds that define the structures of the bases themselves. Such weak bonds are crucial to biochemical systems; they are weak enough to be reversibly broken in biochemical processes, yet they are strong enough, when many form simultaneously, to help stabilize specific structures such as the double helix.

FIGURE 1.2 The double helix. The double-helical structure of DNA proposed by Watson and Crick. The sugar-phosphate backbones of the two chains are shown in red and blue and the bases are shown in green, purple, orange, and yellow.

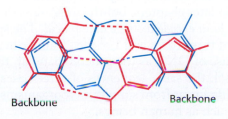

Adenine (A)   **Thymine (T)**   **Guanine (G)**   **Cytosine (C)**

FIGURE 1.3 Watson-Crick base pairs. Adenine pairs with thymine (A–T), and guanine with cytosine (G–C). The dashed lines represent hydrogen bonds.

The structure proposed by Watson and Crick has two properties of central importance to the role of DNA as the hereditary material. First, the structure is compatible with *any sequence of bases*. The base pairs have essentially the same shape (Figure 1.4) and thus fit equally well into the center of the double-helical structure. Second, because of base-pairing, *the sequence of bases along one strand completely determines the sequence along the other strand*. As Watson and Crick so coyly wrote: "It has not escaped our notice that the specific pairing we have postulated immediately suggests a possible copying mechanism for the genetic material." Thus, if the DNA double helix is separated into two single strands, each strand can act as a template for the generation of its partner strand through specific base-pair formation (Figure 1.5). *The three-dimensional structure of DNA beautifully illustrates the close connection between molecular form and function.*

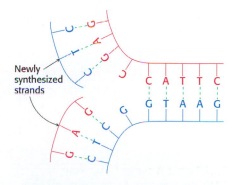

Newly synthesized strands

FIGURE 1.5 DNA replication. If a DNA molecule is separated into two strands, each strand can act as the template for the generation of its partner strand.

FIGURE 1.4 Base-pairing in DNA. The base-pairs A–T (blue) and C–G (red) are shown overlaid. The Watson-Crick base-pairs have the same overall size and shape, allowing them to fit neatly within the double helix.

Backbone   Backbone

Ribose

Uracil (U)

1.1.3 RNA Is an Intermediate in the Flow of Genetic Information

An important nucleic acid in addition to DNA is *ribonucleic acid (RNA)*. Some viruses use RNA as the genetic material, and even those organisms that employ DNA must first convert the genetic information into RNA for the information to be accessible or functional. Structurally, RNA is quite similar to DNA. It is a linear polymer made up of a limited number of repeating monomers, each composed of a sugar, a phosphate, and a base. The sugar is ribose instead of deoxyribose (hence, *RNA*) and one of the bases is uracil (U) instead of thymine (T). Unlike DNA, an RNA molecule usually exists as a single strand, although significant segments within an RNA molecule may be double stranded, with G pairing primarily with C and A pairing with U. This intrastrand base-pairing generates RNA molecules with complex structures and activities, including catalysis.

RNA has three basic roles in the cell. First, it serves as the intermediate in the flow of information from DNA to protein, the primary functional molecules of the cell. The DNA is copied, or *transcribed*, into messenger RNA (mRNA), and the mRNA is *translated* into protein. Second, RNA molecules serve as adaptors that translate the information in the nucleic acid sequence of mRNA into information designating the sequence of constituents that make up a protein. Finally, RNA molecules are important functional components of the molecular machinery, called ribosomes, that carries out the translation process. As will be discussed in Chapter 2, the unique position of RNA between the storage of genetic information in DNA and the functional expression of this information as protein as well as its potential to combine genetic and catalytic capabilities are indications that RNA played an important role in the evolution of life.

1.1.4 Proteins, Encoded by Nucleic Acids, Perform Most Cell Functions

A major role for many sequences of DNA is to encode the sequences of *proteins*, the workhorses within cells, participating in essentially all processes. Some proteins are key structural components, whereas others are specific catalysts (termed *enzymes*) that promote chemical reactions. Like DNA and RNA, proteins are linear polymers. However, proteins are more complicated in that they are formed from a selection of 20 building blocks, called *amino acids*, rather than 4.

The functional properties of proteins, like those of other biomolecules, are determined by their three-dimensional structures. Proteins possess an extremely important property: a protein spontaneously folds into a well-defined and elaborate three-dimensional structure that is dictated entirely by the sequence of amino acids along its chain (Figure 1.6). *The self-folding nature of proteins constitutes the transition from the one-dimensional world of sequence information to the three-dimensional world of biological function.* This marvelous ability of proteins to self assemble into complex structures is responsible for their dominant role in biochemistry.

How is the sequence of bases along DNA translated into a sequence of amino acids along a protein chain? We will consider the details of this process in later chapters, but the important finding is that *three bases along a DNA chain encode a single amino acid.* The specific correspondence between a set of three bases and 1 of the 20 amino acids is called the *genetic code*. Like the use of DNA as the genetic material, the genetic code is essentially universal; the same sequences of three bases encode the same amino acids in all life forms from simple microorganisms to complex, multicellular organisms such as human beings.

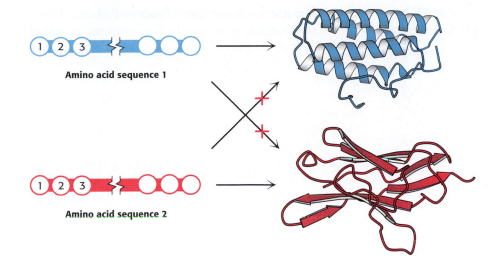

Amino acid sequence 1

Amino acid sequence 2

FIGURE 1.6 Folding of a protein. The three-dimensional structure of a protein, a linear polymer of amino acids, is dictated by its amino acid sequence.

Knowledge of the functional and structural properties of proteins is absolutely essential to understanding the significance of the human genome sequence. For example, the sequence at the beginning of this chapter corresponds to a region of the genome that differs in people who have the genetic disorder *cystic fibrosis*. The most common mutation causing cystic fibrosis, the loss of three consecutive Ts from the gene sequence, leads to the loss of a single amino acid within a protein chain of 1480 amino acids. This seemingly slight difference—a loss of 1 amino acid of nearly 1500—creates a life-threatening condition. What is the normal function of the protein encoded by this gene? What properties of the encoded protein are compromised by this subtle defect? Can this knowledge be used to develop new treatments? These questions fall in the realm of biochemistry. Knowledge of the human genome sequence will greatly accelerate the pace at which connections are made between DNA sequences and disease as well as other human characteristics. However, these connections will be nearly meaningless without the knowledge of biochemistry necessary to interpret and exploit them.

Cystic fibrosis—

A disease that results from a decrease in fluid and salt secretion by a transport protein referred to as the cystic fibrosis transmembrane conductance regulator (CFTR). As a result of this defect, secretion from the pancreas is blocked, and heavy, dehydrated mucus accumulates in the lungs, leading to chronic lung infections.

1.2 BIOCHEMICAL UNITY UNDERLIES BIOLOGICAL DIVERSITY

The stunning variety of living systems (Figure 1.7) belies a striking similarity. The common use of DNA and the genetic code by all organisms underlies one of the most powerful discoveries of the past century—namely, that *organisms are remarkably uniform at the molecular level*. All organisms are built from similar molecular components distinguishable by relatively minor variations. *This uniformity reveals that all organisms on Earth have*

FIGURE 1.7 The diversity of living systems. The distinct morphologies of the three organisms shown—a plant (the false hellebora, or Indian poke) and two animals (sea urchins and a common house cat)—might suggest that they have little in common. Yet biochemically they display a remarkable commonality that attests to a common ancestry. [(Left and right) John Dudak/Phototake. (Middle) Jeffrey L. Rotman/Peter Arnold.]

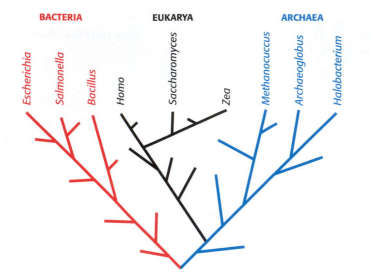

FIGURE 1.8 The tree of life. A possible evolutionary path from a common ancestral cell to the diverse species present in the modern world can be deduced from DNA sequence analysis.

arisen from a common ancestor. A core of essential biochemical processes, common to all organisms, appeared early in the evolution of life. The diversity of life in the modern world has been generated by evolutionary processes acting on these core processes through millions or even billions of years. As we will see repeatedly, the generation of diversity has very often resulted from the adaptation of existing biochemical components to new roles rather than the development of fundamentally new biochemical technology. The striking uniformity of life at the molecular level affords the student of biochemistry a particularly clear view into the essence of biological processes that applies to all organisms from human beings to the simplest microorganisms.

On the basis of their biochemical characteristics, the diverse organisms of the modern world can be divided into three fundamental groups called *domains: Eukarya* (eukaryotes), *Bacteria* (formerly Eubacteria), and *Archaea* (formerly Archaebacteria). Eukarya comprise all macroscopic organisms, including human beings as well as many microscopic, unicellular organisms such as yeast. The defining characteristic of *eukaryotes* is the presence of a well-defined nucleus within each cell. Unicellular organisms such as bacteria, which lack a nucleus, are referred to as *prokaryotes*. The prokaryotes were reclassified as two separate domains in response to Carl Woese's discovery in 1977 that certain bacteria-like organisms are biochemically quite distinct from better-characterized bacterial species. These organisms, now recognized as having diverged from bacteria early in evolution, are archaea. Evolutionary paths from a common ancestor to modern organisms can be developed and analyzed on the basis of biochemical information. One such path is shown in Figure 1.8.

By examining biochemistry in the context of the tree of life, we can often understand how particular molecules or processes helped organisms adapt to specific environments or life styles. We can ask not only *what* biochemical processes take place, but also *why* particular strategies appeared in the course of evolution. In addition to being sources of historical insights, *the answers to such questions are often highly instructive with regard to the biochemistry of contemporary organisms.*

1.3 CHEMICAL BONDS IN BIOCHEMISTRY

The essence of biological processes—the basis of the uniformity of living systems—is in its most fundamental sense molecular interactions; in other words, the chemistry that takes place between molecules. Biochemistry is the *chemistry* that takes place within living systems. To truly understand biochemistry, we need to understand chemical bonding. We review here the types of chemical bonds that are important for biochemicals and their transformations.

The strongest bonds that are present in biochemicals are *covalent bonds*, such as the bonds that hold the atoms together within the individual bases shown in Figure 1.3. A covalent bond is formed by the sharing of a pair of electrons between adjacent atoms. A typical carbon–carbon (C–C) covalent bond has a bond length of 1.54 Å and bond energy of 85 kcal mol^{-1} (356 kJ mol^{-1}). Because this energy is relatively high, considerable energy must be expended to break covalent bonds. More than one electron pair can be shared between two atoms to form a multiple covalent bond. For example, three of the bases in Figure 1.4 include carbon–oxygen (C=O) double bonds. These bonds are even stronger than C–C single bonds, with energies near 175 kcal mol^{-1} (732 kJ mol^{-1}).

For some molecules, more than one pattern of covalent bonding can be written. For example, benzene can be written in two equivalent ways called *resonance structures*. Benzene's true structure is a composite of its two resonance structures. A molecule that can be written as several resonance structures of approximately equal energies has greater stability than does a molecule without multiple resonance structures. Thus, because of its resonance structures, benzene is unusually stable.

Chemical reactions entail the breaking and forming of covalent bonds. The flow of electrons in the course of a reaction can be depicted by curved arrows, a method of representation called "arrow pushing." Each arrow represents an electron pair.

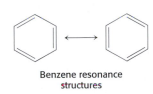

Benzene resonance
structures

1.3.1 Reversible Interactions of Biomolecules Are Mediated by Three Kinds of Noncovalent Bonds

Readily reversible, noncovalent molecular interactions are key steps in the dance of life. Such weak, noncovalent forces play essential roles in the faithful replication of DNA, the folding of proteins into intricate three-dimensional forms, the specific recognition of substrates by enzymes, and the detection of molecular signals. Indeed, all biological structures and processes depend on the interplay of noncovalent interactions as well as covalent ones. The three fundamental noncovalent bonds are *electrostatic interactions, hydrogen bonds,* and *van der Waals interactions.* They differ in geometry, strength, and specificity. Furthermore, these bonds are greatly affected in different ways by the presence of water. Let us consider the characteristics of each:

1. *Electrostatic interactions.* An electrostatic interaction depends on the electric charges on atoms. The energy of an electrostatic interaction is given by *Coulomb's law:*

$$E = kq_1q_2/Dr$$

where E is the energy, q_1 and q_2 are the charges on the two atoms (in units of the electronic charge), r is the distance between the two atoms (in angstroms), D is the dielectric constant (which accounts for the effects of the intervening medium), and k is a proportionality constant ($k = 332$, to give energies in units of kilocalories per mole, or 1389, for energies in kilojoules per mole). Thus, the electrostatic interaction between two atoms bearing single opposite charges separated by 3 Å in water (which has a dielectric constant of 80) has an energy of 1.4 kcal mol^{-1} (5.9 kJ mol^{-1}).

2. *Hydrogen bonds.* Hydrogen bonds are relatively weak interactions, which nonetheless are crucial for biological macromolecules such as DNA and proteins. These interactions are also responsible for many of the properties of water that make it such a special solvent. The hydrogen atom in a hydrogen bond is partly shared between two relatively electronegative atoms such as nitrogen or oxygen. The *hydrogen-bond donor* is the group that includes both the atom to which the hydrogen is more tightly linked and the hydrogen atom itself, whereas the *hydrogen-bond acceptor* is the atom less tightly linked to the hydrogen atom (Figure 1.9). Hydrogen bonds are fundamentally

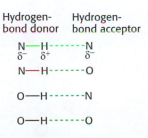

Hydrogen-   Hydrogen-
bond donor   bond acceptor

FIGURE 1.9 Hydrogen bonds that include nitrogen and oxygen atoms. The positions of the partial charges (δ^+ and δ^-) are shown.

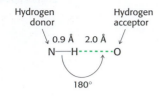

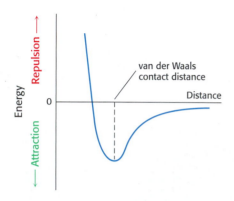

FIGURE 1.10 Energy of a van der Waals interaction as two atoms approach one another. The energy is most favorable at the van der Waals contact distance. The energy rises rapidly owing to electron–electron repulsion as the atoms move closer together than this distance.

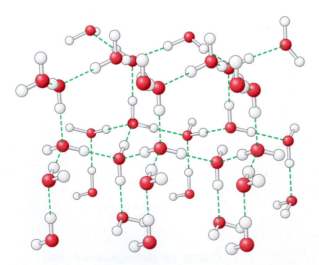

FIGURE 1.11 Structure of ice. Hydrogen bonds (shown as dashed lines) are formed between water molecules.

electrostatic interactions. The relatively electronegative atom to which the hydrogen atom is covalently bonded pulls electron density away from the hydrogen atom so that it develops a partial positive charge (δ^+). Thus, it can interact with an atom having a partial negative charge (δ^-) through an electrostatic interaction.

Hydrogen bonds are much weaker than covalent bonds. They have energies of 1–3 kcal mol^{-1} (4–13 kJ mol^{-1}) compared with approximately 100 kcal mol^{-1} (418 kJ mol^{-1}) for a carbon–hydrogen covalent bond. Hydrogen bonds are also somewhat longer than are covalent bonds; their bond distances (measured from the hydrogen atom) range from 1.5 to 2.6 Å; hence, distances ranging from 2.4 to 3.5 Å separate the two nonhydrogen atoms in a hydrogen bond. The strongest hydrogen bonds have a tendency to be approximately straight, such that the hydrogen-bond donor, the hydrogen atom, and the hydrogen-bond acceptor lie along a straight line.

3. *van der Waals interactions.* The basis of a van der Waals interaction is that the distribution of electronic charge around an atom changes with time. At any instant, the charge distribution is not perfectly symmetric. This transient asymmetry in the electronic charge around an atom acts through electrostatic interactions to induce a complementary asymmetry in the electron distribution around its neighboring atoms. The resulting attraction between two atoms increases as they come closer to each other, until they are separated by the van der Waals *contact distance* (Figure 1.10). At a shorter distance, very strong repulsive forces become dominant because the outer electron clouds overlap.

Energies associated with van der Waals interactions are quite small; typical interactions contribute from 0.5 to 1.0 kcal mol^{-1} (from 2 to 4 kJ mol^{-1}) per atom pair. When the surfaces of two large molecules come together, however, a large number of atoms are in van der Waals contact, and the net effect, summed over many atom pairs, can be substantial.

1.3.2 The Properties of Water Affect the Bonding Abilities of Biomolecules

Weak interactions are the key means by which molecules interact with one another—enzymes with their substrates, hormones with their receptors, antibodies with their antigens. The strength and specificity of weak interactions are highly dependent on the medium in which they take place, and the majority of biological interactions take place in water. Two properties of water are especially important biologically:

1. *Water is a polar molecule.* The water molecule is bent, not linear, and so the distribution of charge is asymmetric. The oxygen nucleus draws electrons away from the hydrogen nuclei, which leaves the region around the hydrogen nuclei with a net positive charge. The water molecule is thus an electrically polar structure.

2. *Water is highly cohesive.* Water molecules interact strongly with one another through hydrogen bonds. These interactions are apparent in the structure of ice (Figure 1.11). Networks of hydrogen bonds hold the structure together; similar interactions link molecules in liquid water and account for the cohesion of liquid water, although, in the liquid state, some of the hydrogen bonds are broken. The highly cohesive nature of water dramatically affects the interactions between molecules in aqueous solution.

What is the effect of the properties of water on the weak interactions discussed in Section 1.3.1? The polarity and hydrogen-bonding capability of water make it a highly interacting molecule. Water is an excellent solvent for polar molecules. The reason is that water greatly weakens electrostatic forces and hydrogen bonding between polar molecules by competing for their attractions. For example, consider the effect of water on hydrogen bonding between a carbonyl group and the NH group of an amide.

A hydrogen atom of water can replace the amide hydrogen atom as a hydrogen-bond donor, whereas the oxygen atom of water can replace the carbonyl oxygen atom as a hydrogen-bond acceptor. Hence, a strong hydrogen bond between a CO group and an NH group forms only if water is excluded.

The dielectric constant of water is 80, so water diminishes the strength of electrostatic attractions by a factor of 80 compared with the strength of those same interactions in a vacuum. The dielectric constant of water is unusually high because of its polarity and capacity to form oriented solvent shells around ions. These oriented solvent shells produce electric fields of their own, which oppose the fields produced by the ions. Consequently, the presence of water markedly weakens electrostatic interactions between ions.

The existence of life on Earth depends critically on the capacity of water to dissolve a remarkable array of polar molecules that serve as fuels, building blocks, catalysts, and information carriers. High concentrations of these polar molecules can coexist in water, where they are free to diffuse and interact with one another. However, the excellence of water as a solvent poses a problem, because it also weakens interactions between polar molecules. *The presence of water-free microenvironments within biological systems largely circumvents this problem.* We will see many examples of these specially constructed niches in protein molecules. Moreover, the presence of water with its polar nature permits another kind of weak interaction to take place, one that drives the folding of proteins (Section 1.3.4) and the formation of cell boundaries (Section 12.4).

The essence of these interactions, like that of all interactions in biochemistry, is energy. To understand much of biochemistry—bond formation, molecular structure, enzyme catalysis—we need to understand energy. Thermodynamics provides a valuable tool for approaching this topic. We will revisit this topic in more detail when we consider enzymes (Chapter 8) and the basic concepts of metabolism (Chapter 14).

1.3.3 Entropy and the Laws of Thermodynamics

The highly structured, organized nature of living organisms is apparent and astonishing. This organization extends from the organismal through the cellular to the molecular level. Indeed, biological processes can seem magical in that the well-ordered structures and patterns emerge from the chaotic and disordered world of inanimate objects. However, the organization visible in a cell or a molecule arises from biological events that are subject to the same physical laws that govern all processes—in particular, the *laws of thermodynamics.*

How can we understand the creation of order out of chaos? We begin by noting that the laws of thermodynamics make a distinction between a system and its surroundings. A *system* is defined as the matter within a defined region of space. The matter in the rest of the universe is called the *surroundings. The First Law of Thermodynamics states that the total energy of a system and its surroundings is constant.* In other words, the energy content of the universe is constant; energy can be neither created nor destroyed. Energy can take different forms, however. Heat, for example, is one form of energy. Heat is a manifestation of the *kinetic energy* associated with the random motion of molecules. Alternatively, energy can be present as *potential energy,* referring to the ability of energy to be released on the occurrence of some process. Consider, for example, a ball held at the top of a tower. The ball has considerable potential energy because, when it is released, the ball will develop kinetic energy associated with its motion as it falls. Within chemical systems, potential energy is related to the likelihood that atoms can react with one another. For instance, a mixture of gasoline and oxygen has much potential energy because these molecules may react to form carbon dioxide and release energy as heat. The First Law requires that any energy released in the formation of chemical bonds be used to break other bonds, be released as heat, or be stored in some other form.

Another important thermodynamic concept is that of *entropy.* Entropy is a measure of the level of randomness or disorder in a system. *The Second Law of Thermodynamics states that the total entropy of a system and its surroundings always increases for a spontaneous process.* At first glance, this law appears to contradict much common experience, particularly about biological systems. Many biological processes, such as the generation of a well-defined structure such as a leaf from carbon dioxide gas and other nutrients, clearly increase the level of order and hence decrease entropy. Entropy may be decreased locally in the formation of such ordered structures only if the entropy of other parts of the universe is increased by an equal or greater amount.

An example may help clarify the application of the laws of thermodynamics to a chemical system. Consider a container with 2 moles of hydrogen gas on one side of a divider and 1 mole of oxygen gas on the other (Figure 1.12). If the divider is removed, the gases will intermingle spontaneously to form a uniform mixture. The process of mixing increases entropy as an ordered arrangement is replaced by a randomly distributed mixture.

Other processes within this system can decrease the entropy locally while increasing the entropy of the universe. A spark applied to the mixture initiates a chemical reaction in which hydrogen and oxygen combine to form water:

$$2\ H_2 + O_2 \longrightarrow 2\ H_2O$$

If the temperature of the system is held constant, the entropy of the system decreases because 3 moles of two differing reactants have been combined to form 2 moles of a single product. The gas now consists of a uniform set of indistinguishable molecules. However, the reaction releases a significant amount of heat into the surroundings, and this heat will increase the entropy of the surrounding molecules by increasing their random movement.

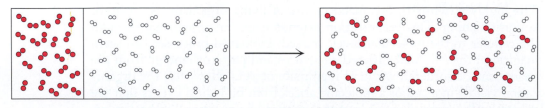

FIGURE 1.12 From order to disorder. The spontaneous mixing of gases is driven by an increase in entropy.

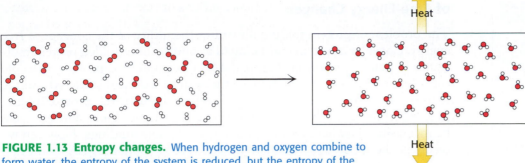

FIGURE 1.13 Entropy changes. When hydrogen and oxygen combine to form water, the entropy of the system is reduced, but the entropy of the universe is increased owing to the release of heat to the surroundings.

The entropy increase in the surroundings is enough to allow water to form spontaneously from hydrogen and oxygen (Figure 1.13).

The change in the entropy of the surroundings will be proportional to the amount of heat transferred from the system and inversely proportional to the temperature of the surroundings, because an input of heat leads to a greater increase in entropy at lower temperatures than at higher temperatures. In biological systems, T [in kelvin (K), absolute temperature] is assumed to be constant. If we define the heat content of a system as *enthalpy (H)*, then we can express the relation linking the entropy (S) of the surroundings to the transferred heat and temperature as a simple equation:

$$\Delta S_{surroundings} = -\Delta H_{system}/T \qquad (1)$$

The total entropy change is given by the expression

$$\Delta S_{total} = \Delta S_{system} + \Delta S_{surroundings} \qquad (2)$$

Substituting equation 1 into equation 2 yields

$$\Delta S_{total} = \Delta S_{system} - \Delta H_{system}/T \qquad (3)$$

Multiplying by $-T$ gives

$$-T\Delta S_{total} = \Delta H_{system} - T\Delta S_{system} \qquad (4)$$

The function $-T\Delta S$ has units of energy and is referred to as *free energy* or *Gibbs free energy*, after Josiah Willard Gibbs, who developed this function in 1878:

$$\Delta G = \Delta H_{system} - T\Delta S_{system} \qquad (5)$$

The free-energy change, ΔG, will be used throughout this book to describe the energetics of biochemical reactions.

Recall that the Second Law of Thermodynamics states that, for a reaction to be spontaneous, the entropy of the universe must increase. Examination of equation 3 shows that the total entropy will increase if and only if

$$\Delta S_{system} > \Delta H_{system}/T \qquad (6)$$

Rearranging gives $T\Delta S_{system} > \Delta H$, or entropy will increase if and only if

$$\Delta G = \Delta H_{system} - T\Delta S_{system} < 0 \qquad (7)$$

In other words, *the free-energy change must be negative for a reaction to be spontaneous.* A negative free-energy change occurs with an increase in the overall entropy of the universe. Thus, we need to consider only one term, the free energy of the system, to decide whether a reaction can occur spontaneously; any effects of the changes within the system on the rest of the universe are automatically taken into account.

Unfolded ensemble

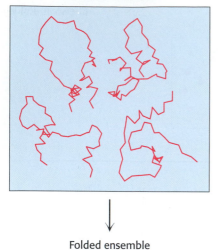

Folded ensemble

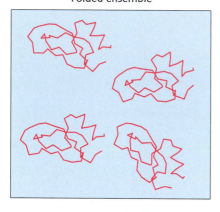

FIGURE 1.14 Protein folding. Protein folding entails the transition from a disordered mixture of unfolded molecules to a relatively uniform solution of folded protein molecules.

1.3.4 Protein Folding Can Be Understood in Terms of Free-Energy Changes

The problem of protein folding illustrates the utility of the concept of free energy. Consider a system consisting of a solution of unfolded protein molecules in aqueous solution (Figure 1.14). Each unfolded protein molecule can adopt a unique conformation, so the system is quite disordered and the entropy of the collection of molecules is relatively high. Yet, protein folding proceeds spontaneously under appropriate conditions. Thus, entropy must be increasing elsewhere in the system or in the surroundings. How can we reconcile the apparent contradiction that proteins spontaneously assume an ordered structure, and yet entropy increases? The entropy decrease in the system on folding is not as large as it appears to be, because of the properties of water. Molecules in aqueous solution interact with water molecules through the formation of hydrogen and ionic interactions. However, some molecules (termed *nonpolar molecules*) cannot participate in hydrogen or ionic interactions. The interactions of nonpolar molecules with water are not as favorable as are interactions between the water molecules themselves. The water molecules in contact with these nonpolar surfaces form "cages" around the nonpolar molecule, becoming more well ordered (and, hence, lower in entropy) than water molecules free in solution. As two such nonpolar molecules come together, some of the water molecules are released, and so they can interact freely with bulk water (Figure 1.15). Hence, nonpolar molecules have a tendency to aggregate in water because the entropy of the water is increased through the release of water molecules. This phenomenon, termed the *hydrophobic effect,* helps promote many biochemical processes.

How does the hydrophobic effect favor protein folding? Some of the amino acids that make up proteins have nonpolar groups. These nonpolar amino acids have a strong tendency to associate with one another inside the interior of the folded protein. The increased entropy of water resulting from the interaction of these hydrophobic amino acids helps to compensate for the entropy losses inherent in the folding process.

Hydrophobic interactions are not the only means of stabilizing protein structure. Many weak bonds, including hydrogen bonds and van der Waals interactions, are formed in the protein-folding process, and heat is released into the surroundings as a consequence. Although these interactions replace interactions with water that take place in the unfolded protein, the net result is the release of heat to the surroundings and thus a negative (favorable) change in enthalpy for the system.

The folding process can occur when the combination of the entropy associated with the hydrophobic effect and the enthalpy change associated with hydrogen bonds and van der Waals interactions makes the overall free energy negative.

FIGURE 1.15 The hydrophobic effect. The aggregation of nonpolar groups in water leads to an increase in entropy owing to the release of water molecules into bulk water.

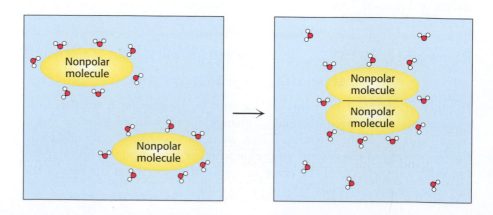

1.4 BIOCHEMISTRY AND HUMAN BIOLOGY

Our understanding of biochemistry has had and will continue to have extensive effects on many aspects of human endeavor. *First, biochemistry is an intrinsically beautiful and fascinating body of knowledge.* We now know the essence and many of the details of the most fundamental processes in biochemistry, such as how a single molecule of DNA replicates to generate two identical copies of itself and how the sequence of bases in a DNA molecule determines the sequence of amino acids in an encoded protein. Our ability to describe these processes in detailed, mechanistic terms places a firm chemical foundation under other biological sciences. Moreover, the realization that we can understand essential life processes, such as the transmission of hereditary information, as chemical structures and their reactions has significant philosophical implications. What does it mean, biochemically, to be human? What are the biochemical differences between a human being, a chimpanzee, a mouse, and a fruit fly? Are we more similar than we are different?

Second, biochemistry is greatly influencing medicine and other fields. The molecular lesions causing sickle-cell anemia, cystic fibrosis, hemophilia, and many other genetic diseases have been elucidated at the biochemical level. Some of the molecular events that contribute to cancer development have been identified. An understanding of the underlying defects opens the door to the discovery of effective therapies. Biochemistry makes possible the rational design of new drugs, including specific inhibitors of enzymes required for the replication of viruses such as human immunodeficiency virus (HIV). Genetically engineered bacteria or other organisms can be used as "factories" to produce valuable proteins such as insulin and stimulators of blood-cell development. Biochemistry is also contributing richly to clinical diagnostics. For example, elevated levels of telltale enzymes in the blood reveal whether a patient has recently had a myocardial infarction (heart attack). DNA probes are coming into play in the precise diagnosis of inherited disorders, infectious diseases, and cancers. Agriculture, too, is benefiting from advances in biochemistry with the development of more effective, environmentally safer herbicides and pesticides and the creation of genetically engineered plants that are, for example, more resistant to insects. All of these endeavors are being accelerated by the advances in genomic sequencing.

Third, advances in biochemistry are enabling researchers to tackle some of the most exciting questions in biology and medicine. How does a fertilized egg give rise to cells as different as those in muscle, brain, and liver? How do the senses work? What are the molecular bases for mental disorders such as Alzheimer disease and schizophrenia? How does the immune system distinguish between self and nonself? What are the molecular mechanisms of short-term and long-term memory? The answers to such questions, which once seemed remote, have been partly uncovered and are likely to be more thoroughly revealed in the near future.

Because all living organisms on Earth are linked by a common origin, evolution provides a powerful organizing theme for biochemistry. This book is organized to emphasize the unifying principles revealed by evolutionary considerations. We begin in the next chapter with a brief tour along a plausible evolutionary path from the formation of some of the chemicals that we now associate with living organisms through the evolution of the processes essential for the development of complex, multicellular organisms. The remainder of Part I of the book more fully introduces the most important classes of biochemicals as well as catalysis and regulation. Part II, Transducing and Storing Energy, describes how energy from chemicals or from sunlight is converted into usable forms and how this conversion is regulated. As we will see, a small set of molecules such as adenosine triphosphate

(ATP) act as energy currencies that allow energy, however captured, to be utilized in a variety of biochemical processes. This part of the text examines the important pathways for the conversion of environmental energy into molecules such as ATP and uncovers many unifying principles. Part III, Synthesizing the Molecules of Life, illustrates the use of the molecules discussed in Part II to synthesize key molecular building blocks, such as the bases of DNA and amino acids, and then shows how these precursors are assembled into DNA, RNA, and proteins. In Parts II and III, we will highlight the relation between the reactions within each pathway and between those in different pathways so as to suggest how these individual reactions may have combined early in evolutionary history to produce the necessary molecules. From the student's perspective, the existence of features common to several pathways enables material mastered in one context to be readily applied to new contexts. Part IV, Responding to Environmental Changes, explores some of the mechanisms that cells and multicellular organisms have evolved to detect and respond to changes in the environment. The topics range from general mechanisms, common to all organisms, for regulating the expression of genes to the sensory systems used by human beings and other complex organisms. In many cases, we can now see how these elaborate systems evolved from pathways that existed earlier in evolutionary history. Many of the sections in Part IV link biochemistry with other fields such as cell biology, immunology, and neuroscience. We are now ready to begin our journey into biochemistry with events that took place more than 3 billion years ago.

APPENDIX: DEPICTING MOLECULAR STRUCTURES

The authors of a biochemistry text face the problem of trying to present three-dimensional molecules in the two dimensions available on the printed page. The interplay between the three-dimensional structures of biomolecules and their biological functions will be discussed extensively throughout this book. Toward this end, we will frequently use representations that, although of necessity are rendered in two dimensions, emphasize the three-dimensional structures of molecules.

Stereochemical Renderings

Most of the chemical formulas in this text are drawn to depict the geometric arrangement of atoms, crucial to chemical bonding and reactivity, as accurately as possible. For example, the carbon atom of methane is sp^3 hybridized and tetrahedral, with

H–C–H angles of 109.5 degrees while the carbon atom in formaldehyde is sp^2 hybridized with bond angles of 120 degrees.

Methane   **Formaldehyde**

To illustrate the correct *stereochemistry* about carbon atoms, wedges will be used to depict the direction of a bond into or out of the plane of the page. A solid wedge with the broad end away from the carbon denotes a bond coming toward the viewer out of the plane. A dashed wedge, with the broad end of the bond at the carbon represents a bond going away from the viewer into the plane of the page. The remaining two bonds are depicted as straight lines.

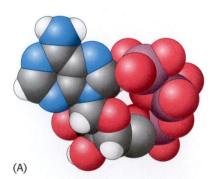

(A)

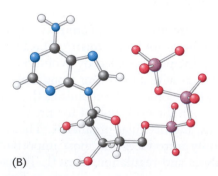

(B)

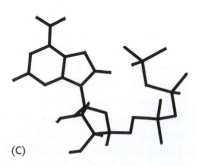

(C)

FIGURE 1.16 Molecular representations. Comparison of (A) space-filling, (B) ball-and-stick, and (C) skeletal models of ATP.

Fischer Projections

Although more representative of the actual structure of a compound, stereochemical structures are often difficult to draw quickly. An alternative method of depicting structures with tetrahedral carbon centers relies on the use of *Fischer projections*.

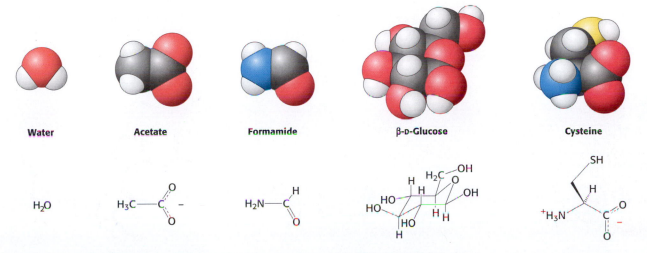

Fischer projection   **Stereochemical rendering**

In a Fischer projection, the bonds to the central carbon are represented by horizontal and vertical lines from the substituent atoms to the carbon atom, which is assumed to be at the center of the cross. By convention, the horizontal bonds are assumed to project out of the page toward the viewer, whereas the vertical bonds are assumed to project into the page away from the viewer. The Glossary of Compounds found at the back of the book is a structural glossary of the key molecules in biochemistry, each presented in two forms: with stereochemically accurate bond angles and as a Fisher projection.

For depicting molecular architecture in more detail, five types of models will be used: space filling, ball and stick, skeletal, ribbon, and surface representations (Figure 1.16). The first three types show structures at the atomic level.

1. *Space-filling models.* The space-filling models are the most realistic. The size and position of an atom in a space-filling model are determined by its bonding properties and van der Waals radius, or contact distance (Section 1.3.1). A van der Waals radius describes how closely two atoms can approach each other when they are not linked by a covalent bond. The colors of the model are set by convention.

Carbon, black	Hydrogen, white	Nitrogen, blue
Oxygen, red	Sulfur, yellow	Phosphorus, purple

Space-filling models of several simple molecules are shown in Figure 1.17.

2. *Ball-and-stick models.* Ball-and-stick models are not as realistic as space-filling models, because the atoms are depicted as spheres of radii smaller than their van der Waals radii. However, the bonding arrangement is easier to see because the bonds are explicitly represented as sticks. In an illustration, the taper of a stick, representing parallax, tells which of a pair of bonded atoms is closer to the reader. A ball-and-stick model reveals a complex structure more clearly than a space-filling model does.

3. *Skeletal models.* An even simpler image is achieved with a skeletal model, which shows only the molecular framework. In skeletal models, atoms are not shown explicitly. Rather, their positions are implied by the junctions and ends of bonds. Skeletal models are frequently used to depict larger, more complex structures.

As biochemistry has advanced, more attention has been focused on the structures of biological macromolecules and their complexes. These structures comprise thousands or even tens of thousands of atoms. Although these structures can be depicted at the atomic level, it is difficult to discern the relevant structural features because of the large number of atoms. Thus, more schematic representations—ribbon diagrams and surface representations—have been developed for the depiction of macromolecular structures in which atoms are not shown explicitly (Figure 1.18).

4. *Ribbon diagrams.* These diagrams are highly schematic and most commonly used to accent a few dramatic aspects of protein structure, such as the α helix (a coiled ribbon), the β strand (a broad arrow), and loops (simple lines), so as to provide simple and clear views of the folding patterns of proteins.

5. *Surface representations.* Often, the interactions between macromolecules take place exclusively at their surfaces. Surface representations have been developed to better visualize macromolecular surfaces. These representations display the overall shapes of macromolecules and can be shaded or colored to indicate particular features such as surface topography or the distribution of electric charges.

Water   **Acetate**   **Formamide**   **β-D-Glucose**   **Cysteine**

FIGURE 1.17 Space-filling models. Structural formulas and space-filling representations of selected molecules are shown.

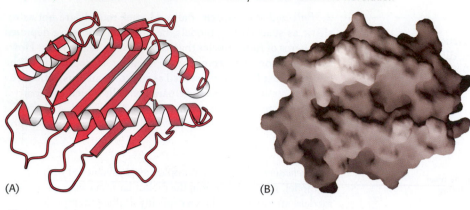

(A)

(B)

FIGURE 1.18 Alternative representations of protein structure. A ribbon diagram (A) and a surface representation (B) of a key protein from the immune system emphasize different aspects of structure.

KEY TERMS

deoxyribonucleic acid (DNA) (p. 4)

double helix (p. 5)

ribonucleic acid (RNA) (p. 6)

protein (p. 6)

amino acid (p. 6)

genetic code (p. 6)

Eukarya (p. 8)

Bacteria (p. 8)

Archaea (p. 8)

eukaryote (p. 8)

prokaryote (p. 8)

covalent bond (p. 8)

resonance structure (p. 9)

electrostatic interaction (p. 9)

hydrogen bond (p. 9)

van der Waals interaction (p. 9)

entropy (p. 12)

enthalpy (p. 13)

free energy (p. 13)

hydrophobic effect (p. 14)

stereochemistry (p. 16)

Fischer projection (p. 17)

space-filling model (p. 17)

ball-and stick-model (p. 17)

skeletal model (p. 17)

ribbon diagram (p. 17)

surface presentation (p. 17)

CHAPTER 2

Biochemical Evolution

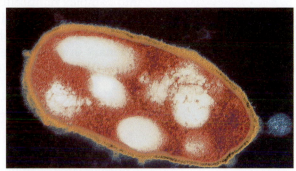

Natural selection, one of the key forces powering evolution, opens an array of improbable ecological niches to species that can adapt biochemically. (Left) Salt pools, where the salt concentration can be greater than 1.5 M, would seem to be highly inhospitable environments for life. Yet certain halophilic archaea, such as *Haloferax mediteranei* (right), possess biochemical adaptations that enable them to thrive under these harsh conditions. [(Left) Kaj R. Svensson/Science Photo Library/Photo Researchers; (right) Wanner/Eye of Science/Photo Researchers.]

Earth is approximately 4.5 billion years old. Remarkably, there is convincing fossil evidence that organisms morphologically (and very probably biochemically) resembling certain modern bacteria were in existence 3.5 billion years ago. With the use of the results of directed studies and accidental discoveries, it is now possible to construct a hypothetical yet plausible evolutionary path from the prebiotic world to the present. A number of uncertainties remain, particularly with regard to the earliest events. Nonetheless, a consideration of the steps along this path and the biochemical problems that had to be solved provides a useful perspective from which to regard the processes found in modern organisms. *These evolutionary connections make many aspects of biochemistry easier to understand.*

We can think of the path leading to modern living species as consisting of stages, although it is important to keep in mind that these stages were almost certainly not as distinct as presented here. The first stage was the initial generation of some of the key molecules of life—nucleic acids, proteins, carbohydrates, and lipids—by nonbiological processes. The second stage was fundamental—the transition from prebiotic chemistry to replicating systems. With the passage of time, these systems became increasingly sophisticated, enabling the formation of living cells. In the third stage, mechanisms evolved for interconverting energy from chemical sources and sunlight into forms that can be utilized to drive biochemical reactions. Intertwined with these energy-conversion processes are pathways for synthesizing the components of nucleic acids, proteins, and other key substances from simpler molecules. With the development of energy-conversion processes and biosynthetic pathways, a wide variety of unicellular organisms evolved. The fourth stage was the evolution of mechanisms that allowed cells to adjust their biochemistry to different, and often changing, environments. Organisms with these capabilities could form colonies comprising groups of interacting cells, and some eventually evolved into complex multicellular organisms.

This chapter introduces key challenges posed in the evolution of life, whose solutions are elaborated in later chapters. Exploring a possible evolutionary origin for these fundamental processes makes their use, in contrast with that of potential alternatives, more understandable.

2.1 KEY ORGANIC MOLECULES ARE USED BY LIVING SYSTEMS

Approximately 1 billion years after Earth's formation, life appeared, as already mentioned. Before life could exist, though, another major process needed to have taken place—the synthesis of the organic molecules required for living systems from simpler molecules found in the environment. The components of nucleic acids and proteins are relatively complex organic molecules, and one might expect that only sophisticated synthetic routes could produce them. However, this requirement appears not to have been the case. How did the building blocks of life come to be?

2.1.1 Many Components of Biochemical Macromolecules Can Be Produced in Simple, Prebiotic Reactions

Among several competing theories about the conditions of the *prebiotic world,* none is completely satisfactory or problem-free. One theory holds that Earth's early atmosphere was highly reduced, rich in methane (CH_4), ammonia (NH_3), water (H_2O), and hydrogen (H_2), and that this atmosphere was subjected to large amounts of solar radiation and lightning. For the sake of argument, we will assume that these conditions were indeed those of prebiotic Earth. Can complex organic molecules be synthesized under these conditions? In the 1950s, Stanley Miller and Harold Urey set out to answer this question. An electric discharge, simulating lightning, was passed through a mixture of methane, ammonia, water, and hydrogen (Figure 2.1). Remarkably, these experiments yielded a highly nonrandom mixture of organic compounds, including amino acids and other substances fundamental to biochemistry. The procedure produces the amino acids glycine and alanine in approximately 2% yield, depending on the amount of carbon supplied as methane. More complex amino acids such as glutamic acid and leucine are produced in smaller amounts (Figure 2.2). Hydrogen cyanide (HCN), another likely component of the early atmosphere, will con-

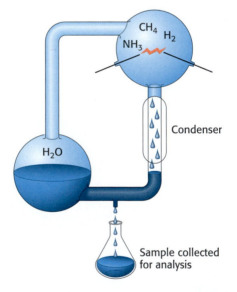

FIGURE 2.1 The Urey-Miller experiment. An electric discharge (simulating lightning) passed through an atmosphere of CH_4, NH_3, H_2O, and H_2 leads to the generation of key organic compounds such as amino acids.

FIGURE 2.2 Products of prebiotic synthesis. Amino acids produced in the Urey-Miller experiment.

Glycine   Alanine   Glutamic acid   Leucine

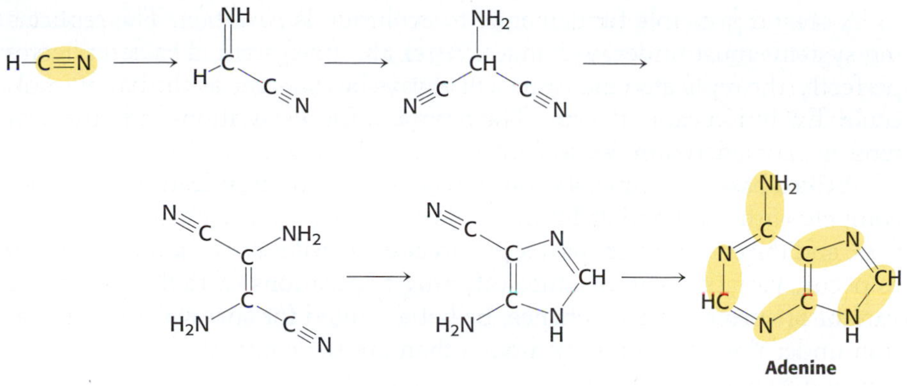

FIGURE 2.3 Prebiotic synthesis of a nucleic acid component. Adenine can be generated by the condensation of HCN.

dense on exposure to heat or light to produce adenine, one of the four nucleic acid bases (Figure 2.3). Other simple molecules combine to form the remaining bases. A wide array of sugars, including ribose, can be formed from formaldehyde under prebiotic conditions.

2.1.2 Uncertainties Obscure the Origins of Some Key Biomolecules

The preceding observations suggest that many of the building blocks found in biology are unusually easy to synthesize and that significant amounts could have accumulated through the action of nonbiological processes. However, it is important to keep in mind that there are many uncertainties. For instance, ribose is just one of many sugars formed under prebiotic conditions. In addition, ribose is rather unstable under possible prebiotic conditions. Futhermore, ribose occurs in two mirror-image forms, only one of which occurs in modern RNA. To circumvent those problems, the first nucleic acid-like molecules have been suggested to have been bases attached to a different backbone and only later in evolutionary time was ribose incorporated to form nucleic acids as we know them today. Despite these uncertainties, an assortment of prebiotic molecules did arise in some fashion, and from this assortment *those with properties favorable for the processes that we now associate with life began to interact and to form more complicated compounds.* The processes through which modern organisms synthesize molecular building blocks will be discussed in Chapters 24, 25, and 26.

2.2 EVOLUTION REQUIRES REPRODUCTION, VARIATION, AND SELECTIVE PRESSURE

Once the necessary building blocks were available, how did a living system arise and evolve? Before the appearance of life, simple molecular systems must have existed that subsequently evolved into the complex chemical systems that are characteristic of organisms. To address how this evolution occurred, we need to consider the *process* of evolution. There are several basic principles common to evolving systems, whether they are simple collections of molecules or competing populations of organisms. First, the most fundamental property of evolving systems is their ability to *replicate* or *reproduce*. Without this ability of *reproduction*, each "species" of molecule that might appear is doomed to extinction as soon as all its individual molecules degrade. For example, individual molecules of biological polymers such as ribonucleic acid are degraded by hydrolysis reactions and other processes. However, *molecules that can replicate will continue to be represented in the population even if the lifetime of each individual molecule remains short.*

A second principle fundamental to evolution is *variation*. The replicating systems must undergo changes. After all, if a system always replicates perfectly, the replicated molecule will always be the same as the parent molecule. Evolution cannot occur. The nature of these variations in living systems are considered in Section 2.2.5.

A third basic principle of evolution is *competition*. Replicating molecules compete with one another for available resources such as chemical precursors, and the competition allows the process of *evolution by natural selection* to occur. Variation will produce differing populations of molecules. Some variant offspring may, by chance, be better suited for survival and replication under the prevailing conditions than are their parent molecules. The prevailing conditions exert a *selective pressure* that gives an advantage to one of the variants. Those molecules that are best able to survive and to replicate themselves will increase in relative concentration. Thus, new molecules arise that are better able to replicate under the conditions of their environment. The same principles hold true for modern organisms. Organisms reproduce, show variation among individual organisms, and compete for resources; those variants with a selective advantage will reproduce more successfully. The changes leading to variation still take place at the molecular level, but the selective advantage is manifest at the organismal level.

2.2.1 The Principles of Evolution Can Be Demonstrated in Vitro

Is there any evidence that evolution can take place at the molecular level? In 1967, Sol Spiegelman showed that replicating molecules could evolve new forms in an experiment that allowed him to observe molecular evolution in the test tube. He used as his evolving molecules RNA molecules derived from a bacterial virus called bacteriophage $Q\beta$. The genome of bacteriophage $Q\beta$, a single RNA strand of approximately 3300 bases, depends for its replication on the activity of a protein complex termed $Q\beta$ replicase. Spiegelman mixed the replicase with a starting population of $Q\beta$ RNA molecules. Under conditions in which there are ample amounts of precursors, no time constraints, and no other selective pressures, the composition of the population does not change from that of the parent molecules on replication. When selective pressures are applied, however, the composition of the population of molecules can change dramatically. For example, decreasing the time available for replication from 20 minutes to 5 minutes yielded, incrementally over 75 generations, a population of molecules dominated by a single species comprising only 550 bases. This species is replicated 15 times as rapidly as the parental $Q\beta$ RNA (Figure 2.4). Spiegelman applied other selective pressures by, for example, limiting the concentrations of precursors or adding compounds that inhibit the replication process. In each case, new species appeared that replicated more effectively under the conditions imposed.

The process of evolution demonstrated in these studies depended on the existence of machinery for the replication of RNA fragments in the form of the $Q\beta$ replicase. As noted in Chapter 1, one of the most elegant characteristics of nucleic acids is that the mechanism for their replication follows naturally from their molecular structure. This observation suggests that nucleic acids, perhaps RNA, could have become *self-replicating*. Indeed, the results of studies have revealed that single-stranded nucleic acids can serve as templates for the synthesis of their complementary strands and that this synthesis can occur spontaneously—that is, without biologically de-

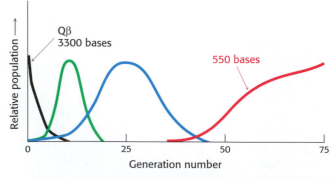

FIGURE 2.4 Evolution in a test tube. Rapidly replicating species of RNA molecules were generated from $Q\beta$ RNA by exerting selective pressure. The green and blue curves correspond to species of intermediate size that accumulated and then became extinct in the course of the experiment.

rived replication machinery. However, investigators have not yet found conditions in which an RNA molecule is fully capable of independent self-replication from simple starting materials.

2.2.2 RNA Molecules Can Act As Catalysts

The development of capabilities beyond simple replication required the generation of specific catalysts. A *catalyst* is a molecule that accelerates a particular chemical reaction without itself being chemically altered in the process. The properties of catalysts will be discussed in detail in Chapters 8 and 9. Some catalysts are highly specific; they promote certain reactions without substantially affecting closely related processes. Such catalysts allow the reactions of specific pathways to take place in preference to those of potential alternative pathways. Until the 1980s, all biological catalysts, termed *enzymes*, were believed to be proteins. Then, Tom Cech and Sidney Altman independently discovered that certain RNA molecules can be effective catalysts. These RNA catalysts have come to be known as *ribozymes*. The discovery of ribozymes suggested the possibility that catalytic RNA molecules could have played fundamental roles early in the evolution of life.

The catalytic ability of RNA molecules is related to their ability to adopt specific yet complex structures. This principle is illustrated by a "hammerhead" ribozyme, an RNA structure first identified in plant viruses (Figure 2.5). This RNA molecule promotes the cleavage of specific RNA molecules at specific sites; this cleavage is necessary for certain aspects of the viral life cycle. The ribozyme, which requires mg^{2+} ion or other ions for the cleavage step to take place, forms a complex with its substrate RNA molecule that can adopt a reactive conformation.

The existence of RNA molecules that possess specific binding and catalytic properties makes plausible the idea of an early *"RNA world"* inhabited by life forms dependent on RNA molecules to play all major roles, including those important in heredity, the storage of information, and the promotion of specific reactions—that is, biosynthesis and energy metabolism.

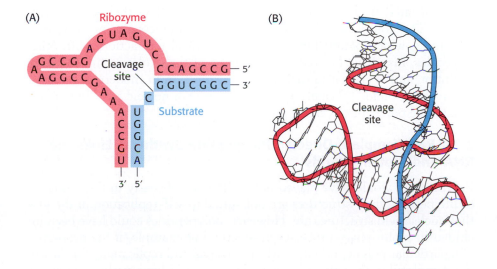

(A)

(B)

FIGURE 2.5 Catalytic RNA.
(A) The base-pairing pattern of a "hammerhead" ribozyme and its substrate. (B) The folded conformation of the complex. The ribozyme cleaves the bond at the cleavage site. The paths of the nucleic acid backbones are highlighted in red and blue.

2.2.3 Amino Acids and Their Polymers Can Play Biosynthetic and Catalytic Roles

In the early RNA world, the increasing populations of replicating RNA molecules would have consumed the building blocks of RNA that had been generated over long periods of time by prebiotic reactions. A shortage of these compounds would have favored the evolution of alternative mechanisms

FIGURE 2.6 Biosynthesis of RNA bases.
Amino acids are building blocks for the
biosynthesis of purines and pyrimidines.

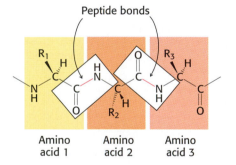

Peptide bonds

**FIGURE 2.7 An alternative functional
polymer.** Proteins are built of amino acids
linked by peptide bonds.

for their synthesis. A large number of pathways are possible. Examining the
biosynthetic routes utilized by modern organisms can be a source of insight
into which pathways survived. A striking observation is that simple amino
acids are used as building blocks for the RNA bases (Figure 2.6). For both
purines (adenine and guanine) and pyrimidines (uracil and cytosine), an
amino acid serves as a core onto which the remainder of the base is elaborated. In addition, nitrogen atoms are donated by the amino group of the
amino acid aspartic acid and by the amide group of the glutamine side chain.

Amino acids are chemically more versatile than nucleic acids because
their side chains carry a wider range of chemical functionality. Thus, amino
acids or short polymers of amino acids linked by *peptide bonds*, called
polypeptides (Figure 2.7), may have functioned as components of ribozymes
to provide a specific reactivity. Furthermore, longer polypeptides are capable of spontaneously folding to form well-defined three-dimensional structures, dictated by the sequence of amino acids along their polypeptide
chains. The ability of polypeptides to fold spontaneously into elaborate
structures, which permit highly specific chemical interactions with other
molecules, may have favored the expansion of their roles in the course of
evolution and is crucial to their dominant position in modern organisms.
Today, most biological catalysts (enzymes) are not nucleic acids but are instead large polypeptides called *proteins*.

2.2.4 RNA Template-Directed Polypeptide Synthesis Links the RNA and Protein Worlds

Polypeptides would have played only a limited role early in the evolution
of life because their structures are not suited to self-replication in the way
that nucleic acid structures are. However, polypeptides could have been included in evolutionary processes indirectly. For example, if the properties
of a particular polypeptide favored the survival and replication of a class of
RNA molecules, then these RNA molecules could have evolved ribozyme
activities that promoted the synthesis of that polypeptide. This method of
producing polypeptides with specific amino acid sequences has several limitations. First, it seems likely that only relatively short specific polypeptides
could have been produced in this manner. Second, it would have been difficult to accurately link the particular amino acids in the polypeptide in a
reproducible manner. Finally, a different ribozyme would have been required for each polypeptide. A critical point in evolution was reached when

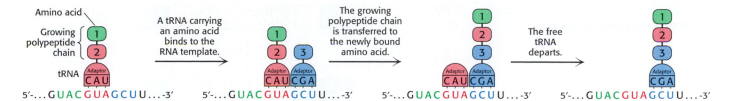

FIGURE 2.8 Linking the RNA and protein worlds. Polypeptide synthesis is directed by an RNA template. Adaptor RNA molecules, with amino acids attached, sequentially bind to the template RNA to facilitate the formation of a peptide bond between two amino acids. The growing polypeptide chain remains attached to an adaptor RNA until the completion of synthesis.

an apparatus for polypeptide synthesis developed that allowed *the sequence of bases in an RNA molecule to directly dictate the sequence of amino acids in a polypeptide.* A code evolved that established a relation between a specific sequence of three bases in RNA and an amino acid. We now call this set of three-base combinations, each encoding an amino acid, the *genetic code.* A decoding, or *translation,* system exists today as the *ribosome* and associated factors that are responsible for essentially all polypeptide synthesis from RNA templates in modern organisms. The essence of this mode of polypeptide synthesis is illustrated in Figure 2.8.

An RNA molecule (*messenger RNA,* or *mRNA*), containing in its base sequence the information that specifies a particular protein, acts as a template to direct the synthesis of the polypeptide. Each amino acid is brought to the template attached to an adapter molecule specific to that amino acid. These adapters are specialized RNA molecules (called *transfer RNAs* or *tRNAs*). After initiation of the polypeptide chain, a tRNA molecule with its associated amino acid binds to the template through specific Watson-Crick base-pairing interactions. Two such molecules bind to the ribosome and peptide-bond formation is catalyzed by an RNA component (called *ribosomal RNA* or *rRNA*) of the ribosome. The first RNA departs (with neither the polypeptide chain nor an amino acid attached) and another tRNA with its associated amino acid bonds to the ribosome. The growing polypeptide chain is transferred to this newly bound amino acid with the formation of a new peptide bond. This cycle then repeats itself. This scheme allows the sequence of the RNA template to encode the sequence of the polypeptide and thereby makes possible the production of long polypeptides with specified sequences. The mechanism of protein synthesis will be discussed in Chapter 29. Importantly, the ribosome is composed largely of RNA and is a highly sophisticated ribozyme, suggesting that it might be a surviving relic of the RNA world.

2.2.5 The Genetic Code Elucidates the Mechanisms of Evolution

The sequence of bases that encodes a functional protein molecule is called a *gene.* The genetic code—that is, the relation between the base sequence of a gene and the amino acid sequence of the polypeptide whose synthesis the gene directs—applies to all modern organisms with only very minor exceptions. This universality reveals that the genetic code was fixed early in the course of evolution and has been maintained to the present day.

We can now examine the mechanisms of evolution. Earlier, we considered how variation is required for evolution. We can now see that such variations in living systems are changes that alter the meaning of the genetic message. These variations are called *mutations.* A mutation can be as simple as a change in a single nucleotide (called a *point mutation*), such that a

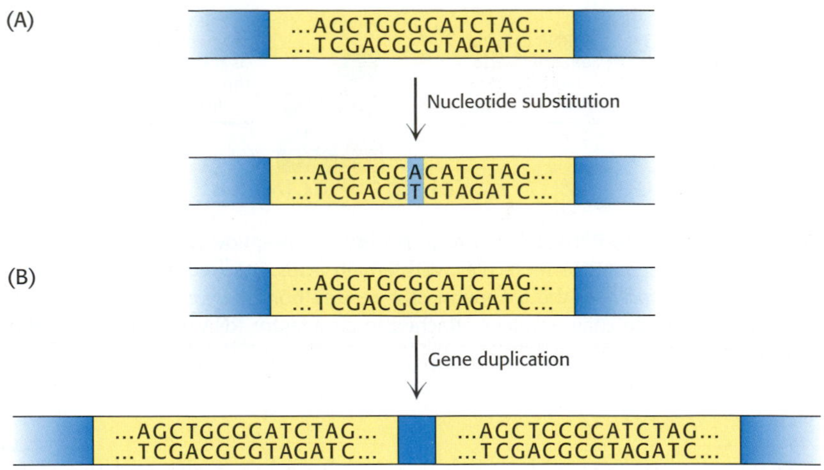

(A)

...AGCTGCGCATCTAG...
...TCGACGCGTAGATC...

↓ Nucleotide substitution

...AGCTGCACATCTAG...
...TCGACGTGTAGATC...

(B)

...AGCTGCGCATCTAG...
...TCGACGCGTAGATC...

↓ Gene duplication

...AGCTGCGCATCTAG...   ...AGCTGCGCATCTAG...
...TCGACGCGTAGATC...   ...TCGACGCGTAGATC...

FIGURE 2.9 Mechanisms of evolution. A change in a gene can be (A) as simple as a single base change or (B) as dramatic as partial or complete gene duplication.

sequence of bases that encoded a particular amino acid may now encode another (Figure 2.9A). A mutation can also be the insertion or deletion of several nucleotides.

Other types of alteration permit the more rapid evolution of new biochemical activities. For instance, entire sections of the coding material can be duplicated, a process called *gene duplication* (Figure 2.9B). One of the duplication products may accumulate mutations and eventually evolve into a gene with a different, but related, function. Furthermore, parts of a gene may be duplicated and added to parts of another to give rise to a completely new gene, which encodes a protein with properties associated with each parent gene. Higher organisms contain many large families of enzymes and other macromolecules that are clearly related to one another in the same manner. Thus, gene duplication followed by specialization has been a crucial process in evolution. It allows the generation of macromolecules having particular functions without the need to start from scratch. The accumulation of genes with subtle and large differences allows for the generation of more complex biochemical processes and pathways and thus more complex organisms.

2.2.6 Transfer RNAs Illustrate Evolution by Gene Duplication

Transfer RNA molecules are the adaptors that associate an amino acid with its correct base sequence. Transfer RNA molecules are structurally similar to one another: each adopts a three-dimensional cloverleaf pattern of base-paired groups (Figure 2.10). Subtle differences in structure enable the protein-synthesis machinery to distinguish transfer RNA molecules with different amino acid specificities.

This family of related RNA molecules likely was generated by gene duplication followed by specialization. A nucleic acid sequence encoding one member of the family was duplicated, and the two copies evolved independently to generate molecules with specificities for different amino acids. This process was repeated, starting from one primordial transfer RNA gene until the 20 (or more) distinct members of the transfer RNA family present in modern organisms arose.

2.2.7 DNA Is a Stable Storage Form for Genetic Information

It is plausible that RNA was utilized to store genetic information early in the history of life. However, in modern organisms (with the exception of some viruses), the RNA derivative DNA (*deoxy*ribonucleic acid) performs this function (Sections 1.1.1 and 1.1.3). The 2′-hydroxyl group in the

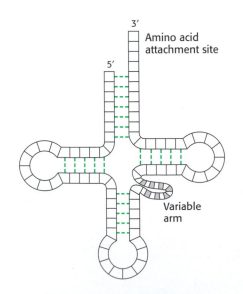

FIGURE 2.10 Cloverleaf pattern of tRNA. The pattern of base-pairing interactions observed for all transfer RNA molecules reveals that these molecules had a common evolutionary origin.

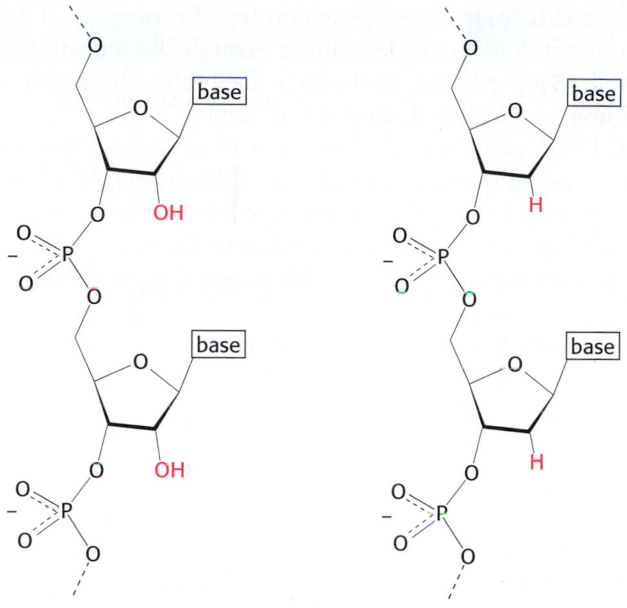

FIGURE 2.11 RNA and DNA compared. Removal of the 2′-hydroxyl group from RNA to form DNA results in a backbone that is less susceptible to cleavage by hydrolysis and thus enables more-stable storage of genetic information.

ribose unit of the RNA backbone is replaced by a hydrogen atom in DNA (Figure 2.11).

What is the selective advantage of DNA over RNA as the genetic material? The genetic material must be extremely stable so that sequence information can be passed on from generation to generation without degradation. RNA itself is a remarkably stable molecule; negative charges in the sugar-phosphate backbone protect it from attack by hydroxide ions that would lead to hydrolytic cleavage. However, the 2′-hydroxyl group makes the RNA susceptible to base-catalyzed hydrolysis. The removal of the 2′-hydroxyl group from the ribose decreases the rate of hydrolysis by approximately 100-fold under neutral conditions and perhaps even more under extreme conditions. Thus, the conversion of the genetic material from RNA into DNA would have substantially increased its chemical stability.

The evolutionary transition from RNA to DNA is recapitulated in the biosynthesis of DNA in modern organisms. In all cases, the building blocks used in the synthesis of DNA are synthesized from the corresponding building blocks of RNA by the action of enzymes termed *ribonucleotide reductases*. These enzymes convert ribonucleotides (a base and phosphate groups linked to a *ribose* sugar) into deoxyribonucleotides (a base and phosphates linked to *deoxyribose* sugar).

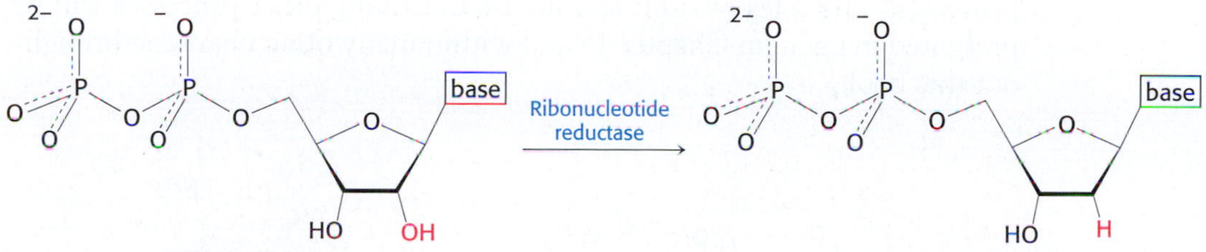

The properties of the ribonucleotide reductases vary substantially from species to species, but evidence suggests that they have a common mechanism of action and appear to have evolved from a common primordial enzyme.

The covalent structures of RNA and DNA differ in one other way. Whereas RNA contains *uracil*, DNA contains a methylated uracil derivative termed *thymine*. This modification also serves to protect the integrity of the genetic sequence, although it does so in a less direct manner. As we will see in Chapter 27, the methyl group present in thymine facilitates the repair of damaged DNA, providing an additional selective advantage.

Although DNA replaced RNA in the role of storing the genetic information, RNA maintained many of its other functions. RNA still provides the template that directs polypeptide synthesis, the adaptor molecules, the catalytic activity of the ribosomes, and other functions. Thus, the genetic message is *transcribed* from DNA into RNA and then *translated* into protein.

$$\text{DNA} \xrightarrow{\text{Transcription}} \text{RNA} \xrightarrow{\text{Translation}} \text{polypeptide} \xrightarrow{\text{Folding}} \begin{array}{c}\text{functional} \\ \text{protein}\end{array}$$

| Linear nucleic acid | Linear nucleic acid | Linear amino acid sequence | Three-dimensional structure |

This flow of sequence information from DNA to RNA to protein (to be considered in detail in Chapters 5, 28, and 29) applies to all modern organisms (with minor exceptions for certain viruses).

2.3 ENERGY TRANSFORMATIONS ARE NECESSARY TO SUSTAIN LIVING SYSTEMS

Most of the reactions that lead to the biosynthesis of nucleic acids and other biomolecules are not thermodynamically favorable under most conditions; they require an input of energy to proceed. Thus, they can proceed only if they are coupled to processes that release energy. How can energy-requiring and energy-releasing reactions be linked? How is energy from the environment transformed into a form that living systems can use? Answering these questions fundamental to biochemistry is the objective of much of this book.

2.3.1 ATP, a Common Currency for Biochemical Energy, Can Be Generated Through the Breakdown of Organic Molecules

Just as most economies simplify trade by using currency rather than bartering, biochemical systems have evolved common currencies for the exchange of energy. The most important of these currencies are molecules related to *adenosine triphosphate (ATP)* that contain an array of three linked phosphates (Figure 2.12). The bonds linking the phosphates persist in solution under a variety of conditions, but, when they are broken, an unusually large amount of energy is released that can be used to promote other processes. The roles of ATP and its use in driving other processes will be presented in detail in Chapter 14 and within many other chapters throughout this book.

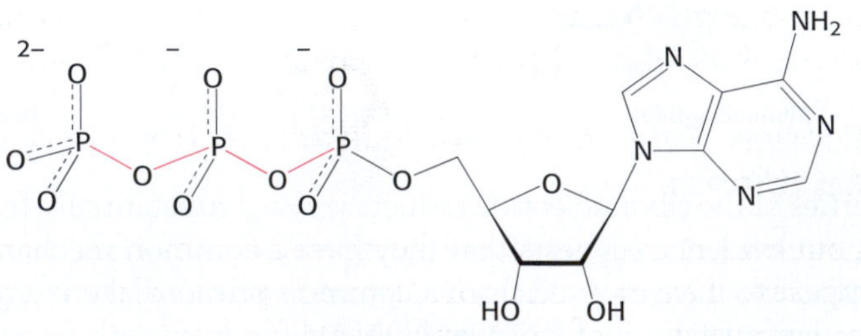

FIGURE 2.12 ATP, the energy currency of living systems. The phosphodiester bonds (red) release considerable energy when cleaved by hydrolysis or other processes.

Adenosine triphosphate (ATP)

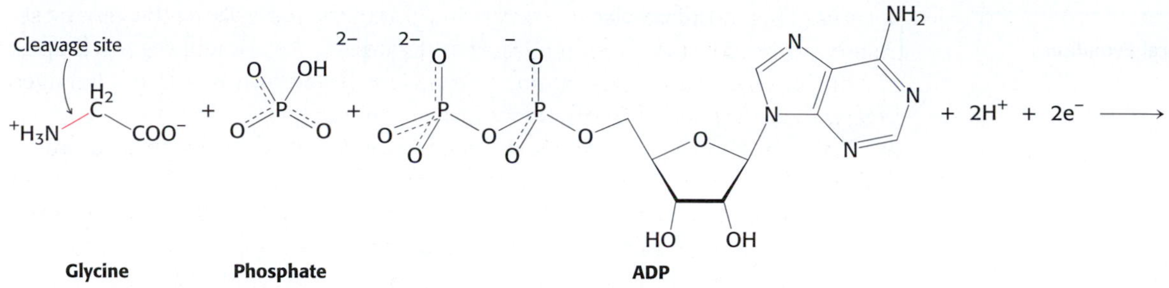

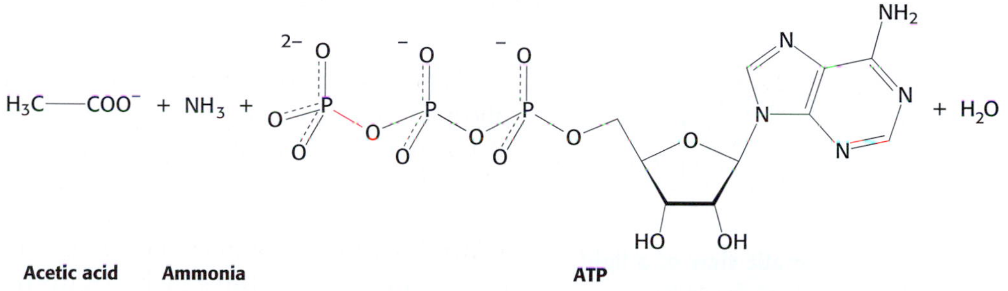

FIGURE 2.13 A possible early method for generating ATP. The synthesis of ATP might have been driven by the degradation of glycine.

ATP must be generated in appropriate quantities to be available for such reactions. The energy necessary for the synthesis of ATP can be obtained by the breakdown of other chemicals. Specific enzymes have evolved to couple these degradative processes to the phosphorylation of adenosine diphosphate (ADP) to yield ATP. Amino acids such as glycine, which were probably present in relatively large quantities in the prebiotic world and early in evolution, were likely sources of energy for ATP generation. The degradation of glycine to acetic acid may be an ATP-generation system that functioned early in evolution (Figure 2.13). In this reaction, the carbon–nitrogen bond in glycine is cleaved by reduction (the addition of electrons), and the energy released from the cleavage of this bond drives the coupling of ADP and orthophosphate (P_i) to produce ATP.

Amino acids are still broken down to produce ATP in modern organisms. However, sugars such as glucose are a more commonly utilized energy source because they are more readily metabolized and can be stored. The most important process for the direct synthesis of ATP in modern organisms is *glycolysis*, a complex process that derives energy from glucose.

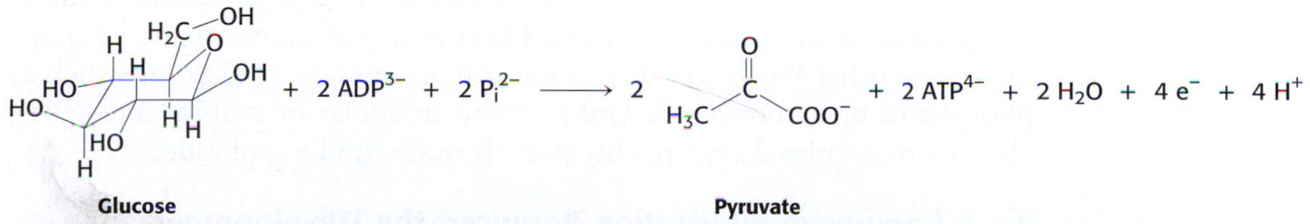

Glycolysis presumably evolved as a process for ATP generation after carbohydrates such as glucose were being produced in significant quantities by other pathways. Glycolysis will be discussed in detail in Chapter 16.

2.3.2 Cells Were Formed by the Inclusion of Nucleic Acids Within Membranes

Modern organisms are made up of *cells*. A cell is composed of nucleic acids, proteins, and other biochemicals surrounded by a *membrane* built from lipids. These membranes completely enclose their contents, and so cells have a defined inside and outside. A typical membrane-forming lipid is phosphatidyl choline.

Hydrophobic   Hydrophilic

Phosphatidyl choline

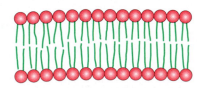

FIGURE 2.14 Schematic view of a lipid bilayer. These structures define the boundaries of cells.

The most important feature of membrane-forming molecules such as phosphatidyl choline is that they are *amphipathic*—that is, they contain both *hydrophilic* (water-loving) and *hydrophobic* (water-avoiding) components. Membrane-forming molecules consist of fatty acids, whose long alkyl groups are hydrophobic, connected to shorter hydrophilic "head groups." When such lipids are in contact with water, they spontaneously aggregate to form specific structures such that the hydrophobic parts of the molecules are packed together away from water, whereas the hydrophilic parts are exposed to the aqueous solution. The structure that is important for membrane formation is the *lipid bilayer* (Figure 2.14). A bilayer is formed from two layers of lipids arranged such that the fatty acid tails of each layer interact with each other to form a hydrophobic interior while the hydrophilic head groups interact with the aqueous solution on each side. Such bilayer structures can fold onto themselves to form hollow spheres having interior compartments filled with water. The hydrophobic interior of the bilayer serves as a barrier between two aqueous phases. If such structures are formed in the presence of other molecules such as nucleic acids and proteins, these molecules can become trapped inside, thus forming cell-like structures. The structures of lipids and lipid bilayers will be considered in detail in Chapter 12.

At some stage in evolution, sufficient quantities of appropriate amphipathic molecules must have accumulated from biosynthetic or other processes to allow some nucleic acids to become entrapped and cell-like organisms to form. Such compartmentalization has many advantages. When the components of a cell are enclosed in a membrane, the products of enzymatic reactions do not simply diffuse away into the environment but instead are contained where they can be used by the cell that produced them. The containment is aided by the fact that nearly all biosynthetic intermediates and other biochemicals include one or more charged groups such as phosphates or carboxylates. Unlike more nonpolar or neutral molecules, charged molecules do not readily pass through lipid membranes.

2.3.3 Compartmentalization Required the Development of Ion Pumps

Despite its many advantages, the enclosure of nucleic acids and proteins within membranes introduced several complications. Perhaps the most significant were the effects of *osmosis*. Membranes are somewhat permeable to water and small nonpolar molecules, whereas they are impermeable to macromolecules such as nucleic acids. When macromolecules are concentrated inside a compartment surrounded by such a semipermeable membrane, osmotic forces drive water through the membrane into the compartment. Without counterbalancing effects, the flow of water will burst the cell (Figure 2.15).

Osmosis—

The movement of a solvent across a membrane in the direction that tends to equalize concentrations of solute on the two sides of the membrane.

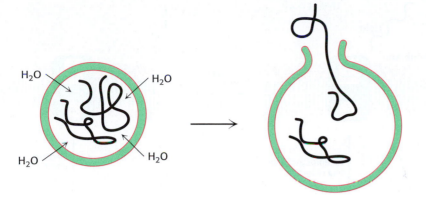

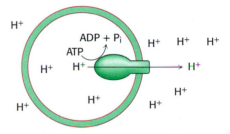

FIGURE 2.15 The "osmotic crisis."
A cell consisting of macromolecules surrounded by a semipermeable membrane will take up water from outside the cell and burst.

Modern cells have two distinct mechanisms for resisting these osmotic forces. One mechanism is to toughen the cell membrane by the introduction of an additional structure such as a cell wall. However, such a chemically elaborate structure may not have evolved quickly, especially because it must completely surround a cell to be effective. The other mechanism is the use of *energy-dependent ion pumps*. These pumps can lower the concentration of ions inside a cell relative to the outside, favoring the flow of water molecules from inside to outside. The resulting unequal distribution of ions across an inherently impermeable membrane is called an *ion gradient*. Appropriate ion gradients can balance the osmotic forces and maintain a cell at a constant volume. Membrane proteins such as ion pumps will be considered in Chapter 13.

Ion gradients can prevent osmotic crises, but they require energy to be produced. Most likely, an ATP-driven proton pump was the first existing component of the machinery for generating an ion gradient (Figure 2.16). Such pumps, which are found in essentially all modern cells, hydrolyze ATP to ADP and inorganic phosphate and utilize the energy released to transport protons from the inside to the outside of a cell. The pump thus establishes a proton gradient that, in turn, can be coupled to other membrane-transport processes such as the removal of sodium ions from the cell. The proton gradient and other ion gradients generated from it act together to counteract osmotic effects and prevent the cell from swelling and bursting.

FIGURE 2.16 Generating an ion gradient. ATP hydrolysis can be used to drive the pumping of protons (or other ions) across a membrane.

2.3.4 Proton Gradients Can Be Used to Drive the Synthesis of ATP

Enzymes act to accelerate reactions, but they cannot alter the position of chemical equilibria. An enzyme that accelerates a reaction in the forward direction must also accelerate the reaction to the same extent in the reverse direction. Thus, the existence of an enzyme that utilized the hydrolysis of ATP to generate a proton gradient presented a tremendous opportunity for the evolution of alternative systems for generating ATP. Such an enzyme could synthesize ATP by reversing the process that produces the gradient. Enzymes, now called *ATP synthases,* do in fact use proton gradients to drive the bonding of ADP and P_i to form ATP (Figure 2.17). These proteins will be considered in detail in Chapter 18.

Organisms have evolved a number of elaborate mechanisms for the generation of proton gradients across membranes. An example is *photosynthesis,* a process first used by bacteria and now also used by plants to harness the light energy from the sun. The essence of photosynthesis is the light-driven transfer of an electron across a membrane. The fundamental processes are illustrated in Figure 2.18.

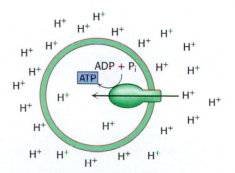

FIGURE 2.17 Use of proton gradients to synthesize ATP. ATP can be synthesized by the action of an ATP-driven proton pump running in reverse.

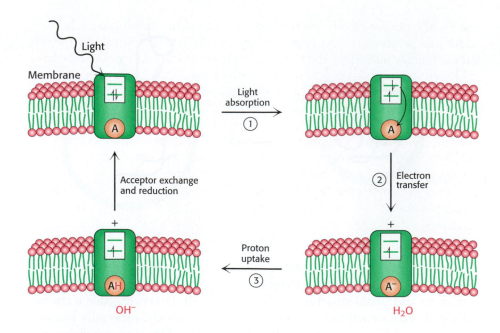

FIGURE 2.18 Photosynthesis.
Absorption of light (1) leads to electron transfer across a membrane (2). For each electron transfer, one excess hydroxide ion is generated inside the cell (3). The process produces a proton gradient across the membrane that can drive ATP synthesis.

The photosynthetic apparatus, which is embedded in a membrane, contains pigments that efficiently absorb light from the sun. The absorbed light provides the energy to promote an electron in the pigment molecule to an excited state. The high-energy electron can then jump to an appropriate acceptor molecule located in the part of the membrane facing the inside of the cell. The acceptor molecule, now reduced, binds a proton from a water molecule, generating an hydroxide ion inside the cell. The electronic "hole" left in the pigment on the outside of the membrane can then be filled by the donation of an electron from a suitable reductant on the outside of the membrane. Because the generation of an hydroxide ion inside the cell is equivalent to the generation of a proton outside the cell, a proton gradient develops across the membrane. Protons flow down this gradient through ATP synthases to generate ATP.

Photosynthesis is but one of a range of processes in different organisms that lead to ATP synthesis through the action of proteins evolutionarily related to the primordial ATP-driven pumps. In animals, the degradation of carbohydrates and other organic compounds is the source of the electron flow across membranes that can be used to develop proton gradients. The formation of ATP-generating proton gradients by fuel metabolism will be considered in Chapter 18 and by light absorption in Chapter 19.

2.3.5 Molecular Oxygen, a Toxic By-Product of Some Photosynthetic Processes, Can Be Utilized for Metabolic Purposes

As stated earlier, photosynthesis generates electronic "holes" in the photosynthetic apparatus on the outside of the membrane. These holes are powerful oxidizing agents; that is, they have very high affinities for electrons and can pull electrons from many types of molecules. They can even oxidize water. Thus, for many photosynthetic organisms, the electron donor that completes the photosynthetic cycle is water. The product of water oxidation is oxygen gas—that is, molecular oxygen (O_2).

$$2 \ H_2O \longrightarrow O_2 + 4 \ e^- + 4 \ H^+$$

The use of water as the electron donor significantly increases the efficiency of photosynthetic ATP synthesis because the generation of one molecule of oxygen is accompanied not only by the release of four electrons (e^-), but also by the release of four protons on one side of the membrane. Thus, an additional proton is released for each proton equivalent produced by the initial electron-transfer process, so twice as many protons are available to drive ATP synthesis. Oxygen generation will be considered in Chapter 19.

Oxygen was present in only small amounts in the atmosphere before organisms evolved that could oxidize water. The "pollution" of the air with oxygen produced by photosynthetic organisms greatly affected the course of evolution. Oxygen is quite reactive and thus extremely toxic to many organisms. Many biochemical processes have evolved to protect cells from the deleterious effects of oxygen and other reactive species that can be generated from this molecule. Subsequently, organisms evolved mechanisms for taking advantage of the high reactivity of oxygen to promote favorable processes. Most important among these mechanisms are those for the oxidation of organic compounds such as glucose. Through the action of oxygen, a glucose molecule can be completely converted into carbon dioxide and water, releasing enough energy to synthesize approximately 30 molecules of ATP.

$$\text{Glucose} + 6\,O_2 \longrightarrow 6\,CO_2 + 6\,H_2O + \text{energy}$$

This number represents a 15-fold increase in ATP yield compared with the yield from the breakdown of glucose in the absence of oxygen in the process of glycolysis. This increased efficiency is apparent in everyday life; our muscles exhaust their fuel supply and tire quickly if they do not receive enough oxygen and are forced to use glycolysis as the sole ATP source. The role of oxygen in the extraction of energy from organic molecules will be considered in Chapter 18.

Glucose

2.4 CELLS CAN RESPOND TO CHANGES IN THEIR ENVIRONMENTS

The environments in which cells grow often change rapidly. For example, cells may consume all of a particular food source and must utilize others. To survive in a changing world, cells evolved mechanisms for adjusting their biochemistry in response to signals indicating environmental change. The adjustments can take many forms, including changes in the activities of preexisting enzyme molecules, changes in the rates of synthesis of new enzyme molecules, and changes in membrane-transport processes.

Initially, the detection of environmental signals occurred inside cells. Chemicals that could pass into cells, either by diffusion through the cell membrane or by the action of transport proteins, and could bind directly to proteins inside the cell and modulate their activities. An example is the use of the sugar arabinose by the bacterium *Escherichia coli* (Figure 2.19). *E. coli* cells are normally unable to use arabinose efficiently as a source of

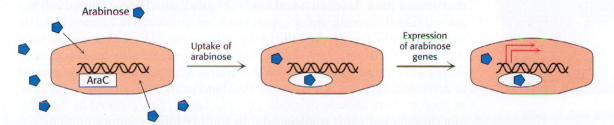

FIGURE 2.19 Responding to environmental conditions. In *E. coli* cells, the uptake of arabinose from the environment triggers the production of enzymes necessary for its utilization.

energy. However, if arabinose is their only source of carbon, *E. coli* cells synthesize enzymes that catalyze the conversion of this sugar into useful forms. This response is mediated by arabinose itself. If present in sufficient quantity outside the cell, arabinose can enter the cell through transport proteins. Once inside the cell, arabinose binds to a protein called AraC. This binding alters the structure of AraC so that it can now bind to specific sites in the bacterial DNA and increase RNA transcription from genes encoding enzymes that metabolize arabinose. The mechanisms of gene regulation will be considered in Chapter 31.

Subsequently, mechanisms appeared for detecting signals at the cell surface. Cells could thus respond to signaling molecules even if those molecules did not pass into the cell. Receptor proteins evolved that, embedded in the membrane, could bind chemicals present in the cellular environment. Binding produced changes in the protein structure that could be detected at the inside surface of the cell membrane. By this means, chemicals outside the cell could influence events inside the cell. Many of these *signal-transduction pathways* make use of substances such as cyclic adenosine monophosphate (cAMP) and calcium ion as "second messengers" that can diffuse throughout the cell, spreading the word of environmental change.

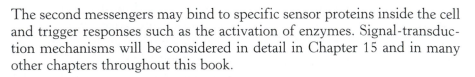

Cyclic AMP (cAMP)   **Calcium ion in water**

The second messengers may bind to specific sensor proteins inside the cell and trigger responses such as the activation of enzymes. Signal-transduction mechanisms will be considered in detail in Chapter 15 and in many other chapters throughout this book.

2.4.1 Filamentous Structures and Molecular Motors Enable Intracellular and Cellular Movement

The development of the ability to move was another important stage in the evolution of cells capable of adapting to a changing environment. Without this ability, nonphotosynthetic cells might have starved after consuming the nutrients available in their immediate vicinity.

Bacteria swim through the use of filamentous structures termed *flagella* that extend from their cell membranes (Figure 2.20). Each bacterial cell has several flagella, which, under appropriate conditions, form rotating bundles that efficiently propel the cell through the water. These flagella are long polymers consisting primarily of thousands of identical protein subunits. At the base of each flagellum are assemblies of proteins that act as motors to drive its rotation. The rotation of the flagellar motor is driven by the flow of protons from outside to inside the cell. Thus, energy stored in the form of a proton gradient is transduced into another form, rotatory motion.

Other mechanisms for motion, also depending on filamentous structures, evolved in other cells. The most important of these structures are *microfilaments* and *microtubules*. Microfilaments are polymers of the protein

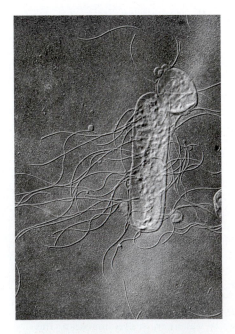

FIGURE 2.20 Bacteria with flagella. A bacterium (*Proteus mirabilis*) swims through the rotation of filamentous structures called flagella. [Fred E. Hossler/ Visuals Unlimited.]

actin, and microtubules are polymers of two closely related proteins termed α- and β-*tubulin.* Unlike a bacterial flagellum, these filamentous structures are highly dynamic: they can rapidly increase or decrease in length through the addition or subtraction of component protein molecules. Microfilaments and microtubules also serve as tracks on which other proteins move, driven by the hydrolysis of ATP. Cells can change shape through the motion of *molecular motor proteins* along such filamentous structures that are changing in shape as a result of dynamic polymerization (Figure 2.21). Coordinated shape changes can be a means of moving a cell across a surface and are crucial to cell division. The motor proteins are also responsible for the transport of organelles and other structures within eukaryotic cells. Molecular motors will be considered in Chapter 34.

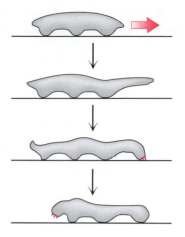

FIGURE 2.21 Alternative movement. Cell mobility can be achieved by changes in cell shape.

2.4.2 Some Cells Can Interact to Form Colonies with Specialized Functions

Early organisms lived exclusively as single cells. Such organisms interacted with one another only indirectly by competing for resources in their environments. Certain of these organisms, however, developed the ability to form colonies comprising many interacting cells. In such groups, the environment of a cell is dominated by the presence of surrounding cells, which may be in direct contact with one another. These cells communicate with one another by a variety of signaling mechanisms and may respond to signals by altering enzyme activity or levels of gene expression. One result may be *cell differentiation;* differentiated cells are genetically identical but have different properties because their genes are expressed differently.

Several modern organisms are able to switch back and forth from existence as independent single cells to existence as multicellular colonies of differentiated cells. One of the most well characterized is the slime mold *Dictyostelium.* In favorable environments, this organism lives as individual cells; under conditions of starvation, however, the cells come together to form a cell aggregate. This aggregate, sometimes called a *slug,* can move as a unit to a potentially more favorable environment where it then forms a multicellular structure, termed a *fruiting body,* that rises substantially above the surface on which the cells are growing. Wind may carry cells released from the top of the fruiting body to sites where the food supply is more plentiful. On arriving in a well-stocked location, the cells grow, reproduce, and live as individual cells until the food supply is again exhausted (Figure 2.22).

The transition from unicellular to multicellular growth is triggered by cell–cell communication and reveals much about signaling processes between and within cells. Under starvation conditions, *Dictyostelium* cells release the signal molecule cyclic AMP. This molecule signals surrounding cells by binding to a membrane-bound protein receptor on the cell surface.

FIGURE 2.22 Unicellular to multicellular transition in *Dictyostelium.* This scanning electron migrograph shows the transformation undergone by the slime mold *Dictyostelium.* Hundreds of thousands of single cells aggregate to form a migrating slug, seen in the lower left. Once the slug comes to a stop, it gradually elongates to form the fruiting body. [Courtesy of M. J. Grimsom and R. L. Blanton, Texas Tech University.]

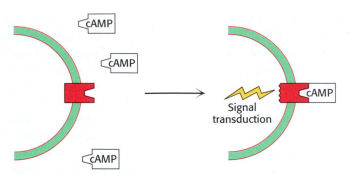

FIGURE 2.23 Intracellular signaling. Cyclic AMP, detected by cell-surface receptors, initiates the formation of aggregates in *Dictyostelium*.

FIGURE 2.24 Cell differentiation in ***Dictyostelium.*** The colors represent the distribution of cell types expressing similar sets of genes in the *Dictyostelium* fruiting body.

FIGURE 2.25 The nematode ***Caenor-*** ***habditis elegans.*** This organism serves as a useful model for development. [Sinclair Stammers Science Photo Library/Photo Researchers.]

The binding of cAMP molecules to these receptors triggers several responses, including movement in the direction of higher cAMP concentration, as well as the generation and release of additional cAMP molecules (Figure 2.23).

The cells aggregate by following cAMP gradients. Once in contact, they exchange additional signals and then differentiate into distinct *cell types,* each of which expresses the set of genes appropriate for its eventual role in forming the fruiting body (Figure 2.24). The life cycles of organisms such as *Dictyostelium* foreshadow the evolution of organisms that are multicellular throughout their lifetimes. It is also interesting to note the cAMP signals starvation in many organisms, including human beings.

2.4.3 The Development of Multicellular Organisms Requires the Orchestrated Differentiation of Cells

The fossil record indicates that macroscopic, multicellular organisms appeared approximately 600 million years ago. Most of the organisms familiar to us consist of many cells. For example, an adult human being contains approximately 100,000,000,000,000 cells. The cells that make up different organs are distinct and, even within one organ, many different cell types are present. Nonetheless, the DNA sequence in each cell is identical. The differences between cell types are the result of differences in how these genes are expressed.

Each multicellular organism begins as a single cell. For this cell to develop into a complex organism, the embryonic cells must follow an intricate program of regulated gene expression, cell division, and cell movement. The developmental program relies substantially on the responses of cells to the environment created by neighboring cells. Cells in specific positions within the developing embryo divide to form particular tissues, such as muscle. Developmental pathways have been extensively studied in a number of organisms, including the nematode *Caenorhabditis elegans* (Figure 2.25), a 1-mm-long worm containing 959 cells. A detailed map describing the fate of each cell in *C. elegans* from the fertilized egg to the adult is shown in Figure 2.26. Interestingly, proper development requires not only cell division but also the death of specific cells at particular points in time through a process called programmed cell death or *apoptosis.*

Investigations of genes and proteins that control development in a wide range of organisms have revealed a great many common features. Many of the molecules that control human development are evolutionarily related to those in relatively simple organisms such as *C. elegans.* Thus, solutions to the problem of controlling development in multicellular organisms arose early in evolution and have been adapted many times in the course of evolution, generating the great diversity of complex organisms.

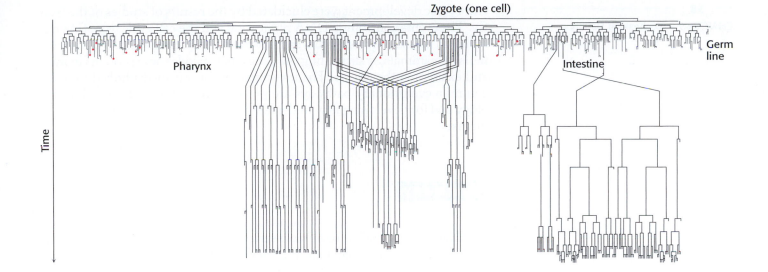

Zygote (one cell)

Time

Pharynx

Intestine

Germ line

FIGURE 2.26 Developmental pathways of *C. elegans.* The nematode develops from a single cell, called a zygote, into a complex organism. The fate of each individual cell in *C. elegans* is known and can be followed by referring to the cell-lineage diagram. The labels indicate cells that form specific organs. Cells that undergo programmed cell death are shown in red.

2.4.4 The Unity of Biochemistry Allows Human Biology to Be Effectively Probed Through Studies of Other Organisms

All organisms on Earth have a common origin (Figure 2.27). How could complex organisms such as human beings have evolved from the simple organisms that existed at life's start? The path outlined in this chapter reveals that most of the fundamental processes of biochemistry were largely fixed early in the history of life. The complexity of organisms such as human beings is manifest, at a biochemical level, in the interactions between overlapping and competing pathways, which lead to the generation of intricately connected groups of specialized cells. The evolution of biochemical and physiological complexity is made possible by the effects of gene duplication followed by specialization. Paradoxically, the reliance on gene duplication also makes this complexity easier to comprehend. Consider, for example, the protein kinases—enzymes that transfer phosphoryl groups from ATP to specific amino acids in proteins. These enzymes play essential roles in many signal-transduction pathways and in the control of cell growth and differentiation. The human genome encodes approximately 500 proteins of this class; even a relatively simple, unicellular organism such as brewer's yeast has more than 100 protein kinases. Yet each of these enzymes is the evolutionary descendant of a common ancestral enzyme. Thus, *we can learn much about the essential behavior of this large collection of proteins through studies of a single family member.* After the essential behavior is understood, we can evaluate the specific adaptations that allow each family member to perform its particular biological functions.

Most central processes in biology have been characterized first in relatively simple organisms, often through a combination of genetic, physiological, and biochemical studies. Many of the processes controlling early

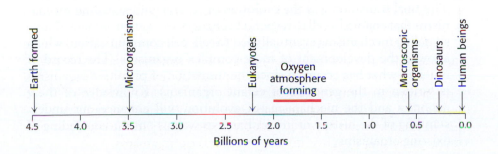

Earth formed

Microorganisms

Eukaryotes

Oxygen atmosphere forming

Macroscopic organisms

Dinosaurs

Human beings

4.5 4.0 3.5 3.0 2.5 2.0 1.5 1.0 0.5 0.0

Billions of years

FIGURE 2.27 A possible time line for biochemical evolution. Key events are indicated.

embryonic development were elucidated by the results of studies of the fruit fly. The events controlling DNA replication and the cell cycle were first deciphered in yeast. Investigators can now test the functions of particular proteins in mammals by disrupting the genes that encode these proteins in mice and examining the effects. The investigations of organisms linked to us by common evolutionary pathways are powerful tools for exploring all of biology and for developing new understanding of normal human function and disease.

SUMMARY

Key Organic Molecules Are Used by Living Systems

The evolution of life required a series of transitions, beginning with the generation of organic molecules that could serve as the building blocks for complex biomolecules. How these molecules arose is a matter of conjecture, but experiments have established that they could have formed under hypothesized prebiotic conditions.

Evolution Requires Reproduction, Variation, and Selective Pressure

The next major transition in the evolution of life was the formation of replicating molecules. Replication, coupled with variation and selective pressure, marked the beginning of evolution. Variation was introduced by a number of means, from simple base substitutions to the duplication of entire genes. RNA appears to have been an early replicating molecule. Furthermore, some RNA molecules possess catalytic activity. However, the range of reactions that RNA is capable of catalyzing is limited. With time, the catalytic activity was transferred to proteins— linear polymers of the chemically versatile amino acids. RNA directed the synthesis of these proteins and still does in modern organisms through the development of a genetic code, which relates base sequence to amino acid sequence. Eventually, RNA lost its role as the gene to the chemically similar but more stable nucleic acid DNA. In modern organisms, RNA still serves as the link between DNA and protein.

Energy Transformations Are Necessary to Sustain Living Systems

Another major transition in evolution was the ability to transform environmental energy into forms capable of being used by living systems. ATP serves as the cellular energy currency that links energy-yielding reactions with energy-requiring reactions. ATP itself is a product of the oxidation of fuel molecules, such as amino acids and sugars. With the evolution of membranes—hydrophobic barriers that delineate the borders of cells—ion gradients were required to prevent osmotic crises. These gradients were formed at the expense of ATP hydrolysis. Later, ion gradients generated by light or the oxidation of fuel molecules were used to synthesize ATP.

Cells Can Respond to Changes in Their Environments

The final transition was the evolution of sensing and signaling mechanisms that enabled a cell to respond to changes in its environment. These signaling mechanisms eventually led to cell–cell communication, which allowed the development of more-complex organisms. The record of much of what has occurred since the formation of primitive organisms is written in the genomes of extant organisms. Knowledge of these genomes and the mechanisms of evolution will enhance our understanding of the history of life on Earth as well as our understanding of existing organisms.

KEY TERMS

prebiotic world (p. 20)
reproduction (p. 21)
variation (p. 22)
competition (p. 22)
selective pressure (p. 22)
catalyst (p. 23)
enzyme (p. 23)
ribozyme (p. 23)

RNA world (p. 23)
proteins (p. 24)
genetic code (p. 25)
translation (p. 25)
gene (p. 25)
mutation (p. 25)
gene duplication (p. 26)
ATP (adenosine triphosphate) (p. 28)

membrane (p. 29)
ion pump (p. 31)
ion gradient (p. 31)
photosynthesis (p. 31)
signal transduction pathway (p. 34)
molecular motor protein (p. 35)
cell differentiation (p. 35)
unity of biochemistry (p. 37)

SELECTED READINGS

Where to start

Pace, N. R., 2000. The universal nature of biochemistry. *Proc. Natl. Acad. Sci. USA* 98:805–808.

Orgel, L. E., 1987. Evolution of the genetic apparatus: A review. *Cold Spring Harbor Symp. Quant. Biol.* 52:9–16.

Lazcano, A., and Miller, S. L., 1996. The origin and early evolution of life: Prebiotic chemistry, the pre-RNA world, and time. *Cell* 85:793–798.

Orgel, L. E. 1998. The origin of life: A review of facts and speculations. *Trends Biochem. Sci.* 23:491–495.

Books

Darwin, C., 1975. *On the Origin of Species, a Facsimile of the First Edition.* Harvard University Press.

Gesteland, R. F., Cech, T., and Atkins, J. F., 1999. *The RNA World.* Cold Spring Harbor Laboratory Press.

Dawkins, R., 1996. *The Blind Watchmaker.* Norton.

Smith, J. M., and Szathmáry, E., 1995. *The Major Transitions in Evolution.* W. H. Freeman and Company.

Prebiotic chemistry

Miller, S. L., 1987. Which organic compounds could have occurred on the prebiotic earth? *Cold Spring Harbor Symp. Quant. Biol.* 52:17–27.

Westheimer, F. H., 1987. Why nature chose phosphates. *Science* 235:1173–1178.

Levy, M., and Miller, S. L., 1998. The stability of the RNA bases: Implications for the origin of life. *Proc. Natl. Acad. Sci. USA* 95:7933–7938.

Sanchez, R., Ferris, J., and Orgel, L. E., 1966. Conditions for purine synthesis: Did prebiotic synthesis occur at low temperatures? *Science* 153:72–73.

In vitro evolution

Mills, D. R., Peterson, R. L., and Spiegelman, S., 1967. An extracellular Darwinian experiment with a self-duplicating nucleic acid molecule. *Proc. Natl. Acad. Sci. USA* 58:217–224.

Levisohn, R., and Spiegelman, S., 1969. Further extracellular Darwinian experiments with replicating RNA molecules: Diverse variants isolated under different selective conditions. *Proc. Natl. Acad. Sci. USA* 63:805–811.

Wilson, D. S., and Szostak, J. W., 1999. In vitro selection of functional nucleic acids. *Annu. Rev. Biochem.* 68:611–647.

Replication and catalytic RNA

Cech, T. R., 1993. The efficiency and versatility of catalytic RNA: Implications for an RNA world. *Gene* 135:33–36.

Orgel, L. E., 1992. Molecular replication. *Nature* 358:203–209.

Zielinski, W. S., and Orgel, L. E., 1987. Autocatalytic synthesis of a tetranucleotide analogue. *Nature* 327:346–347.

Nelson, K. E., Levy, M., and Miller, S. L., 2000. Peptide nucleic acids rather than RNA may have been the first genetic molecule. *Proc. Natl. Acad. Sci. USA* 97:3868–3871.

Transition from RNA to DNA

Reichard, P., 1997. The evolution of ribonucleotide reduction. *Trends Biochem. Sci.* 22:81–85.

Jordan, A., and Reichard, P., 1998. Ribonucleotide reductases. *Annu. Rev. Biochem.* 67:71–98.

Membranes

Wilson, T. H., and Maloney, P. C., 1976. Speculations on the evolution of ion transport mechanisms. *Fed. Proc.* 35:2174–2179.

Wilson, T. H., and Lin, E. C., 1980. Evolution of membrane bioenergetics. *J. Supramol. Struct.* 13:421–446.

Multicellular organisms and development

Mangiarotti, G., Bozzaro, S., Landfear, S., and Lodish, H. F., 1983. Cell–cell contact, cyclic AMP, and gene expression during development of *Dictyostelium discoideum*. *Curr. Top. Dev. Biol.* 18:117–154.

Kenyon, C., 1988. The nematode *Caenorhabditis elegans*. *Science* 240:1448–1453.

Hodgkin, J., Plasterk, R. H., and Waterston, R. H., 1995. The nematode *Caenorhabditis elegans* and its genome. *Science* 270:410–414.

PROBLEMS

1. *Finding the fragments.* Identify the likely source (CH_4, NH_3, H_2O, or H_2) of each atom in alanine generated in the Miller-Urey experiment.

2. *Following the populations.* In an experiment analogous to the Spiegelman experiment, suppose that a population of RNA molecules consists of 99 identical molecules, each of which replicates once in 15 minutes, and 1 molecule that replicates once in 5 minutes. Estimate the composition of the population after 1, 10, and 25 "generations" if a generation is defined as 15 minutes of replication. Assume that all necessary components are readily available.

3. *Selective advantage.* Suppose that a replicating RNA molecule has a mutation (genotypic change) and the phenotypic result is that it binds nucleotide monomers more tightly than do other RNA molecules in its population. What might the selective advantage of this mutation be? Under what conditions would you expect this selective advantage to be most important?

4. *Opposite of randomness.* Ion gradients prevent osmotic crises, but they require energy to be produced. Why does the formation of a gradient require an energy input?

5. *Coupled gradients.* How could a proton gradient with a higher concentration of protons inside a cell be used to pump ions out of a cell?

6. *Proton counting.* Consider the reactions that take place across a photosynthetic membrane. On one side of the membrane, the following reaction takes place:

$$4\,e^- + 4\,A^- + 4\,H_2O \longrightarrow 4\,AH + 4\,OH^-$$

whereas, on the other side of the membrane, the reaction is:

$$2\,H_2O \longrightarrow O_2 + 4\,e^- + 4\,H^+$$

How many protons are made available to drive ATP synthesis for each reaction cycle?

7. *An alternative pathway.* To respond to the availability of sugars such as arabinose, a cell must have at least two types of proteins: a transport protein to allow the arabinose to enter the cell and a gene-control protein, which binds the arabinose and modifies gene expression. To respond to the availability of some very hydrophobic molecules, a cell requires only one protein. Which one and why?

8. *How many divisions?* In the development pathway of *C. elegans,* cell division is initially synchronous—that is, all cells divide at the same rate. Later in development, some cells divide more frequently than do others. How many times does each cell divide in the synchronous period? Refer to Figure 2.26.

CHAPTER 3

Protein Structure and Function

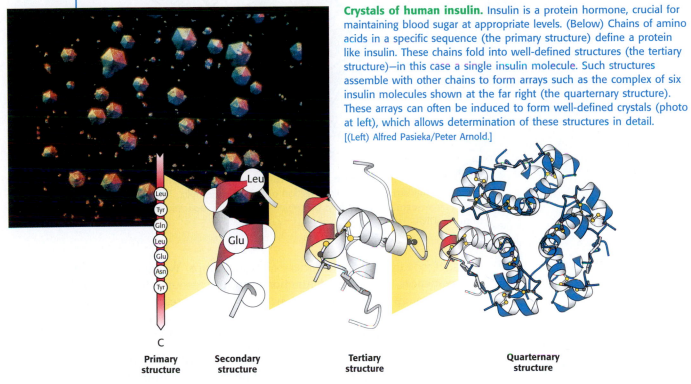

Crystals of human insulin. Insulin is a protein hormone, crucial for maintaining blood sugar at appropriate levels. (Below) Chains of amino acids in a specific sequence (the primary structure) define a protein like insulin. These chains fold into well-defined structures (the tertiary structure)—in this case a single insulin molecule. Such structures assemble with other chains to form arrays such as the complex of six insulin molecules shown at the far right (the quarternary structure). These arrays can often be induced to form well-defined crystals (photo at left), which allows determination of these structures in detail. [(Left) Alfred Pasieka/Peter Arnold.]

| Primary structure | Secondary structure | Tertiary structure | Quarternary structure |

Proteins are the most versatile macromolecules in living systems and serve crucial functions in essentially all biological processes. They function as catalysts, they transport and store other molecules such as oxygen, they provide mechanical support and immune protection, they generate movement, they transmit nerve impulses, and they control growth and differentiation. Indeed, much of this text will focus on understanding what proteins do and how they perform these functions.

Several key properties enable proteins to participate in such a wide range of functions.

1. *Proteins are linear polymers built of monomer units called amino acids.* The construction of a vast array of macromolecules from a limited number of monomer building blocks is a recurring theme in biochemistry. Does protein function depend on the linear sequence of amino acids? The function of a protein is directly dependent on its three-dimensional structure (Figure 3.1). Remarkably, proteins spontaneously fold up into three-dimensional structures that are determined by the sequence of amino acids in the protein polymer. Thus, *proteins are the embodiment of the transition from the one-dimensional world of sequences to the three-dimensional world of molecules capable of diverse activities.*

2. *Proteins contain a wide range of functional groups.* These functional groups include alcohols, thiols, thioethers, carboxylic

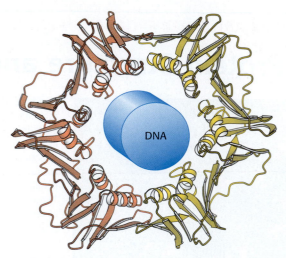

FIGURE 3.1 Structure dictates function. A protein component of the DNA replication machinery surrounds a section of DNA double helix. The structure of the protein allows large segments of DNA to be copied without the replication machinery dissociating from the DNA.

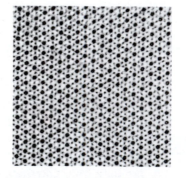

FIGURE 3.2 A complex protein assembly. An electron micrograph of insect flight tissue in cross section shows a hexagonal array of two kinds of protein filaments. [Courtesy of Dr. Michael Reedy.]

acids, carboxamides, and a variety of basic groups. When combined in various sequences, this array of functional groups accounts for the broad spectrum of protein function. For instance, the chemical reactivity associated with these groups is essential to the function of *enzymes,* the proteins that catalyze specific chemical reactions in biological systems (see Chapters 8–10).

3. *Proteins can interact with one another and with other biological macromolecules to form complex assemblies.* The proteins within these assemblies can act synergistically to generate capabilities not afforded by the individual component proteins (Figure 3.2). These assemblies include macromolecular machines that carry out the accurate replication of DNA, the transmission of signals within cells, and many other essential processes.

4. *Some proteins are quite rigid, whereas others display limited flexibility.* Rigid units can function as structural elements in the cytoskeleton (the internal scaffolding within cells) or in connective tissue. Parts of proteins with limited flexibility may act as hinges, springs, and levers that are crucial to protein function, to the assembly of proteins with one another and with other molecules into complex units, and to the transmission of information within and between cells (Figure 3.3).

FIGURE 3.3 Flexibility and function. Upon binding iron, the protein lactoferrin undergoes conformational changes that allow other molecules to distinguish between the iron-free and the iron-bound forms.

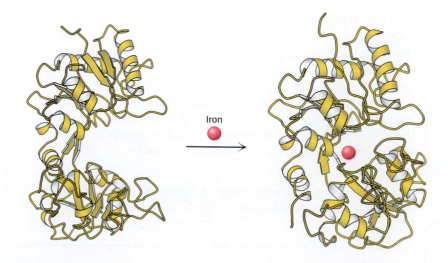

Iron

3.1 PROTEINS ARE BUILT FROM A REPERTOIRE OF 20 AMINO ACIDS

Amino acids are the building blocks of proteins. An α-*amino acid* consists of a central carbon atom, called the α *carbon*, linked to an amino group, a carboxylic acid group, a hydrogen atom, and a distinctive R group. The R group is often referred to as the *side chain*. With four different groups connected to the tetrahedral α-carbon atom, α-amino acids are *chiral;* the two mirror-image forms are called the L isomer and the D isomer (Figure 3.4).

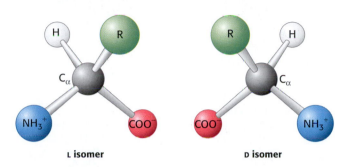

L isomer   **D isomer**

FIGURE 3.4 The L and D isomers of amino acids. R refers to the side chain. The L and D isomers are mirror images of each other.

Only L amino acids are constituents of proteins. For almost all amino acids, the L isomer has *S* (rather than *R*) absolute configuration (Figure 3.5). Although considerable effort has gone into understanding why amino acids in proteins have this absolute configuration, no satisfactory explanation has been arrived at. It seems plausible that the selection of L over D was arbitrary but, once made, was fixed early in evolutionary history.

Amino acids in solution at neutral pH exist predominantly as *dipolar ions* (also called *zwitterions*). In the dipolar form, the amino group is protonated ($-NH_3^+$) and the carboxyl group is deprotonated ($-COO^-$). The ionization state of an amino acid varies with pH (Figure 3.6). In acid solution (e.g., pH 1), the amino group is protonated ($-NH_3^+$) and the carboxyl group is not dissociated (–COOH). As the pH is raised, the carboxylic acid is the first group to give up a proton, inasmuch as its pK_a is near 2. The dipolar form persists until the pH approaches 9, when the protonated amino group

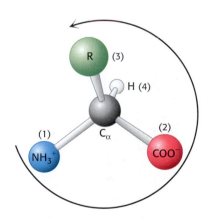

FIGURE 3.5 Only L amino acids are found in proteins. Almost all L amino acids have an *S* absolute configuration (from the Latin *sinister* meaning "left"). The counterclockwise direction of the arrow from highest- to lowest-priority substituents indicates that the chiral center is of the *S* configuration.

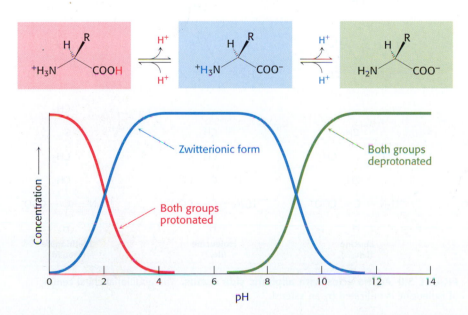

Zwitterionic form

Both groups deprotonated

Both groups protonated

Concentration

pH

FIGURE 3.6 Ionization state as a function of pH. The ionization state of amino acids is altered by a change in pH. The zwitterionic form predominates near physiological pH.

FIGURE 3.7 Structures of glycine and alanine. (Top) Ball-and-stick models show the arrangement of atoms and bonds in space. (Middle) Stereochemically realistic formulas show the geometrical arrangement of bonds around atoms (see Chapter 1 Appendix). (Bottom) Fischer projections show all bonds as being perpendicular for a simplified representation (see Chapter 1 Appendix).

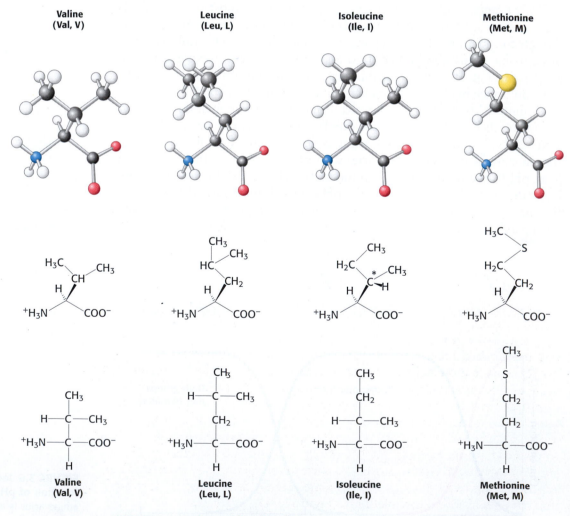

FIGURE 3.8 Amino acids with aliphatic side chains. The additional chiral center of isoleucine is indicated by an asterisk.

loses a proton. For a review of acid–base concepts and pH, see the appendix to this chapter.

Twenty kinds of side chains varying in *size, shape, charge, hydrogen-bonding capacity, hydrophobic character,* and *chemical reactivity* are commonly found in proteins. Indeed, all proteins in all species—bacterial, archaeal, and eukaryotic—are constructed from the same set of 20 amino acids. This fundamental alphabet of proteins is several billion years old. The remarkable range of functions mediated by proteins results from the diversity and versatility of these 20 building blocks. Understanding how this alphabet is used to create the intricate three-dimensional structures that enable proteins to carry out so many biological processes is an exciting area of biochemistry and one that we will return to in Section 3.6.

Let us look at this set of amino acids. The simplest one is *glycine*, which has just a hydrogen atom as its side chain. With two hydrogen atoms bonded to the α-carbon atom, glycine is unique in being achiral. *Alanine*, the next simplest amino acid, has a methyl group (–CH_3) as its side chain (Figure 3.7).

Larger hydrocarbon side chains are found in *valine, leucine,* and *isoleucine* (Figure 3.8). *Methionine* contains a largely *aliphatic* side chain that includes a *thioether* (–S–) group. The side chain of isoleucine includes an additional chiral center; only the isomer shown in Figure 3.8 is found in proteins. The larger aliphatic side chains are *hydrophobic*—that is, they tend to cluster together rather than contact water. The three-dimensional structures of water-soluble proteins are stabilized by this tendency of hydrophobic groups to come together, called *the hydrophobic effect* (see Section 1.3.4). The different sizes and shapes of these hydrocarbon side chains enable them to pack together to form compact structures with few holes. *Proline* also has an aliphatic side chain, but it differs from other members of the set of 20 in that its side chain is bonded to both the nitrogen and the α-carbon atoms (Figure 3.9). Proline markedly influences protein architecture because its ring structure makes it more conformationally restricted than the other amino acids.

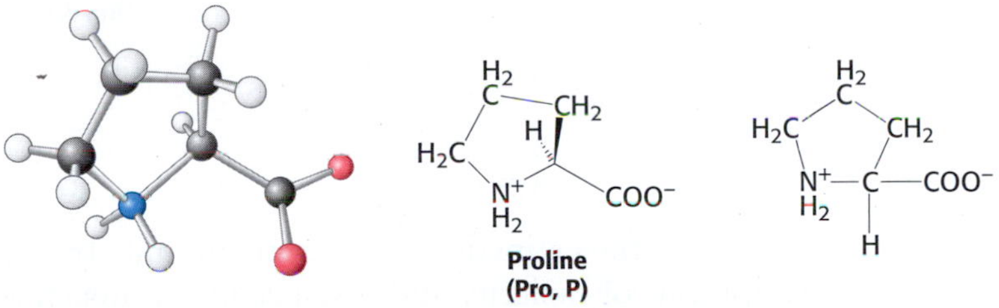

Proline (Pro, P)

FIGURE 3.9 Cyclic structure of proline. The side chain is joined to both the α carbon and the amino group.

Three amino acids with relatively simple *aromatic side chains* are part of the fundamental repertoire (Figure 3.10). *Phenylalanine*, as its name indicates, contains a phenyl ring attached in place of one of the hydrogens of alanine. The aromatic ring of *tyrosine* contains a hydroxyl group. This hydroxyl group is reactive, in contrast with the rather inert side chains of the other amino acids discussed thus far. *Tryptophan* has an indole ring joined to a methylene (–CH_2–) group; the indole group comprises two fused rings and an NH group. Phenylalanine is purely hydrophobic, whereas tyrosine and tryptophan are less so because of their hydroxyl and NH groups. The aromatic rings of tryptophan and tyrosine contain delocalized π electrons that strongly absorb ultraviolet light (Figure 3.11).

A compound's *extinction coefficient* indicates its ability to absorb light. Beer's law gives the absorbance (A) of light at a given wavelength:

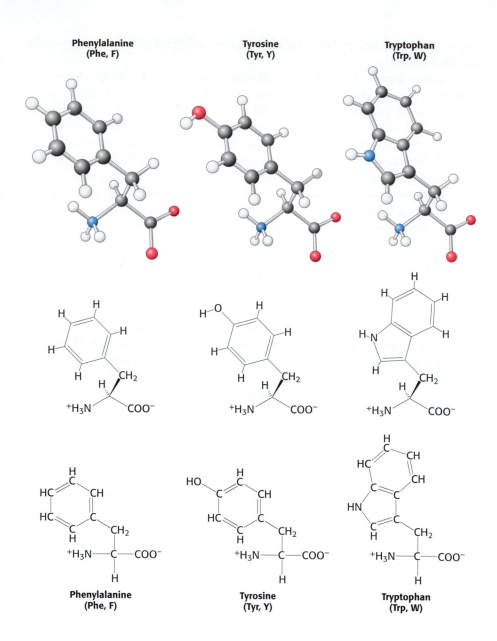

FIGURE 3.10 **Amino acids with aromatic side chains.** Phenylalanine, tyrosine, and tryptophan have hydrophobic character. Tyrosine and tryptophan also have hydrophilic properties because of their –OH and –NH– groups, respectively.

$$A = \epsilon c l \qquad \text{Beer's law}$$

where ϵ is the extinction coefficient [in units that are the reciprocals of molarity and distance in centimeters (M^{-1} cm^{-1})], c is the concentration of the absorbing species (in units of molarity, M), and l is the length through which the light passes (in units of centimeters). For tryptophan, absorption is maximum at 280 nm and the extinction coefficient is 3400 M^{-1} cm^{-1} whereas, for tyrosine, absorption is maximum at 276 nm and the extinction coefficient is a less-intense 1400 M^{-1} cm^{-1}. Phenylalanine absorbs light less strongly and at shorter wavelengths. The absorption of light at 280 nm can be used to estimate

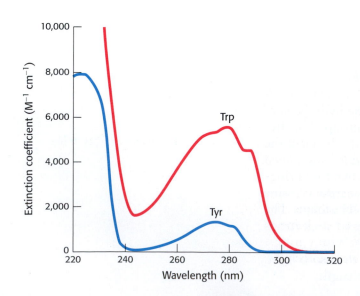

FIGURE 3.11 **Absorption spectra of the aromatic amino acids tryptophan (red) and tyrosine (blue).** Only these amino acids absorb strongly near 280 nm. [Courtesy of Greg Gatto].

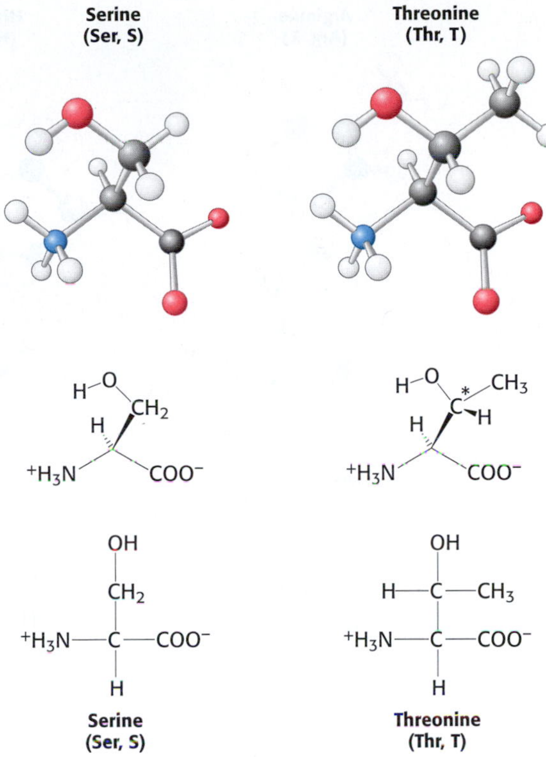

Serine
(Ser, S)

Threonine
(Thr, T)

Serine
(Ser, S)

Threonine
(Thr, T)

FIGURE 3.12 Amino acids containing aliphatic hydroxyl groups. Serine and threonine contain hydroxyl groups that render them hydrophilic. The additional chiral center in threonine is indicated by an asterisk.

the concentration of a protein in solution if the number of tryptophan and tyrosine residues in the protein is known.

Two amino acids, *serine* and *threonine*, contain aliphatic *hydroxyl groups* (Figure 3.12). Serine can be thought of as a hydroxylated version of alanine, whereas threonine resembles valine with a hydroxyl group in place of one of the valine methyl groups. The hydroxyl groups on serine and threonine make them much more *hydrophilic* (water loving) and *reactive* than alanine and valine. Threonine, like isoleucine, contains an additional asymmetric center; again only one isomer is present in proteins.

Cysteine is structurally similar to serine but contains a *sulfhydryl*, or *thiol* (–SH), group in place of the hydroxyl (–OH) group (Figure 3.13). The sulfhydryl group is much more reactive. Pairs of sulfhydryl groups may come together to form disulfide bonds, which are particularly important in stabilizing some proteins, as will be discussed shortly.

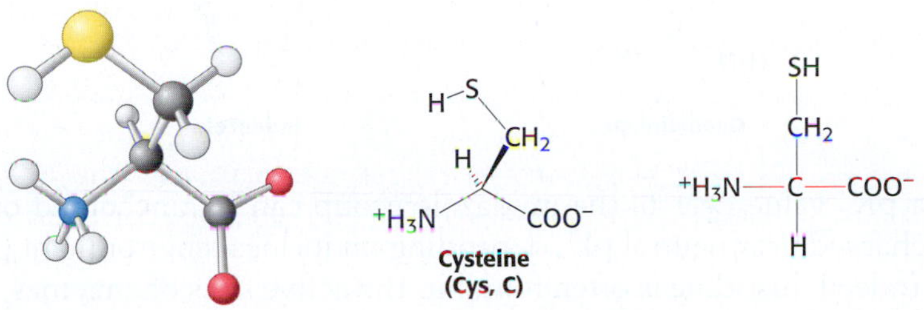

Cysteine
(Cys, C)

FIGURE 3.13 Structure of cysteine.

We turn now to amino acids with very polar side chains that render them highly hydrophilic. *Lysine* and *arginine* have relatively long side chains that terminate with groups that are *positively charged* at neutral pH. Lysine is capped by a primary amino group and arginine by a guanidinium group. *Histidine* contains an imidazole group, an aromatic ring that also can be positively charged (Figure 3.14).

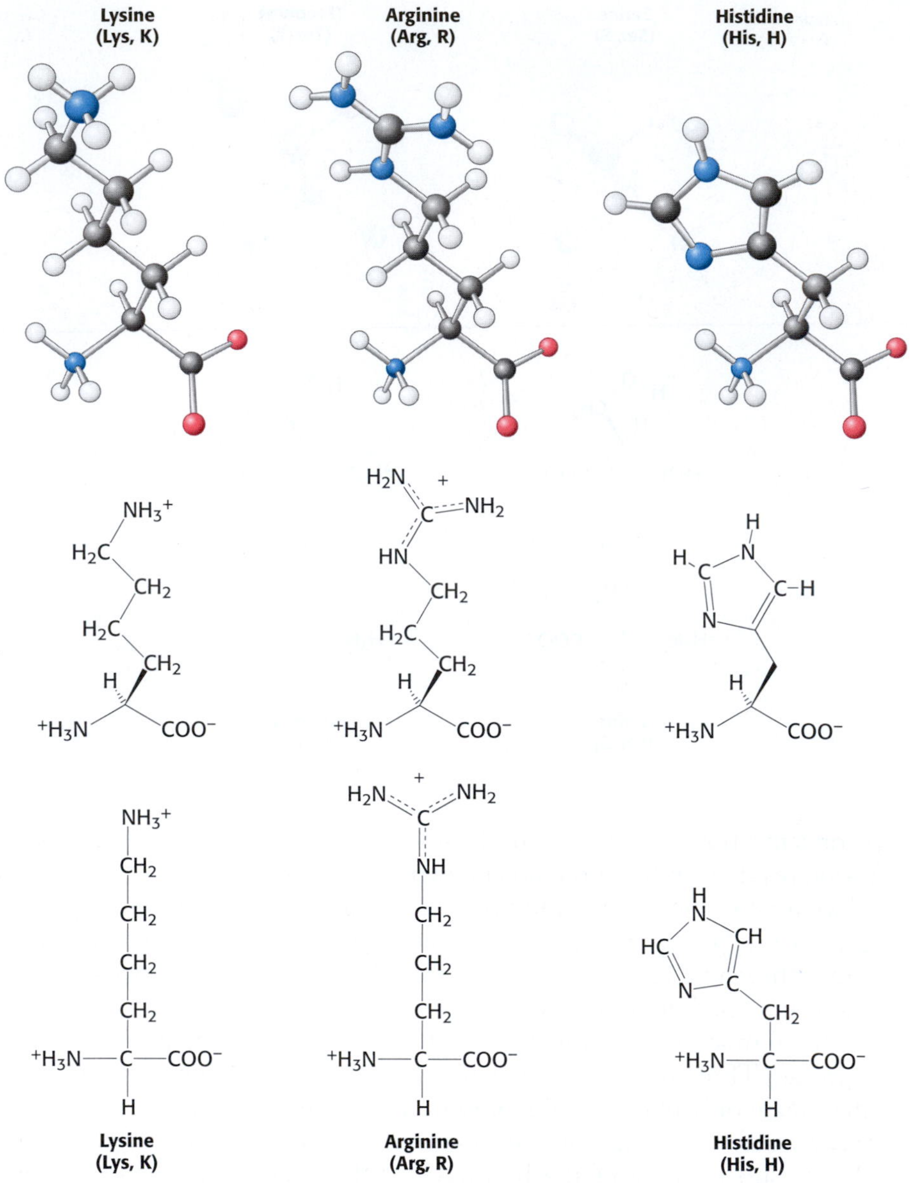

**Lysine
(Lys, K)**

**Arginine
(Arg, R)**

**Histidine
(His, H)**

FIGURE 3.14 The basic amino acids lysine, arginine, and histidine.

**Lysine
(Lys, K)**

**Arginine
(Arg, R)**

**Histidine
(His, H)**

Guanidinium

Imidazole

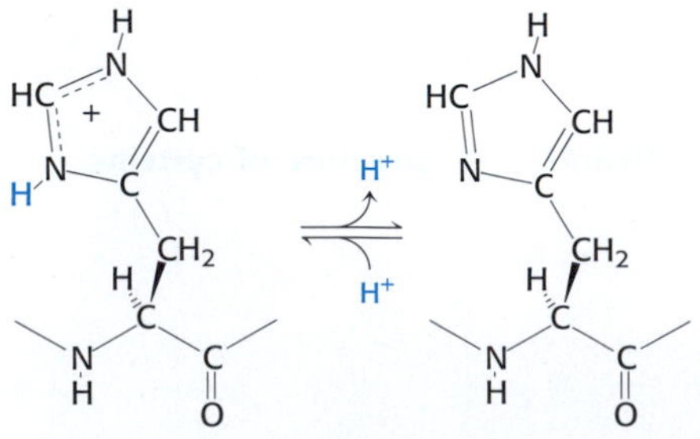

FIGURE 3.15 Histidine ionization.

Histidine can bind or release protons near physiological pH.

With a pK_a value near 6, the imidazole group can be uncharged or positively charged near neutral pH, depending on its local environment (Figure 3.15). Indeed, histidine is often found in the active sites of enzymes, where the imidazole ring can bind and release protons in the course of enzymatic reactions.

The set of amino acids also contains two with *acidic side chains: aspartic acid* and *glutamic acid* (Figure 3.16). These amino acids are often called *aspartate* and *glutamate* to emphasize that their side chains are usually negatively charged at physiological pH. Nonetheless, in some proteins these side chains do accept protons, and this ability is often functionally important. In addition, the set includes uncharged derivatives of aspartate and

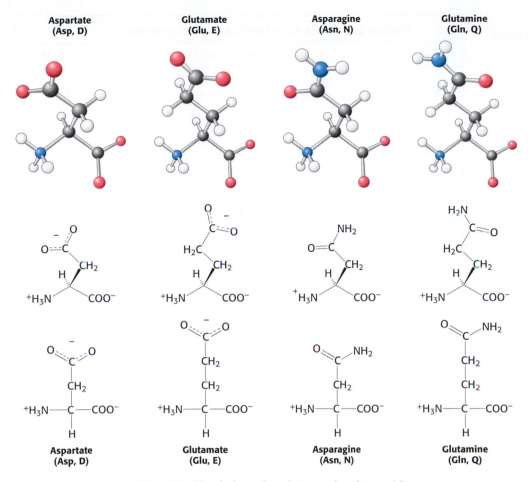

Aspartate (Asp, D)   **Glutamate (Glu, E)**   **Asparagine (Asn, N)**   **Glutamine (Gln, Q)**

FIGURE 3.16 Amino acids with side-chain carboxylates and carboxamides.

glutamate—*asparagine* and *glutamine*—each of which contains a terminal *carboxamide* in place of a carboxylic acid (Figure 3.16).

Seven of the 20 amino acids have readily ionizable side chains. These 7 amino acids are able to donate or accept protons to facilitate reactions as well as to form ionic bonds. Table 3.1 gives equilibria and typical pK_a values for ionization of the side chains of tyrosine, cysteine, arginine, lysine, histidine, and aspartic and glutamic acids in proteins. Two other groups in proteins—the terminal α-amino group and the terminal α-carboxyl group—can be ionized, and typical pK_a values are also included in Table 3.1.

Amino acids are often designated by either a three-letter abbreviation or a one-letter symbol (Table 3.2). The abbreviations for amino acids are the first three letters of their names, except for asparagine (Asn), glutamine (Gln), isoleucine (Ile), and tryptophan (Trp). The symbols for many amino acids are the first letters of their names (e.g., G for glycine and L for leucine); the other symbols have been agreed on by convention. These abbreviations and symbols are an integral part of the vocabulary of biochemists.

How did this particular set of amino acids become the building blocks of proteins? First, as a set, they are diverse; their structural and chemical properties span a wide range, endowing proteins with the versatility to assume many functional roles. Second, as noted in Section 2.1.1, many of these amino acids were probably available from prebiotic reactions. Finally, excessive intrinsic reactivity may have eliminated other possible amino

TABLE 3.1 **Typical pK_a values of ionizable groups in proteins**

Group	Acid	⇌	Base	Typical pK_a*
Terminal α-carboxyl group				3.1
Aspartic acid Glutamic acid				4.1
Histidine				6.0
Terminal α-amino group				8.0
Cysteine				8.3
Tyrosine				10.9
Lysine				10.8
Arginine				12.5

*pK_a values depend on temperature, ionic strength, and the microenvironment of the ionizable group.

TABLE 3.2 **Abbreviations for amino acids**

Amino acid	Three-letter abbreviation	One-letter abbreviation	Amino acid	Three-letter abbreviation	One-letter abbreviation
Alanine	Ala	A	Methionine	Met	M
Arginine	Arg	R	Phenylalanine	Phe	F
Asparagine	Asn	N	Proline	Pro	P
Aspartic Acid	Asp	D	Serine	Ser	S
Cysteine	Cys	C	Threonine	Thr	T
Glutamine	Gln	Q	Tryptophan	Trp	W
Glutamic Acid	Glu	E	Tyrosine	Tyr	Y
Glycine	Gly	G	Valine	Val	V
Histidine	His	H	Asparagine or aspartic acid	Asx	B
Isoleucine	Ile	I			
Leucine	Leu	L	Glutamine or glutamic acid	Glx	Z
Lysine	Lys	K			

acids. For example, amino acids such as homoserine and homocysteine tend to form five-membered cyclic forms that limit their use in proteins; the alternative amino acids that are found in proteins—serine and cysteine—do not readily cyclize, because the rings in their cyclic forms are too small (Figure 3.17).

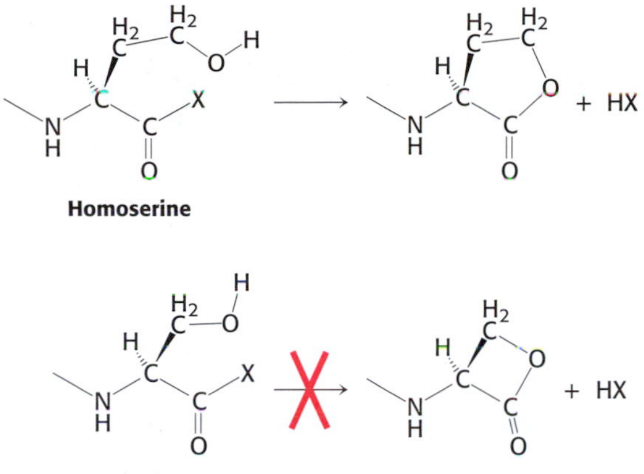

Homoserine

Serine

FIGURE 3.17 Undesirable reactivity in amino acids. Some amino acids are unsuitable for proteins because of undesirable cyclization. Homoserine can cyclize to form a stable, five-membered ring, potentially resulting in peptide-bond cleavage. Cyclization of serine would form a strained, four-membered ring and thus is unfavored. X can be an amino group from a neighboring amino acid or another potential leaving group.

3.2 PRIMARY STRUCTURE: AMINO ACIDS ARE LINKED BY PEPTIDE BONDS TO FORM POLYPEPTIDE CHAINS

Proteins are *linear polymers* formed by linking the α-carboxyl group of one amino acid to the α-amino group of another amino acid with a *peptide bond* (also called an *amide bond*). The formation of a dipeptide from two amino acids is accompanied by the loss of a water molecule (Figure 3.18). The equilibrium of this reaction lies on the side of hydrolysis rather than synthesis. Hence, the biosynthesis of peptide bonds requires an input of free energy. Nonetheless, peptide bonds are quite *stable kinetically;* the lifetime of a peptide bond in aqueous solution in the absence of a catalyst approaches 1000 years.

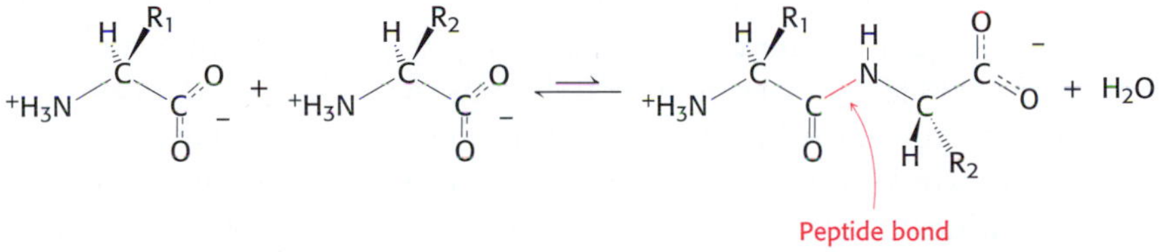

Peptide bond

FIGURE 3.18 Peptide-bond formation. The linking of two amino acids is accompanied by the loss of a molecule of water.

A series of amino acids joined by peptide bonds form a *polypeptide chain,* and each amino acid unit in a polypeptide is called a *residue. A polypeptide chain has polarity* because its ends are different, with an α-amino group at one end and an α-carboxyl group at the other. By convention, *the amino end is taken to be the beginning of a polypeptide chain,* and so the sequence of amino acids in a polypeptide chain is written starting with the amino-terminal residue. Thus, in the pentapeptide Tyr-Gly-Gly-Phe-Leu (YGGFL), tyrosine is the amino-terminal (N-terminal) residue and leucine is the carboxyl-terminal (C-terminal) residue (Figure 3.19). Leu-Phe-Gly-Gly-Tyr (LFGGY) is a different pentapeptide, with different chemical properties.

A polypeptide chain consists of a regularly repeating part, called the *main chain* or *backbone,* and a variable part, comprising the distinctive *side chains* (Figure 3.20). The polypeptide backbone is rich in hydrogen-bonding potential. Each residue contains a carbonyl group, which is a good hydrogen-bond acceptor and, with the exception of proline, an NH group, which is a

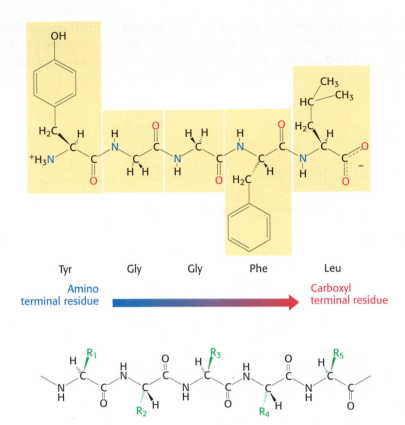

Tyr · Gly · Gly · Phe · Leu

Amino terminal residue → Carboxyl terminal residue

FIGURE 3.19 Amino acid sequences have direction. This illustration of the pentapeptide Try-Gly-Gly-Phe-Leu (YGGFL) shows the sequence from the amino terminus to the carboxyl terminus. This pentapeptide, Leu-enkephalin, is an opioid peptide that modulates the perception of pain. The reverse pentapeptide, Leu-Phe-Gly-Gly-Tyr (LFGGY), is a different molecule and shows no such effects.

FIGURE 3.20 Components of a polypeptide chain. A polypeptide chain consists of a constant backbone (shown in black) and variable side chains (shown in green).

Dalton—
A unit of mass very nearly equal to that of a hydrogen atom. Named after John Dalton (1766–1844), who developed the atomic theory of matter.

Kilodalton (kd)—
A unit of mass equal to 1000 daltons.

good hydrogen-bond donor. These groups interact with each other and with functional groups from side chains to stabilize particular structures, as will be discussed in detail.

Most natural polypeptide chains contain between 50 and 2000 amino acid residues and are commonly referred to as *proteins*. Peptides made of small numbers of amino acids are called *oligopeptides* or simply *peptides*. The mean molecular weight of an amino acid residue is about 110, and so the molecular weights of most proteins are between 5500 and 220,000. We can also refer to the mass of a protein, which is expressed in units of daltons; one *dalton* is equal to one atomic mass unit. A protein with a molecular weight of 50,000 has a mass of 50,000 daltons, or 50 kd (kilodaltons).

In some proteins, the linear polypeptide chain is cross-linked. The most common cross-links are *disulfide bonds*, formed by the oxidation of a pair of cysteine residues (Figure 3.21). The resulting unit of linked cysteines is

Cysteine

Cysteine

$\xrightarrow{\text{Oxidation}}_{\text{Reduction}}$

$+ \ 2\,H^+ \ + \ 2\,e^-$

Cystine

FIGURE 3.21 Cross-links. The formation of a disulfide bond from two cysteine residues is an oxidation reaction.

called *cystine*. Extracellular proteins often have several disulfide bonds, whereas intracellular proteins usually lack them. Rarely, nondisulfide cross-links derived from other side chains are present in some proteins. For example, collagen fibers in connective tissue are strengthened in this way, as are fibrin blood clots.

3.2.1 Proteins Have Unique Amino Acid Sequences That Are Specified by Genes

In 1953, Frederick Sanger determined the amino acid sequence of insulin, a protein hormone (Figure 3.22). *This work is a landmark in biochemistry because it showed for the first time that a protein has a precisely defined amino acid sequence.* Moreover, it demonstrated that insulin consists only of L amino acids linked by peptide bonds between α-amino and α-carboxyl groups. This accomplishment stimulated other scientists to carry out sequence studies of a wide variety of proteins. Indeed, the complete amino acid sequences of more than 100,000 proteins are now known. *The striking fact is that each protein has a unique, precisely defined amino acid sequence.* The amino acid sequence of a protein is often referred to as its *primary structure*.

A series of incisive studies in the late 1950s and early 1960s revealed that the amino acid sequences of proteins are genetically determined. The sequence of nucleotides in DNA, the molecule of heredity, specifies a complementary sequence of nucleotides in RNA, which in turn specifies the amino acid sequence of a protein. In particular, each of the 20 amino acids of the repertoire is encoded by one or more specific sequences of three nucleotides (Section 5.5).

Knowing amino acid sequences is important for several reasons. First, knowledge of the sequence of a protein is usually essential to elucidating its mechanism of action (e.g., the catalytic mechanism of an enzyme). Moreover, proteins with novel properties can be generated by varying the sequence of known proteins. Second, amino acid sequences determine the three-dimensional structures of proteins. Amino acid sequence is the link between the genetic message in DNA and the three-dimensional structure that performs a protein's biological function. Analyses of relations between amino acid sequences and three-dimensional structures of proteins are uncovering the rules that govern the folding of polypeptide chains. Third, sequence determination is a component of molecular pathology, a rapidly growing area of medicine. Alterations in amino acid sequence can produce abnormal function and disease. Severe and sometimes fatal diseases, such as sickle-cell anemia and cystic fibrosis, can result from a change in a single amino acid within a protein. Fourth, the sequence of a protein reveals much about its evolutionary history (see Chapter 7). Proteins resemble one another in amino acid sequence only if they have a common ancestor. Consequently, molecular events in evolution can be traced from amino acid sequences; molecular paleontology is a flourishing area of research.

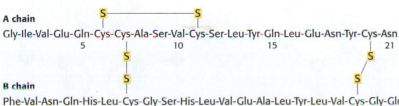

A chain

Gly-Ile-Val-Glu-Gln-Cys-Cys-Ala-Ser-Val-Cys-Ser-Leu-Tyr-Gln-Leu-Glu-Asn-Tyr-Cys-Asn

B chain

Phe-Val-Asn-Gln-His-Leu-Cys-Gly-Ser-His-Leu-Val-Glu-Ala-Leu-Tyr-Leu-Val-Cys-Gly-Glu-Arg-Gly-Phe-Phe-Tyr-Thr-Pro-Lys-Ala

FIGURE 3.22 Amino acid sequence of bovine insulin.

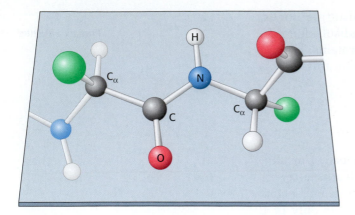

FIGURE 3.23 **Peptide bonds are planar.** In a pair of linked amino acids, six atoms (C_α, C, O, N, H, and C_α) lie in a plane. Side chains are shown as green balls.

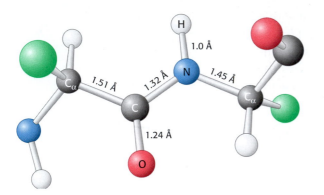

FIGURE 3.24 **Typical bond lengths within a peptide unit.** The peptide unit is shown in the trans configuration.

3.2.2 Polypeptide Chains Are Flexible Yet Conformationally Restricted

Examination of the geometry of the protein backbone reveals several important features. First, *the peptide bond is essentially planar* (Figure 3.23). Thus, for a pair of amino acids linked by a peptide bond, six atoms lie in the same plane: the α-carbon atom and CO group from the first amino acid and the NH group and α-carbon atom from the second amino acid. The nature of the chemical bonding within a peptide explains this geometric preference. The peptide bond has considerable *double-bond character*, which prevents rotation about this bond.

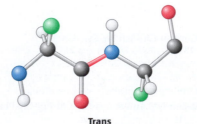

Peptide bond resonance structures

The inability of the bond to rotate constrains the conformation of the peptide backbone and accounts for the bond's planarity. This double-bond character is also expressed in the length of the bond between the CO and NH groups. The C–N distance in a peptide bond is typically 1.32 Å, which is between the values expected for a C–N single bond (1.49 Å) and a C=N double bond (1.27 Å), as shown in Figure 3.24. Finally, the peptide bond is uncharged, allowing polymers of amino acids linked by peptide bonds to form tightly packed globular structures.

Two configurations are possible for a planar peptide bond. In the trans configuration, the two α-carbon atoms are on opposite sides of the peptide bond. In the cis configuration, these groups are on the same side of the peptide bond. *Almost all peptide bonds in proteins are trans.* This preference for trans over cis can be explained by the fact that steric clashes between groups attached to the α-carbon atoms hinder formation of the cis form but do not occur in the trans configuration (Figure 3.25). By far the most common cis peptide bonds are X–Pro linkages. Such bonds show less preference for the trans configuration because the nitrogen of proline is bonded to two tetrahedral carbon atoms, limiting the steric differences between the trans and cis forms (Figure 3.26).

In contrast with the peptide bond, the bonds between the amino group and the α-carbon atom and between the α-carbon atom and the carbonyl group are pure single bonds. The two adjacent rigid peptide units may rotate about these bonds, taking on various orientations. *This freedom of rotation*

FIGURE 3.25 **Trans and cis peptide bonds.** The trans form is strongly favored because of steric clashes that occur in the cis form.

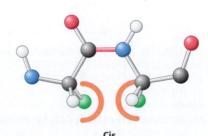

Trans   **Cis**

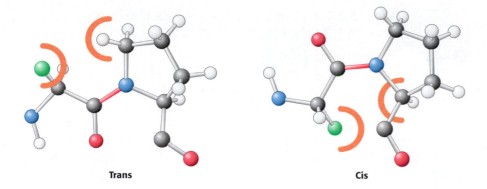

Trans   **Cis**

FIGURE 3.26 Trans and cis X–Pro bonds. The energies of these forms are relatively balanced because steric clashes occur in both forms.

about two bonds of each amino acid allows proteins to fold in many different ways. The rotations about these bonds can be specified by dihedral angles (Figure 3.27). The angle of rotation about the bond between the nitrogen and the α-carbon atoms is called *phi* (ϕ). The angle of rotation about the bond between the α-carbon and the carbonyl carbon atoms is called *psi* (ψ). A clockwise rotation about either bond as viewed from the front of the back group corresponds to a positive value. The ϕ and ψ angles determine the path of the polypeptide chain.

Are all combinations of ϕ and ψ possible? G. N. Ramachandran recognized that many combinations are forbidden because of steric collisions between atoms. The allowed values can be visualized on a two-dimensional plot called a *Ramachandran diagram* (Figure 3.28). Three-quarters of the possible (ϕ, ψ) combinations are excluded simply by local steric clashes. *Steric exclusion, the fact that two atoms cannot be in the same place at the same time, can be a powerful organizing principle.*

The ability of biological polymers such as proteins to fold into well-defined structures is remarkable thermodynamically. Consider the equilibrium between an unfolded polymer that exists as a random coil—that is, as a mixture of many possible conformations—and the folded form that adopts a unique conformation. The favorable entropy associated with the large number of conformations in the unfolded form opposes folding and must be overcome by interactions favoring the folded form. Thus, highly flexible polymers with a large number of possible conformations do not fold into unique structures. *The rigidity of the peptide unit and the restricted set of allowed ϕ and ψ angles limits the number of structures accessible to the unfolded form sufficiently to allow protein folding to occur.*

Dihedral angle—
A measure of the rotation about a bond, usually taken to lie between $-180°$ and $+180°$. Dihedral angles are sometimes called torsion angles.

(A)   (B)   (C)

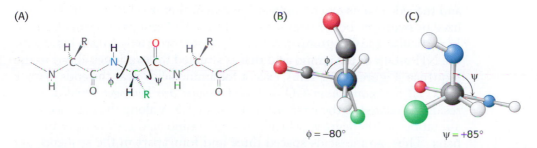

$\phi = -80°$ $\psi = +85°$

FIGURE 3.27 Rotation about bonds in a polypeptide. The structure of each amino acid in a polypeptide can be adjusted by rotation about two single bonds. (A) Phi (ϕ) is the angle of rotation about the bond between the nitrogen and the α-carbon atoms, whereas psi (ψ) is the angle of rotation about the bond between the α-carbon and the carbonyl carbon atoms. (B) A view down the bond between the nitrogen and the α-carbon atoms, showing how ϕ is measured. (C) A view down the bond between the α-carbon and the carbonyl carbon atoms, showing how ψ is measured.

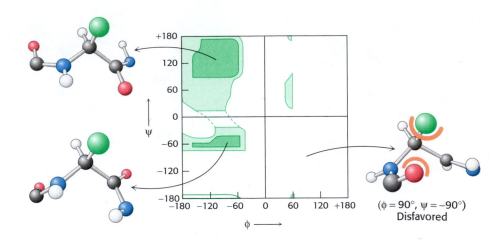

FIGURE 3.28 A Ramachandran diagram showing the values of φ and ψ. Not all φ and ψ values are possible without collisions between atoms. The most favorable regions are shown in dark green; borderline regions are shown in light green. The structure on the right is disfavored because of steric clashes.

($φ = 90°$, $ψ = -90°$)
Disfavored

3.3 SECONDARY STRUCTURE: POLYPEPTIDE CHAINS CAN FOLD INTO REGULAR STRUCTURES SUCH AS THE ALPHA HELIX, THE BETA SHEET, AND TURNS AND LOOPS

Can a polypeptide chain fold into a regularly repeating structure? In 1951, Linus Pauling and Robert Corey proposed two periodic structures called the *α helix* (alpha helix) and the *β pleated sheet* (beta pleated sheet). Subsequently, other structures such as the *β turn* and *omega (Ω) loop* were identified. Although not periodic, these common turn or loop structures are well defined and contribute with α helices and β sheets to form the final protein structure.

STRUCTURAL INSIGHTS, **Elements of Protein Structure** *provides interactive representations of some of the important elements of protein architecture described in this chapter, including a summary of secondary structure motifs.*

3.3.1 The Alpha Helix Is a Coiled Structure Stabilized by Intrachain Hydrogen Bonds

In evaluating potential structures, Pauling and Corey considered which conformations of peptides were sterically allowed and which most fully exploited the hydrogen-bonding capacity of the backbone NH and CO groups. The first of their proposed structures, the *α helix*, is a rodlike structure (Figure 3.29). A tightly coiled backbone forms the inner part of the rod and the side chains extend outward in a helical array. The α helix is stabilized by hydrogen bonds between the NH and CO groups of the main chain. In particular, the CO group of each amino acid forms a hydrogen bond with the NH group of the amino acid that is situated four residues ahead in the sequence (Figure 3.30). Thus, except for amino acids near the ends of an α helix, all *the main-chain CO and NH groups are hydrogen bonded.* Each residue is related to the next one by a rise of 1.5 Å along the helix axis and a rotation of 100 degrees, which gives 3.6 amino acid residues per turn of helix. Thus, amino acids spaced three and four apart in the sequence are spatially quite close to one another in an α helix. In contrast, amino acids two apart in the sequence are situated on opposite sides of the helix and so are unlikely to make contact. The *pitch* of the α helix, which is equal to the product of the translation (1.5 Å) and the number of residues per turn (3.6), is 5.4 Å. The *screw sense* of a helix can be right-handed (clockwise) or left-handed (counterclockwise). The Ramachandran diagram reveals that both

STRUCTURAL INSIGHTS, appearing throughout the book, are molecular modeling-based tutorials that enable you to review structure and learn what the latest research tells us about the workings of the molecule. To access, go to the Web site: www.whfreeman.com/biochem5, and select the chapter, Structural Insights, and the title.

Screw sense—
Describes the direction in which a helical structure rotates with respect to its axis. If, viewed down the axis of a helix, the chain turns in a clockwise direction, it has a right-handed screw sense. If the turning is counterclockwise, the screw sense is left-handed.

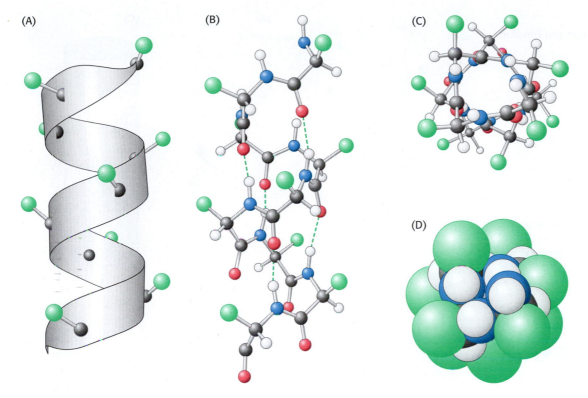

FIGURE 3.29 **Structure of the** α **helix.** (A) A ribbon depiction with the α-carbon atoms and side chains (green) shown. (B) A side view of a ball-and-stick version depicts the hydrogen bonds (dashed lines) between NH and CO groups. (C) An end view shows the coiled backbone as the inside of the helix and the side chains (green) projecting outward. (D) A space-filling view of part C shows the tightly packed interior core of the helix.

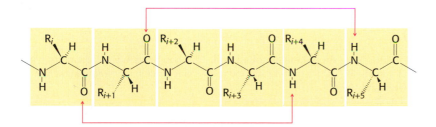

FIGURE 3.30 **Hydrogen-bonding scheme for an** α **helix.** In the α helix, the CO group of residue n forms a hydrogen bond with the NH group of residue $n + 4$.

the right-handed and the left-handed helices are among allowed conformations (Figure 3.31). However, right-handed helices are energetically more favorable because there is less steric clash between the side chains and the backbone. *Essentially all* α *helices found in proteins are right-handed.* In schematic diagrams of proteins, α helices are depicted as twisted ribbons or rods (Figure 3.32).

Pauling and Corey predicted the structure of the α helix 6 years before it was actually seen in the x-ray reconstruction of the structure of myoglobin. *The elucidation of the structure of the* α *helix is a landmark in biochemistry because it demonstrated that the conformation of a polypeptide chain can be predicted if the properties of its components are rigorously and precisely known.*

The α-helical content of proteins ranges widely, from nearly none to almost 100%. For example, about 75% of the residues in ferritin, a protein that helps store iron, are in α helices (Figure 3.33). Single α helices are usually less than 45 Å long. However, two or more α helices can entwine to form a very stable structure, which can have a length of 1000 Å (100 nm, or 0.1 μm)

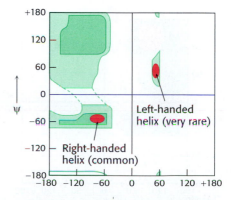

FIGURE 3.31 **Ramachandran diagram for helices.** Both right- and left-handed helices lie in regions of allowed conformations in the Ramachandran diagram. However, essentially all α helices in proteins are right-handed.

(A)   (B)   (C)

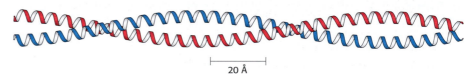

FIGURE 3.32 **Schematic views of α helices.** (A) A ball-and-stick model. (B) A ribbon depiction. (C) A cylindrical depiction.

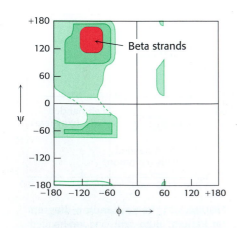

FIGURE 3.33 **A largely α helical protein.** Ferritin, an iron-storage protein, is built from a bundle of α helices.

or more (Figure 3.34). Such *α-helical coiled coils* are found in myosin and tropomyosin in muscle, in fibrin in blood clots, and in keratin in hair. The helical cables in these proteins serve a mechanical role in forming stiff bundles of fibers, as in porcupine quills. The cytoskeleton (internal scaffolding) of cells is rich in so-called intermediate filaments, which also are two-stranded α-helical coiled coils. Many proteins that span biological membranes also contain α helices.

20 Å

FIGURE 3.34 **An α-helical coiled coil.** The two helices wind around one another to form a superhelix. Such structures are found in many proteins including keratin in hair, quills, claws, and horns.

3.3.2 Beta Sheets Are Stabilized by Hydrogen Bonding Between Polypeptide Strands

Pauling and Corey discovered another periodic structural motif, which they named the *β pleated sheet* (β because it was the second structure that they elucidated, the α helix having been the first). The β pleated sheet (or, more simply, the β sheet) differs markedly from the rodlike α helix. A polypeptide chain, called a *β strand,* in a β sheet is almost fully extended rather than being tightly coiled as in the α helix. A range of extended structures are sterically allowed (Figure 3.35).

The distance between adjacent amino acids along a β strand is approximately 3.5 Å, in contrast with a distance of 1.5 Å along an α helix. The side chains of adjacent amino acids point in opposite directions (Figure 3.36). A β sheet is formed by linking two or more β strands by hydrogen bonds. Adjacent chains in a β sheet can run in opposite directions (antiparallel β sheet) or in the same direction (parallel β sheet). In the antiparallel arrangement, the NH group and the CO group of each amino acid are respectively hydrogen bonded to the CO group and the NH group of a partner on the adjacent chain (Figure 3.37). In the parallel arrangement, the hydrogen-bonding scheme is slightly more complicated. For each amino acid, the NH group is hydrogen

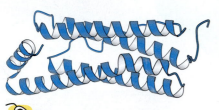

FIGURE 3.35 **Ramachandran diagram for β strands.** The red area shows the sterically allowed conformations of extended, β-strand-like structures.

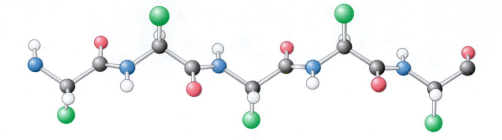

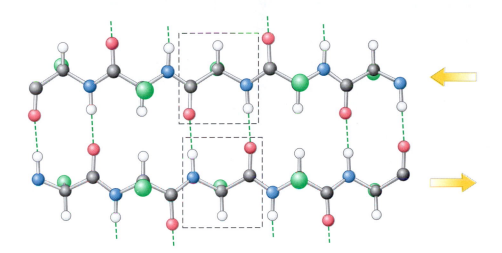

FIGURE 3.36 Structure of a β strand. The side chains (green) are alternately above and below the plane of the strand.

FIGURE 3.37 An antiparallel β sheet. Adjacent β strands run in opposite directions. Hydrogen bonds between NH and CO groups connect each amino acid to a single amino acid on an adjacent strand, stabilizing the structure.

bonded to the CO group of one amino acid on the adjacent strand, whereas the CO group is hydrogen bonded to the NH group on the amino acid two residues farther along the chain (Figure 3.38). Many strands, typically 4 or 5 but as many as 10 or more, can come together in β sheets. Such β sheets can be purely antiparallel, purely parallel, or mixed (Figure 3.39).

In schematic diagrams, β strands are usually depicted by broad arrows pointing in the direction of the carboxyl-terminal end to indicate the type of β sheet formed—parallel or antiparallel. More structurally diverse than α helices, β sheets can be relatively flat but most adopt a somewhat twisted shape (Figure 3.40). The β sheet is an important structural element in many proteins. For example, fatty acid-binding proteins, important for lipid metabolism, are built almost entirely from β sheets (Figure 3.41).

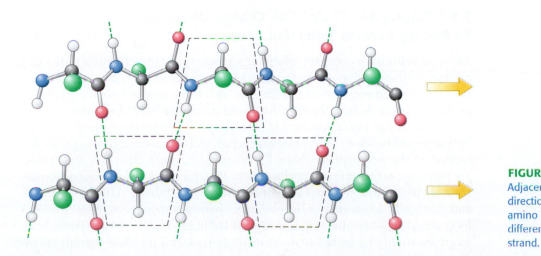

FIGURE 3.38 A parallel β sheet. Adjacent β strands run in the same direction. Hydrogen bonds connect each amino acid on one strand with two different amino acids on the adjacent strand.

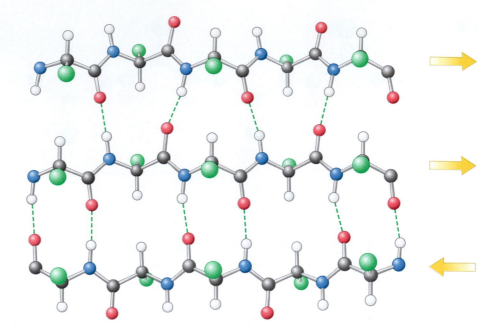

FIGURE 3.39 Structure of a mixed β sheet.

(A)

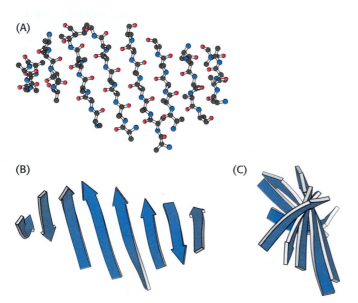

(B)

(C)

FIGURE 3.40 A twisted β sheet. (A) A ball-and-stick model. (B) A schematic model. (C) The schematic view rotated by 90 degrees to illustrate the twist more clearly.

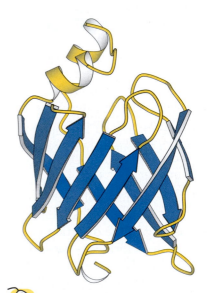

FIGURE 3.41 A protein rich in β sheets. The structure of a fatty acid-binding protein.

3.3.3 Polypeptide Chains Can Change Direction by Making Reverse Turns and Loops

Most proteins have compact, globular shapes, requiring reversals in the direction of their polypeptide chains. Many of these reversals are accomplished by a common structural element called the *reverse turn* (also known as the *β turn* or *hairpin bend*), illustrated in Figure 3.42. In many reverse turns, the CO group of residue i of a polypeptide is hydrogen bonded to the NH group of residue $i + 3$. This interaction stabilizes abrupt changes in direction of the polypeptide chain. In other cases, more elaborate structures are responsible for chain reversals. These structures are called *loops* or sometimes Ω *loops* (omega loops) to suggest their overall shape. Unlike α helices and β strands, loops do not have regular, periodic structures. Nonetheless, loop structures are often rigid and well defined (Figure 3.43). Turns and loops invariably lie on the surfaces of proteins and thus often participate in

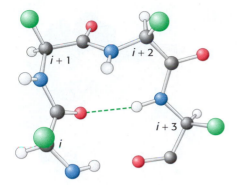

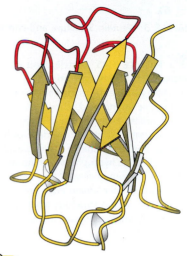

FIGURE 3.42 **Structure of a reverse turn.** The CO group of residue *i* of the polypeptide chain is hydrogen bonded to the NH group of residue *i* + 3 to stabilize the turn.

interactions between proteins and other molecules. The distribution of α helices, β strands, and turns along a protein chain is often referred to as its *secondary structure*.

FIGURE 3.43 **Loops on a protein surface.** A part of an antibody molecule has surface loops (shown in red) that mediate interactions with other molecules.

3.4 TERTIARY STRUCTURE: WATER-SOLUBLE PROTEINS FOLD INTO COMPACT STRUCTURES WITH NONPOLAR CORES

Let us now examine how amino acids are grouped together in a complete protein. X-ray crystallographic and nuclear magnetic resonance studies (Section 4.5) have revealed the detailed three-dimensional structures of thousands of proteins. We begin here with a preview of *myoglobin,* the first protein to be seen in atomic detail.

Myoglobin, the oxygen carrier in muscle, is a single polypeptide chain of 153 amino acids (see also Chapters 7 and 10). The capacity of myoglobin to bind oxygen depends on the presence of *heme,* a nonpolypeptide *prosthetic (helper) group* consisting of protoporphyrin IX and a central iron atom. *Myoglobin is an extremely compact molecule.* Its overall dimensions are 45 × 35 × 25 Å, an order of magnitude less than if it were fully stretched out (Figure 3.44).

(A)
(B)

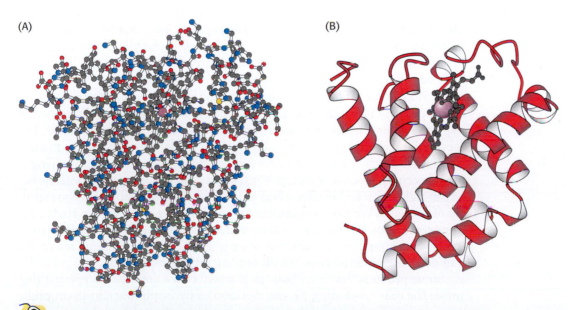

FIGURE 3.44 **Three-dimensional structure of myoglobin.** (A) This ball-and-stick model shows all nonhydrogen atoms and reveals many interactions between the amino acids. (B) A schematic view shows that the protein consists largely of α helices. The heme group is shown in black and the iron atom is shown as a purple sphere.

FIGURE 3.45 Distribution of amino acids in myoglobin. (A) A space-filling model of myoglobin with hydrophobic amino acids shown in yellow, charged amino acids shown in blue, and others shown in white. The surface of the molecule has many charged amino acids, as well as some hydrophobic amino acids. (B) A cross-sectional view shows that mostly hydrophobic amino acids are found on the inside of the structure, whereas the charged amino acids are found on the protein surface.

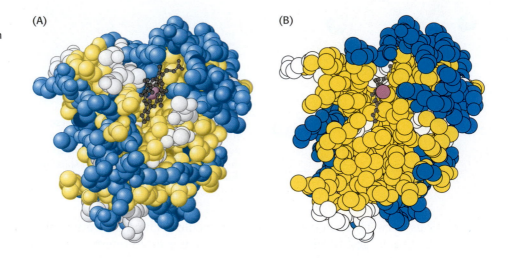

About 70% of the main chain is folded into eight α helices, and much of the rest of the chain forms turns and loops between helices.

The folding of the main chain of myoglobin, like that of most other proteins, is complex and devoid of symmetry. The overall course of the polypeptide chain of a protein is referred to as its *tertiary structure*. A unifying principle emerges from the distribution of side chains. The striking fact is that *the interior consists almost entirely of nonpolar residues* such as leucine, valine, methionine, and phenylalanine (Figure 3.45). Charged residues such as aspartate, glutamate, lysine, and arginine are absent from the inside of myoglobin. The only polar residues inside are two histidine residues, which play critical roles in binding iron and oxygen. The outside of myoglobin, on the other hand, consists of both polar and nonpolar residues. The space-filling model shows that there is very little empty space inside.

This contrasting distribution of polar and nonpolar residues reveals a key facet of protein architecture. In an aqueous environment, protein folding is driven by the strong tendency of hydrophobic residues to be excluded from water (see Section 1.3.4). Recall that a system is more thermodynamically stable when hydrophobic groups are clustered rather than extended into the aqueous surroundings. *The polypeptide chain therefore folds so that its hydrophobic side chains are buried and its polar, charged chains are on the surface.* Many α helices and β strands are amphipathic; that is, the α helix or β strand has a hydrophobic face, which points into the protein interior, and a more polar face, which points into solution. The fate of the main chain accompanying the hydrophobic side chains is important, too. An unpaired peptide NH or CO group markedly prefers water to a nonpolar milieu. The secret of burying a segment of main chain in a hydrophobic environment is pairing all the NH and CO groups by hydrogen bonding. This pairing is neatly accomplished in an α helix or β sheet. Van der Waals interactions between tightly packed hydrocarbon side chains also contribute to the stability of proteins. We can now understand why the set of 20 amino acids contains several that differ subtly in size and shape. They provide a palette from which to choose to fill the interior of a protein neatly and thereby maximize van der Waals interactions, which require intimate contact.

Some proteins that span biological membranes are "the exceptions that prove the rule" regarding the distribution of hydrophobic and hydrophilic amino acids throughout three-dimensional structures. For example, consider porins, proteins found in the outer membranes of many bacteria (Figure 3.46). The permeability barriers of membranes are built largely of alkane chains that are quite hydrophobic (Section 12.4). Thus, porins are

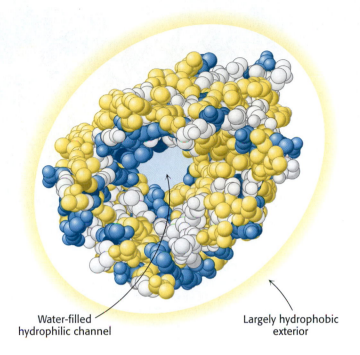

Water-filled
hydrophilic channel

Largely hydrophobic
exterior

**FIGURE 3.46 "Inside out"
amino acid distribution in porin.**
The outside of porin (which contacts
hydrophobic groups in membranes) is
covered largely with hydrophobic residues,
whereas the center includes a water-filled
channel lined with charged and polar
amino acids.

covered on the outside largely with hydrophobic residues that interact with
the neighboring alkane chains. In contrast, the center of the protein contains many charged and polar amino acids that surround a water-filled
channel running through the middle of the protein. Thus, because porins
function in hydrophobic environments, they are "inside out" relative to proteins that function in aqueous solution.

Some polypeptide chains fold into two or more compact regions that may
be connected by a flexible segment of polypeptide chain, rather like pearls
on a string. These compact globular units, called *domains,* range in size from
about 30 to 400 amino acid residues. For example, the extracellular part of
CD4, the cell-surface protein on certain cells of the immune system to which
the human immunodeficiency virus (HIV) attaches itself, comprises four
similar domains of approximately 100 amino acids each (Figure 3.47). Often, proteins are found to have domains in common even if their overall tertiary structures are different.

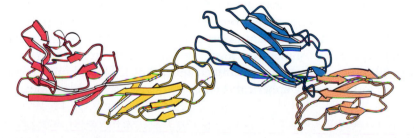

FIGURE 3.47 Protein domains.
The cell-surface protein CD4 consists of
four similar domains.

3.5 QUATERNARY STRUCTURE: POLYPEPTIDE CHAINS CAN ASSEMBLE INTO MULTISUBUNIT STRUCTURES

Four levels of structure are frequently cited in discussions of protein architecture. So far, we have considered three of them. *Primary structure* is the
amino acid sequence. *Secondary structure* refers to the spatial arrangement
of amino acid residues that are nearby in the sequence. Some of these
arrangements are of a regular kind, giving rise to a periodic structure. The
α helix and β strand are elements of secondary structure. *Tertiary structure*

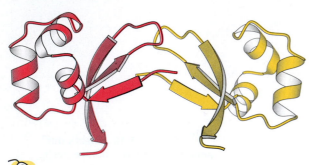

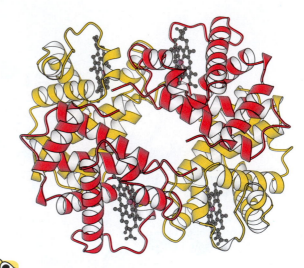

FIGURE 3.48 Quaternary structure. The Cro protein of bacteriophage λ is a dimer of identical subunits.

FIGURE 3.49 The $\alpha_2\beta_2$ tetramer of human hemoglobin. The structure of the two identical α subunits (red) is similar to but not identical with that of the two identical β subunits (yellow). The molecule contains four heme groups (black with the iron atom shown in purple).

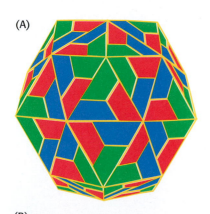

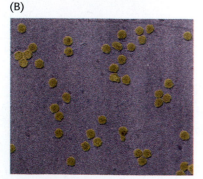

FIGURE 3.50 Complex quaternary structure. The coat of rhinovirus comprises 60 copies of each of four subunits. (A) A schematic view depicting the three types of subunits (shown in red, blue, and green) visible from outside the virus. (B) An electron micrograph showing rhinovirus particles. [Courtesy of Norm Olson, Dept. of Biological Sciences, Purdue University.]

refers to the spatial arrangement of amino acid residues that are far apart in the sequence and to the pattern of disulfide bonds. We now turn to proteins containing more than one polypeptide chain. Such proteins exhibit a fourth level of structural organization. Each polypeptide chain in such a protein is called a *subunit. Quaternary structure* refers to the spatial arrangement of subunits and the nature of their interactions. The simplest sort of quaternary structure is a *dimer,* consisting of two identical subunits. This organization is present in the DNA-binding protein Cro found in a bacterial virus called λ (Figure 3.48). More complicated quaternary structures also are common. More than one type of subunit can be present, often in variable numbers. For example, human hemoglobin, the oxygen-carrying protein in blood, consists of two subunits of one type (designated α) and two subunits of another type (designated β), as illustrated in Figure 3.49. Thus, the hemoglobin molecule exists as an $\alpha_2\beta_2$ tetramer. Subtle changes in the arrangement of subunits within the hemoglobin molecule allow it to carry oxygen from the lungs to tissues with great efficiency (Section 10.2).

Viruses make the most of a limited amount of genetic information by forming coats that use the same kind of subunit repetitively in a symmetric array. The coat of rhinovirus, the virus that causes the common cold, includes 60 copies each of four subunits (Figure 3.50). The subunits come together to form a nearly spherical shell that encloses the viral genome.

3.6 THE AMINO ACID SEQUENCE OF A PROTEIN DETERMINES ITS THREE-DIMENSIONAL STRUCTURE

How is the elaborate three-dimensional structure of proteins attained, and how is the three-dimensional structure related to the one-dimensional amino acid sequence information? The classic work of Christian Anfinsen in the 1950s on the enzyme ribonuclease revealed the relation between the amino acid sequence of a protein and its conformation. Ribonuclease is a single polypeptide chain consisting of 124 amino acid residues cross-linked by four disulfide bonds (Figure 3.51). Anfinsen's plan was to destroy the

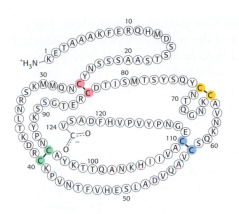

FIGURE 3.51 Amino acid sequence of bovine ribonuclease. The four disulfide bonds are shown in color. [After C. H. W. Hirs, S. Moore, and W. H. Stein, *J. Biol. Chem.* 235 (1960):633.]

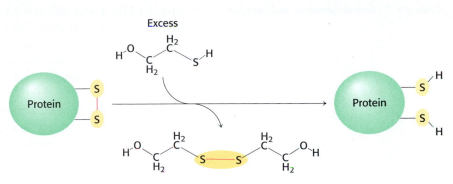

FIGURE 3.52 Role of β-mercaptoethanol in reducing disulfide bonds. Note that, as the disulfides are reduced, the β-mercaptoethanol is oxidized and forms dimers.

three-dimensional structure of the enzyme and to then determine what conditions were required to restore the structure.

Agents such as urea or guanidinium chloride effectively disrupt the noncovalent bonds, although the mechanism of action of these agents is not fully understood. The disulfide bonds can be cleaved reversibly by reducing them with a reagent such as *β-mercaptoethanol* (Figure 3.52). In the presence of a large excess of β-mercaptoethanol, a protein is produced in which the disulfides (cystines) are fully converted into sulfhydryls (cysteines).

Most polypeptide chains devoid of cross-links assume a *random-coil conformation* in 8 M urea or 6 M guanidinium chloride, as evidenced by physical properties such as viscosity and optical activity. When ribonuclease was treated with β-mercaptoethanol in 8 M urea, the product was a fully reduced, randomly coiled polypeptide chain *devoid of enzymatic activity*. In other words, ribonuclease was *denatured* by this treatment (Figure 3.53).

Anfinsen then made the critical observation that the denatured ribonuclease, freed of urea and β-mercaptoethanol by dialysis, slowly regained enzymatic activity. He immediately perceived the significance of this chance finding: the sulfhydryl groups of the denatured enzyme became oxidized by air, and the enzyme spontaneously refolded into a catalytically active form. Detailed studies then showed that nearly all the original enzymatic activity was regained if the sulfhydryl groups were oxidized under suitable conditions. All the measured physical and chemical properties of the refolded enzyme were virtually identical with those of the native enzyme. These experiments showed that *the information needed to specify the catalytically active structure of ribonuclease is contained in its amino acid sequence.* Subsequent studies have established the generality of this central principle of biochemistry: *sequence specifies conformation.* The dependence of conformation on sequence is especially significant because of the intimate connection between conformation and function.

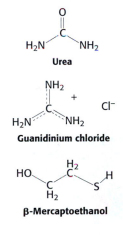

Urea

Guanidinium chloride

β-Mercaptoethanol

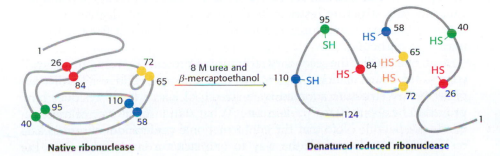

Native ribonuclease

Denatured reduced ribonuclease

8 M urea and β-mercaptoethanol

FIGURE 3.53 Reduction and denaturation of ribonuclease.

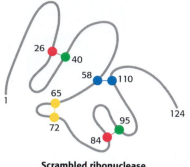

Scrambled ribonuclease

Trace of
β-mercaptoethanol

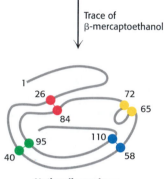

Native ribonuclease

FIGURE 3.54 Reestablishing correct disulfide pairing. Native ribonuclease can be reformed from scrambled ribonuclease in the presence of a trace of β-mercaptoethanol.

A quite different result was obtained when reduced ribonuclease was reoxidized while it was still in 8 M urea and the preparation was then dialyzed to remove the urea. Ribonuclease reoxidized in this way had only 1% of the enzymatic activity of the native protein. Why were the outcomes so different when reduced ribonuclease was reoxidized in the presence and absence of urea? The reason is that the wrong disulfides formed pairs in urea. There are 105 different ways of pairing eight cysteine molecules to form four disulfides; only one of these combinations is enzymatically active. The 104 wrong pairings have been picturesquely termed "scrambled" ribonuclease. Anfinsen found that scrambled ribonuclease spontaneously converted into fully active, native ribonuclease when trace amounts of β-mercaptoethanol were added to an aqueous solution of the protein (Figure 3.54). The added β-mercaptoethanol catalyzed the rearrangement of disulfide pairings until the native structure was regained in about 10 hours. *This process was driven by the decrease in free energy as the scrambled conformations were converted into the stable, native conformation of the enzyme.* The native disulfide pairings of ribonuclease thus contribute to the stabilization of the thermodynamically preferred structure.

Similar refolding experiments have been performed on many other proteins. In many cases, the native structure can be generated under suitable conditions. For other proteins, however, refolding does not proceed efficiently. In these cases, the unfolding protein molecules usually become tangled up with one another to form aggregates. Inside cells, proteins called *chaperones* block such illicit interactions (Section 11.3.6).

3.6.1 Amino Acids Have Different Propensities for Forming Alpha Helices, Beta Sheets, and Beta Turns

How does the amino acid sequence of a protein specify its three-dimensional structure? How does an unfolded polypeptide chain acquire the form of the native protein? These fundamental questions in biochemistry can be approached by first asking a simpler one: What determines whether a particular sequence in a protein forms an α helix, a β strand, or a turn? Examining the frequency of occurrence of particular amino acid residues in these secondary structures (Table 3.3) can be a source of insight into this determination. Residues such as alanine, glutamate, and leucine tend to be present in α helices, whereas valine and isoleucine tend to be present in β strands. Glycine, asparagine, and proline have a propensity for being in turns.

The results of studies of proteins and synthetic peptides have revealed some reasons for these preferences. The α helix can be regarded as the default conformation. Branching at the β-carbon atom, as in valine, threonine, and isoleucine, tends to destabilize α helices because of steric clashes. These residues are readily accommodated in β strands, in which their side chains project out of the plane containing the main chain. Serine, aspartate, and asparagine tend to disrupt α helices because their side chains contain hydrogen-bond donors or acceptors in close proximity to the main chain, where they compete for main-chain NH and CO groups. Proline tends to disrupt both α helices and β strands because it lacks an NH group and because its ring structure restricts its φ value to near −60 degrees. Glycine readily fits into all structures and for that reason does not favor helix formation in particular.

Can one predict the secondary structure of proteins by using this knowledge of the conformational preferences of amino acid residues? Predictions of secondary structure adopted by a stretch of six or fewer residues have proved to be about 60 to 70% accurate. What stands in the way of more accurate prediction? Note that the conformational preferences of amino acid residues are not tipped all the way to one structure (see Table 3.3). For

TABLE 3.3 Relative frequencies of amino acid residues in secondary structures

Amino acid	α helix	β sheet	Turn
Ala	**1.29**	0.90	0.78
Cys	**1.11**	0.74	0.80
Leu	**1.30**	1.02	0.59
Met	**1.47**	0.97	0.39
Glu	**1.44**	0.75	1.00
Gln	**1.27**	0.80	0.97
His	**1.22**	1.08	0.69
Lys	**1.23**	0.77	0.96
Val	0.91	**1.49**	0.47
Ile	0.97	**1.45**	0.51
Phe	1.07	**1.32**	0.58
Tyr	0.72	**1.25**	1.05
Trp	0.99	**1.14**	0.75
Thr	0.82	**1.21**	1.03
Gly	0.56	0.92	**1.64**
Ser	0.82	0.95	**1.33**
Asp	1.04	0.72	**1.41**
Asn	0.90	0.76	**1.28**
Pro	0.52	0.64	**1.91**
Arg	0.96	0.99	0.88

Note: The amino acids are grouped according to their preference for α helices (top group), β sheets (second group), or turns (third group). Arginine shows no significant preference for any of the structures.
After T. E. Creighton, *Proteins: Structures and Molecular Properties,* 2d ed. (W. H. Freeman and Company, 1992), p. 256.

example, glutamate, one of the strongest helix formers, prefers α helix to β strand by only a factor of two. The preference ratios of most other residues are smaller. Indeed, some penta- and hexapeptide sequences have been found to adopt one structure in one protein and an entirely different structure in another (Figure 3.55). Hence, some amino acid sequences do not uniquely determine secondary structure. Tertiary interactions—interactions between residues that are far apart in the sequence—may be decisive in specifying the secondary structure of some segments. *The context is often crucial in determining the conformational outcome.* The conformation of a protein evolved to work in a particular environment or context.

Pathological conditions can result if a protein assumes an inappropriate conformation for the context. Striking examples are *prion diseases,* such as Creutzfeldt-Jacob disease, kuru, and mad cow disease. These conditions result when a brain protein called a prion converts from its normal conformation (designated PrP^C) to an altered one (PrP^{SC}). This conversion is self-propagating, leading to large aggregates of PrP^{SC}. The role of these aggregates in the generation of the pathological conditions is not yet understood.

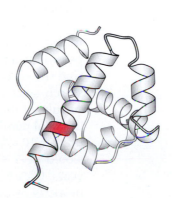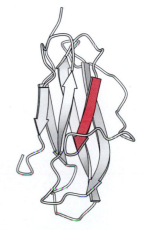

FIGURE 3.55 Alternative conformations of a peptide sequence. Many sequences can adopt alternative conformations in different proteins. Here the sequence VDLLKN shown in red assumes an α helix in one protein context (left) and a β strand in another (right).

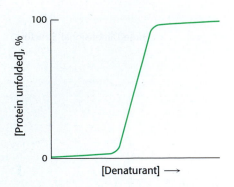

FIGURE 3.56 Transition from folded to unfolded state. Most proteins show a sharp transition from the folded to unfolded form on treatment with increasing concentrations of denaturants.

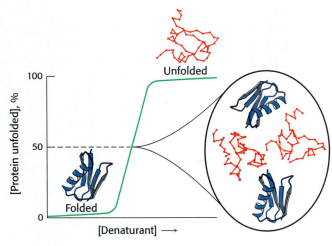

FIGURE 3.57 Components of a partially denatured protein solution. In a half-unfolded protein solution, half the molecules are fully folded and half are fully unfolded.

3.6.2 Protein Folding Is a Highly Cooperative Process

As stated earlier, proteins can be denatured by heat or by chemical denaturants such as urea or guanidium chloride. For many proteins, a comparison of the degree of unfolding as the concentration of denaturant increases has revealed a relatively sharp transition from the folded, or native, form to the unfolded, or denatured, form, suggesting that only these two conformational states are present to any significant extent (Figure 3.56). A similar sharp transition is observed if one starts with unfolded proteins and removes the denaturants, allowing the proteins to fold.

Protein folding and unfolding is thus largely an *"all or none" process* that results from a *cooperative transition*. For example, suppose that a protein is placed in conditions under which some part of the protein structure is thermodynamically unstable. As this part of the folded structure is disrupted, the interactions between it and the remainder of the protein will be lost. The loss of these interactions, in turn, will destabilize the remainder of the structure. Thus, conditions that lead to the disruption of any part of a protein structure are likely to unravel the protein completely. The structural properties of proteins provide a clear rationale for the cooperative transition.

The consequences of cooperative folding can be illustrated by considering the contents of a protein solution under conditions corresponding to the middle of the transition between the folded and unfolded forms. Under these conditions, the protein is "half folded." Yet the solution will contain no half-folded molecules but, instead, will be a 50/50 mixture of fully folded and fully unfolded molecules (Figure 3.57). Structures that are partly intact and partly disrupted are not thermodynamically stable and exist only transiently. Cooperative folding ensures that partly folded structures that might interfere with processes within cells do not accumulate.

3.6.3 Proteins Fold by Progressive Stabilization of Intermediates Rather Than by Random Search

The cooperative folding of proteins is a thermodynamic property; its occurrence reveals nothing about the kinetics and mechanism of protein folding. How does a protein make the transition from a diverse ensemble of unfolded structures into a unique conformation in the native form? One possibility a priori would be that all possible conformations are tried out to find the energetically most favorable one. How long would such a random search take? Consider a small protein with 100 residues. Cyrus Levinthal calculated that, if each residue can assume three different conformations, the total number of structures would be 3^{100}, which is equal to 5×10^{47}. If it takes 10^{-13} s to convert one structure into another, the total search time would be $5 \times 10^{47} \times 10^{-13}$ s, which is equal to 5×10^{34} s, or 1.6×10^{27} years. Clearly, it would take much too long for even a small protein to fold properly by randomly trying out all possible conformations. The enormous difference between calculated and actual folding times is called *Levinthal's paradox.*

The way out of this dilemma is to recognize the power of *cumulative selection*. Richard Dawkins, in *The Blind Watchmaker,* asked how long it would take a monkey poking randomly at a typewriter to reproduce Hamlet's remark to Polonius, "Methinks it is like a weasel" (Figure 3.58). An astronomically large number of keystrokes, of the order of 10^{40}, would be required. However, suppose that we preserved each correct character and allowed the monkey to retype only the wrong ones. In this case, only a few thousand keystrokes, on average, would be needed. The crucial difference between these cases is that the first employs a completely random search, whereas, in the second, *partly correct intermediates are retained.*

The essence of protein folding is the retention of partly correct intermediates. However, the protein-folding problem is much more difficult than the one presented to our simian Shakespeare. First, the criterion of correctness is not a residue-by-residue scrutiny of conformation by an omniscient observer but rather the total free energy of the transient species. Second, proteins are only marginally stable. The free-energy difference between the folded and the unfolded states of a typical 100-residue protein is 10 kcal mol^{-1} (42 kJ mol^{-1}), and thus each residue contributes on average only 0.1 kcal mol^{-1} (0.42 kJ mol^{-1}) of energy to maintain the folded state. This amount is less than that of thermal energy, which is 0.6 kcal mol^{-1} (2.5 kJ mol^{-1}) at room temperature. This meager stabilization energy means that correct intermediates, especially those formed early in folding, can be lost. The analogy is that the monkey would be somewhat free to undo its correct keystrokes. Nonetheless, the interactions that lead to cooperative folding can stabilize intermediates as structure builds up. Thus, local regions, which have significant structural preference, though not necessarily stable on their own, will tend to adopt their favored structures and, as they form, can interact with one other, leading to increasing stabilization.

3.6.4 Prediction of Three-Dimensional Structure from Sequence Remains a Great Challenge

The amino acid sequence completely determines the three-dimensional structure of a protein. However, the prediction of three-dimensional structure from sequence has proved to be extremely difficult. As we have seen, the local sequence appears to determine only between 60% and 70% of the secondary structure; long-range interactions are required to fix the full secondary structure and the tertiary structure.

Investigators are exploring two fundamentally different approaches to predicting three-dimensional structure from amino acid sequence. The first is *ab initio prediction,* which attempts to predict the folding of an amino acid sequence without any direct reference to other known protein structures. Computer-based calculations are employed that attempt to minimize the free energy of a structure with a given amino acid sequence or to simulate the folding process. The utility of these methods is limited by the vast number of possible conformations, the marginal stability of proteins, and the subtle energetics of weak interactions in aqueous solution. The second approach takes advantage of our growing knowledge of the three-dimensional structures of many proteins. In these *knowledge-based methods,* an amino acid sequence of unknown structure is examined for compatibility with any known protein structures. If a significant match is detected, the known structure can be used as an initial model. Knowledge-based methods have been a source of many insights into the three-dimensional conformation of proteins of known sequence but unknown structure.

3.6.5 Protein Modification and Cleavage Confer New Capabilities

Proteins are able to perform numerous functions relying solely on the versatility of their 20 amino acids. However, many proteins are covalently modifed, through the attachment of groups other than amino acids, to augment their functions (Figure 3.59). For example, *acetyl groups* are attached to the amino termini of many proteins, a modification that

```
 200  ?T(\G{+s x[A.N5~,#ATxSGpn`e□@
 400  oDr'Jh7s DFR:W4l'u+^v6zpJseOi
 600  e2ih'8zs n527x8l8d_ih=Hldseb.
 800  S#dh>}/s ]tZqC%1P%DK<|!^aseZ.
1000  VOth>nLs ut/isjl_kwojjwMasef.
1200  juth+nvs it is[lukh?SCw=ase5.
1400  Iithdn4s it isOl/ks/IxwLase~.
1600  M?thinrs it is lXk?T"_woasel.
1800  MSthinWs it is lwkN7□Kw(asel.
2000  Mhthin`s it is likv,aww_asel.
2200  MMthinns it is lik+5avwlasel.
2400  MethinXs it is likydaqw)asel.
2600  Methin4s it is lik2dasweasel.
2800  MethinHs it is like□aTweasel.
2883  Methinks it is like a weasel.
```

```
 200  )z~hg)W4{{cu!kO{d6jS!NlEyUx}p
 400  "W hi\kR.<&CfA%4-Y1G!iT$6({|6
 600  .L=hinkm4(uMGP^lAWoE6klwW=yiS
 800   AthinkaPa_vYH liR\Hb,Uo4\-"(
1000  OFthinksP)@fZO li8v] /+Eln26B
1200  6ithinksMVt -V likm+gl#K~}BFk
1400  vxthinksaEt □w like.S1Geutks.
1600  :Othinks<it MC likesN2[eaVe4.
1800  uxthinksqit 0r likeQh)weaoeW.
2000  Y/thinks it id like7alwea)e&.
2200  Methinks it iW like a[weaWel.
2400  Methinks it is like a;weasel.
2431  Methinks it is like a weasel.
```

FIGURE 3.58 Typing monkey anology. A monkey randomly poking a typewriter could write a line from Shakespeare's *Hamlet,* provided that correct keystrokes were retained. In the two computer simulations shown, the cumulative number of keystrokes is given at the left of each line.

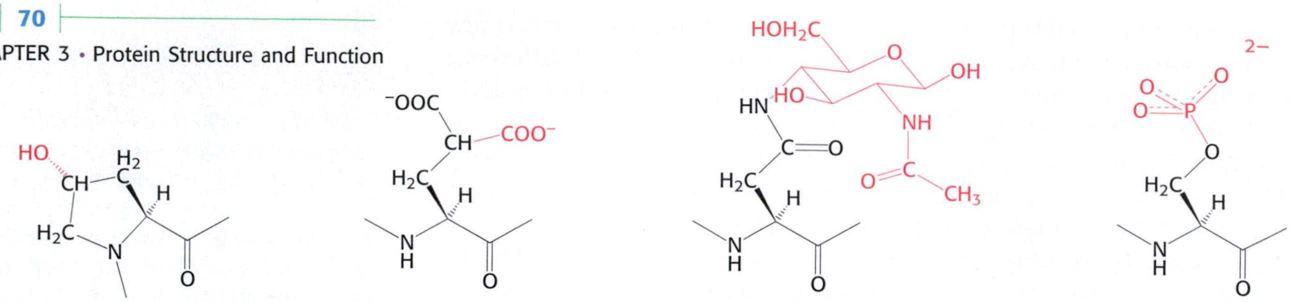

Hydroxyproline   **γ-Carboxyglutamate**   **Carbohydrate–asparagine adduct**   **Phosphoserine**

FIGURE 3.59 Finishing touches.
Some common and important covalent modifications of amino acid side chains are shown.

makes these proteins more resistant to degradation. The addition of *hydroxyl groups* to many proline residues stabilizes fibers of newly synthesized collagen, a fibrous protein found in connective tissue and bone. The biological significance of this modification is evident in the disease scurvy: a deficiency of vitamin C results in insufficient hydroxylation of collagen and the abnormal collagen fibers that result are unable to maintain normal tissue strength. Another specialized amino acid produced by a finishing touch is *γ-carboxyglutamate*. In vitamin K deficiency, insufficient carboxylation of glutamate in prothrombin, a clotting protein, can lead to hemorrhage. Many proteins, especially those that are present on the surfaces of cells or are secreted, acquire *carbohydrate units* on specific asparagine residues. The addition of sugars makes the proteins more hydrophilic and able to participate in interactions with other proteins. Conversely, the addition of a *fatty acid* to an α-amino group or a cysteine sulfhydryl group produces a more hydrophobic protein.

Many hormones, such as epinephrine (adrenaline), alter the activities of enzymes by stimulating the phosphorylation of the hydroxyl amino acids serine and threonine; *phosphoserine* and *phosphothreonine* are the most ubiquitous modified amino acids in proteins. Growth factors such as insulin act by triggering the phosphorylation of the hydroxyl group of tyrosine residues to form *phosphotyrosine*. The phosphoryl groups on these three modified amino acids are readily removed; thus they are able to act as reversible switches in regulating cellular processes. The roles of phosphorylation in signal transduction will be discussed extensively in Chapter 15.

The preceding modifications consist of the addition of special groups to amino acids. Other special groups are generated by chemical rearrangements of side chains and, sometimes, the peptide backbone. For example, certain jellyfish produce a fluorescent green protein (Figure 3.60). The source of the fluorescence is a group formed by the spontaneous rearrangement and oxidation of the sequence Ser-Tyr-Gly within the center of the protein. This protein is of great utility to researchers as a marker within cells (Section 4.3.5).

Finally, many proteins are cleaved and trimmed after synthesis. For example, digestive enzymes are synthesized as inactive precursors that can be stored safely in the pancreas. After release into the intestine, these precursors become activated by peptide-bond cleavage. In blood clotting, peptide-bond cleavage converts soluble fibrinogen into insoluble fibrin. A number of polypeptide hormones, such as adrenocorticotropic hormone, arise from the splitting of a single large precursor protein. Likewise, many virus proteins are produced by the cleavage of large polyprotein precursors. We shall encounter many more examples of modification and cleavage as essential features of protein formation and function. Indeed, these finishing touches account for much of the versatility, precision, and elegance of protein action and regulation.

(A)

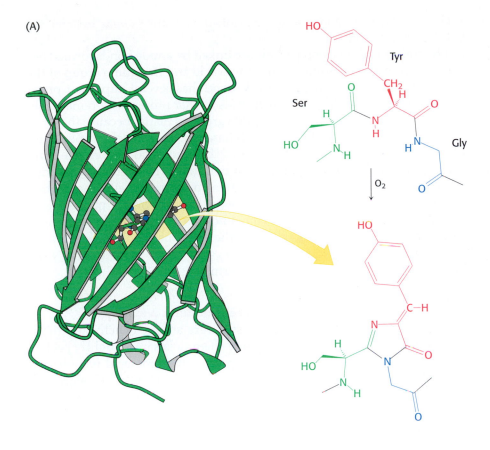

(B)

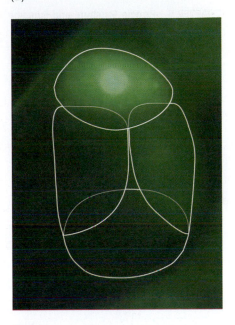

FIGURE 3.60 Chemical rearrangement in GFP. (A) The structure of green fluorescent protein (GFP). The rearrangement and oxidation of the sequence Ser-Try-Gly is the source of fluorescence. (B) Fluorescence micrograph of a four-cell embryo (cells are outlined) from the roundworm *C. elegans* containing a protein, PIE-1, labeled with GFP. The protein is expressed only in the cell (top) that will give rise to the germline. [(B) Courtesy of Geraldine Seydoux.]

SUMMARY

Proteins are the workhorses of biochemistry, participating in essentially all cellular processes. Protein structure can be described at four levels. The primary structure refers to the amino acid sequence. The secondary structure refers to the conformation adopted by local regions of the polypeptide chain. Tertiary structure describes the overall folding of the polypeptide chain. Finally, quaternary structure refers to the specific association of multiple polypeptide chains to form multisubunit complexes.

Proteins Are Built from a Repertoire of 20 Amino Acids

Proteins are linear polymers of amino acids. Each amino acid consists of a central tetrahedral carbon atom linked to an amino group, a carboxylic acid group, a distinctive side chain, and a hydrogen. These tetrahedral centers, with the exception of that of glycine, are chiral; only the L isomer exists in natural proteins. All natural proteins are constructed from the same set of 20 amino acids. The side chains of these 20 building blocks vary tremendously in size, shape, and the presence of functional groups. They can be grouped as follows: (1) aliphatic side chains—glycine, alanine, valine, leucine, isoleucine, methionine, and proline; (2) aromatic side chains—phenylalanine, tyrosine, and tryptophan; (3) hydroxyl-containing aliphatic side chains—serine and threonine; (4) sulfhydryl-containing cysteine; (5) basic side chains—lysine, arginine, and histidine; (6) acidic side chains—aspartic acid and glutamic acid; and (7) carboxamide-containing side chains—asparagine and glutamine. These groupings are somewhat arbitrary and many other sensible groupings are possible.

● **Primary Structure: Amino Acids Are Linked by Peptide Bonds to Form Polypeptide Chains**

The amino acids in a polypeptide are linked by amide bonds formed between the carboxyl group of one amino acid and the amino group of the next. This linkage, called a peptide bond, has several important properties. First, it is resistant to hydrolysis so that proteins are remarkably stable kinetically. Second, the peptide group is planar because the C–N bond has considerable double-bond character. Third, each peptide bond has both a hydrogen-bond donor (the NH group) and a hydrogen-bond acceptor (the CO group). Hydrogen bonding between these backbone groups is a distinctive feature of protein structure. Finally, the peptide bond is uncharged, which allows proteins to form tightly packed globular structures having significant amounts of the backbone buried within the protein interior. Because they are linear polymers, proteins can be described as sequences of amino acids. Such sequences are written from the amino to the carboxyl terminus.

● **Secondary Structure: Polypeptide Chains Can Fold into Regular Structures Such as the Alpha Helix, the Beta Sheet, and Turns and Loops**

Two major elements of secondary structure are the α helix and the β strand. In the α helix, the polypeptide chain twists into a tightly packed rod. Within the helix, the CO group of each amino acid is hydrogen bonded to the NH group of the amino acid four residues along the polypeptide chain. In the β strand, the polypeptide chain is nearly fully extended. Two or more β strands connected by NH-to-CO hydrogen bonds come together to form β sheets.

● **Tertiary Structure: Water-Soluble Proteins Fold into Compact Structures with Nonpolar Cores**

The compact, asymmetric structure that individual polypeptides attain is called tertiary structure. The tertiary structures of water-soluble proteins have features in common: (1) an interior formed of amino acids with hydrophobic side chains and (2) a surface formed largely of hydrophilic amino acids that interact with the aqueous environment. The driving force for the formation of the tertiary structure of water-soluble proteins is the hydrophobic interactions between the interior residues. Some proteins that exist in a hydrophobic environment, in membranes, display the inverse distribution of hydrophobic and hydrophilic amino acids. In these proteins, the hydrophobic amino acids are on the surface to interact with the environment, whereas the hydrophilic groups are shielded from the environment in the interior of the protein.

● **Quaternary Structure: Polypeptide Chains Can Assemble into Multisubunit Structures**

Proteins consisting of more than one polypeptide chain display quaternary structure, and each individual polypeptide chain is called a subunit. Quaternary structure can be as simple as two identical subunits or as complex as dozens of different subunits. In most cases, the subunits are held together by noncovalent bonds.

● **The Amino Acid Sequence of a Protein Determines Its Three-Dimensional Structure**

The amino acid sequence completely determines the three-dimensional structure and, hence, all other properties of a protein. Some proteins can be unfolded completely yet refold efficiently when placed under conditions in which the folded form of the protein is stable. The amino acid sequence of a protein is determined by the sequences of bases in a DNA molecule. This one-dimensional sequence information is extended into the three-dimensional world by the ability of proteins to fold spontaneously. Protein folding is a highly cooperative process; structural intermediates between the unfolded and folded forms do not accumulate.

The versatility of proteins is further enhanced by covalent modifications. Such modifications can incorporate functional groups not present in the 20 amino acids. Other modifications are important to the regulation of protein activity. Through their structural stability, diversity, and chemical reactivity, proteins make possible most of the key processes associated with life.

APPENDIX: ACID–BASE CONCEPTS

Ionization of Water

Water dissociates into hydronium (H_3O^+) and hydroxyl (OH^-) ions. For simplicity, we refer to the hydronium ion as a hydrogen ion (H^+) and write the equilibrium as

$$H_2O \rightleftharpoons H^+ + OH^-$$

The equilibrium constant K_{eq} of this dissociation is given by

$$K_{eq} = [H^+][OH^-]/[H_2O] \qquad (1)$$

in which the terms in brackets denote molar concentrations. Because the concentration of water (55.5 M) is changed little by ionization, expression 1 can be simplified to give

$$K_W = [H^+][OH^-] \qquad (2)$$

in which K_w is the ion product of water. At 25°C, K_w is 1.0×10^{-14}.

Note that the concentrations of H^+ and OH^- are reciprocally related. If the concentration of H^+ is high, then the concentration of OH^- must be low, and vice versa. For example, if $[H^+] = 10^{-2}$ M, then $[OH^-] = 10^{-12}$ M.

Definition of Acid and Base

An acid is a proton donor. A base is a proton acceptor.

$$\text{Acid} \rightleftharpoons H^+ + \text{base}$$

$$\underset{\text{Acetic acid}}{CH_3COOH} \rightleftharpoons H^+ + \underset{\text{Acetate}}{CH_3COO^-}$$

$$\underset{\text{Ammonium ion}}{NH_4^+} \rightleftharpoons H^+ + \underset{\text{Ammonia}}{NH_3}$$

The species formed by the ionization of an acid is its conjugate base. Conversely, protonation of a base yields its conjugate acid. Acetic acid and acetate ion are a conjugate acid–base pair.

Definition of pH and p*K*

The pH of a solution is a measure of its concentration of H^+. The pH is defined as

$$pH = \log_{10}(1/[H^+]) = -\log_{10}[H^+] \qquad (3)$$

The ionization equilibrium of a weak acid is given by

$$HA \rightleftharpoons H^+ + A^-$$

The apparent equilibrium constant K_a for this ionization is

$$K_a = [H^+][A^-]/[HA] \qquad (4)$$

The pK_a of an acid is defined as

$$pK_a = -\log K_a = \log(1/K_a) \qquad (5)$$

Inspection of equation 4 shows that the pK_a of an acid is the pH at which it is half dissociated, when $[A^-] = [HA]$.

Henderson-Hasselbalch Equation

What is the relation between pH and the ratio of acid to base? A useful expression can be derived from equation 4. Rearrangement of that equation gives

$$1/[H^+] = 1/K_a[A^-]/[HA] \qquad (6)$$

Taking the logarithm of both sides of equation 6 gives

$$\log(1/[H^+]) = \log(1/K_a) + \log([A^-]/[HA]) \qquad (7)$$

Substituting pH for log $1/[H^+]$ and pK_a for log $1/K_a$ in equation 7 yields

$$pH = pK_a + \log([A^-]/[HA]) \qquad (8)$$

which is commonly known as the Henderson-Hasselbalch equation.

The pH of a solution can be calculated from equation 8 if the molar proportion of A^- to HA and the pK_a of HA are known. Consider a solution of 0.1 M acetic acid and 0.2 M acetate ion. The pK_a of acetic acid is 4.8. Hence, the pH of the solution is given by

$$pH = 4.8 + \log(0.2/0.1) = 4.8 + \log 2.0 = 4.8 + 0.3 = 5.1$$

Conversely, the pK_a of an acid can be calculated if the molar proportion of A^- to HA and the pH of the solution are known.

Buffers

An acid–base conjugate pair (such as acetic acid and acetate ion) has an important property: it resists changes in the pH of a solution. In other words, it acts as a *buffer*. Consider the addition of OH^- to a solution of acetic acid (HA):

$$HA + OH^- \rightleftharpoons A^- + H_2O$$

A plot of the dependence of the pH of this solution on the amount of OH^- added is called a *titration curve* (Figure 3.61). Note that there is an inflection point in the curve at pH 4.8, which is the pK_a of acetic acid. In the vicinity of this pH, a relatively large amount of OH^- produces little change in pH. In other words, the buffer maintains the value of pH near a given value, despite the addition of other either protons or hydroxide

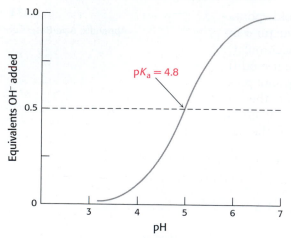

FIGURE 3.61 Titration curve of acetic acid.

TABLE 3.4   **pK_a values of some amino acids**

Amino acid	pK_a values (25°C)		
	α-COOH group	α-NH$_3^+$ group	Side chain
Alanine	2.3	9.9	
Glycine	2.4	9.8	
Phenylalanine	1.8	9.1	
Serine	2.1	9.2	
Valine	2.3	9.6	
Aspartic acid	2.0	10.0	3.9
Glutamic acid	2.2	9.7	4.3
Histidine	1.8	9.2	6.0
Cysteine	1.8	10.8	8.3
Tyrosine	2.2	9.1	10.9
Lysine	2.2	9.2	10.8
Arginine	1.8	9.0	12.5

After J. T. Edsall and J. Wyman, *Biophysical Chemistry* (Academic Press, 1958), Chapter 8.

ions. In general, a weak acid is most effective in buffering against pH changes in the vicinity of its pK_a value.

pK_a Values of Amino Acids

An amino acid such as glycine contains two ionizable groups: an α-carboxyl group and a protonated α-amino group. As base is added, these two groups are titrated (Figure 3.62). The pK_a of the α-COOH group is 2.4, whereas that of the α-NH$_3^+$ group is 9.8. The pK_a values of these groups in other amino acids are similar (Table 3.4). Some amino acids, such as aspartic acid, also contain an ionizable side chain. The pK_a values of ionizable side chains in amino acids range from 3.9 (aspartic acid) to 12.5 (arginine).

FIGURE 3.62 Titration of the α-carboxyl and α-amino groups of an amino acid.

KEY TERMS

side chain (R group) (p. 43)
L amino acid (p. 43)
dipolar ion (zwitterion) (p. 43)
peptide bond (amide bond) (p. 51)
disulfide bond (p. 52)
primary structure (p. 53)
phi (φ) angle (p. 55)

psi (ψ) angle (p. 55)
Ramachandran diagram (p. 55)
α helix (p. 56)
β pleated sheet (p. 58)
β strand (p. 58)
reverse turn (β turn; hairpin turn) (p. 60)
secondary structure (p. 61)

tertiary structure (p. 62)
domain (p. 63)
subunit (p. 64)
quaternary structure (p. 64)
cooperative transition (p. 68)

SELECTED READINGS

Where to start

Richardson, J. S., 1981. The anatomy and taxonomy of protein structure. *Adv. Protein Chem.* 34:167–339.
Doolittle, R. F., 1985. Proteins. *Sci. Am.* 253(4):88–99.
Richards, F. M., 1991. The protein folding problem. *Sci. Am.* 264(1): 54–57.
Weber, A. L., and Miller, S. L., 1981. Reasons for the occurrence of the twenty coded protein amino acids. *J. Mol. Evol.* 17:273–284.

Books

Branden, C., Tooze, J., 1999. *Introduction to Protein Structure* (2d ed.). Garland.

Perutz, M. F., 1992. *Protein Structure: New Approaches to Disease and Therapy.* W. H. Freeman and Company.
Creighton, T. E., 1992. *Proteins: Structures and Molecular Principles* (2d ed.). W. H. Freeman and Company.
Schultz, G. E., and Schirmer, R. H., 1979. *Principles of Protein Structure.* Springer-Verlag.

Conformation of proteins

Richardson, J. S., Richardson, D. C., Tweedy, N. B., Gernert, K. M., Quinn, T. P., Hecht, M. H., Erickson, B. W., Yan, Y., McClain, R. D., Donlan, M. E., and Suries, M. C., 1992. Looking at proteins: Representations, folding, packing, and design. *Biophys. J.* 63: 1186–1220.

Chothia, C., and Finkelstein, A. V., 1990. The classification and origin of protein folding patterns. *Annu. Rev. Biochem.* 59:1007–1039.

Alpha helices, beta sheets, and loops

O'Neil, K. T., and DeGrado, W. F., 1990. A thermodynamic scale for the helix-forming tendencies of the commonly occurring amino acids. *Science* 250:646–651.

Zhang, C., and Kim, S. H., 2000. The anatomy of protein beta-sheet topology. *J. Mol. Biol.* 299:1075–1089.

Regan, L., 1994. Protein structure: Born to be beta. *Curr. Biol.* 4:656–658.

Leszczynski, J. F., and Rose, G. D., 1986. Loops in globular proteins: A novel category of secondary structure. *Science* 234:849–855.

Srinivasan, R., and Rose, G. D., 1999. A physical basis for protein secondary structure. *Proc. Natl. Acad. Sci. USA* 96:14258–14263.

Domains

Bennett, M. J., Choe, S., and Eisenberg, D., 1994. Domain swapping: Entangling alliances between proteins. *Proc. Natl. Acad. Sci. USA* 91:3127–3131.

Bergdoll, M., Eltis, L. D., Cameron, A. D., Dumas, P., and Bolin, J. T., 1998. All in the family: Structural and evolutionary relationships among three modular proteins with diverse functions and variable assembly. *Protein Sci.* 7:1661–1670.

Hopfner, K. P., Kopetzki, E., Kresse, G. B., Bode, W., Huber, R., and Engh, R. A., 1998. New enzyme lineages by subdomain shuffling. *Proc. Natl. Acad. Sci. USA* 95:9813–9818.

Ponting, C. P., Schultz, J., Copley, R. R., Andrade, M. A., and Bork, P., 2000. Evolution of domain families. *Adv. Protein Chem.* 54:185–244.

Protein folding

Anfinsen, C. B., 1973. Principles that govern the folding of protein chains. *Science* 181:223–230.

Baldwin, R. L., and Rose, G. D., 1999. Is protein folding hierarchic? I. Local structure and peptide folding. *Trends Biochem. Sci.* 24:26–33.

Baldwin, R. L., and Rose, G. D., 1999. Is protein folding hierarchic? II. Folding intermediates and transition states. *Trends Biochem. Sci.* 24:77–83.

Staley, J. P., and Kim, P. S., 1990. Role of a subdomain in the folding of bovine pancreatic trypsin inhibitor. *Nature* 344:685–688.

Neira, J. L., and Fersht, A. R., 1999. Exploring the folding funnel of a polypeptide chain by biophysical studies on protein fragments. *J. Mol. Biol.* 285:1309–1333.

Covalent modification of proteins

Krishna, R. G., and Wold, F., 1993. Post-translational modification of proteins. *Adv. Enzymol. Relat. Areas. Mol. Biol.* 67:265–298.

Aletta, J. M., Cimato, T. R., and Ettinger, M. J., 1998. Protein methylation: A signal event in post-translational modification. *Trends Biochem. Sci.* 23:89–91.

Glazer, A. N., DeLange, R. J., and Sigman, D. S., 1975. *Chemical Modification of Proteins.* North-Holland.

Tsien, R. Y., 1998. The green fluorescent protein. *Annu. Rev. Biochem.* 67:509–544.

Molecular graphics

Kraulis, P., 1991. MOLSCRIPT: A program to produce both detailed and schematic plots of protein structures. *J. Appl. Cryst.* 24:946–950.

Ferrin, T., Huang, C., Jarvis, L., and Langridge, R., 1988. The MIDAS display system. *J. Mol. Graphics* 6:13–27.

Richardson, D. C., and Richardson, J. S., 1994. Kinemages: Simple macromolecular graphics for interactive teaching and publication. *Trends Biochem. Sci.* 19:135–138.

PROBLEMS

1. *Shape and dimension.* (a) Tropomyosin, a 70-kd muscle protein, is a two-stranded α-helical coiled coil. Estimate the length of the molecule? (b) Suppose that a 40-residue segment of a protein folds into a two-stranded antiparallel β structure with a 4-residue hairpin turn. What is the longest dimension of this motif?

2. *Contrasting isomers.* Poly-L-leucine in an organic solvent such as dioxane is α helical, whereas poly-L-isoleucine is not. Why do these amino acids with the same number and kinds of atoms have different helix-forming tendencies?

3. *Active again.* A mutation that changes an alanine residue in the interior of a protein to valine is found to lead to a loss of activity. However, activity is regained when a second mutation at a different position changes an isoleucine residue to glycine. How might this second mutation lead to a restoration of activity?

4. *Shuffle test.* An enzyme that catalyzes disulfide–sulfhydryl exchange reactions, called protein disulfide isomerase (PDI), has been isolated. PDI rapidly converts inactive scrambled ribonuclease into enzymatically active ribonuclease. In contrast, insulin is rapidly inactivated by PDI. What does this important observation imply about the relation between the amino acid sequence of insulin and its three-dimensional structure?

5. *Stretching a target.* A protease is an enzyme that catalyzes the hydrolysis of the peptide bonds of target proteins. How might a protease bind a target protein so that its main chain becomes fully extended in the vicinity of the vulnerable peptide bond?

6. *Often irreplaceable.* Glycine is a highly conserved amino acid residue in the evolution of proteins. Why?

7. *Potential partners.* Identify the groups in a protein that can form hydrogen bonds or electrostatic bonds with an arginine side chain at pH 7.

8. *Permanent waves.* The shape of hair is determined in part by the pattern of disulfide bonds in keratin, its major protein. How can curls be induced?

9. *Location is everything.* Proteins that span biological membranes often contain α helices. Given that the insides of membranes are highly hydrophobic (Section 12.2.1), predict what type of amino acids would be in such a helix. Why is an α helix particularly suited to exist in the hydrophobic environment of the interior of a membrane?

10. *Issues of stability.* Proteins are quite stable. The lifetime of a peptide bond in aqueous solution is nearly 1000 years. However, the $\Delta G°'$ of hydrolysis of proteins is negative and quite large. How can you account for the stability of the peptide bond in light of the fact that hydrolysis releases much energy?

11. *Minor species.* For an amino acid such as alanine, the major species in solution at pH 7 is the zwitterionic form. Assume a pK_a value of 8 for the amino group and a pK_a value of 3 for the

carboxylic acid and estimate the ratio of the concentration of neutral amino acid species (with the carboxylic acid protonated and the amino group neutral) to that of the zwitterionic species at pH 7.

12. *A matter of convention.* All L amino acids have an *S* absolute configuration except L-cysteine, which has the *R* configuration. Explain why L-cysteine is designated as the *R* absolute configuration.

13. *Hidden message.* Translate the following amino acid sequence into one-letter code: Leu-Glu-Ala-Arg-Asn-Ile-Asn-Gly-Ser-Cys-Ile-Glu-Cys-Glu-Ile-Ser-Gly-Arg-Glu-Ala-Thr.

14. *Who goes first?* Would you expect Pro–X peptide bonds to tend to have cis conformations like those of X–Pro bonds? Why or why not?

15. *Matching.* For each of the amino acid derivatives shown below (A–E), find the matching set of φ and ψ values (a–e).

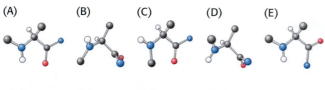

(A) (B) (C) (D) (E)

(a)
φ = 120°,
ψ = 120°

(b)
φ = 180°,
ψ = 0°

(c)
φ = 180°,
ψ = 180°

(d)
φ = 0°,
ψ = 180°

(e)
φ = −60°,
ψ = −40°

16. *Concentrate of the concentration.* A solution of a protein whose sequence includes three tryptophan residues, no tyrosine residues, and no phenylalanine residues has an absorbance of 0.1 at 280 nm in a cell with a path length of 1 cm. Estimate the concentration of the protein in units of molarity. If the protein has a molecular mass of 100 kd, estimate the concentration in units of milligrams of protein per milliliter of solution.

 Media Problem

You can use the **Structural Insights** and **Conceptual Insights** as visual aids to help you answer Media Problems. Go to the Website: www.whfreeman.com/biochem5, and select the applicable module.

17. *Inside-out, back-to-front.* In the Media Problem section of the **Structural Insights** module on protein structure, you can examine molecular models of four putative protein structures. One of the four structures has been determined by x-ray crystallography. The other three have been made-up, and in fact are very unlikely to occur. Which are the structures that are unlikely to occur and why?

CHAPTER 4

Exploring Proteins

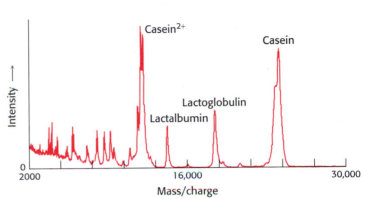

Milk, a source of nourishment for all mammals, is composed, in part, of a variety of proteins. The protein components of milk are revealed by the technique of MALDI–TOF mass spectrometry, which separates molecules on the basis of their mass to charge ratio. [(Left) Jean Paul Iris/FPG (Right) courtesy of Brian Chait.]

In the preceding chapter, we saw that proteins play crucial roles in nearly all biological processes—in catalysis, signal transmission, and structural support. This remarkable range of functions arises from the existence of thousands of proteins, each folded into a distinctive three-dimensional structure that enables it to interact with one or more of a highly diverse array of molecules. A major goal of biochemistry is to determine how amino acid sequences specify the conformations of proteins. Other goals are to learn how individual proteins bind specific substrates and other molecules, mediate catalysis, and transduce energy and information.

The purification of the protein of interest is the indispensable first step in a series of studies aimed at exploring protein function. Proteins can be separated from one another on the basis of solubility, size, charge, and binding ability. When a protein has been purified, the amino acid sequence can be determined. The strategy is to divide and conquer, to obtain specific fragments that can be readily sequenced. Automated peptide sequencing and the application of recombinant DNA methods are providing a wealth of amino acid sequence data that are opening new vistas. To understand the physiological context of a protein, antibodies are choice probes for locating proteins in vivo and measuring their quantities. Monoclonal antibodies able to probe for specific proteins can be obtained in large amounts. The synthesis of peptides is possible, which makes feasible the synthesis of new drugs, functional protein fragments, and antigens for inducing the formation of specific antibodies. Nuclear magnetic resonance (NMR) spectroscopy and x-ray crystallography are the principal techniques for elucidating three-dimensional structure, the key determinant of function.

The exploration of proteins by this array of physical and chemical techniques has greatly enriched our understanding of the molecular basis of life and makes it possible to tackle some of the most challenging questions of biology in molecular terms.

4.0.1 The Proteome Is the Functional Representation of the Genome

Many organisms are yielding their DNA base sequences to gene sequencers, including several metazoans. The roundworm *Caenorhabditis elegans* has a genome of 97 million bases and about 19,000 protein-encoding genes, whereas that of the fruit fly *Drosophilia melanogaster* contains 180 million bases and about 14,000 genes. The incredible progress being made in gene sequencing has already culminated in the elucidation of the complete sequence of the human genome, all 3 billion bases with an estimated 40,000 genes. But this genomic knowledge is analogous to a list of parts for a car—it does not explain how the parts work together. A new word has been coined, the *proteome,* to signify a more complex level of information content, the level of *functional information,* which encompasses the type, functions, and interactions of proteins that yield a functional unit.

The term proteome is derived from *prote*ins expressed by the gen*ome.* Whereas the genome tells us what is possible, the proteome tells us what is functionally present—for example, which proteins interact to form a signal-transduction pathway or an ion channel in a membrane. The proteome is not a fixed characteristic of the cell. Rather, because it represents the functional expression of information, it varies with cell type, developmental stage, and environmental conditions, such as the presence of hormones. The proteome is much larger than the genome because of such factors as alternatively spliced RNA, the posttranslational modification of proteins, the temporal regulation of protein synthesis, and varying protein–protein interactions. Unlike the genome, the proteome is not static.

An understanding of the proteome is acquired by investigating, characterizing, and cataloging proteins. An investigator often begins the process by separating a particular protein from all other biomolecules in the cell.

4.1 THE PURIFICATION OF PROTEINS IS AN ESSENTIAL FIRST STEP IN UNDERSTANDING THEIR FUNCTION

An adage of biochemistry is, Never waste pure thoughts on an impure protein. Starting from pure proteins, we can determine amino acid sequences and evolutionary relationships between proteins in diverse organisms and we can investigate a protein's biochemical function. Moreover, crystals of the protein may be grown from pure protein, and from such crystals we can obtain x-ray data that will provide us with a picture of the protein's tertiary structure—the actual *functional* unit.

4.1.1 The Assay: How Do We Recognize the Protein That We Are Looking For?

Purification should yield a sample of protein containing only one type of molecule, the protein in which the biochemist is interested. This protein sample may be only a fraction of 1% of the starting material, whether that starting material consists of cells in culture or a particular organ from a plant or animal. How is the biochemist able to isolate a particular protein from a complex mixture of proteins?

The biochemist needs a test, called an *assay*, for some unique identifying property of the protein so that he or she can tell when the protein is present. Determining an effective assay is often difficult; but the more specific the assay, the more effective the purification. For enzymes, which are protein catalysts (Chapter 8), the assay is usually based on the reaction that the enzyme catalyzes in the cell. Consider the enzyme lactate dehydrogenase, an important player in the anaerobic generation of energy from glucose as well as in the synthesis of glucose from lactate. Lactate dehydrogenase carries out the following reaction:

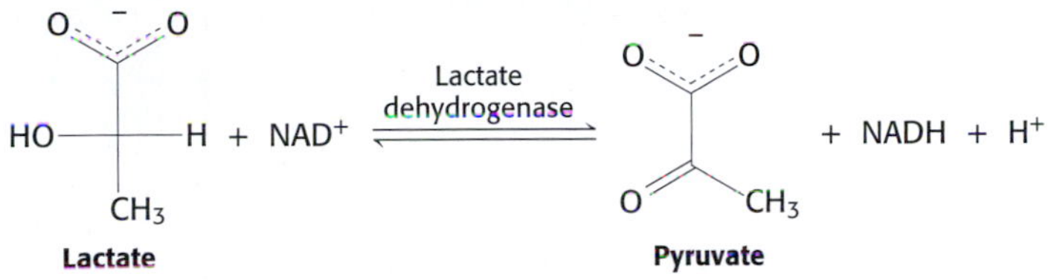

Nicotinamide adenine dinucleotide [reduced (NADH); Section 14.3.1] is distinguishable from the other components of the reaction by its ability to absorb light at 340 nm. Consequently, we can follow the progress of the reaction by examining how much light the reaction mixture absorbs at 340 nm in unit time—for instance, within 1 minute after the addition of the enzyme. Our assay for enzyme activity during the purification of lactate dehydrogenase is thus the increase in absorbance of light at 340 nm observed in 1 minute.

To be certain that our purification scheme is working, we need one additional piece of information—the amount of protein present in the mixture being assayed. There are various rapid and accurate means of determining protein concentration. With these two experimentally determined numbers—enzyme activity and protein concentration—we then calculate the *specific activity*, the ratio of enzyme activity to the amount of protein in the enzyme assay. The specific activity will rise as the purification proceeds and the protein mixture being assayed consists to a greater and greater extent of lactate dehydrogenase. In essence, the point of the purification is to maximize the specific activity.

4.1.2 Proteins Must Be Released from the Cell to Be Purified

Having found an assay and chosen a source of protein, we must now fractionate the cell into components and determine which component is enriched in the protein of interest. Such fractionation schemes are developed by trial and error, on the basis of previous experience. In the first step, a *homogenate* is formed by disrupting the cell membrane, and the mixture is fractionated by centrifugation, yielding a dense pellet of heavy material at the bottom of the centrifuge tube and a lighter supernatant above (Figure 4.1). The supernatant is again centrifuged at a greater force to yield yet another pellet and supernatant. The procedure, called *differential centrifugation*, yields several fractions of decreasing density, each still containing hundreds of different proteins, which are subsequently assayed for the activity being purified. Usually, one fraction will be enriched for such activity, and it then serves as the source of material to which more discriminating purification techniques are applied.

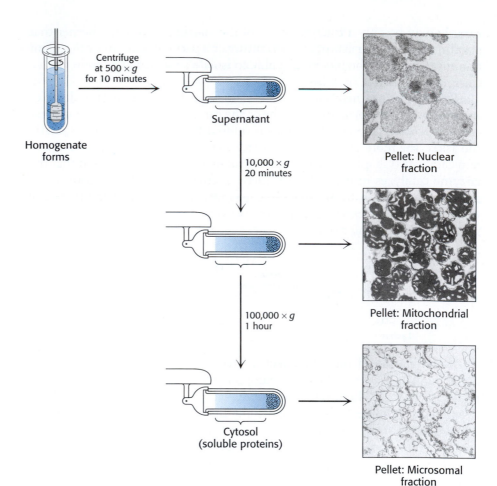

FIGURE 4.1 Differential centrifugation. Cells are disrupted in a homogenizer and the resulting mixture, called the homogenate, is centrifuged in a step-by-step fashion of increasing centrifugal force. The denser material will form a pellet at lower centrifugal force than will the less-dense material. The isolated fractions can be used for further purification. [Photographs courtesy of S. Fleischer and B. Fleischer.]

4.1.3 Proteins Can Be Purified According to Solubility, Size, Charge, and Binding Affinity

Several thousand proteins have been purified in active form on the basis of such characteristics as *solubility, size, charge,* and *specific binding affinity.* Usually, protein mixtures are subjected to a series of separations, each based on a different property to yield a pure protein. At each step in the purification, the preparation is assayed and the protein concentration is determined. Substantial quantities of purified proteins, of the order of many milligrams, are needed to fully elucidate their three-dimensional structures and their mechanisms of action. Thus, the overall yield is an important feature of a purification scheme. A variety of purification techniques are available.

Salting Out. Most proteins are less soluble at high salt concentrations, an effect called *salting out.* The salt concentration at which a protein precipitates differs from one protein to another. Hence, salting out can be used to fractionate proteins. For example, 0.8 M ammonium sulfate precipitates fibrinogen, a blood-clotting protein, whereas a concentration of 2.4 M is needed to precipitate serum albumin. Salting out is also useful for concentrating dilute solutions of proteins, including active fractions obtained from other purification steps. Dialysis can be used to remove the salt if necessary.

Dialysis. Proteins can be separated from small molecules by *dialysis* through a semipermeable membrane, such as a cellulose membrane with pores (Figure 4.2). Molecules having dimensions significantly greater than the pore diameter are retained inside the dialysis bag, whereas smaller molecules and ions traverse the pores of such a membrane and emerge in the dialysate outside the bag. This technique is useful for removing a salt or other small molecule, but it will not distinguish between proteins effectively.

Gel-Filtration Chromatography. More discriminating separations on the basis of size can be achieved by the technique of *gel-filtration chromatography* (Figure 4.3). The sample is applied to the top of a column consisting of porous beads made of an insoluble but highly hydrated polymer such as dextran or agarose (which are carbohydrates) or polyacrylamide. Sephadex, Sepharose, and Biogel are commonly used commercial preparations of these beads, which are typically 100 μm (0.1 mm) in diameter. Small molecules can enter these beads, but large ones cannot. The result is that small molecules are distributed in the aqueous solution both inside the beads and between them, whereas large molecules are located only in the solution between the beads. *Large molecules flow more rapidly through this column and emerge first because a smaller volume is accessible to them.* Molecules that are of a size to occasionally enter a bead will flow from the column at an intermediate position, and small molecules, which take a longer, tortuous path, will exit last.

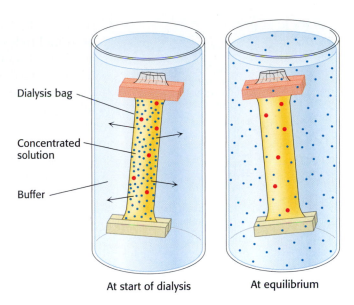

FIGURE 4.2 Dialysis. Dialysis bag — Protein molecules (red) are retained within the dialysis bag, Concentrated solution — whereas small molecules (blue) diffuse Buffer — into the surrounding medium.

At start of dialysis   At equilibrium

FIGURE 4.2 Dialysis. Protein molecules (red) are retained within the dialysis bag, whereas small molecules (blue) diffuse into the surrounding medium.

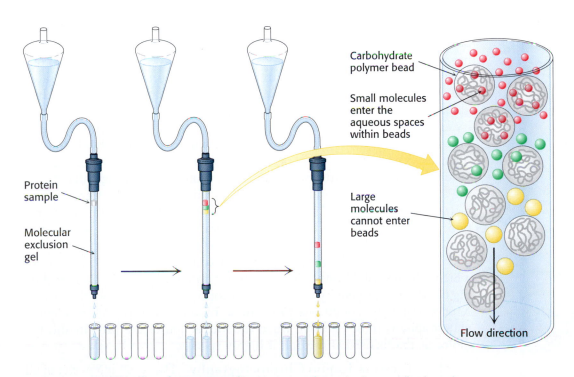

Protein sample

Molecular exclusion gel

Carbohydrate polymer bead

Small molecules enter the aqueous spaces within beads

Large molecules cannot enter beads

Flow direction

FIGURE 4.3 Gel filtration chromatography. A mixture of proteins in a small volume is applied to a column filled with porous beads. Because large proteins cannot enter the internal volume of the beads, they emerge sooner than do small ones.

Ion-Exchange Chromatography. Proteins can be separated on the basis of their net charge by *ion-exchange chromatography.* If a protein has a net positive charge at pH 7, it will usually bind to a column of beads containing carboxylate groups, whereas a negatively charged protein will not

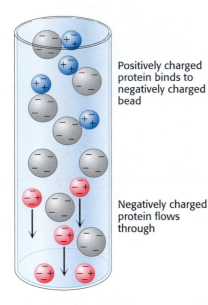

FIGURE 4.4 Ion-exchange chromatography. This technique separates proteins mainly according to their net charge.

Positively charged protein binds to negatively charged bead

Negatively charged protein flows through

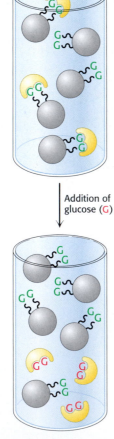

Glucose-binding protein attaches to glucose residues (G) on beads

Addition of glucose (G)

Glucose-binding proteins are released on addition of glucose

FIGURE 4.5 Affinity chromatography. Affinity chromatography of concanavalin A (shown in yellow) on a solid support containing covalently attached glucose residues (G).

(Figure 4.4). A positively charged protein bound to such a column can then be eluted (released) by increasing the concentration of sodium chloride or another salt in the eluting buffer because sodium ions compete with positively charged groups on the protein for binding to the column. Proteins that have a low density of net positive charge will tend to emerge first, followed by those having a higher charge density. Positively charged proteins (cationic proteins) can be separated on negatively charged carboxymethyl–cellulose (CM–cellulose) columns. Conversely, negatively charged proteins (anionic proteins) can be separated by chromatography on positively charged diethylaminoethyl–cellulose (DEAE–cellulose) columns.

Cellulose or agarose

Carboxymethyl (CM) group
(ionized form)

Cellulose or agarose

Diethylaminoethyl (DEAE) group
(protonated form)

Affinity Chromatography. *Affinity chromatography* is another powerful and generally applicable means of purifying proteins. This technique takes advantage of the high affinity of many proteins for specific chemical groups. For example, the plant protein concanavalin A can be purified by passing a crude extract through a column of beads containing covalently attached glucose residues. Concanavalin A binds to such a column because it has affinity for glucose, whereas most other proteins do not. The bound concanavalin A can then be released from the column by adding a concentrated solution of glucose. The glucose in solution displaces the column-attached glucose residues from binding sites on concanavalin A (Figure 4.5). Affinity chromatography is a powerful means of isolating transcription factors, proteins that regulate gene expression by binding to specific DNA sequences. A protein mixture is percolated through a column containing specific DNA sequences attached to a matrix; proteins with a high affinity for the sequence will bind and be retained. In this instance, the transcription factor is released by washing with a solution containing a high concentration of salt. In general, affinity chromatography can be effectively used to isolate a protein that recognizes group X by (1) covalently attaching X or a derivative of it to a column, (2) adding a mixture of proteins to this column, which is then washed with buffer to remove unbound proteins, and (3) eluting the desired protein by adding a high concentration of a soluble form of X or altering the conditions to decrease binding affinity. Affinity chromatography is most effective when the interaction of the protein and the molecule that is used as the bait is highly specific.

High-Pressure Liquid Chromatography. The resolving power of all of the column techniques can be improved substantially through the use of a technique called *high-pressure liquid chromatography (HPLC)*, which is an enhanced version of the column techniques already discussed. The column materials themselves are much more finely divided and, as a consequence, there are more interaction sites and thus greater resolving power. Because the column is made of finer material, pressure must be applied to the column to obtain adequate flow rates. The net result is high resolution as well as rapid separation (Figure 4.6).

4.1.4 Proteins Can Be Separated by Gel Electrophoresis and Displayed

How can we tell whether a purification scheme is effective? One way is to ascertain that the specific activity rises with each purification step. Another is to visualize the effectiveness by displaying the proteins present at each step. The technique of electrophoresis makes the latter method possible.

Gel Electrophoresis. A molecule with a net charge will move in an electric field. This phenomenon, termed *electrophoresis*, offers a powerful means of separating proteins and other macromolecules, such as DNA and RNA. The velocity of migration (v) of a protein (or any molecule) in an electric field depends on the electric field strength (E), the net charge on the protein (z), and the frictional coefficient (f).

$$v = \frac{Ez}{f} \tag{1}$$

The electric force Ez driving the charged molecule toward the oppositely charged electrode is opposed by the viscous drag fv arising from friction between the moving molecule and the medium. The frictional coefficient f depends on both the mass and shape of the migrating molecule and the viscosity (η) of the medium. For a sphere of radius r,

$$f = 6\pi\eta r \tag{2}$$

Electrophoretic separations are nearly always carried out in gels (or on solid supports such as paper) because the gel serves as a molecular sieve that enhances separation (Figure 4.7). Molecules that are small compared with the pores in the gel readily move through the gel, whereas molecules much larger than the pores are almost immobile. Intermediate-size molecules move through the gel with various degrees of facility. Electrophoresis is performed

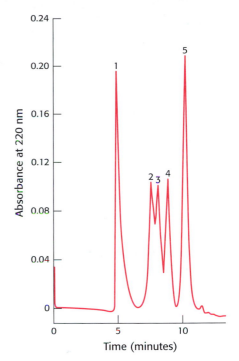

FIGURE 4.6 High-pressure liquid chromatography (HPLC). Gel filtration by HPLC clearly defines the individual proteins because of its greater resolving power: (1) thyroglobulin (669 kd), (2) catalase (232 kd), (3) bovine serum albumin (67 kd), (4) ovalbumin (43 kd), and (5) ribonuclease (13.4 kd). [After K. J. Wilson and T. D. Schlabach. In *Current Protocols in Molecular Biology*, vol. 2, suppl. 41, F. M. Ausbel, R. Brent, R. E. Kingston, D. D. Moore, J. G. Seidman, J. A. Smith, and K. Struhl, Eds. (Wiley, 1998), p. 10.14.1.]

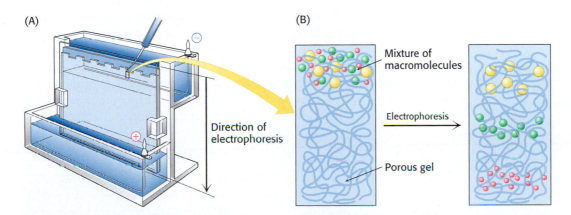

FIGURE 4.7 Polyacrylamide gel electrophoresis. (A) Gel electrophoresis apparatus. Typically, several samples undergo electrophoresis on one flat polyacrylamide gel. A microliter pipette is used to place solutions of proteins in the wells of the slab. A cover is then placed over the gel chamber and voltage is applied. The negatively charged SDS (sodium dodecyl sulfate)–protein complexes migrate in the direction of the anode, at the bottom of the gel. (B) The sieving action of a porous polyacrylamide gel separates proteins according to size, with the smallest moving most rapidly.

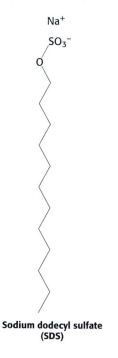

FIGURE 4.8 Formation of a polyacrylamide gel. A three-dimensional mesh is formed by co-polymerizing activated monomer (blue) and cross-linker (red).

Acrylamide   **Methylenebisacrylamide**

$S_2O_8^{2-}$ (persulfate)

$2\ SO_4^{-\cdot}$ (sulfate radical, initiates polymerization)

$CONH_2$ $CONH_2$ $CONH_2$ $CONH_2$

$CONH_2$ $CONH_2$ $CONH_2$ $CONH_2$

Na^+

SO_3^-

Sodium dodecyl sulfate (SDS)

in a thin, vertical slab of polyacrylamide. The direction of flow is from top to bottom. Polyacrylamide gels, formed by the polymerization of acrylamide and cross-linked by methylenebisacrylamide, are choice supporting media for electrophoresis because they are chemically inert and are readily formed (Figure 4.8). Electrophoresis is the opposite of gel filtration in that all of the molecules, regardless of size, are forced to move through the same matrix. The gel behaves as one bead of a gel-filtration column.

Proteins can be separated largely on the basis of mass by electrophoresis in a polyacrylamide gel under denaturing conditions. The mixture of proteins is first dissolved in a solution of sodium dodecyl sulfate (SDS), an anionic detergent that disrupts nearly all noncovalent interactions in native proteins. Mercaptoethanol (2-thioethanol) or dithiothreitol also is added to reduce disulfide bonds. Anions of SDS bind to main chains at a ratio of about one SDS anion for every two amino acid residues. This complex of SDS with a denatured protein has a large net negative charge that is roughly proportional to the mass of the protein. The negative charge acquired on binding SDS is usually much greater than the charge on the native protein; this native charge is thus rendered insignificant. The SDS–protein complexes are then subjected to electrophoresis. When the electrophoresis is complete, the proteins in the gel can be visualized by staining them with silver or a dye such as Coomassie blue, which reveals a series of bands (Figure 4.9). Radioactive

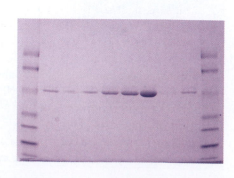

FIGURE 4.9 Staining of proteins after electrophoresis. Proteins subjected to electrophoresis on an SDS–polyacrylamide gel can be visualized by staining with Coomassie blue. [Courtesy of Kodak Scientific Imaging Systems.]

labels can be detected by placing a sheet of x-ray film over the gel, a procedure called *autoradiography*.

Small proteins move rapidly through the gel, whereas large proteins stay at the top, near the point of application of the mixture. The mobility of most polypeptide chains under these conditions is linearly proportional to the logarithm of their mass (Figure 4.10). Some carbohydrate-rich proteins and membrane proteins do not obey this empirical relation, however. SDS–polyacrylamide gel electrophoresis (SDS-PAGE) is rapid, sensitive, and capable of a high degree of resolution. As little as 0.1 μg ($\sim$2 pmol) of a protein gives a distinct band when stained with Coomassie blue, and even less ($\sim$0.02 μg) can be detected with a silver stain. Proteins that differ in mass by about 2% (e.g., 40 and 41 kd, arising from a difference of about 10 residues) can usually be distinguished.

We can examine the efficacy of our purification scheme by analyzing a part of each fraction by SDS-PAGE. The initial fractions will display dozens to hundreds of proteins. As the purification progresses, the number of bands will diminish, and the prominence of one of the bands should increase. This band will correspond to the protein of interest.

Isoelectric Focusing. Proteins can also be separated electrophoretically on the basis of their relative contents of acidic and basic residues. The *isoelectric point* (pl) of a protein is the pH at which its net charge is zero. At this pH, its electrophoretic mobility is zero because z in equation 1 is equal to zero. For example, the pI of cytochrome c, a highly basic electron-transport protein, is 10.6, whereas that of serum albumin, an acidic protein in blood, is 4.8. Suppose that a mixture of proteins undergoes electrophoresis in a pH gradient in a gel in the absence of SDS. Each protein will move until it reaches a position in the gel at which the pH is equal to the pI of the protein. This method of separating proteins according to their isoelectric point is called *isoelectric focusing*. The pH gradient in the gel is formed first by subjecting a mixture of *polyampholytes* (small multicharged polymers) having many pI values to electrophoresis. Isoelectric focusing can readily resolve proteins that differ in pI by as little as 0.01, which means that proteins differing by one net charge can be separated (Figure 4.11).

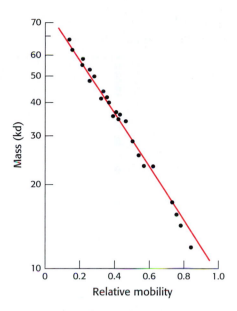

FIGURE 4.10 Electrophoresis can determine mass. The electrophoretic mobility of many proteins in SDS–polyacrylamide gels is inversely proportional to the logarithm of their mass. [After K. Weber and M. Osborn, *The Proteins*, vol. 1, 3d ed. (Academic Press, 1975), p. 179.]

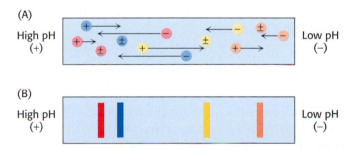

FIGURE 4.11 The principle of isoelectric focusing. A pH gradient is established in a gel before loading the sample. (A) The sample is loaded and voltage is applied. The proteins will migrate to their isoelectric pH, the location at which they have no net charge. (B) The proteins form bands that can be excised and used for further experimentation.

Two-Dimensional Electrophoresis. Isoelectric focusing can be combined with SDS–PAGE to obtain very high resolution separations. A single sample is first subjected to isoelectric focusing. This single-lane gel is then placed horizontally on top of an SDS–polyacrylamide slab. The proteins are thus spread across the top of the polyacrylamide gel according to how far they migrated during isoelectric focusing. They then undergo electrophoresis again in a perpendicular direction (vertically) to yield a two-dimensional pattern of spots. In such a gel, proteins have been separated in the horizontal direction on the basis of isoelectric point and in the vertical direction on the basis of mass. It is remarkable that more than a thousand different proteins in the bacterium *Escherichia coli* can be resolved in a single experiment by two-dimensional electrophoresis (Figure 4.12).

(A)

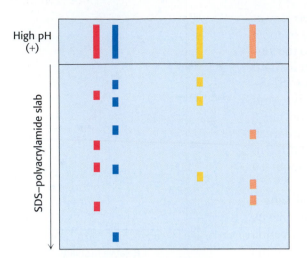

High pH (+)

SDS–polyacrylamide slab

(B) Isoelectric focusing

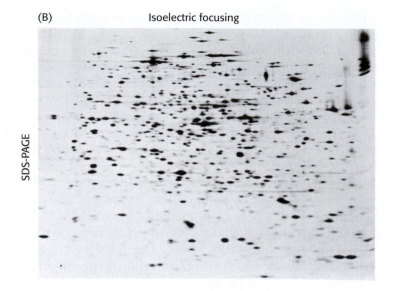

SDS-PAGE

FIGURE 4.12 Two-dimensional gel electrophoresis. (A) A protein sample is initially fractionated in one dimension by isoelectric focusing as described in Figure 4.11. The isoelectric focusing gel is then attached to an SDS–polyacrylamide gel, and electrophoresis is performed in the second dimension, perpendicular to the original separation. Proteins with the same pI are now separated on the basis of mass. (B) Proteins from *E. coli* were separated by two-dimensional gel electrophoresis, resolving more than a thousand different proteins. The proteins were first separated according to their isoelectric pH in the horizontal direction and then by their apparent mass in the vertical direction. [(B) Courtesy of Dr. Patrick H. O'Farrell.]

Proteins isolated from cells under different physiological conditions can be subjected to two-dimensional electrophoresis, followed by an examination of the intensity of the signals. In this way, particular proteins can be seen to increase or decrease in concentration in response to the physiological state. How can we tell what protein is being regulated? A former drawback to the power of the two-dimensional gel is that, although many proteins are displayed, they are not identified. It is now possible to identify proteins by coupling two-dimensional gel electrophoresis with mass spectrometric techniques. We will consider these techniques when we examine how the mass of a protein is determined (Section 4.1.7).

4.1.5 A Protein Purification Scheme Can Be Quantitatively Evaluated

To determine the success of a protein purification scheme, we monitor the procedure at each step by determining specific activity and by performing an SDS-PAGE analysis. Consider the results for the purification of a fictitious protein, summarized in Table 4.1 and Figure 4.13. At each step, the following parameters are measured:

Total protein. The quantity of protein present in a fraction is obtained by determining the protein concentration of a part of each fraction and multiplying by the fraction's total volume.

Total activity. The enzyme activity for the fraction is obtained by measuring the enzyme activity in the volume of fraction used in the assay and multiplying by the fraction's total volume.

TABLE 4.1 Quantification of a purification protocol for a fictitious protein

Step	Total protein (mg)	Total activity (units)	Specific activity, (units mg^{-1})	Yield (%)	Purification level
Homogenization	15,000	150,000	10	100	1
Salt fractionation	4,600	138,000	30	92	3
Ion-exchange chromatography	1,278	115,500	90	77	9
Molecular exclusion chromatography	68.8	75,000	1,100	50	110
Affinity chromatography	1.75	52,500	30,000	35	3,000

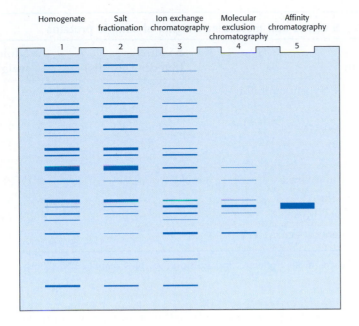

FIGURE 4.13 Electrophoretic analysis of a protein purification. The purification scheme in Table 4.1 was analyzed by SDS-PAGE. Each lane contained 50 μg of sample. The effectiveness of the purification can be seen as the band for the protein of interest becomes more prominent relative to other bands.

Specific activity. This parameter is obtained by dividing total activity by total protein.

Yield. This parameter is a measure of the activity retained after each purification step as a percentage of the activity in the crude extract. The amount of activity in the initial extract is taken to be 100%.

Purification level. This parameter is a measure of the increase in purity and is obtained by dividing the specific activity, calculated after each purification step, by the specific activity of the initial extract.

As we see in Table 4.1, the first purification step, salt fractionation, leads to an increase in purity of only 3-fold, but we recover nearly all the target protein in the original extract, given that the yield is 92%. After dialysis to lower the high concentration of salt remaining from the salt fractionation, the fraction is passed through an ion-exchange column. The purification now increases to 9-fold compared with the original extract, whereas the yield falls to 77%. Molecular exclusion chromatography brings the level of purification to 100-fold, but the yield is now at 50%. The final step is affinity chromatography with the use of a ligand specific for the target enzyme. This step, the most powerful of these purification procedures, results in a purification level of 3000-fold, while lowering the yield to 35%. The SDS-PAGE in Figure 4.13 shows that, if we load a constant amount of protein onto each lane after each step, the number of bands decreases in proportion to the level of purification, and the amount of protein of interest increases as a proportion of the total protein present.

A good purification scheme takes into account both purification levels and yield. A high degree of purification and a poor yield leave little protein with which to experiment. A high yield with low purification leaves many contaminants (proteins other than the one of interest) in the fraction and complicates the interpretation of experiments.

4.1.6 Ultracentrifugation Is Valuable for Separating Biomolecules and Determining Their Masses

We have already seen that centrifugation is a powerful and generally applicable method for separating a crude mixture of cell components, but it is also useful for separating and analyzing biomolecules themselves. With this

TABLE 4.2   S values and molecular weights of sample proteins

Protein	S value (Svedberg units)	Molecular weight
Pancreatic trypsin inhibitor	1	6,520
Cytochrome c	1.83	12,310
Ribonuclease A	1.78	13,690
Myoglobin	1.97	17,800
Trypsin	2.5	23,200
Carbonic anhydrase	3.23	28,800
Concanavlin A	3.8	51,260
Malate dehydrogenase	5.76	74,900
Lactate dehydrogenase	7.54	146,200

From T. Creighton, *Proteins,* 2nd Edition (W. H. Freeman and Company, 1993), Table 7.1.

technique, we can determine such parameters as mass and density, learn something about the shape of a molecule, and investigate the interactions between molecules. To deduce these properties from the centrifugation data, we need a mathematical description of how a particle behaves in a centrifugal force.

A particle will move through a liquid medium when subjected to a centrifugal force. A convenient means of quantifying the rate of movement is to calculate the sedimentation coefficient, *s*, of a particle by using the following equation:

$$s = m(1 - \bar{v}\rho)/f$$

where *m* is the mass of the particle, $\bar{v}$ is the partial specific volume (the reciprocal of the particle density), ρ is the density of the medium and *f* is the frictional coefficient (a measure of the shape of the particle). The $(1 - \bar{v}\rho)$ term is the buoyant force exerted by liquid medium.

Sedimentation coefficients are usually expressed in *Svedberg units (S)*, equal to 10^{-13} s. The smaller the S value, the slower a molecule moves in a centrifugal field. The S values for a number of biomolecules and cellular components are listed in Table 4.2 and Figure 4.14.

Several important conclusions can be drawn from the preceding equation:

1. The sedimentation velocity of a particle depends in part on its mass. A more massive particle sediments more rapidly than does a less massive particle of the same shape and density.

2. Shape, too, influences the sedimentation velocity because it affects the viscous drag. The frictional coefficient *f* of a compact particle is smaller than that of an extended particle of the same mass. Hence, elongated particles sediment more slowly than do spherical ones of the same mass.

3. A dense particle moves more rapidly than does a less dense one because the opposing buoyant force $(1 - \bar{v}\rho)$ is smaller for the denser particle.

4. The sedimentation velocity also depends on the density of the solution. (ρ). Particles sink when $\bar{v}\rho < 1$, float when $\bar{v}\rho > 1$, and do not move when $\bar{v}\rho = 1$.

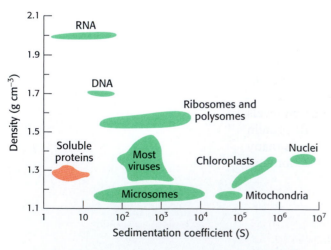

FIGURE 4.14 Density and sedimentation coefficients of cellular components. [After L. J. Kleinsmith and V. M. Kish, *Principles of Cell and Molecular Biology,* 2d ed. (Harper Collins, 1995), p. 138.]

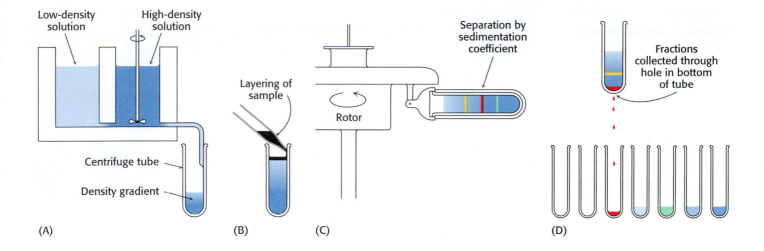

Low-density
solution

High-density
solution

Centrifuge tube

Density gradient

(A)

Layering of
sample

(B)

Rotor

(C)

Separation by
sedimentation
coefficient

Fractions
collected through
hole in bottom
of tube

(D)

A technique called *zonal, band,* or most commonly *gradient* centrifugation can be used to separate proteins with different sedimentation coefficients. The first step is to form a density gradient in a centrifuge tube. Differing proportions of a low-density solution (such as 5% sucrose) and a high-density solution (such as 20% sucrose) are mixed to create a linear gradient of sucrose concentration ranging from 20% at the bottom of the tube to 5% at the top (Figure 4.15). The role of the gradient is to prevent connective flow. A small volume of a solution containing the mixture of proteins to be separated is placed on top of the density gradient. When the rotor is spun, proteins move through the gradient and separate according to their sedimentation coefficients. The time and speed of the centrifugation is determined empirically. The separated bands, or zones, of protein can be harvested by making a hole in the bottom of the tube and collecting drops. The drops can be measured for protein content and catalytic activity or another functional property. This sedimentation-velocity technique readily separates proteins differing in sedimentation coefficient by a factor of two or more.

The mass of a protein can be directly determined by *sedimentation equilibrium,* in which a sample is centrifuged at relatively low speed so that sedimentation is counterbalanced by diffusion. *The sedimentation-equilibrium technique for determining mass is very accurate and can be applied under nondenaturing conditions in which the native quaternary structure of multimeric proteins is preserved.* In contrast, SDS–polyacrylamide gel electrophoresis (Section 4.1.4) provides an *estimate* of the mass of dissociated polypeptide chains under *denaturing* conditions. Note that, if we know the mass of the dissociated components of a multimeric protein as determined by SDS–polyacrylamide analysis and the mass of the intact multimeric protein as determined by sedimentation equilibrium analysis, we can determine how many copies of each polypeptide chain is present in the multimeric protein.

4.1.7 The Mass of a Protein Can Be Precisely Determined by Mass Spectrometry

Mass spectrometry has been an established analytical technique in organic chemistry for many years. Until recently, however, the very low volatility of proteins made mass spectrometry useless for the investigation of these molecules. This difficulty has been circumvented by the introduction of techniques for effectively dispersing proteins and other macromolecules into the gas phase. These methods are called *matrix-assisted laser desorption–ionization (MALDI)* and *electrospray spectrometry.* We will focus on

FIGURE 4.15 Zonal centrifugation. The steps are as follows: (A) form a density gradient, (B) layer the sample on top of the gradient, (C) place the tube in a swinging-bucket rotor and centrifuge it, and (D) collect the samples. [After D. Freifelder, *Physical Biochemistry,* 2d ed. (W. H. Freeman and Company, 1982), p. 397.]

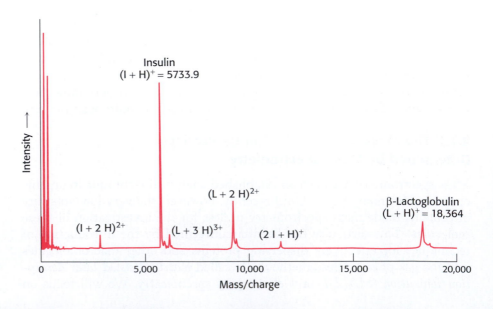

(1) Protein sample is ionized

(2) Electrical field accelerates ions

Beam splitter

Laser

Trigger

(4) Laser triggers a clock

Transient recorder

Ion source

Flight tube

(3) Lightest ions arrive at the detector first

Detector

Laser beam

Matrix

Sample

Protein

FIGURE 4.16 MALDI-TOF mass spectrometry. (1) The protein sample, embedded in an appropriate matrix, is ionized by the application of a laser beam. (2) An electrical field accelerates the ions formed through the flight tube toward the detector. (3) The lightest ions arrive first. (4) The ionizing laser pulse also triggers a clock that measures the time of flight (TOF) for the ions. [After J. T. Watson, *Introduction to Mass Spectrometry*, 3d ed. (Lippincott-Raven, 1997), p. 279.]

MALDI spectrometry. In this technique, protein ions are generated and then accelerated through an electrical field (Figure 4.16). They travel through the flight tube, with the smallest traveling fastest and arriving at the detector first. Thus, the *time of flight (TOF)* in the electrical field is a measure of the mass (or, more precisely, the mass/charge ratio). Tiny amounts of biomolecules, as small as a few picomoles (pmol) to femtomoles (fmol), can be analyzed in this manner. A MALDI-TOF mass spectrum for a mixture of the proteins insulin and β-lactoglobulin is shown in Figure 4.17. The masses determined by MALDI-TOF are 5733.9 and 18,364, respectively, compared with calculated values of 5733.5 and 18,388. MALDI-TOF is indeed an accurate means of determining protein mass.

FIGURE 4.17 MALDI-TOF mass spectrum of insulin and β-lactoglobulin. A mixture of 5 pmol each of insulin (I) and β-lactoglobulin (L) was ionized by MALDI, which produces predominately singly charged molecular ions from peptides and proteins ($I + H^+$ for insulin and $L + H^+$ for lactoglobulin). However, molecules with multiple charges as well as small quantities of a singly charged dimer of insulin, $(2 I + H)^+$, also are produced. [After J. T. Watson, *Introduction to Mass Spectrometry*, 3d ed. (Lippincott-Raven, 1997), p. 282.]

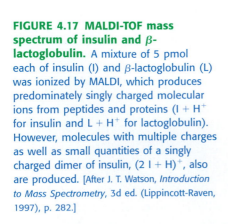

Insulin $(I + H)^+ = 5733.9$

Intensity →

$(I + 2 H)^{2+}$

$(L + 3 H)^{3+}$

$(L + 2 H)^{2+}$

$(2 I + H)^+$

β-Lactoglobulin $(L + H)^+ = 18,364$

0   5,000   10,000   15,000   20,000

Mass/charge

Mass spectrometry has permitted the development of *peptide mass fingerprinting*. This technique for identifying peptides has greatly enhanced the utility of two-dimensional gels. Two-dimensional electrophoresis is performed as described in Section 4.1.4. The sample of interest is extracted and cleaved *specifically* by chemical or enzymatic means. The masses of the protein fragments are then determined with the use of mass spectrometry. Finally, the peptide masses, or *fingerprint*, are matched against the fingerprint found in databases of proteins that have been "electronically cleaved" by a computer simulating the same fragmentation technique used for the experimental sample. This technique has provided some outstanding results. For example, of 150 yeast proteins analyzed with the use of two-dimensional gels, peptide mass fingerprinting unambiguously identified 80%. Mass spectrometry has provided name tags for many of the proteins in two-dimensional gels.

4.2 AMINO ACID SEQUENCES CAN BE DETERMINED BY AUTOMATED EDMAN DEGRADATION

The protein of interest having been purified and its mass determined, the next analysis usually performed is to determine the protein's amino acid sequence, or primary structure. As stated previously (Section 3.2.1), a wealth of information about a protein's function and evolutionary history can often be obtained from the primary structure. Let us examine first how we can sequence a simple peptide, such as

Ala-Gly-Asp-Phe-Arg-Gly

The first step is to determine the *amino acid composition* of the peptide. The peptide is hydrolyzed into its constituent amino acids by heating it in 6 N HCl at 110°C for 24 hours. Amino acids in hydrolysates can be separated by ion-exchange chromatography on columns of sulfonated polystyrene. The identity of the amino acid is revealed by its elution volume, which is the volume of buffer used to remove the amino acid from the column (Figure 4.18), and quantified by reaction with *ninhydrin*. Amino acids treated with ninhydrin give an intense blue color, except for proline, which gives a yellow color because it contains a secondary amino group. The concentration of an amino acid in a solution, after heating with ninhydrin, is proportional to the optical absorbance of the solution. This technique can detect a microgram (10 nmol) of an amino acid, which is about the amount

Ninhydrin

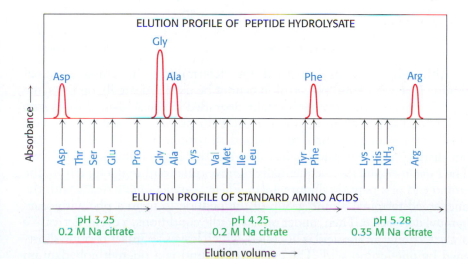

FIGURE 4.18 Determination of amino acid composition. Different amino acids in a peptide hydrolysate can be separated by ion-exchange chromatography on a sulfonated polystyrene resin (such as Dowex-50). Buffers (in this case, sodium citrate) of increasing pH are used to elute the amino acids from the column. The amount of each amino acid present is determined from the absorbance. Aspartate, which has an acidic side chain, is first to emerge, whereas arginine, which has a basic side chain, is the last. The original peptide is revealed to be composed of one aspartate, one alanine, one phenylalanine, one arginine, and two glycine residues.

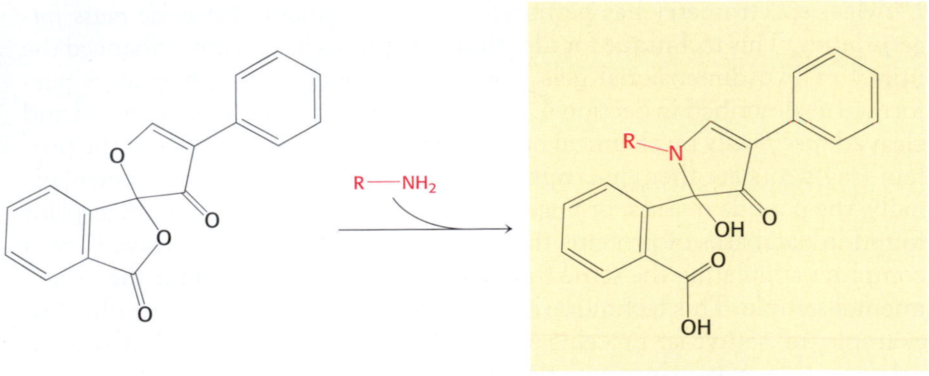

FIGURE 4.19 Fluorescent derivatives of amino acids. Fluorescamine reacts with the α-amino group of an amino acid to form a fluorescent derivative.

Fluorescamine

Amine derivative

present in a thumbprint. As little as a nanogram (10 pmol) of an amino acid can be detected by replacing ninhydrin with *fluorescamine,* which reacts with the α-amino group to form a highly fluorescent product (Figure 4.19). A comparison of the chromatographic patterns of our sample hydrolysate with that of a standard mixture of amino acids would show that the amino acid composition of the peptide is

$$(Ala, Arg, Asp, Gly_2, Phe)$$

The parentheses denote that this is the amino acid composition of the peptide, not its sequence.

The next step is often to identify the N-terminal amino acid by labeling it with a compound that forms a stable covalent bond. *Fluorodinitrobenzene* (FDNB) was first used for this purpose by Frederick Sanger. *Dabsyl chloride* is now commonly used because it forms fluorescent derivatives that can be detected with high sensitivity. It reacts with an uncharged α-NH_2 group to form a sulfonamide derivative that is stable under conditions that hydrolyze peptide bonds (Figure 4.20). Hydrolysis of our sample dabsyl–peptide in 6 N HCl would yield a dabsyl–amino acid, which could be identified as dabsyl–alanine by its chromatographic properties. *Dansyl chloride,* too, is a valuable labeling reagent because it forms fluorescent sulfonamides.

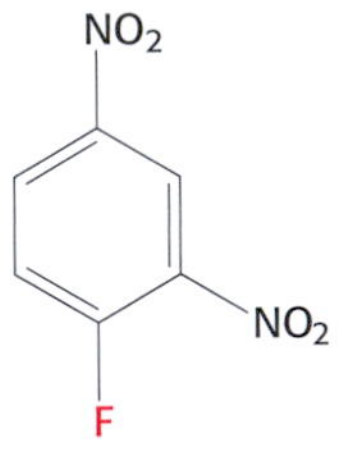

Fluorodinitrobenzene

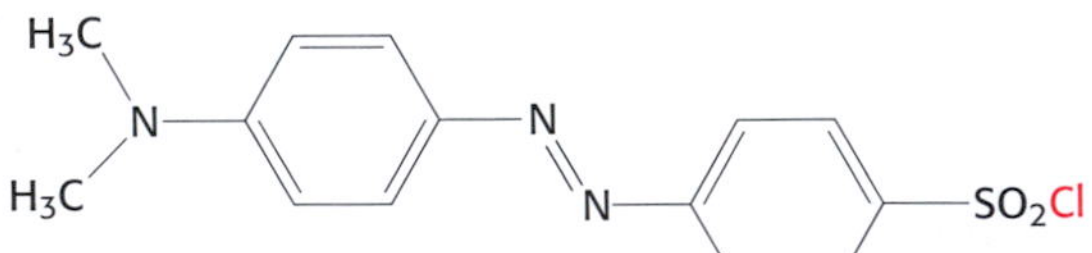

Dabsyl chloride

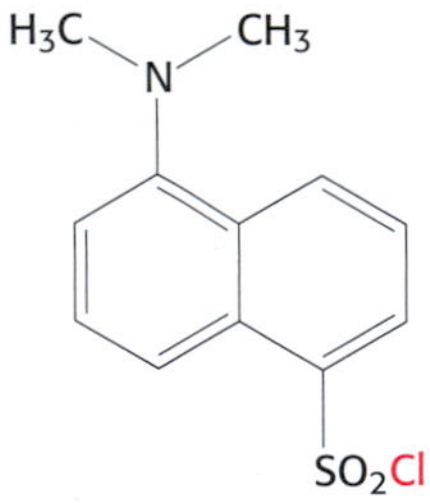

Dansyl chloride

Although the dabsyl method for determining the amino-terminal residue is sensitive and powerful, it cannot be used repeatedly on the same peptide, because the peptide is totally degraded in the acid-hydrolysis step and thus all sequence information is lost. Pehr Edman devised a method for labeling the amino-terminal residue and cleaving it from the peptide without disrupting the peptide bonds between the other amino acid residues. The *Edman degradation* sequentially removes one residue at a time from the amino end of a peptide (Figure 4.21). *Phenyl isothiocyanate* reacts with the uncharged terminal amino group of the peptide to form a phenylthiocarbamoyl derivative. Then, under mildly acidic conditions, a cyclic derivative of the terminal amino acid is liberated, which leaves an intact peptide shortened by one amino acid. The cyclic compound is a phenylthiohydantoin

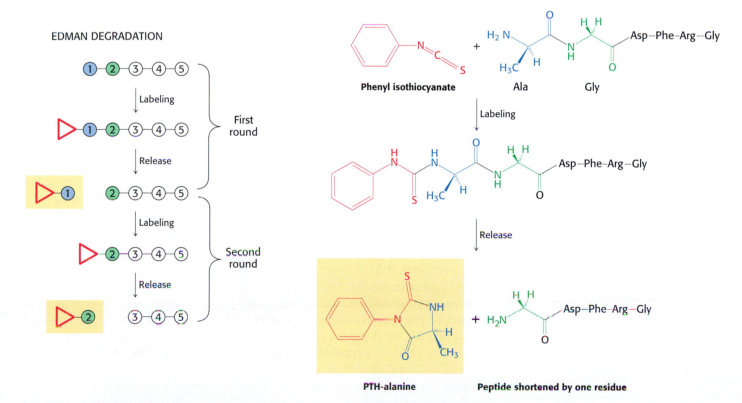

FIGURE 4.20 Determination of the amino-terminal residue of a peptide. Dabsyl chloride labels the peptide, which is then hydrolyzed with the use of hydrochloric acid. The dabsyl–amino acid (dabsyl–alanine in this example) is identified by its chromatographic characteristics.

FIGURE 4.21 The Edman degradation. The labeled amino-terminal residue (PTH–alanine in the first round) can be released without hydrolyzing the rest of the peptide. Hence, the amino-terminal residue of the shortened peptide (Gly-Asp-Phe-Arg-Gly) can be determined in the second round. Three more rounds of the Edman degradation reveal the complete sequence of the original peptide.

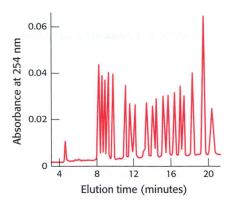

FIGURE 4.22 Separation of PTH–amino acids. PTH–amino acids can be rapidly separated by high-pressure liquid chromatography (HPLC). In this HPLC profile, a mixture of PTH–amino acids is clearly resolved into its components. An unknown amino acid can be identified by its elution position relative to the known ones.

(PTH)–amino acid, which can be identified by chromatographic procedures. The Edman procedure can then be repeated on the shortened peptide, yielding another PTH–amino acid, which can again be identified by chromatography. Three more rounds of the Edman degradation will reveal the complete sequence of the original peptide pentapeptide.

The development of automated sequencers has markedly decreased the time required to determine protein sequences. One cycle of the Edman degradation—the cleavage of an amino acid from a peptide and its identification—is carried out in less than 1 hour. By repeated degradations, the amino acid sequence of some 50 residues in a protein can be determined. High-pressure liquid chromatography provides a sensitive means of distinguishing the various amino acids (Figure 4.22). Gas-phase sequenators can analyze picomole quantities of peptides and proteins. This high sensitivity makes it feasible to analyze the sequence of a protein sample eluted from a single band of an SDS–polyacrylamide gel.

4.2.1 Proteins Can Be Specifically Cleaved into Small Peptides to Facilitate Analysis

In principle, it should be possible to sequence an entire protein by using the Edman method. In practice, the peptides cannot be much longer than about 50 residues. This is so because the reactions of the Edman method, especially the release step, are not 100% efficient, and so not all peptides in the reaction mixture release the amino acid derivative at each step. For instance, if the efficiency of release for each round were 98%, the proportion of "correct" amino acid released after 60 rounds would be (0.98^{60}), or 0.3—a hopelessly impure mix. This obstacle can be circumvented by cleaving the original protein at specific amino acids into smaller peptides that can be sequenced. In essence, the strategy is to *divide and conquer.*

Specific cleavage can be achieved by chemical or enzymatic methods. For example, *cyanogen bromide* (CNBr) splits polypeptide chains only on the carboxyl side of methionine residues (Figure 4.23).

FIGURE 4.23 Cleavage by cyanogen bromide. Cyanogen bromide cleaves polypeptides on the carboxyl side of methionine residues.

A protein that has 10 methionine residues will usually yield 11 peptides on cleavage with CNBr. Highly specific cleavage is also obtained with *trypsin*, a proteolytic enzyme from pancreatic juice. Trypsin cleaves polypeptide chains on the carboxyl side of arginine and lysine residues (Figure 4.24 and Section 9.1.4). A protein that contains 9 lysine and 7 arginine residues will usually yield 17 peptides on digestion with trypsin. Each of these tryptic

FIGURE 4.24 Cleavage by trypsin. Trypsin hydrolyzes polypeptides on the carboxyl side of arginine and lysine residues.

TABLE 4.3 Specific cleavage of polypeptides

Reagent	Cleavage site
Chemical cleavage	
Cyanogen bromide	Carboxyl side of methionine residues
O-Iodosobenzoate	Carboxyl side of tryptophan residues
Hydroxylamine	Asparagine–glycine bonds
2-Nitro-5-thiocyanobenzoate	Amino side of cysteine residues
Enzymatic cleavage	
Trypsin	Carboxyl side of lysine and arginine residues
Clostripain	Carboxyl side of arginine residues
Staphylococcal protease	Carboxyl side of aspartate and glutamate residues (glutamate only under certain conditions)
Thrombin	Carboxyl side of arginine
Chymotrypsin	Carboxyl side of tyrosine, tryptophan, phenylalanine, leucine, and methionine
Carboxypeptidase A	Amino side of C-terminal amino acid (not arginine, lysine, or proline)

peptides, except for the carboxyl-terminal peptide of the protein, will end with either arginine or lysine. Table 4.3 gives several other ways of specifically cleaving polypeptide chains.

The peptides obtained by specific chemical or enzymatic cleavage are separated by some type of chromatography. The sequence of each purified peptide is then determined by the Edman method. At this point, the amino acid sequences of segments of the protein are known, but the order of these segments is not yet defined. How can we order the peptides to obtain the primary structure of the original protein? The necessary additional information is obtained from *overlap peptides* (Figure 4.25). A second enzyme is used to split the polypeptide chain at different linkages. For example, chymotrypsin cleaves preferentially on the carboxyl side of aromatic and some other bulky nonpolar residues (Section 9.1.3). Because these chymotryptic peptides overlap two or more tryptic peptides, they can be used to establish the order of the peptides. The entire amino acid sequence of the polypeptide chain is then known.

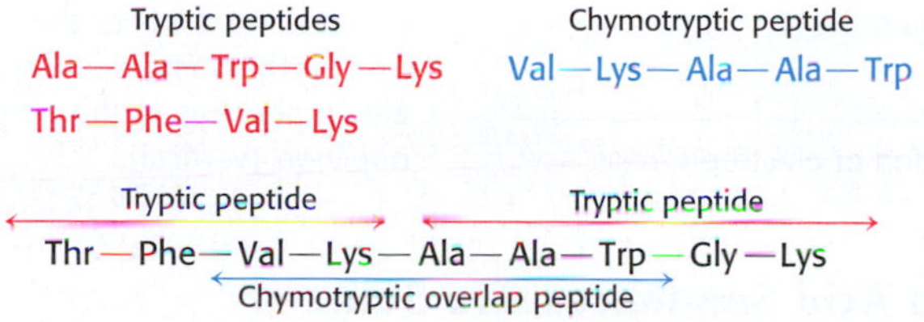

Tryptic peptides   Chymotryptic peptide

Ala—Ala—Trp—Gly—Lys   Val—Lys—Ala—Ala—Trp

Thr—Phe—Val—Lys

Tryptic peptide   Tryptic peptide

Thr—Phe—Val—Lys—Ala—Ala—Trp—Gly—Lys

Chymotryptic overlap peptide

FIGURE 4.25 Overlap peptides. The peptide obtained by chymotrypic digestion overlaps two tryptic peptides, establishing their order.

Additional steps are necessary if the initial protein sample is actually several polypeptide chains. SDS–gel electrophoresis under reducing conditions should display the number of chains. Alternatively, the number of distinct N-terminal amino acids could be determined. For a protein made up of two or more polypeptide chains held together by noncovalent bonds, denaturing agents, such as urea or guanidine hydrochloride, are used to dissociate the chains from one another. The dissociated chains must be separated from one another before sequence determination of the individual chains can begin. Polypeptide chains linked by disulfide bonds are separated by reduction with thiols such as β-mercaptoethanol or dithiothreitol. To prevent the cysteine residues from recombining, they are then alkylated with iodoacetate to form stable S-carboxymethyl derivatives (Figure 4.26). Sequencing can then be performed as heretofore described.

To complete our understanding of the protein's structure, we need to determine the positions of the original disulfide bonds. This information can be obtained by using a *diagonal electrophoresis* technique to isolate the peptide sequences containing such bonds (Figure 4.27). First, the protein is specifically cleaved into peptides under conditions in which the disulfides remain intact. The mixture of peptides is applied to a corner of a sheet of paper and subjected to electrophoresis in a single lane along one side. The resulting sheet is exposed to vapors of performic acid, which cleaves disulfides and converts them into cysteic acid residues. Peptides originally linked by disulfides are now independent and more acidic because of the formation of an SO_3^- group.

FIGURE 4.26 Disulfide-bond reduction. Polypeptides linked by disulfide bonds can be separated by reduction with dithiothreitol followed by alkylation to prevent reformation.

This mixture is subjected to electrophoresis in the perpendicular direction under the same conditions as those of the first electrophoresis. Peptides that were devoid of disulfides will have the same mobility as before, and consequently all will be located on a single diagonal line. In contrast, the newly formed peptides containing cysteic acid will usually migrate differently from their parent disulfide-linked peptides and hence will lie off the diagonal. These peptides can then be isolated and sequenced, and the location of the disulfide bond can be established.

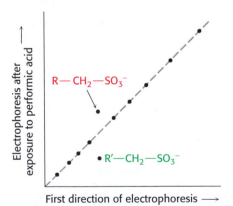

FIGURE 4.27 Diagonal electrophoresis. Peptides joined together by disulfide bonds can be detected by diagonal electrophoresis. The mixture of peptides is subjected to electrophoresis in a single lane in one direction (horizontal) and then treated with performic acid, which cleaves and oxidizes the disulfide bonds. The sample is then subjected to electrophoresis in the perpendicular direction (vertical).

4.2.2 Amino Acid Sequences Are Sources of Many Kinds of Insight

A protein's amino acid sequence, once determined, is a valuable source of insight into the protein's function, structure, and history.

1. *The sequence of a protein of interest can be compared with all other known sequences to ascertain whether significant similarities exist. Does this protein belong to one of the established families?* A search for kinship between a newly sequenced protein and the thousands of previously sequenced ones takes

only a few seconds on a personal computer (Section 7.2). If the newly isolated protein is a member of one of the established classes of protein, we can begin to infer information about the protein's function. For instance, chymotrypsin and trypsin are members of the serine protease family, a clan of proteolytic enzymes that have a common catalytic mechanism based on a reactive serine residue (Section 9.1.4). If the sequence of the newly isolated protein shows sequence similarity with trypsin or chymotrypsin, the result suggests that it may be a serine protease.

2. *Comparison of sequences of the same protein in different species yields a wealth of information about evolutionary pathways.* Genealogical relations between species can be inferred from sequence differences between their proteins. We can even estimate the time at which two evolutionary lines diverged, thanks to the clocklike nature of random mutations. For example, a comparison of serum albumins found in primates indicates that human beings and African apes diverged 5 million years ago, not 30 million years ago as was once thought. Sequence analyses have opened a new perspective on the fossil record and the pathway of human evolution.

3. *Amino acid sequences can be searched for the presence of internal repeats.* Such internal repeats can reveal information about the history of an individual protein itself. Many proteins apparently have arisen by duplication of a primordial gene followed by its diversification. For example, calmodulin, a ubiquitous calcium sensor in eukaryotes, contains four similar calcium-binding modules that arose by gene duplication (Figure 4.28).

4. *Many proteins contain amino acid sequences that serve as signals designating their destinations or controlling their processing.* A protein destined for export from a cell or for location in a membrane, for example, contains a signal sequence, a stretch of about 20 hydrophobic residues near the amino terminus that directs the protein to the appropriate membrane. Another protein may contain a stretch of amino acids that functions as a nuclear localization signal, directing the protein to the nucleus.

5. *Sequence data provide a basis for preparing antibodies specific for a protein of interest.* Careful examination of the amino acid sequence of a protein can reveal which sequences will be most likely to elicit an antibody when injected into a mouse or rabbit. Peptides with these sequences can be synthesized and used to generate antibodies to the protein. These specific antibodies can be very useful in determining the amount of a protein present in solution or in the blood, ascertaining its distribution within a cell, or cloning its gene (Section 4.3.3).

6. *Amino acid sequences are valuable for making DNA probes that are specific for the genes encoding the corresponding proteins* (Section 6.1.4). Knowledge of a protein's primary structure permits the use of reverse genetics. DNA probes that correspond to a part of the amino acid sequence can be constructed on the basis of the genetic code. These probes can be used to isolate the gene of the protein so that the entire sequence of the protein can be determined. The gene in turn can provide valuable information about the physiological regulation of the protein. Protein sequencing is an integral part of molecular genetics, just as DNA cloning is central to the analysis of protein structure and function.

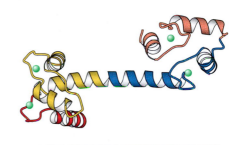

FIGURE 4.28 Repeating motifs in a protein chain. Calmodulin, a calcium sensor, contains four similar units in a single polypeptide chain shown in red, yellow, blue, and orange. Each unit binds a calcium ion (shown in green).

4.2.3 Recombinant DNA Technology Has Revolutionized Protein Sequencing

Hundreds of proteins have been sequenced by Edman degradation of peptides derived from specific cleavages. Nevertheless, heroic effort is required to elucidate the sequence of large proteins, those with more than 1000 residues. For sequencing such proteins, a complementary experimental approach based on recombinant DNA technology is often more efficient. As will be discussed in Chapter 6, long stretches of DNA can be cloned and sequenced, and the nucleotide sequence directly reveals the amino acid sequence of the protein encoded by the gene (Figure 4.29). Recombinant DNA technology is producing a wealth of amino acid sequence information at a remarkable rate.

DNA sequence   GGG | TTC | TTG | GGA | GCA | GCA | GGA | AGC | ACT | ATG | GGC | GCA |

Amino acid sequence   Gly   Phe   Leu   Gly   Ala   Ala   Gly   Ser   Thr   Met   Gly   Ala

FIGURE 4.29 DNA sequence yields the amino acid sequence. The complete nucleotide sequence of HIV-1 (human immunodeficiency virus), the cause of AIDS (acquired immune deficiency syndrome), was determined within a year after the isolation of the virus. A part of the DNA sequence specified by the RNA genome of the virus is shown here with the corresponding amino acid sequence (deduced from a knowledge of the genetic code).

Even with the use of the DNA base sequence to determine primary structure, there is still a need to work with isolated proteins. The amino acid sequence deduced by reading the DNA sequence is that of the *nascent* protein, the direct product of the translational machinery. Many proteins are modified after synthesis. Some have their ends trimmed, and others arise by cleavage of a larger initial polypeptide chain. Cysteine residues in some proteins are oxidized to form disulfide links, connecting either parts within a chain or separate polypeptide chains. Specific side chains of some proteins are altered. Amino acid sequences derived from DNA sequences are rich in information, but they do not disclose such posttranslational modifications. Chemical analyses of proteins in their final form are needed to delineate the nature of these changes, which are critical for the biological activities of most proteins. *Thus, genomic and proteomic analyses are complementary approaches to elucidating the structural basis of protein function.*

4.3 IMMUNOLOGY PROVIDES IMPORTANT TECHNIQUES WITH WHICH TO INVESTIGATE PROTEINS

Immunological methods have become important tools used to purify a protein, locate it in the cell, or quantify how much of the protein is present. These methods are predicated on the exquisite specificity of antibodies for their target proteins. Labeled antibodies provide a means to tag a specific protein so that it can be isolated, quantified, or visualized.

4.3.1 Antibodies to Specific Proteins Can Be Generated

Immunological techniques begin with the generation of antibodies to a particular protein. An *antibody* (also called an *immunoglobulin*, Ig) is a protein synthesized by an animal in response to the presence of a foreign substance, called an *antigen*, and normally functions to protect the animal from infection (Chapter 33). Antibodies have specific and high affinity for the antigens that elicited their synthesis. Proteins, polysaccharides, and nucleic acids can be effective antigens. An antibody recognizes a specific group or cluster of amino acids on a large molecule called an *antigenic determinant,* or *epitope* (Figures 4.30 and 4.31). Small foreign molecules, such as synthetic peptides, also can elicit antibodies, provided that the small molecule contains a recognized epitope and is attached to a macromolecular carrier. The small foreign molecule itself is called a *hapten*. Animals have a very

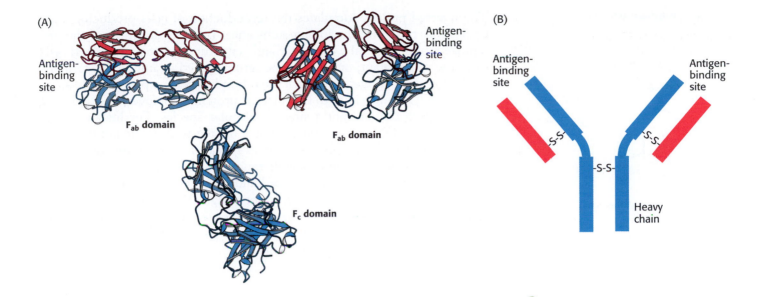

(A)

Antigen-binding site

F_{ab} domain

F_{ab} domain

Antigen-binding site

F_c domain

(B)

Antigen-binding site

Antigen-binding site

-S-S-

-S-S-

-S-S-

Heavy chain

large repertoire of antibody-producing cells, each producing an antibody of a single specificity. An antigen acts by stimulating the proliferation of the small number of cells that were already forming an antibody capable of recognizing the antigen (Chapter 33).

Immunological techniques depend on our being able to generate antibodies to a specific antigen. To obtain antibodies that recognize a particular protein, a biochemist injects the protein into a rabbit twice, 3 weeks apart.

FIGURE 4.30 Antibody structure.
(A) IgG antibodies consist of four chains, two heavy chains (blue) and two light chains (red), linked by disulfide bonds. The heavy and light chains come together to form F_{ab} domains, which have the antigen-binding sites at the ends. The two heavy chains form the F_c domain. The F_{ab} domains are linked to the F_c domain by flexible linkers. (B) A more schematic representation of an IgG molecule.

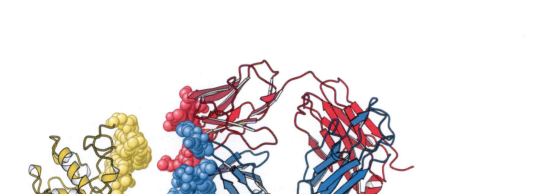

Antigen

Antibody

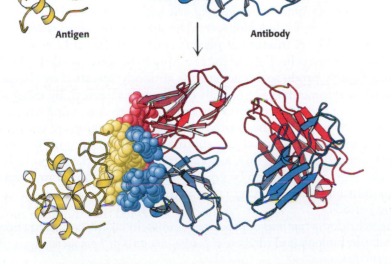

FIGURE 4.31 Antigen–antibody interactions. A protein antigen, in this case lysozyme, binds to the end of an F_{ab} domain from an antibody. The end of the antibody and the antigen have complementary shapes, allowing a large amount of surface to be buried on binding.

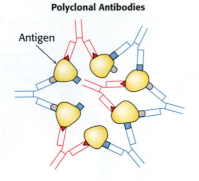

Polyclonal Antibodies

Antigen

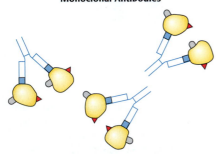

Monoclonal Antibodies

FIGURE 4.32 Polyclonal and monoclonal antibodies. Most antigens have several epitopes. Polyclonal antibodies are heterogeneous mixtures of antibodies, each specific for one of the various epitopes on an antigen. Monoclonal antibodies are all identical, produced by clones of a single antibody-producing cell. They recognize one specific epitope. [After R. A. Goldsby, T. J. Kindt, B. A. Osborne, *Kuby Immunology,* 4th ed. (W. H. Freeman and Company, 2000), p. 154.]

The injected protein stimulates the reproduction of cells producing antibodies that recognize the foreign substance. Blood is drawn from the immunized rabbit several weeks later and centrifuged to separate blood cells from the supernatant, or serum. The serum, called an *antiserum*, contains antibodies to all antigens to which the rabbit has been exposed. Only some of them will be antibodies to the injected protein. Moreover, antibodies of a given specificity are not a single molecular species. For instance, 2,4-dinitrophenol (DNP) has been used as a hapten to generate antibodies to DNP. Analyses of anti-DNP antibodies revealed a wide range of binding affinities—the dissociation constants ranged from about 0.1 nM to 1 μM. Correspondingly, a large number of bands were evident when anti-DNP antibody was subjected to isoelectric focusing. These results indicate that cells are producing many different antibodies, each recognizing a different surface feature of the same antigen. The antibodies are heterogeneous, or *polyclonal* (Figure 4.32). This heterogeneity is a barrier, which can complicate the use of these antibodies.

4.3.2 Monoclonal Antibodies with Virtually Any Desired Specificity Can Be Readily Prepared

The discovery of a means of producing *monoclonal antibodies* of virtually any desired specificity was a major breakthrough that intensified the power of immunological approaches. Just as working with impure proteins makes it difficult to interpret data and understand function, so too does working with an impure mixture of antibodies. The ideal would be to isolate a clone of cells that produce only a single antibody. The problem is that antibody-producing cells isolated from an organism die in a short time.

Immortal cell lines that produce monoclonal antibodies do exist. These cell lines are derived from a type of cancer, *multiple myeloma*, a malignant disorder of antibody-producing cells. In this cancer, a single transformed plasma cell divides uncontrollably, generating a very large number of *cells of a single kind*. They are a *clone* because they are descended from the same cell and have identical properties. The identical cells of the myeloma secrete large amounts of normal *immunoglobulin of a single kind* generation after generation. A myeloma can be transplanted from one mouse to another, where it continues to proliferate. These antibodies were useful for elucidating antibody structure, but nothing is known about their specificity and so they are useless for the immunological methods described in the next pages.

Cesar Milstein and Georges Köhler discovered that *large amounts of homogeneous antibody of nearly any desired specificity could be obtained by fusing a short-lived antibody-producing cell with an immortal myeloma cell.* An antigen is injected into a mouse, and its spleen is removed several weeks later (Figure 4.33). A mixture of plasma cells from this spleen is fused in vitro with myeloma cells. Each of the resulting hybrid cells, called *hybridoma cells,* indefinitely produces homogeneous antibody specified by the parent cell from the spleen. Hybridoma cells can then be screened, by using some sort of assay for the antigen–antibody interaction, to determine which ones produce antibody having the desired specificity. Collections of cells shown to produce the desired antibody are subdivided and reassayed. This process is repeated until a pure cell line, a clone producing a single antibody, is isolated. These positive cells can be grown in culture medium or injected into mice to induce myelomas. Alternatively, the cells can be frozen and stored for long periods.

The hybridoma method of producing monoclonal antibodies has opened new vistas in biology and medicine. *Large amounts of homogeneous antibod-*

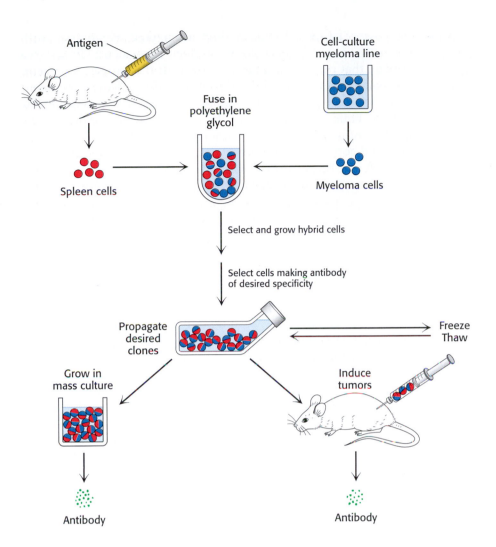

Antigen

Cell-culture
myeloma line

Fuse in
polyethylene
glycol

Spleen cells

Myeloma cells

Select and grow hybrid cells

Select cells making antibody
of desired specificity

Propagate
desired
clones

Freeze
Thaw

Grow in
mass culture

Induce
tumors

Antibody

Antibody

FIGURE 4.33 Preparation of monoclonal antibodies. Hybridoma cells are formed by fusion of antibody-producing cells and myeloma cells. The hybrid cells are allowed to proliferate by growing them in selective medium. They are then screened to determine which ones produce antibody of the desired specificity. [After C. Milstein. Monoclonal antibodies. Copyright © 1980 by Scientific American, Inc. All rights reserved.]

ies with tailor-made specificities can be readily prepared. They are sources of insight into relations between antibody structure and specificity. Moreover, monoclonal antibodies can serve as precise analytical and preparative reagents. For example, a pure antibody can be obtained against an antigen that has not yet been isolated (Section 4.4). Proteins that guide development have been identified with the use of monoclonal antibodies as tags (Figure 4.34). Monoclonal antibodies attached to solid supports can be used as affinity columns to purify scarce proteins. This method has been used to purify interferon (an antiviral protein) 5000-fold from a crude mixture. Clinical laboratories are using monoclonal antibodies in many assays. For example, the detection in blood of isozymes that are normally localized in the heart points to a myocardial infarction (heart attack). Blood transfusions have been made safer by antibody screening of donor blood for viruses that cause AIDS (acquired immune deficiency syndrome), hepatitis, and other infectious diseases. Monoclonal antibodies are also being evaluated for use as therapeutic agents, as in the treatment of cancer. Furthermore, the vast repertoire of antibody specificity can be tapped to generate catalytic antibodies having novel features not found in naturally occurring enzymes.

4.3.3 Proteins Can Be Detected and Quantitated by Using an Enzyme-Linked Immunosorbent Assay

Antibodies can be used as exquisitely specific analytic reagents to quantify the amount of a protein or other antigen. The technique is the *enzyme-linked*

FIGURE 4.34 Fluorescence micrograph of a developing *Drosophila* embryo. The embryo was stained with a fluorescent-labeled monoclonal antibody for the DNA-binding protein encoded by *engrailed*, an essential gene in specifying the body plan. [Courtesy of Dr. Nipam Patel and Dr. Corey Goodman.]

immunosorbent assay (ELISA). In this method, an enzyme, which reacts with a colorless substrate to produce a colored product, is covalently linked to a specific antibody that recognizes a target antigen. If the antigen is present, the antibody–enzyme complex will bind to it, and the enzyme component of the antibody–enzyme complex will catalyze the reaction generating the colored product. Thus, the presence of the colored product indicates the presence of the antigen. Such an enzyme-linked immunosorbent assay, which is rapid and convenient, can detect less than a nanogram (10^{-9} g) of a protein. ELISA can be performed with either polyclonal or monoclonal antibodies, but the use of monoclonal antibodies yields more reliable results.

We will consider two among the several types of ELISA. *The indirect ELISA is used to detect the presence of antibody* and is the basis of the test for HIV infection. In that test, viral core proteins (the antigen) are absorbed to the bottom of a well. Antibodies from a patient are then added to the coated well and allowed to bind to the antigen. Finally, enzyme-linked antibodies to human antibodies (for instance, goat antibodies that recognize human antibodies) are allowed to react in the well and unbound antibodies are removed by washing. Substrate is then applied. An enzyme reaction suggests that the enzyme-linked antibodies were bound to human antibodies, which in turn implies that the patient had antibodies to the viral antigen (Figure 4.35).

The sandwich ELISA allows both the detection and the quantitation of antigen. Antibody to a particular antigen is first absorbed to the bottom of a well. Next, the antigen (or blood or urine containing the antigen) is added to the well and binds to the antibody. Finally, a second, different antibody to the antigen is added. This antibody is enzyme linked and is processed as described for indirect ELISA. In this case, the extent of reaction is directly proportional to the amount of antigen present. Consequently, it permits the measurement of small quantities of antigen (see Figure 4.35).

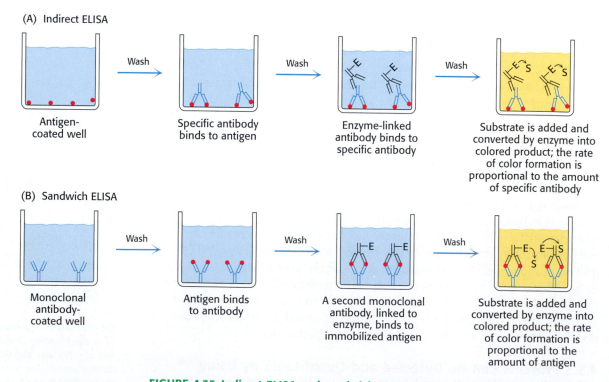

(A) Indirect ELISA

Antigen-coated well →(Wash)→ Specific antibody binds to antigen →(Wash)→ Enzyme-linked antibody binds to specific antibody →(Wash)→ Substrate is added and converted by enzyme into colored product; the rate of color formation is proportional to the amount of specific antibody

(B) Sandwich ELISA

Monoclonal antibody-coated well →(Wash)→ Antigen binds to antibody →(Wash)→ A second monoclonal antibody, linked to enzyme, binds to immobilized antigen →(Wash)→ Substrate is added and converted by enzyme into colored product; the rate of color formation is proportional to the amount of antigen

FIGURE 4.35 Indirect ELISA and sandwich ELISA (A) In indirect ELISA, the production of color indicates the amount of an antibody to a specific antigen. (B) In sandwich ELISA, the production of color indicates the quantity of antigen. [After R. A. Goldsby, T. J. Kindt, B. A. Osborne, *Kuby Immunology*, 4th ed. (W. H. Freeman and Company, 2000), p. 162.]

4.3.4 Western Blotting Permits the Detection of Proteins Separated by Gel Electrophoresis

Often it is necessary to detect small quantities of a particular protein in the presence of many other proteins, such as a viral protein in the blood. Very small quantities of a protein of interest in a cell or in body fluid can be detected by an immunoassay technique called *Western blotting* (Figure 4.36). A sample is subjected to electrophoresis on an SDS–polyacrylamide gel. Blotting (or more typically electroblotting) transfers the resolved proteins on the gel to the surface of a polymer sheet to make them more accessible for reaction. An antibody that is specific for the protein of interest is added to the sheet and reacts with the antigen. The antibody–antigen complex on the sheet then can be detected by rinsing the sheet with a second antibody specific for the first (e.g., goat antibody that recognizes mouse antibody). A radioactive label on the second antibody produces a dark band on x-ray film (an autoradiogram). Alternatively, an enzyme on the second antibody generates a colored product, as in the ELISA method. Western blotting makes it possible to find a protein in a complex mixture, the proverbial needle in a haystack. It is the basis for the test for infection by hepatitis C, where it is used to detect a core protein of the virus. This technique is also very useful in the cloning of genes.

FIGURE 4.36 Western blotting. Proteins on an SDS–polyacrylamide gel are transferred to a polymer sheet and stained with radioactive antibody. A band corresponding to the protein to which the antibody binds appears in the autoradiogram.

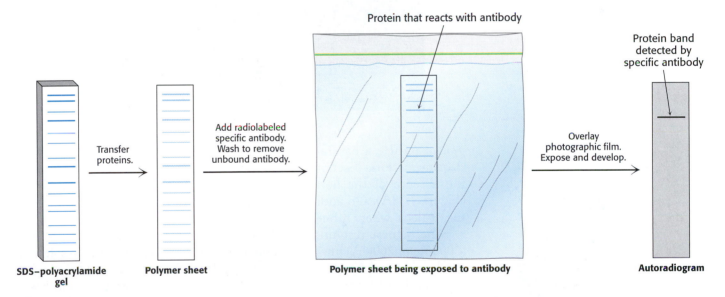

Protein that reacts with antibody

Protein band detected by specific antibody

Transfer proteins.

Add radiolabeled specific antibody. Wash to remove unbound antibody.

Overlay photographic film. Expose and develop.

SDS–polyacrylamide gel

Polymer sheet

Polymer sheet being exposed to antibody

Autoradiogram

4.3.5 Fluorescent Markers Make Possible the Visualization of Proteins in the Cell

Biochemistry is often performed in test tubes or polyacrylamide gels. However, most proteins function in the context of a cell. Fluorescent markers provide a powerful means of examining proteins in their biological context. For instance, cells can be stained with fluorescence-labeled antibodies or other fluorescent proteins and examined by *fluorescence microscopy* to reveal the location of a protein of interest. Arrays of parallel bundles are evident in cells stained with antibody specific for actin, a protein that polymerizes into filaments (Figure 4.37). Actin filaments are constituents of the cytoskeleton, the internal scaffolding of cells that controls their shape and movement. By tracking protein location, fluorescent markers also provide clues to protein function. For instance, the glucocorticoid receptor protein is a transcription factor that controls gene expression in response to the steroid hormone cortisone. The receptor was linked to *green fluorescent*

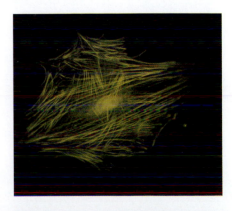

FIGURE 4.37 Actin filaments. Fluorescence micrograph of actin filaments in a cell stained with an antibody specific to actin. [Courtesy of Dr. Elias Lazarides.]

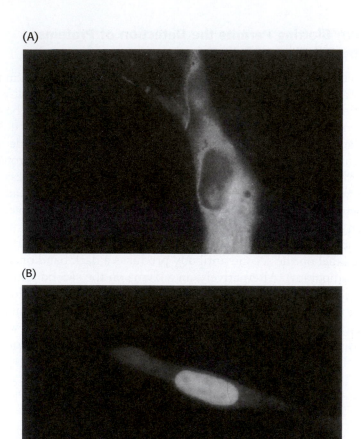

(A)

(B)

FIGURE 4.38 Nuclear localization of a steroid receptor. (A) The receptor, made visible by attachment of the green fluorescent protein, is located predominantly in the cytoplasm of the cultured cell. (B) Subsequent to the addition of corticosterone (a glucocorticoid steroid), the receptor moves into the nucleus. [Courtesy of Professor William B. Pratt/Department of Pharmacology, University of Michigan.]

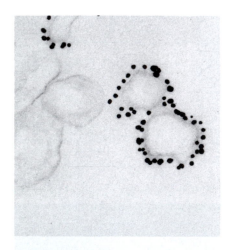

FIGURE 4.39 Immunoelectron microscopy. The opaque particles (150-Å, or 15-nm, diameter) in this electron micrograph are clusters of gold atoms bound to antibody molecules. These membrane vesicles from the synapses of neurons contain a channel protein that is recognized by the specific antibody. [Courtesy of Dr. Peter Sargent.]

protein (GFP), a naturally fluorescent protein isolated from the jellyfish *Aequorea victoria* (Section 3.6.5). Fluorescence microscopy revealed that, in the absence of the hormone, the receptor is located in the cytoplasm (Figure 4.38A). On addition of the steroid, the receptor is translocated to the nucleus, where it binds to DNA (Figure 4.38B).

The highest resolution of fluorescence microscopy is about 0.2 μm (200 nm, or 2000 Å), the wavelength of visible light. Finer spatial resolution can be achieved by electron microscopy by using antibodies tagged with electron-dense markers. For example, ferritin conjugated to an antibody can be readily visualized by electron microscopy because it contains an electron-dense core rich in iron. Clusters of gold also can be conjugated to antibodies to make them highly visible under the electron microscope. *Immunoelectron microscopy* can define the position of antigens to a resolution of 10 nm (100 Å) or finer (Figure 4.39).

4.4 PEPTIDES CAN BE SYNTHESIZED BY AUTOMATED SOLID-PHASE METHODS

The ability to synthesize peptides of defined sequence is a powerful technique for extending biochemical analysis for several reasons.

1. *Synthetic peptides can serve as antigens to stimulate the formation of specific antibodies.* For instance, as discussed earlier, it is often more efficient to obtain a protein sequence from a nucleic acid sequence than by sequencing the protein itself (see also Chapter 6). Peptides can be synthesized on the basis of the nucleic acid sequence, and antibodies can be raised that

target these peptides. These antibodies can then be used to isolate the intact protein from the cell.

2. *Synthetic peptides can be used to isolate receptors for many hormones and other signal molecules.* For example, white blood cells are attracted to bacteria by formylmethionyl (fMet) peptides released in the breakdown of bacterial proteins. Synthetic formylmethionyl peptides have been useful in identifying the cell-surface receptor for this class of peptide. Moreover, synthetic peptides can be attached to agarose beads to prepare affinity chromatography columns for the purification of receptor proteins that specifically recognize the peptides.

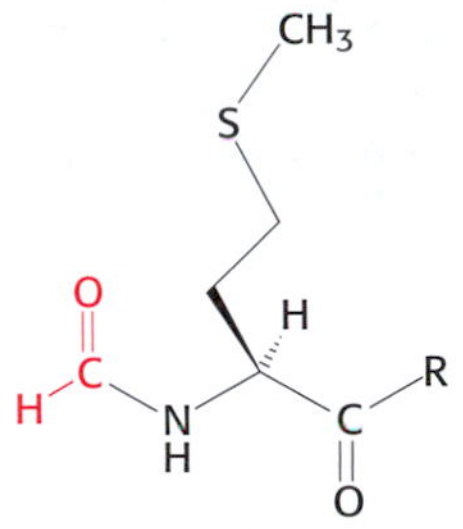

fMet peptide

3. *Synthetic peptides can serve as drugs.* Vasopressin is a peptide hormone that stimulates the reabsorption of water in the distal tubules of the kidney, leading to the formation of more concentrated urine. Patients with diabetes insipidus are deficient in *vasopressin* (also called *antidiuretic hormone*), and so they excrete large volumes of urine (more than 5 liters per day) and are continually thirsty. This defect can be treated by administering 1-desamino-8-D-arginine vasopressin, a synthetic analog of the missing hormone (Figure 4.40). This synthetic peptide is degraded in vivo much more slowly than vasopressin and, additionally, does not increase the blood pressure.

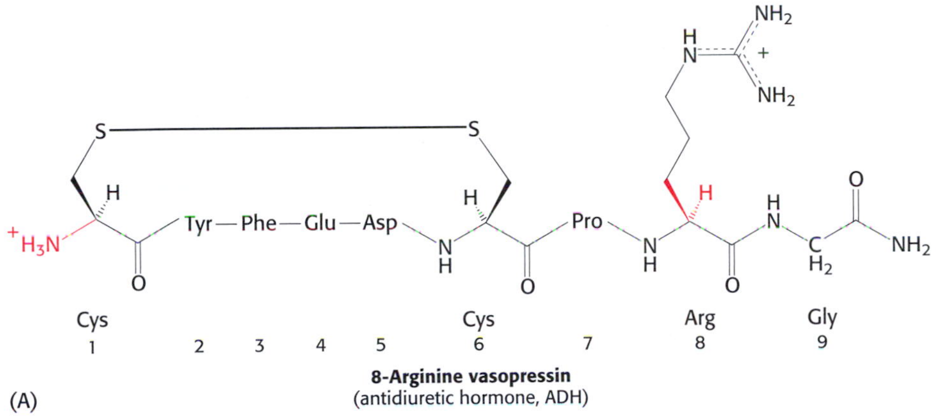

(A)
8-Arginine vasopressin
(antidiuretic hormone, ADH)

(B)
1-Desamino-8-D-arginine vasopressin

FIGURE 4.40 Vasopressin and synthetic vasopressin. Structural formulas of (A) vasopressin, a peptide hormone that stimulates water resorption, and (B) 1-desamino-8-D-arginine vasopressin, a more stable synthetic analog of this antidiuretic hormone.

4. Finally, *studying synthetic peptides can help define the rules governing the three-dimensional structure of proteins.* We can ask whether a particular sequence by itself folds into an α helix, β strand, or hairpin turn or behaves as a random coil.

How are these peptides constructed? The amino group of one amino acid is linked to the carboxyl group of another. However, a unique product is formed only if a single amino group and a single carboxyl group are available for reaction. Therefore, it is necessary to block some groups and to

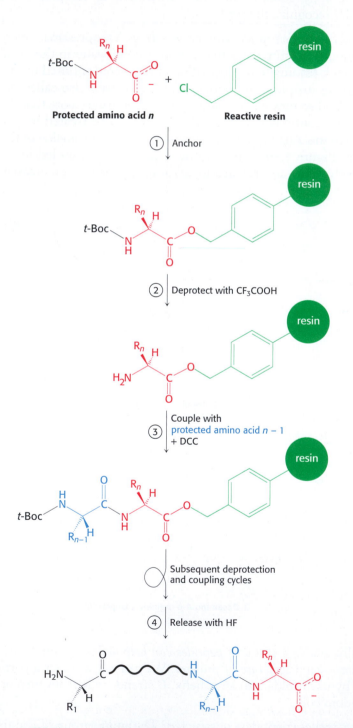

activate others to prevent unwanted reactions. The α-amino group of the first amino acid of the desired peptide is blocked with a *tert*-butyloxycarbonyl (*t*-Boc) group, yielding a *t*-Boc amino acid. The carboxyl group of this same amino acid is activated by reacting it with a reagent such as *dicyclohexylcarbodiimide* (DCC), as illustrated in Figure 4.41. The free amino group of the next amino acid to be linked attacks the activated carboxyl, leading to the formation of a peptide bond and the release of dicyclohexylurea. The carboxyl group of the resulting dipeptide is activated with DCC and reacted with the free amino group of the amino acid that will be the third residue

FIGURE 4.41 Amino acid activation. Dicyclohexylcarbodiimide is used to activate carboxyl groups for the formation of peptide bonds.

FIGURE 4.42 Solid-phase peptide synthesis. The sequence of steps in solid-phase synthesis is: (1) anchoring of the C-terminal amino acid, (2) deprotection of the amino terminus, and (3) coupling of the next residue. Steps 2 and 3 are repeated for each added amino acid. Finally, in step 4, the completed peptide is released from the resin.

in the peptide. This process is repeated until the desired peptide is synthesized. Exposing the peptide to dilute acid removes the *t*-Boc protecting group from the first amino acid while leaving peptide bonds intact.

Peptides containing more than 100 amino acids can be synthesized by sequential repetition of the preceding reactions. Linking the growing peptide chain to an insoluble matrix, such as polystyrene beads, further enhances efficiency. A major advantage of this *solid-phase method* is that the desired product at each stage is bound to beads that can be rapidly filtered and washed, and so there is no need to purify intermediates. All reactions are carried out in a single vessel, eliminating losses caused by repeated transfers of products. The carboxyl-terminal amino acid of the desired peptide sequence is first anchored to the polystyrene beads (Figure 4.42). The *t*-Boc protecting group of this amino acid is then removed. The next amino acid (in the protected *t*-Boc form) and dicyclohexylcarbodiimide, the coupling agent, are added together. After the peptide bond forms, excess reagents and dicyclohexylurea are washed away, leaving the desired dipeptide product attached to the beads. Additional amino acids are linked by the same sequence of reactions. At the end of the synthesis, the peptide is released from the beads by adding hydrofluoric acid (HF), which cleaves the carboxyl ester anchor without disrupting peptide bonds. Protecting groups on potentially reactive side chains, such as that of lysine, also are removed at this time. This cycle of reactions can be readily automated, which makes it feasible to routinely synthesize peptides containing about 50 residues in good yield and purity. In fact, the solid-phase method has been used to synthesize interferons (155 residues) that have antiviral activity and ribonuclease (124 residues) that is catalytically active.

4.5 THREE-DIMENSIONAL PROTEIN STRUCTURE CAN BE DETERMINED BY NMR SPECTROSCOPY AND X-RAY CRYSTALLOGRAPHY

A crucial question is, What does the three-dimensional structure of a specific protein look like? Protein structure determines function, given that the specificity of active sites and binding sites depends on the precise three-dimensional conformation. Nuclear magnetic resonance spectroscopy and x-ray crystallography are two of the most important techniques for elucidating the conformation of proteins.

4.5.1 Nuclear Magnetic Resonance Spectroscopy Can Reveal the Structures of Proteins in Solution

Nuclear magnetic resonance (NMR) spectroscopy is unique in being able to reveal the *atomic structure* of macromolecules *in solution*, provided that highly concentrated solutions ($\sim$1 mM, or 15 mg ml^{-1} for a 15-kd protein) can be obtained. This technique depends on the fact that certain atomic nuclei are intrinsically magnetic. Only a limited number of isotopes display this property, called *spin*, and the ones most important to biochemistry are listed in Table 4.4. The simplest example is the hydrogen nucleus (^{1}H), which is a proton. The spinning of a proton generates a magnetic moment. This moment can take either of two orientations, or spin states (called α and β), when an external magnetic field is applied (Figure 4.43). The energy difference between these states is proportional to the strength of the imposed magnetic field. The α state has a slightly lower energy and hence is slightly more populated (by a factor of the order of 1.00001 in a typical experiment) because it is aligned with the field. A spinning proton in an α state can be

TABLE 4.4 Biologically important nuclei giving NMR signals

Nucleus	Natural abundance (% by weight of the element)
^{1}H	99.984
^{2}H	0.016
^{13}C	1.108
^{14}N	99.635
^{15}N	0.365
^{17}O	0.037
^{23}Na	100.0
^{25}Mg	10.05
^{31}P	100.0
^{35}Cl	75.4
^{39}K	93.1

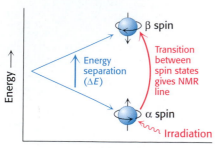

FIGURE 4.43 Basis of NMR spectroscopy. The energies of the two orientations of a nucleus of spin ½ (such as ^{31}P and ^1H) depend on the strength of the applied magnetic field. Absorption of electromagnetic radiation of appropriate frequency induces a transition from the lower to the upper level.

FIGURE 4.44 One-dimensional NMR spectra. (A) ^1H-NMR spectrum of ethanol (CH_3CH_2OH) shows that the chemical shifts for the hydrogen are clearly resolved. (B) ^1H-NMR spectrum from a 55 amino acid fragment of a protein with a role in RNA splicing shows a greater degree of complexity. A large number of peaks are present and many overlap. [(A) After C. Branden and J. Tooze, *Introduction to Protein Structure* (Garland, 1991), p. 280; (B) courtesy of Barbara Amann and Wesley McDermott.]

raised to an excited state (β state) by applying a pulse of electromagnetic radiation (a radio-frequency, or RF, pulse), provided the frequency corresponds to the energy difference between the α and the β states. In these circumstances, the spin will change from α to β; in other words, *resonance* will be obtained. A resonance spectrum for a molecule can be obtained by varying the magnetic field at a constant frequency of electromagnetic radiation or by keeping the magnetic field constant and varying electromagnetic radiation.

These properties can be used to examine the chemical surroundings of the hydrogen nucleus. The flow of electrons around a magnetic nucleus generates a small local magnetic field that opposes the applied field. The degree of such shielding depends on the surrounding electron density. Consequently, nuclei in different environments will change states, or resonate, at slightly different field strengths or radiation frequencies. The nuclei of the perturbed sample absorb electromagnetic radiation at a frequency that can be measured. The different frequencies, termed *chemical shifts*, are expressed in fractional units δ (parts per million, or ppm) relative to the shifts of a standard compound, such as a water-soluble derivative of tetramethylsilane, that is added with the sample. For example, a $-CH_3$ proton typically exhibits a chemical shift (δ) of 1 ppm, compared with a chemical shift of 7 ppm for an aromatic proton. The chemical shifts of most protons in protein molecules fall between 0 and 9 ppm (Figure 4.44). It is possible to resolve most protons in many proteins by using this technique of *one-dimensional NMR*. With this information, we can then deduce changes to a particular chemical group under different conditions, such as the conformational change of a protein from a disordered structure to an α helix in response to a change in pH.

We can garner even more information by examining how the spins on different protons affect their neighbors. By inducing a transient magnetization in a sample through the application a radio-frequency pulse, it is possible to alter the spin on one nucleus and examine the effect on the spin of a neighboring nucleus. Especially revealing is a *two-dimensional spectrum obtained by nuclear Overhauser enhancement spectroscopy* (NOESY), *which graphically displays pairs of protons that are in close proximity*, even if they are not close together in the primary structure. The basis for this technique is the *nuclear Overhauser effect* (NOE), an interaction between nuclei that is proportional to the inverse sixth power of the distance between them. Magnetization is transferred from an excited nucleus to an unexcited one if they are less than about 5 Å apart (Figure 4.45A). In other words, the effect provides a means of detecting the location of atoms relative to one another in the three-dimensional structure of the protein. The diagonal of a NOESY spectrum corresponds to a one-dimensional spectrum. The off-diagonal peaks provide crucial new information: *they identify pairs of protons*

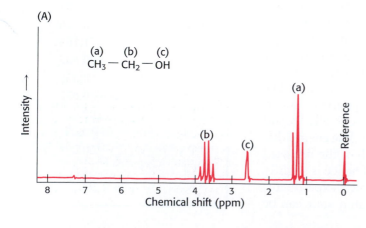

(A)

(a) (b) (c)
$CH_3 - CH_2 - OH$

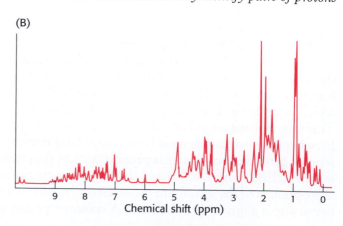

(B)

(A)

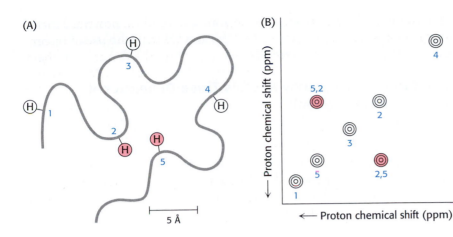

(B)

FIGURE 4.45 The nuclear Overhauser effect. The nuclear Overhauser effect (NOE) identifies pairs of protons that are in close proximity. (A) Schematic representation of a polypeptide chain highlighting five particular protons. Protons 2 and 5 are in close proximity (~4 Å apart), whereas other pairs are farther apart. (B) A highly simplified NOESY spectrum. The diagonal shows five peaks corresponding to the five protons in part A. The peaks above the diagonal and the symmetrically related one below reveal that proton 2 is close to proton 5.

that are less than 5 Å apart (Figure 4.45B). A two-dimensional NOESY spectrum for a protein comprising 55 amino acids is shown in Figure 4.46. The large number of off-diagonal peaks reveals short proton–proton distances. The three-dimensional structure of a protein can be reconstructed with the use of such proximity relations. Structures are calculated such that protons that must be separated by less than 5 Å on the basis of NOESY spectra are close to one another in the three-dimensional structure (Figure 4.47). If a sufficient number of distance constraints are applied, the three-dimensional structure can be determined nearly uniquely. A family of related structures is generated for three reasons (Figure 4.48). First, not enough constraints may be experimentally accessible to fully specify the structure. Second, the distances obtained from analysis of the NOESY spectrum are only approximate. Finally, the experimental observations are made not on single molecules but on a large number of molecules in solution that may have slightly different structures at any given moment. Thus, the family of structures generated from NMR structure analysis indicates the range of conformations for the protein in solution. At present, NMR spectroscopy can determine the structures of only relatively small proteins (<40 kd), but its revolving power is certain to increase. The power of

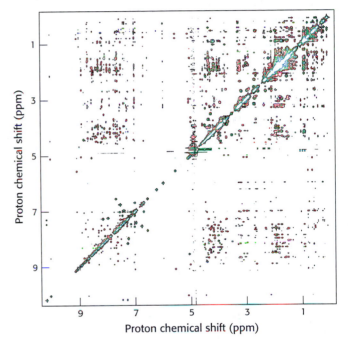

FIGURE 4.46 Detecting short proton–proton distances. A NOESY spectrum for a 55 amino acid domain from a protein having a role in RNA splicing. Each off-diagonal peak corresponds to a short proton–proton separation. This spectrum reveals hundreds of such short proton–proton distances, which can be used to determine the three-dimensional structure of this domain. [Courtesy of Barbara Amann and Wesley McDermott.]

(A)

(B)

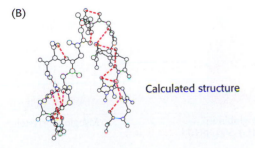

Calculated structure

FIGURE 4.47 Structures calculated on the basis of NMR constraints. (A) NOESY observations show that protons (connected by dotted red lines) are close to one another in space. (B) A three-dimensional structure calculated with these proton pairs constrained to be close together.

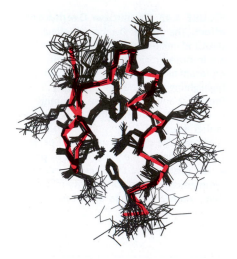

FIGURE 4.48 A family of structures. A set of 25 structures for a 28 amino acid domain from a zinc-finger-DNA-binding protein. The red line traces the average course of the protein backbone. Each of these structures is consistent with hundreds of constraints derived from NMR experiments. The differences between the individual structures are due to a combination of imperfections in the experimental data and the dynamic nature of proteins in solution. [Courtesy of Barbara Amann.]

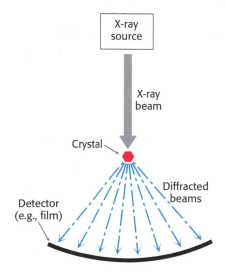

FIGURE 4.49 Essence of an x-ray crystallographic experiment: an x-ray beam, a crystal, and a detector.

NMR has been greatly enhanced by the ability to produce proteins labeled uniformly or at specific sites with ^{13}C, ^{15}N, and ^2H with the use of recombinant DNA technology (Chapter 6).

4.5.2 X-Ray Crystallography Reveals Three-Dimensional Structure in Atomic Detail

X-ray crystallography provides the finest visualization of protein structure currently available. This technique can reveal the precise three-dimensional positions of most atoms in a protein molecule. The use of x-rays provides the best resolution because the wavelength of x-rays is about the same length as that of a covalent bond. The three components in an x-ray crystallographic analysis are a *protein crystal*, a *source of x-rays*, and a *detector* (Figure 4.49).

The technique requires that all molecules be precisely oriented, so the first step is to obtain crystals of the protein of interest. Slowly adding ammonium sulfate or another salt to a concentrated solution of protein to reduce its solubility favors the formation of highly ordered crystals. This is the process of salting out discussed in Section 4.1.3. For example, myoglobin crystallizes in 3 M ammonium sulfate (Figure 4.50). Some proteins crystallize readily, whereas others do so only after much effort has been expended in identifying the right conditions. Crystallization is an art; the best practitioners have great perseverance and patience. Increasingly large and complex proteins are being crystallized. For example, poliovirus, an 8500-kd assembly of 240 protein subunits surrounding an RNA core, has been crystallized and its structure solved by x-ray methods. Crucially, protein crystals frequently display their biological activity, indicating that the proteins have crystallized in their biologically active configuration. For instance, enzyme crystals may display catalytic activity if the crystals are suffused with substrate.

Next, a source of x-rays is required. A beam of x-rays of wavelength 1.54 Å is produced by accelerating electrons against a copper target. A narrow beam of x-rays strikes the protein crystal. Part of the beam goes straight through the crystal; the rest is *scattered* in various directions. Finally, these scattered, or *diffracted*, x-rays are detected by x-ray film, the blackening of the emulsion being proportional to the intensity of the scattered x-ray beam, or by a solid-state electronic detector. The scattering pattern provides abundant information about protein structure. The basic physical principles underlying the technique are:

1. *Electrons scatter x-rays.* The amplitude of the wave scattered by an atom is proportional to its number of electrons. Thus, a carbon atom scatters six times as strongly as a hydrogen atom does.

2. *The scattered waves recombine.* Each atom contributes to each scattered beam. The scattered waves reinforce one another at the film or detector if they are in phase (in step) there, and they cancel one another if they are out of phase.

3. *The way in which the scattered waves recombine depends only on the atomic arrangement.*

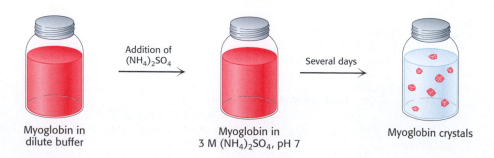

FIGURE 4.50 Crystallization of myoglobin.

Myoglobin in dilute buffer

Addition of $(NH_4)_2SO_4$

Myoglobin in 3 M $(NH_4)_2SO_4$, pH 7

Several days

Myoglobin crystals

The protein crystal is mounted and positioned in a precise orientation with respect to the x-ray beam and the film. The crystal is rotated so that the beam can strike the crystal from many directions. This rotational motion results in an x-ray photograph consisting of a regular array of spots called *reflections*. The x-ray photograph shown in Figure 4.51 is a two-dimensional section through a three-dimensional array of 25,000 spots. The intensity of each spot is measured. These *intensities and their positions* are the basic experimental data of an x-ray crystallographic analysis. The next step is to reconstruct an image of the protein from the observed intensities. In light microscopy or electron microscopy, the diffracted beams are focused by lenses to directly form an image. However, appropriate lenses for focusing x-rays do not exist. Instead, the image is formed by applying a mathematical relation called a Fourier transform. For each spot, this operation yields a wave of electron density whose amplitude is proportional to the square root of the observed intensity of the spot. Each wave also has a *phase*—that is, the timing of its crests and troughs relative to those of other waves. The phase of each wave determines whether the wave reinforces or cancels the waves contributed by the other spots. These phases can be deduced from the well-understood diffraction patterns produced by electron-dense heavy-atom reference markers such as uranium or mercury at specific sites in the protein.

The stage is then set for the calculation of an electron-density map, which gives the density of electrons at a large number of regularly spaced points in the crystal. This three-dimensional electron-density distribution is represented by a series of parallel sections stacked on top of one another. Each section is a transparent plastic sheet (or, more recently, a layer in a computer image) on which the electron-density distribution is represented by contour lines (Figure 4.52), like the contour lines used in geological survey maps to depict altitude (Figure 4.53). The next step is to interpret the electron-density map. A critical factor is the *resolution* of the x-ray analysis, which is determined by the number of scattered intensities used in the Fourier synthesis. The fidelity of the image depends on the resolution of the

(A)

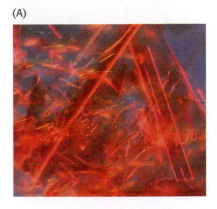

(B)

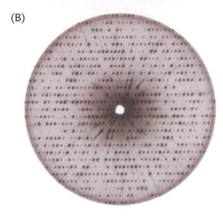

FIGURE 4.51 Myoglobin crystal and x-ray. (A) Crystal of myoglobin. (B) X-ray precession photograph of a myoglobin crystal. [(A) Mel Pollinger/Fran Heyl Associates.]

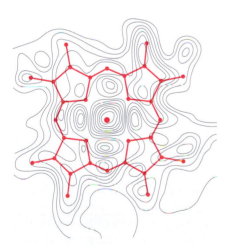

FIGURE 4.52 Section of the electron-density map of myoglobin. This section of the electron-density map shows the heme group. The peak of the center of this section corresponds to the position of the iron atom. [From J. C. Kendrew. The three-dimensional structure of a protein molecule. Copyright © 1961 by Scientific American, Inc. All rights reserved.]

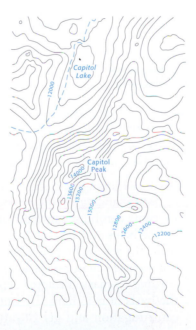

FIGURE 4.53 Section of a U.S. Geological Survey map. Capitol Peak Quadrangle, Colorado.

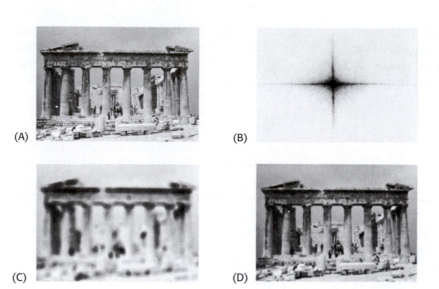

(A)   (B)

(C)   (D)

FIGURE 4.54 Resolution affects the quality of an image. The effect of resolution on the quality of a reconstructed image is shown by an optical analog of x-ray diffraction: (A) a photograph of the Parthenon; (B) an optical diffraction pattern of the Parthenon; (C and D) images reconstructed from the pattern in part B. More data were used to obtain image D than image C, which accounts for the higher quality of image D. [(A) Courtesy of Dr. Thomas Steitz. (B) Courtesy of Dr. David DeRosier).]

Fourier synthesis, as shown by the optical analogy in Figure 4.54. A resolution of 6 Å reveals the course of the polypeptide chain but few other structural details. The reason is that polypeptide chains pack together so that their centers are between 5 Å and 10 Å apart. Maps at higher resolution are needed to delineate groups of atoms, which lie between 2.8 Å and 4.0 Å apart, and individual atoms, which are between 1.0 Å and 1.5 Å apart. The ultimate resolution of an x-ray analysis is determined by the degree of perfection of the crystal. For proteins, this limiting resolution is usually about 2 Å.

The structures of more than 10,000 proteins had been elucidated by NMR and x-ray crystallography by mid-2000, and several new structures are now determined each day. The coordinates are collected at the Protein Data Bank (http://www.rcsb.org/pdb) and the structures can be accessed for visualization and analysis. Knowledge of the detailed molecular architecture of proteins has been a source of insight into how proteins recognize and bind other molecules, how they function as enzymes, how they fold, and how they evolved. This extraordinarily rich harvest is continuing at a rapid pace and is greatly influencing the entire field of biochemistry.

SUMMARY

The rapid progress in gene sequencing has advanced another goal of biochemistry—elucidation of the proteome. The proteome is the complete set of proteins expressed and includes information about how they are modified, how they function, and how they interact with other molecules.

The Purification of Proteins Is an Essential Step in Understanding Their Function

Proteins can be separated from one another and from other molecules on the basis of such characteristics as solubility, size, charge, and binding affinity. SDS–polyacrylamide gel electrophoresis separates the polypeptide chains of proteins under denaturing conditions largely according to mass. Proteins can also be separated electrophoretically on the basis of net charge by isoelectric focusing in a pH gradient. Ultracentrifugation and gel-filtration chromatography resolve proteins according to size, whereas ion-exchange chromatography separates them mainly on the basis of net charge. The high affinity of many proteins for specific chemical groups is exploited in affinity chromatography, in which proteins bind to columns containing beads bearing covalently linked substrates, inhibitors, or other specifically recognized groups. The mass of a protein can be precisely determined by sedimentation equilibrium measurements or by mass spectrometry.

Amino Acid Sequences Can Be Determined by Automated Edman Degradation

The amino acid composition of a protein can be ascertained by hydrolyzing it into its constituent amino acids in 6 N HCl at 110°C. The amino acids can be separated by ion-exchange chromatography and quantitated by reacting them with ninhydrin or fluorescamine. Amino acid sequences can be determined by Edman degradation, which removes one amino acid at a time from the amino end of a peptide. Phenyl isothiocyanate reacts with the terminal amino group to form a phenyl-thiocarbamoyl derivative, which cyclizes under mildly acidic conditions to give a phenylthiohydantoin–amino acid and a peptide shortened by one residue. Automated repeated Edman degradations by a sequenator can analyze sequences of about 50 residues. Longer polypeptide chains are broken into shorter ones for analysis by specifically cleaving them with a reagent such as cyanogen bromide, which splits peptide bonds on the carboxyl side of methionine residues. Enzymes such as trypsin, which cleaves on the carboxyl side of lysine and arginine residues, also are very useful in splitting proteins. Amino acid sequences are rich in information concerning the kinship of proteins, their evolutionary relations, and diseases produced by mutations. Knowledge of a sequence provides valuable clues to conformation and function.

Immunology Provides Important Techniques with Which to Investigate Proteins

Proteins can be detected and quantitated by highly specific antibodies; monoclonal antibodies are especially useful because they are homogeneous. Enzyme-linked immunosorbent assays and Western blots of SDS–polyacrylamide gels are used extensively. Proteins can also be localized within cells by immunofluorescence microscopy and immunoelectron microscopy.

Peptides Can Be Synthesized by Automated Solid-Phase Methods

Polypeptide chains can be synthesized by automated solid-phase methods in which the carboxyl end of the growing chain is linked to an insoluble support. The α-carboxyl group of the incoming amino acid is activated by dicyclohexylcarbodiimide and joined to the α-amino group of the growing chain. Synthetic peptides can serve as drugs and as antigens to stimulate the formation of specific antibodies. They can also be sources of insight into relations between amino acid sequence and conformation.

Three-Dimensional Protein Structure Can Be Determined by NMR Spectroscopy and X-Ray Crystallography

Nuclear magnetic resonance spectroscopy and x-ray crystallography have greatly enriched our understanding of how proteins fold, recognize other molecules, and catalyze chemical reactions. Nuclear magnetic resonance spectroscopy reveals the structure and dynamics of proteins in solution. The chemical shift of nuclei depends on their local environment. Furthermore, the spins of neighboring nuclei interact with each other in ways that provide definitive structural information.

X-ray crystallography is possible because electrons scatter x-rays; the way in which the scattered waves recombine depends only on the atomic arrangement. The three-dimensional structures of thousands of proteins are now known in atomic detail.

KEY TERMS

proteome (p. 78)

assay (p. 79)

homogenate (p. 79)

salting out (p. 80)

dialysis (p. 80)

gel-filtration chromatography (p. 81)

ion-exchange chromatography (p. 81)

affinity chromatography (p. 82)

high-pressure liquid chromatography (HPLC) (p. 82)

gel electrophoresis (p. 83)

isoelectric point (p. 85)

isoelectric focusing (p. 85)

two-dimensional electrophoresis (p. 85)

sedimentation coefficient (Svedberg units, S) (p. 88)

matrix-assisted laser desorption–ionization–time of flight spectrometry (MALDI-TOF) (p. 89)

dabsyl chloride (p. 92)

dansyl chloride (p. 92)

Edman degradation (p. 92)

phenyl isothiocyanate (p. 92)

cyanogen bromide (CNBr) (p. 94)

overlap peptides (p. 95)

diagonal electrophoresis (p. 96)

antibody (p. 98)

antigen (p. 98)

antigenic determinant (epitope) (p. 98)

monoclonal antibodies (p. 100)

enzyme-linked immunosorbent assay (ELISA) (p. 101)

Western blotting (p. 103)

fluorescence microscopy (p. 103)

green fluorescent protein (GFP) (p. 103)

solid-phase method (p. 107)

nuclear magnetic resonance (NMR) spectroscopy (p. 107)

x-ray crystallography (p. 110)

SELECTED READINGS

Where to start

Hunkapiller, M. W., and Hood, L. E., 1983. Protein sequence analysis: Automated microsequencing. *Science* 219:650–659.

Merrifield, B., 1986. Solid phase synthesis. *Science* 232:341–347.

Sanger, F., 1988. Sequences, sequences, sequences. *Annu. Rev. Biochem.* 57:1–28.

Milstein, C., 1980. Monoclonal antibodies. *Sci. Am.* 243(4):66–74.

Moore, S., and Stein, W. H., 1973. Chemical structures of pancreatic ribonuclease and deoxyribonuclease. *Science* 180:458–464.

Books

Creighton, T. E., 1993. *Proteins: Structure and Molecular Properties* (2d ed.). W. H. Freeman and Company.

Kyte, J., 1994. *Structure in Protein Chemistry*. Garland.

Van Holde, K. E., Johnson, W. C., and Ho, P.-S., 1998. *Principles of Physical Biochemistry*. Prentice Hall.

Methods in Enzymology. Academic Press. [The more than 200 volumes of this series are a treasure house of experimental procedures.]

Cantor, C. R., and Schimmel, P. R., 1980. *Biophysical Chemistry*. W. H. Freeman and Company.

Freifelder, D., 1982. *Physical Biochemistry: Applications to Biochemistry and Molecular Biology*. W. H. Freeman and Company.

Johnstone, R. A. W., 1996. *Mass Spectroscopy for Chemists and Biochemists* (2d ed.). Cambridge University Press.

Wilkins, M. R., Williams, K. L., Appel, R. D., and Hochstrasser, D. F., 1997. *Proteome Research: New Frontiers in Functional Genomics (Principles and Practice)*. Springer Verlag

Protein purification and analysis

Deutscher, M. (Ed.), 1997. *Guide to Protein Purification*. Academic Press.

Scopes, R. K., and Cantor, C., 1994. *Protein Purification: Principles and Practice* (3d ed.). Springer Verlag.

Dunn, M. J., 1997. Quantitative two-dimensional gel electrophoresis: From proteins to proteomes. *Biochem. Soc. Trans.* 25:248–254.

Aebersold, R., Pipes, G. D., Wettenhall, R. E., Nika, H., and Hood, L. E., 1990. Covalent attachment of peptides for high sensitivity solid-phase sequence analysis. *Anal. Biochem.* 187:56–65.

Blackstock, W. P., and Weir, M. P., 1999. Proteomics: Quantitative and physical mapping of cellular proteins. *Trends Biotechnol.* 17:121–127.

Dutt, M. J., and Lee, K. H., 2000. Proteomic analysis. *Curr. Opin. Biotechnol.* 11:176–179.

Pandey, A., and Mann, M., 2000. Proteomics to study genes and genomes. *Nature* 405:837–846.

Ultracentrifugation and mass spectrometry

Schuster, T. M., and Laue, T. M., 1994. *Modern Analytical Ultracentrifugation*. Springer Verlag.

Arnott, D., Shabanowtiz, J., and Hunt, D. F., 1993. Mass spectrometry of proteins and peptides: Sensitive and accurate mass measurement and sequence analysis. *Clin. Chem.* 39:2005–2010.

Chait, B. T., and Kent, S. B. H., 1992. Weighing naked proteins: Practical, high-accuracy mass measurement of peptides and proteins. *Science* 257:1885–1894.

Jardine, I., 1990. Molecular weight analysis of proteins. *Methods Enzymol.* 193:441–455.

Edmonds, C. G., Loo, J. A., Loo, R. R., Udseth, H. R., Barinaga, C. J., and Smith, R. D., 1991. Application of electrospray ionization mass spectrometry and tandem mass spectrometry in combination with capillary electrophoresis for biochemical investigations. *Biochem. Soc. Trans.* 19:943–947.

Li, L., Garden, R. W., and Sweedler, J. V., 2000. Single-cell MALDI: A new tool for direct peptide profiling. *Trends Biotechnol.* 18:51–160.

Pappin, D. J., 1997. Peptide mass fingerprinting using MALDI-TOF mass spectrometry. *Methods Mol. Biol.* 64:165–173.

Yates, J. R., 3rd, 1998. Mass spectrometry and the age of the proteome. *J. Mass Spectrom.* 33:1–19.

X-ray crystallography and spectroscopy

Glusker, J. P., 1994. X-ray crystallography of proteins. *Methods Biochem. Anal.* 37:1–72.

Wery, J. P., and Schevitz, R. W., 1997. New trends in macromolecular x-ray crystallography. *Curr. Opin. Chem. Biol.* 1:365–369.

Brunger, A. T., 1997. X-ray crystallography and NMR reveal complementary views of structure and dynamics. *Nat. Struct. Biol.* 4 (suppl.):862–865.

Wüthrich, K., 1989. Protein structure determination in solution by nuclear magnetic resonance spectroscopy. *Science* 243:45–50.

Clore, G. M., and Gronenborn, A. M., 1991. Structures of larger proteins in solution: Three- and four-dimensional heteronuclear NMR spectroscopy. *Science* 252:1390–1399.

Wüthrich, K., 1986. *NMR of Proteins and Nucleic Acids*. Wiley-Interscience.

Monoclonal antibodies and fluorescent molecules

Köhler, G., and Milstein, C., 1975. Continuous cultures of fused cells secreting antibody of predefined specificity. *Nature* 256:495–497.

Goding, J. W., 1996. *Monoclonal Antibodies: Principles and Practice.* Academic Press.

Immunology Today, 2000. Volume 21, issue 8.

Tsien, R. Y., 1998. The green fluorescent protein. *Annu. Rev. Biochem.* 67:509–544.

Kendall, J. M., and Badminton, M. N., 1998. *Aequorea victoria* bioluminescence moves into an exciting era. *Trends Biotechnol.* 16:216–234.

Chemical synthesis of proteins

Mayo, K. H., 2000. Recent advances in the design and construction of synthetic peptides: For the love of basics or just for the technology of it. *Trends Biotechnol.* 18:212–217.

Borgia, J. A., and Fields, G. B., 2000. Chemical synthesis of proteins. *Trends Biotechnol.* 18:243–251.

─ PROBLEMS ├──

1. *Valuable reagents.* The following reagents are often used in protein chemistry:

CNBr	Performic acid	Phenyl isothiocyanate
Urea	Dabsyl chloride	Chymotrypsin
Mercaptoethanol	6 N HCl	
Trypsin	Ninhydrin	

Which one is the best suited for accomplishing each of the following tasks?

(a) Determination of the amino acid sequence of a small peptide.
(b) Identification of the amino-terminal residue of a peptide (of which you have less than 0.1 μg).
(c) Reversible denaturation of a protein devoid of disulfide bonds. Which additional reagent would you need if disulfide bonds were present?
(d) Hydrolysis of peptide bonds on the carboxyl side of aromatic residues.
(e) Cleavage of peptide bonds on the carboxyl side of methionines.
(f) Hydrolysis of peptide bonds on the carboxyl side of lysine and arginine residues.

2. *Finding an end.* Anhydrous hydrazine ($H_2N—NH_2$) has been used to cleave peptide bonds in proteins. What are the reaction products? How might this technique be used to identify the carboxyl-terminal amino acid?

3. *Crafting a new breakpoint.* Ethyleneimine reacts with cysteine side chains in proteins to form S-aminoethyl derivatives. The peptide bonds on the carboxyl side of these modified cysteine residues are susceptible to hydrolysis by trypsin. Why?

4. *Spectrometry.* The absorbance A of a solution is defined as

$$A = \log_{10}(I_0/I)$$

in which I_0 is the incident light intensity and I is the transmitted light intensity. The absorbance is related to the molar absorption coefficient (extinction coefficient) ϵ (in $M^{-1}\,cm^{-1}$), concentration c (in M), and path length l (in cm) by

$$A = \epsilon l c$$

The absorption coefficient of myoglobin at 580 nm is 15,000 $M^{-1}\,cm^{-1}$. What is the absorbance of a 1 mg ml^{-1} solution across a 1-cm path? What percentage of the incident light is transmitted by this solution?

5. *A slow mover.* Tropomyosin, a 93-kd muscle protein, sediments more slowly than does hemoglobin (65 kd). Their sedimentation coefficients are 2.6S and 4.31S, respectively. Which structural feature of tropomyosin accounts for its slow sedimentation?

6. *Sedimenting spheres.* What is the dependence of the sedimentation coefficient S of a spherical protein on its mass? How much more rapidly does an 80-kd protein sediment than does a 40-kd protein?

7. *Size estimate.* The relative electrophoretic mobilities of a 30-kd protein and a 92-kd protein used as standards on an SDS–polyacrylamide gel are 0.80 and 0.41, respectively. What is the apparent mass of a protein having a mobility of 0.62 on this gel?

8. *A new partnership?* The gene encoding a protein with a single disulfide bond undergoes a mutation that changes a serine residue into a cysteine residue. You want to find out whether the disulfide pairing in this mutant is the same as in the original protein. Propose an experiment to directly answer this question.

9. *Sorting cells.* Fluorescence-activated cell sorting (FACS) is a powerful technique for separating cells according to their content of particular molecules. For example, a fluorescence-labeled antibody specific for a cell-surface protein can be used to detect cells containing such a molecule. Suppose that you want to isolate cells that possess a receptor enabling them to detect bacterial degradation products. However, you do not yet have an antibody directed against this receptor. Which fluorescence-labeled molecule would you prepare to identify such cells?

10. *Column choice.* (a) The octapeptide AVGWRVKS was digested with the enzyme trypsin. Would ion exchange or molecular exclusion be most appropriate for separating the products? Explain. (b) Suppose that the peptide was digested with chymotrypsin. What would be the optimal separation technique? Explain.

11. *Making more enzyme?* In the course of purifying an enzyme, a researcher performs a purification step that results in an *increase* in the total activity to a value greater than that present in the original crude extract. Explain how the amount of total activity might increase.

12. *Protein purification problem.* Complete the table below.

Purification procedure	Total protein (mg)	Total activity (units)	Specific activity (units mg^{-1})	Purification level	Yield (%)
Crude extract	20,000	4,000,000		1	100
(NH)$_4$SO$_4$ precipitation	5,000	3,000,000			
DEAE–cellulose chromatography	1,500	1,000,000			
Size-exclusion chromatography	500	750,000			
Affinity chromatography	45	675,000			

Chapter Integration Problems

13. *Quaternary structure.* A protein was purified to homogeneity. Determination of the molecular weight by molecular exclusion chromatography yields 60 kd. Chromatography in the presence of 6 M urea yields a 30-kd species. When the chromatography is repeated in the presence of 6 M urea and 10 mM β-mercaptoethanol, a single molecular species of 15 kd results. Describe the structure of the molecule.

14. *Helix–coil transitions.*

(a) NMR measurements have shown that poly-L-lysine is a random coil at pH 7 but becomes α helical as the pH is raised above 10. Account for this pH-dependent conformational transition.
(b) Predict the pH dependence of the helix–coil transition of poly-L-glutamate.

15. *Peptides on a chip.* Large numbers of different peptides can be synthesized in a small area on a solid support. This high-density array can then be probed with a fluorescence-labeled protein to find out which peptides are recognized. The binding of an antibody to an array of 1024 different peptides occupying a total area the size of a thumbnail is shown in the figure below. How would you synthesize such a peptide array? [Hint: Use light instead of acid to deprotect the terminal amino group in each round of synthesis.]

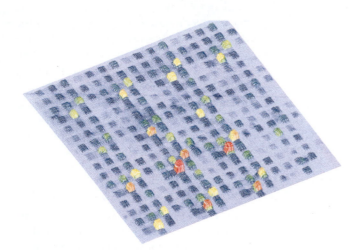

Data Interpretation Problems

16. *Protein sequencing I.* Determine the sequence of hexapeptide based on the following data. Note: When the sequence is not known, a comma separates the amino acids. (See Table 4.3)

Amino acid composition: (2R,A,S,V,Y)
N-terminal analysis of the hexapeptide: A
Trypsin digestion: (R,A,V) and (R,S,Y)
Carboxypeptidase digestion: No digestion.
Chymotrypsin digestion: (A,R,V,Y) and (R,S)

17. *Protein sequencing II.* Determine the sequence of a peptide consisting of 14 amino acids on the basis of the following data.

Amino acid composition: (4S,2L,F,G,I,K,M,T,W,Y)
N-terminal analysis: S
Carboxypeptidase digestion: L
Trypsin digestion: (3S,2L,F,I,M,T,W) (G,K,S,Y)
Chymotrypsin digestion:
  (F,I,S) (G,K,L) (L,S) (M,T) (S,W) (S,Y)
N-terminal analysis of (F,I,S) peptide: S
Cyanogen bromide treatment:
  (2S,F,G,I,K,L,M*,T,Y) (2S,L,W)

   M*, methionine detected as homoserine

18. *Edman degradation.* Alanine amide was treated with phenyl isothiocyanate to form PTH–alanine. Write a mechanism for this reaction.

Fluorescence scan of an array of 1024 peptides in a 1.6-cm² area. Each synthesis site is a 400-μm square. A fluorescently labeled monoclonal antibody was added to the array to identify peptides that are recognized. The height and color of each square denote the fluorescence intensity. [After S. P. A. Fodor, J. O. Read, M. C. Pirrung, L. Stryer, A. T. Lu, and D. Solas. *Science* 251(1991):767.]

ANIMATED TECHNIQUES

Visit www.whfreeman.com/biochem5 to see animations of SDS Gel Electrophoresis, Immunoblotting, Preparing Monoclonal Antibodies, Reporter Constructs (GFP). [Courtesy Lodish et al., 2000, *Molecular Cell Biology*, 4th ed., W. H. Freeman and Company.]

CHAPTER 5

DNA, RNA, and the Flow of Genetic Information

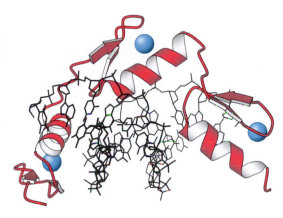

Having genes in common accounts for the resemblance of a mother and her daughters. Genes must be expressed to exert an effect, and proteins regulate such expression. One such regulatory protein, a zinc-finger protein (zinc ion is blue, protein is red), is shown bound to a control or promoter region of DNA (black). [(Left) Barnaby Hall/Photonica.]

DNA and RNA are long linear polymers, called nucleic acids, that carry information in a form that can be passed from one generation to the next. These macromolecules consist of a large number of linked nucleotides, each composed of a sugar, a phosphate, and a base. Sugars linked by phosphates form a common backbone, whereas the bases vary among four kinds. *Genetic information is stored in the sequence of bases along a nucleic acid chain.* The bases have an additional special property: they form specific pairs with one another that are stabilized by hydrogen bonds. The base pairing results in the formation of a double helix, a helical structure consisting of two strands. *These base pairs provide a mechanism for copying the genetic information in an existing nucleic acid chain to form a new chain.* Although RNA probably functioned as the genetic material very early in evolutionary history, the genes of all modern cells and many viruses are made of DNA. DNA is replicated by the action of DNA polymerase enzymes. These exquisitely specific enzymes copy sequences from nucleic acid templates with an error rate of less than 1 in 100 million nucleotides.

Genes specify the kinds of proteins that are made by cells, but DNA is not the direct template for protein synthesis. Rather, the templates for protein synthesis are RNA (ribonucleic acid) molecules. In particular, a class of RNA molecules called *messenger RNA* (mRNA) are the information-carrying intermediates in protein synthesis. Other RNA molecules, such as *transfer RNA* (tRNA) and *ribosomal RNA* (rRNA), are part of the protein-synthesizing machinery. All forms of cellular RNA are synthesized by RNA polymerases that take instructions from DNA templates. This process of *transcription* is followed by *translation*, the synthesis of proteins according to instructions given by mRNA templates. Thus, the flow of genetic information, or *gene expression*, in normal cells is:

$$\text{DNA} \xrightarrow{\text{Transcription}} \text{RNA} \xrightarrow{\text{Translation}} \text{Protein}$$

This flow of information is dependent on the genetic code, which defines the relation between the sequence of bases in DNA (or its mRNA transcript) and the sequence of amino acids in a protein. The code is nearly the same in all organisms: a sequence of three bases, called a *codon*, specifies an amino acid. Codons in mRNA are read sequentially by tRNA molecules, which serve as adaptors in protein synthesis. Protein synthesis takes place on ribosomes, which are complex assemblies of rRNAs and more than 50 kinds of proteins.

The last theme to be considered is the interrupted character of most eukaryotic genes, which are mosaics of nucleic acid sequences called *introns* and *exons*. Both are transcribed, but introns are cut out of newly synthesized RNA molecules, leaving mature RNA molecules with continuous exons. The existence of introns and exons has crucial implications for the evolution of proteins.

5.1 A NUCLEIC ACID CONSISTS OF FOUR KINDS OF BASES LINKED TO A SUGAR-PHOSPHATE BACKBONE

The nucleic acids DNA and RNA are well suited to function as the carriers of genetic information by virtue of their covalent structures. These macromolecules are *linear polymers* built up from similar units connected end to end (Figure 5.1). Each monomer unit within the polymer consists of three components: a sugar, a phosphate, and a base. The sequence of bases uniquely characterizes a nucleic acid and represents a form of linear information.

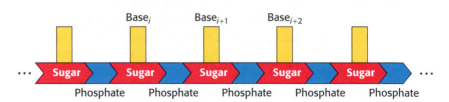

FIGURE 5.1 Polymeric structure of nucleic acids.

5.1.1 RNA and DNA Differ in the Sugar Component and One of the Bases

The sugar in *deoxyribonucleic acid (DNA)* is *deoxyribose*. The deoxy prefix indicates that the 2′ carbon atom of the sugar lacks the oxygen atom that is linked to the 2′ carbon atom of *ribose* (the sugar in *ribonucleic acid*, or *RNA*), as shown in Figure 5.2. The sugars in nucleic acids are linked to one another by phosphodiester bridges. Specifically, the 3′-hydroxyl (3′-OH) group of the sugar moiety of one nucleotide is esterified to a phosphate group, which is, in turn, joined to the 5′-hydroxyl group of the adjacent sugar. The chain of sugars linked by phosphodiester bridges is referred to as the *backbone* of the nucleic acid (Figure 5.3). Whereas the backbone is constant in DNA and RNA, the bases vary from one monomer to the next. Two of the bases

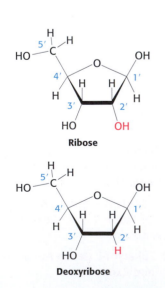

FIGURE 5.2 Ribose and deoxyribose.
Atoms are numbered with primes to distinguish them from atoms in bases (see Figure 5.4).

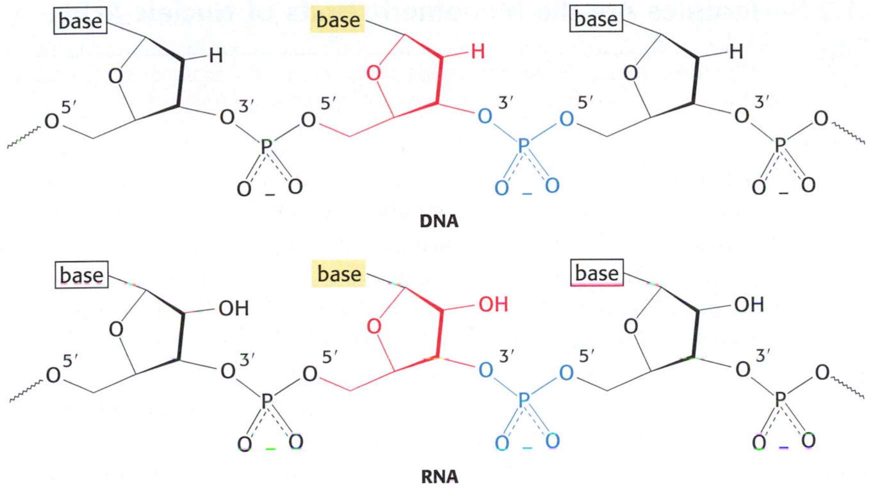

FIGURE 5.3 Backbones of DNA and RNA. The backbones of these nucleic acids are formed by 3'-to-5' phosphodiester linkages. A sugar unit is highlighted in red and a phosphate group in blue.

are derivatives of *purine*—adenine (A) and guanine (G)—and two of *pyrimidine*—cytosine (C) and thymine (T, DNA only) or uracil (U, RNA only), as shown in Figure 5.4.

RNA, like DNA, is a long unbranched polymer consisting of nucleotides joined by $3' \rightarrow 5'$ phosphodiester bonds (see Figure 5.3). The covalent structure of RNA differs from that of DNA in two respects. As stated earlier and as indicated by its name, the sugar units in RNA are riboses rather than deoxyriboses. Ribose contains a 2'-hydroxyl group not present in deoxyribose. As a consequence, in addition to the standard $3' \rightarrow 5'$ linkage, a $2' \rightarrow 5'$ linkage is possible for RNA. This later linkage is important in the removal of introns and the joining of exons for the formation of mature RNA (Section 28.3.4). The other difference, as already mentioned, is that one of the four major bases in RNA is uracil (U) instead of thymine (T).

Note that each phosphodiester bridge has a negative charge. This negative charge repels nucleophilic species such as hydroxide ion; consequently, phosphodiester linkages are much less susceptible to hydrolytic attack than are other esters such as carboxylic acid esters. This resistance is crucial for maintaining the integrity of information stored in nucleic acids. The absence of the 2'-hydroxyl group in DNA further increases its resistance to hydrolysis. The greater stability of DNA probably accounts for its use rather than RNA as the hereditary material in all modern cells and in many viruses.

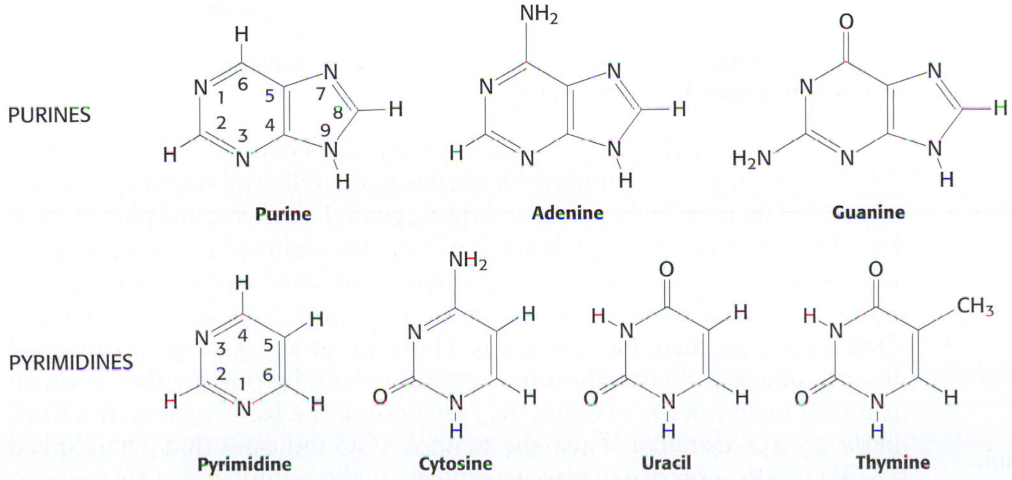

FIGURE 5.4 Purines and pyrimidines. Atoms within bases are numbered without primes. Uracil instead of thymine is used in RNA.

5.1.2 Nucleotides Are the Monomeric Units of Nucleic Acids

STRUCTURAL INSIGHTS, Nucleic Acids *offers a three-dimensional perspective on nucleotide structure, base pairing, and other aspects of DNA and RNA structure.*

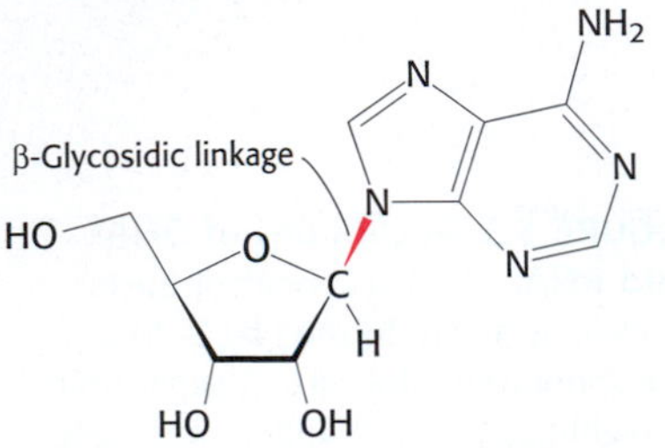

A unit consisting of a base bonded to a sugar is referred to as a *nucleoside*. The four nucleoside units in RNA are called *adenosine, guanosine, cytidine,* and *uridine,* whereas those in DNA are called *deoxyadenosine, deoxyguanosine, deoxycytidine,* and *thymidine.* In each case, N-9 of a purine or N-1 of a pyrimidine is attached to C-1' of the sugar (Figure 5.5). The base lies above the plane of sugar when the structure is written in the standard orientation; that is, the configuration of the N-glycosidic linkage is β. A *nucleotide* is a nucleoside joined to one or more phosphate groups by an ester linkage. The most common site of esterification in naturally occurring nucleotides is the hydroxyl group attached to C-5' of the sugar. A compound formed by the attachment of a phosphate group to the C-5' of a nucleoside sugar is called a *nucleoside 5'-phosphate* or a *5'-nucleotide.* For example, ATP is *adenosine 5'-triphosphate.* Another nucleotide is deoxyguanosine 3'-monophosphate (3'-dGMP; Figure 5.6). This nucleotide differs from ATP in that it contains guanine rather than adenine, contains deoxyribose rather than ribose (indicated by the prefix "d"), contains one rather than three phosphates, and has the phosphate esterified to the hydroxyl group in the 3' rather than the 5' position. Nucleotides are the monomers that are linked to form RNA and DNA. The four nucleotide units in DNA are called *deoxyadenylate, deoxyguanylate, deoxycytidylate,* and *thymidylate.* Note that thymidylate contains deoxyribose; by convention, the prefix deoxy is not added because thymine-containing nucleotides are only rarely found in RNA.

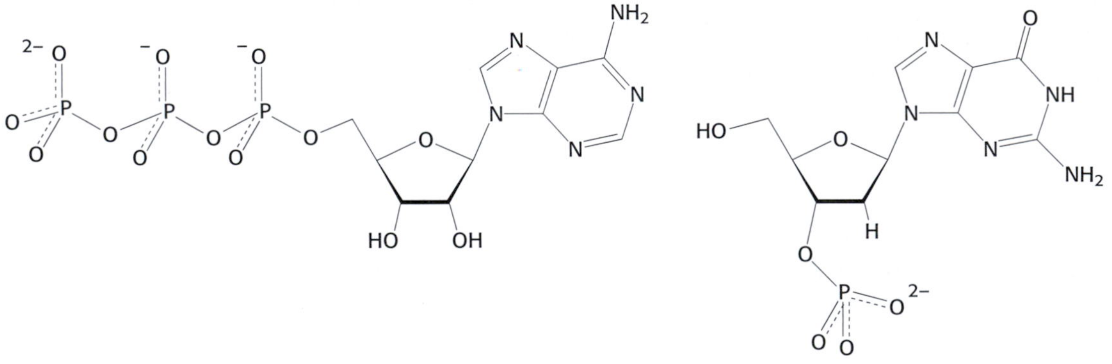

FIGURE 5.5 β-Glycosidic linkage in a nucleoside.

FIGURE 5.6 Nucleotides adenosine 5'-triphosphate (5'-ATP) and deoxyguanosine 3'-monophosphate (3'-dGMP).

The abbreviated notations pApCpG or pACG denote a trinucleotide of DNA consisting of the building blocks deoxyadenylate monophosphate, deoxycytidylate monophosphate, and deoxyguanylate monophosphate linked by a phosphodiester bridge, where "p" denotes a phosphate group (Figure 5.7). The 5' end will often have a phosphate attached to the 5'-OH group. Note that, like a polypeptide (see Section 3.2), *a DNA chain has polarity.* One end of the chain has a free 5'-OH group (or a 5'-OH group attached to a phosphate), whereas the other end has a 3'-OH group, neither of which is linked to another nucleotide. By convention, *the base sequence is written in the 5'-to-3' direction.* Thus, the symbol ACG indicates that the unlinked 5'-OH group is on deoxyadenylate, whereas the unlinked 3'-OH group is on deoxyguanylate. Because of this polarity, ACG and GCA correspond to different compounds.

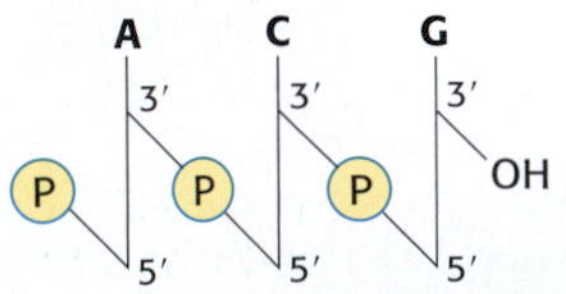

FIGURE 5.7 Structure of a DNA chain. The chain has a 5' end, which is usually attached to a phosphate, and a 3' end, which is usually a free hydroxyl group.

A striking characteristic of naturally occurring DNA molecules is their length. A DNA molecule must comprise many nucleotides to carry the genetic information necessary for even the simplest organisms. For example, the DNA of a virus such as polyoma, which can cause cancer in certain organisms, is 5100 nucleotides in length. We can quantify the information carrying capacity of nucleic acids in the following way. Each position can be one of four bases, corresponding to two bits of information ($2^2 = 4$). Thus, a chain of 5100 nucleotides corresponds to $2 \times 5100 = 10,200$ bits, or 1275 bytes (1 byte = 8 bits). The *E. coli* genome is a single DNA molecule consisting of two chains of 4.6 million nucleotides, corresponding to 9.2 million bits, or 1.15 megabytes, of information (Figure 5.8).

DNA molecules from higher organisms can be much larger. The human genome comprises approximately 3 billion nucleotides, divided among 24 distinct DNA molecules (22 autosomes, x and y sex chromosomes) of different sizes. One of the largest known DNA molecules is found in the Indian muntjak, an Asiatic deer; its genome is nearly as large as the human genome but is distributed on only 3 chromosomes (Figure 5.9). The largest of these chromosomes has chains of more than 1 billion nucleotides. If such a DNA molecule could be fully extended, it would stretch more than 1 foot in length. Some plants contain even larger DNA molecules.

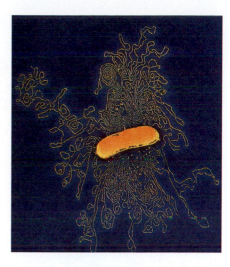

FIGURE 5.8 Electron micrograph of part of the *E. coli* genome. [Dr. Gopal Murti/Science Photo Library/Photo Researchers.]

 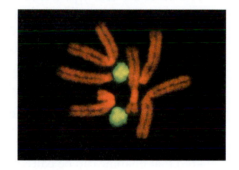

FIGURE 5.9 The Indian muntjak and its chromosomes. Cells from a female Indian muntjak (right) contain three pairs of very large chromosomes (stained orange). The cell shown is a hybrid containing a pair of human chromosomes (stained green) for comparison. [(Left) M. Birkhead, OSF/Animals Animals. (Right) J–Y Lee, M. Koi, E. J. Stanbridge, M. Oshimura, A. T. Kumamoto, and A. P. Feinberg. *Nature Genetics* 7(1994):30.]

5.2 A PAIR OF NUCLEIC ACID CHAINS WITH COMPLEMENTARY SEQUENCES CAN FORM A DOUBLE-HELICAL STRUCTURE

The covalent structure of nucleic acids accounts for their ability to carry information in the form of a sequence of bases along a nucleic acid chain. Other features of nucleic acid structure facilitate the process of *replication*—that is, the generation of two copies of a nucleic acid from one. These features depend on the ability of the bases found in nucleic acids to form *specific base pairs* in such a way that a helical structure consisting of two strands is formed. The double-helical structure of DNA facilitates the replication of the genetic material (Section 5.2.2).

5.2.1 The Double Helix Is Stabilized by Hydrogen Bonds and Hydrophobic Interactions

The existence of specific base-pairing interactions was discovered in the course of studies directed at determining the three-dimensional structure of DNA. Maurice Wilkins and Rosalind Franklin obtained x-ray diffraction photographs of fibers of DNA (Figure 5.10). The characteristics of these diffraction patterns indicated that DNA was formed of two chains that wound in a regular helical structure. From these and other data, James Watson and Francis Crick inferred a structural model for DNA that accounted

3.4-Å spacing

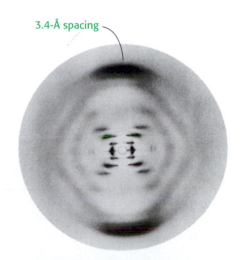

FIGURE 5.10 X-ray diffraction photograph of a hydrated DNA fiber. The central cross is diagnostic of a helical structure. The strong arcs on the meridian arise from the stack of nucleotide bases, which are 3.4 Å apart. [Courtesy of Dr. Maurice Wilkins.]

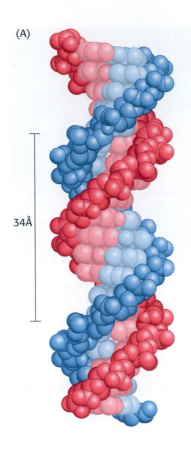

34Å

(B)

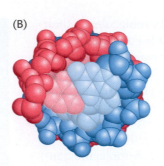

FIGURE 5.11 Watson-Crick model of double-helical DNA.
One polynucleotide chain is shown in blue and the other in red. The purine and pyrimidine bases are shown in lighter colors than the sugar-phosphate backbone. (A) Axial view. The structure repeats along the helical axis (vertical) at intervals of 34 Å, which corresponds to 10 nucleotides on each chain. (B) Radial view, looking down the helix axis.

for the diffraction pattern and was also the source of some remarkable insights into the functional properties of nucleic acids (Figure 5.11).

The features of the Watson-Crick model of DNA deduced from the diffraction patterns are:

1. Two helical polynucleotide chains are coiled around a common axis. The chains run in opposite directions.

2. The sugar-phosphate backbones are on the outside and, therefore, the purine and pyrimidine bases lie on the inside of the helix.

3. The bases are nearly perpendicular to the helix axis, and adjacent bases are separated by 3.4 Å. The helical structure repeats every 34 Å, so there are 10 bases (= 34 Å per repeat/3.4 Å per base) per turn of helix. There is a rotation of 36 degrees per base (360 degrees per full turn/10 bases per turn).

4. The diameter of the helix is 20 Å.

How is such a regular structure able to accommodate an arbitrary sequence of bases, given the different sizes and shapes of the purines and pyrimidines? In attempting to answer this question, Watson and Crick discovered that guanine can be paired with cytosine and adenine with thymine to form base pairs that have essentially the same shape (Figure 5.12). These base pairs are held together by specific hydrogen bonds. This base-pairing scheme was supported by earlier studies of the base composition of DNA from different species. In 1950, Erwin Chargaff reported that the ratios of adenine to thymine and of guanine to cytosine were nearly the same in all species studied. Note in Table 5.1 that all the adenine:thymine and guanine:cytosine ratios are close to 1, whereas the adenine-to-guanine ratio varies considerably. The meaning of these equivalences was not evident until the Watson-Crick model was proposed, when it became clear that they represent an essential facet of DNA structure.

The spacing of approximately 3.4 Å between nearly parallel base pairs is readily apparent in the DNA diffraction pattern (see Figure 5.10). The

Guanine   Cytosine

Adenine   Thymine

FIGURE 5.12 Structures of the base pairs proposed by Watson and Crick.

TABLE 5.1   Base compositions experimentally determined for a variety of organisms

Species	A:T	G:C	A:G
Human being	1.00	1.00	1.56
Salmon	1.02	1.02	1.43
Wheat	1.00	0.97	1.22
Yeast	1.03	1.02	1.67
Escherichia coli	1.09	0.99	1.05
Serratia marcescens	0.95	0.86	0.70

stacking of bases one on top of another contributes to the stability of the double helix in two ways (Figure 5.13). First, adjacent base pairs attract one another through van der Waals forces (Section 1.3.1). Energies associated with van der Waals interactions are quite small, such that typical interactions contribute from 0.5 to 1.0 kcal mol^{-1} per atom pair. In the double helix, however, a large number of atoms are in van der Waals contact, and the net effect, summed over these atom pairs, is substantial. In addition, the double helix is stabilized by the hydrophobic effect (Section 1.3.4): base stacking, or hydrophobic interactions between the bases, results in the exposure of the more polar surfaces to the surrounding water. This arrangement is reminiscent of protein folding, where hydrophobic amino acids are interior in the protein and hydrophilic are exterior (Section 3.4). Base stacking in DNA is also favored by the conformations of the relatively rigid five-membered rings of the backbone sugars. The sugar rigidity affects both the single-stranded and the double-helical forms.

5.2.2 The Double Helix Facilitates the Accurate Transmission of Hereditary Information

The double-helical model of DNA and the presence of specific base pairs immediately suggested how the genetic material might replicate. The sequence of bases of one strand of the double helix precisely determines the sequence of the other strand; a guanine base on one strand is always paired with a cytosine base on the other strand, and so on. Thus, separation of a double helix into its two component chains would yield two single-stranded templates onto which new double helices could be constructed, each of which would have the same sequence of bases as the parent double helix. Consequently, as DNA is replicated, one of the chains of each daughter DNA molecule would be newly synthesized, whereas the other would be passed unchanged from the parent DNA molecule. This distribution of parental atoms is achieved by *semiconservative replication.*

Matthew Meselson and Franklin Stahl carried out a critical test of this hypothesis in 1958. They labeled the parent DNA with ^{15}N, a heavy isotope of nitrogen, to make it denser than ordinary DNA. The labeled DNA was generated by growing *E. coli* for many generations in a medium that contained ^{15}NH$_4$Cl as the sole nitrogen source. After the incorporation of heavy nitrogen was complete, the bacteria were abruptly transferred to a medium that contained ^{14}N, the ordinary isotope of nitrogen. The question asked was: What is the distribution of ^{14}N and ^{15}N in the DNA molecules after successive rounds of replication?

The distribution of ^{14}N and ^{15}N was revealed by the technique of *density-gradient equilibrium sedimentation.* A small amount of DNA was dissolved in a concentrated solution of cesium chloride having a density close to that of the DNA (1.7 g cm^{-3}). This solution was centrifuged until it was nearly at equilibrium. The opposing processes of sedimentation and diffusion created a gradient in the concentration of cesium chloride across the centrifuge cell. The result was a stable density gradient, ranging from 1.66 to 1.76 g cm^{-3}. The DNA molecules in this density gradient were driven by centrifugal force into the region where the solution's density was equal to their own. The genomic DNA yielded a narrow band that was detected by its absorption of ultraviolet light. A mixture of ^{14}N DNA and ^{15}N DNA molecules gave clearly separate bands because they differ in density by about 1% (Figure 5.14).

DNA was extracted from the bacteria at various times after they were transferred from a ^{15}N to a ^{14}N medium and centrifuged. Analysis of these samples showed that there was a single band of DNA after one generation. The density of this band was precisely halfway between the densities of the

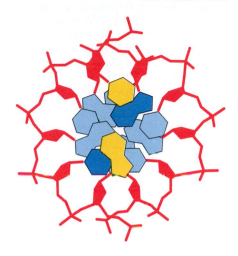

FIGURE 5.13 Axial view of DNA. Base pairs are stacked nearly one on top of another in the double helix.

(A)

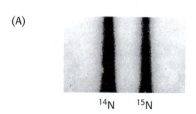

^{14}N ^{15}N

(B)

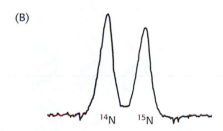

^{14}N ^{15}N

FIGURE 5.14 Resolution of ^{14}N DNA and ^{15}N DNA by density-gradient centrifugation. (A) Ultraviolet absorption photograph of a centrifuge cell showing the two distinct bands of DNA. (B) Densitometric tracing of the absorption photograph. [From M. Meselson and F. W. Stahl. *Proc. Natl. Acad. Sci. U.S.A.* 44(1958):671.]

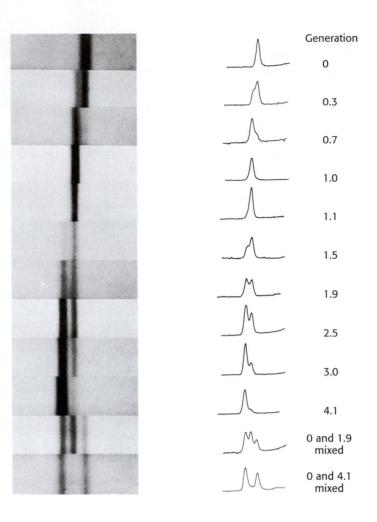

Generation

0

0.3

0.7

1.0

1.1

1.5

1.9

2.5

3.0

4.1

0 and 1.9 mixed

0 and 4.1 mixed

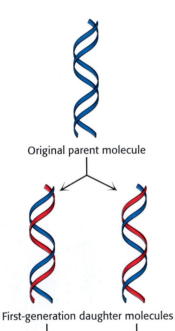

Original parent molecule

First-generation daughter molecules

Second-generation daughter molecules

FIGURE 5.16 Diagram of semiconservative replication. Parental DNA is shown in blue and newly synthesized DNA in red. [After M. Meselson and F. W. Stahl. *Proc. Natl. Acad. Sci. U.S.A.* 44(1958):671.]

^{14}N DNA and ^{15}N DNA bands (Figure 5.15). The absence of ^{15}N DNA indicated that parental DNA was not preserved as an intact unit after replication. The absence of ^{14}N DNA indicated that all the daughter DNA derived some of their atoms from the parent DNA. This proportion had to be half because the density of the hybrid DNA band was halfway between the densities of the ^{14}N DNA and ^{15}N DNA bands.

After two generations, there were equal amounts of two bands of DNA. One was hybrid DNA, and the other was ^{14}N DNA. Meselson and Stahl concluded from these incisive experiments "that the nitrogen in a DNA molecule is divided equally between two physically continuous subunits; that following duplication, each daughter molecule receives one of these; and that the subunits are conserved through many duplications." Their results agreed perfectly with the Watson-Crick model for DNA replication (Figure 5.16).

5.2.3 The Double Helix Can Be Reversibly Melted

During DNA replication and other processes, the two strands of the double helix must be separated from one another, at least in a local region. In the laboratory, the double helix can be disrupted by heating a solution of DNA. The heating disrupts the hydrogen bonds between base pairs and thereby causes the strands to separate. The dissociation of the double helix is often called *melting* because it occurs relatively abruptly at a certain temperature. The *melting temperature* (T_m) is defined as the temperature at which half the helical structure is lost. Strands may also be separated by adding acid or alkali to ionize the nucleotide bases and disrupt base pairing.

Stacked bases in nucleic acids absorb less ultraviolet light than do unstacked bases, an effect called *hypochromism*. Thus, the melting of nucleic

(A)

(B)

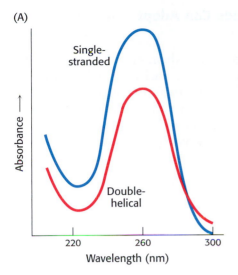

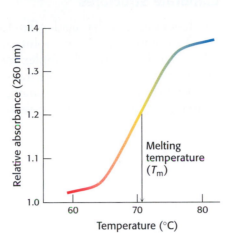

FIGURE 5.17 Hypochromism.
(A) Single-stranded DNA absorbs light more effectively than does double-helical DNA. (B) The absorbance of a DNA solution at a wavelength of 260 nm increases when the double helix is melted into single strands.

acids is easily followed by monitoring their absorption of light, which peaks at a wavelength of 260 nm (Figure 5.17).

Separated complementary strands of nucleic acids spontaneously reassociate to form a double helix when the temperature is lowered below T_m. This renaturation process is sometimes called *annealing*. The facility with which double helices can be melted and then reassociated is crucial for the biological functions of nucleic acids. Of course, inside cells, the double helix is not melted by the addition of heat. Instead, proteins called *helicases* use chemical energy (from ATP) to disrupt the structure of double-stranded nucleic acid molecules.

The ability to melt and reanneal DNA reversibly in the laboratory provides a powerful tool for investigating sequence similarity as well as gene structure and expression. For instance, DNA molecules from two different organisms can be melted and allowed to reanneal or *hybridize* in the presence of each other. If the sequences are similar, hybrid DNA duplexes, with DNA from each organism contributing a strand of the double helix, can form. Indeed, the degree of hybridization is an indication of the relatedness of the genomes and hence the organisms. Similar hybridization experiments with RNA and DNA can locate genes in a cell's DNA that correspond to a particular RNA. We will return to this important technique in Chapter 6.

5.2.4 Some DNA Molecules Are Circular and Supercoiled

The DNA molecules in human chromosomes are linear. However, electron microscopic and other studies have shown that intact DNA molecules from some other organisms are circular (Figure 5.18A). The term *circular* refers to the continuity of the DNA chains, not to their geometrical form. DNA molecules inside cells necessarily have a very compact shape. Note that the *E. coli* chromosome, fully extended, would be about 1000 times as long as the greatest diameter of the bacterium.

A new property appears in the conversion of a linear DNA molecule into a closed circular molecule. The axis of the double helix can itself be twisted into a *superhelix* (Figure 5.18B). A circular DNA molecule without any superhelical turns is known as a *relaxed molecule*. Supercoiling is biologically important for two reasons. First, *a supercoiled DNA molecule has a more compact shape than does its relaxed counterpart. Second, supercoiling may hinder or favor the capacity of the double helix to unwind and thereby affects the interactions between DNA and other molecules.* These topological features of DNA will be considered further in Section 27.3.

(A)

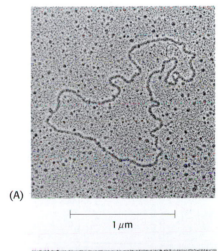

$\vdash$ 1 μm $\dashv$

(B)

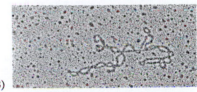

FIGURE 5.18 Electron micrographs of circular DNA from mitochondria.
(A) Relaxed form. (B) Supercoiled form.
[Courtesy of Dr. David Clayton.]

5.2.5 Single-Stranded Nucleic Acids Can Adopt Elaborate Structures

Single-stranded nucleic acids often fold back on themselves to form well-defined structures. Early in evolutionary history, nucleic acids, particularly RNA, may have adopted complex and diverse structures both to store genetic information and to catalyze its transmission (Section 2.2.2). Such structures are also important in all modern organisms in entities such as the ribosome, a large complex of RNAs and proteins on which proteins are synthesized.

FIGURE 5.19 Stem-loop structures.
Stem-loop structures may be formed from single-stranded DNA and RNA molecules.

The simplest and most common structural motif formed is a *stem-loop*, created when two complementary sequences within a single strand come together to form double-helical structures (Figure 5.19). In many cases, these double helices are made up entirely of Watson-Crick base pairs. In other cases, however, the structures include mismatched or unmatched (bulged) bases. Such mismatches destabilize the local structure but introduce deviations from the standard double-helical structure that can be important for higher-order folding and for function (Figure 5.20).

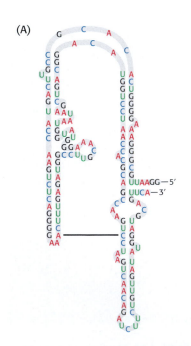

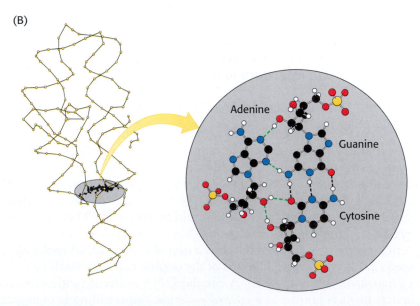

FIGURE 5.20 Complex structure of an RNA molecule. A single-stranded RNA molecule may fold back on itself to form a complex structure. (A) The nucleotide sequence showing Watson-Crick base pairs and other nonstandard base pairings in stem-loop structures. (B) The three-dimensional structure and one important long-range interaction between three bases. Hydrogen bonds within the Watson-Crick base pair are shown as dashed black lines; additional hydrogen bonds are shown as dashed green lines.

Single-stranded nucleic acids can adopt structures more complex than simple stem-loops through the interaction of more widely separated bases. Often, three or more bases may interact to stabilize these structures. In such cases, hydrogen-bond donors and acceptors that do not participate in Watson-Crick base pairs may participate in hydrogen bonds of nonstandard pairings. Metal ions such as magnesium ion (Mg^{2+}) often assist in the stabilization of these more elaborate structures.

5.3 DNA IS REPLICATED BY POLYMERASES THAT TAKE INSTRUCTIONS FROM TEMPLATES

We now turn to the molecular mechanism of DNA replication. The full replication machinery in cells comprises more than 20 proteins engaged in intricate and coordinated interplay. In 1958, Arthur Kornberg and his colleagues isolated the first known of the enzymes, called DNA polymerases, that promote the formation of the bonds joining units of the DNA backbone.

5.3.1 DNA Polymerase Catalyzes Phosphodiester-Bond Formation

DNA polymerases catalyze the step-by-step addition of deoxyribonucleotide units to a DNA chain (Figure 5.21). Importantly, *the new DNA chain is assembled directly on a preexisting DNA template*. The reaction catalyzed, in its simplest form, is:

$$(DNA)_n + dNTP \rightleftharpoons (DNA)_{n+1} + PP_i$$

where dNTP stands for any deoxyribonucleotide and PP_i is a pyrophosphate molecule. The template can be a single strand of DNA or a double strand with one of the chains broken at one or more sites. If single stranded, the template DNA must be bound to a *primer* strand having a free 3'-hydroxyl group. The reaction also requires all four activated precursors—that is, the deoxynucleoside 5'-triphosphates dATP, dGTP, TTP, and dCTP—as well as Mg^{2+} ion.

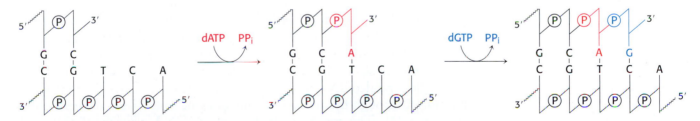

FIGURE 5.21 Polymerization reaction catalyzed by DNA polymerases.

The chain-elongation reaction catalyzed by DNA polymerases is a nucleophilic attack by the 3'-hydroxyl group of the primer on the innermost phosphorus atom of the deoxynucleoside triphosphate (Figure 5.22). A phosphodiester bridge forms with the concomitant release of pyrophosphate. The subsequent hydrolysis of pyrophosphate by pyrophosphatase, a ubiquitous enzyme, helps drive the polymerization forward. Elongation of the DNA chain proceeds in the 5'-to-3' direction.

DNA polymerases catalyze the formation of a phosphodiester bond efficiently only if the base on the incoming nucleoside triphosphate is complementary to the base on the template strand. Thus, DNA polymerase is

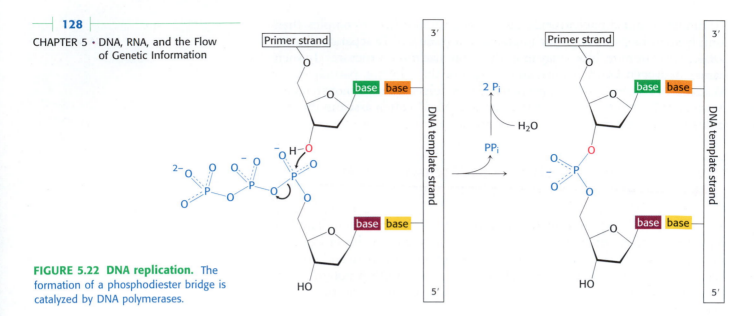

FIGURE 5.22 DNA replication. The formation of a phosphodiester bridge is catalyzed by DNA polymerases.

a *template-directed enzyme* that synthesizes a product with a base sequence complementary to that of the template. Many DNA polymerases also have a separate nuclease activity that allows them to correct mistakes in DNA by using a different reaction to remove mismatched nucleotides. These properties of DNA polymerases contribute to the remarkably high fidelity of DNA replication, which has an error rate of less than 10^{-8} per base pair.

5.3.2 The Genes of Some Viruses Are Made of RNA

Genes in all cellular organisms are made of DNA. The same is true for some viruses, but for others the genetic material is RNA. Viruses are genetic elements enclosed in protein coats that can move from one cell to another but are not capable of independent growth. One well-studied example of an RNA virus is the tobacco mosaic virus, which infects the leaves of tobacco plants. This virus consists of a single strand of RNA (6930 nucleotides) surrounded by a protein coat of 2130 identical subunits. An RNA-directed RNA polymerase catalyzes the replication of this viral RNA.

Another important class of RNA virus comprises the *retroviruses*, so called because the genetic information flows from RNA to DNA rather than from DNA to RNA. This class includes human immunodeficiency virus 1 (HIV-1), the cause of AIDS, as well as a number of RNA viruses that produce tumors in susceptible animals. Retrovirus particles contain two copies of a single-stranded RNA molecule. On entering the cell, the RNA is copied into DNA through the action of a viral enzyme called *reverse transcriptase* (Figure 5.23). The resulting double-helical DNA version of the viral

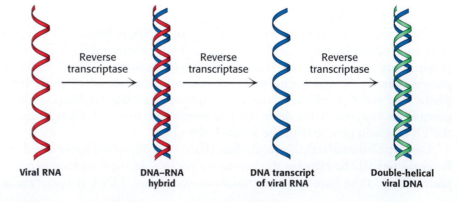

Viral RNA → Reverse transcriptase → DNA–RNA hybrid → Reverse transcriptase → DNA transcript of viral RNA → Reverse transcriptase → Double-helical viral DNA

FIGURE 5.23 Flow of information from RNA to DNA in retroviruses. The RNA genome of a retrovirus is converted into DNA by reverse transcriptase, an enzyme brought into the cell by the infecting virus particle. Reverse transcriptase catalyzes the synthesis of a complementary DNA strand, the digestion of the RNA, and the subsequent synthesis of the DNA strand.

genome can become incorporated into the chromosomal DNA of the host and is replicated along with the normal cellular DNA. At a later time, the integrated viral genome is expressed to form viral RNA and viral proteins, which assemble into new virus particles.

Note that RNA viruses are not vestiges of the RNA world. Instead, fragments of RNA in these viruses have evolved to encode their protein coats and other structures needed for transferring from cell to cell and replicating.

5.4 GENE EXPRESSION IS THE TRANSFORMATION OF DNA INFORMATION INTO FUNCTIONAL MOLECULES

The information stored as DNA becomes useful when it is expressed in the production of RNA and proteins. This rich and complex topic is the subject of several chapters later in this book, but here we introduce the basics of gene expression. DNA can be thought of as archival information, stored and manipulated judiciously to minimize damage (mutations). It is expressed in two steps. First, an RNA copy is made. An RNA molecule that encodes proteins can be thought of as a photocopy of the original information—it can be made in multiple copies, used, and then disposed of. Second, an RNA molecule can be further thought of as encoding directions for protein synthesis that must be translated to be of use. The information in messenger RNA is translated into a functional protein. Other types of RNA molecules exist to facilitate this translation. We now examine the transcription of DNA information into RNA, the translation of RNA information into protein, and the genetic code that links nucleotide sequence with amino acid sequence.

5.4.1 Several Kinds of RNA Play Key Roles in Gene Expression

Cells contain several kinds of RNA (Table 5.2).

1. *Messenger RNA* is the template for protein synthesis or *translation*. An mRNA molecule may be produced for each gene or group of genes that is to be expressed in *E. coli*, whereas a distinct mRNA is produced for each gene in eukaryotes. Consequently, mRNA is a heterogeneous class of molecules. In *E. coli*, the average length of an mRNA molecule is about 1.2 kilobases (kb).

2. *Transfer RNA* carries amino acids in an activated form to the ribosome for peptide-bond formation, in a sequence dictated by the mRNA template. There is at least one kind of tRNA for each of the 20 amino acids. Transfer RNA consists of about 75 nucleotides (having a mass of about 25 kd), which makes it one of the smallest of the RNA molecules.

Kilobase (kb)—

A unit of length equal to 1000 base pairs of a double-stranded nucleic acid molecule (or 1000 bases of a single-stranded molecule).

One kilobase of double-stranded DNA has a contour length of 0.34 μm and a mass of about 660 kd.

TABLE 5.2   RNA molecules in *E. coli*

Type	Relative amount (%)	Sedimentation coefficient (S)	Mass (kd)	Number of nucleotides
Ribosomal RNA (rRNA)	80	23	1.2×10^3	3700
		16	0.55×10^3	1700
		5	3.6×10^1	120
Transfer RNA (tRNA)	15	4	2.5×10^1	75
Messenger RNA (mRNA)	5		Heterogeneous	

3. *Ribosomal RNA (rRNA),* the major component of ribosomes, plays both a catalytic and a structural role in protein synthesis (Section 29.3.1). In *E. coli,* there are three kinds of rRNA, called *23S, 16S,* and *5S RNA* because of their sedimentation behavior. One molecule of each of these species of rRNA is present in each ribosome.

Ribosomal RNA is the most abundant of the three types of RNA. Transfer RNA comes next, followed by messenger RNA, which constitutes only 5% of the total RNA. Eukaryotic cells contain additional small RNA molecules. *Small nuclear RNA (snRNA)* molecules, for example, participate in the splicing of RNA exons. A small RNA molecule in the cytosol plays a role in the targeting of newly synthesized proteins to intracellular compartments and extracellular destinations.

5.4.2 All Cellular RNA Is Synthesized by RNA Polymerases

The synthesis of RNA from a DNA template is called *transcription* and is catalyzed by the enzyme *RNA polymerase* (Figure 5.24). RNA polymerase requires the following components:

1. *A template.* The preferred template is *double-stranded DNA.* Single-stranded DNA also can serve as a template. RNA, whether single or double stranded, is not an effective template; nor are RNA–DNA hybrids.

2. *Activated precursors.* All four *ribonucleoside triphosphates*—ATP, GTP, UTP, and CTP—are required.

3. *A divalent metal ion.* Mg^{2+} or Mn^{2+} are effective.

FIGURE 5.24 RNA polymerase. A large enzyme comprising many subunits including β (red) and β′ (yellow), which form a "claw" that holds the DNA to be transcribed. The active site includes a Mg^{2+} ion (red ball) at the center of the structure.

RNA polymerase catalyzes the initiation and elongation of RNA chains. The reaction catalyzed by this enzyme is:

$$(\text{RNA})_{n\text{ residues}} + \text{ribonucleoside triphosphate} \rightleftharpoons (\text{RNA})_{n+1\text{ residues}} + \text{PP}_i$$

The synthesis of RNA is like that of DNA in several respects (Figure 5.25). First, the direction of synthesis is $5' \rightarrow 3'$. Second, the mechanism of elongation is similar: the 3′-OH group at the terminus of the growing chain

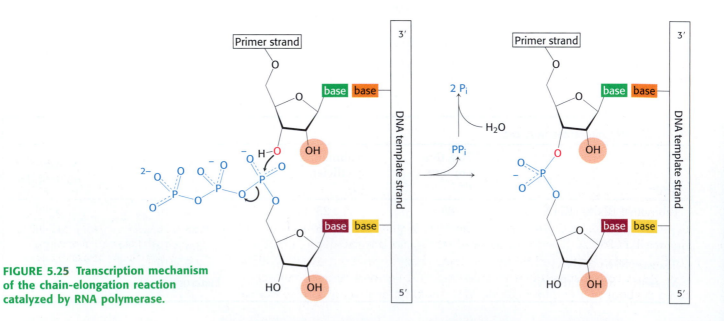

FIGURE 5.25 Transcription mechanism of the chain-elongation reaction catalyzed by RNA polymerase.

makes a nucleophilic attack on the innermost phosphate of the incoming nucleoside triphosphate. Third, the synthesis is driven forward by the hydrolysis of pyrophosphate. In contrast with DNA polymerase, however, RNA polymerase does not require a primer. In addition, RNA polymerase lacks the nuclease capability used by DNA polymerase to excise mismatched nucleotides.

All three types of cellular RNA—mRNA, tRNA, and rRNA—are synthesized in *E. coli* by the same RNA polymerase according to instructions given by a DNA template. In mammalian cells, there is a division of labor among several different kinds of RNA polymerases. We shall return to these RNA polymerases in Chapter 28.

5.4.3 RNA Polymerases Take Instructions from DNA Templates

RNA polymerase, like the DNA polymerases described earlier, takes instructions from a DNA template. The earliest evidence was the finding that the *base composition* of newly synthesized RNA is the complement of that of the DNA template strand, as exemplified by the RNA synthesized from a template of single-stranded ϕX174 DNA (Table 5.3). *Hybridization experiments* also revealed that RNA synthesized by RNA polymerase is complementary to its DNA template. In these experiments, DNA is melted and allowed to reassociate in the presence of mRNA. RNA–DNA hybrids will form if the RNA and DNA have complementary sequences. The strongest evidence for the fidelity of transcription came from base-sequence studies showing that the RNA sequence is the precise complement of the DNA template sequence (Figure 5.26).

TABLE 5.3 Base composition (percentage) of RNA synthesized from a viral DNA template

DNA template (plus strand of ϕX174)		RNA product	
A	25	25	U
T	33	32	A
G	24	23	C
C	18	20	G

5′—GCGGCGACGCGCAGUUAAUCCCACAGCCGCCAGUUCCGCUGGCGGCAUUUU—3' **mRNA**
3′—CGCCGCTGCGCGTCAATTAGGGTGTCGGCGGTCAAGGCGACCGCCGTAAAA—5' **Template strand of DNA**
5′—GCGGCGACGCGCAGTTAATCCCACAGCCGCCAGTTCCGCTGGCGGCATTTT—3' **Coding strand of DNA**

FIGURE 5.26 Complementarity between mRNA and DNA. The base sequence of mRNA (red) is the complement of that of the DNA template strand (blue). The sequence shown here is from the tryptophan operon, a segment of DNA containing the genes for five enzymes that catalyze the synthesis of tryptophan. The other strand of DNA (black) is called the coding strand because it has the same sequence as the RNA transcript except for thymine (T) in place of uracil (U).

5.4.4 Transcription Begins near Promoter Sites and Ends at Terminator Sites

RNA polymerase must detect and transcribe discrete genes from within large stretches of DNA. What marks the beginning of a transcriptional unit? DNA templates contain regions called *promoter sites* that specifically bind RNA polymerase and determine where transcription begins. In bacteria, two sequences on the 5′ (upstream) side of the first nucleotide to be transcribed function as promoter sites (Figure 5.27A). One of them, called the *Pribnow box*, has the consensus sequence TATAAT and is centered at −10 (10 nucleotides on the 5′ side of the first nucleotide transcribed, which is denoted by +1). The other, called the −35 *region*, has the consensus sequence TTGACA. The first nucleotide transcribed is usually a purine.

Eukaryotic genes encoding proteins have promoter sites with a TATAAA consensus sequence, called a *TATA box* or a *Hogness box*, centered at about −25 (Figure 5.27B). Many eukaryotic promoters also have a

Consensus sequence—
The base sequences of promoter sites are not all identical. However, they do possess common features, which can be represented by an idealized consensus sequence. Each base in the consensus sequence TATAAT is found in a majority of prokaryotic promoters. Nearly all promoter sequences differ from this consensus sequence at only one or two bases.

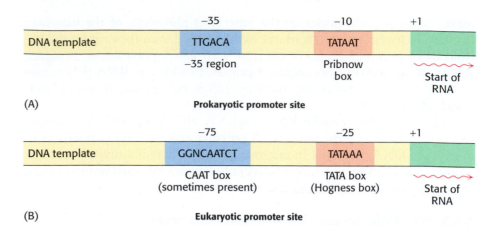

FIGURE 5.27 **Promoter sites for transcription.** Promoter sites are required for the initiation of transcription in both (A) prokaryotes and (B) eukaryotes. Consensus sequences are shown. The first nucleotide to be transcribed is numbered +1. The adjacent nucleotide on the 5′ side is numbered −1. The sequences shown are those of the coding strand of DNA.

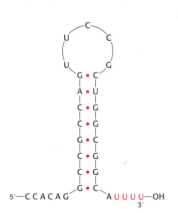

FIGURE 5.28 **Base sequence of the 3′ end of an mRNA transcript in *E. coli*.** A stable hairpin structure is followed by a sequence of uridine (U) residues.

CAAT box with a GG*NCAATCT* consensus sequence centered at about −75. Transcription of eukaryotic genes is further stimulated by *enhancer sequences,* which can be quite distant (as many as several kilobases) from the start site, on either its 5′ or its 3′ side.

RNA polymerase proceeds along the DNA template, transcribing one of its strands until it reaches a terminator sequence. This sequence encodes a termination signal, which in *E. coli* is a *base-paired hairpin* on the newly synthesized RNA molecule (Figure 5.28). This hairpin is formed by base pairing of self-complementary sequences that are rich in G and C. Nascent RNA spontaneously dissociates from RNA polymerase when this hairpin is followed by a string of U residues. Alternatively, RNA synthesis can be terminated by the action of *rho,* a protein. Less is known about the termination of transcription in eukaryotes. A more detailed discussion of the initiation and termination of transcription will be given in Chapter 28. The important point now is that *discrete start and stop signals for transcription are encoded in the DNA template.*

In eukaryotes, the mRNA is modified after transcription (Figure 5.29). A "cap" structure is attached to the 5′ end, and a sequence of adenylates, the poly(A) tail, is added to the 3′ end. These modifications will be presented in detail in Section 28.3.1.

FIGURE 5.29 **Modification of mRNA.** Messenger RNA in eukaryotes is modified after transcription. A nucleotide "cap" structure is added to the 5′ end, and a poly(A) tail is added at the 3′ end.

5.4.5 Transfer RNA Is the Adaptor Molecule in Protein Synthesis

We have seen that mRNA is the template for protein synthesis. How then does it direct amino acids to become joined in the correct sequence to form a protein? In 1958, Francis Crick wrote:

> RNA presents mainly a sequence of sites where hydrogen bonding could occur. One would expect, therefore, that whatever went onto the template in a *specific* way did so by forming hydrogen bonds. It is therefore a natural hypothesis that the amino acid is carried to the template by an adaptor molecule, and that the adaptor is the part that actually fits onto the RNA. In its simplest form, one would require twenty adaptors, one for each amino acid.

This highly innovative hypothesis soon became established as fact. *The adaptor in protein synthesis is transfer RNA.* The structure and reactions of these remarkable molecules will be considered in detail in Chapter 29. For the moment, it suffices to note that tRNA contains an *amino acid-attachment site* and a *template-recognition site*. A tRNA molecule carries a specific amino acid in an activated form to the site of protein synthesis. The carboxyl group of this amino acid is esterified to the 3'- or 2'-hydroxyl group of the ribose unit at the 3' end of the tRNA chain (Figure 5.30). The joining of an amino acid to a tRNA molecule to form an *aminoacyl-tRNA* is catalyzed by a specific enzyme called an *aminoacyl-tRNA synthetase*. This esterification reaction is driven by ATP cleavage. There is at least one specific synthetase for each of the 20 amino acids. The template-recognition site on tRNA is a sequence of three bases called an *anticodon* (Figure 5.31). The anticodon on tRNA recognizes a complementary sequence of three bases, called a *codon*, on mRNA.

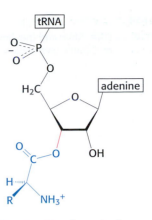

FIGURE 5.30 Attachment of an amino acid to a tRNA molecule. The amino acid (shown in blue) is esterified to the 3'-hydroxyl group of the terminal adenosine of tRNA.

5.5 AMINO ACIDS ARE ENCODED BY GROUPS OF THREE BASES STARTING FROM A FIXED POINT

The *genetic code* is the relation between the sequence of bases in DNA (or its RNA transcripts) and the sequence of amino acids in proteins. Experiments by Francis Crick, Sydney Brenner, and others established the following features of the genetic code by 1961:

1. *Three nucleotides encode an amino acid.* Proteins are built from a basic set of 20 amino acids, but there are only four bases. Simple calculations show that a minimum of three bases is required to encode at least 20 amino acids. Genetic experiments showed that *an amino acid is in fact encoded by a group of three bases, or codon.*

2. *The code is nonoverlapping.* Consider a base sequence ABCDEF. In an overlapping code, ABC specifies the first amino acid, BCD the next, CDE the next, and so on. In a nonoverlapping code, ABC designates the first amino acid, EFG the second, and so forth. Genetics experiments again established the code to be nonoverlapping.

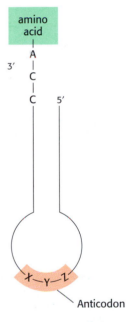

FIGURE 5.31 Symbolic diagram of an aminoacyl-tRNA. The amino acid is attached at the 3' end of the RNA. The anticodon is the template-recognition site.

| ABC | DEF | GHI | JKL | **Base sequence** |

| aa₁ | aa₂ | aa₃ | aa₄ | **Amino acid sequence** |

3. *The code has no punctuation.* In principle, one base (denoted as Q) might serve as a "comma" between groups of three bases.

. . . QABCQDEFQGHIQJKLQ . . .

This is not the case. Rather, *the sequence of bases is read sequentially from a fixed starting point,* without punctuation.

Start

ABC | DEF | GHI | JKL | MNO

aa₁ — aa₂ — aa₃ — aa₄ — aa₅

4. *The genetic code is degenerate.* Some amino acids are encoded by more than one codon, inasmuch as there are 64 possible base triplets and only 20 amino acids. In fact, 61 of the 64 possible triplets specify particular amino acids and 3 triplets (called stop codons) designate the termination of translation. Thus, *for most amino acids, there is more than one code word.*

5.5.1 Major Features of the Genetic Code

All 64 codons have been deciphered (Table 5.4). Because the code is highly degenerate, only tryptophan and methionine are encoded by just one triplet each. The other 18 amino acids are each encoded by two or more. Indeed, leucine, arginine, and serine are specified by six codons each. The number of codons for a particular amino acid correlates with its frequency of occurrence in proteins.

Codons that specify the same amino acid are called *synonyms*. For example, CAU and CAC are synonyms for histidine. Note that synonyms are not distributed haphazardly throughout the genetic code (depicted in Table 5.4). An amino acid specified by two or more synonyms occupies a single box (unless it is specified by more than four synonyms). The amino acids in a box are specified by codons that have the same first two bases but differ in the third base, as exemplified by GUU, GUC, GUA, and GUG. Thus, *most synonyms differ only in the last base of the triplet.* Inspection of the code shows that XYC and XYU always encode the same amino acid, whereas XYG and XYA usually encode the same amino acid. The structural basis for these equivalences of codons will become evident when we consider the nature of the anticodons of tRNA molecules (Section 29.3.9).

What is the biological significance of the extensive degeneracy of the genetic code? If the code were not degenerate, 20 codons would designate amino acids and 44 would lead to chain termination. The probability of mutating to chain termination would therefore be much higher with a nondegenerate code. Chain-termination mutations usually lead to inactive proteins, whereas substitutions of one amino acid for another are usually rather

TABLE 5.4   The genetic code

First position (5′ end)	Second position				Third position (3′ end)
	U	C	A	G	
U	Phe	Ser	Tyr	Cys	U
	Phe	Ser	Tyr	Cys	C
	Leu	Ser	Stop	Stop	A
	Leu	Ser	Stop	Trp	G
C	Leu	Pro	His	Arg	U
	Leu	Pro	His	Arg	C
	Leu	Pro	Gln	Arg	A
	Leu	Pro	Gln	Arg	G
A	Ile	Thr	Asn	Ser	U
	Ile	Thr	Asn	Ser	C
	Ile	Thr	Lys	Arg	A
	Met	Thr	Lys	Arg	G
G	Val	Ala	Asp	Gly	U
	Val	Ala	Asp	Gly	C
	Val	Ala	Glu	Gly	A
	Val	Ala	Glu	Gly	G

Note: This table identifies the amino acid encoded by each triplet. For example, the codon 5′ AUG 3′ on mRNA specifies methionine, whereas CAU specifies histidine. UAA, UAG, and UGA are termination signals. AUG is part of the initiation signal, in addition to coding for internal methionine residues.

harmless. Thus, *degeneracy minimizes the deleterious effects of mutations.* Degeneracy of the code may also be significant in permitting DNA base composition to vary over a wide range without altering the amino acid sequence of the proteins encoded by the DNA. The G + C content of bacterial DNA ranges from less than 30% to more than 70%. DNA molecules with quite different G + C contents could encode the same proteins if different synonyms of the genetic code were consistently used.

5.5.2 Messenger RNA Contains Start and Stop Signals for Protein Synthesis

Messenger RNA is translated into proteins on *ribosomes,* large molecular complexes assembled from proteins and ribosomal RNA. How is mRNA interpreted by the translation apparatus? As already mentioned, *UAA, UAG, and UGA designate chain termination.* These codons are read not by tRNA molecules but rather by specific proteins called *release factors* (Section 29.4.4). Binding of the release factors to the ribosomes releases the newly synthesized protein. The start signal for protein synthesis is more complex. Polypeptide chains in bacteria start with a modified amino acid—namely, formylmethionine (fMet). A specific tRNA, the initiator tRNA, carries fMet. This fMet-tRNA recognizes the codon AUG or, less frequently, GUG. However, AUG is also the codon for an internal methionine residue, and GUG is the codon for an internal valine residue. Hence, the signal for the first amino acid in a prokaryotic polypeptide chain must be more complex than that for all subsequent ones. *AUG (or GUG) is only part of the initiation signal* (Figure 5.32). In bacteria, the initiating AUG (or GUG) codon is preceded several nucleotides away by a purine-rich sequence that base-pairs with a complementary sequence in a ribosomal RNA molecule (Section 29.3.4). In eukaryotes, the AUG closest to the 5′ end of an mRNA molecule is usually the start signal for protein synthesis. This particular AUG is read by an initiator tRNA conjugated to methionine. Once the initiator AUG is located, the *reading frame* is established—groups of three nonoverlapping nucleotides are defined, beginning with the initiator AUG codon.

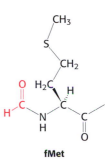

fMet

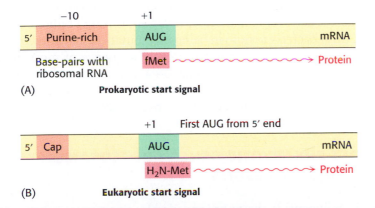

(A) **Prokaryotic start signal**

(B) **Eukaryotic start signal**

FIGURE 5.32 Initiation of protein synthesis. Start signals are required for the initiation of protein synthesis in (A) prokaryotes and (B) eukaryotes.

5.5.3 The Genetic Code Is Nearly Universal

Is the genetic code the same in all organisms? The base sequences of many wild-type and mutant genes are known, as are the amino acid sequences of their encoded proteins. In each case, the nucleotide change in the gene and the amino acid change in the protein are as predicted by the genetic code. Furthermore, mRNAs can be correctly translated by the protein-synthesizing machinery of very different species. For example, human hemoglobin mRNA is correctly translated by a wheat germ extract, and bacteria efficiently express recombinant DNA molecules encoding human

TABLE 5.5   Distinctive codons of human mitochondria

Codon	Standard code	Mitochondrial code
UGA	Stop	Trp
UGG	Trp	Trp
AUA	Ile	Met
AUG	Met	Met
AGA	Arg	Stop
AGG	Arg	Stop

proteins such as insulin. These experimental findings strongly suggested that the genetic code is universal.

A surprise was encountered when the sequence of human mitochondrial DNA became known. Human mitochondria read UGA as a codon for tryptophan rather than as a stop signal (Table 5.5). Furthermore, AGA and AGG are read as stop signals rather than as codons for arginine, and AUA is read as a codon for methionine instead of isoleucine. Mitochondria of other species, such as those of yeast, also have genetic codes that differ slightly from the standard one. The genetic code of mitochondria can differ from that of the rest of the cell because mitochondrial DNA encodes a distinct set of tRNAs. Do any cellular protein-synthesizing systems deviate from the standard genetic code? Ciliated protozoa differ from most organisms in reading UAA and UAG as codons for amino acids rather than as stop signals; UGA is their sole termination signal. Thus, *the genetic code is nearly but not absolutely universal.* Variations clearly exist in mitochondria and in species, such as ciliates, that branched off very early in eukaryotic evolution. It is interesting to note that two of the codon reassignments in human mitochondria diminish the information content of the third base of the triplet (e.g., both AUA and AUG specify methionine). Most variations from the standard genetic code are in the direction of a simpler code.

Why has the code remained nearly invariant through billions of years of evolution, from bacteria to human beings? A mutation that altered the reading of mRNA would change the amino acid sequence of most, if not all, proteins synthesized by that particular organism. Many of these changes would undoubtedly be deleterious, and so there would be strong selection against a mutation with such pervasive consequences.

5.6 MOST EUKARYOTIC GENES ARE MOSAICS OF INTRONS AND EXONS

In bacteria, polypeptide chains are encoded by a continuous array of triplet codons in DNA. For many years, genes in higher organisms also were assumed to be continuous. This view was unexpectedly shattered in 1977, when investigators in several laboratories discovered that several genes are *discontinuous.* The mosaic nature of eukaryotic genes was revealed by electron microscopic studies of hybrids formed between mRNA and a segment of DNA containing the corresponding gene (Figure 5.33). For example, the gene for the β chain of hemoglobin is interrupted within its amino acid-coding sequence by a long *intervening sequence* of 550 base pairs and a short one of 120 base pairs. Thus, the *β-globin gene is split into three coding sequences.*

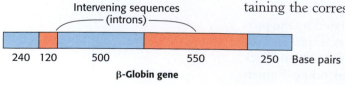

Intervening sequences
(introns)

240 120   500   550   250   Base pairs

β-Globin gene

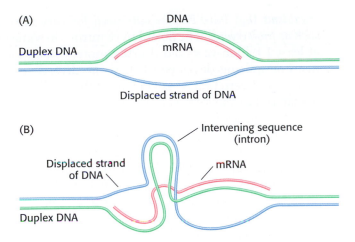

(A)

Duplex DNA

DNA

mRNA

Displaced strand of DNA

(B)

Intervening sequence
(intron)

Displaced strand
of DNA

mRNA

Duplex DNA

FIGURE 5.33 Detection of intervening sequences by electron microscopy. An mRNA molecule (shown in red) is hybridized to genomic DNA containing the corresponding gene. (A) A single loop of single-stranded DNA (shown in blue) is seen if the gene is continuous. (B) Two loops of single-stranded DNA (blue) and a loop of double-stranded DNA (blue and green) are seen if the gene contains an intervening sequence. Additional loops are evident if more than one intervening sequence is present.

5.6.1 RNA Processing Generates Mature RNA

At what stage in gene expression are intervening sequences removed? Newly synthesized RNA chains (pre-mRNA) isolated from nuclei are much larger than the mRNA molecules derived from them; in the case of β-globin RNA, the former sediment at 15S in zonal centrifugation experiments (Section 4.1.6) and the latter at 9S. In fact, the primary transcript of the β-globin gene contains two regions that are not present in the mRNA. *These intervening sequences in the 15S primary transcript are excised, and the coding sequences are simultaneously linked by a precise splicing enzyme to form the mature 9S mRNA* (Figure 5.34). Regions that are removed from the primary transcript are called *introns* (for *intervening sequences*), whereas those that are retained in the mature RNA are called *exons* (for *expressed regions*). A common feature in the expression of split genes is that their exons are ordered in the same sequence in mRNA as in DNA. Thus, split genes, like continuous genes, are colinear with their polypeptide products.

Splicing is a facile complex operation that is carried out by *spliceosomes*, which are assemblies of proteins and small RNA molecules (Section 28.3.4). This enzymatic machinery recognizes signals in the nascent RNA that specify the splice sites. *Introns nearly always begin with GU and end with an AG that is preceded by a pyrimidine-rich tract* (Figure 5.35). *This consensus sequence is part of the signal for splicing.*

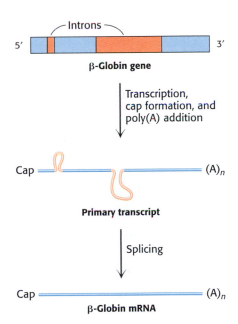

Introns

5′   3′

β-Globin gene

Transcription, cap formation, and poly(A) addition

Cap   (A)$_n$

Primary transcript

Splicing

Cap   (A)$_n$

β-Globin mRNA

FIGURE 5.34 Transcription and processing of the β-globin gene. The gene is transcribed to yield the primary transcript, which is modified by cap and poly(A) addition. The intervening sequences in the primary RNA transcript are removed to form the mRNA.

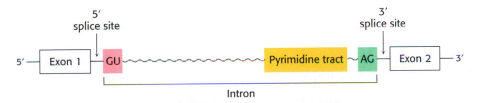

5′
splice site

3′
splice site

5′ — Exon 1 — GU ⌇⌇⌇⌇⌇⌇ Pyrimidine tract ⌇ AG — Exon 2 — 3′

Intron

FIGURE 5.35 Consensus sequence for the splicing of mRNA precursors.

5.6.2 Many Exons Encode Protein Domains

Most genes of higher eukaryotes, such as birds and mammals, are split. Lower eukaryotes, such as yeast, have a much higher proportion of continuous genes. In prokaryotes, split genes are extremely rare. Have introns been inserted into genes in the evolution of higher organisms? Or have introns been removed from genes to form the streamlined genomes of prokaryotes and simple eukaryotes? Comparisons of the DNA sequences of genes encoding proteins that are highly conserved in evolution suggest that *introns were present in ancestral genes and were lost in the evolution of*

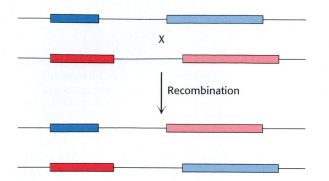

FIGURE 5.36 Exon shuffling. Exons can be readily shuffled by recombination of DNA to expand the genetic repertoire.

Membrane-bound antibody molecule

Extracellular side

Cell membrane

Cytosol

Membrane-anchoring unit encoded by a separate exon

(A)

Soluble antibody molecule

Alternative splicing of RNA excludes membrane-anchoring domain

Secreted into extracellular medium

(B)

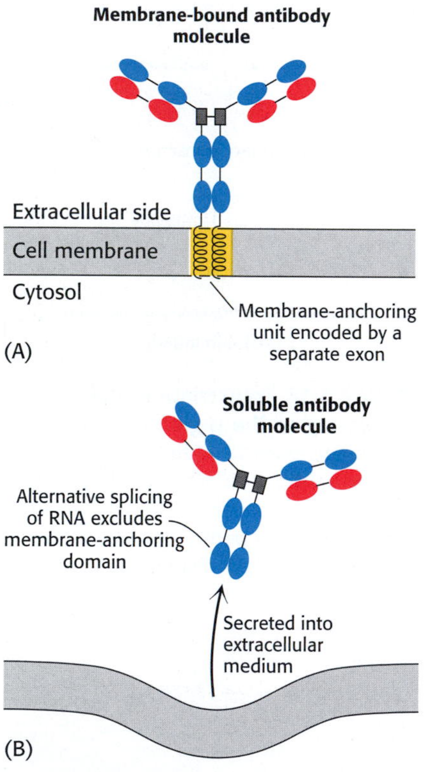

FIGURE 5.37 Alternative splicing. Alternative splicing generates mRNAs that are templates for different forms of a protein: (A) a membrane-bound antibody on the surface of a lymphocyte, and (B) its soluble counterpart, exported from the cell. The membrane-bound antibody is anchored to the plasma membrane by a helical segment (highlighted in yellow) that is encoded by its own exon.

organisms that have become optimized for very rapid growth, such as prokaryotes. The positions of introns in some genes are at least 1 billion years old. Furthermore, a common mechanism of splicing developed before the divergence of fungi, plants, and vertebrates, as shown by the finding that mammalian cell extracts can splice yeast RNA. *Many exons encode discrete structural and functional units of proteins.* An attractive hypothesis is that *new proteins arose in evolution by the rearrangement of exons encoding discrete structural elements, binding sites, and catalytic sites,* a process called *exon shuffling.* Because it preserves functional units but allows them to interact in new ways, exon shuffling is a rapid and efficient means of generating novel genes (Figure 5.36). Introns are extensive regions in which DNA can break and recombine with no deleterious effect on encoded proteins. In contrast, the exchange of sequences between different exons usually leads to loss of function.

Another advantage conferred by split genes is the potentiality for generating a series of related proteins by splicing a nascent RNA transcript in different ways. For example, a precursor of an antibody-producing cell forms an antibody that is anchored in the cell's plasma membrane (Figure 5.37). Stimulation of such a cell by a specific foreign antigen that is recognized by the attached antibody leads to cell differentiation and proliferation. The activated antibody-producing cells then splice their nascent RNA transcript in an alternative manner to form soluble antibody molecules that are secreted rather than retained on the cell surface. We see here a clear-cut example of a benefit conferred by the complex arrangement of introns and exons in higher organisms. *Alternative splicing is a facile means of forming a set of proteins that are variations of a basic motif according to a developmental program without requiring a gene for each protein.*

SUMMARY

A Nucleic Acid Consists of Four Kinds of Bases Linked to a Sugar-Phosphate Backbone

DNA and RNA are linear polymers of a limited number of monomers. In DNA, the repeating units are nucleotides with the sugar being a deoxyribose and the bases being adenine (A), thymine (T), guanine (G), and cytosine (C). In RNA, the sugar is a ribose and the base uracil (U) is used in place of thymine. DNA is the molecule of heredity in all prokaryotic and eukaryotic organisms. In viruses, the genetic material is either DNA or RNA.

A Pair of Nucleic Acid Chains with Complementary Sequences Can Form a Double-Helical Structure

All cellular DNA consists of two very long, helical polynucleotide chains coiled around a common axis. The sugar-phosphate backbone of each strand is on the outside of the double helix, whereas the purine and pyrimidine bases are on the inside. The two chains are held together by hydrogen bonds between pairs of bases: adenine is always paired with thymine, and guanine is always paired with cytosine. Hence, one strand of a double helix is the complement of the other. The two strands of the double helix run in opposite directions. Genetic information is encoded in the precise sequence of bases along a strand. Most RNA molecules are single stranded, but many contain extensive double-helical regions that arise from the folding of the chain into hairpins.

DNA Is Replicated by Polymerases That Take Instructions from Templates

In the replication of DNA, the two strands of a double helix unwind and separate as new chains are synthesized. Each parent strand acts as a template for the formation of a new complementary strand. Thus, the replication of DNA is semiconservative—each daughter molecule receives one strand from the parent DNA molecule. The replication of DNA is a complex process carried out by many proteins, including several DNA polymerases. The activated precursors in the synthesis of DNA are the four deoxyribonucleoside 5'-triphosphates. The new strand is synthesized in the $5' \rightarrow 3'$ direction by a nucleophilic attack by the 3'-hydroxyl terminus of the primer strand on the innermost phosphorus atom of the incoming deoxyribonucleoside triphosphate. Most important, DNA polymerases catalyze the formation of a phosphodiester bond only if the base on the incoming nucleotide is complementary to the base on the template strand. In other words, DNA polymerases are template-directed enzymes. The genes of some viruses, such as tobacco mosaic virus, are made of single-stranded RNA. An RNA-directed RNA polymerase mediates the replication of this viral RNA. Retroviruses, exemplified by HIV-1, have a single-stranded RNA genome that is transcribed into double-stranded DNA by reverse transcriptase, an RNA-directed DNA polymerase.

Gene Expression Is the Transformation of DNA Information into Functional Molecules

The flow of genetic information in normal cells is from DNA to RNA to protein. The synthesis of RNA from a DNA template is called transcription, whereas the synthesis of a protein from an RNA template is termed translation. Cells contain several kinds of RNA: messenger RNA (mRNA), transfer RNA (tRNA), and ribosomal RNA (rRNA), which vary in size from 75 to more than 5000 nucleotides. All cellular RNA is synthesized by RNA polymerases according to instructions given by DNA templates. The activated intermediates are ribonucleoside triphosphates and the direction of synthesis, like that of DNA, is $5' \rightarrow 3'$. RNA polymerase differs from DNA polymerase in not requiring a primer.

Amino Acids Are Encoded by Groups of Three Bases Starting from a Fixed Point

The genetic code is the relation between the sequence of bases in DNA (or its RNA transcript) and the sequence of amino acids in proteins. Amino acids are encoded by groups of three bases (called codons) starting from a fixed point. Sixty-one of the 64 codons specify particular amino acids, whereas the other 3 codons (UAA, UAG, and UGA) are signals for chain termination. Thus, for most amino acids, there is more than one code word. In other words, the code is degenerate. The genetic code is nearly the same in all organisms. Natural mRNAs contain start and stop signals for translation, just as genes do for directing where transcription begins and ends.

Most Eukaryotic Genes Are Mosaics of Introns and Exons

Most genes in higher eukaryotes are discontinuous. Coding sequences (exons) in these split genes are separated by intervening sequences (introns), which are removed in the conversion of the primary transcript into mRNA and other functional mature RNA molecules. Split genes, like continuous genes, are colinear with their polypeptide products. A striking feature of many exons is that they encode functional domains in proteins. New proteins probably arose in the course of evolution by the shuffling of exons. Introns may have been present in primordial genes but were lost in the evolution of such fast-growing organisms as bacteria and yeast.

KEY TERMS

deoxyribonucleic acid (DNA) (p. 118)
deoxyribose (p. 118)
ribose (p. 118)
ribonucleic acid (RNA) (p. 118)
purine (p. 119)
pyrimidine (p. 119)
nucleoside (p. 120)
nucleotide (p. 120)
replication (p. 121)
double helix (p. 121)
semiconservative replication (p. 123)

DNA polymerase (p. 127)
template (p. 127)
primer (p. 127)
reverse transcriptase (p. 128)
messenger RNA (mRNA) (p. 129)
translation (p. 129)
transfer RNA (tRNA) (p. 129)
ribosomal RNA (rRNA) (p. 130)
small nuclear RNA (snRNA) (p. 130)
transcription (p. 130)
RNA polymerase (p. 130)

promoter site (p. 131)
codon (p. 133)
genetic code (p. 133)
ribosome (p. 135)
intron (p. 137)
exon (p. 137)
splicing (p. 137)
spliceosomes (p. 137)
exon shuffling (p. 138)
alternative splicing (p. 138)

SELECTED READINGS

Where to start

Felsenfeld, G., 1985. DNA. *Sci. Am.* 253(4):58–67.
Darnell, J. E., Jr., 1985. RNA. *Sci. Am.* 253(4):68–78.
Dickerson, R. E., 1983. The DNA helix and how it is read. *Sci. Am.* 249(6):94–111.
Crick, F. H. C., 1954. The structure of the hereditary material. *Sci. Am.*191(4): 54–61.
Chambon, P., 1981. Split genes. *Sci. Am.* 244(5):60–71.
Watson, J. D., and Crick, F. H. C., 1953. Molecular structure of nucleic acids. A structure for deoxyribose nucleic acid. *Nature* 171:737–738.
Watson, J. D., and Crick, F. H. C., 1953. Genetic implications of the structure of deoxyribonucleic acid. *Nature* 171:964–967.
Meselson, M., and Stahl, F. W., 1958. The replication of DNA in *Escherichia coli. Proc. Natl. Acad. Sci. USA* 44:671–682.

Books

Bloomfield, V. A., Crothers, D. M., Tinoco, I., and Hearst, J., 2000. *Nucleic Acids: Structures, Properties, and Functions.* University Science Books.
Singer, M., Berg, P., 1991. *Genes and Genomes: A Changing Perspective.* University Science Books.
Lodish, H., Berk, A., Zipursky, L., and Matsudaira, P., 1999. *Molecular Cell Biology* (4th ed.). W. H. Freeman and Company.
Lewin, B., 2000. *Genes VII.* Oxford University Press.
Watson, J. D., Hopkins, N. H., Roberts, J. W., Steitz, J. A., and Weiner, A. M., 2000. *Molecular Biology of the Gene* (5th ed.). Benjamin Cummings.

DNA structure

Saenger, W., 1984. *Principles of Nucleic Acid Structure.* Springer Verlag.
Dickerson, R. E., Drew, H. R., Conner, B. N., Wing, R. M., Fratini, A. V., and Kopka, M. L., 1982. The anatomy of A-, B-, and Z-DNA. *Science* 216:475–485.
Sinden, R. R., 1994. *DNA structure and function.* Academic Press.

DNA replication

Kornberg, A., and Baker, T. A., 1992. *DNA Replication* (2d ed.). W. H. Freeman and Company.
Hübscher, U., Nasheuer, H.-P., and Syväoja, J. E. 2000. Eukaryotic DNA polymerases: A growing family. *Trends Biochem. Sci.* 25:143–147.
Brautigam, C. A., and Steitz, T. A., 1998. Structural and functional insights provided by crystal structures of DNA polymerases and their substrate complexes. *Curr. Opin. Struct. Biol.* 8:54–63.

Discovery of messenger RNA

Jacob, F., and Monod, J., 1961. Genetic regulatory mechanisms in the synthesis of proteins. *J. Mol. Biol.* 3:318–356.

Brenner, S., Jacob, F., and Meselson, M., 1961. An unstable intermediate carrying information from genes to ribosomes for protein synthesis. *Nature* 190:576–581.
Hall, B. D., and Spiegelman, S., 1961. Sequence complementarity of T2-DNA and T2-specific RNA. *Proc. Natl. Acad. Sci. USA* 47:137–146.

Genetic code

Crick, F. H. C., Barnett, L., Brenner, S., and Watts-Tobin, R. J., 1961. General nature of the genetic code for proteins. *Nature* 192: 1227–1232.
Nirenberg, M., 1968. The genetic code. In *Nobel Lectures: Physiology or Medicine* (1963–1970), pp. 372–395. American Elsevier (1973).
Crick, F. H. C., 1958. On protein synthesis. *Symp. Soc. Exp. Biol.* 12:138–163.
Woese, C. R., 1967. *The Genetic Code.* Harper & Row.
Knight, R. D., Freeland, S. J., and Landweber L. F., 1999. Selection, history and chemistry: The three faces of the genetic code. *Trends Biochem. Sci.* 24(6):241–247

Introns, exons, and split genes

Sharp, P. A., 1988. RNA splicing and genes. *J. Am. Med. Assoc.* 260: 3035–3041.
Dorit, R. L., Schoenbach, L., and Gilbert, W., 1990. How big is the universe of exons? *Science* 250:1377–1382.
Cochet, M., Gannon, F., Hen, R., Maroteaux, L., Perrin, F., and Chambon, P., 1979. Organization and sequence studies of the 17-piece chicken conalbumin gene. *Nature* 282:567–574.
Tilghman, S. M., Tiemeier, D. C., Seidman, J. G., Peterlin, B. M., Sullivan, M., Maizel, J. V., and Leder, P., 1978. Intervening sequence of DNA identified in the structural portion of a mouse β-globin gene. *Proc. Natl. Acad. Sci. USA* 75:725–729.

Reminiscences and historical accounts

Watson, J. D., 1968. *The Double Helix.* Atheneum.
McCarty, M., 1985. *The Transforming Principle: Discovering That Genes Are Made of DNA.* Norton.
Cairns, J., Stent, G. S., and Watson, J. D., 2000. *Phage and the Origins of Molecular Biology.* Cold Spring Harbor Laboratory.
Olby, R., 1974. *The Path to the Double Helix.* University of Washington Press.
Portugal, F. H., and Cohen, J. S., 1977. *A Century of DNA: A History of the Discovery of the Structure and Function of the Genetic Substance.* MIT Press.
Judson, H., 1996. *The Eighth Day of Creation.* Cold Spring Harbor Laboratory.
Sayre, A. 2000. *Rosalind Franklin and DNA.* Norton.

PROBLEMS

1. *Complements.* Write the complementary sequence (in the standard $5' \rightarrow 3'$ notation) for (a) GATCAA, (b) TCGAAC, (c) ACGCGT, and (d) TACCAT.

2. *Compositional constraint.* The composition (in mole-fraction units) of one of the strands of a double-helical DNA molecule is [A] = 0.30 and [G] = 0.24. (a) What can you say about [T] and [C] for the same strand? (b) What can you say about [A], [G], [T], and [C] of the complementary strand?

3. *Lost DNA.* The DNA of a deletion mutant of λ bacteriophage has a length of 15 μm instead of 17 μm. How many base pairs are missing from this mutant?

4. *An unseen pattern.* What result would Meselson and Stahl have obtained if the replication of DNA were conservative (i.e., the parental double helix stayed together)? Give the expected distribution of DNA molecules after 1.0 and 2.0 generations for conservative replication.

5. *Tagging DNA.* (a) Suppose that you want to radioactively label DNA but not RNA in dividing and growing bacterial cells. Which radioactive molecule would you add to the culture medium? (b) Suppose that you want to prepare DNA in which the backbone phosphorus atoms are uniformly labeled with ^{32}P. Which precursors should be added to a solution containing DNA polymerase I and primed template DNA? Specify the position of radioactive atoms in these precursors.

6. *Finding a template.* A solution contains DNA polymerase I and the Mg^{2+} salts of dATP, dGTP, dCTP, and TTP. The following DNA molecules are added to aliquots of this solution. Which of them would lead to DNA synthesis? (a) A single-stranded closed circle containing 1000 nucleotide units. (b) A double-stranded closed circle containing 1000 nucleotide pairs. (c) A single-stranded closed circle of 1000 nucleotides base-paired to a linear strand of 500 nucleotides with a free 3'-OH terminus. (d) A double-stranded linear molecule of 1000 nucleotide pairs with a free 3'-OH group at each end.

7. *The right start.* Suppose that you want to assay reverse transcriptase activity. If polyriboadenylate is the template in the assay, what should you use as the primer? Which radioactive nucleotide should you use to follow chain elongation?

8. *Essential degradation.* Reverse transcriptase has ribonuclease activity as well as polymerase activity. What is the role of its ribonuclease activity?

9. *Virus hunting.* You have purified a virus that infects turnip leaves. Treatment of a sample with phenol removes viral proteins. Application of the residual material to scraped leaves results in the formation of progeny virus particles. You infer that the infectious substance is a nucleic acid. Propose a simple and highly sensitive means of determining whether the infectious nucleic acid is DNA or RNA.

10. *Mutagenic consequences.* Spontaneous deamination of cytosine bases in DNA occurs at low but measurable frequency. Cytosine is converted into uracil by loss of its amino group. After this conversion, which base pair occupies this position in each of the daughter strands resulting from one round of replication? Two rounds of replication?

11. *Information content.* (a) How many different 8-mer sequences of DNA are there? (Hint: There are 16 possible dinucleotides and 64 possible trinucleotides.) (b) How many bits of information are stored in an 8-mer DNA sequence? In the *E. coli* genome? In the human genome? (c) Compare each of these values with the amount of information that can be stored on a personal computer diskette. A byte is equal to 8 bits.

12. *Key polymerases.* Compare DNA polymerase I and RNA polymerase from *E. coli* in regard to each of the following features: (a) activated precursors, (b) direction of chain elongation, (c) conservation of the template, and (d) need for a primer.

13. *Encoded sequences.* (a) Write the sequence of the mRNA molecule synthesized from a DNA template strand having the sequence

$$5'\text{-ATCGTACCGTTA-}3'$$

(b) What amino acid sequence is encoded by the following base sequence of an mRNA molecule? Assume that the reading frame starts at the 5' end.

$$5'\text{-UUGCCUAGUGAUUGGAUG-}3'$$

(c) What is the sequence of the polypeptide formed on addition of poly(UUAC) to a cell-free protein-synthesizing system?

14. *A tougher chain.* RNA is readily hydrolyzed by alkali, whereas DNA is not. Why?

15. *A potent blocker.* How does cordycepin (3'-deoxyadenosine) block the synthesis of RNA?

16. *Silent RNA.* The code word GGG cannot be deciphered in the same way as can UUU, CCC, and AAA, because poly(G) does not act as a template. Poly(G) forms a triple-stranded helical structure. Why is it an ineffective template?

17. *Two from one.* Synthetic RNA molecules of defined sequence were instrumental in deciphering the genetic code. Their synthesis first required the synthesis of DNA molecules to serve as a template. H. Gobind Khorana synthesized, by organic-chemical methods, two complementary deoxyribonucleotides, each with nine residues: d(TAC)₃ and d(GTA)₃. Partly overlapping duplexes that formed on mixing these oligonucleotides then served as templates for the synthesis by DNA polymerase of long, repeating double-helical DNA chains. The next step was to obtain long polyribonucleotide chains with a sequence complementary to only one of the two DNA strands. How did he obtain only poly(UAC)? Only poly(GUA)?

18. *Overlapping or not.* In a nonoverlapping triplet code, each group of three bases in a sequence ABCDEF ... specifies only one amino acid—ABC specifies the first, DEF the second, and so forth—whereas, in a completely overlapping triplet code, ABC specifies the first amino acid, BCD the second, CDE the third, and so forth. Assume that you can mutate an individual nucleotide of a codon and detect the mutation in the amino acid sequence. Design an experiment that would establish whether the genetic code is overlapping or nonoverlapping.

19. *Triple entendre.* The RNA transcript of a region of T4 phage DNA contains the sequence 5'-AAAUGAGGA-3'. This sequence encodes three different polypeptides. What are they?

20. *Valuable synonyms.* Proteins generally have low contents of Met and Trp, intermediate ones of His and Cys, and high ones of Leu and Ser. What is the relation between the number of codons of an amino acid and its frequency of occurrence in proteins? What might be the selective advantage of this relation?

21. *A new translation.* A transfer RNA with a UGU anticodon is enzymatically conjugated to ^{14}C-labeled cysteine. The cysteine unit is then chemically modified to alanine (with the use of Raney nickel, which removes the sulfur atom of cysteine). The altered aminoacyl-tRNA is added to a protein-synthesizing system containing normal components except for this tRNA. The mRNA added to this mixture contains the following sequence:

5'-UUUUGCCAUGUUUGUGCU-3'

What is the sequence of the corresponding radiolabeled peptide?

Chapter Integration Problems

22. *Eons ago.* The atmosphere of the primitive Earth before the emergence of life contained N_2, NH_3, H_2, HCN, CO, and H_2O. Which of these compounds is a likely precursor of most of the atoms in adenine? Why?

23. *Back to the bench.* A protein chemist told a molecular geneticist that he had found a new mutant hemoglobin in which aspartate replaced lysine. The molecular geneticist expressed surprise and sent his friend scurrying back to the laboratory. (a) Why did the molecular geneticist doubt the reported amino acid substitution? (b) Which amino acid substitutions would have been more palatable to the molecular geneticist?

24. *Eons apart.* The amino acid sequences of a yeast protein and a human protein carrying out the same function are found to be 60% identical. However, the corresponding DNA sequences are only 45% identical. Account for this differing degree of identity.

 Media Problem

25. *More than one way to pair a base.* Genetic mutations can arise due to nonstandard base pairing during DNA replication. Such mispairing can be made more likely by the chemical modification of bases (which is how mutagens work). One example is oxidation of guanine to 8-oxoguanine. An effect of this modification is to introduce some steric strain into the *anti* configuration of the glycosylic bond, making the *syn* configuration more favorable than usual. Look at the Media Problem section of the **Structural Insights** module on nucleic acids and explain why 8-oxoguanine often mispairs with adenine.

CHAPTER 6

Exploring Genes

Processes such as development from a caterpillar into a butterfly involve dramatic changes in patterns of gene expression. The expression levels of thousands of genes can be monitored through the use of DNA arrays. At right, a GeneChip® reveals the expression levels of more than 12,000 human genes; the brightness of each spot indicates the expression level of the corresponding gene. [(Left) Roger Hart/Rainbow. (Right) GeneChip courtesy of Affymetrix.]

Recombinant DNA technology has revolutionized biochemistry since it came into being in the 1970s. The genetic endowment of organisms can now be precisely changed in designed ways. Recombinant DNA technology is a fruit of several decades of basic research on DNA, RNA, and viruses. It depends, first, on having enzymes that can cut, join, and replicate DNA and reverse transcribe RNA. Restriction enzymes cut very long DNA molecules into specific fragments that can be manipulated; DNA ligases join the fragments together. The availability of many kinds of restriction enzymes and DNA ligases makes it feasible to treat DNA sequences as modules that can be moved at will from one DNA molecule to another. Thus, recombinant DNA technology is based on nucleic acid enzymology.

A second foundation is the base-pairing language that allows complementary sequences to recognize and bind to each other. Hybridization with complementary DNA or RNA probes is a sensitive and powerful means of detecting specific nucleotide sequences. In recombinant DNA technology, base-pairing is used to construct new combinations of DNA as well as to detect and amplify particular sequences. This revolutionary technology is also critically dependent on our understanding of viruses, the ultimate parasites. Viruses efficiently deliver their own DNA (or RNA) into hosts, subverting them either to replicate the viral genome and produce viral proteins or to incorporate viral DNA into the host genome. Likewise, plasmids, which are accessory chromosomes found in bacteria, have been indispensable in recombinant DNA technology.

These new methods have wide-ranging benefits. Entire genomes, including the human genome, are being deciphered. New insights are emerging, for example, into the regulation of gene expression in cancer and development and the evolutionary history of proteins as well as organisms. New proteins can be created by altering genes in specific ways to provide detailed views into protein function. Clinically useful proteins, such as hormones, are now synthesized by recombinant DNA techniques. Crops are being generated to resist pests and harsh conditions. The new opportunities opened by recombinant DNA technology promise to have broad effects.

6.1 THE BASIC TOOLS OF GENE EXPLORATION

The rapid progress in biotechnology—indeed its very existence—is a result of a relatively few techniques.

1. *Restriction-enzyme analysis.* Restriction enzymes are precise, molecular scalpels that allow the investigator to manipulate DNA segments.

2. *Blotting techniques.* The Southern and Northern blots are used to separate and characterize DNA and RNA, respectively. The Western blot, which uses antibodies to characterize proteins, was described in Section 4.3.4.

3. *DNA sequencing.* The precise nucleotide sequence of a molecule of DNA can be determined. Sequencing has yielded a wealth of information concerning gene architecture, the control of gene expression, and protein structure.

4. *Solid-phase synthesis of nucleic acids.* Precise sequences of nucleic acids can be synthesized de novo and used to identify or amplify other nucleic acids.

5. *The polymerase chain reaction (PCR).* The polymerase chain reaction leads to a billionfold amplification of a segment of DNA. One molecule of DNA can be amplified to quantities that permit characterization and manipulation. This powerful technique is being used to detect pathogens and genetic diseases, to determine the source of a hair left at the scene of a crime, and to resurrect genes from fossils.

A final tool, the use of which will be highlighted in the next chapter, is the computer. Without the computer, it would be impossible to catalog, access, and characterize the abundant information, especially DNA sequence information, that the techniques just outlined are rapidly generating.

6.1.1 Restriction Enzymes Split DNA into Specific Fragments

Restriction enzymes, also called *restriction endonucleases,* recognize specific base sequences in double-helical DNA and cleave, at specific places, both strands of a duplex containing the recognized sequences. To biochemists, these exquisitely precise scalpels are marvelous gifts of nature. They are indispensable for analyzing chromosome structure, sequencing very long DNA molecules, isolating genes, and creating new DNA molecules that can be cloned. Werner Arber and Hamilton Smith discovered restriction enzymes, and Daniel Nathans pioneered their use in the late 1960s.

Restriction enzymes are found in a wide variety of prokaryotes. Their biological role is to cleave foreign DNA molecules. The cell's own DNA is not degraded, because the sites recognized by its own restriction enzymes are methylated. Many restriction enzymes recognize specific sequences of four to eight base pairs and hydrolyze a phosphodiester bond in each strand in this region. A striking characteristic of these cleavage sites is that they almost always possess *twofold rotational symmetry.* In other words, the recognized sequence is *palindromic,* or an inverted repeat, and the cleavage sites

Palindrome—

A word, sentence, or verse that reads the same from right to left as it does from left to right.

Radar
Madam, I'm Adam
Able was I ere I saw Elba
Roma tibi subito motibus ibit amor

Derived from the Greek *palindromos,* "running back again."

are symmetrically positioned. For example, the sequence recognized by a restriction enzyme from *Streptomyces achromogenes* is:

Cleavage site

5′ C—C—G—C—G—G 3′
3′ G—G—C—G—C—C 5′

Cleavage site   Symmetry axis

In each strand, the enzyme cleaves the C–G phosphodiester bond on the 3′ side of the symmetry axis. As we shall see in Chapter 9, this symmetry reflects that of structures of the restriction enzymes themselves.

More than 100 restriction enzymes have been purified and characterized. Their names consist of a three-letter abbreviation for the host organism (e.g., *Eco* for *Escherichia coli, Hin* for *Haemophilus influenzae, Hae* for *Haemophilus aegyptius*) followed by a strain designation (if needed) and a roman numeral (if more than one restriction enzyme from the same strain has been identified). The specificities of several of these enzymes are shown in Figure 6.1. Note that the cuts may be staggered or even.

Restriction enzymes are used to cleave DNA molecules into specific fragments that are more readily analyzed and manipulated than the entire parent molecule. For example, the 5.1-kb circular duplex DNA of the tumor-producing SV40 virus is cleaved at 1 site by *Eco*RI, 4 sites by *Hpa*I, and 11 sites by *Hind*III. A piece of DNA produced by the action of one restriction enzyme can be specifically cleaved into smaller fragments by another restriction enzyme. The pattern of such fragments can serve as a *fingerprint* of a DNA molecule, as will be discussed shortly. Indeed, complex chromosomes containing hundreds of millions of base pairs can be mapped by using a series of restriction enzymes.

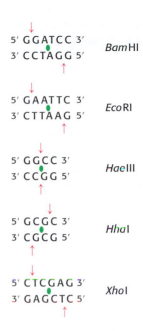

5′ GGATCC 3′
3′ CCTAGG 5′   *Bam*HI

5′ GAATTC 3′
3′ CTTAAG 5′   *Eco*RI

5′ GGCC 3′
3′ CCGG 5′   *Hae*III

5′ GCGC 3′
3′ CGCG 5′   *Hha*I

5′ CTCGAG 3′
3′ GAGCTC 5′   *Xho*I

FIGURE 6.1 Specificities of some restriction endonucleases. The base-pair sequences that are recognized by these enzymes contain a twofold axis of symmetry. The two strands in these regions are related by a 180-degree rotation about the axis marked by the green symbol. The cleavage sites are denoted by red arrows. The abbreviated name of each restriction enzyme is given at the right of the sequence that it recognizes.

6.1.2 Restriction Fragments Can Be Separated by Gel Electrophoresis and Visualized

Small differences between related DNA molecules can be readily detected because their restriction fragments can be separated and displayed by gel electrophoresis. In many types of gels, the electrophoretic mobility of a DNA fragment is inversely proportional to the logarithm of the number of base pairs, up to a certain limit. Polyacrylamide gels are used to separate fragments containing about as many as 1000 base pairs, whereas more porous agarose gels are used to resolve mixtures of larger fragments (about as many as 20 kb). An important feature of these gels is their high resolving power. In certain kinds of gels, *fragments differing in length by just one nucleotide of several hundred can be distinguished*. Moreover, entire chromosomes containing millions of nucleotides can be separated on agarose gels by applying pulsed electric fields (pulsed-field gel electrophoresis, PFGE) in different directions. This technique depends on the differential stretching and relaxing of large DNA molecules as an electric field is turned off and on at short intervals. Bands or spots of radioactive DNA in gels can be visualized by autoradiography (Section 4.1.4). Alternatively, a gel can be stained with ethidium bromide, which fluoresces an intense orange when bound to double-helical DNA molecule (Figure 6.2). A band containing only 50 ng of DNA can be readily seen.

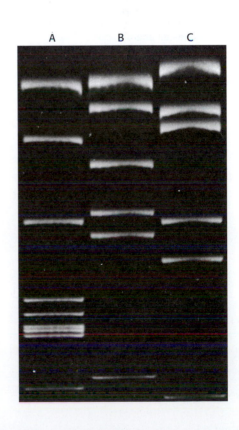

A   B   C

FIGURE 6.2 Gel electrophoresis pattern of a restriction digest. This gel shows the fragments produced by cleaving SV40 DNA with each of three restriction enzymes. These fragments were made fluorescent by staining the gel with ethidium bromide. [Courtesy of Dr. Jeffrey Sklar.]

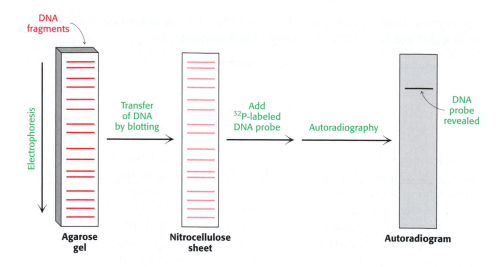

FIGURE 6.3 Southern blotting. A DNA fragment containing a specific sequence can be identified by separating a mixture of fragments by electrophoresis, transferring them to nitrocellulose, and hybridizing with a ^{32}P-labeled probe complementary to the sequence. The fragment containing the sequence is then visualized by autoradiography.

Restriction-fragment-length polymorphism (RFLP)—

Southern blotting can be used to follow the inheritance of selected genes. Mutations within restriction sites change the sizes of restriction fragments and hence the positions of bands in Southern-blot analyses. The existence of genetic diversity in a population is termed polymorphism. The detected mutation may itself cause disease or it may be closely linked to one that does. Genetic diseases such as sickle-cell anemia, cystic fibrosis, and Huntington chorea can be detected by RFLP analyses.

A restriction fragment containing a specific base sequence can be identified by hybridizing it with a labeled complementary DNA strand (Figure 6.3). A mixture of restriction fragments is separated by electrophoresis through an agarose gel, denatured to form single-stranded DNA, and transferred to a nitrocellulose sheet. The positions of the DNA fragments in the gel are preserved on the nitrocellulose sheet, where they are exposed to a ^{32}P-labeled single-stranded *DNA probe*. The probe hybridizes with a restriction fragment having a complementary sequence, and autoradiography then reveals the position of the restriction-fragment-probe duplex. A particular fragment in the midst of a million others can be readily identified in this way, like finding a needle in a haystack. This powerful technique is known as *Southern blotting* because it was devised by Edwin Southern.

Similarly, RNA molecules can be separated by gel electrophoresis, and specific sequences can be identified by hybridization subsequent to their transfer to nitrocellulose. This analogous technique for the analysis of RNA has been whimsically termed *Northern blotting*. A further play on words accounts for the term *Western blotting*, which refers to a technique for detecting a particular protein by staining with specific antibody (Section 4.3.4). Southern, Northern, and Western blots are also known respectively as *DNA, RNA,* and *protein blots*.

6.1.3 DNA Is Usually Sequenced by Controlled Termination of Replication (Sanger Dideoxy Method)

The analysis of DNA structure and its role in gene expression also have been markedly facilitated by the development of powerful techniques for the *sequencing* of DNA molecules. The key to DNA sequencing is the generation of DNA fragments whose length depends on the last base in the sequence. Collections of such fragments can be generated through the *controlled interruption of enzymatic replication*, a method developed by Frederick Sanger and coworkers. This technique has superseded alternative methods because of its simplicity. The same procedure is performed on four reaction mixtures at the same time. In all these mixtures, a DNA polymerase is used to make the complement of a particular sequence within a single-stranded DNA molecule. The synthesis is primed by a fragment, usually obtained by chemical synthetic methods described in Section 6.1.4, that is complementary to a part of the sequence known from other studies. In addition to the four deoxyribonucleoside triphosphates (radioactively labeled), each reaction mixture contains a small amount of the *2′,3′-dideoxy analog* of *one* of the nucleotides, a different nucleotide for each reaction mixture.

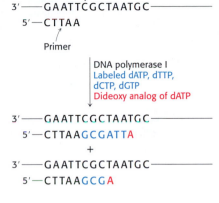

2', 3'-Dideoxy analog

DNA to be sequenced

3' ——— G A A T T C G C T A A T G C ———————

5' — C T T A A

Primer

DNA polymerase I
Labeled dATP, dTTP,
dCTP, dGTP
Dideoxy analog of dATP

3' ——— G A A T T C G C T A A T G C ———————

5' — C T T A A G C G A T T A

+

3' ——— G A A T T C G C T A A T G C ———————

5' — C T T A A G C G A

New DNA strands are separated
and electrophoresed

FIGURE 6.4 Strategy of the chain-termination method for sequencing DNA. Fragments are produced by adding the 2',3'-dideoxy analog of a dNTP to each of four polymerization mixtures. For example, the addition of the dideoxy analog of dATP (shown in red) results in fragments ending in A. The dideoxy analog cannot be extended.

The incorporation of this analog blocks further growth of the new chain because it lacks the 3'-hydroxyl terminus needed to form the next phosphodiester bond. The concentration of the dideoxy analog is low enough that chain termination will take place only occasionally. The polymerase will sometimes insert the correct nucleotide and other times the dideoxy analog, stopping the reaction. For instance, if the dideoxy analog of dATP is present, fragments of various lengths are produced, but all will be terminated by the dideoxy analog (Figure 6.4). Importantly, this dideoxy analog of dATP will be inserted only where a T was located in the DNA being sequenced. Thus, the fragments of different length will correspond to the positions of T. Four such sets of *chain-terminated fragments* (one for each dideoxy analog) then undergo electrophoresis, and the base sequence of the new DNA is read from the autoradiogram of the four lanes.

Fluorescence detection is a highly effective alternative to autoradiography. A fluorescent tag is attached to an oligonucleotide priming fragment—a differently colored one in each of the four chain-terminating reaction mixtures (e.g., a blue emitter for termination at A and a red one for termination at C). The reaction mixtures are combined and subjected to electrophoresis together. The separated bands of DNA are then detected by their fluorescence as they emerge from the gel; the sequence of their colors directly gives the base sequence (Figure 6.5). Sequences of as many as 500 bases can be determined in this way. Alternatively, the dideoxy analogs can be labeled, each with a specific fluorescent label. When this method is used, all four terminators can be placed in a single tube, and only one reaction is necessary. Fluorescence detection is attractive because it eliminates the use of radioactive reagents and can be readily automated.

Sanger and coworkers determined the complete sequence of the 5386 bases in the DNA of the φX174 DNA virus in 1977, just a quarter century after Sanger's pioneering elucidation of the amino acid sequence of a protein. This accomplishment is a landmark in molecular biology because it revealed the total information content of a DNA genome. This tour de force was followed several years later by the determination of the sequence of human mitochondrial DNA, a double-stranded circular DNA molecule containing 16,569 base pairs. It encodes 2 ribosomal RNAs, 22 transfer RNAs, and 13 proteins. In

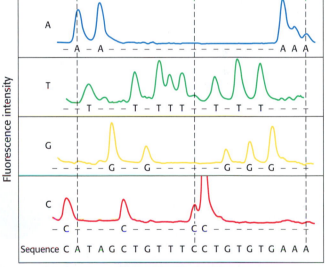

Fluorescence intensity

Oligonucleotide length

FIGURE 6.5 Fluorescence detection of oligonucleotide fragments produced by the dideoxy method. Each of the four chain-terminating mixtures is primed with a tag that fluoresces at a different wavelength (e.g., blue for A). The sequence determined by fluorescence measurements at four wavelengths is shown at the bottom. [From L. M. Smith, J. Z. Sanders, R. J. Kaiser, P. Hughes, C. Dodd, C. R. Connell, C. Heiner, S. B. H. Kent, and L. E. Hood. *Nature* 321(1986):674.]

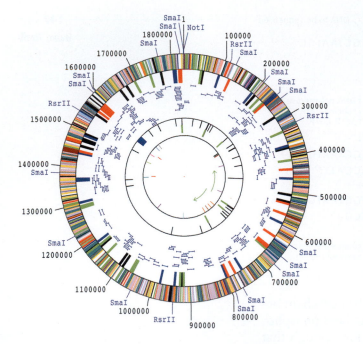

FIGURE 6.6 A complete genome. The diagram depicts the genome of *Haemophilus influenzae,* the first complete genome of a free-living organism to be sequenced. The genome encodes more than 1700 proteins and 70 RNA molecules. The likely function of approximately one-half of the proteins was determined by comparisons with sequences from proteins previously characterized in other species. [From R. D. Fleischmann et al., *Science* 269(1995):496; scan courtesy of TIGR.]

recent years, the complete genomes of free-living organisms have been sequenced. The first such sequence to be completed was that of the bacterium *Haemophilus influenzae.* Its genome comprises 1,830,137 base pairs and encodes approximately 1740 proteins (Figure 6.6).

Many other bacterial and archaeal genomes have since been sequenced. The first eukaryotic genome to be completely sequenced was that of baker's yeast, *Saccharomyces cerevisiae,* which comprises approximately 12 million base pairs, distributed on 16 chromosomes, and encodes more than 6000 proteins. This achievement was followed by the first complete sequencing of the genome of a multicellular organism, the nematode *Caenorhabditis elegans,* which contains nearly 100 million base pairs. The human genome is considerably larger at more than 3 billion base pairs, but it has been essentially completely sequenced. *The ability to determine complete genome sequences has revolutionized biochemistry and biology.*

6.1.4 DNA Probes and Genes Can Be Synthesized by Automated Solid-Phase Methods

DNA strands, like polypeptides (Section 4.4), can be synthesized by the sequential addition of activated monomers to a growing chain that is linked to an insoluble support. The activated monomers are protonated *deoxyribonucleoside 3'-phosphoramidites.* In step 1, the 3' phosphorus atom of this incoming unit becomes joined to the 5' oxygen atom of the growing chain to form a *phosphite triester* (Figure 6.7). The 5'-OH group of the activated

FIGURE 6.7 Solid-phase synthesis of a DNA chain by the phosphite triester method. The activated monomer added to the growing chain is a deoxyribonucleoside 3'-phosphoramidite containing a DMT protecting group on its 5' oxygen atom, a β-cyanoethyl (βCE) protecting group on its 3' phosphoryl oxygen, and a protecting group on the base.

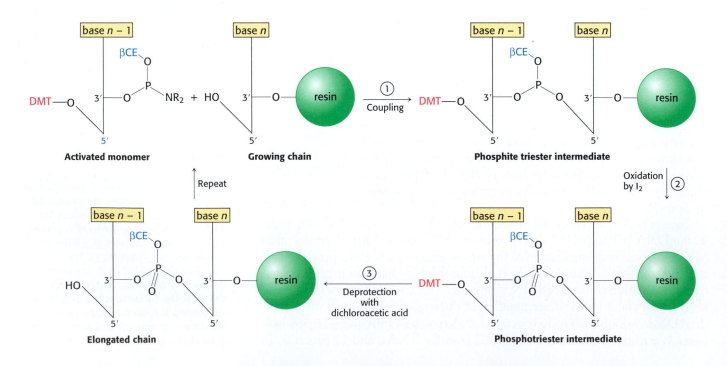

monomer is unreactive because it is blocked by a dimethoxytrityl (DMT) protecting group, and the 3′-phosphoryl group is rendered unreactive by attachment of the β-cyanoethyl (βCE) group. Likewise, amino groups on the purine and pyrimidine bases are blocked.

Coupling is carried out under anhydrous conditions because water reacts with phosphoramidites. In step 2, the phosphite triester (in which P is trivalent) is oxidized by iodine to form a *phosphotriester* (in which P is pentavalent). In step 3, the DMT protecting group on the 5′-OH of the growing chain is removed by the addition of dichloro-acetic acid, which leaves other protecting groups intact. The DNA chain is now elongated by one unit and ready for another cycle of addition. Each cycle takes only about 10 minutes and elongates more than 98% of the chains.

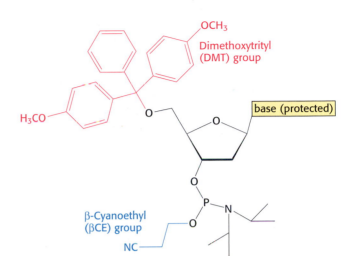

A deoxyribonucleoside 3′-phosphoramidite with DMT and βCE attached

This solid-phase approach is ideal for the synthesis of DNA, as it is for polypeptides, because the desired product stays on the insoluble support until the final release step. All the reactions take place in a single vessel, and excess soluble reagents can be added to drive reactions to completion. At the end of each step, soluble reagents and by-products are washed away from the glass beads that bear the growing chains. At the end of the synthesis, NH_3 is added to remove all protecting groups and release the oligonucleotide from the solid support. Because elongation is never 100% complete, the new DNA chains are of diverse lengths—the desired chain is the longest one. The sample can be purified by high-pressure liquid chromatography or by electrophoresis on polyacrylamide gels. DNA chains of as many as 100 nucleotides can be readily synthesized by this automated method.

The ability to rapidly synthesize DNA chains of any selected sequence opens many experimental avenues. For example, synthesized oligonucleotide labeled at one end with ^{32}P or a fluorescent tag can be used to search for a complementary sequence in a very long DNA molecule or even in a genome consisting of many chromosomes. The use of labeled oligonucleotides as DNA probes is powerful and general. For example, a DNA probe that can base-pair to a known complementary sequence in a chromosome can serve as the starting point of an exploration of adjacent uncharted DNA. Such a probe can be used as a *primer* to initiate the replication of neighboring DNA by DNA polymerase. One of the most exciting applications of the solid-phase approach is the *synthesis of new tailor-made genes*. New proteins with novel properties can now be produced in abundance by expressing synthetic genes. *Protein engineering* has become a reality.

6.1.5 Selected DNA Sequences Can Be Greatly Amplified by the Polymerase Chain Reaction

In 1984, Kary Mullis devised an ingenious method called the *polymerase chain reaction (PCR)* for amplifying specific DNA sequences. Consider a DNA duplex consisting of a target sequence surrounded by nontarget DNA. Millions of the target sequences can be readily obtained by PCR if the flanking sequences of the target are known. PCR is carried out by adding the following components to a solution containing the target sequence: (1) a pair of primers that hybridize with the flanking sequences of the target, (2) all four deoxyribonucleoside triphosphates (dNTPs), and (3) a heat-stable DNA polymerase. A PCR cycle consists of three steps (Figure 6.8).

1. *Strand separation*. The two strands of the parent DNA molecule are separated by heating the solution to 95°C for 15 s.

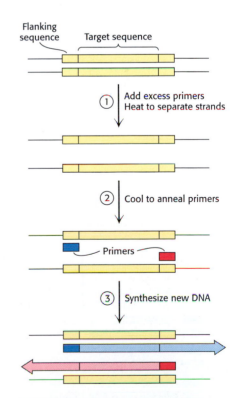

FIGURE 6.8 The first cycle in the polymerase chain reaction (PCR). A cycle consists of three steps: strand separation, hybridization of primers, and extension of primers by DNA synthesis.

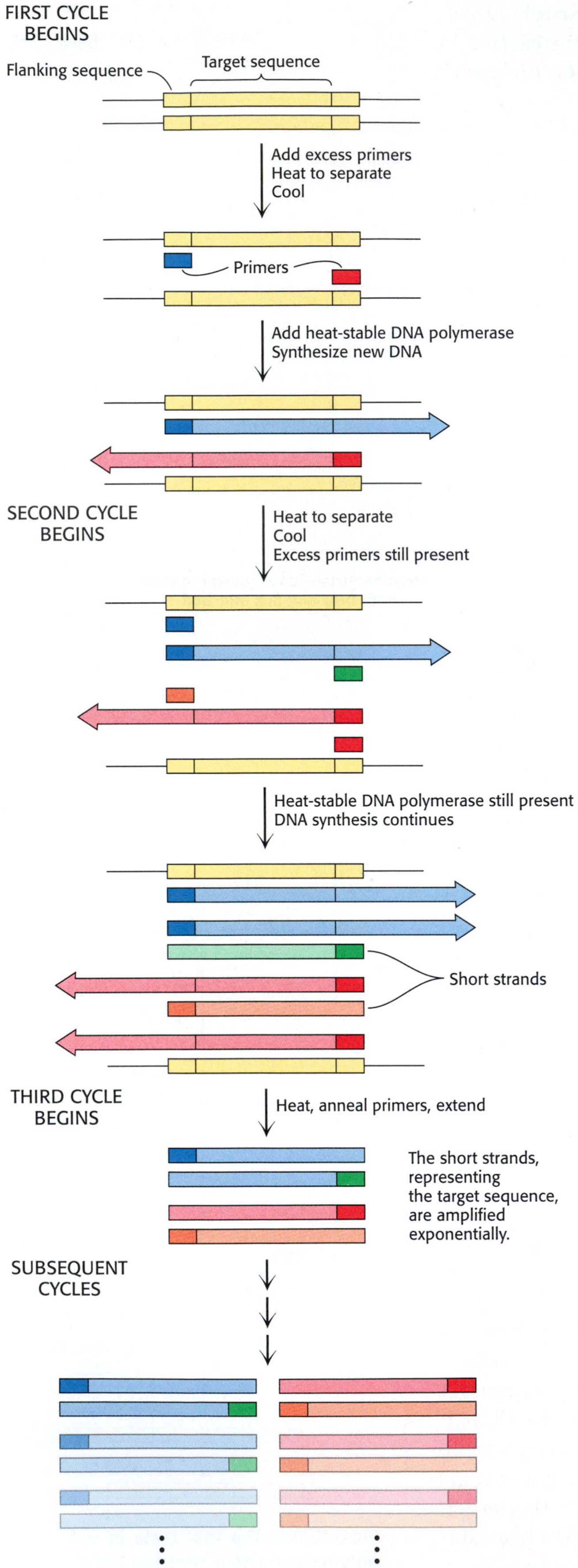

FIRST CYCLE BEGINS

Flanking sequence — Target sequence

Add excess primers
Heat to separate
Cool

Primers

Add heat-stable DNA polymerase
Synthesize new DNA

SECOND CYCLE BEGINS

Heat to separate
Cool
Excess primers still present

Heat-stable DNA polymerase still present
DNA synthesis continues

Short strands

THIRD CYCLE BEGINS

Heat, anneal primers, extend

The short strands, representing the target sequence, are amplified exponentially.

SUBSEQUENT CYCLES

FIGURE 6.9 Multiple cycles of the polymerase chain reaction. The two short strands produced at the end of the third cycle (along with longer stands not shown) represent the target sequence. Subsequent cycles will amplify the target sequence exponentially and the parent sequence arithmetically.

2. *Hybridization of primers.* The solution is then abruptly cooled to 54°C to allow each primer to hybridize to a DNA strand. One primer hybridizes to the 3'-end of the target on one strand, and the other primer hybridizes to the 3' end on the complementary target strand. Parent DNA duplexes do not form, because the primers are present in large excess. Primers are typically from 20 to 30 nucleotides long.

3. *DNA synthesis.* The solution is then heated to 72°C, the optimal temperature for *Taq* DNA polymerase. This heat-stable polymerase comes from <u>Thermus aquaticus</u>, a thermophilic bacterium that lives in hot springs. The polymerase elongates both primers in the direction of the target sequence because DNA synthesis is in the 5'-to-3' direction. DNA synthesis takes place on both strands but extends beyond the target sequence.

These three steps—strand separation, hybridization of primers, and DNA synthesis—constitute one cycle of the PCR amplification and can be carried out repetitively just by changing the temperature of the reaction mixture. The thermostability of the polymerase makes it feasible to carry out PCR in a closed container; no reagents are added after the first cycle. The duplexes are heated to begin the second cycle, which produces four duplexes, and then the third cycle is initiated (Figure 6.9). At the end of the third cycle, two short strands appear that constitute only the target sequence—the sequence including and bounded by the primers. Subsequent cycles will amplify the target sequence exponentially. The larger strands increase in number arithmetically and serve as a source for the synthesis of more short strands. Ideally, after n cycles, this sequence is amplified 2^n-fold. The amplification is a millionfold after 20 cycles and a billionfold after 30 cycles, which can be carried out in less than an hour.

Several features of this remarkable method for amplifying DNA are noteworthy. First, the sequence of the target need not be known. All that is required is knowledge of the flanking sequences. Second, the target can be much larger than the primers. Targets larger than 10 kb have been amplified by PCR. Third, primers do not have to be perfectly matched to flanking sequences to amplify targets. With the use of primers derived from a gene of known sequence, it is possible to search for variations on the theme. In this way, families of genes are being discovered by PCR. Fourth, PCR is highly specific because of the stringency of hybridization at high temperature (54°C). Stringency is the required closeness of the match between primer and target, which can be controlled by temperature and salt. At high temperatures, the only DNA that is amplified is that situated between primers that have hybridized. A gene constituting less than a millionth of the total DNA of a higher organism is accessible by PCR. Fifth, PCR is exquisitely sensitive. A single DNA molecule can be amplified and detected.

6.1.6 PCR Is a Powerful Technique in Medical Diagnostics, Forensics, and Molecular Evolution

PCR can provide valuable diagnostic information in medicine. Bacteria and viruses can be readily detected with the use of specific primers. For example, PCR can reveal the presence of human immunodeficiency virus in people who have not mounted an immune response to this pathogen and would therefore be missed with an antibody assay. Finding *Mycobacterium tuberculosis* bacilli in tissue specimens is slow and laborious. With PCR, as few as 10 tubercle bacilli per million human cells can be readily detected. PCR is a promising method for the early detection of certain cancers. This technique can identify mutations of certain growth-control genes, such as the *ras* genes (Section 15.4.2). The capacity to greatly amplify selected regions of DNA can also be highly informative in monitoring cancer chemotherapy. Tests using PCR can detect when cancerous cells have been eliminated and treatment can be stopped; they can also detect a relapse and the need to immediately resume treatment. PCR is ideal for detecting leukemias caused by chromosomal rearrangements.

PCR is also having an effect in forensics and legal medicine. An individual DNA profile is highly distinctive because many genetic loci are highly variable within a population. For example, variations at a specific one of these locations determines a person's HLA type (human leukocyte antigen type); organ transplants are rejected when the HLA types of the donor and recipient are not sufficiently matched. PCR amplification of multiple genes is being used to establish biological parentage in disputed paternity and immigration cases. Analyses of blood stains and semen samples by PCR have implicated guilt or innocence in numerous assault and rape cases. The root of a single shed hair found at a crime scene contains enough DNA for typing by PCR (Figure 6.10).

DNA is a remarkably stable molecule, particularly when relatively shielded from air, light, and water. Under such circumstances, large fragments of DNA can remain intact for thousands of years or longer. PCR provides an ideal method for amplifying such ancient DNA molecules so that they can be detected and characterized (Section 7.5.1). PCR can also be used to amplify DNA from microorganisms that have not yet been isolated and cultured. As will be discussed in the next chapter, sequences from these PCR products can be sources of considerable insight into evolutionary relationships between organisms.

6.2 RECOMBINANT DNA TECHNOLOGY HAS REVOLUTIONIZED ALL ASPECTS OF BIOLOGY

The pioneering work of Paul Berg, Herbert Boyer, and Stanley Cohen in the early 1970s led to the development of recombinant DNA technology, which has permitted biology to move from an exclusively analytical science to a synthetic one. New combinations of unrelated genes can be constructed in the laboratory by applying recombinant DNA techniques. These novel combinations can be cloned—amplified manyfold—by introducing them into suitable cells, where they are replicated by the DNA-synthesizing machinery of the host. The inserted genes are often transcribed and translated in their new setting. What is most striking is that the genetic endowment of the host can be permanently altered in a designed way.

FIGURE 6.10 DNA and forensics. DNA analysis can be used to establish guilt in criminal cases. Here, DNA was isolated from bloodstains on the pants and shirt of a defendant and amplified by PCR. The DNA was then compared to the DNA from the victim as well as the defendant using gel electrophoresis and autoradiography. DNA from the bloodstains on the defendant's clothing matched the pattern of the victim, but not that of the defendant. The frequency of a coincidental match of the DNA pattern on the clothing and the victim is approximately 1 in 33 billion. Lanes λ, 1kb, and TS = Control DNA samples; lane D = DNA from the defendant; jeans = DNA isolated from bloodstains on defendent's pants; shirt = DNA isolated from bloodstains of the defendant's shirt (two different amounts analyzed); V = DNA sample from victim's blood. [Courtesy of Cellmark Diagnostics, Germantown MD.]

6.2.1 Restriction Enzymes and DNA Ligase Are Key Tools in Forming Recombinant DNA Molecules

Let us begin by seeing how novel DNA molecules can be constructed in the laboratory. A DNA fragment of interest is covalently joined to a DNA *vector*. The essential feature of a vector is that it can replicate autonomously in an appropriate host. *Plasmids* (naturally occurring circles of DNA that act as accessory chromosomes in bacteria) and bacteriophage λ, a virus, are choice vectors for cloning in *E. coli*. The vector can be prepared for accepting a new DNA fragment by cleaving it at a single specific site with a restriction enzyme. For example, the plasmid pSC101, a 9.9-kb double-helical circular DNA molecule, is split at a unique site by the *Eco*RI restriction enzyme. The staggered cuts made by this enzyme produce *complementary single-stranded ends*, which have specific affinity for each other and hence are known as *cohesive* or *sticky ends*. Any DNA fragment can be inserted into this plasmid if it has the same cohesive ends. Such a fragment can be prepared from a larger piece of DNA by using the same restriction enzyme as was used to open the plasmid DNA (Figure 6.11).

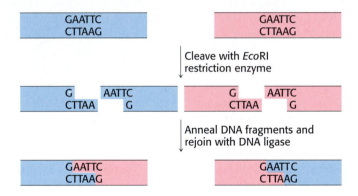

FIGURE 6.11 Joining of DNA molecules by the cohesive-end method. Two DNA molecules, cleaved with a common restriction enzyme such as *Eco*RI, can be ligated to form recombinant molecules.

The single-stranded ends of the fragment are then complementary to those of the cut plasmid. The DNA fragment and the cut plasmid can be annealed and then joined by *DNA ligase*, which catalyzes the formation of a phosphodiester bond at a break in a DNA chain. DNA ligase requires a free 3′-hydroxyl group and a 5′-phosphoryl group. Furthermore, the chains joined by ligase must be in a double helix. An energy source such as ATP or NAD$^+$ is required for the joining reaction, as will be discussed in Chapter 27.

This cohesive-end method for joining DNA molecules can be made general by using a *short, chemically synthesized DNA linker* that can be cleaved by restriction enzymes. First, the linker is covalently joined to the ends of a DNA fragment or vector. For example, the 5′ ends of a decameric linker and a DNA molecule are phosphorylated by polynucleotide kinase and then joined by the ligase from T4 phage (Figure 6.12). This ligase can form a covalent bond between blunt-ended (flush-ended) double-helical DNA molecules. Cohesive ends are produced when these terminal extensions are cut by an appropriate restriction enzyme. Thus, *cohesive ends corresponding to a particular restriction enzyme can be added to virtually any DNA molecule*. We see here the fruits of combining enzymatic and synthetic chemical approaches in crafting new DNA molecules.

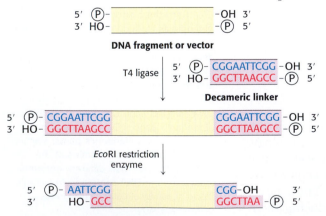

FIGURE 6.12 Formation of cohesive ends. Cohesive ends are formed by the addition and cleavage of a chemically synthesized linker.

6.2.2 Plasmids and Lambda Phage Are Choice Vectors for DNA Cloning in Bacteria

Many plasmids and bacteriophages have been ingeniously modified to enhance the delivery of recombinant DNA molecules into bacteria and to facilitate the selection of bacteria harboring these vectors. Plasmids are circular duplex DNA molecules occurring naturally in some bacteria and ranging in size from 2 to several hundred kilobases. They carry genes for the inactivation of antibiotics, the production of toxins, and the breakdown of natural products. These *accessory chromosomes* can replicate independently of the host chromosome. In contrast with the host genome, they are dispensable under certain conditions. A bacterial cell may have no plasmids at all or it may house as many as 20 copies of a plasmid.

pBR322 Plasmid. One of the most useful plasmids for cloning is *pBR322*, which contains genes for resistance to tetracycline and ampicillin (an antibiotic like penicillin). Different endonucleases can cleave this plasmid at a variety of unique sites, at which DNA fragments can be inserted. Insertion of DNA at the *Eco*RI restriction site does not alter either of the genes for antibiotic resistance (Figure 6.13). However, insertion at the *Hind*III, *Sal*I, or *Bam*HI restriction site inactivates the gene for tetracycline resistance, an effect called *insertional inactivation*. Cells containing pBR322 with a DNA insert at one of these restriction sites are resistant to ampicillin but sensitive to tetracycline, and so they can be readily *selected*. Cells that failed to take up the vector are sensitive to both antibiotics, whereas cells containing pBR322 without a DNA insert are resistant to both.

Lambda (λ) Phage. Another widely used vector, λ *phage*, enjoys a choice of life styles: this bacteriophage can destroy its host or it can become part of its host (Figure 6.14). In the *lytic pathway*, viral functions are fully expressed: viral DNA and proteins are quickly produced and packaged into virus particles, leading to the lysis (destruction) of the host cell and the sudden appearance of about 100 progeny virus particles, or *virions*. In the *lysogenic pathway*, the phage DNA becomes inserted into the host-cell genome and can be replicated together with host-cell DNA for many generations, remaining inactive. Certain environmental changes can trigger the expression of this dormant viral DNA, which leads to the formation of progeny virus and lysis of the host. Large segments of the 48-kb DNA of λ phage are not essential for productive infection and can be replaced by foreign DNA, thus making λ phage an ideal vector.

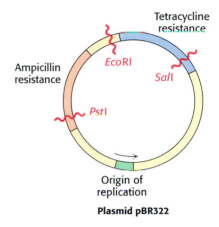

Plasmid pBR322

FIGURE 6.13 Genetic map of the plasmid pBR322. This plasmid carries two genes for antibiotic resistance. Like all other plasmids, it is a circular duplex DNA.

FIGURE 6.14 Alternative infection modes for λ phage. Lambda phage can multiply within a host and lyse it (lytic pathway), or its DNA can become integrated into the host genome (lysogenic pathway), where it is dormant until activated.

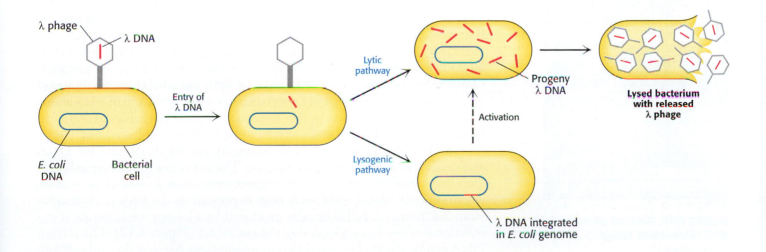

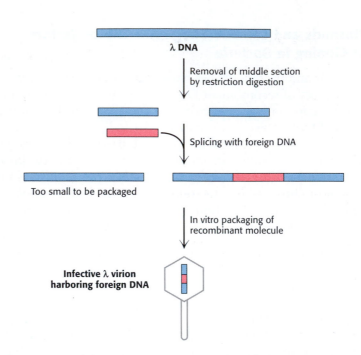

λ DNA

Removal of middle section
by restriction digestion

Splicing with foreign DNA

Too small to be packaged

In vitro packaging of
recombinant molecule

Infective λ virion
harboring foreign DNA

FIGURE 6.15 Mutant λ phage as a cloning vector. The packaging process selects DNA molecules that contain an insert.

FIGURE 6.16 Electron micrograph of M13 filamentous phage. [Courtesy of Dr. Robley Williams.]

Mutant λ phages designed for cloning have been constructed. An especially useful one called λgt-λβ contains only two *Eco*RI cleavage sites instead of the five normally present (Figure 6.15). After cleavage, the middle segment of this λ DNA molecule can be removed. The two remaining pieces of DNA (called arms) have a combined length equal to 72% of a normal genome length. This amount of DNA is too little to be packaged into a λ particle, because only DNA measuring from 75% to 105% of a normal genome in length can be readily packaged. However, *a suitably long DNA insert (such as 10 kb) between the two ends of λ DNA enables such a recombinant DNA molecule (93% of normal length) to be packaged.* Nearly all infective λ particles formed in this way will contain an inserted piece of foreign DNA. Another advantage of using these modified viruses as vectors is that they enter bacteria much more easily than do plasmids. Among the variety of λ mutants that have been constructed for use as cloning vectors, one of them, called a *cosmid*, is essentially a hybrid of λ phage and a plasmid that can serve as a vector for large DNA inserts (as large as 45 kb).

M13 Phage. Another very useful vector for cloning DNA, *M13 phage* is especially useful for sequencing the inserted DNA. This filamentous virus is 900 nm long and only 9 nm wide (Figure 6.16). Its 6.4-kb single-stranded circle of DNA is protected by a coat of 2710 identical protein subunits. M13 enters *E. coli* through the bacterial sex pilus, a protein appendage that permits the transfer of DNA between bacteria. The single-stranded DNA in the virus particle [called the (+) strand] is replicated through an intermediate circular double-stranded replicative form (RF) containing (+) and (−) strands. Only the (+) strand is packaged into new virus particles. About a thousand progeny M13 are produced per generation. A striking feature of M13 is that it does not kill its bacterial host. Consequently, large quantities of M13 can be grown and easily harvested (1 gram from 10 liters of culture fluid).

An M13 vector is prepared for cloning by cutting its circular RF DNA at a single site with a restriction enzyme. The cut is made in a *polylinker* region that contains a series of closely spaced recognition sites for restriction enzymes; only one of each such sites is present in the vector. A double-stranded foreign DNA fragment produced by cleavage with the same restriction enzyme is then ligated to the cut vector (Figure 6.17). The foreign DNA can be inserted in two different orientations because the ends of both

DNA molecules are the same. Hence, half the new (+) strands packaged into virus particles will contain one of the strands of the foreign DNA, and half will contain the other strand. Infection of *E. coli* by a single virus particle will yield a large amount of single-stranded M13 DNA containing the same strand of the foreign DNA. DNA cloned into M13 can be easily sequenced. An oligonucleotide that hybridizes adjacent to the polylinker region is used as a primer for sequencing the insert. This oligomer is called a *universal sequencing primer* because it can be used to sequence *any* insert. M13 is ideal for sequencing but not for long-term propagation of recombinant DNA, because inserts longer than about 1 kb are not stably maintained.

6.2.3 Specific Genes Can Be Cloned from Digests of Genomic DNA

Ingenious cloning and selection methods have made feasible the isolation of a specific DNA segment several kilobases long out of a genome containing more than 3×10^6 kb. Let us see how a gene that is present just once in a human genome can be cloned. A sample containing many molecules of total genomic DNA is first mechanically sheared or partly digested by restriction enzymes into large fragments (Figure 6.18). This nearly random population of overlapping DNA fragments is then separated by gel electrophoresis to isolate a set about 15 kb long. Synthetic linkers are attached to the ends of these fragments, cohesive ends are formed, and the fragments are then inserted into a vector, such as λ phage DNA, prepared with the same cohesive ends. *E. coli* bacteria are then infected with these recombinant phages. The resulting lysate contains fragments of human DNA housed in a sufficiently large number of virus particles to ensure that nearly the entire genome is represented. These phages constitute a *genomic library*. Phages can be propagated indefinitely, and so the library can be used repeatedly over long periods.

This genomic library is then screened to find the very small portion of phages harboring the gene of interest. For the human genome, a calculation shows that a 99% probability of success requires screening about 500,000 clones; hence, a very rapid and efficient screening process is essential. Rapid screening can be accomplished by DNA hybridization.

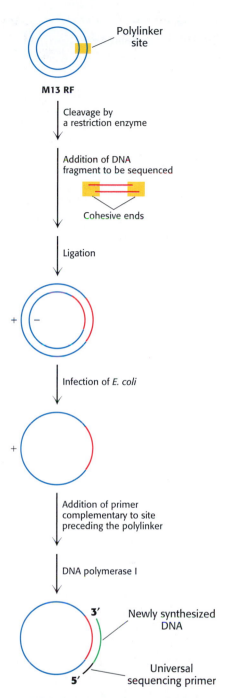

FIGURE 6.17 M13 phage DNA, a cloning and sequencing vector. M13 phage DNA is very useful in sequencing DNA fragments by the dideoxy method. A double-stranded DNA fragment is inserted into M13 RF DNA. Synthesis of new strand is primed by an oligonucleotide that is complementary to a sequence near the inserted DNA.

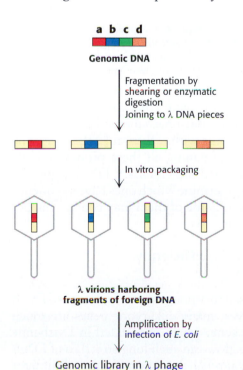

FIGURE 6.18 Creation of a genomic library. A genomic library can be created from a digest of a whole eukaryotic genome.

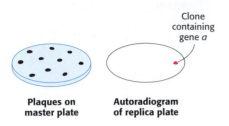

FIGURE 6.19 Screening a genomic library for a specific gene. Here, a plate is tested from plaques containing gene *a* of Figure 6.18.

A dilute suspension of the recombinant phages is first plated on a lawn of bacteria (Figure 6.19). Where each phage particle has landed and infected a bacterium, a *plaque* containing identical phages develops on the plate. A replica of this master plate is then made by applying a sheet of nitrocellulose. Infected bacteria and phage DNA released from lysed cells adhere to the sheet in a pattern of spots corresponding to the plaques. Intact bacteria on this sheet are lysed with NaOH, which also serves to denature the DNA so that it becomes accessible for hybridization with a ^{32}P-labeled probe. *The presence of a specific DNA sequence in a single spot on the replica can be detected by using a radioactive complementary DNA or RNA molecule as a probe.* Autoradiography then reveals the positions of spots harboring recombinant DNA. The corresponding plaques are picked out of the intact master plate and grown. A single investigator can readily screen a million clones in a day.

This method makes it possible to isolate virtually any gene, *provided that a probe is available.* How does one obtain a specific probe? One approach is to *start with the corresponding mRNA from cells in which it is abundant.* For example, precursors of red blood cells contain large amounts of mRNA for hemoglobin, and plasma cells are rich in mRNAs for antibody molecules. The mRNAs from these cells can be fractionated by size to enrich for the one of interest. As will be described shortly, a DNA complementary to this mRNA can be synthesized in vitro and cloned to produce a highly specific probe.

Alternatively, *a probe for a gene can be prepared if part of the amino acid sequence of the protein encoded by the gene is known.* A problem arises because a given peptide sequence can be encoded by a number of oligonucleotides (Figure 6.20). Thus, for this purpose, peptide sequences containing tryptophan and methionine are preferred, because these amino acids are specified by a single codon, whereas other amino acid residues have between two and six codons (Section 5.5.1).

Amino acid sequence	... Cys	Pro	Asn	Lys	Trp	Thr	His ...
Potential oligonucleotide sequences	TG$_\mathrm{T}^\mathrm{C}$	CG$_\mathrm{T}^\mathrm{A\ C\ G}$	AA$_\mathrm{T}^\mathrm{C}$	AA$_\mathrm{G}^\mathrm{A}$	TGG	AC$_\mathrm{T}^\mathrm{A\ C\ G}$	CA$_\mathrm{T}^\mathrm{C}$

FIGURE 6.20 Probes generated from a protein sequence. A probe can be generated by synthesizing all possible oligonucleotides encoding a particular sequence of amino acids. Because of the degeneracy of the genetic code, 256 distinct oligonucleotides must be synthesized to ensure that the probe matching the sequence of seven amino acids is present.

All the DNA sequences (or their complements) that encode the selected peptide sequence are synthesized by the solid-phase method and made radioactive by phosphorylating their 5′ ends with ^{32}P from [^{32}P]-ATP. The replica plate is exposed to a mixture containing all these probes and autoradiographed to identify clones with a complementary DNA sequence. Positive clones are then sequenced to determine which ones have a sequence matching that of the protein of interest. Some of them may contain the desired gene or a significant segment of it.

6.2.4 Long Stretches of DNA Can Be Efficiently Analyzed by Chromosome Walking

A typical genomic DNA library housed in λ phage vectors consists of DNA fragments about 15 kb long. However, many eukaryotic genes are much longer—for example, the *dystrophin* gene, which is mutated in Duchenne muscular dystrophy, is 2000 kb long. How can such long stretches of DNA be analyzed? The development of cosmids helped because these chimeras of plasmids and λ phages can house 45-kb inserts. Much larger pieces of DNA can now be propagated in *bacterial artificial chromosomes (BACs)* or *yeast artificial chromosomes (YACs)*. YACs contain a centromere, an *autonomous replication sequence* (ARS, where replication begins), a pair of telomeres (normal ends of eukaryotic chromosomes), selectable marker genes, and a cloning site (Figure 6.21). Genomic DNA is partly digested by a restriction endonuclease that cuts, on the average, at distant sites. The fragments are then separated by pulsed-field gel electrophoresis, and the large ones (~ 450 kb) are eluted and ligated into YACs. Artificial chromosomes bearing inserts ranging from 100 to 1000 kb are efficiently replicated in yeast cells.

Equally important in analyzing large genes is the capacity to scan long regions of DNA. The principle technique for this purpose makes use of overlaps in the library fragments. The fragments in a cosmid or YAC library are produced by random cleavage of many DNA molecules, and so some of the fragments overlap one another. Suppose that a fragment containing region *A* selected by hybridization with a complementary probe *A'* also contains region *B* (Figure 6.22). A new probe *B'* can be prepared by cleaving this fragment between regions *A* and *B* and subcloning region *B*. If the library is screened again with probe *B'*, new fragments containing region *B* will be found. Some will contain a previously unknown region *C*. Hence, we now have information about a segment of DNA encompassing regions *A*, *B*, and *C*. This process of subcloning and rescreening is called *chromosome walking*. Long stretches of DNA can be analyzed in this way, provided that each of the new probes is complementary to a unique region.

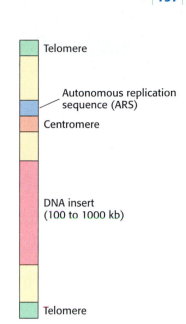

FIGURE 6.21 Diagram of a yeast artificial chromosome (YAC). DNA inserts as large as 1000 kb can be propagated in this vector.

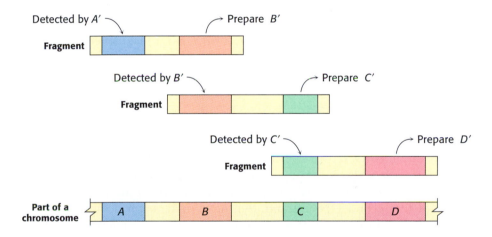

FIGURE 6.22 Chromosome walking. Long regions of unknown DNA can be explored, starting with a known base sequence, by subcloning and rescreening. New probes are designed on the basis of the DNA sequences that have been determined.

6.3 MANIPULATING THE GENES OF EUKARYOTES

Eukaryotic genes, in a simplified form, can be introduced into bacteria, and the bacteria can be used as factories to produce a desired protein product. It is also possible to introduce DNA into higher organisms. In regard to animals, this ability provides a valuable tool for examining gene action, and it will be the basis of gene therapy. In regard to plants, introduced genes may make a plant resistant to pests or capable of growing in harsh conditions or able to carry greater quantities of essential nutrients. The manipulation of eukaryotic genes holds much promise for medical and agricultural benefits, but it is also the source of controversy.

6.3.1 Complementary DNA Prepared from mRNA Can Be Expressed in Host Cells

How can mammalian DNA be cloned and expressed by *E. coli*? Recall that most mammalian genes are mosaics of introns and exons (Section 5.6). These

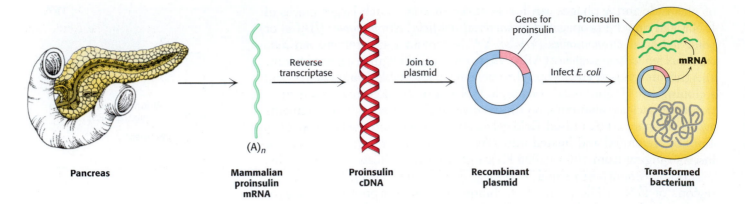

Pancreas → Reverse transcriptase → **Mammalian proinsulin mRNA** (A)$_n$ → **Proinsulin cDNA** → Join to plasmid → **Recombinant plasmid** (Gene for proinsulin) → Infect *E. coli* → **Transformed bacterium** (Proinsulin, mRNA)

FIGURE 6.23 Synthesis of proinsulin by bacteria. Proinsulin, a precursor of insulin, can be synthesized by transformed (genetically altered) clones of *E. coli*. The clones contain the mammalian proinsulin gene.

interrupted genes cannot be expressed by bacteria, which lack the machinery to splice introns out of the primary transcript. However, this difficulty can be circumvented by causing bacteria to take up recombinant DNA that is complementary to mRNA. For example, proinsulin, a precursor of insulin, is synthesized by bacteria harboring plasmids that contain DNA complementary to mRNA for proinsulin (Figure 6.23). Indeed, bacteria produce much of the insulin used today by millions of diabetics.

The key to forming *complementary DNA (cDNA)* is the enzyme *reverse transcriptase*. As discussed in Section 5.3.1, a retrovirus uses this enzyme to form a DNA–RNA hybrid in replicating its genomic RNA. Reverse transcriptase synthesizes a DNA strand complementary to an RNA template if it is provided with a DNA primer that is base-paired to the RNA and contains a free 3′-OH group. We can use a simple sequence of linked thymidine [oligo(T)] residues as the primer. This oligo(T) sequence pairs with the poly(A) sequence at the 3′ end of most eukaryotic mRNA molecules (Section 5.4.4), as shown in Figure 6.24. The reverse transcriptase then synthesizes the rest of the cDNA strand in the presence of the four deoxyribonucleoside triphosphates. The RNA strand of this RNA–DNA hybrid is subsequently hydrolyzed by raising the pH. Unlike RNA, DNA is resistant to alkaline hydrolysis. The single-stranded DNA is converted into double-stranded DNA by creating another primer site. The enzyme *terminal transferase* adds nucleotides—for instance, several residues of dG—to the 3′ end of DNA. Oligo(dC) can bind to dG residues and prime the synthesis of the second DNA strand. Synthetic linkers can be added to this double-helical DNA for ligation to a suitable vector. Complementary DNA for all mRNA that a cell contains can be made, inserted into vectors, and then inserted into bacteria. Such a collection is called a *cDNA library*.

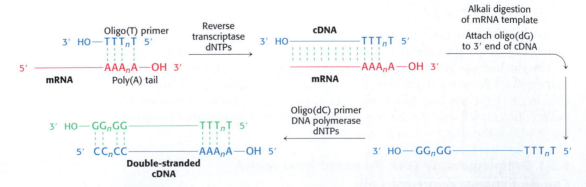

FIGURE 6.24 Formation of a cDNA duplex. A cDNA duplex is created from mRNA by using reverse transcriptase to synthesize a cDNA strand, first along the mRNA template and then, after digestion of the mRNA, along that same newly synthesized cDNA strand.

Complementary DNA molecules can be inserted into vectors that favor their efficient expression in hosts such as *E. coli*. Such plasmids or phages are called *expression vectors*. To maximize transcription, the cDNA is inserted into the vector in the correct reading frame near a strong bacterial promoter site. In addition, these vectors ensure efficient translation by encoding a ribosome-binding site on the mRNA near the initiation codon. Clones of cDNA can be screened on the basis of their capacity to direct the synthesis of a foreign protein in bacteria. A radioactive antibody specific for the protein of interest can be used to identify colonies of bacteria that harbor the corresponding cDNA vector (Figure 6.25). As described in Section 6.2.3, spots of bacteria on a replica plate are lysed to release proteins, which bind to an applied nitrocellulose filter. A ^{125}I-labeled antibody specific for the protein of interest is added, and autoradiography reveals the location of the desired colonies on the master plate. This immunochemical screening approach can be used whenever a protein is expressed and corresponding antibody is available.

6.3.2 Gene-Expression Levels Can Be Comprehensively Examined

Most genes are present in the same quantity in every cell—namely, one copy per haploid cell or two copies per diploid cell. However, the level at which a gene is expressed, as indicated by mRNA quantities, can vary widely, ranging from no expression to hundreds of mRNA copies per cell. Gene-expression patterns vary from cell type to cell type, distinguishing, for example, a muscle cell from a nerve cell. Even within the same cell, gene-expression levels may vary as the cell responds to changes in physiological circumstances.

Using our knowledge of complete genome sequences, it is now possible to analyze the pattern and level of expression of all genes in a particular cell or tissue. One of the most powerful methods developed to date for this purpose is based on hybridization. High-density arrays of oligonucleotides, called *DNA microarrays* or *gene chips*, can be constructed either through light-directed chemical synthesis carried out with photolithographic microfabrication techniques used in the semiconductor industry or by placing very small dots of oligonucleotides or cDNAs on a solid support such as a microscope slide. Fluorescently labeled cDNA is hybridized to the chip to reveal the expression level for each gene, identifiable by its known location on the chip. (Figure 6.26).

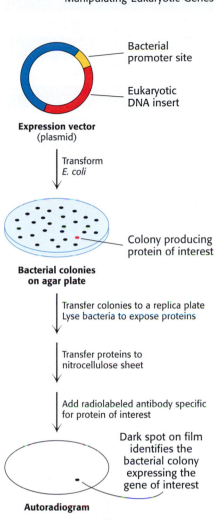

FIGURE 6.25 Screening of cDNA clones. A method of screening for cDNA clones is to identify expressed products by staining with specific antibody.

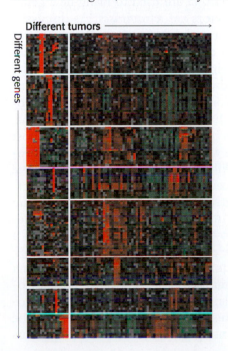

FIGURE 6.26 Gene expression analysis using microarrays. The expression levels of thousands of genes can be simultaneously analyzed using DNA microarrays (gene chips). Here, analysis of 1733 genes in 84 breast tumor samples reveals that the tumors can be divided into distinct classes based on their gene expression patterns. Red corresponds to gene induction and green corresponds to gene repression. [Adapted from C. M. Perou et al., *Nature* 406(2000):747.]

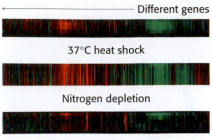

Different genes

37°C heat shock

Nitrogen depletion

Amino acid starvation

FIGURE 6.27 Monitoring changes in yeast gene expression. This microarray analysis shows levels of gene expression for yeast genes under different conditions. [Adapted from Iyer et al., *Nature* 409(2000):533.]

The intensity of the fluorescent spot on the chip reveals the extent of transcription of a particular gene. DNA chips have been prepared that contain oligonucleotides complementary to all known open reading frames, 6200 in number, within the yeast genome (Figure 6.27). An analysis of mRNA pools with the use of these chips revealed, for example, that approximately 50% of all yeast genes are expressed at steady-state levels of between 0.1 and 1.0 mRNA copy per cell. This method readily detected variations in expression levels displayed by specific genes under different growth conditions. These tools will continue to grow in power as genome sequencing efforts continue.

6.3.3 New Genes Inserted into Eukaryotic Cells Can Be Efficiently Expressed

Bacteria are ideal hosts for the amplification of DNA molecules. They can also serve as factories for the production of a wide range of prokaryotic and eukaryotic proteins. However, bacteria lack the necessary enzymes to carry out posttranslational modifications such as the specific cleavage of polypeptides and the attachment of carbohydrate units. Thus, many eukaryotic genes can be correctly expressed only in eukaryotic host cells. The introduction of recombinant DNA molecules into cells of higher organisms can also be a source of insight into how their genes are organized and expressed. How are genes turned on and off in embryological development? How does a fertilized egg give rise to an organism with highly differentiated cells that are organized in space and time? These central questions of biology can now be fruitfully approached by expressing foreign genes in mammalian cells.

Recombinant DNA molecules can be introduced into animal cells in several ways. In one method, foreign DNA molecules precipitated by calcium phosphate are taken up by animal cells. A small fraction of the imported DNA becomes stably integrated into the chromosomal DNA. The efficiency of incorporation is low, but the method is useful because it is easy to apply. In another method, DNA is *microinjected* into cells. A fine-tipped (0.1-μm-diameter) glass micropipet containing a solution of foreign DNA is inserted into a nucleus (Figure 6.28). A skilled investigator can inject hundreds of cells per hour. About 2% of injected mouse cells are viable and contain the new gene. In a third method, *viruses* are used to bring new genes into animal cells. The most effective vectors are *retroviruses*. As discussed in Section 5.3.1, retroviruses replicate through DNA intermediates, the reverse of the normal flow of information. A striking feature of the life cycle of a retrovirus is that the double-helical DNA form of its genome, produced by the action of reverse transcriptase, becomes randomly incorporated into host chromosomal DNA. This DNA version of the viral genome, called *proviral DNA*, can be efficiently expressed by the host cell and replicated along with normal cellular DNA. Retroviruses do not usually kill their hosts. Foreign genes have been efficiently introduced into mammalian cells by infecting them with vectors derived from *Moloney murine leukemia virus*, which can accept inserts as long as 6 kb. Some genes introduced by this retroviral vector into the genome of a transformed host cell are efficiently expressed.

Two other viral vectors are extensively used. *Vaccinia virus*, a large DNA-containing virus, replicates in the cytoplasm of mammalian cells, where it shuts down host-cell protein synthesis. *Baculovirus* infects insect cells, which can be conveniently cultured. Insect larvae infected with this virus

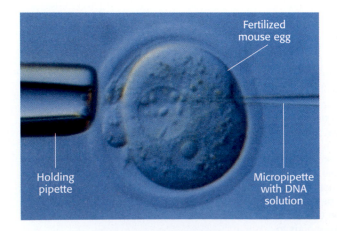

Fertilized mouse egg

Holding pipette

Micropipette with DNA solution

FIGURE 6.28 Microinjection of DNA. Cloned plasmid DNA is being microinjected into the male pronucleus of a fertilized mouse egg.

can serve as efficient protein factories. Vectors based on these large-genome viruses have been engineered to express DNA inserts efficiently.

6.3.4 Transgenic Animals Harbor and Express Genes That Were Introduced into Their Germ Lines

Genetically engineered giant mice illustrate the expression of foreign genes in mammalian cells (Figure 6.29). Giant mice were produced by introducing the gene for rat growth hormone into a fertilized mouse egg. *Growth hormone (somatotropin)*, a 21-kd protein, is normally synthesized by the pituitary gland. A deficiency of this hormone produces dwarfism, and an excess leads to gigantism. The gene for rat growth hormone was placed on a plasmid next to the mouse metallothionein *promoter* (Figure 6.30). This promoter site is normally located on a chromosome, where it controls the transcription of *metallothionein*, a cysteine-rich protein that has high affinity for heavy metals. Metallothionein binds to and sequesters heavy metals, many of which are toxic for metabolic processes (Section 17.3.2). The synthesis of this protective protein by the liver is induced by heavy-metal ions such as cadmium. Hence, if mice contain the new gene, its expression can be initiated by the addition of cadmium to the drinking water.

FIGURE 6.29 Transgenic mice. Injection of the gene for growth hormone into a fertilized mouse egg gave rise to a giant mouse (left), about twice the weight of his silbling (right). [Courtesy of Dr. Ralph Brinster.]

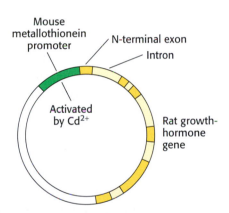

Mouse metallothionein promoter

N-terminal exon

Intron

Activated by Cd^{2+}

Rat growth-hormone gene

FIGURE 6.30 Rat growth hormone–metallothionein gene construct. The gene for rat growth hormone (shown in yellow) was inserted into a plasmid next to the metallothionein promoter, which is activated by the addition of heavy metals, such as cadmium ion.

   Several hundred copies of the plasmid containing the promoter and growth-hormone gene were microinjected into the male pronucleus of a fertilized mouse egg, which was then inserted into the uterus of a foster mother mouse. A number of mice that developed from such microinjected eggs contained the gene for rat growth hormone, as shown by Southern blots of their DNA. These *transgenic mice*, containing multiple copies (~ 30 per cell) of the rat growth-hormone gene, grew much more rapidly than did control mice. In the presence of cadmium, the level of growth hormone in these mice was 500 times as high as in normal mice, and their body weight at maturity was twice normal. The foreign DNA had been transcribed and its five introns correctly spliced out to form functional mRNA. *These experiments strikingly demonstrate that a foreign gene under the control of a new promoter site can be integrated and efficiently expressed in mammalian cells.*

FIGURE 6.31 Gene disruption by homologous recombination. (A) A mutated version of the gene to be disrupted is constructed, maintaining some regions of homology with the normal gene (red). When the foreign mutated gene is introduced into an embryonic stem cell, (B) recombination takes place at regions of homology and (C) the normal (targeted) gene is replaced, or "knocked out," by the foreign gene The cell is inserted into embryos, and mice lacking the gene (knockout mice) are produced.

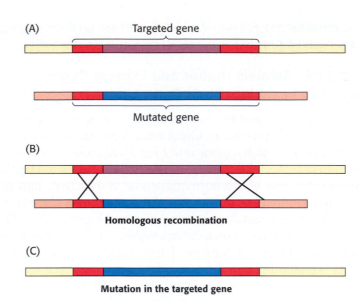

(A)

Targeted gene

Mutated gene

(B)

Homologous recombination

(C)

Mutation in the targeted gene

6.3.5 Gene Disruption Provides Clues to Gene Function

A gene's function can also be probed by inactivating the gene and looking for resulting abnormalities. Powerful methods have been developed for accomplishing *gene disruption* (also called *gene knockout*) in organisms such as yeast and mice. These methods rely on the process of *homologous recombination*. Through this process, regions of strong sequence similarity exchange segments of DNA. Foreign DNA inserted into a cell thus can disrupt any gene that is at least in part homologous by exchanging segments (Figure 6.31). Specific genes can be targeted if their nucleotide sequences are known.

For example, the gene knockout approach has been applied to the genes encoding gene regulatory proteins (also called transcription factors) that control the differentiation of muscle cells. When both copies of the gene for the regulatory protein *myogenin* are disrupted, an animal dies at birth because it lacks functional skeletal muscle. Microscopic inspection reveals that the tissues from which muscle normally forms contain precursor cells that have failed to differentiate fully (Figure 6.32). Heterozygous mice containing one normal myogenin gene and one disrupted gene appear normal, indicating that the level of gene expression is not essential for its function. Analogous studies have probed the function of many other genes to generate animal models for known human genetic diseases.

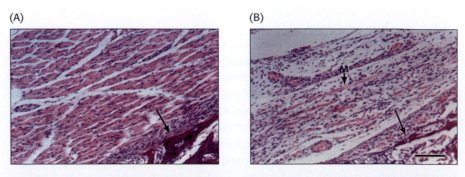

(A)   (B)

FIGURE 6.32 Consequences of gene disruption. Sections of muscle from normal (A) and gene-disrupted (B) mice, as viewed under the light microscope. Muscles do not develop properly in mice having both myogenin genes disrupted. [From P. Hasty, A. Bradley, J. H. Morris, D. G. Edmondson, J. M. Venuti, E. N. Olson, and W. H. Klein, *Nature* 364(1993):501.]

6.3.6 Tumor-Inducing Plasmids Can Be Used to Introduce New Genes into Plant Cells

The common soil bacterium *Agrobacterium tumefaciens* infects plants and introduces foreign genes into plants cells (Figure 6.33). A lump of tumor tissue called a *crown gall* grows at the site of infection. Crown galls synthesize opines, a group of amino acid derivatives that are metabolized by the infecting bacteria. In essence, the metabolism of the plant cell is diverted to satisfy the highly distinctive appetite of the intruder. *Tumor-inducing plasmids (Ti plasmids)* that are carried by *Agrobacterium* carry instructions for the switch to the tumor state and the synthesis of opines. A small part of the Ti plasmid becomes integrated into the genome of infected plant cells; this 20-kb segment is called *T-DNA* (transferred DNA; Figure 6.34).

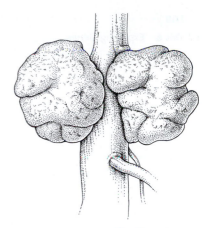

FIGURE 6.33 Tumors in plants. Crown gall, a plant tumor, is caused by a bacterium (*Agrobacterium tumefaciens*) that carries a tumor-inducing plasmid (Ti plasmid).

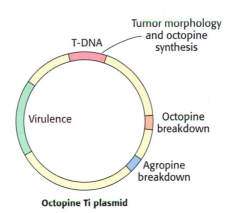

FIGURE 6.34 Ti plasmids. Agrobacteria containing Ti plasmids can deliver foreign genes into some plant cells. [After M. Chilton. A vector for introducing new genes into plants. Copyright ©1983 by Scientific American, Inc. All rights reserved.]

Ti plasmid derivatives can be used as vectors to deliver foreign genes into plant cells. First, a segment of foreign DNA is inserted into the T-DNA region of a small plasmid through the use of restriction enzymes and ligases. This synthetic plasmid is added to *Agrobacterium* colonies harboring naturally occurring Ti plasmids. By recombination, Ti plasmids containing the foreign gene are formed. These Ti vectors hold great promise as tools for exploring the genomes of plant cells and modifying plants to improve their agricultural value and crop yield. However, they are not suitable for transforming all types of plants. Ti-plasmid transfer is effective with dicots (broad-leaved plants such as grapes) and a few kinds of monocots but not with economically important cereal monocots.

Foreign DNA can be introduced into cereal monocots as well as dicots by applying intense electric fields, a technique called *electroporation* (Figure 6.35). First, the cellulose wall surrounding plant cells is removed by adding cellulase; this treatment produces *protoplasts*, plant cells with exposed plasma membranes. Electric pulses are then applied to a suspension of protoplasts and plasmid DNA. Because high electric fields make membranes transiently permeable to large molecules, plasmid DNA molecules enter the cells. The cell wall is then allowed to reform, and the plant cells are again viable. Maize cells and carrot cells have been stably transformed in this way with the use of plasmid DNA that includes genes for resistance to antibiotics. Moreover, the transformed cells efficiently express the plasmid DNA. Electroporation is also an effective means of delivering foreign DNA into animal cells.

The most effective means of transforming plant cells is through the use of "gene guns," or bombardment-mediated transformation. DNA is coated onto 1-μm-diameter tungsten pellets, and these microprojectiles are fired

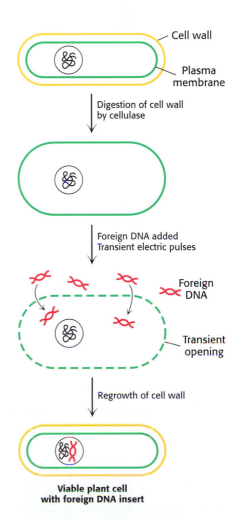

FIGURE 6.35 Electroporation. Foreign DNA can be introduced into plant cells by electroporation, the application of intense electric fields to make their plasma membranes transiently permeable.

at the target cells with a velocity greater than 400 m s^{-1}. Despite its apparent crudeness, this technique is proving to be the most effective way of transforming plants, especially important crop species such as soybean, corn, wheat, and rice. The gene-gun technique affords an opportunity to develop genetically modified organisms (GMOs) with beneficial characteristics. Such characteristics could include the ability to grow in poor soils, resistance to natural climatic variation, resistance to pests, and nutritional fortification. These crops might be most useful in developing countries. The use of genetically modified organisms is highly controversial at this point because of fears of unexpected side effects.

The first GMO to come to market was a tomato characterized by delayed ripening, rendering it ideal for shipment. Pectin is a polysaccharide that gives tomatoes their firmness and is naturally destroyed by the enzyme *polygalacturonase*. As pectin is destroyed, the tomatoes soften, making shipment difficult. DNA was introduced that disrupts the polygalacturonase gene. Less of the enzyme was produced, and the tomatoes stayed fresh longer. However, the tomato's poor taste hindered its commercial success.

6.4 NOVEL PROTEINS CAN BE ENGINEERED BY SITE-SPECIFIC MUTAGENESIS

Much has been learned about genes and proteins by analyzing mutated genes selected from the repertoire offered by nature. In the classic genetic approach, mutations are generated randomly throughout the genome, and those exhibiting a particular phenotype are selected. Analysis of these mutants then reveals which genes are altered, and DNA sequencing identifies the precise nature of the changes. *Recombinant DNA technology now makes it feasible to create specific mutations in vitro.*

6.4.1 Proteins with New Functions Can Be Created Through Directed Changes in DNA

We can construct new genes with designed properties by making three kinds of directed changes: *deletions, insertions,* and *substitutions.*

Deletions. A specific deletion can be produced by cleaving a plasmid at two sites with a restriction enzyme and ligating to form a smaller circle. This simple approach usually removes a large block of DNA. A smaller deletion can be made by cutting a plasmid at a single site. The ends of the linear DNA are then digested with an exonuclease that removes nucleotides from both strands. The shortened piece of DNA is then ligated to form a circle that is missing a short length of DNA about the restriction site.

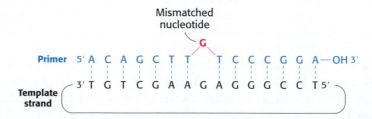

FIGURE 6.36 Oligonucleotide-directed mutagenesis. A primer containing a mismatched nucleotide is used to produce a desired change in the DNA sequence.

Substitutions: Oligonucleotide-Directed Mutagenesis. Mutant proteins with single amino acid substitutions can be readily produced by *oligonucleotide-directed mutagenesis* (Figure 6.36). Suppose that we want to replace a particular serine residue with cysteine. This mutation can be made if (1) we have a plasmid containing the gene or cDNA for the protein and (2) we know the base sequence around the site to be altered. If the serine of interest is encoded by TCT, we need to change the C to a G to get cysteine, which is encoded by TGT. This type of mutation is called a point mutation because only one base is altered. The key to this mutation is to prepare an oligonucleotide primer that is complementary to this region of the gene except that it contains TGT instead of TCT. The two strands of the plasmid are separated, and the primer is then annealed to the complementary strand. The mismatch of 1 base pair of 15 is tolerable if the annealing is carried out at an appropriate temperature. After annealing to the complementary strand, the primer is elongated by DNA polymerase, and the double-stranded circle is closed by adding DNA ligase. Subsequent replication of this duplex yields two kinds of progeny plasmid, half with the original TCT sequence and half with the mutant TGT sequence. Expression of the plasmid containing the new TGT sequence will produce a protein with the desired substitution of serine for cysteine at a unique site. We will encounter many examples of the use of oligonucleotide-directed mutagenesis to precisely alter regulatory regions of genes and to produce proteins with tailor-made features.

Insertions: Cassette Mutagenesis. In another valuable approach, *cassette mutagenesis*, plasmid DNA is cut with a pair of restriction enzymes to remove a short segment (Figure 6.37). A synthetic double-stranded oligonucleotide (the *cassette*) with cohesive ends that are complementary to the ends of the cut plasmid is then added and ligated. Each plasmid now contains the desired mutation. It is convenient to introduce into the plasmid unique restriction sites spaced about 40 nucleotides apart so that mutations can be readily made anywhere in the sequence.

Designer Genes. Novel proteins can also be created by splicing together gene segments that encode domains that are not associated in nature. For example, a gene for an antibody can be joined to a gene for a toxin to produce a chimeric protein that kills cells that are recognized by the antibody. These *immunotoxins* are being evaluated as anticancer agents. Entirely new genes can be synthesized de novo by the solid-phase method. Furthermore, noninfectious coat proteins of viruses can be produced in large amounts by recombinant DNA methods. They can serve as *synthetic vaccines* that are safer than conventional vaccines prepared by inactivating pathogenic viruses. A subunit of the hepatitis B virus produced in yeast is proving to be an effective vaccine against this debilitating viral disease.

6.4.2 Recombinant DNA Technology Has Opened New Vistas

Recombinant DNA technology has revolutionized the analysis of the molecular basis of life. Complex chromosomes are being rapidly mapped and dissected into units that can be manipulated and deciphered. The amplification of genes by cloning has provided abundant quantities of DNA for sequencing. Genes are now open books that can be read. New insights are emerging, as exemplified by the discovery of introns in eukaryotic genes. Central questions of biology, such as the molecular basis of development, are now being fruitfully explored. DNA and RNA sequences provide a wealth of information about evolution. Biochemists now move back and forth between gene and protein and feel at home in both areas of inquiry.

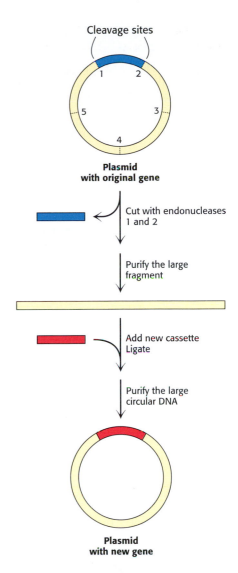

FIGURE 6.37 Cassette mutagenesis. DNA is cleaved at a pair of unique restriction sites by two different restriction endonuclease. A synthetic oligonucleotide with ends that are complementary to these sites (the *cassette*) is then ligated to the cleaved DNA. The method is highly versatile because the inserted DNA can have any desired sequence.

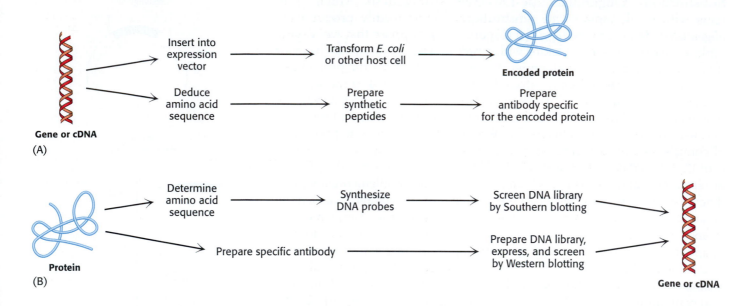

(A)

(B)

FIGURE 6.38 The techniques of protein chemistry and nucleic acid chemistry are mutually reinforcing. (A) From DNA (or RNA) to protein and (B) from protein to DNA.

Analyses of genes and cDNA can reveal the existence of previously unknown proteins, which can be isolated and purified (Figure 6.38A). Conversely, purification of a protein can be the starting point for the isolation and cloning of its gene or cDNA (Figure 6.39B). Very small amounts of protein or nucleic acid suffice because of the sensitivity of recently developed microchemical techniques and the amplification afforded by gene cloning and the polymerase chain reaction. The powerful techniques of protein chemistry, nucleic acid chemistry, immunology, and molecular genetics are highly synergistic.

New kinds of proteins can be created by altering genes in specific ways. Site-specific mutagenesis opens the door to understanding how proteins fold, recognize other molecules, catalyze reactions, and process information. Large amounts of protein can be obtained by expressing cloned genes or cDNAs in bacteria or eukaryotic cells. Hormones, such as insulin, and antiviral agents, such as interferon, are being produced by bacteria. Tissue plasminogen activator, which is administered to a patient after a heart attack, is made in large quantities in mammalian cells. A new pharmacology, using proteins produced by recombinant DNA technology as drugs, is beginning to significantly alter the practice of medicine. Recombinant DNA technology is also providing highly specific diagnostic reagents, such as DNA probes for the detection of genetic diseases, infections, and cancers. Human gene therapy has been successfully initiated. White blood cells deficient in adenosine deaminase, an essential enzyme, are taken from patients and returned after being transformed in vitro to correct the genetic error. Agriculture, too, is benefiting from genetic engineering. Transgenic crops with increased resistance to insects, herbicides, and drought have been produced.

SUMMARY

The Basic Tools of Gene Exploration

The recombinant DNA revolution in biology is rooted in the repertoire of enzymes that act on nucleic acids. Restriction enzymes are a key group among them. These endonucleases recognize specific base sequences in double-helical DNA and cleave both strands of the duplex, forming specific fragments of DNA. These restriction fragments can be separated and displayed by gel electrophoresis. The pattern that they form on the gel is a fingerprint of a DNA molecule. A DNA fragment containing a particular sequence can be identified by hybridizing it with a labeled single-stranded DNA probe (Southern blotting).

Rapid sequencing techniques have been developed to further the analysis of DNA molecules. DNA can be sequenced by controlled interruption of replication (Sanger dideoxy method). The fragments produced are separated by gel electrophoresis and visualized by autoradiography of a ^{32}P label at the 5′ end or by fluorescent tags. The recent sequencing of many complete genomes demonstrates the power of these techniques.

DNA probes for hybridization reactions, as well as new genes, can be synthesized by the automated solid-phase method. The technique is to add deoxyribonucleoside 3′-phosphoramidites to one another to form a growing chain that is linked to an insoluble support. DNA chains a hundred nucleotides long can be readily synthesized by this automated solid-phase method. The polymerase chain reaction makes it possible to greatly amplify specific segments of DNA in vitro. The region amplified is determined by the placement of a pair of primers that are added to the target DNA along with a thermostable DNA polymerase and deoxyribonucleoside triphosphates. The exquisite sensitivity of PCR makes it a choice technique in detecting pathogens and cancer markers, in genotyping, and in reading DNA from fossils that are many thousands of years old.

Recombinant DNA Technology Has Revolutionized All Aspects of Biology

New genes can be constructed in the laboratory, introduced into host cells, and expressed. Novel DNA molecules are made by joining fragments that have complementary cohesive ends produced by the action of a restriction enzyme. DNA ligase seals breaks in DNA chains. Vectors for propagating the DNA include plasmids, λ phage, and yeast artificial chromosomes. Specific genes can be cloned from a genomic library using a DNA or RNA probe. Foreign DNA can be expressed after insertion into prokaryotic and eukaryotic cells by the appropriate vector.

Manipulating the Genes of Eukaryotes

The production of giant mice by injecting the gene for rat growth hormone into fertilized mouse eggs vividly shows that mammalian cells can be genetically altered in a designed way. New DNA can be brought into plant cells by the soil bacterium *Agrobacterium tumefaciens*, which harbors Ti (tumor-inducing) plasmids. DNA can also be introduced into plant cells by applying intense electric fields, which render them transiently permeable to very large molecules, or by bombarding them with DNA-coated microparticles. Gene-expression levels can be examined through the hybridization of cellular mRNA to arrays of oligonucleotides synthesized on solid supports (gene chips). The functions of particular genes can also be investigated by disruption.

Novel Proteins Can Be Engineered by Site-Specific Mutagenesis

Specific mutations can be generated in vitro to engineer novel proteins. A mutant protein with a single amino acid substitution can be produced by priming DNA replication with an oligonucleotide encoding the new amino acid. Plasmids can be engineered to permit the facile insertion of a DNA cassette containing any desired mutation. The techniques of protein and nucleic acid chemistry are highly synergistic. Investigators now move back and forth between gene and protein with great facility. Recombinant DNA technology is beginning to significantly alter the practice of medicine by providing new diagnostic and therapeutic agents and revealing molecular mechanisms of disease.

KEY TERMS

restriction enzyme (p. 144)
palindrome (p. 144)
DNA probe (p. 146)
Southern blotting (p.146)
Northern blotting (p. 146)
controlled termination of replication
(Sanger dideoxy method) (p. 146)
polymerase chain reaction (PCR)
(p. 149)
vector (p. 152)
plasmid (p. 152)

sticky ends (p. 152)
DNA ligase (p. 152)
lambda (λ) phage (p. 153)
genomic library (p. 155)
bacterial artificial chromosome (BAC)
(p. 157)
yeast artificial chromosome (YAC)
(p. 157)
chromosome walking (p. 157)
complementary DNA (cDNA) (p. 158)
reverse transcriptase (p. 158)

cDNA library (p. 158)
expression vector (p. 159)
DNA microarray (gene chip) (p. 159)
transgenic mice (p. 161)
gene disruption (gene knockout) (p. 162)
tumor-inducing plasmid (Ti plasmid)
(p. 163)
oligonucleotide-directed mutagenesis
(p. 165)
cassette mutagenesis (p. 165)

SELECTED READINGS

Where to start

Berg, P., 1981. Dissections and reconstructions of genes and chromosomes. *Science* 213:296–303.
Gilbert, W., 1981. DNA sequencing and gene structure. *Science* 214:1305–1312.
Sanger, F., 1981. Determination of nucleotide sequences in DNA. *Science* 214:1205–1210.
Mullis, K.B., 1990. The unusual origin of the polymerase chain reaction. *Sci. Am.* 262(4):56–65.

Books on recombinant DNA technology

Watson, J. D., Gilman, M., Witkowski, J., and Zoller, M., 1992. *Recombinant DNA* (2d ed.). Scientific American Books.
Grierson, D. (Ed.), 1991. *Plant Genetic Engineering.* Chapman and Hall.
Mullis, K. B., Ferré, F., and Gibbs, R. A. (Eds.), 1994. *The Polymerase Chain Reaction.* Birkhäuser.
Russel, D., Sambrook, J., and Russel, D., 2000. *Molecular Cloning: A Laboratory Manual* (3d ed.). Cold Spring Harbor Laboratory Press.
Ausubel, F. M., Brent, R., Kingston, R. E., and Moore, D. D., (Eds.) 1999. *Short Protocols in Molecular Biology: A Compendium of Methods from Current Protocols in Molecular Biology.* Wiley.
Birren, B., Green, E. D., Klapholz, S., Myers, R. M., Roskams, J., Riethamn, H., and Hieter, P. (Eds.), 1999. *Genome Analysis* (vols. 1–4). Cold Spring Harbor Laboratory Press.
Methods in Enzymology. Academic Press. [Many volumes in this series deal with recombinant DNA technology.]

DNA sequencing and synthesis

Hunkapiller, T., Kaiser, R. J., Koop, B. F., and Hood, L., 1991. Large-scale and automated DNA sequence determination. *Science* 254:59–67.
Sanger, F., Nicklen, S., and Coulson, A. R., 1977. DNA sequencing with chain-terminating inhibitors. *Proc. Natl. Acad. Sci. USA* 74:5463–5467.
Maxam, A. M., and Gilbert, W., 1977. A new method for sequencing DNA. *Proc. Natl. Acad. Sci. USA* 74:560–564.
Smith, L. M., Sanders, J. Z., Kaiser, R. J., Hughes, P., Dodd, C., Connell, C. R., Heiner, C., Kent, S. B. H., and Hood, L. E., 1986. Fluorescence detection in automated DNA sequence analysis. *Nature* 321:674–679.
Pease, A. C., Solas, D., Sullivan, E. J., Cronin, M. T., Holmes, C. P., and Fodor, S. P. A., 1994. Light-generated oligonucleotide arrays for rapid DNA sequence analysis. *Proc. Natl. Acad. Sci. USA* 91:5022–5026.
Venter, J. C., Adams, M. D., Sutton, G. G., Kerlavage, A.R., Smith, H.O., and Hunkapiller, M., 1998. Shotgun sequencing of the human genome. *Science* 280:1540–1542.

Polymerase chain reaction (PCR)

Arnheim, N., and Erlich, H., 1992. Polymerase chain reaction strategy. *Annu. Rev. Biochem.* 61:131–156.
Kirby, L.T. (Ed.), 1997. *DNA Fingerprinting: An Introduction.* Stockton Press.
Eisenstein, B. I., 1990. The polymerase chain reaction: A new method for using molecular genetics for medical diagnosis. *N. Engl. J. Med.* 322:178–183.
Foley, K. P., Leonard, M. W., and Engel, J. D., 1993. Quantitation of RNA using the polymerase chain reaction. *Trends Genet.* 9:380–386.
Pääbo, S., 1993. Ancient DNA. *Sci. Am.* 269(5):86–92.
Hagelberg, E., Gray, I. C., and Jeffreys, A. J., 1991. Identification of the skeletal remains of a murder victim by DNA analysis. *Nature* 352:427–429.
Lawlor, D. A., Dickel, C. D., Hauswirth, W. W., and Parham, P., 1991. Ancient HLA genes from 7500-year-old archaeological remains. *Nature* 349:785–788.
Krings, M., Geisert, H., Schmitz, R. W., Krainitzki, H., and Pääbo, S., 1999. DNA sequence of the mitochondrial hypervariable region II for the Neanderthal type specimen. *Proc. Natl. Acad. Sci. USA* 96:5581–5585.
Ovchinnikov, I. V., Götherström, A., Romanova, G. P., Kharitonov, V. M., Lidén, K., and Goodwin, W., 2000. Molecular analysis of Neanderthal DNA from the northern Caucasus. *Nature* 404:490–493.

DNA arrays

Duggan, D. J., Bittner, J. M., Chen, Y., Meltzer, P., and Trent, J. M., 1999. Expression profiling using cDNA microarrays. *Nat. Genet.* 21:10–14.
Golub, T. R., Slonim, D. K., Tamayo, P., Huard, C., Gaasenbeek, M., Mesirov, J. P., Coller, H., Loh, M. L., Downing, J. R., Caligiuri, M. A., Bloomfield, C. D., and Lander, E. S., 1999. Molecular classification of cancer: Class discovery and class prediction by gene expression monitoring. *Science* 286:531–537.
Perou, C. M., Sørlie, T., Eisen, M. B., van de Rijn, M., Jeffery, S. S., Rees, C. A., Pollack, J. R., Ross, D. T., Johnsen, H., Akslen, L. A., Fluge, Ø., Pergamenschikov, A., Williams, C., Zhu, S. X., Lønning, P. E., Børresen-Dale, A.-L., Brown, P. O., and Botstein, D., 2000. Molecular portraits of human breast tumours. *Nature* 406:747–752.

Introduction of genes into animal cells

Anderson, W. F., 1992. Human gene therapy. *Science* 256:808–813.
Friedmann, T., 1997. Overcoming the obstacles to gene therapy. *Sci. Am.* 277(6):96-101.

Blaese, R. M., 1997. Gene therapy for cancer. *Sci. Am.* 277 (6):111–115.

Brinster, R. L., and Palmiter, R. D., 1986. Introduction of genes into the germ lines of animals. *Harvey Lect.* 80:1–38.

Capecchi, M. R., 1989. Altering the genome by homologous recombination. *Science* 244:1288–1292.

Hasty, P., Bradley, A., Morris, J. H., Edmondson, D. G., Venuti, J. M., Olson, E. N., and Klein, W. H., 1993. Muscle deficiency and neonatal death in mice with a targeted mutation in the myogenin gene. *Nature* 364:501–506

Parkmann, R., Weinberg, K., Crooks, G., Nolta, J., Kapoor, N., and Kohn, D., 2000. Gene therapy for adenosine deaminase deficiency. *Annu. Rev. Med.* 51:33–47.

Genetic engineering of plants

Gasser, C. S., and Fraley, R. T., 1992. Transgenic crops. *Sci. Am.* 266(6):62–69.

Gasser, C. S., and Fraley, R. T., 1989. Genetically engineering plants for crop improvement. *Science* 244:1293–1299.

Shimamoto, K., Terada, R., Izawa, T., and Fujimoto, H., 1989. Fertile transgenic rice plants regenerated from transformed protoplasts. *Nature* 338:274–276.

Chilton, M.-D., 1983. A vector for introducing new genes into plants. *Sci. Am.* 248(6):50.

Hansen, G., Wright, M. S., 1999. Recent advances in the transformation of plants. *Trends Plant Sci.* 4:226–231.

Hammond, J., 1999. Overview: The many uses of transgenic plants. *Curr. Top. Microbiol. Immunol.* 240:1–20.

Finer, J. J., Finer, K. R., and Ponappa, T., 1999. Particle bombardment mediated transformation. *Curr. Top. Microbiol. Immunol.* 240:60–80.

PROBLEMS

1. *Reading sequences.* An autoradiogram of a sequencing gel containing four lanes of DNA fragments is shown in the adjoining illustration. (a) What is the sequence of the DNA fragment? (b) Suppose that the Sanger dideoxy method shows that the template strand sequence is 5′-TGCAATGGC-3′. Sketch the gel pattern that would lead to this conclusion.

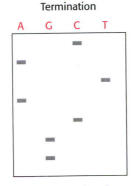

Termination
A   G   C   T

2. *The right template.* Ovalbumin is the major protein of egg white. The chicken ovalbumin gene contains eight exons separated by seven introns. Should one use ovalbumin cDNA or ovalbumin genomic DNA to form the protein in *E. coli?* Why?

3. *Cleavage frequency.* The restriction enzyme *Alu*I cleaves at the sequence 5′-AGCT-3′, and *Not*I cleaves at 5′-GCGGC-CGC-3′. What would be the average distance between cleavage sites for each enzyme on digestion of double-stranded DNA?

4. *The right cuts.* Suppose that a human genomic library is prepared by exhaustive digestion of human DNA with the *Eco*RI restriction enzyme. Fragments averaging about 4 kb in length would be generated.

(a) Is this procedure suitable for cloning large genes? Why?
(b) Is this procedure suitable for mapping extensive stretches of the genome by chromosome walking? Why?

5. *A revealing cleavage.* Sickle-cell anemia arises from a mutation in the gene for the β chain of human hemoglobin. The change from GAG to GTG in the mutant eliminates a cleavage site for the restriction enzyme *Mst*II, which recognizes the target sequence CCTGAGG. These findings form the basis of a diagnostic test for the sickle-cell gene. Propose a rapid procedure for distinguishing between the normal and the mutant gene. Would a positive result prove that the mutant contains GTG in place of GAG?

6. *Many melodies from one cassette.* Suppose that you have isolated an enzyme that digests paper pulp and have obtained its cDNA. The goal is to produce a mutant that is effective at high temperature. You have engineered a pair of unique restriction sites in the cDNA that flank a 30-bp coding region. Propose a rapid technique for generating many different mutations in this region.

7. *A blessing and a curse.* The power of PCR can also create problems. Suppose someone claims to have isolated dinosaur DNA by using PCR. What questions might you ask to determine if it is indeed dinosaur DNA?

8. *Questions of accuracy.* The stringency (Section 6.1.5) of PCR amplification can be controlled by altering the temperature at which the hybridization of the primers to the target DNA occurs. How would altering the temperature of hybridization effect the amplification? Suppose that you have a particular yeast gene *A*, and you wish to see if it has a counterpart in humans. How would controlling the stringency of the hybridization help you?

9. *Terra incognita.* PCR is typically used to amplify DNA that lies between two known sequences. Suppose that you want to explore DNA on both sides of a single known sequence. Devise a variation of the usual PCR protocol that would enable you to amplify entirely new genomic terrain.

10. *A puzzling ladder.* A gel pattern displaying PCR products shows four strong bands. The four pieces of DNA have lengths that are approximately in the ratio of 1:2:3:4. The largest band is cut out of the gel, and PCR is repeated with the same primers. Again, a ladder of four bands is evident in the gel. What does this result reveal about the structure of the encoded protein?

11. *Landmarks in the genome.* Many laboratories throughout the world are mapping the human genome. It is essential that the results be merged at an early stage to provide a working physical map of each chromosome. In particular, it is necessary to know whether a YAC studied in one laboratory overlaps a YAC studied in another when only a small proportion of each (less than 5%) has been sequenced. Propose a simple test for overlap based on the *transfer of information but not of materials* between the two laboratories.

Chapter Integration Problem

12. *Designing primers.* A successful PCR experiment often depends on designing the correct primers. In particular, the T_m for each primer should be approximately the same. What is the basis of this requirement?

Chapter Integration and Data Analysis Problem

13. *Any direction but east.* A series of people are found to have difficulty eliminating certain types of drugs from their bloodstreams. The problem has been linked to a gene *X,* which encodes an enzyme Y. Six people were tested with the use of various techniques of molecular biology. Person A is a normal control, person B is asymptomatic but some of his children have the metabolic problem, and persons C through F display the trait. Tissue samples from each person were obtained. Southern analysis was performed on the DNA after digestion with the restriction enzyme *Hind*III. Northern analysis of mRNA also was done. In both types of analysis, the gels were probed with labeled *X* cDNA. Finally, a Western blot with an enzyme-linked monoclonal antibody was used to test for the presence of protein Y. The results are shown here. Why is person B without symptoms? Suggest possible defects in the other people.

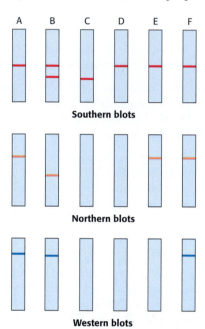

Data Interpretation Problem

14. *DNA diagnostics.* Representations of sequencing gels for variants of the α chain of human hemoglobin are shown here. What is the nature of the amino acid change in each of the variants? The first triplet encodes valine.

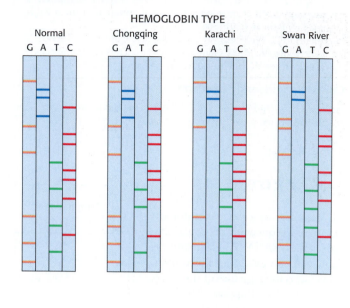

ANIMATED TECHNIQUES

Visit www.whfreeman.com/biochem5 to see animations of Dideoxy Sequencing of DNA, Polymerase Chain Reaction, Synthesizing an Oligonucleotide Array, Screening an Oligonucleotide Array for Patterns of Gene Expression, Plasmid Cloning, In Vitro Mutagenesis of Cloned Genes. Creating a Transgenic Mouse. [Courtesy Lodish et al., 2000, *Molecular Cell Biology,* 4th ed., W. H. Freeman and Company.]

CHAPTER 7

Exploring Evolution

Evolutionary relationships are manifest in protein sequences. The close kinship between human beings and chimpanzees, hinted at by the mutual interest shown by Jane Goodall and a chimpanzee in the photograph, is revealed in the amino acid sequences of myoglobin. The human sequence (red) differs from the chimpanzee sequence (blue) in only one amino acid in a protein chain of 153 residues. [(Left) Kennan Ward/Corbis.]

GLSDGEWQLVLNVWGKVEADIPGHGQEVLIRLFKGHPETLEKFDKFKHLKSEDEMKASEDLKKHGATVLTALGGIL–
GLSDGEWQLVLNVWGKVEADIPGHGQEVLIRLFKGHPETLEKFDKFKHLKSEDEMKASEDLKKHGATVLTALGGIL–

KKKGHHEAEIKPLAQSHATKHKIPVKYLEFISECIIQVLHSKHPGDFGADAQGAMNKALELFRKDMASNYKELGFQG
KKKGHHEAEIKPLAQSHATKHKIPVKYLEFISECIIQVLQSKHPGDFGADAQGAMNKALELFRKDMASNYKELGFQG

Like members of a human family, members of molecular families often have features in common. Such family resemblance is most easily detected by comparing three-dimensional structure, the aspect of a molecule most closely linked to function. Consider, for example, ribonuclease from cows, which was introduced in our consideration of protein folding (Section 3.6). Comparing structures reveals that the three-dimensional structure of this protein and that of a human ribonuclease are quite similar (Figure 7.1). Although this similarity is not unexpected, given the similarity in biological function, similarities revealed by comparisons are sometimes surprising. For example, angiogenin, a protein identified on the basis of its ability to stimulate the growth of new blood vessels, also turns out to be structurally similar to ribonuclease—so similar that it is clear that both angiogenin and ribonuclease are members of the same protein family (Figure 7.2). Angiogenin and ribonuclease must have had a common ancestor at some earlier stage of evolution.

Unfortunately, three-dimensional structures have been determined for only a relatively small number of proteins. In contrast, gene sequences and the corresponding amino acid sequences are available for a great number of proteins, largely owing to the tremendous power of DNA cloning and sequencing techniques. Evolutionary relationships also are manifest in amino acid sequences. For example, comparison of the amino acid sequences of bovine ribonuclease and angiogenin reveals that 35% of amino acids in corresponding

OUTLINE

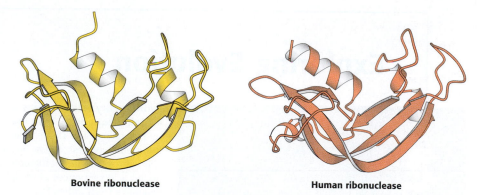

Bovine ribonuclease   **Human ribonuclease**

FIGURE 7.1 Structures of ribonucleases from cows and human beings. Structural similarity often follows functional similarity.

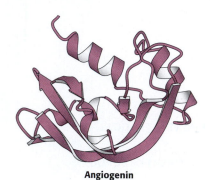

Angiogenin

FIGURE 7.2 Structure of angiogenin. The protein angiogenin, identified on the basis of its ability to stimulate blood-vessel growth, is highly similar to ribonuclease.

positions are identical. Is this level sufficiently high to ensure an evolutionary relationship? If not, what level is required? In this chapter, we shall examine the methods that are used to compare amino acid sequences and to deduce such evolutionary relationships.

Sequence-comparison methods have become a powerful tool in modern biochemistry. Sequence databases can be probed for matches to a newly elucidated sequence in order to identify related molecules. This information can often be a source of considerable insight into the function and mechanism of the newly sequenced molecule. When three-dimensional structures are available, they may be compared to confirm relationships suggested by sequence comparisons and to reveal others that are not readily detected at the level of sequence alone.

By examining the footprints present in modern protein sequences, the biochemist can become a molecular archeologist able to learn about events in the evolutionary past. Sequences comparisons can often reveal both pathways of evolutionary descent and estimated dates of specific evolutionary landmarks. This information can be used to construct evolutionary trees that trace the evolution of a particular protein or nucleic acid in many cases from Archaea and Bacteria through Eukarya, including human beings. Molecular evolution can also be studied experimentally. In some cases, DNA from fossils can be amplified by PCR methods (Section 6.1.5) and sequenced, giving a direct view into the past. In addition, investigators can observe molecular evolution taking place in the laboratory, through experiments based on nucleic acid replication. The results of such studies are revealing more about how evolution proceeds.

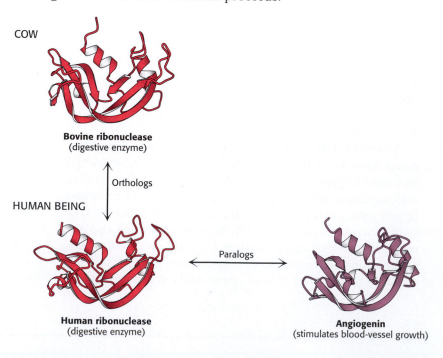

COW

Bovine ribonuclease
(digestive enzyme)

Orthologs

HUMAN BEING

Human ribonuclease
(digestive enzyme)

Paralogs

Angiogenin
(stimulates blood-vessel growth)

FIGURE 7.3 Two classes of homologs. Homologs that perform identical or very similar functions in different organisms are called orthologs, whereas homologs that perform different functions within one organism are called paralogs.

7.1 HOMOLOGS ARE DESCENDED FROM A COMMON ANCESTOR

The exploration of biochemical evolution consists largely of an attempt to determine how proteins, other molecules, and biochemical pathways have been transformed through time. The most fundamental relationship between two entities is *homology;* two molecules are said to be *homologous* if they have been derived from a common ancestor. Homologous molecules, or *homologs,* can be divided into two classes (Figure 7.3). *Paralogs* are homologs that are present within one species. Paralogs often differ in their detailed biochemical functions. *Orthologs* are homologs that are present within different species and have very similar or identical functions. Understanding the homology between molecules can reveal the evolutionary history of the molecules as well as information about their function; if a newly sequenced protein is homologous to an already characterized protein, we have a strong indication of the new protein's biochemical function.

How can we tell whether two human proteins are paralogs or whether a yeast protein is the ortholog of a human protein? As will be discussed in Section 7.2, *homology is often manifested by significant similarity in nucleotide or amino acid sequence and almost always manifested in three-dimensional structure.*

7.2 STATISTICAL ANALYSIS OF SEQUENCE ALIGNMENTS CAN DETECT HOMOLOGY

CONCEPTUAL INSIGHTS, Sequence Analysis, *provides opportunities to interactively explore issues involved in sequence alignment.*

CONCEPTUAL INSIGHTS, appearing throughout the book, are interactive animations that help you build your understanding of key biochemical principles and concepts. To access, go to the Web site: www.whfreeman.com/biochem5, and select the chapter, Conceptual Insights, and the title.

A significant sequence similarity between two molecules implies that they are likely to have the same evolutionary origin and, therefore, the same three-dimensional structure, function, and mechanism. Although both nucleic acid and protein sequences can be compared to detect homology, a comparison of protein sequences is much more effective for several reasons, most notably that proteins are built from 20 different building blocks, whereas RNA and DNA are synthesized from only 4 building blocks.

To illustrate sequence-comparison methods, let us consider a class of proteins called the *globins.* Myoglobin is a protein that binds oxygen in muscle, whereas hemoglobin is the oxygen-carrying protein in blood (Section 10.2). Both proteins cradle a heme group, an iron-containing organic molecule that binds the oxygen. Each human hemoglobin molecule is composed of four heme-containing polypeptide chains, two identical α chains and two identical β chains. Here, we consider only the α chain. We wish to examine the similarity between the amino acid sequence of the human α chain and that of human myoglobin (Figure 7.4). To detect such similarity, methods have been developed for *sequence alignment.*

Human hemoglobin (α chain)

VLSPADKTNVKAAWGKVGAHAGEYGAEALERMFLSFPTTKTYFPHFDLSHG
SAQVKGHGKKVADALTNAVAHVDDMPNALSALSDLHAHKLRVDPVNFKLLS
HCLLVTLAAHLPAEFTPAVHASLDKFLASVSTVLTSKYR

Human myoglobin

GLSDGEWQLVLNWWGKVEADIPGHGQEVLIRLFKGHPETLEKFDKFKHLKS
EDEMKASEDLKKHGATVLTALGGILKKKGHHEAEIKPLAQSHATKHKIPVK
YLEFISECIIQVLQSKHPGDFGADAQGAMNKALELFRKDMASNYKELGFQG

FIGURE 7.4 Amino acid sequences of human hemoglobin (α chain) and human myoglobin. Hemoglobin α is composed of 141 amino acids; myoglobin consists of 153 amino acids. (One-letter abbreviations designating amino acids are used; see Table 3.2.)

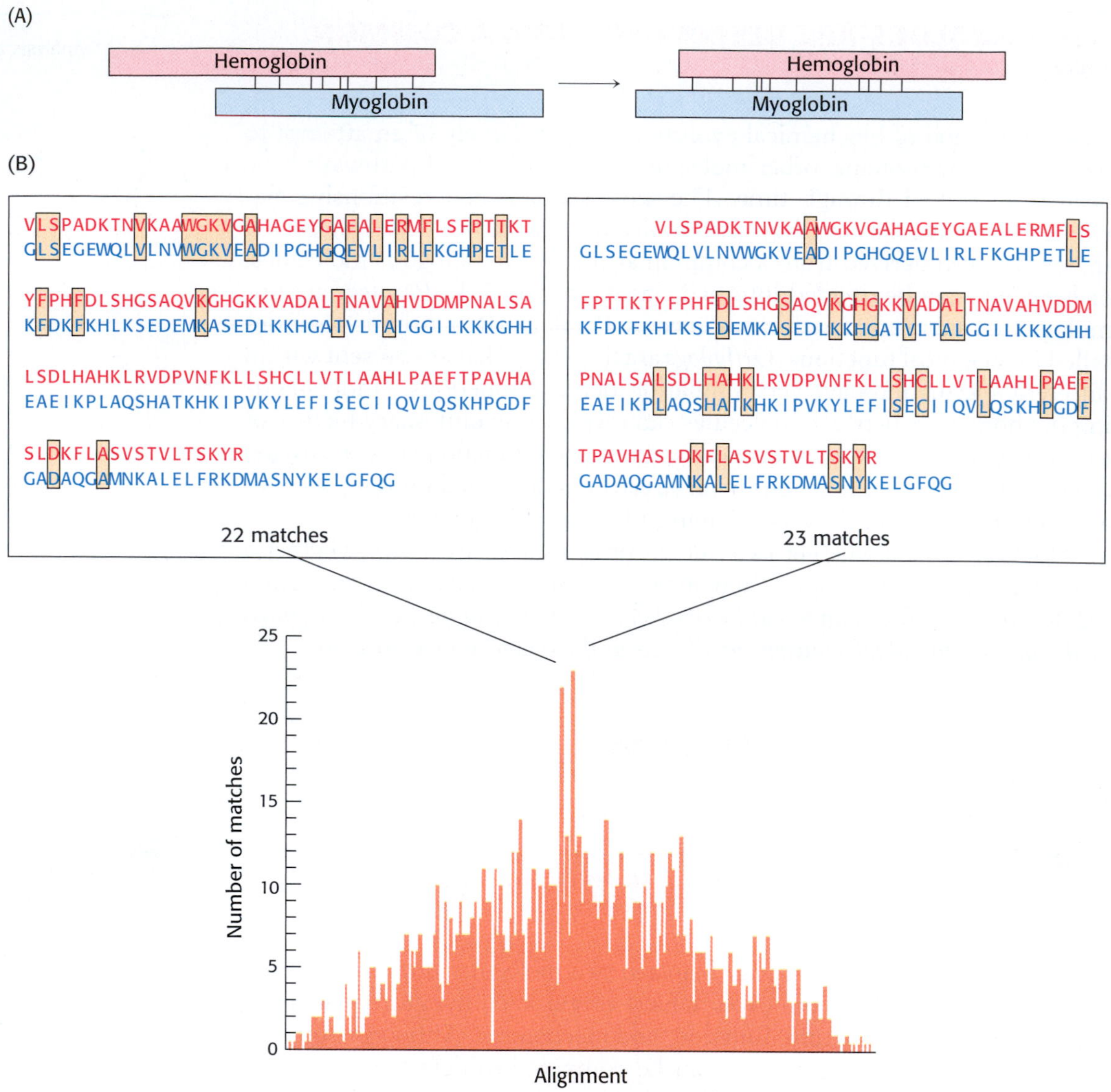

(A)

(B)

22 matches   23 matches

FIGURE 7.5 Comparing the amino acid sequences of hemoglobin α and myoglobin. (A) A comparison is made by sliding the sequences of the two proteins past one another, one amino acid at a time, and counting the number of amino acid identities between the proteins. (B) The two alignments with the largest number of matches are shown above the graph, which plots the matches as a function of alignment.

How can we tell where to align the two sequences? The simplest approach is to compare all possible juxtapositions of one protein sequence with another, in each case recording the number of identical residues that are aligned with one another. This comparison can be accomplished by simply sliding one sequence past the other, one amino acid at a time, and counting the number of matched residues (Figure 7.5).

For hemoglobin α and myoglobin, the best alignment reveals 23 sequence identities, spread throughout the central parts of the sequences. However, a nearby alignment showing 22 identities is nearly as good. In this alignment, the identities are concentrated toward the amino-terminal end of the sequences. The sequences can be aligned to capture most of the identities in *both* alignments by introducing a *gap* into one of the sequences (Figure 7.6). Such gaps must often be inserted to compensate for the insertions or deletions of nucleotides that may have taken place in the gene for one molecule but not the other in the course of evolution.

Hemoglobin α   V L S P A D K T N V K A A W G K V G A H A G E Y G A E A L E R M F L S F P T T K T Y F P H F - - - - - - D
Myoglobin   G L S E G E W Q L V L N V W G K V E A D I P G H G Q E V L I R L F K G H P E T L E K F D K F K H L K S E D

L S H G S A Q V K G H G K K V A D A L T N A V A H V D D M P N A L S A L S D L H A H K L R V D P V N K K L
E M K A S E D L K K H G A T V L T A L G G I L K K K G H H E A E I K P L A Q S H A T K H K I P V K Y L E F

L S H C L L V T L A A H L P A E F T P A V H A S L D K F L A S V S T V L T S K Y R
I S E C I I Q V L Q S K H P G D F G A D A Q G A M N K A L E L F R K D M A S N Y K E L G F Q G

Gap

FIGURE 7.6 Alignment with gap insertion. The alignment of hemoglobin α and myoglobin after a gap has been inserted into the hemoglobin α sequence.

The use of gaps substantially increases the complexity of sequence alignment because, in principle, the insertion of gaps of arbitrary sizes must be considered throughout each sequence. However, methods have been developed for the insertion of gaps in the automatic alignment of sequences. These methods use scoring systems to compare different alignments, and they include penalties for gaps to prevent the insertion of an unreasonable number of them. Here is an example of such a scoring system: each identity between aligned sequences results in +10 points, whereas each gap introduced, regardless of size, results in −25 points. For the alignment shown in Figure 7.6, there are 38 identities and 1 gap, producing a score of (38 × 10 − 1 × 25 = 355). Overall, there are 38 matched amino acids in an average length of 147 residues; so the sequences are 25.9% identical. The next step is to ask, Is this precentage of identity significant?

7.2.1 The Statistical Significance of Alignments Can Be Estimated by Shuffling

The similarities in sequence in Figure 7.5 appear striking, yet there remains the possibility that a grouping of sequence identities has occurred by chance alone. How can we estimate the probability that a specific series of identities is a chance occurrence? To make such an estimate, the amino acid sequence in one of the proteins is "shuffled"—that is, randomly rearranged—and the alignment procedure is repeated (Figure 7.7). This process is repeated to build up a distribution showing, for each possible score, the number of shuffled sequences that received that score.

When this procedure is applied to the sequences of myoglobin and hemoglobin α, the authentic alignment clearly stands out (Figure 7.8). Its score is far above the mean for the alignment scores based on shuffled sequences. The odds of such a deviation occurring owing due to chance alone are

T H I S I S T H E A U T H E N T I C S E Q U E N C E

Shuffling

S N U C S N S E A T E E I T U H E Q I H H T T C E I

FIGURE 7.7 The generation of a shuffled sequence.

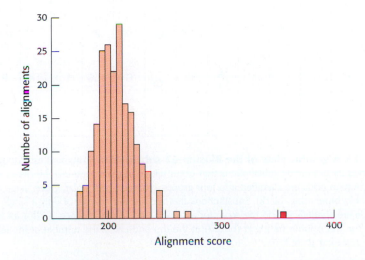

FIGURE 7.8 Statistical comparison of alignment scores. Alignment scores are calculated for many shuffled sequences, and the number of sequences generating a particular score is plotted against the score. The resulting plot is a distribution of alignment scores occurring by chance. The alignment score for hemoglobin α and myoglobin (shown in red) is substantially greater than any of these scores, strongly suggesting that the sequence similarity is significant.

approximately 1 in 10^{20}. Thus, we can comfortably conclude that the two sequences are genuinely similar; the simplest explanation for this similarity is that these sequences are homologous—that is, that the two molecules have descended by divergence from a common ancestor.

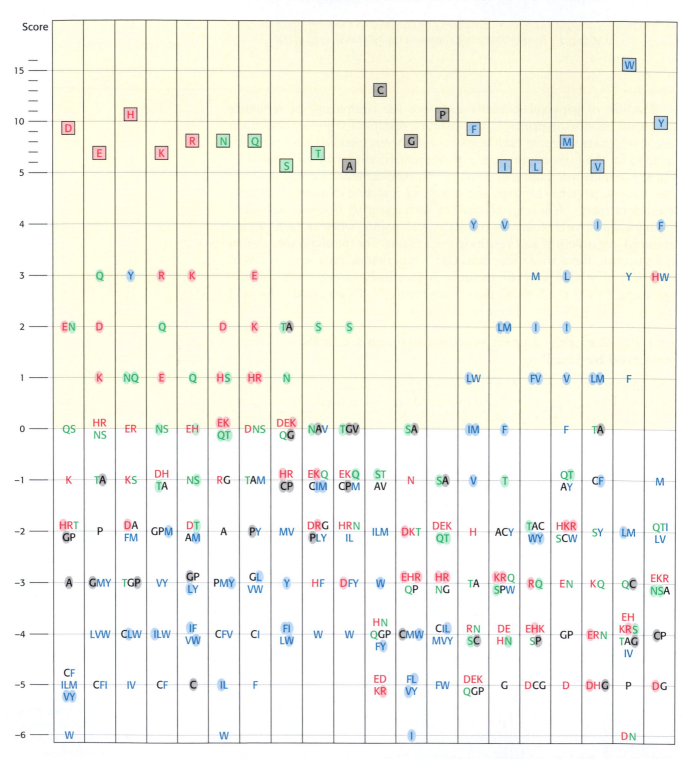

FIGURE 7.9 A graphic view of the Blosum-62 substitution matrix. This scoring scheme was derived by examining substitutions that occur within aligned sequence blocks in related proteins. Amino acids are classified into four groups (charged, red; polar, green; large and hydrophobic, blue; other, black). Substitutions that require the change of only a single nucleotide are shaded. To find the score for a substitution of, for instance, a Y for an H, you find the Y in the column having H (boxed) at the top and check the number at the left. In this case, the resulting score is 3.

7.2.2 Distant Evolutionary Relationships Can Be Detected Through the Use of Substitution Matrices

The scoring scheme in Section 7.2.1 assigns points only to positions occupied by identical amino acids in the two sequences being compared. No credit is given for any pairing that is not an identity. However, not all substitutions are equivalent. Some are structurally *conservative substitutions*, replacing one amino acid with another that is similar in size and chemical properties. Such conservative amino acid substitutions may have relatively minor effects on protein structure and can thus be tolerated without compromising function. In other substitutions, an amino acid replaces one that is dissimilar. Furthermore, some amino acid substitutions result from the replacement of only a single nucleotide in the gene sequence; whereas others require two or three replacements. Conservative and single-nucleotide substitutions are likely to be more common than are substitutions with more radical effects. How can we account for the type of substitution when comparing sequences? We can approach this problem by first examining the substitutions that have actually taken place in evolutionarily related proteins.

From the examination of appropriately aligned sequences, *substitution matrices* can be deduced. In these matrices, a large positive score corresponds to a substitution that occurs relatively frequently, whereas a large negative score corresponds to a substitution that occurs only rarely. The Blosum-62 substitution matrix illustrated in Figure 7.9 is an example. The highest scores in this substitution matrix indicate that amino acids such as cysteine (C) and tryptophan (W) tend to be conserved more than those such as serine (S) and alanine (A). Furthermore, structurally conservative substitutions such as lysine (K) for arginine (R) and isoleucine (I) for valine (V) have relatively high scores. When two sequences are compared, each substitution is assigned a score based on the matrix. In addition, a gap penalty is often assigned according to the size of the gap. For example, the introduction of a gap lowers the alignment score by 12 points and the extension of an existing gap costs 2 points per residue. Using this scoring system, the alignment shown in Figure 7.6 receives a score of 115. In many regions, most substitutions are conservative (defined as those substitutions with scores greater than 0) and relatively few are strongly disfavored types (Figure 7.10).

This scoring system detects homology between less obviously related sequences with greater sensitivity than would a comparison of identities only. Consider, for example, the protein leghemoglobin, an oxygen-binding protein found in the roots of some plants. The amino acid sequence of leghemoglobin from the herb lupine can be aligned with that of human myoglobin and scored by using either the simple scoring scheme based on identities only or the Blosum-62 scoring matrix (see Figure 7.9). Repeated

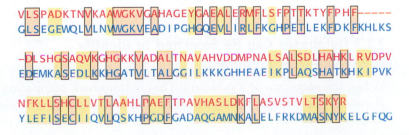

Hemoglobin α
Myoglobin

FIGURE 7.10 Alignment with conservative substitutions noted. The alignment of hemoglobin α and myoglobin with conservative substitutions indicated by yellow shading and identities by orange.

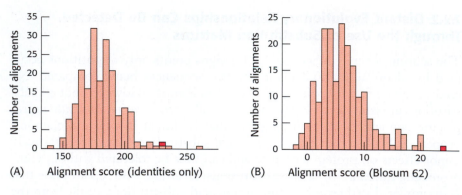

FIGURE 7.11 Alignment of identities only versus the Blosum 62 matrix. Repeated shuffling and scoring reveal the significance of sequence alignment for human myoglobin versus lupine leghemoglobin with the use of either (A) the simple, identity-based scoring system or (B) the Blosum-62 matrix. The scores for the alignment of the authentic sequences are shown in red. The Blosum matrix provides greater statistical power.

shuffling and scoring provides a distribution of alignment scores (Figure 7.11). Scoring based on identities only indicates that the odds of the alignment between myoglobin and leghemoglobin occurring by chance alone are 1 in 20. Thus, although the level of similarity suggests a relationship, there is a 5% chance that the similarity is accidental on the basis of this analysis. In contrast, users of the substitution matrix are able to incorporate the effects of conservative substitutions. From such an analysis, the odds of the alignment occurring by chance are calculated to be approximately 1 in 300. Thus, an analysis performed by using the substitution matrix reaches a much firmer conclusion about the evolutionary relationship between these proteins (Figure 7.12).

Myoglobin
Leghemoglobin

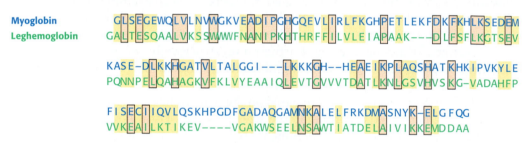

FIGURE 7.12 Alignment of human myoglobin and lupine leghemoglobin. The use of the Blosum-62 substitution matrix yields the alignment shown between human myoglobin and lupine leghemoglobin, illustrating identities (orange) and conservative substitutions (yellow). These sequences are 23% identical.

Experience with sequence analysis has led to the development of simpler rules of thumb. For sequences longer than 100 amino acids, sequence identities greater than 25% are almost certainly not the result of chance alone; such sequences are probably homologous. In contrast, if two sequences are less than 15% identical, pairwise comparison alone is unlikely to indicate statistically significant similarity. For sequences that are between 15% and 25% identical, further analysis is necessary to determine the statistical significance of the alignment. It must be emphasized that *the lack of a statistically significant degree of sequence similarity does not rule out homology.* The sequences of many proteins that have descended from common ancestors have diverged to such an extent that the relationship between the proteins can no longer be detected from their sequences alone. As we will see, such homologous proteins can often be detected by examining three-dimensional structures.

7.2.3 Databases Can Be Searched to Identify Homologous Sequences

When the sequence of a protein is first determined, comparing it with all previously characterized sequences can be a source of tremendous insight into its evolutionary relatives and, hence, its structure and function. *Indeed, an extensive sequence comparison is almost always the first analysis performed on a newly elucidated sequence.* The sequence alignment methods heretofore described are used to compare an individual sequence with all members of a database of known sequences.

In 1995, investigators reported the first complete sequence of the genome of a free-living organism, the bacterium *Haemophilus influenzae*. Of 1743 identified open reading frames (Section 6.3.2), 1007 (58%) could be linked by sequence-comparison methods to some protein of known function that had been previously characterized in another organism. An additional 347 open reading frames could be linked to sequences in the database for which no function had yet been assigned ("hypothetical proteins"). The remaining 389 sequences did not match any sequence present in the database at the time at which the *Haemophilus influenzae* sequence was completed. Thus, investigators were able to identify likely functions for more than half the proteins within this organism solely through the use of sequence-comparison methods.

7.3 EXAMINATION OF THREE-DIMENSIONAL STRUCTURE ENHANCES OUR UNDERSTANDING OF EVOLUTIONARY RELATIONSHIPS

Sequence comparison is a powerful tool for extending our knowledge of protein function and kinship. However, biomolecules generally function as intricate three-dimensional structures rather than as linear polymers. Mutations occur at the level of sequence, but the effects of the mutations are at the level of function, and function is directly related to tertiary structure. Consequently, to gain a deeper understanding of evolutionary relationships between proteins, we must examine three-dimensional structures, especially in conjunction with sequence information. The techniques of structural determination are presented in Chapter 4.

7.3.1 Tertiary Structure Is More Conserved Than Primary Structure

Because three-dimensional structure is much more closely associated with function than is sequence, tertiary structure is more evolutionarily conserved than is primary structure. This conservation is apparent in the tertiary structures of the globins (Figure 7.13), which are extremely similar even though the similarity between human myoglobin and lupine leghemoglobin is just barely detectable at the sequence level and that between human hemoglobin (α chain) and lupine leghemoglobin is not statistically

FIGURE 7.13 Conservation of three-dimensional structure. The tertiary structures of human hemoglobin (α chain), human myoglobin, and lupine leghemoglobin are conserved. Each heme group contains an iron atom to which oxygen binds.

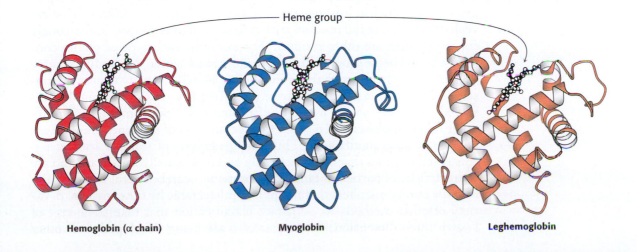

Heme group

Hemoglobin (α chain)   Myoglobin   Leghemoglobin

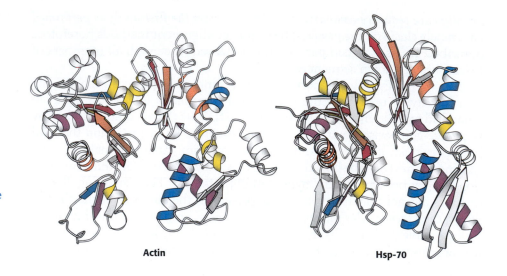

FIGURE 7.14 Structures of actin and the large fragment of heat shock protein 70 (Hsp-70). A comparison of the identically colored elements of secondary structure reveals the overall similarity in structure despite the difference in biochemical activities.

significant (15.6% identity). This structural similarity firmly establishes that the framework that binds the heme group and facilitates the reversible binding of oxygen has been conserved over a long evolutionary period.

Anyone aware of the similar biochemical functions of hemoglobin, myoglobin, and leghemoglobin could expect the structural similarities. In a growing number of other cases, however, a comparison of three-dimensional structures has revealed striking similarities between proteins that were *not* expected to be related. A case in point is the protein actin, a major component of the cytoskeleton, and heat shock protein 70 (Hsp-70), which assists protein folding inside cells. These two proteins were found to be noticeably similar in structure despite only 15.6% sequence identity (Figure 7.14). On the basis of their three-dimensional structures, actin and Hsp-70 are paralogs. The level of structural similarity strongly suggests that, despite their different biological roles in modern organisms, these proteins descended from a common ancestor. As the three-dimensional structures of more proteins are determined, such unexpected kinships are being discovered with increasing frequency. The search for such kinships relies ever more frequently on computer-based search procedures that allow the three-dimensional structure of any protein to be compared with all other known structures.

7.3.2 Knowledge of Three-Dimensional Structures Can Aid in the Evaluation of Sequence Alignments

The sequence-comparison methods described thus far treat all positions within a sequence equally. However, examination of families of homologous proteins for which at least one three-dimensional structure is known has revealed that regions and residues critical to protein function are more strongly conserved than are other residues. For example, each type of globin contains a bound heme group with an iron atom at its center. A histidine residue that interacts directly with this iron (residue 64 in human myoglobin) is conserved in all globins. After we have identified key residues or highly conserved sequences within a family of proteins, we can sometimes identify other family members even when the overall level of sequence similarity is below statistical significance. Thus, the generation of *sequence templates*—conserved residues that are structurally and functionally important and are characteristic of particular families of proteins—can be useful for recognizing new family members that might be undetectable by other means. A variety of other methods for sequence classification that take advantage of known three-dimensional structures also are being developed. Still other methods are able to identify relatively conserved residues within a family of homologous proteins, even without a known three-dimensional structure. These methods are proving to be powerful in identifying distant evolutionary relationships.

7.3.3 Repeated Motifs Can Be Detected by Aligning Sequences with Themselves

More than 10% of all proteins contain sets of two or more domains that are similar to one another. The afore-described sequence search methods can often detect internally repeated sequences that have been characterized in other proteins. Where repeated units do not correspond to previously identified domains, their presence can be detected by attempting to align a given sequence with itself. This alignment is most easily visualized with the use of a *self-diagonal plot*. Here, the protein sequence is displayed on both the vertical and the horizontal axes, running from amino to carboxyl terminus; a dot is placed at each point in the space defined by the axes at which the amino acid directly below along the horizontal axis is the same as that directly across along the vertical axis. The central diagonal represents the sequence aligned with itself. Internal repeats are manifested as lines of dots parallel to the central diagonal, illustrated by the plot in Figure 7.15 prepared for the TATA-box-binding protein, a key protein in the initiation of gene transcription (Section 28.2.3).

The statistical significance of such repeats can be tested by aligning the regions in question as if these regions were sequences from separate proteins. For the TATA-box-binding protein, the alignment is highly significant: 30% of the amino acids are identical over 90 residues (Figure 7.16A). The estimated probability of such an alignment occurring by chance is 1 in 10^{13}. The determination of the three-dimensional structure of the TATA-box-binding protein confirmed the presence of repeated structures; the protein is formed of two nearly identical domains (Figure 7.16B). The evidence is convincing that the gene encoding this protein evolved by duplication of a gene encoding a single domain.

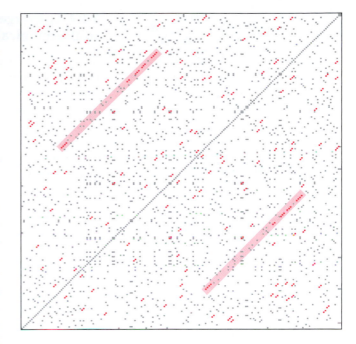

FIGURE 7.15 A self-diagonal plot for the TATA-box-binding protein from the plant *Arabidopsis*. Self-diagonal plots are used to search for amino acid sequence repeats within a protein. The central diagonal is the sequence aligned with itself. Red dots indicating a correspondence of amino acids appear where two or more amino acids in a row match. Lines of dots, highlighted in pink, parallel to the central diagonal suggest an internal repeat.

(A)

1 MTDQGLEGSNPVDLSKHPS

20   GIVPTLQNIVSTVNLDCKLDLKAIALQ–ARNAEYNPKRFAAVIMRIR
110 FKDFKIQNIVGSCDVKFPIRLEGLAYSHAAFSSYEPELFPGLIYRMK

66   EPKTTALIFASGKMVCTGAKSEDFSKMAARKYARIVQKLGFPAK
157 VPKIVLLIFVSGKIVITGAKMRDETYKAFENIYPVLSEFRKIQQ

(B)
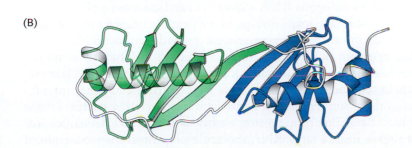

FIGURE 7.16 Sequence alignment of internal repeats. (A) An alignment of the sequences of the two repeats of the TATA-box-binding protein. The amino-terminal repeat is shown in green and the carboxyl-terminal repeat in blue. (B) Structure of the TATA-box-binding protein. The amino-terminal domain is shown in green and the carboxyl-terminal domain in blue.

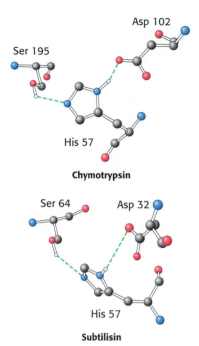

FIGURE 7.17 Convergent evolution of protease active sites. The relative positions of the three key residues shown are nearly identical in the active sites of the serine proteases chymotrypsin and subtilisin.

7.3.4 Convergent Evolution: Common Solutions to Biochemical Challenges

Thus far, we have been exploring proteins derived from common ancestors—that is, through *divergent evolution*. In other cases, clear examples have been found of proteins that are structurally similar in important ways but are not descended from a common ancestor. How might two unrelated proteins come to resemble each other structurally? Two proteins evolving independently may have converged on a similar structure in order to perform a similar biochemical activity. Perhaps that structure was an especially effective solution to a biochemical problem that organisms face. The process by which very different evolutionary pathways lead to the same solution is called *convergent evolution*.

One example of convergent evolution is found among the serine proteases. These enzymes, to be discussed in more detail in Chapter 9, cleave peptide bonds by hydrolysis. Figure 7.17 shows for two such enzymes the structure of the active sites—that is, the sites on the proteins at which the hydrolysis reaction takes place. These active-site structures are remarkably similar. In each case, a serine residue, a histidine residue, and an aspartic acid residue are positioned in space in nearly identical arrangements. As we will see, this is the case because chymotrypsin and subtilisin use the same mechanistic solution to the problem of peptide hydrolysis. At first glance, this similarity might suggest that these proteins are homologous. However, striking differences in the overall structures of these proteins make an evolutionary relationship extremely unlikely (Figure 7.18). Whereas chymotrypsin consists almost entirely of β sheets, subtilisin contains extensive α-helical structure. Moreover, the key serine, histidine, and aspartic acid residues do not occupy similar positions or even appear in the same order within the two sequences. It is extremely unlikely that two proteins evolving from a common ancestor could have retained similar active-site structures while other aspects of the structure changed so dramatically.

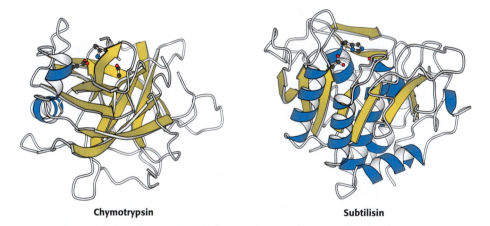

Chymotrypsin

Subtilisin

FIGURE 7.18 Structures of chymotrypsin and subtilisin. The β strands are shown in yellow and α helices in blue. The overall structures are quite dissimilar, in stark contrast with the active sites, shown at the top of each structure.

7.3.5 Comparison of RNA Sequences Can Be a Source of Insight into Secondary Structures

A comparison of homologous RNA sequences can be a source of important insights into evolutionary relationships in a manner similar to that already described. In addition, such comparisons provide clues to the three-dimensional structure of the RNA itself. As noted in Chapter 5, single-stranded nucleic acid molecules fold back on themselves to form elaborate structures held together by Watson-Crick base-pairing and other interactions. In a family of sequences that form such base-paired structures, base sequences may vary, but base-pairing ability is conserved. Consider, for example, a region from a large RNA molecule present in the ribosomes of all organisms (Figures 7.19). In the region shown, the *E. coli* sequence has a guanine (G) residue in position 9 and a cytosine (C) residue in position 22, whereas the human sequence has uracil (U) in position 9 and adenine (A) in position 22. Examination of the six sequences shown in Figure 7.20 (and many others) reveals that the bases in positions 9 and 22 retain the ability to form a Watson-Crick base pair even though the identities of the bases in these positions vary. Base-pairing ability is also conserved in neighboring positions; we can deduce that two segments with such compensating mutations are likely to form a double helix. Where sequences are known for several homologous RNA molecules, this type of sequence analysis can often suggest complete secondary structures as well as some additional interactions.

(A)

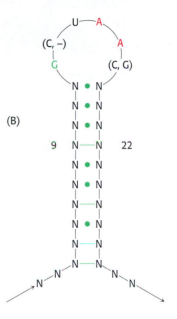
(B)

		↓9	↓22
BACTERIA	*Escherichia coli*	C A C A C G G C G G G U G C U A A C G U C C G U C G U G A A	
	Pseudomonas aeruginosa	A C C A C G G C G G G U G C U A A C G U C C G U C G U G A A	
ARCHAEA	*Halobacterium halobium*	C C G G U G U G C G G G G – U A A G C C U G U G C A C C G U	
	Methanococcus vannielli	G A G G G C A U A C G G G – U A A G C U G U A U G U C C G A	
EUKARYA	*Homo sapiens*	G G G C C A C U U U U G G – U A A G C A G A A C U G G C G C	
	Saccharomyces cerevisiae	G G G C C A U U U U U G G – U A A G C A G A A C U G G C G A	

FIGURE 7.19 Comparison of RNA sequences. (A) A comparison of sequences in a part of ribosomal RNA taken from a variety of species. (B) The implied secondary structure. Bars indicate positions at which Watson-Crick base-pairing is completely conserved in the sequences shown, whereas dots indicate positions at which Watson-Crick base-pairing is conserved in most cases.

7.4 EVOLUTIONARY TREES CAN BE CONSTRUCTED ON THE BASIS OF SEQUENCE INFORMATION

CONCEPTUAL INSIGHTS, Sequence Analysis, *offers insights into evolutionary trees through interactive analysis of simulated evolutionary histories.*

The observation that homology is often manifested as sequence similarity suggests that the evolutionary pathway relating the members of a family of proteins may be deduced by examination of sequence similarity. This approach is based on the notion that sequences that are more similar to one another have had less evolutionary time to diverge from one another than have sequences that are less similar. This method can be illustrated by using the three globin sequences in Figures 7.10 and 7.12, as well as the sequence for the human hemoglobin β chain. These sequences can be aligned with the additional constraint that gaps, if present, should be at the same positions in all of the proteins. These aligned sequences can be used to construct an *evolutionary tree* in which the length of the branch connecting each pair of proteins is proportional to the number of amino acid differences between the sequences (Figure 7.20).

Such comparisons reveal only the relative divergence times—for example, that myoglobin diverged from hemoglobin twice as long ago as the α chain diverged from the β chain. How can we estimate the approximate dates of gene duplications and other evolutionary events? Evolutionary

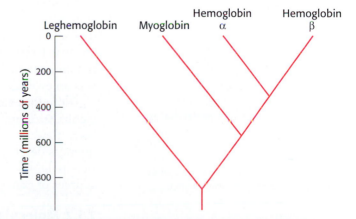

FIGURE 7.20 An evolutionary tree for globins. The branching structure was deduced by sequence comparison, whereas the results of fossil studies provided the overall time scale showing when divergence occurred.

FIGURE 7.21 The lamprey. A jawless fish whose ancestors diverged from bony fish approximately 400 million years ago, the lamprey contains hemoglobin molecules that contain only a single type of polypeptide chain. [Brent P. Kent.]

trees can be calibrated by comparing the deduced branch points with divergence times determined from the fossil record. For example, the duplication leading to the two chains of hemoglobin appears to have occurred 350 million years ago. This estimate is supported by the observation that jawless fish such as the lamprey, which diverged from bony fish approximately 400 million years ago, contain hemoglobins built from a single type of subunit (Figure 7.21).

These methods can be applied to both relatively modern and very ancient molecules, such as the ribosomal RNAs that are found in all organisms. Indeed, it was such an RNA sequence analysis that led to the suggestion that Archaea are a distinct group of organisms that diverged from Bacteria very early in evolutionary history.

7.5 MODERN TECHNIQUES MAKE THE EXPERIMENTAL EXPLORATION OF EVOLUTION POSSIBLE

Two techniques of biochemistry have made it possible to examine the course of evolution more directly and not simply by inference. The polymerase chain reaction (Section 6.1.5) allows the direct examination of ancient DNA sequences, releasing us, at least in some cases, from the constraints of being able to examine existing genomes from living organisms only. Molecular evolution may be investigated through the use of *combinatorial chemistry*, the process of producing large populations of molecules en masse and selecting for a biochemical property. This exciting process provides a glimpse into the types of molecules that may have existed in the RNA world.

7.5.1 Ancient DNA Can Sometimes Be Amplified and Sequenced

The tremendous chemical stability of DNA (Section 2.2.7) makes the molecule well suited to its role as the storage site of genetic information. So stable is the molecule that samples of DNA have survived for many thousands of years under appropriate conditions. With the development of PCR methods, such ancient DNA can sometimes be amplified and sequenced. This approach has been applied to mitochondrial DNA from a Neanderthal fossil estimated at between 30,000 and 100,000 years of age found near Düsseldorf, Germany, in 1856. Investigators managed to identify a total of 379 bases of sequence. Comparison with a number of the corresponding sequences from *Homo sapiens* revealed between 22 and 36 substitutions, considerably fewer than the average of 55 differences between human beings and chimpanzees over the common bases in this region. Further analysis suggested that the common ancestor of modern human beings and Neanderthals lived approximately 600 thousand years ago. An evolutionary tree constructed by using these and other data revealed that the Neanderthal was not an intermediate between chimpanzees and human beings but, instead, was an evolutionary "dead end" that became extinct (Figure 7.22).

Note that earlier studies describing the sequencing of much more ancient DNA such as that found in insects trapped in amber appear to have been flawed; contaminating modern DNA was responsible for the sequences determined. Successful sequencing of ancient DNA requires sufficient DNA for reliable amplification and the rigorous exclusion of all sources of contamination.

Homo sapiens   Chimpanzee
Neanderthal

FIGURE 7.22 Placing Neanderthal on an evolutionary tree. Comparison of DNA sequences revealed that Neanderthal is not on the line of direct descent leading to *Homo sapiens* but, instead, branched off earlier and then became extinct.

7.5.2 Molecular Evolution Can Be Examined Experimentally

Evolution requires three processes: (1) the generation of a diverse population, (2) the selection of members based on some criterion of fitness, and (3) reproduction to enrich the population in more fit members (Section 2.2).

Nucleic acid molecules are capable of undergoing all three processes in vitro under appropriate conditions. The results of such studies enable us to glimpse how evolutionary processes might have generated catalytic activities and specific binding abilities—important biochemical functions in all living systems.

A diverse population of nucleic acid molecules can be synthesized in the laboratory by the process of combinatorial chemistry, which rapidly produces large populations of a particular type of molecule such as a nucleic acid. A population of molecules of a given size can be generated randomly so that many or all possible sequences are present in the mixture. When an initial population has been generated, it is subjected to a selection process that isolates specific molecules with desired binding or reactivity properties. Finally, molecules that have survived the selection process are replicated through the use of PCR; primers are directed toward specific sequences included at the ends of each member of the population.

As an example of this approach, consider an experiment that set a goal of creating an RNA molecule capable of binding adenosine triphosphate and related nucleotides. Such ATP-bonding molecules are of interest because they might have been present in the RNA world. An initial population of RNA molecules 169 nucleotides long was created; 120 of the positions differed randomly, with equimolar mixtures of adenine, cytosine, guanine, and uracil. The initial synthetic pool that was used contained approximately 10^{14} RNA molecules. Note that this number is a very small fraction of the total possible pool of random 120-base sequences. From this pool, those molecules that bound to ATP, which had been immobilized on a column, were selected (Figure 7.23).

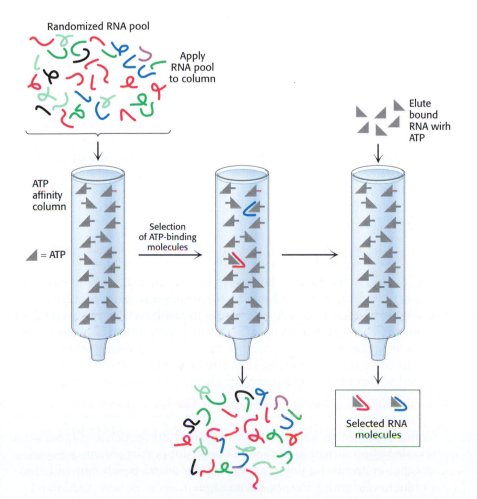

Randomized RNA pool

Apply RNA pool to column

Elute bound RNA wirh ATP

ATP affinity column

▲ = ATP

Selection of ATP-binding molecules

Selected RNA molecules

FIGURE 7.23 Evolution in the laboratory. A collection of RNA molecules of random sequences is synthesized by combinatorial chemistry. This collection is selected for the ability to bind ATP by passing the RNA through an ATP affinity column (Section 4.1.3). The ATP-binding RNA molecules are released from the column by washing with excess ATP, and replicated. The process of selection and replication is then repeated several times. The final RNA products with significant ATP-binding ability are isolated and characterized.

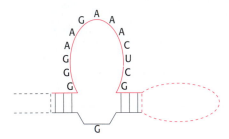

FIGURE 7.24 A conserved secondary structure. The secondary structure shown is common to RNA molecules selected for ATP binding.

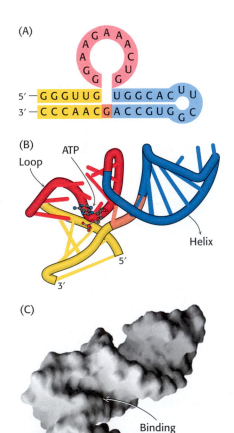

FIGURE 7.25 RNA molecule binds ATP. (A) The Watson-Crick base-pairing pattern, (B) the folding pattern, and (C) a surface representation of an RNA molecule selected to bind adenosine nucleotides. The bound ATP is shown in part B, and the binding site is revealed as a deep pocket in part C.

The collection of molecules that were bound well by the ATP affinity column was allowed to reproduce by reverse transcription into DNA, amplification by PCR, and transcription back into RNA. This new population was subjected to additional rounds of selection for ATP-binding activity. After eight generations, members of the selected population were characterized by sequencing. Seventeen different sequences were obtained, 16 of which could form the structure shown in Figure 7.24. Each of these molecules bound ATP with high affinity, as indicated by dissociation constants less than 50 μM.

The folded structure of the ATP-binding region from one of these RNAs was determined by nuclear magnetic resonance (Section 4.5.1) methods (Figure 7.25). As expected, this 40-nucleotide molecule is composed of two Watson-Crick base-paired helical regions separated by an 11-nucleotide loop. This loop folds back on itself in an intricate way to form a deep pocket into which the adenine ring can fit. Thus, a structure was generated, or evolved, that was capable of a specific interaction.

SUMMARY

Homologs Are Descended from a Common Ancestor

Exploring evolution biochemically often means searching for homology, or relatedness, between molecules, because homologous molecules, or homologs, evolved from a common ancestor. Paralogs are homologous molecules that are found in one species and have acquired different functions through evolutionary time. Orthologs are homologous molecules that are found in different species and have similar or identical functions.

Statistical Analysis of Sequence Alignments Can Detect Homology

Protein and nucleic acid sequences are two of the primary languages of biochemistry. Sequence-alignment methods are the most powerful tools of the evolutionary detective. Sequences can be aligned to maximize their similarity, and the significance of these alignments can be judged by statistical tests. The detection of a statistically significant alignment between two sequences strongly suggests that two sequences are related by divergent evolution from a common ancestor. The use of substitution matrices makes the detection of more distant evolutionary relationships possible. Any sequence can be used to probe sequence databases to identify related sequences present in the same organism or in other organisms.

Examination of Three-Dimensional Structure Enhances Our Understanding of Evolutionary Relationships

The evolutionary kinship between proteins may be even more profoundly evident in the conserved three-dimensional structures. The analysis of three-dimensional structure in combination with analysis of especially conserved sequences has made it possible to determine evolutionary relationships that are not possible to detect by other means. Sequence-comparison methods can also be used to detect imperfectly repeated sequences within a protein, indicative of linked similar domains.

Evolutionary Trees Can Be Constructed on the Basis of Sequence Information

Construction of an evolutionary tree based on sequence comparisons revealed approximate times for the gene duplication events separating myoglobin and hemoglobin as well as the α and β subunits of hemoglobin. Evolutionary trees based on sequences can be compared to those based on fossil records.

- **Modern Techniques Make the Experimental Exploration of Evolution Possible**

The exploration of evolution can also be a laboratory science. In favorable cases, PCR amplification of well-preserved samples allows the determination of the nucleotide sequences from extinct organisms. Sequences so determined can help authenticate aspects of an evolutionary tree constructed by other means. Molecular evolutionary experiments performed in the test tube can examine how molecules such as ligand-binding RNA molecules might have been generated.

KEY TERMS

homolog (p. 173)

paralog (p. 173)

ortholog (p. 173)

sequence alignment (p. 173)

conservative substitution (p. 177)

substitution matrix (p. 177)

sequence template (p. 180)

self-diagonal plot (p. 181)

divergent evolution (p. 182)

convergent evolution (p. 182)

evolutionary tree (p. 183)

combinatorial chemistry (p. 184)

SELECTED READINGS

Book

Doolittle, R. F., 1987. *Of UFS and ORFS.* University Science Books.

Sequence alignment

Henikoff, S., and Henikoff, J. G., 1992. Amino acid substitution matrices from protein blocks. *Proc. Natl. Acad. Sci. USA* 89: 10915–10919.

Johnson, M. S., and Overington, J. P., 1993. A structural basis for sequence comparisons: An evaluation of scoring methodologies. *J. Mol. Biol.* 233:716–738.

Aravind, L., and Koonin, E. V., 1999. Gleaning non-trivial structural, functional and evolutionary information about proteins by iterative database searches. *J. Mol. Biol.* 287:1023–1040.

Altschul, S. F., Madden. T. L., Schaffer, A. A., Zhang, J., Zhang, Z., Miller, W., and Lipman, D. J., 1997. Gapped BLAST and PSI-BLAST: A new generation of protein database search programs. *Nucleic Acids Res.* 25:3389–3402.

Structure comparison

Bashford, D., Chothia, C., and Lesk, A. M., 1987. Determinants of a protein fold: Unique features of the globin amino acid sequences. *J. Mol. Biol.* 196:199–216.

Harutyunyan, E. H., Safonova, T. N., Kuranova, I. P., Popov, A. N., Teplyakov, A. V., Obmolova, G. V., Rusakov, A. A., Vainshtein, B. K., Dodson, G. G., Wilson, J. C., et al., 1995. The structure of deoxy- and oxy-leghemoglobin from lupin. *J. Mol. Biol.* 251: 104–115.

Flaherty, K. M., McKay, D. B., Kabsch, W., and Holmes, K. C., 1991. Similarity of the three-dimensional structures of actin and the ATPase fragment of a 70-kDa heat shock cognate protein. *Proc. Natl. Acad. Sci. U. S. A.* 88:5041–5045.

Murzin, A. G., Brenner, S. E., Hubbard, T., and Chothia, C., 1995. SCOP: A structural classification of proteins database for the investigation of sequences and structures. *J. Mol. Biol.* 247:536–540.

Hadley, C., and Jones, D. T., 1999. A systematic comparison of protein structure classification: SCOP, CATH and FSSP. *Structure Fold. Des.* 7:1099–1112.

Domain detection

Ploegman, J. H., Drent, G., Kalk, K. H., and Hol, W. G., 1978. Structure of bovine liver rhodanese I: Structure determination at 2.5 Å resolution and a comparison of the conformation and sequence of its two domains. *J. Mol. Biol.* 123:557–594.

Nikolov, D. B., Hu, S. H., Lin, J., Gasch, A., Hoffmann, A., Horikoshi, M., Chua, N. H., Roeder, R. G., and Burley, S. K., 1992. Crystal structure of TFIID TATA-box binding protein. *Nature* 360:40–46.

Doolittle, R. F., 1995. The multiplicity of domains in proteins. *Annu. Rev. Biochem.* 64:287–314.

Heger, A., and Holm, L., 2000. Rapid automatic detection and alignment of repeats in protein sequences. *Proteins* 41:224–237.

Evolutionary trees

Doolittle, R. F., 1992. Stein and Moore Award address. Reconstructing history with amino acid sequences. *Protein Sci.* 1:191–200.

Zukerkandl, E., and Pauling, L., 1965. Molecules as documents of evolutionary history. *J. Theor. Biol.* 8:357–366.

Ancient DNA

Krings, M., Stone, A., Schmitz, R. W., Krainitzki, H., Stoneking, M., and Pääbo, S., 1997. Neandertal DNA sequences and the origin of modern humans [see comments]. *Cell* 90:19–30.

Krings, M., Geisert, H., Schmitz, R. W., Krainitzki, H., and Pääbo, S., 1999. DNA sequence of the mitochondrial hypervariable region II from the Neanderthal type specimen. *Proc. Natl. Acad. Sci. USA* 96:5581–5585.

Evolution in the laboratory

Gold, L., Polisky, B., Uhlenbeck, O., and Yarus, M., 1995. Diversity of oligonucleotide functions. *Annu. Rev. Biochem.* 64:763–797.

Wilson, D. S., and Szostak, J. W., 1999. In vitro selection of functional nucleic acids. *Annu. Rev. Biochem.* 68:611–647.

Hermann, T., and Patel, D. J., 2000. Adaptive recognition by nucleic acid aptamers. *Science* 287:820–825.

Web sites

The Protein Databank (PDB) site is the repository for three-dimensional macromolecular structures. It currently contains nearly 14,000 structures. (http://www.rcsb.org/pdb/)

National Center for Biotechnology Information (NCBI) contains molecular biological databases and software for analysis. (http://www.ncbi.nlm.nih.gov/)

PROBLEMS

1. *What's the score?* Using the identity-based scoring system (Section 7.2), calculate the score for the following alignment. Do you think the score is statistically significant?

(1) WYLGKITRMDAEVLLKKPTVRDGHFLVTQCESSPGEF-
(2) WYFGKITRRESERLLLNPENPRGTFLVRESETTKGAY-

SISVRFGDSVQ-----HFKVLRDQNGKYYLWAVK-FN-
CLSVSDFDNAKGLNVKHYKIRKLDSGGFYITSRTQFS-

SLNELVAYHRTASVSRTHTILLSDMNV
SSLQQLVAYYSKHADGLCHRLTNV

2. *Sequence and structure.* A comparison of the aligned amino acid sequences of two proteins each consisting of 150 amino acids reveals them to be only 8% identical. However, their three-dimensional structures are very similar. Are these two proteins related evolutionarily? Explain.

3. *It depends on how you count.* Consider the following two sequence alignments:

(a) A-SNLFDIRLIG
    GSNDFYEVKIMD

(b) ASNLFDIRLI-G
    GSNDFYEVKIMD

Which alignment has a higher score if the identity-based scoring system (Section 7.2) is used? Which alignment has a higher score if the Blosum-62 substitution matrix (Figure 7.9) is used?

4. *Discovering a new base pair.* Examine the ribosomal RNA sequences in Figure 7.19. In sequences that do not contain Watson-Crick base pairs, what base tends to be paired with G? Propose a structure for your new base pair.

5. *Overwhelmed by numbers.* Suppose that you wish to synthesize a pool of RNA molecules that contain all four bases at each of 40 positions. How much RNA must you have in grams if the pool is to have at least a single molecule of each sequence? The average molecular weight of a nucleotide is 330 g mol^{-1}.

6. *Form follows function.* The three-dimensional structure of biomolecules is more conserved evolutionarily than is sequence. Why is this the case?

7. *Shuffling.* Using the identity-based scoring system (Section 7.2), calculate the alignment score for the alignment of the following two short sequences:

(1) ASNFLDKAGK   (2) ATDYLEKAGK

Generate a shuffled version of sequence 2 by randomly reordering these 10 amino acids. Align your shuffled sequence with sequence 1 without allowing gaps, and calculate the alignment score between sequence 1 and your shuffled sequence.

8. *Interpreting the score.* Suppose that the sequences of two proteins each with 200 amino acids are aligned and that the percentage of identical residues has been calculated. How would you interpret each of the following results in regard to the possible divergence of the two proteins from a common ancestor?

(a) 80%, (b) 50%, (c) 20%, (d) 10%

9. *A set of three.* The sequences of three proteins (A, B, and C) are compared with one another, yielding the following levels of identity:

	A	B	C
A	100%	65%	15%
B	65%	100%	55%
C	15%	55%	100%

Assume that the sequence matches are distributed relatively uniformly along each aligned sequence pair. Would you expect protein A and protein C to have similar three-dimensional structures? Explain.

10. *RNA alignment.* Sequences of an RNA fragment from five species have been determined and aligned. Propose a likely secondary structure for these fragments.

(1) UUGGAGAUUCGGUAGAAUCUCCC
(2) GCCGGGAAUCGACAGAUUCCCCG
(3) CCCAAGUCCCGGCAGGGACUUAC
(4) CUCACCUGCCGAUAGGCAGGUCA
(5) AAUACCACCCGGUAGGGUGGUUC

 Media Problem

11. *Evolutionary time machine.* It has been suggested that ancestral protein sequences might be inferred from evolutionary trees of sequences that exist today. The **Conceptual Insights** module on sequence analysis allows you to try your hand at inferring ancestral sequences from model evolutionary trees. Based on this experience, explain why you would *not* expect to be able to successfully infer dinosaur sequences.

CHAPTER 8

Enzymes: Basic Concepts and Kinetics

$$\xrightarrow[\text{Aequorin}]{O_2,\ Ca^{2+}}$$

+ CO_2 + light (466 nm)

The activity of an enzyme is responsible for the glow of the luminescent jellyfish at left. The enzyme aequorin catalyzes the oxidation of a compound by oxygen in the presence of calcium to release CO_2 and light. [(Left) Fred Bavendam/Peter Arnold.]

Enzymes, the catalysts of biological systems, are remarkable molecular devices that determine the patterns of chemical transformations. They also mediate the transformation of one form of energy into another. The most striking characteristics of enzymes are their *catalytic power and specificity*. Catalysis takes place at a particular site on the enzyme called the *active site. Nearly all known enzymes are proteins.* However, proteins do not have an absolute monopoly on catalysis; the discovery of catalytically active RNA molecules provides compelling evidence that RNA was an early biocatalyst (Section 2.2.2).

Proteins as a class of macromolecules are highly effective catalysts for an enormous diversity of chemical reactions because of their capacity *to specifically bind a very wide range of molecules.* By utilizing the full repertoire of intermolecular forces, enzymes bring substrates together in an optimal orientation, the prelude to making and breaking chemical bonds. They catalyze reactions *by stabilizing transition states*, the highest-energy species in reaction pathways. By selectively stabilizing a transition state, an enzyme determines which one of several potential chemical reactions actually takes place.

8.1 ENZYMES ARE POWERFUL AND HIGHLY SPECIFIC CATALYSTS

Enzymes accelerate reactions by factors of as much as a million or more (Table 8.1). Indeed, most reactions in biological systems do not take place at perceptible rates in the absence of enzymes. Even a reaction as simple as the hydration of carbon dioxide is catalyzed by an enzyme—namely, carbonic anhydrase (Section 9.2). The transfer of CO_2 from the tissues into the blood and then to the alveolar air would be less complete in the absence of this enzyme. In fact, carbonic anhydrase is one of the fastest enzymes known. Each enzyme molecule can hydrate 10^6 molecules of CO_2 per second. This catalyzed reaction is 10^7 times as fast as the uncatalyzed one. We will consider the mechanism of carbonic anhydrase catalysis in Chapter 9. Enzymes are highly specific both in the reactions that they catalyze and in their choice of reactants, which are called *substrates*. An enzyme usually catalyzes a single chemical reaction or a set of closely related reactions. Side reactions leading to the wasteful formation of by-products are rare in enzyme-catalyzed reactions, in contrast with uncatalyzed ones.

Let us consider *proteolytic enzymes* as an example. In vivo, these enzymes catalyze *proteolysis*, the hydrolysis of a peptide bond.

Peptide   **Carboxyl component**   **Amino component**

Most proteolytic enzymes also catalyze a different but related reaction in vitro—namely, the hydrolysis of an ester bond. Such reactions are more easily monitored than is proteolysis and are useful in experimental investigations of these enzymes (Section 9.1.2).

Ester   **Acid**   **Alcohol**

Proteolytic enzymes differ markedly in their degree of substrate specificity. Subtilisin, which is found in certain bacteria, is quite undiscriminat-

TABLE 8.1 Rate enhancement by selected enzymes

Enzyme	Nonenzymatic half-life		Uncatalyzed rate (k_{un} s^{-1})	Catalyzed rate (k_{cat} s^{-1})	Rate enhancement (k_{cat}/k_{un})
OMP decarboxylase	78,000,000	years	2.8×10^{-16}	39	1.4×10^{17}
Staphylococcal nuclease	130,000	years	1.7×10^{-13}	95	5.6×10^{14}
AMP nucleosidase	69,000	years	1.0×10^{-11}	60	6.0×10^{12}
Carboxypeptidase A	7.3	years	3.0×10^{-9}	578	1.9×10^{11}
Ketosteroid isomerase	7	weeks	1.7×10^{-7}	66,000	3.9×10^{11}
Triose phosphate isomerase	1.9	days	4.3×10^{-6}	4,300	1.0×10^{9}
Chorismate mutase	7.4	hours	2.6×10^{-5}	50	1.9×10^{6}
Carbonic anhydrase	5	seconds	1.3×10^{-1}	1×10^{6}	7.7×10^{6}

Abbreviations: OMP, orotidine monophosphate; AMP, adenosine monophosphate.
Source: After A. Radzicka and R. Wofenden. *Science* 267 (1995):90–93.

ing: it will cleave any peptide bond with little regard to the identity of the adjacent side chains. Trypsin, a digestive enzyme, is quite specific and catalyzes the splitting of peptide bonds only on the carboxyl side of lysine and arginine residues (Figure 8.1A). Thrombin, an enzyme that participates in blood clotting, is even more specific than trypsin. It catalyzes the hydrolysis of Arg–Gly bonds in particular peptide sequences only (Figure 8.1B).

DNA polymerase I, a template-directed enzyme (Section 27.2), is another highly specific catalyst. It adds nucleotides to a DNA strand that is being synthesized, in a sequence determined by the sequence of nucleotides in another DNA strand that serves as a template. DNA polymerase I is remarkably precise in carrying out the instructions given by the template. It inserts the wrong nucleotide into a new DNA strand less than one in a million times.

The specificity of an enzyme is due to the precise interaction of the substrate with the enzyme. This precision is a result of the intricate three-dimensional structure of the enzyme protein.

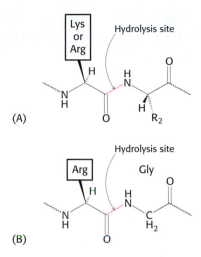

(A)

(B)

FIGURE 8.1 Enzyme specificity. (A) Trypsin cleaves on the carboxyl side of arginine and lysine residues, whereas (B) thrombin cleaves Arg–Gly bonds in particular sequences specifically.

8.1.1 Many Enzymes Require Cofactors for Activity

The catalytic activity of many enzymes depends on the presence of small molecules termed *cofactors*, although the precise role varies with the cofactor and the enzyme. Such an enzyme without its cofactor is referred to as an *apoenzyme*; the complete, catalytically active enzyme is called a *holoenzyme*.

$$\text{Apoenzyme} + \text{cofactor} = \text{holoenzyme}$$

Cofactors can be subdivided into two groups: metals and small organic molecules (Table 8.2). The enzyme carbonic anhydrase, for example, requires Zn^{2+} for its activity (Section 9.2.1). Glycogen phosphorylase (Section 21.1.5), which mobilizes glycogen for energy, requires the small organic molecule pyridoxal phosphate (PLP).

TABLE 8.2 Enzyme cofactors

Cofactor	Enzyme
Coenzyme	
Thiamine pyrophosphate	Pyruvate dehydrogenase
Flavin adenine nucleotide	Monoamine oxidase
Nicotinamide adenine dinucleotide	Lactate dehydrogenase
Pyridoxal phosphate	Glycogen phosphorylase
Coenzyme A (CoA)	Acetyl CoA carboxylase
Biotin	Pyruvate carboxylase
5′-Deoxyadenosyl cobalamin	Methylmalonyl mutase
Tetrahydrofolate	Thymidylate synthase
Metal	
Zn^{2+}	Carbonic anhydrase
Zn^{2+}	Carboxypeptidase
Mg^{2+}	*Eco*RV
Mg^{2+}	Hexokinase
Ni^{2+}	Urease
Mo	Nitrate reductase
Se	Glutathione peroxidase
Mn^{2+}	Superoxide dismutase
K^+	Propionyl CoA carboxylase

Cofactors that are small organic molecules are called *coenzymes*. Often derived from vitamins, coenzymes can be either tightly or loosely bound to the enzyme. If tightly bound, they are called *prosthetic groups*. Loosely associated coenzymes are more like cosubstrates because they bind to and are released from the enzyme just as substrates and products are. The use of the same coenzyme by a variety of enzymes and their source in vitamins sets coenzymes apart from normal substrates, however. Enzymes that use the same coenzyme are usually mechanistically similar. In Chapter 9, we will examine the mechanistic importance of cofactors to enzyme activity. A more detailed discussion of coenzyme vitamins can be found in Section 8.6.

8.1.2 Enzymes May Transform Energy from One Form into Another

In many biochemical reactions, *the energy of the reactants is converted with high efficiency into a different form.* For example, in photosynthesis, light energy is converted into chemical-bond energy through an ion gradient. In mitochondria, the free energy contained in small molecules derived from food is converted first into the free energy of an ion gradient and then into a different currency, the free energy of adenosine triphosphate. Enzymes may then use the chemical-bond energy of ATP in many ways. The enzyme myosin converts the energy of ATP into the mechanical energy of contracting muscles. Pumps in the membranes of cells and organelles, which can be thought of as enzymes that move substrates rather than chemically altering them, create chemical and electrical gradients by using the energy of ATP to transport molecules and ions (Figure 8.2). The molecular mechanisms of these energy-transducing enzymes are being unraveled. We will see in subsequent chapters how unidirectional cycles of discrete steps—binding, chemical transformation, and release—lead to the conversion of one form of energy into another.

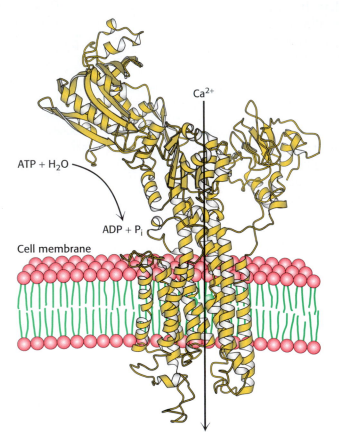

FIGURE 8.2 An energy-transforming enzyme. Ca^{2+} ATPase uses the energy of ATP hydrolysis to transport Ca^{2+} across the membrane, generating a Ca^{2+} gradient.

Ca^{2+}

ATP + H_2O

ADP + P_i

Cell membrane

8.1.3 Enzymes Are Classified on the Basis of the Types of Reactions That They Catalyze

Many enzymes have common names that provide little information about the reactions that they catalyze. For example, a proteolytic enzyme secreted by the pancreas is called trypsin. Most other enzymes are named for their substrates and for the reactions that they catalyze, with the suffix "ase" added. Thus, an ATPase is an enzyme that breaks down ATP, whereas ATP synthase is an enzyme that synthesizes ATP.

To bring some consistency to the classification of enzymes, in 1964 the International Union of Biochemistry established an Enzyme Commission to develop a nomenclature for enzymes. Reactions were divided into six major groups numbered 1 through 6 (Table 8.3). These groups were subdivided and further subdivided, so that a four-digit number preceded by the letters *EC* for Enzyme Commission could precisely identify all enzymes.

Consider as an example nucleoside monophosphate (NMP) kinase, an enzyme that we will examine in detail in the next chapter (Section 9.4). It catalyzes the following reaction:

$$ATP + NMP \rightleftharpoons ADP + NDP$$

TABLE 8.3 Six major classes of enzymes

Class	Type of reaction	Example	Chapter
1. Oxidoreductases	Oxidation-reduction	Lactate dehydrogenase	16
2. Transferases	Group transfer	Nucleoside monophosphate kinase (NMP kinase)	9
3. Hydrolases	Hydrolysis reactions (transfer of functional groups to water)	Chymotrypsin	9
4. Lyases	Addition or removal of groups to form double bonds	Fumarase	18
5. Isomerases	Isomerization (intramolecular group transfer)	Triose phosphate isomerase	16
6. Ligases	Ligation of two substrates at the expense of ATP hydrolysis	Aminoacyl-tRNA synthetase	29

NMP kinase transfers a phosphoryl group from ATP to NMP to form a nucleoside diphosphate (NDP) and ADP. Consequently, it is a transferase, or member of group 2. Many groups in addition to phosphoryl groups, such as sugars and carbon units, can be transferred. Transferases that shift a phosphoryl group are designated 2.7. Various functional groups can accept the phosphoryl group. If a phosphate is the acceptor, the transferase is designated 2.7.4. The final number designates the acceptor more precisely. In regard to NMP kinase, a nucleoside monophosphate is the acceptor, and the enzyme's designation is EC 2.7.4.4. Although the common names are used routinely, the classification number is used when the precise identity of the enzyme might be ambiguous.

8.2 FREE ENERGY IS A USEFUL THERMODYNAMIC FUNCTION FOR UNDERSTANDING ENZYMES

Some of the principles of thermodynamics were introduced in Chapter 1—notably the idea of *free energy (G)*. To fully understand how enzymes operate, we need to consider two thermodynamic properties of the reaction: (1) the free-energy difference (ΔG) between the products and reactants and (2) the energy required to initiate the conversion of reactants to products. The former determines whether the reaction will be spontaneous, whereas the later determines the rate of the reaction. Enzymes affect only the latter. First, we will consider the thermodynamics of reactions and then, in Section 8.3, the rates of reactions.

8.2.1 The Free-Energy Change Provides Information About the Spontaneity but Not the Rate of a Reaction

As stated in Section 1.3.3, the free-energy change of a reaction (ΔG) tells us if the reaction can occur spontaneously:

1. *A reaction can occur spontaneously only if ΔG is negative.* Such reactions are said to be exergonic.

2. A system is at equilibrium and no net change can take place if ΔG is zero.

3. A reaction cannot occur spontaneously if ΔG is positive. An input of free energy is required to drive such a reaction. These reactions are termed endergonic.

Two additional points need to be emphasized. The ΔG of a reaction depends only on the free energy of the products (the final state) minus the free energy of the reactants (the initial state). *The ΔG of a reaction is independent of the path (or molecular mechanism) of the transformation.* The mechanism of a reaction has no effect on ΔG. For example, the ΔG for the oxidation of glucose to CO_2 and H_2O is the same whether it occurs by combustion in vitro or by a series of enzyme-catalyzed steps in a cell. *The ΔG provides no information about the rate of a reaction.* A negative ΔG indicates that a reaction *can* occur spontaneously, but it does not signify whether it will proceed at a perceptible rate. As will be discussed shortly (Section 8.3), the rate of a reaction depends on the *free energy of activation* ($\Delta G^{\ddagger}$), which is largely unrelated to the ΔG of the reaction.

8.2.2 The Standard Free-Energy Change of a Reaction Is Related to the Equilibrium Constant

As for any reaction, we need to be able to determine ΔG for an enzyme-catalyzed reaction in order to know whether the reaction is spontaneous or an input of energy is required. To determine this important thermodynamic parameter, we need to take into account the nature of both the reactants and the products as well as their concentrations.

Consider the reaction

$$A + B \rightleftharpoons C + D$$

The ΔG of this reaction is given by

$$\Delta G = \Delta G^{\circ} + RT \ln \frac{[C][D]}{[A][B]} \tag{1}$$

in which ΔG° is the *standard free-energy change*, R is the gas constant, T is the absolute temperature, and [A], [B], [C], and [D] are the molar concentrations (more precisely, the activities) of the reactants. ΔG° is the free-energy change for this reaction under standard conditions—that is, when each of the reactants A, B, C, and D is present at a concentration of 1.0 M (for a gas, the standard state is usually chosen to be 1 atmosphere). Thus, the ΔG of a reaction depends on the *nature* of the reactants (expressed in the ΔG° term of equation 1) and on their *concentrations* (expressed in the logarithmic term of equation 1).

A convention has been adopted to simplify free-energy calculations for biochemical reactions. The standard state is defined as having a pH of 7. Consequently, when H^+ is a reactant, its activity has the value 1 (corresponding to a pH of 7) in equations 1 and 4 (below). The activity of water also is taken to be 1 in these equations. The *standard free-energy change at pH 7*, denoted by the symbol $\Delta G^{\circ\prime}$ will be used throughout this book. The *kilocalorie* (abbreviated *kcal*) and the *kilojoule (kJ)* will be used as the units of energy. One kilocalorie is equivalent to 4.184 kilojoules.

The relation between the standard free energy and the equilibrium constant of a reaction can be readily derived. This equation is important because it displays the energetic relation between products and reactants in terms of their concentrations. At equilibrium, $\Delta G = 0$. Equation 1 then becomes

$$0 = \Delta G^{\circ\prime} + RT \ln \frac{[C][D]}{[A][B]} \tag{2}$$

and so

$$\Delta G^{\circ\prime} = -RT \ln \frac{[C][D]}{[A][B]} \tag{3}$$

Units of energy—

A *calorie* (cal) is equivalent to the amount of heat required to raise the temperature of 1 gram of water from 14.5°C to 15.5°C.

A *kilocalorie* (kcal) is equal to 1000 cal.

A *joule* (J) is the amount of energy needed to apply a 1-newton force over a distance of 1 meter.

A *kilojoule* (kJ) is equal to 1000 J.

1 kcal = 4.184 kJ

The equilibrium constant under standard conditions, K'_{eq}, is defined as

$$K'_{eq} = \frac{[C][D]}{[A][B]} \qquad (4)$$

Substituting equation 4 into equation 3 gives

$$\Delta G^{\circ\prime} = -RT \ln K'_{eq} \qquad (5)$$

$$\Delta G^{\circ\prime} = -2.303RT \log_{10} K'_{eq} \qquad (6)$$

which can be rearranged to give

$$K'_{eq} = 10^{-\Delta G^{\circ\prime}/(2.303RT)} \qquad (7)$$

Substituting $R = 1.987 \times 10^{-3}$ kcal mol^{-1} deg^{-1} and $T = 298$ K (corresponding to 25°C) gives

$$K'_{eq} = 10^{-\Delta G^{\circ\prime}/1.36} \qquad (8)$$

where $\Delta G^{\circ\prime}$ is here expressed in kilocalories per mole because of the choice of the units for R in equation 7. Thus, the standard free energy and the equilibrium constant of a reaction are related by a simple expression. For example, an equilibrium constant of 10 gives a standard free-energy change of -1.36 kcal mol^{-1} (-5.69 kJ mol^{-1}) at 25°C (Table 8.4). Note that, for each 10-fold change in the equilibrium constant, the $\Delta G^{\circ\prime}$ changes by 1.36 kcal mol^{-1} (5.69 kJ mol^{-1}).

As an example, let us calculate $\Delta G^{\circ\prime}$ and ΔG for the isomerization of dihydroxyacetone phosphate (DHAP) to glyceraldehyde 3-phosphate (GAP). This reaction takes place in glycolysis (Section 16.1.4). At equilibrium, the ratio of GAP to DHAP is 0.0475 at 25°C (298 K) and pH 7. Hence, $K'_{eq} = 0.0475$. The standard free-energy change for this reaction is then calculated from equation 6:

$$\Delta G^{\circ\prime} = -2.303RT \log_{10} K'_{eq}$$
$$= -2.303 \times 1.987 \times 10^{-3} \times 298 \times \log_{10}(0.0475)$$
$$= +1.80 \text{ kcal mol}^{-1}(7.53 \text{ kJ mol}^{-1})$$

Under these conditions, the reaction is endergonic. DHAP will not spontaneously convert to GAP.

Now let us calculate ΔG for this reaction when the initial concentration of DHAP is 2×10^{-4} M and the initial concentration of GAP is 3×10^{-6} M. Substituting these values into equation 1 gives

$$\Delta G = 1.80 \text{ kcal mol}^{-1} + 2.303RT \log_{10} \frac{3 \times 10^{-6} \text{ M}}{2 \times 10^{-4} \text{ M}}$$

$$= 1.80 \text{ kcal mol}^{-1} - 2.49 \text{ kcal mol}^{-1}$$

$$= -0.69 \text{ kcal mol}^{-1} (-2.89 \text{ kJ mol}^{-1})$$

This negative value for the ΔG indicates that the isomerization of DHAP to GAP is exergonic and can occur spontaneously when these species are present at the aforestated concentrations. Note that ΔG for this reaction is negative, although $\Delta G^{\circ\prime}$ is positive. *It is important to stress that whether the ΔG for a reaction is larger, smaller, or the same as $\Delta G^{\circ\prime}$ depends on the concentrations of the reactants and products.* The criterion of spontaneity for a reaction is ΔG, not $\Delta G^{\circ\prime}$. This point is important because reactions that are not spontaneous based on $\Delta G^{\circ\prime}$ can be made spontaneous by adjusting the concentrations of reactants and products. This principle is the basis of the coupling of reactions to form metabolic pathways (Chapter 14).

TABLE 8.4   Relation between $\Delta G^{\circ\prime}$ and K'_{eq} (at 25°C)

K'_{eq}	$\Delta G^{\circ\prime}$ kcal mol^{-1}	$\Delta G^{\circ\prime}$ kJ/mol^{-1}
10^{-5}	6.82	28.53
10^{-4}	5.46	22.84
10^{-3}	4.09	17.11
10^{-2}	2.73	11.42
10^{-1}	1.36	5.69
1	0	0
10	-1.36	-5.69
10^2	-2.73	-11.42
10^3	-4.09	-17.11
10^4	-5.46	-22.84
10^5	-6.82	-28.53

Dihydroxyacetone phosphate (DHAP)

Glyceraldehyde 3-phosphate (GAP)

8.2.3 Enzymes Alter Only the Reaction Rate and Not the Reaction Equilibrium

Because enzymes are such superb catalysts, it is tempting to ascribe to them powers that they do not have. An enzyme cannot alter the laws of thermodynamics and *consequently cannot alter the equilibrium of a chemical reaction.* This inability means that an enzyme accelerates the forward and reverse reactions by precisely the same factor. Consider the interconversion of A and B. Suppose that, in the absence of enzyme, the forward rate constant (k_F) is 10^{-4} s^{-1} and the reverse rate constant (k_R) is 10^{-6} s^{-1}. The equilibrium constant K is given by the ratio of these rate constants:

$$A \underset{10^{-6}\,s^{-1}}{\overset{10^{-4}\,s^{-1}}{\rightleftharpoons}} B$$

$$K = \frac{[B]}{[A]} = \frac{k_F}{k_R} = \frac{10^{-4}}{10^{-6}} = 100$$

The equilibrium concentration of B is 100 times that of A, whether or not enzyme is present. However, it might take considerable time to approach this equilibrium without enzyme, whereas equilibrium would be attained rapidly in the presence of a suitable enzyme. *Enzymes accelerate the attainment of equilibria but do not shift their positions. The equilibrium position is a function only of the free-energy difference between reactants and products.*

8.3 ENZYMES ACCELERATE REACTIONS BY FACILITATING THE FORMATION OF THE TRANSITION STATE

The free-energy difference between reactants and products accounts for the equilibrium of the reaction, but enzymes accelerate how quickly this equilibrium is attained. How can we explain the rate enhancement in terms of thermodynamics? To do so, we have to consider not the end points of the reaction but the chemical pathway between the end points.

A chemical reaction of substrate S to form product P goes through a *transition state* S‡ that has a higher free energy than does either S or P. The double dagger denotes a thermodynamic property of the transition state. The transition state is the most seldom occupied species along the reaction pathway because it is the one with the highest free energy. The difference in free energy between the transition state and the substrate is called the *Gibbs free energy of activation* or simply the *activation energy*, symbolized by $\Delta G^{\ddagger}$, as mentioned in Section 8.2.1 (Figure 8.3).

$$\Delta G^{\ddagger} = G_{S^{\ddagger}} - G_S$$

Note that the energy of activation, or $\Delta G^{\ddagger}$, does not enter into the final ΔG calculation for the reaction, because the energy input required to reach the transition state is returned when the transition state forms the product. The activation-energy barrier immediately suggests how enzymes enhance reaction rate without altering ΔG of the reaction: enzymes function to lower the activation energy, or, in other words, *enzymes facilitate the formation of the transition state.*

One approach to understanding how enzymes achieve this facilitation is to assume that the transition state (S‡) and the substrate (S) are in equilibrium.

$$S \overset{K^{\ddagger}}{\rightleftharpoons} S^{\ddagger} \overset{v}{\longrightarrow} P$$

in which $K^{\ddagger}$ is the equilibrium constant for the formation of S‡, and v is the rate of formation of product from S‡.

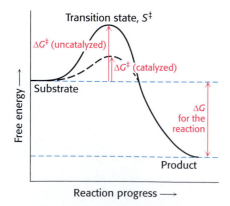

FIGURE 8.3 Enzymes decrease the activation energy. Enzymes accelerate reactions by decreasing $\Delta G^{\ddagger}$, the free energy of activation.

The rate of the reaction is proportional to the concentration of S‡:

$$\text{Rate} \propto \left[\text{S}^{\ddagger}\right]$$

because only S‡ can be converted into product. The concentration of S‡ is in turn related to the free energy difference between S‡ and S, because these two chemical species are assumed to be in equilibrium; the greater the difference between these two states, the smaller the amount of S‡.

Because the reaction rate is proportional to the concentration of S‡, and the concentration of S‡ depends on $\Delta G^{\ddagger}$, the rate of reaction V depends on $\Delta G^{\ddagger}$. Specifically,

$$V = v\left[\text{S}^{\ddagger}\right] = \frac{kT}{h}[\text{S}]e^{-\Delta G^{\ddagger}/RT}$$

In this equation, k is Boltzmann's constant, and h is Planck's constant. The value of kT/h at 25°C is 6.2×10^{12} s^{-1}. Suppose that the free energy of activation is 6.82 kcal mol^{-1} (28.53 kJ mol^{-1}). The ratio [S‡]/[S] is then 10^{-5} (see Table 8.4). If we assume for simplicity's sake that [S] = 1M, then the reaction rate V is 6.2×10^{7} s^{-1}. If $\Delta G^{\ddagger}$ were lowered by 1.36 kcal mol^{-1} (5.69 kJ mol^{-1}), the ratio [S‡]/[S] is then 10^{-4}, and the reaction rate would be 6.2×10^{8} s^{-1}. As Table 8.4 shows, a decrease of 1.36 kcal mol^{-1} in $\Delta G^{\ddagger}$ yields a tenfold larger V. A relatively small decrease in $\Delta G^{\ddagger}$ (20% in this particular reaction) results in a much greater increase in V.

Thus, we see the key to how enzymes operate: *Enzymes accelerate reactions by decreasing* $\Delta G^{\ddagger}$, *the activation energy.* The combination of substrate and enzyme creates a new reaction pathway whose transition-state energy is lower than that of the reaction in the absence of enzyme (see Figure 8.3). The lower activation energy means that more molecules have the required energy to reach the transition state. Decreasing the activation barrier is analogous to lowering the height of a high-jump bar; more athletes will be able to clear the bar. *The essence of catalysis is specific binding of the transition state.*

> "I think that enzymes are molecules that are complementary in structure to the activated complexes of the reactions that they catalyze, that is, to the molecular configuration that is intermediate between the reacting substances and the products of reaction for these catalyzed processes. The attraction of the enzyme molecule for the activated complex would thus lead to a decrease in its energy and hence to a decrease in the energy of activation of the reaction and to an increase in the rate of reaction."
>
> —LINUS PAULING
> *Nature* 161(1948):707

8.3.1 The Formation of an Enzyme–Substrate Complex Is the First Step in Enzymatic Catalysis

Much of the catalytic power of enzymes comes from their bringing substrates together in favorable orientations to promote the formation of the transition states in *enzyme–substrate* (ES) complexes. The substrates are bound to a specific region of the enzyme called the *active site.* Most enzymes are highly selective in the substrates that they bind. Indeed, the catalytic specificity of enzymes depends in part on the specificity of binding.

What is the evidence for the existence of an enzyme–substrate complex?

1. The first clue was the observation that, at a constant concentration of enzyme, the reaction rate increases with increasing substrate concentration until a maximal velocity is reached (Figure 8.4). In contrast, uncatalyzed reactions do not show this saturation effect. *The fact that an enzyme-catalyzed reaction has a maximal velocity suggests the formation of a discrete ES complex.* At a sufficiently high substrate concentration, all the catalytic sites are filled and so the reaction rate cannot increase. Although indirect, this is the most general evidence for the existence of ES complexes.

2. *X-ray crystallography* has provided high-resolution images of substrates and substrate analogs bound to the active sites of many enzymes (Figure 8.5). In Chapter 9, we will take a close look at several of these complexes. X-ray studies carried out at low temperatures (to slow reactions down) are providing revealing views of enzyme–substrate complexes and their subsequent

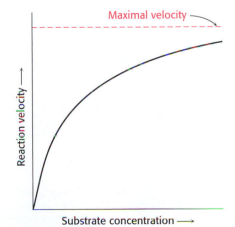

FIGURE 8.4 Reaction velocity versus substrate concentration in an enzyme-catalyzed reaction. An enzyme-catalyzed reaction reaches a maximal velocity.

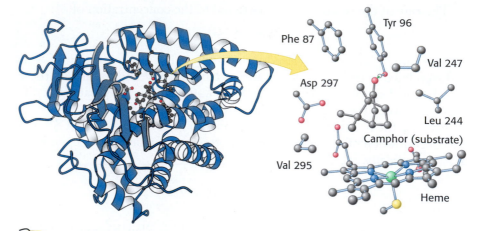

FIGURE 8.5 **Structure of an enzyme-substrate complex.** (Left) The enzyme cytochrome P-450 is illustrated bound to its substrate camphor. (Right) In the active site, the substrate is surrounded by residues from the enzyme. Note also the presence of a heme cofactor.

reactions. A new technique, *time-resolved crystallography*, depends on co-crystallizing a photolabile substrate analog with the enzyme. The substrate analog can be converted to substrate light, and images of the enzyme–substrate complex are obtained in a fraction of a second by scanning the crystal with intense, polychromatic x-rays from a synchrotron.

3. The *spectroscopic characteristics* of many enzymes and substrates change on formation of an ES complex. These changes are particularly striking if the enzyme contains a colored prosthetic group. Tryptophan synthetase, a bacterial enzyme that contains a pyridoxal phosphate (PLP) prosthetic group, provides a nice illustration. This enzyme catalyzes the synthesis of L-tryptophan from L-serine and indole-derivative. The addition of L-serine to the enzyme produces a marked increase in the fluorescence of the PLP group (Figure 8.6). The subsequent addition of indole, the second substrate, quenches this fluorescence to a level even lower than that of the enzyme alone. Thus, fluorescence spectroscopy reveals the existence of an enzyme–serine complex and of an enzyme–serine–indole complex. Other spectroscopic techniques, such as nuclear magnetic resonance and electron spin resonance, also are highly informative about ES interactions.

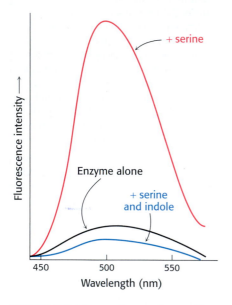

FIGURE 8.6 **Change in spectroscopic characteristics with the formation of an enzyme–substrate complex.** Fluorescence intensity of the pyridoxal phosphate group at the active site of tryptophan synthetase changes on addition of serine and indole, the substrates.

8.3.2 The Active Sites of Enzymes Have Some Common Features

The active site of an enzyme is the region that binds the substrates (and the cofactor, if any). It also contains the residues that directly participate in the making and breaking of bonds. These residues are called the *catalytic groups*. In essence, *the interaction of the enzyme and substrate at the active site promotes the formation of the transition state*. The active site is the region of the enzyme that most directly lowers the $\Delta G^{\ddagger}$ of the reaction, which results in the rate enhancement characteristic of enzyme action. Although enzymes differ widely in structure, specificity, and mode of catalysis, a number of generalizations concerning their active sites can be stated:

1. *The active site is a three-dimensional cleft* formed by groups that come from different parts of the amino acid sequence—indeed, residues far apart in the sequence may interact more strongly than adjacent residues in the amino acid sequence. In lysozyme, an enzyme that degrades the cell walls of some bacteria, the important groups in the active site are contributed by

(A)

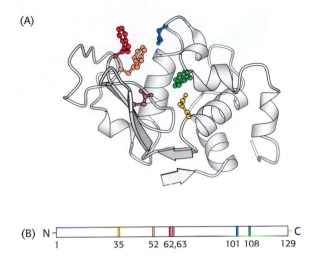

(B) N ⊢—————————————————————⊣ C
     1   35   52 62,63   101 108   129

FIGURE 8.7 Active sites may include distant residues. (A) Ribbon diagram of the enzyme lysozyme with several components of the active site shown in color. (B) A schematic representation of the primary structure of lysozyme shows that the active site is composed of residues that come from different parts of the polypeptide chain.

residues numbered 35, 52, 62, 63, and 101 in the sequence of the 129 amino acids (Figure 8.7).

2. *The active site takes up a relatively small part of the total volume of an enzyme.* Most of the amino acid residues in an enzyme are not in contact with the substrate, which raises the intriguing question of why enzymes are so big. Nearly all enzymes are made up of more than 100 amino acid residues, which gives them a mass greater than 10 kd and a diameter of more than 25 Å. The "extra" amino acids serve as a scaffold to create the three-dimensional active site from amino acids that are far apart in the primary structure. Amino acids near to one another in the primary structure are often sterically constrained from adopting the structural relations necessary to form the active site. In many proteins, the remaining amino acids also constitute regulatory sites, sites of interaction with other proteins, or channels to bring the substrates to the active sites.

3. *Active sites are clefts or crevices.* In all enzymes of known structure, substrate molecules are bound to a cleft or crevice. Water is usually excluded unless it is a reactant. The nonpolar character of much of the cleft enhances the binding of substrate as well as catalysis. Nevertheless, the cleft may also contain polar residues. In the nonpolar microenvironment of the active site, certain of these polar residues acquire special properties essential for substrate binding or catalysis. The internal positions of these polar residues are biologically crucial exceptions to the general rule that polar residues are exposed to water.

4. *Substrates are bound to enzymes by multiple weak attractions.* ES complexes usually have equilibrium constants that range from 10^{-2} to 10^{-8} M, corresponding to free energies of interaction ranging from about -3 to -12 kcal mol^{-1} (from -13 to -50 kJ mol^{-1}). The noncovalent interactions in ES complexes are much weaker than covalent bonds, which have energies between -50 and -110 kcal mol^{-1} (between -210 and -460 kJ mol^{-1}). As discussed in Chapter 1 (Section 1.3.1), electrostatic interactions, hydrogen bonds, van der Waals forces, and hydrophobic interactions mediate reversible interactions of biomolecules. Van der Waals forces become significant in binding only when numerous substrate atoms simultaneously come close to many enzyme atoms. Hence, the enzyme and substrate should have complementary shapes. The directional character of hydrogen bonds between enzyme and substrate often enforces a high degree of specificity, as seen in the RNA-degrading enzyme ribonuclease (Figure 8.8).

FIGURE 8.8 Hydrogen bonds between an enzyme and substrate. The enzyme ribonuclease forms hydrogen bonds with the uridine component of the substrate. [After F. M. Richards, H. W. Wyckoff, and N. Allewel. In *The Neurosciences: Second Study Program,* F. O. Schmidt, Ed. (Rockefeller University Press, 1970), p. 970.]

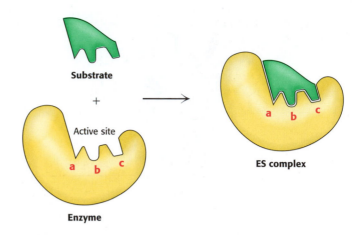

FIGURE 8.9 Lock-and-key model of enzyme–substrate binding. In this model, the active site of the unbound enzyme is complementary in shape to the substrate.

5. *The specificity of binding depends on the precisely defined arrangement of atoms in an active site.* Because the enzyme and the substrate interact by means of short-range forces that require close contact, a substrate must have a matching shape to fit into the site. Emil Fischer's analogy of the lock and key (Figure 8.9), expressed in 1890, has proved to be highly stimulating and fruitful. However, we now know that enzymes are flexible and that the shapes of the active sites can be markedly modified by the binding of substrate, as was postulated by Daniel E. Koshland, Jr., in 1958. The active site of some enzymes assume a shape that is complementary to that of the transition state only *after* the substrate is bound. This process of dynamic recognition is called *induced fit* (Figure 8.10).

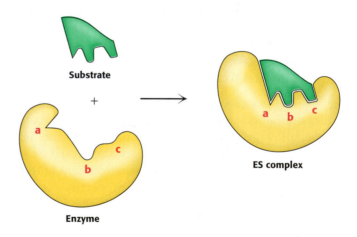

FIGURE 8.10 Induced-fit model of enzyme–substrate binding. In this model, the enzyme changes shape on substrate binding. The active site forms a shape complementary to the substrate only after the substrate has been bound.

8.4 THE MICHAELIS-MENTEN MODEL ACCOUNTS FOR THE KINETIC PROPERTIES OF MANY ENZYMES

The primary function of enzymes is to enhance rates of reactions so that they are compatible with the needs of the organism. To understand how enzymes function, we need a kinetic description of their activity. For many enzymes, the rate of catalysis V_0, which is defined as the number of moles of product formed per second, varies with the substrate concentration [S] in a manner shown in Figure 8.11. The rate of catalysis rises linearly as substrate concentration increases and then begins to level off and approach a maximum at higher substrate concentrations. Before we can accurately interpret

FIGURE 8.11 Michaelis-Menten kinetics. A plot of the reaction velocity (V_0) as a function of the substrate concentration [S] for an enzyme that obeys Michaelis-Menten kinetics shows that the maximal velocity (V_{max}) is approached asymptotically. The Michaelis constant (K_M) is the substrate concentration yielding a velocity of $V_{max}/2$.

this graph, we need to understand how it is generated. Consider an enzyme that catalyzes the S to P by the following pathway:

$$E + S \underset{k_{-1}}{\overset{k_1}{\rightleftharpoons}} ES \underset{k_{-2}}{\overset{k_2}{\rightleftharpoons}} E + P$$

The extent of product formation is determined as a function of time for a series of substrate concentrations (Figure 8.12). As expected, in each case, the amount of product formed increases with time, although eventually a time is reached when there is *no net change* in the concentration of S or P. The enzyme is still actively converting substrate into product and visa versa, but the reaction equilibrium has been attained. Figure 8.13A illustrates the changes in concentration observed in all of the reaction participants with time until equilibrium has been reached.

Enzyme kinetics are more easily approached if we can ignore the back reaction. We define V_0 as the rate of increase in product with time when [P] is low; that is, at times close to zero (hence, V_0) (Figure 8.13B). Thus, for the graph in Figure 8.11, V_0 is determined for each substrate concentration by measuring the rate of product formation at early times before P accumulates (see Figure 8.12).

We begin our kinetic examination of enzyme activity with the graph shown in Figure 8.11. At a fixed concentration of enzyme, V_0 is almost linearly proportional to [S] when [S] is small but is nearly independent of [S] when [S] is large. In 1913, Leonor Michaelis and Maud Menten proposed a simple model to account for these kinetic characteristics. The critical feature in their treatment is that a specific ES complex is a necessary intermediate in catalysis. The model proposed, which is the simplest one that accounts for the kinetic properties of many enzymes, is

$$E + S \underset{k_{-1}}{\overset{k_1}{\rightleftharpoons}} ES \xrightarrow{k_2} E + P \qquad (9)$$

An enzyme E combines with substrate S to form an ES complex, with a rate constant k_1. The ES complex has two possible fates. It can dissociate to E and S, with a rate constant k_{-1}, or it can proceed to form product P, with a rate constant k_2. Again, we assume that almost none of the product reverts to the initial substrate, a condition that holds in the initial stage of a reaction before the concentration of product is appreciable.

We want an expression that relates the rate of catalysis to the concentrations of substrate and enzyme and the rates of the individual steps. Our

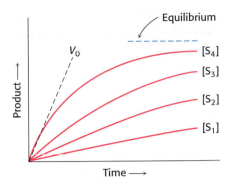

FIGURE 8.12 Determining initial velocity. The amount of product formed at different substrate concentrations is plotted as a function of time. The initial velocity (V_0) for each substrate concentration is determined from the slope of the curve at the beginning of a reaction, when the reverse reaction is insignificant.

FIGURE 8.13 Changes in the concentration of reaction participants of an enzyme-catalyzed reaction with time. Concentration changes under (A) steady-state conditions, and (B) the pre-steady-state conditions.

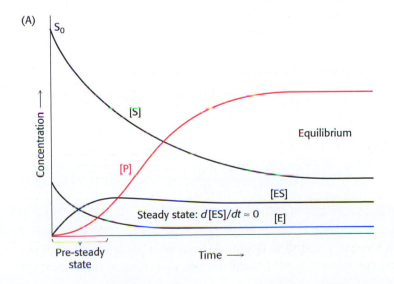

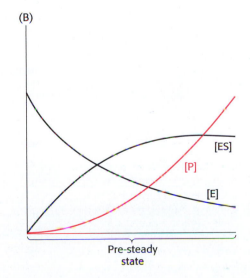

starting point is that the catalytic rate is equal to the product of the concentration of the ES complex and k_2.

$$V_0 = k_2[\text{ES}] \tag{10}$$

Now we need to express [ES] in terms of known quantities. The rates of formation and breakdown of ES are given by:

$$\text{Rate of formation of ES} = k_1[\text{E}][\text{S}] \tag{11}$$

$$\text{Rate of breakdown of ES} = (k_{-1} + k_2)[\text{ES}] \tag{12}$$

To simplify matters, we will work under the *steady-state assumption*. In a steady state, the concentrations of intermediates, in this case [ES], stay the same even if the concentrations of starting materials and products are changing. This occurs when the rates of formation and breakdown of the ES complex are equal. Setting the right-hand sides of equations 11 and 12 equal gives

$$k_1[\text{E}][\text{S}] = (k_{-1} + k_2)[\text{ES}] \tag{13}$$

By rearranging equation 13, we obtain

$$[\text{E}][\text{S}]/[\text{ES}] = (k_{-1} + k_2)/k_1 \tag{14}$$

Equation 14 can be simplified by defining a new constant, K_M, called the *Michaelis constant*:

$$K_\text{M} = \frac{k_{-1} + k_2}{k_1} \tag{15}$$

Note that K_M has the units of concentration. K_M is an important characteristic of enzyme–substrate interactions and is independent of enzyme and substrate concentrations.

Inserting equation 15 into equation 14 and solving for [ES] yields

$$[\text{ES}] = \frac{[\text{E}][\text{S}]}{K_\text{M}} \tag{16}$$

Now let us examine the numerator of equation 16. The concentration of uncombined substrate [S] is very nearly equal to the total substrate concentration, provided that the concentration of enzyme is much lower than that of substrate. The concentration of uncombined enzyme [E] is equal to the total enzyme concentration $[\text{E}]_\text{T}$ minus the concentration of the ES complex.

$$[\text{E}] = [\text{E}]_\text{T} - [\text{ES}] \tag{17}$$

Substituting this expression for [E] in equation 16 gives

$$[\text{ES}] = \frac{([\text{E}]_\text{T} - [\text{ES}])[\text{S}]}{K_\text{M}} \tag{18}$$

Solving equation 18 for [ES] gives

$$[\text{ES}] = \frac{[\text{E}]_\text{T}/K_\text{M}}{1 + [\text{S}]/K_\text{M}} \tag{19}$$

or

$$[\text{ES}] = [\text{E}]_\text{T}\frac{[\text{S}]}{[\text{S}] + K_\text{M}} \tag{20}$$

By substituting this expression for [ES] into equation 10, we obtain

$$V_0 = k_2 [E]_T \frac{[S]}{[S] + K_M} \qquad (21)$$

The maximal rate, V_{max}, is attained when the catalytic sites on the enzyme are saturated with substrate—that is, when $[ES] = [E]_T$. Thus,

$$V_{max} = k_2 [E]_T \qquad (22)$$

Substituting equation 22 into equation 21 yields the *Michaelis-Menten equation*:

$$V_0 = V_{max} \frac{[S]}{[S] + K_M} \qquad (23)$$

This equation accounts for the kinetic data given in Figure 8.11. At very low substrate concentration, when [S] is much less than K_M, $V_0 = (V_{max}/K_M)/[S]$; that is, the rate is directly proportional to the substrate concentration. At high substrate concentration, when [S] is much greater than K_M, $V_0 = V_{max}$; that is, the rate is maximal, independent of substrate concentration.

The meaning of K_M is evident from equation 23. When $[S] = K_M$, then $V_0 = V_{max}/2$. Thus, K_M *is equal to the substrate concentration at which the reaction rate is half its maximal value.* K_M is an important characteristic of an enzyme-catalyzed reaction and is significant for its biological function.

The physiological consequence of K_M is illustrated by the sensitivity of some individuals to ethanol. Such persons exhibit facial flushing and rapid heart rate (tachycardia) after ingesting even small amounts of alcohol. In the liver, alcohol dehydrogenase converts ethanol into acetaldehyde.

$$CH_3CH_2OH + NAD^+ \underset{\text{dehydrogenase}}{\overset{\text{Alchohol}}{\rightleftharpoons}} CH_3CHO + H^+ + NADH$$

Normally, the acetaldehyde, which is the cause of the symptoms when present at high concentrations, is processed to acetate by acetaldehyde dehydrogenase.

$$CH_3CHO + NAD^+ \underset{\text{dehydrogenase}}{\overset{\text{Acetaldehyde}}{\rightleftharpoons}} CH_3COO^- + NADH + 2\,H^+$$

Most people have two forms of the acetaldehyde dehydrogenase, a low K_M mitochondrial form and a high K_M cytosolic form. In susceptible persons, the mitochondrial enzyme is less active due to the substitution of a single amino acid, and acetaldehyde is processed only by the cytosolic enzyme. Because this enzyme has a high K_M, less acetaldehyde is converted into acetate; excess acetaldehyde escapes into the blood and accounts for the physiological effects.

8.4.1 The Significance of K_M and V_{max} Values

CONCEPTUAL INSIGHTS, Steady-State Enzyme Kinetics. *Learn how the kinetic parameters K_M and V_{max} can be determined experimentally using the enzyme kinetics lab simulation in this media module.*

The Michaelis constant, K_M, and the maximal rate, V_{max}, can be readily derived from rates of catalysis measured at a variety of substrate concentrations if an enzyme operates according to the simple scheme given in equation 23. The derivation of K_M and V_{max} is most commonly achieved

TABLE 8.5 K_M **values of some enzymes**

Enzyme	Substrate	$K_M(\mu M)$
Chymotrypsin	Acetyl-L-tryptophanamide	5000
Lysozyme	Hexa-N-acetylglucosamine	6
β-Galactosidase	Lactose	4000
Threonine deaminase	Threonine	5000
Carbonic anhydrase	CO_2	8000
Penicillinase	Benzylpenicillin	50
Pyruvate carboxylase	Pyruvate	400
	HCO_3^-	1000
	ATP	60
Arginine-tRNA synthetase	Arginine	3
	tRNA	0.4
	ATP	300

with the use of curve-fitting programs on a computer (see the appendix to this chapter for alternative means of determining K_M and V_{max}). The K_M values of enzymes range widely (Table 8.5). For most enzymes, K_M lies between 10^{-1} and 10^{-7} M. The K_M value for an enzyme depends on the particular substrate and on environmental conditions such as pH, temperature, and ionic strength. The Michaelis constant, K_M, has two meanings. First, K_M is the concentration of substrate at which half the active sites are filled. Thus, K_M provides a measure of the substrate concentration required for significant catalysis to occur. In fact, for many enzymes, experimental evidence suggests that K_M provides an approximation of substrate concentration in vivo. When the K_M is known, the fraction of sites filled, f_{ES}, at any substrate concentration can be calculated from

$$f_{ES} = \frac{V}{V_{max}} = \frac{[S]}{[S] + K_M} \tag{24}$$

Second, K_M is related to the rate constants of the individual steps in the catalytic scheme given in equation 9. In equation 15, K_M is defined as $(k_{-1} + k_2)/k_1$. Consider a limiting case in which k_{-1} is much greater than k_2. Under such circumstances, the ES complex dissociates to E and S much more rapidly than product is formed. Under these conditions ($k_{-1} \gg k_2$),

$$K_M = \frac{k_{-1}}{k_1} \tag{25}$$

The dissociation constant of the ES complex is given by

$$K_{ES} = \frac{[E][S]}{[ES]} = \frac{k_{-1}}{k_1} \tag{26}$$

In other words, K*M* *is equal to the dissociation constant of the ES complex if* k₂ *is much smaller than* k₋₁. When this condition is met, K_M is a measure of the strength of the ES complex: a high K_M indicates weak binding; a low K_M indicates strong binding. It must be stressed that K_M indicates the affinity of the ES complex only when k_{-1} is much greater than k_2.

The maximal rate, V_{max}, reveals the *turnover number* of an enzyme, which is *the number of substrate molecules converted into product by an enzyme molecule in a unit time when the enzyme is fully saturated with substrate.* It is equal to the kinetic constant k_2, which is also called k_{cat}. The maximal rate, V_{max}, reveals the turnover number of an enzyme if the concentration of active sites

[E]$_T$ is known, because

$$V_{max} = k_2[E]_T$$

and thus

$$k_2 = V_{max}/[E]_T \qquad (27)$$

For example, a 10^{-6} M solution of carbonic anhydrase catalyzes the formation of 0.6 M H_2CO_3 per second when it is fully saturated with substrate. Hence, k_2 is 6×10^5 s^{-1}. This turnover number is one of the largest known. Each catalyzed reaction takes place in a time equal to $1/k_2$, which is 1.7 μs for carbonic anhydrase. The turnover numbers of most enzymes with their physiological substrates fall in the range from 1 to 10^4 per second (Table 8.6).

8.4.2 Kinetic Perfection in Enzymatic Catalysis: The k_{cat}/K_M Criterion

When the substrate concentration is much greater than K_M, the rate of catalysis is equal to k_{cat}, the turnover number, as described in Section 8.4.1. However, most enzymes are not normally saturated with substrate. Under physiological conditions, the $[S]/K_M$ ratio is typically between 0.01 and 1.0. When $[S] << K_M$, the enzymatic rate is much less than k_{cat} because most of the active sites are unoccupied. Is there a number that characterizes the kinetics of an enzyme under these more typical cellular conditions? Indeed there is, as can be shown by combining equations 10 and 16 to give

$$V_0 = \frac{k_{cat}}{K_M}[E][S] \qquad (28)$$

When $[S] << K_M$, the concentration of free enzyme, [E], is nearly equal to the total concentration of enzyme [E$_T$], so

$$V_0 = \frac{k_{cat}}{K_M}[S][E]_T \qquad (29)$$

Thus, when $[S] << K_M$, the enzymatic velocity depends on the values of k_{cat}/K_M, [S], and [E]$_T$. Under these conditions, k_{cat}/K_M is the rate constant for the interaction of S and E and can be used as a measure of catalytic efficiency. For instance, by using k_{cat}/K_M values, one can compare an enzyme's preference for different substrates. Table 8.7 shows the k_{cat}/K_M

TABLE 8.6 Maximum turnover numbers of some enzymes

Enzyme	Turnover number (per second)
Carbonic anhydrase	600,000
3-Ketosteroid isomerase	280,000
Acetylcholinesterase	25,000
Penicillinase	2,000
Lactate dehydrogenase	1,000
Chymotrypsin	100
DNA polymerase I	15
Tryptophan synthetase	2
Lysozyme	0.5

TABLE 8.7 Substrate preferences of chymotrypsin

Amino acid in ester	Amino acid side chain	k_{cat}/K_M (s^{-1} M^{-1})
Glycine	—H	1.3×10^{-1}
Valine	—CH with CH$_3$ and CH$_3$	2.0
Norvaline	—CH$_2$CH$_2$CH$_3$	3.6×10^2
Norleucine	—CH$_2$CH$_2$CH$_2$CH$_3$	3.0×10^3
Phenylalanine	—CH$_2$— (phenyl ring)	1.0×10^5

Source: After A. Fersht, *Structure and Mechanism in Protein Science: A Guide to Enzyme Catalysis and Protein Folding* (W. H. Freeman and Company, 1999), Table 7.3.

values for several different substrates of chymotrypsin (Section 9.1.1). Chymotrypsin clearly has a preference for cleaving next to bulky, hydrophobic side chains.

How efficient can an enzyme be? We can approach this question by determining whether there are any physical limits on the value of k_{cat}/K_M. Note that this ratio depends on k_1, k_{-1}, and k_2, as can be shown by substituting for K_M.

$$\frac{k_{cat}}{K_M} = \frac{k_{cat}}{k_{-1} + k_{cat}/k_1} = \frac{K_{cat}}{k_{cat} + k_{-1}} k_1 < k_1 \qquad (30)$$

Suppose that the rate of formation of product (k_{cat}) is much faster than the rate of dissociation of the ES complex (k_{-1}). The value of k_{cat}/K_M then approaches k_1. Thus, the ultimate limit on the value of k_{cat}/K_M is set by k_1, the rate of formation of the ES complex. *This rate cannot be faster than the diffusion-controlled encounter of an enzyme and its substrate.* Diffusion limits the value of k_1 so that it cannot be higher than between 10^8 and 10^9 s^{-1} M^{-1}. Hence, the upper limit on k_{cat}/K_M is between 10^8 and 10^9 s^{-1} M^{-1}.

The k_{cat}/K_M ratios of the enzymes superoxide dismutase, acetylcholinesterase, and triosephosphate isomerase are between 10^8 and 10^9 s^{-1} M^{-1}. Enzymes such as these that have k_{cat}/K_M ratios at the upper limits have attained *kinetic perfection*. *Their catalytic velocity is restricted only by the rate at which they encounter substrate in the solution* (Table 8.8). Any further gain in catalytic rate can come only by decreasing the time for diffusion. Remember that the active site is only a small part of the total enzyme structure. Yet, for catalytically perfect enzymes, every encounter between enzyme and substrate is productive. In these cases, there may be attractive electrostatic forces on the enzyme that entice the substrate to the active site. These forces are sometimes referred to poetically as *Circe effects*.

The limit imposed by the rate of diffusion in solution can also be partly overcome by confining substrates and products in the limited volume of a multienzyme complex. Indeed, some series of enzymes are associated into organized assemblies (Section 17.1.9) so that the product of one enzyme is very rapidly found by the next enzyme. In effect, products are channeled from one enzyme to the next, much as in an assembly line.

Circe effect—

The utilization of attractive forces to lure a substrate into a site in which it undergoes a transformation of structure, as defined by William P. Jencks, an enzymologist, who coined the term.

A goddess of Greek mythology, Circe lured Odysseus's men to her house and then transformed them into pigs.

TABLE 8.8 Enzymes for which k_{cat}/K_M is close to the diffusion-controlled rate of encounter

Enzyme	k_{cat}/K_M (s^{-1}M^{-1})
Acetylcholinesterase	1.6×10^8
Carbonic anhydrase	8.3×10^7
Catalase	4×10^7
Crotonase	2.8×10^8
Fumarase	1.6×10^8
Triose phosphate isomerase	2.4×10^8
β-Lactamase	1×10^8
Superoxide dismutase	7×10^9

Source: After A. Fersht, *Structure and Mechanism in Protein Science: A Guide to Enzyme Catalysis and Protein Folding* (W. H. Freeman and Company, 1999), Table 4.5.

8.4.3 Most Biochemical Reactions Include Multiple Substrates

Most reactions in biological systems usually include two substrates and two products and can be represented by the bisubstrate reaction:

$$A + B \rightleftharpoons P + Q$$

The majority of such reactions entail the transfer of a functional group, such as a phosphoryl or an ammonium group, from one substrate to the other. In oxidation–reduction reactions, electrons are transferred between substrates. Multiple substrate reactions can be divided into two classes: *sequential displacement* and *double displacement*.

Sequential Displacement. In the sequential mechanism, all substrates must bind to the enzyme before any product is released. Consequently, in a bisubstrate reaction, a *ternary complex* of the enzyme and both substrates forms. Sequential mechanisms are of two types: ordered, in which the substrates bind the enzyme in a defined sequence, and random.

Many enzymes that have NAD^+ or $NADH$ as a substrate exhibit the sequential ordered mechanism. Consider lactate dehydrogenase, an important enzyme in glucose metabolism (Section 16.1.9). This enzyme reduces pyruvate to lactate while oxidizing $NADH$ to NAD^+.

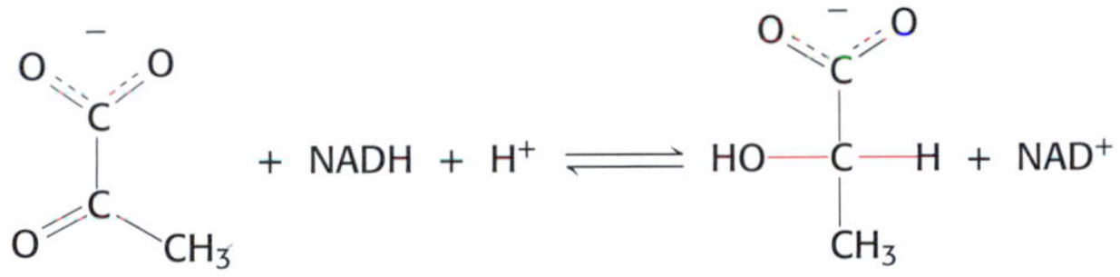

Pyruvate   **Lactate**

In the ordered sequential mechanism, the coenzyme always binds first and the lactate is always released first. This sequence can be represented as follows in a notation developed by W. Wallace Cleland:

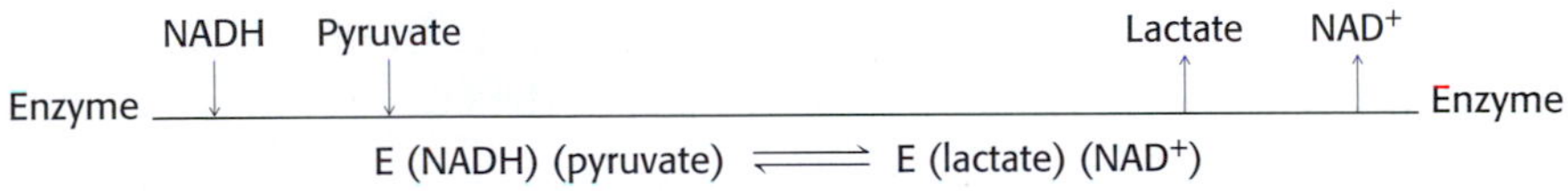

The enzyme exists as a ternary complex: first, consisting of the enzyme and substrates and, after catalysis, the enzyme and products.

In the random sequential mechanism, the order of addition of substrates and release of products is random. Sequential random reactions are illustrated by the formation of phosphocreatine and ADP from ATP and creatine, a reaction catalyzed by creatine kinase (Section 14.1.5).

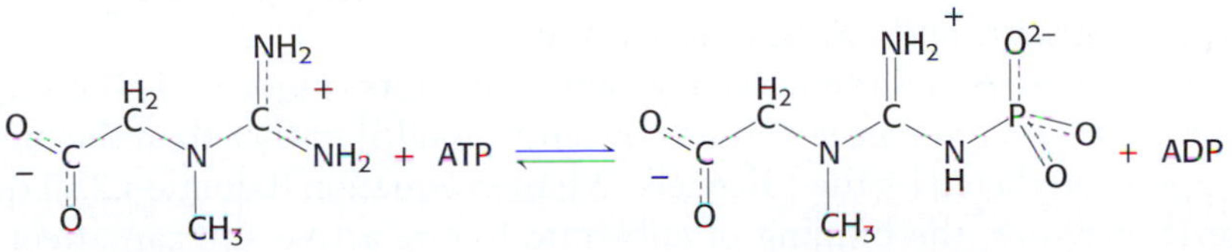

Creatine   **Phosphocreatine**

Phosphocreatine is an important energy source in muscle. Sequential random reactions can also be depicted in the Cleland notation.

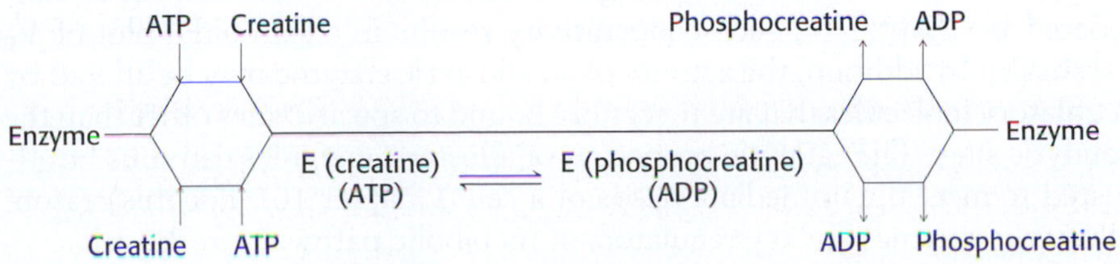

Although the order of certain events is random, the reaction still passes through the ternary complexes including, first, substrates and, then, products.

Double-Displacement (Ping-Pong) Reactions. In double-displacement, or Ping-Pong, reactions, one or more products are released before all substrates bind the enzyme. The defining feature of double-displacement reactions is the existence of a *substituted enzyme intermediate*, in which the enzyme is temporarily modified. Reactions that shuttle amino groups between amino acids and α-keto acids are classic examples of double-displacement mechanisms. The enzyme aspartate aminotransferase (Section 23.3.1) catalyzes the transfer of an amino group from aspartate to α-ketoglutarate.

Aspartate α-Ketoglutarate   Glutamate   Oxaloacetate

The sequence of events can be portrayed as the following diagram.

Aspartate   Oxaloacetate α-Ketoglutarate   Glutamate

Enzyme

E (aspartate)   (E-NH$_3$) (oxaloacetate)   (E-NH$_3$)   (E-NH$_3$) (α-ketoglutarate)   E (glutamate)   Enzyme

After aspartate binds to the enzyme, the enzyme removes aspartate's amino group to form the substituted enzyme intermediate. The first product, oxaloacetate, subsequently departs. The second substrate, α-ketoglutarate, binds to the enzyme, accepts the amino group from the modified enzyme, and is then released as the final product, glutamate. In the Cleland notation, the substrates appear to bounce on and off the enzyme analogously to a Ping-Pong ball bouncing on a table.

8.4.4 Allosteric Enzymes Do Not Obey Michaelis-Menten Kinetics

The Michaelis-Menten model has greatly assisted the development of enzyme chemistry. Its virtues are simplicity and broad applicability. However, the Michaelis-Menten model cannot account for the kinetic properties of many enzymes. An important group of enzymes that do not obey Michaelis-Menten kinetics comprises the *allosteric enzymes*. These enzymes consist of multiple subunits and multiple active sites.

Allosteric enzymes often display sigmoidal plots (Figure 8.14) of the reaction velocity V_0 versus substrate concentration [S], rather than the hyperbolic plots predicted by the Michaelis-Menten equation (equation 23). In allosteric enzymes, the binding of substrate to one active site can affect the properties of other active sites in the same enzyme molecule. A possible outcome of this interaction between subunits is that the binding of substrate becomes *cooperative;* that is, the binding of substrate to one active site of the enzyme facilitates substrate binding to the other active sites. As will be considered in Chapter 10, such cooperativity results in a sigmoidal plot of V_0 versus [S]. In addition, the activity of an allosteric enzyme may be altered by regulatory molecules that are reversibly bound to specific sites other than the catalytic sites. The catalytic properties of allosteric enzymes can thus be adjusted to meet the immediate needs of a cell (Chapter 10). For this reason, allosteric enzymes are key regulators of metabolic pathways in the cell.

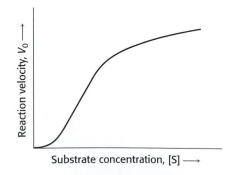

FIGURE 8.14 Kinetics for an allosteric enzyme. Allosteric enzymes display a sigmoidal dependence of reaction velocity on substrate concentration.

8.5 ENZYMES CAN BE INHIBITED BY SPECIFIC MOLECULES

The activity of many enzymes can be inhibited by the binding of specific small molecules and ions. This means of inhibiting enzyme activity serves as a major control mechanism in biological systems. The regulation of allosteric enzymes typifies this type of control. In addition, many drugs and toxic agents act by inhibiting enzymes. Inhibition by particular chemicals can be a source of insight into the mechanism of enzyme action: specific inhibitors can often be used to identify residues critical for catalysis. The value of transition-state analogs as potent inhibitors will be discussed shortly.

Enzyme inhibition can be either reversible or irreversible. An *irreversible inhibitor* dissociates very slowly from its target enzyme because it has become tightly bound to the enzyme, either covalently or noncovalently. Some irreversible inhibitors are important drugs. Penicillin acts by covalently modifying the enzyme transpeptidase, thereby preventing the synthesis of bacterial cell walls and thus killing the bacteria (Section 8.5.5). Aspirin acts by covalently modifying the enzyme cyclooxygenase, reducing the synthesis of inflammatory signals.

Reversible inhibition, in contrast with irreversible inhibition, is characterized by a rapid dissociation of the enzyme–inhibitor complex. In *competitive inhibition*, an enzyme can bind substrate (forming an ES complex) or inhibitor (EI) but not both (ESI). The competitive inhibitor resembles the substrate and binds to the active site of the enzyme (Figure 8.15). The substrate is thereby prevented from binding to the same active site. *A competitive inhibitor diminishes the rate of catalysis by reducing the proportion of enzyme molecules bound to a substrate.* At any given inhibitor concentration, competitive inhibition can be relieved by increasing the substrate concentration. Under these conditions, the substrate "outcompetes" the inhibitor for the active site. Methotrexate is a structural analog of tetrahydrofolate, a coenzyme for the enzyme dihydrofolate reductase, which plays a role in the biosynthesis of purines and pyrimidines (Figure 8.16). It binds to dihydrofolate reductase 1000-fold more tightly than the natural substrate and inhibits nucleotide base synthesis. It is used to treat cancer.

FIGURE 8.15 Distinction between a competitive and a noncompetitive inhibitor. (Top) enzyme–substrate complex; (middle) a competitive inhibitor binds at the active site and thus prevents the substrate from binding; (bottom) a noncompetitive inhibitor does not prevent the substrate from binding.

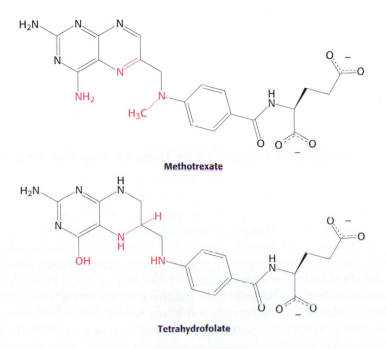

Methotrexate

Tetrahydrofolate

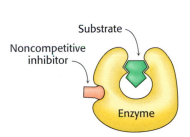

FIGURE 8.16 Enzyme inhibitors. The cofactor tetrahydrofolate and its structural analog methotrexate. Regions with structural differences are shown in red.

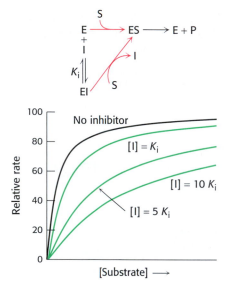

FIGURE 8.17 Kinetics of a competitive inhibitor. As the concentration of a competitive inhibitor increases, higher concentrations of substrate are required to attain a particular reaction velocity. The reaction pathway suggests how sufficiently high concentrations of substrate can completely relieve competitive inhibition.

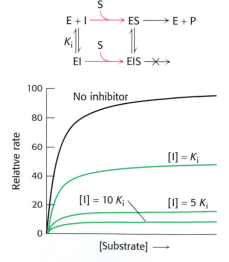

FIGURE 8.18 Kinetics of a non-competitive inhibitor. The reaction pathway shows that the inhibitor binds both to free enzyme and to enzyme complex. Consequently, V_{max} cannot be attained, even at high substrate concentrations.

In *noncompetitive inhibition*, which also is reversible, the inhibitor and substrate can bind simultaneously to an enzyme molecule at different binding sites (see Figure 8.16). A noncompetitive inhibitor acts by decreasing the turnover number rather than by diminishing the proportion of enzyme molecules that are bound to substrate. Noncompetitive inhibition, in contrast with competitive inhibition, cannot be overcome by increasing the substrate concentration. A more complex pattern, called *mixed inhibition*, is produced when a single inhibitor both hinders the binding of substrate and decreases the turnover number of the enzyme.

8.5.1 Competitive and Noncompetitive Inhibition Are Kinetically Distinguishable

How can we determine whether a reversible inhibitor acts by competitive or noncompetitive inhibition? Let us consider only enzymes that exhibit Michaelis-Menten kinetics. Measurements of the rates of catalysis at different concentrations of substrate and inhibitor serve to distinguish the two types of inhibition. In *competitive inhibition*, the inhibitor competes with the substrate for the active site. The dissociation constant for the inhibitor is given by

$$K_i = [E][I]/[EI]$$

Because increasing the amount of substrate can overcome the inhibition, V_{max} can be attained in the presence of a competitive inhibitor (Figure 8.17). *The hallmark of competitive inhibition is that it can be overcome by a sufficiently high concentration of substrate.* However, the apparent value of K_M is altered; the effect of a competitive inhibitor is to increase the apparent value of K_M. This new value of K_M, called K_M^{app}, is numerically equal to

$$K_M^{app} = K_M(1 + [I]/K_i)$$

where [I] is the concentration of inhibitor and K_i is the dissociation constant for the enzyme–inhibitor complex. As the value of [I] increases, the value of K_M^{app} increases (see Figure 8.17). In the presence of a competitive inhibitor, an enzyme will have the same V_{max} as in the absence of an inhibitor.

In *noncompetitive inhibition* (Figure 8.18), substrate can still bind to the enzyme–inhibitor complex. However, the enzyme–inhibitor–substrate complex *does not* proceed to form product. The value of V_{max} is decreased to a new value called V_{max}^{app} while the value of K_M is unchanged. Why is V_{max} lowered while K_M remains unchanged? In essence, the inhibitor simply lowers the concentration of functional enzyme. The remaining enzyme behaves like a more dilute solution of enzyme; V_{max} is lower, but K_M is the same. *Noncompetitive inhibition cannot be overcome by increasing the substrate concentration.*

8.5.2 Irreversible Inhibitors Can Be Used to Map the Active Site

In Chapter 9, we will examine the chemical details of how enzymes function. The first step in obtaining the chemical mechanism of an enzyme is to determine what functional groups are required for enzyme activity. How can we ascertain these functional groups? X-ray crystallography (Section 4.5.2) of the enzyme bound to its substrate provides one approach. Irreversible inhibitors that covalently bond to the enzyme provide an alternative and often complementary means for elucidating functional groups at the enzyme active site because they modify the functional groups, which can then be identified. Irreversible inhibitors can be divided into three categories: group-specific reagents, substrate analogs, and suicide inhibitors.

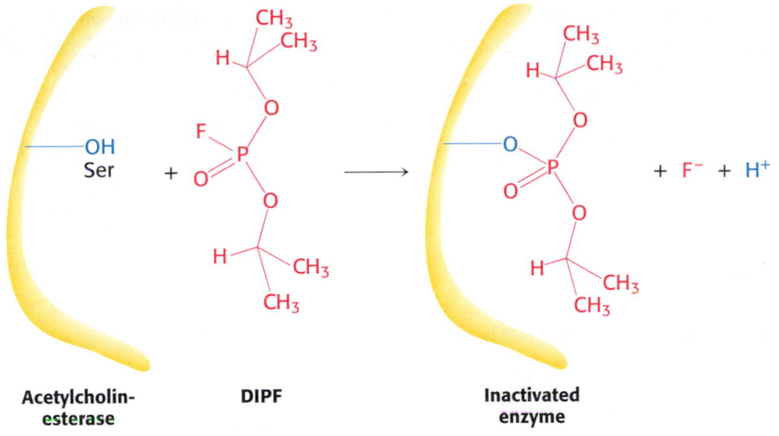

**Acetylcholin-
esterase**   **DIPF**   **Inactivated
enzyme**

FIGURE 8.19 Enzyme inhibition by diisopropylphosphofluoridate (DIPF), a group-specific reagent. DIPF can inhibit an enzyme by covalently modifying a crucial serine residue (Section 9.1.1).

Group-specific reagents react with specific R groups of amino acids. Two examples of group-specific reagents are diisopropylphosphofluoridate (DIPF; Figure 8.19) and iodoactemide (Figure 8.20). DIPF modifies only 1 of the 27 serine residues in the proteolytic enzyme chymotrypsin, implying that this serine residue is especially reactive. As we will see in Chapter 9, it is indeed the case that this serine residue is at the active site. DIPF also revealed a reactive serine residue in acetylcholinesterase, an enzyme important in the transmission of nerve impulses (see Figure 8.19). Thus, DIPF and similar compounds that bind and inactivate acetylcholinesterase are potent nerve gases.

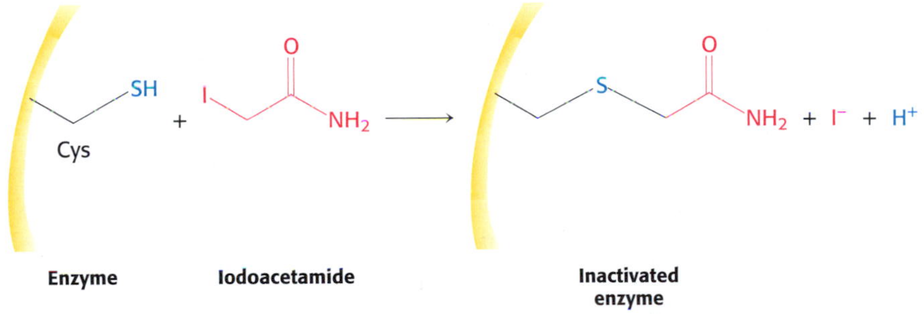

Enzyme   **Iodoacetamide**   **Inactivated
enzyme**

FIGURE 8.20 Enzyme inactivation by iodoacetamide, a group-specific reagent. Iodoacetamide can inactivate an enzyme by reacting with a critical cysteine residue.

Affinity labels are molecules that are structurally similar to the substrate for the enzyme that covalently modify active site residues. They are thus more specific for the enzyme active site than are group-specific reagents. Tosyl-L-phenylalanine chloromethyl ketone (TPCK) is a substrate analog for chymotrypsin (Figure 8.21). TPCK binds at the active site; and then reacts irreversibly with a histidine residue at that site, inhibiting the enzyme. The compound 3-bromoacetol is an affinity label for the enzyme triose phosphate isomerase (TIM). It mimics the normal substrate, dihydroxyacetone phosphate, by binding at the active site; then it covalently modifies the enzyme such that the enzyme is irreversibly inhibited (Figure 8.22).

Suicide inhibitors, or *mechanism-based inhibitors* are modified substrates that provide the most specific means to modify an enzyme active site. The inhibitor binds to the enzyme as a substrate and is initially processed by the normal catalytic mechanism. The mechanism of catalysis then generates a chemically reactive intermediate that inactivates the enzyme through covalent modification. The fact that the enzyme participates in its own irreversible inhibition strongly suggests that the covalently modified group on the enzyme is catalytically vital. One example of such an inhibitor is *N,N*-dimethylpropargylamine. A flavin prosthetic group of monoamine oxidase

FIGURE 8.21 Affinity labeling.
(A) Tosyl-L-phenylalanine chloromethyl ketone (TPCK) is a reactive analog of the normal substrate for the enzyme chymotrypsin. (B) TPCK binds at the active site of chymotrypsin and modifies an essential histidine residue.

FIGURE 8.22 Bromoacetol phosphate, an affinity label for triose phosphate isomerase (TIM). Bromoacetol phosphate, an analog of dihydroxyacetone phosphate, binds at the active site of the enzyme and covalently modifies a glutamic acid residue required for enzyme activity.

FIGURE 8.23 Mechanism-based (suicide) inhibition. Monoamine oxidase, an enzyme important for neurotransmitter synthesis, requires the cofactor FAD (flavin adenine dinucleotide). *N,N*-Dimethylpropargylamine inhibits monoamine oxidase by covalently modifying the flavin prosthetic group only after the inhibitor is first oxidized. The N-5 flavin adduct is stabilized by the addition of a proton.

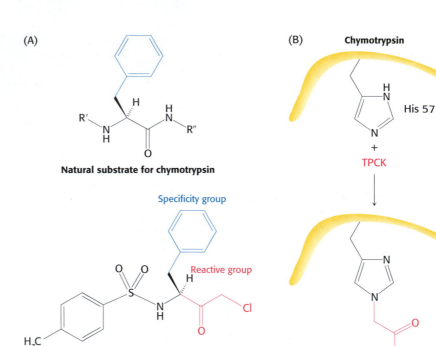

(A)

Natural substrate for chymotrypsin

Specificity group

Reactive group

Tosyl-L-phenylalanine chloromethyl ketone (TPCK)

(B) Chymotrypsin

His 57

TPCK

Triose phosphate isomerase (TIM)

Glu

Bromoacetol phosphate

Inactivated enzyme

Flavin prosthetic group

N,N-Dimethylpropargylamine

Oxidation

Alkylation − H⁺

+ H⁺

Stably modified flavin of inactivated enzyme

(MAO) oxidizes the *N,N*-dimethylpropargylamine, which in turn inactivates the enzyme by covalently modifying the flavin prosthetic group by alkylating N-5 (Figure 8.23). Monoamine oxidase deaminates neurotransmitters such as dopamine and serotonin, lowering their levels in the brain. Parkinson disease is associated with low levels of dopamine, and depression is associated with low levels of serotonin. The drug (−)deprenyl, which is used to treat Parkinson disease and depression, is a suicide inhibitor of monoamine oxidase.

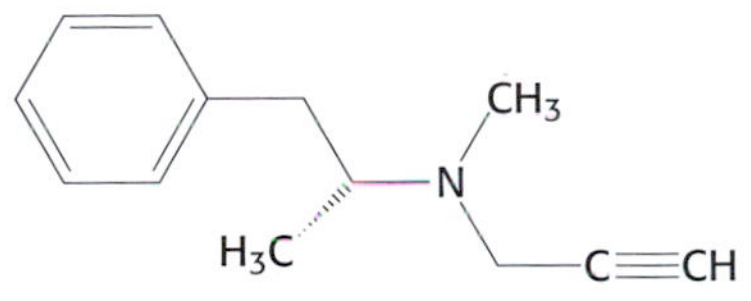

8.5.3 Transition-State Analogs Are Potent Inhibitors of Enzymes

We turn now to compounds that provide the most intimate views of the catalytic process itself. Linus Pauling proposed in 1948 that compounds resembling the transition state of a catalyzed reaction should be very effective inhibitors of enzymes. These mimics are called *transition-state analogs*. The inhibition of proline racemase is an instructive example. *The racemization of proline proceeds through a transition state in which the tetrahedral α-carbon atom has become trigonal by loss of a proton* (Figure 8.24). In the trigonal form, all three bonds are in the same plane; C_α also carries a net negative charge. This symmetric carbanion can be reprotonated on one side to give the L isomer or on the other side to give the D isomer. This picture is supported by the finding that the inhibitor pyrrole 2-carboxylate binds to the racemase 160 times as tightly as does proline. *The α-carbon atom of this inhibitor, like that of the transition state, is trigonal.* An analog that also carries a negative charge on C_α would be expected to bind even more tightly. In general, synthesizing compounds that more closely resemble the transition state than the substrate itself can produce highly potent and specific inhibitors of enzymes. The inhibitory power of transition-state analogs underscores the essence of catalysis: *selective binding of the transition state.*

(−)Deprenyl

(A)

(B)

L-Proline   **Planar transition state**   **D-Proline**   **Pyrrole 2-carboxylic acid** (transition-state analog)

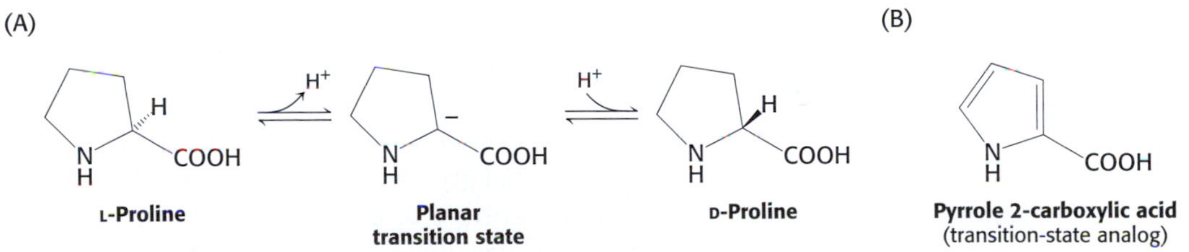

FIGURE 8.24 Inhibition by transition state analogs. (A) The isomerization of L-proline to D-proline by proline racemase, a bacterial enzyme, proceeds through a planar transition state in which the α carbon is trigonal rather than tetrahedral. (B) Pyrrole 2-carboxylate, a transition state analog because of its trigonal geometry, is a potent inhibitor of proline racemase.

8.5.4 Catalytic Antibodies Demonstrate the Importance of Selective Binding of the Transition State to Enzymatic Activity

Antibodies that recognize transition states should function as catalysts, if our understanding of the importance of the transition state to catalysis is correct. The preparation of an antibody that catalyzes the insertion of a metal ion into a porphyrin nicely illustrates the validity of this approach. Ferrochelatase, the final enzyme in the biosynthetic pathway for the production of heme, catalyzes the insertion of Fe^{2+} into protoporphyrin IX. The nearly planar porphyrin must be bent for iron to enter. The recently determined crystal structure of the ferrochelatase bound to a substrate analog confirms that the enzyme does indeed bend one of the pyrole rings, distorting it 36 degrees to insert the iron.

FIGURE 8.25 Use of transition-state analogs to generate catalytic antibodies. The insertion of a metal ion into a porphyrin by ferrochelatase proceeds through a transition state in which the porphyrin is bent. *N*-Methylmesoporphyrin, a bent porphyrin that resembles the transition state of the ferrochelatase-catalyzed reaction, was used to generate an antibody that also catalyzes the insertion of a metal ion into a porphyrin ring.

The problem was to find a transition-state analog for this metallation reaction that could be used as an antigen (immunogen) to generate an antibody. The solution came from the results of studies showing that an alkylated porphyrin, *N*-methylprotoporphyrin, is a potent inhibitor of ferrochelatase. This compound resembles the transition state because N-*alkylation forces the porphyrin to be bent*. Moreover, it was known that *N*-alkylporphyrins chelate metal ions 10^4 times as fast as their unalkylated counterparts do. Bending increases the exposure of the pyrrole nitrogen lone pairs of electrons to solvent, which facilitates metal in binding.

An antibody catalyst was produced with the use of an *N*-alkylporphyrin as the immunogen. The resulting antibody presumably distorts a planar porphyrin (Figure 8.25) to facilitate the entry of a metal. On average, an antibody molecule metallated 80 porphyrin molecules per hour, a rate only 10-fold less than that of ferrochelatase and 2500-fold faster than the uncatalyzed reaction. *Catalytic antibodies (abzymes) can indeed be produced by using transition-state analogs as antigens.* Antibodies catalyzing many other kinds of chemical reactions—exemplified by ester and amide hydrolysis, amide-bond formation, transesterification, photoinduced cleavage, photoinduced dimerization, decarboxylation, and oxidization—have been produced with the use of similar strategies. The results of studies with transition-state analogs provide strong evidence that enzymes can function complementary in structure to the transition state. *The power of transition-state analogs is now evident: (1) they are sources of insight into catalytic mechanisms, (2) they can serve as potent and specific inhibitors of enzymes, and (3) they can be used as immunogens to generate a wide range of novel catalysts.*

8.5.5 Penicillin Irreversibly Inactivates a Key Enzyme in Bacterial Cell-Wall Synthesis

Penicillin, the first antibiotic discovered, consists of a thiazolidine ring fused to a *β-lactam* ring, to which a variable R group is attached by a peptide bond (Figure 8.26A). In benzyl penicillin, for example, R is a benzyl group (Figure 8.26B). This structure can undergo a variety of rearrangements, and, in particular, the β-lactam ring is very labile. Indeed, this instability is closely tied to the antibiotic action of penicillin, as will be evident shortly.

How does penicillin inhibit bacterial growth? In 1957, Joshua Lederberg showed that bacteria ordinarily susceptible to penicillin could be grown in its presence if a hypertonic medium were used. The organisms obtained in this way, called *protoplasts*, are devoid of a cell wall and consequently lyse when transferred to a normal medium. Hence, penicillin was inferred to interfere with the synthesis of the bacterial cell wall. The cell-wall macromolecule, called a *peptidoglycan*, consists of linear polysaccharide chains that

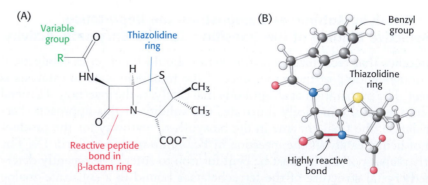

FIGURE 8.26 Structure of penicillin. The reactive site of penicillin is the peptide bond of its β-lactam ring. (A) Structural formula of penicillin. (B) Representation of benzyl penicillin.

are cross-linked by short peptides (Figure 8.27). The enormous bag-shaped peptidoglycan confers mechanical support and prevents bacteria from bursting in response to their high internal osmotic pressure.

In 1965, James Park and Jack Strominger independently deduced that penicillin blocks the last step in cell-wall synthesis—namely, the cross-linking of different peptidoglycan strands. In the formation of the cell wall of *Staphylococcus aureus*, the amino group at one end of a pentaglycine chain attacks the peptide bond between two D-alanine residues in another peptide unit (Figure 8.28). A peptide bond is formed between glycine and one of the D-alanine residues; the other D-alanine residue is released. This cross-linking reaction is catalyzed by *glycopeptide transpeptidase*. Bacterial cell walls are unique in containing D amino acids, which form cross-links by a mechanism different from that used to synthesize proteins.

Penicillin inhibits the cross-linking transpeptidase by the Trojan horse stratagem. The transpeptidase normally forms an *acyl intermediate* with the penultimate D-alanine residue of the D-Ala-D-Ala peptide (Figure 8.29). This covalent acyl-enzyme intermediate then reacts with the amino group of the terminal glycine in another peptide to form the cross-link. Penicillin is welcomed into the active site of the transpeptidase because it mimics the D-Ala-D-Ala moiety of the normal substrate (Figure 8.30). Bound penicillin then forms a covalent bond with a serine residue at the active site of the enzyme. *This penicilloyl-enzyme does not react further. Hence, the transpeptidase is irreversibly inhibited and cell-wall synthesis cannot take place.*

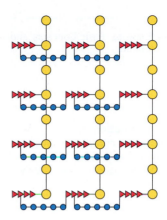

FIGURE 8.27 Schematic representation of the peptidoglycan in *Staphylococcus aureus*. The sugars are shown in yellow, the tetrapeptides in red, and the pentaglycine bridges in blue. The cell wall is a single, enormous, bag-shaped macromolecule because of extensive cross-linking.

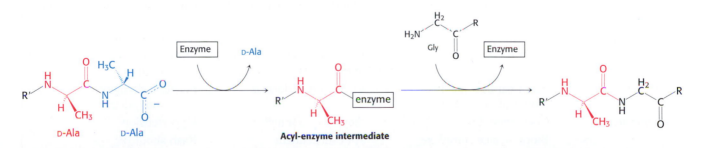

Terminal glycine residue of pentaglycine bridge   **Terminal D-Ala-D-Ala unit**   **Gly-D-Ala crosslink**   **D-Ala**

FIGURE 8.28 Formation of cross-links in *S. aureus* peptidoglycan. The terminal amino group of the pentaglycine bridge in the cell wall attacks the peptide bond between two D-alanine residues to form a cross-link.

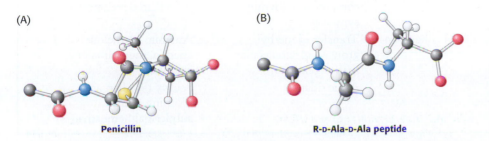

FIGURE 8.29 Transpeptidation reaction. An acyl-enzyme intermediate is formed in the transpeptidation reaction leading to cross-link formation.

(A)   (B)

Penicillin   **R-D-Ala-D-Ala peptide**

FIGURE 8.30 Conformations of penicillin and a normal substrate. The conformation of penicillin in the vicinity of its reactive peptide bond (A) resembles the postulated conformation of the transition state of R-D-Ala-D-Ala (B) in the transpeptidation reaction. [After B. Lee. *J. Mol. Biol.* 61(1971):464.]

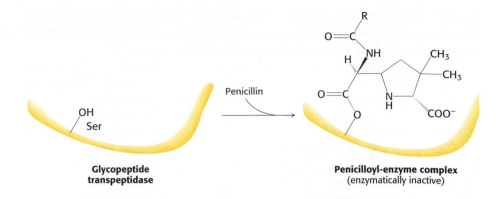

FIGURE 8.31 Formation of a penicilloyl-enzyme complex. Penicillin reacts with the transpeptidase to form an inactive complex, which is indefinitely stable.

Glycopeptide
transpeptidase

Penicilloyl-enzyme complex
(enzymatically inactive)

Why is penicillin such an effective inhibitor of the transpeptidase? The highly strained, four-membered β-lactam ring of penicillin makes it especially reactive. On binding to the transpeptidase, the serine residue at the active site attacks the carbonyl carbon atom of the lactam ring to form the penicilloyl-serine derivative (Figure 8.31). Because the peptidase participates in its own inactivation, penicillin acts as a suicide inhibitor.

8.6 VITAMINS ARE OFTEN PRECURSORS TO COENZYMES

Earlier (Section 8.1.1), we considered the fact that many enzymes require cofactors to be catalytically active. One class of these cofactors, termed coenzymes, consists of small organic molecules, many of which are derived from *vitamins*. Vitamins themselves are organic molecules that are needed in small amounts in the diets of some higher animals. These mole-

TABLE 8.9 Water-Soluble Vitamins

Vitamin	Coenzyme	Typical reaction type	Consequences of deficiency
Thiamine (B_1)	Thiamine pyrophosphate	Aldehyde transfer	Beriberi (weight loss, heart problems, neurological dysfunction)
Riboflavin (B_2)	Flavin adenine dinucleotide (FAD)	Oxidation–reduction	Cheliosis and angular stomatitus (lesions of the mouth), dermatitis
Pyridoxine (B_6)	Pyridoxal phosphate	Group transfer to or from amino acids	Depression, confusion, convulsions
Nicotinic acid (niacin)	Nicotinamide adenine dinucleotide (NAD^+)	Oxidation–reduction	Pellagra (dermatitis, depression, diarrhea)
Pantothenic acid	Coenzyme A	Acyl–group transfer	Hypertension
Biotin	Biotin–lysine complexes (biocytin)	ATP-dependent carboxylation and carboxyl-group transfer	Rash about the eyebrows, muscle pain, fatigue (rare)
Folic acid	Tetrahydrofolate	Transfer of one-carbon components; thymine synthesis	Anemia, neural-tube defects in development
B_{12}	5′-Deoxyadenosyl cobalamin	Transfer of methyl groups; intramolecular rearrangements	Anemia, pernicious anemia, methylmalonic acidosis
C (ascorbic acid)		Antioxidant	Scurvy (swollen and bleeding gums, subdermal hemorrhages)

TABLE 8.10 Fat-soluble vitamins

Vitamin	Function	Deficiency
A	Antioxidant	Inhibition of sperm production; lesions in muscles and nerves (rare)
D	Regulation of calcium and phosphate metabolism	Rickets (children): skeletal deformaties, impaired growth Osteomalacia (adults): soft, bending bones
E	Roles in vision, growth, reproduction	Night blindness, cornea damage, damage to respiratory and gastrointestinal tract
K	Blood coagulation	Subdermal hemorrhaging

cules serve the same roles in nearly all forms of life, but higher animals lost the capacity to synthesize them in the course of evolution. For instance, whereas *E. coli* can thrive on glucose and organic salts, human beings require at least 12 vitamins in the diet. The biosynthetic pathways for vitamins can be complex; thus, it is biologically more efficient to ingest vitamins than to synthesize the enzymes required to construct them from simple molecules. This efficiency comes at the cost of dependence on other organisms for chemicals essential for life. Indeed, vitamin deficiency can generate diseases in all organisms requiring these molecules (Tables 8.9 and 8.10). Vitamins can be grouped according to whether they are soluble in water or in nonpolar solvents.

8.6.1 Water-Soluble Vitamins Function As Coenzymes

Table 8.9 lists the *water-soluble vitamins*—ascorbic acid (vitamin C) and a series known as the vitamin B complex (Figure 8.32). Ascorbate, the ionized form of ascorbic acid, serves as a reducing agent (an antioxidant), as will be discussed shortly. The vitamin B series comprises components of coenzymes. Note that, in all cases except vitamin C, the vitamin must be modified before it can serve its function.

Vitamin deficiencies are capable of causing a variety of pathological conditions (see Table 8.9). However, many of the same symptoms can result from conditions other than lack of a vitamin. For this reason and because vitamins are required in relatively small amounts, pathological conditions resulting from vitamin deficiencies are often difficult to diagnose.

The requirement for vitamin C proved relatively straightforward to demonstrate. This water-soluble vitamin is not used as a coenzyme but is

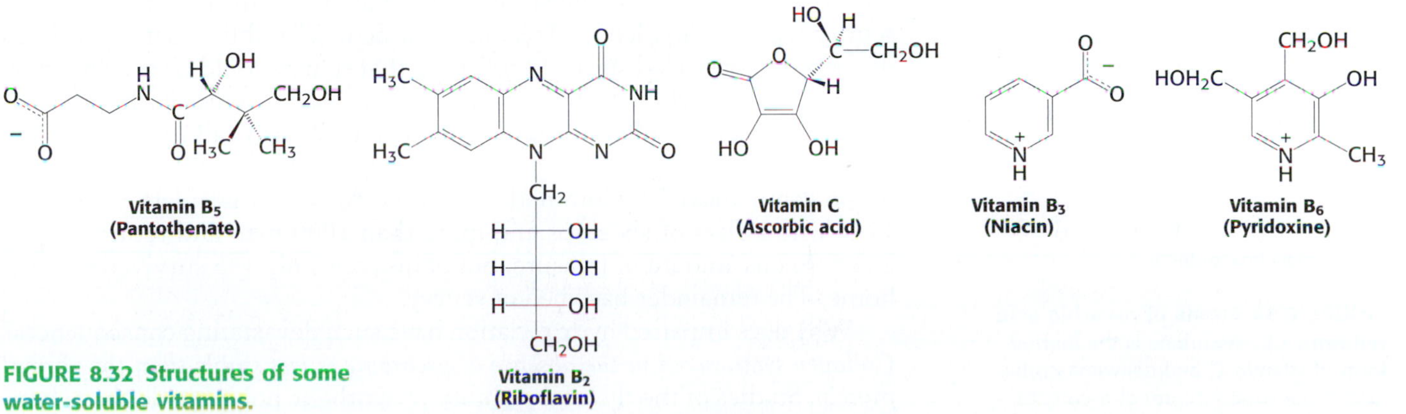

FIGURE 8.32 Structures of some water-soluble vitamins.

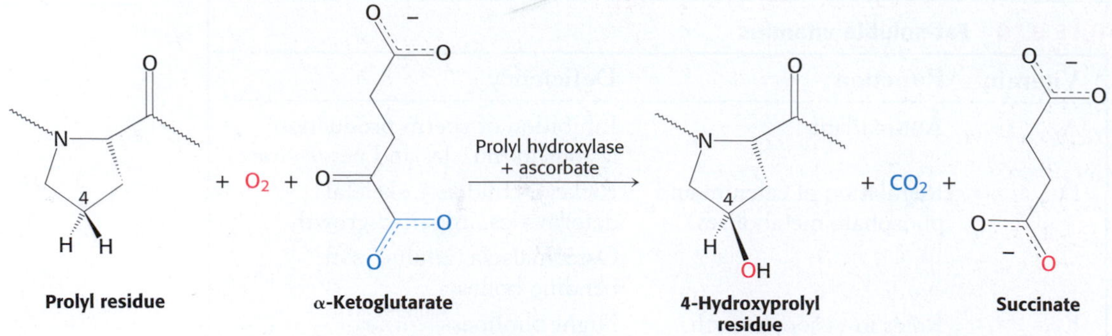

FIGURE 8.33 Formation of 4-hydroxyproline. Proline is hydroxylated at C-4 by the action of prolyl hydroxylase, an enzyme that activates molecular oxygen.

still required for the continued activity of proyl hydroxylase. This enzyme synthesizes 4-hydroxyproline, an amino acid that is required in collagen, the major connective tissue in vertebrates, but is rarely found anywhere else. How is this unusual amino acid formed and what is its role? The results of radioactive-labeling studies showed that proline residues on the amino side of glycine residues in nascent collagen chains become hydroxylated. The oxygen atom that becomes attached to C-4 of proline comes from molecular oxygen, O_2. The other oxygen atom of O_2 is taken up by α-ketoglutarate, which is converted into succinate (Figure 8.33). This complex reaction is catalyzed by *prolyl hydroxylase*, a *dioxygenase*. It is assisted by an Fe^{2+} ion, which is tightly bound to it and needed to activate O_2. The enzyme also converts α-ketoglutarate into succinate without hydroxylating proline. In this partial reaction, an oxidized iron complex is formed, which inactivates the enzyme. How is the active enzyme regenerated? *Ascorbate (vitamin C)* comes to the rescue by reducing the ferric ion of the inactivated enzyme. In the recovery process, ascorbate is oxidized to dehydroascorbic acid (Figure 8.34). Thus, ascorbate serves here as a specific *antioxidant*.

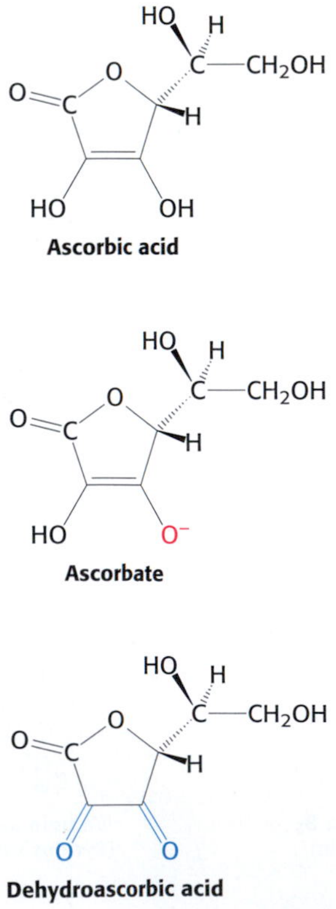

Primates are unable to synthesize ascorbic acid and hence must acquire it from their diets. The importance of ascorbate becomes strikingly evident in *scurvy*. Jacques Cartier in 1536 gave a vivid description of this dietary deficiency disease, which afflicted his men as they were exploring the Saint Lawrence River:

> Some did lose all their strength, and could not stand on their feet. … Others also had all their skins spotted with spots of blood of a purple colour: then did it ascend up to their ankles, knees, thighs, shoulders, arms, and necks. Their mouths became stinking, their gums so rotten, that all the flesh did fall off, even to the roots of the teeth, which did also almost all fall out.

James Lind, a Scottish physician, illuminated the means of preventing scurvy in an article titled "A Treatise of the Scurvy" published in 1747. Lind described a controlled study establishing that scurvy could be prevented by including citrus fruits in the diet. The Royal Navy eventually began issuing lime rations to sailors, from which custom British sailors acquired the nickname "limeys." Lind's research was inspired by the plight of an expedition commanded by Commodore George Anson. Anson left England in 1740 with a fleet of six ships and more than 1000 men and returned with an enormous amount of treasure, but of his crew only 145 survived to reach home. The remainder had died of scurvy.

Why does impaired hydroxylation have such devastating consequences? *Collagen synthesized in the absence of ascorbate is less stable than the normal protein.* Studies of the thermal stability of synthetic polypeptides have been

FIGURE 8.34 Forms of ascorbic acid (vitamin C). Ascorbate is the ionized form of vitamin C, and dehydroascorbic acid is the oxidized form of ascorbate.

especially informative. Hydroxyproline stabilizes the collagen triple helix by forming interstrand hydrogen bonds. The abnormal fibers formed by insufficiently hydroxylated collagen contribute to the skin lesions and blood-vessel fragility seen in scurvy.

8.6.2 Fat-Soluble Vitamins Participate in Diverse Processes Such as Blood Clotting and Vision

Not all vitamins function as coenzymes. The *fat-soluble vitamins*, which are designated by the letters A, D, E, and K (Figure 8.35, Table 8.10), have a diverse array of functions. Vitamin K, which is required for normal blood clotting (*K* from the German *koagulation*), participates in the carboxylation of glutamate residues to γ-carboxyglutamate, which makes modified glutamic acid a much stronger chelator of Ca^{2+} (Section 10.5.7). Vitamin A (retinol) is the precursor of retinal, the light-sensitive group in rhodopsin and other visual pigments (Section 32.3.1). A deficiency of this vitamin leads to night blindness. In addition, young animals require vitamin A for growth. Retinoic acid, which contains a terminal carboxylate in place of the alcohol terminus of retinol, serves as a signal molecule and activates the transcription of specific genes that mediate growth and development (Section 31.3). A metabolite of vitamin D is a hormone that regulates the metabolism of calcium and phosphorus. A deficiency in vitamin D impairs bone formation in growing animals. Infertility in rats is a consequence of vitamin E (α-tocopherol) deficiency. This vitamin reacts with and neutralizes reactive oxygen species such as hydroxyl, radicals before they can oxidize unsaturated membrane lipids, damaging cell structures.

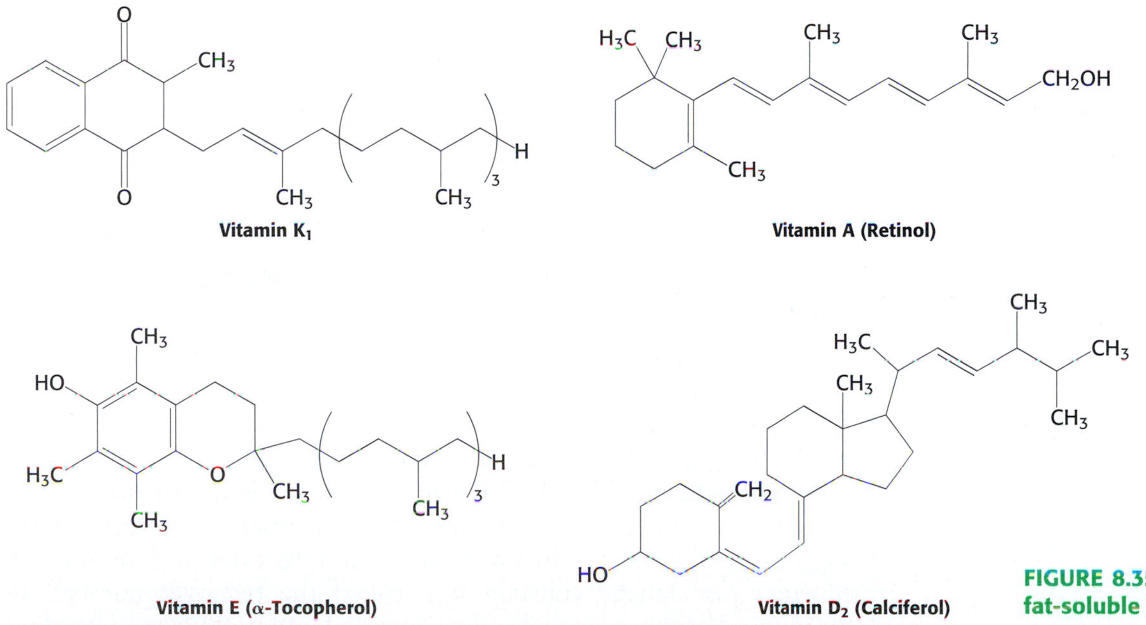

Vitamin K₁

Vitamin A (Retinol)

Vitamin E (α-Tocopherol)

Vitamin D₂ (Calciferol)

FIGURE 8.35 Structures of some fat-soluble vitamins.

SUMMARY

● **Enzymes are Powerful and Highly Specific Catalysts**

The catalysts in biological systems are enzymes, and nearly all enzymes are proteins. Enzymes are highly specific and have great catalytic power. They can enhance reaction rates by factors of 10^6 or more. Many enzymes require cofactors for activity. Such cofactors can be metal ions or small, vitamin-derived organic molecules called coenzymes.

Free Energy Is a Useful Thermodynamic Function for Understanding Enzymes

Free energy (G) is the most valuable thermodynamic function for determining whether a reaction can take place and for understanding the energetics of catalysis. A reaction can occur spontaneously only if the change in free energy (ΔG) is negative. The free-energy change of a reaction that takes place when reactants and products are at unit activity is called the standard free-energy change $(\Delta G°)$. Biochemists usually use $\Delta G°'$, the standard free-energy change at pH 7. Enzymes do not alter reaction equilibria; rather, they increase reaction rates.

Enzymes Accelerate Reactions by Facilitating the Formation of the Transition State

Enzymes serve as catalysts by decreasing the free energy of activation of chemical reactions. Enzymes accelerate reactions by providing a reaction pathway in which the transition state (the highest-energy species) has a lower free energy and hence is more rapidly formed than in the uncatalyzed reaction.

The first step in catalysis is the formation of an enzyme–substrate complex. Substrates are bound to enzymes at active-site clefts from which water is largely excluded when the substrate is bound. The specificity of enzyme–substrate interactions arises mainly from hydrogen bonding, which is directional, and the shape of the active site, which rejects molecules that do not have a sufficiently complementary shape. The recognition of substrates by enzymes is accompanied by conformational changes at active sites, and such changes facilitate the formation of the transition state.

The Michaelis-Menten Model Accounts for the Kinetic Properties of Many Enzymes

The Michaelis-Menten model accounts for the kinetic properties of some enzymes. In this model, an enzyme (E) combines with a substrate (S) to form an enzyme–substrate (ES) complex, which can proceed to form a product (P) or to dissociate into E and S.

$$E + S \underset{k_{-1}}{\overset{k_1}{\rightleftharpoons}} ES \xrightarrow{k_2} E + P$$

The rate V_0 of formation of product is given by the Michaelis-Menten equation:

$$V_0 = V_{max} \frac{[S]}{[S] + K_M}$$

in which V_{max} is the reaction rate when the enzyme is fully saturated with substrate and K_M, the Michaelis constant, is the substrate concentration at which the reaction rate is half maximal. The maximal rate, V_{max}, is equal to the product of k_2 or k_{cat} and the total concentration of enzyme. The kinetic constant k_{cat}, called the turnover number, is the number of substrate molecules converted into product per unit time at a single catalytic site when the enzyme is fully saturated with substrate. Turnover numbers for most enzymes are between 1 and 10^4 per second. The ratio of k_{cat}/K_M provides a penetrating probe into enzyme efficiency.

Allosteric enzymes constitute an important class of enzymes whose catalytic activity can be regulated. These enzymes, which do not conform to Michaelis-Menton kinetics, have multiple active sites. These active sites display cooperativity, as evidenced by a sigmoidal dependence of reaction velocity on substrate concentration.

Enzymes Can Be Inhibited by Specific Molecules

Specific small molecules or ions can inhibit even nonallosteric enzymes. In irreversible inhibition, the inhibitor is covalently linked to the enzyme or bound so tightly that its dissociation from the enzyme is very slow. Covalent inhibitors provide a means of mapping the enzyme's active site. In contrast, reversible inhibition is characterized by a rapid equilibrium between enzyme and inhibitor. A competitive inhibitor prevents the substrate from binding to the active site. It reduces the reaction velocity by diminishing the proportion of enzyme molecules that are bound to substrate. In noncompetitive inhibition, the inhibitor decreases the turnover number. Competitive inhibition can be distinguished from noncompetitive inhibition by determining whether the inhibition can be overcome by raising the substrate concentration.

The essence of catalysis is selective stabilization of the transition state. Hence, an enzyme binds the transition state more tightly than the substrate. Transition-state analogs are stable compounds that mimic key features of this highest-energy species. They are potent and specific inhibitors of enzymes. Proof that transition-state stabilization is a key aspect of enzyme activity comes from the generation of catalytic antibodies. Transition-state analogs are used as antigens, or immunogens, in generating catalytic antibodies.

Vitamins Are Often Precursors to Coenzymes

Vitamins are small biomolecules that are needed in small amounts in the diets of higher animals. The water-soluble vitamins are vitamin C (ascorbate, an antioxidant) and the vitamin B complex (components of coenzymes). Ascorbate is required for the hydroxylation of proline residues in collagen, a key protein of connective tissue. The fat-soluble vitamins are vitamin A (a precursor of retinal), D (a regulator of calcium and phosphorus metabolism), E (an antioxidant in membranes), and K (a participant in the carboxylation of glutamate).

APPENDIX: V_{max} AND K_M CAN BE DETERMINED BY DOUBLE-RECIPROCAL PLOTS

Before the availability of computers, the determination of K_M and V_{max} values required algebraic manipulation of the basic Michaelis-Menten equation. Because V_{max} is approached asymptotically (see Figure 8.11), it is impossible to obtain a definitive value from a typical Michaelis-Menten plot. Because K_M is the concentration of substrate at $V_{max}/2$, it is likewise impossible to determine an accurate value of K_M. However, V_{max} can be accurately determined if the Michaelis-Menten equation is transformed into one that gives a straight-line plot. Taking the reciprocal of both sides of equation 23 gives

$$\frac{1}{V_0} = \frac{1}{V_{max}} + \frac{K_M}{V_{max}} \cdot \frac{1}{[S]} \tag{31}$$

A plot of $1/V_0$ versus $1/[S]$, called a *Lineweaver-Burk* or *double-reciprocal plot*, yields a straight line with an intercept of $1/V_{max}$ and a slope of K_M/V_{max} (Figure 8.36). The intercept on the x-axis is $-1/K_M$.

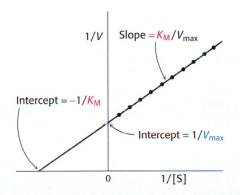

FIGURE 8.36 A double-reciprocal or Lineweaver-Burk plot. A double-reciprocal plot of enzyme kinetics is generated by plotting $1/V_0$ as a function $1/[S]$. The slope is the K_M/V_{max}, the intercept on the vertical axis is $1/V_{max}$, and the intercept on the horizontal axis is $-1/K_M$.

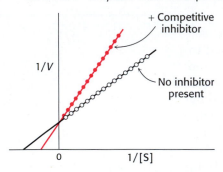

FIGURE 8.37 Competitive inhibition illustrated on a double-reciprocal plot. A double-reciprocal plot of enzyme kinetics in the presence (—•—•—) and absence (—◦—◦—) of a competitive inhibitor illustrates that the inhibitor has no effect on V_{max} but increases K_M.

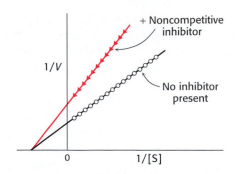

FIGURE 8.38 Noncompetitive inhibition illustrated on a double-reciprocal plot. A double-reciprocal plot of enzyme kinetics in the presence (—•—•—) and absence (—◦—◦—) of a noncompetitive inhibitor shows that K_M is unaltered and V_{max} is decreased.

Double-reciprocal plots are especially useful for distinguishing between competitive and noncompetitive inhibitors. In competitive inhibition, the intercept on the y-axis of the plot of $1/V_0$ versus $1/[S]$ is the same in the presence and in the absence of inhibitor, although the slope is increased (Figure 8.37). That the intercept is unchanged is because a competitive inhibitor does not alter V_{max}. At a sufficiently high concentration, virtually all the active sites are filled by substrate, and the enzyme is fully operative. The increase in the slope of the $1/V_0$ versus $1/[S]$ plot indicates the strength of binding of competitive inhibitor. In the presence of a competitive inhibitor, equation 31 is replaced by

$$\frac{1}{V_0} = \frac{1}{V_{max}} + \frac{K_M}{V_{max}}\left(1 + \frac{[I]}{K_i}\right)\left(\frac{1}{[S]}\right)$$

in which [I] is the concentration of inhibitor and K_i is the dissociation constant of the enzyme–inhibitor complex.

$$E + I \rightleftharpoons EI$$

$$K_i = \frac{[E][I]}{[EI]}$$

In other words, the slope of the plot is increased by the factor $(1 + [I]/K_i)$ in the presence of a competitive inhibitor. Consider an enzyme with a K_M of 10^{-4} M. In the absence of inhibitor, $V_0 = V_{max}/2$ when $[S] = 10^{-4}$ M. In the presence of 2×10^{-3} M competitive inhibitor that is bound to the enzyme with a K_i of 10^{-3} M, the apparent K_M (K_M^{app}) will be equal to $K_M \times (1 + [I]/K_i)$, or 3×10^{-4} M. Substitution of these values into equation 23 gives $V_0 = V_{max}/4$, when $[S] = 10^{-4}$ M. The presence of the competitive inhibitor thus cuts the reaction rate in half at this substrate concentration.

In noncompetitive inhibition (Figure 8.38), the inhibitor can combine with either the enzyme or the enzyme–substrate complex. In pure noncompetitive inhibition, the values of the dissociation constants of the inhibitor and enzyme and of the inhibitor and enzyme–substrate complex are equal (Section 8.5.1). The value of V_{max} is decreased to a new value called V_{max}^{app}, and so the intercept on the vertical axis is increased. The new slope, which is equal to K_M/V_{max}^{app}, is larger by the same factor. In contrast with V_{max}, K_M is not affected by pure noncompetitive inhibition. The maximal velocity in the presence of a pure noncompetitive inhibitor, V_{max}^i, is given by

$$V_{max}^{app} = \frac{V_{max}}{1 + [I]/K_i}$$

KEY TERMS

enzyme (p. 189)
substrate (p. 190)
cofactor (p. 191)
apoenzyme (p. 191)
holoenzyme (p. 191)
coenzyme (p. 192)
prosthetic group (p. 192)
free energy (p. 193)
transition state (p. 196)
free energy of activation (p. 196)
active site (p. 197)

induced fit (p. 200)
K_M (the Michaelis constant) (p. 202)
V_{max} (p. 203)
Michaelis-Menten equation (p. 203)
turnover number (p. 204)
k_{cat}/K_M (p. 205)
sequential displacement reaction (p. 207)
double-displacement (Ping-Pong) reaction (p. 208)

allosteric enzyme (p. 208)
competitive inhibition (p. 209)
noncompetitive inhibition (p. 210)
group-specific reagent (p. 211)
affinity label (p. 211)
mechanism-based (suicide) inhibition (p. 211)
transition-state analog (p. 213)
catalytic antibody (abzyme) (p. 213)
vitamin (p. 216)

—| **SELECTED READINGS** |——————————————————

Where to start

Koshland, D. E., Jr., 1987. Evolution of catalytic function. *Cold Spring Harbor Symp. Quant. Biol.* 52:1–7.

Jencks, W. P., 1987. Economics of enzyme catalysis. *Cold Spring Harbor Symp. Quant. Biol.* 52:65–73.

Lerner, R. A., and Tramontano, A., 1988. Catalytic antibodies. *Sci. Am.* 258(3):58–70.

Books

Fersht, A., 1999. *Structure and Mechanism in Protein Science: A Guide to Enzyme Catalysis and Protein Folding.* W. H. Freeman and Company.

Walsh, C., 1979. *Enzymatic Reaction Mechanisms.* W. H. Freeman and Company.

Page, M. I., and Williams, A. (Eds.), 1987. *Enzyme Mechanisms.* Royal Society of Chemistry.

Bender, M. L., Bergeron, R. J., and Komiyama, M., 1984. *The Bioorganic Chemistry of Enzymatic Catalysis.* Wiley-Interscience.

Abelson, J. N., and Simon, M. I. (Eds.), 1992. *Methods in Enzymology.* Academic Press.

Boyer, P. D. (Ed.), 1970. *The Enzymes* (3d ed.). Academic Press.

Friedmann, H. (ed.), 1981. *Benchmark Papers in Biochemistry.* Vol. 1, *Enzymes.* Hutchinson Ross.

Transition-state stabilization, analogs, and other enzyme inhibitors

Schramm, V. L., 1998. Enzymatic transition states and transition state analog design. *Annu. Rev. Biochem.* 67:693–720.

Pauling, L., 1948. Nature of forces between large molecules of biological interest. *Nature* 161:707–709.

Leinhard, G. E., 1973. Enzymatic catalysis and transition-state theory. *Science* 180:149–154.

Kraut, J., 1988. How do enzymes work? *Science* 242:533–540.

Waxman, D. J., and Strominger, J. L., 1983. Penicillin-binding proteins and the mechanism of action of β-lactam antibiotics. *Annu. Rev. Biochem.* 52:825–869.

Abraham, E. P., 1981. The β-lactam antibiotics. *Sci. Am.* 244:76–86.

Walsh, C. T., 1984. Suicide substrates, mechanism-based enzyme inactivators: Recent developments. *Annu. Rev. Biochem.* 53:493–535.

Catalytic antibodies

Hilvert, D., 2000. Critical analysis of antibody catalysis. *Annu. Rev. Biochem.* 69:751–794.

Wade, H., and Scanlan, T. S., 1997. The structural and functional basis of antibody catalysis. *Annu. Rev. Biophys. Biomol. Struct.* 26:461–493.

Lerner, R. A., Benkovic, S. J., and Schultz, P. G., 1991. At the crossroads of chemistry and immunology: Catalytic antibodies. *Science* 252:659–667.

Cochran, A. G., and Schultz, P. G., 1990. Antibody-catalyzed porphyrin metallation. *Science* 249:781–783.

Enzyme kinetics and mechanisms

Xie, X. S., and Lu, H. P., 1999. Single-molecule enzymology. *J. Biol. Chem.* 274:15967–15970.

Miles, E. W., Rhee, S., and Davies, D. R., 1999. The molecular basis of substrate channeling. *J. Biol. Chem.* 274:12193–12196.

Warshel, A., 1998. Electrostatic origin of the catalytic power of enzymes and the role of preorganized active sites. *J. Biol. Chem.* 273:27035–27038.

Cannon, W. R., and Benkovic, S. J., 1999. Solvation, reorganization energy, and biological catalysis. *J. Biol. Chem.* 273:26257–26260.

Cleland, W. W., Frey, P. A., and Gerlt, J. A., 1998. The low barrier hydrogen bond in enzymatic catalysis. *J. Biol. Chem.* 273:25529–25532.

Romesberg, F. E., Santarsiero, B. D., Spiller, B., Yin, J., Barnes, D., Schultz, P. G., and Stevens, R. C., 1998. Structural and kinetic evidence for strain in biological catalysis. *Biochemistry* 37:14404–14409.

Lu, H. P., Xun, L., and Xie, X. S., 1998. Single-molecule enzymatic dynamics. *Science* 282:1877–1882.

Fersht, A. R., Leatherbarrow, R. J., and Wells, T. N. C., 1986. Binding energy and catalysis: A lesson from protein engineering of the tyrosyl-tRNA synthetase. *Trends Biochem. Sci.* 11:321–325.

Jencks, W. P., 1975. Binding energy, specificity, and enzymic catalysis: The Circe effect. *Adv. Enzymol.* 43:219–410.

Knowles, J. R., and Albery, W. J., 1976. Evolution of enzyme function and the development of catalytic efficiency. *Biochemistry* 15:5631–5640.

—| **PROBLEMS** |——————————————————

1. *Hydrolytic driving force.* The hydrolysis of pyrophosphate to orthophosphate is important in driving forward biosynthetic reactions such as the synthesis of DNA. This hydrolytic reaction is catalyzed in *Escherichia coli* by a pyrophosphatase that has a mass of 120 kd and consists of six identical subunits. For this enzyme, a unit of activity is defined as the amount of enzyme that hydrolyzes 10 μmol of pyrophosphate in 15 minutes at 37°C under standard assay conditions. The purified enzyme has a V_{max} of 2800 units per milligram of enzyme.

(a) How many moles of substrate are hydrolyzed per second per milligram of enzyme when the substrate concentration is much greater than K_M?

(b) How many moles of active site are there in 1 mg of enzyme? Assume that each subunit has one active site.

(c) What is the turnover number of the enzyme? Compare this value with others mentioned in this chapter.

2. *Destroying the Trojan horse.* Penicillin is hydrolyzed and thereby rendered inactive by penicillinase (also known as β-lac

tamase), an enzyme present in some resistant bacteria. The mass of this enzyme in *Staphylococcus aureus* is 29.6 kd. The amount of penicillin hydrolyzed in 1 minute in a 10-ml solution containing 10^{-9} g of purified penicillinase was measured as a function of the concentration of penicillin. Assume that the concentration of penicillin does not change appreciably during the assay.

[Penicillin] (μM)	Amount hydrolyzed (nanomoles)
1	0.11
3	0.25
5	0.34
10	0.45
30	0.58
50	0.61

(a) Plot V_0 versus [S] and $1/V_0$ versus $1/[S]$ for these data. Does penicillinase appear to obey Michaelis-Menten kinetics? If so, what is the value of K_M?

(b) What is the value of V_{max}?

(c) What is the turnover number of penicillinase under these experimental conditions? Assume one active site per enzyme molecule.

3. *Counterpoint.* Penicillinase (β-lactamase) hydrolyzes penicillin. Compare penicillinase with glycopeptide transpeptidase.

4. *Mode of inhibition.* The kinetics of an enzyme are measured as a function of substrate concentration in the presence and in the absence of 2 mM inhibitor (I).

[S] (μM)	Velocity (μmol/minute)	
	No inhibitor	Inhibitor
3	10.4	4.1
5	14.5	6.4
10	22.5	11.3
30	33.8	22.6
90	40.5	33.8

(a) What are the values of V_{max} and K_M in the absence of inhibitor? In its presence?

(b) What type of inhibition is it?

(c) What is the binding constant of this inhibitor?

(d) If [S] = 10 μM and [I] = 2 mM, what fraction of the enzyme molecules have a bound substrate? A bound inhibitor?

(e) If [S] = 30 μM, what fraction of the enzyme molecules have a bound substrate in the presence and in the absence of 2 mM inhibitor? Compare this ratio with the ratio of the reaction velocities under the same conditions.

5. *A different mode.* The kinetics of the enzyme considered in problem 4 are measured in the presence of a different inhibitor. The concentration of this inhibitor is 100 μM.

(a) What are the values of V_{max} and K_M in the presence of this inhibitor? Compare them with those obtained in problem 4.

(b) What type of inhibition is it?

(c) What is the dissociation constant of this inhibitor?

[S] (μM)	Velocity (μmol/minute)	
	No inhibitor	Inhibitor
3	10.4	2.1
5	14.5	2.9
10	22.5	4.5
30	33.8	6.8
90	40.5	8.1

(d) If [S] = 30 μM, what fraction of the enzyme molecules have a bound substrate in the presence and in the absence of 100 μM inhibitor?

6. *A fresh view.* The plot of $1/V_0$ versus $1/[S]$ is sometimes called a Lineweaver-Burk plot. Another way of expressing the

kinetic data is to plot V_0 versus $V_0/[S]$, which is known as an Eadie-Hofstee plot.

(a) Rearrange the Michaelis-Menten equation to give V_0 as a function of $V_0/[S]$.

(b) What is the significance of the slope, the vertical intercept, and the horizontal intercept in a plot of V_0 versus $V_0/[S]$?

(c) Sketch a plot of V_0 versus $V_0/[S]$ in the absence of an inhibitor, in the presence of a competitive inhibitor, and in the presence of a noncompetitive inhibitor.

7. *Potential donors and acceptors.* The hormone progesterone contains two ketone groups. At pH 7, which side chains of the receptor might form hydrogen bonds with progesterone?

8. *Competing substrates.* Suppose that two substrates, A and B, compete for an enzyme. Derive an expression relating the ratio of the rates of utilization of A and B, V_A/V_B, to the concentrations of these substrates and their values of k_2 and K_M. (Hint: Express V_A as a function of k_2/K_M for substrate A, and do the same for V_B.) Is specificity determined by K_M alone?

9. *A tenacious mutant.* Suppose that a mutant enzyme binds a substrate 100-fold as tightly as does the native enzyme. What is the effect of this mutation on catalytic rate if the binding of the transition state is unaffected?

10. *Uncompetitive inhibition.* The following reaction represents the mechanism of action of an uncompetitive inhibitor.

$$E + S \underset{k_{-1}}{\overset{k_1}{\rightleftharpoons}} ES \overset{k_2}{\longrightarrow} E + P$$
$$+$$
$$I$$
$$\Big\updownarrow k_I$$
$$ESI$$

(a) Draw a standard Michaelis-Menton curve in the absence and in the presence of increasing amounts of inhibitor. Repeat for a double-reciprocal plot.

(b) Explain the results obtained in part *a*.

11. *More Michaelis-Menten.* For an enzyme that follows simple Michaelis-Menten kinetics, what is the value of V_{max} if V_0 is equal to 1 μmol/minute at $1/10\ K_M$?

Data Interpretation Problems

12. *Varying the enzyme.* For a one-substrate, enzyme-catalyzed reaction, double-reciprocal plots were determined for three different enzyme concentrations. Which of the following three families of curve would you expect to be obtained? Explain.

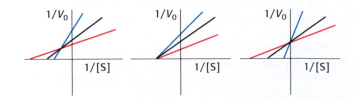

13. *Too much of a good thing.* A simple Michaelis-Menten enzyme, in the absence of any inhibitor, displayed the following kinetic behavior. The expected value of V_{max} is shown on the y-axis.

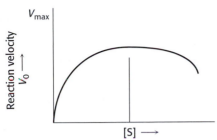

(a) Draw a double-reciprocal plot that corresponds to the velocity-versus-substrate curve.
(b) Provide an explanation for the kinetic results.

14. *Rate-limiting step.* In the conversion of A into D in the following biochemical pathway, enzymes E_A, E_B, and E_C have the K_M values indicated under each enzyme. If all of the substrates and products are present at a concentration of 10^{-4} M, which step will be rate limiting and why?

$$A \underset{}{\overset{E_A}{\rightleftharpoons}} B \underset{}{\overset{E_B}{\rightleftharpoons}} C \underset{}{\overset{E_C}{\rightleftharpoons}} D$$

$$K_M = \quad 10^{-2}\,M \quad\quad 10^{-4}\,M \quad\quad 10^{-4}\,M$$

Chapter Integration Problems

15. *Titration experiment.* The effect of pH on the activity of an enzyme was examined. At its active site, the enzyme has an ionizable group that must be negatively charged for substrate binding and catalysis to take place. The ionizable group has a pK_a of 6.0. The substrate is positively charged throughout the pH range of the experiment.

$$
\begin{array}{c}
E^- + S^+ \rightleftharpoons E^- \!\!-\!\! S^+ \longrightarrow E^- + P^+ \\
+ \\
H^+ \\
\Updownarrow \\
EH
\end{array}
$$

(a) Draw the V_0-versus-pH curve when the substrate concentration is much greater than the enzyme K_M.
(b) Draw the V_0-versus-pH curve when the substrate concentration is much less than the enzyme K_M.
(c) At which pH will the velocity equal one-half of the maximal velocity attainable under these condition?

16. *A question of stability.* Pyridoxal phosphate (PLP) is a coenzyme for the enzyme ornithine aminotransferase. The enzyme was purified from cells grown in PLP-deficient media as well as from cells grown in media that contained pyridoxal phosphate. The stability of the enzyme was then measured by incubating the enzyme at 37°C and assaying for the amount of enzyme activity remaining. The following results were obtained.

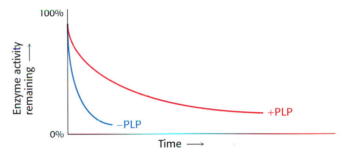

(a) Why does the amount of active enzyme decrease with the time of incubation?
(b) Why does the amount of enzyme from the PLP-deficient cells decline more rapidly?

Media Problem

17. *Not all data points are created equal.* Your lab partner, who is both systematic and frugal, decides to perform a series of enzyme assays at substrate concentrations of 1, 2, 4, and 8 μM. You argue for doing the experiments at [S] = 1, 4, 16, and 100 μM. Try both sets of experiments using the simulated enzyme kinetics lab in the Steady-State Enzyme Kinetics **Conceptual Insights** module. Who had the better idea, and why?

CHAPTER 9

Catalytic Strategies

Strategy and tactics. Chess and enzymes have in common the use of strategy, consciously thought out in the game of chess and selected by evolution for the action of an enzyme. The three amino acid residues at the right, denoted by the white bonds, constitute a catalytic triad found in the active site of a class of enzymes that cleave peptide bonds. The substrate, represented by the molecule with black bonds, is as hopelessly trapped as the king in the photograph of a chess match at the left and is sure to be cleaved. [(Left) Courtesy of Wendie Berg.]

What are the sources of the catalytic power and specificity of enzymes? This chapter presents the catalytic strategies used by four classes of enzymes: the serine proteases, carbonic anhydrases, restriction endonucleases, and nucleoside monophosphate (NMP) kinases. The first three classes of enzymes catalyze reactions that require the addition of water to a substrate. For the serine proteases, exemplified by chymotrypsin, the challenge is to promote a reaction that is almost immeasurably slow at neutral pH in the absence of a catalyst. For carbonic anhydrases, the challenge is to achieve a high absolute rate of reaction, suitable for integration with other rapid physiological processes. For restriction endonucleases such as *Eco*RV, the challenge is to attain a very high level of specificity. Finally, for NMP kinases, the challenge is to transfer a phosphoryl group from ATP to a nucleotide and not to water. The actions of these enzymes illustrate many important principles of catalysis. The mechanisms of these enzymes have been revealed through the use of incisive experimental probes, including the techniques of protein structure determination (Chapter 4) and site-directed mutagenesis (Chapter 6). These mechanisms include the use of binding energy and induced fit as well as several specific catalytic strategies. Properties common to an enzyme family reveal how their enzyme active sites have evolved and been refined. Structural and mechanistic comparisons of enzyme action are thus sources of insight into the evolutionary history of enzymes. These comparisons also reveal particularly effective solutions to biochemical problems that are used repeatedly in biological systems. In addition, our knowledge of catalytic strategies has been used to develop practical applications, including drugs that are potent and specific enzyme inhibitors. Finally, although we shall not consider catalytic RNA molecules (Section 28.4) explicitly in this chapter, the principles apply to these catalysts in addition to protein catalysts.

9.0.1 A Few Basic Catalytic Principles Are Used by Many Enzymes

In Chapter 8, we learned that enzymatic catalysis begins with substrate binding. The *binding energy* is the free energy released in the formation of a large number of weak interactions between the enzyme and the substrate. We can envision this binding energy as serving two purposes: it establishes substrate specificity and increases catalytic efficiency. Only the correct substrate can participate in most or all of the interactions with the enzyme and thus maximize binding energy, accounting for the exquisite substrate specificity exhibited by many enzymes. Furthermore, the full complement of such interactions is formed only when the substrate is in the transition state. Thus, interactions between the enzyme and the substrate not only favor substrate binding but stabilize the transition state, thereby lowering the activation energy. The binding energy can also promote structural changes in both the enzyme and the substrate that facilitate catalysis, a process referred to as *induced fit*.

Enzymes commonly employ one or more of the following strategies to catalyze specific reactions:

1. *Covalent catalysis.* In covalent catalysis, the active site contains a reactive group, usually a powerful nucleophile that becomes temporarily covalently modified in the course of catalysis. The proteolytic enzyme chymotrypsin provides an excellent example of this mechanism (Section 9.1.2).

2. *General acid–base catalysis.* In general acid–base catalysis, a molecule other than water plays the role of a proton donor or acceptor. Chymotrypsin uses a histidine residue as a base catalyst to enhance the nucleophilic power of serine (Section 9.1.3).

3. *Metal ion catalysis.* Metal ions can function catalytically in several ways. For instance, a metal ion may serve as an electrophilic catalyst, stabilizing a negative charge on a reaction intermediate. Alternatively, the metal ion may generate a nucleophile by increasing the acidity of a nearby molecule, such as water in the hydration of CO_2 by carbonic anhydrase (Section 9.2.2). Finally, the metal ion may bind to substrate, increasing the number of interactions with the enzyme and thus the binding energy. This strategy is used by NMP kinases (Section 9.4.2).

4. *Catalysis by approximation.* Many reactions include two distinct substrates. In such cases, the reaction rate may be considerably enhanced by bringing the two substrates together along a single binding surface on an enzyme. NMP kinases bring two nucleotides together to facilitate the transfer of a phosphoryl group from one nucleotide to the other (Section 9.4.3).

9.1 PROTEASES: FACILITATING A DIFFICULT REACTION

Protein turnover is an important process in living systems (Chapter 23). Proteins that have served their purpose must be degraded so that their constituent amino acids can be recycled for the synthesis of new proteins. Proteins ingested in the diet must be broken down into small peptides and amino acids for absorption in the gut. Furthermore, as described in detail in Chapter 10, proteolytic reactions are important in regulating the activity of certain enzymes and other proteins.

Proteases cleave proteins by a hydrolysis reaction—the addition of a molecule of water to a peptide bond:

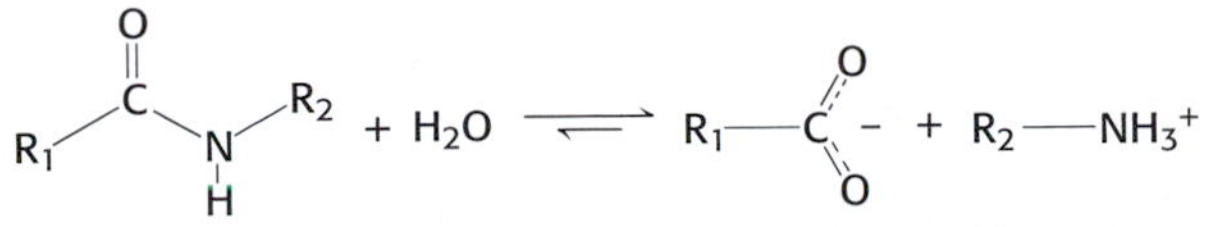

$$R_1 - \overset{O}{\underset{}{C}} - \underset{H}{N} - R_2 + H_2O \rightleftharpoons R_1 - \overset{O}{\underset{O}{C}} - + R_2 - NH_3^+$$

Although the hydrolysis of peptide bonds is thermodynamically favored, such hydrolysis reactions are extremely slow. In the absence of a catalyst, the half-life for the hydrolysis of a typical peptide at neutral pH is estimated to be between 10 and 1000 years. Yet, peptide bonds must be hydrolyzed within milliseconds in some biochemical processes.

The chemical bonding in peptide bonds is responsible for their kinetic stability. Specifically, the resonance structure that accounts for the planarity of a peptide bond (Section 3.2.2) also makes such bonds resistant to hydrolysis. This resonance structure endows the peptide bond with partial double-bond character:

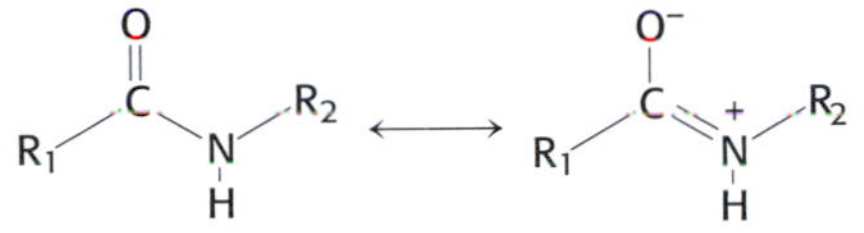

$$R_1 - \overset{O}{\underset{H}{C}} - \underset{H}{N} - R_2 \longleftrightarrow R_1 - \overset{O^-}{\underset{H}{C}} = \overset{+}{\underset{H}{N}} - R_2$$

The carbon–nitrogen bond is strengthened by its double-bond character, and the carbonyl carbon atom is less electrophilic and less susceptible to nucleophilic attack than are the carbonyl carbon atoms in compounds such as carboxylate esters. Consequently, to promote peptide-bond cleavage, an enzyme must facilitate nucleophilic attack at a normally unreactive carbonyl group.

9.1.1 Chymotrypsin Possesses a Highly Reactive Serine Residue

A number of proteolytic enzymes participate in the breakdown of proteins in the digestive systems of mammals and other organisms. One such enzyme, chymotrypsin, cleaves peptide bonds selectively on the carboxyl-terminal side of the large hydrophobic amino acids such as tryptophan, tyrosine, phenylalanine, and methionine (Figure 9.1). Chymotrypsin is a good example of the use of *covalent modification* as a catalytic strategy. The enzyme employs a powerful nucleophile to attack the unreactive carbonyl group of the substrate. This nucleophile becomes covalently attached to the substrate briefly in the course of catalysis.

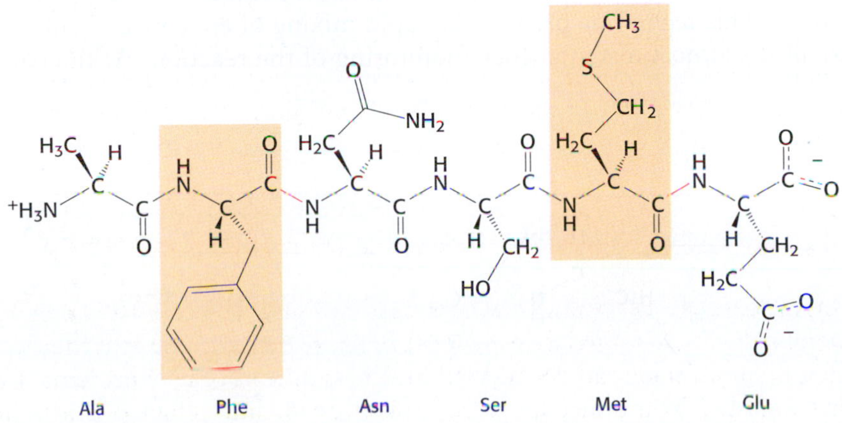

FIGURE 9.1 Specificity of chymotrypsin. Chymotrypsin cleaves proteins on the carboxyl side of aromatic or large hydrophobic amino acids (shaded yellow). The likely bonds cleaved by chymotrypsin are indicated in red.

Ala   Phe   Asn   Ser   Met   Glu

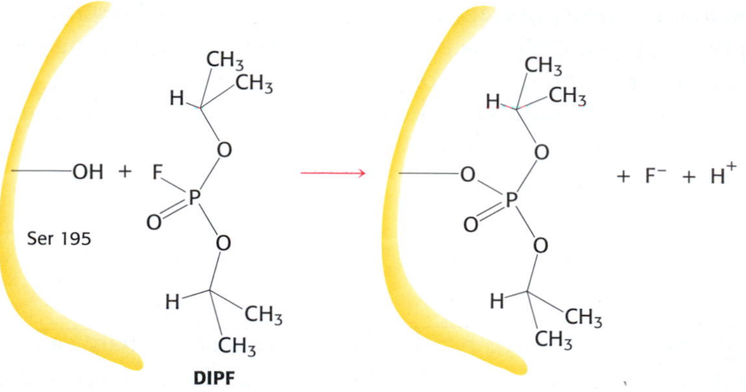

FIGURE 9.2 An unusually reactive serine in chymotrypsin. Chymotrypsin is inactivated by treatment with diisopropylphosphofluoridate (DIPF), which reacts only with serine 195 among 28 possible serine residues.

What is the nucleophile that chymotrypsin employs to attack the substrate carbonyl group? A clue came from the fact that chymotrypsin contains an extraordinarily reactive serine residue. Treatment with organofluorophosphates such as diisopropylphosphofluoridate (DIPF) (Section 8.5.2) was found to inactivate the enzyme irreversibly (Figure 9.2). Despite the fact that the enzyme possesses 28 serine residues, only one, serine 195, was modified, resulting in a total loss of enzyme activity. This *chemical modification reaction* suggested that this unusually reactive serine residue plays a central role in the catalytic mechanism of chymotrypsin.

9.1.2 Chymotrypsin Action Proceeds in Two Steps Linked by a Covalently Bound Intermediate

CONCEPTUAL INSIGHTS, Enzyme Kinetics. *See the section entitled "Pre-Steady-State Kinetics" in the Conceptual Insights module to better understand why a "burst" phase at short reaction times implies the existence of an enzyme-substrate intermediate.*

How can we elucidate the role of serine 195 in chymotrypsin action? A study of the enzyme's kinetics provided a second clue to chymotrypsin's catalytic mechanism and the role of serine 195. The kinetics of enzyme action are often easily monitored by having the enzyme act on a substrate analog that forms a colored product. For chymotrypsin, such a *chromogenic substrate* is N-acetyl-L-phenylalanine p-nitrophenyl ester. This substrate is an ester rather than an amide, but many proteases will also hydrolyze esters. One of the products formed by chymotrypsin's cleavage of this substrate is p-nitrophenolate, which has a yellow color (Figure 9.3). Measurements of the absorbance of light revealed the amount of p-nitrophenolate being produced.

Under steady-state conditions, the cleavage of this substrate obeys Michaelis-Menten kinetics with a K_M of 20 μM and a k_{cat} of 77 s^{-1}. The initial phase of the reaction was examined by using the stopped-flow method. This technique permits the rapid mixing of enzyme and substrate and allows almost instantaneous monitoring of the reaction. At the begin-

FIGURE 9.3 Chromogenic substrate. N-Acetyl-L-phenylalanine p-nitrophenyl ester yields a yellow product, p-nitrophenolate, on cleavage by chymotrypsin. p-Nitrophenolate forms by deprotonation of p-nitrophenol at pH 7.

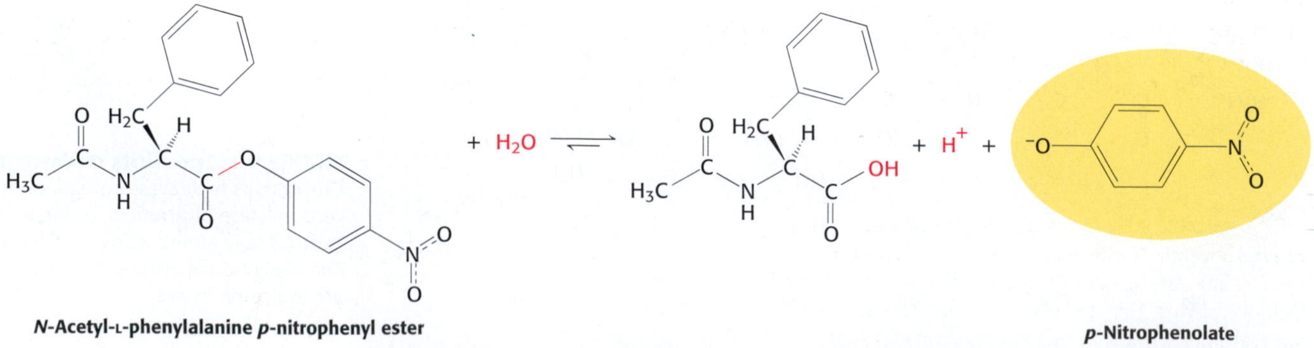

N-Acetyl-L-phenylalanine p-nitrophenyl ester

p-Nitrophenolate

ning of the reaction, this method revealed a "burst" phase during which the colored product was produced rapidly (Figure 9.4). Product was then produced more slowly as the reaction reached the steady state. These results suggest that hydrolysis proceeds in two steps. The burst is observed because, for this substrate, the first step is more rapid than the second step.

The two steps are explained by the reaction of the serine nucleophile with the substrate to form the covalently bound enzyme–substrate intermediate (Figure 9.5). First, the highly reactive serine 195 hydroxyl group attacks the carbonyl group of the substrate to form the acyl-enzyme intermediate, releasing the alcohol p-nitrophenol (or an amine if the substrate is an amide rather than an ester). Second, the acyl-enzyme intermediate is hydrolyzed to release the carboxylic acid component of the substrate and regenerate the free enzyme. Thus, p-nitrophenolate is produced rapidly on the addition of the substrate as the acyl-enzyme intermediate is formed, but it takes longer for the enzyme to be "reset" by the hydrolysis of the acyl-enzyme intermediate.

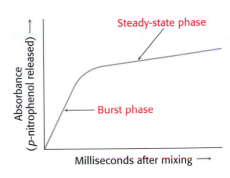

FIGURE 9.4 Kinetics of chymotrypsin catalysis. Two stages are evident in the cleaving of N-acetyl-L-phenylalanine p-nitrophenyl ester by chymotrypsin: a rapid burst phase (pre-steady state) and a steady-state phase.

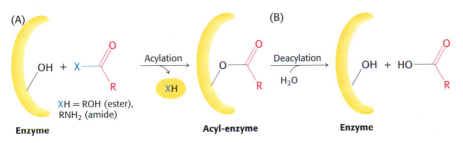

FIGURE 9.5 Covalent catalysis. Hydrolysis by chymotrypsin takes place in two stages: (A) acylation to form the acyl-enzyme intermediate followed by (B) deacylation to regenerate the free enzyme.

9.1.3 Serine is Part of a Catalytic Triad That Also Includes Histidine and Aspartic Acid

STRUCTURAL INSIGHTS, Chymotrypsin: A Serine Protease. *Work with interactive molecular models to learn more about the structural bases of active site specificity and reactivity, and some of the ways in which active site residues can be identified.*

The determination of the three-dimensional structure of chymotrypsin by David Blow in 1967 was a source of further insight into its mechanism of action. Overall, chymotrypsin is roughly spherical and comprises three polypeptide chains, linked by disulfide bonds. It is synthesized as a single polypeptide, termed chymotrypsinogen, which is activated by the proteolytic cleavage of the polypeptide to yield the three chains. The active site of chymotrypsin, marked by serine 195, lies in a cleft on the surface of the enzyme (Figure 9.6). The structural analysis revealed the chemical basis of the special reactivity of serine 195 (Figure 9.7). The side chain of serine 195 is hydrogen bonded to the imidazole ring of histidine 57. The —NH group of this imidazole ring is, in turn, hydrogen

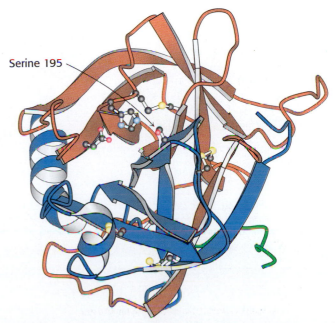

FIGURE 9.6 Three-dimensional structure of chymotrypsin. The three chains are shown in ribbon form in orange, blue, and green. The side chains of the catalytic triad residues, including serine 195, are shown as ball-and-stick representations, as are two intrastrand and interstrand disulfide bonds.

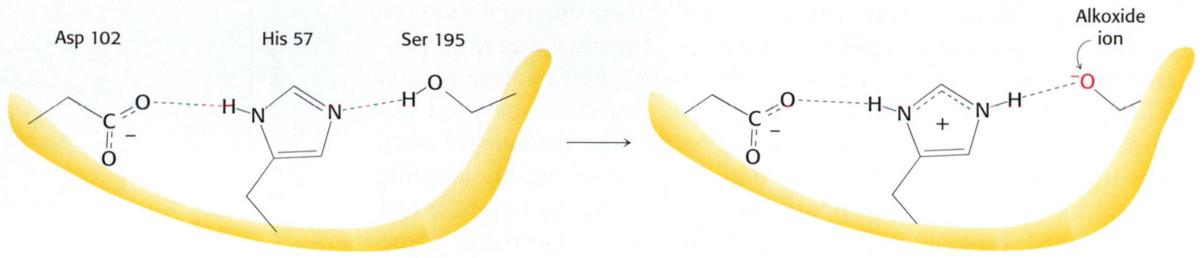

FIGURE 9.7 The catalytic triad. The catalytic triad, shown on the left, converts serine 195 into a potent nucleophile, as illustrated on the right.

FIGURE 9.8 Peptide hydrolysis by chymotrypsin. The mechanism of peptide hydrolysis illustrates the principles of covalent and acid–base catalysis. The dashed green lines indicate favorable interactions between the negatively charged aspartate residue and the positively charged histidine residue, which make the histidine residue a more powerful base.

bonded to the carboxylate group of aspartate 102. This constellation of residues is referred to as the *catalytic triad*. How does this arrangement of residues lead to the high reactivity of serine 195? The histidine residue serves to position the serine side chain and to polarize its hydroxyl group. In doing so, the residue acts as a general base catalyst, a hydrogen ion acceptor, because the polarized hydroxyl group of the serine residue is poised for deprotonation. The withdrawal of the proton from the hydroxyl group generates an alkoxide ion, which is a much more powerful nucleophile than an alcohol is. The aspartate residue helps orient the histidine residue and make it a better proton acceptor through electrostatic effects.

These observations suggest a mechanism for peptide hydrolysis (Figure 9.8). After substrate binding (step 1), the reaction begins with the hydroxyl group of serine 195 making a nucleophilic attack on the carbonyl carbon atom of the substrate (step 2). The nucleophilic attack changes the geometry around this carbon atom from trigonal planar to tetrahedral. The inherently unstable *tetrahedral intermediate* formed bears a formal negative

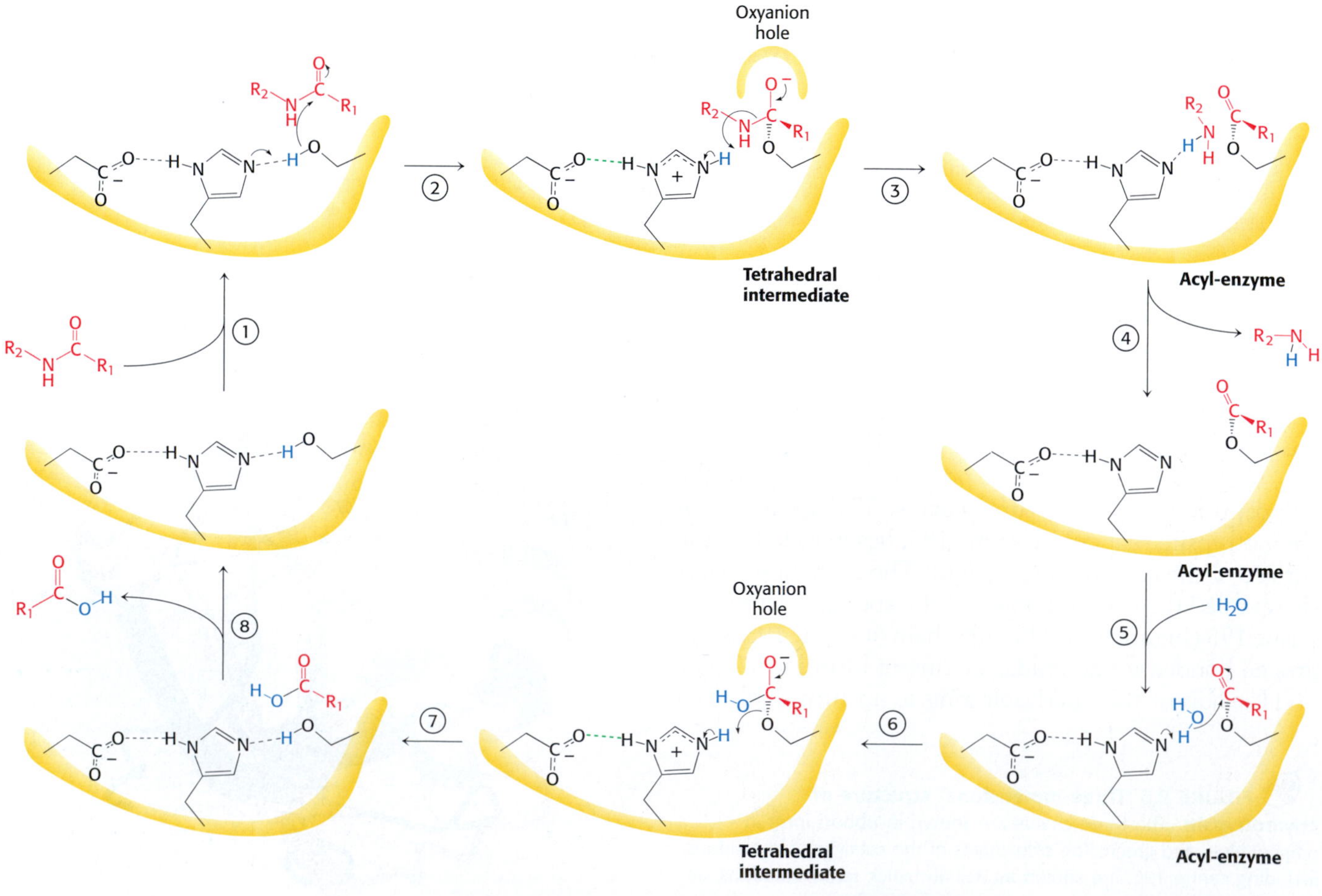

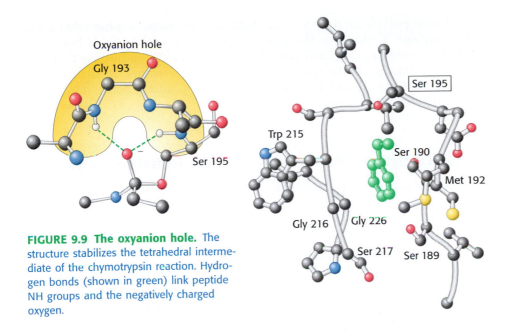

FIGURE 9.9 The oxyanion hole. The structure stabilizes the tetrahedral intermediate of the chymotrypsin reaction. Hydrogen bonds (shown in green) link peptide NH groups and the negatively charged oxygen.

FIGURE 9.10 The hydrophobic pocket of chymotrypsin. The hydrophobic pocket of chymotrypsin is responsible for its substrate specificity. The key amino acids that constitute the binding site are labeled, including the active-site serine residue (boxed). The position of an aromatic ring bound in the pocket is shown in green.

charge on the oxygen atom derived from the carbonyl group. This charge is stabilized by interactions with NH groups from the protein in a site termed the *oxyanion hole* (Figure 9.9). These interactions also help stabilize the transition state that precedes the formation of the tetrahedral intermediate. This tetrahedral intermediate then collapses to generate the acyl-enzyme (step 3). This step is facilitated by the transfer of a proton from the positively charged histidine residue to the amino group formed by cleavage of the peptide bond. The amine component is now free to depart from the enzyme (step 4) and is replaced by a water molecule (step 5). The ester group of the acyl-enzyme is now hydrolyzed by a process that is essentially a repeat of steps 2 through 4. The water molecule attacks the carbonyl group while a proton is concomitantly removed by the histidine residue, which now acts as a general acid catalyst, forming a tetrahedral intermediate (step 6). This structure breaks down to form the carboxylic acid product (step 7). Finally, the release of the carboxylic acid product (step 8) readies the enzyme for another round of catalysis.

This mechanism accounts for all characteristics of chymotrypsin action except the observed preference for cleaving the peptide bonds just past residues with large, hydrophobic side chains. Examination of the three-dimensional structure of chymotrypsin with substrate analogs and enzyme inhibitors revealed the presence of a deep, relatively hydrophobic pocket, called the S_1 pocket, into which the long, uncharged side chains of residues such as phenylalanine and tryptophan can fit. *The binding of an appropriate side chain into this pocket positions the adjacent peptide bond into the active site for cleavage* (Figure 9.10). The specificity of chymotrypsin depends almost entirely on which amino acid is directly on the amino-terminal side of the peptide bond to be cleaved. Other proteases have more-complex specificity patterns, as illustrated in Figure 9.11. Such enzymes have

FIGURE 9.11 Specificity nomenclature for protease–substrate interactions. The potential sites of interaction of the substrate with the enzyme are designated P (shown in red), and corresponding binding sites on the enzyme are designated S. The scissile bond (also shown in red) is the reference point.

additional pockets on their surfaces for the recognition of other residues in the substrate. Residues on the amino-terminal side of the scissile bond (the bond to be cleaved) are labeled P_1, P_2, P_3, and so forth, indicating their positions in relation to the scissile bond. Likewise, residues on the carboxyl side of the scissile bond are labeled P_1', P_2', P_3', and so forth. The corresponding sites on the enzyme are referred to as S_1, S_2 or S_1', S_2', and so forth.

9.1.4 Catalytic Triads Are Found in Other Hydrolytic Enzymes

Many other proteins have subsequently been found to contain catalytic triads similar to that discovered in chymotrypsin. Some, such as trypsin and elastase, are obvious homologs of chymotrypsin. The sequences of these proteins are approximately 40% identical with that of chymotrypsin, and their overall structures are nearly the same (Figure 9.12). These proteins operate by mechanisms identical with that of chymotrypsin. However, they have very different substrate specificities. Trypsin cleaves at the peptide bond after residues with long, positively charged side chains—namely, arginine and lysine—whereas elastase cleaves at the peptide bond after amino acids with small side chains—such as alanine and serine. Comparison of the S_1 pockets of these enzymes reveals the basis of the specificity. In trypsin, an aspartate residue (Asp 189) is present at the bottom of the S_1 pocket in place of a serine residue in chymotrypsin. The aspartate residue attracts and stabilizes a positively charged arginine or lysine residue in the substrate. In elastase, two residues at the top of the pocket in chymotrypsin and trypsin are replaced with valine (Val 190 and Val 216). These residues close off the mouth of the pocket so that only small side chains may enter (Figure 9.13).

Other members of the chymotrypsin family include a collection of proteins that take part in blood clotting, to be discussed in Chapter 10. In addition, a wide range of proteases found in bacteria and viruses also belong to this clan. Furthermore, other enzymes that are not homologs of chymotrypsin have been found to contain very similar active sites. As noted in Chapter 7, the presence of very similar active sites in these different protein families is a consequence of convergent evolution. Subtilisin, a protease in bacteria such as *Bacillus amyloliquefaciens*, is a particularly well characterized example. The active site of this enzyme includes both the catalytic triad and the oxyanion hole. However, one of the NH groups that forms the oxyanion hole comes from the side chain of an asparagine residue rather than from the peptide backbone (Figure 9.14). Subtilisin is the founding member of another large family of proteases that includes representatives from Archaea, Eubacteria, and Eukarya.

Yet another example of the catalytic triad has been found in carboxypeptidase II from wheat. The structure of this enzyme is not significantly similar to either chymotrypsin or subtilisin (Figure 9.15). This protein is a member of an intriguing family of homologous proteins that includes esterases such as acetylcholine esterase and certain lipases. These enzymes all make use of histidine-activated nucleophiles, but the nucleophiles may be cysteine rather than serine. Finally, other proteases have been discovered that contain an active-site serine or threonine residue that is activated not by a histidine–aspartate pair but by a primary amino group from the side chain of lysine or by the N-terminal amino group of the polypeptide chain.

Thus, the catalytic triad in proteases has emerged at least three times in the course of evolution. We can conclude that this catalytic strategy must be an especially effective approach to the hydrolysis of peptides and related bonds.

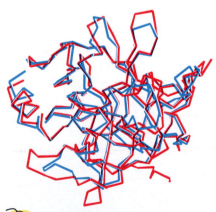

FIGURE 9.12 Structural similarity of trypsin and chymotrypsin. An overlay of the structure of chymotrypsin (red) on that of trypsin (blue) shows the high degree of similarity. Only α-carbon atom positions are shown. The mean deviation in position between corresponding α-carbon atoms is 1.7 Å.

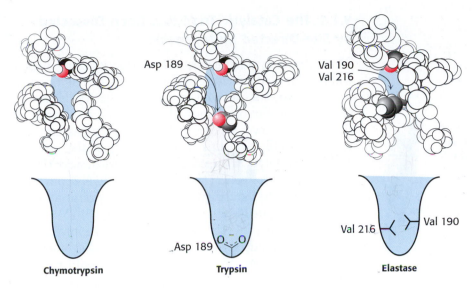

FIGURE 9.13 The S₁ pockets of chymotrypsin, trypsin, and elastase. Certain residues play key roles in determining the specificity of these enzymes. The side chains of these residues, as well as those of the active-site serine residues, are shown in color.

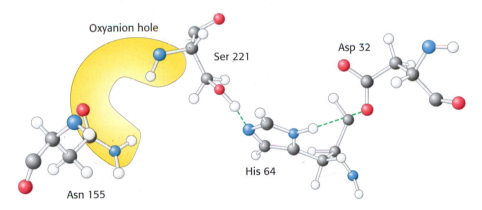

FIGURE 9.14 The catalytic triad and oxyanion hole of subtilisin. The peptide bond attacked by nucleophilic serine 221 of the catalytic triad will develop a negative charge, which is stabilized by enzyme NH groups (both in the backbone and in the side chain of Asn 155) located in the oxyanion hole.

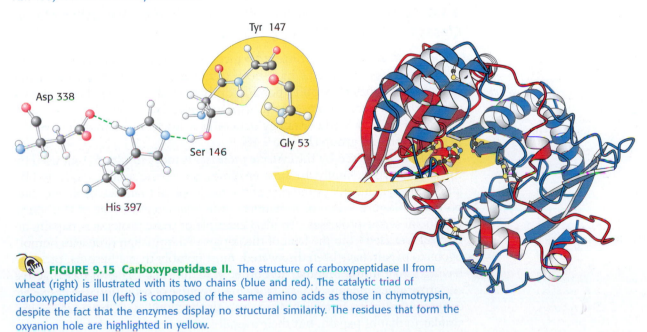

FIGURE 9.15 Carboxypeptidase II. The structure of carboxypeptidase II from wheat (right) is illustrated with its two chains (blue and red). The catalytic triad of carboxypeptidase II (left) is composed of the same amino acids as those in chymotrypsin, despite the fact that the enzymes display no structural similarity. The residues that form the oxyanion hole are highlighted in yellow.

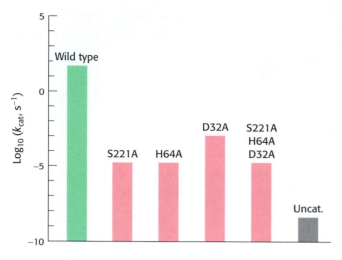

FIGURE 9.16 Site-directed mutagenesis of subtilisin. Residues of the catalytic triad were mutated to alanine, and the activity of the mutated enzyme was measured. Mutations in any component of the catalytic triad cause a dramatic loss of enzyme activity. Note that the activity is displayed on a logarithmic scale. The mutations are identified as follows: the first letter is the one-letter abbreviation for the amino acid being altered; the number identifies the position of the residue in the primary structure; and the second letter is the one-letter abbreviation for the amino acid replacing the original one.

9.1.5 The Catalytic Triad Has Been Dissected by Site-Directed Mutagenesis

The techniques of molecular biology discussed in Chapter 6 have permitted detailed examination of the catalytic triad. In particular, site-directed mutagenesis has been used to test the contribution of individual amino acid residues to the catalytic power of an enzyme. Subtilisin has been extensively studied by this method. Each of the residues within the catalytic triad, consisting of aspartic acid 32, histidine 64, and serine 221, has been individually converted into alanine, and the ability of each mutant enzyme to cleave a model substrate has been examined (Figure 9.16). As expected, the conversion of active-site serine 221 into alanine dramatically reduced catalytic power; the value of k_{cat} fell to less than *one-millionth* of its value for the wild-type enzyme. The value of K_M was essentially unchanged: its increase by no more than a factor of two indicated that substrate binding is not significantly affected. The mutation of histidine 64 to alanine had very similar effects. These observations support the notion that the serine–histidine pair act together to generate a nucleophile of sufficient power to attack the carbonyl group of a peptide bond. The conversion of aspartate 32 into alanine had a smaller effect, although the value of k_{cat} still fell to less than 0.005% of its wild-type value. The simultaneously conversion of all three catalytic triad residues into alanine was no more deleterious than the conversion of serine or histidine alone. Despite the reduction in their catalytic power, the mutated enzymes still hydrolyze peptides a thousand times as rapidly as does buffer at pH 8.6.

Because the oxyanion hole of subtilisin includes a side-chain NH group in addition to backbone NH groups, it is possible to probe the importance of the oxyanion hole for catalysis by site-directed mutagenesis. The mutation of asparagine 155 to glycine reduced the value of k_{cat} to 0.2% of its wild-type value but increased the value of K_M by only a factor of two. These observations demonstrate that the NH group of the asparagine residue plays a significant role in stabilizing the tetrahedral intermediate and the transition state leading to it.

9.1.6 Cysteine, Aspartyl, and Metalloproteases Are Other Major Classes of Peptide-Cleaving Enzymes

Not all proteases utilize strategies based on activated serine residues. Classes of proteins have been discovered that employ three alternative approaches to peptide-bond hydrolysis (Figure 9.17). These classes are the (1) cysteine proteases, (2) aspartyl proteases, and (3) metalloproteases. In each case, the strategy generates a nucleophile that attacks the peptide carbonyl group (Figure 9.18).

The strategy used by the *cysteine proteases* is most similar to that used by the chymotrypsin family. In these enzymes, a cysteine residue, activated by a histidine residue, plays the role of the nucleophile that attacks the peptide bond (see Figure 9.18), in a manner quite analogous to that of the serine residue in serine proteases. An ideal example of these proteins is papain, an enzyme purified from the fruit of the papaya. Mammalian proteases homologous to papain have been discovered, most notably the cathepsins, proteins having a role in the immune and other systems. The cysteine-based active site arose independently at least twice in the course of evolution; the caspases, enzymes that play a major role in apoptosis (Section 2.4.3), have active sites similar to that of papain, but their overall structures are unrelated.

CYSTEINE PROTEASES

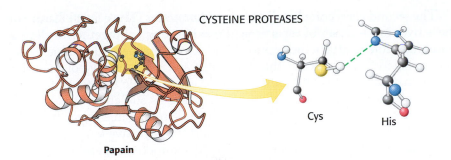

Cys   His

Papain

ASPARTYL PROTEASES

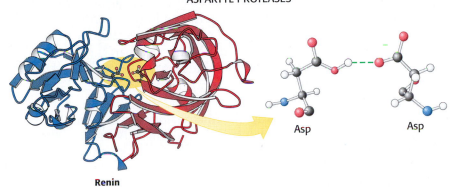

Asp   Asp

Renin

METALLOPROTEASES

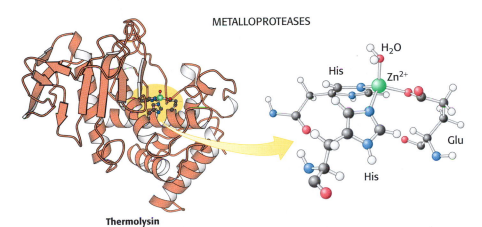

H_2O

His Zn^{2+}

Glu

His

Thermolysin

FIGURE 9.17 Three classes of proteases and their active sites. These examples of a cysteine protease, an aspartyl protease, and a metalloprotease use a histidine-activated cysteine residue, an aspartate-activated water molecule, and a metal-activated water molecule, respectively, as the nucleophile. The two halves of renin are in blue and red to highlight the approximate twofold symmetry of aspartyl proteases.

(A) CYSTEINE PROTEASES

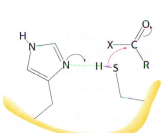

(B) ASPARTYL PROTEASES

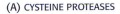

(C) METALLOPROTEASES

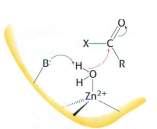

FIGURE 9.18 The activation strategies for three classes of proteases. The peptide carbonyl group is attacked by (A) a histidine-activated cysteine, in the cysteine proteases; (B) an aspartate-activated water molecule, in the aspartyl proteases; and (C) a metal-activated water molecule, in the metalloproteases. For the metalloproteases, the letter B represents a base (often a glutamate) that helps deprotonate the metal-bound water.

The second class comprises the *aspartyl proteases*. The central feature of the active sites is a pair of aspartic acid residues that act together to allow a water molecule to attack the peptide bond. One aspartic acid residue (in its deprotonated form) activates the attacking water molecule by poising it for deprotonation, whereas the other aspartic acid residue (in its protonated form) polarizes the peptide carbonyl, increasing its susceptibility to attack (see Figure 9.18). Members of this class include renin, an enzyme having a role in the regulation of blood pressure, and the digestive enzyme pepsin. These proteins possess approximate twofold symmetry, suggesting that the two halves are evolutionarily related. A likely scenario is that two copies of a gene for the ancestral enzyme fused to form a single gene that encoded a single-chain enzyme. Each copy of the gene would have contributed an aspartate residue to the active site. The human immunodeficiency virus (HIV) and other retroviruses contain an unfused dimeric aspartyl protease that is similar to the fused protein, but the individual chains are not joined to make a single chain (Figure 9.19). This observation is consistent with the idea that the enzyme may have originally existed as separate subunits.

The *metalloproteases* constitute the final major class of peptide-cleaving enzymes. The active site of such a protein contains a bound metal ion, almost always zinc, that activates a water molecule to act as a nucleophile to attack the peptide carbonyl group. The bacterial enzyme thermolysin and the digestive enzyme carboxypeptidase A are classic examples of the zinc proteases. Thermolysin, but not carboxypeptidase A, is a member of a large and diverse family of homologous zinc proteases that includes the matrix metalloproteases, enzymes that catalyze the reactions in tissue remodeling and degradation.

In each of these three classes of enzymes, *the active site includes features that allow for the activation of water or another nucleophile as well as for the polarization of the peptide carbonyl group and subsequent stabilization of a tetrahedral intermediate* (see Figure 9.18).

9.1.7 Protease Inhibitors Are Important Drugs

Compounds that block or modulate the activities of proteases can have dramatic biological effects. Most natural *protease inhibitors* are similar in structure to the peptide substrates of the enzyme that each inhibits (Section 10.5.4). *Several important drugs are protease inhibitors.* For example, captopril, an inhibitor of the metalloprotease angiotensin-converting enzyme (ACE), has been used to regulate blood pressure. Crixivan, an inhibitor of the HIV protease, is used in the treatment of AIDS. This protease cleaves multidomain viral proteins into their active forms; blocking this process completely prevents the virus from being infectious (see Figure 9.19). To prevent unwanted side effects, protease inhibitors used as drugs must be specific for one enzyme without inhibiting other proteins within the body.

Let us examine the interaction of Crixivan with HIV protease in more detail. Crixivan is constructed around an alcohol that mimics the tetrahedral intermediate; other groups are present to bind into the S_2, S_1, S_1', and S_2' recognition sites on the enzyme (Figure 9.20). The results of x-ray crystallographic studies revealed the structure of the enzyme–Crixivan complex, showing that Crixivan adopts a conformation that approximates the twofold symmetry of the enzyme (Figure 9.21). The active site of HIV protease is covered by two apparently flexible flaps that fold down on top of the bound in-

FIGURE 9.19 The structure of HIV protease and its binding pocket. The protease is a dimer of identical subunits, shown in blue and yellow, consisting of 99 amino acids each. The active-site aspartic acid residues, one from each chain, are shown as ball-and-stick structures. The flaps will close down on the binding pocket after substrate has been bound.

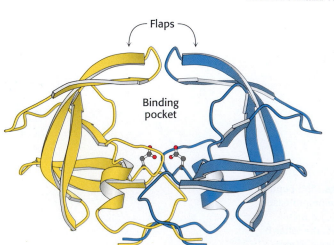

Flaps

Binding pocket

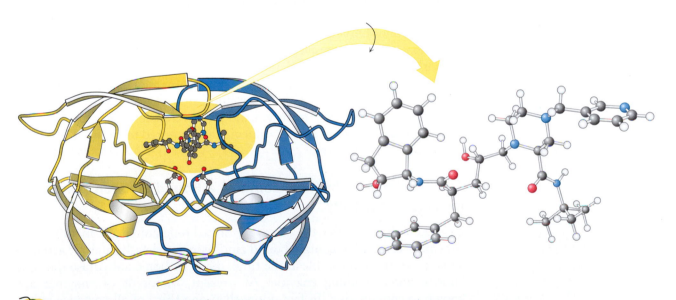

Crixivan

Peptide substrate

FIGURE 9.20 Crixivan, an HIV protease inhibitor. The structure of Crixivan is shown in comparison with that of a peptide substrate of HIV protease. The scissile bond in the substrate is highlighted in red.

hibitor. The hydroxyl group of the central alcohol interacts with two aspartate residues of the active site, one in each subunit. In addition, two carbonyl groups of the inhibitor are hydrogen bonded to a water molecule (not shown), which, in turn, is hydrogen bonded to a peptide NH group in each of the flaps. This interaction of the inhibitor with water and the enzyme is not possible with cellular aspartyl proteases such as renin and thus may contribute to the specificity of Crixivan and other inhibitors for HIV protease.

FIGURE 9.21 HIV protease–Crixivan complex. (Left) The HIV protease is shown with the inhibitor crixivan bound at the active site. (Right) The drug has been rotated to reveal its approximately twofold symmetric conformation.

9.2 MAKING A FAST REACTION FASTER: CARBONIC ANHYDRASES

Carbon dioxide is a major end product of aerobic metabolism. In complex organisms, this carbon dioxide is released into the blood and transported to the lungs for exhalation. While in the blood, carbon dioxide reacts with water. The product of this reaction is a moderately strong acid, carbonic acid ($pK_a = 3.5$), which becomes bicarbonate ion on the loss of a proton.

Even in the absence of a catalyst, this hydration reaction proceeds at a moderate pace. At 37°C near neutral pH, the second-order rate constant k_1 is 0.0027 M^{-1} s^{-1}. This corresponds to an effective first-order rate constant of 0.15 s^{-1} in water ($[H_2O] = 55.5$ M). Similarly, the reverse reaction, the dehydration of bicarbonate, is relatively rapid, with a rate constant of $k_{-1} = 50$ s^{-1}. These rate constants correspond to an equilibrium constant of $K_1 = 5.4 \times 10^{-5}$ and a ratio of $[CO_2]$ to $[H_2CO_3]$ of 340:1.

Despite the fact that CO_2 hydration and HCO_3^- dehydration occur spontaneously at reasonable rates in the absence of catalysts, almost all organisms contain enzymes, referred to as *carbonic anhydrases*, that catalyze these processes. Such enzymes are required because CO_2 hydration and HCO_3^- dehydration are often coupled to rapid processes, particularly transport processes. For example, HCO_3^- in the blood must be dehydrated to form CO_2 for exhalation as the blood passes through the lungs. Conversely, CO_2 must be converted into HCO_3^- for the generation of the aqueous humor of the eye and other secretions. Furthermore, both CO_2 and HCO_3^- are substrates and products for a variety of enzymes, and the rapid interconversion of these species may be necessary to ensure appropriate substrate levels. So important are these enzymes in human beings that mutations in some carbonic anhydrases have been found to cause osteopetrosis (excessive formation of dense bones accompanied by anemia) and mental retardation.

Carbonic anhydrases accelerate CO_2 hydration dramatically. The most active enzymes, typified by human carbonic anhydrase II, hydrate CO_2 at rates as high as $k_{cat} = 10^6$ s^{-1}, or a million times a second. Fundamental physical processes such as diffusion and proton transfer ordinarily limit the rate of hydration, and so special strategies are required to attain such prodigious rates.

9.2.1 Carbonic Anhydrase Contains a Bound Zinc Ion Essential for Catalytic Activity

Less than 10 years after the discovery of carbonic anhydrase in 1932, this enzyme was found to contain bound zinc, associated with catalytic activity. This discovery, remarkable at the time, made carbonic anhydrase the first known zinc-containing enzyme. At present, hundreds of enzymes are known to contain zinc. In fact, more than one-third of all enzymes either contain bound metal ions or require the addition of such ions for activity. The chemical reactivity of metal ions—associated with their positive charges, with their ability to form relatively strong yet kinetically labile bonds, and, in some cases, with their capacity to be stable in more than one oxidation state—explains why catalytic strategies that employ metal ions have been adopted throughout evolution.

The results of x-ray crystallographic studies have supplied the most detailed and direct information about the zinc site in carbonic anhydrase. At least seven carbonic anhydrases, each with its own gene, are present in human beings. They are all clearly homologous, as revealed by substantial levels of sequence identity. Carbonic anhydrase II, present in relatively

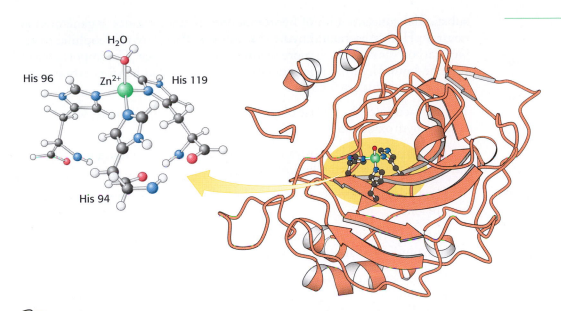

FIGURE 9.22 The structure of human carbonic anhydrase II and its zinc site. (Left) The zinc is bound to the imidazole rings of three histidine residues as well as to a water molecule. (Right) The location of the zinc site in the enzyme.

high concentrations in red blood cells, has been the most extensively studied (Figure 9.22).

Zinc is found only in the + 2 state in biological systems; so we need consider only this oxidation level as we examine the mechanism of carbonic anhydrase. A zinc atom is essentially always bound to four or more ligands; in carbonic anhydrase, three coordination sites are occupied by the imidazole rings of three histidine residues and an additional coordination site is occupied by a water molecule (or hydroxide ion, depending on pH). Because all of the molecules occupying the coordination sites are neutral, the overall charge on the $Zn(His)_3$ unit remains $+2$.

9.2.2 Catalysis Entails Zinc Activation of Water

How does this zinc complex facilitate carbon dioxide hydration? A major clue comes from the pH profile of enzymatically catalyzed carbon dioxide hydration (Figure 9.23). At pH 8, the reaction proceeds near its maximal rate. As the pH decreases, the rate of the reaction drops. The midpoint of this transition is near pH 7, suggesting that a group with $pK_a = 7$ plays an important role in the activity of carbonic anhydrase and that the deprotonated (high pH) form of this group participates more effectively in catalysis. Although some amino acids, notably histidine, have pK_a values near 7, *a variety of evidence suggests that the group responsible for this transition is not an amino acid but is the zinc-bound water molecule.* Thus, the binding of a water molecule to the positively charged zinc center reduces the pK_a of the water molecule from 15.7 to 7 (Figure 9.24). With the lowered pK_a, a

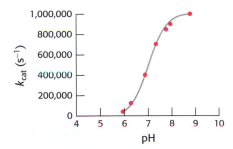

FIGURE 9.23 Effect of pH on carbonic anhydrase activity. Changes in pH alter the rate of carbon dioxide hydration catalyzed by carbonic anhydrase II. The enzyme is maximally active at high pH.

FIGURE 9.24 The pK_a of water-bound zinc. Binding to zinc lowers the pK_a of water from 15.7 to 7.

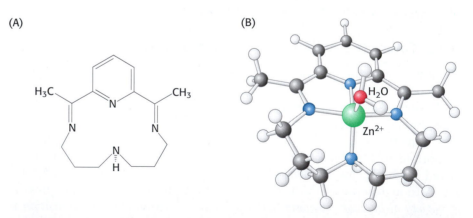

FIGURE 9.25 **Mechanism of carbonic anhydrase.** The zinc-bound hydroxide mechanism for the hydration of carbon dioxide catalyzed by carbonic anhydrase.

substantial concentration of hydroxide ion (bound to zinc) is generated at neutral pH. A zinc-bound hydroxide ion is sufficiently nucleophilic to attack carbon dioxide much more readily than water does. The importance of the zinc-bound hydroxide ion suggests a simple mechanism for carbon dioxide hydration (Figure 9.25).

1. Zinc facilitates the release of a proton from a water molecule, which generates a hydroxide ion.

2. The carbon dioxide substrate binds to the enzyme's active site and is positioned to react with the hydroxide ion.

3. The hydroxide ion attacks the carbon dioxide, converting it into bicarbonate ion.

4. The catalytic site is regenerated with the release of the bicarbonate ion and the binding of another molecule of water.

Thus, the binding of water to zinc favors the formation of the transition state, leading to bicarbonate formation by facilitating proton release and by bringing the two reactants into close proximity. A range of studies supports this mechanism. In particular, studies of a *synthetic analog model system* provide evidence for its plausibility. A simple synthetic ligand binds zinc through four nitrogen atoms (compared with three histidine nitrogen atoms in the enzyme), as shown in Figure 9.26. One water molecule remains bound to the zinc ion in the complex. Direct measurements reveal that this water molecule has a pK_a value of 8.7, not as low as the value for the water molecule in carbonic anhydrase but substantially lower than the value for free water. At pH 9.2, this complex accelerates the hydration of carbon dioxide more than 100-fold. Although catalysis by this synthetic system is much less efficient than catalysis by carbonic anhydrase, the model system strongly suggests that the zinc-bound hydroxide mechanism is likely to be correct. Carbonic anhydrases have evolved to utilize the reactivity intrinsic to a zinc-bound hydroxide ion as a potent catalyst.

(A) (B)

FIGURE 9.26 **A synthetic analog model system for carbonic anhydrase.** (A) An organic compound, capable of binding zinc, was synthesized as a model for carbonic anhydrase. The zinc complex of this ligand accelerates the hydration of carbon dioxide more than 100-fold under appropriate conditions. (B) The structure of the presumed active complex showing zinc bound to the ligand and to one water molecule.

9.2.3 A Proton Shuttle Facilitates Rapid Regeneration of the Active Form of the Enzyme

As noted earlier, some carbonic anhydrases can hydrate carbon dioxide at rates as high as a million times a second (10^6 s^{-1}). The magnitude of this rate can be understood from the following observations. At the conclusion of a carbon dioxide hydration reaction, the zinc-bound water molecule must lose a proton to regenerate the active form of the enzyme (Figure 9.27). The rate of the reverse reaction, the protonation of the zinc-bound hydroxide ion, is limited by the rate of proton diffusion. Protons diffuse very rapidly with second-

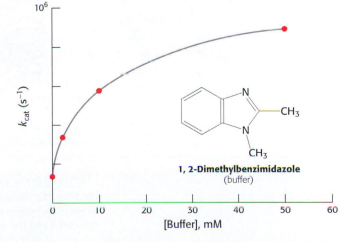

$$K = k_1/k_{-1} = 10^{-7}$$

FIGURE 9.27 Kinetics of water deprotonation. The kinetics of deprotonation and protonation of the zinc-bound water molecule in carbonic anhydrase.

order rate constants near $10^{-11}\,M^{-1}\,s^{-1}$. Thus, the backward rate constant k_{-1} must be less than $10^{11}\,M^{-1}\,s^{-1}$. Because the equilibrium constant K is equal to k_1/k_{-1}, the forward rate constant is given by $k_1 = K \cdot k_{-1}$. Thus, if $k_{-1} \le 10^{11}\,M^{-1}\,s^{-1}$ and $K = 10^{-7}\,M$ (because $pK_a = 7$), then k_1 must be less than or equal to $10^4\,s^{-1}$. In other words, the rate of proton diffusion limits the rate of proton release to less than $10^4\,s^{-1}$ for a group with $pK_a = 7$. However, if carbon dioxide is hydrated at a rate of $10^6\,s^{-1}$, then every step in the mechanism (see Figure 9.25) must take place at least this fast. How can this apparent paradox be resolved?

The answer became clear with the realization that *the highest rates of carbon dioxide hydration require the presence of buffer, suggesting that the buffer components participate in the reaction.* The buffer can bind or release protons. The advantage is that, whereas the concentrations of protons and hydroxide ions are limited to $10^{-7}\,M$ at neutral pH, the concentration of buffer components can be much higher, on the order of several millimolar. If the buffer component BH^+ has a pK_a of 7 (matching that for the zinc-bound water), then the equilibrium constant for the reaction in Figure 9.28 is 1.

$$K = k_1'/k_{-1}' \approx 1$$

FIGURE 9.28 The effect of buffer on deprotonation. The deprotonation of the zinc-bound water molecule in carbonic anhydrase is aided by buffer component B.

The rate of proton abstraction is given by $k_1' \cdot [B]$. The second-order rate constants k_1' and k_{-1}' will be limited by buffer diffusion to values less than approximately $10^9\,M^{-1}\,s^{-1}$. Thus, buffer concentrations greater than $[B] = 10^{-3}\,M$ (1 mM) may be high enough to support carbon dioxide hydration rates of $10^6\,M^{-1}\,s^{-1}$ because $k_1' \cdot [B] = (10^9\,M^{-1}\,s^{-1}) \cdot (10^{-3}\,M) = 10^6\,s^{-1}$. This prediction is confirmed experimentally (Figure 9.29).

The molecular components of many buffers are too large to reach the active site of carbonic anhydrase. Carbonic anhydrase II has evolved a *proton shuttle* to allow buffer components to participate in the reaction from solution. The primary component of this shuttle is histidine 64. This residue transfers protons from the zinc-bound water molecule to the protein surface and then to the buffer (Figure 9.30). *Thus, catalytic function has been enhanced through the evolution of an apparatus for controlling proton transfer from and to the active site.* Because protons participate in many biochemical reactions, the manipulation of the proton inventory within active sites is crucial to the function of many enzymes and explains the prominence of acid–base catalysis.

FIGURE 9.29 The effect of buffer concentration on the rate of carbon dioxide hydration. The rate of carbon dioxide hydration increases with the concentration of the buffer 1,2-dimethylbenzimidazole. The buffer enables the enzyme to achieve its high catalytic rates.

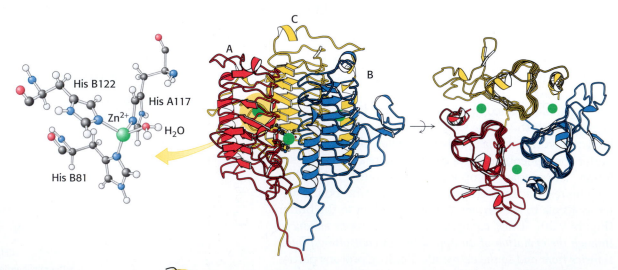

FIGURE 9.30 Histidine proton shuttle.
(1) Histidine 64 abstracts a proton from the zinc bound water molecule, generating a nucleophilic hydroxide ion and a protonated histidine. (2) The buffer (B) removes a proton from the histidine, regenerating the unprotonated form.

9.2.4 Convergent Evolution Has Generated Zinc-Based Active Sites in Different Carbonic Anhydrases

Carbonic anhydrases homologous to the human enzymes, referred to as *α-carbonic anhydrases*, are common in animals and in some bacteria and algae. In addition, two other families of carbonic anhydrases have been discovered. The *β-carbonic anhydrases* are found in higher plants and in many bacterial species, including *E. coli*. These proteins contain the zinc required for catalytic activity but are not significantly similar in sequence to the α-carbonic anhydrases. Furthermore, the β-carbonic anhydrases have only one conserved histidine residue, whereas the α-carbonic anhydrases have three. No three-dimensional structure is yet available, but spectroscopic studies suggest that the zinc is bound by one histidine residue, two cysteine residues (conserved among β-carbonic anhydrases), and a water molecule. A third family, the *γ-carbonic anhydrases*, also has been identified, initially in the archaeon *Methanosarcina thermophila*. The crystal structure of this enzyme reveals three zinc sites extremely similar to those in the α-carbonic anhydrases. In this case, however, the three zinc sites lie at the interfaces between the three subunits of a trimeric enzyme (Figure 9.31). The very striking left-handed β-helix (a β strand twisted into a left-handed helix) structure present in this enzyme has also been found in enzymes that catalyze reactions unrelated to those of carbonic anhydrase. Thus, convergent evolution has generated carbonic anhydrases that rely on coordinated zinc ions at least three times. In each

FIGURE 9.31 γ-Carbonic anhydrase. (Left) The zinc site of γ-carbonic anhydrase. (Middle) The trimeric structure of the protein (individual chains are labeled A, B, and C). (Right) The protein is rotated to show a top-down view that highlights its threefold symmetry and the position of the zinc sites (green) at the interfaces between subunits.

case, the catalytic activity appears to be associated with zinc-bound water molecules.

9.3 RESTRICTION ENZYMES: PERFORMING HIGHLY SPECIFIC DNA-CLEAVAGE REACTIONS

Let us next consider a hydrolytic reaction that results in the cleavage of DNA. Bacteria and archaea have evolved mechanisms to protect themselves from viral infections. Many viruses inject their DNA genomes into cells; once inside, the viral DNA hijacks the cell's machinery to drive the production of viral proteins and, eventually, of progeny virus. Often, a viral infection results in the death of the host. A major protective strategy for the host is to use *restriction endonucleases* (restriction enzymes) to degrade the viral DNA on its introduction into a cell. These enzymes recognize particular base sequences, called *recognition sequences* or *recognition sites,* in their target DNA and cleave that DNA at defined positions. The most well studied class of restriction enzymes comprises the so-called type II restriction enzymes, which cleave DNA *within* their recognition sequences. Other types of restriction enzymes cleave DNA at positions somewhat distant from their recognition sites.

Restriction endonucleases must show tremendous specificity at two levels. First, they must cleave only DNA molecules that contain recognition sites (hereafter referred to as *cognate DNA*) without cleaving DNA molecules that lack these sites. Suppose that a recognition sequence is six base pairs long. Because there are 4^6, or 4096, sequences having six base pairs, the concentration of sites that must not be cleaved will be approximately 5000-fold as high as the concentration of sites that should be cleaved. Thus, to keep from damaging host-cell DNA, endonucleases must cleave cognate DNA molecules much more than 5000 times as efficiently as they cleave nonspecific sites. Second, restriction enzymes must not degrade the host DNA. How do these enzymes manage to degrade viral DNA while sparing their own?

The restriction endonuclease *Eco*RV (from *E. coli*) cleaves double-stranded viral DNA molecules that contain the sequence 5'-GATATC-3' but leaves intact host DNA containing hundreds of such sequences. The host DNA is protected by other enzymes called methylases, which methylate adenine bases within host recognition sequences (Figure 9.32). For each restriction endonuclease, the host cell produces a corresponding methylase that marks the host DNA and prevents its degradation. These pairs of enzymes are referred to as *restriction-modification systems.* We shall return to the mechanism used to achieve the necessary levels of specificity after considering the chemistry of the cleavage process.

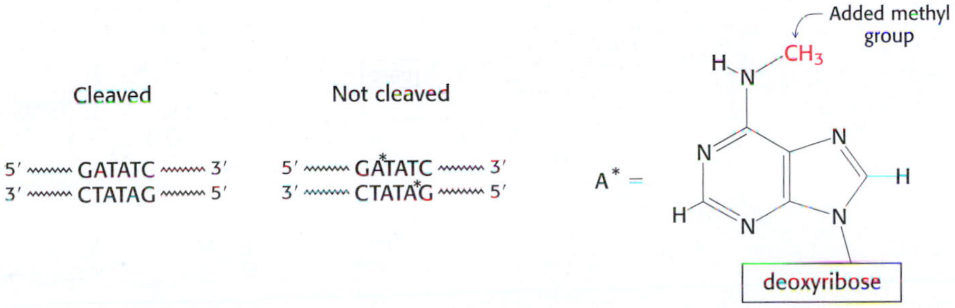

FIGURE 9.32 Protection by methylation. The recognition sequence for *Eco*RV endonuclease (left) and the sites of methylation (right) in DNA protected from the catalytic action of the enzyme.

9.3.1 Cleavage Is by In-Line Displacement of 3′ Oxygen from Phosphorus by Magnesium-Activated Water

The fundamental reaction catalyzed by restriction endonucleases is the hydrolysis of the phosphodiester backbone of DNA. Specifically, the bond between the 3′ oxygen atom and the phosphorus atom is broken. The products of this reaction are DNA strands with a free 3′-hydroxyl group and a 5′-phosphoryl group (Figure 9.33). This reaction proceeds by nucleophilic attack at the phosphorus atom. We will consider two types of mechanism, as suggested by analogy with the proteases. The restriction endonuclease might cleave DNA in mechanism 1 through a covalent intermediate, employing a potent nucleophile (Nu), or in mechanism 2 by direct hydrolysis:

Mechanism Type 1 (covalent intermediate)

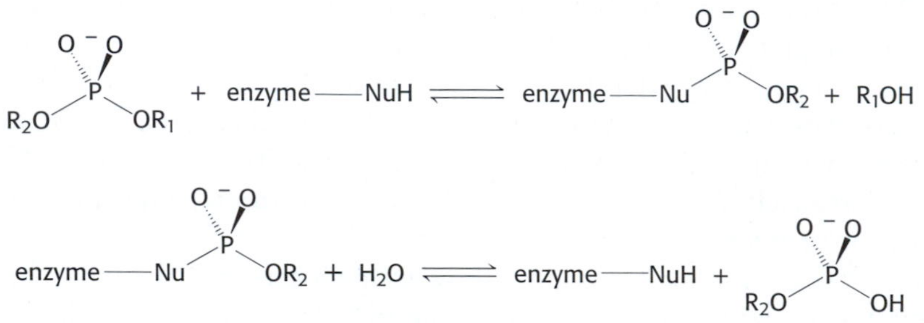

Mechanism Type 2 (direct hydrolysis)

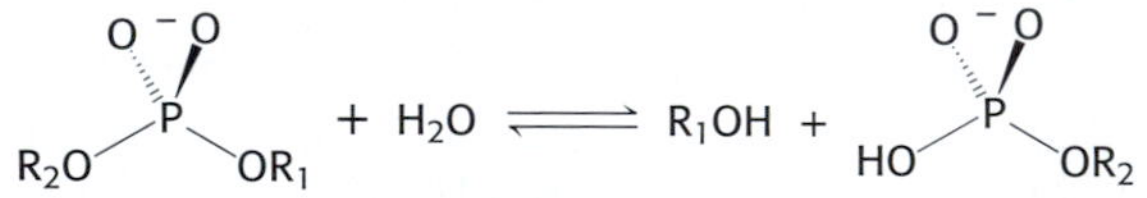

Each postulates a different nucleophile to carry out the attack on the phosphorus. In either case, each reaction takes place by an *in-line displacement* path:

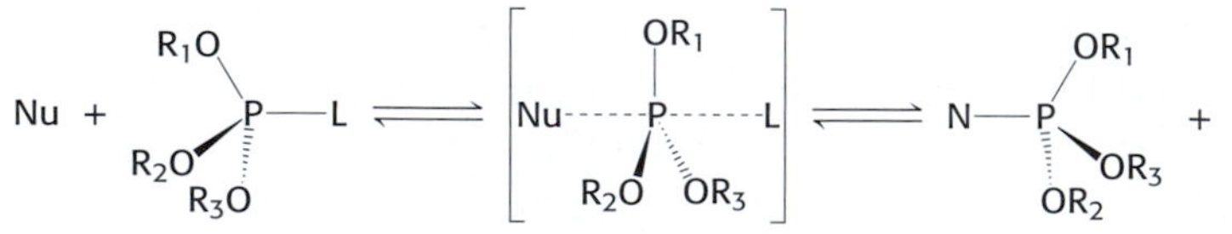

The incoming nucleophile attacks the phosphorus atom, and a pentacoordinate transition state is formed. This species has a trigonal bipyramidal geometry centered at the phosphorus atom, with the incoming nucleophile at one

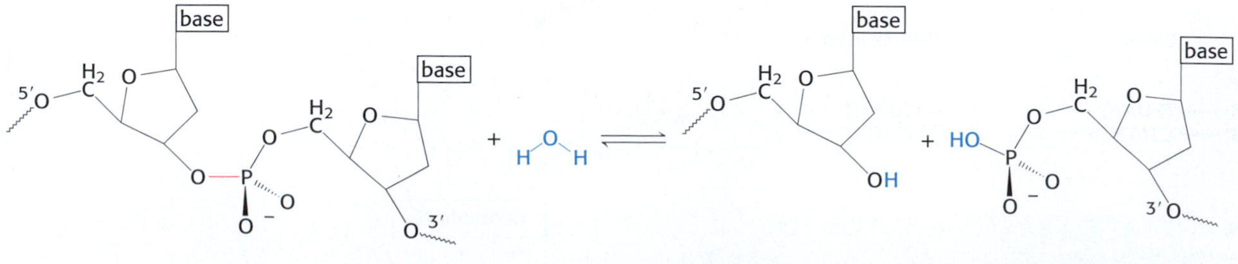

FIGURE 9.33 Hydrolysis of a phosphodiester bond. All restriction enzymes catalyze the hydrolysis of DNA phosphodiester bonds, leaving a phosphoryl group attached to the 5′ end. The bond that is cleaved is shown in red.

apex of the two pyramids and the group that is displaced (the leaving group, L) at the other apex. The two mechanisms differ in the number of times the displacement occurs in the course of the reaction.

In the first type of mechanism, a nucleophile in the enzyme (analogous to serine 195 in chymotrypsin) attacks the phosphoryl group to form a covalent intermediate. In a second step, this intermediate is hydrolyzed to produce the final products. Because two displacement reactions take place at the phosphorus atom in the first mechanism, the stereochemical configuration at the phosphorus atom would be inverted and then inverted again, and the overall configuration would be retained. In the second type of mechanism, analogous to that used by the aspartyl and metalloproteases, an activated water molecule attacks the phosphorus atom directly. In this mechanism, a single displacement reaction takes place at the phosphorus atom. Hence, the stereochemical configuration of the tetrahedral phosphorus atom is inverted each time a displacement reaction takes place. Monitoring the stereochemical changes of the phosphorus could be one approach to determining the mechanism of restriction endonuclease action.

A difficulty is that the phosphorus centers in DNA are not chiral, because two of the groups bound to the phosphorus atom are simple oxygen atoms, identical with each other. This difficulty can be circumvented by preparing DNA molecules that contain chiral phosphoryl groups, made by replacing one oxygen atom with sulfur (called a phosphorothioate). Let us consider *Eco*RV endonuclease. This enzyme cleaves the phosphodiester bond between the T and the A at the center of the recognition sequence 5′-GATATC-3′. The first step in monitoring the activity of the enzyme is to synthesize an appropriate substrate for *Eco*RV containing phosphorothioates at the sites of cleavage (Figure 9.34). The reaction is then performed in water that has been greatly enriched in ^{18}O to allow the incoming oxygen atom to be marked. The location of the ^{18}O label with respect to the sulfur atom indicates whether the reaction proceeds with inversion or retention of stereochemistry. *The analysis revealed that the stereochemical configuration at the phosphorus atom was inverted only once with cleavage.* This result is consistent with a direct attack of water at phosphorus and rules out the formation of any covalently bound intermediate (Figure 9.35).

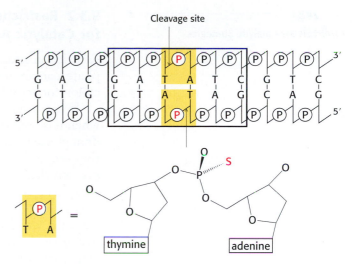

FIGURE 9.34 Labeling with phosphorothioates. Phosphorothioates, groups in which one of the nonbridging oxygen atoms is replaced with a sulfur atom, can be used to label specific sites in the DNA backbone to determine the overall stereochemical course of a displacement reaction. Here, a phosphorothioate is placed at sites that can be cleaved by *Eco*RV endonuclease.

FIGURE 9.35 Stereochemistry of cleaved DNA. Cleavage of DNA by *Eco*RV endonuclease results in overall inversion of the stereochemical configuration at the phosphorus atom, as indicated by the stereochemistry of the phosphorus atom bound to one bridging oxygen atom, one ^{16}O, one ^{18}O, and one sulfur atom. This configuration strongly suggests that the hydrolysis takes place by the direct attack of water on the phosphorus atom.

9.3.2 Restriction Enzymes Require Magnesium for Catalytic Activity

Restriction endonucleases as well as many other enzymes that act on phosphate-containing substrates require Mg^{2+} or some other similar divalent cation for activity. What is the function of this metal?

It has been possible to examine the interactions of the magnesium ion when it is bound to the enzyme. Crystals have been produced of *Eco*RV endonuclease bound to oligonucleotides that contain the appropriate recognition sequences. These crystals are grown in the absence of magnesium to prevent cleavage; then, when produced, the crystals are soaked in solutions containing the metal. No cleavage takes place, allowing the location of the magnesium ion binding sites to be determined (Figure 9.36). The magnesium ion was found to be bound to six ligands: three are water molecules, two are carboxylates of the enzyme's aspartate residues, and one is an oxygen atom of the phosphoryl group at the site of cleavage. The magnesium ion holds a water molecule in a position from which the water molecule can attack the phosphoryl group and, in conjunction with the aspartate residues, helps polarize the water molecule toward deprotonation. Because cleavage does not take place within these crystals, the observed structure cannot be the true catalytic conformation. Additional studies have revealed that a second magnesium ion must be present in an adjacent site for *Eco*RV endonuclease to cleave its substrate.

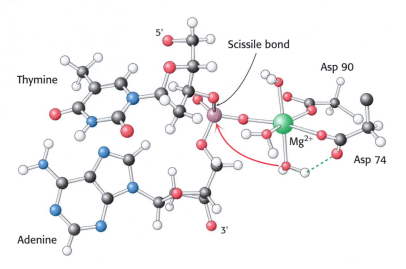

FIGURE 9.36 Magnesium ion binding site in *Eco*RV endonuclease. The magnesium ion helps to activate a water molecule and positions it so that it can attack the phosphate.

9.3.3 The Complete Catalytic Apparatus Is Assembled Only Within Complexes of Cognate DNA Molecules, Ensuring Specificity

We now return to the question of specificity, the defining feature of restriction enzymes. The recognition sequences for most restriction endonucleases are *inverted repeats*. This arrangement gives the three-dimensional structure of the recognition site a *twofold rotational symmetry* (Figure 9.37). The restriction enzymes display a corresponding symmetry to facilitate recognition: they are dimers whose two subunits are related by twofold rotational symmetry. The matching symmetry of the recognition sequence and the enzyme has been confirmed by the determination of the structure of the complex between *Eco*RV endonuclease and DNA fragments containing its recognition sequence (Figure 9.38). The enzyme surrounds the DNA in a tight embrace. Examination of this structure reveals features that are highly significant in determining specificity.

(A)

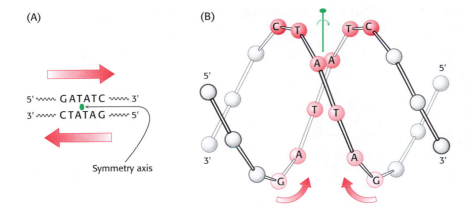

5' ~~~~ G A T A T C ~~~~ 3'
3' ~~~~ C T A T A G ~~~~ 5'

Symmetry axis

(B)

FIGURE 9.37 Structure of the recognition site of *Eco*RV endonuclease. (A) The sequence of the recognition site, which is symmetric around the axis of rotation designated in green. (B) The inverted repeat within the recognition sequence of *Eco*RV (and most other restriction endonucleases) endows the DNA site with twofold rotational symmetry.

A unique set of interactions occurs between the enzyme and a cognate DNA sequence. Within the 5'-GATATC-3' sequence, the G and A bases at the 5' end of each strand and their Watson-Crick partners directly contact the enzyme by hydrogen bonding with residues that are located in two loops, one projecting from the surface of each enzyme subunit (Figure 9.39). The most striking feature of this complex is the *distortion of the DNA*, which is substantially kinked in the center (Figure 9.40). The central two TA base pairs in the recognition sequence play a key role in producing the kink. They do not make contact with the enzyme but appear to be required because of their ease of distortion. 5'-TA-3' sequences are known to be among the most easily deformed base pairs. The distortion of the DNA at this site has severe effects on the specificity of enzyme action.

Specificity is often determined by an enzyme's binding affinity for substrates. In regard to *Eco*RV endonuclease, however, binding studies performed in the absence of magnesium have demonstrated that the enzyme binds to all sequences, both cognate and noncognate, with approximately equal affinity. However, the structures of complexes formed with noncognate DNA fragments are strikingly different from those formed with cognate DNA: the noncognate DNA conformation is not substantially distorted (Figure 9.41). *This lack of distortion has important consequences with regard to catalysis. No phosphate is positioned sufficiently close to the active-site*

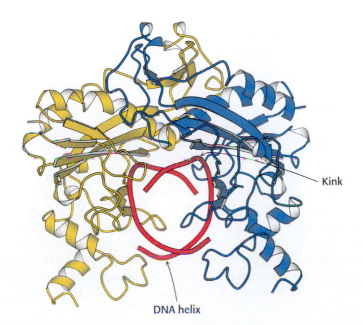

Kink

DNA helix

FIGURE 9.38 Structure of the *Eco*RV–cognate DNA complex. This view of the structure of *Eco*RV endonuclease bound to a cognate DNA fragment is down the helical axis of the DNA. The two protein subunits are in yellow and blue, and the DNA backbone is in red. The twofold axes of the enzyme dimer and the DNA are aligned.

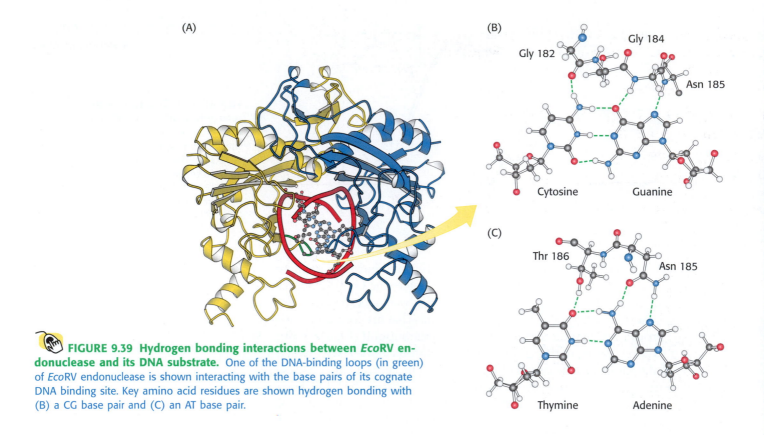

(A)

(B)

Gly 182   Gly 184

Asn 185

Cytosine   Guanine

(C)

Thr 186   Asn 185

Thymine   Adenine

FIGURE 9.39 Hydrogen bonding interactions between *Eco*RV endonuclease and its DNA substrate. One of the DNA-binding loops (in green) of *Eco*RV endonuclease is shown interacting with the base pairs of its cognate DNA binding site. Key amino acid residues are shown hydrogen bonding with (B) a CG base pair and (C) an AT base pair.

aspartate residues to complete a magnesium ion binding site (see Figure 9.36). Hence, the nonspecific complexes do not bind the magnesium ion and the complete catalytic apparatus is never assembled. The distortion of the substrate and the subsequent binding of the magnesium ion account for the catalytic specificity of more than 1,000,000-fold that is observed for *Eco*RV endonuclease despite very little preference at the level of substrate binding.

We can now see the role of binding energy in this strategy for attaining catalytic specificity. In binding to the enzyme, the DNA is distorted in such a way that additional contacts are made between the enzyme and the substrate, increasing the binding energy. However, this increase is canceled by the energetic cost of distorting the DNA from its relaxed conformation (Figure 9.42). Thus, for *Eco*RV endonuclease, there is little difference in binding affinity for cognate and nonspecific DNA fragments. However, the distortion in the cognate complex dramatically affects catalysis by

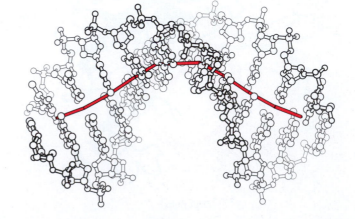

FIGURE 9.40 Distortion of the recognition site. The DNA is represented as a ball-and-stick model. The path of the DNA helical axis, shown in red, is substantially distorted on binding to the enzyme. For the B form of DNA, the axis is straight (not shown).

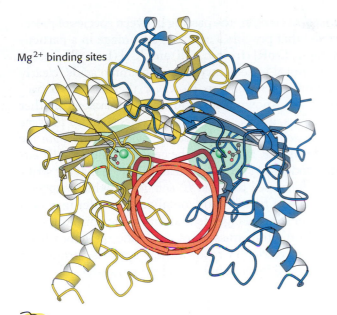

Mg²⁺ binding sites

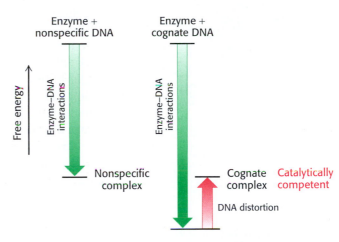

Enzyme + nonspecific DNA

Enzyme + cognate DNA

Free energy

Enzyme–DNA interactions

Enzyme–DNA interactions

Nonspecific complex

Cognate complex

Catalytically competent

DNA distortion

👆 **FIGURE 9.41 Nonspecific and cognate DNA within *Eco*RV endonuclease.** A comparison of the positions of the nonspecific (orange) and the cognate DNA (red) within *Eco*RV reveals that, in the nonspecific complex, the DNA backbone is too far from the enzyme to complete the magnesium ion binding sites.

FIGURE 9.42 Greater binding energy of *Eco*RV endonuclease bound to cognate versus noncognate DNA. The additional interactions between *Eco*RV endonuclease and cognate DNA increase the binding energy, which can be used to drive DNA distortions necessary for forming a catalytically competent complex.

completing the magnesium ion binding site. *This example illustrates how enzymes can utilize available binding energy to deform substrates and poise them for chemical transformation.* Interactions that take place within the distorted substrate complex stabilize the transition state leading to DNA hydrolysis.

The distortion in the DNA explains how methylation blocks catalysis and protects host-cell DNA. When a methyl group is added to the amino group of the adenine nucleotide at the 5′ end of the recognition sequence, the methyl group's presence precludes the formation of a hydrogen bond between the amino group and the side-chain carbonyl group of asparagine 185 (Figure 9.43). This asparagine residue is closely linked to the other amino acids that form specific contacts with the DNA. The absence of the hydrogen bond disrupts other interactions between the enzyme and the DNA substrate, and the distortion necessary for cleavage will not take place.

9.3.4 Type II Restriction Enzymes Have a Catalytic Core in Common and Are Probably Related by Horizontal Gene Transfer

🌿 Type II restriction enzymes are prevalent in Archaea and Eubacteria. What can we tell of the evolutionary history of these enzymes? Comparison of the amino acid sequences of a variety of type II restriction endonucleases did not reveal significant sequence similarity between most pairs of enzymes. However, a careful examination of three-dimensional structures, taking into account the location of the active sites, revealed the presence of a core structure conserved in the different enzymes. This structure includes β strands that contain the aspartate (or, in some cases, glutamate) residues forming the magnesium ion binding sites (Figure 9.44).

These observations indicate that many type II restriction enzymes are indeed evolutionary related. Analyses of the sequences in greater detail suggest that bacteria may have obtained genes encoding these enzymes from

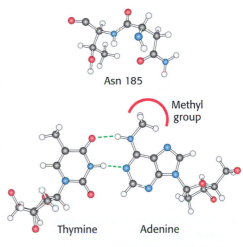

*Eco*RV

Asn 185

Methyl group

Thymine   Adenine

Methylated DNA

FIGURE 9.43 Methylation of adenine. The methylation of adenine blocks the formation of hydrogen bonds between *Eco*RV endonuclease and cognate DNA molecules and prevents their hydrolysis.

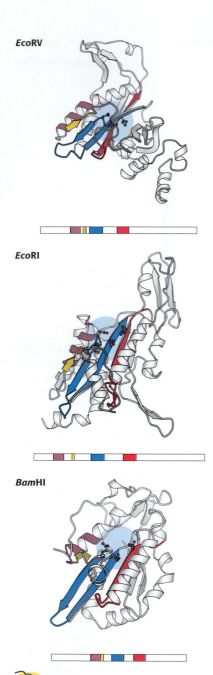

EcoRV

EcoRI

Bam HI

FIGURE 9.44 A conserved structural core in type II restriction enzymes. Four conserved structural elements, including the active-site region (in blue), are highlighted in color in these models of a single monomer from each dimeric enzyme. The positions of the amino acid sequences that form these elements within each overall sequence are represented schematically below each structure.

other species by *horizontal gene transfer,* the passing between species of pieces of DNA (such as plasmids) that provide a selective advantage in a particular environment. For example, *Eco*RI (from *E. coli*) and *Rsr*I (from *Rhodobacter sphaeroides*) are 50% identical in sequence over 266 amino acids, clearly indicative of a close evolutionary relationship. However, these species of bacteria are not closely related, as is known from sequence comparisons of other genes and other evidence. Thus, *it appears that these species obtained the gene for this restriction endonuclease from a common source more recently than the time of their evolutionary divergence.* Moreover, the gene encoding *Eco*RI endonuclease uses particular codons to specify given amino acids that are strikingly different from the codons used by most *E. coli* genes, which suggests that the gene did not originate in *E. coli.* Horizontal gene transfer may be a relatively common event. For example, genes that inactivate antibiotics are often transferred, leading to the transmission of antibiotic resistance from one species to another. For restriction-modification systems, protection against viral infections may have favored horizontal gene transfer.

9.4 NUCLEOSIDE MONOPHOSPHATE KINASES: CATALYZING PHOSPHORYL GROUP EXCHANGE BETWEEN NUCLEOTIDES WITHOUT PROMOTING HYDROLYSIS

The final enzymes that we shall consider are the nucleoside monophosphate kinases (NMP kinases), typified by adenylate kinase. These enzymes catalyze the transfer of the terminal phosphoryl group from a nucleoside triphosphate (NTP), usually ATP, to the phosphoryl group on a nucleoside monophosphate (Figure 9.45). The challenge for NMP kinases is to promote the transfer of the phosphoryl group from NTP to NMP without promoting the competing reaction—the transfer of a phosphoryl group from NTP to water; that is, NTP hydrolysis. We shall see how the use of induced fit by these enzymes is used to solve this problem. Moreover, these enzymes employ metal ion catalysis; but, in this case, the metal forms a complex with the substrate to enhance enzyme–substrate interaction.

FIGURE 9.45 Phosphoryl group transfer by nucleoside monophosphate kinases. These enzymes catalyze the interconversion of nucleoside triphosphate (here, ATP) and a nucleoside monophosphate (NMP) into two nucleoside diphosphates by the transfer of a phosphoryl group (shown in red).

9.4.1 NMP Kinases Are a Family of Enzymes Containing P-Loop Structures

X-ray crystallographic methods have yielded the three-dimensional structures of a number of different NMP kinases, both free and bound to substrates or substrate analogs. Comparison of these structures reveals that these enzymes form a family of homologous proteins (Figure 9.46). In particular, such comparisons reveal the presence of a conserved NTP-binding domain. This domain consists of a central β sheet, surrounded on both sides by α helices (Figure 9.47). A characteristic feature of this domain is a loop between the first β strand and the first helix. This loop, which typically has an amino acid sequence of the form Gly-X-X-X-X-Gly-Lys, is often referred to as the *P-loop* because it interacts with phosphoryl groups on the bound nucleotide (Figure 9.48). As described in Section 9.4.4, similar domains containing P-loops are present in a wide variety of important nucleotide-binding proteins.

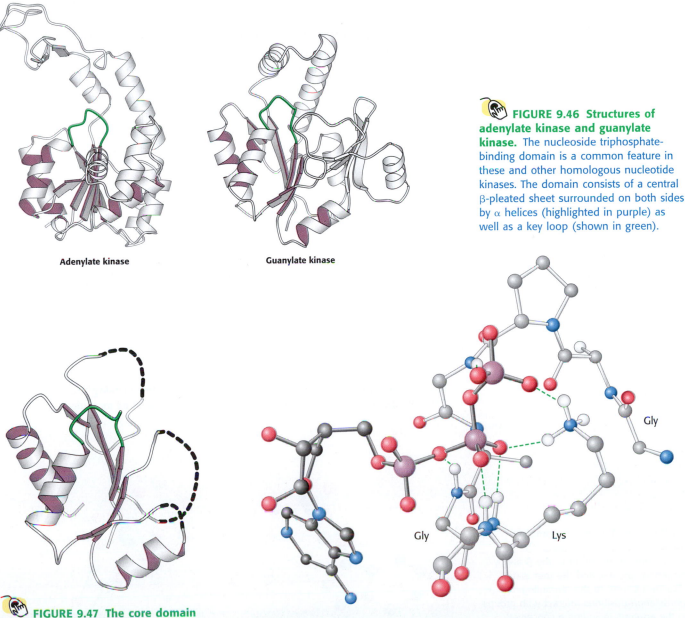

Adenylate kinase

Guanylate kinase

FIGURE 9.46 Structures of adenylate kinase and guanylate kinase. The nucleoside triphosphate-binding domain is a common feature in these and other homologous nucleotide kinases. The domain consists of a central β-pleated sheet surrounded on both sides by α helices (highlighted in purple) as well as a key loop (shown in green).

Gly

Gly

Lys

FIGURE 9.47 The core domain of NMP kinases. The P-loop is shown in green. The dashed lines represent the remainder of the protein structure.

FIGURE 9.48 P-loop interaction with ATP. The P-loop of adenylate kinase interacts with the phosphoryl groups of ATP (shown with dark bonds). Hydrogen bonds (green) link ATP to peptide NH groups as well as a lysine residue conserved among NMP kinases.

9.4.2 Magnesium (or Manganese) Complexes of Nucleoside Triphosphates Are the True Substrates for Essentially All NTP-Dependent Enzymes

Kinetic studies of NMP kinases, as well as many other enzymes having ATP or other nucleoside triphosphates as a substrate, reveal that these enzymes are essentially inactive in the absence of divalent metal ions such as magnesium (Mg^{2+}) or manganese (Mn^{2+}), but acquire activity on the addition of these ions. In contrast with the enzymes discussed so far, the metal is not a component of the active site. Rather, nucleotides such as ATP bind these ions, and *it is the metal ion–nucleotide complex that is the true substrate for the enzymes.* The dissociation constant for the ATP–Mg^{2+} complex is approximately 0.1 mM, and thus, given that intracellular Mg^{2+} concentrations are typically in the millimolar range, essentially all nucleoside triphosphates are present as NTP–Mg^{2+} complexes.

How does the binding of the magnesium ion to the nucleotide affect catalysis? There are a number of related consequences, but all serve to enhance the specificity of the enzyme–substrate interactions by enhancing binding energy. First, the magnesium ion neutralizes some of the negative charges present on the polyphosphate chain, reducing nonspecific ionic interactions between the enzyme and the polyphosphate group of the nucleotide. Second, the interactions between the magnesium ion and the oxygen atoms in the phosphoryl group hold the nucleotide in well-defined conformations that can be specifically bound by the enzyme (Figure 9.49). Magnesium ions are typically coordinated to six groups in an octahedral arrangement. Typically, two oxygen atoms are directly coordinated to a magnesium ion, with the remaining coordination positions often occupied

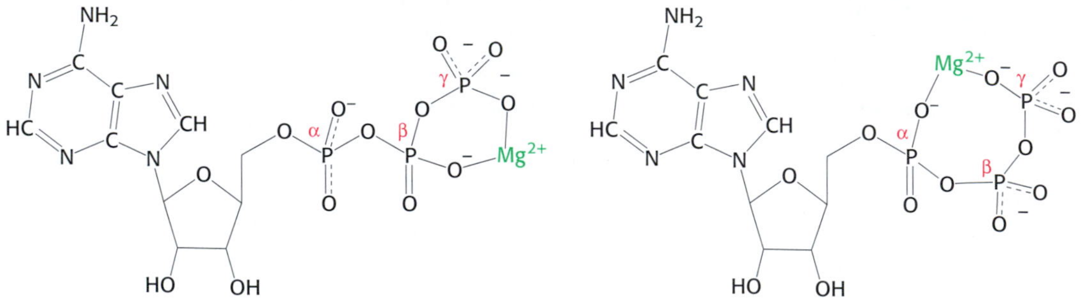

FIGURE 9.49 The structures of two isomeric forms of the ATP–Mg^{2+} complex. Other groups coordinated to the magnesium ion have been omitted for clarity.

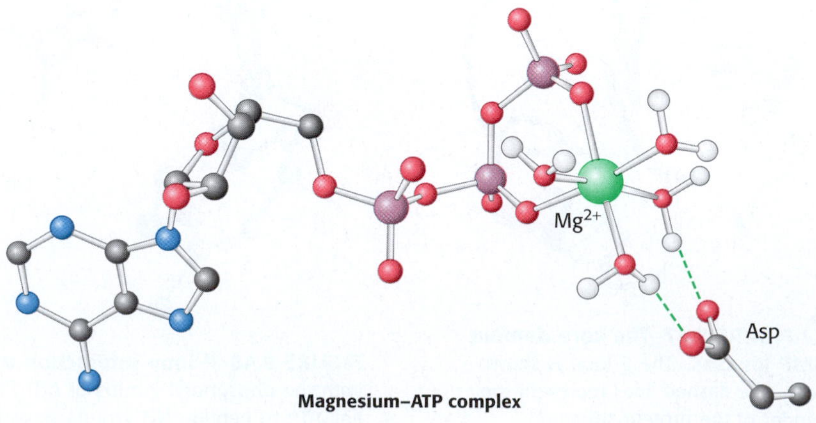

FIGURE 9.50 ATP–Mg^{2+} complex bound to adenylate kinase. The magnesium ion is bound to the β and γ phosphoryl groups, and the four water molecules bound to the remaining coordination positions interact with groups on the enzyme, including a conserved aspartate residue. Other interactions have been omitted for clarity.

Magnesium–ATP complex

by water molecules. Oxygen atoms of the α and β, β and γ, or α and γ phosphoryl groups may contribute, depending on the particular enzyme. In addition, different stereoisomers are produced, depending on exactly which oxygen atoms bind to the metal ion. Third, the magnesium ion provides additional points of interaction between the ATP–Mg^{2+} complex and the enzyme, thus increasing the binding energy. In some cases, such as the DNA polymerases (Section 27.2.2), side chains (often aspartate and glutamate residues) of the enzyme can bind directly to the magnesium ion. In other cases, the enzyme interacts indirectly with the magnesium ion through hydrogen bonds to the coordinated water molecules (Figure 9.50). Such interactions have been observed in adenylate kinases bound to ATP analogs.

9.4.3 ATP Binding Induces Large Conformational Changes

Comparison of the structure of adenylate kinase in the presence and absence of an ATP analog reveals that substrate binding induces large structural changes in the kinase, providing a classic example of the use of induced fit (Figure 9.51). The P-loop closes down on top of the polyphosphate chain, interacting most extensively with the β phosphoryl group. The movement of the P-loop permits the top domain of the enzyme to move down to form a lid over the bound nucleotide. This motion is favored by interactions between basic residues (conserved among the NMP kinases), the peptide backbone NH groups, and the nucleotide. With the ATP nucleotide held in this position, its γ phosphoryl group is positioned next to the binding site for the second substrate, NMP. In sum, the direct interactions with the nucleotide substrate lead to local structural rearrangements (movement of the P-loop) within the enzyme, which in turn allow more extensive changes (the closing down of the top domain) to take place. The binding of the second substrate, NMP, induces additional conformational changes. Both sets of changes ensure that a catalytically competent conformation is formed only when *both* the donor and the acceptor are bound, preventing wasteful transfer of the phosphoryl group to water. The enzyme holds its two substrates close together and appropriately oriented to stabilize the transition state that leads to the transfer of a phosphoryl group from the ATP to the NMP. This is an example of *catalysis by approximation*. We will see such examples of a catalytically competent active site being generated only on substrate binding many times in our study of biochemistry.

9.4.4 P-Loop NTPase Domains Are Present in a Range of Important Proteins

Domains similar (and almost certainly homologous) to those found in NMP kinases are present in a remarkably wide array of proteins, many of which participate in essential biochemical processes. Examples include ATP synthase, the key enzyme responsible for ATP generation; molecular motor proteins such as myosin; signal-transduction proteins such as transducin; proteins essential for translating mRNA into proteins, such as elongation factor Tu; and DNA and RNA unwinding helicases. The wide utility of P-loop NTPase domains is perhaps best explained by their ability to undergo substantial conformational changes on nucleoside triphosphate binding and hydrolysis. We shall encounter these domains (hereafter referred to as P-loop NTPases) throughout the book and shall observe how they function as springs, motors, and clocks. To allow easy recognition of these domains, they, like the binding domains of the NMP kinases, will be depicted with the inner surfaces of the ribbons in a ribbon diagram shown in purple and the P-loop shown in green (Figure 9.52).

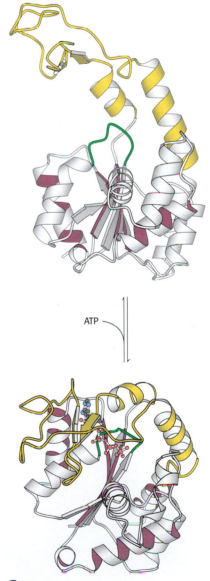

ATP

FIGURE 9.51 Conformational changes in adenylate kinase. Large conformational changes are associated with the binding of ATP by adenylate kinase. The P-loop is shown in green in each structure. The lid domain is highlighted in yellow.

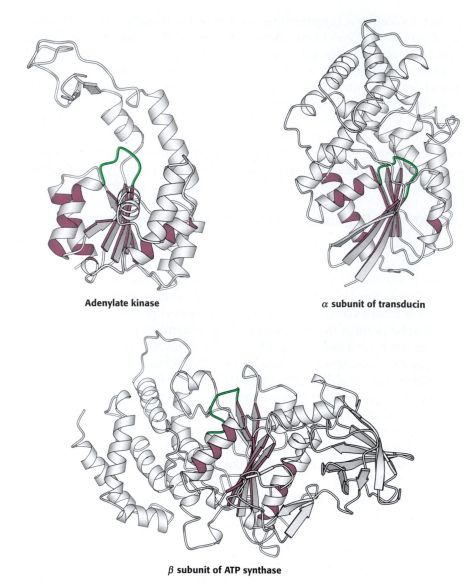

Adenylate kinase

α subunit of transducin

β subunit of ATP synthase

FIGURE 9.52 Three proteins containing P-loop NTPase domains. For the conserved domain, the inner surfaces of the ribbons are purple and the P-loops are green.

SUMMARY

Enzymes adopt conformations that are structurally and chemically complementary to the transition states of the reactions that they catalyze. Sets of interacting amino acid residues make up sites with the special structural and chemical properties necessary to stabilize the transition state. Enzymes use five basic strategies to form and stabilize the transition state: (1) the use of binding energy, (2) covalent catalysis, (3) general acid–base catalysis, (4) metal ion catalysis, and (5) catalysis by approximation. Of the enzymes examined in this chapter, three groups of enzymes catalyze the addition of water to their substrates but have different requirements for catalytic speed and specificity, and a fourth group of enzymes must prevent reaction with water.

Proteases: Facilitating a Difficult Reaction

The cleavage of peptide bonds by chymotrypsin is initiated by the attack of a serine residue on the peptide carbonyl group. The attacking hydroxyl group is activated by interaction with the imidazole group of a histidine residue, which is, in turn, linked to an aspartate residue. This Ser-His-Asp catalytic triad generates a powerful nucleophile. The product of this initial reaction is a covalent intermediate formed by the enzyme and an acyl group derived from the bound substrate. The hydrolysis of this acyl-enzyme intermediate completes the cleavage process. The tetrahedral intermediates for these reactions have a negative charge on the peptide carbonyl oxygen atom. This negative charge is stabilized by interactions with peptide NH groups in a region on the enzyme termed the oxyanion hole.

Other proteases employ the same catalytic strategy. Some of these proteases, such as trypsin and elastase, are homologs of chymotrypsin. In other proteases, such as subtilisin, a very similar catalytic triad has arisen by convergent evolution. Active-site structures that differ from the catalytic triad are present in a number of other classes of proteases. These classes employ a range of catalytic strategies but, in each case, a nucleophile is generated that is sufficiently powerful to attack the peptide carbonyl group. In some enzymes, the nucleophile is derived from a side chain; whereas, in others, an activated water molecule attacks the peptide carbonyl directly.

Carbonic Anhydrases: Making a Fast Reaction Faster

Carbonic anhydrases catalyze the reaction of water with carbon dioxide to generate carbonic acid. The catalysis can be extremely fast: molecules of some carbonic anhydrases hydrate carbon dioxide at rates as high as 1 million times per second. A tightly bound zinc ion is a crucial component of the active sites of these enzymes. Each zinc ion binds a water molecule and promotes its deprotonation to generate a hydroxide ion at neutral pH. This hydroxide attacks carbon dioxide to form bicarbonate ion, HCO_3^-. Because of the physiological roles of carbon dioxide and bicarbonate ions, speed is of the essence for this enzyme. To overcome limitations imposed by the rate of proton transfer from the zinc-bound water molecule, the most active carbonic anhydrases have evolved a proton shuttle to transfer protons to a buffer.

Restriction Enzymes: Performing Highly Specific DNA Cleavage Reactions

A high level of substrate specificity is often the key to biological function. Restriction endonucleases that cleave DNA at specific recognition sequences discriminate between molecules that contain these recognition sequences and those that do not. Within the enzyme–substrate complex, the DNA substrate is distorted in a manner that generates a magnesium ion binding site between the enzyme and DNA. The magnesium ion binds and activates a water molecule, which attacks the phosphodiester backbone.

Some enzymes discriminate between potential substrates by binding them with different affinities. Others may bind many potential substrates but promote chemical reactions efficiently only on specific molecules. Restriction endonucleases such as *Eco*RV endonuclease employ the latter mechanism to achieve levels of discrimination as high as million-fold. Structural studies reveal that these enzymes may bind nonspecific DNA molecules, but such molecules are not distorted in a manner that allows magnesium ion binding and, hence, catalysis. Restriction enzymes are prevented from acting on the DNA of a host cell by the methylation of key sites within their recognition sequences. The added methyl groups block specific interactions between the enzymes and the DNA such that the distortion necessary for cleavage does not take place.

Nucleoside Monophosphate Kinases: Catalyzing Phosphoryl Group Exchange Without Promoting Hydrolysis

Finally, NMP kinases illustrate that induced fit—the alteration of enzyme structure on substrate binding—facilitates phosphoryl transfer between nucleotides rather than to a molecule of water. This class of enzyme displays a structural motif called the P-loop NTPase domain that is present in a wide array of nucleotide-binding proteins. The closing of the P-loop over a bound nucleoside triphosphate substrate permits the top domain of the enzyme to form a lid over the bound nucleotide, positioning the triphosphate near the monophosphate with which it will react, in an example of catalysis by approximation. These enzymes are dependent on metal ions, but the ions bind to substrate instead of directly to the enzyme. The binding of the metal ion to the nucleoside triphosphate enhances the specificity of the enzyme–substrate interactions by holding the nucleotide in a well-defined conformation and providing additional points of interaction, thus increasing binding energy.

KEY TERMS

binding energy (p. 228)

induced fit (p. 228)

covalent catalysis (p. 228)

general acid–base catalysis (p. 228)

metal ion catalysis (p. 228)

catalysis by approximation (p. 228)

chemical modification reaction (p. 230)

catalytic triad (p. 231)

oxyanion hole (p. 233)

protease inhibitor (p. 238)

proton shuttle (p. 243)

recognition sequence (p. 245)

restriction-modification system (p. 245)

in-line displacement (p. 246)

horizontal gene transfer (p. 252)

P-loop (p. 253)

SELECTED READINGS

Where to start

Stroud, R. M., 1974. A family of protein-cutting proteins. *Sci. Am.* 231(1):74–88.

Kraut, J., 1977. Serine proteases: structure and mechanism of catalysis. *Annu. Rev. Biochem.* 46:331–358.

Lindskog, S., 1997. Structure and mechanism of carbonic anhydrase. *Pharmacol. Ther.* 74:1–20.

Jeltsch, A., Alves, J., Maass, G., Pingoud, A., 1992. On the catalytic mechanism of EcoRI and EcoRV: A detailed proposal based on biochemical results, structural data and molecular modelling. *FEBS Lett.* 304:4–8.

Yan, H., and Tsai, M.-D., 1999. Nucleoside monophosphate kinases: Structure, mechanism, and substrate specificity. *Adv. Enzymol. Relat. Areas Mol. Biol.* 73:103–134.

Lolis, E., and Petsko, G. A., 1990. Transition-state analogues in protein crystallography: Probes of the structural source of enzyme catalysis. *Annu. Rev. Biochem.* 59:597–630.

Books

Fersht, A., 1999. *Structure and Mechanism in Protein Science: A Guide to Enzyme Catalysis and Protein Folding.* W. H. Freeman and Company.

Silverman, R. B., 2000. *The Organic Chemistry of Enzyme-Catalyzed Reactions.* Academic Press.

Page, M., and Williams, A., 1997. *Organic and Bio-organic Mechanisms.* Addison Wesley Longman.

Chymotrypsin and other serine proteases

Fastrez, J., and Fersht, A. R., 1973. Demonstration of the acyl-enzyme mechanism for the hydrolysis of peptides and anilides by chymotrypsin. *Biochemistry* 12:2025–2034.

Sigler, P. B., Blow, D. M., Matthews, B. W., and Henderson, R., 1968. Structure of crystalline-chymotrypsin II: A preliminary report including a hypothesis for the activation mechanism. *J. Mol. Biol.* 35:143–164.

Kossiakoff, A. A., and Spencer, S. A., 1981. Direct determination of the protonation states of aspartic acid-102 and histidine-57 in the tetrahedral intermediate of the serine proteases: Neutron structure of trypsin. *Biochemistry* 20:6462–6474.

Carter, P., and Wells, J. A., 1988. Dissecting the catalytic triad of a serine protease. *Nature* 332:564–568.

Carter, P., and Wells, J. A., 1990. Functional interaction among catalytic residues in subtilisin BPN'. *Proteins* 7:335–342.

Koepke, J., Ermler, U., Warkentin, E., Wenzl, G., and Flecker, P., 2000. Crystal structure of cancer chemopreventive Bowman-Birk inhibitor in ternary complex with bovine trypsin at 2.3 Å resolution: Structural basis of Janus-faced serine protease inhibitor specificity. *J. Mol. Biol.* 298:477–491.

Gaboriaud, C., Rossi, V., Bally, I., Arlaud, G. J., and Fontecilla-Camps, J. C., 2000. Crystal structure of the catalytic domain of human complement C1s: A serine protease with a handle. *EMBO J.* 19:1755–1765.

Other proteases

Kamphuis, I. G., Kalk, K. H., Swarte, M. B., and Drenth, J., 1984. Structure of papain refined at 1.65 Å resolution. *J. Mol. Biol.* 179:233–256.

Kamphuis, I. G., Drenth, J., and Baker, E. N., 1985. Thiol proteases: Comparative studies based on the high-resolution structures of papain and actinidin, and on amino acid sequence information for cathepsins B and H, and stem bromelain. *J. Mol. Biol.* 182:317–329.

Sivaraman, J., Nagler, D. K., Zhang, R., Menard, R., and Cygler, M., 2000. Crystal structure of human procathepsin X: A cysteine protease with the proregion covalently linked to the active site cysteine. *J. Mol. Biol.* 295:939–951.

Davies, D. R., 1990. The structure and function of the aspartic proteinases. *Annu. Rev. Biophys. Biophys. Chem.* 19:189–215.

Dorsey, B. D., Levin, R. B., McDaniel, S. L., Vacca, J. P., Guare, J. P., Darke, P. L., Zugay, J. A., Emini, E. A., Schleif, W. A., Quintero, J. C., et al., 1994. L-735,524: The design of a potent and orally bioavailable HIV protease inhibitor. *J. Med. Chem.* 37:3443–3451.

Chen, Z., Li, Y., Chen, E., Hall, D. L., Darke, P. L., Culberson, C., Shafer, J. A., and Kuo, L. C., 1994. Crystal structure at 1.9-Å resolution of human immunodeficiency virus (HIV) II protease complexed with L-735,524, an orally bioavailable inhibitor of the HIV proteases. *J. Biol. Chem.* 269:26344–26348.

Ollis, D. L., Cheah, E., Cygler, M., Dijkstra, B., Frolow, F., Franken, S. M., Harel, M., Remington, S. J., Silman, I., Schrag, J., et al., 1992. The α/β hydrolase fold. *Protein Eng.* 5:197–211.

Carbonic anhydrase

Lindskog, S., and Coleman, J. E., 1973. The catalytic mechanism of carbonic anhydrase. *Proc. Natl. Acad. Sci. USA* 70:2505–2508.

Kannan, K. K., Notstrand, B., Fridborg, K., Lovgren, S., Ohlsson, A., and Petef, M., 1975. Crystal structure of human erythrocyte carbonic anhydrase B: Three-dimensional structure at a nominal 2.2-Å resolution. *Proc. Natl. Acad. Sci. USA* 72:51–55.

Boriack-Sjodin, P. A., Zeitlin, S., Chen, H. H., Crenshaw, L., Gross, S., Dantanarayana, A., Delgado, P., May, J. A., Dean, T., and Christianson, D. W., 1998. Structural analysis of inhibitor binding to human carbonic anhydrase II. *Protein Sci.* 7:2483–2489.

Wooley, P., 1975. Models for metal ion function in carbonic anhydrase. *Nature* 258:677–682.

Jonsson, B. H., Steiner, H., and Lindskog, S., 1976. Participation of buffer in the catalytic mechanism of carbonic anhydrase. *FEBS Lett.* 64:310–314.

Sly, W. S., and Hu, P. Y., 1995. Human carbonic anhydrases and carbonic anhydrase deficiencies. *Annu. Rev. Biochem.* 64:375–401.

Maren, T. H., 1988. The kinetics of HCO_3^- synthesis related to fluid secretion, pH control, and CO_2 elimination. *Annu. Rev. Physiol.* 50:695–717.

Kisker, C., Schindelin, H., Alber, B. E., Ferry, J. G., and Rees, D. C., 1996. A left-hand beta-helix revealed by the crystal structure of a carbonic anhydrase from the archaeon *Methanosarcina thermophila*. *EMBO J.* 15:2323–2330.

Restriction enzymes

Winkler, F. K., Banner, D. W., Oefner, C., Tsernoglou, D., Brown, R. S., Heathman, S. P., Bryan, R. K., Martin, P. D., Petratos, K., and Wilson, K. S., 1993. The crystal structure of EcoRV endonuclease and of its complexes with cognate and non-cognate DNA fragments. *EMBO J.* 12:1781–1795.

Kostrewa, D., and Winkler, F. K., 1995. Mg^{2+} binding to the active site of EcoRV endonuclease: A crystallographic study of complexes with substrate and product DNA at 2 Å resolution. *Biochemistry* 34:683–696.

Athanasiadis, A., Vlassi, M., Kotsifaki, D., Tucker, P. A., Wilson, K. S., and Kokkinidis, M., 1994. Crystal structure of PvuII endonuclease reveals extensive structural homologies to EcoRV. *Nat. Struct. Biol.* 1:469–475.

Sam, M. D., and Perona, J. J., 1999. Catalytic roles of divalent metal ions in phosphoryl transfer by EcoRV endonuclease. *Biochemistry* 38:6576–6586.

Jeltsch, A., and Pingoud, A., 1996. Horizontal gene transfer contributes to the wide distribution and evolution of type II restriction-modification systems. *J. Mol. Evol.* 42:91–96.

NMP kinases

Byeon, L., Shi, Z., and Tsai, M. D., 1995. Mechanism of adenylate kinase: The "essential lysine" helps to orient the phosphates and the active site residues to proper conformations. *Biochemistry* 34:3172–3182.

Dreusicke, D., and Schulz, G. E., 1986. The glycine-rich loop of adenylate kinase forms a giant anion hole. *FEBS Lett.* 208:301–304.

Pai, E. F., Sachsenheimer, W., Schirmer, R. H., and Schulz, G. E., 1977. Substrate positions and induced-fit in crystalline adenylate kinase. *J. Mol. Biol.* 114:37–45.

Schlauderer, G. J., Proba, K., and Schulz, G. E., 1996. Structure of a mutant adenylate kinase ligated with an ATP-analogue showing domain closure over ATP. *J. Mol. Biol.* 256:223–227.

Vonrhein, C., Schlauderer, G. J., and Schulz, G. E., 1995. Movie of the structural changes during a catalytic cycle of nucleoside monophosphate kinases. *Structure* 3:483–490.

Muller-Dieckmann, H. J., and Schulz, G. E., 1994. The structure of uridylate kinase with its substrates, showing the transition state geometry. *J. Mol. Biol.* 236:361–367.

PROBLEMS

1. *No burst.* Examination of the cleavage of the chromogenic *amide* substrate, A, by chymotrypsin with the use of stopped-flow kinetic methods reveals no burst. Why?

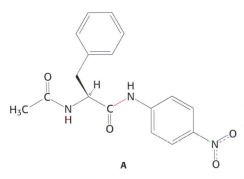

A

2. *Contributing to your own demise.* Consider the subtilisin substrates A and B.

Phe-Ala-Gln-Phe-X   Phe-Ala-His-Phe-X
       A   B

These substrates are cleaved (between Phe and X) by native subtilisin at essentially the same rate. However, the His 64-to-Ala mutant of subtilisin cleaves substrate B more than 1000-fold as rapidly as it cleaves substrate A. Propose an explanation.

3. $1 + 1 \neq 2$. Consider the following argument. In subtilisin, mutation of Ser 221 to Ala results in a 10^6-fold decrease in activity. Mutation of His 64 to Ala results in a similar 10^6-fold decrease. Therefore, simultaneous mutation of Ser 221 to Ala and His 64 to Ala should result in a $10^6 \times 10^6 = 10^{12}$-fold reduction in activity. Is this correct? Why or why not?

4. *Adding a charge.* In chymotrypsin, a mutant was constructed with Ser 189, which is in the bottom of the substrate specificity pocket, changed to Asp. What effect would you predict for this Ser 189 → Asp 189 mutation?

5. *Conditional results.* In carbonic anhydrase II, mutation of the proton-shuttle residue His 64 to Ala was expected to result in a decrease in the maximal catalytic rate. However, in buffers such as imidazole with relatively small molecular components, no rate reduction was observed. In buffers with larger molecular components, significant rate reductions were observed. Propose an explanation.

6. *How many sites?* A researcher has isolated a restriction endonuclease that cleaves at only one particular 10-base-pair site. Would this enzyme be useful in protecting cells from viral infections, given that a typical viral genome is 50,000 base pairs long? Explain.

7. *Is faster better?* Restriction endonucleases are, in general, quite slow enzymes with typical turnover numbers of 1 s^{-1}. Suppose that endonucleases were faster with turnover numbers similar to those for carbonic anhydrase (10^6 s^{-1}). Would this increased rate be beneficial to host cells, assuming that the fast enzymes have similar levels of specificity?

8. *Adopting a new gene.* Suppose that one species of bacteria obtained one gene encoding a restriction endonuclease by horizontal gene transfer. Would you expect this acquisition to be beneficial?

9. *Predict the product.* Adenylate kinase is treated with adenosine disphosphate (ADP).

(a) What products will be generated?

(b) If the initial concentration of ADP is 1 mM, estimate the concentrations of ADP and the products from part *a* after incubation with adenylate kinase for a long time.

10. *Chelation therapy.* Treatment of carbonic anhydrase with high concentrations of the metal chelator EDTA (ethylenediaminetetraacetic acid) results in the loss of enzyme activity. Propose an explanation.

11. *Identify the enzyme.* Consider the structure of molecule A. Which enzyme discussed in this chapter do you think molecule A will most effectively inhibit?

Molecule A

12. *An aldehyde inhibitor.* Elastase is specifically inhibited by an aldehyde derivative of one of its substrates:

(a) Which residue in the active site of elastase is most likely to form a covalent bond with this aldehyde?

(b) What type of covalent link would be formed?

Mechanism Problem

13. *Complete the mechanism.* On the basis of the information provided in Figure 9.18, complete the mechanisms for peptide-bond cleavage by (a) a cysteine protease, (b) an aspartyl protease, and (c) a metalloprotease.

Media Problems

14. *Now you see it, now you don't.* Pre-steady-state experiments using chymotrypsin and a chromogenic substrate (N-acetyl-L-phenylalanine p-nitrophenyl ester) show a "burst" of product at very short times (Figure 9.4). The **Conceptual Insights** module on enzyme kinetics explains this result. What results would you see if the product detected by the assay was the free N-terminal component of the substrate instead of the C-terminal component? (*Hint:* Use the pre-steady-state reaction simulation to simulate the experiment. Select different times following mixing and observe the amount of each product.).

15. *Seeing is disbelieving.* DIPF reacts specifically with serine 195 of chymotrypsin. One hypothesis as to why this is so might be that serine 195 is unusually exposed on the surface of the protein compared to other serines. After looking at the **Structural Insights** module on chymotrypsin, what do you think of this hypothesis?

CHAPTER 10

Regulatory Strategies: Enzymes and Hemoglobin

Like motor traffic, metabolic pathways flow more efficiently when regulated by signals. CTP, the final product of a multistep pathway, controls flux through the pathway by inhibiting the committed step catalyzed by aspartate transcarbamoylase (ATCase). [(Left) Richard Berenholtz/The Stock Market.]

The activity of proteins, including enzymes, often must be regulated so that they function at the proper time and place. The biological activity of proteins is regulated in four principal ways:

1. *Allosteric control.* Allosteric proteins contain distinct regulatory sites and multiple functional sites. Regulation by small signal molecules is a significant means of controlling the activity of many proteins. The binding of these regulatory molecules at sites distinct from the active site triggers conformational changes that are transmitted to the active site. Moreover, allosteric proteins show the property of *cooperativity*: activity at one functional site affects the activity at others. As a consequence, a slight change in substrate concentration can produce substantial changes in activity. Proteins displaying allosteric control are thus information transducers: their activity can be modified in response to signal molecules or to information shared among active sites. This chapter examines two of the best-understood allosteric proteins: the enzyme *aspartate transcarbamoylase* (ATCase) and the oxygen-carrying protein *hemoglobin*. Catalysis by aspartate transcarbamoylase of the first step in pyrimidine biosynthesis is inhibited by cytidine triphosphate, the final product of that biosynthesis, in an example of *feedback inhibition*. The binding of O_2 by hemoglobin is cooperative and is regulated by H^+, CO_2 and 2,3-bisphosphoglycerate (2,3-BPG).

2. *Multiple forms of enzymes.* Isozymes, or isoenzymes, provide an avenue for varying regulation of the same reaction at distinct locations or times. Isozymes are homologous enzymes within a single organism that catalyze the same reaction but differ slightly in structure and more obviously in K_M and V_{max} values, as well as regulatory properties. Often, isozymes are expressed in a distinct tissue or organelle or at a distinct stage of development.

3. *Reversible covalent modification.* The catalytic properties of many enzymes are markedly altered by the covalent attachment of a modifying group, most commonly a phosphoryl group. ATP serves as the phosphoryl donor in these reactions, which are catalyzed by *protein kinases*. The removal of phosphoryl groups by hydrolysis is catalyzed by *protein phosphatases*. This chapter considers the structure, specificity, and control of *protein kinase A* (PKA), a ubiquitous eukaryotic enzyme that regulates diverse target proteins.

4. *Proteolytic activation.* The enzymes controlled by some of these mechanisms cycle between active and inactive states. A different regulatory motif is used to *irreversibly* convert an inactive enzyme into an active one. Many enzymes are activated by the hydrolysis of a few or even one peptide bond in inactive precursors called *zymogens* or *proenzymes*. This regulatory mechanism generates digestive enzymes such as chymotrypsin, trypsin, and pepsin. Caspases, which are proteolytic enzymes that are the executioners in *programmed cell death,* or *apoptosis* (Section 2.4.3), are proteolytically activated from the procaspase form. Blood clotting is due to a remarkable cascade of zymogen activations. Active digestive and clotting enzymes are switched off by the irreversible binding of specific inhibitory proteins that are irresistible lures to their molecular prey.

To begin, we will consider the principles of allostery by examining two proteins: the enzyme aspartate transcarbamoylase and the oxygen-transporting protein hemoglobin.

10.1 ASPARTATE TRANSCARBAMOYLASE IS ALLOSTERICALLY INHIBITED BY THE END PRODUCT OF ITS PATHWAY

Aspartate transcarbamoylase catalyzes the first step in the biosynthesis of pyrimidines, bases that are components of nucleic acids. The reaction catalyzed by this enzyme is the condensation of aspartate and carbamoyl phosphate to form N-carbamoylaspartate and orthophosphate (Figure 10.1).

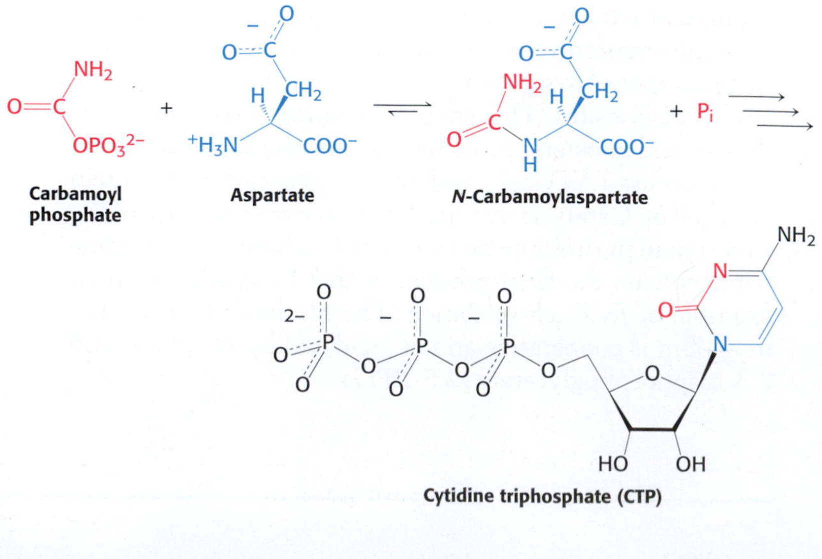

FIGURE 10.1 ATCase reaction.
Aspartate transcarbamoylase catalyzes the committed step, the condensation of aspartate and carbamoyl phosphate to form *N*-carbamoylaspartate, in pyrimidine synthesis.

Carbamoyl phosphate

Aspartate

N-Carbamoylaspartate

Cytidine triphosphate (CTP)

ATCase catalyzes the committed step in the pathway that will ultimately yield pyrimidine nucleotides such as cytidine triphosphate (CTP). How is this enzyme regulated to generate precisely the amount of CTP needed by the cell?

John Gerhart and Arthur Pardee found that ATCase is inhibited by CTP, the final product of the ATCase-controlled pathway. The rate of the reaction catalyzed by ATCase is fast in the absence of high concentrations of CTP but decreases as the CTP concentration increases (Figure 10.2). Thus, more molecules are sent along the pathway to make new pyrimidines until sufficient quantities of CTP have accumulated. The effect of CTP on the enzyme exemplifies the *feedback*, or *end-product*, *inhibition* mentioned earlier. Despite the fact that end-product regulation makes considerable physiological sense, the observation that ATCase is inhibited by CTP is remarkable because *CTP is structurally quite different from the substrates of the reaction* (see Figure 10.1). Owing to this structural dissimilarity, CTP must bind to a site distinct from the active site where substrate binds. Such sites are called *allosteric* (from the Greek *allos*, "other," and *stereos*, "structure") or *regulatory sites*. CTP is an example of an *allosteric inhibitor*. In ATCase (but not all allosterically regulated enzymes), the catalytic sites and the regulatory sites are on separate polypeptide chains.

10.1.1 ATCase Consists of Separable Catalytic and Regulatory Subunits

What is the evidence that ATCase has distinct regulatory and catalytic sites? ATCase can be literally separated into regulatory and catalytic subunits by treatment with a mercurial compound such as *p*-hydroxymercuribenzoate, which reacts with sulfhydryl groups (Figure 10.3). The results of ultracentrifugation studies carried out by Gerhart and Howard Schachman showed that *p*-hydroxymercuribenzoate dissociates ATCase into two kinds of subunits (Figure 10.4). The sedimentation coefficient of the native enzyme is 11.6S, whereas those of the dissociated subunits are 2.8S and 5.8S, indicating subunits of different size. The subunits can be readily separated by ion-exchange chromatography because they differ markedly in charge (Section 4.1.3) or by centrifugation in a sucrose density gradient because they differ in size (Section 4.1.6). Furthermore, the attached *p*-mercuribenzoate groups can be removed from the separated subunits by adding an excess of mercaptoethanol. The isolated subunits provide materials that can be used to investigate and characterize the individual subunits and their interactions with one another.

The larger subunit is called the *catalytic* (or *c*) *subunit*. This subunit displays catalytic activity, but it is not affected by CTP. The isolated smaller subunit can bind CTP, but has no catalytic activity. Hence, that subunit is called the *regulatory* (or *r*) *subunit*. The catalytic subunit, which consists of

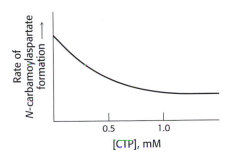

FIGURE 10.2 CTP inhibits ATCase. Cytidine triphosphate, an end product of the pyrimidine synthesis pathway, inhibits aspartate transcarbamoylase despite having little structural similarity to reactants or products.

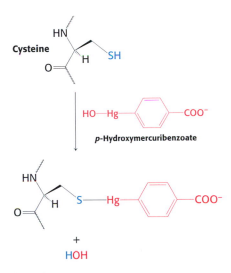

FIGURE 10.3 Modification of cysteine residues. *p*-Hydroxymercuribenzoate reacts with crucial cysteine residues in aspartate transcarbamoylase.

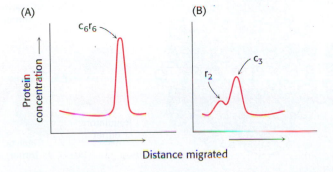

(A)

(B)

FIGURE 10.4 Ultracentrifugation studies of ATCase. Sedimentation velocity patterns of (A) native ATCase and (B) the enzyme after treatment with *p*-hydroxymercuribenzoate show that the enzyme can be dissociated into regulatory and catalytic subunits. [After J. C. Gerhart and H. K. Schachman. *Biochemistry* 4(1965):1054.]

three chains (34 kd each), is referred to as c_3. The regulatory subunit, which consists of two chains (17 kd each), is referred to as r_2. The catalytic and regulatory subunits combine rapidly when they are mixed. The resulting complex has the same structure, c_6r_6, as the native enzyme: two catalytic trimers and three regulatory dimers.

$$2\,c_3 + 3\,r_2 \longrightarrow c_6r_6$$

Furthermore, the reconstituted enzyme has the same allosteric properties as the native enzyme. Thus, ATCase is composed of discrete catalytic and regulatory subunits, which interact in the native enzyme to produce its allosteric behavior.

10.1.2 Allosteric Interactions in ATCase Are Mediated by Large Changes in Quaternary Structure

How can the binding of CTP to a regulatory subunit influence reactions at the active site of a catalytic subunit? Significant clues have been provided by the determination of the three-dimensional structure of ATCase in various forms by x-ray crystallography in the laboratory of William Lipscomb. The structure of the enzyme without any ligands bound to it confirms the overall structure of the enzyme. Two catalytic trimers are stacked one on top of the other, linked by three dimers of the regulatory chains (Figure 10.5). There are significant contacts between the two catalytic trimers: each r chain within a regulatory dimer interacts with a c chain within a catalytic trimer through a structural domain stabilized by a zinc ion bound to four cysteine residues.

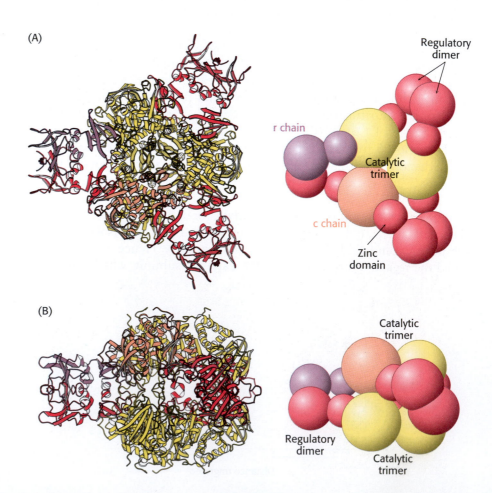

FIGURE 10.5 Structure of ATCase.
(A) The quaternary structure of aspartate transcarbamoylase as viewed from the top. The schematic drawing at the right is a simplified representation of the relationships between subunits. A single trimer [catalytic (c) chains, shown in orange and yellow] is visible; in this view, the second trimer is hidden behind the one visible. (B) A side view of the complex.

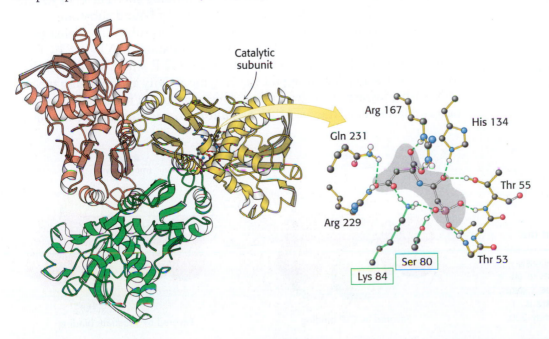

Bound substrates　　　　　**Reaction intermediate**

N-(Phosphonacetyl)-ʟ-aspartate
(PALA)

FIGURE 10.6 PALA, a bisubstrate analog. (Top) Nucleophilic attack by the amino group of aspartate on the carbonyl carbon atom of carbamoyl phosphate generates an intermediate on the pathway to the formation of *N*-carbamoylaspartate. (Bottom) *N*-(Phosphonacetyl)-ʟ-aspartate (PALA) is an analog of the reaction intermediate and a potent competitive inhibitor of aspartate transcarbamoylase.

The ability of *p*-hydroxymercuribenzoate to dissociate the catalytic and regulatory subunits is related to the ability of mercury to bind strongly to the cysteine residues, displacing the zinc and destabilizing this domain.

To understand the mechanism of allosteric regulation, it is crucial to locate each active site and each regulatory site in the three-dimensional structure. To locate the active sites, the enzyme was crystallized in the presence of *N*-(phosphonacetyl)-ʟ-aspartate (PALA), a bisubstrate analog (an analog of the two substrates) that resembles an intermediate along the pathway of catalysis (Figure 10.6). PALA is a potent competitive inhibitor of ATCase; it binds to and blocks the active sites. The structure of the ATCase–PALA complex reveals that PALA binds at sites lying at the boundaries between pairs of c chains within a catalytic trimer (Figure 10.7). Note that, though most of the residues belong to one subunit, several key residues belong to a neighboring subunit. Thus, because the active sites are at the subunit interface, each catalytic trimer contributes three active sites to the complete enzyme. Suitable amino acid residues are available in the active sites for recognizing all features of the bisubstrate analog, including the phosphate and both carboxylate groups.

FIGURE 10.7 The active site of ATCase. Some of the crucial active-site residues are shown binding to the inhibitor PALA. The active site is composed mainly of residues from one subunit, but an adjacent subunit also contributes important residues (boxed in green).

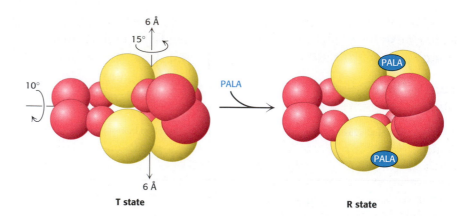

FIGURE 10.8 The T-to-R state transition in ATCase. Aspartate transcarbamoylase exists in two conformations: a compact, relatively inactive form called the tense (T) state and an expanded form called the relaxed (R) state. PALA binding stabilizes the R state.

Further examination of the ATCase–PALA complex reveals a remarkable change in quaternary structure on binding of PALA. The two catalytic trimers move 12 Å farther apart and rotate approximately 10 degrees about their common threefold axis of symmetry. Moreover, the regulatory dimers rotate approximately 15 degrees to accommodate this motion (Figure 10.8). The enzyme literally expands on PALA binding. In essence, ATCase has two distinct quaternary forms: one that predominates in the absence of substrate or substrate analogs and another that predominates when substrates or analogs are bound. These forms will be referred to as the T (for tense) state and the R (for relaxed) state, respectively. The T state has lower affinity for substrates and, hence, lower catalytic activity than does the R state. In the presence of any fixed concentration of aspartate and carbamoyl phosphate, the enzyme exists in equilibrium between the T and the R forms. *The position of the equilibrium depends on the number of active sites that are occupied by substrate.*

Having located the active sites and seen that PALA binding results in substantial structural changes in the entire ATCase molecule, we now turn our attention to the effects of CTP. Where on the regulatory subunit does CTP bind? Determination of the structure of ATCase in the presence of CTP reveals a binding site for this nucleotide in each regulatory chain in a domain that does not interact with the catalytic subunit (Figure 10.9). The question naturally arises as to how CTP can inhibit the catalytic activity of the enzyme when it does not interact with the catalytic chain. Each active site is more than 50 Å from the nearest CTP binding site. The CTP-bound form is in the T quaternary state in the absence of bound substrate.

The quaternary structural changes observed on substrate-analog binding suggest a mechanism for the allosteric regulation of ATCase by CTP (Figure 10.10). The binding of the inhibitor CTP shifts the equilibrium toward the T state, decreasing the net enzyme activity and reducing the rate of *N*-carbamoylaspartate generation. This mechanism for allosteric regulation is referred to as the *concerted mechanism* because the change in the en-

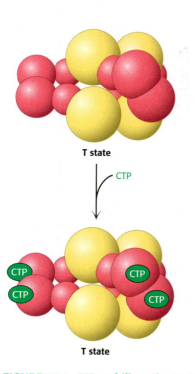

FIGURE 10.9 CTP stabilizes the T state. The binding of CTP to the regulatory subunit of aspartate transcarbamoylase stabilizes the T state.

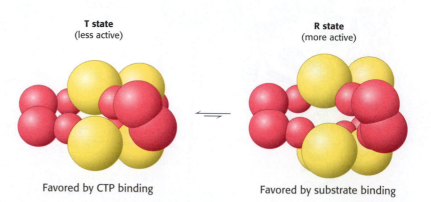

FIGURE 10.10 The R state and the T state are in equilibrium. Even in the absence of any substrate or regulators, aspartate transcarbamoylase exists in an equilibrium between the R and the T states. Under these conditions, the T state is favored by a factor of approximately 200.

zyme is "all or none"; the entire enzyme is converted from T into R, affecting all of the catalytic sites equally. The concerted mechanism stands in contrast with the sequential mechanism, which will be discussed shortly.

10.1.3 Allosterically Regulated Enzymes Do Not Follow Michaelis-Menten Kinetics

Allosteric enzymes are distinguished by their response to substrate concentration in addition to their susceptibility to regulation by other molecules. Examining the rate of product formation as a function of substrate concentration can be a source of further insights into the mechanism of regulation of ATCase (Figure 10.11). The curve differs from that expected for an enzyme that follows Michaelis-Menten kinetics. The observed curve is referred to as sigmoid because it resembles an "S." How can we explain this kinetic behavior in light of the structural observations? In the absence of substrate, the enzyme exists almost entirely in the T state. However, the binding of substrate molecules to the enzyme shifts the enzyme toward the R state. A transition from T to R favored by substrate binding to one site will increase the enzymatic activity of the remaining five sites, leading to an overall increase in enzyme activity. This important property is called *cooperativity* because the subunits cooperate with one another. If one subunit switches conformation, they all do. The sigmoid curve can be pictured as a composite of two Michaelis-Menten curves, one corresponding to the T state and the other to the R state. An increase in substrate concentration favors a transition from the T-state curve to the R-state curve (Figure 10.12).

The importance of the changes in quaternary structure in determining the sigmoidal curve is illustrated nicely by studies of the isolated catalytic trimer, freed by *p*-hydroxymercuribenzoate treatment. The catalytic subunit shows Michaelis-Menton kinetics with kinetic parameters that are indistinguishable from those deduced for the R state. Thus, the term *tense* is apt: in the T state, the regulatory dimers hold the two catalytic trimers sufficiently close to one other that key loops on their surfaces collide and interfere with conformational adjustments necessary for high-affinity substrate binding and catalysis.

10.1.4 Allosteric Regulators Modulate the T-to-R Equilibrium

What is the effect of CTP on the kinetic profile of ATCase? CTP increases the initial phase of the sigmoidal curve (Figure 10.13). As noted earlier, CTP inhibits the activity of ATCase. In the presence of CTP, the enzyme becomes less responsive to the cooperative effects facilitated by substrate binding; more substrate is required to attain a given reaction rate. Interestingly, ATP, too, is an allosteric effector of ATCase. However, the effect of ATP is to *increase* the reaction rate at a given aspartate concentration (Figure 10.14). At high concentrations of ATP, the kinetic profile shows a less-pronounced sigmoidal behavior. Note that such sigmoidal behavior has an

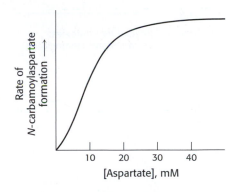

FIGURE 10.11 ATCase displays sigmoidal kinetics. A plot of product formation as a function of substrate concentration produces a sigmoidal curve because the binding of substrate to one active site favors the conversion of the entire enzyme into the R state, increasing the activity at the other active sites. Thus, the active sites show cooperativity.

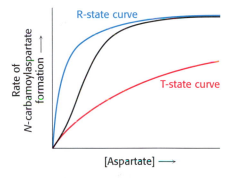

FIGURE 10.12 Basis for the sigmoidal curve. The generation of the sigmoidal curve by the property of cooperativity can be understood by imagining an allosteric enzyme as a mixture of two Michaelis-Menten enzymes, one with a high value of K_m that corresponds to the T state and another with a low value of K_m that corresponds to the R state. As the concentration of substrate is increased, the equilibrium shifts from the T state to the R state, which results in a steep rise in activity with respect to substrate concentration.

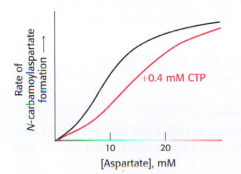

FIGURE 10.13 Effect of CTP on ATCase kinetics. Cytidine triphosphate (CTP) stabilizes the T state of aspartate transcarbamoylase, making it more difficult for substrate binding to convert the enzyme into the R state. As a result, the curve is shifted to the right, as shown in red.

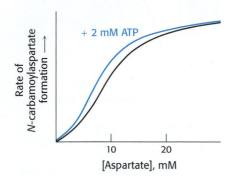

FIGURE 10.14 Effect of ATP on ATCase kinetics. ATP is an allosteric activator of aspartate transcarbamoylase because it stabilizes the R state, making it easier for substrate to bind. As a result, the curve is shifted to the left, as shown in blue.

additional consequence: in the concentration range where the T-to-R transition is taking place, the curve depends quite steeply on the substrate concentration. The effects of substrates on allosteric enzymes are referred to as *homotropic effects* (from the Greek *homós*, "same"). In contrast, the effects of nonsubstrate molecules on allosteric enzymes (such as those of CTP and ATP on ATCase) are referred to as *heterotropic effects* (from the Greek *héteros*, "different").

The increase in ATCase activity in response to increased ATP concentration has two potential physiological explanations. First, high ATP concentration signals a high concentration of purine nucleotides in the cell; the increase in ATCase activity will tend to balance the purine and pyrimidine pools. Second, a high concentration of ATP indicates that there is significant energy stored in the cell to promote mRNA synthesis and DNA replication.

10.1.5 The Concerted Model Can Be Formulated in Quantitative Terms

CONCEPTUAL INSIGHTS, Cooperative Binding and Kinetics. *Interactive graphing activities allow you to experiment with changes in the parameters and conditions of the MWC model in order to increase your understanding of the model and its implications for cooperative binding and kinetics.*

The concerted model was first proposed by Jacques Monod, Jeffries Wyman, and Jean-Pierre Changeux; hence, it is often referred to as the MWC model. This model can be formulated in quantitative terms. Consider an enzyme with n identical active sites. Suppose that the enzyme exists in equilibrium between a T form with a low affinity for its substrate and an R form with a high affinity for the substrate. We can define L as the equilibrium constant between the R and the T forms; c as the ratio of the affinities of the two forms for the substrate, S, measured as dissociation constants; and α as the ratio of substrate concentration to the dissociation constant K_R.

$$R \rightleftharpoons T \qquad\qquad L = \frac{[T]}{[R]}$$

$$Site_R + S \rightleftharpoons Site_R - S \qquad K_R = \frac{[Site_R][S]}{[Site_R - S]}$$

$$Site_T + S \rightleftharpoons Site_T - S \qquad K_T = \frac{[Site_T][S]}{[Site_T - S]}$$

Define

$$c = \frac{K_R}{K_T} \quad \text{and} \quad \alpha = \frac{[S]}{K_R}$$

The fraction of active sites bound to substrate (fractional saturation, Y_S) is given by

$$Y_S = \frac{\alpha(1 - \alpha)^{n-1} + Lc\alpha(1 + c\alpha)^{n-1}}{(1 + \alpha)^n + L(1 + c\alpha)^n}$$

where n is the number of sites in the enzyme.

This quantitative model can be used to examine the data from ATCase, for which $n = 6$. Excellent agreement with experimental data is obtained with $L \approx 200$ and $c \approx 0.1$. Thus, in the absence of bound substrate, the equilibrium favors the T form by a factor of 200 (i.e., only 1 in 200 molecules is in the R form), and the affinity of the R form for substrate is approximately 10 times as high as that of the T form. As substrate binds to each active site, the equilibrium shifts toward the R form. For example, with these parameters, when half the active sites (three of six) are occupied by substrate, the equilibrium has shifted so that the ratio of T to R is now 1 to 5; that is, nearly all the molecules are in the R form.

The effects of CTP and ATP can be modeled simply by changing the value of L. For the CTP-saturated form, the value of L increases to 1250. Thus, it takes more substrate to shift the equilibrium appreciably to the R form. For the ATP saturated form, the value of L decreases to 70 (Figure 10.15).

10.1.6 Sequential Models Also Can Account for Allosteric Effects

In the concerted model, an allosteric enzyme can exist in one of only two states, T and R; no intermediate states are allowed. An alternative, first proposed by Daniel Koshland, posits that *sequential* changes in structure take place within an oligomeric enzyme as active sites are occupied. The binding of substrate to one site influences the substrate affinity of neighboring active sites *without necessarily inducing a transition encompassing the entire enzyme* (Figure 10.16). An important feature of sequential in contrast with concerted models is that the former can account for *negative cooperativity*, in which the binding of substrate to one active site *decreases* the affinity of other sites for substrate. The results of studies of a number of allosteric proteins suggest that most behave according to some combination of the sequential and cooperative models.

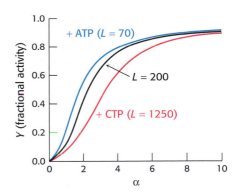

FIGURE 10.15 Quantitative description of the MWC model. Fractional activity, *Y*, is the fraction of active sites bound to substrate and is directly proportional to reaction velocity; α is the ratio of [S] to the dissociation constant of S with the enzyme in the R state; *L* is the ratio of the concentration of enzyme in the T state to that in the R state. The binding of the regulators ATP and CTP to ATCase changes the value of *L* and thus the response to substrate concentration.

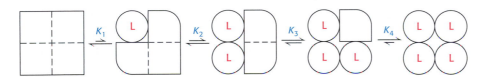

FIGURE 10.16 Simple sequential model for a tetrameric allosteric enzyme. The binding of a ligand (L) to a subunit changes the conformation of that particular subunit from the T (square) to the R (circle) form. This transition affects the affinity of the other subunits for the ligand.

10.2 HEMOGLOBIN TRANSPORTS OXYGEN EFFICIENTLY BY BINDING OXYGEN COOPERATIVELY

Allostery is a property not limited to enzymes. The basic principles of allostery are also well illustrated by the oxygen-transport protein hemoglobin (Section 7.2). The binding of oxygen to hemoglobin isolated from red blood cells displays marked sigmoidal behavior (similar to that observed for the activity of ATCase, as a function of substrate concentration), which is indicative of cooperation between subunits (Figure 10.17). What is the physiological significance of the cooperative binding of oxygen by hemoglobin? Oxygen must be transported in the blood from the lungs, where the partial pressure of oxygen (pO_2) is relatively high (approximately 100 torr), to the tissues, where the partial pressure of oxygen is much lower (typically 20 torr). Let us consider how the cooperative behavior represented by the sigmoidal

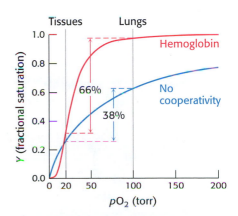

FIGURE 10.17 Cooperativity enhances oxygen delivery by hemoglobin. Because of cooperativity between O_2-binding sites, hemoglobin delivers more O_2 to tissues than would a noncooperative protein (pO_2, partial pressure of oxygen.)

Torr—

A unit of pressure equal to that exerted by a column of mercury 1 mm high at 0°C and standard gravity (1 mm Hg). Named after Evangelista Torricelli (1608–1647), inventor of the mercury barometer.

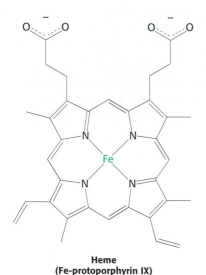

**Heme
(Fe-protoporphyrin IX)**

curve leads to efficient oxygen transport. In the lungs, hemoglobin becomes nearly saturated with oxygen such that 98% of the oxygen-binding sites are occupied. When hemoglobin moves to the tissues, the saturation level drops to 32%. Thus, a total of $98 - 32 = 66\%$ of the potential oxygen-binding sites contribute to oxygen transport. In comparison, for a hypothetical noncooperative transport protein, the most oxygen that can be transported from a region in which pO_2 is 100 torr to one in which it is 20 torr is $63 - 25 = 38\%$ (see Figure 10.17). Thus, *the cooperative binding of oxygen by hemoglobin enables it to deliver 1.7 times as much oxygen as it would if the sites were independent.* The homotropic regulation of hemoglobin by its ligand oxygen dramatically increases its physiological oxygen-carrying capacity.

10.2.1 Oxygen Binding Induces Substantial Structural Changes at the Iron Sites in Hemoglobin

The results of structural studies pioneered by Max Perutz revealed the structure of hemoglobin in various forms. Human hemoglobin A, present in adults, consists of four subunits: two α subunits and two β subunits. The α and β subunits are homologous and have similar three-dimensional structures (Section 7.4). The capacity of hemoglobin to bind oxygen depends on the presence of a bound prosthetic group called *heme*. The heme group is responsible for the distinctive red color of blood. The heme group consists of an organic component and a central iron atom. The organic component, called *protoporphyrin*, is made up of four pyrrole rings linked by methene bridges to form a tetrapyrrole ring. Four methyl groups, two vinyl groups, and two propionate side chains are attached.

The iron atom lies in the center of the protoporphyrin, bonded to the four pyrrole nitrogen atoms. Under normal conditions, the iron is in the ferrous (Fe^{2+}) oxidation state. The iron ion can form two additional bonds, one on each side of the heme plane. These binding sites are called the fifth and sixth coordination sites. In hemoglobin, the fifth coordination site is occupied by the imidazole ring of a histidine residue from the protein. In deoxyhemoglobin, the sixth coordination site remains unoccupied. The iron ion lies approximately 0.4 Å outside the porphyrin plane because iron, in this form, is slightly too large to fit into the well-defined hole within the porphyrin ring (Figure 10.18).

The binding of the oxygen molecule at the sixth coordination site of the iron ion substantially rearranges the electrons within the iron so that the ion becomes effectively smaller, allowing it to move into the plane of the porphyrin (Figure 10.19). This change in electronic structure is paralleled by

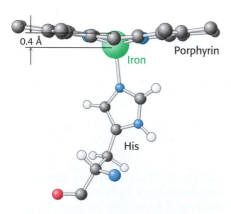

FIGURE 10.18 Position of iron in deoxyhemoglobin. The iron ion lies slightly outside the plane of the porphyrin in heme.

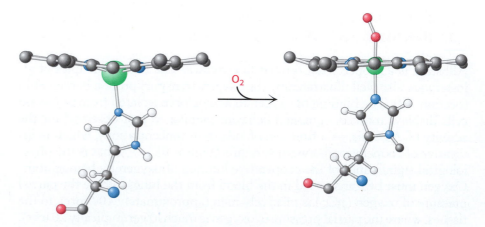

FIGURE 10.19 Oxygen binding initiates structural changes. The iron ion moves into the plane of the heme on oxygenation. The proximal histidine is pulled along with the iron ion.

changes in the magnetic properties of hemoglobin, which are the basis for functional magnetic resonance imaging (fMRI; Section 32.1.3). Indeed, the structural changes that take place on oxygen binding were anticipated by Linus Pauling, based on magnetic measurements in 1936, nearly 25 years before the three-dimensional structure of hemoglobin was elucidated.

10.2.2 Oxygen Binding Markedly Changes the Quaternary Structure of Hemoglobin

The three-dimensional structure of hemoglobin is best described as a pair of identical $\alpha\beta$ dimers ($\alpha_1\beta_1$ and $\alpha_2\beta_2$) that associate to form the hemoglobin tetramer (Figure 10.20). In deoxyhemoglobin, these $\alpha\beta$ dimers are linked by an extensive interface, which includes, among other regions, the carboxyl terminus of each chain. The heme groups are well separated in the tetramer with iron–iron distances ranging from 24 to 40 Å. The deoxy form corresponds to the T state in the context of either the concerted or the sequential model for hemoglobin cooperativity. On oxygen binding, there are substantial changes in quaternary structure that correspond to the T-to-R state transition (Figure 10.21).

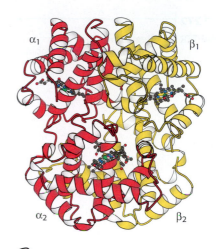

FIGURE 10.20 **Quaternary structure of hemoglobin.** Hemoglobin, which is composed of two α chains and two β chains, functions as a pair of $\alpha\beta$ dimers.

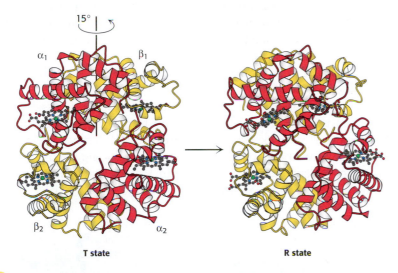

T state   R state

FIGURE 10.21 **Transition from T to R state in hemoglobin.** On oxygenation, one pair of $\alpha\beta$ subunits shifts with respect to the other by a rotation of 15 degrees.

The $\alpha_1\beta_1$ and $\alpha_2\beta_2$ dimers rotate approximately 15 degrees with respect to one another. The dimers themselves are relatively unchanged, although localized conformational shifts do occur. Thus, the interface between the $\alpha_1\beta_1$ and $\alpha_2\beta_2$ dimers is most effected by this structural transition.

How does oxygen binding lead to the structural transition from the T state to the R state? When the iron ion moves into the plane of the porphyrin, the histidine residue bound in the fifth coordination site moves with it. This histidine residue is part of an α helix, which also moves (Figure 10.22). The carboxyl terminal end of this α helix lies in the interface between the two $\alpha\beta$ dimers. Consequently, the structural transition at the iron ion is directly transmitted to the other subunits. The rearrangement of the dimer interface provides a pathway for communication between subunits, enabling the cooperative binding of oxygen.

Is the cooperative binding of oxygen by hemoglobin best described by the concerted or the sequential model? Neither model in its pure form fully accounts for the behavior of hemoglobin. Instead, a combined model

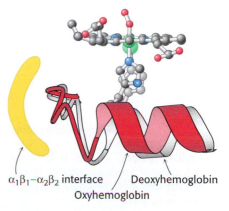

$\alpha_1\beta_1$–$\alpha_2\beta_2$ interface   Deoxyhemoglobin
Oxyhemoglobin

FIGURE 10.22 **Conformational changes in hemoglobin.** The movement of the iron ion on oxygenation brings the iron-associated histidine residue toward the porphyrin ring. The corresponding movement of the histidine-containing α helix alters the interface between the $\alpha\beta$ pairs, instigating other structural changes. For comparison, the deoxyhemoglobin structure is shown in gray behind the oxyhemoglobin structure in color.

is required. Hemoglobin behavior is concerted in that hemoglobin with three sites occupied by oxygen is in the quaternary structure associated with the R state. The remaining open binding site has an affinity for oxygen more than 20-fold as great as that of fully deoxygenated hemoglobin binding its first oxygen. However, the behavior is not fully concerted, because hemoglobin with oxygen bound to only one of four sites remains in the T-state quaternary structure. Yet, this molecule binds oxygen 3 times as strongly as does fully deoxygenated hemoglobin, an observation consistent only with a sequential model. These results highlight the fact that the concerted and sequential models represent idealized limiting cases, which real systems may approach but rarely attain.

10.2.3 Tuning the Oxygen Affinity of Hemoglobin: The Effect of 2,3-Bisphosphoglycerate

Examination of the oxygen binding of hemoglobin fully purified from red blood cells revealed that the oxygen affinity of purified hemoglobin is much greater than that for hemoglobin within red blood cells. This dramatic difference is due to the presence within these cells of 2,3-bisphosphoglycerate (2,3-BPG) (also known as 2,3-diphosphoglycerate or 2,3-DPG). This highly anionic compound is present in red blood cells at approximately the same concentration as that of hemoglobin (~ 2 mM). Without 2,3-BPG, hemoglobin would be an extremely inefficient oxygen transporter, releasing only 8% of its cargo in the tissues.

How does 2,3-BPG affect oxygen affinity so significantly? Examination of the crystal structure of deoxyhemoglobin in the presence of 2,3-BPG reveals that a single molecule of 2,3-BPG binds in a pocket, present only in the T form, in the center of the hemoglobin tetramer (Figure 10.23). On T-to-R transition, this pocket collapses. Thus, 2,3-BPG binds preferentially to deoxyhemoglobin and stabilizes it, effectively reducing the oxygen affinity. In order for the structural transition from T to R to take place, the bonds between hemoglobin and 2,3-BPG must be broken and 2,3-BPG must be expelled.

**2,3-Bisphosphoglycerate
(2,3-BPG)**

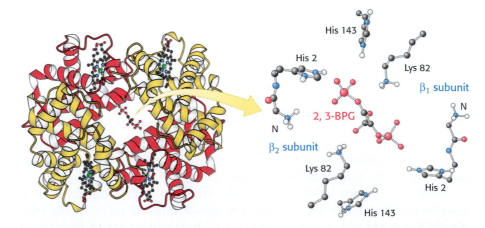

FIGURE 10.23 Mode of binding of 2,3-BPG to human deoxyhemoglobin. 2,3-BPG binds in the central cavity of deoxyhemoglobin. There, it interacts with three positively charged groups on each β chain.

2,3-BPG binding to hemoglobin has other crucial physiological consequences. The globin gene expressed by fetuses differs from that expressed by human adults; fetal hemoglobin tetramers include two α chains and two γ chains. The γ chain, a result of another gene duplication, is 72% identical in amino acid sequence with the β chain. One noteworthy change is the substitution of a serine residue for His 143 in the β chain of the 2,3-BPG-binding site. This change removes two positive charges from the 2,3-BPG-binding site (one from each chain) and reduces the affinity of 2,3-BPG for fetal hemoglobin, thereby increasing the oxygen-binding affinity of fetal hemoglobin relative to that of maternal (adult) hemoglobin (Figure 10.24). This difference in oxygen affinity allows oxygen to be effectively transferred from maternal to fetal red cells. We see again an example of where gene duplication and specialization produced a ready solution to a biological challenge—in this case, the transport of oxygen from mother to fetus.

10.2.4 The Bohr Effect: Hydrogen Ions and Carbon Dioxide Promote the Release of Oxygen

Rapidly metabolizing tissues, such as contracting muscle, have a high need for oxygen and generate large amounts of hydrogen ions and carbon dioxide as well (Sections 16.1.9 and 17.1). Both of these species are heterotropic effectors of hemoglobin that enhance oxygen release. The oxygen affinity of hemoglobin decreases as pH decreases from the value of 7.4 found in the lungs (Figure 10.25). Thus, as hemoglobin moves into a region of low pH, its tendency to release oxygen increases. For example, transport from the lungs, with pH 7.4 and an oxygen partial pressure of 100 torr, to active muscle, with a pH of 7.2 and an oxygen partial pressure of 20 torr, results in a release of oxygen amounting to 77% of total carrying capacity. Recall that only 66% of the oxygen would be released in the absence of any change in pH. In addition, hemoglobin responds to carbon dioxide with a decrease in oxygen affinity, thus facilitating the release of oxygen in tissues with a high carbon dioxide concentration. In the presence of carbon dioxide at a partial pressure of 40 torr, the amount of oxygen released approaches 90% of the maximum carrying capacity. Thus, *the heterotropic regulation of hemoglobin by hydrogen ions and carbon dioxide further increases the oxygen-transporting efficiency of this magnificent allosteric protein.*

The regulation of oxygen binding by hydrogen ions and carbon dioxide is called the *Bohr effect* after Christian Bohr, who described this phenomenon in 1904. The results of structural and chemical studies have revealed much about the chemical basis of the Bohr effect. At least two sets of chemical groups are responsible for the effect of protons: the amino termini and the side chains of histidines β146 and α122, which have pK_a values near pH 7. Consider histidine β146. In deoxyhemoglobin, the terminal carboxylate group of β146 forms a salt bridge with a lysine residue in the α subunit of the other αβ dimer. This interaction locks the side chain of histidine β146 in a position where it can participate in a salt bridge with negatively charged aspartate 94 in the same chain, provided that the imidazole group of the histidine residue is protonated (Figure 10.26). At high pH, the side chain of histidine β146 is not protonated and the salt bridge does not form. As the pH drops, however, the side chain of histidine β146 becomes protonated, the

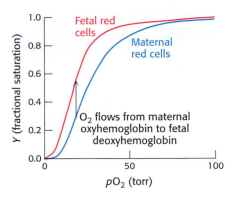

FIGURE 10.24 Oxygen affinity of fetal red blood cells. Fetal red blood cells have a higher oxygen affinity than that of maternal red blood cells because fetal hemoglobin does not bind 2,3-BPG as well as maternal hemoglobin does.

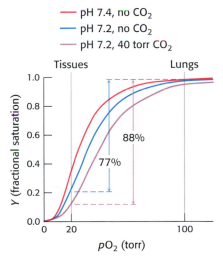

FIGURE 10.25 Effect of pH and CO_2 concentration on the oxygen affinity of hemoglobin. Lowering the pH from 7.4 (red curve) to 7.2 (blue curve) results in the release of O_2 from oxyhemoglobin. Raising the CO_2 partial pressure from 0 to 40 torr (purple curve) also favors the release of O_2 from oxyhemoglobin.

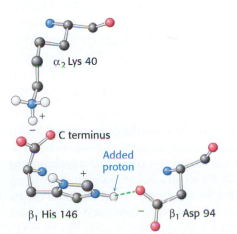

FIGURE 10.26 Chemical basis of the Bohr effect. In deoxyhemoglobin, shown here, three amino acid residues form two salt bridges that stabilize the T quaternary structure. The formation of one of the salt bridges depends on the presence of an added proton on histidine β146. The proximity of the negative charge on aspartate 94 favors protonation of histidine β146 in deoxyhemoglobin.

$$R-N(H)(H) + C(=O)(=O) \rightleftharpoons R-N(H)-C(=O)(O^-) + H^+$$

Carbamate

salt bridge with aspartate β94 forms, and the quaternary structure characteristic of deoxyhemoglobin is stabilized, leading to a greater tendency for oxygen to be released at actively metabolizing tissues. No significant change takes place in oxyhemoglobin over the same pH range.

Carbon dioxide also stabilizes deoxyhemoglobin by reacting with the terminal amino groups to form carbamate groups, which are negatively charged, in contrast with the neutral or positive charges on the free amino groups. The amino termini lie at the interface between the αβ dimers, and these negatively charged carbamate groups participate in salt-bridge interactions, characteristic of the T-state structure, which stabilize deoxyhemoglobin's structure and favor the release of oxygen.

Hemoglobin with bound carbon dioxide and hydrogen ions is carried in the blood back to the lungs, where it releases the hydrogen ions and carbon dioxide and rebinds oxygen. Thus, hemoglobin helps to transport hydrogen ions and carbon dioxide in addition to transporting oxygen. However, transport by hemoglobin accounts for only about 14% of the total transport of these species; both hydrogen ions and carbon dioxide are also transported in the blood as bicarbonate (HCO_3^-) formed spontaneously or through the action of carbonic anhydrase (Section 9.2), an abundant enzyme in red blood cells.

10.3 ISOZYMES PROVIDE A MEANS OF REGULATION SPECIFIC TO DISTINCT TISSUES AND DEVELOPMENTAL STAGES

Isozymes or *isoenzymes,* are enzymes that differ in amino acid sequence yet catalyze the same reaction. Usually, these enzymes display different kinetic parameters, such as K_M, or different regulatory properties. They are encoded by different genetic loci, which usually arise through gene duplication and divergence (Section 2.2.5). Isozymes differ from allozymes, which are enzymes that arise from allelic variation at one gene locus. Isozymes can often be distinguished from one another by biochemical properties such as electrophoretic mobility.

The existence of isozymes permits the fine-tuning of metabolism to meet the particular needs of a given tissue or developmental stage. Consider the example of lactate dehydrogenase (LDH), an enzyme that functions in anaerobic glucose metabolism and glucose synthesis. Human beings have two isozymic polypeptide chains for this enzyme: the H isozyme highly expressed in heart and the M isozyme found in skeletal muscle. The amino acid sequences are 75% identical. The functional enzyme is tetrameric, and many different combinations of the two subunits are possible. The H_4 isozyme, found in the heart, has a higher affinity for substrates than does the M_4 isozyme. The two isozymes also differ in that high levels of pyruvate allosterically inhibit the H_4 but not the M_4 isozyme. The other combinations, such as H_3M, have intermediate properties depending on the ratio of the two kinds of chains. We will consider these isozymes in their biological context in Chapter 16.

The M_4 isozyme functions optimally in an anaerobic environment, whereas the H_4 isozyme does so in an aerobic environment. Indeed, the proportions of these isozymes are altered in the course of development of the rat heart as the tissue switches from an anaerobic environment to an aerobic one (Figure 10.27A). Figure 10.27B shows the tissue-specific forms of lactate dehydrogenase in adult rat tissues. The appearance of some

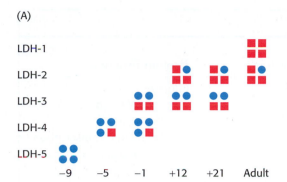

(A)

	−9	−5	−1	+12	+21	Adult
LDH-1						
LDH-2						
LDH-3						
LDH-4						
LDH-5						

(B)

	Heart	Kidney	Red blood cell	Brain	Leukocyte	Muscle	Liver
H_4	▬	▬	▬	▬	▬	—	—
H_3M	▬	▬	▬	▬	▬		▬
H_2M_2	—	▬	—	▬	▬	▬	
HM_3	—				▬		
M_4	—	—	—	—	—	▬	▬

FIGURE 10.27 Isozymes of lactate dehydrogenase. (A) The rat heart LDH isozyme profile changes in the course of development. The H isozyme is represented by squares and the M isozyme by circles. The negative and positive numbers denote the days before and after birth, respectively. (B) LDH isozyme content varies by tissue. [(A) After W.-H. Li, *Molecular Evolution* (Sinauer, 1997), p. 283; (B) After K. Urich, *Comparative Animal Biochemistry* (Springer Verlag, 1990), p. 542.]

isozymes in the blood is indicative of tissue damage and can be used for clinical diagnosis. For instance, an increase in serum levels of H_4 relative to H_3M is an indication that a myocardial infarction, or heart attack, has occurred.

10.4 COVALENT MODIFICATION IS A MEANS OF REGULATING ENZYME ACTIVITY

The covalent attachment of another molecule can modify the activity of enzymes and many other proteins. In these instances, a donor molecule provides a functional moiety that modifies the properties of the enzyme. Most modifications are reversible. Phosphorylation and dephosphorylation are the most common but not the only means of covalent modification. Histones—proteins that assist in the packaging of DNA into chromosomes as well as in gene regulation—are rapidly acetylated and deacetylated in vivo (Section 31.3.4). More heavily acetylated histones are associated with genes that are being actively transcribed. The acetyltransferase and deacetylase enzymes are themselves regulated by phosphorylation, showing that the covalent modification of histones may be controlled by the covalent modification of the modifying enzymes.

Modification is not readily reversible in some cases. Some proteins in signal-transduction pathways, such as Ras and Src (a protein tyrosine kinase), are localized to the cytoplasmic face of the plasma membrane by the irreversible attachment of a lipid group (Section 12.5.3). Fixed in this location, the proteins are better able to receive and transmit information that is being passed along their signaling pathways (Chapter 15). The attachment of ubiquitin, a protein comprising 72 amino acids, is a signal that a protein is to be destroyed, the ultimate means of regulation (Chapter 23). Cyclin, an important protein in cell-cycle regulation, must be ubiquitinated and destroyed before a cell can enter anaphase and proceed through the cell cycle (Table 10.1).

Virtually all the metabolic processes that we will examine are regulated in part by covalent modification. Indeed, the allosteric properties of many enzymes are modified by covalent modification. Table 10.1 lists some of the common covalent modifications.

Acetylated lysine

TABLE 10.1   Common covalent modifications of protein activity

Modification	Donor molecule	Example of modified protein	Protein function
Phosphorylation	ATP	Glycogen phosphorylase	Glucose homeostasis; energy transduction
Acetylation	Acetyl CoA	Histones	DNA packing; transcription
Myristoylation	Myristoyl CoA	Src	Signal transduction
ADP-ribosylation	NAD	RNA polymerase	Transcription
Farnesylation	Farnesyl pyrophosphate	Ras	Signal transduction
γ-Carboxylation	HCO_3^-	Thrombin	Blood clotting
Sulfation	3′-Phosphoadenosine-5′-phosphosulfate	Fibrinogen	Blood-clot formation
Ubiquitination	Ubiquitin	Cyclin	Control of cell cycle

10.4.1 Phosphorylation Is a Highly Effective Means of Regulating the Activities of Target Proteins

The activities of many enzymes, membrane channels, and other target proteins are regulated by phosphorylation, the most prevalent reversible covalent modification. Indeed, we will see this regulatory mechanism in virtually every metabolic process in eukaryotic cells. The enzymes catalyzing phosphorylation reactions are called *protein kinases*, which constitute one of the largest protein families known, with more than 100 homologous enzymes in yeast and more than 550 in human beings. This multiplicity of enzymes allows regulation to be fine-tuned according to a specific tissue, time, or substrate.

The terminal (γ) phosphoryl group of ATP is transferred to specific *serine* and *threonine* residues by one class of protein kinases and to specific *tyrosine* residues by another.

TABLE 10.2 Examples of serine and threonine kinases and their activating signals

Signal	Enzyme
Cyclic nucleotides	Cyclic AMP-dependent protein kinase
	Cyclic GMP-dependent protein kinase
Ca^{2+} and calmodulin	Ca^{2+}-calmodulin protein kinase
	Phosphorylase kinase or glycogen synthase kinase 2
AMP	AMP-activated kinase
Diacylglycerol	Protein kinase C
Metabolic intermediates and other "local" effectors	Many target specific enzymes, such as pyruvate dehydrogenase kinase and branched-chain ketoacid dehydrogenase kinase

Source: After D. Fell, *Understanding the Control of Metabolism* (Portland Press, 1997), Table 7.2.

The acceptors in protein phosphorylation reactions are located inside cells, where the phosphoryl-group donor ATP is abundant. Proteins that are entirely extracellular are not regulated by reversible phosphorylation. Table 10.2 lists a few of the known protein kinases. *Protein phosphatases* reverse the effects of kinases by catalyzing the hydrolytic removal of phosphoryl groups attached to proteins.

Phosphorylated protein + H_2O →(Protein phosphatase) + **Orthophosphate (P_i)**

The unmodified hydroxyl-containing side chain is regenerated and orthophosphate (P_i) is produced.

It is important to note that phosphorylation and dephosphorylation are not the reverse of one another; each is essentially irreversible under physiological conditions. Furthermore, both reactions take place at negligible rates in the absence of enzymes. Thus, phosphorylation of a protein substrate will take place only through the action of a specific protein kinase and at the expense of ATP cleavage, and dephosphorylation will result only through the action of a phosphatase. The rate of cycling between the phosphorylated and the dephosphorylated states depends on the relative activities of kinases and phosphatases. Note that the net outcome of the two reactions is the hydrolysis of ATP to ADP and P_i, which has a ΔG of -12 kcal mol^{-1} (-50 kJ mol^{-1}) under cellular conditions (Section 14.1.2). This highly favorable free-energy change ensures that target proteins cycle unidirectionally between unphosphorylated and phosphorylated forms.

Phosphorylation is a highly effective means of controlling the activity of proteins for structural, thermodynamic, kinetic, and regulatory reasons:

1. A phosphoryl group adds two negative charges to a modified protein. Electrostatic interactions in the unmodified protein can be disrupted and new electrostatic interactions can be formed. Such structural changes can markedly alter substrate binding and catalytic activity.

Free energy

Protein–OH + ATP

Protein–OPO$_3^{2-}$ + ADP

–H_2O

Protein–OH + HOPO$_3^{2-}$

2. A phosphate group can form three or more hydrogen bonds. The tetrahedral geometry of the phosphoryl group makes these hydrogen bonds highly directional, allowing for specific interactions with hydrogen-bond donors.

3. The free energy of phosphorylation is large. Of the -12 kcal mol^{-1} (-50 kJ mol^{-1}) provided by ATP, about half is consumed in making phosphorylation irreversible; the other half is conserved in the phosphorylated protein. Recall that a free-energy change of 1.36 kcal mol^{-1} (5.69 kJ mol^{-1}) corresponds to a factor of 10 in an equilibrium constant (Section 14.1.3). Hence, phosphorylation can change the conformational equilibrium between different functional states by a large factor, of the order of 10^4.

4. Phosphorylation and dephosphorylation can take place in less than a second or over a span of hours. The kinetics can be adjusted to meet the timing needs of a physiological process.

5. Phosphorylation often evokes *highly amplified* effects. A single activated kinase can phosphorylate hundreds of target proteins in a short interval. Further amplification can take place because the target proteins may be enzymes, each of which can then transform a large number of substrate molecules.

6. ATP is the cellular energy currency (Chapter 14). The use of this compound as a phosphoryl group donor links the energy status of the cell to the regulation of metabolism.

Protein kinases vary in their degree of specificity. *Dedicated protein kinases* phosphorylate a single protein or several closely related ones. *Multifunctional protein kinases* modify many different targets; they have a wide reach and can coordinate diverse processes. Comparisons of amino acid sequences of many phosphorylation sites show that a multifunctional kinase recognizes related sequences. For example, the *consensus sequence* recognized by protein kinase A is Arg-Arg-X-*Ser*-Z or Arg-Arg-X-*Thr*-Z, in which X is a small residue, Z is a large hydrophobic one, and *Ser* or *Thr* is the site of phosphorylation. It should be noted that this sequence is not absolutely required. Lysine, for example, can substitute for one of the arginine residues but with some loss of affinity. Short synthetic peptides containing a consensus motif are nearly always phosphorylated by serine-threonine protein kinases. Thus, *the primary determinant of specificity is the amino acid sequence surrounding the serine or threonine phosphorylation site*. However, distant residues can contribute to specificity. For instance, changes in protein conformation may alter the accessibility of a possible phosphorylation site.

10.4.2 Cyclic AMP Activates Protein Kinase A by Altering the Quaternary Structure

Protein kinases modulate the activity of many proteins, but what leads to the activation of a kinase? Activation is often a multistep process initiated by hormones (Chapter 15). In some cases, the hormones trigger the formation of cyclic AMP, a molecule formed by cyclization of ATP. Cyclic AMP serves as an intracellular messenger in mediating the physiological actions of hormones, as will be discussed in Chapter 15. The striking finding is that *most effects of cAMP in eukaryotic cells are achieved through the activation by cAMP of a single protein kinase. This key enzyme is called protein kinase A or PKA.* The kinase alters the activities of target proteins by phosphorylating specific serine or threonine residues. As we shall see, PKA provides a clear example of the integration of allosteric regulation and phosphorylation.

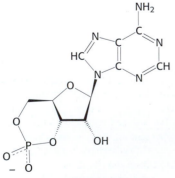

**Cyclic adenosine monophosphate
(cAMP)**

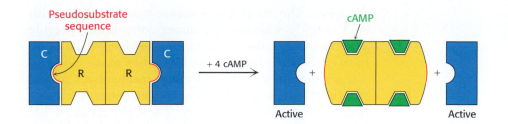

Pseudosubstrate sequence

cAMP

C R R C + 4 cAMP → C + R R + C

Active   Active

FIGURE 10.28 Regulation of protein kinase A. The binding of four molecules of cAMP activates protein kinase A by dissociating the inhibited holoenzyme (R_2C_2) into a regulatory subunit (R_2) and two catalytically active subunits (C).

PKA is activated by cAMP concentrations of the order of 10 nM. The activation mechanism is reminiscent of that of aspartate transcarbamoylase. Like that enzyme, PKA in muscle consists of two kinds of subunits: a 49-kd regulatory (R) subunit, which has high affinity for cAMP, and a 38-kd catalytic (C) subunit. In the absence of cAMP, the regulatory and catalytic subunits form an R_2C_2 complex that is enzymatically inactive (Figure 10.28). The binding of two molecules of cAMP to each of the regulatory subunits leads to the dissociation of R_2C_2 into an R_2 subunit and two C subunits. These free catalytic subunits are then enzymatically active. Thus, *the binding of cAMP to the regulatory subunit relieves its inhibition of the catalytic subunit.* PKA and most other kinases exist in isozymic forms for fine-tuning regulation to meet the needs of a specific cell or developmental stage.

How does the binding of cAMP activate the kinase? Each R chain contains the sequence Arg-Arg-Gly-*Ala*-Ile, which matches the consensus sequence for phosphorylation except for the presence of alanine in place of serine. In the R_2C_2 complex, this *pseudosubstrate sequence* of R occupies the catalytic site of C, thereby preventing the entry of protein substrates (see Figure 10.28). The binding of cAMP to the R chains allosterically moves the pseudosubstrate sequences out of the catalytic sites. The released C chains are then free to bind and phosphorylate substrate proteins.

10.4.3 ATP and the Target Protein Bind to a Deep Cleft in the Catalytic Subunit of Protein Kinase A

The three-dimensional structure of the catalytic subunit of PKA containing a bound 20-residue peptide inhibitor was determined by x-ray crystallography. The 350-residue catalytic subunit has two lobes (Figure 10.29). ATP and part of the inhibitor fill a deep cleft between

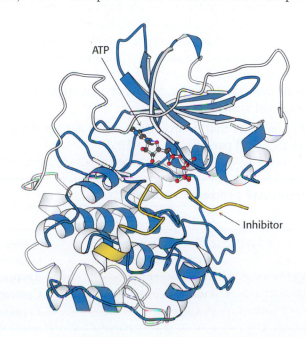

ATP

Inhibitor

FIGURE 10.29 Protein kinase A bound to an inhibitor. Three-dimensional structure of a complex of the catalytic subunit of protein kinase A and an inhibitor bearing a pseudosubstrate sequence. The inhibitor (yellow) binds in a cleft between the domains of the enzyme. The bound ATP is adjacent to the active site.

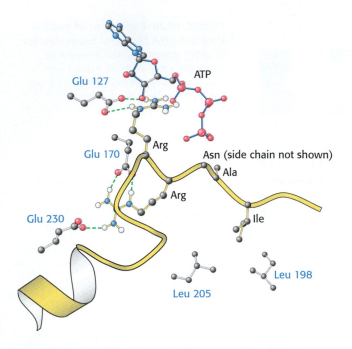

FIGURE 10.30 Binding of pseudosubstrate to protein kinase A. The two arginine side chains of the pseudosubstrate form salt bridges with three glutamate carboxylates. Hydrophobic interactions are also important in the recognition of substrate. The isoleucine residue of the pseudosubstrate is in contact with a pair of leucine residues of the enzyme.

the lobes. The smaller lobe makes many contacts with ATP–Mg^{2+}, whereas the larger lobe binds peptide and contributes the key catalytic residues. Like other kinases (Section 16.1.1), the two lobes move closer to one another on substrate binding; mechanisms for restricting this domain closure provides a means for regulating protein kinase activity. *The PKA structure has broad significance because residues 40 to 280 constitute a conserved catalytic core that is common to essentially all known protein kinases.* We see here an example of a successful biochemical solution to the problem of protein phosphorylation being employed many times in the course of evolution.

The bound peptide in this crystal occupies the active site because it contains the pseudosubstrate sequence Arg-Arg-Asn-*Ala*-Ile (Figure 10.30). The structure of the complex reveals the basis for the consensus sequence. The guanidinium group of the first arginine residue forms an ion pair with the carboxylate side chain of a glutamate residue (Glu 127) of the enzyme. The second arginine likewise interacts with two other carboxylates. The nonpolar side chain of isoleucine, which matches Z in the consensus sequence, fits snugly in a hydrophobic groove formed by two leucine residues of the enzyme.

10.5 MANY ENZYMES ARE ACTIVATED BY SPECIFIC PROTEOLYTIC CLEAVAGE

We turn now to a different mechanism of enzyme regulation. Many enzymes acquire full enzymatic activity as they spontaneously fold into their characteristic three-dimensional forms. In contrast, other enzymes are synthesized as inactive precursors that are subsequently activated by cleavage of one or a few specific peptide bonds. The inactive precursor is called a *zymogen* (or a *proenzyme*). An energy source (ATP) is not needed for cleavage. Therefore, in contrast with reversible regulation by phosphorylation, even proteins located outside cells can be activated by this means. Another noteworthy difference is that proteolytic activation, in contrast with allosteric control and reversible covalent modification, occurs just once in the life of an enzyme molecule.

Specific proteolysis is a common means of activating enzymes and other proteins in biological systems. For example:

1. The *digestive enzymes* that hydrolyze proteins are synthesized as zymogens in the stomach and pancreas (Table 10.3).

TABLE 10.3   Gastric and pancreatic zymogens

Site of synthesis	Zymogen	Active enzyme
Stomach	Pepsinogen	Pepsin
Pancreas	Chymotrypsinogen	Chymotrypsin
Pancreas	Trypsinogen	Trypsin
Pancreas	Procarboxypeptidase	Carboxypeptidase
Pancreas	Proelastase	Elastase

2. *Blood clotting* is mediated by a cascade of proteolytic activations that ensures a rapid and amplified response to trauma.

3. Some protein hormones are synthesized as inactive precursors. For example, *insulin* is derived from *proinsulin* by proteolytic removal of a peptide.

4. The fibrous protein *collagen*, the major constituent of skin and bone, is derived from *procollagen*, a soluble precursor.

5. Many *developmental processes* are controlled by the activation of zymogens. For example, in the metamorphosis of a tadpole into a frog, large amounts of collagen are resorbed from the tail in the course of a few days. Likewise, much collagen is broken down in a mammalian uterus after delivery. The conversion of *procollagenase* into *collagenase*, the active protease, is precisely timed in these remodeling processes.

6. *Programmed cell death*, or *apoptosis*, is mediated by proteolytic enzymes called *caspases*, which are synthesized in precursor form as *procaspases*. When activated by various signals, caspases function to cause cell death in most organisms, ranging from *C. elegans* (Section 2.4.3) to human beings. Apoptosis provides a means sculpting the shapes of body parts in the course of development and a means of eliminating cells producing anti-self antibodies or infected with pathogens as well as cells containing large amounts of damaged DNA.

We next examine the activation and control of zymogens, using as examples several digestive enzymes as well as blood-clot formation.

10.5.1 Chymotrypsinogen Is Activated by Specific Cleavage of a Single Peptide Bond

Chymotrypsin is a digestive enzyme that hydrolyzes proteins in the small intestine. Its mechanism of action was discussed in detail in Chapter 9. Its inactive precursor, *chymotrypsinogen*, is synthesized in the pancreas, as are several other zymogens and digestive enzymes. Indeed, the pancreas is one of the most active organs in synthesizing and secreting proteins. The enzymes and zymogens are synthesized in the acinar cells of the pancreas and stored inside membrane-bounded granules (Figure 10.31). The zymogen granules accumulate at the apex of the acinar cell; when the cell is stimulated by a hormonal signal or a nerve impulse, the contents of the granules are released into a duct leading into the duodenum.

Chymotrypsinogen, a single polypeptide chain consisting of 245 amino acid residues, is virtually devoid of enzymatic activity. It is converted into a fully active enzyme when the peptide bond joining arginine 15 and isoleucine 16 is cleaved by trypsin (Figure 10.32). The resulting active enzyme, called π-chymotrypsin, then acts on other π-chymotrypsin molecules. Two dipeptides are removed to yield α-chymotrypsin, the stable form of the enzyme. The three resulting chains in α-chymotrypsin remain linked to one another by two interchain disulfide bonds. The striking feature of this activation process is that *cleavage of a single specific peptide bond transforms the protein from a catalytically inactive form into one that is fully active.*

10.5.2 Proteolytic Activation of Chymotrypsinogen Leads to the Formation of a Substrate-Binding Site

How does cleavage of a single peptide bond activate the zymogen? Key conformational changes, which were revealed by the elucidation of the three-dimensional structure of chymotrypsinogen, result from the cleavage of the peptide bond between amino acids 15 and 16.

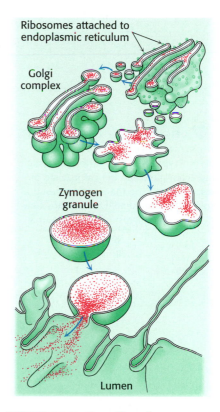

FIGURE 10.31 Secretion of zymogens by an acinar cell of the pancreas.

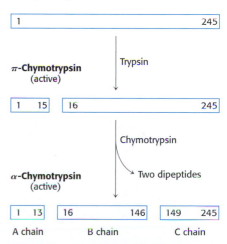

FIGURE 10.32 Proteolytic activation of chymotrypsinogen. The three chains of α-chymotrypsin are linked by two interchain disulfide bonds (A to B, and B to C).

1. The newly formed *amino-terminal group of isoleucine 16 turns inward and forms an ionic bond with aspartate 194* in the interior of the chymotrypsin molecule (Figure 10.33). Protonation of this amino group stabilizes the active form of chymotrypsin.

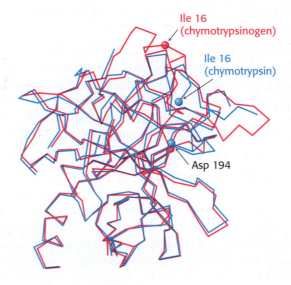

FIGURE 10.33 Conformations of chymotrypsinogen (red) and chymotrypsin (blue). The electrostatic interaction between the carboxylate of aspartate 194 and the α-amino group of isoleucine 16, essential for the structure of active chymotrypsin, is possible only in chymotrypsin.

2. This electrostatic interaction triggers a number of conformational changes. Methionine 192 moves from a deeply buried position in the zymogen to the surface of the active enzyme, and residues 187 and 193 become more extended. These changes result in the formation of the *substrate-specificity site* for aromatic and bulky nonpolar groups. One side of this site is made up of residues 189 through 192. *This cavity for binding part of the substrate is not fully formed in the zymogen.*

3. The tetrahedral transition state in catalysis by chymotrypsin is stabilized by hydrogen bonds between the negatively charged carbonyl oxygen atom of the substrate and two NH groups of the main chain of the enzyme (Section 9.1.3). One of these NH groups is not appropriately located in chymotrypsinogen, and so *the oxyanion hole is incomplete in the zymogen.*

4. The conformational changes elsewhere in the molecule are very small. Thus, *the switching on of enzymatic activity in a protein can be accomplished by discrete, highly localized conformational changes that are triggered by the hydrolysis of a single peptide bond.*

10.5.3 The Generation of Trypsin from Tryspinogen Leads to the Activation of Other Zymogens

The structural changes accompanying the activation of trypsinogen, the precursor of the proteolytic enzyme trypsin, are somewhat different from those in the activation of chymotrypsinogen. X-ray analyses have shown that the conformation of four stretches of polypeptide, constituting about 15% of the molecule, changes markedly on activation. *These regions, called the activation domain, are very flexible in the zymogen, whereas they have a well-defined conformation in trypsin.* Furthermore, the oxyanion hole (Section 9.1.3) in trypsinogen is too far from histidine 57 to promote the formation of the tetrahedral transition state.

The digestion of proteins in the duodenum requires the concurrent action of several proteolytic enzymes, because each is specific for a limited number of side chains. Thus, the zymogens must be switched on at the same

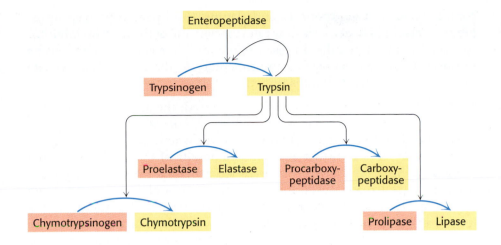

FIGURE 10.34 Zymogen activation by proteolytic cleavage. Enteropeptidase initiates the activation of the pancreatic zymogens by activating trypsin, which then activates other zymogens. Active enzymes are shown in yellow; zymogens are shown in orange.

time. Coordinated control is achieved by the action of *trypsin as the common activator of all the pancreatic zymogens*—trypsinogen, chymotrypsinogen, proelastase, procarboxypeptidase, and prolipase, a lipid degrading enzyme. To produce active trypsin, the cells that line the duodenum secrete an enzyme, *enteropeptidase*, that hydrolyzes a unique lysine–isoleucine peptide bond in trypsinogen as the zymogen enters the duodenum from the pancreas. The small amount of trypsin produced in this way activates more trypsinogen and the other zymogens (Figure 10.34). Thus, *the formation of trypsin by enteropeptidase is the master activation step.*

10.5.4 Some Proteolytic Enzymes Have Specific Inhibitors

The conversion of a zymogen into a protease by cleavage of a single peptide bond is a precise means of switching on enzymatic activity. However, this activation step is irreversible, and so a different mechanism is needed to stop proteolysis. Specific protease inhibitors accomplish this task. For example, *pancreatic trypsin inhibitor*, a 6-kd protein, inhibits trypsin by binding very tightly to its active site. The dissociation constant of the complex is 0.1 pM, which corresponds to a standard free energy of binding of about -18 kcal mol^{-1} (-75 kJ mol^{-1}). In contrast with nearly all known protein assemblies, this complex is not dissociated into its constituent chains by treatment with denaturing agents such as 8 M urea or 6 M guanidine hydrochloride.

The reason for the exceptional stability of the complex is that pancreatic trypsin inhibitor is a very effective substrate analog. X-ray analyses showed that the inhibitor lies in the active site of the enzyme, positioned such that the side chain of lysine 15 of this inhibitor interacts with the aspartate side chain in the specificity pocket of trypsin. In addition, there are many hydrogen bonds between the main chain of trypsin and that of its inhibitor. Furthermore, the carbonyl group of lysine 15 and the surrounding atoms of the inhibitor fit snugly in the active site of the enzyme. Comparison of the structure of the inhibitor bound to the enzyme with that of the free inhibitor reveals that *the structure is essentially unchanged on binding to the enzyme* (Figure 10.35). Thus, the inhibitor is preorganized into a structure that is highly complementary to the enzyme's active site. Indeed, the

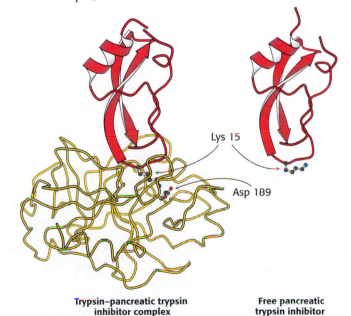

Lys 15

Asp 189

Trypsin–pancreatic trypsin
inhibitor complex

Free pancreatic
trypsin inhibitor

FIGURE 10.35 Interaction of trypsin with its inhibitor. Structure of a complex of trypsin (yellow) and pancreatic trypsin inhibitor (red). Lysine 15 of the inhibitor penetrates into the active site of the enzyme and forms a salt bridge with aspartate 189 in the active site. The bound inhibitor and the free inhibitor are almost identical in structure.

peptide bond between lysine 15 and alanine 16 in pancreatic trypsin inhibitor is cleaved but at a very slow rate: the half-life of the trypsin–inhibitor complex is several months. In essence, the inhibitor is a substrate, but its intrinsic structure is so nicely complementary to the enzyme's active site that it binds very tightly and is turned over slowly.

Why does trypsin inhibitor exist? Indeed, the amount of trypsin is much greater that that of the inhibitor. Under what circumstances is it beneficial to inhibit trypsin? Recall that trypsin activates other zymogens. Consequently, it is vital that even small amounts of trypsin be prevented from initiating the cascade prematurely. Trypsin molecules activated in the pancreas or pancreatic ducts could severely damage those tissues, leading to acute pancreatitis. Tissue necrosis may result from the activation of proteolytic enzymes (as well as prolipases) by trypsin, and hemorrhaging may result from its activation of elastase. We see here the physiological need for the tight binding of the inhibitor to trypsin.

Pancreatic trypsin inhibitor is not the only important protease inhibitor. α_1-*Antitrypsin* (also called α_1-*antiproteinase*), a 53-kd plasma protein, protects tissues from digestion by elastase, a secretory product of neutrophils (white blood cells that engulf bacteria). *Antielastase* would be a more accurate name for this inhibitor, because it blocks elastase much more effectively than it blocks trypsin. Like pancreatic trypsin inhibitor, α_1-antitrypsin blocks the action of target enzymes by binding nearly irreversibly to their active sites. Genetic disorders leading to a deficiency of α_1-antitrypsin show that this inhibitor is physiologically important. For example, the substitution of lysine for glutamate at residue 53 in the type Z mutant slows the secretion of this inhibitor from liver cells. Serum levels of the inhibitor are about 15% of normal in people homozygous for this defect. The consequence is that excess elastase destroys alveolar walls in the lungs by digesting elastic fibers and other connective-tissue proteins.

The resulting clinical condition is called *emphysema* (also known as *destructive lung disease*). People with emphysema must breathe much harder than normal people to exchange the same volume of air, because their alveoli are much less resilient than normal. Cigarette smoking markedly increases the likelihood that even a type Z heterozygote will develop emphysema. The reason is that smoke oxidizes methionine 358 of the inhibitor (Figure 10.36), a residue essential for binding elastase. Indeed, this methionine side chain is the bait that selectively traps elastase. The *methionine sulfoxide* oxidation product, in contrast, does not lure elastase, a striking consequence of the insertion of just one oxygen atom into a protein. We will consider another protease inhibitor, antithrombin III, when we examine the control of blood clotting.

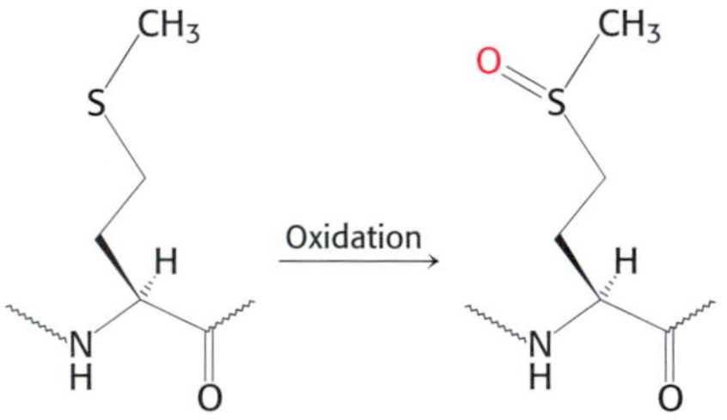

FIGURE 10.36 Oxidation of methionine to the sulfoxide.

10.5.5 Blood Clotting Is Accomplished by a Cascade of Zymogen Activations

Enzymatic cascades are often employed in biochemical systems to achieve a rapid response. In a cascade, an initial signal institutes a series of steps, each of which is catalyzed by an enzyme. At each step, the signal is amplified. For instance, if a signal molecule activates an enzyme that in turn activates 10 enzymes and each of the 10 enzymes in turn activates 10 additional enzymes, after four steps the original signal will have been amplified 10,000-fold. Blood clots are formed by a *cascade of zymogen activations*: the activated form of one clotting factor catalyzes the activation of the next (Figure 10.37). Thus, very small amounts of the initial factors suffice to trigger the cascade, ensuring a rapid response to trauma.

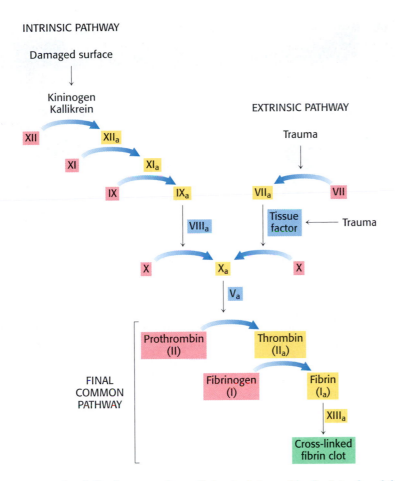

FIGURE 10.37 Blood-clotting cascade. A fibrin clot is formed by the interplay of the intrinsic, extrinsic, and final common pathways. The intrinsic pathway begins with the activation of factor XII (Hageman factor) by contact with abnormal surfaces produced by injury. The extrinsic pathway is triggered by trauma, which activates factor VII and releases a lipoprotein, called tissue factor, from blood vessels. Inactive forms of clotting factors are shown in red; their activated counterparts (indicated by the subscript "a") are in yellow. Stimulatory proteins that are not themselves enzymes are shown in blue. A striking feature of this process is that the activated form of one clotting factor catalyzes the activation of the next factor.

There are two means of initiating clotting: the intrinsic and extrinsic pathways. The *intrinsic clotting pathway* is activated by exposure of anionic surfaces on rupture of the endothelial lining of the blood vessels. These surfaces serve as binding sites for factors in the clotting cascade. Substances that are released from tissues as a consequence of trauma to them trigger the *extrinsic clotting pathway*. The extrinsic and intrinsic pathways converge on a common sequence of final steps to form a clot composed of the protein fibrin. The two pathways interact with each other in vivo. Indeed, both are needed for proper clotting, as evidenced by clotting disorders caused by a deficiency of a single protein in one of the pathways. Note that the active forms of the clotting factors are designated with a subscript "a."

10.5.6 Fibrinogen Is Converted by Thrombin into a Fibrin Clot

The best-characterized part of the clotting process is the final step in the cascade: the conversion of fibrinogen into fibrin by thrombin, a proteolytic enzyme. Fibrinogen is made up of three globular units connected by two

(A)

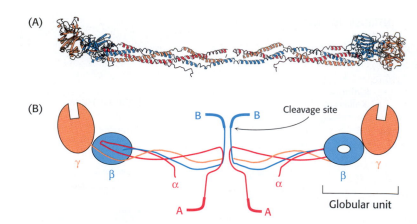

(B)

FIGURE 10.38 Structure of a fibrinogen molecule. (A) A ribbon diagram. The two rod regions are α-helical coiled coils, connected to a globular region at each end. (B) A schematic representation showing the positions of the fibrinopeptides A and B.

rods (Figure 10.38). This 340-kd protein consists of six chains: two each of Aα, Bβ, and γ. The rod regions are triple-stranded α-helical coiled coils, a recurring motif in proteins. Thrombin cleaves four *arginine–glycine peptide bonds* in the central globular region of fibrinogen. On cleavage, an A peptide of 18 residues from each of the two Aα chains and a B peptide of 20 residues from each of the two Bβ chains are released from the globular region. These A and B peptides are called *fibrinopeptides*. A fibrinogen molecule devoid of these fibrinopeptides is called a *fibrin monomer* and has the subunit structure $(\alpha\beta\gamma)_2$.

Fibrin monomers spontaneously assemble into ordered fibrous arrays called *fibrin*. Electron micrographs and low-angle x-ray patterns show that fibrin has a periodic structure that repeats every 23 nm (Figure 10.39).

FIGURE 10.39 Electron micrograph of fibrin. The 23-nm period along the fiber axis is half the length of a fibrinogen molecule. [Courtesy of Dr. Henry Slayter.]

Higher-resolution images reveal the detailed structure of the fibrin monomer, how the fibrin monomers come together, and how this assembly is facilitated by removal of the fibrinopeptides. The homologous β and γ chains have globular domains at the carboxyl-terminal ends (Figure 10.40). These domains have binding "holes" that interact with peptides. The β domain is specific for sequences of the form H_3N^+-Gly-His-Arg-, whereas the γ domain binds H_3N^+-Gly-Pro-Arg-. Exactly these sequences (some-

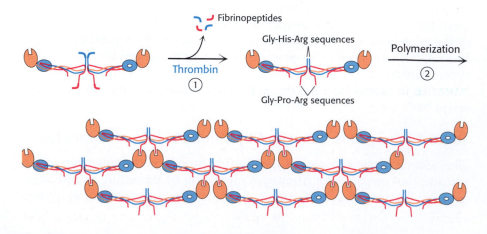

FIGURE 10.40 Formation of a fibrin clot. (1) Thrombin cleaves fibrinopeptides A and B from the central globule of fibrinogen. (2) Globular domains at the carboxyl-terminal ends of the β and γ chains interact with "knobs" exposed at the amino-terminal ends of the β and α chains to form clots.

times called "knobs") are exposed at the amino-terminal ends of the β and α chains, respectively, on thrombin cleavage. The knobs of the α subunits fit into the holes on the γ subunits of another monomer to form a protofibril. This protofibril is extended when the knobs of the β subunits fit into holes of β subunits of other protofibrils. Thus, analogous to the activation of chymotrypsinogen, peptide-bond cleavage exposes new amino termini that can participate in specific interactions. The newly formed clot is stabilized by the formation of amide bonds between the side chains of lysine and glutamine residues in different monomers.

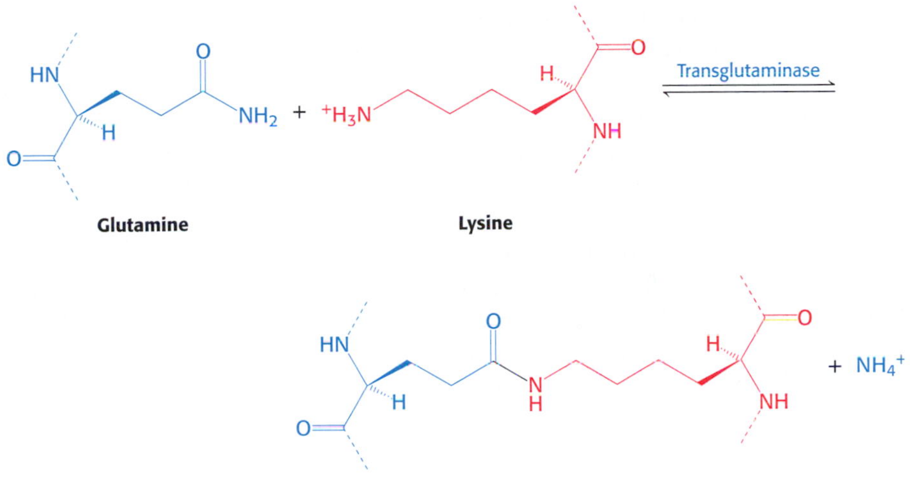

Glutamine   **Lysine**

Cross-link

This cross-linking reaction is catalyzed by *transglutaminase (factor XIII$_a$)*, which itself is activated from the protransglutaminase form by thrombin.

10.5.7 Prothrombin Is Readied for Activation by a Vitamin K-Dependent Modification

Thrombin is synthesized as a zymogen called *prothrombin*, which comprises four major domains, with the serine protease domain at its carboxyl terminus. The first domain is called a *gla domain*, whereas domains 2 and 3 are called *kringle domains* (Figure 10.41). These domains work in concert to keep prothrombin in an inactive form and to target it to appropriate sites for its activation by factor X$_a$ (a serine protease) and factor V$_a$ (a stimulatory protein). Activation is accomplished by proteolytic cleavage of the bond between arginine 274 and threonine 275 to release a fragment containing the first three domains and by cleavage of the bond between arginine 323 and isoleucine 324 (analogous to the key bond in chymotrypsinogen) to yield active thrombin.

Cleavage sites

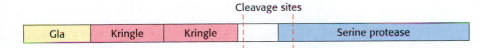

FIGURE 10.41 Modular structure of prothrombin. Cleavage of two peptide bonds yields thrombin. All the γ-carboxyglutamate residues are in the gla domain.

Vitamin K (Section 8.6.2 and Figure 10.42) has been known for many years to be essential for the synthesis of prothrombin and several other clotting factors. The results of studies of the abnormal prothrombin synthesized in the absence of vitamin K or in the presence of vitamin K antagonists, such as dicoumarol, revealed the mode of action of this vitamin. *Dicoumarol* is found in spoiled sweet clover and causes a fatal

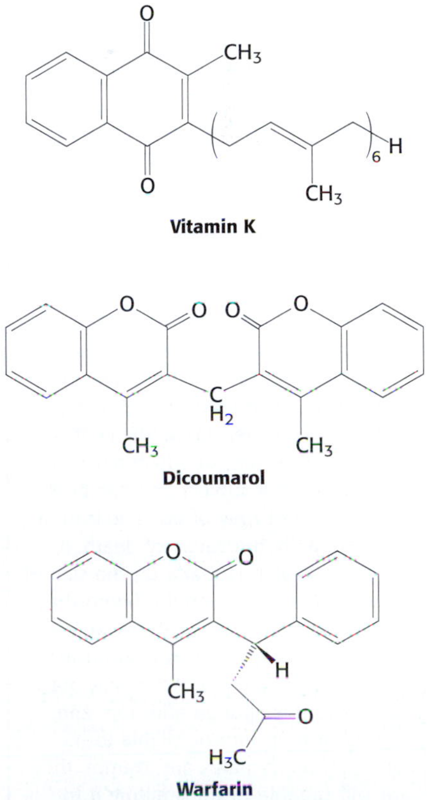

Vitamin K

Dicoumarol

Warfarin

FIGURE 10.42 Structures of vitamin K and two antagonists, dicoumarol and warfarin.

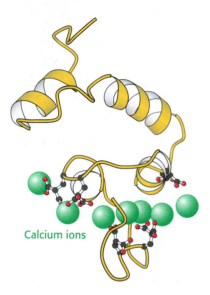

Calcium ions

FIGURE 10.43 The calcium-binding region of prothrombin. Prothrombin binds calcium ions with the modified amino acid γ-carboxyglutamate (red).

hemorrhagic disease in cattle fed on this hay. This coumarin derivative is used clinically as an *anticoagulant* to prevent thromboses in patients prone to clot formation. Dicoumarol and such related vitamin K antagonists as *warfarin* also serve as effective rat poisons. Cows fed dicoumarol synthesize an abnormal prothrombin that does not bind Ca^{2+}, in contrast with normal prothrombin. This difference was puzzling for some time because abnormal prothrombin has the same number of amino acid residues and gives the same amino acid analysis after acid hydrolysis as does normal prothrombin.

Nuclear magnetic resonance studies revealed that normal prothrombin contains γ-*carboxyglutamate*, a previously unknown residue that evaded detection because its second carboxyl group is lost on acid hydrolysis. The abnormal prothrombin formed subsequent to the administration of anticoagulants lacks this modified amino acid. In fact, the first 10 glutamate residues in the amino-terminal region of prothrombin are carboxylated to γ-carboxyglutamate by a vitamin K-dependent enzyme system (Figure

γ-**Carboxyglutamate residue**

10.43). *The vitamin K-dependent carboxylation reaction converts glutamate, a weak chelator of Ca^{2+}, into γ-carboxyglutamate, a much stronger chelator.* Prothrombin is thus able to bind Ca^{2+}, but what is the effect of this binding? The binding of Ca^{2+} by prothrombin anchors the zymogen to phospholipid membranes derived from blood platelets after injury. The binding of prothrombin to phospholipid surfaces is crucial because it brings prothrombin into close proximity to two clotting proteins that catalyze its conversion into thrombin. The proteolytic activation of prothrombin removes the calcium-binding domain and frees thrombin from the membrane so that it can cleave fibrinogen and other targets.

10.5.8 Hemophilia Revealed an Early Step in Clotting

Some important breakthroughs in the elucidation of clotting path ways have come from studies of patients with bleeding disorders. *Classic hemophilia,* or *hemophilia A,* the best-known clotting defect, is genetically transmitted as a sex-linked recessive characteristic. *In classic hemophilia, factor VIII (antihemophilic factor) of the intrinsic pathway is missing or has markedly reduced activity.* Although factor VIII is not itself a protease, it markedly stimulates the activation of factor X, the final protease of the intrinsic pathway, by factor IX_a, a serine protease (Figure 10.44). Thus, activation of the intrinsic pathway is severely impaired in hemophilia.

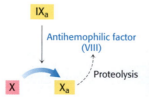

FIGURE 10.44 Action of antihemophilic factor. Antihemophilic factor (VIII) stimulates the activation of factor X by factor IX_a. It is interesting to note that the activity of factor VIII is markedly increased by limited proteolysis by thrombin and factor X_a. This positive feedback amplifies the clotting signal and accelerates clot formation after a threshold has been reached.

In the past, hemophiliacs were treated with transfusions of a concentrated plasma fraction containing factor VIII. This therapy carried the risk of infection. Indeed, many hemophiliacs contracted hepatitis and AIDS. A safer preparation of factor VIII was urgently needed. With the use of biochemical purification and recombinant DNA techniques, the gene for factor VIII was isolated and expressed in cells grown in culture. Recombinant factor VIII purified from these cells has largely replaced plasma concentrates in treating hemophilia.

10.5.9 The Clotting Process Must Be Precisely Regulated

There is a fine line between hemorrhage and thrombosis. Clots must form rapidly yet remain confined to the area of injury. What are the mechanisms that normally limit clot formation to the site of injury? The lability of clotting factors contributes significantly to the control of clotting. Activated factors are short-lived because they are diluted by blood flow, removed by the liver, and degraded by proteases. For example, the stimulatory proteins factors V_a and VIII_a are digested by protein C, a protease that is switched on by the action of thrombin. *Thus, thrombin has a dual function: it catalyzes the formation of fibrin and it initiates the deactivation of the clotting cascade.*

Specific inhibitors of clotting factors are also critical in the termination of clotting. The most important one is *antithrombin III,* a plasma protein that inactivates thrombin by forming an irreversible complex with it. Antithrombin III resembles α_1-antitrypsin except that it inhibits thrombin much more strongly than it inhibits elastase. Antithrombin III also blocks other serine proteases in the clotting cascade—namely, factors XII_a, XI_a, IX_a, and X_a. The inhibitory action of antithrombin III is enhanced by *heparin,* a negatively charged polysaccharide found in mast cells near the walls of blood vessels and on the surfaces of endothelial cells (Figure 10.45). Heparin acts as an *anticoagulant* by increasing the rate of formation of irreversible complexes between antithrombin III and the serine protease clotting factors. Antitrypsin and antithrombin are *serpins,* a family of *ser*ine protease *in*hibitors.

The importance of the ratio of thrombin to antithrombin is illustrated in the case in which a 14-year-old boy died of a bleeding disorder because of a mutation in his α_1-antitrypsin, which normally inhibits elastase (Section 10.5.4). Methionine 358 in α_1-antitrypsin's binding pocket for elastase was replaced by arginine, resulting in a change in specificity from an elastase inhibitor to a thrombin inhibitor. α_1-Antitrypsin activity normally increases markedly after injury to counteract excess elastase arising from stimulated neutrophils. The mutant α_1-antitrypsin caused the patient's thrombin activity to drop to such a low level that hemorrhage ensued. *We see here a striking example of how a change of a single residue in a protein can dramatically alter specificity and an example of the critical importance of having the right amount of a protease inhibitor.*

Antithrombin limits the extent of clot formation, but what happens to the clots themselves? Clots are not permanent structures but are designed to be dissolved when the structural integrity of damaged areas is restored. Fibrin is split by *plasmin,* a serine protease that hydrolyzes peptide bonds in the coiled-coil regions. Plasmin molecules can diffuse through aqueous channels in the porous fibrin clot to cut the accessible connector rods. Plasmin is formed by proteolytic activation of *plasminogen,* an inactive precursor that has a high affinity for the fibrin clots. This conversion is carried out by *tissue-type plasminogen activator* (TPA), a 72-kd protein that has a domain structure closely related to that of prothrombin (Figure 10.46).

Fibrin binding	Kringle	Kringle		Serine protease

FIGURE 10.46 Modular structure of tissue-type plasminogen activator (TPA).

However, a domain that targets TPA to fibrin clots replaces the membrane-targeting gla domain of prothrombin. The TPA bound to fibrin clots swiftly activates adhering plasminogen. In contrast, TPA activates free plasminogen very slowly. The gene for TPA has been cloned and expressed in cultured mammalian cells. The results of clinical studies have shown that TPA administered intravenously within an hour of the formation of a blood clot in a coronary artery markedly increases the likelihood of surviving a heart attack (Figure 10.47).

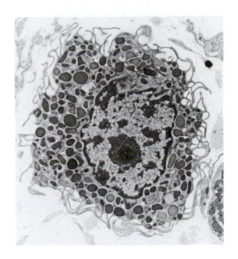

FIGURE 10.45 Electron micrograph of a mast cell. Heparin and other molecules in the dense granules are released into the extracellular space when the cell is triggered to secrete. [Courtesy of Lynne Mercer.]

(A)

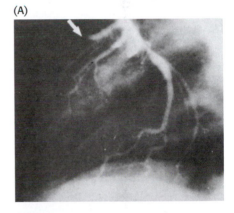

(B)

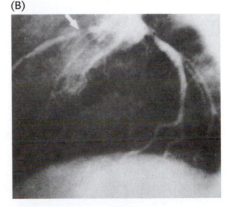

FIGURE 10.47 The effect of tissue-type plasminogen factor. TPA leads to the dissolution of blood clots, as shown by x-ray images of blood vessels in the heart (A) before and (B) 3 hours after the administration of TPA. The position of the clot is marked by the arrow in part A. [After F. Van de Werf, P. A. Ludbrook, S. R. Bergmann, A. J. Tiefenbrunn, K. A. A. Fox, H. de Geest, M. Verstraete, D. Collen, and B. E. Sobel. *New Engl. J. Med.* 310(1984):609.]

SUMMARY

Aspartate Transcarbamoylase Is Allosterically Inhibited by the End Product of Its Pathway

Allosteric proteins constitute an important class of proteins whose biological activity can be regulated. Specific regulatory molecules can modulate the activity of allosteric proteins by binding to distinct regulatory sites, separate from the functional site. These proteins have multiple functional sites, which display cooperation as evidenced by a sigmoidal dependence of function on substrate concentration. Aspartate transcarbamoylase (ATCase), one of the best-understood allosteric enzymes, catalyzes the synthesis of N-carbamoylaspartate, the first intermediate in the synthesis of pyrimidines. ATCase is feedback inhibited by cytidine triphosphate(CTP), the final product of the pathway. ATP reverses this inhibition. ATCase consists of separable catalytic (c_3) subunits (which bind the substrates) and regulatory (r_2) subunits (which bind CTP and ATP). The inhibitory effect of CTP, the stimulatory action of ATP, and the cooperative binding of substrates are mediated by large changes in quaternary structure. On binding substrates, the c_3 subunits of the c_6r_6 enzyme move apart and reorient themselves. This allosteric transition is highly concerted, as postulated by the Monod-Wyman-Changeux (MWC) model. All subunits of an ATCase molecule simultaneously interconvert from the T (low-affinity) to the R (high-affinity) state.

Hemoglobin Transports Oxygen Efficiently by Binding Oxygen Cooperatively

Hemoglobin, the oxygen carrier in the blood, is an allosteric protein. Hemoglobin consists of four polypeptide chains, each with a heme group—a substituted porphyrin with a central iron. Hemoglobin A, the predominant hemoglobin in adults, has the subunit structure $\alpha_2\beta_2$. Hemoglobin transports H^+ and CO_2 in addition to O_2. Hemoglobin exhibits three kinds of allosteric effects. First, the oxygen-binding curve of hemoglobin is sigmoidal, which indicates that the binding of oxygen is cooperative. The binding of oxygen to one heme group facilitates the binding of oxygen to the other heme groups in the same molecule. Second, the binding of H^+ and CO_2 promotes the release of O_2 from hemoglobin, an effect that is physiologically important in enhancing the release of O_2 in metabolically active tissues such as muscle. These allosteric linkages between the binding of H^+, CO_2, and O_2 are known as the Bohr effect. Third, the affinity of hemoglobin for O_2 is further regulated by 2,3-bisphosphoglycerate (2,3-BPG), a small molecule with a very high density of negative charge. 2,3-Bisphosphoglycerate binds tightly to deoxyhemoglobin but not to oxyhemoglobin. Hence, 2,3-BPG lowers the oxygen affinity of hemoglobin. Fetal hemoglobin ($\alpha_2\gamma_2$) has a higher oxygen affinity than human adult hemoglobin because fetal hemoglobin binds 2,3-BPG less tightly. Neither the sequential nor the concerted model completely describes the allosteric behavior of hemoglobin. Rather, the behavior of hemoglobin is best described by a combined model that employs features of both models.

Isozymes Provide a Means of Regulation Specific to Distinct Tissues and Developmental Stages

Isozymes differ in structural characteristics but catalyze the same reaction. They provide a means of fine-tuning metabolism to meet the needs of a given tissue or developmental stage. The results of gene-duplication events provide the means for subtle regulation of enzyme function.

• Covalent Modification Is a Means of Regulating Enzyme Activity

Covalent modification of proteins is a potent means of controlling the activity of enzymes and other proteins. Phosphorylation is the most common type of reversible covalent modification. Signals can be highly amplified by phosphorylation because a single kinase can act on many target molecules. The regulatory actions of protein kinases are reversed by protein phosphatases, which catalyze the hydrolysis of attached phosphoryl groups.

Cyclic AMP serves as an intracellular messenger in the transduction of many hormonal and sensory stimuli. Cyclic AMP switches on protein kinase A (PKA), a major multifunctional kinase, by binding to the regulatory subunit of the enzyme, thereby releasing the active catalytic subunits of PKA. In the absence of cAMP, the catalytic sites of PKA are occupied by pseudosubstrate sequences of the regulatory subunit.

• Many Enzymes Are Activated by Specific Proteolytic Cleavage

The activation of an enzyme by proteolytic cleavage of one or a few peptide bonds is a recurring control mechanism seen in processes as diverse as the activation of digestive enzymes and blood clotting. The inactive precursor is called a zymogen (or a proenzyme). Trypsinogen is activated by enteropeptidase or trypsin, and trypsin then activates a host of other zymogens, leading to the digestion of foodstuffs. For instance, trypsin converts chymotrypsinogen, a zymogen, into active chymotrypsin by hydrolyzing a single peptide bond.

A striking feature of the clotting process is that it is accomplished by a cascade of zymogen conversions, in which the activated form of one clotting factor catalyzes the activation of the next precursor. Many of the activated clotting factors are serine proteases. In the final step of clot formation, fibrinogen, a highly soluble molecule in the plasma, is converted by thrombin into fibrin by the hydrolysis of four arginine–glycine bonds. The resulting fibrin monomer spontaneously forms long, insoluble fibers called fibrin. Zymogen activation is also essential in the lysis of clots. Plasminogen is converted into plasmin, a serine protease that cleaves fibrin, by tissue-type plasminogen activator (TPA). Although zymogen activation is irreversible, specific inhibitors of some proteases exert control. The irreversible protein inhibitor antithrombin III holds blood clotting in check in the clotting cascade.

KEY TERMS

feedback (end-product) inhibition (p. 263)
regulatory site (p. 263)
concerted mechanism (p. 266)
cooperativity (p. 267)
homotropic effect (p. 268)
heterotropic effect (p. 268)
sequential model (p. 269)

heme (p. 270)
protoporphyrin (p. 270)
Bohr effect (p. 273)
isozyme (isoenzyme) (p. 274)
covalent modification (p. 275)
protein kinase (p. 276)
protein phosphatase (p. 277)

consensus sequence (p. 278)
protein kinase A (PKA) (p. 278)
pseudosubstrate sequence (p. 279)
zymogen (proenzyme) (p. 280)
enzymatic cascade (p. 284)
intrinsic clotting pathway (p. 285)
extrinsic clotting pathway (p. 285)

SELECTED READINGS

Where to start

Kantrowitz, E. R., and Lipscomb, W. N., 1990. *Escherichia coli* aspartate transcarbamoylase: The molecular basis for a concerted allosteric transition. *Trends Biochem. Sci.* 15:53–59.

Schachman, H. K., 1988. Can a simple model account for the allosteric transition of aspartate transcarbamoylase? *J. Biol. Chem.* 263: 18583–18586.
Dickerson, R. E., and Geis, I.,1983. *Hemoglobin: Structure, Function, Evolution and Pathology.* Benjamin Cummings.

Neurath, H., 1989. Proteolytic processing and physiological regulation. *Trends Biochem. Sci.* 14:268–271.

Bode, W., and Huber, R., 1992. Natural protein proteinase inhibitors and their interaction with proteinases. *Eur. J. Biochem.* 204: 433–451.

Aspartate transcarbamoylase and allosteric interactions

Endrizzi, J. A., Beernink, P. T., Alber, T., and Schachman, H. K., 2000. Binding of bisubstrate analog promotes large structural changes in the unregulated catalytic trimer of aspartate transcarbamoylase: Implications for allosteric regulation. *Proc. Natl. Acad. Sci. U. S. A.* 97:5077–5082.

Beernink, P. T., Endrizzi, J. A., Alber, T., and Schachman, H. K., 1999. Assessment of the allosteric mechanism of aspartate transcarbamoylase based on the crystalline structure of the unregulated catalytic subunit. *Proc. Natl. Acad. Sci. U. S. A.* 96:5388–5393.

Wales, M. E., Madison, L. L., Glaser, S. S., and Wild, J. R., 1999. Divergent allosteric patterns verify the regulatory paradigm for aspartate transcarbamylase. *J. Mol. Biol.* 294:1387–1400.

LiCata, V. J., Burz, D. S., Moerke, N. J., and Allewell, N. M., 1998. The magnitude of the allosteric conformational transition of aspartate transcarbamylase is altered by mutations. *Biochemistry* 37: 17381–17385.

Eisenstein, E., Markby, D. W., and Schachman, H. K., 1990. Heterotropic effectors promote a global conformational change in aspartate transcarbamoylase. *Biochemistry* 29:3724–3731.

Werner, W. E., and Schachman, H. K., 1989. Analysis of the ligand-promoted global conformational change in aspartate transcarbamoylase: Evidence for a two-state transition from boundary spreading in sedimentation velocity experiments. *J. Mol. Biol.* 206:221–230.

Newell, J. O., Markby, D. W., and Schachman, H. K., 1989. Cooperative binding of the bisubstrate analog *N*-(phosphonacetyl)-L-aspartate to aspartate transcarbamoylase and the heterotropic effects of ATP and CTP. *J. Biol. Chem.* 264:2476–2481.

Stevens, R. C., Reinisch, K. M., and Lipscomb, W. N., 1991. Molecular structure of *Bacillus subtilis* aspartate transcarbamoylase at 3.0 Å resolution. *Proc. Natl. Acad. Sci. U. S. A.* 88:6087–6091.

Stevens, R. C., Gouaux, J. E., and Lipscomb, W. N., 1990. Structural consequences of effector binding to the T state of aspartate carbamoyltransferase: Crystal structures of the unligated and ATP- and CTP-complexed enzymes at 2.6-Å resolution. *Biochemistry* 29:7691–7701.

Gouaux, J. E., and Lipscomb, W. N., 1990. Crystal structures of phosphonoacetamide ligated T and phosphonoacetamide and malonate ligated R states of aspartate carbamoyltransferase at 2.8-Å resolution and neutral pH. *Biochemistry* 29:389–402.

Labedan, B., Boyen, A., Baetens, M., Charlier, D., Chen, P., Cunin, R., Durbeco, V., Glansdorff, N., Herve, G., Legrain, C., Liang, Z., Purcarea, C., Roovers, M., Sanchez, R., Toong, T. L., Van de Casteele, M., van Vliet, F., Xu, Y., and Zhang, Y. F., 1999. The evolutionary history of carbamoyltransferases: A complex set of paralogous genes was already present in the last universal common ancestor. *J. Mol. Evol.* 49:461–473.

Hemoglobin

Perutz, M. F., Wilkinson, A. J., Paoli, M., and Dodson, G. G., 1998. The stereochemical mechanism of the cooperative effects in hemoglobin revisited. *Annu. Rev. Biophys. Biomol. Struct.* 27:1–34.

Ackers, G. K., 1998. Deciphering the molecular code of hemoglobin allostery. *Adv. Protein Chem.* 51:185–253.

Ackers, G. K., Doyle, M. L., Myers, D., and Daugherty, M. A., 1992. Molecular code for cooperativity in hemoglobin. *Science* 255:54–63.

Fermi, G., Perutz, M. F., Shaanan, B., and Fourme, R., 1984. The crystal structure of human deoxyhaemoglobin at 1.74 Å resolution. *J. Mol. Biol.* 175:159–174.

Kilmartin, J. V. and Rossi-Bernardi, L., 1973. Interaction of hemoglobin with hydrogen ions, carbon dioxide, and organic phosphates. *Physiol. Rev.* 53:836–890.

Covalent modification

Johnson, L. N., and Barford, D., 1993. The effects of phosphorylation on the structure and function of proteins. *Annu. Rev. Biophys. Biomol. Struct.* 22:199–232.

Ziegler, M., 2000. New functions of a long-known molecule: Emerging roles of NAD in cellular signaling. *Eur. J. Biochem.* 267:1550–1564.

Ng, H. H. and Bird, A., 2000. Histone deacetylases: Silencers for hire. *Trends Biochem. Sci.* 25:121–126.

Raju, R. V., Kakkar, R., Radhi, J. M., and Sharma, R. K., 1997. Biological significance of phosphorylation and myristoylation in the regulation of cardiac muscle proteins. *Mol. Cell. Biochem.* 176:135–143.

Jacobson, M. K., and Jacobson, E. L., 1999. Discovering new ADP-ribose polymer cycles: Protecting the genome and more. *Trends Biochem. Sci.* 24:415–417.

Barford, D., Das, A. K., and Egloff, M. P., 1998. The structure and mechanism of protein phosphatases: Insights into catalysis and regulation. *Annu. Rev. Biophys. Biomol. Struct.* 27:133–164.

Protein kinase A

Taylor, S. S., Knighton, D. R., Zheng, J., Sowadski, J. M., Gibbs, C. S., and Zoller, M. J., 1993. A template for the protein kinase family. *Trends Biochem. Sci.* 18:84–89.

Gibbs, C. S., Knighton, D. R., Sowadski, J. M., Taylor, S. S., and Zoller, M. J., 1992. Systematic mutational analysis of cAMP-dependent protein kinase identifies unregulated catalytic subunits and defines regions important for the recognition of the regulatory subunit. *J. Biol. Chem.* 267:4806–4814.

Knighton, D. R., Zheng, J. H., TenEyck, L., Ashford, V. A., Xuong, N. H., Taylor, S. S., and Sowadski, J. M., 1991. Crystal structure of the catalytic subunit of cyclic adenosine monophosphate-dependent protein kinase. *Science* 253:407–414.

Knighton, D. R., Zheng, J. H., TenEyck, L., Xuong, N. H., Taylor, S. S., and Sowadski, J. M., 1991. Structure of a peptide inhibitor bound to the catalytic subunit of cyclic adenosine monophosphate-dependent protein kinase. *Science* 253:414–420.

Adams, S. R., Harootunian, A. T., Buechler, Y. J., Taylor, S. S., and Tsien, R. Y., 1991. Fluorescence ratio imaging of cyclic AMP in single cells. *Nature* 349:694–697.

Zymogen activation

Neurath, H., 1986. The versatility of proteolytic enzymes. *J. Cell. Biochem.* 32:35–49.

Bode, W., and Huber, R., 1986. Crystal structure of pancreatic serine endopeptidases. In *Molecular and Cellular Basis of Digestion* (pp. 213–234), edited by P. Desnuelle, H. Sjostrom, and O. Noren. Elsevier.

Huber, R., and Bode, W., 1978. Structural basis of the activation and action of trypsin. *Acc. Chem. Res.* 11:114–122.

Stroud, R. M., Kossiakoff, A. A., and Chambers, J. L., 1977. Mechanism of zymogen activation. *Annu. Rev. Biophys. Bioeng.* 6:177–193.

Sielecki, A. R., Fujinaga, M., Read, R. J., and James, M. N., 1991. Refined structure of porcine pepsinogen at 1.8 Å resolution. *J. Mol. Biol.* 219:671–692.

Protease inhibitors

Carrell, R., and Travis, J., 1985. α_1-Antitrypsin and the serpins: Variation and countervariation. *Trends Biochem. Sci.* 10:20–24.

Carp, H., Miller, F., Hoidal, J. R., and Janoff, A., 1982. Potential mechanism of emphysema: α_1-Proteinase inhibitor recovered from lungs of cigarette smokers contains oxidized methionine and has decreased elastase inhibitory capacity. *Proc. Natl. Acad. Sci. U. S. A.* 79:2041–2045.

Owen, M. C., Brennan, S. O., Lewis, J. H., and Carrell, R. W., 1983. Mutation of antitrypsin to antithrombin. *New Engl. J. Med.* 309:694–698.

Travis, J., and Salvesen, G. S., 1983. Human plasma proteinase inhibitors. *Annu. Rev. Biochem.* 52:655–709.

Clotting cascade

Fuentes-Prior, P., Iwanaga, Y., Huber, R., Pagila, R., Rumennik, G., Seto, M., Morser, J., Light, D. R., and Bode, W., 2000. Structural basis for the anticoagulant activity of the thrombin-thrombomodulin complex. *Nature.* 404:518–525.

Herzog, R. W., and High, K. A., 1998. Problems and prospects in gene therapy for hemophilia. *Curr. Opin. Hematol.* 5:321–326.

Doolittle, R. F., 1981. Fibrinogen and fibrin. *Sci. Am.* 245(12):126–135.

Lawn, R. M., and Vehar, G. A., 1986. The molecular genetics of hemophilia. *Sci. Am.* 254(3):48–65.

Brown, J. H., Volkmann, N., Jun, G., Henschen-Edman, A. H. and Cohen, C., 2000. The crystal structure of modified bovine fibrinogen. *Proc. Natl. Acad. Sci. U. S. A.* 97:85–90.

Stubbs, M. T., Oschkinat, H., Mayr, I., Huber, R., Angliker, H., Stone, S. R., and Bode, W., 1992. The interaction of thrombin with fibrinogen: A structural basis for its specificity. *Eur. J. Biochem.* 206:187–195.

Rydel, T. J., Tulinsky, A., Bode, W., and Huber, R., 1991. Refined structure of the hirudin-thrombin complex. *J. Mol. Biol.* 221: 583–601.

PROBLEMS

1. *Activity profile.* A histidine residue in the active site of aspartate transcarbamoylase is thought to be important in stabilizing the transition state of the bound substrates. Predict the pH dependence of the catalytic rate, assuming that this interaction is essential and dominates the pH-activity profile of the enzyme.

2. *Allosteric switching.* A substrate binds 100 times as tightly to the R state of an allosteric enzyme as to its T state. Assume that the concerted (MWC) model applies to this enzyme.

(a) By what factor does the binding of one substrate molecule per enzyme molecule alter the ratio of the concentrations of enzyme molecules in the R and T states?

(b) Suppose that L, the ratio of [T] to [R] in the absence of substrate, is 10^7 and that the enzyme contains four binding sites for substrate. What is the ratio of enzyme molecules in the R state to that in the T state in the presence of saturating amounts of substrate, assuming that the concerted model is obeyed?

3. *Allosteric transition.* Consider an allosteric protein that obeys the concerted model. Suppose that the ratio of T to R formed in the absence of ligand is 10^5, $K_T = 2$ mM, and $K_R = 5$ μM. The protein contains four binding sites for ligand. What is the fraction of molecules in the R form when 0, 1, 2, 3, and 4 ligands are bound?

4. *Negative cooperativity.* You have isolated a dimeric enzyme that contains two identical active sites. The binding of substrate to one active site decreases the substrate affinity of the other active site. Which allosteric model best accounts for this negative cooperativity?

5. *Paradoxical at first glance.* Recall that phosphonacetyl-L-aspartate (PALA) is a potent inhibitor of ATCase because it mimics the two physiological substrates. However, low concentrations of this unreactive bisubstrate analog *increase* the reaction velocity. On the addition of PALA, the reaction rate increases until an average of three molecules of PALA are bound per molecule of enzyme. This maximal velocity is 17-fold as great as it is in the absence of PALA. The reaction rate then decreases to nearly zero on the addition of three more molecules of PALA per molecule of enzyme. Why do low concentrations of PALA activate ATCase?

6. *R versus T.* An allosteric enzyme that follows the concerted mechanism has a T/R ratio of 300 in the absence of substrate. Suppose that a mutation reversed the ratio. How would this mutation affect the relation between the rate of the reaction and substrate concentration?

7. *Tuning oxygen affinity.* What is the effect of each of the following treatments on the oxygen affinity of hemoglobin A in vitro?

(a) Increase in pH from 7.2 to 7.4.
(b) Increase in pCO_2 from 10 to 40 torr.
(c) Increase in [2,3-BPG] from 2×10^{-4} to 8×10^{-4} M.
(d) Dissociation of $\alpha_2\beta_2$ into monomeric subunits.

8. *Avian and reptilian counterparts.* The erythrocytes of birds and turtles contain a regulatory molecule different from 2,3-BPG. This substance is also effective in reducing the oxygen affinity of human hemoglobin stripped of 2,3-BPG. Which of the following substances would you predict to be most effective in this regard?

(a) Glucose 6-phosphate (d) Malonate
(b) Inositol hexaphosphate (e) Arginine
(c) HPO_4^{2-} (f) Lactate

9. *Tuning proton affinity.* The pK of an acid depends partly on its environment. Predict the effect of the following environmental changes on the pK of a glutamic acid side chain.

(a) A lysine side chain is brought into close proximity.
(b) The terminal carboxyl group of the protein is brought into close proximity.
(c) The glutamic acid side chain is shifted from the outside of the protein to a nonpolar site inside.

10. *Zymogen activation.* When very low concentrations of pepsinogen are added to acidic media, how does the half-time for activation depend on zymogen concentration?

11. *A revealing assay.* Suppose that you have just examined a young boy with a bleeding disorder highly suggestive of classic hemophilia (factor VIII deficiency). Because of the late hour, the laboratory that carries out specialized coagulation assays is closed. However, you happen to have a sample of blood from a classic hemophiliac whom you admitted to the hospital an hour earlier. What is the simplest and most rapid test that you can perform to determine whether your present patient also is deficient in factor VIII activity?

12. *Counterpoint.* The synthesis of factor X, like that of prothrombin, requires vitamin K. Factor X also contains γ-carboxyglutamate residues in its amino-terminal region. However, activated factor X, in contrast with thrombin, retains this region of the molecule. What is a likely functional consequence of this difference between the two activated species?

13. *A discerning inhibitor.* Antithrombin III forms an irreversible complex with thrombin but not with prothrombin. What is the most likely reason for this difference in reactivity?

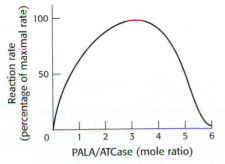

Effect of PALA on ATCase rate.

14. *Repeating heptads.* Each of the three types of fibrin chains contains repeating heptapeptide units *(abcdefg)* in which residues *a* and *d* are hydrophobic. Propose a reason for this regularity.

15. *Drug design.* A drug company has decided to use recombinant DNA methods to prepare a modified α_1-antitrypsin that will be more resistant to oxidation than is the naturally occurring inhibitor. Which single amino acid substitution would you recommend?

Data Interpretation Problems

16. *Distinguishing between models.* The graph below shows the fraction of an allosteric enzyme in the R state (f_R) and the fraction of active sites bound to substrate (Y) as a function of substrate concentration. Which model, the concerted or sequential, best explains these results?

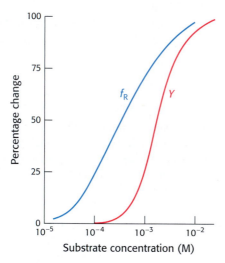

[From M. W. Kirschner and H. K. Schachman, *Biochemistry* 12 (1966):2997.]

17. *Reporting live from ATCase.* ATCase was reacted with nitromethane to form a colored nitrotyrosine group ($\lambda_{max} =$ 430 nm) in each of its catalytic chains. The absorption by this reporter group depends on its immediate environment. An essential lysine residue at each catalytic site also was modified to block the binding of substrate. Catalytic trimers from this doubly modified enzyme were then combined with native trimers to form a hybrid enzyme. The absorption by the nitrotyrosine group was measured on addition of the substrate analog succinate. What is the significance of the alteration in the absorbance at 430 nm?

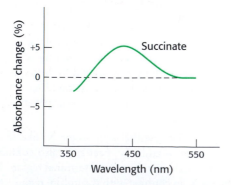

[After H. K. Schachman. *J. Biol. Chem.* 263 (1988):18583]

The binding of succinate, an unreactive analog of aspartate, to one trimer changed the visible absorption spectrum of nitrotyrosine residues attached to the other trimer.

18. *Reporting live from ATCase 2.* A different ATCase hybrid was constructed to test the effects of allosteric activators and inhibitors. Normal regulatory subunits were combined with nitrotyrosine-containing catalytic subunits. The addition of ATP in the absence of substrate increased the absorbance at 430 nm, the same change elicited by the addition of succinate (see the graph in Problem 17). Conversely, CTP in the absence of substrate decreased the absorbance at 430 nm. What is the significance of the changes in absorption of the reporter groups?.

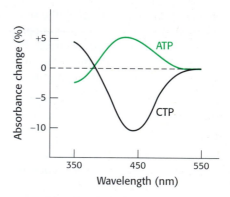

[After H. K. Schachman. *J. Biol. Chem.* 263(1988):18583.]

Chapter Integration Problem

19. *Sticky patches.* The substitution of valine for glutamate at position 6 of the β chains of hemoglobin places a nonpolar residue on the outside of hemoglobin S, the version of hemoglobin responsible for sickle-cell anemia. The oxygen affinity and allosteric properties of hemoglobin are virtually unaffected by this change. However, the valine side chain of hemoglobin S interacts with a complementary sticky patch (formed by phenylalanine β85 and leucine β88) on another hemoglobin molecule—a patch that is exposed in deoxygenated but not in oxygenated hemoglobin. What is the chemical basis for the interaction between the hemoglobin molecules? What would be the effect of this interaction?

Mechanism Problems

20. *Aspartate transcarbamoylase.* Write the mechanism (in detail) for the conversion of aspartate and carbamoyl phosphate into N-carbamoylaspartate. Include a role for the residue histidine present in the active site.

21. *Protein kinases.* Write a mechanism (in detail) for the phosphorylation of a serine residue by ATP catalyzed by a protein kinase. What groups might you expect to find in the enzyme active site?

Media Problem

22. *Cooperative, more or less.* (a) Suppose you purify the same enzyme from two different species and find that one is a tetramer ($n = 4$) and the other a hexamer ($n = 6$). All other things being equal, which would you expect would show the greater cooperativity? (b) To your surprise, both enzymes show similar cooperativity. Explain how differences in the MWC parameter c could explain this. (*Hint:* Make sure you do the exercise in Section 4 of the **Conceptual Insights** module on cooperative binding and kinetics.)

CHAPTER 11

Carbohydrates

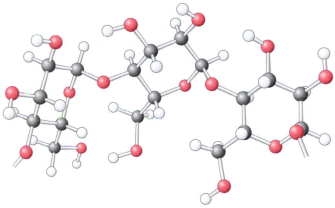

Carbohydrates in food are important sources of energy. Starch, found in plant-derived food such as pasta, consists of chains of linked glucose molecules. These chains are broken down into individual glucose molecules for eventual use in generation of ATP and building blocks for other molecules. [(Left) Superstock.]

Let us take an overview of carbohydrates, one of the four major classes of biomolecules along with proteins, nucleic acids, and lipids. Carbohydrates are aldehyde or ketone compounds with multiple hydroxyl groups. They make up most of the organic matter on Earth because of their extensive roles in all forms of life. First, carbohydrates serve as *energy stores, fuels, and metabolic intermediates.* Second, ribose and deoxyribose sugars form part of the *structural framework of RNA and DNA.* Third, polysaccharides are *structural elements in the cell walls of bacteria and plants.* In fact, cellulose, the main constituent of plant cell walls, is one of the most abundant organic compounds in the biosphere. Fourth, carbohydrates are *linked to many proteins and lipids,* where they play key roles in mediating interactions among cells and interactions between cells and other elements in the cellular environment.

A key related property of carbohydrates in their role as mediators of cellular interactions is the tremendous *structural diversity* possible within this class of molecules. Carbohydrates are built from monosaccharides, small molecules that typically contain from three to nine carbon atoms and vary in size and in the stereochemical configuration at one or more carbon centers. These monosaccharides may be linked together to form a large variety of oligosaccharide structures. The unraveling of these oligosaccharide structures, the discovery of their placement at specific sites within proteins, and the determination of their function are tremendous challenges in the field of proteomics.

11.1 MONOSACCHARIDES ARE ALDEHYDES OR KETONES WITH MULTIPLE HYDROXYL GROUPS

Monosaccharides, the simplest carbohydrates, are aldehydes or ketones that have two or more hydroxyl groups; the empirical formula of many is $(C–H_2O)_n$, literally a "carbon hydrate." Monosaccharides are important fuel molecules as well as building blocks for nucleic acids. The smallest monosaccharides, for which $n = 3$, are dihydroxyacetone and D- and L-glyceraldehyde.

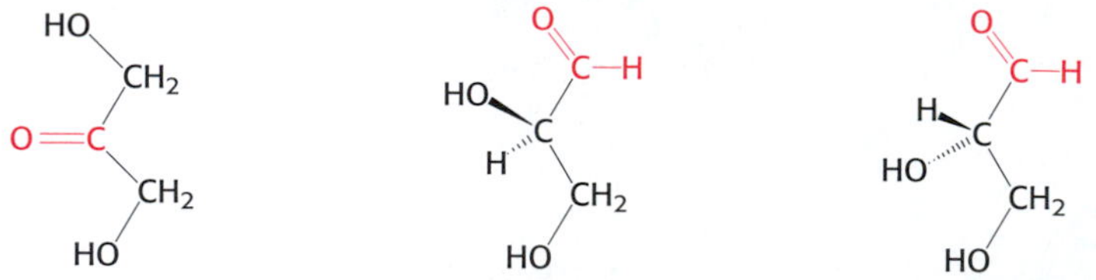

Dihydroxyacetone
(a ketose)

D-Glyceraldehyde
(an aldose)

L-Glyceraldehyde
(an aldose)

They are referred to as *trioses* (tri- for 3). Dihydroxyacetone is called a *ketose* because it contains a keto group, whereas glyceraldehyde is called an *aldose* because it contains an aldehyde group. Glyceraldehyde has a single asymmetric carbon and, thus, there are two stereoisomers of this sugar. D-Glyceraldehyde and L-glyceraldehyde are *enantiomers*, or mirror images of each other. As mentioned in Chapter 3, the prefixes D and L designate the absolute configuration. Monosaccharides and other sugars will often be represented in this book by *Fischer projections* (Figure 11.1). Recall that, in a Fischer projection of a molecule, atoms joined to an asymmetric carbon atom by horizontal bonds are in front of the plane of the page, and those joined by vertical bonds are behind (see the Appendix in Chapter 1). Fischer projections are useful for depicting carbohydrate structures because they provide clear and simple views of the stereochemistry at each carbon center.

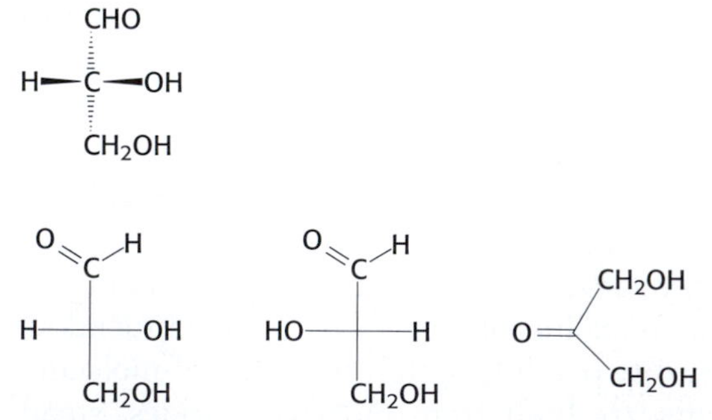

D-Glyceraldehyde   **L-Glyceraldehyde**   **Dihydroxyacetone**

FIGURE 11.1 Fischer projections of trioses. The top structure reveals the stereochemical relations assumed for Fischer projections.

Simple monosaccharides with four, five, six, and seven carbon atoms are called *tetroses*, *pentoses*, *hexoses*, and *heptoses*, respectively. Because these molecules have multiple asymmetric carbons, they exist as *diastereoisomers*, isomers that are not mirror images of each other, as well as enantiomers. In regard to these monosaccharides, the symbols D and L designate the ab-

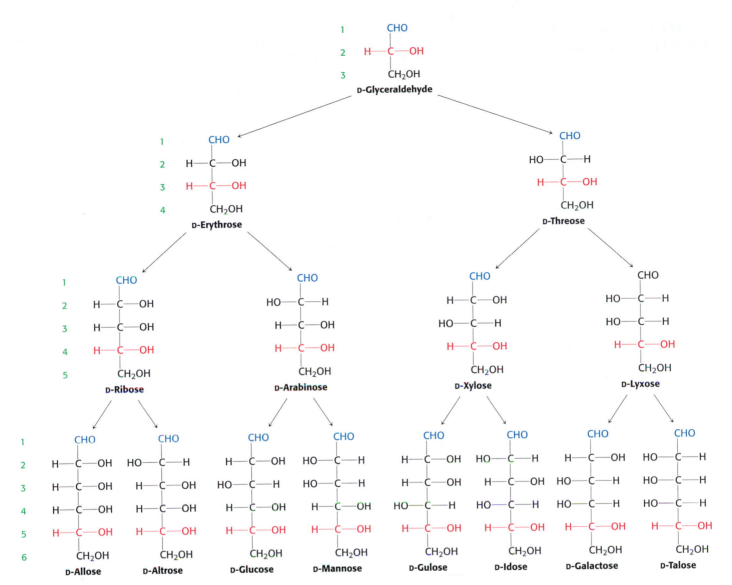

FIGURE 11.2 **D-Aldoses containing three, four, five, and six carbon atoms.**
D-Aldoses contain an aldehyde group (shown in blue) and have the absolute configuration of D-glyceraldehyde at the asymmetric center (shown in red) farthest from the aldehyde group. The numbers indicate the standard designations for each carbon atom.

solute configuration of the asymmetric carbon farthest from the aldehyde or keto group. Figure 11.2 shows the common D-aldose sugars. D-Ribose, the carbohydrate component of RNA, is a five-carbon aldose. D-Glucose, D-mannose, and D-galactose are abundant six-carbon aldoses. Note that D-glucose and D-mannose differ in configuration only at C-2. Sugars differing in configuration at a single asymmetric center are called *epimers*. Thus, D-glucose and D-mannose are epimeric at C-2; D-glucose and D-galactose are epimeric at C-4.

Dihydroxyacetone is the simplest ketose. The stereochemical relation between D-ketoses containing as many as six carbon atoms are shown in Figure 11.3. Note that ketoses have one fewer asymmetric center than do aldoses with the same number of carbons. D-Fructose is the most abundant ketohexose.

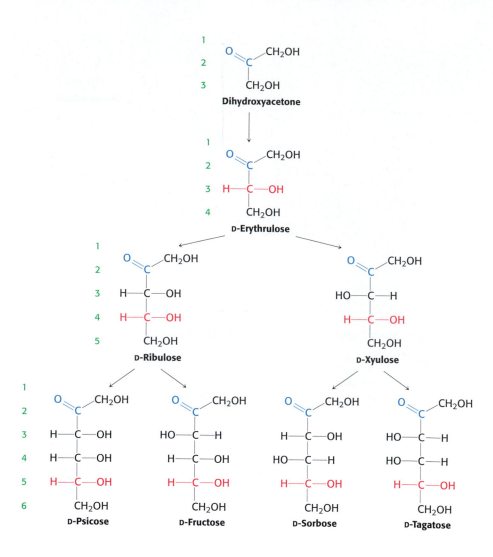

FIGURE 11.3 D-Ketoses containing three- four, five, and six carbon atoms. The keto group is shown in blue. The asymmetric center farthest from the keto group, which determines the D designation, is shown in red.

11.1.1 Pentoses and Hexoses Cyclize to Form Furanose and Pyranose Rings

The predominant forms of ribose, glucose, fructose, and many other sugars in solution are not open chains. Rather, the open-chain forms of these sugars cyclize into rings. In general, an aldehyde can react with an alcohol to form a *hemiacetal*.

Furan

Pyran

For an aldohexose such as glucose, the C-1 aldehyde in the open-chain form of glucose reacts with the C-5 hydroxyl group to form an *intramolecular hemiacetal*. The resulting cyclic hemiacetal, a six-membered ring, is called *pyranose* because of its similarity to *pyran* (Figure 11.4). Similarly, a ketone can react with an alcohol to form a *hemiketal*.

The C-2 keto group in the open-chain form of a ketohexose, such as fructose, can form an *intramolecular hemiketal* by reacting with either the

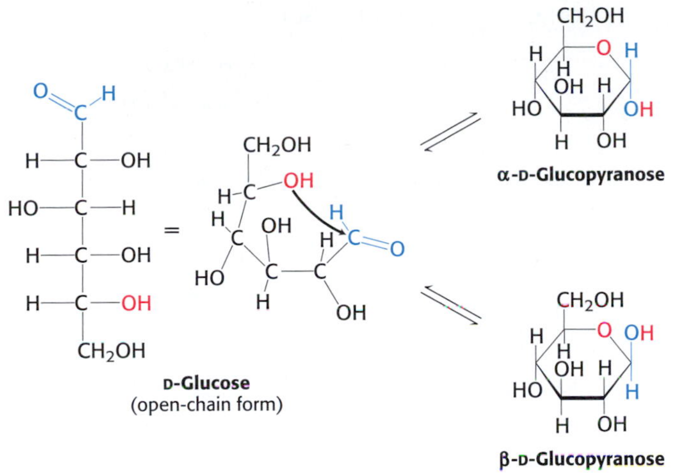

FIGURE 11.4 Pyranose formation. The open-chain form of glucose cyclizes when the C-5 hydroxyl group attacks the oxygen atom of the C-1 aldehyde group to form an intramolecular hemiacetal. Two anomeric forms, designated α and β, can result.

C-6 hydroxyl group to form a six-membered cyclic hemiketal or the C-5 hydroxyl group to form a five-membered cyclic hemiketal (Figure 11.5). The five-membered ring is called a *furanose* because of its similarity to *furan*.

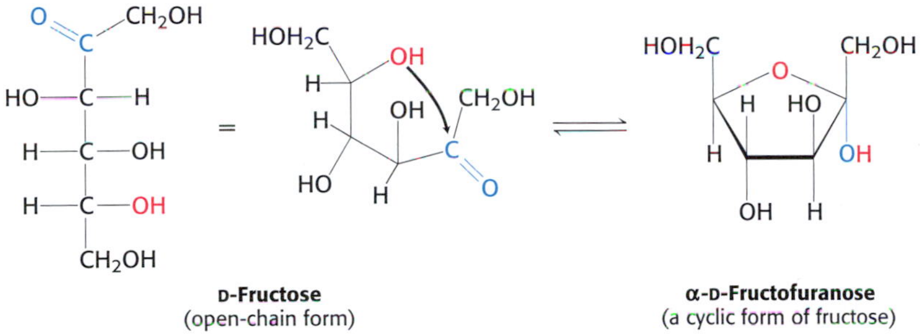

FIGURE 11.5 Furanose formation. The open-chain form of fructose cyclizes to a five-membered ring when the C-5 hydroxyl group attacks the C-2 ketone to form an intramolecular hemiketal. Two anomers are possible, but only the α anomer is shown.

The depictions of glucopyranose and fructofuranose shown in Figures 11.4 and 11.5 are *Haworth projections*. In such projections, the carbon atoms in the ring are not explicitly shown. The approximate plane of the ring is perpendicular to the plane of the paper, with the heavy line on the ring projecting toward the reader. Like Fischer projections, Haworth projections allow easy depiction of the stereochemistry of sugars. We will return to a more structurally realistic view of the conformations of cyclic monosaccharides shortly.

An additional asymmetric center is created when a cyclic hemiacetal is formed. In glucose, C-1, the carbonyl carbon atom in the open-chain form, becomes an asymmetric center. Thus, two ring structures can be formed: α-D-glucopyranose and β-D-glucopyranose (see Figure 11.4). For D sugars drawn as Haworth projections, *the designation α means that the hydroxyl group attached to C-1 is below the plane of the ring; β means that it is above the plane of the ring*. The C-1 carbon atom is called the *anomeric carbon atom*, and the α and β forms are called *anomers*. An equilibrium mixture of glucose contains approximately one-third α anomer, two-thirds β anomer, and <1% of the open-chain form.

The same nomenclature applies to the furanose ring form of fructose, except that α and β refer to the hydroxyl groups attached to C-2, the anomeric carbon atom (see Figure 11.5). Fructose forms both pyranose and furanose rings. The pyranose form predominates in fructose free in solution, and the furanose form predominates in many fructose derivatives (Figure 11.6). Pentoses such as D-ribose and 2-deoxy-D-ribose form furanose rings, as we have seen in the structure of these units in RNA and DNA.

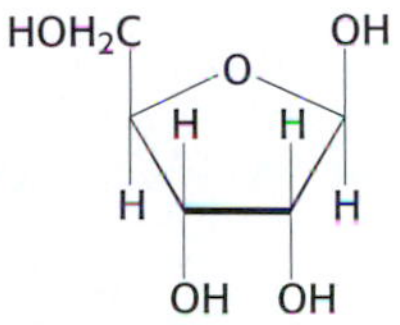

D-Ribose

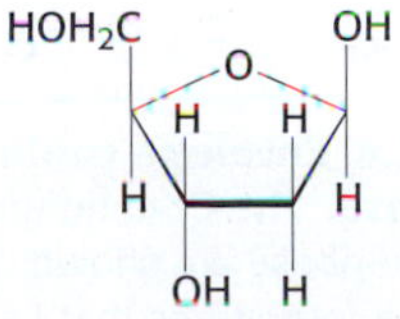

2-Deoxy-D-ribose

α-D-Fructofuranose   β-D-Fructofuranose

α-D-Fructopyranose   β-D-Fructopyranose

FIGURE 11.6 Ring structures of fructose. Fructose can form both five-membered furanose and six-membered pyranose rings. In each case, both α and β anomers are possible.

11.1.2 Conformation of Pyranose and Furanose Rings

The six-membered pyranose ring is not planar, because of the tetrahedral geometry of its saturated carbon atoms. Instead, pyranose rings adopt two classes of conformations, termed *chair* and *boat* because of the resemblance to these objects (Figure 11.7). In the chair form, the substituents on the ring carbon atoms have two orientations: *axial* and *equatorial*. Axial bonds are nearly perpendicular to the average plane of the ring, whereas equatorial bonds are nearly parallel to this plane. Axial substituents sterically hinder each other if they emerge on the same side of the ring (e.g., 1,3-diaxial groups). In contrast, equatorial substituents are less crowded. The chair form of *β*-D-*glucopyranose predominates because all axial positions are occupied by hydrogen atoms.* The bulkier –OH and –CH$_2$OH groups emerge at the less-hindered periphery. The boat form of glucose is disfavored because it is quite sterically hindered.

Steric hindrance

FIGURE 11.7 Chair and boat forms of β-D-glucopyranose. The chair form is more stable because of less steric hindrance as the axial positions are occupied by hydrogen atoms.

Chair form   Boat form

Furanose rings, like pyranose rings, are not planar. They can be puckered so that four atoms are nearly coplanar and the fifth is about 0.5 Å away from this plane (Figure 11.8). This conformation is called an *envelope form* because the structure resembles an opened envelope with the back flap raised. In the ribose moiety of most biomolecules, either C-2 or C-3 is out of the plane on the same side as C-5. These conformations are called C$_2$-endo and C$_3$-endo, respectively.

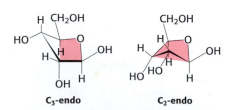

C$_3$-endo   C$_2$-endo

FIGURE 11.8 Envelope conformations of β-D-ribose. The C$_2$-endo and C$_3$-endo forms of β-D-ribose are shown. The color indicates the four atoms that lie approximately in a plane.

11.1.3 Monosaccharides Are Joined to Alcohols and Amines Through Glycosidic Bonds

Monosaccharides can be modified by reaction with alcohols and amines to form *adducts*. For example, D-glucose will react with methanol in an acid-catalyzed process: the anomeric carbon atom reacts with the hydroxyl group of methanol to form two products, methyl α-D-glucopyranoside and methyl β-D-glucopyranoside. These two glucopyranosides differ in the configuration at the anomeric carbon atom. The new bond formed between the anomeric carbon atom of glucose and the hydroxyl oxygen atom of methanol is called a *glycosidic bond*—specifically, an O-*glycosidic bond*. The anomeric carbon atom of a sugar can be linked to the nitrogen atom of an amine to form an N-*glycosidic bond*.

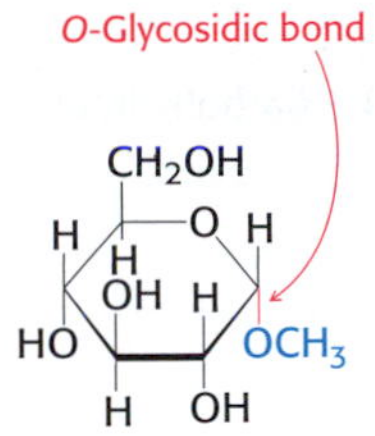

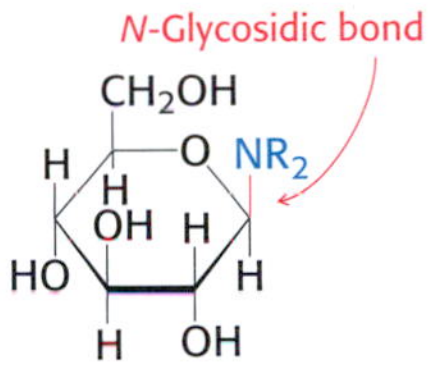

Indeed, we have previously encountered such reaction products; nucleosides are adducts between sugars such as ribose and amines such as adenine (Section 5.1.2). Some other important modified sugars are shown in Figure 11.9. Compounds such as methyl glucopyranoside show differences in reactivity from that of the parent monosaccharide. For example, unmodified glucose reacts with oxidizing agents such as cupric ion (Cu^{2+}) because the open-chain form has a free aldehyde group that is readily oxidized.

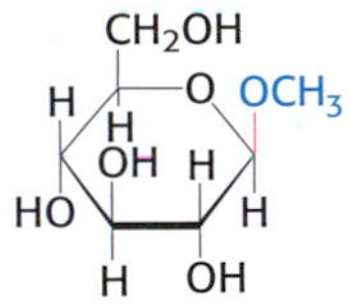

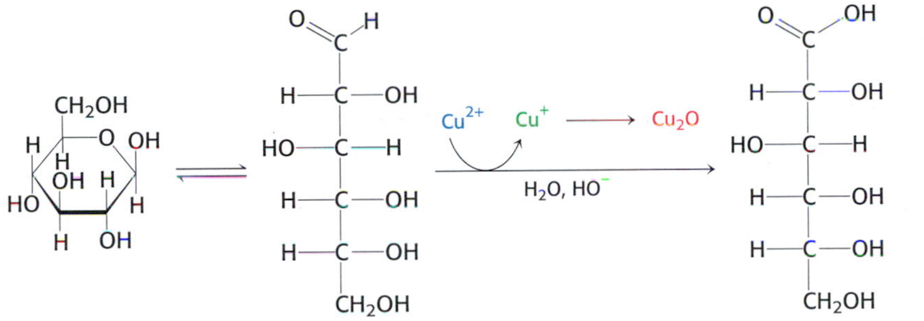

O-Glycosidic bond

Methyl α-D-glucopyranoside

Methyl β-D-glucopyranoside

N-Glycosidic bond

Glycosides such as methyl glucopyranoside do not react, because they are not readily interconverted with a form that includes a free aldehyde group. Solutions of cupric ion (known as Fehling's solution) provide a simple test for sugars such as glucose. Sugars that react are called *reducing sugars*; those that do not are called *nonreducing sugars*.

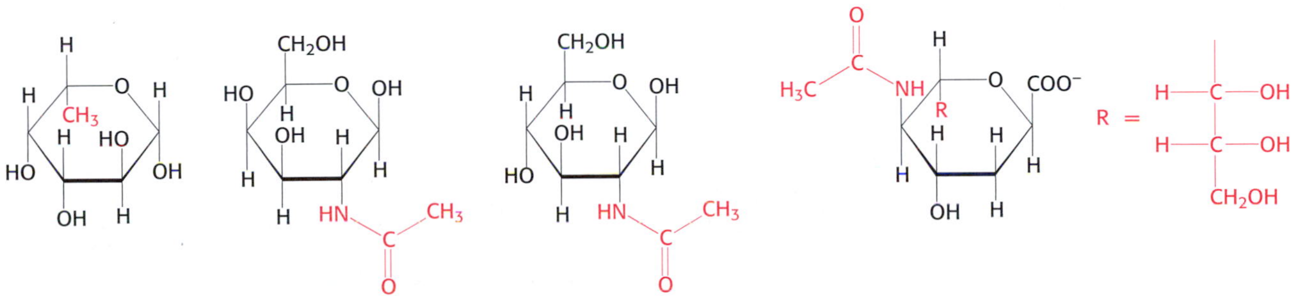

β-L-Fucose (Fuc)	β-D-Acetylgalactosamine (GalNAc)	β-D-Acetylglucosamine (GlcNAc)	Sialic acid (Sia) (N-Acetylneuraminate)

FIGURE 11.9 Modified monosaccharides. Carbohydrates can be modified by the addition of substituents (shown in red) other than hydroxyl groups. Such modified carbohydrates are often expressed on cell surfaces.

11.2 COMPLEX CARBOHYDRATES ARE FORMED BY LINKAGE OF MONOSACCHARIDES

Because sugars contain many hydroxyl groups, glycosidic bonds can join one monosaccharide to another. *Oligosaccharides* are built by the linkage of two or more monosaccharides by O-glycosidic bonds (Figure 11.10). In maltose, for example, two D-glucose residues are joined by a glycosidic linkage between the α-anomeric form of C-1 on one sugar and the hydroxyl oxygen atom on C-4 of the adjacent sugar. Such a linkage is called an α-1,4-glycosidic bond. The fact that monosaccharides have multiple hydroxyl groups means that

Glycosidic bond

FIGURE 11.10 Maltose, a disaccharide. Two molecules of glucose are linked by an α-1,4-glycosidic bond to form the disaccharide maltose.

various glycosidic linkages are possible. Indeed, the wide array of these linkages in concert with the wide variety of monosaccharides and their many isomeric forms makes complex carbohydrates information-rich molecules.

11.2.1 Sucrose, Lactose, and Maltose Are the Common Disaccharides

A *disaccharide* consists of two sugars joined by an *O*-glycosidic bond. Three abundant disaccharides are sucrose, lactose, and maltose (Figure 11.11). *Sucrose* (common table sugar) is obtained commercially from cane or beet. The anomeric carbon atoms of a glucose unit and a fructose unit are joined in this disaccharide; the configuration of this glycosidic linkage is α for glucose and β for fructose. Sucrose can be cleaved into its component monosaccharides by the enzyme *sucrase*.

Lactose, the disaccharide of milk, consists of galactose joined to glucose by a β-1,4-glycosidic linkage. Lactose is hydrolyzed to these monosaccharides by *lactase* in human beings (Section 16.1.12) and by *β-galactosidase* in bacteria. In *maltose*, two glucose units are joined by an α-1,4 glycosidic linkage, as stated earlier. Maltose comes from the hydrolysis of starch and is in turn hydrolyzed to glucose by *maltase*. Sucrase, lactase, and maltase are located on the outer surfaces of epithelial cells lining the small intestine (Figure 11.12).

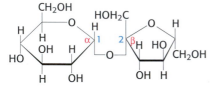

Sucrose
(α-D-Glucopyranosyl-(1→2)-β-D-fructofuranose

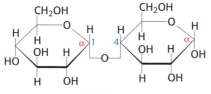

Lactose
(β-D-Galactopyranosyl-(1→4)-α-D-glucopyranose

Maltose
(α-D-Glucopyranosyl-(1→4)-α-D-glucopyranose

FIGURE 11.11 Common disaccharides. Sucrose, lactose, and maltose are common dietary components.

FIGURE 11.12 Electron micrograph of a microvillus. Lactase and other enzymes that hydrolyze carbohydrates are present on microvilli that project from the outer face of the plasma membrane of intestinal epithelial cells. [From M. S. Mooseker and L. G. Tilney, *J. Cell. Biol.* 67(1975):725.]

11.2.2 Glycogen and Starch Are Mobilizable Stores of Glucose

Large polymeric oligosaccharides, formed by the linkage of multiple monosaccharides, are called *polysaccharides*. Polysaccharides play vital roles in energy storage and in maintaining the structural integrity of an organism. If all of the monosaccharides are the same, these polymers are called *homopolymers*. The most common homopolymer in animal cells is *glycogen*, the storage form

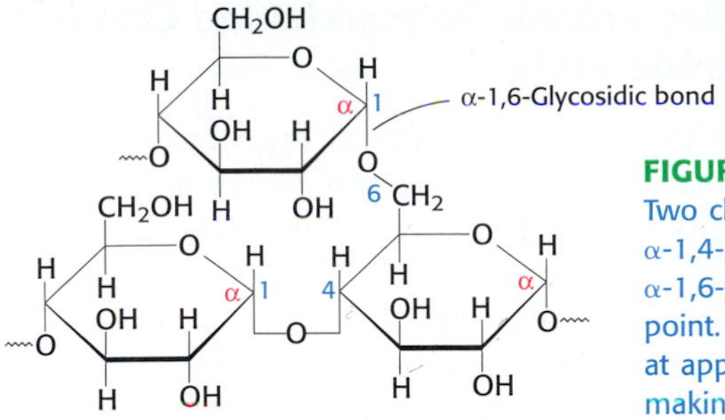

FIGURE 11.13 Branch point in glycogen.
Two chains of glucose molecules joined by α-1,4-glycosidic bonds are linked by an α-1,6-glycosidic bond to create a branch point. Such an α-1,6-glycosidic bond forms at approximately every 10 glucose units, making glycogen a highly branched molecule.

of glucose. As will be considered in detail in Chapter 21, glycogen is a very large, branched polymer of glucose residues. Most of the glucose units in glycogen are linked by α-1,4-glycosidic bonds. The branches are formed by α-1,6-glycosidic bonds, present about once in 10 units (Figure 11.13).

The nutritional reservoir in plants is *starch,* of which there are two forms. *Amylose,* the unbranched type of starch, consists of glucose residues in α-1,4 linkage. *Amylopectin,* the branched form, has about 1 α-1,6 linkage per 30 α-1,4 linkages, in similar fashion to glycogen except for its lower degree of branching. More than half the carbohydrate ingested by human beings is starch. Both amylopectin and amylose are rapidly hydrolyzed by α-amylase, an enzyme secreted by the salivary glands and the pancreas.

11.2.3 Cellulose, the Major Structural Polymer of Plants, Consists of Linear Chains of Glucose Units

Cellulose, the other major polysaccharide of glucose found in plants, serves a structural rather than a nutritional role. *Cellulose is one of the most abundant organic compounds in the biosphere.* Some 10^{15} kg of cellulose is synthesized and degraded on Earth each year. It is an unbranched polymer of glucose residues joined by β-1,4 linkages. The β configuration allows cellulose to form very long, straight chains. Fibrils are formed by parallel chains that interact with one another through hydrogen bonds. The α-1,4 linkages in glycogen and starch produce a very different molecular architecture from that of cellulose. A hollow helix is formed instead of a straight chain (Figure 11.14). These differing consequences of the α and β linkages are biologically important. The straight chain formed by β linkages is optimal for the construction of fibers having a high tensile strength. In contrast, the open helix formed by α linkages is well suited to forming an accessible store of sugar. Mammals lack cellulases and therefore cannot digest wood and vegetable fibers.

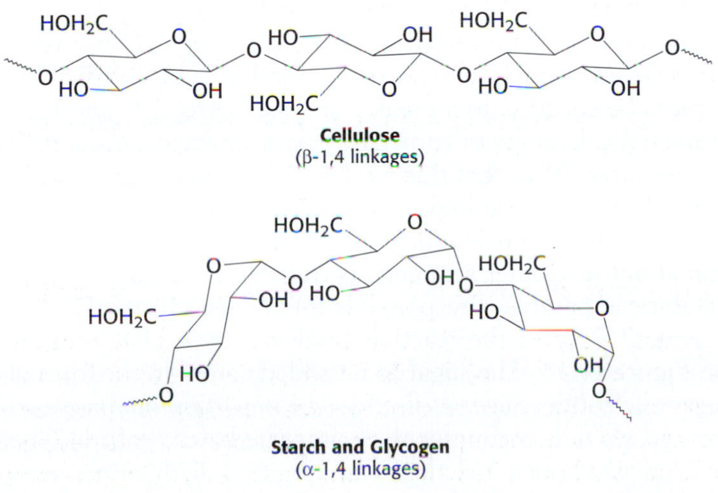

Cellulose
(β-1,4 linkages)

Starch and Glycogen
(α-1,4 linkages)

FIGURE 11.14 Glycosidic bonds determine polysaccharide structure. The β-1,4 linkages favor straight chains, which are optimal for structural purposes. The α-1,4 linkages favor bent structures, which are more suitable for storage.

11.2.4 Glycosaminoglycans Are Anionic Polysaccharide Chains Made of Repeating Disaccharide Units

A different kind of repeating polysaccharide is present on the animal cell surface and in the extracellular matrix. Many *glycosaminoglycans* are made of *disaccharide repeating units* containing a derivative of an *amino sugar,* either glucosamine or galactosamine (Figure 11.15). At least one of the sugars in the repeating unit has a *negatively charged carboxylate or sulfate group.* Chondroitin sulfate, keratan sulfate, heparin, heparan sulfate, dermatan sulfate, and hyaluronate are the major glycosaminoglycans.

Glycosaminoglycans are usually attached to proteins to form *proteoglycans.* Heparin is synthesized in a nonsulfated form, which is then deacetylated and sulfated. Incomplete modification leads to a mixture of variously sulfated sequences. Some of them act as anticoagulants by binding specifically to antithrombin, which accelerates its sequestration of thrombin (Section 10.5.6). Heparan sulfate is like heparin except that it has fewer N- and O-sulfate groups and more acetyl groups.

Proteoglycans resemble polysaccharides more than proteins in as much as the carbohydrate makes up as much as 95% of the biomolecule by weight. Proteoglycans function as lubricants and structural components in connective tissue, mediate adhesion of cells to the extracellular matrix, and bind factors that stimulate cell proliferation.

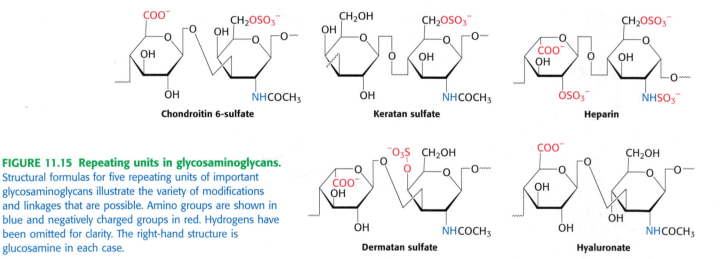

FIGURE 11.15 Repeating units in glycosaminoglycans. Structural formulas for five repeating units of important glycosaminoglycans illustrate the variety of modifications and linkages that are possible. Amino groups are shown in blue and negatively charged groups in red. Hydrogens have been omitted for clarity. The right-hand structure is glucosamine in each case.

11.2.5 Specific Enzymes Are Responsible for Oligosaccharide Assembly

Oligosaccharides are synthesized through the action of specific enzymes, *glycosyltransferases,* which catalyze the formation of glycosidic bonds. Each enzyme must be specific, to a greater or lesser extent, to the sugars being linked. Given the diversity of known glycosidic linkages, many different enzymes are required. Note that this mode of assembly stands in contrast with those used for the other biological polymers heretofore discussed—that is, polypeptides and oligonucleotides. As these polymers are assembled, information about monomer sequence is transferred from a template, and a single catalytic apparatus is responsible for all bond formation.

The general form of the reaction catalyzed by a glycosyltransferase is shown in Figure 11.16. The sugar to be added comes in the form of an activated sugar nucleotide. *Sugar nucleotides are important intermediates in many processes,* and we will encounter these intermediates again in Chapters 16 and 21. Note that such reactions can proceed with either retention or

UDP-glucose

UDP

inversion of configuration at the glycosidic carbon atom at which the new bond is formed; a given enzyme proceeds by one stereochemical path or the other.

The human ABO blood groups illustrate the effects of glycosyltransferases. Carbohydrates are attached to glycoproteins and glycolipids on the surfaces of red blood cells. For one type of blood group, one of the three different structures, termed A, B, and O, may be present (Figure 11.17). These structures have in common an oligosaccharide foundation called the O (or sometimes H) antigen. The A and B antigens differ from the O antigen by the addition of one extra monosaccharide, either N-acetylgalactosamine (for A) or galactose (for B) through an α-1,3 linkage to a galactose moiety of the O antigen.

Specific glycosyltransferases add the extra monosaccharide to the O antigen. Each person inherits the gene for one glycosyltransferase of this type from each parent. The type A transferase specifically adds N-acetylgalactosamine, whereas the type B transferase adds galactose. These enzymes are identical in all but 4 of 354 positions. The O phenotype is the result of a mutation that leads to premature termination of translation and, hence, to the production of no active glycosyltransferase.

These structures have important implications for blood transfusions and other transplantation procedures. If an antigen not normally present in a person is introduced, the person's immune system recognizes it as foreign. Adverse reactions can ensue, initiated by the intravascular destruction of the incompatible red blood cells.

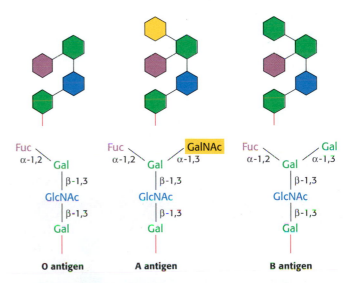

FIGURE 11.17 **Structures of A, B, and O oligosaccharide antigens.** Abbreviations: Fuc, fucose; Gal, galactose; GalNAc, N-acetylgalactosamine; GlcNAc, N-acetylglucosamine.

Why are different blood types present in the human population? Suppose that a pathogenic organism such as a parasite expresses on its cell surface a carbohydrate antigen similar to one of the blood-group

Asn   **Ser**

N-linked GlcNAc   **O-linked GalNAc**

FIGURE 11.18 Glycosidic bonds between proteins and carbohydrates. A glycosidic bond links a carbohydrate to the side chain of asparagine (*N*-linked) or to the side chain of serine or threonine (*O*-linked). The glycosidic bonds are shown in red.

Abbreviations for sugars—

Fuc	Fucose
Gal	Galactose
GalNAc	*N*-Acetylgalactosamine
Glc	Glucose
GlcNAc	*N*-Acetylglucosamine
Man	Mannose
Sia	Sialic acid
NeuNAc	*N*-Acetylneuraminate (sialic acid)

antigens. This antigen may not be readily detected as foreign in a person with the blood type that matches the parasite antigen, and the parasite will flourish. However, other people with different blood types will be protected. Hence, there will be selective pressure on human beings to vary blood type to prevent parasitic mimicry and a corresponding selective pressure on parasites to enhance mimicry. The constant "arms race" between pathogenic microorganisms and human beings drives the evolution of diversity of surface antigens within the human population.

11.3 CARBOHYDRATES CAN BE ATTACHED TO PROTEINS TO FORM GLYCOPROTEINS

Carbohydrate groups are covalently attached to many different proteins to form *glycoproteins*. Carbohydrates are a much smaller percentage of the weight of glycoproteins than of proteoglycans. Many glycoproteins are components of cell membranes, where they play a variety of roles in processes such as cell adhesion and the binding of sperm to eggs.

11.3.1 Carbohydrates May Be Linked to Proteins Through Asparagine (*N*-Linked) or Through Serine or Threonine (*O*-Linked) Residues

In glycoproteins, sugars are attached either to the amide nitrogen atom in the side chain of asparagine (termed an N-*linkage*) or to the oxygen atom in the side chain of serine or threonine (termed an O-*linkage*), as shown in Figure 11.18. An asparagine residue can accept an oligosaccharide only if the residue is part of an Asn-X-Ser or Asn-X-Thr sequence, in which X can be any residue. Thus, *potential glycosylation sites can be detected within amino acid sequences.* However, which of these potential sites is actually glycosylated depends on other aspects of the protein structure and on the cell type in which the protein is expressed. All *N*-linked oligosaccharides have in common a pentasaccharide core consisting of three mannose and two *N*-acetylglucosamine residues. Additional sugars are attached to this core to form the great variety of oligosaccharide patterns found in glycoproteins (Figure 11.19).

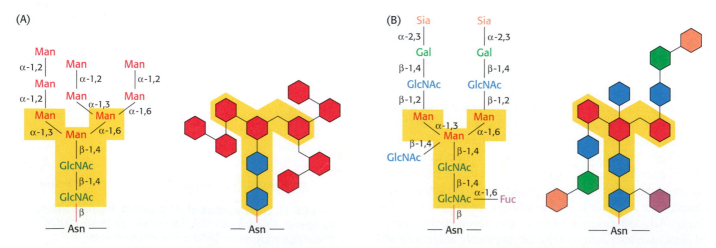

FIGURE 11.19 *N*-linked oligosaccharides. A pentasaccharide core (shaded yellow) is common to all *N*-linked oligosaccharides and serves as the foundation for a wide variety of *N*-linked oligosaccharides, two of which are illustrated: (A) high-mannose type; (B) complex type. Detailed chemical formulas and schematic structures are shown for each type.

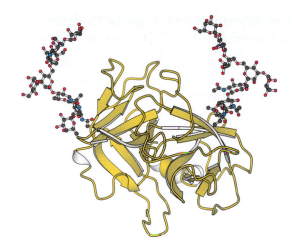

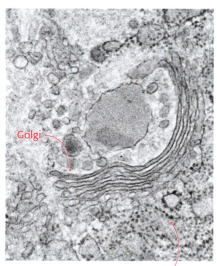

FIGURE 11.20 Elastase, a secreted glycoprotein, showing linked carbohydrates on its surface. Elastase is a protease found in serum. Note that the oligosaccharide chains have substantial size even for this protein, which has a relatively low level of glycosylation.

Carbohydrates are linked to some soluble proteins as well as membrane proteins. In particular, many of the proteins secreted from cells are glycosylated. Most proteins present in the serum component of blood are glycoproteins (Figure 11.20). Furthermore, N-acetylglucosamine residues are O-linked to some intracellular proteins. The role of these carbohydrates, which are dynamically added and removed, is under active investigation.

11.3.2 Protein Glycosylation Takes Place in the Lumen of the Endoplasmic Reticulum and in the Golgi Complex

Protein glycosylation takes place inside the lumen of the *endoplasmic reticulum* (ER) and the *Golgi complex*, organelles that play central roles in protein trafficking (Figure 11.21). One such glycoprotein (depicted in Figure 11.20) is the proteolytic enzyme elastase (Section 9.1.4), which is secreted by the pancreas as a zymogen (Section 10.5). This protein is synthesized by ribosomes attached to the cytoplasmic face of the ER membrane, and the peptide chain is inserted into the lumen of the ER as it grows, guided by a signal sequence of 29 amino acids at the amino terminus. This signal sequence, which directs the protein through a channel in the ER membrane, is cleaved from the protein in the transport process into the ER (Figure 11.22). After the protein has entered the ER, the glycosylation process begins. The N-linked glycosylation begins in the ER and continues in the Golgi complex, whereas the O-linked glycosylation takes place exclusively in the Golgi complex.

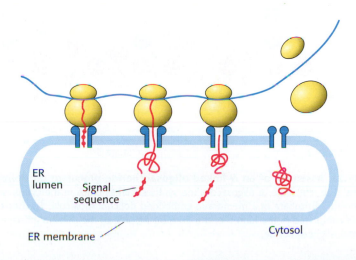

Golgi

Endoplasmic reticulum

FIGURE 11.21 Golgi complex and endoplasmic reticulum. The electron micrograph shows the Golgi complex and adjacent endoplasmic reticulum. The black dots on the cytoplasmic surface of the ER membrane are ribosomes. [Micrograph courtesy of Lynne Mercer.]

ER lumen

Signal sequence

ER membrane

Cytosol

FIGURE 11.22 Transport into the endoplasmic reticulum. As translation takes place, a signal sequence on membrane and secretory proteins directs the nascent protein through channels in the ER membrane and into the lumen. In most cases, the signal sequence is subsequently cleaved and degraded.

11.3.3 *N*-Linked Glycoproteins Acquire Their Initial Sugars from Dolichol Donors in the Endoplasmic Reticulum

A large oligosaccharide destined for attachment to the asparagine residue of a protein is assembled attached to *dolichol phosphate,* a specialized lipid molecule containing as many as 20 isoprene (C_5) units (Section 26.4.8).

$n = 15–19$

Dolichol phosphate

The terminal phosphate group is the site of attachment of the activated oligosaccharide, which is subsequently transferred to the protein acceptor. Dolichol phosphate resides in the ER membrane with its phosphate terminus on the cytoplasmic face.

The assembly process proceeds in three stages. First, 2 N-acetylglucosamine residues and 5 mannose residues are added to the dolichol phosphate through the action of a number of cytoplasmic enzymes that catalyze monosaccharide transfer from sugar nucleotides. Then, in a remarkable (and, as yet, not well understood) process, this large structure is "flipped" through the ER membrane into the lumen of the ER. Finally, additional sugars are added by enzymes in the ER lumen, this time with the use of monosaccharides activated by attachment to dolichol phosphate. This process ends with the formation of a 14-residue oligosaccharide attached to dolichol phosphate (Figure 11.23).

The 14-sugar-residue precursor attached to this dolichol phosphate intermediate is then transferred en bloc to a specific asparagine residue of the growing polypeptide chain. In regard to elastase, oligosaccharides are linked

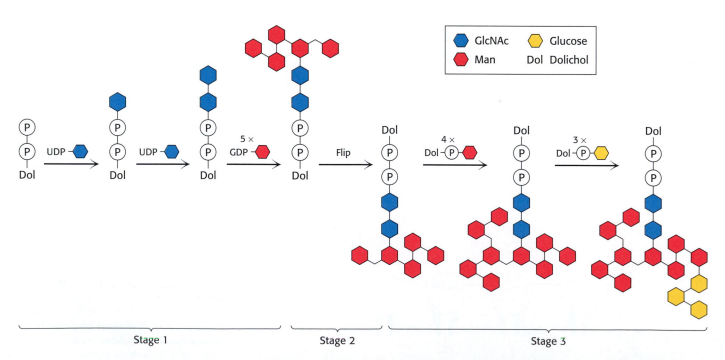

FIGURE 11.23 Assembly of an *N*-linked oligosaccharide precursor on dolichol phosphate. The first stage of oligosaccharide synthesis takes place in the cytoplasm on the exposed phosphate of a membrane-embedded dolichol molecule. Synthesis of the precursor is completed in the lumen of the ER after flipping of the dolichol phosphate and attached oligosaccharide.

to the asparagine residues in the recognition sequences Asn 109-Gly-Ser and Asn 159-Val-Thr. Both the activated sugars and the complex enzyme that is responsible for transferring the oligosaccharide to the protein are located on the lumenal side of the ER, accounting for the fact that proteins in the cytosol are not glycosylated by this pathway. Before the glycoprotein leaves the lumen of the ER, 3 glucose molecules are removed from the 14-residue oligosaccharide. As we will see in Section 11.3.6, the sequential removal of these glucose molecules is a quality-control step that ensures that only properly folded glycoproteins are further processed.

Dolichol pyrophosphate released in the transfer of the oligosaccharide to the protein is recycled to dolichol phosphate by the action of a phosphatase. This hydrolysis is blocked by *bacitracin,* an antibiotic. Another interesting antibiotic inhibitor of *N*-glycosylation is *tunicamycin,* a hydrophobic analog of the sugar nucleotide uridine diphosphate–*N*-acetylglucosamine (UDP-GlcNAc), the activated form of *N*-acetylglucosamine used as a substrate for the enzymes that synthesize the oligosaccharide unit on dolichol phosphate. Tunicamycin blocks the addition of *N*-acetylglucosamine to dolichol phosphate, the first step in the formation of the core oligosaccharide.

11.3.4 Transport Vesicles Carry Proteins from the Endoplasmic Reticulum to the Golgi Complex for Further Glycosylation and Sorting

Proteins in the lumen of the ER and in the ER membrane are transported to the Golgi complex, which is a stack of flattened membranous sacs. The Golgi complex has two principal roles. First, *carbohydrate units of glycoproteins are altered and elaborated in the Golgi complex.* The O-linked sugar units are fashioned there, and the *N*-linked sugars, arriving from the ER as a component of a glycoprotein, are modified in many different ways. Second, *the Golgi complex is the major sorting center of the cell.* Proteins proceed from the Golgi complex to lysosomes, secretory granules (as is the case for the elastase zymogen), or the plasma membrane, according to signals encoded within their amino acid sequences and three-dimensional structures (Figure 11.24).

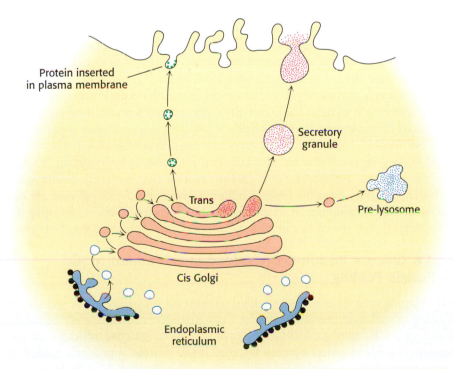

FIGURE 11.24 Golgi complex as sorting center. The Golgi complex is the sorting center in the targeting of proteins to lysosomes, secretory vesicles, and the plasma membrane. The cis face of the Golgi complex receives vesicles from the ER, and the trans face sends a different set of vesicles to target sites. Vesicles also transfer proteins from one compartment of the Golgi complex to another. [Courtesy of Dr. Marilyn Farquhar.]

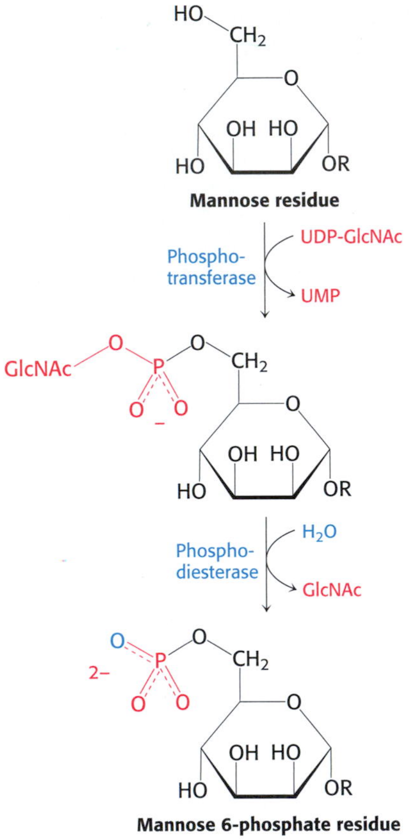

FIGURE 11.25 Formation of a mannose 6-phosphate marker. A glycoprotein destined for delivery to lysosomes acquires a phosphate marker in the cis Golgi compartment in a two-step process. First, a phosphotransferase adds a phospho-N-acetylglucosamine unit to the 6-OH group of a mannose, and then a phosphodiesterase removes the added sugar to generate a mannose 6-phosphate residue in the core oligosaccharide.

The Golgi complex of a typical mammalian cell has 3 or 4 membranous sacs (cisternae), and those of many plant cells have about 20. The Golgi complex is differentiated into (1) a *cis* compartment, the receiving end, which is closest to the ER; (2) *medial* compartments; and (3) a *trans* compartment, which exports proteins to a variety of destinations. These compartments contain different enzymes and mediate distinctive functions.

The N-linked carbohydrate units of glycoproteins are further modified in each of the compartments of the Golgi complex. In the cis Golgi compartment, three mannose residues are removed from the oligosaccharide chains of proteins destined for secretion or for insertion in the plasma membrane. The carbohydrate units of glycoproteins targeted to the lumen of lysosomes are further modified. In the medial Golgi compartments of some cells, two more mannose residues are removed, and two N-acetylglucosamines residues and a fucose residue are added. Finally, in the trans Golgi, another N-acetylglucosamine residue can be added, followed by galactose and sialic acid, to form a complex oligosaccharide unit. The sequence of N-linked oligosaccharide units of a glycoprotein is determined both by (1) the sequence and conformation of the protein undergoing glycosylation and by (2) the glycosyltransferases present in the Golgi compartment in which they are processed. Note that, despite all of this processing, N-glycosylated proteins have in common a pentasaccharide core (see Figure 11.19). Carbohydrate processing in the Golgi complex is called *terminal glycosylation* to distinguish it from *core glycosylation*, which takes place in the ER. Tremendous structural diversification can occur as a result of the terminal glycosylation process.

11.3.5 Mannose 6-phosphate Targets Lysosomal Enzymes to Their Destinations

A carbohydrate marker directs certain proteins from the Golgi complex to lysosomes. A clue to the identity of this marker came from analyses of *I-cell disease* (also called *mucolipidosis II*), a lysosomal storage disease. *Lysosomes* are organelles that degrade and recycle damaged cellular components or material brought into the cell by endocytosis. Patients with I-cell disease suffer severe psychomotor retardation and skeletal deformities. Their lysosomes contain large *inclusions* of undigested glycosaminoglycans (Section 11.2.4) and glycolipids (Section 12.3.3)—hence the "I" in the name of the disease. These inclusions are present because at least eight acid hydrolases required for their degradation are missing from affected lysosomes. In contrast, very high levels of the enzymes are present in the blood and urine. Thus, active enzymes are synthesized, but they are exported instead of being sequestered in lysosomes. In other words, *a whole series of enzymes is mislocated in I-cell disease*. Normally, these enzymes contain a mannose 6-phosphate residue, but, in I-cell disease, the attached mannose is unmodified (Figure 11.25). *Mannose 6-phosphate is in fact the marker that normally directs many hydrolytic enzymes from the Golgi complex to lysosomes. I-cell patients are deficient in the phosphotransferase catalyzing the first step in the addition of the phosphoryl group; the consequence is the mistargeting of eight essential enzymes.*

11.3.6 Glucose Residues Are Added and Trimmed to Aid in Protein Folding

The oligosaccharide precursors added to proteins may play a role in protein folding as well as in protein targeting. As we have seen, before a glycoprotein leaves the ER, two glucosidases cleave the three glucose residues of the oligosaccharide in a step-by-step fashion. If the protein is properly folded, it moves to the Golgi complex for further processing (Section 11.3.3). However,

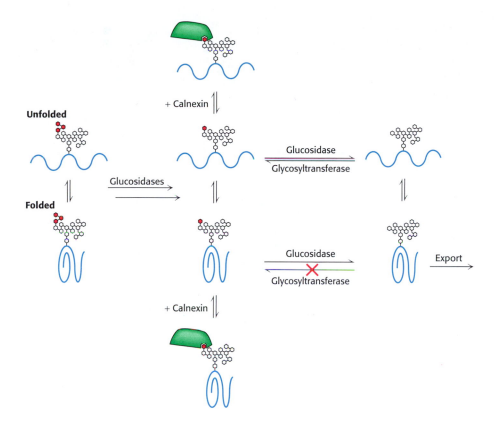

FIGURE 11.26 Quality-control system for protein folding in the ER. A properly folded glycoprotein will move to the Golgi complex after the removal of glucose moieties (shown in red). An unfolded or misfolded protein will receive a glucose residue, through the action of a glucosyltransferase. Such glucosylated glycoproteins bind to calnexin (or the related protein calreticulin), which serves as a chaperone to allow multiple attempts to attain correct folding. Properly folded proteins are not reglucosylated.

if the protein is sufficiently unfolded that the oligosaccharide can act as a substrate for glucosyltransferase, another enzyme residing in the lumen of the ER, a glucose residue will be reattached (Figure 11.26). This residue, in turn, is bound by one of two chaperone proteins called *calnexin* and *calreticulin*. Calnexin, the more fully understood of the two proteins, is membrane bound, whereas calreticulin is a soluble component of the ER lumen. Unfolded proteins held by these carbohydrate-binding proteins (lectins, Section 11.4) cannot leave the ER, giving the unfolded proteins time to fold properly. When a chaperone releases the bound protein, the glucose residue will be cleaved by a glucosidase. If the folding is correct, the protein moves to the Golgi complex. Otherwise, the protein will repeat another cycle of glucose addition and binding until the glucose-free (and, hence, properly folded) protein can be translocated to the Golgi complex. This quality-control system reveals an important principle: *carbohydrates carry information.* Here, the availability of carbohydrates to specific glycosyltransferases conveys information about the folding state of the protein. Moreover, we see the reiteration of a theme in the control of protein folding: other chaperone proteins rely on the same essential mechanism of allowing misfolded proteins multiple attempts to reach a folded state (Section 3.6), even though carbohydrate modification is not a part of their reaction cycles.

11.3.7 Oligosaccharides Can Be "Sequenced"

Given the large diversity of oligosaccharide structures and the many possible points of attachment to most proteins, how is it possible to determine the structure of a glycoprotein? Most approaches are based on the use of enzymes that cleave oligosaccharides at specific types of linkages. For example, N-linked oligosaccharides can be released from proteins by an enzyme such as Peptide N-glycosidase F, which cleaves the N-glycosidic bonds linking the oligosaccharide to the protein. The oligosaccharides can then be isolated and analyzed. Through the use of MALDI-TOF or other mass spectrometric

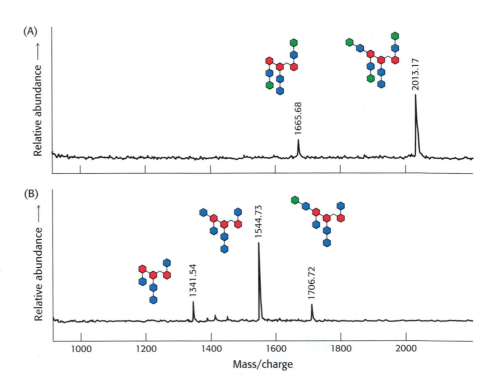

FIGURE 11.27 Mass spectrometric "sequencing" of oligosaccharides. Carbohydrate-cleaving enzymes were used to release and specifically cleave the oligosaccharide component of the glycoprotein fetuin from bovine serum. Parts A and B show the masses obtained with MALDI-TOF spectrometry as well as the corresponding structures of the oligosaccharide digestion products (using the same scheme as that in Figure 11.19): (A) digestion with Peptide *N*-glycosidase F (to release the oligosaccharide from the protein) and neuraminidase; (B) digestion with Peptide *N*-glycosidase F, neuraminidase, and β-1,4-galactosidase. Knowledge of the enzyme specificities and the masses of the products permits the characterization of the oligosaccharide. [After A. Varki, R. Cummings, J. Esko, H. Freeze, G. Hart, and J. Marth (Eds.), *Essentials of Glycobiology* (Cold Spring Harbor Laboratory Press, 1999), p. 596.]

techniques (Section 4.1.7), the mass of an oligosaccharide fragment can be determined. However, given the large number of potential monosaccharide combinations, many possible oligosaccharide structures are consistent with a given mass. More complete information can be obtained by cleaving the oligosaccharide with enzymes of varying specificities. For example, β-1,4-galactosidase cleaves β-glycosidic bonds exclusively at galactose residues. The products can again be analyzed by mass spectrometry (Figure 11.27). The repetition of this process with the use of an array of enzymes of different specificity will eventually reveal the structure of the oligosaccharide.

The points of oligosaccharide attachment can be determined through the use of proteases. Cleavage of a protein by applying specific proteases yields a characteristic pattern of peptide fragments that can be analyzed chromatographically (Section 4.2.1). The chromatographic properties of peptides attached to oligosaccharides will change on glycosidase treatment. Mass spectrometric analysis or direct peptide sequencing can reveal the identity of the peptide in question and, with additional effort, the exact site of oligosaccharide attachment.

Posttranslational modifications such as glycosylation greatly increase the complexity of the proteome. A given protein with several potential glycosidation sites can have many different glycosylated forms (sometimes called *glycoforms*), each of which may be generated only in a specific cell type or developmental stage. Now that the sequencing of the human genome is essentially complete, the characterization of the much more complex proteome, including the biological roles of specifically modified proteins, can begin in earnest.

11.4 LECTINS ARE SPECIFIC CARBOHYDRATE-BINDING PROTEINS

The diverse carbohydrate structures displayed on cell surfaces are well suited to serve as interaction sites between cells and their environments. Proteins termed *lectins* (from the Latin *legere*, "to select") are the partners that bind specific carbohydrate structures. Lectins are ubiquitous, being found in animals, plants, and microorganisms. We have already seen that some lectins, such as calnexin, function as chaperones in protein folding (Section 11.3.6).

11.4.1 Lectins Promote Interactions Between Cells

The chief function of lectins in animals is to facilitate cell–cell contact. A lectin usually contains two or more binding sites for carbohydrate units; some lectins form oligomeric structures with multiple binding sites. The binding sites of lectins on the surface of one cell interact with arrays of carbohydrates displayed on the surface of another cell. Lectins and carbohydrates are linked by a number of relatively weak interactions that ensure specificity yet permit unlinking as needed. The interactions between one cell surface with carbohydrates and another with lectins resemble the action of Velcro; each interaction is relatively weak but the composite is strong.

The exact role of lectins in plants is unclear, although they can serve as potent insecticides. Castor beans contain so much lectin that they are toxic to most organisms. The binding specificities of lectins from plants have been well characterized (Figure 11.28). Bacteria, too, contain lectins. *Escherichia coli* bacteria are able to adhere to epithelial cells of the gastrointestinal tract because lectins on the *E. coli* surface recognize oligosaccharide units on the surfaces of target cells. These lectins are located on slender hairlike appendages called *fimbriae* (*pili*).

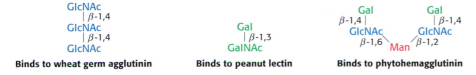

Binds to wheat germ agglutinin **Binds to peanut lectin** **Binds to phytohemagglutinin**

FIGURE 11.28 Binding selectivities of plant lectins. The plant lectins wheat germ agglutinin, peanut lectin, and phytohemagglutinin recognize different oligosaccharides.

Lectins can be divided into classes on the basis of their amino acid sequences and biochemical properties. One large class is the C type (for *calcium-requiring*) found in animals. These proteins have in common a domain of 120 amino acids that is responsible for carbohydrate binding. The structure of one such domain bound to a carbohydrate target is shown in Figure 11.29.

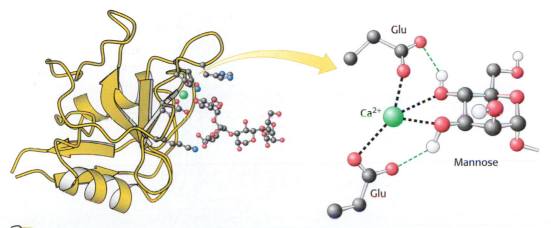

FIGURE 11.29 Structure of a C-type carbohydrate-binding domain from an animal lectin. A calcium ion links a mannose residue to the lectin. Selected interactions are shown, with some hydrogen atoms omitted for clarity.

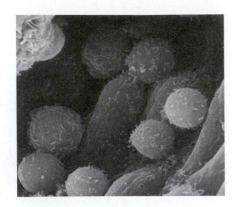

FIGURE 11.30 Selectins mediate cell–cell interactions. The scanning electron micrograph shows lymphocytes adhering to the endothelial lining of a lymph node. The L selectins on the lymphocyte surface bind specifically to carbohydrates on the lining of the lymph-node vessels. [Courtesy of Dr. Eugene Butcher.]

A calcium ion acts as a bridge between the protein and the sugar through direct interactions with sugar hydroxyl groups. In addition, two glutamate residues in the protein bind to both the calcium ion and the sugar, while other protein side chains form hydrogen bonds with other hydroxyl groups on the carbohydrate. Changes in the amino acid residues that interact with the carbohydrate alter the carbohydrate-binding specificity of the lectin.

Proteins termed *selectins* are members of the C-type family. Selectins bind immune-system cells to the sites of injury in the inflammatory response (Figure 11.30). The L, E, and P forms of selectins bind specifically to carbohydrates on *lymph-node* vessels, *endothelium*, or activated blood *platelets*, respectively. New therapeutic agents that control inflammation may emerge from a deeper understanding of how selectins bind and distinguish different carbohydrates.

11.4.2 Influenza Virus Binds to Sialic Acid Residues

The ability of viruses to infect specific cell types is dictated in part by the ability of these viruses to bind to particular structures or receptors on the surfaces of cells. In some cases, these receptors are carbohydrates. For example, influenza virus recognizes sialic acid residues present on cell-surface glycoproteins. The viral protein that binds to these sugars is called *hemagglutinin* (Figure 11.31).

After these surface interactions have taken place and the virus has been taken into the cell, another viral protein, *neuraminidase*, cleaves the glycosidic bonds to the sialic acid residues, freeing the virus to infect the cell. Inhibitors of this enzyme are showing some promise as anti-influenza agents.

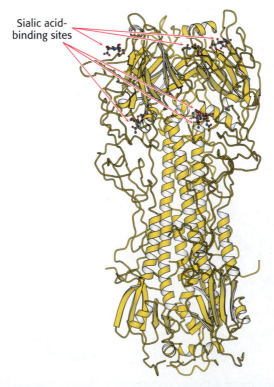

Sialic acid-binding sites

FIGURE 11.31 Structure of a part of influenza hemagglutinin. This viral protein has multiple binding sites for linking to sialic acid residues on the target-cell surface.

SUMMARY

Monosaccharides Are Aldehydes or Ketones with Multiple Hydroxyl Groups

An aldose is a carbohydrate with an aldehyde group (as in glyceraldehyde and glucose), whereas a ketose contains a keto group (as in dihydroxyacetone and fructose). A sugar belongs to the D series if the absolute configuration of its asymmetric carbon farthest from the aldehyde or keto group is the same as that of D-glyceraldehyde. Most naturally occurring sugars belong to the D series. The C-1 aldehyde in the open-chain form of glucose reacts with the C-5 hydroxyl group to form a six-membered pyranose ring. The C-2 keto group in the open-chain form of fructose reacts with the C-5 hydroxyl group to form a five-membered furanose ring. Pentoses such as ribose and deoxyribose also form furanose rings. An additional asymmetric center is formed at the anomeric carbon atom (C-1 in aldoses and C-2 in ketoses) in these cyclizations. The hydroxyl group attached to the anomeric carbon atom is below the plane of the ring (viewed in the standard orientation) in the α anomer, whereas it is above the ring in the β anomer. Not all the atoms in the rings lie in the same plane. Rather, pyranose rings usually adopt the chair conformation, and furanose rings usually adopt the envelope conformation. Sugars are joined to alcohols and amines by glycosidic bonds from the anomeric carbon atom. For example, N-glycosidic bonds link sugars to purines and pyrimidines in nucleotides, RNA, and DNA.

Complex Carbohydrates Are Formed by Linkage of Monosaccharides

Sugars are linked to one another in disaccharides and polysaccharides by O-glycosidic bonds. Sucrose, lactose, and maltose are the common disaccharides. Sucrose (common table sugar), obtained from cane or beet, consists of α-glucose and β-fructose joined by a glycosidic linkage between their anomeric carbon atoms. Lactose (in milk) consists of galactose joined to glucose by a β-1,4 linkage. Maltose (from starch) consists of two glucoses joined by an α-1,4 linkage. Starch is a polymeric form of glucose in plants, and glycogen serves a similar role in animals. Most of the glucose units in starch and glycogen are in α-1,4 linkage. Glycogen has more branch points formed by α-1,6 linkages than does starch, which makes glycogen more soluble. Cellulose, the major structural polymer of plant cell walls, consists of glucose units joined by β-1,4 linkages. These β linkages give rise to long straight chains that form fibrils with high tensile strength. In contrast, the α linkages in starch and glycogen lead to open helices, in keeping with their roles as mobilizable energy stores. Cell surfaces and extracellular matrices of animals contain polymers of repeating disaccharides called glycosaminoglycans. One of the units in each repeat is a derivative of glucosamine or galactosamine. These highly anionic carbohydrates have a high density of carboxylate or sulfate groups. Proteins bearing covalently linked glycosaminoglycans are termed proteoglycans.

Carbohydrates Can Attach to Proteins to Form Glycoproteins

Specific enzymes link the oligosaccharide units on proteins either to the side-chain oxygen atom of a serine or threonine residue or to the side-chain amide nitrogen atom of an asparagine residue. Protein glycosylation takes place in the lumen of the endoplasmic reticulum. The N-linked oligosaccharides are synthesized on dolichol phosphate and subsequently transferred to the protein acceptor. Additional sugars are attached in the Golgi complex to form diverse patterns.

● **Lectins Are Specific Carbohydrate-Binding Proteins**

Carbohydrates are recognized by proteins called lectins, which are found in animals, plants, and microorganisms. In animals, the interplay of lectins and their sugar targets guides cell–cell contact. The viral protein hemagglutinin on the surface of the influenza virus recognizes sialic acid residues on the surfaces of the cells invaded by the virus. A small number of carbohydrate residues can be joined in many different ways to form highly diverse patterns that can be distinguished by the lectin domains of protein receptors.

KEY TERMS

monosaccharide (p. 296)

triose (p. 296)

ketose (p. 296)

aldose (p. 296)

enantiomer (p. 296)

tetrose (p. 296)

pentose (p. 296)

hexose (p. 296)

heptose (p. 296)

diastereoisomer (p. 296)

epimer (p. 297)

hemiacetal (p. 298)

pyranose (p. 298)

hemiketal (p. 298)

furanose (p. 299)

anomer (p. 299)

glycosidic bond (p. 301)

reducing sugar (p. 301)

nonreducing sugar (p. 301)

oligosaccharide (p. 301)

disaccharide (p. 302)

polysaccharide (p. 302)

glycogen (p. 302)

starch (p. 303)

cellulose (p. 303)

glycosaminoglycan (p. 304)

proteoglycan (p. 304)

glycosyltransferase (p. 304)

glycoprotein (p. 306)

endoplasmic reticulum (p. 307)

Golgi complex (p. 307)

dolichol phosphate (p. 308)

lectin (p. 312)

selectin (p. 314)

SELECTED READINGS

Where to start

Sharon, N., and Lis, H., 1993. Carbohydrates in cell recognition. *Sci. Am.* 268(1):82–89.

Lasky, L. A., 1992. Selectins: Interpreters of cell-specific carbohydrate information during inflammation. *Science* 258:964–969.

Weiss, P., and Ashwell, G., 1989. The asialoglycoprotein receptor: Properties and modulation by ligand. *Prog. Clin. Biol. Res.* 300: 169–184.

Sharon, N., 1980. Carbohydrates. *Sci. Am.* 245(5):90–116.

Paulson, J. C.,1989. Glycoproteins: What are the sugar side chains for? *Trends Biochem. Sci.* 14:272–276.

Woods, R. J., 1995. Three-dimensional structures of oligosaccharides. *Curr. Opin. Struct. Biol.* 5:591–598.

Books

Varki, A., Cummings, R., Esko, J., Freeze, H., Hart, G., and Marth, J., 1999. *Essentials of Glycobiology.* Cold Spring Harbor Laboratory Press.

Fukuda, M., and Hindsgaul, O., 2000. *Molecular Glycobiology.* IRL Press at Oxford University Press.

El Khadem, H. S., 1988. *Carbohydrate Chemistry.* Academic Press.

Ginsburg, V., and Robbins, P. W. (Eds.), 1981. *Biology of Carbohydrates* (vols. 1–3). Wiley.

Fukuda, M. (Ed.), 1992. *Cell Surface Carbohydrates and Cell Development.* CRC Press.

Preiss, J. (Ed.), 1988. *The Biochemistry of Plants: A Comprehensive Treatise: Carbohydrates.* Academic Press.

Structure of carbohydrate-binding proteins

Ünligil, U., and Rini, J. M., 2000. Glycosyltransferase structure and mechanism. *Curr. Opin. Struct. Biol.* 10:510–517.

Bouckaert, J., Hamelryck, T., Wyns, L., and Loris, R., 1999. Novel structures of plant lectins and their complexes with carbohydrates. *Curr. Opin. Struct. Biol.* 9:572–577.

Weis, W. I., and Drickamer, K., 1996. Structural basis of lectin-carbohydrate recognition. *Annu. Rev. Biochem.* 65:441–473.

Vyas, N. K., 1991. Atomic features of protein-carbohydrate interactions. *Curr. Opin. Struct. Biol.* 1:732–740.

Weis, W. I., Drickamer, K., and Hendrickson, W. A., 1992. Structure of a C-type mannose-binding protein complexed with an oligosaccharide. *Nature* 360:127–134.

Wright, C. S., 1992. Crystal structure of a wheat germ agglutinin/glycophorin-sialoglycopeptide receptor complex: Structural basis for cooperative lectin-cell binding. *J. Biol. Chem.* 267:14345–14352.

Shaanan, B., Lis, H., and Sharon, N., 1991. Structure of a legume lectin with an ordered *N*-linked carbohydrate in complex with lactose. *Science* 254:862–866.

Glycoproteins

Spiro, R. G., 2000. Glucose residues as key determinants in the biosynthesis and quality control of glycoproteins with *N*-linked oligosaccharides. *J. Biol. Chem.* 275:35657–35660.

Bernfield, M., Götte, M., Park, P. W., Reizes,O., Fitzgerald, M. L., Lincecum, J., and Zako, M.,1999. Functions of cell surface heparan sulfate proteoglycans. *Annu. Rev. Biochem.* 68:729–777.

Iozzo, R. V., 1998. Matrix proteoglycans: From molecular design to cellular function. *Annu. Rev. Biochem.* 67:609–652.

Trombetta, E. S., and Helenius, A., 1998. Lectins as chaperones in glycoprotein folding. *Curr. Opin. Struct. Biol.* 8:587–592.

Yanagishita, M., and Hascall, V. C., 1992. Cell surface heparan sulfate proteoglycans. *J. Biol. Chem.* 267:9451–9454.

Iozzo, R. V., 1999. The biology of small leucine-rich proteoglycans: Functional network of interactive proteins. *J. Biol. Chem.* 274: 18843–18846.

Carbohydrates in recognition processes

Weis, W. I., 1997. Cell-surface carbohydrate recognition by animal and viral lectins. *Curr. Opin. Struct. Biol.* 7:624–630.

Sharon, N., and Lis, H., 1989. Lectins as cell recognition molecules. *Science* 246:227–234.

Turner, M. L., 1992. Cell adhesion molecules: A unifying approach to topographic biology. *Biol. Rev. Camb. Philos. Soc.* 67:359–377.

Feizi, T., 1992. Blood group–related oligosaccharides are ligands in cell-adhesion events. *Biochem. Soc. Trans.* 20:274–278.

Jessell, T. M., Hynes, M. A., and Dodd, J.,1990. Carbohydrates and carbohydrate-binding proteins in the nervous system. *Annu. Rev. Neurosci.* 13:227–255.

Clothia, C., and Jones, E. V., 1997. The molecular structure of cell adhesion molecules. *Annu. Rev. Biochem.* 66:823–862.

Carbohydrate sequencing

Venkataraman, G., Shriver, Z., Raman, R., and Sasisekharan, R., 1999. Sequencing complex polysaccharides. *Science* 286:537–542.

Zhao, Y., Kent, S. B. H., and Chait, B. T., 1997. Rapid, sensitive structure analysis of oligosaccharides. *Proc. Natl. Acad. Sci. USA* 94:1629–1633.

Rudd, P. M., Guile, G. R., Küster, B., Harvey, D. J., Opdenakker, G., and Dwek, R. A. 1997. Oligosaccharide sequencing technology. *Nature* 388:205–207.

──┤ PROBLEMS ├──

1. *Word origin.* Account for the origin of the term *carbohydrate.*

2. *Diversity.* How many different oligosaccharides can be made by linking one glucose, one mannose, and one galactose? Assume that each sugar is in its pyranose form. Compare this number with the number of tripeptides that can be made from three different amino acids.

3. *Couples.* Indicate whether each of the following pairs of sugars consists of anomers, epimers, or an aldose-ketose pair:

(a) D-glyceraldehyde and dihydroxyacetone
(b) D-glucose and D-mannose
(c) D-glucose and D-fructose
(d) α-D-glucose and β-D-glucose
(e) D-ribose and D-ribulose
(f) D-galactose and D-glucose

4. *Tollen's test.* Glucose and other aldoses are oxidized by an aqueous solution of a silver–ammonia complex. What are the reaction products?

5. *Mutarotation.* The specific rotations of the α and β anomers of D-glucose are +112 degrees and +18.7 degrees, respectively. Specific rotation, $[\alpha]_D$, is defined as the observed rotation of light of wavelength 589 nm (the D line of a sodium lamp) passing through 10 cm of a 1 g ml^{-1} solution of a sample. When a crystalline sample of α-D-glucopyranose is dissolved in water, the specific rotation decreases from 112 degrees to an equilibrium value of 52.7 degrees. On the basis of this result, what are the proportions of the α and β anomers at equilibrium? Assume that the concentration of the open-chain form is negligible.

6. *Telltale adduct.* Glucose reacts slowly with hemoglobin and other proteins to form covalent compounds. Why is glucose reactive? What is the nature of the adduct formed?

7. *Periodate cleavage.* Compounds containing hydroxyl groups on adjacent carbon atoms undergo carbon–carbon bond cleavage when treated with periodate ion (IO_4^-). How can this reaction be used to distinguish between pyranosides and furanosides?

8. *Oxygen source.* Does the oxygen atom attached to C-1 in methyl α-D-glucopyranoside come from glucose or methanol?

9. *Sugar lineup.* Identify the following four sugars.

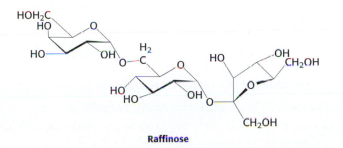

10. *Cellular glue.* A trisaccharide unit of a cell-surface glycoprotein is postulated to play a critical role in mediating cell–cell adhesion in a particular tissue. Design a simple experiment to test this hypothesis.

11. *Mapping the molecule.* Each of the hydroxyl groups of glucose can be methylated with reagents such as dimethylsulfate under basic conditions. Explain how exhaustive methylation followed by compete digestion of a known amount of glycogen would enable you to determine the number of branch points and reducing ends.

12. *Component parts.* Raffinose is a trisaccharide and a minor constituent in sugar beets.

(a) Is raffinose a reducing sugar? Explain.
(b) What are the monosaccharides that compose raffinose?
(c) β-Galactosidase is an enzyme that will remove galactose residues from an oligosaccharide. What are the products of β-galactosidase treatment of raffinose?

Raffinose

13. *Anomeric differences.* α-D-Mannose is a sweet-tasting sugar. β-D-Mannose, on the other hand, tastes bitter. A pure solution of α-D-mannose loses its sweet taste with time as it is converted into the β anomer. Draw the β anomer and explain how it is formed from the α anomer.

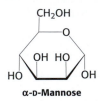

α-D-Mannose

14. *A taste of honey.* Fructose in its β-D-pyranose form accounts for the powerful sweetness of honey. The β-D-furanose form, although sweet, is not as sweet as the pyranose form. The furanose form is the more stable form. Draw the two forms and explain why it may not always be wise to cook with honey.

15. *Making ends meet.* (a) Compare the number of reducing ends to nonreducing ends in a molecule of glycogen. (b) As we will see in Chapter 21, glycogen is an important fuel storage form that is rapidly mobilized. At which end—the reducing or nonreducing—would you expect most metabolism to take place?

16. *Carbohydrates and proteomics.* Suppose that a protein contains six potential N-linked glycosylation sites. How many possible proteins can be generated, depending on which of these sites is actually glycosylated? Do not include the effects of diversity within the carbohydrate added.

Chapter Integration Problem

17. *Stereospecificity.* Sucrose, a major product of photosynthesis in green leaves, is synthesized by a battery of enzymes. The substrates for sucrose synthesis, D-glucose and D-fructose, are a mixture of α and β anomers as well as acyclic compounds in solution. Nonetheless, sucrose consists of α-D-glucose linked by its carbon-1 atom to the carbon-2 atom of β-D-fructose. How can the specificity of sucrose be explained in light of the potential substrates?

CHAPTER 12

Lipids and Cell Membranes

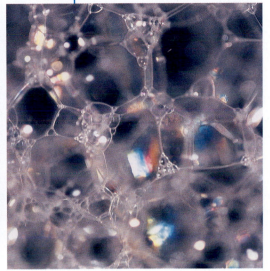

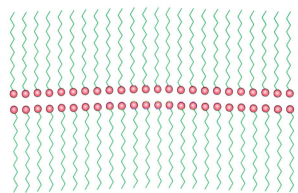

The surface of a soap bubble is a bilayer formed by detergent molecules. The polar heads (red) pack together leaving the hydrophobic groups (green) in contact with air on the inside and outside of the bubble. Other bilayer structures define the boundary of a cell. [(Left) Photonica.]

The boundaries of cells are formed by *biological membranes*, the barriers that *define the inside and the outside of a cell* (Figure 12.1). These barriers prevent molecules generated inside the cell from leaking out and unwanted molecules from diffusing in; yet they also contain transport systems that allow specific molecules to be taken up and unwanted compounds to be removed from the cell. Such transport systems confer on membranes the important property of *selective permeability*.

Membranes are dynamic structures in which proteins float in a sea of lipids. The lipid components of the membrane form the permeability barrier, and protein components act as a transport system of pumps and channels that endow the membrane with selective permeability.

In addition to an external cell membrane (called the plasma membrane), eukaryotic cells also contain internal membranes that form the boundaries of organelles such as mitochondria, chloroplasts, peroxisomes, and lysosomes. Functional specialization in the course of evolution has been closely linked to the formation of such compartments. Specific systems have evolved to allow targeting of selected proteins into or through particular internal membranes and, hence, into specific organelles. External and internal membranes have essential features in common, and these essential features are the subject of this chapter.

OUTLINE

- **12.1 Many Common Features Underlie the Diversity of Biological Membranes**
- **12.2 Fatty Acids Are Key Constituents of Lipids**
- **12.3 There Are Three Common Types of Membrane Lipids**
- **12.4 Phospholipids and Glycolipids Readily Form Bimolecular Sheets in Aqueous Media**
- **12.5 Proteins Carry Out Most Membrane Processes**
- **12.6 Lipids and Many Membrane Proteins Diffuse Rapidly in the Plane of the Membrane**
- **12.7 Eukaryotic Cells Contain Compartments Bounded by Internal Membranes**

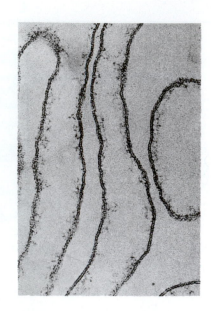

FIGURE 12.1 Red-blood-cell plasma membrane. An electron micrograph of a preparation of plasma membranes from red blood cells showing the membranes as seen "on edge," in cross section. [Courtesy of Dr. Vincent Marchesi.]

Biological membranes serve several additional important functions indispensable for life, such as energy storage and information transduction, that are dictated by the proteins associated with them. In this chapter, we will examine the general properties of membrane proteins—how they can exist in the hydrophobic environment of the membrane while connecting two hydrophilic environments—and delay a discussion of the functions of these proteins to the next and later chapters.

12.1 MANY COMMON FEATURES UNDERLIE THE DIVERSITY OF BIOLOGICAL MEMBRANES

Membranes are as diverse in structure as they are in function. However, they do have in common a number of important attributes:

1. Membranes are *sheetlike structures,* only two molecules thick, that form *closed boundaries* between different compartments. The thickness of most membranes is between 60 Å (6 nm) and 100 Å (10 nm).

2. Membranes consist mainly of *lipids* and *proteins.* Their mass ratio ranges from 1:4 to 4:1. Membranes also contain *carbohydrates* that are linked to lipids and proteins.

3. Membrane lipids are relatively small molecules that have both *hydrophilic* and *hydrophobic* moieties. These lipids spontaneously form *closed bimolecular sheets* in aqueous media. These *lipid bilayers* are barriers to the flow of polar molecules.

4. *Specific proteins mediate distinctive functions of membranes.* Proteins serve as pumps, channels, receptors, energy transducers, and enzymes. Membrane proteins are embedded in lipid bilayers, which create suitable environments for their action.

5. Membranes are *noncovalent assemblies.* The constituent protein and lipid molecules are held together by many noncovalent interactions, which are cooperative.

6. Membranes are *asymmetric.* The two faces of biological membranes always differ from each other.

7. Membranes are *fluid structures.* Lipid molecules diffuse rapidly in the plane of the membrane, as do proteins, unless they are anchored by specific interactions. In contrast, lipid molecules and proteins do not readily rotate across the membrane. Membranes can be regarded as *two-dimensional solutions of oriented proteins and lipids.*

8. Most cell membranes are *electrically polarized,* such that the inside is negative [typically − 60 millivolts (mV)]. Membrane potential plays a key role in transport, energy conversion, and excitability (Chapter 13).

12.2 FATTY ACIDS ARE KEY CONSTITUENTS OF LIPIDS

Among the most biologically significant properties of lipids are their hydrophobic properties. These properties are mainly due to a particular component of lipids: fatty acids, or simply fats. Fatty acids also play important roles in signal-transduction pathways (Sections 15.2 and 22.6.2).

12.2.1 The Naming of Fatty Acids

Fatty acids are hydrocarbon chains of various lengths and degrees of unsaturation that terminate with carboxylic acid groups. The systematic name

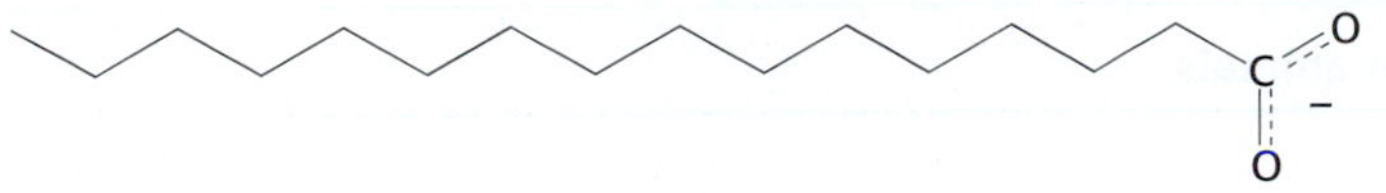

Palmitate
(ionized form of palmitic acid)

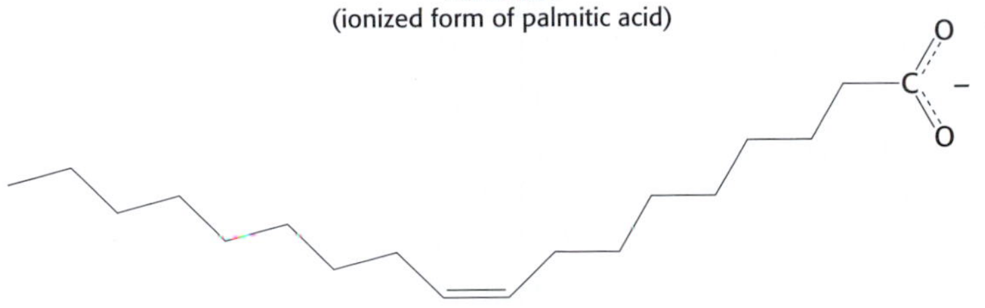

Oleate
(ionized form of oleic acid)

FIGURE 12.2 Structures of two fatty acids. Palmitate is a 16-carbon, saturated fatty acid, and oleate is an 18-carbon fatty acid with a single cis double bond.

for a fatty acid is derived from the name of its parent hydrocarbon by the substitution of *oic* for the final *e*. For example, the C_{18} saturated fatty acid is called *octadecanoic acid* because the parent hydrocarbon is octadecane. A C_{18} fatty acid with one double bond is called octade*cenoic* acid; with two double bonds, octadeca*dienoic* acid; and with three double bonds, octa-deca*trienoic* acid. The notation 18:0 denotes a C_{18} fatty acid with no double bonds, whereas 18:2 signifies that there are two double bonds. The structures of the ionized forms of two common fatty acids—palmitic acid (C_{16}, saturated) and oleic acid (C_{18}, monounsaturated)—are shown in Figure 12.2.

Fatty acid carbon atoms are numbered starting at the carboxyl terminus, as shown in the margin. Carbon atoms 2 and 3 are often referred to as α and β, respectively. The methyl carbon atom at the distal end of the chain is called the ω-*carbon atom*. The position of a double bond is represented by the symbol Δ followed by a superscript number. For example, *cis*-Δ^9 means that there is a cis double bond between carbon atoms 9 and 10; *trans*-Δ^2 means that there is a trans double bond between carbon atoms 2 and 3. Alternatively, the position of a double bond can be denoted by counting from the distal end, with the ω-carbon atom (the methyl carbon) as number 1. An ω-3 fatty acid, for example, has the structure shown in the margin. Fatty acids are ionized at physiological pH, and so it is appropriate to refer to them according to their carboxylate form: for example, palmitate or hexadecanoate.

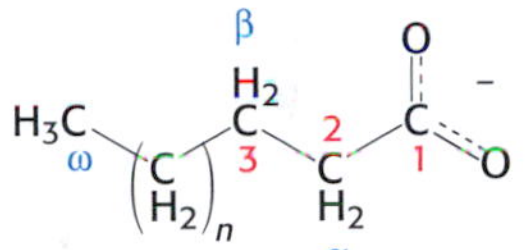

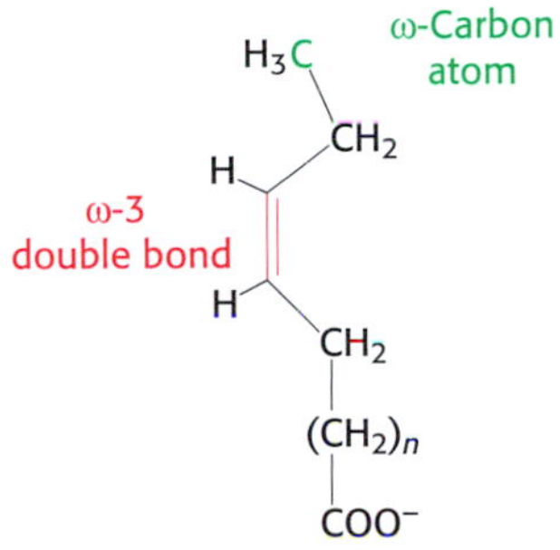

An ω-3 fatty acid

12.2.2 Fatty Acids Vary in Chain Length and Degree of Unsaturation

Fatty acids in biological systems usually contain an even number of carbon atoms, typically between 14 and 24 (Table 12.1). The 16- and 18-carbon fatty acids are most common. The hydrocarbon chain is almost invariably unbranched in animal fatty acids. The alkyl chain may be saturated or it may contain one or more double bonds. The configuration of the double bonds in most unsaturated fatty acids is cis. The double bonds in polyunsaturated fatty acids are separated by at least one methylene group.

The properties of fatty acids and of lipids derived from them are markedly dependent on chain length and degree of saturation. Unsaturated fatty acids have lower melting points than saturated fatty acids of the same length. For example, the melting point of stearic acid is 69.6°C, whereas that of oleic acid (which contains one cis double bond) is 13.4°C. The melting points of polyunsaturated fatty acids of the C_{18} series are even lower. Chain length also affects the melting point, as illustrated by the fact that the melting temperature of palmitic acid (C_{16}) is 6.5 degrees lower than that of stearic acid (C_{18}). Thus, *short chain length and unsaturation enhance the fluidity of fatty acids and of their derivatives.*

TABLE 12.1   Some naturally occurring fatty acids in animals

Number of carbons	Number of double bonds	Common name	Systematic name	Formula
12	0	Laurate	n-Dodecanoate	$CH_3(CH_2)_{10}COO^-$
14	0	Myristate	n-Tetradecanoate	$CH_3(CH_2)_{12}COO^-$
16	0	Palmitate	n-Hexadecanoate	$CH_3(CH_2)_{14}COO^-$
18	0	Stearate	n-Octadecanoate	$CH_3(CH_2)_{16}COO^-$
20	0	Arachidate	n-Eicosanoate	$CH_3(CH_2)_{18}COO^-$
22	0	Behenate	n-Docosanoate	$CH_3(CH_2)_{20}COO^-$
24	0	Lignocerate	n-Tetracosanoate	$CH_3(CH_2)_{22}COO^-$
16	1	Palmitoleate	cis-Δ^9-Hexadecenoate	$CH_3(CH_2)_5CH{=}CH(CH_2)_7COO^-$
18	1	Oleate	cis-Δ^9-Octadecenoate	$CH_3(CH_2)_7CH{=}CH(CH_2)_7COO^-$
18	2	Linoleate	cis,cis-Δ^9,Δ^{12}-Octadecadienoate	$CH_3(CH_2)_4(CH{=}CHCH_2)_2(CH_2)_6COO^-$
18	3	Linolenate	all-cis-$\Delta^9,\Delta^{12},\Delta^{15}$-Octadecatrienoate	$CH_3CH_2(CH{=}CHCH_2)_3(CH_2)_6COO^-$
20	4	Arachidonate	all-cis-$\Delta^5,\Delta^8,\Delta^{11},\Delta^{14}$-Eicosatetraenoate	$CH_3(CH_2)_4(CH{=}CHCH_2)_4(CH_2)_2COO^-$

12.3 THERE ARE THREE COMMON TYPES OF MEMBRANE LIPIDS

Lipids differ markedly from the other groups of biomolecules considered thus far. By definition, *lipids are water-insoluble biomolecules that are highly soluble in organic solvents such as chloroform.* Lipids have a variety of biological roles: they serve as fuel molecules, highly concentrated energy stores, signal molecules, and components of membranes. The first three roles of lipids will be discussed in later chapters. Here, our focus is on lipids as membrane constituents. The three major kinds of membrane lipids are *phospholipids, glycolipids,* and *cholesterol.* We begin with lipids found in eukaryotes and bacteria. The lipids in archaea are distinct, although they have many features related to their membrane-forming function in common with lipids of other organisms.

12.3.1 Phospholipids Are the Major Class of Membrane Lipids

Phospholipids are abundant in all biological membranes. A phospholipid molecule is constructed from four components: fatty acids, a platform to which the fatty acids are attached, a phosphate, and an alcohol attached to the phosphate (Figure 12.3). The fatty acid components provide a hydrophobic barrier, whereas the remainder of the molecule has hydrophilic properties to enable interaction with the environment.

The platform on which phospholipids are built may be *glycerol,* a 3-carbon alcohol, or *sphingosine,* a more complex alcohol. Phospholipids derived from glycerol are called *phosphoglycerides.* A phosphoglyceride consists of a glycerol backbone to which two fatty acid chains (whose characteristics were described in Section 12.2.2) and a phosphorylated alcohol are attached.

In phosphoglycerides, the hydroxyl groups at C-1 and C-2 of glycerol are esterified to the carboxyl groups of the two fatty acid chains. The C-3 hydroxyl group of the glycerol backbone is esterified to phosphoric acid. When no further additions are made, the resulting compound is *phosphatidate (diacylglycerol 3-phosphate),* the simplest phosphoglyceride. Only small

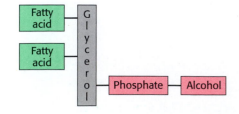

FIGURE 12.3 Schematic structure of a phospholipid.

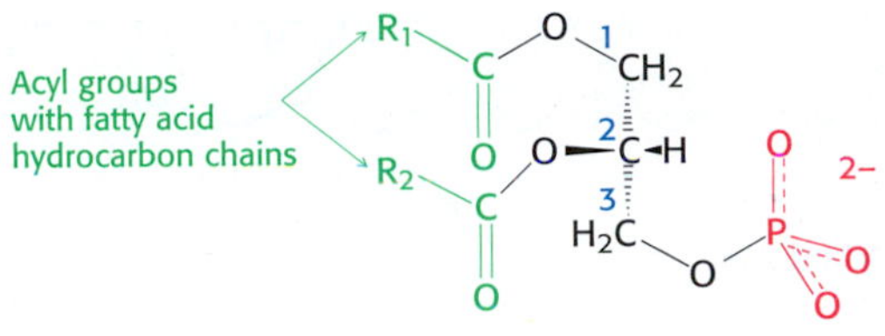

FIGURE 12.4 Structure of phosphatidate (diacylglycerol 3-phosphate). The absolute configuration of the center carbon (C-2) is shown.

Phosphatidate
(Diacylglycerol 3-phosphate)

amounts of phosphatidate are present in membranes. However, the molecule is a key intermediate in the biosynthesis of the other phosphoglycerides (Section 26.1). The absolute configuration of the glycerol 3-phosphate moiety of membrane lipids is shown in Figure 12.4.

The major phosphoglycerides are derived from phosphatidate by the formation of an ester bond between the phosphate group of phosphatidate and the hydroxyl group of one of several alcohols. The common alcohol moieties of phosphoglycerides are the amino acid serine, ethanolamine, choline, glycerol, and the inositol.

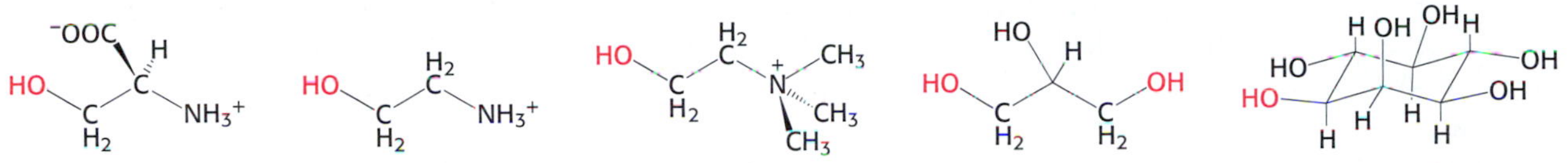

Serine   Ethanolamine   Choline   Glycerol   Inositol

The structural formulas of phosphatidyl choline and the other principal phosphoglycerides—namely, phosphatidyl ethanolamine, phosphatidyl serine, phosphatidyl inositol, and diphosphatidyl glycerol—are given in Figure 12.5.

Sphingomyelin is a phospholipid found in membranes that is not derived from glycerol. Instead, the backbone in sphingomyelin is *sphingosine*, an

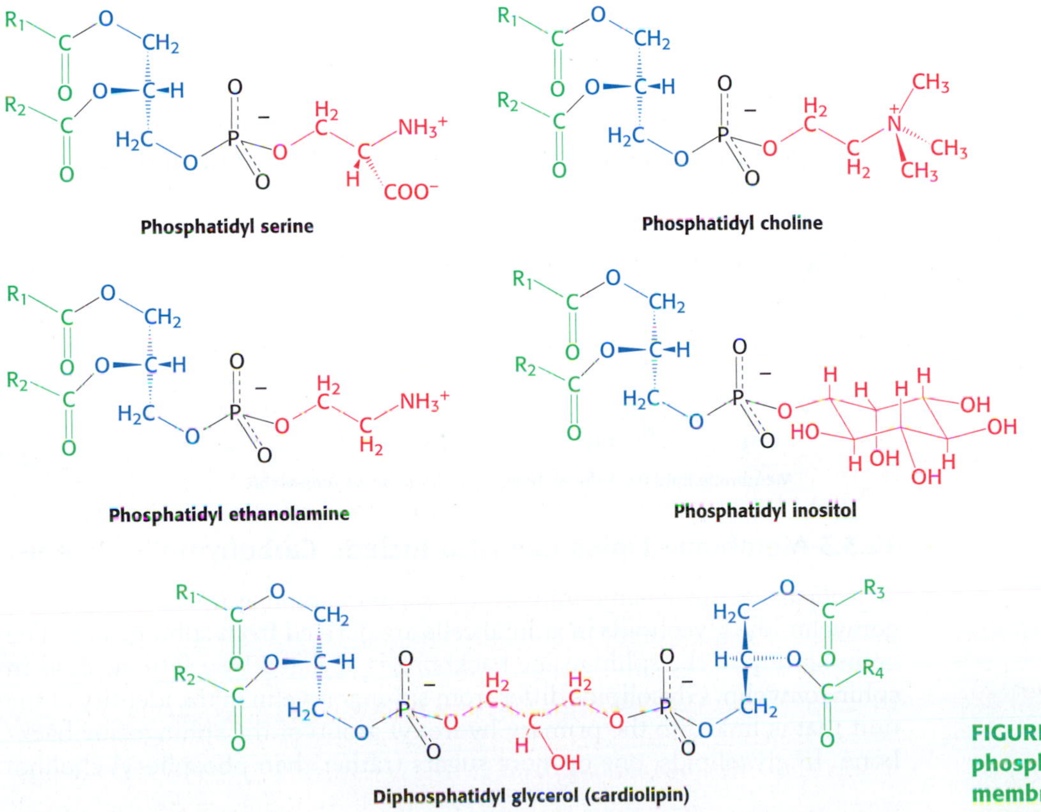

Phosphatidyl serine

Phosphatidyl choline

Phosphatidyl ethanolamine

Phosphatidyl inositol

Diphosphatidyl glycerol (cardiolipin)

FIGURE 12.5 Some common phosphoglycerides found in membranes.

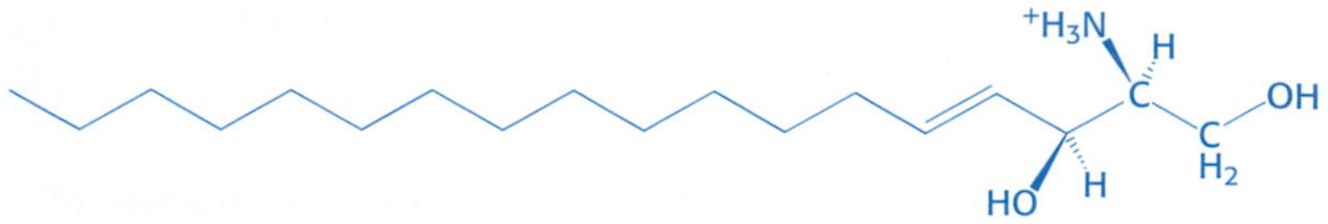

Sphingosine

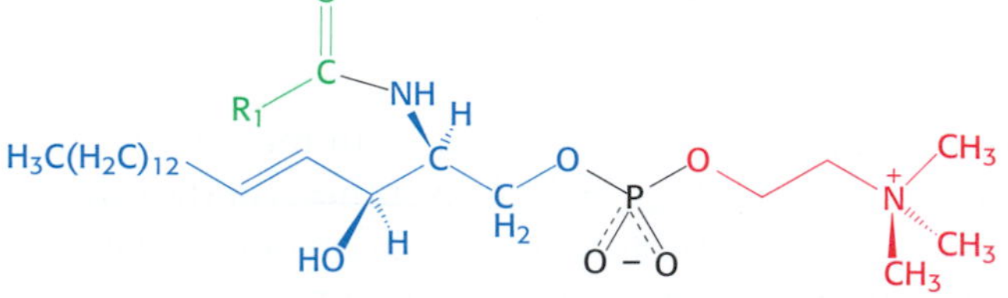

Sphingomyelin

FIGURE 12.6 Structures of sphingosine and sphingomyelin. The sphingosine moiety of sphingomyelin is highlighted in blue.

FIGURE 12.7 An archaeon and its environment. Archaea can thrive in habitats as harsh as a volcanic vent. Here, the archaea form an orange mat surrounded by yellow sulfurous deposits. [Krafft-Explorer/Photo Researchers.]

amino alcohol that contains a long, unsaturated hydrocarbon chain (Figure 12.6). In sphingomyelin, the amino group of the sphingosine backbone is linked to a fatty acid by an amide bond. In addition, the primary hydroxyl group of sphingosine is esterified to phosphoryl choline.

12.3.2 Archaeal Membranes Are Built from Ether Lipids with Branched Chains

The membranes of archaea differ in composition from those of eukaryotes or bacteria in three important ways. Two of these differences clearly relate to the hostile living conditions of many archaea (Figure 12.7). First, the non-polar chains are joined to a glycerol backbone by *ether* rather than ester linkages. The ether linkage is more resistant to hydrolysis. Second, the alkyl chains are *branched* rather than linear. They are built up from repeats of a fully saturated five-carbon fragment. These branched, saturated hydrocarbons are more resistant to oxidation. The ability of archaeal lipids to resist hydrolysis and oxidation may help these organisms to withstand the extreme conditions, such as high temperature, low pH, or high salt concentration, under which some of these archaea grow. Finally, the stereochemistry of the central glycerol is inverted compared with that shown in Figure 12.4.

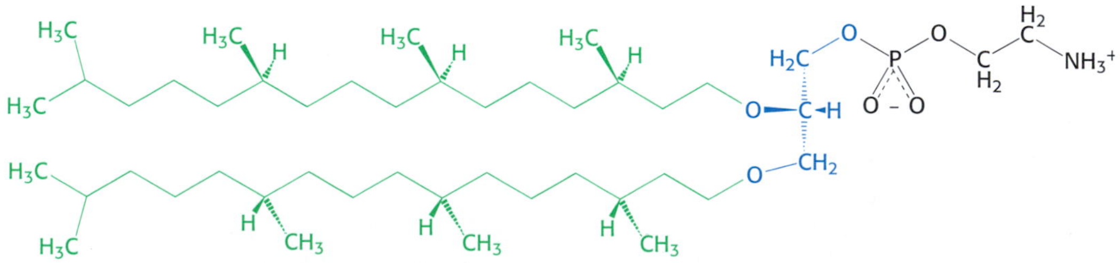

Membrane lipid from the archaeon *Methanococcus jannaschii*

12.3.3 Membrane Lipids Can Also Include Carbohydrate Moieties

Glycolipids, as their name implies, are *sugar-containing lipids*. Like sphingomyelin, the glycolipids in animal cells are derived from sphingosine. The amino group of the sphingosine backbone is acylated by a fatty acid, as in sphingomyelin. Glycolipids differ from sphingomyelin in the identity of the unit that is linked to the primary hydroxyl group of the sphingosine backbone. In glycolipids, one or more sugars (rather than phosphoryl choline) are attached to this group. The simplest glycolipid, called a *cerebroside*, contains a single sugar residue, either glucose or galactose.

Cerebroside
(a glycolipid)

More complex glycolipids, such as *gangliosides*, contain a branched chain of as many as seven sugar residues. Glycolipids are oriented in a completely asymmetric fashion with the *sugar residues always on the extracellular side of the membrane.*

12.3.4 Cholesterol Is a Lipid Based on a Steroid Nucleus

Cholesterol is a lipid with a structure quite different from that of phospholipids. It is a steroid, built from four linked hydrocarbon rings.

Cholesterol

A hydrocarbon tail is linked to the steroid at one end, and a hydroxyl group is attached at the other end. In membranes, the molecule is oriented parallel to the fatty acid chains of the phospholipids, and the hydroxyl group interacts with the nearby phospholipid head groups. Cholesterol is absent from prokaryotes but is found to varying degrees in virtually all animal membranes. It constitutes almost 25% of the membrane lipids in certain nerve cells but is essentially absent from some intracellular membranes.

12.3.5 A Membrane Lipid Is an Amphipathic Molecule Containing a Hydrophilic and a Hydrophobic Moiety

The repertoire of membrane lipids is extensive, perhaps even bewildering, at first sight. However, they possess a critical common structural theme: *membrane lipids are amphipathic molecules* (amphiphilic molecules). A membrane lipid contains both a *hydrophilic* and a *hydrophobic* moiety.

Let us look at a model of a phosphoglyceride, such as phosphatidyl choline. Its overall shape is roughly rectangular (Figure 12.8A). The two hydrophobic fatty acid chains are approximately parallel to each other, whereas the hydrophilic phosphoryl choline moiety points in the opposite direction. Sphingomyelin has a similar conformation, as does the archaeal lipid depicted. Therefore, the following shorthand has been adopted to represent these membrane lipids: the hydrophilic unit, also called the *polar head group*, is represented by a circle, whereas the hydrocarbon tails are depicted by straight or wavy lines (Figure 12.8B).

(A)

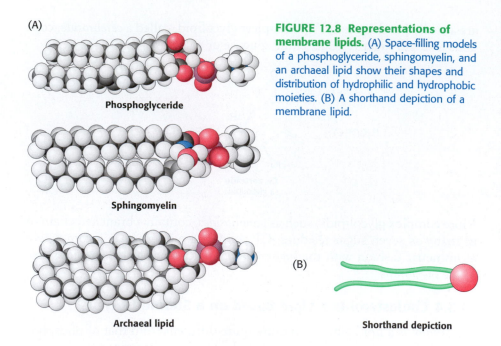

Phosphoglyceride

Sphingomyelin

Archaeal lipid

(B)

Shorthand depiction

FIGURE 12.8 Representations of membrane lipids. (A) Space-filling models of a phosphoglyceride, sphingomyelin, and an archaeal lipid show their shapes and distribution of hydrophilic and hydrophobic moieties. (B) A shorthand depiction of a membrane lipid.

12.4 PHOSPHOLIPIDS AND GLYCOLIPIDS READILY FORM BIMOLECULAR SHEETS IN AQUEOUS MEDIA

What properties enable phospholipids to form membranes? *Membrane formation is a consequence of the amphipathic nature of the molecules.* Their polar head groups favor contact with water, whereas their hydrocarbon tails interact with one another, in preference to water. How can molecules with these preferences arrange themselves in aqueous solutions? One way is to form a *micelle*, a globular structure in which polar head groups are surrounded by water and hydrocarbon tails are sequestered inside, interacting with one another (Figure 12.9).

Alternatively, the strongly opposed preferences of the hydrophilic and hydrophobic moieties of membrane lipids can be satisfied by forming a *lipid bilayer*, composed of two lipid sheets (Figure 12.10). A lipid bilayer is also called a *bimolecular sheet*. The hydrophobic tails of each individual sheet interact with one another, forming a hydrophobic interior that acts as a permeability barrier. The hydrophilic head groups interact with the aqueous medium on each side of the bilayer. The two opposing sheets are called leaflets.

The favored structure for most phospholipids and glycolipids in aqueous media is a bimolecular sheet rather than a micelle. The reason is that the two fatty acyl chains of a phospholipid or a glycolipid are too bulky to fit into the interior of a micelle. In contrast, salts of fatty acids (such as sodium palmitate, a constituent of soap), which contain only one chain, readily form micelles. *The formation of bilayers instead of micelles by phospholipids is of critical biological importance.* A micelle is a limited structure, usually less than 20 nm (200 Å) in diameter. In contrast, a bimolecular sheet can have macroscopic dimensions, such as a millimeter (10^6 nm, or 10^7 Å). Phospholipids and related molecules are important membrane constituents because they readily form extensive bimolecular sheets (Figure 12.11).

The formation of lipid bilayers is a *self-assembly process*. In other words, the structure of a bimolecular sheet is inherent in the structure of the constituent lipid molecules. The growth of lipid bilayers from phospholipids is a rapid and spontaneous process in water. *Hydrophobic interactions are the*

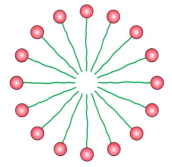

FIGURE 12.9 Diagram of a section of a micelle. Ionized fatty acids readily form such structures, but most phospholipids do not.

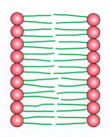

FIGURE 12.10 Diagram of a section of a bilayer membrane.

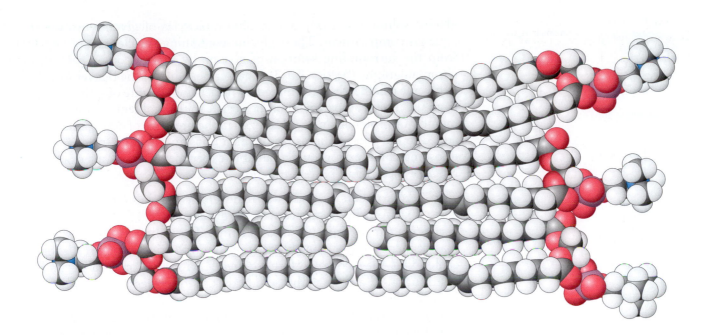

FIGURE 12.11 Space-filling model
of a section of phospholipid bilayer
membrane.

major driving force for the formation of lipid bilayers. Recall that hydrophobic interactions also play a dominant role in the folding of proteins (Sections 1.3.4 and 3.4) and in the stacking of bases in nucleic acids (Section 5.2.1). Water molecules are released from the hydrocarbon tails of membrane lipids as these tails become sequestered in the nonpolar interior of the bilayer. Furthermore, *van der Waals attractive forces between the hydrocarbon tails favor close packing of the tails*. Finally, there are *electrostatic and hydrogen-bonding attractions between the polar head groups and water molecules*. Thus, lipid bilayers are stabilized by the full array of forces that mediate molecular interactions in biological systems.

Because lipid bilayers are held together by many *reinforcing, noncovalent interactions (predominantly hydrophobic)*, they are *cooperative structures*. These hydrophobic interactions have three significant biological consequences: (1) lipid bilayers have an inherent tendency to be *extensive*; (2) lipid bilayers will tend to *close on themselves* so that there are no edges with exposed hydrocarbon chains, and so they form compartments; and (3) lipid bilayers are *self-sealing* because a hole in a bilayer is energetically unfavorable.

12.4.1 Lipid Vesicles Can Be Formed from Phospholipids

The propensity of phospholipids to form membranes has been used to create an important experimental and clinical tool. *Lipid vesicles*, or *liposomes*, aqueous compartments enclosed by a lipid bilayer (Figure 12.12), can be used to study membrane permeability or to deliver chemicals to cells. Liposomes are formed by suspending a suitable lipid, such as phosphatidyl choline, in an aqueous medium, and then *sonicating* (i.e., agitating by high-frequency sound waves) to give a dispersion of closed vesicles that are quite uniform in size. Vesicles formed by these methods are nearly spherical in shape and have a diameter of about 50 nm (500 Å). Larger vesicles (of the order of 1 μm, or 10^4 Å, in diameter) can be prepared by slowly evaporating the organic solvent from a suspension of phospholipid in a mixed solvent system.

Ions or molecules can be trapped in the aqueous compartments of lipid vesicles by forming the vesicles in the presence of these substances (Figure 12.13). For example, 50-nm-diameter vesicles formed in a 0.1 M

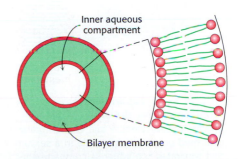

FIGURE 12.12 Liposome. A liposome,
or lipid vesicle, is a small aqueous
compartment surrounded by a lipid bilayer.

Inner aqueous
compartment

Bilayer membrane

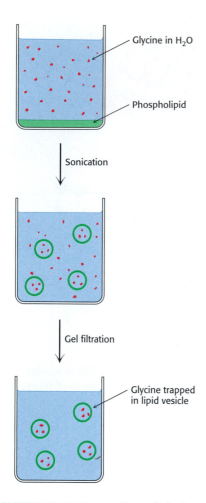

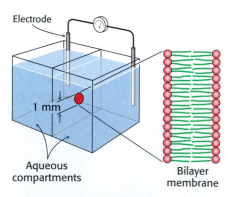

FIGURE 12.13 Preparation of glycine-containing liposomes. Liposomes containing glycine are formed by sonication of phospholipids in the presence of glycine. Free glycine is removed by gel filtration.

FIGURE 12.14 Experimental arrangement for the study of planar bilayer membrane. A bilayer membrane is formed across a 1-mm hole in a septum that separates two aqueous compartments. This arrangement permits measurements of the permeability and electrical conductance of lipid bilayers.

glycine solution will trap about 2000 molecules of glycine in each inner aqueous compartment. These glycine-containing vesicles can be separated from the surrounding solution of glycine by dialysis or by gel-filtration chromatography (Section 4.1.3). The permeability of the bilayer membrane to glycine can then be determined by measuring the rate of efflux of glycine from the inner compartment of the vesicle to the ambient solution. Specific membrane proteins can be solubilized in the presence of detergents and then added to the phospholipids from which liposomes will be formed. Protein–liposome complexes provide valuable experimental tools for examining a range of membrane-protein functions.

Experiments are underway to develop clinical uses for liposomes. For example, liposomes containing drugs or DNA for gene-therapy experiments can be injected into patients. These liposomes fuse with the plasma membrane of many kinds of cells, introducing into the cells the molecules that they contain. Drug delivery with liposomes also alters the distribution of a drug within the body and often lessens its toxicity. For instance, less of the drug is distributed to normal tissues because long-circulating liposomes have been shown to concentrate in regions of increased blood circulation, such as solid tumors and sites of inflammation. Moreover, the selective fusion of lipid vesicles with particular kinds of cells is a promising means of controlling the delivery of drugs to target cells.

Another well-defined synthetic membrane is a *planar bilayer membrane*. This structure can be formed across a 1-mm hole in a partition between two aqueous compartments by dipping a fine paintbrush into a membrane-forming solution, such as phosphatidyl choline in decane. The tip of the brush is then stroked across a hole (1 mm in diameter) in a partition between two aqueous media. The lipid film across the hole thins spontaneously into a lipid bilayer. The electrical conduction properties of this macroscopic bilayer membrane are readily studied by inserting electrodes into each aqueous compartment (Figure 12.14). For example, its permeability to ions is determined by measuring the current across the membrane as a function of the applied voltage.

12.4.2 Lipid Bilayers Are Highly Impermeable to Ions and Most Polar Molecules

The results of permeability studies of lipid vesicles and electrical-conductance measurements of planar bilayers have shown that *lipid bilayer membranes have a very low permeability for ions and most polar molecules.* Water is a conspicuous exception to this generalization; it readily traverses such membranes because of its small size, high concentration, and lack of a complete charge. The range of measured permeability coefficients is very wide (Figure 12.15). For example, Na^+ and K^+ traverse these membranes 10^9

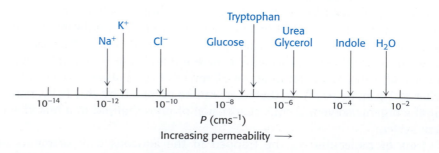

FIGURE 12.15 Permeability coefficients (P) of ions and molecules in a lipid bilayer. The ability of molecules to cross a lipid bilayer spans a wide range of values.

times as slowly as does H_2O. Tryptophan, a zwitterion at pH 7, crosses the membrane 10^3 times as slowly as does indole, a structurally related molecule that lacks ionic groups. In fact, *the permeability coefficients of small molecules are correlated with their solubility in a nonpolar solvent relative to their solubility in water.* This relation suggests that a small molecule might traverse a lipid bilayer membrane in the following way: first, it sheds its solvation shell of water; then, it becomes dissolved in the hydrocarbon core of the membrane; finally, it diffuses through this core to the other side of the membrane, where it becomes resolvated by water. An ion such as Na^+ traverses membranes very slowly because the replacement of its coordination shell of polar water molecules by nonpolar interactions with the membrane interior is highly unfavorable energetically.

12.5 PROTEINS CARRY OUT MOST MEMBRANE PROCESSES

We now turn to membrane proteins, which are responsible for most of the dynamic processes carried out by membranes. Membrane lipids form a permeability barrier and thereby establish compartments, whereas *specific proteins mediate nearly all other membrane functions.* In particular, proteins transport chemicals and information across a membrane. Membrane lipids create the appropriate environment for the action of such proteins.

Membranes differ in their protein content. Myelin, a membrane that serves as an insulator around certain nerve fibers, has a low content of protein (18%). Relatively pure lipids are well suited for insulation. In contrast, the plasma membranes or exterior membranes of most other cells are much more active. They contain many pumps, channels, receptors, and enzymes. The protein content of these plasma membranes is typically 50%. Energy-transduction membranes, such as the internal membranes of mitochondria and chloroplasts, have the highest content of protein, typically 75%.

The protein components of a membrane can be readily visualized by *SDS–polyacrylamide gel electrophoresis.* As discussed earlier (Section 4.1.4), the electrophoretic mobility of many proteins in SDS-containing gels depends on the mass rather than on the net charge of the protein. The gel-electrophoresis patterns of three membranes—the plasma membrane of erythrocytes, the photoreceptor membrane of retinal rod cells, and the sarcoplasmic reticulum membrane of muscle—are shown in Figure 12.16. It is evident that each of these three membranes contains many proteins but has a distinct protein composition. In general, *membranes performing different functions contain different repertoires of proteins.*

12.5.1 Proteins Associate with the Lipid Bilayer in a Variety of Ways

The ease with which a protein can be dissociated from a membrane indicates how intimately it is associated with the membrane. Some membrane proteins can be solubilized by relatively mild means, such as extraction by a solution of high ionic strength (e.g., 1 M NaCl). Other membrane proteins are bound much more tenaciously; they can be solubilized only by using a detergent or an organic solvent. Membrane proteins can be classified as being either *peripheral* or *integral* on the basis of this difference in dissociability (Figure 12.17). *Integral membrane proteins* interact extensively with the hydrocarbon chains of membrane lipids, and so only agents that compete for these nonpolar interactions can release them. In fact, most integral membrane proteins span the lipid bilayer. In contrast, *peripheral membrane proteins* are bound to

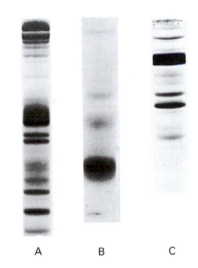

A   B   C

FIGURE 12.16 SDS–acrylamide gel patterns of membrane proteins.
(A) The plasma membrane of erythrocytes. (B) The photoreceptor membranes of retinal rod cells. (C) The sarcoplasmic reticulum membrane of muscle cells. [Courtesy of Dr. Theodore Steck (Part A) and Dr. David MacLennan (Part C).]

FIGURE 12.17 Integral and peripheral membrane proteins. Integral membrane proteins (*a*, *b*, and *c*) interact extensively with the hydrocarbon region of the bilayer. Nearly all known integral membrane proteins traverse the lipid bilayer. Peripheral membrane proteins (*d* and *e*) bind to the surfaces of integral proteins. Some peripheral membrane proteins interact with the polar head groups of the lipids (not shown).

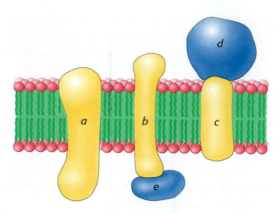

membranes primarily by electrostatic and hydrogen-bond interactions with the head groups of lipids. These polar interactions can be disrupted by adding salts or by changing the pH. Many peripheral membrane proteins are bound to the surfaces of integral proteins, on either the cytosolic or the extracellular side of the membrane. Others are anchored to the lipid bilayer by a covalently attached hydrophobic chain, such as a fatty acid.

12.5.2 Proteins Interact with Membranes in a Variety of Ways

Although membrane proteins are more difficult to purify and crystallize than are water-soluble proteins, researchers using x-ray crystallographic or electron microscopic methods have determined the three-dimensional structures of more than 20 such proteins at sufficiently high resolution to discern the molecular details. As noted in Chapter 3, the structures of membrane proteins differ from those of soluble proteins with regard to the distribution of hydrophobic and hydrophilic groups. We will consider the structures of three membrane proteins in some detail.

Proteins Can Span the Membrane with Alpha Helices. The first membrane protein that we consider is the archaeal protein *bacteriorhodopsin*, shown in Figure 12.18. This protein acts in energy transduction by using light energy to transport protons from inside the cell to outside. The proton gradient generated in this way is used to form ATP. Bacteriorhodopsin is built almost entirely of α helices; seven closely packed α helices, arranged almost perpendicularly to the plane of the membrane, span its 45-Å width. Examination of the primary structure of bacteriorhodopsin reveals that most of the amino acids in these membrane-spanning α helices are nonpolar and only a very few are charged (Figure 12.19). This distribution of nonpolar amino acids is sensible because these residues are either in contact with the

(A)

(B)

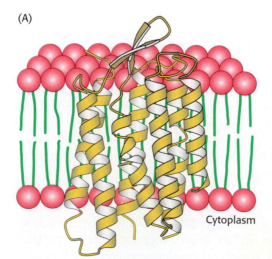

FIGURE 12.18 Structure of bacteriorhodopsin. Bacteriorhodopsin consists largely of membrane-spanning α helices. (A) View through the membrane bilayer. The interior of the membrane is green and the head groups are red. (B) A view from the cytoplasmic side of the membrane.

Cytoplasm

hydrocarbon core of the membrane or with one another. *Membrane-spanning α helices are the most common structural motif in membrane proteins.* As will be discussed in Section 12.5.4, such regions can often be detected from amino acid sequence information alone.

AQITGRPEWIWLALGTALMGLGTLYFLVKGMGVSDPDAKKFYAITTLVPA
IAFTMYLSMLLGYGLTMVPFGGEQNPIYWARYADWLFTTPLLLLDLALLV
DADQGTILALVGADGIMIGTGLVGALTKVYSYRFVWWAISTAAMLYILYV
LFFGFTSKAESMRPEVASTFKVLRNVTVVLWSAYVVVWLIGSEGAGIVPL
NIETLLFMVLDVSAKVGFGLILLRSRAIFGEAEAPEPSADGAAATS
```

**FIGURE 12.19 Amino acid sequence of bacteriorhodopsin.** The seven helical regions are highlighted in yellow and the charged residues in red.

**A Channel Protein Can Be Formed from Beta Strands.** Porin, a protein from the outer membranes of bacteria such as *E. coli* and *Rhodobacter capsulatus*, represents a class of membrane proteins with a completely different type of structure. Structures of this type are built from β strands and contain essentially no α helices (Figure 12.20).

(A)

(B)

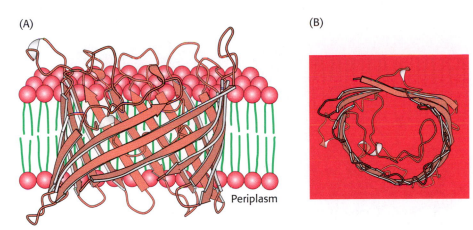

Periplasm

**FIGURE 12.20 Structure of bacterial porin (from *Rhodopseudomonas blastica*).** Porin is a membrane protein built entirely of beta strands. (A) Side view. (B) View from the periplasmic space. Only one monomer of the trimeric protein is shown.

The arrangement of β strands is quite simple: each strand is hydrogen bonded to its neighbor in an antiparallel arrangement, forming a single β sheet. The β sheet curls up to form a hollow cylinder that functions as the active unit. As its name suggests, porin forms pores, or channels, in the membranes. A pore runs through the center of each cylinder-like protein. The outside surface of porin is appropriately nonpolar, given that it interacts with the hydrocarbon core of the membrane. In contrast, the inside of the channel is quite hydrophilic and is filled with water. This arrangement of nonpolar and polar surfaces is accomplished by the alternation of hydrophobic and hydrophilic amino acids along each β strand (Figure 12.21).

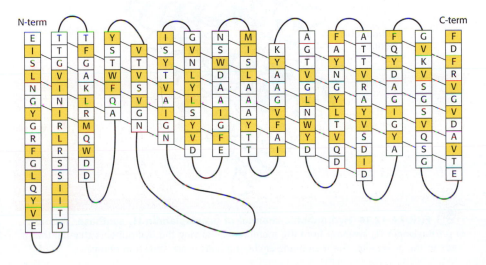

**FIGURE 12.21 Amino acid sequence of a porin.** Some membrane proteins such as porins are built from β strands that tend to have hydrophobic and hydrophilic amino acids in adjacent positions. The secondary structure of *Rhodopseudomonas blastica* is shown, with the diagonal lines indicating the direction of hydrogen bonding along the β sheet. Hydrophobic residues (F, I, L, M, V, W, and Y) are shown in yellow. These residues tend to lie on the outside of the structure, in contact with the hydrophobic core of the membrane.
```

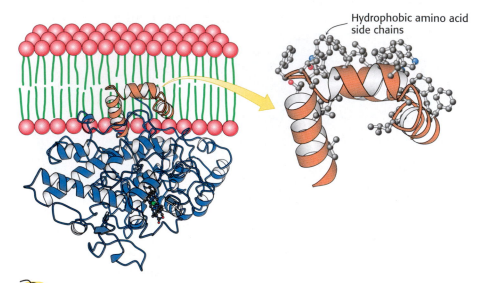

Arachidonate

Cyclooxygenase | 2 O$_2$

Prostaglandin G$_2$

Peroxidase | 2 H$^+$ + 2 e$^-$
→ H$_2$O

Prostaglandin H$_2$

FIGURE 12.22 Formation of prostaglandin H$_2$. Prostaglandin H$_2$ synthase-1 catalyzes the formation of prostaglandin H$_2$ from arachidonic acid in two steps.

Embedding Part of a Protein in a Membrane Can Link It to the Membrane Surface. The structure of the membrane-bound enzyme prostaglandin H$_2$ synthase-1 reveals a rather different role for α helices in protein–membrane associations. This enzyme catalyzes the conversion of arachidonic acid into prostaglandin H$_2$ in two steps: a cyclooxygenase reaction and a peroxidase reaction (Figure 12.22). Prostaglandin H$_2$ promotes inflammation and modulates gastric acid secretion. The enzyme that produces prostaglandin H$_2$ is a homodimer with a rather complicated structure consisting primarily of α helices. Unlike bacteriorhodopsin, this protein is not largely embedded in the membrane but, instead, lies along the outer surface of the membrane firmly bound by a set of α helices with hydrophobic surfaces that extend from the bottom of the protein into the membrane (Figure 12.23). This linkage is sufficiently strong that only the action of detergents can release the protein from the membrane. Thus, this enzyme is classified as an integral membrane protein, although it is not a membrane-spanning protein.

Hydrophobic amino acid side chains

FIGURE 12.23 Attachment of prostaglandin H$_2$ synthase-1 to the membrane. Prostaglandin H$_2$ synthase-1 is held in the membrane by a set of α helices coated with hydrophobic side chains. One monomer of the dimeric enzyme is shown.

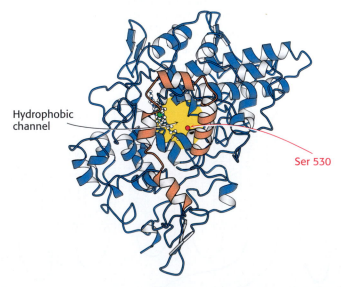

Hydrophobic channel

Ser 530

FIGURE 12.24 Hydrophobic channel of prostaglandin H$_2$ synthase. A view of prostaglandin H$_2$ synthase from the membrane, showing the hydrophobic channel that leads to the active site. The membrane-anchoring helices are shown in orange.

The localization of prostaglandin H_2 synthase-1 in the membrane is crucial to its function. The substrate for this enzyme, arachidonic acid, is a hydrophobic molecule generated by the hydrolysis of membrane lipids. Arachidonic acid reaches the active site of the enzyme from the membrane without entering an aqueous environment by traveling through a hydrophobic channel in the protein (Figure 12.24). Indeed, nearly all of us have experienced the importance of this channel: drugs such as aspirin and ibuprofen block the channel and prevent prostaglandin synthesis by inhibiting the cyclooxygenase activity of the synthase. In the case of aspirin, the drug acts through the transfer of an acetyl group from the aspirin to a serine residue (Ser 530) that lies along the path to the active site (Figure 12.25).

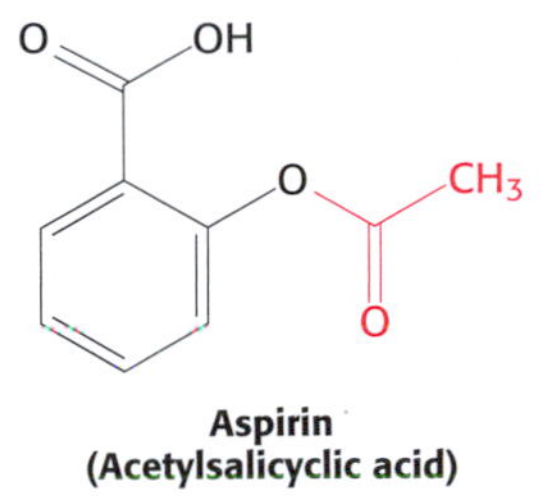

Two important features emerge from our examination of these three examples of membrane protein structure. First, the parts of the protein that interact with the hydrophobic parts of the membrane are coated with nonpolar amino acid side chains, whereas those parts that interact with the aqueous environment are much more hydrophilic. Second, the structures positioned within the membrane are quite regular and, in particular, all backbone hydrogen-bond donors and acceptors participate in hydrogen bonds. *Breaking a hydrogen bond within a membrane is quite unfavorable, because little or no water is present to compete for the polar groups.*

Aspirin
(Acetylsalicyclic acid)

FIGURE 12.25 Aspirin's effects on prostaglandin H_2 synthase-1. Aspirin acts by transferring an acetyl group to a serine residue in prostaglandin H_2 synthase-1.

12.5.3 Some Proteins Associate with Membranes Through Covalently Attached Hydrophobic Groups

The membrane proteins considered thus far associate with the membrane through surfaces generated by hydrophobic amino acid side chains. However, even otherwise soluble proteins can associate with membranes if the association is mediated by hydrophobic groups attached to the proteins. Three such groups are shown in Figure 12.26: (1) a palmitoyl group attached to a specific cysteine residue by a thioester bond, (2) a farnesyl group attached to a cysteine residue at the carboxyl terminus, and (3) a glycolipid structure termed a glycosyl phosphatidyl inositol (GPI) anchor attached to the carboxyl terminus. These modifications are attached by enzyme systems that recognize specific signal sequences near the site of attachment.

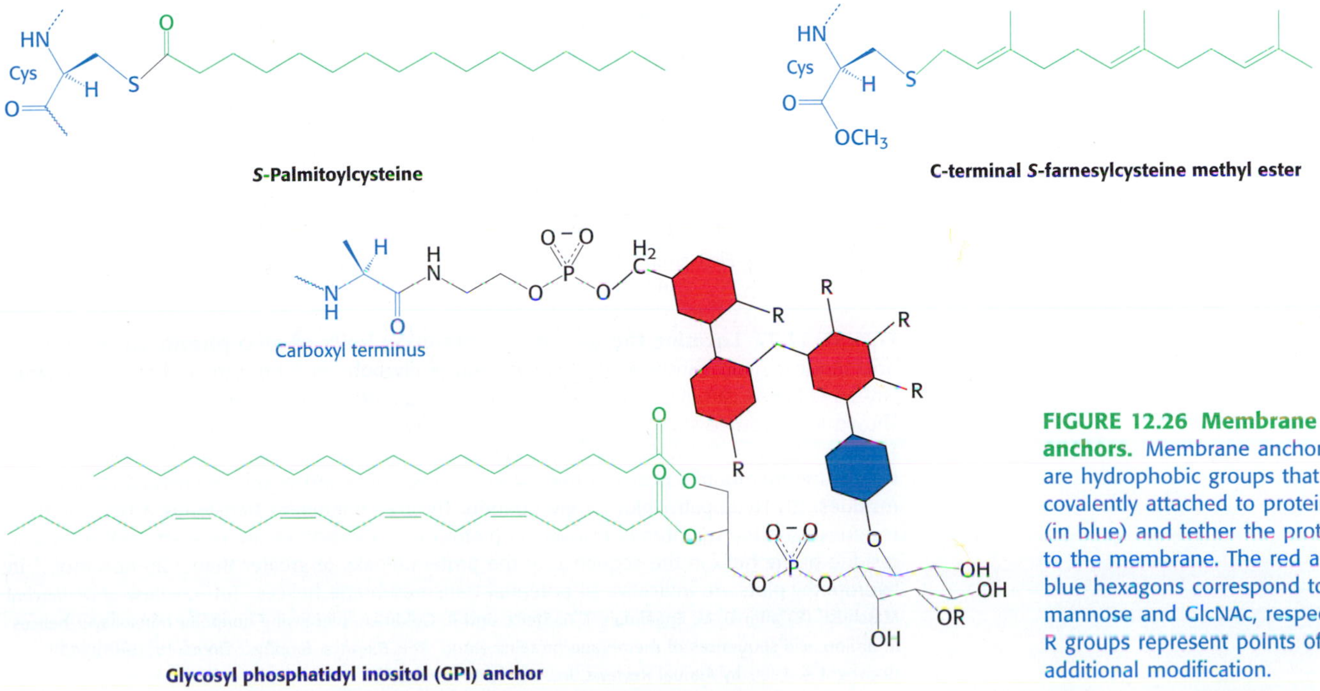

S-Palmitoylcysteine

C-terminal S-farnesylcysteine methyl ester

Glycosyl phosphatidyl inositol (GPI) anchor

FIGURE 12.26 Membrane anchors. Membrane anchors are hydrophobic groups that are covalently attached to proteins (in blue) and tether the proteins to the membrane. The red and blue hexagons correspond to mannose and GlcNAc, respectively. R groups represent points of additional modification.

Amino acid residue	Transfer free energy kcal mol^{-1} (kJ mol^{-1})	
Phe	3.7	(15.5)
Met	3.4	(14.3)
Ile	3.1	(13.0)
Leu	2.8	(11.8)
Val	2.6	(10.9)
Cys	2.0	(8.4)
Trp	1.9	(8.0)
Ala	1.6	(6.7)
Thr	1.2	(5.0)
Gly	1.0	(4.2)
Ser	0.6	(2.5)
Pro	−0.2	(−0.8)
Tyr	−0.7	(−2.9)
His	−3.0	(−12.6)
Gln	−4.1	(−17.2)
Asn	−4.8	(−20.2)
Glu	−8.2	(−34.4)
Lys	−8.8	(−37.0)
Asp	−9.2	(−38.6)
Arg	−12.3	(−51.7)

Source: After D. M. Engelman, T. A. Steitz, and A. Goldman. *Annu. Rev. Biophys. Biophys. Chem.* 15(1986):330. *Note*: The free energies are for the transfer of an amino acid residue in an α helix from the membrane interior (assumed to have a dielectric constant of 2) to water.

12.5.4 Transmembrane Helices Can Be Accurately Predicted from Amino Acid Sequences

Many membrane proteins, like bacteriorhodopsin, employ α helices to span the hydrophobic part of a membrane. As noted earlier, typically most of the residues in these α helices are nonpolar and almost none of them are charged. Can we use this information to identify putative membrane-spanning regions from sequence data alone? One approach to identifying transmembrane helices is to ask whether a postulated helical segment is likely to be most stable in a hydrocarbon milieu or in water. Specifically, we want to estimate the free-energy change when a helical segment is transferred from the interior of a membrane to water. Free-energy changes for the transfer of individual amino acid residues from a hydrophobic to an aqueous environment are given in Table 12.2. For example, the transfer of a poly-L-arginine helix, a homopolymer of a positively charged amino acid, from the interior of a membrane to water would be highly favorable [−12.3 kcal mol^{-1} (−51.5 kJ mol^{-1}) per arginine residue in the helix], whereas the transfer of

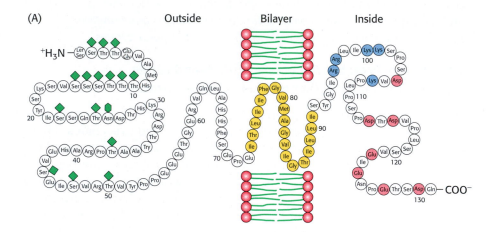

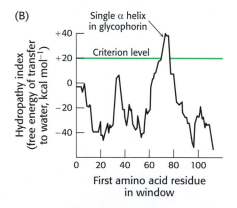

FIGURE 12.27 Locating the membrane-spanning helix of glycophorin. (A) Amino acid sequence and transmembrane disposition of glycophorin A from the red-cell membrane. Fifteen *O*-linked carbohydrate units are shown as diamond shapes, and an *N*-linked unit is shown as a lozenge shape. The hydrophobic residues (yellow) buried in the bilayer form a transmembrane α helix. The carboxyl-terminal part of the molecule, located on the cytosolic side of the membrane, is rich in negatively charged (red) and positively charged (blue) residues. (B) Hydropathy plot for glycophorin. The free energy for transfering a helix of 20 residues from the membrane to water is plotted as a function of the position of the first residue of the helix in the sequence of the protein. Peaks of greater than +20 kcal mol^{-1} in hydropathy plots are indicative of potential transmembrane helices. [(A) Courtesy of Dr. Vincent Marchesi; (B) after D. M. Engelman, T. A. Steitz, and A. Goldman. Identifying nonpolar transbilayer helices in amino acid sequences of membrane proteins. *Annu. Rev. Biophys. Biophys. Chem.* 15(1986):343. Copyright © 1986 by Annual Reviews, Inc. All rights reserved.]

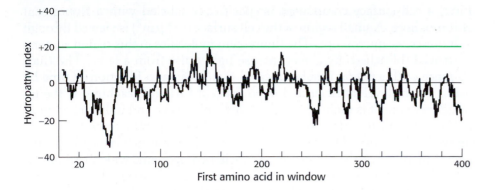

FIGURE 12.28 Hydropathy plot for porin. No strong peaks are observed for this intrinsic membrane protein because it is constructed from membrane-spanning β strands rather than α helices.

a poly-L-phenylalanine helix, a homopolymer of a hydrophobic amino acid, would be unfavorable [$+3.7$ kcal mol^{-1} ($+15.5$ kJ mol^{-1}) per phenylala-nine residue in the helix].

The hydrocarbon core of a membrane is typically 30 Å wide, a length that can be traversed by an α helix consisting of 20 residues. We can take the amino acid sequence of a protein and estimate the free-energy change that takes place when a hypothetical α helix formed of residues 1 through 20 is transferred from the membrane interior to water. The same calculation can be made for residues 2 through 21, 3 through 22, and so forth, until we reach the end of the sequence. The span of 20 residues chosen for this calculation is called a *window*. The free-energy change for each window is plotted against the first amino acid at the window to create a *hydropathy plot*. Empirically, a peak of $+20$ kcal mol^{-1} ($+84$ kJ mol^{-1}) or more in a hydropathy plot based on a window of 20 residues indicates that a polypeptide segment could be a membrane-spanning α helix. For example, glycophorin, a protein found in the membranes of red blood cells, is predicted by this criterion to have one membrane-spanning helix, in agreement with experimental findings (Figure 12.27). It should be noted, however, that a peak in the hydropathy plot does not prove that a segment is a transmembrane helix. Even soluble proteins may have highly nonpolar regions. Conversely, some membrane proteins contain transmembrane structural features (such as membrane-spanning β strands) that escape detection by these plots (Figure 12.28).

12.6 LIPIDS AND MANY MEMBRANE PROTEINS DIFFUSE RAPIDLY IN THE PLANE OF THE MEMBRANE

Biological membranes are not rigid, static structures. On the contrary, lipids and many membrane proteins are constantly in lateral motion, a process called *lateral diffusion*. The rapid lateral movement of membrane proteins has been visualized by means of fluorescence microscopy through the use of the technique of *fluorescence recovery after photobleaching* (FRAP; Figure 12.29).

FIGURE 12.29 Fluorescence recovery after photobleaching (FRAP) technique. (A) The cell-surface fluoresces because of a labeled surface component. (B) The fluorescent molecules of a small part of the surface are bleached by an intense light pulse. (C) The fluorescence intensity recovers as bleached molecules diffuse out of the region and unbleached molecules diffuse into it. (D) The rate of recovery depends on the diffusion coefficient.

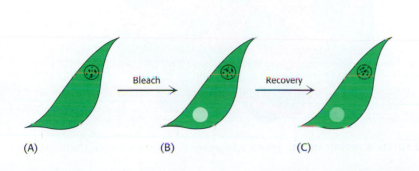

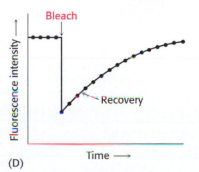

First, a cell-surface component is specifically labeled with a fluorescent chromophore. A small region of the cell surface ($\sim 3~\mu m^2$) is viewed through a fluorescence microscope. The fluorescent molecules in this region are then destroyed (bleached) by a very intense light pulse from a laser. The fluorescence of this region is subsequently monitored as a function of time by using a light level sufficiently low to prevent further bleaching. If the labeled component is mobile, bleached molecules leave and unbleached molecules enter the illuminated region, which results in an increase in the fluorescence intensity. The rate of recovery of fluorescence depends on the lateral mobility of the fluorescence-labeled component, which can be expressed in terms of a diffusion coefficient, D. The average distance s traversed in time t depends on D according to the expression

$$s = (4Dt)^{1/2}$$

The diffusion coefficient of lipids in a variety of membranes is about $1~\mu m^2~s^{-1}$. Thus, a phospholipid molecule diffuses an average distance of $2~\mu m$ in 1 s. This rate means that *a lipid molecule can travel from one end of a bacterium to the other in a second.* The magnitude of the observed diffusion coefficient indicates that the viscosity of the membrane is about 100 times that of water, rather like that of olive oil.

In contrast, proteins vary markedly in their lateral mobility. *Some proteins are nearly as mobile as lipids, whereas others are virtually immobile.* For example, the photoreceptor protein rhodopsin (Section 32.3.1), a very mobile protein, has a diffusion coefficient of $0.4~\mu m^2~s^{-1}$. The rapid movement of rhodopsin is essential for fast signaling. At the other extreme is fibronectin, a peripheral glycoprotein that interacts with the extracellular matrix. For fibronectin, D is less than $10^{-4}~\mu m^2~s^{-1}$. Fibronectin has a very low mobility because it is anchored to actin filaments on the inside of the plasma membrane through *integrin*, a transmembrane protein that links the extracellular matrix to the cytoskeleton.

12.6.1 The Fluid Mosaic Model Allows Lateral Movement but Not Rotation Through the Membrane

On the basis of the dynamic properties of proteins in membranes, S. Jonathan Singer and Garth Nicolson proposed the concept of a *fluid mosaic model* for the overall organization of biological membranes in 1972 (Figure 12.30). The essence of their model is that *membranes are two-dimensional solutions of oriented lipids and globular proteins.* The lipid bilayer has a dual role: it is both a *solvent* for integral membrane proteins and a *permeability*

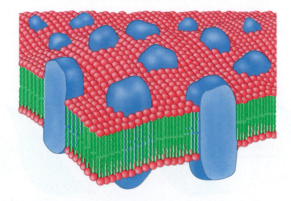

FIGURE 12.30 Fluid mosaic model. [After S. J. Singer and G. L. Nicolson. *Science* 175(1972):723.]

barrier. Membrane proteins are free to diffuse laterally in the lipid matrix unless restricted by special interactions.

Although the lateral diffusion of membrane components can be rapid, the spontaneous rotation of lipids from one face of a membrane to the other is a very slow process. The transition of a molecule from one membrane surface to the other is called *transverse diffusion* or *flip-flop* (Figure 12.31) The flip-flop of phospholipid molecules in phosphatidyl choline vesicles has been directly measured by electron spin resonance techniques, which show that *a phospholipid molecule flip-flops once in several hours.* Thus, a phospholipid molecule takes about 10^9 times as long to flip-flop across a membrane as it takes to diffuse a distance of 50 Å in the lateral direction. The free-energy barriers to flip-flopping are even larger for protein molecules than for lipids because proteins have more extensive polar regions. In fact, the flip-flop of a protein molecule has not been observed. Hence, *membrane asymmetry can be preserved for long periods.*

12.6.2 Membrane Fluidity Is Controlled by Fatty Acid Composition and Cholesterol Content

Many membrane processes, such as transport or signal transduction, depend on the fluidity of the membrane lipids, which in turn depends on the properties of fatty acid chains, which can exist in an ordered, rigid state or in a relatively disordered, fluid state. The transition from the rigid to the fluid state occurs rather abruptly as the temperature is raised above T_m, the melting temperature (Figure 12.32). *This transition temperature depends on the length of the fatty acyl chains and on their degree of unsaturation* (Table 12.3). The presence of saturated fatty acyl residues favors the rigid state because their straight hydrocarbon chains interact very favorably with each other. On the other hand, *a cis double bond produces a bend in the hydrocarbon chain. This bend interferes with a highly ordered packing of fatty acyl chains, and so* T_m *is lowered* (Figure 12.33). The length of the fatty acyl chain also affects the transition temperature. Long hydrocarbon chains interact more strongly than do short ones. Specifically, each additional $-CH_2-$ group makes a favorable contribution of about -0.5 kcal mol^{-1} (-2.1 kJ mol^{-1}) to the free energy of interaction of two adjacent hydrocarbon chains.

Bacteria regulate the fluidity of their membranes by varying the number of double bonds and the length of their fatty acyl chains. For example, the ratio of saturated to unsaturated fatty acyl chains in the *E. coli* membrane decreases from 1.6 to 1.0 as the growth temperature is lowered from 42°C to 27°C. This decrease in the proportion of saturated residues prevents the membrane from becoming too rigid at the lower temperature.

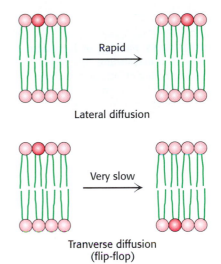

Lateral diffusion

Very slow

Tranverse diffusion
(flip-flop)

FIGURE 12.31 Lipid movement in membranes. Lateral diffusion of lipids is much more rapid than transverse diffusion (flip-flop).

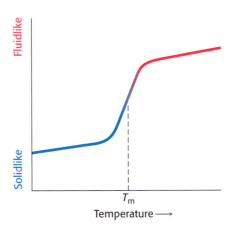

Fluidlike

Solidlike

T_m
Temperature ⟶

FIGURE 12.32 The phase-transition, or melting, temperature (T_m) for a phospholipid membrane. As the temperature is raised, the phospholipid membrane changes from a packed, ordered state to a more random one.

TABLE 12.3 The melting temperature of phosphatidyl choline containing different pairs of identical fatty acid chains

Number of carbons	Number of double bonds	Fatty acid		T_m (°C)
		Common name	Systematic name	
22	0	Behenate	*n*-Docosanote	75
18	0	Stearate	*n*-Octadecanoate	58
16	0	Palmitate	*n*-Hexadecanoate	41
14	0	Myristate	*n*-Tetradecanoate	24
18	1	Oleate	*cis*-Δ^9-Octadecenoate	-22

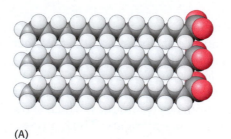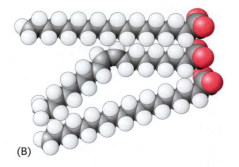

(A)   (B)

FIGURE 12.33 Packing of fatty acid chains in a membrane. The highly ordered packing of fatty acid chains is disrupted by the presence of cis double bonds. The space-filling models show the packing of (A) three molecules of stearate (C_{18}, saturated) and (B) a molecule of oleate (C_{18}, unsaturated) between two molecules of stearate.

In animals, cholesterol is the key regulator of membrane fluidity. Cholesterol contains a bulky steroid nucleus with a hydroxyl group at one end and a flexible hydrocarbon tail at the other end. Cholesterol inserts into bilayers with its long axis perpendicular to the plane of the membrane. The hydroxyl group of cholesterol forms a hydrogen bond with a carbonyl oxygen atom of a phospholipid head group, whereas the hydrocarbon tail of cholesterol is located in the nonpolar core of the bilayer. The different shape of cholesterol compared with phospholipids disrupts the regular interactions between fatty acyl chains. In addition, cholesterol appears to form specific complexes with some phospholipids. Such complexes may concentrate in specific regions within membranes. One result of these interactions is the *moderation of membrane fluidity*, making membranes less fluid but at the same time less subject to phase transitions.

12.6.3 All Biological Membranes Are Asymmetric

Membranes are structurally and functionally asymmetric. The outer and inner surfaces of *all known biological membranes have different components and different enzymatic activities.* A clear-cut example is the pump that regulates the concentration of Na^+ and K^+ ions in cells (Figure 12.34). This transport protein is located in the plasma membrane of nearly all cells in higher organisms. The Na^+-K^+ pump is oriented so that it pumps Na^+ out of the cell and K^+ into it. Furthermore, ATP must be on the inside of the cell to drive the pump. Ouabain, a specific inhibitor of the pump, is effective only if it is located outside.

Membrane proteins have a unique orientation because they are synthesized and inserted into the membrane in an asymmetric manner. This absolute asymmetry is preserved because membrane proteins do not rotate from one side of the membrane to the other and because *membranes are always synthesized by the growth of preexisting membranes.* Lipids, too, are asymmetrically distributed as a consequence of their mode of biosynthesis, but this asymmetry is usually not absolute, except for glycolipids. In the red-blood-cell membrane, sphingomyelin and phosphatidyl choline are preferentially located in the outer leaflet of the bilayer, whereas phosphatidyl ethanolamine and phosphatidyl serine are located mainly in the inner leaflet. Large amounts of cholesterol are present in both leaflets.

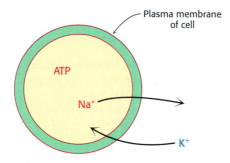

Plasma membrane of cell

ATP

Na^+

K^+

FIGURE 12.34 Asymmetry of the Na^+-K^+ transport system in plasma membranes. The Na^+-K^+ transport system pumps Na^+ out of the cell and K^+ into the cell.

12.7 EUKARYOTIC CELLS CONTAIN COMPARTMENTS BOUNDED BY INTERNAL MEMBRANES

Thus far we have considered only the plasma membrane of cells. Many bacteria such as *E. coli* have two membranes separated by a cell wall (made of proteins, peptides, and carbohydrates) lying in between (Figure 12.35). The inner membrane acts as the permeability barrier, and the outer membrane and

(A)

Outer membrane

Lipoprotein

Cell wall

Inner membrane

Porin

Periplasm

Cytoplasm

(B)

Cell wall

Membrane

FIGURE 12.35 **Cell membranes of prokaryotes.** A schematic view of the membrane in bacterial cells surrounded by (A) two membranes or (B) one membrane.

the cell wall provide additional protection. The outer membrane is quite permeable to small molecules owing to the presence of porins. The region between the two membranes containing the cell wall is called the *periplasm*. Other bacteria and archaea have only a single membrane surrounded by a cell wall.

Eukaryotic cells, with the exception of plant cells, do not have cell walls, and their cell membranes consist of a single lipid bilayer. In plant cells, the cell wall is on the outside of the plasma membrane. Eukaryotic cells are distinguished by the use of membranes inside the cell to form internal compartments (Figure 12.36). For example, peroxisomes, organelles that play a major role in the oxidation of fatty acids for energy conversion, are defined by a single membrane. Mitochondria, the organelles in which ATP is synthesized, are surrounded by two membranes. Much like the case for a bacterium, the outer membrane is quite permeable to small molecules, whereas the inner membrane is not. Indeed, considerable evidence now indicates that mitochondria evolved from bacteria by *endosymbiosis* (Section 18.1.2). A double membrane also surrounds the nucleus. However, the *nuclear envelope* is not continuous but, instead, consists of a set of closed membranes that come together at structures called *nuclear pores*. These pores regulate transport into and out of the nucleus. The nuclear membranes are linked to another membrane-defined structure, the *endoplasmic reticulum*, which plays a host of cellular roles, including drug detoxification and the modification of proteins for secretion (Section 11.3.4). Thus, a eukaryotic cell comprises interacting compartments, and transport into and out of these compartments is essential to many biochemical processes.

12.7.1 Proteins Are Targeted to Specific Compartments by Signal Sequences

The compartmentalization of eukaryotic cells makes possible many processes that must be separated from the remainder of the cellular environment to function properly. Specific proteins are found in peroxisomes, others in mitochondria, and still others in the nucleus. How do proteins end up in the proper compartment? Even for bacteria, some targeting of proteins is required: some proteins are secreted from the cell, whereas others remain in the cytosol.

Proteins include specific sequences that serve as address labels to direct the molecules to the proper location. For example, most peroxisomal proteins end with a sequence, Ser-Lys-Leu-COO$^-$, that acts as an autonomous

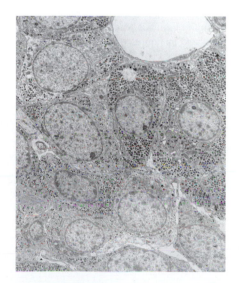

FIGURE 12.36 **Internal membranes of eukaryotes.** Electron micrograph of a thin section of a hormone-secreting cell for the rat pituitary, showing the presence of internal structures bounded by membranes. [Biophoto Associates/Photo Researchers.]

^+H_3N —— M L R T S S L F T R R V Q P S L F R N I L R L Q S T

FIGURE 12.37 A mitochondrial targeting sequence. This sequence is recognized by receptors on the external face of the outer mitochondrial membrane. A protein bearing the sequence will be imported into the mitochondrion. Hydrophobic residues are shown in yellow, basic ones in blue, and serine and threonine in red.

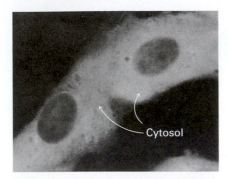

(A)

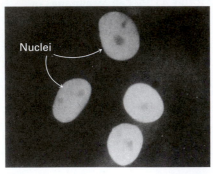

Nuclei

(B)

FIGURE 12.38 Movement of a protein into the nucleus. Localization of (A) unmodified pyruvate kinase, and (B) pyruvate kinase containing a nuclear localization signal sequence attached to its amino terminus. The protein was visualized by fluorescence microscopy after staining with a specific antibody. [From W. D. Richardson, B. L. Roberts, and A. E. Smith. *Cell* 44(1986):79.]

targeting signal. The removal of this sequence from a protein that normally resides in peroxisomes blocks its import into that organelle, whereas the addition of this sequence to a protein that normally resides in the cytosol can direct that protein to peroxisomes. A protein destined to pass through both mitochondrial membranes usually has a targeting sequence at its amino terminus (Figure 12.37). Unlike the peroxisomal targeting sequence, these amino-terminal sequences are highly variable; no clear consensus exists. They are typically from 15 to 35 residues long and rich in positively charged residues and in serines and threonines. Proteins destined for the nucleus have internal targeting sequences. A typical nuclear localization signal contains five consecutive positively charged residues such as Lys-Lys-Lys-Arg-Lys. The addition of such a sequence to a protein not found in the nucleus can direct it to the nucleus (Figure 12.38). Other sequences can direct proteins out of the nucleus. The known targeting sequences are given in Table 12.4.

TABLE 12.4 Targeting sequences

Target	Signal
Nucleus	–KKXK or $-(K/R)_2-X_{10-12}-(K/R)^*$
Peroxisome	$-SKL-COO^-$
Mitochondrion	N-terminal amphipathic helix
Endoplasmic reticulum	$-KDEL-COO^-$(ER retention)

*The "/" means that either K or R is required.

Targeting sequences act by binding to specific proteins associated with each organelle. The determination of the structure of a protein, α-karyopherin, that binds to the nuclear localization signal reveals how the protein recognizes such a targeting sequence (Figure 12.39). A peptide containing the appropriate sequence binds to a specific site on the protein. The target peptide is held in an extended conformation through interactions between the target peptide backbone and asparagine side chains of the α-karyopherin while each of the basic residues lies in a deep pocket near the bottom, lined with negatively charged residues. Proteins that bind to the other targeting signal sequences presumably also have structures that allow recognition of

FIGURE 12.39 Protein targeting signal recognition. The structure of the nuclear localization signal–binding protein α-karyopherin (also known as α-importin) with a nuclear localization signal peptide bound to its major recognition site.

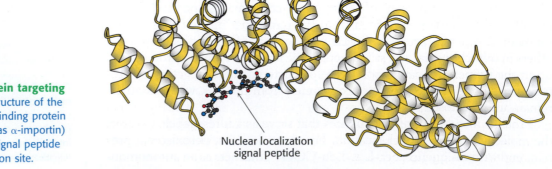

Nuclear localization signal peptide

the required features. Note that we have considered only how proteins are marked for different compartments. Other mechanisms are required for proteins actually to cross membranes (Section 11.3.2).

12.7.2 Membrane Budding and Fusion Underlie Several Important Biological Processes

Membranes must be able to separate or join together to take up, transport, and release molecules. Many take up molecules through the process of *receptor-mediated endocytosis* (Figure 12.40). Here, a protein or larger complex initially binds to a receptor on the cell surface. After the protein is bound, specialized proteins act to cause the membrane in the vicinity of the bound protein to invaginate. The invaginated membrane eventually breaks off and fuses to form a *vesicle*.

Receptor-mediated endocytosis plays a key role in cholesterol metabolism (Section 26.3.3). Some cholesterol in the blood is in the form of a lipid–protein complex called *low-density lipoprotein* (LDL). Low density lipoprotein binds to an LDL receptor, an integral membrane protein. The segment of the plasma membrane containing the LDL–LDL receptor complex then invaginates and buds off from the membrane. The LDL separates from the receptor, which is recycled back to the membrane in a separate vesicle. The vesicle containing the LDL fuses with a *lysosome*, an organelle containing an array of digestive enzymes. The cholesterol is released into the cell for storage or use in membrane biosynthesis, and the remaining protein components are degraded. Various hormones, transport proteins, and antibodies employ receptor-mediated endocytosis to gain entry into a cell. A less advantageous consequence is that this pathway is available to viruses and toxins as a means of entry into cells. The reverse process—the fusion of a vesicle to a membrane—is a key step in the release of neurotransmitters from a neuron into the synaptic cleft (Figure 12.41). Although the processes of budding and fusion appear deceptively simple, the structures of the intermediates in the budding and fusing processes and the detailed mechanisms remain active areas of investigation.

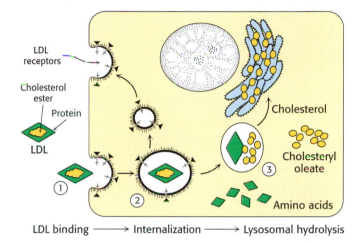

FIGURE 12.40 Receptor-mediated endocytosis. The process of receptor-mediated endocytosis is illustrated for the cholesterol-carrying complex, low-density lipoprotein (LDL): (1) LDL binds to a specific receptor, the LDL receptor; (2) this complex invaginates to form an internal vesicle; (3) after separation from its receptor, the LDL-containing vesicle fuses with a lysosome, leading to degradation of the LDL and release of the cholesterol.

SUMMARY

● **Many Common Features Underlie the Diversity of Biological Membranes**
Biological membranes are sheetlike structures, typically from 60 to 100 Å thick, that are composed of protein and lipid molecules held together by noncovalent interactions. Membranes are highly selective permeability barriers. They create closed compartments, which may be entire cells or organelles within a cell. Proteins in membranes regulate the molecular and ionic compositions of these compartments. Membranes also control the flow of information between cells.

● **Fatty Acids Are Key Constituents of Lipids**
Fatty acids are hydrocarbon chains of various lengths and degrees of unsaturation that terminate with a carboxylic acid group. The fatty acid chains in membranes usually contain between 14 and 24 carbon atoms; they may be saturated or unsaturated. Short chain length and unsaturation enhance the fluidity of fatty acids and their derivatives by lowering the melting temperature.

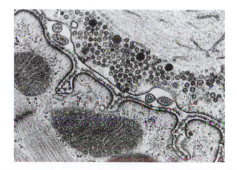

FIGURE 12.41 Neurotransmitter release. Neurotransmitter-containing synaptic vesicles are arrayed near the plasma membrane of a nerve cell. Synaptic vesicles fuse with the plasma membrane, releasing the neurotransmitter into the synaptic cleft. [T. Reese/Don Fawcett/Photo Researchers.]

There Are Three Common Types of Membrane Lipids

The major classes of membrane lipids are phospholipids, glycolipids, and cholesterol. Phosphoglycerides, a type of phospholipid, consist of a glycerol backbone, two fatty acid chains, and a phosphorylated alcohol. Phosphatidyl choline, phosphatidyl serine, and phosphatidyl ethanolamine are major phosphoglycerides. Sphingomyelin, a different type of phospholipid, contains a sphingosine backbone instead of glycerol. Glycolipids are sugar-containing lipids derived from sphingosine. Cholesterol, which modulates membrane fluidity, is constructed from a steroid nucleus. A common feature of these membrane lipids is that they are amphipathic molecules, having hydrophobic and hydrophilic ends.

Phospholipids and Glycolipids Readily Form Bimolecular Sheets in Aqueous Media

Membrane lipids spontaneously form extensive bimolecular sheets in aqueous solutions. The driving force for membrane formation is the hydrophobic interactions among the fatty acid tails of membrane lipids. The hydrophilic head groups interact with the aqueous medium. Lipid bilayers are cooperative structures, held together by many weak bonds. These lipid bilayers are highly impermeable to ions and most polar molecules, yet they are quite fluid, which enables them to act as a solvent for membrane proteins.

Proteins Carry Out Most Membrane Processes

Specific proteins mediate distinctive membrane functions such as transport, communication, and energy transduction. Many integral membrane proteins span the lipid bilayer, whereas others are partly embedded in the membrane. Peripheral membrane proteins are bound to membrane surfaces by electrostatic and hydrogen-bond interactions. Membrane-spanning proteins have regular structures, including β strands, although the α helix is the most common membrane-spanning domain. Indeed, sequences of 20 consecutive nonpolar amino acids can be diagnostic of a membrane-spanning α-helical region of a protein.

Lipids and Many Membrane Proteins Diffuse Rapidly in the Plane of the Membrane

Membranes are structurally and functionally asymmetric, as exemplified by the restriction of sugar residues to the external surface of mammalian plasma membranes. Membranes are dynamic structures in which proteins and lipids diffuse rapidly in the plane of the membrane (lateral diffusion), unless restricted by special interactions. In contrast, the rotation of lipids from one face of a membrane to the other (transverse diffusion, or flip-flop) is usually very slow. Proteins do not rotate across bilayers; hence, membrane asymmetry can be preserved. The degree of fluidity of a membrane partly depends on the chain length of its lipids and the extent to which their constituent fatty acids are unsaturated. In animals, cholesterol content also regulates membrane fluidity.

Eukaryotic Cells Contain Compartments Bounded by Internal Membranes

An extensive array of internal membranes in eukaryotes creates compartments within a cell for distinct biochemical functions. For instance, a double membrane surrounds the nucleus, the location of most of the cell's genetic material, and the mitochondria, the location of most ATP synthesis. A single membrane defines the other internal compartments, such as the endoplasmic reticulum. Some compartments can exchange material by the process of membrane budding and fusion. As with all membranes, the proteins associated with these membranes determine the specific biochemical function. Specific amino acid sequences in the proteins direct these molecules to the appropriate compartment.

KEY TERMS

fatty acid (p. 320)

phospholipid (p. 322)

sphingosine (p. 322)

phosphoglyceride (p. 322)

sphingomyelin (p. 323)

glycolipid (p. 324)

cerebroside (p. 325)

ganglioside (p. 325)

cholesterol (p. 325)

amphipathic molecule (p. 325)

lipid bilayer (p. 326)

liposome (p. 327)

integral membrane protein (p. 329)

peripheral membrane protein (p. 329)

hydropathy plot (p. 335)

lateral diffusion (p. 335)

fluid mosaic model (p. 336)

targeting sequence (p. 340)

receptor-mediated endocytosis (p. 341)

SELECTED READINGS

Where to start

De Weer, P., 2000. A century of thinking about cell membranes. *Annu. Rev. Physiol.* 62:919–926.

Bretscher, M. S., 1985. The molecules of the cell membrane. *Sci. Am.* 253(4):100–108.

Unwin, N., and Henderson, R., 1984. The structure of proteins in biological membranes. *Sci. Am.* 250(2):78–94.

Deisenhofer, J., and Michel, H., 1989. The photosynthetic reaction centre from the purple bacterium *Rhodopseudomonas viridis. EMBO J.* 8:2149–2170.

Singer, S. J., and Nicolson, G. L., 1972. The fluid mosaic model of the structure of cell membranes. *Science* 175:720–731.

Jacobson, K., Sheets, E. D., and Simson, R., 1995. Revisiting the fluid mosaic model of membranes. *Science* 268:1441–1442.

Books

Gennis, R. B., 1989. *Biomembranes: Molecular Structure and Function.* Springer Verlag.

Vance, D. E., and Vance, J. E. (Eds.)., 1996. *Biochemistry of Lipids, Lipoproteins, and Membranes.* Elsevier.

Lipowsky, R., and Sackmann, E., 1995. *The structure and dynamics of membranes.* Elsevier.

Racker, E., 1985. *Reconstitutions of Transporters, Receptors, and Pathological States.* Academic Press.

Tanford, C., 1980. *The Hydrophobic Effect: Formation of Micelles and Biological Membranes* (2d ed.). Wiley-Interscience.

Membrane lipids and dynamics

Saxton, M. J., and Jacobson, K., 1997. Single-particle tracking: Applications to membrane dynamics. *Annu. Rev. Biophys. Biomol. Struct.* 26:373–399.

Bloom, M., Evans, E., Mouritsen, O. G., 1991. Physical properties of the fluid lipid-bilayer component of cell membranes: A perspective. *Q. Rev. Biophys.* 24:293–397.

Elson, E. L., 1986. Membrane dynamics studied by fluorescence correlation spectroscopy and photobleaching recovery. *Soc. Gen. Physiol. Ser.* 40:367–383.

Zachowski, A., and Devaux, P. F., 1990. Transmembrane movements of lipids. *Experientia* 46:644–656.

Devaux, P. F., 1992. Protein involvement in transmembrane lipid asymmetry. *Annu. Rev. Biophys. Biomol. Struct.* 21:417–439.

Silvius, J. R., 1992. Solubilization and functional reconstition of biomembrane components. *Annu. Rev. Biophys. Biomol. Struct.* 21:323–348.

Yeagle, P. L., Albert, A. D., Boeze-Battaglia, K., Young, J., and Frye, J., 1990. Cholesterol dynamics in membranes. *Biophys. J.* 57:413–424.

Nagle, J. F., and Tristram-Nagle, S., 2000. Lipid bilayer structure. *Curr. Opin. Struct. Biol.* 10: 474–480.

Dowhan, W., 1997. Molecular basis for membrane phospholipid diversity: Why are there so many lipids? *Annu. Rev. Biochem.* 66:199–232.

Huijbregts, R. P. H., de Kroon, A. I. P. M., and de Kruijff, B., 1998. Rapid transmembrane movement of newly synthesized phosphatidylethanolamine across the inner membrane of *Escherichia coli. J. Biol.Chem.* 273:18936–18942.

Structure of membrane proteins

Popot, J.-L., and Engleman, D. M., 2000. Helical membrane protein folding, stability and evolution. *Annu. Rev. Biochem.* 69:881–922.

White, S. H., and Wimley, W. C., 1999. Membrane protein folding and stability: Physical principles. *Annu. Rev. Biophys. Biomol. Struct.* 28:319–365.

Marassi, F. M., and Opella, S. J., 1998. NMR structural studies of membrane proteins. *Curr. Opin. Struct. Biol.* 8:640–648.

Lipowsky, R., 1991. The conformation of membranes. *Nature* 349:475–481.

Altenbach, C., Marti, T., Khorana, H. G., and Hubbell, W. L., 1990. Transmembrane protein structure: Spin labeling of bacteriorhodopsin mutants. *Science* 248:1088–1092.

Fasman, G. D., and Gilbert, W. A., 1990. The prediction of transmembrane protein sequences and their conformation: An evaluation. *Trends Biochem. Sci.* 15:89–92.

Jennings, M. L., 1989. Topography of membrane proteins. *Annu. Rev. Biochem.* 58:999–1027.

Engelman, D. M., Steitz, T. A., and Goldman, A., 1986. Identifying non-polar transbilayer helices in amino acid sequences of membrane proteins. *Annu. Rev. Biophys. Biophys. Chem.* 15:321–353.

Undenfriend, S., and Kodukola, K., 1995. How glycosyl-phosphatidyl-inositiol-anchored membrane proteins are made. *Annu. Rev. Biochem.* 64:563–591.

Intracellular membranes

Skehel, J. J., and Wiley, D. C., 2000. Receptor binding and membrane fusion in virus entry: The influenza hemagglutinin. *Annu. Rev. Biochem.* 69:531–569.

Roth, M. G., 1999. Lipid regulators of membrane traffic through the Golgi complex. *Trends Cell Biol.* 9:174–179.

Jahn, R., and Sudhof, T. C., 1999. Membrane fusion and exocytosis. *Annu. Rev. Biochem.* 68:863–911.

Stroud, R. M., and Walter, P., 1999. Signal sequence recognition and protein targeting. *Curr. Opin. Struct. Biol.* 9:754–759.

Teter, S. A., and Klionsky, D. J., 1999. How to get a folded protein across a membrane. *Trends Cell Biol.* 9:428–31.

Hettema, E. H., Distel, B., and Tabak, H. F., 1999. Import of proteins into peroxisomes. *Biochim. Biophys. Acta* 1451:17–34.

PROBLEMS

1. *Population density.* How many phospholipid molecules are there in a 1-μm^2 region of a phospholipid bilayer membrane? Assume that a phospholipid molecule occupies 70 $Å^2$ of the surface area.

2. *Lipid diffusion.* What is the average distance traversed by a membrane lipid in 1 μs, 1 ms, and 1 s? Assume a diffusion coefficient of 10^{-8} cm^{2}s^{-1}.

3. *Protein diffusion.* The diffusion coefficient, D, of a rigid spherical molecule is given by

$$D = kT/6\pi\eta r$$

in which η is the viscosity of the solvent, r is the radius of the sphere, k is the Boltzman constant (1.38×10^{-16} erg/degree), and T is the absolute temperature. What is the diffusion coefficient at 37°C of a 100-kd protein in a membrane that has an effective viscosity of 1 poise (1 poise = 1 erg s^{-1}/cm^{-3})? What is the average distance traversed by this protein in 1 μs, 1 ms, and 1 s? Assume that this protein is an unhydrated, rigid sphere of density 1.35 g cm^{-3}.

4. *Cold sensitivity.* Some antibiotics act as carriers that bind an ion on one side of a membrane, diffuse through the membrane, and release the ion on the other side. The conductance of a lipid-bilayer membrane containing a carrier antibiotic decreased abruptly when the temperature was lowered from 40°C to 36°C. In contrast, there was little change in conductance of the same bilayer membrane when it contained a channel-forming antibiotic. Why?

5. *Flip-flop.* The transverse diffusion of phospholipids in a bilayer membrane was investigated by using a paramagnetic analog of phosphatidyl choline, called *spin-labeled phosphatidyl choline.*

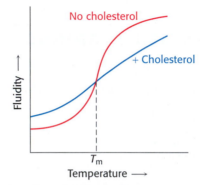

Spin-labeled phosphatidyl choline

The nitroxide (NO) group in spin-labeled phosphatidyl choline gives a distinctive paramagnetic resonance spectrum. This spectrum disappears when nitroxides are converted into amines by reducing agents such as ascorbate.

Lipid vesicles containing phosphatidyl choline (95%) and the spin-labeled analog (5%) were prepared by sonication and purified by gel-filtration chromatography. The outside diameter of these liposomes was about 25 nm. The amplitude of the paramagnetic resonance spectrum decreased to 35% of its initial value within a few minutes of the addition of ascorbate. There was no detectable change in the spectrum within a few minutes after the addition of a second aliquot of ascorbate. However, the amplitude of the residual spectrum decayed exponentially with a half-time of 6.5 hours. How would you interpret these changes in the amplitude of the paramagnetic spectrum?

6. *Flip-flop 2.* Although proteins rarely if ever flip-flop across a membrane, distribution of membrane lipids between the membrane leaflets is not absolute except in the case of glycolipids. Why are glycosylated lipids less likely to flip-flop?

7. *Cis versus trans.* Why are most unsaturated fatty acids found in phospholipids in the cis rather than the trans conformation? Draw the structure of a 16-carbon fatty acid as saturated, trans monounsaturated, and cis monounsaturated.

8. *A question of competition.* Would a homopolymer of alanine be more likely to form an α helix in water or in a hydrophobic medium? Explain.

9. *Maintaining fluidity.* A culture of bacteria growing at 37°C was shifted to 25°C. How would you expect this shift to alter the fatty acid composition of the membrane phospholipids? Explain.

Data Interpretation Problems

10. *Cholesterol effects.* The red line on the following graph shows the fluidity of the fatty acids of a phospholipid bilayer as a function of temperature. The blue line shows the fluidity in the presence of cholesterol.

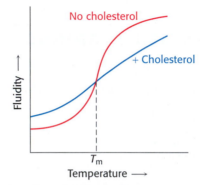

(a) What is the effect of cholesterol?
(b) Why might this effect be biologically important?

11. *Hydropathy plots.* On the basis of the following hydropathy plots for three proteins, predict which would be membrane proteins. What are the ambiguities with respect to using such plots to determine if a protein is a membrane protein?

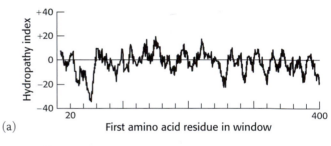

(a)

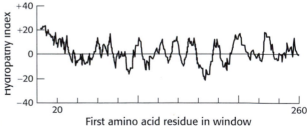

(b)

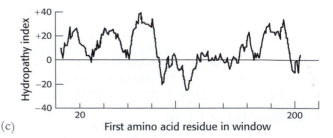

(c)

Chapter Integration Problem

12. *The proper environment.* An understanding of the structure and function of membrane proteins has lagged behind that of other proteins. The primary reason is that membrane proteins are more difficult to purify and crystallize. Why might this be the case?

CHAPTER 13

Membrane Channels and Pumps

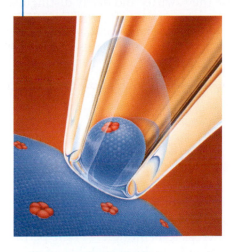

← Closed

← Open

The flow of ions through a single membrane channel (channels are shown in red in the illustration at the left) can be detected by the patch clamp technique, which records current changes as the channel transits between the open and closed states. [(Left) After E. Neher and B. Sakmann. The patch clamp technique. Copyright © 1992 by Scientific American, Inc. All rights reserved. (Right) Courtesy of Dr. Mauricio Montal.]

The lipid bilayer of biological membranes, as discussed in Chapter 12, is intrinsically impermeable to ions and polar molecules. Permeability is conferred by two classes of membrane proteins, *pumps* and *channels.* Pumps use a source of free energy such as ATP or light to drive the thermodynamically uphill transport of ions or molecules. Pump action is an example of *active transport*. Channels, in contrast, enable ions to flow rapidly through membranes in a downhill direction. Channel action illustrates *passive transport*, or *facilitated diffusion*.

Pumps are energy transducers in that they convert one form of free energy into another. Two types of ATP-driven pumps, P-type ATPases and the ATP-binding cassette pumps, undergo conformational changes on ATP binding and hydrolysis that cause a bound ion to be transported across the membrane. Phosphorylation and dephosphorylation of both the Ca^{2+}-ATPase and the Na^{+}-K^{+}-ATPase pumps, which are representative of P-type ATPase, are coupled to changes in orientation and affinity of their ion-binding sites.

A different mechanism of active transport, one that utilizes the gradient of one ion to drive the active transport of another, will be illustrated by the sodium–calcium exchanger. This pump plays an important role in extruding Ca^{2+} from cells.

We begin our examination of channels with the *acetylcholine receptor*, a channel that mediates the transmission of nerve signals across synapses, the functional junctions between neurons. The acetylcholine receptor is a *ligand-gated* channel in that the channel opens in response to the

OUTLINE

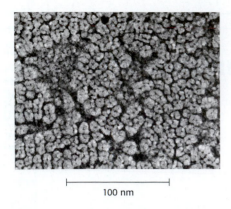

FIGURE 13.1 Acetylcholine receptors.
An electron micrograph shows the densely packed acetylcholine receptors embedded in a postsynaptic membrane. [Courtesy of Dr. John Heuser and Dr. Shelly Salpeter.]

binding of acetylcholine (Figure 13.1). In contrast, the sodium and potassium channels, which mediate action potentials in neuron axon membranes, are opened by membrane depolarization rather than by the binding of an allosteric effector. These channels are *voltage-gated*. These channels are also of interest because they swiftly and deftly distinguish between quite similar ions (e.g., Na^+ and K^+). The flow of ions through a single channel in a membrane can readily be detected by using the *patch-clamp technique*.

The chapter concludes with a view of a different kind of channel—the cell-to-cell channel, or *gap junction*. These channels allow the transport of ions and metabolites between cells.

13.1 THE TRANSPORT OF MOLECULES ACROSS A MEMBRANE MAY BE ACTIVE OR PASSIVE

Before we consider the specifics of membrane-protein function, we will consider some general principles of membrane transport. Two factors determine whether a molecule will cross a membrane: (1) the permeability of the molecule in a lipid bilayer and (2) the availability of an energy source.

13.1.1 Many Molecules Require Protein Transporters to Cross Membranes

As discussed in Chapter 12, some molecules can pass through cell membranes because they dissolve in the lipid bilayer. Such molecules are called *lipophilic molecules*. The steroid hormones provide a physiological example. These cholesterol relatives can pass through a membrane in their path of movement, but what determines the direction in which they will move? Such molecules will pass through a membrane located down their concentration gradient in a process called *simple diffusion*. In accord with the Second Law of Thermodynamics, molecules spontaneously move from a region of higher concentration to one of lower concentration. Thus, in this case, an entropy increase powers transport across the membrane.

Matters become more complicated when the molecule is highly polar. For example, sodium ions are present at 143 mM outside the cell and 14 mM inside the cell, yet sodium does not freely enter the cell because the positively charged ion cannot pass through the hydrophobic membrane interior. In some circumstances, as during a nerve impulse (Section 13.5.3), sodium ions must enter the cell. How are they able to do so? Sodium ions pass through specific channels in the hydrophobic barrier formed by membrane proteins. This means of crossing the membrane is called *facilitated diffusion,* because the diffusion across the membrane is facilitated by the channel. It is also called *passive transport,* because the energy driving the ion movement originates from the ion gradient itself, without any contribution by the transport system. Channels, like enzymes, display substrate specificity.

How is the sodium gradient established in the first place? In this case, sodium must move, or be pumped, *against* a concentration gradient. Because moving the ion from a low concentration to a higher concentration results in a decrease in entropy, it requires an input of free energy. Protein transporters embedded in the membrane are capable of using an energy source to move the molecule up a concentration gradient. Because an input of energy from another source is required, this means of crossing the membrane is called *active transport.*

13.1.2 Free Energy Stored in Concentration Gradients Can Be Quantified

An unequal distribution of molecules is an energy-rich condition because free energy is minimized when all concentrations are equal. Consequently, to attain such an unequal distribution of molecules, called a *concentration gradient,* requires an input of free energy. In fact, the creation of a concentration gradient is the result of active transport. Can we quantify the amount of energy required to generate a concentration gradient (Figure 13.2)? Consider an uncharged solute molecule. The free-energy change in transporting this species from side 1, where it is present at a concentration of c_1, to side 2, where it is present at concentration c_2, is

$$\Delta G = RT \ln(c_2/c_1) = 2.303 RT \log_{10}(c_2/c_1)$$

For a charged species, the unequal distribution across the membrane generates an electrical potential that also must be considered because the ions will be repelled by the like charges. The sum of the concentration and electrical terms is called the electrochemical potential. The free-energy change is then given by

$$\Delta G = RT \ln(c_2/c_1) + ZF\Delta V = 2.303 RT \log_{10}(c_2/c_1) + ZF\Delta V$$

in which Z is the electrical charge of the transported species, ΔV is the potential in volts across the membrane, and F is the faraday [23.1 kcal V^{-1} mol^{-1} (96.5 kJ V^{-1} mol^{-1})].

A transport process must be active when ΔG is positive, whereas it can be passive when ΔG is negative. For example, consider the transport of an uncharged molecule from $c_1 = 10^{-3}$ M to $c_2 = 10^{-1}$ M.

$$\Delta G = 2.303 RT \log_{10}(10^{-1}/10^{-3})$$
$$= 2.303 \times 1.99 \times 298 \times 2$$
$$= +2.7 \text{ kcal mol}^{-1} (+11.3 \text{ kJ mol}^{-1})$$

At 25°C (298 K), ΔG is + 2.7 kcal mol^{-1} (+11.3 kJ mol^{-1}), indicating that this transport process requires an input of free energy. It could be driven, for example, by the hydrolysis of ATP, which yields −12 kcal mol^{-1} (−50.2 kJ mol^{-1}) under typical cellular conditions. If ΔG is negative, the transport process can occur spontaneously without free-energy input.

Ion gradients are important energy storage forms in all biological systems. For instance, bacteriorhodopsin (Section 12.5.2) generates a proton gradient at the expense of light energy, an example of active transport. The energy of the proton gradient in turn can be converted into chemical energy in the form of ATP. This example illustrates the use of membranes and membrane proteins to transform energy: from light energy into an ion gradient into chemical energy.

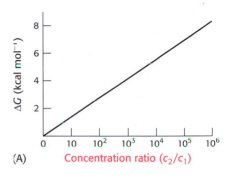

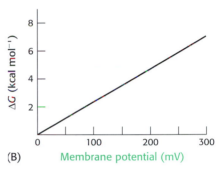

(A) Concentration ratio (c_2/c_1)

(B) Membrane potential (mV)

FIGURE 13.2 Free energy and transport. The free-energy change in transporting (A) an uncharged solute from a compartment at concentration c_1 to one at c_2 and (B) a singly charged species across a membrane to the side having the same charge as that of the transported ion. Note that the free-energy change imposed by a membrane potential of 59 mV is equivalent to that imposed by a concentration ratio of 10 for a singly charged ion at 25°C.

13.2 A FAMILY OF MEMBRANE PROTEINS USES ATP HYDROLYSIS TO PUMP IONS ACROSS MEMBRANES

The extracellular fluid of animal cells has a salt concentration similar to that of sea water. However, cells must control their intracellular salt concentrations to prevent unfavorable interactions with high concentrations of ions such as calcium and to facilitate specific processes. For instance, most animal cells contain a high concentration of K^+ and a low concentration of Na^+ relative to the external medium. These ionic gradients are generated by a specific transport system, an enzyme that is called the Na^+-K^+ *pump* or the

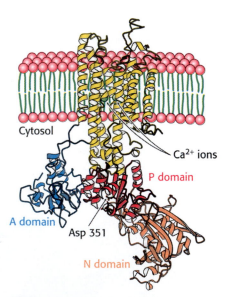

β-Phosphorylaspartate

FIGURE 13.3 Phosphoaspartate.
Phosphoaspartate (also referred to as β-aspartyl phosphate) is a key intermediate in the reaction cycles of P-type ATPases.

FIGURE 13.4 Structure of SR Ca²⁺ ATPase. This enzyme, the calcium pump of the sarcoplasmic reticulum, comprises a membrane-spanning domain of 10 α helices and a cytoplasmic headpiece consisting of three domains (N, P, and A). Two calcium ions (green) bind within the membrane-spanning region. The aspartate residue characteristic of this protein family is indicated.

Na^+-K^+ *ATPase.* The hydrolysis of ATP by the pump provides the energy needed for the active transport of Na^+ out of the cell and K^+ into the cell, generating the gradient. The pump is called the Na^+-K^+ ATPase because the hydrolysis of ATP occurs only when Na^+ and K^+ are bound to the pump. Moreover, this ATPase, like all such enzymes, requires Mg^{2+} (Section 9.4.2). The active transport of Na^+ and K^+ is of great physiological significance. Indeed, more than a third of the ATP consumed by a resting animal is used to pump these ions. The Na^+-K^+ gradient in animal cells controls cell volume, renders neurons and muscle cells electrically excitable, and drives the active transport of sugars and amino acids.

The subsequent purification of other ion pumps has revealed a large family of evolutionarily related ion pumps including proteins from bacteria, archaea, and all eukaryotes. These pumps are specific for an array of ions. Of particular interest are the Ca^{2+} ATPase, the enzyme that transports Ca^{2+} out of the cytoplasm and into the sarcoplasmic reticulum of muscle cells, and the gastric H^+-K^+ ATPase, the enzyme responsible for pumping sufficient protons into the stomach to lower the pH below 1.0. These enzymes and the hundreds of known homologs, including the Na^+-K^+ ATPase, are referred to as *P-type ATPases* because they form a key phosphorylated intermediate. In the formation of this intermediate, a phosphoryl group obtained from the hydrolysis of ATP is linked to the side chain of a specific conserved aspartate residue in the ATPase (Figure 13.3).

13.2.1 The Sarcoplasmic Reticulum Ca²⁺ ATPase Is an Integral Membrane Protein

We will consider the structural and mechanistic features of these enzymes by examining the Ca^{2+} ATPase found in the sarcoplasmic reticulum (SR Ca^{2+} ATPase) of muscle cells. This enzyme, which constitutes 80% of the sarcoplasmic reticulum membrane protein, plays an important role in muscle contraction, which is triggered by an abrupt rise in the cytosolic calcium level. Muscle relaxation depends on the rapid removal of Ca^{2+} from the cytosol into the sarcoplasmic reticulum, a specialized compartment for calcium storage, by the SR Ca^{2+} ATPase. This pump maintains a Ca^{2+} concentration of approximately 0.1 μM in the cytosol compared with 1.5 mM in the sarcoplasmic reticulum.

The SR Ca^{2+} ATPase is a single 110-kd polypeptide with a transmembrane domain consisting of 10 α helices. A large cytoplasmic head piece constitutes nearly half the molecular weight of the protein and consists of three distinct domains (Figure 13.4). The three cytoplasmic domains of the SR Ca^{2+} ATPase have distinct functions. One domain (N) binds the ATP nucleotide, another (P) accepts the phosphoryl group on its conserved aspartate residue, and the third (A) may serve as an actuator for the N domain. The relation between these three domains changes significantly on ATP hydrolysis. The crystal structure in the absence of ATP shows the likely nucleotide-binding site separated by more than 25 Å from the phosphorylation site, suggesting that the N and P domains move toward one another during the catalytic cycle. This closure is facilitated by ATP binding and by the binding of Ca^{2+} to the membrane-spanning helices.

The results of mechanistic studies of the SR Ca^{2+} ATPase and other P-type ATPases have revealed two common features. First, as we have seen, each protein can be phosphorylated on a specific aspartate residue. For the SR Ca^{2+} ATPase, this reaction takes place at Asp 351 only in the presence of relatively high cytosolic concentrations of Ca^{2+}. Second, each pump can interconvert between at least two different conformations, denoted by E_1 and E_2. Thus, at least four conformational states—E_1, E_1-P, E_2-P, and E_2—participate

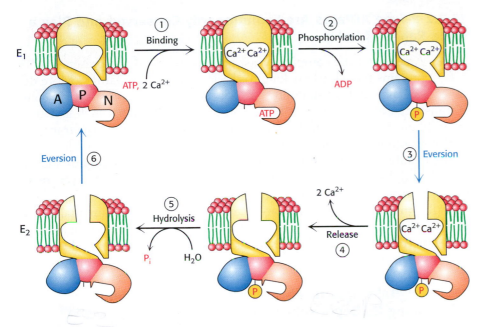

in the transport process. From these four states, it is possible to construct a plausible mechanism of action for these enzymes, although further studies are necessary to confirm the mechanism and provide more details (Figure 13.5):

1. The postulated reaction cycle begins with the binding of ATP and two Ca^{2+} ions to the E_1 state.

2. The enzyme cleaves ATP, transferring the γ-phosphoryl group to the key aspartate residue. Calcium must be bound to the enzyme for the phosphorylation to take place. Phosphorylation shifts the conformational equilibrium of the ATPase toward E_2.

3. The transition from the E_1 to the E_2 state causes the ion-binding sites to "evert" so that the ions can dissociate only to the luminal side of the membrane.

4. In the E_2-P state, the enzyme has low affinity for the Ca^{2+} ions, so they are released.

5. With the release of Ca^{2+}, the phosphoaspartate residue is hydrolyzed, and the phosphate group is released.

6. The enzyme, devoid of a covalently attached phosphoryl group, is not stable in the E_2 form. It everts back to the E_1 form, completing the cycle.

Essentially the same mechanism is employed by the Na^+-K^+ ATPase. The E_1 state binds three Na^+ ions and transports them across the membrane and out of the cell as a result of the protein's phosphorylation and transition to the E_2 state. The three Na^+ ions are released into the extracellular medium. The E_2 state of this enzyme also binds ions—namely, two K^+ ions. These K^+ ions are carried across the membrane in the opposite direction by eversion driven by the hydrolysis of the phosphoaspartate residue and are released into the cytosol.

The change in free energy accompanying the transport of Na^+ and K^+ can be calculated (Section 13.1.1). Suppose that the concentration of Na^+ outside and inside the cell is 143 and 14 mM, respectively, and that of K^+ is 4 and 157 mM. At a membrane potential of -50 mV and a temperature of 37 °C, the free-energy change in transporting 3 moles of Na^+ out of and 2 moles of K^+ into the cell is $+10.0$ kcal ($+41.8$ kJ mol^{-1}). The hydrolysis of a single ATP per transport cycle provides sufficient free energy, about -12 kcal mol^{-1} (-50 kJ mol^{-1}) under cellular conditions, to drive the uphill transport of these ions.

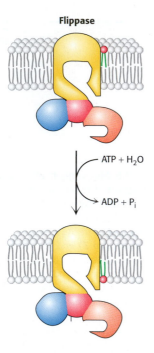

Flippase

ATP + H$_2$O

ADP + P$_i$

FIGURE 13.6 P-type ATPases can transport lipids. Flippases are enzymes that maintain membrane asymmetry by "flipping" phospholipids (displayed with a red head group) from the outer to the inner layer of the membrane.

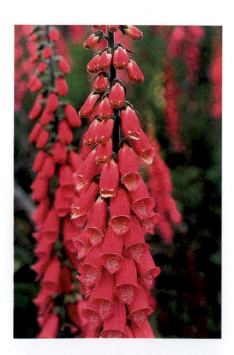

Foxglove (*Digitalis purpurea*) is the source of digitalis, one of the most widely used drugs. [Inga Spence/Visuals Unlimited.]

13.2.2 P-Type ATPases Are Evolutionarily Conserved and Play a Wide Range of Roles

Analysis of the complete yeast genome revealed the presence of 16 proteins that clearly belong to the P-type ATPase family. More detailed sequence analysis suggests that 2 of these proteins transport H$^+$ ions, 2 transport Ca^{2+}, 3 transport Na$^+$, and 2 transport metals such as Cu^{2+}. In addition, 5 members of this family appear to participate in the transport of phospholipids with amino acid head groups. These latter proteins assist in the maintenance of membrane asymmetry by transporting lipids such as phosphatidyl serine from the outer to the inner leaflet of the bilayer membrane (Figure 13.6). Such enzymes have been termed "flippases."

All members of this protein family employ the same fundamental mechanism. The free energy of ATP hydrolysis drives membrane transport by effecting conformational changes associated with the addition and removal of a phosphoryl group at an analogous aspartate site in each protein.

13.2.3 Digitalis Specifically Inhibits the Na$^+$-K$^+$ Pump by Blocking Its Dephosphorylation

Certain steroids derived from plants are potent inhibitors (K$_i$ ≈ 10 nM) of the Na$^+$-K$^+$ pump. Digitoxigenin and ouabain are members of this class of inhibitors, which are known as *cardiotonic steroids* because of their strong effects on the heart (Figure 13.7). These compounds inhibit the dephosphorylation of the E$_2$-P form of the ATPase when applied on the *extracellular* face of the membrane.

(A)

CH$_3$

CH$_3$

OH

HO

H

Digitoxigenin

(B)

$$E_2\text{—}P + H_2O \xrightarrow{\quad\quad} E_2 + P_i$$

K$^+$

Inhibited by cardiotonic steroids

FIGURE 13.7 Digitoxigenin. Cardiotonic steroids such as digitoxigenin inhibit the Na$^+$-K$^+$ pump by blocking the dephosphorylation of E$_2$-P.

Digitalis, a mixture of cardiotonic steroids derived from the dried leaf of the foxglove plant (*Digitalis purpurea*), is of great clinical significance. Digitalis increases the force of contraction of heart muscle, which makes it a choice drug in the treatment of congestive heart failure. Inhibition of the Na$^+$-K$^+$ pump by digitalis leads to a higher level of Na$^+$ inside the cell. The diminished Na$^+$ gradient results in slower extrusion of Ca^{2+} by the sodium–calcium exchanger (Section 13.4). The subsequent increase in the intracellular level of Ca^{2+} enhances the contractility of cardiac muscle. It is interesting to note that digitalis was effectively used long before the discovery of the Na$^+$-K$^+$ ATPase. In 1785, William Withering, a British physician, heard tales of an elderly woman, known as "the old woman of Shropshire," who cured people of "dropsy" (which today would be recognized as congestive heart failure) with an extract of foxglove. Withering conducted the first scientific study of the effects of foxglove on congestive heart failure and documented its effectiveness.

13.3 MULTIDRUG RESISTANCE AND CYSTIC FIBROSIS HIGHLIGHT A FAMILY OF MEMBRANE PROTEINS WITH ATP-BINDING CASSETTE DOMAINS

Tumor cells in culture often become resistant to drugs that were initially quite toxic to the cells. Remarkably, the development of resistance to one drug also makes the cells less sensitive to a range of other compounds. This phenomenon is known as *multidrug resistance*. In a significant discovery, the onset of multidrug resistance was found to correlate with the expression and activity of a membrane protein with an apparent molecular mass of 170 kd. This protein acts as an ATP-dependent pump that extrudes a wide range of small molecules from cells that express it. The protein is called the *multidrug resistance protein (MDR)* or *P-glycoprotein* ("glyco" because it includes a carbohydrate moiety). Thus, when cells are exposed to a drug, the MDR pumps the drug out of the cell before the drug can exert its effects. A related protein was discovered through genetic studies of the hereditary disease *cystic fibrosis* (Section 1.1.4). In one of the first studies leading to the identification of a specific genetic change causing human disease, investigators performed a comparative genetic analysis of many people having this disease and family members who did not have the disease. The gene found to be mutated in the affected persons encodes a protein, now called *cystic fibrosis transmembrane conductance regulator (CFTR)*. CFTR acts as an ATP-regulated chloride channel in the plasma membranes of epithelial cells. As mentioned in Chapter 1, cystic fibrosis results from a decrease in fluid and salt secretion by CFTR. As a consequence of this defect, secretion from the pancreas is blocked and heavy, dehydrated mucus accumulates in the lungs, leading to chronic lung infections.

Analysis of the amino acid sequences of MDR, CFTR, and homologous proteins revealed a common architecture (Figure 13.8). Each protein comprises four domains: two membrane-binding domains of unknown structure and two ATP-binding domains. The ATP-binding domains of these proteins are called *ATP-binding cassettes* (or ABCs) and are homologous to domains in a large family of transport proteins of bacteria and archaea. Indeed, with 79 members, the ABC proteins are the largest single family identified in the *E. coli* genome. The ABC proteins are members of the P-loop NTPase superfamily (Section 9.4.4). Some ABC proteins, particularly those of prokaryotes, are multisubunit proteins constructed such that the membrane-spanning domains and the ABC domains are present on separate polypeptide chains. The consolidation of enzymatic activities of several polypeptide chains in prokaryotes to a single chain in eukaryotes is a theme that we will see again (Section 22.4.4). For example, the histidine permease of *Salmonella typhimurium*, which transports the amino acid histidine into the bacterium, consists of (1) two different protein subunits with membrane-spanning domains (HisQ and HisM) and (2) a homodimeric protein (HisP) with ABC domains (Figure 13.9). A soluble, histidine-binding protein (HisJ) binds histidine after the amino acid enters the cell.

Like other members of the P-loop NTPase superfamily, proteins with ABC domains undergo conformational changes on ATP binding and hydrolysis. These structural changes are coupled within each dimeric transporter unit in a manner that allows these membrane proteins to drive the uptake or efflux of specific compounds or to act as gates for open membrane channels.

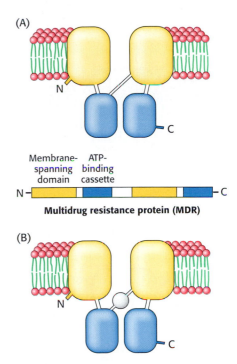

FIGURE 13.8 ABC transporters. The multidrug resistance protein (MDR) and the cystic fibrosis transmembrane regulator (CFTR) are homologous proteins composed of two transmembrane domains and two ATP-binding domains, called ATP-binding cassettes (ABCs).

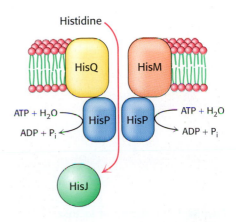

FIGURE 13.9 Histidine permease. In the histidine permease of *S. typhimurium*, the membrane-spanning regions (yellow and orange) and ABC regions (blue) are on separate polypeptide chains (compare with Figure 13.8). ATP hydrolysis drives the transport of histidine into the cell.

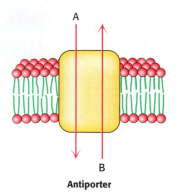

Antiporter

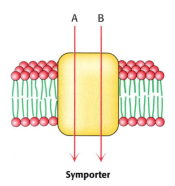

Symporter

FIGURE 13.10 Secondary transporters.
These transporters employ the downhill
flow of one gradient to power the
formation of another gradient. In
antiporters, the chemical species move
in opposite directions. In symporters, the
two species move in the same direction.

13.4 SECONDARY TRANSPORTERS USE ONE CONCENTRATION GRADIENT TO POWER THE FORMATION OF ANOTHER

Many active-transport processes are not directly driven by the hydrolysis of ATP. Instead, the thermodynamically uphill flow of one species of ion or molecule is coupled to the downhill flow of a different species. Membrane proteins that pump ions or molecules uphill by this means are termed *secondary transporters* or *cotransporters*. These proteins can be classified as either *antiporters* or *symporters*. Antiporters couple the downhill flow of one species to the uphill flow of another in the *opposite direction* across the membrane; symporters use the flow of one species to drive the flow of a different species in the *same direction* across the membrane (Figure 13.10).

The *sodium–calcium exchanger* in the plasma membrane of an animal cell is an antiporter that uses the electrochemical gradient of Na^+ to pump Ca^{2+} out of the cell. Three Na^+ ions enter the cell for each Ca^{2+} ion that is extruded. The cost of transport by this exchanger is paid by the Na^+-K^+-ATPase pump, which generates the requisite sodium gradient. Because Ca^{2+} is a vital messenger inside the cell, its concentration must be tightly controlled. *The exchanger has lower affinity for Ca^{2+} than does the Ca^{2+}ATPase (Section 13.2.1), but its capacity to extrude Ca^{2+} is greater.* The exchanger can lower the cytosolic Ca^{2+} level to several micromolar; submicromolar Ca^{2+} levels are attained by the subsequent action of the Ca^{2+} ATPase. The exchanger can extrude about 2000 Ca^{2+} ions per second, compared with only 30 ions per second for the Ca^{2+}-ATPase pump.

Glucose is pumped into some animal cells by a symporter powered by the simultaneous entry of Na^+. The entry of Na^+ provides a free-energy input of 2.2 kcal mol^{-1} (9.2 kJ mol^{-1}) under typical cellular conditions (external $[Na^+]$ = 143 mM, internal $[Na^+]$ = 14 mM, and membrane potential = -50 mV). This free-energy input is sufficient to generate a 66-fold concentration gradient of an uncharged molecule such as glucose.

Secondary transporters are ancient molecular machines, common today in bacteria and archaea as well as in eukaryotes. For example, approximately 160 (of approximately 4000) proteins encoded by the *E. coli* genome appear to be secondary transporters. Sequence comparison and hydropathy analysis suggest that members of the largest family have 12 transmembrane helices that appear to have arisen by duplication and fusion of a membrane protein with 6 transmembrane helices. Included in this family is the lactose permease of *E. coli*. This symporter uses the H^+ gradient across the *E. coli* membrane generated by the oxidation of fuel molecules to drive the uptake of lactose and other sugars against a concentration gradient. The permease has a proton-binding site and a lactose-binding site (Figure 13.11).

FIGURE 13.11 Action of lactose permease. Lactose permease pumps lactose into bacterial cells by drawing on the proton-motive force. The binding sites evert when a lactose molecule (L) and a proton (H^+) are bound to external sites. After these species are released inside the cell, the binding sites again evert to complete the transport cycle. Lactose permease is an example of a symporter.

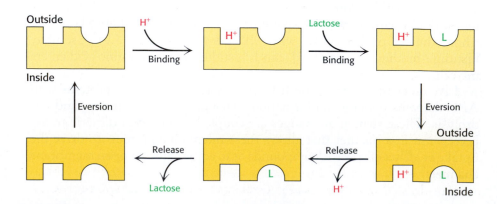

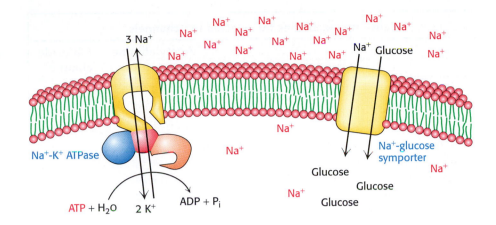

FIGURE 13.12 Energy transduction by membrane proteins. The Na$^+$-K$^+$ ATPase converts the free energy of phosphoryl transfer into the free energy of a Na$^+$ ion gradient. The ion gradient can then be used to pump materials into the cell, through the action of a secondary transporter such as the Na$^+$-glucose symporter.

A proton and a lactose molecule bind to sites facing the outside of the cell. The permease, with both binding sites full, everts, releasing first the proton and then the lactose inside the bacterium. Another eversion places the empty sites on the outside. Thus, the energetically uphill transfer of one lactose molecule is coupled to the downhill transport of one proton. Further analysis of the three-dimensional structures is underway and should provide more information about their mechanisms of action as well as the evolutionary relationships within this large group of ancient proteins.

These observations reveal how different energy currencies are interconverted. A single energy currency, ATP, is used by P-type ATPases to generate gradients of a small number of types of ions, particularly Na$^+$ and H$^+$, across membranes. These gradients then serve as an energy source for the large number of secondary transporters, which allow many different molecules to be taken up or transported out of cells (Figure 13.12).

13.5 SPECIFIC CHANNELS CAN RAPIDLY TRANSPORT IONS ACROSS MEMBRANES

Pumps and secondary transporters can transport ions at rates approaching several thousand ions per second. Other membrane proteins, *ion channels*, which are passive transport systems, are capable of ion-transport rates that are more than 1000 times as high. These rates of transport through ion channels are close to rates expected for ions diffusing freely through aqueous solution. Yet, ion channels are not simply tubes that span membranes through which ions can rapidly flow. Instead, they are highly sophisticated molecular machines that respond to chemical and physical changes in their environments and undergo precisely timed conformational changes that facilitate their roles as essential components of the nervous and other systems.

Several key properties characterize ion channels:

1. *Ion channels can be highly selective for particular ions.* For example, some channels allow the flow of K$^+$ very effectively but do not allow appreciable levels of Na$^+$ to cross the membrane. Other channels transport positively charged ions (cations), but block the flow of negatively charged ions (anions). The selectivities of some ion-channel proteins are shown in Table 13.1.

2. *Ion channels exist in open and closed states.* These channels undergo transitions from the closed state, incapable of supporting ion transport, to the open state, through which ions can flow.

3. *Transitions between the open and the closed states are regulated.* Ion channels are divided into two classes: *ligand-gated channels* and *voltage-gated*

TABLE 13.1   Relative permeabilities for selected ion channels

	Na$^+$ channel	K$^+$ channel	Acetylcholine receptor	Chloride channel
Li$^+$	0.93	< 0.01	0.87	< 0.01
Na$^+$	1.00	< 0.01	1.00	< 0.01
K$^+$	0.09	1.00	1.11	< 0.01
Rb$^+$	< 0.01	0.91		
Cs$^+$	< 0.01	< 0.08	1.42	
NH$_4^+$	0.16	0.13	1.79	
H$_3$NOH$^+$	0.94	< 0.03	1.92	
H$_2$NNH$_3^+$	0.59	< 0.03		
H$_3$CNH$_3^+$	< 0.01	< 0.02		
Cl$^-$	< 0.01	< 0.01	< 0.01	1.00

channels. Ligand-gated channels open and close in response to the binding of specific chemicals, whereas voltage-gated channels open and close in response to the electrical potential across the membrane in which they are found.

4. *Open states of channels often spontaneously convert into inactivated states.* Most ion channels do not remain in an open state indefinitely but, instead, spontaneously transform into inactivated states that do not conduct ions. The spontaneous transitions of ion channels from open to inactivated states act as built-in timers that determine the duration of ion flow.

13.5.1 Patch-Clamp Conductance Measurements Reveal the Activities of Single Channels

The study of ion channels has been revolutionized by the *patch-clamp technique*, which was introduced by Erwin Neher and Bert Sakmann in 1976 (Figure 13.13). This powerful technique enables the measurement of the activity of a single channel to be measured. A clean glass pipette with a tip diameter of about 1 μm is pressed against an intact cell to form a seal. Slight suction leads to the formation of a very tight seal so that the resistance between the inside of the pipette and the bathing solution is many gigaohms (1 gigaohm is equal to 10^9 ohms). Thus, a gigaohm seal (called a *gigaseal*)

FIGURE 13.13 Patch-clamp modes. The patch-clamp technique for monitoring channel activity is highly versatile. A high-resistance seal (gigaseal) is formed between the pipette and a small patch of plasma membrane. This configuration is called *cell attached.* The breaking of the membrane patch by increased suction produces a low-resistance pathway between the pipette and interior of the cell. The activity of the channels in the entire plasma membrane can be monitored in this *whole-cell mode.* To prepare a membrane in the *excised-patch mode,* the pipette is pulled away from the cell. A piece of plasma membrane with its cytosolic side now facing the medium is monitored by the patch pipette.

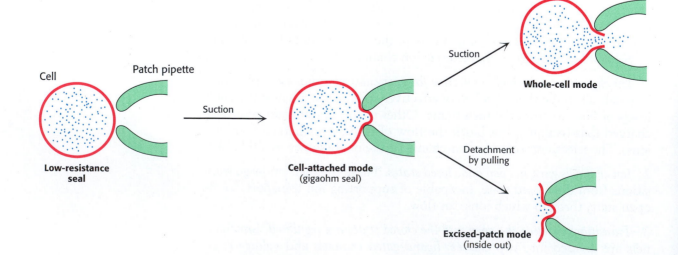

ensures that an electric current flowing through the pipette is identical with the current flowing through the membrane covered by the pipette. The gigaseal makes possible high-resolution current measurements while a known voltage is applied across the membrane. In fact, patch clamping increased the precision of such measurements 100-fold. *The flow of ions through a single channel and transitions between the open and closed states of a channel can be monitored with a time resolution of microseconds.* Furthermore, the activity of a channel in its native membrane environment, even in an intact cell, can be directly observed. Patch-clamp methods provided one of the first views of single biomolecules in action. Subsequently, other methods for observing single molecules were invented, opening new vistas on biochemistry at its most fundamental level.

13.5.2 Ion-Channel Proteins Are Built of Similar Units

How do ion channels, vital to a wide array of biological functions, operate at a molecular level? We will examine three channels important in the propagation of nerve impulses: the ligand-gated channel; the acetylcholine receptor channel, which communicates the nerve impulse between certain neurons; and the voltage-gated Na^+ and K^+ channels, which conduct the nerve impulse down the axon of a neuron.

Nerve impulses are communicated across most synapses by small, diffusible molecules called *neurotransmitters*, of which one is acetylcholine, referred to as a cholinergic neurotransmitter because it is derived from choline (Section 12.3.1). The presynaptic membrane of a synapse is separated from the postsynaptic membrane by a gap of about 50 nm, called the *synaptic cleft*. The end of the presynaptic axon is filled with *synaptic vesicles*, each containing about 10^4 acetylcholine molecules (Figure 13.14). The arrival of a nerve impulse leads to the synchronous export of the contents of some 300 vesicles, which raises the acetylcholine concentration in the cleft from 10 nM to 500 μM in less than a millisecond. The binding of acetylcholine to the postsynaptic membrane markedly changes its ionic permeabilities (Figure 13.15). *The conductance of both Na^+ and K^+ increases greatly within 0.1 ms, leading to a large inward current of Na^+ and a smaller outward current of K^+.* The inward Na^+ current depolarizes the postsynaptic membrane and triggers an action potential (Section 13.5.3). *Acetylcholine opens a single kind of cation channel, which is almost equally permeable to Na^+ and K^+.* This change in ion permeability is mediated by the *acetylcholine receptor*.

The acetylcholine receptor is the best-understood ligand-gated channel. The activity of a single such channel is graphically displayed in patch-clamp

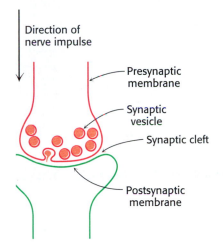

FIGURE 13.14 Schematic representation of a synapse.

Acetylcholine

FIGURE 13.15 Membrane depolarization. Acetylcholine depolarizes the postsynaptic membrane by increasing the conductance of Na^+ and K^+.

Polarized postsynaptic membrane (~ −75 mV)

High [K^+] Low [Na^+]

Acetylcholine

Depolarized postsynaptic membrane (~ 0 mV)

Na^+

K^+

Direction of action potential

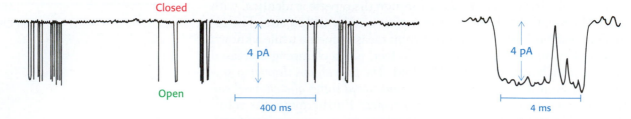

Closed

Open

4 pA

400 ms

4 pA

4 ms

FIGURE 13.16 Patch-clamp recordings of the acetylcholine receptor channel. Patch-clamp recordings illustrate changes in the conductance of an acetylcholine receptor channel in the presence of acetylcholine. The channel undergoes frequent transitions between open and closed states. [Courtesy of Dr. D. Colquhoun and Dr. B. Sakmann.]

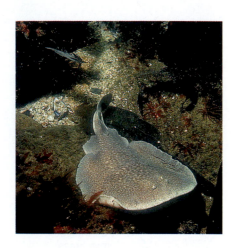

The torpedo (*Torpedo marmorata,* also known as the electric ray) has an electric organ, rich in acetylcholine receptors, that can deliver a shock of as much as 200 V for approximately 1 s. [Yves Gladu/Jacana/Photo Researchers.]

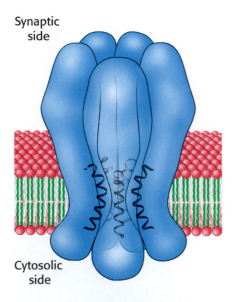

Synaptic side

Cytosolic side

FIGURE 13.17 Schematic representation of the closed form of the acetylcholine receptor channel. In the closed state, the narrowest part of the pore is occluded by side chains coming from five helices. [Courtesy of Dr. Nigel Unwin.]

recordings of postsynaptic membranes of skeletal muscle (Figure 13.16). The addition of acetylcholine is followed by transient openings of the channel. The current, i, flowing through an open channel is 4 pA (picoamperes) when the membrane potential, V, is -100 mV. An ampere is the flow of 6.24×10^{18} charges per second. Hence, 2.5×10^7 ions per second flow through an open channel.

The *electric organ* of *Torpedo marmorata,* an electric fish, is a choice source of acetylcholine receptors for study because its electroplaxes (voltage-generating cells) are very rich in cholinergic postsynaptic membranes. The receptor is very densely packed in these membranes ($\sim 20{,}000/\mu m^2$). An exotic biological material has been invaluable in the isolation of acetylcholine receptors. Snake neurotoxins such as α-*bungarotoxin* (from the venom of a Formosan snake) and *cobratoxin* block the transmission of impulses between nerve and muscle. These small (7-kd) basic proteins bind specifically and very tightly to acetylcholine receptors and hence can be used as tags.

The acetylcholine receptor of the electric organ has been solubilized by adding a nonionic detergent to a postsynaptic membrane preparation and purified by affinity chromatography on a column bearing covalently attached cobratoxin. With the use of techniques presented in Chapter 4, the 268-kd receptor was identified as a pentamer of four kinds of membrane-spanning subunits— α_2, β, γ, and δ—arranged in the form of a ring that creates a pore through the membrane (Figure 13.17). The cloning and sequencing of the cDNAs for the four kinds of subunits (50–58 kd) showed that they have clearly similar sequences; the genes for the α, β, γ, and δ subunits arose by duplication and divergence of a common ancestral gene. Each subunit has a large extracellular domain, followed at the carboxyl end by four predominantly hydrophobic segments that span the bilayer membrane. Acetylcholine binds at the α–γ and α–δ interfaces. Electron microscopic studies of purified acetylcholine receptors demonstrated that the structure has approximately fivefold symmetry, in harmony with the similarity of its five constituent subunits.

What is the basis of channel opening? A comparison of the structures of the closed and open forms of the channel would be highly revealing, but such comparisons have been difficult to obtain. Cryoelectron micrographs indicate that the binding of acetylcholine to the extracellular domain causes a structural alteration, which initiates rotations of the α-helical rods lining the membrane-spanning pore. The amino acid sequences of these helices point to the presence of alternating ridges of small polar or neutral residues (serine, threonine, glycine) and large nonpolar ones (isoleucine, leucine, phenylalanine). In the closed state, the large residues may occlude the channel by forming a tight hydrophobic ring (Figure 13.18). Indeed, each subunit has a bulky

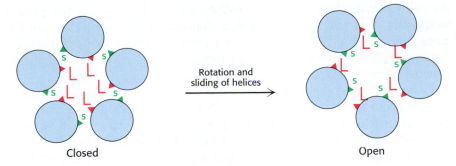

Closed   Open

FIGURE 13.18 Opening of the acetylcholine channel pore. Large hydrophobic side chains (L) occlude the pore of the closed form of the acetylcholine receptor channel. Channel opening is probably mediated by the tilting of helices that line the pore. Large residues move away from the pore and small ones (S) take their place. [After N. Unwin. *Neuron* 3(1989):665.]

leucine residue at a critical position. The binding of acetylcholine could allosterically rotate the membrane-spanning helices so that the pore would be lined by small polar residues rather than by large hydrophobic ones. The wider, more polar pore would then be open to the passage of Na^+ and K^+ ions.

13.5.3 Action Potentials Are Mediated by Transient Changes in Na^+ and K^+ Permeability

We turn now from ligand-gated channels to voltage-gated channels, which are responsible for the propagation of nerve impulses. A *nerve impulse* is an electrical signal produced by the flow of ions across the plasma membrane of a neuron and is the fundamental means of communication in the nervous system. The interior of a neuron, like that of most other cells, has a high concentration of K^+ and a low concentration of Na^+. These ionic gradients are generated by an ATP-driven pump (Section 13.2.1). In the resting state, the membrane potential is -60 mV. A nerve impulse, or *action potential*, is generated when the membrane potential is depolarized beyond a critical threshold value (i.e., from -60 to -40 mV). The membrane potential becomes positive within about a millisecond and attains a value of about $+30$ mV before turning negative again. This amplified depolarization is propagated along the nerve terminal (Figure 13.19)

Ingenious experiments carried out by Alan Hodgkin and Andrew Huxley revealed that *action potentials arise from large, transient changes in the permeability of the axon membrane to Na^+ and K^+ ions* (see Figure 13.19A). Two kinds of voltage-sensitive channels, one selectively permeable to Na^+ and the other to K^+, were defined. The conductance of the membrane to Na^+ changes first. Depolarization of the membrane beyond the threshold level leads to an opening of Na^+ channels. Sodium ions begin to flow into

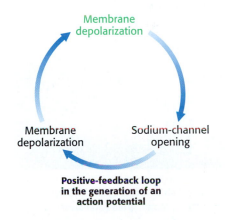

Membrane depolarization

Membrane depolarization   Sodium-channel opening

Positive-feedback loop in the generation of an action potential

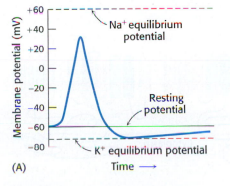

(A)

Membrane potential (mV)

+60 — Na^+ equilibrium potential
+40
+20
0
−20
−40   Resting potential
−60
−80 — K^+ equilibrium potential

Time →

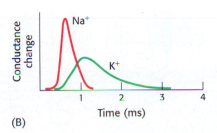

Conductance change

Na^+

K^+

1   2   3   4
Time (ms)

(B)

FIGURE 13.19 Membrane potential. Depolarization of an axon membrane results in an action potential. Time course of (A) the change in membrane potential and (B) the change in Na^+ and K^+ conductances.

the cell because of the large electrochemical gradient across the plasma membrane. The entry of Na^+ further depolarizes the membrane, and so more gates for Na^+ are opened. This positive feedback between depolarization and Na^+ entry leads to a very rapid and large change in membrane potential, from about -60 mV to $+30$ mV in a millisecond.

Sodium channels spontaneously close and potassium channels begin to open at about this time (see Figure 13.19B). Consequently, potassium ions flow outward, and so the membrane potential returns to a negative value. The resting level of -60 mV is restored in a few milliseconds as the K^+ conductance decreases to the value characteristic of the unstimulated state. Only a very small proportion of the sodium and potassium ions in a nerve cell, of the order of one in a million, flows across the plasma membrane during the action potential. Clearly, the action potential is a very efficient means of signaling over large distances.

13.5.4 The Sodium Channel Is an Example of a Voltage-Gated Channel

Like the acetylcholine receptor channel, the sodium channel also was purified on the basis of its ability to bind a specific neurotoxin. Tetrodotoxin, an organic compound isolated from the puffer fish, binds to sodium channels with great avidity ($K_i \approx 1$ nM). The lethal dose of this poison for an adult human being is about 10 ng. The sodium channel was first purified from the electric organ of electric eel, which is a rich source of the protein forming this channel. The isolated protein is a single chain of 260 kd.

The availability of purified protein enabled Shosaku Numa and coworkers to clone and sequence the cDNA for the sodium channel from the electroplax cells of the eel electric organ and then from the rat. Subsequently, a large number of sodium channel cDNAs have been cloned from other sources, and sequence comparisons have been made. The eel and rat cDNA sequences are approximately 61% identical, which indicates that the amino acid sequence of the sodium channel has been conserved over a long evolutionary period. Most interesting, the channel contains four internal repeats, or homology units, having similar amino acid sequences, suggesting that gene duplication and divergence have produced the gene for this channel. Hydrophobicity profiles indicate that each homology unit contains five hydrophobic segments (S1, S2, S3, S5, and S6). Each repeat also contains a highly positively charged S4 segment; arginine or lysine residues are present at nearly every third residue. Numa proposed that segments S1 through S6 are membrane-spanning α helices (Figure 13.20). The positively charged residues in S4 segments act as the voltage sensors of the channel. The purification of calcium channels and the subsequent cloning and sequencing of their cDNAs revealed that these proteins are homologous to the sodium channels and have quite similar architectures; each protein comprises four imperfectly repeated units, each of which has regions corresponding to segments S1 through S6.

Tetrodotoxin

A puffer fish is regarded as a culinary delicacy in Japan. [Fred Bavendam/ Peter Arnold.]

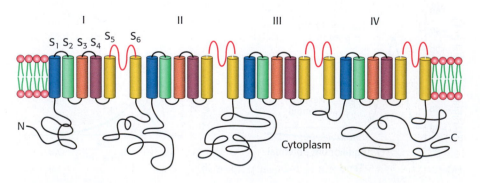

FIGURE 13.20 The sodium channel.
The Na^+ channel is a single polypeptide chain with four repeating units (I–IV). Each repeat probably folds into six transmembrane helices. The loops (shown in red) between helices 5 and 6 of each domain form the pore of the channel.

We can thus note similarities between ligand-gated and voltage-gated channels. Like the acetylcholine receptor, the sodium channel is constructed of similar units. The acetylcholine receptor has five units, whereas the sodium channel has four units that have been fused into a single polypeptide chain. The acetylcholine receptor is composed of similar but noncovalently attached subunits.

13.5.5 Potassium Channels Are Homologous to the Sodium Channel

The purification of potassium channels proved to be much more difficult because of their low abundance and the lack of known high-affinity ligands comparable to tetrodotoxin. The breakthrough came in studies of mutant fruit flies that shake violently when anesthetized with ether. The mapping and cloning of the gene, termed *shaker*, responsible for this defect revealed the amino acid sequence encoded by a potassium-channel gene. The availability of this gene sequence has led to the cloning of potassium-channel cDNAs from many other organisms. *Shaker* cDNA encodes a 70-kd protein that has regions that correspond to one of the homology units of the sodium channel containing the membrane-spanning segments S1 through S6. Thus, a potassium-channel subunit is homologous to one of the repeated homology units of the sodium and calcium channels. Consistent with this hypothesis, four potassium-channel subunits come together to form a functional channel. Subsequently, other potassium channels were discovered, including some from bacteria, which contain only the two membrane-spanning regions corresponding to segments S5 and S6. This and other information *pointed to the region between S5 and S6 as a key component of the ion-channel pore in the potassium channel and in the sodium and calcium channels as well.* The sequence relationships between these ion channels are summarized in Figure 13.21.

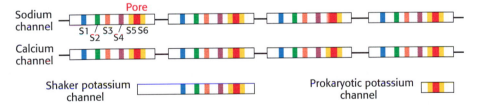

FIGURE 13.21 Sequence relationships of ion channels. Like colors indicate structurally similar regions of the sodium, calcium, and potassium channels. These channels exhibit approximate fourfold symmetry.

13.5.6 The Structure of a Potassium Channel Reveals the Basis of Rapid Ion Flow with Specificity

STRUCTURAL INSIGHTS, The Potassium Channel, *examines the structural basis of the potassium channel's ion specificity and high conductivity in further detail.*

Scientists were slowly discovering the likely structures of ion channels through a combination of patch-clamp methods, site-directed mutagenesis, and other methods. However, progress was limited by the lack of a high-resolution three-dimensional structure. The need was met by the determination of the structure of a bacterial potassium channel by x-ray crystallography in 1998. The resulting structural framework is a source of insight into many aspects of ion-channel function, including specificity and rapidity of ion flow.

As expected, the potassium channel is a tetramer of identical subunits, each of which includes two membrane-spanning α helices. The four subunits come together to form a pore in the shape of a cone that runs through the center of the structure (Figure 13.22). Beginning from the inside of the cell, the pore starts with a diameter of approximately 10 Å and then constricts

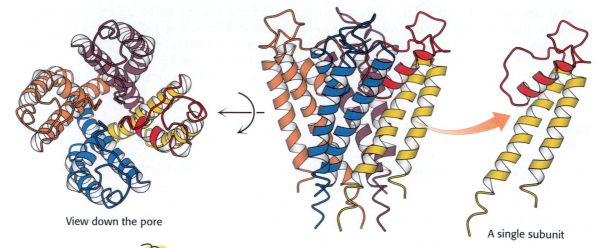

View down the pore

A single subunit

FIGURE 13.22 Structure of the potassium channel. The potassium channel, composed of four identical subunits, is cone shaped, with the larger opening facing the inside of the cell (center). A view down the pore, looking toward the outside of the cell, shows the relations of the individual subunits (left). One of the four identical subunits of the pore is illustrated at the right, with the pore-forming region highlighted in red.

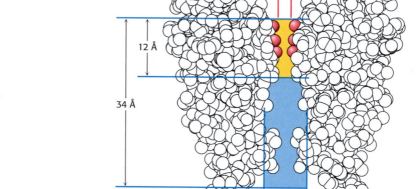

FIGURE 13.23 Path through a channel. A potassium ion entering the potassium channel can pass a distance of 22 Å into the membrane while remaining solvated with water (blue). At this point, the pore diameter narrows to 3 Å (yellow), and potassium must shed its water and interact with carbonyl groups (red) of the pore amino acids.

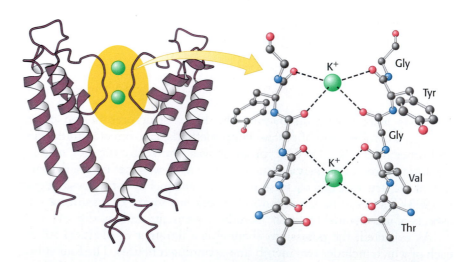

FIGURE 13.24 Selectivity filter of the potassium channel. Potassium ions interact with the carbonyl groups of the TVGYG sequence of the selectivity filter, located at the 3-Å-diameter pore of the potassium channel.

to a smaller cavity with a diameter of 8 Å. Both the opening to the outside and the central cavity of the pore are filled with water, and a K^+ ion can fit in the pore without losing its shell of bound water molecules. Approximately two-thirds of the way through the membrane, the pore becomes more constricted (3-Å diameter). At that point, any K^+ ions must give up their water molecules and interact directly with groups from the protein. The channel structure effectively reduces the thickness of the membrane from 34 Å to 12 Å by allowing the solvated ions to penetrate into the membrane before the ions must directly interact with the channel (Figure 13.23).

For potassium ions to relinquish their water molecules, other polar interactions must replace those with water. The restricted part of the pore is built from residues between the two transmembrane helices (which correspond to segments S5 and S6 in the sodium channel). In particular, a five-amino-acid stretch within this region functions as the *selectivity filter* that determines the preference for K^+ over other ions (Figure 13.24). The stretch has the sequence Thr-Val-Gly-Tyr-Gly, which is nearly completely conserved in all K^+ channels and had already been identified as a signature sequence useful for identifying potential K^+ channels. This region lies in a relatively extended conformation and is oriented such that the peptide carbonyl groups are directed into the channel, facilitating interaction with the potassium ions.

Potassium channels are 100-fold as permeable to K^+ as to Na^+. How is this high degree of selectivity achieved? The narrow diameter (3 Å) of the selectivity filter of the potassium channel enables the filter to reject ions having a radius larger than 1.5 Å. However, a bare Na^+ is small enough (Table 13.2) to pass through the pore. Indeed, the ionic radius of Na^+ is substantially smaller than that of K^+. How then is Na^+ rejected?

We need to consider the free-energy cost of dehydrating the Na^+ and K^+ ions, given that they cannot pass through this part of the channel bearing a retinue of water molecules. The key point is that the free-energy costs of dehydrating these ions are considerable [Na^+, 72 kcal mol^{-1} (301 kJ mol^{-1}), and K^+, 55 kcal mol^{-1} (203 kJ mol^{-1})]. *The channel pays the cost of dehydrating K^+ by providing compensating interactions with the carbonyl oxygen atoms lining the selectivity filter.* However, these oxygen atoms are positioned such that they do not interact very favorably with Na^+, because it is too small (Figure 13.25). The higher cost of dehydrating Na^+ would be unrecovered, and so Na^+ would be rejected. The ionic radii of oxygen, potassium, and sodium are 1.4, 1.33, and 0.95 Å, respectively. Hence a ring of oxygen atoms positioned so that the K^+–O distance is 2.73 Å (1.4 + 1.33 Å) would be optimal for interaction with K^+ compared with the shorter Na^+–O bonds (0.95 + 1.4 = 2.35 Å) optimal for interaction with Na^+. Thus, the potassium channel avoids closely embracing Na^+ ions, which must stay hydrated and hence are impermeant.

TABLE 13.2 Properties of alkali cations

Ion	Ionic radius (Å)	Hydration free energy in kcal mol^{-1} (kJ mol^{-1})
Li^+	0.60	-98 (-410)
Na^+	0.95	-72 (-301)
K^+	1.33	-55 (-230)
Rb^+	1.48	-51 (-213)
Cs^+	1.69	-47 (-197)

FIGURE 13.25 Energetic basis of ion selectivity. The energy cost of dehydrating a potassium ion is compensated by favorable interactions with the selectivity filter. Because sodium is too small to interact favorably with the selectivity filter, the free energy of desolvation cannot be compensated and the sodium does not pass through the channel.

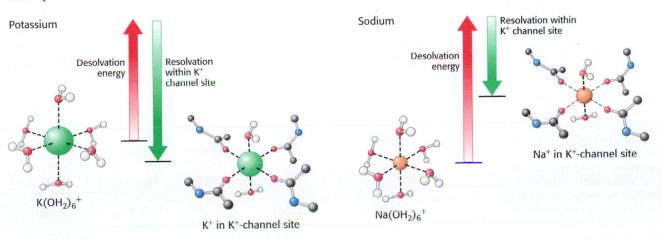

Potassium

Desolvation energy

Resolution within K^+ channel site

$K(OH_2)_6{}^+$

K^+ in K^+-channel site

Sodium

Desolvation energy

Resolution within K^+ channel site

$Na(OH_2)_6{}^+$

Na^+ in K^+-channel site

13.5.7 The Structure of the Potassium Channel Explains Its Rapid Rates of Transport

In addition to selectivity, ion channels display rapid rates of ion transport. A structural analysis provides an appealing explanation for this proficiency. The results of such studies revealed the presence of two potassium-binding sites in the constricted regions of the potassium channel that are crucial for rapid ion flow. Consider the process of ion conductance. One K^+ ion proceeds into the channel and through the relatively unrestricted part of the channel. It then gives up most or all of its coordinated water molecules and binds to the first site in the selectivity filter region, a favorable binding site. It can then jump to the second site, which appears to have comparable binding energy. However, the binding energy of the second site presents a free-energy barrier, or trap, preventing the ion from completing its journey; there is no energetic reason to leave the second ion-binding site. However, if a second ion moves through the channel into the first site, the electrostatic repulsion between the two ions will destabilize the initially bound ion and help push it into solution (Figure 13.26). This mechanism provides a solution to the apparent paradox of high ion selectivity (requiring tight binding sites) and rapid flow.

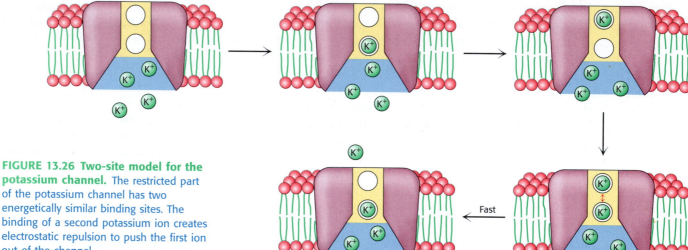

FIGURE 13.26 Two-site model for the potassium channel. The restricted part of the potassium channel has two energetically similar binding sites. The binding of a second potassium ion creates electrostatic repulsion to push the first ion out of the channel.

Fast

The structure determined for K^+ channels is a good start for considering the amino acid sequence similarities, as well as the structural and functional relations, for Na^+ and Ca^{2+} channels because of their homology to K^+ channels. Sequence comparisons and the results of mutagenesis experiments have also implicated the region between segments S5 and S6 in ion selectivity in the Ca^{2+} channels. In Ca^{2+} channels, one glutamate residue of this region in each of the four units plays a major role in determining ion selectivity. The Na^+ channel's selection of Na^+ over K^+ depends on ionic radius; the diameter of the pore is sufficiently restricted that small ions such as Na^+ and Li^+ can pass through the channel, but larger ions such as K^+ are significantly hindered (Figure 13.27).

FIGURE 13.27 Selectivity of the sodium channel. The ionic selectivity of the sodium channel partly depends on steric factors. Sodium and lithium ions, together with a water molecule, fit in the channel, as do hydroxylamine and hydrazine. In contrast, K^+ with a water molecule is too large. [After R. D. Keynes. Ion channels in the nerve-cell membrane. Copyright © 1979 by Scientific American, Inc. All rights reserved.]

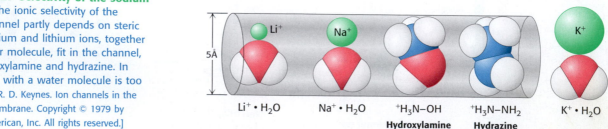

5 Å

| Li^+ • H_2O | Na^+ • H_2O | $^+H_3N–OH$ Hydroxylamine | $^+H_3N–NH_2$ Hydrazine | K^+ • H_2O |

Residues in the positions corresponding to the glutamate residues in Ca^{2+} channels are major components of the selectivity filter of the Na^+ channel. These residues are aspartate, glutamate, lysine, and alanine in units 1, 2, 3, and 4, respectively (the DEKA locus). Thus, the potential fourfold symmetry of the channel is clearly broken in this region, providing one explanation of why Na^+ channels comprise single large polypeptide chains rather than a noncovalent assembly of four identical subunits.

13.5.8 A Channel Can Be Inactivated by Occlusion of the Pore: The Ball-and-Chain Model

The potassium channel and the sodium channel undergo inactivation within milliseconds of channel opening (Figure 13.28). A first clue to the mechanism of inactivation came from exposing the cytoplasmic side of either channel to trypsin; cleavage by trypsin produced a trimmed channel that stayed persistently open after depolarization. A second clue was the finding that alternatively spliced variants of the potassium channel have markedly different inactivation kinetics; these variants differed from one another only near the amino terminus, which is on the cytoplasmic side of the channel. A mutant Shaker channel lacking 42 amino acids near the amino terminus opened in response to depolarization but did not inactivate (see Figure 13.28). Most revealing, inactivation was restored by adding a synthetic peptide corresponding to the first 20 residues of the native channel.

These experiments strongly support the *ball-and-chain model* for channel inactivation that had been proposed years earlier (Figure 13.29). According to this model, the first 20 residues of the potassium channel form a cytoplasmic unit (the *ball*) that is attached to a flexible segment of the polypeptide (the *chain*). When the channel is closed, the ball rotates freely in the aqueous solution. When the channel opens, the ball quickly finds a complementary site in the pore and occludes it. Hence, the channel opens for only a brief interval before it undergoes inactivation by occlusion. Shortening the chain speeds inactivation because the ball finds its target more quickly. Conversely, lengthening the chain slows inactivation. Thus, the duration of the open state can be controlled by the length and flexibility of the tether.

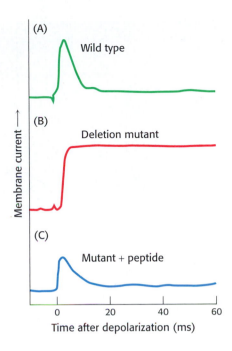

FIGURE 13.28 Inactivation of the potassium channel. The amino-terminal region of the potassium chain is critical for inactivation. (A) The wild-type Shaker potassium channel displays rapid inactivation after opening. (B) A mutant channel lacking residues 6 through 46 does not inactivate. (C) Inactivation can be restored by adding a peptide consisting of residues 1 through 20 at a concentration of 100 μM. [After W. N. Zagotta, T. Hoshi, and R. W. Aldrich. *Science* 250(1990):568.]

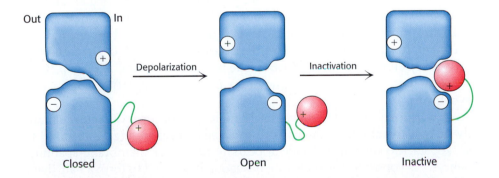

FIGURE 13.29 Ball-and-chain model for channel inactivation. The inactivation domain, or "ball" (red), is tethered to the channel by a flexible "chain" (green). In the closed state, the ball is located in the cytosol. Depolarization opens the channel and creates a negatively charged binding site for the positively charged ball near the mouth of the pore. Movement of the ball into this site inactivates the channel by occluding it. [After C. M. Armstrong and F. Bezanilla. *J. Gen. Physiol.* 70(1977):567.]

13.6 GAP JUNCTIONS ALLOW IONS AND SMALL MOLECULES TO FLOW BETWEEN COMMUNICATING CELLS

The ion channels that we have considered thus far have narrow pores and are moderately to highly selective in regard to which ions are permeant. They are closed in the resting state and have short lifetimes in the open state, typically a millisecond, that enable them to transmit highly frequent neural signals. We turn now to a channel with a very different role. *Gap junctions*, also known as *cell-to-cell channels*, serve as passageways between the

(A)

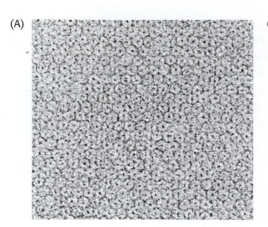

(B)

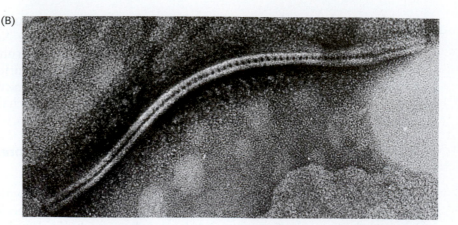

FIGURE 13.30 Gap junctions. (A) This electron micrograph shows a sheet of isolated gap junctions. The cylindrical connexons form a hexagonal lattice having a unit-cell length of 85 Å. The densely stained central hole has a diameter of about 20 Å. (B) Electron micrograph of a tangential view of apposed cell membranes that are joined by gap junctions. [(A) Courtesy of Dr. Nigel Unwin and Dr. Guido Zampighi; (B) from E. L. Hertzberg and N. B. Gilula. *J. Biol. Chem.* 254(1979):2143.]

interiors of contiguous cells. Gap junctions are clustered in discrete regions of the plasma membranes of apposed cells. Electron micrographs of sheets of gap junctions show them tightly packed in a regular hexagonal array (Figure 13.30A). A 20-Å central hole, the lumen of the channel, is prominent in each gap junction. A tangential view (Figure 13.30B) shows that these channels span the intervening space, or gap, between apposed cells (hence, the name *gap junction*). The width of the gap between the cytosols of the two cells is about 35 Å.

Small hydrophilic molecules as well as ions can pass through gap junctions. The pore size of the junctions was determined by microinjecting a series of fluorescent molecules into cells and observing their passage into adjoining cells. All polar molecules with a mass of less than about 1 kd can readily pass through these cell-to-cell channels. Thus, *inorganic ions and most metabolites (e.g., sugars, amino acids, and nucleotides) can flow between the interiors of cells joined by gap junctions.* In contrast, proteins, nucleic acids, and polysaccharides are too large to traverse these channels. *Gap junctions are important for intercellular communication.* Cells in some excitable tissues, such as heart muscle, are coupled by the rapid flow of ions through these junctions, which ensure a rapid and synchronous response to stimuli. Gap junctions are also essential for the nourishment of cells that are distant from blood vessels, as in lens and bone. Moreover, communicating channels are important in development and differentiation. For example, a pregnant uterus is transformed from a quiescent protector of the fetus to a forceful ejector at the onset of labor; the formation of functional gap junctions at that time creates a syncytium of muscle cells that contract in synchrony.

A cell-to-cell channel is made of 12 molecules of *connexin*, one of a family of transmembrane proteins with molecular masses ranging from 30 to 42 kd. Each connexin molecule appears to have four membrane-spanning helices. Six connexin molecules are hexagonally arrayed to form a half channel, called a *connexon* or *hemichannel* (Figure 13.31). Two connexons join end to end in the intercellular space to form a functional channel between the communicating cells. Cell-to-cell channels differ from other membrane channels in three respects: (1) they traverse *two* membranes rather than one; (2) they connect cytosol to cytosol, rather than to the extracellular space or the lumen of an organelle; and (3) the connexons forming a channel are synthesized by different cells. Gap junctions form readily when cells are brought together. A cell-to-cell channel, once formed, tends to stay open for seconds to minutes. They are closed by high concentrations of calcium ion and by low pH. *The closing of gap junctions by Ca^{2+} and H^+ serves to seal normal cells from traumatized or dying neighbors.* Gap junctions are also controlled by membrane potential and by hormone-induced phosphorylation.

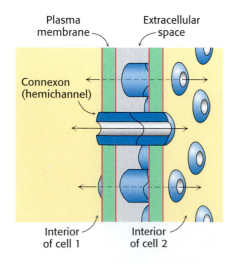

Plasma membrane

Extracellular space

Connexon (hemichannel)

Interior of cell 1

Interior of cell 2

FIGURE 13.31 Schematic representation of a gap junction. [Courtesy of Dr. Werner Loewenstein.]

SUMMARY

The Transport of Molecules Across a Membrane May Be Active or Passive

For a net movement of molecules across a membrane, two features are required: (1) the molecule must be able to cross a hydrophobic barrier and (2) an energy source must power the movement. Lipophilic molecules can pass through a membrane's hydrophobic interior by simple diffusion. These molecules will move down their concentration gradients. Polar or charged molecules require proteins to form passages through the hydrophobic barrier. Passive transport or facilitated diffusion occurs when an ion or polar molecule moves down its concentration gradient. If a molecule moves against a concentration gradient, an external energy source is required; this movement is referred to as active transport and results in the generation of concentration gradients. Concentration gradients are a commonly used form of energy in all organisms.

A Family of Membrane Proteins Uses ATP Hydrolysis to Pump Ions Across Membranes

Active transport is often carried out at the expense of ATP hydrolysis. P-type ATPases pump ions against a concentration gradient and become transiently phosphorylated on an aspartic acid residue in the process of transport. P-type ATPases, which include the sarcoplasmic reticulum Ca^{2+} ATPase and the Na^+-K^+ ATPase, are integral membrane proteins with conserved structures and catalytic mechanisms.

Multidrug Resistance and Cystic Fibrosis Highlight a Family of Membrane Proteins with ATP-Binding Cassette Domains

The membrane proteins with ATP-binding cassette (ABC) domains are complex ATP-dependent pumps. Each pump includes four major domains: two domains span the membrane and two others contain ABC P-loop ATPase structures. The multidrug resistance proteins confer resistance on cancer cells by pumping chemotherapeutic drugs out of a cancer cell before the drugs can exert their effects. Another ABC domain protein is the cystic fibrosis transmembrane conductance regulator (CFTR), an ATP-gated chloride channel. Mutations in CFTR can result in cystic fibrosis.

Secondary Transporters Use One Concentration Gradient to Power the Formation of Another

Many active-transport systems couple the uphill flow of one ion or molecule to the downhill flow of another. These membrane proteins, called secondary transporters or cotransporters, can be classified as antiporters or symporters. Antiporters couple the downhill flow of one type of ion in one direction to the uphill flow of another in the opposite direction. Symporters move both ions in the same direction.

Specific Channels Can Rapidly Transport Ions Across Membranes

Ion channels allow the rapid movement of ions across the hydrophobic barrier of the membrane. Such channels allow ions to flow down their concentration gradients. The channels have several features in common: (1) ion specificity, (2) the existence of open and closed states, (3) regulation by ligands or voltage. Ion channels are exemplified by the Na^+ and K^+ channels responsible for nerve impulses.

Gap Junctions Allow Ions and Small Molecules to Flow Between Communicating Cells

In contrast with many channels, which connect the cell interior with the environment, gap junctions, or cell-to-cell channels, serve to connect the interiors of contiguous groups of cells. A cell-to-cell channel is composed of 12 molecules of connexin, which associate to form two 6-membered connexons.

KEY TERMS

facilitated diffusion (passive transport) (p. 346)

active transport (p. 346)

Na^+-K^+ pump (Na^+-K^+-ATPase) (p. 347)

P-type ATPase (p. 348)

digitalis (p. 350)

multidrug resistance (p. 351)

ATP-binding cassette (ABC) domain (p. 351)

secondary transporter (cotransporter) (p. 352)

antiporter (p. 352)

symporter (p. 352)

ion channel (p. 353)

ligand-gated channel (p. 353)

voltage-gated channel (p. 353)

patch-clamp technique (p. 354)

gigaseal (p. 354)

neurotransmitter (p. 355)

action potential (p. 357)

selectivity filter (p. 361)

gap junction (p. 363)

SELECTED READINGS

Where to start

Welsh, M. J., and Smith, A. E., 1995. Cystic fibrosis. *Sci. Am.* 273(6):52–59.

Unwin, N., 1993. Neurotransmitter action: Opening of ligand-gated ion channels. *Cell* 72:31–41.

Lienhard, G. E., Slot, J. W., James, D. E., and Mueckler, M. M., 1992. How cells absorb glucose. *Sci. Am.* 266(1):86–91.

Neher, E., and Sakmann, B., 1992. The patch clamp technique. *Sci. Am.* 266(3):28–35.

Sakmann, B., 1992. Elementary steps in synaptic transmission revealed by currents through single ion channels. *Science* 256:503–512.

Books

Ashcroft, F. M., 2000. *Ion Channels and Disease.* Academic Press.

Conn, P. M. (Ed.), 1998. *Ion Channels,* vol. 293, *Methods in Enzymology.* Academic Press.

Aidley, D. J., and Stanfield, P. R., 1996. *Ion Channels: Molecules in Action.* Cambridge University Press.

Hille, B., 1992. *Ionic Channels of Excitable Membranes* (2d ed.). Sinauer.

Läuger, P., 1991. *Electrogenic Ion Pumps.* Sinauer.

Stein, W. D., 1990. *Channels, Carriers, and Pumps: An Introduction to Membrane Transport.* Academic Press.

Hodgkin, A., 1992. *Chance and Design: Reminiscences of Science in Peace and War.* Cambridge University Press.

Voltage-gated ion channels

Bezanilla, F., 2000. The voltage sensor in voltage-dependent ion channels. *Physiol. Rev.* 80:555–592.

Biggin, P. C., Roosild, T., and Choe, S., 2000. Potassium channel structure: Domain by domain. *Curr. Opin. Struct. Biol.* 10:456–461.

Sansom, M. S. P., 2000. Potassium channels: Watching a voltage-sensor tilt and twist. *Curr. Biol.* 10:R206–R209.

Shieh, C.-C., Coghlan, M., Sullivan, J. P., and Gopalakrishnan, M., 2000. Potassium channels: Molecular defects, diseases, and therapeutic opportunities. *Pharmacol. Rev.* 52:557–594.

Horn, R., 2000. Conversation between voltage sensors and gates of ion channels. *Biochemistry* 39:15653–15658.

Perozo, E., Cortes, D. M., and Cuello, L. G., 1999. Structural rearrangements underlying K^+-channel activation gating. *Science* 285:73–78.

Doyle, D. A., Morais Cabral, J., Pfuetzner, R. A., Kuo, A., Gulbis, J. M., Cohen, S. L., Chait, B. T., and MacKinnon R., 1998. The structure of the potassium channel: Molecular basis of K^+ conduction and selectivity. *Science* 280:69–77.

Marban, E., Yamagishi, T., and Tomaselli, G. F., 1998. Structure and function of the voltage-gated Na^+ channel. *J. Physiol.* 508:647–657.

Gonzalez, C., Rosenman, E., Bezanilla, F., Alvarez, O., and Latorre, R., 2000. Modulation of the Shaker K^+ channel gating kinetics by the S3-S4 linker. *J. Gen. Physiol.* 114:193–297.

Miller, R. J., 1992. Voltage-sensitive Ca^{2+} channels. *J. Biol. Chem.* 267:1403–1406.

Catterall, W. A., 1991. Excitation-contraction coupling in vertebrate skeletal muscle: A tale of two calcium channels. *Cell* 64:871–874.

Ligand-gated ion channels

Miyazawa, A., Fujiyoshi, Y., Stowell, M., and Unwin, N., 1999. Nicotinic acetylcholine receptor at 4.6 Å resolution: Transverse tunnels in the channel wall. *J. Mol. Biol.* 288:765–786.

Barrantes, F. J., Antollini, S. S., Blanton, M. P., and Prieto, M., 2000. Topography of the nicotinic acetylcholine receptor membrane-embedded domains. *J. Biol. Chem.* 275:37333–37339.

Cordero-Erausquin, M., Marubio, L. M., Klink, R., and Changeux, J. P., 2000. Nicotinic receptor function: New perspectives from knockout mice. *Trends Pharmacol. Sci.* 21:211–217.

Le Novère, N., and Changeux, J. P., 1995. Molecular evolution of the nicotinic acetylcholine receptor: An example of multigene family in excitable cells. *J. Mol. Evol.* 40:155–172.

Kunishima, N., Shimada, Y., Tsuji, Y., Sato, T., Yamamoto, M., Kumasaka, T., Nakanishi, S., Jingami, H., and Morikawa, K., 2000. Structural basis of glutamate recognition by dimeric metabotropic glutamate receptor. *Nature* 407:971–978.

Betz, H., Kuhse, J., Schmieden, V., Laube, B., Kirsch, J., and Harvey, R. J., 1999. Structure and functions of inhibitory and excitatory glycine receptors. *Ann. N. Y. Acad. Sci.* 868:667–676.

Unwin, N., 1995. Acetylcholine receptor channel imaged in the open state. *Nature* 373:37–43.

Colquhoun, D., and Sakmann, B., 1981. Fluctuations in the microsecond time range of the current through single acetylcholine receptor ion channels. *Nature* 294:464–466.

Seeburg, P. H., Wisden, W., Verdoorn, T. A., Pritchett, D. B., Werner, P., Herb, A., Luddens, H., Sprengel, R., and Sakmann, B., 1990. The $GABA_A$ receptor family: Molecular and functional diversity. *Cold Spring Harbor Symp. Quant. Biol.* 55:29–40.

ATP-driven ion pumps

Toyoshima, C., Nakasako, M., Nomura, H., and Ogawa, H., 2000. Crystal structure of the calcium pump of sarcoplasmic reticulum at 2.6 Å resolution. *Nature* 405:647–655.

Auer, M., Scarborough, G. A., and Kuhlbrandt, W., 1998. Three-dimensional map of the plasma membrane H^+-ATPase in the open conformation. *Nature* 392:840–843.

Axelsen, K. B., and Palmgren, M. G., 1998. Evolution of substrate specificities in the P-type ATPase superfamily. *J. Mol. Evol.* 46:84–101.

Pedersen, P. A., Jorgensen, J. R., and Jorgensen, P. L., 2000. Importance of conserved alpha-subunit [709]GDGVND for Mg^{2+} binding, phosphorylation, energy transduction in Na, K-ATPase. *J. Biol. Chem.* 275:37588–37595.

Blanco, G., and Mercer, R. W., 1998. Isozymes of the Na-K-ATPase: Heterogeneity in structure, diversity in function. *Am. J. Physiol.* 275:F633–F650.

MacLennan, D. H., 1990. Molecular tools to elucidate problems in excitation-contraction coupling. *Biophys. J.* 58:1355–1365.

Estes, J. W., and White, P. D., 1965. William Withering and the purple foxglove. *Sci. Am.* 212(6):110–117.

ATP-binding cassette (ABC) proteins

Akabas, M. H., 2000. Cystic fibrosis transmembrane conductance regulator: Structure and function of an epithelial chloride channel. *J. Biol. Chem.* 275:3729–3732.

Chen, J., Sharma, S., Quiocho, F. A., and Davidson, A. L., 2001. Trapping the transition state of an ATP-binding cassette transporter: Evidence for a concerted mechanism of maltose transport. *Proc. Natl. Acad. Sci. USA* 98:1525–1530.

Zhang, Z. R., McDonough, S. I., and McCarty, N. A., 2000. Interaction between permeation and gating in the putative pore domain mutant in the cystic fibrosis transmembrane conductance regulator. *Biophys. J.* 79:298–313.

Sheppard, D. N., and Welsh, M. J., 1999. Structure and function of the CFTR chloride channel. *Physiol. Rev.* 79:S23–S45.

Riordan, J. R., Rommens, J. M., Kerem, B. S., Alon, N., Rozmahel, R., Grzelczak, Z., Zielenski, J., Lok, S., Plavsic, N., Chou, J. L., Drumm, M. T., Iannuzzi, M. C., Collins, F. S., and Tsui, L. C., 1989. Identification of the cystic fibrosis gene: Cloning and characterization of complementary DNA. *Science* 245:1066–1073.

Jones, P. M., and George, A. M., 2000. Symmetry and structure in P-glycoprotein and ABC transporters: What goes around comes around. *Eur. J. Biochem.* 287:5298–5305.

Chen, Y., and Simon, S. M., 2000. In situ biochemical demonstration that P-glycoprotein is a drug efflux pump with broad specificity. *J. Cell Biol.* 148:863–870.

Saier, M. H., Jr., Paulsen, I. T., Sliwinski, M. K., Pao, S. S., Skurray, R. A., and Nikaido, H., 1998. Evolutionary origins of multidrug and drug-specific efflux pumps in bacteria. *FASEB J.* 12:265–74.

Symporters and antiporters

Philipson, K. D., and Nicoll, D. A., 2000. Sodium-calcium exchange: A molecular perspective. *Annu. Rev. Physiol.* 62:111–133.

Green, A. L., Anderson, E. J., and Brooker, R. J., 2000. A revised model for the structure and function of the lactose permease: Evidence that a face on transmembrane segment 2 is important for conformational changes. *J. Biol. Chem.* 275:23240–23246.

Pao, S. S., Paulsen, I. T., and Saier, M. H., Jr., 1998. Major facilitator superfamily. *Microbiol. Mol. Biol. Rev.* 62:1–34.

Wright, E. M., Hirsch, J. R., Loo, D. D., and Zampighi, G. A., 1997. Regulation of Na^+/glucose cotransporters. *J. Exp. Biol.* 200:287–293.

Wright, E. M., Loo, D. D., Turk, E., and Hirayama, B. A., 1996. Sodium cotransporters. *Curr. Opin. Cell Biol.* 8:468–473.

Kaback, H. R., Bibi, E., and Roepe, P. D., 1990. β-Galactoside transport in *E. coli*: A functional dissection of *lac* permease. *Trends Biochem. Sci.* 8:309–314.

Hilgemann, D. W., Nicoll, D. A., and Philipson, K. D., 1991. Charge movement during Na^+ translocation by native and cloned cardiac Na^+/Ca^{2+} exchanger. *Nature* 352:715–718.

Hediger, M. A., Turk, E., and Wright, E. M., 1989. Homology of the human intestinal Na^+/glucose and *Escherichia coli* Na^+/proline cotransporters. *Proc. Natl. Acad. Sci. USA* 86:5748–5752.

Gap junctions

Revilla, A., Bennett, M. V. L., and Barrio, L. C., 2000. Molecular determinants of membrane potential dependence in vertebrate gap junction channels. *Proc. Natl. Acad. Sci. USA* 97:14760–14765.

Unger, V. M., Kumar, N. M., Gilula, N. B., and Yeager, M., 1999. Three-dimensional structure of a recombinant gap junction membrane channel. *Science* 283:1176–1180.

Simon, A. M., 1999. Gap junctions: More roles and new structural data. *Trends Cell Biol.* 9:169–170.

White, T. W., and Paul, D. L., 1999. Genetic diseases and gene knockouts reveal diverse connexin functions. *Annu. Rev. Physiol.* 61:283–310.

PROBLEMS

1. *Concerted opening*. Suppose that a channel obeys the concerted allosteric model (MWC model, Section 10.1.5). The binding of ligand to the R state (the open form) is 20 times as tight as to the T state (the closed form). In the absence of ligand, the ratio of closed to open channels is 10^5. If the channel is a tetramer, what is the fraction of open channels when 1, 2, 3, and 4 ligands are bound?

2. *Respiratory paralysis*. Tabun and sarin have been used as chemical-warfare agents, and parathion has been employed as an insecticide. What is the molecular basis of their lethal actions?

Tabun

Sarin

Parathion

3. *Ligand-induced channel opening*. The ratio of open to closed forms of the acetylcholine receptor channel containing zero, one, and two bound acetylcholine molecules is 5×10^{-6}, 1.2×10^{-3}, and 14, respectively.

(a) By what factor is the open/closed ratio increased by the binding of the first acetylcholine molecule? The second acetylcholine molecule?

(b) What are the corresponding free-energy contributions to channel opening at 25°C?

(c) Can the allosteric transition be accounted for by the MWC concerted model?

4. *Voltage-induced channel opening*. The fraction of open channels at 5 mV increments beginning at -45 mV and ending at $+5$ mV at 20°C is 0.02, 0.04, 0.09, 0.19, 0.37, 0.59, 0.78, 0.89, 0.95, 0.98, and 0.99.

(a) At what voltage are half the channels open?

(b) What is the value of the gating charge?

(c) How much free energy is contributed by the movement of the gating charge in the transition from -45 mV to $+5$ mV?

5. *Different directions*. The potassium channel and the sodium channel have similar structures and are arranged in the same orientation in the cell membrane. Yet, the sodium channel allows sodium ions to flow into the cell and the potassium channel allows potassium ions to flow out of the cell. Explain.

6. *Structure–activity relations.* On the basis of the structure of tetrodotoxin, propose a mechanism by which the toxin inhibits sodium flow through the sodium channel.

7. *A dangerous snail.* Cone snails are carnivores that inject a powerful set of toxins into their prey, leading to rapid paralysis. Many of these toxins are found to bind to specific ion-channel proteins. Why are such molecules so toxic? How might such toxins be useful for biochemical studies?

8. *Only a few.* Why do only a small number of sodium ions need to flow through the sodium channel to significantly change the membrane potential?

9. *Frog poison.* Batrachotoxin (BTX) is a steroidal alkaloid from the skin of *Phyllobates terribilis*, a poisonous Colombian frog (source of the poison used on blowgun darts). In the presence of BTX, sodium channels in an excised patch stay persistently open when the membrane is depolarized. They close when the membrane is repolarized. Which transition is blocked by BTX?

Phyllobates terribilis. [Tom McHugh, Photo Researchers.]

10. *Valium target.* γ-Aminobutyric acid (GABA) opens channels that are specific for chloride ions. The $GABA_A$ receptor channel is pharmacologically important because it is the target of Valium, which is used to diminish anxiety.

(a) The extracellular concentration of Cl^- is 123 mM and the intracellular concentration is 4 mM. In which direction does Cl^- flow through an open channel when the membrane potential is in the -60 mV to $+30$ mV range?
(b) What is the effect of chloride-channel opening on the excitability of a neuron?
(c) The hydropathy profile of the $GABA_A$ receptor resembles that of the acetylcholine receptor. Predict the number of subunits in this chloride channel.

11. *The price of extrusion.* What is the free-energy cost of pumping Ca^{2+} out of a cell when the cytosolic concentration is 0.4 μM, the extracellular concentration is 1.5 mM, and the membrane potential is -60 mV?

12. *Rapid transit.* A channel exhibits current increments of 5 pA at a membrane potential of -50 mV. The mean open time is 1 ms.

(a) What is the conductance of this channel?
(b) How many univalent ions flow through the channel during its mean open time?
(c) How long does it take an ion to pass through the channel?

13. *Pumping protons.* Design an experiment to show that lactose permease can be reversed in vitro to pump protons.

Chapter Integration Problem

14. *Speed and efficiency matter.* Acetylcholine is rapidly destroyed by the enzyme acetylcholinesterase. This enzyme, which has a turnover number of 25,000 per second, has attained catalytic perfection with a k_{cat}/K_M of 2×10^8 $M^{-1}s^{-1}$. Why is it physiologically crucial that this enzyme be so efficient?

Mechanism Problem

15. *Remembrance of mechanisms past.* Acetylcholinesterase converts acetylcholine into acetate and choline. Show the reaction as chemical structures :

$$\text{Acetylcholine} + H_2O \longrightarrow \text{acetate} + \text{choline}$$

Like serine proteases, acetylcholinesterase is inhibited by DIPF. Propose a catalytic mechanism for acetylcholine digestion by acetylcholinesterase.

Data Interpretation Problem

16. *Tarantula toxin.* Acid sensing is associated with pain, tasting, and other biological activities (Chapter 32). Acid sensing is carried out by a ligand-gated channel that permits sodium influx in response to H^+. This family of acid-sensitive ion channels (ASICs) comprises a number of members. Psalmotoxin 1 (PcTX1), a venom from the tarantula, inhibits some members of this family. Below are electrophysiological recordings of cells containing one of several members of the ASIC family made in the presence of the toxin at a concentration of 10 nM. The channels were opened by changing the pH from 7.4 to the indicated values. The PcTX1 was present for a short time (indicated by the black bar above the recordings) after which time it was rapidly washed from the system.

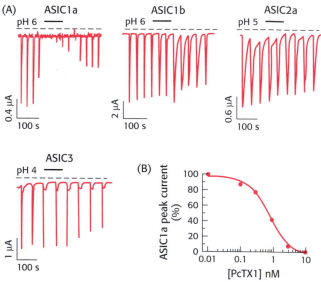

(A) Electrophysiological recordings of cells exposed to tarantula toxin. (B) Plot of peak current of a cell containing the ASIC1a protein versus the toxin concentration. [From P. Escoubas, et al., 2000, *J. Biol. Chem.* 275:25116–215121.]

(a) Which of the ASIC family members—ASIC1a, ASIC1b, ASIC2a, or ASIC3—is most sensitive to the toxin?
(b) Is the effect of the toxin reversible? Explain.
(c) What concentration of PcTX1 yields 50% inhibition of the sensitive channel?

17. *Channel problems 1*. A number of pathological conditions result from mutations in the acetylcholine receptor channel. One such mutation in the β subunit, βV266M, causes muscle weakness and rapid fatigue. An investigation of the acetylcholine-generated currents through the acetylcholine receptor channel for both a control and a patient yielded the following results. What is the effect of the mutation on channel function? Suggest some possible biochemical explanations for the effect.

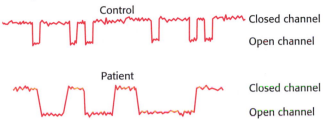

Control

Closed channel

Open channel

Patient

Closed channel

Open channel

18. *Channel problems 2*. The acetylcholine receptor channel can also undergo mutation leading to fast channel syndrome (FCS), with clinical manifestations similar to those of slow channel syndrome (SCS). What would the recordings of ion movement look like in this syndrome? Again, suggest a biochemical explanation.

19. *Transport differences*. The rate of transport of two molecules, indole and glucose, across a cell membrane is shown in the right column. What are the differences between the transport mechanisms of the two molecules? Suppose that ouabain inhibited the transport of glucose. What would this inhibition suggest about the mechanism of transport?

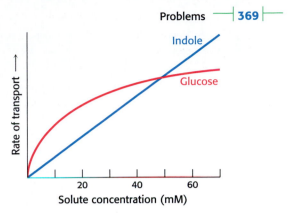

Rate of transport →

Indole

Glucose

20   40   60

Solute concentration (mM)

 **Media Problem**

20. *The merits of inflexibility*. The selectivity filter of potassium channels restricts passage of sodium ions even though sodium ions are smaller than potassium ions. Mutation of the tyrosine that is part of this filter to leucine has been shown to reduce selectivity against sodium in a channel homologous to the one whose structure has been determined [Chapman, M. L., Krovetz, H. S., and VanDongen, A. M. J., 2001. GYGD pore motifs in neighboring potassium channel subunits interact to determine ion selectivity. *J. Physiol.* 530:21–33]. Look at the three-dimensional structure in the Potassium Channel **Structural Insights** module and propose, in general terms, an explanation for the role of this tyrosine in the wild-type protein and the effect of its mutation to leucine.

PART II

Transducing and Storing Energy

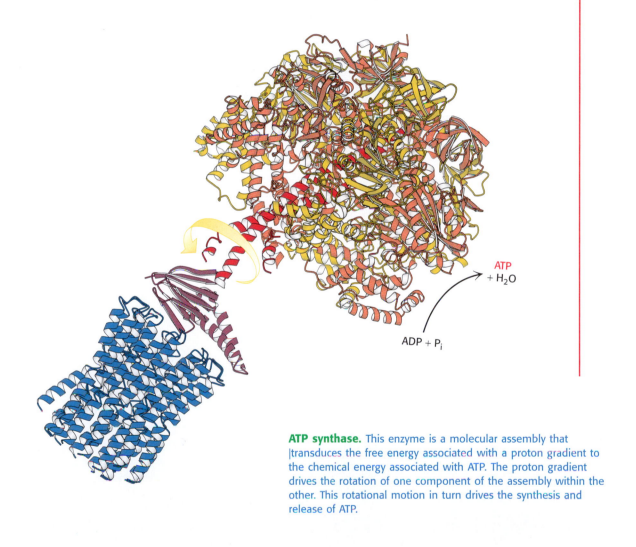

ATP
+ H$_2$O

ADP + P$_i$

ATP synthase. This enzyme is a molecular assembly that transduces the free energy associated with a proton gradient to the chemical energy associated with ATP. The proton gradient drives the rotation of one component of the assembly within the other. This rotational motion in turn drives the synthesis and release of ATP.

CHAPTER 14

Metabolism: Basic Concepts and Design

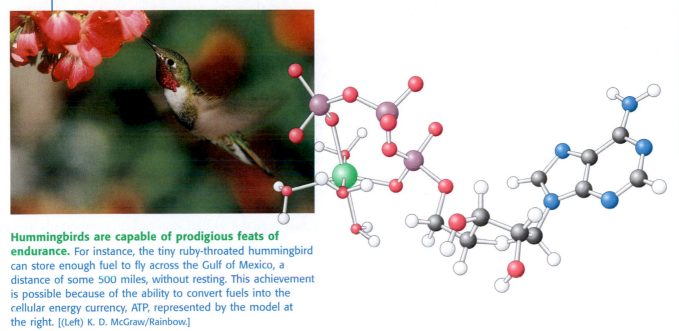

Hummingbirds are capable of prodigious feats of endurance. For instance, the tiny ruby-throated hummingbird can store enough fuel to fly across the Gulf of Mexico, a distance of some 500 miles, without resting. This achievement is possible because of the ability to convert fuels into the cellular energy currency, ATP, represented by the model at the right. [(Left) K. D. McGraw/Rainbow.]

The concepts of conformation and dynamics developed in Part I—especially those dealing with the specificity and catalytic power of enzymes, the regulation of their catalytic activity, and the transport of molecules and ions across membranes—enable us to now ask questions fundamental to biochemistry:

1. *How does a cell extract energy and reducing power from its environment?*

2. *How does a cell synthesize the building blocks of its macromolecules and then the macromolecules themselves?*

These processes are carried out by a highly integrated network of chemical reactions that are collectively known as *metabolism.*

More than a thousand chemical reactions take place in even as simple an organism as *Escherichia coli*. The array of reactions may seem overwhelming at first glance. However, closer scrutiny reveals that metabolism has a *coherent design containing many common motifs*. These motifs include the use of an energy currency and the repeated appearance of a limited number of activated intermediates. In fact, a group of about 100 molecules play central roles in all forms of life. Furthermore, although the number of reactions in metabolism is large, the number of *kinds* of reactions is small and the mechanisms of these reactions are usually quite simple. Metabolic pathways are also regulated in common ways. The purpose of this chapter is to introduce some general principles and motifs of metabolism to provide a foundation for the more detailed studies to follow.

14.0.1 Cells Transform Different Types of Energy

Living organisms require a continual input of free energy for three major purposes: (1) the performance of mechanical work in muscle contraction and other cellular movements, (2) the active transport of molecules and ions, and (3) the synthesis of macromolecules and other biomolecules from simple precursors. The free energy used in these processes, which maintain an organism in a state that is far from equilibrium, is derived from the environment.

The First Law of Thermodynamics states that energy can be neither created nor destroyed. The amount of energy in the universe is constant. Nevertheless, energy can be converted from one form into another.

Photosynthetic organisms, or *phototrophs*, use the energy of sunlight to convert simple energy-poor molecules into more-complex energy-rich molecules that serve as fuels. In other words, photosynthetic organisms transform light energy into chemical energy. Indeed, this transformation is ultimately the primary source of chemical energy for the vast majority of organisms, human beings included. *Chemotrophs*, which include animals, obtain chemical energy through the oxidation of foodstuffs generated by phototrophs.

Chemical energy obtained from the oxidation of carbon compounds may be transformed into the unequal distribution of ions across a membrane, resulting in an ion gradient. This gradient, in turn, is an energy source that can be used to move molecules across membranes, that can be converted into yet other types of chemical energy, or that can convey information in the form of nerve impulses. In addition, chemical energy can be transduced into mechanical energy. We convert the chemical energy of a fuel into structural alterations of contractile proteins that result in muscle contraction and movement. Finally, chemical energy powers the reactions that result in the synthesis of biomolecules.

At any given instant in a cell, thousands of energy transformations are taking place. Energy is being extracted from fuels and used to power biosynthetic processes. These transformations are referred to as *metabolism* or *intermediary metabolism*.

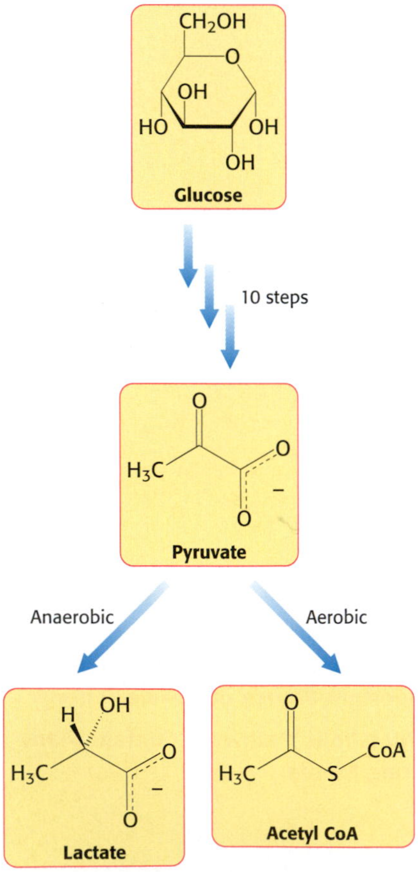

14.1 METABOLISM IS COMPOSED OF MANY COUPLED, INTERCONNECTING REACTIONS

Metabolism is essentially a linked series of chemical reactions that begins with a particular molecule and converts it into some other molecule or molecules in a carefully defined fashion (Figure 14.1). There are many such defined pathways in the cell (Figure 14.2), and we will examine a few of them in some detail later. These pathways are interdependent, and their activity is coordinated by exquisitely sensitive means of communication in which allosteric enzymes are predominant (Section 10.1). We will consider the principles of this communication in Chapter 15.

We can divide metabolic pathways into two broad classes: (1) those that convert energy into biologically useful forms and (2) those that require inputs of energy to proceed. Although this division is often imprecise, it is nonetheless a useful distinction in an examination of metabolism. Those reactions that transform fuels into cellular energy are called *catabolic reactions* or, more generally, *catabolism*.

$$\text{Fuels (carbohydrates, fats)} \xrightarrow{\text{Catabolism}} CO_2 + H_2O + \text{useful energy}$$

Those reactions that require energy—such as the synthesis of glucose, fats, or DNA—are called *anabolic reactions* or *anabolism*. The useful forms of energy that are produced in catabolism are employed in anabolism to gen-

FIGURE 14.1 Glucose metabolism. Glucose is metabolized to pyruvate in 10 linked reactions. Under anaerobic conditions, pyruvate is metabolized to lactate and, under aerobic conditions, to acetyl CoA. The glucose-derived carbons are subsequently oxidized to CO_2.

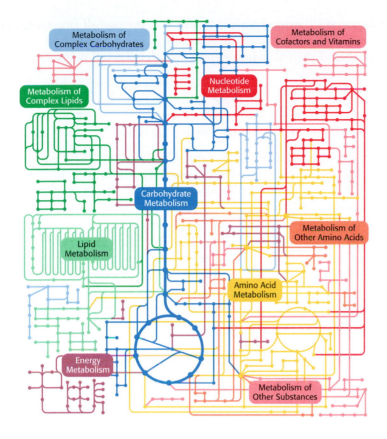

FIGURE 14.2 Metabolic pathways.
[From the Kyoto Encyclopedia of Genes and Genomes (www.genome.ad.jp/kegg).]

erate complex structures from simple ones, or energy-rich states from energy-poor ones.

$$\text{Useful energy} + \text{small molecules} \xrightarrow{\text{Anabolism}} \text{complex molecules}$$

Some pathways can be either anabolic or catabolic, depending on the energy conditions in the cell. They are referred to as *amphibolic pathways*.

14.1.1 A Thermodynamically Unfavorable Reaction Can Be Driven by a Favorable Reaction

CONCEPTUAL INSIGHTS, Energetic Coupling, *offers a graphical presentation of how enzymatic coupling enables a favorable reaction to drive an unfavorable reaction.*

How are specific pathways constructed from individual reactions? A pathway must satisfy minimally two criteria: (1) the individual reactions must be *specific* and (2) the entire set of reactions that constitute the pathway must be *thermodynamically favored*. A reaction that is specific will yield only one particular product or set of products from its reactants. As discussed in Chapter 8, a function of enzymes is to provide this specificity. The thermodynamics of metabolism is most readily approached in terms of free energy, which was discussed in Sections 1.3.3, 8.2.1, and 8.2.2. A reaction can occur spontaneously only if ΔG, the change in free energy, is negative. Recall that ΔG for the formation of products C and D from substrates A and B is given by

$$\Delta G = \Delta G^{\circ\prime} + RT \ln \frac{[\text{C}][\text{D}]}{[\text{A}][\text{B}]}$$

Thus, the ΔG of a reaction depends on the *nature* of the reactant products (expressed by the $\Delta G^{\circ\prime}$ term, the standard free-energy change) and on their *concentrations* (expressed by the second term).

An important thermodynamic fact is that *the overall free-energy change for a chemically coupled series of reactions is equal to the sum of the free-energy changes of the individual steps.* Consider the following reactions:

$$A \rightleftharpoons B + C \qquad \Delta G^{\circ\prime} = +5 \text{ kcal mol}^{-1} (+21 \text{ kJ mol}^{-1})$$
$$B \rightleftharpoons D \qquad \Delta G^{\circ\prime} = -8 \text{ kcal mol}^{-1} (-34 \text{ kJ mol}^{-1})$$
$$\overline{A \rightleftharpoons C + D \qquad \Delta G^{\circ\prime} = -3 \text{ kcal mol}^{-1} (-13 \text{ kJ mol}^{-1})}$$

Under standard conditions, A cannot be spontaneously converted into B and C, because ΔG is positive. However, the conversion of B into D under standard conditions is thermodynamically feasible. Because free-energy changes are additive, the conversion of A into C and D has a $\Delta G^{\circ\prime}$ of -3 kcal mol^{-1} (-13 kJ mol^{-1}), which means that it can occur spontaneously under standard conditions. Thus, *a thermodynamically unfavorable reaction can be driven by a thermodynamically favorable reaction to which it is coupled.* In this example, the chemical intermediate B, common to both reactions, couples the reactions. Thus, metabolic pathways are formed by the coupling of enzyme-catalyzed reactions such that the overall free energy of the pathway is negative.

14.1.2 ATP Is the Universal Currency of Free Energy in Biological Systems

Just as commerce is facilitated by the use of a common currency, the commerce of the cell—metabolism—is facilitated by the use of a common energy currency, *adenosine triphosphate* (ATP). Part of the free energy derived from the oxidation of foodstuffs and from light is transformed into this highly accessible molecule, which acts as the free-energy donor in most energy-requiring processes such as motion, active transport, or biosynthesis.

ATP is a nucleotide consisting of an adenine, a ribose, and a triphosphate unit (Figure 14.3). The active form of ATP is usually a complex of ATP with Mg^{2+} or Mn^{2+} (Section 9.4.2). In considering the role of ATP as an energy carrier, we can focus on its triphosphate moiety. *ATP is an energy-rich molecule because its triphosphate unit contains two phosphoanhydride bonds.* A large amount of free energy is liberated when ATP is hydrolyzed to adenosine diphosphate (ADP) and orthophosphate (P_i) or when ATP is hydrolyzed to adenosine monophosphate (AMP) and pyrophosphate (PP_i).

$$ATP + H_2O \rightleftharpoons ADP + P_i$$
$$\Delta G^{\circ\prime} = -7.3 \text{ kcal mol}^{-1} (-30.5 \text{ kJ mol}^{-1})$$
$$ATP + H_2O \rightleftharpoons AMP + PP_i$$
$$\Delta G^{\circ\prime} = -10.9 \text{ kcal mol}^{-1} (-45.6 \text{ kJ mol}^{-1})$$

The precise $\Delta G^{\circ\prime}$ for these reactions depends on the ionic strength of the medium and on the concentrations of Mg^{2+} and other metal ions. Under typical cellular concentrations, the actual ΔG for these hydrolyses is approximately -12 kcal mol^{-1} (-50 kJ mol^{-1}).

The free energy liberated in the hydrolysis of ATP is harnessed to drive reactions that require an input of free energy, such as muscle contraction. In turn, ATP is formed from ADP and P_i when fuel molecules are oxidized in chemotrophs or when light is trapped by phototrophs. *This ATP–ADP cycle is the fundamental mode of energy exchange in biological systems.*

Some biosynthetic reactions are driven by hydrolysis of nucleoside triphosphates that are analogous to ATP—namely, guanosine triphosphate (GTP), uridine triphosphate (UTP), and cytidine triphosphate (CTP). The diphosphate forms of these nucleotides are denoted by GDP, UDP, and CDP, and the monophosphate forms by GMP, UMP, and CMP. Enzymes

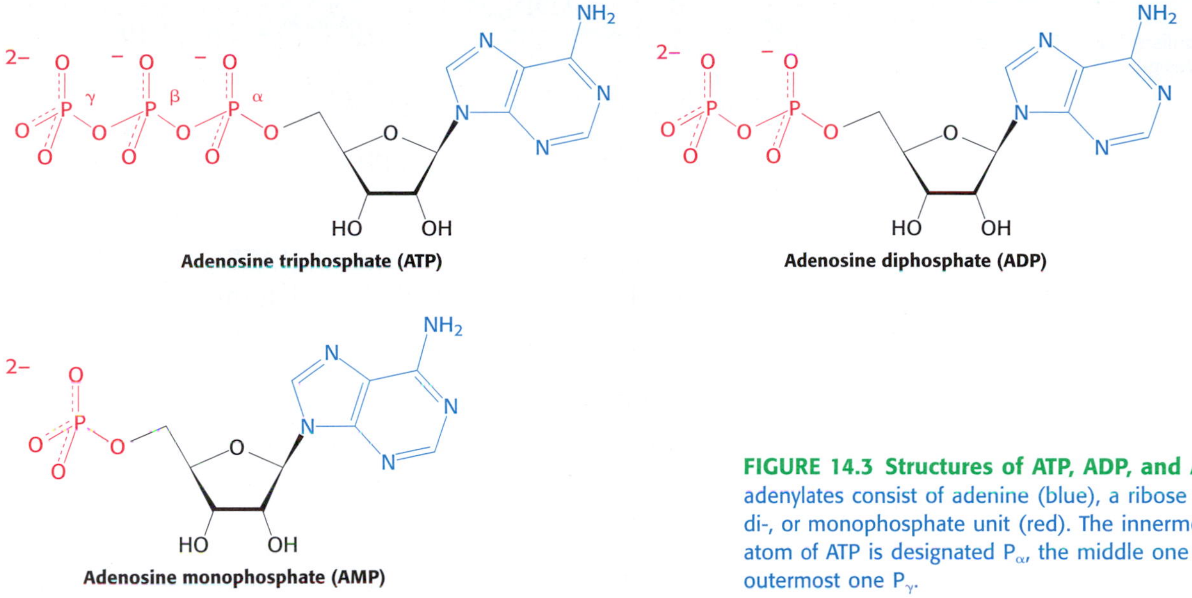

Adenosine triphosphate (ATP)

Adenosine diphosphate (ADP)

Adenosine monophosphate (AMP)

FIGURE 14.3 Structures of ATP, ADP, and AMP. These adenylates consist of adenine (blue), a ribose (black), and a tri-, di-, or monophosphate unit (red). The innermost phosphorus atom of ATP is designated P_α, the middle one P_β, and the outermost one P_γ.

can catalyze the transfer of the terminal phosphoryl group from one nucleotide to another. The phosphorylation of nucleoside monophosphates is catalyzed by a family of nucleoside monophosphate kinases, as discussed in Section 9.4.1. The phosphorylation of nucleoside diphosphates is catalyzed by nucleoside diphosphate kinase, an enzyme with broad specificity. It is intriguing to note that, although all of the nucleotide triphosphates are energetically equivalent, ATP is nonetheless the primary cellular energy carrier. In addition, two important electron carriers, NAD^+ and FAD, are derivatives of ATP. *The role of ATP in energy metabolism is paramount.*

14.1.3 ATP Hydrolysis Drives Metabolism by Shifting the Equilibrium of Coupled Reactions

How does coupling to ATP hydrolysis make possible an otherwise unfavorable reaction? Consider a chemical reaction that is thermodynamically unfavorable without an input of free energy, a situation common to many biosynthetic reactions. Suppose that the standard free energy of the conversion of compound A into compound B is $+4.0$ kcal mol^{-1} ($+13$ kJ mol^{-1}):

$$A \rightleftharpoons B \qquad \Delta G^{\circ\prime} = +4.0 \text{ kcal mol}^{-1} (+13 \text{ kJ mol}^{-1})$$

The equilibrium constant K'_{eq} of this reaction at 25°C is related to $\Delta G^{\circ\prime}$ (in units of kilocalories per mole) by

$$K'_{eq} = [B]_{eq}/[A]_{eq} = 10^{-\Delta G^{\circ\prime}/1.36} = 1.15 \times 10^{-3}$$

Thus, net conversion of A into B cannot occur when the molar ratio of B to A is equal to or greater than 1.15×10^{-3}. However, A can be converted into B under these conditions if the reaction is coupled to the hydrolysis of ATP. The new overall reaction is

$$A + ATP + H_2O \rightleftharpoons B + ADP + P_i + H^+$$
$$\Delta G^{\circ\prime} = -3.3 \text{ kcal mol}^{-1} (-13.8 \text{ kJ mol}^{-1})$$

Its standard free-energy change of -3.3 kcal mol^{-1} (-13.8 kJ mol^{-1}) is the sum of the value of $\Delta G^{\circ\prime}$ for the conversion of A into B [$+4.0$ kcal mol^{-1} ($+12.6$ kJ mol^{-1})] and the value of $\Delta G^{\circ\prime}$ for the hydrolysis of ATP [-7.3 kcal mol^{-1} (-30.5 kJ mol^{-1})]. At pH 7, the equilibrium constant of this coupled reaction is

$$K'_{eq} = \frac{[B]_{eq}}{[A]_{eq}} \times \frac{[ADP]_{eq}[P_i]_{eq}}{[ATP]_{eq}} = 10^{3.3/1.36} = 2.67 \times 10^2$$

At equilibrium, the ratio of [B] to [A] is given by

$$\frac{[B]_{eq}}{[A]_{eq}} = K'_{eq} \frac{[ATP]_{eq}}{[ADP]_{eq}[P_i]_{eq}}$$

The ATP-generating system of cells maintains the [ATP]/[ADP][P_i] ratio at a high level, typically of the order of 500 M^{-1}. For this ratio,

$$\frac{[B]_{eq}}{[A]_{eq}} = 2.67 \times 10^2 \times 500 = 1.34 \times 10^5$$

which means that the hydrolysis of ATP enables A to be converted into B until the [B]/[A] ratio reaches a value of 1.34×10^5. This equilibrium ratio is strikingly different from the value of 1.15×10^{-3} for the reaction A → B in the absence of ATP hydrolysis. In other words, coupling the hydrolysis of ATP with the conversion of A into B has changed the equilibrium ratio of B to A by a factor of about 10^8.

We see here the thermodynamic essence of ATP's action as an *energy-coupling agent*. Cells maintain a high level of ATP by using oxidizable substrates or light as sources of free energy. The hydrolysis of an ATP molecule in a coupled reaction then changes the equilibrium ratio of products to reactants by a very large factor, of the order of 10^8. More generally, the hydrolysis of *n* ATP molecules changes the equilibrium ratio of a coupled reaction (or sequence of reactions) by a factor of 10^{8n}. For example, the hydrolysis of three ATP molecules in a coupled reaction changes the equilibrium ratio by a factor of 10^{24}. Thus, *a thermodynamically unfavorable reaction sequence can be converted into a favorable one by coupling it to the hydrolysis of a sufficient number of ATP molecules in a new reaction.* It should also be emphasized that A and B in the preceding coupled reaction may be interpreted very generally, not only as different chemical species. For example, A and B may represent activated and unactivated conformations of a protein; in this case, phosphorylation with ATP may be a means of conversion into an activated conformation. Such a conformation can store free energy, which can then be used to drive a thermodynamically unfavorable reaction. Through such changes in conformation, molecular motors such as myosin, kinesin, and dynein convert the chemical energy of ATP into mechanical energy (Chapter 34). Indeed, this conversion is the basis of muscle contraction.

Alternatively, A and B may refer to the concentrations of an ion or molecule on the outside and inside of a cell, as in the active transport of a nutrient. The active transport of Na^+ and K^+ across membranes is driven by the phosphorylation of the sodium–potassium pump by ATP and its subsequent dephosphorylation (Section 13.2.1).

14.1.4 Structural Basis of the High Phosphoryl Transfer Potential of ATP

As illustrated by molecular motors (Chapter 34) and ion pumps (Section 13.2), phosphoryl transfer is a common means of energy coupling. Furthermore, as we shall see in Chapter 15, phosphoryl transfer is also widely used in the intracellular transmission of information. What makes ATP a particularly efficient phosphoryl-group donor? Let us compare the standard free energy of hydrolysis of ATP with that of a phosphate ester, such as glycerol 3-phosphate:

$$\text{ATP} + H_2O \rightleftharpoons \text{ADP} + P_i + H^+$$
$$\Delta G^{\circ\prime} = -7.3 \text{ kcal mol}^{-1} (-30.5 \text{ kJ mol}^{-1})$$

$$\text{Glycerol 3-phosphate} + H_2O \rightleftharpoons \text{glycerol} + P_i$$
$$\Delta G^{\circ\prime} = -2.2 \text{ kcal mol}^{-1} (-9.2 \text{ kJ mol}^{-1})$$

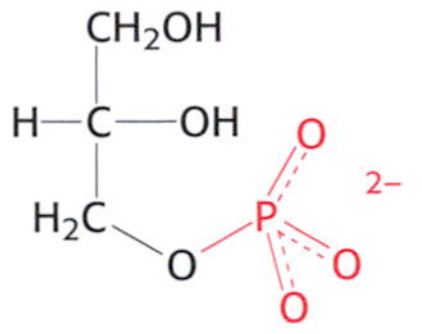

The magnitude of $\Delta G^{\circ\prime}$ for the hydrolysis of glycerol 3-phosphate is much smaller than that of ATP, which means that ATP has a stronger tendency to transfer its terminal phosphoryl group to water than does glycerol 3-phosphate. In other words, ATP has a higher *phosphoryl transfer potential (phosphoryl-group transfer potential)* than does glycerol 3-phosphate.

What is the structural basis of the high phosphoryl transfer potential of ATP? Because $\Delta G^{\circ\prime}$ depends on the *difference* in free energies of the products and reactants, the structures of both ATP and its hydrolysis products, ADP and P_i, must be examined to answer this question. Three factors are important: *resonance stabilization, electrostatic repulsion,* and *stabilization due to hydration.* ADP and, particularly, P_i, have greater resonance stabilization than does ATP. Orthophosphate has a number of resonance forms of similar energy (Figure 14.4), whereas the γ-phosphoryl group of ATP has a smaller number.

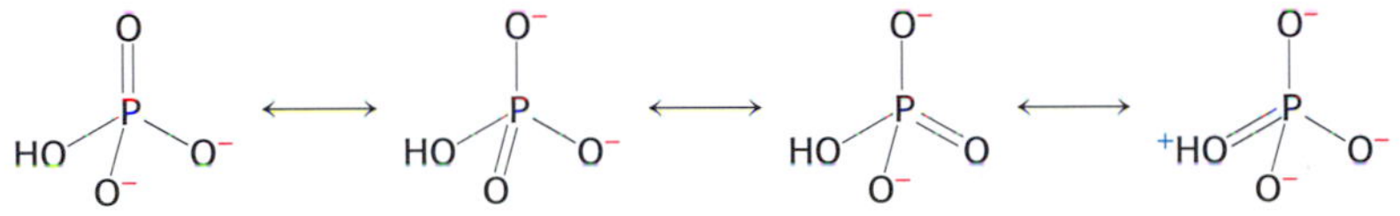

Glycerol 3-phosphate

FIGURE 14.4 Resonance structures of orthophosphate.

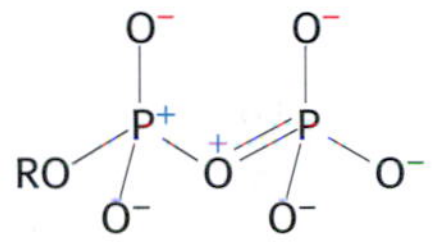

FIGURE 14.5 Improbable resonance structure. The structure contributes little to the terminal part of ATP, because two positive charges are placed adjacent to each other.

Forms like that shown in Figure 14.5 are unfavorable because a positively charged oxygen atom is adjacent to a positively charged phosphorus atom, an electrostatically unfavorable juxtaposition. Furthermore, at pH 7, the triphosphate unit of ATP carries about four negative charges. These charges repel one another because they are in close proximity. The repulsion between them is reduced when ATP is hydrolyzed. Finally, water can bind more effectively to ADP and P_i than it can to the phosphoanhydride part of ATP, stabilizing the ADP and P_i by hydration.

ATP is often called a high-energy phosphate compound, and its phosphoanhydride bonds are referred to as high-energy bonds. Indeed, a "squiggle" ($\sim$P) is often used to indicate such a bond. Nonetheless, there is nothing special about the bonds themselves. *They are high-energy bonds in the sense that much free energy is released when they are hydrolyzed,* for the aforegiven reasons.

14.1.5 Phosphoryl Transfer Potential Is an Important Form of Cellular Energy Transformation

The standard free energies of hydrolysis provide a convenient means of comparing the phosphoryl transfer potential of phosphorylated compounds. Such comparisons reveal that ATP is not the only compound with a high phosphoryl transfer potential. In fact, some compounds in biological systems have a higher phosphoryl transfer potential than that of ATP. These compounds include phosphoenolpyruvate (PEP), 1,3-bisphosphoglycerate (1,3-BPG), and creatine phosphate (Figure 14.6). Thus, PEP can transfer its phosphoryl group to ADP to form ATP. Indeed, this is one of the ways in which ATP is generated in the breakdown of sugars (Sections 14.2.1, 16.1.6, and 16.1.7). It is significant that ATP has a phosphoryl transfer potential that is intermediate among the biologically important phosphorylated molecules (Table 14.1). *This intermediate position enables ATP to function efficiently as a carrier of phosphoryl groups.*

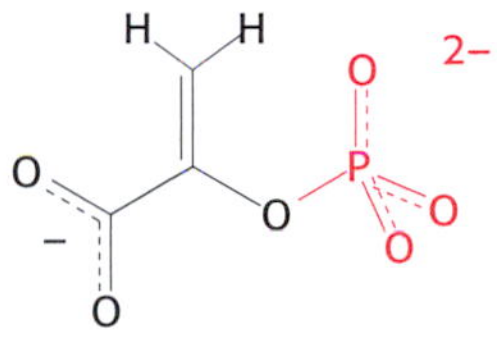

Phosphoenolpyruvate

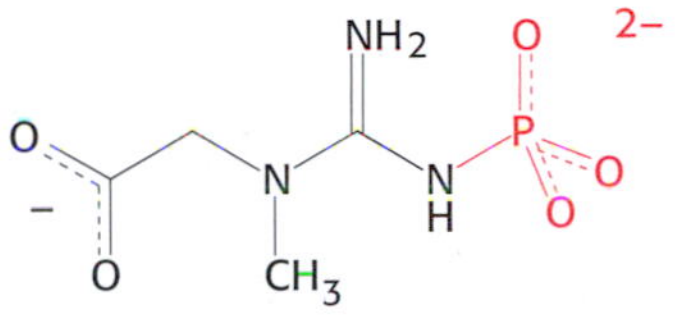

Creatine phosphate

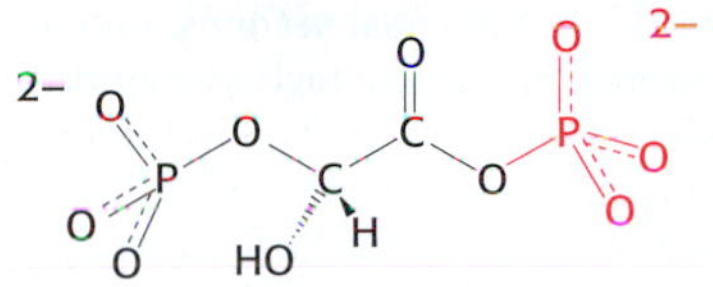

1,3-Bisphosphoglycerate

FIGURE 14.6 High phosphoryl transfer potential compounds. These compounds have a higher phosphoryl transfer potential than that of ATP and can be used to phosphorylate ADP to form ATP.

TABLE 14.1 **Standard free energies of hydrolysis of some phosphorylated compounds**

Compound	kcal mol^{-1}	kJ mol^{-1}
Phosphoenolpyruvate	-14.8	-61.9
1,3-Bisphosphoglycerate	-11.8	-49.4
Creatine phosphate	-10.3	-43.1
ATP (to ADP)	-7.3	-30.5
Glucose 1-phosphate	-5.0	-20.9
Pyrophosphate	-4.6	-19.3
Glucose 6-phosphate	-3.3	-13.8
Glycerol 3-phosphate	-2.2	-9.2

Creatine phosphate in vertebrate muscle serves as a reservoir of high-potential phosphoryl groups that can be readily transferred to ATP. Indeed, we use creatine phosphate to regenerate ATP from ADP every time we exercise strenuously. This reaction is catalyzed by *creatine kinase*.

$$\text{Creatine phosphate} + \text{ADP} + \text{H}^+ \overset{\text{Creatine kinase}}{\rightleftharpoons} \text{ATP} + \text{creatine}$$

At pH 7, the standard free energy of hydrolysis of creatine phosphate is -10.3 kcal mol^{-1} (-43.1 kJ mol^{-1}), compared with -7.3 kcal mol^{-1} (-30.5 kJ mol^{-1}) for ATP. Hence, the standard free-energy change in forming ATP from creatine phosphate is -3.0 kcal mol^{-1} (-12.6 kJ mol^{-1}), which corresponds to an equilibrium constant of 162.

$$K_{eq} = \frac{[\text{ATP}][\text{creatine}]}{[\text{ADP}][\text{creatine phosphate}]} = 10^{-\Delta G^{\circ\prime}/(2.303RT)} = 10^{3/1.36} = 162$$

In resting muscle, typical concentrations of these metabolites are [ATP] = 4 mM, [ADP] = 0.013 mM, [creatine phosphate] = 25 mM, and [creatine] = 13 mM. The amount of ATP in muscle suffices to sustain contractile activity for less than a second. The abundance of creatine phosphate and its high phosphoryl transfer potential relative to that of ATP make it a highly effective phosphoryl buffer. Indeed, creatine phosphate is the major source of phosphoryl groups for ATP regeneration for a runner during the first 4 seconds of a 100-meter sprint. After that, ATP must be generated through metabolism (Figure 14.7).

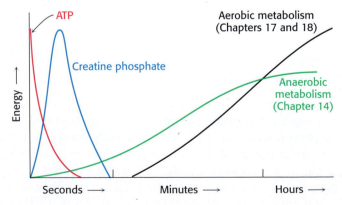

FIGURE 14.7 Sources of ATP during exercise. In the initial seconds, exercise is powered by existing high phosphoryl transfer compounds (ATP and creatine phosphate). Subsequently, the ATP must be regenerated by metabolic pathways.

14.2 THE OXIDATION OF CARBON FUELS IS AN IMPORTANT SOURCE OF CELLULAR ENERGY

ATP serves as the principal *immediate donor of free energy* in biological systems rather than as a long-term storage form of free energy. In a typical cell, an ATP molecule is consumed within a minute of its formation. Although the total quantity of ATP in the body is limited to approximately 100 g, *the turnover of this small quantity of ATP is very high*. For example, a resting human being consumes about 40 kg of ATP in 24 hours. During strenuous exertion, the rate of utilization of ATP may be as high as 0.5 kg/minute. For a 2-hour run, 60 kg (132 pounds) of ATP is utilized. Clearly, it is vital to have mechanisms for regenerating ATP. Motion, active transport, signal amplification, and biosynthesis can occur only if ATP is continually re-generated from ADP (Figure 14.8). The generation of ATP is one of the primary roles of catabolism. The carbon in fuel molecules—such as glucose and fats—is oxidized to CO_2, and the energy released is used to regenerate ATP from ADP and P_i.

In aerobic organisms, the ultimate electron acceptor in the oxidation of carbon is O_2 and the oxidation product is CO_2. Consequently, the more re-duced a carbon is to begin with, the more exergonic its oxidation will be. Figure 14.9 shows the $\Delta G°'$ of oxidation for one-carbon compounds.

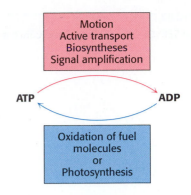

FIGURE 14.8 ATP–ADP cycle. This cycle is the fundamental mode of energy exchange in biological systems.

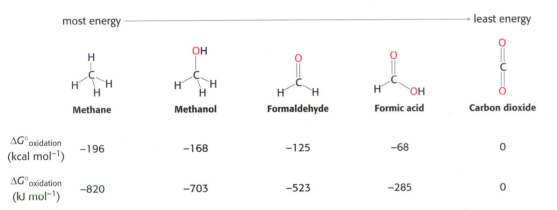

	Methane	Methanol	Formaldehyde	Formic acid	Carbon dioxide
$\Delta G°_{\text{oxidation}}$ (kcal mol^{-1})	−196	−168	−125	−68	0
$\Delta G°_{\text{oxidation}}$ (kJ mol^{-1})	−820	−703	−523	−285	0

FIGURE 14.9 Free energy of oxidation of single-carbon compounds.

Although fuel molecules are more complex (Figure 14.10) than the single-carbon compounds depicted in Figure 14.9, when a fuel is oxidized, the oxidation takes place one carbon at a time. The carbon oxidation energy is used in some cases to create a compound with high phosphoryl transfer po-tential and in other cases to create an ion gradient. In either case, the end point is the formation of ATP.

FIGURE 14.10 Prominent fuels. Fats are a more efficient fuel source than carbohydrates such as glucose because the carbon in fats is more reduced.

14.2.1 High Phosphoryl Transfer Potential Compounds Can Couple Carbon Oxidation to ATP Synthesis

How is the energy released in the oxidation of a carbon compound converted into ATP? As an example, consider glyceraldehyde 3-phosphate (shown in the margin), which is a metabolite of glucose formed in the oxidation of that sugar. The C-1 carbon (shown in red) is a component of an aldehyde and is not in its most oxidized state. Oxidation of the aldehyde to an acid will re-lease energy.

Glyceraldehyde 3-phosphate (GAP)

Glyceraldehyde 3-phosphate → (Oxidation) → **3-Phosphoglyceric acid**

However, the oxidation does not take place directly. Instead, the carbon oxidation generates an acyl phosphate, 1,3-bisphosphoglycerate. The electrons released are captured by NAD^+, which we will consider shortly.

Glyceraldehyde 3-phosphate (GAP) $+ NAD^+ + HPO_4^{2-}$ ⟶ **1,3-Bisphosphoglycerate (1,3-BPG)** $+ NADH + H^+$

For reasons similar to those discussed for ATP (Section 14.1.4), 1,3-bisphosphoglycerate has a high phosphoryl transfer potential. Thus, the cleavage of 1,3-BPG can be coupled to the synthesis of ATP.

1,3-Bisphosphoglycerate $+ ADP$ ⟶ **3-Phosphoglyceric acid** $+ ATP$

The energy of oxidation is initially trapped as a high-energy phosphate compound and then used to form ATP. The oxidation energy of a carbon atom is transformed into phosphoryl transfer potential, first as 1,3-bisphosphoglycerate and ultimately as ATP. We will consider these reactions in mechanistic detail in Section 16.1.5.

14.2.2 Ion Gradients Across Membranes Provide an Important Form of Cellular Energy That Can Be Coupled to ATP Synthesis

The electrochemical potential of *ion gradients across membranes*, produced by the oxidation of fuel molecules or by photosynthesis, ultimately powers the synthesis of most of the ATP in cells. In general, ion gradients are versatile means of coupling thermodynamically unfavorable reactions to favorable ones. Indeed, in animals, *proton gradients* generated by the oxidation of carbon fuels account for more than 90% of ATP generation (Figure 14.11). This process is called *oxidative phosphorylation* (Chapter 18). ATP hydrolysis can then be used to form ion gradients of different types and functions. The electrochemical potential of a Na^+ gradient, for example, can be tapped to pump Ca^{2+} out of cells (Section 13.4) or to transport nutrients such as sugars and amino acids into cells.

14.2.3 Stages in the Extraction of Energy from Foodstuffs

Let us take an overall view of the processes of energy conversion in higher organisms before considering them in detail in subsequent chapters. Hans Krebs described three stages in the generation of energy from the oxidation of foodstuffs (Figure 14.12).

In the first stage, large molecules in food are broken down into smaller units. Proteins are hydrolyzed to their 20 kinds of constituent amino acids, poly-

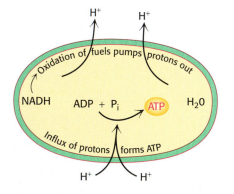

FIGURE 14.11 Proton gradients. The oxidation of fuels can power the formation of proton gradients. These proton gradients can in turn drive the synthesis of ATP.

saccharides are hydrolyzed to simple sugars such as glucose, and fats are hydrolyzed to glycerol and fatty acids. This stage is strictly a preparation stage; no useful energy is captured in this phase.

In the second stage, these numerous small molecules are degraded to a few simple units that play a central role in metabolism. In fact, most of them—sugars, fatty acids, glycerol, and several amino acids—are converted into the acetyl unit of acetyl CoA (Section 14.3.1). Some ATP is generated in this stage, but the amount is small compared with that obtained in the third stage.

In the third stage, ATP is produced from the complete oxidation of the acetyl unit of acetyl CoA. The third stage consists of the citric acid cycle and oxidative phosphorylation, which are the final common pathways in the oxidation of fuel molecules. Acetyl CoA brings acetyl units into the citric acid cycle [also called the tricarboxylic acid (TCA) cycle or Krebs cycle], where they are completely oxidized to CO_2. Four pairs of electrons are transferred (three to NAD^+ and one to FAD) for each acetyl group that is oxidized. Then, a proton gradient is generated as electrons flow from the reduced forms of these carriers to O_2, and this gradient is used to synthesize ATP.

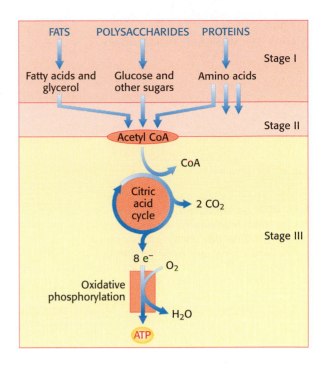

FIGURE 14.12 Stages of catabolism. The extraction of energy from fuels can be divided into three stages.

14.3 METABOLIC PATHWAYS CONTAIN MANY RECURRING MOTIFS

At first glance, metabolism appears intimidating because of the sheer number of reactants and reactions. Nevertheless, there are unifying themes that make the comprehension of this complexity more manageable. These unifying themes include common metabolites, reactions, and regulatory schemes that stem from a common evolutionary heritage.

14.3.1 Activated Carriers Exemplify the Modular Design and Economy of Metabolism

We have seen that phosphoryl transfer can be used to drive otherwise endergonic reactions, alter the energy of conformation of a protein, or serve as a signal to alter the activity of a protein. The phosphoryl-group donor in all of these reactions is ATP. In other words, *ATP is an activated carrier of phosphoryl groups because phosphoryl transfer from ATP is an exergonic process.* The use of activated carriers is a recurring motif in biochemistry, and we will consider several such carriers here.

1. *Activated carriers of electrons for fuel oxidation.* In aerobic organisms, the ultimate electron acceptor in the oxidation of fuel molecules is O_2. However, electrons are not transferred directly to O_2. Instead, fuel molecules transfer electrons to special carriers, which are either *pyridine nucleotides* or *flavins*. The reduced forms of these carriers then transfer their high-potential electrons to O_2.

Nicotinamide adenine dinucleotide is a major electron carrier in the oxidation of fuel molecules (Figure 14.13). The reactive part of NAD^+ is its nicotinamide ring, a pyridine derivative synthesized from the vitamin niacin. *In the oxidation of a substrate, the nicotinamide ring of NAD^+ accepts a hydrogen ion and two electrons, which are equivalent to a hydride ion.*

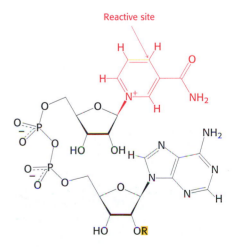

FIGURE 14.13 Structures of the oxidized forms of nicotinamide-derived electron carriers. Nicotinamide adenine dinucleotide (NAD^+) and nicotinamide adenine dinucleotide phosphate ($NADP^+$) are prominent carriers of high-energy electrons. In NAD^+, R = H; in $NADP^+$, R = PO_3^{2-}.

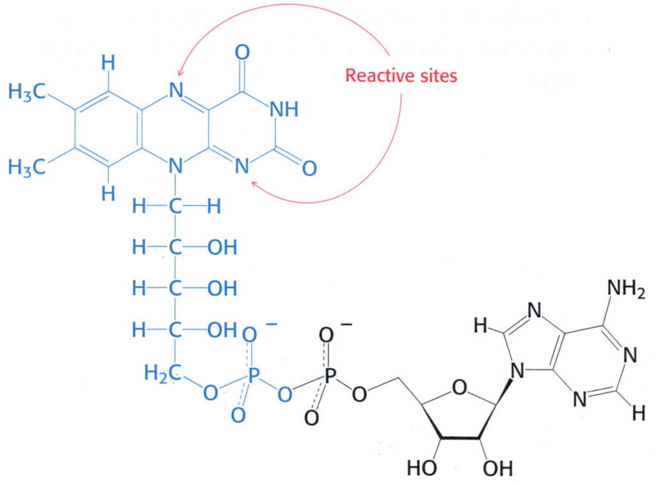

FIGURE 14.14 Structure of the oxidized form of flavin adenine dinucleotide (FAD). This electron carrier consists of a flavin mononucleotide (FMN) unit (shown in blue) and an AMP unit (shown in black).

The reduced form of this carrier is called *NADH*. In the oxidized form, the nitrogen atom carries a positive charge, as indicated by NAD^+. NAD^+ is the electron acceptor in many reactions of the type

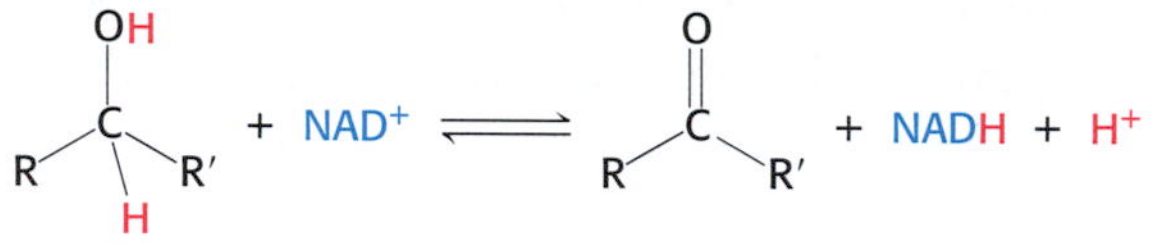

In this dehydrogenation, one hydrogen atom of the substrate is directly transferred to NAD^+, whereas the other appears in the solvent as a proton. Both electrons lost by the substrate are transferred to the nicotinamide ring.

The other major electron carrier in the oxidation of fuel molecules is the coenzyme *flavin adenine dinucleotide* (Figure 14.14). The abbreviations for the oxidized and reduced forms of this carrier are FAD and $FADH_2$, respectively. FAD is the electron acceptor in reactions of the type

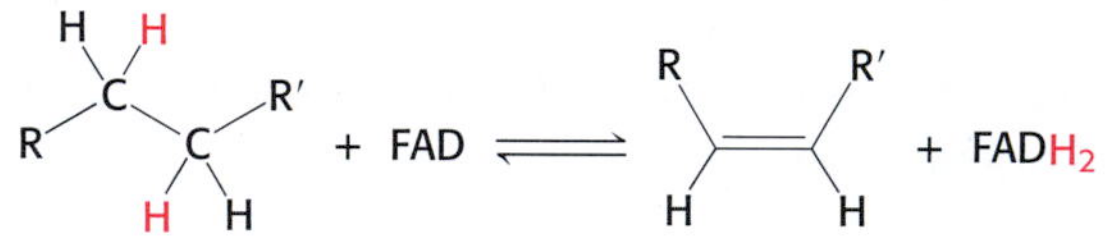

The reactive part of FAD is its isoalloxazine ring, a derivative of the vitamin riboflavin (Figure 14.15). FAD, like NAD^+, can accept two electrons. In doing so, FAD, unlike NAD^+, takes up two protons. These carriers of high-potential electrons as well as flavin mononucleotide (FMN), an electron carrier related to FAD, will be considered further in Chapter 18.

2. *An activated carrier of electrons for reductive biosynthesis.* High-potential electrons are required in most biosyntheses because the precursors are more oxidized than the products. Hence, reducing power is needed in addition to

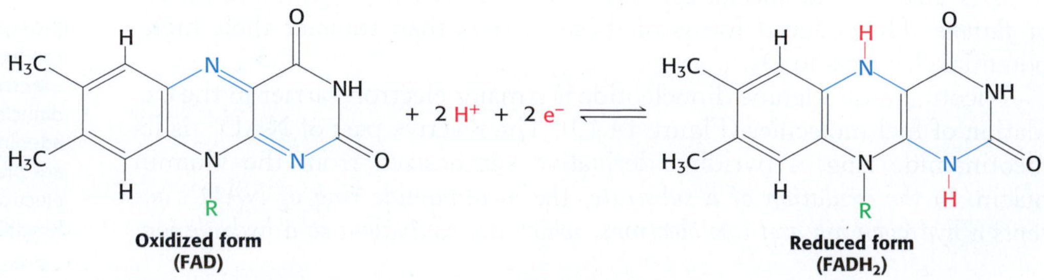

FIGURE 14.15 Structures of the reactive parts of FAD and $FADH_2$. The electrons and protons are carried by the isoalloxazine ring component of FAD and $FADH_2$.

Oxidized form (FAD)

$$+ 2\,H^+ + 2\,e^- \rightleftharpoons$$

Reduced form ($FADH_2$)

ATP. For example, in the biosynthesis of fatty acids, the keto group of an added two-carbon unit is reduced to a methylene group in several steps. This sequence of reactions requires an input of four electrons.

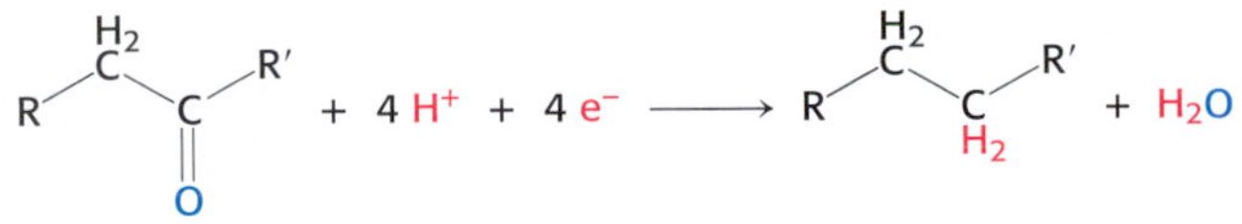

$$R\overset{\overset{\displaystyle H_2}{C}}{}\overset{\displaystyle C}{\underset{\displaystyle O}{}}R' + 4\,H^+ + 4\,e^- \longrightarrow R\overset{\overset{\displaystyle H_2}{C}}{}\overset{\displaystyle C}{\underset{\displaystyle H_2}{}}R' + H_2O$$

The electron donor in most reductive biosyntheses is NADPH, the reduced form of nicotinamide adenine dinucleotide phosphate ($NADP^+$; see Figure 14.13). NADPH differs from NADH in that the $2'$-hydroxyl group of its adenosine moiety is esterified with phosphate. NADPH carries electrons in the same way as NADH. However, *NADPH is used almost exclusively for reductive biosyntheses, whereas NADH is used primarily for the generation of ATP*. The extra phosphoryl group on NADPH is a tag that enables enzymes to distinguish between high-potential electrons to be used in anabolism and those to be used in catabolism.

3. *An activated carrier of two-carbon fragments.* Coenzyme A, another central molecule in metabolism, is a carrier of acyl groups (Figure 14.16). Acyl groups are important constituents both in catabolism, as in the oxidation of fatty acids, and in anabolism, as in the synthesis of membrane lipids. The terminal sulfhydryl group in CoA is the reactive site. Acyl groups are linked to CoA by thioester bonds. The resulting derivative is called an *acyl CoA*. An acyl group often linked to CoA is the acetyl unit; this derivative is called *acetyl CoA*. The $\Delta G°'$ for the hydrolysis of acetyl CoA has a large negative value:

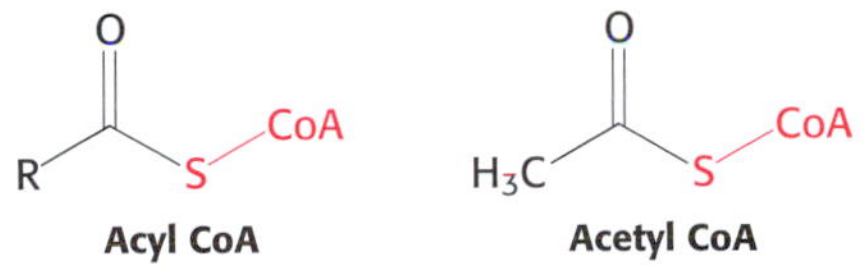

$$\text{Acetyl CoA} + H_2O \rightleftharpoons \text{acetate} + \text{CoA} + H^+$$
$$\Delta G°' = -7.5 \text{ kcal mol}^{-1} \, (-31.4 \text{ kJ mol}^{-1})$$

The hydrolysis of a thioester is thermodynamically more favorable than that of an oxygen ester because the electrons of the $C{=}O$ bond cannot form resonance structures with the C–S bond that are as stable as those that they can form with the C–O bond. Consequently, *acetyl CoA has a high acetyl potential (acetyl group-transfer potential) because transfer of the acetyl group is exergonic.* Acetyl CoA carries an activated acetyl group, just as ATP carries an activated phosphoryl group.

The use of activated carriers illustrates two key aspects of metabolism. First, NADH, NADPH, and $FADH_2$ react slowly with O_2 in the absence of a catalyst. Likewise, ATP and acetyl CoA are hydrolyzed slowly (in times of many hours or even days) in the absence of a catalyst. These molecules are kinetically quite stable in the face of a large thermodynamic driving force for reaction with O_2 (in regard to the electron carriers) and H_2O (in regard to ATP and acetyl CoA). *The kinetic stability of these molecules in the absence of specific catalysts is essential for their biological function because it enables enzymes to control the flow of free energy and reducing power.*

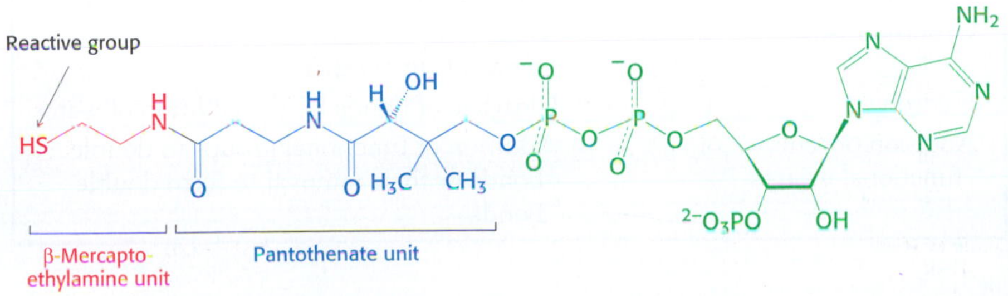

Acyl CoA   **Acetyl CoA**

FIGURE 14.16 Structure of coenzyme A (CoA).

TABLE 14.2 Some activated carriers in metabolism

Carrier molecule in activated form	Group carried	Vitamin precursor
ATP	Phosphoryl	
NADH and NADPH	Electrons	Nicotinate (niacin)
FADH$_2$	Electrons	Riboflavin (vitamin B$_2$)
FMNH$_2$	Electrons	Riboflavin (vitamin B$_2$)
Coenzyme A	Acyl	Pantothenate
Lipoamide	Acyl	
Thiamine pyrophosphate	Aldehyde	Thiamine (vitamin B$_1$)
Biotin	CO$_2$	Biotin
Tetrahydrofolate	One-carbon units	Folate
S-Adenosylmethionine	Methyl	
Uridine diphosphate glucose	Glucose	
Cytidine diphosphate diacylglycerol	Phosphatidate	
Nucleoside triphosphates	Nucleotides	

Note: Many of the activated carriers are coenzymes that are derived from water-soluble vitamins (Section 8.6.1).

Second, *most interchanges of activated groups in metabolism are accomplished by a rather small set of carriers* (Table 14.2). The existence of a recurring set of activated carriers in all organisms is one of the unifying motifs of biochemistry. Furthermore, it illustrates the modular design of metabolism. A small set of molecules carries out a very wide range of tasks. Metabolism is readily comprehended because of the economy and elegance of its underlying design.

14.3.2 Key Reactions Are Reiterated Throughout Metabolism

Just as there is an economy of design in the use of activated carriers, so is there an economy of design in biochemical reactions. The thousands of metabolic reactions, bewildering at first in their variety, can be subdivided into just six types (Table 14.3). Specific reactions of each type appear repeatedly, further reducing the number of reactions necessary for the student to learn.

TABLE 14.3 Types of chemical reactions in metabolism

Type of reaction	Description
Oxidation–reduction	Electron transfer
Ligation requiring ATP cleavage	Formation of covalent bonds (i.e., carbon–carbon bonds)
Isomerization	Rearrangement of atoms to form isomers
Group transfer	Transfer of a functional group from one molecule to another
Hydrolytic	Cleavage of bonds by the addition of water
Addition or removal of functional groups	Addition of functional groups to double bonds or their removal to form double bonds

1. *Oxidation–reduction reactions* are essential components of many pathways. Useful energy is often derived from the oxidation of carbon compounds. Consider the following two reactions:

Succinate   **Fumarate**   (1)

Malate   **Oxaloacetate**   (2)

These two oxidation–reduction reactions are components of the citric acid cycle (Chapter 17), which completely oxidizes the activated two-carbon fragment of acetyl CoA to two molecules of CO_2. In reaction 1, $FADH_2$ carries the electrons, whereas, in reaction 2, electrons are carried by NADH. In biosynthetic oxidation–reduction reactions, NADPH is the reductant.

2. *Ligation reactions* form bonds by using free energy from ATP cleavage. Reaction 3 illustrates the ATP-dependent formation of a carbon–carbon bond, necessary to combine smaller molecules to form larger ones. Oxaloacetate is formed from pyruvate and CO_2.

Pyruvate   + CO_2 + ATP + H_2O ⇌

+ ADP + P_i + H^+ (3)

Oxaloacetate

The oxaloacetate can be used in the citric acid cycle or converted into amino acids such as aspartic acid.

3. *Isomerization reactions* rearrange particular atoms within the molecule. Their role is often to prepare a molecule for subsequent reactions such as the oxidation–reduction reactions described in point 1.

Citrate   **Isocitrate**   (4)

Reaction 4 is, again, a component of the citric acid cycle. This isomerization prepares the molecule for subsequent oxidation and decarboxylation by moving the hydroxyl group of citrate from a tertiary to a secondary position.

4. *Group-transfer reactions* play a variety of roles. Reaction 5 is representative of such a reaction. A phosphoryl group is transferred from the activated phosphoryl-group carrier, ATP, to glucose. This reaction traps glucose in the cell so that further catabolism can take place.

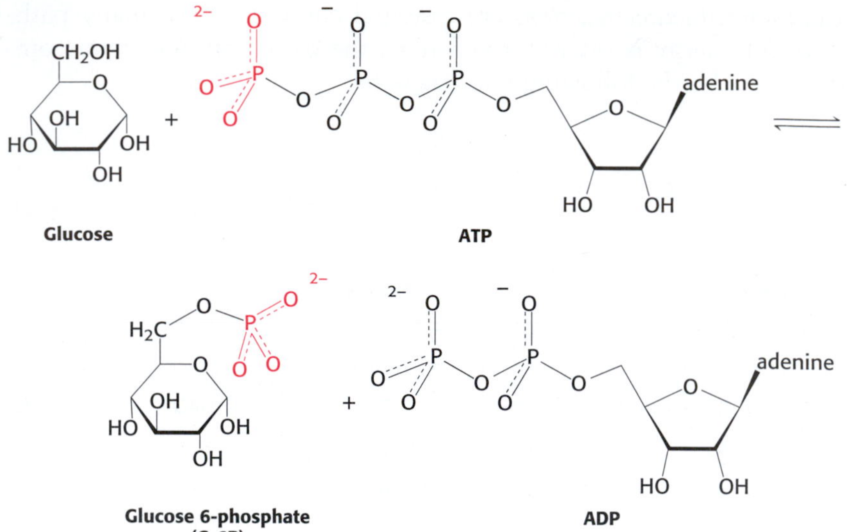

Glucose   **ATP**

Glucose 6-phosphate   **ADP**   (5)
(G-6P)

We saw earlier that group-transfer reactions are used to synthesize ATP (14.2.1). We will also see examples of their use in signaling pathways (Chapter 15).

5. *Hydrolytic reactions* cleave bonds by the addition of water. Hydrolysis is a common means employed to break down large molecules, either to facilitate further metabolism or to reutilize some of the components for biosynthetic purposes. Proteins are digested by hydrolytic cleavage (Chapters 9 and 10). Reaction 6 illustrates the hydrolysis of a peptide to yield two smaller peptides.

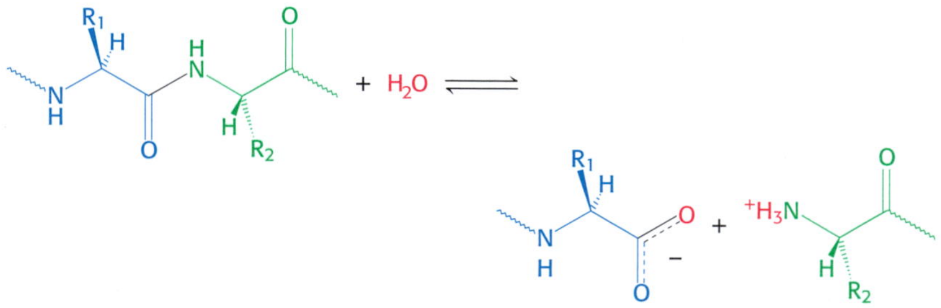

(6)

6. *The addition of functional groups to double bonds or the removal of groups to form double bonds* constitutes the final class of reactions. The enzymes that catalyze these types of reaction are classified as lyases (Section 8.1.3). An important example, illustrated in reaction 7, is the conversion of the six-carbon molecule fructose 1,6-bisphosphate (F-1, 6-BP) into 2 three-carbon fragments: dihydroxyacetone phosphate and glyceraldehyde 3-phosphate.

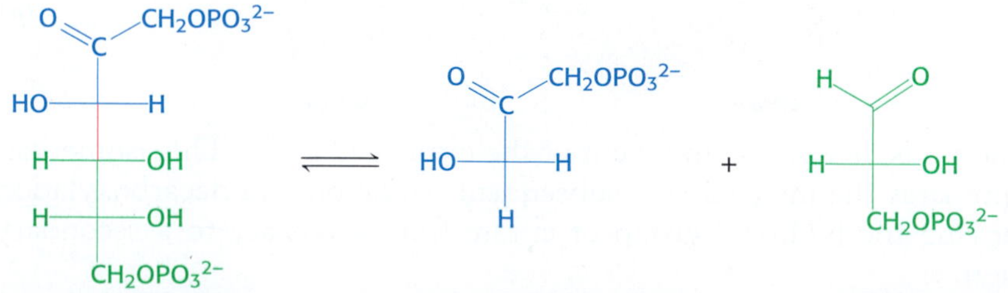

Fructose 1,6-bisphosphate   **Dihydroxyacetone phosphate**   **Glyceraldehyde 3-phosphate**
(F-1,6-BP)   **(DHAP)**   **(GAP)**

                                                                                                 (7)

This reaction is a key step in glycolysis, a key pathway for extracting energy from glucose (Section 16.1.3). Dehydrations to form double bonds, such as the formation of phosphoenolpyruvate (Table 14.1) from 2-phos-phoglycerate (reaction 8), are important reactions of this type.

$$\text{2-Phosphoglycerate} \rightleftharpoons \text{Phosphoenolpyruvate (PEP)} + H_2O \qquad (8)$$

The dehydration sets up the next step in the pathway, a group-transfer reaction that uses the that uses the high phosphoryl transfer potential of the product PEP to form ATP from ADP.

These six fundamental reaction types are the basis of metabolism. Remember that all six types can proceed in either direction, depending on the standard free energy for the specific reaction and the concentration of the reactants and products inside the cell. As an example of how simple themes are reiterated, consider the reactions shown in Figure 14.17. The same sequence of reactions is employed in the citric acid cycle, fatty acid degradation, the degradation of the amino acid lysine, and (in reverse) the biosynthesis of fatty acids. An effective way to learn is to look for commonalties in the diverse metabolic pathways that we will be studying. There is a chemical logic that, when exposed, renders the complexity of the chemistry of living systems more manageable and reveals its elegance.

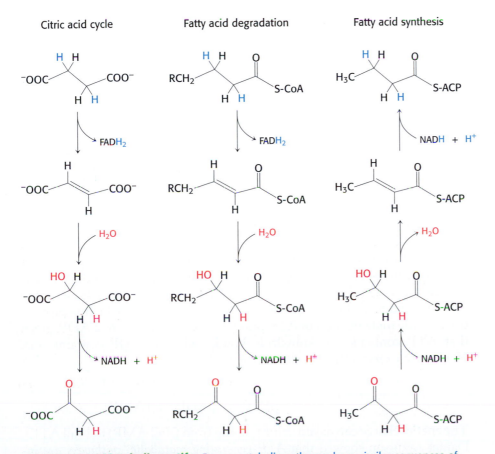

FIGURE 14.17 Metabolic motifs. Some metabolic pathways have similar sequences of reactions in common—in this case, an oxidation, the addition of a functional group (from a water molecule) to a double bond, and another oxidation. ACP designates acyl carrier protein.

14.3.3 Metabolic Processes Are Regulated in Three Principal Ways

It is evident that the complex network of reactions constituting intermediary metabolism must be rigorously regulated. At the same time, metabolic control must be flexible, because the external environments of cells are not constant. Metabolism is regulated by controlling (1) *the amounts of enzymes*, (2) *their catalytic activities*, and (3) *the accessibility of substrates*. The amount of a particular enzyme depends on both its rate of synthesis and its rate of degradation. The level of most enzymes is adjusted primarily by changing the *rate of transcription* of the genes encoding them. In *E. coli*, the presence of lactose, for example, induces within minutes a more than 50-fold increase in the rate of synthesis of β-galactosidase, an enzyme required for the breakdown of this disaccharide.

The catalytic activity of enzymes is controlled in several ways. *Reversible allosteric control* is especially important. For example, the first reaction in many biosynthetic pathways is allosterically inhibited by the ultimate product of the pathway. The inhibition of aspartate transcarbamoylase by cytidine triphosphate (Section 10.1) is a well-understood example of *feedback inhibition*. This type of control can be almost instantaneous. Another recurring mechanism is *reversible covalent modification*. For example, glycogen phosphorylase, the enzyme catalyzing the breakdown of glycogen, a storage form of sugar, is activated by phosphorylation of a particular serine residue when glucose is scarce (Section 21.2.1).

Hormones coordinate metabolic relations between different tissues, often by regulating the reversible modification of key enzymes. Hormones such as epinephrine trigger signal transduction cascades that lead to highly amplified changes in metabolic patterns in target tissues such as the muscle (Section 15.0.1). The hormone insulin promotes the entry of glucose into many kinds of cells. As will be discussed again in Chapter 15, many hormones act through intracellular messengers, such as cyclic AMP and calcium ion, that coordinate the activities of many target proteins.

Controlling the *flux of substrates* also regulates metabolism. The transfer of substrates from one compartment of a cell to another (e.g., from the cytosol to mitochondria) can serve as a control point.

An important general principle of metabolism is that *biosynthetic and degradative pathways are almost always distinct*. This separation is necessary for energetic reasons, as will be evident in subsequent chapters. It also facilitates the control of metabolism. In eukaryotes, metabolic regulation and flexibility also are enhanced by compartmentalization. For example, fatty acid oxidation takes place in mitochondria, whereas fatty acid synthesis takes place in the cytosol. *Compartmentalization segregates opposed reactions.*

Many reactions in metabolism are controlled by the *energy status* of the cell. One index of the energy status is the *energy charge*, which is proportional to the mole fraction of ATP plus half the mole fraction of ADP, given that ATP contains two anhydride bonds, whereas ADP contains one. Hence, the energy charge is defined as

$$\text{Energy charge} = \frac{[\text{ATP}] + \frac{1}{2}[\text{ADP}]}{[\text{ATP}] + [\text{ADP}] + [\text{AMP}]}$$

The energy charge can have a value ranging from 0 (all AMP) to 1 (all ATP). Daniel Atkinson showed that *ATP-generating (catabolic) pathways are inhibited by an energy charge, whereas ATP-utilizing (anabolic) pathways are stimulated by a high-energy charge.* In plots of the reaction rates of such pathways versus the energy charge, the curves are steep near an energy charge of 0.9, where they usually intersect (Figure 14.18). It is evident that the control

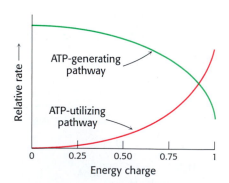

FIGURE 14.18 Energy charge regulates metabolism. High concentrations of ATP inhibit the relative rates of a typical ATP-generating (catabolic) pathway and stimulate the typical ATP-utilizing (anabolic) pathway.

of these pathways has evolved to maintain the energy charge within rather narrow limits. In other words, *the energy charge, like the pH of a cell, is buffered*. The energy charge of most cells ranges from 0.80 to 0.95. An alternative index of the energy status is the *phosphorylation potential*, which is defined as

$$\text{Phosphorylation potential} = \frac{[\text{ATP}]}{[\text{ADP}][\text{P}_i]}$$

The phosphorylation potential, in contrast with the energy charge, depends on the concentration of P_i and is directly related to the free energy-storage available from ATP.

14.3.4 Evolution of Metabolic Pathways

How did the complex pathways that constitute metabolism evolve? This question, a difficult one to address, was approached in Chapter 2. The current thinking is that RNA was an early biomolecule and that, in an early RNA world, RNA served as catalysts and information-storage molecules (Section 2.2.2).

Why do activated carriers such as ATP, NADH, FADH$_2$, and coenzyme A contain adenosine diphosphate units (Figure 14.19)? A possible explanation is that these molecules evolved from the early RNA catalysts. Non-RNA units such as the isoalloxazine ring may have been recruited to serve as efficient carriers of activated electrons and chemical units, a function not readily performed by RNA itself. We can picture the adenine ring of FADH$_2$ binding to a uracil unit in a niche of an RNA enzyme (ribozyme) by base-pairing, whereas the isoalloxazine ring protrudes and functions as an electron carrier. When the more versatile proteins replaced RNA as the major catalysts, the ribonucleotide coenzymes stayed essentially unchanged because they were already well suited to their metabolic roles. The nicotinamide unit of NADH, for example, can readily transfer electrons irrespective of whether the adenine unit interacts with a base in an RNA enzyme or with amino acid residues in a protein enzyme. With the advent of protein enzymes, these important cofactors evolved as free molecules without losing the adenosine diphosphate vestige of their RNA-world ancestry. That molecules and motifs of metabolism are common to all forms of life testifies to their common origin and to the retention of functioning modules through billions of years of evolution. Our understanding of metabolism, like that of other biological processes, is enriched by inquiry into how these beautifully integrated patterns of reactions came into being.

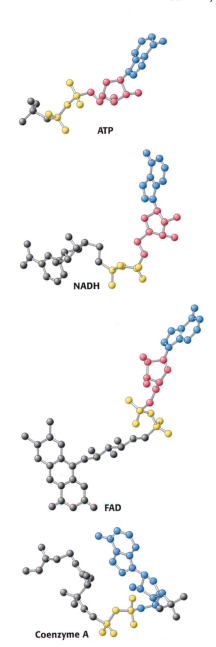

FIGURE 14.19 Adenosine diphosphate (ADP) is an ancient module in metabolism. This fundamental building block is present in key molecules such as ATP, NADH, FAD, and coenzyme A. The adenine unit is shown in blue, the ribose unit in red, and the diphosphate unit in yellow.

SUMMARY

All cells transform energy. They extract energy from their environment and use this energy to convert simple molecules into cellular components.

Metabolism Is Composed of Many Coupled, Interconnecting Reactions

The process of energy transduction takes place through a highly integrated network of chemical reactions called metabolism. Metabolism can be subdivided into catabolism (reactions employed to extract energy from fuels) and anabolism (reactions that use this energy for biosynthesis). The most valuable thermodynamic concept for understanding bioenergetics is free energy. A reaction can occur spontaneously only if the change in free energy (ΔG) is negative. A thermodynamically unfavorable reaction can be driven by a thermodynamically favorable one, which in many cases is the hydrolysis of ATP. The hydrolysis of ATP shifts the equilibrium of a coupled reaction by a factor of about 10^8. ATP, the universal currency of energy in biological systems, is an energy-rich molecule because it contains two phosphoanhydride bonds.

The Oxidation of Carbon Fuels Is an Important Source of Cellular Energy

ATP formation is coupled to the oxidation of carbon fuels, either directly or through the formation of ion gradients. Photosynthetic organisms can use light to generate such gradients. ATP is consumed in muscle contraction and other motions of cells, in active transport, in signal transduction processes, and in biosyntheses. There are three stages in the extraction of energy from foodstuffs by aerobic organisms. In the first stage, large molecules are broken down into smaller ones, such as amino acids, sugars, and fatty acids. In the second stage, these small molecules are degraded to a few simple units that have a pervasive role in metabolism. One of them is the acetyl unit of acetyl CoA, a carrier of activated acyl groups. The third stage of metabolism is the citric acid cycle and oxidative phosphorylation, in which ATP is generated as electrons flow to O_2, the ultimate electron acceptor, and fuels are completely oxidized to CO_2.

Metabolic Pathways Contain Many Recurring Motifs

Metabolism is characterized by common motifs. A small number of activated carriers, such as ATP, NADH, and acetyl CoA, are used in many metabolic pathways. NADPH, which carries two electrons at a high potential, provides reducing power in the biosynthesis of cell components from more-oxidized precursors. ATP and NADPH are continually generated and consumed. Most transfers of activated groups in metabolism are mediated by a recurring set of carriers. Moreover, key reaction types are used repeatedly in metabolic pathways.

Metabolism is regulated in a variety of ways. The amounts of some critical enzymes are controlled by regulation of the rate of protein synthesis and degradation. In addition, the catalytic activities of many enzymes are regulated by allosteric interactions (as in feedback inhibition) and by covalent modification. The movement of many substrates into cells and subcellular compartments also is controlled. Distinct pathways for biosynthesis and degradation contribute to metabolic regulation. The energy charge, which depends on the relative amounts of ATP, ADP, and AMP, plays a role in metabolic regulation. A high-energy charge inhibits ATP-generating (catabolic) pathways, whereas it stimulates ATP-utilizing (anabolic) pathways.

KEY TERMS

phototroph (p. 374)

chemotroph (p. 374)

metabolism or intermediary metabolism (p. 374)

catabolism (p. 374)

anabolism (p. 374)

amphibolic pathway (p. 375)

coupled reaction (p. 377)

phosphoryl transfer potential (p. 379)

oxidative phosphorylation (p. 382)

activated carrier (p. 383)

oxidation–reduction reaction (p. 387)

ligation reaction (p. 387)

isomerization reaction (p. 387)

group-transfer reaction (p. 387)

hydrolytic reaction (p. 388)

addition to or formation of double-bond reaction (p. 388)

energy charge (p. 390)

phosphorylation potential (p. 391)

SELECTED READINGS

Where to start

McGrane, M. M., Yun, J. S., Patel, Y. M., and Hanson, R. W., 1992. Metabolic control of gene expression: In vivo studies with transgenic mice. *Trends Biochem. Sci.* 17:40–44.

Books

Harold, F. M., 1986. *The Vital Force: A Study of Bioenergetics.* W. H. Freeman and Company.

Krebs, H. A., and Kornberg, H. L., 1957. *Energy Transformations in Living Matter.* Springer Verlag.

Linder, M. C. (Ed.), 1991. *Nutritional Biochemistry and Metabolism* (2d ed.). Elsevier.

Gottschalk, G., 1986. *Bacterial Metabolism* (2d ed.). Springer Verlag.

Nicholls, D. G., and Ferguson, S. J., 1997. *Bioenergetics 2* (2d ed.) Academic Press.

Martin, B. R., 1987. *Metabolic Regulation: A Molecular Approach.* Blackwell Scientific.

Frayn, K. N., 1996. *Metabolic Regulation: A Human Perspective.* Portland Press.

Fell, D., 1997. *Understanding the Control of Metabolism.* Portland Press.

Harris, D. A., 1995. *Bioenergetics at a Glance.* Blackwell Scientific.

Thermodynamics

Von Baeyer, H. C., 1999. *Warmth Disperses and Time Passes: A History of Heat.* Modern Library.

Edsall, J. T., and Gutfreund, H., 1983. *Biothermodynamics: The Study of Biochemical Processes at Equilibrium.* Wiley.

Klotz, I. M., 1967. *Energy Changes in Biochemical Reactions.* Academic Press.

Hill, T. L., 1977. *Free Energy Transduction in Biology.* Academic Press.

Alberty, R. A., 1993. Levels of thermodynamic treatment of biochemical reaction systems. *Biophys. J.* 65:1243–1254.

Alberty, R. A., and Goldberg, R. N., 1992. Standard thermodynamic formation properties for the adenosine 5'-triphosphate series. *Biochemistry* 31:10610–10615.

Alberty, R. A., 1968. Effect of pH and metal ion concentration on the equilibrium hydrolysis of adenosine triphosphate to adenosine diphosphate. *J. Biol. Chem.* 243:1337–1343.

Goldberg, R. N., 1984. *Compiled Thermodynamic Data Sources for Aqueous and Biochemical Systems: An Annotated Bibliography (1930–1983).* National Bureau of Standards Special Publication 685, U.S. Government Printing Office.

Frey, P. A., and Arabshahi, A., 1995. Standard free energy change for the hydrolysis of the α, β-phosphoanhydride bridge in ATP. *Biochemistry* 34:11307–11310.

Westheimer, F. H., 1987. Why nature chose phosphates. *Science* 235:1173–1178.

Bioenergetics and metabolism

Schilling, C. H., Letscher, D., and Palsson, B. O., 2000. Theory for the systemic definition of metabolic pathways and their use in interpreting metabolic function from a pathway-oriented perspective. *J. Theor. Biol.* 203:229–248.

DeCoursey, T. E., and Cherny, V. V., 2000. Common themes and problems of bioenergetics and voltage-gated proton channels. *Biochim. Biophys. Acta.* 1458:104–119.

Giersch, C., 2000. Mathematical modelling of metabolism. *Curr. Opin. Plant Biol.* 3:249–253.

Rees, D. C., and Howard, J. B., 1999. Structural bioenergetics and energy transduction mechanisms. *J. Mol. Biol.* 293:343–350.

Regulation of metabolism

Kemp, G. J., 2000. Studying metabolic regulation in human muscle. *Biochem. Soc. Trans.* 28:100–103.

Towle, H. C., Kaytor, E. N., and Shih, H. M., 1996. Metabolic regulation of hepatic gene expression. *Biochem. Soc. Trans.* 24:364–368.

Hofmeyr, J. H., 1995. Metabolic regulation: A control analytic perspective. *J. Bioenerg. Biomembr.* 27:479–490.

Atkinson, D. E., 1977. *Cellular Energy Metabolism and Its Regulation.* Academic Press.

Erecińska, M., and Wilson, D. F., 1978. Homeostatic regulation of cellular energy metabolism. *Trends Biochem. Sci.* 3:219–223.

Historical aspects

Kalckar, H. M., 1991. 50 years of biological research: From oxidative phosphorylation to energy requiring transport regulation. *Annu. Rev. Biochem.* 60:1–37.

Kalckar, H. M. (Ed.), 1969. *Biological Phosphorylations.* Prentice Hall.

Fruton, J. S., 1972. *Molecules and Life.* Wiley-Interscience.

Lipmann, F., 1971. *Wanderings of a Biochemist.* Wiley-Interscience.

———| PROBLEMS |———

1. *Energy flow.* What is the direction of each of the following reactions when the reactants are initially present in equimolar amounts? Use the data given in Table 14.1.

(a) ATP + creatine $\rightleftharpoons$ creatine phosphate + ADP
(b) ATP + glycerol $\rightleftharpoons$ glycerol 3-phosphate + ADP
(c) ATP + pyruvate $\rightleftharpoons$ phosphoenolpyruvate + ADP
(d) ATP + glucose $\rightleftharpoons$ glucose 6-phosphate + ADP

2. *A proper inference.* What information do the $\Delta G°'$ data given in Table 14.1 provide about the relative rates of hydrolysis of pyrophosphate and acetyl phosphate?

3. *A potent donor.* Consider the following reaction:

$$\text{ATP} + \text{pyruvate} \rightleftharpoons \text{phosphoenolpyruvate} + \text{ADP}$$

(a) Calculate $\Delta G°'$ and K'_{eq} at 25°C for this reaction, by using the data given in Table 14.1.
(b) What is the equilibrium ratio of pyruvate to phosphoenolpyruvate if the ratio of ATP to ADP is 10?

4. *Isomeric equilibrium.* Calculate $\Delta G°'$ for the isomerization of glucose 6-phosphate to glucose 1-phosphate. What is the equilibrium ratio of glucose 6-phosphate to glucose 1-phosphate at 25°C?

5. *Activated acetate.* The formation of acetyl CoA from acetate is an ATP-driven reaction:

$$\text{Acetate} + \text{ATP} + \text{CoA} \rightleftharpoons \text{acetyl CoA} + \text{AMP} + \text{PP}_i$$

(a) Calculate $\Delta G°'$ for this reaction by using data given in this chapter.
(b) The PP$_i$ formed in the preceding reaction is rapidly hydrolyzed in vivo because of the ubiquity of inorganic pyrophosphatase. The $\Delta G°'$ for the hydrolysis of PP$_i$ is -4.6 kcal mol^{-1}. Calculate the $\Delta G°'$ for the overall reaction. What effect does the hydrolysis of PP$_i$ have on the formation of acetyl CoA?

6. *Acid strength.* The pK of an acid is a measure of its proton-group-transfer potential.

(a) Derive a relation between $\Delta G°'$ and pK.
(b) What is the $\Delta G°'$ for the ionization of acetic acid, which has a pK of 4.8?

7. *Raison d'être.* The muscles of some invertebrates are rich in *arginine phosphate* (phosphoarginine). Propose a function for this amino acid derivative. How would you test your hypothesis?

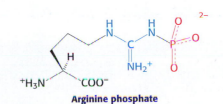

Arginine phosphate

8. *Recurring motif.* What is the structural feature common to ATP, FAD, NAD^+, and CoA?

9. *Ergogenic help or hindrance?* Creatine is a popular, but untested, dietary supplement.

(a) What is the biochemical rationale for the use of creatine?
(b) What type of exercise would most benefit from creatine supplementation?

10. *Standard conditions versus real life.* The enzyme aldolase catalyzes the following reaction in the glycolytic pathway:

Fructose 1, 6 bisphosphate $\xrightleftharpoons{\text{Aldolase}}$
  dihydroxyacetone phosphate + glyceraldehyde 3–phosphate

The $\Delta G°'$ for the reaction is $+5.7$ kcal mol^{-1}, whereas the ΔG in the cell is -0.3 kcal mol^{-1}. Calculate the ratio of reactants to products under equilibrium and intracellular conditions. Using your results, explain how the reaction can be endergonic under standard conditions and exergonic under intracellular conditions.

11. *Not all alike.* The concentrations of ATP, ADP, and P_i differ with cell type. Consequently, the release of free energy with the hydrolysis of ATP will vary with cell type. Using the following table, calculate the ΔG for the hydrolysis of ATP in muscle, liver, and brain cells. In which cell type is the free energy of ATP hydrolysis greatest?

	ATP (mM)	ADP (mM)	P_i (mM)
Liver	3.5	1.8	5.0
Muscle	8.0	0.9	8.0
Brain	2.6	0.7	2.7

12. *Running downhill.* Glycolysis is a series of 10 linked reactions that convert one molecule of glucose into two molecules of pyruvate with the concomitant synthesis of two molecules of ATP (Chapter 16). The $\Delta G°'$ for this set of reactions is -8.5

kcal mol^{-1} (-35.6 kJ mol^{-1}), whereas the ΔG is -18.3 kcal mol^{-1} (-76.6 kJ mol^{-1}). Explain why the free-energy release is so much greater under intracellular conditions than under standard conditions.

Chapter Integration Problem

13. *Activated sulfate.* Fibrinogen contains tyrosine-*O*-sulfate. Propose an activated form of sulfate that could react in vivo with the aromatic hydroxyl group of a tyrosine residue in a protein to form tyrosine-*O*-sulfate.

Data Interpretation

14. *Opposites attract.* The following graph shows how the ΔG for the hydrolysis of ATP varies as a function of the Mg^{2+} concentration ($pMg = \log 1/[Mg^{2+}]$).

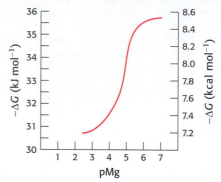

(a) How does decreasing $[Mg^{2+}]$ affect the ΔG of hydrolysis for ATP?
(b) How can you explain this effect?

👆 Media Problem

15. *Coupled reactions.* The most obvious role of enzymes is to accelerate reactions, but a second critical role is to couple reactions that would ordinarily be unrelated. From what you have learned from the text and the **Conceptual Insights** module on energetic coupling, explain why the coupling of chemical reactions in a single enzyme is critical for life.

CHAPTER 15

Signal-Transduction Pathways: An Introduction to Information Metabolism

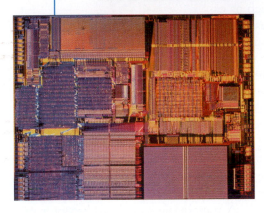

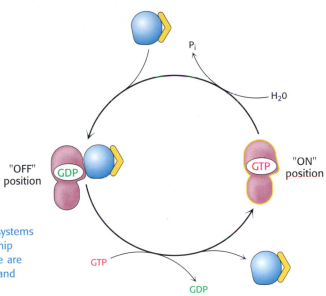

"OFF" position

"ON" position

Molecular switches. Signal transduction circuits in biological systems have molecular on/off switches that, like those in a computer chip (above), transmit information when "on." Common among these are G proteins (right), which transmit a signal when bound to GTP and are silent when bound to GDP. [(Left) Courtesy of Intel.]

A cell is highly responsive to specific chemicals in its environment. Hormones are chemical signals that tell a cell to respond to a change in conditions. Molecules in food or aromas communicate taste and smell through their interaction with specialized sensory cells. This chapter provides an overview of *information metabolism*—how cells receive, process, and respond to information from the environment. The results of genome-sequencing efforts have underscored how widespread and diverse these information-processing circuits are. For example, approximately half of the 25 largest protein families encoded by the human genome deal primarily with information processing.

Signal-transduction cascades mediate the sensing and processing of stimuli. These molecular circuits detect, amplify, and integrate diverse external signals to generate responses such as changes in enzyme activity, gene expression, or ion-channel activity. This chapter is an introduction to some of the most important classes of molecules that participate in common signal-transduction pathways. We will encounter many specific pathways in their biochemical contexts in later chapters. In this chapter, we will also consider the consequences of defects in these pathways, particularly those leading to cancer.

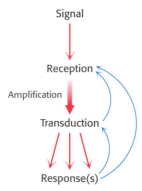

FIGURE 15.1 Principles of signal transduction. An environmental signal, such as a hormone, is first received by interaction with a cellular component, most often a cell-surface receptor. The information that the signal has arrived is then converted into other chemical forms, or *transduced*. The signal is often amplified before evoking a response. Feedback pathways regulate the entire signaling process.

15.0.1 Signal Transduction Depends on Molecular Circuits: An Overview

Signal-transduction pathways follow a broadly similar course that can be viewed as a molecular circuit (Figure 15.1). We begin by examining the challenges posed by transferring extracellular information to a cell's interior.

1. *Membrane receptors transfer information from the environment to the cell's interior.* A few nonpolar signal molecules such as estrogens and other steroid hormones are able to diffuse through the cell membranes and, hence, enter the cell. Once inside the cell, these molecules can bind to proteins that interact directly with DNA and modulate gene transcription. Thus, a chemical signal enters the cell and directly alters gene-expression patterns. These important signaling systems will be discussed in Chapter 31. However, most signal molecules are too large and too polar to pass through the membrane, and no appropriate transport systems are present. Thus, *the information that signal molecules are present must be transmitted across the cell membrane without the molecules themselves entering the cell.* A membrane-associated receptor protein often performs the function of information transfer across the membrane.

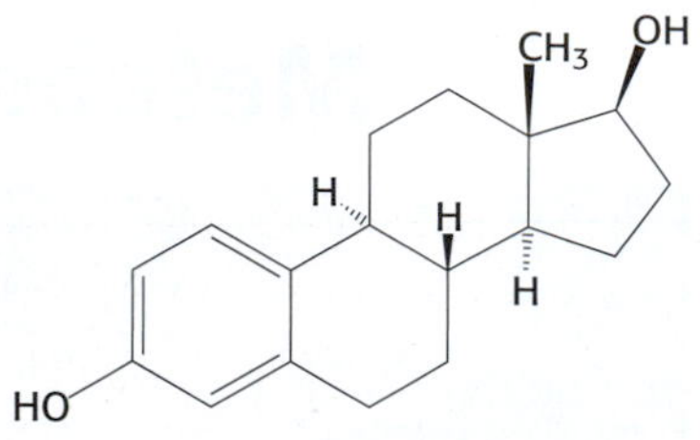

Estrogen (estradiol)

Such a receptor is an intrinsic membrane protein that has both extracellular and intracellular domains. A binding site on the extracellular domain specifically recognizes the signal molecule (often referred to as the *ligand*). Such binding sites are analogous to enzyme active sites except that no catalysis takes place within them. The interaction of the ligand and the receptor alters the tertiary or quaternary structure of the receptor, including the intracellular domain. These structural changes are not sufficient to yield an appropriate response, because they are restricted to a small number of receptor molecules in the cell membrane. The information embodied by the presence of the ligand, often called the *primary messenger,* must be transduced into other forms that can alter the biochemistry of the cell.

2. *Second messengers relay information from the receptor–ligand complex.* Changes in the concentration of small molecules, called *second messengers,* constitute the next step in the molecular information circuit. Particularly important second messengers include cyclic AMP and cyclic GMP, calcium ion, inositol 1,4,5-trisphosphate, (IP$_3$), and diacylglycerol (DAG; Figure 15.2).

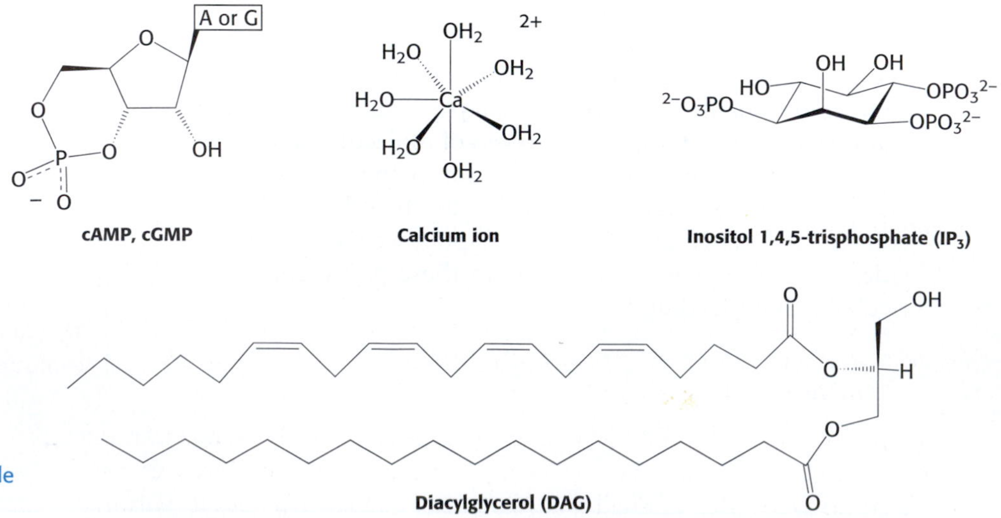

cAMP, cGMP

Calcium ion

Inositol 1,4,5-trisphosphate (IP$_3$)

FIGURE 15.2 Common second messengers. Second messengers are intracellular molecules that change in concentration in response to environmental signals. That change in concentration conveys information inside the cell.

Diacylglycerol (DAG)

The use of second messengers has several consequences. First, second messengers are often free to diffuse to other compartments of the cell, such as the nucleus, where they can influence gene expression and other processes. Second, the signal may be amplified significantly in the generation of second messengers. Enzymes or membrane channels are almost always activated in second-messenger generation; each activated macromolecule can lead to the generation of many second messengers within the cell. Thus, *a low concentration of signal in the environment, even as little as a single molecule, can yield a large intracellular signal and response.* Third, the use of common second messengers in multiple signaling pathways creates both opportunities and potential problems. Input from several signaling pathways, often called *cross talk,* may affect the concentrations of common second messengers. Cross talk permits more finely tuned regulation of cell activity than would the action of individual independent pathways. However, inappropriate cross talk can cause second messengers to be misinterpreted.

3. *Protein phosphorylation is a common means of information transfer.* Many second messengers elicit responses by activating *protein kinases.* These enzymes transfer phosphoryl groups from ATP to specific serine, threonine, and tyrosine residues in proteins.

We previously encountered the cAMP-dependent protein kinase in Section 10.4.2. This protein kinase and others are the link that transduces changes in the concentrations of free second messengers into changes in the covalent structures of proteins. Although these changes are less transient than the changes in secondary-messenger concentrations, protein phosphorylation is not irreversible. Indeed, *protein phosphatases* are enzymes that hydrolytically remove specific phosphoryl groups from modified proteins.

4. *The signal is terminated.* Protein phosphatases are one mechanism for the termination of a signaling process. After a signaling process has been initiated and the information has been transduced to affect other cellular processes, the signaling processes must be terminated. Without such termination, cells lose their responsiveness to new signals. Moreover, signaling processes that fail to be terminated properly may lead to uncontrolled cell growth and the possibility of cancer.

Essentially every biochemical process presented in the remainder of this book either is a component of a signal-transduction pathway or can be affected by one. As we shall see, the use of *protein modules* in various combinations is a clear, even dominant, theme in the construction of signal-transduction proteins. Signal-transduction proteins have evolved by the addition of such ancillary modules to core domains to facilitate interactions with other proteins or cell membranes. By controlling which proteins interact with one another, these modules play important roles in determining the wiring diagrams for signal-transduction circuits.

We begin by considering the largest and one of the most important classes of receptor, the seven-transmembrane-helix receptors.

15.1 SEVEN-TRANSMEMBRANE-HELIX RECEPTORS CHANGE CONFORMATION IN RESPONSE TO LIGAND BINDING AND ACTIVATE G PROTEINS

CONCEPTUAL INSIGHTS, Signaling Pathways: Response and Recovery. *Animations in this media module show how 7TM receptors and G proteins are employed in two sensory-system signaling pathways.*

The *seven-transmembrane-helix (7TM) receptors* are responsible for transmitting information initiated by signals as diverse as photons, odorants, tastants, hormones, and neurotransmitters (Table 15.1). Several thousand such receptors are known, and the list continues to grow. As the name indicates, these receptors contain seven helices that span the membrane bilayer. The receptors are sometimes referred to as serpentine receptors because the single polypeptide chain "snakes" through the membrane seven times (Figure 15.3A). A well-characterized member of this family is *rhodopsin*. The "ligand" for this protein, which plays an essential role in vision, is a photon (Section 32.3.1). An example of a receptor that responds to chemical signals is the *β-adrenergic receptor*. This protein binds epinephrine (also called *adrenaline*), a hormone responsible for the "fight or flight" response. We will address the biochemical roles of this hormone in more detail later (Section 21.3.1).

Recently, the three-dimensional structure of bovine rhodopsin was determined in its unactivated form (Figure 15.3B). A variety of evidence reveals that the 7TM receptors, particularly their cytoplasmic loops and

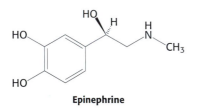

Epinephrine

TABLE 15.1 Biological functions mediated by 7TM receptors

- Smell
- Taste
- Vision
- Neurotransmission
- Hormone secretion
- Chemotaxis
- Exocytosis
- Control of blood pressure
- Embryogenesis
- Cell growth and differentiation
- Development
- Viral infection
- Carcinogenesis

Source: After J. S. Gutkind, *J. Biol. Chem.* 273(1998):1839.

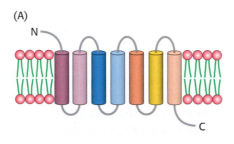

(A)

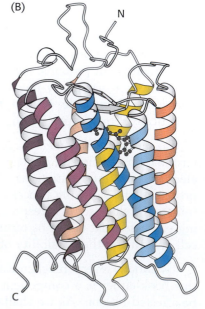

(B)

FIGURE 15.3 A 7TM receptor.
(A) Schematic representation of a 7TM receptor showing how it passes through the membrane seven times. (B) Three-dimensional structure of rhodopsin, a 7TM receptor taking part in visual signal transduction. As the first 7TM receptor whose structure has been determined, its structure provides a framework for understanding other 7TM receptors. A linked photoreceptor molecule, retinal, is present in the position where, in at least other 7TM receptors, ligands likely bind.

their carboxyl termini, change conformation in response to ligand binding, although the details of these conformational changes remain to be established. Thus, *the binding of a ligand from outside the cell induces a conformational change in the 7TM receptor that can be detected inside the cell.* Even though vision and response to hormones would seem to have little in common, a comparison of the amino acid sequences of rhodopsin and the β-adrenergic receptor clearly reveals homology. On the basis of this sequence comparison, the β-adrenergic receptor is expected to have a structure quite similar to that of rhodopsin. As we shall see, these receptors also have in common the next step in their signaling transduction cascades.

15.1.1 Ligand Binding to 7TM Receptors Leads to the Activation of G Proteins

What is the next step in the pathway after the binding of epinephrine by the β-adrenergic receptor? An important clue was Martin Rodbell's finding that GTP in addition to hormone is essential for signal transduction to proceed. Equally revealing was the observation that hormone binding stimulates GTP hydrolysis. These findings led to the discovery by Alfred Gilman that *a guanyl nucleotide-binding protein is an intermediary in signal transduction from the 7TM receptors.* This signal-coupling protein is termed a *G protein* (G for guanyl nucleotide). The activated G protein stimulates the activity of *adenylate cyclase,* an enzyme that increases the concentration of cAMP by forming it from ATP (Figure 15.4).

15.1.2 G Proteins Cycle Between GDP- and GTP-Bound Forms

How do these G proteins operate? *In the unactivated state, the guanyl nucleotide bound to the G protein is GDP.* In this form, the G protein exists as a *heterotrimer* consisting of α, β, and γ subunits; the α subunit (referred to as G_α) binds the nucleotide (Figure 15.5). The α subunit is a member of the P-loop NTPases family (Section 9.4.1) and the P-loop that participates in nucleotide binding. The β subunit contains a seven-bladed propeller structure, and the γ subunit comprises a pair of α helices that wrap around the

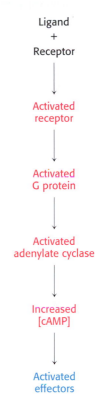

Ligand
+
Receptor

↓

Activated
receptor

↓

Activated
G protein

↓

Activated
adenylate cyclase

↓

Increased
[cAMP]

↓

Activated
effectors

FIGURE 15.4 The β-adrenergic receptor signal-transduction pathway. On binding of ligand, the receptor activates a G protein that in turn activates the enzyme adenylate cyclase. Adenylate cyclase generates the second messenger cAMP. The increase in cAMP results in a biochemical response to the initial signal.

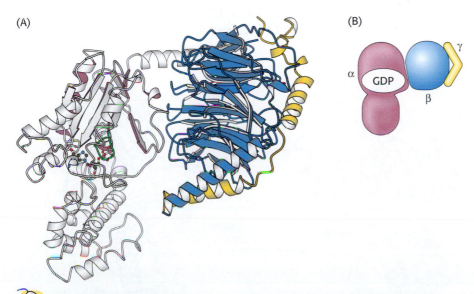

(A)   (B)

FIGURE 15.5 A heterotrimeric G protein. (A) A ribbon diagram shows the relation between the three subunits. In this complex, the α subunit (gray and purple) is bound to GDP. (B) A schematic representation of the heterotrimeric G protein.

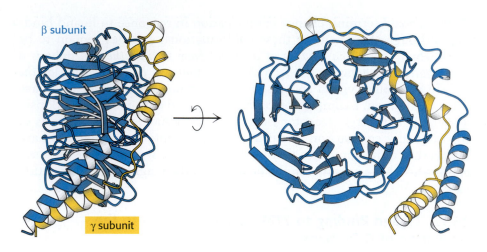

β subunit

γ subunit

FIGURE 15.6 The βγ subunits of the heterotrimeric G protein. Two views illustrate the interaction between the β and the γ subunits. The helices of the γ subunit (yellow) wrap around the β subunit (blue). The seven-bladed propeller structure of the β subunit is readily apparent in the representation on the right.

β subunit (Figure 15.6). The α and γ subunits are usually anchored to the membrane by covalently attached fatty acids. *The role of the hormone-bound receptor is to catalyze the exchange of GTP for bound GDP.* The hormone–receptor complex interacts with the heterotrimeric G protein and opens the nucleotide-binding site so that GDP can depart and GTP from solution can bind. The α subunit simultaneously dissociates from the βγ dimer ($G_{\beta\gamma}$). The structure of the G_α subunit conforms tightly to the GTP molecule; in particular, three stretches of polypeptide (termed switch I, switch II, and switch III) interact either directly or indirectly with the γ phosphate of GTP (Figure 15.7). These structural changes are responsible for the reduced affinity of G_α for $G_{\beta\gamma}$. *The dissociation of the G-protein heterotrimer into G_α and $G_{\beta\gamma}$ units transmits the signal that the receptor has bound its ligand.* Moreover, the surfaces of G_α and $G_{\beta\gamma}$ that had formed the trimer interface are now exposed to interact with other proteins.

A single hormone–receptor complex can stimulate nucleotide exchange in many G-protein heterotrimers. Thus, hundreds of G_α molecules are converted from their GDP into their GTP forms for each bound molecule of hormone, giving an *amplified response*. All 7TM receptors appear to be coupled to G pro-

FIGURE 15.7 Conformational changes in G_α on nucleotide exchange. (Left) Prior to activation, G_α binds GDP. (Right) On GTP for GDP exchanges, the three switch regions (shown in blue) close upon the nucleoside triphosphate, generating the active conformation.

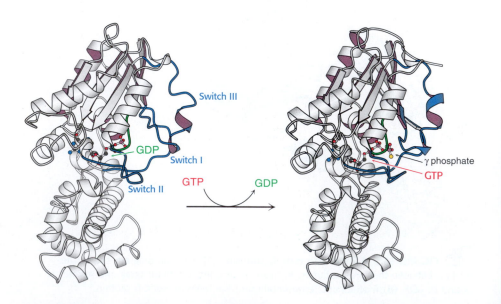

Switch III

GDP

Switch I

GTP   GDP

Switch II

γ phosphate

GTP

TABLE 15.2   G-protein families and their functions

G_α class	Initiating signal	Downstream signal
$G_{\alpha s}$	β-Adrenergic amines, glucagon, parathyroid hormone, many others	Stimulates adenylate cyclase
$G_{\alpha i}$	Acetylcholine, α-adrenergic amines, many neurotransmitters	Inhibits adenylate cyclase
$G_{\alpha t}$	Photons	Stimulates cGMP phosphodiesterase
$G_{\alpha q}$	Acetylcholine, α-adrenergic amines, many neurotransmitters	Increases IP_3 and intracellular calcium
$G_{\alpha 13}$	Thrombin, other agonists	Stimulates Na^+ and H^+ exchange

Source: Z. Farfel, H. R. Bourne, and T. Iiri. *N. Engl. J. Med.* 340(1999):1012.

teins, and so the 7TM receptors are sometimes referred to as *G-protein-coupled receptors* or *GPCRs.* Do all of the signals that function by means of 7TM receptors funnel through the same G protein? Indeed not. Different G proteins exist, and they can affect downstream targets in different ways when activated. For example, in regard to the G protein coupled to the β-adrenergic receptor, the α subunit binds to adenylate cyclase and stimulates its enzymatic activity. This subunit is referred to as $G_{\alpha s}$, the *s* in the subscript indicating the subunit's *stimulatory* role. The human genome contains more than 15 genes encoding the α subunits, 5 encoding the β subunits, and 10 encoding the γ subunits. Thus, in principle, there could be more than a thousand heterotrimeric G proteins; however, the number of combinations that actually exists is not known. Selected members of this family are shown in Table 15.2. Only a small subset of these proteins is expressed in a particular cell.

15.1.3 Activated G Proteins Transmit Signals by Binding to Other Proteins

The adenylate cyclase enzyme that is activated by the epinephrine–β-adrenergic receptor complex is a membrane protein that contains 12 presumed membrane-spanning helices. The enzymatically active part of the protein is formed from two large intracellular domains: one is located between transmembrane helices 6 and 7 and the other after the last transmembrane helix. The structure was determined for a complex formed between $G_{\alpha s}$ bound to a GTP analog and protein fragments corresponding to the active adenylate cyclase (Figure 15.8). As expected, the G protein binds to adenylate cyclase

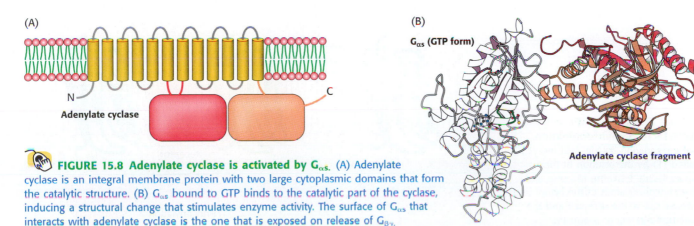

(A)

N

C

Adenylate cyclase

(B)

$G_{\alpha s}$ (GTP form)

Adenylate cyclase fragment

FIGURE 15.8 Adenylate cyclase is activated by $G_{\alpha s}$. (A) Adenylate cyclase is an integral membrane protein with two large cytoplasmic domains that form the catalytic structure. (B) $G_{\alpha s}$ bound to GTP binds to the catalytic part of the cyclase, inducing a structural change that stimulates enzyme activity. The surface of $G_{\alpha s}$ that interacts with adenylate cyclase is the one that is exposed on release of $G_{\beta\gamma}$.

through the surface that had bound the βγ dimer when the G protein was in its GDP form. The activation of the G protein exposes this surface and subtly changes it so that it now binds the surface of adenylate cyclase in preference to $G_{\beta\gamma}$. The interaction of $G_{\alpha s}$ with adenylate cyclase favors a more catalytically active conformation of the enzyme, thus stimulating cAMP production. *The net result is that the binding of epinephrine to the receptor on the cell surface increases the rate of cAMP production inside the cell.*

15.1.4 G Proteins Spontaneously Reset Themselves Through GTP Hydrolysis

The ability to shut down signal-transduction pathways is as critical as the ability to turn them on. How is the signal initiated by activated 7TM receptors switched off? *G_α subunits have intrinsic GTPase activity*, hydrolyzing bound GTP to GDP and P_i. This hydrolysis reaction is slow, however, requiring from seconds to minutes and thus allowing the GTP form of G_α to activate downstream components of the signal-transduction pathway before GTP hydrolysis deactivates the subunit. In essence, *the bound GTP acts as a built-in clock that spontaneously resets the G_α subunit after a short time period.* After GTP hydrolysis and the release of P_i, the GDP-bound form of G_α then reassociates with $G_{\beta\gamma}$ to reform the heterotrimeric protein (Figure 15.9).

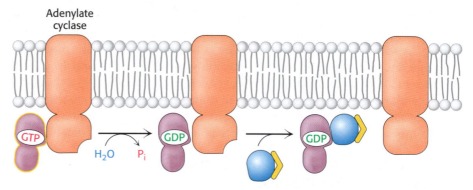

FIGURE 15.9 Resetting G_α. On hydrolysis of the bound GTP by the intrinsic GTPase activity of G_α, G_α reassociates with the βγ subunits to form the heterotrimeric G protein, thereby terminating the activation of adenyl cyclase.

The hormone-bound activated receptor must be reset as well to prevent the continuous activation of G proteins. This resetting is accomplished by two processes (Figure 15.10). First, the hormone dissociates, returning the receptor to its initial, unactivated state. The likelihood that the receptor remains in its unbound state depends on the concentration of hormone. Second, the hormone–receptor complex is deactivated by the phosphorylation of serine and threonine residues in the carboxyl-terminal tail. In the example

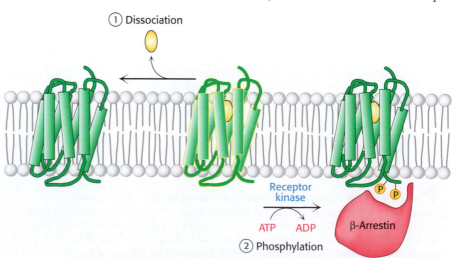

FIGURE 15.10 Signal termination.
Signal transduction by the 7TM receptor is halted (1) by dissociation of the signal molecule from the receptor and (2) by phosphorylation of the cytoplasmic C-terminal tail of the receptor and the subsequent binding of β-arrestin.

under consideration, *β-adrenergic receptor kinase* phosphorylates the carboxyl-terminal tail of the hormone–receptor complex but not the unoccupied receptor. Finally, the binding of *β-arrestin,* binds to the phosphorylated receptor and further diminishes its G-protein-activating ability. Phosphorylation and the binding of β-arrestin account for the *desensitization (adaptation)* of the receptor subsequent to prolonged exposure to epinephrine. The epinephrine-initiated cascade, like many other signal-transduction processes, has evolved to *respond to changes in the strength of stimuli rather than to their absolute level.* Adaptation is advantageous because it enables receptors to respond to changes in the level of stimuli over a wide range of background levels.

15.1.5 Cyclic AMP Stimulates the Phosphorylation of Many Target Proteins by Activating Protein Kinase A

Let us continue to follow the information flow down this signal-transduction pathway. The increased concentration of cAMP can affect a wide range of cellular processes. For example, it enhances the degradation of storage fuels, increases the secretion of acid by the gastric mucosa, leads to the dispersion of melanin pigment granules, diminishes the aggregation of blood platelets, and induces the opening of chloride channels. How does cAMP influence so many cellular processes? Is there a common denominator for its diverse effects? Indeed there is. *Most effects of cyclic AMP in eukaryotic cells are mediated by activation of a single protein kinase.* This key enzyme is called *protein kinase A (PKA).* As discussed in Section 10.4.2, PKA consists of two regulatory (R) chains and two catalytic (C) chains. In the absence of cAMP, the R_2C_2 complex is catalytically inactive. The binding of cAMP to the regulatory chains releases the catalytic chains, which are enzymatically active on their own. Activated PKA then phosphorylates specific serine and threonine residues in many targets to alter their activity. The significance and far reach of the *adenylate cyclase cascade* are seen in the following examples:

1. In *glycogen metabolism* (Section 21.5), PKA phosphorylates two enzymes that lead to the breakdown of this polymeric store of glucose and the inhibition of further glycogen synthesis.

2. PKA stimulates the *expression of specific genes* by phosphorylating a transcriptional activator called the cAMP-response element binding (CREB) protein (Section 31.3.6). This activity of PKA illustrates that signal-transduction pathways can extend into the nucleus to alter gene expression.

3. *Synaptic transmission* between pairs of neurons in *Aplysia* (a marine snail) is enhanced by serotonin, a neurotransmitter that is released by adjacent interneurons. Serotonin binds to a 7TM receptor to trigger an adenylate cyclase cascade. The rise in cAMP level activates PKA, which facilitates the closing of potassium channels by phosphorylating them. Closure of potassium channels increases the excitability of the target cell.

Thus, signal-transduction pathways that include 7TM receptors, the activation of adenylate cyclase, and the activation of PKA can modulate enzyme activities, gene-expression patterns, and membrane excitability.

15.2 THE HYDROLYSIS OF PHOSPHATIDYL INOSITOL BISPHOSPHATE BY PHOSPHOLIPASE C GENERATES TWO MESSENGERS

Cyclic AMP is not the only second messenger employed by 7TM receptors and the G proteins. We turn now to another ubiquitous second-messenger cascade that is used by many hormones to evoke a variety of responses. The

OPO$_3^{2-}$
OH
$^{2-}$O$_3$PO
OH
HO
O
P
O   O

O
O

O
H
O
O
O

Phosphatidyl inositol 4,5-bisphosphate (PIP$_2$)

Cleavage
site

phosphoinositide cascade, like the adenylate cyclase cascade, converts extracellular signals into intracellular ones. The intracellular messengers formed
by activation of this pathway arise from the cleavage *phosphatidyl inositol
4,5-bisphosphate* (PIP$_2$), a phospholipid present in cell membranes. The
binding of a hormone such as vasopressin to a 7TM receptor leads to the
activation of the β isoform of *phospholipase C.* The G$_\alpha$ protein that activates
phospholipase C is called G$_{\alpha q}$. The activated enzyme then hydrolyzes the
phosphodiester bond linking the phosphorylated inositol unit to the acylated glycerol moiety. The cleavage of PIP$_2$ produces two messengers:
inositol 1,4,5-trisphosphate, a soluble molecule that can diffuse from the
membrane, and diacylglycerol, which stays in the membrane.

Comparison of the amino acid sequences of different isoforms of phospholipase C as well as examination of the known three-dimensional structures of phospholipase components reveal an intriguing modular structure
(Figure 15.11).

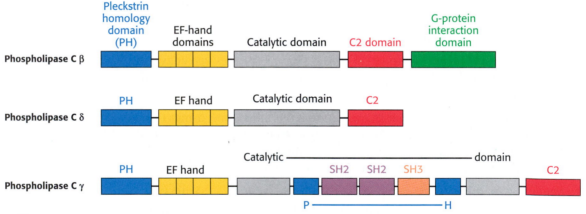

FIGURE 15.11 Modular structure of phospholipase C. The domain structures of three
isoforms of phospholipase C reveal similarities and differences among the isoforms. Only
the β isoform, with the G-protein-binding domain, can be stimulated directly by G proteins.
For phospholipase Cγ, the insertion of two SH2 (Src homology 2) domains and one SH3
(Src homology 3) domain splits the catalytic domain and a PH domain into two parts.

**Pleckstrin
homology
domain**

PIP$_2$

This analysis reveals the basis for both phospholipase enzymatic activity
and its regulation by signal-transduction pathways. The catalytic core of
these enzymes has an αβ barrel structure similar to the catalytic core of triose
phosphate isomerase and other enzymes (Section 16.1.4). This domain is
flanked by domains that interact with membrane components. At the amino
terminus is a pleckstrin homology (PH) domain. This ~120-residue domain
binds a lipid head group such as that of PIP$_2$ (Figure 15.12). The PH

FIGURE 15.12 Pleckstrin homology domain. PH domains facilitate the binding
of proteins to membrane lipids, particularly PIP$_2$. In regard to phospholipase C, the PH
domains help to localize the enzyme near its substrate, PIP$_2$.

domain is joined to the catalytic domain by a set of four EF-hand domains. Although EF-hand domains often take part in calcium-binding (Section 15.3.2), the EF-hand domains of phospholipase C lack many of the calcium-binding residues. On the carboxyl-terminal side of the catalytic domain is a C2 domain (for protein kinase C domain 2). This ~130-residue domain is a member of the immunoglobulin domain super-family (Chapter 33) and plays a role in binding phospholipid headgroups. Such interactions, often but not always, require the presence of bound calcium ions.

The binding of a G protein brings the enzyme into a catalytically active position. The β isoform of phospholipase C has an additional domain at its carboxyl terminus—a domain that interacts with the α subunit of G_q in its GTP form. Because this G protein is linked to the membrane by its fatty acid anchor, this interaction helps pull the β isoform of phospholipase to the membrane. This interaction acts in concert with the binding of the PH and C2 domains of phospholipase C to membrane components to bring the active site in the catalytic core into a position against a membrane surface that is favorable for cleavage of the phosphodiester bond of PIP_2 (Figure 15.13). Some of these interactions and the enzymatic reaction itself also depend on the presence of calcium ion. Phospholipase isoforms that lack the carboxyl-terminal regulatory domain do not respond to these signal-transduction pathways. The two products of the cleavage reaction, inositol 1,4,5-trisphosphate and diacylglycerol, each trigger additional steps in the signal-transduction cascades.

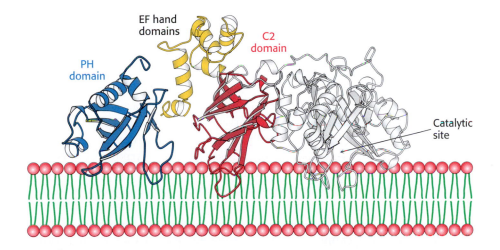

FIGURE 15.13 Phospholipase C acts at the membrane surface. The PH and C2 domains of phospholipase help to position the enzyme's catalytic site for ready access to the phosphodiester bond of the membrane lipid substrate, PIP_2.

15.2.1 Inositol 1,4,5-trisphosphate Opens Channels to Release Calcium Ions from Intracellular Stores

What are the biochemical effects of the second messenger inositol 1,4,5-trisphosphate? These effects were delineated by microinjecting IP_3 molecules into cells or by allowing IP_3 molecules to diffuse into cells whose plasma membranes had been made permeable. Michael Berridge and coworkers found that *IP_3 causes the rapid release of Ca^{2+} from intracellular stores—the endoplasmic reticulum and, in smooth muscle cells, the sarcoplasmic reticulum.* The elevated level of Ca^{2+} in the cytosol then triggers processes such as smooth muscle contraction, glycogen breakdown, and vesicle release. In *Xenopus* oocytes, the injection of IP_3 suffices to activate many of the early events of fertilization.

Inositol 1,4,5-trisphosphate is able to increase Ca^{2+} concentration by associating with a membrane protein called the *IP_3-gated channel* or *IP_3 receptor*. This receptor, which is composed of four large, identical subunits,

Inositol

Inositol 1,4,5-trisphosphate

Inositol 1,3,4,5-tetrakisphosphate

Inositol 1,3,4-trisphosphate

Inositol

FIGURE 15.14 Metabolism of IP₃. The IP₃ signal is terminated by the metabolism of the compound into derivatives lacking second-messenger capabilities.

forms an ion channel. The ion-conducting channel itself is likely similar to the structurally characterized K^+ channel (Section 13.5.6). At least three molecules of IP₃ must bind to sites on the cytosolic side of the membrane protein to open the channel and release Ca^{2+}. *The highly cooperative opening of calcium channels by nanomolar concentrations of IP₃ enables cells to detect and amplify very small changes in the concentration of this messenger.* The cooperativity of IP₃ binding and channel opening is another example of the role of allosteric interactions.

How is the IP₃-initiated signal turned off? *Inositol 1,4,5-trisphosphate is a short-lived messenger because it is rapidly converted into derivatives that do not open the channel* (Figure 15.14). Its lifetime in most cells is less than a few seconds. Inositol 1,4,5-trisphosphate can be degraded to inositol by the sequential action of phosphatases or it can be phosphorylated to inositol 1,3,4,5-tetrakisphosphate, which is then converted into inositol by an alternative route. Lithium ion, widely used to treat bipolar affective disorder, may act by inhibiting the recycling of inositol 1,3,4-trisphosphate.

15.2.2 Diacylglycerol Activates Protein Kinase C, Which Phosphorylates Many Target Proteins

Diacylglycerol, the other molecule formed by the receptor-triggered hydrolysis of PIP₂, also is a second messenger and it, too, activates a wide array of targets. We will examine how diacylglycerol activates *protein kinase C (PKC)*, a protein kinase that phosphorylates serine and threonine residues in many target proteins. The amino acid sequences and the results of three-dimensional structural studies of protein kinase C isozymes reveal an elegant modular architecture (Figure 15.15). The α, β, and γ isozymes of protein kinase C have in common a structurally similar catalytic domain, homologous to that in PKA, at the carboxyl terminus. Adjacent to the catalytic domain is a C2 domain that is related to the C2 domain of phospholipase C and that interacts with membrane phospholipids. On the other side of the C2 domain is a pair of C1 domains, each organized around two bound zinc ions. These two domains, particularly the second (C1B) domain, bind diacylglycerol (Figure 15.16A). Finally, at the amino terminus is a sequence of the form -A-R-K-G-**A**-L-R-Q-K-. This sequence is striking because the consensus sequence for PKC substrates is X-R-X-X-(S,T)-Hyd-R-X in which Hyd refers to a large, hydrophobic residue. This sequence from PKC is referred to as a *pseudosubstrate sequence* because it resembles the substrate sequence except that it has an alanine residue in place of the serine or threonine residue; so it cannot be phosphorylated. Recall that an analogous pseudosubstrate sequence is present in the regulatory chain of PKA and that it prevents activity by occluding the active site in the R_2C_2 tetramer of PKA (Section 10.4.2). Similarly, the pseudosubstrate sequence of PKC binds to that enzyme's active site and prevents substrate binding.

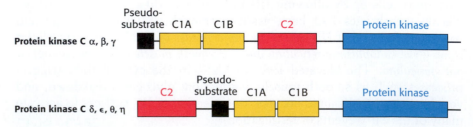

FIGURE 15.15 Modular structure of protein kinase C. Seven isozymes of PKC can be divided into two classes on the basis of their domain organization.

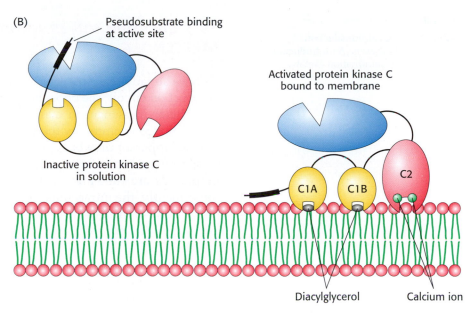

(A)

Zinc ion

Diacylglycerol-binding site

(B)

Pseudosubstrate binding
at active site

Inactive protein kinase C
in solution

Activated protein kinase C
bound to membrane

C2

C1A   C1B

Diacylglycerol   Calcium ion

FIGURE 15.16 Protein kinase C activation. (A) The C1 domain of PKC, structurally organized around two bound zinc ions, binds diacylglycerol. (B) When the C1 domains bind to diacylglycerol in the membrane, the pseudosubstrate is pulled from the active site, permitting catalysis. Calcium-binding C2 domains help to localize PKC to the membrane.

With this structure in mind, we can now understand how PKC is activated on PIP$_2$ hydrolysis (Figure 15.16B). Before activation, PKC is free in solution. On PIP$_2$ hydrolysis in the membrane by phospholipase C, the C1B domain of PKC binds to diacylglycerol. This binding and the interaction of the C2 domain with membrane phospholipids anchors the enzyme to the membrane. The interaction between the C2 domain and the phospholipids, especially phosphatidyl serine, requires calcium ions. The binding of the C1A domain to diacylglycerol pulls the pseudosubstrate out of the active site; the released pseudosubstrate, which is quite positively charged, probably interacts with the negatively charged membrane surface as well. The result of the conformational transitions is the binding of PKC to diacylglycerol-rich regions of the membrane in an active state, ready to phosphorylate serine and threonine residues in appropriate sequence contexts. Note that diacylglycerol and IP$_3$ work in tandem: IP$_3$ increases the Ca^{2+} concentration, and Ca^{2+} facilitates PKC activation as well as phospholipase C activity.

Diacylglycerol, like IP$_3$, acts transiently because it is rapidly metabolized. It can be phosphorylated to phosphatidate (Section 26.1) or it can be hydrolyzed to glycerol and its constituent fatty acids (Figure 15.17). Arachidonate, the C$_{20}$ polyunsaturated fatty acid that usually occupies the position 2 on the glycerol moiety of PIP$_2$, is the precursor of a series of 20-carbon hormones, including the prostaglandins (Section 22.6.2). Thus, *the phosphoinositide pathway gives rise to many molecules that have signaling roles.*

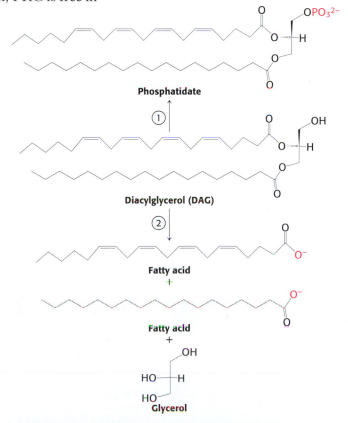

Phosphatidate

①

Diacylglycerol (DAG)

②

Fatty acid
+

Fatty acid
+

Glycerol

FIGURE 15.17 Metabolism of diacylglycerol. Diacylglycerol may be (1) phosphorylated to phosphatidate or (2) hydrolyzed to glycerol and fatty acids.

15.3 CALCIUM ION IS A UBIQUITOUS CYTOSOLIC MESSENGER

We have already seen that Ca^{2+} is an important component of one signal-transduction circuit, the phosphoinositide cascade. Indeed, Ca^{2+} is itself an intracellular messenger in many eukaryotic signal-transducing pathways. Why is this ion commonly found to mediate so many signaling processes? First, an apparent drawback is in fact an advantage: calcium complexes of phosphorylated and carboxylated compounds are often insoluble, but such compounds are fundamental to many biochemical processes in the cell. Consequently, the intracellular levels of Ca^{2+} must be kept low to prevent precipitation of these compounds. These low levels are maintained by transport systems for the extrusion of Ca^{2+}. In eukaryotic cells, two in particular—the Ca^{2+} ATPase (Section 13.2.1) and the sodium–calcium exchanger—are especially important. Because of their actions, the cytosolic level of Ca^{2+} in unexcited cells is typically 100 nM, several orders of magnitude lower than the concentration in the blood, which is approximately 5 mM. This steep concentration gradient presents cells with a matchless opportunity: *the cytosolic Ca^{2+} concentration can be abruptly raised for signaling purposes by transiently opening calcium channels in the plasma membrane or in an intracellular membrane.*

A second property of Ca^{2+} that makes it a highly suitable intracellular messenger is that it can bind tightly to proteins (Figure 15.18). Negatively charged oxygen atoms (from the side chains of glutamate and aspartate) and uncharged oxygen atoms (main-chain carbonyl groups and side-chain oxygen atoms from glutamine and asparagine) bind well to Ca^{2+}. *The capacity of Ca^{2+} to be coordinated to multiple ligands—from six to eight oxygen atoms— enables it to cross-link different segments of a protein and induce significant conformational changes.*

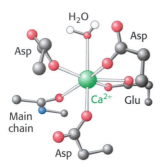

FIGURE 15.18 Mode of binding of Ca^{2+} to calmodulin. Calcium is coordinated to six oxygen atoms from the protein and one (top) of water.

15.3.1 Ionophores Allow the Visualization of Changes in Calcium Concentration

Our understanding of the role of calcium in cellular processes has been greatly enhanced by the use of calcium-specific reagents. *Ionophores* such as *A23187* and *ionomycin* can traverse a lipid bilayer because they have a hydrophobic periphery.

A23187

Ionomycin

They can be used to introduce Ca^{2+} into cells and organelles. Many physiological responses that are normally triggered by the binding of hormones to cell-surface receptors can also be elicited by using calcium ionophores to raise the cytosolic calcium level. Conversely, the concentration of unbound calcium in a cell can be made very low (nanomolar or less) by introducing a calcium-specific chelator such as EGTA. The concentration of free Ca^{2+} in intact cells can be monitored by using polycyclic chelators such as Fura-2.

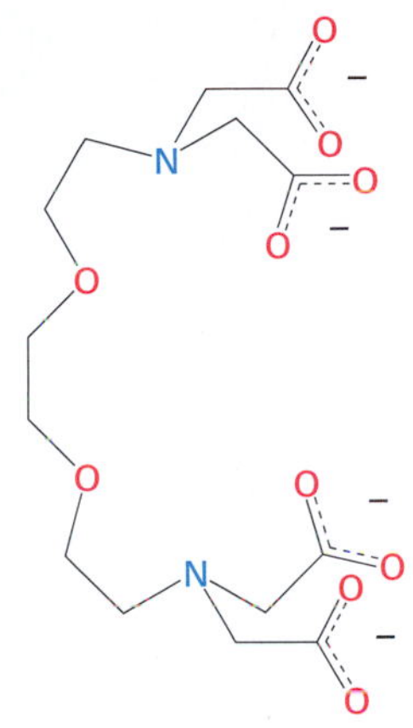

EGTA
[Ethylene glycol bis(β-aminoethyl ether)-
_N,N,N',N'_-tetraacetate]

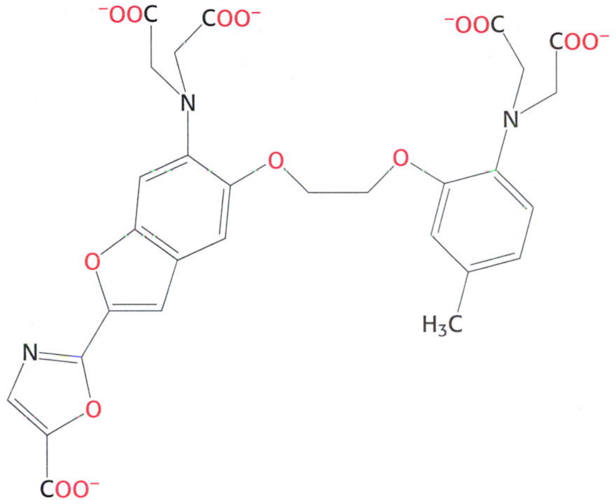

Fura-2

The fluorescence properties of this and related indicators change markedly when Ca^{2+} is bound (Figure 15.19). These compounds can be introduced into cells to allow measurement of the available Ca^{2+} concentration in real time through the use of fluorescence microscopy. Such methods allow the direct detection of calcium fluxes and diffusion within living cells in response to the activation of specific signal-transduction pathways (Figure 15.20).

FIGURE 15.19 Calcium indicator. The fluorescence spectra of the calcium-binding dye Fura-2 can be used to measure available calcium ion concentrations in solution and in cells. [After S. J. Lippard and J. M. Berg, _Principles of Bioinorganic Chemistry._ (University Science Books, 1994), p. 193.]

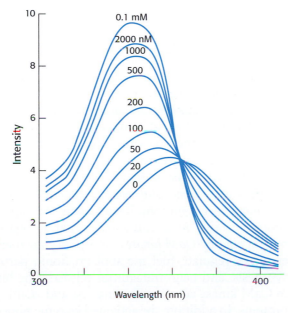

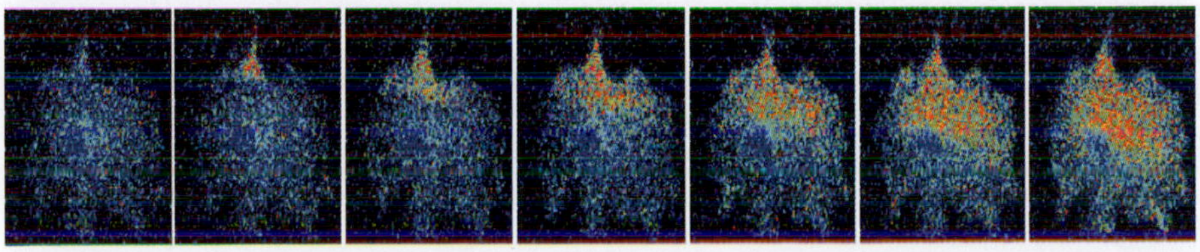

FIGURE 15.20 Calcium imaging. A series of images shows Ca^{2+} spreading across a cell. These images were obtained through the use of a fluorescent calcium-binding dye. The images are false colored: red represents high Ca^{2+} concentrations and blue low Ca^{2+} concentrations. [Courtesy of Dr. Masashi Isshiki, Dept. of Nephrology, University of Tokyo, and Dr. G. W. Anderson, Dept. of Cell Biology, University of Texas Southwestern Medical School.]

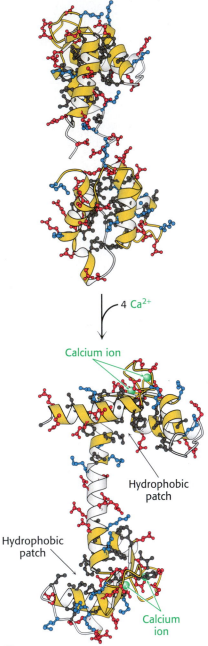

4 Ca²⁺

Calcium ion

Hydrophobic
patch

Hydrophobic
patch

Calcium
ion

FIGURE 15.22 Conformational changes in calmodulin on calcium binding. In the absence of calcium (top), the EF hands have hydrophobic cores. On binding of a calcium ion (green sphere) to each EF hand, structural changes expose hydrophobic patches on the calmodulin surface. These patches serve as docking regions for target proteins. Acidic residues are shown in red, basic residues in blue, and hydrophobic residues in black. The central helix in calmodulin remains somewhat flexible, even in the calcium-bound state.

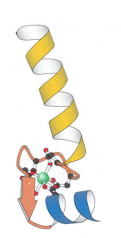
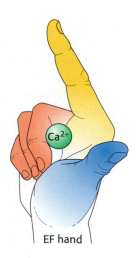

FIGURE 15.21 EF hand. Formed by a helix-loop-helix unit, an EF hand is a binding site for calcium in many calcium sensing proteins. Here, the E helix is yellow, the F helix is blue, and calcium is represented by the green sphere.

EF hand

15.3.2 Calcium Activates the Regulatory Protein Calmodulin, Which Stimulates Many Enzymes and Transporters

Calmodulin (CaM), a 17-kd protein with four calcium-binding sites, serves as a calcium sensor in nearly all eukaryotic cells. Calmodulin is activated by the binding of Ca²⁺ when the cytosolic calcium level is raised above about 500 nM. Calmodulin is a member of the *EF-hand protein family.* The *EF hand* is a Ca²⁺-binding motif that consists of a helix, a loop, and a second helix. This motif, originally discovered in the protein parvalbumin, was named the EF hand because helices designated E and F in parvalbumin form the calcium-binding motif and because the two helices are positioned like the forefinger and thumb of the right hand (Figure 15.21). Seven oxygen atoms are coordinated to each Ca^{2+}: six from the protein and one from a bound water molecule.

Let us consider how calmodulin changes conformation in response to Ca^{2+} binding and how these conformational changes enable the calmodulin to interact with other proteins (Figure 15.22). In the absence of bound Ca^{2+}, calmodulin consists of two domains, each consisting of a pair of EF-hand motifs joined by a flexible helix. Many of the residues that typically participate in calcium binding are on the surface of these domains, oriented in a manner inappropriate for Ca^{2+} binding. On the binding of one Ca^{2+} to each EF hand, these units change conformation: the Ca^{2+}-binding sites turn inward to bind the Ca^{2+}, moving hydrophobic residues from the inside to the outside of the domains. These conformational changes generate hydrophobic patches on the surface of each domain that are suitable for interacting with other proteins.

The Ca^{2+}–calmodulin complex stimulates a wide array of enzymes, pumps, and other target proteins. Two targets are especially noteworthy: one that propagates the signal and another that abrogates it. *Calmodulin-dependent protein kinases (CaM kinases)* phosphorylate many different proteins. These enzymes regulate fuel metabolism, ionic permeability, neurotransmitter synthesis, and neurotransmitter release. The binding of Ca^{2+}-calmodulin to CaM kinase activates the enzyme and enables it to phosphorylate target proteins. In addition, the activated enzyme phosphorylates itself and is thus partly active even after Ca^{2+} concentration falls and calmodulin is released from the kinase. In essence, autophosphorylation of CaM kinase is the memory of a previous calcium pulse. The plasma membrane Ca^{2+}-*ATPase pump* is another important target of Ca^{2+}-calmodulin. Stimulation of the pump by Ca^{2+}-calmodulin drives the calcium level down to restore the low-calcium basal state, thus helping to terminate the signal.

Are there structural features common to all calmodulin target proteins? Comparisons of the amino acid sequences of calmodulin-binding domains

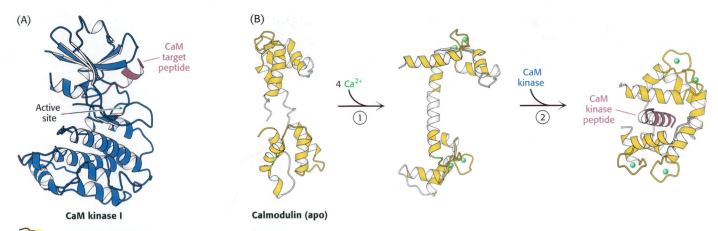

(A)

CaM
target
peptide

Active
site

CaM kinase I

(B)

4 Ca²⁺ ①

CaM
kinase ②

CaM
kinase
peptide

Calmodulin (apo)

FIGURE 15.23 Calmodulin binds to amphipathic α helices. (A) An amphipatic α helix (purple) in CaM kinase I is a target for calmodulin. (B) After calcium binding (1), the two halves of calmodulin clamp down around the target helix (2), binding it through hydrophobic and ionic interactions. In CaM kinase I, this interaction extracts a C-terminal α helix, allowing the enzyme to adopt an active conformation.

of target proteins suggests that *calmodulin recognizes positively charged, amphipathic α helices*. The results of structural studies fully support this conclusion and reveal the details of calmodulin–target interactions (Figure 15.23). The two domains of calmodulin surround the amphipathic helix and are linked to it by extensive hydrophobic and ionic interactions. The α helix linking the two EF hands folds back onto itself to facilitate binding to the target helix. How does this binding of calmodulin to its target lead to enzyme activation? In regard to CaM kinase I, calmodulin targets a peptide present at the carboxyl terminus of the kinase. This region interacts with a loop that is crucial for ATP binding and holds it in a conformation inappropriate for ATP binding. Calmodulin surrounds this peptide and extracts it, freeing the kinase active site to bind ATP and phosphorylate appropriate substrates.

15.4 SOME RECEPTORS DIMERIZE IN RESPONSE TO LIGAND BINDING AND SIGNAL BY CROSS-PHOSPHORYLATION

The 7TM receptors initiate signal-transduction pathways through changes in tertiary structure that are induced by ligand binding. A fundamentally different mechanism is utilized by a number of other classes of receptors. For these receptors, ligand binding leads to changes in quaternary structures—specifically, the formation of receptor dimers. As we shall see, receptor dimerization is crucial because protein kinase domains associated with the intracellular domains of the receptors are brought together in such a way that they can phosphorylate one another. Such *cross-phosphorylation* initiates further signaling.

We consider human growth hormone and its receptor as our first example. Growth hormone is a monomeric protein of 217 amino acids that forms a compact four-helix bundle structure (Figure 15.24). The growth-hormone receptor comprises 638 amino acids, divided into an extracellular domain of 250 amino acids, a single membrane-spanning helix, and an intracellular domain of 350 amino acids. In the absence of bound hormone, the receptor is present as a monomer. Growth hormone binds to the extracellular domain of the receptor. Remarkably, each monomeric hormone binds to two receptor molecules, thus promoting the formation of a dimer

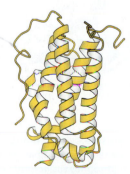

Human growth hormone

FIGURE 15.24 Human growth hormone structure. Human growth hormone forms a four-helix bundle.

Growth
hormone

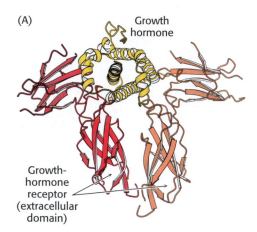

Growth-
hormone
receptor
(extracellular
domain)

(B)

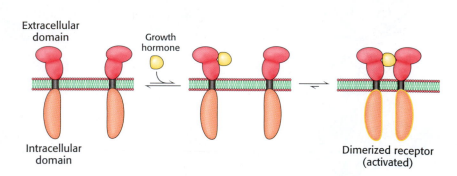

Extracellular
domain

Growth
hormone

Intracellular
domain

Dimerized receptor
(activated)

FIGURE 15.25 Binding of growth hormone leads to receptor dimerization.
(A) A single growth-hormone molecule (yellow) interacts with the extracellular domain of
two receptors (red and orange). (B) The binding of one hormone molecule to two
receptors leads to the formation of a receptor dimer. Dimerization is a key step in this
signal-transduction pathway.

of the receptor (Figure 15.25). The binding between the hormone and the
receptors is highly cooperative; once a hormone has bound to a single receptor molecule, the binding of the second receptor is highly favored.

This process takes place outside the cell. Dimerization of the extracellular domains of the receptor brings together the intracellular domains as
well. Associated with each intracellular domain is a molecule of a protein
kinase termed *Janus kinase 2 (JAK2)* in an unactivated form. Janus kinases
have modular structures consisting of four previously described domains
(Figure 15.26).

ERM   SH2   Protein kinase-like   Protein kinase

FIGURE 15.26 Janus kinase domain structure. A Janus kinase (JAK) includes four
recognized domains: an ERM domain that favors interactions with membranes, an SH2
domain that binds phosphotyrosine-containing peptides, and two domains homologous to
protein kinases. Only the second protein kinase domain appears to be enzymatically functional.

The carboxyl terminus is a protein kinase domain; its amino acid sequence
and known biochemical properties suggest that this domain functions as a
tyrosine kinase. Adjacent to this domain is a second region that is clearly
homologous to a protein kinase, although several key residues have been
changed and the biochemical activity of this domain is not well established.
Indeed, this pair of kinase-like domains accounts for the name of these proteins; Janus is the two-headed Roman god of gateways. At the amino terminus is a 300-amino-acid domain, called the ERM domain, that helps anchor JAK2 to membranes. In between this domain and the kinase domains
is an *SH2 domain* (SH for Src homology; Section 15.4). SH2 domains are
100-amino-acid domains that bind peptides containing phosphotyrosine
(Figure 15.27). The domains bind phosphotyrosine through interactions
with conserved arginine residues among other residues.

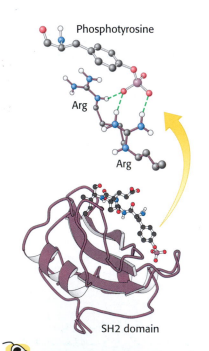

Phosphotyrosine

Arg

Arg

SH2 domain

**FIGURE 15.27 Recognition of
phosphotyrosine by SH2 domains.**
The structure of an SH2 domain (purple)
bound to a phosphotyrosine-containing
peptide. The hydrogen-bonding
interactions between the phosphotyrosine
residue and two arginine residues are
shown; interactions with other residues
are omitted for clarity.

**STRUCTURAL INSIGHTS, SH2 Domains: An Example of Modular Regulatory
Domains,** *takes a close look at phosphotyrosine–SH2 domain interactions and the
diverse ways they can affect protein function.*

Dimerization of the growth-hormone receptors brings together the
JAK2 proteins associated with each intracellular domain, apparently delivering a key loop (termed the *activation loop*) of one kinase domain into the
active site of the kinase bound to the other receptor, which results in cross-

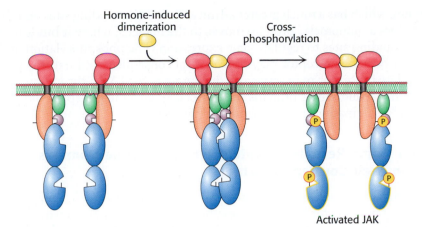

Hormone-induced
dimerization

Cross-
phosphorylation

Activated JAK

FIGURE 15.28 Cross-phosphorylation of JAKs induced by receptor dimerization. The binding of growth hormone leads to receptor dimerization, which brings two JAKs together in such a way that each phosphorylates key residues on the other. The activated JAKs remain bound to the receptor.

phosphorylation (Figure 15.28). How phosphorylation of JAK2 leads to its activation has not been directly established. However, the results of studies of other kinases reveal that the activation loop is in a conformation unsuitable for catalysis in the unphosphorylated form but changes to an active conformation when phosphoryl groups are added to key sites (Figure 15.29).

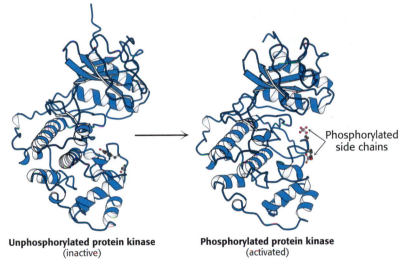

Phosphorylated
side chains

Unphosphorylated protein kinase
(inactive)

Phosphorylated protein kinase
(activated)

FIGURE 15.29 Activation of a protein kinase by phosphorylation. In the unphosphorylated state, a key loop is in a conformation unsuitable for catalysis. Phosphorylation (at two sites in the case shown) stabilizes an active conformation.

When activated by cross-phosphorylation, JAK2 can phosphorylate other substrates. In the present case, at least two important proteins are phosphorylated—a regulator of gene expression called STAT5 (STAT for *signal transducers and activators of transcription*) and the growth-hormone receptor itself. What are the consequences of the phosphorylation of these two proteins? STAT5 is phosphorylated on a tyrosine residue near the carboxyl terminus of the protein. The phosphotyrosine residue binds to an SH2 domain of another STAT5 molecule. Reciprocal interactions lead to the formation of a stable STAT dimer (Figure 15.30). The dimerized STAT

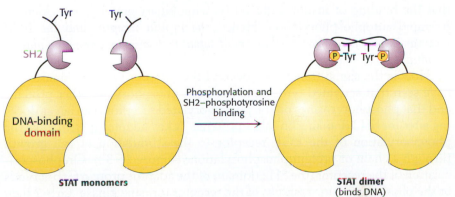

Tyr   Tyr

SH2

Phosphorylation and
SH2–phosphotyrosine
binding

P-Tyr Tyr-P

DNA-binding
domain

STAT monomers

STAT dimer
(binds DNA)

FIGURE 15.30 Phosphorylation-induced dimerization of STAT proteins. The phosphorylation of a key tyrosine residue on each STAT protein leads to an interaction between the phosphotyrosine and an SH2 domain on another STAT monomer. The STAT dimer produced by these reciprocal interactions has a high affinity for specific DNA sequences and is able to alter gene expression after binding to DNA.

Epidermal growth factor (EGF)

FIGURE 15.31 Structure of epidermal growth factor. This protein growth factor is stabilized by three disulfide bonds.

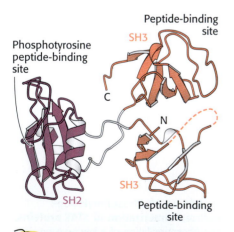

FIGURE 15.32 Structure of Grb-2, an adaptor protein. Grb-2 consists of two SH3 domains and a central SH2 domain. The SH2 domain binds to phosphotyrosine resides on an activated receptor while the SH3 domains bind proline-rich regions on other proteins such as Sos.

protein, which has a much greater affinity for specific binding sites on DNA than does a monomeric protein, moves to the nucleus, where it binds to the DNA binding sites to regulate gene expression. The phosphorylation of the growth-hormone receptor may have several consequences. First, the phosphorylated receptor may serve as a docking site for JAK2 through its SH2 domain. Second, other proteins may associate with the phosphorylated receptor, participating in other signaling pathways.

15.4.1 Some Receptors Contain Tyrosine Kinase Domains Within Their Covalent Structures

Growth factors such as *insulin, epidermal growth factor (EGF),* and *platelet-derived growth factor* bind to the extracellular domains of transmembrane receptors that have tyrosine kinase domains present within their intracellular domains. For these proteins, which are found in multicellular organisms but not in yeast, genes encoding extracellular domains and the signaling kinases fused in the course of evolution. These *receptor tyrosine kinases (RTKs)* signal by mechanisms quite similar to those discussed for the pathway initiated by the growth-hormone receptor.

Consider, for example, *epidermal growth factor,* a 6-kd polypeptide that stimulates the growth of epidermal and epithelial cells (Figure 15.31). This 53-residue growth factor is produced by the cleavage of an *EGF precursor,* a large transmembrane protein. Such processing, which is common for growth factors and hormones, is reminiscent of the processing of zymogens into active enzymes (Section 10.5). The first step in the signal-transduction pathway is the binding of EGF to the *epidermal growth factor receptor,* a single polypeptide chain consisting of 1186 residues. The receptor tyrosine kinase is monomeric and enzymatically inactive in the absence of the growth factor. *The binding of EGF to the extracellular domain causes the receptor to dimerize and undergo cross-phosphorylation and activation.*

The insulin receptor is a disulfide-bonded dimer of αβ pairs even when insulin is not bound. Nevertheless, insulin is still required for the activation of the kinase, demonstrating that dimerization is necessary but not sufficient for activation. The binding of the growth factor must convert the subunits of the dimer into a conformation that brings appropriate tyrosine residues from one chain into the active site of the other chain so that cross-phosphorylation can take place.

An elegant experiment demonstrated the commonality of the receptor tyrosine kinase signaling mechanism. The EGF receptor and the insulin receptor both contain intrinsic tyrosine kinases. Do these receptors transfer information across the membrane in the same way? This question was answered by synthesizing a gene that encoded a *chimeric receptor*—the extracellular part came from the insulin receptor, and the membrane-spanning and cytosolic parts came from the EGF receptor. The striking result was that the binding of insulin induced tyrosine kinase activity, as evidenced by rapid autophosphorylation. Hence, *the insulin receptor and the EGF receptor employ a common mechanism of signal transmission across the plasma membrane.*

How is the signal transferred beyond the receptor tyrosine kinase? We have seen that activated tyrosine kinases can phosphorylate other proteins and that phosphotyrosines on the phosphorylated receptors can act as docking sites for SH2 domains on other proteins. A key *adaptor protein* links the phosphorylation of the EGF receptor to the stimulation of cell growth through a chain of protein phosphorylations (Figure 15.32). On phosphorylation of the receptor, the SH2 domain of the adaptor protein Grb-2 binds to the phosphotyrosine residues of the receptor tyrosine kinase. Grb-2 then recruits a protein called Sos, which interacts with Grb-2 through two *SH3 domains*, domains that bind proline-rich stretches of polypeptide and, like SH2 domains, are recurring domains that mediate protein–protein interactions. Sos, in turn, binds to and activates Ras, a very prominent signal-transduction component that we will consider in Section 15.4.2. Finally, Ras, in its activated form, binds to other components of the molecular circuitry leading to the activation of the specific serine-threonine protein kinases that phosphorylate specific targets that promote cell growth. We see here another example of how a signal-transduction pathway is constructed. Specific protein–protein interactions (through SH2, SH3, and other domains not considered here) link the original ligand–binding event to the final result—stimulation of cell growth.

15.4.2 Ras, Another Class of Signaling G Protein

We now turn our attention to another important family of signal proteins, the *small G proteins,* or small GTPases. This large superfamily of proteins—grouped into subfamilies called Ras, Rho, Arf, Rab, and Ran—plays a major role in a host of cell functions including growth, differentiation, cell motility, cytokinesis, and transport of materials throughout the cell (Table 15.3). Like their relatives the heterotrimeric G proteins (Section 15.1.2), the small G proteins cycle between an active GTP-bound form and an inactive GDP-bound form. They differ from the heterotrimeric G proteins in being smaller (20–25 kd versus 30–35 kd) and monomeric. Nonetheless, the two families are related by divergent evolution, and small G proteins have many key mechanistic and structural motifs in common with the G_α subunit of the heterotrimeric G proteins.

In their activated GTP-bound form, small G proteins such as Ras stimulate cell growth and differentiation. Recall that Sos is the immediate upstream link to Ras in the circuit conveying the EGF signal. How does Sos activate Ras? Sos binds to Ras, reaches into the nucleotide-binding pocket, and opens it up, allowing GDP to escape and GTP to enter in its place (Figure 15.33). This process is

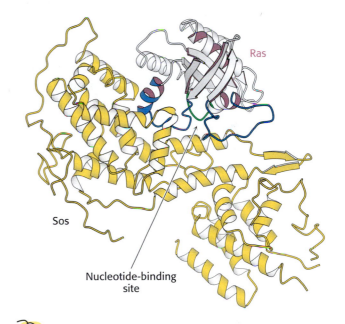

FIGURE 15.33 Structure of Sos, a guanine-nucleotide exchange factor. Sos (yellow) binds to Ras and opens up its nucleotide-binding site, allowing GDP to escape and GTP to bind. In the GTP-bound form, Ras can bind to and activate other proteins, including protein kinases.

TABLE 15.3   Ras superfamily of GTPases

Subfamily	Function
Ras	Regulates cell growth through serine-threonine protein kinases
Rho	Reorganizes cytoskeleton through serine-threonine protein kinases
Arf	Activates the ADP-ribosyltransferase of the cholera toxin A subunit; regulates vesicular trafficking pathways; activates phospholipase D
Rab	Plays a key role in secretory and endocytotic pathways
Ran	Functions in the transport of RNA and protein into and out of the nucleus

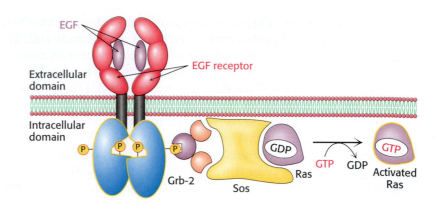

FIGURE 15.34 EGF signaling pathway.
The binding of epidermal growth factor (EGF) to its receptor leads to cross-phosphorylation of the receptor. The phosphorylated receptor binds Grb-2, which, in turn, binds Sos. Sos stimulates the exchange of GTP for GDP in Ras. Activated Ras binds to and stimulates protein kinases (not shown).

presumably analogous to the stimulation of nucleotide exchange in heterotrimeric G proteins by activated 7TM receptors, a process for which structural details are not yet available. Sos is referred to as a *guanine-nucleotide exchange factor (GEF)*. Thus, the binding of EGF to the EGF receptor leads to the conversion of Ras into its GTP form through the intermediacy of Grb-2 and Sos (Figure 15.34).

Like the G_α protein, Ras possesses an intrinsic GTPase activity, which serves to terminate the signal and return the system to the inactive state. This activity is slow but is augmented by helper proteins termed *GTPase-activating proteins (GAPs)*. The guanine-nucleotide exchange factors and the GTPase-activating proteins allow the G-protein cycle to proceed with rates appropriate for a balanced level of downstream signaling.

15.5 DEFECTS IN SIGNALING PATHWAYS CAN LEAD TO CANCER AND OTHER DISEASES

In light of their complexity, it comes as no surprise that signal-transduction pathways occasionally fail, leading to pathological or disease states. Cancer, a set of diseases characterized by uncontrolled or inappropriate cell growth, is strongly associated with defects in signal-transduction proteins. Indeed, the study of cancer, particularly cancer caused by certain viruses, has contributed greatly to our understanding of signal-transduction proteins and pathways.

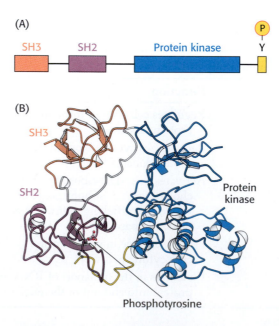

FIGURE 15.35 Src structure.
(A) Cellular Src includes an SH3 domain, an SH2 domain, a protein kinase domain, and a carboxyl-terminal tail that includes a key tyrosine residue. (B) Structure of c-Src in an inactivated form with the key tyrosine residue phosphorylated. The phosphotyrosine residue is bound in the SH2 domain; the linker between the SH2 domain and the protein kinase domain is bound by the SH3 domain. These interactions hold the kinase domain in an inactive conformation.

For example, Rous sarcoma virus is a retrovirus that causes sarcoma (a cancer of tissues of mesodermal origin such as muscle or connective tissue) in chickens. In addition to the genes necessary for viral replication, this virus carries a gene termed v-src. The v-src gene is an *oncogene*; it leads to the transformation of susceptible cell types. The protein encoded by v-src is a protein tyrosine kinase that includes SH2 and SH3 domains (Figure 15.35). Indeed, the names of these domains derive from the fact that they are Src *homology* domains. The v-Src protein is similar in amino acid sequence to a protein normally found in chicken muscle cells referred to as c-Src (for cellular Src). The c-src gene does not induce cell transformation and is termed a *proto-oncogene*. The protein that it encodes is a signal-transduction protein that regulates cell growth. As we shall see, small differences in the amino acid sequences between the proteins encoded by the proto-oncogene and the oncogene are responsible for the oncogene product being trapped in the "on" position.

An examination of the structure of c-Src in an inactive conformation reveals an intricate relation between the three major domains. The SH3 domain lies nearest the amino terminus, followed by the SH2 domain and then the kinase domain. There is also an extended carboxyl-terminal stretch that includes a phosphotyrosine residue. The phosphotyrosine residue is bound within the SH2 domain, whereas the linker between the SH2 domain and the kinase domain is bound by the SH3 domain. These interactions hold the kinase domain in an inactive conformation. The Src protein in this form can be activated by three distinct processes (Figure 15.36): (1) the phosphotyrosine residue bound in the SH2 pocket can be displaced by another phosphotyrosine-containing polypeptide with a higher affinity for this SH2 domain, (2) the phosphoryl group on the tyrosine residue can be removed by a phosphatase, and (3) the linker can be displaced from the SH3 domain by a polypeptide with a higher affinity for this SH3 domain. Thus, Src responds to the presence of one of a set of distinct signals. The amino acid sequence of the viral oncogene is more that 90% identical with its cellular counterpart. Why does it have such a different biological activity? The C-terminal 19 amino acids of c-Src are replaced by a completely different stretch of 11 amino acids, and this region lacks the key tyrosine residue that is phosphorylated in inactive c-Src. Since the discovery of Src, many other mutated protein kinases have been identified as oncogenes.

How did the Rous sarcoma virus acquire the mutated version of *src*? In an infection, a viral genome may pick up a gene from its host in such a way that the region encoding the last few amino acids is missing.

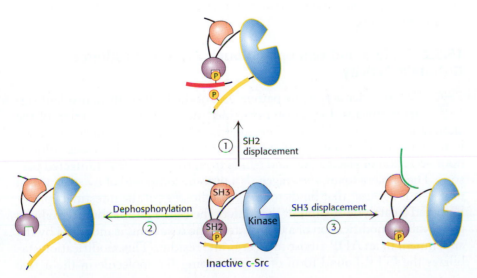

① SH2 displacement

Dephosphorylation ②

SH3 displacement ③

SH3
SH2
Kinase

Inactive c-Src

FIGURE 15.36 Activation pathways for c-Src. Inactive c-Src can be activated by one of at least three distinct pathways: (1) displacement of the SH2 domain, (2) dephosphorylation, or (3) displacement of the SH3 domain.

Such a modified gene may have given the Rous sarcoma virus a selective advantage because it will have favored viral growth when introduced with the virus into a host cell.

Impaired GTPase activity in a regulatory protein also can lead to cancer. Indeed, *ras* is one of the genes most commonly mutated in human tumors. Mammalian cells contain three 21-kd Ras proteins (H-, K-, and N-Ras) that cycle between GTP and GDP forms. The most common mutations in tumors lead to a loss of the ability to hydrolyze GTP. Thus, the Ras protein is trapped in the "on" position and continues to stimulate cell growth.

15.5.1 Protein Kinase Inhibitors May Be Effective Anticancer Drugs

The widespread occurrence of over active protein kinases in cancer cells suggests that molecules that inhibit these enzymes might act as antitumor agents. Recent results have dramatically supported this concept. More that 90% of patients with chronic myologenous leukemia (CML) show a specific chromosomal defect in affected cells (Figure 15.37). The translocation of genetic material between chromosomes 9 and 22 causes the c-*abl* gene, which encodes a tyrosine kinase, to be inserted into the *bcr* gene on chromosome 22. The result is the production of a fusion protein called Bcr-Abl that consists primarily of sequences for the c-Abl kinase. However, the *bcr-abl* gene is not regulated appropriately; it is expressed at higher levels than the gene encoding the normal c-Abl kinase. In addition, the Bcr-Abl protein may have regulatory properties that are subtly different from those of the c-Abl kinase itself. Thus, leukemia cells express a unique target for chemotherapy. Recent clinical trials of a specific inhibitor of the Bcr-Abl kinase have shown dramatic results; more than 90% of patients responded well to the treatment. This approach to cancer chemotherapy is fundamentally distinct from most approaches, which target cancer cells solely on the basis of their rapid growth, leading to side effects because normal rapidly growing cells also are affected. *Thus, our understanding of signal-transduction pathways is leading to conceptually new disease treatments.*

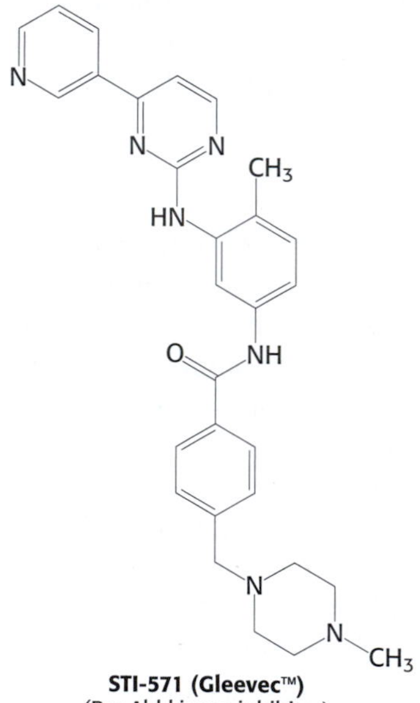

STI-571 (Gleevec™)
(Bcr-Abl kinase inhibitor)

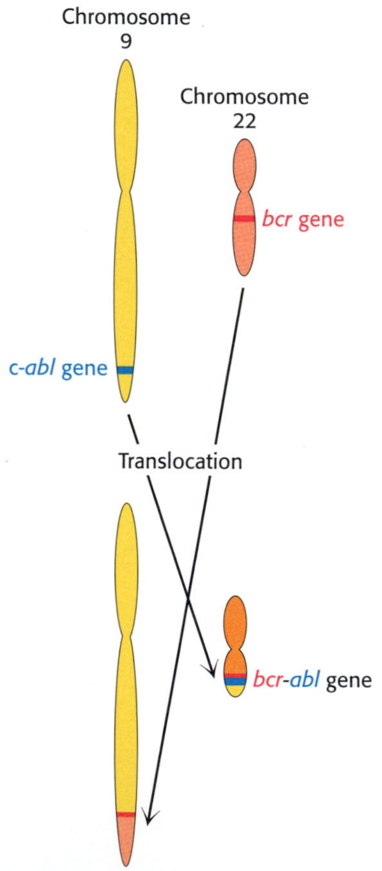

15.5.2 Cholera and Whooping Cough Are Due to Altered G-Protein Activity

We consider here some pathologies of the G-protein-dependent signal pathways. Let us first consider the mechanism of action of the cholera toxin, secreted by the intestinal bacterium *Vibrio cholera*. Cholera is an acute diarrheal disease that can be life threatening. It causes voluminous secretion of electrolytes and fluids from the intestines of infected persons. The cholera toxin, *choleragen*, is a protein composed of two functional units—a B subunit that binds to G_{M1} gangliosides of the intestinal epithelium and a catalytic A subunit that enters the cell. The A subunit catalyzes the covalent modification of a $G_{\alpha s}$ protein: the α subunit is modified by the attachment of an ADP-ribose to an arginine residue. This modification stabilizes the GTP-bound form of $G_{\alpha s}$, trapping the molecule in the active

FIGURE 15.37 Formation of the *bcr-abl* gene by translocation. In chronic myologenous leukemia, parts of chromosomes 9 and 22 are reciprocally exchanged, causing the *bcr* and *abl* genes to fuse. The protein kinase encoded by the *bcr-abl* gene is expressed at higher levels in cells having this translocation than is the c-*abl* gene in normal cells.

TABLE 15.4   Diseases of heterotrimeric G proteins

Disease
Excessive signaling
Cholera
Cancer (adenoma) of pituitary and thyroid
Cancer (adenoma) of adrenal and ovary
Essential hypertension
Deficient signaling
Night blindness
Pseudohypoparathyroidism type Ib
Pertussis

Source: After Z. Farfel, H. R. Bourne, and T. Iiri. *N. Engl. J. Med.* 340(1999):1012.

conformation. The active G protein, in turn, continuously activates protein kinase A. PKA opens a chloride channel (a CFTR channel; Section 13.3) and inhibits the Na^+–H^+ exchanger by phosphorylation. The net result of the phosphorylation of these channels is an excessive loss of NaCl and the loss of large amounts of water into the intestine. Patients suffering from cholera for 4 to 6 days may pass as much as twice their body weight in fluid. Treatment consists of rehydration with a glucose–electrolyte solution.

Whereas cholera is a result of a G protein trapped in the active conformation, causing the signal-transduction pathway to be perpetually stimulated, pertussis, or whooping cough, is a result of the opposite situation. Pertussis toxin also adds an ADP-ribose moiety,— in this case, to a $G_{\alpha i}$ protein, a G_α protein that inhibits adenyl cyclase, closes Ca^{2+} channels, and opens K^+ channels. The effect of this modification, however, is to lower the G protein's affinity for GTP, effectively trapping it in the "off" conformation. The pulmonary symptoms have not yet been traced to a particular target of the $G_{\alpha i}$ protein. Pertussis toxin is secreted by *Bordetella pertussis,* the bacterium responsible for whooping cough.

Cholera and pertussis are but two examples of diseases caused by defects in G proteins. Table 15.4 lists others. In light of the fact that G proteins relay signals for more than 500 receptors, it is likely that this list will continue to grow.

15.6 RECURRING FEATURES OF SIGNAL-TRANSDUCTION PATHWAYS REVEAL EVOLUTIONARY RELATIONSHIPS

Many features of signal-transduction pathways are ancient. For example, cAMP signals the need for energy in prokaryotes as well as eukaryotes, although the mechanisms for detecting cAMP are different. Similarly, the GTP-binding proteins—the G_α subunits of the heterotrimeric G proteins and the members of the Ras family—are part of an ancient superfamily of evolutionarily related proteins. Other members of this superfamily are proteins that cycle between ATP- and ADP-bound forms; these proteins function in ATP synthesis (Section 18.4.5) and in generating molecular motion (Chapter 34). The superfamily also includes proteins taking part in protein synthesis (Section 29.4.2). The key feature of these proteins is that they undergo significant conformational changes on binding nucleoside triphosphates and hydrolyzing them to nucleoside

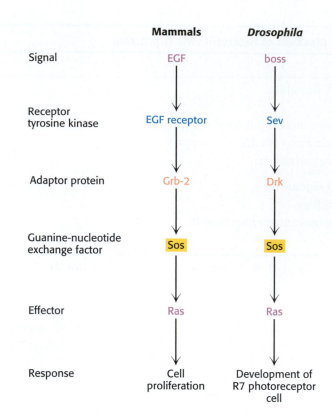

FIGURE 15.38 Pathway conservation. A pathway homologous to the mammalian EGF signal-transduction pathway functions in *Drosophila* to control the development of a specific photoreceptor cell in the eye.

diphosphates. These proteins can thus function as molecular "on–off" switches. A domain with this ability must have arisen early in evolution and been adapted to meet a range of biochemical needs since.

Other proteins crucial to signal-transduction pathways arose much later. For example, the eukaryotic protein kinases are one of the largest protein families in all eukaryotes and yet appear to be absent in prokaryotes. The evolution of the eukaryotic protein kinase domain appears to have been an important biochemical step in the appearance of eukaryotes and the subsequent development of multicellular organisms.

Entire signaling pathways have been conserved between organisms. For example, a key pathway in eye development in *Drosophila* is completely analogous to the EGF pathways in human beings (Figure 15.38). Thus, the wiring of this growth-control pathway is at least 800 million years old.

SUMMARY

The highly specific binding of signal molecules, many of which are hormones and growth factors, to receptor molecules initiates the signal-transduction cascade. Secondary messengers carry the signal inside the cell and often use protein phosphorylation as a signaling device.

Seven-Transmembrane-Helix Receptors Change Conformation in Response to Ligand Binding and Activate G Proteins

Seven-transmembrane-helix receptors operate in conjunction with heterotrimeric G proteins. The binding of hormone to a 7TM receptor triggers the exchange of GTP for GDP bound to the α subunit of the

G protein. G_α proteins can transmit information in a number of ways. $G_{\alpha s}$-GTP activates adenylate cyclase, an integral membrane protein that catalyzes the synthesis of cAMP. Cyclic AMP then activates protein kinase A by binding to its regulatory subunit, thus unleashing its catalytic chains. PKA, a multifunctional kinase, alters the activity of many target proteins by phosphorylating serine and threonine residues.

The Hydrolysis of Phosphatidyl Inositol Bisphosphate by Phospholipase C Generates Two Messengers

The phosphoinositide cascade is mediated by 7TM receptors and $G_{\alpha q}$ proteins. The receptor-triggered activation of phospholipase C generates two intracellular messengers by hydrolysis of phosphatidyl inositol 4,5-bisphosphate. Inositol trisphosphate opens calcium channels in the endoplasmic and sarcoplasmic reticulum membranes, leading to an elevated level of Ca^{2+} in the cytosol. Diacylglycerol activates protein kinase C, which phosphorylates serine and threonine residues in target proteins.

Calcium Ion Is a Ubiquitous Cytosolic Messenger

Calcium ion acts by binding to calmodulin and other calcium sensors. Calmodulin contains four calcium-binding modules called EF hands that recur in other proteins. Ca^{2+}-calmodulin activates target proteins by binding to positively charged amphipathic helices.

Some Receptors Dimerize in Response to Ligand Binding and Signal by Cross-Phosphorylation

Some ligands induce dimerization of the receptors to which they bind. Such a receptor contains an extracellular domain that binds the ligand, a transmembrane region, and a cytosolic domain that either binds or contains a protein kinase. The growth-hormone receptor participates in an example of this type of signal-transduction pathway. Dimerization of the receptor activates Janus kinase 2, a protein kinase associated with the intracellular part of the receptor. The kinase, in turn, phosphorylates and activates a transcription factor called STAT5.

Intrinsic tyrosine kinases are covalently incorporated in the intracellular domains of some receptors, such as epidermal growth factor receptor and the insulin receptor. When such receptor tyrosine kinases dimerize, cross-phosphorylation occurs. The phosphorylated tyrosines in activated receptor tyrosine kinases serve as docking sites for SH2 domains present in numerous signaling proteins and permit further propagation of the signal. A prominent component of such pathways is the small GTPase Ras. The Ras protein, like the G_α subunit, cycles between an inactive form bound to GDP and an active form bound to GTP.

Defects in Signaling Pathways Can Lead to Cancer and Other Diseases

If the genes encoding components of the signal-transduction pathways are altered by mutation, pathological conditions, most notably cancer, may result. In their mutated form, these genes are called oncogenes. The normal counterparts are called proto-oncogenes and function in pathways that control cell growth and replication. Mutated versions of *ras* are frequently found in human cancers.

Recurring Features of Signal-Transduction Pathways Reveal Evolutionary Relationships

Many aspects of signal transduction are ancient and appear throughout the kingdoms of life. G proteins of various classes are employed as molecular switches in a host of biochemical processes.

KEY TERMS

ligand (p. 396)

primary messenger (p. 396)

second messenger (p. 396)

protein kinase (p. 397)

protein phosphatase (p. 397)

seven-transmembrane-helix (7TM) receptor (p. 398)

G protein (p. 399)

adenylate cyclase (p. 399)

G_{α} (p. 399)

$G_{\beta\gamma}$ (p. 400)

G-protein-coupled receptor (GCPR) (p. 401)

desensitization (adaptation) (p. 403)

protein kinase A (PKA) (p. 403)

phosphoinositide cascade (p. 404)

phospholipase C (p. 404)

protein kinase C (PKC) (p. 406)

pseudosubstrate sequence (p. 406)

calmodulin (CaM) (p. 409)

EF hand (p. 409)

calmodulin-dependent protein (CaM) kinase (p. 410)

SH2 domain (p. 412)

receptor tyrosine kinase (RTK) (p. 414)

SH3 domain (p. 415)

small G protein (p. 415)

oncogene (p. 417)

proto-oncogene (p. 417)

SELECTED READINGS

Where to start

Scott, J. D., and Pawson, T., 2000. Cell communication: The inside story. *Sci. Am.* 282(6):7279.

Pawson, T., 1995. Protein modules and signalling networks. *Nature* 373:573–580.

Hurley, J. H., and Grobler, J. A., 1997. Protein kinase C and phospholipase C: Bilayer interactions and regulation. *Curr. Opin. Struct. Biol.* 7:557–565.

Okada, T., Ernst, O. P., Palczewski, K., and Hofmann, K. P., 2001. Activation of rhodopsin: New insights from structural and biochemical studies. *Trends Biochem. Sci.* 26:318–324.

Tsien, R. Y., 1992. Intracellular signal transduction in four dimensions: From molecular design to physiology. *Am. J. Physiol.* 263:C723–C728.

Loewenstein, W. R., 1999. *Touchstone of Life : Molecular Information, Cell Communication, and the Foundations of Life.* Oxford University Press.

G proteins and 7TM receptors

Palczewski, K., Kumasaka, T., Hori, T., Behnke, C. A., Motoshima, H., Fox, B. A., Le Trong, I., Teller, D. C., Okada, T., Stenkamp, R. E., Yamamoto, M., and Miyano, M., 2000. Crystal structure of rhodopsin: A G protein-coupled receptor. *Science* 289:739–745.

Lefkowitz, R. J., 2000. The superfamily of heptahelical receptors. *Nat. Cell Biol.* 2:E133–E136.

Bourne, H. R., Sanders, D. A., and McCormick, F., 1991. The GTPase superfamily: Conserved structure and molecular mechanism. *Nature* 349:117–127.

Lambright, D. G., Noel, J. P., Hamm, H. E., and Sigler, P. B., 1994. Structural determinants for activation of the alpha-subunit of a heterotrimeric G protein. *Nature* 369:621–628.

Noel, J. P., Hamm, H. E., and Sigler, P. B., 1993. The 2.2 Å crystal structure of transducin-alpha complexed with GTP gamma S. *Nature* 366:654–663.

Sondek, J., Lambright, D. G., Noel, J. P., Hamm, H. E., and Sigler, P. B., 1994. GTPase mechanism of G proteins from the 1.7-Å crystal structure of transducin alpha-GDP-AIF$_4^-$. *Nature* 372:276–279.

Sondek, J., Bohm, A., Lambright, D. G., Hamm, H. E., and Sigler, P. B., 1996. Crystal structure of a G-protein beta gamma dimer at 2.1 Å resolution. *Nature* 379:369–374.

Wedegaertner, P. B., Wilson, P. T., and Bourne, H. R., 1995. Lipid modifications of trimeric G proteins. *J. Biol. Chem.* 270:503–506.

Farfel, Z., Bourne, H. R., and Iiri, T., 1999. The expanding spectrum of G protein diseases. *N. Engl. J. Med.* 340:1012–1020.

Bockaert, J., and Pin, J. P., 1999. Molecular tinkering of G protein-coupled receptors: An evolutionary success. *EMBO J.* 18:1723–1729.

cAMP cascade

Hurley, J. H., 1998. The adenylyl and guanylyl cyclase superfamily. *Curr. Opin. Struct. Biol.* 8:770–777.

Hurley, J. H., 1999. Structure, mechanism, and regulation of mammalian adenylyl cyclase. *J. Biol. Chem.* 274:7599–7602.

Tesmer, J. J., Sunahara, R. K., Gilman, A. G., and Sprang, S. R., 1997. Crystal structure of the catalytic domains of adenylyl cyclase in a complex with $G_{s\alpha} \cdot$ GTPγS. *Science* 278:1907–1916.

Smith, C. M., Radzio-Andzelm, E., Madhusudan, Akamine, P., and Taylor, S. S., 1999. The catalytic subunit of cAMP-dependent protein kinase: Prototype for an extended network of communication. *Prog. Biophys. Mol. Biol.* 71:313–341.

Taylor, S. S., Buechler, J. A., and Yonemoto, W., 1990. cAMP-dependent protein kinase: Framework for a diverse family of regulatory enzymes. *Annu. Rev. Biochem.* 59:971–1005.

Phosphoinositide cascade

Berridge, M. J., and Irvine, R. F., 1989. Inositol phosphates and cell signalling. *Nature* 341:197–205.

Berridge, M. J., 1993. Inositol trisphosphate and calcium signalling. *Nature* 361:315–325.

Essen, L. O., Perisic, O., Cheung, R., Katan, M., and Williams, R. L., 1996. Crystal structure of a mammalian phosphoinositide-specific phospholipase C δ. *Nature* 380:595–602.

Ferguson, K. M., Lemmon, M. A., Schlessinger, J., and Sigler, P. B., 1995. Structure of the high affinity complex of inositol trisphosphate with a phospholipase C pleckstrin homology domain. *Cell* 83:1037–1046.

Baraldi, E., Carugo, K. D., Hyvonen, M., Surdo, P. L., Riley, A. M., Potter, B. V., O'Brien, R., Ladbury, J. E., and Saraste, M., 1999. Structure of the PH domain from Bruton's tyrosine kinase in complex with inositol 1,3,4,5-tetrakisphosphate. *Structure Fold Des.* 7:449–460.

Calcium

Ikura, M., Clore, G. M., Gronenborn, A. M., Zhu, G., Klee, C. B., and Bax, A., 1992. Solution structure of a calmodulin-target peptide complex by multidimensional NMR. *Science* 256:632–638.

Kuboniwa, H., Tjandra, N., Grzesiek, S., Ren, H., Klee, C. B., and Bax, A., 1995. Solution structure of calcium-free calmodulin. *Nat. Struct. Biol.* 2:768–776.

Grynkiewicz, G., Poenie, M., and Tsien, R. Y., 1985. A new generation of Ca^{2+} indicators with greatly improved fluorescence properties. *J. Biol. Chem.* 260:3440–3450.

Kerr, R., Lev-Ram, V., Baird, G., Vincent, P., Tsien, R. Y., and Schafer, W. R., 2000. Optical imaging of calcium transients in neurons and pharyngeal muscle of *C. elegans*. *Neuron* 26:583–594.

Chin, D., and Means, A. R., 2000. Calmodulin: a prototypical calcium sensor. *Trends Cell Biol.* 10:322–328.

Dawson, A. P., 1997. Calcium signalling: How do IP3 receptors work? *Curr. Biol.* 7:R544–R547.

Protein kinases, including receptor tyrosine kinases

Riedel, H., Dull, T. J., Honegger, A. M., Schlessinger, J., and Ullrich, A., 1989. Cytoplasmic domains determine signal specificity, cellular routing characteristics and influence ligand binding of epidermal growth factor and insulin receptors. *EMBO J.* 8:2943–2954.

Taylor, S. S., Knighton, D. R., Zheng, J., Sowadski, J. M., Gibbs, C. S., and Zoller, M. J., 1993. A template for the protein kinase family. *Trends Biochem. Sci.* 18:84–89.

Sicheri, F., Moarefi, I., and Kuriyan, J., 1997. Crystal structure of the Src family tyrosine kinase Hck. *Nature* 385:602–609.

Waksman, G., Shoelson, S. E., Pant, N., Cowburn, D., and Kuriyan, J., 1993. Binding of a high affinity phosphotyrosyl peptide to the Src SH2 domain: Crystal structures of the complexed and peptide-free forms. *Cell* 72:779–790.

Schlessinger, J., 2000. Cell signaling by receptor tyrosine kinases. *Cell* 103:211–225.

Simon, M. A., 2000. Receptor tyrosine kinases: Specific outcomes from general signals. *Cell* 103:13–15.

Robinson, D. R., Wu, Y. M., and Lin, S. F., 2000. The protein tyrosine kinase family of the human genome. *Oncogene* 19:5548–5557.

Hubbard, S. R., 1999. Structural analysis of receptor tyrosine kinases. *Prog. Biophys. Mol. Biol.* 71:343–358.

Carter-Su, C., and Smit, L. S., 1998. Signaling via JAK tyrosine kinases: Growth hormone receptor as a model system. *Recent Prog. Horm. Res.* 53:61–82.

Ras

Milburn, M. V., Tong, L., deVos, A. M., Brunger, A., Yamaizumi, Z., Nishimura, S., and Kim, S. H., 1990. Molecular switch for signal transduction: Structural differences between active and inactive forms of protooncogenic Ras proteins. *Science* 247:939–945.

Boriack-Sjodin, P. A., Margarit, S. M., Bar-Sagi, D., and Kuriyan, J., 1998. The structural basis of the activation of Ras by Sos. *Nature* 394:337–343.

Maignan, S., Guilloteau, J. P., Fromage, N., Arnoux, B., Becquart, J., and Ducruix, A., 1995. Crystal structure of the mammalian Grb2 adaptor. *Science* 268:291–293.

Takai, Y., Sasaki, T., and Matozaki, T., 2001. Small GTP-binding proteins. *Physiol. Rev.* 81:153–208.

Cancer

Druker, B. J., Sawyers, C. L., Kantarjian, H., Resta, D. J., Reese, S. F., Ford, J. M., Capdeville, R., and Talpaz, M., 2001. Activity of a specific inhibitor of the BCR-ABL tyrosine kinase in the blast crisis of chronic myeloid leukemia and acute lymphoblastic leukemia with the Philadelphia chromosome. *N. Engl. J. Med.* 344:1038–1042.

Vogelstein, B., and Kinzler, K. W., 1993. The multistep nature of cancer. *Trends Genet.* 9:138–141.

Ellis, C. A., and Clark, G., 2000. The importance of being K-Ras. *Cell Signal.* 12:425–434.

Hanahan, D., and Weinberg, R. A., 2000. The hallmarks of cancer. *Cell* 100:57–70.

McCormick, F., 1999. Signalling networks that cause cancer. *Trends Cell Biol.* 9:M53–M56.

PROBLEMS

1. *Levels of amplification.* At which stages in the signaling pathway from epinephrine to cAMP does a significant amount of amplification occur? Answer the same question for the signaling pathways from human growth hormone to STAT5 and from EGF to Ras.

2. *Active mutants.* Some protein kinases are inactive unless they are phosphorylated on key serine or threonine residues. In some cases, active enzymes can be generated by mutating these serine or threonine residues to glutamate. Propose an explanation.

3. *In the pocket.* SH2 domains bind phosphotyrosine residues in deep pockets on their surfaces. Would you expect SH2 domains to bind phosphoserine or phosphothreonine with high affinity? Why or why not?

4. *Inhibitory constraints.* Many proteins in signal-transduction pathways are activated by the removal of an inhibitory constraint. Give two examples of this recurring mechanism.

5. *Pseudosubstrate mutant.* A mutated form of a protein kinase C isozyme that has three changes in amino acids in the pseudosubstrate sequence is isolated. What properties would you expect this mutated form to have?

6. *Antibodies mimicking hormones.* Antibodies have two identical antigen-binding sites. Remarkably, antibodies to the extracellular parts of growth-factor receptors often lead to the same cellular effects as does exposure to growth factors. Explain this observation.

7. *Facile exchange.* A mutated form of the α subunit of the heterotrimeric G protein has been identified; this form readily exchanges nucleotides even in the absence of an activated receptor. What effect would you expect this mutated α subunit to have on its signaling pathway?

8. *Blocking one side.* Human growth hormone binds simultaneously to two growth-hormone receptors through two different surfaces. Suppose sequence changes in a mutated form of growth hormone disrupt one interaction surface but do not affect the other. What properties would such a mutated hormone have? Why might it be useful?

9. *Diffusion rates.* Normally, rates of diffusion vary inversely with molecular weights; so smaller molecules diffuse faster than do larger ones. In cells, however, calcium ion diffuses more slowly than does cAMP. Propose a possible explanation.

10. *Awash with glucose.* Glucose is mobilized for ATP generation in muscle in response to epinephrine, which activates $G_{\alpha s}$. Cyclic AMP phosphodiesterase is an enzyme that converts cAMP into AMP. How would inhibitors of cAMP phosphodiesterase affect glucose mobilization in muscle?

Chapter Integration Problem

11. *Nerve growth factor pathway.* Nerve growth factor (NGF) binds to a protein tyrosine kinase receptor. The amount of diacylglycerol in the plasma membrane increases in cells express-

ing this receptor when treated with NGF. Propose a simple signaling pathway and identify the isoform of any participating enzymes. Would you expect the concentrations of any other common second messengers to increase on NGF treatment?

Mechanism Problem

12. *Distant relatives.* The structure of adenylate cyclase is similar to the structures of some types of DNA polymerases, suggesting that these enzymes derived from a common ancestor. Compare the reactions catalyzed by these two enzymes. In what ways are they similar?

Data Interpretation Problems

13. *Establishing specificity.* You wish to determine the hormone-binding specificity of a newly identified membrane receptor. Three different hormones, X, Y, and Z, were mixed with the receptor in separate experiments, and the percentage of binding capacity of the receptor was determined as a function of hormone concentration, as shown in graph A.

(a) What concentrations of each hormone yield 50% maximal binding?
(b) Which hormone shows the highest binding affinity for the receptor?

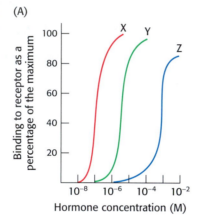

(A)

You next wish to determine whether the hormone–receptor complex stimulates the adenylate cyclase cascade. To do so, you measure adenylate cyclase activity as a function of hormone concentration, as shown in graph B.

(c) What is the relation between the binding affinity of the hormone–receptor complex and the ability of the hormone to enhance adenylate cyclase activity? What can you conclude about the mechanism of action of the hormone–receptor complex?
(d) Suggest experiments that would determine whether a $G_{\alpha s}$ protein is a component of the signal-transduction pathway.

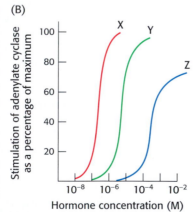

(B)

14. *Binding issues.* A scientist wishes to determine the number of receptors specific for a ligand X, which he has in both radioactive and nonradioactive form. In one experiment, he adds increasing amounts of the radioactive X and measures how much of it is bound to the cells. The result is shown as total activity in the adjoining graph. Next, he performs the same experiment, except that he includes a several hundredfold excess of nonradioactive X. This result is shown as nonspecific binding. The difference between the two curves is the specific binding.

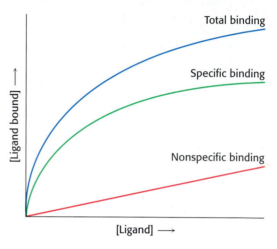

(a) Why is the total binding not an accurate representation of the number of receptors on the cell surface?
(b) What is the purpose of performing the experiment in the presence of excess nonradioactive ligand?
(c) What is the significance of the fact that specific binding attains a plateau?

15. *Counting receptors.* With the use of experiments such as those described in problems 13 and 14, it is possible to calculate the number of receptors in the cell membrane. Suppose that the specific activity of the ligand is 10^{12} cpm per millimole and that the maximal specific binding is 10^4 cpm per milligram of membrane protein. There are 10^{10} cells per milligram of membrane protein. Assume that one ligand binds per receptor. Calculate the number of receptor molecules present per cell.

Media Problem

16. *Grubby binding.* The **Structural Insights** module on SH2 domains describes some of the determinants of SH2 specificity and the ways in which SH2-phosphotyrosine binding can affect protein function. Given that the Src kinase SH2 domain binds Src phosphotyrosine 527, what effect do you think mutation of Glu 529 to Asn would have on the protein kinase activity of Src? Suppose you now obtained a second mutation within Src that reversed the effect of the first. Can you predict what that second mutation might be?

CHAPTER 16

Glycolysis and Gluconeogenesis

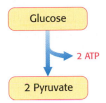

Glycolysis produces energy. Michael Johnson sprints to another victory in the 200-meter semifinals of the Olympics. Johnson, like anyone who sprints, requires a source of energy that can be rapidly accessed. The anaerobic metabolism of glucose—the process of glycolysis—provides such a source of energy for short, intense bouts of exercise. [(Left) Simon Bruty/Allsport.]

The first metabolic pathway that we encounter is *glycolysis*, an ancient pathway employed by a host of organisms. *Glycolysis is the sequence of reactions that metabolizes one molecule of glucose to two molecules of pyruvate with the concomitant net production of two molecules of ATP.* This process is anaerobic (i.e., it does not require O_2) inasmuch as it evolved before the accumulation of substantial amounts of oxygen in the atmosphere. Pyruvate can be further processed anaerobically (fermented) to lactate (*lactic acid fermentation*) or ethanol (*alcoholic fermentation*). Under aerobic conditions, pyruvate can be completely oxidized to CO_2, generating much more ATP, as will be discussed in Chapters 17 and 18.

Glucose can be synthesized from noncarbohydrate precursors, such as pyruvate and lactic acid, in the process of *gluconeogenesis*. Although glycolysis and gluconeogenesis have some of the same enzymes in common, the two pathways are not simply the reverse of each other. In particular, the highly exergonic, irreversible steps of glycolysis are bypassed in gluconeogenesis. Both pathways are stringently controlled by intercellular and intracellular signals, and they are reciprocally regulated so that glycolysis and gluconeogenesis do not take place simultaneously in the same cell to a significant extent.

Our understanding of glucose metabolism, especially glycolysis, has a rich history. Indeed, the development of biochemistry and the delineation of glycolysis went hand in hand. A key discovery was made by Hans Buchner and Eduard Buchner in 1897, quite by accident. The Buchners

OUTLINE

- **16.1 Glycolysis Is an Energy-Conversion Pathway in Many Organisms**
- **16.2 The Glycolytic Pathway Is Tightly Controlled**
- **16.3 Glucose Can Be Synthesized from Noncarbohydrate Precursors**
- **16.4 Gluconeogenesis and Glycolysis Are Reciprocally Regulated**

Glycolysis—
Derived from the Greek stem *glyk-*, "sweet," and the word *lysis*, "dissolution."

were interested in manufacturing cell-free extracts of yeast for possible therapeutic use. These extracts had to be preserved without the use of antiseptics such as phenol, and so they decided to try sucrose, a commonly used preservative in kitchen chemistry. They obtained a startling result: sucrose was rapidly fermented into alcohol by the yeast juice. The significance of this finding was immense. *The Buchners demonstrated for the first time that fermentation could take place outside living cells.* The accepted view of their day, asserted by Louis Pasteur in 1860, was that fermentation is inextricably tied to living cells. The chance discovery of the Buchners refuted this vitalistic dogma and opened the door to modern biochemistry. *Metabolism became chemistry.*

Studies of muscle extracts then showed that many of the reactions of lactic acid fermentation were the same as those of alcoholic fermentation. *This exciting discovery revealed an underlying unity in biochemistry.* The complete glycolytic pathway was elucidated by 1940, largely through the pioneering contributions of Gustav Embden, Otto Meyerhof, Carl Neuberg, Jacob Parnas, Otto Warburg, Gerty Cori, and Carl Cori. Glycolysis is also known as the *Embden-Meyerhof pathway.*

In our consideration of the glycolytic and gluconeogenic pathways, we shall examine the mechanisms of selected enzymes in some detail. Of particular interest will be the enzymes that play the most central roles in converting one type of chemical energy into another.

16.0.1 Glucose Is an Important Fuel for Most Organisms

Glucose is an important and common fuel. In mammals, glucose is the only fuel that the brain uses under nonstarvation conditions and the only fuel that red blood cells can use at all. Indeed, almost all organisms use glucose, and most that do process it in a similar fashion. Recall from Chapter 11 that there are many carbohydrates. Why is glucose instead of some other monosaccharide such a prominent fuel? We can speculate on the reasons. First, glucose is one of the monosaccharides formed from formaldehyde under prebiotic conditions, so it may have been available as a fuel source for primitive biochemical systems. Second, glucose has a low tendency, relative to other monosaccharides, to nonenzymatically glycosylate proteins. In their open-chain (carbonyl) forms, monosaccharides can react with the amino groups of proteins to form Schiff bases, which rearrange to form a more stable amino ketone linkage. Such nonspecifically modified proteins often do not function effectively. Glucose has a strong tendency to exist in the ring formation and, consequently, relatively little tendency to modify proteins. Recall that all the hydroxyl groups in the ring conformation of β-glucose are equatorial, contributing to this high relative stability (Section 11.12.12).

16.0.2 Fermentations Provide Usable Energy in the Absence of Oxygen

Although glycolysis is a nearly universal process, the fate of its end product, pyruvate, may vary in different organisms or even in different tissues. In the presence of oxygen, the most common situation in multicellular organisms and many unicellular ones, pyruvate is metabolized to carbon dioxide and water through the citric acid cycle and the electron-transport chain. In the absence of oxygen, fermentation generates a lesser amount of energy; pyruvate is converted, or fermented, into lactic acid in lactic acid fermentation or into ethanol in alcoholic fermentation (Figure 16.1). Lactic acid production takes place in skeletal muscle when energy needs outpace the ability to transport oxygen. Although we will consider only these two fer-

---

Enzyme—

A term coined by Friedrich Whilhelm Kühne in 1878 to designate catalytically active substances that had previously been called ferments. Derived from the Greek words *en,* "in," and *zyme,* "leaven."

Fermentation—

An ATP-generating process in which organic compounds act as both donors and acceptors of electrons. Fermentation can take place in the absence of O_2. Discovered by Louis Pasteur, who described fermentation as "la vie sans l'air" ("life without air").

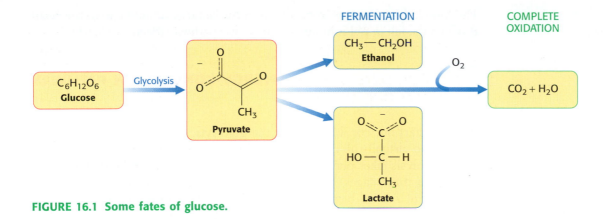

FIGURE 16.1 **Some fates of glucose.**

mentations, microorganisms are capable of generating a wide array of molecules as end points to fermentation (Table 16.1). Indeed, many food products are the result of fermentations. These foods include sour cream, yogurt, various cheeses, beer, wine, and sauerkraut.

Fermentations yield only a fraction of the energy available from the complete combustion of glucose. Why is a relatively inefficient metabolic pathway so extensively used? The fundamental reason is that oxygen is not required. The ability to survive without oxygen affords a host of living accommodations such as soils, deep water, and skin pores. Some organisms, called *obligate anaerobes*, cannot survive in the presence of O_2, a highly reactive compound. The bacterium *Clostridium perfringens*, the cause of gangrene, is an example of an obligate anaerobe. Other pathogenic obligate anaerobes are listed in Table 16.2.

Facultative anaerobes can function in the presence or absence of oxygen. For instance, organisms that live in the intertidal zone, such as the bivalve *Mytilus* (Figure 16.2), can function aerobically, using gills when they are under water and anaerobically when exposed to the air. Such organisms display habitat-dependent anaerobic functioning, or *habitat-dependent anaerobiosis*. Muscles in most animals display *activity-dependent anaerobiosis*, meaning that they can function anaerobically for short periods. For example, when animals perform bursts of intense exercise, their ATP needs rise faster than the ability of the body to provide oxygen to the muscle.

TABLE 16.1 Starting and ending points of various fermentations

Glucose	⟶ lactate
Lactate	⟶ acetate
Glucose	⟶ ethanol
Ethanol	⟶ acetate
Arginine	⟶ carbon dioxide
Pyrimidines	⟶ carbon dioxide
Purines	⟶ formate
Ethylene glycol	⟶ acetate
Threonine	⟶ propionate
Leucine	⟶ 2-alkylacetate
Phenylalanine	⟶ propionate

Note: The products of some fermentations are the substrates for others.

TABLE 16.2 Examples of pathogenic obligate anaerobes

Bacterium	Results of infection
Clostridium tetani	Tetanus (lockjaw)
Clostridium botulinum	Botulism (an especially severe type of food poisoning)
Clostridium perfringens	Gas gangrene (gas is produced as an end point of the fermentation, distorting and destroying the tissue)
Bartonella hensela	Cat scratch fever (flulike symptoms)
Bacteroides fragilis	Abdominal, pelvic, pulmonary, and blood infections

FIGURE 16.2 **The bivalve *Mytilus*.** These mussels, inhabitants of the intertidal zone, display habitat-dependent anaerobiosis. [Ed Reschke/Peter Arnold.]

The muscle functions anaerobically until the lactic acid builds up to the point at which the fall in pH inhibits the anaerobic pathway (Section 16.2.1).

16.1 GLYCOLYSIS IS AN ENERGY-CONVERSION PATHWAY IN MANY ORGANISMS

We now start our consideration of the glycolytic pathway. This pathway is common to virtually all cells, both prokaryotic and eukaryotic. In eukaryotic cells, glycolysis takes place in the cytosol. This pathway can be thought of as comprising three stages (Figure 16.3). Stage 1, which is the conversion of glucose into fructose 1,6-bisphosphate, consists of three steps: a phosphorylation, an isomerization, and a second phosphorylation reaction. *The strategy of these initial steps in glycolysis is to trap the glucose in the cell and form a compound that can be readily cleaved into phosphorylated three-carbon units.* Stage 2 is the cleavage of the fructose 1,6-bisphosphate into two three-carbon fragments. These resulting three-carbon units are readily interconvertible. In stage 3, ATP is harvested when the three-carbon fragments are oxidized to pyruvate.

16.1.1 Hexokinase Traps Glucose in the Cell and Begins Glycolysis

Glucose enters cells through specific transport proteins (Section 16.2.4) and has one principal fate: *it is phosphorylated by ATP to form glucose 6-phosphate.* This step is notable for two reasons: (1) glucose 6-phosphate cannot diffuse through the membrane, because of its negative charges, and (2) the addition of the phosphoryl group begins to destabilize glucose, thus facilitating its further metabolism. The transfer of the phosphoryl group from ATP to the hydroxyl group on carbon 6 of glucose is catalyzed by *hexokinase.*

Phosphoryl transfer is a fundamental reaction in biochemistry and is one that was discussed in mechanistic and structural detail earlier (Section 9.4). *Kinases* are enzymes that catalyze the transfer of a phosphoryl group from ATP to an acceptor. Hexokinase, then, catalyzes the transfer of a phosphoryl group from ATP to a variety of six-carbon sugars (*hexoses*), such as glucose and mannose. *Hexokinase, like adenylate kinase (Section 9.4.2) and all other kinases, requires Mg^{2+} (or another divalent metal ion such as Mn^{2+}) for activity.* The divalent metal ion forms a complex with ATP.

The results of x-ray crystallographic studies of yeast hexokinase revealed that the binding of glucose induces a large conformational change in the enzyme, analogous to the conformational changes undergone by NMP kinases on substrate binding (Section 9.4.3). Hexokinase consists of two lobes, which move toward each other when glucose is bound (Figure 16.4). On glucose binding, one lobe rotates 12 degrees with respect to the other, resulting in movements of the polypeptide backbone of as much as 8 Å. The cleft between the lobes closes, and the bound glucose becomes surrounded by protein, except for the hydroxyl group of carbon 6, which will accept the phosphoryl group from ATP. The closing of the cleft in hexokinase is a striking example of the role of *induced fit* in enzyme action (Section 8.3.2).

Stage 1 of glycolysis. The three steps of stage 1 begin with the phosphorylation of glucose by hexokinase.

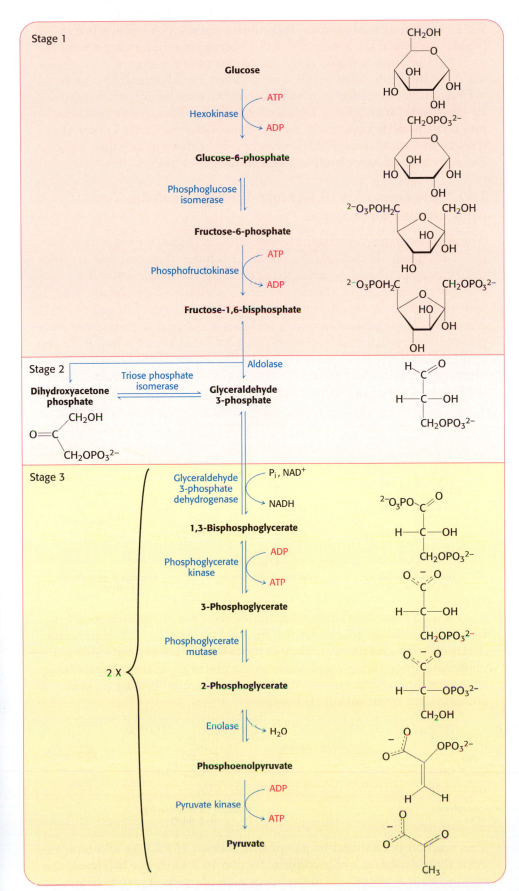

Stage 1

Glucose

ATP

Hexokinase

ADP

Glucose-6-phosphate

Phosphoglucose isomerase

Fructose-6-phosphate

ATP

Phosphofructokinase

ADP

Fructose-1,6-bisphosphate

Stage 2

Aldolase

Triose phosphate isomerase

Dihydroxyacetone phosphate ⇌ **Glyceraldehyde 3-phosphate**

Stage 3

2 X

Glyceraldehyde 3-phosphate dehydrogenase

P_i, NAD^+

NADH

1,3-Bisphosphoglycerate

ADP

Phosphoglycerate kinase

ATP

3-Phosphoglycerate

Phosphoglycerate mutase

2-Phosphoglycerate

Enolase

H_2O

Phosphoenolpyruvate

ADP

Pyruvate kinase

ATP

Pyruvate

FIGURE 16.3 Stages of glycolysis. The glycolytic pathway can be divided into three stages: (1) glucose is trapped and destabilized; (2) two interconvertible three-carbon molecules are generated by cleavage of six-carbon fructose; and (3) ATP is generated.

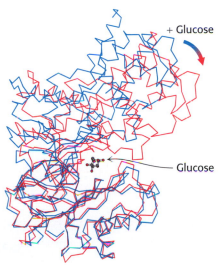

FIGURE 16.4 Induced fit in hexokinase. As shown in blue, the two lobes of hexokinase are separated in the absence of glucose. The conformation of hexokinase changes markedly on binding glucose, as shown in red. The two lobes of the enzyme come together and surround the substrate. [Courtesy of Dr. Thomas Steitz.]

The glucose-induced structural changes are significant in two respects. First, the environment around the glucose becomes much more nonpolar, which favors the donation of the terminal phosphoryl group of ATP. Second, as noted in Section 9.4.3, the substrate-induced conformational changes within the kinase enables it to discriminate against H_2O as a substrate. If hexokinase were rigid, a molecule of H_2O occupying the binding site for the $-CH_2OH$ of glucose would attack the γ phosphoryl group of ATP, forming ADP and P_i. In other words, a rigid kinase would necessarily also be an ATPase. It is interesting to note that other kinases taking part in glycolysis—pyruvate kinase, phosphoglycerate kinase, and phosphofructokinase—also contain clefts between lobes that close when substrate is bound, although the structures of these enzymes are different in other regards. *Substrate-induced cleft closing is a general feature of kinases.*

16.1.2 The Formation of Fructose 1,6-bisphosphate from Glucose 6-phosphate

The next step in glycolysis is the *isomerization of glucose 6-phosphate to fructose 6-phosphate.* Recall that the open-chain form of glucose has an aldehyde group at carbon 1, whereas the open-chain form of fructose has a keto group at carbon 2. Thus, the isomerization of glucose 6-phosphate to fructose 6-phosphate is a *conversion of an aldose into a ketose.* The reaction catalyzed by *phosphoglucose isomerase* includes additional steps because both glucose 6-phosphate and fructose 6-phosphate are present primarily in the cyclic forms. The enzyme must first open the six-membered ring of glucose 6-phosphate, catalyze the isomerization, and then promote the formation of the five-membered ring of fructose 6-phosphate.

Glucose 6-phosphate (G-6P) ⇌ **Glucose 6-phosphate** (Open chain form) ⇌ **Fructose 6-phosphate** (Open chain form) ⇌ **Fructose 6-phosphate** (F-6P)

A second phosphorylation reaction follows the isomerization step. *Fructose 6-phosphate is phosphorylated by ATP to fructose 1,6-bisphosphate* (F-1,6-BP). The prefix *bis-* in bisphosphate means that two separate monophosphate groups are present, whereas the prefix *di-* in diphosphate (as in adenosine diphosphate) means that two phosphate groups are present and are connected by an anhydride bond.

Fructose 6-phosphate (F-6P) + ATP —[Phosphofructokinase]→ **Fructose 1,6-bisphosphate** (F-1, 6-BP) + ADP + H^+

This reaction is catalyzed by *phosphofructokinase (PFK),* an allosteric enzyme that sets the pace of glycolysis (Section 16.2.1). As we will learn, this enzyme plays a central role in the integration of much of metabolism.

16.1.3 The Six-Carbon Sugar Is Cleaved into Two Three-Carbon Fragments by Aldolase

The second stage of glycolysis begins with the splitting of fructose 1,6-bisphosphate into *glyceraldehyde 3-phosphate (GAP)* and *dihydroxyacetone phosphate (DHAP)*. The products of the remaining steps in glycolysis consist of three-carbon units rather than six-carbon units.

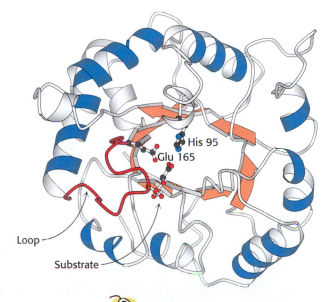

Stage 2 of glycolysis. Two three-carbon fragments are produced from one six-carbon sugar.

This reaction is catalyzed by *aldolase*. This enzyme derives its name from the nature of the reverse reaction, an aldol condensation. The reaction catalyzed by aldolase is readily reversible under intracellular conditions.

16.1.4 Triose Phosphate Isomerase Salvages a Three-Carbon Fragment

Glyceraldehyde 3-phosphate is on the direct pathway of glycolysis, whereas dihydroxyacetone phosphate is not. Unless a means exists to convert dihydroxyacetone phosphate into glyceraldehyde 3-phosphate, a three-carbon fragment useful for generating ATP will be lost. These compounds are isomers that can be readily interconverted: dihydroxyacetone phosphate is a ketose, whereas glyceraldehyde 3-phosphate is an aldose. The isomerization of these three-carbon phosphorylated sugars is catalyzed by *triose phosphate isomerase (TIM;* Figure 16.5). This reaction is rapid and reversible. At equilibrium, 96% of the triose phosphate is dihydroxyacetone phosphate. However, the reaction proceeds readily from dihydroxyacetone phosphate to glyceraldehyde 3-phosphate because the subsequent reactions of glycolysis remove this product.

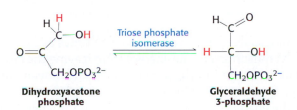

FIGURE 16.5 Structure of triose phosphate isomerase. This enzyme consists of a central core of eight parallel β strands (orange) surrounded by eight α helices (blue). This structural motif, called an αβ barrel, is also found in the glycolytic enzymes aldolase, enolase, and pyruvate kinase. Histidine 95 and glutamate 165, essential components of the active site of triose phosphate isomerase, are located in the barrel. A loop (red) closes off the active site on substrate binding.

Much is known about the catalytic mechanism of triose phosphate isomerase. TIM catalyzes the transfer of a hydrogen atom from carbon 1 to carbon 2 in converting dihydroxyacetone phosphate into glyceraldehyde 3-phosphate, an intramolecular oxidation–reduction. This isomerization of a ketose into an aldose proceeds through an *enediol intermediate* (Figure 16.6).

X-ray crystallographic and other studies showed that glutamate 165 (see Figure 16.5) plays the role of a general acid–base catalyst. However, this

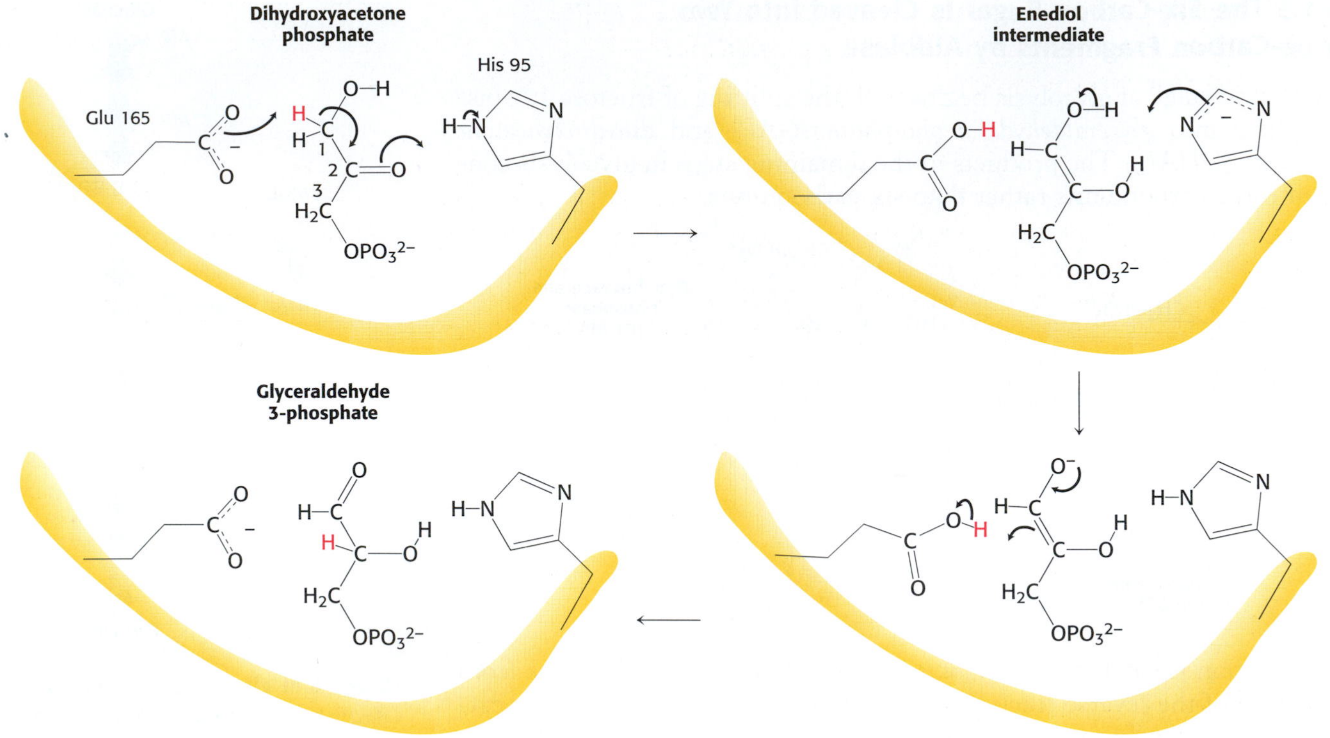

Dihydroxyacetone
phosphate

Enediol
intermediate

His 95

Glu 165

Glyceraldehyde
3-phosphate

FIGURE 16.6 Catalytic mechanism of triose phosphate isomerase. Glutamate 165 transfers a proton between carbons with the assistance of histidine 95, which shuttles between the neutral and relatively rare negatively charged form. The latter is stabilized by interactions with other parts of the enzyme.

carboxylate group by itself is not basic enough to pull a proton away from a carbon atom adjacent to a carbonyl group. Histidine 95 assists catalysis by donating a proton to stabilize the negative charge that develops on the C-2 carbonyl group.

Two features of this enzyme are noteworthy. First, TIM displays great catalytic prowess. It accelerates isomerization by a factor of 10^{10} compared with the rate obtained with a simple base catalyst such as acetate ion. Indeed, the k_{cat}/K_M ratio for isomerization of glyceraldehyde 3-phosphate is $2 \times 10^8 \, M^{-1} \, s^{-1}$, which is close to the diffusion-controlled limit. In other words, the rate-limiting step in catalysis is the diffusion-controlled encounter of substrate and enzyme. TIM is an example of a *kinetically perfect enzyme* (Section 8.2.5). Second, TIM suppresses an undesired side reaction, the decomposition of the enediol intermediate into methyl glyoxal and orthophosphate.

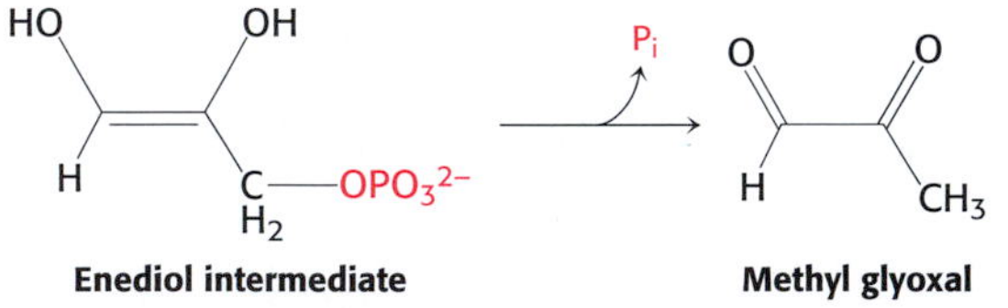

Enediol intermediate

Methyl glyoxal

In solution, this physiologically useless reaction is 100 times as fast as isomerization. Hence, TIM must prevent the enediol from leaving the enzyme. This labile intermediate is trapped in the active site by the movement of a loop of 10 residues (see Figure 16.5). This loop serves as a lid on the active site, shutting it when the enediol is present and reopening it when isomerization is completed. *We see here a striking example not only of catalytic perfection, but also of the acceleration of a desirable reaction so that it takes place much faster than an undesirable alternative reaction.* Thus, two molecules of glyceraldehyde 3-phosphate are formed from one molecule of fructose 1,6-bisphosphate by the sequential action of aldolase and triose phosphate isomerase. The economy of metabolism is evident in this reaction sequence. The isomerase funnels dihydroxyacetone phosphate into the main glycolytic pathway—a separate set of reactions is not needed.

16.1.5 Energy Transformation: Phosphorylation Is Coupled to the Oxidation of Glyceraldehyde 3-phosphate by a Thioester Intermediate

The preceding steps in glycolysis have transformed one molecule of glucose into two molecules of glyceraldehyde 3-phosphate, but no energy has yet been extracted. On the contrary, thus far two molecules of ATP have been invested. We come now to a series of steps that harvest some of the energy contained in glyceraldehyde 3-phosphate. The initial reaction in this sequence is the *conversion of glyceraldehyde 3-phosphate into 1,3-bisphosphoglycerate (1,3-BPG), a reaction catalyzed by glyceraldehyde 3-phosphate dehydrogenase* (Figure 16.7).

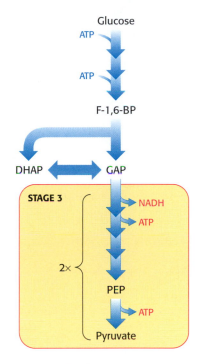

Stage 3 of glycolysis. The oxidation of three-carbon fragments yields ATP.

$$\text{Glyceraldehyde 3-phosphate (GAP)} + NAD^+ + P_i \rightleftharpoons \text{1,3-Bisphosphoglycerate (1,3-BPG)} + NADH + H^+$$

(glyceraldehyde 3-phosphate dehydrogenase)

1,3-Bisphosphoglycerate is an acyl phosphate. Such compounds have a high phosphoryl-transfer potential; one of its phosphoryl groups is transferred to ADP in the next step in glycolysis. The reaction catalyzed by glyceraldehyde 3-phosphate dehydrogenase is really the sum of two processes: the *oxidation* of the aldehyde to a carboxylic acid by NAD^+ and the *joining* of the carboxylic acid and orthophosphate to form the acyl-phosphate product.

$$\text{(aldehyde)} + NAD^+ + H_2O \xrightarrow{\text{Oxidation}} \text{(carboxylic acid)} + NADH + H^+$$

$$\text{(carboxylic acid)} + P_i \xrightarrow{\text{Acyl-phosphate formation (dehydration)}} \text{(acyl phosphate)} + H_2O$$

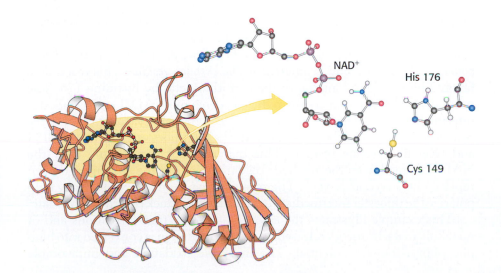

FIGURE 16.7 Structure of glyceraldehyde 3-phosphate dehydrogenase. The active site includes a cysteine residue and a histidine residue adjacent to a bound NAD^+.

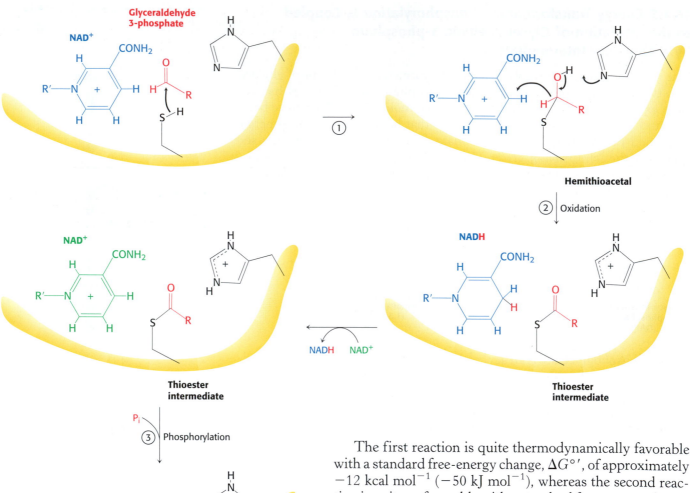

FIGURE 16.8 Catalytic mechanism of glyceraldehyde 3-phosphate dehydrogenase. The reaction proceeds through a thioester intermediate, which allows the oxidation of glyceraldehyde to be coupled to the phosphorylation of 3-phosphoglycerate.

The first reaction is quite thermodynamically favorable with a standard free-energy change, $\Delta G°'$, of approximately -12 kcal mol^{-1} (-50 kJ mol^{-1}), whereas the second reaction is quite unfavorable with a standard free-energy change of the same magnitude but the opposite sign. If these two reactions simply took place in succession, the second reaction would have a very large activation energy and thus not take place at a biologically significant rate. These two processes *must be coupled* so that the favorable aldehyde oxidation can be used to drive the formation of the acyl phosphate. How are these reactions coupled? The key is an intermediate, formed as a result of the aldehyde oxidation, that is higher in free energy than the free carboxylic acid is. This intermediate reacts with orthophosphate to form the acyl-phosphate product.

Let us consider the mechanism of glyceraldehyde 3-phosphate dehydrogenase in detail (Figure 16.8). In step 1, the aldehyde substrate reacts with the sulfhydryl group of cysteine 149 on the enzyme to form a hemithioacetal. Step 2 is the *transfer of a hydride ion to a molecule of NAD$^+$ that is tightly bound to the enzyme and is adjacent to the cysteine residue.* This reaction is favored by the deprotonation of the hemithioacetal by histidine 176. The products of this reaction are the reduced coenzyme NADH and a thioester intermediate. *This thioester intermediate has a free energy close to that of the reactants.* In step 3, orthophosphate attacks the thioester to form 1,3-BPG and free the cysteine residue. This displacement occurs only after the NADH formed from the aldehyde oxidation has left the enzyme and been replaced by a second NAD$^+$. The positive charge on the NAD$^+$ may help polarize the thioester intermediate to facilitate the attack by orthophosphate.

This example illustrates the essence of energy transformations and of metabolism itself: energy released by carbon oxidation is converted into high phosphoryl-transfer potential. The favorable oxidation and unfavorable phosphorylation reactions are *coupled by the thioester intermediate,* which preserves much of the free energy released in the oxidation reaction. We see here the *use of a covalent enzyme-bound intermediate as a mechanism of energy coupling.* A free-energy profile of the glyceraldehyde 3-phosphate dehydrogenase reaction, compared with a hypothetical process in which the reaction proceeds without this intermediate, reveals how this intermediate allows a favorable process to drive an unfavorable one (Figure 16.9).

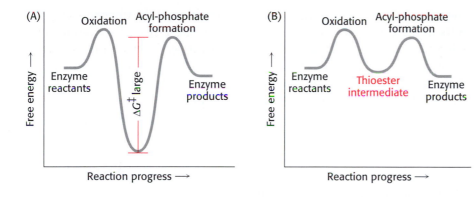

FIGURE 16.9 Free-energy profiles for glyceraldehyde oxidation followed by acyl-phosphate formation. (A) A hypothetical case with no coupling between the two processes. The second step must have a large activation barrier, making the reaction very slow. (B) The actual case with the two reactions coupled through a thioester intermediate.

16.1.6 The Formation of ATP from 1,3-Bisphosphoglycerate

The final stage in glycolysis is the generation of ATP from the phosphorylated three-carbon metabolites of glucose. *Phosphoglycerate kinase* catalyzes the transfer of the phosphoryl group from the acyl phosphate of 1,3-bisphosphoglycerate to ADP. ATP and 3-phosphoglycerate are the products.

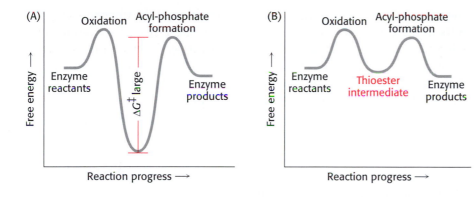

The formation of ATP in this manner is referred to as *substrate-level phosphorylation* because the phosphate donor, 1,3-BPG, is a substrate with high phosphoryl-transfer potential. We will contrast this manner of ATP formation with that in which ATP is formed from ionic gradients in Chapters 18 and 19.

Thus, the outcomes of the reactions catalyzed by glyceraldehyde 3-phosphate dehydrogenase and phosphoglycerate kinase are:

1. Glyceraldehyde 3-phosphate, an aldehyde, is oxidized to 3-phosphoglycerate, a carboxylic acid.

2. NAD^+ is concomitantly reduced to NADH.

3. ATP is formed from P_i and ADP at the expense of carbon oxidation energy.

Keep in mind that, because of the actions of aldolase and triose phosphate isomerase, two molecules of glyceraldehyde 3-phosphate were formed and hence two molecules of ATP were generated. These ATP molecules make up for the two molecules of ATP consumed in the first stage of glycolysis.

16.1.7 The Generation of Additional ATP and the Formation of Pyruvate

In the remaining steps of glycolysis, 3-phosphoglycerate is converted into pyruvate with the concomitant conversion of ADP into ATP.

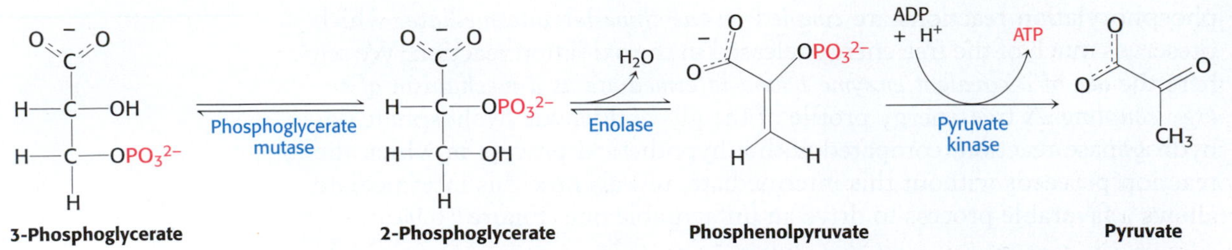

3-Phosphoglycerate — Phosphoglycerate mutase → **2-Phosphoglycerate** — Enolase → **Phosphenolpyruvate** — Pyruvate kinase → **Pyruvate**

The first reaction is a rearrangement. The position of the phosphoryl group shifts in the conversion of *3-phosphoglycerate into 2-phosphoglycerate*, a reaction catalyzed by *phosphoglycerate mutase*. In general, a *mutase* is an enzyme that catalyzes the intramolecular shift of a chemical group, such as a phosphoryl group. The phosphoglycerate mutase reaction has an interesting mechanism: the phosphoryl group is not simply moved from one carbon to another. This enzyme requires catalytic amounts of 2,3-bisphosphoglycerate to maintain an active-site histidine residue in a phosphorylated form.

Enz-His + 2,3 bisphosphoglycerate ⇌

Enz-His-phosphate + 2-phosphoglycerate

Enz-His-phosphate + 3-phosphoglycerate ⇌

Enz-His + 2,3-bisphosphoglycerate

The sum of these reactions yields the mutase reaction:

3-Phosphoglycerate ⇌ 2-phosphoglycerate

Examination of the first partial reaction reveals that the mutase functions as a phosphatase—it converts 2,3-bisphosphoglycerate into 2-phosphoglycerate. However, the phosphoryl group remains linked to the enzyme. This phosphoryl group is then transferred to 3-phosphoglycerate to reform 2,3-bisphosphoglycerate.

In the next reaction, an *enol* is formed by the dehydration of 2-phosphoglycerate. *Enolase* catalyzes the formation of *phosphoenolpyruvate* (PEP). This dehydration markedly elevates the transfer potential of the phosphoryl group. An *enol phosphate* has a high phosphoryl-transfer potential,

TABLE 16.3 **Reactions of glycolysis**

Step	Reaction
1	Glucose + ATP ⟶ glucose 6-phosphate + ADP + H^+
2	Glucose 6-phosphate ⇌ fructose 6-phosphate
3	Fructose 6-phosphate + ATP ⟶ fructose 1,6-bisphosphate + ADP + H^+
4	Fructose 1,6-bisphosphate ⇌ dihydroxyacetone phosphate + glyceraldehyde 3-phosphate
5	Dihydroxyacetone phosphate ⇌ glyceraldehyde 3-phosphate
6	Glyceraldehyde 3-phosphate + P_i + NAD^+ ⇌ 1,3-bisphosphoglycerate + NADH + H^+
7	1,3-Bisphosphoglycerate + ADP ⇌ 3-phosphoglycerate + ATP
8	3-Phosphoglycerate ⇌ 2-phosphoglycerate
9	2-Phosphoglycerate ⇌ phosphoenolpyruvate + H_2O
10	Phosphoenolpyruvate + ADP + H^+ ⟶ pyruvate + ATP

Note: ΔG, the actual free-energy change, has been calculated from $\Delta G^{\circ\prime}$ and known concentrations of reactants under typical physiologic conditions. Glycolysis can proceed only if the ΔG values of all reactions are negative. The small positive ΔG values of three of the

whereas the phosphate ester, such as 2-phosphoglycerate, of an ordinary alcohol has a low one. The $\Delta G°'$ of the hydrolysis of a phosphate ester of an ordinary alcohol is -3 kcal mol^{-1} (-13 kJ mol^{-1}), whereas that of phosphoenolpyruvate is -14.8 kcal mol^{-1} (-62 kJ mol^{-1}). Why does phosphoenolpyruvate have such a high phosphoryl-transfer potential? The phosphoryl group traps the molecule in its unstable enol form. When the phosphoryl group has been donated to ATP, the enol undergoes a conversion into the more stable ketone—namely, pyruvate.

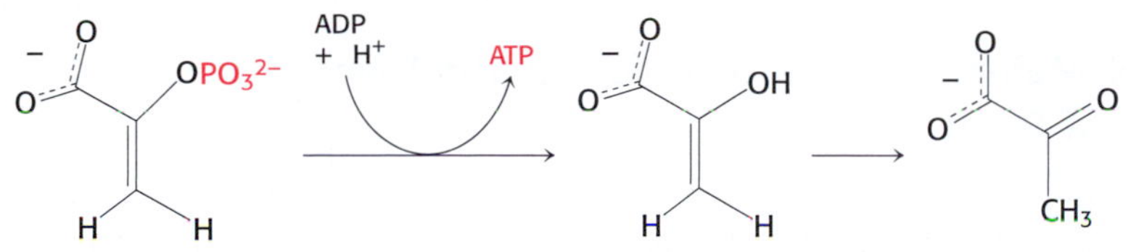

Phosphenolpyruvate   **Pyruvate**
(enol form)   **Pyruvate**

Thus, *the high phosphoryl-transfer potential of phosphoenolpyruvate arises primarily from the large driving force of the subsequent enol–ketone conversion.* Hence, pyruvate is formed, and ATP is generated concomitantly. The virtually irreversible transfer of a phosphoryl group from phosphoenolpyruvate to ADP is catalyzed by *pyruvate kinase*. Because the molecules of ATP used in forming fructose 1,6-bisphosphate have already been regenerated, the two molecules of ATP generated from phosphoenolpyruvate are "profit."

16.1.8 Energy Yield in the Conversion of Glucose into Pyruvate

The net reaction in the transformation of glucose into pyruvate is:

$$\text{Glucose} + 2\,P_i + 2\,\text{ADP} + 2\,\text{NAD}^+ \longrightarrow$$
$$2\,\text{pyruvate} + 2\,\text{ATP} + 2\,\text{NADH} + 2\,\text{H}^+ + 2\,\text{H}_2\text{O}$$

Thus, *two molecules of ATP are generated in the conversion of glucose into two molecules of pyruvate.* The reactions of glycolysis are summarized in Table 16.3.

Enzyme	Reaction type	$\Delta G°'$ in kcal mol^{-1} (kJ mol^{-1})	ΔG in kcal mol^{-1} (kJ mol^{-1})
Hexokinase	Phosphoryl transfer	$-4.0\ (-16.7)$	$-8.0\ (-33.5)$
Phosphoglucose isomerase	Isomerization	$+0.4\ (+1.7)$	$-0.6\ (-2.5)$
Phosphofructokinase	Phosphoryl transfer	$-3.4\ (-14.2)$	$-5.3\ (-22.2)$
Aldolase	Aldol cleavage	$+5.7\ (+23.8)$	$-0.3\ (-1.3)$
Triose phosphate isomerase	Isomerization	$+1.8\ (+7.5)$	$+0.6\ (+2.5)$
Glyceraldehyde 3-phosphate dehydrogenase	Phosphorylation coupled to oxidation	$+1.5\ (+6.3)$	$+0.6\ (+2.5)$
Phosphoglycerate kinase	Phosphoryl transfer	$-4.5\ (-18.8)$	$+0.3\ (+1.3)$
Phosphoglycerate mutase	Phosphoryl shift	$+1.1\ (+4.6)$	$+0.2\ (+0.8)$
Enolase	Dehydration	$+0.4\ (+1.7)$	$-0.8\ (-3.3)$
Pyruvate kinase	Phosphoryl transfer	$-7.5\ (-31.4)$	$-4.0\ (-16.7)$

above reactions indicate that the concentrations of metabolites in vivo in cells undergoing glycolysis are not precisely known.

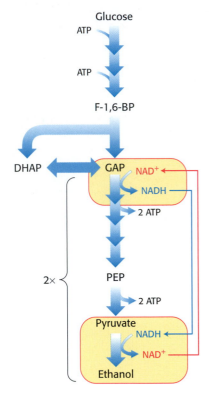

Location of redox balance steps. The generation and consumption of NADH, located within the glycolytic pathway.

Note that the energy released in the anaerobic conversion of glucose into two molecules of pyruvate is -47 kcal mol^{-1} (-197 kJ mol^{-1}). We shall see in Chapters 17 and 18 how much more energy can be released from glucose in the presence of oxygen.

> **CONCEPTUAL INSIGHTS, Energetics of Glucose Metabolism.** *See the section on the energetics of glycolysis in the Conceptual Insights module for a graphical representation of free energy differences among glycolytic metabolites, and how these differences are used to drive ATP and NADH synthesis in coupled reactions.*

16.1.9 Maintaining Redox Balance: The Diverse Fates of Pyruvate

The conversion of glucose into two molecules of pyruvate has resulted in the net synthesis of ATP. However, an energy-converting pathway that stopped at pyruvate would not proceed for long, because redox balance has not been maintained. As we have seen, the activity of glyceraldehyde 3-phosphate dehydrogenase, in addition to generating a compound with high phosphoryl-transfer potential, of necessity leads to the reduction of NAD$^+$ to NADH. There are limited amounts of NAD$^+$ in the cell, which is derived from the vitamin *niacin*, a dietary requirement in human beings. Consequently, NAD$^+$ must be regenerated for glycolysis to proceed. Thus, the final process in the pathway is the regeneration of NAD$^+$ through the metabolism of pyruvate. The sequence of reactions from glucose to pyruvate is similar in most organisms and most types of cells. In contrast, the fate of pyruvate is variable. Three reactions of pyruvate are of prime importance: conversion into ethanol, lactic acid, or carbon dioxide (Figure 16.10).

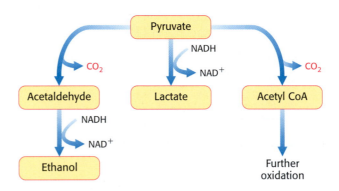

FIGURE 16.10 Diverse fates of pyruvate. Ethanol and lactate can be formed by reactions involving NADH. Alternatively, a two-carbon unit from pyruvate can be coupled to coenzyme A (see Section 17.1.1) to form acetyl CoA.

1. *Ethanol* is formed from pyruvate in yeast and several other microorganisms. The first step is the decarboxylation of pyruvate. This reaction is catalyzed by *pyruvate decarboxylase*, which requires the coenzyme thiamine pyrophosphate. This coenzyme, derived from the vitamin thiamine (B$_1$), also participates in reactions catalyzed by other enzymes (Section 17.1.1). The second step is the reduction of acetaldehyde to ethanol by NADH, in a reaction catalyzed by *alcohol dehydrogenase*. This process regenerates NAD$^+$.

The active site of alcohol dehydrogenase contains a zinc ion that is coordinated to the sulfur atoms of two cysteine residues and a nitrogen atom of histidine (Figure 16.11). This zinc ion polarizes the carbonyl group of the substrate to favor the transfer of a hydride from NADH.

The conversion of glucose into ethanol is an example of *alcoholic fermentation*. The net result of this anaerobic process is:

$$\text{Glucose} + 2\ P_i + 2\ ADP + 2\ H^+ \longrightarrow$$
$$2\ \text{ethanol} + 2\ CO_2 + 2\ ATP + 2\ H_2O$$

Note that NAD^+ and NADH do not appear in this equation, even though they are crucial for the overall process. NADH generated by the oxidation of glyceraldehyde 3-phosphate is consumed in the reduction of acetaldehyde to ethanol. Thus, *there is no net oxidation–reduction in the conversion of glucose into ethanol* (Figure 16.12). The ethanol formed in alcoholic fermentation provides a key ingredient for brewing and winemaking.

2. *Lactate* is formed from pyruvate in a variety of microorganisms in a process called *lactic acid fermentation*. The reaction also takes place in the cells of higher organisms when the amount of oxygen is limiting, as in muscle during intense activity. The reduction of pyruvate by NADH to form lactate is catalyzed by *lactate dehydrogenase*.

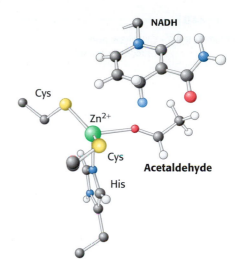

FIGURE 16.11 **Active site of alcohol dehydrogenase.** The active site contains a zinc ion bound to two cysteine residues and one histidine residue. The zinc ion binds the acetaldehyde substrate through its oxygen atom, polarizing it so that it more easily accepts a hydride (light blue) from NADH. Only the nicotinamide ring of NADH is shown.

The overall reaction in the conversion of glucose into lactate is:

$$\text{Glucose} + 2\ P_i + 2\ ADP \longrightarrow 2\ \text{lactate} + 2\ ATP + 2\ H_2O$$

As in alcoholic fermentation, there is no net oxidation–reduction. The NADH formed in the oxidation of glyceraldehyde 3-phosphate is consumed in the reduction of pyruvate. *The regeneration of NAD^+ in the reduction of pyruvate to lactate or ethanol sustains the continued operation of glycolysis under anaerobic conditions.*

3. Only a fraction of the energy of glucose is released in its anaerobic conversion into ethanol or lactate. Much more energy can be extracted aerobically by means of the citric acid cycle and the electron-transport chain. The entry point to this oxidative pathway is *acetyl coenzyme A* (acetyl CoA), which is formed inside mitochondria by the oxidative decarboxylation of pyruvate.

$$\text{Pyruvate} + NAD^+ + CoA \longrightarrow \text{acetyl CoA} + CO_2 + NADH$$

FIGURE 16.12 **Maintaining redox balance.** The NADH produced by the glyceraldehyde 3-phosphate dehydrogenase reaction must be reoxidized to NAD^+ for the glycolytic pathway to continue. In alcoholic fermentation, alcohol dehydrogenase oxidizes NADH and generates ethanol. In lactic acid fermentation (not shown), lactate dehydrogenase oxidizes NADH while generating lactic acid.

This reaction, which is catalyzed by the pyruvate dehydrogenase complex, will be discussed in detail in Chapter 18. The NAD^+ required for this reaction and for the oxidation of glyceraldehyde 3-phosphate is regenerated when NADH ultimately transfers its electrons to O_2 through the electron-transport chain in mitochondria.

16.1.10 The Binding Site for NAD^+ Is Similar in Many Dehydrogenases

Although the enzymes taking part in glycolysis and the subsequent conversion of pyruvate are structurally diverse, the three dehydrogenases—glyceraldehyde 3-phosphate dehydrogenase, alcohol dehydrogenase, and lactate dehydrogenase—have in common a domain for NAD^+ binding (Figure 16.13). This nucleotide-binding region is made up of four α helices and a sheet of six parallel β strands. Moreover, in all cases, the bound NAD^+ displays nearly the same conformation. This common structural domain, one of the first recurring structural domains to be discovered, is often called a *Rossmann fold* after Michael Rossmann, who first recognized it. This fold likely represents a primordial dinucleotide-binding domain that recurs in the dehydrogenases of glycolysis and other enzymes because of their descent from a common ancestor.

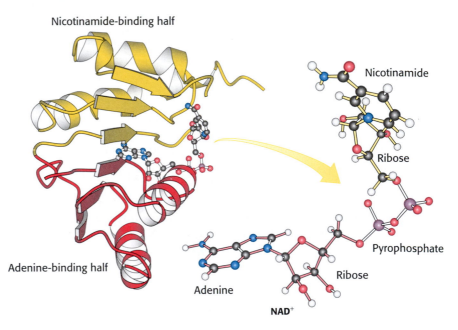

FIGURE 16.13 NAD^+-binding region in dehydrogenases. The nicotinamide-binding half (yellow) is structurally similar to the adenine-binding half (red). The two halves together form a structural motif called a Rossmann fold. The NAD^+ molecule binds in an extended conformation.

16.1.11 The Entry of Fructose and Galactose into Glycolysis

Although glucose is the most widely used monosaccharide, others also are important fuels. Let us consider how two abundant sugars—fructose and galactose—can be funneled into the glycolytic pathway (Figure 16.14). Much of the ingested fructose is metabolized by the liver, using the *fructose 1-phosphate pathway* (Figure 16.15). The first step is the phosphorylation of *fructose* to *fructose 1-phosphate* by *fructokinase*. Fructose 1-phosphate is then split into *glyceraldehyde* and *dihydroxyacetone phosphate*, an intermediate in glycolysis. This aldol cleavage is catalyzed by a specific *fructose 1-phosphate aldolase*. Glyceraldehyde is then phosphorylated to *glyceraldehyde 3-phosphate*, a glycolytic intermediate, by *triose kinase*. Alternatively, *fructose can be phosphorylated to fructose 6-phosphate by hexokinase*. However, the affinity of hexokinase for glucose is 20 times as great as it is for fructose. Little fructose 6-phosphate is formed in the liver because glucose is so much more abundant in this organ. Moreover, glucose, as the preferred fuel, is also trapped in the muscle by the hexokinase reaction. Because liver and muscle phosphorylate glucose rather than fructose, adipose tissue is exposed to more fructose than glucose. Hence, the formation of fructose 6-phosphate is not competitively inhibited to a biologically significant extent, and most of the fructose in adipose tissue is metabolized through fructose 6-phosphate.

There are no catabolic pathways to metabolize *galactose*, so the strategy is to convert galactose into a metabolite of glucose. *Galactose is converted into glucose 6-phosphate* in four steps. The first reaction in the *galactose–glucose interconversion pathway* is the phosphorylation of galactose to galactose 1-phosphate by *galactokinase*.

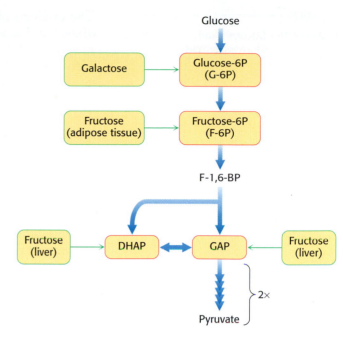

FIGURE 16.14 Entry points in glycolysis for galactose and fructose.

Galactose 1-phosphate then acquires a uridyl group from uridine diphosphate glucose (UDP-glucose), an intermediate in the synthesis of glycosidic linkages (Section 21.4.2).

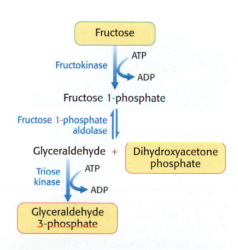

FIGURE 16.15 Fructose metabolism. Fructose enters the glycolytic pathway in the liver through the fructose 1-phosphate pathway.

The products of this reaction, which is catalyzed by *galactose 1-phosphate uridyl transferase,* are UDP-galactose and glucose 1-phosphate. The galactose moiety of UDP-galactose is then epimerized to glucose. The configuration of the hydroxyl group at carbon 4 is inverted by *UDP-galactose 4-epimerase.*

The sum of the reactions catalyzed by galactokinase, the transferase, and the epimerase is:

$$\text{Galactose} + \text{ATP} \longrightarrow \text{glucose 1-phosphate} + \text{ADP} + \text{H}^+$$

Note that UDP-glucose is not consumed in the conversion of galactose into glucose, because it is regenerated from UDP-galactose by the epimerase. This reaction is reversible, and the product of the reverse direction also is important. *The conversion of UDP-glucose into UDP-galactose is essential for the synthesis of galactosyl residues in complex polysaccharides and glycoproteins if the amount of galactose in the diet is inadequate to meet these needs.*

Finally, glucose 1-phosphate, formed from galactose, is isomerized to glucose 6-phosphate by *phosphoglucomutase.* We shall return to this reaction when we consider the synthesis and degradation of glycogen, which proceeds through glucose 1-phosphate, in Chapter 21.

16.1.12 Many Adults Are Intolerant of Milk Because They Are Deficient in Lactase

Many adults are unable to metabolize the milk sugar lactose and experience gastrointestinal disturbances if they drink milk. *Lactose intolerance,* or hypolactasia, is most commonly caused by a deficiency of the enzyme lactase, which cleaves lactose into glucose and galactose.

Lactose $+ \text{H}_2\text{O}$ Lactase   **Galactose** $+$ **Glucose**

"Deficiency" is not quite the appropriate term, because a decrease in lactase is normal during development in all mammals. As children are weaned and milk becomes less prominent in their diets, lactase activity normally declines to about 5 to 10% of the level at birth. This decrease is not as pronounced with some groups of people, most notably Northern Europeans, and people from these groups can continue to ingest milk without gastrointestinal difficulties. With the appearance of milk-producing domesticated animals, a human being with a genetic alteration endowing high levels of lactase activity in adulthood would hypothetically have a selective advantage in being able to consume calories from the readily available milk.

What happens to the lactose in the intestine of a lactase-deficient person? The lactose is a good energy source for microorganisms in the colon, and they ferment it to lactic acid while also generating methane (CH_4) and hydrogen gas (H_2). The gas produced creates the uncomfortable feeling of gut distention and the annoying problem of flatulence. The lactic acid produced by the microorganisms is osmotically active and draws water into the intestine, as does any undigested lactose, resulting in diarrhea. If severe enough, the gas and diarrhea hinder the absorption of other nutrients such as fats and proteins. The simplest treatment is to avoid the consumption of products containing much lactose. Alternatively, the enzyme lactase can be ingested with milk products.

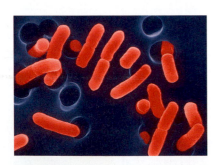

Scanning electron micrograph of *Lactobacillus.* The anaerobic bacteria *Lactobacillus* is shown here (artificially colored) at a magnification of 22, 245×. As suggested by its name, this genus of bacteria ferments glucose into lactic acid, and is widely used in the food industry. Lactobacillus is also a component of the normal human bacterial flora of the urogenital tract where, because of its ability to generate an acidic environment, it prevents growth of harmful bacteria. [Dr. Dennis Kunkel/PhotoTake.]

16.1.13 Galactose Is Highly Toxic If the Transferase Is Missing

Less common than lactose intolerance are disorders that interfere with the metabolism of galactose. The disruption of galactose metabolism is referred to as *galactosemia*. The most common form, called classic galactosemia, is an inherited deficiency in galactose 1-phosphate uridyl transferase activity. Afflicted infants fail to thrive. They vomit or have diarrhea after consuming milk, and enlargement of the liver and jaundice are common, sometimes progressing to cirrhosis. Cataracts will form, and lethargy and retarded mental development also are common. The blood-galactose level is markedly elevated, and galactose is found in the urine. The absence of the transferase in red blood cells is a definitive diagnostic criterion.

The most common treatment is to remove galactose (and lactose) from the diet. The enigma of galactosemia is that, although elimination of galactose from the diet prevents liver disease and cataract development, the majority of patients still suffer from central nervous system malfunction, most commonly a delayed acquisition of language skills. Females will also display ovarian failure.

Cataract formation is better understood. A cataract is the clouding of the normally clear lens of the eye. If the transferase is not active in the lens of the eye, the presence of aldose reductase causes the accumulating galactose to be reduced to galactitol.

Galactose → **Galactitol** (via Aldose reductase, NADPH + H⁺ → NADP⁺)

Galactitol is osmotically active, and water will diffuse into the lens, instigating the formation of cataracts. In fact, there is a high incidence of cataract formation with age in populations that consume substantial amounts of milk into adulthood.

16.2 THE GLYCOLYTIC PATHWAY IS TIGHTLY CONTROLLED

The flux through the glycolytic pathway must be adjusted in response to conditions both inside and outside the cell. The rate of conversion of glucose into pyruvate is regulated to meet two major cellular needs: (1) the production of ATP, generated by the degradation of glucose, and (2) the provision of building blocks for synthetic reactions, such as the formation of fatty acids. *In metabolic pathways, enzymes catalyzing essentially irreversible reactions are potential sites of control.* In glycolysis, the reactions catalyzed by hexokinase, phosphofructokinase, and pyruvate kinase are virtually irreversible; hence, these enzymes would be expected to have regulatory as well as catalytic roles. In fact, each of them serves as a control site. Their activities are regulated by the reversible binding of allosteric effectors or by covalent modification. In addition, the amounts of these important enzymes are varied by the regulation of transcription to meet changing metabolic needs. The time required for reversible allosteric control, regulation by phosphorylation, and transcriptional control is typically in milliseconds, seconds, and hours, respectively.

16.2.1 Phosphofructokinase Is the Key Enzyme in the Control of Glycolysis

Phosphofructokinase is the most important control element in the mammalian glycolytic pathway (Figure 16.16). High levels of ATP allosterically inhibit the enzyme in the liver (a 340-kd tetramer), thus lowering its affinity for fructose 6-phosphate. A high concentration of ATP converts the hyperbolic binding curve of fructose 6-phosphate into a sigmoidal one (Figure 16.17). ATP elicits this effect by binding to a specific regulatory site that is distinct from the catalytic site. AMP reverses the inhibitory action of ATP, and so *the activity of the enzyme increases when the ATP/AMP ratio is lowered*. In other words, *glycolysis is stimulated as the energy charge falls*. A fall in pH also inhibits phosphofructokinase activity. The inhibition of phosphofructokinase by H^+ prevents excessive formation of lactic acid (Section 16.1.9) and a precipitous drop in blood pH (acidosis).

FIGURE 16.16 Structure of phosphofructokinase. Phosphofructokinase in the liver is a tetramer of four identical subunits. The positions of the catalytic and allosteric sites are indicated.

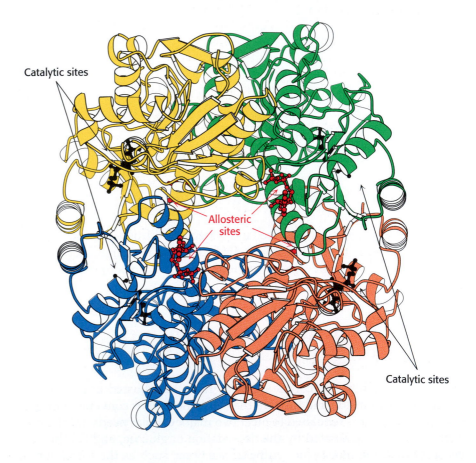

Catalytic sites

Allosteric sites

Catalytic sites

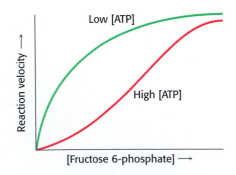

FIGURE 16.17 Allosteric regulation of phosphofructokinase. A high level of ATP inhibits the enzyme by decreasing its affinity for fructose 6-phosphate. AMP diminishes and citrate enhances the inhibitory effect of ATP.

Why is AMP and not ADP the positive regulator of phosphofructokinase? When ATP is being utilized rapidly, the enzyme *adenylate kinase* (Section 9.4) can form ATP from ADP by the following reaction:

$$ADP + ADP \rightleftharpoons ATP + AMP$$

Thus, some ATP is salvaged from ADP, and AMP becomes the signal for the low-energy state. Moreover, the use of AMP as an allosteric regulator

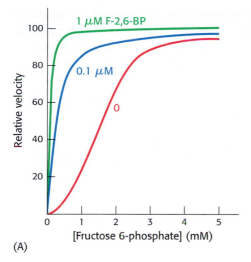

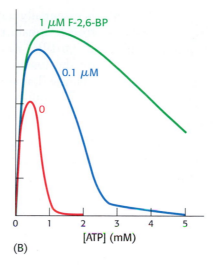

(A) (B)

FIGURE 16.18 Activation of phosphofructokinase by fructose 2,6-bisphosphate. (A) The sigmoidal dependence of velocity on substrate concentration becomes hyperbolic in the presence of 1 μM fructose 2,6-bisphosphate. (B) ATP, acting as a substrate, initially stimulates the reaction. As the concentration of ATP increases, it acts as an allosteric inhibitor. The inhibitory effect of ATP is reversed by fructose 2,6-bisphosphate. [After E. Van Schaftingen, M. F. Jett, L. Hue, and H. G. Hers. *Proc. Natl. Acad. Sci.* 78(1981):3483.]

provides an especially sensitive control. We can understand why by considering, first, that the total adenylate pool ([ATP], [ADP], [AMP]) in a cell is constant over the short term and, second, that the concentration of ATP is greater than that of ADP and the concentration of ADP is, in turn, greater than that of AMP. Consequently, small-percentage changes in [ATP] result in larger-percentage changes in the concentrations of the other adenylate nucleotides. This magnification of small changes in [ATP] to larger changes in [AMP] leads to tighter control by increasing the range of sensitivity of phosphofructokinase.

Glycolysis also furnishes carbon skeletons for biosyntheses, and so a signal indicating whether building blocks are abundant or scarce should also regulate phosphofructokinase. Indeed, *phosphofructokinase is inhibited by citrate*, an early intermediate in the citric acid cycle (Section 17.1.3). A high level of citrate means that biosynthetic precursors are abundant and additional glucose should not be degraded for this purpose. Citrate inhibits phosphofructokinase by enhancing the inhibitory effect of ATP.

In 1980, *fructose 2,6-bisphosphate (F-2,6-BP)* was identified as a potent activator of phosphofructokinase. Fructose 2,6-bisphosphate activates phosphofructokinase by increasing its affinity for fructose 6-phosphate and diminishing the inhibitory effect of ATP (Figure 16.18). In essence, *Fructose 2,6-bisphosphate is an allosteric activator that shifts the conformational equilibrium of this tetrameric enzyme from the T state to the R state.*

16.2.2 A Regulated Bifunctional Enzyme Synthesizes and Degrades Fructose 2,6-bisphosphate

How is the concentration of fructose 2,6-bisphosphate appropriately controlled? Two enzymes regulate the concentration of this important regulator of glycolysis by phosphorylating fructose 6-phosphate and dephosphorylating fructose 2,6-bisphosphate. Fructose 2,6-bisphosphate is formed in a reaction catalyzed by *phosphofructokinase 2 (PFK2)*, a different enzyme from phosphofructokinase. Fructose 2,6-bisphosphate is hydrolyzed to fructose 6-phosphate by a specific phosphatase, *fructose bisphosphatase 2 (FBPase2)*. The striking finding is that *both PFK2 and FBPase2 are present in a single 55-kd polypeptide chain* (Figure 16.19). This *bifunctional enzyme* contains an N-terminal *regulatory domain*, followed by a *kinase domain* and a *phosphatase domain*. PFK2 resembles adenylate kinase in having a P-loop NTPase domain (Sections 9.4.1 and 9.4.3), whereas FBPase2 resembles phosphoglycerate mutase (Section 16.1.7). Recall that the mutase

$$^{2-}O_3POH_2C \qquad OPO_3^{2-}$$

Fructose 2,6-bisphosphate (F-2,6-BP)

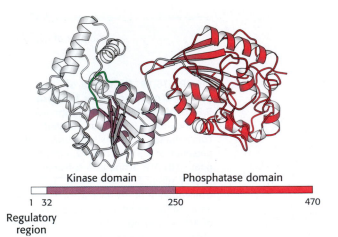

Kinase domain | Phosphatase domain

1 32   250   470

Regulatory
region

FIGURE 16.19 Domain structure of the bifunctional enzyme phosphofructokinase 2. The kinase domain (purple) is fused to the phosphatase domain (red). The kinase domain is a P-loop NTP hydrolase domain, as indicated by the purple shading (Section 9.4.4). The bar represents the amino acid sequence of the enzyme.

FIGURE 16.20 Control of the synthesis and degradation of fructose 2,6-bisphosphate. A low blood-glucose level as signaled by glucagon leads to the phosphorylation of the bifunctional enzyme and hence to a lower level of fructose 2,6-bisphosphate, slowing glycolysis. High levels of fructose 6-phosphate accelerate the formation of fructose 2,6-bisphosphate by facilitating the dephosphorylation of the bifunctional enzyme.

is essentially a phosphatase. In the bifunctional enzyme, the phosphatase activity evolved to become specific for F-2,6-BP. The bifunctional enzyme itself probably arose by the fusion of genes encoding the kinase and phosphatase domains.

The bifunctional enzyme exists in five isozymic forms (isoforms) that differ in size and kinetics as well as immunological and regulatory properties. Recall that isoenzymes, or isozymes, have essentially the same architectural plan and catalytic properties but differ in how they are regulated. The L isoform, which predominates in the liver, and the M isoform, found in muscle are generated by alternative splicing (Section 28.3.6) of the transcription product of a single gene. The L isoform helps to maintain blood-glucose homeostasis. In the liver, the concentration of fructose 6-phosphate rises when blood-glucose concentration is high, and the abundance of fructose 6-phosphate accelerates the synthesis of F-2,6-BP. Hence, *an abundance of fructose 6-phosphate leads to a higher concentration of F-2,6-BP, which in turn stimulates phosphofructokinase.* Such a process is called *feedforward stimulation.* What controls whether PFK2 or FBPase2 dominates the bifunctional enzyme's activities in the liver? The activities of PFK2 and FBPase2 are reciprocally controlled by *phosphorylation of a single serine residue.* When glucose is scarce, a rise in the blood level of the hormone glucagon triggers a cyclic AMP cascade, through its 7TM receptor and $G_s\alpha$ (Section 15.1), leading to the phosphorylation of this bifunctional enzyme by protein kinase A (Figure 16.20). This covalent modification activates FBPase2 and inhibits PFK2, lowering the level of F-2,6-BP. Thus, glucose metabolism by the liver is curtailed. Conversely, when glucose is abundant, the enzyme loses its attached phosphate group. This covalent modification activates PFK2 and inhibits FBPase2, raising the level of F-2,6-BP and accelerating glycolysis. This coordinated control is facilitated by the location of the kinase and phosphatase domains on the same polypeptide chain as the regulatory domain. We shall return to this elegant switch when we consider the integration of carbohydrate metabolism (Section 16.4).

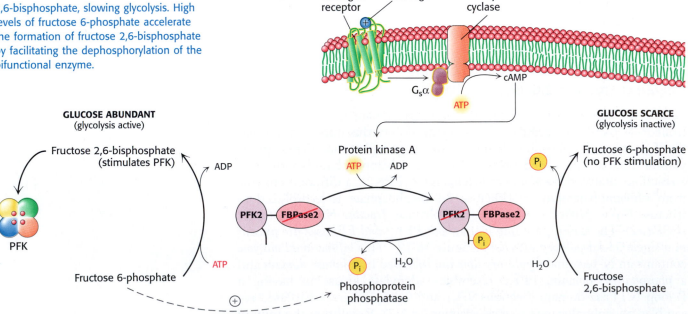

Glucagon receptor | Glucagon | Adenylate cyclase

$G_s\alpha$
ATP
cAMP

GLUCOSE ABUNDANT
(glycolysis active)

Fructose 2,6-bisphosphate
(stimulates PFK)

ADP

PFK

ATP

Fructose 6-phosphate

Protein kinase A
ATP   ADP

PFK2 — FBPase2   PFK2 — FBPase2

P_i H_2O

Phosphoprotein phosphatase

$+$

P_i

P_i

H_2O

GLUCOSE SCARCE
(glycolysis inactive)

Fructose 6-phosphate
(no PFK stimulation)

Fructose 2,6-bisphosphate

16.2.3 Hexokinase and Pyruvate Kinase Also Set the Pace of Glycolysis

Phosphofructokinase is the most prominent regulatory enzyme in glycolysis, but it is not the only one. Hexokinase, the enzyme catalyzing the first step of glycolysis, is inhibited by its product, glucose 6-phosphate. High concentrations of this molecule signal that the cell no longer requires glucose for energy, for storage in the form of glycogen, or as a source of biosynthetic precursors, and the glucose will be left in the blood. For example, when phosphofructokinase is inactive, the concentration of fructose 6-phosphate rises. In turn, the level of glucose 6-phosphate rises because it is in equilibrium with fructose 6-phosphate. Hence, *the inhibition of phosphofructokinase leads to the inhibition of hexokinase.* However, the liver, in keeping with its role as monitor of blood-glucose levels, possesses a specialized isozyme of hexokinase called *glucokinase* that is not inhibited by glucose 6-phosphate. Glucokinase phosphorylates glucose only when it is abundant because it has about a 50-fold lower affinity for glucose than does hexokinase. The role of glucokinase is to provide glucose 6-phosphate for the synthesis of glycogen, a storage form of glucose (Section 21.4), and for the formation of fatty acids (Section 22.1). The low glucose affinity of glucokinase in the liver gives the brain and muscles first call on glucose when its supply is limited, whereas it ensures that glucose will not be wasted when it is abundant.

Why is phosphofructokinase rather than hexokinase the pacemaker of glycolysis? The reason becomes evident on noting that glucose 6-phosphate is not solely a glycolytic intermediate. Glucose 6-phosphate can also be converted into glycogen or it can be oxidized by the pentose phosphate pathway (Section 20.3) to form NADPH. The first irreversible reaction unique to the glycolytic pathway, the *committed step*, (Section 10.2), is the phosphorylation of fructose 6-phosphate to fructose 1,6-bisphosphate. Thus, it is highly appropriate for phosphofructokinase to be the primary control site in glycolysis. In general, *the enzyme catalyzing the committed step in a metabolic sequence is the most important control element in the pathway.*

Pyruvate kinase, the enzyme catalyzing the third irreversible step in glycolysis, controls the outflow from this pathway. This final step yields ATP and pyruvate, a central metabolic intermediate that can be oxidized further or used as a building block. Several isozymic forms of pyruvate kinase (a tetramer of 57-kd subunits) encoded by different genes are present in mammals: the L type predominates in liver, and the M type in muscle and brain. The L and M forms of pyruvate kinase have many properties in common. Both bind phosphoenolpyruvate cooperatively. Fructose 1,6-bisphosphate, the product of the preceding irreversible step in glycolysis, activates both isozymes to enable them to keep pace with the oncoming high flux of intermediates. ATP allosterically inhibits both the L and the M forms of pyruvate kinase to slow glycolysis when the energy charge is high. Finally, alanine (synthesized in one step from pyruvate, Section 24.2.2) also allosterically inhibits the pyruvate kinases—in this case, to signal that building blocks are abundant.

The isozymic forms differ in their susceptibility to covalent modification. The catalytic properties of the L form—but not of the M form—are also controlled by reversible phosphorylation (Figure 16.21). When the blood-glucose level is low, the glucagon-triggered cyclic AMP cascade (Section 15.1.5) leads to the phosphorylation of pyruvate kinase, which diminishes its activity. *These hormone-triggered phosphorylations, like that of the bifunctional enzyme controlling the levels of fructose 2,6-bisphosphate, prevent the liver from consuming glucose when it is more urgently needed by brain*

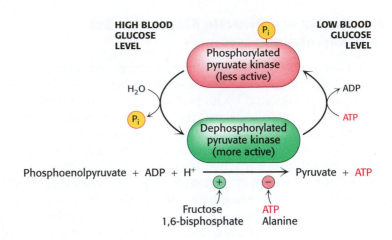

FIGURE 16.21 Control of the catalytic activity of pyruvate kinase. Pyruvate kinase is regulated by allosteric effectors and covalent modification.

and muscle (Section 30.3). We see here a clear-cut example of how isoenzymes contribute to the metabolic diversity of different organs. We will return to the control of glycolysis after considering gluconeogenesis.

16.2.4 A Family of Transporters Enables Glucose to Enter and Leave Animal Cells

Several glucose transporters mediate the thermodynamically downhill movement of glucose across the plasma membranes of animal cells. Each member of this protein family, named GLUT1 to GLUT5, consists of a single polypeptide chain about 500 residues long (Table 16.4). The common structural theme is the presence of 12 transmembrane segments (Figure 16.22).

The members of this family have distinctive roles:

1. GLUT1 and GLUT3, present in nearly all mammalian cells, are responsible for basal glucose uptake. Their K_M value for glucose is about 1 mM, significantly less than the normal serum-glucose level, which typically ranges from 4 mM to 8 mM. Hence, GLUT1 and GLUT3 continually transport glucose into cells at an essentially constant rate.

TABLE 16.4   Family of glucose transporters

Name	Tissue location	K_m	Comments
GLUT1	All mammalian tissues	1 mM	Basal glucose uptake
GLUT2	Liver and pancreatic β cells	15–20 mM	In the pancreas, plays a role in regulation of insulin In the liver, removes excess glucose from the blood
GLUT3	All mammalian tissues	1 mM	Basal glucose uptake
GLUT4	Muscle and fat cells	5 mM	Amount in muscle plasma membrane increases with endurance training
GLUT5	Small intestine	—	Primarily a fructose transporter

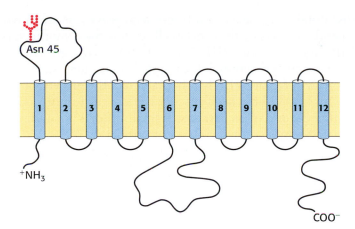

FIGURE 16.22 Model of a mammalian glucose transporter. The hydrophobicity profile of the protein indicates 12 transmembrane α helices. [From M. Muekler, C. Caruso, S. A. Baldwin, M. Panico, M. Blench, H. R. Morris, W. J. Allard, G. E. Lienhard, and H. F. Lodish. *Science* 229(1985):941.]

2. GLUT2, present in liver and pancreatic β cells, is distinctive in having a very high K_M value for glucose (15–20 mM). Hence, glucose enters these tissues at a biologically significant rate only when there is much glucose in the blood. The pancreas can thereby sense the glucose level and accordingly adjust the rate of insulin secretion. Insulin signals the need to remove glucose from the blood for storage as glycogen or conversion into fat (Section 30.3). The high K_M value of GLUT2 also ensures that glucose rapidly enters liver cells only in times of plenty.

3. GLUT4, which has a K_M value of 5 mM, transports glucose into muscle and fat cells. The presence of insulin, which signals the fed state, leads to a rapid increase in the number of GLUT4 transporters in the plasma membrane. Hence, insulin promotes the uptake of glucose by muscle and fat. The amount of this transporter present in muscle membranes increases in response to endurance exercise training.

4. GLUT5, present in the small intestine, functions primarily as a fructose transporter.

This family of transporters vividly illustrates how isoforms of a single protein can significantly shape the metabolic character of cells and contribute to their diversity and functional specialization. The transporters are members of a superfamily of transporters called the major facilitator (MF) superfamily. Members of this family transport sugars in organisms as diverse as *E. coli*, *Trypansoma brucei* (a parasitic protozoan that causes sleeping sickness), and human beings. An elegant solution to the problem of fuel transport evolved early and has been tailored to meet the needs of different organisms and even different tissues.

16.2.5 Cancer and Glycolysis

It has been known for decades that tumors display enhanced rates of glucose uptake and glycolysis. We now know that these enhanced rates of glucose processing are not fundamental to the development of cancer, but we can ask what selective advantage they might confer on cancer cells.

Cancer cells grow more rapidly than the blood vessels to nourish them; thus, as solid tumors grow, they are unable to obtain oxygen efficiently. In other words, they begin to experience *hypoxia*. Under these conditions, glycolysis leading to lactic acid fermentation becomes the primary source of ATP. Glycolysis is made more efficient in hypoxic tumors by the action of a transcription factor, *hypoxia-inducible transcription factor (HIF-1)*. In the absence of oxygen, HIF-1 increases the expression of most glycolytic enzymes and the glucose transporters GLUT1 and GLUT3 (Table 16.5). In

TABLE 16.5 Proteins in glucose metabolism encoded by genes regulated by hypoxia-inducible factor

GLUT1
GLUT3
Hexokinase
Phosphofructokinase
Aldolase
Glyceraldehyde 3-phosphate dehydrogenase
Phosphoglycerate kinase
Enolase
Pyruvate kinase
Lactate dehydrogenase

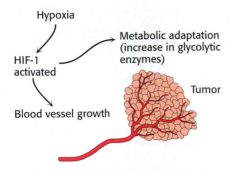

FIGURE 16.23 Alteration of gene expression in tumors due to hypoxia. The hypoxic conditions inside a tumor mass lead to the activation of the hypoxia-inducible transcription factor (HIF-1), which induces metabolic adaptation (increase in glycolytic enzymes) and activates angiogenic factors that stimulate the growth of new blood vessels. [Adapted from C. V. Dang and G. L. Semenza. *Trends Biochem. Sci.* 24(1999):68–72.]

fact, glucose uptake correlates with tumor aggressiveness and a poor prognosis. These adaptations by the cancer cells enable the tumor to survive until vascularization can occur. HIF-1 also stimulates the growth of new tumors by increasing the expression of signal molecules, such as vascular endothelial growth factor (VEGF), that facilitate the growth of blood vessels (Figure 16.23). Without such vascularization, the tumor would cease to grow and either die or remain harmlessly small. Efforts are underway to develop drugs that inhibit vascularization of tumors.

16.3 GLUCOSE CAN BE SYNTHESIZED FROM NONCARBOHYDRATE PRECURSORS

We now turn to the *synthesis of glucose from noncarbohydrate precursors*, a process called *gluconeogenesis*. This metabolic pathway is important because the brain depends on glucose as its primary fuel and red blood cells use only glucose as a fuel. The daily glucose requirement of the brain in a typical adult human being is about 120 g, which accounts for most of the 160 g of glucose needed daily by the whole body. The amount of glucose present in body fluids is about 20 g, and that readily available from glycogen, a storage form of glucose (Section 21.1), is approximately 190 g. Thus, the direct glucose reserves are sufficient to meet glucose needs for about a day. During a longer period of starvation, glucose must be formed from noncarbohydrate sources (Section 30.3.1).

The gluconeogenic pathway converts pyruvate into glucose. Noncarbohydrate precursors of glucose are first converted into pyruvate or enter the pathway at later intermediates such as oxaloacetate and dihydroxyacetone phosphate (Figure 16.24). The major noncarbohydrate precursors are *lactate, amino acids,* and *glycerol*. Lactate is formed by active skeletal muscle when the rate of glycolysis exceeds the rate of oxidative metabolism. Lactate is readily converted into pyruvate by the action of lactate dehydrogenase (Section 16.1.9). Amino acids are derived from proteins in the diet and, during starvation, from the breakdown of proteins in skeletal muscle (Section 30.3.1). The hydrolysis of triacylglycerols (Section 22.2.1) in fat cells yields glycerol and fatty acids. Glycerol is a precursor of glucose, but animals cannot convert fatty acids into glucose, for reasons that will be discussed later (Section 22.3.7). Glycerol may enter either the gluconeogenic or the glycolytic pathway at dihydroxyacetone phosphate.

The major site of gluconeogenesis is the *liver*, with a small amount also taking place in the *kidney*. Little gluconeogenesis takes place in the brain, skeletal muscle, or heart muscle. Rather, *gluconeogenesis in the liver and kidney helps to maintain the glucose level in the blood so that brain and muscle can extract sufficient glucose from it to meet their metabolic demands.*

16.3.1 Gluconeogenesis Is Not a Reversal of Glycolysis

In glycolysis, glucose is converted into pyruvate; in gluconeogenesis, pyruvate is converted into glucose. However, *gluconeogenesis is not a reversal of*

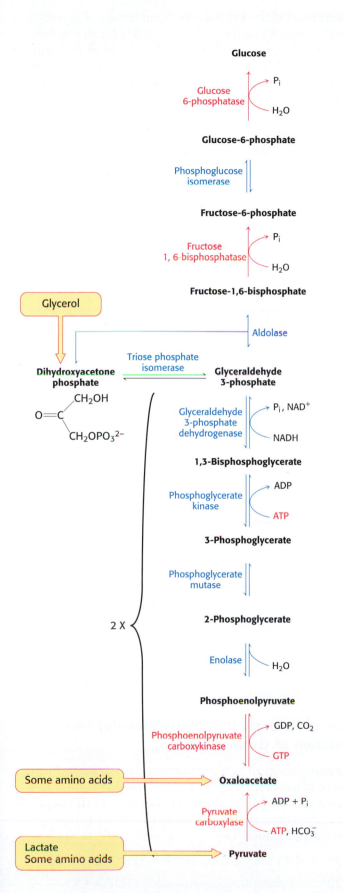

Glucose

Glucose 6-phosphatase → P_i, H_2O

Glucose-6-phosphate

Phosphoglucose isomerase

Fructose-6-phosphate

Fructose 1, 6-bisphosphatase → P_i, H_2O

Fructose-1,6-bisphosphate

Glycerol

Aldolase

Triose phosphate isomerase

Dihydroxyacetone phosphate ⇌ **Glyceraldehyde 3-phosphate**

Glyceraldehyde 3-phosphate dehydrogenase → P_i, NAD^+, $NADH$

1,3-Bisphosphoglycerate

Phosphoglycerate kinase → ADP, ATP

3-Phosphoglycerate

Phosphoglycerate mutase

2-Phosphoglycerate

Enolase → H_2O

Phosphoenolpyruvate

Phosphoenolpyruvate carboxykinase → GDP, CO_2, GTP

Some amino acids → **Oxaloacetate**

Pyruvate carboxylase → ADP + P_i, ATP, HCO_3^-

Lactate
Some amino acids → **Pyruvate**

2 X

FIGURE 16.24 Pathway of gluconeo-genesis. The distinctive reactions and enzymes of this pathway are shown in red. The other reactions are common to glycolysis. The enzymes for gluconeo-genesis are located in the cytosol, except for pyruvate carboxylase (in the mitochondria) and glucose 6-phosphatase (membrane bound in the endoplasmic reticulum). The entry points for lactate, glycerol, and amino acids are shown.

glycolysis. Several reactions must differ because the equilibrium of glycolysis lies far on the side of pyruvate formation. The actual ΔG for the formation of pyruvate from glucose is about -20 kcal mol^{-1} (-84 kJ mol^{-1}) under typical cellular conditions. Most of the decrease in free energy in glycolysis takes place in the three essentially irreversible steps catalyzed by hexokinase, phosphofructokinase, and pyruvate kinase.

$$\text{Glucose} + \text{ATP} \xrightarrow{\text{Hexokinase}} \text{glucose 6-phosphate} + \text{ADP}$$
$$\Delta G = -8.0 \text{ kcal mol}^{-1} (-33 \text{ kJ mol}^{-1})$$

$$\text{Fructose 6-phosphate} + \text{ATP} \xrightarrow{\text{Phosphofructokinase}}$$
$$\text{fructose 1,6-bisphosphate} + \text{ADP}$$
$$\Delta G = -5.3 \text{ kcal mol}^{-1} (-22 \text{ kJ mol}^{-1})$$

$$\text{Phosphoenolpyruvate} + \text{ADP} \xrightarrow{\text{Pyruvate kinase}} \text{pyruvate} + \text{ATP}$$
$$\Delta G = -4.0 \text{ kcal mol}^{-1} (-17 \text{ kJ mol}^{-1})$$

In gluconeogenesis, the following new steps bypass these virtually irreversible reactions of glycolysis:

1. *Phosphoenolpyruvate is formed from pyruvate by way of oxaloacetate* through the action of pyruvate carboxylase and phosphoenolpyruvate carboxykinase.

$$\text{Pyruvate} + CO_2 + \text{ATP} + H_2O \xrightarrow{\text{Pyruvate carboxylase}}$$
$$\text{oxaloacetate} + \text{ADP} + P_i + 2 \text{ H}^+$$

$$\text{Oxaloacetate} + \text{GTP} \xrightarrow{\text{Phosphoenolpyruvate carboxykinase}}$$
$$\text{phosphoenolpyruvate} + \text{GDP} + CO_2$$

2. *Fructose 6-phosphate is formed from fructose 1,6-bisphosphate by hydrolysis of the phosphate ester at carbon 1.* Fructose 1,6-bisphosphatase catalyzes this exergonic hydrolysis.

$$\text{Fructose 1,6-bisphosphate} + H_2O \longrightarrow \text{fructose 6-phosphate} + P_i$$

3. *Glucose is formed by hydrolysis of glucose 6-phosphate* in a reaction catalyzed by glucose 6-phosphatase.

$$\text{Glucose 6-phosphate} + H_2O \longrightarrow \text{glucose} + P_i$$

We will examine each of these steps in turn.

16.3.2 The Conversion of Pyruvate into Phosphoenolpyruvate Begins with the Formation of Oxaloacetate

The first step in gluconeogenesis is the carboxylation of pyruvate to form oxaloacetate at the expense of a molecule of ATP. Then, oxaloacetate is decarboxylated and phosphorylated to yield phosphoenolpyruvate, at the expense of the high phosphoryl-transfer potential of GTP. Both of these reactions take place inside the mitochondria.

$$\text{Pyruvate} + CO_2 + \text{ATP} + H_2O \rightleftharpoons \text{oxaloacetate} + \text{ADP} + P_i + 2 \text{ H}^+$$
$$\text{Oxaloacetate} + \text{GTP} \rightleftharpoons \text{phosphoenolpyruvate} + \text{GDP} + CO_2$$

The first reaction is catalyzed by *pyruvate carboxylase* and the second by *phosphoenolpyruvate carboxykinase*. The sum of these reactions is:

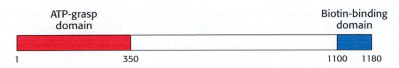

FIGURE 16.25 **Domain structure of pyruvate carboxylase.** The ATP-grasp domain activates HCO_3^- and transfers CO_2 to the biotin-binding domain. From there, the CO_2 is transferred to pyruvate generated in the central domain.

$$\text{Pyruvate} + \text{ATP} + \text{GTP} + \text{H}_2\text{O} \rightleftharpoons$$
$$\text{phosphoenolpyruvate} + \text{ADP} + \text{GDP} + \text{P}_i + 2\,\text{H}^+$$

Pyruvate carboxylase is of special interest because of its structural, catalytic, and allosteric properties. The N-terminal 300 to 350 amino acids form an *ATP-grasp domain* (Figure 16.25), which is a widely used ATP-activating domain to be discussed in more detail when we investigate nucleotide biosynthesis (Section 25.1.1). The C-terminal 80 amino acids constitute a biotin-binding domain (Figure 16.26) that we will see again in fatty acid synthesis (Section 22.4.1). *Biotin* is a covalently attached prosthetic group, which serves as a *carrier of activated CO_2*. The carboxylate group of biotin is linked to the ε-amino group of a specific lysine residue by an amide bond (Figure 16.27). Note that biotin is attached to pyruvate carboxylase by a *long, flexible chain.*

The carboxylation of pyruvate takes place in three stages:

$$\text{HCO}_3^- + \text{ATP} \rightleftharpoons \text{HOCO}_2\text{-PO}_3^{2-} + \text{ADP}$$

$$\text{Biotin-enzyme} + \text{HOCO}_2\text{-PO}_3^{2-} \rightleftharpoons \text{CO}_2\text{-biotin-enzyme} + \text{P}_i$$

$$\text{CO}_2\text{-biotin-enzyme} + \text{pyruvate} \rightleftharpoons \text{biotin-enzyme} + \text{oxaloacetate}$$

Recall that, in aqueous solutions, CO_2 exists as HCO_3^- with the aid of carbonic anhydrase (Section 9.2). The HCO_3^- is activated to carboxyphosphate. This activated CO_2 is subsequently bonded to the N-1 atom of the biotin ring to form the carboxybiotin-enzyme intermediate (see Figure 16.27). The CO_2 attached to the biotin is quite activated. The $\Delta G°'$ for its cleavage

$$\text{CO}_2\text{-biotin-enzyme} + \text{H}^+ \longrightarrow \text{CO}_2 + \text{biotin-enzyme}$$

is -4.7 kcal mol^{-1} (-20 kJ mol^{-1}). This negative $\Delta G°'$ indicates that carboxybiotin is able to transfer CO_2 to acceptors without the input of additional free energy.

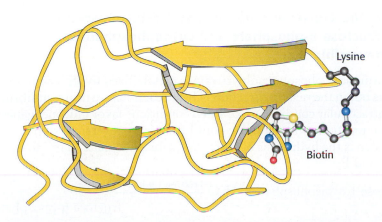

FIGURE 16.26 **Biotin-binding domain of pyruvate carboxylase.** This likely structure is based on the structure of the homologous domain from the enzyme acetyl CoA carboxylase (Section 22.4.1). The biotin is on a flexible tether, allowing it to move between the ATP-bicarbonate site and the pyruvate site.

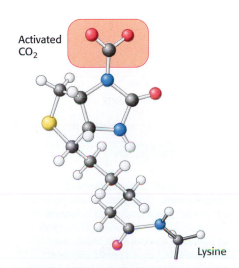

FIGURE 16.27 **Structure of carboxybiotin.**

The activated carboxyl group is then transferred from carboxybiotin to pyruvate to form oxaloacetate. The long, flexible link between biotin and the enzyme enables this prosthetic group to rotate from one active site of the enzyme (the ATP-bicarbonate site) to the other (the pyruvate site).

The first partial reaction of pyruvate carboxylase, the formation of carboxybiotin, depends on the presence of acetyl CoA. *Biotin is not carboxylated unless acetyl CoA is bound to the enzyme.* Acetyl CoA has no effect on the second partial reaction. The allosteric activation of pyruvate carboxylase by acetyl CoA is an important physiological control mechanism that will be discussed in Section 17.3.1.

16.3.3 Oxaloacetate Is Shuttled into the Cytosol and Converted into Phosphoenolpyruvate

Pyruvate carboxylase is a mitochondrial enzyme, whereas the other enzymes of gluconeogenesis are cytoplasmic. Oxaloacetate, the product of the pyruvate carboxylase reaction, is reduced to malate inside the mitochondrion for transport to the cytosol. The reduction is accomplished by an NADH-linked malate dehydrogenase. When malate has been transported across the mitochondrial membrane, it is reoxidized to oxaloacetate by an NAD^+-linked malate dehydrogenase in the cytosol (Figure 16.28).

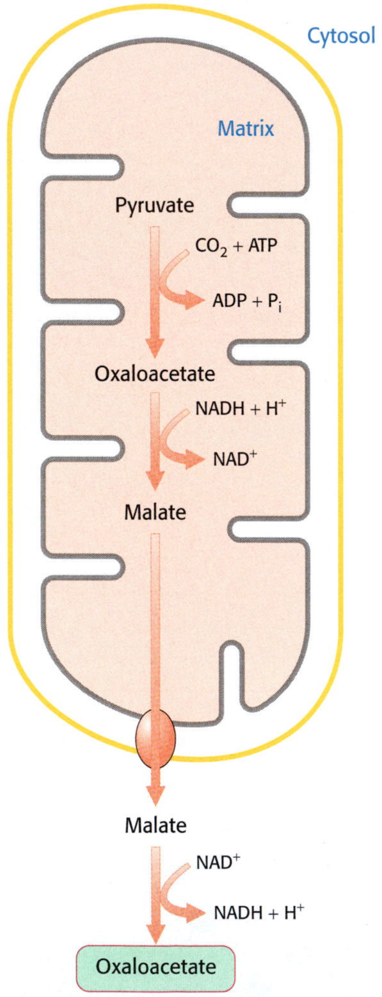

Finally, oxaloacetate is simultaneously *decarboxylated* and *phosphorylated* by phosphoenolpyruvate carboxykinase in the cytosol. The CO_2 that was added to pyruvate by pyruvate carboxylase comes off in this step. Recall that, in glycolysis, the presence of a phosphoryl group traps the unstable enol isomer of pyruvate as phosphoenolpyruvate (Section 16.1.7). In gluconeogenesis, the formation of the unstable enol is driven by decarboxylation—the oxidation of the carboxylic acid to CO_2—and trapped by the addition of a phosphate to carbon 2 from GTP. The two-step pathway for the formation of phosphoenolpyruvate from pyruvate has a $\Delta G^{\circ\prime}$ of $+0.2$ kcal mol^{-1} ($+0.13$ kJ mol^{-1}) in contrast with $+7.5$ kcal mol^{-1} ($+31$ kJ mol^{-1}) for the reaction catalyzed by pyruvate kinase. The much more favorable $\Delta G^{\circ\prime}$ for the two-step pathway results from the use of a molecule of ATP to add a molecule of CO_2 in the carboxylation step that can be removed to power the formation of phosphoenolpyruvate in the decarboxylation step. *Decarboxylations often drive reactions otherwise highly endergonic.* This metabolic motif is used in the citric acid cycle (Section 17.1), the pentose phosphate pathway (Section 20.3.1), and fatty acid synthesis (Section 22.4.3).

16.3.4 The Conversion of Fructose 1,6-bisphosphate into Fructose 6-phosphate and Orthophosphate Is an Irreversible Step

On formation, phosphoenolpyruvate is metabolized by the enzymes of glycolysis but in the reverse direction. These reactions are near equilibrium under intracellular conditions; so, when conditions favor gluconeogenesis, the reverse reactions will take place until the next irreversible step is reached. This step is the hydrolysis of fructose 1,6-bisphosphate to fructose 6-phosphate and P_i.

$$\text{Fructose 1,6-bisphosphate} + H_2O \xrightarrow{\text{Fructose 1,6-bisphosphatase}} \text{fructose 6-phosphate} + P_i$$

The enzyme responsible for this step is fructose 1,6-bisphosphatase. Like its glycolytic counterpart, it is an allosteric enzyme that participates in the regulation of gluconeogenesis. We will return to its regulatory properties later in the chapter.

FIGURE 16.28 Compartmental cooperation. Oxaloacetate utilized in the cytosol for gluconeogenesis is formed in the mitochondrial matrix by carboxylation of pyruvate. Oxaloacetate leaves the mitochondrion by a specific transport system (not shown) in the form of malate, which is reoxidized to oxaloacetate in the cytosol.

16.3.5 The Generation of Free Glucose Is an Important Control Point

The fructose 6-phosphate generated by fructose 1,6-bisphosphatase is readily converted into glucose 6-phosphate. In most tissues, gluconeogenesis ends here. Free glucose is not generated; rather, the glucose 6-phosphate is processed in some other fashion, notably to form glycogen. One advantage to ending gluconeogenesis at glucose 6-phosphate is that, unlike free glucose, the molecule cannot diffuse out of the cell. To keep glucose inside the cell, the generation of free glucose is controlled in two ways. First, the enzyme responsible for the conversion of glucose 6-phosphate into glucose, *glucose 6-phosphatase*, is regulated. Second, the enzyme is present only in tissues whose metabolic duty is to maintain blood-glucose homeostasis— tissues that release glucose into the blood. These tissues are the liver and to a lesser extent the kidney.

This final step in the generation of glucose does not take place in the cytosol. Rather, glucose 6-phosphate is transported into the lumen of the endoplasmic reticulum, where it is hydrolyzed to glucose by glucose 6-phosphatase, which is bound to the membrane (Figure 16.29). An associated Ca^{2+}-binding stabilizing protein is essential for phosphatase activity. Glucose and P_i are then shuttled back to the cytosol by a pair of transporters. The glucose transporter in the endoplasmic reticulum membrane is like those found in the plasma membrane (Section 16.2.4). It is striking that five proteins are needed to transform cytosolic glucose 6-phosphate into glucose.

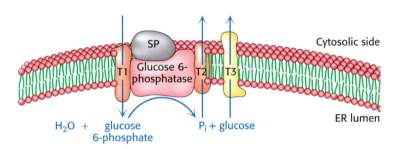

$H_2O +$ glucose 6-phosphate $\quad P_i +$ glucose

FIGURE 16.29 Generation of glucose from glucose 6-phosphate. Several endoplasmic reticulum (ER) proteins play a role in the generation of glucose from glucose 6-phosphate. T1 transports glucose 6-phosphate into the lumen of the ER, whereas T2 and T3 transport P_i and glucose, respectively, back into the cytosol. Glucose 6-phosphatase is stabilized by a Ca^{2+}-binding protein (SP). [After A. Buchell and I. D. Waddel. *Biochem. Biophys. Acta* 1092(1991):129.]

16.3.6 Six High-Energy Phosphate Bonds Are Spent in Synthesizing Glucose from Pyruvate

CONCEPTUAL INSIGHTS, Energetics of Glucose Metabolism. *See the section on gluconeogenesis in the Conceptual Insights module to review why and how gluconeogenesis must differ from the reversal of glycolysis.*

The stoichiometry of gluconeogenesis is:

$$2 \text{ Pyruvate} + 4 \text{ ATP} + 2 \text{ GTP} + 2 \text{ NADH} + 6 \text{ H}_2\text{O} \longrightarrow$$
$$\text{glucose} + 4 \text{ ADP} + 2 \text{ GDP} + 6 \text{ P}_i + 2 \text{ NAD}^+ + 2 \text{ H}^+$$
$$\Delta G^{\circ\prime} = -9 \text{ kcal mol}^{-1} (-38 \text{ kJ mol}^{-1})$$

In contrast, the stoichiometry for the reversal of glycolysis is:

$$2 \text{ Pyruvate} + 2 \text{ ATP} + 2 \text{ NADH} + 2 \text{ H}_2\text{O} \longrightarrow$$
$$\text{glucose} + 2 \text{ ADP} + 2 \text{ P}_i + 2 \text{ NAD}^+$$
$$\Delta G^{\circ\prime} = +20 \text{ kcal mol}^{-1} (+84 \text{ kJ mol}^{-1})$$

Note that *six* nucleotide triphosphate molecules are hydrolyzed to synthesize glucose from pyruvate in gluconeogenesis, whereas only *two* molecules of ATP are generated in glycolysis in the conversion of glucose into pyruvate. Thus, the extra cost of gluconeogenesis is four high phosphoryl-transfer potential molecules per molecule of glucose synthesized from pyruvate. The four additional high phosphoryl-transfer potential molecules are needed to turn an energetically unfavorable process (the reversal of glycolysis, $\Delta G^{\circ\prime} = +20$ kcal mol^{-1} [+84 kJ mol^{-1}]) into a favorable one (gluconeogenesis, $\Delta G^{\circ\prime} = -9$ kcal mol^{-1} [−38 kJ mol^{-1}]). This is a clear example of the coupling of reactions: ATP hydrolysis is used to power an energetically unfavorable reaction.

16.4 GLUCONEOGENESIS AND GLYCOLYSIS ARE RECIPROCALLY REGULATED

Gluconeogenesis and glycolysis are coordinated so that within a cell one pathway is relatively inactive while the other is highly active. If both sets of reactions were highly active at the same time, the net result would be the hydrolysis of four nucleotide triphosphates (two ATP plus two GTP) per reaction cycle. Both glycolysis and gluconeogenesis are highly exergonic under cellular conditions, and so there is no thermodynamic barrier to such simultaneous activity. However, the *amounts* and *activities* of the distinctive enzymes of each pathway are controlled so that both pathways are not highly active at the same time. The rate of glycolysis is also determined by the concentration of glucose, and the rate of gluconeogenesis by the concentrations of lactate and other precursors of glucose.

The interconversion of fructose 6-phosphate and fructose 1,6-bisphosphate is stringently controlled (Figure 16.30). As discussed in Section 16.2.1, AMP stimulates phosphofructokinase, whereas ATP and citrate inhibit it. Fructose 1,6-bisphosphatase, on the other hand, is inhibited by AMP and activated by citrate. A high level of AMP indicates that the energy charge is low and signals the need for ATP generation. Conversely,

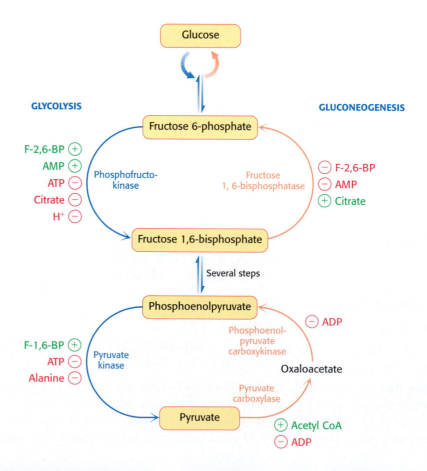

FIGURE 16.30 Reciprocal regulation of gluconeogenesis and glycolysis in the liver. The level of fructose 2,6-bisphosphate is high in the fed state and low in starvation. Another important control is the inhibition of pyruvate kinase by phosphorylation during starvation.

high levels of ATP and citrate indicate that the energy charge is high and that biosynthetic intermediates are abundant. Under these conditions, glycolysis is nearly switched off and gluconeogenesis is promoted.

Phosphofructokinase and fructose 1,6-bisphosphatase are also reciprocally controlled by *fructose 2,6-bisphosphate in the liver* (Section 16.2.2). The level of F-2,6-BP is low during starvation and high in the fed state, because of the antagonistic effects of glucagon and insulin on the production and degradation of this signal molecule. *Fructose 2,6-bisphosphate strongly stimulates phosphofructokinase and inhibits fructose 1,6-bisphosphatase.* Hence, glycolysis is accelerated and gluconeogenesis is diminished in the fed state. During starvation, gluconeogenesis predominates because the level of F-2,6-BP is very low. Glucose formed by the liver under these conditions is essential for the viability of brain and muscle.

The interconversion of phosphoenolpyruvate and pyruvate also is precisely regulated. Recall that pyruvate kinase is controlled by allosteric effectors and by phosphorylation (Section 16.2.3). High levels of ATP and alanine, which signal that the energy charge is high and that building blocks are abundant, inhibit the enzyme in liver. Conversely, pyruvate carboxylase, which catalyzes the first step in gluconeogenesis from pyruvate, is activated by acetyl CoA and inhibited by ADP. Likewise, ADP inhibits phosphoenolpyruvate carboxykinase. Hence, gluconeogenesis is favored when the cell is rich in biosynthetic precursors and ATP.

The amounts and the activities of these essential enzymes also are regulated. The regulators in this case are hormones. Hormones affect gene expression primarily by changing the rate of transcription, as well as by regulating the degradation of mRNA. Insulin, which rises subsequent to eating, stimulates the expression of phosphofructokinase, pyruvate kinase, and the bifunctional enzyme that makes and degrades F-2,6-BP. Glucagon, which rises during starvation, inhibits the expression of these enzymes and stimulates instead the production of two key gluconeogenic enzymes, phosphoenolpyruvate carboxykinase and fructose 1,6-bisphosphatase. Transcriptional control in eukaryotes is much slower than allosteric control; it takes hours or days in contrast with seconds to minutes. The richness and complexity of hormonal control are graphically displayed by the promoter of the phosphoenolpyruvate carboxykinase gene, which contains regulatory sequences that respond to insulin, glucagon, glucocorticoids, and thyroid hormone (Figure 16.31).

FIGURE 16.31 The promoter of the phosphoenolpyruvate carboxykinase gene. This promoter is approximately 500 bp in length and contains regulatory sequences (response elements) that mediate the action of several hormones. IRE, insulin response element; GRE, glucocorticoid response element; TRE, thyroid hormone response element; CREI and CREII, cAMP response elements. [After M. M. McGrane, J. S Jun, Y. M. Patel, and R. W. Hanson. *Trends Biochem. Sci.* 17(1992):40.]

16.4.1 Substrate Cycles Amplify Metabolic Signals and Produce Heat

A pair of reactions such as the phosphorylation of fructose 6-phosphate to fructose 1,6-bisphosphate and its hydrolysis back to fructose 6-phosphate is called a *substrate cycle*. As already mentioned, both reactions are not simultaneously fully active in most cells, because of reciprocal allosteric

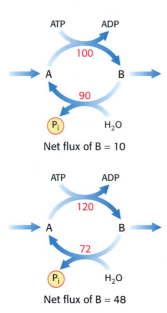

ATP   ADP

100

A   B

90

P_i H_2O

Net flux of B = 10

ATP   ADP

120

A   B

72

P_i H_2O

Net flux of B = 48

FIGURE 16.32 Substrate cycle. This ATP-driven cycle operates at two different rates. A small change in the rates of the two opposing reactions results in a large change in the *net* flux of product B.

controls. However, the results of isotope-labeling studies have shown that some fructose 6-phosphate is phosphorylated to fructose 1,6-bisphosphate in gluconeogenesis. There also is a limited degree of cycling in other pairs of opposed irreversible reactions. This cycling was regarded as an imperfection in metabolic control, and so substrate cycles have sometimes been called *futile cycles*. Indeed, there are pathological conditions, such as malignant hyperthermia, in which control is lost and both pathways proceed rapidly with the concomitant generation of heat by the rapid, uncontrolled hydrolysis of ATP.

Despite such extraordinary circumstances, it now seems likely that substrate cycles are biologically important. One possibility is that *substrate cycles amplify metabolic signals.* Suppose that the rate of conversion of A into B is 100 and of B into A is 90, giving an initial net flux of 10. Assume that an allosteric effector increases the A ⟶ B rate by 20% to 120 and reciprocally decreases the B ⟶ A rate by 20% to 72. The new net flux is 48, and so a 20% change in the rates of the opposing reactions has led to a 380% increase in the net flux. In the example shown in Figure 16.32, this amplification is made possible by the rapid hydrolysis of ATP. It has been suggested that the flux down the glycolytic pathway may increase 1000-fold at the initiation of intense exercise. Because it seems unlikely that allosteric activation of enzymes alone could explain this increased flux, the existence of substrate cycles may partly account for the rapid rise in the rate of glycolysis.

The other potential biological role of substrate cycles is the *generation of heat produced by the hydrolysis of ATP.* A striking example is provided by bumblebees, which must maintain a thoracic temperature of about 30°C to fly. A bumblebee is able to maintain this high thoracic temperature and forage for food even when the ambient temperature is only 10°C because phosphofructokinase and fructose 1,6-bisphosphatase in its flight muscle are simultaneously highly active; the continuous hydrolysis of ATP generates heat. This bisphosphatase is not inhibited by AMP, which suggests that the enzyme is specially designed for the generation of heat. In contrast, the honeybee has almost no fructose 1,6-bisphosphatase activity in its flight muscle and consequently cannot fly when the ambient temperature is low.

16.4.2 Lactate and Alanine Formed by Contracting Muscle Are Used by Other Organs

Lactate produced by active skeletal muscle and erythrocytes is a source of energy for other organs. Erythrocytes lack mitochondria and can never oxidize glucose completely. In contracting skeletal muscle during vigorous exercise, the rate at which glycolysis produces pyruvate exceeds the rate at which the citric acid cycle oxidizes it. Under these conditions, moreover, the rate of formation of NADH by glycolysis is greater than the rate of its oxidation by aerobic metabolism. Continued glycolysis depends on the availability of NAD^+ for the oxidation of glyceraldehyde 3-phosphate. The accumulation of NADH is reversed by lactate dehydrogenase, which oxidizes NADH to NAD^+ as it reduces pyruvate to lactate (Section 16.1.7). However, lactate is a dead end in metabolism. It must be converted back into pyruvate before it can be metabolized. The only role of the reduction of pyruvate to lactate is to regenerate NAD^+ so that glycolysis can proceed in active skeletal muscle and erythrocytes. *The formation of lactate buys time and shifts part of the metabolic burden from muscle to other organs.*

The plasma membranes of some cells, particularly cells in cardiac muscle, contain carriers that make them highly permeable to lactate and pyruvate. Both substances diffuse out of active skeletal muscle into the blood and then into these permeable cells. Once inside these well-oxygenated cells,

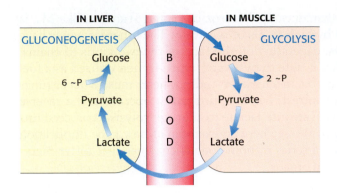

FIGURE 16.33 The Cori cycle. Lactate formed by active muscle is converted into glucose by the liver. This cycle shifts part of the metabolic burden of active muscle to the liver.

lactate can be reverted back to pyruvate and metabolized through the citric acid cycle and oxidative phosphorylation to generate ATP. The use of lactate in place of glucose by these cells makes more circulating glucose available to the active muscle cells. Excess lactate enters the liver and is converted first into pyruvate and then into glucose by the gluconeogenic pathway. *Thus, the liver restores the level of glucose necessary for active muscle cells, which derive ATP from the glycolytic conversion of glucose into lactate. Contracting skeletal muscle supplies lactate to the liver, which uses it to synthesize glucose. These reactions constitute the Cori cycle* (Figure 16.33). Studies have shown that alanine, like lactate, is a major precursor of glucose. In muscle, alanine is formed from pyruvate by transamination (Section 24.2.2); the reverse reaction takes place in the liver. This process also helps maintain nitrogen balance. The interplay between glycolysis and gluconeogenesis is summized in Figure 16.34, which shows how these pathways help meet the energy needs of different cell types.

Isozymic forms of lactate dehydrogenase in different tissues catalyze the interconversions of pyruvate and lactate. Lactate dehydrogenase is a tetramer of two kinds of 35-kd subunits encoded by similar genes: the H type predominates in the heart, and the homologous M type in skeletal muscle and the liver. These subunits associate to form five types of tetramers: H_4, H_3M_1, H_2M_2, H_1M_3, and M_4. The H_4 isozyme (type 1) has higher

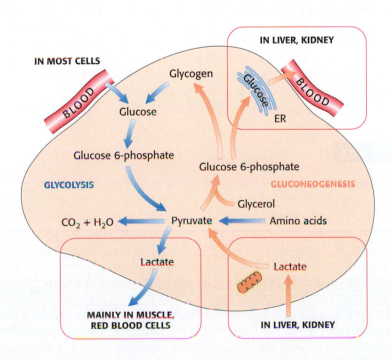

FIGURE 16.34 Cooperation between glycolysis and gluconeogenesis. Glycolysis and gluconeogenesis are coordinated, in a tissue-specific fashion, to ensure that the glucose-dependent energy needs of all cells are met.

affinity for substrates than does the M_4 isozyme (type 5) and, unlike M_4, is allosterically inhibited by high levels of pyruvate. The other isozymes have intermediate properties, depending on the ratio of the two kinds of chains. H_4 is designed to oxidize lactate to pyruvate, which is then utilized as a fuel by the heart through aerobic metabolism. Indeed, heart muscle never functions anaerobically. In contrast, M_4 is optimized to operate in the reverse direction, to convert pyruvate into lactate to allow glycolysis to proceed under anaerobic conditions. We see here an example of how gene duplication and divergence generate a series of homologous enzymes that foster metabolic cooperation between organs.

16.4.3 Glycolysis and Gluconeogenesis Are Evolutionarily Intertwined

The metabolism of glucose has ancient origins. Organisms living in the early biosphere depended on the anaerobic generation of energy until significant amounts of oxygen began to accumulate 2 billion years ago. The fact that glycolytic enzymes with similar properties do not have similar amino acid sequences also provides a clue to how the pathway originated. Although there are four kinases and two isomerases in the pathway, both sequence and structural comparisons do not suggest that these sets of enzymes are related to one another by divergent evolution. The absence of such similarities implies that glycolytic enzymes were derived independently rather than by gene duplication. The common dinucleotide-binding domain found in the dehydrogenases (Section 16.1.10) and the $\alpha\beta$ barrels are the only major recurring elements.

We can speculate on the relationship between glycolysis and gluconeogenesis if we think of glycolysis as consisting of two segments: the metabolism of hexoses (the upper segment) and the metabolism of trioses (the lower segment). The enzymes from the upper segment are different in some species and are missing entirely in some archaea, whereas enzymes from the lower segment are quite conserved. In fact, four enzymes of the lower segment are present in all species. *This lower part of the pathway is common to glycolysis and gluconeogenesis.* This common part of the two pathways may be the oldest part, constituting the core to which the other steps were added. The upper part would vary according to the sugars that were available to evolving organisms in particular niches. Interestingly, this core part of carbohydrate metabolism can generate triose precursors for ribose sugars, a component of RNA and a critical requirement for the RNA world. Thus, we are left with the unanswered question, Was the original core pathway used for energy conversion or biosynthesis?

SUMMARY

● **Glycolysis Is an Energy-Conversion Pathway in Many Organisms**

Glycolysis is the set of reactions that converts glucose into pyruvate. The 10 reactions of glycolysis take place in the cytosol. In the first stage, glucose is converted into fructose 1,6-bisphosphate by a phosphorylation, an isomerization, and a second phosphorylation reaction. Two molecules of ATP are consumed per molecule of glucose in these reactions, which are the prelude to the net synthesis of ATP. In the second stage, fructose 1,6-bisphosphate is cleaved by aldolase into dihydroxyacetone phosphate and glyceraldehyde 3-phosphate, which are readily interconvertible. In the third stage, ATP is generated. Glyceraldehyde 3-phosphate is oxidized and phosphorylated to form 1,3-bisphosphoglycerate, an acyl phosphate with a high phosphoryl-transfer potential. This molecule transfers a phosphoryl group to ADP to form ATP and 3-phosphoglycerate. A phosphoryl shift and a dehydration form phosphoenolpyruvate, a second intermediate with a high phosphoryl-transfer potential. Another molecule of ATP is generated as phosphoenolpyruvate is converted into pyruvate. There is a net gain of two molecules of ATP in the formation of two molecules of pyruvate from one molecule of glucose.

The electron acceptor in the oxidation of glyceraldehyde 3-phosphate is NAD^+, which must be regenerated for glycolysis to continue. In aerobic organisms, the NADH formed in glycolysis transfers its electrons to O_2 through the electron-transport chain, which thereby regenerates NAD^+. Under anaerobic conditions and in some microorganisms, NAD^+ is regenerated by the reduction of pyruvate to lactate. In other microorganisms, NAD^+ is regenerated by the reduction of pyruvate to ethanol. These two processes are examples of fermentations.

The Glycolytic Pathway Is Tightly Controlled

The glycolytic pathway has a dual role: it degrades glucose to generate ATP, and it provides building blocks for the synthesis of cellular components. The rate of conversion of glucose into pyruvate is regulated to meet these two major cellular needs. Under physiologic conditions, the reactions of glycolysis are readily reversible except for the ones catalyzed by hexokinase, phosphofructokinase, and pyruvate kinase. Phosphofructokinase, the most important control element in glycolysis, is inhibited by high levels of ATP and citrate, and it is activated by AMP and fructose 2,6-bisphosphate. In the liver, this bisphosphate signals that glucose is abundant. Hence, phosphofructokinase is active when either energy or building blocks are needed. Hexokinase is inhibited by glucose 6-phosphate, which accumulates when phosphofructokinase is inactive. ATP and alanine allosterically inhibit pyruvate kinase, the other control site, and fructose 1,6-bisphosphate activates the enzyme. Consequently, pyruvate kinase is maximally active when the energy charge is low and glycolytic intermediates accumulate.

Glucose Can Be Synthesized from Noncarbohydrate Precursors

Gluconeogenesis is the synthesis of glucose from noncarbohydrate sources, such as lactate, amino acids, and glycerol. Several of the reactions that convert pyruvate into glucose are common to glycolysis. Gluconeogenesis, however, requires four new reactions to bypass the essential irreversibility of three reactions in glycolysis. In two of the new reactions, pyruvate is carboxylated in mitochondria to oxaloacetate, which in turn is decarboxylated and phosphorylated in the cytosol to phosphoenolpyruvate. Two high phosphoryl-transfer potential molecules are consumed in these reactions, which are catalyzed by pyruvate carboxylase and phosphoenolpyruvate carboxykinase. Pyruvate carboxylase contains a biotin prosthetic group. The other distinctive reactions of gluconeogenesis are the hydrolyses of fructose 1,6-bisphosphate and glucose 6-phosphate, which are catalyzed by specific phosphatases. The major raw materials for gluconeogenesis by the liver are lactate and alanine produced from pyruvate by active skeletal muscle. The formation of lactate during intense muscular activity buys time and shifts part of the metabolic burden from muscle to the liver.

Gluconeogenesis and Glycolysis Are Reciprocally Regulated

Gluconeogenesis and glycolysis are reciprocally regulated so that one pathway is relatively inactive while the other is highly active. Phosphofructokinase and fructose 1,6-bisphosphatase are key control points.

Fructose 2,6-bisphosphate, an intracellular signal molecule present at higher levels when glucose is abundant, activates glycolysis and inhibits gluconeogenesis by regulating these enzymes. Pyruvate kinase and pyruvate carboxylase are regulated by other effectors so that both are not maximally active at the same time. Allosteric regulation and reversible phosphorylation, which are rapid, are complemented by transcriptional control, which takes place in hours or days.

KEY TERMS

glycolysis (p. 425)

lactic acid fermentation (p. 425)

alcoholic fermentation (p. 425)

gluconeogenesis (p. 425)

obligate anaerobe (p. 427)

facultative anaerobe (p. 427)

hexokinase (p. 428)

kinase (p. 428)

phosphofructokinase (PFK) (p. 430)

thioester intermediate (p. 434)

substrate-level phosphorylation (p. 435)

mutase (p. 436)

enol phosphate (p. 436)

pyruvate kinase (p. 437)

bifunctional enzyme (p. 445)

feedforward stimulation (p. 446)

committed step (p. 447)

pyruvate carboxylase (p. 452)

biotin (p. 453)

glucose 6-phosphatase (p. 455)

substrate cycle (p. 457)

Cori cycle (p. 459)

SELECTED READINGS

Where to start

Knowles, J. R., 1991. Enzyme catalysis: Not different, just better. *Nature* 350:121–124.

Granner, D., and Pilkis, S., 1990. The genes of hepatic glucose metabolism. *J. Biol. Chem.* 265:10173–10176.

McGrane, M. M., Yun, J. S., Patel, Y. M., and Hanson, R. W., 1992. Metabolic control of gene expression: In vivo studies with transgenic mice. *Trends Biochem. Sci.* 17:40–44.

Pilkis, S. J., and Granner, D. K., 1992. Molecular physiology of the regulation of hepatic gluconeogenesis and glycolysis. *Annu. Rev. Physiol.* 54:885–909.

Books

Fell, D., 1997. *Understanding the Control of Metabolism.* Portland.

Fersht, A., 1999. *Structure and Mechanism in Protein Science: A Guide to Enzyme Catalysis and Protein Folding.* W. H. Freeman and Company.

Frayn, K. N., 1996. *Metabolic Regulation: A Human Perspective.* Portland.

Hargreaves, M., and Thompson, M. (Eds.) 1999. *Biochemistry of Exercise X.* Human Kinetics.

Structure of glycolytic and gluconeogenic enzymes

Aleshin, A. E., Kirby, C., Liu, X., Bourenkov, G. P., Bartunik, H. D., Fromm, H. J., and Honzatko, R. B., 2000. Crystal structures of mutant monomeric hexokinase I reveal multiple ADP binding sites and conformational changes relevant to allosteric regulation. *J. Mol. Biol.* 296:1001–1015.

Jeffery, C. J., Bahnson, B. J., Chien, W., Ringe, D., and Petsko, G. A., 2000. Crystal structure of rabbit phosphoglucose isomerase, a glycolytic enzyme that moonlights as neuroleukin, autocrine motility factor, and differentiation mediator. *Biochemistry* 39:955–964.

Schirmer, T., and Evans, P. R., 1990. Structural basis of the allosteric behaviour of phosphofructokinase. *Nature* 343:140–145.

Cooper, S. J., Leonard, G. A., McSweeney, S. M., Thompson, A. W., Naismith, J. H., Qamar, S., Plater, A., Berry, A., and Hunter, W. N., 1996. The crystal structure of a class II fructose-1,6-bisphosphate aldolase shows a novel binuclear metal-binding active site embedded in a familiar fold. *Structure* 4:1303–1315.

Davenport, R. C., Bash, P. A., Seaton, B. A., Karplus, M., Petsko, G. A., and Ringe, D., 1991. Structure of the triosephosphate isomerase–phosphoglycolohydroxamate complex: An analogue of the intermediate on the reaction pathway. *Biochemistry* 30:5821–5826.

Skarzynski, T., Moody, P. C., and Wonacott, A. J., 1987. Structure of holo-glyceraldehyde-3-phosphate dehydrogenase from *Bacillus stearothermophilus* at 1.8 Å resolution. *J. Mol. Biol.* 193:171–187.

Bernstein, B. E., and Hol, W. G., 1998. Crystal structures of substrates and products bound to the phosphoglycerate kinase active site reveal the catalytic mechanism. *Biochemistry* 37:4429–4436.

Rigden, D. J., Alexeev, D., Phillips, S. E. V., and Fothergill-Gilmore, L. A., 1998. The 2.3 Å X-ray crystal structure of *S. cerevisiae* phosphoglycerate mutase. *J. Mol. Biol.* 276:449–459.

Zhang, E., Brewer, J. M., Minor, W., Carreira, L. A., and Lebioda, L., 1997. Mechanism of enolase: The crystal structure of asymmetric dimer enolase-2-phospho-D-glycerate/enolase-phosphoenolpyruvate at 2.0 Å resolution. *Biochemistry* 36:12526–12534.

Mattevi, A., Valentini, G., Rizzi, M., Speranza, M. L., Bolognesi, M., and Coda, A., 1995. Crystal structure of *Escherichia coli* pyruvate kinase type I: Molecular basis of the allosteric transition. *Structure* 3:729–741.

Hasemann, C. A., Istvan E. S., Uyeda, K., and Deisenhofer, J., 1996. The crystal structure of the bifunctional enzyme 6-phosphofructo-2-kinase/fructose-2,6-biphosphatase reveals distinct domain homologies. *Structure* 4:1017–1029.

Tari, L. W., Matte, A., Pugazhenthi, U., Goldie, H., and Delbaere, L. T. J., 1996. Snapshot of an enzyme reaction intermediate in the structure of the ATP-Mg^{2+}-oxalate ternary complex of *Escherichia coli* PEP carboxykinase. *Nat. Struct. Biol.* 3:355–363.

Catalytic mechanisms

Soukri, A., Mougin, A., Corbier, C., Wonacott, A., Branlant, C., and Branlant, G., 1989. Role of the histidine 176 residue in glyceraldehyde-3-phosphate dehydrogenase as probed by site-directed mutagenesis. *Biochemistry* 28:2586–2592.

Bash, P. A., Field, M. J., Davenport, R. C., Petsko, G. A., Ringe, D., and Karplus, M., 1991. Computer simulation and analysis of the reaction pathway of triosephosphate isomerase. *Biochemistry* 30:5826–5832.

Knowles, J. R., and Albery, W. J., 1977. Perfection in enzyme catalysis: The energetics of triosephosphate isomerase. *Acc. Chem. Res.* 10:105–111.

Rose, I. A., 1981. Chemistry of proton abstraction by glycolytic enzymes (aldolase, isomerases, and pyruvate kinase). *Philos. Trans. R. Soc. Lond.: Series B, Biol. Sci.* 293:131–144.

Regulation

Dang, C. V., and Semenza, G. L., 1999. Oncogenic alterations of metabolism. *Trends Biochem. Sci.* 24:68–72.

Depre, C., Rider, M. H., and Hue, L., 1998. Mechanisms of control of heart glycolysis. *Eur. J. Biochem.* 258:277–290.

Gleeson, T. T., 1996. Post-exercise lactate metabolism: A comparative review of sites, pathways, and regulation. *Annu. Rev. Physiol* 58:556–581.

Hers, H. -G., and Van Schaftingen, E., 1982. Fructose 2,6-bisphosphate two years after its discovery. *Biochem. J.* 206:1–12.

Middleton, R. J., 1990. Hexokinases and glucokinases. *Biochem. Soc. Trans.* 18:180–183.

Nordlie, R. C., Foster, J. D., and Lange, A. J., 1999. Regulation of glucose production by the liver. *Annu. Rev. Nutr.* 19:379–406.

Jitrapakdee, S., and Wallace, J. C., 1999. Structure, function and regulation of pyruvate carboxylase. *Biochem. J.* 340:1–16.

Pilkis, S. J., and Claus, T. H., 1991. Hepatic gluconeogenesis/glycolysis: Regulation and structure/function relationships of substrate cycle enzymes. *Annu. Rev. Nutr.* 11:465–515.

Plaxton, W. C., 1996. The organization and regulation of plant glycolysis. *Annu. Rev. Plant Physiol. Plant Mol. Biol.* 47:185–214.

van de Werve, G., Lange, A., Newgard, C., Mechin, M. C., Li, Y., and Berteloot, A., 2000. New lessons in the regulation of glucose metabolism taught by the glucose 6-phosphatase system. *Eur. J. Biochem.* 267:1533–1549.

Sugar transporters

Czech, M. P., and Corvera, S., 1999. Signaling mechanisms that regulate glucose transport. *J. Biol. Chem.* 274:1865–1868.

Silverman, M., 1991. Structure and function of hexose transporters. *Annu. Rev. Biochem.* 60:757–794.

Thorens, B., Charron, M. J., and Lodish, H. F., 1990. Molecular physiology of glucose transporters. *Diabetes Care* 13:209–218.

Genetic diseases

Scriver, C. R., Beaudet, A. L., Sly, W. S., and Valle, D. (Eds.), 1995. *The Metabolic and Molecular Basis of Inherited Disease* (7th ed.). McGraw-Hill.

Evolution

Dandekar, T., Schuster, S., Snel, B., Huynen, M., and Bork, P., 1999. Pathway alignment: Application to the comparative analysis of glycolytic enzymes. *Biochem. J.* 343:115–124.

Heinrich, R., Melendez-Hevia, E., Montero, F., Nuno, J. C., Stephani, A., and Waddell, T. G., 1999. The structural design of glycolysis: An evolutionary approach. *Biochem. Soc. Trans.* 27:294–298.

Walmsley, A. R., Barrett, M. P., Bringaud, F., and Gould, G. W., 1998. Sugar transporters from bacteria, parasites and mammals: Structure-activity relationships. *Trends Biochem. Sci.* 23:476–480.

Maes, D., Zeelen, J. P., Thanki, N., Beaucamp, N., Alvarez, M., Thi, M. H., Backmann, J., Martial, J. A., Wyns, L., Jaenicke, R., and Wierenga, R. K., 1999. The crystal structure of triosephosphate isomerase (TIM) from *Thermotoga maritima*: A comparative thermostability structural analysis of ten different TIM structures. *Proteins* 37:441–453.

Historical aspects

Fruton, J. S., 1972. *Molecules and Life: Historical Essays on the Interplay of Chemistry and Biology.* Wiley Interscience.

Kalckar, H. M. (Ed.), 1969. *Biological Phosphorylations: Development of Concepts.* Prentice Hall.

PROBLEMS

1. *Kitchen chemistry.* Sucrose is commonly used to preserve fruits. Why is glucose not suitable for preserving foods?

2. *Tracing carbon atoms I.* Glucose labeled with ^{14}C at C-1 is incubated with the glycolytic enzymes and necessary cofactors.

(a) What is the distribution of ^{14}C in the pyruvate that is formed? (Assume that the interconversion of glyceraldehyde 3-phosphate and dihydroxyacetone phosphate is very rapid compared with the subsequent step.)

(b) If the specific activity of the glucose substrate is 10 mCi mM^{-1}, what is the specific activity of the pyruvate that is formed?

3. *Lactic acid fermentation.* (a) Write a balanced equation for the conversion of glucose into lactate. (b) Calculate the standard free-energy change of this reaction by using the data given in Table 16.3 and the fact that $\Delta G^{\circ\prime}$ is -6 kcal for the following reaction:

$$\text{Pyruvate} + \text{NADH} + \text{H}^+ \rightleftharpoons \text{lactate} + \text{NAD}^+$$

What is the free-energy change (ΔG, not $\Delta G^{\circ\prime}$) of this reaction when the concentrations of reactants are: glucose, 5 mM; lactate, 0.05 mM; ATP, 2 mM; ADP, 0.2 mM; and P_i, 1 mM?

4. *High potential.* What is the equilibrium ratio of phosphoenolpyruvate to pyruvate under standard conditions when [ATP]/[ADP] = 10?

5. *Hexose–triose equilibrium.* What are the equilibrium concentrations of fructose 1,6-bisphosphate, dihydroxyacetone phosphate, and glyceraldehyde 3-phosphate when 1 mM fructose 1,6-bisphosphate is incubated with aldolase under standard conditions?

6. *Double labeling.* 3-Phosphoglycerate labeled uniformly with ^{14}C is incubated with 1,3-BPG labeled with ^{32}P at C-1. What is the radioisotope distribution of the 2,3-BPG that is formed on addition of BPG mutase?

7. *An informative analog.* Xylose has the same structure as that of glucose except that it has a hydrogen atom at C-6 in place of a hydroxymethyl group. The rate of ATP hydrolysis by hexokinase is markedly enhanced by the addition of xylose. Why?

8. *Distinctive sugars.* The intravenous infusion of fructose into healthy volunteers leads to a two- to fivefold increase in the level of lactate in the blood, a far greater increase than that observed after the infusion of the same amount of glucose.

(a) Why is glycolysis more rapid after the infusion of fructose?

(b) Fructose has been used in place of glucose for intravenous feeding. Why is this use of fructose unwise?

9. *Metabolic mutants.* Predict the effect of each of the following mutations on the pace of glycolysis in liver cells:

(a) Loss of the allosteric site for ATP in phosphofructokinase.

(b) Loss of the binding site for citrate in phosphofructokinase.

(c) Loss of the phosphatase domain of the bifunctional enzyme that controls the level of fructose 2,6-bisphosphate.
(d) Loss of the binding site for fructose 1,6-bisphosphate in pyruvate kinase.

10. *Metabolic mutant.* What are the likely consequences of a genetic disorder rendering fructose 1,6-bisphosphatase in liver less sensitive to regulation by fructose 2,6-bisphosphate?

11. *Biotin snatcher.* Avidin, a 70-kd protein in egg white, has very high affinity for biotin. In fact, it is a highly specific inhibitor of biotin enzymes. Which of the following conversions would be blocked by the addition of avidin to a cell homogenate?

(a) Glucose $\longrightarrow$ pyruvate
(b) Pyruvate $\longrightarrow$ glucose
(c) Oxaloacetate $\longrightarrow$ glucose
(d) Malate $\longrightarrow$ oxaloacetate
(e) Pyruvate $\longrightarrow$ oxaloacetate
(f) Glyceraldehyde 3-phosphate $\longrightarrow$ fructose 1,6-bisphosphate

12. *Tracing carbon atoms II.* If cells synthesizing glucose from lactate are exposed to CO_2 labeled with ^{14}C, what will be the distribution of label in the newly synthesized glucose?

13. *Arsenate poisoning.* Arsenate (AsO_4^{3-}) closely resembles P_i in structure and reactivity. In the reaction catalyzed by glyceraldehyde 3-phosphate dehydrogenase, arsenate can replace phosphate in attacking the energy-rich thioester intermediate. The product of this reaction, 1-arseno-3-phosphoglycerate, is unstable. It and other acyl arsenates are rapidly and spontaneously hydrolyzed. What is the effect of arsenate on energy generation in a cell?

14. *Reduce, reuse, recycle.* In the conversion of glucose into two molecules of lactate, the NADH generated earlier in the pathway is oxidized to NAD^+. Why is it not to the cell's advantage to simply make more NAD^+ so that the regeneration would not be necessary? After all, the cell would save much energy because it would no longer need to synthesize lactic acid dehydrogenase.

15. *Adenylate kinase again.* Adenylate kinase, an enzyme discussed in great detail in Chapter 9, is responsible for interconverting the adenylate nucleotide pool:

$$ADP + ADP \xrightleftharpoons{\text{Adenylate kinase}} ATP + AMP$$

The equilibrium constant for this reaction is close to 1, inasmuch as the number of phosphoanhydride bonds is the same on each side of the equation. Using the equation for the equilibrium constant for this reaction, show why changes in [AMP] are a more effective indicator of the adenylate pool than [ATP].

16. *Working at cross-purposes?* Gluconeogenesis takes place during intense exercise, which seems counterintuitive. Why would an organism synthesize glucose and at the same time use glucose to generate energy?

17. *Powering pathways.* Compare the stoichiometries of glycolysis and gluconeogenesis (Sections 16.1.8 and 16.3.6). Recall that the input of one ATP equivalent changes the equilibrium constant of a reaction by a factor of about 10^8 (Section 14.1.3). By what factor do the additional high phosphoryl-transfer potential compounds alter the equilibrium constant of gluconeogenesis?

Mechanism Problem

18. *Argument by analogy.* Propose a mechanism for the conversion of glucose 6-phosphate into fructose 6-phosphate by phosphoglucose isomerase based on the mechanism of triose phosphate isomerase.

Chapter Integration Problem

19. *Not just for energy.* People with galactosemia display central nervous system abnormalities even if galactose is eliminated from the diet. The precise reason for it is not known. Suggest a plausible explanation.

Data Interpretation Problem

20. *Now, that's unusual.* Phosphofructokinase has recently been isolated from the hyperthermophilic archaeon *Pyrococcus furiosus*. It was subjected to standard biochemical analysis to determine basic catalytic parameters. The processes under study were of the form:

Fructose 6-phosphate + $(x - P_i)\longrightarrow$
$$\text{fructose 1,6-bisphosphate} + (x)$$

The assay measured the increase in fructose 1,6-bisphosphate. Selected results are shown in the adjoining graph.

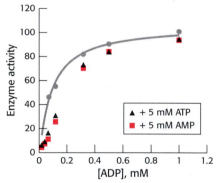

[Data after: J. E. Tuininga et al. (1999) *J. Biol Chem.* 274:21023–21028.]

(a) How does the *P. furiosus* phosphofructokinase differ from the phosphofructokinase discussed in this chapter?
(b) What effects do AMP and ATP have on the reaction with ADP?

Media Problems

21. *No free lunch.* Suppose a microorganism is discovered to be lacking pyruvate kinase. Nevertheless, the organism does metabolize glucose to pyruvate. Surprisingly, it does so with a net gain of four molecules of ATP (or ATP equivalents) per molecule of metabolized glucose. Looking at the **Conceptual Insights** module on the energetics of glucose metabolism, suggest a biochemical pathway for metabolism of glucose in this organism. Why might evolution have favored the normal glycolytic pathway rather than this alternative?

22. *Oxidation in the absence of oxygen.* The energetics of glycolysis section of the **Conceptual Insights** module shows that there is a large free energy drop upon oxidation of glyceraldehyde 3-phosphate to 1,3-bisphosphoglycerate (reaction 6). In the presence of oxygen, some to this energy is ultimately converted into ATP production. However, no such conversion happens under anaerobic conditions. Explain why.

CHAPTER 17

The Citric Acid Cycle

Roundabouts, or traffic circles, function as hubs to facilitate traffic flow.
The citric acid cycle is the biochemical hub of the cell, oxidizing carbon fuels, usually in the form of acetyl CoA, as well as serving as a source of precursors for biosynthesis. [(Above) Chris Warren/International Stock.]

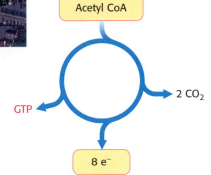

From Chapter 16, we know that glucose can be metabolized to pyruvate anaerobically to synthesize ATP through the glycolytic pathway. Glycolysis, however, harvests but a fraction of the ATP available from glucose. We now begin an exploration of the aerobic processing of glucose, which is the source of most of the ATP generated in metabolism. The aerobic processing of glucose starts with the complete oxidation of glucose derivatives to carbon dioxide. This oxidation takes place in the *citric acid cycle,* a series of reactions also known as the *tricarboxylic acid (TCA) cycle* or the *Krebs cycle.* The citric acid cycle is the *final common pathway for the oxidation of fuel molecules*—amino acids, fatty acids, and carbohydrates. Most fuel molecules enter the cycle as acetyl coenzyme A.

OUTLINE

- **17.1 The Citric Acid Cycle Oxidizes Two-Carbon Units**

- **17.2 Entry to the Citric Acid Cycle and Metabolism Through It Are Controlled**

- **17.3 The Citric Acid Cycle Is a Source of Biosynthetic Precursors**

- **17.4 The Glyoxylate Cycle Enables Plants and Bacteria to Grow on Acetate**

Acetyl coenzyme A (Acetyl CoA)

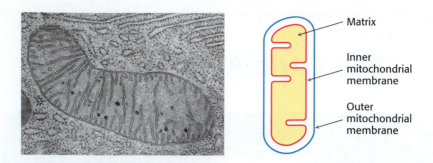

FIGURE 17.1 Mitochondrion. The double membrane of the mitochondrion is evident in this electron micrograph. The numerous invaginations of the inner mitochondrial membrane are called cristae. The oxidative decarboxylation of pyruvate and the sequence of reactions in the citric acid cycle take place within the matrix. [(Left) Omikron/Photo Researchers.]

Under aerobic conditions, the pyruvate generated from glucose is oxidatively decarboxylated to form acetyl CoA. In eukaryotes, the reactions of the citric acid cycle take place inside mitochondria, in contrast with those of glycolysis, which take place in the cytosol (Figure 17.1).

17.0.1 An Overview of the Citric Acid Cycle

The citric acid cycle is the central metabolic hub of the cell. It is the gateway to the aerobic metabolism of any molecule that can be transformed into an acetyl group or dicarboxylic acid. The cycle is also an important source of precursors, not only for the storage forms of fuels, but also for the building blocks of many other molecules such as amino acids, nucleotide bases, cholesterol, and porphyrin (the organic component of heme).

What is the function of the citric acid cycle in transforming fuel molecules into ATP? Recall that fuel molecules are carbon compounds that are capable of being oxidized—of losing electrons (Chapter 14). The citric acid cycle includes a series of oxidation–reduction reactions that result in the oxidation of an acetyl group to two molecules of carbon dioxide.

The overall pattern of the citric acid cycle is shown in Figure 17.2. A four- carbon compound (oxaloacetate) condenses with a two-carbon acetyl unit to yield a six-carbon tricarboxylic acid (citrate). An isomer of citrate is then oxidatively decarboxylated. The resulting five-carbon compound (α-ketoglutarate) also is oxidatively decarboxylated to yield a four-carbon compound (succinate). Oxaloacetate is then regenerated from succinate. Two carbon atoms enter the cycle as an acetyl unit and two carbon atoms leave the cycle in the form of two molecules of carbon dioxide. Three hydride ions (hence, six electrons) are transferred to three molecules of nicotinamide adenine dinucleotide (NAD^+), whereas one pair of hydrogen atoms (hence, two electrons) is transferred to one molecule of flavin adenine dinucleotide (FAD). *The function of the citric acid cycle is the harvesting of high-energy electrons from carbon fuels.* Note that the citric acid cycle itself nei-

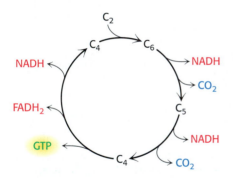

FIGURE 17.2 Overview of the citric acid cycle. The citric acid cycle oxidizes two-carbon units, producing two molecules of CO_2, one molecule of GTP, and high-energy electrons in the form of NADH and $FADH_2$.

FIGURE 17.3 Cellular respiration. The citric acid cycle constitutes the first stage in cellular respiration, the removal of high-energy electrons from carbon fuels (left). These electrons reduce O_2 to generate a proton gradient (middle), which is used to synthesize ATP (right). The reduction of O_2 and the synthesis of ATP constitute oxidative phosphorylation.

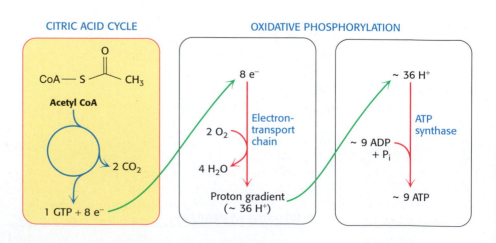

ther generates a large amount of ATP nor includes oxygen as a reactant (Figure 17.3). Instead, the citric acid cycle removes electrons from acetyl CoA and uses these electrons to form NADH and $FADH_2$. In *oxidative phosphorylation* (Chapter 18), electrons released in the reoxidation of NADH and $FADH_2$ flow through a series of membrane proteins (referred to as the *electron-transport chain*) to generate a proton gradient across the membrane. These protons then flow through ATP synthase to generate ATP from ADP and inorganic phosphate. Oxygen is required for the citric acid cycle indirectly inasmuch as it is the electron acceptor at the end of the electron-transport chain, necessary to regenerate NAD^+ and FAD.

The citric acid cycle, in conjunction with oxidative phosphorylation, provides the vast majority of energy used by aerobic cells—in human beings, greater than 95%. It is highly efficient because a limited number of molecules can generate large amounts of NADH and $FADH_2$. Note in Figure 17.2 that the four-carbon molecule, oxaloacetate, that initiates the first step in the citric acid cycle is regenerated at the end of one passage through the cycle. The oxaloacetate acts catalytically: it participates in the oxidation of the acetyl group but is itself regenerated. Thus, one molecule of oxaloacetate is capable of participating in the oxidation of many acetyl molecules.

17.1 THE CITRIC ACID CYCLE OXIDIZES TWO-CARBON UNITS

Acetyl CoA is the fuel for the citric acid cycle. This important molecule is formed from the breakdown of glycogen (the storage form of glucose), fats, and many amino acids. Indeed, as we will see in Chapter 22, fats contain strings of reduced two-carbon units that are first oxidized to acetyl CoA and then completely oxidized to CO_2 by the citric acid cycle.

17.1.1 The Formation of Acetyl Coenzyme A from Pyruvate

The formation of acetyl CoA from carbohydrates is less direct than from fat. Recall that carbohydrates, most notably glucose, are processed by glycolysis into pyruvate (Chapter 16). Under anaerobic conditions, the pyruvate is converted into lactic acid or ethanol, depending on the organism. Under aerobic conditions, the pyruvate is transported into mitochondria in exchange for OH^- by the pyruvate carrier, an antiporter (Section 13.4). In the mitochondrial matrix, pyruvate is oxidatively decarboxylated by the *pyruvate dehydrogenase complex* to form acetyl CoA.

$$\text{Pyruvate} + \text{CoA} + NAD^+ \longrightarrow \text{acetyl CoA} + CO_2 + \text{NADH}$$

This irreversible reaction is the link between glycolysis and the citric acid cycle. (Figure 17.4) Note that, in the preparation of the glucose derivative pyruvate for the citric acid cycle, an oxidative decarboxylation takes place and high-transfer-potential electrons in the form of NADH are captured. Thus, the pyruvate dehydrogenase reaction has many of the key features of the reactions of the citric acid cycle itself.

The pyruvate dehydrogenase complex is a large, highly integrated complex of three kinds of enzymes (Table 17.1). Pyruvate dehydrogenase is a member of a family of homologous complexes that includes the citric acid cycle enzyme α-ketoglutarate dehydrogenase (Section 17.1.6), a branched-chain α-ketoacid dehydrogenase, and acetoin dehydrogenase, found in certain prokaryotes. These complexes are giant, with molecular masses ranging from 4 to 10 million daltons (Figure 17.5). As we will see, their elaborate structures allow groups to travel from one active site to another, connected

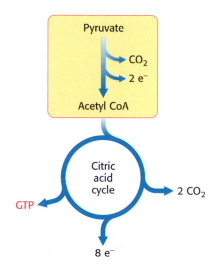

FIGURE 17.4 The link between glycolysis and the citric acid cycle. Pyruvate produced by glycolysis is converted into acetyl CoA, the fuel of the citric acid cycle.

FIGURE 17.5 Electron micrograph of the pyruvate dehydrogenase complex from *E. coli.* [Courtesy of Dr. Lester Reed.]

TABLE 17.1 **Pyruvate dehydrogenase complex of *E. coli***

Enzyme	Abbreviation	Number of chains	Prosthetic group	Reaction catalyzed
Pyruvate dehydrogenase component	E_1	24	TPP	Oxidative decarboxylation of pyruvate
Dihydrolipoyl transacetylase	E_2	24	Lipoamide	Transfer of the acetyl group to CoA
Dihydrolipoyl dehydrogenase	E_3	12	FAD	Regeneration of the oxidized form of lipoamide

by tethers to the core of the structure. The mechanism of the pyruvate dehydrogenase reaction is wonderfully complex, more so than is suggested by its relatively simple stoichiometry. The reaction requires the participation of the three enzymes of the pyruvate dehydrogenase complex, each composed of several polypeptide chains, and five coenzymes: *thiamine pyrophosphate (TPP), lipoic acid,* and *FAD* serve as catalytic cofactors, and CoA and NAD^+ are stoichiometric cofactors.

Thiamine pyrophosphate (TPP)   **Lipoic acid**

At least two additional enzymes regulate the activity of the complex.

The conversion of pyruvate into acetyl CoA consists of three steps: decarboxylation, oxidation, and transfer of the resultant acetyl group to CoA.

Pyruvate   **Acetyl CoA**

These steps must be coupled to preserve the free energy derived from the decarboxylation step to drive the formation of NADH and acetyl CoA. First, pyruvate combines with TPP and is then decarboxylated (Figure 17.6). This reaction is catalyzed by the *pyruvate dehydrogenase component* (E_1) of the multienzyme complex. A key feature of TPP, the prosthetic group of the pyruvate dehydrogenase component, is that the carbon atom between the nitrogen and sulfur atoms in the thiazole ring is much more acidic than most $=CH-$ groups, with a pK_a value near 10. This center ionizes to form a *carbanion,* which readily adds to the carbonyl group of pyruvate.

TPP   **Carbanion of TPP**

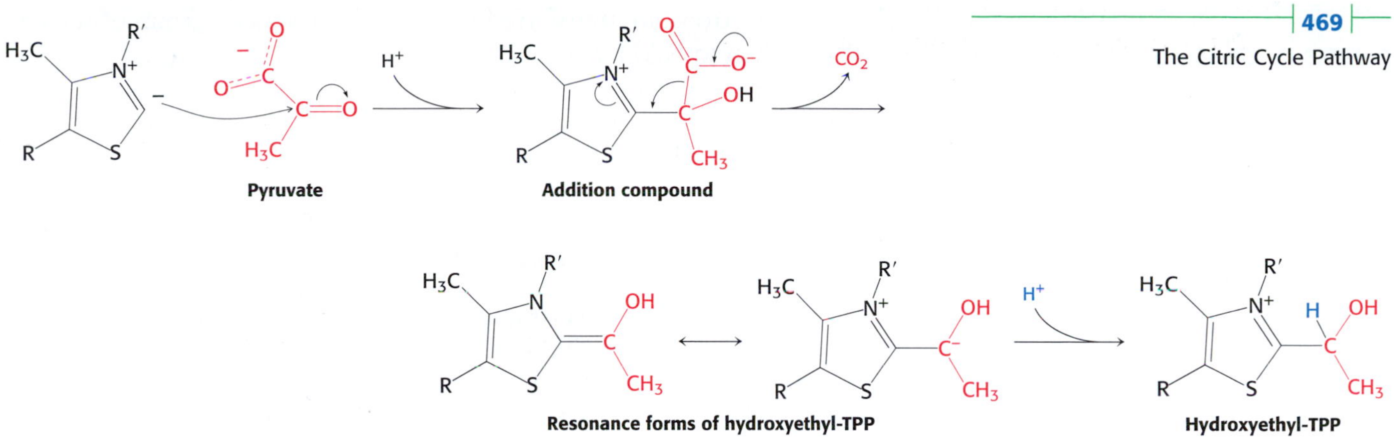

FIGURE 17.6 Mechanism of the decarboxylation reaction of E_1, the pyruvate dehydrogenase component of the pyruvate dehydrogenese complex.

This addition is followed by the decarboxylation of pyruvate. The positively charged ring of TPP acts as an electron sink that stabilizes the negative charge that is transferred to the ring as part of the decarboxylation. Protonation yields hydroxyethyl-TPP.

Second, the hydroxyethyl group attached to TPP is *oxidized* to form an acetyl group and concomitantly *transferred* to lipoamide, a derivative of lipoic acid that is linked to the side chain of a lysine residue by an amide linkage.

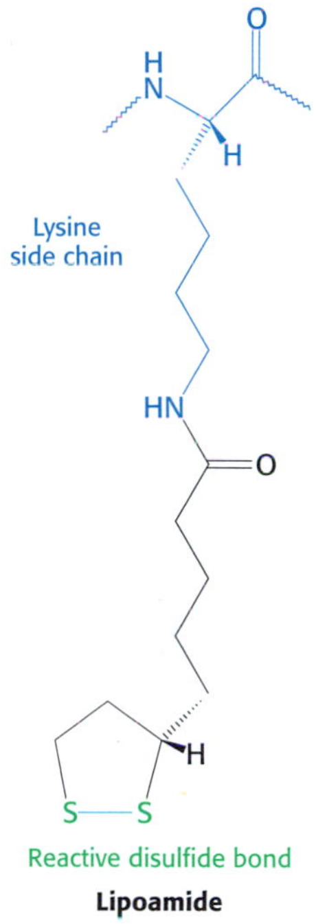

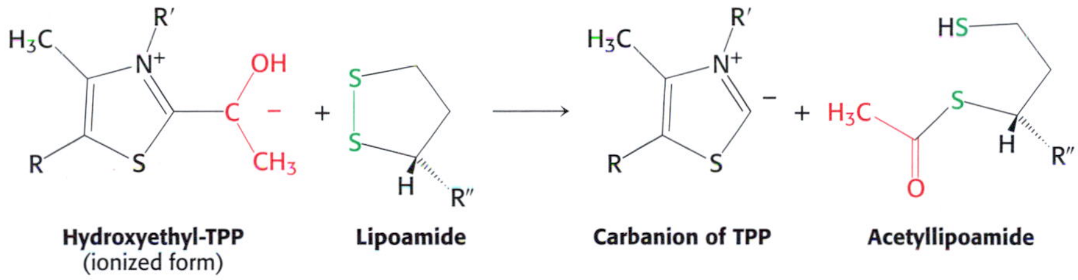

The oxidant in this reaction is the disulfide group of lipoamide, which is reduced to its disulfhydryl form. This reaction, also catalyzed by the pyruvate dehydrogenase component E_1, yields *acetyllipoamide*.

Third, *the acetyl group is transferred from acetyllipoamide to CoA to form acetyl CoA.*

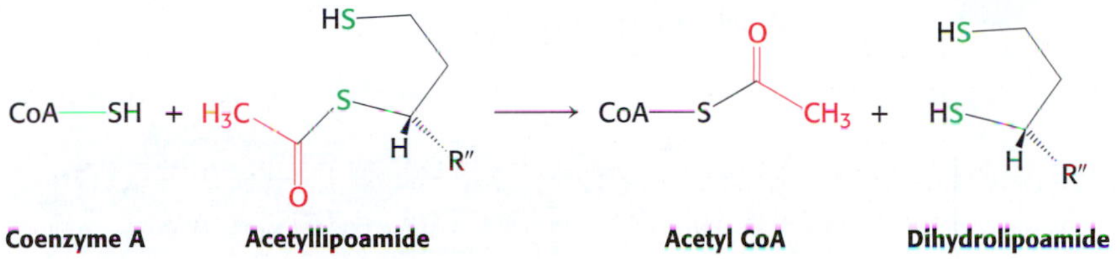

Dihydrolipoyl transacetylase (E_2) catalyzes this reaction. The energy-rich thioester bond is preserved as the acetyl group is transferred to CoA. Recall that CoA serves as a carrier of many activated acyl groups, of which acetyl is the simplest (Section 14.3.1). Acetyl CoA, the fuel for the citric acid cycle, has now been generated from pyruvate.

The pyruvate dehydrogenase complex cannot complete another catalytic cycle until the dihydrolipoamide is oxidized to lipoamide. In a fourth step, *the oxidized form of lipoamide is regenerated by dihydrolipoyl dehydrogenase*

(E_3). Two electrons are transferred to an FAD prosthetic group of the enzyme and then to NAD^+.

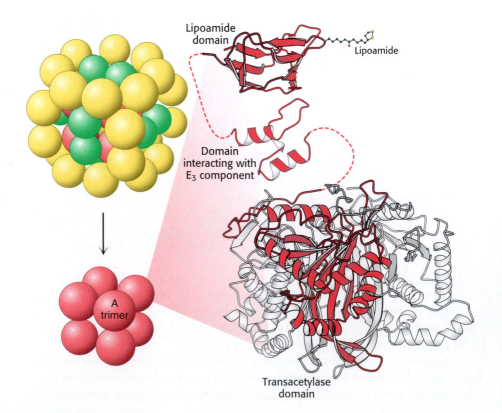

This electron transfer to FAD is unusual, because the common role for FAD is to receive electrons from NADH. The electron transfer potential of FAD is altered by its association with the enzyme and enables it to transfer electrons to NAD^+. Proteins tightly associated with FAD or flavin mononucleotide (FMN) are called *flavoproteins*.

17.1.2 Flexible Linkages Allow Lipoamide to Move Between Different Active Sites

Although the structure of an intact member of the pyruvate dehydrogenase complex family has not yet been determined in atomic detail, the structures of all of the component enzymes are now known, albeit from different complexes and species. Thus, it is now possible to construct an atomic model of the complex to understand its activity (Figure 17.7).

The core of the complex is formed by E_2. Transacetylase consists of eight catalytic trimers assembled to form a hollow cube. Each of the three subunits forming a trimer has three major domains (Figure 17.8). At the amino terminus is a small domain that contains a bound lipoamide cofactor attached to a lysine residue. This domain is homologous to biotin-binding domains such as that of pyruvate carboxylase (see Figure 16.26). The lipoamide domain is followed by a small domain that interacts with E_3 within the complex. A larger transacetylase domain completes an E_2 subunit. E_1 is an $\alpha_2 \beta_2$ tetramer, and E_3 is a $\alpha\beta$ dimer. Twenty-four copies of

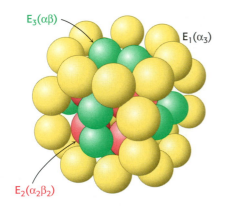

FIGURE 17.7 Schematic representation of the pyruvate dehydrogenase complex. The transacetylase core (E_2) is shown in red, the pyruvate dehydrogenase component (E_1) in yellow, and the dihydrolipoyl dehydrogenase (E_3) in green.

FIGURE 17.8 Structure of the transacetylase (E_2) core. Each red ball represents a trimer of three E_2 subunits. Each subunit consists of three domains: a lipoamide-binding domain, a small domain for interaction with E_3, and a large transacetylase catalytic domain. All three subunits of the transacetylase domain are shown in the ribbon representation, with one depicted in red.

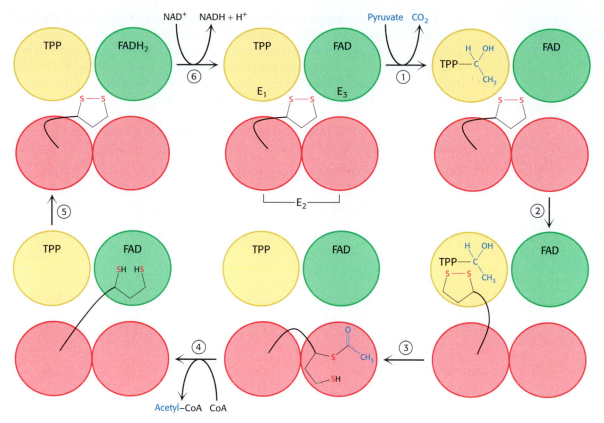

FIGURE 17.9 Reactions of the pyruvate dehydrogenase complex. At the top (center), the enzyme (represented by a yellow, a blue, and two red spheres) is unmodified and ready for a catalytic cycle. (1) Pyruvate is decarboxylated to form the hydroxyethyl TPP. (2) The dihydrolipoyl arm of E_2 moves into the active site of E_1. (3) E_1 catalyzes the transfer of the two-carbon group to the dihydrolipoyl group to form the acetyl-lipoyl complex. (4) E_2 catalyzes the transfer of the acetyl moiety to CoA to form the product acetyl CoA. The disulfhydryl lipoyl arm then swings to the active site of E_3. E_3 catalyzes (5) the reduction of the lipoic acid and (6) the transfer of the protons and electrons to NAD^+ to complete the reaction cycle.

E_1 and 12 copies of E_3 surround the E_2 core. How do the three distinct active sites work in concert (Figure 17.9)?

1. Pyruvate is decarboxylated at the active site of E_1, forming the substituted TPP intermediate, and CO_2 leaves as the first product. This active site lies within the E_1 complex, connected to the enzyme surface by a 20-Å-long hydrophobic channel.

2. E_2 inserts the lipoyl-lysine arm of the lipoamide domain into the channel in E_1.

3. E_1 catalyzes the transfer of the acetyl group to the lipoamide. The acetylated lipoyl-lysine arm then leaves E_1 and enters the E_2 cube through 30 Å windows on the sides of the cube to visit the active site of E_2, located deep in the cube at the subunit interface.

4. The acetyl moiety is then transferred to CoA, and the second product, acetyl CoA, leaves the cube. The reduced lipoyl-lysine arm then swings to the active site of the E_3 flavoprotein.

5. At the E_3 active site, the lipoamide acid is oxidized by coenzyme FAD.

6. The final product, NADH, is produced with the reoxidation of $FADH_2$, and the reactivated lipoamide is ready to begin another reaction cycle.

The structural integration of three kinds of enzymes makes the coordinated catalysis of a complex reaction possible. The proximity of one enzyme to another *increases the overall reaction rate* and *minimizes side reactions.* All the intermediates in the oxidative decarboxylation of pyruvate are tightly bound to the complex and are readily transferred because of the ability of the lipoyl-lysine arm of E_2 to call on each active site in turn.

17.1.3 Citrate Synthase Forms Citrate from Oxaloacetate and Acetyl Coenzyme A

The citric acid cycle begins with the condensation of a four-carbon unit, oxaloacetate, and a two-carbon unit, the acetyl group of acetyl CoA. Oxaloacetate reacts with acetyl CoA and H_2O to yield citrate and CoA.

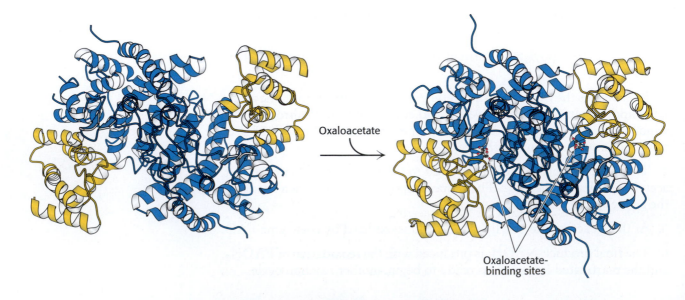

Oxaloacetate   Acetyl CoA   Citryl CoA   Citrate

Synthase—
An enzyme catalyzing a synthetic reaction in which two units are joined without the direct participation of ATP (or another nucleoside triphosphate).

This reaction, which is an aldol condensation followed by a hydrolysis, is catalyzed by *citrate synthase.* Oxaloacetate first condenses with acetyl CoA to form *citryl CoA,* which is then hydrolyzed to citrate and CoA. The hydrolysis of citryl CoA, a high-energy thioester intermediate, drives the overall reaction far in the direction of the synthesis of citrate. In essence, the hydrolysis of the thioester powers the synthesis of a new molecule from two precursors. Because this reaction initiates the cycle, it is very important that side reactions be minimized. Let us briefly consider the how citrate synthase prevents wasteful processes such as the hydrolysis of acetyl CoA.

Mammalian citrate synthase is a dimer of identical 49-kd subunits. Each active site is located in a cleft between the large and small domains of a subunit, adjacent to the subunit interface. The results of x-ray crystallographic studies of citrate synthase and its complexes with several substrates and inhibitors revealed that the enzyme undergoes large conformational changes in the course of catalysis. Citrate synthase exhibits sequential, ordered

FIGURE 17.10 Conformational changes in citrate synthase on binding oxaloacetate. The small domain of each subunit of the homodimer is shown in yellow; the large domain is shown in blue. (Left) Open form of enzyme alone. (Right) Closed form of the liganded enzyme.

kinetics: oxaloacetate binds first, followed by acetyl CoA. The reason for the ordered binding is that *oxaloacetate induces a major structural rearrangement leading to the creation of a binding site for acetyl CoA*. The open form of the enzyme observed in the absence of ligands is converted into a closed form by the binding of oxaloacetate (Figure 17.10). In each subunit, the small domain rotates 19 degrees relative to the large domain. *Movements as large as 15 Å are produced by the rotation of α helices elicited by quite small shifts of side chains around bound oxaloacetate*. This conformational transition is reminiscent of the cleft closure in hexokinase induced by the binding of glucose (Section 16.1.1).

Citrate synthase catalyzes the condensation reaction by bringing the substrates into close proximity, orienting them, and polarizing certain bonds. Two histidine residues and an aspartate residue are important players (Figure 17.11). One of the histidine residues (His 274) donates a proton to the carbonyl oxygen of acetyl CoA to promote the removal of a methyl proton by Asp 375. Oxaloacetate is activated by the transfer of a proton from His 320 to its carbonyl carbon atom. The concomitant attack of the enol of acetyl CoA on the carbonyl carbon of oxaloacetate results in the formation of a carbon–carbon bond. The newly formed citryl CoA induces additional structural changes in the enzyme. The active site becomes completely enclosed. His 274 participates again as a proton donor to hydrolyze the thioester. Coenzyme A leaves the enzyme, followed by citrate, and the enzyme returns to the initial open conformation.

We can now understand how the wasteful hydrolysis of acetyl CoA is prevented. Citrate synthase is well suited to hydrolyze *citryl* CoA but not *acetyl* CoA. How is this discrimination accomplished? First, acetyl CoA does not bind to the enzyme until oxaloacetate is bound and ready for condensation. Second, the catalytic residues crucial for hydrolysis of the thioester linkage are not appropriately positioned *until citryl CoA is formed*. As with hexokinase (Section 16.1.1) and triose phosphate isomerase (Section 16.1.4), *induced fit prevents an undesirable side reaction*.

FIGURE 17.11 Mechanism of synthesis of citryl CoA by citrate synthase. The condensation of oxaloacetate and acetyl CoA proceeds through an enol intermediate. The subsequent hydrolysis of citryl CoA yields citrate and CoA.

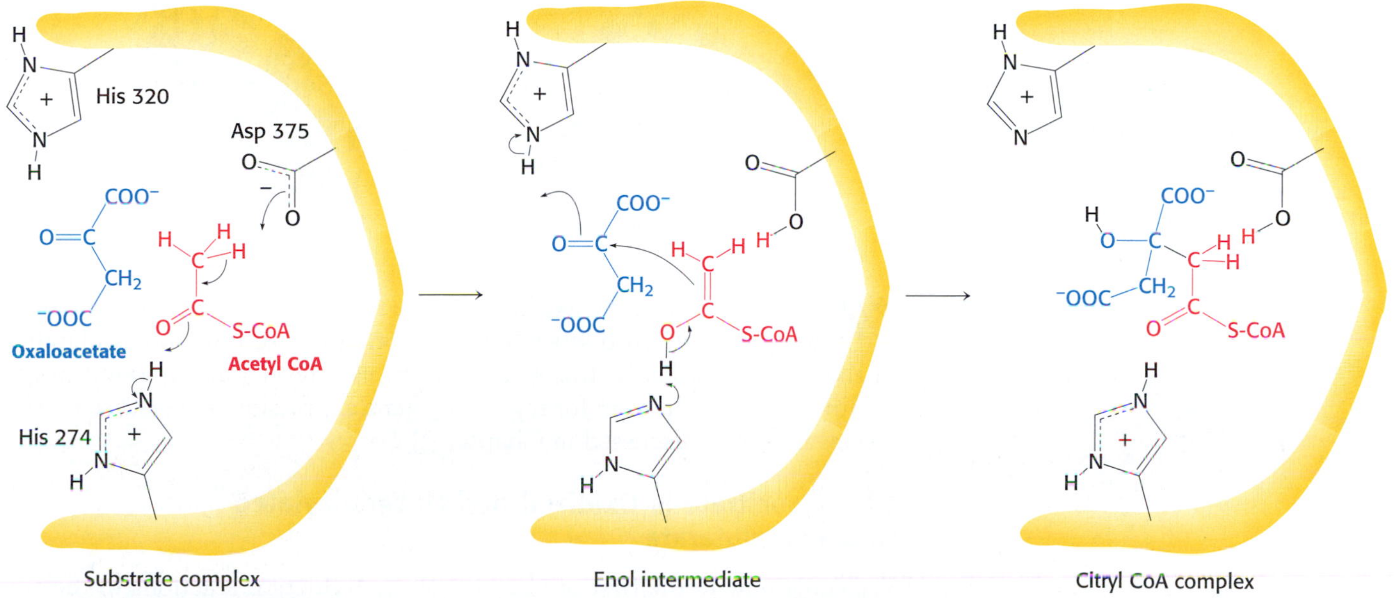

Substrate complex   Enol intermediate   Citryl CoA complex

17.1.4 Citrate Is Isomerized into Isocitrate

The tertiary hydroxyl group is not properly located in the citrate molecule for the oxidative decarboxylations that follow. Thus, citrate is isomerized into isocitrate to enable the six-carbon unit to undergo oxidative decarboxylation.

The isomerization of citrate is accomplished by a *dehydration* step followed by a *hydration* step. The result is an interchange of a hydrogen atom and a hydroxyl group. The enzyme catalyzing both steps is called *aconitase* because *cis-aconitate* is an intermediate.

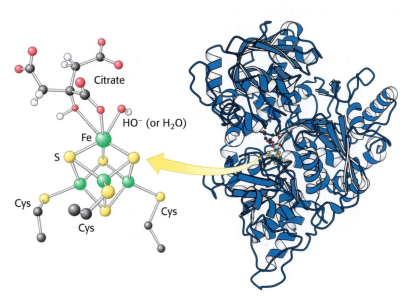

Aconitase is an *iron–sulfur protein*, or *nonheme iron protein*. It contains four iron atoms that are not incorporated as part of a heme group. The four iron atoms are complexed to four inorganic sulfides and three cysteine sulfur atoms, leaving one iron atom available to bind citrate and then isocitrate through their carboxylate and hydroxyl groups (Figure 17.12). This iron center, in conjunction with other groups on the enzyme, facilitates the dehydration and rehydration reactions. We will consider the role of these iron–sulfur clusters in the electron-transfer reactions of oxidative phosphorylation subsequently (Section 18.3.1).

FIGURE 17.12 Binding of citrate to the iron–sulfur complex of aconitase. A 4Fe-4S iron–sulfur cluster is a component of the active site of aconitase. One of the iron atoms of the cluster binds to the carboxylate and hydroxyl groups of citrate.

The iron–sulfur cluster in aconitase is somewhat unstable, so one or more iron atoms dissociate under conditions of low iron availability in the cell. Remarkably, this sensitivity to iron level was exploited in the evolution of a mechanism for regulating gene expression in response to iron levels, as will be discussed in Chapter 31.

17.1.5 Isocitrate Is Oxidized and Decarboxylated to α-Ketoglutarate

We come now to the first of four oxidation–reduction reactions in the citric acid cycle. The oxidative decarboxylation of isocitrate is catalyzed by *isocitrate dehydrogenase*.

$$\text{Isocitrate} + NAD^+ \longrightarrow \alpha\text{-ketoglutarate} + CO_2 + NADH$$

The intermediate in this reaction is oxalosuccinate, an unstable β-ketoacid. While bound to the enzyme, it loses CO_2 to form α-ketoglutarate.

Isocitrate → **Oxalosuccinate** → **α-Ketoglutarate**

The rate of formation of α-ketoglutarate is important in determining the overall rate of the cycle, as will be discussed in Section 17.2.2. This oxidation generates the first high-transfer-potential electron carrier NADH in the cycle.

17.1.6 Succinyl Coenzyme A Is Formed by the Oxidative Decarboxylation of α-Ketoglutarate

The conversion of isocitrate into α-ketoglutarate is followed by a second oxidative decarboxylation reaction, the formation of succinyl CoA from α-ketoglutarate.

α-Ketoglutarate + NAD$^+$ + CoA ⟶ **Succinyl CoA** + CO_2 + NADH

The oxidative decarboxylation of α-ketoglutarate closely resembles that of pyruvate, also an α-ketoacid.

$$\text{Pyruvate} + \text{CoA} + \text{NAD}^+ \longrightarrow \text{acetyl CoA} + CO_2 + \text{NADH}$$

Both reactions include the decarboxylation of an α-ketoacid and the subsequent formation of a high-transfer-potential thioester linkage with CoA. The complex that catalyzes the oxidative decarboxylation of α-ketoglutarate is homologous to the pyruvate dehydrogenase complex, and the reaction mechanism is entirely analogous. The *α-ketoglutarate dehydrogenase* component (E_2) and transsuccinylase (E_1) are different from but homologous to the corresponding enzymes in the pyruvate dehydrogenase complex, whereas the dihydrolipoyl dehydrogenase components (E_3) of the two complexes are identical.

17.1.7 A High Phosphoryl-Transfer Potential Compound Is Generated from Succinyl Coenzyme A

Succinyl CoA is an energy-rich thioester compound. The $\Delta G°'$ for the hydrolysis of succinyl CoA is about -8 kcal mol^{-1} (-33.5 kJ mol^{-1}), which is comparable to that of ATP (-7.3 kcal mol^{-1}, or -30.5 kJ mol^{-1}). In the citrate synthase reaction, the cleavage of the thioester bond powers the synthesis of the six-carbon citrate from the four-carbon oxaloacetate and the two-carbon fragment. *The cleavage of the thioester bond of succinyl CoA is coupled to the phosphorylation of a purine nucleoside diphosphate, usually GDP.* This reaction is catalyzed by *succinyl CoA synthetase* (succinate thiokinase). This enzyme is an $\alpha_2\beta_2$ heterodimer; the functional unit is one αβ pair. The mechanism is a clear example of energy transformations: energy inherent in

Succinyl CoA + P_i + GDP ⟶ **Succinate** + CoA + GTP

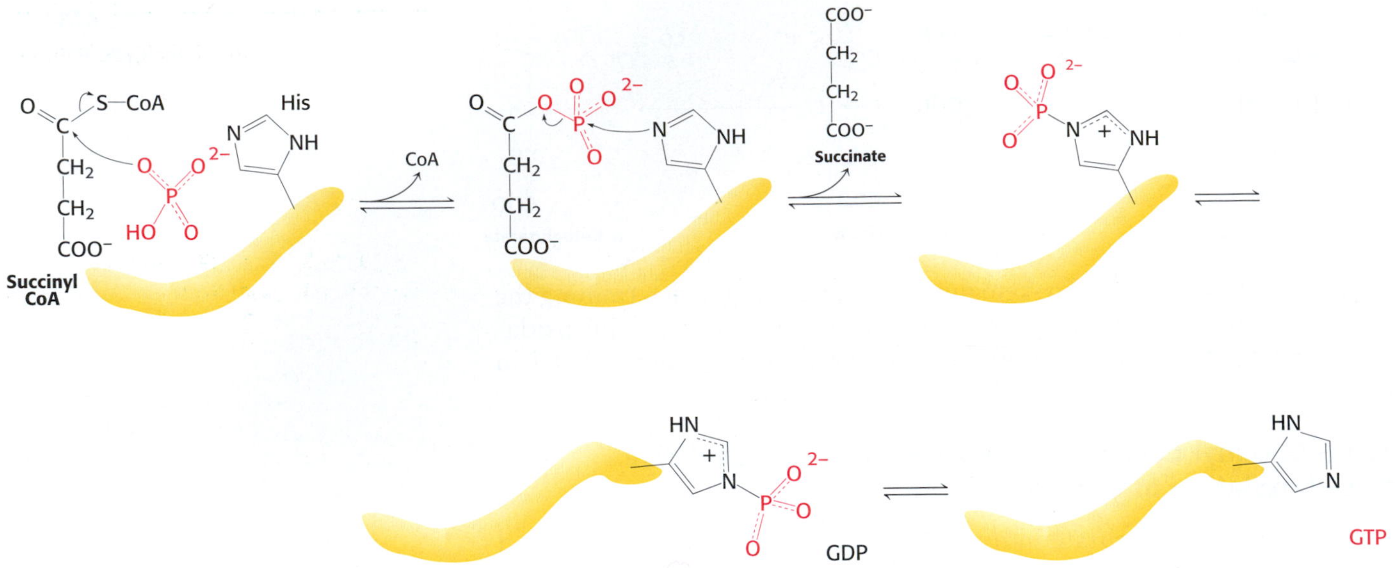

FIGURE 17.13 Reaction mechanism of succinyl CoA synthetase. The formation of GTP at the expense of succinyl CoA is an example of substrate-level phosphorylation. The reaction proceeds through a phosphorylated enzyme intermediate.

the thioester molecule is transformed into phosphoryl-group transfer potential (Figure 17.13). The first step is the displacement of coenzyme A by orthophosphate, which generates another energy-rich compound, succinyl phosphate. A histidine residue of the α subunit removes the phosphoryl group with the concomitant generation of succinate and phosphohistidine. The phosphohistidine residue then swings over to a bound nucleoside diphosphate and the phosphoryl group is transferred to form the nucleoside triphosphate. The participation of high-energy compounds in all the steps is attested to by the fact that the reaction is readily reversible: $\Delta G°' = -0.8$ kcal mol^{-1} (-3.4 kJ mol^{-1}). This is the only step in the citric acid cycle that directly yields a compound with high phosphoryl transfer potential through a substrate-level phosphorylation. Some mammalian succinyl CoA synthetases are specific for GDP and others for ADP. The *E. coli* enzyme uses either GDP or ADP as the phosphoryl-group acceptor. We have already seen that GTP is an important component of signal-transduction systems (Chapter 15). Alternatively, its γ-phosphoryl group can be readily transferred to ADP to form ATP, in a reaction catalyzed by *nucleoside diphosphokinase*.

$$GTP + ADP \rightleftharpoons GDP + ATP$$

The mechanism of succinyl CoA synthetase reveals that a phosphoryl group is transferred first to succinyl CoA bound in the α subunit and then to a nucleoside diphosphate bound in the β subunit. Examination of the three-dimensional structure of succinyl CoA synthetase shows that each subunit comprises two domains (Figure 17.14). The carboxyl-terminal domains of the two subunits are similar to one another, whereas the amino-terminal domains have different structures, each characteristic of its role in the mechanism. The amino-terminal domain of the α subunit forms a Rossmann fold (Section 16.1.10), which binds the ADP component of succinyl CoA, whereas the amino-terminal domain of the β subunit is an ATP-grasp domain, a nucleotide-activating domain found in many enzymes, especially those catalyzing purine biosynthesis (Section 16.3.2 and Chapter 25). Succinyl CoA synthetase has evolved by adopting these domains and harnessing them to allow the capture of the energy associated with succinyl CoA cleavage to drive the generation of a nucleoside triphosphate.

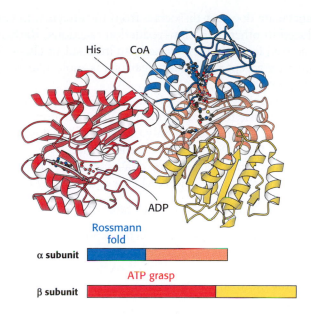

FIGURE 17.14 Structure of succinyl CoA synthetase. The enzyme is composed of two subunits. The α subunit contains a Rossmann fold that binds the ADP component of CoA, and the β subunit contains a nucleotide-activating region called the ATP-grasp domain. The ATP-grasp domain is shown here binding a molecule of ADP. The histidine residue picks up the phosphoryl group from near the CoA and swings over to transfer it to the nucleotide bound in the ATP-grasp domain.

17.1.8 Oxaloacetate Is Regenerated by the Oxidation of Succinate

Reactions of four-carbon compounds constitute the final stage of the citric acid cycle: the regeneration of oxaloacetate.

Succinate	Fumarate	Malate	Oxaloacetate

The reactions constitute a metabolic motif that we will see again in fatty acid synthesis and degradation as well as in the degradation of some amino acids (see Figure 14.17). A methylene group (CH_2) is converted into a carbonyl group ($C=O$) in three steps: an oxidation, a hydration, and a second oxidation reaction. Not only is oxaloacetate thereby regenerated for another round of the cycle, but also more energy is extracted in the form of $FADH_2$ and NADH.

Succinate is oxidized to fumarate by *succinate dehydrogenase*. The hydrogen acceptor is FAD rather than NAD^+, which is used in the other three oxidation reactions in the cycle. In succinate dehydrogenase, the isoalloxazine ring of FAD is covalently attached to a histidine side chain of the enzyme (denoted E-FAD).

$$E\text{-}FAD + \text{succinate} \rightleftharpoons E\text{-}FADH_2 + \text{fumarate}$$

FAD is the hydrogen acceptor in this reaction because the free-energy change is insufficient to reduce NAD^+. FAD is nearly always the electron acceptor in oxidations that remove two hydrogen *atoms* from a substrate.

Succinate dehydrogenase, like aconitase, is an iron–sulfur protein. Indeed, succinate dehydrogenase contains three different kinds of iron–sulfur clusters, 2Fe-2S (two iron atoms bonded to two inorganic sulfides), 3Fe-4S, and 4Fe-4S. Succinate dehydrogenase—which consists of two subunits, one 70 kd and the other 27 kd—differs from other enzymes in the citric acid cycle in being embedded in the inner mitochondrial membrane. In fact, *succinate dehydrogenase is directly associated with the electron-transport chain, the link between the citric acid cycle and ATP formation.* $FADH_2$ produced by the

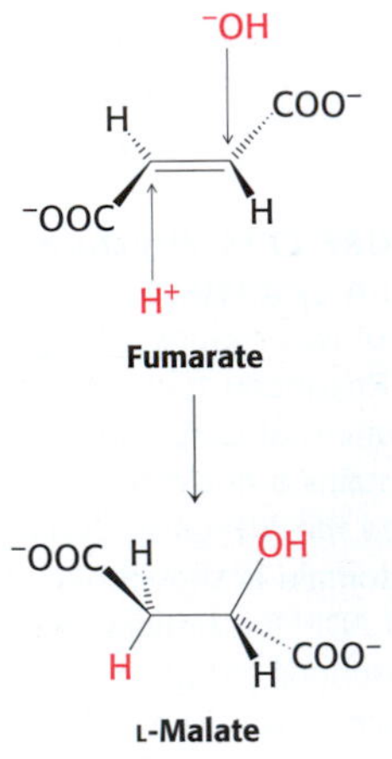

Fumarate

↓

L-Malate

oxidation of succinate does not dissociate from the enzyme, in contrast with NADH produced in other oxidation–reduction reactions. Rather, two electrons are transferred from $FADH_2$ directly to iron–sulfur clusters of the enzyme. The ultimate acceptor of these electrons is molecular oxygen, as we shall see in Chapter 18.

The next step is the hydration of fumarate to form L-malate. *Fumarase* catalyzes a stereospecific trans addition of a hydrogen atom and a hydroxyl group. The hydroxyl group adds to only one side of the double bond of fumarate; hence, only the L isomer of malate is formed.

Finally, malate is oxidized to form oxaloacetate. This reaction is catalyzed by *malate dehydrogenase*, and NAD^+ is again the hydrogen acceptor.

$$\text{Malate} + \text{NAD}^+ \rightleftharpoons \text{oxaloacetate} + \text{NADH} + \text{H}^+$$

Note that the standard free energy for this reaction, unlike that for the other steps in the citric acid cycle, is significantly positive. The oxidation of malate is driven by the utilization of the products—oxaloacetate by citrate synthase and NADH by the electron-transport chain.

17.1.9 Stoichiometry of the Citric Acid Cycle

The net reaction of the citric acid cycle is:

$$\text{Acetyl CoA} + 3\,\text{NAD}^+ + \text{FAD} + \text{GDP} + \text{P}_i + 2\,\text{H}_2\text{O} \longrightarrow$$
$$2\,\text{CO}_2 + 3\,\text{NADH} + \text{FADH}_2 + \text{GTP} + 2\,\text{H}^+ + \text{CoA}$$

Let us recapitulate the reactions that give this stoichiometry (Figure 17.15 and Table 17.2):

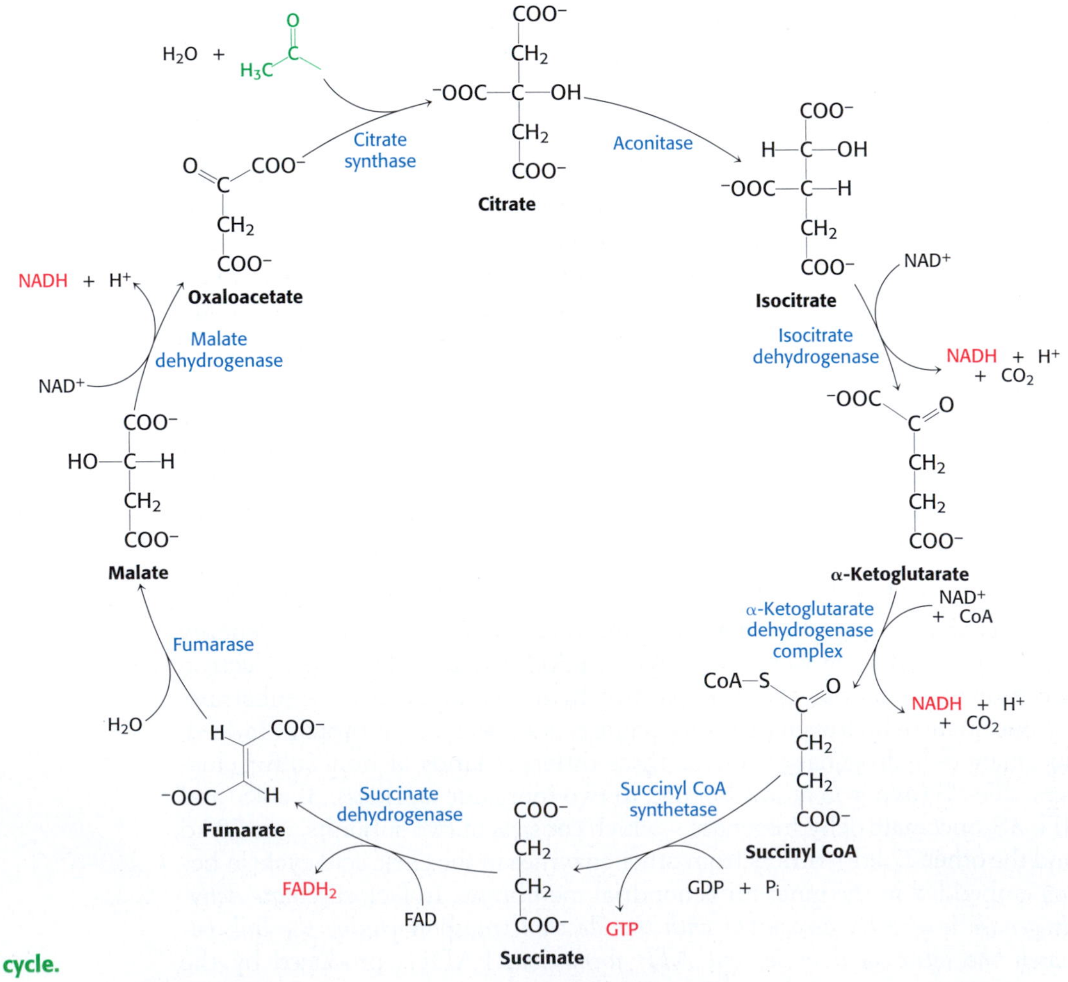

FIGURE 17.15 The citric acid cycle.

TABLE 17.2   Citric acid cycle

Step	Reaction	Enzyme	Prosthetic group	Type*	$\Delta G^{\circ\prime}$ kcal mol^{-1}	$\Delta G^{\circ\prime}$ kJ mol^{-1}
1	Acetyl CoA + oxaloacetate + $H_2O \longrightarrow$ citrate + CoA + H^+	Citrate synthase		a	-7.5	-31.4
2a	Citrate $\rightleftharpoons$ cis-aconitate + H_2O	Aconitase	Fe–S	b	$+2.0$	$+8.4$
2b	cis-Aconitate + $H_2O \rightleftharpoons$ isocitrate	Aconitase	Fe–S	c	-0.5	-2.1
3	Isocitrate + NAD$^+$ $\rightleftharpoons$ α-ketoglutarate + CO_2 + NADH	Isocitrate dehydrogenase		d + e	-2.0	-8.4
4	α-Ketoglutarate + NAD$^+$ + CoA $\rightleftharpoons$ succinyl CoA + CO_2 + NADH	α-Ketoglutarate dehydrogenase complex	Lipoic acid, FAD, TPP	d + e	-7.2	-30.1
5	Succinyl CoA + P_i + GDP $\rightleftharpoons$ succinate + GTP + CoA	Succinyl CoA synthetase		f	-0.8	-3.3
6	Succinate + FAD (enzyme-bound) $\rightleftharpoons$ fumarate + FADH$_2$ (enzyme-bound)	Succinate dehydrogenase	FAD, Fe–S	e	~ 0	0
7	Fumarate + $H_2O \rightleftharpoons$ L-malate	Fumarase		c	-0.9	-3.8
8	L-Malate + NAD$^+$ $\rightleftharpoons$ oxaloacetate + NADH + H^+	Malate dehydrogenase		e	$+7.1$	$+29.7$

*Reaction type: (a) condensation; (b) dehydration; (c) hydration; (d) decarboxylation; (e) oxidation; (f) substrate-level phosphorylation.

1. Two carbon atoms enter the cycle in the condensation of an acetyl unit (from acetyl CoA) with oxaloacetate. Two carbon atoms leave the cycle in the form of CO_2 in the successive decarboxylations catalyzed by isocitrate dehydrogenase and α-ketoglutarate dehydrogenase. Interestingly, the results of isotope-labeling studies revealed that the two carbon atoms that enter each cycle are not the ones that leave.

2. Four pairs of hydrogen atoms leave the cycle in four oxidation reactions. Two molecules of NAD$^+$ are reduced in the oxidative decarboxylations of isocitrate and α-ketoglutarate, one molecule of FAD is reduced in the oxidation of succinate, and one molecule of NAD$^+$ is reduced in the oxidation of malate.

3. One compound with high phosphoryl transfer potential, usually GTP, is generated from the cleavage of the thioester linkage in succinyl CoA.

4. Two molecules of water are consumed: one in the synthesis of citrate by the hydrolysis of citryl CoA and the other in the hydration of fumarate.

Recall also that NADH is generated in the formation of acetyl CoA from pyruvate by the pyruvate dehydrogenase reaction.

The efficiency of the citric acid cycle may be enhanced by the arrangement of the constituent enzymes. Evidence is accumulating that the enzymes are physically associated with one another to facilitate substrate channeling between active sites. The word *metabolon* has been suggested as the name for such multienzyme complexes.

As will be discussed in Chapter 18, the electron-transport chain oxidizes the NADH and FADH$_2$ formed in the citric acid cycle. The transfer of electrons from these carriers to O_2, the ultimate electron acceptor, leads to the generation of a proton gradient across the inner mitochondrial membrane. This proton-motive force then powers the generation of ATP; the net stoichiometry is about 2.5 ATP per NADH, and 1.5 ATP per $FADH_2$. Consequently, 9 high-transfer-potential phosphoryl groups are generated when the electron-transport chain oxidizes 3 molecules of NADH and 1 molecule of $FADH_2$, and 1 high-transfer-potential phosphoryl group per acetyl unit is directly formed in the citric acid cycle. Thus, 1 acetate unit generates approximately 10 molecules of ATP. In dramatic contrast, only 2 molecules of ATP are generated per molecule of glucose (which generates 2 molecules of acetyl CoA) by anaerobic glycolysis.

Recall that molecular oxygen does not participate directly in the citric acid cycle. However, the cycle operates only under aerobic conditions because NAD^+ and FAD can be regenerated in the mitochondrion only by the transfer of electrons to molecular oxygen. *Glycolysis has both an aerobic and an anaerobic mode, whereas the citric acid cycle is strictly aerobic.* Glycolysis can proceed under anaerobic conditions because NAD^+ is regenerated in the conversion of pyruvate into lactate.

17.2 ENTRY TO THE CITRIC ACID CYCLE AND METABOLISM THROUGH IT ARE CONTROLLED

The citric acid cycle is the final common pathway for the aerobic oxidation of fuel molecules. Moreover, as we will see shortly (Section 17.3) and repeatedly elsewhere in our study of biochemistry, the cycle is an important source of building blocks for a host of important biomolecules. As befits its role as the metabolic hub of the cell, entry into the cycle and the rate of the cycle itself are controlled at several stages.

17.2.1 The Pyruvate Dehydrogenase Complex Is Regulated Allosterically and by Reversible Phosphorylation

As we saw earlier, glucose can be formed from pyruvate (Section 16.3). *However, the formation of acetyl CoA from pyruvate is an irreversible step in animals and thus they are unable to convert acetyl CoA back into glucose.* The oxidative decarboxylation of pyruvate to acetyl CoA commits the carbon atoms of glucose to two principal fates: oxidation to CO_2 by the citric acid cycle, with the concomitant generation of energy, or incorporation into lipid (Figure 17.16). As expected of an enzyme at a critical branch point in metabolism, the activity of the pyruvate dehydrogenase complex is stringently controlled by several means (Figure 17.17). High concentrations of reaction products of the complex inhibit the reaction: acetyl CoA inhibits the transacetylase component (E_2), whereas NADH inhibits the dihydrolipoyl

FIGURE 17.16 From glucose to acetyl CoA. The synthesis of acetyl CoA by the pyruvate dehydrogenase complex is a key irreversible step in the metabolism of glucose.

FIGURE 17.17 Regulation of the pyruvate dehydrogenase complex. The complex is inhibited by its immediate products, NADH and acetyl CoA. The pyruvate dehydrogenase component is also regulated by covalent modification. A specific kinase phosphorylates and inactivates pyruvate dehydrogenase, and a phosphatase actives the dehydrogenase by removing the phosphoryl. The kinase and the phosphatase also are highly regulated enzymes.

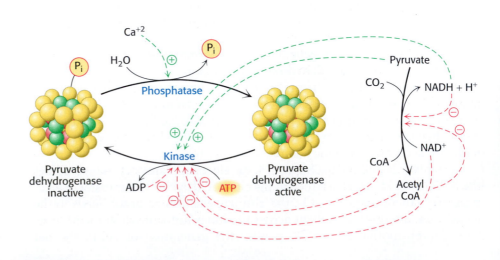

dehydrogenase (E_3). However, the key means of regulation in eukaryotes is covalent modification of the pyruvate dehydrogenase component. *Phosphorylation of the pyruvate dehydrogenase component (E_1) by a specific kinase switches off the activity of the complex. Deactivation is reversed by the action of a specific phosphatase.* The site of phosphorylation is the transacetylase component (E_2), again highlighting the structural and mechanistic importance of this core. Increasing the $NADH/NAD^+$, acetyl CoA/CoA, or ATP/ADP ratio promotes phosphorylation and, hence, deactivation of the complex. In other words, high concentrations of immediate (acetyl CoA and NADH) and ultimate (ATP) products inhibit the activity. Thus, *pyruvate dehydrogenase is switched off when the energy charge is high and biosynthetic intermediates are abundant.* On the other hand, pyruvate as well as ADP (a signal of low energy charge) activate the dehydrogenase by inhibiting the kinase.

In contrast, α_1-adrenergic agonists and hormones such as vasopressin stimulate pyruvate dehydrogenase by triggering a rise in the cytosolic Ca^{2+} level (Section 15.3.2), which in turn elevates the mitochondrial Ca^{2+} level. The rise in mitochondrial Ca^{2+} activates the pyruvate dehydrogenase complex by stimulating the phosphatase. Insulin also accelerates the conversion of pyruvate into acetyl CoA by stimulating the dephosphorylation of the complex. In turn, glucose is funneled into pyruvate.

The importance of this covalent control is illustrated in people with a phosphatase deficiency. Because pyruvate dehydrogenase is always phosphorylated and thus inactive, glucose is processed to lactic acid. This condition results in unremitting lactic acidosis (high blood levels of lactic acid), which leads to the malfunctioning of many tissues, most notably the central nervous system (Section 17.3.2).

17.2.2 The Citric Acid Cycle Is Controlled at Several Points

The rate of the citric acid cycle is precisely adjusted to meet an animal cell's needs for ATP (Figure 17.18). The primary control points are the allosteric enzymes isocitrate dehydrogenase and α-ketoglutarate dehydrogenase.

Isocitrate dehydrogenase is allosterically stimulated by ADP, which enhances the enzyme's affinity for substrates. The binding of isocitrate, NAD^+, Mg^{2+}, and ADP is mutually cooperative. In contrast, NADH inhibits isocitrate dehydrogenase by directly displacing NAD^+. ATP, too, is inhibitory. It is important to note that several steps in the cycle require NAD^+ or FAD, which are abundant only when the energy charge is low.

A second control site in the citric acid cycle is α-ketoglutarate dehydrogenase. Some aspects of this enzyme's control are like those of the pyruvate dehydrogenase complex, as might be expected from the homology of the two enzymes. α-Ketoglutarate dehydrogenase is inhibited by succinyl CoA and NADH, the products of the reaction that it catalyzes. In addition, α-ketoglutarate dehydrogenase is inhibited by a high energy charge. Thus, *the rate of the cycle is reduced when the cell has a high level of ATP.*

In many bacteria, the funneling of two-carbon fragments into the cycle also is controlled. *The synthesis of citrate from oxaloacetate and acetyl CoA carbon units is an important control point in these organisms.* ATP is an allosteric inhibitor of citrate synthase. The effect of ATP is to increase the value of K_M for acetyl CoA. Thus, as the level of ATP increases, less of this enzyme is saturated with acetyl CoA and so less citrate is formed.

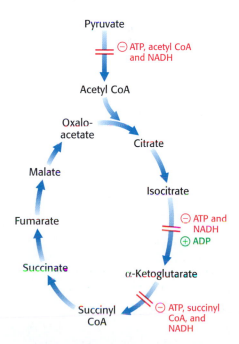

FIGURE 17.18 Control of the citric acid cycle. The citric acid cycle is regulated primarily by the concentration of ATP and NADH. The key control points are the enzymes isocitrate dehydrogenase and α-ketoglutarate dehydrogenase.

17.3 THE CITRIC ACID CYCLE IS A SOURCE OF BIOSYNTHETIC PRECURSORS

Thus far, discussion has focused on the citric acid cycle as the *major degradative pathway for the generation of ATP.* As a major metabolic hub of the cell, the citric acid cycle also *provides intermediates for biosyntheses* (Figure 17.19). For example, most of the carbon atoms in porphyrins come from *succinyl CoA.* Many of the amino acids are derived from *α-ketoglutarate* and *oxaloacetate.* These biosynthetic processes will be discussed in subsequent chapters.

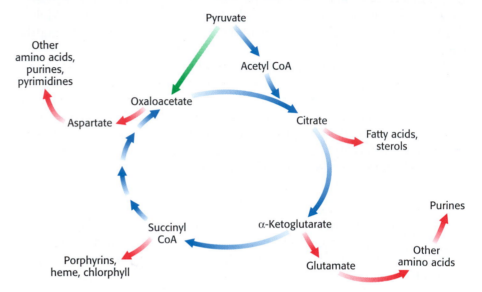

FIGURE 17.19 Biosynthetic roles of the citric acid cycle. Intermediates drawn off for biosyntheses (shown by red arrows) are replenished by the formation of oxaloacetate from pyruvate.

17.3.1 The Citric Acid Cycle Must Be Capable of Being Rapidly Replenished

The important point now is that *citric acid cycle intermediates must be replenished if any are drawn off for biosyntheses.* Suppose that much oxaloacetate is converted into amino acids for protein synthesis and, subsequently, the energy needs of the cell rise. The citric acid cycle will operate to a reduced extent unless new oxaloacetate is formed, because acetyl CoA cannot enter the cycle unless it condenses with oxaloacetate. Even though oxaloacetate is recycled, a minimal level must be maintained to allow the cycle to function.

How is oxaloacetate replenished? Mammals lack the enzymes for the net conversion of acetyl CoA into oxaloacetate or any other citric acid cycle intermediate. Rather, oxaloacetate is formed by the carboxylation of pyruvate, in a reaction catalyzed by the biotin-dependent enzyme *pyruvate carboxylase.*

$$\text{Pyruvate} + CO_2 + ATP + H_2O \longrightarrow \text{oxaloacetate} + ADP + P_i + 2\,H^+$$

Recall that this enzyme plays a crucial role in gluconeogenesis (Section 16.3.2). It is active only in the presence of acetyl CoA, which signifies the need for more oxaloacetate. If the energy charge is high, oxaloacetate is converted into glucose. If the energy charge is low, oxaloacetate replenishes the citric acid cycle. The synthesis of oxaloacetate by the carboxylation of pyruvate is an example of an *anaplerotic reaction* (of Greek origin, meaning to "fill up"), a reaction that leads to the net synthesis, or replenishment, of pathway components. Note that, because the citric acid cycle is a cycle, it can be replenished by the generation of any of the intermediates.

17.3.2 The Disruption of Pyruvate Metabolism Is the Cause of Beriberi and Poisoning by Mercury and Arsenic

Beriberi, a neurologic and cardiovascular disorder, is caused by a dietary deficiency of thiamine (also called *vitamin B₁*). The disease has been and continues to be a serious health problem in the Far East because rice, the major food, has a rather low content of thiamine. This deficiency is partly ameliorated if the whole rice grain is soaked in water before milling—some of the thiamine in the husk then leaches into the rice kernel. The problem is exacerbated if the rice is polished, because only the outer layer contains significant amounts of thiamine. Beriberi is also occasionally seen in alcoholics who are severely malnourished and thus thiamine deficient. The disease is characterized by neurologic and cardiac symptoms. Damage to the peripheral nervous system is expressed as pain in the limbs, weakness of the musculature, and distorted skin sensation. The heart may be enlarged and the cardiac output inadequate.

Which biochemical processes might be affected by a deficiency of thiamine? *Thiamine pyrophosphate is the prosthetic group of three important enzymes: pyruvate dehydrogenase, α-ketoglutarate dehydrogenase, and transketolase.* Transketolase functions in the pentose phosphate pathway, which will be discussed in Chapter 20. *The common feature of enzymatic reactions utilizing TPP is the transfer of an activated aldehyde unit.* In beriberi, the levels of pyruvate and α-ketoglutarate in the blood are higher than normal. The increase in the level of pyruvate in the blood is especially pronounced after the ingestion of glucose. A related finding is that the activities of the pyruvate and α-ketoglutarate dehydrogenase complexes in vivo are abnormally low. The low transketolase activity of red cells in beriberi is an easily measured and reliable diagnostic indicator of the disease.

Why does TPP deficiency lead primarily to neurological disorders? The nervous system relies essentially on glucose as its only fuel. In contrast, most other tissues can use fats as a source of fuel for the citric acid cycle. The product of aerobic glycolysis, pyruvate, can enter the citric acid cycle only through the pyruvate dehydrogenase complex.

Symptoms similar to those of beriberi arise if an organism is exposed to mercury or arsenite (AsO_3^{3-}). Both elements have a high affinity for neighboring sulfhydryls, such as those in the reduced dihydrolipoyl groups of the dihydrolipoyl dehydrogenase component of the pyruvate dehydrogenase complex (Figure 17.20). The binding of mercury or arsenite to the dihydrolipoyl groups inhibits the complex and leads to central nervous system

Beriberi—
A vitamin-deficiency disease first described in 1630 by Jacob Bonitus, a Dutch physician working in Java:

"A certain very troublesome affliction, which attacks men, is called by the inhabitants Beriberi (which means sheep). I believe those, whom this same disease attacks, with their knees shaking and the legs raised up, walk like sheep. It is a kind of paralysis, or rather Tremor: for it penetrates the motion and sensation of the hands and feet indeed sometimes of the whole body."

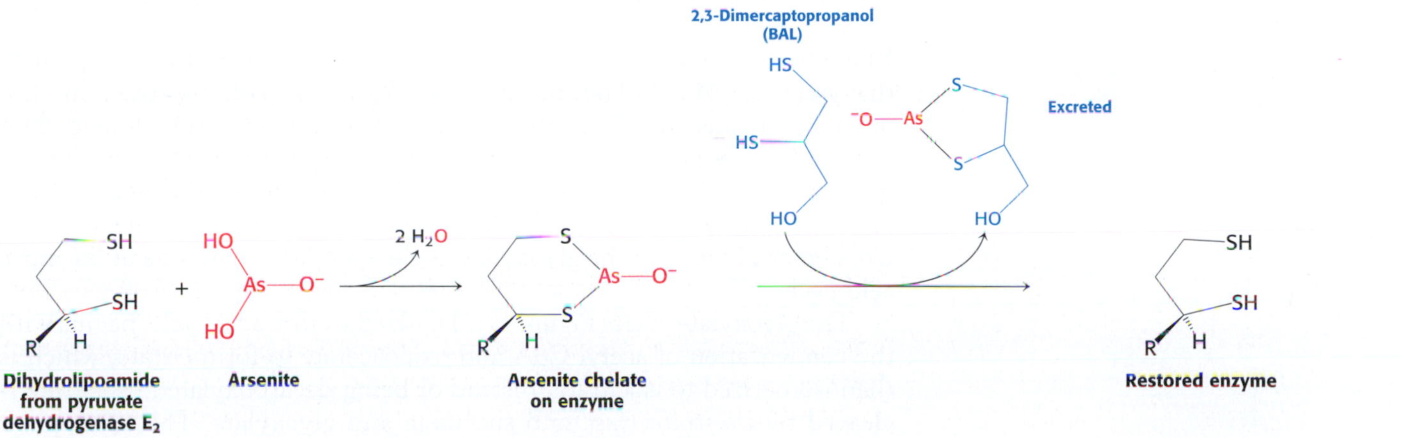

FIGURE 17.20 Arsenite poisoning. Arsenite inhibits the pyruvate dehydrogenase complex by inactivating the dihydrolipoamide component of the transacetylase. Some sulfhydryl reagents, such as 2,3-dimercaptoethanol, relieve the inhibition by forming a complex with the arsenite that can be excreted.

Dihydrolipoamide from pyruvate dehydrogenase E₂   Arsenite   Arsenite chelate on enzyme   2,3-Dimercaptopropanol (BAL)   Excreted   Restored enzyme

[The Granger Collection.]

pathologies. The proverbial phrase "mad as a hatter" refers to the strange behavior of poisoned hat makers who used mercury nitrate to soften and shape animal furs. This form of mercury is absorbed through the skin. Similar problems afflicted the early photographers, who used vaporized mercury to create daguerreotypes.

Treatment for these poisons is the administration of sulfhydryl reagents with adjacent sulfhydryl groups to compete with the dihydrolipoyl residues for binding with the metal ion, which is then excreted in the urine. Indeed, 2,3-dimercaptopropanol (see Figure 17.20) was developed after World War I as an antidote to lewisite, an arsenic-based chemical weapon. This compound was initially called BAL, for British anti-lewisite.

17.3.3 Speculations on the Evolutionary History of the Citric Acid Cycle

How did the citric acid cycle come into being? Although definitive answers are elusive, it is nevertheless instructive to speculate how this complicated central hub of metabolism developed. We can perhaps begin to comprehend how evolution might work at the level of biochemical pathways.

It is most likely that the citric acid cycle was assembled from preexisting reaction pathways. As noted earlier, many of the intermediates formed in the citric acid cycle are used in biosynthetic pathways to generate amino acids and porphyrins. Thus, compounds such as pyruvate, α-ketoglutarate, and oxaloacetate were likely present early in evolution for biosynthetic purposes. The oxidative decarboxylation of these α-ketoacids is quite favorable thermodynamically. The elegant modular structures of the pyruvate and α-ketoglutarate dehydrogenase complexes reveal how three reactions (decarboxylation, oxidation, and thioester formation) can be linked to harness the free energy associated with decarboxylation to drive the synthesis of both acyl CoA derivatives and NADH. These reactions almost certainly formed the core of processes that preceded the citric acid cycle evolutionarily. Interestingly, α-ketoglutarate can be directly converted into oxaloacetate by transamination of the respective amino acids by aspartate aminotransferase, another key biosynthetic enzyme. Thus, cycles comprising smaller numbers of intermediates could have existed before the present form evolved to harvest the electrons from pyruvate or other compounds more efficiently.

17.4 THE GLYOXYLATE CYCLE ENABLES PLANTS AND BACTERIA TO GROW ON ACETATE

Many bacteria and plants are able to subsist on acetate or other compounds that yield acetyl CoA. They make use of a metabolic pathway absent in most other organisms that converts two-carbon acetyl units into four-carbon units (succinate) for energy production and biosyntheses. This reaction sequence, called the *glyoxylate cycle*, bypasses the two decarboxylation steps of the citric acid cycle. Another key difference is that two molecules of acetyl CoA enter per turn of the glyoxylate cycle, compared with one in the citric acid cycle.

The glyoxylate cycle (Figure 17.21), like the citric acid cycle, begins with the condensation of acetyl CoA and oxaloacetate to form citrate, which is then isomerized to isocitrate. Instead of being decarboxylated, isocitrate is cleaved by *isocitrate lyase* into succinate and glyoxylate. The subsequent steps regenerate oxaloacetate from glyoxylate. Acetyl CoA condenses with

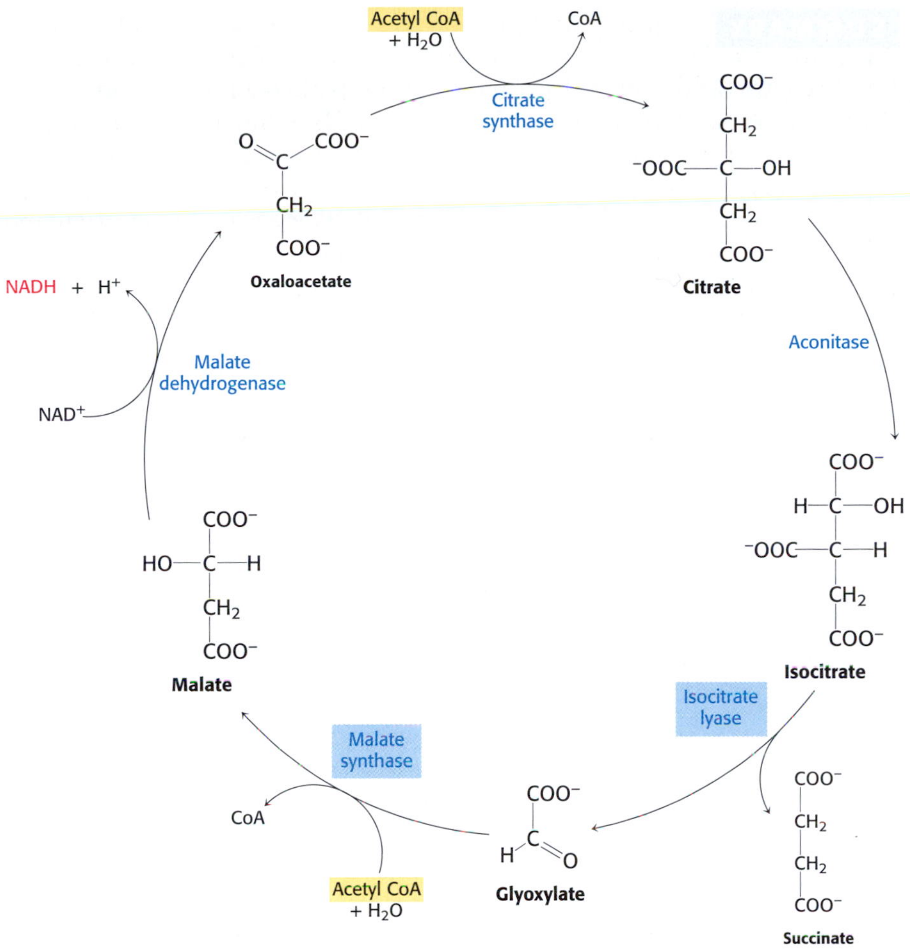

FIGURE 17.21 The glyoxylate pathway. The glyoxylate cycle allows plants and some microorganisms to grow on acetate because the cycle bypasses the decarboxylation steps of the citric acid cycle. The enzymes that permit the conversion of acetate into succinate—isocitrate lyase and malate synthase—are boxed in blue.

glyoxylate to form malate in a reaction catalyzed by *malate synthase*, which resembles citrate synthase. Finally, malate is oxidized to oxaloacetate, as in the citric acid cycle. The sum of these reactions is:

$$2 \text{ Acetyl CoA} + \text{NAD}^+ + 2 \text{ H}_2\text{O} \longrightarrow$$
$$\text{succinate} + 2 \text{ CoA} + \text{NADH} + 2 \text{ H}^+$$

In plants, these reactions take place in organelles called *glyoxysomes*. Succinate, released midcycle, can be converted into carbohydrates by a combination of the citric acid cycle and gluconeogenesis. Thus, organisms with the glyoxylate cycle gain a metabolic versatility.

Bacteria and plants can synthesize acetyl CoA from acetate and CoA by an ATP-driven reaction that is catalyzed by *acetyl CoA synthetase*.

$$\text{Acetate} + \text{CoA} + \text{ATP} \longrightarrow \text{acetyl CoA} + \text{AMP} + \text{PP}_\text{i}$$

Pyrophosphate is then hydrolyzed to orthophosphate, and so the equivalents of two compounds having high phosphoryl transfer potential are consumed in the activation of acetate. We will return to this type of activation reaction in fatty acid degradation (Section 22.2.2), where it is used to form fatty acyl CoA, and in protein synthesis, where it is used to link amino acids to transfer RNAs (Section 29.2.1).

SUMMARY

The citric acid cycle is the final common pathway for the oxidation of fuel molecules. It also serves as a source of building blocks for biosyntheses. Most fuel molecules enter the cycle as acetyl CoA. The link between glycolysis and the citric acid cycle is the oxidative decarboxylation of pyruvate to form acetyl CoA. In eukaryotes, this reaction and those of the cycle take place inside mitochondria, in contrast with glycolysis, which takes place in the cytosol.

The Citric Acid Cycle Oxidizes Two-Carbon Units

The cycle starts with the condensation of oxaloacetate (C_4) and acetyl CoA (C_2) to give citrate (C_6), which is isomerized to isocitrate (C_6). Oxidative decarboxylation of this intermediate gives α-ketoglutarate (C_5). The second molecule of carbon dioxide comes off in the next reaction, in which α-ketoglutarate is oxidatively decarboxylated to succinyl CoA (C_4). The thioester bond of succinyl CoA is cleaved by inorthophosphate to yield succinate, and a high phosphoryl transfer potential compound in the form of GTP is concomitantly generated. Succinate is oxidized to fumarate (C_4), which is then hydrated to form malate (C_4). Finally, malate is oxidized to regenerate oxaloacetate (C_4). Thus, two carbon atoms from acetyl CoA enter the cycle, and two carbon atoms leave the cycle as CO_2 in the successive decarboxylations catalyzed by isocitrate dehydrogenase and α-ketoglutarate dehydrogenase. In the four oxidation–reduction reactions in the cycle, three pairs of electrons are transferred to NAD^+ and one pair to FAD. These reduced electron carriers are subsequently oxidized by the electron-transport chain to generate approximately 9 molecules of ATP. In addition, 1 molecule of a compound having a high phosphoryl transfer potential is directly formed in the citric acid cycle. Hence, a total of 10 molecules of compounds having high phosphoryl transfer potential are generated for each two-carbon fragment that is completely oxidized to H_2O and CO_2.

Entry to the Citric Acid Cycle and Metabolism Through It Are Controlled

The citric acid cycle operates only under aerobic conditions because it requires a supply of NAD^+ and FAD. The irreversible formation of acetyl CoA from pyruvate is an important regulatory point for the entry of glucose-derived pyruvate into the citric acid cycle. The activity of the pyruvate dehydrogenase complex is stringently controlled by reversible phosphorylation. The electron acceptors are regenerated when NADH and $FADH_2$ transfer their electrons to O_2 through the electron-transport chain, with the concomitant production of ATP. Consequently, the rate of the citric acid cycle depends on the need for ATP. In eukaryotes, the regulation of two enzymes in the cycle also is important for control. A high energy charge diminishes the activities of isocitrate dehydrogenase and α-ketoglutarate dehydrogenase. These mechanisms complement each other in reducing the rate of formation of acetyl CoA when the energy charge of the cell is high and when biosynthetic intermediates are abundant.

The Citric Acid Cycle Is a Source of Biosynthetic Precursors

When the cell has adequate energy available, the citric acid cycle can also provide a source of building blocks for a host of important biomolecules, such as nucleotide bases, proteins, and heme groups. This use depletes the cycle of intermediates. When the cycle again needs to metabolize fuel, anaplerotic reactions replenish the cycle intermediates.

● **The Glyoxylate Cycle Enables Plants and Bacteria to Grow on Acetate**

The glyoxylate cycle enhances the metabolic versatility of many plants and bacteria. This cycle, which uses some of the reactions of the citric acid cycle, enables these organisms to subsist on acetate because it bypasses the two decarboxylation steps of the citric acid cycle.

KEY TERMS

citric acid (tricarboxylic acid, TCA; Krebs) cycle (p. 465)

oxidative phosphorylation (p. 467)

acetyl CoA (p. 467)

pyruvate dehydrogenase complex (p. 467)

flavoprotein (p. 470)

citrate synthase (p. 472)

iron–sulfur (nonheme iron) protein (p. 474)

isocitrate dehydrogenase (p. 474)

α-ketoglutarate dehydrogenase (p. 475)

metabolon (p. 479)

anaplerotic reaction (p. 482)

beriberi (p. 483)

glyoxylate cycle (p. 484)

isocitrate lyase (p. 484)

malate synthase (p. 485)

glyoxysome (p. 485)

SELECTED READINGS

Where to start

Reed, L. J., and Hackert, M. L., 1990. Structure-function relationships in dihydrolipoamide acyltransferases. *J. Biol. Chem.* 265:8971–8974.

Mattevi, A., Obmolova, G., Schulze, E., Kalk, K. H., Westphal, A. H., De Kok, A., and Hol, W. G., 1992. Atomic structure of the cubic core of the pyruvate dehydrogenase multienzyme complex. *Science* 255:1544–1550.

Pyruvate dehydrogenase complex

Izard, T., Ævarsson, A., Allen, M. D., Westphal, A. H., Perham, R. N., De Kok, A., and Hol, W. G., 1999. Principles of quasi-equivalence and Euclidean geometry govern the assembly of cubic and dodecahedral cores of pyruvate dehydrogenase complexes. *Proc. Natl. Acad. Sci. USA* 96:1240–1245.

Ævarsson, A., Seger, K., Turley, S., Sokatch, J. R., and Hol, W. M. J., 1999. Crystal structure of 2-oxoisovalerate and dehydrogenase and the architecture of 2-oxo acid dehydrogenase multiple enzyme complexes. *Nat. Struct. Biol.* 6:785–792.

Domingo, G. J., Chauhan, H. J., Lessard, I. A., Fuller, C., and Perham, R. N., 1999. Self-assembly and catalytic activity of the pyruvate dehydrogenase multienzyme complex from *Bacillus stearothermophilus*. *Eur. J. Biochem.* 266:1136–1146.

Jones, D. D., Horne, H. J., Reche, P. A., and Perham, R. N., 2000. Structural determinants of post-translational modification and catalytic specificity for the lipoyl domains of the pyruvate dehydrogenase multienzyme complex of *Escherichia coli*. *J. Mol. Biol.* 295:289–306.

McCartney, R. G., Rice, J. E., Sanderson, S. J., Bunik, V., Lindsay, H., and Lindsay, J. G., 1998. Subunit interactions in the mammalian alpha-ketoglutarate dehydrogenase complex: Evidence for direct association of the alpha-ketoglutarate dehydrogenase and dihydrolipoamide dehydrogenase components. *J. Biol. Chem.* 273:24158–24164.

Structure of citric acid cycle enzymes

Chapman, A. D., Cortes, A., Dafforn, T. R., Clarke, A. R., and Brady, R. L., 1999. Structural basis of substrate specificity in malate dehydrogenases: Crystal structure of a ternary complex of porcine cytoplasmic malate dehydrogenase, alpha-ketomalonate and tetrahydoNAD. *J. Mol. Biol.* 285:703–712.

Fraser, M. E., James, M. N., Bridger, W. A., and Wolodko, W. T., 1999. A detailed structural description of *Escherichia coli* succinyl-CoA

synthetase. *J. Mol. Biol.* 285:1633–1653. [Published erratum appears in May 7, 1999, issue of *J. Mol. Biol.* 288(3):501.]

Lloyd, S. J., Lauble, H., Prasad, G. S., and Stout, C. D., 1999. The mechanism of aconitase: 1.8 Å resolution crystal structure of the S642a:citrate complex. *Protein Sci.* 8:2655–2662.

Remington, S. J., 1992. Structure and mechanism of citrate synthase. *Curr. Top. Cell. Regul.* 33:209–229.

Rose, I. A., 1998. How fumarase recycles after the malate ⟶ fumarate reaction: Insights into the reaction mechanism. *Biochemistry* 37:17651–17658.

Johnson, J. D., Muhonen, W. W., and Lambeth, D. O., 1998. Characterization of the ATP- and GTP-specific succinyl-CoA synthetases in pigeon: The enzymes incorporate the same subunit. *J. Biol. Chem.* 273:27573–27579.

Karpusas, M., Branchaud, B., and Remington, S. J., 1990. Proposed mechanism for the condensation reaction of citrate synthase: 1.9-Å structure of the ternary complex with oxaloacetate and carboxymethyl coenzyme A. *Biochemistry* 29:2213–2219.

Lauble, H., Kennedy, M. C., Beinert, H., and Stout, C. D., 1992. Crystal structures of aconitase with isocitrate and nitroisocitrate bound. *Biochemistry* 31:2735–2748.

Organization of the citric acid cycle

Velot, C., Mixon, M. B., Teige, M., and Srere, P. A., 1997. Model of a quinary structure between Krebs TCA cycle enzymes: A model for the metabolon. *Biochemistry* 36:14271–14276.

Barnes, S. J., and Weitzman, P. D., 1986. Organization of citric acid cycle enzymes into a multienzyme cluster. *FEBS Lett.* 201:267–270.

Haggie, P. M., and Brindle, K. M., 1999. Mitochondrial citrate synthase is immobilized in vivo. *J. Biol. Chem.* 274:3941–3945.

Morgunov, I., and Srere, P. A., 1998. Interaction between citrate synthase and malate dehydrogenase: Substrate channeling of oxaloacetate. *J. Biol. Chem.* 273:29540–29544.

Regulation

Huang, B., Gudi, R., Wu, P., Harris, R. A., Hamilton, J., and Popov, K. M., 1998. Isoenzymes of pyruvate dehydrogenase phosphatase: DNA-derived amino acid sequences, expression, and regulation. *J. Biol. Chem.* 273:17680–17688.

Bowker-Kinley, M., and Popov, K. M., 1999. Evidence that pyruvate dehydrogenase kinase belongs to the ATPase/kinase superfamily. *Biochem. J.* 1:47–53.

Jitrapakdee, S., and Wallace, J. C., 1999. Structure, function and regulation of pyruvate carboxylase. *Biochem. J.* 340:1–16.

Hurley, J. H., Dean, A. M., Sohl, J. L., Koshland, D. J., and Stroud, R. M., 1990. Regulation of an enzyme by phosphorylation at the active site. *Science* 249:1012–1016.

Evolutionary aspects

Meléndez-Hevia, E., Waddell, T. G., and Cascante, M., 1996. The puzzle of the Krebs citric acid cycle: Assembling the pieces of chemically feasible reactions, and opportunism in the design of metabolic pathways in evolution. *J. Mol. Evol.* 43:293–303.

Baldwin, J. E., and Krebs, H., 1981. The evolution of metabolic cycles. *Nature* 291:381–382.

Gest, H., 1987. Evolutionary roots of the citric acid cycle in prokaryotes. *Biochem. Soc. Symp.* 54:3–16.

Weitzman, P. D. J., 1981. Unity and diversity in some bacterial citric acid cycle enzymes. *Adv. Microbiol. Physiol.* 22:185–244.

Discovery of the citric acid cycle

Krebs, H. A., and Johnson, W. A., 1937. The role of citric acid in intermediate metabolism in animal tissues. *Enzymologia* 4:148–156.

Krebs, H. A., 1970. The history of the tricarboxylic acid cycle. *Perspect. Biol. Med.* 14:154–170.

Krebs, H. A., and Martin, A., 1981. *Reminiscences and Reflections.* Clarendon Press.

PROBLEMS

1. *Flow of carbon atoms.* What is the fate of the radioactive label when each of the following compounds is added to a cell extract containing the enzymes and cofactors of the glycolytic pathway, the citric acid cycle, and the pyruvate dehydrogenase complex? (The ^{14}C label is printed in red.)

(a) (b) (c) (d)

(e) Glucose 6-phosphate labeled at C-1.

2. $C_2 + C_2 \longrightarrow C_4$.

(a) Which enzymes are required to get *net synthesis* of oxaloacetate from acetyl CoA?
(b) Write a balanced equation for the net synthesis.
(c) Do mammalian cells contain the requisite enzymes?

3. *Driving force.* What is the $\Delta G°'$ for the complete oxidation of the acetyl unit of acetyl CoA by the citric acid cycle?

4. *Acting catalytically.* The citric acid cycle itself, which is composed of enzymatically catalyzed steps, can be thought of essentially as the product of a supramolecular enzyme. Explain.

5. *Probing stereospecificity.* A sample of deuterated reduced NAD was prepared by incubating $H_3C–CD_2–OH$ and NAD^+ with alcohol dehydrogenase. This reduced coenzyme was added to a solution of 1,3-BPG and glyceraldehyde 3-phosphate dehydrogenase. The NAD^+ formed by this second reaction contained one atom of deuterium, whereas glyceraldehyde 3-phosphate, the other product, contained none. What does this experiment reveal about the stereospecificity of glyceraldehyde 3-phosphate dehydrogenase?

6. *A potent inhibitor.* Thiamine thiazolone pyrophosphate binds to pyruvate dehydrogenase about 20,000 times as strongly

as does thiamine pyrophosphate, and it competitively inhibits the enzyme. Why?

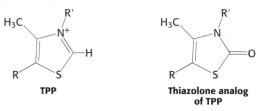

TPP **Thiazolone analog of TPP**

7. *Lactic acidosis.* Patients in shock will often suffer from lactic acidosis due to a deficiency of O_2. Why does a lack of O_2 lead to lactic acid accumulation? One treatment for shock is to administer dichloroacetate, which inhibits the kinase associated with the pyruvate dehydrogenase complex. What is the biochemical rationale for this treatment?

8. *Coupling reactions.* The oxidation of malate by NAD^+ to form oxaloacetate is a highly endergonic reaction under standard conditions [$\Delta G°' = + 7$ kcal mol^{-1} ($+ 29$ kJ mol^{-1})]. The reaction proceeds readily under physiological conditions.

(a) Why?
(b) Assuming an [NAD^+]/[$NADH$] ratio of 8 and a pH of 7, what is the lowest [malate]/[oxaloacetate] ratio at which oxaloacetate can be formed from malate?

9. *Synthesizing α-ketoglutarate.* It is possible, with the use of the reactions and enzymes discussed in this chapter, to convert pyruvate into α-ketoglutarate without depleting any of the citric acid cycle components. Write a balanced reaction scheme for this conversion, showing cofactors and identifying the required enzymes.

Chapter Integration Problem

10. *Fats into glucose?* Fats are usually metabolized into acetyl CoA and then further processed through the citric acid cycle. In Chapter 16, we learned that glucose could be synthesized from oxaloacetate, a citric acid cycle intermediate. Why, then, after a long bout of exercise depletes our carbohydrate stores, do we need to replenish those stores by eating carbohydrates? Why do we not simply replace them by converting fats into carbohydrates?

Mechanism Problems

11. *Theme and variation.* Propose a reaction mechanism for the condensation of acetyl CoA and glyoxylate in the glyoxylate cycle of plants and bacteria.

12. *Symmetry problems.* In experiments carried out in 1941 to investigate the citric acid cycle, oxaloacetate labeled with ^{14}C in the carboxyl carbon atom farthest from the keto group was introduced to an active preparation of mitochondria.

Oxaloacetate

Analysis of the α-ketoglutarate formed showed that none of the radioactive label had been lost. Decarboxylation of α-ketoglutarate then yielded succinate devoid of radioactivity. All the label was in the released CO_2. Why were the early investigators of the citric acid cycle surprised that *all* the label emerged in the CO_2?

13. *Symmetric molecules reacting asymmetrically.* The interpretation of the experiments described in problem 12 was that citrate (or any other symmetric compound) cannot be an intermediate in the formation of α-ketoglutarate, because of the asymmetric fate of the label. This view seemed compelling until Alexander Ogston incisively pointed out in 1948 that "it is possible that *an asymmetric enzyme which attacks a symmetrical compound can distinguish between its identical groups.*" For simplicity, consider a molecule in which two hydrogen atoms, a group X, and a different group Y are bonded to a tetrahedral carbon atom as a model for citrate. Explain how a symmetric molecule can react with an enzyme in an asymmetric way.

Data Interpretation

14. *A little goes a long way.* As will become clearer in Chapter 18, the activity of the citric acid cycle can be monitored by monitoring the amount of O_2 consumed. The greater the rate of O_2 consumption, the faster the rate of the cycle. Hans Krebs used this assay to investigate the cycle in 1937. He used as his experimental system minced pigeon breast muscle, which is rich in mitochondria. In one set of experiments, Krebs measured the O_2 consumption in the presence of carbohydrate only and in the presence of carbohydrate and citrate. The results are shown in the following table.

Effect of citrate on oxygen consumption by minced pigeon breast muscle

Time (min)	Micromoles of oxygen consumed	
	Carbohydrate only	Carbohydrate plus 3 μmol of citrate
10	26	28
60	43	62
90	46	77
150	49	85

(a) How much O_2 would be absorbed if the added citrate were completely oxidized to H_2O and CO_2?
(b) Based on your answer to part *a*, what do the results given in the table suggest?

15. *Arsenite poisoning.* The effect of arsenite on the experimental system of problem 14 was then examined. Experimental data (not presented here) showed that the amount of citrate present did not change in the course of the experiment in the absence of arsenite. However, if arsenite was added to the system, different results were obtained, as shown in the following table.

Disappearance of citric acid in pigeon breast muscle in the presence of arsenite

Micromoles of citrate added	Micromoles of citrate found after 40 minutes	Micromoles of citrate used
22	0.6	21
44	20	24
90	56	34

(a) What is the effect of arsenite on the disappearance of citrate?
(b) How is the arsenite's action altered by the addition of more citrate?
(c) What do these data suggest about the site of action of arsenite?

16. *Isocitrate lyase and tuberculosis.* The bacterium *Mycobacterium tuberculosis*, the cause of tuberculosis, can invade the lungs and persist in a latent state for years. During this time, the bacteria reside in granulomas—nodular scars containing bacteria and host-cell debris in the center and surrounded by immune cells. The granulomas are lipid-rich, oxygen-poor environments. How these bacteria manage to persist is something of a mystery. The results of recent research suggest that the glyoxylate cycle is required for the persistence. The following data show the amount of bacteria [presented as colony-forming units (cfu)] in mice lungs in the weeks after an infection.

In graph A, the black circles represent the results for wild-type bacteria and the red circles represent the results for bacteria from which the gene for isocitrate lyase was deleted.

(A)

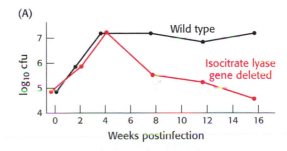

(a) What is the effect of the absence of isocitrate lyase?

The techniques discussed in Chapter 6 were used to reinsert the gene encoding isocitrate lyase into bacteria from which it had previously been deleted.

In graph B, black circles represent bacteria into which the gene was reinserted and red circles bacteria from which the gene was still missing.

(b) Do these results support those obtained in part *a*?

(c) What is the purpose of the experiment in part *b*?

(d) Why do these bacteria perish in the absence of the glyoxylate cycle?

Need extra help? Purchase chapters of the *Student Companion* with complete solutions online at www.whfreeman.com/biochem5.

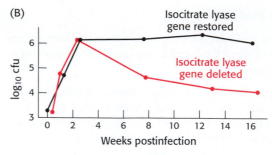

(B)

[Data after McKinney et al., 2000. *Nature* 406:735–738.]

CHAPTER 18

Oxidative Phosphorylation

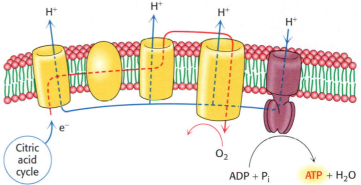

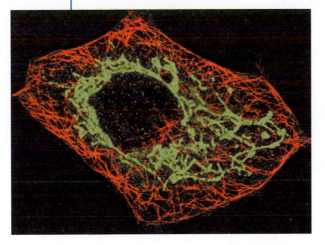

Mitochondria, stained green, form a network inside a fibroblast cell (left). Mitochondria oxidize carbon fuels to form cellular energy. This transformation requires electron transfer through several large protein complexes (above), some of which pump protons, forming a proton gradient that powers the synthesis of ATP. [(Left) Courtesy of Michael P. Yaffee, Dept. of Biology, University of California at San Diego.]

The NADH and FADH$_2$ formed in glycolysis, fatty acid oxidation, and the citric acid cycle are energy-rich molecules because each contains a pair of electrons having a high transfer potential. When these electrons are used to reduce molecular oxygen to water, a large amount of free energy is liberated, which can be used to generate ATP. *Oxidative phosphorylation is the process in which ATP is formed as a result of the transfer of electrons from NADH or FADH$_2$ to O$_2$ by a series of electron carriers.* This process, which takes place in mitochondria, is the major source of ATP in aerobic organisms (Figure 18.1). For example, oxidative phosphorylation generates 26 of the 30 molecules of ATP that are formed when glucose is completely oxidized to CO$_2$ and H$_2$O.

Oxidative phosphorylation is conceptually simple and mechanistically complex. Indeed, the unraveling of the mechanism of oxidative phosphorylation has been one of the most challenging problems of biochemistry. The flow of electrons from NADH or FADH$_2$ to O$_2$ through protein complexes located in the mitochondrial inner membrane leads to the pumping of protons out of the mitochondrial matrix. The resulting uneven distribution of protons generates a pH gradient and a transmembrane electrical potential that creates a *proton-motive force*. ATP is synthesized when protons flow back to the mitochondrial matrix

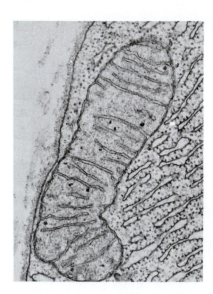

FIGURE 18.1 Electron micrograph of a mitochondrion. [Courtesy of Dr. George Palade.]

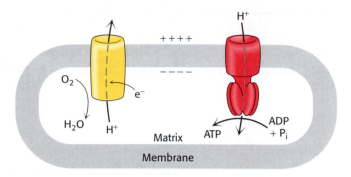

FIGURE 18.2 Essence of oxidative phosphorylation. Oxidation and ATP synthesis are coupled by transmembrane proton fluxes.

through an enzyme complex. Thus, *the oxidation of fuels and the phosphorylation of ADP are coupled by a proton gradient across the inner mitochondrial membrane* (Figure 18.2).

Oxidative phosphorylation is the culmination of a series of energy transformations that are called *cellular respiration* or simply *respiration* in their entirety. First, carbon fuels are oxidized in the citric acid cycle to yield electrons with high transfer potential. Then, this electron-motive force is converted into a proton-motive force and, finally, the proton-motive force is converted into phosphoryl transfer potential. The conversion of electron-motive force into proton-motive force is carried out by three electron-driven proton pumps—NADH-Q oxidoreductase, Q-cytochrome *c* oxidoreductase, and cytochrome *c* oxidase. These large transmembrane complexes contain multiple oxidation–reduction centers, including quinones, flavins, iron–sulfur clusters, hemes, and copper ions. The final phase of oxidative phosphorylation is carried out by *ATP synthase,* an ATP-synthesizing assembly that is driven by the flow of protons back into the mitochondrial matrix. Components of this remarkable enzyme rotate as part of its catalytic mechanism. Oxidative phosphorylation vividly shows that *proton gradients are an interconvertible currency of free energy in biological systems.*

Respiration—

An ATP-generating process in which an inorganic compound (such as molecular oxygen) serves as the ultimate electron acceptor. The electron donor can be either an organic compound or an inorganic one.

18.1 OXIDATIVE PHOSPHORYLATION IN EUKARYOTES TAKES PLACE IN MITOCHONDRIA

Mitochondria are oval-shaped organelles, typically about 2 μm in length and 0.5 μm in diameter, about the size of a bacterium. Eugene Kennedy and Albert Lehninger discovered a half-century ago that *mitochondria contain the respiratory assembly, the enzymes of the citric acid cycle, and the enzymes of fatty acid oxidation.*

18.1.1 Mitochondria Are Bounded by a Double Membrane

Electron microscopic studies by George Palade and Fritjof Sjöstrand revealed that mitochondria have two membrane systems: an *outer membrane* and an extensive, highly folded *inner membrane.* The inner membrane is folded into a series of internal ridges called *cristae.* Hence, there are two compartments in mitochondria: (1) the *intermembrane space* between the outer and the inner membranes and (2) the *matrix,* which is bounded by the inner membrane (Figure 18.3). Oxidative phosphor-

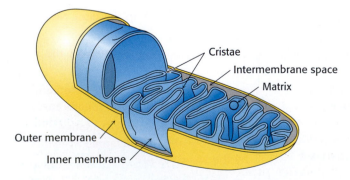

FIGURE 18.3 Diagram of a mitochondrion. [After *Biology of the Cell* by Stephen L. Wolfe. © 1972 by Wadsworth Publishing Company, Inc., Belmost, California 94002. Adapted by permission of the publisher.]

ylation takes place in the inner mitochondrial membrane, in contrast with most of the reactions of the citric acid cycle and fatty acid oxidation, which take place in the matrix.

The outer membrane is quite permeable to most small molecules and ions because it contains many copies of *mitochondrial porin*, a 30–35 kd pore-forming protein also known as VDAC, for voltage-dependent anion channel. VDAC plays a role in the regulated flux of metabolites—usually anionic species such as phosphate, chloride, organic anions, and the adenine nucleotides—across the outer membrane. VDAC appears to form an open β-barrel structure similar to that of the bacterial porins (Section 12.5.2), although mitochondrial porins and bacterial porins may have evolved independently. Some cytoplasmic kinases bind to VDAC, thereby obtaining preferential access to the exported ATP. In contrast, the inner membrane is intrinsically impermeable to nearly all ions and polar molecules. A large family of transporters shuttles metabolites such as ATP, pyruvate, and citrate across the inner mitochondrial membrane. The two faces of this membrane will be referred to as the *matrix side* and the *cytosolic side* (the latter because it is freely accessible to most small molecules in the cytosol). They are also called the *N* and *P* sides, respectively, because the membrane potential is negative on the matrix side and positive on the cytosolic side.

In prokaryotes, the electron-driven proton pumps and ATP-synthesizing complex are located in the cytoplasmic membrane, the inner of two membranes. The outer membrane of bacteria, like that of mitochondria, is permeable to most small metabolites because of the presence of porins.

18.1.2 Mitochondria Are the Result of an Endosymbiotic Event

Mitochondria are semiautonomous organelles that live in an endosymbiotic relation with the host cell. These organelles contain their own DNA, which encodes a variety of different proteins and RNAs. The genomes of mitochondrial range broadly in size across species. The mitochondrial genome of the protist *Plasmodium falciparum* consists of fewer than 6000 base pairs (6 kbp), whereas those of some land plants comprise more than 200 kbp (Figure 18.4). Human mitochondrial DNA comprises 16,569 bp and encodes 13 respiratory-chain proteins as well as the small and large ribosomal RNAs and enough tRNAs to translate all codons. However, mitochondria also contain many proteins encoded by nuclear DNA. Cells that contain mitochondria depend on these organelles for oxidative phosphorylation, and the mitochondria in turn depend on the cell for their very existence. How did this intimate symbiotic relation come to exist?

An *endosymbiotic event* is thought to have occurred whereby a free-living organism capable of oxidative phosphorylation was engulfed by another cell. The double membrane, circular DNA (with some exceptions), and mitochondrial-specific transcription and translation machinery all point to this conclusion. Thanks to the rapid accumulation of sequence data for mitochondrial and bacterial genomes, it is now possible to speculate on the origin of the "original" mitochondrion with some authority. The most mitochondrial-like bacterial genome is that of *Rickettsia prowazekii*, the cause of louse-borne typhus. The genome for this organism is more than 1 million base pairs in size and contains 834 protein-encoding genes. Sequence data suggest that all extant mitochondria are derived from an ancestor of *R. prowazekii* as the result of a single endosymbiotic event.

The evidence that modern mitochondria result from a single event comes from examination of the most bacteria-like mitochondrial genome, that of the protozoan *Reclinomonas americana*. Its genome contains 97 genes, of which 62 specify proteins that include all of the protein-coding genes found

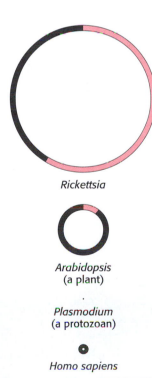

Rickettsia

Arabidopsis
(a plant)

Plasmodium
(a protozoan)

Homo sapiens

FIGURE 18.4 Sizes of mitochondrial genomes. The sizes of three mitochondrial genomes compared with the genome of *Rickettsia*, a relative of the the presumed ancestor of all mitochondria. For genomes of more than 60 kbp, the DNA coding region for genes with known function is shown in red.

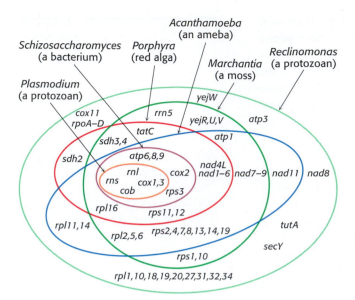

FIGURE 18.5 Overlapping gene complements of mitochondria. The genes present within each oval are those present within the organism represented by the oval. Only rRNA- and protein-coding genes are shown. The genome of *Reclinomonas* contains all the protein-coding genes found in all the sequenced mitochondrial genomes. [After M. W. Gray, G. Burger, and B. F. Lang. *Science* 283(1999): 1476–1481.]

in all of the sequenced mitochondrial genomes (Figure 18.5). Yet, this genome encodes less than 2% of the protein-coding genes in the bacterium *E. coli*. It seems unlikely that mitochondrial genomes resulting from several endosymbiotic events could have been independently reduced to the same set of genes found in *R. americana*.

Note that transient engulfment of prokaryotic cells by larger cells is not uncommon in the microbial world. In regard to mitochondria, such a transient relation became permanent as the bacterial cell lost DNA, making it incapable of independent living, and the host cell became dependent on the ATP generated by its tenant.

18.2 OXIDATIVE PHOSPHORYLATION DEPENDS ON ELECTRON TRANSFER

In Chapter 17, the primary function of the citric acid cycle was identified as the generation of NADH and $FADH_2$ by the oxidation of acetyl CoA. In oxidative phosphorylation, NADH and $FADH_2$ are used to reduce molecular oxygen to water. The highly exergonic reduction of molecular oxygen by NADH and $FADH_2$ occurs in a number of electron-transfer reactions, taking place in a set of membrane proteins known as the *electron-transport chain*.

18.2.1 High-Energy Electrons: Redox Potentials and Free-Energy Changes

High-energy electrons and redox potentials are of fundamental importance in oxidative phosphorylation. In oxidative phosphorylation, the *electron transfer potential* of NADH or $FADH_2$ is converted into the *phosphoryl transfer potential* of ATP. We need quantitative expressions for these forms of free energy. The measure of phosphoryl transfer potential is already familiar to us: it is given by $\Delta G^{\circ\prime}$ for the hydrolysis of the activated phosphate compound. The corresponding expression for the electron transfer potential is E_0', the *reduction potential* (also called the *redox potential* or *oxidation–reduction potential*).

The reduction potential is an electrochemical concept. Consider a substance that can exist in an oxidized form X and a reduced form X^-. Such a pair is called a *redox couple*. The reduction potential of this couple can be determined by measuring the electromotive force generated by a *sample half-cell* connected to a *standard reference half-cell* (Figure 18.6). The sample half-cell consists of an electrode immersed in a solution of 1 M oxidant (X) and 1 M reductant (X^-). The standard reference half-cell consists of an electrode immersed in a 1 M H^+ solution that is in equilibrium with H_2 gas at 1 atmosphere pressure. The electrodes are connected to a voltmeter, and an agar bridge establishes electrical continuity between the half-cells. Electrons then flow from one half-cell to the other. If the reaction proceeds in the direction

$$X^- + H^+ \longrightarrow X + \tfrac{1}{2}H_2$$

the reactions in the half-cells (referred to as *half-reactions* or *couples*) must be

$$X^- \longrightarrow X + e^- \qquad H^+ + e^- \longrightarrow \tfrac{1}{2}H_2$$

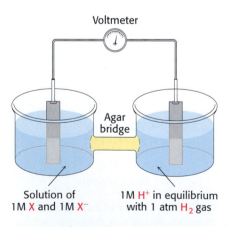

Voltmeter

Agar bridge

Solution of 1M X and 1M X⁻

1M H⁺ in equilibrium with 1 atm H₂ gas

FIGURE 18.6 Measurement of redox potential. Apparatus for the measurement of the standard oxidation–reduction potential of a redox couple. Electrons, but not X or X⁻, can flow through the agar bridge.

Thus, electrons flow from the sample half-cell to the standard reference half-cell, and the sample-cell electrode is taken to be negative with respect to the standard-cell electrode. *The reduction potential of the $X:X^-$ couple is the observed voltage at the start of the experiment* (when X, X^-, and H^+ are 1 M). *The reduction potential of the $H^+:H_2$ couple is defined to be 0 volts.*

The meaning of the reduction potential is now evident. A negative reduction potential means that the reduced form of a substance has lower affinity for electrons than does H_2, as in the preceding example. A positive reduction potential means that the reduced form of a substance has higher affinity for electrons than does H_2. These comparisons refer to standard conditions—namely, 1 M oxidant, 1 M reductant, 1 M H^+, and 1 atmosphere H_2. Thus, *a strong reducing agent (such as NADH) is poised to donate electrons and has a negative reduction potential, whereas a strong oxidizing agent (such as O_2) is ready to accept electrons and has a positive reduction potential.*

The reduction potentials of many biologically important redox couples are known (Table 18.1). Table 18.1 is like those presented in chemistry texts except that a hydrogen ion concentration of 10^{-7} M (pH 7) instead of 1 M (pH 0) is the standard state adopted by biochemists. This difference is denoted by the prime in E_0'. Recall that the prime in $\Delta G^{\circ\prime}$ denotes a standard free-energy change at pH 7.

The standard free-energy change $\Delta G^{\circ\prime}$ is related to the change in reduction potential $\Delta E_0'$ by

$$\Delta G^{\circ\prime} = -nF\Delta E_0'$$

in which n is the number of electrons transferred, F is a proportionality constant called the *faraday* [23.06 kcal mol^{-1} V^{-1} (96.48 kJ mol^{-1} V^{-1})], $\Delta E_0'$ is in volts, and $\Delta G^{\circ\prime}$ is in kilocalories or kilojoules per mole.

The free-energy change of an oxidation–reduction reaction can be readily calculated from the reduction potentials of the reactants. For example,

TABLE 18.1 **Standard reduction potentials of some reactions**

Oxidant	Reductant	n	E_0' (V)
Succinate + CO_2	α-Ketoglutarate	2	-0.67
Acetate	Acetaldehyde	2	-0.60
Ferredoxin (oxidized)	Ferredoxin (reduced)	1	-0.43
2 H^+	H_2	2	-0.42
NAD^+	$NADH + H^+$	2	-0.32
$NADP^+$	$NADPH + H^+$	2	-0.32
Lipoate (oxidized)	Lipoate (reduced)	2	-0.29
Glutathione (oxidized)	Glutathione (reduced)	2	-0.23
FAD	$FADH_2$	2	-0.22
Acetaldehyde	Ethanol	2	-0.20
Pyruvate	Lactate	2	-0.19
Fumarate	Succinate	2	0.03
Cytochrome b (+3)	Cytochrome b (+2)	1	0.07
Dehydroascorbate	Ascorbate	2	0.08
Ubiquinone (oxidized)	Ubiquinone (reduced)	2	0.10
Cytochrome c (+3)	Cytochrome c (+2)	1	0.22
Fe (+3)	Fe (+2)	1	0.77
$\frac{1}{2}O_2 + 2 H^+$	H_2O	2	0.82

Note: E_0' is the standard oxidation–reduction potential (pH 7, 25°C) and n is the number of electrons transferred. E_0' refers to the partial reaction written as

Oxidant + e$^-$ ⟶ reductant

consider the reduction of pyruvate by NADH, catalyzed by lactate dehydrogenase.

$$\text{Pyruvate} + \text{NADH} + \text{H}^+ \rightleftharpoons \text{lactate} + \text{NAD}^+ \qquad \text{(a)}$$

The reduction potential of the NAD^+:NADH couple, or half-reaction, is -0.32 V, whereas that of the pyruvate: lactate couple is -0.19 V. By convention, reduction potentials (as in Table 18.1) refer to partial reactions written as reductions: oxidant $+ e^- \longrightarrow$ reductant. Hence,

$$\text{Pyruvate} + 2\,\text{H}^+ + 2\,e^- \longrightarrow \text{lactate} \qquad E_0' = -0.19 \text{ V} \qquad \text{(b)}$$

$$\text{NAD}^+ + \text{H}^+ + 2\,e^- \longrightarrow \text{NADH} \qquad E_0' = -0.32 \text{ V} \qquad \text{(c)}$$

To obtain reaction a from reactions b and c, we need to reverse the direction of reaction c so that NADH appears on the left side of the arrow. In doing so, the sign of E_0' must be changed.

$$\text{Pyruvate} + 2\,\text{H}^+ + 2\,e^- \longrightarrow \text{lactate} \qquad E_0' = -0.19 \text{ V} \qquad \text{(b)}$$

$$\text{NADH} \longrightarrow \text{NAD}^+ + \text{H}^+ + 2\,e^- \qquad E_0' = +0.32 \text{ V} \qquad \text{(d)}$$

For reaction b, the free energy can be calculated with $n = 2$.

$$\Delta G^{\circ\prime} = -2 \times 23.06 \text{ kcal mol}^{-1}\,\text{V}^{-1} \times -0.19 \text{ V}$$
$$= +8.8 \text{ kcal mol}^{-1}\,(36.7 \text{ kJ mol}^{-1})$$

Likewise, for reaction d,

$$\Delta G^{\circ\prime} = -2 \times 23.06 \text{ kcal mol}^{-1}\,\text{V}^{-1} \times +0.32 \text{ V}$$
$$= -14.8 \text{ kcal mol}^{-1}\,(61.7 \text{ kJ mol}^{-1})$$

Thus, the free energy for reaction a is given by

$$\Delta G^{\circ\prime} = \Delta G^{\circ\prime} \text{ (for reaction b)} + \Delta G^{\circ\prime} \text{ (for reaction d)}$$
$$= +8.8 + (-14.8)$$
$$= -6.0 \text{ kcal mol}^{-1}\,(-25.1 \text{ kJ mol}^{-1})$$

18.2.2 A 1.14-Volt Potential Difference Between NADH and O₂ Drives Electron Transport Through the Chain and Favors the Formation of a Proton Gradient

The driving force of oxidative phosphorylation is the electron transfer potential of NADH or FADH_2 relative to that of O_2. How much energy is released by the reduction of O_2 with NADH? Let us calculate $\Delta G^{\circ\prime}$ for this reaction. The pertinent half-reactions are

$$\tfrac{1}{2}\,O_2 + 2\,\text{H}^+ + 2\,e^- \longrightarrow \text{H}_2\text{O} \qquad E_0' = +0.82 \text{ V} \qquad \text{(a)}$$

$$\text{NAD}^+ + \text{H}^+ + 2\,e^- \longrightarrow \text{NADH} \qquad E_0' = -0.32 \text{ V} \qquad \text{(b)}$$

Subtracting reaction b from reaction a yields

$$\tfrac{1}{2}\,O_2 + \text{NADH} + \text{H}^+ \longrightarrow \text{H}_2\text{O} + \text{NAD}^+ \qquad \text{(c)}$$

The standard free energy for this reaction is then given by

$$\Delta G^{\circ\prime} = -2 \times 23.06 \text{ kcal mol}^{-1}\,\text{V}^{-1} \times +0.82 \text{ V} -$$
$$(-2 \times 23.06 \text{ kcal mol}^{-1}\,\text{V}^{-1} \times -0.32 \text{ V})$$
$$= -37.8 \text{ kcal mol}^{-1} - (14.8 \text{ kcal mol}^{-1})$$
$$= -52.6 \text{ kcal mol}^{-1}\,(-220.1 \text{ kJ mol}^{-1})$$

This is a substantial release of free energy. Recall that $\Delta G^{\circ\prime} = -7.5$ kcal mol^{-1} ($-31.4 \text{ kJ mol}^{-1}$) for the hydrolysis of ATP. The released energy is used initially to generate a proton gradient that is then used for the synthesis of ATP and the transport of metabolites across the mitochondrial membrane.

How can the energy associated with a proton gradient be quantified? Recall that the free-energy change for a species moving from one side of a membrane where it is at concentration c_1 to the other side where it is at a concentration c_2 is given by

$$\Delta G = RT \ln (c_2/c_1) + ZF\Delta V = 2.303RT \log_{10}(c_2/c_1) + ZF\Delta V$$

in which Z is the electrical charge of the transported species and ΔV is the potential in volts across the membrane (Section 13.1.2). Under typical conditions for the inner mitochondrial membrane, the pH outside is 1.4 units lower than inside [corresponding to $\log_{10}(c_2/c_1)$ of 1.4] and the membrane potential is 0.14 V, the outside being positive. Because $Z = +1$ for protons, the free-energy change is $(2.303 \times 1.98 \times 10^{-3}$ kcal mol^{-1} K$^{-1} \times 310$ K $\times$ 1.4) + ($+1 \times 23.06$ kcal mol^{-1} V$^{-1} \times 0.14$ V) = 5.2 kcal mol^{-1} (21.8 kJ mol^{-1}). Thus, each proton that is transported out of the matrix to the cytosolic side corresponds to 5.2 kcal mol^{-1} of free energy.

18.2.3 Electrons Can Be Transferred Between Groups That Are Not in Contact

As will be discussed shortly, the electron-carrying groups in the protein constituents of the electron-transport chain are *flavins, iron–sulfur clusters, quinones, hemes,* and *copper ions.* How are electrons transferred between electron-carrying groups that are frequently buried in the interior of a protein in fixed positions and are therefore not directly in contact? Electrons can move through space, even through a vacuum. However, the rate of electron transfer through space falls off rapidly as the electron donor and electron acceptor move apart from each other, decreasing by a factor of 10 for each increase in separation of 0.8 Å. The protein environment provides more-efficient pathways for electron conduction: typically, the rate of electron transfer decreases by a factor of 10 every 1.7 Å (Figure 18.7). For groups in contact, electron-transfer reactions can be quite fast with rates of approximately 10^{13} s^{-1}. Within proteins in the electron-transport chain, electron-carrying groups are typically separated by 15 Å beyond their van der Waals contact distance. For such separations, we expect electron-transfer rates of approximately 10^4 s^{-1} (i.e., electron transfer in less than 1 ms), assuming that all other factors are optimal. Without the mediation of the protein, an electron transfer over this distance would take approximately 1 day.

Another important factor in determining the rate of electron transfer is the driving force, the free-energy change associated with the reaction (Figure 18.8). Like the rates of most reactions, those of electron-transfer reactions tend to increase as the free-energy change for the reaction becomes more favorable. Interestingly, however, each electron-transfer reaction has an optimal driving force; making the reaction more favorable beyond this point decreases the rate of the electron-transfer process. This so-called *inverted region* is of tremendous importance for the light reactions of photosynthesis, to be discussed in Chapter 19. For the purposes of the electron-

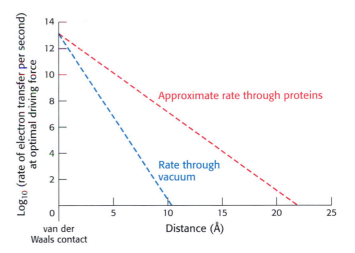

FIGURE 18.7 Distance dependence of electron-transfer rate. The rate of electron transfer decreases as the electron donor and the electron acceptor move apart. In a vacuum, the rate decreases by a factor of 10 for every increase of 0.8 Å. In proteins, the rate decreases more gradually, by a factor of 10 for every increase of 1.7 Å. This rate is only approximate because variations in the structure of the intervening protein medium can affect the rate.

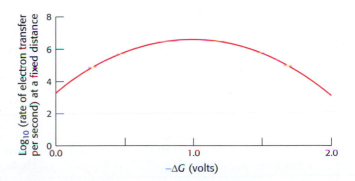

FIGURE 18.8 Free-energy dependence of electron-transfer rate. The rate of an electron-transfer reaction at first increases as the driving force for the reaction increases. The rate reaches a maximum and then decreases at very large driving forces.

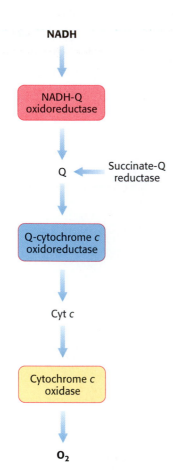

NADH

NADH-Q
oxidoreductase

Q ← Succinate-Q
reductase

Q-cytochrome c
oxidoreductase

Cyt c

Cytochrome c
oxidase

O_2

FIGURE 18.9 Sequence of electron carriers in the respiratory chain.

transport chain, the effects of distance and driving force combine to determine which pathway, among the set of those possible, will be used at each stage in the course of a reaction.

18.3 THE RESPIRATORY CHAIN CONSISTS OF FOUR COMPLEXES: THREE PROTON PUMPS AND A PHYSICAL LINK TO THE CITRIC ACID CYCLE

Electrons are transferred from NADH to O_2 through a chain of three large protein complexes called *NADH-Q oxidoreductase, Q-cytochrome* c *oxidoreductase,* and *cytochrome* c *oxidase* (Figure 18.9 and Table 18.2). *Electron flow within these transmembrane complexes leads to the transport of protons across the inner mitochondrial membrane.* Electrons are carried from NADH-Q oxidoreductase to Q-cytochrome c oxidoreductase, the second complex of the chain, by the reduced form of *coenzyme Q (Q),* also known as *ubiquinone* because it is a *ubiquitous quinone* in biological systems. Ubiquinone is a hydrophobic quinone that diffuses rapidly within the inner mitochondrial membrane. Ubiquinone also carries electrons from $FADH_2$, generated in succinate dehydrogenase in the citric acid cycle, to Q-cytochrome c oxidoreductase, generated through *succinate-Q reductase.* Cytochrome c, a small, soluble protein, shuttles electrons from Q-cytochrome c oxidoreductase to cytochrome c oxidase, the final component in the chain and the one that catalyzes the reduction of O_2. NADH-Q oxidoreductase, succinate-Q reductase, Q-cytochrome c oxidoreductase, and cytochrome c oxidase are also called *Complex I, II, III,* and *IV,* respectively. Succinate-Q reductase (Complex II), in contrast with the other complexes, does not pump protons.

Coenzyme Q is a quinone derivative with a long isoprenoid tail. The number of five-carbon isoprene units in coenzyme Q depends on the species. The most common form in mammals contains 10 isoprene units (coenzyme Q_{10}). For simplicity, the subscript will be omitted from this abbreviation because all varieties function in an identical manner. Quinones can exist in three oxidation states (Figure 18.10). In the fully oxidized state (Q), coenzyme Q has two keto groups. The addition of one electron and one

TABLE 18.2   Components of the mitochondrial electron-transport chain

| Enzyme complex | Mass (kd) | Subunits | Prosthetic group | Oxidant or reductant | | |
				Matrix side	Membrane core	Cytosolic side
NADH-Q oxidoreductase	880	≥ 34	FMN Fe-S	NADH	Q	
Succinate-Q reductase	140	4	FAD Fe-S	Succinate	Q	
Q-cytochrome c oxidoreductase	250	10	Heme b_H Heme b_L Heme c_1 Fe-S		Q	Cytochrome c
Cytochrome c oxidase	160	10	Heme a Heme a_3 Cu_A and Cu_B			Cytochrome c

Sources: J. W. DePierre and L. Ernster, *Annu. Rev. Biochem.* 46(1977):215; Y. Hatefi, *Annu Rev. Biochem.* 54(1985);1015; and J. E. Walker, *Q. Rev. Biophys.* 25(1992):253.

proton results in the semiquinone form (QH·). The semiquinone form is relatively easily deprotonated to form a semiquinone radical anion (Q·⁻). The addition of a second electron and proton generates ubiquinol (QH₂), the fully reduced form of coenzyme Q, which holds its protons more tightly. Thus, *for quinones, electron-transfer reactions are coupled to proton binding and release,* a property that is key to transmembrane proton transport.

FIGURE 18.10 Oxidation states of quinones. The reduction of ubiquinone (Q) to ubiquinol (QH₂) proceeds through a semiquinone anion intermediate (Q·⁻).

(QH·)

H⁺

Oxidized form of coenzyme Q
(Q, ubiquinone)

Semiquinone intermediate
(Q·⁻)

Reduced form of coenzyme Q
(QH₂, ubiquinol)

18.3.1 The High-Potential Electrons of NADH Enter the Respiratory Chain at NADH-Q Oxidoreductase

The electrons of NADH enter the chain at *NADH-Q oxidoreductase* (also called *NADH dehydrogenase*), an enormous enzyme (880 kd) consisting of at least 34 polypeptide chains. The construction of this proton pump, like that of the other two in the respiratory chain, is a cooperative effort of genes residing in both the mitochondria and the nucleus. The structure of this enzyme has been determined only at moderate resolution (Figure 18.11). NADH-Q oxidoreductase is L-shaped, with a horizontal arm lying in the membrane and a vertical arm that projects into the matrix. Although a detailed understanding of the mechanism is likely to require higher-resolution structural information, some aspects of the mechanism have been established.

The reaction catalyzed by this enzyme appears to be

$$NADH + Q + 5\,H^+_{matrix} \longrightarrow NAD^+ + QH_2 + 4\,H^+_{cytosol}$$

The initial step is the binding of NADH and the transfer of its two high-potential electrons to the *flavin mononucleotide (FMN)* prosthetic group of this complex to give the reduced form, FMNH₂. Like quinones, flavins bind protons when they are reduced. FMN can also accept one electron instead of two (or FMNH₂ can donate one electron) by forming a semiquinone radical intermediate (Figure 18.12). The electron acceptor of FMN, the isoalloxazine ring, is identical with that of FAD. Electrons are then transferred from FMNH₂ to a series of *iron–sulfur clusters,* the second type of prosthetic group in NADH-Q oxidoreductase.

Fe-S clusters in *iron–sulfur proteins* (also called *nonheme iron proteins*) play a critical role in a wide range of reduction reactions in biological systems. Several types of Fe-S clusters are known (Figure 18.13). In the simplest

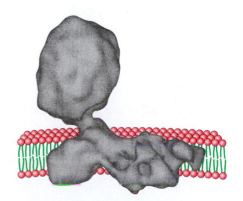

FIGURE 18.11 Structure of NADH-Q oxidoreductase (Complex I). The structure, determined by electron microscopy at 22-Å resolution, consists of a membrane-spanning part and a long arm that extends into the matrix. NADH is oxidized in the arm, and the electrons are transferred to reduce Q in the membrane. [After N. Grigorieff, *J. Mol. Biol.* 277(1998):1033–1048.]

Flavin mononucleotide (oxidized)
(FMN)

Semiquinone intermediate

Flavin mononucleotide (reduced)
(FMNH₂)

FIGURE 18.12 Oxidation states of flavins. The reduction of flavin mononucleotide (FMN) to $FMNH_2$ proceeds through a semiquinone intermediate.

kind, a single iron ion is tetrahedrally coordinated to the sulfhydryl groups of four cysteine residues of the protein. A second kind, denoted by 2Fe-2S, contains two iron ions and two inorganic sulfides. Such clusters are usually coordinated by four cysteine residues, although exceptions exist, as we shall see when we consider Q-cytochrome c oxidoreductase. A third type, designated 4Fe-4S, contains four iron ions, four inorganic sulfides, and four cysteine residues. We encountered a variation of this type of cluster in aconitase in Section 17.1.4. NADH-Q oxidoreductase contains both 2Fe-2S and 4Fe-4S clusters. Iron ions in these Fe-S complexes cycle between Fe^{2+} (reduced) or Fe^{3+} (oxidized) states. Unlike quinones and flavins, iron–sulfur clusters generally undergo oxidation-reduction reactions without releasing or binding protons.

Electrons in the iron–sulfur clusters of NADH-Q oxidoreductase are shuttled to coenzyme Q. *The flow of two electrons from NADH to coenzyme Q through NADH-Q oxidoreductase leads to the pumping of four hydrogen ions out of the matrix of the mitochondrion.* The details of this process remain the subject of active investigation. However, the coupled electron–proton transfer reactions of Q are crucial. NADH binds to a site on the vertical arm and transfers its electrons to FMN. These electrons flow within the vertical unit to three 4Fe-4S centers and then to a bound Q. *The reduction of Q to QH₂ results in the uptake of two protons from the matrix* (Figure 18.14). The pair of electrons on bound QH_2 are transferred to a 4Fe-4S center and the protons are released on the cytosolic side. Finally, these elections are transferred to a *mobile Q* in the hydrophobic core of the membrane, re-

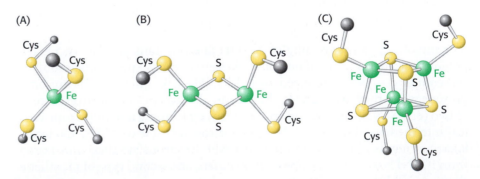

FIGURE 18.13 Iron–sulfur clusters. (A) A single iron ion bound by four cysteine residues. (B) 2Fe-2S cluster with iron ions bridged by sulfide ions. (C) 4Fe-4S cluster. Each of these clusters can undergo oxidation–reduction reactions.

sulting in the uptake of two additional protons from the matrix. The challenge is to delineate the binding events and conformational changes induced by these electron transfers and learn how the uptake and release of protons from the appropriate sides of the membrane is facilitated.

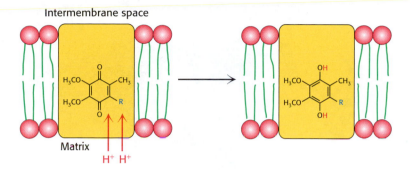

Intermembrane space

Matrix

H⁺ H⁺

FIGURE 18.14 Coupled electron–proton transfer reactions. The reduction of a quinone (Q) to QH_2 in an appropriate site can result in the uptake of two protons from the mitochondrial matrix.

18.3.2 Ubiquinol Is the Entry Point for Electrons from FADH₂ of Flavoproteins

The citric acid cycle enzyme succinate dehydrogenase, which generates $FADH_2$ with the oxidation of succinate to fumarate (Section 17.1.8), is part of the *succinate-Q reductase complex (Complex II)*, an integral membrane protein of the inner mitochondrial membrane. $FADH_2$ does not leave the complex. Rather, its electrons are transferred to Fe-S centers and then to Q for entry into the electron-transport chain. Two other enzymes that we will encounter later, *glycerol phosphate dehydrogenase* (Section 18.5.1) and *fatty acyl CoA dehydrogenase* (Section 22.2.4), likewise transfer their high-potential electrons from $FADH_2$ to Q to form ubiquinol (QH_2), the reduced state of ubiquinone. The succinate-Q reductase complex and other enzymes that transfer electrons from $FADH_2$ to Q, in contrast with NADH-Q oxidoreductase, do not transport protons. Consequently, less ATP is formed from the oxidation of $FADH_2$ than from NADH.

18.3.3 Electrons Flow from Ubiquinol to Cytochrome *c* Through Q-Cytochrome *c* Oxidoreductase

The second of the three proton pumps in the respiratory chain is *Q-cytochrome c oxidoreductase* (also known as *Complex III* and cytochrome reductase). A cytochrome is an electron-transferring protein that contains a heme prosthetic group. The iron ion of a cytochrome alternates between a reduced ferrous (+2) state and an oxidized ferric (+3) state during electron transport. The function of Q-cytochrome *c* oxidoreductase is to catalyze the transfer of electrons from QH_2 to oxidized *cytochrome c (cyt c)*, a water-soluble protein, and concomitantly pump protons out of the mitochondrial matrix.

$$QH_2 + 2\,\text{Cyt}\,c_{ox} + 2\,H^+_{matrix} \longrightarrow Q + 2\,\text{Cyt}\,c_{red} + 4\,H^+_{cytosol}$$

Q-cytochrome *c* oxidoreductase is a dimer with each monomer containing 11 subunits (Figure 18.15). Q-cytochrome *c* oxidoreductase itself contains a total of three hemes, contained within two cytochrome subunits: two *b*-type hemes, termed heme b_L (L for low affinity) and heme b_H (H for high affinity), within cytochrome *b*, and one *c*-type heme within cytochrome c_1. The prosthetic group of the heme in cytochromes *b*, c_1, and *c* is iron-protoporphyrin IX, the same heme as in myoglobin and hemoglobin (Section 10.2.1). The hemes in cytochromes *c* and c_1, in contrast with those in

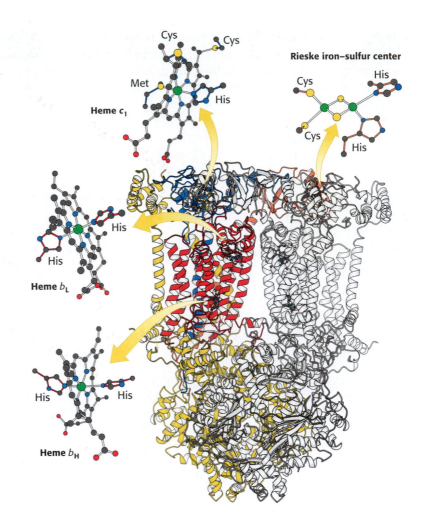

FIGURE 18.15 Structure of Q-cytochrome c oxidoreductase (cytochrome bc_1). This enzyme is a homodimer with 11 distinct polypeptide chains. The major prosthetic groups, three hemes and a 2Fe-2S cluster, mediate the electron-transfer reactions between quinones in the membrane and cytochrome c in the intermembrane space.

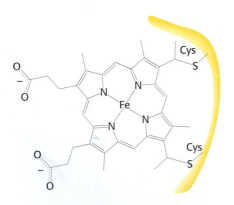

FIGURE 18.16 Attachment of c-type cytochromes. A heme group is covalently attached to a protein through thioether linkages formed by the addition of sulfhydryl groups of cysteine residues to vinyl groups on protoporphyrin.

cytochrome b, are covalently attached to the protein (Figure 18.16). The linkages are thioethers formed by the addition of the sulfhydryl groups of two cysteine residues to the vinyl groups of the heme. Because of these groups, this enzyme is also known as cytochrome bc_1. In addition to the hemes, the enzyme also contains an iron–sulfur protein with an 2Fe-2S center. This center, termed the *Rieske center,* is unusual in that one of the iron ions is coordinated by two histidine residues rather than two cysteine residues. This coordination stabilizes the center in its reduced form, raising its reduction potential. Finally, Q-cytochrome c oxidoreductase contains two distinct binding sites for ubiquinone termed Q_o and Q_i, with the Q_i site lying closer to the inside of the matrix.

18.3.4 Transmembrane Proton Transport: The Q Cycle

The mechanism for the coupling of electron transfer from Q to cytochrome c to transmembrane proton transport is known as the *Q cycle* (Figure 18.17). The Q cycle also facilitates the switch from the two-electron carrier ubiquinol to the one-electron carrier cytochrome c. The cycle begins as ubiquinol (QH_2) binds in the Q_o site. Ubiquinol transfers its electrons, one at a time. One electron flows first to the Rieske 2Fe-2S cluster, then to cytochrome c_1, and finally to a molecule of oxidized cytochrome c, converting it into its reduced form. The reduced cytochrome c molecule is free to diffuse away from the enzyme. The second electron is transferred first to cytochrome b_L, then to cytochrome b_H, and finally to an oxidized uniquinone bound in the Q_i site. This quinone (Q) molecule is reduced to a semiquinone anion $(Q \cdot ^-)$. Importantly, as the QH_2 in the Q_o site is oxidized to Q, its protons

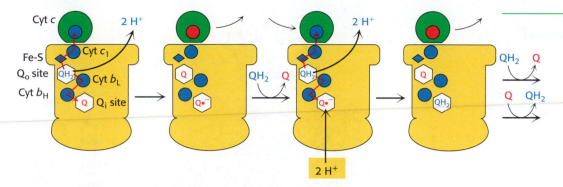

FIGURE 18.17 Q cycle. The two electrons of a bound QH_2 are transferred, one to cytochrome c and the other to a bound Q to form the semiquinone $Q\cdot^-$. The newly formed Q dissociates and is replaced by a second QH_2, which also gives up its electrons, one to a second molecule of cytochrome c and the other to reduce $Q\cdot^-$ to QH_2. This second electron transfer results in the uptake of two protons from the matrix. Prosthetic groups are shown in their oxidized forms in blue and in their reduced forms in red.

are released to the cytosolic side of the membrane. This Q molecule in the Q_o site is free to diffuse out into the ubiquinone pool.

At this point, $Q\cdot^-$ resides in the Q_i site. A second molecule of QH_2 binds to the Q_o site and reacts in the same way as the first. One of its electrons is transferred through the Rieske center and cytochrome c_1 to reduce a second molecule of cytochrome c. The other electron goes through cytochromes b_L and b_H to $Q\cdot^-$ bound in the Q_i site. On the addition of the second electron, this quinone radical anion takes up two protons from the matrix side to form QH_2. *The removal of these two protons from the matrix contributes to the formation of the proton gradient.* At the end of the Q cycle, two molecules of QH_2 are oxidized to form two molecules of Q, and one molecule of Q is reduced to QH_2, two molecules of cytochrome c are reduced, four protons are released on the cytoplasmic side, and two protons are removed from the mitochondrial matrix.

18.3.5 Cytochrome c Oxidase Catalyzes the Reduction of Molecular Oxygen to Water

The final stage of the electron-transport chain is the oxidation of the reduced cytochrome c generated by Complex III, which is coupled to the reduction of O_2 to two molecules of H_2O. This reaction is catalyzed by *cytochrome c oxidase (Complex IV)*. The four-electron reduction of oxygen directly to water without the release of intermediates poses a challenge. Nevertheless, this reaction is quite thermodynamically favorable. From the reduction potentials in Table 18.1, the standard free-energy change for this reaction is calculated to be $\Delta G^{\circ\prime} = -55.4$ kcal mol^{-1} (-231.8 kJ mol^{-1}). As much of this free energy as possible must be captured in the form of a proton gradient for subsequent use in ATP synthesis.

Bovine cytochrome c oxidase is reasonably well understood at the structural level (Figure 18.18). It consists of 13 subunits, of which 3 (called *subunits I, II,* and *III*) are encoded by the mitochondrial genome. Cytochrome c oxidase contains two heme A groups and three copper ions, arranged as two copper centers, designated A and B. One center, Cu_A/Cu_A, contains two copper ions linked by two bridging cysteine residues. This center initially accepts electrons from reduced cytochrome c. The remaining copper ion, Cu_B, is coordinated by three histidine residues, one of which is modified by covalent linkage to a tyrosine residue. Heme A differs from the heme

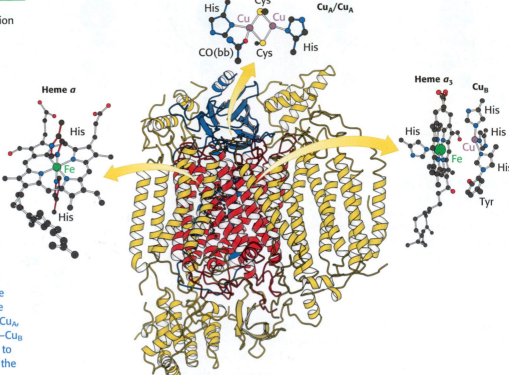

FIGURE 18.18 Structure of cytochrome c oxidase. This enzyme consists of 13 polypeptide chains. The major prosthetic groups include Cu_A/Cu_A, heme a, and heme a_3-Cu_B. Heme a_3–Cu_B is the site of the reduction of oxygen to water. CO(bb) is a carbonyl group of the peptide backbone.

in cytochrome c and c_1 in three ways: (1) a formyl group replaces a methyl group, (2) a C_{15} hydrocarbon chain replaces one of the vinyl groups, and (3) the heme is not covalently attached to the protein.

Heme A

The two heme A molecules, termed *heme* a and *heme* a₃, have distinct properties because they are located in different environments within cytochrome c oxidase. Heme a functions to carry electrons from Cu_A/Cu_A, whereas heme a_3 passes electrons to Cu_B, to which it is directly adjacent. Together, *heme a_3 and Cu_B form the active center at which O_2 is reduced to H_2O.*

The catalytic cycle begins with the enzyme in its fully oxidized form (Figure 18.19). One molecule of reduced cytochrome c transfers an electron, initially to Cu_A/Cu_A. From there, the electron moves to heme a, then to heme a_3, and finally to Cu_B, which is reduced from the Cu^{2+} (cupric) form to the Cu^+ (cuprous) form. A second molecule of cytochrome c introduces a second electron that flows down the same path, stopping at heme a_3, which is reduced to the Fe^{2+} form. Recall that the iron in hemoglobin is in the Fe^{2+} form when it binds oxygen (Section 10.2.1). Thus, at this stage, cy-

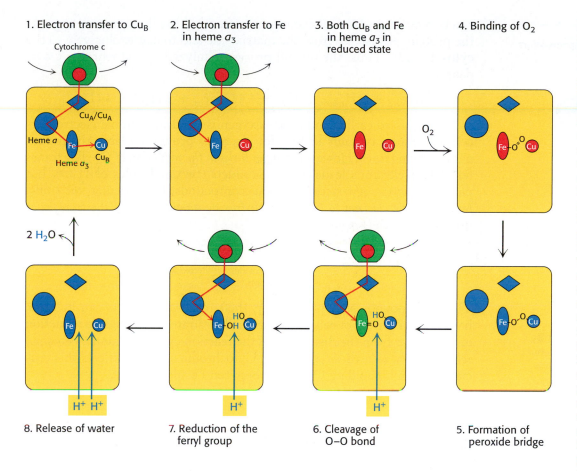

1. Electron transfer to Cu_B

Cytochrome c

Cu_A/Cu_A

Heme a

Fe — Cu

Cu_B

Heme a_3

2. Electron transfer to Fe in heme a_3

Fe   Cu

3. Both Cu_B and Fe in heme a_3 in reduced state

Fe   Cu

O_2

4. Binding of O_2

Fe-O-O Cu

5. Formation of peroxide bridge

Fe-O-O-Cu

6. Cleavage of O–O bond

HO
Fe=O Cu

H^+

7. Reduction of the ferryl group

HO
Fe-OH Cu

H^+

8. Release of water

Fe   Cu

H^+ H^+

2 H_2O

FIGURE 18.19
Cytochrome oxidase mechanism. The cycle begins with all prosthetic groups in their oxidized forms (shown in blue). Reduced cytochrome c introduces an electron that reduces Cu_B. A second reduced cytochrome c then reduces the iron in heme a_3. This Fe^{2+} center then binds oxygen. Two electrons are transferred to the bound oxygen to form peroxide, which bridges between the iron and Cu_B. The introduction of an additional electron by a third molecule of reduced cytochrome c cleaves the O–O bond and results in the uptake of a proton from the matrix. The introduction of a final electron and three more protons generates two molecules of H_2O, which are released from the enzyme to regenerate the initial state. The four protons found in the water molecules come from the matrix.

tochrome c oxidase is poised to bind oxygen and does so. The proximity of Cu_B in its reduced form to the heme a_3–oxygen complex allows the oxygen to be reduced to peroxide (O_2^{2-}), which forms a bridge between the Fe^{3+} in heme a_3 and Cu_B^{2+} (Figure 18.20). The addition of a third electron from cytochrome c as well as a proton results in the cleavage of the O–O bond, yielding a ferryl group, $Fe^{4+}=O$, at heme a_3 and Cu_B^{2+}–OH. The addition of the final electron from cytochrome c and a second proton reduces the ferryl group to Fe^{3+}–OH. Reaction with two additional protons allows the release of two molecules of water and resets the enzyme to its initial, fully oxidized form.

This reaction can be summarized as

$$4\ Cyt\ c_{red} + 4\ H^+_{matrix} + O_2 \longrightarrow 4\ Cyt\ c_{ox} + 2\ H_2O$$

The four protons in this reaction come exclusively from the matrix. Thus, the consumption of these four protons contributes directly to the proton gradient. Recall that each proton contributes 5.2 kcal mol^{-1} (21.8 kJ mol^{-1}) to the free energy associated with the proton gradient; so these four protons contribute 20.8 kcal mol^{-1} (87.2 kJ mol^{-1}), an amount substantially less than the free energy available from the reduction of oxygen to water. Remarkably, *cytochrome c oxidase evolved to pump four additional protons from the matrix to the cytoplasmic side of the membrane in the course of each reaction cycle for a total of eight protons removed from the matrix* (Figure 18.21). The details of how these protons are transported through the protein is still under study. However, two effects contribute to the mechanism. First, charge neutrality tends to be maintained in the interior of proteins. Thus, the addition of an electron to a site inside a protein tends to favor the binding of a proton to a nearby site. Second, conformational changes take place, particularly around the heme a_3–Cu_B center, in the course of the

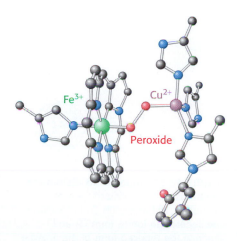

Fe^{3+} Cu^{2+}

Peroxide

FIGURE 18.20 Peroxide bridge. The oxygen bound to heme a_3 is reduced to peroxide by the presence of Cu_B.

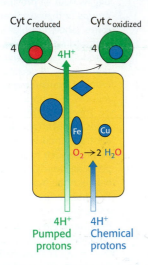

FIGURE 18.21 Proton transport by cytochrome c oxidase. Four "chemical" protons are taken up from the matrix side to reduce one molecule of O_2 to two molecules of H_2O. Four additional "pumped" protons are transported out of the matrix and released on the cytosolic side in the course of the reaction. The pumped protons double the efficiency of free-energy storage in the form of a proton gradient for this final step in the electron-transport chain.

Dismutation—

A reaction in which a single reactant is converted into two different products.

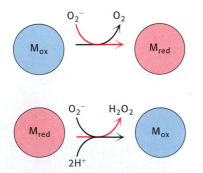

FIGURE 18.22 Superoxide dismutase mechanism. The oxidized form of superoxide dismutase (M_{ox}) reacts with one superoxide ion to form O_2 and generate the reduced form of the enzyme (M_{red}). The reduced form then reacts with a second superoxide and two protons to form hydrogen peroxide and regenerate the oxidized form of the enzyme.

reaction cycle, and these changes must be used to allow protons to enter the protein exclusively from the matrix side and to exit exclusively to the cytosolic side. Thus, the overall process catalyzed by cytochrome c oxidase is

$$4 \text{ Cyt } c_{\text{red}} + 8 \text{ H}^+_{\text{matrix}} + O_2 \longrightarrow 4 \text{ Cyt } c_{\text{ox}} + 2 \text{ H}_2\text{O} + 4 \text{ H}^+_{\text{cytosol}}$$

As discussed earlier, molecular oxygen is an ideal terminal electron acceptor, because its high affinity for electrons provides a large thermodynamic driving force. However, *danger lurks in the reduction of O_2*. The transfer of four electrons leads to safe products (two molecules of H_2O), but partial reduction generates hazardous compounds. In particular, *the transfer of a single electron to O_2 forms superoxide anion, whereas the transfer of two electrons yields peroxide.*

$$O_2 \xrightarrow{e^-} O_2^{\bullet -} \xrightarrow{e^-} O_2^{2-}$$
$$\text{Superoxide} \qquad \text{Peroxide}$$
$$\text{anion}$$

These compounds and, particularly, their reaction products can be quite harmful to a variety of cell components. The strategy for the safe reduction of O_2 is clear from the discussion of the reaction cycle: *the catalyst does not release partly reduced intermediates.* Cytochrome c oxidase meets this crucial criterion by holding O_2 tightly between Fe and Cu ions.

18.3.6 Toxic Derivatives of Molecular Oxygen Such as Superoxide Radical Are Scavenged by Protective Enzymes

Although cytochrome *c* oxidase and other proteins that reduce O_2 are remarkably successful in not releasing intermediates, small amounts of superoxide anion and hydrogen peroxide are unavoidably formed. Superoxide, hydrogen peroxide, and species that can be generated from them such as OH· are collectively referred to as *reactive oxygen species* or *ROS*.

What are the cellular defense strategies against oxidative damage by ROS? Chief among them is the enzyme *superoxide dismutase*. This enzyme scavenges superoxide radicals by catalyzing the conversion of two of these radicals into hydrogen peroxide and molecular oxygen.

$$2 \text{ O}_2^{\bullet -} + 2 \text{ H}^+ \underset{\text{dismutase}}{\overset{\text{Superoxide}}{\rightleftharpoons}} O_2 + \text{H}_2\text{O}_2$$

Eukaryotes contain two forms of this enzyme, a manganese-containing version located in mitochondria and a copper-zinc-dependent cytosolic form. These enzymes perform the dismutation reaction by a similar mechanism (Figure 18.22). The oxidized form of the enzyme is reduced by superoxide to form oxygen. The reduced form of the enzyme, formed in this reaction, then reacts with a second superoxide ion to form peroxide, which takes up two protons along the reaction path to yield hydrogen peroxide.

The hydrogen peroxide formed by superoxide dismutase and by other processes is scavenged by *catalase*, a ubiquitous heme protein that catalyzes the dismutation of hydrogen peroxide into water and molecular oxygen.

$$2 \text{ H}_2\text{O}_2 \overset{\text{Catalase}}{\rightleftharpoons} O_2 + 2 \text{ H}_2\text{O}$$

Superoxide dismutase and catalase are remarkably efficient, performing their reactions at or near the diffusion-limited rate (Section 8.4.2). Other cellular defenses against oxidative damage include the antioxidant vitamins, vitamins E and C. Because it is lipophilic, vitamin E is especially useful in protecting membranes from lipid peroxidation.

The importance of the cell's defense against ROS is demonstrated by the presence of superoxide dismutase in all aerobic organisms. *Escherichia coli* mutants lacking this enzyme are highly vulnerable to oxidative damage. Moreover, oxidative damage is believed to cause, at least in part, a growing number of diseases (Table 18.3).

18.3.7 The Conformation of Cytochrome *c* Has Remained Essentially Constant for More Than a Billion Years

Cytochrome c is present in all organisms having mitochondrial respiratory chains: plants, animals, and eukaryotic microorganisms. This electron carrier evolved more than 1.5 billion years ago, before the divergence of plants and animals. Its function has been conserved throughout this period, as evidenced by the fact that *the cytochrome* c *of any eukaryotic species reacts in vitro with the cytochrome* c *oxidase of any other species tested thus far.* Finally, some prokaryotic cytochromes, such as cytochrome c_2 from a photosynthetic bacterium and cytochrome c_{550} from a denitrifying bacterium, closely resemble cytochrome *c* from tuna heart mitochondria (Figure 18.23). This evidence attests that the structural and functional characteristics of cytochrome *c* present an efficient evolutionary solution to electron transfer.

TABLE 18.3 Pathological and other conditions that may entail free-radical injury

Atherogenesis
Emphysema; bronchitis
Parkinson disease
Duchenne muscular dystrophy
Cervical cancer
Alcoholic liver disease
Diabetes
Acute renal failure
Down syndrome
Retrolental fibroplasia
Cerebrovascular disorders
Ischemia; reperfusion injury

Source: After D.B. Marks, A.D. Marks, and C.M. Smith, *Basic Medical Biochemistry: A Clinical Approach* (Williams & Wilkins, 1996, p. 331).

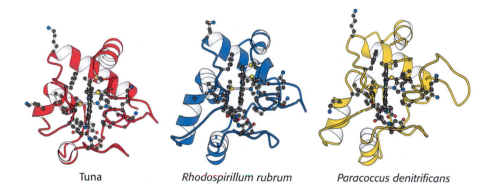

Tuna   *Rhodospirillum rubrum*   *Paracoccus denitrificans*

FIGURE 18.23 Conservation of the three-dimensional structure of cytochrome c. The side chains are shown for the 21 conserved amino acids and the heme.

The resemblance among cytochrome *c* molecules extends to the level of amino acid sequence. Because of the molecule's relatively small size and ubiquity, the amino acid sequences of cytochrome *c* from more than 80 widely ranging eukaryotic species were determined by direct protein sequencing by Emil Smith, Emanual Margoliash, and others. Comparison of these sequences revealed that *26 of 104 residues have been invariant for more than one and a half billion years of evolution.* A phylogenetic tree, constructed from the amino acid sequences of cytochrome *c*, reveals the evolutionary relationships between many animal species (Figure 18.24).

18.4 A PROTON GRADIENT POWERS THE SYNTHESIS OF ATP

CONCEPTUAL INSIGHTS, Energy Transformations in Oxidative Phosphorylation. *View this media module for an animated, interactive summary of how electron transfer potential is converted into proton-motive force and, finally, phosphoryl transfer potential in oxidative phosphorylation.*

Thus far, we have considered the flow of electrons from NADH to O_2, an exergonic process.

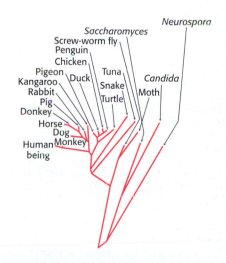

FIGURE 18.24 Evolutionary tree constructed from sequences of cytochrome c. Branch lengths are proportional to the number of amino acid changes that are believed to have occurred. This drawing is an adaptation of the work of Walter M. Fitch and Emanuel Margoliash.

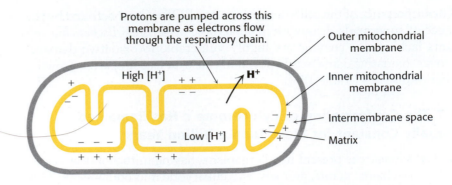

Protons are pumped across this
membrane as electrons flow
through the respiratory chain.

Outer mitochondrial
membrane

High [H$^+$]

H$^+$

Inner mitochondrial
membrane

Intermembrane space

Low [H$^+$]

Matrix

FIGURE 18.25 Chemiosmotic hypothesis. Electron transfer through the respiratory chain leads to the pumping of protons from the matrix to the cytosolic side of the inner mitochondrial membrane. The pH gradient and membrane potential constitute a proton-motive force that is used to drive ATP synthesis.

$$NADH + \tfrac{1}{2}O_2 + H^+ \rightleftharpoons H_2O + NAD^+$$
$$\Delta G°' = -52.6 \text{ kcal mol}^{-1} (-220.1 \text{ kJ mol}^{-1})$$

Next, we consider how this process is coupled to the synthesis of ATP, an endergonic process.

$$ADP + P_i + H^+ \rightleftharpoons ATP + H_2O$$
$$\Delta G°' = +7.3 \text{ kcal mol}^{-1} (+30.5 \text{ kJ mol}^{-1})$$

A molecular assembly in the inner mitochondrial membrane carries out the synthesis of ATP. This enzyme complex was originally called the *mitochondrial ATPase* or *F_1F_0ATPase* because it was discovered through its catalysis of the reverse reaction, the hydrolysis of ATP. *ATP synthase*, its preferred name, emphasizes its actual role in the mitochondrion. It is also called *Complex V*.

How is the oxidation of NADH coupled to the phosphorylation of ADP? It was first suggested that electron transfer leads to the formation of a covalent high-energy intermediate that serves as a high phosphoryl transfer potential compound or to the formation of an activated protein conformation, which then drives ATP synthesis. The search for such intermediates for several decades proved fruitless.

In 1961, Peter Mitchell proposed that electron transport and ATP synthesis are coupled by *a proton gradient across the inner mitochondrial membrane* rather than by a covalent high-energy intermediate or an activated protein conformation. In his model, the transfer of electrons through the respiratory chain leads to the pumping of protons from the matrix to the cytosolic side of the inner mitochondrial membrane. The H$^+$ concentration becomes lower in the matrix, and an electrical field with the matrix side negative is generated (Figure 18.25). Mitchell's idea, called *the chemiosmotic hypothesis*, was that this proton-motive force drives the synthesis of ATP by ATP synthase. Mitchell's highly innovative hypothesis that oxidation and phosphorylation are coupled by a proton gradient is now supported by a wealth of evidence. Indeed, electron transport does generate a proton gradient across the inner mitochondrial membrane. The pH outside is 1.4 units lower than inside, and the membrane potential is 0.14 V, the outside being positive. As we calculated in Section 18.2.2, this membrane potential corresponds to a free energy of 5.2 kcal (21.8 kJ) per mole of protons.

An artificial system was created to elegantly demonstrate the basic principle of the chemiosmotic hypothesis. Synthetic vesicles containing bacteriorhodopsin, a purple-membrane protein from halobacteria that pumps protons when illuminated, and mitochondrial ATP synthase purified from beef heart were created (Figure 18.26). When the vesicles were exposed to light, ATP was formed. This key experiment clearly showed that *the respiratory chain and ATP synthase are biochemically separate systems, linked only by a proton-motive force.*

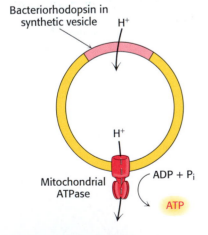

Bacteriorhodopsin in
synthetic vesicle

H$^+$

H$^+$

Mitochondrial
ATPase

ADP + P$_i$

ATP

FIGURE 18.26 Testing the chemiosmotic hypothesis. ATP is synthesized when reconstituted membrane vesicles containing bacteriorhodopsin (a light-driven proton pump) and ATP synthase are illuminated. The orientation of ATP synthase in this reconstituted membrane is the reverse of that in the mitochondrion.

18.4.1 ATP Synthase Is Composed of a Proton-Conducting Unit and a Catalytic Unit

Biochemical, electron microscopic, and crystallographic studies of ATP synthase have revealed many details of its structure (Figure 18.27). It is a large, complex membrane-embedded enzyme that looks like a ball on a stick. The 85-Å-diameter ball, called the F_1 subunit, protrudes into the mitochondrial matrix and contains the catalytic activity of the synthase. In fact, isolated F_1 subunits display ATPase activity. The F_1 subunit consists of five types of polypeptide chains (α_3, β_3, γ, δ, and ϵ) with the indicated stoichiometry. The α and β subunits, which make up the bulk of the F_1, are arranged alternately in a hexameric ring; they are homologous to one another and are members of the P-loop NTPase family (Section 9.4.1). Both bind nucleotides but only the β subunits participate directly in catalysis. The central stalk consists of two proteins: γ and ϵ. The γ subunit includes a long α-helical coiled coil that extends into the center of the $\alpha_3\beta_3$ hexamer. *The γ subunit breaks the symmetry of the $\alpha_3\beta_3$ hexamer: each of the β subunits is distinct by virtue of its interaction with a different face of γ.* Distinguishing the three β subunits is crucial for the mechanism of ATP synthesis.

The F_0 subunit is a hydrophobic segment that spans the inner mitochondrial membrane. *F_0 contains the proton channel of the complex.* This channel consists of a ring comprising from 10 to 14 **c** subunits that are embedded in the membrane. A single **a** subunit binds to the outside of this ring. The proton channel depends on both the **a** subunit and the **c** ring. The F_0 and F_1 subunits are connected in two ways, by the central $\gamma\epsilon$ stalk and by an exterior column. The exterior column consists of one **a** subunit, two **b** subunits, and the δ subunit. As will be discussed shortly, we can think of the enzyme as consisting of two functional components: (1) a moving unit, or *rotor,* consisting of the **c** ring and the $\gamma\epsilon$ stalk, and (2) a stationary unit, or *stator,* composed of the remainder of the molecule.

18.4.2 Proton Flow Through ATP Synthase Leads to the Release of Tightly Bound ATP: The Binding-Change Mechanism

CONCEPTUAL INSIGHTS, ATP Synthase as Motor Protein, *looks further into the chemistry and mechanics of ATP synthase rotation.*

ATP synthase catalyzes the formation of ATP from ADP and orthophosphate.

$$ADP^{3-} + HPO_4^{2-} + H^+ \rightleftharpoons ATP^{4-} + H_2O$$

The actual substrates are Mg^{2+} complexes of ADP and ATP, as in all known phosphoryl transfer reactions with these nucleotides. A terminal oxygen atom of ADP attacks the phosphorus atom of P_i to form a pentacovalent intermediate, which then dissociates into ATP and H_2O (Figure 18.28). The attacking oxygen atom of ADP and the departing oxygen atom of P_i occupy the apices of a trigonal bipyramid.

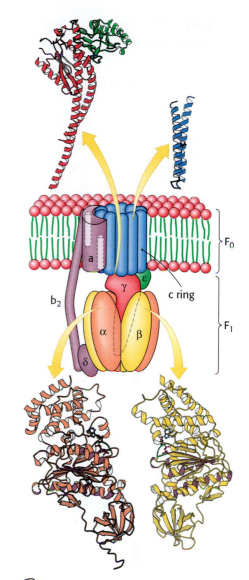

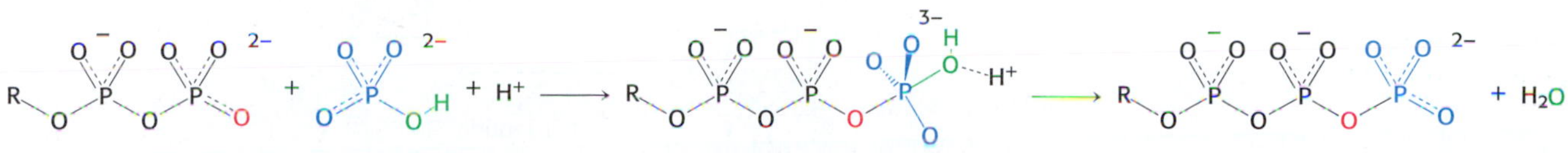

FIGURE 18.27 Structure of ATP synthase. A schematic structure is shown along with detailed structures of the components for which structures have been determined to high resolution. The P-loop NTPase domains of the α and β subunits are indicated by purple shading.

FIGURE 18.28 ATP synthesis mechanism. One of the oxygen atoms of ADP attacks the phosphorus atom of P_i to form a pentacovalent intermediate, which then forms ATP and releases a molecule of H_2O.

How does the flow of protons drive the synthesis of ATP? The results of isotopic-exchange experiments unexpectedly revealed that *enzyme-bound ATP forms readily in the absence of a proton-motive force.* When ADP and P_i were added to ATP synthase in $H_2^{18}O$, ^{18}O became incorporated into P_i through the synthesis of ATP and its subsequent hydrolysis (Figure 18.29).

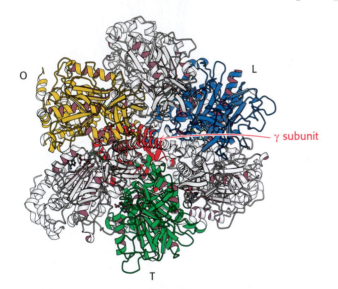

FIGURE 18.29 ATP forms without a proton-motive force but is not released. The results of isotope-exchange experiments indicate that enzyme-bound ATP is formed from ADP and P_i in the absence of a proton-motive force.

The rate of incorporation of ^{18}O into P_i showed that about equal amounts of bound ATP and ADP are in equilibrium at the catalytic site, even in the absence of a proton gradient. However, ATP does not leave the catalytic site unless protons flow through the enzyme. Thus, *the role of the proton gradient is not to form ATP but to release it from the synthase.*

On the basis of these and other observations, Paul Boyer proposed a *binding-change mechanism* for proton-driven ATP synthesis. This proposal states that changes in the properties of the three β subunits allow sequential ADP and P_i binding, ATP synthesis, and ATP release. The concepts of this initial proposal refined by more recent crystallographic and other data yield a satisfying mechanism for ATP synthesis. As already noted, interactions with the γ subunit make the three β subunits inequivalent (Figure 18.30). One β subunit can be in the T, or tight, conformation. This conformation binds ATP with great avidity. Indeed, its affinity for ATP is so high that it will convert bound ADP and P_i into ATP with an equilibrium constant near 1, as indicated by the aforediscussed isotopic-exchange experiments. However, the conformation of this subunit is sufficiently constrained that it cannot release ATP. A second subunit will then be in the L, or loose, conformation. This conformation binds ADP and P_i. It, too, is sufficiently constrained that it cannot release bound nucleotides. The final subunit will be in the O, or open, form. This form can exist with a bound nucleotide in a structure that is similar to those of the T and L forms, but it can also convert to form a more open conformation and release a bound nucleotide (Figure 18.31). This structure, with one of the three β subunits in an open, nucleotide-free state, as well as one with one of the β subunits in a nucleotide-bound O conformation, have been observed crystallographically.

FIGURE 18.30 ATP synthase nucleotide-binding sites are not equivalent. The γ subunit passes through the center of the $\alpha_3\beta_3$ hexamer and makes the nucleotide-binding sites in the β subunits distinct from one another.

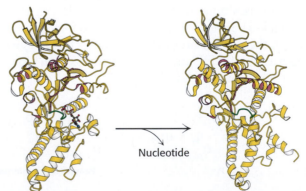

FIGURE 18.31 ATP release from the β subunit in the open form. Unlike the tight and loose forms, the open form of the β subunit can change conformation sufficiently to release bound nucleotides.

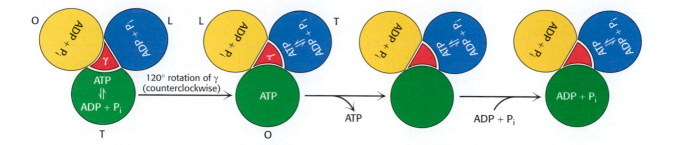

FIGURE 18.32 Binding-change mechanism for ATP synthase. The rotation of the γ subunit interconverts the three β subunits. The subunit in the T (tight) form, which contains newly synthesized ATP that cannot be released, is converted into the O (open) form. In this form, it can release ATP and then bind ADP and P_i to begin a new cycle.

The interconversion of these three forms can be driven by rotation of the γ subunit (Figure 18.32). Suppose the γ subunit is rotated 120 degrees in a counterclockwise direction (as viewed from the top). This rotation will change the subunit in the T conformation into the O conformation, allowing the subunit to release the ATP that has been formed within it. The subunit in the L conformation will be converted into the T conformation, allowing the transition of bound ADP + P_i into ATP. Finally, the subunit in the O conformation will be converted into the L conformation, trapping the bound ADP and P_i so that they cannot escape. The binding of ADP and P_i to the subunit now in the O conformation completes the cycle. This mechanism suggests that ATP can be synthesized by driving the rotation of the γ subunit in the appropriate direction. Likewise, this mechanism suggests that the hydrolysis of ATP by the enzyme should drive the rotation of the γ subunit in the opposite direction.

18.4.3 The World's Smallest Molecular Motor: Rotational Catalysis

Is it possible to observe the proposed rotation directly? Elegant experiments were performed with the use of a simple experimental system consisting of cloned $\alpha_3\beta_3\gamma$ subunits only (Figure 18.33). The β subunits were engineered to contain amino-terminal polyhistidine tags, which have a high affinity for nickel ions. This property of the tags allowed the $\alpha_3\beta_3$ assembly to be immobilized on a glass surface that had been coated with nickel ions. The γ subunit was linked to a fluorescently labeled actin filament to provide a long segment that could be observed under a fluorescence microscope. Remarkably, the addition of ATP caused the actin filament to rotate unidirectionally in a counterclockwise direction. *The γ subunit was rotating, being driven by the hydrolysis of ATP.* Thus, the catalytic activity of an individual molecule could be observed. The counterclockwise rotation is consistent with the predicted mechanism for hydrolysis because the molecule was viewed from below relative to the view shown in Figure 18.32.

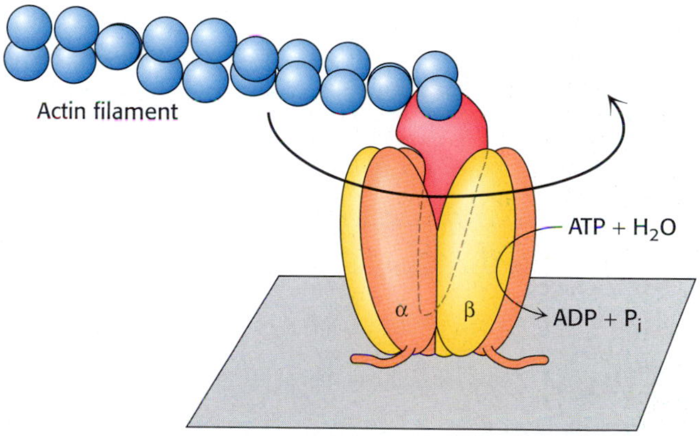

More detailed analysis in the presence of lower concentrations of ATP revealed that the γ subunit rotates in 120-degree increments, with each step corresponding to the hydrolysis of a single ATP molecule. In addition, from the results obtained by varying the length of the actin filament and measuring the rate of rotation, the enzyme appears to operate near 100% efficiency; that is, essentially all of the energy released by ATP hydrolysis is converted into rotational motion.

FIGURE 18.33 Direct observation of ATP-driven rotation in ATP synthase. The $\alpha_3\beta_3$ hexamer of ATP synthase is fixed to a surface, with the γ subunit projecting upward and linked to a fluorescently labeled actin filament. The addition and subsequent hydrolysis of ATP result in the counterclockwise rotation of the γ subunit, which can be directly seen under a fluorescence microscope.

18.4.4 Proton Flow Around the c Ring Powers ATP Synthesis

The direct observation of rotary motion of the γ subunit is strong evidence for the rotational mechanism for ATP synthesis. The last remaining

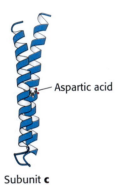

Aspartic acid

Subunit **c**

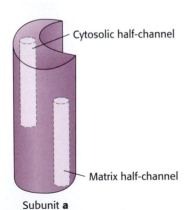

Cytosolic half-channel

Matrix half-channel

Subunit **a**

FIGURE 18.34 Components of the proton-conducting unit of ATP synthase. The **c** subunit consists of two α helices that span the membrane. An aspartic acid residue in the second helix lies on the center of the membrane. The structure of the **a** subunit has not yet been directly observed, but it appears to include two half-channels that allow protons to enter and pass partway but not completely through the membrane.

question is: How does proton flow through F_0 drive the rotation of the γ subunit? Howard Berg and George Oster proposed an elegant mechanism that provides a clear answer to this question. The mechanism depends on the structures of the **a** and **c** subunits of F_0 (Figure 18.34). The structure of the **c** subunit was determined both by NMR methods and by x-ray crystallography. Each polypeptide chain forms a pair of α helices that span the membrane. An aspartic acid residue (Asp 61) is found in the middle of the second helix. When Asp 61 is in contact with the hydrophobic part of the membrane, the residue must be in the neutral aspartic acid form, rather than in the charged, aspartate form. From 9 to 12 **c** subunits assemble into a symmetric membrane-spanning ring. Although the structure of the **a** subunit has not yet been experimentally determined, a variety of evidence is consistent with a structure that includes two proton half-channels that do not span the membrane (see Figure 18.34). Thus, protons can pass into either of these channels, but they cannot move completely across the membrane. The **a** subunit directly abuts the ring comprising the **c** subunits, with each half-channel directly interacting with one **c** subunit.

With this structure in mind, we can see how a proton gradient can drive rotation of the **c** ring. Suppose that the Asp 61 residues of the two **c** subunits that are in contact with a half-channel have given up their protons so that they are in the charged aspartate form (Figure 18.35), which is possible because they are in relatively hydrophilic environments inside the half-channel. The **c** ring cannot rotate in either direction, because such a rotation would move a charged aspartate residue into the hydrophobic part of the membrane. A proton can move through either half-channel to protonate one of the aspartate residues. However, it is much more likely to pass through the channel that is connected to the cytosolic side of the membrane because the proton concentration is more than 25 times as high on this side as on the matrix side, owing to the action of the electron-transport-chain proteins. The entry of protons into the cytosolic half-channel is further facilitated by the membrane potential of $+0.14$ V (positive on the cytoplasmic side), which increases the concentration of protons near the mouth of the cytosolic half-channel. *If the aspartate residue is protonated to its neutral form, the **c** ring can now rotate, but only in a clockwise direction.* Such a rotation moves the newly protonated aspartic acid residue into contact with the membrane, moves the charged aspartate residue from contact with the matrix half-channel to the cytosolic half-channel, and moves a different protonated aspartic acid residue from contact with the membrane to the matrix half-channel. The proton can then dissociate from aspartic acid and move through the half-channel into the proton-poor matrix to restore the initial state. This dissociation is favored by the positive charge on a conserved arginine residue (Arg 210) in the **a** subunit. Thus, *the difference in proton concentration and potential on the two sides of the membrane leads to different probabilities of protonation through the two half-channels, which yields directional rotational motion.* Each proton moves through the membrane by riding around on the rotating **c** ring to exit through the matrix half-channel (Figure 18.36).

The **c** ring is tightly linked to the γ and ε subunits. Thus, as the **c** ring turns, these subunits are turned inside the $\alpha_3\beta_3$ hexamer unit of F_1. The exterior column formed by the two **b** chains and the δ subunit prevent the $\alpha_3\beta_3$ hexamer from rotating. Thus, the proton-gradient-driven rotation of the **c** ring drives the rotation of the γ subunit, which in turn promotes the synthesis of ATP through the binding-change mechanism. Recall that the number of **c** subunits in the **c** ring appears to range between 10 and 14. This number is significant because it determines the number of protons that must be transported to generate a molecule of ATP. Each 360-degree rotation of the γ subunit leads to the synthesis and release of three molecules of ATP. Thus, if there are 10 **c** subunits in the ring (as was observed in a crystal

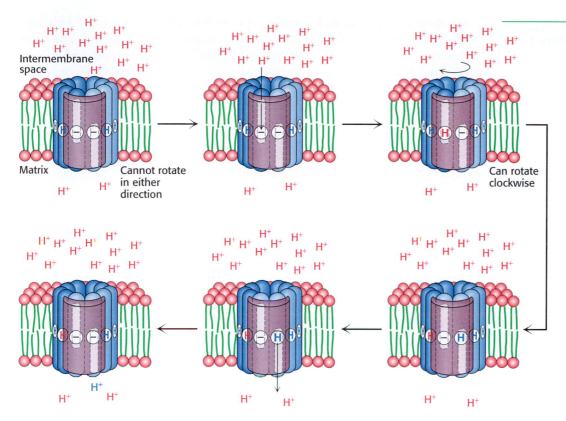

FIGURE 18.35 Proton motion across the membrane drives rotation of the c ring. A proton enters from the intermembrane space into the cytosolic half-channel to neutralize the charge on an aspartate residue in a **c** subunit. With this charge neutralized, the **c** ring can rotate clockwise by one **c** subunit, moving an aspartic acid residue out of the membrane into the matrix half-channel. This proton can move into the matrix, resetting the system to its initial state.

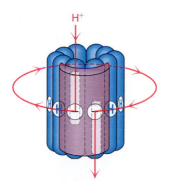

FIGURE 18.36 Proton path through the membrane. Each proton enters the cytosolic half-channel, follows a complete rotation of the **c** ring, and exits through the other half-channel into the matrix.

structure of yeast mitochondrial ATP synthase), each ATP generated requires the transport of $10/3 = 3.33$ protons. For simplicity, we will assume that 3 protons must flow into the matrix for each ATP formed, but we must keep in mind that the true value may differ.

18.4.5 ATP Synthase and G Proteins Have Several Common Features

The α and β subunits of ATP synthase are members of the P-loop NTPase family of proteins. In Chapter 15, we learned that the signaling properties of other members of this family, the G proteins, depend on their ability to bind nucleoside triphosphates and nucleoside diphosphates with great kinetic tenacity. They do not exchange nucleotides unless they are stimulated to do so by interaction with other proteins. The binding-change mechanism of ATP synthase is a variation on this theme. The three different faces of the γ subunit of ATP synthase interact with the P-loop regions of the β subunits to favor the structures of either the NDP- or NTP-binding forms or to facilitate nucleotide release. The conformational changes take place in an orderly way, driven by the rotation of the γ subunit.

18.5 MANY SHUTTLES ALLOW MOVEMENT ACROSS THE MITOCHONDRIAL MEMBRANES

The inner mitochondrial membrane must be impermeable to most molecules, yet much exchange has to take place between the cytosol and the mitochondria. This exchange is mediated by an array of membrane-spanning transporter proteins (Section 13.4).

18.5.1 Electrons from Cytosolic NADH Enter Mitochondria by Shuttles

Recall that the glycolytic pathway generates NADH in the cytosol in the oxidation of glyceraldehyde 3-phosphate, and NAD^+ must be regenerated for glycolysis to continue. How is cytosolic NADH reoxidized under aerobic conditions? NADH cannot simply pass into mitochondria for oxidation by the respiratory chain, because the inner mitochondrial membrane is impermeable to NADH and NAD^+. The solution is that *electrons from NADH*, rather than NADH itself, are carried across the mitochondrial membrane. One of several means of introducing electrons from NADH into the electron transport chain is the *glycerol 3-phosphate shuttle* (Figure 18.37). The first step in this shuttle is the transfer of a pair of electrons from NADH to dihydroxyacetone phosphate, a glycolytic intermediate, to form glycerol 3-phosphate. This reaction is catalyzed by a glycerol 3-phosphate dehydrogenase in the cytosol. Glycerol 3-phosphate is reoxidized to dihydroxyacetone phosphate on the outer surface of the inner mitochondrial membrane by a membrane-bound isozyme of glycerol 3-phosphate dehydrogenase. An electron pair from glycerol 3-phosphate is transferred to a FAD prosthetic group in this enzyme to form $FADH_2$. This reaction also regenerates dihydroxyacetone phosphate.

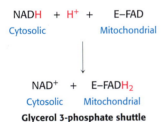

NAD**H** + **H⁺** + E–FAD
Cytosolic   Mitochondrial

NAD⁺ + E–FAD**H₂**
Cytosolic   Mitochondrial
Glycerol 3-phosphate shuttle

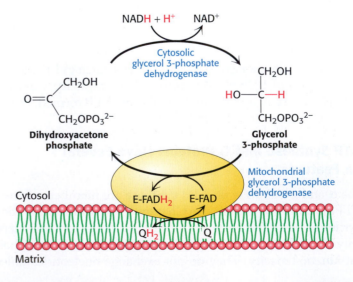

FIGURE 18.37 Glycerol 3-phosphate shuttle. Electrons from NADH can enter the mitochondrial electron transport chain by being used to reduce dihydroxyacetone phosphate to glycerol 3-phosphate. Glycerol 3-phosphate is reoxidized by electron transfer to an FAD prosthetic group in a membrane-bound glycerol 3-phosphate dehydrogenase. Subsequent electron transfer to Q to form QH₂ allows these electrons to enter the electron-transport chain.

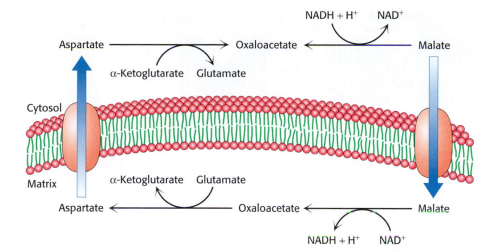

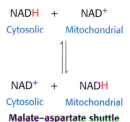

FIGURE 18.38 Malate–aspartate shuttle.

The reduced flavin transfers its electrons to the electron carrier Q, which then enters the respiratory chain as QH_2. *When cytosolic NADH transported by the glycerol 3-phosphate shuttle is oxidized by the respiratory chain, 1.5 rather than 2.5 ATP are formed.* The yield is lower because FAD rather than NAD^+ is the electron acceptor in mitochondrial glycerol 3-phosphate dehydrogenase. The use of FAD enables electrons from cytosolic NADH to be transported into mitochondria against an NADH concentration gradient. The price of this transport is one molecule of ATP per two electrons. This glycerol 3-phosphate shuttle is especially prominent in muscle and enables it to sustain a very high rate of oxidative phosphorylation. Indeed, some insects lack lactate dehydrogenase and are completely dependent on the glycerol 3-phosphate shuttle for the regeneration of cytosolic NAD^+.

In the heart and liver, electrons from cytosolic NADH are brought into mitochondria by the *malate–aspartate shuttle*, which is mediated by two membrane carriers and four enzymes (Figure 18.38). Electrons are transferred from NADH in the cytosol to oxaloacetate, forming malate, which traverses the inner mitochondrial membrane and is then reoxidized by NAD^+ in the matrix to form NADH in a reaction catalyzed by the citric acid cycle enzyme malate dehydrogenase. The resulting oxaloacetate does not readily cross the inner mitochondrial membrane, and so a transamination reaction (Section 23.3.1) is needed to form aspartate, which can be transported to the cytosolic side. Mitochondrial glutamate donates an amino group, forming aspartate and α-ketoglutarate. In the cytoplasm, aspartate is then deaminated to form oxaloacetate and the cycle is restarted. This shuttle, in contrast with the glycerol 3-phosphate shuttle, is readily reversible. Consequently, NADH can be brought into mitochondria by the malate–aspartate shuttle only if the $NADH/NAD^+$ ratio is higher in the cytosol than in the mitochondrial matrix. This versatile shuttle also facilitates the exchange of key intermediates between mitochondria and the cytosol.

18.5.2 The Entry of ADP into Mitochondria Is Coupled to the Exit of ATP by ATP-ADP Translocase

The major function of oxidative phosphorylation is to generate ATP from ADP. However, ATP and ADP do not diffuse freely across the inner mitochondrial membrane. How are these highly charged molecules moved across the inner membrane into the cytosol? A specific transport protein, *ATP-ADP translocase* (also called *adenine nucleotide translocase* or *ANT*), enables these molecules to traverse this permeability barrier. Most

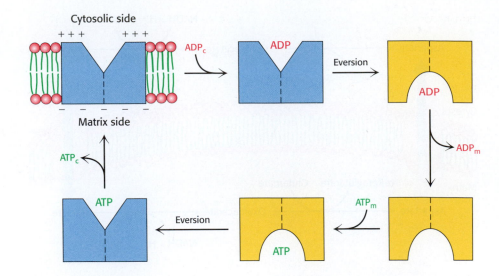

FIGURE 18.39 Mechanism of mitochondrial ATP-ADP translocase. The translocase catalyzes the coupled entry of ADP and exit of ATP into and from the matrix. The reaction cycle is driven by membrane potential. The actual conformational change corresponding to eversion of the binding site could be quite small.

important, the flows of ATP and ADP are coupled. *ADP enters the mitochondrial matrix only if ATP exits, and vice versa.* The reaction catalyzed by the translocase, which acts as an antiporter, is

$$\text{ADP}^{3-}_{\text{cytosol}} + \text{ATP}^{4-}_{\text{matrix}} \longrightarrow \text{ADP}^{3-}_{\text{matrix}} + \text{ATP}^{4-}_{\text{cytosol}}$$

ATP-ADP translocase is highly abundant, constituting about 14% of the protein in the inner mitochondrial membrane. The translocase, a dimer of identical 30-kd subunits, contains a single nucleotide-binding site that alternately faces the matrix and cytosolic sides of the membrane (Figure 18.39). ATP and ADP (both devoid of Mg^{2+}) are bound with nearly the same affinity. In the presence of a positive membrane potential (as would be the case for an actively respiring mitochondrion), *the rate of binding-site eversion from the matrix to the cytosolic side is more rapid for ATP than for ADP because ATP has one more negative charge.* Hence, ATP is transported out of the matrix about 30 times as rapidly as is ADP, which leads to a higher phosphoryl transfer potential on the cytosolic side than on the matrix side. The translocase does not evert at an appreciable rate unless a molecule of ADP is bound at the open, cytosolic site, which then everts to the mitochondrial matrix side. This feature ensures that the entry of ADP into the matrix is precisely coupled to the exit of ATP. The other side of the coin is that *the membrane potential and hence the proton-motive force are decreased by the exchange of ATP for ADP, which results in a net transfer of one negative charge out of the matrix.* ATP–ADP exchange is energetically expensive; about a quarter of the energy yield from electron transfer by the respiratory chain is consumed to regenerate the membrane potential that is tapped by this exchange process. The inhibition of this process leads to the subsequent inhibition of cellular respiration as well (Section 18.6.3).

18.5.3 Mitochondrial Transporters for Metabolites Have a Common Tripartite Motif

ATP-ADP translocase is but one of many mitochondrial transporters for ions and charged metabolites (Figure 18.40). For historical reasons, these transmembrane proteins are sometimes called carriers. Recall that some of them function as symporters and others as antiporters (Section 13.4). The *phosphate carrier,* which works in concert with ATP-ADP translocase, mediates the electroneutral exchange of H_2PO_4^- for OH^- (or, indistinguishably, the electroneutral symport of H_2PO_4^- and H^+). The combined action of these two transporters leads to the exchange of cytosolic ADP and P_i for matrix ATP at the cost of an influx of one H^+. The *dicarboxylate carrier*

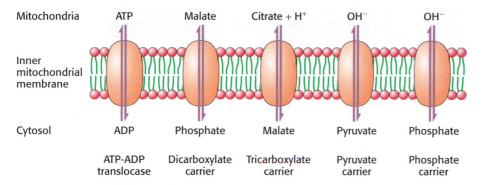

FIGURE 18.40 **Mitochondrial transporters.** Transporters (also called carriers) are transmembrane proteins that move ions and charged metabolites across the inner mitochondrial membrane.

enables malate, succinate, and fumarate to be exported from mitochondria in exchange for P_i. The *tricarboxylate carrier* transports citrate and H^+ in exchange for malate. Pyruvate in the cytosol enters the mitochondrial matrix in exchange for OH^- (or together with H^+) by means of the *pyruvate carrier*. These mitochondrial transporters and more than five others have a common structural motif. They are constructed from three tandem repeats of a 100-residue module, each containing two putative transmembrane segments (Figure 18.41).

18.6 THE REGULATION OF CELLULAR RESPIRATION IS GOVERNED PRIMARILY BY THE NEED FOR ATP

Because ATP is the end product of cellular respiration, its concentration is the ultimate determinant of the rate of all of the components of respiratory pathways.

18.6.1 The Complete Oxidation of Glucose Yields About 30 Molecules of ATP

We can now estimate how many molecules of ATP are formed when glucose is completely oxidized to CO_2. The number of ATP (or GTP) molecules formed in glycolysis and the citric acid cycle is unequivocally known because it is determined by the stoichiometries of chemical reactions. In contrast, the ATP yield of oxidative phosphorylation is less certain because the stoichiometries of proton pumping, ATP synthesis, and metabolite transport processes need not be integer numbers or even have fixed values. As discussed earlier, the best current estimates for the number of protons pumped out of the matrix by NADH-Q oxidoreductase, Q-cytochrome *c* oxidoreductase, and cytochrome *c* oxidase per electron pair are four, two, and four, respectively. The synthesis of a molecule of ATP is driven by the flow of about three protons through ATP synthase. An additional proton is consumed in transporting ATP from the matrix to the cytosol. Hence, about 2.5 molecules of cytosolic ATP are generated as a result of the flow of a pair of electrons from NADH to O_2. For electrons that enter at the level of Q-cytochrome *c* oxidoreductase, such as those from the oxidation of succinate or cytosolic NADH, the yield is about 1.5 molecules of ATP per electron pair. Hence, as tallied in Table 18.4, *about 30 molecules of ATP are formed when glucose is completely oxidized to CO_2;* this value supersedes the traditional estimate of 36 molecules of ATP. Most of the ATP, 26 of 30 molecules formed, is generated by oxidative phosphorylation. Recall that the anaerobic metabolism of glucose yields only 2 molecules of ATP.

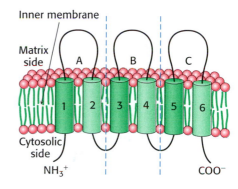

FIGURE 18.41 **Structure of mitochondrial transporters.** Many mitochondrial transporters consist of three similar 100-residue units. These proteins contain six putative membrane-spanning segments. [After J. E. Walker. *Curr. Opin. Struct. Biol.* 2(1992):519.]

TABLE 18.4   ATP yield from the complete oxidation of glucose

Reaction sequence	ATP yield per glucose molecule
Glycolysis: Conversion of glucose into pyruvate (in the cytosol)	
Phosphorylation of glucose	− 1
Phosphorylation of fructose 6-phosphate	− 1
Dephosphorylation of 2 molecules of 1,3-BPG	+ 2
Dephosphorylation of 2 molecules of phosphoenolpyruvate	+ 2
2 molecules of NADH are formed in the oxidation of 2 molecules of glyceraldehyde 3-phosphate	
Conversion of pyruvate into acetyl CoA (inside mitochondria)	
2 molecules of NADH are formed	
Citric acid cycle (inside mitochondria)	
2 molecules of guanosine triphosphate are formed from 2 molecules of succinyl CoA	+ 2
6 molecules of NADH are formed in the oxidation of 2 molecules each of isocitrate, α-ketoglutarate, and malate	
2 molecules of $FADH_2$ are formed in the oxidation of 2 molecules of succinate	
Oxidative phosphorylation (inside mitochondria)	
2 molecules of NADH formed in glycolysis; each yields 1.5 molecules of ATP (assuming transport of NADH by the glycerol 3-phosphate shuttle)	+ 3
2 molecules of NADH formed in the oxidative decarboxylation of pyruvate; each yields 2.5 molecules of ATP	+ 5
2 molecules of $FADH_2$ formed in the citric acid cycle; each yields 1.5 molecules of ATP	+ 3
6 molecules of NADH formed in the citric acid cycle; each yields 2.5 molecules of ATP	+ 15
NET YIELD PER MOLECULE OF GLUCOSE	+ 30

Source: The ATP yield of oxidative phosphorylation is based on values given in P. C. Hinkle, M. A. Kumar, A. Resetar, and D. L. Harris, *Biochemistry* 30(1991):3576. *Note*: The current value of 30 molecules of ATP per molecule of glucose supersedes the earlier one of 36 molecules of ATP. The stoichiometries of proton pumping, ATP synthesis, and metabolite transport should be regarded as estimates. About two more molecules of ATP are formed per molecule of glucose oxidized when the malate–aspartate shuttle rather than the glycerol 3-phosphate shuttle is used.

18.6.2 The Rate of Oxidative Phosphorylation Is Determined by the Need for ATP

How is the rate of the electron-transport chain controlled? Under most physiological conditions, electron transport is tightly coupled to phosphorylation. *Electrons do not usually flow through the electron-transport chain to O_2 unless ADP is simultaneously phosphorylated to ATP.* Oxidative phosphorylation requires a supply of NADH (or other source of electrons at high potential), O_2, ADP, and P_i. The most important factor in determining the rate of oxidative phosphorylation is the *level of ADP.* The rate of oxygen consumption by mitochondria increases markedly when ADP is added and then returns to its initial value when the added ADP has been converted into ATP (Figure 18.42).

FIGURE 18.42 Respiratory control. Electrons are transferred to O_2 only if ADP is concomitantly phosphorylated to ATP.

The regulation of the rate of oxidative phosphorylation by the ADP level is called *respiratory control* or *acceptor control*. The level of ADP likewise affects the rate of the citric acid cycle because of its need for NAD^+ and FAD. The physiological significance of this regulatory mechanism is evident. The ADP level increases when ATP is consumed, and so oxidative phosphorylation is coupled to the utilization of ATP. *Electrons do not flow from fuel molecules to O_2 unless ATP needs to be synthesized.* We see here another example of the regulatory significance of the energy charge.

18.6.3 Oxidative Phosphorylation Can Be Inhibited at Many Stages

Oxidative phosphorylation is susceptible to inhibition at all stages of the process. Specific inhibitors of electron transport were invaluable in revealing the sequence of electron carriers in the respiratory chain. For example, *rotenone* and *amytal* block electron transfer in NADH-Q oxidoreductase and thereby prevent the utilization of NADH as a substrate (Figure 18.43). In contrast, electron flow resulting from the oxidation of succinate is unimpaired, because these electrons enter through QH_2, beyond the block. *Antimycin A* interferes with electron flow from cytochrome b_H in Q-cytochrome c oxidoreductase. Furthermore, electron flow in cytochrome c oxidase can be blocked by *cyanide* (CN^-), *azide* (N^{3-}), and *carbon monoxide* (CO). Cyanide and azide react with the ferric form of heme a_3, whereas carbon monoxide inhibits the ferrous form. Inhibition of the electron-transport chain also inhibits ATP synthesis because the proton-motive force can no longer be generated.

ATP synthase also can be inhibited. Oligomycin and dicyclohexylcarbodiimide (DCCD) prevent the influx of protons through ATP synthase. If actively respiring mitochondria are exposed to an inhibitor of ATP synthase, the electron-transport chain ceases to operate. Indeed, this observation clearly illustrates that electron transport and ATP synthesis are normally tightly coupled.

This tight coupling of electron transport and phosphorylation in mitochondria can be disrupted (uncoupled) by 2,4-dinitrophenol (Figure 18.44) and certain other acidic aromatic compounds. These substances carry protons across the inner mitochondrial membrane. In the presence of these uncouplers, electron transport from NADH to O_2 proceeds in a normal fashion, but ATP is not formed by mitochondrial ATP synthase because the proton-motive force across the inner mitochondrial membrane is dissipated. This loss of respiratory control leads to increased oxygen consumption and oxidation of NADH. Indeed, in the accidental ingestion of uncouplers, large amounts of metabolic fuels are consumed, but no energy is stored as ATP. Rather, energy is released as heat. DNP and other uncouplers are very useful in metabolic studies because of their specific effect on oxidative phosphorylation. The regulated uncoupling of oxidative phosphorylation is a biologically useful means of generating heat.

ATP-ADP translocase is specifically inhibited by very low concentrations of *atractyloside* (a plant glycoside) or *bongkrekic acid* (an antibiotic from a mold). Atractyloside binds to the translocase when its nucleotide site faces the cytosol, whereas bongkrekic acid binds when this site faces the mitochondrial matrix. Oxidative phosphorylation stops soon after either inhibitor is added, showing that ATP-ADP translocase is essential.

18.6.4 Regulated Uncoupling Leads to the Generation of Heat

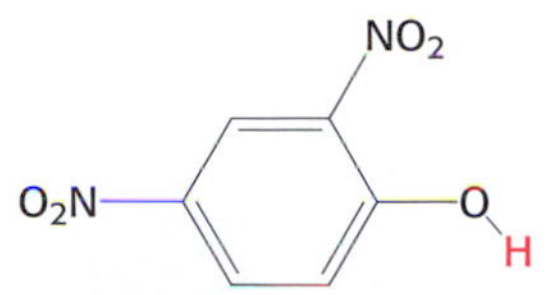

The uncoupling of oxidative phosphorylation is a means of generating heat to maintain body temperature in hibernating animals, in some newborn animals

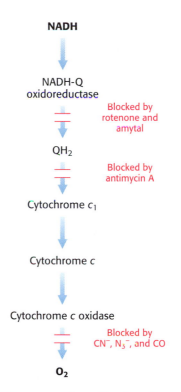

FIGURE 18.43 Sites of action of some inhibitors of electron transport.

FIGURE 18.44 Uncoupler of oxidative phosphorylation. 2,4-Dinitrophenol, a lipid-soluble substance, can carry protons across the inner mitochondrial membrane. The dissociable proton is shown in red.

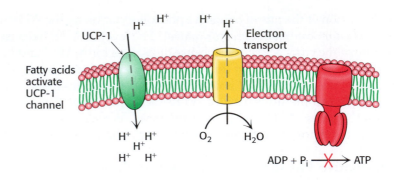

FIGURE 18.45 Action of an uncoupling protein. Uncoupling protein-1 (UCP-1) generates heat by permitting the influx of protons into the mitochondria without the synthesis of ATP.

(including human beings), and in mammals adapted to cold. Brown adipose tissue, which is very rich in mitochondria (often referred to as brown fat mitochondria), is specialized for this process of *nonshivering thermogenesis*. The inner mitochondrial membrane of these mitochondria contains a large amount of *uncoupling protein (UCP)*, here UCP-1, or *thermogenin*, a dimer of 33-kd subunits that resembles ATP-ADP translocase. UCP-1 forms a pathway for the flow of protons from the cytosol to the matrix. In essence, *UCP-1 generates heat by short-circuiting the mitochondrial proton battery*. This dissipative proton pathway is activated by free fatty acids liberated from triacylglycerols in response to hormonal signals, such as β-adrenergic agonists (Figure 18.45).

In addition to UCP-1, two other uncoupling proteins have been identified. UCP-2, which is 56% identical in sequence with UCP-1, is found in a wide variety of tissues. UCP-3 (57% identical with UCP-1 and 73% identical with UCP-2) is localized to skeletal muscle and brown fat. This family of uncoupling proteins, especially UCP-2 and UCP-3, may play a role in energy homeostasis. In fact, the genes for UCP-2 and UCP-3 map to regions of the human and mouse chromosomes that have been linked to obesity, substantiating the notion that they function as a means of regulating body weight. The use of uncoupling proteins is not limited to animals, however. The skunk cabbage uses an analogous mechanism to heat its floral spikes, increasing the evaporation of odoriferous molecules that attract insects to fertilize its flowers.

18.6.5 Mitochondrial Diseases Are Being Discovered

As befitting an organelle that is so central to energy metabolism, mitochondrial malfunction can lead to pathological conditions. The number of diseases that can be attributed to mitochondrial mutations is steadily growing in step with our growing understanding of the biochemistry and genetics of mitochondria. The first mitochondrial disease to be understood was Leber hereditary optic neuropathy (LHON), a form of blindness that strikes in midlife as a result of mutations to the NADH-Q oxidoreductase component of Complex I. Some of these mutations impair NADH utilization, whereas others block electron transfer to Q. The accumulation of mutations in mitochondrial genes in the course of several decades may contribute to aging, degenerative disorders, and cancer.

A human egg harbors several hundred thousand molecules of mitochondrial DNA, whereas a sperm contributes only a few hundred and thus has little effect on the mitochondrial genotype. Because the maternally inherited mitochondria are present in large numbers and not all of the mitochondria may be affected, the pathologies of mitochondrial mutants can be quite complex.

Even within a single family carrying an identical mutation, chance fluctuations in the percentage of mitochondria with the mutation lead to large variations in the nature and severity of the symptoms of the pathological condition as well as the time of onset. As the percentage of defective mitochondria increases, energy-generating capacity diminishes until, at some threshold, the cell can no longer function properly. Defects in cellular respiration are doubly dangerous. Not only does energy transduction decrease, but also the likelihood that reactive oxygen species will be generated increases. Organs that are highly dependent on oxidative phosphorylation, such as the nervous system and the heart, are most vulnerable to mutations in mitochondrial DNA.

18.6.6 Mitochondria Play a Key Role in Apoptosis

In the course of development or in cases of significant cell damage, individual cells within multicellular organisms undergo *programmed cell death,* or *apoptosis.* Mitochondria act as control centers regulating this process. Although the details have not yet been established, a pore called the mitochondrial permeability transition pore (mtPTP) forms in damaged mitochondria. This pore appears to consist of VDAC (the adenine nucleotide translocator) and several other mitochondrial proteins, including members of a family of proteins (Bcl family) that were initially discovered because of their role in cancer. One of the most potent activators of apoptosis is cytochrome *c.* Its presence in the cytosol activates a cascade of proteolytic enzymes called *caspases.* These cysteine proteases (Section 9.1.6) are conserved in evolution, being found in organisms ranging from hydra to human beings. Cytochrome *c,* in conjunction with other proteins, initiates the cascade by activating procaspase 9 to form caspase 9, which then activates other caspases. Activation of the caspase cascade does not lead to generalized protein destruction. Rather, the caspases have particular targets. For instance, the proteins that maintain cell structure are destroyed. Another example is the degradation of a protein that inhibits an enzyme that destroys DNA (caspase-activated DNAse, CAD), freeing CAD to cleave the genetic material. This cascade of proteolytic enzymes has been called "death by a thousand tiny cuts."

18.6.7 Power Transmission by Proton Gradients: A Central Motif of Bioenergetics

The main concept presented in this chapter is that mitochondrial electron transfer and ATP synthesis are linked by a transmembrane proton gradient. ATP synthesis in bacteria and chloroplasts (Section 19.4) also is driven by proton gradients. In fact, proton gradients power a variety of energy-requiring processes such as the active transport of calcium ions by mitochondria, the entry of some amino acids and sugars into bacteria, the rotation of bacterial flagella, and the transfer of electrons from $NADP^+$ to NADPH. Proton gradients can also be used to generate heat, as in hibernation. It is evident that *proton gradients are a central interconvertible currency of free energy in biological systems* (Figure 18.46). Mitchell noted that the proton-motive force is a marvelously simple and effective store of free energy because it requires only a thin, closed lipid membrane between two aqueous phases.

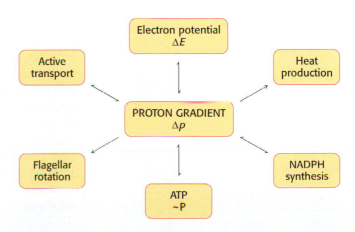

FIGURE 18.46 The proton gradient is an interconvertible form of free energy.

SUMMARY

Oxidative Phosphorylation in Eukaryotes Takes Place in Mitochondria

Mitochondria generate most of the ATP required by aerobic cells by a joint endeavor of the reactions of citric acid cycle, which take place in the mitochondrial matrix, and oxidative phosphorylation, which takes place in the inner mitochondrial membrane. Mitochondria are descendents of a free-living bacterium that established a symbiotic relation with another cell.

Oxidative Phosphorylation Depends on Electron Transfer

In oxidative phosphorylation, the synthesis of ATP is coupled to the flow of electrons from NADH or $FADH_2$ to O_2 by a proton gradient across the inner mitochondrial membrane. Electron flow through three asymmetrically oriented transmembrane complexes results in the pumping of protons out of the mitochondrial matrix and the generation of a membrane potential. ATP is synthesized when protons flow back to the matrix through a channel in an ATP-synthesizing complex, called ATP synthase (also known as F_0F_1-ATPase). Oxidative phosphorylation exemplifies a fundamental theme of bioenergetics: the transmission of free energy by proton gradients.

The Respiratory Chain Consists of Four Complexes: Three Proton Pumps and a Physical Link to the Citric Acid Cycle

The electron carriers in the respiratory assembly of the inner mitochondrial membrane are quinones, flavins, iron–sulfur complexes, heme groups of cytochromes, and copper ions. Electrons from NADH are transferred to the FMN prosthetic group of NADH-Q oxidoreductase (Complex I), the first of four complexes. This oxidoreductase also contains Fe-S centers. The electrons emerge in QH_2, the reduced form of ubiquinone (Q). The citric acid cycle enzyme succinate dehydrogenase is a component of the succinate-Q reductase complex (Complex II), which donates electrons from $FADH_2$ to Q to form QH_2. This highly mobile hydrophobic carrier transfers its electrons to Q-cytochrome c oxidoreductase (Complex III), a complex that contains cytochromes b and c_1 and an Fe-S center. This complex reduces cytochrome c, a water-soluble peripheral membrane protein. Cytochrome c, like Q, is a mobile carrier of electrons, which it then transfers to cytochrome c oxidase (Complex IV). This complex contains cytochromes a and a_3 and three copper ions. A heme iron ion and a copper ion in this oxidase transfer electrons to O_2, the ultimate acceptor, to form H_2O.

A Proton Gradient Powers the Synthesis of ATP

The flow of electrons through Complexes I, III, and IV leads to the transfer of protons from the matrix side to the cytosolic side of the inner mitochondrial membrane. A proton-motive force consisting of a pH gradient (matrix side basic) and a membrane potential (matrix side negative) is generated. The flow of protons back to the matrix side through ATP synthase drives ATP synthesis. The enzyme complex is a molecular motor made of two operational units: a rotating component and a stationary component. The rotation of the γ subunit induces structural changes in the β subunit that result in the synthesis and release of ATP from the enzyme. Proton influx provides the force for the rotation.

The flow of two electrons through NADH-Q oxidoreductase, Q-cytochrome c oxidoreductase, and cytochrome c oxidase generates a gradient sufficient to synthesize 1, 0.5, and 1 molecule of ATP, respectively. Hence, 2.5 molecules of ATP are formed per molecule of NADH oxidized in the mitochondrial matrix, whereas only 1.5 molecules of ATP are made per molecule of $FADH_2$ oxidized because its electrons enter the chain at QH_2, after the first proton-pumping site.

Many Shuttles Allow Movement Across the Mitochondrial Membranes
Mitochondria employ a host of carriers, or transporters, to move molecules across the inner mitochondrial membrane. The electrons of cytoplasmic NADH are transferred into the mitochondria by the glycerol phosphate shuttle to form $FADH_2$ from FAD. The entry of ADP into the mitochondrial matrix is coupled to the exit of ATP by ATP-ADP translocase, a transporter driven by membrane potential.

The Regulation of Oxidative Phosphorylation Is Governed Primarily by the Need for ATP
About 30 molecules of ATP are generated when a molecule of glucose is completely oxidized to CO_2 and H_2O. Electron transport is normally tightly coupled to phosphorylation. NADH and $FADH_2$ are oxidized only if ADP is simultaneously phosphorylated to ATP, a form of regulation called acceptor or respiratory control. Uncouplers such as DNP can disrupt this coupling; they dissipate the proton gradient by carrying protons across the inner mitochondrial membrane. Proteins have been identified that uncouple electron transport and ATP synthesis for the generation of heat.

KEY TERMS

oxidative phosphorylation (p. 491)

proton-motive force (p. 491)

cellular respiration (p. 492)

electron-transport chain (p. 494)

reduction (redox, oxidation–reduction, E'_0) potential (p. 494)

inverted region (p. 497)

coenzyme Q (Q, ubiquinone) (p. 498)

NADH-Q oxidoreductase (Complex I) (p. 499)

flavin mononucleotide (FMN) (p. 499)

iron–sulfur (nonheme iron) protein (p. 499)

succinate-Q reductase (Complex II) (p. 501)

Q-cytochrome c oxidoreductase (Complex III) (p. 501)

cytochrome c (cyt c) (p. 501)

Rieske center (p. 502)

Q cycle (p. 502)

cytochrome c oxidase (Complex IV) (p. 503)

superoxide dismutase (p. 506)

catalase (p. 506)

ATP synthase (Complex V, F_1F_0 ATPase) (p. 508)

glycerol 3-phosphate shuttle (p. 514)

malate–aspartate shuttle (p. 515)

ATP-ADP translocase (adenine nucleotide translocase, ANT) (p. 515)

respiratory (acceptor) control (p. 519)

uncoupling protein (UCP) (p. 520)

programmed cell death (apoptosis) (p. 521)

caspase (p. 521)

SELECTED READINGS

Where to start

Gray, M. W., Burger, G., and Lang, B. F., 1999. Mitochondrial evolution. *Science* 283:1476–1481.

Wallace, D. C., 1997. Mitochondrial DNA in aging and disease. *Sci. Am.* 277(2):40–47.

Saraste, M., 1999. Oxidative phosphorylation at the *fin de siècle*. *Science* 283:1488–1493.

Shultz, B. E., and Chan, S. I., 2001. Structures and proton-pumping strategies of mitochondrial respiratory enzymes. *Ann. Rev. Biophys. Biomol. Struct.* 30:23–65.

Moser, C. C., Keske, J. M., Warncke, K., Farid, R. S., and Dutton, P. L., 1992. Nature of biological electron transfer. *Nature* 355: 796–802.

Books

Scheffler, I. E., 1999. *Mitochondria*. Wiley.

Nicholls, D. G., and Ferguson, S. J., 1997. *Bioenergetics 2*. Academic Press.

Harold, F. M., 1986. *The Vital Force: A Study of Bioenergetics*. W. H. Freeman and Company.

Ernster, L. (Ed.), 1992. *Molecular Mechanisms in Bioenergetics*. Elsevier.

Electron-transport chain

Zaslavsky, D., and Gennis, R. B., 2000. Proton pumping by cytochrome oxidase: Progress, problems and postulates. *Biochim. Biophys. Acta* 1458:164–179.

Grigorieff, N., 1999. Structure of the respiratory NADH:ubiquinone oxidoreductase (complex I). *Curr. Opin. Struct. Biol.* 9:476–483.

Ackrell, B. A., 2000. Progress in understanding structure–function relationships in respiratory chain complex II. *FEBS Lett.* 466:1–5.

Grigorieff, N., 1998. Three-dimensional structure of bovine NADH: ubiquinone oxidoreductase (complex I) at 22 Å in ice. *J. Mol. Biol.* 277:1033–1046.

Dutton, P. L., Moser, C. C., Sled, V. D., Daldal, F., and Ohnishi, T., 1998. A reductant-induced oxidation mechanism for complex I. *Biochim. Biophys. Acta* 1364:245–257.

Zhang, Z., Huang, L., Shulmeister, V. M., Chi, Y. I., Kim, K. K., Hung, L. W., Crofts, A. R., Berry, E. A., and Kim, S. H., 1998. Electron transfer by domain movement in cytochrome bc₁. *Nature* 392:677–684.

Xia, D., Yu, C. A., Kim, H., Xia, J. Z., Kachurin, A. M., Zhang, L., Yu, L., and Deisenhofer, J., 1997. Crystal structure of the cytochrome bc₁ complex from bovine heart mitochondria. *Science* 277:60–66.

Michel, H., Behr, J., Harrenga, A., and Kannt, A., 1998. Cytochrome c oxidase: Structure and spectroscopy. *Annu. Rev. Biophys. Biomol. Struct.* 27:329–356.

Verkhovsky, M. I., Jasaitis, A., Verkhovskaya, M. L., Morgan, J. E., and Wikstrom, M., 1999. Proton translocation by cytochrome c oxidase. *Nature* 400:480–483.

Wikstrom, M., and Morgan, J. E., 1992. The dioxygen cycle: Spectral, kinetic, and thermodynamic characteristics of ferryl and peroxy intermediates observed by reversal of the cytochrome oxidase reaction. *J. Biol. Chem.* 267:10266–10273.

ATP synthase

Noji, H., and Yoshida, M., 2001. The rotary machine in the cell: ATP synthase. *J. Biol. Chem.* 276:1665–1668.

Yasuda, R., Noji, H., Kinosita, K., Jr., and Yoshida, M., 1998. F1-ATPase is a highly efficient molecular motor that rotates with discrete 120 degree steps. *Cell* 93:1117–1124.

Kinosita, K., Jr., Yasuda, R., Noji, H., Ishiwata, S., and Yoshida, M., 1998. F1-ATPase: A rotary motor made of a single molecule. *Cell* 93:21–24.

Noji, H., Yasuda, R., Yoshida, M., and Kinosita, K., Jr., 1997. Direct observation of the rotation of F1-ATPase. *Nature* 386:299–302.

Tsunoda, S. P., Aggeler, R., Yoshida, M., and Capaldi, R. A., 2001. Rotation of the c subunit oligomer in fully functional F₁F₀ ATP synthase. *Proc. Natl. Acad. Sci. USA* 987:898–902.

Gibbons, C., Montgomery, M. G., Leslie, A. G. W., and Walker, J., 2000. The structure of the central stock in F₁-ATPase at 2.4 Å resolution. *Nat. Struct. Biol.* 7:1055–1061.

Boyer, P. D., 2000. Catalytic site forms and controls in ATP synthase catalysis. *Biochim. Biophys. Acta* 1458:252–262.

Stock, D., Leslie, A. G., and Walker, J. E., 1999. Molecular architecture of the rotary motor in ATP synthase. *Science* 286:1700–1705.

Sambongi, Y., Iko, Y., Tanabe, M., Omote, H., Iwamoto-Kihara, A., Ueda, I., Yanagida, T., Wada, Y., and Futai, M., 1999. Mechanical rotation of the c subunit oligomer in ATP synthase (F₀F₁): Direct observation. *Science* 286:1722–1724.

Abrahams, J. P., Leslie, A. G., Lutter, R., and Walker, J. E., 1994. Structure at 2.8 Å resolution of F1-ATPase from bovine heart mitochondria. *Nature* 370:621–628.

Bianchet, M. A., Hullihen, J., Pedersen, P. L., and Amzel, L. M., 1998. The 2.8-Å structure of rat liver F1-ATPase: Configuration of a critical intermediate in ATP synthesis/hydrolysis. *Proc. Natl. Acad. Sci. USA* 95:11065–11070.

Translocators

Klingenberg, M., and Huang, S. G., 1999. Structure and function of the uncoupling protein from brown adipose tissue. *Biochim. Biophys. Acta* 1415:271–296.

Nicholls, D. G., and Rial, E., 1999. A history of the first uncoupling protein, UCP1. *J. Bioenerg. Biomembr.* 31:399–406.

Ricquier, D., and Bouillaud, F., 2000. The uncoupling protein homologues: UCP1, UCP2, UCP3, StUCP and AtUCP. *Biochem. J.* 345:161–179.

Walker, J. E., 1992. The mitochondrial transporter family. *Curr. Opin. Struct. Biol.* 2:519–526.

Klingenberg, M., 1992. Structure-function of the ADP/ATP carrier. *Biochem. Soc. Trans.* 20:547–550.

Superoxide dismutase and catalase

Culotta, V. C., 2000. Superoxide dismutase, oxidative stress, and cell metabolism. *Curr. Top. Cell Regul.* 36:117–132.

Morrison, B. M., Morrison, J. H., and Gordon, J. W., 1998. Superoxide dismutase and neurofilament transgenic models of amyotrophic lateral sclerosis. *J. Exp. Zool.* 282:32–47.

Tainer, J. A., Getzoff, E. D., Richardson, J. S., and Richardson, D. C., 1983. Structure and mechanism of copper, zinc superoxide dismutase. *Nature* 306:284–287.

Reid, T. J., Murthy, M. R., Sicignano, A., Tanaka, N., Musick, W. D., and Rossmann, M. G., 1981. Structure and heme environment of beef liver catalase at 2.5 Å resolution. *Proc. Natl. Acad. Sci. USA* 78:4767–4771.

Stallings, W. C., Pattridge, K. A., Strong, R. K., and Ludwig, M. L., 1984. Manganese and iron superoxide dismutases are structural homologs. *J. Biol. Chem.* 259:10695–10699.

Hsieh, Y., Guan, Y., Tu, C., Bratt, P. J., Angerhofer, A., Lepock, J. R., Hickey, M. J., Tainer, J. A., Nick, H. S., and Silverman, D. N., 1998. Probing the active site of human manganese superoxide dismutase: The role of glutamine 143. *Biochemistry* 37:4731–4739.

Mitochondrial diseases

Smeitink, J., van den Heuvel, L., and DiMauro, S., 2001. The genetics and pathology of oxidative phosphorylation. *Nature Rev. Genet* 2:342–352.

Wallace, D. C., 1999. Mitochondrial diseases in man and mouse. *Science* 283:1482–1488.

Wallace, D. C., 1992. Diseases of the mitochondrial DNA. *Annu. Rev. Biochem.* 61:1175–1212.

Benecke, R., Strumper, P., and Weiss, H., 1992. Electron transfer Complex I defect in idiopathic dystonia. *Ann. Neurol.* 32:683–686.

Apoptosis

Joza, N., Susin, S. A., Daugas, E., Stanford, W. L., Cho, S. K., Li, C. Y. J., Sasaki, T., Elia, A. J., Cheng, H.-Y. M., Ravagnan, L., Ferri, K. F., Zamzami, N., Wakeham, A., Hakem, R., Yoshida, H., Kong, Y.-Y., Mak, T. W., Zúñiga-Pflücker, J. C., Kroemer, G., and Penninger, J. M., 2001. Essential role of the mitochondrial *apoptosis*-inducing factor in programmed cell death. *Nature* 410:549–554.

Desagher, S., and Martinou, J. C., 2000. Mitochondria as the central control point of apoptosis. *Trends Cell Biol.* 10:369–377.

Hengartner, M. O., 2000. The biochemistry of apoptosis. *Nature* 407:770–776.

Earnshaw, W. C., Martins, L. M., and Kaufmann, S. H., 1999. Mammalian caspases: Structure, activation, substrates, and functions during apoptosis. *Annu. Rev. Biochem.* 68:383–424.

Wolf, B. B., and Green, D. R., 1999. Suicidal tendencies: Apoptotic cell death by caspase family proteinases. *J. Biol. Chem.* 274:20049–20052.

Historical aspects

Mitchell, P., 1979. Keilin's respiratory chain concept and its chemiosmotic consequences. *Science* 206:1148–1159.

Mitchell, P., 1976. Vectorial chemistry and the molecular mechanics of chemiosmotic coupling: Power transmission by proticity. *Biochem. Soc. Trans.* 4:399–430.

Racker, E., 1980. From Pasteur to Mitchell: A hundred years of bioenergetics. *Fed. Proc.* 39:210–215.

Keilin, D., 1966. *The History of Cell Respiration and Cytochromes.* Cambridge University Press.

Kalckar, H. M. (Ed.), 1969. *Biological Phosphorylations: Development of Concepts.* Prentice Hall.

Kalckar, H. M., 1991. Fifty years of biological research: From oxidative phosphorylation to energy requiring transport and regulation. *Annu. Rev. Biochem.* 60:1–37.

Fruton, J. S., 1972. *Molecules and Life: Historical Essays on the Interplay of Chemistry and Biology.* Wiley-Interscience.

PROBLEMS

1. *Energy harvest.* What is the yield of ATP when each of the following substrates is completely oxidized to CO_2 by a mammalian cell homogenate? Assume that glycolysis, the citric acid cycle, and oxidative phosphorylation are fully active.

(a) Pyruvate
(b) Lactate
(c) Fructose 1,6-bisphosphate
(d) Phosphoenolpyruvate
(e) Galactose
(f) Dihydroxyacetone phosphate

2. *Reference states.* The standard oxidation–reduction potential for the reduction of O_2 to H_2O is given as 0.82 V in Table 18.1. However, the value given in textbooks of chemistry is 1.23 V. Account for this difference.

3. *Potent poisons.* What is the effect of each of the following inhibitors on electron transport and ATP formation by the respiratory chain?

(a) Azide
(b) Atractyloside
(c) Rotenone
(d) DNP
(e) Carbon monoxide
(f) Antimycin A

4. *A question of coupling.* What is the mechanistic basis for the observation that the inhibitors of ATP synthase also lead to an inhibition of the electron-transport chain?

5. *O_2 consumption.* Using the axes in the adjoining illustration, draw an oxygen-uptake curve ([O_2] versus time) for a suspension of isolated mitochondria when the following compounds are added in the indicated order. With the addition of each compound, all of the previously added compounds remain present. The experiment starts with the amount of oxygen indicated by the arrow on the y-axis. [O_2] can only decrease or be unaffected.

(a) Glucose
(b) ADP + P_i
(c) Citrate
(d) Oligomycin
(e) Succinate
(f) Dinitrophenol
(g) Rotenone
(h) Cyanide

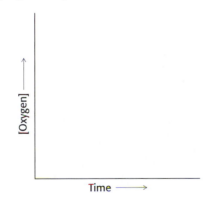

6. *P:O ratios.* The number of molecules of inorganic phosphate incorporated into organic form per atom of oxygen consumed, termed the *P:O ratio*, was frequently used as an index of oxidative phosphorylation.

(a) What is the relation of the P:O ratio to the ratio of the number of protons translocated per electron pair ($H^+/2\ e^-$) and the ratio of the number of protons needed to synthesize ATP and transport it to the cytosol (P/H^+)?

(b) What are the P:O ratios for electrons donated by matrix NADH and by succinate?

7. *Thermodynamic constraint.* Compare the $\Delta G^{\circ\prime}$ values for the oxidation of succinate by NAD^+ and by FAD. Use the data given in Table 18.1, and assume that E'_0 for the FAD–$FADH_2$ redox couple is nearly 0 V. Why is FAD rather than NAD^+ the electron acceptor in the reaction catalyzed by succinate dehydrogenase?

8. *Cyanide antidote.* The immediate administration of nitrite is a highly effective treatment for cyanide poisoning. What is the basis for the action of this antidote? (Hint: Nitrite oxidizes ferrohemoglobin to ferrihemoglobin.)

9. *Currency exchange.* For a proton-motive force of 0.2 V (matrix negative), what is the maximum [ATP]/[ADP][P_i] ratio compatible with ATP synthesis? Calculate this ratio three times, assuming that the number of protons translocated per ATP formed is two, three, and four and that the temperature is 25°C.

10. *Runaway mitochondria 1.* Suppose that the mitochondria of a patient oxidizes NADH irrespective of whether ADP is present. The P:O ratio for oxidative phosphorylation by these mitochondria is less than normal. Predict the likely symptoms of this disorder.

11. *An essential residue.* Conduction of protons by the F_0 unit of ATP synthase is blocked by modification of a single side chain by dicyclohexylcarbodiimide. What are the most likely targets of action of this reagent? How might you use site-specific mutagenesis to determine whether this residue is essential for proton conduction?

12. *Recycling device.* The cytochrome b component of Q-cytochrome c oxidoreductase enables both electrons of QH_2 to be effectively utilized in generating a proton-motive force. Cite another recycling device in metabolism that brings a potentially dead end reaction product back into the mainstream.

13. *Crossover point.* The precise site of action of a respiratory-chain inhibitor can be revealed by the *crossover technique*. Britton Chance devised elegant spectroscopic methods for determining the proportions of the oxidized and reduced forms of each carrier. This determiniation is feasible because the forms have distinctive absorption spectra, as illustrated in the adjoining graph for cytochrome c. You are given a new inhibitor and find that its addition to respiring mitochondria causes the carriers between NADH and QH_2 to become more reduced and those between cytochrome c and O_2 to become more oxidized. Where does your inhibitor act?

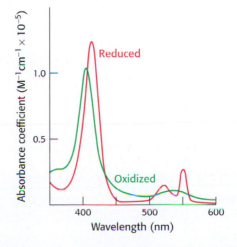

14. *Runaway mitochondria 2.* Years ago, it was suggested that un-couplers would make wonderful diet drugs. Explain why this idea was proposed and why it was rejected. Why might the producers of antiperspirants be supportive of the idea?

15. *Identifying the inhibition.* You are asked to determine whether a chemical is an electron-transport-chain inhibitor or an inhibitor of ATP synthase. Design an experiment to determine this.

Chapter Integration Problem

16. *No exchange.* Mice that are completely lacking adenine nucleotide translocase (ANT^-/ANT^-) can be made by the "knockout" technique. Remarkably, these mice are viable but have the following pathological conditions: (1) high serum levels of lactate, alanine, and succinate; (2) there is little electron transport; and (3) the levels of mitochondrial H_2O_2 are increased six- to eightfold compared with normal mice. Provide a possible biochemical explanation for each of these conditions.

Data Interpretation Problem

17. *Mitochondrial disease.* A mutation in a mitochondrial gene encoding a component of ATP synthase has been identified. People who have this mutation suffer from muscle weakness, ataxia, and retinitis pigmentosa. A tissue biopsy was performed on each patient, and submitochondrial particles were isolated that were capable of succinate-sustained ATP synthesis. First, the activity of the ATP synthase was measured on the addition of succinate and the following results were obtained.

	ATP synthase activity (nmol of ATP formed min^{-1} mg^{-1})
Controls	3.0
Patient 1	0.25
Patient 2	0.11
Patient 3	0.17

(a) What was the purpose of the addition of succinate?

(b) What is the effect of the mutation on succinate-coupled ATP synthesis?

Next, the ATPase activity of the enzyme was measured by incubating the submitochondrial particle with ATP in the absence of succinate.

	ATP hydrolysis (nmol of ATP hydrolyzed min^{-1} mg^{-1})
Controls	33
Patient 1	30
Patient 2	25
Patient 3	31

(c) Why was succinate omitted from the reaction?

(d) What is the effect of the mutation on ATP hydrolysis?

(e) What do these results, in conjunction with those obtained in the first experiment, tell you about the nature of the mutation?

Mechanism Problem

18. *Chiral clue.* ATPγS, a slowly hydrolyzed analog of ATP, can be used to probe the mechanism of phosphoryl transfer reactions. Chiral ATPγS has been synthesized containing ^{18}O in a specific γ position and ordinary ^{16}O elsewhere in the molecule. Hydrolysis of this chiral molecule by ATP synthase in ^{17}O-enriched water yields inorganic $[^{16}O,^{17}O,^{18}O]$thiophosphate having the following absolute configuration. In contrast, hydrolysis of this chiral ATPγS by a calcium-pumping ATPase from muscle gives thiophosphate of the opposite configuration. What is the simplest interpretation of these data?

ATPγS   Thiophosphate

Media Problem

19. *Queueing up?* The **Conceptual Insights** module on oxidative phophorylation summarizes the steps in the electron transport chain and can help you answer the following questions. Suppose you have isolated actively respiring mitochondria. What do you think might happen to ATP production in the short-term if you were to add large amounts of coenzyme Q? If you were to add large amounts of QH_2? How might the long-term responses differ from the short-term?

CHAPTER 19

The Light Reactions of Photosynthesis

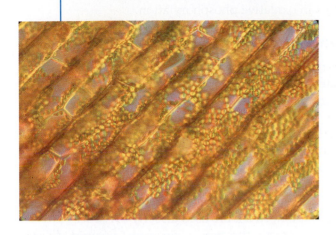

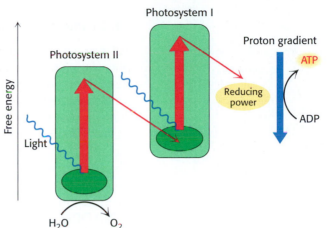

Chloroplasts (left) convert light energy into chemical energy. High-energy electrons in chloroplasts are transported through two photosystems (right). During this transit, which culminates in the generation of reducing power, ATP is synthesized in a manner analogous to mitochondrial ATP synthesis. Unlike as in mitochondrial electron transport, however, electrons in chloroplasts are energized by light. [(Left) Herb Charles Ohlmeyer/Fran Heyl Associates.]

Essentially all free energy utilized by biological systems arises from solar energy that is trapped by the process of photosynthesis. The basic equation of photosynthesis is deceptively simple. Water and carbon dioxide combine to form carbohydrates and molecular oxygen.

$$CO_2 + H_2O \xrightarrow{\text{Light}} (CH_2O) + O_2$$

In this equation, (CH_2O) represents carbohydrate, primarily sucrose and starch. The mechanism of photosynthesis is complex and requires the interplay of many proteins and small molecules. Photosynthesis in green plants takes place in *chloroplasts* (Figure 19.1). *The energy of light captured by pigment molecules, called chlorophylls, in chloroplasts is used to generate high-energy electrons with great reducing potential.* These electrons are used to produce NADPH as well as ATP in a series of reactions called the *light reactions* because they require light. NADPH and ATP formed by the action of light then reduce carbon dioxide and convert it into *3-phosphoglycerate* by a series of reactions called the *Calvin cycle* or the *dark reactions.* The Calvin cycle will be discussed in Chapter 20. The amount of energy stored by photosynthesis is enormous. More than 10^{17} kcal (4.2×10^{17} kJ) of free energy is stored annually by photosynthesis on Earth, which corresponds to the assimilation of more than 10^{10} tons of carbon into carbohydrate and other forms of organic matter.

FIGURE 19.1 Electron micrograph of a chloroplast from a spinach leaf. The thylakoid membranes pack together to form grana. [Courtesy of Dr. Kenneth Miller.]

Photosynthetic yield—

"If a year's yield of photosynthesis were amassed in the form of sugar cane, it would form a heap over two miles high and with a base 43 square miles."

—G. E. FOGGE

If all of this sugar cane were converted into sugar cubes (0.5 inch on a side) and stacked end to end, the sugar cubes would extend 1.6×10^{10} miles, or to the planet Pluto.

Photosynthetic catastrophe—

If photosynthesis were to cease, all higher forms of life would be extinct in about 25 years. A milder version of such a catastrophe ended the Cretaceous period 65.1 million years ago when a large asteroid struck the Yucatan Peninsula of Mexico. Enough dust was sent into the atmosphere that photosynthetic capacity was greatly diminished, which apparently led to the disappearance of the dinosaurs and allowed the mammals to rise to prominence.

As animals ourselves, we perhaps easily overlook the ultimate primacy of photosynthesis for our biosphere. Photosynthesis is the source of essentially all the carbon compounds and all the oxygen that makes aerobic metabolism possible. Moreover, as we shall see, there are considerable mechanistic and evolutionary parallels between the light reactions of photosynthesis and steps in oxidative phosphorylation.

19.0.1 Photosynthesis: An Overview

We can use our understanding of the citric acid cycle and oxidative phosphorylation to anticipate the processes required of photosynthesis. The citric acid cycle oxidizes carbon fuels to CO_2 to generate high-energy electrons, notably in the form of NADH. The flow of these high-energy electrons generates a proton-motive force through the action of the electron-transport chain. This proton-motive force is then transduced by ATP synthase to form ATP. A principal difference between oxidative phosphorylation and photosynthesis is the source of the high-energy electrons. The light reactions of photosynthesis use energy from *photons* to generate high-energy electrons (Figure 19.2). These electrons are used directly to reduce $NADP^+$ to NADPH and are used indirectly through an electron-transport chain to generate a proton-motive force across a membrane. A side product of these reactions is O_2. The proton-motive force drives ATP synthesis through the action of an ATP synthase, homologous to that in oxidative phosphorylation. In the dark reactions, the NADPH and ATP formed by the action of light drive the reduction of CO_2 to more-useful organic compounds.

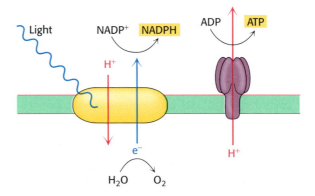

FIGURE 19.2 The light reactions of photosynthesis. Light is absorbed and the energy is used to drive electrons from water to generate NADPH and to drive protons across a membrane. These protons return through ATP synthase to make ATP.

19.1 PHOTOSYNTHESIS TAKES PLACE IN CHLOROPLASTS

In Chapter 18, we saw that oxidative phosphorylation, the predominant means of generating ATP from fuel molecules, was compartmentalized into mitochondria. Likewise, photosynthesis, the means of converting light into chemical energy, is sequestered into organelles called *chloroplasts*, typically 5 μm long. Like a mitochondrion, a chloroplast has an outer membrane and an inner membrane, with an intervening intermembrane space (Figure 19.3). The inner membrane surrounds a *stroma*, which is the site of the carbon chemistry of photosynthesis (Section 20.1). In the stroma are membranous structures called *thylakoids*, which are flattened sacs, or discs. The thylakoid sacs are stacked to form a *granum*. Different grana are linked by regions of thylakoid membrane called *stroma lamellae*. The thylakoid membranes separate the thylakoid space from the stroma space.

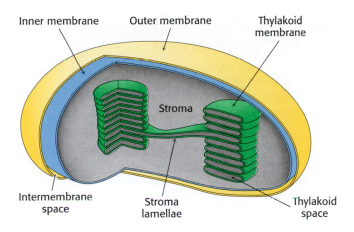

FIGURE 19.3 Diagram of a chloroplast.
[After S. L. Wolfe, *Biology of the Cell,* p. 130. ©
1972 by Wadsworth Publishing Company, Inc.
Adapted by permission of the publisher.]

Thus, chloroplasts have three different membranes (*outer, inner,* and *thylakoid membranes*) and three separate spaces (*intermembrane, stroma,* and *thylakoid spaces*). In developing chloroplasts, thylakoids are believed to arise from invaginations of the inner membrane, and so they are analogous to the mitochondrial cristae. Like the mitochondrial cristae, they are the site of coupled oxidation–reduction reactions that generate the proton-motive force.

19.1.1 The Primary Events of Photosynthesis Take Place in Thylakoid Membranes

The thylakoid membranes contain the energy-transducing machinery: light-harvesting proteins, reaction centers, electron-transport chains, and ATP synthase. They have nearly equal amounts of lipids and proteins. The lipid composition is highly distinctive: about 40% of the total lipids are *galactolipids* and 4% are *sulfolipids,* whereas only 10% are phospholipids. The thylakoid membrane and the inner membrane, like the inner mitochondrial membrane, are impermeable to most molecules and ions. The outer membrane of a chloroplast, like that of a mitochondrion, is highly permeable to small molecules and ions. The stroma contains the soluble enzymes that utilize the NADPH and ATP synthesized by the thylakoids to convert CO_2 into sugar. Plant leaf cells contain between 1 and 100 chloroplasts, depending on the species, cell type, and growth conditions.

19.1.2 The Evolution of Chloroplasts

Chloroplasts contain their own DNA and the machinery for replicating and expressing it. However, chloroplasts are not autonomous: they also contain many proteins encoded by nuclear DNA. How did the intriguing relation between the cell and its chloroplasts develop? We now believe that, in a manner analogous to the evolution of mitochondria (Section 18.1.2), chloroplasts are the result of endosymbiotic events in which a photosynthetic microorganism, most likely an ancestor of a cyanobacterium (Figure 19.4), was engulfed by a eukaryotic host. Evidence suggests that chloroplasts in higher plants and green algae are derived from a single endosymbiotic event, whereas those in red and brown algae arose from at least one additional event.

The chloroplast genome is smaller than that of a cyanobacterium; however, like that of a cyanobacterium, it is circular with a single start site for DNA replication, and its genes are arranged in operons—sequences of functionally related genes under common control (Chapter 31). In the course of evolution, many of the genes of the chloroplast ancestor were transferred to the plant cell's nucleus or, in some cases, lost entirely, thus establishing a fully dependent relation.

FIGURE 19.4 Cyanobacteria. A colony of the photosynthetic filamentous cyanobacteria *Anabaena* shown at 450× magnification. Ancestors of these bacteria are thought to have evolved into present-day chloroplasts. [James W. Richardson/Visuals Unlimited.]

19.2 LIGHT ABSORPTION BY CHLOROPHYLL INDUCES ELECTRON TRANSFER

The trapping of light energy is the key to photosynthesis. The first event is the absorption of light by a photoreceptor molecule. The principal photoreceptor in the chloroplasts of most green plants is *chlorophyll* a, a substituted tetrapyrrole (Figure 19.5). The four nitrogen atoms of the pyrroles are coordinated to a magnesium ion. Unlike a porphyrin such as heme, chlorophyll has a reduced pyrrole ring. Another distinctive feature of chlorophyll is the presence of *phytol*, a highly hydrophobic 20-carbon alcohol, esterified to an acid side chain.

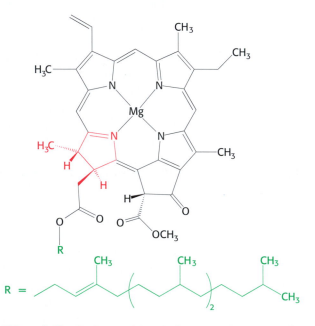

FIGURE 19.5 Chlorophyll. Like heme, chlorophyll *a* is a cyclic tetrapyrrole. One of the pyrrole rings (shown in red) is reduced. A phytol chain (shown in green) is connected by an ester linkage. Magnesium ion binds at the center of the structure.

Chlorophylls are very effective photoreceptors because they contain networks of alternating single and double bonds. Such compounds are called *polyenes*. They have very strong absorption bands in the visible region of the spectrum, where the solar output reaching Earth also is maximal (Figure 19.6). The peak molar absorption coefficient (ϵ, Section 3.1) of chlorophyll *a* is higher than 10^5 M^{-1} cm^{-1}, among the highest observed for organic compounds.

What happens when light is absorbed by a molecule such as chlorophyll? The energy from the light excites an electron from its ground energy level to an excited energy level (Figure 19.7). This high-energy electron can have several fates. For most compounds that absorb light, the electron simply returns to the ground state and the absorbed energy is converted into heat. However, if a suitable electron acceptor is nearby, the excited electron can move from the initial molecule to the acceptor (Figure 19.8). This process results in the formation of a positive charge on the initial molecule (due to the loss of an electron) and a negative charge on the acceptor and is, hence, referred to as *photoinduced charge separation*. The site where the separational change occurs is called the *reaction center*. We shall see how the photosynthetic apparatus is arranged to make photoinduced charge separation extremely efficient. The electron, extracted from its initial site by absorption of light, can reduce other species to store the light energy in chemical forms.

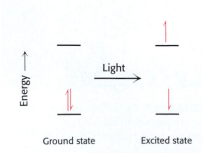

FIGURE 19.6 Light absorption by chlorophyll *a*. Chlorophyll *a* absorbs visible light efficiently as judged by the extinction coefficients near 10^5 M^{-1} cm^{-1}.

FIGURE 19.7 Light absorption. The absorption of light leads to the excitation of an electron from its ground state to a higher energy level.

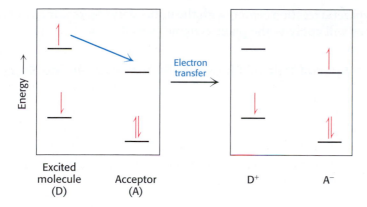

FIGURE 19.8 Photoinduced charge separation. If a suitable electron acceptor is nearby, an electron that has been moved to a high energy level by light absorption can move from the excited molecule to the acceptor.

19.2.1 Photosynthetic Bacteria and the Photosynthetic Reaction Centers of Green Plants Have a Common Core

Photosynthesis in green plants is mediated by two kinds of membrane-bound, light-sensitive complexes—*photosystem I (PS I)* and *photosystem II (PS II)*. Photosystem I typically includes 13 polypeptide chains, more than 60 chlorophyll molecules, a quinone (vitamin K_1), and three 4Fe-4S clusters. The total molecular mass is more than 800 kd. Photosystem II is only slightly less complex with at least 10 polypeptide chains, more than 30 chlorophyll molecules, a nonheme iron ion, and four manganese ions. Photosynthetic bacteria such as *Rhodopseudomonas viridis* contain a simpler, single type of photosynthetic reaction center, the structure of which was revealed at atomic resolution. The bacterial reaction center consists of four polypeptides: L (31 kd), M (36 kd), and H (28 kd) subunits and C, a *c*-type cytochrome (Figure 19.9). *The results of sequence comparisons and low-resolution structural studies of photosystems I and II revealed that the bacterial reaction center is homologous to the more complex plant systems.* Thus, we begin our consideration of the mechanisms of the light reactions within the bacterial

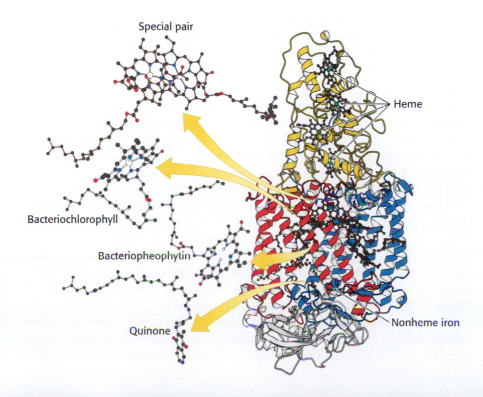

FIGURE 19.9 Bacterial photosynthetic reaction center. The core of the reaction center from *Rhodopseudomonas viridis* consists of two similar chains: L (red) and M (blue). An H chain (white) and a cytochrome subunit (yellow) complete the structure. A chain of prosthetic groups, beginning with a special pair of bacteriochlorophylls and ending at a bound quinone, runs through the structure from top to bottom in this view.

**Bacteriochlorophyll *b*
(BChl-*b*)**

**Bacteriopheophytin
(BPh)**

photosynthetic reaction center, with the understanding that many of our observations will apply to the plant systems as well.

19.2.2 A Special Pair of Chlorophylls Initiates Charge Separation

The L and M subunits form the structural and functional core of the bacterial photosynthetic reaction center (see Figure 19.9). Each of these homologous subunits contains five transmembrane helices. The H subunit, which has only one transmembrane helix, lies on the cytoplasmic side of the membrane. The cytochrome subunit, which contains four *c*-type hemes, lies on the opposite periplasmic side. Four bacteriochlorophyll *b* (BChl-*b*) molecules, two bacteriopheophytin *b* (BPh) molecules, two quinones (Q_A and Q_B), and a ferrous ion are associated with the L and M subunits.

Bacteriochlorophylls are similar to chlorophylls, except for the reduction of an additional pyrrole ring and some other minor differences that shift their absorption maxima to the near infrared, to wavelengths as long as 1000 nm. *Bacteriopheophytin* is the term for a bacteriochlorophyll that has two protons instead of a magnesium ion at its center.

The reaction begins with light absorption by a dimer of BChl-*b* molecules that lie near the periplasmic side of the membrane. This dimer, called a *special pair* because of its fundamental role in photosynthesis, absorbs light maximally at 960 nm, in the infrared near the edge of the visible region. For this reason, the special pair is often referred to as *P960* (P stands for pigment). Excitation of the special pair leads to the ejection of an electron, which is transferred through another molecule of BChl-*b* to the bacteriopheophytin in the L subunit (Figure 19.10, steps 1 and 2). This initial charge separation, which yields a positive charge on the special pair ($P960^+$) and a negative charge on BPh, occurs in less than 10 picoseconds (10^{-11} seconds). Interestingly, only one of the two possible paths within the nearly symmetric L-M dimer is utilized. In their high-energy states, $P960^+$ and BPh^- could undergo charge recombination; that is, the electron on BPh^- could move back to neutralize the positive charge on the special pair. Its return to the special pair would waste a valuable high-energy electron and simply convert the absorbed light energy into heat. Three factors in the structure of the reaction center work together to suppress charge recombination nearly completely (Figure 19.10, steps 3 and 4). First, another electron acceptor, a tightly bound quinone (Q_A), is less than 10 Å away from BPh^-, and so the electron is rapidly transferred farther away from the special pair. Recall that electron-transfer rates depend strongly on distance (Section 18.2.3). Second, one of the hemes of the cytochrome subunit is less than 10 Å away from the special pair, and so the positive charge is neutralized by the transfer of an electron from the reduced cytochrome. Finally, the electron transfer from BPh^- to the positively charged special pair is especially slow: the transfer is so thermodynamically favorable that it takes place in the inverted region where electron-transfer rates become slower (Section 18.2.3). Thus, electron transfer proceeds efficiently from BPh^- to Q_A.

From Q_A, the electron moves to a more loosely associated quinone, Q_B. The absorption of a second photon and the movement of a second electron down the path from the special pair completes the two-electron reduction of Q_B from Q to QH_2. Because the Q_B-binding site lies near the cytoplasmic side of the membrane, two protons are taken up from the cytoplasm, contributing to the development of a proton gradient across the cell membrane (Figure 19.10, steps 5, 6, and 7).

How does the cytochrome subunit of the reaction center regain an electron to complete the cycle? The reduced quinone (QH_2) is reoxidized to Q by complex III of the respiratory electron-transport chain (Section 18.3.3). The

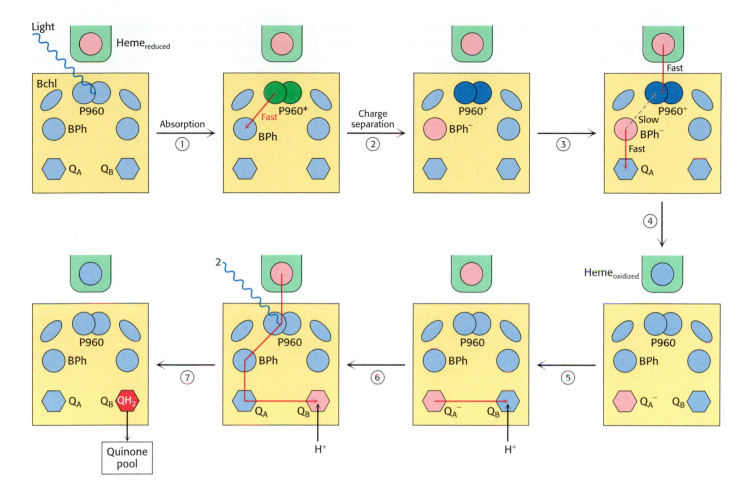

FIGURE 19.10 Electron chain in the photosynthetic bacterial reaction center. The absorption of light by the special pair (P960) results in the rapid transfer of an electron from this site to a bacteriopheophytin (BPh), creating a photoinduced charge separation (steps 1 and 2). (The asterisk on P960 stands for excited state.) The possible return of the electron from the pheophytin to the oxidized special pair is suppressed by the "hole" in the special pair being refilled with an electron from the cytochrome subunit and the electron from the pheophytin being transferred to a quinone (Q_A) that is farther away from the special pair (steps 3 and 4). The reduction of a quinone (Q_B) on the periplasmic side of the membrane results in the uptake of two protons from the periplasmic space (steps 5 and 6). The reduced quinone can move into the quinone pool in the membrane (step 7).

electrons from the reduced quinone are transferred through a soluble cytochrome c intermediate, called cytochrome c_2, in the periplasm to the cytochrome subunit of the reaction center. The flow of electrons is thus cyclic. The proton gradient generated in the course of this cycle drives the generation of ATP through the action of ATP synthase.

19.3 TWO PHOTOSYSTEMS GENERATE A PROTON GRADIENT AND NADPH IN OXYGENIC PHOTOSYNTHESIS

Photosynthesis by oxygen-evolving organisms depends on the interplay of two photosystems, linked by common intermediates (Figure 19.11). These two systems were discovered because of slight differences in the wave-

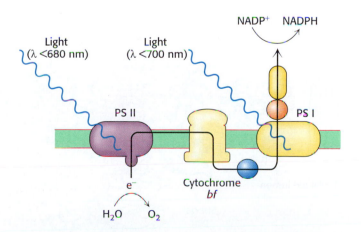

FIGURE 19.11 Two photosystems. The absorption of photons by two distinct photosytems (PS I and PS II) is required for complete electron flow from water to $NADP^+$.

lengths of light to which they respond. Photosystem I responds to light with wavelengths shorter than 700 nm, whereas photosystem II responds to wavelengths shorter than 680 nm. Under normal conditions, electrons flow first through photosystem II, then through cytochrome *bf*, a membrane-bound complex homologous to Q–cytochrome *c* oxidoreductase from oxidative phosphorylation (Section 18.3.3), and then through photosystem I. The electrons are derived from water: two molecules of H_2O are oxidized to generate a molecule of O_2 for every four electrons sent through this electron-transport chain. The electrons end up reducing $NADP^+$ to NADPH, a versatile reagent for driving biosynthetic processes. These processes generate a proton gradient across the thylakoid membrane that drives the formation of ATP.

19.3.1 Photosystem II Transfers Electrons from Water to Plastoquinone and Generates a Proton Gradient

Photosystem II of green plants is reasonably similar to the bacterial reaction center (Figure 19.12). The core of photosystem II is formed by D1 and D2, a pair of similar 32-kd subunits that span the thylakoid membrane. These subunits are homologous to the L and M chains of the bacterial reaction center. Unlike the bacterial system, photosystem II contains a large number of additional subunits that bind additional chlorophylls and increase the efficiency with which light energy is absorbed and transferred to the reaction center (Section 19.5).

The overall reaction catalyzed by photosystem II is:

$$2\,Q + 2\,H_2O \xrightarrow{\text{Light}} O_2 + 2\,QH_2$$

in which Q represents plastoquinone and QH_2 represents plastoquinol. This reaction is similar to one catalyzed by the bacterial system in that a quinone is converted from its oxidized into its reduced form. However, instead of obtaining the electrons for this reduction from a reduced cytochrome *c* molecule, photosystem II extracts the electrons from water, generating molecular oxygen. This remarkable reaction takes place at a special center containing four manganese ions.

Plastoquinone
(oxidized form, Q)

$(n = 6 \text{ to } 10)$

Plastoquinol
(reduced form, QH_2)

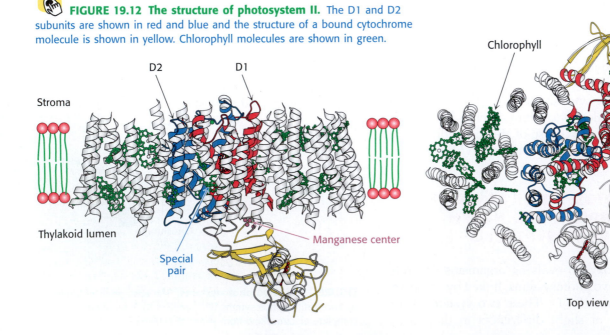

FIGURE 19.12 The structure of photosystem II. The D1 and D2 subunits are shown in red and blue and the structure of a bound cytochrome molecule is shown in yellow. Chlorophyll molecules are shown in green.

Stroma

D2   D1

Thylakoid lumen

Special pair

Manganese center

Chlorophyll

Top view

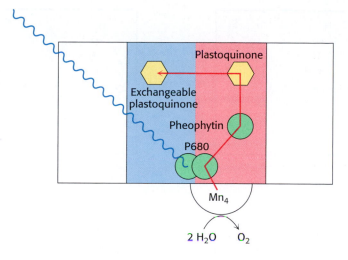

FIGURE 19.13 Electron flow through photosystem II. Light absorption induces electron transfer from P680 down an electron-transfer pathway to an exchangeable plastoquinone. The positive charge on P680 is neutralized by electron flow from water molecules bound at the manganese center.

The photochemistry of photosystem II begins with excitation of a special pair of chlorophyll molecules that are bound by the D1 and D2 subunits (Figure 19.13). This pair of molecules is analogous to the special pair in the bacterial reaction center, but it absorbs light at shorter wavelengths (maximum absorbance at 680 nm) because it consists of chlorophyll *a* molecules rather than bacteriochlorophyll. The special pair is often called *P680*. On excitation, P680 rapidly transfers an electron to a nearby pheophytin (chlorophyll with two H^+ ions in place of the central Mg^{2+} ion). From there, the electron is transferred first to a tightly bound plastoquinone at site Q_A and then to an exchangeable plastoquinone at site Q_B. This electron flow is entirely analogous to that in the bacterial system. With the arrival of a second electron and the uptake of two protons, the exchangeable plastoquinone is reduced to QH_2.

The major difference between the bacterial system and photosystem II is the source of the electrons that are used to neutralize the positive charge formed on the special pair. *$P680^+$, a very strong oxidant, extracts electrons from water molecules bound at the manganese center.* The structure of this center, which includes four manganese ions, a calcium ion, a chloride ion, and a tyrosine residue that forms a radical, has not been fully established, although the results of extensive spectroscopic studies and a recent X-ray crystallographic study at moderate resolution have provided many constraints. Manganese was apparently evolutionarily selected for this role because of its ability to exist in multiple oxidation states (Mn^{2+}, Mn^{3+}, Mn^{4+}, Mn^{5+}) and to form strong bonds with oxygen-containing species. The manganese center, in its reduced form, oxidizes two molecules of water to form a single molecule of oxygen. Each time the absorbance of a photon kicks an electron out of P680, the positively charged special pair extracts an electron from the manganese center (Figure 19.14). Thus four photochemical steps are

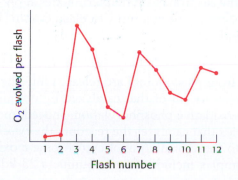

FIGURE 19.14 Four photons are required to generate one oxygen molecule. When dark-adapted chloroplasts are exposed to a brief flash of light, one electron passes through photosystem II. Monitoring the O_2 released after each flash reveals that four flashes are required to generate each O_2 molecule. The peaks in O_2 release occur after the 3rd, 7th, and 11th flashes because the dark-adapted chloroplasts start in the S_1 state—that is, the one-electron reduced state.

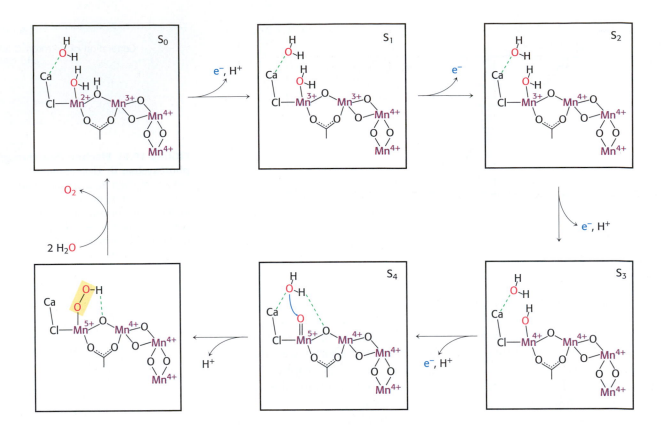

FIGURE 19.15 A plausible scheme for oxygen evolution from the manganese center. A possible partial structure for the manganese center is shown. The center is oxidized, one electron at a time, until two bound H_2O molecules are linked to form a molecule of O_2, which is then released from the center. A tyrosine residue (not shown) also participates in the coupled proton–electron transfer steps. The structures are designated S_0 through S_4 to indicate the number of electrons that have been removed.

required to extract the electrons and reduce the manganese center (Figure 19.15). The four electrons harvested from water are used to reduce two molecules of Q to QH_2.

Photosystem II spans the thylakoid membrane such that the site of quinone reduction is on the side of the stroma, whereas the manganese center and, hence, the site of water oxidation lies in the thylakoid lumen. Thus, the two protons that are taken up with the reduction of each molecule of plastoquinone come from the stroma, and the four protons that are liberated in the course of water oxidation are released into the lumen. This distribution of protons generates a proton gradient across the thylakoid membrane characterized by an excess of protons in the thylakoid lumen compared with the stroma (Figure 19.16). Thus, the direction of the proton gradient is the reverse of that generated during oxidative phosphorylation, which depletes the mitochondrial matrix of protons. As we shall see, this difference is consistent with the reversed orientations of other membrane proteins, including ATP synthase.

19.3.2 Cytochrome *bf* Links Photosystem II to Photosystem I

The plastoquinol (QH_2) produced by photosystem II contributes its electrons to continue the electron chain that terminates at photosystem I. These electrons are transferred, one at a time, to plastocyanin (Pc), a copper protein in the thylakoid lumen.

$$QH_2 + 2\,Pc(Cu^{2+}) \longrightarrow Q + 2\,Pc(Cu^+) + 2\,H^+_{thylakoid\ lumen}$$

The two protons from plastoquinol are released into the thylakoid lumen. This reaction is reminiscent of that catalyzed by ubiquinol cytochrome *c* oxidoreductase in oxidative phosphorylation. Indeed, most components of the enzyme complex that catalyzes the reaction, the *cytochrome* bf complex, are homologous to those of ubiquinol cytochrome *c* oxidoreductase. The cytochrome *bf* complex includes four subunits: a 23-kd cytochrome with

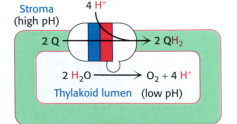

FIGURE 19.16 Proton-gradient direction. Photosystem II releases protons into the thylakoid lumen and takes them up from the stroma. The result is a pH gradient across the thylakoid membrane with an excess of protons (low pH) inside.

two b-type hemes, a 20-kd Rieske-type Fe-S protein, a 33-kd cytochrome f with a c-type cytochrome, and a 17-kd chain.

This complex catalyzes the reaction through the Q cycle (Section 18.3.4). In the first half of the Q cycle, plastoquinol is oxidized to plastoquinone, one electron at a time. The electrons from plastoquinol flow through the Fe-S protein to convert oxidized plastocyanin into its reduced form. Plastocyanin is a small, soluble protein with a single copper ion bound by a cysteine residue, two histidine residues, and a methionine residue in a distorted tetrahedral arrangement (Figure 19.17). This geometry facilitates the interconversion between the Cu^{2+} and the Cu^{+} states and sets the reduction potential at an appropriate value relative to that of plastoquinol. Plastocyanin is intensely blue in color in its oxidized form, marking it as a member of the "blue copper protein," or type I copper protein family.

The oxidation of plastoquinol results in the release of two protons into the thylakoid lumen. In the second half of the Q cycle (Section 18.3.4), cytochrome bf reduces a second molecule of plastoquinone from the Q pool to plastoquinol, taking up two protons from one side of the membrane, and then reoxidizes plastoquinol to release these protons on the other side. The enzyme is oriented so that protons are released into the thylakoid lumen and taken up from the stroma, contributing further to the proton gradient across the thylakoid membrane (Figure 19.18).

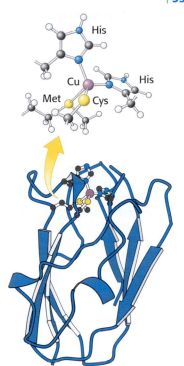

FIGURE 19.17 Structure of plastocyanin. Two histadine residues, a cysteine residue, and a methionine residue coordinate a copper ion in a distorted tetrahedral manner in this protein, which carries electrons from cytochrome bf complex to photosystem I.

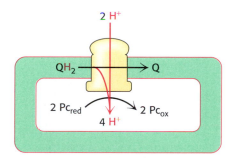

FIGURE 19.18 Cytochrome bf contribution to proton gradient. The cytochrome bf complex oxidizes QH₂ to Q through the Q cycle. Four protons are released into the thylakoid lumen in each cycle.

19.3.3 Photosystem I Uses Light Energy to Generate Reduced Ferredoxin, a Powerful Reductant

The final stage of the light reactions is catalyzed by photosystem I (Figure 19.19). The core of this system is a pair of similar subunits psaA (83 kd) and psaB (82 kd). These subunits are quite a bit larger than the core subunits

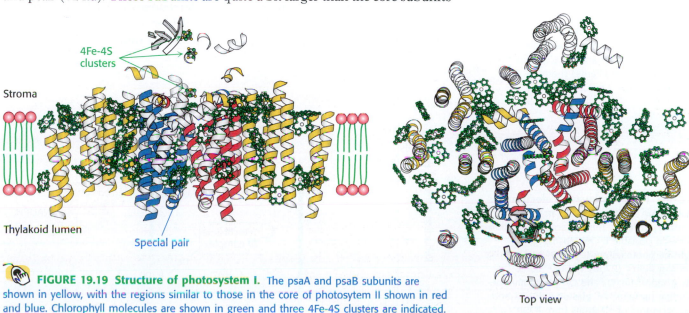

FIGURE 19.19 Structure of photosystem I. The psaA and psaB subunits are shown in yellow, with the regions similar to those in the core of photosytem II shown in red and blue. Chlorophyll molecules are shown in green and three 4Fe-4S clusters are indicated.

Top view

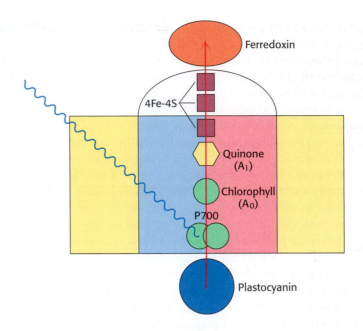

FIGURE 19.20 Electron flow through photosystem I to ferredoxin. Light absorption induces electron transfer from P700 down an electron-transfer pathway that includes a chlorophyll molecule, a quinone molecule, and three 4Fe-4S clusters to reach ferredoxin. The positive charge left on P700 is neutralized by electron transfer from reduced plastocyanin.

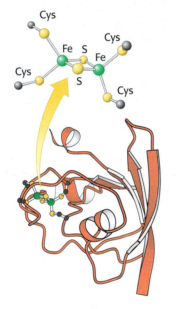

FIGURE 19.21 Structure of ferredoxin. In plants, ferredoxin contains a 2Fe-2S cluster. This protein accepts electrons from photosystem I and carries them to ferredoxin-NADP$^+$ reductase.

of photosystem II and the bacterial reaction center. Nonetheless, they appear to be homologous; the terminal 40% of each subunit is similar to a corresponding subunit of photosystem II. A special pair of chlorophyll *a* molecules lies at the center of the structure and absorb light maximally at 700 nm. This center, *P700*, initiates photoinduced charge separation (Figure 19.20). The electron is transferred down a pathway through chlorophyll at site A_0 and quinone at site A_1 to a set of 4Fe-4S clusters. From there, the electron is transferred to ferredoxin (Fd), a soluble protein containing a 2Fe-2S cluster coordinated to four cysteine residues (Figure 19.21). The positive charge of P700$^+$ is neutralized by the transfer of an electron from reduced plastocyanin. Thus, the overall reaction catalyzed by photosystem I is a simple one-electron oxidation–reduction reaction.

$$Pc(Cu^+) + Fd_{ox} \xrightarrow{\text{Light}} Pc(Cu^{2+}) + Fd_{red}$$

Given that the reduction potentials for plastocyanin and ferredoxin are $+0.37$ V and -0.45 V, respectively, the standard free energy for the oxidation of reduced plastocyanin by oxidized ferredoxin is $+18.9$ kcal mol^{-1} ($+79.1$ kJ mol^{-1}). This uphill reaction is driven by the absorption of a 700-nm photon which has an energy of 40.9 kcal mol^{-1} (171 kJ mol^{-1}).

FIGURE 19.22 Pathway of electron flow from H$_2$O to NADP$^+$ in photosynthesis. This endergonic reaction is made possible by the absorption of light by photosystem II (P680) and photosystem I (P700). Abbreviations: Ph, pheophytin; Q_A and Q_B, plastoquinone-binding proteins; Pc, plastocyanin; A_0 and A_1, acceptors of electrons from P700*; Fd, ferredoxin; Mn, manganese.

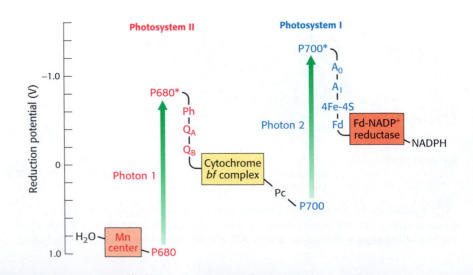

The cooperation between photosystem I and photosystem II creates electron flow from H_2O to $NADP^+$. The pathway of electron flow is called the *Z scheme of photosynthesis* because the redox diagram from P680 to P700* looks like the letter Z (Figure 9.22).

19.3.4 Ferredoxin-NADP$^+$ Reductase Converts NADP$^+$ into NADPH

Although reduced ferredoxin is a strong reductant, it is not useful for driving many reactions, in part because ferredoxin carries only one available electron. In contrast, NADPH, a two-electron reductant, is widely used in biosynthetic processes, including the reactions of the Calvin cycle (Chapter 20). How can reduced ferredoxin be used to drive the reduction of $NADP^+$ to NADPH? This reaction is catalyzed by *ferredoxin-NADP$^+$ reductase*, a flavoprotein (Figure 19.23A). The bound FAD moiety in this enzyme accepts electrons, one at a time, from two molecules of reduced ferredoxin as it proceeds from its oxidized form, through a semiquinone intermediate, to its fully reduced form (Figure 19.23B). The enzyme then transfers a hydride ion to $NADP^+$ to form NADPH. Note that this reaction takes place on the stromal side of the membrane. Hence, the uptake of a proton in the reduction of $NADP^+$ further contributes to the proton gradient across the thylakoid membrane.

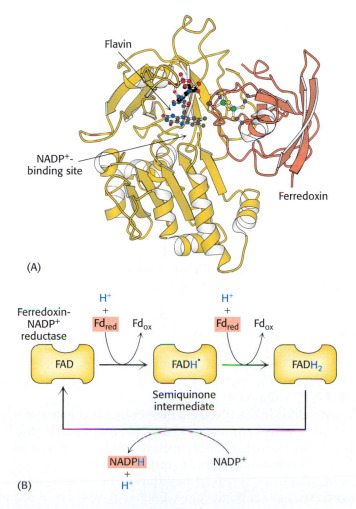

FIGURE 19.23 Structure and function of ferredoxin-NADP$^+$ reductase. (A) Structure of ferredoxin-NADP$^+$ reductase. This enzyme accepts electrons, one at a time, from ferredoxin (shown in orange). (B) Ferredoxin-NADP$^+$ reductase first accepts one electron from reduced ferredoxin to form a flavin semiquinone intermediate. The enzyme then accepts a second electron to form FADH$_2$, which then transfers two electrons and a proton to NADP$^+$ to form NADPH.

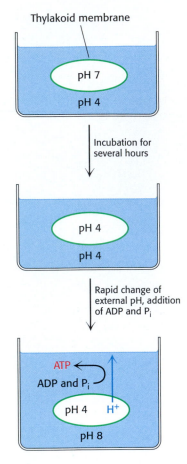

Thylakoid membrane

FIGURE 19.24 Jagendorg's demonstration. Chloroplasts synthesize ATP after the imposition of a pH gradient.

19.4 A PROTON GRADIENT ACROSS THE THYLAKOID MEMBRANE DRIVES ATP SYNTHESIS

In 1966, André Jagendorf showed that chloroplasts synthesize ATP in the dark when an artificial pH gradient is imposed across the thylakoid membrane. To create this transient pH gradient, he soaked chloroplasts in a pH 4 buffer for several hours and then rapidly mixed them with a pH 8 buffer containing ADP and P_i. The pH of the stroma suddenly increased to 8, whereas the pH of the thylakoid space remained at 4. *A burst of ATP synthesis then accompanied the disappearance of the pH gradient across the thylakoid membrane* (Figure 19.24). This incisive experiment was one of the first to unequivocally support the hypothesis put forth by Peter Mitchell that ATP synthesis is driven by proton-motive force.

The principles by which ATP synthesis takes place in chloroplasts are nearly identical with those for oxidative phosphorylation. We have seen how light induces electron transfer through photosystems II and I and the cytochrome *bf* complex. At various stages in this process, protons are released into the thylakoid lumen or taken up from the stroma, generating a proton gradient. Such a gradient can be maintained because the thylakoid membrane is essentially impermeable to protons. *The thylakoid space becomes markedly acidic, with the pH approaching 4. The light-induced transmembrane proton gradient is about 3.5 pH units.* As discussed in Section 18.4, energy inherent in the proton gradient, called the *proton-motive force (Δp)*, is described as the sum of two components: a charge gradient and a chemical gradient. In chloroplasts, nearly all of Δp arises from the pH gradient, whereas, in mitochondria, the contribution from the membrane potential is larger. The reason for this difference is that the thylakoid membrane is quite permeable to Cl^- and Mg^{2+}. The light-induced transfer of H^+ into the thylakoid space is accompanied by the transfer of either Cl^- in the same direction or Mg^{2+} (1 Mg^{2+} per 2 H^+) in the opposite direction. Consequently, electrical neutrality is maintained and no membrane potential is generated. A pH gradient of 3.5 units across the thylakoid membrane corresponds to a proton-motive force of 0.20 V or a ΔG of -4.8 kcal mol^{-1} (-20.0 kJ mol^{-1}).

19.4.1 The ATP Synthase of Chloroplasts Closely Resembles Those of Mitochondria and Prokaryotes

The proton-motive force generated by the light reactions is converted into ATP by the *ATP synthase* of chloroplasts, also called the *CF_1-CF_0 complex* (*C* stands for chloroplast and *F* for factor). CF_1-CF_0 ATP synthase closely resembles the F_1-F_0 complex of mitochondria (Section 18.4.1). CF_0 conducts protons across the thylakoid membrane, whereas CF_1 catalyzes the formation of ATP from ADP and P_i.

CF_0 is embedded in the thylakoid membrane. It consists of four different polypeptide chains known as I (17 kd), II (16.5 kd), III (8 kd), and IV (27 kd) having an estimated stoichiometry of 1:2:12:1. Subunits I, II, and III correspond to subunits **a, b,** and **c,** respectively, of the mitochondrial F_0 subunit, and subunit IV is similar in sequence to subunit **a.** CF_1, the site of ATP synthesis, has a subunit composition $\alpha_3\beta_3\gamma\delta\epsilon$. The β subunits contain the catalytic sites, similar to the F_1 subunit of mitochondrial ATP synthase. Remarkably, β subunits of corn chloroplast ATP synthase are more than 60% identical in amino acid sequence with those of human ATP synthase, despite the passage of approximately 1 billion years since the separation of the plant and animal kingdoms.

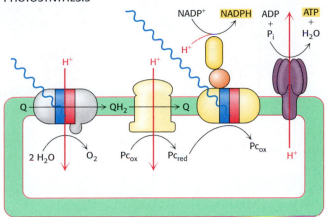

OXIDATIVE PHOSPHORYLATION

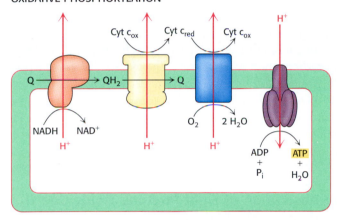

FIGURE 19.25 Comparison of photosynthesis and oxidative phosphorylation. The light-induced electron transfer in photosynthesis drives protons into the thylakoid lumen. The excess protons flow out of the lumen through ATP synthase to generate ATP in the stroma. In oxidative phosphorylation, electron flow down the electron-transport chain pumps protons out of the mitochondrial matrix. Excess protons from the intermembrane space flow into the matrix through ATP synthase to generate ATP in the matrix.

Significantly, the membrane orientation of CF_1-CF_0 is reversed compared with that of the mitochondrial ATP synthase (Figure 19.25). Thus, protons flow *out* of the thylakoid lumen through ATP synthase into the stroma. Because CF_1 is on the stromal surface of the thylakoid membrane, the newly synthesized ATP is released directly into the stromal space. Recall that NADPH formed through the action of photosystem I and ferredoxin-$NADP^+$ reductase also is released into the stromal space. Thus, *ATP and NADPH, the products of the light reactions of photosynthesis, are appropriately positioned for the subsequent dark reactions, in which CO_2 is converted into carbohydrate.*

19.4.2 Cyclic Electron Flow Through Photosystem I Leads to the Production of ATP Instead of NADPH

An alternative pathway for electrons arising from P700, the reaction center of photosystem I, contributes to the versatility of photosynthesis. The electron in reduced ferredoxin can be transferred to the cytochrome *bf* complex rather than to $NADP^+$. This electron then flows back through the cytochrome *bf* complex to reduce plastocyanin, which can then be reoxidized

FIGURE 19.26 Cyclic photo-phosphorylation. In this pathway, electrons from reduced ferredoxin are transferred to the cytochrome *bf* complex rather than to ferredoxin-NADP$^+$ reductase. The flow of electrons through cytochrome *bf* pumps protons into the thylakoid lumen. These protons flow through ATP synthase to generate ATP. Neither NADPH nor O$_2$ is generated by this pathway.

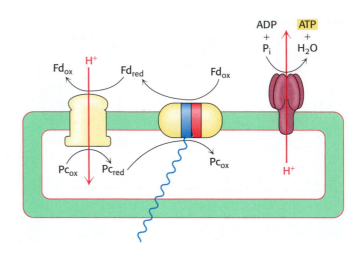

by P700$^+$ to complete a cycle. The net outcome of this cyclic flow of electrons is the pumping of protons by the cytochrome *bf* complex. The resulting proton gradient then drives the synthesis of ATP. In this process, called *cyclic photophosphorylation, ATP is generated without the concomitant formation of NADPH* (Figure 19.26). Photosystem II does not participate in cyclic photophosphorylation, and so O$_2$ is not formed from H$_2$O. Cyclic photophosphorylation takes place when NADP$^+$ is unavailable to accept electrons from reduced ferredoxin, because of a very high ratio of NADPH to NADP$^+$.

19.4.3 The Absorption of Eight Photons Yields One O$_2$, Two NADPH, and Three ATP Molecules

We can now estimate the overall stoichiometry for the light reactions. The absorption of 4 photons by photosystem II generates 1 molecule of O$_2$ and releases 4 protons into the thylakoid lumen. The 2 molecules of plastoquinol are oxidized by the Q cycle of the cytochrome *bf* complex to release 8 protons into the lumen. Finally, the electrons from 4 molecules of reduced plastocyanin are driven to ferredoxin by the absorption of 4 additional photons. The 4 molecules of reduced ferredoxin generate 2 molecules of NADPH. Thus, the overall reaction is:

$$2\,H_2O + 2\,NADP^+ + 10\,H^+_{stroma} \longrightarrow O_2 + 2\,NADPH + 12\,H^+_{lumen}$$

The 12 protons released in the lumen can then flow through ATP synthase. Given the apparent stoichiometry of 12 subunit III components in CF$_0$, we expect that 12 protons must pass through CF$_0$ to complete one full rotation of CF$_1$ and, hence, generate and release 3 molecules of ATP. Given this ratio, the overall reaction is

$$2\,H_2O + 2\,NADP^+ + 10\,H^+_{stroma} \longrightarrow O_2 + 2\,NADPH + 12\,H^+_{lumen}$$
$$\underline{3\,ADP^{3-} + 3\,P_i^{2-} + 3\,H^+ + 12\,H^+_{lumen} \longrightarrow 3\,ATP^{4-} + 3\,H_2O + 12\,H^+_{stroma}}$$
$$2\,NADP^+ + 3\,ADP^{3-} + 3\,P_i^{2-} + H^+ \longrightarrow O_2 + 2\,NADPH + 3\,ATP^{4-} + H_2O$$

Cyclic photophosphorylation is somewhat more productive with regard to ATP synthesis. The absorption of 4 photons by photosystem I results in the release of 8 protons into the lumen by the cytochrome *bf* system. These protons flow through ATP synthase to yield 2 molecules of ATP (assuming the same ratio of ATP molecules generated per proton). Thus, each 2 absorbed photons yield 1 molecule of ATP. No NADPH is produced.

19.5 ACCESSORY PIGMENTS FUNNEL ENERGY INTO REACTION CENTERS

A light-harvesting system that relied only on the chlorophyll *a* molecules of the special pair would be rather inefficient for two reasons. First, chlorophyll *a* molecules absorb light only at specific wavelengths (see Figure 19.6). A large gap is present in the middle of the visible regions between approximately 450 and 650 nm. This gap corresponds to the peak of the solar spectrum, so failure to collect this light would constitute a considerable lost opportunity. Second, even in spectral regions where chlorophyll *a* absorbs light, many photons would pass through without being absorbed, owing to the relatively low density of chlorophyll *a* molecules in a reaction center. Accessory pigments, both additional chlorophylls as well as other classes of molecules, are closely associated with reaction centers. *These pigments absorb light and funnel the energy to the reaction center for conversion into chemical forms.* Indeed, experiments in 1932 by Robert Emerson and William Arnold on *Chlorella* cells (unicellular green algae) demonstrated that only 1 molecule of O_2 was produced for 2500 chlorophyll molecules excited.

19.5.1 Resonance Energy Transfer Allows Energy to Move from the Site of Initial Absorbance to the Reaction Center

How is energy funneled from an associated pigment to a reaction center? We have already seen how the absorption of a photon can lead to electron excitation and transfer (Section 19.2). Another reaction to photon absorption, not leading to electron transfer, is more common. Through electromagnetic interactions through space, the excitation energy can be transferred from one molecule to a nearby molecule (Figure 19.27). The rate of this process, called *resonance energy transfer*, depends strongly on the distance between the energy donor and the energy acceptor molecules; an increase in the distance between the donor and the acceptor by a factor of two typically results in a decrease in the energy-transfer rate by a factor of $2^6 = 64$. For reasons of conservation of energy, energy transfer must be from a donor in the excited state to an acceptor of equal or lower energy. *The excited state of the special pair of chlorophyll molecules is lower in energy than that for single chlorophyll molecules, allowing reaction centers to trap the energy transferred from other molecules* (Figure 19.28).

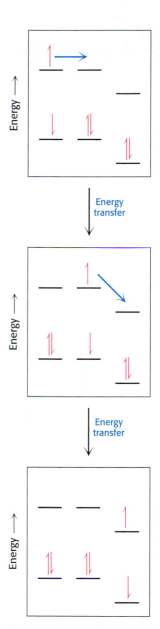

FIGURE 19.27 Resonance energy transfer. Energy, absorbed by one molecule, can be transferred to nearby molecules with excited states of equal or lower energy.

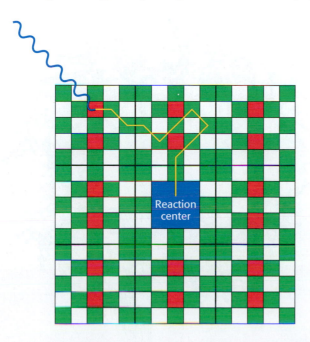

FIGURE 19.28 Energy transfer from accessory pigments to reaction centers. Light energy absorbed by accessory chlorophyll molecules or other pigments can be transferred to reaction centers, where it drives photoinduced charge separation. The green squares represent accessory chlorophyll molecules and the red squares carotenoid molecules; the white squares designate protein.

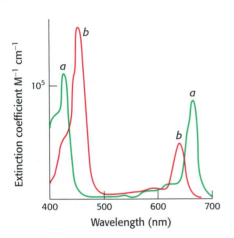

Chlorophyll b

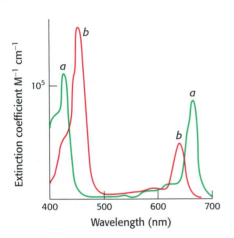

FIGURE 19.29 Absorption spectra of
chlorophylls *a* and *b*.

19.5.2 Light-Harvesting Complexes Contain Additional Chlorophylls and Carotinoids

Chlorophyll b and *carotenoids* are important light-harvesting molecules that funnel energy to the reaction center. Chlorophyll *b* differs from chlorophyll *a* in having a formyl group in place of a methyl group. This small difference moves its two major absorption peaks toward the center of the visible region. In particular, chlorophyll *b* efficiently absorbs light with wavelengths between 450 and 500 nm (Figure 19.29).

Carotenoids are extended polyenes that also absorb light between 400 and 500 nm.

Lycopene

β-Carotene

The carotenoids are responsible for most of the yellow and red colors of fruits and flowers, and they provide the brilliance of fall, when the chlorophyll molecules are degraded to reveal the carotenoids.

In addition to their role in transferring energy to reaction centers, the carotenoids serve a safeguarding function. Carotenoids suppress damaging photochemical reactions, particularly those including oxygen, that can be induced by bright sunlight. Indeed, plants lacking carotenoids are quickly killed on exposure to light and oxygen.

The accessory pigments are arranged in numerous *light-harvesting complexes* that completely surround the reaction center. The 26-kd subunit of light-harvesting complex II (LHC-II) is the most abundant membrane protein in chloroplasts. This subunit binds seven chlorophyll *a* molecules, six chlorophyll *b* molecules, and two carotenoid molecules. Similar light-harvesting assemblies exist in photosynthetic bacteria (Figure 19.30).

FIGURE 19.30 Structure of a light-harvesting complex. Eight polypeptides, each of which binds three chlorophyll molecules (green) and a carotenoid molecule (red), surround a central cavity that contains the reaction center.

19.5.3 Phycobilisomes Serve as Molecular Light Pipes in Cyanobacteria and Red Algae

Little blue or red light reaches algae living at a depth of a meter or more in seawater, because such light is absorbed by water and by chlorophyll molecules in organisms lying above. How can photosynthetic organisms survive under such conditions? Cyanobacteria (blue-green algae) and red algae contain large protein assemblies called *phycobilisomes* that enable them to harvest the green and yellow light that penetrates to their ecological niche. Phycobilisomes are bound to the outer face of the thylakoid membrane, where they serve as light-absorbing antennas to funnel excitation energy into the reaction centers of photosystem II. They absorb maximally in the 470- to 650-nm region, in the valley between the blue and far-red absorption peaks of chlorophyll *a*. Phycobilisomes are very large assemblies (several million daltons) of many *phycobiliprotein* subunits, each

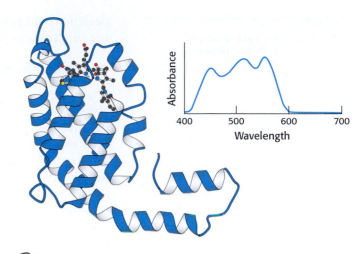

FIGURE 19.31 Structure of a phycobilisome subunit. This protein, a phycoerythrin, contains a phycoerythrobilin linked to a cysteine residue. The inset shows the absorption spectrum of a phycoerythrin.

containing many covalently attached *bilin* prosthetic groups, as well as linker polypeptides (Figure 19.31). Phycobilisomes contain hundreds of bilins. Phycocyanobilin and phycoerythrobilin are the two most common ones.

Peptide-linked phycocyanobilin

Peptide-linked phycoerythrobilin

The geometrical arrangement of phycobiliproteins in phycobilisomes (Figure 19.32), as well as their spectral properties, contributes to the efficiency of energy transfer, which is greater than 95%. Excitation energy absorbed by phycoerythrin subunits at the periphery of these antennas appears at the reaction center in less than 100 ps. Phycobilisomes are elegantly

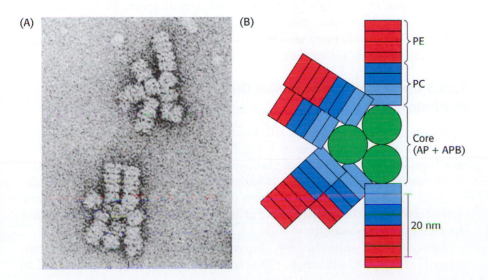

(A)

(B)

PE

PC

Core
(AP + APB)

20 nm

FIGURE 19.32 Structure of a phycobilisome. (A) Electron micrograph of phycobilisomes from a cyanobacterium (*Synechocystis*). (B) Schematic representation of a phycobilisome from the cyanobacterium *Synechocystis* 6701. Rods containing phycoerythrin (PE) and phycocyanin (PC) emerge from a core made of allophycocyanin (AP) and allophycocyanin B (APB). The core region binds to the thylakoid membrane. [(A) Courtesy of Dr. Robley Williams and Dr. Alexander Glazer; (B) after a drawing kindly provided by Dr. Alexander Glazer.]

designed light pipes that enable algae to occupy ecological niches that would not support organisms relying solely on chlorophyll for the trapping of light.

19.5.4 Components of Photosynthesis Are Highly Organized

The complexity of photosynthesis, seen already in the elaborate interplay of complex components, extends even to the placement of the components in the thylakoid membranes. *Thylakoid membranes of most plants are differentiated into stacked (appressed) and unstacked (nonappressed) regions* (see Figures 19.1 and 19.3). Stacking increases the amount of thylakoid membrane in a given chloroplast volume. Both regions surround a common internal thylakoid space, but only unstacked regions make direct contact with the chloroplast stroma. Stacked and unstacked regions differ in the nature of their photosynthetic assemblies (Figure 19.33). Photosystem I and ATP synthase are located almost exclusively in unstacked regions, whereas photosystem II is present mostly in stacked regions. The cytochrome *bf* complex is found in both regions. Indeed, this complex rapidly moves back and forth between the stacked and unstacked regions. Plastoquinone and plastocyanin are the mobile carriers of electrons between assemblies located in different regions of the thylakoid membrane. A common internal thylakoid space enables protons liberated by photosystem II in stacked membranes to be utilized by ATP synthase molecules that are located far away in unstacked membranes.

FIGURE 19.33 Location of photosynthesis components.
Photosynthetic assemblies are differentially distributed in the stacked (appressed) and unstacked (nonappressed) regions of thylakoid membranes. [After a drawing kindly provided by Dr. Jan M. Anderson and Dr. Bertil Andersson.]

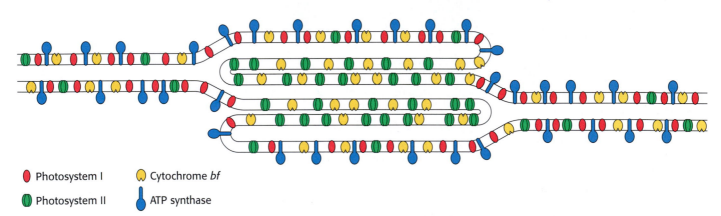

- 🔴 Photosystem I
- 🟢 Photosystem II
- 🟡 Cytochrome *bf*
- 🔵 ATP synthase

What is the functional significance of this lateral differentiation of the thylakoid membrane system? The positioning of photosystem I in the unstacked membranes also gives it direct access to the stroma for the reduction of $NADP^+$. ATP synthase, too, is located in the unstacked region to provide space for its large CF_1 globule and to give access to ADP. In contrast, the tight quarters of the appressed region pose no problem for photosystem II, which interacts with a small polar electron donor (H_2O) and a highly lipid soluble electron carrier (plastoquinone).

19.5.5 Many Herbicides Inhibit the Light Reactions of Photosynthesis

Many commercial herbicides kill weeds by interfering with the action of photosystem II or photosystem I. Inhibitors of photosystem II block electron flow, whereas inhibitors of photosystem I divert electrons from the terminal part of this photosystem. Photosystem II inhibitors include urea derivatives such as *diuron* and triazine derivatives such as *atrazine*. These chemicals bind to the Q_B site of the D1 subunit of photosystem II and block the formation of plastoquinol (QH_2).

Paraquat (1,1'-dimethyl-4-4'-bipyridinium) is an inhibitor of photosystem I. Paraquat, a dication, can accept electrons from photosystem I to

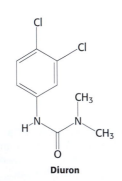

Diuron

Atrazine

become a radical. This radical reacts with O_2 to produce reactive oxygen species such as superoxide and hydroxyl radical ($OH\cdot$). Such reactive oxygen species react with double bonds in membrane lipids, damaging the membrane (Section 18.3.6).

19.6 THE ABILITY TO CONVERT LIGHT INTO CHEMICAL ENERGY IS ANCIENT

The ability to convert light energy into chemical energy is a significant evolutionary advantage. Indeed, photosynthesis arose early in the history of life on Earth, which began about 3.5 billion years ago. Geological evidence suggests that oxygenic photosynthesis became important approximately 2 billion years ago. Anoxygenic photosynthetic systems existed even earlier (Table 19.1). The photosynthetic system of the nonsulfur purple

TABLE 19.1 Major Groups of Photosynthetic Prokaryotes

Bacteria	Photosynthetic electron donor	O_2 use
Green sulfur	H_2, H_2S, S	Anoxygenic
Green nonsulfur	Variety of amino acids and organic acids	Anoxygenic
Purple sulfur	H_2, H_2S, S	Anoxygenic
Purple nonsulfur	Usually organic molecules	Anoxygenic
Cyanobacteria	H_2O	Oxygenic

bacterium *Rhodospeudomonas viridis* has most features common to oxygenic photosynthetic systems and clearly predates them. Green sulfur bacteria such as *Chlorobium thiosulfatophilum* carry out a reaction that also seems to have appeared before oxygenic photosynthesis and is even more similar to oxygenic photosynthesis than that of *R. viridis*. Reduced sulfur species such as H_2S are electron donors in the overall photosynthetic reaction:

$$CO_2 + 2\,H_2S \xrightarrow{\text{Light}} (CH_2O) + 2\,S + H_2O$$

Nonetheless, photosynthesis did not evolve immediately at the origin of life. The failure to discover photosynthesis in the domain of Archaea implies that photosynthesis evolved exclusively in the domain of Bacteria. Eukaryotes appropriated through endosymbiosis the basic photosynthetic units that were the products of bacterial evolution. All domains of life do have electron-transport chains in common, however. As we have seen, components such as the ubiquinone–cytochrome c oxidoreductase and cytochrome bf family are present in both respiratory and photosynthetic electron-transport chains. These components were the foundations on which light-energy-capturing systems evolved.

SUMMARY

Photosynthesis Takes Place in Chloroplasts

The proteins that participate in the light reactions of photosynthesis are located in the thylakoid membranes of chloroplasts. The light reactions result in (1) the creation of reducing power for the production of NADPH, (2) the generation of a transmembrane proton gradient for the formation of ATP, and (3) the production of O_2.

Light Absorption by Chlorophyll Induces Electron Transfer

Chlorophyll molecules—tetrapyrroles with a central magnesium ion—absorb light quite efficiently because they are polyenes. An electron excited to a high-energy state by the absorption of a photon can move to nearby electron acceptors. In photosynthesis, an excited electron leaves a pair of associated chlorophyll molecules known as the special pair. The functional core of photosynthesis, a reaction center, from a photosynthetic bacterium has been studied in great detail. In this system, the electron moves from the special pair (containing bacteriochlorophyll) to a bacteriopheophytin (a bacteriochlorophyll lacking the central magnesium ion) to quinones. The reduction of quinones leads to the generation of a proton gradient, which drives ATP synthesis in a manner analogous to that of oxidative phosphorylation.

Two Photosystems Generate a Proton Gradient and NADPH in Oxygenic Photosynthesis.

Photosynthesis in green plants is mediated by two linked photosystems. In photosystem II, excitation of P680, a special pair of chlorophyll molecules located at the interface of two similar subunits, leads to electron transfer to plastoquinone in a manner analogous to that for the bacterial reaction center. The electrons are replenished by the extraction of electrons from water at a center containing four manganese ions. One molecule of O_2 is generated at this center for each four electrons transferred. The plastoquinol produced at photosystem II is reoxidized by the cytochrome *bf* complex, which transfers the electrons to plastocyanin, a soluble copper protein. From plastocyanin, the electrons enter photosystem I. In photosytem I, excitation of the special pair P700 releases electrons that flow to ferredoxin, a powerful reductant. Ferredoxin-$NADP^+$ reductase, a flavoprotein located on the stromal side of the membrane, then catalyzes the formation of NADPH. A proton gradient is generated as electrons pass through photosystem II, through the cytochrome *bf* complex, and through ferredoxin-$NADP^+$ reductase.

A Proton Gradient Across the Thylakoid Membrane Drives ATP Synthesis

The proton gradient across the thylakoid membrane creates a proton-motive force, used by ATP synthase to form ATP. The ATP synthase of chloroplasts (also called CF_0-CF_1) closely resembles the ATP-synthesizing assemblies of bacteria and mitochondria (F_0-F_1). If the $NADPH$:$NADP^+$ ratio is high, electrons transferred to ferredoxin by photosystem I can reenter the cytochrome *bf* complex. This process, called cyclic photophosphorylation, leads to the generation of a proton gradient by the cytochrome *bf* complex without the formation of NADPH or O_2.

Accessory Pigments Funnel Energy into Reaction Centers

Light-harvesting complexes that surround the reaction centers contain additional molecules of chlorophyll *a*, as well as carotenoids and chlorophyll *b* molecules, which absorb light in the center of the visible spectrum. These accessory pigments increase efficiency in light capture by absorbing light and transferring the energy to reaction centers through resonance energy transfer. In blue-green and red algae, phycobilisomes—large protein assemblies with bound pigments called bilins—act as light-absorbing antennas.

The Ability to Convert Light into Chemical Energy Is Ancient

The photosystems have structural features in common that suggest a common evolutionary origin. Similarities in organization and molecular structure to those of oxidative phosphorylation suggest that the photosynthetic apparatus evolved from an early energy-transduction system.

KEY TERMS

light reaction (p. 527)

chloroplast (p. 528)

stroma (p. 528)

thylakoid (p. 528)

granum (p. 528)

chlorophyll *a* (p. 530)

photoinduced charge separation (p. 530)

reaction center (p. 530)

photosystem I (PS I) (p. 531)

photosystem II (PS II) (p. 531)

special pair (p. 532)

P960 (p. 532)

P680 (p. 535)

manganese center (p. 535)

cytochrome *bf* (p. 536)

P700 (p. 538)

Z scheme of photosynthesis (p. 539)

proton-motive force (Δp) (p. 540)

ATP synthase (CF_1-CF_0 complex) (p. 540)

cyclic photophosphorylation (p. 542)

carotenoid (p. 544)

light-harvesting complex (p. 544)

phycobilisome (p. 545)

SELECTED READINGS

Where to start

Huber, R., 1989. A structural basis of light energy and electron transfer in biology. *EMBO J.* 8:2125–2147.

Deisenhofer, J., and Michel, H., 1989. The photosynthetic reaction centre from the purple bacterium *Rhodopseudomonas viridis*. *EMBO J.* 8:2149–2170.

Barber, J., and Andersson, B., 1994. Revealing the blueprint of photosynthesis. *Nature* 370:31–34.

Books and general reviews

Raghavendra, A. S., 1998. *Photosynthesis: A Comprehensive Treatise*. 1998. Cambridge University Press.

Cramer, W. A., and Knaff, D. B., 1991. *Energy Transduction in Biological Membranes: A Textbook of Bioenergetics*. Springer Verlag.

Nicholls, D. G., and Ferguson, S. J., 1997. *Bioenergetics* (2d ed.). Academic Press.

Harold, F. M., 1986. *The Vital Force: A Study of Bioenergetics*. W. H. Freeman and Company.

Electron-transfer mechanisms

Beratan, D., and Skourtis, S., 1998. Electron transfer mechanisms. *Curr. Opin. Chem. Biol.* 2:235–243.

Moser, C. C., Keske, J. M., Warncke, K., Farid, R. S., and Dutton, P. L., 1992. Nature of biological electron transfer. *Nature* 355:796–802.

Boxer, S. G., 1990. Mechanisms of long-distance electron transfer in proteins: Lessons from photosynthetic reaction centers. *Annu. Rev. Biophys. Biophys. Chem.* 19:267–299.

Photosystem II

Zouni, A., Witt, H. T., Kern, J., Fromme, P., Krauss, N., Saenger, W., and Orth, P., 2001. Crystal structure of photosystem II from *Synechococcus elongatus* at 3.8 Å resolution. *Nature* 409:739–743.

Rhee, K. H., Morris, E. P., Barber, J., and Kuhlbrandt, W., 1998. Three-dimensional structure of the plant photosystem II reaction centre at 8 Å resolution. *Nature* 396:283–286.

Morris, E. P., Hankamer, B., Zheleva, D., Friso, G., and Barber, J., 1997. The three-dimensional structure of a photosystem II core complex determined by electron crystallography. *Structure* 5:837–849.

Deisenhofer, J., and Michel, H., 1991. High-resolution structures of photosynthetic reaction centers. *Annu. Rev. Biophys. Biophys. Chem.* 20:247–266.

Vermaas, W., 1993. Molecular-biological approaches to analyze photosystem II structure and function. *Annu. Rev. Plant Physiol. Plant Mol. Biol.* 44:457–481.

Oxygen evolution

Hoganson, C. W., and Babcock, G. T., 1997. A metalloradical mechanism for the generation of oxygen from water in photosynthesis. *Science* 277:1953–1956.

Yamachandra, V. K., DeRose, V. J., Latimer, M. J., Mukerji, I., Sauer, K., and Klein, M. P., 1993. Where plants make oxygen: A structural model for the photosynthetic oxygen-evolving manganese complex. *Science* 260:675–679.

Brudvig, G. W., Beck, W. F., and de Paula, J. C., 1989. Mechanism of photosynthetic water oxidation. *Annu. Rev. Biophys. Biophys. Chem.* 18:25–46.

Peloquin, J. M., and Britt, R. D., 2001. EPR/ENDOR characterization of the physical and electronic structure of the OEC Mn cluster. *Biochim. Biophys. Acta* 1503:96–111.

Photosystem I and cytochrome *bf*

Schubert, W. D., Klukas, O., Saenger, W., Witt, H. T., Fromme, P., and Krauss, N., 1998. A common ancestor for oxygenic and anoxygenic photosynthetic systems: A comparison based on the structural model of photosystem I. *J. Mol. Biol.* 280:297–314.

Fotiadis, D., Muller, D. J., Tsiotis, G., Hasler, L., Tittmann, P., Mini, T., Jeno, P., Gross, H., and Engel, A., 1998. Surface analysis of the photosystem I complex by electron and atomic force microscopy. *J. Mol. Biol.* 283:83–94.

Klukas, O., Schubert, W. D., Jordan, P., Krauss, N., Fromme, P., Witt, H. T., and Saenger, W., 1999. Photosystem I, an improved model of the stromal subunits PsaC, PsaD, and PsaE. *J. Biol. Chem.* 274:7351–7360.

Jensen, P. E., Gilpin, M., Knoetzel, J., and Scheller, H. V., 2000. The PSI-K subunit of photosystem I is involved in the interaction between light-harvesting complex I and the photosystem I reaction center core. *J. Biol. Chem.* 275:24701–24708.

Kitmitto, A., Mustafa, A. O., Holzenburg, A., and Ford, R. C., 1998. Three-dimensional structure of higher plant photosystem I determined by electron crystallography. *J. Biol. Chem.* 273:29592–29599.

Krauss, N., Hinrichs, W., Witt, I., Fromme, P., Pritzkow, W., Dauter, Z., Betzel, C., Wilson, K. S., Witt, H. T., and Saenger, W., 1993. Three-dimensional structure of system I photosynthesis at 6 Å resolution. *Nature* 361:326–331.

Malkin, R., 1992. Cytochrome bc_1 and $b_6 f$ complexes of photosynthetic membranes. *Photosynth. Res.* 33:121–136.

Karplus, P. A., Daniels, M. J., and Herriott, J. R., 1991. Atomic structure of ferredoxin-$NADP^+$ reductase: Prototype for a structurally novel flavoenzyme family. *Science* 251:60–66.

ATP synthase

Richter, M. L., Hein, R., and Huchzermeyer, B., 2000. Important subunit interactions in the chloroplast ATP synthase. *Biochim. Biophys. Acta* 1458:326–329.

Oster, G., and Wang, H., 1999. ATP synthase: Two motors, two fuels. *Structure* 7:R67–R72.

Junge, W., Lill, H., and Engelbrecht, S., 1997. ATP synthase: An electrochemical transducer with rotatory mechanics. *Trends Biochem. Sci.* 22:420–423.

Weber, J., and Senior, A. E., 2000. ATP synthase: What we know about ATP hydrolysis and what we do not know about ATP synthesis. *Biochim. Biophys. Acta* 1458:300–309.

Light-harvesting assemblies

Conroy, M. J., Westerhuis, W. H., Parkes Loach, P. S., Loach, P. A., Hunter, C. N., and Williamson, M. P., 2000. The solution structure of *Rhodobacter sphaeroides* LH1beta reveals two helical domains separated by a more flexible region: Structural consequences for the LH1 complex. *J. Mol. Biol.* 298:83–94.

Koepke, J., Hu, X., Muenke, C., Schulten, K., and Michel, H., 1996. The crystal structure of the light-harvesting complex II (B800–850) from *Rhodospirillum molischianum*. *Structure* 4:581–597.

Grossman, A. R., Bhaya, D., Apt, K. E., and Kehoe, D. M., 1995. Light-harvesting complexes in oxygenic photosynthesis: Diversity, control, and evolution. *Annu. Rev. Genet.* 29:231–288.

Kühlbrandt, W., Wang, D.-N., and Fujiyoshi, Y., 1994. Atomic model of plant light-harvesting complex by electron crystallography. *Nature* 367:614–621.

Glazer, A. N., 1983. Comparative biochemistry of photosynthetic light-harvesting systems. *Annu. Rev. Biochem.* 52:125–157.

Evolution

Green, B. R., 2001. Was "molecular opportunism" a factor in the evolution of different light-harvesting photosynthetic light-harvesting pigment systems? *Proc. Natl. Acad. Sci. USA* 98:2119–2121.

Dismukes, G. C., Klimov, V. V., Baranov, S. V., Nozlov, Y. N., DasGupta, J., and Tyryshkin, A., 2001. The origin of atmospheric oxygen on earth: The innovation of oxygenic photosynthesis. *Proc. Natl. Acad. Sci. USA* 98:2170–2175.

Moreira, D., Le Guyader, H., and Phillippe, H., 2000. The origin of red algae and the evolution of chloroplasts. *Nature* 405:69–72.

Cavalier-Smith, T., 2000. Membrane heredity and early chloroplast evolution. *Trends Plant Sci.* 5:174–182.

Blankenship, R. E., and Hartman, H., 1998. The origin and evolution of oxygenic photosynthesis. *Trends Biochem. Sci.* 23:94–97.

PROBLEMS

1. *Electron transfer.* Calculate the $\Delta E'_0$ and $\Delta G°'$ for the reduction of $NADP^+$ by ferredoxin. Use data given in Table 18.1 in Section 18.2.1.

2. *To boldly go.* (a) It can be argued that, if life were to exist elsewhere in the universe, it would require some process like photosynthesis. Why is this a reasonable argument? (b) If the Enterprise were to land on a distant plant and find no measurable oxygen in the atmosphere, could the crew conclude that photosynthesis is not taking place?

3. *Weed killer 1.* Dichlorophenyldimethylurea (DCMU), a herbicide, interferes with photophosphorylation and O_2 evolution. However, it does not block O_2 evolution in the presence of an artificial electron acceptor such as ferricyanide. Propose a site for the inhibitory action of DCMU.

4. *Weed killer 2.* Predict the effect of the herbicide dichlorophenyldimethylurea (DCMU) on a plant's ability to perform cyclic photophosphorylation.

5. *Infrared harvest.* Consider the relation between the energy of a photon and its wavelength.

(a) Some bacteria are able to harvest 1000-nm light. What is the energy (in kilocalories or kilojoules) of a mole (also called an einstein) of 1000-nm photons?
(b) What is the maximum increase in redox potential that can be induced by a 1000-nm photon?
(c) What is the minimum number of 1000-nm photons needed to form ATP from ADP and P_i? Assume a ΔG of 12 kcal mol^{-1} (50 kJ mol^{-1}) for the phosphorylation reaction.

6. *Missing acceptors.* Suppose that a bacterial reaction center containing only the special pair and the quinones was prepared. Given the separation of 22 Å between the special pair and the closest quinone, estimate the rate of electron transfer between the excited special pair and this quinone.

7. *Self-preservation.* Blue-green bacteria deprived of a source of nitrogen digest their least-essential proteins. Which phycobilisome component is likely to be degraded first under starvation conditions?

8. *Close approach.* Suppose that energy transfer between two chlorophyll *a* molecules separated by 10 Å occurs in 10 picoseconds. Suppose that this distance is increased to 20 Å with all other factors remaining the same. How long would energy transfer take?

Mechanism Problem

9. *Hill reaction.* In 1939, Robert Hill discovered that chloroplasts evolve O_2 when they are illuminated in the presence of an artificial electron acceptor such as ferricyanide $[Fe^{3+}(CN)_6]^{3-}$. Ferricyanide is reduced to ferrocyanide $[Fe^{2+}(CN)_6]^{4-}$ in this process. No NADPH or reduced plastocyanin is produced. Propose a mechanism for the Hill reaction.

Data Interpretation and Chapter Integration Problem

10. *The same, but different.* The $\alpha_3\beta_3\gamma$ complex of mitochondrial or chloroplast ATP synthase will function as an ATPase in vitro. The chloroplast enzyme (both synthase and ATPase activity) is sensitive to redox control, whereas the mitochondrial enzyme is not. To determine where the enzymes differ, a segment of the mitochondrial γ subunit was removed and replaced with the equivalent segment from the chloroplast γ subunit. The ATPase activity of the modified enzyme was then measured as a function of redox conditions.

(a) What is the redox regulator of the ATP synthase in vivo? The graph below shows the ATPase activity of modified and control enzymes under various redox conditions.

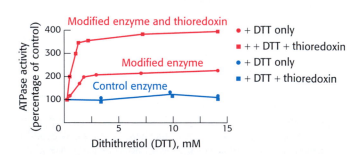

[Data after O. Bald, et al., 2000. *J Biol. Chem.*, 275: 12757–12762.]

(b) What is the effect of increasing the reducing power of the reaction mixture for the control and modified enzymes?
(c) What was the effect of the addition of thioredoxin? How did these results differ from those in the presence of DTT alone? Suggest a possible explanation for the difference.
(d) Did the researchers succeed in identifying the region of the γ subunit responsible for redox regulation?
(e) What is the biological rationale of regulation by high concentrations of reducing agents?
(f) What amino acids in the γ subunit are most likely affected by the reducing conditions?
(g) What experiments might confirm your answer to part *e*?

CHAPTER 20

The Calvin Cycle and the Pentose Phosphate Pathway

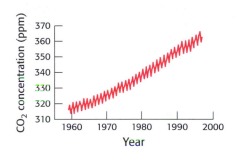

Atmospheric carbon dioxide measurements at Mauna Loa, Hawaii. These measurements show annual cycles resulting from seasonal variation in carbon dioxide fixation by the Calvin cycle in terrestrial plants. Much of this fixation takes place in rain forests, which account for approximately 50% of terrestrial fixation. [Dennis Potokar/Photo Researchers.]

Photosynthesis proceeds in two parts: the light reactions and the dark reactions. The light reactions, which were discussed in Chapter 19, transform light energy into ATP and biosynthetic reducing power, NADPH. The dark reactions, which constitute the Calvin cycle, named after Melvin Calvin, the biochemist who elucidated the pathway, reduce carbon atoms from their fully oxidized state as carbon dioxide to the more reduced state as a hexose. The components of the Calvin cycle and called the dark reactions because, in contrast with the light reactions, these reactions do not directly depend on the presence of light.

In addition to ATP, the dark reactions require reducing power in the form of *NADPH, the currency of readily available reducing power in cells.* The phosphoryl group on the 2'-hydroxyl group of one of the ribose units of NADPH distinguishes NADPH from NADH. There is a *fundamental distinction between NADPH and NADH in biochemistry: NADH is oxidized by the respiratory chain to generate ATP, whereas NADPH serves as a reductant in biosynthetic processes.*

The second half of this chapter examines a pathway common to all organisms, known variously as the pentose phosphate pathway, the hexose monophosphate pathway, the phosphogluconate pathway, or the pentose shunt. The pathway provides a means by which glucose can be oxidized to generate NADPH and is the source of much of the NADPH that is needed for the biosynthesis of many biomolecules, most notably fats. We will observe the use of

OUTLINE

- **20.1 The Calvin Cycle Synthesizes Hexoses from Carbon Dioxide and Water**

- **20.2 The Activity of the Calvin Cycle Depends on Environmental Conditions**

- **20.3 The Pentose Phosphate Pathway Generates NADPH and Synthesizes Five-Carbon Sugars**

- **20.4 The Metabolism of Glucose 6-phosphate by the Pentose Phosphate Pathway Is Coordinated with Glycolysis**

- **20.5 Glucose 6-phosphate Dehydrogenase Plays a Key Role in Protection Against Reactive Oxygen Species**

NADPH in many of the biosynthetic reactions considered in Part III of this text. The pentose phosphate pathway can also be used for the catabolism of pentose sugars from the diet, the synthesis of pentose sugars for nucleotide biosynthesis, and the catabolism and synthesis of less common four- and seven-carbon sugars. The pentose phosphate pathway and the Calvin cycle have in common several enzymes and intermediates that attest to an evolutionary kinship. Like glycolysis and gluconeogenesis (Chapter 16), these pathways are mirror images of one another: the Calvin cycle uses NADPH to reduce carbon dioxide to generate hexoses, whereas the pentose phosphate pathway breaks down glucose into carbon dioxide to generate NADPH. The Calvin cycle is sometimes referred to as the *reductive pentose phosphate pathway*.

20.1 THE CALVIN CYCLE SYNTHESIZES HEXOSES FROM CARBON DIOXIDE AND WATER

We saw in Chapter 16 that glucose could be formed from noncarbohydrate precursors, such as lactate and amino acids, by gluconeogenesis. The synthesis of glucose from these compounds is simplified because the carbons are already incorporated into relatively complex organic molecules. In contrast, the source of the carbon atoms in the Calvin cycle is the simple molecule carbon dioxide. In this extremely important process, *carbon dioxide gas is trapped in a form that is useful for many processes*. The Calvin cycle brings into living systems the carbon atoms that will become constituents of nucleic acids, proteins, and fats. Photosynthetic organisms are called *autotrophs* (literally "self-feeders") because they can synthesize glucose from carbon dioxide and water, by using sunlight as an energy source, and then recover some of this energy from the synthesized glucose through the glycolytic pathway and aerobic metabolism. Organisms that obtain energy from chemical fuels only are called *heterotrophs*, which ultimately depend on autotrophs for their fuel. The Calvin cycle also differs from gluconeogenesis in where it takes place in photosynthetic eukaryotes. Whereas gluconeogenesis takes place in the cytoplasm, the Calvin cycle takes place in the stroma of chloroplasts, the photosynthetic organelles.

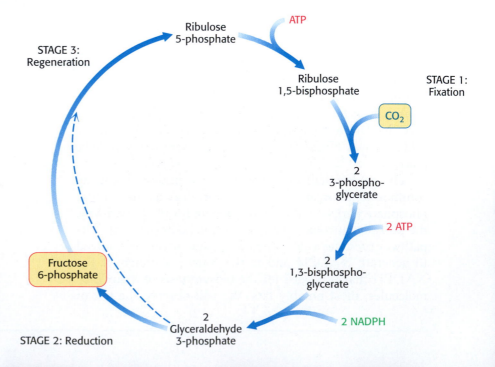

FIGURE 20.1 Calvin cycle. The Calvin cycle consists of three stages. Stage 1 is the fixation of carbon by the carboxylation of ribulose 1,5-bisphosphate. Stage 2 is the reduction of the fixed carbon to begin the synthesis of hexose. Stage 3 is the regeneration of the starting compound, ribulose 1,5-bisphosphate.

The Calvin cycle comprises three stages (Figure 20.1):

1. The fixation of CO_2 by ribulose 1,5-bisphosphate to form two molecules of 3-phosphoglycerate.

2. The reduction of 3-phosphoglycerate to form hexose sugars.

3. The regeneration of ribulose 1,5-bisphosphate so that more CO_2 can be fixed.

Although we will focus on the Calvin cycle, other means of fixing carbon dioxide into hexose sugars exist in the photosynthetic world, notably a version of the citric acid cycle running in reverse.

20.1.1 Carbon Dioxide Reacts with Ribulose 1,5-bisphosphate to Form Two Molecules of 3-Phosphoglycerate

The first step in the Calvin cycle is the fixation of CO_2. The CO_2 molecule condenses with ribulose 1,5-bisphosphate to form an unstable six-carbon compound, which is rapidly hydrolyzed to two molecules of 3-phosphoglycerate.

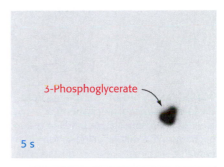

3-Phosphoglycerate

5 s

60 s

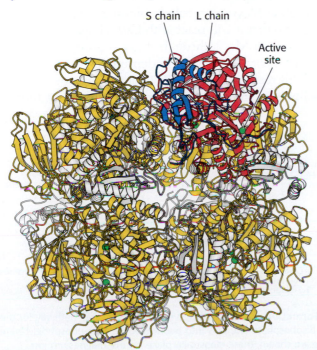

The initial incorporation of CO_2 into 3-phosphoglycerate was revealed through the use of a carbon-14 radioactive tracer (Figure 20.2). This highly exergonic reaction $[\Delta G°' = -12.4 \text{ kcal mol}^{-1} (-51.9 \text{ kJ mol}^{-1})]$ is catalyzed by *ribulose 1,5-bisphosphate carboxylase/oxygenase* (usually called *rubisco*), an enzyme located on the stromal surface of the thylakoid membranes of chloroplasts. This important reaction is the rate-limiting step in hexose synthesis. Rubisco in chloroplasts consists of eight large (L, 55-kd) subunits and eight small (S, 13-kd) ones (Figure 20.3). Each L chain contains a catalytic site and a regulatory site. The S chains enhance the catalytic

S chain   L chain

Active site

FIGURE 20.2 Tracing the fate of carbon dioxide. Radioactivity from $^{14}CO_2$ is incorporated into 3-phosphoglycerate within 5 s in irradiated cultures of algae. After 60 s, the radioactivity appears in many compounds, the intermediates within the Calvin cycle. [Courtesy of Dr. J. A. Bassham.]

FIGURE 20.3 Structure of rubisco. The enzyme ribulose 1,5-bisphosphate carboxylase/oxygenase (rubisco) comprises eight large subunits (one shown in red and the others in yellow) and eight small subunits (one shown in blue and the others in white). The active sites lie in the large subunits.

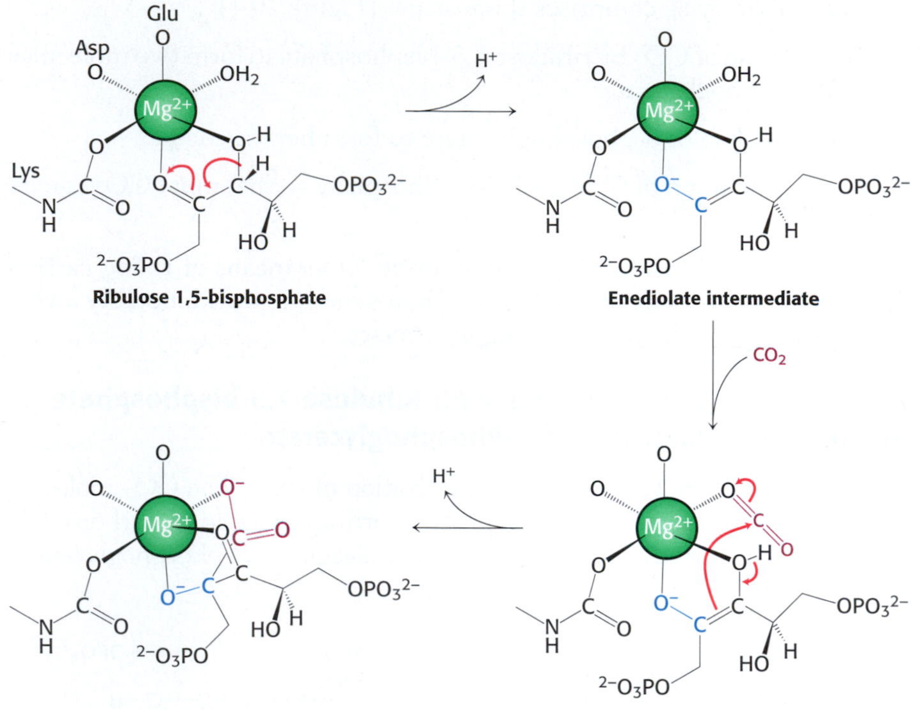

FIGURE 20.4 Role of the magnesium ion in the rubisco mechanism. Ribulose 1,5-bisphosphate binds to a magnesium ion that is linked to rubisco through a glutamate residue, an aspartate residue, and the lysine carbamate. The coordinated ribulose 1,5-bisphosphate gives up a proton to form a reactive enediolate species that reacts with CO_2 to form a new carbon–carbon bond.

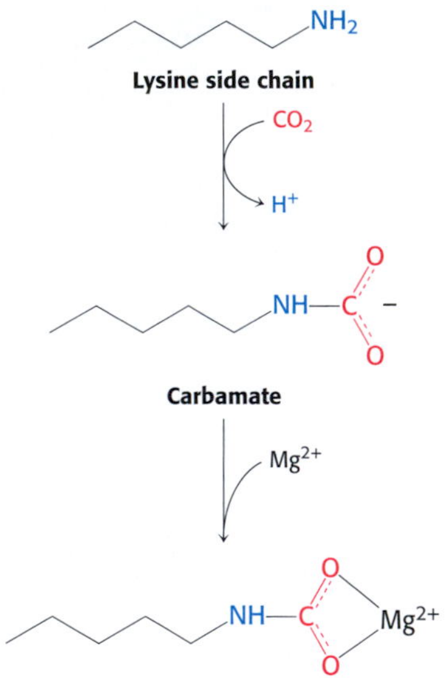

activity of the L chains. This enzyme is very abundant in chloroplasts, constituting more than 16% of their total protein. In fact, rubisco is the most abundant enzyme and probably the most abundant protein in the biosphere. Large amounts are present because rubisco is a slow enzyme; its maximal catalytic rate is only $3 \ s^{-1}$.

Rubisco requires a bound divalent metal ion for activity, usually magnesium ion. Like the zinc ion in the active site of carbonic anhydrase (Section 9.2.1), this metal ion serves to activate a bound substrate molecule by stabilizing a negative charge. Interestingly, a CO_2 molecule other than the substrate is required to complete the assembly of the Mg^{2+} binding site in rubisco. This CO_2 molecule adds to the uncharged ϵ-amino group of lysine 201 to form a *carbamate*. This negatively charged adduct then binds the Mg^{2+} ion. The formation of the carbamate is facilitated by the enzyme *rubisco activase*, although it will also form spontaneously at a lower rate.

The metal center plays a key role in binding ribulose 1,5-bisphosphate and activating it so that it will react with CO_2 (Figure 20.4). Ribulose 1,5-bisphosphate binds to Mg^{2+} through its keto group and an adjacent hydroxyl group. This complex is readily deprotonated to form an enediolate intermediate. This reactive species, analogous to the zinc–hydroxide species in carbonic anhydrase (Section 9.2.2), couples with CO_2, forming the new

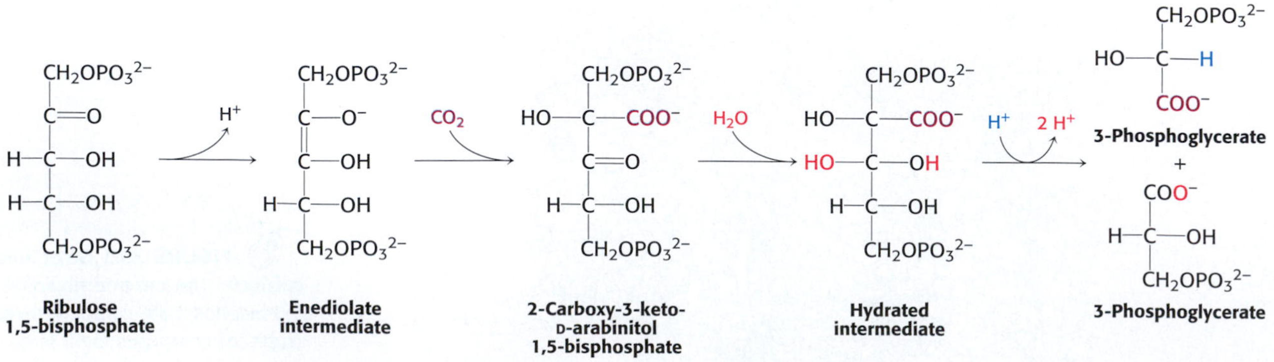

FIGURE 20.5 Formation of 3-phosphoglycerate. The overall pathway for the conversion of ribulose 1,5 bisphosphate and CO_2 into two molecules of 3-phosphoglycerate. Although the free species are shown, these steps take place on the magnesium ion.

carbon–carbon bond. The resulting product is coordinated to the Mg^{2+} ion through three groups, including the newly formed carboxylate. A molecule of H_2O is then added to this β-ketoacid to form an intermediate that cleaves to form two molecules of 3-phosphoglycerate (Figure 20.5).

20.1.2 Catalytic Imperfection: Rubisco Also Catalyzes a Wasteful Oxygenase Reaction

The reactive intermediate generated on the Mg^{2+} ion sometimes reacts with O_2 instead of CO_2. Thus, rubisco also catalyzes a deleterious oxygenase reaction. The products of this reaction are *phosphoglycolate* and *3-phosphoglycerate* (Figure 20.6). The rate of the carboxylase reaction is four times

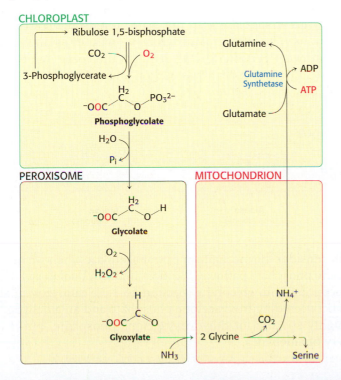

FIGURE 20.6 A wasteful side reaction. The reactive enediolate intermediate on rubisco also reacts with molecular oxygen to form a hydroperoxide intermediate, which then proceeds to form one molecule of 3-phosphoglycerate and one molecule of phosphoglycolate.

that of the oxygenase reaction under normal atmospheric conditions at 25°C; the stromal concentration of CO_2 is then 10 μM and that of O_2 is 250 μM. The oxygenase reaction, like the carboxylase reaction, requires that lysine 201 be in the carbamate form. Because this carbamate forms only in the presence of CO_2, this property would prevent rubisco from catalyzing the oxygenase reaction exclusively when CO_2 is absent.

Phosphoglycolate is not a versatile metabolite. A salvage pathway recovers part of its carbon skeleton (Figure 20.7). A specific phosphatase converts phosphoglycolate into *glycolate*, which enters *peroxisomes* (also called

FIGURE 20.7 Photorespiratory reactions. Phosphoglycolate is formed as a product of the oxygenase reaction in chloroplasts. After dephosphorylation, glycolate is transported into peroxisomes where it is converted into glyoxylate and then glycine. In mitochondria, two glycines are converted into serine, after losing a carbon as CO_2 and ammonia. The ammonia is salvaged in chloroplasts.

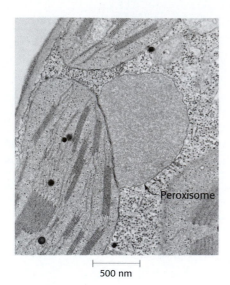

FIGURE 20.8 Electron micrograph of a peroxisome nestled between two chloroplasts. [Courtesy of Dr. Sue Ellen Frederick.]

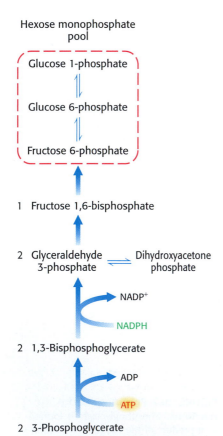

FIGURE 20.9 Hexose phosphate formation. 3-Phosphoglycerate is converted into fructose 6-phosphate in a pathway parallel to that of glyconeogenesis.

microbodies; Figure 20.8). Glycolate is then oxidized to *glyoxylate* by glycolate oxidase, an enzyme with a flavin mononucleotide prosthetic group. The H_2O_2 produced in this reaction is cleaved by catalase to H_2O and O_2. Transamination of glyoxylate then yields *glycine*. Two glycine molecules can be used to form serine, a potential precursor of glucose, with the release of CO_2 and ammonia (NH_4^+). The ammonia, used in the synthesis of nitrogen-containing compounds, is salvaged by glutamine synthetase reaction.

This salvage pathway serves to recycle three of the four carbon atoms of two molecules of glycolate. However, one carbon atom is lost as CO_2. This process is called *photorespiration* because O_2 is consumed and CO_2 is released. Photorespiration is wasteful because organic carbon is converted into CO_2 without the production of ATP, NADPH, or another energy-rich metabolite. Moreover, the oxygenase activity increases more rapidly with temperature than the carboxylase activity, presenting a problem for tropical plants (Section 20.2.3). Evolutionary processes have presumably enhanced the preference of rubisco for carboxylation. For instance, the rubisco of higher plants is eightfold as specific for carboxylation as that of photosynthetic bacteria.

20.1.3 Hexose Phosphates Are Made from Phosphoglycerate, and Ribulose 1,5-bisphosphate Is Regenerated

The 3-phosphoglycerate product of rubisco is next converted into three forms of hexose phosphate: glucose 1-phosphate, glucose 6-phosphate, and fructose 6-phosphate. Recall that these isomers are readily interconvertible (Sections 16.1.2 and 16.1.11). The steps in this conversion (Figure 20.9) are like those of the gluconeogenic pathway (Section 16.3.1), except that glyceraldehyde 3-phosphate dehydrogenase in chloroplasts, which generates glyceraldehyde 3-phosphate (GAP), is specific for NADPH rather than NADH. Alternatively, the glyceraldehyde 3-phosphate can be transported to the cytosol for glucose synthesis. These reactions and that catalyzed by rubisco bring CO_2 to the level of a hexose, converting CO_2 into a chemical fuel at the expense of NADPH and ATP generated from the light reactions.

The third phase of the Calvin cycle is the regeneration of ribulose 1,5-bisphosphate, the acceptor of CO_2 in the first step. The problem is to construct a five-carbon sugar from six-carbon and three-carbon sugars. A transketolase and an aldolase play the major role in the rearrangement of the carbon atoms. The *transketolase*, which we will see again in the pentose phosphate pathway (Section 20.2.3), requires the coenzyme thiamine pyrophosphate (TPP) to transfer a two-carbon unit (CO–CH_2OH) from a ketose to an aldose.

We will consider the mechanism of transketolase when we meet it again in the pentose phosphate pathway (Section 20.3.2). *Aldolase*, which we have already encountered in glycolysis (Section 16.1.3), catalyzes an aldol condensation between dihydroxyacetone phosphate and an aldehyde. This enzyme is highly specific for dihydroxyacetone phosphate, but it accepts a wide variety of aldehydes.

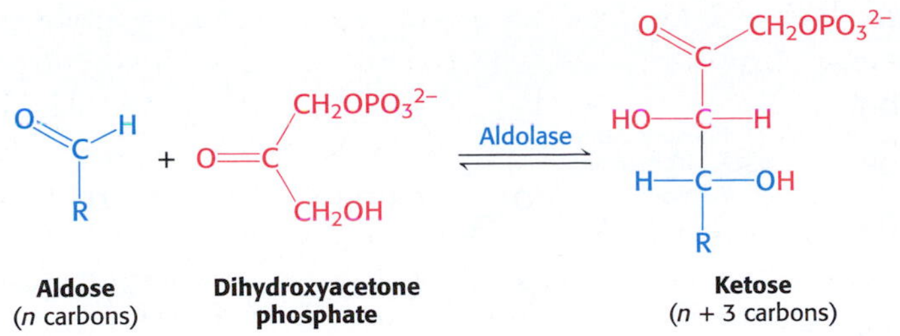

Aldose (n carbons) + **Dihydroxyacetone phosphate** —Aldolase→ **Ketose** (n + 3 carbons)

With these enzymes, the construction of the five-carbon sugar proceeds as shown in Figure 20.10.

Finally, ribose-5-phosphate is converted into ribulose 5-phosphate by *phosphopentose isomerase* while xylulose 5-phosphate is converted into ribulose 5-phosphate by *phosphopentose epimerase*. Ribulose 5-phosphate is converted into ribulose 1,5-bisphosphate through the action of *phosphoribulose kinase* (Figure 20.11). The sum of these reactions is

$$\text{Fructose 6-phosphate} + 2 \text{ glyceraldehyde 3-phosphate}$$
$$+ \text{ dihydroxyacetone phosphate} + 3\,\text{ATP} \longrightarrow$$
$$3 \text{ ribulose 1,5-bisphosphate} + 3\,\text{ADP}$$

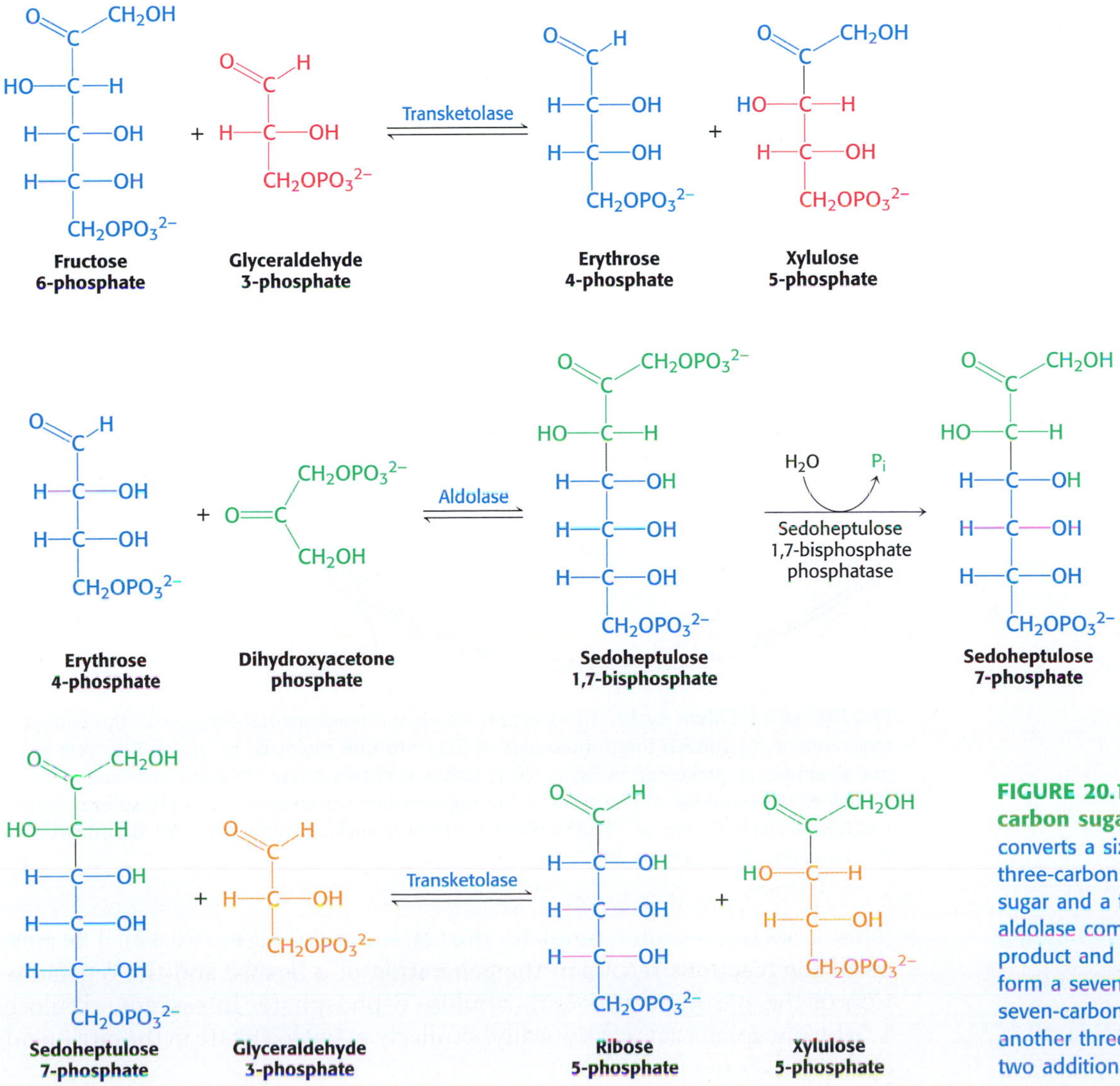

Fructose 6-phosphate + **Glyceraldehyde 3-phosphate** ⇌Transketolase⇌ **Erythrose 4-phosphate** + **Xylulose 5-phosphate**

Erythrose 4-phosphate + **Dihydroxyacetone phosphate** ⇌Aldolase⇌ **Sedoheptulose 1,7-bisphosphate** —(H_2O, P_i) Sedoheptulose 1,7-bisphosphate phosphatase→ **Sedoheptulose 7-phosphate**

Sedoheptulose 7-phosphate + **Glyceraldehyde 3-phosphate** ⇌Transketolase⇌ **Ribose 5-phosphate** + **Xylulose 5-phosphate**

FIGURE 20.10 Formation of five-carbon sugars. First, transketolase converts a six-carbon sugar and a three-carbon sugar into a four-carbon sugar and a five-carbon sugar. Then, aldolase combines the four-carbon product and a three-carbon sugar to form a seven-carbon sugar. Finally, this seven-carbon fragment combines with another three-carbon fragment to form two additional five-carbon sugars.

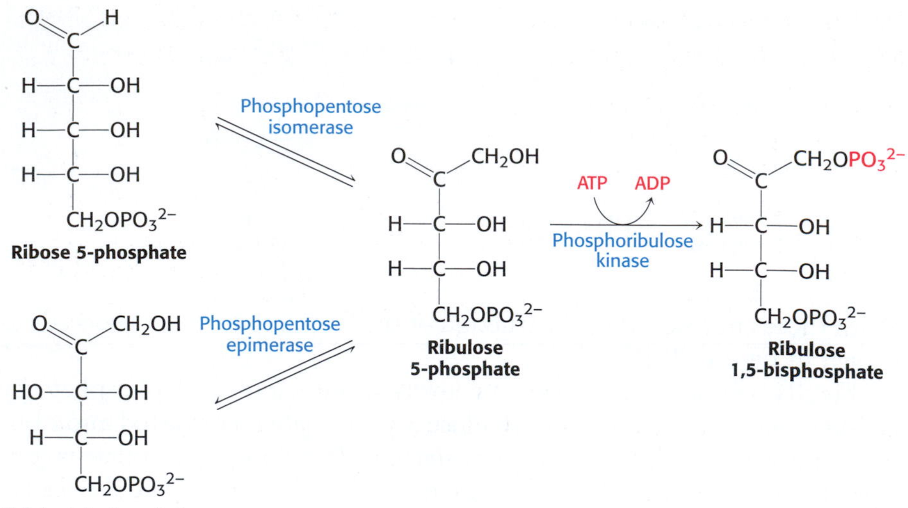

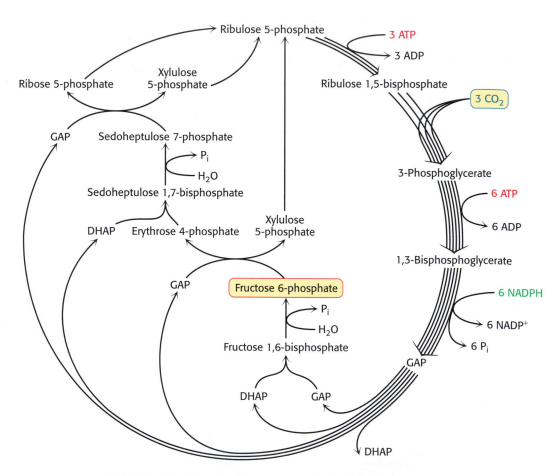

FIGURE 20.11 Regeneration of ribulose 1,5-bisphosphate. Both ribose 5-phosphate and xylulose 5-phosphate are converted into ribulose 5-phosphate, which is then phosphorylated to complete the regeneration of ribulose 1,5-bisphosphate.

FIGURE 20.12 Calvin cycle. The diagram shows the reactions necessary with the correct stoichiometry to convert three molecules of CO_2 into one molecule of DHAP. The cycle is not as simple as presented in Figure 20.1; rather, it entails many reactions that lead ultimately to the synthesis of glucose and the regeneration of ribulose 1,5-bisphosphate. [After J. R. Bowyer and R. C. Leegood. "Photosynthesis," in *Plant Biochemistry*, P. M. Dey and J. B. Harborne, Eds. (Academic Press, 1997), p. 85.]

This series of reactions completes the Calvin cycle (Figure 20.12). The sum of all the reactions results in the generation of a hexose and the regeneration of the starting compound, ribulose 5-phosphate. In essence, ribulose 1,5-bisphosphate acts catalytically, similarly to oxaloacetate in the citric acid cycle.

20.1.4 Three Molecules of ATP and Two Molecules of NADPH Are Used to Bring Carbon Dioxide to the Level of a Hexose

What is the energy expenditure for synthesizing a hexose? Six rounds of the Calvin cycle are required, because one carbon atom is reduced in each round. Twelve molecules of ATP are expended in phosphorylating 12 molecules of 3-phosphoglycerate to 1,3-bisphosphoglycerate, and 12 molecules of NADPH are consumed in reducing 12 molecules of 1,3-bisphosphoglycerate to glyceraldehyde 3-phosphate. An additional six molecules of ATP are spent in regenerating ribulose 1,5-bisphosphate. We can now write a balanced equation for the net reaction of the Calvin cycle.

$$6\,CO_2 + 18\,ATP + 12\,NADPH + 12\,H_2O \longrightarrow$$
$$C_6H_{12}O_6 + 18\,ADP + 18\,P_i + 12\,NADP^+ + 6\,H^+$$

Thus, three molecules of ATP and two molecules of NADPH are consumed in incorporating a single CO_2 molecule into a hexose such as glucose or fructose.

20.1.5 Starch and Sucrose Are the Major Carbohydrate Stores in Plants

Plants contain two major storage forms of sugar: *starch* and *sucrose*. Starch, like its animal counterpart glycogen, is a polymer of glucose residues, but it is less branched than glycogen because it contains a smaller proportion of α-1,6-glycosidic linkages (Section 11.2.2). Another difference is that ADP-glucose, not UDP-glucose, is the activated precursor. Starch is synthesized and stored in chloroplasts.

In contrast, sucrose (common table sugar), a disaccharide, is synthesized in the cytosol. Plants lack the ability to transport hexose phosphates across the chloroplast membrane, but an abundant phosphate translocator mediates the transport of triose phosphates from chloroplasts to the cytosol in exchange for phosphate. Fructose 6-phosphate formed from triose phosphates joins the glucose unit of UDP-glucose to form sucrose 6-phosphate (Figure 20.13). Hydrolysis of the phosphate ester yields sucrose, a readily

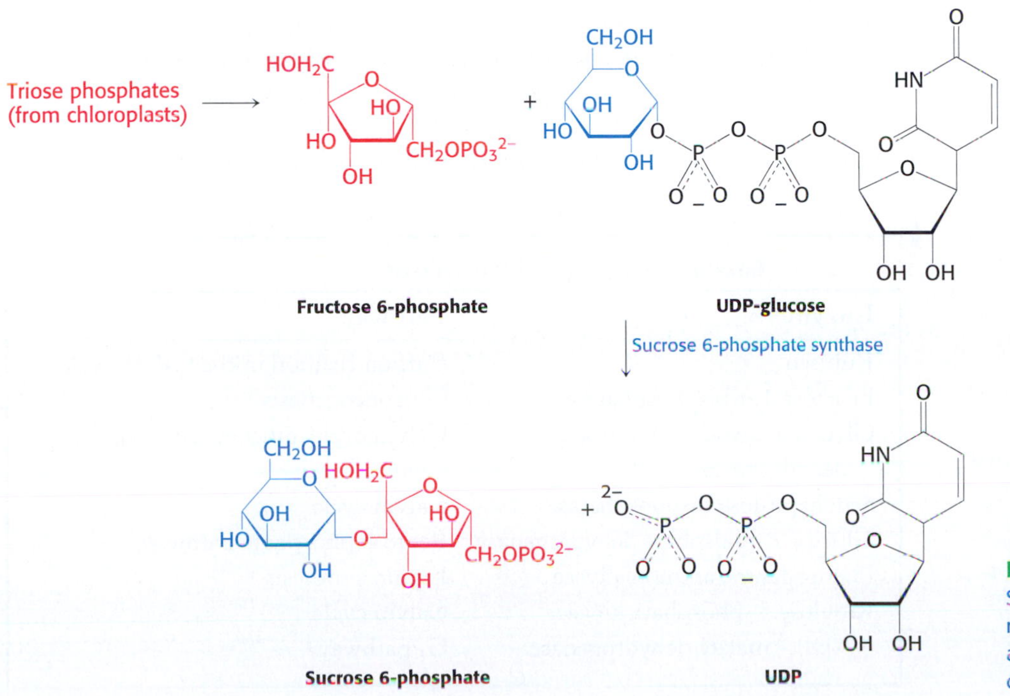

FIGURE 20.13 Synthesis of sucrose. Sucrose 6-phosphate is formed by the reaction between fructose 6-phosphate and the activated intermediate uridine diphosphate glucose (UDP-glucose).

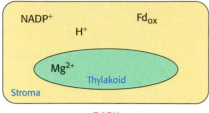

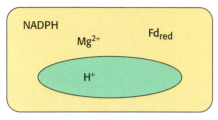

FIGURE 20.14 Light regulation of the Calvin cycle. The light reactions of photosynthesis transfer electrons out of the thylakoid lumen into the stroma and they transfer protons from the stroma into the thylakoid lumen. As a consequence of these processes, the concentrations of NADPH, reduced ferredoxin (Fd), and Mg^{2+} in the stroma are higher in the light than in the dark, whereas the concentration of H^+ is lower in the dark. Each of these concentration changes helps couple the Calvin cycle reactions to the light reactions.

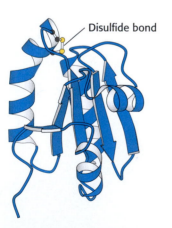

Disulfide bond

FIGURE 20.15 Thioredoxin. The oxidized form of thioredoxin contains a disulfide bond. When thioredoxin is reduced by reduced ferredoxin, the disulfide bond is converted into two free sulfhydryl groups. Reduced thioredoxin can cleave disulfide bonds in enzymes, activating certain Calvin cycle enzymes and inactivating some degradative enzymes.

transportable and mobilizable sugar that is stored in many plant cells, as in sugar beets and sugar cane.

20.2 THE ACTIVITY OF THE CALVIN CYCLE DEPENDS ON ENVIRONMENTAL CONDITIONS

Carbon dioxide assimilation by the Calvin cycle operates during the day, whereas carbohydrate degradation to yield energy takes place primarily at night. How are synthesis and degradation coordinately controlled? The light reactions lead to changes in the stroma—namely, an increase in pH and in Mg^{2+}, NADPH, and reduced ferredoxin concentration—all of which contribute to the activation of certain Calvin cycle enzymes (Figure 20.14).

20.2.1 Rubisco Is Activated by Light-Driven Changes in Proton and Magnesium Ion Concentrations

As stated earlier, the rate-limiting step in the Calvin cycle is the carboxylation of ribulose 1,5-bisphosphate to form two molecules of 3-phosphoglycerate. *The activity of rubisco increases markedly on illumination.* The addition of CO_2 to lysine 201 of rubisco to form the carbamate is essential for Mg^{2+} coordination and, hence, catalytic activity (Section 20.1.1). Carbamate formation is favored by alkaline pH and high concentrations of Mg^{2+} ion in the stroma, both of which are consequences of the light-driven pumping of protons from the stroma into the thylakoid space. Magnesium ion concentration rises because Mg^{2+} ions from the thylakoid space are released into the stroma to compensate for the influx of protons.

20.2.2 Thioredoxin Plays a Key Role in Regulating the Calvin Cycle

Light-driven reactions lead to electron transfer from water to ferredoxin and, eventually, to NADPH. Both reduced ferredoxin and NADPH regulate enzymes from the Calvin cycle. One key protein in these regulatory processes is *thioredoxin*, a 12-kd protein containing neighboring cysteine residues that cycle between a reduced sulfhydryl and an oxidized disulfide form (Figure 20.15). The reduced form of thioredoxin activates many biosynthetic enzymes by reducing disulfide bridges that control their activity and inhibits several degradative enzymes by the same means (Table 20.1). In chloroplasts, oxidized thioredoxin is reduced by ferredoxin in a reaction catalyzed by *ferredoxin-thioredoxin reductase*. This enzyme contains

TABLE 20.1 Enzymes regulated by thioredoxin

Enzyme	Pathway
Rubisco	Carbon fixation in the Calvin cycle
Fructose 1,6-bisphosphatase	Gluconeogenesis
Glyceraldehyde 3-phosphate dehydrogenase	Calvin cycle, gluconeogenesis, glycolysis
Sedoheptulose bisphosphatase	Calvin cycle
Glucose 6-phosphate dehydrogenase	Pentose phosphate pathway
Phenylalanine ammonia lyase	Lignin synthesis
Ribulose 5′-phosphate kinase	Calvin cycle
$NADP^+$-malate dehydrogenase	C_4 pathway

a 4Fe-4S cluster that couples two one-electron oxidations of reduced ferredoxin to the two-electron reduction of thioredoxin. Thus, *the activities of the light and dark reactions of photosynthesis are coordinated through electron transfer from reduced ferredoxin to thioredoxin and then to component enzymes containing regulatory disulfide bonds* (Figure 20.16). We shall return to thioredoxin when we consider the reduction of ribonucleotides (Section 25.3).

Other means of control also exist. For instance, phosphoribulose kinase and glyceraldehyde 3-phosphate dehydrogenase also are regulated by NADPH directly. In the dark, these enzymes associate with a small protein called CP12 to form a large complex in which the enzymes are inactivated. NADPH generated in the light reactions binds to this complex, leading to the release of the enzymes. Thus, the activity of these enzymes depends first on reduction by thioredoxin and then on the NADPH-mediated release from CP12.

20.2.3 The C₄ Pathway of Tropical Plants Accelerates Photosynthesis by Concentrating Carbon Dioxide

Recall that the oxygenase activity of rubisco increases more rapidly with temperature than does its carboxylase activity. How then do plants, such as sugar cane, that grow in hot climates prevent very high rates of wasteful photorespiration? Their solution to this problem is to achieve a high local concentration of CO_2 at the site of the Calvin cycle in their photosynthetic cells. The essence of this process, which was elucidated by M. D. Hatch and C. R. Slack, is that *four-carbon (C₄) compounds such as oxaloacetate and malate carry CO_2 from mesophyll cells, which are in contact with air, to bundle-sheath cells, which are the major sites of photosynthesis* (Figure 20.17). Decarboxylation of the four-carbon compound in a bundle-sheath cell maintains a high concentration of CO_2 at the site of the Calvin cycle. The three-carbon compound pyruvate returns to the mesophyll cell for another round of carboxylation.

The *C₄ pathway* for the transport of CO_2 starts in a mesophyll cell with the condensation of CO_2 and phosphoenolpyruvate to form *oxaloacetate*, in a reaction catalyzed by *phosphoenolpyruvate carboxylase*. In some species, oxaloacetate is converted into *malate* by an $NADP^+$-linked malate dehydrogenase. Malate goes into the bundle-sheath cell and is oxidatively decarboxylated within the chloroplasts by an $NADP^+$-linked malate dehydrogenase. The released CO_2 enters the Calvin cycle in the usual way by condensing with ribulose 1,5-bisphosphate. Pyruvate formed in this decarboxylation reaction returns to the mesophyll cell. Finally, phosphoenolpyruvate is formed from pyruvate by *pyruvate-P_i dikinase*.

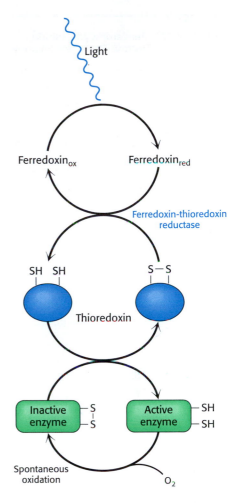

FIGURE 20.16 Enzyme activation by thioredoxin. Reduced thioredoxin activates certain Calvin cycle enzymes by cleaving regulatory disulfide bonds.

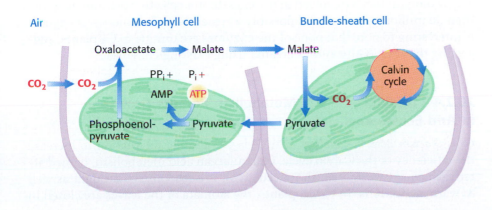

FIGURE 20.17 C₄ pathway. Carbon dioxide is concentrated in bundle-sheath cells by the expenditure of ATP in mesophyll cells.

The net reaction of this C_4 pathway is

$$CO_2 \text{ (in mesophyll cell)} + ATP + H_2O \longrightarrow$$
$$CO_2 \text{ (in bundle-sheath cell)} + AMP + 2\,P_i + H^+$$

Thus, *the energetic equivalent of two ATP molecules is consumed in transporting CO_2 to the chloroplasts of the bundle-sheath cells.* In essence, this process is active transport: the pumping of CO_2 into the bundle-sheath cell is driven by the hydrolysis of one molecule of ATP to one molecule of AMP and two molecules of orthophosphate. The CO_2 concentration can be 20-fold as great in the bundle-sheath cells as in the mesophyll cells.

When the C_4 pathway and the Calvin cycle operate together, the net reaction is

$$6\,CO_2 + 30\,ATP + 12\,NADPH + 12\,H_2O \longrightarrow$$
$$C_6H_{12}O_6 + 30\,ADP + 30\,P_i + 12\,NADP^+ + 18\,H^+$$

Note that 30 molecules of ATP are consumed per hexose molecule formed when the C_4 pathway delivers CO_2 to the Calvin cycle, in contrast with 18 molecules of ATP per hexose molecule in the absence of the C_4 pathway. The high concentration of CO_2 in the bundle-sheath cells of C_4 plants, which is due to the expenditure of the additional 12 molecules of ATP, is critical for their rapid photosynthetic rate, because CO_2 is limiting when light is abundant. A high CO_2 concentration also minimizes the energy loss caused by photorespiration.

Tropical plants with a C_4 pathway do little photorespiration because the high concentration of CO_2 in their bundle-sheath cells accelerates the carboxylase reaction relative to the oxygenase reaction. This effect is especially important at higher temperatures. The geographic distribution of plants having this pathway (C_4 plants) and those lacking it (C_3 plants) can now be understood in molecular terms. *C_4 plants* have the advantage in a hot environment and under high illumination, which accounts for their prevalence in the tropics. *C_3 plants*, which consume only 18 molecules of ATP per hexose molecule formed in the absence of photorespiration (compared with 30 molecules of ATP for C_4 plants), are more efficient at temperatures of less than about 28°C, and so they predominate in temperate environments.

Rubisco is found in bacteria, eukaryotes, and even archaea, though other photosynthetic components have not been found in archaea. Thus, rubisco emerged early in evolution, when the atmosphere was rich in CO_2 and almost devoid of O_2. The enzyme was not originally selected to operate in an environment like the present one, which is almost devoid of CO_2 and rich in O_2. Photorespiration became significant about 60 million years ago, when the CO_2 concentration fell to present levels. The C_4 pathway is thought to have evolved in response to this selective pressure no more than 30 million years ago and possibly as recently as 7 million years ago. It is interesting to note that none of the enzymes are unique to C_4 plants, suggesting that this pathway was created using existing enzymes.

20.2.4 Crassulacean Acid Metabolism Permits Growth in Arid Ecosystems

Crassulacean acid metabolism (CAM) is yet another adaptation to increase the efficiency of the Calvin cycle. Crassulacean acid metabolism, named after the genus *Crassulacea* (the succulents), is a response to drought as well as warm conditions. In CAM plants, the stomata of the leaves are closed in the heat of the day to prevent water loss (Figure 20.18). As a consequence, CO_2 cannot be absorbed during the daylight hours when it is needed for glucose synthesis. When the stomata open at the cooler temperatures of night, CO_2 is fixed by the C_4 pathway into malate, which is stored in vacuoles. During the day, malate is decarboxylated and the CO_2 becomes available to the Calvin cycle. In contrast with C_4 plants, CO_2 accumulation is separated from CO_2 utilization temporally in CAM plants rather than spatially.

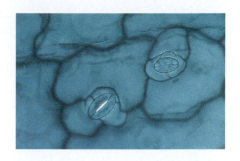

20.3 THE PENTOSE PHOSPHATE PATHWAY GENERATES NADPH AND SYNTHESIZES FIVE-CARBON SUGARS

The pentose phosphate pathway meets the need of all organisms for a source of NADPH to use in reductive biosynthesis (Table 20.2). This pathway consists of two phases: the oxidative generation of NADPH and the nonoxidative interconversion of sugars (Figure 20.19). In the oxidative phase, NADPH is generated when glucose 6-phosphate is oxidized to ribose 5-phosphate. This five-carbon sugar and its derivatives are components of RNA and DNA, as well as ATP, NADH, FAD, and coenzyme A.

$$\text{Glucose 6-phosphate} + 2\,\text{NADP}^+ + H_2O \longrightarrow$$
$$\text{ribose 5-phosphate} + 2\,\text{NADPH} + 2\,H^+ + CO_2$$

In the nonoxidative phase, the pathway catalyzes the interconversion of three-, four-, five-, six-, and seven-carbon sugars in a series of nonoxidative reactions that can result in the synthesis of five-carbon sugars for nucleotide biosynthesis or the conversion of excess five-carbon sugars into intermediates of the glycolytic pathway. All these reactions take place in the cytosol. These interconversions rely on the same reactions that lead to the regeneration of ribulose 1,5-bisphosphate in the Calvin cycle.

TABLE 20.2 Pathways requiring NADPH
Synthesis
Fatty acid biosynthesis
Cholesterol biosynthesis
Neurotransmitter biosynthesis
Nucleotide biosynthesis
Detoxification
Reduction of oxidized glutathione
Cytochrome P450 monooxygenases

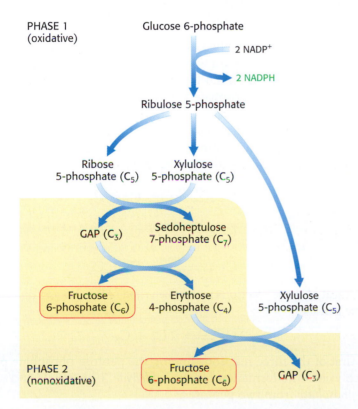

FIGURE 20.19 Pentose phosphate pathway. The pathway consists of (1) an oxidative phase that generates NADPH and (2) a nonoxidative phase that interconverts phosphorylated sugars.

20.3.1 Two Molecules of NADPH Are Generated in the Conversion of Glucose 6-phosphate into Ribulose 5-phosphate

The oxidative phase of the pentose phosphate pathway starts with the dehydrogenation of glucose 6-phosphate at carbon 1, a reaction catalyzed by *glucose 6-phosphate dehydrogenase* (Figure 20.20). This enzyme is highly specific for $NADP^+$; the K_M for NAD^+ is about a thousand times as great as that for $NADP^+$. The product is *6-phosphoglucono-δ-lactone*, which is an intramolecular ester between the C-1 carboxyl group and the C-5 hydroxyl group. The next step is the hydrolysis of 6-phosphoglucono-δ-lactone by a specific *lactonase* to give *6-phosphogluconate*. This six-carbon sugar is then oxidatively decarboxylated by *6-phosphogluconate dehydrogenase* to yield *ribulose 5-phosphate*. $NADP^+$ is again the electron acceptor. The final step in the synthesis of ribose 5-phosphate is the isomerization of ribulose 5-phosphate by phosphopentose isomerase (see Figure 20.11)

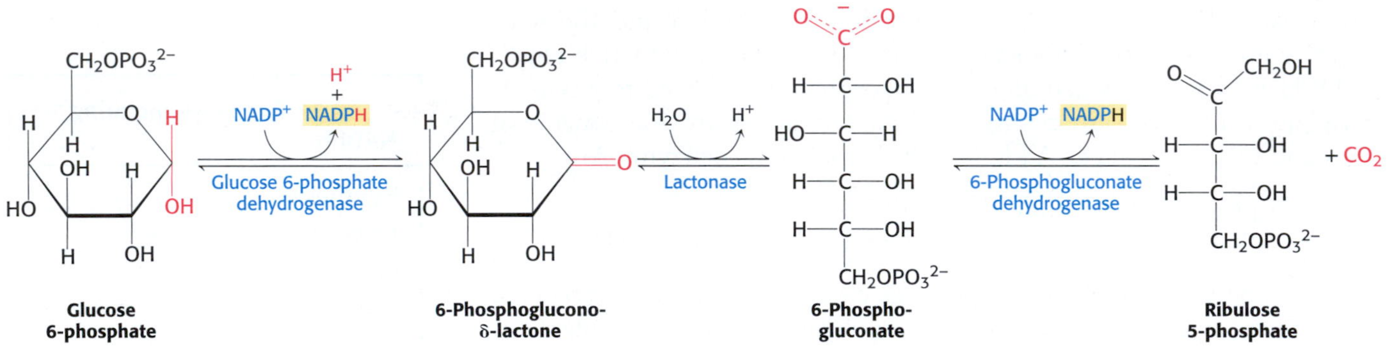

FIGURE 20.20 Oxidative phase of the pentose phosphate pathway. Glucose 6-phosphate is oxidized to 6-phosphoglucono-δ-lactone to generate one molecule of NADPH. The lactone product is hydrolyzed to 6-phosphogluconate, which is oxidatively decarboxylated to ribulose 5-phosphate with the generation of a second molecule of NADPH.

20.3.2 The Pentose Phosphate Pathway and Glycolysis Are Linked by Transketolase and Transaldolase

The preceding reactions yield two molecules of NADPH and one molecule of ribose 5-phosphate for each molecule of glucose 6-phosphate oxidized. However, many cells need NADPH for reductive biosyntheses much more than they need ribose 5-phosphate for incorporation into nucleotides and nucleic acids. In these cases, ribose 5-phosphate is converted into glyceraldehyde 3-phosphate and fructose 6-phosphate by *transketolase* and *transaldolase*. *These enzymes create a reversible link between the pentose phosphate pathway and glycolysis by catalyzing these three successive reactions.*

$$C_5 + C_5 \xrightleftharpoons{\text{Transketolase}} C_3 + C_7$$

$$C_3 + C_7 \xrightleftharpoons{\text{Transaldolase}} C_6 + C_4$$

$$C_4 + C_5 \xrightleftharpoons{\text{Transketolase}} C_6 + C_3$$

The net result of these reactions is the *formation of two hexoses and one triose from three pentoses*:

$$3\,C_5 \rightleftharpoons 2\,C_6 + C_3$$

The first of the three reactions linking the pentose phosphate pathway and glycolysis is the formation of *glyceraldehyde 3-phosphate* and *sedoheptulose 7-phosphate* from two pentoses.

| Xylulose 5-phosphate | Ribose 5-phosphate | Transketolase | Glyceraldehyde 3-phosphate | Sedoheptulose 7-phosphate |

The donor of the two-carbon unit in this reaction is xylulose 5-phosphate, an epimer of ribulose 5-phosphate. A ketose is a substrate of transketolase only if its hydroxyl group at C-3 has the configuration of xylulose rather than ribulose. Ribulose 5-phosphate is converted into the appropriate epimer for the transketolase reaction by *phosphopentose epimerase* (see Figure 20.11) in the reverse reaction of that which occurs in the Calvin cycle.

Glyceraldehyde 3-phosphate and sedoheptulose 7-phosphate generated by the transketolase then react to form *fructose 6-phosphate* and *erythrose 4-phosphate*.

| Glyceraldehyde 3-phosphate | Sedoheptulose 7-phosphate | Transaldolase | Fructose 6-phosphate | Erythrose 4-phosphate |

This synthesis of a four-carbon sugar and a six-carbon sugar is catalyzed by *transaldolase*.

In the third reaction, transketolase catalyzes the synthesis of *fructose 6-phosphate* and *glyceraldehyde 3-phosphate* from erythrose 4-phosphate and xylulose 5-phosphate.

| Erythrose 4-phosphate | Xylulose 5-phosphate | Transketolase | Fructose 6-phosphate | Glyceraldehyde 3-phosphate |

The sum of these reactions is

2 Xylulose 5-phosphate + ribose 5-phosphate $\rightleftharpoons$
2 fructose 6-phosphate + glyceraldehyde 3-phosphate

Xylulose 5-phosphate can be formed from ribose 5-phosphate by the sequential action of phosphopentose isomerase and phosphopentose epimerase, and so the net reaction starting from ribose 5-phosphate is

3 Ribose 5-phosphate $\rightleftharpoons$
2 fructose 6-phosphate + glyceraldehyde 3-phosphate

TABLE 20.3 Pentose phosphate pathway

Reaction	Enzyme
Oxidative phase	
Glucose 6-phosphate + NADP$^+$ $\longrightarrow$ 6-phosphoglucono-δ-lactone + NADPH + H$^+$	Glucose 6-phosphate dehydrogenase
6-Phosphoglucono-δ-lactone + H$_2$O $\longrightarrow$ 6-phosphogluconate + H$^+$	Lactonase
6-Phosphogluconate + NADP$^+$ $\longrightarrow$ ribulose 5-phosphate + CO$_2$ + NADPH	6-Phosphogluconate dehydrogenase
Nonoxidative Phase	
Ribulose 5-phosphate $\rightleftharpoons$ ribose 5-phosphate	Phosphopentose isomerase
Ribulose 5-phosphate $\rightleftharpoons$ xylulose 5-phosphate	Phosphopentose epimerase
Xylulose 5-phosphate + ribose 5-phosphate $\rightleftharpoons$ sedoheptulose 7-phosphate + glyceraldehyde 3-phosphate	Transketolase
Sedoheptulose 7-phosphate + glyceraldehyde 3-phosphate $\rightleftharpoons$ fructose 6-phosphate + erythrose 4-phosphate	Transaldolase
Xylulose 5-phosphate + erythrose 4-phosphate $\rightleftharpoons$ fructose 6-phosphate + glyceraldehyde 3-phosphate	Transketolase

Thus, *excess ribose 5-phosphate formed by the pentose phosphate pathway can be completely converted into glycolytic intermediates.* Moreover, any ribose ingested in the diet can be processed into glycolytic intermediates by this pathway. It is evident that the carbon skeletons of sugars can be extensively rearranged to meet physiologic needs (Table 20.3).

20.3.3 Transketolase and Transaldolase Stabilize Carbanionic Intermediates by Different Mechanisms

The reactions catalyzed by transketolase and transaldolase are distinct yet similar in many ways. One difference is that transketolase transfers a two-carbon unit, whereas transaldolase transfers a three-carbon unit. Each of these units is transiently attached to the enzyme in the course of the reaction. In transketolase, the site of addition of the unit is the thiazole ring of the required coenzyme thiamine pyrophosphate. Transketolase is homologous to the E$_1$ subunit of the pyruvate dehydrogenase complex (Section 17.1.1) and the reaction mechanism is similar (Figure 20.21).

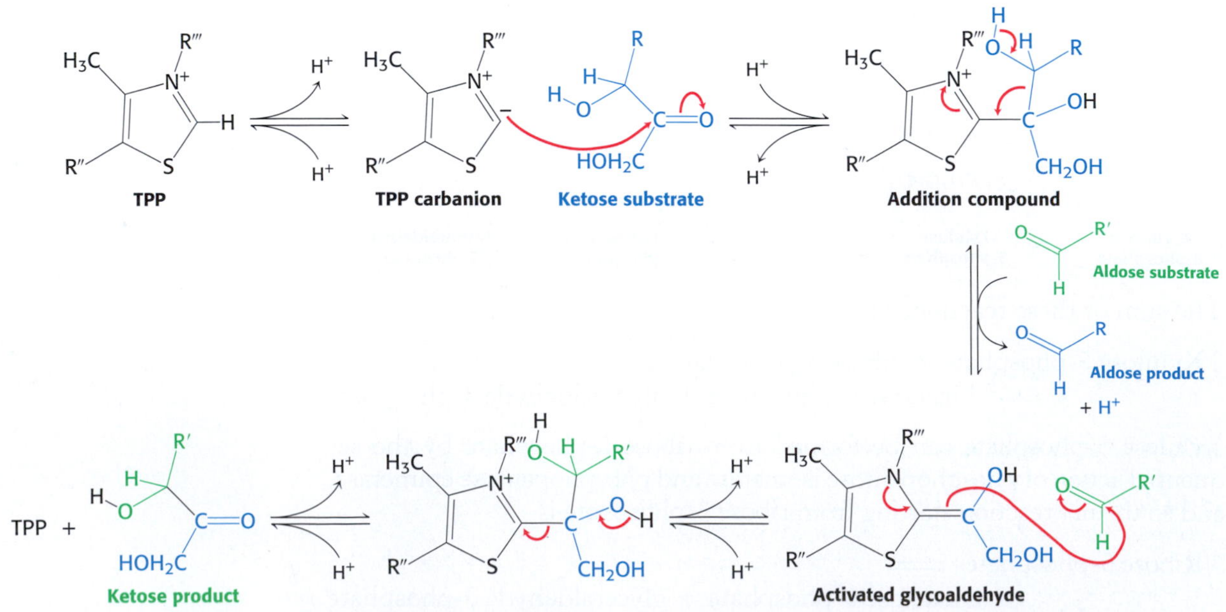

FIGURE 20.21 Transketolase mechanism. The carbanion of thiamine pyrophosphate (TPP) attacks the ketose substrate. Cleavage of a carbon–carbon bond frees the aldose product and leaves a two-carbon fragment joined to TPP. This activated glycoaldehyde intermediate attacks the aldose substrate to form a new carbon–carbon bond. The ketose product is released, freeing the TPP for the next reaction cycle.

The C-2 carbon atom of bound TPP readily ionizes to give a *carbanion*. The negatively charged carbon atom of this reactive intermediate attacks the carbonyl group of the ketose substrate. The resulting addition compound releases the aldose product to yield an *activated glycoaldehyde unit*. The positively charged nitrogen atom in the thiazole ring acts as an *electron sink* in the development of this activated intermediate. The carbonyl group of a suitable aldose acceptor then condenses with the activated glycoaldehyde unit to form a new ketose, which is released from the enzyme.

Transaldolase transfers a three-carbon *dihydroxyacetone* unit from a ketose donor to an aldose acceptor. Transaldolase, in contrast with transketolase, does not contain a prosthetic group. Rather, *a Schiff base is formed between the carbonyl group of the ketose substrate and the ε-amino group of a lysine residue at the active site of the enzyme* (Figure 20.22). This kind of covalent enzyme–substrate intermediate is like that formed in fructose 1,6-bisphosphate aldolase in the glycolytic pathway (Section 16.1.3) and, indeed, the enzymes are homologous. The Schiff base becomes protonated, the bond between C-3 and C-4 is split, and an aldose is released. The negative charge on the Schiff-base carbanion moiety is stabilized by resonance. The positively charged nitrogen atom of the protonated Schiff base acts as an electron sink. The Schiff-base adduct is stable until a suitable aldose becomes bound. The dihydroxyacetone moiety then reacts with the carbonyl group of the aldose. The ketose product is released by hydrolysis of the Schiff base. The nitrogen atom of the protonated Schiff base plays the same role in transaldolase as the thiazole-ring nitrogen atom does in transketolase. In each enzyme, a group within an intermediate reacts like

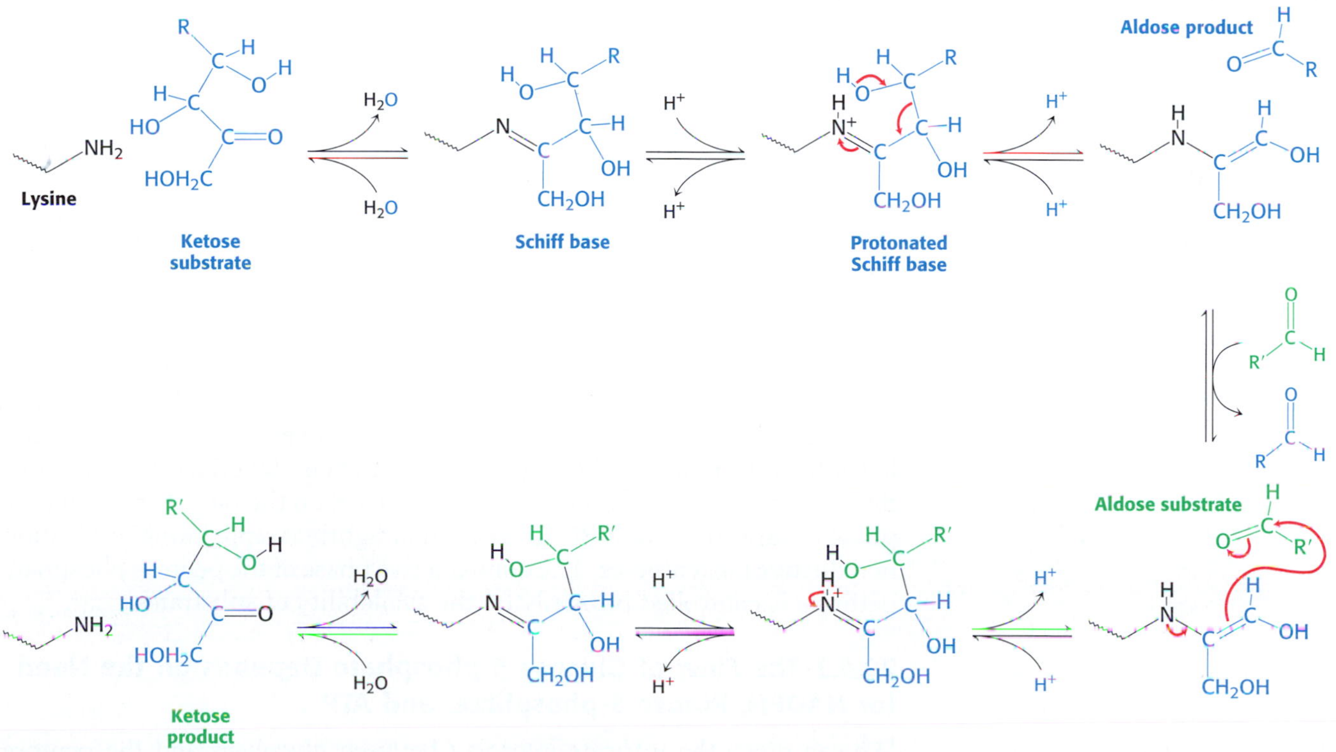

FIGURE 20.22 Transaldolase mechanism. The reaction begins with the formation of a Schiff base between a lysine residue in transaldolase and the ketose substrate. Protonation of the Schiff base leads to release of the aldose product, leaving a three-carbon fragment attached to the lysine residue. This intermediate adds to the aldose substrate to form a new carbon–carbon bond. The reaction cycle is completed by release of the ketose product from the lysine side chain.

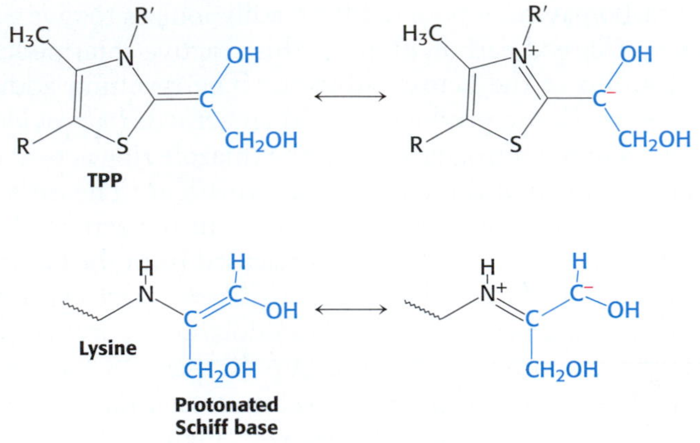

FIGURE 20.23 Carbanion intermediates.
For transketolase and transaldolase, a
carbanion intermediate is stabilized by
resonance. In transketolase, TPP stabilizes
this intermediate; in transaldolase, a
protonated Schiff base plays this role.

**CONCEPTUAL INSIGHTS,
Overview of Carbohydrate and Fatty
Acid Metabolism.** *View this media
module to gain a "bigger picture" understanding of the roles of the pentose
phosphate pathway in the context of
other metabolic pathways (glycolysis,
citric acid cycle, glycogen and fatty
acid metabolism).*

a carbanion in attacking a carbonyl group to form a new carbon–carbon
bond. In each case, the charge on the carbanion is stabilized by resonance
(Figure 20.23).

20.4 THE METABOLISM OF GLUCOSE 6-PHOSPHATE BY THE PENTOSE PHOSPHATE PATHWAY IS COORDINATED WITH GLYCOLYSIS

Glucose 6-phosphate is metabolized by both the glycolytic pathway (Chapter 16) and the pentose phosphate pathway. How is the processing of this
important metabolite partitioned between these two metabolic routes? The
cytoplasmic concentration of $NADP^+$ plays a key role in determining the
fate of glucose 6-phosphate.

20.4.1 The Rate of the Pentose Phosphate Pathway Is Controlled by the Level of $NADP^+$

The first reaction in the oxidative branch of the pentose phosphate pathway, the dehydrogenation of glucose 6-phosphate, is essentially irreversible.
In fact, this reaction is rate limiting under physiological conditions and
serves as the control site. The most important regulatory factor is the level
of $NADP^+$, the electron acceptor in the oxidation of glucose 6-phosphate
to 6-phosphoglucono-δ-lactone. The inhibitory effect of low levels of
$NADP^+$ is exacerbated by the fact that NADPH competes with $NADP^+$
in binding to the enzyme. The ratio of $NADP^+$ to NADPH in the cytosol
of a liver cell from a well-fed rat is about 0.014, several orders of magnitude
lower than the ratio of NAD^+ to NADH, which is 700 under the same conditions. The marked effect of the $NADP^+$ level on the rate of the oxidative
phase ensures that NADPH generation is tightly coupled to its utilization
in reductive biosyntheses. The nonoxidative phase of the pentose phosphate
pathway is controlled primarily by the availability of substrates.

20.4.2 The Flow of Glucose 6-phosphate Depends on the Need for NADPH, Ribose 5-phosphate, and ATP

We can grasp the intricate interplay between glycolysis and the pentose
phosphate pathway by examining the metabolism of glucose 6-phosphate
in four different metabolic situations (Figure 20.24).

Mode 1. *Much more ribose 5-phosphate than NADPH is required.* For example, rapidly dividing cells need ribose 5-phosphate for the synthesis of
nucleotide precursors of DNA. Most of the glucose 6-phosphate is converted

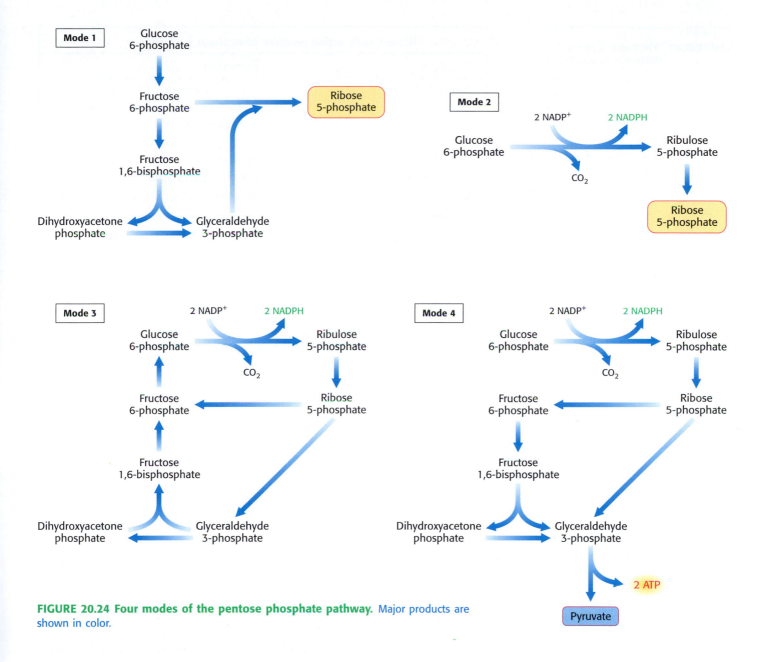

FIGURE 20.24 Four modes of the pentose phosphate pathway. Major products are shown in color.

into fructose 6-phosphate and glyceraldehyde 3-phosphate by the glycolytic pathway. Transaldolase and transketolase then convert two molecules of fructose 6-phosphate and one molecule of glyceraldehyde 3-phosphate into three molecules of ribose 5-phosphate by a reversal of the reactions described earlier. The stoichiometry of mode 1 is

$$5 \text{ Glucose 6-phosphate} + \text{ATP} \longrightarrow 6 \text{ ribose 5-phosphate} + \text{ADP} + \text{H}^+$$

Mode 2. *The needs for NADPH and ribose 5-phosphate are balanced.* The predominant reaction under these conditions is the formation of two molecules of NADPH and one molecule of ribose 5-phosphate from one molecule of glucose 6-phosphate in the oxidative phase of the pentose phosphate pathway. The stoichiometry of mode 2 is

$$\text{Glucose 6-phosphate} + 2 \text{ NADP}^+ + \text{H}_2\text{O} \longrightarrow$$
$$\text{ribose 5-phosphate} + 2 \text{ NADPH} + 2 \text{ H}^+ + \text{CO}_2$$

Mode 3. *Much more NADPH than ribose 5-phosphate is required.* For example, adipose tissue requires a high level of NADPH for the synthesis of

TABLE 20.4 Tissues with active pentose phosphate pathways

Tissue	Function
Adrenal gland	Steroid synthesis
Liver	Fatty acid and cholesterol synthesis
Testes	Steroid synthesis
Adipose tissue	Fatty acid synthesis
Ovary	Steroid synthesis
Mammary gland	Fatty acid synthesis
Red blood cells	Maintenance of reduced glutathione

fatty acids (Table 20.4). In this case, glucose 6-phosphate is completely oxidized to CO_2. Three groups of reactions are active in this situation. First, the oxidative phase of the pentose phosphate pathway forms two molecules of NADPH and one molecule of ribose 5-phosphate. Then, ribose 5-phosphate is converted into fructose 6-phosphate and glyceraldehyde 3-phosphate by transketolase and transaldolase. Finally, glucose 6-phosphate is resynthesized from fructose 6-phosphate and glyceraldehyde 3-phosphate by the gluconeogenic pathway. The stoichiometries of these three sets of reactions are

$$6 \text{ Glucose 6-phosphate} + 12 \text{ NADP}^+ + 6 \text{ H}_2\text{O} \longrightarrow$$
$$6 \text{ ribose 5-phosphate} + 12 \text{ NADPH} + 12 \text{ H}^+ + 6 \text{ CO}_2$$

$$6 \text{ Ribose 5-phosphate} \longrightarrow$$
$$4 \text{ fructose 6-phosphate} + 2 \text{ glyceraldehyde 3-phosphate}$$

$$4 \text{ Fructose 6-phosphate} + 2 \text{ glyceraldehyde 3-phosphate} + \text{H}_2\text{O} \longrightarrow$$
$$5 \text{ glucose 6-phosphate} + \text{P}_i$$

The sum of the mode 3 reactions is

$$\text{Glucose 6-phosphate} + 12 \text{ NADP}^+ + 7 \text{ H}_2\text{O} \longrightarrow$$
$$6 \text{ CO}_2 + 12 \text{ NADPH} + 12 \text{ H}^+ + \text{P}_i$$

Thus, *the equivalent of glucose 6-phosphate can be completely oxidized to CO_2 with the concomitant generation of NADPH.* In essence, ribose 5-phosphate produced by the pentose phosphate pathway is recycled into glucose 6-phosphate by transketolase, transaldolase, and some of the enzymes of the gluconeogenic pathway.

Mode 4. *Both NADPH and ATP are required.* Alternatively, ribose 5-phosphate formed by the oxidative phase of the pentose phosphate pathway can be converted into pyruvate. Fructose 6-phosphate and glyceraldehyde 3-phosphate derived from ribose 5-phosphate enter the glycolytic pathway rather than reverting to glucose 6-phosphate. In this mode, *ATP and NADPH are concomitantly generated, and five of the six carbons of glucose 6-phosphate emerge in pyruvate.*

$$3 \text{ Glucose 6-phosphate} + 6 \text{ NADP}^+ + 5 \text{ NAD}^+ + 5 \text{ P}_i + 8 \text{ ADP} \longrightarrow$$
$$5 \text{ pyruvate} + 3 \text{ CO}_2 + 6 \text{ NADPH} + 5 \text{ NADH}$$
$$+ 8 \text{ ATP} + 2 \text{ H}_2\text{O} + 8 \text{ H}^+$$

Pyruvate formed by these reactions can be oxidized to generate more ATP or it can be used as a building block in a variety of biosyntheses.

20.4.3 Through the Looking Glass: The Calvin Cycle and the Pentose Phosphate Pathway

The complexities of the Calvin cycle and the pentose phosphate pathway are easier to comprehend if we consider them mirror images of each other. The Calvin cycle begins with the fixation of CO_2 and goes on to use NADPH in the synthesis of glucose. The pentose phosphate pathway begins with the oxidation of a glucose-derived carbon atom to CO_2 and concomitantly generates NADPH. The regeneration phase of the Calvin cycle converts C_6 and C_3 molecules back into the starting material—the C_5 molecule ribulose 1,5-bisphosphate. The pentose phosphate pathway converts a C_5 molecule, ribose 5-phosphate, into C_6 and C_3 intermediates of the glycolytic pathway. Not surprisingly, in photosynthetic organisms, many enzymes are common to the two pathways. We see the economy of evolution: the use of identical enzymes for similar reactions with different ends.

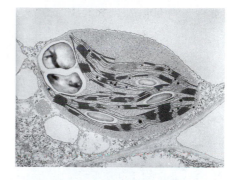

Electron micrograph of a chloroplast. The thylakoid membranes course throughout the stroma of a chloroplast from a cell of *Phleum pratense*, a grass. The dark areas of stacked thylakoid membrane are grana. Several large starch granules, which store the newly synthesized glucose, are also obvious. [Biophoto Associates/Photo Researchers.]

20.5 GLUCOSE 6-PHOSPHATE DEHYDROGENASE PLAYS A KEY ROLE IN PROTECTION AGAINST REACTIVE OXYGEN SPECIES

Reactive oxygen species (ROS) generated in oxidative metabolism inflict damage on all classes of macromolecules and can ultimately lead to cell death. Indeed, ROS are implicated in a number of human diseases (Section 18.3.6). Reduced *glutathione* (GSH), a tripeptide with a free sulfhydryl group, is required to combat oxidative stress and maintain the normal reduced state in the cell. Oxidized glutathione (GSSG) is reduced by NADPH generated by glucose 6-phosphate dehydrogenase in the pentose phosphate pathway. Indeed, cells with reduced levels of glucose 6-phosphate dehydrogenase are especially sensitive to oxidative stress. This stress is most acute in red blood cells because, lacking mitochondria, they have no alternative means of generating reducing power.

Glutathione (reduced)
(γ-Glutamylcysteinylglycine)

20.5.1 Glucose 6-phosphate Dehydrogenase Deficiency Causes a Drug-Induced Hemolytic Anemia

At its introduction in 1926, an antimalarial drug, pamaquine, was associated with the appearance of severe and mysterious ailments. Most patients tolerated the drug well, but a few developed severe symptoms within a few days after therapy was started. The urine turned black, jaundice developed, and the hemoglobin content of the blood dropped sharply. In some cases, massive destruction of red blood cells caused death.

This drug-induced hemolytic anemia was shown 30 years later to be caused by a *deficiency of glucose 6-phosphate dehydrogenase*, the enzyme catalyzing the first step in the oxidative branch of the pentose phosphate pathway. This defect, which is inherited on the X chromosome, is the most common enzymopathy, affecting hundreds of millions of people. The major role of NADPH in red cells is to reduce the disulfide form of glutathione to the sulfhydryl form. The enzyme that catalyzes the regeneration of reduced glutathione, the flavoprotein *glutathione reductase*, a dimer of 50-kd subunits, is homologous to ferredoxin-$NADP^+$ reductase, which we encountered in photosynthesis (Section 19.3.4). The reduced form of glutathione serves as a *sulfhydryl buffer* that maintains the cysteine residues of hemoglobin and other red-blood-cell proteins in the reduced state. Normally, the ratio of the reduced to oxidized forms of glutathione in red blood cells is 500.

How is GSH regenerated from GSSG and NADPH by glutathione reductase? The electrons from NADPH are not directly transferred to the

Vicia faba. The Mediterranean plant *Vicia faba* is a source of fava beans that contain the purine glycoside vicine. [Inga Spence/Visuals Unlimited.]

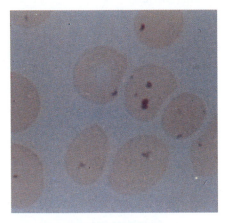

FIGURE 20.25 Red blood cells with Heinz bodies. The light micrograph shows red blood cells obtained from a person deficient in glucose 6-phosphate dehydrogenase. The dark particles, called Heinz bodies, inside the cells are clumps of denatured protein that adhere to the plasma membrane and stain with basic dyes. Red blood cells in such people are highly susceptible to oxidative damage. [Courtesy of Dr. Stanley Schrier.]

disulfide bond in oxidized glutathione. Rather, they are transferred from NADPH to a tightly bound flavin adenine dinucleotide (FAD) on the reductase, then to a disulfide bridge between two cysteine residues in the enzyme subunit, and finally to oxidized glutathione.

Reduced glutathione is essential for maintaining the normal structure of red blood cells and for keeping hemoglobin in the ferrous state. The reduced form also plays a role in detoxification by reacting with hydrogen peroxide and organic peroxides.

$$2\,GSH + ROOH \longrightarrow GSSG + H_2O + ROH$$

Cells with a lowered level of reduced glutathione are more susceptible to hemolysis. How can we explain this phenomenon biochemically? The presence of pamaquine, a purine glycoside of fava beans, or other nonenzymatic oxidative agents leads to the generation of peroxides, reactive oxygen species that can damage membranes as well as other biomolecules. Peroxides are normally eliminated by glutathione peroxidase with the use of glutathione as a reducing agent (Section 24.4.1). Moreover, in the absence of the enzyme, the hemoglobin sulfhydryl groups can no longer be maintained in the reduced form and hemoglobin molecules then cross-link with one another to form aggregates called *Heinz bodies* on cell membranes (Figure 20.25). Membranes damaged by the Heinz bodies and reactive oxygen species become deformed and the cell is likely to undergo lysis. In the absence of oxidative stress, however, the deficiency is quite benign. The occurrence of this dehydrogenase deficiency also clearly demonstrates that *atypical reactions to drugs may have a genetic basis.*

20.5.2 A Deficiency of Glucose 6-phosphate Dehydrogenase Confers an Evolutionary Advantage in Some Circumstances

The incidence of the most common form of glucose 6-phosphate dehydrogenase deficiency, characterized by a tenfold reduction in enzymatic activity in red blood cells, is 11% among Americans of African heritage. This high frequency suggests that the deficiency may be advantageous under certain environmental conditions. Indeed, *glucose 6-phosphate dehydrogenase deficiency protects against falciparum malaria.* The parasites causing this disease require reduced glutathione and the products of the pentose phosphate pathway for optimal growth. Thus, glucose 6-phosphate dehydrogenase deficiency is a mechanism of protection against malaria, which accounts for its high frequency in malaria-infested regions of the world. We see here once again the interplay of heredity and environment in the production of disease.

SUMMARY

● **The Calvin Cycle Synthesizes Hexoses from Carbon Dioxide and Water**

ATP and NADPH formed in the light reactions of photosynthesis are used to convert CO_2 into hexoses and other organic compounds. The dark phase of photosynthesis, called the Calvin cycle, starts with the reaction of CO_2 and ribulose 1,5-bisphosphate to form two molecules of 3-phosphoglycerate. The steps in the conversion of 3-phosphoglycerate into fructose 6-phosphate and glucose 6-phosphate are like those of gluconeogenesis, except that glyceraldehyde 3-phosphate dehydrogenase in chloroplasts is specific for NADPH rather than NADH. Ribulose 1,5-bisphosphate is regenerated from fructose 6-phosphate, glyceraldehyde 3-phosphate, and dihydroxyacetone phosphate by a complex series of reactions. Several of the steps in the regeneration of ribulose 1,5-bisphosphate are like those of the pentose phosphate pathway. Three molecules of ATP and two molecules of NADPH are consumed for each molecule of CO_2 converted into a hexose. Starch in chloroplasts and sucrose in the cytosol are the major carbohydrate stores in plants.

The Activity of the Calvin Cycle Depends on Environmental Conditions

Reduced thioredoxin formed by the light-driven transfer of electrons from ferredoxin activates enzymes of the Calvin cycle by reducing disulfide bridges. The light-induced increase in pH and Mg^{2+} level of the stroma is important in stimulating the carboxylation of ribulose 1,5-bisphosphate by ribulose 1,5-bisphosphate carboxylase. This enzyme also catalyzes a competing oxygenase reaction, which produces phosphoglycolate and 3-phosphoglycerate. The recycling of phosphoglycolate leads to the release of CO_2 and further consumption of O_2 in a process called photorespiration. This wasteful side reaction is minimized in tropical plants, which have an accessory pathway—called the C_4 pathway—for concentrating CO_2 at the site of the Calvin cycle. This pathway enables tropical plants to take advantage of high levels of light and minimize the oxygenation of ribulose 1,5-bisphosphate. Plants in arid ecosystems employ Crassulacean acid metabolism (CAM) to prevent dehydration. In CAM plants, the C_4 pathway is active during the night when the plant exchanges gases with the air. During the day, gas exchange is eliminated and CO_2 is generated from malate stored in vacuoles.

The Pentose Phosphate Pathway Generates NADPH and Synthesizes Five-Carbon Sugars

Whereas the Calvin cycle is present only in photosynthetic organisms, the pentose phosphate pathway is present in all organisms. The pentose phosphate pathway generates NADPH and ribose 5-phosphate in the cytosol. NADPH is used in reductive biosyntheses, whereas ribose 5-phosphate is used in the synthesis of RNA, DNA, and nucleotide coenzymes. The pentose phosphate pathway starts with the dehydrogenation of glucose 6-phosphate to form a lactone, which is hydrolyzed to give 6-phosphogluconate and then oxidatively decarboxylated to yield ribulose 5-phosphate. $NADP^+$ is the electron acceptor in both of these oxidations. The last step is the isomerization of ribulose 5-phosphate (a ketose) to ribose 5-phosphate (an aldose). A different mode of the pathway is active when cells need much more NADPH than ribose 5-phosphate. Under these conditions, ribose 5-phosphate is converted into glyceraldehyde 3-phosphate and fructose 6-phosphate by transketolase and transaldolase. These two enzymes create a reversible link between the pentose phosphate pathway and glycolysis. Xylulose 5-phosphate, sedoheptulose 7-phosphate, and erythrose 4-phosphate are intermediates in these interconversions. In this way, 12 molecules of NADPH can be generated for each molecule of glucose 6-phosphate that is completely oxidized to CO_2.

The Metabolism of Glucose 6-phosphate by the Pentose Phosphate Pathway Is Coordinated with Glycolysis

Only the nonoxidative branch of the pathway is significantly active when much more ribose 5-phosphate than NADPH needs to be synthesized. Under these conditions, fructose 6-phosphate and glyceraldehyde 3-phosphate (formed by the glycolytic pathway) are converted into ribose 5-phosphate without the formation of NADPH. Alternatively, ribose 5-phosphate formed by the oxidative branch can be converted into pyruvate through fructose 6-phosphate and glyceraldehyde 3-phosphate. In this mode, ATP and NADPH are generated, and five of the six carbons of glucose 6-phosphate emerge in pyruvate. The interplay of the glycolytic and pentose phosphate pathways enables the levels of NADPH, ATP, and building blocks such as ribose 5-phosphate and pyruvate to be continuously adjusted to meet cellular needs.

• Glucose 6-phosphate Dehydrogenase Plays a Key Role in Protection Against Reactive Oxygen Species

NADPH generated by glucose 6-phosphate dehydrogenase maintains the appropriate levels of reduced glutathione required to combat oxidative stress and maintain the proper reducing environment in the cell. Cells with diminished glucose 6-phosphate dehydrogenase activity are especially sensitive to oxidative stress.

KEY TERMS

Calvin cycle (dark reactions) (p. 552)

autotroph (p. 552)

heterotroph (p. 552)

rubisco (ribulose 1,5-bisphosphate carboxylase/oxygenase) (p. 553)

peroxisome (p. 555)

photorespiration (p. 556)

starch (p. 559)

sucrose (p. 559)

thioredoxin (p. 560)

C_4 pathway (p. 561)

C_4 plant (p. 562)

C_3 plant (p. 562)

Crassulacean acid metabolism (CAM) (p. 562)

pentose phosphate pathway (p. 563)

glucose 6-phosphate dehydrogenase (p. 564)

transketolase (p. 564)

transaldolase (p. 564)

glutathione (p. 571)

SELECTED READINGS

Where to start

Horecker, B. L., 1976. Unravelling the pentose phosphate pathway. In *Reflections on Biochemistry* (pp. 65–72), edited by A. Kornberg, L. Cornudella, B. L. Horecker, and J. Oro, Pergamon.

Levi, P., 1984. Carbon. In *The Periodic Table.* Random House.

Melendez-Hevia, E., and Isidoro, A., 1985. The game of the pentose phosphate cycle. *J. Theor. Biol.* 117:251–263.

Barber, J., and Andersson, B., 1994. Revealing the blueprint of photosynthesis. *Nature* 370:31–34.

Rawsthorne, S., 1992. Towards an understanding of C_3-C_4 photosynthesis. *Essays Biochem.* 27:135–146.

Books and general reviews

Wood, T., 1985. *The Pentose Phosphate Pathway.* Academic Press.

Buchanan, B. B., Gruissem, W., and Jones, R. L., 2000. *Biochemistry and Molecular Biology of Plants.* American Society of Plant Physiologists.

Schneider, G., Lindqvist, Y., and Branden, C., 1992. Rubisco: Structure and mechanism. *Annu. Rev. Biophys. Biomol. Struct.* 21:119–143.

Spreitzer, R. J., 1993. Genetic dissection of rubisco structure and function. *Annu. Rev. Plant Physiol. Plant Mol. Biol.* 44:411–434.

Enzymes and reaction mechanisms

Harrison, D. H., Runquist, J. A., Holub, A., and Miziorko, H. M., 1998. The crystal structure of phosphoribulokinase from *Rhodobacter sphaeroides* reveals a fold similar to that of adenylate kinase. *Biochemistry* 37:5074–5085.

Miziorko, H. M., 2000. Phosphoribulokinase: Current perspectives on the structure/function basis for regulation and catalysis. *Adv. Enzymol. Relat. Areas Mol. Biol.* 74:95–127.

Thorell, S., Gergely, P., Jr., Banki, K., Perl, A., and Schneider, G., 2000. The three-dimensional structure of human transaldolase. *FEBS Lett.* 475:205–208.

Lindqvist, Y., Schneider, G., Ermler, U., and Sundstrom, M., 1992. Three-dimensional structure of transketolase, a thiamine diphosphate dependent enzyme, at 2.5 Å resolution. *EMBO J.* 11:2373–2379.

Robinson, B. H., and Chun, K., 1993. The relationships between transketolase, yeast pyruvate decarboxylase and pyruvate dehydrogenase of the pyruvate dehydrogenase complex. *FEBS Lett.* 328:99–102.

Carbon dioxide fixation and rubisco

Sugawara, H., Yamamoto, H., Shibata, N., Inoue, T., Okada, S., Miyake, C., Yokota, A., and Kai, Y., 1999. Crystal structure of carboxylase reaction-oriented ribulose 1,5-bisphosphate carboxylase/oxygenase from a thermophilic red alga, *Galdieria partita. J. Biol. Chem.* 274:15655–15661.

Hansen, S., Vollan, V. B., Hough, E., and Andersen, K., 1999. The crystal structure of rubisco from *Alcaligenes eutrophus* reveals a novel central eight-stranded beta-barrel formed by beta-strands from four subunits. *J. Mol. Biol.* 288:609–621.

Knight, S., Andersson, I., and Branden, C. I., 1990. Crystallographic analysis of ribulose 1,5-bisphosphate carboxylase from spinach at 2.4 Å resolution: Subunit interactions and active site. *J. Mol. Biol.* 215:113–160.

Taylor, T. C., and Andersson, I., 1997. The structure of the complex between rubisco and its natural substrate ribulose 1,5-bisphosphate. *J. Mol. Biol.* 265:432–444.

Cleland, W. W., Andrews, T. J., Gutteridge, S., Hartman, F. C., and Lorimer, G. H., 1998. Mechanism of rubisco: The carbamate as general base. *Chem. Rev.* 98:549–561.

Buchanan, B. B., 1992. Carbon dioxide assimilation in oxygenic and anoxygenic photosynthesis. *Photosynth. Res.* 33:147–162.

Hatch, M. D., 1987. C_4 photosynthesis: A unique blend of modified biochemistry, anatomy, and ultrastructure. *Biochim. Biophys. Acta* 895:81–106.

Regulation

Rokka, A., Zhang, L., and Aro, E.-M., 2001. Rubisco activase: An enzyme with a temperature-dependent dual function? *Plant J.* 25:463–472.

Zhang, N., and Portis, A. R., Jr., 1999. Mechanism of light regulation of rubisco: A specific role for the larger rubisco activase isoform involving reductive activation by thioredoxin-f. *Proc. Natl. Acad. Sci. USA* 96:9438–9443.

Wedel, N., Soll, J., and Paap, B. K., 1997. CP12 provides a new mode of light regulation of Calvin cycle activity in higher plants. *Proc. Natl. Acad. Sci. USA* 94:10479–10484.

Avilan, L., Lebreton, S., and Gontero, B., 2000. Thioredoxin activation of phosphoribulokinase in a bi-enzyme complex from *Chlamydomonas reinhardtii* chloroplasts. *J. Biol. Chem.* 275:9447–9451.

Irihimovitch, V., and Shapira, M., 2000. Glutathione redox potential modulated by reactive oxygen species regulates translation of rubisco large subunit in the chloroplast. *J. Biol. Chem.* 275:16289–16295.

Glucose 6-phosphate dehydrogenase

Au, S. W., Gover, S., Lam, V. M., and Adams, M. J., 2000. Human glucose-6-phosphate dehydrogenase: The crystal structure reveals a structural NADP(+) molecule and provides insights into enzyme deficiency. *Structure Fold. Des.* 8:293–303.

Salvemini, F., Franze, A., Iervolino, A., Filosa, S., Salzano, S., and Ursini, M. V., 1999. Enhanced glutathione levels and oxidoresistance mediated by increased glucose-6-phosphate dehydrogenase expression. *J. Biol. Chem.* 274:2750–2757.

Tian, W. N., Braunstein, L. D., Apse, K., Pang, J., Rose, M., Tian, X., and Stanton, R. C., 1999. Importance of glucose-6-phosphate dehydrogenase activity in cell death. *Am. J. Physiol.* 276:C1121–C1131.

Tian, W. N., Braunstein, L. D., Pang, J., Stuhlmeier, K. M., Xi, Q. C., Tian, X., and Stanton, R. C., 1998. Importance of glucose-6-phosphate dehydrogenase activity for cell growth. *J. Biol. Chem.* 273:10609–10617.

Ursini, M. V., Parrella, A., Rosa, G., Salzano, S., and Martini, G., 1997. Enhanced expression of glucose-6-phosphate dehydrogenase in human cells sustaining oxidative stress. *Biochem. J.* 323:801–806.

Evolution

Coy, J. F., Dubel, S., Kioschis, P., Thomas, K., Micklem, G., Delius, H., and Poustka, A., 1996. Molecular cloning of tissue-specific transcripts of a transketolase-related gene: Implications for the evolution of new vertebrate genes. *Genomics* 32:309–316.

Schenk, G., Layfield, R., Candy, J. M., Duggleby, R. G., and Nixon, P. F., 1997. Molecular evolutionary analysis of the thiamine-diphosphate-dependent enzyme, transketolase. *J. Mol. Evol.* 44: 552–572.

Notaro, R., Afolayan, A., and Luzzatto L., 2000. Human mutations in glucose 6-phosphate dehydrogenase reflect evolutionary history. *FASEB J.* 14:485–494.

Wedel, N., and Soll, J., 1998. Evolutionary conserved light regulation of Calvin cycle activity by NADPH-mediated reversible phosphoribulokinase/CP12/glyceraldehyde-3-phosphate dehydrogenase complex dissociation. *Proc. Natl. Acad. Sci. USA* 95:9699–9704.

Martin, W., and Schnarrenberger, C., 1997. The evolution of the Calvin cycle from prokaryotic to eukaryotic chromosomes: A case study of functional redundancy in ancient pathways through endosymbiosis. *Curr. Genet.* 32:1–18.

Ku, M. S., Kano-Murakami, Y., and Matsuoka, M., 1996. Evolution and expression of C4 photosynthesis genes. *Plant Physiol.* 111:949–957.

Pereto, J. G., Velasco, A. M., Becerra, A., and Lazcano, A., 1999. Comparative biochemistry of CO_2 fixation and the evolution of autotrophy. *Int. Microbiol.* 2:3–10.

PROBLEMS

1. *Variation on a theme.* Sedoheptulose 1,7-bisphosphate is an intermediate in the Calvin cycle but not in the pentose phosphate pathway. What is the enzymatic basis of this difference?

2. *Total eclipse.* An illuminated suspension of *Chlorella* is actively carrying out photosynthesis. Suppose that the light is suddenly switched off. How would the levels of 3-phosphoglycerate and ribulose 1,5-bisphosphate change in the next minute?

3. *CO_2 deprivation.* An illuminated suspension of *Chlorella* is actively carrying out photosynthesis in the presence of 1% CO_2. The concentration of CO_2 is abruptly reduced to 0.003%. What effect would this reduction have on the levels of 3-phosphoglycerate and ribulose 1,5-bisphosphate in the next minute?

4. *A potent analog.* 2-Carboxyarabinitol 1,5-bisphosphate (CABP) has been useful in studies of rubisco.
(a) Write the structural formula of CABP.
(b) Which catalytic intermediate does it resemble?
(c) Predict the effect of CABP on rubisco.

5. *Salvage operation.* Write a balanced equation for the transamination of glyoxylate to yield glycine.

6. *When one equals two.* In the C4 pathway, one ATP molecule is used in combining the CO_2 with phosphoenolpyruvate to form oxaloacetate (Figure 20.17), but, in the computation of energetics bookkeeping, two ATP molecules are said to be consumed. Explain.

7. *Dog days of August.* Before the days of pampered lawns, most homeowners practiced horticultural Darwinism. A result was that the lush lawns of early summer would often convert into robust cultures of crabgrass in the dog days of August. Provide a possible biochemical explanation for this transition.

8. *Global warming.* C3 plants are most common in higher latitudes and become less common at latitudes near the equator. The reverse is true of C4 plants. How might global warming affect this distribution?

9. *Tracing glucose.* Glucose labeled with ^{14}C at C-6 is added to a solution containing the enzymes and cofactors of the oxidative phase of the pentose phosphate pathway. What is the fate of the radioactive label?

10. *Recurring decarboxylations.* Which reaction in the citric acid cycle is most analogous to the oxidative decarboxylation of 6-phosphogluconate to ribulose 5-phosphate? What kind of enzyme-bound intermediate is formed in both reactions?

11. *Carbon shuffling.* Ribose 5-phosphate labeled with ^{14}C at C-1 is added to a solution containing transketolase, transaldolase, phosphopentose epimerase, phosphopentose isomerase, and glyceraldehyde 3-phosphate. What is the distribution of the radioactive label in the erythrose 4-phosphate and fructose 6-phosphate that are formed in this reaction mixture?

12. *Synthetic stoichiometries.* What is the stoichiometry of synthesis of (a) ribose 5-phosphate from glucose 6-phosphate with-

out the concomitant generation of NADPH; (b) NADPH from glucose 6-phosphate without the concomitant formation of pentose sugars?

13. *Trapping a reactive lysine.* Design a chemical experiment to identify the lysine residue that forms a Schiff base at the active site of transaldolase.

14. *Reductive power.* What ratio of NADPH to $NADP^+$ is required to sustain [GSH] = 10 mM and [GSSG] = 1 mM? Use the redox potentials given in Table 18.1.

Mechanism Problems

15. *An alternative approach.* The mechanisms of some aldolases do not include Schiff-base intermediates. Instead, these enzymes require bound metal ions. Propose such a mechanism for the conversion of dihydroxyacetone phosphate and glyceraldehyde 3-phosphate into fructose 1,6-bisphosphate.

16. *A recurring intermediate.* Phosphopentose isomerase interconverts the aldose ribose 5-phosphate and the ketose ribulose 5-phosphate. Propose a mechanism.

Chapter Integration Problems

17. *Catching carbons.* Radioactive-labeling experiments can yield estimates of how much glucose 6-phosphate is metabolized by the pentose phosphate pathway and how much is metabolized by the combined action of glycolysis and the citric acid cycle. Suppose that you have samples of two different tissues as well as two radioactively labeled glucose samples, one with glucose labeled with ^{14}C at C-1 and the other with glucose labeled with ^{14}C at C-6. Design an experiment that would enable you to determine the relative activity of the aerobic metabolism of glucose compared with metabolism by the pentose phosphate pathway.

18. *Photosynthetic efficiency.* Use the following information to estimate the efficiency of photosynthesis.

The $\Delta G°'$ for the reduction of CO_2 to the level of hexose is +114 kcal mol^{-1} (+477 kJ mol^{-1}).

A mole of 600-nm photons has an energy content of 47.6 kcal (199 kJ).

Assume that the proton gradient generated in producing the required NADPH is sufficient to drive the synthesis of the required ATP.

Date Interpretation Problem

19. Graph A shows the photosynthetic activity of two species of plant, one a C_4 plant and the other a C_3 plant, as a function of leaf temperature.

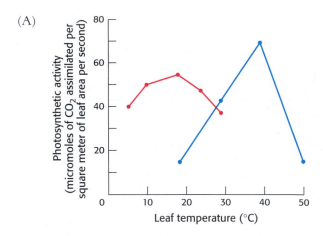

(A)

(a) Which data were most likely generated by the C_4 plant and which by the C_3 plant? Explain.
(b) Suggest some possible explanations for why the photosynthetic activity falls at higher temperatures.

Graph B illustrates how the photosynthetic activity of C_3 and C_4 plants varies with CO_2 concentration when temperature (30°C) and light intensity (high) are constant.

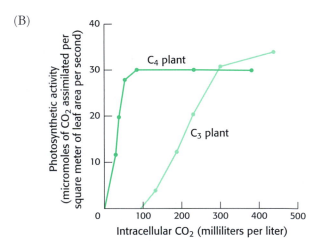

(B)

(c) Why can C_4 plants thrive at CO_2 concentrations that do not support the growth of C_3 plants?
(d) Suggest a plausible explanation for why C_3 plants continue to increase photosynthetic activity at higher CO_2 concentrations, whereas C_4 plants reach a plateau.

CHAPTER 21

Glycogen Metabolism

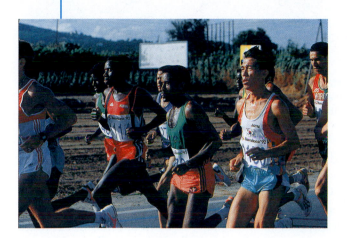

Signal cascades lead to the mobilization of glycogen to produce glucose, an energy source for runners.
[(Left) Mike Powell/Allsport.]

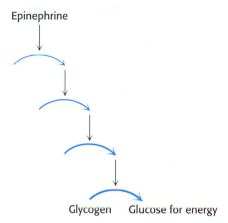

Glycogen is a *readily mobilized storage form of glucose*. It is a very large, branched polymer of glucose residues (Figure 21.1) that can be broken down to yield glucose molecules when energy is needed. Most of the glucose residues in glycogen are linked by α-1,4-glycosidic bonds. Branches at about every tenth residue are created by α-1,6-glycosidic bonds. Recall that α-glycosidic linkages form open helical polymers, whereas β linkages produce nearly straight strands that form structural fibrils, as in cellulose (Section 11.2.3).

Glycogen is not as reduced as fatty acids are and consequently not as energy rich. Why do animals store any energy as glycogen? Why not convert all excess fuel into fatty acids? Glycogen is an important fuel reserve for several reasons. The controlled breakdown of glycogen and release of glucose increase the amount of glucose that is available between meals. Hence, glycogen serves as a buffer to maintain blood-glucose levels. Glycogen's role in maintaining blood-glucose levels is especially important because glucose is virtually the only fuel used by the brain, except during prolonged starvation. Moreover, the glucose from glycogen is readily mobilized and is therefore a good source of energy for sudden, strenuous activity. Unlike fatty acids, the released glucose can provide energy in the absence of oxygen and can thus supply energy for anaerobic activity.

The two major sites of glycogen storage are the liver and skeletal muscle. The concentration of glycogen is higher in the liver than in muscle (10% versus 2% by weight), but

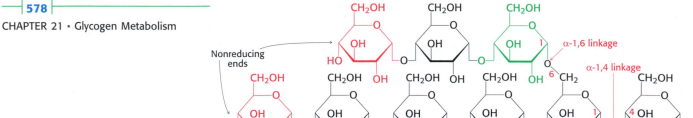

FIGURE 21.1 Glycogen structure. In this structure of two outer branches of a glycogen molecule, the residues at the nonreducing ends are shown in red and residue that starts a branch is shown in green. The rest of the glycogen molecule is represented by R.

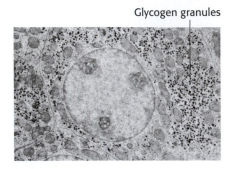

FIGURE 21.2 Electron micrograph of a liver cell. The dense particles in the cytoplasm are glycogen granules. [Courtesy of Dr. George Palade.]

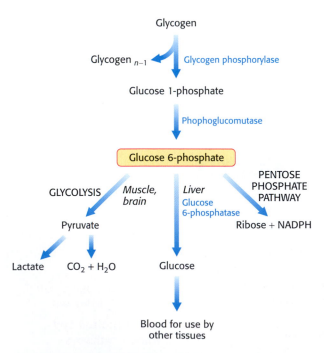

FIGURE 21.3 Fates of glucose 6-phosphate. Glucose 6-phosphate derived from glycogen can (1) be used as a fuel for anaerobic or aerobic metabolism as in, for instance, muscle; (2) be converted into free glucose in the liver and subsequently released into the blood; (3) be processed by the pentose phosphate pathway to generate NADPH or ribose in a variety of tissues.

more glycogen is stored in skeletal muscle overall because of its much greater mass. Glycogen is present in the cytosol in the form of granules ranging in diameter from 10 to 40 nm (Figure 21.2). In the liver, glycogen synthesis and degradation are regulated to maintain blood-glucose levels as required to meet the needs of the organism as a whole. In contrast, in muscle, these processes are regulated to meet the energy needs of the muscle itself.

21.0.1 An Overview of Glycogen Metabolism

Glycogen degradation and synthesis are relatively simple biochemical processes. Glycogen degradation consists of three steps: (1) the release of glucose 1-phosphate from glycogen, (2) the remodeling of the glycogen substrate to permit further degradation, and (3) the conversion of glucose 1-phosphate into glucose 6-phosphate for further metabolism. The glucose 6-phosphate derived from the breakdown of glycogen has three fates (Figure 21.3): (1) It is the initial substrate for glycolysis, (2) it can be processed by the pentose phosphate pathway to yield NADPH and ribose derivatives; and (3) it can be converted into free glucose for release into the bloodstream. This conversion takes place mainly in the liver and to a lesser extent in the intestines and kidneys.

Glycogen synthesis requires an activated form of glucose, uridine diphosphate glucose (UDP-glucose), which is formed by the reaction of UTP and glucose 1-phosphate. UDP-glucose is added to the nonreducing end of glycogen molecules. As is the case for glycogen degradation, the glycogen molecule must be remodeled for continued synthesis.

The regulation of these processes is quite complex. Several enzymes taking part in glycogen metabolism allosterically respond to metabolites that signal the energy needs of the cell. *These allosteric responses allow the adjustment of enzyme activity to meet the needs of the cell in which the enzymes are expressed.* Glycogen metabolism is also regulated by hormonally stimulated cascades that lead to the reversible phosphorylation of enzymes, which alters their kinetic properties. *Regulation by hormones allows glycogen metabolism to adjust to the needs of the entire organism.* By both these mechanisms, glycogen degradation is integrated with glycogen synthesis. We will first examine the metabolism, followed by enzyme regulation and then the elaborate integration of control mechanisms.

21.1 GLYCOGEN BREAKDOWN REQUIRES THE INTERPLAY OF SEVERAL ENZYMES

The efficient breakdown of glycogen to provide glucose 6-phosphate for further metabolism requires four enzyme activities: one to degrade glycogen, two to remodel glycogen so that it remains a substrate for degradation, and one to convert the product of glycogen breakdown into a form suitable for further metabolism. We will examine each of these activities in turn.

21.1.1 Phosphorylase Catalyzes the Phosphorolytic Cleavage of Glycogen to Release Glucose 1-phosphate

Glycogen phosphorylase, the key enzyme in glycogen breakdown, cleaves its substrate by the addition of orthophosphate (P_i) to yield *glucose 1-phosphate.* The cleavage of a bond by the addition of orthophosphate is referred to as *phosphorolysis.*

$$\text{Glycogen} + P_i \rightleftharpoons \text{glucose 1-phosphate} + \text{glycogen}$$
$$(n \text{ residues}) \qquad\qquad\qquad\qquad (n-1 \text{ residues})$$

Phosphorylase catalyzes the sequential removal of glycosyl residues from the nonreducing ends of the glycogen molecule (the ends with a free 4-OH groups; Section 11.1.3). Orthophosphate splits the glycosidic linkage between C-1 of the terminal residue and C-4 of the adjacent one. Specifically, it cleaves the bond between the C-1 carbon atom and the glycosidic oxygen atom, and the α configuration at C-1 is retained.

Glycogen (n residues) → Glucose 1-phosphate + Glycogen (n – 1 residues)

Glucose 1-phosphate released from glycogen can be readily converted into glucose 6-phosphate (Section 21.1.3), an important metabolic intermediate, by the enzyme phosphoglucomutase.

The reaction catalyzed by phosphorylase is readily reversible in vitro. At pH 6.8, the equilibrium ratio of orthophosphate to glucose 1-phosphate is 3.6. The value of $\Delta G^{\circ\prime}$ for this reaction is small because a glycosidic bond is replaced by a phosphoryl ester bond that has a nearly equal transfer potential. However, phosphorolysis proceeds far in the direction of glycogen breakdown in vivo because the [P_i]/[glucose 1-phosphate] ratio is usually greater than 100, substantially favoring phosphorolysis. We see here an example of how the cell can alter the free-energy change of a reaction to favor the reaction's occurrence by altering the ratio of substrate and product.

The phosphorolytic cleavage of glycogen is energetically advantageous because the released sugar is already phosphorylated. In contrast, a hydrolytic cleavage would yield glucose, which would then have to be phosphorylated at the expense of the hydrolysis of a molecule of ATP to enter the glycolytic pathway. An additional advantage of phosphorolytic cleavage for muscle cells is that glucose 1-phosphate, negatively charged under physiological conditions, cannot diffuse out of the cell.

21.1.2 A Debranching Enzyme Also Is Needed for the Breakdown of Glycogen

Glycogen phosphorylase, the key enzyme in glycogen breakdown, can carry out this process by itself only to a limited extent before encountering an obstacle. The α-1,6-glycosidic bonds at the branch points are not susceptible to cleavage by phosphorylase. Indeed, phosphorylase stops cleaving α-1,4 linkages when it reaches a terminal residue four residues away from a branch point. Because about 1 in 10 residues is branched, glycogen degradation by the phosphorylase alone would come to a halt after the release of six glucose molecules per branch.

How can the remainder of the glycogen molecule be mobilized for use as a fuel? Two additional enzymes, a *transferase* and *α-1,6-glucosidase*, remodel the glycogen for continued degradation by the phosphorylase (Figure 21.4). *The transferase shifts a block of three glycosyl residues from one outer branch to the other.* This transfer exposes a single glucose residue joined by an α-1,6-glycosidic linkage. α-1,6-Glucosidase, also known as the debranching enzyme, hydrolyzes the α-1,6-glycosidic bond, resulting in the release of a free glucose molecule.

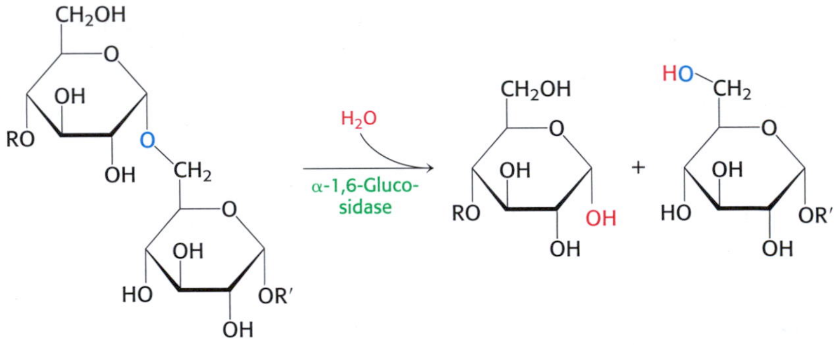

Glycogen (*n* residues)   **Glucose**   **Glycogen** (*n* − 1 residues)

This free glucose molecule is phosphorylated by the glycolytic enzyme hexokinase. Thus, the transferase and α-1,6-glucosidase convert the branched structure into a linear one, which paves the way for further cleavage by phosphorylase. It is noteworthy that, in eukaryotes, the transferase and the α-1,6-glucosidase activities are present in a single 160-kd polypeptide chain,

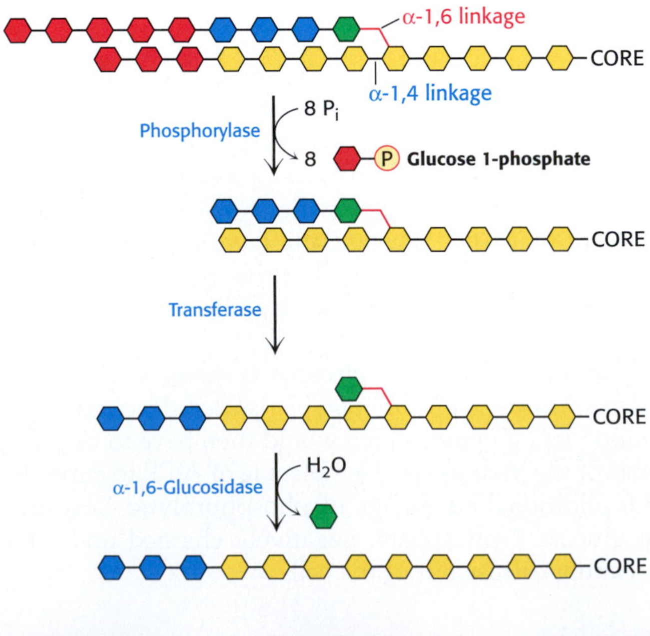

FIGURE 21.4 Glycogen remodeling. First, α-1,4-glycosidic bonds on each branch are cleaved by phosphorylase, leaving four residues along each branch. The transferase shifts a block of three glycosyl residues from one outer branch to the other. In this reaction, the α-1,4-glycosidic link between the blue and the green residues is broken and a new α-1,4 link between the blue and the yellow residues is formed. The green residue is then removed by α-1,6-glucosidase, leaving a linear chain with all α-1,4 linkages, suitable for further cleavage by phosphorylase.

providing yet another example of a bifunctional enzyme (Section 16.2.2). Furthermore, these enzymes may have additional features in common (Section 21.4.3).

21.1.3 Phosphoglucomutase Converts Glucose 1-phosphate into Glucose 6-phosphate

Glucose 1-phosphate formed in the phosphorolytic cleavage of glycogen must be converted into glucose 6-phosphate to enter the metabolic mainstream. This shift of a phosphoryl group is catalyzed by *phosphoglucomutase*. Recall that this enzyme is also used in galactose metabolism (Section 16.1.11). To effect this shift, the enzyme engages in the exchange of a phosphoryl group with the substrate (Figure 21.5).

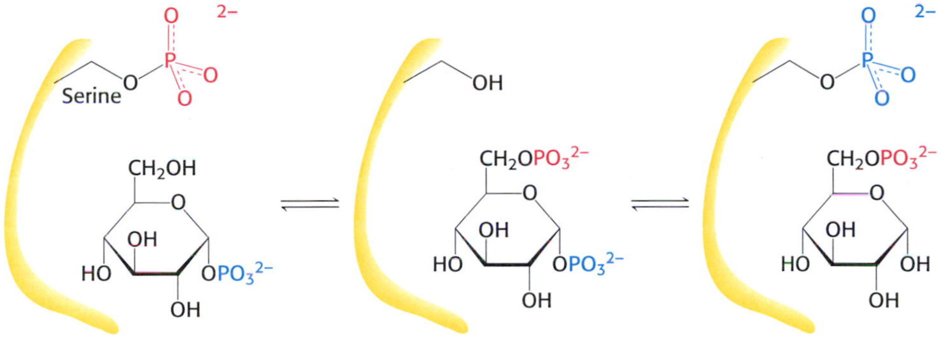

FIGURE 21.5 Reaction catalyzed by phosphoglucomutase. A phosphoryl group is transferred from the enzyme to the substrate, and a different phosphoryl group is transferred back to restore the enzyme to its initial state.

Glucose
1-phosphate

Glucose
1,6-bisphosphate

Glucose
6-phosphate

The catalytic site of an active mutase molecule contains a phosphorylated serine residue. The phosphoryl group is transferred from the serine residue to the C-6 hydroxyl group of glucose 1-phosphate to form glucose 1,6-bisphosphate. The C-1 phosphoryl group of this intermediate is then shuttled to the same serine residue, resulting in the formation of glucose 6-phosphate and the regeneration of the phosphoenzyme.

These reactions are like those of *phosphoglycerate mutase*, a glycolytic enzyme (Section 16.1.7). The role of glucose 1,6-bisphosphate in the interconversion of the phosphoglucoses is like that of 2,3-bisphosphoglycerate (2,3-BPG) in the interconversion of 2-phosphoglycerate and 3-phosphoglycerate in glycolysis. A phosphoenzyme intermediate participates in both reactions.

21.1.4 Liver Contains Glucose 6-phosphatase, a Hydrolytic Enzyme Absent from Muscle

A major function of the liver is to maintain a near constant level of glucose in the blood. The liver releases glucose into the blood during muscular activity and between meals to be taken up primarily by the brain and skeletal muscle. However, the phosphorylated glucose produced by glycogen breakdown, in contrast with glucose, is not readily transported out of cells. The liver contains a hydrolytic enzyme, *glucose 6-phosphatase*, which cleaves the phosphoryl group to form free glucose and orthophosphate. This glucose 6-phosphatase, located on the lumenal side of the smooth endoplasmic reticulum membrane, is the same enzyme that releases free glucose at the conclusion of gluconeogenesis. Recall that glucose 6-phosphate is transported into the endoplasmic reticulum; glucose and orthophosphate formed by hydrolysis are then shuttled back into the cytosol (Section 16.3.5).

$$\text{Glucose 6-phosphate} + H_2O \longrightarrow \text{glucose} + P_i$$

Glucose 6-phosphatase is absent from most other tissues. Consequently, glucose 6-phosphate is retained for the generation of ATP. In contrast, glucose is not a major fuel for the liver.

21.1.5 Pyridoxal Phosphate Participates in the Phosphorolytic Cleavage of Glycogen

Let us now examine the catalytic mechanism of glycogen phosphorylase, which is a dimer of two identical 97-kd subunits. Each subunit is compactly folded into an *amino-terminal domain* (480 residues) containing a *glycogen-binding site* and a *carboxyl-terminal domain* (360 residues; Figure 21.6). The catalytic site is located in a deep crevice formed by residues from amino- and carboxyl-terminal domains. The special challenge faced by phosphorylase is to cleave glycogen phosphorolytically rather than hydrolytically to save the ATP required to phosphorylate free glucose. This cleavage requires that water be excluded from the active site. Several clues provide us with information about the mechanism by which phosphorylase achieves the exclusion of water. First, both the glycogen substrate and the glucose 1-phosphate product have an α configuration at C-1 (the designation α means that the oxygen atom attached to C-1 is below the plane of the ring; Section 11.1.3). A direct attack of phosphate on C-1 of a sugar would invert the configuration at this carbon because the reaction would proceed through a pentacovalent transition state. Because the resulting glucose 1-phosphate has an α rather than a β configuration, an even number of steps (most simply, two) is required. The most likely explanation for these results is that a *carbonium ion intermediate* is formed.

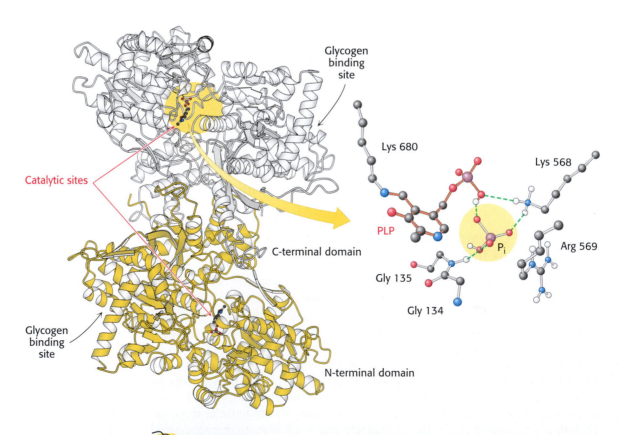

FIGURE 21.6 Structure of glycogen phosphorylase. This enzyme forms a homodimer: one subunit is shown in white and the other in yellow. Each catalytic site includes a pyridoxal-phosphate (PLP) group, linked to lysine 680 of the enzyme. The binding site for the phosphate (P_i) substrate is shown.

A second clue to the catalytic mechanism of phosphorylase is its requirement for *pyridoxal phosphate (PLP)*, a derivative of pyridoxine (vitamin B_6, Section 8.6.1). The aldehyde group of this coenzyme forms a Schiff base with a specific lysine side chain of the enzyme (Figure 21.7). The results of structural studies indicate that the reacting orthophosphate group takes a position between the 5′-phosphate group of PLP and the glycogen substrate (Figure 21.8). *The 5′-phosphate group of PLP acts in tandem with orthophosphate by serving as a proton donor and then as a proton acceptor (that is, as a general acid–base catalyst).* Orthophosphate (in the HPO_4^{2-} form) donates a proton to the oxygen atom attached to carbon 4 of the departing glycogen chain and simultaneously acquires a proton from PLP. The carbonium ion intermediate formed in this step is then attacked by orthophosphate to form α-glucose 1-phosphate, with the concomitant return of a hydrogen atom to pyridoxal phosphate. The requirement that water be excluded from the active site calls for the special role of pyridoxal phosphate in facilitating the phosphorolytic cleavage.

The glycogen-binding site is 30 Å away from the catalytic site (see Figure 21.6), but it is connected to the catalytic site by a narrow crevice able to accommodate four or five glucose units. The large separation between the binding site and the catalytic site enables the enzyme to phosphorolyze many residues without having to dissociate and reassociate after each catalytic cycle. An enzyme that can catalyze many reactions without having to dissociate and reassociate after each catalytic step is said to be *processive*—a property of enzymes that synthesize and degrade large polymers. We will see such enzymes again when we consider DNA and RNA synthesis.

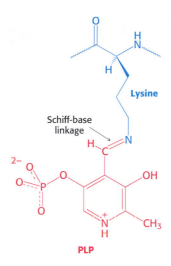

FIGURE 21.7 PLP–Schiff-base linkage. A pyridoxal phosphate group (red) forms a Schiff base with a lysine residue (blue) at the active site of phosphorylase.

Carbocation intermediate

FIGURE 21.8 Phosphorylase mechanism. A bound HPO_4^{2-} group (red) favors the cleavage of the glycosidic bond by donating a proton to the departing glucose (black). This reaction results in the formation of a carbocation and is favored by the transfer of a proton from the protonated phosphate group of the bound pyridoxal phosphate PLP group (blue). The combination of the carbocation and the orthophosphate results in the formation of glucose 1-phosphate.

21.2 PHOSPHORYLASE IS REGULATED BY ALLOSTERIC INTERACTIONS AND REVERSIBLE PHOSPHORYLATION

Glycogen metabolism is precisely controlled by multiple interlocking mechanisms, and the focus of this control is glycogen phosphorylase. *Phosphorylase is regulated by several allosteric effectors that signal the energy state of the cell as well as by reversible phosphorylation, which is responsive to hormones such as insulin, epinephrine, and glucagon.* We will examine the differences in the control of glycogen metabolism in two tissues: skeletal muscle and liver. These differences are due to the fact that *the muscle uses glucose to produce energy for itself, whereas the liver maintains glucose homeostasis of the organism as a whole.*

STRUCTURAL INSIGHTS, Glycogen Phosphorylase, *looks closely at the structural mechanisms of phosphorylase regulation, examining the effects of allosteric effectors and serine phosphorylation.*

21.2.1 Muscle Phosphorylase Is Regulated by the Intracellular Energy Charge

We begin by considering the glycogen phosphorylase from muscle. The dimeric skeletal muscle phosphorylase exists in two interconvertible forms:

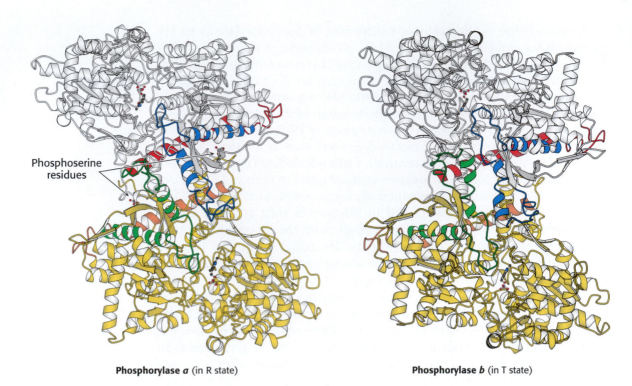

Phosphorylase *a* (in R state)   **Phosphorylase *b*** (in T state)

Phosphoserine residues

FIGURE 21.9 Structures of phosphorylase *a* and phosphorylase *b*. Phosphorylase *a* is phosphorylated on serine 14 of each subunit. This modification favors the structure of the more active R state. One subunit is shown in white, with helices and loops important for regulation shown in blue and red. The other subunit is shown in yellow, with the regulatory structures shown in orange and green. Phosphorylase *b* is not phosphorylated and exists predominantly in the T state.

a *usually active* phosphorylase *a* and a *usually inactive* phosphorylase *b* (Figure 21.9). Each of these two forms exists in equilibrium between an active relaxed (R) state and a much less active tense (T) state, but the equilibrium for phosphorylase *a* favors the R state whereas the equilibrium for phosphorylase *b* favors the T state (Figure 21.10). Phosphorylase *a* and phosphorylase *b* differ by a single phosphoryl group in each subunit. Phosphorylase *b* is converted into phosphorylase *a* when it is phosphorylated at a single serine residue (serine 14) in each subunit. The regulatory enzyme *phosphorylase kinase* catalyzes this covalent modification. As will be described, increased levels of epinephrine (resulting from fear or from the excitement of exercise) and the electrical stimulation of muscle result in phosphorylation of the enzyme to the phosphorylase *a* form.

FIGURE 21.10 Phosphorylase regulation. Both phosphorylase *b* and phosphorylase *a* exist as equilibria between an active R state and a less-active T state. Phosphorylase *b* is usually inactive because the equilibrium favors the T state. Phosphorylase *a* is usually active because the equilibrium favors the R state. Regulatory structures are shown in blue and green.

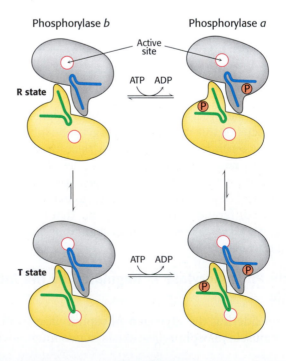

Comparison of the structures of phosphorylase *a* and phosphorylase *b* reveals that subtle structural changes at the subunit interfaces are transmitted to the active sites (see Figure 21.9). The transition from the T state (represented by phosphorylase *b*) to the R state (represented by phosphorylase *a*) entails a 10-degree rotation around the twofold axis of the dimer. Most importantly, this transition is associated with structural changes in α helices that move a loop out of the active site of each subunit. Thus, the T state is less active because the catalytic site is partly blocked. In the R state, the catalytic site is more accessible and a binding site for orthophosphate is well organized.

The position of the equilibrium of phosphorylase *b* between the T and the R form is responsive to conditions in the cell. Muscle phosphorylase *b* is active only in the presence of high concentrations of AMP, which binds to a nucleotide-binding site and stabilizes the conformation of phosphorylase *b* in the R state (Figure 21.11). ATP acts as a negative allosteric effector by competing with AMP and so favors the T state. Thus, *the transition of phosphorylase* b *between the T and the R state is controlled by the energy charge of the muscle cell.* Glucose 6-phosphate also favors the T state of phosphorylase *b*, an example of feedback inhibition.

Under most physiological conditions, *phosphorylase* b *is inactive because of the inhibitory effects of ATP and glucose 6-phosphate.* In contrast, *phosphorylase* a *is fully active,* regardless of the levels of AMP, ATP, and glucose 6-phosphate. In resting muscle, nearly all the enzyme is in the inactive *b* form. When exercise commences, the elevated level of AMP leads to the activation of phosphorylase *b*. Exercise will also result in hormone release that generates the phosphorylated *a* form of the enzyme. The absence of glucose 6-phosphatase in muscle ensures that glucose 6-phosphate derived from glycogen remains within the cell for energy transformation.

21.2.2 Liver Phosphorylase Produces Glucose for Use by Other Tissues

The regulation of liver glycogen phosphorylase differs markedly from that of muscle, a consequence of the role of the liver in glucose homeostasis for the organism as a whole. In human beings, liver phosphorylase and muscle phosphorylase are approximately 90% identical in amino acid sequence. The differences result in subtle but important shifts in the stability of various forms of the enzyme. In contrast with the muscle enzyme, liver phosphorylase *a* but not *b* exhibits the most responsive T-to-R transition. The binding of glucose shifts the allosteric equilibrium of the *a* form from the R to the T state, deactivating the enzyme (Figure 21.12). Why would glucose function as a negative regulator of liver phosphorylase *a*? The role of glycogen degradation in the liver is to form glucose for *export to other tissues* when the blood-glucose level is low. Hence, if free glucose is present from some other source such as diet, there is no need to mobilize glycogen. Unlike the enzyme in muscle, the liver phosphorylase is insensitive to regulation by AMP because the liver does not undergo the dramatic changes in energy charge seen in a contracting muscle. We see here a clear example of the use of isozymic forms of the same enzyme to establish the tissue-specific biochemical properties of muscle and the liver.

Phosphorylase *b*
(muscle)

Nucleotide-binding sites

AMP

ATP

Glucose 6-phosphate

T state   R state

FIGURE 21.11 Allosteric regulation of muscle phosphorylase. A low energy charge, represented by high concentrations of AMP, favors the transition to the R state.

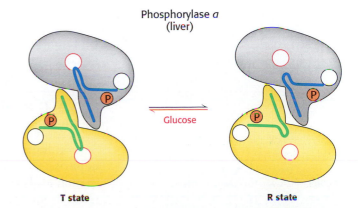

Phosphorylase *a*
(liver)

Glucose

T state   R state

FIGURE 21.12 Allosteric regulation of liver phosphorylase. The binding of glucose to phosphorylase *a* shifts the equilibrium to the T state and inactivates the enzyme. Thus, glycogen is not mobilized when glucose is already abundant.

21.2.3 Phosphorylase Kinase Is Activated by Phosphorylation and Calcium Ions

The next upstream component of this signal-transduction pathway is the enzyme that covalently modifies phosphorylase. This enzyme is *phosphorylase kinase*, a very large protein with a subunit composition in skeletal muscle of $(\alpha\beta\gamma\delta)_4$ and a mass of 1200 kd. The catalytic activity resides in the γ subunit, whereas the other subunits serve a regulatory function. This kinase is under dual control. Like its own substrate, phosphorylase kinase is regulated by phosphorylation: the kinase is converted from *a low-activity form into a high-activity one by phosphorylation of its β subunit*. The enzyme catalyzing the activation of phosphorylase kinase is *protein kinase A (PKA)*, which is switched on by a second messenger, cyclic AMP (Sections 10.4.2 and 15.1.5). As will be discussed, hormones such as epinephrine induce the breakdown of glycogen by activating a cyclic AMP cascade (Figure 21.13).

Phosphorylase kinase can also be partly activated by Ca^{2+} levels of the order of 1 μM. Its δ subunit is *calmodulin*, a calcium sensor that stimulates many enzymes in eukaryotes (Section 15.3.2). This mode of activation of the kinase is important in muscle, where contraction is triggered by the release of Ca^{2+} from the sarcoplasmic reticulum. Phosphorylase kinase attains maximal activity only after both phosphorylation of the β subunit and activation of the δ subunit by Ca^{2+} binding.

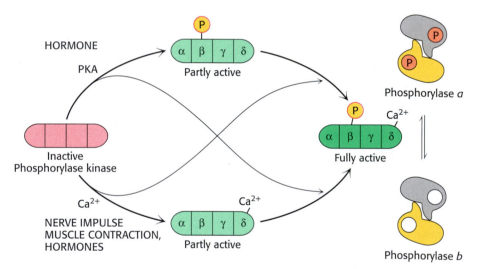

FIGURE 21.13 Activation of phosphorylase kinase. Phosphorylase kinase is activated by hormones that lead to the phosphorylation of the β subunit and by Ca^{2+} binding of the δ subunit. Both types of stimulation are required for maximal enzyme activity.

21.3 EPINEPHRINE AND GLUCAGON SIGNAL THE NEED FOR GLYCOGEN BREAKDOWN

Protein kinase A activates phosphorylase kinase, which in turn activates glycogen phosphorylase. What activates protein kinase A? What is the signal that ultimately triggers an increase in glycogen breakdown?

21.3.1 G Proteins Transmit the Signal for the Initiation of Glycogen Breakdown

Several hormones greatly affect glycogen metabolism. Glucagon and epinephrine trigger the breakdown of glycogen. Muscular activity or its anticipation leads to the release of *epinephrine (adrenaline)*, a catecholamine derived from tyrosine, from the adrenal medulla. Epinephrine markedly stimulates glycogen breakdown in muscle and, to a lesser extent, in the liver. The liver is more responsive to *glucagon*, a polypeptide hormone that is secreted by the α cells of the pancreas when the blood-sugar level is low. Physiologically, glucagon signifies the starved state.

Epinephrine

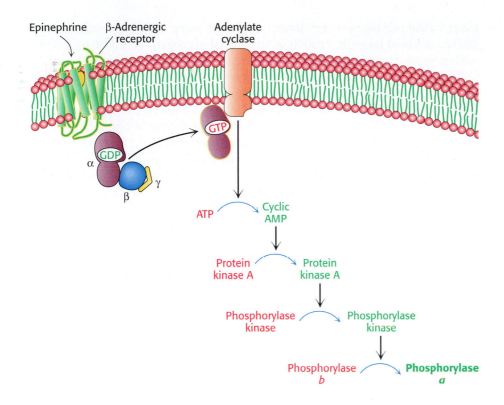

FIGURE 21.14 Regulatory cascade for glycogen breakdown. Glycogen degradation is stimulated by hormone binding to 7TM receptors. Hormone binding initiates a G-protein–dependent signal-transduction pathway that results in the phosphorylation and activation of glycogen phosphorylase.

How do hormones trigger the breakdown of glycogen? We will briefly review this signal-transduction cascade, already discussed in Section 15.1 (Figure 21.14).

$$^+H_3N-His-Ser-Glu-Gly-\overset{5}{}Thr-Phe-Thr-Ser-Asp-\overset{10}{Tyr}-$$

$$-Ser-\overset{15}{Lys}-Tyr-Leu-Asp-Ser-Arg-\overset{20}{Arg}-Ala-Gln-$$

$$-Asp-\overset{25}{Phe}-Val-Gln-Trp-Leu-\overset{29}{Met}-Asn-Thr-COO^-$$

Glucagon

1. The signal molecules epinephrine and glucagon bind to specific 7TM receptors in the plasma membranes of muscle and liver cells, respectively. Epinephrine binds to the β-adrenergic receptor in muscle, whereas glucagon binds to the glucagon receptor. These binding events activate the α subunit of the heteromeric G_s protein. *A specific external signal has been transmitted into the cell through structural changes,* first in the receptor and then in the G protein.

2. The GTP-bound form of the α subunit of G_s activates adenylate cyclase, a transmembrane protein that catalyzes the formation of the secondary messenger cyclic AMP from ATP.

3. The elevated cytosolic level of cyclic AMP activates *protein kinase A* through the binding of cyclic AMP to the regulatory subunits, which then dissociate from the catalytic subunits. The free catalytic subunits are now active.

4. Protein kinase A phosphorylates the β subunit of phosphorylase kinase, which subsequently activates glycogen phosphorylase.

The cyclic AMP cascade highly amplifies the effects of hormones. Hence, the binding of a small number of hormone molecules to cell-surface receptors leads to the release of a very large number of sugar units. Indeed, the amplification is so large that much of the stored glycogen would be mobilized within seconds were it not for a counterregulatory system (Section 21.3.2).

The signal-transduction processes in the liver are more complex than those in muscle. Epinephrine can also elicit glycogen degradation in the liver. However, in addition to binding to the β-adrenergic receptor, it binds to the 7TM α-adrenergic receptor, which then activates phospholipase C and, hence, initiates the *phosphoinositide cascade* (Section 15.2). The consequent rise in the level of inositol 1,4,5-trisphosphate induces the release of Ca^{2+} from endoplasmic reticulum stores. Recall that the δ subunit of phosphorylase kinase is the Ca^{2+} sensor calmodulin. Binding of Ca^{2+} to calmodulin leads to a partial activation of phosphorylase kinase. Stimulation by both glucagon and epinephrine leads to maximal mobilization of liver glycogen.

21.3.2 Glycogen Breakdown Must Be Capable of Being Rapidly Turned Off

There must be a way to shut down the high-gain system of glycogen breakdown quickly to prevent the wasteful depletion of glycogen after energy needs have been met. Indeed, another cascade leads to the dephosphorylation and inactivation of phosphorylase kinase and glycogen phosphorylase. Simultaneously, glycogen synthesis is activated.

The signal-transduction pathway leading to the activation of glycogen phosphorylase is shut down by the process already described for such pathways employing G proteins and cyclic AMP. The inherent GTPase activity of the G protein converts the bound GTP into GDP, thereby halting signal transduction. Cells also contain phosphodiesterase activity that converts cyclic AMP into AMP. Protein kinase A sets the stage for the shutdown of glycogen degradation by adding a phosphoryl group to the α subunit of phosphorylase kinase after first phosphorylating the β subunit. This addition of a phosphoryl group renders the enzyme a better substrate for dephosphorylation and consequent inactivation by the enzyme *protein phosphatase 1 (PP1)*. Protein phosphatase 1 also removes the phosphoryl group from glycogen phosphorylase, converting the enzyme into the usually inactive *b* form.

21.3.3 The Regulation of Glycogen Phosphorylase Became More Sophisticated as the Enzyme Evolved

Analyses of the primary structures of glycogen phosphorylase from human beings, rats, *Dictyostelium* (slime mold), yeast, potatoes, and *E. coli* have enabled inferences to be made about the evolution of this important enzyme. The 16 residues that come into contact with glucose at the active site are identical in nearly all the enzymes. There is more variation but still substantial conservation of the 15 residues at the pyridoxal phosphate-binding site. Likewise, the glycogen-binding site is relatively well conserved in all the enzymes. The high degree of similarity of the active site, the pyridoxal phosphate-binding site and the glycogen-binding site shows that the catalytic mechanism has been maintained throughout evolution.

Differences arise, however, when we compare the regulatory sites. The simplest type of regulation would be feedback inhibition by glucose 6-phosphate. Indeed, the glucose 6-phosphate regulatory site is highly conserved among the majority of the phosphorylases. The crucial amino acid residues that participate in regulation by phosphorylation and nucleotide binding are well conserved only in the mammalian enzymes. Thus, this level of regulation was a later evolutionary acquisition.

21.4 GLYCOGEN IS SYNTHESIZED AND DEGRADED BY DIFFERENT PATHWAYS

CONCEPTUAL INSIGHTS, Overview of Carbohydrate and Fatty Acid Metabolism. *View this media module to better understand how glycogen metabolism fits in with other energy storage and utilization pathways (glycolysis, citric acid cycle, pentose phosphate pathway, and fatty acid metabolism).*

As we have seen in glycolysis and gluconeogenesis, biosynthetic and degradative pathways rarely operate by precisely the same reactions in the forward and reverse directions. Glycogen metabolism provided the first known example of this important principle. *Separate pathways afford much greater flexibility, both in energetics and in control.*

In 1957, Luis Leloir and his coworkers showed that glycogen is synthesized by a pathway that utilizes *uridine diphosphate glucose (UDP-glucose)* rather than glucose 1-phosphate as the activated glucose donor.

$$\text{Synthesis: Glycogen}_n + \text{UDP-glucose} \longrightarrow \text{glycogen}_{n+1} + \text{UDP}$$

$$\text{Degradation: Glycogen}_{n+1} + \text{P}_i \longrightarrow \text{glycogen}_n + \text{glucose 1-phosphate}$$

21.4.1 UDP-Glucose Is an Activated Form of Glucose

UDP-glucose, the glucose donor in the biosynthesis of glycogen, is an *activated form of glucose,* just as ATP and acetyl CoA are activated forms of orthophosphate and acetate, respectively. The C-1 carbon atom of the glucosyl unit of UDP-glucose is activated because its hydroxyl group is esterified to the diphosphate moiety of UDP.

UDP-glucose is synthesized from glucose 1-phosphate and uridine triphosphate (UTP) in a reaction catalyzed by *UDP-glucose pyrophosphorylase.* The pyrophosphate liberated in this reaction comes from the outer two phosphoryl residues of UTP.

Uridine diphosphate glucose (UDP-glucose)

Glucose 1-phosphate   UTP   UDP-glucose

This reaction is readily reversible. However, pyrophosphate is rapidly hydrolyzed in vivo to orthophosphate by an inorganic pyrophosphatase. The essentially irreversible hydrolysis of pyrophosphate drives the synthesis of UDP-glucose.

$$\text{Glucose 1-phosphate} + \text{UTP} \rightleftharpoons \text{UDP-glucose} + \text{PP}_i$$

$$\text{PP}_i + \text{H}_2\text{O} \longrightarrow 2\,\text{P}_i$$

$$\overline{\text{Glucose 1-phosphate} + \text{UTP} + \text{H}_2\text{O} \longrightarrow \text{UDP-glucose} + 2\,\text{P}_i}$$

The synthesis of UDP-glucose exemplifies another recurring theme in biochemistry: *many biosynthetic reactions are driven by the hydrolysis of pyrophosphate.*

21.4.2 Glycogen Synthase Catalyzes the Transfer of Glucose from UDP-Glucose to a Growing Chain

New glucosyl units are added to the nonreducing terminal residues of glycogen. The activated glucosyl unit of UDP-glucose is transferred to the hydroxyl group at a C-4 terminus of glycogen to form an α-1,4-glycosidic linkage. In elongation, UDP is displaced by the terminal hydroxyl group of the growing glycogen molecule. This reaction is catalyzed by *glycogen synthase, the key regulatory enzyme in glycogen synthesis.*

UDP-glucose

Glycogen
(*n* residues)

UTP

Glycogen
(*n* + 1 residues)

Glycogen synthase can add glucosyl residues only if the polysaccharide chain already contains more than four residues. Thus, glycogen synthesis requires a *primer*. This priming function is carried out by *glycogenin*, a protein composed of two identical 37-kd subunits, each bearing an oligosaccharide of α-1,4-glucose units. Carbon 1 of the first unit of this chain, the reducing end, is covalently attached to the phenolic hydroxyl group of a specific tyrosine in each glycogenin subunit. How is this chain formed? Each subunit of glycogenin catalyzes the addition of eight glucose units to its partner in the glycogenin dimer. UDP-glucose is the donor in this autoglycosylation. At this point, glycogen synthase takes over to extend the glycogen molecule.

21.4.3 A Branching Enzyme Forms α-1,6 Linkages

Glycogen synthase catalyzes only the synthesis of α-1,4 linkages. Another enzyme is required to form the α-1,6 linkages that make glycogen a branched polymer. Branching occurs after a number of glucosyl residues are joined in α-1,4 linkage by glycogen synthase. A branch is created by the breaking of an α-1,4 link and the formation of an α-1,6 link: this reaction is different from debranching. A block of residues, typically 7 in number, is transferred to a more interior site. The *branching enzyme* that catalyzes this reaction is quite exacting. The block of 7 or so residues must include the nonreducing terminus and come from a chain at least 11 residues long. In addition, the new branch point must be at least 4 residues away from a preexisting one.

Branching is important because it increases the solubility of glycogen. Furthermore, branching creates a large number of terminal residues, the sites of action of glycogen phosphorylase and synthase (Figure 21.15). Thus, *branching increases the rate of glycogen synthesis and degradation.*

Glycogen branching requires a single transferase activity. Glycogen debranching requires two enzyme activities: a transferase and an α-1,6 glucosidase. Sequence analysis suggests that the two transferases and, perhaps, the α-1,6 glucosidase are members of the same enzyme family, termed the *α-amylase family.* Such an enzyme catalyzes a reaction by forming a covalent intermediate attached to a conserved aspartate residue (Figure 21.16). Thus, the branching enzyme appears to function through the transfer of a chain of glucose molecules from an α-1,4 linkage to an

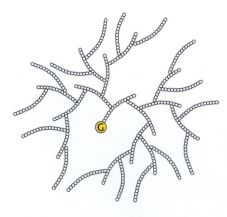

FIGURE 21.15 Cross section of a glycogen molecule. The component labeled G is glycogenin.

aspartate residue on the enzyme and then from this site to a more interior location on the glycogen molecule to form an α-1,6 linkage.

21.4.4 Glycogen Synthase Is the Key Regulatory Enzyme in Glycogen Synthesis

The activity of glycogen synthase, like that of phosphorylase, is regulated by covalent modification. Glycogen synthase is phosphorylated at multiple sites by protein kinase A and several other kinases. The resulting alteration of the charges in the protein lead to its inactivation (Figure 21.17). *Phosphorylation has opposite effects on the enzymatic activities of glycogen synthase and phosphorylase.* Phosphorylation converts the active *a* form of the synthase into a usually inactive *b* form. The phosphorylated *b* form requires a high level of the allosteric activator glucose 6-phosphate for activity, whereas the *a* form is active whether or not glucose 6-phosphate is present.

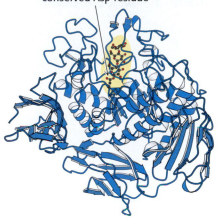

Carbohydrate linked to conserved Asp residue

FIGURE 21.16 Structure of glycogen transferase. A conserved aspartate residue forms a covalent intermediate with a chain of glucose molecules.

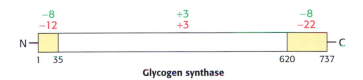

Glycogen synthase

FIGURE 21.17 Charge distribution of glycogen synthase. Glycogen synthase has a highly asymmetric charge distribution. Phosphorylation markedly changes the net charge of the amino- and carboxyl-terminal regions (yellow) of the enzyme. The net charge of these regions and the interior of the enzyme before and after complete phosphorylation are shown in green and red, respectively. [After M. F. Browner, K. Nakano, A. G. Bang, and R. J. Fletterick. *Proc. Natl. Acad. Sci. USA* 86(1989):1443.]

21.4.5 Glycogen Is an Efficient Storage Form of Glucose

What is the cost of converting glucose 6-phosphate into glycogen and back into glucose 6-phosphate? The pertinent reactions have already been described, except for reaction 5, which is the regeneration of UTP. ATP phosphorylates UDP in a reaction catalyzed by *nucleoside diphosphokinase.*

$$\text{Glucose 6-phosphate} \longrightarrow \text{glucose 1-phosphate} \tag{1}$$

$$\text{Glucose 1-phosphate} + \text{UTP} \longrightarrow \text{UDP-glucose} + \text{PP}_i \tag{2}$$

$$\text{PP}_i + \text{H}_2\text{O} \longrightarrow 2\,\text{P}_i \tag{3}$$

$$\text{UDP-glucose} + \text{glycogen}_n \longrightarrow \text{glycogen}_{n+1} + \text{UDP} \tag{4}$$

$$\text{UDP} + \text{ATP} \longrightarrow \text{UTP} + \text{ADP} \tag{5}$$

Sum: Glucose 6-phosphate + ATP + glycogen$_n$ + H$_2$O $\longrightarrow$
$$\text{glycogen}_{n+1} + \text{ADP} + 2\,\text{P}_i$$

Thus, one ATP is hydrolyed incorporating glucose 6-phosphate into glycogen. The energy yield from the breakdown of glycogen is highly efficient. About 90% of the residues are phosphorolytically cleaved to glucose 1-phosphate, which is converted at no cost into glucose 6-phosphate. The other 10% are branch residues, which are hydrolytically cleaved. One molecule of ATP is then used to phosphorylate each of these glucose molecules to glucose 6-phosphate. The complete oxidation of glucose 6-phosphate yields about 31 molecules of ATP, and storage consumes slightly more than one molecule of ATP per molecule of glucose 6-phosphate; so *the overall efficiency of storage is nearly 97%.*

21.5 GLYCOGEN BREAKDOWN AND SYNTHESIS ARE RECIPROCALLY REGULATED

We now return to the regulation of glycogen metabolism with a knowledge of both degradation and synthesis. *Glycogen breakdown and synthesis are reciprocally regulated by a hormone-triggered cAMP cascade acting through protein kinase A* (Figure 21.18). In addition to phosphorylating and activating phosphorylase kinase, protein kinase A adds a phosphoryl group to glycogen synthase, which leads to a *decrease* in enzymatic activity. This important control mechanism prevents glycogen from being synthesized at the same time that it is being broken down. How is the enzymatic activity reversed so that glycogen breakdown halts and glycogen synthesis begins?

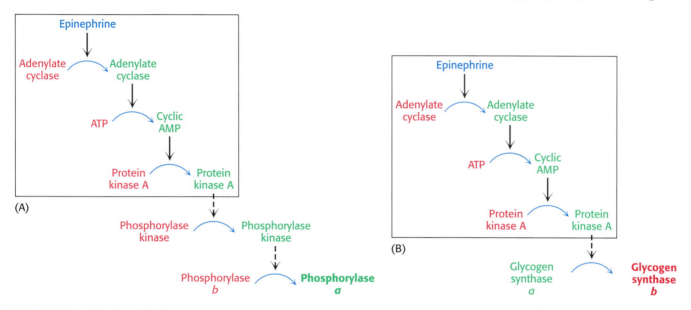

FIGURE 21.18 Coordinate control of glycogen metabolism. Glycogen metabolism is regulated, in part, by hormone-triggered cyclic AMP cascades: (A) glycogen degradation; (B) glycogen synthesis. Inactive forms are shown in red, and active ones in green. The sequence of reactions leading to the activation of protein kinase A is the same in the regulation of glycogen degradation and synthesis. Phosphorylase kinase also inactivates glycogen synthase.

21.5.1 Protein Phosphatase 1 Reverses the Regulatory Effects of Kinases on Glycogen Metabolism

The changes in enzymatic activity produced by protein kinases are reversed by *protein phosphatases*. The hydrolysis of phosphorylated serine and threonine residues in proteins is catalyzed by protein phosphatases. One enzyme, termed *protein phosphatase 1, plays key roles in regulating glycogen metabolism.* PP1 inactivates phosphorylase kinase and phosphorylase *a* by dephosphorylating these enzymes. PP1 decreases the rate of glycogen breakdown: it reverses the effects of the phosphorylation cascade. Moreover, PP1 also removes the phosphoryl group from glycogen synthase *b* to convert it into the much more active *a* form. Hence, PP1 accelerates glycogen synthesis. PP1 is yet another molecular device for coordinating carbohydrate storage.

The complete complex of PP1 consists of three components: PP1 itself, a 37-kd catalytic subunit; a 123-kd R_{Gl} subunit that confers high affinity for glycogen; and inhibitor 1, a small regulatory subunit that, when phosphorylated, inhibits PP1. The importance of the R_{Gl} subunit is that it brings PP1, which is active only when associated with glycogen molecules, into proximity with its substrates.

How is the phosphatase activity of PP1 itself regulated? Consider the case in which glycogen degradation is predominant (Figure 21.19). In this case, PKA is active. Two components of PP1 are themselves substrates for protein kinase A. Phosphorylation of the R_{Gl} component by protein kinase A prevents R_{Gl} from binding the catalytic subunit of PP1. Consequently,

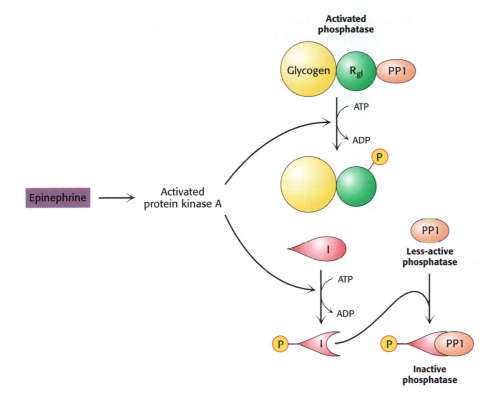

FIGURE 21.19 Regulation of protein phosphatase 1 (PP1). Phosphorylation of R_{GI} by protein kinase A dissociates the catalytic subunit from the glycogen particle and hence the PP1 substrates. Inhibition is complete when the inhibitor subunit (I) is phosphorylated and binds to PP1 to inactivate it.

activation of the cAMP cascade leads to the inactivation of PP1 because it can no longer bind its substrates. Phosphorylation of inhibitor 1 by protein kinase A blocks catalysis by PP1. Thus, when glycogen degradation is switched on by cAMP, the accompanying phosphorylation of inhibitor 1 keeps phosphorylase in the active *a* form and glycogen synthase in the inactive *b* form. The epinephrine-induced phosphorylation of the R_{GI} subunit and inhibitor 1 are complementary devices for sustaining glycogen degradation.

21.5.2 Insulin Stimulates Glycogen Synthesis by Activating Protein Phosphatase 1

How is glycogen synthesis stimulated? As stated earlier, the presence of glucagon signifies the starved state and initiates glycogen breakdown while inhibiting glycogen synthesis. When blood-glucose levels are high, *insulin stimulates the synthesis of glycogen by triggering a pathway that activates protein phosphatase 1* (Figure 21.20). The first step in the action of insulin is its binding to a receptor tyrosine kinase in the plasma membrane. Multiple

FIGURE 21.20 Insulin activates protein phosphatase 1. Insulin triggers a cascade leading to the activation of protein phosphatase 1, which results in the stimulation of glycogen synthesis and inhibition of its breakdown. The activated receptor tyrosine kinase switches on a putative master kinase that phosphorylates the insulin-sensitive protein kinase. In turn, the glycogen-targeting subunit (R_{GI} subunit) of the phosphatase is phosphorylated, which activates the enzyme. [After P. Dent, A. Lavoinne, S. Nakielny, F. B. Caudwell, P. Watt, and P. Cohen. *Nature* 348(1990):306.]

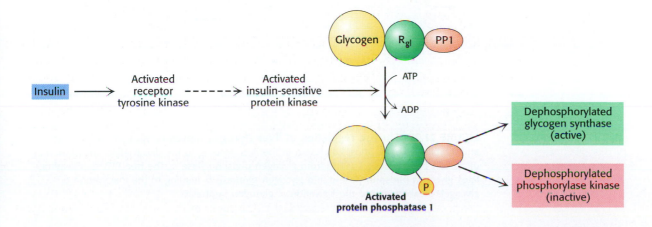

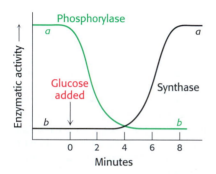

FIGURE 21.21 Blood glucose regulates liver glycogen metabolism. The infusion of glucose into the bloodstream leads to the inactivation of phosphorylase, followed by the activation of glycogen synthase, in the liver. [After W. Stalmans, H. De Wulf, L. Hue, and H.-G. Hers. *Eur. J. Biochem.* 41(1974):127.]

phosphorylations again serve as the instigation for a regulatory wave of *dephosphorylations*. The binding of insulin to its receptor leads to the activation of an *insulin-sensitive protein kinase* that phosphorylates the R_{GI} subunit of PP1 at a site *different* from that modified by protein kinase A. This phosphorylation leads to the association of the R_{GI} subunit with PP1 and the glycogen molecule. The consequent dephosphorylation of glycogen synthase, phosphorylase kinase, and phosphorylase promotes glycogen synthesis and blocks its degradation. Once again we see that *glycogen synthesis and breakdown are coordinately controlled.*

21.5.3 Glycogen Metabolism in the Liver Regulates the Blood-Glucose Level

After a meal rich in carbohydrates, blood-glucose levels rise, leading to an increase in glycogen synthesis in the liver. Although insulin is the primary signal for glycogen synthesis, other, nonhormonal mechanisms also function in the liver. One signal is the concentration of glucose in the blood, which normally ranges from about 80 to 120 mg per 100 ml (4.4 – 6.7 mM). The liver senses the concentration of glucose in the blood and takes up or releases glucose accordingly. The amount of liver phosphorylase *a* decreases rapidly when glucose is infused (Figure 21.21). After a lag period, the amount of glycogen synthase *a* increases, which results in the synthesis of glycogen. In fact, *phosphorylase a is the glucose sensor in liver cells.* The binding of glucose to phosphorylase *a* shifts its allosteric equilibrium from the active R form to the inactive T form. This conformational change *renders the phosphoryl group on serine 14 a substrate for protein phosphatase 1.* It is significant that PP1 binds tightly to phosphorylase *a* but acts catalytically only when glucose induces the transition to the T form. Recall that the R ⟷ T transition of muscle phosphorylase *a* is unaffected by glucose and is thus unaffected by the rise in blood-glucose levels (Section 21.2.2).

How does glucose activate glycogen synthase? Phosphorylase *b*, in contrast with phosphorylase *a*, does not bind the phosphatase. Consequently, the conversion of *a* into *b* is accompanied by the *release of PP1, which is then free to activate glycogen synthase* (Figure 21.22). Removal of the phosphoryl group of inactive glycogen synthase *b* converts it into the active *a* form. Initially, there are about 10 phosphorylase *a* molecules per molecule of phosphatase. Hence, *the activity of glycogen synthase begins to increase*

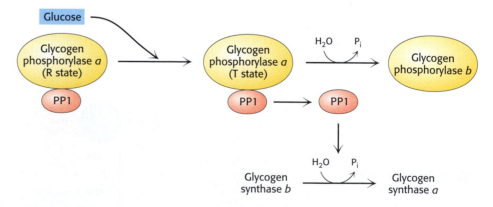

FIGURE 21.22 Glucose regulation of liver glycogen metabolism. Glucose binds to and inhibits glycogen phosphorylase *a* in the liver, leading to the dissociation and activation of protein phosphatase 1 (PP1) from glycogen phosphorylase *a*. The free PP1 dephosphorylates glycogen phosphorylase *a* and glycogen synthase *b*, leading to the inactivation of glycogen breakdown and the activation of glycogen synthesis.

only after most of phosphorylase a is converted into b. This remarkable glucose-sensing system depends on three key elements: (1) communication between the serine phosphate and the allosteric site for glucose, (2) the use of PP1 to inactivate phosphorylase and activate glycogen synthase, and (3) the binding of the phosphatase to phosphorylase a to prevent the premature activation of glycogen synthase.

21.5.4 A Biochemical Understanding of Glycogen-Storage Diseases Is Possible

Edgar von Gierke described the first glycogen-storage disease in 1929. A patient with this disease has a huge abdomen caused by a *massive enlargement of the liver*. There is a pronounced *hypoglycemia* between meals. Furthermore, the blood-glucose level does not rise on administration of epinephrine and glucagon. An infant with this glycogen-storage disease may have convulsions because of the low blood-glucose level.

The enzymatic defect in von Gierke disease was elucidated in 1952 by Carl and Gerty Cori. They found that *glucose 6-phosphatase is missing from the liver of a patient with this disease*. This was the first demonstration of an inherited deficiency of a liver enzyme. The liver glycogen is normal in structure but present in abnormally large amounts. The absence of glucose 6-phosphatase in the liver causes hypoglycemia because glucose cannot be formed from glucose 6-phosphate. This phosphorylated sugar does not leave the liver, because it cannot cross the plasma membrane. The presence of excess glucose 6-phosphate triggers an increase in glycolysis in the liver, leading to a high level of lactate and pyruvate in the blood. Patients who have von Gierke disease also have an increased dependence on fat metabolism. This disease can also be produced by a mutation in the gene that encodes the *glucose 6-phosphate transporter*. Recall that glucose 6-phosphate must be transported into the lumen of the endoplasmic reticulum to be hydrolyzed by phosphatase (Section 16.3.5). Mutations in the other three essential proteins of this system can likewise lead to von Gierke disease.

Seven other glycogen-storage diseases have been characterized (Table 21.1). In Pompe disease (type II), lysosomes become engorged with glycogen because they lack α-1,4-glucosidase, a hydrolytic enzyme confined to these organelles (Figure 21.23). The Coris elucidated the biochemical defect in another glycogen-storage disease (type III), which cannot be distinguished from von Gierke disease (type I) by physical examination alone. In type III disease, the structure of liver and muscle glycogen is abnormal and the amount is markedly increased. Most striking, the outer branches of the glycogen are very short. *Patients having this type lack the debranching enzyme (α-1,6-glucosidase)*, and so only the outermost branches of glycogen can be effectively utilized. Thus, only a small fraction of this abnormal glycogen is functionally active as an accessible store of glucose.

A defect in glycogen metabolism confined to muscle is found in McArdle disease (type V). *Muscle phosphorylase activity is absent*, and the patient's capacity to perform strenuous exercise is limited because of painful muscle cramps. The patient is otherwise normal and well developed. Thus, effective utilization of muscle glycogen is not essential for life. The results of phosphorus-31 nuclear magnetic resonance studies of these patients have been very informative. The pH of skeletal muscle cells of normal people drops during strenuous exercise because of the production of lactate. In contrast, the muscle cells of patients with McArdle disease become more alkaline during exercise because of the breakdown of creatine phosphate (Section 14.1.5). Lactate does not accumulate in these patients because the

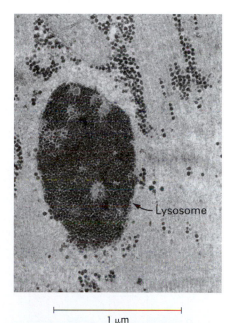

1 μm

FIGURE 21.23 Glycogen-engorged lysosome. This electron micrograph shows skeletal muscle from an infant with type II glycogen-storage disease (Pompe disease). The lysosomes are filled with glycogen because of a deficiency in α-1,4-glucosidase, a hydrolytic enzyme confined to lysosomes. The amount of glycogen in the cytosol is normal. [From H.-G. Hers and F. Van Hoof, Eds. *Lysosomes and Storage Diseases* (Academic Press, 1973), p. 205.]

TABLE 21.1   Glycogen-storage diseases

Type	Defective enzyme	Organ affected	Glycogen in the affected organ	Clinical features
I Von Gierke disease	Glucose 6-phosphatase or transport system	Liver and kidney	Increased amount; normal structure.	Massive enlargement of the liver. Failure to thrive. Severe hypoglycemia, ketosis, hyperuricemia, hyperlipemia.
II Pompe disease	α-1,4-Glucosidase (lysosomal)	All organs	Massive increase in amount; normal structure.	Cardiorespiratory failure causes death, usually before age 2.
III Cori disease	Amylo-1,6-glucosidase (debranching enzyme)	Muscle and liver	Increased amount; short outer branches.	Like type I, but milder course.
IV Andersen disease	Branching enzyme (α-1,4 $\longrightarrow$ α-1,6)	Liver and spleen	Normal amount; very long outer branches.	Progressive cirrhosis of the liver. Liver failure causes death, usually before age 2.
V McArdle disease	Phosphorylase	Muscle	Moderately increased amount; normal structure.	Limited ability to perform strenuous exercise because of painful muscle cramps. Otherwise patient is normal and well developed.
VI Hers disease	Phosphorylase	Liver	Increased amount.	Like type I, but milder course.
VII	Phosphofructokinase	Muscle	Increased amount; normal structure.	Like type V.
VIII	Phosphorylase kinase	Liver	Increased amount; normal structure.	Mild liver enlargement. Mild hypoglycemia.

Note: Types I through VII are inherited as autosomal recessives. Type VIII is sex linked.

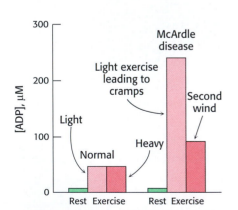

FIGURE 21.24 NMR study of human arm muscle. The level of ADP during exercise increases much more in a patient with McArdle glycogen-storage disease (type V) than in normal controls. [After G. K. Radda. *Biochem. Soc. Trans.* 14(1986):522.]

glycolytic rate of their muscle is much lower than normal; their glycogen cannot be mobilized. The results of NMR studies have also shown that the painful cramps in this disease are correlated with high levels of ADP (Figure 21.24). NMR spectroscopy is a valuable, noninvasive technique for assessing dietary and exercise therapy for this disease.

SUMMARY

Glycogen, a readily mobilized fuel store, is a branched polymer of glucose residues. Most of the glucose units in glycogen are linked by α-1,4 glycosidic bonds. At about every tenth residue, a branch is created by an α-1,6-glycosidic bond. Glycogen is present in large amounts in muscle cells and in liver cells, where it is stored in the cytoplasm in the form of hydrated granules.

Glycogen Breakdown Requires the Interplay of Several Enzymes
Most of the glycogen molecule is degraded to glucose 1-phosphate by the action of glycogen phosphorylase, the key enzyme in glycogen breakdown. The glycosidic linkage between C-1 of a terminal residue and C-4 of the adjacent one is split by orthophosphate to give glucose

1-phosphate, which can be reversibly converted into glucose 6-phosphate. Branch points are degraded by the concerted action of an oligosaccharide transferase and an α-1,6-glucosidase.

Phosphorylase Is Regulated by Allosteric Interactions and Reversible Phosphorylation

Phosphorylase is regulated by allosteric effectors and reversible covalent modifications. Phosphorylase *b*, which is usually inactive, is converted into active phosphorylase *a* by the phosphorylation of a single serine residue in each subunit. This reaction is catalyzed by phosphorylase kinase. The *b* form in muscle can also be activated by the binding of AMP, an effect antagonized by ATP and glucose 6-phosphate. The *a* form in the liver is inhibited by glucose. The AMP-binding sites and phosphorylation sites are located at the subunit interface. In muscle, phosphorylase is activated to generate glucose for use inside the cell as a fuel for contractile activity. In contrast, liver phosphorylase is activated to liberate glucose for export to other organs, such as skeletal muscle and the brain.

Epinephrine and Glucagon Signal the Need for Glycogen Breakdown

Epinephrine and glucagon stimulate glycogen breakdown through specific TTM receptors. Muscle is the primary target of epinephrine, whereas the liver is responsive to glucagon. Both signal molecules initiate a kinase cascade that leads to the activation of glycogen phosphorylase.

Glycogen Is Synthesized and Degraded by Different Pathways

Glycogen is synthesized by a different pathway from that of glycogen breakdown. UDP-glucose, the activated intermediate in glycogen synthesis, is formed from glucose 1-phosphate and UTP. Glycogen synthase catalyzes the transfer of glucose from UDP-glucose to the C-4 hydroxyl group of a terminal residue in the growing glycogen molecule. Synthesis is primed by glycogenin, an autoglycosylating protein that contains a covalently attached oligosaccharide unit on a specific tyrosine residue. A branching enzyme converts some of the α-1,4 linkages into α-1,6 linkages to increase the number of ends so that glycogen can be made and degraded more rapidly.

Glycogen Breakdown and Synthesis Are Reciprocally Regulated

Glycogen synthesis and degradation are coordinated by several amplifying reaction cascades. Epinephrine and glucagon stimulate glycogen breakdown and inhibit its synthesis by increasing the cytosolic level of cyclic AMP, which activates protein kinase A. Elevated cytosolic Ca^{2+} levels stimulate glycogen degradation by activating phosphorylase kinase. Hence, muscle contraction and calcium-mobilizing hormones promote glycogen breakdown.

The glycogen-mobilizing actions of PKA are reversed by protein phosphatase 1, which is regulated by several hormones. Epinephrine inhibits this phosphatase by blocking its attachment to glycogen molecules and by turning on an inhibitor. Insulin, in contrast, activates this phosphatase by triggering a cascade that phosphorylates the glycogen-targeting subunit of this enzyme. Hence, glycogen synthesis is decreased by epinephrine and increased by insulin. Glycogen synthase and phosphorylase are also regulated by noncovalent allosteric interactions. In fact, phosphorylase is a key part of the glucose-sensing system of liver cells. Glycogen metabolism exemplifies the power and precision of reversible phosphorylation in regulating biological processes.

KEY TERMS

glycogen phosphorylase (p. 579)

phosphorolysis (p. 579)

pyridoxal phosphate (p. 583)

phosphorylase kinase (p. 584)

protein kinase A (p. 586)

calmodulin (p. 586)

epinephrine (adrenaline) (p. 586)

glucagon (p. 586)

uridine diphosphate glucose (UDP-glucose) (p. 589)

glycogen synthase (p. 589)

glycogenin (p. 590)

protein phosphatase 1 (p. 592)

insulin (p. 593)

SELECTED READINGS

Where to start

Krebs, E. G., 1993. Protein phosphorylation and cellular regulation I. *Biosci. Rep.* 13:127–142.

Fischer, E. H., 1993. Protein phosphorylation and cellular regulation II. *Angew. Chem. Int. Ed.* 32:1130–1137.

Johnson, L. N., 1992. Glycogen phosphorylase: Control by phosphorylation and allosteric effectors. *FASEB J.* 6:2274–2282.

Browner, M. F., and Fletterick, R. J., 1992. Phosphorylase: A biological transducer. *Trends Biochem. Sci.* 17:66–71.

Books and general reviews

Shulman, R. G., and Rothman, D. L., 1996. Enzymatic phosphorylation of muscle glycogen synthase: A mechanism for maintenance of metabolic homeostasis. *Proc. Natl. Acad. Sci. USA* 93:7491–7495.

Roach, P. J., Cao, Y., Corbett, C. A., DePaoli, R. A., Farkas, I., Fiol, C. J., Flotow, H., Graves, P. R., Hardy, T. A., and Hrubey, T. W., 1991. Glycogen metabolism and signal transduction in mammals and yeast. *Adv. Enzyme Regul.* 31:101–120.

Shulman, G. I., and Landau, B. R., 1992. Pathways of glycogen repletion. *Physiol. Rev.* 72:1019–1035.

X-ray crystallographic studies

Lowe, E. D., Noble, M. E., Skamnaki, V. T., Oikonomakos, N. G., Owen, D. J., and Johnson, L. N., 1997. The crystal structure of a phosphorylase kinase peptide substrate complex: Kinase substrate recognition. *EMBO J.* 16:6646–6658.

Barford, D., Hu, S. H., and Johnson, L. N., 1991. Structural mechanism for glycogen phosphorylase control by phosphorylation and AMP. *J. Mol. Biol.* 218:233–260.

Sprang, S. R., Withers, S. G., Goldsmith, E. J., Fletterick, R. J., and Madsen, N. B., 1991. Structural basis for the activation of glycogen phosphorylase *b* by adenosine monophosphate. *Science* 254:1367–1371.

Johnson, L. N., and Barford, D., 1990. Glycogen phosphorylase: The structural basis of the allosteric response and comparison with other allosteric proteins. *J. Biol. Chem.* 265:2409–2412.

Browner, M. F., Fauman, E. B., and Fletterick, R. J., 1992. Tracking conformational states in allosteric transitions of phosphorylase. *Biochemistry* 31:11297–11304.

Martin, J. L., Johnson, L. N., and Withers, S. G., 1990. Comparison of the binding of glucose and glucose 1-phosphate derivatives to T-state glycogen phosphorylase *b*. *Biochemistry* 29:10745–10757.

Johnson, L. N., Acharya, K. R., Jordan, M. D., and McLaughlin, P. J., 1990. Refined crystal structure of the phosphorylase-heptulose-2-phosphate-oligosaccharide-AMP complex. *J. Mol. Biol.* 211:645–661.

Priming of glycogen synthesis

Alonso, M. D., Lomako, J., Lomako, W. M., and Whelan, W. J., 1995. A new look at the biogenesis of glycogen. *FASEB J.* 9:1126–1137.

Lin, A., Mu, J., Yang, J., and Roach, P. J., 1999. Self-glucosylation of glycogenin, the initiator of glycogen biosynthesis, involves an inter-subunit reaction. *Arch. Biochem. Biophys.* 363:163–170.

Roach, P. J., and Skurat, A. V., 1997. Self-glucosylating initiator proteins and their role in glycogen biosynthesis. *Prog. Nucleic Acid Res. Mol. Biol.* 57:289–316.

Smythe, C., and Cohen, P., 1991. The discovery of glycogenin and the priming mechanism for glycogen biogenesis. *Eur. J. Biochem.* 200:625–631.

Catalytic mechanisms

Skamnaki, V. T., Owen, D. J., Noble, M. E., Lowe, E. D., Lowe, G., Oikonomakos, N. G., and Johnson, L. N., 1999. Catalytic mechanism of phosphorylase kinase probed by mutational studies. *Biochemistry* 38:14718–14730.

Buchbinder, J. L., and Fletterick, R. J., 1996. Role of the active site gate of glycogen phosphorylase in allosteric inhibition and substrate binding. *J. Biol. Chem.* 271:22305–22309.

Palm, D., Klein, H. W., Schinzel, R., Buehner, M., and Helmreich, E. J. M., 1990. The role of pyridoxal 5′-phosphate in glycogen phosphorylase catalysis. *Biochemistry* 29:1099–1107.

Regulation of glycogen metabolism

Pederson, B. A., Cheng, C., Wilson, W. A., and Roach, P. J., 2000. Regulation of glycogen synthase: Identification of residues involved in regulation by the allosteric ligand glucose-6-P and by phosphorylation. *J. Biol. Chem.* 275:27753–27761.

Melendez, R., Melendez-Hevia, E., and Canela, E. I., 1999. The fractal structure of glycogen: A clever solution to optimize cell metabolism. *Biophys. J.* 77:1327–1332.

Franch, J., Aslesen, R., and Jensen, J., 1999. Regulation of glycogen synthesis in rat skeletal muscle after glycogen-depleting contractile activity: Effects of adrenaline on glycogen synthesis and activation of glycogen synthase and glycogen phosphorylase. *Biochem. J.* 344 (pt.1):231–235.

Aggen, J. B., Nairn, A. C., and Chamberlin, R., 2000. Regulation of protein phosphatase-1. *Chem. Biol.* 7:R13–R23.

Egloff, M. P., Johnson, D. F., Moorhead, G., Cohen, P. T., Cohen, P., and Barford, D., 1997. Structural basis for the recognition of regulatory subunits by the catalytic subunit of protein phosphatase 1. *EMBO J.* 16:1876–1887.

Ragolia, L., and Begum, N., 1998. Protein phosphatase-1 and insulin action. *Mol. Cell. Biochem.* 182:49–58.

Wu, J., Liu, J., Thompson, I., Oliver, C. J., Shenolikar, S., and Brautigan, D. L., 1998. A conserved domain for glycogen binding in protein phosphatase-1 targeting subunits. *FEBS Lett.* 439:185–191.

Genetic diseases

Chen, Y.-T., and Burchell, A., 1995. Glycogen storage diseases. In *The Metabolic Basis of Inherited Diseases* (7th ed., pp. 935–965) edited by C. R. Striver., A. L. Beaudet, W. S. Sly, D. Valle, J. B. Stanbury, J. B. Wyngaarden, and D. S. Fredrickson, McGraw-Hill.

Burchell, A., and Waddell, I. D., 1991. The molecular basis of the hepatic microsomal glucose-6-phosphatase system. *Biochim. Biophys. Acta* 1092:129–137.

Lei, K. J., Shelley, L. L., Pan, C. J., Sidbury, J. B., and Chou, J. Y., 1993. Mutations in the glucose-6-phosphatase gene that cause glycogen storage disease type Ia. *Science* 262:580–583.

Ross, B. D., Radda, G. K., Gadian, D. G., Rocker, G., Esiri, M., and Falconer-Smith, J., 1981. Examination of a case of suspected McArdle's syndrome by ^{31}P NMR. *N. Engl. J. Med.* 304:1338–1342.

Evolution

Holm, L., and Sander, C., 1995. Evolutionary link between glycogen phosphorylase and a DNA modifying enzyme. *EMBO J.* 14:1287–1293.

Hudson, J. W., Golding, G. B., and Crerar, M. M., 1993. Evolution of allosteric control in glycogen phosphorylase. *J. Mol. Biol.* 234:700–721.

Rath, V. L., and Fletterick, R. J., 1994. Parallel evolution in two homologues of phosphorylase. *Nat. Struct. Biol.* 1:681–690.

Melendez, R., Melendez-Hevia, E., and Cascante, M., 1997. How did glycogen structure evolve to satisfy the requirement for rapid mobilization of glucose? A problem of physical constraints in structure building. *J. Mol. Evol.* 45:446–455.

Rath, V. L., Lin, K.,Hwang, P. K., and Fletterick, R. J., 1996. The evolution of an allosteric site in phosphorylase. *Structure* 4:463–473.

PROBLEMS

1. *Carbohydrate conversion.* Write a balanced equation for the formation of glycogen from galactose.

2. *If a little is good, a lot is better.* α-Amylose is an unbranched glucose polymer. Why would this polymer not be as effective a storage form for glucose as glycogen?

3. *Telltale products.* A sample of glycogen from a patient with liver disease is incubated with orthophosphate, phosphorylase, the transferase, and the debranching enzyme (α-1,6-glucosidase). The ratio of glucose 1-phosphate to glucose formed in this mixture is 100. What is the most likely enzymatic deficiency in this patient?

4. *Excessive storage.* Suggest an explanation for the fact that the amount of glycogen in type I glycogen-storage disease (von Gierke disease) is increased.

5. *A shattering experience.* Crystals of phosphorylase *a* grown in the presence of glucose shatter when a substrate such as glucose 1-phosphate is added. Why?

6. *Recouping an essential phosphoryl.* The phosphoryl group on phosphoglucomutase is slowly lost by hydrolysis. Propose a mechanism that utilizes a known catalytic intermediate for restoring this essential phosphoryl group. How might this phosphoryl donor be formed?

7. *Hydrophobia.* Why is water excluded from the active site of phosphorylase? Predict the effect of a mutation that allows water molecules to enter.

8. *Removing all traces.* In human liver extracts, the catalytic activity of glycogenin was detectable only after treatment with α-amylase (Section 11.2.2). Why was α-amylase necessary to reveal the glycogenin activity?

9. *Two in one.* A single polypeptide chain houses the transferase and debranching enzyme. Cite a potential advantage of this arrangement.

10. *How did they do that?* A strain of mice has been developed that lack the enzyme phosphorylase kinase. Yet, after strenuous exercise, the glycogen stores of a mouse of this strain are depleted. Explain how this is possible.

11. *Metabolic mutants.* Predict the major consequence of each of the following mutations:

(a) Loss of the AMP-binding site in muscle phosphorylase.
(b) Mutation of Ser 14 to Ala 14 in liver phosphorylase.
(c) Overexpression of phosphorylase kinase in the liver.
(d) Loss of the gene that encodes inhibitor 1 of protein phosphatase 1.

(e) Loss of the gene that encodes the glycogen-targeting subunit of protein phosphatase 1.
(f) Loss of the gene that encodes glycogenin.

12. *More metabolic mutants.* Briefly, predict the major consequences of each of the following mutations affecting glycogen utilization.

(a) Loss of GTPase activity of the G protein α subunit.
(b) Loss of the gene that encodes inhibitor 1 of protein phosphatase 1.
(c) Loss of phosphodiesterase activity.

13. *Multiple phosphorylation.* Protein kinase A activates muscle phosphorylase kinase by rapidly phosphorylating its β subunits. The α subunits of phosphorylase kinase are then slowly phosphorylated, which makes the α and β subunits susceptible to the action of protein phosphatase 1. What is the functional significance of the slow phosphorylation of α?

14. *The wrong switch.* What would be the consequences to glycogen mobilization of a mutation in phosphorylase kinase that leads to the phosphorylation of the α-subunit prior to that of the β subunit?

Mechanism Problem

15. *Family resemblance.* Propose mechanisms for the two enzymes catalyzing steps in glycogen debranching based on their potential membership in the α-amylase family.

Chapter Integration and Data Interpretation Problems

16. *Glycogen isolation 1.* The liver is a major storage site for glycogen. Purified from two samples of human liver, glycogen was either treated or not treated with α-amylase and subsequently analyzed by SDS-PAGE and Western blotting with the use of antibodies to glycogenin. The results are presented in the adjoining illustration.

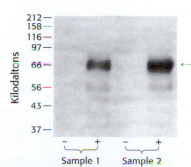

Glycogen isolation 1.
[Courtesy of Dr. Peter J. Roach, Indiana University School of Medicine.]

(a) Why are no proteins visible in the lanes without amylase treatment?

(b) What is the effect of treating the samples with α-amylase? Explain the results.

(c) List other proteins that you might expect to be associated with glycogen. Why are other proteins not visible?

17. *Glycogen isolation 2.* The gene for glycogenin was transfected into a cell line that normally stores only small amounts of glycogen. The cells were then manipulated according to the following protocol, and glycogen was isolated and analyzed by SDS-PAGE and Western blotting using an antibody to glycogenin with and without α-amylase treatment. The results are presented in the adjoining illustration.

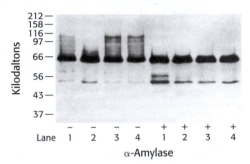

Glycogen isolation 2. [Courtesy of Dr. Peter J. Roach, Indiana University School of Medicine.]

The protocol: Cells cultured in growth medium and 25 mM glucose (lane 1) were switched to medium containing no glucose for 24 hours (lane 2). Glucose-starved cells were re-fed with medium containing 25 mM glucose for 1 hour (lane 3) or 3 hours (lane 4). Samples (12 μg of protein) were either treated or not treated with α-amylase, as indicated, before being loaded on the gel.

(a) Why did the Western analysis produce a "smear"—that is, the high-molecular-weight staining in lane 1(−)?

(b) What is the significance of the decrease in high-molecular-weight staining in lane 2(−)?

(c) What is the significance of the difference between lanes 2(−) and 3(−)?

(d) Suggest a plausible reason why there is essentially no difference between lanes 3(−) and 4(−)?

(e) Why are the bands at 66 kd the same in the lanes treated with amylase, despite the fact that the cells were treated differently?

 **Media Problem**

18. *Either/or.* Mammalian glycogen phosphorylase is activated by phosphorylation of serine 14, and deactivated by dephosphorylation. The binding of glucose to glycogen phosphorylase *a* inhibits the enzyme by promoting the protein phosphatase–catalyzed dephosphorylation of serine 14, which converts the *a* form of the enzyme to the *b* form. From the information presented in the **Structural Insights** module on glycogen phosphorylase, provide an explanation how glucose binding might promote dephosphorylation of serine 14.

CHAPTER 22

Fatty Acid Metabolism

Fats provide an efficient means for storing energy for later use. (Right) The processes of fatty acid synthesis (preparation for energy storage) and fatty acid degradation (preparation for energy use) are, in many ways, the reverse of each other. (Above) Studies of mice are revealing the interplay between these pathways and the biochemical bases of appetite and weight control. [(Above) © Jackson/Visuals Unlimited.]

We turn now from the metabolism of carbohydrates to that of fatty acids. A fatty acid contains a long hydrocarbon chain and a terminal carboxylate group. Fatty acids have four major physiological roles. First, *fatty acids are building blocks of phospholipids and glycolipids.* These amphipathic molecules are important components of biological membranes, as we discussed in Chapter 12. Second, many proteins are modified by the *covalent attachment of fatty acids, which targets them to membrane locations* (Section 12.5.3). Third, *fatty acids are fuel molecules.* They are stored as *triacylglycerols* (also called *neutral fats* or *triglycerides*), which are uncharged esters of fatty acids with glycerol (Figure 22.1). Fatty acids mobilized from triacylglycerols are oxidized to meet the energy needs of a cell or organism. Fourth, *fatty acid derivatives serve as hormones and intracellular messengers.* In this chapter, we will focus on the oxidation and synthesis of fatty acids, processes that are reciprocally regulated in response to hormones.

22.0.1 An Overview of Fatty Acid Metabolism

Fatty acid degradation and synthesis are relatively simple processes that are essentially the reverse of each other. The process of degradation converts an aliphatic compound into a set of activated acetyl units (acetyl CoA) that can be processed by the citric acid cycle (Figure 22.2). An activated

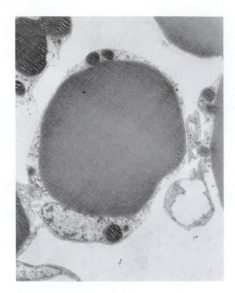

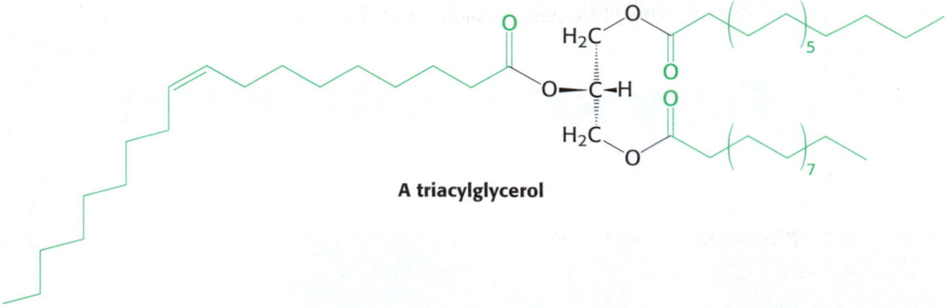

A triacylglycerol

fatty acid is oxidized to introduce a double bond; the double bond is hydrated to introduce an oxygen; the alcohol is oxidized to a ketone; and, finally, the four carbon fragment is cleaved by coenzyme A to yield acetyl CoA and a fatty acid chain two carbons shorter. If the fatty acid has an even number of carbon atoms and is saturated, the process is simply repeated until the fatty acid is completely converted into acetyl CoA units.

Fatty acid synthesis is essentially the reverse of this process. Because the result is a polymer, the process starts with monomers—in this case with activated acyl group (most simply, an acetyl unit) and malonyl units (see Figure 22.2). The malonyl unit is condensed with the acetyl unit to form a four-carbon fragment. To produce the required hydrocarbon chain, the carbonyl must be reduced. The fragment is reduced, dehydrated, and reduced

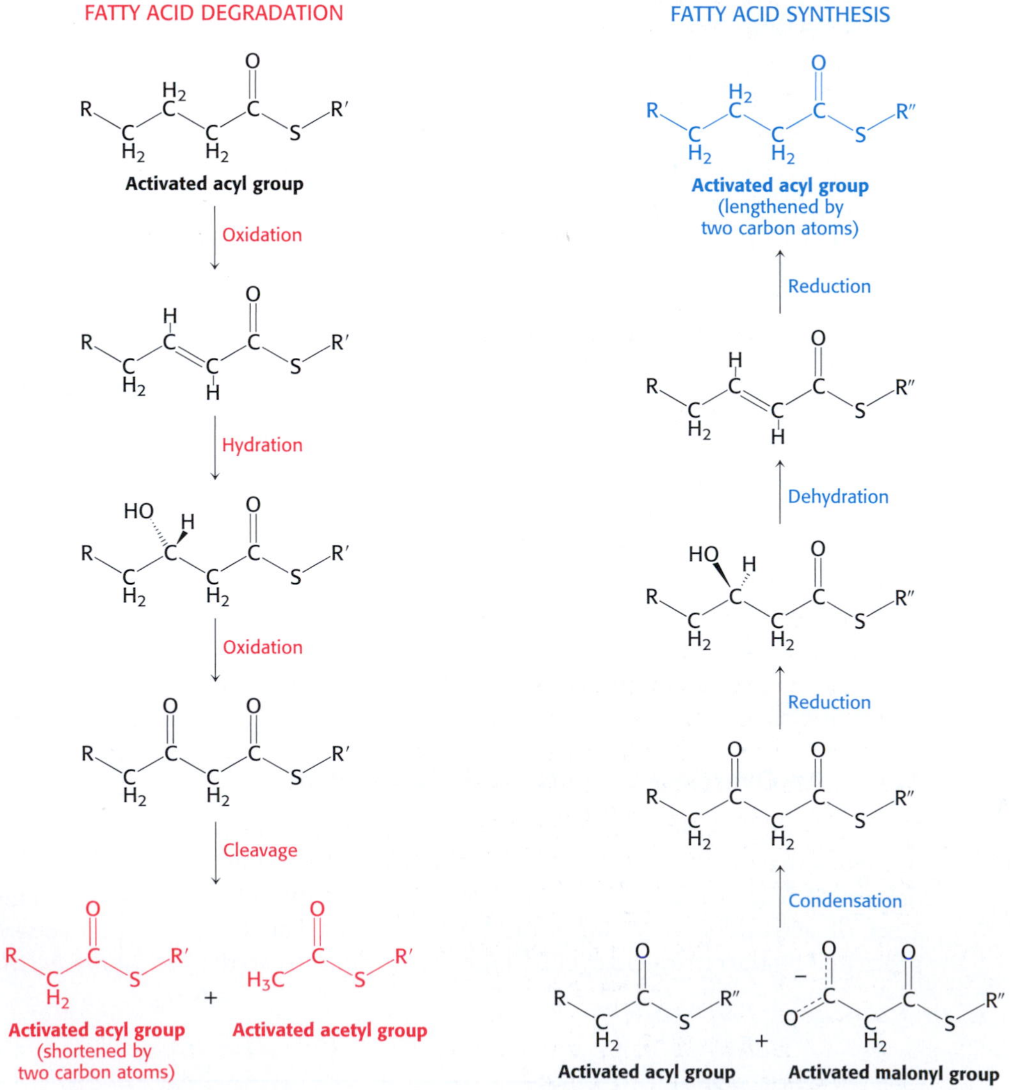

FIGURE 22.1 Electron micrograph of an adipocyte. A small band of cytoplasm surrounds the large deposit of triacylglycerols. [Biophoto Associates/ Photo Researchers.]

FATTY ACID DEGRADATION **FATTY ACID SYNTHESIS**

FIGURE 22.2 Steps in fatty acid degradation and synthesis. The two processes are in many ways mirror images of each other.

again, exactly the opposite of degradation, to bring the carbonyl group to the level of a methylene group with the formation of butyryl CoA. Another activated malonyl group condenses with the butyryl unit and the process is repeated until a C_{16} fatty acid is synthesized.

22.1 TRIACYLGLYCEROLS ARE HIGHLY CONCENTRATED ENERGY STORES

Triacylglycerols are highly concentrated stores of metabolic energy because they are *reduced* and *anhydrous*. The yield from the complete oxidation of fatty acids is about 9 kcal g^{-1} (38 kJ g^{-1}), in contrast with about 4 kcal g^{-1} (17 kJ g^{-1}) for carbohydrates and proteins. The basis of this large difference in caloric yield is that fatty acids are much more reduced. Furthermore, triacylglycerols are nonpolar, and so they are stored in a nearly anhydrous form, whereas much more polar proteins and carbohydrates are more highly hydrated. In fact, 1 g of dry glycogen binds about 2 g of water. Consequently, *a gram of nearly anhydrous fat stores more than six times as much energy as a gram of hydrated glycogen*, which is likely the reason that triacylglycerols rather than glycogen were selected in evolution as the major energy reservoir. Consider a typical 70-kg man, who has fuel reserves of 100,000 kcal (420,000 kJ) in triacylglycerols, 25,000 kcal (100,000 kJ) in protein (mostly in muscle), 600 kcal (2500 kJ) in glycogen, and 40 kcal (170 kJ) in glucose. Triacylglycerols constitute about 11 kg of his total body weight. If this amount of energy were stored in glycogen, his total body weight would be 55 kg greater. The glycogen and glucose stores provide enough energy to sustain biological function for about 24 hours, whereas the triacylglycerol stores allow survival for several weeks.

In mammals, the major site of accumulation of triacylglycerols is the cytoplasm of *adipose cells (fat cells)*. Droplets of triacylglycerol coalesce to form a large globule, which may occupy most of the cell volume (see Figure 22.1). Adipose cells are specialized for the synthesis and storage of triacylglycerols and for their mobilization into fuel molecules that are transported to other tissues by the blood.

The utility of triacylglycerols as an energy source is dramatically illustrated by the abilities of migratory birds, which can fly great distances without eating. Examples are the American golden plover and the ruby-throated sparrow. The golden plover flies from Alaska to the southern tip of South America; a large segment of the flight (3800 km, or 2400 miles) is over open ocean, where the birds cannot feed. The ruby-throated hummingbird can fly nonstop across the Gulf of Mexico. Fatty acids provide the energy source for both these prodigious feats.

Triacylglycerols fuel the long migration flights of the American Golden Plover (*Pluvialis dominica*). [Gerard Fuehrer/Visuals Unlimited.]

22.1.1 Dietary Lipids Are Digested by Pancreatic Lipases

Most lipids are ingested in the form of triacylglycerols but must be degraded to fatty acids for absorption across the intestinal epithelium. Recall that lipids are not easily solubilized, yet they must be in order to be degraded. Triacylglycerols in the intestinal lumen are incorporated into micelles formed with the aid of *bile salts* (Figure 22.3), amphipathic molecules synthesized from cholesterol in the liver and secreted from the gall bladder. Incorporation of lipids into micelles orients the ester bonds of the lipid toward the surface of the micelle,

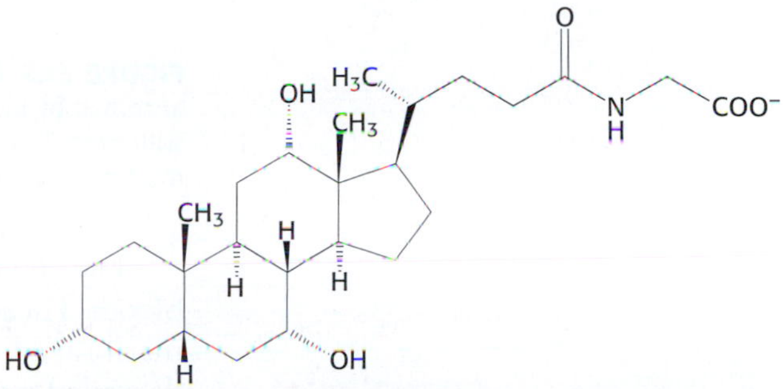

Glycocholate

FIGURE 22.3 Glycocholate. Bile salts, such as glycocholate, facilitate lipid digestion in the intestine.

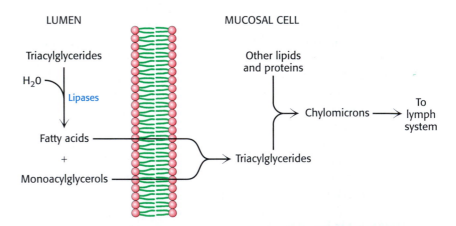

Triacylglyceride   **Diacylglyceride**   **Monoacylglycerol**

FIGURE 22.4 Action of pancreatic lipases. Lipases secreted by the pancreas convert triacylglycerols into fatty acids and monoacylglycerol for absorption into the intestine.

rendering the bonds more susceptible to digestion by pancreatic lipases that are in aqueous solution. If the production of bile salts is inadequate owing to liver disease, large amounts of fats (as much as 30 g day^{-1}) are excreted in the feces. This condition is referred to as steatorrhea, from the Greek *steato*, "fat."

The lipases digest the triacylglycerols into free fatty acids and monoacylglycerol (Figure 22.4). These digestion products are carried in micelles to the intestinal epithelium where they are absorbed across the plasma membrane.

22.1.2 Dietary Lipids Are Transported in Chylomicrons

In the intestinal mucosal cells, the triacylglycerols are resynthesized from fatty acids and monoacylglycerols and then packaged into lipoprotein transport particles called *chylomicrons*, stable particles ranging from approximately 180 to 500 nm in diameter (Figure 22.5). These particles are composed mainly of triacylglycerols, with apoprotein B-48 as the main protein component. Protein constituents of lipoprotein particles are called *apolipoproteins*. Chylomicrons also function in the transport of fat-soluble vitamins and cholesterol.

LUMEN   **MUCOSAL CELL**

Triacylglycerides

H_2O

Lipases

Fatty acids

+

Monoacylglycerols

Other lipids and proteins

Triacylglycerides

Chylomicrons ⟶ To lymph system

FIGURE 22.5 Chylomicron formation. Free fatty acids and monoacylglycerols are absorbed by intestinal epithelial cells. Triacylglycerols are resynthesized and packaged with other lipids and apoprotein B-48 to form chylomicrons, which are then released into the lymph system.

The chylomicrons are released into the lymph system and then into the blood. These particles bind to membrane-bound lipoprotein lipases, primarily at adipose tissue and muscle, where the triacylglycerols are once again degraded into free fatty acids and monoacylglycerol for transport into the tissue. The triacylglycerols are then resynthesized inside the cell and stored. In the muscle, they can be oxidized to provide energy, as will be discussed shortly.

22.2 THE UTILIZATION OF FATTY ACIDS AS FUEL REQUIRES THREE STAGES OF PROCESSING

Peripheral tissues gain access to the lipid energy reserves stored in adipose tissue through three stages of processing. First, the lipids must be mobilized. In this process, triacylglycerols are degraded to fatty acids and glycerol, which are released from the adipose tissue and transported to the energy-requiring tissues. Second, at these tissues, the fatty acids must be activated and transported into mitochondria for degradation. Third, the fatty acids are broken down in a step-by-step fashion into acetyl CoA, which is then processed in the citric acid cycle.

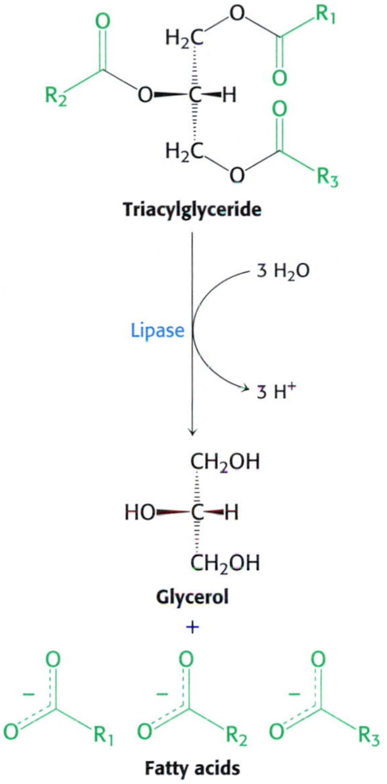

22.2.1 Triacylglycerols Are Hydrolyzed by Cyclic AMP-Regulated Lipases

The initial event in the utilization of fat as an energy source is the hydrolysis of triacylglycerols by lipases, an event referred to as *lipolysis*. The lipase of adipose tissue are activated on treatment of these cells with the hormones epinephrine, norepinephrine, glucagon, and adrenocorticotropic hormone. In adipose cells, these hormones trigger 7TM receptors that activate adenylate cyclase (Section 15.1.3). The increased level of cyclic AMP then stimulates protein kinase A, which activates the lipases by phosphorylating them. Thus, *epinephrine, norepinephrine, glucagon, and adrenocorticotropic hormone induce lipolysis* (Figure 22.6). In contrast, *insulin inhibits lipolysis*. The released fatty acids are not soluble in blood plasma, and so, on release, serum albumin binds the fatty acids and serves as a carrier. By these means, free fatty acids are made accessible as a fuel in other tissues.

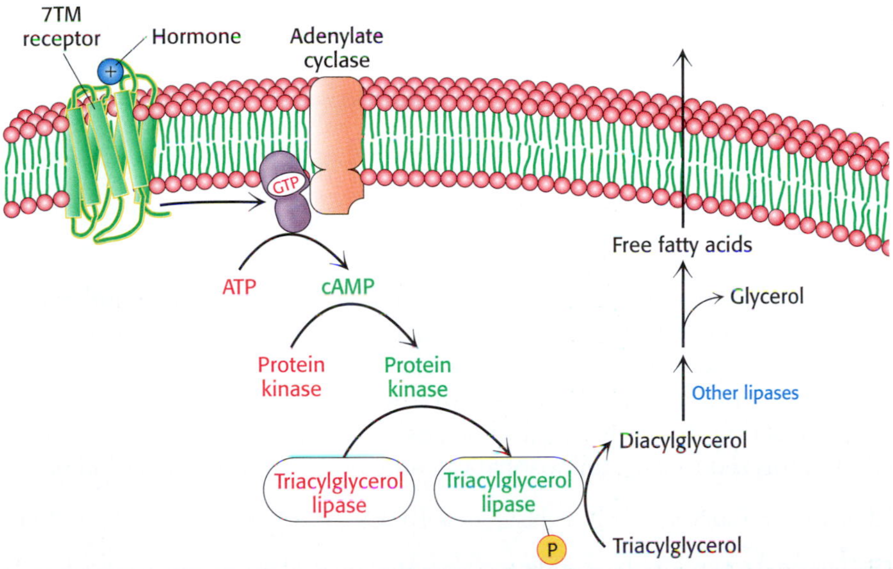

FIGURE 22.6 Mobilization of triacylglycerols. Triacylglycerols in adipose tissue are converted into free fatty acids and glycerol for release into the bloodstream in response to hormonal signals. A hormone-sensitive lipase initiates the process.

Glycerol formed by lipolysis is absorbed by the liver and phosphorylated, oxidized to dihydroxyacetone phosphate, and then isomerized to glyceraldehyde 3-phosphate. This molecule is an intermediate in both the glycolytic and the gluconeogenic pathways.

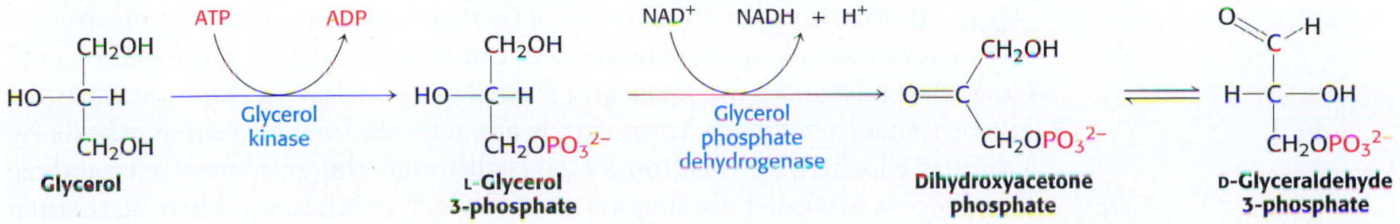

Hence, glycerol can be converted into pyruvate or glucose in the liver, which contains the appropriate enzymes. The reverse process can take place by the reduction of dihydroxyacetone phosphate to glycerol 3-phosphate. Hydrolysis by a phosphatase then gives glycerol. Thus, glycerol and glycolytic intermediates are readily interconvertible.

22.2.2 Fatty Acids Are Linked to Coenzyme A Before They Are Oxidized

Eugene Kennedy and Albert Lehninger showed in 1949 that fatty acids are oxidized in mitochondria. Subsequent work demonstrated that they are activated before they enter the mitochondrial matrix. Adenosine triphosphate (ATP) drives the formation of a thioester linkage between the carboxyl group of a fatty acid and the sulfhydryl group of CoA. This activation reaction takes place on the outer mitochondrial membrane, where it is catalyzed by *acyl CoA synthetase* (also called *fatty acid thiokinase*).

Paul Berg showed that the activation of a fatty acid is accomplished in two steps. First, the fatty acid reacts with ATP to form an *acyl adenylate*. In this mixed anhydride, the carboxyl group of a fatty acid is bonded to the phosphoryl group of AMP. The other two phosphoryl groups of the ATP substrate are released as pyrophosphate. The sulfhydryl group of CoA then attacks the acyl adenylate, which is tightly bound to the enzyme, to form acyl CoA and AMP.

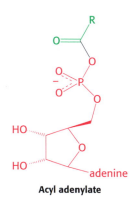

Acyl adenylate

These partial reactions are freely reversible. In fact, the equilibrium constant for the sum of these reactions is close to 1. One high-transfer-potential compound is cleaved (between PP_i and AMP) and one high-transfer-potential compound is formed (the thioester acyl CoA). How is the overall reaction driven forward? The answer is that pyrophosphate is rapidly hydrolyzed by a pyrophosphatase, and so the complete reaction is

$$RCOO^- + CoA + ATP + H_2O \longrightarrow RCO-CoA + AMP + P_i + 2\,H^+$$

This reaction is quite favorable because the equivalent of two molecules of ATP is hydrolyzed, whereas only one high-transfer-potential compound is formed. We see here another example of a recurring theme in biochemistry: *many biosynthetic reactions are made irreversible by the hydrolysis of inorganic pyrophosphate.*

Another motif recurs in this activation reaction. The enzyme-bound acyl-adenylate intermediate is not unique to the synthesis of acyl CoA. *Acyl adenylates are frequently formed when carboxyl groups are activated in biochemical reactions.* Amino acids are activated for protein synthesis by a similar mechanism (Section 29.2.1), although the enzymes that catalyze this process are not homologous to acyl CoA synthetase. Thus, *activation by adenylation recurs in part because of convergent evolution.*

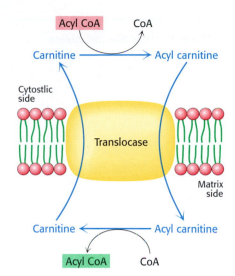

22.2.3 Carnitine Carries Long-Chain Activated Fatty Acids into the Mitochondrial Matrix

Fatty acids are activated on the outer mitochondrial membrane, whereas they are oxidized in the mitochondrial matrix. A special transport mechanism is needed to carry long-chain acyl CoA molecules across the inner mitochondrial membrane. Activated long-chain fatty acids are transported across the membrane by conjugating them to *carnitine*, a zwitterionic alcohol. The acyl group is transferred from the sulfur atom of CoA to the hydroxyl group of carnitine to form *acyl carnitine*. This reaction is catalyzed by *carnitine acyltransferase I* (also called *carnitine palmitoyl transferase I*), which is bound to the outer mitochondrial membrane.

Acyl CoA + **Carnitine** ⇌ **Acyl carnitine** + HS—CoA

Acyl carnitine is then shuttled across the inner mitochondrial membrane by a translocase (Figure 22.7). The acyl group is transferred back to CoA on the matrix side of the membrane. This reaction, which is catalyzed by *carnitine acyltransferase II (carnitine palmitoyl transferase II)*, is simply the reverse of the reaction that takes place in the cytosol. Normally, the transfer of an acyl group from an alcohol to a sulfhydryl group is thermodynamically unfavorable. However, the equilibrium constant for this reaction for carnitine is near 1, apparently because carnatine and its esters are solvated differently from most other alcohols and their esters because of the zwitterionic nature of carnitine. As a result, the O-acyl link in carnitine has a high group-transfer potential. Finally, the translocase returns carnitine to the cytosolic side in exchange for an incoming acyl carnitine.

A number of diseases have been traced to a deficiency of carnitine, the transferase or the translocase. The symptoms of carnitine deficiency range from mild muscle cramping to severe weakness and even death. The muscle, kidney, and heart are the tissues primarily affected. Muscle weakness during prolonged exercise is an important characteristic of a deficiency of carnitine acyl transferases because muscle relies on fatty acids as a long-term source of energy. Medium-chain (C_8–C_{10}) fatty acids, which do not require carnitine to enter the mitochondria, are oxidized normally in these patients. These diseases illustrate that the impaired flow of a metabolite from one compartment of a cell to another can lead to a pathological condition.

22.2.4 Acetyl CoA, NADH, and FADH₂ Are Generated in Each Round of Fatty Acid Oxidation

A saturated acyl CoA is degraded by a recurring sequence of four reactions: oxidation by flavin adenine dinucleotide (FAD), hydration, oxidation by NAD^+, and thiolysis by CoA (Figure 22.8). The fatty acyl chain is

Acyl CoA

FAD → FADH₂ | Oxidation

trans-Δ²-Enoyl CoA

H₂O | Hydration

L-3-Hydroxyacyl CoA

NAD^+ → H^+ + NADH | Oxidation

3-Ketoacyl CoA

HS—CoA | Thiolysis

Acyl CoA (shortened by two carbon atoms) + **Acetyl CoA**

FIGURE 22.8 **Reaction sequence for the degradation of fatty acids.** Fatty acids are degraded by the repetition of a four-reaction sequence consisting of oxidation, hydration, oxidation, and thiolysis.

shortened by two carbon atoms as a result of these reactions, and $FADH_2$, NADH, and acetyl CoA are generated. Because oxidation is on the β carbon, this series of reactions is called the *β-oxidation pathway*.

The first reaction in each round of degradation is the *oxidation* of acyl CoA by an *acyl CoA dehydrogenase* to give an enoyl CoA with a trans double bond between C-2 and C-3.

$$\text{Acyl CoA} + \text{E-FAD} \longrightarrow \textit{trans-}\Delta^2\text{-enoyl CoA} + \text{E-FADH}_2$$

As in the dehydrogenation of succinate in the citric acid cycle, FAD rather than NAD^+ is the electron acceptor because the value of ΔG for this reaction is insufficient to drive the reduction of NAD^+. Electrons from the $FADH_2$ prosthetic group of the reduced acyl CoA dehydrogenase are transferred to a second flavoprotein called *electron-transferring flavoprotein* (ETF). In turn, ETF donates electrons to *ETF:ubiquinone reductase*, an iron–sulfur protein. Ubiquinone is thereby reduced to ubiquinol, which delivers its high-potential electrons to the second proton-pumping site of the respiratory chain (Section 18.3.3). Consequently, 1.5 molecules of ATP are generated per molecule of $FADH_2$ formed in this dehydrogenation step, as in the oxidation of succinate to fumarate (Section 18.3.2).

R—CH₂—CH₂—R' E-FAD ETF-FADH₂ Fe-S (oxidized) Ubiquinol (QH₂)
R—CH=CH—R' E-FADH₂ ETF-FAD Fe-S (reduced) Ubiquinone (Q)

The next step is the *hydration* of the double bond between C-2 and C-3 by *enoyl CoA hydratase*.

$$\textit{trans-}\Delta^2\text{-Enoyl CoA} + \text{H}_2\text{O} \rightleftharpoons \text{L-3-hydroxyacyl CoA}$$

The hydration of enoyl CoA is stereospecific. Only the L isomer of 3-hydroxyacyl CoA is formed when the trans-Δ^2 double bond is hydrated. The enzyme also hydrates a cis-Δ^2 double bond, but the product then is the D isomer. We shall return to this point shortly in considering how unsaturated fatty acids are oxidized.

The hydration of enoyl CoA is a prelude to the second *oxidation* reaction, which converts the hydroxyl group at C-3 into a keto group and generates NADH. This oxidation is catalyzed by *L-3-hydroxyacyl CoA dehydrogenase*, which is specific for the L isomer of the hydroxyacyl substrate.

$$\text{L-3-Hydroxyacyl CoA} + \text{NAD}^+ \rightleftharpoons \text{3-ketoacyl CoA} + \text{NADH} + \text{H}^+$$

TABLE 22.1 Principal reactions in fatty acid oxidation

Step	Reaction	Enzyme
1	Fatty acid + CoA + ATP $\rightleftharpoons$ acyl CoA + AMP + PP$_i$	Acyl CoA synthetase [also called fatty acid thiokinase and fatty acid:CoA ligase (AMP)]
2	Carnitine + acyl CoA $\rightleftharpoons$ acyl carnitine + CoA	Carnitine acyltransferase (also called carnitine palmitoyl transferase)
3	Acyl CoA + E-FAD $\longrightarrow$ *trans-*Δ^2-enoyl CoA + E-FADH$_2$	Acyl CoA dehydrogenases (several isozymes having different chain-length specificity)
4	*trans-*Δ^2-Enoyl CoA + H$_2$O $\rightleftharpoons$ L-3-hydroxyacyl CoA	Enoyl CoA hydratase (also called crotonase or 3-hydroxyacyl CoA hydrolyase)
5	L-3-Hydroxyacyl CoA + NAD$^+$ $\rightleftharpoons$ 3-ketoacyl CoA + NADH + H$^+$	L-3-Hydroxyacyl CoA dehydrogenase
6	3-Ketoacyl CoA + CoA $\rightleftharpoons$ acetyl CoA + acyl CoA (shortened by C$_2$)	β-Ketothiolase (also called thiolase)

The preceding reactions have oxidized the methylene group at C-3 to a keto group. The final step is the cleavage of 3-ketoacyl CoA by the thiol group of a second molecule of CoA, which yields acetyl CoA and an acyl CoA shortened by two carbon atoms. This thiolytic cleavage is catalyzed by β-ketothiolase.

$$3\text{-Ketoacyl CoA} + \text{HS—CoA} \rightleftharpoons \text{acetyl CoA} + \text{acyl CoA}$$
$$(n \text{ carbons}) \qquad\qquad\qquad (n-2 \text{ carbons})$$

Table 22.1 summarizes the reactions in fatty acid oxidation.

The shortened acyl CoA then undergoes another cycle of oxidation, starting with the reaction catalyzed by acyl CoA dehydrogenase (Figure 22.9). Fatty acyl chains containing from 12 to 18 carbon atoms are oxidized by the long-chain acyl CoA dehydrogenase. The medium-chain acyl CoA dehydrogenase oxidizes fatty acyl chains having from 14 to 4 carbons, whereas the short-chain acyl CoA dehydrogenase acts only on 4- and 6- carbon acyl chains. In contrast, β-ketothiolase, hydroxyacyl dehydrogenase, and enoyl CoA hydratase have broad specificity with respect to the length of the acyl group.

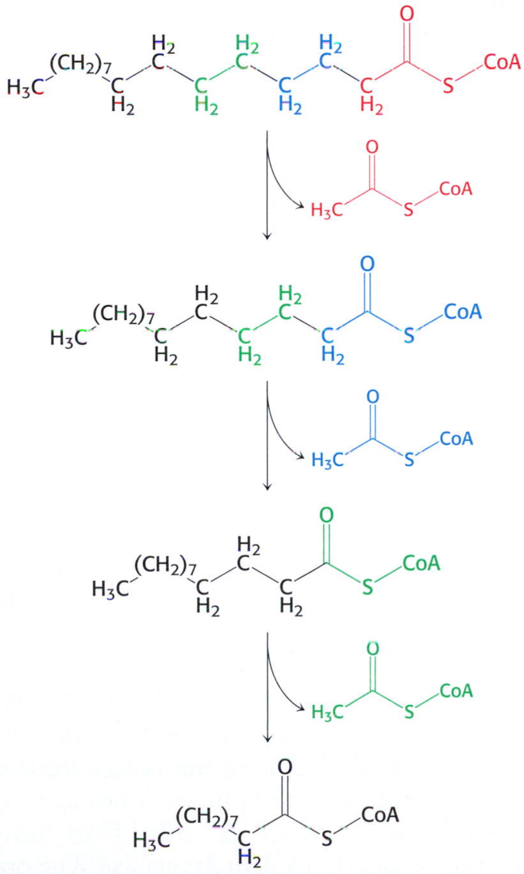

FIGURE 22.9 First three rounds in the degradation of palmitate. Two-carbon units are sequentially removed from the carboxyl end of the fatty acid.

22.2.5 The Complete Oxidation of Palmitate Yields 106 Molecules of ATP

We can now calculate the energy yield derived from the oxidation of a fatty acid. In each reaction cycle, an acyl CoA is shortened by two carbon atoms, and one molecule each of $FADH_2$, NADH, and acetyl CoA is formed.

$$C_n\text{-acyl CoA} + \text{FAD} + \text{NAD}^+ + H_2O + \text{CoA} \longrightarrow$$
$$C_{n-2}\text{-acyl CoA} + \text{FADH}_2 + \text{NADH} + \text{acetyl CoA} + H^+$$

The degradation of palmitoyl CoA (C_{16}-acyl CoA) requires seven reaction cycles. In the seventh cycle, the C_4-ketoacyl CoA is thiolyzed to two molecules of acetyl CoA. Hence, the stoichiometry of oxidation of palmitoyl CoA is

$$\text{Palmitoyl CoA} + 7\,\text{FAD} + 7\,\text{NAD}^+ + 7\,\text{CoA} + 7\,\text{H}_2\text{O} \longrightarrow$$
$$8\,\text{acetyl CoA} + 7\,\text{FADH}_2 + 7\,\text{NADH} + 7\,\text{H}^+$$

Approximately 2.5 molecules of ATP are generated when the respiratory chain oxidizes each of the 7 molecules of NADH, whereas 1.5 molecules of ATP are formed for each of the 7 molecules of FADH_2 because their electrons enter the chain at the level of ubiquinol. Recall that the oxidation of acetyl CoA by the citric acid cycle yields 10 molecules of ATP. Hence, the number of ATP molecules formed in the oxidation of palmitoyl CoA is 10.5 from the 7 molecules of FADH_2, 17.5 from the 7 molecules of NADH, and 80 from the 8 molecules of acetyl CoA, which gives a total of 108. The equivalent of 2 molecules of ATP is consumed in the activation of palmitate, in which ATP is split into AMP and 2 molecules of P_i. Thus, *the complete oxidation of a molecule of palmitate yields 106 molecules of ATP.*

22.3 CERTAIN FATTY ACIDS REQUIRE ADDITIONAL STEPS FOR DEGRADATION

The β-oxidation pathway accomplishes the complete degradation of saturated fatty acids having an even number of carbon atoms. Most fatty acids have such structures because of their mode of synthesis (Section 22.4.3). However, not all fatty acids are so simple. The oxidation of fatty acids containing double bonds requires additional steps. Likewise, fatty acids containing an odd number of carbon atoms yield a propionyl CoA at the final thiolysis step that must be converted into an easily usable form by additional enzyme reactions.

22.3.1 An Isomerase and a Reductase Are Required for the Oxidation of Unsaturated Fatty Acids

The oxidation of unsaturated fatty acids presents some difficulties, yet many such fatty acids are available in the diet. Most of the reactions are the same as those for saturated fatty acids. In fact, only two additional enzymes—an isomerase and a reductase—are needed to degrade a wide range of unsaturated fatty acids.

Consider the oxidation of palmitoleate. This C_{16} unsaturated fatty acid, which has one double bond between C-9 and C-10, is activated and transported across the inner mitochondrial membrane in the same way as saturated fatty acids. Palmitoleoyl CoA then undergoes three cycles of degradation, which are carried out by the same enzymes as in the oxidation of saturated fatty acids. However, the cis-Δ^3-enoyl CoA formed in the third round is not a substrate for acyl CoA dehydrogenase. The presence of a double bond between C-3 and C-4 prevents the formation of another double bond between C-2 and C-3. This impasse is resolved by a new reaction that shifts the position and configuration of the cis-Δ^3 double bond. *An isomerase converts this double bond into a trans-Δ^2 double bond.* The subsequent reactions are those of the saturated fatty acid oxidation pathway, in which the trans-Δ^2-enoyl CoA is a regular substrate.

Another problem arises with the oxidation of polyunsaturated fatty acids. Consider linoleate, a C_{18} polyunsaturated fatty acid with cis-Δ^9 and cis-Δ^{12} double bonds (Figure 22.10). The cis-Δ^3 double bond formed after three rounds of β oxidation is converted into a trans-Δ^2 double bond by the aforementioned isomerase. The acyl CoA produced by another round of β

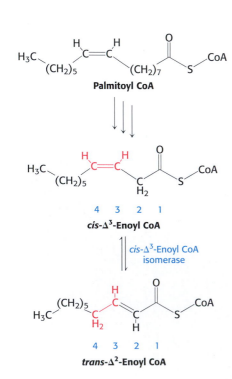

Palmitoyl CoA

cis-Δ^3-Enoyl CoA

cis-Δ^3-Enoyl CoA isomerase

trans-Δ^2-Enoyl CoA

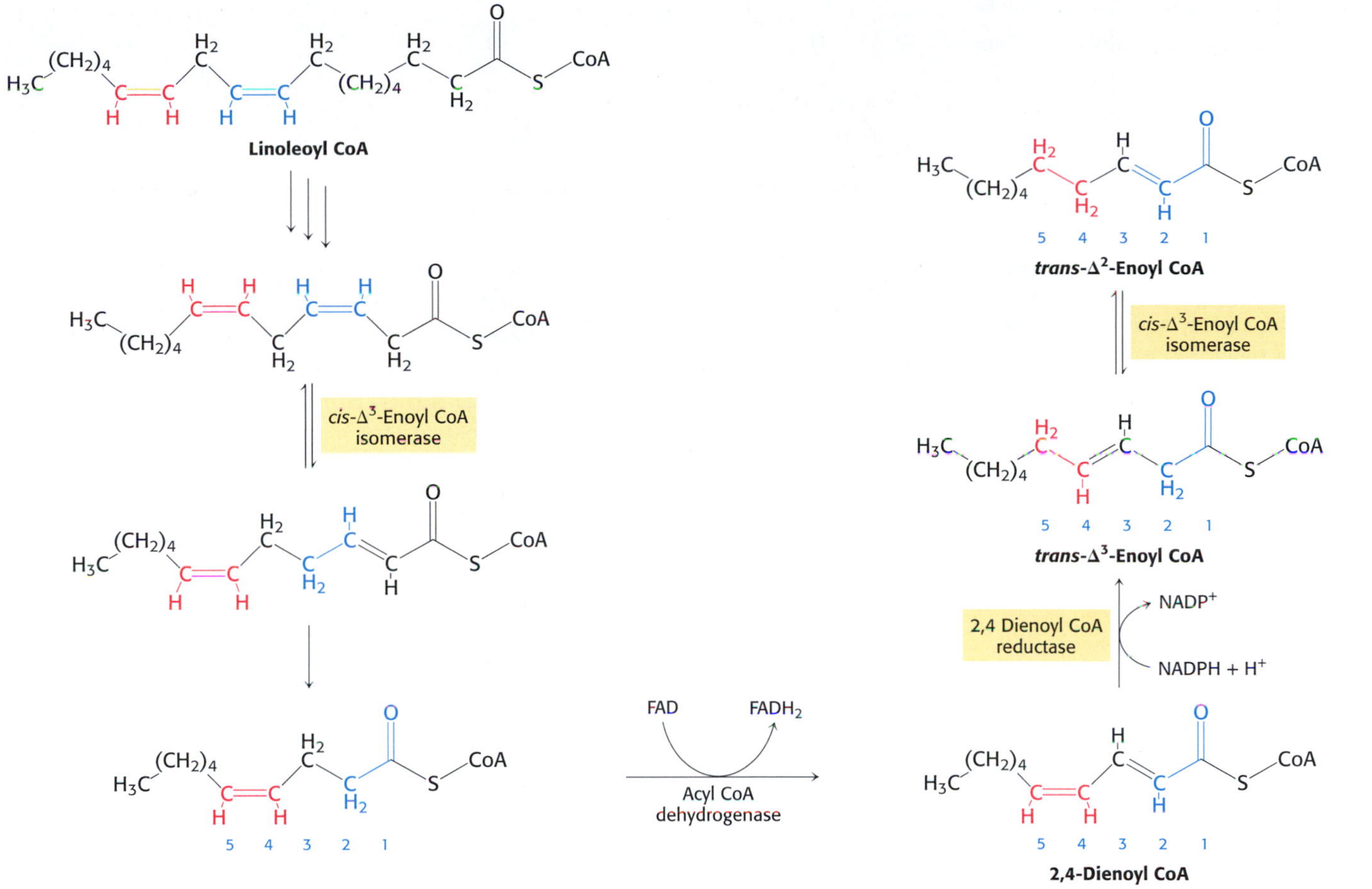

FIGURE 22.10 Oxidation of linoleoyl CoA. The complete oxidation of the diunsaturated fatty acid linoleate is facilitated by the activity of enoyl CoA isomerase and 2,4-dienoyl CoA reductase.

oxidation contains a cis-Δ^4 double bond. Dehydrogenation of this species by acyl CoA dehydrogenase yields a *2,4-dienoyl intermediate*, which is not a substrate for the next enzyme in the β-oxidation pathway. This impasse is circumvented by *2,4-dienoyl CoA reductase*, an enzyme that uses NADPH to reduce the 2,4-dienoyl intermediate to *trans*-Δ^3-enoyl CoA. *cis*-Δ^3-Enoyl CoA isomerase then converts *trans*-Δ^3-enoyl CoA into the trans-Δ^2 form, a customary intermediate in the β-oxidation pathway. These catalytic strategies are elegant and economical. Only two extra enzymes are needed for the oxidation of *any* polyunsaturated fatty acid. *Odd-numbered double bonds are handled by the isomerase, and even-numbered ones by the reductase and the isomerase.*

22.3.2 Odd-Chain Fatty Acids Yield Propionyl Coenzyme A in the Final Thiolysis Step

Fatty acids having an odd number of carbon atoms are minor species. They are oxidized in the same way as fatty acids having an even number, except that propionyl CoA and acetyl CoA, rather than two molecules of acetyl CoA, are produced in the final round of degradation. The activated three-carbon unit in propionyl CoA enters the citric acid cycle after it has been converted into succinyl CoA.

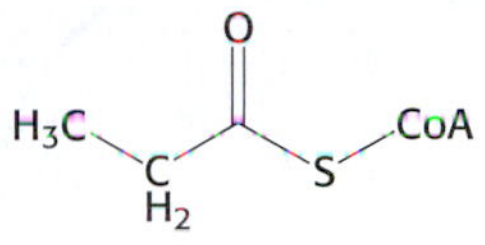

22.3.3 Propionyl CoA Is Converted into Succinyl CoA in a Reaction That Requires Vitamin B$_{12}$

The pathway from propionyl CoA to succinyl CoA is especially interesting because it entails a rearrangement that requires *vitamin B$_{12}$* (also known as

Propionyl CoA

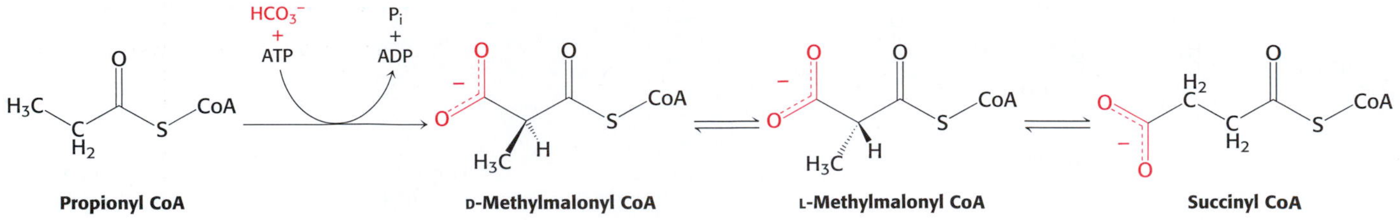

FIGURE 22.11 Conversion of propionyl CoA into succinyl CoA. Propionyl CoA, generated from fatty acids with an odd number of carbons as well as some amino acids, is converted into the citric acid cycle intermediate succinyl CoA.

FIGURE 22.12 Structure of coenzyme B$_{12}$ (5′-deoxyadenosylcobalamin). Substitution of cyano and methyl groups create cyanocobalamin and methylcobalamin, respectively.

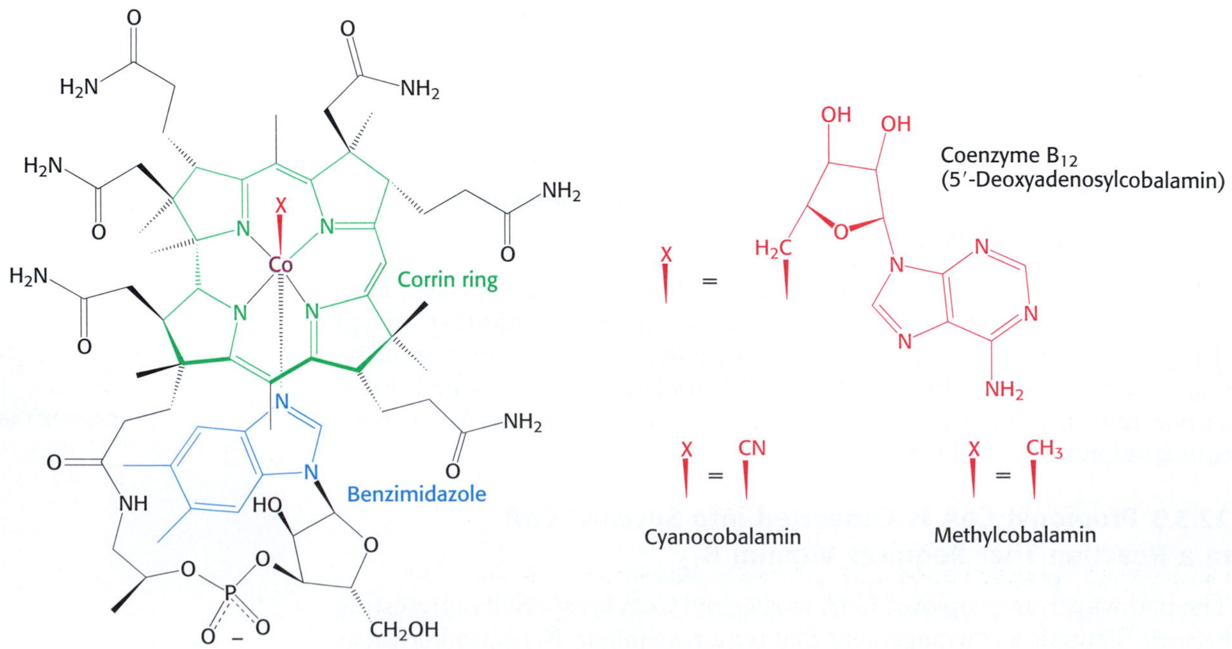

cobalamin). Propionyl CoA is carboxylated at the expense of the hydrolysis of an ATP to yield the D isomer of methylmalonyl CoA (Figure 22.11). This carboxylation reaction is catalyzed by *propionyl CoA carboxylase*, a biotin enzyme that is homologous to and has a catalytic mechanism like that of pyruvate carboxylase (Section 16.3.2). The D isomer of methylmalonyl CoA is racemized to the L isomer, the substrate for a mutase that converts it into *succinyl CoA* by an *intramolecular rearrangement*. The –CO–S–CoA group migrates from C-2 to C-3 in exchange for a hydrogen atom. This very unusual isomerization is catalyzed by *methylmalonyl CoA mutase*, which contains a derivative of vitamin B$_{12}$, cobalamin, as its coenzyme.

Cobalamin enzymes, which are present in most organisms, catalyze three types of reactions: (1) *intramolecular rearrangements;* (2) *methylations,* as in the synthesis of methionine (Section 24.2.7); and (3) *reduction of ribonucleotides to deoxyribonucleotides* (Section 25.3). In mammals, the conversion of L-methylmalonyl CoA into succinyl CoA and the formation of methionine by methylation of homocysteine are the only reactions that are known to require coenzyme B$_{12}$. The latter reaction is especially important because methionine is required for the generation of coenzymes that participate in the synthesis of purines and thymine, which are needed for nucleic acid synthesis.

The core of cobalamin consists of a *corrin ring with a central cobalt atom* (Figure 22.12). The corrin ring, like a porphyrin, has *four pyrrole units*. Two of them are directly bonded to each other, whereas methene bridges, as in porphyrins, join the others. The corrin ring is more reduced than that of porphyrins and the substituents are different. A cobalt atom is bonded to the four pyrrole nitrogens. Linked to the corrin ring is a derivative of *dimethylbenzimidazole* that contains ribose 3-phosphate and aminoisopropanol.

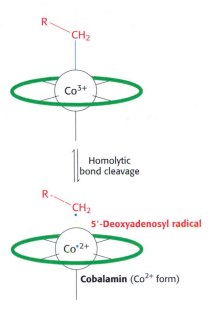

FIGURE 22.13 Rearrangement reaction catalyzed by cobalamin enzymes. The R group can be an amino group, a hydroxyl group, or a substituted carbon.

In free cobalamin, one of the nitrogen atoms of dimethylbenzimidazole is the *fifth substituent* linked to the cobalt atom. In coenzyme B_{12}, *the sixth substituent* linked to the cobalt atom is a 5′-*deoxyadenosyl unit*. This position can also be occupied by a cyano group, a methyl group, or other ligands. In these compounds, the cobalt is in the +3 oxidation state.

The rearrangement reactions catalyzed by coenzyme B_{12} are exchanges of two groups attached to adjacent carbon atoms (Figure 22.13). A hydrogen atom migrates from one carbon atom to the next, and an R group (such as the –CO–S–CoA group of methylmalonyl CoA) concomitantly moves in the reverse direction. The first step in these intramolecular rearrangements is the cleavage of the carbon–cobalt bond of 5′-deoxyadenosylcobalamin to form coenzyme B_{12} (Co^{2+}) and a 5′-deoxyadenosyl radical, –CH_2·) (Figure 22.14). In this *homolytic cleavage reaction*, one electron of the Co–C bond stays with Co (reducing it from the +3 to the +2 oxidation state) while the other stays with the carbon atom, generating a free radical. In contrast, nearly all other cleavage reactions in biological systems are *heterolytic*—an electron *pair* is transferred to one of the two atoms that were bonded together.

What is the role of this very unusual –CH_2· radical? This highly reactive species abstracts a *hydrogen atom* from the substrate to form 5′-deoxyadenosine and a substrate radical (Figure 22.15). This substrate radical spontaneously rearranges: the carbonyl CoA group migrates to the position formerly occupied by H on the neighboring carbon atom to produce a different radical. This product radical abstracts a hydrogen atom from the methyl group of 5′-deoxyadenosine to complete the rearrangement and return the deoxyadenosyl unit to the radical form. *The role of coenzyme B_{12} in such intramolecular migrations is to serve as a source of free radicals for the abstraction of hydrogen atoms.*

An essential property of coenzyme B_{12} is the weakness of its cobalt–carbon bond, the facile cleavage of which generates a radical. To facilitate

FIGURE 22.14 Formation of a 5′-deoxyadenosyl radical. The methylmalonyl CoA mutase reaction begins with the homolytic cleavage of the bond joining Co^{3+} to a carbon of the ribose of the adenosine moiety. The cleavage generates a 5′-deoxyadenosyl radical and leads to the reduction of Co^{3+} to Co^{2+}.

L-Methylmalonyl CoA — 5′-Deoxyadenosyl radical — 5′-Deoxyadenosine — Succinyl CoA

FIGURE 22.15 Formation of succinyl CoA by a rearrangement reaction. A free radical abstracts a hydrogen atom in the rearrangement of methylmalonyl CoA to succinyl CoA.

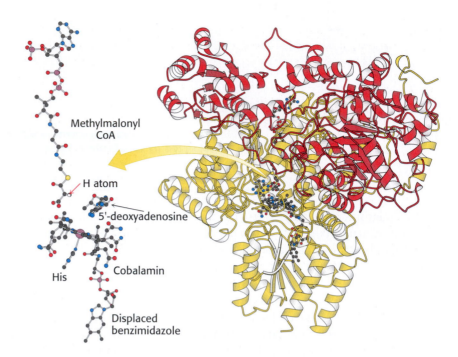

Methylmalonyl CoA

H atom

5'-deoxyadenosine

His

Cobalamin

Displaced benzimidazole

FIGURE 22.16 Active site of methylmalonyl CoA mutase. The arrangement of substrate and coenzyme in the active site facilitates the cleavage of the cobalt–carbon bond and the subsequent abstraction of a hydrogen atom from the substrate.

the cleavage of this bond, enzymes such as methylmalonyl CoA mutase displace the benzamidazole group from the cobalt and coordinate the cobalamin through a histidine residue (Figure 22.16). The steric crowding around the cobalt–carbon bond within the corrin ring system contributes to the bond weakness.

22.3.4 Fatty Acids Are Also Oxidized in Peroxisomes

Although most fatty acid oxidation takes place in mitochondria, some oxidation takes place in cellular organelles called *peroxisomes* (Figure 22.17). These organelles are characterized by high concentrations of the enzyme catalase, which catalyzes the dismutation of hydrogen peroxide into water and molecular oxygen (Section 18.3.6). Fatty acid oxidation in these organelles, which halts at octanyl CoA, may serve to shorten long chains to make them better substrates of β oxidation in mitochondria. Peroxisomal oxidation differs from β oxidation in the initial dehydrogenation reaction (Figure 22.18). In peroxisomes, a flavoprotein dehydrogenase transfers electrons to O_2 to yield H_2O_2 instead of capturing the high-energy electrons as $FADH_2$, as occurs in mitochondrial β oxidation. Catalase is needed to con-

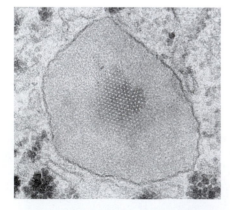

FIGURE 22.17 Electron micrograph of a peroxisome in a liver cell. A crystal of urate oxidase is present inside the organelle, which is bounded by a single bilayer membrane. The dark granular structures outside the peroxisome are glycogen particles. [Courtesy of Dr. George Palade.]

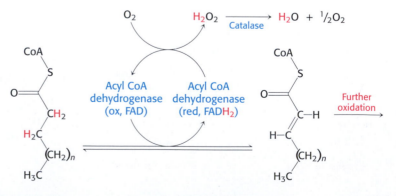

FIGURE 22.18 Initiation of peroxisomal fatty acid degradation. The first dehydration in the degradation of fatty acids in peroxisomes requires a flavoprotein dehydrogenase that transfers electrons to O_2 to yield H_2O_2.

vert the hydrogen peroxide produced in the initial reaction into water and oxygen. Subsequent steps are identical with their mitochondrial counterparts, although they are carried out by different isoforms of the enzymes.

Zellweger syndrome, which results from the absence of functional peroxisomes, is characterized by liver, kidney, and muscle abnormalities and usually results in death by age six. The syndrome is caused by a defect in the import of enzymes into the peroxisomes. Here we see a pathological condition resulting from an inappropriate cellular distribution of enzymes.

22.3.5 Ketone Bodies Are Formed from Acetyl Coenzyme A When Fat Breakdown Predominates

The acetyl CoA formed in fatty acid oxidation enters the citric acid cycle only if fat and carbohydrate degradation are appropriately balanced. The reason is that the entry of acetyl CoA into the citric acid cycle depends on the availability of oxaloacetate for the formation of citrate, but the concentration of oxaloacetate is lowered if carbohydrate is unavailable or improperly utilized. Recall that oxaloacetate is normally formed from pyruvate, the product of glycolysis, by pyruvate carboxylase (Section 16.3.1). This is the molecular basis of the adage that *fats burn in the flame of carbohydrates.*

In fasting or diabetes, oxaloacetate is consumed to form glucose by the gluconeogenic pathway (Section 16.3.2) and hence is unavailable for condensation with acetyl CoA. Under these conditions, acetyl CoA is diverted to the formation of acetoacetate and D-3-hydroxybutyrate. Acetoacetate, D-3-hydroxybutyrate, and acetone are often referred to as *ketone bodies.* Abnormally high levels of ketone bodies are present in the blood of untreated diabetics (Section 22.3.6).

Acetoacetate is formed from acetyl CoA in three steps (Figure 22.19). Two molecules of acetyl CoA condense to form acetoacetyl CoA. This reaction, which is catalyzed by thiolase, is the reverse of the thiolysis step in the oxidation of fatty acids. Acetoacetyl CoA then reacts with acetyl CoA and water to give 3-hydroxy-3-methylglutaryl CoA (HMG-CoA) and CoA. This condensation resembles the one catalyzed by citrate synthase (Section 17.1.3). This reaction, which has a favorable equilibrium owing to the hydrolysis of a thioester linkage, compensates for the unfavorable equilibrium in the formation of acetoacetyl CoA. 3-Hydroxy-3-methylglutaryl CoA is then cleaved to acetyl CoA and acetoacetate. The sum of these reactions is

$$2 \text{ Acetyl CoA} + H_2O \longrightarrow \text{acetoacetate} + 2 \text{ CoA} + H^+$$

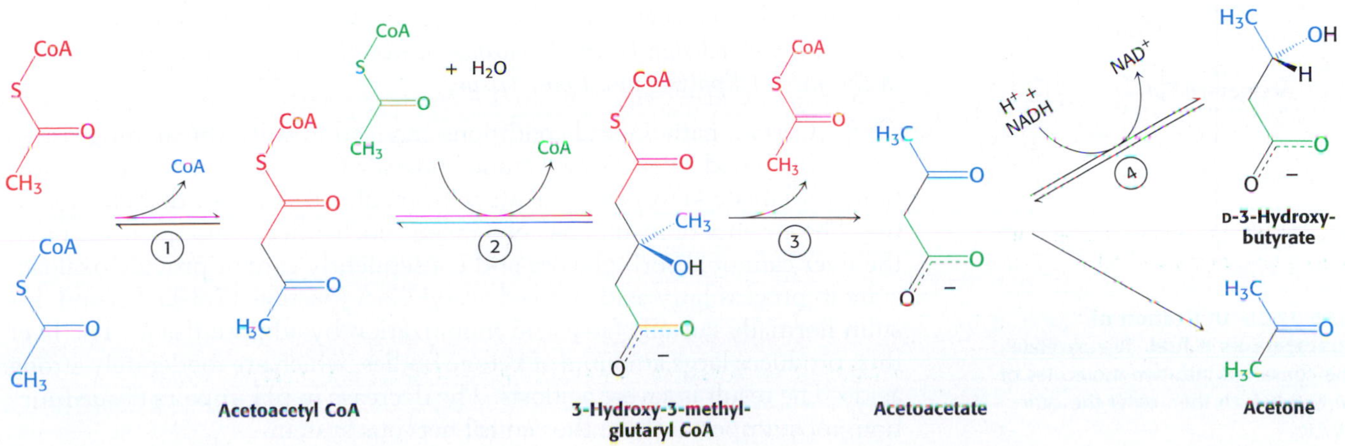

FIGURE 22.19 Formation of ketone bodies. The ketone bodies— acetoacetate, D-3-hydroxybutyrate, and acetone from acetyl CoA—are formed primarily in the liver. Enzymes catalyzing these reactions are (1) 3-ketothiolase, (2) hydroxymethylglutaryl CoA synthase, (3) hydroxymethylglutaryl CoA cleavage enzyme, and (4) D-3-hydroxybutyrate dehydrogenase. Acetoacetate spontaneously decarboxylates to form acetone.

Acetoacetyl CoA   3-Hydroxy-3-methyl-glutaryl CoA   Acetoacetate   D-3-Hydroxy-butyrate   Acetone

D-3-Hydroxybutyrate is formed by the reduction of acetoacetate in the mitochondrial matrix by D-3-hydroxybutyrate dehydrogenase. The ratio of hydroxybutyrate to acetoacetate depends on the $NADH/NAD^+$ ratio inside mitochondria.

Because it is a β-ketoacid, acetoacetate also undergoes a slow, spontaneous decarboxylation to acetone. The odor of acetone may be detected in the breath of a person who has a high level of acetoacetate in the blood.

22.3.6 Ketone Bodies Are a Major Fuel in Some Tissues

The major site of production of acetoacetate and 3-hydroxybutyrate is the liver. These substances diffuse from the liver mitochondria into the blood and are transported to peripheral tissues. These ketone bodies were initially regarded as degradation products of little physiological value. However, the results of studies by George Cahill and others revealed that these derivatives of acetyl CoA are important molecules in energy metabolism. *Acetoacetate and 3-hydroxybutyrate are normal fuels of respiration and are quantitatively important as sources of energy.* Indeed, heart muscle and the renal cortex use acetoacetate in preference to glucose. In contrast, glucose is the major fuel for the brain and red blood cells in well-nourished people on a balanced diet. However, the brain adapts to the utilization of acetoacetate during starvation and diabetes (Sections 30.3.1 and 30.3.2). In prolonged starvation, 75% of the fuel needs of the brain are met by ketone bodies.

3-Hydroxybutyrate is oxidized to produce acetoacetate as well as NADH for use in oxidative phosphorylation.

Acetoacetate can be activated by the transfer of CoA from succinyl CoA in a reaction catalyzed by a specific CoA transferase. Acetoacetyl CoA is then cleaved by thiolase to yield two molecules of acetyl CoA, which can then enter the citric acid cycle (Figure 22.20). The liver has acetoacetate available to supply to other organs because it lacks this particular CoA transferase.

Ketone bodies can be regarded as a water-soluble, transportable form of acetyl units. Fatty acids are released by adipose tissue and converted into acetyl units by the liver, which then exports them as acetoacetate. As might be expected, acetoacetate also has a regulatory role. *High levels of acetoacetate in the blood signify an abundance of acetyl units and lead to a decrease in the rate of lipolysis in adipose tissue.*

Certain pathological conditions can lead to a life-threatening rise in the blood levels of the ketone bodies. Most common of these conditions is diabetic ketosis in patients with insulin-dependent diabetes mellitus. The absence of insulin has two major biochemical consequences. First, the liver cannot absorb glucose and consequently cannot provide oxaloacetate to process fatty acid-derived acetyl CoA (Section 17.3.1). Second, insulin normally curtails fatty acid mobilization by adipose tissue. The liver thus produces large amounts of ketone bodies, which are moderately strong acids. The result is severe acidosis. The decrease in pH impairs tissue function, most importantly in the central nervous system.

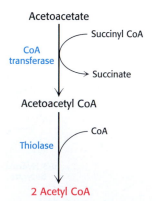

FIGURE 22.20 Utilization of acetoacetate as a fuel. Acetoacetate can be converted into two molecules of acetyl CoA, which then enter the citric acid cycle.

D-3-Hydroxybutyrate   Acetoacetate

22.3.7 Animals Cannot Convert Fatty Acids into Glucose

It is important to note that *animals are unable to effect the net synthesis of glucose from fatty acids*. Specifically, acetyl CoA cannot be converted into pyruvate or oxaloacetate in animals. The two carbon atoms of the acetyl group of acetyl CoA enter the citric acid cycle, but two carbon atoms leave the cycle in the decarboxylations catalyzed by isocitrate dehydrogenase and α-ketoglutarate dehydrogenase. Consequently, oxaloacetate is regenerated, but it is not formed de novo when the acetyl unit of acetyl CoA is oxidized by the citric acid cycle. In contrast, plants have two additional enzymes enabling them to convert the carbon atoms of acetyl CoA into oxaloacetate (Section 17.4.).

22.4 FATTY ACIDS ARE SYNTHESIZED AND DEGRADED BY DIFFERENT PATHWAYS

CONCEPTUAL INSIGHTS, Overview of Carbohydrate and Fatty Acid Metabolism, *will help you understand how fatty acid metabolism fits in with other energy storage and utilization pathways (glycolysis, citric acid cycle, pentose phosphate pathway, glycogen metabolism), with a focus on carbon and energy flux.*

Fatty acid synthesis is not simply a reversal of the degradative pathway. Rather, it consists of a new set of reactions, again exemplifying the principle that *synthetic and degradative pathways are almost always distinct.* Some important differences between the pathways are:

1. Synthesis takes place in the *cytosol*, in contrast with degradation, which takes place primarily in the mitochondrial matrix.

2. Intermediates in fatty acid synthesis are covalently linked to the sulfhydryl groups of an *acyl carrier protein* (ACP), whereas intermediates in fatty acid breakdown are covalently attached to the sulfhydryl group of coenzyme A.

3. The enzymes of fatty acid synthesis in higher organisms are joined in a *single polypeptide chain* called *fatty acid synthase.* In contrast, the degradative enzymes do not seem to be associated.

4. The growing fatty acid chain is elongated by the *sequential addition of two-carbon units* derived from acetyl CoA. The activated donor of two-carbon units in the elongation step is *malonyl ACP.* The elongation reaction is driven by the release of CO_2.

5. The reductant in fatty acid synthesis is *NADPH*, whereas the oxidants in fatty acid degradation are NAD^+ and *FAD*.

6. Elongation by the fatty acid synthase complex stops on formation of *palmitate* (C_{16}). Further elongation and the insertion of double bonds are carried out by other enzyme systems.

22.4.1 The Formation of Malonyl Coenzyme A Is the Committed Step in Fatty Acid Synthesis

Fatty acid synthesis starts with the carboxylation of acetyl CoA to *malonyl CoA*. This irreversible reaction is the committed step in fatty acid synthesis.

Acetyl CoA   **Malonyl CoA**

The synthesis of malonyl CoA is catalyzed by *acetyl CoA carboxylase*, which contains a biotin prosthetic group. The carboxyl group of biotin is covalently attached to the ϵ amino group of a lysine residue, as in pyruvate carboxylase (Section 16.3.2) and propionyl CoA carboxylase (Section 22.3.3). As with these other enzymes, a carboxybiotin intermediate is formed at the expense of the hydrolysis a molecule of ATP. The activated CO_2 group in this intermediate is then transferred to acetyl CoA to form malonyl CoA.

$$\text{Biotin-enzyme} + \text{ATP} + \text{HCO}_3^- \rightleftharpoons \text{CO}_2\text{-biotin-enzyme} + \text{ADP} + \text{P}_i$$

$$\text{CO}_2\text{-biotin-enzyme} + \text{acetyl CoA} \longrightarrow \text{malonyl CoA} + \text{biotin-enzyme}$$

This enzyme is also the essential regulatory enzyme for fatty acid metabolism (Section 22.5).

22.4.2 Intermediates in Fatty Acid Synthesis Are Attached to an Acyl Carrier Protein

The intermediates in fatty acid synthesis are linked to an acyl carrier protein. Specifically, they are linked to the sulfhydryl terminus of a phosphopantetheine group, which is, in turn, attached to a serine residue of the acyl carrier protein (Figure 22.21). Recall that, in the degradation of fatty acids, a phosphopantetheine group is present as part of CoA instead (Section 22.2.2). ACP, a single polypeptide chain of 77 residues, can be regarded as a giant prosthetic group, a "macro CoA."

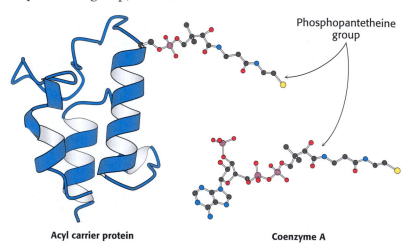

Phosphopantetheine group

Acyl carrier protein

Coenzyme A

FIGURE 22.21 Phosphopantetheine. Both acyl carrier protein and CoA include phosphopantetheine as their reactive units.

22.4.3 The Elongation Cycle in Fatty Acid Synthesis

The enzyme system that catalyzes the synthesis of saturated long-chain fatty acids from acetyl CoA, malonyl CoA, and NADPH is called the *fatty acid synthase*. The constituent enzymes of bacterial fatty acid synthases are dissociated when the cells are broken apart. The availability of these isolated enzymes has facilitated the elucidation of the steps in fatty acid synthesis (Table 22.2). In fact, the reactions leading to fatty acid synthesis in higher organisms are very much like those of bacteria.

The elongation phase of fatty acid synthesis starts with the formation of acetyl ACP and malonyl ACP. *Acetyl transacylase* and *malonyl transacylase* catalyze these reactions.

$$\text{Acetyl CoA} + \text{ACP} \rightleftharpoons \text{acetyl ACP} + \text{CoA}$$
$$\text{Malonyl CoA} + \text{ACP} \rightleftharpoons \text{malonyl ACP} + \text{CoA}$$

Malonyl transacylase is highly specific, whereas acetyl transacylase can transfer acyl groups other than the acetyl unit, though at a much slower rate. Fatty

TABLE 22.2   Principal reactions in fatty acid synthesis in bacteria

Step	Reaction	Enzyme
1	Acetyl CoA + HCO_3^- + ATP $\longrightarrow$ malonyl CoA + ADP + P_i + H^+	Acetyl CoA carboxylase
2	Acetyl CoA + ACP $\rightleftharpoons$ acetyl ACP + CoA	Acetyl transacylase
3	Malonyl CoA + ACP $\rightleftharpoons$ malonyl ACP + CoA	Malonyl transacylase
4	Acetyl ACP + malonyl ACP $\longrightarrow$ acetoacetyl ACP + ACP + CO_2	Acyl-malonyl ACP condensing enzyme
5	Acetoacetyl ACP + NADPH + H^+ $\rightleftharpoons$ D-3-hydroxybutyryl ACP + $NADP^+$	β-Ketoacyl ACP reductase
6	D-3-Hydroxybutyryl ACP $\rightleftharpoons$ crotonyl ACP + H_2O	3-Hydroxyacyl ACP dehydratase
7	Crotonyl ACP + NADPH + H^+ $\longrightarrow$ butyryl ACP + $NADP^+$	Enoyl ACP reductase

acids with an odd number of carbon atoms are synthesized starting with propionyl ACP, which is formed from propionyl CoA by acetyl transacylase.

Acetyl ACP and malonyl ACP react to form acetoacetyl ACP (Figure 22.22). The *acyl-malonyl ACP condensing enzyme* catalyzes this condensation reaction.

Acetyl ACP + malonyl ACP $\longrightarrow$ acetoacetyl ACP + ACP + CO_2

In the condensation reaction, a four-carbon unit is formed from a two-carbon unit and a three-carbon unit, and CO_2 is released. Why is the four-carbon unit not formed from 2 two-carbon units? In other words, why are the reactants acetyl ACP and malonyl ACP rather than two molecules of acetyl ACP? The answer is that the equilibrium for the synthesis of acetoacetyl ACP from two molecules of acetyl ACP is highly unfavorable. In contrast, *the equilibrium is favorable if malonyl ACP is a reactant because its decarboxylation contributes a substantial decrease in free energy.* In effect, ATP drives the condensation reaction, though ATP does not directly participate in the condensation reaction. Rather, ATP is used to carboxylate acetyl CoA to malonyl CoA. The free energy thus stored in malonyl CoA is released in the decarboxylation accompanying the formation of acetoacetyl ACP. Although HCO_3^- is required for fatty acid synthesis, its carbon atom does not appear in the product. Rather, *all the carbon atoms of fatty acids containing an even number of carbon atoms are derived from acetyl CoA.*

The next three steps in fatty acid synthesis reduce the keto group at C-3 to a methylene group (see Figure 22.22). First, acetoacetyl ACP is reduced to D-3-hydroxybutyryl ACP. This reaction differs from the corresponding one in fatty acid degradation in two respects: (1) the D rather than the L isomer is formed; and (2) NADPH is the reducing agent, whereas NAD^+ is the oxidizing agent in β oxidation. This difference exemplifies the general principle that *NADPH is consumed in biosynthetic reactions, whereas NADH is generated in energy-yielding reactions.* Then D-3-hydroxybutyryl ACP is dehydrated to form crotonyl ACP, which is a *trans-Δ^2-enoyl ACP.* The final step in the cycle *reduces* crotonyl ACP to butyryl ACP. NADPH is again the reductant, whereas FAD is the oxidant in the corresponding reaction in β-oxidation. The enzyme that catalyzes this step, *enoyl ACP reductase*, is

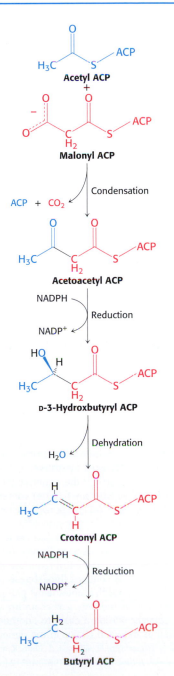

FIGURE 22.22 Fatty acid synthesis. Fatty acids are synthesized by the repetition of the following reaction sequence: condensation, reduction, dehydration, and reduction. The intermediates shown here are produced in the first round of synthesis.

inhibited by triclosan, a broad-spectrum antibacterial agent. Triclosan is used in a variety of products such as toothpaste, soaps, and skin creams. These last three reactions—a reduction, a dehydration, and a second reduction—convert acetoacetyl ACP into butyryl ACP, which completes the first elongation cycle.

In the second round of fatty acid synthesis, butyryl ACP condenses with malonyl ACP to form a C_6-β-ketoacyl ACP. This reaction is like the one in the first round, in which acetyl ACP condenses with malonyl ACP to form a C_4-β-ketoacyl ACP. Reduction, dehydration, and a second reduction convert the C_6-β-ketoacyl ACP into a C_6-acyl ACP, which is ready for a third round of elongation. The elongation cycles continue until C_{16}-acyl ACP is formed. This intermediate is a good substrate for a thioesterase that hydrolyzes C_{16}-acyl ACP to yield palmitate and ACP. *The thioesterase acts as a ruler to determine fatty acid chain length.* The synthesis of longer-chain fatty acids is discussed in Section 22.6.

22.4.4 Fatty Acids Are Synthesized by a Multifunctional Enzyme Complex in Eukaryotes

Although the basic biochemical reactions in fatty acid synthesis are very similar in *E. coli* and eukaryotes, the structure of the synthase varies considerably. The fatty acid synthases of eukaryotes, in contrast with those of *E. coli*, have the component enzymes linked in a large polypeptide chain.

Mammalian fatty acid synthase is a dimer of identical 260-kd subunits. Each chain is folded into three domains joined by flexible regions (Figure 22.23). *Domain 1, the substrate entry and condensation unit*, contains acetyl transferase, malonyl transferase, and β-ketoacyl synthase (condensing enzyme). *Domain 2, the reduction unit*, contains the acyl carrier protein, β-ketoacyl reductase, dehydratase, and enoyl reductase. *Domain 3, the palmitate release unit*, contains the thioesterase. Thus, *seven different catalytic sites are present on a single polypeptide chain.* It is noteworthy that many eukaryotic multienzyme complexes are multifunctional proteins in which different enzymes are linked covalently. An advantage of this arrangement is that the synthetic activity of different enzymes is coordinated. In addition, a multienzyme complex consisting of covalently joined enzymes is more stable than one formed by noncovalent attractions. Furthermore, intermediates can be

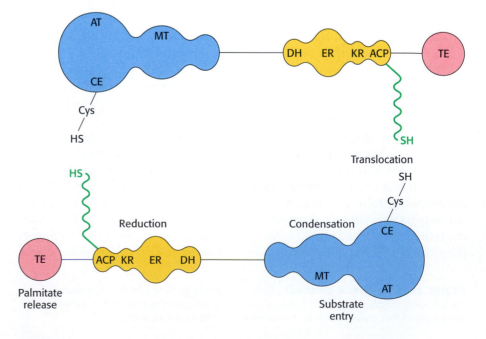

FIGURE 22.23 Schematic representation of animal fatty acid synthase. Each of the identical chains in the dimer contains three domains. Domain 1 (blue) contains acetyl transferase (AT), malonyl transferase (MT), and condensing enzyme (CE). Domain 2 (yellow) contains acyl carrier protein (ACP), β-ketoacyl reductase (KR), dehydratase (DH), and enoyl reductase (ER). Domain 3 (red) contains thioesterase (TE). The flexible phosphopantetheinyl group (green) carries the fatty acyl chain from one catalytic site on a chain to another, as well as between chains in the dimer. [After Y. Tsukamoto, H. Wong, J. S. Mattick, and S. J. Wakil. *J. Biol. Chem.* 258(1983):15312.]

efficiently handed from one active site to another without leaving the assembly. It seems likely that multifunctional enzymes such as fatty acid synthase arose in eukaryotic evolution by exon shuffling (Section 5.6.2), because each of the component enzymes is recognizably homologous to its bacterial counterpart.

22.4.5 The Flexible Phosphopantetheinyl Unit of ACP Carries Substrate from One Active Site to Another

We next examine the coordinated functioning of the mammalian fatty acid synthase. Fatty acid synthesis begins with the transfer of the acetyl group of acetyl CoA first to a serine residue in the active site of acetyl transferase and then to the sulfur atom of a cysteine residue in the active site of the condensing enzyme on one chain of the dimeric enzyme. Similarly, the malonyl group is transferred from malonyl CoA first to a serine residue in the active site of malonyl transferase and then to the sulfur atom of the phosphopantetheinyl group of the acyl carrier protein on the *other* chain in the dimer. Domain 1 of each chain of this dimer interacts with domains 2 and 3 of the other chain. Thus, each of the two functional units of the synthase consists of domains formed by different chains. Indeed, the arenas of catalytic action are the interfaces between domains on opposite chains.

Elongation begins with the joining of the acetyl unit on the condensing enzyme (CE) to a two-carbon part of the malonyl unit on ACP (Figure 22.24).

CO_2 is released and an acetoacetyl-S-phosphopantetheinyl unit is formed on ACP. The active-site sulfhydryl group on the condensing enzyme is restored. The acetoacetyl group is then delivered to three active sites in domain 2 of the opposite chain to reduce it to a butyryl unit. This saturated

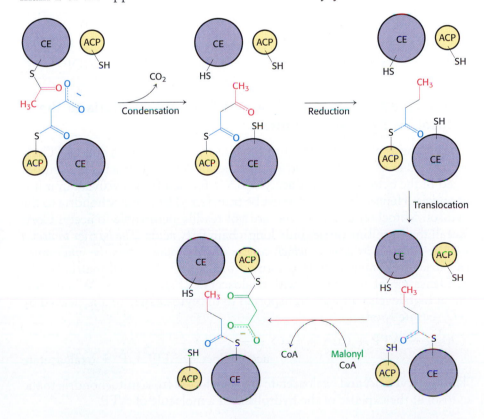

FIGURE 22.24 Reactions of fatty acid synthase. Translocations of the elongating fatty acyl chain between the cysteine sulfhydryl group of the condensing enzyme (CE, blue) and the phosphopantetheine sulfhydryl group of the acyl carrier protein (ACP, yellow) lead to the growth of the fatty acid chain. The reactions are repeated until the palmitoyl product is synthesized.

C_4 unit then migrates from the phosphopantetheinyl sulfur atom on ACP to the cysteine sulfur atom on the condensing enzyme. The synthase is now ready for another round of elongation. The butyryl unit on the condensing enzyme becomes linked to a two-carbon part of the malonyl unit on ACP to form a six-carbon unit on ACP, which undergoes reduction. Five more rounds of condensation and reduction produce a palmitoyl (C_{16}) chain on the condensing enzyme, which is hydrolyzed to palmitate by the thioesterase on domain 3 of the opposite chain. The migration of the growing fatty acyl chain back and forth between ACP and the condensing enzyme in each round of elongation is analogous to the translocations of growing peptide chains that take place in protein synthesis (Section 29.3.7).

The flexibility and 20-Å maximal length of the phosphopantetheinyl moiety are critical for the function of this multienzyme complex. The enzyme subunits need not undergo large structural rearrangements to interact with the substrate. Instead, the substrate is on a long, flexible arm that can reach each of the numerous active sites. Recall that biotin and lipoamide also are on long, flexible arms in their multienzyme complexes. The organization of the fatty acid synthases of higher organisms enhances the efficiency of the overall process because intermediates are directly transferred from one active site to the next.

22.4.6 The Stoichiometry of Fatty Acid Synthesis

The stoichiometry of the synthesis of palmitate is

$$\text{Acetyl CoA} + 7\,\text{malonyl CoA} + 14\,\text{NADPH} + 20\,\text{H}^+ \longrightarrow$$
$$\text{palmitate} + 7\,CO_2 + 14\,\text{NADP}^+ + 8\,\text{CoA} + 6\,H_2O$$

The equation for the synthesis of the malonyl CoA used in the preceding reaction is

$$7\,\text{Acetyl CoA} + 7\,CO_2 + 7\,\text{ATP} \longrightarrow$$
$$7\,\text{malonyl CoA} + 7\,\text{ADP} + 7\,P_i + 14\,\text{H}^+$$

Hence, the overall stoichiometry for the synthesis of palmitate is

$$8\,\text{Acetyl CoA} + 7\,\text{ATP} + 14\,\text{NADPH} + 6\,\text{H}^+ \longrightarrow$$
$$\text{palmitate} + 14\,\text{NADP}^+ + 8\,\text{CoA} + 6\,H_2O + 7\,\text{ADP} + 7\,P_i$$

22.4.7 Citrate Carries Acetyl Groups from Mitochondria to the Cytosol for Fatty Acid Synthesis

The synthesis of palmitate requires the input of 8 molecules of acetyl CoA, 14 molecules of NADPH, and 7 molecules of ATP. Fatty acids are synthesized in the cytosol, whereas acetyl CoA is formed from pyruvate in mitochondria. Hence, acetyl CoA must be transferred from mitochondria to the cytosol. Mitochondria, however, are not readily permeable to acetyl CoA. Recall that carnitine carries only long-chain fatty acids. *The barrier to acetyl CoA is bypassed by citrate, which carries acetyl groups across the inner mitochondrial membrane.* Citrate is formed in the mitochondrial matrix by the condensation of acetyl CoA with oxaloacetate (Figure 22.25). When present at high levels, citrate is transported to the cytosol, where it is cleaved by *ATP-citrate lyase.*

$$\text{Citrate} + \text{ATP} + \text{CoA} + H_2O \rightleftharpoons$$
$$\text{acetyl CoA} + \text{ADP} + P_i + \text{oxaloacetate}$$

Thus, acetyl CoA and oxaloacetate are transferred from mitochondria to the cytosol at the expense of the hydrolysis of a molecule of ATP.

Lyases—

Enzymes catalyzing the cleavage of C–C, C–O, or C–N bonds by elimination. A double bond is formed in these reactions.

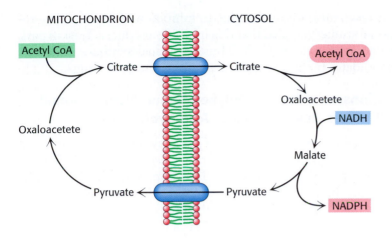

22.4.8 Sources of NADPH for Fatty Acid Synthesis

Oxaloacetate formed in the transfer of acetyl groups to the cytosol must now be returned to the mitochondria. The inner mitochondrial membrane is impermeable to oxaloacetate. Hence, a series of bypass reactions are needed. Most important, these reactions generate much of the NADPH needed for fatty acid synthesis. First, oxaloacetate is reduced to malate by NADH. This reaction is catalyzed by a *malate dehydrogenase* in the cytosol.

$$\text{Oxaloacetate} + \text{NADH} + \text{H}^+ \rightleftharpoons \text{malate} + \text{NAD}^+$$

Second, malate is oxidatively decarboxylated by an *$NADP^+$-linked malate enzyme* (also called *malic enzyme*).

$$\text{Malate} + \text{NADP}^+ \rightleftharpoons \text{pyruvate} + \text{CO}_2 + \text{NADPH}$$

The pyruvate formed in this reaction readily enters mitochondria, where it is carboxylated to oxaloacetate by pyruvate carboxylase.

$$\text{Pyruvate} + \text{CO}_2 + \text{ATP} + \text{H}_2\text{O} \rightleftharpoons$$
$$\text{oxaloacetate} + \text{ADP} + \text{P}_i + 2\,\text{H}^+$$

The sum of these three reactions is

$$\text{NADP}^+ + \text{NADH} + \text{ATP} + \text{H}_2\text{O} \rightleftharpoons$$
$$\text{NADPH} + \text{NAD}^+ + \text{ADP} + \text{P}_i + \text{H}^+$$

Thus, *one molecule of NADPH is generated for each molecule of acetyl CoA that is transferred from mitochondria to the cytosol.* Hence, eight molecules of NADPH are formed when eight molecules of acetyl CoA are transferred to the cytosol for the synthesis of palmitate. *The additional six molecules of NADPH required for this process come from the pentose phosphate pathway* (Section 20.3.1).

The accumulation of the precursors for fatty acid synthesis is a wonderful example of the coordinated use of multiple processes to fulfill a biochemical need. The citric acid cycle, subcellular compartmentalization, and the pentose phosphate pathway provide the carbon atoms and reducing power, whereas glycolysis and oxidative phosphorylation provide the ATP to meet the needs for fatty acid synthesis.

22.4.9 Fatty Acid Synthase Inhibitors May Be Useful Drugs

Fatty acid synthase is overexpressed in some breast cancers. Researchers intrigued by this observation have tested inhibitors of fatty acid synthase on mice to see how the inhibitors affect tumor growth. A startling observation was made: *mice treated with inhibitors of the condensing enzyme showed remarkable weight loss* due to inhibition of feeding. The results of additional studies revealed that this inhibition is due, at least in part, to the accumulation of malonyl CoA. Thus, fatty acid synthase inhibitors are exciting candidates both as antitumor and as antiobesity drugs.

22.4.10 Variations on a Theme: Polyketide and Nonribosomal Peptide Synthetases Resemble Fatty Acid Synthase

The mammalian multifunctional fatty acid synthase is a member of a large family of complex enzymes termed *megasynthases* that participate in step-by-step synthetic pathways. Two important classes of compounds that are synthesized by such enzymes are the *polyketides* and the *nonribosomal peptides*. The antibiotic erythromycin is an example of a polyketide, whereas penicillin (Section 8.5.5) is a nonribosomal peptide.

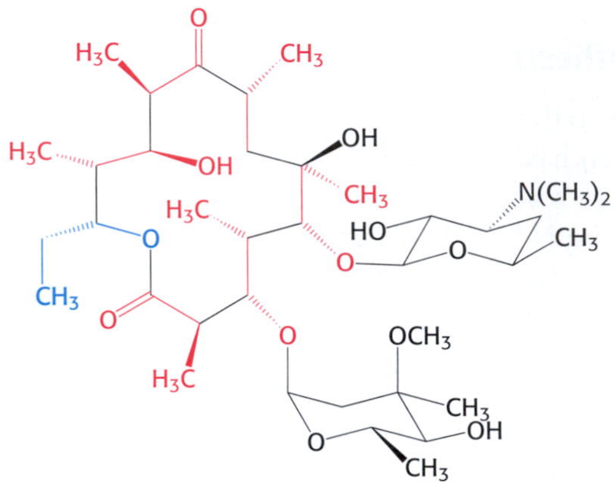

Erythromycin

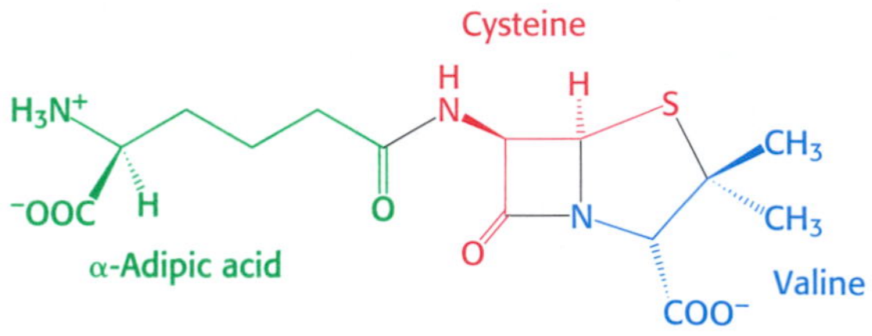

Penicillin

The core of erythromycin (deoxyerythronolide B, or Deb) is synthesized by the following reaction:

Propionyl CoA + 6 methylmalonyl CoA + 6 NADPH + 6 H$^+$ $\longrightarrow$
Deb + 7 CoA + 6 NADP$^+$ + 2 H$_2$O

This reaction is accomplished by three megasynthases consisting of 3491, 3567, and 3172 amino acids. The synthesis of deoxyerythronolide B begins with propionyl CoA linked to a phosphopantetheine chain connected to an acyl carrier protein domain. Similarly, the precursor of penicillin [Δ-(L-aminoadipyl)-L-cysteinyl-D-valine, or ACV] is generated by the following reaction:

L-Aminoadipic acid + L-Cys + L-Val + 3 ATP + H$_2$O $\longrightarrow$
ACV + 3 ADP + 3 P$_i$

which is catalyzed by a megasynthase consisting of 3791 amino acids. Each amino acid is activated by a specific adenylation domain within the enzyme—a domain that is homologous to acyl CoA synthase. Additional domains are responsible for peptide-bond formation and for the epimerization of the valine residue. Again, during synthesis, the components are linked to phosphopantetheine chains. Members of this remarkably modular enzyme family generate many of the natural products that have proved to be useful as drugs.

22.5 ACETYL COENZYME A CARBOXYLASE PLAYS A KEY ROLE IN CONTROLLING FATTY ACID METABOLISM

Fatty acid metabolism is stringently controlled so that synthesis and degradation are highly responsive to physiological needs. Fatty acid synthesis is maximal when carbohydrate and energy are plentiful and when fatty acids are scarce. *Acetyl CoA carboxylase plays an essential role in regulating fatty acid synthesis and degradation.* Recall that this enzyme catalyzes the committed step in fatty acid synthesis: the production of malonyl CoA (the activated two-carbon donor). The carboxylase is controlled by three global signals—glucagon, epinephrine, and insulin—that correspond to the overall energy status of the organism. *Insulin stimulates fatty acid synthesis by activating the carboxylase, whereas glucagon and epinephrine have the reverse effect.* The levels of citrate, palmitoyl CoA, and AMP within a cell also exert control. *Citrate,* a signal that building blocks and energy are abundant, activates the carboxylase. Palmitoyl CoA and AMP, in contrast, lead to the inhibition of the carboxylase. Thus, this important enzyme is subject to both global and local regulation. We will examine each of these levels of regulation in turn.

Global Regulation. Global regulation is carried out by means of reversible phosphorylation. *Acetyl CoA carboxylase is switched off by phosphorylation* and activated by dephosphorylation (Figure 22.26). Modification of a single serine residue by an *AMP-dependent protein kinase (AMPK)* converts the carboxylase into an inactive form. The phosphoryl group on the inhibited carboxylase is removed by *protein phosphatase 2A.* The proportion of carboxylase in the active dephosphorylated form depends on the relative rates of these opposing enzymes.

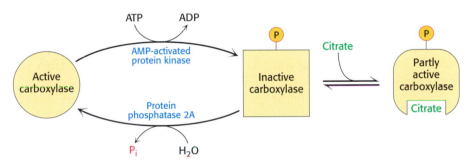

FIGURE 22.26 Control of acetyl CoA carboxylase. Acetyl CoA carboxylase is inhibited by phosphorylation and activated by the binding of citrate.

How is the formation of the inactive, phosphorylated form of the enzyme regulated? AMPK, the enzyme that phosphorylates the carboxylase, is essentially a fuel gauge—it is activated by AMP and inhibited by ATP. Thus, the carboxylase is inactivated when the energy charge is low. This kinase is conserved among eukaryotes. Homologs found in yeast and plants also play roles in sensing the energy status of the cell. The inhibition of phosphatase 2A is necessary to maintain acetyl CoA carboxylase in the phosphorylated state. Epinephrine and glucagon activate protein kinase A, which in turn inhibits the phosphatase by phosphorylating it. Hence, *these catabolic hormones switch off fatty acid synthesis by keeping the carboxylase in the inactive phosphorylated state.*

How is the enzyme dephosphorylated and activated? *Insulin stimulates the carboxylase by causing its dephosphorylation.* It is not clear which of the phosphatases activates the carboxylase in response to insulin. The hormonal control of acetyl CoA carboxylase is reminiscent of that of glycogen synthase (Section 21.5.2).

Local Regulation. Acetyl CoA carboxylase is also under local control. *This enzyme is allosterically stimulated by citrate.* Specifically, citrate partly reverses the inhibition produced by phosphorylation. It acts in an unusual

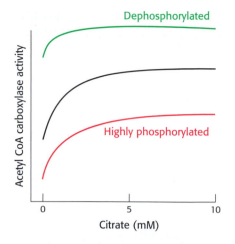

FIGURE 22.27 Dependence of the catalytic activity of acetyl CoA carboxylase on the concentration of citrate. The dephosphorylated form of the carboxylase is highly active even when citrate is absent. Citrate partly overcomes the inhibition produced by phosphorylation. [After G. M. Mabrouk, I. M. Helmy, K. G. Thampy, and S. J. Wakil. *J. Biol. Chem.* 265(1990):6330.]

manner on inactive acetyl CoA carboxylase, which exists as an octamer (Figure 22.27). *Citrate facilitates the polymerization of the inactive octamers into active filaments* (Figure 22.28). The level of citrate is high when both acetyl CoA and ATP are abundant. Recall that mammalian isocitrate dehydrogenase is inhibited by a high energy charge (Section 17.2.2). Hence, a high level of citrate signifies that two-carbon units and ATP are available for fatty acid synthesis. The stimulatory effect of citrate on the carboxylase is antagonized by *palmitoyl CoA*, which is abundant when there is an excess of fatty acids. Palmitoyl CoA causes the filaments to disassemble into the inactive octamers. Palmitoyl CoA also inhibits the translocase that transports citrate from mitochondria to the cytosol, as well as glucose 6-phosphate dehydrogenase, which generates NADPH in the pentose phosphate pathway.

Response to Diet. Fatty acid synthesis and degradation are reciprocally regulated so that both are not simultaneously active. *In starvation, the level of free fatty acids rises because hormones such as epinephrine and glucagon stimulate adipose-cell lipase. Insulin, in contrast, inhibits lipolysis.* Acetyl CoA carboxylase also plays a role in the regulation of fatty acid degradation. Malonyl CoA, the product of the carboxylase reaction, is present at a high level when fuel molecules are abundant. *Malonyl CoA inhibits carnitine acyltransferase I, preventing access of fatty acyl CoAs to the mitochondrial matrix in times of plenty.* Malonyl CoA is an especially effective inhibitor of carnitine acyltransferase I in heart and muscle, tissues that have little fatty acid synthesis capacity of their own. In these tissues, acetyl CoA carboxylase may be a purely regulatory enzyme. Finally, two enzymes in the β-oxidation pathway are markedly inhibited when the energy charge is high. NADH inhibits 3-hydroxyacyl CoA dehydrogenase, and acetyl CoA inhibits thiolase.

Long-term control is mediated by changes in the rates of synthesis and degradation of the enzymes participating in fatty acid synthesis. Animals that have fasted and are then fed high-carbohydrate, low-fat diets show marked increases in their amounts of acetyl CoA carboxylase and fatty acid synthase within a few days. This type of regulation is known as *adaptive control.*

22.6 ELONGATION AND UNSATURATION OF FATTY ACIDS ARE ACCOMPLISHED BY ACCESSORY ENZYME SYSTEMS

The major product of the fatty acid synthase is palmitate. In eukaryotes, longer fatty acids are formed by elongation reactions catalyzed by enzymes on the cytosolic face of the *endoplasmic reticulum membrane*. These reactions add two-carbon units sequentially to the carboxyl ends of both saturated and unsaturated fatty acyl CoA substrates. Malonyl CoA is the two-carbon donor in the elongation of fatty acyl CoAs. Again, condensation is driven by the decarboxylation of malonyl CoA.

22.6.1 Membrane-Bound Enzymes Generate Unsaturated Fatty Acids

Endoplasmic reticulum systems also introduce double bonds into long-chain acyl CoAs. For example, in the conversion of stearoyl CoA into oleoyl CoA, a cis-Δ^9 double bond is inserted by an oxidase that employs *molecular oxygen* and *NADH* (or *NADPH*).

$$\text{Stearoyl CoA} + \text{NADH} + \text{H}^+ + \text{O}_2 \longrightarrow$$
$$\text{oleoyl CoA} + \text{NAD}^+ + 2\,\text{H}_2\text{O}$$

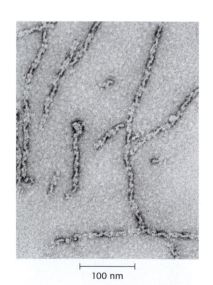

100 nm

FIGURE 22.28 Filaments of acetyl CoA carboxylase. The electron micrograph shows the enzymatically active filamentous form of acetyl CoA carboxylase from chicken liver. The inactive form is an octamer of 265-kd subunits. [Courtesy of Dr. M. Daniel Lane.]

$$H^+ + NADH \rightarrow \boxed{E\text{-FAD}} \leftrightarrow \boxed{Fe^{2+}} \rightarrow \boxed{Fe^{3+}} \leftrightarrow \text{Oleoyl CoA} + 2\,H_2O$$
$$NAD^+ \leftarrow \boxed{E\text{-FADH}_2} \rightarrow \boxed{Fe^{3+}} \leftrightarrow \boxed{Fe^{2+}} \rightarrow \text{Stearoyl CoA} + O_2$$

NADH-cytochrome b_5 **reductase**   **Cytochrome** b_5 **Desaturase**

FIGURE 22.29 Electron-transport chain in the desaturation of fatty acids.

This reaction is catalyzed by a complex of three membrane-bound enzymes: *NADH-cytochrome* b₅ *reductase, cytochrome* b₅, and a *desaturase* (Figure 22.29). First, electrons are transferred from NADH to the FAD moiety of NADH-cytochrome b_5 reductase.

The heme iron atom of cytochrome b_5 is then reduced to the Fe^{2+} state. The nonheme iron atom of the desaturase is subsequently converted into the Fe^{2+} state, which enables it to interact with O_2 and the saturated fatty acyl CoA substrate. A double bond is formed and two molecules of H_2O are released. Two electrons come from NADH and two from the single bond of the fatty acyl substrate.

A variety of unsaturated fatty acids can be formed from oleate by a combination of elongation and desaturation reactions. For example, oleate can be elongated to a 20:1 cis-Δ^{11} fatty acid. Alternatively, a second double bond can be inserted to yield an 18:2 cis-Δ^6,Δ^9 fatty acid. Similarly, palmitate (16:0) can be oxidized to palmitoleate (16:1 cis-Δ^9), which can then be elongated to *cis*-vaccenate (18:1 cis-Δ^{11}).

Unsaturated fatty acids in mammals are derived from either palmitoleate (16:1), oleate (18:1), linoleate (18:2), or linolenate (18:3). The number of carbon atoms from the ω end of a derived unsaturated fatty acid to the nearest double bond identifies its precursor.

Mammals lack the enzymes to introduce double bonds at carbon atoms beyond C-9 in the fatty acid chain. Hence, mammals cannot synthesize linoleate (18:2 cis-Δ^9, Δ^{12}) and linolenate (18:3 cis-Δ^9, Δ^{12}, Δ^{15}). *Linoleate and linolenate are the two essential fatty acids.* The term *essential* means that they must be supplied in the diet because they are required by an organism and cannot be endogenously synthesized. Linoleate and linolenate furnished by the diet are the starting points for the synthesis of a variety of other unsaturated fatty acids.

Precursor	Formula
Linolenate (ω-3)	
	$CH_3-(CH_2)_1-CH=CH-R$
Linoleate (ω-6)	
	$CH_3-(CH_2)_4-CH=CH-R$
Palmitoleate (ω-7)	
	$CH_3-(CH_2)_5-CH=CH-R$
Oleate (ω-9)	
	$CH_3-(CH_2)_7-CH=CH-R$

22.6.2 Eicosanoid Hormones Are Derived from Polyunsaturated Fatty Acids

Arachidonate, a 20:4 fatty acid derived from linoleate, is the major precursor of several classes of signal molecules: prostaglandins, prostacyclins, thromboxanes, and leukotrienes (Figure 22.30).

A prostaglandin is a 20-carbon fatty acid containing a 5-carbon ring (Figure 22.31). A series of prostaglandins is fashioned by reductases and isomerases. The major classes are designated PGA through PGI; a subscript denotes the number of carbon–carbon double bonds outside the ring. Prostaglandins with two double bonds, such as PGE₂, are derived from arachidonate; the other two double bonds of this precursor are lost in forming a five-membered ring. *Prostacyclin* and *thromboxanes* are related compounds that arise from a nascent prostaglandin. They are generated by *prostacylin synthase* and *thromboxane synthase* respectively. Alternatively, arachidonate can be converted into *leukotrienes* by the action of

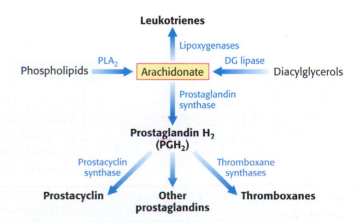

FIGURE 22.30 Arachidonate is the major precursor of eicosanoid hormones. Prostaglandin synthase catalyzes the first step in a pathway leading to prostaglandins, prostacyclins, and thromboxanes. Lipoxygenase catalyzes the initial step in a pathway leading to leukotrienes.

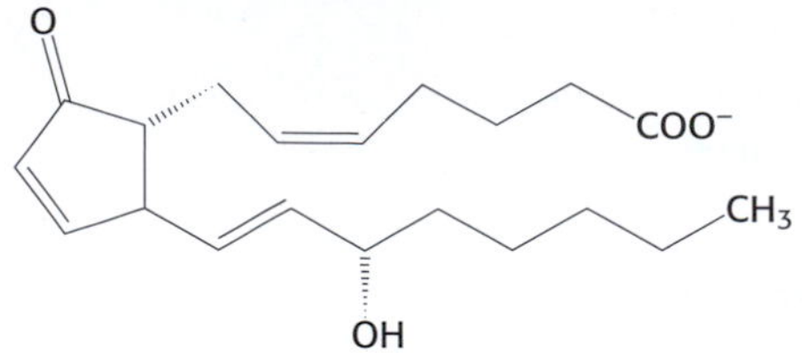

Prostaglandin A_2

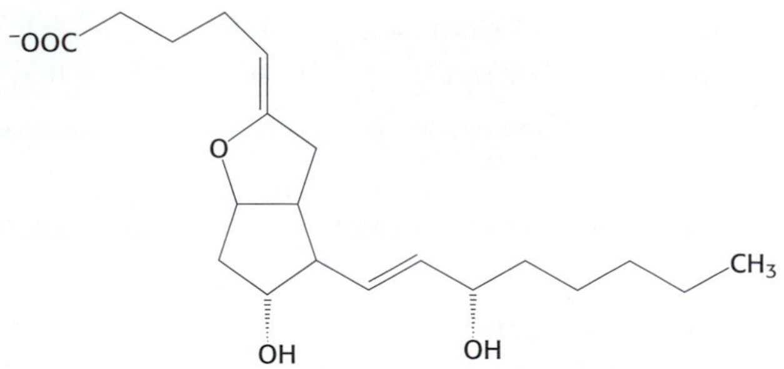

Prostacylin (PGI_2)

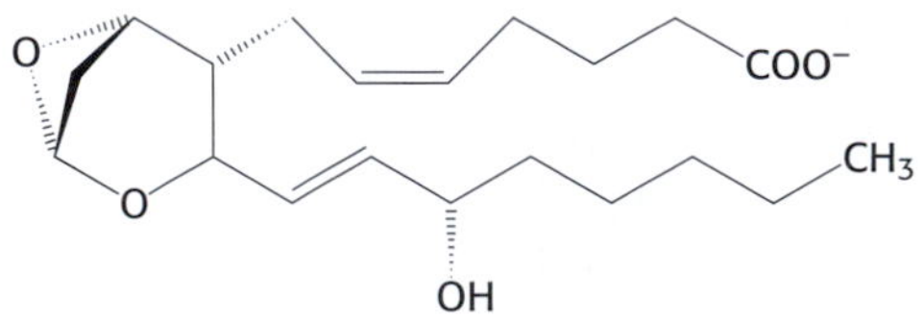

Thromboxane A_2 (TXA_2)

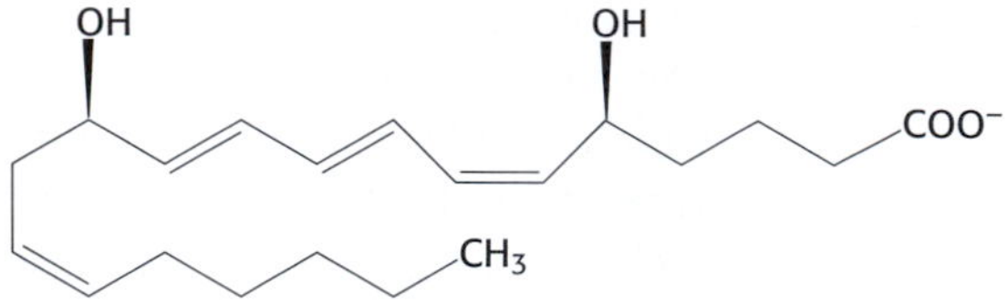

Leukotriene B_4

FIGURE 22.31 Structures of several eicosanoids.

lipoxygenase. These compounds, first found in leukocytes, contain three conjugated double bonds—hence, the name. Prostaglandins, prostacyclin, thromboxanes, and leukotrienes are called *eicosanoids* (from the Greek *eikosi,* "twenty") because they contain 20 carbon atoms.

Prostaglandins and other eicosanoids are *local hormones* because they are short-lived. They alter the activities both of the cells in which they are synthesized and of adjoining cells by binding to 7TM receptors. The nature of these effects may vary from one type of cell to another, in contrast with the more uniform actions of global hormones such as insulin and glucagon. Prostaglandins stimulate inflammation, regulate blood flow to particular organs, control ion transport across membranes, modulate synaptic transmission, and induce sleep.

Recall that aspirin blocks access to the active site of the enzyme that converts arachidonate into prostaglandin H_2 (Section 12.5.2). Because arachidonate is the precursor of other prostaglandins, prostacyclin, and thromboxanes, blocking this step affects many signaling pathways. It accounts for the wide-ranging effects that aspirin and related compounds have on inflammation, fever, pain, and blood clotting.

SUMMARY

Triacylglycerols Are Highly Concentrated Energy Stores

Fatty acids are physiologically important as (1) components of phospholipids and glycolipids, (2) hydrophilic modifiers of proteins, (3) fuel molecules, and (4) hormones and intracellular messengers. They are stored in adipose tissue as triacylglycerols (neutral fat).

The Utilization of Fatty Acids as Fuel Requires Three Stages of Processing

Triacylglycerols can be mobilized by the hydrolytic action of lipases that are under hormonal control. Fatty acids are activated to acyl CoAs, transported across the inner mitochondrial membrane by carnitine, and degraded in the mitochondrial matrix by a recurring sequence of four reactions: oxidation by FAD, hydration, oxidation by NAD^+, and thiolysis by CoA. The $FADH_2$ and NADH formed in the oxidation steps transfer their electrons to O_2 by means of the respiratory chain, whereas the acetyl CoA formed in the thiolysis step normally enters the citric acid cycle by condensing with oxaloacetate. Mammals are unable to convert fatty acids into glucose, because they lack a pathway for the net production of oxaloacetate, pyruvate, or other gluconeogenic intermediates from acetyl CoA.

Certain Fatty Acids Require Additional Steps for Degradation

Fatty acids that contain double bonds or odd numbers of carbon atoms require ancillary steps to be degraded. An isomerase and a reductase are required for the oxidation of unsaturated fatty acids, whereas propionyl CoA derived from chains with odd numbers of carbons requires a vitamin B_{12}-dependent enzyme to be converted into succinyl CoA.

Fatty Acids Are Synthesized and Degraded by Different Pathways

Fatty acids are synthesized in the cytosol by a different pathway from that of β oxidation. Synthesis starts with the carboxylation of acetyl CoA to malonyl CoA, the committed step. This ATP-driven reaction is catalyzed by acetyl CoA carboxylase, a biotin enzyme. The intermediates in fatty acid synthesis are linked to an acyl carrier protein. Acetyl ACP is formed from acetyl CoA, and malonyl ACP is formed from malonyl CoA. Acetyl ACP and malonyl ACP condense to form acetoacetyl ACP, a reaction driven by the release of CO_2 from the activated malonyl unit. A reduction, a dehydration, and a second reduction follow. NADPH is the reductant in these steps. The butyryl ACP formed in this way is ready for a second round of elongation, starting with the addition of a two-carbon unit from malonyl ACP. Seven rounds of elongation yield palmitoyl ACP, which is hydrolyzed to palmitate. In higher organisms, the enzymes carrying out fatty acid synthesis are covalently linked in a multifunctional enzyme complex. A reaction cycle based on the formation and cleavage of citrate carries acetyl groups from mitochondria to the cytosol. NADPH needed for synthesis is generated in the transfer of reducing equivalents from mitochondria by the malate–pyruvate shuttle and by the pentose phosphate pathway.

Acetyl Coenzyme A Carboxylase Plays a Key Role in Controlling Fatty Acid Metabolism

Fatty acid synthesis and degradation are reciprocally regulated so that both are not simultaneously active. Acetyl CoA carboxylase, the essential control site, is stimulated by insulin and inhibited by glucagon and epinephrine. These hormonal effects are mediated by changes in the amounts of the active dephosphorylated and inactive phosphorylated forms of the carboxylase. Citrate, which signals an abundance of building blocks and energy, allosterically stimulates the carboxylase. Glucagon and epinephrine stimulate triacylglycerol breakdown by activating the lipase. Insulin, in contrast, inhibits lipolysis. In times of plenty, fatty acyl CoAs do not enter the mitochondrial matrix because malonyl CoA inhibits carnitine acyltransferase I.

Elongation and Unsaturation of Fatty Acids Are Accomplished by Accessory Enzyme Systems

Fatty acids are elongated and desaturated by enzyme systems in the endoplasmic reticulum membrane. Desaturation requires NADH and O_2 and is carried out by a complex consisting of a flavoprotein, a cytochrome, and a nonheme iron protein. Mammals lack the enzymes to introduce double bonds distal to C-9, and so they require linoleate and linolenate in their diets.

Arachidonate, an essential precursor of prostaglandins and other signal molecules, is derived from linoleate. This 20:4 polyunsaturated fatty acid is the precursor of several classes of signal molecules—prostaglandins, prostacyclins, thromboxanes, and leukotrienes—that act as messengers and local hormones because of their transience. They are called eicosanoids because they contain 20 carbon atoms. Prostaglandin synthase, the enzyme catalyzing the first step in the synthesis of all eicosanoids except the leukotrienes, consists of a cyclooxygenase and a hydroperoxidase. Aspirin (acetylsalicylate), an anti-inflammatory and antithrombotic drug, irreversibly blocks the synthesis of these eicosanoids.

KEY TERMS

triacylglycerol (neutral fat, triacylglyceride) (p. 603)

bile salt (p. 603)

chylomicron (p. 604)

acyl adenylate (p. 606)

carnitine (p. 607)

β-oxidation pathway (p. 608)

vitamin B_{12} (cobalamin) (p. 611)

peroxisome (p. 614)

ketone body (p. 615)

acyl carrier protein (ACP) (p. 617)

fatty acid synthase (p. 617)

malonyl CoA (p. 617)

acetyl CoA carboxylase (p. 618)

megasynthase (p. 624)

polyketide (p. 624)

nonribosomal peptide (p. 624)

AMP-dependent protein kinase (AMPK) (p. 625)

protein phosphatase 2A (p. 625)

arachidonate (p. 627)

prostaglandin (p. 627)

eicosanoid (p. 628)

SELECTED READINGS

Where to start

Wakil, S. J., 1989. Fatty acid synthase, a proficient multifunctional enzyme. *Biochemistry* 28:4523–4530.

Rasmussen, B. B., and Wolfe, R. R., 1999. Regulation of fatty acid oxidation in skeletal muscle. *Annu. Rev. Nutr.* 19: 463–484.

Semenkovich, C. F., 1997. Regulation of fatty acid synthase (FAS). *Prog. Lipid Res.* 36:43–53.

Sul, H. S., Smas, C. M., Wang, D., and Chen, L., 1998. Regulation of fat synthesis and adipose differentiation. *Prog. Nucleic Acid Res. Mol. Biol.* 60:317–345.

Wolf, G., 1996. Nutritional and hormonal regulation of fatty acid synthase. *Nutr. Rev.* 54:122–123.

Munday, M. R., and Hemingway, C. J., 1999. The regulation of acetyl-CoA carboxylase: A potential target for the action of hypolipidemic agents. *Adv. Enzyme Regul.* 39:205–234.

Books

Vance, D. E., and Vance, J. E. (Eds.), 1996. *Biochemistry of Lipids, Lipoproteins, and Membranes.* Elsevier.

Boyer, P. D. (Ed.), 1983. *The Enzymes* (3d ed.). Vol. 16, *Lipid Enzymology.* Academic Press.

Numa, S. (Ed.), 1984. *Fatty Acid Metabolism and Its Regulation.* Elsevier.

Fatty acid oxidation

Barycki, J. J., O'Brien, L. K., Strauss, A. W., and Banaszak, L. J., 2000. Sequestration of the active site by interdomain shifting: Crystallographic and spectroscopic evidence for distinct conformations of L-3-hydroxyacyl-CoA dehydrogenase. *J. Biol. Chem.* 275:27186–27196.

Ramsay, R. R., 2000. The carnitine acyltransferases: Modulators of acyl-CoA-dependent reactions. *Biochem. Soc. Trans.* 28:182–186.

Eaton, S., Bartlett, K., and Pourfarzam, M., 1996. Mammalian mitochondrial beta-oxidation. *Biochem. J.* 320:345–357.

Thorpe, C., and Kim, J. J., 1995. Structure and mechanism of action of the acyl-CoA dehydrogenases. *FASEB J.* 9:718–725.

Foster, D. W., 1984. From glycogen to ketones—and back. *Diabetes* 33:1188–1199.

McGarry, J. D., and Foster, D. W., 1980. Regulation of hepatic fatty acid oxidation and ketone body production. *Annu. Rev. Biochem.* 49:395–420.

Fatty acid synthesis

Zhang, Y.-M., Rao, M. S., Heath, R. J., Price, A. C., Olson, A. J., Rock, C. O., and White, S. W., 2001. Identification and analysis of the acyl carrier protein (ACP) docking site on beta-ketoacyl-ACP synthase III. *J. Biol. Chem.* 276:8231–8238.

Davies, C., Heath, R. J., White, S. W., and Rock, C. O., 2000. The 1.8 Å crystal structure and active-site architecture of beta-ketoacyl-acyl carrier protein synthase III (FabH) from *Escherichia coli. Structure Fold Des.* 8:185–195.

Denton, R. M., Heesom, K. J., Moule, S. K., Edgell, N. J., and Burnett, P., 1997. Signalling pathways involved in the stimulation of fatty acid synthesis by insulin. *Biochem. Soc. Trans.* 25:1238–1242.

Stoops, J. K., Kolodziej, S. J., Schroeter, J. P., Bretaudiere, J. P., and Wakil, S. J., 1992. Structure-function relationships of the yeast fatty acid synthase: Negative-stain, cryo-electron microscopy, and image analysis studies of the end views of the structure. *Proc. Natl. Acad. Sci. USA* 89:6585–6589.

Loftus, T. M., Jaworsky, D. E., Frehywot, G. L., Townsend, C. A., Ronnett, G. V., Lane, M. D., and Kuhajda, F. P., 2000. Reduced food intake and body weight in mice treated with fatty acid synthase inhibitors. *Science* 288:2379–2381.

Acetyl CoA carboxylase

Thoden, J. B., Blanchard, C. Z., Holden, H. M., and Waldrop, G. L., 2000. Movement of the biotin carboxylase B-domain as a result of ATP binding. *J. Biol. Chem.* 275:16183–16190.

Kim, K. H., 1997. Regulation of mammalian acetyl-coenzyme A carboxylase. *Annu. Rev. Nutr.* 17:77–99.

Mabrouk, G. M., Helmy, I. M., Thampy, K. G., and Wakil, S. J., 1990. Acute hormonal control of acetyl–CoA carboxylase: The roles of insulin, glucagon, and epinephrine. *J. Biol. Chem.* 265:6330–6338.

Kim, K. H., Lopez, C. F., Bai, D. H., Luo, X., and Pape, M. E., 1989. Role of reversible phosphorylation of acetyl–CoA carboxylase in long-chain fatty acid synthesis. *FASEB J.* 3:2250–2256.

Witters, L. A., and Kemp, B. E., 1992. Insulin activation of acetyl–CoA carboxylase accompanied by inhibition of the 5′-AMP-activated protein kinase. *J. Biol. Chem.* 267:2864–2867.

Cohen, P., and Hardie, D. G., 1991. The actions of cyclic AMP on biosynthetic processes are mediated indirectly by cyclic AMP-dependent protein kinases. *Biochim. Biophys. Acta* 1094: 292–299.

Moore, F., Weekes, J., and Hardie, D. G., 1991. Evidence that AMP triggers phosphorylation as well as direct allosteric activation of rat liver AMP-activated protein kinase: A sensitive mechanism to protect the cell against ATP depletion. *Eur. J. Biochem.* 199:691–697.

Eicosanoids

Malkowski, M. G., Ginell, S. L., Smith, W. L., and Garavito, R. M., 2000. The productive conformation of arachidonic acid bound to prostaglandin synthase. *Science* 289:1933–1937.

Smith, T., McCracken, J., Shin, Y.-K., and DeWitt, D., 2000. Arachidonic acid and nonsteroidal anti-inflammatory drugs induce conformational changes in the human prostaglandin endoperoxide H2 synthase-2 (cyclooxygenase-2). *J. Biol. Chem.* 275:40407–40415.

Kalgutkar, A. S., Crews, B. C., Rowlinson, S. W., Garner, C., Seibert, K., and Marnett L. J., 1998. Aspirin-like molecules that covalently inactivate cyclooxygenase-2. *Science* 280:1268–1270.

Lands, W. E., 1991. Biosynthesis of prostaglandins. *Annu. Rev. Nutr.* 11:41–60.

Sigal, E., 1991. The molecular biology of mammalian arachidonic acid metabolism. *Am. J. Physiol.* 260:L13–L28.

Weissmann, G., 1991. Aspirin. *Sci. Am.* 264(1):84–90.

Vane, J. R., Flower, R. J., and Botting, R. M., 1990. History of aspirin and its mechanism of action. *Stroke* (12 suppl.):IV12–IV23.

Genetic diseases

Brivet, M., Boutron, A., Slama, A., Costa, C., Thuillier, L., Demaugre, F., Rabier, D., Saudubray, J. M., and Bonnefont, J. P., 1999. Defects in activation and transport of fatty acids. *J. Inherited Metab. Dis.* 22:428–441.

Wanders, R. J., van Grunsven, E. G., and Jansen, G. A., 2000. Lipid metabolism in peroxisomes: Enzymology, functions and dysfunctions of the fatty acid alpha- and beta-oxidation systems in humans. *Biochem. Soc. Trans.* 28:141–149.

Wanders, R. J., Vreken, P., den Boer, M. E., Wijburg, F. A., van Gennip, A. H., and IJlst, L., 1999. Disorders of mitochondrial fatty acyl-CoA β-oxidation. *J. Inherited Metab. Dis.* 22:442–487.

Kerner, J., and Hoppel, C., 1998. Genetic disorders of carnitine metabolism and their nutritional management. *Annu. Rev. Nutr.* 18:179–206.

Bartlett, K., and Pourfarzam, M., 1998. Recent developments in the detection of inherited disorders of mitochondrial β-oxidation. *Biochem. Soc. Trans.* 26:145–152.

Pollitt, R. J, 1995. Disorders of mitochondrial long-chain fatty acid oxidation. *J. Inherited Metab. Dis.* 18:473–490.

Roe, C. R., and Coates, P. M., 1995. Mitochondrial fatty acid oxidation disorders. In *The Metabolic Basis of Inherited Diseases* (7th ed., pp. 1501–1534), edited by C. R. Striver, A. L. Beaudet, W. S. Sly, D. Valle, J. B. Stanbury, J. B. Wyngaarden, and D. S. Frederickson. McGraw-Hill.

—| PROBLEMS |—

1. *After lipolysis.* Write a balanced equation for the conversion of glycerol into pyruvate. Which enzymes are required in addition to those of the glycolytic pathway?

2. *From fatty acid to ketone body.* Write a balanced equation for the conversion of stearate into acetoacetate.

3. *Counterpoint.* Compare and contrast fatty acid oxidation and synthesis with respect to

(a) site of the process.
(b) acyl carrier.
(c) reductants and oxidants.
(d) stereochemistry of the intemediates.
(e) direction of synthesis or degradation.
(f) organization of the enzyme system.

4. *Sources.* For each of the following unsaturated fatty acids, indicate whether the biosynthetic precursor in animals is palmitoleate, oleate, linoleate, or linolenate.

(a) 18:1 cis-Δ^{11}
(b) 18:3 cis-$\Delta^6,\Delta^9,\Delta^{12}$
(c) 20:2 cis-Δ^{11},Δ^{14}
(d) 20:3 cis-$\Delta^5,\Delta^8,\Delta^{11}$
(e) 22:1 cis-Δ^{13}
(f) 22:6 cis-$\Delta^4,\Delta^7,\Delta^{10},\Delta^{13},\Delta^{16},\Delta^{19}$

5. *Tracing carbons.* Consider a cell extract that actively synthesizes palmitate. Suppose that a fatty acid synthase in this preparation forms one molecule of palmitate in about 5 minutes. A large amount of malonyl CoA labeled with ^{14}C in each carbon of its malonyl unit is suddenly added to this system, and fatty acid synthesis is stopped a minute later by altering the pH. The fatty acids in the supernatant are analyzed for radioactivity. Which carbon atom of the palmitate formed by this system is more radioactive—C-1 or C-14?

6. *Driven by decarboxylation.* What is the role of decarboxylation in fatty acid synthesis? Name another key reaction in a metabolic pathway that employs this mechanistic motif.

7. *Kinase surfeit.* Suppose that a promoter mutation leads to the overproduction of protein kinase A in adipose cells. How might fatty acid metabolism be altered by this mutation?

8. *An unaccepting mutant.* The serine residue in acetyl CoA carboxylase that is the target of the AMP-dependent protein kinase is mutated to alanine. What is a likely consequence of this mutation?

9. *Blocked assets.* The presence of a fuel molecule in the cytosol does not ensure that it can be effectively used. Give two examples of how impaired transport of metabolites between compartments leads to disease.

10. *Elegant inversion.* Peroxisomes have an alternative pathway for oxidizing polyunsaturated fatty acids. They contain a hydratase that converts D-3-hydroxyacyl CoA into *trans*-Δ^2-enoyl CoA. How can this enzyme be used to oxidize CoAs containing a cis double bond at an even-numbered carbon atom (e.g., the cis-Δ^{12} double bond of linoleate)?

11. *Covalent catastrophe.* What is a potential disadvantage of having many catalytic sites together on one very long polypeptide chain?

12. *Missing acyl CoA dehydrogenases.* A number of genetic deficiencies in acyl CoA dehydrogenases have been described. This deficiency presents early in life after a period of fasting. Symptoms include vomiting, lethargy, and sometimes coma. Not only

are blood levels of glucose low (hypoglycemia), but starvation-induced ketosis is absent. Provide a biochemical explanation for these last two observations.

13. *Effects of clofibrate.* High blood levels of triacylglycerides are associated with heart attacks and strokes. Clofibrate, a drug that increases the activity of peroxisomes, is sometimes used to treat patients with such a condition. What is the biochemical basis for this treatment?

14. *A different kind of enzyme.* Figure 22.27 shows the response of acetyl CoA carboxylase to varying amounts of citrate. Explain this effect in light of the allosteric effects that citrate has on the enzyme. Predict the effects of increasing concentrations of palmitoyl CoA.

Mechanism Problems

15. *Variation on a theme.* Thiolase is homologous in structure to the condensing enzyme. On the basis of this observation, propose a mechanism for the cleavage of 3-ketoacyl CoA by CoA.

16. *Two plus three to make four.* Propose a reaction mechanism for the condensation of an acetyl unit with a malonyl unit to form an acetoacetyl unit in fatty acid synthesis.

Chapter Integration Problems

17. *Ill-advised diet.* Suppose that, for some bizarre reason, you decided to exist on a diet of whale and seal blubber, exclusively.

(a) How would lack of carbohydrates affect your ability to utilize fats?
(b) What would your breath smell like?
(c) One of your best friends, after trying unsuccessfully to convince you to abandon this diet, makes you promise to consume a healthy dose of odd-chain fatty acids. Does your friend have your best interests at heart? Explain.

18. *Fats to glycogen.* An animal is fed stearic acid that is radioactively labeled with $[^{14}C]$carbon. A liver biopsy reveals the presence of ^{14}C-labeled glycogen. How is this possible in light of the fact that animals cannot convert fats into carbohydrates?

Data Interpretation Problem

19. *Mutant enzyme.* Carnitine palmitoyl transferase I (CPTI) catalyzes the conversion of long-chain acyl CoA into acyl carnitine, a prerequisite for transport into mitochondria and subse-

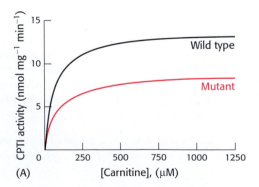

(A)

quent degradation. A mutant enzyme was constructed with a single amino acid change at position 3 of glutamic acid for alanine. Figures A through C show data from studies performed to identify the effect of the mutation [data from J. Shi, H. Zhu, D. N. Arvidson, and G. J. Wodegiorgis. *J. Biol. Chem.* 274(1999): 9421–9426].

(a) What is the effect of the mutation on enzyme activity when the concentration of carnitine is varied? What are the K_M and V_{max} values for the wild-type and mutant enzymes?

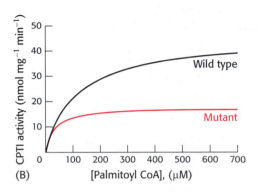

(B)

(b) What is the effect when the experiment is repeated with varying concentrations of palmitoyl CoA? What are the K_M and V_{max} values for the wild-type and mutant enzymes?

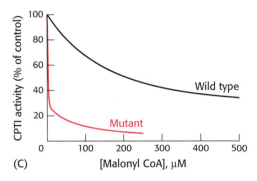

(C)

(c) Figure C shows the inhibitory effect of malonyl CoA on the wild-type and mutant enzymes. Which enzyme is more sensitive to malonyl CoA inhibition?
(d) Suppose that palmitoyl CoA = 100 μM, carnitine = 100 μM, and malonyl CoA = 10 μM. Under these conditions, what is the most prominent effect of the mutation on the properties of the enzyme?
(e) What can you conclude about the role of glutamate 3 in carnitine acyltransferase I function?

 Media Problem

20. *E pluribus unum.* The storage of chemical energy in the form of fatty acids requires carbon building blocks in the form of acetyl-CoA and energy in the form of ATP and NADPH. Using the **Conceptual Insights** overview of carbohydrate and fatty acid metabolism, estimate how many glucose molecules would be required to form one molecule of palmitate (C_{16}) if glucose were the sole carbon and energy source.

CHAPTER 23

Protein Turnover and Amino Acid Catabolism

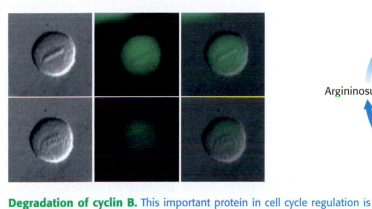

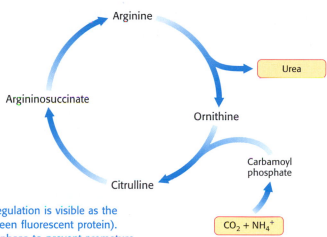

Degradation of cyclin B. This important protein in cell cycle regulation is visible as the green areas in the images above (the protein was fused with green fluorescent protein). Cyclin B is prominent during metaphase, but is degraded in anaphase to prevent premature initiation of another cell cycle. A large protease complex called the proteasome digests the protein into amino acids. These are either reused, or further processed by the urea cycle, which removes the nitrogen as urea. [(Above left) Courtesy of Jonathan Pines, University of Cambridge, Wellcome/CRC Institute of Cancer and Developmental Biology.]

The assembly of new proteins requires a source of amino acids. These building blocks are generated by the digestion of proteins in the intestine and the degradation of proteins within the cell. Many cellular proteins are constantly degraded and resynthesized. To facilitate this recycling, a complex system for the controlled turnover of proteins has evolved. Damaged or unneeded proteins are marked for destruction by the covalent attachment of chains of a small protein, *ubiquitin*. Polyubiquitinated proteins are subsequently degraded by a large, ATP-dependent complex called the *proteasome*.

The primary uses of amino acids are as building blocks for protein and peptide synthesis and as a source of nitrogen for the synthesis of other amino acids and other nitrogenous compounds such as nucleotide bases. Amino acids in excess of those needed for biosynthesis cannot be stored, in contrast with fatty acids and glucose, nor are they excreted. Rather, surplus amino acids are used as metabolic fuel. *The α-amino group is removed, and the resulting carbon skeleton is converted into a major metabolic intermediate.* Most of the amino groups of surplus amino acids are converted into urea through the *urea cycle*, whereas their carbon skeletons are transformed into acetyl CoA, acetoacetyl CoA, pyruvate, or one of the intermediates of the citric acid cycle. Hence, *fatty acids, ketone bodies, and glucose can be formed from amino acids.*

Several coenzymes play key roles in amino acid degradation, foremost among them is *pyridoxal phosphate*. This coenzyme forms Schiff-base intermediates that allow α-amino groups to be shuttled between amino acids and ketoacids. We will consider several genetic errors of amino acid degradation that lead to brain damage and mental retardation unless remedial action is initiated soon after birth. *Phenylketonuria,* which is caused by a block in the conversion of phenylalanine into tyrosine, is readily diagnosed and can be treated by removing phenylalanine from the diet. The study of amino acid metabolism is especially rewarding because it is rich in connections between basic biochemistry and clinical medicine.

23.1 PROTEINS ARE DEGRADED TO AMINO ACIDS

Dietary protein is a vital source of amino acids. Proteins ingested in the diet are digested into amino acids or small peptides that can be absorbed by the intestine and transported in the blood. Another source of amino acids is the degradation of defective or unneeded cellular proteins.

23.1.1 The Digestion and Absorption of Dietary Proteins

Protein digestion begins in the stomach, where the acidic environment favors protein denaturation. Denatured proteins are more accessible as substrates for proteolysis than are native proteins. The primary proteolytic enzyme of the stomach is *pepsin,* a nonspecific protease that, remarkably, is maximally active at pH 2. Thus, pepsin can be active in the highly acidic environment of the stomach, even though other proteins undergo denaturation there.

Protein degradation continues in the lumen of the intestine owing to the activity of proteolytic enzymes secreted by the pancreas. These proteins, introduced in Chapters 9 and 10 (Sections 9.1 and 10.5), are secreted as inactive zymogens and then converted into active enzymes. The battery of enzymes displays a wide array of specificity, and so the substrates are degraded into free amino acids as well as di- and tripeptides. Digestion is further enhanced by proteases, such as aminopeptidase N, that are located in the plasma membrane of the intestinal cells. Aminopeptidases digest proteins from the amino-terminal end. Single amino acids, as well as di- and tripeptides, are transported into the intestinal cells from the lumen and subsequently released into the blood for absorption by other tissues (Figure 23.1).

FIGURE 23.1 Digestion and absorption of proteins. Protein digestion is primarily a result of the activity of enzymes secreted by the pancreas. Aminopeptidases associated with the intestinal epithelium further digest proteins. The amino acids and di- and tripeptides are absorbed into the intestinal cells by specific transporters. Free amino acids are then released into the blood for use by other tissues.

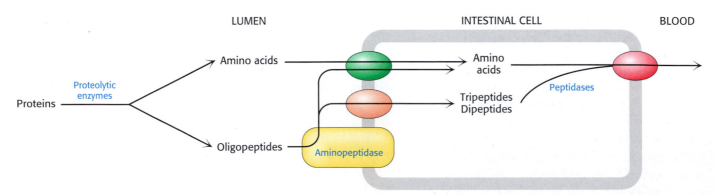

23.1.2 Cellular Proteins Are Degraded at Different Rates

Protein turnover—the degradation and resynthesis of proteins—takes place constantly in cells. Although some proteins are very stable, many proteins are short lived, particularly those that are important in metabolic regulation. Altering the amounts of these proteins can rapidly change metabolic patterns. In addition, cells have mechanisms for detecting and removing damaged proteins. A significant proportion of newly synthesized protein molecules are defective because of errors in translation. Even proteins that are normal when first synthesized may undergo oxidative damage or be altered in other ways with the passage of time.

The half-lives of proteins range over several orders of magnitude (Table 23.1). Ornithine decarboxylase, at approximately 11 minutes, has one of the shortest half-lives of any mammalian protein. This enzyme participates in the synthesis of polyamines, which are cellular cations essential for growth and differentiation. The life of hemoglobin, on the other hand, is limited only by the life of the red blood cell, and the lens protein, crystallin, by the life of the organism.

23.2 PROTEIN TURNOVER IS TIGHTLY REGULATED

How can a cell distinguish proteins that are meant for degradation? *Ubiquitin*, a small (8.5-kd) protein present in all eukaryotic cells, is the tag that marks proteins for destruction (Figure 23.2). Ubiquitin is the cellular equivalent of the "black spot" of Robert Louis Stevenson's *Treasure Island*: the signal for death.

23.2.1 Ubiquitin Tags Proteins for Destruction

Ubiquitin is highly conserved in eukaryotes: yeast and human ubiquitin differ at only 3 of 76 residues. The carboxyl-terminal glycine residue of ubiquitin (Ub) becomes covalently attached to the ϵ-amino groups of several lysine residues on a protein destined to be degraded. The energy for the formation of these *isopeptide bonds* (*iso* because ϵ- rather than α-amino groups are targeted) comes from ATP hydrolysis.

Three enzymes participate in the attachment of ubiquitin to each protein (Figure 23.3): ubiquitin-activating enzyme, or E1, ubiquitin-conjugating enzyme, or E2, and ubiquitin–protein ligase, or E3. First, the terminal carboxylate group of ubiquitin becomes linked to a sulfhydryl group of E1 by a thioester bond. This ATP-driven reaction is

TABLE 23.1 Dependence of the half-lives of cytosolic yeast proteins on the nature of their amino-terminal residues

Highly stabilizing residues $(t_{1/2} > 20 \text{ hours})$			
Ala	Cys	Gly	Met
Pro	Ser	Thr	Val
Intrinsically destabilizing residues $(t_{1/2} = 2 \text{ to } 30 \text{ minutes})$			
Arg	His	Ile	Leu
Lys	Phe	Trp	Tyr
Destabilizing residues after chemical modification $(t_{1/2} = 3 \text{ to } 30 \text{ minutes})$			
Asn	Asp	Gln	Glu

Source: J. W. Tobias, T. E. Schrader, G. Rocap, and A. Varshavsky. *Science* 254(1991):1374.

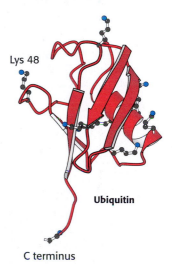

FIGURE 23.2 Ubiquitin. The structure of ubiquitin reveals an extended carboxyl terminus that is activated and linked to other proteins. Lysine residues also are shown, including lysine 48, the major site for linking additional ubiquitin molecules.

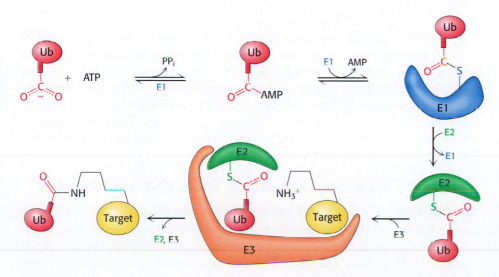

FIGURE 23.3 Ubiquitin conjugation. The ubiquitin-activating enzyme E1 adenylates ubiquitin (Ub) and transfers the ubiquitin to one of its own cysteine residues. Ubiquitin is then transferred to a cysteine residue in the ubiquitin-conjugating enzyme E2. Finally, the ubiquitin–protein ligase E3 transfers the ubiquitin to a lysine residue on the target protein.

reminiscent of fatty acid activation (Section 22.4.1). An adenylate is linked to the C-terminal carboxylate of ubiquitin with the release of pyrophosphate, and the ubiquitin is transferred to a sulfhydryl group of a key cysteine residue in E1. Second, activated ubiquitin is shuttled to a sulfhydryl group of E2. Finally, E3 catalyzes the transfer of ubiquitin from E2 to an ϵ-amino group on the target protein.

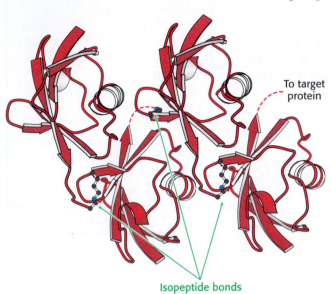

To target
protein

Isopeptide bonds

FIGURE 23.4 Tetraubiquitin. Four ubiquitin molecules are linked by isopeptide bonds. This unit is the primary signal for degradation when linked to a target protein.

The attachment of a single molecule of ubiquitin is only a weak signal for degradation. However, the ubiquitination reaction is processive: chains of ubiquitin can be generated by the linkage of the ϵ-amino group of lysine residue 48 of one ubiquitin molecule to the terminal carboxylate of another. Chains of four or more ubiquitin molecules are particularly effective in signaling degradation (Figure 23.4). The use of chains of ubiquitin molecules may have at least two advantages. First, the ubiquitin molecules interact with one another to form a binding surface distinct from that created by a single ubiquitin molecule. Second, individual ubiquitin molecules can be cleaved off without loss of the degradation signal.

Although most eukaryotes have only one or a small number of distinct E1 enzymes, all eukaryotes have many distinct E2 and E3 enzymes. Moreover, there appears to be only a single family of evolutionarily related E2 proteins but many distinct families of E3 proteins. Although the E3 component provides most of the substrate specificity for ubiquitination, the multiple combinations of the E2–E3 complex allow for more finely tuned substrate discrimination.

What determines whether a protein becomes ubiquitinated? One signal turns out to be unexpectedly simple. *The half-life of a cytosolic protein is determined to a large extent by its amino-terminal residue* (see Table 23.1). This dependency is referred to as the *N-terminal rule.* A yeast protein with methionine at its N terminus typically has a half-life of more than 20 hours, whereas one with arginine at this position has a half-life of about 2 minutes. A highly destabilizing N-terminal residue such as arginine or leucine favors rapid ubiquitination, whereas a stabilizing residue such as methionine or proline does not. E3 enzymes are the readers of N-terminal residues. Other signals thought to identify proteins for degradation include *cyclin destruction boxes,* which are amino acid sequences that mark cell-cycle proteins for destruction, and proteins rich in proline, glutamic acid, serine, and threonine (PEST sequences).

Some pathological conditions vividly illustrate the importance of the regulation of protein turnover. For example, human papilloma virus (HPV) encodes a protein that activates a specific E3 enzyme. The enzyme ubiquitinates the tumor suppressor p53 and other proteins that control DNA repair, which are then destroyed. The activation of this E3 enzyme is observed in more than 90% of cervical carcinomas. Thus, the inappropriate marking of key regulatory proteins for destruction can trigger further events, leading to tumor formation.

23.2.2 The Proteasome Digests the Ubiquitin-Tagged Proteins

If ubiquitin is the mark of death, what is the executioner? A large protease complex called the *proteasome* or the *26S proteasome* digests the ubiquitinated proteins. This ATP-driven multisubunit protease spares ubiquitin, which is then recycled. The 26S proteasome is a complex of two

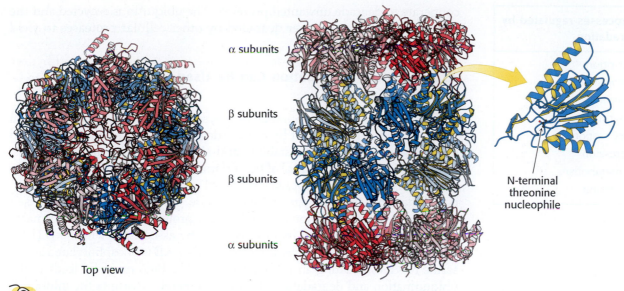

α subunits

β subunits

β subunits

α subunits

Top view

N-terminal
threonine
nucleophile

FIGURE 23.5 20S proteasome. The 20S proteasome comprises 28 homologous subunits (α, red; β, blue), arranged in four rings of 7 subunits each. Some of the β subunits (highlighted in yellow) include protease active sites at the amino termini. The top view shows the approximate seven-fold symmetry of the structure.

components: a 20S proteasome, which contains the catalytic activity, and a 19S regulatory subunit.

The 20S complex is constructed from two copies each of 14 subunits and has a mass of 700 kd (Figure 23.5). All 14 subunits are homologous and adopt the same overall structure. The subunits are arranged in four rings of 7 subunits that stack to form a structure resembling a barrel. The components of the two rings at the ends of the barrel are called the α subunits and those of the two central rings the β subunits. The active sites of the protease are located at the N-termini of certain β subunits on the interior of the barrel—specifically, those β chains having an N-terminal threonine or serine residue. The hydroxyl groups of these amino acids are converted into nucleophiles with the assistance of their own amino groups. These nucleophilic groups then attack the carbonyl groups of peptide bonds and form acyl-enzyme intermediates (Section 9.1.2). The structure of the complex sequesters the proteolytic active sites from potential substrates until they are directed into the barrel. Substrates are degraded in a processive manner without the release of degradation intermediates, until the substrate is reduced to peptides ranging in length from seven to nine residues.

The 20S proteasome is a sealed barrel. Access to its interior is controlled by a 19S regulatory complex, itself a 700-kd complex made up of 20 subunits. This complex binds to both ends of the 20S proteasome core to form the complete 26S proteasome (Figure 23.6). The 19S subunit binds specifically to polyubiquitin chains. Key components of the 19S complex are six distinct ATPases of the AAA class (ATPase associated with various cellular activities) characterized by a conserved 230 amino acid ATP-binding domain of the P-loop NTPase family. This class of ATPase, found in all kingdoms, is associated with a variety of cell functions including cell-cycle regulation and organelle biogenesis. Although the exact role of the ATPase remains uncertain, ATP hydrolysis may assist the 19S complex to unfold the substrate and induce conformational changes in the 20S proteasome so that the substrate can be passed into the center of the complex. Finally, the 19S subunit also contains an isopeptidase that cleaves off intact ubiquitin molecules. Thus, the ubiquitinization pathway and the proteasome

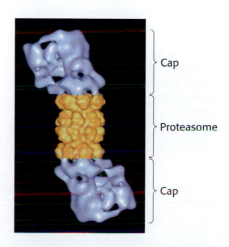

Cap

Proteasome

Cap

FIGURE 23.6 26S proteasome. A 19S cap is attached to each end of the 20S catalytic unit. [From W. Baumeister, J. Walz, F. Zuhl, and E. Seemuller. *Cell* 92(1998):367. Courtesy of Dr. Wolfgang Baumeister.]

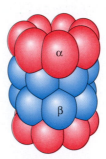

Archaeal proteasome

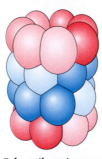

Eukaryotic proteasome

FIGURE 23.7 Proteasome evolution. The archaeal proteasome consists of 14 identical α subunits and 14 identical β subunits. In the eukaryotic proteasome, gene duplication and specialization has led to 7 distinct subunits of each type. The overall architecture of the proteasome is conserved.

cooperate to degrade unwanted proteins. The ubiquitin is recycled and the peptide products are further degraded by other cellular proteases to yield individual amino acids.

23.2.3 Protein Degradation Can Be Used to Regulate Biological Function

Table 23.2 lists a number of physiological processes that are controlled at least in part by protein degradation. This control is exerted by dynamically altering the stability and abundance of regulatory proteins. Consider, for example, control of the inflammatory response. A transcription factor called *NF-κB* (NF for nuclear factor) initiates the expression of a number of the genes that take part in this response. This important factor is itself activated by the degradation of an inhibitory protein, *I-κB*. NF-κB is maintained in the cytoplasm in its inactive state by association with I-κB (I for inhibitor). In response to inflammatory signals, I-κβ is phosphorylated at two serine residues, creating an E3 binding site. The binding of E3 leads to the ubiquitination and degradation of I-κB and thereby disrupts the inhibitor's association with NF-κB. The liberated transcription factor migrates to the nucleus to stimulate transcription of the target genes. The NF-κB–I-κB system illustrates the interplay of several key regulatory motifs: receptor-mediated signal transduction, phosphorylation, compartmentalization, controlled and specific degradation, and selective gene expression.

23.2.4 The Ubiquitin Pathway and the Proteasome Have Prokaryotic Counterparts

Both the ubiquitin pathway and the proteasome appear to be present in all eukaryotes. Homologs of the proteasome are found in prokaryotes, although the physiological roles of these homologs have not been well established. The proteasomes of some archaea are quite similar in overall structure to their eukaryotic counterparts and similarly have 28 subunits (Figure 23.7). In the archaeal proteasome, however, all α subunits and all β subunits are identical; in eukaryotes, each α or β subunit is one of seven different isoforms. This specialization provides distinct substrate specificity.

Ubiquitin, however, has not been found in prokaryotes. Indeed, the high level of sequence similarity between the human and yeast proteins suggests that ubiquitin, in its present form, diverged relatively recently in evolutionary terms. Ubiquitin's molecular ancestors were recently identified in prokaryotes. Remarkably, these proteins take part not in protein modification but in coenzyme biosynthesis (Figure 23.8). The biosynthesis of thiamine (Section 8.6.1) begins with a sulfide ion derived from cysteine. This sulfide is added to the C-terminal carboxylate of the protein ThiS, which had been activated as an acyl adenylate. The activation of ThiS and the addi-

FIGURE 23.8 Biosynthesis of thiamine. The biosynthesis of thiamine begins with the addition of sulfide to the carboxyl terminus of the protein ThiS. This protein is activated by adenylation and conjugated in a manner analogous to the first steps in the ubiquitin pathway.

tion of sulfide are catalyzed by the enzyme ThiF. Human E1 includes two tandem regions of 160 amino acids that are 28% identical in amino acid sequence with a region of ThiF from *E. coli*. The evolutionary relationships between these two pathways were cemented by the determination of the three-dimensional structure of ThiS, which revealed a structure very similar to that of ubiquitin, despite being only 14% identical in amino acid sequence (Figure 23.9). Thus, a eukaryotic system for protein modification evolved from a preexisting prokaryotic pathway for coenzyme biosynthesis.

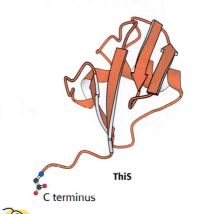

ThiS
C terminus

FIGURE 23.9 Structure of ThiS. The determination of the structure of ThiS revealed it to be structurally similar to ubiquitin despite only 14% sequence identity. This observation suggests that a prokaryotic protein such as ThiS evolved into ubiquitin.

23.3 THE FIRST STEP IN AMINO ACID DEGRADATION IS THE REMOVAL OF NITROGEN

What is the fate of amino acids released on protein digestion or turnover? Any not needed as building blocks are degraded to specific compounds. The major site of amino acid degradation in mammals is the liver. The amino group must be removed, inasmuch as there are no nitrogenous compounds in energy-transduction pathways. The α-ketoacids that result from the deamination of amino acids are metabolized so that the carbon skeletons can enter the metabolic mainstream as precursors to glucose or citric acid cycle intermediates. The fate of the α-amino group will be considered first, followed by that of the carbon skeleton (Section 23.5).

23.3.1 Alpha-Amino Groups Are Converted into Ammonium Ions by the Oxidative Deamination of Glutamate

The α-amino group of many amino acids is transferred to α-ketoglutarate to form *glutamate,* which is then oxidatively deaminated to yield ammonium ion (NH_4^+).

$$\text{Amino acid} \longrightarrow \text{Glutamate} \longrightarrow NH_4^+$$

Aminotransferases catalyze the transfer of an α-amino group from an α-amino acid to an α-ketoacid. These enzymes, also called *transaminases,* generally funnel α-amino groups from a variety of amino acids to α-ketoglutarate for conversion into NH_4^+.

Aspartate aminotransferase, one of the most important of these enzymes, catalyzes the transfer of the amino group of aspartate to α-ketoglutarate.

$$\text{Aspartate} + \alpha\text{-ketoglutarate} \rightleftharpoons \text{oxaloacetate} + \text{glutamate}$$

Alanine aminotransferase catalyzes the transfer of the amino group of alanine to α-ketoglutarate.

$$\text{Alanine} + \alpha\text{-ketoglutarate} \rightleftharpoons \text{pyruvate} + \text{glutamate}$$

These transamination reactions are reversible and can thus be used to synthesize amino acids from α-ketoacids, as we shall see in Chapter 24.

The nitrogen atom that is transferred to α-ketoglutarate in the transamination reaction is converted into free ammonium ion by oxidative deamination. This reaction is catalyzed by *glutamate dehydrogenase*. This enzyme is unusual in being able to utilize either NAD^+ or $NADP^+$, at least in some species. The reaction proceeds by dehydrogenation of the C–N bond, followed by hydrolysis of the resulting Schiff base.

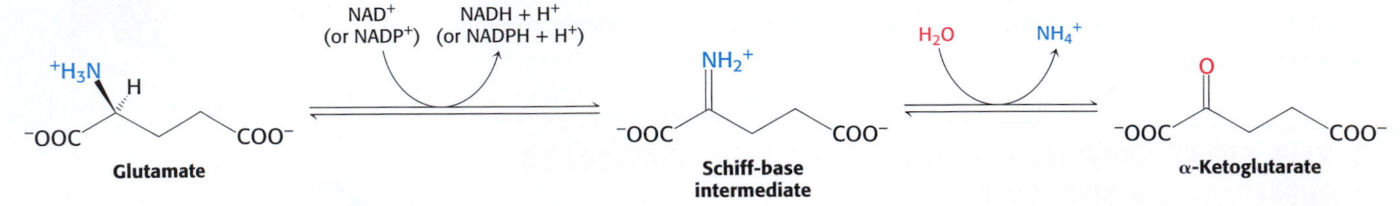

The equilibrium for this reaction favors glutamate; the reaction is driven by the consumption of ammonia. Glutamate dehydrogenase is located in mitochondria, as are some of the other enzymes required for the production of urea. This compartmentalization sequesters free ammonia, which is toxic.

In vertebrates, the activity of glutamate dehydrogenase is allosterically regulated. The enzyme consists of six identical subunits. Guanosine triphosphate and adenosine triphosphate are allosteric inhibitors, whereas guanosine diphosphate and adenosine diphosphate are allosteric activators. Hence, *a lowering of the energy charge accelerates the oxidation of amino acids.*

The sum of the reactions catalyzed by aminotransferases and glutamate dehydrogenase is

$$\text{α-Amino acid} + NAD^+ \rightleftharpoons \text{α-ketoacid} + NH_4^+ + NADH + H^+$$
$$(\text{or } NADP^+) \qquad\qquad (\text{or } NADPH)$$

In most terrestrial vertebrates, NH_4^+ is converted into urea, which is excreted.

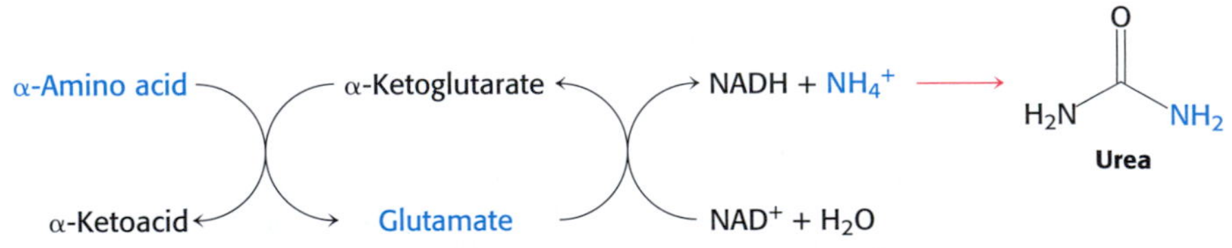

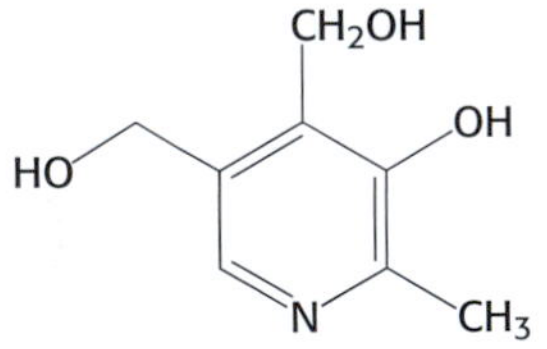

23.3.2 Pyridoxal Phosphate Forms Schiff-Base Intermediates in Aminotransferases

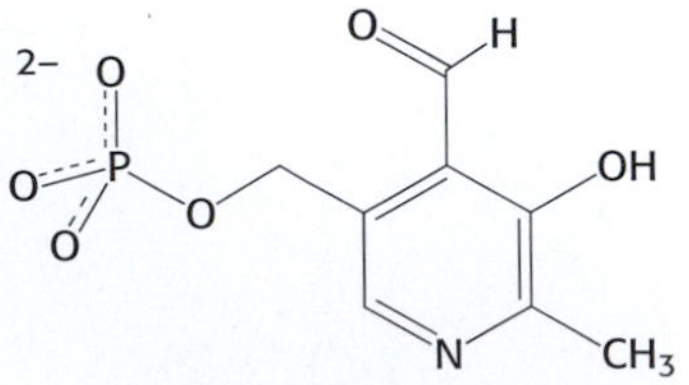

Pyridoxine
(Vitamin B₆)

Pyridoxal phosphate
(PLP)

All aminotransferases contain the prosthetic group *pyridoxal phosphate* (PLP), which is derived from *pyridoxine (vitamin B₆)*. Pyridoxal phosphate includes a pyridine ring that is slightly basic as well as a phenolic hydroxyl group that is slightly acidic. Thus, pyridoxal phosphate derivatives can form stable tautomeric forms in which the pyridine nitrogen atom is protonated and, hence, positively charged while the hydroxyl group is deprotonated, forming a phenolate.

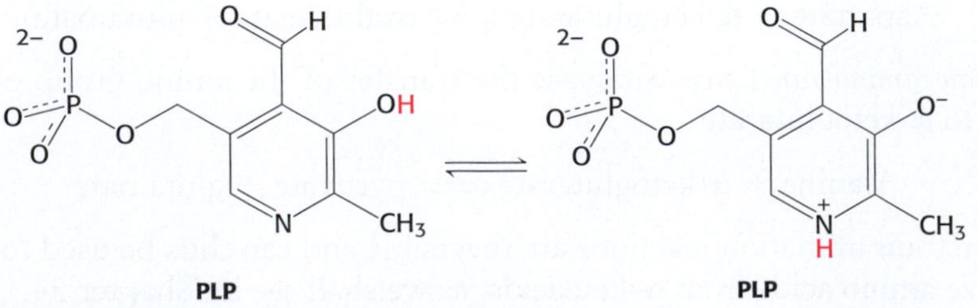

PLP   PLP

The most important functional group on PLP is the aldehyde. This group allows PLP to form covalent Schiff-base intermediates with amino acid substrates. Indeed, even in the absence of substrate, the aldehyde group of PLP usually forms a Schiff-base linkage with the ϵ-amino group of a specific lysine residue of the enzyme. A new Schiff-base linkage is formed on addition of an amino acid substrate. These Schiff-base linkages are often protonated, with the positive charge stabilized by interaction with the negatively charged phenolate group of PLP.

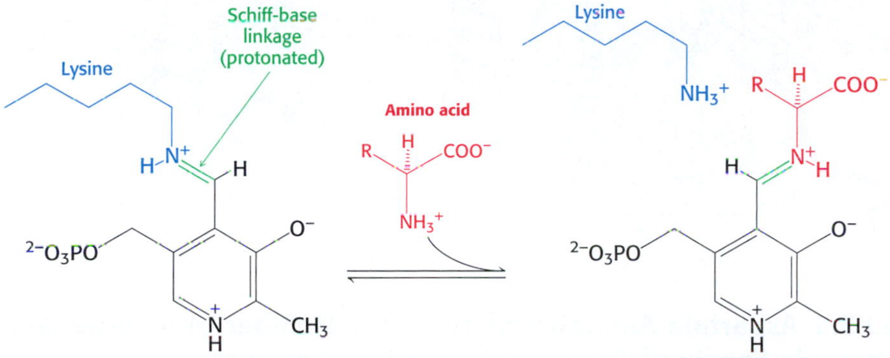

Internal aldimine　　　　　　**External aldimine**

The α-amino group of the amino acid substrate displaces the ϵ-amino group of the active-site lysine residue. In other words, an *internal* aldimine becomes an *external* aldimine. The amino acid–PLP Schiff base that is formed remains tightly bound to the enzyme by multiple noncovalent interactions.

The Schiff base between the amino acid substrate and PLP, the *external aldimine,* loses a proton from the α-carbon atom of the amino acid to form a *quinonoid* intermediate (Figure 23.10).

FIGURE 23.10 Transamination mechanism. The external aldimine loses a proton to form a quinonoid intermediate. Reprotonation of this intermediate at the aldehyde carbon atom yields a ketimine. This intermediate is hydrolyzed to generate the α-ketoacid product and pyridoxamine phosphate.

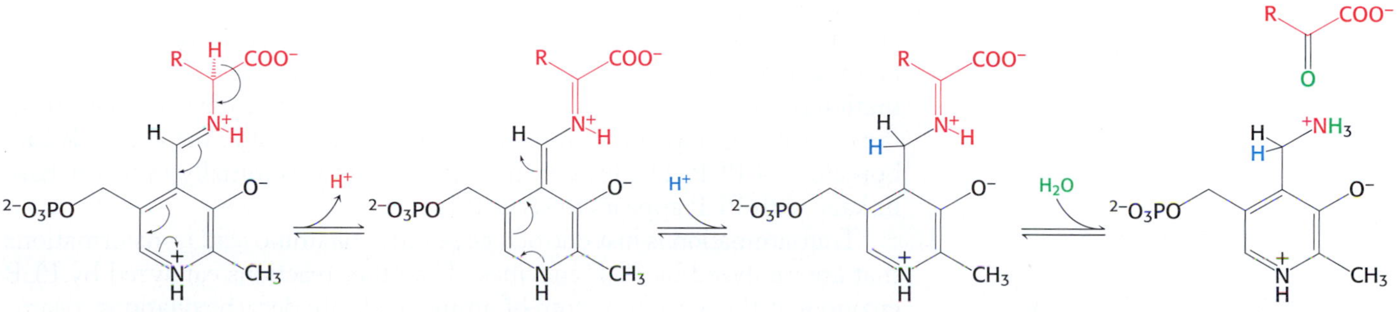

Aldimine　　　　　**Quinonoid intermediate**　　　　　**Ketimine**　　　　　**Pyridoxamine phosphate (PMP)**

The negative charge that is left on the amino acid is stabilized by delocalization into the pyridinium ring. Reprotonation of this intermediate at the aldehyde carbon atom yields a *ketimine.* The ketimine is then hydrolyzed to an α-ketoacid and *pyridoxamine phosphate (PMP).* These steps constitute half of the transamination reaction.

$$\text{Amino acid}_1 + \text{E-PLP} \rightleftharpoons \alpha\text{-ketoacid}_1 + \text{E-PMP}$$

The second half takes place by the reverse of the preceding pathway. A second α-ketoacid reacts with the enzyme–pyridoxamine phosphate complex (E-PMP) to yield a second amino acid and regenerate the enzyme–pyridoxal phosphate complex (E-PLP).

$$\alpha\text{-Ketoacid}_2 + \text{E-PMP} \rightleftharpoons \text{amino acid}_2 + \text{E-PLP}$$

The sum of these partial reactions is

$$\text{Amino acid}_1 + \alpha\text{-ketoacid}_2 \rightleftharpoons \text{amino acid}_2 + \alpha\text{-ketoacid}_1$$

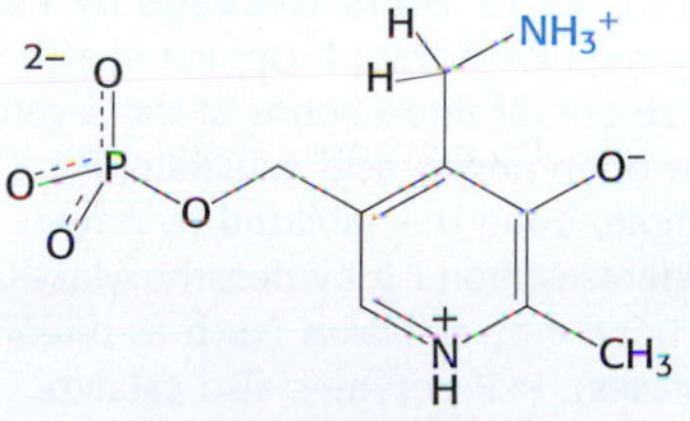

Pyridoxamine phosphate (PMP)

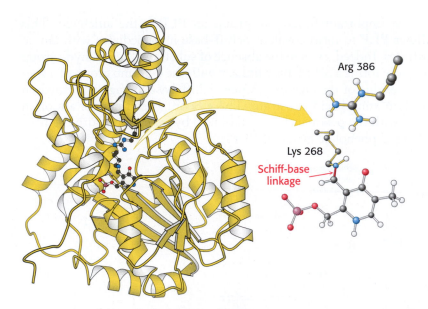

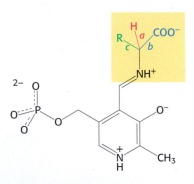

FIGURE 23.11 Aspartate aminotransferase. The active site of this prototypical PLP-dependent enzyme includes pyridoxal phosphate attached to the enzyme by a Schiff-base linkage with lysine 258. An arginine residue in the active site helps orient substrates by binding to their α-carboxylate groups.

Arg 386

Lys 268

Schiff-base
linkage

23.3.3 Aspartate Aminotransferase Is a Member of a Large and Versatile Family of Pyridoxal-Dependent Enzymes

The mitochondrial enzyme aspartate aminotransferase provides an especially well studied example of PLP as a coenzyme for transamination reactions (Figure 23.11). The results of X-ray crystallographic studies provided detailed views of how PLP and substrates are bound and confirmed much of the proposed catalytic mechanism. Each of the identical 45-kd subunits of this dimer consists of a large domain and a small one. PLP is bound to the large domain, in a pocket near the subunit interface. In the absence of substrate, the aldehyde group of PLP is in a Schiff-base linkage with lysine 258, as anticipated. Adjacent to the coenzyme's binding site is a conserved arginine residue that interacts with the α-carboxylate group of the substrate, helping to orient the substrate appropriately in the active site. The transamination reaction (see Figure 23.10) requires a base to remove a proton from the α-carbon group of the amino acid and to transfer it to the aldehyde carbon atom of PLP. The lysine amino group that was initially in Schiff-base linkage with PLP appears to serve this role.

Transamination is just one of a wide range of amino acid transformations that are catalyzed by PLP enzymes. The other reactions catalyzed by PLP enzymes at the α-carbon atom of amino acids are decarboxylations, deaminations, racemizations, and aldol cleavages (Figure 23.12). In addition, PLP enzymes catalyze elimination and replacement reactions at the β-carbon atom (e.g., tryptophan synthetase; Section 24.2.11) and the γ-carbon atom (e.g., cytathionine β-synthase, Section 24.2.9) of amino acid substrates. Three common features of PLP catalysis underlie these diverse reactions.

1. A Schiff base is formed by the amino acid substrate (the amine component) and PLP (the carbonyl component).

2. The protonated form of PLP acts as an *electron sink* to stabilize catalytic intermediates that are negatively charged. Electrons from these intermediates can be transferred into the pyridine ring to neutralize the positive charge on the pyridinium nitrogen. In other words, PLP is an *electrophilic catalyst*.

3. The product Schiff base is cleaved at the completion of the reaction.

Many of the enzymes that catalyze these reactions, such as serine hydroxymethyltransferase, which converts serine into glycine, have the same fold as that of aspartate aminotransferase and are clearly related by

FIGURE 23.12 Bond cleavage by PLP enzymes. Pyridoxal phosphate enzymes labilize one of three bonds at the α-carbon atom of an amino acid substrate. For example, bond *a* is labilized by aminotransferases, bond *b* by decarboxylases, and bond *c* by aldolases (such as threonine aldolases). PLP enzymes also catalyze reactions at the β- and γ-carbon atoms of amino acids.

divergent evolution. Others, such as tryptophan synthetase, have quite different overall structures. Nonetheless, the active sites of these enzymes are remarkably similar to that of aspartate aminotransferase, revealing the effects of convergent evolution.

How does a particular enzyme selectively favor the cleavage of one of three bonds at the α-carbon atom of an amino acid substrate? An important principle is that *the bond being broken must be perpendicular to the π orbitals of the electron sink* (Figure 23.13). An aminotransferase, for example, achieves this goal by binding the amino acid substrate so that the C_α–H bond is perpendicular to the PLP ring (Figure 23.14). In serine hydroxymethyltransferase, the N–C_α bond is rotated so that the C_α–C_β bond is most nearly perpendicular to the plane of the PLP ring, favoring its cleavage. This means of choosing one of several possible catalytic outcomes is called *stereoelectronic control.*

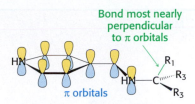

FIGURE 23.13 Stereoelectronic effects. The orientation about the NH–C_α bond determines the most favored reaction catalyzed by a pyridoxal phosphate enzyme. The bond that is most nearly perpendicular to the π orbitals of the pyridoxal phosphate electron sink is most easily cleaved.

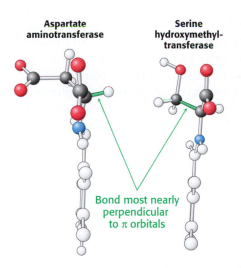

FIGURE 23.14 Reaction choice. In aspartate aminotransferase, the C_α–H bond is most nearly perpendicular to the π-orbital system and is cleaved. In serine hydroxymethyltransferase, a small rotation about the N–C_α bond places the C_α–C_β bond perpendicular to the π system, favoring its cleavage.

23.3.4 Serine and Threonine Can Be Directly Deaminated

Although the nitrogen atoms of most amino acids are transferred to α-ketoglutarate before removal, the α-amino groups of serine and threonine can be directly converted into NH_4^+. These direct deaminations are catalyzed by *serine dehydratase* and *threonine dehydratase*, in which PLP is the prosthetic group.

$$\text{Serine} \longrightarrow \text{pyruvate} + NH_4^+$$

$$\text{Threonine} \longrightarrow \alpha\text{-ketobutyrate} + NH_4^+$$

These enzymes are called *dehydratases* because *dehydration precedes deamination.* Serine loses a hydrogen ion from its α-carbon atom and a hydroxide ion group from its β-carbon atom to yield aminoacrylate. This unstable compound reacts with H_2O to give pyruvate and NH_4^+. Thus, the presence of a hydroxyl group attached to the β-carbon atom in each of these amino acids permits the direct deamination.

23.3.5 Peripheral Tissues Transport Nitrogen to the Liver

Most amino acid degradation takes place in tissues other than the liver. For instance, muscle uses amino acids as a source of fuel during prolonged exercise and fasting. How is the nitrogen processed in these other tissues? As in the liver, the first step is the removal of the nitrogen from the amino acid. However, muscle lacks the enzymes of the urea cycle, so the nitrogen must be released in a form that can be absorbed by the liver and converted into urea.

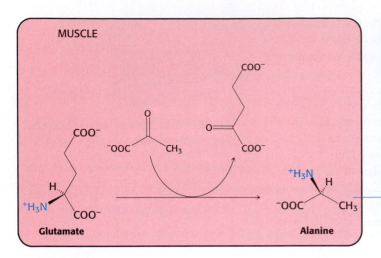
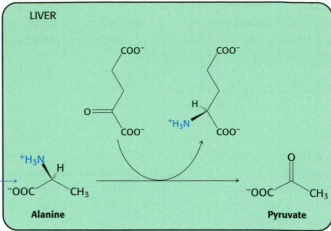

FIGURE 23.15 The alanine cycle. Glutamate in muscle is transaminated to alanine, which is released into the bloodstream. In the liver, alanine is taken up and converted into pyruvate for subsequent metabolism.

Nitrogen is transported from muscle to the liver in two principal transport forms. Glutamate is formed by transamination reactions, but the nitrogen is then transferred to pyruvate to form alanine, which is released into the blood (Figure 23.15). The liver takes up the alanine and converts it back into pyruvate by transamination. The pyruvate can be used for gluconeogenesis and the amino group eventually appears as urea. This transport is referred to as the *alanine cycle*. It is reminiscent of the Cori cycle discussed earlier (Section 16.4.2) and again illustrates the ability of the muscle to shift some of its metabolic burden to the liver.

Nitrogen can also be transported as glutamine. Glutamine synthetase (Section 24.1.2) catalyzes the synthesis of glutamine from glutamate and NH_4^+ in an ATP-dependent reaction:

$$NH_4^+ + \text{glutamate} + ATP \xrightarrow{\text{Glutamine synthetase}} \text{glutamine} + ADP + P_i$$

The nitrogens of glutamine can be converted into urea in the liver.

23.4 AMMONIUM ION IS CONVERTED INTO UREA IN MOST TERRESTRIAL VERTEBRATES

Some of the NH_4^+ formed in the breakdown of amino acids is consumed in the biosynthesis of nitrogen compounds. In most terrestrial vertebrates, the excess NH_4^+ is converted into *urea* and then excreted. Such organisms are referred to as *ureotelic*.

In terrestrial vertebrates, urea is synthesized by the *urea cycle* (Figure 23.16). The urea cycle, proposed by Hans Krebs and Kurt Henseleit in 1932, was the first cyclic metabolic pathway to be discovered. One of the nitrogen atoms of the urea is transferred from an amino acid, aspartate. The other nitrogen atom is derived directly from free NH_4^+, and the carbon atom comes from HCO_3^- (derived by hydration of CO_2; see Section 9.2).

FIGURE 23.16 The urea cycle.

23.4.1 The Urea Cycle Begins with the Formation of Carbamoyl Phosphate

The urea cycle begins with the coupling of free NH_4^+ with HCO_3^- to form carbamoyl phosphate. The synthesis of carbamoyl phosphate, though a simple molecule, is complex, comprising three steps, all catalyzed by *carbamoyl phosphate synthetase*.

The reaction begins with the phosphorylation of HCO_3^- to form carboxyphosphate, which then reacts with ammonium ion to form carbamic acid. Finally, a second molecule of ATP phosphorylates carbamic acid to carbamoyl phosphate. The structure and mechanism of the fascinating enzyme that catalyzes these reactions will be discussed in Chapter 24. The consumption of two molecules of ATP makes this synthesis of carbamoyl phosphate essentially irreversible. The mammalian enzyme requires *N-acetyl-glutamate* for activity, as will be discussed shortly.

The carbamoyl group of carbamoyl phosphate, which has a high transfer potential because of its anhydride bond, is transferred to *ornithine* to form *citrulline*, in a reaction catalyzed by *ornithine transcarbamoylase*.

Ornithine and citrulline are amino acids, but they are not used as building blocks of proteins. The formation of NH_4^+ by glutamate dehydrogenase, its incorporation into carbamoyl phosphate, and the subsequent synthesis of citrulline take place in the mitochondrial matrix. In contrast, the next three reactions of the urea cycle, which lead to the formation of urea, take place in the cytosol.

Citrulline is transported to the cytoplasm where it condenses with aspartate, the donor of the second amino group of urea. This synthesis of *argininosuccinate*, catalyzed by *argininosuccinate synthetase*, is driven by the cleavage of ATP into AMP and pyrophosphate and by the subsequent hydrolysis of pyrophosphate.

Argininosuccinase cleaves argininosuccinate into *arginine* and *fumarate*. Thus, the carbon skeleton of aspartate is preserved in the form of fumarate.

Argininosuccinate $\xrightarrow{\text{Argininosuccinase}}$ **Arginine** + **Fumarate**

Finally, arginine is hydrolyzed to generate urea and ornithine in a reaction catalyzed by *arginase*. Ornithine is then transported back into the mitochondrion to begin another cycle. The urea is excreted. Indeed, human beings excrete about 10 kg (22 pounds) of urea per year.

Arginine $\xrightarrow[\text{Arginase}]{\text{H}_2\text{O}}$ **Ornithine** + **Urea**

23.4.2 The Urea Cycle Is Linked to the Citric Acid Cycle

The stoichiometry of urea synthesis is

$$CO_2 + NH_4^+ + 3\,ATP + aspartate + 2\,H_2O \longrightarrow$$
$$urea + 2\,ADP + 2\,P_i + AMP + PP_i + fumarate$$

Pyrophosphate is rapidly hydrolyzed, and so the equivalent of four molecules of ATP are consumed in these reactions to synthesize one molecule of urea. *The synthesis of fumarate by the urea cycle is important because it links the urea cycle and the citric acid cycle* (Figure 23.17). Fumarate is hydrated to malate, which is in turn oxidized to oxaloacetate. Oxaloacetate has several possible fates: (1) transamination to aspartate, (2) conversion into glucose by the gluconeogenic pathway, (3) condensation with acetyl CoA to form citrate, or (4) conversion into pyruvate.

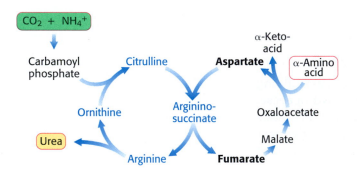

FIGURE 23.17 Metabolic integration of nitrogen metabolism. The urea cycle, the citric acid cycle, and the transamination of oxaloacetate are linked by fumarate and aspartate.

23.4.3 The Evolution of the Urea Cycle

We have previously encountered carbamoyl phosphate as a substrate for aspartate transcarbamoylase, the enzyme that catalyzes the first step in pyrimidine biosynthesis (Section 10.1). Carbamoyl phosphate synthetase generates carbamoyl phosphate for both this pathway and the urea cycle. In mammals, two distinct carbamoyl phosphate synthetase isozymes are present. As discussed earlier, the mitochondrial enzyme uses NH_4^+ as the nitrogen source, as is appropriate for its role in the urea cycle. In pyrimidine biosynthesis, carbamoyl phosphate synthetase differs in two important ways (Section 25.1.1). First, this enzyme utilizes glutamine as a nitrogen source. The side chain amide is hydrolyzed within one domain of the enzyme and the ammonia generated moves through a tunnel in the enzyme to react with carboxyphosphate. Second, this enzyme is part of a large polypeptide called CAD that comprises three distinct enzymes: *c*arbamoyl phosphate synthetase, *a*spartate transcarbamoylase, and *d*ihydroorotase, all of which catalyze steps in pyrimidine biosynthesis (Section 25.1). Interestingly, the domain in which glutamine hydrolysis takes place is largely preserved in the urea-cycle enzyme, although that domain is catalytically inactive. This site binds *N*-acetylglutamate, an allosteric activator of the enzyme. *N*-Acetylglutamate is synthesized only if free amino acids are present, an indication that any ammonia generated must be disposed of. *A catalytic site in one isozyme has been adapted to act as an allosteric site in another isozyme having a different physiological role.*

N-Acetylglutamate

What about the other enzymes in the urea cycle? Ornithine transcarbamoylase is homologous to aspartate transcarbamoylase and the structures of their catalytic subunits are quite similar (Figure 23.18). Thus, two consecutive steps in the pyrimidine biosynthetic pathway were adapted for urea synthesis. The next step in the urea cycle is the addition of aspartate to citrulline to form argininosuccinate, and the subsequent step is the removal of fumarate. These two steps together accomplish the net addition of an amino group to citrulline to form arginine. Remarkably, these steps are analogous to two consecutive steps in the purine biosynthetic pathway (Section 25.2.3).

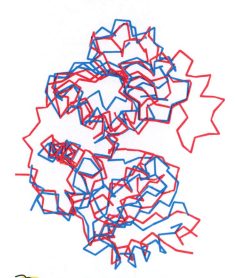

FIGURE 23.18 Homologous enzymes. The structure of the catalytic subunit of ornithine transcarbamoylase (blue) is quite similar to that of the catalytic subunit of aspartate transcarbamoylase (red), indicating that these two enzymes are homologs.

Aspartate

Fumarate

P-ribose   P-ribose   P-ribose

The enzymes that catalyze these steps are homologous to argininosuccinate synthetase and argininosuccinase, respectively. Thus, four of the five enzymes in the urea cycle were adapted from enzymes taking part in nucleotide biosynthesis. The remaining enzyme, arginase, appears to be an ancient enzyme found in all domains of life.

23.4.4 Inherited Defects of the Urea Cycle Cause Hyperammonemia and Can Lead to Brain Damage

The synthesis of urea in the liver is the major route of removal of NH_4^+. A blockage of carbamoyl phosphate synthesis or of any of the four steps of the urea cycle has devastating consequences because there is no alternative pathway for the synthesis of urea. *All defects in the urea cycle lead to an elevated level of NH_4^+ in the blood (hyperammonemia).* Some of these genetic defects become evident a day or two after birth, when the afflicted infant becomes lethargic and vomits periodically. Coma and irreversible brain damage may soon follow. Why are high levels of NH_4^+ toxic? The answer to this question is not yet known. One possibility is that elevated levels of glutamine, formed from NH_4^+ and glutamate (Section 23.3.5), produce osmotic effects that lead directly to brain swelling.

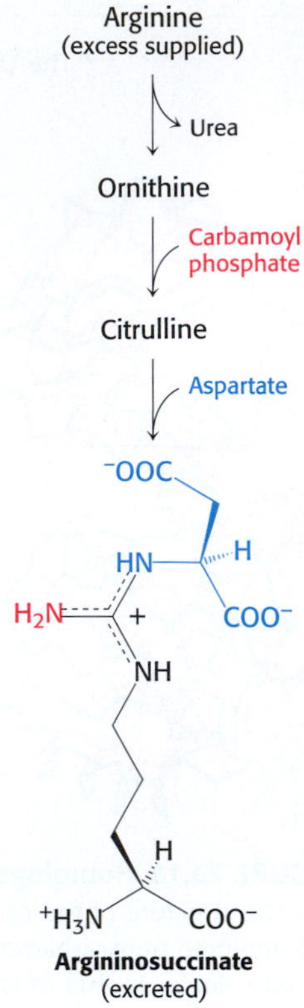

FIGURE 23.19 Treatment of argininosuccinase deficiency. Argininosuccinase deficiency can be managed by supplementing the diet with arginine. Nitrogen is excreted in the form of argininosuccinate.

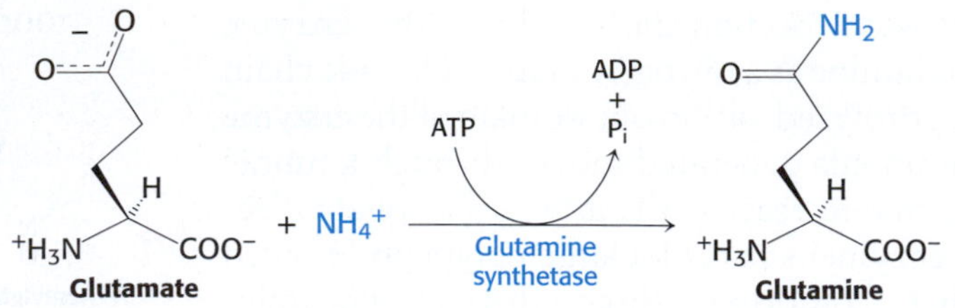

Ingenious strategies for coping with deficiencies in urea synthesis have been devised on the basis of a thorough understanding of the underlying biochemistry. Consider, for example, *argininosuccinase deficiency*. This defect can be partly bypassed by *providing a surplus of arginine in the diet and restricting the total protein intake*. In the liver, arginine is split into urea and ornithine, which then reacts with carbamoyl phosphate to form citrulline (Figure 23.19). This urea-cycle intermediate condenses with aspartate to yield argininosuccinate, which is then excreted. Note that two nitrogen atoms—one from carbamoyl phosphate and another from aspartate—are eliminated from the body per molecule of arginine provided in the diet. In essence, *argininosuccinate substitutes for urea in carrying nitrogen out of the body*.

The treatment of *carbamoyl phosphate synthetase deficiency* or *ornithine transcarbamoylase deficiency* illustrates a different strategy for circumventing a metabolic block. Citrulline and argininosuccinate cannot be used to dispose of nitrogen atoms, because their formation is impaired. Under these conditions, excess nitrogen accumulates in glycine and glutamine. The challenge then is to rid the body of these two amino acids, which is accomplished by supplementing a protein-restricted diet with *large amounts of benzoate and phenylacetate*. Benzoate is activated to benzoyl CoA, which reacts with glycine to form hippurate (Figure 23.20). Likewise, phenylacetate is activated to phenylacetyl CoA, which reacts with glutamine to form phenylacetylglutamine. These conjugates substitute for urea in the disposal of nitrogen. Thus, *latent biochemical pathways can be activated to partly bypass a genetic defect*.

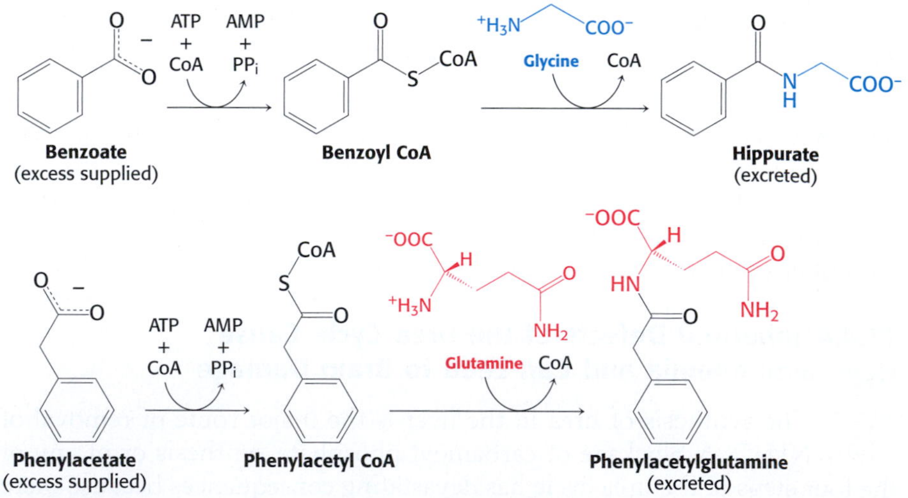

FIGURE 23.20 Treatment of carbamoyl phosphate synthetase and ornithine transcarbamoylase deficiencies. Both deficiencies can be treated by supplementing the diet with benzoate and phenylacetate. Nitrogen is excreted in the form of hippurate and phenylacetylglutamine.

23.4.5 Urea Is Not the Only Means of Disposing of Excess Nitrogen

As discussed earlier, most terrestrial vertebrates are ureotelic; they excrete excess nitrogen as urea. However, urea is not the only excretable form of nitrogen. *Ammoniotelic organisms, such as aquatic vertebrates and invertebrates, release nitrogen as* NH_4^+ *and rely on the aqueous environment to dilute this toxic substance. Interestingly, lungfish, which are normally ammoniotelic, become ureotelic in time of drought, when they live out of the water.

Both ureotelic and ammoniotelic organisms depend on sufficient water, to varying degrees, for nitrogen excretion. *In contrast, uricotelic organisms, which secrete nitrogen as the purine uric acid, require little water.* Disposal of excess nitrogen as uric acid is especially valuable in animals, such as birds, that produce eggs having impermeable membranes that accumulate waste products. The pathway for nitrogen excretion developed in the course of evolution clearly depends on the habitat of the organism.

23.5 CARBON ATOMS OF DEGRADED AMINO ACIDS EMERGE AS MAJOR METABOLIC INTERMEDIATES

We now turn to the fates of the carbon skeletons of amino acids after the removal of the α-amino group. *The strategy of amino acid degradation is to transform the carbon skeletons into major metabolic intermediates that can be converted into glucose or oxidized by the citric acid cycle.* The conversion pathways range from extremely simple to quite complex. The carbon skeletons of the diverse set of 20 fundamental amino acids are funneled into only seven molecules: *pyruvate, acetyl CoA, acetoacetyl CoA, α-ketoglutarate, succinyl CoA, fumarate,* and *oxaloacetate.* We see here a striking example of the remarkable economy of metabolic conversions, as well as an illustration of the importance of certain metabolites.

Amino acids that are degraded to acetyl CoA or acetoacetyl CoA are termed *ketogenic* amino acids because they can give rise to ketone bodies or fatty acids. Amino acids that are degraded to pyruvate, α-ketoglutarate, succinyl CoA, fumarate, or oxaloacetate are termed *glucogenic* amino acids. The net synthesis of glucose from these amino acids is feasible because these citric acid cycle intermediates and pyruvate can be converted into phosphoenolpyruvate and then into glucose (Section 16.3.2). Recall that mammals lack a pathway for the net synthesis of glucose from acetyl CoA or acetoacetyl CoA.

Of the basic set of 20 amino acids, only leucine and lysine are solely ketogenic (Figure 23.21). Isoleucine, phenylalanine, tryptophan, and tyrosine

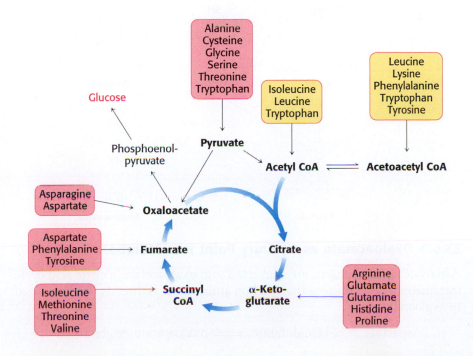

FIGURE 23.21 Fates of the carbon skeletons of amino acids. Glucogenic amino acids are shaded red, and ketogenic amino acids are shaded yellow. Most amino acids are both glucogenic and ketogenic.

are both ketogenic and glucogenic. Some of their carbon atoms emerge in acetyl CoA or acetoacetyl CoA, whereas others appear in potential precursors of glucose. The other 14 amino acids are classed as solely glucogenic. This classification is not universally accepted, because different quantitative criteria are applied. Whether an amino acid is regarded as being glucogenic, ketogenic, or both depends partly on the eye of the beholder. We will identify the degradation pathways by the entry point into metabolism.

23.5.1 Pyruvate as an Entry Point into Metabolism

Pyruvate is the entry point of the three-carbon amino acids—alanine, serine, and cysteine—into the metabolic mainstream (Figure 23.22). The transamination of alanine directly yields pyruvate.

$$\text{Alanine} + \alpha\text{-ketoglutarate} \rightleftharpoons \text{pyruvate} + \text{glutamate}$$

As mentioned previously (Section 23.3.1), glutamate is then oxidatively deaminated, yielding NH_4^+ and regenerating α-ketoglutarate. The sum of these reactions is

$$\text{Alanine} + NAD(P)^+ + H_2O \longrightarrow \text{pyruvate} + NH_4^+ + NAD(P)H + H^+$$

Another simple reaction in the degradation of amino acids is the *deamination of serine to pyruvate* by *serine dehydratase* (Section 23.3.4).

$$\text{Serine} \longrightarrow \text{pyruvate} + NH_4^+$$

Cysteine can be converted into pyruvate by several pathways, with its sulfur atom emerging in H_2S, SCN^-, or SO_3^{2-}.

The carbon atoms of three other amino acids can be converted into pyruvate. *Glycine* can be converted into serine by enzymatic addition of a hydroxymethyl group or it can be cleaved to give CO_2, NH_4^+, and an activated one-carbon unit (Section 24.2.6). *Threonine* can give rise to pyruvate through the intermediate aminoacetone. Three carbon atoms of *tryptophan* can emerge in alanine, which can be converted into pyruvate.

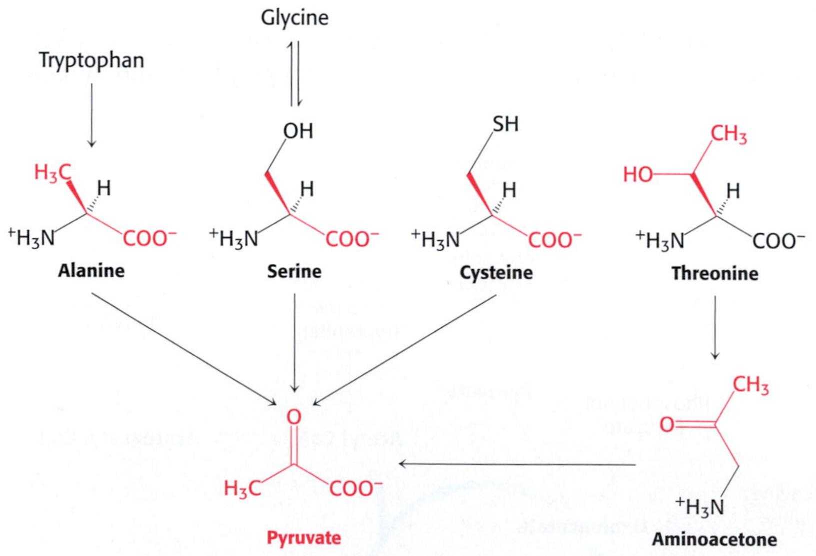

FIGURE 23.22 Pyruvate formation from amino acids. Pyruvate is the point of entry for alanine, serine, cysteine, glycine, threonine, and tryptophan.

23.5.2 Oxaloacetate as an Entry Point into Metabolism

Aspartate and asparagine are converted into oxaloacetate, a citric acid cycle intermediate. *Aspartate*, a four-carbon amino acid, is directly *transaminated to oxaloacetate*.

$$\text{Aspartate} + \alpha\text{-ketoglutarate} \rightleftharpoons \text{oxaloacetate} + \text{glutamate}$$

Asparagine is hydrolyzed by *asparaginase* to NH_4^+ and aspartate, which is then transaminated.

Recall that aspartate can also be converted into *fumarate* by the urea cycle (Section 23.4.2). Fumarate is also a point of entry for half the carbon atoms of tyrosine and phenylalanine, as will be discussed shortly.

23.5.3 Alpha-Ketoglutarate as an Entry Point into Metabolism

The carbon skeletons of several five-carbon amino acids enter the citric acid cycle at α-ketoglutarate. These amino acids are first converted into *glutamate,* which is then oxidatively deaminated by glutamate dehydrogenase to yield α-ketoglutarate (Figure 23.23).

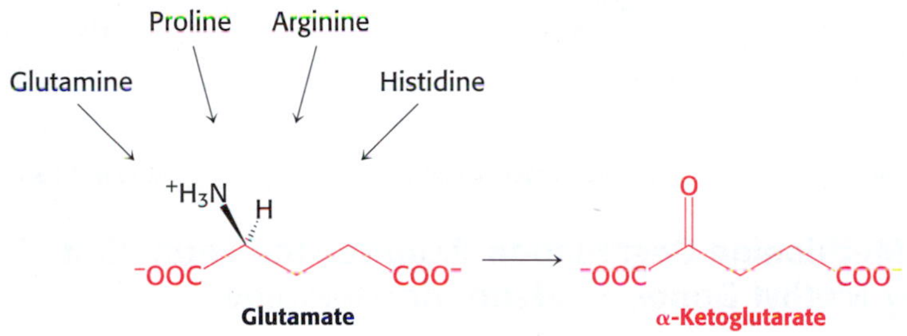

FIGURE 23.23 α-Ketoglutarate formation from amino acids.
α-Ketoglutarate is the point of entry of several five-carbon amino acids that are first converted into glutamate.

Histidine is converted into 4-imidazolone 5-propionate (Figure 23.24). The amide bond in the ring of this intermediate is hydrolyzed to the *N*-formimino derivative of glutamate, which is then converted into glutamate by transfer of its formimino group to tetrahydrofolate, a carrier of activated one-carbon units (Section 24.2.6).

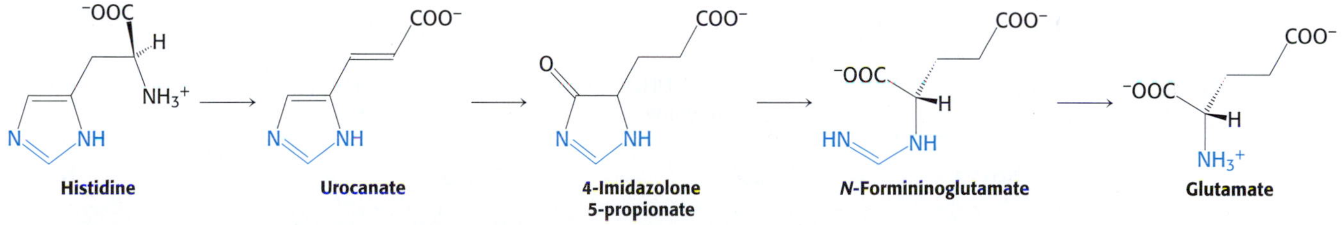

Glutamine is hydrolyzed to glutamate and NH_4^+ by *glutaminase. Proline* and *arginine* are each converted into glutamate γ-semialdehyde, which is then oxidized to glutamate (Figure 23.25).

FIGURE 23.24 Histidine degradation.
Conversion of histidine into glutamate.

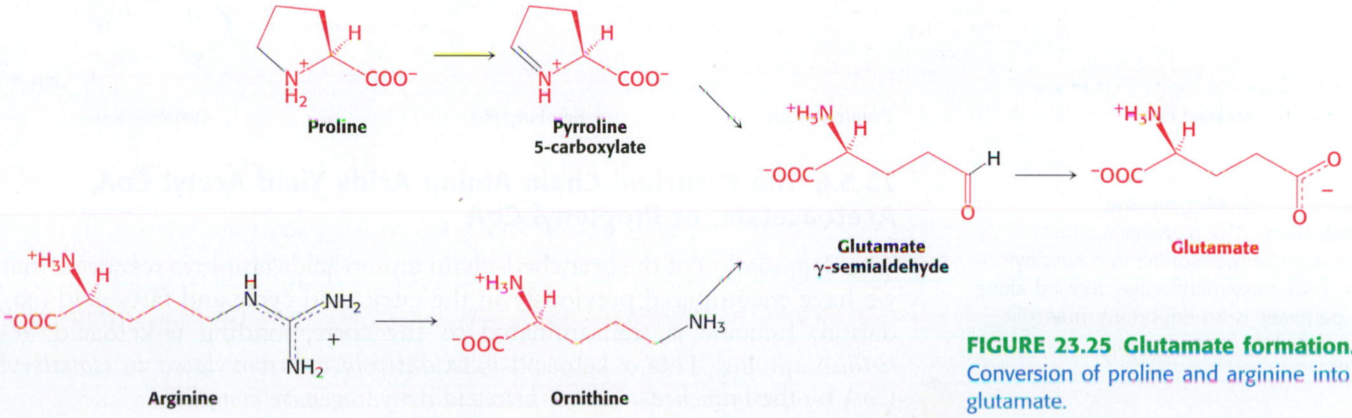

FIGURE 23.25 Glutamate formation.
Conversion of proline and arginine into glutamate.

23.5.4 Succinyl Coenzyme A Is a Point of Entry for Several Nonpolar Amino Acids

Succinyl CoA is a point of entry for some of the carbon atoms of methionine, isoleucine, and valine. Propionyl CoA and then methylmalonyl CoA are intermediates in the breakdown of these three nonpolar amino acids (Figure 23.26). The mechanism for the interconversion of propionyl CoA and methylmalonyl CoA was presented in Section 22.3.3. This pathway from propionyl CoA to succinyl CoA is also used in the oxidation of fatty acids that have an odd number of carbon atoms (Section 22.3.2).

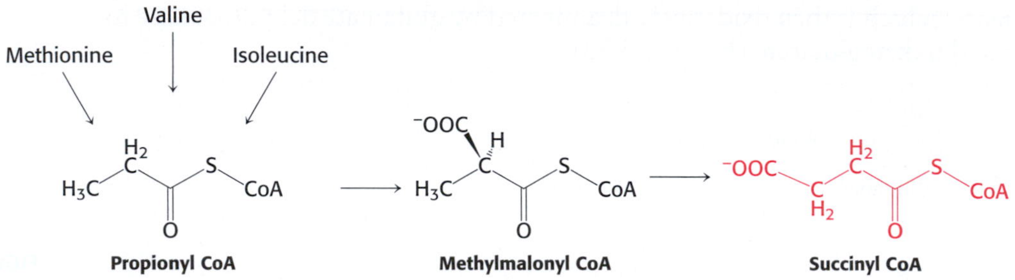

FIGURE 23.26 Succinyl CoA formation. Conversion of methionine, isoleucine, and valine into succinyl CoA.

23.5.5 Methionine Degradation Requires the Formation of a Key Methyl Donor, S-Adenosylmethionine

Methionine is converted into succinyl CoA in nine steps (Figure 24.2.7). The first step is the adenylation of methionine to form S-adenosylmethionine (SAM), a common methyl donor in the cell (Section 24.2.7). Methyl donation and deadenylation yield homocysteine, which is eventually processed to α-ketobutyrate. The enzyme α-ketoacid dehydrogenase complex oxidatively decarboxylates α-ketobutyrate to propionyl CoA, which is processed to succinyl CoA as described in Section 23.3.3.

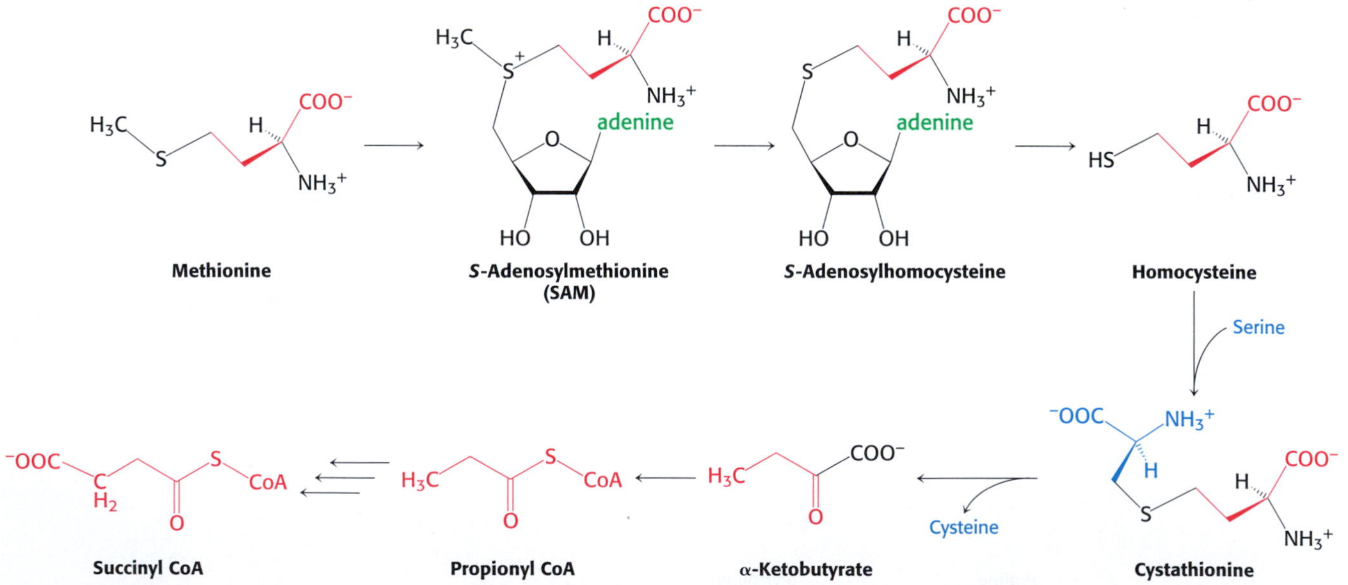

FIGURE 23.27 Methionine metabolism. The pathway for the conversion of methionine into succinyl CoA. S-Adenosylmethionine, formed along this pathway, is an important molecule for transferring methyl groups.

23.5.6 The Branched-Chain Amino Acids Yield Acetyl CoA, Acetoacetate, or Propionyl CoA

The degradation of the branched-chain amino acids employs reactions that we have encountered previously in the citric acid cycle and fatty acid oxidation. Leucine is transaminated to the corresponding α-ketoacid, α-ketoisocaproate. This α-ketoacid is *oxidatively decarboxylated to isovaleryl CoA by the branched-chain α-ketoacid dehydrogenase complex.*

Leucine → **α-Ketoisocaproate** → CoA **Isovaleryl CoA**

The α-ketoacids of valine and isoleucine, the other two branched-chain aliphatic amino acids, as well as α-ketobutyrate derived from methionine also are substrates. The oxidative decarboxylation of these α-ketoacids is analogous to that of pyruvate to acetyl CoA and of α-ketoglutarate to succinyl CoA. The branched-chain α-ketoacid dehydrogenase, a multienzyme complex, is a homolog of pyruvate dehydrogenase (Section 17.1.1) and α-ketoglutarate dehydrogenase (Section 17.1.6). Indeed, the E3 components of these enzymes, which regenerate the oxidized form of lipoamide, are identical.

The isovaleryl CoA derived from leucine is *dehydrogenated* to yield *β-methylcrotonyl CoA*. This oxidation is catalyzed by *isovaleryl CoA dehydrogenase*. The hydrogen acceptor is FAD, as in the analogous reaction in fatty acid oxidation that is catalyzed by acyl CoA dehydrogenase. *β-Methylglutaconyl CoA* is then formed by the *carboxylation* of β-methylcrotonyl CoA at the expense of the hydrolysis of a molecule of ATP. As might be expected, the carboxylation mechanism of β-methylcrotonyl CoA carboxylase is similar to that of pyruvate carboxylase and acetyl CoA carboxylase.

Isovaleryl CoA — FAD → FADH₂ — **β-Methylcrotonyl CoA** — CO₂ + ATP → ADP + Pᵢ — **β-Methylglutaconyl CoA**

β-Methylglutaconyl CoA is then *hydrated* to form *3-hydroxy-3-methylglutaryl CoA*, which is cleaved into *acetyl CoA* and *acetoacetate*. This reaction has already been discussed in regard to the formation of ketone bodies from fatty acids (Section 22.3.5).

β-Methylglutaconyl CoA — H₂O → **3-Hydroxy-3-methylglutaryl CoA** → **Acetyl CoA** + **Acetoacetate**

The degradative pathways of valine and isoleucine resemble that of leucine. After transamination and oxidative decarboxylation to yield a CoA derivative, the subsequent reactions are like those of fatty acid oxidation. Isoleucine yields acetyl CoA and propionyl CoA, whereas valine yields CO_2 and propionyl CoA. The degradation of leucine, valine, and isoleucine validate a point made earlier (Chapter 14): the number of reactions in metabolism is large, but the number of *kinds* of reactions is relatively small. The degradation of leucine, valine, and isoleucine provides a striking illustration of the underlying simplicity and elegance of metabolism.

23.5.7 Oxygenases Are Required for the Degradation of Aromatic Amino Acids

The degradation of the aromatic amino acids is not as straightforward as that of the amino acids previously discussed, although the final products—acetoacetate, fumarate, and pyruvate—are common intermediates. For the aromatic amino acids, *molecular oxygen is used to break an aromatic ring*.

The degradation of phenylalanine begins with its hydroxylation to tyrosine, a reaction catalyzed by *phenylalanine hydroxylase*. This enzyme is called a *monooxygenase* (or *mixed-function oxygenase*) because *one atom of O_2 appears in the product and the other in H_2O*.

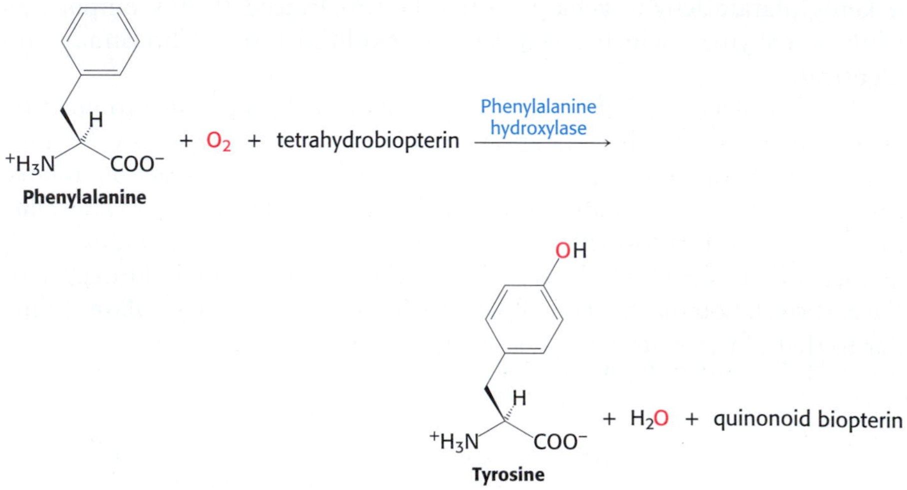

The reductant here is *tetrahydrobiopterin,* an electron carrier that has not been previously discussed and is derived from the cofactor *biopterin*. Because biopterin is synthesized in the body, it is not a vitamin. Tetrahydrobiopterin is initially formed by the reduction of dihydrobiopterin by NADPH in a reaction catalyzed by *dihydrofolate reductase* (Figure 23.28).

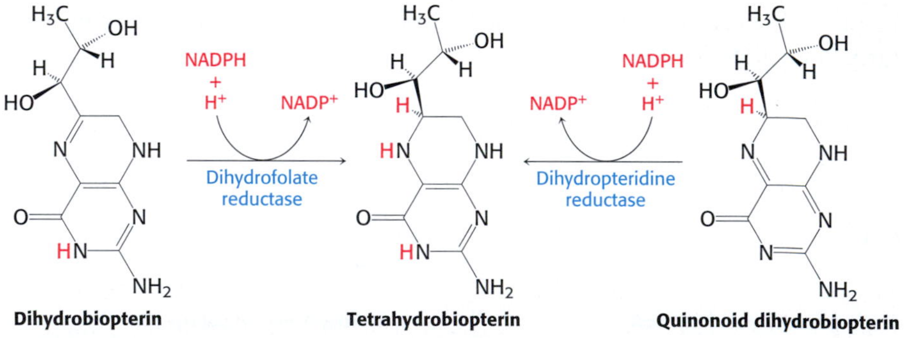

FIGURE 23.28 Formation of tetrahydrobiopterin, an important electron carrier. Tetrahydrobiopterin can be formed by the reduction of either of two forms of dihydrobiopterin.

NADH reduces the quinonoid form of dihydrobiopterin produced in the hydroxylation of phenylalanine back to tetrahydrobiopterin in a reaction catalyzed by *dihydropteridine reductase*. The sum of the reactions catalyzed by phenylalanine hydroxylase and dihydropteridine reductase is

$$\text{Phenylalanine} + O_2 + \text{NADH} + H^+ \longrightarrow \text{tyrosine} + \text{NAD}^+ + H_2O$$

Note that these reactions can also be used to synthesize tyrosine from phenylalanine.

The next step in the degradation of phenylalanine and tyrosine is the transamination of tyrosine to p-*hydroxyphenylpyruvate* (Figure 23.29). This

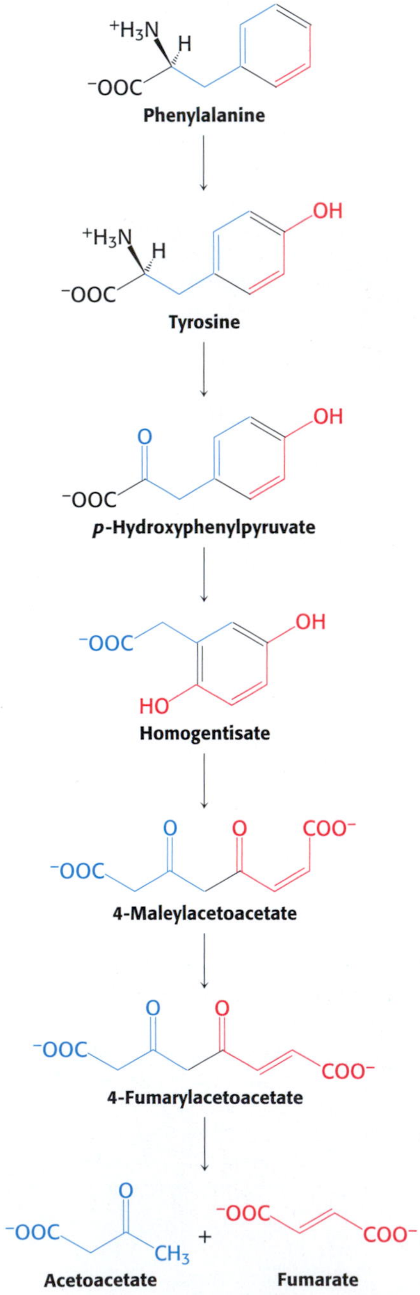

FIGURE 23.29 Phenylalanine and tyrosine degradation. The pathway for the conversion of phenylalanine into acetoacetate and fumarate.

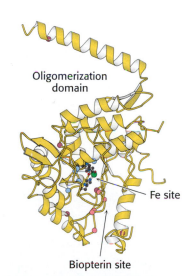

FIGURE 23.30 Tryptophan degradation. The pathway for the conversion of tryptophan into alanine and acetoacetate.

α-ketoacid then reacts with O_2 to form *homogentisate.* The enzyme catalyzing this complex reaction, p-*hydroxyphenylpyruvate hydroxylase,* is called a *dioxygenase* because *both atoms of O_2 become incorporated into the product,* one on the ring and one in the carboxyl group. The aromatic ring of homogentisate is then cleaved by O_2, which yields 4-maleylacetoacetate. This reaction is catalyzed by *homogentisate oxidase,* another dioxygenase. 4-Maleylacetoacetate is then isomerized to *4-fumarylacetoacetate* by an enzyme that uses glutathione as a cofactor. Finally, 4-fumarylacetoacetate is hydrolyzed to *fumarate* and *acetoacetate.*

Tryptophan degradation requires several oxygenases (Figure 23.30). Tryptophan 2,3-dioxygenase cleaves the pyrrole ring, and kynureinine 3-monooxygenase hydroxylates the remaining benzene ring, a reaction similar to the hydroxylation of phenylalanine to form tyrosine. Alanine is removed and the 3-hydroxyanthranilic acid is cleaved with another dioxygenase and subsequently processed to acetoacetyl CoA (Figure 23.31). *Nearly all cleavages of aromatic rings in biological systems are catalyzed by dioxygenases,* a class of enzymes discovered by Osamu Hayaishi. The active sites of these enzymes contain iron that is not part of heme or an iron–sulfur cluster.

FIGURE 23.31 Structure of one subunit of phenylalanine hydroxylase. Mutations in the genes encoding this enzyme cause phenylketonuria. More than 200 point mutations have been identified in these genes. The positions of five mutations affecting the active site (blue), the biopterin-binding site (red), and other regions of the protein (purple) are indicated as colored spheres.

23.6 INBORN ERRORS OF METABOLISM CAN DISRUPT AMINO ACID DEGRADATION

Errors in amino acid metabolism provided some of the first correlations between biochemical defects and pathological conditions. For instance, *alcaptonuria* is an inherited metabolic disorder caused by the absence of homogentisate oxidase. In 1902, Archibald Garrod showed that alcaptonuria is transmitted as a single recessive Mendelian trait. Furthermore, he recognized that homogentisate is a normal intermediate in the

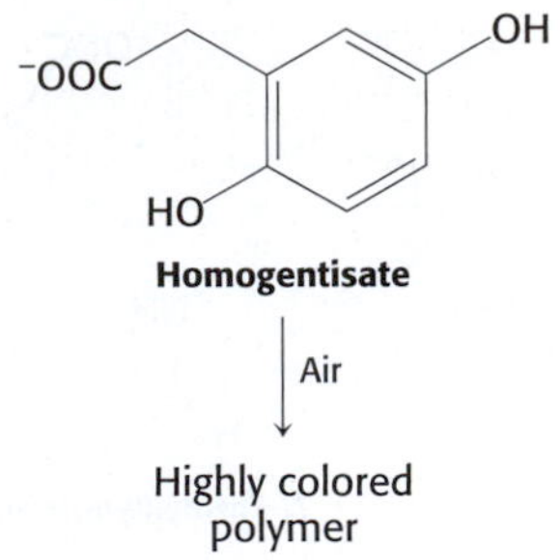

Homogentisate

Air

Highly colored
polymer

degradation of phenylalanine and tyrosine (see Figure 23.29) and that it accumulates in alcaptonuria because its degradation is blocked. He concluded that "the splitting of the benzene ring in normal metabolism is the work of a special enzyme, that in congenital alcaptonuria this enzyme is wanting." Homogentisate accumulates and is excreted in the urine, which turns dark on standing as homogentisate is oxidized and polymerized to a melanin-like substance.

Although alcaptonuria is a relatively harmless condition, such is not the case with other errors in amino acid metabolism. In *maple syrup urine disease,* the oxidative decarboxylation of α-ketoacids derived from valine, isoleucine, and leucine is blocked because the branched-chain dehydrogenase is missing or defective. Hence, the levels of these α-ketoacids and the branched-chain amino acids that give rise to them are markedly elevated in both blood and urine. Indeed, the urine of patients has the odor of maple syrup—hence the name of the disease (also called *branched-chain keto-aciduria*). Maple syrup urine disease usually leads to mental and physical retardation unless the patient is placed on a diet low in valine, isoleucine, and leucine early in life. The disease can be readily detected in newborns by screening urine samples with 2,4-dinitrophenylhydrazine, which reacts with α-ketoacids to form 2,4-dinitrophenylhydrazone derivatives. A definitive diagnosis can be made by mass spectrometry.

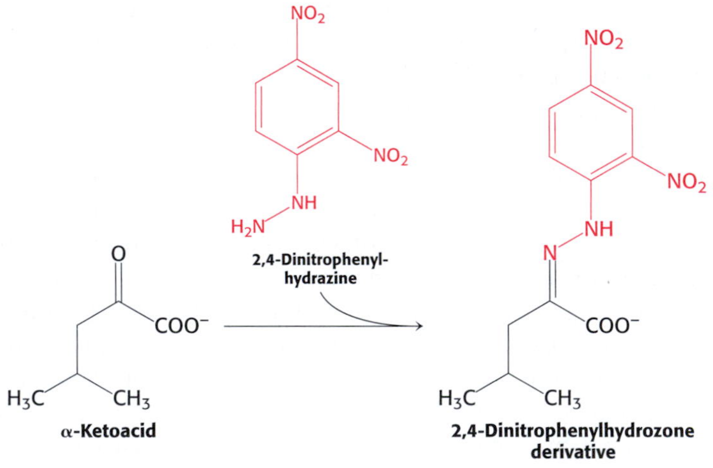

α-Ketoacid → 2,4-Dinitrophenyl-hydrazine → 2,4-Dinitrophenylhydrozone derivative

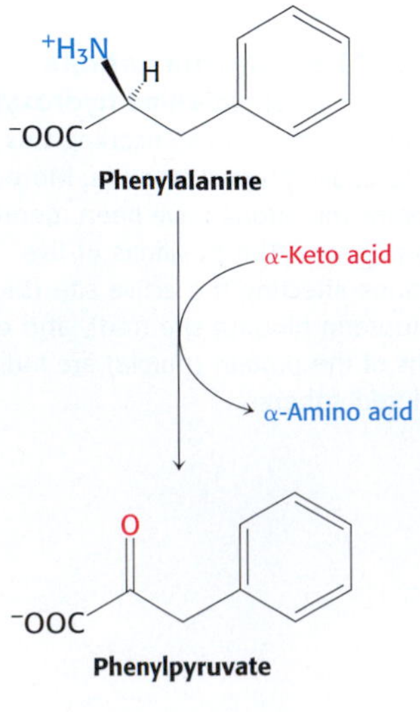

Phenylalanine

α-Keto acid

α-Amino acid

Phenylpyruvate

Phenylketonuria is perhaps the best known of the diseases of amino acid metabolism. Phenylketonuria is caused by an *absence or deficiency of phenylalanine hydroxylase* or, more rarely, of its tetrahydrobiopterin cofactor. *Phenylalanine accumulates in all body fluids because it cannot be converted into tyrosine.* Normally, three-quarters of the phenylalanine is converted into tyrosine, and the other quarter becomes incorporated into proteins. Because the major outflow pathway is blocked in phenylketonuria, the blood level of phenylalanine is typically at least 20-fold as high as in normal people. Minor fates of phenylalanine in normal people, such as the formation of phenylpyruvate, become major fates in phenylketonurics.

Indeed, the initial description of phenylketonuria in 1934 was made by observing the reaction of phenylpyruvate with $FeCl_3$, which turns the urine olive green. *Almost all untreated phenylketonurics are severely mentally retarded.* In fact, about 1% of patients in mental institutions have phenylketonuria. The brain weight of these people is below normal, myelination of their nerves is defective, and their reflexes are hyperactive. The life expectancy of untreated phenylketonurics is drastically shortened. Half are dead by age 20 and three-quarters by age 30. *The biochemical basis of their mental retardation is an enigma.*

Phenylketonurics appear normal at birth, but are severely defective by age 1 if untreated. The therapy for phenylketonuria is a *low phenylalanine diet*. The aim is to provide just enough phenylalanine to meet the needs for growth and replacement. Proteins that have a low content of phenylalanine, such as casein from milk, are hydrolyzed and phenylalanine is removed by adsorption. A low phenylalanine diet must be started very soon after birth to prevent irreversible brain damage. In one study, the average IQ of phenylketonurics treated within a few weeks after birth was 93; a control group treated starting at age 1 had an average IQ of 53.

Early diagnosis of phenylketonuria is essential and has been accomplished by mass screening programs. The phenylalanine level in the blood is the preferred diagnostic criterion because it is more sensitive and reliable than the $FeCl_3$ test. Prenatal diagnosis of phenylketonuria with DNA probes has become feasible because the gene has been cloned and many mutations have been pinpointed to sites in the protein (see Figure 23.31). Interestingly, whereas some mutations affect the activity of the enzyme, others do not affect the activity itself but, instead, decrease the enzyme concentration. These mutations lead to degradation of the enzyme, at least in part by the ubiquitin–proteasome pathway.

The incidence of phenylketonuria is about 1 in 20,000 newborns. The disease is inherited in an *autosomal recessive* manner. Heterozygotes, who make up about 1.5% of a typical population, appear normal. Carriers of the phenylketonuria gene have a reduced level of phenylalanine hydroxylase, as indicated by an increased level of phenylalanine in the blood. However, this criterion is not absolute, because the blood levels of phenylalanine in carriers and normal people overlap to some extent. The measurement of the kinetics of the disappearance of intravenously administered phenylalanine is a more definitive test for the carrier state. It should be noted that a high blood level of phenylalanine in a pregnant woman can result in abnormal development of the fetus. This is a striking example of maternal–fetal relationships at the molecular level. Table 23.3 lists some other diseases of amino acid metabolism.

TABLE 23.3 Inborn errors of amino acid metabolism

Disease	Enzyme deficiency	Symptoms
Citrullinema	Arginosuccinate lyase	Lethergy, siezures, reduced muscle tension
Tyrosinemia	Various enzymes of tyrosine degradation	Weakness, self-mutilation, liver damage, mental retardation
Albinism	Tyrosinase	Absence of pigmentation
Homocystinuria	Cystathionine β-synthase	Scoliosis, muscle weakness, mental retardation, thin blond hair
Hyperlysinemia	α-Aminoadipic semialdehyde dehydrogenase	Seizures, mental retardation, lack of muscle tone, ataxia

SUMMARY

Proteins Are Degraded to Amino Acids

Dietary protein is digested in the intestine, producing amino acids that are transported throughout the body. Cellular proteins are degraded at widely variable rates, ranging from minutes to the life of the organism.

Protein Turnover Is Tightly Regulated

The turnover of cellular proteins is a regulated process requiring complex enzyme systems. Proteins to be degraded are conjugated with ubiquitin, a small conserved protein, in a reaction driven by ATP hydrolysis. The ubiquitin conjugating system is composed of three distinct enzymes. A large, barrel-shaped complex called the proteasome digests the ubiquitinated proteins. The proteasome also requires ATP hydrolysis to function. The resulting amino acids provide a source of precursors for protein, nucleotide bases, and other nitrogenous compounds.

The First Step in Amino Acid Degradation Is the Removal of Nitrogen

Surplus amino acids are used as metabolic fuel. The first step in their degradation is the removal of their α-amino groups by transamination to an α-ketoacid. Pyridoxal phosphate is the coenzyme in all aminotransferases and in many other enzymes catalyzing amino acids transformations. The α-amino group funnels into α-ketoglutarate to form glutamate, which is then oxidatively deaminated by glutamate dehydrogenase to give NH_4^+ and α-ketoglutarate. NAD^+ or $NADP^+$ is the electron acceptor in this reaction.

Ammonium Ion Is Converted into Urea in Most Terrestrial Vertebrates

The first step in the synthesis of urea is the formation of carbamoyl phosphate, which is synthesized from CO_2, NH_4^+, and two molecules of ATP by carbamoyl phosphate synthetase. Ornithine is then carbamoylated to citrulline by orthinine transcarbamoylase. These two reactions take place in mitochondria. Citrulline leaves the mitochondrion and condenses with aspartate to form argininosuccinate, which is cleaved into arginine and fumarate. The other nitrogen atom of urea comes from aspartate. Urea is formed by the hydrolysis of arginine, which also regenerates ornithine. Some enzymatic deficiencies of the urea cycle can be bypassed by supplementing the diet with arginine or compounds that form conjugates with glycine and glutamine.

Carbon Atoms of Degraded Amino Acids Emerge as Major Metabolic Intermediates

The carbon atoms of degraded amino acids are converted into pyruvate, acetyl CoA, acetoacetate, or an intermediate of the citric acid cycle. Most amino acids are solely glucogenic, two are solely ketogenic, and a few are both ketogenic and glucogenic. Alanine, serine, cysteine, glycine, threonine, and tryptophan are degraded to pyruvate. Asparagine and aspartate are converted into oxaloacetate. α-Ketoglutarate is the point of entry for glutamate and four amino acids (glutamine, histidine, proline, and arginine) that can be converted into glutamate. Succinyl CoA is the point of entry for some of the carbon atoms of three amino acids (methionine, isoleucine, and valine) that are degraded through the intermediate methylmalonyl CoA. Leucine is degraded to acetoacetyl CoA and acetyl CoA. The breakdown of valine and isoleucine is like that of leucine. Their α-ketoacid derivatives are oxidatively decarboxylated by the branched-chain α-ketoacid dehydrogenase.

The rings of aromatic amino acids are degraded by oxygenases. Phenylalanine hydroxylase, a monooxygenase, uses tetrahydrobiopterin as the reductant. One of the oxygen atoms of O_2 emerges in tyrosine and the other in water. Subsequent steps in the degradation of these aromatic amino acids are catalyzed by dioxygenases, which catalyze the insertion of both atoms of O_2 into organic products. Four of the carbon atoms of phenylalanine and tyrosine are converted into fumarate, and four emerge in acetoacetate.

● Inborn Errors of Metabolism Can Disrupt Amino Acid Degradation

Errors in amino acid metabolism served as sources of some of the first insights into the correlation between pathology and biochemistry. Although there are many hereditary errors of amino acid metabolism, phenylketonuria is the best known. This condition is the result of the accumulation of high levels of phenylalanine in the body fluids. By unknown mechanisms, this accumulation results in mental retardation unless the afflicted are placed on low phenylalanine diets immediately after birth.

KEY TERMS

ubiquitin (p. 635)

proteasome (p. 636)

aminotransferase (transaminase) (p. 639)

glutamate dehydrogenase (p. 640)

pyridoxal phosphate (PLP) (p. 640)

pyridoxamine phosphate (PMP) (p. 641)

alanine cycle (p. 644)

urea cycle (p. 644)

carbamoyl phosphate synthetase (p. 645)

N-acetylglutamate (p. 645)

ketogenic amino acid (p. 649)

glucogenic amino acid (p. 649)

biopterin (p. 654)

phenylketonuria (p. 656)

SELECTED READINGS

Where to start

Torchinsky, Y. M., 1989. Transamination: Its discovery, biological and chemical aspects. *Trends Biochem. Sci.* 12:115–117.

Eisensmith, R. C., and Woo, S. L. C., 1991. Phenylketonuria and the phenylalanine hydroxylase gene. *Mol. Biol. Med.* 8:3–18.

Levy, H. L., 1989. Nutritional therapy for selected inborn errors of metabolism. *J. Am. Coll. Nutr.* 8:54S–60S.

Schwartz, A. L., and Ciechanover, A., 1999. The ubiquitin-proteasome pathway and pathogenesis of human diseases. *Annu. Rev. Med.* 50: 57–74.

Books

Bender, D. A., 1985. *Amino Acid Metabolism* (2d ed.). Wiley.

Lippard, S. J., and Berg, J. M., 1994. *Principles of Bioinorganic Chemistry.* University Science Books.

Schauder, P., Wahren, J., Paoletti, R., Bernardi, R., and Rinetti, M. (Eds.), 1992. *Branched-Chain Amino Acids: Biochemistry, Physiopathology, and Clinical Sciences.* Raven Press.

Grisolia, S., Báguena, R., and Mayor, F. (Eds.), 1976. *The Urea Cycle.* Wiley.

Walsh, C., 1979. *Enzymatic Reaction Mechanisms.* W. H. Freeman and Company.

Christen, P., and Metzler, D. E., 1985. *Transaminases.* Wiley.

Ubiquitin and the proteasome

Bochtler, M., Ditzel, L., Groll, M., Hartmann, C., and Huber, R., 1999. The proteasome. *Annu. Rev. Biophys. Biomol. Struct.* 28:295–317.

Thrower, J. S., Hoffman, L., Rechsteiner, M., and Pickart, C. M., 2000. Recognition of the polyubiquitin proteolytic signal. *EMBO J.* 19:94–102.

Hochstrasser, M., 2000. Evolution and function of ubiquitin-like protein-conjugation systems. *Nat. Cell Biol.* 2:E153–E157.

Jentsch, S., and Pyrowolakis, G., 2000. Ubiquitin and its kin: How close are the family ties? *Trends Cell Biol.* 10:335–342.

Laney, J. D., and Hochstrasser, M., 1999. Substrate targeting in the ubiquitin system. *Cell* 97:427–430.

Cook, W. J., Jeffrey, L. C., Kasperek, E., and Pickart, C. M., 1994. Structure of tetraubiquitin shows how multiubiquitin chains can be formed. *J. Mol. Biol.* 236:601–609.

Hartmann-Petersen, R., Tanaka, K., and Hendil, K. B., 2001. Quaternary structure of the ATPase complex of human 26S proteasomes determined by chemical cross-linking. *Arch. Biochem. Biophys.* 386: 89–94.

Pyridoxal phosphate-dependent enzymes

Mehta, P. K., and Christen, P., 2000. The molecular evolution of pyridoxal-5'-phosphate-dependent enzymes. *Adv. Enzymol. Relat. Areas Mol. Biol.* 74:129–184.

Schneider, G., Kack, H., and Lindqvist, Y., 2000. The manifold of vitamin B_6 dependent enzymes. *Structure Fold Des.* 8:R1–R6.

Jager, J., Moser, M., Sauder, U., and Jansonius, J. N., 1994. Crystal structures of *Escherichia coli* aspartate aminotransferase in two conformations: Comparison of an unliganded open and two liganded closed forms. *J. Mol. Biol.* 239:285–305.

Malashkevich, V. N., Toney, M. D., and Jansonius, J. N., 1993. Crystal structures of true enzymatic reaction intermediates: Aspartate and glutamate ketimines in aspartate aminotransferase. *Biochemistry* 32:13451–13462.

McPhalen, C. A., Vincent, M. G., Picot, D., Jansonius, J. N., Lesk, A. M., and Chothia, C., 1992. Domain closure in mitochondrial aspartate aminotransferases. *J. Mol. Biol.* 227:197–213.

Urea-cycle enzymes

Huang, X., and Raushel, F. M., 2000. Restricted passage of reaction intermediates through the ammonia tunnel of carbamoyl phosphate synthetase. *J. Biol. Chem.* 275:26233–26240.

Lawson, F. S., Charlebois, R. L., and Dillon, J. A., 1996. Phylogenetic analysis of carbamoylphosphate synthetase genes: Complex evolutionary history includes an internal duplication within a gene which can root the tree of life. *Mol. Biol. Evol.* 13:970–977.

McCudden, C. R., and Powers-Lee, S. G., 1996. Required allosteric effector site for N-acetylglutamate on carbamoyl-phosphate synthetase I. *J. Biol. Chem.* 271:18285–18294.

Kanyo, Z. F., Scolnick, L. R., Ash, D. E., and Christianson, D. W., 1996. Structure of a unique binuclear manganese cluster in arginase. *Nature* 383:554–557.

Turner, M. A., Simpson, A., McInnes, R. R., and Howell, P. L., 1997. Human argininosuccinate lyase: A structural basis for intragenic complementation. *Proc. Natl. Acad. Sci. USA* 94:9063–9068.

Amino acid degradation

Fusetti, F., Erlandsen, H., Flatmark, T., and Stevens, R. C., 1998. Structure of tetrameric human phenylalanine hydroxylase and its implications for phenylketonuria. *J. Biol. Chem.* 273:16962–16967.

Sugimoto, K., Senda, T., Aoshima, H., Masai, E., Fukuda, M., and Mitsui, Y., 1999. Crystal structure of an aromatic ring opening dioxygenase LigAB, a protocatechuate 4,5-dioxygenase, under aerobic conditions. *Structure Fold Des.* 7:953–965.

Titus, G. P., Mueller, H. A., Burgner, J., Rodriguez De Cordoba, S., Penalva, M. A., and Timm, D. E., 2000. Crystal structure of human homogentisate dioxygenase. *Nat. Struct. Biol.* 7:542–546.

Erlandsen, H., and Stevens, R. C., 1999. The structural basis of phenylketonuria. *Mol. Genet. Metab.* 68:103–125.

Genetic diseases

Striver, C. R., Beaudet, A. L., Sly, W. S., Valle, D., Stanbury, J. B., Wyngaarden, J. B., and Fredrickson, D. S. (Eds.), 1995. *The Metabolic Basis of Inherited Diseases* (7th ed.). McGraw-Hill.

Nyhan, W. L. (Ed.), 1984. *Abnormalities in Amino Acid Metabolism in Clinical Medicine.* Appleton-Century-Crofts.

Historical aspects and the process of discovery

Cooper, A. J. L., and Meister, A., 1989. An appreciation of Professor Alexander E. Braunstein: The discovery and scope of enzymatic transamination. *Biochimie* 71:387–404.

Garrod, A. E., 1909. *Inborn Errors in Metabolism.* Oxford University Press (reprinted in 1963 with a supplement by H. Harris).

Childs, B., 1970. Sir Archibald Garrod's conception of chemical individuality: A modern appreciation. *N. Engl. J. Med.* 282:71–78.

Holmes, F. L., 1980. Hans Krebs and the discovery of the ornithine cycle. *Fed. Proc.* 39:216–225.

PROBLEMS

1. *Wasted energy?* Protein hydrolysis is an exergonic process, yet the 26S proteasome is dependent on ATP hydrolysis for activity.

(a) Although the exact function of the ATPase activity is not known, suggest some likely functions.
(b) Small-peptides can be hydrolyzed without the expenditure of ATP. How does this information concur with your answer to part *a*?

2. *Keto counterparts.* Name the α-ketoacid that is formed by transamination of each of the following amino acids:

(a) Alanine (d) Leucine
(b) Aspartate (e) Phenylalanine
(c) Glutamate (f) Tyrosine

3. *A versatile building block.* (a) Write a balanced equation for the conversion of aspartate into glucose through the intermediate oxaloacetate. Which coenzymes participate in this transformation? (b) Write a balanced equation for the conversion of aspartate into oxaloacetate through the intermediate fumarate.

4. *The benefits of specialization.* The archaeal proteasome contains 14 identical active β subunits, whereas the eukaryotic proteasome has 7 distinct β subunits. What are the potential benefits of having several distinct active subunits?

5. *Propose a structure.* The 19S subunit of the proteasome contains 6 subunits that are members of the AAA ATPase family. Other members of this large family are associated into homohexamers with six-fold symmetry. Propose a structure for the AAA ATPases within the 19S proteasome. How might you test and refine your prediction?

6. *Effective electron sinks.* Pyridoxal phosphate stabilizes carbanionic intermediates by serving as an electron sink. Which other prosthetic group catalyzes reactions in this way?

7. *Helping hand.* Propose a role for the positively charged guanidinium nitrogen atom in the cleavage of argininosuccinate into arginine and fumarate.

8. *Completing the cycle.* Four high-transfer-potential phosphoryl groups are consumed in synthesizing urea according to the stoichiometry given in Section 23.4.2. In this reaction, aspartate is converted into fumarate. Suppose that fumarate is converted back into aspartate. What is the resulting stoichiometry of urea synthesis? How many high-transfer-potential phosphoryl groups are spent?

9. *Inhibitor design.* Compound A has been synthesized as a potential inhibitor of a urea-cycle enzyme. Which enzyme do you think compound A might inhibit?

Compound A

10. *Ammonia toxicity.* Glutamate is an important neurotransmitter whose levels must be carefully regulated in the brain. Explain how a high concentration of ammonia might disrupt this regulation. How might a high concentration of ammonia alter the citric acid cycle?

11. *A precise diagnosis.* The urine of an infant gives a positive reaction with 2,4-dinitrophenylhydrazine. Mass spectrometry shows abnormally high blood levels of pyruvate, α-ketoglutarate, and the α-ketoacids of valine, isoleucine, and leucine. Identify a likely molecular defect and propose a definitive test of your diagnosis.

12. *Therapeutic design.* How would you treat an infant who is deficient in argininosuccinate synthetase? Which molecules would carry nitrogen out of the body?

13. *Sweet hazard.* Why should phenylketonurics avoid using aspartame, an artificial sweetener? (Hint: Aspartame is L-aspartyl-L-phenylalanine methyl ester.)

14. *Déjà vu.* N-Acetylglutamate is required as a cofactor in the synthesis of carbamoyl phosphate. How might N-acetylglutamate be synthesized from glutamate?

Mechanism Problems

15. *Serine dehydratase.* Write out a complete mechanism for the conversion of serine into aminoacrylate catalyzed by serine dehydratase.

16. *Serine racemase.* The nervous system contains a substantial amount of D-serine, which is generated from L-serine by serine racemase, a PLP-dependent enzyme. Propose a mechanism for this reaction. What is the equilibrium constant for the reaction L-serine $\rightleftharpoons$ D-serine?

Chapter Integration Problems

17. *Double duty.* Degradation signals are commonly located in protein regions that also facilitate protein–protein interactions. Explain why this coexistence of two functions in the same domain might be useful.

18. *Fuel choice.* Within a few days after a fast begins, nitrogen excretion accelerates to a higher-than-normal level. After a few weeks, the rate of nitrogen excretion falls to a lower level and continues at this low rate. However, after the fat stores have been depleted, nitrogen excretion rises to a high level.

(a) What events trigger the initial surge of nitrogen excretion?
(b) Why does nitrogen excretion fall after several weeks of fasting?
(c) Explain the increase in nitrogen excretion when the lipid stores have been depleted.

19. *Isoleucine degradation.* Isoleucine is degraded to acetyl CoA and succinyl CoA. Suggest a plausible reaction sequence, based on reactions discussed in the text, for this degradation pathway.

Data Interpretation Problem

20. *Another helping hand.* In eukaryotes, the 20S proteasome in conjunction with the 19S component degrades ubiquitinated proteins with the hydrolysis of a molecule of ATP. Archaea lack ubiquitin and the 26S proteasome but do contain a 20S proteasome. Some archaea also contain an ATPase that is homologous to the ATPases of the eukaryotic 19S component. This archaeal ATPase activity was isolated as a 650-kd complex (called PAN) from the archaeon *Thermoplasma*, and experiments were performed to determine if PAN could enhance the activity of the 20S proteasome from *Thermoplasma* as well as other 20S proteasomes.

Protein degradation was measured as a function of time and in the presence of various combinations of components. Graph A shows the results.

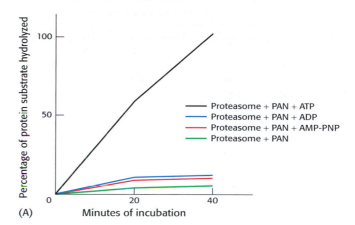

(A)

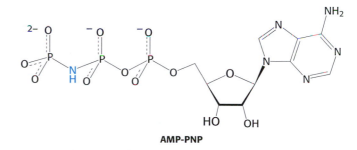

AMP-PNP

(a) What is the effect of PAN on archaeal proteasome activity in the absence of nucleotides?
(b) What is the nucleotide requirement for protein digestion?
(c) What evidence suggests that ATP hydrolysis, and not just the presence of ATP, is required for digestion?

A similar experiment was performed with a small peptide as a substrate for the proteasome instead of a protein. The results obtained are shown in graph B.

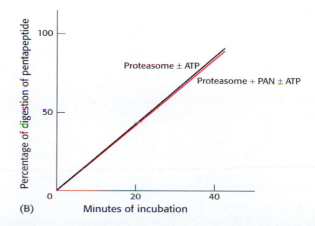

(B)

(d) How do the requirements for peptide digestion differ from those of protein digestion?
(e) Suggest some reasons for the difference.

The ability of PAN from the archaeon *Thermoplasma* to support protein degradation by the 20S proteasomes from the archeaon *Methanosarcina* and rabbit muscle was then examined.

	Percentage of digestion of protein substrate (20S proteasome source)		
Additions	*Thermoplasma*	*Methanosarcina*	Rabbit-muscle
None	11	10	10
PAN	8	8	8
PAN + ATP	100	40	30
PAN + ADP	12	9	10

(f) Can the *Thermoplasma* PAN augment protein digestion by the proteasomes from other organisms?

(g) What is the significance of the stimulation of rabbit muscle proteasome by *Thermoplasma* PAN?

[Data from P. Zwickl, D. Ng, K. M. Woo, H.-P. Klenk, and A. L. Goldberg. An archaebacterial ATPase, homologous to ATPase in the eukaryotic 26S proteasome, activates protein breakdown by 20S proteasomes. *J. Biol. Chem.* 274(1999): 26008–26014.]

Synthesizing the Molecules of Life

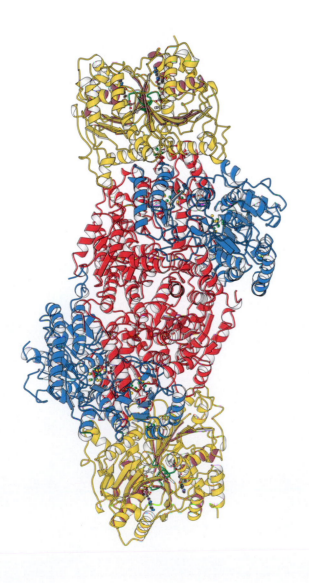

Nitrogenase complex. Nitrogen is an essential component of many biochemical building blocks. The enzyme complex shown here converts nitrogen gas, an abundant but inert compound, into a form that can be used for synthesizing amino acids, nucleotides, and other biochemicals.

CHAPTER 24

The Biosynthesis of Amino Acids

Glutamate

Nitrogen is a key component of amino acids. The atmosphere is rich in nitrogen gas (N_2), a very unreactive molecule. Certain organisms such as bacteria that live in the root nodules of yellow clover (photo at left) can convert nitrogen gas into ammonia. Ammonia can then be used to synthesize first glutamate and then other amino acids. [(Left) Runk/Schoenberger from Grant Heilman]

The assembly of biological molecules, including proteins and nucleic acids, requires the generation of appropriate starting materials. We have already considered the assembly of carbohydrates in regard to the Calvin cycle and the pentose phosphate pathway (Chapter 20). The present chapter and the next two examine the assembly of the other important building blocks—namely, amino acids, nucleotides, and lipids.

The pathways for the biosynthesis of these molecules are extremely ancient, going back to the last common ancestor of all living things. Indeed, these pathways probably predate many of the pathways of energy transduction discussed in Part II and may have provided key selective advantages in early evolution. Many of the intermediates present in energy-transduction pathways play a role in biosynthesis as well. These common intermediates allow efficient interplay between energy-transduction (catabolic) and biosynthetic (anabolic) pathways. Thus, cells are able to balance the degradation of compounds for energy mobilization and the synthesis of starting materials for macromolecular construction.

We begin our consideration of biosynthesis with amino acids. Amino acids are the building blocks of proteins and the nitrogen source for many other important molecules, including nucleotides, neurotransmitters, and prosthetic groups such as porphyrins.

Anabolism—
Biosynthetic processes.

Catabolism—
Degradative processes.
Derived from the Greek *ana,* "up";
kata, "down"; *ballein,* "to throw."

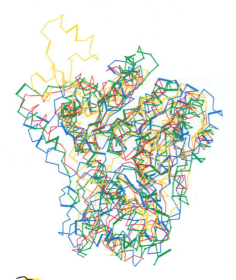

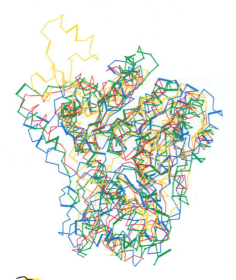
FIGURE 24.1 Aminotransferase protein family. A superposition of the structures of four aminotransferases taking part in amino acid biosynthesis reveals a common fold, indicating that these proteins descended from a common ancestor.

24.0.1 An Overview of Amino Acid Synthesis

A fundamental problem for biological systems is to obtain nitrogen in an easily usable form. This problem is solved by certain microorganisms capable of reducing the inert N≡N molecule (nitrogen gas) to two molecules of ammonia in one of the most remarkable reactions in biochemistry. Nitrogen in the form of ammonia is the source of nitrogen for all the amino acids. The carbon backbones come from the glycolytic pathway, the pentose phosphate pathway, or the citric acid cycle.

In amino acid production, we encounter an important problem in biosynthesis—namely, stereochemical control. Because all amino acids except glycine are chiral, biosynthetic pathways must generate the correct isomer with high fidelity. In each of the 19 pathways for the generation of chiral amino acids, the stereochemistry at the α-carbon atom is established by a transamination reaction that involves pyridoxal phosphate. Almost all the transaminases that catalyze these reactions descend from a common ancestor, illustrating once again that effective solutions to biochemical problems are retained throughout evolution (Figure 24.1).

Biosynthetic pathways are often highly regulated such that building blocks are synthesized only when supplies are low. Very often, a high concentration of the final product of a pathway inhibits the activity of enzymes that function early in the pathway. Often present are allosteric enzymes capable of sensing and responding to concentrations of regulatory species. These enzymes are similar in functional properties to aspartate transcarbamylase and its regulators (Section 10.1). Feedback and allosteric mechanisms ensure that all twenty amino acids are maintained in sufficient amounts for protein synthesis and other processes.

24.1 NITROGEN FIXATION: MICROORGANISMS USE ATP AND A POWERFUL REDUCTANT TO REDUCE ATMOSPHERIC NITROGEN TO AMMONIA

The nitrogen in amino acids, purines, pyrimidines, and other biomolecules ultimately comes from atmospheric nitrogen, N_2. The biosynthetic process starts with the reduction of N_2 to NH_3 (ammonia), a process called *nitrogen fixation.* Although higher organisms are unable to fix nitrogen, this conversion is carried out by some bacteria and archaea. Symbiotic *Rhizobium* bacteria invade the roots of leguminous plants and form root nodules in which they fix nitrogen, supplying both the bacteria and the plants. The amount of N_2 fixed by *diazotrophic (nitrogen-fixing) microorganisms* has been estimated to be 10^{11} kilograms per year, about 60% of Earth's newly fixed nitrogen. Lightning and ultraviolet radiation fix another 15%; the other 25% is fixed by industrial processes. The industrial process for nitrogen fixation devised by Fritz Haber in 1910 is still being used in fertilizer factories.

$$N_2 + 3\,H_2 \rightleftharpoons 2\,NH_3$$

The fixation of N_2 is typically carried out by mixing with H_2 gas over an iron catalyst at about 500°C and a pressure of 300 atmospheres. The extremely strong N≡N bond, which has a bond energy of 225 kcal mol^{-1}, is highly resistant to chemical attack. Indeed, Lavoisier named nitrogen gas "azote," meaning "without life" because it is so unreactive. Nevertheless, the conversion of nitrogen and hydrogen to form ammonia is thermodynamically favorable; the reaction is difficult kinetically because intermediates along the reaction pathway are unstable.

To meet the kinetic challenge, the biological process of nitrogen fixation requires a complex enzyme with multiple redox centers. The *nitrogenase complex,* which carries out this fundamental transformation, consists of two proteins: a *reductase,* which provides electrons with high reducing power, and *nitrogenase,* which uses these electrons to reduce N_2 to NH_3. The transfer of electrons from the reductase to the nitrogenase component is coupled to the hydrolysis of ATP by the reductase (Figure 24.2). The nitrogenase complex is exquisitely sensitive to inactivation by O_2. Leguminous plants maintain a very low concentration of free O_2 in their root nodules by binding O_2 to *leghemoglobin,* a homolog of hemoglobin (Section 7.3.1).

In principle, the reduction of N_2 to NH_3 is a six-electron process.

$$N_2 + 6\,e^- + 6\,H^+ \rightleftharpoons 2\,NH_3$$

FIGURE 24.2 Nitrogen fixation. Electrons flow from ferredoxin to the reductase (iron protein, or Fe protein) to nitrogenase (molybdenum–iron protein, or MoFe protein) to reduce nitrogen to ammonia. ATP hydrolysis within the reductase drives conformational changes necessary for the efficient transfer of electrons.

However, the biological reaction always generates at least 1 mol of H_2 in addition to 2 mol of NH_3 for each mole of $N{\equiv}N$. Hence, an input of two additional electrons is required.

$$N_2 + 8\,e^- + 8\,H^+ \rightleftharpoons 2\,NH_3 + H_2$$

In most nitrogen-fixing microorganisms, *the eight high-potential electrons come from reduced ferredoxin,* generated by photosynthesis or oxidative processes. Two molecules of ATP are hydrolyzed for each electron transferred. Thus, *at least 16 molecules of ATP are hydrolyzed for each molecule of N_2 reduced.*

$$N_2 + 8\,e^- + 8\,H^+ + 16\,ATP + 16\,H_2O \rightleftharpoons$$
$$2\,NH_3 + H_2 + 16\,ADP + 16\,P_i$$

Again, ATP hydrolysis is not required to make nitrogen reduction favorable thermodynamically. Rather, it is essential to reduce the heights of activation barriers along the reaction pathway, thus making the reaction kinetically feasible.

24.1.1 The Iron–Molybdenum Cofactor of Nitrogenase Binds and Reduces Atmospheric Nitrogen

Both the reductase and the nitrogenase components of the complex are *iron–sulfur proteins,* in which iron is bonded to the sulfur atom of a cysteine residue and to inorganic sulfide. The *reductase* (also called the *iron protein* or the *Fe protein*) is a dimer of identical 30-kd subunits bridged by a 4Fe-4S cluster (Figure 24.3).

The role of the reductase is to transfer electrons from a suitable donor, such as reduced ferredoxin, to the nitrogenase component. The binding and hydrolysis of ATP triggers a conformational change that moves the reductase closer to the nitrogenase component from whence it is able to transfer its electron to the center of nitrogen reduction. The structure of the ATP-binding region reveals it to be a member of the P-loop NTPase family (Section 9.4.1) that is clearly related to the nucleotide-binding regions found in G proteins and related proteins (Section 15.1.2). Thus, we see another example of how this domain has been recruited in evolution because of its ability to couple nucleoside triphosphate hydrolysis to conformational changes.

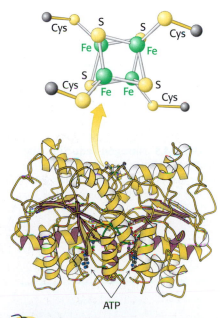

FIGURE 24.3 Fe protein. This protein is a dimer composed of two polypeptide chains linked by a 4Fe-4S cluster. Each monomer is a member of the P-loop NTPase family and contains an ATP-binding site.

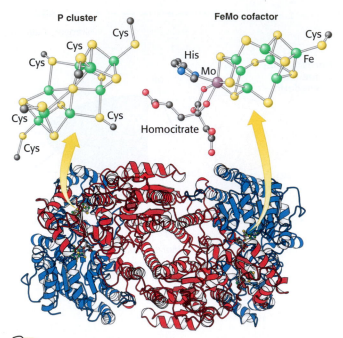

P cluster

Cys

Cys

Cys

Cys

Cys

Cys

FeMo cofactor

Cys

His

Mo

Fe

Homocitrate

👆 **FIGURE 24.4 MoFe protein.** This protein is an heterotetramer composed of two α subunits (red) and two β subunits (blue). The protein contains two copies each of two types of clusters: P clusters and FeMo cofactors. Each P cluster contains eight iron atoms and seven sulfides linked to the protein by six cysteinate residues. Each FeMo cofactor contains one molybdenum atom, seven iron atoms, nine sulfides, and a homocitrate, and is linked to the protein by one cysteinate residue and one histidine residue.

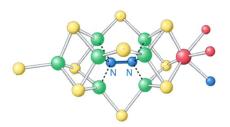

N   N

FIGURE 24.5 Nitrogen-reduction site. The FeMo cofactor contains an open center that is the likely site of nitrogen binding and reduction.

The nitrogenase component is an $\alpha_2\beta_2$ tetramer (240 kd), in which the α and β subunits are homologous to each other and structurally quite similar (Figure 24.4). Electrons enter at the *P clusters*, which are located at the α–β interface. These clusters are each composed of eight iron atoms and seven sulfide ions. In the reduced form, each cluster takes the form of two 4Fe-3S partial cubes linked by a central sulfide ion. Each cluster is linked to the protein through six cysteinate residues. Electrons flow from the P cluster to the *FeMo cofactor*, a very unusual redox center. Because molybdenum is present in this cluster, the nitrogenase component is also called the *molybdenum–iron protein (MoFe protein)*. The FeMo cofactor consists of two M-3Fe-3S clusters, in which molybdenum occupies the M site in one cluster and iron occupies it in the other. The two clusters are joined by three sulfide ions. The FeMo cofactor is also coordinated to a homocitrate moiety and to the α sub-unit through one histidine residue and one cysteinate residue. This cofactor is distinct from the molybdenum-containing cofactor found in sulfite oxidase and apparently all other molybdenum-containing enzymes except nitrogenase.

The FeMo cofactor is the site of nitrogen fixation. Note that each of the six central iron atoms is linked to only three atoms, leaving open a binding opportunity for N_2. It seems likely that N_2 binds in the central cavity of this cofactor (Figure 24.5). The formation of multiple Fe–N interactions in this complex weakens the N≡N bond and thereby lowers the activation barrier for reduction.

24.1.2 Ammonium Ion Is Assimilated into an Amino Acid Through Glutamate and Glutamine

The next step in the assimilation of nitrogen into biomolecules is the entry of NH_4^+ into amino acids. *Glutamate* and *glutamine* play pivotal roles in this regard. The α-amino group of most amino acids comes from the α-amino group of glutamate by transamination (Section 23.3.1). Glutamine, the other major nitrogen donor, contributes its side-chain nitrogen atom in the biosynthesis of a wide range of important compounds, including the amino acids tryptophan and histidine.

Glutamate is synthesized from NH_4^+ and α-ketoglutarate, a citric acid cycle intermediate, by the action of *glutamate dehydrogenase*. We have already encountered this enzyme in the degradation of amino acids (Section 23.3.1). Recall that NAD^+ is the oxidant in catabolism, whereas NADPH is the reductant in biosyntheses. Glutamate dehydrogenase is unusual in that it does not discriminate between NADH and NADPH, at least in some species.

$$NH_4^+ + \alpha\text{-ketoglutarate} + NADPH + H^+ \rightleftharpoons$$
$$glutamate + NADP^+ + H_2O$$

The reaction proceeds in two steps. First, a Schiff base forms between ammonia and α-ketoglutarate. The formation of a Schiff base between an amine and a carbonyl compound is a key reaction that takes place at many stages of amino acid biosynthesis and degradation.

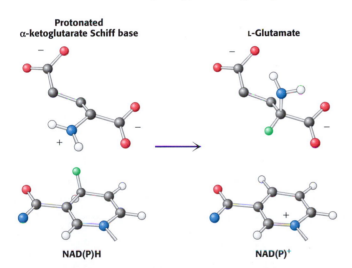

Schiff bases can be easily protonated. With glutamate dehydrogenase, the protonated Schiff base is reduced by the transfer of a hydride ion from NADPH to form glutamate.

This reaction is crucial because it establishes the stereochemistry of the α-carbon atom (S absolute configuration) in glutamate. The enzyme binds the α-ketoglutarate substrate in such a way that hydride transferred from NAD(P)H is added to form the L isomer of glutamate (Figure 24.6). As we shall see, this stereochemistry is established for other amino acids by transamination reactions that rely on pyridoxal phosphate.

Protonated
α-ketoglutarate Schiff base

L-Glutamate

NAD(P)H

NAD(P)⁺

FIGURE 24.6 Establishment of chirality. In the active site of glutamate dehydrogenase, hydride transfer from NAD(P)H to a specific face of the achiral protonated Schiff base of α-ketoglutarate establishes the L configuration of glutamate.

A second ammonium ion is incorporated into glutamate to form glutamine by the action of *glutamine synthetase*. This amidation is driven by the hydrolysis of ATP. ATP participates directly in the reaction by phosphorylating the side chain of glutamate to form an acyl-phosphate intermediate, which then reacts with ammonia to form glutamine.

Glutamate

Acyl-phosphate intermediate

Glutamine

A high-affinity ammonia-binding site is formed only after the formation of the acyl-phosphate intermediate. A specific site for ammonia binding is required to prevent attack by water from hydrolyzing the intermediate and wasting a molecule of ATP. The regulation of glutamine synthetase plays a critical role in controlling nitrogen metabolism (Section 24.3.2).

Glutamate dehydrogenase and glutamine synthetase are present in all organisms. Most prokaryotes also contain an evolutionarily unrelated enzyme, *glutamate synthase*, which catalyzes the reductive amination of α-ketoglutarate with the use of glutamine as the nitrogen donor.

$$\alpha\text{-Ketoglutarate} + \text{glutamine} + \text{NADPH} + \text{H}^+ \rightleftharpoons$$
$$2\,\text{glutamate} + \text{NADP}^+$$

The side-chain amide of glutamine is hydrolyzed to generate ammonia within the enzyme, a recurring theme throughout nitrogen metabolism. *When NH_4^+ is limiting, most of the glutamate is made by the sequential action of glutamine synthetase and glutamate synthase.* The sum of these reactions is

$$\text{NH}_4^+ + \alpha\text{-ketoglutarate} + \text{NADPH} + \text{ATP} \longrightarrow$$
$$\text{glutamate} + \text{NADP}^+ + \text{ADP} + \text{P}_i$$

Note that this stoichiometry differs from that of the glutamate dehydrogenase reaction in that ATP is hydrolyzed. Why do prokaryotes sometimes use this more expensive pathway? The answer is that the value of K_M of glutamate dehydrogenase for NH_4^+ is high (≈ 1 mM), and so this enzyme is not saturated when NH_4^+ is limiting. In contrast, glutamine synthetase has very high affinity for NH_4^+. Thus, ATP hydrolysis is required to capture ammonia when it is scarce.

24.2 AMINO ACIDS ARE MADE FROM INTERMEDIATES OF THE CITRIC ACID CYCLE AND OTHER MAJOR PATHWAYS

Thus far, we have considered the conversion of N_2 into NH_4^+ and the assimilation of NH_4^+ into glutamate and glutamine. We turn now to the biosynthesis of the other amino acids. The pathways for the biosynthesis of amino acids are diverse. However, they have an important common feature: *their carbon skeletons come from intermediates of glycolysis, the pentose phosphate pathway, or the citric acid cycle.* On the basis of these starting materials, amino acids can be grouped into six biosynthetic families (Figure 24.7).

FIGURE 24.7 Biosynthetic families of amino acids in bacteria and plants. Major metabolic precursors are shaded blue. Amino acids that give rise to other amino acids are shaded yellow. Essential amino acids are in boldface type.

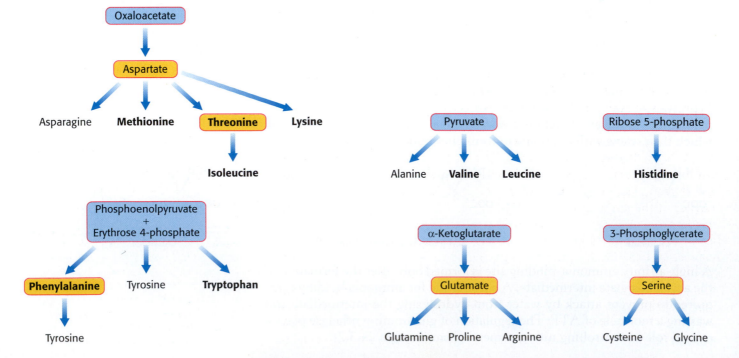

24.2.1 Human Beings Can Synthesize Some Amino Acids but Must Obtain Others from the Diet

Most microorganisms such as *E. coli* can synthesize the entire basic set of 20 amino acids, whereas human beings cannot make 9 of them. The amino acids that must be supplied in the diet are called *essential amino acids,* whereas the others are termed *nonessential amino acids* (Table 24.1). These designations refer to the needs of an organism under a particular set of conditions. For example, enough arginine is synthesized by the urea cycle to meet the needs of an adult but perhaps not those of a growing child. A deficiency of even one amino acid results in a *negative nitrogen balance*. In this state, more protein is degraded than is synthesized, and so more nitrogen is excreted than is ingested.

The nonessential amino acids are synthesized by quite simple reactions, whereas the pathways for the formation of the essential amino acids are quite complex. For example, the nonessential amino acids *alanine* and *aspartate* are synthesized in a single step from pyruvate and oxaloacetate, respectively. In contrast, the pathways for the essential amino acids require from 5 to 16 steps (Figure 24.8). The sole exception to this pattern is arginine, inasmuch as the synthesis of this nonessential amino acid de novo requires 10 steps. Typically, though, it is made in only 3 steps from ornithine as part of the urea cycle. Tyrosine, classified as a nonessential amino acid because it can be synthesized in 1 step from phenylalanine, requires 10 steps to be synthesized from scratch and is essential if phenylalanine is not abundant. We begin with the biosynthesis of nonessential amino acids.

24.2.2 A Common Step Determines the Chirality of All Amino Acids

Three α-ketoacids—α-ketoglutarate, oxaloacetate, and pyruvate—can be converted into amino acids in one step through the addition of an amino group. We have seen that α-ketoglutarate can be converted into glutamate by reductive amination (Section 24.1.2). The amino group from glutamate can be transferred to other α-ketoacids by transamination reactions. Thus, aspartate and alanine can be made from the addition of an amino group to oxaloacetate and pyruvate, respectively.

$$\text{Oxaloacetate} + \text{glutamate} \rightleftharpoons \text{aspartate} + \alpha\text{-ketoglutarate}$$

$$\text{Pyruvate} + \text{glutamate} \rightleftharpoons \text{alanine} + \alpha\text{-ketoglutarate}$$

These reactions are carried out by *pyridoxal phosphate-dependent transaminases*. Transamination reactions participate in the synthesis of most amino acids. We shall review the transaminase mechanism (Section 23.3.1) as it applies to amino acid biosynthesis (see Figure 23.10).

The reaction pathway begins with pyridoxal phosphate in a Schiff-base linkage with lysine at the transaminase active site, forming an internal aldimine (Figure 24.9). An external aldimine forms between PLP and the amino-group donor, glutamate, which displaces the lysine residue. The transfer of the amino group from glutamate to pyridoxal phosphate forms pyridoxamine phosphate, the actual amino donor. When the amino group has been incorporated into pyridoxamine, the reaction pathway proceeds in reverse, and the amino group is transferred to an α-ketoacid to form an amino acid.

Aspartate aminotransferase is the prototype of a large family of PLP-dependent enzymes. Comparisons of amino acid sequences as well as several three-dimensional structures reveal that almost all transaminases having roles in amino acid biosynthesis are related to aspartate aminotransferase

TABLE 24.1	Basic set of 20 amino acids
Nonessential	Essential
Alanine	Histidine
Arginine	Isoleucine
Asparagine	Leucine
Aspartate	Lysine
Cysteine	Methionine
Glutamate	Phenylalanine
Glutamine	Threonine
Glycine	Tryptophan
Proline	Valine
Serine	
Tyrosine	

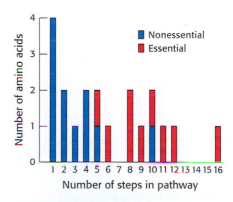

FIGURE 24.8 Essential and nonessential amino acids. Some amino acids are nonessential to human beings because they can be biosynthesized in a small number of steps. Those amino acids requiring a large number of steps for their synthesis are essential in the diet because some of the enzymes for these steps have been lost in the course of evolution.

Lysine

Internal aldimine

Glu → α-Keto-glutarate

Pyridoxamine phosphate (PMP)

H_2O

Ketimine

H^+

L-Amino acid

External aldimine

H^+

Quinonoid intermediate

FIGURE 24.9 Amino acid biosynthesis by transamination. Within a transaminase, the internal aldimine is converted into pyridoxamine phosphate (PMP) by reaction with glutamate. PMP then reacts with an α-ketoacid to generate a ketimine. This intermediate is converted into a quinonoid intermediate, which in turn yields an external aldimine. The aldimine is cleaved to release the newly formed amino acid to complete the cycle.

by divergent evolution. An examination of the aligned amino acid sequences reveals that two residues are completely conserved. These residues are the lysine residue that forms the Schiff base with the pyridoxal phosphate cofactor (lysine 258 in aspartate aminotransferase) and an arginine residue that interacts with the α-carboxylate group of the ketoacid (see Figure 23.11).

An essential step in the transamination reaction is the protonation of the quinonoid intermediate to form the external aldimine. *The chirality of the amino acid formed is determined by the direction from which this proton is added to the quinonoid form.* (Figure 24.10). This protonation step determines the L configuration of the amino acids produced. The interaction between the conserved arginine residue and the α-carboxylate group helps orient the substrate so that, when the lysine residue transfers a proton to the face of the quinonoid intermediate, it generates an aldimine with an L configuration at the C_α center.

FIGURE 24.10 Stereochemistry of proton addition. In a transaminase active site, the addition of a proton from the lysine residue to the bottom face of the quinonoid intermediate determines the L configuration of the amino acid product. The conserved arginine residue interacts with the α-carboxylate group and helps establish the appropriate geometry of the quinonoid intermediate.

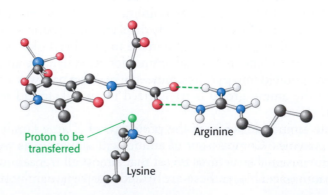

Proton to be transferred

Arginine

Lysine

The formation of asparagine from aspartate is chemically analogous to the formation of glutamine from glutamate. Both transformations are amidation reactions and both are driven by the hydrolysis of ATP. The actual reactions are different, however. In bacteria, the reaction for the asparagine synthesis is

$$\text{Aspartate} + NH_4^+ + ATP \longrightarrow \text{asparagine} + AMP + PP_i + H^+$$

Thus, the products of ATP hydrolysis are AMP and PP_i rather than ADP and P_i. Aspartate is activated by adenylation rather than by phosphorylation.

Aspartate   **Acyl-adenylate intermediate**   **Asparagine**

We have encountered this mode of activation in fatty acid degradation and will see it again in lipid and protein synthesis.

In mammals, the nitrogen donor for asparagine is glutamine rather than ammonia as in bacteria. Ammonia is generated by hydrolysis of the side chain of glutamine and directly transferred to activated aspartate, bound in the active site. An advantage is that the cell is not directly exposed to NH_4^+, which is toxic at high levels to human beings and other mammals. *The use of glutamine hydrolysis as a mechanism for generating ammonia for use within the same enzyme is a motif common throughout biosynthetic pathways.*

24.2.4 Glutamate Is the Precursor of Glutamine, Proline, and Arginine

The synthesis of glutamate by the reductive amination of α-ketoglutarate has already been discussed, as has the conversion of glutamate into glutamine (Section 24.1.2). Glutamate is the precursor of two other nonessential amino acids: *proline* and *arginine*. First, the γ-carboxyl group of glutamate reacts with ATP to form an acyl phosphate. This mixed anhydride is then reduced by NADPH to an aldehyde.

Glutamate   **Acyl-phosphate intermediate**   **Glutamic γ-semialdehyde**

Glutamic γ-semialdehyde cyclizes with a loss of H_2O in a nonenzymatic process to give Δ^1-pyrroline-5-carboxylate, which is reduced by NADPH to proline. Alternatively, the semialdehyde can be transaminated to ornithine, which is converted in several steps into arginine (Section 23.4.1).

The chemical structures at the top of the page show the conversion of Glutamate through Glutamic γ-semialdehyde to Ornithine, and via Δ¹-Pyrroline-5-carboxylate to Proline.

- Glutamate → (ATP + NADPH → ADP + P_i + NADP$^+$) → Glutamic γ-semialdehyde
- Glutamic γ-semialdehyde → (α-Ketoglutarate, Glutamate) → Ornithine (NH$_3^+$)
- Glutamic γ-semialdehyde → (H_2O) → Δ¹-Pyrroline-5-carboxylate → (H$^+$ + NADPH → NADP$^+$) → Proline

Glutamate ... **Glutamic γ-semialdehyde** ... **Ornithine** ... **Δ¹-Pyrroline-5-carboxylate** ... **Proline**

24.2.5 Serine, Cysteine, and Glycine Are Formed from 3-Phosphoglycerate

Serine is synthesized from 3-phosphoglycerate, an intermediate in glycolysis. The first step is an oxidation to 3-phosphohydroxypyruvate. This α-ketoacid is transaminated to 3-phosphoserine, which is then hydrolyzed to serine.

The chemical pathway shows:
- 3-Phosphoglycerate → (NAD$^+$ → H$^+$ + NADH) → 3-Phosphohydroxypyruvate → (Glutamate → α-Ketoglutarate) → 3-Phosphoserine → (H_2O → P_i) → Serine

3-Phospho-glycerate ... **3-Phosphohydroxy-pyruvate** ... **3-Phospho-serine** ... **Serine**

Serine is the precursor of *glycine* and *cysteine*. In the formation of glycine, the side-chain methylene group of serine is transferred to *tetrahydrofolate*, a carrier of one-carbon units that will be discussed shortly.

$$\text{Serine + tetrahydrofolate} \longrightarrow \text{glycine + methylenetetrahydrofolate} + H_2O$$

This interconversion is catalyzed by *serine transhydroxymethylase*, another PLP enzyme that is homologous to aspartate aminotransferase (Figure 24.11). The bond between the α- and β-carbon atoms of serine is labilized by the formation of a Schiff base between serine and PLP (Section 23.3.3). The side-chain methylene group of serine is then transferred to tetrahydrofolate. The conversion of serine into cysteine requires the substitution of a sulfur atom derived from methionine for the side-chain oxygen atom (Section 24.2.8).

24.2.6 Tetrahydrofolate Carries Activated One-Carbon Units at Several Oxidation Levels

Tetrahydrofolate (also called *tetrahydropteroylglutamate*), a highly versatile carrier of activated one-carbon units, consists of three groups: a substituted pteridine, *p*-aminobenzoate, and a chain of one or more glutamate residues (Figure 24.12). Mammals can synthesize the pteridine ring, but they are unable to conjugate it to the other two units. They obtain tetrahydrofolate from their diets or from microorganisms in their intestinal tracts.

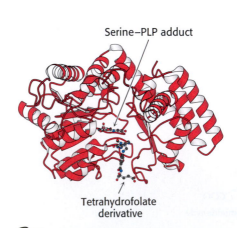

Serine–PLP adduct

Tetrahydrofolate derivative

FIGURE 24.11 Structure of serine hydroxymethyltransferase. This enzyme transfers a one-carbon unit from the side chain of serine to tetrahydrofolate. One subunit of the dimeric enzyme is shown.

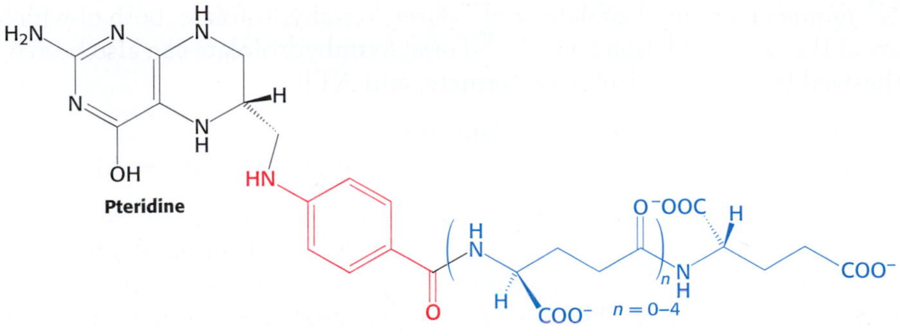

FIGURE 24.12 Tetrahydrofolate. This cofactor includes three components: a pteridine ring, *p*-aminobenzoate, and one or more glutamate residues.

The one-carbon group carried by tetrahydrofolate is bonded to its N-5 or N-10 nitrogen atom (denoted as N^5 and N^{10}) or to both. This unit can exist in three oxidation states (Table 24.2). The most-reduced form carries a *methyl* group, whereas the intermediate form carries a *methylene* group. More-oxidized forms carry a *formyl*, *formimino*, or *methenyl* group. The fully oxidized one-carbon unit, CO_2, is carried by biotin rather than by tetrahydrofolate.

The one-carbon units carried by tetrahydrofolate are interconvertible (Figure 24.13). N^5,N^{10}-*Methylene*tetrahydrofolate can be reduced to N^5-*methyl*tetrahydrofolate or oxidized to N^5,N^{10}-*methenyl*tetrahydrofolate. N^5, N^{10}-*Methenyl*tetrahydrofolate can be converted into

TABLE 24.2 One-carbon groups carried by tetrahydrofolate

Oxidation state	Group	
Most reduced (= methanol)	$-CH_3$	Methyl
Intermediate (= formaldehyde)	$-CH_2-$	Methylene
Most oxidized (= formic acid)	$-CHO$	Formyl
	$-CHNH$	Formimino
	$-CH=$	Methenyl

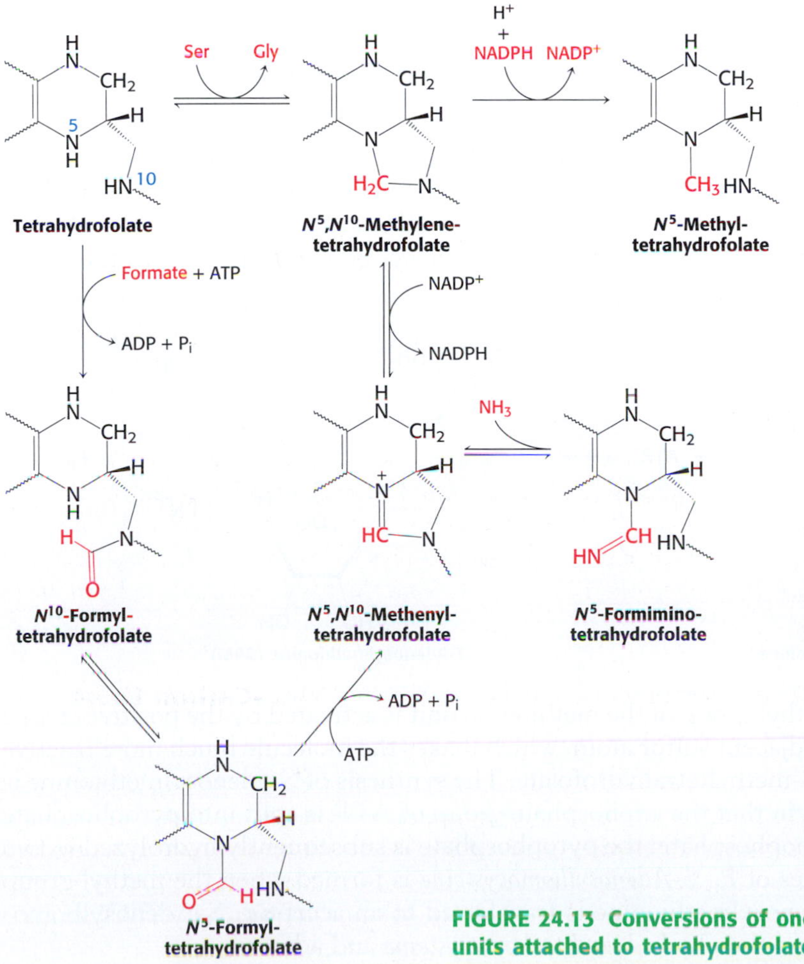

FIGURE 24.13 Conversions of one-carbon units attached to tetrahydrofolate.

N^5-*formiminotetrahydrofolate* or N^{10}-*formyltetrahydrofolate*, both of which are at the same oxidation level. N^{10}-*Formyltetrahydrofolate* can also be synthesized from tetrahydrofolate, formate, and ATP.

$$\text{Formate} + \text{ATP} + \text{tetrahydrofolate} \longrightarrow$$
$$N^{10}\text{-formyltetrahydrofolate} + \text{ADP} + \text{P}_i$$

N^5-*Formyltetrahydrofolate* can be reversibly isomerized to N^{10}-*formyl* tetrahydrofolate or it can be converted into N^5,N^{10}-*methenyltetrahydrofolate*.

These tetrahydrofolate derivatives serve as donors of one-carbon units in a variety of biosyntheses. Methionine is regenerated from homocysteine by transfer of the methyl group of N^5-methyltetrahydrofolate, as will be discussed shortly. We shall see in Chapter 25 that some of the carbon atoms of *purines* are acquired from derivatives of N^{10}-formyltetrahydrofolate. The methyl group of *thymine*, a pyrimidine, comes from N^5,N^{10}-methylenetetrahydrofolate. This tetrahydrofolate derivative can also donate a one-carbon unit in an alternative synthesis of *glycine* that starts with CO_2 and NH_4^+, a reaction catalyzed by *glycine synthase* (called the *glycine cleavage enzyme* when it operates in the reverse direction).

$$CO_2 + NH_4^+ + N^5,N^{10}\text{-methylenetetrahydrofolate} + \text{NADH} \rightleftharpoons$$
$$\text{glycine} + \text{tetrahydrofolate} + \text{NAD}^+$$

Thus, one-carbon units at each of the three oxidation levels are utilized in biosyntheses. Furthermore, *tetrahydrofolate serves as an acceptor of one-carbon units in degradative reactions.* The major source of one-carbon units is the facile conversion of serine into glycine, which yields N^5,N^{10}-methylenetetrahydrofolate. Serine can be derived from 3-phosphoglycerate, and so *this pathway enables one-carbon units to be formed de novo from carbohydrates.*

24.2.7 S-Adenosylmethionine Is the Major Donor of Methyl Groups

Tetrahydrofolate can carry a methyl group on its N-5 atom, but its transfer potential is not sufficiently high for most biosynthetic methylations. Rather, the activated methyl donor is usually S-*adenosylmethionine* (SAM), which is synthesized by the transfer of an adenosyl group from ATP to the sulfur atom of methionine.

Methionine   *S*-Adenosylmethionine (SAM)

The methyl group of the methionine unit is activated by the positive charge on the adjacent sulfur atom, which makes the molecule much more reactive than N^5-methyltetrahydrofolate. The synthesis of S-adenosylmethionine is unusual in that the triphosphate group of ATP is split into pyrophosphate and orthophosphate; the pyrophosphate is subsequently hydrolyzed to two molecules of P_i. S-*Adenosylhomocysteine* is formed when the methyl group of S-adenosylmethionine is transferred to an acceptor. S-Adenosylhomocysteine is then hydrolyzed to *homocysteine* and adenosine.

S-Adenosylmethionine (SAM)   **S-Adenosylhomocysteine**   **Homocysteine**

Methionine can be regenerated by the transfer of a methyl group to homocysteine from N^5-methyltetrahydrofolate, a reaction catalyzed by *methionine synthase* (also known as *homocysteine methyltransferase*).

Homocysteine N^5-**Methyltetrahydrofolate**   **Methionine**   **Tetrahydrofolate**

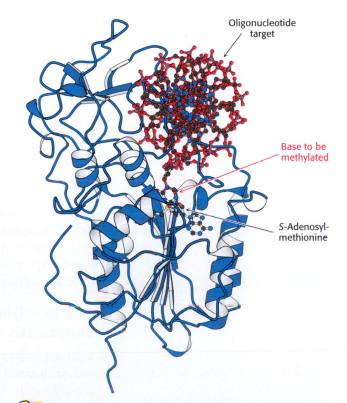

FIGURE 24.14 Activated methyl cycle. The methyl group of methionine is activated by the formation of *S*-adenosylmethionine.

The coenzyme that mediates this transfer of a methyl group is *methylcobalamin*, derived from vitamin B_{12}. In fact, this reaction and the rearrangement of L-methylmalonyl CoA to succinyl CoA (Section 23.5.4), catalyzed by a homologous enzyme, are the only two B_{12}-dependent reactions known to take place in mammals. Another enzyme that converts homocysteine into methionine without vitamin B_{12} also is present in many organisms.

These reactions constitute the *activated methyl cycle* (Figure 24.14). Methyl groups enter the cycle in the conversion of homocysteine into methionine and are then made highly reactive by the addition of adenosyl groups, which make the sulfur atoms positively charged and the methyl groups much more electrophilic. The high transfer potential of the *S*-methyl group enables it to be transferred to a wide variety of acceptors.

Among the acceptors modified by *S*-adenosylmethionine are specific bases in DNA. The methylation of DNA protects bacterial DNA from cleavage by restriction enzymes (Section 9.3). The base to be methylated is flipped out of the DNA double helix into the active site where it can accept a methyl group from *S*-adenosylmethionine (Figure 24.15). A recurring *S*-adenosylmethionine-binding domain is present in many SAM-dependent methylases.

S-Adenosylmethionine is also the precursor of *ethylene*, a gaseous plant hormone that induces the ripening of fruit. *S*-Adenosylmethionine is cyclized to a cyclopropane derivative that is then oxidized to form ethylene. The Greek philosopher Theophrastus recognized more than 2000 years ago that sycamore figs do not ripen unless they are scraped with an iron claw. The reason is now

FIGURE 24.15 DNA methylation. The structure of a DNA methylase bound to an oligonucleotide target shows that the base to be methylated is flipped out of the DNA helix into the active site of a SAM-dependent methylase.

known: *wounding triggers ethylene production, which in turn induces ripening.* Much effort is being made to understand this biosynthetic pathway because ethylene is a culprit in the spoilage of fruit.

$$\text{S-Adenosylmethionine} \xrightarrow{\text{ACC synthase}} \begin{array}{c} H_2C\!-\!CH_2 \\ {}^+H_3N \quad COO^- \end{array} \xrightarrow{\text{ACC oxidase}} H_2C\!=\!CH_2$$

1-Aminocyclopropane-
1-carboxylate (ACC)

Ethylene

24.2.8 Cysteine Is Synthesized from Serine and Homocysteine

In addition to being a precursor of methionine in the activated methyl cycle, homocysteine is an intermediate in the synthesis of cysteine. Serine and homocysteine condense to form *cystathionine*. This reaction is catalyzed by *cystathionine β-synthase*. Cystathionine is then deaminated and cleaved to cysteine and α-ketobutyrate by *cystathioninase*. Both of these enzymes utilize PLP and are homologous to aspartate aminotransferase. The net reaction is

$$\text{Homocysteine} + \text{serine} \rightleftharpoons \text{cysteine} + \alpha\text{-ketobutyrate} + NH_4^+$$

Note that the sulfur atom of cysteine is derived from homocysteine, whereas the carbon skeleton comes from serine.

Homocysteine + Serine → Cystathionine → α-Ketobutyrate + Cysteine

24.2.9 High Homocysteine Levels Are Associated with Vascular Disease

People with elevated serum levels of homocysteine or the disulfide-linked dimer homocystine have an unusually high risk for coronary heart disease and arteriosclerosis. The most common genetic cause of high homocysteine levels is a mutation within the gene encoding cystathionine β-synthase. The molecular basis of homocysteine's action has not been clearly identified, although it appears to damage cells lining blood vessels and to increase the growth of vascular smooth muscle. The amino acid raises oxidative stress as well. Vitamin treatments are effective in reducing homocysteine levels in some people. Treatment with vitamins maximizes the activity of the two major metabolic pathways processing homocysteine. Pyridoxal phosphate, a vitamin B_6 derivative, is necessary for the activity of cystathionine β-synthase, which converts homocysteine into cystathione; tetrahydrofolate, and vitamin B_{12}, supports the methylation of homocysteine to methionine.

24.2.10 Shikimate and Chorismate Are Intermediates in the Biosynthesis of Aromatic Amino Acids

We turn now to the biosynthesis of essential amino acids. These amino acids are synthesized by plants and microorganisms, and those in the human diet are ultimately derived primarily from plants. The essential amino acids are formed by much more complex routes than are the nonessential amino acids. The pathways for the synthesis of aromatic amino acids in bacteria have been selected for discussion here because they are well understood and exemplify recurring mechanistic motifs.

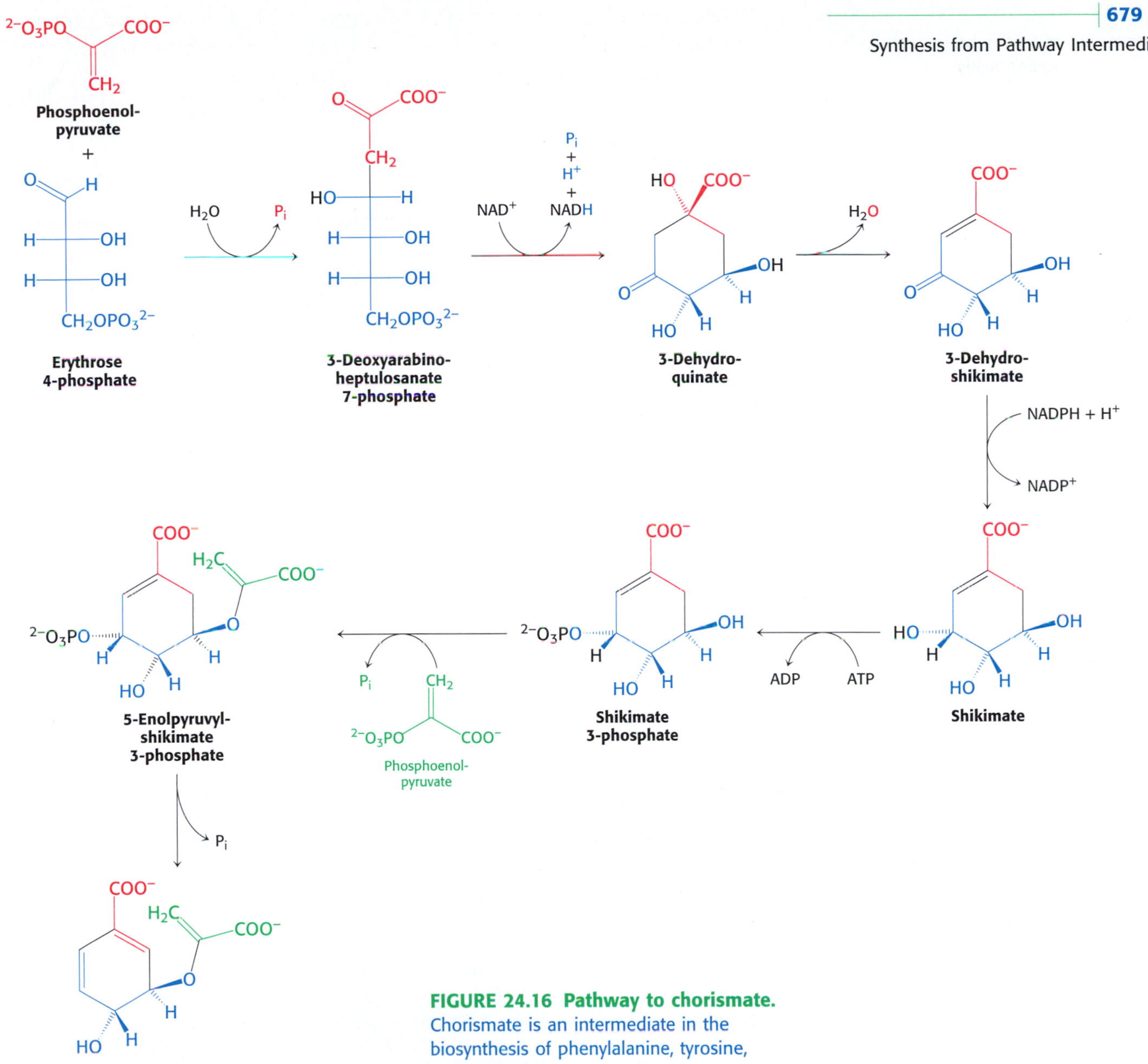

FIGURE 24.16 Pathway to chorismate.
Chorismate is an intermediate in the
biosynthesis of phenylalanine, tyrosine,
and tryptophan.

Phenylalanine, tyrosine, and tryptophan are synthesized by a common pathway in *E. coli* (Figure 24.16). The initial step is the condensation of phosphoenolpyruvate (a glycolytic intermediate) with erythrose 4-phosphate (a pentose phosphate pathway intermediate). The resulting seven-carbon open-chain sugar is oxidized, loses its phosphoryl group, and cyclizes to 3-dehydroquinate. Dehydration then yields 3-dehydroshikimate, which is reduced by NADPH to shikimate. Phosphorylation of shikimate by ATP gives shikimate 3-phosphate, which condenses with a second molecule of phosphoenolpyruvate. This 5-enolpyruvyl intermediate loses its phosphoryl group, yielding chorismate, the common precursor of all three aromatic amino acids. The importance of this pathway is revealed by the effectiveness of glyphosate (Roundup), a broad-spectrum herbicide. This compound inhibits the enzyme that produces 5-enolpyruvylshikimate 3-phosphate and, hence, blocks aromatic amino acid biosynthesis in plants. Because animals lack this enzyme, the herbicide is fairly nontoxic.

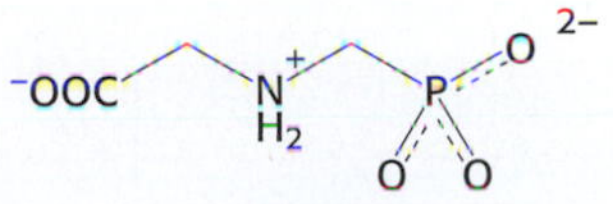

Glyphosate
(Roundup)

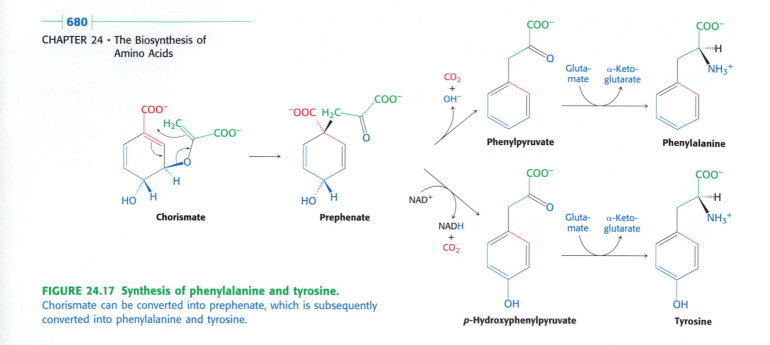

FIGURE 24.17 Synthesis of phenylalanine and tyrosine.
Chorismate can be converted into prephenate, which is subsequently
converted into phenylalanine and tyrosine.

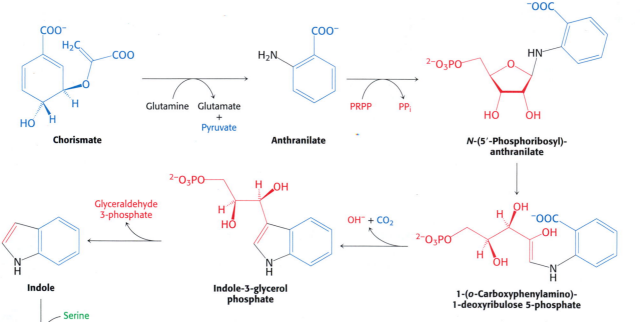

FIGURE 24.18 Synthesis of tryptophan. Chorismate can be converted into anthranilate,
which is subsequently converted into tryptophan.

The pathway bifurcates at chorismate. Let us first follow the *prephenate branch* (Figure 24.17). A mutase converts chorismate into prephenate, the immediate precursor of the aromatic ring of phenylalanine and tyrosine. This fascinating conversion is a rare example of an electrocyclic reaction in biochemistry, mechanistically similar to the well-known Diels-Alder reaction from organic chemistry. Dehydration and decarboxylation yield *phenylpyruvate*. Alternatively, prephenate can be oxidatively decarboxylated to p-*hydroxyphenylpyruvate*. These α-ketoacids are then transaminated to form *phenylalanine* and *tyrosine*.

The branch starting with *anthranilate* leads to the synthesis of tryptophan (Figure 24.18). Chorismate acquires an amino group derived from the hydrolysis of the side chain of glutamine and releases pyruvate to form anthranilate. Then anthranilate condenses with *5-phosphoribosyl-1-pyrophos-phate (PRPP), an activated form of ribose phosphate.* PRPP is also an important intermediate in the synthesis of histidine, purine nucleotides, and pyrimidine nucleotides (Sections 25.1.4 and 25.2.2). The C-1 atom of ribose 5-phosphate becomes bonded to the nitrogen atom of anthranilate in a reaction that is driven by the release and hydrolysis of pyrophosphate. The ribose moiety of phosphoribosylanthranilate undergoes rearrangement to yield 1-(*o*-carboxyphenylamino)-1-deoxyribulose 5-phosphate. This intermediate is dehydrated and then decarboxylated to indole-3-glycerol phosphate, which is cleaved to indole. Then indole reacts with serine to form tryptophan. In these final steps, which are catalyzed by tryptophan synthetase, the side chain of indole-3-glycerol phosphate is removed as glyceraldehyde 3-phosphate and replaced by the carbon skeleton of serine.

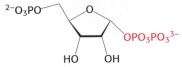

5-Phosphoribosyl-1-pyrophosphate (PRPP)

24.2.11 Tryptophan Synthetase Illustrates Substrate Channeling in Enzymatic Catalysis

Indole

Schiff base of aminoacrylate
(derived from serine)

Tryptophan synthetase of *E. coli*, an $\alpha_2\beta_2$ tetramer, can be dissociated into two α subunits and a β_2 subunit (Figure 24.19). The α subunit catalyzes the formation of indole from indole-3-glycerol phosphate, whereas each β subunit has a PLP-containing active site that catalyzes the condensation of indole and serine to form tryptophan. The overall three-dimensional structure of this enzyme is distinct from that of aspartate aminotransferase and the other PLP enzymes already discussed. Serine forms a Schiff base with this PLP, which is then dehydrated to give the *Schiff base of aminoacrylate.* This reactive intermediate is attacked by indole to give tryptophan.

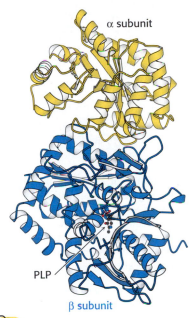

FIGURE 24.19 Structure of tryptophan synthetase. The structure of the complex formed by one α subunit and one β subunit. PLP is bound to the β subunit.

The synthesis of tryptophan poses a challenge. Indole, a hydrophobic molecule, readily traverses membranes and would be lost from the cell if it were allowed to diffuse away from the enzyme. This problem is solved in an ingenious way. A 25-Å-long channel connects the active site of the α subunit with that of the adjacent β subunit in the $\alpha_2\beta_2$ tetramer (Figure 24.20). Thus, indole can diffuse from one active site to the other without being released into bulk solvent. Indeed, the results of isotopic-labeling experiments showed that indole formed by the α subunit does not leave the enzyme when serine is present. Furthermore, the two partial reactions are coordinated. Indole is not formed by the α subunit until the highly reactive aminoacrylate is ready and waiting in the β subunit. We see here a clear-cut example of *substrate channeling* in catalysis by a multienzyme complex. Channeling substantially increases the catalytic rate. Furthermore, a deleterious side reaction—in this case, the potential loss of an intermediate—is prevented. We shall encounter other examples of substrate channeling in Chapter 25.

24.3 AMINO ACID BIOSYNTHESIS IS REGULATED BY FEEDBACK INHIBITION

The rate of synthesis of amino acids depends mainly on the *amounts* of the biosynthetic enzymes and on their *activities*. We now consider the control of enzymatic activity. The regulation of enzyme synthesis will be discussed in Chapter 31.

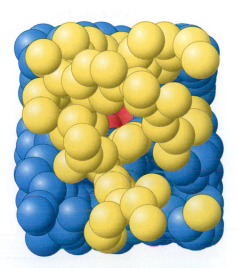

FIGURE 24.20 Substrate channeling. A 25-Å tunnel runs from the active site of the α subunit of tryptophan synthetase (yellow) to the PLP cofactor (red) in the active site of the β subunit (blue).

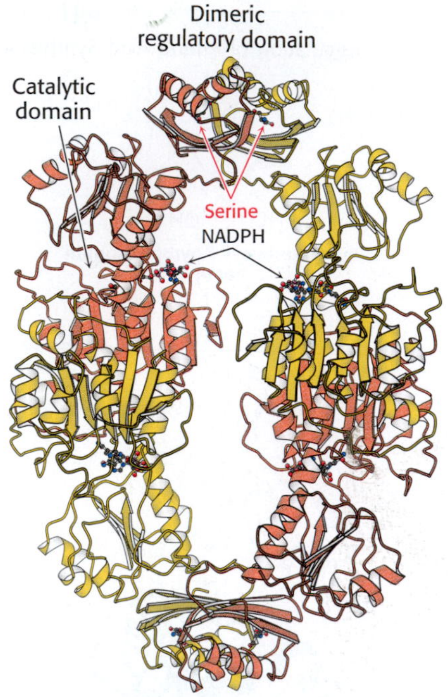

In a biosynthetic pathway, the first irreversible reaction, called the *committed step*, is usually an important regulatory site. *The final product of the pathway (Z) often inhibits the enzyme that catalyzes the committed step (A → B).*

This kind of control is essential for the conservation of building blocks and metabolic energy. Consider the biosynthesis of serine (Section 24.2.5). The committed step in this pathway is the oxidation of 3-phosphoglycerate, catalyzed by the enzyme *3-phosphoglycerate dehydrogenase*. The *E. coli* enzyme is a tetramer of four identical subunits, each comprising a catalytic domain and a serine-binding regulatory domain (Figure 24.21). The regulatory domains of two subunits interact to form a dimeric serine-binding regulatory unit so that the tetrameric enzyme contains two such regulatory units. Each unit is capable of binding two serine molecules. The binding of serine to a regulatory site reduces the value of V_{max} for the enzyme; an enzyme bound to four molecules of serine is essentially inactive. Thus, if serine is abundant in the cell, the enzyme activity is inhibited, and so 3-phosphoglycerate, a key building block that can be used for other processes, is not wasted.

24.3.1 Branched Pathways Require Sophisticated Regulation

The regulation of branched pathways is more complicated because the concentration of two products must be accounted for. In fact, several intricate feedback mechanisms have been found in branched biosynthetic pathways.

Feedback Inhibition and Activation. Consider, for example, the biosynthesis of the amino acids valine, leucine, and isoleucine. A common intermediate, hydroxyethyl thiamine pyrophosphate (hydroxyethyl-TPP; Section 17.1.1), initiates the pathways leading to all three of these amino acids. Hydroxyethyl-TPP can react with α-ketobutyrate in the initial step for the synthesis of isoleucine. Alternatively, hydroxyethyl-TPP can react with pyruvate in the committed step for the pathways leading to valine and leucine. Thus, the relative concentrations of α-ketobutyrate and pyruvate determine how much isoleucine is produced compared with valine and leucine. *Threonine deaminase*, the PLP enzyme that catalyzes the formation of α-ketobutyrate, is allosterically inhibited by isoleucine (Figure 24.22). This enzyme is also allosterically activated by valine. Thus, this enzyme is inhibited by the product of the pathway that it initiates and is activated by the end product of a competitive pathway. This mechanism balances the amounts of different amino acids that are synthesized.

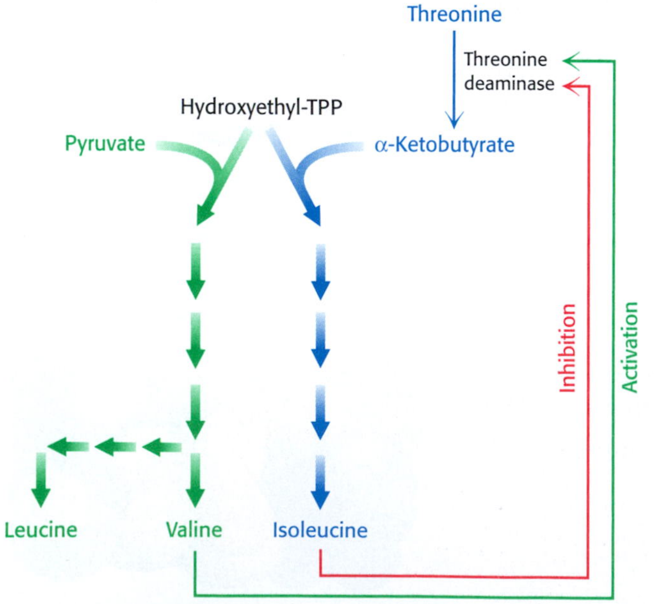

The regulatory domain in threonine deaminase is very similar in structure to the dimeric regulatory domain in 3-phosphoglycerate dehydrogenase (Figure 24.23). In this case, the two half regulatory domains are fused into a single unit with two differentiated amino acid-binding sites, one for isoleucine and the other for valine. Sequence analysis shows that similar regulatory domains are present in other amino acid biosynthetic

FIGURE 24.21 Structure of 3-phosphoglycerate dehydrogenase. This enzyme, which catalyzes the committed step in the serine biosynthetic pathway, includes a serine-binding regulatory domain. Serine binding to this domain reduces the activity of the enzyme.

FIGURE 24.22 Regulation of threonine deaminase. Threonine is converted into α-ketobutyrate in the committed step leading to the synthesis of isoleucine. The enzyme that catalyzes this step, threonine deaminase, is inhibited by isoleucine and activated by valine, the product of a parallel pathway.

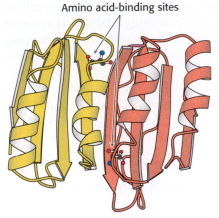
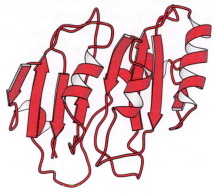

Amino acid-binding sites

Phosphoglycerate dehydrogenase
(dimeric regulatory domain)

Threonine deaminase
(single-chain regulatory domain)

FIGURE 24.23 A recurring regulatory domain. The regulatory domain formed by two subunits of 3-phosphoglycerate dehydrogenase is structurally related to the single-chain regulatory domain of threonine deaminase. Sequence analyses have revealed this amino acid-binding regulatory domain to be present in other enzymes as well.

enzymes. *The similarities suggest that feedback-inhibition processes may have evolved by the linkage of specific regulatory domains to the catalytic domains of biosynthetic enzymes.*

Enzyme Multiplicity. Sophisticated regulation can also evolve by duplication of the genes encoding the biosynthetic enzymes. For example, the phosphorylation of aspartate is the committed step in the biosynthesis of threonine, methionine, and lysine. Three distinct aspartokinases catalyze this reaction in *E. coli*, an example of a regulatory mechanism called *enzyme multiplicity.* (Figure 24.24). The catalytic domains of these enzymes show approximately 30% sequence identity. Although the mechanisms of catalysis are essentially identical, their activities are regulated differently: one enzyme is not subject to feedback inhibition, another is inhibited by threonine, and the third is inhibited by lysine.

Cumulative Feedback Inhibition. The regulation of glutamine synthetase in *E. coli* is a striking example of *cumulative feedback inhibition.* Recall that glutamine is synthesized from glutamate, NH_4^+, and ATP. *Glutamine synthetase* consists of 12 identical 50-kd subunits arranged in two hexagonal rings that face each other (Figure 24.25). Earl Stadtman showed

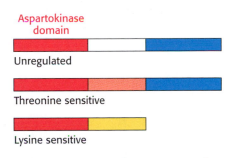

Aspartokinase domain

Unregulated

Threonine sensitive

Lysine sensitive

FIGURE 24.24 Domain structures of three aspartokinases. Each catalyzes the committed step in the biosynthesis of a different amino acid: (top) methionine, (middle) threonine, and (bottom) lysine. They have a catalytic domain in common but differ in their regulatory domains.

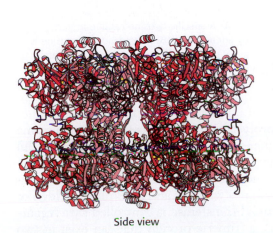

Side view

Top view

FIGURE 24.25 Structure of glutamine synthetase. Glutamine synthetase consists of 12 identical subunits arranged in two rings of six subunits. The active sites are indicated by the presence of manganese ions (two yellow spheres).

that this enzyme regulates the flow of nitrogen and hence plays a key role in controlling bacterial metabolism. The amide group of glutamine is a source of nitrogen in the biosyntheses of a variety of compounds, such as tryptophan, histidine, carbamoyl phosphate, glucosamine 6-phosphate, cytidine triphosphate, and adenosine monophosphate. Glutamine synthetase is cumulatively inhibited by each of these final products of glutamine metabolism, as well as by alanine and glycine. *In cumulative inhibition, each inhibitor can reduce the activity of the enzyme, even when other inhibitors are bound at saturating levels.* The enzymatic activity of glutamine synthetase is switched off almost completely when all final products are bound to the enzyme.

24.3.2 The Activity of Glutamine Synthetase Is Modulated by an Enzymatic Cascade

The activity of glutamine synthetase is also controlled by *reversible covalent modification*—the attachment of an *AMP unit* by a phosphodiester bond to the hydroxyl group of a specific tyrosine residue in each subunit (Figure 24.26). *This adenylylated enzyme is less active and more susceptible to cumulative feedback inhibition than is the deadenylylated form.* The covalently attached AMP unit is removed from the adenylylated enzyme by phosphorolysis. The attachment of an AMP unit is the final step in an enzymatic cascade that is initiated several steps back by reactants and immediate products in glutamine synthesis.

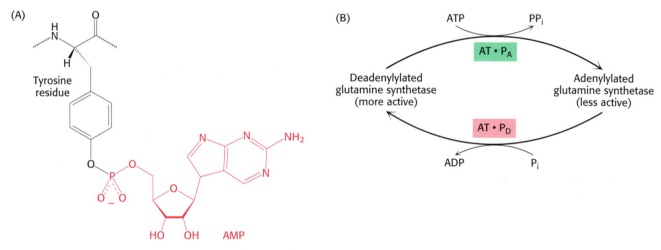

FIGURE 24.26 **Regulation by adenylation.** (A) A specific tyrosine residue in each subunit in glutamine synthetase is modified by adenylation. (B) Adenylation of tyrosine is catalyzed by a complex of adenylyl transferase (AT) and one form of a regulatory protein (P$_A$). The same enzyme catalyzes deadenylation when it is complexed with the other form (P$_D$) of the regulatory protein.

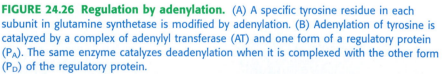

FIGURE 24.27 **Structure of the regulatory protein P.** This trimeric regulatory protein controls the modification of glutamine synthetase.

The adenylation and phosphorolysis reactions are catalyzed by the same enzyme, *adenylyl transferase*. Sequence analysis indicates that this adenylyl transferase comprises two homologous halves, suggesting that one half catalyzes the adenylation reaction and the other half the phospholytic deadenylation reaction. What determines whether an AMP unit is added or removed? The specificity of adenylyl transferase is controlled by a *regulatory protein* (designated P or P$_{II}$), a trimeric protein that can exist in two forms, P$_A$ and P$_D$ (Figure 24.27). The complex of P$_A$ and adenylyl transferase catalyzes the attachment of an AMP unit to glutamine synthetase, which reduces its activity. Conversely, the complex of P$_D$ and adenylyl transferase removes AMP from the adenylylated enzyme.

This brings us to another level of reversible covalent modification. P_A is converted into P_D by the attachment of uridine monophosphate to a specific tyrosine residue (Figure 24.28). This reaction, which is catalyzed by *uridylyl transferase*, is stimulated by ATP and α-ketoglutarate, whereas it is inhibited by glutamine. In turn, the UMP units on P_D are removed by hydrolysis, a reaction promoted by glutamine and inhibited by α-ketoglutarate. These opposing catalytic activities are present on a single polypeptide chain, homologous to adenylyl transferase, and are controlled so that the enzyme does not simultaneously catalyze uridylylation and hydrolysis.

Why is an enzymatic cascade used to regulate glutamine synthetase? One advantage of a cascade is that it *amplifies signals,* as in blood clotting and the control of glycogen metabolism (Sections 10.5.5 and 21.3.1). Another advantage is that the *potential for allosteric control is markedly increased when each enzyme in the cascade is an independent target for regulation.* The integration of nitrogen metabolism in a cell requires that a large number of input signals be detected and processed. In addition, the regulatory protein P also participates in regulating the transcription of genes for glutamine synthetase and other enzymes taking part in nitrogen metabolism. The evolution of a cascade provided many more regulatory sites and made possible a finer tuning of the flow of nitrogen in the cell.

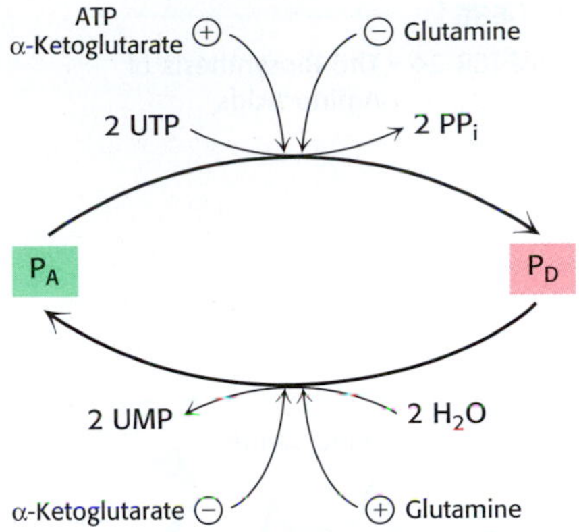

FIGURE 24.28 A higher level in the regulatory cascade of glutamine synthetase. P_A and P_D, the regulatory proteins that control the specificity of adenylyl transferase, are interconvertible. P_A is converted into P_D by uridylylation, which is reversed by hydrolysis. The enzymes catalyzing these reactions are regulated by the concentrations of metabolic intermediates.

24.4 AMINO ACIDS ARE PRECURSORS OF MANY BIOMOLECULES

In addition to being the building blocks of proteins and peptides, amino acids serve as precursors of many kinds of small molecules that have important and diverse biological roles. Let us briefly survey some of the biomolecules that are derived from amino acids (Figure 24.29).

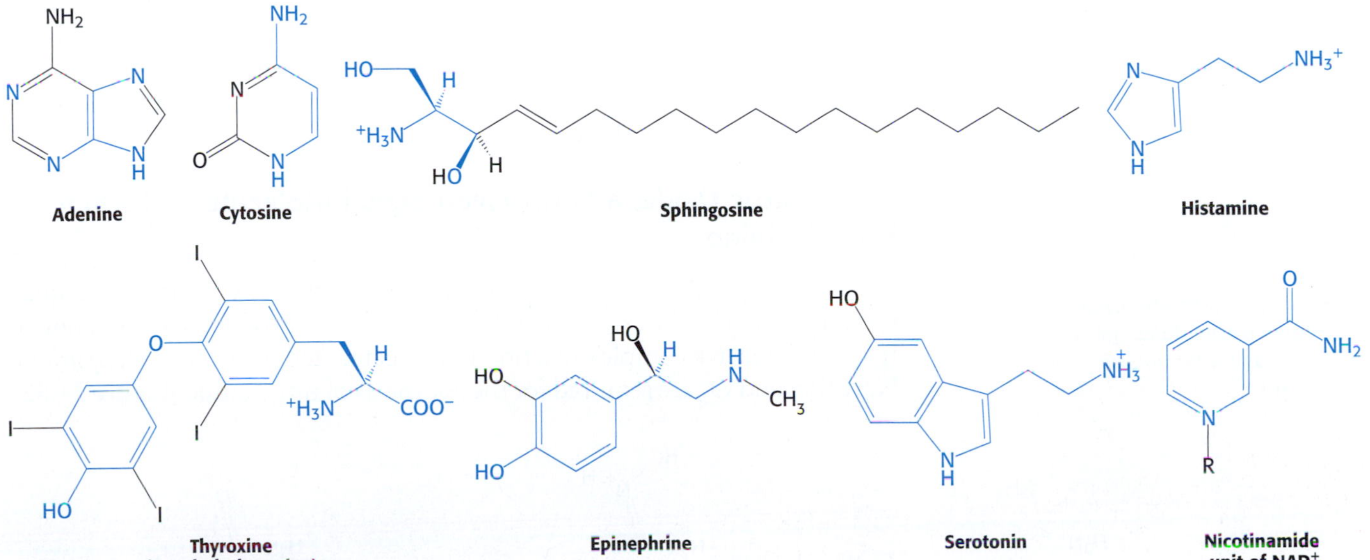

Adenine   **Cytosine**   **Sphingosine**   **Histamine**

Thyroxine (Tetraiodothyronine)   **Epinephrine**   **Serotonin**   **Nicotinamide unit of NAD⁺**

FIGURE 24.29 Selected biomolecules derived from amino acids. The atoms contributed by amino acids are shown in blue.

Purines and *pyrimidines* are derived largely from amino acids. The biosynthesis of these precursors of DNA, RNA, and numerous coenzymes will be discussed in detail in Chapter 25. The reactive terminus of *sphingosine*, an intermediate in the synthesis of sphingolipids, comes from serine. *Histamine,* a potent vasodilator, is derived from histidine by decarboxylation. Tyrosine is a precursor of the hormones *thyroxine* (tetraiodothyronine) and *epinephrine* and of *melanin,* a complex polymeric pigment. The neurotransmitter *serotonin* (5-hydroxytryptamine) and the *nicotinamide ring*

FIGURE 24.30 Glutathione. This
tripeptide consists of a cysteine
residue flanked by a glycine residue
and a glutamate residue that is
linked to cysteine by an isopeptide
bond between glutamate's side-
chain carboxylate group and
cysteine's amino group.

γ-Glutamate   Cysteine   Glycine

of NAD^+ are synthesized from tryptophan. Let us now consider in more detail three particularly important biochemicals derived from amino acids.

24.4.1 Glutathione, a Gamma-Glutamyl Peptide, Serves as a Sulfhydryl Buffer and an Antioxidant

Glutathione, a tripeptide containing a sulfhydryl group, is a highly distinctive amino acid derivative with several important roles (Figure 24.30). For example, glutathione, present at high levels (≈ 5 mM) in animal cells, protects red cells from oxidative damage by serving as a sulfhydryl buffer (Section 20.5.1). It cycles between a reduced thiol form (GSH) and an oxidized form (GSSG) in which two tripeptides are linked by a disulfide bond. GSSG is reduced to GSH by *glutathione reductase*, a flavoprotein that uses NADPH as the electron source. The ratio of GSH to GSSG in most cells is greater than 500. *Glutathione plays a key role in detoxification by reacting with hydrogen peroxide and organic peroxides, the harmful by-products of aerobic life.*

$$2\,GSH + RO{-}OH \rightleftharpoons GSSG + H_2O + ROH$$

Glutathione peroxidase, the enzyme catalyzing this reaction, is remarkable in having a modified amino acid containing a *selenium* (Se) atom (Figure 24.31). Specifically, its active site contains the selenium analog of cysteine, in which selenium has replaced sulfur. The selenolate ($E\text{-}Se^-$) form of this residue reduces the peroxide substrate to an alcohol and is in turn oxidized to selenenic acid ($E\text{-}SeOH$). Glutathione then comes into action by forming a selenosulfide adduct ($E\text{-}Se\text{-}S\text{-}G$). A second molecule of glutathione then regenerates the active form of the enzyme by attacking the selenosulfide to form oxidized glutathione (Figure 24.32).

24.4.2 Nitric Oxide, a Short-Lived Signal Molecule, Is Formed from Arginine

Nitric oxide (NO) is an important messenger in many vertebrate signal-transduction processes. This free-radical gas is produced endogenously from *arginine* in a complex reaction that is catalyzed by *nitric oxide synthase*. NADPH and O_2 are required for the synthesis of nitric oxide (Figure 24.33).

Selenocysteine

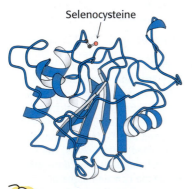

FIGURE 24.31 Structure of glutathione peroxidase. This enzyme, which has a role in peroxide detoxification, contains a selenocysteine residue in its active site.

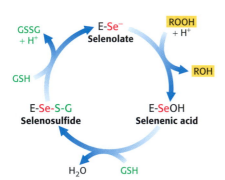

FIGURE 24.32 Catalytic cycle of glutathione peroxidase. [After O. Epp, R. Ladenstein, and A. Wendel. *Eur. J. Biochem.* 133(1983):51.]

Arginine

H⁺ + O₂ + NADPH → H₂O + NADP⁺

N-ω-Hydroxy-arginine

O₂ + NADPH → H₂O + NADP⁺

Citrulline   + ·NO   Nitric oxide

FIGURE 24.33 Formation of nitric oxide. NO is generated by the oxidation of arginine.

Nitric oxide acts by binding to and activating soluble guanylate cyclase, an important enzyme in signal transduction (Section 32.3.3). This enzyme is homologous to adenylate cyclase but includes a heme-containing domain that binds NO.

24.4.3 Mammalian Porphyrins Are Synthesized from Glycine and Succinyl Coenzyme A

The involvement of an amino acid in the biosynthesis of the porphyrin rings of hemes and chlorophylls was first revealed by the results of isotopic-labeling experiments carried out by David Shemin and his colleagues. In 1945, they showed that the nitrogen atoms of heme were labeled after the feeding of [^{15}N]glycine to human subjects (of whom Shemin was the first), whereas the ingestion of [^{15}N]glutamate resulted in very little labeling.

Using ^{14}C, which had just become available, they discovered that 8 of the carbon atoms of heme in nucleated duck erythrocytes are derived from the α-carbon atom of glycine and none from the carboxyl carbon atom. The results of subsequent studies demonstrated that the other 26 carbon atoms of heme can arise from acetate. Moreover, the ^{14}C in methyl-labeled acetate emerged in 24 of these carbons, whereas the ^{14}C in carboxyl-labeled acetate appeared only in the other 2 (Figure 24.34).

This highly distinctive labeling pattern led Shemin to propose that a heme precursor is formed by the condensation of glycine with an activated succinyl compound. In fact, *the first step in the biosynthesis of porphyrins in mammals is the condensation of glycine and succinyl CoA to form δ-aminolevulinate.*

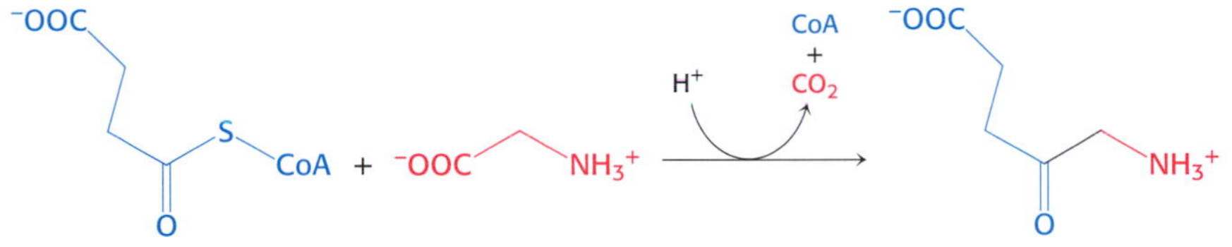

Succinyl CoA **Glycine** **δ-Aminolevulinate**

This reaction is catalyzed by *δ-aminolevulinate synthase,* a PLP enzyme present in mitochondria. Two molecules of δ-aminolevulinate condense to form *porphobilinogen,* the next intermediate. Four molecules of porphobilinogen then condense head to tail to form a linear *tetrapyrrole* in a reaction catalyzed by *porphobilinogen deaminase.* The enzyme-bound linear tetrapyrrole then cyclizes to form uroporphyrinogen III, which has an asymmetric arrangement of side chains. This reaction requires a *cosynthase.* In the presence of synthase alone, uroporphyrinogen I, the nonphysiologic symmetric isomer, is produced. Uroporphyrinogen III is also a key intermediate in the synthesis of vitamin B_{12} by bacteria and that of chlorophyll by bacteria and plants (Figure 24.35).

The porphyrin skeleton is now formed. Subsequent reactions alter the side chains and the degree of saturation of the porphyrin ring (see Figure 24.34). *Coproporphyrinogen III* is formed by the decarboxylation of the acetate side chains. The desaturation of the porphyrin ring and the conversion of two of the propionate side chains into vinyl groups yield *protoporphyrin IX.* The chelation of iron finally gives *heme,* the prosthetic group of proteins such as myoglobin, hemoglobin, catalase, peroxidase, and cytochrome

^{15}N labeling: A pioneer's account—
"Myself as a Guinea Pig
"... in 1944, I undertook, together with David Rittenberg, an investigation on the turnover of blood proteins of man. To this end I synthesized 66 g of glycine labeled with 35 percent ^{15}N at a cost of \$1000 for the ^{15}N. On 12 February 1945, I started the ingestion of the labeled glycine. Since we did not know the effect of relatively large doses of the stable isotope of nitrogen and since we believed that the maximum incorporation into the proteins could be achieved by the administration of glycine in some continual manner, I ingested 1 g samples of glycine at hourly intervals for the next 66 hours.... At stated intervals, blood was withdrawn and after proper preparation the ^{15}N concentrations of different blood proteins were determined."

—David Shemin
Bioessays 10(1989):30

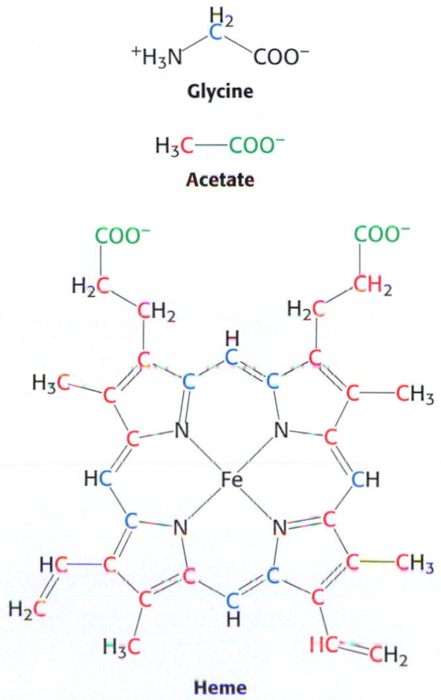

FIGURE 24.34 Heme labeling. The origins of atoms in heme revealed by the results of isotopic labeling studies.

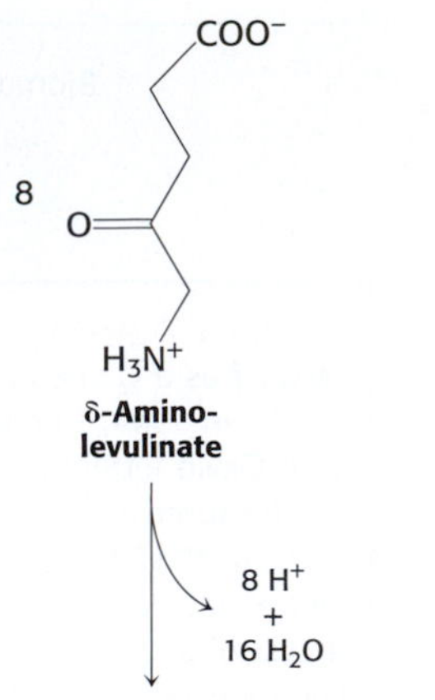

δ-Amino-
levulinate

8 H$^+$
+
16 H$_2$O

Acetate (A)
COO$^-$

Propionate (P)
COO$^-$

H$_3$N$^+$

Porphobilogen

Enz

A P A P A P A P

Linear tetrapyrrole

V = Vinyl

M = Methyl

Protoporphyrin IX

Coproporphyrinogen III

Uroporphyrinogen III

Iron

Heme

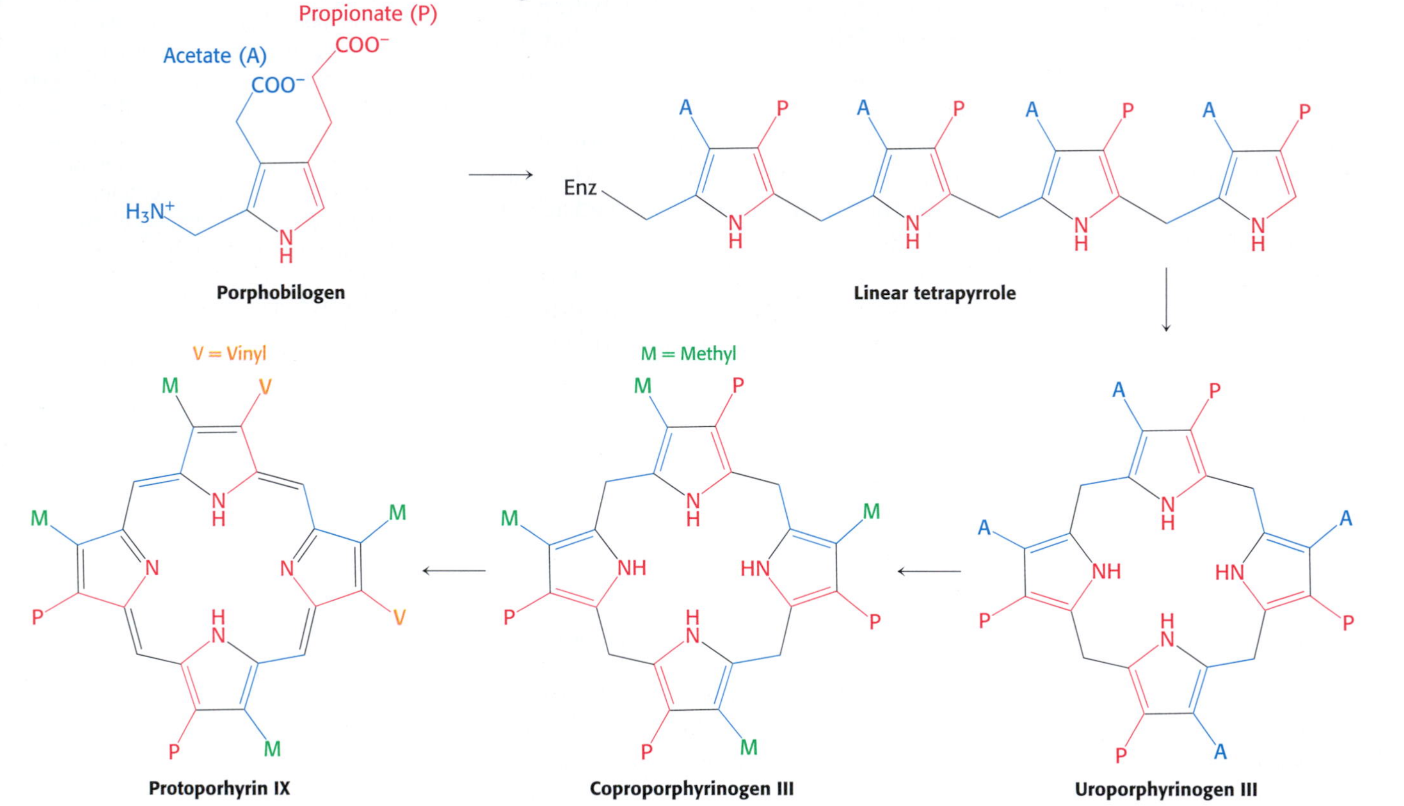

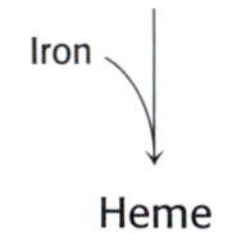

FIGURE 24.35 Heme biosynthetic pathway. The pathway for the formation of heme starts with eight molecules of δ-aminolevulinate.

c. The insertion of the *ferrous* form of iron is catalyzed by *ferrochelatase.* Iron is transported in the plasma by *transferrin,* a protein that binds two ferric ions, and stored in tissues inside molecules of *ferritin.* The large internal cavity (≈ 80 Å in diameter) of ferritin can hold as many as 4500 ferric ions (Section 31.4.2).

The normal human erythrocyte has a life span of about 120 days, as was first shown by the time course of ^{15}N in Shemin's own hemoglobin after he ingested ^{15}N-labeled glycine. The first step in the degradation of the heme group is the cleavage of its α-methene bridge to form the green pigment *biliverdin,* a linear tetrapyrrole. The central methene bridge of biliverdin is then reduced by *biliverdin reductase* to form *bilirubin,* a red pigment (Figure 24.36). The changing color of a bruise is a highly graphic indicator of these degradative reactions.

24.4.4 Porphyrins Accumulate in Some Inherited Disorders of Porphyrin Metabolism

Porphyrias are inherited or acquired disorders caused by a deficiency of enzymes in the heme biosynthetic pathway. Porphyrin is synthesized in both the erythroblasts and the liver, and either one may be the site of a disorder. *Congenital erythropoietic porphyria,* for example, prematurely destroys eythrocytes. This disease results from insufficient cosynthase. In this porphyria, the synthesis of the required amount of uroporphyrinogen III is accompanied by the formation of very large quantities of uroporphyrinogen I, the useless symmetric isomer. Uroporphyrin I, coproporphyrin I, and other symmetric derivatives also accumulate. The urine of patients having this disease is red because of the excretion of large amounts of uroporphyrin I. Their teeth exhibit a strong red fluorescence under ultraviolet light because of the deposition of porphyrins. Furthermore, their *skin*

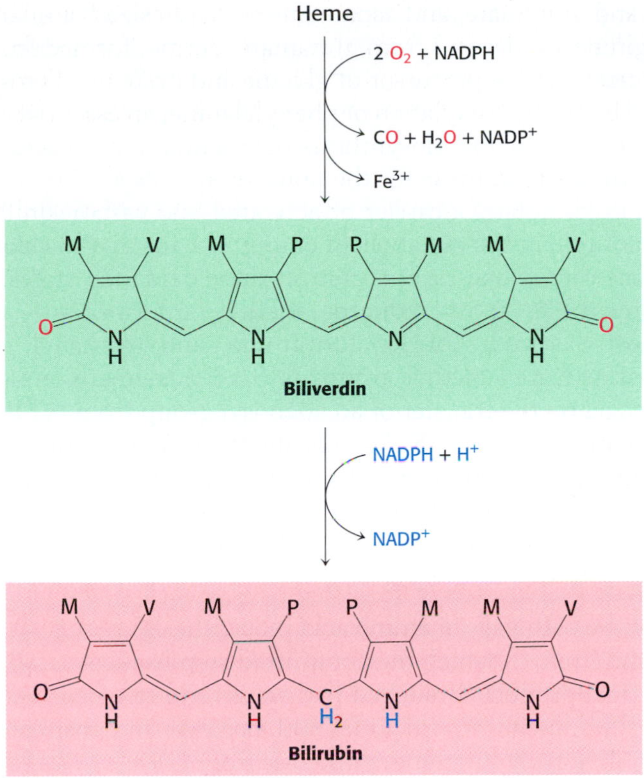

Heme

2 O_2 + NADPH

CO + H_2O + $NADP^+$

Fe^{3+}

Biliverdin

NADPH + H^+

$NADP^+$

Bilirubin

is usually very sensitive to light because photoexcited porphyrins are quite reactive. *Acute intermittent porphyria* is the most prevalent of the porphyrias affecting the liver. This porphyria is characterized by the overproduction of porphobilinogen and δ-aminolevulinate, which results in severe abdominal pain and neurological dysfunction. The "madness" of George III, King of England during the American Revolution, is believed to have been due to this porphyria.

SUMMARY

- ### Nitrogen Fixation: Microorganisms Use ATP and a Powerful Reductant to Reduce Atmospheric Nitrogen to Ammonia

  Microorganisms use ATP and reduced ferredoxin, a powerful reductant, to reduce N_2 to NH_3. An iron–molybdenum cluster in nitrogenase deftly catalyzes the fixation of N_2, a very inert molecule. Higher organisms consume the fixed nitrogen to synthesize amino acids, nucleotides, and other nitrogen-containing biomolecules. The major points of entry of NH_4^+ into metabolism are glutamine or glutamate.

- ### Amino Acids Are Made from Intermediates of the Citric Acid Cycle and Other Major Pathways

  Human beings can synthesize 11 of the basic set of 20 amino acids. These amino acids are called nonessential, in contrast with the essential amino acids, which must be supplied in the diet. The pathways for the synthesis of nonessential amino acids are quite simple. Glutamate dehydrogenase catalyzes the reductive amination of α-ketoglutarate to glutamate. A transamination reaction takes place in the synthesis of most amino acids. At this step, the chirality of the amino acid is established. Alanine and aspartate are synthesized by the transamination of pyruvate and oxaloacetate, respectively. Glutamine is synthesized from

NH_4^+ and glutamate, and asparagine is synthesized similarly. Proline and arginine are derived from glutamate. Serine, formed from 3-phosphoglycerate, is the precursor of glycine and cysteine. Tyrosine is synthesized by the hydroxylation of phenylalanine, an essential amino acid. The pathways for the biosynthesis of essential amino acids are much more complex than those for the nonessential ones.

Tetrahydrofolate, a carrier of activated one-carbon units, plays an important role in the metabolism of amino acids and nucleotides. This coenzyme carries one-carbon units at three oxidation states, which are interconvertible: most reduced—methyl; intermediate—methylene; and most oxidized—formyl, formimino, and methenyl. The major donor of activated methyl groups is S-adenosylmethionine, which is synthesized by the transfer of an adenosyl group from ATP to the sulfur atom of methionine. S-Adenosylhomocysteine is formed when the activated methyl group is transferred to an acceptor. It is hydrolyzed to adenosine and homocysteine, the latter of which is then methylated to methionine to complete the activated methyl cycle.

Amino Acid Biosynthesis Is Regulated by Feedback Inhibition

Most of the pathways of amino acid biosynthesis are regulated by feedback inhibition, in which the committed step is allosterically inhibited by the final product. Branched pathways require extensive interaction among the branches that includes both negative and positive regulation. The regulation of glutamine synthetase from *E. coli* is a striking demonstration of cumulative feedback inhibition and of control by a cascade of reversible covalent modifications.

Amino Acids Are Precursors of Many Biomolecules

Amino acids are precursors of a variety of biomolecules. Glutathione (γ-Glu-Cys-Gly) serves as a sulfhydryl buffer and detoxifying agent. Glutathione peroxidase, a selenoenzyme, catalyzes the reduction of hydrogen peroxide and organic peroxides by glutathione. Nitric oxide, a short-lived messenger, is formed from arginine. Porphyrins are synthesized from glycine and succinyl CoA, which condense to give δ-aminolevulinate. Two molecules of this intermediate become linked to form porphobilinogen. Four molecules of porphobilinogen combine to form a linear tetrapyrrole, which cyclizes to uroporphyrinogen III. Oxidation and side-chain modifications lead to the synthesis of protoporphyrin IX, which acquires an iron atom to form heme.

KEY TERMS

nitrogen fixation (p. 666)

nitrogenase complex (p. 667)

essential amino acids (p. 671)

nonessential amino acids (p. 671)

pyridoxal phosphate (p. 671)

tetrahydrofolate (p. 674)

S-adenosylmethionine (SAM) (p. 676)

activated methyl cycle (p. 677)

substrate channeling (p. 681)

committed step (p. 682)

enzyme multiplicity (p. 683)

cumulative feedback inhibition (p. 683)

SELECTED READINGS

Where to start

Kim, J., and Rees, D. C., 1989. Nitrogenase and biological nitrogen fixation. *Biochemistry* 33:389–397.

Christen, P., Jaussi, R., Juretic, N.,Mehta, P. K., Hale, T. I., and Ziak, M., 1990. Evolutionary and biosynthetic aspects of aspartate aminotransferase isoenzymes and other aminotransferases. *Ann. N. Y. Acad. Sci.* 585:331–338.

Schneider, G., Kack, H., and Lindqvist, Y., 2000. The manifold of vitamin B6 dependent enzymes. *Structure Fold Des.* 8:R1–R6.

Rhee, S. G., Chock, P. B., and Stadtman, E. R., 1989. Regulation of *Escherichia coli* glutamine synthetase. *Adv. Enzymol. Mol. Biol.* 62: 37–92.

Shemin, D., 1989. An illustration of the use of isotopes: The biosynthesis of porphyrins. *Bioessays* 10:30–35.

Books

Bender, D. A., 1985. *Amino Acid Metabolism* (2d ed.). Wiley.

Jordan, P. M. (Ed.), 1991. *Biosynthesis of Tetrapyrroles.* Elsevier.

Scriver, C. R., Beaudet, A. L., Sly, W. S., and Valle, D. (Eds.), 1995. *The Metabolic Basis of Inherited Disease* (7th ed.). McGraw-Hill

Meister, A., 1965. *Biochemistry of the Amino Acids* (vols. 1 and 2, 2d ed.). Academic Press.

Blakley, R. L., and Benkovic, S. J., 1989. *Folates and Pterins* (vol. 2). Wiley.

Walsh, C., 1979. *Enzymatic Reaction Mechanisms.* W. H. Freeman and Company.

Nitrogen fixation

Halbleib, C. M., and Ludden, P. W., 2000. Regulation of biological nitrogen fixation. *J. Nutr.* 130:1081–1084.

Mayer, S. M., Lawson, D. M., Gormal, C. A., Roe, S. M., and Smith, B. E., 1999. New insights into structure-function relationships in nitrogenase: A 1.6 Å resolution X-ray crystallographic study of *Klebsiella pneumoniae* MoFe-protein. *J. Mol. Biol.* 292:871–891.

Peters, J. W., Fisher, K., and Dean, D. R., 1995. Nitrogenase structure and function: A biochemical-genetic perspective. *Annu. Rev. Microbiol.* 49:335–366.

Leigh, G. J., 1995. The mechanism of dinitrogen reduction by molybdenum nitrogenases. *Eur. J. Biochem.* 229:14–20.

Chan, M. K., Kim, J., and Rees, D. C., 1993. The nitrogenase FeMo-cofactor and P-cluster pair: 2.2 Å resolution studies. *Science* 260:792–794.

Georgiadis, M. M., Komiya, H., Chakrabarti, P., Woo, D., Kornuc, J. J., and Rees, D. C., 1992. Crystallographic structure of the nitrogenase iron protein from *Azotobacter vinelandii. Science* 257:1653–1659.

Regulation of amino acid biosynthesis

Eisenberg, D., Gill, H. S., Pfluegl, G. M., and Rotstein, S. H., 2000. Structure-function relationships of glutamine synthetases. *Biochim. Biophys. Acta* 1477:122–145.

Purich, D. L., 1998. Advances in the enzymology of glutamine synthesis. *Adv. Enzymol. Relat. Areas Mol. Biol.* 72:9–42.

Yamashita, M. M., Almassy, R. J., Janson, C. A., Cascio, D., and Eisenberg, D., 1989. Refined atomic model of glutamine synthetase at 3.5 Å resolution. *J. Biol. Chem.* 264:17681–17690.

Schuller, D. J., Grant, G. A., and Banaszak, L. J., 1995. The allosteric ligand site in the V_{max}-type cooperative enzyme phosphoglycerate dehydrogenase. *Nat. Struct. Biol.* 2:69–76.

Rhee, S. G., Park, R., Chock, P. B., and Stadtman, E. R., 1978. Allosteric regulation of monocyclic interconvertible enzyme cascade systems: Use of *Escherichia coli* glutamine synthetase as an experimental model. *Proc. Natl. Acad. Sci. USA* 75:3138–3142.

Wessel, P. M., Graciet, E., Douce, R., and Dumas, R., 2000. Evidence for two distinct effector-binding sites in threonine deaminase by site-directed mutagenesis, kinetic, and binding experiments. *Biochemistry* 39: 15136–15143.

Xu, Y., Carr, P. D., Huber, T., Vasudevan, S. G., and Ollis, D. L., 2001. The structure of the PII-ATP complex. *Eur. J. Biochem.* 268:2028–2037.

Aromatic amino acid biosynthesis

Pan, P., Woehl, E., and Dunn, M. F., 1997. Protein architecture, dynamics and allostery in tryptophan synthase channeling. *Trends Biochem. Sci.* 22:22–27.

Sachpatzidis, A., Dealwis, C., Lubetsky, J. B., Liang, P. H., Anderson, K. S., and Lolis, E., 1999. Crystallographic studies of phosphonate-based alpha-reaction transition-state analogues complexed to tryptophan synthase. *Biochemistry* 38:12665–12674.

Weyand, M., and Schlichting, I., 1999. Crystal structure of wild-type tryptophan synthase complexed with the natural substrate indole-3-glycerol phosphate. *Biochemistry* 38:16469–16480.

Crawford, I. P., 1989. Evolution of a biosynthetic pathway: The tryptophan paradigm. *Annu. Rev. Microbiol.* 43:567–600.

Carpenter, E. P., Hawkins, A. R., Frost, J. W., and Brown, K. A., 1998. Structure of dehydroquinate synthase reveals an active site capable of multistep catalysis. *Nature* 394:299–302.

Schlichting, I., Yang, X. J., Miles, E. W., Kim, A. Y., and Anderson, K. S., 1994. Structural and kinetic analysis of a channel-impaired mutant of tryptophan synthase. *J. Biol. Chem.* 269:26591–26593.

Glutathione

Edwards, R., Dixon, D. P., and Walbot, V., 2000. Plant glutathione S-transferases: Enzymes with multiple functions in sickness and in health. *Trends Plant Sci.* 5:193–198.

Lu, S. C., 2000. Regulation of glutathione synthesis. *Curr. Top. Cell Regul.* 36:95–116.

Schulz, J. B., Lindenau, J., Seyfried, J., and Dichgans, J., 2000. Glutathione, oxidative stress and neurodegeneration. *Eur. J. Biochem.* 267:4904–4911.

Lu, S. C., 1999. Regulation of hepatic glutathione synthesis: Current concepts and controversies. *FASEB J.* 13:1169–1183.

Salinas, A. E., and Wong, M. G., 1991. Glutathione S-transferases: A review. *Curr. Med. Chem.* 6:279–309.

Ethylene and nitric oxide

Haendeler, J., Zeiher, A. M., and Dimmeler, S., 1999. Nitric oxide and apoptosis. *Vitam. Horm.* 57:49–77.

Capitani, G., Hohenester, E., Feng, L., Storici, P., Kirsch, J. F., and Jansonius, J. N., 1999. Structure of 1-aminocyclopropane-1-carboxylate synthase, a key enzyme in the biosynthesis of the plant hormone ethylene. *J. Mol. Biol.* 294:745–756.

Hobbs, A. J., Higgs, A., and Moncada, S., 1999. Inhibition of nitric oxide synthase as a potential therapeutic target. *Annu. Rev. Pharmacol. Toxicol.* 39:191–220.

Stuehr, D. J., 1999. Mammalian nitric oxide synthases. *Biochim. Biophys. Acta* 1411:217–230.

Chang, C., and Shockey, J. A., 1999. The ethylene-response pathway: Signal perception to gene regulation. *Curr. Opin. Plant Biol.* 2:352–358.

Johnson, P. R., and Ecker, J. R., 1998. The ethylene gas signal transduction pathway: A molecular perspective. *Annu. Rev. Genet.* 32:227–254.

Theologis, A., 1992. One rotten apple spoils the whole bushel: The role of ethylene in fruit ripening. *Cell* 70:181–184.

Biosynthesis of porphyrins

Leeper, F. J., 1989. The biosynthesis of porphyrins, chlorophylls, and vitamin B_{12}. *Nat. Prod. Rep.* 6:171–199.

Porra, R. J., and Meisch, H.-U., 1984. The biosynthesis of chlorophyll. *Trends Biochem. Sci.* 9:99–104.

PROBLEMS

1. *From sugar to amino acid.* Write a balanced equation for the synthesis of alanine from glucose.

2. *From air to blood.* What are the intermediates in the flow of nitrogen from N_2 to heme?

3. *One-carbon transfers.* Which derivative of folate is a reactant in the conversion of

(a) glycine into serine?

(b) homocysteine into methionine?

4. *Telltale tag.* In the reaction catalyzed by glutamine synthetase, an oxygen atom is transferred from the side chain of glutamate to orthophosphate, as shown by the results of ^{18}O-labeling studies. Account for this finding.

5. *Therapeutic glycine.* Isovaleric acidemia is an inherited disorder of leucine metabolism caused by a deficiency of isovaleryl CoA dehydrogenase. Many infants having this disease die in the first month of life. The administration of large amounts of glycine sometimes leads to marked clinical improvement. Propose a mechanism for the therapeutic action of glycine.

6. *Deprived algae.* Blue-green algae (cyanobacteria) form *heterocysts* when deprived of ammonia and nitrate. In this form, the algae lack nuclei and are attached to adjacent vegetative cells. Heterocysts have photosystem I activity but are entirely devoid of photosystem II activity. What is their role?

7. *Cysteine and cystine.* Most cytosolic proteins lack disulfide bonds, whereas extracellular proteins usually contain them. Why?

8. *To and fro.* The synthesis of δ-aminolevulinate takes place in the mitochondrial matrix, whereas the formation of porphobilinogen takes place in the cytosol. Propose a reason for the mitochondrial location of the first step in heme synthesis.

9. *Direct synthesis.* Which of the 20 amino acids can be synthesized directly from a common metabolic intermediate by a transamination reaction?

10. *Lines of communication.* For the following example of a branched pathway, propose a feedback inhibition scheme that would result in the production of equal amounts of Y and Z.

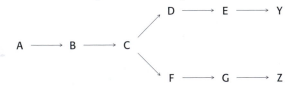

11. *Cumulative feedback inhibition.* The first common step (A → B) is partly inhibited by both of the final products, each acting independently of the other. Suppose that a high level of Y alone decreased the rate of the A → B step from 100 to 60 s^{-1} and that a high level of Z alone decreased the rate from 100 to 40 s^{-1}. What would the rate be in the presence of high levels of both Y and Z?

Mechanism Problems

12. *Ethylene formation.* Propose a mechanism for the conversion of S-adenosylmethionine into 1-aminocyclopropane-1-carboxylate (ACC) by ACC synthase, a PLP enzyme. What is the other product?

13. *Mirror-image serine.* Brain tissue contains substantial amounts of D-serine which is generated from L-serine by serine racemase, a PLP enzyme. Propose a mechanism for the interconversion of L- and D-serine. What is the equilibrium constant for the reaction L-serine ⇌ D-serine?

Chapter Integration Problems

14. *Connections.* How might increased synthesis of aspartate and glutamate affect energy production in a cell? How would the cell respond to such an effect?

15. *Protection required.* Suppose that a mutation in bacteria resulted in diminished activity of methionine adenosyltransferase, the enzyme responsible for the synthesis of SAM from methionine and ATP. Predict how this might affect the stability of the mutated bacteria's DNA.

Chapter Integration and Data Interpretation Problem

16. *Light effects.* The adjoining graph shows the concentration of several free amino acids in light- and dark-adapted plants.

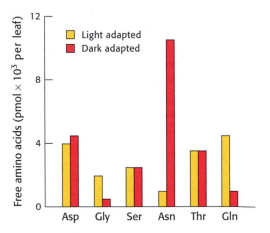

[After B. B. Buchanan, W. Gruissem, and R. L. Jones, *Biochemistry and Molecular Biology of Plants.* (American Society of Plant Physiology, 2000), Figure 8.3, p. 363.]

(a) Of the amino acids shown, which are most affected by light–dark adaptation?
(b) Suggest a plausible biochemical explanation for the difference observed.
(c) White asparagus, a culinary delicacy, is the result of growing asparagus plants in the dark. What chemical might you think enhances the taste of white asparagus?

CHAPTER 25

Nucleotide Biosynthesis

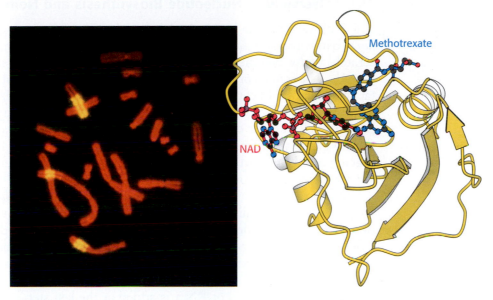

Methotrexate

NAD

Nucleotides are required for cell growth and replication. A key enzyme for the synthesis of one nucleotide is dihydrofolate reductase (right). Cells grown in the presence of methotrexate, a reductase inhibitor, respond by increasing the number of copies of the reductase gene. The bright yellow regions visible on three of the chromosomes in the fluorescence micrograph (left), which were grown in the presence of methotrexate, contain hundreds of copies of the reductase gene. [(Left) Courtesy of Dr. Barbara Trask and Dr. Joyce Hamlin.]

An ample supply of nucleotides is essential for many life processes. First, nucleotides are the *activated precursors of nucleic acids*. As such, they are necessary for the replication of the genome and the transcription of the genetic information into RNA. Second, an adenine nucleotide, ATP, is *the universal currency of energy*. A guanine nucleotide, GTP, also serves as an energy source for a more select group of biological processes. Third, nucleotide derivatives such as UDP-glucose *participate in biosynthetic processes* such as the formation of glycogen. Fourth, nucleotides are *essential components of signal-transduction pathways*. Cyclic nucleotides such as cyclic AMP and cyclic GMP are second messengers that transmit signals both within and between cells. ATP acts as the donor of phosphoryl groups transferred by protein kinases.

In this chapter, we continue along the path begun in Chapter 24, which described the incorporation of nitrogen into amino acids from inorganic sources such as nitrogen gas. The amino acids glycine and aspartate are the scaffolds on which the ring systems present in nucleotides are assembled. Furthermore, aspartate and the side chain of glutamine serve as sources of NH_2 groups in the formation of nucleotides.

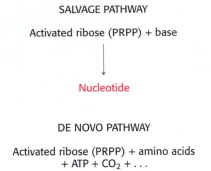

SALVAGE PATHWAY

Activated ribose (PRPP) + base

↓

Nucleotide

DE NOVO PATHWAY

Activated ribose (PRPP) + amino acids
+ ATP + CO₂ + . . .

↓

Nucleotide

FIGURE 25.1 Salvage and de novo pathways. In a salvage pathway, a base is reattached to a ribose, activated in the form of 5-phosphoribosyl-1-pyrophosphate (PRPP). In de novo synthesis, the base itself is synthesized from simpler starting materials, including amino acids. ATP hydrolysis is required for de novo synthesis.

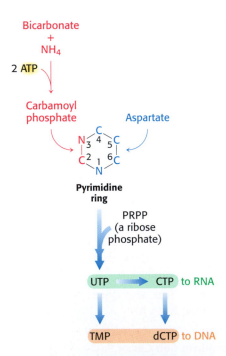

FIGURE 25.2 De novo pathway for pyrimidine nucleotide synthesis. The C-2 and N-3 atoms in the pyrimidine ring come from carbamoyl phosphate, whereas the other atoms of the ring come from aspartate.

Nucleotide biosynthetic pathways are tremendously important as intervention points for therapeutic agents. Many of the most widely used drugs in the treatment of cancer block steps in nucleotide biosynthesis, particularly steps in the synthesis of DNA precursors.

25.0.1 Overview of Nucleotide Biosynthesis and Nomenclature

The pathways for the biosynthesis of nucleotides fall into two classes: *de novo* pathways and *salvage* pathways (Figure 25.1). In de novo (from scratch) pathways, the nucleotide bases are assembled from simpler compounds. The framework for a *pyrimidine* base is assembled first and then attached to ribose. In contrast, the framework for a *purine* base is synthesized piece by piece directly onto a ribose-based structure. These pathways comprise a small number of elementary reactions that are repeated with variation to generate different nucleotides, as might be expected for pathways that appeared very early in evolution. In salvage pathways, preformed bases are recovered and reconnected to a ribose unit.

Both de novo and salvage pathways lead to the synthesis of *ribo*nucleotides. However, DNA is built from *deoxyribo*nucleotides. Consistent with the notion that RNA preceded DNA in the course of evolution, all deoxyribonucleotides are synthesized from the corresponding ribonucleotides. The deoxyribose sugar is generated by the reduction of ribose within a fully formed nucleotide. Furthermore, the methyl group that distinguishes the thymine of DNA from the uracil of RNA is added at the last step in the pathway.

The nomenclature of nucleotides and their constituent units was presented earlier (Section 5.1.2). Recall that a *nucleoside* consists of a purine or pyrimidine base linked to a sugar and that a *nucleotide* is a phosphate ester of a nucleoside. The names of the major bases of RNA and DNA, and of their nucleoside and nucleotide derivatives, are given in Table 25.1.

TABLE 25.1   Nomenclature of bases, nucleosides, and nucleotides

RNA		
Base	Ribonucleoside	Ribonucleotide (5′-monophosphate)
Adenine (A)	Adenosine	Adenylate (AMP)
Guanine (G)	Guanosine	Guanylate (GMP)
Uracil (U)	Uridine	Uridylate (UMP)
Cytosine (C)	Cytidine	Cytidylate (CMP)
DNA		
Base	Deoxyribonucleoside	Deoxyribonucleotide (5′-monophosphate)
Adenine (A)	Deoxyadenosine	Deoxyadenylate (dAMP)
Guanine (G)	Deoxyguanosine	Deoxyguanylate (dGMP)
Thymine (T)	Thymidine	Thymidylate (TMP)
Cytosine (C)	Deoxycytidine	Deoxycytidylate (dCMP)

25.1 IN DE NOVO SYNTHESIS, THE PYRIMIDINE RING IS ASSEMBLED FROM BICARBONATE, ASPARTATE, AND GLUTAMINE

In de novo synthesis of pyrimidines, the ring is synthesized first and then it is attached to ribose to form a *pyrimidine nucleotide* (Figure 25.2). Pyrimidine rings are assembled from bicarbonate, aspartic acid, and ammonia.

Although ammonia can be used directly, it is usually produced from the hydrolysis of the side chain of glutamine.

25.1.1 Bicarbonate and Other Oxygenated Carbon Compounds Are Activated by Phosphorylation

The first step in de novo pyrimidine biosynthesis is the synthesis of *carbamoyl phosphate* from bicarbonate and ammonia in a multistep process, requiring the cleavage of two molecules of ATP. This reaction is catalyzed by *carbamoyl phosphate synthetase (CPS)* (Section 23.4.1). Analysis of the structure of CPS reveals two homologous domains, each of which catalyzes an ATP-dependent step (Figure 25.3).

In the first step of the carbamoyl phosphate synthesis pathway, bicarbonate is phosphorylated by ATP to form carboxyphosphate and ADP. Ammonia then reacts with carboxyphosphate to form carbamic acid and inorganic phosphate.

Bicarbonate → (ATP, ADP) → **Carboxyphosphate** → (NH_3, P_i) → **Carbamic acid**

The active site for this reaction lies in a domain formed by the amino-terminal third of CPS. This domain forms a structure, called an *ATP-grasp fold*, that surrounds ATP and holds it in an orientation suitable for nucleophilic attack at the γ phosphoryl group. Proteins containing ATP-grasp folds catalyze the formation of carbon–nitrogen bonds through acyl-phosphate intermediates and are widely used in nucleotide biosynthesis. In the final step catalyzed by carbamoyl phosphate synthetase, carbamic acid is phosphorylated by another molecule of ATP to form carbamoyl phosphate.

Carbamic acid → (ATP, ADP) → **Carbamoyl phosphate**

This reaction takes place in a second ATP-grasp domain within the enzyme. The active sites leading to carbamic acid formation and carbamoyl phosphate formation are very similar, revealing that this enzyme evolved by a gene duplication event. Indeed, duplication of a gene encoding an ATP-grasp domain followed by specialization was central to the evolution of nucleotide biosynthetic processes (Section 25.2.3).

25.1.2 The Side Chain of Glutamine Can Be Hydrolyzed to Generate Ammonia

Carbamoyl phosphate synthetase primarily uses glutamine as a source of ammonia. In this case, a second polypeptide component of the carbamoyl phosphate synthetase enzyme hydrolyzes glutamine to form ammonia and glutamate. The active site of the glutamine-hydrolyzing component of carbamoyl phosphate synthetase contains a catalytic dyad comprising a cysteine and a histidine residue (Figure 25.4). Such a catalytic dyad, reminiscent of the active site of cysteine proteases (Section 9.1.6), is conserved in a

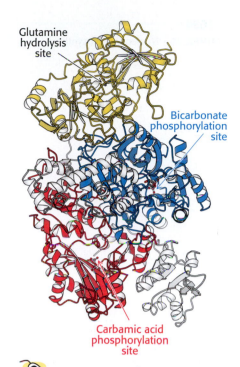

FIGURE 25.3 Structure of carbamoyl phosphate synthetase. This enzyme consists of two chains. The smaller chain (yellow) contains a site for glutamine hydrolysis to generate ammonia. The larger chain includes two ATP-grasp domains (blue and red). In one ATP-grasp domain (blue), bicarbonate is phosphorylated to carboxyphosphate, which then reacts with ammonia to generate carbamic acid. In the other ATP-grasp domain, the carbamic acid is phosphorylated to produce carbamoyl phosphate.

Glutamine hydrolysis site

Bicarbonate phosphorylation site

Carbamic acid phosphorylation site

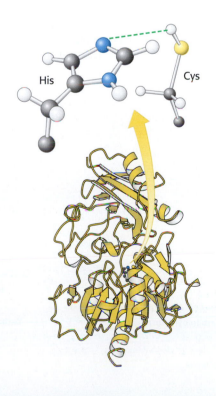

His   Cys

FIGURE 25.4 Ammonia-generation site. The smaller domain of carbamoyl phosphate synthetase contains an active site for the hydrolysis of the side chain carboxamide of glutamine to generate ammonia. Key residues in this active site include a cysteine residue and a histidine residue.

family of amidotransferases, including CTP synthetase (Section 25.1.6) and GMP synthetase (Section 25.2.4).

25.1.3 Intermediates Can Move Between Active Sites by Channeling

Carbamoyl phosphate synthetase contains three different active sites (see Figure 25.3), separated from one another by a total of 80 Å (Figure 25.5). Intermediates generated at one site move to the next without leaving the enzyme; that is, they move by means of substrate channeling, similar to the process described for tryptophan synthetase (Section 24.2.11). The ammonia generated in the glutamine-hydrolysis active site travels 45 Å through a channel within the enzyme to reach the site at which carboxyphosphate has been generated. The carbamic acid generated at this site diffuses an additional 35 Å through an extension of the channel to reach the site at which carbamoyl phosphate is generated. This channeling serves two roles: (1) intermediates generated at one active site are captured with no loss caused by diffusion; and (2) labile intermediates, such as carboxyphosphate and carbamic acid (which decompose in less than 1 s at pH 7), are protected from hydrolysis.

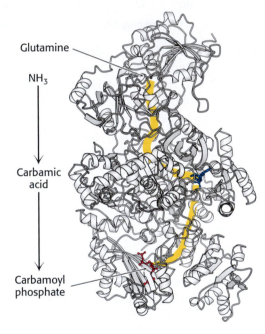

Glutamine

NH_3

Carbamic acid

Carbamoyl phosphate

FIGURE 25.5 Substrate channeling. The three active sites of carbamoyl phosphate synthetase are linked by a channel (yellow) through which intermediates pass. Glutamine enters one active site, and carbamoyl phosphate, which includes the nitrogen atom from the glutamine side chain, leaves another 80 Å away.

25.1.4 Orotate Acquires a Ribose Ring from PRPP to Form a Pyrimidine Nucleotide and Is Converted into Uridylate

Carbamoyl phosphate reacts with aspartate to form carbamoylaspartate in a reaction catalyzed by *aspartate transcarbamoylase* (Section 10.1). Carbamoylaspartate then cyclizes to form dihydroorotate which is then oxidized by NAD^+ to form orotate.

Carbamoyl phosphate — Aspartate, P_i → Carbamoylaspartate — H^+, H_2O → Dihydroorotate — NAD^+, NADH + H^+ → Orotate

At this stage, orotate couples to ribose, in the form of *5-phosphoribosyl-1-pyrophosphate (PRPP)*, a form of ribose activated to accept nucleotide bases. PRPP is synthesized from ribose-5-phosphate, formed by the pentose phosphate pathway, by the addition of pyrophosphate from ATP. Orotate reacts with PRPP to form *orotidylate*, a pyrimidine nucleotide. This reaction is driven by the hydrolysis of pyrophosphate. The enzyme that catalyzes this addition, *pyrimidine phosphoribosyltransferase*, is homologous to a number of other phosphoribosyltransferases that add different groups to PRPP to form the other nucleotides. Orotidylate is then decarboxylated to form *uridylate (UMP),* a major pyrimidine nucleotide that is a precursor to RNA. This reaction is catalyzed by *orotidylate decarboxylase.*

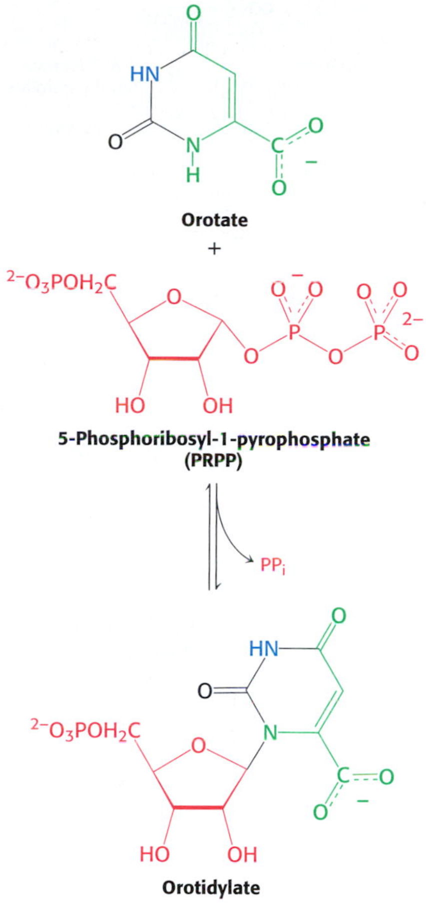

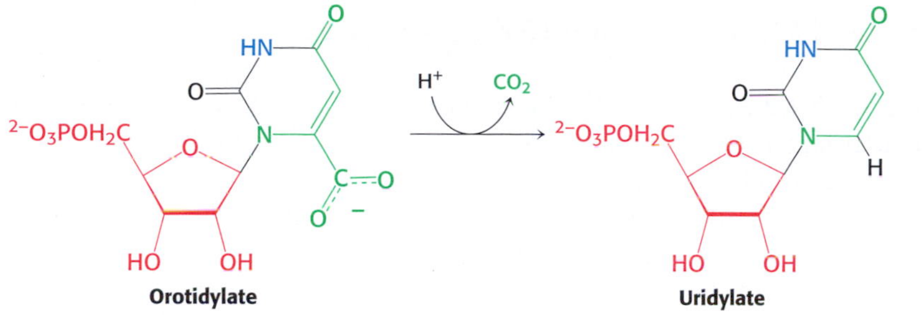

Orotate
+

5-Phosphoribosyl-1-pyrophosphate (PRPP)

PP$_i$

Orotidylate → **Uridylate**

Orotidylate

This enzyme is one of the most proficient enzymes known. In its absence, decarboxylation is extremely slow and is estimated to take place once every 78 million years; with the enzyme present, it takes place approximately once per second, a rate enhancement of 10^{17}-fold!

25.1.5 Nucleotide Mono-, Di-, and Triphosphates Are Interconvertible

How is the other major pyrimidine ribonucleotide, cytidine, formed? It is synthesized from the uracil base of UMP, but UMP is converted into UTP before the synthesis can take place. Recall that the diphosphates and triphosphates are the active forms of nucleotides in biosynthesis and energy conversions. Nucleoside monophosphates are converted into nucleoside triphosphates in stages. First, nucleoside monophosphates are converted into diphosphates by specific *nucleoside monophosphate kinases* that utilize ATP as the phosphoryl-group donor (Section 9.4). For example, UMP is phosphorylated to UDP by *UMP kinase.*

$$\text{UMP} + \text{ATP} \rightleftharpoons \text{UDP} + \text{ADP}$$

Nucleoside diphosphates and triphosphates are interconverted by *nucleoside diphosphate kinase,* an enzyme that has broad specificity, in contrast with the monophosphate kinases. X and Y can represent any of several ribonucleosides or even deoxyribonucleosides.

$$\text{XDP} + \text{YTP} \rightleftharpoons \text{XTR} + \text{YDP}$$

25.1.6 CTP Is Formed by Amination of UTP

After uridine triphosphate has been formed, it can be transformed into *cytidine triphosphate* by the replacement of a carbonyl group by an amino group.

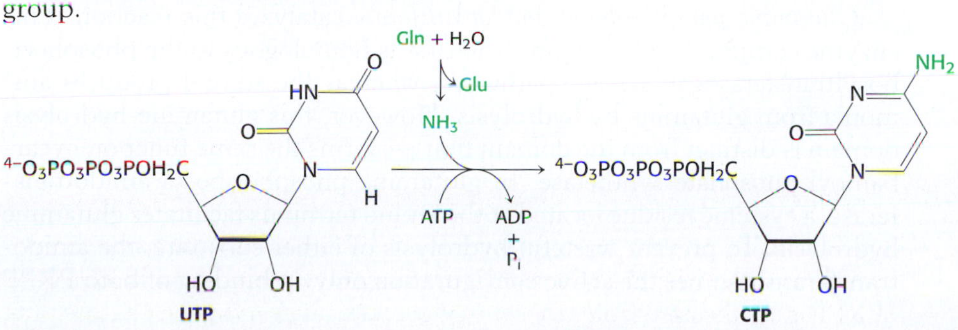

Gln + H$_2$O
Glu
NH$_3$
ATP   ADP + P$_i$

UTP → **CTP**

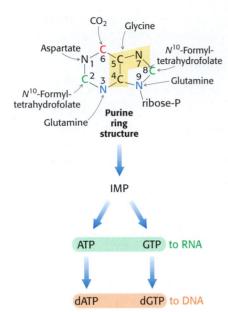

FIGURE 25.6 De novo pathway for purine nucleotide synthesis. The origins of the atoms in the purine ring are indicated.

Like the synthesis of carbamoyl phosphate, this reaction requires ATP and uses glutamine as the source of the amino group. The reaction proceeds through an analogous mechanism in which the O-4 atom is phosphorylated to form a reactive intermediate, and then the phosphate is displaced by ammonia, freed from glutamine by hydrolysis. CTP can then be used in many biochemical processes, including RNA synthesis.

25.2 PURINE BASES CAN BE SYNTHESIZED DE NOVO OR RECYCLED BY SALVAGE PATHWAYS

Purine nucleotides can be synthesized in two distinct pathways. First, purines are synthesized de novo, beginning with simple starting materials such as amino acids and bicarbonate (Figure 25.6). Unlike the case for pyrimidines, the purine bases are assembled already attached to the ribose ring. Alternatively, purine bases, released by the hydrolytic degradation of nucleic acids and nucleotides, can be salvaged and recycled. Purine salvage pathways are especially noted for the energy that they save and the remarkable effects of their absence (Section 25.6.2).

25.2.1 Salvage Pathways Economize Intracellular Energy Expenditure

Free purine bases, derived from the turnover of nucleotides or from the diet, can be attached to PRPP to form purine nucleoside monophosphates, in a reaction analogous to the formation of orotidylate. Two salvage enzymes with different specificities recover purine bases. *Adenine phosphoribosyltransferase* catalyzes the formation of adenylate

$$\text{Adenine} + \text{PRPP} \longrightarrow \text{adenylate} + \text{PP}_i$$

whereas *hypoxanthine-guanine phosphoribosyltransferase* (*HGPRT*) catalyzes the formation of guanylate as well as *inosinate* (inosine monophosphate, IMP), a precursor of guanylate and adenylate (Section 25.2.4).

$$\text{Guanine} + \text{PRPP} \longrightarrow \text{guanylate} + \text{PP}_i$$

$$\textit{Hypoxanthine} + \text{PRPP} \longrightarrow \text{inosinate} + \text{PP}_i$$

Similar salvage pathways exist for pyrimidines. Pyrimidine phosphoribosyltransferase will reconnect uracil, but not cytosine, to PRPP.

25.2.2 The Purine Ring System Is Assembled on Ribose Phosphate

De novo purine biosynthesis, like pyrimidine biosynthesis, requires PRPP, but for purines, PRPP provides the foundation on which the bases are constructed step by step. The initial committed step is the displacement of pyrophosphate by ammonia, rather than by a preassembled base, to produce *5-phosphoribosyl-1-amine*, with the amine in the β configuration.

Glutamine phosphoribosyl amidotransferase catalyzes this reaction. This enzyme comprises two domains: the first is homologous to the phosphoribosyltransferases in salvage pathways, whereas the second produces ammonia from glutamine by hydrolysis. However, this glutamine-hydrolysis domain is distinct from the domain that performs the same function in carbamoyl phosphate synthetase. In glutamine phosphoribosyl amidotransferase, a cysteine residue located at the amino terminus facilitates glutamine hydrolysis. To prevent wasteful hydrolysis of either substrate, the amidotransferase assumes the active configuration only on binding of both PRPP and glutamine. As is the case with carbamoyl phosphate synthetase, the ammonia generated at the glutamine-hydrolysis active site passes through a channel to reach PRPP without being released into solution.

25.2.3 The Purine Ring Is Assembled by Successive Steps of Activation by Phosphorylation Followed by Displacement

Nine additional steps are required to assemble the purine ring. Remarkably, the first six steps are analogous reactions. Most of these steps are catalyzed by enzymes with ATP-grasp domains that are homologous to those in carbamoyl phosphate synthetase. *Each step consists of the activation of a carbon-bound oxygen atom (typically a carbonyl oxygen atom) by phosphorylation, followed by the displacement of a phosphoryl group by ammonia or an amine group acting as a nucleophile (Nu).*

De novo purine biosynthesis proceeds as follows (Figure 25.7).

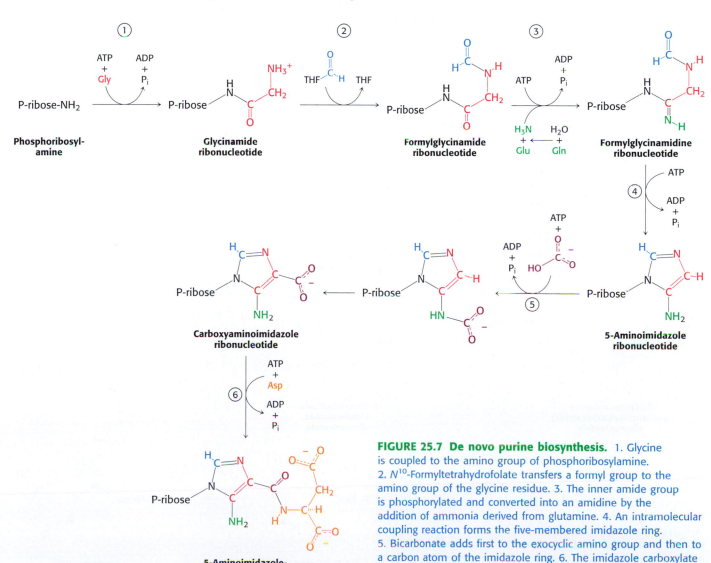

FIGURE 25.7 De novo purine biosynthesis. 1. Glycine is coupled to the amino group of phosphoribosylamine. 2. N^{10}-Formyltetrahydrofolate transfers a formyl group to the amino group of the glycine residue. 3. The inner amide group is phosphorylated and converted into an amidine by the addition of ammonia derived from glutamine. 4. An intramolecular coupling reaction forms the five-membered imidazole ring. 5. Bicarbonate adds first to the exocyclic amino group and then to a carbon atom of the imidazole ring. 6. The imidazole carboxylate is phosphorylated, and the phosphate is displaced by the amino group of aspartate.

1. The carboxylate group of a glycine residue is activated by phosphorylation and then coupled to the amino group of phosphoribosylamine. A new amide bond is formed while the amino group of glycine is free to act as a nucleophile in the next step.

2. Formate is activated and then added to this amino group to form formylglycinamide ribonucleotide. In some organisms, two distinct enzymes can catalyze this step. One enzyme transfers the formyl group from N^{10}-formyltetrahydrofolate (Section 24.2.6). The other enzyme activates formate as formyl phosphate, which is added directly to the glycine amino group.

3. The inner amide group is activated and then converted into an amidine by the addition of ammonia derived from glutamine.

4. The product of this reaction, formylglycinamidine ribonucleotide, cyclizes to form the five-membered imidazole ring found in purines. Although this cyclization is likely to be favorable thermodynamically, a molecule of ATP is consumed to ensure irreversibility. The familiar pattern is repeated: a phosphoryl group from the ATP molecule activates the carbonyl group and is displaced by the nitrogen atom attached to the ribose molecule. Cyclization is thus an intramolecular reaction in which the nucleophile and phosphate-activated carbon atom are present within the same molecule.

5. Bicarbonate is activated by phosphorylation and then attacked by the exocyclic amino group. The product of the reaction in step 5 rearranges to transfer the carboxylate group to the imidazole ring. Interestingly, mammals do not require ATP for this step; bicarbonate apparently attaches directly to the exocyclic amino group and is then transferred to the imidazole ring.

6. The imidazole carboxylate group is phosphorylated again and the phosphate group is displaced by the amino group of aspartate. Thus, a six-step process links glycine, formate, ammonia, bicarbonate, and aspartate to form an intermediate that contains all but two of the atoms necessary for the formation of the purine ring.

Three more steps complete the ring construction (Figure 25.8). Fumarate, an intermediate in the citric acid cycle, is eliminated, leaving the

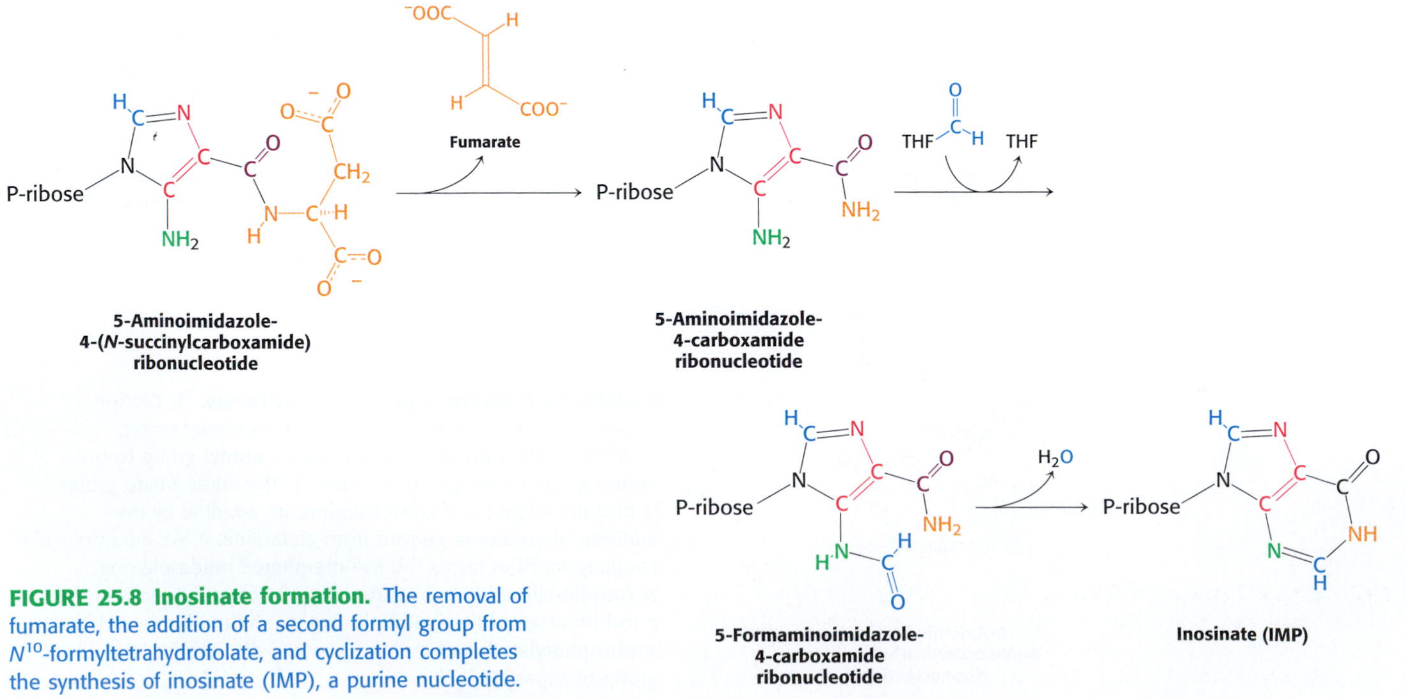

5-Aminoimidazole-4-(N-succinylcarboxamide) ribonucleotide

5-Aminoimidazole-4-carboxamide ribonucleotide

5-Formaminoimidazole-4-carboxamide ribonucleotide

Inosinate (IMP)

FIGURE 25.8 Inosinate formation. The removal of fumarate, the addition of a second formyl group from N^{10}-formyltetrahydrofolate, and cyclization completes the synthesis of inosinate (IMP), a purine nucleotide.

nitrogen atom from aspartate joined to the imidazole ring. The use of aspartate as an amino-group donor and the concomitant release of fumarate are reminiscent of the conversion of citrulline into arginine in the urea cycle and these steps are catalyzed by homologous enzymes in the two pathways (Section 23.4.2). A formyl group from N^{10}-formyltetrahydrofolate is added to this nitrogen atom to form a final intermediate that cyclizes with the loss of water to form inosinate.

25.2.4 AMP and GMP Are Formed from IMP

A few steps convert inosinate into either AMP or GMP (Figure 25.9). *Adenylate* is synthesized from inosinate by the substitution of an amino group for the carbonyl oxygen atom at C-6. Again, the addition of aspartate followed by the elimination of fumarate contributes the amino group. GTP, rather than ATP, is the phosphoryl-group donor in the synthesis of the adenylosuccinate intermediate from inosinate and aspartate. In accord with the use of GTP, the enzyme that promotes this conversion, *adenylsuccinate synthase,* is structurally related to the G-protein family and does not contain an ATP-grasp domain. The same enzyme catalyzes the removal of fumarate from adenylosuccinate in the synthesis of adenylate and from 5-aminoimidazole-4-N-succinocarboxamide ribonucleotide in the synthesis of inosinate.

Guanylate (GMP) is synthesized by the oxidation of inosinate to xanthylate (XMP), followed by the incorporation of an amino group at C-2. NAD^+ is the hydrogen acceptor in the oxidation of inosinate. Xanthylate is activated by the transfer of an AMP group (rather than a phosphoryl group) from ATP to the oxygen atom in the newly formed carbonyl group. Ammonia, generated by the hydrolysis of glutamine, then displaces the AMP group to form guanylate, in a reaction catalyzed by *GMP synthetase.* Note that the synthesis of adenylate requires GTP, whereas the synthesis of guanylate requires ATP. This reciprocal use of nucleotides by the pathways creates an important regulatory opportunity (Section 25.4).

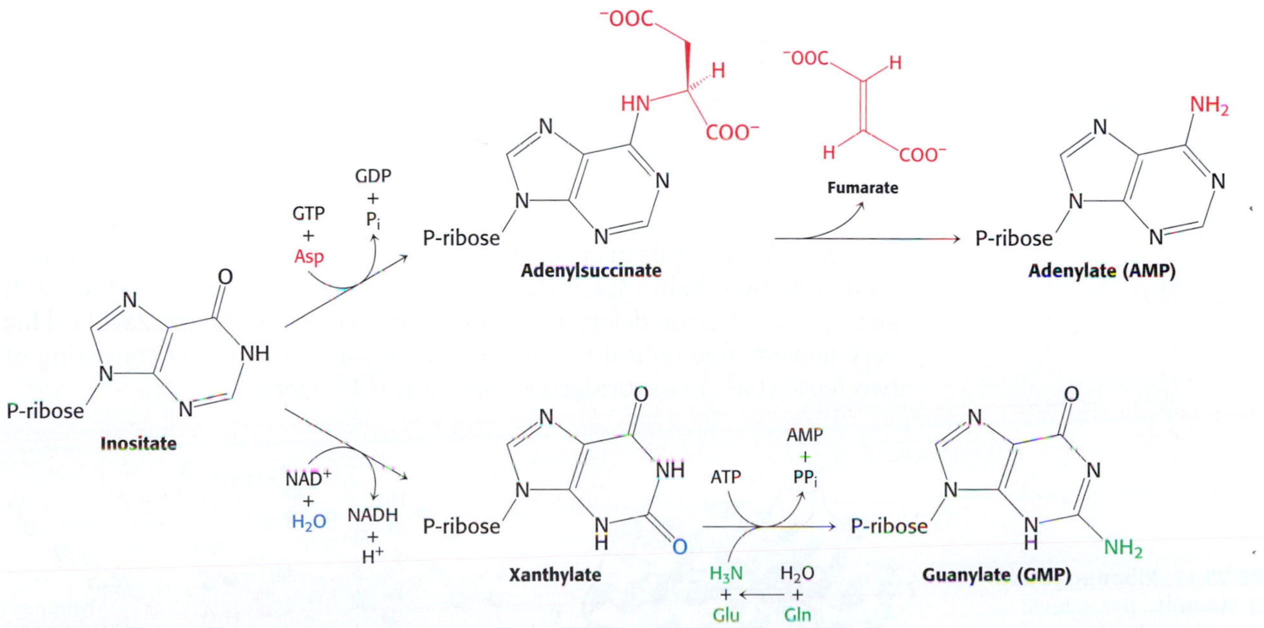

FIGURE 25.9 Generating AMP and GMP. Inosinate is the precursor of AMP and GMP. AMP is formed by the addition of aspartate followed by the release of fumarate. GMP is generated by the addition of water, dehydrogenation by NAD^+, and the replacement of the carbonyl oxygen atom by $-NH_2$ derived by the hydrolysis of glutamine.

25.3 DEOXYRIBONUCLEOTIDES SYNTHESIZED BY THE REDUCTION OF RIBONUCLEOTIDES THROUGH A RADICAL MECHANISM

We turn now to the synthesis of deoxyribonucleotides. These precursors of DNA are formed by the reduction of ribonucleotides—specifically, the 2'-hydroxyl group on the ribose moiety is replaced by a hydrogen atom. The substrates are ribonucleoside diphosphates or triphosphates, and the ultimate reductant is NADPH. The enzyme *ribonucleotide reductase* is responsible for the reduction reaction for all four ribonucleotides. The ribonucleotide reductases of different organisms are a remarkably diverse set of enzymes. The results of detailed studies have revealed that they have a common reaction mechanism, and their three-dimensional structural features indicate that these enzymes are homologous. We will focus on the best understood of these enzymes, that of *E. coli* living aerobically. This ribonucleotide reductase consists of two subunits: R1 (an 87-kd dimer) and R2 (a 43-kd dimer).

The R1 subunit contains the active site as well as two allosteric control sites (Section 25.4). This subunit includes three conserved cysteine residues and a glutamate residue, all four of which participate in the reduction of ribose to deoxyribose (Figure 25.10).

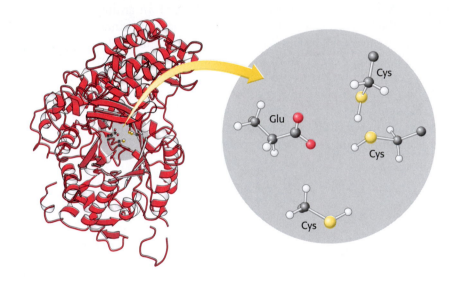

FIGURE 25.10 Ribonucleotide reductase R1 subunit. Ribonucleotide reductase reduces ribonucleotides to deoxyribonucleotides in its R1 subunit in an active site that contains three key cysteine residues and one glutamate residue. Two R1 subunits come together to form a dimer.

The R2 subunit's role in catalysis is to generate a remarkable free radical in each of its two chains. Each R2 chain contains a stable *tyrosyl radical* with an unpaired electron delocalized onto its aromatic ring (Figure 25.11). This very unusual free radical is generated by a nearby *iron center* consisting of two ferric (Fe^{3+}) ions bridged by an oxide (O^{2-}) ion.

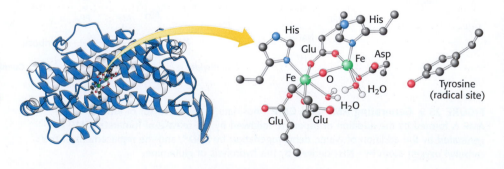

FIGURE 25.11 Ribonucleotide reductase R2 subunit. This subunit contains a stable free radical on a tyrosine residue. This radical is generated by the reaction of oxygen at a nearby site containing two iron atoms. Two R2 subunits come together to form a dimer.

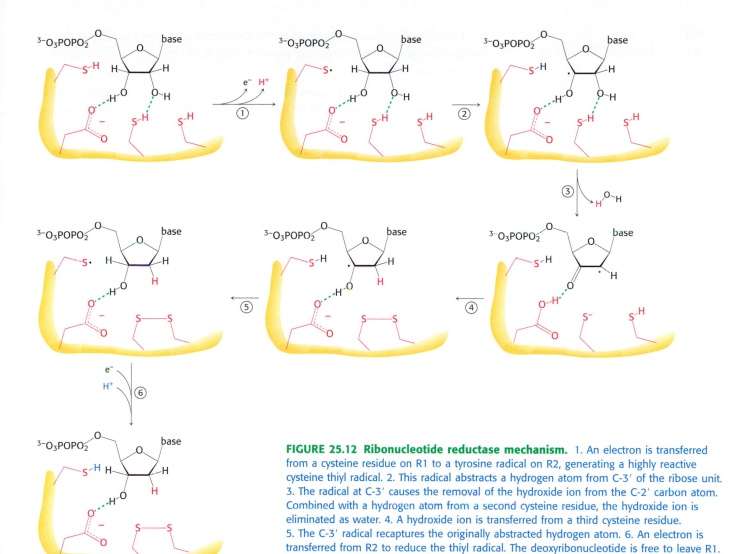

FIGURE 25.12 Ribonucleotide reductase mechanism. 1. An electron is transferred from a cysteine residue on R1 to a tyrosine radical on R2, generating a highly reactive cysteine thiyl radical. 2. This radical abstracts a hydrogen atom from C-3′ of the ribose unit. 3. The radical at C-3′ causes the removal of the hydroxide ion from the C-2′ carbon atom. Combined with a hydrogen atom from a second cysteine residue, the hydroxide ion is eliminated as water. 4. A hydroxide ion is transferred from a third cysteine residue. 5. The C-3′ radical recaptures the originally abstracted hydrogen atom. 6. An electron is transferred from R2 to reduce the thiyl radical. The deoxyribonucleotide is free to leave R1. The disulfide formed in the active site must be reduced to begin another reaction cycle.

In the synthesis of a deoxyribonucleotide, the hydroxyl group bonded to C-2′ of the ribose ring is replaced by H, with retention of the configuration at the C-2′ carbon atom (Figure 25.12).

1. The reaction begins with the transfer of an electron from a cysteine residue on R1 to the tyrosyl radical on R2. The loss of an electron generates a highly reactive *cysteine thiyl radical* within the active site of R1.

2. This radical then abstracts a hydrogen atom from C-3′ of the ribose unit, generating a radical at that carbon atom.

3. The radical at C-3′ promotes the release of the hydroxide ion on the carbon-2 atom. Protonated by a second cysteine residue, the departing hydroxide ion leaves as a water molecule.

4. A hydride ion (a proton on two electrons) is then transferred from a third cysteine residue to complete the reduction of the C-2′ position, form a disulfide bond, and reform a C-3′ radical.

5. This C-3′ radical recaptures the same hydrogen atom originally abstracted by the first cysteine residue, and the deoxyribonucleotide is free to leave the enzyme.

6. The disulfide bond generated in the enzyme's active site is then reduced by specific disulfide-containing proteins, such as thioredoxin, to regenerate the active enzyme.

To complete the overall reaction, the oxidized thioredoxin generated by this process is reduced by NADH in a reaction catalyzed by *thioredoxin reductase*.

Ribonucleotide reductases that do not contain tyrosyl radicals have been characterized in other organisms. Instead, these enzymes contain other stable radicals that are generated by other processes. For example, in one class of reductases, the coenzyme adenosylcobalamin is the radical source. Despite differences in the stable radical employed, the active sites of these enzymes are similar to that of the *E. coli* ribonucleotide reductase, and they appear to act by the same mechanism, based on the exceptional reactivity of cysteine radicals. Thus, these enzymes have a common ancestor but evolved a range of mechanisms for generating stable radical species that function well under different growth conditions. It appears that the primordial enzymes were inactivated by oxygen, whereas enzymes such as the *E. coli* enzyme make use of oxygen to generate the initial tyrosyl radical. Note that the reduction of ribonucleotides to deoxyribonucleotides is a difficult reaction chemically, likely to require a sophisticated catalyst. The existence of a common protein enzyme framework for this process strongly suggests that proteins joined the RNA world before the evolution of DNA as a stable storage form for genetic information.

25.3.1 Thymidylate Is Formed by the Methylation of Deoxyuridylate

Uracil, produced by the pyrimidine synthesis pathway, is not a component of DNA. Rather, DNA contains *thymine,* a methylated analog of uracil. Another step is required to generate thymidylate from uracil. *Thymidylate synthase* catalyzes this finishing touch: deoxyuridylate (dUMP) is methylated to thymidylate (TMP). As will be discussed in Chapter 27, the methylation of this nucleotide facilitates the identification of DNA damage for repair and, hence, helps preserve the integrity of the genetic information stored in DNA. The methyl donor in this reaction is N^5,N^{10}-methylenetetrahydrofolate rather than S-adenosylmethionine.

The methyl group becomes attached to the C-5 atom within the aromatic ring of dUMP, but this carbon atom is not a good nucleophile and cannot itself attack the appropriate group on the methyl donor. Thymidylate synthase promotes the methylation by adding a thiolate from a cysteine side chain to this ring to generate a nucleophilic species that can attack the methylene group of N^5,N^{10}-methylenetetrahydrofolate (Figure 25.13). This methylene group, in turn, is activated by distortions imposed by the enzyme that favor opening the open five-membered ring. The activated UMP's attack on the methylene group forms the new carbon–carbon bond. The intermediate formed is then converted into product: a hydride ion is transferred from the tetrahydrofolate ring to transform the methylene group into a methyl group, and a proton is abstracted from the carbon atom bearing the methyl group to eliminate the cysteine and regenerate the aromatic ring. Thus, the tetrahydrofolate derivative loses both its methylene group and a hydride ion and, hence, is oxidized to dihydrofolate. For the synthesis of more thymidylate, tetrahydrofolate must be regenerated.

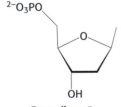

Deoxyribose-P

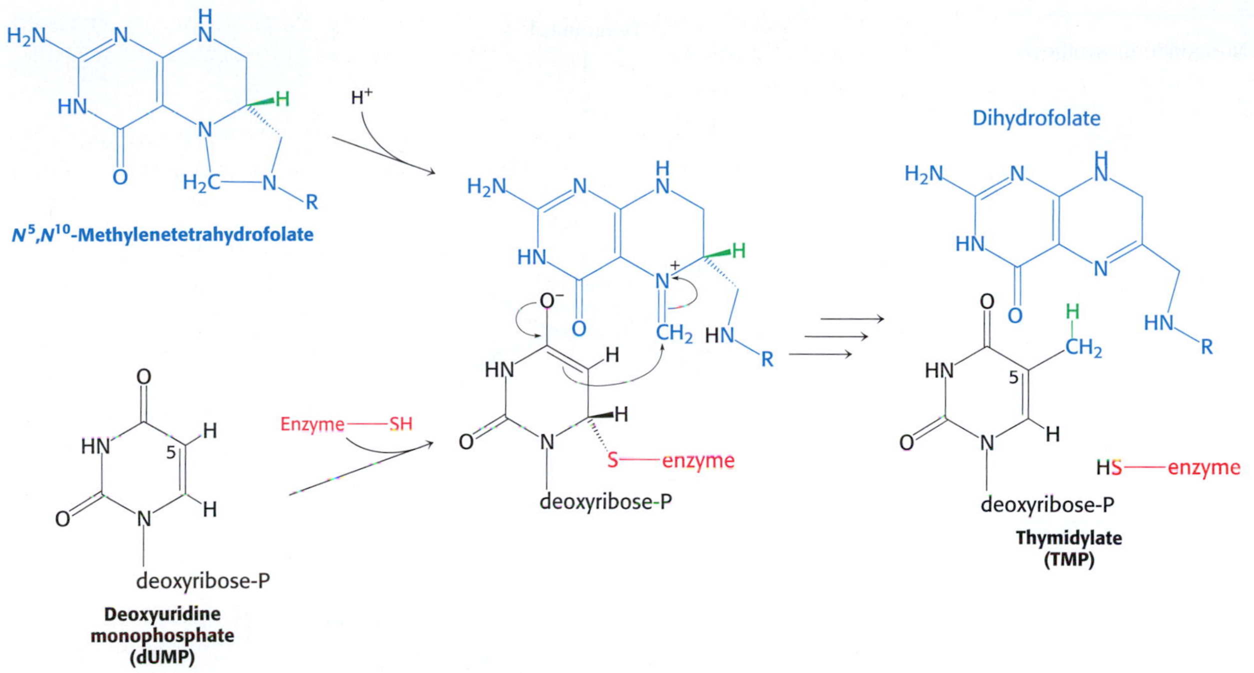

FIGURE 25.13 Thymidylate synthesis. Thymidylate synthase catalyzes the addition of a methyl group (derived from N^5,N^{10}-methylenetertahydrofolate to dUMP to form TMP. The addition of a thiolate from the enzyme activates dUMP. Opening the five-membered ring of the THF derivative prepares the methylene group for nucleophilic attack by the activated dUMP. The reaction is completed by the transfer of a hydride ion to form dihydrofolate.

25.3.2 Dihydrofolate Reductase Catalyzes the Regeneration of Tetrahydrofolate, a One-Carbon Carrier

Tetrahydrofolate is regenerated from the dihydrofolate that is produced in the synthesis of thymidylate. This regeneration is accomplished by *dihydrofolate reductase* with the use of NADPH as the reductant.

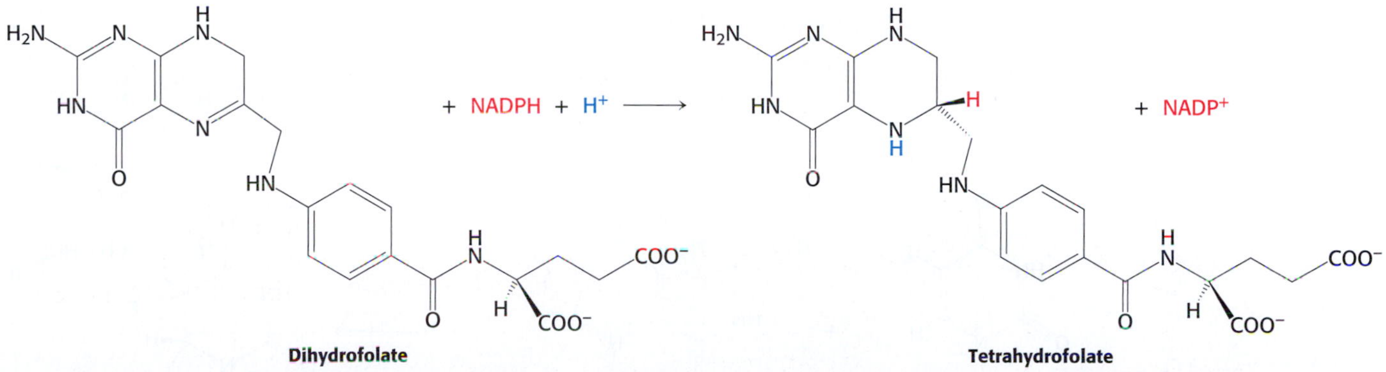

A hydride ion is directly transferred from the nicotinamide ring of NADPH to the pteridine ring of dihydrofolate. The bound dihydrofolate and NADPH are held in close proximity to facilitate the hydride transfer.

25.3.3 Several Valuable Anticancer Drugs Block the Synthesis of Thymidylate

Rapidly dividing cells require an abundant supply of thymidylate for the synthesis of DNA. The vulnerability of these cells to the inhibition of TMP synthesis has been exploited in cancer chemotherapy.

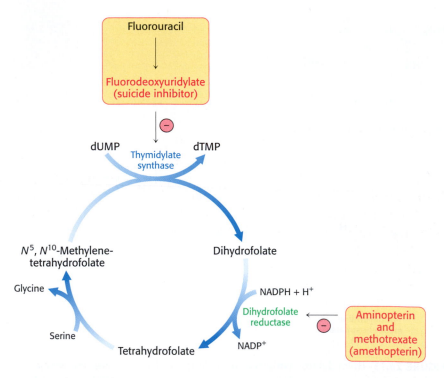

FIGURE 25.14 Anticancer drug targets. Thymidylate synthase and dihydrofolate reductase are choice targets in cancer chemotherapy because the generation of large quantities of precursors for DNA synthesis is required for rapidly dividing cancer cells.

Thymidylate synthase and dihydrofolate reductase are choice targets of chemotherapy (Figure 25.14).

Fluorouracil, a clinically useful anticancer drug, is converted in vivo into *fluorodeoxyuridylate* (F-dUMP). This analog of dUMP irreversibly inhibits thymidylate synthase after acting as a normal substrate through part of the catalytic cycle. Recall that the formation of TMP requires the removal of a proton (H^+) from C-5 of the bound nucleotide (see Figure 25.13). However, the enzyme cannot abstract F^+ from F-dUMP, and so catalysis is blocked at the stage of the covalent complex formed by F-dUMP, methylenetetrahydrofolate, and the sulfhydryl group of the enzyme (Figure 25.15). We see here an example of *suicide inhibition,* in which an enzyme converts a substrate into a reactive inhibitor that halts the enzyme's catalytic activity (Section 8.5.2).

Fluorodeoxyuridylate   **N⁵,N¹⁰-Methylene-tetrahydrofolate**   **Stable adduct**

FIGURE 25.15 Suicide inhibition. Fluorodeoxyuridylate (generated from fluorouracil) traps thymidylate synthase in a form that cannot proceed down the reaction pathway.

The synthesis of TMP can also be blocked by inhibiting the regeneration of tetrahydrofolate. Analogs of dihydrofolate, such as *aminopterin* and *methotrexate* (amethopterin), are potent competitive inhibitors ($K_i < 1$ nM) of dihydrofolate reductase.

Aminopterin (R = H) or methotrexate (R = CH₃)

Methotrexate is a valuable drug in the treatment of many rapidly growing tumors, such as those in acute leukemia and choriocarcinoma, a cancer derived from placental cells. However, methotrexate kills rapidly replicating cells whether they are malignant or not. Stem cells in bone marrow, epithelial cells of the intestinal tract, and hair follicles are vulnerable to the action of this folate antagonist, accounting for its toxic side effects, which include weakening of the immune system, nausea, and hair loss.

Folate analogs such as *trimethoprim* have potent antibacterial and antiprotozoal activity. Trimethoprim binds 10^5-fold less tightly to mammalian dihydrofolate reductase than it does to reductases of susceptible microorganisms. Small differences in the active-site clefts of these enzymes account for its highly selective antimicrobial action. The combination of trimethoprim and sulfamethoxazole (an inhibitor of folate synthesis) is widely used to treat infections.

Trimethoprim

25.4 KEY STEPS IN NUCLEOTIDE BIOSYNTHESIS ARE REGULATED BY FEEDBACK INHIBITION

Nucleotide biosynthesis is regulated by feedback inhibition in a manner similar to the regulation of amino acid biosynthesis (Section 24.3). Indeed, aspartate transcarbamoylase, one of the key enzymes for the regulation of pyrimidine biosynthesis in bacteria, was described in detail in Chapter 10. Recall that *ATCase is inhibited by CTP, the final product of pyrimidine biosynthesis*, and stimulated by ATP. Carbamoyl phosphate synthetase is a site of feedback inhibition in both prokaryotes and eukaryotes.

The synthesis of purine nucleotides is controlled by feedback inhibition at several sites (Figure 25.16).

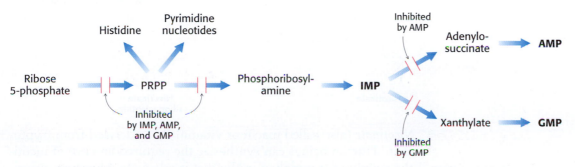

FIGURE 25.16 Control of purine biosynthesis. Feedback inhibition controls both the overall rate of purine biosynthesis and the balance between AMP and GMP production.

1. The committed step in purine nucleotide biosynthesis is the conversion of PRPP into phosphoribosylamine by *glutamine phosphoribosyl amidotransferase*. This important enzyme is feedback-inhibited by many purine ribonucleotides. It is noteworthy that AMP and GMP, the final products of the pathway, are synergistic in inhibiting the amidotransferase.

2. Inosinate is the branch point in the synthesis of AMP and GMP. *The reactions leading away from inosinate are sites of feedback inhibition.* AMP inhibits the conversion of inosinate into adenylosuccinate, its immediate precursor. Similarly, GMP inhibits the conversion of inosinate into xanthylate, its immediate precursor.

3. As already noted, GTP is a substrate in the synthesis of AMP, whereas ATP is a substrate in the synthesis of GMP. This *reciprocal substrate relation* tends to balance the synthesis of adenine and guanine ribonucleotides.

The reduction of ribonucleotides to deoxyribonucleotides is precisely controlled by allosteric interactions. Each polypeptide of the R1 subunit of the aerobic *E. coli* ribonucleotide reductase contains two allosteric sites: one of them controls the *overall activity* of the enzyme, whereas the other regulates *substrate specificity.* The overall catalytic activity of ribonucleotide reductase is diminished by the binding of dATP, which signals an abundance of deoxyribonucleotides. The binding of ATP reverses this feedback inhibition. The binding of dATP or ATP to the substrate-specificity control sites enhances the reduction of UDP and CDP, the pyrimidine nucleotides. The binding of thymidine triphosphate (TTP) promotes the reduction of GDP and inhibits the further reduction of pyrimidine ribonucleotides. The subsequent increase in the level of dGTP stimulates the reduction of ATP to dATP. This complex pattern of regulation supplies the appropriate balance of the four deoxyribonucleotides needed for the synthesis of DNA.

25.5 NAD$^+$, FAD, AND COENZYME A ARE FORMED FROM ATP

Nucleotides are important constituents not only of RNA and DNA, but also of a number of key biomolecules considered many times in our study of biochemistry. NAD$^+$ and NADP$^+$, coenzymes that function in oxidation-reduction reactions, are metabolites of ATP. The first step in the synthesis of *nicotinamide adenine dinucleotide* (NAD$^+$) is the formation of *nicotinate ribonucleotide* from nicotinate and PRPP.

Nicotinate   **Nicotinate ribonucleotide**

Nicotinate (also called niacin or vitamin B$_6$) is derived from tryptophan. Human beings can synthesize the required amount of nicotinate if the supply of tryptophan in the diet is adequate. However, nicotinate must be obtained directly if the dietary intake of tryptophan is low. A dietary deficiency of tryptophan and nicotinate can lead to pellagra, a disease characterized by dermatitis, diarrhea, and dementia. An endocrine tumor that consumes large amounts of tryptophan in synthesizing the hormone and neurotransmitter serotonin (5-hydroxytryptamine) can lead to pellagra-like symptoms.

An AMP moiety is transferred from ATP to nicotinate ribonucleotide to form *desamido*-NAD$^+$. The final step is the transfer of the ammonia generated from the amide group of glutamine to the nicotinate carboxyl group to form NAD^+.

Nicotinate ribonucleotide → **Desamido-NAD$^+$** → **NAD$^+$**

$NADP^+$ is derived from NAD^+ by phosphorylation of the 2′-hydroxyl group of the adenine ribose moiety. This transfer of a phosphoryl group from ATP is catalyzed by *NAD$^+$ kinase*.

Flavin adenine dinucleotide (FAD) is synthesized from riboflavin and two molecules of ATP. Riboflavin is phosphorylated by ATP to give *riboflavin 5′-phosphate* (also called flavin mononucleotide, FMN). FAD is then formed from FMN by the transfer of an AMP moiety from a second molecule of ATP.

$$\text{Riboflavin} + \text{ATP} \longrightarrow \text{riboflavin 5′-phosphate} + \text{ADP}$$
$$\text{Riboflavin 5′-phosphate} + \text{ATP} \longrightarrow \text{flavin adenine dinucleotide} + \text{PP}_i$$

The AMP moiety of coenzyme A also comes from ATP. A common feature of the biosyntheses of NAD^+, FAD, and CoA is the *transfer of the AMP moiety of ATP to the phosphate group of a phosphorylated intermediate.* The pyrophosphate formed in these condensations is then hydrolyzed to orthophosphate. As in many other biosyntheses, *much of the thermodynamic driving force comes from the hydrolysis of the released pyrophosphate.*

Flavin adenine dinucleotide (FAD)

25.6 DISRUPTIONS IN NUCLEOTIDE METABOLISM CAN CAUSE PATHOLOGICAL CONDITIONS

Nucleotides are vital to a host of biochemical processes. It is not surprising, then, that disruption in nucleotide metabolism would have a variety of physiological effects.

25.6.1 Purines Are Degraded to Urate in Human Beings

The nucleotides of a cell undergo continual turnover. Nucleotides are hydrolytically degraded to nucleosides by *nucleotidases*. The phosphorolytic cleavage of nucleosides to free bases and ribose 1-phosphate (or deoxyribose 1-phosphate) is catalyzed by *nucleoside phosphorylases*. Ribose 1-phosphate is isomerized by *phosphoribomutase* to ribose 5-phosphate, a substrate in the synthesis of PRPP. Some of the bases are reused to form nucleotides by salvage pathways. Others are degraded to products that are excreted (Figure 25.17). For example, AMP is degraded to the free base hypoxanthine through deamination and hydrolytic cleavage of the glycosidic bond. *Xanthine oxidase*, a molybdenum- and iron-containing flavoprotein, oxidizes hypoxanthine to *xanthine* and then to *uric acid*. Molecular oxygen, the oxidant in both reactions, is reduced to H_2O_2, which is decomposed to H_2O and O_2 by catalase. Uric acid loses a proton at physiological pH to form *urate. In human beings, urate is the final product of purine degradation and is excreted in the urine.* High serum levels of urate induce gout, a disease in

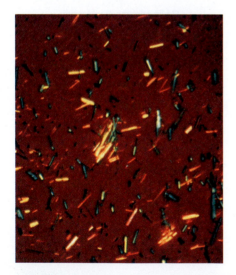

FIGURE 25.17 Purine catabolism.
Purine bases are converted first into xanthine and then into urate for excretion. Xanthine oxidase catalyzes two steps in this process.

Hypoxanthine

Xanthine

Guanine

$$O_2 + H_2O \quad H_2O_2$$

Xanthine oxidase

$$H_2O + O_2$$

Xanthine oxidase

$$H_2O_2$$

Urate

Uric acid

$$H^+$$

AMP

ribose-P

IMP

ribose-P

Allopurinol

FIGURE 25.18 Urate crystals.
Micrograph of sodium urate crystals. Joints and kidneys are damaged by these crystals in gout. [Courtesy of Dr. James McGuire.]

which salts of urate crystallize and damage joints and kidneys (Figure 25.18). Allopurinol, an inhibitor of xanthine oxidase, is used to treat gout in some cases.

The average serum level of urate in humans is close to the solubility limit. In contrast, prosimians (such as lemurs) have tenfold lower levels. A striking increase in urate levels occurred in the evolution of primates. What is the selective advantage of a urate level so high that it teeters on the brink of gout in many people? It turns out that urate has a markedly beneficial action. Urate is a highly effective scavenger of reactive oxygen species. Indeed, urate is about as effective as ascorbate (vitamin C) as an antioxidant. The increased level of urate is humans compared with prosimians and other lower primates may contribute significantly to the longer life span of humans and to lowering the incidence of human cancer.

25.6.2 Lesch-Nyhan Syndrome Is a Dramatic Consequence of Mutations in a Salvage-Pathway Enzyme

Mutations in genes that encode nucleotide biosynthetic enzymes can reduce levels of needed nucleotides and can lead to an accumulation of intermediates. A nearly total absence of hypoxanthine-guanine phosphoribosyltransferase has unexpected and devastating consequences. The most striking expression of this inborn error of metabolism, called the *Lesch-Nyhan syndrome*, is *compulsive self-destructive behavior*. At age 2 or 3, children with this disease begin to bite their fingers and lips and will chew them off if unrestrained. These children also behave aggressively toward others. *Mental deficiency* and *spasticity* are other characteristics of the Lesch-Nyhan syndrome. Elevated levels of urate in the serum lead to the formation of kidney stones early in life, followed by the symptoms of gout years later. The disease is inherited as a sex-linked recessive disorder.

The biochemical consequences of the virtual absence of hypoxanthine-guanine phosphoribosyl transferase are *an elevated concentration of PRPP, a marked increase in the rate of purine biosynthesis by the de novo pathway, and an overproduction of urate.* The relation between the absence of the transferase and the bizarre neurologic signs is an enigma. Specific cells in the brain may be dependent on the salvage pathway for the synthesis of

IMP and GMP. Indeed, transporters of the neurotransmitter dopamine are present at lower levels in affected individuals. Alternatively, cells may be damaged by the accumulation of intermediates to abnormal levels. The Lesch-Nyhan syndrome demonstrates that the salvage pathway for the synthesis of IMP and GMP is not gratuitous. Moreover, the Lesch-Nyhan syndrome reveals that *abnormal behavior such as self-mutilation and extreme hostility can be caused by the absence of a single enzyme.* Psychiatry will no doubt benefit from the unraveling of the molecular basis of such mental disorders.

SUMMARY

In de Novo Synthesis, the Pyrimidine Ring Is Assembled from Bicarbonate, Aspartate, and Glutamine

The pyrimidine ring is assembled first and then linked to ribose phosphate to form a pyrimidine nucleotide. PRPP is the donor of the ribose phosphate moiety. The synthesis of the pyrimidine ring starts with the formation of carbamoylaspartate from carbamoyl phosphate and aspartate, a reaction catalyzed by aspartate transcarbamoylase. Dehydration, cyclization, and oxidation yield orotate, which reacts with PRPP to give orotidylate. Decarboxylation of this pyrimidine nucleotide yields UMP. CTP is then formed by the amination of UTP.

Purines Bases Can Be Synthesized de Novo or Recycled by Salvage Pathways

The purine ring is assembled from a variety of precursors: glutamine, glycine, aspartate, N^{10}-formyltetrahydrofolate, and CO_2. The committed step in the de novo synthesis of purine nucleotides is the formation of 5-phosphoribosylamine from 5-phosphoribosyl-1-pyrophosphate (PRPP) and glutamine. The purine ring is assembled on ribose phosphate, in contrast with the de novo synthesis of pyrimidine nucleotides. The addition of glycine, followed by formylation, amination, and ring closure, yields 5-aminoimidazole ribonucleotide. This intermediate contains the completed five-membered ring of the purine skeleton. The addition of CO_2, the nitrogen atom of aspartate, and a formyl group, followed by ring closure, yields inosinate (IMP), a purine ribonucleotide. AMP and GMP are formed from IMP. Purine ribonucleotides can also be synthesized by a salvage pathway in which a preformed base reacts directly with PRPP.

Deoxyribonucleotides Are Synthesized by the Reduction of Ribonucleotides Through a Radical Mechanism

Deoxyribonucleotides, the precursors of DNA, are formed in *E. coli* by the reduction of ribonucleoside diphosphates. These conversions are catalyzed by ribonucleotide reductase. Electrons are transferred from NADPH to sulfhydryl groups at the active sites of this enzyme by thioredoxin or glutaredoxin. A tyrosyl free radical generated by an iron center in the reductase initiates a radical reaction on the sugar, leading to the exchange of H for OH at C-2'. TMP is formed by methylation of dUMP. The donor of a methylene group and a hydride in this reaction is N^5,N^{10}-methylenetetrahydrofolate, which is converted into dihydrofolate. Tetrahydrofolate is regenerated by the reduction of dihydrofolate by NADPH. Dihydrofolate reductase, which catalyzes this reaction, is inhibited by folate analogs such as aminopterin and methotrexate. These compounds and fluorouracil, an inhibitor of thymidylate synthase, are used as anticancer drugs.

● **Key Steps in Nucleotide Biosynthesis Are Regulated by Feedback Inhibition**

Pyrimidine biosynthesis in *E. coli* is regulated by the feedback inhibition of aspartate transcarbamoylase, the enzyme that catalyzes the committed step. CTP inhibits and ATP stimulates this enzyme. The feedback inhibition of glutamine-PRPP amidotransferase by purine nucleotides is important in regulating their biosynthesis.

● **NAD$^+$, FAD, and Coenzyme A Are Formed from ATP**

Nucleotides are important constituents not only of RNA and DNA, but also of a number of other key biomolecules. Coenzymes NAD$^+$ and FAD, prominent in oxidation–reduction reactions, have ADP as an important constituent. The acyl-group activation compound, coenzyme A, is also derived from ATP.

● **Disruptions in Nucleotide Metabolism Can Cause Pathological Conditions**

Purines are degraded to urate in human beings. Gout, a disease that affects joints and leads to arthritis, is associated with the excessive accumulation of urate. The Lesch-Nyhan syndrome, a genetic disease characterized by self-mutilation, mental deficiency, and gout, is caused by the absence of hypoxanthine-guanine phosphoribosyltransferase. This enzyme is essential for the synthesis of purine nucleotides by the salvage pathway.

KEY TERMS

nucleoside (p. 694)

nucleotide (p. 694)

pyrimidine (p. 694)

carbamoyl phosphate synthetase (CSP) (p. 695)

ATP-grasp fold (p. 695)

5-phosphoribosyl-1-pyrophosphate (PRPP) (p. 696)

orotidylate (p. 697)

purine (p. 698)

salvage pathway (p. 698)

inosinate (p. 698)

hypoxanthine (p. 698)

glutamine phosphoribosyl amidotransferase (p. 698)

ribonucleotide reductase (p. 702)

thymidylate synthase (p. 704)

dihydrofolate reductase (p. 705)

SELECTED READINGS

Where to start

Galperin, M. Y., and Koonin, E. V., 1997. A diverse superfamily of enzymes with ATP-dependent carboxylate-amine/thiol ligase activity. *Protein Sci.* 6:2639–2643.

Jordan, A., and Reichard, P., 1998. Ribonucleotide reductases. *Annu. Rev. Biochem.* 67:71–98.

Seegmiller, J. E., 1989. Contributions of Lesch-Nyhan syndrome to the understanding of purine metabolism. *J. Inherited Metab. Dis.* 12:184–196.

Pyrimidine biosynthesis

Raushel, F. M., Thoden, J. B., Reinhart, G. D., and Holden, H. M., 1998. Carbamoyl phosphate synthetase: A crooked path from substrates to products. *Curr. Opin. Chem. Biol.* 2:624–632.

Begley, T. P., Appleby, T. C., and Ealick, S. E., 2000. The structural basis for the remarkable proficiency of orotidine 5′-monophosphate decarboxylase. *Curr. Opin. Struct. Biol.* 10:711–718.

Traut, T. W., and Temple, B. R., 2000. The chemistry of the reaction determines the invariant amino acids during the evolution and divergence of orotidine 5′-monophosphate decarboxylase. *J. Biol. Chem.* 275:28675–28681.

Lee, L., Kelly, R. E., Pastra-Landis, S. C., and Evans, D. R., 1985. Oligomeric structure of the multifunctional protein CAD that initiates pyrimidine biosynthesis in mammalian cells. *Proc. Natl. Acad. Sci. USA* 82:6802–6806.

Purine biosynthesis

Thoden, J. B., Firestine, S., Nixon, A., Benkovic, S. J., and Holden, H. M., 2000. Molecular structure of *Escherichia coli* PurT-encoded glycinamide ribonucleotide transformylase. *Biochemistry* 39: 8791–8802.

McMillan, F. M., Cahoon, M., White, A., Hedstrom, L., Petsko, G. A., and Ringe, D., 2000. Crystal structure at 2.4 Å resolution of *Borrelia burgdorferi* inosine 5′-monophosphate dehydrogenase: Evidence of a substrate-induced hinged-lid motion by loop 6. *Biochemistry* 39:4533–4542.

Mueller, E. J., Oh, S., Kavalerchik, E., Kappock, T. J., Meyer, E., Li, C., Ealick, S. E., and Stubbe, J., 1999. Investigation of the ATP binding site of *Escherichia coli* aminoimidazole ribonucleotide synthetase using affinity labeling and site-directed mutagenesis. *Biochemistry* 38:9831–9839.

Levdikov, V. M., Barynin, V. V., Grebenko, A. I., Melik-Adamyan, W. R., Lamzin, V. S., and Wilson, K. S., 1998. The structure of SAICAR synthase: An enzyme in the de novo pathway of purine nucleotide biosynthesis. *Structure* 6:363–376.

Smith, J. L., Zaluzec, E. J., Wery, J. P., Niu, L., Switzer, R. L., Zalkin, H., and Satow, Y., 1994. Structure of the allosteric regulatory enzyme of purine biosynthesis. *Science* 264:1427–1433.

Weber, G., Nagai, M., Natsumeda, Y., Ichikawa, S., Nakamura, H., Eble, J. N., Jayaram, H. N., Zhen, W. N., Paulik, E., and Hoffman,

R., 1991. Regulation of de novo and salvage pathways in chemotherapy. *Adv. Enzyme Regul.* 31:45–67.

Ribonucleotide reductases

Reichard, P., 1997. The evolution of ribonucleotide reduction. *Trends Biochem. Sci.* 22:81–85.

Stubbe, J., 2000. Ribonucleotide reductases: The link between an RNA and a DNA world? *Curr. Opin. Struct. Biol.* 10:731–736.

Logan, D. T., Andersson, J., Sjoberg, B. M., and Nordlund, P., 1999. A glycyl radical site in the crystal structure of a class III ribonucleotide reductase. *Science* 283:1499–1504.

Tauer, A., and Benner, S. A., 1997. The B₁₂-dependent ribonucleotide reductase from the archaebacterium *Thermoplasma acidophila:* An evolutionary solution to the ribonucleotide reductase conundrum. *Proc. Natl. Acad. Sci. USA* 94:53–58.

Jordan, A., Torrents, E., Jeanthon, C., Eliasson, R., Hellman, U., Wernstedt, C., Barbe, J., Gibert, I., and Reichard, P., 1997. B₁₂-dependent ribonucleotide reductases from deeply rooted eubacteria are structurally related to the aerobic enzyme from *Escherichia coli.* *Proc. Natl. Acad. Sci. USA* 94:13487–13492.

Stubbe, J., and Riggs-Gelasco, P., 1998. Harnessing free radicals: Formation and function of the tyrosyl radical in ribonucleotide reductase. *Trends Biochem. Sci.* 23:438–443.

Stubbe, J. A., 1989. Protein radical involvement in biological catalysis? *Annu. Rev. Biochem* 58:257–285.

Thymidylate synthase and dihydrofolate reductase

Li, R., Sirawaraporn, R., Chitnumsub, P., Sirawaraporn, W., Wooden, J., Athappilly, F., Turley, S., and Hol, W. G.,2000. Three-dimensional structure of *M. tuberculosis* dihydrofolate reductase reveals opportunities for the design of novel tuberculosis drugs. *J. Mol. Biol.* 295:307–323.

Liang, P. H., and Anderson, K. S., 1998. Substrate channeling and domain-domain interactions in bifunctional thymidylate synthase-dihydrofolate reductase. *Biochemistry* 37:12195–12205.

Miller, G. P., and Benkovic, S. J., 1998. Stretching exercises: Flexibility in dihydrofolate reductase catalysis. *Chem. Biol.* 5:R105–R113.

Blakley, R. L., 1995. Eukaryotic dihydrofolate reductase. *Adv. Enzymol. Relat. Areas Mol. Biol.* 70:23–102.

Carreras, C. W., and Santi, D. V., 1995. The catalytic mechanism and structure of thymidylate synthase. *Annu. Rev. Biochem.* 64:721–762.

Schweitzer, B. I., Dicker, A. P., and Bertino, J. R., 1990. Dihydrofolate reductase as a therapeutic target. *FASEB J.* 4:2441–2452.

Brown, K. A., and Kraut, J., 1992. Exploring the molecular mechanism of dihydrofolate reductase. *Faraday Discuss.* 1992:217–224.

Bystroff, C., Oatley, S. J., and Kraut, J., 1990. Crystal structures of *Escherichia coli* dihydrofolate reductase: The NADP⁺ holoenzyme and the folate NADP⁺ ternary complex—Substrate binding and a model for the transition state. *Biochemistry* 29:3263–3277.

Genetic diseases

Striver, C. R., Beaudet, A. L., Sly, W. S., Valle, D., Stanbury, J. B., Wyngaarden, J. B., and Fredrickson, D. S. (Eds.), 1995. *The Metabolic Basis of Inherited Diseases* (7th ed., pp. 1655–1840). McGraw-Hill.

Nyhan, W. L., 1997. The recognition of Lesch-Nyhan syndrome as an inborn error of purine metabolism. *J. Inherited Metab. Dis.* 20:171–178.

Wong, D. F., Harris, J. C., Naidu, S., Yokoi, F., Marenco, S., Dannals, R. F., Ravert, H. T., Yaster, M., Evans, A., Rousset, O., Bryan, R. N., Gjedde, A., Kuhar, M. J., and Breese, G. R., 1996. Dopamine transporters are markedly reduced in Lesch-Nyhan disease in vivo. *Proc. Natl. Acad. Sci. USA* 93:5539–5543.

Resta, R., and Thompson, L. F., 1997. SCID: The role of adenosine deaminase deficiency. *Immunol. Today* 18:371–374.

Davidson, B. L., Pashmforoush, M., Kelley, W. N., and Palella, T. D., 1989. Human hypoxanthine-guanine phosphoribosyltransferase deficiency: The molecular defect in a patient with gout (HPRTAshville). *J. Biol. Chem.* 264:520–525.

Sculley, D. G., Dawson, P. A., Emerson, B. T., and Gordon, R. B., 1992. A review of the molecular basis of hypoxanthine-guanine phosphoribosyltransferase (HPRT) deficiency. *Hum. Genet.* 90: 195–207.

PROBLEMS

1. *Activated ribose phosphate.* Write a balanced equation for the synthesis of PRPP from glucose through the oxidative branch of the pentose phosphate pathway.

2. *Making a pyrimidine.* Write a balanced equation for the synthesis of orotate from glutamine, CO_2, and aspartate.

3. *Identifying the donor.* What is the activated reactant in the biosynthesis of each of these compounds?

(a) Phosphoribosylamine (d) Nicotinate ribonucleotide
(b) Carbamoylaspartate (e) Phosphoribosylanthranilate
(c) Orotidylate (from orotate)

4. *Inhibiting purine biosynthesis.* Amidotransferases are inhibited by the antibiotic azaserine (*O*-diazoacetyl-L-serine), which is an analog of glutamine.

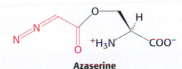

Azaserine

Which intermediates in purine biosynthesis would accumulate in cells treated with azaserine?

5. *The price of methylation.* Write a balanced equation for the synthesis of TMP from dUMP that is coupled to the conversion of serine into glycine.

6. *Sulfa action.* Bacterial growth is inhibited by sulfanilamide and related sulfa drugs, and there is a concomitant accumulation of 5-aminoimidazole-4-carboxamide ribonucleotide. This inhibition is reversed by the addition of *p*-aminobenzoate.

$$H_2N-\!\!\!\!\bigcirc\!\!\!\!-SO_2NH_2$$

Sulfanilamide

Propose a mechanism for the inhibitory effect of sulfanilamide.

7. *A generous donor.* What major biosynthetic reactions utilize PRPP?

8. *HAT medium.* Mutant cells unable to synthesize nucleotides by salvage pathways are very useful tools in molecular and cell biology. Suppose that cell A lacks thymidine kinase, the enzyme catalyzing the phosphorylation of thymidine to thymidylate, and that cell B lacks hypoxanthine-guanine phosphoribosyl transferase.

(a) Cell A and cell B do not proliferate in a *HAT* medium containing *hypoxanthine*, *aminopterin* or *amethopterin* (methotrexate), and *thymine*. However, cell C formed by the fusion of cells A and B grows in this medium. Why?

(b) Suppose that you wanted to introduce foreign genes into cell A. Devise a simple means of distinguishing between cells that have taken up foreign DNA and those that have not.

9. *Adjunct therapy.* Allopurinol is sometimes given to patients with acute leukemia who are being treated with anticancer drugs. Why is allopurinol used?

10. *A hobbled enzyme.* Both side-chain oxygen atoms of aspartate 27 at the active site of dihydrofolate reductase form hydrogen bonds with the pteridine ring of folates. The importance of this interaction was assessed by studying two mutants at this position, Asn 27 and Ser 27. The dissociation constant of methotrexate was 0.07 nM for the wild type, 1.9 nM for the Asn 27 mutant, and 210 nM for the Ser 27 mutant, at 25°C. Calculate the standard free energy of binding of methotrexate by these three proteins. What is the decrease in binding energy resulting from each mutation?

11. *Correcting deficiencies.* Suppose that a person is found who is deficient in an enzyme required for IMP synthesis. How might this person be treated?

12. *Labeled nitrogen.* Purine biosynthesis is allowed to take place in the presence of [^{15}N]aspartate, and the newly synthesized GTP and ATP are isolated. What positions are labeled in the two nucleotides?

13. *Changed inhibitor.* Xanthine oxidase treated with allopurinol results in the formation of a new compound that is an extremely potent inhibitor of the enzyme. Propose a structure for this compound.

Mechanism Problems

14. *The same but not the same.* Write out mechanisms for the conversion of phosphoribosylamine into glycinamide ribonucleotide and of xanthylate into guanylate.

15. *Closing the ring.* Propose a mechanism for the conversion of 5-formamidoimidazole-4-carboxamide ribonucleotide into inosinate.

Chapter Integration Problems

16. *They're everywhere!* Nucleotides play a variety of roles in the cell. Give an example of a nucleotide that acts in each of the following roles or processes.

(a) Second messenger   (e) Transfer of electrons
(b) Phosphoryl-group transfer   (f) DNA sequencing
(c) Activation of carbohydrates   (g) Chemotherapy
(d) Activation of acetyl groups   (h) Allosteric effector

17. *Pernicious anemia.* Purine biosynthesis is impaired by vitamin B_{12} deficiency. Why? How might fatty acid and amino acid metabolism also be affected by a vitamin B_{12} deficiency?

18. *Hyperuricemia.* Many patients with glucose 6-phosphatase deficiency have high serum levels of urate. Hyperuricemia can be induced in normal people by the ingestion of alcohol or by strenuous exercise. Propose a common mechanism that accounts for these findings.

19. *Labeled carbon.* Succinate uniformly labeled with ^{14}C is added to cells actively engaged in pyrimidine biosynthesis. Propose a mechanism by which carbon atoms from succinate could be incorporated into a pyrimidine. At what positions is the pyrimidine labeled?

20. *Exercising muscle.* Some interesting reactions take place in muscle tissue to facilitate the generation of ATP for contraction.

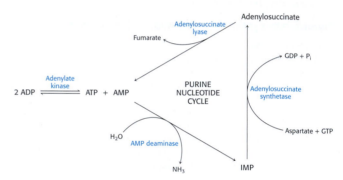

In muscle contraction, ATP is converted into ADP. Adenylate kinase converts two molecules of ADP into a molecule of ATP and AMP.

(a) Why is this reaction beneficial to contracting muscle?
(b) Why is the equilibrium for the adenylate kinase approximately equal to 1?

Muscle can metabolize AMP by using the purine nucleotide cycle. The initial step in this cycle, catalyzed by AMP deaminase, is the conversion of AMP into IMP.

(c) Why might the deamination of AMP facilitate ATP formation in muscle?
(d) How does the purine nucleotide cycle assist the aerobic generation of ATP?

CHAPTER 26

The Biosynthesis of Membrane Lipids and Steroids

Fats such as the triacylglycerol molecule (below) are widely used to store excess energy for later use and to fulfill other purposes, illustrated by the insulating blubber of whales. The natural tendency of fats to exist in nearly water-free forms makes these molecules well-suited for these roles. [(Left) François Cohier/Photo Researchers.]

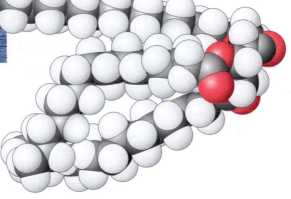

This chapter examines the biosynthesis of three important components of biological membranes—phospholipids, sphingolipids, and cholesterol (Chapter 12). Triacylglycerols also are considered here because the pathway for their synthesis overlaps that of phospholipids. Cholesterol is of interest both as a membrane component and as a precursor of many signal molecules, including the steroid hormones progesterone, testosterone, estrogen, and cortisol. The biosynthesis of cholesterol exemplifies a fundamental mechanism for the assembly of extended carbon skeletons from five-carbon units.

The transport of cholesterol in blood by the low-density lipoprotein and its uptake by a specific receptor on the cell surface vividly illustrate a recurring mechanism for the entry of metabolites and signal molecules into cells. The absence of this receptor in people with *familial hypercholesterolemia*, a genetic disease, leads to markedly elevated cholesterol levels in the blood, cholesterol deposits on blood vessels, and childhood heart attacks. Indeed, cholesterol is implicated in the development of atherosclerosis in individuals without genetic defects. Thus, the regulation of cholesterol synthesis and transport can be a source of especially clear insight into the role that our understanding of biochemistry plays in medicine.

OUTLINE

- **26.1 Phosphatidate Is a Common Intermediate in the Synthesis of Phospholipids and Triacylglycerols**

- **26.2 Cholesterol Is Synthesized from Acetyl Coenzyme A in Three Stages**

- **26.3 The Complex Regulation of Cholesterol Biosynthesis Takes Place at Several Levels**

- **26.4 Important Derivatives of Cholesterol Include Bile Salts and Steroid Hormones**

26.1 PHOSPHATIDATE IS A COMMON INTERMEDIATE IN THE SYNTHESIS OF PHOSPHOLIPIDS AND TRIACYLGLYCEROLS

The first step in the synthesis of both phospholipids for membranes and triacylglycerols for energy storage is the synthesis of *phosphatidate* (diacylglycerol 3-phosphate). In mammalian cells, phosphatidate is synthesized in the endoplasmic reticulum and the outer mitochondrial membrane. It is formed by the addition of two fatty acids to *glycerol 3-phosphate*, which in turn is formed primarily by the reduction of dihydroxyacetone phosphate, a glycolytic intermediate, and to a lesser extent by the phosphorylation of glycerol. Glycerol 3-phosphate is acylated by acyl CoA to form *lysophosphatidate*, which is again acylated by acyl CoA to yield phosphatidate.

Glycerol 3-phosphate → **Lysophosphatidate** → **Phosphatidate**

These acylations are catalyzed by *glycerol phosphate acyltransferase*. In most phosphatidates, the fatty acyl chain attached to the C-1 atom is saturated, whereas the one attached to the C-2 atom is unsaturated.

The pathways diverge at phosphatidate. In the synthesis of triacylglycerols, phosphatidate is hydrolyzed by a specific phosphatase to give a *diacylglycerol (DAG)*. This intermediate is acylated to a *triacylglycerol* in a reaction that is catalyzed by *diglyceride acyltransferase*. Both enzymes are associated in a *triacylglycerol synthetase complex* that is bound to the endoplasmic reticulum membrane.

Phosphatidate → **Diacylglycerol (DAG)** → **Triacylglycerol**

The liver is the primary site of triacylglycerol synthesis. From the liver, the triacylglycerols are transported to the muscles for energy conversion or to the adipocytes for storage.

26.1.1 The Synthesis of Phospholipids Requires an Activated Intermediate

Phospholipid synthesis requires the combination of a diacylglyceride with an alcohol. As in most anabolic reactions, one of the components must be activated. In this case, either of the two components may be activated, depending on the source of the reactants.

Synthesis from an Activated Diacylglycerol. The de novo pathway starts with the reaction of phosphatidate with cytidine triphosphate (CTP) to form *cytidine diphosphodiacylglycerol (CDP-diacylglycerol)* (Figure 26.1). This reaction, like those of many biosyntheses, is driven forward by the hydrolysis of pyrophosphate.

Phosphatidate → (CTP, PP$_i$) → **CDP-diacylglycerol**

The activated phosphatidyl unit then reacts with the hydroxyl group of an alcohol to form a phosphodiester linkage. If the alcohol is serine, the products are *phosphatidyl serine* and cytidine monophosphate (CMP).

CDP-diacylglycerol + **Serine** →

Phosphatidyl serine + **CMP**

Likewise, phosphatidyl inositol is formed by the transfer of a diacylglycerol phosphate unit from CDP-diacylglycerol to inositol. Subsequent phosphorylations catalyzed by specific kinases lead to the synthesis of *phosphatidyl inositol 4,5-bisphosphate*, an important molecule in signal transduction. Recall that hormonal and sensory stimuli activate *phospholipase C*, an enzyme that hydrolyzes this phospholipid to form two intracellular messengers—diacylglycerol and inositol 1,4,5-trisphosphate (Section 15.2).

The fatty acid components of a phospholipid may vary, and thus phosphatidyl serine, as well as most other phospholipids, represents a class of molecules rather than a single species. As a result, a single mammalian cell may contain thousands of distinct phospholipids. Phosphatidyl inositol is unusual in that it has a nearly fixed fatty acid composition. Stearic acid usually occupies the C-1 position and arachidonic acid (Section 22.6.2) the C-2 position.

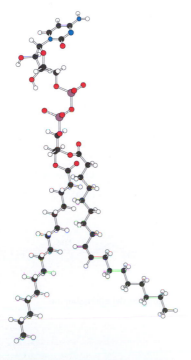

FIGURE 26.1 Structure of CDP-diacylglycerol. A key intermediate in the synthesis of phospholipids consists of phosphatidate and CMP joined by a pyrophosphate linkage.

In bacteria, the decarboxylation of phosphatidyl serine by a pyridoxal phosphate-dependent enzyme yields *phosphatidyl ethanolamine*, another common phospholipid. The amino group of this phosphoglyceride is then methylated three times to form *phosphatidyl choline*. S-*Adenosylmethionine* is the methyl donor.

In mammals, phosphatidyl ethanolamine can be formed from phosphatidyl serine by the enzyme-catalyzed exchange of ethanolamine for the serine moiety of the phospholipid.

Synthesis from an Activated Alcohol. In mammals, phosphatidyl ethanolamine can also be synthesized from ethanolamine through the formation of CDP-ethanolamine. In this case, the alcohol ethanolamine is phosphorylated by ATP to form the precursor, *phosphorylethanolamine*. This precursor then reacts with CTP to form the activated alcohol, *CDP-ethanolamine*. The phosphorylethanolamine unit of CDP-ethanolamine is then transferred to a diacylglycerol to form *phosphatidyl ethanolamine*.

In mammals, a pathway that utilizes choline obtained from the diet ends in the synthesis of phosphatidyl choline, the most common phospholipid in these organisms. In this case, choline is activated in a series of reactions analogous to those in the activation of ethanolamine. Interestingly, the liver possesses an enzyme, *phosphatidyl ethanolamine methyltransferase*, that synthesizes phosphatidyl choline from phosphatidyl ethanolamine, through the successive methylation of ethanolamine. Thus, phosphatidyl choline can be produced by two distinct pathways, ensuring that this phospholipid can be synthesized even if the components for one pathway are in limited supply.

Note that a cytidine nucleotide plays the same role in the synthesis of these phosphoglycerides as a uridine nucleotide does in the formation of glycogen (Section 21.4.1). In all of these biosyntheses, an activated intermediate (UDP-glucose, CDP-diacylglycerol, or CDP-alcohol) is formed from a phosphorylated substrate (glucose 1-phosphate, phosphatidate, or a phosphorylalcohol) and a nucleoside triphosphate (UTP or CTP). The activated intermediate then reacts with a hydroxyl group (the terminus of glycogen, the side chain of serine, or a diacylglycerol).

26.1.2 Plasmalogens and Other Ether Phospholipids Are Synthesized from Dihydroxyacetone Phosphate

Glyceryl ether phospholipids contain an ether linkage instead of an acyl linkage at C-1 and are synthesized starting with dihydroxyacetone phosphate rather than glycerol 3-phosphate (Figure 26.2). Acylation by a fatty acyl CoA yields a 1-acyl derivative that reacts with a long-chain alcohol to form an ether at C-1. NADPH reduces the keto group at C-2, and the resulting alcohol is acylated by a long-chain acyl CoA. Removal of the 3-phosphate group yields 1-alkyl-2-acylglycerol, which reacts with CDP-choline to form the ether analog of phosphatidyl choline.

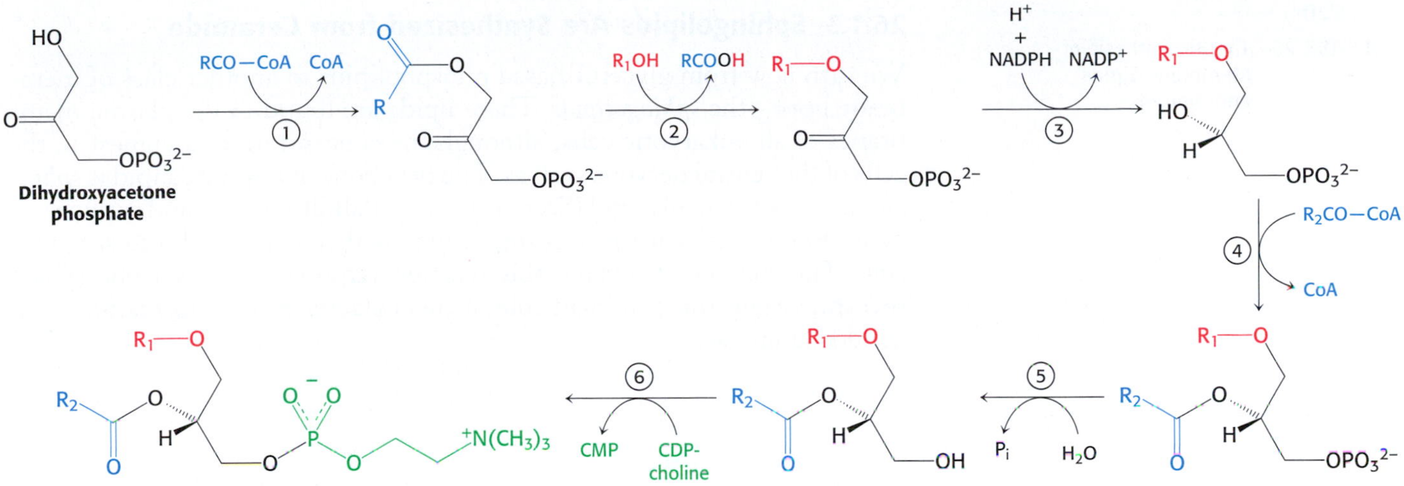

1-Alkyl-2-acyl-phosphatidyl choline

FIGURE 26.2 Synthesis of an ether phospholipid. Steps in the synthesis include (1) acylation of dihydroxyacetone phosphate by acyl CoA, (2) exchange of an alcohol for the carboxylic acid, (3) reduction by NADPH, (4) acylation by a second acyl CoA, (5) hydrolysis of the phosphate ester, and (6) transfer of a phosphocholine moiety.

Platelet-activating factor (PAF) is an ether phospholipid implicated in a number of allergic and inflammatory responses.

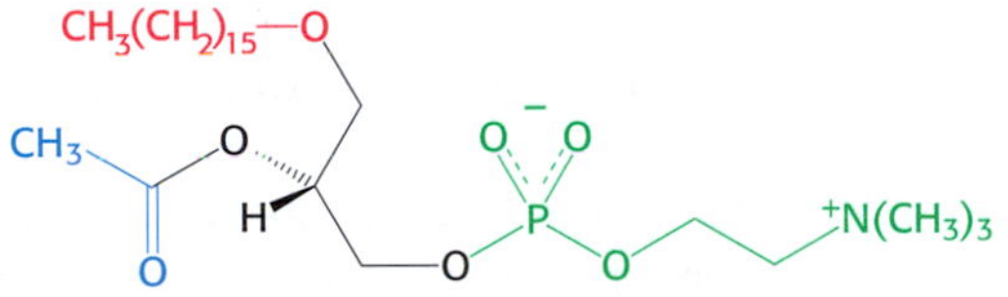

Platelet activating factor (PAF)

Subnanomolar concentrations of this 1-alkyl-2-acetyl ether analog of phosphatidyl choline induce the aggregation of blood platelets, smooth muscle contraction, and the activation of cells of the immune system. It is also a mediator of anaphylactic shock, a severe and often fatal allergic response. The presence of an acetyl group rather than a long-chain acyl group at
C-2 increases the water solubility of this lipid, enabling it to function in the aqueous environment of the blood. PAF functions through a 7TM receptor.

Plasmalogens are phospholipids containing an α,β-unsaturated ether at C-1. Phosphatidal choline, the plasmalogen corresponding to phosphatidyl choline, is formed by desaturation of a 1-alkyl precursor.

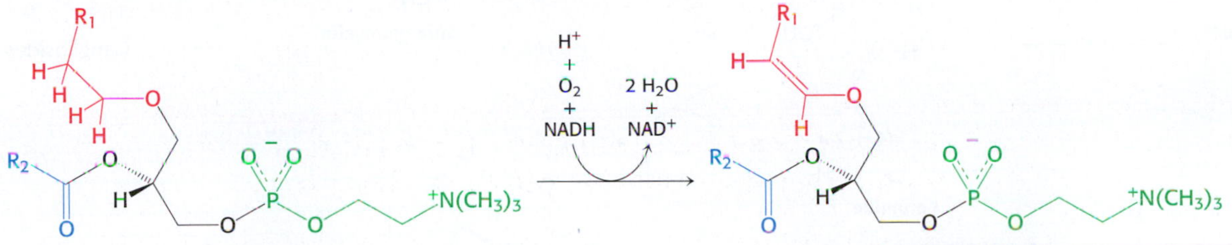

1-Alkyl precursor

Phosphatidal choline (a plasmalogen)

The desaturase catalyzing this final step in the synthesis of a plasmalogen is an endoplasmic reticulum enzyme akin to the one that introduces double bonds into long-chain fatty acyl CoA molecules. In both cases, O_2 and NADH are reactants, and cytochrome b_5 participates in catalysis (Section 22.6).

26.1.3 Sphingolipids Are Synthesized from Ceramide

We turn now from glycerol-based phospholipids to another class of membrane lipid—the *sphingolipids*. These lipids are found in the plasma membranes of all eukaryotic cells, although the concentration is highest in the cells of the central nervous system. The backbone of a sphingolipid is *sphingosine*, rather than glycerol (Section 12.3.1). Palmitoyl CoA and serine condense to form dehydrosphingosine, which is then converted into sphingosine. The enzyme catalyzing this reaction requires pyridoxal phosphate, revealing again the dominant role of this cofactor in transformations that include amino acids.

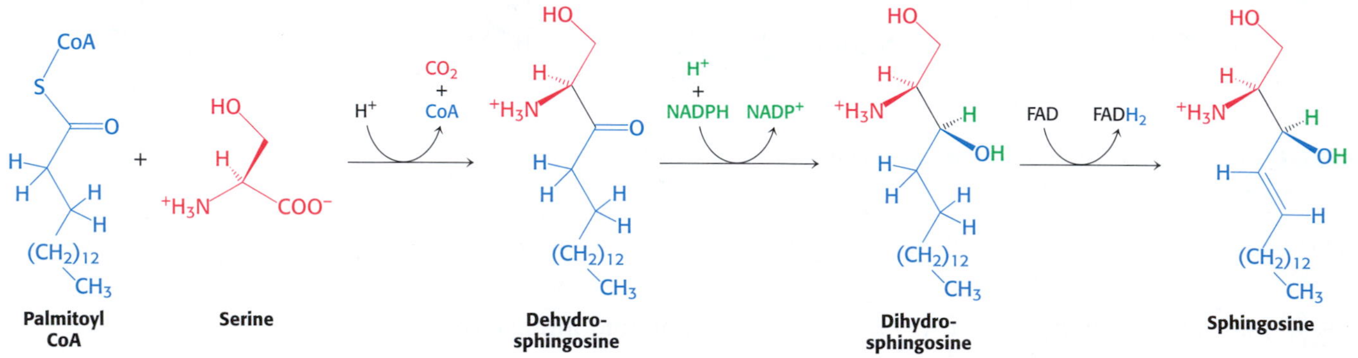

In all sphingolipids, the amino group of sphingosine is acylated: a long-chain acyl CoA reacts with sphingosine to form *ceramide (N-acyl sphingosine)* (Figure 26.3). The terminal hydroxyl group also is substituted. In *sphingomyelin*, a component of the myelin sheath covering many nerve fibers, the substituent is phosphorylcholine, which comes from phosphatidyl choline. In a *cerebroside*, the substituent is glucose or galactose.

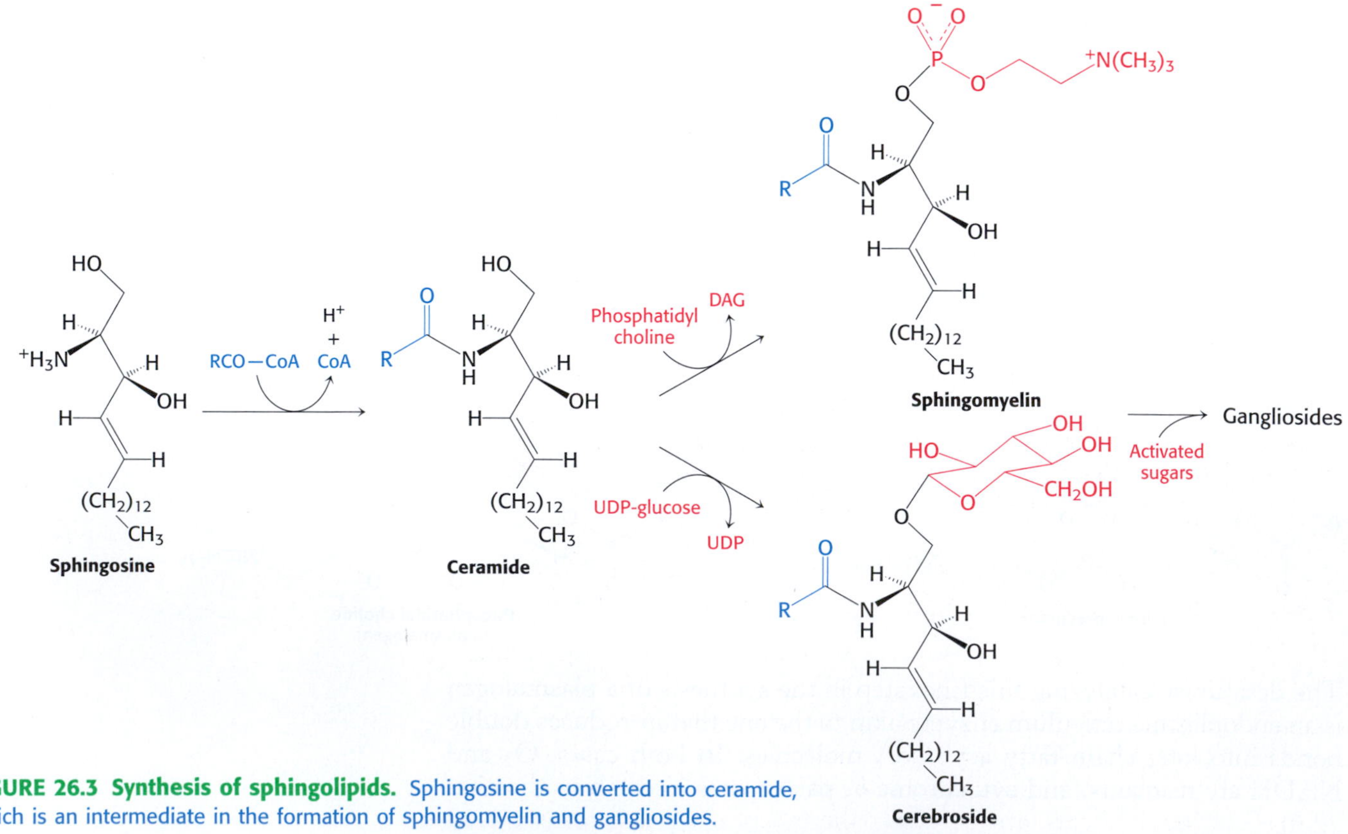

FIGURE 26.3 Synthesis of sphingolipids. Sphingosine is converted into ceramide, which is an intermediate in the formation of sphingomyelin and gangliosides.

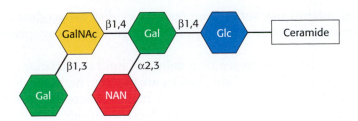

FIGURE 26.4 Ganglioside G$_{M1}$. This ganglioside consists of five monosaccharides linked to ceramide: one glucose (Glc) molecule, two galactose (Gal) molecules, one *N*-acetylgalactosamine (GalNAc) molecule, and one *N*-acetylneuraminate (NAN) molecule. The structures of the linkages are indicated.

UDP-glucose or UDP-galactose is the sugar donor. In a *ganglioside,* an oligosaccharide is linked to the terminal hydroxyl group of ceramide by a glucose residue (Figure 26.4).

26.1.4 Gangliosides Are Carbohydrate-Rich Sphingolipids That Contain Acidic Sugars

In gangliosides, the most complex sphingolipids, an *oligosaccharide chain* attached to the ceramide contains at least one acidic sugar. The acidic sugar is *N-acetylneuraminate* or *N-glycolylneuraminate*. These acidic sugars are called *sialic acids.* Their nine-carbon backbones are synthesized from phospho-enolpyruvate (a three-carbon unit) and *N*-acetylmannosamine 6-phosphate (a six-carbon unit).

Gangliosides are synthesized by the ordered, step-by-step addition of sugar residues to ceramide. The synthesis of these complex lipids requires the activated sugars UDP-glucose, UDP-galactose, and UDP-N-acetyl-galactosamine, as well as the CMP derivative of *N*-acetylneuraminate. CMP-*N*-acetylneuraminate is synthesized from CTP and *N*-acetylneur-aminate. The structure of the resulting ganglioside is determined by the specificity of the glycosyltransferases in the cell. More than 60 different gangliosides have been characterized (see Figure 26.4 for the structure of ganglioside G$_{M1}$).

R$_2$ = H, *N*-acetylneuraminate
R$_2$ = OH, *N*-glycolylneuraminate

26.1.5 Sphingolipids Confer Diversity on Lipid Structure and Function

The structures of sphingolipids and the more abundant glycerophospho-lipids are very similar. Given the structural similarity of these two types of lipids, why are sphingolipids required at all? Indeed, the prefix "sphingo" was applied to capture the "sphinxlike" properties of this enigmatic class of lipids. Although the precise role of sphingolipids is not firmly established, progress toward solving the riddle of their function is being made. The most notable function attributed to sphingolipids is their role as a source of second messengers. For instance, ceramide derived from a sphingolipid may initiate programmed cell death in some cell types.

26.1.6 Respiratory Distress Syndrome and Tay-Sachs Disease Result from the Disruption of Lipid Metabolism

Respiratory distress syndrome is a pathological condition resulting from a failure in the biosynthetic pathway for dipalmitoyl phosphatidyl choline. This phospholipid, in conjunction with specific proteins and other phospholipids, is found in the extracellular fluid that surrounds the alveoli of the lung, where it decreases the surface tension of the fluid to prevent lung collapse at the end of the expiration phase of breathing.

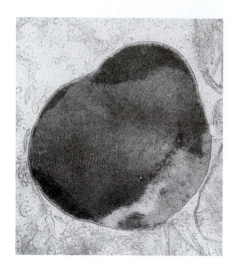

FIGURE 26.5 Lysosome with lipids.
An electron micrograph of a lysosome containing an abnormal amount of lipid. [Courtesy of Dr. George Palade.]

Premature infants may suffer from respiratory distress syndrome because their immature lungs do not synthesize enough dipalmitoyl phosphatidyl choline.

Tay-Sachs disease is caused by a failure of lipid degradation: an inability to degrade gangliosides. Gangliosides are found in highest concentration in the nervous system, particularly in gray matter, where they constitute 6% of the lipids. Gangliosides are normally degraded inside lysosomes by the sequential removal of their terminal sugars but, in Tay-Sachs disease, this degradation does not occur. As a consequence, neurons become significantly swollen with lipid-filled lysosomes (Figure 26.5). An affected infant displays weakness and retarded psychomotor skills before 1 year of age. The child is demented and blind by age 2 and usually dead before age 3.

The ganglioside content of the brain of an infant with Tay-Sachs disease is greatly elevated. *The concentration of ganglioside G_{M2} is many times as high as normal because its terminal N-acetylgalactosamine residue is removed very slowly or not at all.* The missing or deficient enzyme is a specific β-N-acetylhexosaminidase.

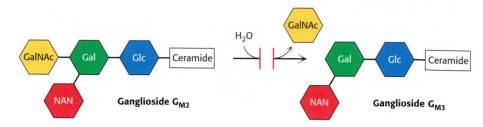

Tay-Sachs disease can be diagnosed in the course of fetal development. Amniotic fluid is obtained by amniocentesis and assayed for β-*N*-acetylhexosaminidase activity.

26.2 CHOLESTEROL IS SYNTHESIZED FROM ACETYL COENZYME A IN THREE STAGES

We now turn our attention to the synthesis of the fundamental lipid *cholesterol*. This steroid modulates the fluidity of animal cell membranes (Section 12.6.2) and is the precursor of steroid hormones such as progesterone, testosterone, estradiol, and cortisol. *All 27 carbon atoms of cholesterol are derived from acetyl CoA in a three-stage synthetic process* (Figure 26.6).

1. Stage one is the synthesis of isopentenyl pyrophosphate, an activated isoprene unit that is the key building block of cholesterol.

2. Stage two is the condensation of six molecules of isopentenyl pyrophosphate to form squalene.

FIGURE 26.6 Labeling of cholesterol. The results of isotope-labeling experiments reveal the source of carbon atoms in cholesterol synthesized from acetate labeled in its methyl group (blue) or carboxylate atom (red).

3. In stage three, squalene cyclizes in an astounding reaction and the tetra-cyclic product is subsequently converted into cholesterol.

26.2.1 The Synthesis of Mevalonate, Which Is Activated as Isopentenyl Pyrophosphate, Initiates the Synthesis of Cholesterol

The first stage in the synthesis of cholesterol is the formation of isopentenyl pyrophosphate from acetyl CoA. This set of reactions, which takes place in the cytosol, starts with the formation of 3-hydroxy-3-methylglutaryl CoA (HMG CoA) from acetyl CoA and acetoacetyl CoA. This intermediate is reduced to *mevalonate* for the synthesis of cholesterol (Figure 26.7). Recall that mitochondrial 3-hydroxy-3-methylglutaryl CoA is processed to form ketone bodies (Section 22.3.5).

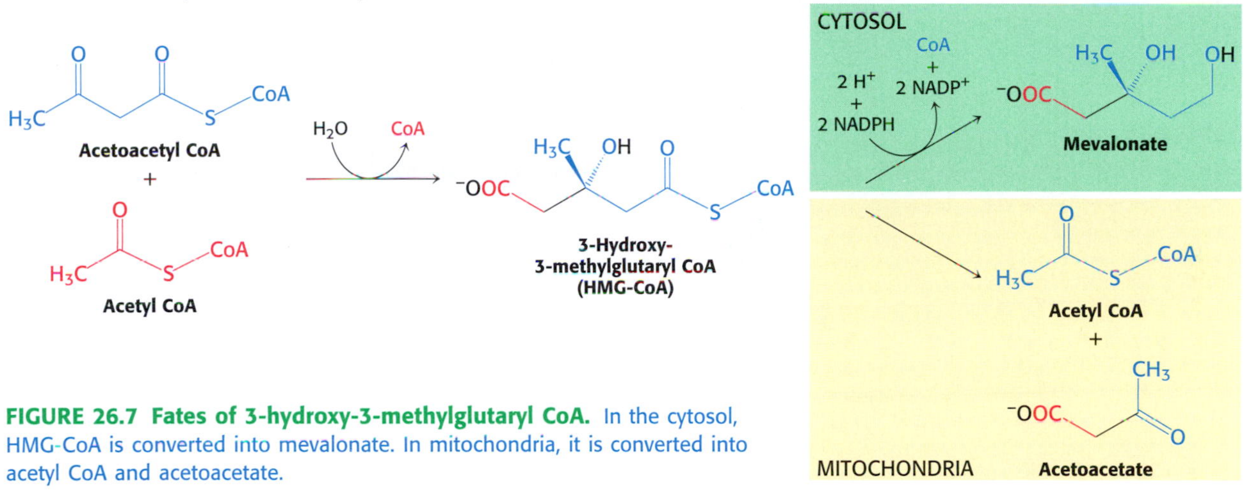

FIGURE 26.7 Fates of 3-hydroxy-3-methylglutaryl CoA. In the cytosol, HMG-CoA is converted into mevalonate. In mitochondria, it is converted into acetyl CoA and acetoacetate.

The synthesis of mevalonate is the committed step in cholesterol formation. The enzyme catalyzing this irreversible step, *3-hydroxy-3-methylglutaryl CoA reductase (HMG-CoA reductase)*, is an important control site in cholesterol biosynthesis, as will be discussed shortly.

3-Hydroxy-3-methylglutaryl CoA + 2 NADPH + 2 H$^+$ $\longrightarrow$
mevalonate + 2 NADP$^+$ + CoA

HMG-CoA reductase is an integral membrane protein in the endoplasmic reticulum.

Mevalonate is converted into *3-isopentenyl pyrophosphate* in three consecutive reactions requiring ATP (Figure 26.8). Decarboxylation yields isopentenyl pyrophosphate, an activated isoprene unit that is a key building block for many important biomolecules throughout the kingdoms of life. We will return to a discussion of this molecule later in the chapter.

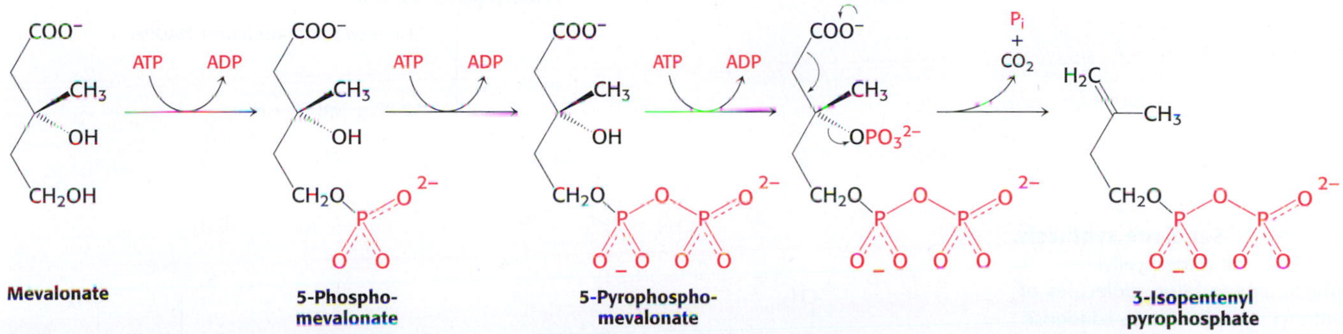

FIGURE 26.8 Synthesis of isopentenyl pyrophosphate. This activated intermediate is formed from mevalonate in three steps, the last of which includes a decarboxylation.

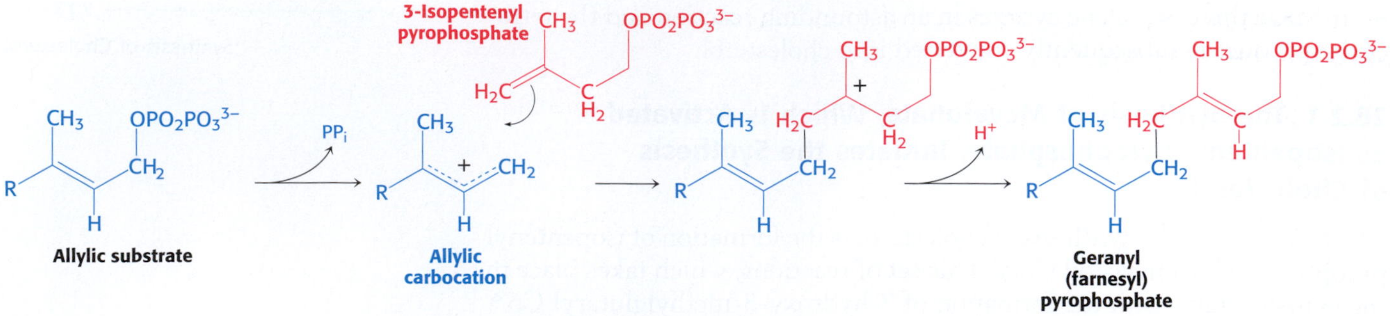

FIGURE 26.9 Condensation mechanism in cholesterol synthesis. The mechanism for joining dimethylallyl pyrophosphate and isopentenyl pyrophosphate to form geranyl pyrophosphate. The same mechanism is used to add an additional isopentenyl pyrophosphate to form farnesyl pyrophosphate.

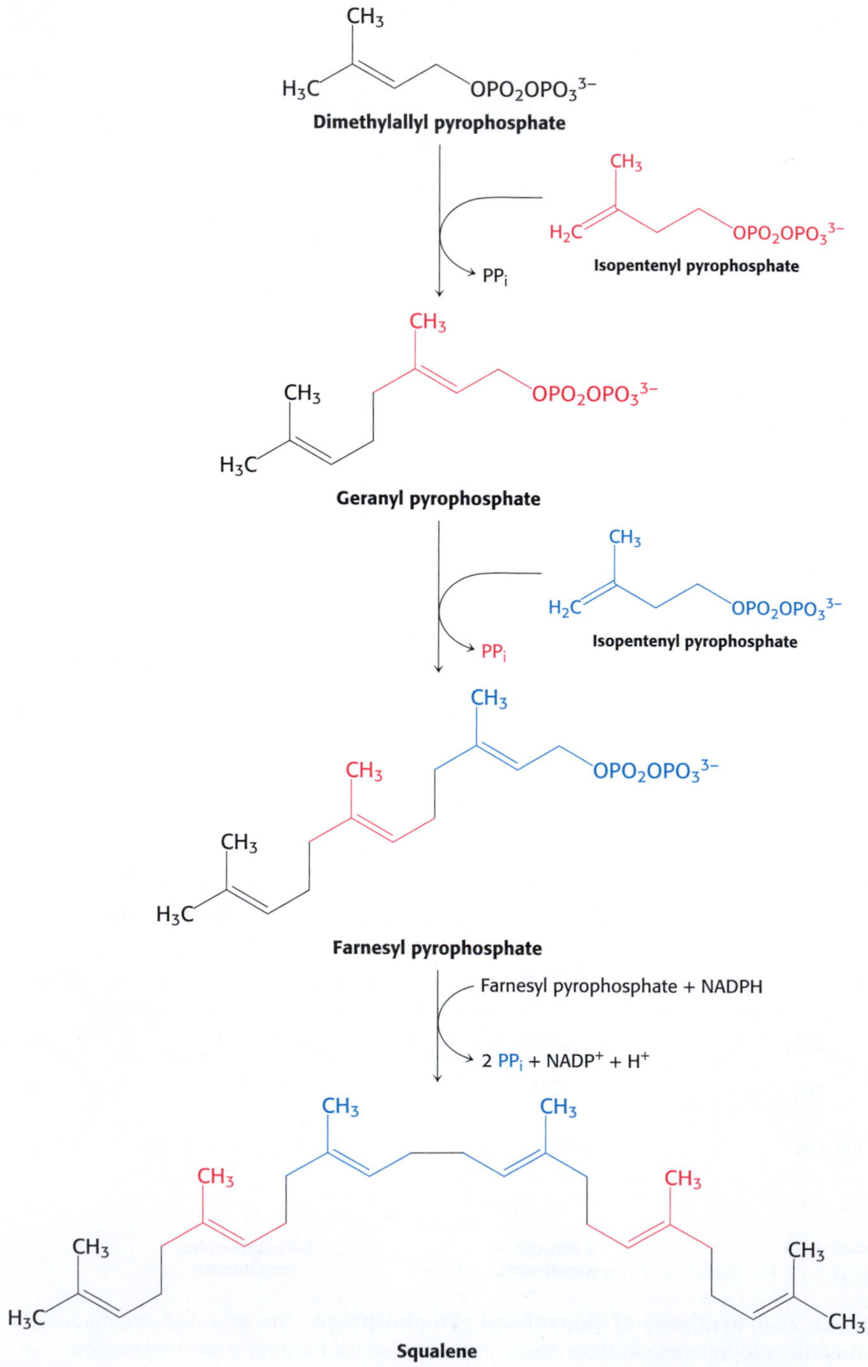

FIGURE 26.10 Squalene synthesis.
One molecule of dimethyallyl pyrophosphate and two molecules of isopentenyl pyrophosphate condense to form farnesyl pyrophosphate. The tail-to-tail coupling of two molecules of farnesyl pyrophosphate yields squalene.

26.2.2 Squalene (C$_{30}$) Is Synthesized from Six Molecules of Isopentenyl Pyrophosphate (C$_5$)

Squalene is synthesized from isopentenyl pyrophosphate by the reaction sequence

$$C_5 \longrightarrow C_{10} \longrightarrow C_{15} \longrightarrow C_{30}$$

This stage in the synthesis of cholesterol starts with the isomerization of *isopentenyl pyrophosphate* to *dimethylallyl pyrophosphate*.

Isopentenyl pyrophosphate   **Dimethylallyl pyrophosphate**

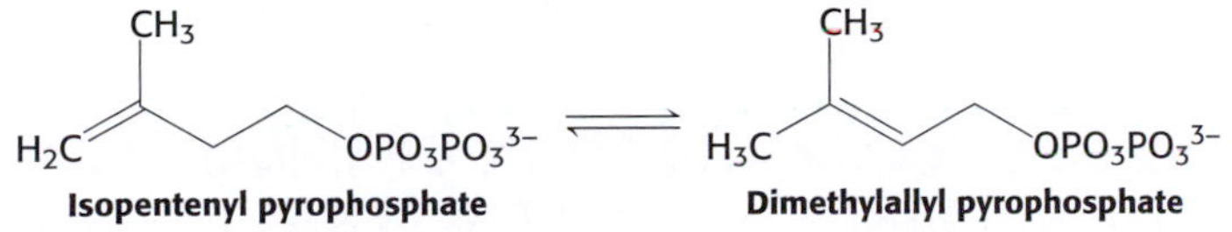

These isomeric C$_5$ units condense to form a C$_{10}$ compound: isopentenyl pyrophosphate attacks an allylic carbonium ion formed from dimethylallyl pyrophosphate to yield *geranyl pyrophosphate* (Figure 26.9). The same kind of reaction takes place again: geranyl pyrophosphate is converted into an allylic carbonium ion, which is attacked by isopentenyl pyrophosphate. The resulting C$_{15}$ compound is called *farnesyl pyrophosphate*. The same enzyme, *geranyl transferase*, catalyzes each of these condensations.

The last step in the synthesis of *squalene* is a reductive tail-to-tail condensation of two molecules of farnesyl pyrophosphate catalyzed by the endoplasmic reticulum enzyme *squalene synthase*.

2 Farnesyl pyrophosphate (C$_{15}$) + NADPH $\longrightarrow$ squalene (C$_{30}$) + 2 PP$_i$ + NADP$^+$ + H$^+$

The reactions leading from C$_5$ units to squalene, a C$_{30}$ isoprenoid, are summarized in Figure 26.10.

26.2.3 Squalene Cyclizes to Form Cholesterol

The final stage of cholesterol biosynthesis starts with the cyclization of squalene (Figure 26.11). Squalene is first activated by conversion into squalene

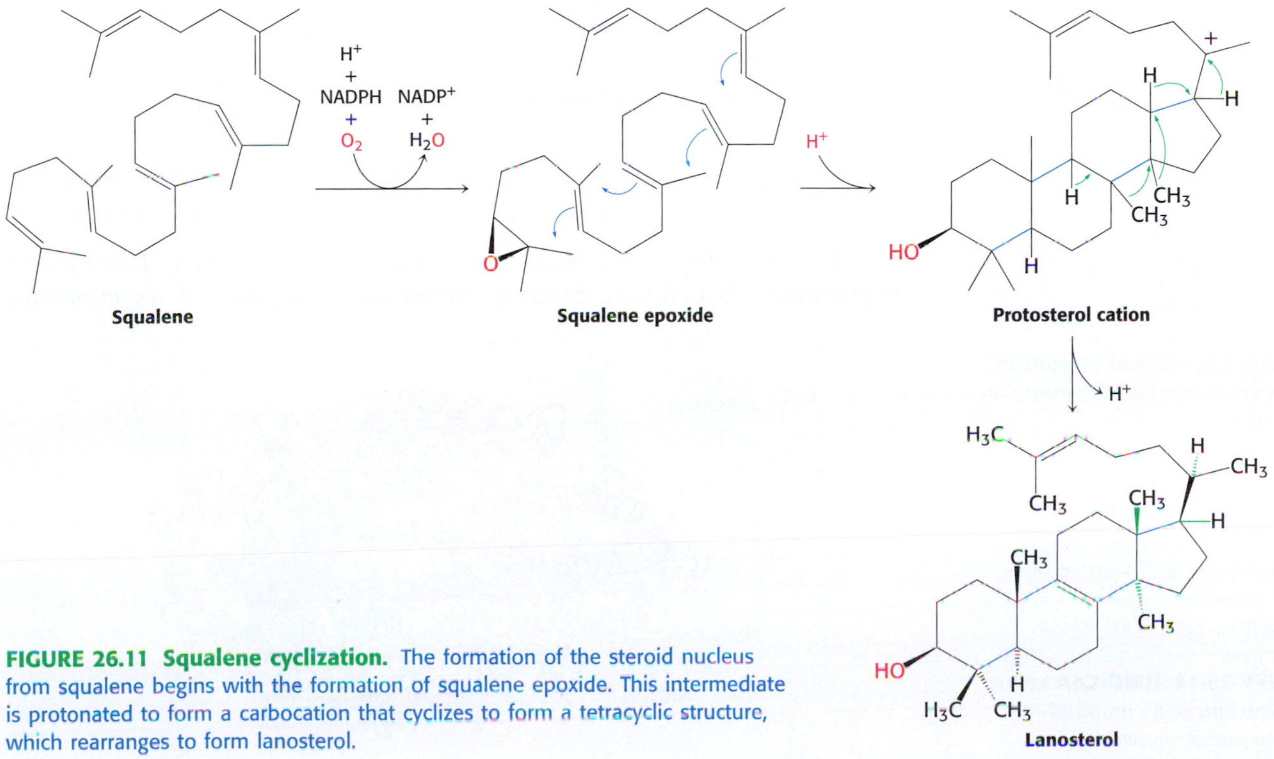

FIGURE 26.11 Squalene cyclization. The formation of the steroid nucleus from squalene begins with the formation of squalene epoxide. This intermediate is protonated to form a carbocation that cyclizes to form a tetracyclic structure, which rearranges to form lanosterol.

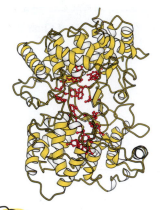

FIGURE 26.12 Oxidosqualene cyclase. The structure of an enzyme homologous to oxidosqualene cyclase shows a central cavity lined primarily with hydrophobic side chains (shown in red) in which the cyclization reaction takes place.

Lanosterol

19 steps → HCOOH + 2 CO_2

Cholesterol

FIGURE 26.13 Cholesterol formation. Lanosterol is converted into cholesterol in a complex process.

epoxide (2,3-oxidosqualene) in a reaction that uses O_2 and NADPH. Squalene epoxide is then cyclized to *lanosterol* by *oxidosqualene cyclase* (Figure 26.12). This remarkable transformation proceeds in a concerted fashion. The enzyme holds squalene epoxide in an appropriate conformation and initiates the reaction by protonating the epoxide oxygen. The carbocation formed spontaneously rearranges to produce lanosterol. Lanosterol is converted into cholesterol in a multistep process by the removal of three methyl groups, the reduction of one double bond by NADPH, and the migration of the other double bond (Figure 26.13).

26.3 THE COMPLEX REGULATION OF CHOLESTEROL BIOSYNTHESIS TAKES PLACE AT SEVERAL LEVELS

Cholesterol can be obtained from the diet or it can be synthesized de novo. An adult on a low-cholesterol diet typically synthesizes about 800 mg of cholesterol per day. The liver is the major site of cholesterol synthesis in mammals, although the intestine also forms significant amounts. The rate of cholesterol formation by these organs is highly responsive to the cellular level of cholesterol. *This feedback regulation is mediated primarily by changes in the amount and activity of 3-hydroxy-3-methylglutaryl CoA reductase* (Figure 26.14). As discussed in Section 26.2.1, this enzyme catalyzes the formation of mevalonate, the committed step in cholesterol biosynthesis. HMG CoA reductase is controlled in multiple ways:

1. The rate of *synthesis of reductase mRNA* is controlled by the *sterol regulatory element binding protein (SREBP)*. This transcription factor binds to a short DNA sequence called the *sterol regulatory element (SRE)* on the 5′ side of the reductase gene. In its inactive state, the SREBP is anchored to the endoplasmic reticulum or nuclear membrane. When cholesterol levels fall, the amino-terminal domain is released from its association with the membrane by two specific proteolytic cleavages. The released protein migrates to the nucleus and binds the SRE of the HMG-CoA reductase gene, as well as several other genes in the cholesterol biosynthetic pathway, to enhance transcription. When cholesterol levels rise, the proteolytic release of the SREBP is blocked, and the SREBP in the nucleus is rapidly degraded. These two events halt the transcription of the genes of the cholesterol biosynthetic pathways.

2. The rate of *translation of reductase mRNA* is inhibited by nonsterol metabolites derived from mevalonate as well as by dietary cholesterol.

3. The *degradation of the reductase* is stringently controlled. The enzyme is bipartite: its cytosolic domain carries out catalysis and *its membrane*

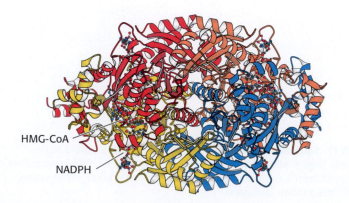

HMG-CoA

NADPH

FIGURE 26.14 HMG-CoA reductase. The structure of a portion of the tetrameric enzyme is shown.

domain senses signals that lead to its degradation. The membrane domain may undergo a change in its oligomerization state *in response to increasing concentrations of sterols such as cholesterol,* making the enzyme more susceptible to proteolysis. Homologous sterol-sensing regions are present in the protease that activates SREBP. The reductase may be further degraded by ubiquitination and targeting to the 26S proteasome under some conditions. A combination of these three regulatory devices can regulate the amount of enzyme over a 200-fold range.

4. *Phosphorylation decreases the activity of the reductase.* This enzyme, like acetyl CoA carboxylase (which catalyzes the committed step in fatty acid synthesis, Section 22.5), is switched off by an AMP-activated protein kinase. Thus, cholesterol synthesis ceases when the ATP level is low.

As we will see shortly, all four regulatory mechanisms are modulated by receptors that sense the presence of cholesterol in the blood.

26.3.1 Lipoproteins Transport Cholesterol and Triacylglycerols Throughout the Organism

Cholesterol and triacylglycerols are transported in body fluids in the form of *lipoprotein particles.* Each particle consists of a core of hydrophobic lipids surrounded by a shell of more polar lipids and proteins. The protein components of these macromolecular aggregates have two roles: *they solubilize hydrophobic lipids and contain cell-targeting signals.* Lipoprotein particles are classified according to increasing density (Table 26.1): *chylomicrons, chylomicron remnants, very low density lipoproteins (VLDL), intermediate-density lipoproteins (IDL), low-density lipoproteins (LDL),* and *high-density lipoproteins (HDL).* Ten principal apoproteins have been isolated and characterized. They are synthesized and secreted by the liver and the intestine.

Triacylglycerols, cholesterol, and other lipids obtained from the diet are carried away from the intestine in the form of large *chylomicrons* (180–500 nm in diameter; Section 22.1.2). These particles have a very low density ($d < 0.94 \text{ g cm}^{-3}$) because triacylglycerols constitute ~99% of their content. Apolipoprotein B-48 (apo B-48), a large protein (240 kd), forms an amphipathic spherical shell around the fat globule; the external face of this shell is hydrophilic. The triacylglycerols in chylomicrons are released through hydrolysis by *lipoprotein lipases.* These enzymes are located on the lining of blood vessels in muscle and other tissues that use fatty acids as fuels and in

TABLE 26.1 Properties of plasma lipoproteins

Lipoproteins	Major core lipids	Apoproteins	Mechanism of lipid delivery
Chylomicron	Dietary triacylglycerols	B-48, C, E	Hydrolysis by lipoprotein lipase
Chylomicron remnant	Dietary cholesterol esters	B-48, E	Receptor-mediated endocytosis by liver
Very low density lipoprotein (VLDL)	Endogenous triacylglycerols	B-100, C, E	Hydrolysis by lipoprotein lipase
Intermediate-density lipoprotein (IDL)	Endogenous cholesterol esters	B-100, E	Receptor-mediated endocytosis by liver and conversion into LDL
Low-density lipoprotein (LDL)	Endogenous cholesterol esters	B-100	Receptor-mediated endocytosis by liver and other tissues
High-density lipoprotein (HDL)	Endogenous cholesterol esters	A	Transfer of cholesterol esters to IDL and LDL

Source: After M. S. Brown and J. L. Goldstein, *The Pharmacological Basis of Therapeutics.* 7th ed., A. G. Gilman, L. S. Goodman, T. W. Rall, and F. Murad, Eds. (Macmillan, 1985), p. 828.

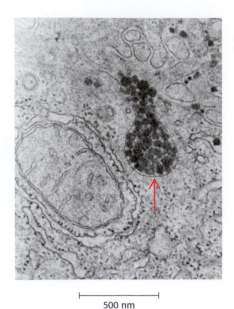

FIGURE 26.15 Site of cholesterol synthesis. Electron micrograph of a part of a liver cell actively engaged in the synthesis and secretion of very low density lipoprotein (VLDL). The arrow points to a vesicle that is releasing its content of VLDL particles. [Courtesy of Dr. George Palade.]

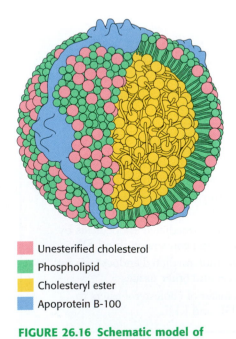

- Unesterified cholesterol
- Phospholipid
- Cholesteryl ester
- Apoprotein B-100

FIGURE 26.16 Schematic model of low-density lipoprotein. The LDL particle is approximately 22 nm (220 Å) in diameter.

the synthesis of fat. The liver then takes up the cholesterol-rich residues, known as *chylomicron remnants.*

The liver is a major site of triacylglycerol and cholesterol synthesis (Figure 26.15). Triacylglycerols and cholesterol in excess of the liver's own needs are exported into the blood in the form of very low density lipoproteins ($d < 1.006$ g cm^{-3}). These particles are stabilized by two lipoproteins—apo B-100 and apo E (34 kd). Apo B-100, one of the largest proteins known (513 kd), is a longer version of apo B-48. Both apo B proteins are encoded by the same gene and produced from the same initial RNA transcript. In the intestine, RNA editing (Section 28.3.2) modifies the transcript to generate the mRNA for apo B-48, the truncated form. Triacylglycerols in very low density lipoproteins, as in chylomicrons, are hydrolyzed by lipases on capillary surfaces. The resulting remnants, which are rich in cholesteryl esters, are called *intermediate-density lipoproteins* ($1.006 < d < 1.019$ g cm^{-3}). These particles have two fates. Half of them are taken up by the liver for processing, and half are converted into low-density lipoprotein ($1.019 < d < 1.063$ g cm^{-3}) by the removal of more triacylglycerol.

Low-density lipoprotein is the major carrier of cholesterol in blood. This lipoprotein particle has a diameter of 22 nm and a mass of about 3 million daltons (Figure 26.16). It contains a core of some 1500 esterified cholesterol molecules; the most common fatty acyl chain in these esters is linoleate, a polyunsaturated fatty acid. A shell of phospholipids and unesterified cholesterols surrounds this highly hydrophobic core. The shell also contains a single copy of apo B-100, which is recognized by target cells. *The role of LDL is to transport cholesterol to peripheral tissues and regulate de novo cholesterol synthesis at these sites,* as described in Section 26.3.3. A different purpose is served by *high-density lipoprotein* ($1.063 < d < 1.21$ g cm^{-3}), which picks up cholesterol released into the plasma from dying cells and from membranes undergoing turnover. An acyltransferase in HDL esterifies these cholesterols, which are then either rapidly shuttled to VLDL or LDL by a specific transfer protein or returned by HDL to the liver.

26.3.2 The Blood Levels of Certain Lipoproteins Can Serve Diagnostic Purposes

High serum levels of cholesterol cause disease and death by contributing to the formation of atherosclerotic plaques in arteries throughout the body. This excess cholesterol is present in the form of the low density lipoprotein particle, so-called "bad cholesterol." The ratio of cholesterol in the form of high density lipoprotein, sometimes referred to as "good cholesterol," to that in the form of LDL can be used to evaluate susceptibility to the development of heart disease. For a healthy person, the LDL/HDL ratio is 3.5.

High-density lipoprotein functions as a shuttle that moves cholesterol throughout the body. HDL binds and esterifies cholesterol released from the peripheral tissues and then transfers cholesteryl esters to the liver or to tissues that use cholesterol to synthesize steroid hormones. A specific receptor mediates the docking of the HDL to these tissues. The exact nature of the protective effect of HDL levels is not known; however, a possible mechanism is discussed in Section 26.3.5.

26.3.3 Low-Density Lipoproteins Play a Central Role in Cholesterol Metabolism

Cholesterol metabolism must be precisely regulated to prevent atherosclerosis. The mode of control in the liver, the primary site of cholesterol synthesis, has already been discussed: dietary cholesterol reduces the activity and amount of 3-hydroxy-3-methylglutaryl CoA reductase, the enzyme catalyzing the committed step. The results of studies by Michael Brown and Joseph Goldstein are sources of insight into the control of cholesterol metabolism in nonhepatic cells. In general, cells outside the liver and intestine obtain cholesterol from the plasma rather than synthesizing it de novo. Specifically, *their primary source of cholesterol is the low-density lipoprotein.* The process of LDL uptake, called *receptor-mediated endocytosis,* serves as a paradigm for the uptake of many molecules.

The steps in the receptor-mediated endocytosis of LDL are as follows (see Figure 12.40).

1. Apolipoprotein B-100 on the surface of an LDL particle binds to a specific receptor protein on the plasma membrane of nonhepatic cells. The receptors for LDL are localized in specialized regions called *coated pits,* which contain a specialized protein called *clathrin.*

2. The receptor–LDL complex is internalized by *endocytosis,* that is, the plasma membrane in the vicinity of the complex invaginates and then fuses to form an endocytic vesicle (Figure 26.17).

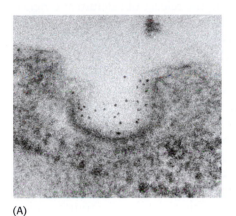

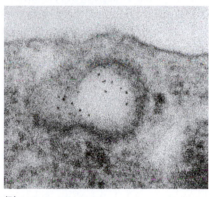

(A)   (B)

FIGURE 26.17 Endocytosis of LDL bound to its receptor. (A) Electron micrograph showing LDL (conjugated to ferritin for visualization, dark spots) bound to a coated-pit region on the surface of a cultured human fibroblast cell. (B) Micrograph showing this region invaginating and fusing to form an endocytic vesicle. [From R. G. W. Anderson, M. S. Brown, and J. L. Goldstein. *Cell* 10 (1977): 351.]

3. These vesicles, containing LDL, subsequently fuse with *lysosomes,* acidic vesicles that carry a wide array of degradative enzymes. The protein component of the LDL is hydrolyzed to free amino acids. The cholesteryl esters in the LDL are hydrolyzed by a lysosomal acid lipase. The LDL receptor itself usually returns unscathed to the plasma membrane. The round-trip time for a receptor is about 10 minutes; in its lifetime of about a day, it may bring many LDL particles into the cell.

4. *The released unesterified cholesterol can then be used for membrane biosynthesis.* Alternatively, it can be *reesterified for storage inside the cell.* In fact, free cholesterol activates *acyl CoA:cholesterol acyltransferase (ACAT),* the enzyme catalyzing this reaction. Reesterified cholesterol contains mainly oleate and palmitoleate, which are monounsaturated fatty acids, in contrast with the cholesterol esters in LDL, which are rich in linoleate, a polyunsaturated fatty acid (see Table 24.1). It is imperative that the cholesterol be reesterified. High concentrations of unesterified cholesterol disrupt the integrity of cell membranes.

The synthesis of LDL receptor is itself subject to feedback regulation. The results of studies of cultured fibroblasts show that, *when cholesterol is abundant inside the cell, new LDL receptors are not synthesized, and so the uptake of additional cholesterol from plasma LDL is blocked.* The gene for the LDL receptor, like that for the reductase, is regulated by SREBP, which binds to a sterol regulatory element that controls the rate of mRNA synthesis.

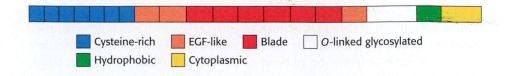

Cysteine-rich   EGF-like   Blade   *O*-linked glycosylated

Hydrophobic   Cytoplasmic

26.3.4 The LDL Receptor Is a Transmembrane Protein Having Five Different Functional Regions

The amino acid sequence of the human LDL receptor reveals the mosaic structure of this 115-kd protein, which is composed of six different types of domain (Figure 26.18). The amino-terminal region of the mature receptor consists of a cysteine-rich sequence of about 40 residues that is repeated, with some variation, seven times to form the LDL-binding domain (Figure 26.19). A set of conserved acidic side chains in this domain bind calcium ion; this metal ion lies at the center of each domain and, along with disulfide bonds formed from the conserved cysteine residues, stabilizes the three-dimensional structure. Protonation of these glutamate and aspartate side chains of the receptor in lysosomes leads to the release of calcium and hence to structural disruption and the release of LDL from its receptor. A second region of the LDL receptor includes two types of recognizable domains, three domains homologous to epidermal growth factor and six repeats that are similar to the blades of the transducin β subunit (Section 15.2.2). The six repeats form a propeller-like structure that packs against one of the EGF-like domains (Figure 26.20). An aspartate residue forms hydrogen bonds that hold each blade to the rest of the structure. These interactions, too, would most likely be disrupted at the low pH in the lysosome.

The third region contains a single domain that is very rich in serine and threonine residues and contains *O*-linked sugars. These oligosaccharides may function as struts to keep the receptor extended from the membrane so that the LDL-binding domain is accessible to LDL. The fourth region contains the fifth type of domain, which consists of 22 hydrophobic residues that span the plasma membrane. The final region contains the sixth type of domain; it consists of 50 residues and emerges on the cytosolic side of the membrane, where it controls the interaction of the receptor with coated pits and participates in endocytosis. The gene for the LDL receptor consists of 18 exons, which correspond closely to the structural units of the protein. *The LDL receptor is a striking example of a mosaic protein encoded by a gene that was assembled by exon shuffling.*

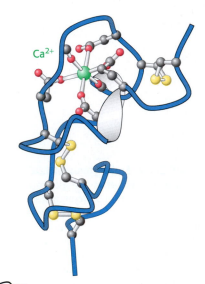

Ca²⁺

FIGURE 26.19 **Structure of cysteine-rich domain.** This calcium-binding cysteine-rich domain is repeated seven times at the amino terminus of the LDL receptor.

26.3.5 The Absence of the LDL Receptor Leads to Hypercholesteremia and Atherosclerosis

The results of Brown and Goldstein's pioneering studies of *familial hypercholesterolemia* revealed the physiologic importance of the LDL receptor. The total concentration of cholesterol and LDL in the plasma is markedly elevated in this genetic disorder, which results from a mutation at a single autosomal locus. The cholesterol level in the plasma of homozygotes is typically 680 mg dl^{-1}, compared with 300 mg dl^{-1} in heterozygotes (clinical assay results are often expressed in milligrams per deciliter, which is equal to milligrams per 100 milliliters). A value of < 200 mg dl^{-1} is regarded as desirable, but many people have higher levels. *In familial hypercholesterolemia, cholesterol is deposited in various tissues because of the high concentration of LDL cholesterol in the plasma.* Nodules of cholesterol called *xanthomas* are prominent in skin and tendons. Of particular concern is the oxidation of the excess blood LDL to form oxidized LDL (oxLDL). The oxLDL is taken up by immune-system cells called macrophages, which

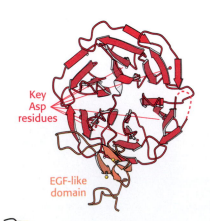

Key Asp residues

EGF-like domain

FIGURE 26.20 **Structure of propeller domain.** The six-bladed propeller domain and an adjacent EGF-like domain of the LDL receptor.

become engorged to form foam cells. These foam cells become trapped in the walls of the blood vessels and contribute to the formation of atherosclerotic plaques that cause arterial narrowing and lead to heart attacks (Figure 26.21). In fact, *most homozygotes die of coronary artery disease in childhood.* The disease in heterozygotes (1 in 500 people) has a milder and more variable clinical course. A serum esterase that degrades oxidized lipids is found in association with HDL. Possibly, the HDL-associated protein destroys the oxLDL, accounting for HDL's ability to protect against coronary disease.

The molecular defect in most cases of familial hypercholesterolemia is an absence or deficiency of functional receptors for LDL. Receptor mutations that disrupt each of the stages in the endocytotic pathway have been identified. Homozygotes have almost no functional receptors for LDL, whereas heterozygotes have about half the normal number. Consequently, the entry of LDL into liver and other cells is impaired, leading to an increased plasma level of LDL. Furthermore, less IDL enters liver cells because IDL entry, too, is mediated by the LDL receptor. Consequently, IDL stays in the blood longer in familial hypercholesterolemia, and more of it is converted into LDL than in normal people. All deleterious consequences of an absence or deficiency of the LDL receptor can be attributed to the ensuing elevated level of LDL cholesterol in the blood.

26.3.6 The Clinical Management of Cholesterol Levels Can Be Understood at a Biochemical Level

Homozygous familial hypercholesterolemia can be treated only by a liver transplant. A more generally applicable therapy is available for heterozygotes and others with high levels of cholesterol. *The goal is to reduce the amount of cholesterol in the blood by stimulating the single normal gene to produce more than the customary number of LDL receptors.* We have already observed that the production of LDL receptors is controlled by the cell's need for cholesterol. Therefore, in essence, the strategy is to deprive the cell of ready sources of cholesterol. When cholesterol is required, the amount of mRNA for the LDL receptor rises and more receptor is found on the cell surface. This state can be induced by a two-pronged approach. First, the intestinal reabsorption of bile salts is inhibited. Bile salts are cholesterol derivatives that promote the absorption of dietary cholesterol and dietary fats (Section 22.1.1). Second, de novo synthesis of cholesterol is blocked.

The reabsorption of bile is impeded by oral administration of positively charged polymers, such as cholestyramine, that bind negatively charged bile salts and are not themselves absorbed. Cholesterol synthesis can be effectively blocked by a class of compounds called *statins* (e.g., lovastatin, which is also called mevacor; Figure 26.22). These compounds are potent competitive inhibitors (K_i = 1 nM) of HMG-CoA reductase, the essential control point in the biosynthetic pathway. Plasma cholesterol levels decrease by 50% in many patients given both lovastatin and inhibitors of bile-salt reabsorption. Lovastatin and other inhibitors of HMG-CoA reductase are widely used to lower the plasma cholesterol level in people who have atherosclerosis, which is the leading cause of death in industrialized societies.

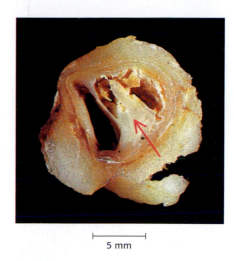

|— 5 mm —|

FIGURE 26.21 An atherosclerotic plaque. A plaque (marked by an arrow) blocks most of the lumen of this blood vessel. The plaque is rich in cholesterol. [Courtesy of Dr. Jeffrey Sklar.]

Lovastatin

FIGURE 26.22 Lovastatin, a competitive inhibitor of HMG-CoA reductase. The part of the structure that resembles the 3-hydroxy-3-methylglutaryl moiety is shown in red.

26.4 IMPORTANT DERIVATIVES OF CHOLESTEROL INCLUDE BILE SALTS AND STEROID HORMONES

Cholesterol is a precursor for other important steroid molecules: the bile salts, steroid hormones, and vitamin D.

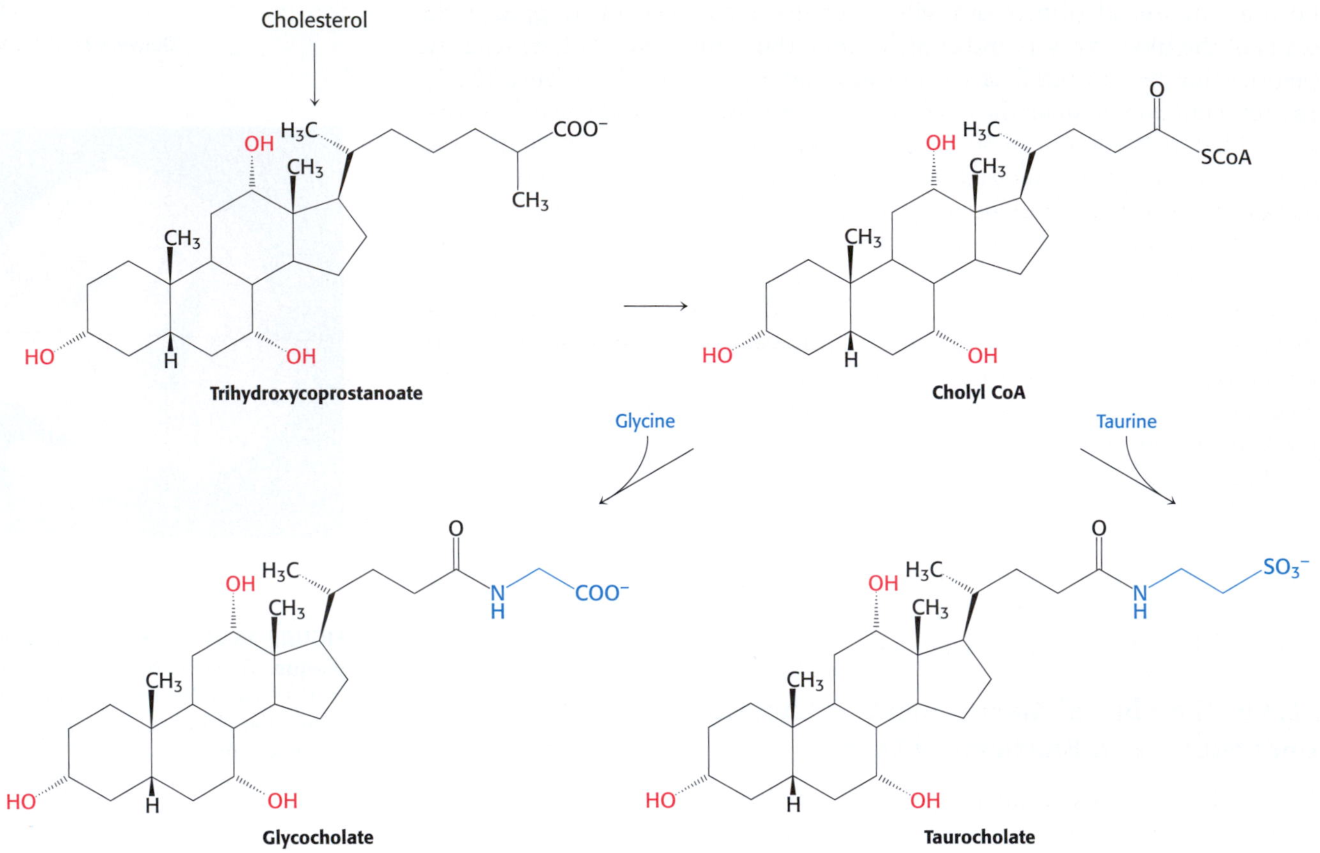

FIGURE 26.23 Synthesis of bile salts. Pathways for the formation of bile salts from cholesterol.

Bile Salts. As polar derivatives of cholesterol, *bile salts* are highly effective *detergents* because they contain both polar and nonpolar regions. Bile salts are synthesized in the liver, stored and concentrated in the gall bladder, and then released into the small intestine. Bile salts, the major constituent of bile, *solubilize dietary lipids* (Section 22.1.1). Solubilization increases in the effective surface area of lipids with two consequences: more surface area is exposed to the digestive action of lipases and lipids are more readily absorbed by the intestine. Bile salts are also the major breakdown products of cholesterol.

Cholesterol is converted into trihydroxycoprostanoate and then into *cholyl CoA*, the activated intermediate in the synthesis of most bile salts (Figure 26.23). The activated carboxyl carbon of cholyl CoA then reacts with the amino group of glycine to form *glycocholate* or it reacts with the amino group of taurine ($H_2NCH_2CH_2SO_3^-$), derived from cysteine, to form *taurocholate*. *Glycocholate is the major bile salt.*

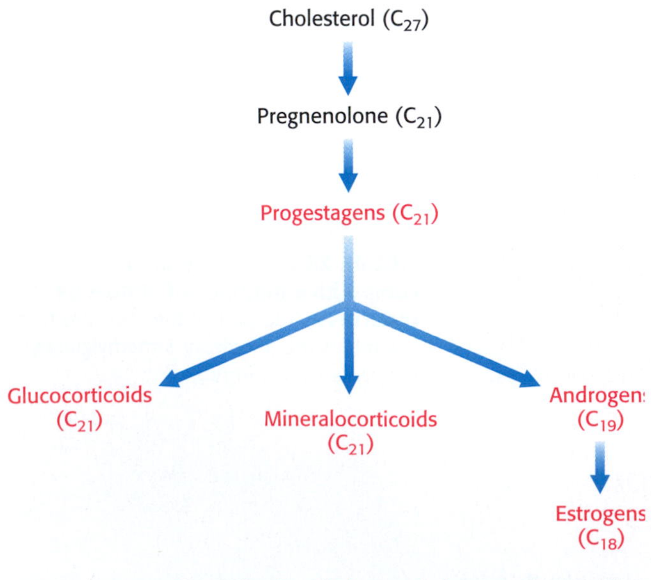

Steroid Hormones. Cholesterol is the precursor of the five major classes of *steroid hormones*: progestagens, glucocorticoids, mineralocorticoids, androgens, and estrogens (Figure 26.24). These hormones are powerful signal molecules that regulate a host of organismal functions. *Progesterone*, a *progestagen*, prepares the lining of the uterus for implantation of an ovum. Progesterone is also essential for the maintenance of pregnancy. *Androgens* (such as *testosterone*) are responsible for the development of male secondary sex characteristics, whereas *estrogens* (such as *estrone*) are required for the development of

FIGURE 26.24 Biosynthetic relations of classes of steroid hormones and cholesterol.

female secondary sex characteristics. Estrogens, along with progesterone, also participate in the ovarian cycle. *Glucocorticoids* (such as *cortisol*) promote gluconeogenesis and the formation of glycogen, enhance the degradation of fat and protein, and inhibit the inflammatory response. They enable animals to respond to stress—indeed, the absence of glucocorticoids can be fatal. *Mineralocorticoids* (primarily *aldosterone*) act on the distal tubules of the kidney to increase the reabsorption of Na^+ and the excretion of K^+ and H^+, which leads to an increase in blood volume and blood pressure. The major sites of synthesis of these classes of hormones are the corpus luteum, for progestagens; the ovaries, for estrogens; the testes, for androgens; and the adrenal cortex, for glucocorticoids and mineralocorticoids.

Steroid hormones bind to and activate receptor molecules that serve as transcription factors to regulate gene expression (Chapter 31.3.1). These small, relatively similar molecules are able to have greatly differing effects because the slight structural differences among them allow interactions with specific receptor molecules.

26.4.1 The Nomenclature of Steroid Hormones

Carbon atoms in steroids are numbered as shown for cholesterol in (Figure 26.25). The rings in steroids are denoted by the letters A, B, C, and D. Cholesterol contains two angular methyl groups: the C-19 methyl group is attached to C-10, and the C-18 methyl group is attached to C-13. The C-18 and C-19 methyl groups of cholesterol lie *above* the plane containing the four rings. A substituent that is above the plane is termed β *oriented*, whereas a substituent that is below the plane is α *oriented*.

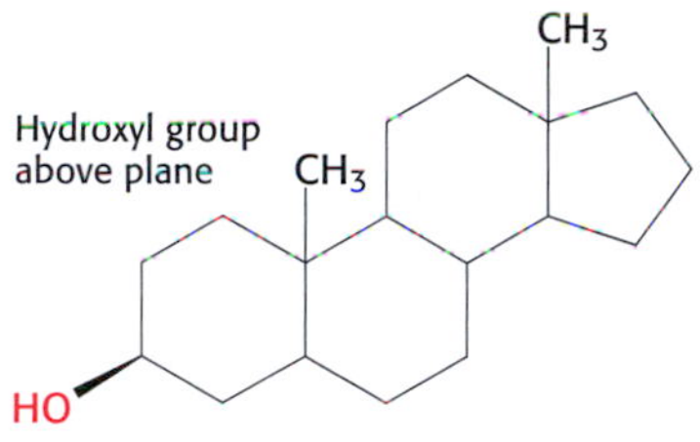

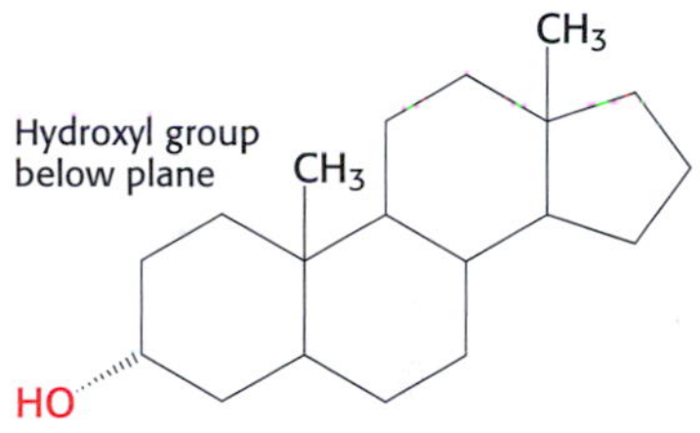

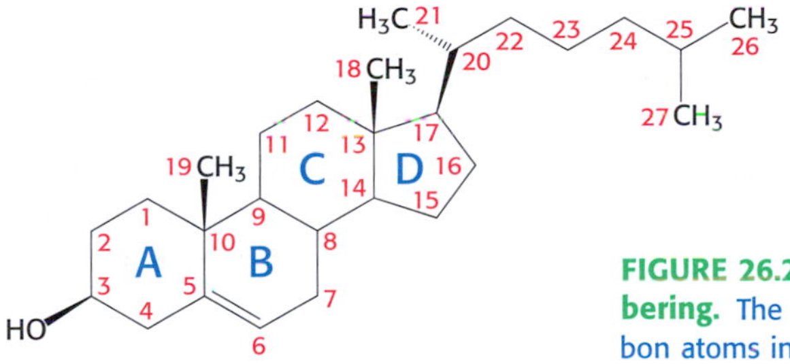

FIGURE 26.25 Cholesterol carbon numbering. The numbering scheme for the carbon atoms in cholesterol and other steroids.

If a hydrogen atom is attached to C-5, it can be either α or β oriented. The A and B steroid rings are fused in a *trans* conformation if the C-5 hydrogen is α oriented, and *cis* if it is β oriented. The absence of a Greek letter for the C-5 hydrogen atom on the steroid nucleus implies a trans fusion. The C-5 hydrogen atom is α oriented in all steroid hormones that contain a hydrogen atom in that position. In contrast, bile salts have a β-oriented hydrogen atom at C-5. Thus, *a cis fusion is characteristic of the bile salts, whereas a trans fusion is characteristic of all steroid hormones that possess a hydrogen atom at C-5.* A trans fusion yields a nearly planar structure, whereas a cis fusion gives a buckled structure.

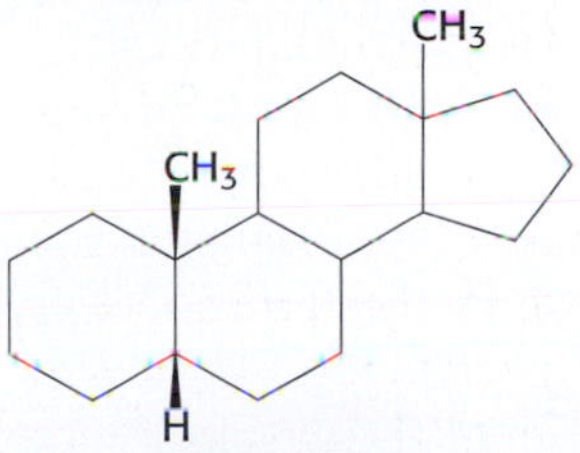

5β-Hydrogen (cis fusion)

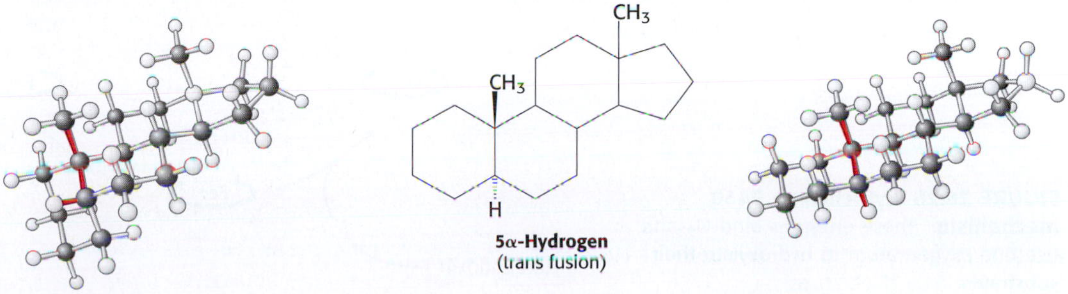

5α-Hydrogen (trans fusion)

26.4.2 Steroids Are Hydroxylated by Cytochrome P450 Monooxygenases That Utilize NADPH and O_2

Hydroxylation reactions play a very important role in the synthesis of cholesterol from squalene and in the conversion of cholesterol into steroid hormones and bile salts. All these hydroxylations require *NADPH and O_2*. The oxygen atom of the incorporated hydroxyl group comes from O_2 rather than from H_2O. While one oxygen atom of the O_2 molecule goes into the substrate, the other is reduced to water. The enzymes catalyzing these reactions are called *monooxygenases* (or *mixed-function oxygenases*). Recall that a monooxygenase also participates in the hydroxylation of aromatic amino acids (Section 23.5.7).

$$RH + O_2 + NADPH + H^+ \longrightarrow ROH + H_2O + NADP^+$$

Hydroxylation requires the activation of oxygen. In the synthesis of steroid hormones and bile salts, activation is accomplished by a cytochrome P450, a family of cytochromes that absorb light maximally at 450 nm when complexed in vitro with exogenous carbon monoxide. These membrane-anchored proteins ($\sim$50 kd) contain a heme prosthetic group. Because the hydroxylation reactions promoted by P450 enzymes are oxidation reactions, it is at first glance surprising that they also consume the reductant NADPH. NADPH transfers its high-potential electrons to a flavoprotein, which transfers them, one at a time, to *adrenodoxin*, a nonheme iron protein. Adrenodoxin transfers one electron to reduce the ferric (Fe^{3+}) form of P450 to the ferrous (Fe^{2+}) form (Figure 26.26). Without the addition of this electron, P450 will not bind oxygen. Recall that only the ferrous form of hemoglobin binds oxygen (Section 10.2.1). The binding of O_2 to the heme is followed by the acceptance of a second electron from adrenodoxin. The acceptance of this second electron leads to cleavage of the O–O bond. One of the oxygen atoms is then protonated and released as water. The remaining oxygen atom forms a highly reactive ferryl ($Fe = O$) intermediate. This intermediate abstracts a hydrogen atom from the substrate RH to form R•. This transient free radical captures the OH group from the iron atom to form ROH, the hydroxylated product, returning the iron atom to the ferric state.

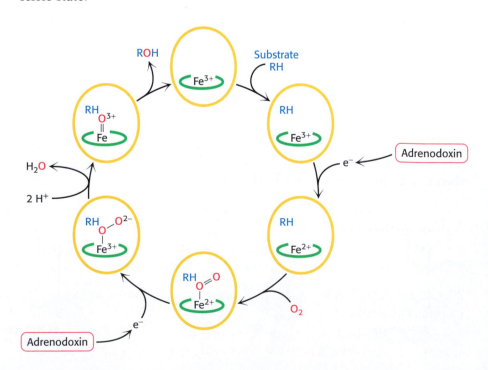

FIGURE 26.26 Cytochrome P450 mechanism. These enzymes bind O_2 and use one oxygen atom to hydroxylate their substrates.

26.4.3 The Cytochrome P450 System Is Widespread and Performs a Protective Function

The cytochrome P450 system, which in mammals is located primarily in the endoplasmic reticulum of the liver and small intestine, is also important in the *detoxification of foreign substances* (xenobiotic compounds) by oxidative metabolism. For example, the hydroxylation of phenobarbital, a barbiturate, *increases its solubility* and *facilitates its excretion.* Likewise, polycyclic aromatic hydrocarbons are hydroxylated by P450, providing sites for conjugation with highly polar units (e.g., glucuronate or sulfate), which markedly increase the solubility of the modified aromatic molecule. One of the most relevant functions of the cytochrome P450 system to human beings is its role in drug metabolism. Drugs such as caffeine and ibuprofen are oxidatively metabolized by these monooxygenases. Indeed, the duration of action of many medications depends on their rate of inactivation by the P450 system. Despite its general protective role in the removal of foreign chemicals, the action of the P450 system is not always beneficial. *Some of the most powerful carcinogens are generated from harmless compounds by the P450 system in vivo* in the process of *metabolic activation.* In plants, the cytochrome P450 system plays a role in the synthesis of toxic compounds as well as the pigments of flowers.

The cytochrome P450 system is a ubiquitous superfamily of monooxygenases that is present in plants, animals, and prokaryotes. The human genome encodes more than 50 members of the family, whereas the genome of the plant *Arabidopsis* encodes more than 250 members. All members of this large family arose by gene duplication followed by subsequent divergence that generated a range of substrate specificity. Indeed, the specificity of these enzymes is encoded in delimited regions of the primary structure, and the substrate specificity of closely related members is often defined by a few critical residues or even a single amino acid.

26.4.4 Pregnenolone, a Precursor for Many Other Steroids, Is Formed from Cholesterol by Cleavage of Its Side Chain

Steroid hormones contain 21 or fewer carbon atoms, whereas cholesterol contains 27. Thus, the first stage in the synthesis of steroid hormones is the removal of a six-carbon unit from the side chain of cholesterol to form *pregnenolone.* The side chain of cholesterol is hydroxylated at C-20 and then at C-22, and the bond between these carbon atoms is subsequently cleaved by *desmolase.* Three molecules of NADPH and three molecules of O_2 are consumed in this remarkable six-electron oxidation.

Cholesterol   20α,22β-Dihydroxycholesterol   Pregnenolone

Adrenocorticotropic hormone (ACTH, or corticotropin), a polypeptide synthesized by the anterior pituitary gland, stimulates the conversion of cholesterol into pregnenolone, the precursor of all steroid hormones.

26.4.5 The Synthesis of Progesterone and Corticosteroids from Pregnenolone

Progesterone is synthesized from pregnenolone in two steps. The 3-hydroxyl group of pregnenolone is oxidized to a 3-keto group, and the Δ^5 double bond

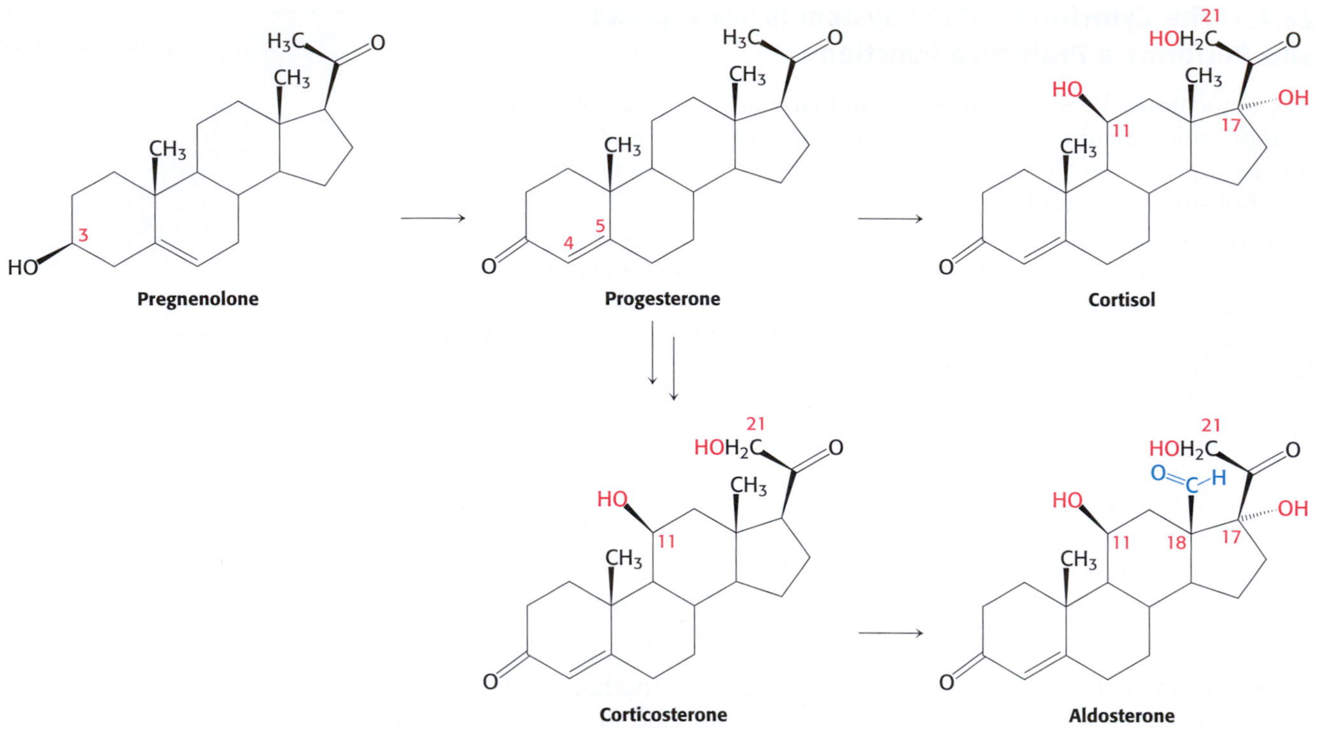

FIGURE 26.27 Pathways for the formation of progesterone, cortisol, and aldosterone.

is isomerized to a Δ^4 double bond (Figure 26.27). *Cortisol,* the major glucocorticoid, is synthesized from progesterone by hydroxylations at C-17, C-21, and C-11; C-17 must be hydroxylated before C-21 is, whereas C-11 can be hydroxylated at any stage. The enzymes catalyzing these hydroxylations are highly specific, as shown by some inherited disorders. The initial step in the synthesis of *aldosterone,* the major mineralocorticoid, is the hydroxylation of progesterone at C-21. The resulting deoxycorticosterone is hydroxylated at C-11. The oxidation of the C-18 angular methyl group to an aldehyde then yields aldosterone.

26.4.6 The Synthesis of Androgens and Estrogens from Pregnenolone

Androgens and estrogens also are synthesized from pregnenolone through the intermediate progesterone. Androgens contain 19 carbon atoms. The synthesis of androgens (Figure 26.28) starts with the hydroxylation of progesterone at C-17. The side chain consisting of C-20 and C-21 is then cleaved to yield *androstenedione,* an androgen. *Testosterone,* another androgen, is formed by the reduction of the 17-keto group of androstenedione. Testosterone, through its actions in the brain, is paramount in the development of male sexual behavior. It is also important for maintenance of the testes and development of muscle mass. Owing to the latter activity, testosterone is referred to as an *anabolic steroid.* Testosterone is reduced by *5α-reductase* to yield *dihydrotestosterone (DHT),* a powerful embryonic androgen that instigates the development and differentiation of the male phenotype. Estrogens are synthesized from androgens by the loss of the C-19 angular methyl group and the formation of an aromatic A ring. *Estrone,* an estrogen, is derived from androstenedione, whereas *estradiol,* another estrogen, is formed from testosterone.

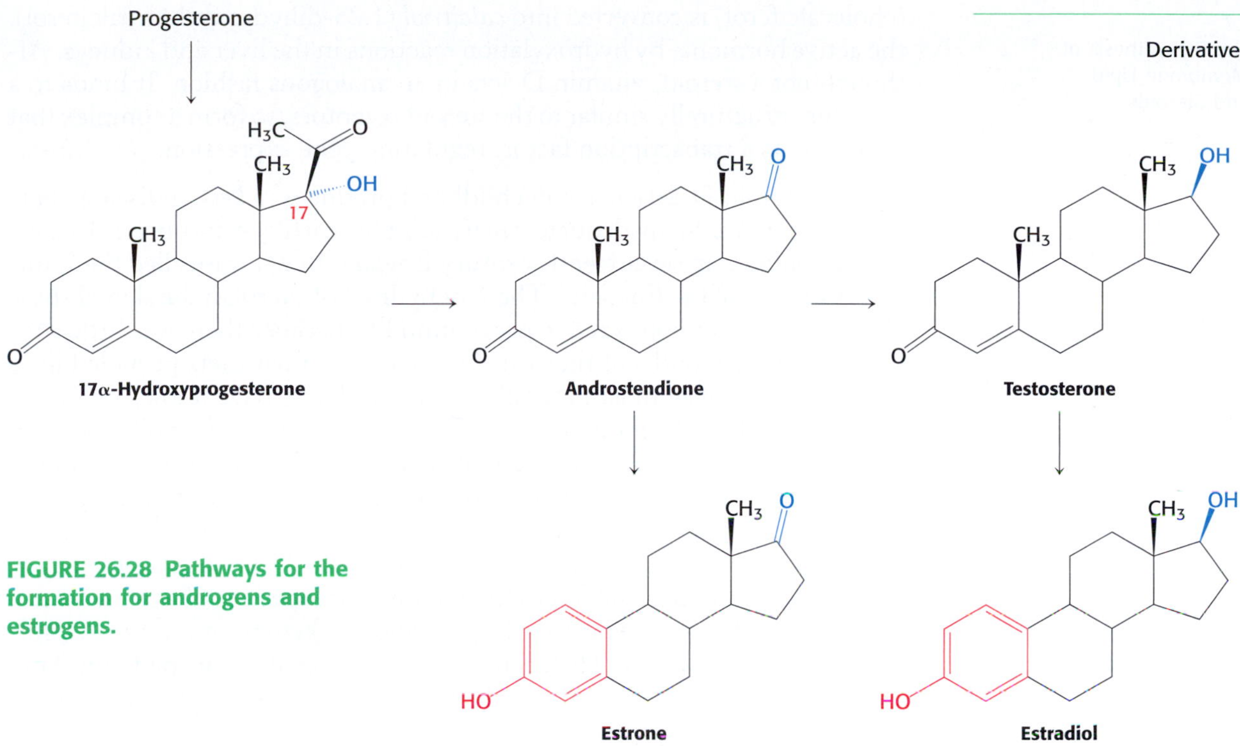

Progesterone

17α-Hydroxyprogesterone

Androstendione

Testosterone

Estrone

Estradiol

FIGURE 26.28 Pathways for the formation for androgens and estrogens.

26.4.7 Vitamin D Is Derived from Cholesterol by the Ring-Splitting Activity of Light

Cholesterol is also the precursor of vitamin D, which plays an essential role in the control of calcium and phosphorus metabolism. *7-Dehydrocholesterol (provitamin D₃) is photolyzed by the ultraviolet light of sunlight to previtamin D₃, which spontaneously isomerizes to vitamin D₃* (Figure 26.29). Vitamin D₃

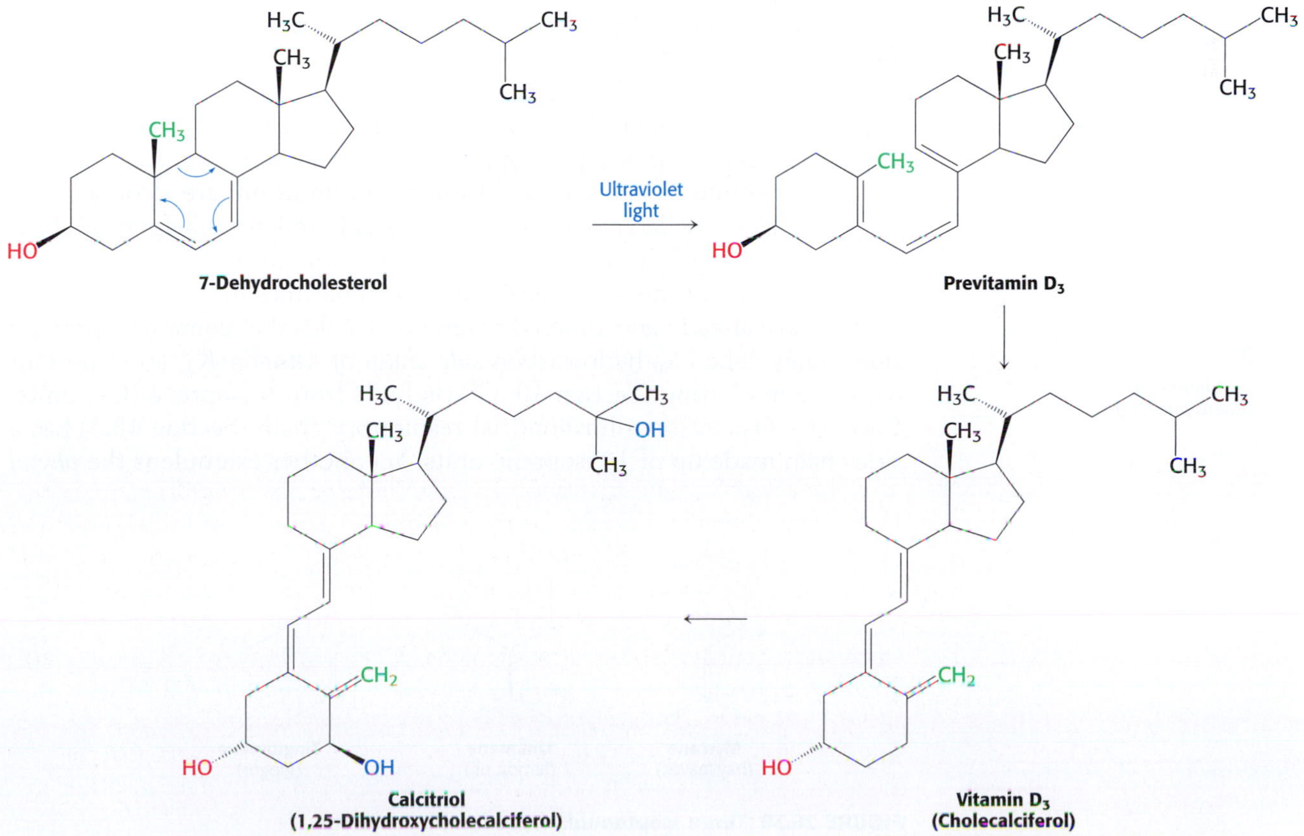

FIGURE 26.29 Vitamin D synthesis. The pathway for the conversion of 7-dehydrocholesterol into vitamin D₃ and then into calcitriol, the active hormone.

7-Dehydrocholesterol

Ultraviolet light

Previtamin D₃

Calcitriol
(1,25-Dihydroxycholecalciferol)

Vitamin D₃
(Cholecalciferol)

(cholecalciferol) is converted into *calcitriol* (1,25-dihydroxycholecalciferol), the active hormone, by hydroxylation reactions in the liver and kidneys. Although not a steroid, vitamin D acts in an analogous fashion. It binds to a receptor, structurally similar to the steroid receptors, to form a complex that functions as a transcription factor, regulating gene expression.

Vitamin D deficiency in childhood produces *rickets*, a disease characterized by inadequate calcification of cartilage and bone. Rickets was so common in seventeenth-century England that it was called the "children's disease of the English." The 7-dehydrocholesterol in the skin of these children was not photolyzed to previtamin D_3, because there was little sunlight for many months of the year. Furthermore, their diets provided little vitamin D, because most naturally occurring foods have a low content of this vitamin. Fish-liver oils are a notable exception. Cod-liver oil, abhorred by generations of children because of its unpleasant taste, was used in the past as a rich source of vitamin D. Today, the most reliable dietary sources of vitamin D are fortified foods. Milk, for example, is fortified to a level of 400 international units per quart (10 μg per quart). The recommended daily intake of vitamin D is 400 international units, irrespective of age. In adults, vitamin D deficiency leads to softening and weakening of bones, a condition called *osteomalacia*. The occurrence of osteomalacia in Bedouin Arab women who are clothed so that only their eyes are exposed to sunlight is a striking reminder that vitamin D is needed by adults as well as by children.

26.4.8 Isopentenyl Pyrophosphate Is a Precursor for a Wide Variety of Biomolecules

Before this chapter ends, we will revisit isopentenyl pyrophosphate, the activated precursor of cholesterol. The combination of isopentenyl pyrophosphate (C_5) units to form squalene (C_{30}) exemplifies a fundamental mechanism for the assembly of carbon skeletons of biomolecules. *A remarkable array of compounds is formed from isopentenyl pyrophosphate, the basic five-carbon building block.* The fragrances of many plants arise from volatile C_{10} and C_{15} compounds, which are called *terpenes*. For example, myrcene ($C_{10}H_{16}$) from bay leaves consists of two isoprene units, as does limonene ($C_{10}H_{15}$) from lemon oil (Figure 26.30). Zingiberene ($C_{15}H_{24}$), from the oil of ginger, is made up of three isoprene units. Some terpenes, such as geraniol from geraniums and menthol from peppermint oil, are alcohols; others, such as citronellal, are aldehydes. We shall see later (Chapter 32) how specialized sets of 7-TM receptors are responsible for the diverse and delightful odor and taste sensations that these molecules induce.

We have already encountered several molecules that contain isoprenoid side chains. The C_{30} hydrocarbon *side chain of vitamin K_2*, an important molecule in clotting (Section 10.5.7), is built from 6 isoprene (C_5) units. *Coenzyme Q_{10}* in the mitochondrial respiratory chain (Section 18.3) has a side chain made up of 10 isoprene units. Yet another example is the *phytol*

Isoprene unit

Vitamin K₂

Isoprene unit

Ubiquinone (Coenzyme Q₁₀)

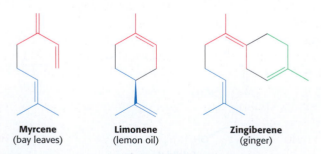

Myrcene
(bay leaves)

Limonene
(lemon oil)

Zingiberene
(ginger)

FIGURE 26.30 Three isoprenoids from familiar sources.

side chain of chlorophyll (Section 19.2), which is formed from 4 isoprene units. Many proteins are targeted to membranes by the covalent attachment of a farnesyl (C_{15}) or a geranylgeranyl (C_{20}) unit to the carboxyl-terminal cysteine residue of the protein (Section 12.5.3). *The attachment of isoprenoid side chains confers hydrophobic character.*

Isoprenoids can delight by their color as well as by their fragrance. The color of tomatoes and carrots comes from *carotenoids*. These compounds absorb light because they contain extended networks of single and double bonds and are important pigments in photosynthesis (Section 19.5.2). Their C_{40} carbon skeletons are built by the successive addition of C_5 units to form *geranylgeranyl pyrophosphate, a C_{20} intermediate,* which then condenses tail-to-tail with another molecule of geranylgeranyl pyrophosphate.

$$C_5 \longrightarrow C_{10} \longrightarrow C_{15} \longrightarrow C_{20} \longrightarrow C_{40} \text{ (phytoene)}$$

Phytoene, the C_{40} condensation product, is dehydrogenated to yield lycopene. Cyclization of both ends of lycopene gives β-carotene, which is the precursor of retinal, the chromophore in all known visual pigments (Section 32.3.2). *These examples illustrate the fundamental role of isopentenyl pyrophosphate in the assembly of extended carbon skeletons of biomolecules.* It is evident that isoprenoids are ubiquitous in nature and have diverse significant roles, including the enhancement of the sensuality of life.

SUMMARY

● **Phosphatidate Is a Common Intermediate in the Synthesis of Phospholipids and Triacylglycerols**

Phosphatidate is formed by successive acylations of glycerol 3-phosphate by acyl CoA. Hydrolysis of its phosphoryl group followed by acylation yields a triacylglycerol. CDP-diacylglycerol, the activated intermediate in the de novo synthesis of several phosphoglycerides, is formed from phosphatidate and CTP. The activated phosphatidyl unit is then transferred to the hydroxyl group of a polar alcohol, such as serine, to form a phospholipid such as phosphatidyl serine. In bacteria, decarboxylation of this phosphoglyceride yields phosphatidyl ethanolamine, which is methylated by S-adenosylmethionine to form phosphatidyl choline. In mammals, this phosphoglyceride is usually synthesized by a pathway that utilizes dietary choline. CDP-choline is the activated intermediate in this route.

Sphingolipids are synthesized from ceramide, which is formed by the acylation of sphingosine. Gangliosides are sphingolipids that contain an oligosaccharide unit having at least one residue of N-acetylneuraminate or a related sialic acid. They are synthesized by the step-by-step addition of activated sugars, such as UDP-glucose, to ceramide.

● **Cholesterol Is Synthesized from Acetyl Coenzyme A in Three Stages**

Cholesterol is a steroid component of animal membranes and a precursor of steroid hormones. The committed step in its synthesis is the formation of mevalonate from 3-hydroxy-3-methylglutaryl CoA (derived from acetyl CoA and acetoacetyl CoA). Mevalonate is converted into isopentenyl pyrophosphate (C_5), which condenses with its isomer, dimethylallyl pyrophosphate (C_5), to form geranyl pyrophosphate (C_{10}). The addition of a second molecule of isopentenyl pyrophosphate yields farnesyl pyrophosphate (C_{15}), which condenses with itself to form squalene (C_{30}). This intermediate cyclizes to lanosterol (C_{30}), which is modified to yield cholesterol (C_{27}).

● **The Complex Regulation of Cholesterol Biosynthesis Takes Place at Several Levels**

In the liver, cholesterol synthesis is regulated by changes in the amount and activity of 3-hydroxy-3-methylglutaryl CoA reductase. Transcription of the gene, translation of the mRNA, and degradation of the enzyme are stringently controlled. In addition, the activity of the reductase is regulated by phosphorylation.

Triacylglycerols exported by the intestine are carried by chylomicrons and then hydrolyzed by lipases lining the capillaries of target tissues. Cholesterol and other lipids in excess of those needed by the liver are exported in the form of very low density lipoprotein. After delivering its content of triacylglycerols to adipose tissue and other peripheral tissue, VLDL is converted into intermediate-density lipoprotein and then into low-density lipoprotein. IDL and LDL carry cholesteryl esters, primarily cholesteryl linoleate. Liver and peripheral tissue cells take up LDL by receptor-mediated endocytosis. The LDL receptor, a protein spanning the plasma membrane of the target cell, binds LDL and mediates its entry into the cell. Absence of the LDL receptor in the homozygous form of familial hypercholesterolemia leads to a markedly elevated plasma level of LDL-cholesterol, the deposition of cholesterol on blood-vessel walls, and heart attacks in childhood. Apolipoprotein B, a very large protein, is a key structural component of chylomicrons, VLDL, and LDL.

● **Important Derivatives of Cholesterol Include Bile Salts and Steroid Hormones**

In addition to bile salts, which facilitate the digestion of lipids, five major classes of steroid hormones are derived from cholesterol: progestagens, glucocorticoids, mineralocorticoids, androgens, and estrogens. Hydroxylations by P450 monooxygenases that use NADPH and O_2 play an important role in the synthesis of steroid hormones and bile salts from cholesterol. P450 enzymes, a large superfamily, also participate in the detoxification of drugs and other foreign substances.

Pregnenolone (C_{21}) is an essential intermediate in the synthesis of steroids. This steroid is formed by scission of the side chain of cholesterol. Progesterone (C_{21}), synthesized from pregnenolone, is the precursor of cortisol and aldosterone. Hydroxylation of progesterone and cleavage of its side chain yields androstenedione, an androgen (C_{19}). Estrogens (C_{18}) are synthesized from androgens by the loss of an angular methyl group and the formation of an aromatic A ring. Vitamin D, which is important in the control of calcium and phosphorus metabolism, is formed from a derivative of cholesterol by the action of light.

In addition to cholesterol and its derivatives, a remarkable array of biomolecules are synthesized from isopentenyl pyrophosphate, the basic five-carbon building block. The hydrocarbon side chains of vitamin K_2, coenzyme Q_{10}, and chlorophyll are extended chains constructed from this activated C_5 unit. Prenyl groups that are derived from this activated intermediate target many proteins to membranes.

KEY TERMS

phosphatidate (p. 716)

triacylglycerol (p. 716)

phospholipid (p. 716)

cytidine diphosphodiacylglycerol (CDP-diacylglycerol) (p. 717)

glyceryl ether phospholipid (p. 718)

sphingolipid (p. 720)

ceramide (*N*-acyl sphingosine) (p. 720)

cerebroside (p. 720)

ganglioside (p. 721)

cholesterol (p. 722)

mevalonate (p. 723)

3-hydroxy-3-methylglutaryl CoA reductase (HMG-CoA reductase) (p. 723)

——| **SELECTED READINGS** |——

Where to start

Vance, D. E. and Van den Bosch, H., 2000. Cholesterol in the year 2000. *Biochim. Biophys. Acta* 1529:1–8.

Brown, M. S., and Goldstein, J. L., 1986. A receptor-mediated pathway for cholesterol homeostasis. *Science* 232:34–47.

Brown, M. S., and Goldstein, J. L., 1984. How LDL receptors influence cholesterol and atherosclerosis. *Sci. Am.* 25l(5):58–66.

Chan, L., 1992. Apolipoprotein B, the major protein component of triglyceride-rich and low density lipoproteins. *J. Biol. Chem.* 267: 25621–25624.

Endo, A., 1992. The discovery and development of HMG-CoA reductase inhibitors. *J. Lipid Res.* 33:1569–1582.

Hakomori. S., 1986. Glycosphingolipids. *Sci. Am.* 254(5):44–53.

Books

Vance, D. E., and Vance, J. E. (Eds.), 1996. *Biochemistry of Lipids, Lipoproteins and Membranes.* Elsevier.

Striver, C. R., Beaudet, A. L., Sly, W. S., Valle, D., Stanbury, J. B., Wyngaarden, J. B., and Fredrickson, D. S. (Eds.), 1995. *The Metabolic Basis of Inherited Diseases* (7th ed.). McGraw-Hill.

Phospholipids and sphingolipids

Huwilera, A., Kolterb, T., Pfeilschiftera, J., and Sandhoffb, K., 2000. Physiology and pathophysiology of sphingolipid metabolism and signaling. *Biochim. Biophys. Acta* 1485:63–99.

Lykidis, A., and Jackowski, S., 2000. Regulation of mammalian cell membrane biosynthesis. *Prog. Nucleic Acid Res. Mol. Biol.* 65: 361–393.

Carman, G. M., and Zeimetz, G. M., 1996. Regulation of phospholipid biosynthesis in the yeast *Saccharomyces cerevisiae. J. Biol. Chem.* 271:13293–13296.

Henry, S. A., and Patton-Vogt, J. L., 1998. Genetic regulation of phospholipid metabolism: Yeast as a model eukaryote. *Prog. Nucleic Acid Res. Mol. Biol.* 61:133–179.

Kent, C., 1995. Eukaryotic phospholipid biosynthesis. *Annu. Rev. Biochem.* 64:315–343.

Prescott, S. M., Zimmerman, G. A., Stafforini, D. M., and McIntyre, T. M., 2000. Platelet-activating factor and related lipid mediators. *Annu. Rev. Biochem.* 69:419–445.

Biosynthesis of cholesterol and steroids

Goldstein, J. L., and Brown, M. S., 1990. Regulation of the mevalonate pathway. *Nature* 343:425–430.

Gardner R. G., Shan, H., Matsuda, S. P. T., and Hampton, R. Y., 2001. An oxysterol-derived positive signal for 3-hydroxy-3-methylglutaryl-CoA reductase degradation in yeast. *J. Biol. Chem.* 276: 8681–8694.

Istvan, E. S., and Deisenhofer, J., 2001. Structural mechanism for statin inhibition of HMG-CoA reductase. *Science* 292:1160–1164.

Ness, G. C., and Chambers, C. M., 2000. Feedback and hormonal regulation of hepatic 3-hydroxy-3-methylglutaryl coenzyme A reductase: The concept of cholesterol buffering capacity. *Proc. Soc. Exp. Biol. Med.* 224:8–19.

Libby, P., Aikawa, M., and Schonbeck, U., 2000. Cholesterol and atherosclerosis. *Biochim. Biophys. Acta* 1529:299–309.

Yokoyama, S., 2000. Release of cellular cholesterol: Molecular mechanism for cholesterol homeostasis in cells and in the body. *Biochim. Biophys. Acta* 1529:231–244.

Cronin, S. R., Khoury, A., Ferry, D. K., and Hampton, R. Y., 2000. Regulation of HMG-CoA reductase degradation requires the P-type ATPase Cod1p/Spf1p. *J. Cell Biol.* 148:915–924.

Edwards, P. A., Tabor, D., Kast, H. R., and Venkateswaran, A., 2000. Regulation of gene expression by SREBP and SCAP. *Biochim. Biophys. Acta* 1529:103–113.

Istvan, E. S., Palnitkar, M., Buchanan, S. K., and Deisenhofer, J., 2000. Crystal structure of the catalytic portion of human HMG-CoA reductase: Insights into regulation of activity and catalysis. *EMBO J.* 19:819–830.

Tabernero, L., Bochar, D. A., Rodwell, V. W., and Stauffacher, C. V., 1999. Substrate-induced closure of the flap domain in the ternary complex structures provides insights into the mechanism of catalysis by 3-hydroxy-3-methylglutaryl-CoA reductase. *Proc. Natl. Acad. Sci. USA.* 96:7167–7171.

Fass, D., Blacklow, S., Kim, P. S., and Berger, J. M., 1997. Molecular basis of familial hypercholesterolaemia from structure of LDL receptor module. *Nature* 388:691–693.

Jeon, H., Meng, W., Takagi, J., Eck, M. J., Springer, T. A., and Blacklow, S. C., 2001. Implications for familial hypercholesterolemia from the structure of the LDL receptor YWTD-EGF domain pair. *Nat. Struct. Biol.* 8:499–504.

Lipoproteins and their receptors

Brouillette, C. G., Anantharamaiah, G. M., Engler, J. A., and Borhani, D. W., 2001. Structural models of human apolipoprotein A-I: A critical analysis and review. *Biochem. Biophys. Acta* 1531:4–46.

Hevonoja, T., Pentikainen, M. O., Hyvonen, M. T., Kovanen, P. T., and Ala-Korpela, M., 2000. Structure of low density lipoprotein (LDL) particles: Basis for understanding molecular changes in modified LDL. *Biochim. Biophys. Acta* 1488:189–210.

Silver, D. L., Jiang, X. C., Arai, T., Bruce, C., and Tall, A. R., 2000. Receptors and lipid transfer proteins in HDL metabolism. *Ann. N. Y. Acad. Sci.* 902:103–111.

Nimpf, J., and Schneider, W. J., 2000. From cholesterol transport to signal transduction: Low density lipoprotein receptor, very low density lipoprotein receptor, and apolipoprotein E receptor-2. *Biochim. Biophys. Acta* 1529:287–298.

Borhani, D. W., Rogers, D. P., Engler, J. A., and Brouillette, C. G., 1997. Crystal structure of truncated human apolipoprotein A-I suggests a lipid-bound conformation. *Proc. Natl. Acad. Sci. USA.* 94:12291–12296.

Wilson, C., Wardell, M. R., Weisgraber, K. H., Mahley, R. W., and Agard, D. A., 1991. Three-dimensional structure of the LDL receptor-binding domain of human apolipoprotein E. *Science* 252:1817–1822.

Plump, A. S., Smith, J. D., Hayek, T., Aalto-Setälä, K., Walsh, A., Verstuyft, J. G., Rubin, E. M., and Breslow, J. L., 1992. Severe hypercholesterolemia and atherosclerosis in apolipoprotein E-deficient mice created by homologous recombination in ES cells. *Cell* 71:343–353.

Sudhof, T. C., Goldstein, J. L., Brown, M. S., and Russell, D. W., 1985. The LDL receptor gene: A mosaic of exons shared with different proteins. *Science* 228:815–822.

Oxygen activation and P450 catalysis

Ingelman-Sundberg, M., Oscarson, M., and McLellan, R. A., 1999. Polymorphic human cytochrome P450 enzymes: An opportunity for individualized drug treatment. *Trends Pharmacol. Sci.* 20:342–349.

Nelson, D. R., 1999. Cytochrome P450 and the individuality of species. *Arch. Biochem. Biophys.* 369:1–10.

Wong, L. L., 1998. Cytochrome P450 monooxygenases. *Curr. Opin. Chem. Biol.* 2:263–268.

Denison, M. S., and Whitlock, J. P., 1995. Xenobiotic-inducible transcription of cytochrome P450 genes. *J. Biol. Chem.* 270: 18175–18178.

Poulos, T. L., 1995. Cytochrome P450. *Curr. Opin. Struct. Biol.* 5: 767–774.

Vaz, A. D., and Coon, M. J., 1994. On the mechanism of action of cytochrome P450: Evaluation of hydrogen abstraction in oxygen-dependent alcohol oxidation. *Biochemistry* 33:6442–6449.

Gonzalez, F. J., and Nebert, D. W., 1990. Evolution of the P450 gene superfamily: Animal-plant "warfare," molecular drive and human genetic differences in drug oxidation. *Trends Genet.* 6: 182–186.

PROBLEMS

1. *Making fat.* Write a balanced equation for the synthesis of a triacylglycerol, starting from glycerol and fatty acids.

2. *Making phospholipid.* Write a balanced equation for the synthesis of phosphatidyl serine by the de novo pathway, starting from serine, glycerol, and fatty acids.

3. *Activated donors.* What is the activated reactant in each of the following biosyntheses?

(a) Phosphatidyl serine from serine
(b) Phosphatidyl ethanolamine from ethanolamine
(c) Ceramide from sphingosine
(d) Sphingomyelin from ceramide
(e) Cerebroside from ceramide
(f) Ganglioside G_{M1} from ganglioside G_{M2}
(g) Farnesyl pyrophosphate from geranyl pyrophosphate

4. *Telltale labels.* What is the distribution of isotopic labeling in cholesterol synthesized from each of the following precursors?

(a) Mevalonate labeled with ^{14}C in its carboxyl carbon atom
(b) Malonyl CoA labeled with ^{14}C in its carboxyl carbon atom

5. *Familial hypercholesterolemia.* Several classes of LDL-receptor mutations have been identified as causes of this disease. Suppose that you have been given cells from patients with different mutations, an antibody specific for the LDL receptor that can be seen with an electron microscope, and access to an electron microscope. What differences in antibody distribution might you expect to find in the cells from different patients?

6. *RNA editing.* A shortened version (apo B-48) of apolipoprotein B is formed by the intestine, whereas the full-length protein (apo B-100) is synthesized by the liver. A glutamine codon (CAA) is changed into a stop codon. Propose a simple mechanism for this change.

7. *Inspiration for drug design.* Some actions of androgens are mediated by dihydrotestosterone, which is formed by the reduction of testosterone. This finishing touch is catalyzed by an NADPH-dependent 5α-reductase.

5α-Dihydrotestosterone

Chromosomal XY males with a genetic deficiency of this reductase are born with a male internal urogenital tract but predominantly female external genitalia. These people are usually reared as girls. At puberty, they masculinize because the testos-

terone level rises. The testes of these reductase-deficient men are normal, whereas their prostate glands remain small. How might this information be used to design a drug to treat *benign prostatic hypertrophy*, a common consequence of the normal aging process in men? A majority of men older than age 55 have some degree of prostatic enlargement, which often leads to urinary obstruction.

8. *Drug idiosyncrasies.* Debrisoquine, a β-adrenergic blocking agent, has been used to treat hypertension. The optimal dose varies greatly (20–400 mg daily) in a population of patients. The urine of most patients taking the drug contains a high level of 4-hydroxydebrisoquine. However, those most sensitive to the drug (about 8% of the group studied) excrete debrisoquine and very little of the 4-hydroxy derivative. Propose a molecular basis for this drug idiosyncrasy. Why should caution be exercised in giving other drugs to patients who are very sensitive to debrisoquine?

Debrisoquine

9. *Removal of odorants.* Many odorant molecules are highly hydrophobic and concentrate within the olfactory epithelium. They would give a persistent signal independent of their concentration in the environment if they were not rapidly modified. Propose a mechanism for converting hydrophobic odorants into water-soluble derivatives that can be rapidly eliminated.

10. *Development difficulties.* Propecia (finasteride) is a synthetic steroid that functions as a competitive and specific inhibitor of 5α-reductase, the enzyme responsible for the synthesis of dihydrotestosterone from testosterone.

Finasteride

It is now widely used to retard the development of male pattern hair loss. Pregnant women are advised to avoid handling this drug. Why is it vitally important that pregnant women avoid contact with Propecia?

11. *Life-style consequences.* Human beings and the plant *Arabidopsis* evolved from the same distant ancestor possessing a

small number of cytochrome P450 genes. Human beings have approximately 50 such genes, whereas *Arabidopsis* has more than 250 of them. Propose a role for the large number of P450 isozymes in plants.

12. *Personalized medicine.* The cytochrome P450 system metabolizes many medicinally useful drugs. Although all human beings have the same number of P450 genes, individual polymorphisms exist that alter the specificity and efficiency of the proteins encoded by the genes. How could knowledge of individual polymorphisms be useful clinically?

Mechanism Problem

13. *An interfering phosphate.* In the course of the overall reaction catalyzed by HMG-CoA reductase, a histidine residue protonates a coenzyme A thiolate, $CoAS^-$, generated in a previous step.

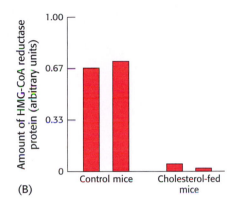

The nearby serine residue can be phosphorylated by AMP-dependent kinase, which results in a loss of activity. Propose an explanation for why phosphorylation of the serine residue inhibits enzyme activity.

14. *Demethylation.* Methyl amines are often demethylated by cytochrome P450 enzymes. Propose a mechanism for the formation of methylamine from dimethylamine catalyzed by cytochrome P450. What is the other product?

Data Interpretation and Chapter Integration Problems

15. *Cholesterol feeding.* Mice were divided into four groups, two of which were fed a normal diet and two of which were fed a cholesterol-rich diet. HMG-CoA reductase mRNA and protein from liver were then isolated and quantified. Graph A shows the results of the mRNA isolation.

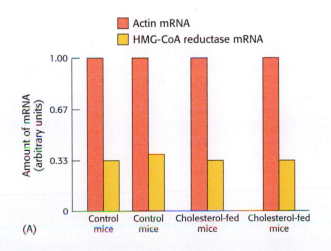

(A)

(a) What is the effect of cholesterol feeding on the amount of HMG-CoA reductase mRNA?

(b) What is the purpose of also isolating the mRNA for the protein actin, which is not under the control of the sterol response element?

HMG-CoA reductase protein was isolated by precipitation with a monoclonal antibody to HMG-CoA reductase. The amount of HMG-CoA protein in each group is shown in graph B.

(B)

(c) What is the effect of the cholesterol diet on the amount of HMG-CoA reductase protein?

(d) Why is this result surprising in light of the results in graph A?

(e) Suggest possible explanations for the results shown in graph B.

CHAPTER 27

DNA Replication, Recombination, and Repair

Faithful copying is essential to the storage of genetic information. With the precision of a diligent monk copying an illuminated manuscript, a DNA polymerase (below) copies DNA strands, preserving the precise sequence of bases with very few errors. [(Left) The Pierpont Morgan Library/ Art Resource.]

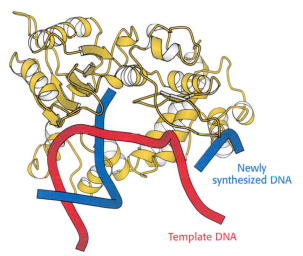

Newly synthesized DNA

Template DNA

Perhaps the most exciting aspect of the structure of DNA deduced by Watson and Crick was, as expressed in their words, that the "specific pairing we have postulated immediately suggests a possible copying mechanism for the genetic material." A double helix separated into two single strands can be replicated because each strand serves as a template on which its complementary strand can be assembled (Figure 27.1). Although this notion of how DNA is replicated is absolutely correct, the double-helical structure of DNA poses a number of challenges to replication, as does the need for extremely faithful copying of the genetic information.

1. The two strands of the double helix have a tremendous affinity for one another, created by the cooperative effects of the many hydrogen bonds that hold adjacent base pairs together. Thus, a mechanism is required for separating the strands in a local region to provide access to the bases that act as templates. Specific proteins melt the double helix at specific sites to initiate DNA replication, and other enzymes, termed *helicases*, use the free energy of ATP hydrolysis to move this melted region along the double helix as replication progresses.

2. The DNA helix must be unwound to separate the two strands. The local unwinding in one region leads to stressful overwinding in surrounding regions (Figure 27.2). Enzymes termed *topoisomerases* introduce supercoils that release the strain caused by overwinding.

FIGURE 27.2 **Consequences of strand separation.** DNA must be locally unwound to expose single-stranded templates for replication. This unwinding puts a strain on the molecule by causing the overwinding of nearby regions.

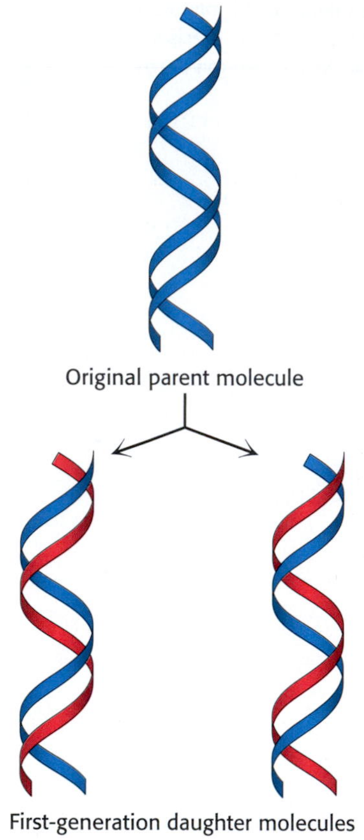

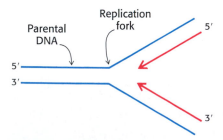

Original parent molecule

First-generation daughter molecules

FIGURE 27.1 **DNA replication.** The double-helical structure immediately suggests how DNA is replicated.

FIGURE 27.3 **DNA replication at low resolution.** On cursory examination, both strands of a DNA template appear to replicate continuously in the same direction.

3. DNA replication must be highly accurate. As noted in Chapter 5, the free energies associated with base pairing within the double helix suggest that approximately 1 in 10^4 bases incorporated will be incorrect. Yet, DNA replication has an error rate estimated to be 1 per 10^{10} nucleotides. As we shall see, additional mechanisms allow proofreading of the newly formed double helix.

4. DNA replication must be very rapid, given the sizes of the genomes and the rates of cell division. The *E. coli* genome contains 4.8 million base pairs and is copied in less than 40 minutes. Thus, 2000 bases are incorporated per second. We shall examine some of the properties of the macromolecular machines that replicate DNA with such high accuracy and speed.

5. The enzymes that copy DNA polymerize nucleotides in the $5' \rightarrow 3'$ direction. The two polynucleotide strands of DNA run in opposite directions, yet both strands appear to grow in the same direction (Figure 27.3). Further analysis reveals that one strand is synthesized in a continuous fashion, whereas the opposite strand is synthesized in fragments in a discontinuous fashion. The synthesis of each fragment must be initiated in an independent manner, and then the fragments must be linked together. The DNA replication apparatus includes enzymes for these priming and ligation reactions.

6. The replication machinery alone cannot replicate the ends of linear DNA molecules, so a mechanism is required to prevent the loss of sequence information with each replication. Specialized structures called *telomeres* are added by another enzyme to maintain the information content at chromosome ends.

7. Most components of the DNA replication machinery serve to preserve the integrity of a DNA sequence to the maximum possible extent, yet a variety of biological processes require DNA formed by the exchange of material between two parent molecules. These processes range from the development of diverse antibody sequences in the immune system (Chapter 33) to the integration of viral genomes into host DNA. Specific enzymes, termed *recombinases*, facilitate these rearrangements.

8. After replication, ultraviolet light and a range of chemical species can damage DNA in a variety of ways. All organisms have enzymes for detecting and repairing harmful DNA modifications. Agents that introduce chemical lesions into DNA are key factors in the development of cancer, as are defects in the repair systems that correct these lesions.

We begin with a review of the structural properties of the DNA double helix.

27.1 DNA CAN ASSUME A VARIETY OF STRUCTURAL FORMS

The double-helical structure of DNA deduced by Watson and Crick immediately suggested how genetic information is stored and replicated. As was discussed earlier (Section 5.2.1), the essential features of their model are:

1. Two polynucleotide chains running in opposite directions coil around a common axis to form a right-handed double helix.

2. Purine and pyrimidine bases are on the inside of the helix, whereas phosphate and deoxyribose units are on the outside.

3. Adenine (A) is paired with thymine (T), and guanine (G) with cytosine (C). An A–T base pair is held together by two hydrogen bonds, and that of a G–C base pair by three such bonds.

27.1.1 A-DNA Is a Double Helix with Different Characteristics from Those of the More Common B-DNA

Watson and Crick based their model (known as the *B-DNA helix*) on x-ray diffraction patterns of DNA fibers, which provided information about properties of the double helix that are averaged over its constituent residues. The results of x-ray diffraction studies of dehydrated DNA fibers revealed a different form called *A-DNA*, which appears when the relative humidity is reduced to less than about 75%. A-DNA, like B-DNA, is a right-handed double helix made up of antiparallel strands held together by Watson-Crick base-pairing. The A helix is wider and shorter than the B helix, and its base pairs are tilted rather than perpendicular to the helix axis (Figure 27.4).

Many of the structural differences between B-DNA and A-DNA arise from different puckerings of their ribose units (Figure 27.5). In A-DNA, C-3′ lies out of the plane (a conformation referred to as C-3′-endo) formed by the other four atoms of the furanose ring; in B-DNA, C-2′ lies out of the

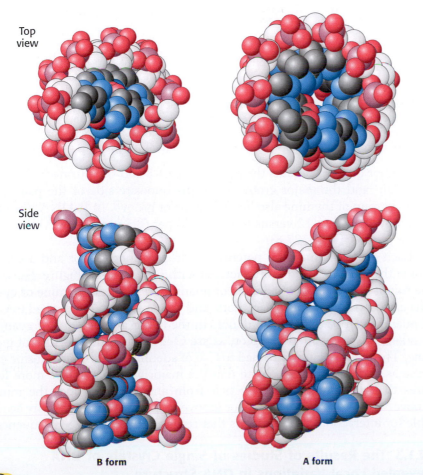

B form   **A form**

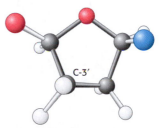

C-3′ endo (A form)

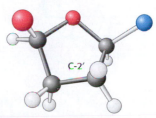

C-2′ endo (B form)

FIGURE 27.5 Sugar puckers. In A-form DNA, the C-3′ carbon atom lies above the approximate plane defined by the four other sugar nonhydrogen atoms (called C-3′ endo). In B-form DNA, each ribose is in a C-2′-endo conformation.

FIGURE 27.4 B-form and A-form DNA. Space-filling models of ten base pairs of B-form and A-form DNA depict their right-handed helical structures. The B-form helix is longer and narrower than the A-form helix. The carbon atoms of the backbone are shown in white.

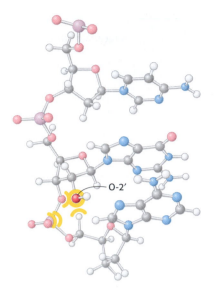

FIGURE 27.6 Steric clash. The introduction of a 2′-hydroxyl group into a B-form structure leads to several steric clashes with nearby atoms.

plane (a conformation called C-2′-endo). The C-3′-endo puckering in A-DNA leads to a 19-degree tilting of the base pairs away from the normal to the helix. The phosphates and other groups in the A helix bind fewer H_2O molecules than do those in B-DNA. Hence, dehydration favors the A form.

The A helix is not confined to dehydrated DNA. *Double-stranded regions of RNA and at least some RNA–DNA hybrids adopt a double-helical form very similar to that of A-DNA.* The position of the 2′-hydroxyl group of ribose prevents RNA from forming a classic Watson-Crick B helix because of steric hindrance (Figure 27.6): the 2′-oxygen atom would come too close to three atoms of the adjoining phosphate group and one atom in the next base. In an A-type helix, in contrast, the 2′-oxygen projects outward, away from other atoms.

27.1.2 The Major and Minor Grooves Are Lined by Sequence-Specific Hydrogen-Bonding Groups

Double-helical nucleic acid molecules contain two grooves, called the *major groove* and the *minor groove*. These grooves arise because the glycosidic bonds of a base pair are not diametrically opposite each other (Figure 27.7).

FIGURE 27.7 Major- and minor-groove sides. Because the two glycosidic bonds are not diametrically opposite each other, each base pair has a larger side that defines the major groove and a smaller side that defines the minor groove. The grooves are lined by potential hydrogen-bond donors (blue) and acceptors (red).

The minor groove contains the pyrimidine O-2 and the purine N-3 of the base pair, and the major groove is on the opposite side of the pair. The methyl group of thymine also lies in the major groove. In B-DNA, the major groove is wider (12 versus 6 Å) and deeper (8.5 versus 7.5 Å) than the minor groove (Figure 27.8).

Each groove is lined by potential hydrogen-bond donor and acceptor atoms that enable specific interactions with proteins (see Figure 27.7). In the minor groove, N-3 of adenine or guanine and O-2 of thymine or cytosine can serve as hydrogen acceptors, and the amino group attached to C-2 of guanine can be a hydrogen donor. In the major groove, N-7 of guanine or adenine is a potential acceptor, as are O-4 of thymine and O-6 of guanine. The amino groups attached to C-6 of adenine and C-4 of cytosine can serve as hydrogen donors. Note that the major groove displays more features that distinguish one base pair from another than does the minor groove. The larger size of the major groove in B-DNA makes it more accessible for interactions with proteins that recognize specific DNA sequences.

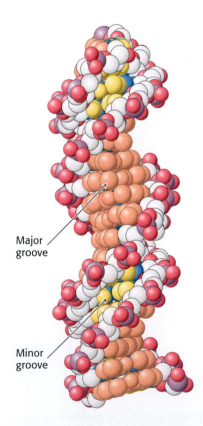

FIGURE 27.8 Major and minor grooves in B-form DNA. The major groove is depicted in orange, and the minor groove is depicted in yellow. The carbon atoms of the backbone are shown in white.

27.1.3 The Results of Studies of Single Crystals of DNA Revealed Local Variations in DNA Structure

X-ray analyses of single crystals of DNA oligomers had to await the development of techniques for synthesizing large amounts of DNA fragments

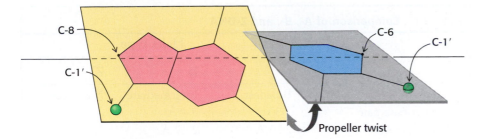

C-8

C-1'

C-6

C-1'

Propeller twist

FIGURE 27.9 Propeller twist. The bases of a DNA base pair are often not precisely coplanar. They are twisted with respect to each other, like the blades of a propeller.

with defined base sequences. X-ray analyses of single crystals of DNA at atomic resolution revealed that *DNA exhibits much more structural variability and diversity than formerly envisaged.*

The x-ray analysis of a crystallized DNA dodecamer by Richard Dickerson and his coworkers revealed that its overall structure is very much like a B-form Watson-Crick double helix. However, the dodecamer differs from the Watson-Crick model in not being uniform; there are rather large local deviations from the average structure. The Watson-Crick model has 10 residues per complete turn, and so a residue is related to the next along a chain by a rotation of 36 degrees. In Dickerson's dodecamer, the rotation angles range from 28 degrees (less tightly wound) to 42 degrees (more tightly wound). Furthermore, the two bases of many base pairs are not perfectly coplanar (Figure 27.9). Rather, they are arranged like the blades of a propeller. This deviation from the idealized structure, called *propeller twisting*, enhances the stacking of bases along a strand. These and other local variations of the double helix depend on base sequence. A protein searching for a specific target sequence in DNA may sense its presence through its effect on the precise shape of the double helix.

27.1.4 Z-DNA Is a Left-Handed Double Helix in Which Backbone Phosphates Zigzag

Alexander Rich and his associates discovered a third type of DNA helix when they solved the structure of dCGCGCG. They found that this hexanucleotide forms a duplex of antiparallel strands held together by Watson-Crick base-pairing, as expected. What was surprising, however, was that this double helix was *left-handed*, in contrast with the *right-handed* screw sense of the A and B helices. Furthermore, the phosphates in the backbone *zigzagged*; hence, they called this new form *Z-DNA* (Figure 27.10).

The Z-DNA form is adopted by short oligonucleotides that have *sequences of alternating pyrimidines and purines*. High salt concentrations are required to minimize electrostatic repulsion between the backbone

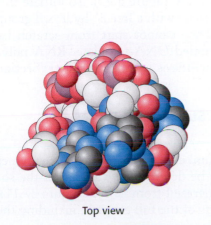

Top view

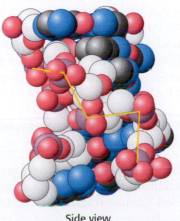

Side view

FIGURE 27.10 Z-DNA. DNA oligomers such as dCGCGCG adopt an alternative conformation under some conditions. This conformation is called Z-DNA because the phosphate groups zigzag along the backbone.

TABLE 27.1   Comparison of A-, B-, and Z-DNA

	Helix type		
	A	**B**	**Z**
Shape	Broadest	Intermediate	Narrowest
Rise per base pair	2.3 Å	3.4 Å	3.8 Å
Helix diameter	25.5 Å	23.7 Å	18.4 Å
Screw sense	Right-handed	Right-handed	Left-handed
Glycosidic bond	*anti*	*anti*	alternating *anti and syn*
Base pairs per turn of helix	11	10.4	12
Pitch per turn of helix	25.3 Å	35.4 Å	45.6 Å
Tilt of base pairs from normal to helix axis	19°	1°	9°
Major groove	Narrow and very deep	Wide and quite deep	Flat
Minor groove	Very broad and shallow	Narrow and quite deep	Very narrow and deep

Primer—

The initial segment of a polymer that is to be extended on which elongation depends.

Template—

A sequence of DNA or RNA that directs the synthesis of a complementary sequence.

phosphates, which are closer to each other than in A- and B-DNA. Under physiological conditions, *most DNA is in the B form*. Although the biological role of Z-DNA is still under investigation, its existence graphically shows that DNA is a flexible, dynamic molecule. The properties of A-, B-, and Z-DNA are compared in Table 27.1.

27.2 DNA POLYMERASES REQUIRE A TEMPLATE AND A PRIMER

DNA polymerases catalyze the formation of polynucleotide chains through the addition of successive nucleotides derived from deoxynucleoside triphosphates. The polymerase reaction takes place only in the presence of an appropriate DNA template. Each incoming nucleoside triphosphate first forms an appropriate base pair with a base in this template. Only then does the DNA polymerase link the incoming base with the predecessor in the chain. Thus, *DNA polymerases are template-directed enzymes.*

DNA polymerases add nucleotides to the 3′ end of a polynucleotide chain. The polymerase catalyzes the nucleophilic attack of the 3′-hydroxyl group terminus of the polynucleotide chain on the α-phosphate group of the nucleoside triphosphate to be added (see Figure 5.22). To initiate this reaction, DNA polymerases require a *primer* with a free 3′-hydroxyl group already base-paired to the template. They cannot start from scratch by adding nucleotides to a free single-stranded DNA template. RNA polymerase, in contrast, can initiate RNA synthesis without a primer (Section 28.1.4).

27.2.1 All DNA Polymerases Have Structural Features in Common

The three-dimensional structures of a number of DNA polymerase enzymes are known. The first such structure to be determined was that of the so-called Klenow fragment of DNA polymerase I from *E. coli* (Figure 27.11). This fragment comprises two main parts of the full enzyme, including the

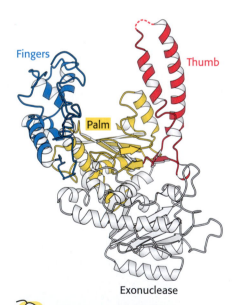

Fingers

Thumb

Palm

Exonuclease

FIGURE 27.11 DNA polymerase structure. The first DNA polymerase structure determined was that of a fragment of *E. coli* DNA polymerase I called the Klenow fragment. Like other DNA polymerases, the polymerase unit resembles a right hand with fingers (blue), palm (yellow), and thumb (red). The Klenow fragment also includes an exonuclease domain.

polymerase unit. This unit approximates the shape of a right hand with domains that are referred to as the fingers, the thumb, and the palm. In addition to the polymerase, the Klenow fragment includes a domain with $3' \rightarrow 5'$ *exonuclease* activity that participates in proofreading and correcting the polynucleotide product (Section 27.2.4).

DNA polymerases are remarkably similar in overall shape, although they differ substantially in detail. At least five structural classes have been identified; some of them are clearly homologous, whereas others are probably the products of convergent evolution. In all cases, the finger and thumb domains wrap around DNA and hold it across the enzyme's active site, which comprises residues primarily from the palm domain. Furthermore, all the polymerases catalyze the same polymerase reaction, which is dependent on two metal ions.

27.2.2 Two Bound Metal Ions Participate in the Polymerase Reaction

Like all enzymes with nucleoside triphosphate substrates, DNA polymerases require metal ions for activity. Examination of the structures of DNA polymerases with bound substrates and substrate analogs reveals the presence of two metal ions in the active site. One metal ion binds both the deoxynucleoside triphosphate (dNTP) and the 3'-hydroxyl group of the primer, whereas the other interacts only with the 3'-hydroxyl group (Figure 27.12). The two metal ions are bridged by the carboxylate groups of two aspartate residues in the palm domain of the polymerase. These side chains hold the metal ions in the proper position and orientation. The metal ion bound to the primer activates the 3'-hydroxyl group of the primer, facilitating its attack on the α-phosphate group of the dNTP substrate in the active site. The two metal ions together help stabilize the negative charge that accumulates on the pentacoordinate transition state. The metal ion initially bound to dNTP stabilizes the negative charge on the pyrophosphate product.

27.2.3 The Specificity of Replication Is Dictated by Hydrogen Bonding and the Complementarity of Shape Between Bases

DNA must be replicated with high fidelity. Each base added to the growing chain should with high probability be the Watson-Crick complement of the base in the corresponding position in the template strand. The binding of the NTP containing the proper base is favored by the formation of a base pair, which is stabilized by specific hydrogen bonds. The binding of a noncomplementary base is unlikely, because the interactions are unfavorable. The hydrogen bonds linking two complementary bases make a significant contribution to the fidelity of DNA replication. However, DNA polymerases replicate DNA more faithfully than these interactions alone can account for.

The examination of the crystal structures of various DNA polymerases indicated several additional mechanisms by which replication fidelity is improved. First, residues of the enzyme form hydrogen bonds with the minor-groove side of the base pair in the active site (Figure 27.13). In the minor groove, hydrogen-bond acceptors are present in the same positions for all Watson-Crick base pairs. These interactions act as a "ruler" that measures whether a properly spaced base pair has formed in the active site. Second, DNA polymerases close down around the incoming NTP (Figure 27.14). The binding of a nucleoside triphosphate into the active site of a DNA polymerase triggers a conformational change: the finger domain

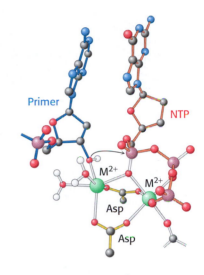

FIGURE 27.12 DNA polymerase mechanism. Two metal ions (typically, Mg^{2+}) participate in the DNA polymerase reaction. One metal ion coordinates the 3'-hydroxyl group of the primer, whereas the phosphate group of the nucleoside triphosphate bridges between the two metal ions. The hydroxyl group of the primer attacks the phosphate group to form a new O—P bond.

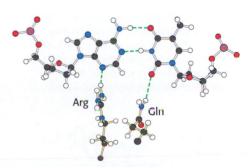

FIGURE 27.13 Minor-groove interactions. DNA polymerases donate two hydrogen bonds to base pairs in the minor groove. Hydrogen-bond acceptors are present in these two positions for all Watson-Crick base pairs including the A–T base pair shown.

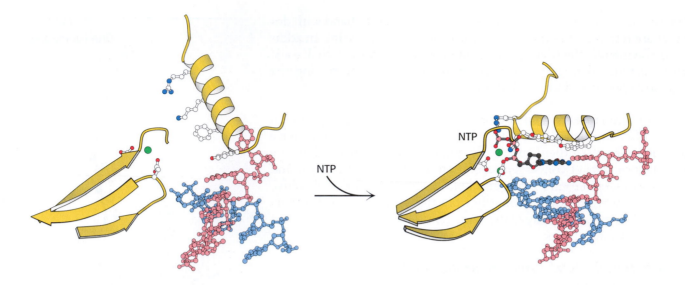

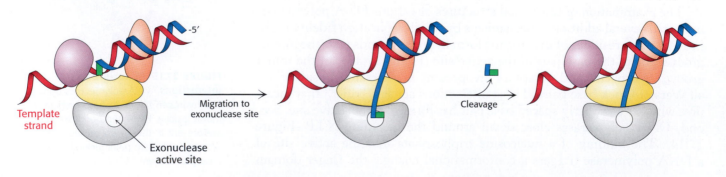

Template strand

Exonuclease active site

Migration to exonuclease site

Cleavage

FIGURE 27.14 Shape selectivity. The binding of a nucleoside triphosphate (NTP) to DNA polymerase induces a conformational change, generating a tight pocket for the base pair consisting of the NTP and its partner on the template strand. Such a conformational change is possible only when the NTP corresponds to the Watson-Crick partner of the template base.

FIGURE 27.15 Proofreading. The growing polynucleotide chain occasionally leaves the polymerase site of DNA polymerase I and migrates to the exonuclease site. There, the last nucleotide added is removed by hydrolysis. Because mismatched bases are more likely to leave the polymerase site, this process serves to proofread the sequence of the DNA being synthesized.

rotates to form a tight pocket into which only a properly shaped base pair will readily fit. The mutation of a conserved tyrosine residue at the top of the pocket results in a polymerase that is approximately 40 times as error prone as the parent polymerase.

27.2.4 Many Polymerases Proofread the Newly Added Bases and Excise Errors

Many polymerases further enhance the fidelity of replication by the use of proofreading mechanisms. As already noted, the Klenow fragment of *E. coli* DNA polymerase I includes an exonuclease domain that does not participate in the polymerization reaction itself. Instead, this domain removes mismatched nucleotides from the 3' end of DNA by hydrolysis. The exonuclease active site is 35 Å from the polymerase active site, yet it can be reached by the newly synthesized polynucleotide chain under appropriate conditions. The proofreading mechanism relies on the increased probability that the end of a growing strand with an incorrectly incorporated nucleotide will leave the polymerase site and transiently move to the exonuclease site (Figure 27.15).

How does the enzyme sense whether a newly added base is correct? First, an incorrect base will not pair correctly with the template strand. Its greater structural fluctuation, permitted by the weaker hydrogen bonding, will frequently bring the newly synthesized strand to the exonuclease site. Second, after the addition of a new nucleotide, the DNA translocates by one base pair into the enzyme. The newly formed base pair must be of the proper dimensions to fit into a tight binding site and participate in hydrogen-bonding interactions in the minor groove similar to those in the polymerization site itself (see Figure 27.13). Indeed, the duplex DNA within the enzyme adopts an A-form structure, allowing clear access to the minor groove. If an incorrect base is incorporated, the enzyme stalls, and the pause provides additional time for the strand to migrate to the exonuclease site. There is a cost to this editing function, however: DNA polymerase I removes approximately 1 correct nucleotide in 20 by hydrolysis. Although the removal of correct nucleotides is slightly wasteful energetically, proofreading increases the accuracy of replication by a factor of approximately 1000.

27.2.5 The Separation of DNA Strands Requires Specific Helicases and ATP Hydrolysis

For a double-stranded DNA molecule to replicate, the two strands of the double helix must be separated from each other, at least locally. This separation allows each strand to act as a template on which a new polynucleotide chain can be assembled. For long double-stranded DNA molecules, the rate of spontaneous strand separation is negligibly low under physiological conditions. Specific enzymes, termed *helicases*, utilize the energy of ATP hydrolysis to power strand separation.

The detailed mechanisms of helicases are still under active investigation. However, the determination of the three-dimensional structures of several helicases has been a source of insight. For example, a bacterial helicase called *PcrA* comprises four domains, hereafter referred to as domains A1, A2, B1, and B2 (Figure 27.16). Domain A1 contains a P-loop NTPase fold, as was expected from amino acid sequence analysis. This domain participates in ATP binding and hydrolysis. Domain B1 is homologous to domain A1 but lacks a P-loop. Domains A2 and B2 have unique structures.

From an analysis of a set of helicase crystal structures bound to nucleotide analogs and appropriate double- and single-stranded DNA molecules, a mechanism for the action of these enzymes was proposed (Figure 27.17). Domains A1 and B1 are capable of binding single-stranded DNA. In the absence of bound ATP, both domains are bound to DNA. The binding of ATP triggers conformational changes in the P-loop and adjacent regions that lead to the closure of the cleft between these two domains. To achieve this movement, domain A1 releases the DNA and slides along the DNA strand, moving closer to domain B1. The enzyme then catalyzes the hydrolysis of ATP to form ADP and orthophosphate. On product release, the cleft between domains A and B springs open. In this state, however, domain A1 has a tighter grip on the DNA than does domain B1, so the DNA is pulled across domain B1 toward domain A1. The result is the translocation of the enzyme along the DNA strand in a manner similar to the way in which an inchworm moves. In regard to PcrA, the enzyme translocates in

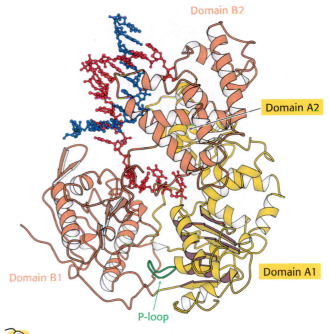

FIGURE 27.16 Helicase structure. The bacterial helicase PcrA comprises four domains: A1, A2, B1, and B2. The A1 domain includes a P-loop NTPase fold, whereas the B1 domain has a similar overall structure but lacks a P-loop and does not bind nucleotides. Single-stranded DNA binds to the A1 and B1 domains near the interfaces with domains A2 and B2.

Domain B2

Domain A2

Domain A1

Domain B1

P-loop

FIGURE 27.17 Helicase mechanism. Initially, both domains A1 and B1 of PcrA bind single-stranded DNA. On binding of ATP, the cleft between these domains closes and domain A1 slides along the DNA. On ATP hydrolysis, the cleft opens up, pulling the DNA from domain B1 toward domain A1. As this process is repeated, double-stranded DNA is unwound.

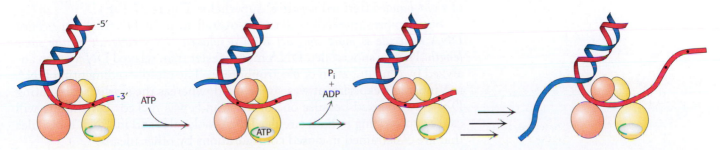

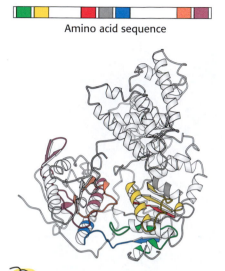

Amino acid sequence

the $3' \rightarrow 5'$ direction. When the helicase encounters a region of double-stranded DNA, it continues to move along one strand and displaces the opposite DNA strand as it progresses. Interactions with specific pockets on the helicase help destabilize the DNA duplex, aided by ATP-induced conformational changes.

Helicases constitute a large and diverse class of enzymes. Some of these enzymes move in a $5' \rightarrow 3'$ direction, whereas others unwind RNA rather than DNA and participate in processes such as RNA splicing and the initiation of mRNA translation. A comparison of the amino acid sequences of hundreds of these enzymes reveals seven regions of striking conservation (Figure 27.18). Mapping these regions onto the PcrA structure shows that they line the ATP-binding site and the cleft between the two domains, consistent with the notion that other helicases undergo conformational changes analogous to those found in PcrA. However, whereas PcrA appears to function as a monomer, other members of the helicase class function as oligomers. The hexameric structures of one important group are similar to that of the F_1 component of ATP synthase (Section 18.4.1), suggesting potential mechanistic similarities.

27.3 DOUBLE-STRANDED DNA CAN WRAP AROUND ITSELF TO FORM SUPERCOILED STRUCTURES

The separation of the two strands of DNA in replication requires the local unwinding of the double helix. This local unwinding must lead either to the overwinding of surrounding regions of DNA or to supercoiling. To prevent the strain induced by overwinding, a specialized set of enzymes is present to introduce supercoils that favor strand separation.

27.3.1 The Linking Number of DNA, a Topological Property, Determines the Degree of Supercoiling

In 1963, Jerome Vinograd found that circular DNA from polyoma virus separated into two distinct species when it was centrifuged. In pursuing this puzzle, he discovered an important property of circular DNA not possessed by linear DNA with free ends. Consider a linear 260-bp DNA duplex in the B-DNA form (Figure 27.19). Because the number of residues per turn in an unstressed DNA molecule is 10.4, this linear DNA molecule has 25 (260/10.4) turns. The ends of this helix can be joined to produce a *relaxed* circular DNA (Figure 27.19B). A different circular DNA can be formed by unwinding the linear duplex by two turns before joining its ends (Figure 27.19C). What is the structural consequence of unwinding before ligation? Two limiting conformations are possible: the DNA can either fold into a structure containing 23 turns of B helix and an unwound loop (Figure 27.19D) or adopt a *supercoiled* structure with 25 turns of B helix and 2 turns of *right-handed* (termed *negative*) superhelix (Figure 27.19E).

Supercoiling markedly alters the overall form of DNA. *A supercoiled DNA molecule is more compact than a relaxed DNA molecule of the same length.* Hence, supercoiled DNA moves faster than relaxed DNA when analyzed by centrifugation or electrophoresis. The rapidly sedimenting DNA in Vinograd's experiment was supercoiled, whereas the slowly sedimenting DNA was relaxed because one of its strands was nicked. Unwinding will cause supercoiling in both circular DNA molecules and in DNA molecules that are constrained in closed configurations by other means.

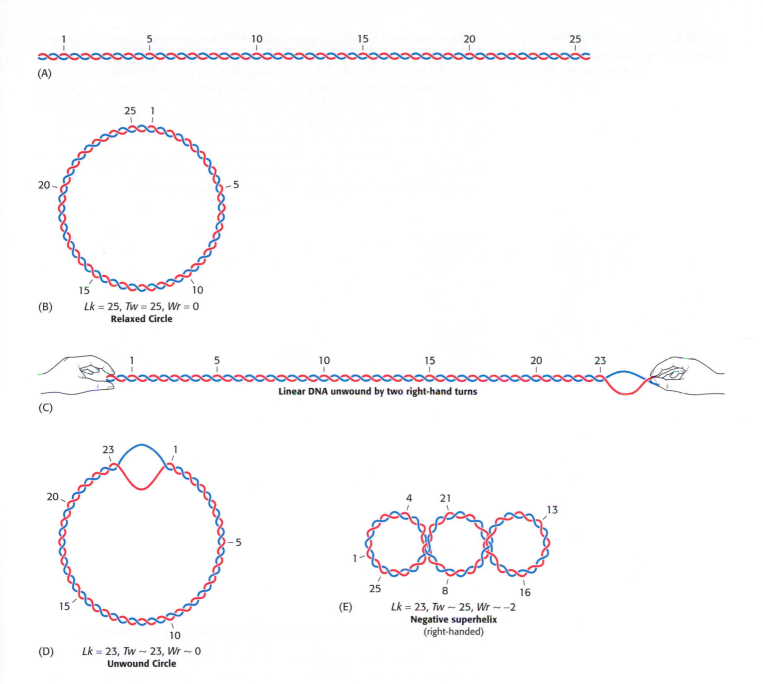

(A)

(B) $Lk = 25$, $Tw = 25$, $Wr = 0$
Relaxed Circle

(C) Linear DNA unwound by two right-hand turns

(D) $Lk = 23$, $Tw \sim 23$, $Wr \sim 0$
Unwound Circle

(E) $Lk = 23$, $Tw \sim 25$, $Wr \sim -2$
Negative superhelix
(right-handed)

27.3.2 Helical Twist and Superhelical Writhe Are Correlated with Each Other Through the Linking Number

Our understanding of the conformation of DNA is enriched by concepts drawn from topology, a branch of mathematics dealing with structural properties that are unchanged by deformations such as stretching and bending. A key topological property of a circular DNA molecule is its *linking number* (Lk), which is equal to the number of times that a strand of DNA winds in the right-handed direction around the helix axis when the axis is constrained to lie in a plane. For the relaxed DNA shown in Figure 27.19B, $Lk = 25$. For the partly unwound molecule shown in part D and the supercoiled one shown in part E, $Lk = 23$ because the linear duplex was unwound two complete turns *before* closure. Molecules differing only in linking number are *topological isomers (topoisomers)* of one another. *Topoisomers of DNA can be interconverted only by cutting one or both DNA strands and then rejoining them.*

FIGURE 27.19 Linking number. The relations between the linking number *(Lk)*, twisting number *(Tw)*, and writhing number *(Wr)* of a circular DNA molecule revealed schematically. [After W. Saenger, *Principles of Nucleic Acid Structure* (Springer-Verlag, 1984), p. 452.]

The unwound DNA and supercoiled DNA shown in Figure 27.19D and E are topologically identical but geometrically different. They have the same value of Lk but differ in Tw (twist) and Wr (writhe). Although the rigorous definitions of twist and writhe are complex, twist is a measure of the helical winding of the DNA strands around each other, whereas writhe is a measure of the coiling of the axis of the double helix, which is called *supercoiling*. A right-handed coil is assigned a negative number (negative supercoiling) and a left-handed coil is assigned a positive number (positive supercoiling). Is there a relation between Tw and Wr? Indeed, there is. Topology tells us that the sum of Tw and Wr is equal to Lk.

$$Lk = Tw + Wr$$

In Figure 27.19, the partly unwound circular DNA has $Tw \sim 23$ and $Wr \sim 0$, whereas the supercoiled DNA has $Tw \sim 25$ and $Wr \sim -2$. These forms can be interconverted without cleaving the DNA chain because they have the same value of Lk; namely, 23. The partitioning of Lk (which must be an integer) between Tw and Wr (which need not be integers) is determined by energetics. The free energy is minimized when about 70% of the change in Lk is expressed in Wr and 30% is expressed in Tw. Hence, the most stable form would be one with $Tw = 24.4$ and $Wr = -1.4$. Thus, *a lowering of* Lk *causes both right-handed (negative) supercoiling of the DNA axis and unwinding of the duplex.* Topoisomers differing by just 1 in Lk, and consequently by 0.7 in Wr, can be readily separated by agarose gel electrophoresis because their hydrodynamic volumes are quite different—*supercoiling condenses DNA* (Figure 27.20). Most naturally occurring DNA molecules are negatively supercoiled. What is the basis for this prevalence? As already stated, negative supercoiling arises from the unwinding or underwinding of the DNA. In essence, negative supercoiling prepares DNA for processes requiring separation of the DNA strands, such as replication or transcription. Positive supercoiling condenses DNA as effectively, but it makes strand separation more difficult.

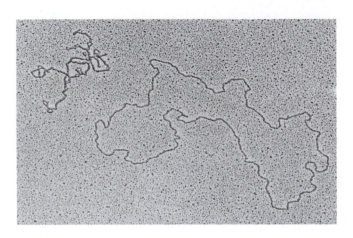

FIGURE 27.20 Topoisomers. An electron micrograph showing negatively supercoiled and relaxed DNA. [Courtesy of Dr. Jack Griffith.]

27.3.3 Type I Topoisomerases Relax Supercoiled Structures

The interconversion of topoisomers of DNA is catalyzed by enzymes called *topoisomerases* which were discovered by James Wang and Martin Gellert. These enzymes alter the linking number of DNA by catalyzing a three-step process: (1) the *cleavage* of one or both strands of DNA, (2) the *passage* of a segment of DNA through this break, and (3) the *resealing* of the DNA break. Type I topoisomerases cleave just one strand of DNA, whereas type II enzymes cleave both strands. Both type I and type II topoisomerases play important roles in DNA replication and in transcription and recombination.

Type I topoisomerases catalyze the relaxation of supercoiled DNA, a thermodynamically favorable process. *Type II topoisomerases* utilize free energy from ATP hydrolysis to add negative supercoils to DNA. The two types of enzymes have several common features, including the use of key tyrosine residues to form covalent links to the polynucleotide backbone that is transiently broken.

The three-dimensional structures of several type I topoisomerases have been determined (Figure 27.21). These structures reveal many features of the reaction mechanism. Human type I topoisomerase comprises four domains, which are arranged around a central cavity having a diameter of 20 Å, just the correct size to accommodate a double-stranded DNA

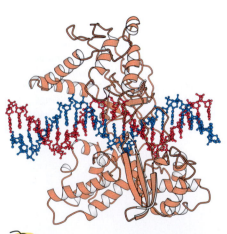

FIGURE 27.21 Structure of a topoisomerase. The structure of a complex between a fragment of human topoisomerase I and DNA.

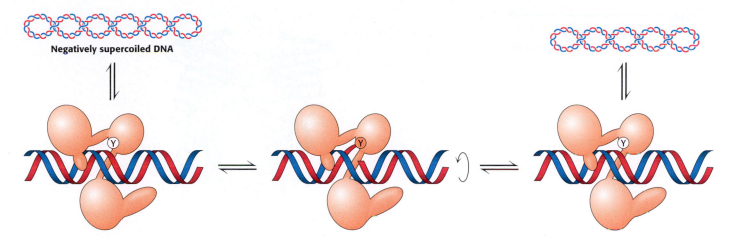

FIGURE 27.22 Topoisomerase I mechanism. On binding to DNA, topoisomerase I cleaves one strand of the DNA through a tyrosine (Y) residue attacking a phosphate. When the strand has been cleaved, it rotates in a controlled manner around the other strand. The reaction is completed by religation of the cleaved strand. This process results in partial or complete relaxation of a supercoiled plasmid.

molecule. This cavity also includes a tyrosine residue (Tyr 723), which acts as a nucleophile to cleave the DNA backbone in the course of catalysis.

From analyses of these structures and the results of other studies, the relaxation of negatively supercoiled DNA molecules are known to proceed in the following manner (Figure 27.22). First, the DNA molecule binds inside the cavity of the topoisomerase. The hydroxyl group of tyrosine 723 attacks a phosphate group on one strand of the DNA backbone to form a phosphodiester linkage between the enzyme and the DNA, cleaving the DNA and releasing a free 5′-hydroxyl group.

With the backbone of one strand cleaved, the DNA can now rotate around the remaining strand, driven by the release of the energy stored because of the supercoiling. The rotation of the DNA unwinds supercoils. The enzyme controls the rotation so that the unwinding is not rapid. The free hydroxyl group of the DNA attacks the phosphotyrosine residue to reseal the backbone and release tyrosine. The DNA is then free to dissociate from the enzyme. Thus, reversible cleavage of one strand of the DNA allows controlled rotation to partly relax supercoiled DNA.

27.3.4 Type II Topoisomerases Can Introduce Negative Supercoils Through Coupling to ATP Hydrolysis

Supercoiling requires an input of energy because a supercoiled molecule, in contrast with its relaxed counterpart, is torsionally stressed. The introduction of an additional supercoil into a 3000-bp plasmid typically requires about 7 kcal mol^{-1}.

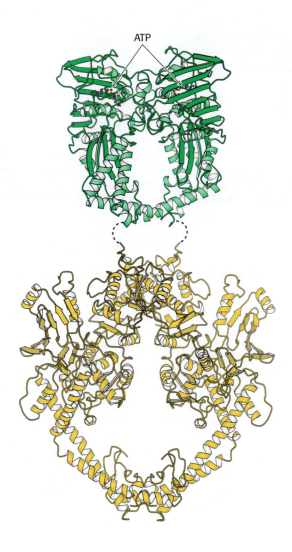

ATP

FIGURE 27.23 Structure of topoisomerase II. A composite structure of topoisomerase II formed from the amino-terminal ATP-binding domain of *E. coli* topoisomerase II (green) and the carboxyl-terminal fragment from yeast topoisomerase II (yellow). Both units form dimeric structures as shown.

Supercoiling is catalyzed by type II topoisomerases. These elegant molecular machines couple the binding and hydrolysis of ATP to the directed passage of one DNA double helix through another that has been temporarily cleaved. These enzymes have several mechanistic features in common with the type I topoisomerases.

The topoisomerase II from yeast is a heart-shaped dimer with a large central cavity (Figure 27.23). This cavity has gates at both the top and the bottom that are crucial to topoisomerase action. The reaction begins with the binding of one double helix (hereafter referred to as the G, for gate, segment) to the enzyme (Figure 27.24). Each strand is positioned next to a tyrosine residue, one from each monomer, capable of forming a covalent linkage with the DNA backbone. This complex then loosely binds a second DNA double helix (hereafter referred to as the T, for transported, segment). Each monomer of the enzyme has a domain that binds ATP; this ATP binding leads to a conformational change that strongly favors the coming together of the two domains. As these domains come closer together, they trap the bound T segment. This conformational change also forces the separation and cleavage of the two strands of the G segment. Each strand is joined to the enzyme by a tyrosine–phosphodiester linkage. Unlike the type I enzymes, the type II topoisomerases hold the DNA tightly so that it cannot rotate. The T segment then passes through the cleaved G segment and into the large central cavity. The ligation of the G segment leads to release of the T segment through the gate at the bottom of the enzyme. The hydrolysis of ATP and the release of ADP and orthophosphate allow the ATP-

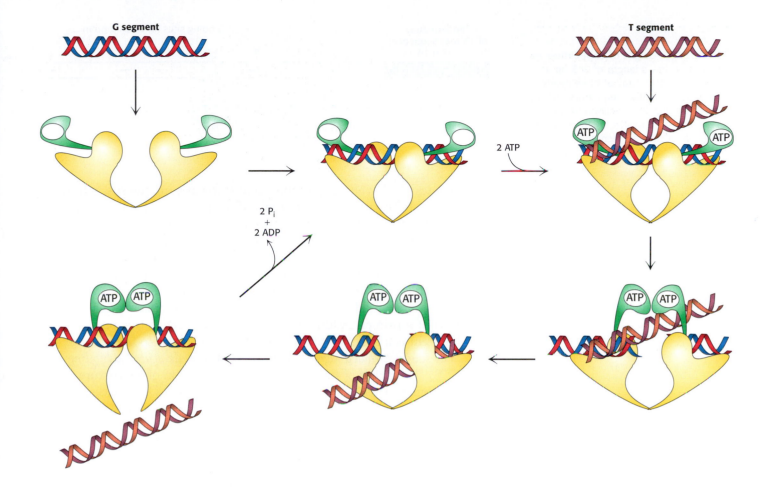

G segment

T segment

2 ATP

2 P$_i$
+
2 ADP

ATP   ATP

ATP   ATP

ATP   ATP

binding domains to separate, preparing the enzyme to bind another T segment. The overall process leads to a decrease in the linking number by two.

The degree of supercoiling of DNA is thus determined by the opposing actions of two enzymes. Negative supercoils are introduced by topoisomerase II and are relaxed by topoisomerase I. The amounts of these enzymes and their activities are regulated to maintain an appropriate degree of negative supercoiling.

The bacterial topoisomerase II (often called DNA gyrase) is the target of several antibiotics that inhibit the prokaryotic enzyme much more than the eukaryotic one. *Novobiocin* blocks the binding of ATP to gyrase. *Nalidixic acid* and *ciprofloxacin,* in contrast, interfere with the breakage and rejoining of DNA chains. These two gyrase inhibitors are widely used to treat urinary tract and other infections. *Camptothecin,* an antitumor agent, inhibits human topoisomerase I by stabilizing the form of the enzyme covalently linked to DNA.

Nalidixic acid

FIGURE 27.24 Topoisomerase II mechanism. Topoisomerase II first binds one DNA duplex termed the G (for gate) segment. The binding of ATP to the two N-terminal domains brings these two domains together. This conformational change leads to the cleavage of both strands of the G segment and the binding of an additional DNA duplex, the T segment. This T segment then moves through the break in the G segment and out the bottom of the enzyme. The hydrolysis of ATP resets the enzyme with the G segment still bound.

27.4 DNA REPLICATION OF BOTH STRANDS PROCEEDS RAPIDLY FROM SPECIFIC START SITES

So far, we have met many of the key players in DNA replication. Here, we ask, Where on the DNA molecule does replication begin, and how is the double helix manipulated to allow the simultaneous use of the two strands

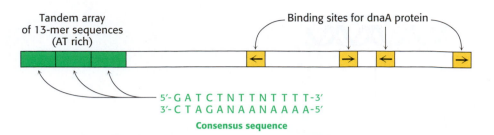

FIGURE 27.25 Origin of replication in *E. coli*. OriC has a length of 245 bp. It contains a tandem array of three nearly identical 13-nucleotide sequences (green) and four binding sites (yellow) for the dnaA protein. The relative orientations of the four dnaA sites are denoted by arrows.

Tandem array
of 13-mer sequences
(AT rich)

Binding sites for dnaA protein

5′- G A T C T N T T N T T T T -3′
3′- C T A G A N A A N A A A A -5′
Consensus sequence

as templates? In *E. coli*, DNA replication starts at a unique site within the entire 4.8×10^6 bp genome. This *origin of replication*, called the oriC *locus*, is a 245-bp region that has several unusual features (Figure 27.25). The oriC locus contains four repeats of a sequence that together act as a binding site for an initiation protein called *dnaA*. In addition, the locus contains a tandem array of 13-bp sequences that are rich in A–T base pairs.

The binding of the dnaA protein to the four sites initiates an intricate sequence of steps leading to the unwinding of the template DNA and the synthesis of a primer. Additional proteins join dnaA in this process. The *dnaB* protein is a helicase that utilizes ATP hydrolysis to unwind the duplex. *The single-stranded regions are trapped by a single-stranded binding protein* (SSB). The result of this process is the generation of a structure called the *prepriming complex*, which makes single-stranded DNA accessible for other enzymes to begin synthesis of the complementary strands.

27.4.1 An RNA Primer Synthesized by Primase Enables DNA Synthesis to Begin

Even with the DNA template exposed, new DNA cannot be synthesized until a primer is constructed. Recall that all known DNA polymerases require a primer with a free 3′-hydroxyl group for DNA synthesis. How is this primer formed? An important clue came from the observation that RNA synthesis is essential for the initiation of DNA synthesis. In fact, *RNA primes the synthesis of DNA*. A specialized RNA polymerase called *primase* joins the prepriming complex in a multisubunit assembly called the *primosome*. Primase synthesizes a short stretch of RNA (~5 nucleotides) that is complementary to one of the template DNA strands (Figure 27.26). The primer is RNA rather than DNA because DNA polymerases cannot start chains de novo. Recall that, to ensure fidelity, DNA polymerase tests the correctness of the preceding base pair before forming a new phosphodiester bond (Section 27.2.4). RNA polymerases can start chains de novo because they do not examine the preceding base pair. Consequently, their error rates are orders of magnitude as high as those of DNA polymerases. The ingenious solution is to start DNA synthesis with a low-fidelity stretch of polynucleotide but mark it "temporary" by placing ribonucleotides in it. The RNA primer is removed by hydrolysis by a $5' \rightarrow 3'$ exonuclease; in *E. coli*, the exonuclease is present as an additional domain of DNA polymerase I, rather than being present in the Klenow fragment. Thus, the complete polymerase I has three distinct active sites: a $3' \rightarrow 5'$ exonuclease proofreading activity, a polymerase activity, and a $5' \rightarrow 3'$ exonuclease activity.

27.4.2 One Strand of DNA Is Made Continuously, Whereas the Other Strand Is Synthesized in Fragments

Both strands of parental DNA serve as templates for the synthesis of new DNA. The site of DNA synthesis is called the *replication fork* because the complex formed by the newly synthesized daughter strands arising from the

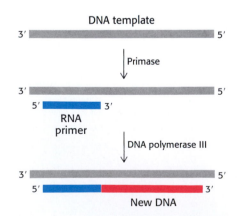

DNA template

Primase

RNA primer

DNA polymerase III

New DNA

FIGURE 27.26 Priming. DNA replication is primed by a short stretch of DNA that is synthesized by primase, an RNA polymerase. The RNA primer is removed at a later stage of replication.

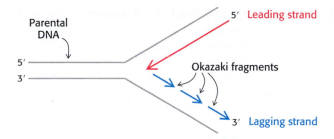

FIGURE 27.27 Okazaki fragments. At a replication fork, both strands are synthesized in a 5' → 3' direction. The leading strand is synthesized continuously, whereas the lagging strand is synthesized in short pieces termed Okazaki fragments.

parental duplex resembles a two-pronged fork. Recall that the two strands are antiparallel; that is, they run in opposite directions. As shown in Figure 27.3, both daughter strands appear to grow in the same direction on cursory examination. However, all known DNA polymerases synthesize DNA in the 5' → 3' direction but not in the 3' → 5' direction. How then does one of the daughter DNA strands appear to grow in the 3' → 5' direction?

This dilemma was resolved by Reiji Okazaki, who found that *a significant proportion of newly synthesized DNA exists as small fragments*. These units of about a thousand nucleotides (called *Okazaki fragments*) are present briefly in the vicinity of the replication fork (Figure 27.27). As replication proceeds, these fragments become covalently joined through the action of DNA ligase (Section 27.4.3) to form one of the daughter strands. The other new strand is synthesized continuously. The strand formed from Okazaki fragments is termed the *lagging strand,* whereas the one synthesized without interruption is the *leading strand.* Both the Okazaki fragments and the leading strand are synthesized in the 5' → 3' direction. *The discontinuous assembly of the lagging strand enables 5' → 3' polymerization at the nucleotide level to give rise to overall growth in the 3' → 5' direction.*

27.4.3 DNA Ligase Joins Ends of DNA in Duplex Regions

The joining of Okazaki fragments requires an enzyme that catalyzes the joining of the ends of two DNA chains. The existence of circular DNA molecules also points to the existence of such an enzyme. In 1967, scientists in several laboratories simultaneously discovered *DNA ligase. This enzyme catalyzes the formation of a phosphodiester bond between the 3' hydroxyl group at the end of one DNA chain and the 5'-phosphate group at the end of the other* (Figure 27.28). An energy source is required to drive this thermodynamically uphill reaction. In eukaryotes and archaea, ATP is the energy source. In bacteria, NAD^+ typically plays this role. We shall examine the mechanistic features that allow these two molecules to power the joining of two DNA chains.

FIGURE 27.28 DNA ligase reaction. DNA ligase catalyzes the joining of one DNA strand with a free 3'-hydroxyl group to another with a free 5'-phosphate group. In eukaryotes and archaea, ATP is cleaved to AMP and PP_i to drive this reaction. In bacteria, NAD^+ is cleaved to AMP and nicotinamide mononucleotide (NMN).

$$E + \text{ATP (or NAD}^+) \rightleftharpoons E\text{-AMP} + PP_i \text{ (or NMN)}$$

$$E\text{-AMP} + \textcircled{P}\text{—5'-DNA} \rightleftharpoons E + \text{AMP}\text{—}\textcircled{P}\text{—5'-DNA}$$

$$\text{DNA-3'-OH} + \text{AMP}\text{—}\textcircled{P}\text{—5'-DNA} \rightleftharpoons \text{DNA-3'-O}\text{—}\textcircled{P}\text{—5'-DNA} + \text{AMP}$$

$$\text{DNA-3'-OH} + \textcircled{P}\text{—5'-DNA} + \text{ATP (or NAD}^+) \rightleftharpoons \text{DNA-3'-O}\text{—}\textcircled{P}\text{—5'-DNA} + \text{AMP} + PP_i \text{ (or NMN)}$$

FIGURE 27.29 DNA ligase mechanism. DNA ligation proceeds by the transfer of an AMP unit first to a lysine side chain on DNA ligase and then to the 5'-phosphate group of the substrate. The AMP unit is released on formation of the phosphodiester linkage in DNA.

DNA ligase cannot link two molecules of single-stranded DNA or circularize single-stranded DNA. Rather, *ligase seals breaks in double-stranded DNA molecules.* The enzyme from *E. coli* ordinarily forms a phosphodiester bridge only if there are at least several base pairs near this link. Ligase encoded by T4 bacteriophage can link two blunt-ended double-helical fragments, a capability that is exploited in recombinant DNA technology.

Let us look at the mechanism of joining, which was elucidated by I. Robert Lehman (Figure 27.29). ATP donates its activated AMP unit to DNA ligase to form a *covalent enzyme–AMP (enzyme–adenylate) complex* in which AMP is linked to the ε-amino group of a lysine residue of the enzyme through a phosphoamide bond. Pyrophosphate is concomitantly released. The activated AMP moiety is then transferred from the lysine residue to the phosphate group at the 5' terminus of a DNA chain, forming a *DNA–adenylate complex.* The final step is a nucleophilic attack by the 3' hydroxyl group at the other end of the DNA chain on this activated 5' phosphorus atom.

In bacteria, NAD$^+$ instead of ATP functions as the AMP donor. NMN is released instead of pyrophosphate. Two high-transfer-potential phosphoryl groups are spent in regenerating NAD$^+$ from NMN and ATP when NAD$^+$ is the adenylate donor. Similarly, two high-transfer-potential phosphoryl groups are spent by the ATP-utilizing enzymes because the pyrophosphate released is hydrolyzed. The results of structural studies revealed that the ATP- and NAD$^+$-utilizing enzymes are homologous even though this homology could not be deduced from their amino acid sequences alone.

27.4.4 DNA Replication Requires Highly Processive Polymerases

Enzyme activities must be highly coordinated to replicate entire genomes precisely and rapidly. A prime example is provided by *DNA polymerase III holoenzyme,* the enzyme responsible for DNA replication in *E. coli.* The hallmarks of this multisubunit assembly are its *very high catalytic potency, fidelity, and processivity. Processivity* refers to the ability of an enzyme to catalyze many consecutive reactions without releasing its substrate. The holoenzyme catalyzes the formation of many thousands of phosphodiester bonds before releasing its template, compared with only 20 for DNA polymerase I. DNA polymerase III holoenzyme has evolved to grasp its template and not let go until the template has been completely replicated. A second distinctive feature of the holoenzyme is its catalytic prowess: 1000 nucleotides are added per second compared with only 10 per second for DNA polymerase I. This acceleration is accomplished with no loss of accuracy. The greater catalytic prowess of polymerase III is largely due to its processivity; no time is lost in repeatedly stepping on and off the template.

Processive enzyme—

From the Latin *procedere,* "to go forward."
An enzyme that catalyzes multiple rounds of elongation or digestion of a polymer while the polymer stays bound. A *distributive enzyme,* in contrast, releases its polymeric substrate between successive catalytic steps.

These striking features of DNA polymerase III do not come cheaply. The holoenzyme consists of 10 kinds of polypeptide chains and has a mass of ~900 kd, nearly an order of magnitude as large as that of a single-chain DNA polymerase, such as DNA polymerase I. This replication complex is an *asymmetric dimer* (Figure 27.30). The holoenzyme is structured as a dimer to enable it to replicate both strands of parental DNA in the same place at the same time. It is asymmetric because the leading and lagging strands are synthesized differently. A τ_2 subunit is associated with one branch of the holoenzyme; γ_2 and $(\delta\delta'\chi\psi)_2$ are associated with the other. The core of each branch is the same, an $\alpha\epsilon\theta$ complex. The α subunit is the polymerase, and the ϵ subunit is the proofreading $3' \rightarrow 5'$ exonuclease. Each core is catalytically active but not processive. Processivity is conferred by β_2 and τ_2.

The source of the processivity was revealed by the determination of the three-dimensional structure of the β_2 subunit (Figure 27.31). This unit has the form of a star-shaped ring. A 35-Å-diameter hole in its center can readily accommodate a duplex DNA molecule, yet leaves enough space between the DNA and the protein to allow rapid sliding and turning during replication. A catalytic rate of 1000 nucleotides polymerized per second requires the sliding of 100 turns of duplex DNA (a length of 3400 Å, or 0.34 μm) through the central hole of β_2 per second. Thus, β_2 *plays a key role in replication by serving as a sliding DNA clamp.*

27.4.5 The Leading and Lagging Strands Are Synthesized in a Coordinated Fashion

The holoenzyme synthesizes the leading and lagging strands simultaneously at the replication fork (Figure 27.32). DNA polymerase III begins the synthesis of the leading strand by using the RNA primer formed by primase. The duplex DNA ahead of the polymerase is unwound by an ATP-driven helicase. Single-stranded binding protein again keeps the strands separated so that both strands can serve as templates. The leading strand is synthesized continuously by polymerase III, which does not release the template until replication has been completed. Topoisomerases II (DNA gyrase) concurrently introduces right-handed (negative) supercoils to avert a topological crisis.

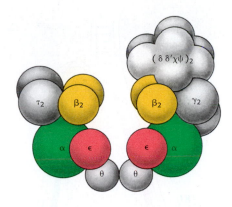

FIGURE 27.30 Proposed architecture of DNA polymerase III holoenzyme. [After A. Kornberg and T. Baker, *DNA Replication*, 2d ed. (W. H. Freeman and Company, 1992).]

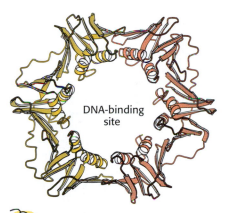

FIGURE 27.31 Structure of the sliding clamp. The dimeric β_2 subunit of DNA polymerase III forms a ring that surrounds the DNA duplex. It allows the polymerase enzyme to move without falling off the DNA substrate.

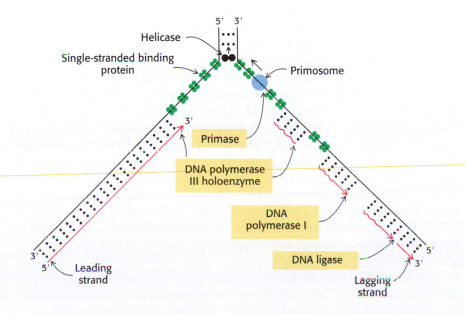

FIGURE 27.32 Replication fork. Schematic representation of the enzymatic events at a replication fork in *E. coli*. Enzymes shaded in yellow catalyze chain initiation, elongation, and ligation. The wavy lines on the lagging strand denote RNA primers. [After A. Kornberg and T. Baker, *DNA Replication*, 2d ed. (W. H. Freeman and Company, 1992).]

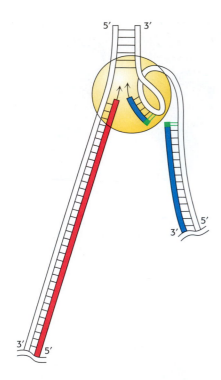

FIGURE 27.33 Coordination between the leading and the lagging strands. The looping of the template for the lagging strand enables a dimeric DNA polymerase III holoenzyme to synthesize both daughter strands. The leading strand is shown in red, the lagging strand in blue, and the RNA primers in green. [Courtesy of Dr. Arthur Kornberg.]

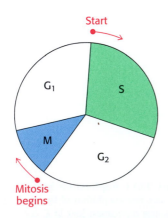

FIGURE 27.34 Eukaryotic cell cycle. DNA replication and cell division must take place in a highly coordinated fashion in eukaryotes. Mitosis (M) takes place only after DNA synthesis (S). Two gaps (G_1 and G_2) in time separate the two processes.

The mode of synthesis of the lagging strand is necessarily more complex. As mentioned earlier, the lagging strand is synthesized in fragments so that $5' \rightarrow 3'$ polymerization leads to overall growth in the $3' \rightarrow 5'$ direction. A looping of the template for the lagging strand places it in position for $5' \rightarrow 3'$ polymerization (Figure 27.33). The looped lagging-strand template passes through the polymerase site in one subunit of a dimeric polymerase III in the same direction as that of the leading-strand template in the other subunit. DNA polymerase III lets go of the lagging-strand template after adding about 1000 nucleotides. A new loop is then formed, and primase again synthesizes a short stretch of RNA primer to initiate the formation of another Okazaki fragment.

The gaps between fragments of the nascent lagging strand are then filled by DNA polymerase I. This essential enzyme also uses its $5' \rightarrow 3'$ exonuclease activity to remove the RNA primer lying ahead of the polymerase site. The primer cannot be erased by DNA polymerase III, because the enzyme lacks $5' \rightarrow 3'$ editing capability. Finally, DNA ligase connects the fragments.

27.4.6 DNA Synthesis Is More Complex in Eukaryotes Than in Prokaryotes

Replication in eukaryotes is mechanistically similar to replication in prokaryotes but is more challenging for a number of reasons. One of them is sheer size: *E. coli* must replicate 4.8 million base pairs, whereas a human diploid cell must replicate 6 billion base pairs. Second, the genetic information for *E. coli* is contained on 1 chromosome, whereas, in human beings, 23 pairs of chromosomes must be replicated. Finally, whereas the *E. coli* chromosome is circular, human chromosomes are linear. Unless countermeasures are taken (Section 27.4.7), linear chromosomes are subject to shortening with each round of replication.

The first two challenges are met by the use of multiple origins of replication, which are located between 30 and 300 kbp apart. In human beings, replication requires about 30,000 origins of replication, with each chromosome containing several hundred. Each origin of replication represents a replication unit, or *replicon*. The use of multiple origins of replication requires mechanisms for ensuring that each sequence is replicated once and only once. The events of eukaryotic DNA replication are linked to the eukaryotic *cell cycle* (Figure 27.34). In the cell cycle, the processes of DNA synthesis and cell division (mitosis) are coordinated so that the replication of all DNA sequences is complete before the cell progresses into the next phase of the cycle. This coordination requires several *checkpoints* that control the progression along the cycle.

The origins of replication have not been well characterized in higher eukaryotes but, in yeast, the DNA sequence is referred to as an *autonomously replicating sequence* (ARS) and is composed of an AT-rich region made up of discrete sites. The ARS serves as a docking site for the *origin of replication complex* (ORC). The ORC is composed of six proteins with an overall mass of ~400 kd. The ORC recruits other proteins to form the prereplication complex. Several of the recruited proteins are called *licensing factors* because they permit the formation of the initiation complex. These proteins serve to ensure that each replicon is replicated once and only once in a cell cycle. How is this regulation achieved? After the licensing factors have established the initiation complex, these factors are marked for destruction by the attachment of ubiquitin and subsequently destroyed by proteasomal digestion (Section 23.2.2).

DNA helicases separate the parental DNA strands, and the single strands are stabilized by the binding of *replication protein A*, a single-stranded–

DNA–binding protein. Replication begins with the binding of *DNA polymerase α*, which is the initiator polymerase. This enzyme has primase activity, used to synthesize RNA primers, as well as DNA polymerase activity, although it possesses no exonuclease activity. After a stretch of about 20 deoxynucleotides have been added to the primer, another replication protein, called *protein replication factor C (RFC)*, displaces DNA polymerase α and attracts *proliferating cell nuclear antigen (PCNA)*. Homologous to the β_2 subunit of *E. coli* polymerase III, PCNA then binds to *DNA polymerase δ*. The association of polymerase δ with PCNA renders the enzyme highly processive and suitable for long stretches of replication. This process is called *polymerase switching* because polymerase δ has replaced polymerase α. Polymerase δ has $3' \rightarrow 5'$ exonuclease activity and can thus edit the replicated DNA. Replication continues in both directions from the origin of replication until adjacent replicons meet and fuse. RNA primers are removed and the DNA fragments are ligated by DNA ligase.

27.4.7 Telomeres Are Unique Structures at the Ends of Linear Chromosomes

Whereas the genomes of essentially all prokaryotes are circular, the chromosomes of human beings and other eukaryotes are linear. The free ends of linear DNA molecules introduce several complications that must be resolved by special enzymes. In particular, it is difficult to fully replicate DNA ends, because polymerases act only in the $5' \rightarrow 3'$ direction. The lagging strand would have an incomplete 5' end after the removal of the RNA primer. Each round of replication would further shorten the chromosome.

The first clue to how this problem is resolved came from sequence analyses of the ends of chromosomes, which are called *telomeres* (from the Greek *telos*, "an end"). Telomeric DNA contains hundreds of tandem repeats of a hexanucleotide sequence. One of the strands is G rich at the 3' end, and it is slightly longer than the other strand. In human beings, the repeating G-rich sequence is AGGGTT.

The structure adopted by telomeres has been extensively investigated. Recent evidence suggests that they may form large duplex loops (Figure 27.35). The single-stranded region at the very end of the structure has been proposed to loop back to form a DNA duplex with another part of the repeated sequence, displacing a part of the original telomeric duplex. This loop-like structure is formed and stabilized by specific telomere-binding proteins. Such structures would nicely protect and mask the end of the chromosome.

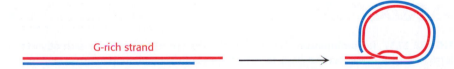

G-rich strand

FIGURE 27.35 Proposed model for telomeres. A single-stranded segment of the G-rich strand extends from the end of the telomere. In one model for telomeres, this single-stranded region invades the duplex to form a large duplex loop.

27.4.8 Telomeres Are Replicated by Telomerase, a Specialized Polymerase That Carries Its Own RNA Template

How are the repeated sequences generated? An enzyme, termed *telomerase*, that executes this function has been purified and characterized. When a primer ending in GGTT is added to the human enzyme in the presence of deoxynucleoside triphosphates, the sequences GGTTAGGGTT and GGTTAGGGTTAGGGTT, as well as longer products, are generated.

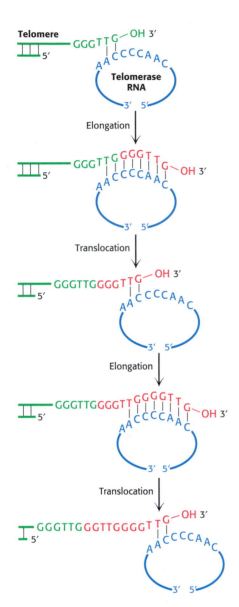

FIGURE 27.36 Telomere formation.
Mechanism of synthesis of the G-rich strand of telomeric DNA. The RNA template of telomerase is shown in blue and the nucleotides added to the G-rich strand of the primer are shown in red. [After E. H. Blackburn. *Nature* 350(1991):569.]

Elizabeth Blackburn and Carol Greider discovered that the enzyme contains an RNA molecule that serves as the template for elongation of the G-rich strand (Figure 27.36). Thus, the enzyme carries the information necessary to generate the telomere sequences. The exact number of repeated sequences is not crucial.

Subsequently, a protein component of telomerases also was identified. From its amino acid sequence, this component is clearly related to reverse transcriptases, enzymes first discovered in retroviruses that copy RNA into DNA. Thus, *telomerase is a specialized reverse transcriptase that carries its own template*. Telomeres may play important roles in cancer-cell biology and in cell aging.

27.5 DOUBLE-STRANDED DNA MOLECULES WITH SIMILAR SEQUENCES SOMETIMES RECOMBINE

Most processes associated with DNA replication function to copy the genetic message as faithfully as possible. However, several biochemical processes require the *recombination* of genetic material between two DNA molecules. In genetic recombination, two daughter molecules are formed by the exchange of genetic material between two parent molecules (Figure 27.37).

1. In meiosis, the limited exchange of genetic material between paired chromosomes provides a simple mechanism for generating genetic diversity in a population.

2. As we shall see in Chapter 33, recombination plays a crucial role in generating molecular diversity for antibodies and some other molecules in the immune system.

3. Some viruses utilize recombination pathways to integrate their genetic material into the DNA of the host cell.

4. Recombination is used to manipulate genes in, for example, the generation of "gene knockout" mice (Section 6.3.5).

Recombination is most efficient between DNA sequences that are similar in sequence. Such processes are often referred to as *homologous recombination* reactions.

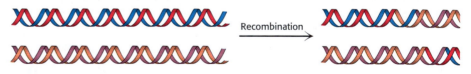

Recombination

FIGURE 27.37 Recombination. Two DNA molecules can recombine with each other to form new DNA molecules that have segments from both parental molecules.

27.5.1 Recombination Reactions Proceed Through Holliday Junction Intermediates

The **STRUCTURAL INSIGHTS** *module for this chapter shows how a recombinase forms a Holliday junction from two DNA duplexes and suggests how this intermediate is resolved to produce recombinants.*

Enzymes called *recombinases* catalyze the exchange of genetic material that takes place in recombination. By what pathway do these enzymes catalyze this exchange? An appealing scheme was proposed by Robin Holliday in 1964. A key intermediate in this mechanism is a crosslike structure, known as a *Holliday junction*, formed by four polynucleotide chains. Such

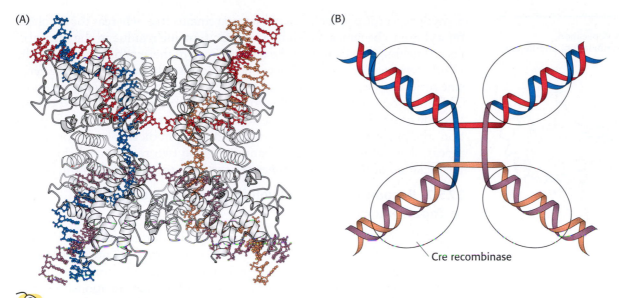

(A)

(B)

Cre recombinase

👆 **FIGURE 27.38 Holliday junction.** (A) The structure of a Holliday junction bound by Cre recombinase (gray), a bacteriophage protein. (B) A schematic view of a Holliday junction.

intermediates have been characterized by a wide range of techniques including x-ray crystallography (Figure 27.38). Note that such intermediates can form only when the nucleotide sequences of the two parental duplexes are very similar or identical in the region of recombination because specific base pairs must form between the bases of the two parental duplexes.

How are such intermediates formed from the parental duplexes and resolved to form products? Many details for this process are now available, based largely on the results of studies of Cre recombinase from bacteriophage P1. This mechanism begins with the recombinase binding to the DNA substrates (Figure 27.39). Four molecules of the enzyme and their associated DNA molecules come together to form a *recombination synapse*. The reaction begins with the cleavage of one strand from each duplex. The

FIGURE 27.39 Recombination mechanism. Recombination begins as two DNA molecules come together to form a recombination synapse. One strand from each duplex is cleaved by the recombinase enzyme; the 3′ end of each of the cleaved strands is linked to a tyrosine (Y) residue on the recombinase enzyme. New phosphodiester bonds are formed when a 5′ end of the other cleaved strand in the complex attacks these tyrosine–DNA adducts. After isomerization, these steps are repeated to form the recombined products.

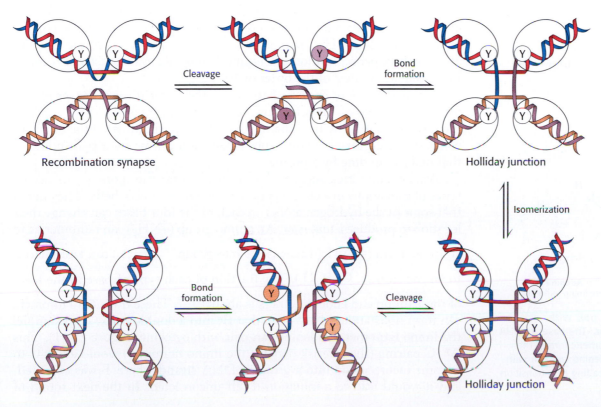

Recombination synapse — Cleavage — Bond formation — Holliday junction

Isomerization

Bond formation — Cleavage — Holliday junction

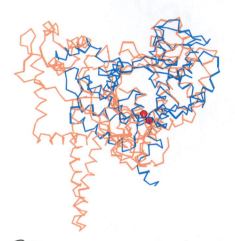

FIGURE 27.40 Recombinases and topoisomerase I. A superposition of Cre recombinase (blue) and topoisomerase I (orange) reveals that these two enzymes have a common structural core. The positions of the tyrosine residues that participate in DNA cleavage reactions are shown as red spheres for both enzymes.

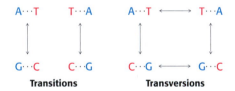

Transitions   **Transversions**

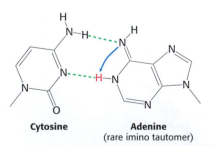

Cytosine   Adenine
(rare imino tautomer)

FIGURE 27.41 Base pair with mutagenic tautomer. The bases of DNA can exist in rare tautomeric forms. The imino tautomer of adenine can pair with cytosine, eventually leading to a transition from A–T to G–C.

5′-hydroxyl group of each cleaved strand remains free, whereas the 3′-phosphoryl group becomes linked to a specific tyrosine residue in the recombinase. The free 5′ ends invade the other duplex in the synapse and attack the DNA–tyrosine units to form new phosphodiester-bonds and free the tyrosine residues. These reactions result in the formation of a Holliday junction. This junction can then isomerize to form a structure in which the polynucleotide chains in the center of the structure are reoriented. From this junction, the processes of strand cleavage and phosphodiester-bond formation repeat. The result is a synapse containing the two recombined duplexes. Dissociation of this complex generates the final recombined products.

27.5.2 Recombinases Are Evolutionarily Related to Topoisomerases

The intermediates that form in recombination reactions, with their tyrosine adducts possessing 3′-phosphoryl groups, are reminiscent of the intermediates that form in the reactions catalyzed by topoisomerases. This mechanistic similarity reflects deeper evolutionary relationships. Examination of the three-dimensional structures of recombinases and type I topoisomerases reveals that these proteins are related by divergent evolution despite little amino acid sequence similarity (Figure 27.40). From this perspective, the action of a recombinase can be viewed as an intermolecular topoisomerase reaction. In each case, a tyrosine–DNA adduct is formed. In a topoisomerase reaction, this adduct is resolved when the 5′-hydroxyl group of the same duplex attacks to reform the same phosphodiester bond that was initially cleaved. In a recombinase reaction, the attacking 5′-hydroxyl group comes from a DNA chain that was not initially linked to the phosphoryl group participating in the phosphodiester bond.

27.6 MUTATIONS INVOLVE CHANGES IN THE BASE SEQUENCE OF DNA

We now turn from DNA replication to DNA mutations and repair. Several types of mutations are known: (1) the *substitution* of one base pair for another, (2) the *deletion* of one or more base pairs, and (3) the *insertion* of one or more base pairs. The spontaneous mutation rate of T4 phage is about 10^{-7} per base per replication. *E. coli* and *Drosophila melanogaster* have much lower mutation rates, of the order of 10^{-10}.

The substitution of one base pair for another is the a common type of mutation. Two types of substitutions are possible. A *transition* is the replacement of one purine by the other or that of one pyrimidine by the other. In contrast, a *transversion* is the replacement of a purine by a pyrimidine or that of a pyrimidine by a purine.

Watson and Crick suggested a mechanism for the spontaneous occurrence of transitions in a classic paper on the DNA double helix. They noted that some of the hydrogen atoms on each of the four bases can change their location to produce a *tautomer*. An amino group (–NH_2) can tautomerize to an imino form (=NH). Likewise, a keto group ($C=O$) can tautomerize to an enol form ($C-OH$). The fraction of each type of base in the form of these imino and enol tautomers is about 10^{-4}. These transient tautomers can form nonstandard base pairs that fit into a double helix. For example, the imino tautomer of adenine can pair with cytosine (Figure 27.41). This A*–C pairing (the asterisk denotes the imino tautomer) would allow C to become incorporated into a growing DNA strand where T was expected, and it would lead to a mutation if left uncorrected. In the next round of replication, A* will probably retautomerize to the standard form, which pairs as usual with thymine, but the cytosine residue will pair with guanine. Hence, one of the daughter DNA molecules will contain a G–C base pair in place of the normal A–T base pair.

27.6.1 Some Chemical Mutagens Are Quite Specific

Base analogs such as 5-bromouracil and 2-aminopurine can be incorporated into DNA and are even more likely than normal nucleic acid bases to form transient tautomers that lead to transition mutations. 5-Bromouracil, an analog of thymine, normally pairs with adenine. However, the proportion of 5-bromouracil in the enol tautomer is higher than that of thymine because the bromine atom is more electronegative than is a methyl group on the C-5 atom. Thus, the incorporation of 5-bromouracil is especially likely to cause altered base-pairing in a subsequent round of DNA replication (Figure 27.42).

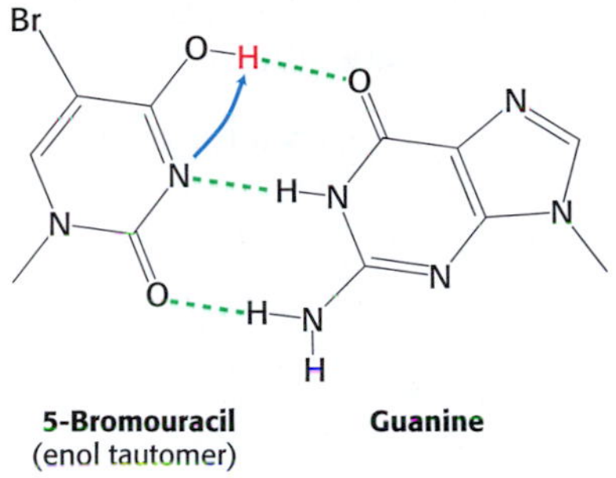

Other mutagens act by chemically modifying the bases of DNA. For example, nitrous acid (HNO_2) reacts with bases that contain amino groups. Adenine is oxidatively deaminated to hypoxanthine, cytosine to uracil, and guanine to xanthine. Hypoxanthine pairs with cytosine rather than with thymine (Figure 27.43). Uracil pairs with adenine rather than with guanine. Xanthine, like guanine, pairs with cytosine. Consequently, nitrous acid causes A–T $\leftrightarrow$ G–C transitions.

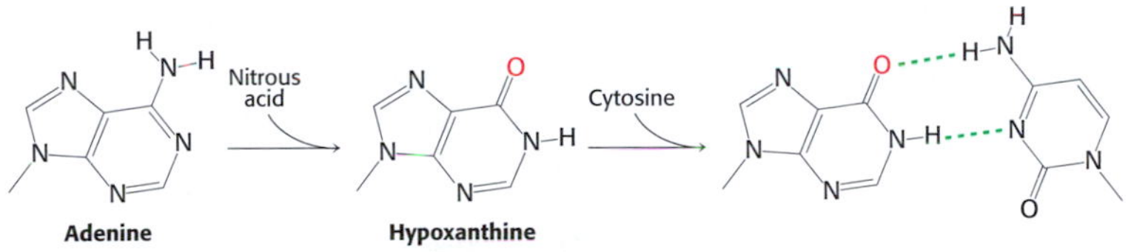

Tautomerization—
The interconversion of two isomers that differ only in the position of protons (and, often, double bonds).

5-Bromouracil (enol tautomer) **Guanine**

FIGURE 27.42 Base pair with 5-bromouracil. This analog of thymine has a higher tendency to form an enol tautomer than does thymine itself. The pairing of the enol tautomer of 5-bromouracil with guanine will lead to a transition from T–A to C–G.

Adenine **Hypoxanthine**

FIGURE 27.43 Chemical mutagenesis. Treatment of DNA with nitrous acid results in the conversion of adenine into hypoxanthine. Hypoxanthine pairs with cytosine, inducing a transition from A–T to G–C.

A different kind of mutation is produced by flat aromatic molecules such as the acridines (Figure 27.44). These compounds intercalate in DNA—that is, they slip in between adjacent base pairs in the DNA double helix. Consequently, they lead to the insertion or deletion of one or more base pairs. The effect of such mutations is to alter the reading frame in translation, unless an integral multiple of three base pairs is inserted or deleted. In fact, the analysis of such mutants contributed greatly to the revelation of the triplet nature of the genetic code.

FIGURE 27.44 Acridines. Acridine dyes induce frameshift mutations by intercalating into the DNA, leading to the incorporation of an additional base on the opposite strand.

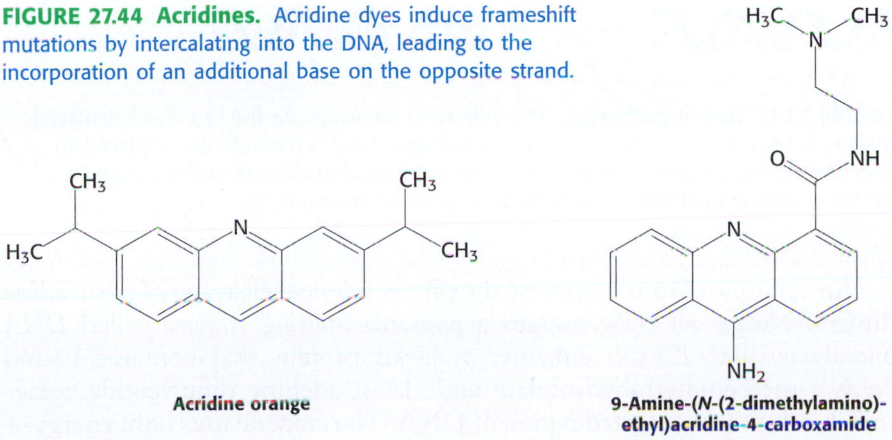

Acridine orange

9-Amine-(*N*-(2-dimethylamino)-ethyl)acridine-4-carboxamide

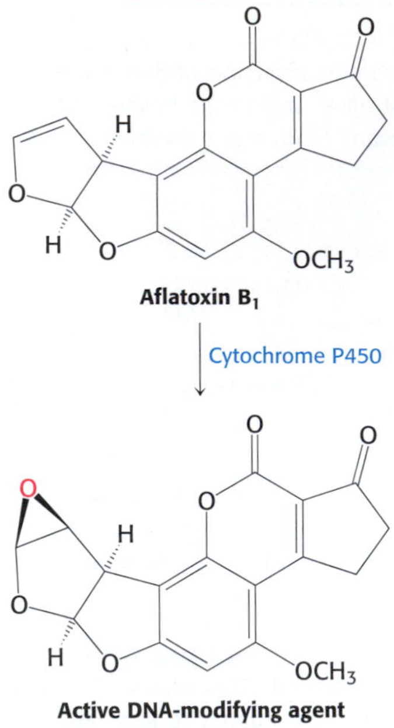

Aflatoxin B₁

Cytochrome P450

↓

Active DNA-modifying agent

FIGURE 27.45 Aflatoxin reaction. The compound, produced by molds that grow on peanuts, is activated by cytochrome P450 to form a highly reactive species that modifies bases such as guanine in DNA, leading to mutations.

Some compounds are converted into highly active mutagens through the action of enzymes that normally play a role in detoxification. A striking example is aflatoxin B_1, a compound produced by molds that grows on peanuts and other foods. A cytochrome P450 enzyme (Section 26.4.3) converts this compound into a highly reactive epoxide (Figure 27.45). This agent reacts with the N-7 atom of guanosine to form an adduct that frequently leads to a G–C-to-T–A transversion.

27.6.2 Ultraviolet Light Produces Pyrimidine Dimers

The ultraviolet component of sunlight is a ubiquitous DNA-damaging agent. Its major effect is to covalently link adjacent pyrimidine residues along a DNA strand (Figure 27.46). Such a pyrimidine dimer cannot fit into a double helix, and so replication and gene expression are blocked until the lesion is removed.

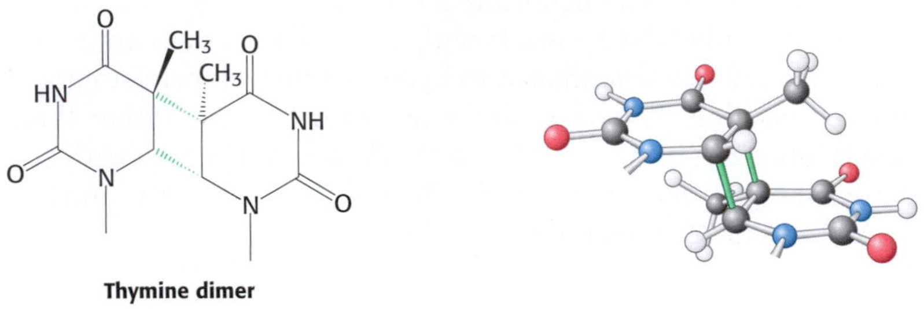

Thymine dimer

FIGURE 27.46 Cross-linked dimer of two thymine bases. Ultraviolet light induces cross-links between adjacent pyrimidines along one strand of DNA.

27.6.3 A Variety of DNA-Repair Pathways Are Utilized

The maintenance of the integrity of the genetic message is key to life. Consequently, all cells possess mechanisms to repair damaged DNA. Three types of repair pathways are direct repair, base-excision repair, and nucleotide-excision repair (Figure 27.47).

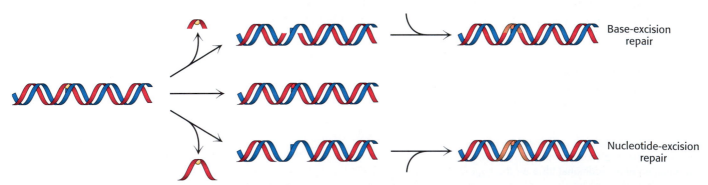

Base-excision repair

Nucleotide-excision repair

FIGURE 27.47 Repair pathways. Three different pathways are used to repair damaged regions in DNA. In base-excision repair, the damaged base is removed and replaced. In direct repair, the damaged region is corrected in place. In nucleotide-excision repair, a stretch of DNA around the site of damage is removed and replaced.

An example of *direct repair* is the photochemical cleavage of pyrimidine dimers. Nearly all cells contain a *photoreactivating enzyme* called *DNA photolyase*. The *E. coli* enzyme, a 35-kd protein that contains bound N^5,N^{10}-methenyltetrahydrofolate and flavin adenine dinucleotide cofactors, binds to the distorted region of DNA. The enzyme uses light energy—

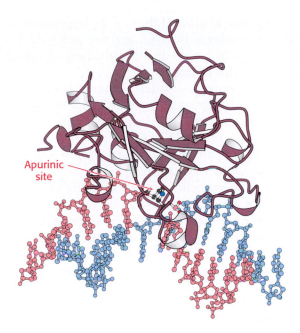

Apurinic
site

FIGURE 27.48 Structure of DNA-repair enzyme. A complex between the DNA-repair enzyme AlkA and an analog of an apurinic site. Note that the damaged base is flipped out of the DNA double helix into the active site of the enzyme for excision.

specifically, the absorption of a photon by the N^5,N^{10}-methenyltetrahydrofolate coenzyme—to form an excited state that cleaves the dimer into its original bases.

The excision of modified bases such as 3-methyladenine by the *E. coli* enzyme *AlkA* is an example of *base-excision repair*. The binding of this enzyme to damaged DNA flips the affected base out of the DNA double helix and into the active site of the enzyme (Figure 27.48). Base flipping also occurs in the enzymatic addition of methyl groups to DNA bases (Section 24.2.7). The enzyme then acts as a glycosylase, cleaving the glycosidic bond to release the damaged base. At this stage, the DNA backbone is intact, but a base is missing. This hole is called an *AP site* because it is apurinic (devoid of A or G) or apyrimidinic (devoid of C or T). An AP endonuclease recognizes this defect and nicks the backbone adjacent to the missing base. *Deoxyribose phosphodiesterase* excises the residual deoxyribose phosphate unit, and DNA polymerase I inserts an undamaged nucleotide, as dictated by the base on the undamaged complementary strand. Finally, the repaired strand is sealed by DNA ligase.

One of the best-understood examples of *nucleotide-excision repair* is the excision of a pyrimidine dimer. Three enzymatic activities are essential for this repair process in *E. coli* (Figure 27.49). First, an enzyme complex consisting of the proteins encoded by the *uvrABC* genes detects the distortion produced by the pyrimidine dimer. A specific uvrABC enzyme then cuts the damaged DNA strand at two sites, 8 nucleotides away from the dimer on the 5' side and 4 nucleotides away on the 3' side. The 12-residue oligonucleotide excised by this highly specific *excinuclease* (from the Latin *exci,* "to cut out") then diffuses away. DNA polymerase I enters the gap to carry out repair synthesis. The 3' end of the nicked strand is the primer, and the intact complementary strand is the template. Finally, the 3' end of the newly synthesized stretch of DNA and the original part of the DNA chain are joined by *DNA ligase*.

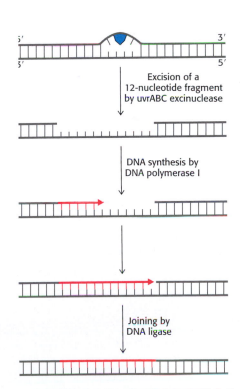

Excision of a
12-nucleotide fragment
by uvrABC excinuclease

DNA synthesis by
DNA polymerase I

Joining by
DNA ligase

FIGURE 27.49 Excision repair. Repair of a region of DNA containing a thymine dimer by the sequential action of a specific excinuclease, a DNA polymerase, and a DNA ligase. The thymine dimer is shown in blue, and the new region of DNA is in red. [After P. C. Hanawalt. *Endevour* 31(1982):83.]

27.6.4 The Presence of Thymine Instead of Uracil in DNA Permits the Repair of Deaminated Cytosine

The presence in DNA of thymine rather than uracil was an enigma for many years. Both bases pair with adenine. The only difference between them is a

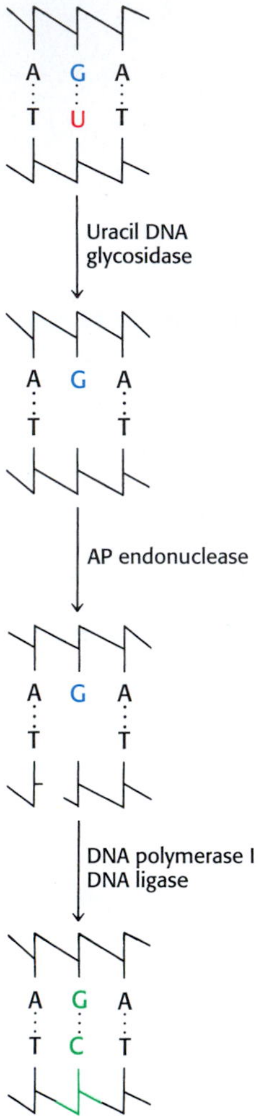

FIGURE 27.50 Uracil repair. Uracil bases in DNA, formed by the deamination of cytosine, are excised and replaced by cytosine.

methyl group in thymine in place of the C-5 hydrogen atom in uracil. Why is a methylated base employed in DNA and not in RNA? The existence of an active repair system to correct the deamination of cytosine provides a convincing solution to this puzzle.

Cytosine in DNA spontaneously deaminates at a perceptible rate to form uracil. The deamination of cytosine is potentially mutagenic because uracil pairs with adenine, and so one of the daughter strands will contain an U–A base pair rather than the original C–G base pair (Figure 27.50). This mutation is prevented by a repair system that recognizes uracil to be foreign to DNA. This enzyme, *uracil DNA glycosylase,* is homologous to AlkA. The enzyme hydrolyzes the glycosidic bond between the uracil and deoxyribose moieties but does not attack thymine-containing nucleotides. The AP site generated is repaired to reinsert cytosine. Thus, *the methyl group on thymine is a tag that distinguishes thymine from deaminated cytosine.* If thymine were not used in DNA, uracil correctly in place would be indistinguishable from uracil formed by deamination. The defect would persist unnoticed, and so a C–G base pair would necessarily be mutated to U–A in one of the daughter DNA molecules. This mutation is prevented by a repair system that searches for uracil and leaves thymine alone. *Thymine is used instead of uracil in DNA to enhance the fidelity of the genetic message.* In contrast, RNA is not repaired, and so uracil is used in RNA because it is a less-expensive building block.

27.6.5 Many Cancers Are Caused by Defective Repair of DNA

As discussed in Chapter 15, cancers are caused by mutations in genes associated with growth control. Defects in DNA-repair systems are expected to increase the overall frequency of mutations and, hence, the likelihood of a cancer-causing mutation. *Xeroderma pigmentosum,* a rare human skin disease, is genetically transmitted as an autosomal recessive trait. The skin in an affected homozygote is extremely sensitive to sunlight or ultraviolet light. In infancy, severe changes in the skin become evident and worsen with time. The skin becomes dry, and there is a marked atrophy of the dermis. Keratoses appear, the eyelids become scarred, and the cornea ulcerates. Skin cancer usually develops at several sites. Many patients die before age 30 from metastases of these malignant skin tumors.

Ultraviolet light produces pyrimidine dimers in human DNA, as it does in *E. coli* DNA. Furthermore, the repair mechanisms are similar. Studies of skin fibroblasts from patients with xeroderma pigmentosum have revealed a biochemical defect in one form of this disease. In normal fibroblasts, half the pyrimidine dimers produced by ultraviolet radiation are excised in less than 24 hours. In contrast, almost no dimers are excised in this time interval in fibroblasts derived from patients with xeroderma pigmentosum. The results of these studies show that *xeroderma pigmentosum can be produced by a defect in the excinuclease that hydrolyzes the DNA backbone near a pyrimidine dimer. The drastic clinical consequences of this enzymatic defect emphasize the critical importance of DNA-repair processes.* The disease can also be caused by mutations in eight other genes for DNA repair, which attests to the complexity of repair processes.

Defects in other repair systems can increase the frequency of other tumors. For example, *hereditary nonpolyposis colorectal cancer (HNPCC, or Lynch syndrome)* results from defective DNA mismatch repair. HNPCC is not rare—as many as 1 in 200 people will develop this form of cancer. Mutations in two genes, called *hMSH2* and *hMLH1,* account for most cases of this hereditary predisposition to cancer. The striking finding is that these genes encode the human counterparts of MutS and MutL of *E. coli.* The

MutS protein binds to mismatched base pairs (e.g., G–T) in DNA. An MutH protein, together with MutL, participates in cleaving one of the DNA strands in the vicinity of this mismatch to initiate the repair process (Figure 27.51). It seems likely that mutations in *hMSH2* and *hMLH1* lead to the accumulation of mutations throughout the genome. In time, genes important in controlling cell proliferation become altered, resulting in the onset of cancer.

27.6.6 Some Genetic Diseases Are Caused by the Expansion of Repeats of Three Nucleotides

Some genetic diseases are caused by the presence of DNA sequences that are inherently prone to errors in the course of replication. A particularly important class of such diseases are characterized by the presence of long tandem arrays of repeats of three nucleotides. An example is *Huntington disease*, an autosomal dominant neurological disorder with a variable age of onset. The mutated gene in this disease expresses a protein called huntingtin, which is expressed in the brain and contains a stretch of consecutive glutamine residues. These glutamine residues are encoded by a tandem array of CAG sequences within the gene. In unaffected persons, this array is between 6 and 31 repeats, whereas, in those with the disease, the array is between 36 and 82 repeats or longer. Moreover, the array tends to become longer from one generation to the next. The consequence is a phenomenon called *anticipation*: the children of an affected parent tend to show symptoms of the disease at an earlier age than did the parent.

The tendency of these *trinucleotide repeats* to expand is explained by the formation of alternative structures in DNA replication (Figure 27.52). Part of the array within the daughter strand can loop out without disrupting base-pairing outside this region. DNA polymerase extends this strand through the remainder of the array, leading to an increase in the number of copies of the trinucleotide sequence.

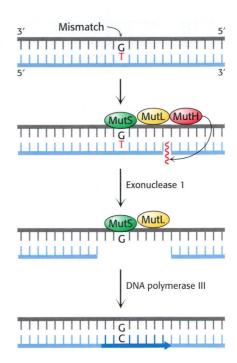

FIGURE 27.51 Mismatch repair. DNA mismatch repair in *E. coli* is initiated by the interplay of MutS, MutL, and MutH proteins. A G–T mismatch is recognized by MutS. MutH cleaves the backbone in the vicinity of the mismatch. A segment of the DNA strand containing the erroneous T is removed by exonuclease I and synthesized anew by DNA polymerase III. [After R. F. Service. *Science* 263(1994):1559.]

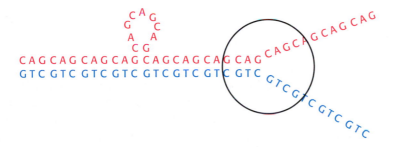

FIGURE 27.52 Triplet repeat expansion. Sequences containing tandem arrays of repeated triplet sequences can be expanded to include more repeats by the looping out of some of the repeats before replication.

A number of other neurological diseases are characterized by expanding arrays of trinucleotide repeats. How do these long stretches of repeated amino acids cause disease? For huntingtin, it appears that the polyglutamine stretches become increasingly prone to aggregate as their length increases; the additional consequences of such aggregation are still under active investigation.

27.6.7 Many Potential Carcinogens Can Be Detected by Their Mutagenic Action on Bacteria

Many human cancers are caused by exposure to chemicals. These chemical carcinogens usually cause mutations, which suggests that *damage to DNA*

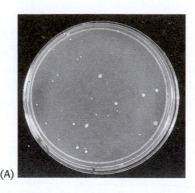 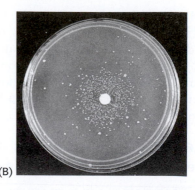

(A)   (B)

FIGURE 27.53 Ames test. (A) A petri dish containing about 10^9 *Salmonella* bacteria that cannot synthesize histidine and (B) a petri dish containing a filter-paper disc with a mutagen, which produces a large number of revertants that can synthesize histidine. After 2 days, the revertants appear as rings of colonies around the disc. The small number of visible colonies in plate A are spontaneous revertants. [From B. N. Ames, J. McCann, and E. Yamasaki. *Mutat. Res.* 31(1975):347.]

is a fundamental event in the origin of mutations and cancer. It is important to identify these compounds and ascertain their potency so that human exposure to them can be minimized. Bruce Ames devised a simple and sensitive test for detecting chemical mutagens. In the *Ames test,* a thin layer of agar containing about 10^9 bacteria of a specially constructed tester strain of *Salmonella* is placed on a petri dish. These bacteria are unable to grow in the absence of histidine, because a mutation is present in one of the genes for the biosynthesis of this amino acid. The addition of a chemical mutagen to the center of the plate results in many new mutations. A small proportion of them reverse the original mutation, and histidine can be synthesized. These *revertants* multiply in the absence of an external source of histidine and appear as discrete colonies after the plate has been incubated at 37°C for 2 days (Figure 27.53). For example, 0.5 μg of 2-aminoanthracene gives 11,000 revertant colonies, compared with only 30 spontaneous revertants in its absence. A series of concentrations of a chemical can be readily tested to generate a dose–response curve. These curves are usually linear, which suggests that there is no threshold concentration for mutagenesis.

Some of the tester strains are responsive to *base-pair substitutions,* whereas others detect *deletions or additions of base pairs (frameshifts).* The sensitivity of these specially designed strains has been enhanced by the genetic deletion of their excision-repair systems. Potential mutagens enter the tester strains easily because the lipopolysaccharide barrier that normally coats the surface of *Salmonella* is incomplete in these strains.

A key feature of this detection system is the inclusion of a *mammalian liver homogenate* (Section 4.1.2). Recall that some potential carcinogens such as aflatoxin are converted into their active forms by enzyme systems in the liver or other mammalian tissues (Section 27.6.1). Bacteria lack these enzymes, and so the test plate requires a few milligrams of a liver homogenate to activate this group of mutagens.

The *Salmonella* test is extensively used to help evaluate the mutagenic and carcinogenic risks of a large number of chemicals. This rapid and inexpensive bacterial assay for mutagenicity complements epidemiological surveys and animal tests that are necessarily slower, more laborious, and far more expensive. The *Salmonella* test for mutagenicity is an outgrowth of studies of gene–protein relations in bacteria. It is a striking example of how fundamental research in molecular biology can lead directly to important advances in public health.

SUMMARY

DNA Can Assume a Variety of Structural Forms

DNA is a structurally dynamic molecule that can exist in a variety of helical forms: A-DNA, B-DNA (the classic Watson-Crick helix), and Z-DNA. DNA can be bent, kinked, and unwound. In A-, B-, and Z-DNA, two antiparallel chains are held together by Watson-Crick base pairs and stacking interactions between bases in the same strand. The sugar-phosphate backbone is on the outside, and the bases are inside the double helix. A- and B-DNA are right-handed helices. In B-DNA, the base pairs are nearly perpendicular to the helix axis. In A–DNA, the bases are tilted rather than perpendicular. An important structural feature of the B helix is the presence of major and minor grooves, which display different potential hydrogen-bond acceptors and donors according to the base sequence. X-ray analysis of a single crystal of B-DNA reveals that the structure is much more variable than was originally imagined. Dehydration induces the transition from B- to A-DNA. Z-DNA is a left-handed helix. It can be formed in regions of DNA in which purines alternate with pyrimidines, as in CGCG or CACA. Most of the DNA in a cell is in the B-form.

DNA Polymerases Require a Template and a Primer

DNA polymerases are template-directed enzymes that catalyze the formation of phosphodiester bonds by the 3'-hydroxyl group's nucleophilic attack on the innermost phosphorus atom of a deoxyribonucleoside 5'-triphosphate. They cannot start chains de novo; a primer with a free 3'-hydroxyl group is required. DNA polymerases from a variety of sources have important structural features in common as well as a catalytic mechanism requiring the presence of two metal ions. Many DNA polymerases proofread the nascent product; their $3' \rightarrow 5'$ exonuclease activity potentially edits the outcome of each polymerization step. A mispaired nucleotide is excised before the next step proceeds. In *E. coli*, DNA polymerase I repairs DNA and participates in replication. Fidelity is further enhanced by an induced fit that results in a catalytically active conformation only when the complex of enzyme, DNA, and correct dNTP is formed. Helicases prepare the way for DNA replication by using ATP hydrolysis to separate the strands of the double helix.

Double-Stranded DNA Can Wrap Around Itself to Form Supercoiled Structures

A key topological property of DNA is its linking number *(Lk)*, which is defined as the number of times one strand of DNA winds around the other in the right-hand direction when the DNA axis is constrained to lie in a plane. Molecules differing in linking number are topoisomers of one another and can be interconverted only by cutting one or both DNA strands; these reactions are catalyzed by topoisomerases. Changes in linking number generally lead to changes in both the number of turns of double helix and the number of turns of superhelix. Topoisomerase II (DNA gyrase) catalyzes the ATP-driven introduction of negative supercoils, which leads to the compaction of DNA and renders it more susceptible to unwinding. Supercoiled DNA is relaxed by topoisomerase I. The unwinding of DNA at the replication fork is catalyzed by an ATP-driven helicase.

• **DNA Replication of Both Strands Proceeds Rapidly from Specific Start Sites**

DNA replication in *E. coli* starts at a unique origin (*oriC*) and proceeds sequentially in opposite directions. More than 20 proteins are required for replication. An ATP-driven helicase unwinds the *oriC* region to create a replication fork. At this fork, both strands of parental DNA serve as templates for the synthesis of new DNA. A short stretch of RNA formed by primase, an RNA polymerase, primes DNA synthesis. One strand of DNA (the leading strand) is synthesized continuously, whereas the other strand (the lagging strand) is synthesized discontinuously, in the form of 1-kb fragments (Okazaki fragments). Both new strands are formed simultaneously by the concerted actions of the highly processive DNA polymerase III holoenzyme, an asymmetric dimer. The discontinuous assembly of the lagging strand enables $5' \rightarrow 3'$ polymerization at the atomic level to give rise to overall growth of this strand in the $3' \rightarrow 5'$ direction. The RNA primer stretch is hydrolyzed by the $5' \rightarrow 3'$ nuclease activity of DNA polymerase I, which also fills gaps. Finally, nascent DNA fragments are joined by DNA ligase in a reaction driven by ATP or NAD^+ cleavage.

DNA synthesis in eukaryotes is more complex than in prokaryotes. Eukaryotes require thousands of origins of replication to complete replication in a timely fashion. A special RNA-dependent DNA polymerase called telomerase is responsible for the replication of the ends of linear chromosomes.

• **Double-Stranded DNA Molecules with Similar Sequences Sometimes Recombine**

DNA molecules that are similar in nucleotide sequences in a local region can recombine to form DNA duplexes that begin with the sequence of one molecule and continue with the sequence of the other. Recombinases catalyze the formation of these products through the formation and resolution of Holliday junctions. In these structures, four DNA strands come together to form a crosslike structure.

Recombinases cleave DNA strands and form specific adducts in which a tyrosine residue of the enzyme is linked to a 3'-phosphoryl group of the DNA. These intermediates then react with 5'-hydroxyl groups of other DNA strands to form the new phosphodiester bonds that are present in the recombination products. The reaction mechanisms of recombinases are similar to those of type I topoisomerases.

• **Mutations Are Produced by Several Types of Changes in the Base Sequence of DNA**

Mutations are produced by mistakes in base-pairing, by the covalent modification of bases, and by the deletion and insertion of bases. The $3' \rightarrow 5'$ exonuclease activity of DNA polymerases is critical in lowering the spontaneous mutation rate, which arises in part from mispairing by tautomeric bases. Lesions in DNA are continually being repaired. Multiple repair processes utilize information present in the intact strand to correct the damaged strand. For example, pyrimidine dimers formed by the action of ultraviolet light are excised by uvrABC excinuclease, an enzyme that removes a 12-nucleotide region containing the dimer. Xeroderma pigmentosum, a genetically transmitted disease, is caused by the defective repair of lesions in DNA, such as pyrimidine dimers; patients with this disease usually develop skin cancers. Many cancers of the colon are caused by defective DNA mismatch repair arising from mutations in human genes that have been highly conserved in evolution. Damage to DNA is a fundamental event in both carcinogenesis and mutagenesis. Many potential carcinogens can be detected by their mutagenic action on bacteria.

KEY TERMS

B-DNA helix (p. 747)

A-DNA helix (p. 747)

major groove (p. 748)

minor groove (p. 748)

Z-DNA helix (p. 749)

DNA polymerase (p. 750)

template (p. 750)

primer (p. 750)

exonuclease (p. 751)

helicase (p. 753)

supercoil (p. 754)

linking number (Lk) (p. 755)

topoisomer (p. 755)

twist (Tw) (p. 756)

writhe (Wr) (p. 756)

topoisomerase (p. 756)

origin of replication (p. 760)

primase (p. 760)

replication fork (p. 760)

Okazaki fragment (p. 761)

lagging strand (p. 761)

leading stand (p. 761)

DNA ligase (p. 761)

processivity (p. 762)

cell cycle (p. 764)

telomere (p. 765)

telomerase (p. 765)

recombinase (p. 766)

Holliday junction (p. 766)

recombination synapse (p. 767)

transition (p. 768)

transversion (p. 768)

mutagen (p. 769)

direct repair (p. 770)

base-excision repair (p. 771)

nucleotide-excision repair (p. 771)

trinucleotide repeat (p. 773)

Ames test (p. 774)

SELECTED READINGS

Where to begin

Kornberg, A., 1988. DNA replication. *J. Biol. Chem.* 263:1–4.

Dickerson, R. E., 1983. The DNA helix and how it is read. *Sci. Am.* 249(6):94–111.

Wang, J. C., 1982. DNA topoisomerases. *Sci. Am.* 247(1):94–109.

Lindahl, T., 1993. Instability and decay of the primary structure of DNA. *Nature* 362:709–715.

Greider, C. W., and Blackburn, E. H., 1996. Telomeres, telomerase, and cancer. *Sci. Am.* 274(2):92–97.

Books

Kornberg, A., and Baker, T. A., 1992. *DNA Replication* (2d ed.). W. H. Freeman and Company.

Bloomfield, V. A., Crothers, D., Tinoco, I., and Hearst, J., 2000. *Nucleic Acids: Structures, Properties and Functions.* University Science Books.

Friedberg, E. C., Walker, G. C., Siede, W., 1995. *DNA Repair and Mutagenesis.* American Society for Microbiology.

Cozzarelli, N. R., and Wang, J. C. (Eds.), 1990. *DNA Topology and Its Biological Effects.* Cold Spring Harbor Laboratory Press.

DNA structure

Chiu, T. K., and Dickerson, R. E., 2000. 1 Å crystal structures of B-DNA reveal sequence-specific binding and groove-specific bending of DNA by magnesium and calcium. *J. Mol. Biol.* 301:915–945.

Herbert, A., and Rich, A., 1999. Left-handed Z-DNA: Structure and Function. *Genetica* 106:37–47.

Dickerson, R. E., 1992. DNA Structure from A to Z. *Methods Enzymol.* 211:67–111.

Quintana, J. R., Grzeskowiak, K., Yanagi, K., and Dickerson, R. E., 1992. Structure of a B-DNA decamer with a central T·A step: C-G-A-T-T-A-A-T-C-G. *J. Mol. Biol.* 225:379–395.

Verdaguer, N., Aymami, J., Fernandez, F. D., Fita, I., Coll, M., Huynh, D. T., Igolen, J., and Subirana, J. A., 1991. Molecular structure of a complete turn of A-DNA. *J. Mol. Biol.* 221:623–635.

DNA topology and topoisomerases

Sikder, D., Unniraman, S., Bhaduri, T., and Nagaraja, V., 2001. Functional cooperation between topoisomerase I and single strand DNA-binding protein. *J. Mol. Biol.* 306:669–679.

Yang, Z., and Champoux, J. J., 2001. The role of histidine 632 in catalysis by human topoisomerase I. *J. Biol. Chem.* 276:677–685.

Fortune, J. M., and Osheroff, N., 2000. Topoisomerase II as a target for anticancer drugs: When enzymes stop being nice. *Prog. Nucleic Acid Res. Mol. Biol.* 64:221–253.

Isaacs, R. J., Davies, S. L., Sandri, M. I., Redwood, C., Wells, N. J., and Hickson, I. D., 1998. Physiological regulation of eukaryotic topoisomerase II. *Biochim. Biophys. Acta* 1400:121–137.

Wang, J. C., 1996. DNA topoisomerases. *Annu. Rev. Biochem.* 65:635–692.

Wang, J. C., 1998. Moving one DNA double helix through another by a type II DNA topoisomerase: The story of a simple molecular machine. *Q. Rev. Biophys.* 31:107–144.

Baird, C. L., Harkins, T. T., Morris, S. K., and Lindsley, J. E., 1999. Topoisomerase II drives DNA transport by hydrolyzing one ATP. *Proc. Natl. Acad. Sci. USA* 96:13685–13690.

Vologodskii, A. V., Levene, S. D., Klenin, K. V., Frank, K. M., and Cozzarelli, N. R., 1992. Conformational and thermodynamic properties of supercoiled DNA. *J. Mol. Biol.* 227:1224–1243.

Fisher, L. M., Austin, C. A., Hopewell, R., Margerrison, M., Oram, M., Patel, S., Wigley, D. B., Davies, G. J., Dodson, E. J., Maxwell, A., and Dodson, G., 1991. Crystal structure of an N-terminal fragment of the DNA gyrase B protein. *Nature* 351:624–629.

Mechanism of replication

Davey, M. J., and O'Donnell, M., 2000. Mechanisms of DNA replication. *Curr. Opin. Chem. Biol.* 4:581–586.

Keck, J. L., and Berger, J. M., 2000. DNA replication at high resolution. *Chem. Biol.* 7:R63–R71.

Kunkel, T. A., and Bebenek, K., 2000. DNA replication fidelity. *Annu. Rev. Biochem.* 69:497–529.

Waga, S., and Stillman, B., 1998. The DNA replication fork in eukaryotic cells. *Annu. Rev. Biochem.* 67:721–751.

Marians, K. J., 1992. Prokaryotic DNA replication. *Annu. Rev. Biochem.* 61:673–719.

DNA polymerases and other enzymes of replication

Hubscher, U., Nasheuer, H. P., and Syvaoja, J. E., 2000. Eukaryotic DNA polymerases: A growing family. *Trends Biochem. Sci.* 25:143–147.

Doublié, S., Tabor, S., Long, A. M., Richardson, C. C., and Ellenberger, T., 1998. Crystal structure of a bacteriophage T7 DNA replication complex at 2.2 Å resolution. *Nature* 391:251–258.

Arezi, B., and Kuchta, R. D., 2000. Eukaryotic DNA primase. *Trends Biochem. Sci.* 25:572–576.

Jager, J., and Pata, J. D., 1999. Getting a grip: Polymerases and their substrate complexes. *Curr. Opin. Struct. Biol.* 9:21–28.

Steitz, T. A., 1999. DNA polymerases: Structural diversity and common mechanisms. *J. Biol. Chem.* 274:17395–17398.

Beese, L. S., Derbyshire, V., and Steitz, T. A., 1993. Structure of DNA polymerase I Klenow fragment bound to duplex DNA. *Science* 260:352–355.

McHenry, C. S., 1991. DNA polymerase III holoenzyme: Components, structure, and mechanism of a true replicative complex. *J. Biol. Chem.* 266:19127–19130.

Kong, X. P., Onrust, R., O'Donnell, M., and Kuriyan, J., 1992. Three-dimensional structure of the beta subunit of *E. coli* DNA polymerase III holoenzyme: A sliding DNA clamp. *Cell* 69:425–437.

Polesky, A. H., Steitz, T. A., Grindley, N. D., and Joyce, C. M., 1990. Identification of residues critical for the polymerase activity of the Klenow fragment of DNA polymerase I from *Escherichia coli*. *J. Biol. Chem.* 265:14579–14591.

Lee, J. Y., Chang, C., Song, H. K., Moon, J., Yang, J. K., Kim, H. K., Kwon, S. T., and Suh, S. W., 2000. Crystal structure of NAD(+)-dependent DNA ligase: Modular architecture and functional implications. *EMBO J.* 19:1119–1129.

Timson, D. J., and Wigley, D. B., 1999. Functional domains of an NAD$^+$-dependent DNA ligase. *J. Mol. Biol.* 285:73–83.

Doherty, A. J., and Wigley, D. B., 1999. Functional domains of an ATP-dependent DNA ligase. *J. Mol. Biol.* 285:63–71.

von Hippel, P. H., and Delagoutte, E., 2001. A general model for nucleic acid helicases and their "coupling" within macromolecular machines. *Cell* 104:177–190.

Tye, B. K., and Sawyer, S., 2000. The hexameric eukaryotic MCM helicase: Building symmetry from nonidentical parts. *J. Biol. Chem.* 275:34833–34836.

Marians, K. J., 2000. Crawling and wiggling on DNA: Structural insights to the mechanism of DNA unwinding by helicases. *Structure Fold Des.* 5:R227–R235.

Soultanas, P., and Wigley, D. B., 2000. DNA helicases: 'Inching forward.' *Curr. Opin. Struct. Biol.* 10:124–128.

Bachand, F., and Autexier, C., 2001. Functional regions of human telomerase reverse transcriptase and human telomerase RNA required for telomerase activity and RNA-protein interactions. *Mol. Cell Biol.* 21:1888–1897.

Bryan, T. M., and Cech, T. R., 1999. Telomerase and the maintenance of chromosome ends. *Curr. Opin. Cell Biol.* 11:318–324.

Griffith, J. D., Comeau, L., Rosenfield, S., Stansel, R. M., Bianchi, A., Moss, H., and de Lange, T., 1999. Mammalian telomeres end in a large duplex loop. *Cell* 97:503–514.

McEachern, M. J., Krauskopf, A., and Blackburn, E. H., 2000. Telomeres and their control. *Annu. Rev. Genet.* 34:331–358.

Recombinases

Van Duyne, G. D., 2001. A structural view of cre-loxp site-specific recombination. *Annu. Rev. Biophys. Biomol. Struct.* 30:87–104.

Chen, Y., Narendra, U., Iype, L. E., Cox, M. M., and Rice, P. A., 2000. Crystal structure of a Flp recombinase-Holliday junction complex: Assembly of an active oligomer by helix swapping. *Mol. Cell* 6:885–897.

Craig, N. L., 1997. Target site selection in transposition. *Annu. Rev. Biochem.* 66:437–474.

Gopaul, D. N., Guo, F., and Van Duyne, G. D., 1998. Structure of the Holliday junction intermediate in Cre-loxP site-specific recombination. *EMBO J.* 17:4175–4187.

Gopaul, D. N., and Duyne, G. D., 1999. Structure and mechanism in site-specific recombination. *Curr. Opin. Struct. Biol.* 9:14–20.

Mutations and DNA repair

Michelson, R. J., and Weinert, T., 2000. Closing the gaps among a web of DNA repair disorders. *Bioessays* 22:966–969.

Aravind, L., Walker, D. R., and Koonin, E. V., 1999. Conserved domains in DNA repair proteins and evolution of repair systems. *Nucleic Acids Res.* 27:1223–1242.

Mol, C. D., Parikh, S. S., Putnam, C. D., Lo, T. P., and Tainer, J. A., 1999. DNA repair mechanisms for the recognition and removal of damaged DNA bases. *Annu. Rev. Biophys. Biomol. Struct.* 28:101–128.

Parikh, S. S., Mol, C. D., and Tainer, J. A., 1997. Base excision repair enzyme family portrait: Integrating the structure and chemistry of an entire DNA repair pathway. *Structure* 5:1543–1550.

Vassylyev, D. G., and Morikawa, K., 1997. DNA-repair enzymes. *Curr. Opin. Struct. Biol.* 7:103–109.

Verdine, G. L., and Bruner, S. D., 1997. How do DNA repair proteins locate damaged bases in the genome? *Chem. Biol.* 4:329–334.

Bowater, R. P., and Wells, R. D., 2000. The intrinsically unstable life of DNA triplet repeats associated with human hereditary disorders. *Prog. Nucleic Acid Res. Mol. Biol.* 66:159–202.

Cummings, C. J., and Zoghbi, H. Y., 2000. Fourteen and counting: Unraveling trinucleotide repeat diseases. *Hum. Mol. Genet.* 9:909–916.

Defective DNA repair and cancer

Berneburg, M., and Lehmann, A. R., 2001. Xeroderma pigmentosum and related disorders: Defects in DNA repair and transcription. *Adv. Genet.* 43:71–102.

Lambert, M. W., and Lambert, W. C., 1999. DNA repair and chromatin structure in genetic diseases. *Prog. Nucleic Acid Res. Mol. Biol.* 63:257–310.

Buys, C. H., 2000. Telomeres, telomerase, and cancer. *N. Engl. J. Med.* 342:1282–1283.

Urquidi, V., Tarin, D., and Goodison, S., 2000. Role of telomerase in cell senescence and oncogenesis. *Annu. Rev. Med.* 51:65–79.

Lynch, H. T., Smyrk, T. C., Watson, P., Lanspa, S. J., Lynch, J. F., Lynch, P. M., Cavalieri, R. J., and Boland, C. R., 1993. Genetics, natural history, tumor spectrum, and pathology of hereditary nonpolyposis colorectal cancer: An updated review. *Gastroenterology* 104:1535–1549.

Fishel, R., Lescoe, M. K., Rao, M. R. S., Copeland, N. G., Jenkins, N. A., Garber, J., Kane, M., and Kolodner, R., 1993. The human mutator gene homolog *MSH2* and its association with hereditary nonpolyposis colon cancer. *Cell* 75:1027–1038.

Ames, B. N., and Gold, L. S., 1991. Endogenous mutagens and the causes of aging and cancer. *Mutat. Res.* 250:3–16.

Ames, B. N., 1979. Identifying environmental chemicals causing mutations and cancer. *Science* 204:587–593.

PROBLEMS

1. *Activated intermediates.* DNA polymerase I, DNA ligase, and topoisomerase I catalyze the formation of phosphodiester bonds. What is the activated intermediate in the linkage reaction catalyzed by each of these enzymes? What is the leaving group?

2. *Fuel for a new ligase.* Whether the joining of two DNA chains by known DNA ligases is driven by NAD$^+$ or ATP depends on the species. Suppose that a new DNA ligase requiring a different energy donor is found. Propose a plausible substitute for NAD$^+$ or ATP in this reaction.

3. *Life in a hot tub.* An archaeon (*Sulfolobus acidocaldarius*) found in acidic hot springs contains a topoisomerase that catalyzes the ATP-driven introduction of positive supercoils into DNA. How might this enzyme be advantageous to this unusual organism?

4. *A cooperative transition.* The transition from B-DNA to Z-DNA occurs over a small change in the superhelix density, which shows that the transition is highly cooperative.

(a) Consider a DNA molecule at the midpoint of this transition. Are B- and Z-DNA regions frequently intermingled or are there long stretches of each?

(b) What does this finding reveal about the energetics of forming a junction between the two kinds of helices?

(c) Would you expect the transition from B- to A-DNA to be more or less cooperative than the one from B- to Z-DNA? Why?

5. *Molecular motors in replication.* (a) How fast does template DNA spin (expressed in revolutions per second) at an *E. coli* replication fork? (b) What is the velocity of movement (in micrometers per second) of DNA polymerase III holoenzyme relative to the template?

6. *Wound tighter than a drum.* Why would replication come to a halt in the absence of topoisomerase II?

7. *Telomeres and cancer.* Telomerase is not active in most human cells. Some cancer biologists have suggested that activation of the telomerase gene would be a requirement for a cell to become cancerous. Explain why this might be the case.

8. *Nick translation.* Suppose that you wish to make a sample of DNA duplex highly radioactive to use as a DNA probe. You have a DNA endonuclease that cleaves the DNA internally to generate 3'-OH and 5'-phosphate groups, intact DNA polymerase I, and radioactive dNTPs. Suggest a means for making the DNA radioactive.

9. *Revealing tracks.* Suppose that replication is initiated in a medium containing *moderately* radioactive tritiated thymine. After a few minutes of incubation, the bacteria are transferred to a medium containing *highly* radioactive tritiated thymidine. Sketch the autoradiographic pattern that would be seen for (a) undirectional replication and (b) bidirectional replication, from a single origin.

10. *Mutagenic trail.* Suppose that the single-stranded RNA from tobacco mosaic virus was treated with a chemical mutagen, that mutants were obtained having serine or leucine instead of proline at a specific position, and that further treatment of these mutants with the same mutagen yielded phenylalanine at this position.

$$ \text{Pro} \xrightarrow{\text{Ser}} \text{Phe} \xrightarrow{\text{Leu}} $$

(a) What are the plausible codon assignments for these four amino acids?

(b) Was the mutagen 5-bromouracil, nitrous acid, or an acridine dye?

11. *Induced spectrum.* DNA photolyases convert the energy of light in the near ultraviolet or visible region (300–500 nm) into chemical energy to break the cyclobutane ring of pyrimidine dimers. In the absence of substrate, these photoreactivating enzymes do not absorb light of wavelengths longer than 300 nm. Why is the substrate-induced absorption band advantageous?

Mechanism Problems

12. *AMP-induced relaxation.* DNA ligase from *E. coli* relaxes supercoiled circular DNA in the presence of AMP but not in its absence. What is the mechanism of this reaction, and why is it dependent on AMP?

13. *A revealing analog.* AMP-PNP, the β,γ-imido analog of ATP, is hydrolyzed very slowly by most ATPases.

AMP-PNP

The addition of AMP-PNP to topoisomerase II and circular DNA leads to the negative supercoiling of a single molecule of DNA per enzyme. DNA remains bound to the enzyme in the presence of this analog. What does this finding reveal about the catalytic mechanism?

Data Interpretation and Chapter Integration Problems

14. *Like a ladder.* Circular DNA from SV40 virus was isolated and subjected to gel electrophoresis. The results are shown in lane A (the control) of the adjoining gel patterns.

A B C

(a) Why does the DNA separate in agarose gel electrophoresis? How does the DNA in each band differ?

The DNA was then incubated with topoisomerase I for 5 minutes and again analyzed by gel electrophoresis with the results shown in lane B.

(b) What types of DNA do the various bands represent?

Another sample of DNA was incubated with topoisomerase I for 30 minutes and again analyzed as shown in lane C.

(c) What is the significance of the fact that more of the DNA is in slower moving forms?

15. *Ames test.* The adjoining illustration shows four petri plates used for the Ames test. A piece of filter paper (white circle in the center of each plate) was soaked in one of four preparations and then placed on a petri plate. The four preparations contained

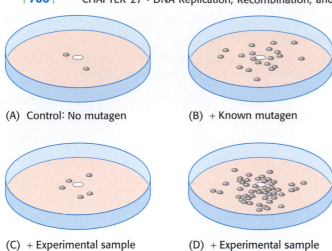

(A) Control: No mutagen

(B) + Known mutagen

(C) + Experimental sample

(D) + Experimental sample
after treatment with
liver homogenate

(A) purified water (control), (B) a known mutagen, (C) a chemical whose mutagenicity is under investigation, and (D) the same chemical after treatment with liver homogenate. The number of revertants, visible as colonies on the petri plates, was determined in each case.

(a) What was the purpose of the control plate, which was exposed only to water?

(b) Why was it wise to use a known mutagen in the experimental system?

(c) How would you interpret the results obtained with the experimental compound?

(d) What liver components would you think are responsible for the effects observed part D?

 **Media Problem**

16. *Cre-ative duplexes.* Site specific recombinases like Cre require that the sequence between the recombinase binding sites be identical in the two DNA molecules that are to be recombined. Examine the structures of the various Cre-DNA complexes in the **Structural Insights** module on Cre recombinase. How might the recombination mechanism check these sequences to ensure they are identical in the parent DNA molecules?

CHAPTER 28

RNA Synthesis and Splicing

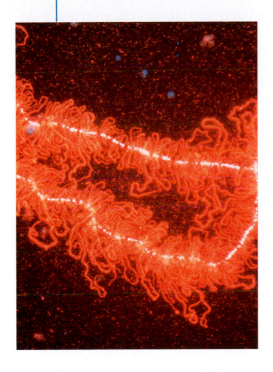

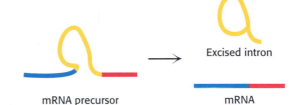

mRNA precursor → Excised intron / mRNA

RNA synthesis is a key step in the expression of genetic information. For eukaryotic cells, the initial RNA transcript (the mRNA precursor) is often spliced, removing introns that do not encode protein sequences. Often, the same pre-mRNA is spliced differently in different cell types or at different developmental stages. In the image at the left, proteins associated with RNA splicing (stained with a fluorescent antibody) highlight regions of the newt genome that are being actively transcribed. [(Left) courtesy of Mark B. Roth and Joseph G. Gall.]

DNA stores genetic information in a stable form that can be readily replicated. However, the expression of this genetic information requires its flow from DNA to RNA to protein, as was introduced in Chapter 5. The present chapter deals with how RNA is synthesized and spliced. We begin with transcription in *Escherichia coli* and focus on three questions: What are the properties of promoters (the DNA sites at which RNA transcription is initiated), and how do the promoters function? How do RNA polymerase, the DNA template, and the nascent RNA chain interact with one another? How is transcription terminated?

We then turn to transcription in eukaryotes, beginning with promoter structure and the transcription-factor proteins that regulate promoter activity. A distinctive feature of eukaryotic DNA templates is the presence of enhancer sequences that can stimulate transcriptional initiation more than a thousand base pairs away from the start site. Primary transcripts in eukaryotes are extensively modified, as exemplified by the capping of the 5′ end of an mRNA precursor and the addition of a long poly(A) tail to its 3′ end. Most striking is the splicing of mRNA precursors, which is catalyzed by spliceosomes consisting of small nuclear ribonucleoprotein particles (snRNPs). The small nuclear RNA (snRNA) molecules in these complexes play a key role in directing the alignment of splice sites and in mediating catalysis. Indeed, some RNA molecules can splice themselves in the absence of protein. This landmark discovery by Thomas Cech and Sidney Altman revealed that RNA molecules can serve as catalysts and greatly influenced our view of molecular evolution.

RNA splicing is not merely a curiosity. Approximately 15% of all genetic diseases are caused by mutations that affect RNA splicing. Morever, the same pre-mRNA can be spliced differently in various cell types, at different stages of development, or in response to other biological signals. In addition, individual bases in some pre-mRNA molecules are changed, in a process called *RNA editing*. One of the biggest surprises of the sequencing of the human genome was that only approximately 40,000 genes were identified compared with previous estimates of 100,000 or more. The ability of one gene to encode more than one distinct mRNA and, hence, more than one protein may play a key role in expanding the repertoire of our genomes.

28.0.1 An Overview of RNA Synthesis

RNA synthesis, or *transcription*, is the process of transcribing DNA nucleotide sequence information into RNA sequence information. RNA synthesis is catalyzed by a large enzyme called *RNA polymerase*. The basic biochemistry of RNA synthesis is common to prokaryotes and eukaryotes, although its regulation is more complex in eukaryotes. The close connection between prokaryotic and eukaryotic transcription has been beautifully illustrated by the recently determined three-dimensional structures of representative RNA polymerases from prokaryotes and eukaryotes (Figure 28.1). Despite substantial differences in size and number of polypeptide subunits, the overall structures of these enzymes are quite similar, revealing a common evolutionary origin.

RNA synthesis, like nearly all biological polymerization reactions, takes place in three stages: *initiation*, *elongation*, and *termination*. RNA polymerase performs multiple functions in this process:

1. It searches DNA for initiation sites, also called *promoter sites* or simply *promoters*. For instance, *E. coli* DNA has about 2000 promoter sites in its 4.8×10^6 bp genome. Because these sequences are on the *same* molecule of DNA as the genes being transcribed, they are called *cis-acting elements*.

2. It unwinds a short stretch of double-helical DNA to produce a single-stranded DNA template from which it takes instructions.

FIGURE 28.1 RNA polymerase structures. The three-dimensional structures of RNA polymerases from a prokaryote *(Thermus aquaticus)* and a eukaryote *(Saccharoromyces cerevisiae).* The two largest subunits for each structure are shown in dark red and dark blue. The similarity of these structures reveals that these enzymes have the same evolutionary origin and have many mechanistic features in common.

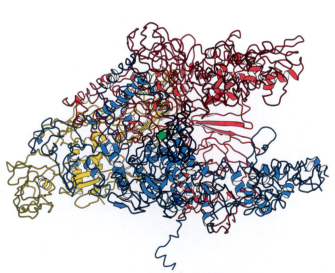

Prokaryotic RNA polymerase

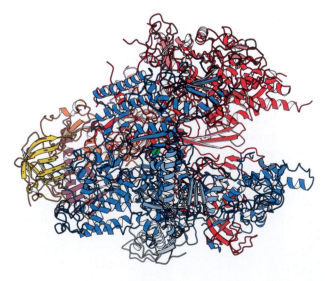

Eukaryotic RNA polymerase

3. It selects the correct ribonucleoside triphosphate and catalyzes the formation of a phosphodiester bond. This process is repeated many times as the enzyme moves unidirectionally along the DNA template. RNA polymerase is completely processive—a transcript is synthesized from start to end by a single RNA polymerase molecule.

4. It detects termination signals that specify where a transcript ends.

5. It interacts with activator and repressor proteins that modulate the rate of transcription initiation over a wide dynamic range. These proteins, which play a more prominent role in eukaryotes than in prokaryotes, are called *transcription factors* or *trans-acting elements*. Gene expression is controlled mainly at the level of transcription, as will be discussed in detail in Chapter 31.

The fundamental reaction of RNA synthesis is the formation of a phosphodiester bond. The 3'-hydroxyl group of the last nucleotide in the chain nucleophilically attacks the α-phosphate group of the incoming nucleoside triphosphate with the concomitant release of a pyrophosphate (see Figure 5.25). This reaction is thermodynamically favorable, and the subsequent degradation of the pyrophosphate to orthophosphate locks the reaction in the direction of RNA synthesis.

The chemistry of RNA synthesis is identical for all forms of RNA, including messenger RNA, transfer RNA, and ribosomal RNA. The basic steps just outlined also apply to all forms. Their synthetic processes differ mainly in regulation, posttranscriptional processing, and the specific polymerase that participates.

28.1 TRANSCRIPTION IS CATALYZED BY RNA POLYMERASE

We begin our consideration of transcription by examining the process in bacteria such as *E. coli*. RNA polymerase from *E. coli* is a very large (~400 kd) and complex enzyme consisting of four kinds of subunits (Table 28.1). The subunit composition of the entire enzyme, called the *holoenzyme*, is $\alpha_2\beta\beta'\sigma$. The σ subunit helps find a promoter site where transcription begins, participates in the initiation of RNA synthesis, and then dissociates from the rest of the enzyme. RNA polymerase without this subunit ($\alpha_2\beta\beta'$) is called the *core enzyme*. The core enzyme contains the catalytic site.

This catalytic site resembles that of DNA polymerase (Secion 27.2.2) in that it includes two metal ions in its active form (Figure 28.2). One metal ion remains bound to

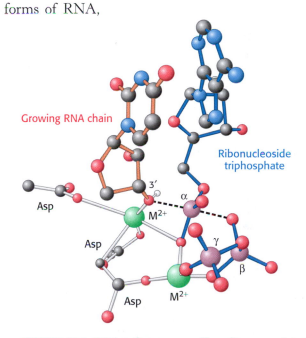

Growing RNA chain

Ribonucleoside triphosphate

FIGURE 28.2 **RNA polymerase active site.** A model of the transition state for phosphodiester-bond formation in the active site of RNA polymerase. The 3'-hydroxyl group of the growing RNA chain attacks the α-phosphate of the incoming nucleoside triphosphate. This transition state is structurally similar to that in DNA polymerase (see Figure 27.12).

TABLE 28.1	Subunits of RNA polymerase from *E. coli*		
Subunit	**Gene**	**Number**	**Mass (kd)**
α	*rpoA*	2	37
β	*rpoB*	1	151
β'	*rpoC*	1	155
σ^{70}	*rpoD*	1	70

the enzyme, whereas the other appears to come in with the nucleoside triphosphate and leave with the pyrophosphate. Three conserved aspartate residues of the enzyme participate in binding these metal ions. Note that the overall structures of DNA polymerase and RNA polymerase are quite different; their similar active sites are the products of convergent evolution.

28.1.1 Transcription Is Initiated at Promoter Sites on the DNA Template

Transcription starts at *promoters* on the DNA template. *Promoters are sequences of DNA that direct the RNA polymerase to the proper initiation site for transcription.* Promoter sites can be identified and characterized by a combination of techniques. One powerful technique for characterizing these and other protein-binding sites on DNA is called *footprinting* (Figure 28.3). First, one of the strands of a DNA fragment under investigation is labeled on one end with ^{32}P. RNA polymerase is added to the labeled DNA, and *the complex is digested with DNAse just long enough to make an average of one cut in each chain.* A part of the radioactive DNA is treated in the same way but without the addition of RNA polymerase to serve as a control. The resulting DNA fragments are separated according to size by electrophoresis. The gel pattern is highly revealing: a series of bands present in the control sample is absent from the sample containing RNA polymerase. These bands are missing because RNA polymerase shields DNA from cleavages that would give rise to the corresponding fragments.

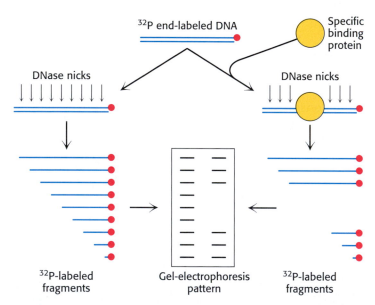

FIGURE 28.3 Footprinting. One end of a DNA chain is labeled with ^{32}P (shown as a red circle). This labeled DNA is then treated with DNAse I such that each fragment is cut only once. The same cleavage is carried out after a protein that binds to specific sites on the DNA has been added. The bound protein protects a segment on the DNA from the action of DNAse I. Hence, certain fragments present in the reaction without protein will be missing. These missing bands in the gel pattern identify the binding site on DNA.

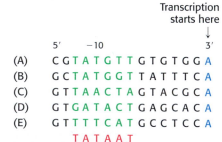

	5′	−10		3′
(A)	C G	T A T G T T	G T G T G G	A
(B)	G C	T A T G G T	T A T T T C	A
(C)	G T	T A A C T A	G T A C G C	A
(D)	G T	G A T A C T	G A G C A C	A
(E)	G T	T T T C A T	G C C T C C	A
		T A T A A T		

FIGURE 28.4 Prokaryotic promoter sequences. A comparison of five sequences from prokaryotic promoters reveals a recurring sequence of TATAAT centered on position −10. The −10 consensus sequence (in red) was deduced from a large number of promoter sequences. The sequences are from the (A) *lac,* (B) *gal,* and (C) *trp* operons of *E. coli;* (D) from λ phage; and (E) from φX174 phage.

A striking pattern is evident when the sequences of many prokaryotic promoters are compared. *Two common motifs are present on the 5′ (upstream) side of the start site.* They are known as the *−10 sequence* and the *−35 sequence* because they are centered at about 10 and 35 nucleotides upstream of the start site. These sequences are each 6 bp long. Their *consensus (average) sequences,* deduced from analyses of many promoters (Figure 28.4), are

```
      -35              -10         +1
5'~~~TTGACA~~~~~~~~TATAAT~~~ Start
                              site
```

The first nucleotide (the start site) of a transcribed DNA sequence is denoted as +1 and the second one as +2; the nucleotide preceding the start site is denoted as −1. These designations refer to the coding strand of DNA. Recall that the sequence of the *template strand of DNA* is the *complement* of that of the RNA transcript (see Figure 5.26). In contrast, the *coding strand of DNA* has the *same* sequence as that of the RNA transcript except for thymine (T) in place of uracil (U). The coding strand is also known as the *sense* (+) *strand*, and the template strand as the *antisense* (−) *strand*.

Promoters differ markedly in their efficacy. Genes with strong promoters are transcribed frequently—as often as every 2 seconds in *E. coli*. In contrast, genes with very weak promoters are transcribed about once in 10 minutes. The −10 and −35 regions of most strong promoters have sequences that correspond closely to the consensus sequences, whereas weak promoters tend to have multiple substitutions at these sites. Indeed, mutation of a single base in either the −10 sequence or the −35 sequence can diminish promoter activity. The distance between these conserved sequences also is important; a separation of 17 nucleotides is optimal. Thus, *the efficiency or strength of a promoter sequence serves to regulate transcription.* Regulatory proteins that bind to specific sequences near promoter sites and interact with RNA polymerase (Chapter 31) also markedly influence the frequency of transcription of many genes.

28.1.2 Sigma Subunits of RNA Polymerase Recognize Promoter Sites

The $\alpha_2\beta\beta'$ core of RNA polymerase is unable to start transcription at promoter sites. Rather, the complete $\alpha_2\beta\beta'\sigma$ holoenzyme is essential for initiation at the correct start site. The σ *subunit* contributes to specific initiation in two ways. First, *it decreases the affinity of RNA polymerase for general regions of DNA by a factor of 10^4.* In its absence, the core enzyme binds DNA indiscriminately and tightly. Second, the σ subunit enables *RNA polymerase to recognize promoter sites.* A large fragment of a σ subunit was found to have an α helix on its surface; this helix has been implicated in recognizing the 5'-TATAAT sequence of the −10 region (Figure 28.5). The holoenzyme binds to duplex DNA and moves along the double helix in search of a promoter, forming transient hydrogen bonds with exposed hydrogen-donor and -acceptor groups on the base pairs. The search is rapid because RNA polymerase slides along DNA instead of repeatedly binding and dissociating from it. *In other words, the promoter site is encountered by a random walk in one dimension rather than in three dimensions.* The observed rate constant for the binding of RNA polymerase holoenzyme to promoter sequences is 10^{10} M^{-1} s^{-1}, more than 100 times as fast as could be accomplished by repeated encounters moving on and off the DNA. The σ subunit is released when the nascent RNA chain reaches nine or ten nucleotides in length. After its release, it can assist initiation by another core enzyme. Thus, the σ *subunit acts catalytically.*

E. coli contains multiple σ factors to recognize several types of promoter sequences contained in *E. coli* DNA. The type that recognizes the consensus sequences described earlier is called σ^{70} because it has a mass of 70 kd. A different σ factor comes into play when the temperature is raised abruptly. *E. coli* responds by synthesizing σ^{32}, which recognizes the promoters of *heat-shock genes.* These promoters exhibit −10 sequences that are somewhat

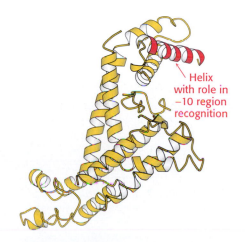

Helix with role in −10 region recognition

FIGURE 28.5 Structure of the σ subunit. The structure of a fragment from the *E. coli* subunit σ^{70} reveals the position of an α helix on the protein surface; this helix plays an important role in binding to the −10 TATAAT sequence.

```
                  −35                    −10
        5′~~~T T G A C A~~~~~~~~~T A T A A T~~~~3′     Standard promoter
5′~~~T N N C N C N C T T G A A~~~~~~~~C C C A T N~~~~3′     Heat-shock promoter
        5′~~~C T G G G N A~~~~~~~~~T T G C A~~~~3′     Nitrogen-starvation promoter
```

FIGURE 28.6 Alternative promoter sequences. A comparison of the consensus sequences of standard promoters, heat-shock promoters, and nitrogen-starvation promoters of *E. coli*. These promoters are recognized by σ^{70}, σ^{32}, and σ^{54}, respectively.

different from the −10 sequence for standard promoters (Figure 28.6). The increased transcription of heat-shock genes leads to the coordinated synthesis of a series of protective proteins. Other σ factors respond to environmental conditions, such as nitrogen starvation. These findings demonstrate that *σ plays a key role in determining where RNA polymerase initiates transcription.*

28.1.3 RNA Polymerase Must Unwind the Template Double Helix for Transcription to Take Place

Although RNA polymerase can search for promoter sites when bound to double-helical DNA, a segment of the helix must be unwound before synthesis can begin. A region of duplex DNA must be unpaired so that nucleotides on one of its strands become accessible for base-pairing with incoming ribonucleoside triphosphates. The DNA template strand selects the correct ribonucleoside triphosphate by forming a Watson-Crick base pair with it (Section 5.2.1), as in DNA synthesis.

How much of the template DNA is unwound by the polymerase? Because unwinding increases the negative supercoiling of the DNA (Section 27.3.2), this question was answered by analyzing the supercoiling of a circular duplex DNA exposed to varying amounts of RNA polymerase. Topoisomerase I, an enzyme catalyzing the concerted cleavage and resealing of duplex DNA (Section 27.3.3), was then added to relax the part of circular DNA not in contact with polymerase molecules. These DNA samples were analyzed by gel electrophoresis after the removal of bound protein. *The degree of negative supercoiling increased in proportion to the number of RNA polymerase molecules bound per template DNA, showing that the enzyme unwinds DNA. Each bound polymerase molecule unwinds a 17-bp segment of DNA, which corresponds to 1.6 turns of B-DNA helix* (Figure 28.7).

Negative supercoiling of circular DNA favors the transcription of genes because it facilitates unwinding (Section 27.3.2). Thus, the introduction of negative supercoils into DNA by topoisomerase II can increase the efficiency of promoters located at distant sites. However, not all promoter sites are stimulated by negative supercoiling. The promoter site for topoisomerase II itself is a noteworthy exception. Negative supercoiling decreases the rate of transcription of this gene, an elegant feedback control ensuring that DNA does not become excessively supercoiled. Negative supercoiling could decrease the efficiency of this promoter by changing the structural relation of the −10 and −35 regions.

The transition from the *closed promoter complex* (in which DNA is double helical) to the *open promoter complex* (in which a DNA segment is unwound) is an essential event in transcription. The stage is now set for the formation of the first phosphodiester bond of the new RNA chain.

28.1.4 RNA Chains Are Formed de Novo and Grow in the 5′-to-3′ Direction

In contrast with DNA synthesis, *RNA synthesis can start de novo, without the requirement for a primer.* Most newly synthesized RNA chains carry a

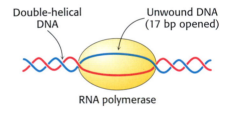

Double-helical DNA   Unwound DNA (17 bp opened)

RNA polymerase

FIGURE 28.7 DNA unwinding. RNA polymerase unwinds about 17 base pairs of template DNA.

highly distinctive tag on the 5′ end: the first base at that end is either *pppG* or *pppA*.

The presence of the triphosphate moiety suggests that RNA synthesis starts at the 5′ end. The results of labeling experiments with γ-^{32}P substrates confirmed that RNA chains, like DNA chains, grow in the 5′ → 3′ direction.

5′ → 3′ growth

28.1.5 Elongation Takes Place at Transcription Bubbles That Move Along the DNA Template

The elongation phase of RNA synthesis begins after the formation of the first phosphodiester bond. An important change is the loss of σ; recall that the core enzyme without σ binds more strongly to the DNA template. Indeed, RNA polymerase stays bound to its template until a termination signal is reached. The region containing RNA polymerase, DNA, and nascent RNA is called a *transcription bubble* because it contains a locally melted "bubble" of DNA (Figure 28.8). The newly synthesized RNA forms a hybrid helix with the template DNA strand. This RNA–DNA helix is about 8 bp long, which corresponds to nearly one turn of a double helix (Section 27.1.3). The 3′-hydroxyl group of the RNA in this hybrid helix is positioned so that it can attack the α-phosphorus atom of an incoming ribonucleoside triphosphate. The core enzyme also contains a binding site for the other DNA strand. About 17 bp of DNA are unwound throughout the elongation phase, as in the initiation phase. The transcription bubble moves a distance of 170 Å (17 nm) in a second, which corresponds to a rate of elongation of about 50 nucleotides per second. Although rapid, it is much slower than the rate of DNA synthesis, which is 800 nucleotides per second.

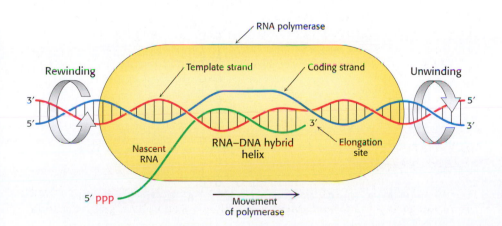

FIGURE 28.8 Transcription bubble. A schematic representation of a transcription bubble in the elongation of an RNA transcript. Duplex DNA is unwound at the forward end of RNA polymerase and rewound at its rear end. The RNA–DNA hybrid rotates during elongation.

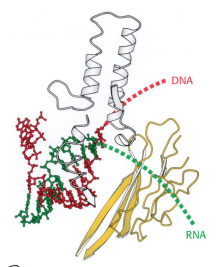

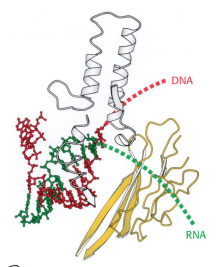 **FIGURE 28.9 RNA–DNA hybrid separation.** A structure within RNA polymerase forces the separation of the RNA–DNA hybrid, allowing the DNA strand to exit in one direction and the RNA product to exit in another.

The lengths of the RNA–DNA hybrid and of the unwound region of DNA stay rather constant as RNA polymerase moves along the DNA template. This finding indicates that DNA is rewound at about the same rate at the rear of RNA polymerase as it is unwound at the front of the enzyme. The RNA–DNA hybrid must also rotate each time a nucleotide is added so that the 3'-OH end of the RNA stays at the catalytic site. The length of the RNA–DNA hybrid is determined by a structure within the enzyme that forces the RNA–DNA hybrid to separate, allowing the RNA chain to exit from the enzyme and the DNA chain to rejoin its DNA partner (Figure 28.9).

It is noteworthy that RNA polymerase lacks nuclease activity. Thus, in contrast with DNA polymerase, *RNA polymerase does not correct the nascent polynucleotide chain. Consequently, the fidelity of transcription is much lower than that of replication.* The error rate of RNA synthesis is of the order of one mistake per 10^4 or 10^5 nucleotides, about 10^5 times as high as that of DNA synthesis. The much lower fidelity of RNA synthesis can be tolerated because mistakes are not transmitted to progeny. For most genes, many RNA transcripts are synthesized, a few defective transcripts are unlikely to be harmful.

28.1.6 An RNA Hairpin Followed by Several Uracil Residues Terminates the Transcription of Some Genes

The termination of transcription is as precisely controlled as its initiation. In the termination phase of transcription, the formation of phosphodiester bonds ceases, the RNA–DNA hybrid dissociates, the melted region of DNA rewinds, and RNA polymerase releases the DNA. What determines where transcription is terminated? *The transcribed regions of DNA templates contain stop signals.* The simplest one is a *palindromic GC-rich region followed by an AT-rich region.* The RNA transcript of this DNA palindrome is self-complementary (Figure 28.10). Hence, its bases can pair to form a hairpin structure with a stem and loop, a structure favored by its high content of G and C residues. Guanine–cytosine base pairs are more stable than adenine–thymine pairs because of the extra hydrogen bond in the base pair. This stable hairpin is followed by a sequence of four or more uracil residues, which also are crucial for termination. The RNA transcript ends within or just after them.

How does this combination hairpin-oligo(U) structure terminate transcription? First, it seems likely that RNA polymerase pauses immediately after it has synthesized a stretch of RNA that folds into a hairpin. Furthermore, the RNA–DNA hybrid helix produced after the hairpin is unstable

FIGURE 28.10 Termination signal. A termination signal found at the 3' end of an mRNA transcript consists of a series of bases that form a stable stem-loop structure and a series of U residues.

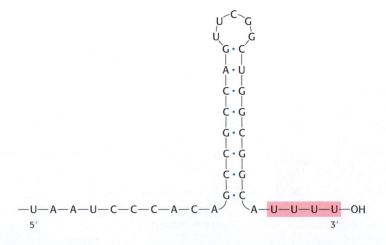

because its rU–dA base pairs are the weakest of the four kinds. Hence, the pause in transcription caused by the hairpin permits the weakly bound *nascent RNA to dissociate from the DNA template and then from the enzyme.* The solitary DNA template strand rejoins its partner to re-form the DNA duplex, and the transcription bubble closes.

28.1.7 The Rho Protein Helps Terminate the Transcription of Some Genes

RNA polymerase needs no help to terminate transcription at a hairpin followed by several U residues. At other sites, however, termination requires the participation of an additional factor. This discovery was prompted by the observation that some RNA molecules synthesized in vitro by RNA polymerase acting alone are *longer* than those made in vivo. The missing factor, a protein that caused the correct termination, was isolated and named *rho* (ρ). Additional information about the action of ρ was obtained by adding this termination factor to an incubation mixture at various times after the initiation of RNA synthesis (Figure 28.11). RNAs with sedimentation coefficients of 10S, 13S, and 17S were obtained when ρ was added at initiation, a few seconds after initiation, and 2 minutes after initiation, respectively. If no ρ was added, transcription yielded a 23S RNA product. It is evident that the template contains at least three termination sites that respond to ρ (yielding 10S, 13S, and 17S RNA) and one termination site that does not (yielding 23S RNA). Thus, specific termination at a site producing 23S RNA can occur in the absence of ρ. However, ρ detects additional termination signals that are not recognized by RNA polymerase alone.

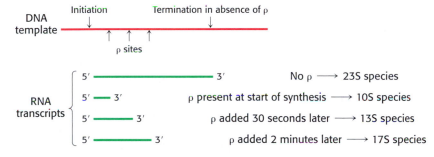

FIGURE 28.11 Effect of ρ protein on the size of RNA transcripts.

How does ρ provoke the termination of RNA synthesis? A key clue is the finding that *ρ hydrolyzes ATP in the presence of single-stranded RNA but not in the presence of DNA or duplex RNA.* Hexameric ρ, which is structurally similar to and homologous to ATP synthase (Section 18.4.1), specifically binds single-stranded RNA; a stretch of 72 nucleotides is bound in such a way that the RNA passes through the center of the structure (Figure 28.12). *Rho is brought into action by sequences located in the nascent RNA that are rich in cytosine and poor in guanine.* The ATPase activity of ρ enables the protein to pull the nascent RNA while pursuing RNA polymerase. When ρ catches RNA polymerase at the transcription bubble, it breaks the RNA–DNA hybrid helix by functioning as an RNA–DNA helicase. Given the structural and evolutionary connection, it is possible that the mechanism of action of ρ is similar to that of ATP synthase, with the single-stranded RNA playing the role of the γ subunit.

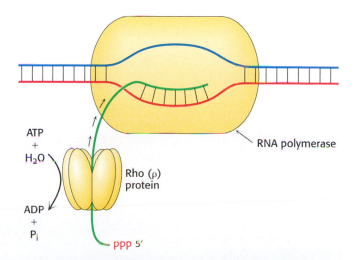

FIGURE 28.12 Mechanism for the termination of transcription by ρ protein. This protein is an ATP-dependent helicase that binds the nascent RNA chain and pulls it away from RNA polymerase and the DNA template.

Proteins in addition to ρ mediate and modulate termination. For example, the *nusA protein* enables RNA polymerase in *E. coli* to recognize a characteristic class of termination sites. In *E. coli*, specialized termination signals called *attenuators* are regulated to meet the nutritional needs of the cell (Section 31.4.1). *A common feature of protein-independent and protein-dependent termination is that the functioning signals lie in newly synthesized RNA rather than in the DNA template.*

28.1.8 Precursors of Transfer and Ribosomal RNA Are Cleaved and Chemically Modified After Transcription

In prokaryotes, messenger RNA molecules undergo little or no modification after synthesis by RNA polymerase. Indeed, many mRNA molecules are translated while they are being transcribed. In contrast, *transfer RNA and ribosomal RNA molecules are generated by cleavage and other modifications of nascent RNA chains.* For example, in *E. coli*, three kinds of rRNA molecules and a tRNA molecule are excised from a single primary RNA transcript that also contains spacer regions (Figure 28.13). Other transcripts contain arrays of several kinds of tRNA or of several copies of the same tRNA. The nucleases that cleave and trim these precursors of rRNA and tRNA are highly precise. *Ribonuclease P*, for example, generates the correct 5′ terminus of all tRNA molecules in *E. coli*. This interesting enzyme contains a catalytically active RNA molecule. *Ribonuclease III* excises 5S, 16S, and 23S rRNA precursors from the primary transcript by cleaving double-helical hairpin regions at specific sites.

A second type of processing is the *addition of nucleotides to the termini of some RNA chains.* For example, CCA, a terminal sequence required for the function of all tRNAs, is added to the 3′ ends of tRNA molecules that do not already possess this terminal sequence. A third type of processing is the *modification of bases and ribose units* of ribosomal RNAs. In prokaryotes, some bases of rRNA are methylated. Unusual bases are found in all tRNA molecules (Section 29.1.2). They are formed by the enzymatic modification of a standard ribonucleotide in a tRNA precursor. For example, uridylate residues are modified after transcription to form *ribothymidylate* and *pseudouridylate*. These modifications generate diversity, allowing greater structural and functional versatility.

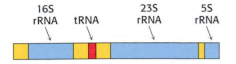

FIGURE 28.13 Primary transcript. Cleavage of this transcript produces 5S, 16S, and 23S rRNA molecules and a tRNA molecule. Spacer regions are shown in yellow.

2′-*O*-Methylribose
(in eukaryotes)

6-Dimethyladenosine
(in prokaryotes)

Uridylate

Ribothymidylate

Pseudouridylate

28.1.9 Antibiotic Inhibitors of Transcription

Many antibiotics are highly specific inhibitors of biological processes. Rifampicin and actinomycin are two antibiotics that inhibit transcription, although in quite different ways. *Rifampicin* is a semisynthetic derivative of *rifamycins*, which are derived from a strain of *Streptomyces*.

Rifampicin

This antibiotic *specifically inhibits the initiation of RNA synthesis*. Rifampicin does not block the binding of RNA polymerase to the DNA template; rather, it interferes with the formation of the first few phosphodiester bonds in the RNA chain. The structure of a complex between a prokaryotic RNA polymerase and rifampicin reveals that the antibiotic blocks the channel into which the RNA–DNA hybrid generated by the enzyme must pass (Figure 28.14). The binding site is 12 Å from the active site itself. Rifampicin does not hinder chain elongation once initiated, because the RNA–DNA hybrid present in the enzyme prevents the antibiotic from binding.

Actinomycin D, a polypeptide-containing antibiotic from a different strain of *Streptomyces*, inhibits transcription by an entirely different mechanism. *Actinomycin D binds tightly and specifically to double-helical DNA and thereby prevents it from being an effective template for RNA synthesis.* It does not bind to single-stranded DNA or RNA, double-stranded RNA, or RNA–DNA hybrids. The results of spectroscopic and hydrodynamic studies of complexes of actinomycin D and DNA suggested that the phenoxazone ring of actinomycin slips in between neighboring base pairs in DNA.

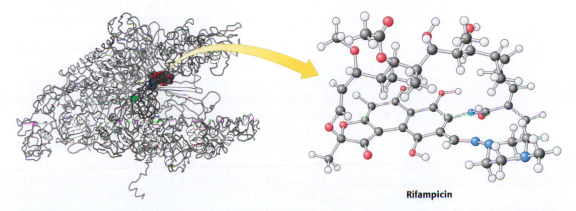

Rifampicin

FIGURE 28.14 Antibiotic action. Rifampicin binds to a pocket in the channel that is normally occupied by the newly formed RNA–DNA hybrid. Thus the antibiotic blocks elongation after only two or three nucleotides have been added.

This mode of binding is called *intercalation*. At low concentrations, actinomycin D inhibits transcription without significantly affecting DNA replication or protein synthesis. Hence, *actinomycin D is extensively used as a highly specific inhibitor of the formation of new RNA in both prokaryotic and eukaryotic cells.* Its ability to inhibit the growth of rapidly dividing cells makes it an effective therapeutic agent in the treatment of some cancers.

28.2 EUKARYOTIC TRANSCRIPTION AND TRANSLATION ARE SEPARATED IN SPACE AND TIME

We turn now to transcription in eukaryotes, a much more complex process than in prokaryotes. *In eukaryotes, transcription and translation take place in different cellular compartments:* transcription takes place in the membrane-bounded nucleus, whereas translation takes place outside the nucleus in the cytoplasm. In prokaryotes, the two processes are closely coupled (Figure 28.15). Indeed, the translation of bacterial mRNA begins while the transcript is still being synthesized. *The spatial and temporal separation of transcription and translation enables eukaryotes to regulate gene expression in much more intricate ways, contributing to the richness of eukaryotic form and function.*

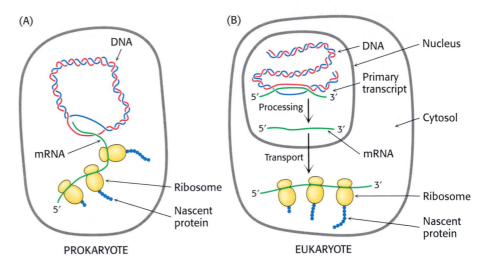

FIGURE 28.15 Transcription and translation. These two processes are closely coupled in prokaryotes, whereas they are spacially and temporally separate in eukaryotes. (A) In prokaryotes, the primary transcript serves as mRNA and is used immediately as the template for protein synthesis. (B) In eukaryotes, mRNA precursors are processed and spliced in the nucleus before being transported to the cytosol for translation into protein. [After J. Darnell, H. Lodish, and D. Baltimore. *Molecular Cell Biology*, 2d ed. (Scientific American Books, 1990), p. 230.]

A second major difference between prokaryotes and eukaryotes is the extent of RNA processing. Although both prokaryotes and eukaryotes modify tRNA and rRNA, *eukaryotes very extensively process nascent RNA destined to become mRNA.* Primary transcripts (pre-mRNA molecules), the products of RNA polymerase action, acquire a cap at their 5′ ends and a poly(A) tail at their 3′ ends. Most importantly, *nearly all mRNA precursors in higher eukaryotes are spliced* (Section 5.6.1). Introns are precisely excised from primary transcripts, and exons are joined to form mature mRNAs with continuous messages. Some mRNAs are only a tenth the size of their precursors, which can be as large as 30 kb or more. The pattern of splicing can be regulated in the course of development to generate variations on a theme, such as membrane-bound and secreted forms of antibody molecules. Alternative splicing enlarges the repertoire of proteins in eukaryotes and is a clear illustration of why the proteome is more complex than the genome.

28.2.1 RNA in Eukaryotic Cells Is Synthesized by Three Types of RNA Polymerase

In prokaryotes, RNA is synthesized by a single kind of polymerase. In contrast, the nucleus of a eukaryote contains three types of RNA polymerase differing in template specificity, location in the nucleus, and susceptibility to inhibitors (Table 28.2). All these polymerases are large proteins, containing from 8 to 14 subunits and having a total molecular mass greater than 500 kd. *RNA polymerase I* is located in nucleoli, where it transcribes the tandem array of genes for 18S, 5.8S, and 28S ribosomal RNA (Section 29.3.1). The other ribosomal RNA molecule (5S rRNA, Section 29.3.1) and all the transfer RNA molecules (Section 29.1.2) are synthesized by *RNA polymerase III*, which is located in the nucleoplasm rather than in nucleoli. *RNA polymerase II*, which also is located in the nucleoplasm, synthesizes the precursors of messenger RNA as well as several small RNA molecules, such as those of the splicing apparatus (Section 28.3.5). Although all eukaryotic RNA polymerases are homologous to one another and to prokaryotic RNA polymerase, RNA polymerase II contains a unique carboxyl-terminal domain on the 220-kd subunit; this domain is unusual because it contains multiple repeats of a YSPTSPS consensus sequence. The activities of RNA polymerase II are regulated by phosphorylation on the serine and threonine residues of the carboxyl-terminal domain. Another major distinction among the polymerases lies in their responses to the toxin α-*amanitin*, a cyclic octapeptide that contains several modified amino acids.

α-Amanitin

α-Amanitin is produced by the poisonous mushroom *Amanita phalloides*, which is also called the *death cup* or the *destroying angel* (Figure 28.16). More than a hundred deaths result worldwide each year from the ingestion of poisonous mushrooms. α-Amanitin binds very tightly ($K_d = 10$ nM) to

FIGURE 28.16 RNA polymerase poison. *Amanita phalloides*, a poisonous mushroom that produces α-amanitin. [After G. Lincoff and D. H. Mitchel, *Toxic and Hallucinogenic Mushroom Poisoning* (Van Nostrand Reinhold, 1977), p. 30.]

TABLE 28.2   Eukaryotic RNA polymerases

Type	Location	Cellular transcripts	Effects of α-amanitin
I	Nucleolus	18S, 5.8S, and 28S rRNA	Insensitive
II	Nucleoplasm	mRNA precursors and snRNA	Strongly inhibited
III	Nucleoplasm	tRNA and 5S rRNA	Inhibited by high concentrations

RNA polymerase II and thereby blocks the elongation phase of RNA synthesis. Higher concentrations of α-amanitin (1 μM) inhibit polymerase III, whereas polymerase I is insensitive to this toxin. This pattern of sensitivity is highly conserved throughout the animal and plant kingdoms.

28.2.2 Cis- And Trans-Acting Elements: Locks and Keys of Transcription

Eukaryotic genes, like their prokaryotic counterparts, require promoters for transcription initiation. Each of the three types of polymerase has distinct promoters. RNA polymerase I transcribes from a single type of promoter, present only in rRNA genes, that encompasses the initiation site. In some genes, RNA polymerase III responds to promoters located in the normal, upstream position; in other genes, it responds to promoters imbedded in the genes, downstream of the initiation site. Promoters for RNA polymerase II can be simple or complex (Section 28.2.3). As is the case for prokaryotes, promoters are always on the same molecule of DNA as the gene they regulate. Consequently, promoters are referred to as *cis-acting elements*.

However, promoters are not the only types of cis-acting DNA sequences. Eukaryotes and their viruses also contain *enhancers*. These DNA sequences, although not promoters themselves, can enormously increase the effectiveness of promoters. Interestingly, the positions of enhancers relative to promoters are not fixed; they can vary substantially. Enhancers play key roles in regulating gene expression in a specific tissue or developmental stage (Section 31.2.4).

The DNA sequences of cis-acting elements are binding sites for proteins called *transcription factors*. Such a protein is sometimes called a *trans-acting factor* because it may be encoded by a gene on a DNA molecule other than that containing the gene being regulated. The binding of a transcription factor to its cognate DNA sequence enables the RNA polymerase to locate the proper initiation site. We will continue our investigation of transcription by examining these cis- and trans-acting elements in turn.

28.2.3 Most Promoters for RNA Polymerase II Contain a TATA Box Near the Transcription Start Site

Promoters for RNA polymerase II, like those for bacterial polymerases, are located on the 5′ side of the start site for transcription. The results of mutagenesis experiments, footprinting studies, and comparisons of many higher eukaryotic genes have demonstrated the importance of several upstream regions. For most genes transcribed by RNA polymerase II, the most important cis-acting element is called the *TATA box* on the basis of its consensus sequence (Figure 28.17). The TATA box is usually centered between positions −30 and −100. Note that the eukaryotic TATA box closely resembles the prokaryotic −10 sequence (TATAAT) but is farther from the start site. The mutation of a single base in the TATA box markedly impairs promoter activity. Thus, the precise sequence, not just a high content of AT pairs, is essential.

The TATA box is necessary but not sufficient for strong promoter activity. Additional elements are located between −40 and −150. Many promoters contain a *CAAT box*, and some contain a *GC box* (Figure 28.18). Constitutive genes (genes that are continuously expressed rather than regulated) tend to have GC boxes in their promoters. The positions of these upstream sequences vary from one promoter to another, in contrast with the quite constant location of the −35 region in prokaryotes. Another difference is that the CAAT box and the GC box can be effective when present

5′ T_{82} A_{97} T_{93} A_{85} A_{63} A_{88} A_{50} 3′
TATA box

FIGURE 28.17 TATA box. Comparisons of the sequences of more than 100 eukaryotic promoters led to the consensus sequence shown. The subscripts denote the frequency (%) of the base at that position.

5′ G G N C A A T C T 3′
CAAT box

5′ G G G C G G 3′
GC box

FIGURE 28.18 CAAT box and GC box. Consensus sequences for the CAAT and GC boxes of eukaryotic promoters for mRNA precursors.

on the template (antisense) strand, unlike the −35 region, which must be present on the coding (sense) strand. These differences between prokaryotes and eukaryotes reflect fundamentally different mechanisms for the recognition of cis-acting elements. The −10 and −35 sequences in prokaryotic promoters correspond to binding sites for RNA polymerase and its associated σ factor. In contrast, the TATA, CAAT, and GC boxes and other cis-acting elements in eukaryotic promoters are recognized by proteins other than RNA polymerase itself.

28.2.4 The TATA-Box-Binding Protein Initiates the Assembly of the Active Transcription Complex

Cis-acting elements constitute only part of the puzzle of eukaryotic gene expression. Transcription factors that bind to these elements also are required. For example, RNA polymerase II is guided to the start site by a set of transcription factors known collectively as *TFII* (*TF* stands for transcription factor, and *II* refers to RNA polymerase II). Individual TFII factors are called TFIIA, TFIIB, and so on. Initiation begins with the binding of TFIID to the TATA box (Figure 28.19).

The key initial event is the recognition of the TATA box by the TATA-box-binding protein (TBP), a 30-kd component of the 700-kd TFIID complex. TBP binds 10^5 times as tightly to the TATA box as to noncognate sequences; the dissociation constant of the specific complex is approximately 1 nM. TBP is a saddle-shaped protein consisting of two similar domains (Section 7.3.3; Figure 28.20). The TATA box of DNA binds to the concave surface of TBP. This binding induces large conformational changes in the bound DNA. The double helix is substantially unwound to widen its *minor groove*, enabling it to make extensive contact with the antiparallel β strands on the concave side of TBP. Hydrophobic interactions are prominent at this interface. Four phenylalanine residues, for example, are intercalated between base pairs of the TATA box. The flexibility of AT-rich sequences is generally exploited here in bending the DNA. Immediately outside the TATA box, classical B-DNA resumes. This complex is distinctly asymmetric. The asymmetry is crucial for specifying a unique start site and ensuring that transcription proceeds unidirectionally.

TBP bound to the TATA box is the heart of the initiation complex (see Figure 28.19). The surface of the TBP saddle provides docking sites for the binding of other components (Figure 28.21). Additional transcription factors assemble on this nucleus in a defined sequence. TFIIA is recruited, followed by TFIIB and then TFIIF—an ATP-dependent helicase that initially separates the DNA duplex for the polymerase. Finally, RNA polymerase II and then TFIIE join the other factors to form a complex called the *basal transcription apparatus*. Sometime in the formation of this complex, the carboxyl-terminal domain of the polymerase is phosphorylated on the serine and threonine residues, a process required for successful initiation. The importance of the carboxyl-terminal domain is highlighted by the finding that yeast containing mutant polymerase II with fewer than 10 repeats is not viable. Most of the factors are released before the polymerase leaves the promoter and can then participate in another round of initiation.

Although bacteria lack TBP, archaea utilize a TBP molecule that is structurally quite similar to the eukaryotic protein. In fact, transcriptional control processes in archaea are, in general, much more similar to those in eukaryotes than are the processes in bacteria. Many components of the eukaryotic transcriptional machinery evolved from an ancestor of archaea.

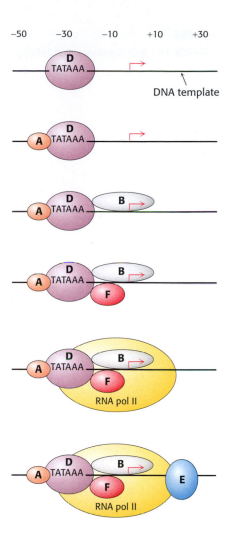

FIGURE 28.19 Transcription initation. Transcription factors TFIIA, B, D, E, and F are essential in initiating transcription by RNA polymerase II. The step-by-step assembly of these general transcription factors begins with the binding of TFIID (purple) to the TATA box. The arrow marks the transcription start site. [After L. Guarente. *Trends Genet.* 8(1992):28.]

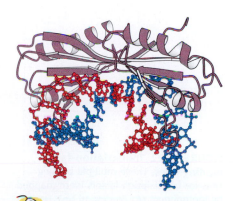

FIGURE 28.20 Complex formed by TATA-box-binding protein and DNA. The saddlelike structure of the protein sits atop a DNA fragment that is both significantly unwound and bent.

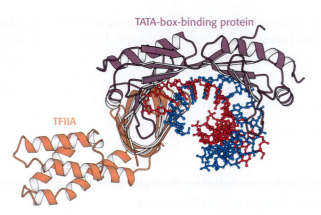

FIGURE 28.21 Assembly of the initiation complex. A ternary complex between the TATA-box-binding protein (purple), TFIIA (orange), and DNA. TFIIA interacts primarily with the other protein.

28.2.5 Multiple Transcription Factors Interact with Eukaryotic Promoters

The basal transcription complex described in Section 28.2.4. initiates transcription at a relatively low frequency. Additional transcription factors that bind to other sites are required to achieve a high rate of mRNA synthesis and to selectively stimulate specific genes. Upstream stimulatory sites in eukaryotic genes are diverse in sequence and variable in position. Their variety suggests that they are recognized by many different specific proteins. Indeed, many transcription factors have been isolated, and their binding sites have been identified by footprinting experiments (Figure 28.22). For example, *Sp1*, an ~100-kd protein from mammalian cells, binds to promoters that contain GC boxes. The duplex DNA of SV40 virus (a cancer-producing virus that infects monkey cells) contains five GC boxes from 50 to 100 bp upstream or downstream of start sites. The *CCAAT-binding transcription factor* (CTF; also called NF1), a 60-kd protein from mammalian cells, binds to the CAAT box. *A heat-shock transcription factor* (HSTF) is expressed in *Drosophila* after an abrupt increase in temperature. This 93-kd DNA-binding protein binds to the consensus sequence

$$5'\text{-CNNGAANNTCCNNG-}3'$$

Several copies of this sequence, known as the *heat-shock response element*, are present starting at a site 15 bp upstream of the TATA box. HSTF differs from σ^{32}, a heat-shock protein of *E. coli* (Section 28.1.2), in binding directly to response elements in heat-shock promoters rather than first becoming associated with RNA polymerase.

FIGURE 28.22 Transcription-factor-binding sites. These multiple binding sites for transcription factors were mapped by footprinting. (A) Binding of Sp1 (green) to the SV40 viral promoter and to the dihydrofolate reductase (DHFR) promoter. (B) Binding of HSTF (blue) to a *Drosophila* heat-shock promoter. [After W. S. Dynan and R. Tjian. *Nature* 316(1985):774.]

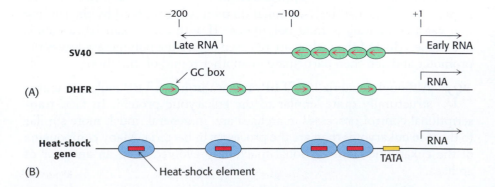

28.2.6 Enhancer Sequences Can Stimulate Transcription at Start Sites Thousands of Bases Away

The activities of many promoters in higher eukaryotes are greatly increased by another type of cis-acting element called an *enhancer*. Enhancers' sequences have no promoter activity of their own *yet can exert their stimulatory actions over distances of several thousand base pairs. They can be upstream, downstream, or even in the midst of a transcribed gene.* Moreover, enhancers are effective when present on *either DNA strand* (equivalently, in either orientation). Enhancers in yeast are known as *upstream activator sequences* (UASs).

A particular enhancer is effective only in certain cells. For example, the immunoglobulin enhancer functions in B lymphocytes but not elsewhere. Cancer can result if the relation between genes and enhancers is disrupted. In Burkitt lymphoma and B-cell leukemia, a chromosomal translocation brings the proto-oncogene *myc* (a transcription factor itself) under the control of a powerful immunoglobin enhancer. The consequent dysregulation of the *myc* gene is believed to play a role in the progression of the cancer.

Transcription factors and other proteins that bind to regulatory sites on DNA can be regarded as passwords that cooperatively open multiple locks, giving RNA polymerase access to specific genes. The discovery of promoters and enhancers has opened the door to understanding how genes are selectively expressed in eukaryotic cells. The regulation of gene transcription, discussed in Chapter 31, is the fundamental means of controlling gene expression.

28.3 THE TRANSCRIPTION PRODUCTS OF ALL THREE EUKARYOTIC POLYMERASES ARE PROCESSED

Virtually all the initial products of transcription are further processed in eukaryotes. For example, tRNA precursors are converted into mature tRNAs by a series of alterations: cleavage of a 5′ leader sequence, splicing to remove an intron, replacement of the 3′-terminal UU by CCA, and modification of several bases (Figure 28.23). A series of enzymes may act on the ribonucleic acid chain or its constituent bases to achieve the final product.

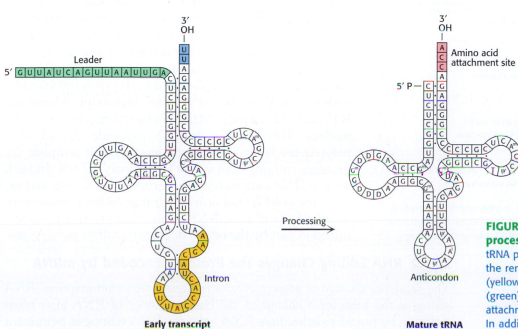

Early transcript → Processing → **Mature tRNA**

FIGURE 28.23 Transfer RNA precursor processing. The conversion of a yeast tRNA precursor into a mature tRNA requires the removal of a 14-nucleotide intron (yellow), the cleavage of a 5′ leader (green), and the removal of UU and the attachment of CCA at the 3′ end (red). In addition, several bases are modified.

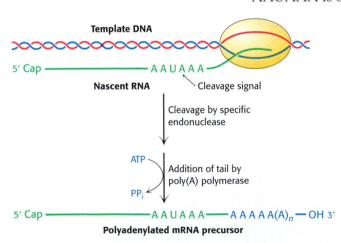

FIGURE 28.24 **Capping the 5′ end.** Caps at the 5′ end of eukaryotic mRNA include 7-methylguanylate (red) attached by a triphosphate linkage to the ribose at the 5′ end. None of the riboses are methylated in cap 0, one is methylated in cap 1, and both are methylated in cap 2.

28.3.1 The Ends of the Pre-mRNA Transcript Acquire a 5′ Cap and a 3′ Poly(A) Tail

Perhaps the most extensively modified transcription product is the product of RNA polymerase II: the majority of this RNA will be processed to mRNA. The immediate product of an RNA polymerase is sometimes referred to as *pre-mRNA*. Most pre-mRNA molecules are spliced to remove the introns. Moreover, both the 5′ and the 3′ ends are modified, and both modifications are retained as the pre-mRNA is converted into mRNA (Section 28.3.3). As in prokaryotes, eukaryotic transcription usually begins with A or G. However, the 5′ triphosphate end of the nascent RNA chain is immediately modified. First, a phosphate is released by hydrolysis. The diphosphate 5′ end then attacks the α-phosphorus atom of GTP to form a very unusual 5′–5′ triphosphate linkage. This distinctive terminus is called a *cap* (Figure 28.24). The N-7 nitrogen of the terminal guanine is then methylated by *S*-adenosylmethionine to form *cap 0*. The adjacent riboses may be methylated to form *cap 1* or *cap 2*. Transfer RNA and ribosomal RNA molecules, in contrast with messenger RNAs and small RNAs that participate in splicing, do not have caps. Caps contribute to the stability of mRNAs by protecting their 5′ ends from phosphatases and nucleases. In addition, caps enhance the translation of mRNA by eukaryotic protein-synthesizing systems (Section 29.5).

As mentioned earlier, pre-mRNA is also modified at the 3′ end. *Most eukaryotic mRNAs contain a polyadenylate, poly(A), tail at that end*, added after transcription has ended. Thus, DNA does not encode this poly(A) tail. Indeed, the nucleotide preceding poly(A) is not the last nucleotide to be transcribed. Some primary transcripts contain hundreds of nucleotides beyond the 3′ end of the mature mRNA.

How is the 3′ end of the pre-mRNA given its final form? *Eukaryotic primary transcripts are cleaved by a specific endonuclease that recognizes the sequence AAUAAA* (Figure 28.25). Cleavage does not occur if this sequence or a segment of some 20 nucleotides on its 3′ side is deleted. The presence of internal AAUAAA sequences in some mature mRNAs indicates that AAUAAA is only part of the cleavage signal; its context also is important.

After cleavage by the endonuclease, a *poly(A) polymerase* adds about 250 adenylate residues to the 3′ end of the transcript; ATP is the donor in this reaction.

The role of the poly(A) tail is still not firmly established despite much effort. However, evidence that it enhances translation efficiency and the stability of mRNA is accumulating. Blocking the synthesis of the poly(A) tail by exposure to *3′-deoxyadenosine (cordycepin)* does not interfere with the synthesis of the primary transcript. Messenger RNA devoid of a poly(A) tail can be transported out of the nucleus. However, an mRNA molecule devoid of a poly(A) tail is usually a much less effective template for protein synthesis than is one with a poly(A) tail. Indeed, some mRNAs are stored in an unadenylated form and receive the poly(A) tail only when translation is imminent. The half-life of an mRNA molecule may also be determined in part by the rate of degradation of its poly(A) tail.

FIGURE 28.25 **Polyadenylation of a primary transcript.** A specific endonuclease cleaves the RNA downstream of AAUAAA. Poly(A) polymerase then adds about 250 adenylate residues.

28.3.2 RNA Editing Changes the Proteins Encoded by mRNA

The sequence content of some mRNAs is altered after transcription. *RNA editing* is the term for a change in the base sequence of RNA after transcription by processes other than RNA splicing. RNA editing is prominent in some systems already discussed. *Apolipoprotein B (apo B)* plays an important role in the transport of triacylglycerols and cholesterol by forming an amphipathic spherical shell around the lipids carried in lipoprotein particles (Section 26.3.1). Apo B exists in two forms, a 512-kd *apo B-100* and a 240-kd *apo B-48*. The larger form, synthesized by the liver, participates in the transport of lipids synthesized in the cell. The smaller form, synthesized by the small intestine, carries dietary fat in the form of chylomicrons. Apo B-48 contains the 2152 N-terminal residues of the 4536-residue apo B-100. This truncated molecule can form lipoprotein particles but cannot bind to the low-density-lipoprotein receptor on cell surfaces. What is the biosynthetic relation of these two forms of apo B? One possibility a priori is that apo B-48 is produced by proteolytic cleavage of apo B-100, and another is that the two forms arise from alternative splicing (see Section 28.3.6). The results of experiments show that neither occurs. A totally unexpected and new mechanism for generating diversity is at work: *the changing of the nucleotide sequence of mRNA after its synthesis* (Figure 28.26). *A specific cytidine residue of mRNA is deaminated to uridine, which changes the codon at residue 2153 from CAA (Gln) to UAA (stop).* The deaminase that catalyzes this reaction is present in the small intestine, but not in the liver, and is expressed only at certain developmental stages.

RNA editing is not confined to apolipoprotein B. Glutamate opens cation-specific channels in the vertebrate central nervous system by binding to receptors in postsynaptic membranes. RNA editing changes a single glutamine codon (CAG) in the mRNA for the glutamate receptor to the codon for arginine (read as CGG). The substitution of Arg for Gln in the receptor prevents Ca^{2+}, but not Na^+, from flowing through the channel. RNA editing is likely much more common than was previously thought. The chemical reactivity of nucleotide bases, including the susceptibility to deamination that necessitates complex DNA-repair mechanisms (Section 27.6.3), has been harnessed as an engine for generating molecular diversity at the RNA and, hence, protein levels.

In trypanosomes (parasitic protozoans), a different kind of RNA editing markedly changes several mitochondrial mRNAs. Nearly half the uridine residues in these mRNAs are *inserted* by RNA editing. A *guide RNA molecule* identifies the sequences to be modified, and a *poly(U) tail* on the guide donates uridine residues to the mRNAs undergoing editing. It is evident that DNA sequences do not always faithfully disclose the sequence of encoded proteins—functionally crucial changes to mRNA can take place.

28.3.3 Splice Sites in mRNA Precursors Are Specified by Sequences at the Ends of Introns

Most genes in higher eukaryotes are composed of exons and introns. The introns must be excised and the exons linked to form the final mRNA in a process called *splicing*. This splicing must be exquisitely sensitive: a one-nucleotide slippage in a splice point would shift the reading frame on the 3′ side of the splice to give an entirely different amino acid sequence. Thus, the correct splice site must be clearly marked. Does a particular sequence denote the splice site? The base sequences of thousands of intron–exon junctions within RNA transcripts are known. In eukaryotes from yeast to mammals, these sequences have a common structural motif: *the base sequence of an intron begins with GU and ends with AG.* The consensus sequence at the 5′ splice in vertebrates is AGGUAAGU (Figure 28.27). At the 3′ end of an

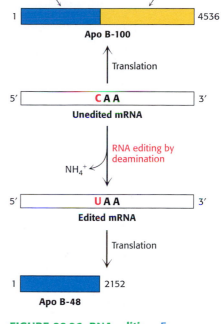

FIGURE 28.26 RNA editing. Enzyme-catalyzed deamination of a specific cytidine residue in the mRNA for apolipoprotein B-100 changes a codon for glutamine (CAA) to a stop codon (UAA). Apolipoprotein B-48, a truncated version of the protein lacking the LDL receptor-binding domain, is generated by this posttranscriptional change in the mRNA sequence. [After P. Hodges and J. Scott. *Trends Biochem. Sci.* 17(1992):77.]

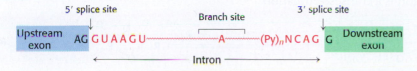

FIGURE 28.27 Splice sites. Consensus sequences for the 5′ splice site and the 3′ splice site are shown. Py stands for pyrimidine.

intron, the consensus sequence is a stretch of *10 pyrimidines* (U or C), followed by any base and then by C, and ending with the invariant AG. Introns also have an important internal site located between 20 and 50 nucleotides upstream of the 3′ splice site; it is called the *branch site* for reasons that will be evident shortly. In yeast, the branch site sequence is nearly always UACUAAC, whereas in mammals a variety of sequences are found.

Parts of introns other than the 5′ and 3′ splice sites and the branch site appear to be less important in determining where splicing takes place. The length of introns ranges from 50 to 10,000 nucleotides. Much of an intron can be deleted without altering the site or efficiency of splicing. Likewise, splicing is unaffected by the insertion of long stretches of DNA into the introns of genes. Moreover, chimeric introns crafted by recombinant DNA methods from the 5′ end of one intron and the 3′ end of a very different intron are properly spliced, provided that the splice sites and branch site are unaltered. In contrast, mutations in each of these three critical regions lead to aberrant splicing.

Despite our knowledge of splice-site sequences, predicting splicing patterns from genomic DNA sequence information remains a challenge. Other information that contributes to splice-site selection is present in DNA sequences, but it is more loosely distributed than are the splice-site sequences themselves.

Aberrant splicing causes some forms of thalassemia, a group of hereditary anemias characterized by the defective synthesis of hemoglobin. In one patient, a mutation of G to A 19 nucleotides away from the normal 3′ splice site of the first intron created a new 3′ splice site (Figure 28.28). The resulting mRNA contains a series of codons not normally present. The sixth codon after the splice is a stop signal for protein synthesis, and so the aberrant protein ends prematurely. *Mutations affecting splice sites have been estimated to cause 15% of all genetic diseases.*

Normal 3′ end
of intron

Normal   5′ CCTATT**GG**TCTATTTTCCACCC**TTAG**GCTGCTG 3′

β-Thalassemia   5′ CCTA**TTAG**TCTATTTTCCACCCTTAGGCTGCTG 3′

FIGURE 28.28 Splicing defects. Mutation of a single base (G to A) in an intron of the β-globin gene leads to thalassemia. This mutation generates a new 3′ splice site (blue) akin to the normal one (yellow) but farther upstream.

28.3.4 Splicing Consists of Two Transesterification Reactions

The splicing of nascent mRNA molecules is a complicated process. It requires the cooperation of several small RNAs and proteins that form a large complex called a *spliceosome*. However, the chemistry of the splicing process is simple. Splicing begins with the cleavage of the phosphodiester bond between the upstream exon (exon 1) and the 5′ end of the intron (Figure 28.29). The attacking group in this reaction is the 2′-hydroxyl group of an adenylate residue in the branch site. A 2′,5′-phosphodiester bond is formed between this A residue and the 5′ terminal phosphate of the intron. This reaction is a transesterification.

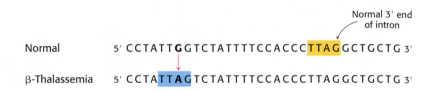

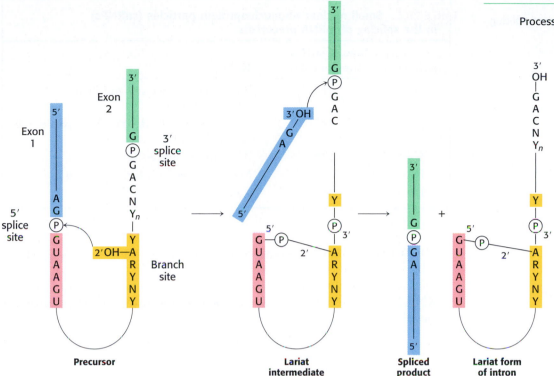

FIGURE 28.29 Splicing mechanism used for mRNA precursors. The upstream (5′) exon is shown in blue, the downstream (3′) exon in green, and the branch site in yellow. Y stands for a purine nucleotide, R for a pyrimidine nucleotide, and N for any nucleotide. The 5′ splice site is attacked by the 2′-OH group of the branch-site adenosine residue. The 3′ splice site is attacked by the newly formed 3′-OH group of the upstream exon. The exons are joined, and the intron is released in the form of a lariat. [After P. A. Sharp. *Cell* 2(1985):3980.]

Note that this adenylate residue is also joined to two other nucleotides by normal 3′,5′-phosphodiester bonds (Figure 28.30). Hence a *branch* is generated at this site, and a *lariat intermediate* is formed.

The 3′-OH terminus of exon 1 then attacks the phosphodiester bond between the intron and exon 2. Exons 1 and 2 become joined, and the intron is released in lariat form. Again, this reaction is a transesterification. Splicing is thus accomplished by two *transesterification reactions* rather than by hydrolysis followed by ligation. The first reaction generates a free 3′-hydroxyl group at the 3′ end of exon 1, and the second reaction links this group to the 5′-phosphate of exon 2. *The number of phosphodiester bonds stays the same during these steps*, which is crucial because it allows the splicing reaction itself to proceed without an energy source such as ATP or GTP.

28.3.5 Small Nuclear RNAs in Spliceosomes Catalyze the Splicing of mRNA Precursors

The nucleus contains many types of small RNA molecules with fewer than 300 nucleotides, referred to as *snRNAs* (*small nuclear RNAs*). A few of them—designated U1, U2, U4, U5, and U6—are essential for splicing mRNA precursors. The secondary structures of these RNAs are highly conserved in organisms ranging from yeast to human beings. These RNA molecules are associated with specific proteins to form complexes termed *snRNPs* (*small nuclear ribonucleoprotein particles*); investigators often speak of them as "snurps." Spliceosomes are large (60S), dynamic assemblies composed of snRNPs, other proteins called *splicing factors*, and the mRNA precursors being processed (Table 28.3).

FIGURE 28.30 Splicing branch point. The structure of the branch point in the lariat intermediate in which the adenylate residue is joined to three nucleotides by phosphodiester bonds. The new 2′-to-5′ linkage is shown in red, and the usual 3′-to-5′ linkages are shown in blue.

TABLE 28.3	Small nuclear ribonucleoprotein particles (snRNPs) in the splicing of mRNA precursors	
snRNP	Size of snRNA (nucleotides)	Role
U1	165	Binds the 5' splice site and then the 3' splice site
U2	185	Binds the branch site and forms part of the catalytic center
U5	116	Binds the 5' splice site
U4	145	Masks the catalytic activity of U6
U6	106	Catalyzes splicing

In mammalian cells, splicing begins with the recognition of the 5' splice site by U1 snRNP (Figure 28.31). In fact, U1 RNA contains a highly conserved six-nucleotide sequence that base pairs to the 5' splice site of the pre-mRNA. This binding initiates spliceosome assembly on the pre-mRNA molecule.

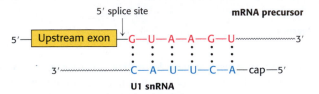

U2 snRNP then binds the branch site in the intron by base-pairing between a highly conserved sequence in U2 snRNA and the pre-mRNA. U2 snRNP binding requires ATP hydrolysis. A preassembled U4-U5-U6 complex joins this complex of U1, U2, and the mRNA precursor to form a complete spliceosome. This association also requires ATP hydrolysis.

A revealing view of the interplay of RNA molecules in this assembly came from examining the pattern of cross-links formed by *psoralen*, a photoactivable reagent that joins neighboring pyrimidines in base-paired regions. These cross-links suggest that splicing takes place in the following way. First, U5 interacts with exon sequences in the 5' splice site and sub-

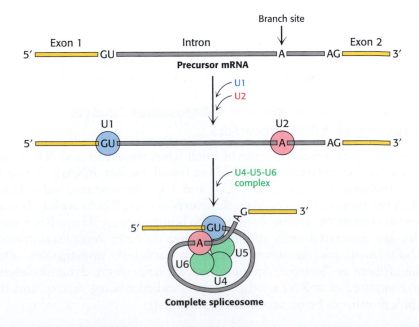

FIGURE 28.31 Spliceosome assembly.
U1 (blue) binds the 5' splice site and U2 (red) to the branch point. A preformed U4-U5-U6 complex then joins the assembly to form the complete spliceosome.

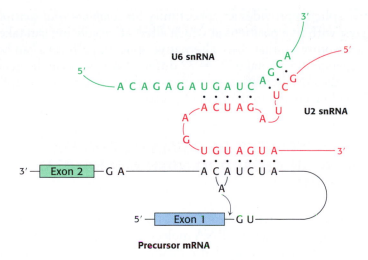

FIGURE 28.32 Splicing catalytic center.
The catalytic center of the spliceosome is
formed by U2 snRNA (red) and U6 snRNA
(green), which are base paired. U2 is also
base paired to the branch site of the mRNA
precursor. [After H. D. Madhani and C. Guthrie.
Cell 71(1992):803.]

sequently with the 3′ exon. Next, U6 disengages from U4 and undergoes
an intramolecular rearrangement that permits base-pairing with U2 and displaces U1 from the spliceosome by interacting with the 5′ end of the intron.
The U2·U6 helix is indispensable for splicing, suggesting that *U2 and U6
snRNAs probably form the catalytic center of the spliceosome* (Figure 28.32).
U4 serves as an inhibitor that masks U6 until the specific splice sites are
aligned. These rearrangements result in the first transesterification reaction,
generating the lariat intermediate and a cleaved 5′ exon.

Further rearrangements of RNA in the spliceosome facilitate the second
transesterification. These rearrangements align the free 5′ exon with the 3′
exon such that the 3′-hydroxyl group of the 5′ exon is positioned to nucleophilically attack the 3′ splice site to generate the spliced product. U2, U5,
and U6 bound to the excised lariat intron are released to complete the splicing reaction.

Many of the steps in the splicing process require ATP hydrolysis. How
is the free energy associated with ATP hydrolysis used to power splicing?
To achieve the well-ordered rearrangements necessary for splicing, ATP-
powered RNA helicases must unwind RNA helices and allow alternative
base-pairing arrangements to form. Thus, two features of the splicing
process are noteworthy. First, *RNA molecules play key roles in directing the
alignment of splice sites and in carrying out catalysis.* Second, *ATP-powered
helicases unwind RNA duplex intermediates that facilitate catalysis and induce the release of snRNPs from the mRNA.*

28.3.6 Some Pre-mRNA Molecules Can Be Spliced in Alternative Ways to Yield Different mRNAs

Alternative splicing is a widespread mechanism for generating protein diversity. The differential inclusion of exons into a mature RNA, alternative
splicing may be regulated to produce distinct forms of a protein for specific
tissues or developmental stages (Figure 28.33). A sample of the growing list
of proteins known to result from alternative splicing is presented in Table
28.4, and recent estimates suggest that the RNA products of 30% of human
genes are alternatively spliced. *Alternative splicing provides a powerful
mechanism for expanding the versatility of genomic sequences.* Suppose,
for example, that it is beneficial to have two forms of a protein with somewhat different properties that are expressed in different tissues. The evolution of an alternative splicing pathway provides a route to meeting this need
by means other than gene duplication and specialization. Furthermore,

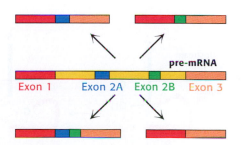

**FIGURE 28.33 Alternative splicing
patterns.** A pre-mRNA with multiple exons
is sometimes spliced in different ways.
Here, with two alternative exons (exons
2A and 2B) present, the mRNA can be
produced with neither, either, or both
exons included. More complex alternative
splicing patterns also are possible.

TABLE 28.4   Selected proteins exhibiting alternative RNA splicing

Actin
Alcohol dehydrogenase
Aldolase
K-ras
Calcitonin
Fibrinogen
Fibronectin
Myosin
Nerve growth factor
Tropomyosin
Troponin

Source: R. E. Breitbart, A. Andreadis, and B. Nadal-Ginard. *Annu. Rev. Biochem.* 56(1987):467–495.

alternative splicing provides an opportunity for *combinatorial control*. Consider a gene with five positions at which alternative splicing can take place. With the assumption that these alternative splicing pathways can be regulated independently, a total of $2^5 = 32$ different mRNAs can be generated. Further studies of alternative splicing and the mechanisms of splice-site selection will be crucial to the field of proteomics.

28.4 THE DISCOVERY OF CATALYTIC RNA WAS REVEALING IN REGARD TO BOTH MECHANISM AND EVOLUTION

RNAs form a surprisingly versatile class of molecules. As we have seen, splicing is catalyzed largely by RNA molecules, with proteins playing a secondary role. Another enzyme that contains a key RNA component is ribonuclease P (RNAse P), which catalyzes the maturation of tRNA by removing nucleotides from the 5′ end of the precursor molecule (Section 28.1.8). Finally, as we shall see in Chapter 29, the RNA component of ribosomes is the catalyst that carries out protein synthesis.

The versatility of RNA first became clear from observations regarding the processing carried out on ribosomal RNA in a single-cell eukaryote. In *Tetrahymena* (a ciliated protozoan), a 414-nucleotide intron is removed from a 6.4-kb precursor to yield the mature 26S rRNA molecule (Figure 28.34). In an elegant series of studies of this splicing reaction, Thomas Cech and his coworkers established that the RNA spliced itself to precisely excise the 414-nucleotide intron. These remarkable experiments demonstrated that an RNA molecule can *splice itself* in the absence of protein and, indeed, can have highly specific catalytic activity.

The *self-splicing* reaction requires an added guanosine nucleotide. Nucleotides were originally included in the reaction mixture because it was thought that ATP or GTP might be needed as an energy source. Instead, the nucleotides were found to be necessary as cofactors. The required cofactor proved to be a guanosine unit, in the form of guanosine, GMP, GDP, or GTP. G (denoting any one of these species) serves not as an energy source but as an attacking group that becomes transiently incorporated into the RNA (see Figure 28.34). G binds to the RNA and then attacks the 5′ splice

FIGURE 28.34 Self-splicing. A ribosomal RNA precursor from *Tetrahymena* splices itself in the presence of a guanosine cofactor (G, shown in green). A 414-nucleotide intron (red) is released in the first splicing reaction. This intron then splices itself twice again to produce a linear RNA that has lost a total of 19 nucleotides. This L19 RNA is catalytically active. [After T. Cech. RNA as an enzyme. Copyright © 1986 by Scientific American, Inc. All rights reserved.]

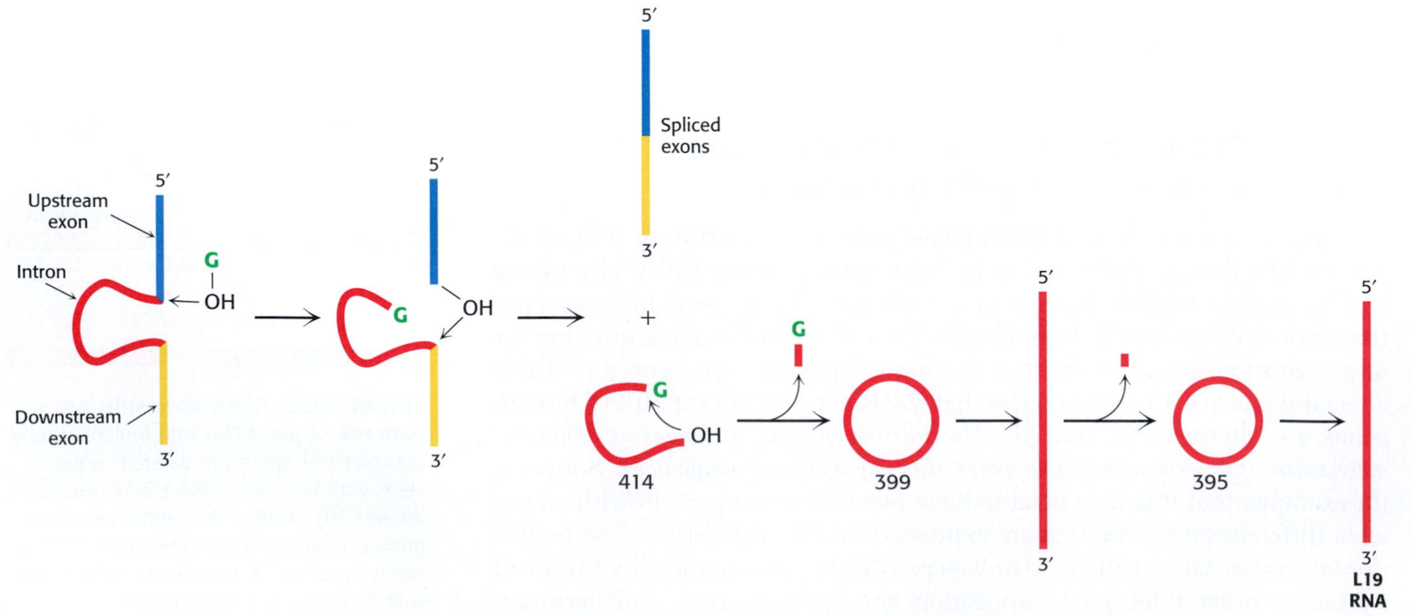

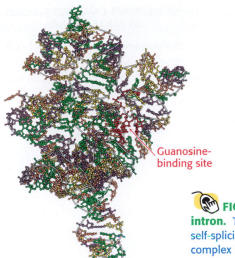

Guanosine-
binding site

FIGURE 28.35 Structure of a self-splicing intron. The structure of a large fragment of the self-splicing intron from *Tetrahymena* reveals a complex folding pattern of helices and loops. Bases are shown in green, A; yellow, C; purple, G; and orange, U.

site to form a phosphodiester bond with the 5′ end of the intron. This transesterification reaction generates a 3′-OH group at the end of the upstream exon. This newly attached 3′-OH group then attacks the 3′ splice. This second transesterification reaction joins the two exons and leads to the release of the 414-nucleotide intron.

Self-splicing depends on the structural integrity of the rRNA precursor. Much of the intron is needed for self-splicing. This molecule, like many RNAs, has a folded structure formed by many double-helical stems and loops (Figure 28.35). Examination of the three-dimensional structure determined by x-ray crystallography reveals a compact folding structure that is in many ways analogous to the structures of protein enzymes. A well-defined pocket for binding the guanosine is formed within the structure.

Analysis of the base sequence of the rRNA precursor suggested that the 5′ splice site is aligned with the catalytic residues by base-pairing between a *pyrimidine-rich region* (CUCUCU) of the upstream exon and a *purine-rich guide sequence* (GGGAGG) within the intron (Figure 28.36). The intron brings together the guanosine cofactor and the 5′ splice site so that the 3′-OH group of G can nucleophilically attack the phosphorus atom at this splice site. Another part of the intron then holds the downstream exon in position for attack by the newly formed 3′-OH group of the upstream exon. A phosphodiester bond is formed between the two exons, and the intron is released as a linear molecule. Like catalysis by protein enzymes, self-catalysis of bond formation and breakage in this rRNA precursor is highly specific.

The finding of enzymatic activity in the self-splicing intron and in the RNA component of RNAse P has opened new areas of inquiry and changed the way in which we think about molecular evolution. The discovery that RNA can be a catalyst as well as an information carrier suggests that an RNA world may have existed early in the evolution of life, before the appearance of DNA and protein (Section 2.2.2).

Messenger RNA precursors in the mitochondria of yeast and fungi also undergo self-splicing, as do some RNA precursors in the chloroplasts of unicellular organisms such as *Chlamydomonas*. Self-splicing reactions can be classified according to the nature of the unit that attacks the upstream splice site. Group I self-splicing is mediated by a guanosine cofactor, as in *Tetrahymena*. The attacking moiety in group II splicing is the 2′-OH group of a specific adenylate of the intron (Figure 28.37).

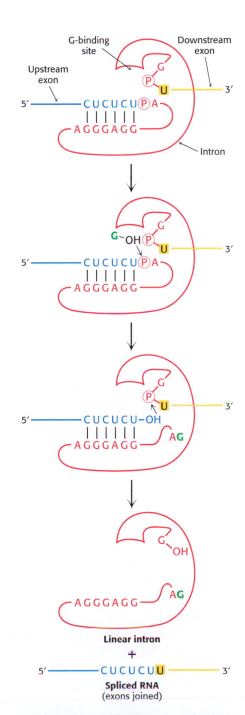

FIGURE 28.36 Self-splicing mechanism. The catalytic mechanism of the self-splicing intron from *Tetrahymena* includes a series of transesterification reactions. [After T. Cech. RNA as an enzyme. Copyright © 1986 by Scientific American, Inc. All rights reserved.]

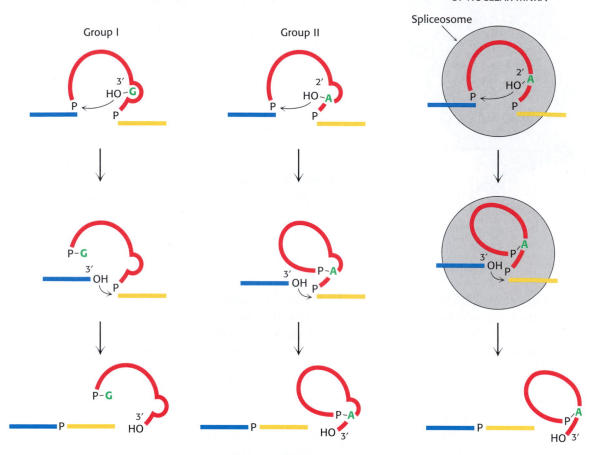

SELF-SPLICING INTRONS

Group I

Group II

SPLICEOSOME-CATALYZED SPLICING OF NUCLEAR mNRA

Spliceosome

FIGURE 28.37 Comparison of splicing pathways. The exons being joined are shown in blue and yellow and the attacking unit is shown in green. The catalytic site is formed by the intron itself (red) in group I and group II splicing. In contrast, the splicing of nuclear mRNA precursors is catalyzed by snRNAs and their associated proteins in the spliceosome. [After P. A. Sharp. *Science* 235(1987):769.]

Group I and group II self-splicing resembles spliceosome-catalyzed splicing in two respects. First, in initial step, a ribose hydroxyl group attacks the 5′ splice site. The newly formed 3′-OH terminus of the upstream exon then attacks the 3′ splice site to form a phosphodiester bond with the downstream exon. Second, both reactions are transesterifications in which the phosphate moieties at each splice site are retained in the products. The number of phosphodiester bonds stays constant. Group II splicing is like the spliceosome-catalyzed splicing of mRNA precursors in several additional ways. The attack at the 5′ splice site is carried out by a part of the intron itself (the 2′-OH group of adenosine) rather than by an external cofactor (G). In both cases, the intron is released in the form of a lariat. Moreover, in some instances, the group II intron is transcribed in pieces that assemble through hydrogen bonding to from the catalytic intron, in a manner analogous to the assembly of the snRNAs in the spliceosome.

These similarities have led to the suggestion that the spliceosome-catalyzed splicing of mRNA precursors evolved from RNA-catalyzed self-splicing. Group II splicing may well be an intermediate between group I splicing and the splicing in the nuclei of higher eukaryotes. *A major step in this transition was the transfer of catalytic power from the intron itself to other molecules.* The formation of spliceosomes gave genes a new freedom because introns were no longer constrained to provide the catalytic center for splicing. Another advantage of external catalysts for splicing is that they can be more readily regulated. However, it is important to note that similarities do not establish ancestry. The similarities between group II introns and mRNA splicing may be a result of convergent evolution. Perhaps there are only a limited number of ways to carry out efficient, specific intron excision. The determination of whether these similarities stem from ancestry or from chemistry will require expanding our understanding of RNA biochemistry.

SUMMARY

Transcription Is Catalyzed by RNA Polymerase

All cellular RNA molecules are synthesized by RNA polymerases according to instructions given by DNA templates. The activated monomer substrates are ribonucleoside triphosphates. The direction of RNA synthesis is $5' \rightarrow 3'$, as in DNA synthesis. RNA polymerases, unlike DNA polymerases, do not need a primer and do not possess proofreading nuclease activity.

RNA polymerase in *E. coli* is a multisubunit enzyme. The subunit composition of the $\sim$500-kd holoenzyme is $\alpha_2\beta\beta'\sigma$ and that of the core enzyme is $\alpha_2\beta\beta'$. Transcription is initiated at promoter sites consisting of two sequences, one centered near -10 and the other near -35; that is, 10 and 35 nucleotides away from the start site in the 5' (upstream) direction. The consensus sequence of the -10 region is TATAAT. The σ subunit enables the holoenzyme to recognize promoter sites. When the growth temperature is raised, *E. coli* expresses a special σ that selectively binds the distinctive promoter of heat-shock genes. RNA polymerase must unwind the template double helix for transcription to take place. Unwinding exposes some 17 bases on the template strand and sets the stage for the formation of the first phosphodiester bond. RNA chains usually start with pppG or pppA. The σ subunit dissociates from the holoenzyme after the initiation of the new chain. Elongation takes place at transcription bubbles that move along the DNA template at a rate of about 50 nucleotides per second. The nascent RNA chain contains stop signals that end transcription. One stop signal is an RNA hairpin, which is followed by several U residues. A different stop signal is read by the *rho* protein, an ATPase. In *E. coli*, precursors of transfer RNA and ribosomal RNA are cleaved and chemically modified after transcription, whereas mRNA is used unchanged as a template for protein synthesis.

Eukaryotic Transcription and Translation Are Separated in Space and Time

RNA synthesis in eukaryotes takes place in the nucleus, whereas protein synthesis takes place in the cytoplasm. There are three types of RNA polymerase in the nucleus: RNA polymerase I makes ribosomal RNA precursors, II makes messenger RNA precursors, and III makes transfer RNA precursors. Eukaryotic promoters are complex, being composed of several different elements. Promoters for RNA polymerase II are located on the 5' side of the start site for transcription. Each consists of a TATA box centered between -30 and -100 and additional upstream sequences. They are recognized by proteins called transcription factors rather than by RNA polymerase II. The saddle-shaped TATA-box-binding protein unwinds and sharply bends DNA at TATA-box sequences and serves as a focal point for the assembly of transcription complexes. The TATA-box-binding protein initiates the assembly of the active transcription complex. The activity of many promoters is greatly increased by enhancer sequences that have no promoter activity of their own. Enhancer sequences can act over distances of several kilobases, and they can be located either upstream or downstream of a gene.

The Transcription Products of All Three Eukaryotic Polymerases Are Processed

The 5′ ends of mRNA precursors become capped and methylated in the course of transcription. A 3′ poly(A) tail is added to most mRNA precursors after the nascent chain has been cleaved by an endonuclease. RNA editing processes alter the nucleotide sequence of some mRNAs, such as the one for apolipoprotein B.

The splicing of mRNA precursors is carried out by spliceosomes, which consist of small nuclear ribonucleoprotein particles (snRNPs). Splice sites in mRNA precursors are specified by sequences at ends of introns and by branch sites near their 3′ ends. The 2′-OH group of an adenosine residue in the branch site attacks the 5′ splice site to form a lariat intermediate. The newly generated 3′-OH terminus of the upstream exon then attacks the 3′ splice site to become joined to the downstream exon. Splicing thus consists of two transesterification reactions, with the number of phosphodiester bonds remaining constant during reactions. Small nuclear RNAs in spliceosomes catalyze the splicing of mRNA precursors. In particular, U2 and U6 snRNAs form the active centers of spliceosomes.

The Discovery of Catalytic RNA Was Revealing with Regard to Both Mechanism And Evolution

Some RNA molecules, such as the 26S ribosomal RNA precursor from *Tetrahymena*, undergo self-splicing in the absence of protein. A self-modified version of this rRNA intron displays true catalytic activity. Spliceosome-catalyzed splicing may have evolved from self-splicing. The discovery of catalytic RNA has opened new vistas in our exploration of early stages of molecular evolution.

KEY TERMS

transcription (p. 782)

RNA polymerase (p. 782)

promoter sites (p. 782)

transcription factor (p. 783)

footprinting (p. 784)

consensus sequence (p. 784)

sigma subunit (p. 785)

transcription bubble (p. 787)

rho (ρ) protein (p. 789)

TATA box (p. 794)

enhancer (p. 797)

pre-mRNA (p. 798)

5′ cap (p. 798)

poly(A) tail (p. 798)

RNA editing (p. 798)

RNA splicing (p. 799)

spliceosome (p. 800)

small nuclear RNA (snRNA) (p. 801)

alternative splicing (p. 803)

catalytic RNA (p. 804)

self-splicing (p. 804)

SELECTED READINGS

Where to begin

Woychik, N. A., 1998. Fractions to functions: RNA polymerase II thirty years later. *Cold Spring Harbor Symp. Quant. Biol.* 63: 311–317.

Losick, R., 1998. Summary: Three decades after sigma. *Cold Spring Harbor Symp. Quant. Biol.* 63:653–666.

Darnell, J. E., Jr. 1985. RNA. *Sci. Am.* 253(4):68–78.

Cech, T. R., 1986. RNA as an enzyme. *Sci. Am.* 255(5):64–75.

Sharp, P. A., 1994. Split genes and RNA splicing. *Cell* 77:805–815.

Cech, T. R., 1990. Self-splicing and enzymatic activity of an intervening sequence RNA from *Tetrahymena*. *Biosci. Rep.* 10:239–261.

Guthrie, C., 1991. Messenger RNA splicing in yeast: Clues to why the spliceosome is a ribonucleoprotein. *Science* 253:157–163.

Books

Lewin, B., 2000. *Genes* (7th ed.). Oxford University Press.

Kornberg, A., and Baker, T. A., 1992. *DNA Replication* (2d ed.). W. H. Freeman and Company.

Lodish, H., Berk, A., Zipursky, S. L., Matsudaira, P., Baltimore, D., and Darnell, J., 2000. *Molecular Cell Biology* (4th ed.). W. H. Freeman and Company.

Watson, J. D., Hopkins, N. H., Roberts, J. W., Steitz, J. A., and Weiner, A. M., 1987. *Molecular Biology of the Gene* (4th ed.). Benjamin Cummings.

Gesteland, R. F., Cech, T., and Atkins, J. F. (Eds.), 1999. *The RNA World* (2d ed.). Cold Spring Harbor Laboratory Press.

RNA polymerases

Cramer, P., Bushnell, D. A., and Kornberg, R. D., 2001. Structural basis of transcription: RNA polymerase II at 2.8 Å resolution. *Science* 292:1863–1875.

Gnatt, A. L., Cramer, P., Fu, J., Bushnell, D. A., and Kornberg, R. D., 2001. Structural basis of transcription: An RNA polymerase II elongation complex at 3.3 Å resolution. *Science* 292:1876–1882.

Zhang, G., Campbell, E. A., Minakhin, L., Richter, C., Severinov, K., and Darst, S. A., 1999. Crystal structure of *Thermus aquaticus* core RNA polymerase at 3.3 Å resolution. *Cell* 98:811–824.

Campbell, E. A., Korzheva, N., Mustaev, A., Murakami, K., Nair, S., Goldfarb, A., and Darst, S. A., 2001. Structural mechanism for rifampicin inhibition of bacterial RNA polymerase. *Cell* 104:901–912.

Cheetham, G. M., and Steitz, T. A., 1999. Structure of a transcribing T7 RNA polymerase initiation complex. *Science* 286:2305–2309.

Ebright, R. H., 2000. RNA polymerase: Structural similarities between bacterial RNA polymerase and eukaryotic RNA polymerase II. *J. Mol. Biol.* 304:687–698.

Paule, M. R., and White, R. J., 2000. Survey and summary: Transcription by RNA polymerases I and III. *Nucleic Acids Res.* 28: 1283–1298.

Initiation and elongation

Buratowski, S., 2000. Snapshots of RNA polymerase II transcription initiation. *Curr. Opin. Cell Biol.* 12:320–325.

Conaway, J. W., and Conaway, R. C., 1999. Transcription elongation and human disease. *Annu. Rev. Biochem.* 68:301–319.

Conaway, J. W., Shilatifard, A., Dvir, A., and Conaway, R. C., 2000. Control of elongation by RNA polymerase II. *Trends Biochem. Sci.* 25:375–380.

Korzheva, N., Mustaev, A., Kozlov, M., Malhotra, A., Nikiforov, V., Goldfarb, A., and Darst, S. A., 2000. A structural model of transcription elongation. *Science* 289:619–625.

Reines, D., Conaway, R. C., and Conaway, J. W., 1999. Mechanism and regulation of transcriptional elongation by RNA polymerase II. *Curr. Opin. Cell Biol.* 11:342–346.

Promoters, enhancers, and transcription factors

Merika, M., and Thanos, D., 2001. Enhanceosomes. *Curr. Opin. Genet. Dev.* 11:205–208.

Park, J. M., Gim, B. S., Kim, J. M., Yoon, J. H., Kim, H. S., Kang, J. G., and Kim, Y. J., 2001. *Drosophila* mediator complex is broadly utilized by diverse gene-specific transcription factors at different types of core promoters. *Mol. Cell. Biol.* 21:2312–2323.

Fiering, S., Whitelaw, E., and Martin, D. I., 2000. To be or not to be active: The stochastic nature of enhancer action. *Bioessays* 22:381–387.

Hampsey, M., and Reinberg, D., 1999. RNA polymerase II as a control panel for multiple coactivator complexes. *Curr. Opin. Genet. Dev.* 9:132–139.

Chen, L., 1999. Combinatorial gene regulation by eukaryotic transcription factors. *Curr. Opin. Struct. Biol.* 9:48–55.

Muller, C. W., 2001. Transcription factors: Global and detailed views. *Curr. Opin. Struct. Biol.* 11:26–32.

Sakurai, H., and Fukasawa, T., 2000. Functional connections between mediator components and general transcription factors of *Saccharomyces cerevisiae*. *J. Biol. Chem.* 275:37251–37256.

Droge, P., and Muller-Hill, B., 2001. High local protein concentrations at promoters: Strategies in prokaryotic and eukaryotic cells. *Bioessays* 23:179–183.

Fickett, J. W., and Hatzigeorgiou, A. G., 1997. Eukaryotic promoter recognition. *Genome Res.* 7:861–878.

Smale, S. T., Jain, A., Kaufmann, J., Emami, K. H., Lo, K., and Garraway, I. P., 1998. The initiator element: A paradigm for core promoter heterogeneity within metazoan protein-coding genes. *Cold Spring Harbor Symp. Quant. Biol.* 63:21–31.

Kim, Y., Geiger, J. H., Hahn, S., and Sigler, P. B., 1993. Crystal structure of a yeast TBP/TATA-box complex. *Nature* 365:512–520.

Kim, J. L., Nikolov, D. B., and Burley, S. K., 1993. Co-crystal structure of TBP recognizing the minor groove of a TATA element. *Nature* 365:520–527.

White, R. J., and Jackson, S. P., 1992. The TATA-binding protein: A central role in transcription by RNA polymerases I, II and III. *Trends Genet.* 8:284–288.

Termination

Burgess, B. R., and Richardson, J. P., 2001. RNA passes through the hole of the protein hexamer in the complex with *Escherichia coli* Rho factor. *J. Biol. Chem.* 276:4182–4189.

Yu, X., Horiguchi, T., Shigesada, K., and Egelman, E. H., 2000. Three-dimensional reconstruction of transcription termination factor rho: Orientation of the N-terminal domain and visualization of an RNA-binding site. *J. Mol. Biol.* 299:1279–1287.

Stitt, B. L., 2001. *Escherichia coli* transcription termination factor Rho binds and hydrolyzes ATP using a single class of three sites. *Biochemistry* 40:2276–2281.

Henkin, T. M., 2000. Transcription termination control in bacteria. *Curr. Opin. Microbiol.* 3:149–153.

Gusarov, I., and Nudler, E., 1999. The mechanism of intrinsic transcription termination. *Mol. Cell.* 3:495–504.

5′-Cap formation and polyadenylation

Shatkin, A. J., and Manley, J. L., 2000. The ends of the affair: Capping and polyadenylation. *Nat. Struct. Biol.* 7:838–842.

Ro-Choi, T. S., 1999. Nuclear snRNA and nuclear function (discovery of 5′ cap structures in RNA). *Crit. Rev. Eukaryotic Gene Expr.* 9:107–158.

Shuman, S., Liu, Y., and Schwer, B., 1994. Covalent catalysis in nucleotidyl transfer reactions: Essential motifs in *Saccharomyces cerevisiae* RNA capping enzyme are conserved in *Schizosaccharomyces pombe* and viral capping enzymes and among polynucleotide ligases. *Proc. Natl. Acad. Sci. U. S. A.* 91:12046–12050.

Bard, J., Zhelkovsky, A. M., Helmling, S., Earnest, T. N., Moore, C. L., and Bohm, A., 2000. Structure of yeast poly(A) polymerase alone and in complex with 3′-dATP. *Science* 289:1346–1349.

Martin, G., Keller, W., and Doublie, S., 2000. Crystal structure of mammalian poly(A) polymerase in complex with an analog of ATP. *EMBO J.* 19:4193–4203.

Zhao, J., Hyman, L., and Moore, C., 1999. Formation of mRNA 3′ ends in eukaryotes: Mechanism, regulation, and interrelationships with other steps in mRNA synthesis. *Microbiol. Mol. Biol. Rev.* 63: 405–445.

Minvielle-Sebastia, L., and Keller, W., 1999. mRNA polyadenylation and its coupling to other RNA processing reactions and to transcription. *Curr. Opin. Cell Biol.* 11:352–357.

Wahle, E., and Kuhn, U., 1997. The mechanism of 3′ cleavage and polyadenylation of eukaryotic pre-mRNA. *Prog. Nucleic Acid Res. Mol. Biol.* 57:41–71.

RNA editing

Gott, J. M., and Emeson, R. B., 2000. Functions and mechanisms of RNA editing. *Annu. Rev. Genet.* 34:499–531.

Simpson, L., Thiemann, O. H., Savill, N. J., Alfonzo, J. D., and Maslov, D. A., 2000. Evolution of RNA editing in trypanosome mitochondria. *Proc. Natl. Acad. Sci. USA* 97:6986–6993.

Chester, A., Scott, J., Anant, S., and Navaratnam, N., 2000. RNA editing: Cytidine to uridine conversion in apolipoprotein B mRNA. *Biochim. Biophys. Acta* 1494:1–3.

Maas, S., and Rich, A., 2000. Changing genetic information through RNA editing. *Bioessays* 22:790–802.

Splicing of mRNA precursors

Stark, H., Dube, P., Luhrmann, R., and Kastner, B., 2001. Arrangement of RNA and proteins in the spliceosomal U1 small nuclear ribonucleoprotein particle. *Nature* 409:539–542.

Strehler, E. E., and Zacharias, D. A., 2001. Role of alternative splicing in generating isoform diversity among plasma membrane calcium pumps. *Physiol. Rev.* 81:21–50.

Graveley, B. R., 2001. Alternative splicing: Increasing diversity in the proteomic world. *Trends Genet.* 17:100–107.

Newman, A., 1998. RNA splicing. *Curr. Biol.* 8:R903–R905.

Reed, R., 2000. Mechanisms of fidelity in pre-mRNA splicing. *Curr. Opin. Cell Biol.* 12:340–345.

Sleeman, J. E., and Lamond, A. I., 1999. Nuclear organization of pre-mRNA splicing factors. *Curr. Opin. Cell Biol.* 11:372–377.

Black, D. L., 2000. Protein diversity from alternative splicing: A challenge for bioinformatics and post-genome biology. *Cell* 103:367–370.

Collins, C. A., and Guthrie, C., 2000. The question remains: Is the spliceosome a ribozyme? *Nat. Struct. Biol.* 7:850–854.

Self-splicing and RNA catalysis

Carola, C., and Eckstein, F., 1999. Nucleic acid enzymes. *Curr. Opin. Chem. Biol.* 3:274–283.

Doherty, E. A., and Doudna, J. A., 2000. Ribozyme structures and mechanisms. *Annu. Rev. Biochem.* 69:597–615.

Fedor, M. J., 2000. Structure and function of the hairpin ribozyme. *J. Mol. Biol.* 297:269–291.

Hanna, R., and Doudna, J. A., 2000. Metal ions in ribozyme folding and catalysis. *Curr. Opin. Chem. Biol.* 4:166–170.

Scott, W. G., 1998. RNA catalysis. *Curr. Opin. Struct. Biol.* 8:720–726.

Scott, W. G., and Klug, A., 1996. Ribozymes: Structure and mechanism in RNA catalysis. *Trends Biochem. Sci.* 21:220–224.

Cech, T. R., Herschlag, D., Piccirilli, J. A., and Pyle, A. M., 1992. RNA catalysis by a group I ribozyme: Developing a model for transition state stabilization. *J. Biol. Chem.* 267:17479–17482.

Herschlag, D., and Cech, T. R., 1990. Catalysis of RNA cleavage by the *Tetrahymena thermophila* ribozyme 1: Kinetic description of the reaction of an RNA substrate complementary to the active site. *Biochemistry* 29:10159–10171.

Piccirilli, J. A., Vyle, J. S., Caruthers, M. H., and Cech, T. R., 1993. Metal ion catalysis in the *Tetrahymena* ribozyme reaction. *Nature* 361:85–88.

Wang, J. F., Downs, W. D., and Cech, T. R., 1993. Movement of the guide sequence during RNA catalysis by a group I ribozyme. *Science* 260:504–508.

PROBLEMS

1. *Complements.* The sequence of part of an mRNA is

5′-AUGGGGAACAGCAAGAGUGGGGGCCCUGUCCAAGGAG-3′

What is the sequence of the DNA coding strand? Of the DNA template strand?

2. *Checking for errors.* Why is RNA synthesis not as carefully monitored for errors as is DNA synthesis?

3. *Speed is not of the essence.* Why is it advantageous for DNA synthesis to be more rapid than RNA synthesis?

4. *Potent inhibitor.* Heparin inhibits transcription by binding to RNA polymerase. What properties of heparin allow it to bind so effectively to RNA polymerase?

5. *A loose cannon.* Sigma protein by itself does not bind to promoter sites. Predict the effect of a mutation enabling σ to bind to the −10 region in the absence of other subunits of RNA polymerase.

6. *Stuck sigma.* What would be the likely effect of a mutation that would prevent σ from dissociating from the RNA polymerase core?

7. *Transcription time.* What is the minimum length of time required for the synthesis by *E. coli* polymerase of an mRNA encoding a 100-kd protein?

8. *Between bubbles.* How far apart are transcription bubbles on *E. coli* genes that are being transcribed at a maximal rate?

9. *A revealing bubble.* Consider the synthetic RNA–DNA transcription bubble illustrated here. Let us refer to the coding DNA strand, the template strand, and the RNA strand as strands 1, 2, and 3, respectively.

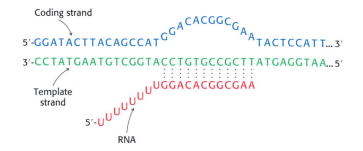

(a) Suppose that strand 3 is labeled with ^{32}P at its 5′ end and that polyacrylamide gel electrophoresis is carried out under nondenaturing conditions. Predict the autoradiographic pattern for (i) strand 3 alone, (ii) strands 1 and 3, (iii) strands 2 and 3, (iv) strands 1, 2, and 3, and (v) strands 1, 2, and 3 and core RNA polymerase.

(b) What is the likely effect of rifampicin on RNA synthesis in this system?

(c) Heparin blocks elongation of the RNA primer if it is added to core RNA polymerase before the onset of transcription but not if added after transcription starts. Account for this difference.

(d) Suppose that synthesis is carried out in the presence of ATP, CTP, and UTP. Compare the length of the longest product obtained with that expected when all four ribonucleoside triphosphates are present.

10. *Abortive cycling.* Di- and trinucleotides are occasionally released from RNA polymerase at the very start of transcription, a process called abortive cycling. This process requires the restart of transcription. Suggest a plausible explanation for abortive cycling.

11. *Polymerase inhibition.* Cordycepin inhibits poly(A) synthesis at low concentrations and RNA synthesis at higher concentrations.

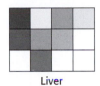

Cordycepin (3'-deoxyadenosine)

(a) What is the basis of inhibition by cordycepin?
(b) Why is poly(A) synthesis more sensitive to the presence of cordycepin?
(c) Does cordycepin need to be modified to exert its effect?

12. *An extra piece.* What is the amino acid sequence of the extra segment of protein synthesized in a thalassemic patient having a mutation leading to aberrant splicing (see Figure 28.28)? The reading frame after the splice site begins with TCT.

13. *A long-tailed messenger.* Another thalassemic patient had a mutation leading to the production of an mRNA for the β chain of hemoglobin that was 900 nucleotides longer than the normal one. The poly(A) tail of this mutant mRNA was located a few nucleotides after the only AAUAAA sequence in the additional sequence. Propose a mutation that would lead to the production of this altered mRNA.

Mechanism Problem

14. *RNA editing.* Many uridine molecules are inserted into some mitochondrial mRNAs in trypanosomes. The uridine residues come from the poly(U) tail of a donor strand. Nucleoside triphosphates do not participate in this reaction. Propose a reaction mechanism that accounts for these findings. (Hint: Relate RNA editing to RNA splicing.)

Chapter Integration Problems

15. *Proteome complexity.* What processes considered in this chapter make the proteome more complex than the genome? What processes might further enhance this complexity?

16. *Separation technique.* Suggest a means by which you could separate mRNA from the other types of RNA in a eukaryotic cell.

Data Interpretation Problems

17. *Run-off experiment.* Nuclei were isolated from brain, liver, and muscle. The nuclei were then incubated with α-^{32}P-UTP under conditions that allow RNA synthesis, except that an inhibitor of RNA initiation was present. The radioactive RNA was isolated and annealed to various DNA sequences that had been attached to a gene chip. In the adjoining graphs, the intensity of the shading indicates roughly how much mRNA was attached to each RNA sequence.

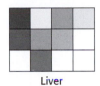

| Liver | Muscle | Brain |

(a) Why does the intensity of hybridization differ between genes?
(b) What is the significance of the fact that some of the RNA molecules display different hybridization patterns in different tissues?
(c) Some genes are expressed in all three tissues. What would you guess is the nature of these genes?
(d) Suggest a reason why an initiation inhibitor was included in the reaction mixture.

18. *Christmas trees.* The adjoining autoradiograph depicts several bacterial genes undergoing transcription. Identify the DNA. What are the strands of increasing length? Where is the beginning of transcription? The end of transcription? On the page, what is the direction of RNA synthesis? What can you conclude about the number of enzymes participating in RNA synthesis on a given gene?

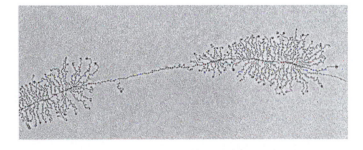

CHAPTER 29

Protein Synthesis

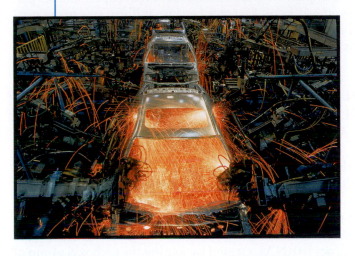

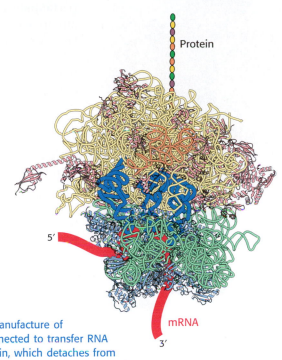

Protein

5'

mRNA

3'

Protein Assembly. The ribosome, shown at the right, is a factory for the manufacture of polypeptides. Amino acids are carried into the ribosome, one at a time, connected to transfer RNA molecules (blue). Each amino acid is joined to the growing polypeptide chain, which detaches from the ribosome only once it is completed. This assembly line approach allows even very long polypeptide chains to be assembled rapidly and with impressive accuracy. [(Left) Doug Martin/Photo Researchers.]

Genetic information is most important because of the proteins that it encodes, in that proteins play most of the functional roles in cells. In Chapters 27 and 28, we examined how DNA is replicated and how DNA is transcribed into RNA. We now turn to the mechanism of protein synthesis, a process called *translation* because the four-letter alphabet of nucleic acids is translated into the entirely different twenty-letter alphabet of proteins. Translation is a conceptually more complex process than either replication or transcription, both of which take place within the framework of a common base-pairing language. Befitting its position linking the nucleic acid and protein languages, the process of protein synthesis depends critically on both nucleic acid and protein factors. Protein synthesis takes place on *ribosomes*—enormous complexes containing three large RNA molecules and more than 50 proteins. One of the great triumphs in biochemistry in recent years has been the determination of the structure of the ribosome and its components so that its function can be examined in atomic detail (Figure 29.1). Perhaps the most significant conclusion from these studies is that *the ribosome is a ribozyme;* that is, the RNA components play the most fundamental roles. These observations strongly support the notion that the ribosome is a surviving inhabitant of the RNA world. As such, the ribosome is rich in information regarding very early steps in evolution.

OUTLINE

- **29.1 Protein Synthesis Requires the Translation of Nucleotide Sequences into Amino Acid Sequences**

- **29.2 Aminoacyl-Transfer RNA Synthetases Read the Genetic Code**

- **29.3 A Ribosome Is a Ribonucleoprotein Particle (70S) Made of a Small (30S) and a Large (50S) Subunit**

- **29.4 Protein Factors Play Key Roles in Protein Synthesis**

- **29.5 Eukaryotic Protein Synthesis Differs from Prokaryotic Protein Synthesis Primarily in Translation Initiation**

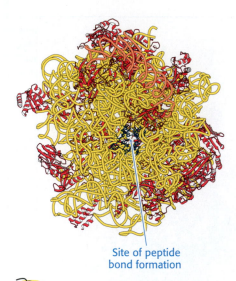

Site of peptide
bond formation

FIGURE 29.1 Ribosome structure.
The structure of a part of the ribosome showing the site at which peptide-bond formation takes place. This site contains only RNA (shown in yellow), with no protein (red) within 20 Å.

Transfer RNA molecules (tRNAs), messenger RNA, and many proteins participate in protein synthesis along with ribosomes. The link between amino acids and nucleic acids is first made by enzymes called aminoacyl-tRNA synthetases. By specifically linking a particular amino acid to each tRNA, these enzymes implement the genetic code. This chapter focuses primarily on protein synthesis in prokaryotes because it illustrates many general principles and is relatively well understood. Some distinctive features of protein synthesis in eukaryotes also are presented.

29.1 PROTEIN SYNTHESIS REQUIRES THE TRANSLATION OF NUCLEOTIDE SEQUENCES INTO AMINO ACID SEQUENCES

The basics of protein synthesis are the same across all kingdoms of life, attesting to the fact that the protein-synthesis system arose very early in evolution. A protein is synthesized in the amino-to-carboxyl direction by the sequential addition of amino acids to the carboxyl end of the growing peptide chain (Figure 29.2). The amino acids arrive at the growing chain in activated form as aminoacyl-tRNAs, created by joining the carboxyl group of an amino acid to the 3′ end of a transfer RNA molecule. The linking of an amino acid to its corresponding tRNA is catalyzed by an *aminoacyl-tRNA synthetase*. ATP cleavage drives this activation reaction. For each amino acid, there is usually one activating enzyme and at least one kind of tRNA.

FIGURE 29.2 Polypeptide-chain growth.
Proteins are synthesized by the successive addition of amino acids to the carboxyl terminus.

29.1.1 The Synthesis of Long Proteins Requires a Low Error Frequency

The process of transcription is analogous to copying, word for word, a page from a book. There is no change of alphabet or vocabulary; so the likelihood of a change in meaning is small. Translating the base sequence of an mRNA molecule into a sequence of amino acids is similar to translating the page of a book into another language. Translation is a complex process, entailing many steps and dozens of molecules. The potential for error exists at each step. The complexity of translation creates a conflict between two requirements: the process must be not only accurate, but also fast enough to meet a cell's needs. How fast is "fast enough"? In *E.coli*, translation takes place at a rate of 40 amino acids per second, a truly impressive speed considering the complexity of the process.

How accurate must protein synthesis be? Let us consider error rates. The probability p of forming a protein with no errors depends on n, the num-

TABLE 29.1 Accuracy of protein synthesis

Frequency of inserting an incorrect amino acid	Probability of synthesizing an error-free protein		
	Number of amino acid residues		
	100	300	1000
10^{-2}	0.366	0.049	0.000
10^{-3}	0.905	0.741	0.368
10^{-4}	0.990	0.970	0.905
10^{-5}	0.999	0.997	0.990

ber of amino acid residues, and ϵ, the frequency of insertion of a wrong amino acid:

$$p = (1 - \epsilon)^n$$

As Table 29.1 shows, an error frequency of 10^{-2} would be intolerable, even for quite small proteins. An ϵ value of 10^{-3} would usually lead to the error-free synthesis of a 300-residue protein (~33 kd) but not of a 1000-residue protein (~110 kd). Thus, the error frequency must not exceed approximately 10^{-4} to produce the larger proteins effectively. Lower error frequencies are conceivable; however, except for the largest proteins, they will not dramatically increase the percentage of proteins with accurate sequences. In addition, such lower error rates are likely to be possible only by a reduction in the rate of protein synthesis because additional time for proofreading will be required. *In fact, the observed values of ϵ are close to 10^{-4}.* An error frequency of about 10^{-4} per amino acid residue was selected in the course of evolution to accurately produce proteins consisting of as many as 1000 amino acids while maintaining a remarkably rapid rate for protein synthesis.

29.1.2 Transfer RNA Molecules Have a Common Design

The fidelity of protein synthesis requires the accurate recognition of three-base *codons* on messenger RNA. Recall that the genetic code relates each amino acid to a three-letter codon (Section 5.5.1). An amino acid cannot itself recognize a codon. Consequently, an amino acid is attached to a specific tRNA molecule that can recognize the codon by Watson-Crick base pairing. *Transfer RNA serves as the adapter molecule that binds to a specific codon and brings with it an amino acid for incorporation into the polypeptide chain.*

Robert Holley first determined the base sequence of a tRNA molecule in 1965, as the culmination of 7 years of effort. Indeed, his study of yeast alanyl-tRNA provided the first complete sequence of any nucleic acid. This adapter molecule is a single chain of 76 ribonucleotides (Figure 29.3). The 5′ terminus is phosphorylated (pG), whereas the 3′ terminus has a free hydroxyl group. The *amino acid attachment site* is the 3′-hydroxyl group of the adenosine residue at the 3′ terminus of the molecule. The sequence IGC in the middle of the molecule is the *anticodon*. It is complementary to GCC, one of the codons for alanine.

The sequences of several other tRNA molecules were determined a short time later. Hundreds of sequences are now known. The striking finding is that all of them can be arranged in a cloverleaf pattern in which about half

Anticodon

3′   5′
—Ċ—Ġ—İ—
 ⋮   ⋮   ⋮
—Ġ—Ċ—Ċ—
5′   3′

Codon

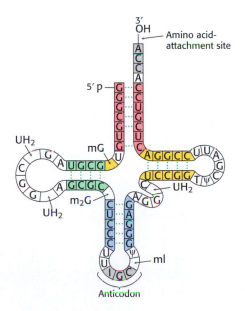

FIGURE 29.3 Alanine-tRNA sequence. The base sequence of yeast alanyl-tRNA and the deduced cloverleaf secondary structure are shown. Modified nucleosides are abbreviated as follows: methylinosine (mI), dihydrouridine (UH₂), ribothymidine (T), pseudouridine (Ψ), methylguanosine (mG), and dimethylguanosine (m₂G). Inosine (I), another modified nucleoside, is part of the anticodon.

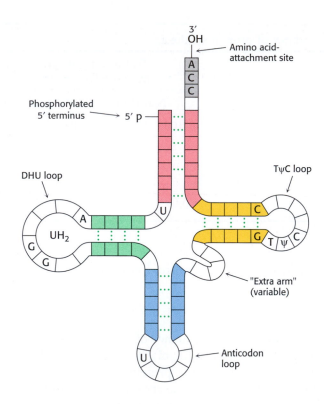

FIGURE 29.4 General structure of tRNA molecules. Comparison of the base sequences of many tRNAs reveals a number of conserved features.

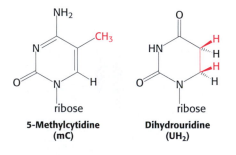

5-Methylcytidine
(mC)

Dihydrouridine
(UH₂)

the residues are base-paired (Figure 29.4). Hence, *tRNA molecules have many common structural features.* This finding is not unexpected, because all tRNA molecules must be able to interact in nearly the same way with the ribosomes, mRNAs, and protein factors that participate in translation.

All known transfer RNA molecules have the following features:

1. Each is a single chain containing between *73 and 93 ribonucleotides* (~25 kd).

2. They contain *many unusual bases*, typically between 7 and 15 per molecule. Some are methylated or dimethylated derivatives of A, U, C, and G formed by enzymatic modification of a precursor tRNA (Section 28.1.8). Methylation prevents the formation of certain base pairs, thereby rendering some of the bases accessible for other interactions. In addition, methylation imparts a hydrophobic character to some regions of tRNAs, which may be important for their interaction with synthetases and ribosomal proteins. Other modifications alter codon recognition, as will be discussed shortly.

3. About half the nucleotides in tRNAs are base-paired to form double helices. Five groups of bases are not base paired in this way: the 3′ *CCA terminal region*, which is part of a region called the *acceptor stem;* the *TψC loop*, which acquired its name from the sequence ribothymine-pseudouracil-cytosine; the "extra arm," which contains a variable number of residues; the *DHU loop*, which contains several dihydrouracil residues; and the *anticodon loop*. The structural diversity generated by this combination of helices and loops containing modified bases ensures that the tRNAs can be uniquely distinguished, though structurally similar overall.

4. The 5′ end of a tRNA is phosphorylated. The 5′ terminal residue is usually pG.

5. The activated amino acid is attached to a hydroxyl group of the adenosine residue located at the end of the 3′ CCA component of the acceptor stem. This region is single stranded at the 3′ end of mature rRNAs.

6. The anticodon is present in a loop near the center of the sequence.

29.1.3 The Activated Amino Acid and the Anticodon of tRNA Are at Opposite Ends of the L-Shaped Molecule

The three-dimensional structure of a tRNA molecule was first determined in 1974 through x-ray crystallographic studies carried out in the laboratories of Alexander Rich and Aaron Klug. The structure determined, that of yeast phenylalanyl-tRNA, is highly similar to all structures subsequently determined for other tRNA molecules. The most important properties of the tRNA structure are:

1. The molecule is *L-shaped* (Figure 29.5).

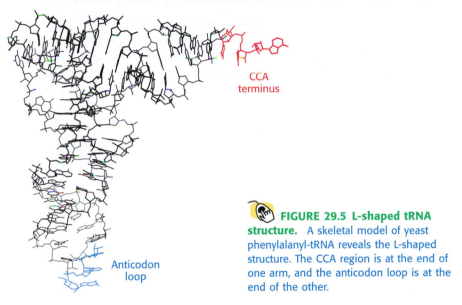

CCA
terminus

Anticodon
loop

FIGURE 29.5 L-shaped tRNA structure. A skeletal model of yeast phenylalanyl-tRNA reveals the L-shaped structure. The CCA region is at the end of one arm, and the anticodon loop is at the end of the other.

2. There are *two* apparently continuous segments of double helix. These segments are like A-form DNA, as expected for an RNA helix (Section 27.1.1). The base-pairing predicted from the sequence analysis is correct. The helix containing the 5′ and 3′ ends stacks on top of the helix that ends in the TψC loop to form one arm of the L; the remaining two helices stack to form the other (Figure 29.6).

3. Most of the bases in the nonhelical regions participate in hydrogen-bonding interactions, even if the interactions are not like those in Watson-Crick base pairs.

4. The CCA terminus containing the *amino acid attachment site* extends from one end of the L. This single-stranded region can change conformation during amino acid activation and protein synthesis.

5. The anticodon loop is at the other end of the L, making accessible the three bases that make up the anticodon.

Thus, the architecture of the tRNA molecule is well suited to its role as adaptor; the anticodon is available to interact with an appropriate codon on mRNA while the end that is linked to an activated amino acid is well positioned to participate in peptide-bond formation.

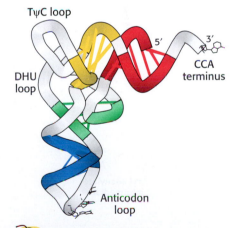

TψC loop

DHU
loop

5′

3′

CCA
terminus

Anticodon
loop

FIGURE 29.6 Helix stacking in tRNA. The four helices of the secondary structure of tRNA (see Figure 29.4) stack to form an L-shaped structure.

29.2 AMINOACYL-TRANSFER RNA SYNTHETASES READ THE GENETIC CODE

The linkage of an amino acid to a tRNA is crucial for two reasons. *First, the attachment of a given amino acid to a particular tRNA establishes the*

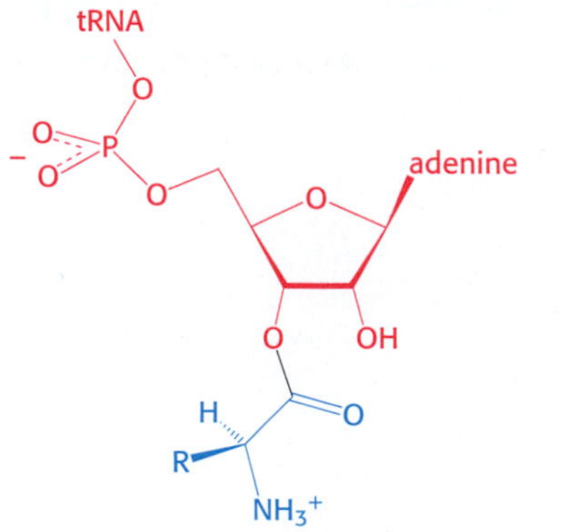

Aminoacyl-tRNA

FIGURE 29.7 Aminoacyl-tRNA. Amino acids are coupled to tRNAs through ester linkages to either the 2′- or the 3′-hydroxyl group of the 3′-adenosine residue. A linkage to the 3′-hydroxyl group is shown.

genetic code. When an amino acid has been linked to a tRNA, it will be incorporated into a growing polypeptide chain at a position dictated by the anticodon of the tRNA. Second, the formation of a peptide bond between free amino acids is not thermodynamically favorable. The amino acid must first be activated for protein synthesis to proceed. *The activated intermediates in protein synthesis are amino acid esters,* in which the carboxyl group of an amino acid is linked to either the *2′-* or the *3′*-hydroxyl group of the ribose unit at the 3′ end of tRNA. An amino acid ester of tRNA is called an *aminoacyl-tRNA* or sometimes a *charged tRNA* (Figure 29.7).

29.2.1 Amino Acids Are First Activated by Adenylation

The activation reaction is catalyzed by specific *aminoacyl-tRNA synthetases,* which are also called *activating enzymes.* The first step is the formation of an *aminoacyl adenylate* from an amino acid and ATP. This activated species is a mixed anhydride in which the carboxyl group of the amino acid is linked to the phosphoryl group of AMP; hence, it is also known as *aminoacyl-AMP.*

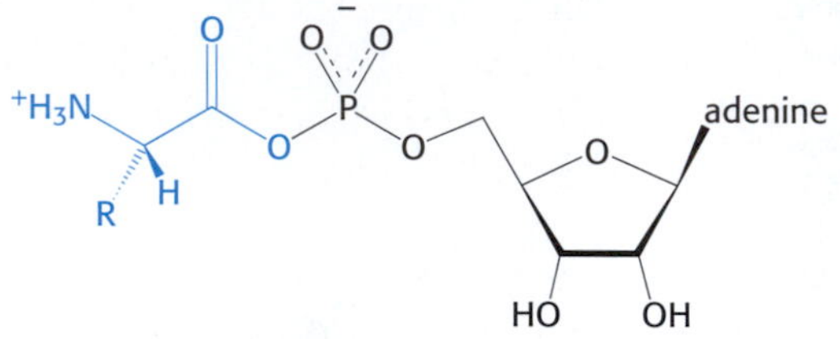

Aminoacyl adenylate

The next step is the transfer of the aminoacyl group of aminoacyl-AMP to a particular tRNA molecule to form *aminoacyl-tRNA.*

$$\text{Aminoacyl-AMP} + \text{tRNA} \rightleftharpoons \text{aminoacyl-tRNA} + \text{AMP}$$

The sum of these activation and transfer steps is

$$\text{Amino acid} + \text{ATP} + \text{tRNA} \rightleftharpoons \text{aminoacyl-tRNA} + \text{AMP} + \text{PP}_i$$

The $\Delta G°'$ of this reaction is close to 0, because the free energy of hydrolysis of the ester bond of aminoacyl-tRNA is similar to that for the hydrolysis of ATP to AMP and PP_i. As we have seen many times, the reaction is driven by the hydrolysis of pyrophosphate. The sum of these three reactions is highly exergonic:

$$\text{Amino acid} + \text{ATP} + \text{tRNA} + \text{H}_2\text{O} \longrightarrow$$
$$\text{aminoacyl-tRNA} + \text{AMP} + 2\,\text{P}_i$$

Thus, *the equivalent of two molecules of ATP are consumed in the synthesis of each aminoacyl-tRNA.* One of them is consumed in forming the ester linkage of aminoacyl-tRNA, whereas the other is consumed in driving the reaction forward.

The activation and transfer steps for a particular amino acid are catalyzed by the same aminoacyl-tRNA synthetase. Indeed, *the aminoacyl-AMP intermediate does not dissociate from the synthetase.* Rather, it is tightly bound to the active site of the enzyme by noncovalent interactions. Aminoacyl-AMP is normally a transient intermediate in the synthesis of aminoacyl-tRNA, but it is relatively stable and readily isolated if tRNA is absent from the reaction mixture.

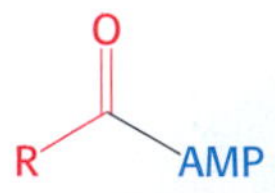

We have already encountered an acyl adenylate intermediate in fatty acid activation (Section 22.2.2). The major difference between these reactions is that the acceptor of the acyl group is CoA in fatty acid activation and tRNA in amino acid activation. The energetics of these biosyntheses are very similar: both are made irreversible by the hydrolysis of pyrophosphate.

Acyl adenylate intermediate

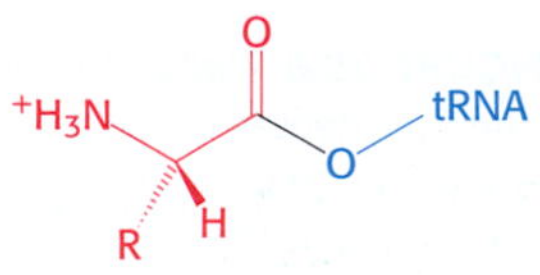

Aminoacyl-tRNA

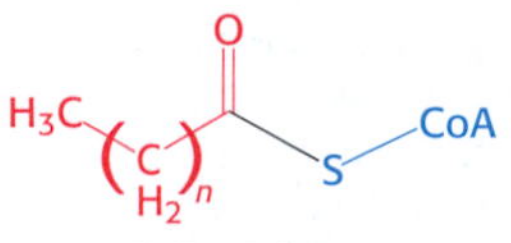

Fatty acyl CoA

29.2.2 Aminoacyl-tRNA Synthetases Have Highly Discriminating Amino Acid Activation Sites

Each aminoacyl-tRNA synthetase is highly specific for a given amino acid. Indeed, a synthetase will incorporate the incorrect amino acid only once in 10^4 or 10^5 catalytic reactions. How is this level of specificity achieved? Each aminoacyl-tRNA synthetase takes advantage of the properties of its amino acid substrate. Let us consider the challenge faced by threonyl-tRNA synthetase. Threonine is particularly similar to two other amino acids—namely, valine and serine. Valine has almost exactly the same shape as threonine, except that it has a methyl group in place of a hydroxyl group. Like threonine, serine has a hydroxyl group but lacks the methyl group. How can the threonyl-tRNA synthetase avoid coupling these incorrect amino acids to threonyl-tRNA?

The structure of the amino acid-binding site of threonyl-tRNA synthetase reveals how valine is avoided (Figure 29.8). The enzyme contains a zinc ion, bound to the enzyme by two histidine residues and one cysteine residue. Like carbonic anhydrase (Section 9.2.1), the remaining coordination sites are available for substrate binding. Threonine coordinates to the zinc ion through its amino group and its side-chain hydroxyl group. The side-chain hydroxyl group is further recognized by an aspartate residue that hydrogen bonds to it. The methyl group present in valine in place of this hydroxyl group cannot participate in these interactions; it is excluded from this active site and, hence, does not become adenylated and transferred to threonyl-tRNA (abbreviated tRNA^{Thr}). Note that the carboxylate group of the amino acid is available to attack the α-phosphate group of ATP to form the aminoacyl adenylate. Other aminoacyl-tRNA synthetases have different strategies for recognizing their cognate amino acids; the use of a zinc ion appears to be unique to threonyl-tRNA synthetase.

The zinc site is less well suited to discrimination against serine because this amino acid does have a hydroxyl group that can bind to the zinc. Indeed, with only this mechanism available, threonyl-tRNA synthetase does mistakenly couple serine to threonyl-tRNA at a rate 10^{-2} to 10^{-3} times that for threonine. As noted in Section 29.1.1, this error rate is likely to lead to many translation errors. How is a higher level of specificity achieved?

Threonine

Valine

Serine

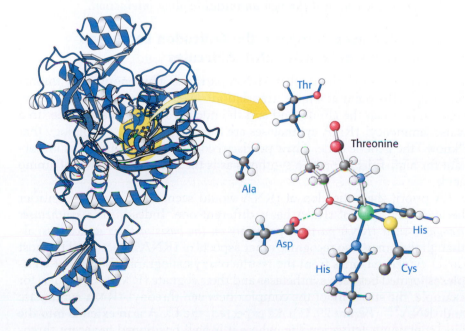

FIGURE 29.8 Structure of threonyl-tRNA synthetase. The structure of a large fragment of threonyl-tRNA synthetase reveals that the amino acid-binding site includes a zinc ion that coordinates threonine through its amino and hydroxyl groups. Only one subunit of the dimeric enzyme is shown in this and subsequent figures.

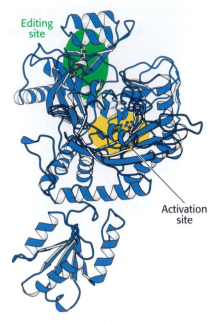

FIGURE 29.9 Editing site. The results of mutagenesis studies revealed the position of the editing site (shown in green) in threonyl-tRNA synthetase.

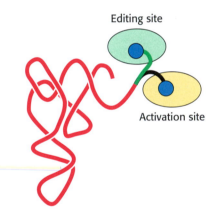

FIGURE 29.10 Editing of aminoacyl-tRNA. The flexible CCA arm of an aminoacyl-tRNA can move the amino acid between the activation site and the editing site. If the amino acid fits well into the editing site, the amino acid is removed by hydrolysis.

29.2.3 Proofreading by Aminoacyl-tRNA Synthetases Increases the Fidelity of Protein Synthesis

Threonyl-tRNA synthetase can be incubated with tRNA^{Thr} that has been covalently linked with serine (Ser-tRNA^{Thr}); the tRNA has been "mischarged." The reaction is immediate: a rapid hydrolysis of the aminoacyl-tRNA forms serine and free tRNA. In contrast, incubation with correctly charged Thr-tRNA^{Thr} results in no reaction. Thus, threonyl-tRNA synthetase contains an additional functional site that hydrolyzes Ser-tRNA^{Thr} but not Thr-tRNA^{Thr}. This editing site provides an opportunity for the synthetase to correct its mistakes and improve its fidelity to less than one mistake in 10^4. The results of structural and mutagenesis studies revealed that the editing site is more than 20 Å from the activation site (Figure 29.9). This site readily accepts and cleaves Ser-tRNA^{Thr} but does not cleave Thr-tRNA^{Thr}. The discrimination of serine from threonine is relatively easy because threonine contains an *extra* methyl group; a site that conforms to the structure of serine will sterically exclude threonine.

Most aminoacyl-tRNA synthetases contain editing sites in addition to acylation sites. These complementary pairs of sites function as a *double sieve* to ensure very high fidelity. In general, the acylation site rejects amino acids that are *larger* than the correct one because there is insufficient room for them, whereas the hydrolytic site cleaves activated species that are *smaller* than the correct one.

The structure of the complex between threonyl-tRNA synthetase and its substrate reveals that the aminoacylated-CCA can swing out of the activation site and into the editing site (Figure 29.10). Thus, the aminoacyl-tRNA can be edited without dissociating from the synthetase. This proofreading, which depends on the conformational flexibility of a short stretch of polynucleotide sequence, is entirely analogous to that of DNA polymerase (Section 27.2.4). In both cases, editing without dissociation significantly improves fidelity with only modest costs in time and energy.

A few synthetases achieve high accuracy without editing. For example, tyrosyl-tRNA synthetase has no difficulty discriminating between tyrosine and phenylalanine; the hydroxyl group on the tyrosine ring enables tyrosine to bind to the enzyme 10^4 times as strongly as phenylalanine. *Proofreading has been selected in evolution only when fidelity must be enhanced beyond what can be obtained through an initial binding interaction.*

29.2.4 Synthetases Recognize the Anticodon Loops and Acceptor Stems of Transfer RNA Molecules

How do synthetases choose their tRNA partners? This enormously important step is the point at which "translation" takes place—at which the correlation between the amino acid and the nucleic acid worlds is made. In a sense, aminoacyl-tRNA synthetases are the only molecules in biology that "know" the genetic code. Their precise recognition of tRNAs is as important for high-fidelity protein synthesis as is the accurate selection of amino acids.

A priori, the anticodon of tRNA would seem to be a good identifier because each type of tRNA has a different one. Indeed, *some synthetases recognize their tRNA partners primarily on the basis of their anticodons*, although they may also recognize other aspects of tRNA structure. The most direct evidence comes from the results of crystallographic studies of complexes formed between synthetases and their cognate tRNAs. Consider, for example, the structure of the complex between threonyl-tRNA synthetase and tRNA^{Thr} (Figure 29.11). As expected, the CCA arm extends into the zinc-containing activation site, where it is well positioned to accept threo-

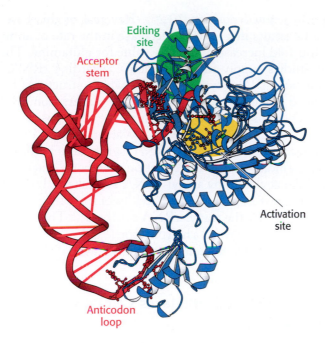

FIGURE 29.11 Threonyl-tRNA synthetase complex. The structure of the complex between threonyl-tRNA synthetase and tRNA^{Thr} reveals that the synthetase binds to both the acceptor stem and the anticodon loop.

nine from threonyl adenylate. The enzyme interacts extensively not only with the acceptor stem of the tRNA, but also with the anticodon loop. The interactions with the anticodon loop are particularly revealing. The bases within the sequence CGU of the anticodon each participate in hydrogen bonds with the enzyme; those in which G and U take part appear to be more important because the C can be replaced by G or U with no loss of acylation efficiency. The importance of the anticodon bases is further underscored by studies of tRNA^{Met}. Changing the anticodon sequence of this tRNA from CAU to GGU allows tRNA^{Met} to be aminoacylated by threonyl-tRNA synthetase nearly as well as tRNA^{Thr}, despite considerable differences in sequence elsewhere in the structure.

The structure of another complex between a tRNA and an aminoacyl-tRNA synthetase, that of glutaminyl-tRNA synthetase, again reveals extensive interactions with both the anticodon loop and the acceptor stem (Figure 29.12). In addition, contacts are made near the "elbow" of the tRNA molecule, particularly with the base pair formed by G in position 10 and C

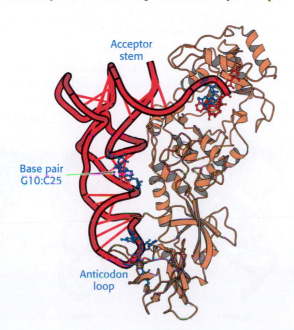

FIGURE 29.12 Glutaminyl-tRNA synthetase complex. The structure of this complex reveals that the synthetase interacts with base pair G10:C25 in addition to the acceptor step and anticodon loop.

```
        A 76
        C
        C
        A
   1  G · C
      G · C
   3  G · U  70
      G · C
      C · G
      U · A
      A · U  66
    U       C 13
   A         U
    G   C
   10
```

FIGURE 29.13 Microhelix recognized by alanyl-tRNA synthetase. A stem-loop containing just 24 nucleotides corresponding to the acceptor stem is aminoacylated by alanyl-tRNA synthetase.

in position 25 (denoted position 10:25). Reversal of this base pair from G · C to C · G results in a fourfold decrease in the rate of aminoacylation as well as a fourfold increase in the K_M value for glutamine. The results of mutagenesis studies supply further evidence regarding tRNA specificity, even for aminoacyl-tRNA synthetases for which structures have not yet been determined. For example, *E. coli* tRNA^{Cys} differs from tRNA^{Ala} at 40 positions and contains a C · G base pair at the 3:70 position. When this C · G base pair is changed to the non-Watson-Crick G · U base pair, tRNA^{Cys} is recognized by alanyl-tRNA synthetase as though it were tRNA^{Ala}. This finding raised the question whether a fragment of tRNA suffices for aminoacylation by alanyl-tRNA synthetase. Indeed, a "microhelix" containing just 24 of the 76 nucleotides of the native tRNA is specifically aminoacylated by the alanyl-tRNA synthetase. This microhelix contains only the acceptor stem and a hairpin loop (Figure 29.13). Thus, specific aminoacylation is possible for some synthetases even if the anticodon loop is completely lacking.

29.2.5 Aminoacyl-tRNA Synthetases Can Be Divided into Two Classes

STRUCTURAL INSIGHTS, Aminoacyl-tRNA Synthetases. *The first parts of the tutorial focus on the structural differences that distinguish class I and class II aminoacyl-tRNA synthetases. The final section of the tutorial looks at the editing process that most tRNA synthetases use to correct tRNA acylation errors.*

At least one aminoacyl-tRNA synthetase exists for each amino acid. The diverse sizes, subunit composition, and sequences of these enzymes were bewildering for many years. Could it be that essentially all synthetases evolved independently? The determination of the three-dimensional structures of several synthetases followed by more-refined sequence comparisons revealed that different synthetases are, in fact, related. Specifically, synthetases fall into two classes, termed *class I* and *class II*, each of which includes enzymes specific for 10 of the 20 amino acids (Table 29.2). Glutaminyl-tRNA synthetase is a representative of class I. The activation domain for class I has a Rossmann fold (Section 16.1.10). Threonyl-tRNA synthetase (see Figure 29.11) is a representative of class II. The activation domain for class II consists largely of β strands. Intriguingly, synthetases from the two classes bind to different faces of the tRNA molecule (Figure 29.14). The CCA arm of tRNA adopts different conformations to accommodate these interactions; the arm is in the helical conformation observed for free tRNA (see Figures 29.5 and 29.6) for class II enzymes and in a hairpin conformation for class I enzymes. These two classes also differ in other ways.

TABLE 29.2 Classification and subunit structure of aminoacyl-tRNA synthetases in *E. coli*

Class I	Class II
Arg (α)	Ala (α_4)
Cys (α)	Asn (α_2)
Gln (α)	Asp (α_2)
Glu (α)	Gly ($\alpha_2\beta_2$)
Ile (α)	His (α_2)
Leu (α)	Lys (α_2)
Met (α)	Phe ($\alpha_2\beta_2$)
Trp (α_2)	Ser (α_2)
Tyr (α_2)	Pro (α_2)
Val (α)	Thr (α_2)

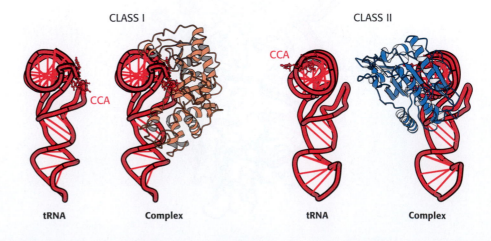

CLASS I CLASS II

tRNA Complex tRNA Complex

FIGURE 29.14 Classes of aminoacyl-tRNA synthetases. Class I and class II synthetases recognize different faces of the tRNA molecule. The CCA arm of tRNA adopts different conformations in complexes with the two classes of synthetase.

1. Class I enzymes acylate the 2′-hydroxyl group of the terminal adenosine of tRNA, whereas class II enzymes (except the enzyme for Phe-tRNA) acylate the 3′-hydroxyl group.

2. These two classes bind ATP in different conformations.

3. Most class I enzymes are monomeric, whereas most class II enzymes are dimeric.

Why did two distinct classes of aminoacyl-tRNA synthetases evolve? The observation that the two classes bind to distinct faces of tRNA suggests at least two possibilities. First, recognition sites on both faces of tRNA may have been required to allow the recognition of 20 different tRNAs. Second, it appears possible that, in some cases, a class I enzyme and a class II enzyme can bind to a tRNA molecule simultaneously without colliding with each other. In this way, enzymes from the two classes could work together to modify specific tRNA molecules.

29.3 A RIBOSOME IS A RIBONUCLEOPROTEIN PARTICLE (70S) MADE OF A SMALL (30S) AND A LARGE (50S) SUBUNIT

We turn now to ribosomes, the molecular machines that coordinate the interplay of charged tRNAs, mRNA, and proteins that leads to protein synthesis. An *E. coli* ribosome is a ribonucleoprotein assembly with a mass of about 2700 kd, a diameter of approximately 200 Å, and a sedimentation coefficient of 70S. The 20,000 ribosomes in a bacterial cell constitute nearly a fourth of its mass.

A ribosome can be dissociated into a *large subunit (50S)* and a *small subunit (30S)* (Figure 29.15). These subunits can be further split into their constituent proteins and RNAs. The 30S subunit contains 21 different proteins (referred to as S1 through S21) and a 16S RNA molecule. The 50S subunit contains 34 different proteins (L1 through L34) and two RNA molecules, a 23S and a 5S species. A ribosome contains one copy of each RNA molecule, two copies of the L7 and L12 proteins, and one copy of each of the other proteins. The L7 protein is identical with L12 except that its amino terminus is acetylated. Only one protein is common to both subunits: S20 is identical with L26. Both the 30S and the 50S subunits can be reconstituted in vitro from their constituent proteins and RNA, as was first achieved by Masayasu Nomura in 1968. *This reconstitution is an outstanding example of the principle that supramolecular complexes can form spontaneously from their macromolecular constituents.*

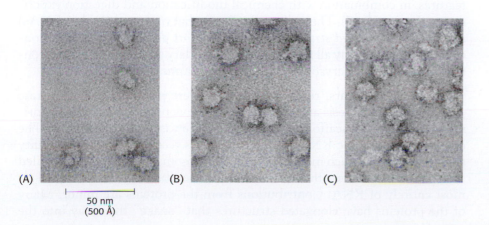

(A) (B) (C)

50 nm
(500 Å)

FIGURE 29.15 Ribosomes at low resolution. Electron micrographs of (A) 30S subunits, (B) 50S subunits, and (C) 70S ribosomes. [Courtesy of Dr. James Lake.]

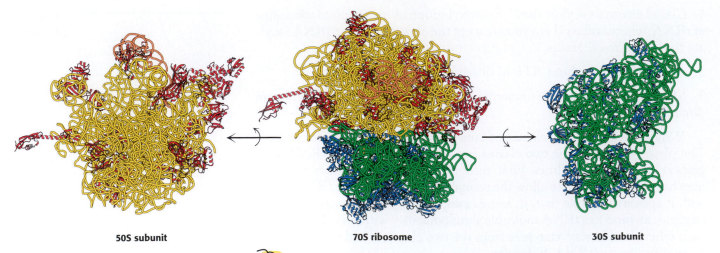

50S subunit   **70S ribosome**   **30S subunit**

FIGURE 29.16 The ribosome at high resolution. Detailed models of the ribosome based on the results of x-ray crystallographic studies of the 70S ribosome and the 30S and 50S subunits. 23S RNA is shown in yellow, 5S RNA in orange, 16S RNA in green, proteins of the 50S subunit in red, and proteins of the 30S subunit in blue.

Electron microscopic studies of the ribosome at increasingly high resolution provided views of the overall structure and revealed the positions of tRNA-binding sites. Astounding progress on the structure of the ribosome has been made by x-ray crystallographic methods, after the pioneering work by Ada Yonath. The structures of both the 30S and the 50S subunits have been determined at or close to atomic resolution, and the elucidation of the structure of intact 70S ribosomes at a similar resolution is following rapidly (Figure 29.16). The determination of this structure requires the positioning of more than 100,000 atoms. The features of these structures are in remarkable agreement with interpretations of less-direct experimental probes. These structures provide an invaluable framework for examining the mechanism of protein synthesis.

29.3.1 Ribosomal RNAs (5S, 16S, and 23S rRNA) Play a Central Role in Protein Synthesis

The prefix *ribo* in the name *ribosome* is apt, because RNA constitutes nearly two-thirds of the mass of these large molecular assemblies. The three RNAs present—5S, 16S, and 23S—are critical for ribosomal architecture and function. They are formed by cleavage of primary 30S transcripts and further processing. The base-pairing patterns of these molecules were deduced by comparing the nucleotide sequences of many species to detect conserved features, in combination with chemical modification and digestion experiments (Figure 29.17). The striking finding is that *ribosomal RNAs* (rRNAs) *are folded into defined structures with many short duplex regions.* This conclusion and essentially all features of the secondary structure have been confirmed by the x-ray crystallographically determined structures.

For many years, ribosomal proteins were presumed to orchestrate protein synthesis and ribosomal RNAs were presumed to serve primarily as structural scaffolding. The current view is almost the reverse. The discovery of catalytic RNA made biochemists receptive to the possibility that RNA plays a much more active role in ribosomal function. The detailed structures make it clear that the key sites in the ribosome are composed almost entirely of RNA. Contributions from the proteins are minor. Many of the proteins have elongated structures that "snake" their way into the

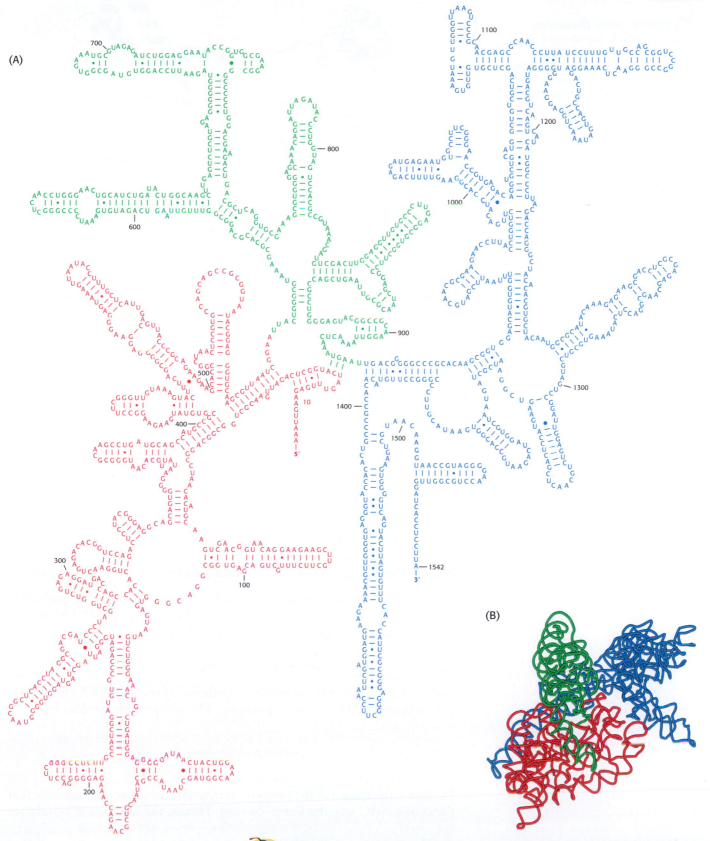

(A)

(B)

FIGURE 29.17 Ribosomal RNA folding pattern. (A) The secondary structure of 16S ribosomal RNA deduced from sequence comparison and the results of chemical studies. (B) The tertiary structure of 16S RNA determined by x-ray crystallography. [Part A courtesy of Dr. Bryn Weiser and Dr. Harry Noller.]

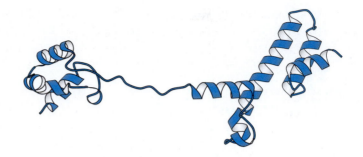

FIGURE 29.18 Ribosomal protein structure. The structure of ribosomal protein L19 of the 50S ribosomal subunit reveals a long segment of extended structure that fits through some of the cavities within the 23S RNA molecule.

RNA matrix (Figure 29.18). The almost inescapable conclusion is that the ribosome initially consisted only of RNA and that the proteins were added later to fine tune its functional properties. This conclusion has the pleasing consequence of dodging a "chicken and egg" question—namely, How can complex proteins be synthesized if complex proteins are required for protein synthesis?

29.3.2 Proteins Are Synthesized in the Amino-to-Carboxyl Direction

Before the mechanism of protein synthesis could be examined, several key facts had to be established. The results of pulse-labeling studies by Howard Dintzis established that protein synthesis proceeds sequentially from the amino terminus. Reticulocytes (young red blood cells) that were actively synthesizing hemoglobin were treated with [³H] leucine. In a period of time shorter than that required to synthesize a complete chain, samples of hemoglobin were taken, separated into α and β chains, and analyzed for the distribution of ³H within their sequences. In the earliest samples, only regions near the carboxyl ends contained radioactivity. In later samples, radioactivity was present closer to the amino terminus as well. This distribution is the one expected if the amino-terminal regions of some chains had already been partly synthesized before the addition of the radioactive amino acid. Thus, *protein synthesis begins at the amino terminus and extends toward the carboxyl terminus.*

29.3.3 Messenger RNA Is Translated in the 5′-to-3′ Direction

The sequence of amino acids in a protein is translated from the nucleotide sequence in mRNA. In which direction is the message read? The answer was established by using the synthetic polynucleotide

$$\overset{5'}{A}—A—A\,\text{+}\,(A—A—A)_n\,A—A—\overset{3'}{C}$$

as the template in a cell-free protein-synthesizing system. AAA encodes lysine, whereas AAC encodes asparagine. The polypeptide product was

$$^+H_3N-Lys-(Lys)_n\!-Asn-\overset{O}{\underset{O}{C}}-$$

Because asparagine was the carboxyl-terminal residue, we can conclude that the codon AAC was the last to be read. Hence, *the direction of translation is 5′ → 3′*.

The direction of translation has important consequences. Recall that transcription also occurs in the 5′ → 3′ direction (Section 28.1.4). If the direction of translation were opposite that of transcription, only fully synthesized mRNA could be translated. In contrast, because the directions are the same, mRNA can be translated while it is being synthesized. In prokaryotes, almost no time is lost between transcription and translation. The 5′ end of mRNA interacts with ribosomes very soon after it is made, much before the 3′ end of the mRNA molecule is finished. *An important feature of prokaryotic gene expression is that translation and transcription are closely coupled in space and time.* Many ribosomes can be translating an mRNA molecule simultaneously. This parallel synthesis markedly increases the efficiency of mRNA translation. The group of ribosomes bound to an mRNA molecule is called a *polyribosome* or a *polysome* (Figure 29.19).

29.3.4 The Start Signal Is AUG (or GUG) Preceded by Several Bases That Pair with 16S rRNA

How does protein synthesis start? The simplest possibility would be for the first three nucleotides of each mRNA to serve as the first codon; no special start signal would then be needed. However, the experimental fact is that translation does not begin immediately at the 5′ terminus of mRNA. Indeed, the first translated codon is nearly always more than 25 nucleotides away from the 5′ end. Furthermore, in prokaryotes, many mRNA molecules are *polycistronic*, or polygenic—that is, they encode two or more polypeptide chains. For example, a single mRNA molecule about 7000 nucleotides long specifies five enzymes in the biosynthetic pathway for tryptophan in *E. coli*. Each of these five proteins has its own start and stop signals on the mRNA. In fact, *all known mRNA molecules contain signals that define the beginning and end of each encoded polypeptide chain.*

A clue to the mechanism of initiation was the finding that nearly half the amino-terminal residues of proteins in *E. coli* are methionine. In fact, the initiating codon in mRNA is AUG (methionine) or, much less frequently, GUG (valine). What additional signals are necessary to specify a translation start site? The first step toward answering this question was the isolation of initiator regions from a number of mRNAs. This isolation was accomplished by using pancreatic ribonuclease to digest mRNA–ribosome complexes (formed under conditions of chain initiation but not elongation). In each case, a sequence of about 30 nucleotides was protected from digestion. As expected, each initiator region displays an AUG (or GUG) codon (Figure 29.20). In addition, each initiator region contains a purine-rich sequence centered about 10 nucleotides on the 5′ side of the initiator codon.

The role of this purine-rich region, called the *Shine-Dalgarno sequence*, became evident when the sequence of 16S rRNA was elucidated. The 3′ end of this rRNA component of the 30S subunit contains a sequence of several bases that is complementary to the purine-rich region in the initiator sites of mRNA. Mutagenesis of the CCUCC sequence near the 3′ end of 16S rRNA to ACACA markedly interferes with the recognition of start sites

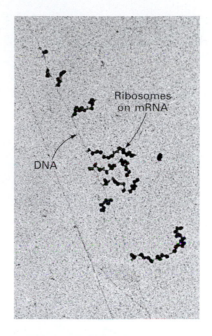

FIGURE 29.19 Polysomes. Transcription of a segment of DNA from *E. coli* generates mRNA molecules that are immediately translated by multiple ribosomes. [From O. L. Miller, Jr., B. A. Hamkalo, and C. A. Thomas, Jr. *Science* 169(1970):392.]

```
5′                                      3′
AGCACGAGGGGAAAUCUGAUGGAACGCUAC    E. coli trpA
UUUGGAUGGAGUGAAACGAUGGCGAUUGCA    E. coli araB
GGUAACCAGGUAACAACCAUGCGAGUGUUG    E. coli thrA
CAAUUCAGGGUGGUGAAUGUGAAACCAGUA    E. coli lacI
AAUCUUGGAGGCUUUUUUUAUGGUUCGUUCU   ϕX174 phage A protein
UAACUAAGGAUGAAAUGCAUGUCUAAGACA    Qβ phage replicase
UCCUAGGAGGUUUGACCUAUGCGAGCUUUU    R17 phage A protein
AUGUACUAAGGAGGUUGUAUGGAACAACGC    λ phage cro
```

Pairs with 16S rRNA   Pairs with initiator tRNA

FIGURE 29.20 Initiation sites. Sequences of mRNA initiation sites for protein synthesis in some bacterial and viral mRNA molecules. Comparison of these sequences reveals some recurring features.

tRNA$_f$
+
Methionine

$\downarrow$ Synthetase

**Methionyl-tRNA$_f$
(Met-tRNA$_f$)**

N^{10}-Formyl-
tetrahydrofolate

$\downarrow$ Transformylase

Tetrahydrofolate

**Formylmethionyl-tRNA$_f$
(fMet-tRNA$_f$)**

FIGURE 29.21 Formylation of methionyl-tRNA. Initiator tRNA (tRNA$_f$) is first charged with methionine, and then a formyl group is transferred to the methionyl-tRNA$_f$ from N^{10}-formyltetrahydrofolate.

FIGURE 29.22 Transfer RNA-binding sites. (A) Three tRNA-binding sites are present on the 70S ribosome. They are called the A (for aminoacyl), P (for peptidyl), and E (for exit) sites. Each tRNA molecule contacts both the 30S and the 50S subunit. (B) The tRNA molecules in sites A and P are base paired with mRNA.

in mRNA. This and other evidence shows that the initiator region of mRNA binds to the 16S rRNA very near its 3′ end. The number of base pairs linking mRNA and 16S rRNA ranges from three to nine. Thus, *two kinds of interactions determine where protein synthesis starts: (1) the pairing of mRNA bases with the 3′ end of 16S rRNA and (2) the pairing of the initiator codon on mRNA with the anticodon of an initiator tRNA molecule.*

29.3.5 Bacterial Protein Synthesis Is Initiated by Formylmethionyl Transfer RNA

The methionine residue found at the amino-terminal end of *E. coli* proteins is usually modified. In fact, *protein synthesis in bacteria starts with N-formylmethionine* (fMet). A special tRNA brings formylmethionine to the ribosome to initiate protein synthesis. This *initiator tRNA* (abbreviated as tRNA$_f$) differs from the one that inserts methionine in internal positions (abbreviated as tRNA$_m$). The subscript "f" indicates that methionine attached to the initiator tRNA can be formylated, whereas it cannot be formylated when attached to tRNA$_m$. In approximately one-half of *E. coli* proteins, N-formylmethionine is removed when the nascent chain is 10 amino acids long.

Methionine is linked to these two kinds of tRNAs by the same aminoacyl-tRNA synthetase. A specific enzyme then formylates the amino group of methionine that is attached to tRNA$_f$ (Figure 29.21). The activated formyl donor in this reaction is N^{10}-formyltetrahydrofolate (Section 24.2.6). It is significant that free methionine and methionyl-tRNA$_m$ are not substrates for this transformylase.

29.3.6 Ribosomes Have Three tRNA-Binding Sites That Bridge the 30S and 50S Subunits

A snapshot of a significant moment in protein synthesis was obtained by determining the structure of the 70S ribosome bound to three tRNA molecules and a fragment of mRNA (Figure 29.22). As expected, the mRNA fragment is bound within the 30S subunit. Each of the tRNA molecules bridges between the 30S and 50S subunits. At the 30S end, two of the three tRNA molecules are bound to the mRNA fragment through anticodon–codon base pairs. These binding sites are called the A site (for *a*minoacyl) and the P site (for *p*eptidyl). The third tRNA molecule is bound to an adjacent site called the E site (for *e*xit).

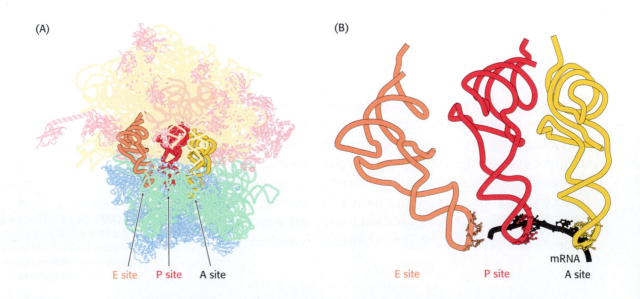

(A)

E site   P site   A site

(B)

mRNA

E site   P site   A site

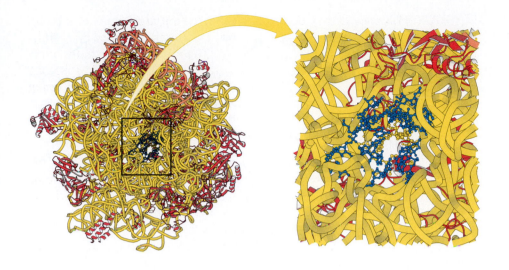

FIGURE 29.23 Polypeptide escape path. A tunnel passes through the 50S subunit beginning at the site of peptide-bond formation (shown in blue). The growing polypeptide chain passes through this tunnel.

The other end of each tRNA molecule interacts with the 50S subunit. The acceptor stems of the tRNA molecules occupying the A site and the P site converge at a site where a peptide bond is formed. Further examination of this site reveals that a tunnel connects this site to the back of the ribosome (Figure 29.23). *The polypeptide chain passes through this tunnel during synthesis.*

29.3.7 The Growing Polypeptide Chain Is Transferred Between tRNAs on Peptide-Bond Formation

Protein synthesis begins with the interaction of the 30S subunit and mRNA through the Shine-Delgarno sequence. On formation of this complex, the initiator tRNA charged with formylmethionine binds to the initiator AUG codon, and the 50S subunit binds to the 30S subunit to form the complete 70S ribosome. How does the polypeptide chain increase in length (Figure 29.24)? The three sites in our snapshot of protein synthesis provide a clue.

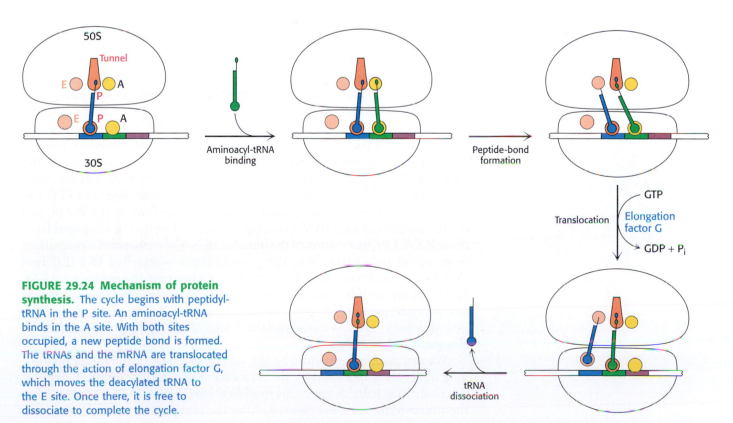

FIGURE 29.24 Mechanism of protein synthesis. The cycle begins with peptidyl-tRNA in the P site. An aminoacyl-tRNA binds in the A site. With both sites occupied, a new peptide bond is formed. The tRNAs and the mRNA are translocated through the action of elongation factor G, which moves the deacylated tRNA to the E site. Once there, it is free to dissociate to complete the cycle.

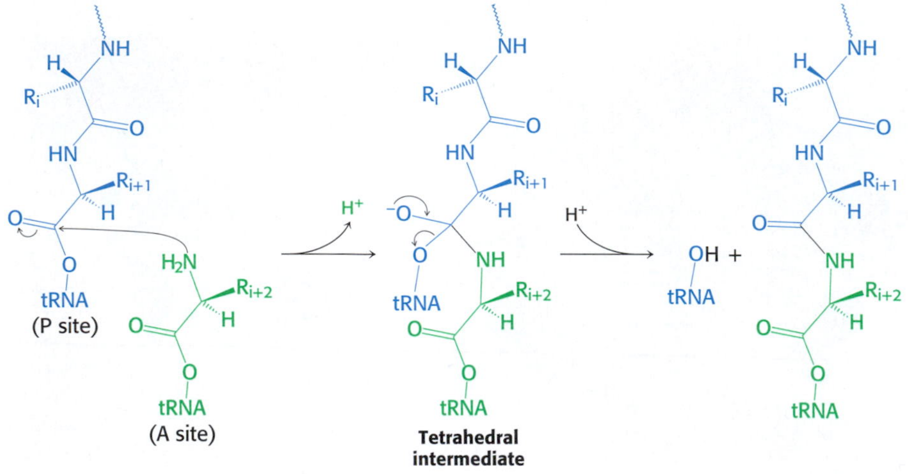

FIGURE 29.25 Peptide-bond formation. The amino group of the aminoacyl-tRNA attacks the carbonyl group of the ester linkage of the peptidyl-tRNA to form a tetrahedral intermediate. This intermediate collapses to form the peptide bond and release the deacylated tRNA.

The initiator tRNA is bound in the P site on the ribosome. A charged tRNA with an anticodon complementary to the codon in the A site then binds. The stage is set for the formation of a peptide bond: the formylmethionine molecule linked to the initiator tRNA will be transferred to the amino group of the amino acid in the A site. The transfer takes place in a ribosome site called the *peptidyl transferase center*.

The amino group of the aminoacyl-tRNA in the A site is well positioned to attack the ester linkage between the initiator tRNA and the formylmethionine molecule (Figure 29.25). The peptidyl transferase center includes bases that promote this reaction by helping to form an $-NH_2$ group on the A site aminoacyl-tRNA and by helping to stabilize the tetrahedral intermediate that forms. This reaction is, in many ways, analogous to the reverse of the reaction catalyzed by serine proteases such as chymotrypsin (Section 9.1.2). The peptidyl-tRNA is analogous to the acyl-enzyme form of a serine protease. In a serine protease, the acyl-enzyme is generated with the use of the free energy associated with cleaving an amide bond. In the ribosome, the free energy necessary to form the analogous species, an amino-acyl-tRNA, comes from the ATP that is cleaved by the aminoacyl-tRNA synthetase before the arrival of the tRNA at the ribosome.

With the peptide bond formed, the peptide chain is now attached to the tRNA in the A site on the 30S subunit while a change in the interaction with the 50S subunit has placed that tRNA and its peptide in the P site of the large subunit. The tRNA in the P site of the 30S subunit is now uncharged. For translation to proceed, the mRNA must be moved (or *translocated*) so that the codon for the next amino acid to be added is in the A site. This translocation takes place through the action of a protein enzyme called *elongation factor G* (Section 29.4.3), driven by the hydrolysis of GTP. On completion of this step, the peptidyl-tRNA is now fully in the P site, and the uncharged initiator tRNA is in the E site and has been disengaged from the mRNA. On dissociation of the initiator tRNA, the ribosome has returned to its initial state except that the peptide chain is attached to a different tRNA, the one corresponding to the first codon past the initiating AUG. Note that *the peptide chain remains in the P site on the 50S subunit throughout this cycle*, presumably growing into the tunnel. This cycle is repeated as new aminoacyl-tRNAs move into the A site, allowing the polypeptide to be elongated indefinitely.

We can now understand why the amino terminus of the initial methionine molecule is modified by the attachment of a formyl group. Chemical reactivity may have dictated this modification (Figure 29.26). Suppose that the amino-terminus is not blocked. After the first peptidyl-transfer reaction, a dipeptide is linked to the tRNA in the P site. If a free amino group is present in the terminal amino acid, this amino group can attack the carbonyl group of the ester linkage to the tRNA, forming a very stable six-membered ring and terminating translation.

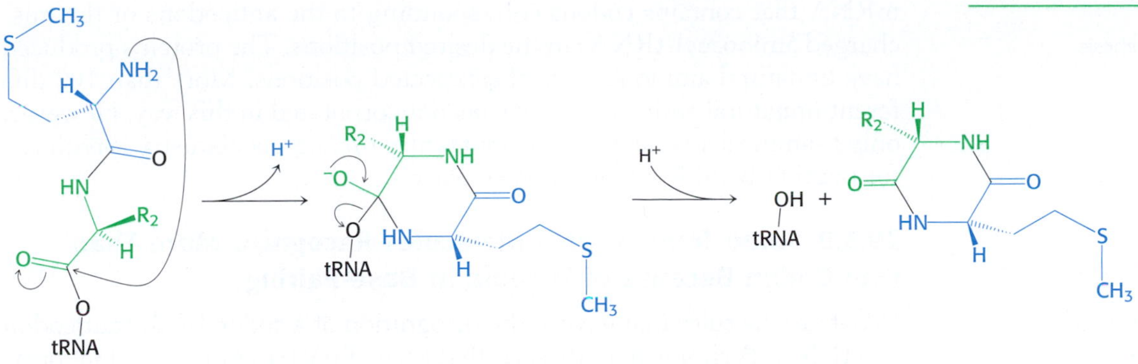

FIGURE 29.26 A role for formylation. With a free terminal amino group, dipeptidyl-tRNA can cyclize to cleave itself from tRNA. Formylation of the amino terminus blocks this reaction.

29.3.8 Only the Codon–Anticodon Interactions Determine the Amino Acid That Is Incorporated

On the basis of the mechanism described in Section 29.3.7, the base-pairing interaction between the anticodon on the incoming tRNA and the codon in the A site on mRNA determines which amino acid is added to the polypeptide chain. Does the amino acid attached to the tRNA play any role in this process? This question was answered in the following way. First, cysteine was attached to its cognate tRNA. The attached cysteine unit was then converted into alanine by adding Raney nickel to Cys-tRNA^{Cys}; the reaction removed the sulfur atom from the cysteine residue without affecting its linkage to tRNA. Thus, a *mischarged aminoacyl-tRNA* was produced in which alanine was covalently attached to a tRNA specific for cysteine.

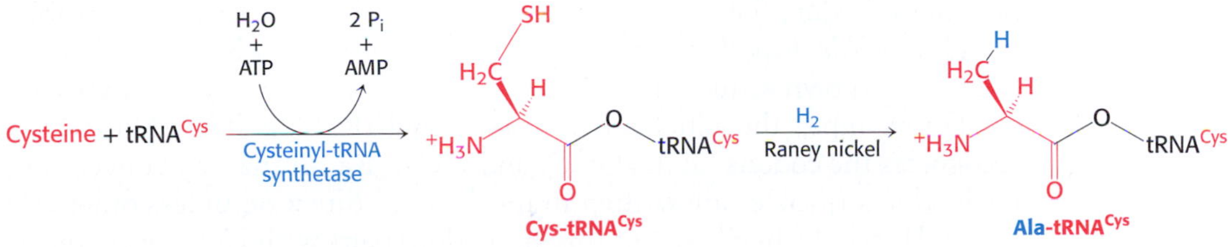

Does this mischarged tRNA recognize the codon for cysteine or for alanine? The answer came when the tRNA was added to a cell-free protein-synthesizing system. The template was a random copolymer of U and G in the ratio of 5:1, which normally incorporates cysteine (encoded by UGU) but not alanine (encoded by GCN). However, alanine was incorporated into a polypeptide when Ala-tRNA^{Cys} was added to the incubation mixture. The same result was obtained when mRNA for hemoglobin served as the template and [^{14}C]alanyl-tRNA^{Cys} was used as the mischarged aminoacyl-tRNA. The only radioactive tryptic peptide produced was one that normally contained cysteine but not alanine. Thus, *the amino acid in aminoacyl-tRNA does not play a role in selecting a codon.*

In recent years, the ability of mischarged tRNAs to transfer their amino acid cargo to a growing polypeptide chain has been used to synthesize peptides with amino acids not found in proteins incorporated into specific sites in a protein. Aminoacyl-tRNAs are first linked to these unnatural amino acids by chemical methods. These mischarged aminoacyl-tRNAs are added to a cell-free protein-synthesizing system along with specially engineered mRNA that contains codons corresponding to the anticodons of the mischarged aminoacyl-tRNAs in the desired positions. The proteins produced have unnatural amino acids in the expected positions. More than 100 different unnatural amino acids have been incorporated in this way. However, only L-amino acids can be used; apparently this stereochemistry is required for peptide-bond formation to take place.

29.3.9 Some Transfer RNA Molecules Recognize More Than One Codon Because of Wobble in Base-Pairing

What are the rules that govern the recognition of a codon by the anticodon of a tRNA? A simple hypothesis is that each of the bases of the codon forms a Watson-Crick type of base pair with a complementary base on the anticodon. The codon and anticodon would then be lined up in an antiparallel fashion. In the diagram in the margin, the prime denotes the complementary base. Thus X and X′ would be either A and U (or U and A) or G and C (or C and G). According to this model, a particular anticodon can recognize only one codon.

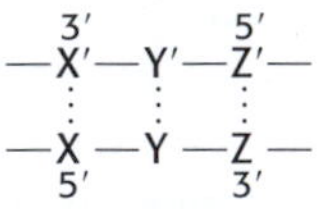

The facts are otherwise. As found experimentally, *some pure tRNA molecules can recognize more than one codon.* For example, the yeast alanyl-tRNA binds to *three* codons: GCU, GCC, and GCA. The first two bases of these codons are the same, whereas the third is different. Could it be that recognition of the third base of a codon is sometimes less discriminating than recognition of the other two? The pattern of degeneracy of the genetic code indicates that this might be so. XYU and XYC always encode the same amino acid; XYA and XYG usually do. Francis Crick surmised from these data that the steric criteria might be less stringent for pairing of the third base than for the other two. Models of various base pairs were built to determine which ones are similar to the standard A · U and G · C base pairs with regard to the distance and angle between the glycosidic bonds. Inosine was included in this study because it appeared in several anticodons. With the assumption of some steric freedom ("wobble") in the pairing of the third base of the codon, the combinations shown in Table 29.3 seemed plausible.

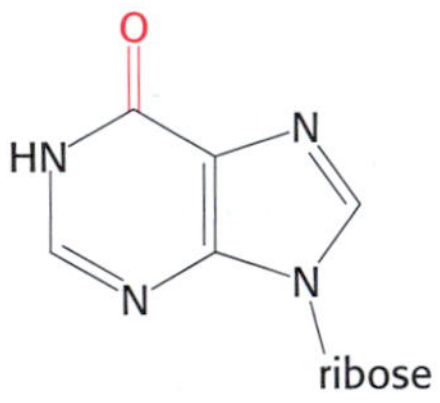

The *wobble hypothesis* is now firmly established. The anticodons of tRNAs of known sequence bind to the codons predicted by this hypothesis. For example, the anticodon of yeast alanyl-tRNA is IGC. This tRNA recognizes the codons GCU, GCC, and GCA. Recall that, by convention, nucleotide sequences are written in the 5′ → 3′ direction unless otherwise noted. Hence, I (the 5′ base of this anticodon) pairs with U, C, or A (the 3′ base of the codon), as predicted.

Anticodon

$$-\overset{3'}{X'}-Y'-\overset{5'}{Z'}-$$
$$-\overset{5'}{X}-Y-\overset{3'}{Z}-$$

Codon

Inosine

TABLE 29.3 Allowed pairings at the third base of the codon according to the wobble hypothesis

First base of anticodon	Third base of codon
C	G
A	U
U	A or G
G	U or C
I	U, C, or A

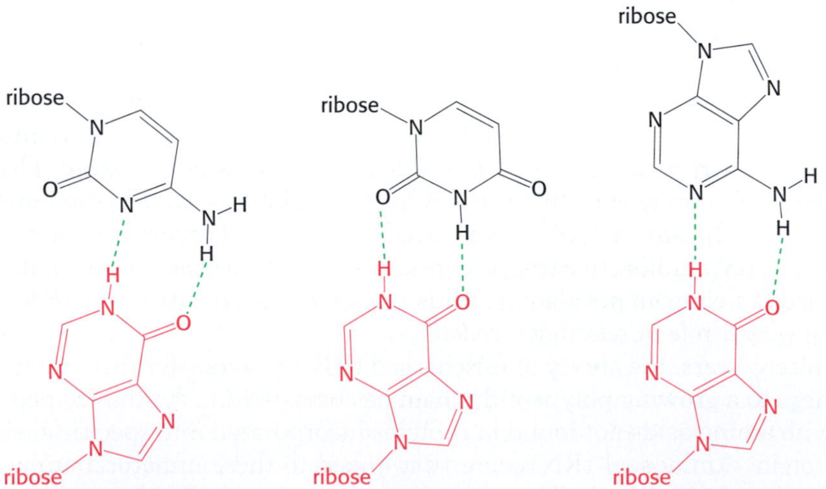

Inosine–cytidine base pair

Inosine–uridine base pair

Inosine–adenosine base pair

Two generalizations concerning the codon–anticodon interaction can be made:

1. The first two bases of a codon pair in the standard way. Recognition is precise. Hence, *codons that differ in either of their first two bases must be recognized by different tRNAs*. For example, both UUA and CUA encode leucine but are read by different tRNAs.

2. The first base of an anticodon determines whether a particular tRNA molecule reads one, two, or three kinds of codons: C or A (one codon), U or G (two codons), or I (three codons). Thus, *part of the degeneracy of the genetic code arises from imprecision (wobble) in the pairing of the third base of the codon with the first base of the anticodon*. We see here a strong reason for the frequent appearance of inosine, one of the unusual nucleosides, in anticodons. *Inosine maximizes the number of codons that can be read by a particular tRNA molecule*. The inosines in tRNA are formed by deamination of adenosine after synthesis of the primary transcript.

Why is wobble tolerated in the third position of the codon but not in the first two? The 30S subunit has two adenine bases (A1492 and A1493 in the 16S RNA) that form hydrogen bonds on the minor-groove side of the codon–anticodon duplex. These interactions serve to check whether Watson-Crick base pairs are present in the first two positions of the codon–anticodon duplex. No such inspection device is present for the third position so more-varied base pairs are tolerated. This mechanism for ensuring fidelity is analogous to the minor-groove interactions utilized by DNA polymerase for a similar purpose (Section 27.2.3). *Thus, the ribosome plays an active role in decoding the codon–anticodon interactions*.

29.4 PROTEIN FACTORS PLAY KEY ROLES IN PROTEIN SYNTHESIS

Although rRNA is paramount in the process of translation, protein factors also are required for the efficient synthesis of a protein. Protein factors participate in the initiation, elongation, and termination of protein synthesis. P-loop NTPases of the G-protein family play particularly important roles. Recall that these proteins serve as molecular switches as they cycle between a GTP-bound form and a GDP-bound form (Chapter 15.1.2).

29.4.1 Formylmethionyl-tRNA$_f$ Is Placed in the P Site of the Ribosome During Formation of the 70S Initiation Complex

Messenger RNA and formylmethionyl-tRNA$_f$ must be brought to the ribosome for protein synthesis to begin. How is this accomplished? Three protein *initiation factors* (IF1, IF2, and IF3) are essential. The 30S ribosomal subunit first forms a complex with IF1 and IF3 (Figure 29.27). The binding of these factors to the 30S subunit prevents it from prematurely joining the 50S subunit to form a dead-end 70S complex, devoid of mRNA and fMet-tRNA$_f$. Initiation factor 2, a member of the G-protein family, binds GTP, and the concomitant conformational change enables IF$_2$ to associate with formylmethionyl-tRNA$_f$. The IF2–GTP–initiator tRNA complex binds with mRNA (correctly positioned by the Shine-Dalgarno sequence interaction with the 16S rRNA) and the 30S subunit to form the *30S initiation complex*. The hydrolysis of GTP bound to IF2 on entry of the 50S subunit leads to the release of the initiation factors. The result is a *70S initiation complex*.

When the 70S initiation complex has been formed, the ribosome is ready for the elongation phase of protein synthesis. The fMet-tRNA$_f$ molecule

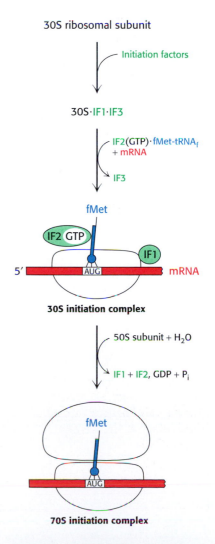

FIGURE 29.27 Translation initiation in prokaryotes. Initiation factors aid the assembly first of the 30S initiation complex and then of the 70S initiation complex.

occupies the P site on the ribosome. The other two sites for tRNA molecules, the A site and the E site, are empty. Formylmethionyl-tRNA$_f$ is positioned so that its anticodon pairs with the initiating AUG (or GUG) codon on mRNA. This interaction sets the reading frame for the translation of the entire mRNA.

29.4.2 Elongation Factors Deliver Aminoacyl-tRNA to the Ribosome

The second phase of protein synthesis is the elongation cycle. This phase begins with the insertion of an aminoacyl-tRNA into the empty A site on the ribosome. The particular species inserted depends on the mRNA codon in the A site. The cognate aminoacyl-tRNA does not simply leave the synthetase and diffuse to the A site. Rather, it is delivered to the A site in association with a 43-kd protein called *elongation factor Tu (EF-Tu)*. Elongation factor Tu, another member of the G-protein family, binds aminoacyl-tRNA only in the GTP form (Figure 29.28). The binding of EF-Tu to aminoacyl-tRNA serves two functions. First, EF-Tu protects the delicate ester linkage in aminoacyl-tRNA from hydrolysis. Second, the GTP in EF-Tu is hydrolyzed to GDP when an appropriate complex between the EF-Tu–aminoacyl-tRNA complex and the ribosome has formed. If the anticodon is not properly paired with the codon, hydrolysis does not take place and the aminoacyl-tRNA is not transferred to the ribosome. This mechanism allows the free energy of GTP hydrolysis to contribute to the fidelity of protein synthesis.

How is EF-Tu in the GDP form reset to bind another aminoacyl-tRNA? *Elongation Factor Ts*, a second elongation factor, joins the EF-Tu complex and induces the dissociation of GDP. Finally, GTP binds to EF-Tu, and EF-Ts is concomitantly released. It is noteworthy that *EF-Tu does not interact with fMet-tRNA$_f$*. Hence, this initiator tRNA is not delivered to the A site. In contrast, Met-tRNA$_m$, like all other aminoacyl-tRNAs, does bind to EF-Tu. These findings account for the fact that *internal AUG codons are not read by the initiator tRNA*. Conversely, initiation factor 2 recognizes fMet-tRNA$_f$ but no other tRNA.

This GTP–GDP cycle of EF-Tu is reminiscent of those of the heterotrimeric G proteins in signal transduction (Section 15.1.2) and the Ras proteins in growth control (Section 15.4.2). This similarity is due to their evolutionary heritage, inasmuch as the amino-terminal domain of EF-Tu is homologous to the P-loop NTPase domains in the other G proteins. The other two domains of the tripartite EF-Tu are distinctive; they mediate interactions with aminoacyl-tRNA and the ribosome. In all these related enzymes, the change in conformation between the GTP and the GDP forms leads to a change in interaction partners. A further similarity is the requirement that an additional protein catalyze the exchange of GTP for GDP; an activated receptor plays the role of EF-Ts for a heterotrimeric G protein, as does Sos for Ras.

29.4.3 The Formation of a Peptide Bond Is Followed by the GTP-Driven Translocation of tRNAs and mRNA

After the correct aminoacyl-tRNA has been placed in the A site, the transfer of the polypeptide chain from the tRNA in the P site is a spontaneous process, driven by the formation of the stronger peptide bond in place of the ester linkage. However, protein synthesis cannot continue without the translocation of the mRNA and the tRNAs within the ribosome. The mRNA must move by a distance of three nucleotides as the deacylated tRNA moves out of the P site into the E site on the 30S subunit and the

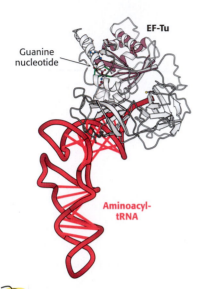

FIGURE 29.28 Structure of elongation factor Tu. The structure of a complex between elongation factor Tu (EF-Tu) and an aminoacyl-tRNA. The amino-terminal domain of EF-Tu is a P-loop NTPase domain similar to those in other G proteins.

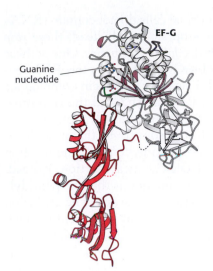

EF-G

Guanine
nucleotide

FIGURE 29.29 Molecular mimicry.
The structure of elongation factor G (EF-G) is
remarkably similar in shape to that of the
EF-Tu–tRNA complex (see Figure 29.28). The
amino-terminal region of EF-G is homologous to
EF-Tu, and the carboxyl-terminal region (shown
in red) comprises a set of protein domains that
adopted the shape of a tRNA molecule over the
course of evolution.

peptidyl-tRNA moves out of the A site into the P site on the 30S subunit.
The result is that the next codon is positioned in the A site for interaction
with the incoming aminoacyl-tRNA.

   Translocation is mediated by *elongation factor G (EF-G,* also called
translocase). The structure of EF-G is exceptional in revealing some aspects
of its mode of action (Figure 29.29). The structure of EF-G closely resembles that of the complex between EF-Tu and tRNA. This is an example of
molecular mimicry; a protein domain evolved so that it mimics the shape of
a tRNA molecule. This structural similarity, as well as other experimental
data, suggests a mechanism for the translocation process (Figure 29.30).
First, EF-G in the GTP form binds to the ribosome, primarily through the
interaction of its EF-Tu-like domain with the 50S subunit. The binding site
includes proteins L11 and the L7-L12 dimer. The tRNA-like domain of
EF-G interacts with the 30S subunit. The binding of EF-G to the ribosome
in this manner stimulates the GTPase activity of EF-G. On GTP hydrolysis, EF-G undergoes a conformational change that forces its arm deeper
into the A site on the 30S subunit. To accommodate this domain, the peptidyl-tRNA in the A site moves to the P site, carrying the mRNA and the
deacylated tRNA with it. The ribosome may be prepared for these rearrangements by the initial binding of EF-G as well. The dissociation of
EF-G leaves the ribosome ready to accept the next aminoacyl-tRNA into
the A site.

**FIGURE 29.30 Translocation
mechanism.** In the GTP form, EF-G
binds to the EF-Tu-binding site on the 50S
subunit. This stimulates GTP hydrolysis,
inducing a conformational change in EF-G,
and driving the stem of EF-G into the A
site on the 30S subunit. To accommodate
this domain, the tRNAs and mRNA move
through the ribosome by a distance
corresponding to one codon.

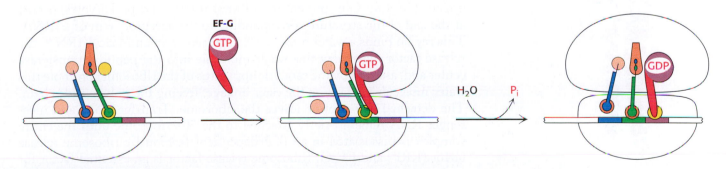

EF-G

GTP

GTP

H₂O   Pᵢ

GDP

29.4.4 Protein Synthesis Is Terminated by Release Factors That Read Stop Codons

The final phase of translation is termination. How does the synthesis of a
polypeptide chain come to an end when a stop codon is encountered?
Aminoacyl-tRNA does not normally bind to the A site of a ribosome if the

codon is UAA, UGA, or UAG, because normal cells do not contain tRNAs with anticodons complementary to these stop signals. Instead, these *stop codons are recognized by release factors* (RFs), *which are proteins*. One of these release factors, RF1, recognizes UAA or UAG. A second factor, RF2, recognizes UAA or UGA. A third factor, RF3, another G protein homologous to EF-Tu, mediates interactions between RF1 or RF2 and the ribosome.

Release factors use a Trojan horse strategy to free the polypeptide chain. One of the most impressive properties of the ribosome is *not* that it catalyzes peptide-bond formation; the formation of a peptide bond by the reaction between an amino group and an ester is a facile chemical reaction. Instead, a more impressive feature crucial to ribosome function is that the peptidyl-tRNA ester linkage is not broken by premature hydrolysis. The exclusion of water from the peptidyl transferase center is crucial in preventing such hydrolysis, which would lead to release of the polypeptide chain. The structure of a prokaryotic release factor has not yet been determined. However, the structure of a eukaryotic release factor, though probably not truly homologous to its prokaryotic counterpart, reveals the strategy (Figure 29.31).

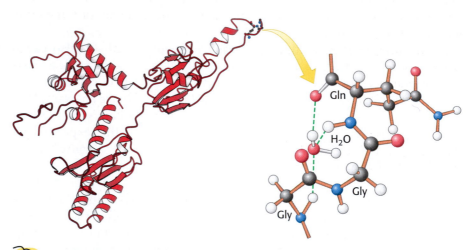

🖱 **FIGURE 29.31 Structure of a release factor.** The structure of a eukaryotic release factor reveals a tRNA-like fold. The acceptor-stem mimic includes the sequence Gly-Gly-Gln at its tip. This region appears to bind a water molecule, which may be brought into the peptidyl transferase center. There it can participate in the cleavage of the peptidyl-tRNA ester bond, with the aid of the glutamine residue and the ribosomal catalytic apparatus.

The structure resembles that of a tRNA by molecular mimicry. The sequence Gly-Gly-Gln, present in both eukaryotes and prokaryotes, occurs at the end of the structure corresponding to the acceptor stem of a tRNA. This region binds a water molecule. Disguised as an aminoacyl-tRNA, the release factor may carry this water molecule into the peptidyl transferase center and, assisted by the catalytic apparatus of the ribosome, promote this water molecule's attack on the ester linkage, freeing the polypeptide chain. The detached polypeptide leaves the ribosome. Transfer RNA and messenger RNA remain briefly attached to the 70S ribosome until the entire complex is dissociated in a GTP-dependent fashion by ribosome release factor (RRF) and EF-G. Ribosome release factor is an essential factor for prokaryotic translation.

The structure of RRF, too, resembles tRNA (Figure 29.32). However, the known tRNA-mimicking structures of RRF, EF-G, and the release factors are distinct; they do not appear to have been generated from a common ancestor. Thus, convergent evolution has provided a similar solution—looking sufficiently like a tRNA to interact with the tRNA-

🖱 **FIGURE 29.32 Structure of ribosome release factor (RRF).** RRF is another protein that resembles tRNA. The α helices of this protein mimic the tRNA structure. In contrast, in EF-G, β strands are the mimics, revealing an independent evolutionary origin.

binding sites on the ribosome—to several problems. The effects of *divergent* evolution are evident in the protein factors that participate in translation, most notably in the form of the homologous G proteins, EF-Tu, EF-G, IF2, and RF3.

29.5 EUKARYOTIC PROTEIN SYNTHESIS DIFFERS FROM PROKARYOTIC PROTEIN SYNTHESIS PRIMARILY IN TRANSLATION INITIATION

The basic plan of protein synthesis in eukaryotes and archaea is similar to that in bacteria. The major structural and mechanistic themes recur in all domains of life. However, eukaryotic protein synthesis entails more protein components than does prokaryotic protein synthesis, and some steps are more intricate. Some noteworthy similarities and differences are as follows:

1. *Ribosomes.* Eukaryotic ribosomes are larger. They consist of a 60S large subunit and a 40S small subunit, which come together to form an 80S particle having a mass of 4200 kd, compared with 2700 kd for the prokaryotic 70S ribosome. The 40S subunit contains an 18S RNA that is homologous to the prokaryotic 16S RNA. The 60S subunit contains three RNAs: the 5S and 28S RNAs are the counterparts of the prokaryotic 5S and 23S molecules; its 5.8S RNA is unique to eukaryotes.

2. *Initiator tRNA.* In eukaryotes, the initiating amino acid is methionine rather than *N*-formylmethionine. However, as in prokaryotes, a special tRNA participates in initiation. This aminoacyl-tRNA is called Met-tRNA$_i$ or Met-tRNA$_f$ (the subscript "i" stands for initiation, and "f" indicates that it can be formylated in vitro).

3. *Initiation.* The initiating codon in eukaryotes is always AUG. Eukaryotes, in contrast with prokaryotes, do not use a specific purine-rich sequence on the 5′ side to distinguish initiator AUGs from internal ones. Instead, the AUG nearest the 5′ end of mRNA is usually selected as the start site. A 40S ribosome attaches to the cap at the 5′ end of eukaryotic mRNA (Section 28.3.1) and searches for an AUG codon by moving step-by-step in the 3′ direction (Figure 29.33). This scanning process in eukaryotic protein synthesis is powered by helicases that hydrolyze ATP. Pairing of the anticodon of Met-tRNA$_i$ with the AUG codon of mRNA signals that the target has been found. In almost all cases, eukaryotic mRNA has only one start site and hence is the template for a single protein. In contrast, a prokaryotic mRNA can have multiple Shine-Dalgarno sequences and, hence, start sites, and it can serve as a template for the synthesis of several proteins.

Eukaryotes utilize many more initiation factors than do prokaryotes, and their interplay is much more intricate. The prefix *eIF* denotes a eukaryotic initiation factor. For example, eIF-4E is a protein that binds directly to the 7-methylguanosine cap (Section 28.3.1), whereas eIF-4A is a helicase. The difference in initiation mechanism between prokaryotes and eukaryotes is, in part, a consequence of the difference in RNA processing. The 5′ end of mRNA is readily available to ribosomes immediately after transcription in prokaryotes. In contrast, pre-mRNA must be processed and transported to the cytoplasm in eukaryotes before translation is initiated. Thus, there is ample opportunity for the formation of complex secondary structures that must be removed to expose signals in the mature mRNA. The 5′ cap provides an easily recognizable starting point. In addition, the complexity of eukaryotic translation initiation provides another mechanism for gene expression that we shall explore further in Chapter 31.

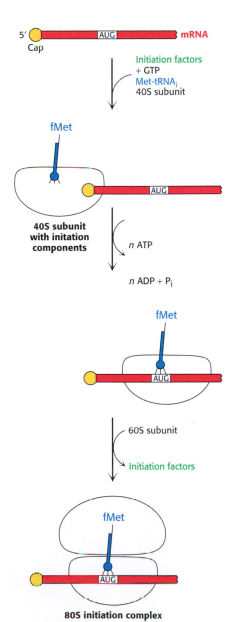

FIGURE 29.33 Eukaryotic translation initiation. In eukaryotes, translation initiation starts with the assembly of a complex on the 5′ cap that includes the 40S subunit and Met-tRNA$_i$. Driven by ATP hydrolysis, this complex scans the mRNA until the first AUG is reached. The 60S subunit is then added to form the 80S initiation complex.

4. *Elongation and termination.* Eukaryotic elongation factors EF1α and EF1βγ are the counterparts of prokaryotic EF-Tu and EF-Ts. The GTP form of EF1α delivers aminoacyl-tRNA to the A site of the ribosome, and EF1βγ catalyzes the exchange of GTP for bound GDP. Eukaryotic EF2 mediates GTP-driven translocation in much the same way as does prokaryotic EF-G. Termination in eukaryotes is carried out by a single release factor, eRF1, compared with two in prokaryotes. Finally, eIF3, like its prokaryotic counterpart IF3, prevents the reassociation of ribosomal subunits in the absence of an initiation complex.

29.5.1 Many Antibiotics Work by Inhibiting Protein Synthesis

The differences between eukaryotic and prokaryotic ribosomes can be exploited for the development of antibiotics (Table 29.4). For example, the antibiotic *puromycin* inhibits protein synthesis by causing nascent prokaryotic polypeptide chains to be released before their synthesis is completed. Puromycin is an analog of the terminal aminoacyl-adenosine part of aminoacyl-tRNA (Figure 29.34).

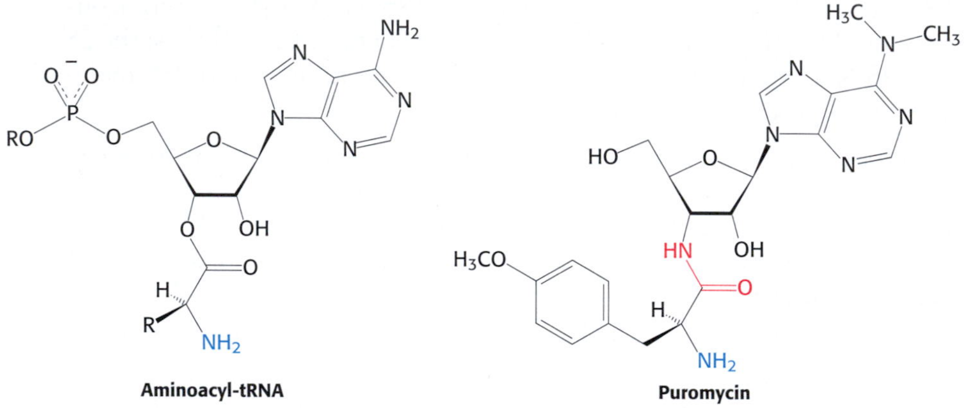

FIGURE 29.34 Antibiotic action of puromycin. Puromycin resembles the aminoacyl terminus of an aminoacyl-tRNA. Its amino group joins the carbonyl group of the growing polypeptide chain to form an adduct that dissociates from the ribosome. This adduct is stable because puromycin has an amide (shown in red) rather than an ester linkage.

Aminoacyl-tRNA

Puromycin

It binds to the A site on the ribosome and inhibits the entry of aminoacyl-tRNA. Furthermore, puromycin contains an α-amino group. This amino group, like the one on aminoacyl-tRNA, forms a peptide bond with the carboxyl group of the growing peptide chain. The product, a peptide having a covalently attached puromycin residue at its carboxyl end, dissociates from the ribosome.

TABLE 29.4 Antibiotic inhibitors of protein synthesis

Antibiotic	Action
Streptomycin and other aminoglycosides	Inhibit initiation and cause misreading of mRNA (prokaryotes)
Tetracycline	Binds to the 30S subunit and inhibits binding of aminoacyl-tRNAs (prokaryotes)
Chloramphenicol	Inhibits the peptidyl transferase activity of the 50S ribosomal subunit (prokaryotes)
Cycloheximide	Inhibits the peptidyl transferase activity of the 60S ribosomal subunit (eukaryotes)
Erythromycin	Binds to the 50S subunit and inhibits translocation (prokaryotes)
Puromycin	Causes premature chain termination by acting as an analog of aminoacyl-tRNA (prokaryotes and eukaryotes)

Streptomycin, a highly basic trisaccharide, interferes with the binding of formylmethionyl-tRNA to ribosomes and thereby prevents the correct initiation of protein synthesis. Other *aminoglycoside antibiotics* such as neomycin, kanamycin, and gentamycin interfere with the *decoding site* located near nucleotide 1492 in 16S rRNA of the 30S subunit (Section 29.3.9). *Chloramphenicol* acts by inhibiting peptidyl transferase activity. *Erythromycin* binds to the 50S subunit and blocks translocation. Finally, *cyclohexamide* blocks peptidyl transferase activity in eukaryotic ribosomes, making a useful laboratory tool for blocking protein synthesis in eukaryotic cells.

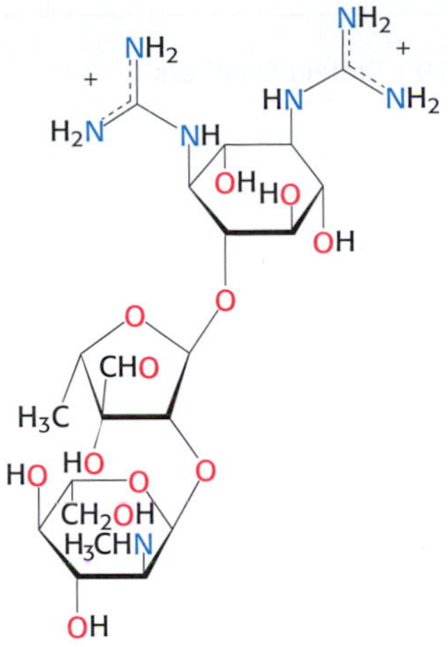

Streptomycin

29.5.2 Diphtheria Toxin Blocks Protein Synthesis in Eukaryotes by Inhibiting Translocation

Diphtheria was a major cause of death in childhood before the advent of effective immunization. The lethal effects of this disease are due mainly to a protein toxin produced by *Corynebacterium diphtheriae*, a bacterium that grows in the upper respiratory tract of an infected person. The gene that encodes the toxin comes from a lysogenic phage that is harbored by some strains of *C. diphtheriae*. A few micrograms of diphtheria toxin is usually lethal in an unimmunized person because it inhibits protein synthesis. The toxin is cleaved shortly after entering a target cell into a 21-kd A fragment and a 40-kd B fragment. *The A fragment of the toxin catalyzes the covalent modification of an important component of the protein-synthesizing machinery, whereas the B fragment enables the A fragment to enter the cytosol of its target cell.*

A single A fragment of the toxin in the cytosol can kill a cell. Why is it so lethal? The target of the A fragment is EF2, the elongation factor catalyzing translocation in eukaryotic protein synthesis. EF2 contains *diphthamide*, an unusual amino acid residue of unknown function that is formed by posttranslational modification of histidine. The A fragment catalyzes the transfer of the adenosine diphosphate ribose unit of NAD^+ to a nitrogen atom of the diphthamide ring (Figure 29.35). *This ADP-ribosylation of a single side chain of EF2 blocks its capacity to carry out translocation of the growing polypeptide chain.* Protein synthesis ceases, accounting for the remarkable toxicity of diphtheria toxin.

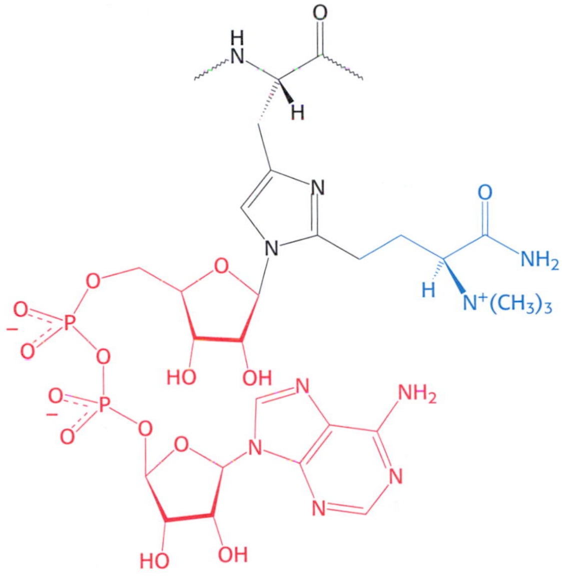

ADP-ribose

FIGURE 29.35 Blocking of translocation by diphtheria toxin. Diphtheria toxin blocks protein synthesis in eukaryotes by catalyzing the transfer of an ADP-ribose unit from NAD^+ to diphthalamide, a modified amino acid residue in elongation factor 2 (translocase). Diphthalamide is formed by a posttranslational modification (blue) of a histidine residue.

SUMMARY

Protein Synthesis Requires the Translation of Nucleotide Sequences into Amino Acid Sequences

Protein synthesis is called translation because information present as a nucleic acid sequence is translated into a different language, the sequence of amino acids in a protein. This complex process is mediated by the coordinated interplay of more than a hundred macromolecules, including mRNA, rRNAs, tRNAs, aminoacyl-tRNA synthetases, and protein factors. Given that proteins typically comprise from 100 to 1000 amino acids, the frequency at which an incorrect amino acid is incorporated in the course of protein synthesis must be less than 10^{-4}. Transfer RNAs are the adaptors that make the link between a nucleic acid and an amino acid. These molecules, single chains of about 80 nucleotides, have an L-shaped structure.

Aminoacyl-Transfer-RNA Synthetases Read the Genetic Code

Each amino acid is activated and linked to a specific transfer RNA by an enzyme called an aminoacyl-tRNA synthetase. Such an enzyme links the carboxyl group of an amino acid to the 2'- or 3'-hydroxyl group of the adenosine unit of a CCA sequence at the 3' end of the tRNA by an ester linkage. There is at least one specific aminoacyl-tRNA synthetase and at least one specific tRNA for each amino acid. A synthetase utilizes both the functional groups and the shape of its cognate amino acid to prevent the attachment of an the incorrect amino acid to a tRNA. Some synthetases have a separate active site at which incorrectly linked amino acids are removed by hydrolysis. A synthetase recognizes the anticodon, the acceptor stem, and sometimes other parts of its tRNA substrate. By specifically recognizing both amino acids and tRNAs, aminoacyl-tRNA synthetases implement the instruction of the genetic code. There exist two evolutionary distinct classes of synthetases, each recognizing 10 amino acids. The two classes recognize opposite faces of tRNA molecules.

A Ribosome Is a Ribonucleoprotein Particle (70S) Made of a Small (30S) and a Large (50S) Subunit

Protein synthesis takes place on ribosomes—ribonucleoprotein particles (about two-thirds RNA and one-third protein) consisting of large and small subunits. In *E. coli*, the 70S ribosome (2700 kd) is made up of 30S and 50S subunits. The 30S subunit consists of 16S ribosomal RNA and 21 different proteins; the 50S subunit consists of 23S and 5S rRNA and 34 different proteins. The structure of almost all components of the ribosome have now been determined at or near atomic resolution.

Proteins are synthesized in the amino-to-carboxyl direction, and mRNA is translated in the $5' \longrightarrow 3'$ direction. The start signal on prokaryotic mRNA is AUG (or GUG) preceded by a purine-rich sequence that can base-pair with 16S rRNA. In prokaryotes, transcription and translation are closely coupled. Several ribosomes can simultaneously translate an mRNA, forming a polysome.

The ribosome includes three sites for tRNA binding called the A (aminoacyl) site, the P (peptidyl) site, and the E (exit) site. With a tRNA attached to the growing peptide chain in the P site, an aminoacyl-tRNA binds to the A site. A peptide bond is formed when the amino group of the aminoacyl-tRNA nucleophically attacks the ester carbonyl group of the peptidyl-tRNA. On peptide-bond formation, the tRNAs and mRNA must be translocated for the next cycle to begin. The deacylated tRNA moves to the E site and then leaves the ribosome, and the peptidyl-tRNA moves from the A site into the P site.

The codons of messenger RNA recognize the anticodons of transfer RNAs rather than the amino acids attached to the tRNAs. A codon on mRNA forms base pairs with the anticodon of the tRNA. Some tRNAs are recognized by more than one codon because pairing of the third base of a codon is less crucial than that of the other two (the wobble mechanism).

Protein Factors Play Key Roles in Protein Synthesis

Protein synthesis takes place in three phases: initiation, elongation, and termination. In prokaryotes, mRNA, formylmethionyl-tRNA$_f$ (the special initiator tRNA that recognizes AUG), and a 30S ribosomal subunit come together with the assistance of initiation factors to form a 30S initiation complex. A 50S ribosomal subunit then joins this complex to form a 70S initiation complex, in which fMet-tRNA$_f$ occupies the P site of the ribosome.

Elongation factor Tu delivers the appropriate aminoacyl-tRNA to the ribosome A (aminoacyl) site as an EF-Tu · aminoacyl-tRNA · GTP ternary complex. EF-Tu serves both to protect the aminoacyl-tRNA from premature cleavage and to increase the fidelity of protein synthesis by ensuring that the correct codon–anticodon pairing has taken place before hydrolyzing GTP and releasing aminoacyl-tRNA into the A site. Elongation factor G uses the free energy of GTP hydrolysis to drive translocation. Protein synthesis is terminated by release factors, which recognize the termination codons UAA, UGA, and UAG and cause the hydrolysis of the ester bond between the polypeptide and tRNA.

• Eukaryotic Protein Synthesis Differs from Prokaryotic Protein Synthesis Primarily in Translation Initiation

The basic plan of protein synthesis in eukaryotes is similar to that of prokaryotes, but there are some significant differences between them. Eukaryotic ribosomes (80S) consist of a 40S small subunit and a 60S large subunit. The initiating amino acid is again methionine, but it is not formylated. The initiation of protein synthesis is more complex in eukaryotes than in prokaryotes. The AUG closest to the 5′ end of mRNA is nearly always the start site. The 40S ribosome finds this site by binding to the 5′ cap and then scanning the RNA until AUG is reached. The regulation of translation in eukaryotes provides a means for regulating gene expression. Many antibiotics act by blocking prokaryotic gene expression.

KEY TERMS

translation (p. 813)

ribosome (p. 813)

transfer RNA (tRNA) (p. 814)

codon (p. 815)

anticodon (p. 815)

aminoacyl-tRNA synthetase (p. 817)

50S subunit (p. 823)

30S subunit (p. 823)

polysome (p. 827)

Shine-Dalgarno sequence (p. 827)

peptidyl transferase center (p. 830)

wobble hypothesis (p. 832)

initiation factor (p. 833)

elongation factor (p. 834)

elongation factor Tu (EF-Tu) (p. 834)

elongation factor Ts (EF-Ts) (p. 834)

elongation factor G (EF-G) (p. 835)

molecular mimicry (p. 835)

release factor (p. 835)

SELECTED READINGS

Where to start

Dahlberg, A. E., 2001. Ribosome structure: The ribosome in action. *Science* 292:868–869.

Ibba, M., Curnow, A. W., and Söll, D., 1997. Aminoacyl-tRNA synthesis: Divergent routes to a common goal. *Trends Biochem. Sci.* 22:39–42.

Davis, B. K., 1999. Evolution of the genetic code. *Prog. Biophys. Mol. Biol.* 72:157–243.

Schimmel, P., and Ribas de Pouplana, L., 2000. Footprints of aminoacyl-tRNA synthetases are everywhere. *Trends Biochem. Sci.* 25:207–209.

Books

Gesteland, R. F., Cech, T., and Atkins, J. F. (Eds.), 1999. *The RNA World.* Cold Spring Harbor Laboratory Press.

Garret, R., Douthwaite, S. R., Liljas, A., Matheson, A. T, Moore, P. B., and Noller, H. F., 2000. *The Ribosome: Structure, Function, Antibiotics and Cellular Interactions.* The American Society for Microbiology.

Aminoacyl-tRNA synthetases

Ibba, M., and Söll, D., 2000. Aminoacyl-tRNA synthesis. *Annu. Rev. Biochem.* 69:617–650.

Sankaranarayanan, R., Dock-Bregeon, A. C., Rees, B., Bovee, M., Caillet, J., Romby, P., Francklyn, C. S., and Moras, D., 2000. Zinc ion mediated amino acid discrimination by threonyl-tRNA synthetase. *Nat. Struct. Biol.* 7:461–465.

Sankaranarayanan, R., Dock-Bregeon, A. C., Romby, P., Caillet, J., Springer, M., Rees, B., Ehresmann, C., Ehresmann, B., and Moras, D., 1999. The structure of threonyl-tRNA synthetase-tRNA(Thr) complex enlightens its repressor activity and reveals an essential zinc ion in the active site. *Cell* 97:371–381.

Dock-Bregeon, A., Sankaranarayanan, R., Romby, P., Caillet, J., Springer, M., Rees, B., Francklyn, C. S., Ehresmann, C., and Moras, D., 2000. Transfer RNA-mediated editing in threonyl-tRNA synthetase: The class II solution to the double discrimination problem. *Cell* 103:877–884.

Serre, L., Verdon, G., Choinowski, T., Hervouet, N., Risler, J. L., and Zelwer, C., 2001. How methionyl-tRNA synthetase creates its amino acid recognition pocket upon L-methionine binding. *J. Mol. Biol.* 306:863–876.

Beuning, P. J., and Musier-Forsyth, K., 2000. Hydrolytic editing by a class II aminoacyl-tRNA synthetase. *Proc. Natl. Acad. Sci. USA* 97:8916–8920.

Bovee, M. L., Yan, W., Sproat, B. S., and Francklyn, C. S., 1999. tRNA discrimination at the binding step by a class II aminoacyl-tRNA synthetase. *Biochemistry* 38:13725–13735.

Fukai, S., Nureki, O., Sekine, S., Shimada, A., Tao, J., Vassylyev, D. G., and Yokoyama, S., 2000. Structural basis for double-sieve discrimination of L-valine from L-isoleucine and L-threonine by the complex of tRNA(Val) and valyl-tRNA synthetase. *Cell* 103:793–803.

de Pouplana, L. R., and Schimmel, P., 2000. A view into the origin of life: Aminoacyl-tRNA synthetases. *Cell. Mol. Life Sci.* 57:865–870.

Carter, C. W., Jr.,1993. Cognition, mechanism, and evolutionary relationships in aminoacyl-tRNA synthetases. *Annu. Rev. Biochem.* 62:715–748.

Transfer RNA

Ibba, M., Becker, H. D., Stathopoulos, C., Tumbula, D. L., and Söll, D., 2000. The adaptor hypothesis revisited.*Trends Biochem. Sci.* 25:311–316.

Weisblum, B., 1999. Back to Camelot: Defining the specific role of tRNA in protein synthesis.*Trends Biochem. Sci.* 24:247–250.

Normanly, J., and Abelson, J., 1989. Transfer RNA identity. *Annu. Rev. Biochem.* 58:1029–1049.

Basavappa, R., and Sigler, P. B., 1991. The 3 Å crystal structure of yeast initiator tRNA: Functional implications in initiator/elongator discrimination. *EMBO J.* 10:3105–3111.

Ribosomes and ribosomal RNAs

Moore, P. B., 2001. The ribosome at atomic resolution. *Biochemistry* 40:3243–3250.

Yonath, A., and Franceschi, F., 1998. Functional universality and evolutionary diversity: Insights from the structure of the ribosome. *Structure* 6:679–684.

Yusupov, M. M., Yusupova, G. Z., Baucom, A., Lieberman, K., Earnest, T. N., Cate, J. H., and Noller, H. F., 2001. Crystal structure of the ribosome at 5.5 Å resolution. *Science* 292:883–896.

Ban, N., Nissen, P., Hansen, J., Moore, P. B., and Steitz, T. A., 2000. The complete atomic structure of the large ribosomal subunit at 2.4 Å resolution. *Science* 289:905–920.

Carter, A. P., Clemons, W. M., Brodersen, D. E., Morgan-Warren, R. J., Wimberly, B. T., and Ramakrishnan, V., 2000. Functional insights from the structure of the 30S ribosomal subunit and its interactions with antibiotics. *Nature* 407:340–348.

Wimberly, B. T., Brodersen, D. E., Clemons, W. M., Morgan-Warren, R. J., Carter, A. P., Vonrhein, C., Hartsch, T., and Ramakrishnan, V., 2000. Structure of the 30S ribosomal subunit. *Nature* 407:327–339.

Agalarov, S. C., Sridhar Prasad, G., Funke, P. M., Stout, C. D., and Williamson, J. R., 2000. Structure of the S15,S6,S18-rRNA complex: Assembly of the 30S ribosome central domain. *Science* 288:107–113.

Frank, J., 2000. The ribosome: A macromolecular machine par excellence. *Chem. Biol.* 7:R133–R141.

Woodson, S. A., and Leontis, N. B., 1998. Structure and dynamics of ribosomal RNA. *Curr. Opin. Struct. Biol.* 8:294–300.

Yonath, A., and Wittmann, H. G., 1988. Approaching the molecular structure of ribosomes. *Biophys. Chem.* 29:17–29.

Initiation factors

Carter, A. P., Clemons, W. M., Jr., Brodersen, D. E., Morgan-Warren, R. J., Hartsch, T., Wimberly, B. T., and Ramakrishnan, V., 2001. Crystal structure of an initiation factor bound to the 30S ribosomal subunit. *Science* 291:498–501.

Guenneugues, M., Caserta, E., Brandi, L., Spurio, R., Meunier, S., Pon, C. L., Boelens, R., and Gualerzi, C. O., 2000. Mapping the fMet-tRNA(f)(Met) binding site of initiation factor IF2. *EMBO J.* 19:5233–5240.

Lee, J. H., Choi, S. K., Roll-Mecak, A., Burley, S. K., and Dever, T. E., 1999. Universal conservation in translation initiation revealed by human and archaeal homologs of bacterial translation initiation factor IF2. *Proc Natl. Acad. Sci. USA* 96:4342–4347.

Meunier, S., Spurio, R., Czisch, M., Wechselberger, R., Guenneugues, M., Gualerzi, C. O., and Boelens, R., 2000. Structure of the fMet-tRNA(fMet)-binding domain of *B. stearothermophilus* initiation factor IF2. *EMBO J.* 19:1918–1926.

Elongation factors

Stark, H., Rodnina, M. V., Wieden, H. J., van Heel, M., and Wintermeyer, W., 2000. Large-scale movement of elongation factor G and extensive conformational change of the ribosome during translocation. *Cell* 100:301–309.

Baensch, M., Frank, R., and Kohl, J., 1998. Conservation of the amino-terminal epitope of elongation factor Tu in eubacteria and Archaea. *Microbiology* 144:2241–2246.

Krasny, L., Mesters, J. R., Tieleman, L. N., Kraal, B., Fucik, V., Hilgenfeld, R., and Jonak, J., 1998. Structure and expression of elongation factor Tu from *Bacillus stearothermophilus*. *J. Mol. Biol.* 283:371–381.

Pape, T., Wintermeyer, W., and Rodnina, M. V., 1998. Complete kinetic mechanism of elongation factor Tu-dependent binding of aminoacyl-tRNA to the A site of the *E. coli* ribosome. *EMBO J.* 17:7490–7497.

Piepenburg, O., Pape, T., Pleiss, J. A., Wintermeyer, W., Uhlenbeck, O. C., and Rodnina, M. V., 2000. Intact aminoacyl-tRNA is required to trigger GTP hydrolysis by elongation factor Tu on the ribosome. *Biochemistry* 39:1734–1738.

Peptide-bond formation and translocation

Yarus, M., and Welch, M., 2000. Peptidyl transferase: Ancient and exiguous. *Chem. Biol.* 7:R187-R190.

Rodriguez-Fonseca, C., Phan, H., Long, K. S., Porse, B. T., Kirillov, S. V., Amils, R., and Garrett, R. A., 2000. Puromycin-rRNA interaction sites at the peptidyl transferase center. *RNA* 6:744–754.

Vladimirov, S. N., Druzina, Z., Wang, R., and Cooperman, B. S., 2000. Identification of 50S components neighboring 23S rRNA nucleotides A2448 and U2604 within the peptidyl transferase center of *Escherichia coli* ribosomes. *Biochemistry* 39:183–193.

Frank, J., and Agrawal, R. K., 2000. A ratchet-like inter-subunit reorganization of the ribosome during translocation. *Nature* 406:318–322.

Termination

Fujiwara, T., Ito, K., and Nakamura, Y., 2001. Functional mapping of ribosome-contact sites in the ribosome recycling factor: A structural view from a tRNA mimic. *RNA* 7:64–70.

Kim, K. K., Min, K., and Suh, S. W., 2000. Crystal structure of the ribosome recycling factor from *Escherichia coli*. *EMBO J.* 19:2362–2370.

Freistroffer, D. V., Kwiatkowski, M., Buckingham, R. H., and Ehrenberg, M., 2000. The accuracy of codon recognition by polypeptide release factors. *Proc. Natl. Acad. Sci. USA* 97:2046–2051.

Heurgue-Hamard, V., Karimi, R., Mora, L., MacDougall, J., Leboeuf, C., Grentzmann, G., Ehrenberg, M., and Buckingham, R. H., 1998. Ribosome release factor RF4 and termination factor RF3 are involved in dissociation of peptidyl-tRNA from the ribosome. *EMBO J.* 17:808–816.

Kisselev, L. L., and Buckingham, R. H., 2000. Translational termination comes of age. *Trends Biochem. Sci.* 25:561–566.

Fidelity and proofreading

Ibba, M., and Söll, D., 1999. Quality control mechanisms during translation. *Science* 286:1893–1897.

Rodnina, M. V., and Wintermeyer, W., 2001. Ribosome fidelity: tRNA discrimination, proofreading and induced fit. *Trends Biochem. Sci.* 26:124–130.

Kurland, C. G., 1992. Translational accuracy and the fitness of bacteria. *Annu. Rev. Genet.* 26:29–50.

Fersht, A., 1999. *Structure and Mechanism in Protein Science: A Guide to Enzyme Catalysis and Protein Folding.* W. H. Freeman and Company.

Eukaryotic protein synthesis

Kozak, M., 1999. Initiation of translation in prokaryotes and eukaryotes. *Gene* 234:187–208.

Negrutskii, B. S., and El'skaya, A. V., 1998. Eukaryotic translation elongation factor 1 alpha: Structure, expression, functions, and possible role in aminoacyl-tRNA channeling. *Prog. Nucleic Acid Res. Mol. Biol.* 60:47–78.

Preiss, T., and Hentze, M. W., 1999. From factors to mechanisms: Translation and translational control in eukaryotes. *Curr. Opin. Genet. Dev.* 9:515–521.

Bushell, M., Wood, W., Clemens, M. J., and Morley, S. J., 2000. Changes in integrity and association of eukaryotic protein synthesis initiation factors during apoptosis. *Eur. J. Biochem.* 267: 1083–1091.

Das, S., Ghosh, R., and Maitra, U., 2001. Eukaryotic translation initiation factor 5 functions as a GTPase-activating protein. *J. Biol. Chem.* 276:6720–6726.

Lee, J. H., Choi, S. K., Roll-Mecak, A., Burley, S. K., and Dever, T. E., 1999. Universal conservation in translation initiation revealed by human and archaeal homologs of bacterial translation initiation factor IF2. *Proc. Natl. Acad. Sci. USA* 96:4342–4347.

Pestova, T. V., and Hellen, C. U., 2000. The structure and function of initiation factors in eukaryotic protein synthesis. *Cell Mol. Life Sci.* 57:651–674.

Antibiotics and toxins

Belova, L., Tenson, T., Xiong, L., McNicholas, P. M., and Mankin, A. S., 2001. A novel site of antibiotic action in the ribosome: Interaction of evernimicin with the large ribosomal subunit. *Proc. Natl. Acad. Sci. USA* 98:3726–3731.

Brodersen, D. E., Clemons, W. M., Jr., Carter, A. P., Morgan-Warren, R. J., Wimberly, B. T., and Ramakrishnan, V., 2000. The structural basis for the action of the antibiotics tetracycline, pactamycin, and hygromycin B on the 30S ribosomal subunit. *Cell* 103:1143–1154.

Porse, B. T., and Garrett, R. A., 1999. Ribosomal mechanics, antibiotics, and GTP hydrolysis. *Cell* 97:423–426.

Eisenberg, D., 1992. The crystal structure of diphtheria toxin. *Nature* 357:216–222.

──┤ PROBLEMS ├──

1. *Synthetase mechanism.* The formation of isoleucyl-tRNA proceeds through the reversible formation of an enzyme-bound Ile-AMP intermediate. Predict whether ^{32}P-labeled ATP is formed from $^{32}PP_i$ when each of the following sets of components is incubated with the specific activating enzyme:

(a) ATP and $^{32}PP_i$
(b) tRNA, ATP, and $^{32}PP_i$
(c) Isoleucine, ATP, and $^{32}PP_i$

2. *Light and heavy ribosomes.* Ribosomes were isolated from bacteria grown in a "heavy" medium (^{13}C and ^{15}N) and from bacteria grown in a "light" medium (^{12}C and ^{14}N). These 60S ribosomes were added to an in vitro system actively engaged in protein synthesis. An aliquot removed several hours later was analyzed by density-gradient centrifugation. How many bands of 70S ribosomes would you expect to see in the density gradient?

3. *The price of protein synthesis.* What is the smallest number of molecules of ATP and GTP consumed in the synthesis of a 200-residue protein, starting from amino acids? Assume that the hydrolysis of PP_i is equivalent to the hydrolysis of ATP for this calculation.

4. *Contrasting modes of elongation.* The two basic mechanisms for the elongation of biomolecules are represented in the adjoining illustration. In type 1, the activating group (X) is released from the growing chain. In type 2, the activating group is released from the incoming unit as it is added to the growing chain. Indicate whether each of the following biosyntheses is by means of a type 1 or a type 2 mechanism:

(a) Glycogen synthesis
(b) Fatty acid synthesis
(c) $C_5 \longrightarrow C_{10} \longrightarrow C_{15}$ in cholesterol synthesis
(d) DNA synthesis
(e) RNA synthesis
(f) Protein synthesis

5. *Suppressing frameshifts.* The insertion of a base in a coding sequence leads to a shift in the reading frame, which in most cases produces a nonfunctional protein. Propose a mutation in a tRNA that might suppress frameshifting.

6. *Tagging a ribosomal site.* Design an affinity-labeling reagent for one of the tRNA binding sites in *E. coli* ribosomes.

7. *Viral mutation.* An mRNA transcript of a T7 phage gene contains the base sequence

$$\downarrow$$
5'-AACUGCACGAGGUAACACAAGAUGGCU-3'

Predict the effect of a mutation that changes the G marked by an arrow to A.

8. *Two synthetic modes.* Compare and contrast protein synthesis by ribosomes with protein synthesis by the solid-phase method (see Section 4.4).

9. *Enhancing fidelity.* Compare the accuracy of (a) DNA replication, (b) RNA synthesis, and (c) protein synthesis. Which mechanisms are used to ensure the fidelity of each of these processes?

10. *Triggered GTP hydrolysis.* Ribosomes markedly accelerate the hydrolysis of GTP bound to the complex of EF-Tu and aminoacyl-tRNA. What is the biological significance of this enhancement of GTPase activity by ribosomes?

11. *Blocking translation.* Devise an experimental strategy for switching off the expression of a specific mRNA without changing the gene encoding the protein or the gene's control elements.

Type 1   Type 2

12. *Directional problem.* Suppose that you have a protein synthesis system that is actively synthesizing a protein designated A. Furthermore, you know that protein A has four trypsin-sensitive sites, equally spaced in the protein, that, on digestion with trypsin, yield the peptides A_1, A_2, A_3, A_4, and A_5. Peptide A_1 is the amino-terminal peptide, and A_5 is the carboxyl peptide. Finally, you know that your system requires 4 minutes to synthesize a complete protein A. At $t = 0$, you add all 20 amino acids, each carrying a ^{14}C label.

(a) At $t = 1$ minute, you isolate intact protein A from the system, cleave it with trypsin, and isolate the five peptides. Which peptide is most heavily labeled?

(b) At $t = 3$ minutes, what will be the order of labeling of peptides from greatest to least?

(c) What does this experiment tell you about the direction of protein synthesis?

13. *Translator.* Aminoacyl-tRNA synthetases are the only component of gene expression that decodes the genetic code. Explain.

14. *A timing device.* EF-Tu, a member of the G-protein family, plays a crucial role in the elongation process of translation. Suppose that a slowly hydrolyzable analog of GTP were added to an elongating system. What would be the effect on rate of protein synthesis?

Mechanism Problems

15. *Molecular attack.* What is the nucleophile in the reaction catalyzed by peptidyl transferase? Write out a plausible mechanism for this reaction.

16. *Evolutionary amino acid choice.* Ornithine is structurally similar to lysine except ornithine's side chain is one methylene group shorter than that of lysine. Attempts to chemically synthesize and isolate ornithinyl-tRNA proved unsuccessful. Propose a mechanistic explanation. (Hint: Six-membered rings are more stable than seven-membered rings).

Chapter Integration Problems

17. *Déjà vu.* Which protein in G-protein cascades plays a role similar to that of elongation factor Ts?

18. *Family resemblance.* Eukaryotic elongation factor 2 is inhibited by ADP ribosylation catalyzed by diphtheria toxin. What other G proteins are sensitive to this mode of inhibition?

Data Interpretation Problem

19. *Helicase helper.* The initiation factor eIF4 displays ATP-dependent RNA helicase activity. Another initiation factor, eIF4H, has been proposed to assist the action of eIF4. Graph A shows some of the experimental results from an assay that can measure the activity of eIF4 helicase in the presence of eIF4H.

(a) What are the effects on eIF4 helicase activity in the presence of eIF4H?

(b) Why did measuring the helicase activity of eIF4H alone serve as an important control?

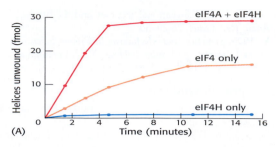

(A)

(c) The initial rate of helicase activity of 0.2 μM of eIF4 was then measured with varying amounts of eIF4H (graph B). What ratio of eIF4H to eIF4 yielded optimal activity?

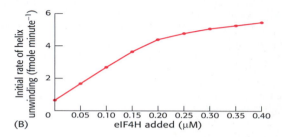

(B)

(d) Next, the effect of RNA–RNA helix stability on the initial rate of unwinding in the presence and absence of eIF4H was tested (graph C). How does the effect of eIF4H vary with helix stability?

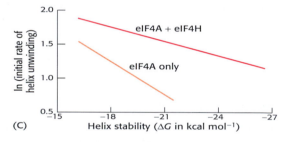

(C)

(e) How might eIF4H affect the helicase activity of eIF4A?

[Data after N. J. Richter, G. W. Rodgers, Jr., J. O. Hensold, and W. C. Merrick, 1999. Further biochemical and kinetic characterization of human eukaryotic initiation factor 4H. *J. Biol. Chem.* 274:35415–35424.]

 Media Problem

20. *Same difference?* Thr-tRNA synthetase (class II) contains an editing site that recognizes and hydrolyzes misacylated Ser-tRNA^{Thr}. This proofreading ability is needed because serine can fit into and make almost as many favorable interactions with the aminoacylation site as threonine, allowing Thr-tRNA synthetase to mistakenly couple serine to threonyl-tRNA. Which aminacyl-tRNA synthetase is most likely to have a comparable need to proofread and edit tRNAs misacylated with valine? Given the class of aminoacyl–tRNA synthetases to which this enzyme belongs, how similar do you expect its editing site to be to that of Thr-tRNA synthetase?

CHAPTER 30

The Integration of Metabolism

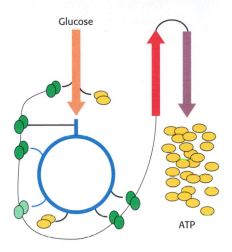

Glucose

ATP

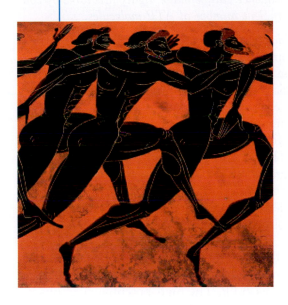

Interplay of metabolic pathways for energy production. At left, the image shows a detail of runners on a Greek amphora painted in the sixth century, B.C. Athletic feats, and others as seemingly simple as maintenance of blood glucose levels, require elaborate metabolic integration. The schematic above represents the oxidation of glucose to yield ATP in a process requiring interplay among glycolysis, the citric acid cycle, and oxidative phosphorylation. These are a few of the many metabolic pathways that must be coordinated to meet the demands of living. [(Left) Metropolitan Museum of Art, Rogers Fund, 1914 (14.130.12). Copyright © 1977 by the Metropolitan Museum of Art.]

We have been examining the biochemistry of metabolism one pathway at a time, but in living systems many pathways are operating simultaneously. Each pathway must be able to sense the status of the others to function optimally to meet the needs of an organism. How is the intricate network of reactions in metabolism coordinated? This chapter presents some of the principles underlying the *integration of metabolism* in mammals. We begin with a recapitulation of the strategy of metabolism and of recurring motifs in its regulation. We then turn to the interplay of different pathways in regard to the flow of molecules at three key crossroads: glucose 6-phosphate, pyruvate, and acetyl CoA. We consider the differences in the metabolic patterns of the brain, muscle, adipose tissue, kidney, and liver. Finally, we examine how the interplay between these tissues is altered in a variety of metabolic perturbations. These discussions will illustrate how biochemical knowledge illuminates the functioning of the organism.

30.1 METABOLISM CONSIST OF HIGHLY INTERCONNECTED PATHWAYS

The basic strategy of catabolic metabolism is to form ATP, reducing power, and building blocks for biosyntheses. Let us briefly review these central themes:

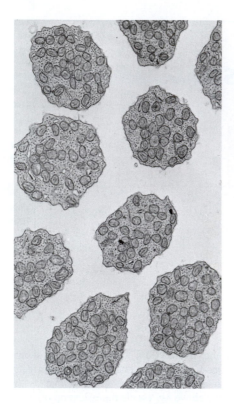

FIGURE 30.1 Electron micrograph of mitochondria. Numerous mitochondria occupy the inner segment of retinal rod cells. These photoreceptor cells generate large amounts of ATP and are highly dependent on a continuous supply of O_2. [Courtesy of Dr. Michael Hogan.]

"To every thing there is a season, and a time to every purpose under the heaven:

A time to be born, and a time to die; a time to plant, and a time to pluck up that which is planted;

A time to kill, and a time to heal; a time to break down, and a time to build up."

ECCLESIASTES 3:1–3

Pasteur effect—

The inhibition of glycolysis by respiration, discovered by Louis Pasteur in studying fermentation by yeast. The consumption of carbohydrate is about sevenfold lower under aerobic conditions than under anaerobic ones. The inhibition of phosphofructokinase by citrate and ATP accounts for much of the Pasteur effect.

1. *ATP is the universal currency of energy.* The high phosphoryl transfer potential of ATP enables it to serve as the energy source in muscle contraction, active transport, signal amplification, and biosyntheses. The hydrolysis of an ATP molecule changes the equilibrium ratio of products to reactants in a coupled reaction by a factor of about 10^8. Hence, *a thermodynamically unfavorable reaction sequence can be made highly favorable by coupling it to the hydrolysis of a sufficient number of ATP molecules.*

2. *ATP is generated by the oxidation of fuel molecules such as glucose, fatty acids, and amino acids.* The common intermediate in most of these oxidations is acetyl CoA. The carbon atoms of the acetyl unit are completely oxidized to CO_2 by the citric acid cycle with the concomitant formation of NADH and $FADH_2$. These electron carriers then transfer their high-potential electrons to the respiratory chain. The subsequent flow of electrons to O_2 leads to the pumping of protons across the inner mitochondrial membrane (Figure 30.1). This proton gradient is then used to synthesize ATP. Glycolysis also generates ATP, but the amount formed is much smaller than that in oxidative phosphorylation. The oxidation of glucose to pyruvate yields only 2 molecules of ATP, whereas the complete oxidation of glucose to CO_2 yields 30 molecules of ATP.

3. *NADPH is the major electron donor in reductive biosyntheses.* In most biosyntheses, the products are more reduced than the precursors, and so reductive power is needed as well as ATP. The high-potential electrons required to drive these reactions are usually provided by NADPH. The pentose phosphate pathway supplies much of the required NADPH.

4. *Biomolecules are constructed from a small set of building blocks.* The highly diverse molecules of life are synthesized from a much smaller number of precursors. The metabolic pathways that generate ATP and NADPH also provide building blocks for the biosynthesis of more-complex molecules. For example, acetyl CoA, the common intermediate in the breakdown of most fuels, supplies a two-carbon unit in a wide variety of biosyntheses, such as those leading to fatty acids, prostaglandins, and cholesterol. Thus, *the central metabolic pathways have anabolic as well as catabolic roles.*

5. *Biosynthetic and degradative pathways are almost always distinct.* For example, the pathway for the synthesis of fatty acids is different from that of their degradation. This separation enables both biosynthetic and degradative pathways to be thermodynamically favorable at all times. A biosynthetic pathway is made exergonic by coupling it to the hydrolysis of a sufficient number of ATP molecules. The separation of biosynthetic and degradative pathways contributes greatly to the effectiveness of metabolic control.

30.1.1 Recurring Motifs in Metabolic Regulation

Anabolism and catabolism must be precisely coordinated. Metabolic networks sense and respond to information on the status of their component pathways. The information is received and metabolism is controlled in several ways:

1. *Allosteric interactions.* The flow of molecules in most metabolic pathways is determined primarily by the activities of certain enzymes rather than by the amount of substrate available. Enzymes that catalyze essentially irreversible reactions are likely control sites, and the first irreversible reaction in a pathway (the committed step) is nearly always tightly controlled. Enzymes

catalyzing committed steps are allosterically regulated, as exemplified by phosphofructokinase in glycolysis and acetyl CoA carboxylase in fatty acid synthesis. Allosteric interactions enable such enzymes to rapidly detect diverse signals and to adjust their activity accordingly.

2. *Covalent modification.* Some regulatory enzymes are controlled by covalent modification in addition to allosteric interactions. For example, the catalytic activity of glycogen phosphorylase is enhanced by phosphorylation, whereas that of glycogen synthase is diminished. Specific enzymes catalyze the addition and removal of these modifying groups (Figure 30.2). Why is covalent modification used in addition to noncovalent allosteric control? The covalent modification of an essential enzyme in a pathway is often the final step in an amplifying cascade and allows metabolic pathways to be rapidly switched on or off by very low concentration of triggering signals. In addition, covalent modifications usually last longer (from seconds to minutes) than do reversible allosteric interactions (from milliseconds to seconds).

3. *Enzyme levels.* The amounts of enzymes, as well as their activities, are controlled. The rates of synthesis and degradation of many regulatory enzymes are altered by hormones. The basics of this control were considered in Chapter 28; we will return to the topic in Chapter 31.

4. *Compartmentation.* The metabolic patterns of eukaryotic cells are markedly affected by the presence of compartments (Figure 30.3). The fates of certain molecules depend on whether they are in the cytosol or in mitochondria, and so their flow across the inner mitochondrial membrane is often regulated. For example, fatty acids are transported into mitochondria for degradation only when energy is required, whereas fatty acids in the cytosol are esterified or exported.

5. *Metabolic specializations of organs.* Regulation in higher eukaryotes is enhanced by the existence of organs with different metabolic roles. Metabolic specialization is the result of differential gene expression.

30.1.2 Major Metabolic Pathways and Control Sites

Let us now review the roles of the major pathways of metabolism and the principal sites for their control:

1. *Glycolysis.* This sequence of reactions in the cytosol converts one molecule of glucose into two molecules of pyruvate with the concomitant generation of two molecules each of ATP and NADH. The NAD^+ consumed in the reaction catalyzed by glyceraldehyde 3-phosphate dehydrogenase must be regenerated for glycolysis to proceed. Under anaerobic conditions, as in highly active skeletal muscle, this regeneration is accomplished by the reduction of pyruvate to lactate. Alternatively, under aerobic conditions, NAD^+ is regenerated by the transfer of electrons from NADH to O_2 through the electron-transport chain. Glycolysis serves two main purposes: it degrades glucose to generate ATP, and it provides carbon skeletons for biosyntheses.

Phosphofructokinase, which catalyzes the committed step in glycolysis, is the most important control site. ATP is both a substrate in the phosphoryl transfer reaction and a regulatory molecule. A high level of ATP inhibits phosphofructokinase—the regulatory sites are distinct from the substrate-binding sites and have a lower affinity for the nucleotide. This inhibitory effect is enhanced by citrate and reversed by AMP (Figure 30.4). Thus, the rate of glycolysis depends on the need for ATP, as signaled by the

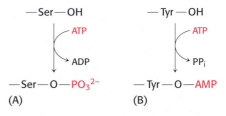

FIGURE 30.2 Covalent modifications. Covalent modifications. Examples of reversible covalent modifications of proteins: (A) phosphorylation, (B) adenylation.

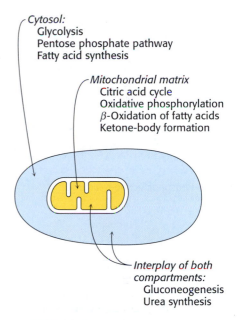

FIGURE 30.3 Compartmentation of the major pathways of metabolism.

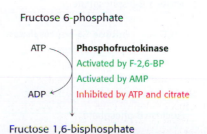

FIGURE 30.4 Regulation of glycolysis. Phosphofructokinase is the key enzyme in the regulation of glycolysis.

ATP/AMP ratio, and on the availability of building blocks, as signaled by the level of citrate. *In liver, the most important regulator of phosphofructo-kinase activity is fructose 2,6-bisphosphate* (F-2,6-BP). Recall that the level of F-2,6-BP is determined by the activity of the kinase that forms it from fructose 6-phosphate and of the phosphatase that hydrolyzes the 2-phosphoryl group (Section 16.2.2). When the blood-glucose level is low, a glucagon-triggered cascade leads to activation of the phosphatase and inhibition of the kinase in the liver. The level of F-2,6-BP declines and, consequently, so does phosphofructokinase activity. Hence, glycolysis is slowed, and the spared glucose is released into the blood for use by other tissues.

2. *Citric acid cycle and oxidative phosphorylation.* The reactions of this common pathway for the oxidation of fuel molecules—carbohydrates, amino acids, and fatty acids—take place inside mitochondria. Most fuels enter the cycle as acetyl CoA. The complete oxidation of an acetyl unit by the citric acid cycle generates one molecule of GTP and four pairs of electrons in the form of three molecules of NADH and one molecule $FADH_2$. These electrons are transferred to O_2 through the electron-transport chain, which results in the formation of a proton gradient that drives the synthesis of nine molecules of ATP. The electron donors are oxidized and recycled back to the citric acid cycle only if ADP is simultaneously phosphorylated to ATP. *This tight coupling, called respiratory control, ensures that the rate of the citric acid cycle matches the need for ATP.* An abundance of ATP also diminishes the activities of two enzymes in the cycle—*isocitrate dehydrogenase* and α-*ketoglutarate dehydrogenase.* The citric acid cycle has an anabolic role as well. In concert with *pyruvate carboxylase,* the citric acid cycle provides intermediates for biosyntheses, such as succinyl CoA for the formation of porphyrins and citrate for the formation of fatty acids.

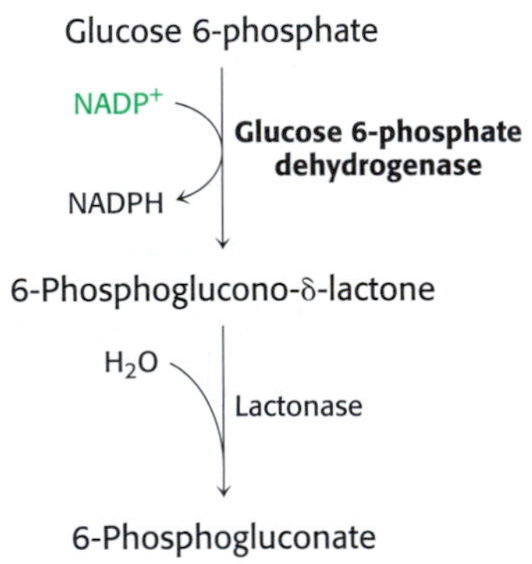

3. *Pentose phosphate pathway.* This series of reactions, which takes place in the cytosol, consists of two stages. The first stage is the oxidative decarboxylation of glucose 6-phosphate. Its purpose is the production of NADPH for reductive biosyntheses and the formation of ribose 5-phosphate for the synthesis of nucleotides. Two molecules of NADPH are generated in the conversion of glucose 6-phosphate into ribose 5-phosphate. The dehydrogenation of glucose 6-phosphate is the committed step in this pathway. This reaction is controlled by the level of $NADP^+$, the electron acceptor (Figure 30.5).

The second stage of the pentose phosphate pathway is the nonoxidative, reversible metabolism of five-carbon phosphosugars into phosphorylated three-carbon and six-carbon glycolytic intermediates. Thus, the nonoxidative branch can either introduce riboses into glycolysis for catabolism or generate riboses from glycolytic intermediates for biosyntheses.

4. *Gluconeogenesis.* Glucose can be synthesized by the liver and kidneys from noncarbohydrate precursors such as lactate, glycerol, and amino acids. The major entry point of this pathway is pyruvate, which is carboxylated to oxaloacetate in mitochondria. Oxaloacetate is then metabolized in the cytosol to form phosphoenolpyruvate. The other distinctive means of gluconeogenesis are two hydrolytic steps that bypass the irreversible reactions of glycolysis. *Gluconeogenesis and glycolysis are usually reciprocally regulated so that one pathway is minimally active while the other is highly active.* For example, AMP inhibits and citrate activates fructose 1,6-bisphosphatase, an essential enzyme in gluconeogenesis, whereas these molecules have opposite effects on phosphofructokinase, the pacemaker of glycolysis (Figure 30.6). Fructose-2,6-bisphosphate also coordinates these processes by inhibiting

FIGURE 30.5 Regulation of the pentose phosphate pathway. The dehydrogenation of glucose 6-phosphate is the committed step in the pentose phosphate pathway.

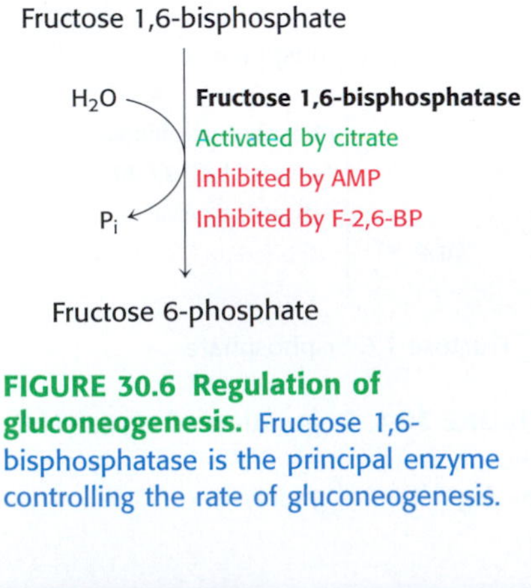

FIGURE 30.6 Regulation of gluconeogenesis. Fructose 1,6-bisphosphatase is the principal enzyme controlling the rate of gluconeogenesis.

fructose 1,6-bisphosphatase. Hence, when glucose is abundant, the high level of F-2,6-BP inhibits gluconeogenesis and activates glycolysis.

5. *Glycogen synthesis and degradation.* Glycogen, a readily mobilizable fuel store, is a branched polymer of glucose residues (Figure 30.7). In glycogen degradation, a phosphorylase catalyzes the cleavage of glycogen by orthophosphate to yield glucose 1-phosphate, which is rapidly converted into glucose 6-phosphate for further metabolism. In glycogen synthesis, the activated intermediate is UDP-glucose, which is formed from glucose 1-phosphate and UTP. Glycogen synthase catalyzes the transfer of glucose from UDP-glucose to the terminal glucose residue of a growing strand. *Glycogen degradation and synthesis are coordinately controlled by a hormone-triggered amplifying cascade so that the phosphorylase is active when synthase is inactive and vice versa.* Phosphorylation and noncovalent allosteric interactions (Section 21.5) regulate these enzymes.

6. *Fatty acid synthesis and degradation.* Fatty acids are synthesized in the cytosol by the addition of two-carbon units to a growing chain on an acyl carrier protein. Malonyl CoA, the activated intermediate, is formed by the carboxylation of acetyl CoA. Acetyl groups are carried from mitochondria to the cytosol as citrate by the citrate–malate shuttle. In the cytosol, citrate is cleaved to yield acetyl CoA. In addition to transporting acetyl CoA, *citrate in the cytosol stimulates acetyl CoA carboxylase, the enzyme catalyzing the committed step. When ATP and acetyl CoA are abundant, the level of citrate increases, which accelerates the rate of fatty acid synthesis* (Figure 30.8).

A different pathway in a different compartment degrades fatty acids. Carnitine transports fatty acids into mitochondria, where they are degraded to acetyl CoA in the mitochondrial matrix by β-oxidation. The acetyl CoA then enters the citric acid cycle if the supply of oxaloacetate is sufficient. Alternatively, acetyl CoA can give rise to ketone bodies. The $FADH_2$ and NADH formed in the β-oxidation pathway transfer their electrons to O_2 through the electron-transport chain. Like the citric acid cycle, β-oxidation can continue only if NAD^+ and FAD are regenerated. Hence, *the rate of fatty acid degradation also is coupled to the need for ATP.* Malonyl CoA, the precursor for fatty acid synthesis, inhibits fatty acid degradation by inhibiting the formation of acyl carnitine by carnitine acyl transferase 1, thus preventing the translocation of fatty acids into mitochondria (Figure 30.9).

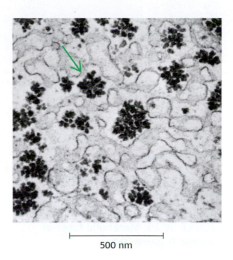

FIGURE 30.7 Glycogen granules. The electron micrograph shows part of a liver cell containing glycogen particles. [Courtesy of Dr. George Palade.]

Acetyl CoA

$HCO_3^- + ATP$ — **Acetyl CoA carboxylase**

Activated by citrate

$ATP + P_i$ — *Inhibited by palmitoyl CoA*

Malonyl CoA

FIGURE 30.8 Regulation of fatty acid synthesis. Acetyl CoA carboxylase is the key control site in fatty acid synthesis.

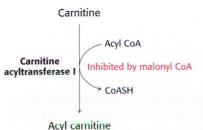

Carnitine

Acyl CoA

Carnitine acyltransferase I — *Inhibited by malonyl CoA*

CoASH

Acyl carnitine

FIGURE 30.9 Control of fatty acid degradation. Malonyl CoA inhibits fatty acid degradation by inhibiting the formation of acyl carnitine.

30.1.3 Key Junctions: Glucose 6-phosphate, Pyruvate, and Acetyl CoA

The factors governing the flow of molecules in metabolism can be further understood by examining three important molecules: glucose 6-phosphate, pyruvate, and acetyl CoA. Each of these molecules has several contrasting fates:

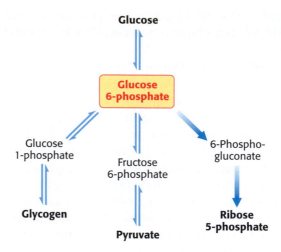

**FIGURE 30.10 Metabolic fates of
glucose 6-phosphate.**

1. *Glucose 6-phosphate.* Glucose entering a cell is rapidly phosphorylated to glucose 6-phosphate and is subsequently stored as glycogen, degraded to pyruvate, or converted into ribose 5-phosphate (Figure 30.10). Glycogen is formed when glucose 6-phosphate and ATP are abundant. In contrast, glucose 6-phosphate flows into the glycolytic pathway when ATP or carbon skeletons for biosyntheses are required. Thus, the conversion of glucose 6-phosphate into pyruvate can be anabolic as well as catabolic. The third major fate of glucose 6-phosphate, to flow through the pentose phosphate pathway, provides NADPH for reductive biosyntheses and ribose 5-phosphate for the synthesis of nucleotides. Glucose 6-phosphate can be formed by the mobilization of glycogen or it can be synthesized from pyruvate and glucogenic amino acids by the gluconeogenic pathway.

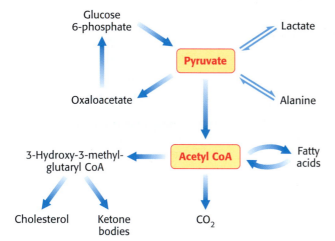

**FIGURE 30.11 Major metabolic fates of pyruvate and
acetyl CoA in mammals.**

2. *Pyruvate.* This three-carbon α-ketoacid is another major metabolic junction (Figure 30.11). Pyruvate is derived primarily from glucose 6-phosphate, alanine, and lactate. Pyruvate can be reduced to lactate by lactate dehydrogenase to regenerate NAD$^+$. This reaction enables glycolysis to proceed transiently under anaerobic conditions in active tissues such as contracting muscle. The lactate formed in active tissue is subsequently oxidized back to pyruvate, in other tissues. The essence of this interconversion buys time and shifts part of the metabolic burden of active muscle to other tissues. Another readily reversible reaction in the cytosol is the transamination of pyruvate, an α-ketoacid, to alanine, the corresponding amino acid. Conversely, several amino acids can be converted into pyruvate. Thus, *transamination is a major link between amino acid and carbohydrate metabolism.*

A third fate of pyruvate is its carboxylation to oxaloacetate inside mitochondria, the first step in gluconeogenesis. This reaction and the subsequent conversion of oxaloacetate into phosphoenolpyruvate bypass an irreversible step of glycolysis and hence enable glucose to be synthesized from pyruvate. The carboxylation of pyruvate is also important for replenishing intermediates of the citric acid cycle. Acetyl CoA activates pyruvate carboxylase, enhancing the synthesis of oxaloacetate, when the citric acid cycle is slowed by a paucity of this intermediate.

A fourth fate of pyruvate is its oxidative decarboxylation to acetyl CoA. *This irreversible reaction inside mitochondria is a decisive reaction in metabolism: it commits the carbon atoms of carbohydrates and amino acids to oxidation by the citric acid cycle or to the synthesis of lipids.* The pyruvate dehydrogenase complex, which catalyzes this irreversible funneling, is stringently regulated by multiple allosteric interactions and covalent modifications. Pyruvate is rapidly converted into acetyl CoA only if ATP is needed or if two-carbon fragments are required for the synthesis of lipids.

3. *Acetyl CoA.* The major sources of this activated two-carbon unit are the oxidative decarboxylation of pyruvate and the β-oxidation of fatty acids (see Figure 30.11). Acetyl CoA is also derived from ketogenic amino acids. The fate of acetyl CoA, in contrast with that of many molecules in metabolism, is quite restricted. The acetyl unit can be completely oxidized to CO_2 by the citric acid cycle. Alternatively, 3-hydroxy-3-methylglutaryl CoA can be formed from three molecules of acetyl CoA. This six-carbon unit is a precursor of cholesterol and of *ketone bodies*, which are transport forms of acetyl units released from the liver for use by some peripheral tissues. A third major fate of acetyl CoA is its export to the cytosol in the form of citrate for the synthesis of fatty acids.

30.2 EACH ORGAN HAS A UNIQUE METABOLIC PROFILE

The metabolic patterns of the brain, muscle, adipose tissue, kidney, and liver are strikingly different. Let us consider how these organs differ in their use of fuels to meet their energy needs:

1. *Brain. Glucose is virtually the sole fuel for the human brain, except during prolonged starvation.* The brain lacks fuel stores and hence requires a continuous supply of glucose. It consumes about 120 g daily, which corresponds to an energy input of about 420 kcal (1760 kJ), accounting for some 60% of the utilization of glucose by the whole body in the resting state. Much of the energy, estimates suggest from 60% to 70%, is used to power transport mechanisms that maintain the Na^+-K^+ membrane potential required for the transmission of the nerve impulses. The brain must also synthesize neurotransmitters and their receptors to propagate nerve impulses. Overall, glucose metabolism remains unchanged during mental activity, although local increases are detected when a subject performs certain tasks.

Glucose is transported into brain cells by the glucose transporter GLUT3. This transporter has a low value of K_M for glucose (1.6 mM), which means that it is saturated under most conditions. Thus, the brain is usually provided with a constant supply of glucose. Noninvasive ^{13}C nuclear magnetic resonance measurements have shown that the concentration of glucose in the brain is about 1 mM when the plasma level is 4.7 mM (84.7 mg/dl), a normal value. Glycolysis slows down when the glucose level approaches the K_M value of hexokinase ($\sim 50 \ \mu M$), the enzyme that traps glucose in the cell (Section 16.1.1). This danger point is reached when the plasma-glucose level drops below about 2.2 mM (39.6 mg/dl) and thus approaches the K_M value of GLUT3.

Fatty acids do not serve as fuel for the brain, because they are bound to albumin in plasma and so do not traverse the blood–brain barrier. In starvation, *ketone bodies generated by the liver partly replace glucose as fuel for the brain.*

2. *Muscle. The major fuels for muscle are glucose, fatty acids, and ketone bodies.* Muscle differs from the brain in having a large store of glycogen (1200 kcal, or 5000 kJ). In fact, about three-fourths of all the glycogen in the body is stored in muscle (Table 30.1). This glycogen is readily converted into glucose 6-phosphate for use within muscle cells. Muscle, like the brain, lacks glucose 6-phosphatase, and so it does not export glucose. Rather, *muscle retains glucose, its preferred fuel for bursts of activity.*

TABLE 30.1 Fuel reserves in a typical 70-kg man

Organ	Available energy in kcal (kJ)		
	Glucose or glycogen	Triacylglycerols	Mobilizable proteins
Blood	60 (250)	45 (200)	0 (0)
Liver	4 (1700)	450 (2000)	400 (1700)
Brain	8 (30)	0 (0)	0 (0)
Muscle	1,200 (5000)	450 (2000)	24,000 (100,000)
Adipose tissue	80 (330)	135,000 (560,000)	40 (170)

Source: After G. F. Cahill, Jr. *Clin. Endocrinol. Metab.* 5(1976):398.

In actively contracting skeletal muscle, the rate of glycolysis far exceeds that of the citric acid cycle, and much of the pyruvate formed is reduced to lactate, some of which flows to the liver, where it is converted into glucose (Figure 30.12).

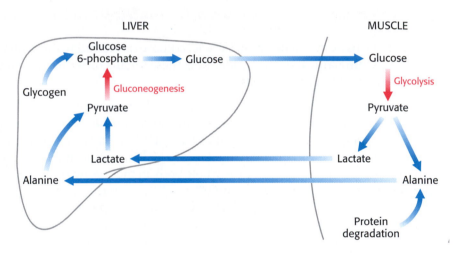

FIGURE 30.12 Metabolic interchanges between muscle and liver.

These interchanges, known as the Cori cycle (Section 16.4.2), shift part of the metabolic burden of muscle to the liver. In addition, a large amount of alanine is formed in active muscle by the transamination of pyruvate. Alanine, like lactate, can be converted into glucose by the liver. Why does the muscle release alanine? Muscle can absorb and transaminate branched-chain amino acids; however, it cannot form urea. Consequently, the nitrogen is released into the blood as alanine. The liver absorbs the alanine, removes the nitrogen for disposal as urea, and processes the pyruvate to glucose or fatty acids. The metabolic pattern of resting muscle is quite different. *In resting muscle, fatty acids are the major fuel, meeting 85% of the energy needs.*

Unlike skeletal muscle, heart muscle functions almost exclusively aerobically, as evidenced by the density of mitochondria in heart muscle. Moreover, the heart has virtually no glycogen reserves. Fatty acids are the heart's main source of fuel, although ketone bodies as well as lactate can serve as fuel for heart muscle. In fact, heart muscle consumes acetoacetate in preference to glucose.

3. *Adipose tissue. The triacylglycerols stored in adipose tissue are an enormous reservoir of metabolic fuel* (see Table 30.1). In a typical 70-kg man, the 15 kg of triacylglycerols have an energy content of 135,000 kcal (565,000 kJ). Adipose tissue is specialized for the esterification of fatty acids and for their release from triacylglycerols. In human beings, the liver is the major site of

fatty acid synthesis. Recall that these fatty acids are esterified in the liver to glycerol phosphate to form triacylglycerol and are transported to the adipose tissue in lipoprotein particles, such as very low density lipoproteins (Section 26.3.1). Triacylglycerols are not taken up by adipocytes; rather, they are first hydrolyzed by an extracellular lipoprotein lipase for uptake. This lipase is stimulated by processes initiated by insulin. After the fatty acids enter the cell, the principal task of adipose tissue is to activate these fatty acids and transfer the resulting CoA derivatives to glycerol in the form of glycerol 3-phosphate. This essential intermediate in lipid biosynthesis comes from the reduction of the glycolytic intermediate dihydroxyacetone phosphate. Thus, *adipose cells need glucose for the synthesis of triacylglycerols* (Figure 30.13).

Triacylglycerols are hydrolyzed to fatty acids and glycerol by intracellular lipases. The release of the first fatty acid from a triacylglycerol, the rate-limiting step, is catalyzed by a hormone-sensitive lipase that is reversibly phosphorylated. The hormone epinephrine stimulates the formation of cyclic AMP, the intracellular messenger in the amplifying cascade, which activates a protein kinase—a recurring theme in hormone action. Triacylglycerols in adipose cells are continually being hydrolyzed and resynthesized. Glycerol derived from their hydrolysis is exported to the liver. Most of the fatty acids formed on hydrolysis are reesterified if glycerol 3-phosphate is abundant. In contrast, they are released into the plasma if glycerol 3-phosphate is scarce because of a paucity of glucose. Thus, *the glucose level inside adipose cells is a major factor in determining whether fatty acids are released into the blood.*

4. *The kidney. The major purpose of the kidney is to produce urine,* which serves as a vehicle for excreting metabolic waste products and for maintaining the osmolarity of the body fluids. The blood plasma is filtered nearly 60 times each day in the renal tubules. Most of the material filtered out of the blood is reabsorbed; so only 1 to 2 liters of urine is produced. Water-soluble materials in the plasma, such as glucose, and water itself are reabsorbed to prevent wasteful loss. The kidneys require large amounts of energy to accomplish the reabsorption. Although constituting only 0.5% of body mass, kidneys consume 10% of the oxygen used in cellular respiration. Much of the glucose that is reabsorbed is carried into the kidney cells by the sodium–glucose cotransporter. Recall that this transporter is powered by the Na^+-K^+ gradient, which is itself maintained by the Na^+-K^+ ATPase (Section 13.4). During starvation, the kidney becomes an important site of gluconeogenesis and may contribute as much as half of the blood glucose.

5. *Liver. The metabolic activities of the liver are essential for providing fuel to the brain, muscle, and other peripheral organs.* Indeed, the liver, which can be from 2% to 4% of body weight, is an organism's metabolic hub (Figure 30.14). Most compounds absorbed by the intestine first pass through the liver, which is thus able to regulate the level of many metabolites in the blood.

Let us first consider how the liver metabolizes carbohydrates. The liver removes two-thirds of the glucose from the blood and all of the remaining monosaccharides. Some glucose is left in the blood for use by other tissues. The absorbed glucose is converted into glucose 6-phosphate by hexokinase and the liver-specific glucokinase. Glucose 6-phosphate, as already stated, has a variety of fates, although the liver uses little of it to meet its own energy needs. Much of the glucose 6-phosphate is converted into glycogen. As much as 400 kcal (1700 kJ) can be stored in this form in the liver. Excess glucose 6-phosphate is metabolized to acetyl CoA, which is used to

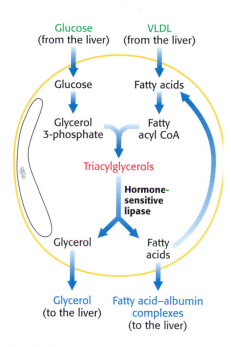

FIGURE 30.13 Synthesis and degradation of triacylglycerols by adipose tissue. Fatty acids are delivered to adipose cells in the form of triacylglycerols contained in very low density lipoproteins (VLDLs).

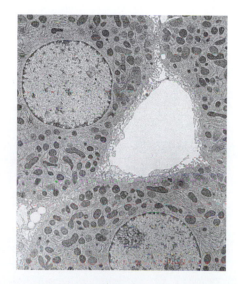

FIGURE 30.14 Electron micrograph of liver cells. The liver plays an essential role in the integration of metabolism. [Courtesy of Dr. Ann Hubbard.]

form fatty acids, cholesterol, and bile salts. The pentose phosphate pathway, another means of processing glucose 6-phosphate, supplies the NADPH for these reductive biosyntheses. The liver can produce glucose for release into the blood by breaking down its store of glycogen and by carrying out gluconeogenesis. The main precursors for gluconeogenesis are lactate and alanine from muscle, glycerol from adipose tissue, and glucogenic amino acids from the diet.

The liver also plays a central role in the regulation of lipid metabolism. When fuels are abundant, fatty acids derived from the diet or synthesized by the liver are esterified and secreted into the blood in the form of very low density lipoprotein (see Figure 30.15). However, in the fasting state, the liver converts fatty acids into ketone bodies. How is the fate of liver fatty acids determined? The selection is made according to whether the fatty acids enter the mitochondrial matrix. Recall that long-chain fatty acids traverse the inner mitochondrial membrane only if they are esterified to carnitine. Carnitine acyltransferase I (also known as carnitine palmitoyl transferase I), which catalyzes the formation of acyl carnitine, is inhibited by malonyl CoA, the committed intermediate in the synthesis of fatty acids (see Figure 30.9). Thus, *when malonyl CoA is abundant, long-chain fatty acids are prevented from entering the mitochondrial matrix, the compartment of β-oxidation and ketone-body formation. Instead, fatty acids are exported to adipose tissue for incorporation into triacylglycerols.* In contrast, the level of malonyl CoA is low when fuels are scarce. Under these conditions, fatty acids liberated from adipose tissues enter the mitochondrial matrix for conversion into ketone bodies.

The liver also plays an essential role in dietary amino acid metabolism. The liver absorbs the majority of amino acids, leaving some in the blood for peripheral tissues. The priority use of amino acids is for protein synthesis rather than catabolism. By what means are amino acids directed to protein synthesis in preference to use as a fuel? The K_M value for the aminoacyl-tRNA synthetases is lower than that of the enzymes taking part in amino acid catabolism. Thus, amino acids are used to synthesize aminoacyl-tRNAs before they are catabolized. When catabolism does take place, the first step is the removal of nitrogen, which is subsequently processed to urea. The liver secretes from 20 to 30 g of urea a day. The α-ketoacids are then used for gluconeogenesis or fatty acid synthesis. Interestingly, the liver cannot remove nitrogen from the branch-chain amino acids (leucine, isoleucine, and valine). Transamination takes place in the muscle.

How does the liver meet its own energy needs? α-Ketoacids derived from the degradation of amino acids are the liver's own fuel. In fact, the main role of glycolysis in the liver is to form building blocks for biosyntheses. Furthermore, the liver cannot use acetoacetate as a fuel, because it has little of the transferase needed for acetoacetate's activation to acetyl CoA. Thus, the liver eschews the fuels that it exports to muscle and the brain.

Fatty acyl carnitine

30.3 FOOD INTAKE AND STARVATION INDUCE METABOLIC CHANGES

We shall now consider the biochemical responses to a series of physiological conditions. Our first example is the *starved–fed cycle*, which we all experience in the hours after an evening meal and through the night's fast. This nightly starved–fed cycle has three stages: the postabsorptive state after a meal, the early fasting during the night, and the refed state after breakfast. A major goal of the many biochemical alterations in this period is to maintain *glucose homeostasis*—that is, a constant blood-glucose level.

1. *The well-fed, or postabsorptive, state.* After we consume and digest an evening meal, glucose and amino acids are transported from the intestine to the blood. The dietary lipids are packaged into chylomicrons and transported to the blood by the lymphatic system. This fed condition leads to the secretion of insulin, which is one of the two most important regulators of fuel metabolism, the other regulator being glucagon. The secretion of the hormone *insulin* by the β cells of the pancreas is stimulated by glucose and the parasympathetic nervous system (Figure 30.15). *In essence, insulin signals the fed state—it stimulates the storage of fuels and the synthesis of proteins in a variety of ways.* For instance, insulin initiates protein kinase cascades—it stimulates glycogen synthesis in both muscle and the liver and suppresses gluconeogenesis by the liver. Insulin also accelerates glycolysis in the liver, which in turn increases the synthesis of fatty acids.

The liver helps to limit the amount of glucose in the blood during times of plenty by storing it as glycogen so as to be able to release glucose in times of scarcity. How is the excess blood glucose present after a meal removed? Insulin accelerates the uptake of blood glucose into the liver by GLUT2. The level of glucose 6-phosphate in the liver rises because only then do the catalytic sites of glucokinase become filled with glucose. Recall that glucokinase is active only when blood-glucose levels are high. Consequently, *the liver forms glucose 6-phosphate more rapidly as the blood-glucose level rises. The increase in glucose 6-phosphate coupled with insulin action leads to a buildup of glycogen stores.* The hormonal effects on glycogen synthesis and storage are reinforced by a direct action of glucose itself. *Phosphorylase a is a glucose sensor in addition to being the enzyme that cleaves glycogen.* When the glucose level is high, the binding of glucose to phosphorylase *a* renders the enzyme susceptible to the action of a phosphatase that converts it into phosphorylase *b*, which does not readily degrade glycogen. Thus, *glucose allosterically shifts the glycogen system from a degradative to a synthetic mode.*

The high insulin level in the fed state also promotes *the entry of glucose into muscle and adipose tissue.* Insulin stimulates the synthesis of glycogen by muscle as well as by the liver. The entry of glucose into adipose tissue provides glycerol 3-phosphate for the synthesis of triacylglycerols. The action of insulin also extends to amino acid and protein metabolism. Insulin promotes the uptake of branched-chain amino acids (valine, leucine, and isoleucine) by muscle. Indeed, insulin has a general stimulating effect on protein synthesis, which favors a building up of muscle protein. In addition, it inhibits the intracellular degradation of proteins.

2. *The early fasting state.* The blood-glucose level begins to drop several hours after a meal, leading to a decrease in insulin secretion and a rise in *glucagon* secretion; glucagon is secreted by the α cells of the pancreas in response to a *low blood-sugar level in the fasting state.* Just as insulin signals the fed state, glucagon signals the starved state. It serves to mobilize glycogen stores when there is no dietary intake of glucose. *The main target organ of glucagon is the liver.* Glucagon stimulates glycogen breakdown and inhibits glycogen synthesis by triggering the cyclic AMP cascade leading to the phosphorylation and activation of phosphorylase and the inhibition of glycogen synthase (Section 21.5). Glucagon also inhibits fatty acid synthesis by diminishing the production of pyruvate and by lowering the activity of acetyl CoA carboxylase by maintaining it in an unphosphorylated state. In addition, glucagon stimulates gluconeogenesis in the liver and blocks glycolysis by lowering the level of F-2,6-BP.

All known actions of glucagon are mediated by protein kinases that are activated by cyclic AMP. The activation of the cyclic AMP cascade results in a higher level of phosphorylase *a* activity and a lower level of glycogen

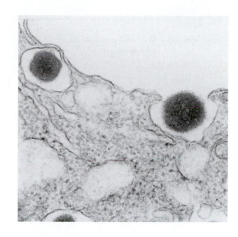

FIGURE 30.15 Insulin secretion. The electron micrograph shows the release of insulin from a pancreatic β cell. One secretory granule is on the verge of fusing with the plasma membrane and releasing insulin into the extracellular space, and the other has already released the hormone. [Courtesy of Dr. Lelio Orci. L. Orci, J.-D. Vassalli, and A. Perrelet. *Sci. Am.* 259 (September 1988):85–94.]

synthase *a* activity. Glucagon's effect on this cascade is reinforced by the diminished binding of glucose to phosphorylase *a*, which makes the enzyme less susceptible to the hydrolytic action of the phosphatase. Instead, the phosphatase remains bound to phosphorylase *a*, and so the synthase stays in the in-active phosphorylated form. Consequently, there is a rapid mobilization of glycogen.

The large amount of glucose formed by the hydrolysis of glucose 6-phosphate derived from glycogen is then released from the liver into the blood. The entry of glucose into muscle and adipose tissue decreases in response to a low insulin level. The diminished utilization of glucose by muscle and adipose tissue also contributes to the maintenance of the blood-glucose level. The net result of these actions of glucagon is to *markedly increase the release of glucose by the liver.*

Both muscle and liver use fatty acids as fuel when the blood-glucose level drops. Thus, *the blood-glucose level is kept at or above 80 mg/dl by three major factors: (1) the mobilization of glycogen and the release of glucose by the liver, (2) the release of fatty acids by adipose tissue, and (3) the shift in the fuel used from glucose to fatty acids by muscle and the liver.*

What is the result of depletion of the liver's glycogen stores? Gluconeogenesis from lactate and alanine continues, but this process merely replaces glucose that had already been converted into lactate and alanine by the peripheral tissues. Moreover, the brain oxidizes glucose completely to CO_2 and H_2O. Thus, for the net synthesis of glucose to occur, another source of carbons is required. Glycerol released from adipose tissue on lipolysis provides some of the carbons, with the remaining carbons coming from the hydrolysis of muscle proteins.

3. *The refed state.* What are the biochemical responses to a hearty breakfast? Fat is processed exactly as it is processed in the normal fed state. However, this is not the case for glucose. The liver does not initially absorb glucose from the blood, but rather leaves it for the peripheral tissues. Moreover, the liver remains in a gluconeogenic mode. Now, however, the newly synthesized glucose is used to replenish the liver's glycogen stores. As the blood-glucose levels continue to rise, the liver completes the replenishment of its glycogen stores and begins to process the remaining excess glucose for fatty acid synthesis.

30.3.1 Metabolic Adaptations in Prolonged Starvation Minimize Protein Degradation

What are the adaptations if fasting is prolonged to the point of starvation? A typical well-nourished 70-kg man has fuel reserves totaling about 161,000 kcal (670,000 kJ; see Table 30.1). The energy need for a 24-hour period ranges from about 1600 kcal (6700 kJ) to 6000 kcal (25,000 kJ), depending on the extent of activity. Thus, stored fuels suffice to meet caloric needs in starvation for 1 to 3 months. However, the carbohydrate reserves are exhausted in only a day.

Even under starvation conditions, the blood-glucose level must be maintained above 2.2 mM (40 mg/dl). *The first priority of metabolism in starvation is to provide sufficient glucose to the brain and other tissues (such as red blood cells) that are absolutely dependent on this fuel.* However, precursors of glucose are not abundant. Most energy is stored in the fatty acyl moieties of triacylglycerols. Recall that fatty acids cannot be converted into glucose, because acetyl CoA cannot be transformed into pyruvate (Section 22.3.7). The glycerol moiety of triacylglycerol can be converted into glucose, but only a limited amount is available. The only other potential source of glu-

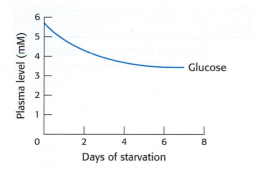

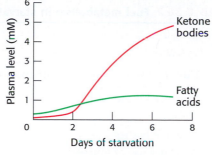

FIGURE 30.16 Fuel choice during starvation. The plasma levels of fatty acids and ketone bodies increase in starvation, whereas that of glucose decreases.

cose is amino acids derived from the breakdown of proteins. However, proteins are not stored, and so any breakdown will necessitate a loss of function. Thus, *the second priority of metabolism in starvation is to preserve protein, which is accomplished by shifting the fuel being used from glucose to fatty acids and ketone bodies* (Figure 30.16).

The metabolic changes on the first day of starvation are like those after an overnight fast. The low blood-sugar level leads to decreased secretion of insulin and increased secretion of glucagon. *The dominant metabolic processes are the mobilization of triacylglycerols in adipose tissue and gluconeogenesis by the liver. The liver obtains energy for its own needs by oxidizing fatty acids released from adipose tissue.* The concentrations of acetyl CoA and citrate consequently increase, which switches off glycolysis. The uptake of glucose by muscle is markedly diminished because of the low insulin level, whereas fatty acids enter freely. Consequently, *muscle shifts almost entirely from glucose to fatty acids for fuel.* The β-oxidation of fatty acids by muscle halts the conversion of pyruvate into acetyl CoA, because acetyl CoA stimulates the phosphorylation of the pyruvate dehydrogenase complex, which renders it inactive (Section 17.2.1). Hence, pyruvate, lactate, and alanine are exported to the liver for conversion into glucose. Glycerol derived from the cleavage of triacylglycerols is another raw material for the synthesis of glucose by the liver.

Proteolysis also provides carbon skeletons for gluconeogenesis. During starvation, degraded proteins are not replenished and serve as carbon sources for glucose synthesis. Initial sources of protein are those that turn over rapidly, such as proteins of the intestinal epithelium and the secretions of the pancreas. Proteolysis of muscle protein provides some of three-carbon precursors of glucose. However, survival for most animals depends on being able to move rapidly, which requires a large muscle mass, and so muscle loss must be minimized.

How is the loss of muscle curtailed? After about 3 days of starvation, the liver forms large amounts of acetoacetate and D-3-hydroxybutyrate (ketone bodies; Figure 30.17). Their synthesis from acetyl CoA increases markedly because the citric acid cycle is unable to oxidize all the acetyl units generated by the degradation of fatty acids. Gluconeogenesis depletes the supply of oxaloacetate, which is essential for the entry of acetyl CoA into the citric acid cycle. Consequently, the liver produces large quantities of ketone bodies, which are released into the blood. At this time, *the brain begins to consume appreciable amounts of acetoacetate in place of glucose.* After 3 days of starvation, about a third of the energy needs of the brain are met by ketone bodies (Table 30.2). The heart also uses ketone bodies as fuel.

After several weeks of starvation, ketone bodies become the major fuel of the brain. Acetoacetate is activated by the transfer of CoA from succinyl

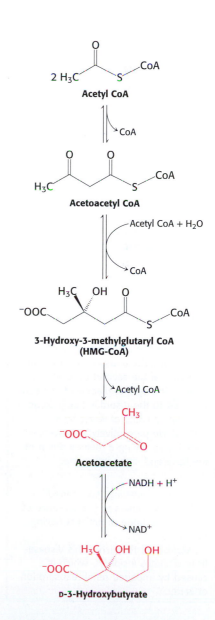

FIGURE 30.17 Synthesis of ketone bodies by the liver.

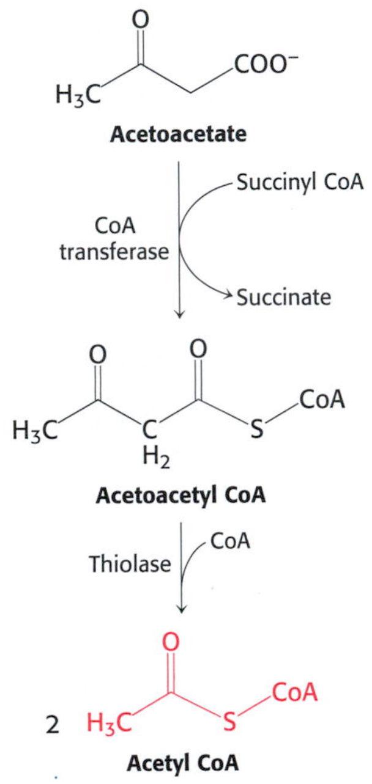

FIGURE 30.18 Entry of ketone bodies into the citric acid cycle.

TABLE 30.2 Fuel metabolism in starvation

Fuel exchanges and consumption	Amount formed or consumed in 24 hours (grams) 3d day	40th day
Fuel use by the brain		
Glucose	100	40
Ketone bodies	50	100
All other use of glucose	50	40
Fuel mobilization		
Adipose-tissue lipolysis	180	180
Muscle-protein degradation	75	20
Fuel output of the liver		
Glucose	150	80
Ketone bodies	150	150

CoA to give acetoacetyl CoA (Figure 30.18). Cleavage by thiolase then yields two molecules of acetyl CoA, which enter the citric acid cycle. In essence, *ketone bodies are equivalents of fatty acids that can pass through the blood–brain barrier*. Only 40 g of glucose is then needed per day for the brain, compared with about 120 g in the first day of starvation. *The effective conversion of fatty acids into ketone bodies by the liver and their use by the brain markedly diminishes the need for glucose. Hence, less muscle is degraded than in the first days of starvation.* The breakdown of 20 g of muscle daily compared with 75 g early in starvation is most important for survival. A person's survival time is mainly determined by the size of the triacylglycerol depot.

What happens after depletion of the triacylglycerol stores? The only source of fuel that remains is proteins. Protein degradation accelerates, and death inevitably results from a loss of heart, liver, or kidney function.

30.3.2 Metabolic Derangements in Diabetes Result from Relative Insulin Insufficiency and Glucagon Excess

We now consider *diabetes mellitus*, a complex disease characterized by grossly abnormal fuel usage: *glucose is overproduced by the liver and underutilized by other organs*. The incidence of diabetes mellitus (usually referred to simply as *diabetes*) is about 5% of the population. Indeed, diabetes is the most common serious metabolic disease in the world; it affects hundreds of millions. *Type I diabetes, or insulin-dependent diabetes mellitus* (IDDM), is caused by autoimmune destruction of the insulin-secreting β cells in the pancreas and usually begins before age 20. The term insulin-dependent means that the individual requires insulin to live. Most diabetics, in contrast, have a normal or even higher level of insulin in their blood, but they are quite unresponsive to the hormone. This form of the disease—known as *type II, or non-insulin-dependent, diabetes mellitus* (NIDDM)—typically arises later in life than does the insulin-dependent form.

In type I diabetes, insulin is absent and consequently glucagon is present at higher-than-normal levels. In essence, the diabetic person is in biochemical starvation mode despite a high concentration of blood glucose. Because insulin is deficient, *the entry of glucose into cells is impaired*. The liver becomes stuck in a gluconeogenic and ketogenic state. The excessive

Diabetes—

Named for the excessive urination in the disease. Aretaeus, a Cappadocian physician of the second century A.D., wrote: "The epithet diabetes has been assigned to the disorder, being something like passing of water by a siphon." He perceptively characterized diabetes as "being a melting-down of the flesh and limbs into urine."

Mellitus—

From Latin, meaning "sweetened with honey." Refers to the presence of sugar in the urine of patients having the disease.

Mellitus distinguishes this disease from diabetes *insipidus*, which is caused by impaired renal reabsorption of water.

level of glucagon relative to insulin leads to a decrease in the amount of F-2,6-BP in the liver. Hence, glycolysis is inhibited and gluconeogenesis is stimulated because of the opposite effects of F-2,6-BP on phosphofructo-kinase and fructose-1,6-bisphosphatase (Section 16.4; see also Figures 30.4 and 30.6). The high glucagon/insulin ratio in diabetes also promotes glyco-gen breakdown. Hence, *an excessive amount of glucose is produced by the liver and released into the blood*. Glucose is excreted in the urine (hence the name *mellitus*) when its concentration in the blood exceeds the reabsorp-tive capacity of the renal tubules. Water accompanies the excreted glucose, and so an untreated diabetic in the acute phase of the disease is hungry and thirsty.

Because carbohydrate utilization is impaired, a lack of insulin leads to the uncontrolled breakdown of lipids and proteins. Large amounts of acetyl CoA are then produced by β-oxidation. However, much of the acetyl CoA cannot enter the citric acid cycle, because there is insufficient oxaloacetate for the condensation step. Recall that mammals can synthe-size oxaloacetate from pyruvate, a product of glycolysis, but not from acetyl CoA; instead, they generate ketone bodies. *A striking feature of diabetes is the shift in fuel usage from carbohydrates to fats; glucose, more abundant than ever, is spurned*. In high concentrations, ketone bodies overwhelm the kidney's capacity to maintain acid–base balance. The un-treated diabetic can go into a coma because of a lowered blood pH level and dehydration.

Type II, or non-insulin-dependent, diabetes accounts for more than 90% of the cases and usually develops in middle-aged, obese people. The exact cause of type II diabetes remains to be elucidated, although a genetic basis seems likely.

30.3.3 Caloric Homeostasis: A Means of Regulating Body Weight

In the United States, obesity has become an epidemic, with nearly 20% of adults classified as obese. Obesity is identified as a risk fac-tor in a host of pathological conditions including diabetes mellitus, hyper-tension, and cardiovascular disease. The cause of obesity is quite simple in the vast majority of cases—more food is consumed than is needed, and the excess calories are stored as fat.

Although the proximal cause of obesity is simple, the biochemical means by which caloric homeostasis and apetite control are usually maintained is enormously complex, but two important signal molecules are insulin and lep-tin. A protein consisting of 146 amino acids, *leptin* is a hormone secreted by adipocytes in direct proportion to fat mass. Leptin acts through a membrane receptor (related in structure and mechanism of action to the growth-hormone receptor; Section 15.4) in the hypothalamus to generate satiation sig-nals. During periods when more energy is expended than ingested (the starved state), adipose tissue loses mass. Under these conditions, the secretion of both leptin and insulin declines, fuel utilization is increased, and energy stores are used. The converse is true when calories are consumed in excess.

The importance of leptin to obesity is dramatically illustrated in mice. Mice lacking leptin are obese and will lose weight if given leptin. Mice that lack the leptin receptor are insensitive to leptin administration. Preliminary evidence indicates that leptin and its receptor play a role in human obesity, but the results are not as clear-cut as in the mouse. The interplay of genes and their products to control caloric homeostasis will be an exciting area of research for some time to come.

30.4 FUEL CHOICE DURING EXERCISE IS DETERMINED BY INTENSITY AND DURATION OF ACTIVITY

The fuels used in anaerobic exercises—sprinting, for example—differ from those used in aerobic exercises—such as distance running. The selection of fuels during these different forms of exercise illustrates many important facets of energy transduction and metabolic integration. ATP directly powers myosin, the protein immediately responsible for converting chemical energy into movement (Chapter 34). However, the amount of ATP in muscle is small. Hence, the power output and, in turn, the velocity of running depend on the rate of ATP production from other fuels. As shown in Table 30.3, *creatine phosphate* (phosphocreatine) can swiftly transfer its high-potential phosphoryl group to ADP to generate ATP (Section 14.1.5). However, the amount of creatine phosphate, like that of ATP itself, is limited. Creatine phosphate and ATP can power intense muscle contraction for 5 to 6 s. Maximum speed in a sprint can thus be maintained for only 5 to 6 s (see Figure 14.7). Thus, the winner in a 100-meter sprint is the runner who slows down the least.

A 100-meter sprint is powered by stored ATP, creatine phosphate, and anaerobic glycolysis of muscle glycogen. The conversion of muscle glycogen into lactate can generate a good deal more ATP, but the rate is slower than that of phosphoryl-group transfer from creatine phosphate. During a ~10-second sprint, the ATP level in muscle drops from 5.2 to 3.7 mM, and that of creatine phosphate decreases from 9.1 to 2.6 mM. The essential role of anaerobic glycolysis is manifested in the elevation of the blood-lactate level from 1.6 to 8.3 mM. The release of H^+ from the intensely active muscle concomitantly lowers the blood pH from 7.42 to 7.24. This pace cannot be sustained in a 1000-meter run (~132 s) for two reasons. First, creatine phosphate is consumed within a few seconds. Second, the lactate produced would cause acidosis. Thus, alternative fuel sources are needed.

The complete oxidation of muscle glycogen to CO_2 substantially increases the energy yield, but this aerobic process is a good deal slower than anaerobic glycolysis. However, as the distance of a run increases, aerobic respiration, or oxidative phosphorylation, becomes increasingly important. For instance, *part of the ATP consumed in a 1000-meter run must come from*

TABLE 30.3 Fuel sources for muscle contraction

Fuel source	Maximal rate of ATP production (mmol/s)	Total ~P available (mmol)
Muscle ATP		223
Creatine phosphate	73.3	446
Conversion of muscle glycogen into lactate	39.1	6,700
Conversion of muscle glycogen into CO_2	16.7	84,000
Conversion of liver glycogen into CO_2	6.2	19,000
Conversion of adipose-tissue fatty acids into CO_2	6.7	4,000,000

Note: Fuels stored are estimated for a 70-kg person having a muscle mass of 28 kg.
Source: After E. Hultman and R. C. Harris. In *Principles of Exercise Biochemistry*, J. R. Poortmans (Ed.). (Karger, 1988), pp. 78–119.

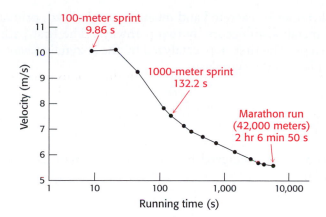

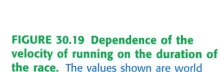

FIGURE 30.19 Dependence of the velocity of running on the duration of the race. The values shown are world track records .

oxidative phosphorylation. Because ATP is produced more slowly by oxidative phosphorylation than by glycolysis (see Table 30.3), the pace is necessarily slower than in a 100-meter sprint. The championship velocity for the 1000-meter run is about 7.6 m/s, compared with approximately 10.2 m/s for the 100-meter event (Figure 30.19).

The running of a marathon (26 miles 385 yards, or 42,200 meters), requires a different selection of fuels and is characterized by cooperation between muscle, liver, and adipose tissue. Liver glycogen complements muscle glycogen as an energy store that can be tapped. However, the total body glycogen stores (103 mol of ATP at best) are insufficient to provide the 150 mol of ATP needed for this grueling ~2-hour event. Much larger quantities of ATP can be obtained by the oxidation of fatty acids derived from the breakdown of *fat in adipose tissue,* but the maximal rate of ATP generation is slower yet than that of glycogen oxidation and is more than tenfold slower than that with creatine phosphate. Thus, *ATP is generated much more slowly from high-capacity stores than from limited ones,* accounting for the different velocities of anaerobic and aerobic events.

ATP generation from fatty acids is essential for distance running. However, a marathon would take about 6 hours to run if all the ATP came from fatty acid oxidation, because it is much slower than glycogen oxidation. Elite runners consume about equal amounts of glycogen and fatty acids during a marathon to achieve a mean velocity of 5.5 m/s, about half that of a 100-meter sprint. How is an optimal mix of these fuels achieved? *A low blood-sugar level leads to a high glucagon/insulin ratio, which in turn mobilizes fatty acids from adipose tissue.* Fatty acids readily enter muscle, where they are degraded by β oxidation to acetyl CoA and then to CO_2. The elevated acetyl CoA level decreases the activity of the pyruvate dehydrogenase complex to block the conversion of pyruvate into acetyl CoA. Hence, fatty acid oxidation decreases the funneling of sugar into the citric acid cycle and oxidative phosphorylation. Glucose is spared so that just enough remains available at the end of the marathon. The simultaneous use of both fuels gives a higher mean velocity than would be attained if glycogen were totally consumed before the start of fatty acid oxidation.

30.5 ETHANOL ALTERS ENERGY METABOLISM IN THE LIVER

 Ethanol has been a part of the human diet for centuries. However, its consumption in excess can result in a number of health problems, most notably liver damage. What is the biochemical basis of these health problems?

Ethanol cannot be excreted and must be metabolized, primarily by the liver. This metabolism occurs by two pathways. The first pathway comprises two steps. The first step, catalyzed by the enzyme *alcohol dehydrogenase*, takes place in the cytoplasm:

$$CH_3CH_2OH + NAD^+ \xrightarrow[\text{dehydrogenase}]{\text{Alcohol}} CH_3CHO + NADH + H^+$$

$$\underset{\text{Ethanol}}{} \qquad\qquad\qquad\qquad \underset{\text{Acetaldehyde}}{}$$

The second step, catalyzed by *aldehyde dehydrogenase*, takes place in mitochondria:

$$CH_3CHO + NAD^+ + H_2O \xrightarrow[\text{dehydrogenase}]{\text{Aldehyde}} CH_3COO^- + NADH + H^+$$

$$\underset{\text{Acetaldehyde}}{} \qquad\qquad\qquad\qquad\qquad \underset{\text{Acetate}}{}$$

Note that *ethanol consumption leads to an accumulation of NADH*. This high concentration of NADH inhibits gluconeogenesis by preventing the oxidation of lactate to pyruvate. In fact, the high concentration of NADH will cause the reverse reaction to predominate, and lactate will accumulate. The consequences may be hypoglycemia and lactic acidosis.

The NADH glut also inhibits fatty acid oxidation. The metabolic purpose of fatty acid oxidation is to generate NADH for ATP generation by oxidative phosphorylation, but an alcohol consumer's NADH needs are met by ethanol metabolism. In fact, the excess NADH signals that conditions are right for fatty acid synthesis. Hence, triacylglycerols accumulate in the liver, leading to a condition known as "fatty liver."

The second pathway for ethanol metabolism is called the ethanol-inducible *microsomal ethanol-oxidizing system* (MEOS). This cytochrome P450-dependent pathway (Section 26.4.2) generates acetaldehyde and subsequently acetate while oxidizing biosynthetic reducing power, NADPH, to $NADP^+$. Because it uses oxygen, this pathway generates free radicals that damage tissues. Moreover, because the system consumes NADPH, the antioxidant glutathione cannot be regenerated (Section 20.5), exacerbating the oxidative stress.

What are the effects of the other metabolites of ethanol? Liver mitochondria can convert acetate into acetyl CoA in a reaction requiring ATP. The enzyme is the thiokinase that normally activates short-chain fatty acids.

$$\text{Acetate} + \text{coenzyme A} + \text{ATP} \longrightarrow \text{acetyl CoA} + \text{AMP} + \text{PP}_i$$

$$\text{PP}_i \longrightarrow 2\,\text{P}_i$$

However, further processing of the acetyl CoA by the citric acid cycle is blocked, because NADH inhibits two important regulatory enzymes—isocitrate dehydrogenase and α-ketoglutarate dehydrogenase. The accumulation of acetyl CoA has several consequences. First, ketone bodies will form and be released into the blood, exacerbating the acidic condition already resulting from the high lactate concentration. The processing of the acetate in the liver becomes inefficient, leading to a buildup of acetaldehyde. This very reactive compound forms covalent bonds with many important functional groups in proteins, impairing protein function. If ethanol is consistently consumed at high levels, the acetaldehyde can significantly damage the liver, eventually leading to cell death.

Liver damage from excessive ethanol consumption occurs in three stages. The first stage is the aforementioned development of fatty liver. In the second stage—alcoholic hepatitis—groups of cells die and inflammation results. This stage can itself be fatal. In stage three—cirrhosis—fibrous structure and scar tissue are produced around the dead cells. Cirrhosis impairs many of the liver's biochemical functions. The cirrhotic liver is unable to convert ammonia into urea, and blood levels of ammonia rise. Ammonia is toxic to the nervous system and can cause coma and death. Cirrhosis of the liver arises in about 25% of alcoholics, and about 75% of all cases of liver cirrhosis are the result of alcoholism. Viral hepatitis is a nonalcoholic cause of liver cirrhosis.

SUMMARY

Metabolism Consists of Highly Interconnected Pathways

The basic strategy of metabolism is simple: form ATP, reducing power, and building blocks for biosyntheses. This complex network of reactions is controlled by allosteric interactions and reversible covalent modifications of enzymes and changes in their amounts, by compartmentation, and by interactions between metabolically distinct organs. The enzyme catalyzing the committed step in a pathway is usually the most important control site. Opposing pathways such as gluconeogenesis and glycolysis are reciprocally regulated so that one pathway is usually quiescent when the other is highly active.

Each Organ Has a Unique Metabolic Profile

The metabolic patterns of the brain, muscle, adipose tissue, kidney, and liver are very different. Glucose is essentially the sole fuel for the brain in a well-fed person. During starvation, ketone bodies (acetoacetate and 3-hydroxybutyrate) become the predominant fuel of the brain. Adipose tissue is specialized for the synthesis, storage, and mobilization of triacylglycerols. The kidney produces urine and reabsorbs glucose. The diverse metabolic activities of the liver support the other organs. The liver can rapidly mobilize glycogen and carry out gluconeogenesis to meet the glucose needs of other organs. It plays a central role in the regulation of lipid metabolism. When fuels are abundant, fatty acids are synthesized, esterified, and sent from the liver to adipose tissue. In the fasting state, however, fatty acids are converted into ketone bodies by the liver.

Food Intake and Starvation Induce Metabolic Changes

Insulin signals the fed state: it stimulates the formation of glycogen and triacylglycerols and the synthesis of proteins. In contrast, glucagon signals a low blood-glucose level: it stimulates glycogen breakdown and gluconeogenesis by the liver and triacylglycerol hydrolysis by adipose tissue. After a meal, the rise in the blood-glucose level leads to increased secretion of insulin and decreased secretion of glucagon. Consequently, glycogen is synthesized in muscle and the liver. When the blood-glucose level drops several hours later, glucose is then formed by the degradation of glycogen and by the gluconeogenic pathway, and fatty acids are released by the hydrolysis of triacylglycerols. The liver and muscle then use fatty acids instead of glucose to meet their own energy needs so that glucose is conserved for use by the brain.

The metabolic adaptations in starvation serve to minimize protein degradation. Large amounts of ketone bodies are formed by the liver from fatty acids and released into the blood within a few days after the onset of starvation. After several weeks of starvation, ketone bodies become the major fuel of the brain. The diminished need for glucose decreases the rate of muscle breakdown, and so the likelihood of survival is enhanced.

Diabetes mellitus, the most common serious metabolic disease, is due to metabolic derangements resulting in an insufficiency of insulin and an excess of glucagon relative to the needs of the individual. The result is an elevated blood-glucose level, the mobilization of triacylglycerols, and excessive ketone-body formation. Accelerated ketone-body formation can lead to acidosis, coma, and death in untreated insulin-dependent diabetics.

Fuel Choice During Exercise Is Determined by Intensity and Duration of Activity

Sprinting and marathon running are powered by different fuels to maximize power output. The 100-meter sprint is powered by stored ATP, creatine phosphate, and anaerobic glycolysis. In contrast, the oxidation of both muscle glycogen and fatty acids derived from adipose tissue is essential in the running of a marathon, a highly aerobic process.

Ethanol Alters Energy Metabolism in the Liver

The oxidation of ethanol results in an unregulated overproduction of NADH, which has several consequences. A rise in the blood levels of lactic acid and ketone bodies causes a fall in blood pH, or acidosis. The liver is damaged because the excess NADH causes excessive fat formation as well as the generation of acetaldehyde, a reactive molecule. Severe liver damage can result.

KEY TERMS

allosteric interaction (p. 846)

covalent modification (p. 847)

glycolysis (p. 847)

phosphofructokinase (p. 847)

citric acid cycle (p. 848)

oxidative phosphorylation (p. 848)

pentose phosphate pathway (p. 848)

gluconeogenesis (p. 848)

glycogen synthesis (p. 849)

glycogen degradation (p. 849)

glucose 6-phosphate (p. 850)

pyruvate (p. 850)

acetyl CoA (p. 851)

ketone body (p. 851)

starved–fed cycle (p. 854)

glucose homeostasis (p. 854)

insulin (p. 855)

glucagon (p. 855)

caloric homeostasis (p. 859)

leptin (p. 859)

creatine phosphate (p. 860)

SELECTED READINGS

Where to start

Kemp, G. J., 2000. Studying metabolic regulation in human muscle. *Biochem. Soc. Trans.* 28:100–103.

Lienhard, G. E., Slot, J. W., James, D. E., and Mueckler, M. M., 1992. How cells absorb glucose. *Sci. Am.* 266(1):86–91.

Books

Fell, D., 1997. *Understanding the Control of Metabolism.* Portland Press.

Frayn, K. N., 1996. *Metabolic Regulation: A Human Perspective.* Portland Press.

Hargreaves, M., and Thompson, M. (Eds.) 1999. *Biochemistry of Exercise X.* Human Kinetics.

Harris, R. A., and Crabb, D. W. 1997. Metabolic interrelationships. In *Textbook of Biochemistry with Clinical Correlations* (pp. 525–562), edited by T. M. Devlin. Wiley-Liss.

Fuel metabolism

Rolland, F., Winderickx, J., and Thevelein, J. M., 2001. Glucose-sensing mechanism in eukaryotic cells. *Trends Biochem. Sci.* 26:310–317.

Rasmussen, B. B., and Wolfe, R. R., 1999. Regulation of fatty acid oxidation in skeletal muscle. *Annu. Rev. Nutr.* 19:463–484.

Hochachka, P. W., 2000. Oxygen, homeostasis, and metabolic regulation. *Adv. Exp. Med. Biol.* 475:311–335.

Holm, E., Sedlaczek, O., and Grips, E., 1999. Amino acid metabolism in liver disease. *Curr. Opin. Clin. Nutr. Metab. Care* 2:47–53.

Wagenmakers, A. J., 1998. Protein and amino acid metabolism in human muscle. *Adv. Exp. Med. Biol.* 441:307–319.

Metabolic adaptations in starvation

Baverel, G., Ferrier, B., and Martin, M., 1995. Fuel selection by the kidney: Adaptation to starvation. *Proc. Nutr. Soc.* 54:197–212.

MacDonald, I. A., and Webber, J., 1995. Feeding, fasting and starvation: Factors affecting fuel utilization. *Proc. Nutr. Soc.* 54:267–274.

Cahill, G. F., Jr., 1976. Starvation in man. *Clin. Endocrinol. Metab.* 5:397–415.

Sugden, M. C., Holness, M. J., and Palmer, T. N., 1989. Fuel selection and carbon flux during the starved-to-fed transition. *Biochem. J.* 263:313–323.

Diabetes mellitus

Rutter, G. A., 2000. Diabetes: The importance of the liver. *Curr. Biol.* 10:R736–R738.

Saltiel, A. R., 2001. New perspectives into the molecular pathogenesis and treatment of type 2 diabetes. *Cell* 104:517–529.

Bell, G. I., Pilikis, S. J., Weber, I. T., and Polonsky, K. S., 1996. Glucokinase mutations, insulin secretion, and diabetes mellitus. *Annu. Rev. Physiol.* 58:171–186.

Withers, D. J., and White, M., 2000. Perspective: The insulin signaling system—a common link in the pathogenesis of type 2 diabetes. *Endocrinology* 141:1917–1921.

Taylor, S. I., 1995. Diabetes mellitus. In *The Metabolic Basis of Inherited Diseases* (7th ed.; pp. 935–936), edited by C. R. Striver, A. L. Beaudet, W. S. Sly, D. Valle, J. B. Stanbury, J. B. Wyngaarden, and D. S. Fredrickson. McGraw-Hill.

Exercise metabolism

Shulman, R. G., and Rothman, D. L., 2001. The "glycogen shunt" in exercising muscle: A role for glycogen in muscle energetics and fatigue. *Proc. Natl. Acad. Sci. USA* 98:457–461.

Gleason, T. 1996. Post-exercise lactate metabolism: A comparative review of sites, pathways, and regulation. *Annu. Rev. Physiol.* 58:556–581.

Holloszy, J. O., and Kohrt, W. M., 1996. Regulation of carbohydrate and fat metabolism during and after exercise. *Annu. Rev. Nutr.* 16:121–138.

Hochachka, P. W., and McClelland, G. B., 1997. Cellular metabolic homeostasis during large-scale change in ATP turnover rates in muscles. *J. Exp. Biol.* 200:381–386.

Horowitz, J. F., and Klein, S., 2000. Lipid metabolism during endurance exercise. *Am. J. Clin. Nutr.* 72:558S–563S.

Wagenmakers, A. J., 1999. Muscle amino acid metabolism at rest and during exercise. *Diabetes Nutr. Metab.* 12:316–322.

Ethanol metabolism

Stewart, S., Jones, D., and Day, C.P., 2001. Alcoholic liver disease: New insights into mechanisms and preventive strategies. *Trends Mol. Med.* 7:408–413.

Lieber, C. S., 2000. Alcohol: Its metabolism and interaction with nutrients. *Annu. Rev. Nutr.* 20:395–430.

Niemela, O., 1999. Aldehyde-protein adducts in the liver as a result of ethanol-induced oxidative stress. *Front. Biosci.* 1:D506–D513.

Riveros-Rosas, H., Julian-Sanchez, A., and Pina, E., 1997. Enzymology of ethanol and acetaldehyde metabolism in mammals. *Arch. Med. Res.* 28:453–471.

Diamond, I., and Gordon, A. S., 1997. Cellular and molecular neuroscience of alcoholism. *Physiol. Rev.* 77:1–20.

PROBLEMS

1. *Distinctive organs.* What are the key enzymatic differences between liver, kidney, muscle, and brain that account for their differing utilization of metabolic fuels?

2. *Missing enzymes.* Predict the major consequence of each of the following enzymatic deficiencies:

(a) Hexokinase in adipose tissue
(b) Glucose 6-phosphatase in liver
(c) Carnitine acyltransferase I in skeletal muscle
(d) Glucokinase in liver
(e) Thiolase in brain
(f) Kinase in liver that synthesizes fructose 2,6-bisphosphate

3. *Contrasting milieux.* Cerebrospinal fluid has a low content of albumin and other proteins compared with plasma.

(a) What effect does this lower content have on the concentration of fatty acids in the extracellular medium of the brain?
(b) Propose a plausible reason for the selection by the brain of glucose rather than fatty acids as the prime fuel.
(c) How does the fuel preference of muscle complement that of the brain?

4. *Metabolic energy and power.* The rate of energy expenditure of a typical 70-kg person at rest is about 70 watts (W), like a light bulb.

(a) Express this rate in kilojoules per second and in kilocalories per second.
(b) How many electrons flow through the mitochondrial electron-transport chain per second under these conditions?
(c) Estimate the corresponding rate of ATP production.
(d) The total ATP content of the body is about 50 g. Estimate how often an ATP molecule turns over in a person at rest.

5. *Respiratory quotient (RQ).* This classic metabolic index is defined as the volume of CO_2 released divided by the volume of O_2 consumed.

(a) Calculate the RQ values for the complete oxidation of glucose and of tripalmitoylglycerol.

(b) What do RQ measurements reveal about the contributions of different energy sources during intense exercise? (Assume that protein degradation is negligible.)

6. *Camel's hump.* Compare the H_2O yield from the complete oxidation of 1 g of glucose with that of 1 g of tripalmitoylglycerol. Relate these values to the evolutionary selection of the contents of a camel's hump.

7. *The wages of sin.* How long does one have to jog to offset the calories obtained from eating 10 macadamia nuts (18 kcal/nut)? [Assume an incremental power consumption of 400 W.]

8. *Sweet hazard.* Ingesting large amounts of glucose before a marathon might seem to be a good way of increasing the fuel stores. However, experienced runners do not ingest glucose before a race. What is the biochemical reason for their avoidance of this potential fuel? (Hint: Consider the effect of glucose ingestion on the level of insulin.)

9. *An effect of diabetes.* Insulin-dependent diabetes is often accompanied by hypertriglyceridemia, which is an excess blood level of triacylglycerides in the form of very low density lipoproteins. Suggest a biochemical explanation.

10. *Sharing the wealth.* The hormone glucagon signifies the starved state, yet it inhibits glycolysis in the liver. How does this inhibition of an energy-production pathway benefit the organism?

11. *Compartmentation.* Glycolysis takes place in the cytoplasm, whereas fatty acid degradation takes place in mitochondria. What metabolic pathways depend on the interplay of reactions that take place in both compartments?

12. *Kwashiorkor.* The most common form of malnutrition in children in the world, kwashiorkor is caused by a diet having ample calories but little protein. The high levels of carbohydrate result in high levels of insulin. What is the effect of high levels of insulin on

(a) lipid utilization?
(b) protein metabolism?

(c) Children suffering from kwashiorkor often have large distended bellies caused by water from the blood leaking into extracellular spaces. Suggest a biochemical basis for this condition.

13. *Oxygen deficit*. After light exercise, the oxygen consumed in recovery is approximately equal to the oxygen deficit, which is the amount of additional oxygen that would have been consumed had oxygen consumption reached steady state immediately. How is the oxygen consumed in recovery used?

14. *Excess post-exercise oxygen consumption*. The oxygen consumed after strenuous exercise stops is significantly greater than the oxygen deficit and is termed *excess post-exercise oxygen consumption* (EPOC). Why is so much more oxygen required after intense exercise?

15. *Psychotropic effects*. Ethanol is unusual in that it is freely soluble in both water and lipids. Thus, it has access to all regions of the highly vascularized brain. Although the molecular basis of ethanol action in the brain is not clear, it is evident that ethanol influences a number of neurotransmitter receptors and ion channels. Suggest a biochemical explanation for the diverse effects of ethanol.

16. *Fiber type*. Skeletal muscle has several distinct fiber types. Type I is used primarily for aerobic activity, whereas type II is specialized for short, intense bursts of activity. How could you distinguish between these types of muscle fiber if you viewed them with an electron microscope?

17. *Tour de France*. Cyclists in the Tour de France (more than 2000 miles in 3 weeks) require about 200,000 kcal of energy, or 10,000 kcal day^{-1} (a resting male requires $\approx$2000 kcal day^{-1}).

(a) With the assumptions that the energy yield of ATP is about 12 kcal mol^{-1} and that ATP has a molecular weight of 503 gm mol^{-1}, how much ATP would be expended by a Tour de France cyclist?

(b) Pure ATP can be purchased at the cost of approximately $150 per gram. How much would it cost to power a cyclist through the Tour de France if the ATP had to be purchased?

CHAPTER 31

The Control of Gene Expression

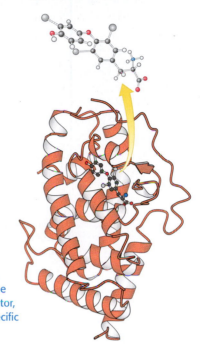

Programming Gene Expression. Complex biological processes often involve coordinated control of the expression of many genes. The maturation of a tadpole into a frog is largely controlled by thyroid hormone. This hormone regulates gene expression by binding to a protein, the thyroid hormone receptor, shown at the right. In response to hormone binding, this protein binds to specific DNA sites in the genome and modulates the expression of nearby genes.
[(Left) Shanon Cummings/Dembinsky Photo Associates.]

Bacteria are highly versatile and responsive organisms: the rate of synthesis of some proteins in bacteria may vary more than a 1000-fold in response to the supply of nutrients or to environmental challenges. Cells of multicellular organisms also respond to varying conditions. Such cells exposed to hormones and growth factors will change substantially in shape, growth rate, and other characteristics. Moreover, many different *cell types* are present in multicellular organisms. For example, cells from muscle and nerve tissue show strikingly different morphologies and other properties, yet they contain exactly the same DNA. These diverse properties are the result of differences in gene expression.

Gene expression is the combined process of the transcription of a gene into mRNA, the processing of that mRNA, and its translation into protein (for protein-encoding genes). A comparison of the gene-expression patterns of cells from the pancreas, which secretes digestive enzymes, and the liver, the site of lipid transport and energy transduction, reveals marked differences in the genes that are highly expressed (Table 31.1), a difference consistent with the physiological roles of these tissues.

How is gene expression controlled? Gene activity is controlled first and foremost at the level of transcription. Much of this control is achieved through the interplay between proteins that bind to specific DNA sequences and their DNA-binding sites. In this chapter, we shall see how signals from the environment of a cell can alter this interplay to induce changes in gene expression. We first consider

TABLE 31.1 Highly expressed protein-encoding genes of the pancreas and liver (as percentage of total mRNA pool)

Rank	Pancreas	%	Liver	%
1	Procarboxypeptidase A1	7.6	Albumin	3.5
2	Pancreatic trypsinogen 2	5.5	Apolipoprotein A-I	2.8
3	Chymotrypsinogen	4.4	Apolipoprotein C-I	2.5
4	Pancreatic trypsin 1	3.7	Apolipoprotein C-III	2.1
5	Elastase IIIB	2.4	ATPase 6/8	1.5
6	Protease E	1.9	Cytochrome oxidase 3	1.1
7	Pancreatic lipase	1.9	Cytochrome oxidase 2	1.1
8	Procarboxypeptidase B	1.7	α-1-Antitrypsin	1.0
9	Pancreatic amylase	1.7	Cytochrome oxidase 1	0.9
10	Bile salt-stimulated lipase	1.4	Apolipoprotein E	0.9

Sources: Data for pancreas from V. E. Velculescu, L. Zhang, B. Vogelstein, and K. W. Kinzler, Serial analysis of gene expression, *Science* 270(1995):484–487. Data for liver from T. Yamashita, S. Hashimoto, S. Kaneko, S. Nagai, N. Toyoda, T. Suzuki, K. Kobayashi, and K. Matsushima, Comprehensive gene expression profile of a normal human liver, *Biochem. Biophys. Res. Commun.* 269(2000):110–116.

gene-regulation mechanisms in prokaryotes and particularly in *E. coli*, because these processes have been extensively investigated in this organism. We then turn to eukaryotic gene regulation. In the chapter's final section, we explore mechanisms for regulating gene expression past the level of transcription.

31.1 PROKARYOTIC DNA-BINDING PROTEINS BIND SPECIFICALLY TO REGULATORY SITES IN OPERONS

Bacteria such as *E. coli* usually rely on glucose as their source of carbon and energy. However, when glucose is scarce, *E. coli* can use lactose as their carbon source even though this disaccharide does not lie on any major metabolic pathways. An essential enzyme in the metabolism of lactose is β-*galactosidase*, which hydrolyzes lactose into galactose and glucose. These products are then metabolized by pathways discussed in Chapter 16.

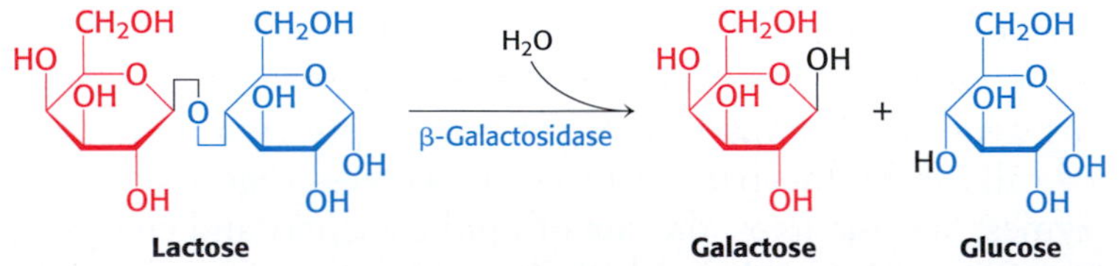

This reaction can be conveniently followed in the laboratory through the use of alternative galactoside substrates that form colored products such as X-Gal (Figure 31.1) An *E. coli* cell growing on a carbon source such as glucose or glycerol contains fewer than 10 molecules of β-galactosidase. In contrast, the same cell will contain several thousand molecules of the enzyme when grown on lactose (Figure 31.2). The presence of lactose in the culture medium induces a large increase in the amount of β-galactosidase by eliciting the synthesis of new enzyme molecules rather than by activating a preexisting but inactive precursor.

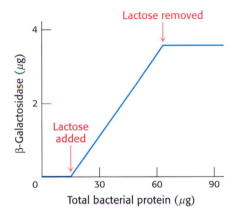

FIGURE 31.1 Following the β-galactosidase reaction. The galactoside substrate X-Gal produces a colored product on cleavage by β-galactosidase. The appearance of this colored product provides a convenient means for monitoring the amount of the enzyme both in vitro and in vivo.

A crucial clue to the mechanism of gene regulation was the observation that two other proteins are synthesized in concert with β-galactosidase— namely, *galactoside permease* and *thiogalactoside transacetylase*. The permease is required for the transport of lactose across the bacterial cell membrane. The transacetylase is not essential for lactose metabolism but appears to play a role in the detoxification of compounds that also may be transported by the permease. Thus, *the expression levels of a set of enzymes that all contribute to the adaptation to a given change in the environment change together*. Such a coordinated unit of gene expression is called an *operon*.

31.1.1 An Operon Consists of Regulatory Elements and Protein-Encoding Genes

The parallel regulation of β-galactosidase, the permease, and the transacetylase suggested that the expression of genes encoding these enzymes is controlled by a common mechanism. Francois Jacob and Jacques Monod proposed the *operon model* to account for this parallel regulation as well as the results of other genetic experiments (Figure 31.3). The genetic elements of the model are a *regulator gene*, a regulatory DNA sequence called *an operator site*, and a *set of structural genes*.

The regulator gene encodes a *repressor* protein that binds to the operator site. The binding of the repressor to the operator prevents transcription of the structural genes. The operator and its associated structural genes constitute the operon. For the *lactose* (lac) *operon*, the *i* gene encodes the repressor, *o* is the operator site, and the *z*, *y*, and *a* genes are the structural genes for β-galactosidase, the permease, and the transacetylase, respectively. The operon also contains a promoter site (denoted by *p*), which directs the RNA polymerase to the correct transcription initiation site. The *z*, *y*, and *a* genes are transcribed to give a single mRNA molecule that encodes all three proteins. An mRNA molecule encoding more than one protein is known as a *polygenic* or *polycistronic* transcript.

FIGURE 31.2 β-Galactosidase induction. The addition of lactose to an *E. coli* culture causes the production of β-galactosidase to increase from very low amounts to much larger amounts. The increase in the amount of enzyme parallels the increase in the number of cells in the growing culture. β-Galactosidase constitutes 6.6% of the total protein synthesized in the presence of lactase.

FIGURE 31.3 Operons. (A) The general structure of an operon as conceived by Jacob and Monod. (B) The structure of the lactose operon. In addition to the promoter (*p*) in the operon, a second promoter is present in front of the regulator gene (*i*) to drive the synthesis of the regulator.

31.1.2 The *lac* Operator Has a Symmetric Base Sequence

The operator site of the *lac* operon has been extensively studied (Figure 31.4). The nucleotide sequence of the operator site shows a nearly perfect inverted repeat, indicating that the DNA in this region has an approximate twofold axis of symmetry. Recall that cleavage sites for restriction enzymes such as *Eco*RV have similar symmetry properties (Section 9.3.3). Symmetry in the operator site usually corresponds to symmetry in the repressor protein that binds the operator site. *Symmetry matching is a recurring theme in protein–DNA interactions.*

5'-...TGTGTGGAATTGTGAGCGGATAACAATTTCACACA...3'
3'-...ACACACCTTAACACTCGCCTAATTGTTAAAGTGTGT...5'

FIGURE 31.4 The *lac* operator. The nucleotide sequence of the *lac* operator shows a nearly perfect inverted repeat, corresponding to twofold rotational symmetry in the DNA. Parts of the sequences that are related by this symmetry are shown in the same color.

31.1.3 The *lac* Repressor Protein in the Absence of Lactose Binds to the Operator and Blocks Transcription

How does the *lac* repressor inhibit the expression of the *lac* operon? The *lac* repressor can exist as a dimer of 37-kd subunits, and two dimers often come together to form a tetramer. In the absence of lactose, the repressor binds very tightly and rapidly to the operator. When the *lac* repressor is bound to DNA, it prevents bound RNA polymerase from locally unwinding the DNA to expose the bases that will act as the template for the synthesis of the RNA strand (Section 28.1.3).

How does the *lac* repressor locate the operator site in the *E. coli* chromosome? The *lac* repressor binds 4×10^6 times as strongly to operator DNA as it does to random sites in the genome. This high degree of selectivity allows the repressor to find the operator efficiently even with a large excess (4.6×10^6) of other sites within the *E. coli* genome. The dissociation constant for the repressor–operator complex is approximately 0.1 pM (10^{-13} M). The rate constant for association ($\approx 10^{10} \text{ M}^{-1} \text{ s}^{-1}$) is strikingly high, indicating that the repressor finds the operator by diffusing along a DNA molecule (a one-dimensional search) rather than encountering it from the aqueous medium (a three-dimensional search).

Inspection of the complete *E. coli* genome sequence reveals two sites within 500 bp of the primary operator site that approximate the sequence of the operator. Other *lac* repressor dimers can bind to these sites, particularly when aided by cooperative interactions with the *lac* repressor dimer at the primary operator site. No other sites that closely match sequence of the *lac* operator site are present in the rest of the *E. coli* genome sequence. Thus, *the DNA-binding specificity of the* lac *repressor is sufficient to specify a nearly unique site within the* E. coli *genome.*

The three-dimensional structure of the *lac* repressor has been determined in various forms. Each monomer consists of a small amino-terminal domain that binds DNA and a larger domain that mediates the formation of the dimer and the tetramer (Figure 31.5). A pair of the amino-terminal domains come together to form the functional DNA-binding unit. Complexes between the *lac* repressor and oligonucleotides that contain the *lac* operator sequence have been structurally characterized. The *lac* repressor binds DNA by inserting an α helix into the major groove of DNA and making a series of contacts with the edges of the base pairs as well as with the phosphodiester backbone (Figure 31.6). For example, an arginine residue in the α helix forms a pair of hydrogen bonds with a guanine residue within the operator. Other bases are not directly contacted but may still be im-

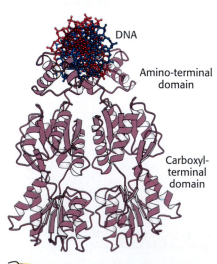

DNA

Amino-terminal domain

Carboxyl-terminal domain

FIGURE 31.5 Structure of the *lac* repressor. A *lac* repressor dimer is shown bound to DNA. A part of the structure that mediates the formation of *lac* repressor tetramers is not shown.

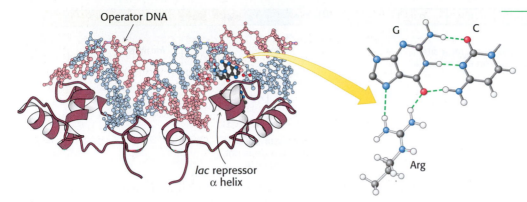

FIGURE 31.6 *Lac* **repressor–DNA interactions.** The *lac* repressor DNA-binding domain inserts an α helix into the major groove of operator DNA. A specific contact between an arginine residue of the repressor and a G–C base pair is shown at the right.

portant for binding by virtue of their effects on local DNA structure. As expected, the twofold axis of the operator coincides with a twofold axis that relates the two DNA-binding domains.

31.1.4 Ligand Binding Can Induce Structural Changes in Regulatory Proteins

How does the presence of lactose trigger expression from the *lac* operon? Interestingly, lactose itself does not have this effect; rather, *allolactose*, a combination of galactose and glucose with an α-1,6 rather than an α-1,4 linkage, does. Allolactose is thus referred to as the *inducer* of the *lac* operon. Allolactose is a side product of the β-galactosidase reaction produced at low levels by the few molecules of β-galactosidase that are present before induction. Some other β-galactosides such as *isopropylthiogalactoside (IPTG)* are potent inducers of β-galactosidase expression, although they are not substrates of the enzyme. IPTG is useful in the laboratory as a tool for inducing gene expression.

How does the presence of the inducer modulate gene expression? *When the* lac *repressor is bound to the inducer, the repressor's affinity for operator DNA is greatly reduced.* The inducer binds in the center of the large domain within each monomer. This binding leads to local conformational changes that are transmitted to the interface with the DNA-binding domains (Figure 31.7).

1,6-Allolactose

Isopropylthiogalactoside (IPTG)

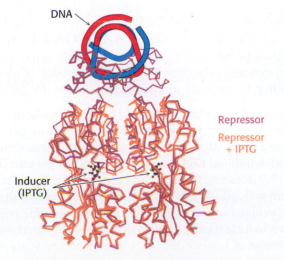

Repressor

Repressor
+ IPTG

Inducer
(IPTG)

FIGURE 31.7 Effects of IPTG on *lac* repressor structure. The structure of the *lac* repressor bound to the inducer isopropylthiogalactoside (IPTG), shown in orange, is superimposed on the structure of the *lac* repressor bound to DNA, shown in purple. The binding of IPTG induces structural changes that alter the relation between the two DNA-binding domains so that they cannot interact effectively with DNA. The DNA-binding domains of the *lac* repressor bound to IPTG are not shown, because these regions are not well ordered in the crystals studied.

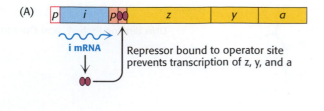

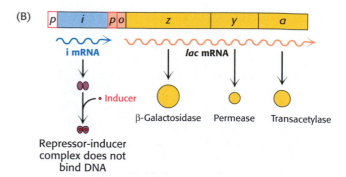

FIGURE 31.8 Induction of the *lac* operon. (A) In the absence of lactose, the *lac* repressor binds DNA and represses transcription from the *lac* operon. (B) Allolactose or another inducer binds to the *lac* repressor, leading to its dissociation from DNA and to the production of *lac* mRNA.

The relation between the two small DNA-binding domains is modified so that they cannot easily contact DNA simultaneously, leading to a dramatic reduction in DNA-binding affinity.

Let us recapitulate the processes that regulate gene expression in the lactose operon (Figure 31.8). In the absence of inducer, the *lac* repressor is bound to DNA in a manner that blocks RNA polymerase from transcribing the *z*, *y*, and *a* genes. Thus, very little β-galactosidase, permease, or transacetylase are produced. The addition of lactose to the environment leads to the formation of allolactose. This inducer binds to the *lac* repressor, leading to conformational changes and the release of DNA by the *lac* repressor. With the operator site unoccupied, RNA polymerase can then transcribe the other *lac* genes and the bacterium will produce the proteins necessary for the efficient utilization of lactose.

The structure of the large domain of the *lac* repressor is similar to those of a large class of proteins that are present in *E. coli* and other bacteria. This family of homologous proteins binds ligands such as sugars and amino acids at their centers. Remarkably, domains of this family are utilized by eukaryotes in taste proteins and in neurotransmitter receptors, as will be discussed in Chapter 32.

31.1.5 The Operon Is a Common Regulatory Unit in Prokaryotes

Many other gene-regulatory networks function in ways analogous to those of the *lac* operon. For example, genes taking part in purine and, to a lesser degree, pyrimidine biosynthesis are repressed by the pur *repressor*. This dimeric protein is 31% identical in sequence with the *lac* repressor and has a similar three-dimensional structure. However, the behavior of the *pur* repressor is opposite that of the *lac* repressor: whereas the *lac* repressor is *released* from DNA by binding to a small molecule, *the* pur *repressor binds DNA specifically only when bound to a small molecule.* Such a small molecule is called a *corepressor*. For the *pur* repressor, the corepressor can be either guanine or hypoxanthine. The dimeric *pur* repressor binds to inverted repeat DNA sites of the form 5′-A**NGCAANCGNTT**N**CNT**-3′, in which the bases shown in boldface type are particularly important. Examination of the *E. coli* genome sequence reveals the presence of more than 20 such sites, regulating 19 operons including more than 25 genes (Figure 31.9).

Because the DNA binding sites for these regulatory proteins are relatively short, it is likely that they evolved independently from one another and are not related by divergence from an ancestral regulatory site. Once a ligand-regulated DNA-binding protein is present in a cell, binding sites may evolve adjacent to additional genes, allowing them to become regulated in a physiologically appropriate manner. Binding sites for the *pur* repressor have evolved in the regulatory regions of a wide range of genes taking part in nucleotide biosynthesis. All such genes can then be regulated in a concerted manner.

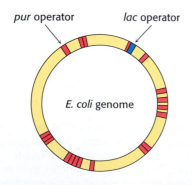

FIGURE 31.9 Binding-site distributions. The *E. coli* genome contains only a single region that closely matches the sequence of the *lac* operator (shown in blue). In contrast, 20 sites match the sequence of the *pur* operator (shown in red). Thus, the *pur* repressor regulates the expression of many more genes than does the *lac* repressor.

31.1.6 Transcription Can Be Stimulated by Proteins That Contact RNA Polymerase

All the DNA-binding proteins discussed thus far function by inhibiting transcription until some environmental condition, such as the presence of lactose, is met. There are also DNA-binding proteins that stimulate transcription. One particularly well studied example is the *catabolite activator protein (CAP)*, which is also known as the cAMP response protein (CRP). When bound to cAMP, CAP, which also is a sequence-specific DNA-binding protein, stimulates the transcription of lactose- and arabinose-catabolizing genes. Within the *lac* operon, CAP binds to an inverted repeat that is centered near position −61 relative to the start site for transcription (Figure 31.10). CAP functions as a dimer of identical subunits.

The CAP–cAMP complex stimulates the initiation of transcription by approximately a factor of 50. A major factor in this stimulation is the recruitment of RNA polymerase to promoters to which CAP is bound. Studies have been undertaken to localize the surfaces on CAP and on the α subunit of RNA polymerase that participate in these interactions (Figure 31.11). These energetically favorable protein–protein contacts increase the likelihood that transcription will be initiated at sites to which the CAP–cAMP complex is bound. Thus, in regard to the *lac* operon, gene expression is maximal when the binding of allolactose relieves the inhibition by the *lac* repressor, and the CAP–cAMP complex stimulates the binding of RNA polymerase.

The *E. coli* genome contains many CAP-binding sites in positions appropriate for interactions with RNA polymerase. Thus, an increase in the cAMP level inside an *E. coli* bacterium results in the formation of CAP–cAMP complexes that bind to many promoters and stimulate the transcription of genes encoding a variety of catabolic enzymes. When grown on glucose, *E. coli* have a very low level of catabolic enzymes such as β-galactosidase. Clearly, it would be wasteful to synthesize these enzymes when glucose is abundant. The inhibitory effect of glucose, called *catabolite repression*, is due to the ability of glucose to lower the intracellular concentration of cyclic AMP.

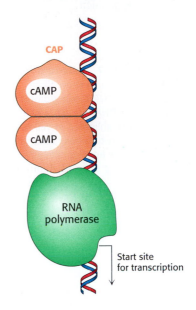

FIGURE 31.10 Binding site for catabolite activator protein (CAP). This protein binds as a dimer to an inverted repeat that is at the position −61 relative to the start site of transcription. The CAP binding site on DNA is adjacent to the position at which RNA polymerase binds.

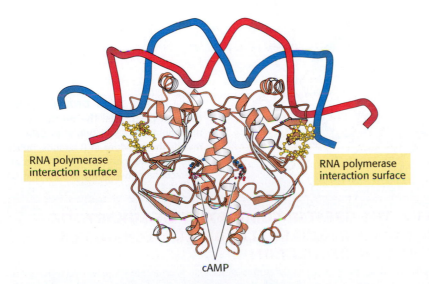

FIGURE 31.11 Structure of a dimer of CAP bound to DNA. The residues in each CAP monomer that have been implicated in direct interactions with RNA polymerase are shown in yellow.

31.1.7 The Helix-Turn-Helix Motif Is Common to Many Prokaryotic DNA-Binding Proteins

The structures of many prokaryotic DNA-binding proteins have now been determined, and amino acid sequences are known for many more. Strikingly,

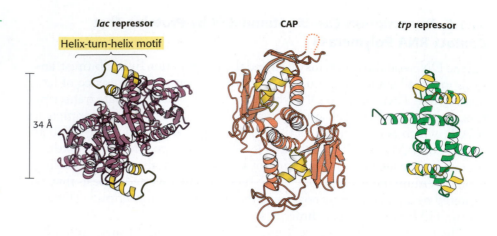

lac repressor   **CAP**   **trp repressor**

Helix-turn-helix motif

34 Å

👆 **FIGURE 31.12 Helix-turn-helix motif.** These structures show three sequence-specific DNA-binding proteins that interact with DNA through a helix-turn-helix motif (highlighted in yellow). In each case, the helix-turn-helix units within a protein dimer are approximately 34 Å apart, corresponding to one full turn of DNA.

the DNA-binding surfaces of many (but not all) of these proteins consist of a pair of α helices separated by a tight turn (Figure 31.12). This *helix-turn-helix motif* is present in the *lac* repressor family, CAP, and many other gene-regulatory proteins. In complexes with DNA, the second of these two helices (often called the *recognition helix*) lies in the major groove, where amino acid side chains make contact with the edges of base pairs, whereas residues of the first helix participate primarily in contacts with the DNA backbone. Helix-turn-helix motifs are very frequently present on proteins that bind DNA as dimers, and thus two of the units will be present, one on each monomer. In this case, the two helix-turn-helix units are related by twofold symmetry along the DNA double helix.

Although the helix-turn-helix motif is the most commonly observed DNA-binding unit in prokaryotes, not all regulatory proteins bind DNA through such units. A striking example is provided by the *E. coli* methionine repressor (Figure 31.13). This protein binds DNA through the insertion of a pair of β strands into the major groove. We shall shortly encounter a variety of other DNA-binding motifs found in eukaryotic cells.

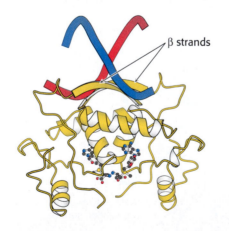

β strands

👆 **FIGURE 31.13 DNA recognition through β strands.** The structure of the methionine repressor bound to DNA reveals that residues in β strands, rather than α helices, participate in the crucial interactions between the protein and DNA.

31.2 THE GREATER COMPLEXITY OF EUKARYOTIC GENOMES REQUIRES ELABORATE MECHANISMS FOR GENE REGULATION

Gene regulation is significantly more complex in eukaryotes than in prokaryotes for a number of reasons. First, the genome being regulated is significantly larger. The *E. coli* genome consists of a single, circular chromosome containing 4.6 Mb. This genome encodes approximately 2000 proteins. In comparison, one of the simplest eukaryotes, *Saccharomyces cerevisiae* (baker's yeast), contains 16 chromosomes ranging in size from 0.2 to 2.2 Mb (Figure 31.14).

Megabase (Mb)—

A length of DNA consisting of 10^6 base pairs (if double stranded) or 10^6 bases (if single stranded).

1 Mb = 10^3 kb = 10^6 bases

The yeast genome totals 17 Mb and encodes approximately 6000 proteins. The genome within a human cell contains 23 pairs of chromosomes ranging in size from 50 to 250 Mb. Approximately 40,000 genes are present within the 3000 Mb of human DNA. It would be very difficult for a DNA-binding protein to recognize a unique site in this vast array of DNA sequences. Consequently, more-elaborate mechanisms are required to achieve specificity.

Another source of complexity in eukaryotic gene regulation is the many different cell types present in most eukaryotes. Liver and pancreatic cells, for example, differ dramatically in the genes that are highly expressed (see Table 31.1). Moreover, eukaryotic genes are not generally organized into operons. Instead, genes that encode proteins for steps within a given pathway are often spread widely across the genome. Finally, transcription and translation are uncoupled in eukaryotes, eliminating some potential gene-regulatory mechanisms.

31.2.1 Nucleosomes Are Complexes of DNA and Histones

The DNA in eukaryotic chromosomes is not bare. Rather eukaryotic DNA is tightly bound to a group of small basic proteins called *histones*. In fact, histones constitute half the mass of a eukaryotic chromosome. The entire complex of a cell's DNA and associated protein is called *chromatin*. Five major histones are present in chromatin: four histones, called H2A, H2B, H3, and H4, associate with one another; the other histone is called H1. Histones have strikingly basic properties because a quarter of the residues in each histone is either arginine or lysine.

In 1974, Roger Kornberg proposed that *chromatin is made up of repeating units, each containing 200 bp of DNA and two copies each of H2A, H2B, H3, and H4*, called the *histone octomer*. These repeating units are known as *nucleosomes*. Strong support for this model comes from the results of a variety of experiments, including observations of appropriately prepared samples of chromatin viewed by electron microscopy (Figure 31.15). Chromatin viewed with the electron microscope has the appearance of beads on a string; each bead has a diameter of approximately 100 Å. Partial digestion of chromatin with DNAse yields the isolated beads. These particles consist of fragments of DNA ≈200 bp in length bound to the eight histones. More extensive digestion yields a reduced DNA fragment of 145 bp bound to the histone octamer. This smaller complex of the histone octamer and the 145-bp DNA fragment is the *nucleosome core particle*. The DNA connecting core particles in undigested chromatin is called *linker DNA*. Histone H1 binds, in part, to the linker DNA.

Size (Mb)

— 1.6
— 2.2

— 1.0

— 0.2

Direction of electrophoresis

FIGURE 31.14 Yeast chromosomes. Pulsed-field electrophoresis allows the separation of 16 yeast chromosomes. [From G. Chu, D. Wollrath, and R. W. Davis. *Science* 234(1986):1583.]

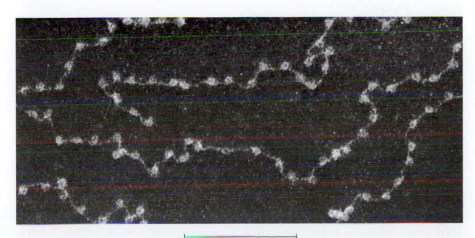

100 nm

FIGURE 31.15 Chromatin structure. An electron micrograph of chromatin showing its "beads on a string" character. [Courtesy of Dr. Ada Olins and Dr. Donald Olins.]

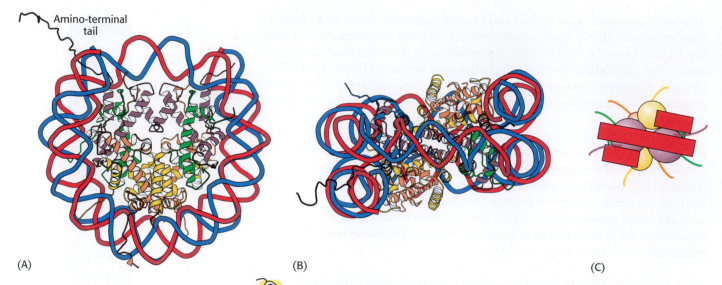

Amino-terminal
tail

(A) (B) (C)

FIGURE 31.16 Nucleosome core particle. The structure consists of a core of eight histone proteins surrounded by DNA. (A) A view showing the DNA wrapping around the histone core. (B) A view related to that in part A by a 90-degree rotation shows that the DNA forms a left-handed superhelix as it wraps around the core. (C) A schematic view.

31.2.2 Eukaryotic DNA Is Wrapped Around Histones to Form Nucleosomes

The overall structure of the nucleosome was revealed through electron microscopic and x-ray crystallographic studies pioneered by Aaron Klug and his colleagues. More recently, the three-dimensional structure of a reconstituted nucleosome core (Figure 31.16) was determined to relatively high resolution by x-ray diffraction methods. As was shown by Evangelos Moudrianakis, the four types of histone that make up the protein core are homologous and similar in structure (Figure 31.17). The eight histones in the core are arranged into a $(H3)_2(H4)_2$ tetramer and a pair of H2A–H2B dimers. The tetramer and dimers come together to form a left-handed superhelical ramp around which the DNA wraps. In addition, each histone has an amino-terminal tail that extends out from the core structure. These tails are flexible and contain a number of lysine and arginine residues. As

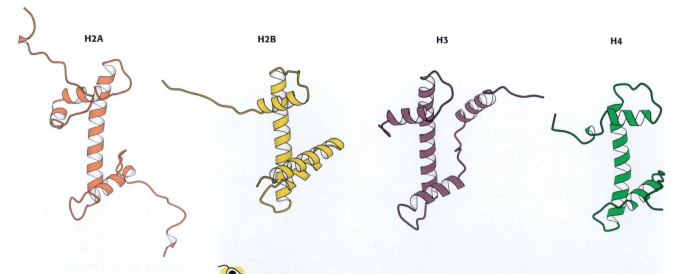

H2A H2B H3 H4

FIGURE 31.17 Homologous histones. Histones H2A, H2B, H3, and H4 each adopt a similar three-dimensional structure as a consequence of common ancestry. Some parts of the tails present at the termini of the proteins are not shown.

we shall see, *covalent modifications of these tails play an essential role in modulating the affinity of the histones for DNA and other properties.*

The DNA forms a left-handed superhelix as it wraps around the outside of the histone octamer. The protein core forms contacts with the inner surface of the superhelix at many points, particularly along the phosphodiester backbone and the minor groove of the DNA. Nucleosomes will form on almost all DNA sites, although some sequences are preferred because the dinucleotide steps are properly spaced to favor bending around the histone core. Histone H1, which has a different structure from the other histones, seals off the nucleosome at the location at which the linker DNA enters and leaves the nucleosome. The amino acid sequences of histones, including their amino- terminal tails, are remarkably conserved from yeast through human beings.

The winding of DNA around the nucleosome core contributes to DNA's packing by decreasing its linear extent. An extended 200-bp stretch of DNA would have a length of about 680 Å. Wrapping this DNA around the histone octamer reduces the length to approximately 100 Å along the long dimension of the nucleosome. Thus the DNA is compacted by a factor of seven. However, human chromosomes in metaphase, which are highly condensed, are compacted by a factor of 10^4. Clearly, the nucleosome is just the first step in DNA compaction. What is the next step? The nucleosomes themselves are arranged in a helical array approximately 360 Å across, forming a series of stacked layers approximately 110 Å apart (Figure 31.18). The folding of these fibers of nucleosomes into loops further compacts DNA.

The writhing of DNA around the histone core in a left-handed helical manner also stores negative supercoils; if the DNA in a nucleosome is straightened out, it will be underwound (Section 23.3.2). This underwinding is exactly what is needed to separate the two DNA strands during replication and transcription (Sections 27.5 and 28.1.5).

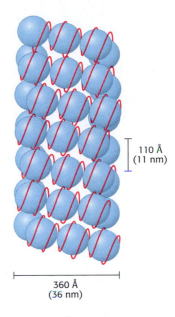

110 Å
(11 nm)

360 Å
(36 nm)

FIGURE 31.18 Higher-order chromatin structure. A proposed model for chromatin arranged in a helical array consisting of six nucleosomes per turn of helix. The DNA double helix (shown in red) is wound around each histone octamer (shown in blue). [After J. T. Finch and A. Klug. *Proc. Natl. Acad. Sci. USA* 73(1976):1900.]

31.2.3 The Control of Gene Expression Requires Chromatin Remodeling

Does chromatin structure play a role in the control of gene expression? Early observations suggested that it does indeed. The treatment of cell nuclei with the nonspecific DNA-cleaving enzyme DNAse I revealed that regions adjacent to genes that are being actively transcribed are more sensitive to cleavage than are other sites in the genome, suggesting that the DNA in these regions is less compacted than it is elsewhere in the genome and more accessible to proteins. In addition, some sites, usually within 1 kb of the start site of an active gene, are exquisitely sensitive to DNAse I and other nucleases. These *hypersensitive sites* correspond to regions that have few nucleosomes or have nucleosomes in an altered conformational state. *Hypersensitive sites are cell-type specific and developmentally regulated.* For example, globin genes in the precursors of erythroid cells from 20-hour-old chicken embryos are insensitive to DNAse I. However, when hemoglobin synthesis begins at 35 hours, regions adjacent to these genes become highly susceptible to digestion. In tissues such as the brain that produce no hemoglobin, the globin genes remain resistant to DNAse I throughout development and into adulthood. The results of these studies suggest that a prerequisite for gene expression is a relaxing of the chromatin structure.

Recent experiments even more clearly revealed the role of chromatin structure in regulating access to DNA binding sites. Genes required for galactose utilization in yeast are activated by a DNA-binding protein called GAL4, which recognizes DNA binding sites with two 5'-GCC-3' sequences separated by 11 base pairs (Figure 31.19). Approximately 4000

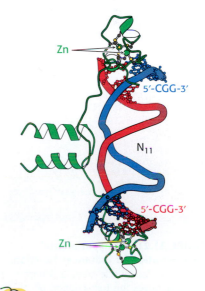

Zn

5'-CGG-3'

N₁₁

5'-CGG-3'

Zn

FIGURE 31.19 GAL4 binding sites. The yeast transcription factor GAL4 binds to DNA sequences of the form 5'-CGG(N)₁₁CCG-3'. Two zinc-based domains are present in the DNA-binding region of this protein. These domains contact the 5'-CGG-3' sequences, leaving the center of the site uncontacted.

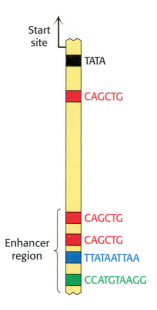

FIGURE 31.20 Enhancer binding sites.
A schematic structure for the region 1 kb
upstream of the start site for the muscle
creatine kinase gene. One binding site of
the form 5'-CAGCTG-3' is present near the
TATA box. The enhancer region farther
upstream contains two binding sites for
the same protein and two additional
binding sites for other proteins.

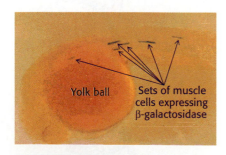

**FIGURE 31.21 An experimental
demonstration of enhancer function.**
A promoter for muscle creatine kinase
artificially drives the transcription of
β-galactosidase in a zebrafish embryo.
Only specific sets of muscle cells produce
β-galactosidase, as visualized by the
formation of the blue product on treatment
of the embryo with X-Gal. [From F. Müller,
D. W. Williamson, J. Kobolák, L. Gauvry,
G. Goldspink, L. Orbán, and N. MacLean. *Molecular
Reproduction and Development* 47(1997): 404.]

potential GAL4 binding sites of the form 5'-CGG(N)₁₁CCG-3' are present in the yeast genome, but only 10 of them regulate genes necessary for galactose metabolism. What fraction of the potential binding sites are actually bound by GAL4? This question is addressed through the use of a technique called *chromatin immunoprecipitation (ChIP)*. GAL4 is first cross-linked to the DNA to which it is bound in chromatin. The DNA is then fragmented into small pieces, and antibodies to GAL4 are used to isolate the chromatin fragments containing GAL4. The cross-linking is reversed, and the DNA is isolated and characterized. The results of these studies reveal that only approximately 10 of the 4000 potential GAL4 sites are occupied by GAL4 when the cells are growing on galactose; more than 99% of the sites appear to be blocked. Thus, whereas in prokaryotes all sites appear to be equally accessible, chromatin structure shields a large number of the potential binding sites in eukaryotic cells. GAL4 is thereby prevented from binding to sites that are unimportant in galactose metabolism.

These lines of evidence and others reveal that chromatin structure is altered in active genes compared with inactive ones. How is chromatin structure modified? As we shall see in Section 31.3.4, specific covalent modifications of histone proteins are crucial. In addition, the binding of specific proteins to DNA sequences called *enhancers* at specific sites in the genome plays a role.

31.2.4 Enhancers Can Stimulate Transcription by Perturbing Chromatin Structure

We can now understand the action of *enhancers*, already introduced in Section 28.2.6. Recall that these DNA sequences, although they have no promoter activity of their own, greatly increase the activities of many promoters in eukaryotes, even when the enhancers are located at a distance of several thousand base pairs from the gene being expressed.

Enhancers function by serving as binding sites for specific regulatory proteins. (Figure 31.20). An enhancer is effective only in the specific cell types in which appropriate regulatory proteins are expressed. In many cases, these DNA-binding proteins influence transcription initiation by perturbing the local chromatin structure to expose a gene or its regulatory sites rather than by direct interactions with RNA polymerase. This mechanism accounts for the ability of enhancers to act at a distance.

The properties of enhancers are illustrated by studies of the enhancer controlling the muscle isoform of creatine kinase (Section 14.1.5). The results of mutagenesis and other studies revealed the presence of an enhancer located between 1350 and 1050 base pairs upstream of the start site of the gene for this enzyme. Experimentally inserting this enhancer near a gene not normally expressed in muscle cells is sufficient to cause the gene to be expressed at high levels in muscle cells, but not other cells (Figure 31.21).

31.2.5 The Modification of DNA Can Alter Patterns of Gene Expression

The modification of DNA provides another mechanism, in addition to packaging with histones, for inhibiting inappropriate gene expression in specific cell types. Approximately 70% of the 5'-CpG-3' sequences in mammalian genomes are methylated at the C-5 position of cytosine by specific methyltransferases. However, the distribution of these methylated cytosines varies, depending on the cell type. Consider, again, the globin genes. In cells that are actively expressing hemoglobin, the region from approximately 1 kb upstream of the start site of the β-globin gene to approximately 100 bp downstream of the start site contains fewer 5-methyl-cytosine residues than does the corresponding region in cells that do not express these genes. The relative absence of 5-methylcytosines near the start site is referred to as *hypomethylation*. The methyl group of 5-methylcytosine protrudes into the major groove where it could easily interfere with the binding of proteins that stimulate transcription.

The distribution of CpG sequences in mammalian genomes is not uniform. The deamination of 5-methylcytosine produces thymine; so CpG sequences are subject to mutation to TpG. Many CpG sequences have been converted into TpG through this mechanism. However, sites near the 5′ ends of genes have been maintained because of their role in gene expression. Thus, most genes are found in *CpG islands*, regions of the genome that contain approximately four times as many CpG sequences as does the remainder of the genome. Note that methylation is not a universal regulatory device, even in multicellular eukaryotes. For example, *Drosophila* DNA is not methylated at all.

31.3 TRANSCRIPTIONAL ACTIVATION AND REPRESSION ARE MEDIATED BY PROTEIN–PROTEIN INTERACTIONS

We have seen how interactions between DNA-binding proteins such as CAP and RNA polymerase can activate transcription in prokaryotic cells (Section 31.1.6). Such protein–protein interactions play a dominant role in eukaryotic gene regulation. In contrast with those of prokaryotic transcription, few eukaryotic transcription factors have any effect on transcription on their own. Instead, *each factor recruits other proteins to build up large complexes that interact with the transcriptional machinery to activate or repress transcription.*

A major advantage of this mode of regulation is that a given regulatory protein can have different effects, depending on what other proteins are present in the same cell. This phenomenon, called *combinatorial control*, is crucial to multicellular organisms that have many different cell types. Even in unicellular eukaryotes such as yeast, combinatorial control allows the generation of distinct cell types.

31.3.1 Steroids and Related Hydrophobic Molecules Pass Through Membranes and Bind to DNA-Binding Receptors

Just as prokaryotes can adjust their patterns of gene expression in response to chemicals in their environment, eukaryotes have many systems for responding to specific molecules with which they come in contact. We first consider a system that detects and responds to estrogens. Synthesized and released by the ovaries, *estrogens*, such as estrone, are cholesterol-derived, steroid hormones (Section 26.4). They are required for the development of female secondary sex characteristics and, along with progesterone, participate in the ovarian cycle.

Estrone
(an estrogen)

Because they are hydrophobic molecules, estrogens easily diffuse across cell membranes. When inside a cell, estrogens bind to highly specific, soluble receptor proteins. Estrogen receptors are members of a large family of proteins that act as receptors for a wide range of hydrophobic molecules, including other steroid hormones, thyroid hormones, and retinoids.

All-*trans*-retinoic acid
(a retinoid)

Thyroxine
(L-3,5,3′,5′-Tetraiodothyronine)
(a thyroid hormone)

These receptors all have a similar mode of action. On binding of the signal molecule (called, generically, a *ligand*), the ligand–receptor complex modifies the expression of specific genes by binding to control elements in the DNA. The human genome encodes approximately 50 members of this family, often referred to as *nuclear hormone receptors*. The genomes of other multicellular eukaryotes encode similar numbers of nuclear hormone receptors, although they are absent in yeast. A comparison of the amino acid sequences of members of this family reveals two highly conserved domains: a DNA-binding domain and a ligand-binding domain (Figure 31.22). The DNA-binding domain lies toward the center of the molecule and includes nine conserved cysteine residues. This domain provides these receptors with sequence-specific DNA activity. Eight of the cysteine residues are conserved because of their role in binding zinc ions: the first four cysteine residues bind one zinc ion, and the second four bind a second zinc ion (see Figure 31.22). The zinc ions stabilize the structure of this small domain; without the bound zinc ions, the domains unfold. Such domains are often referred to as *zinc finger domains*.

The structure of the zinc-binding region of a steroid receptor includes an α helix that begins at the end of the first zinc finger domain. This helix lies in the major groove in the specific DNA complexes formed by estrogen re-

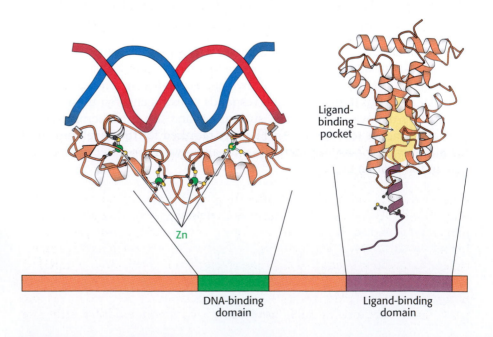

FIGURE 31.22 Structure of two nuclear hormone receptor domains. Nuclear hormone receptors contain two crucial conserved domains: (1) a DNA-binding domain toward the center of the sequence and (2) a ligand-binding domain toward the carboxyl terminus. The structure of a dimer of the DNA-binding domain bound to DNA is shown, as is one monomer of the normally dimeric ligand-binding domain.

Ligand-binding pocket

Zn

DNA-binding domain

Ligand-binding domain

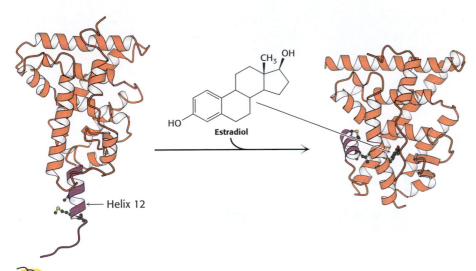

FIGURE 31.23 Ligand binding to nuclear hormone receptor. The ligand lies completely surrounded within a pocket in the ligand-binding domain. The last α helix, helix 12 (shown in purple), folds into a groove on the side of the structure on ligand binding.

ceptors and binds with specific DNA sequences, analogously to prokaryotic DNA-binding proteins. Estrogen receptors bind to specific DNA sites (referred to as *estrogen response elements* or *EREs*) that contain the consensus sequence 5'-**AGGTCA**NNN**TGACCT**-3'. As expected from the symmetry of this sequence, an estrogen receptor binds to such sites as a dimer.

The second highly conserved region of the nuclear receptor proteins lies near the carboxyl terminus and is the ligand-binding site. This domain folds into a structure that consists almost entirely of α helices, arranged in three layers. The ligand binds in a hydrophobic pocket that lies in the center of this array of helices (Figure 31.23). The ligand-binding domain also participates in receptor dimerization.

A comparison of the structures of the ligand-binding domains with and without bound ligand reveals that ligand binding leads to substantial structural rearrangement. In particular, the last α helix (referred to as helix 12), which has hydrophobic residues lining one face but extends out from the receptor in the ligand-free form, folds into a shallow groove on the side of the receptor on ligand binding (see Figure 31.23). How does ligand binding lead to changes in gene expression? The simplest model would have the binding of ligand alter the DNA-binding properties of the receptor, analogously to the *lac* repressor in prokaryotes. However, the results of experiments with purified nuclear hormone receptors revealed that ligand binding does *not* significantly alter DNA-binding affinity and specificity. Another mechanism must be operative.

31.3.2 Nuclear Hormone Receptors Regulate Transcription by Recruiting Coactivators and Corepressors to the Transcription Complex

Because ligand binding does not alter the ability of nuclear hormone receptors to bind DNA, investigators sought to determine whether specific proteins might bind to the nuclear hormone receptors only in the presence of ligand. Such searches led to the identification of several related proteins called *coactivators*, such as SRC-1 (*steroid receptor coactivator*-1), GRIP-1 (*glucocorticoid receptor interacting protein*-1), and NcoA-1 (*nuclear hormone receptor coactivator*-1). These coactivators, referred to as the p160 family because of their size, have a common modular structure

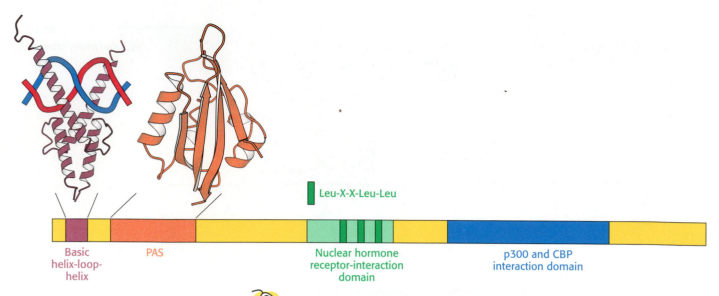

Leu-X-X-Leu-Leu

Basic helix-loop-helix   PAS   Nuclear hormone receptor-interaction domain   p300 and CBP interaction domain

FIGURE 31.24 Coactivator structure. The p160 family of coactivators includes a number of domains that can be recognized at the amino acid sequence level, including a basic helix-loop-helix domain that takes part in DNA binding, a PAS domain that participates in protein–protein interactions, a central domain that contacts the ligand-binding domain of the nuclear hormone receptors, and a domain that interacts with additional coactivators such as p300 and CREB-binding protein (CBP). (CREB stands for cyclic AMP-response element binding protein.) The nuclear hormone receptor interaction domain includes three Leu-X-X-Leu-Leu sequences.

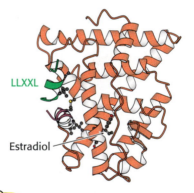

LLXXL

Estradiol

FIGURE 31.25 Coactivator–nuclear hormone receptor interactions. The structure of a complex between the ligand-binding domain of the estrogen receptor with estradiol bound and a peptide from a coactivator reveals that the Leu-X-X-Leu-Leu (LXXLL) sequence forms a helix that binds in a groove on the surface of the ligand-binding domain.

(Figure 31.24). Each coactivator protein contains three sequences of the form Leu-X-X-Leu-Leu within a central region of 200 amino acids. These sequences form short α helices that bind to a hydrophobic patch on the surface of the ligand-binding domains of a nuclear hormone receptor (Figure 31.25). The binding site for the coactivator is fully formed only when ligand is bound, inasmuch as it is adjacent to helix 12. It is likely that a coactivator molecule binds to the ligand-binding domains of a receptor dimer through two of its three Leu-X-X-Leu-Leu sequences. Thus, the binding of ligand to the receptor induces a conformational change that allows the recruitment of a coactivator (Figure 31.26).

Some members of the nuclear hormone receptor family, such as the receptors for thyroid hormone and retinoic acid, repress transcription in the absence of ligand. This repression also is mediated by the ligand-binding domain. In their unbound forms, the ligand-binding domains of these receptors bind to *corepressor proteins*. Members of this family of proteins in-

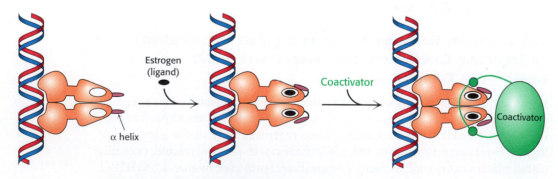

Estrogen (ligand)

Coactivator

α helix

Coactivator

FIGURE 31.26 Coactivator recruitment. The binding of ligand to a nuclear hormone receptor induces a conformational change in the ligand-binding domain. This change in conformation generates favorable sites for the binding of a coactivator.

clude SMRT (Silencing *m*ediator for *r*etinoid and *t*hyroid hormone receptors) and N-Cor (*n*uclear hormone receptor *cor*epressor). Such a corepressor binds to a site in the ligand-binding domain that overlaps the coactivator binding site; ligand binding triggers the release of the corepressor and frees the ligand-binding domain for binding to a coactivator.

31.3.3 Steroid-Hormone Receptors Are Targets for Drugs

Molecules such as estradiol that bind to a receptor and trigger signaling pathways are called *agonists*. Athletes sometimes take natural and synthetic agonists of the androgen receptor, a member of the nuclear hormone receptor family, because their binding to the androgen receptor stimulates the expression of genes that enhance the development of lean muscle mass.

Androstendione
(a natural androgen)

Dianabol
(methandrostenolone)
(a synthetic androgen)

Referred to as *anabolic steroids*, such compounds used in excess are not without side effects. In men, excessive use leads to a decrease in the secretion of testosterone, to testicular atrophy, and sometimes to breast enlargement (gynecomastia) if some of the excess androgen is converted into estrogen. In women, excess testosterone causes a decrease in ovulation and estrogen secretion; it also causes breast regression and growth of facial hair.

Other molecules bind to nuclear hormone receptors but do not effectively trigger signaling pathways. Such compounds are called *antagonists* and are, in many ways, like competitive inhibitors of enzymes. Some important drugs are antagonists that target the estrogen receptor.. For example, *tamoxifen* and *raloxifene* are used in the treatment and prevention of breast cancer, because some breast tumors rely on estrogen-mediated pathways for growth. These compounds are sometimes called *selective estrogen receptor modulators (SERMs)*.

Tamoxifen

Raloxifene

The determination of the structures of complexes between the estrogen receptor and these drugs revealed the basis for their antagonist effect (Figure 31.27). Tamoxifen binds to the same site as estradiol. However, tamoxifen (and other antagonists) have groups that extend out of the normal ligand-binding pocket. These groups prevent helix 12 from binding in its usual position; instead, this helix binds to the site normally occupied by the coactivator. Tamoxifen blocks the binding of coactivators and thus inhibits the activation of gene expression.

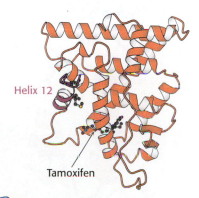

Helix 12

Tamoxifen

FIGURE 31.27 Estrogen receptor–tamoxifen complex. Tamoxifen binds in the pocket normally occupied by estrogen. However, part of the tamoxifen structure extends from this pocket, and so helix 12 cannot pack in its usual position. Instead, this helix blocks the coactivator-binding site.

31.3.4 Chromatin Structure Is Modulated Through Covalent Modifications of Histone Tails

We have seen that nuclear receptors respond to signal molecules by recruiting coactivators and corepressors to the chromatin. Now we can ask, How do coactivators and corepressors modulate transcriptional activity? Much of their effectiveness appears to result from their ability to covalently modify the amino-terminal tails of histones and perhaps other proteins. Some of the p160 coactivators and, in addition, the proteins that they recruit catalyze the transfer of acetyl groups from acetyl CoA to specific lysine residues in the amino-terminal tails of histones.

Lysine in histone tail   **Acetyl CoA**

Enzymes that catalyze such reactions are called *histone acetyltransferases (HATs)*. The histone tails are readily extended; so they can fit into the HAT active site and become acetylated (Figure 31.28).

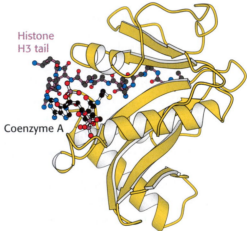

FIGURE 31.28 Structure of histone acetyltransferase. The amino-terminal tail of histone H3 extends into a pocket in which a lysine side chain can accept an acetyl group from acetyl CoA bound in an adjacent site.

What are the consequences of histone acetylation? Lysine bears a positively charged ammonium group at neutral pH. The addition of an acetyl group generates an uncharged amide group. This change dramatically reduces the affinity of the tail for DNA and modestly decreases the affinity of the entire histone complex for DNA. The loosening of the histone complex from the DNA exposes additional DNA regions to the transcription machinery. In addition, the acetylated lysine residues interact with a specific *acetyllysine-binding domain* that is present in many proteins that regulate eukaryotic transcription. This domain, termed a *bromodomain*, comprises approximately 110 amino acids that form a four-helix bundle containing a peptide-binding site at one end (Figure 31.29).

Bromodomain-containing proteins are components of two large complexes essential for transcription. One is a complex of more than 10 polypeptides that binds to the *TATA-box-binding protein*. Recall that the TATA-

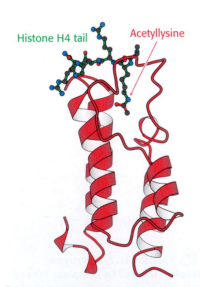

FIGURE 31.29 Structure of a bromodomain. This four-helix-bundle domain binds peptides containing acetyllysine. An acetylated peptide of histone H4 is bound in the structure shown.

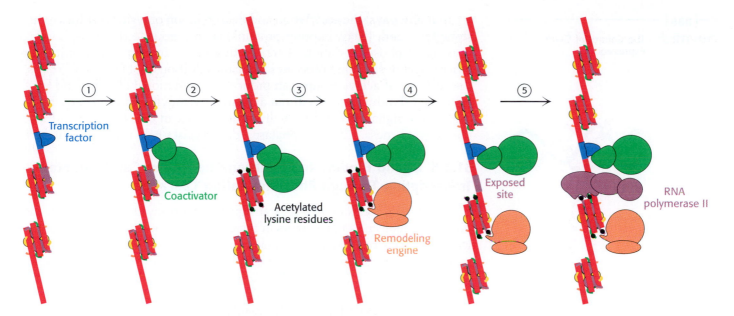

FIGURE 31.30 Chromatin remodeling. Eukaryotic gene regulation begins with an activated transcription factor bound to a specific site on DNA. One scheme for the initiation of transcription by RNA polymerase II requires five steps: (1) recruitment of a coactivator, (2) acetylation of lysine residues in the histone tails, (3) binding of a remodeling engine complex to the acetylated lysine residues, (4) ATP-dependent remodeling of the chromatin structure to expose a binding site for RNA polymerase or for other factors, and (5) recruitment of RNA polymerase. Only two subunits are shown for each complex, although the actual complexes are much larger. Other schemes are possible.

box-binding protein is an essential transcription factor for many genes (Section 28.2.4). Proteins that bind to the TATA-box-binding protein are called *TAFs* (for *TATA-box-binding protein associated factors*). In particular, TAFII250 (named for its participation in RNA polymerase II transcription and its apparent molecular weight of 250 kd) contains a pair of bromodomains near its carboxyl terminus. The two domains are oriented so that each can bind one of two acetyllysine residues at positions 5 and 12 in the histone H4 tail. Thus, *acetylation of the histone tails provides a mechanism for recruiting other components of the transcriptional machinery.*

Bromodomains are also present in some components of large complexes known as *chromatin-remodeling engines*. These complexes, which also contain domains homologous to those of helicases (Section 27.2.5), utilize the free energy of ATP hydrolysis to shift the positions of nucleosomes along the DNA and to induce other conformational changes in chromatin (Figure 31.30). Histone acetylation can lead to reorganization of the chromatin structure, potentially exposing binding sites for other factors. *Thus, histone acetylation can activate transcription through a combination of three mechanisms: by reducing the affinity of the histones for DNA, by recruiting other components of the transcriptional machinery, and by initiating the active remodeling of the chromatin structure.*

31.3.5 Histone Deacetylases Contribute to Transcriptional Repression

Just as in prokaryotes, some changes in a cell's environment lead to the repression of genes that had been active. The modification of histone tails again plays an important role. However, in repression, a key reaction appears to be the deacetylation of acetylated lysine, catalyzed by specific *histone deacetylase* enzymes.

In many ways, the acetylation and deacetylation of lysine residues in histone tails (and, likely, in other proteins) is analogous to the phosphorylation and dephosphorylation of serine, threonine, and tyrosine residues in other stages of signaling processes. Like the addition of phosphoryl groups, the addition of acetyl groups can induce conformational changes and generate novel binding sites. Without a means of removing these groups, however, these signaling switches will become stuck in one position and lose their effectiveness. Like phosphatases, deacetylases help reset the switches.

31.3.6 Ligand Binding to Membrane Receptors Can Regulate Transcription Through Phosphorylation Cascades

In Chapter 15, we examined several signaling pathways that begin with the binding of molecules to receptors in the cell membrane. Some of these pathways lead to the regulation of gene expression. Let us review the pathway initiated by epinephrine. The binding of epinephrine to a 7TM receptor results in the activation of a G protein. The activated G protein, in turn, binds to and activates adenylate cyclase, increasing the intracellular concentration of cAMP. This cAMP binds to the regulatory subunit of protein kinase A (PKA), activating the enzyme. We have previously examined the role of phosphorylation by PKA of a variety of enzymes—for example, those controlling glycogen metabolism. PKA also phosphorylates the *cyclic AMP-response element binding protein (CREB)*, a transcription factor that binds specific DNA sequences as a dimer. Each monomer contributes a long α helix; together, the two helices grab the DNA in the manner of a pair of chopsticks (Figure 31.31).

How does the phosphorylation of CREB affect its ability to activate transcription? Phosphorylation does not appear to alter the DNA-binding properties of this protein. Instead, phosphorylated CREB binds a coactivator protein termed CBP, for *CREB-binding protein*. CBP possesses a highly revealing domain structure (Figure 31.32).

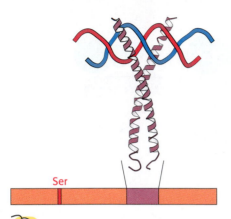

FIGURE 31.31 Cyclic AMP-response element binding protein (CREB). Each of two CREB subunits contributes a long α helix. The two helices coil around each other to form a dimeric DNA-binding unit. CREB is phosphorylated on a specific serine residue by protein kinase A.

Ser

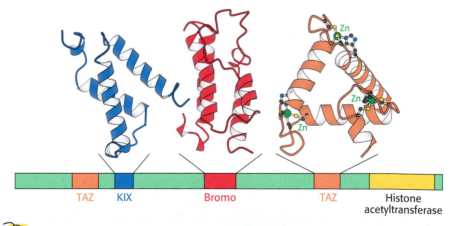

TAZ   KIX   Bromo   TAZ   Histone
                                       acetyltransferase

FIGURE 31.32 Domain structure of CREB-binding protein (CBP). The CREB-binding protein includes at least three types of protein–protein interaction domains in addition to a histone acetyltransferase domain that lies near the carboxyl terminus. The kinase-inducible interaction (KIX) domain interacts specifically with a region of CREB in its phosphorylated form.

Its domains include a KIX domain (for *kinase-inducible interaction*) that binds to the phosphorylated region of CREB (Figure 31.33); a bromodomain that binds acetylated histone tails, and two TAZ domains, zinc-binding domains that facilitate the binding of CBP to a variety of proteins through a remarkable triangular structure (see Figure 31.32). Thus, the pathway initiated by epinephrine binding induces the phosphorylation of a transcription factor, the recruitment of a coactivator, and the assembly of complexes that participate in chromatin remodeling and transcription initiation.

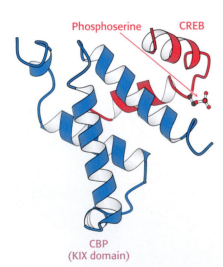

Phosphoserine   CREB

CBP
(KIX domain)

FIGURE 31.33 Interaction between CBP and CREB. The KIX domain of CBP binds a region of CREB in its phosphorylated form.

31.3.7 Chromatin Structure Effectively Decreases the Size of the Genome

The transcriptional regulatory mechanisms utilized by prokaryotes and eukaryotes have some significant differences, many of which are related to the significant difference in genome sizes between these classes of organisms. However, much of the DNA in a eukaryotic cell is stably assembled into chromatin. The packaging of DNA with chromatin renders many potential binding sites for transcription factors inaccessible—in effect, reducing the size of the genome. Thus, rather than scanning through the entire genome, *a eukaryotic DNA-binding protein scans a set of accessible binding sites that is close in size to the genome of a prokaryote.* The cell type is determined by the genes that are accessible.

31.4 GENE EXPRESSION CAN BE CONTROLLED AT POSTTRANSCRIPTIONAL LEVELS

Modulation of the rate of transcriptional initiation is the most common mechanism of gene regulation. However, other stages of transcription also are targets for regulation in some cases. In addition, the process of translation provides other points of intervention for regulating the level of a protein produced in a cell. These mechanisms are quite distinct in prokaryotic and eukaryotic cells because prokaryotes and eukaryotes differ greatly in how transcription and translation are coupled and in how translation is initiated. We shall consider two important examples of posttranscriptional regulation: one from prokaryotes and the other from eukaryotes. In both examples, regulation depends on the formation of distinct secondary structures in mRNA.

31.4.1 Attenuation Is a Prokaryotic Mechanism for Regulating Transcription Through Modulation of Nascent RNA Secondary Structure

A novel mechanism for regulating transcription in bacteria was discovered by Charles Yanofsky and his colleagues as a result of their studies of the tryptophan operon. The 7-kb mRNA transcript from this operon encodes five enzymes that convert chorismate into tryptophan (Section 24.2.10). The mode of regulation of this operon is called *attenuation*, and it depends on features at the 5′ end of the mRNA product (Figure 31.34). Upstream of the coding regions for the enzymes responsible for tryptophan biosynthesis lies a short open reading frame encoding a 14-amino-acid leader peptide. Following this open reading frame is a region of RNA, called an *attenuator*, that is capable of forming several alternative structures. Recall that transcription and translation are tightly coupled in bacteria. Thus, the translation of the *trp* mRNA begins soon after the ribosome-binding site has been synthesized.

FIGURE 31.34 Leader region of *trp* mRNA. (A) The nucleotide sequence of the 5′ end of *trp* mRNA includes a short open reading frame that encodes a peptide comprising 14 amino acids; the peptide has two tryptophan residues and an untranslated attenuator region (blue and red nucleotides). (B and C) The atenuator region can adopt two distinct stem-loop structures.

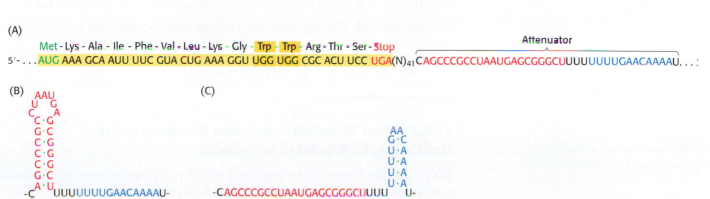

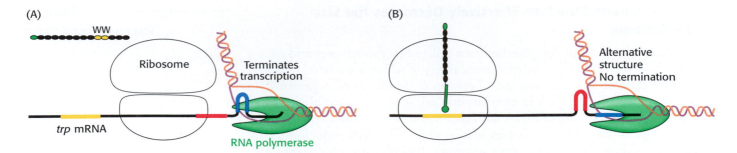

FIGURE 31.35 Attenuation. (A) In the presence of adequate concentrations of tryptophan (and, hence, Trp-tRNA), translation proceeds rapidly and an RNA structure forms that terminates transcription. (B) At low concentrations of tryptophan, translation stalls awaiting Trp-tRNA, giving time for an alternative RNA structure to form that does not terminate transcription efficiently.

A ribosome is able to translate the leader region of the mRNA product only in the presence of adequate concentrations of tryptophan. When enough tryptophan is present, a stem-loop structure forms in the attenuator region, which leads to the release of RNA polymerase from the DNA (Figure 31.35). However, when tryptophan is scarce, transcription is terminated less frequently. How does the level of tryptophan alter transcription of the *trp* operon? An important clue was the finding that the 14-amino-acid leader peptide includes two adjacent tryptophan residues. When tryptophan is scarce, little tryptophanyl-tRNA is present. Thus, the ribosome stalls at the tandem UGG codons encoding tryptophan. This delay leaves the adjacent region of the mRNA exposed as transcription continues. An alternative RNA structure that does not function as a terminator is formed and transcription continues into and through the coding regions for the enzymes. Thus, attenuation provides an elegant means of sensing the supply of tryptophan required for protein synthesis.

Several other operons for the biosynthesis of amino acids in *E. coli* also are regulated by attenuator sites. The leader peptide of each contains an abundance of the amino acid residues of the type controlled by the operon (Figure 31.36). For example, the leader peptide for the phenylalanine operon includes 7 phenylalanine residues among 15 residues. The threonine operon encodes enzymes required for the synthesis of both threonine and isoleucine; the leader peptide contains 8 threonine and 4 isoleucine residues in a 16-residue sequence. The leader peptide for the histidine operon includes 7 histidine residues in a row. In each case, low levels of the corresponding charged tRNA causes the ribosome to stall, trapping the nascent mRNA in a state that can form a structure that allows RNA polymerase to read through the attenuator site.

(A)

Met - Lys - Arg - Ile - Ser - Thr - Thr - Ile - Thr - Thr - Thr - Ile - Thr - Ile - Thr - Thr -

5′   AUG AAA CGC AUU AGC ACC ACC AUU ACC ACC ACC AUC ACC AUU ACC ACA   3′

(B)

Met - Lys - His - Ile - Pro - Phe - Phe - Phe - Ala - Phe - Phe - Phe - Thr - Phe - Pro - Stop

5′   AUG AAA CAC AUA CCG UUU UUC UUC GCA UUC UUU UUU ACC UUC CCC UGA   3′

(C)

Met - Thr - Arg - Val - Gln - Phe - Lys - His - His - His - His - His - His - His - Pro - Asp -

5′   AUG ACA CGC GUU CAA UUU AAA CAC CAC CAU CAU CAC CAU CAU CCU GAC   3′

FIGURE 31.36 Leader peptide sequences. Amino acid sequences and the corresponding mRNA nucleotide sequences of the (A) threonine operon, (B) phenylalanine operon, and (C) histidine operon. In each case, an abundance of one amino acid in the leader peptide sequence leads to attenuation.

31.4.2 Genes Associated with Iron Metabolism Are Translationally Regulated in Animals

RNA secondary structure plays a role in the regulation of iron metabolism in eukaryotes. Iron is an essential nutrient, required for the synthesis of hemoglobin, cytochromes, and many other proteins. However, excess iron can be quite harmful because, untamed by a suitable protein environment, iron can initiate a range of free-radical reactions that damage proteins, lipids, and nucleic acids. Animals have evolved sophisticated systems for the accumulation of iron in times of scarcity and for the safe storage of excess iron for later use. Key proteins include *transferrin*, a transport protein that carries iron in the serum, *transferrin receptor*, a membrane protein that binds iron-loaded transferrin and initiates its entry into cells, and *ferritin*, an impressively efficient iron-storage protein found primarily in the liver and kidneys. Twenty-four ferritin polypeptides form a nearly spherical shell that encloses as many as 2400 iron atoms, a ratio of one iron atom per amino acid (Figure 31.37).

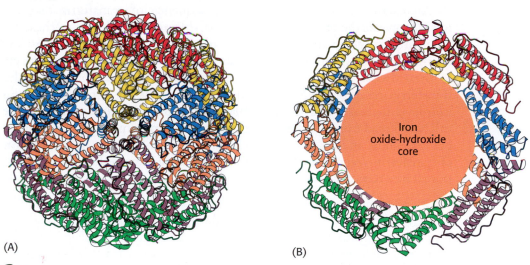

(A)   (B)

FIGURE 31.37 Structure of ferritin. (A) Twenty-four ferritin polypeptides form a nearly spherical shell. (B) A cut-away view reveals the core that stores iron as an iron oxide–hydroxide complex.

Ferritin and transferrin receptor expression levels are reciprocally related in their responses to changes in iron levels. When iron is scarce, the amount of transferrin receptor increases and little or no new ferritin is synthesized. Interestingly, the extent of mRNA synthesis for these proteins does not change correspondingly. Instead, regulation takes place at the level of translation.

Consider ferritin first. Ferritin mRNA includes a stem-loop structure termed an *iron-response element (IRE)* in its 5′ untranslated region (Figure 31.38). This stem-loop binds a 90-kd protein, called an *IRE-binding protein (IRE-BP)*, that blocks the initiation of translation. When the iron level increases, the IRE-BP binds iron as a 4Fe-4S cluster. The IRE-BP bound to iron cannot bind RNA, because the binding sites for iron and RNA

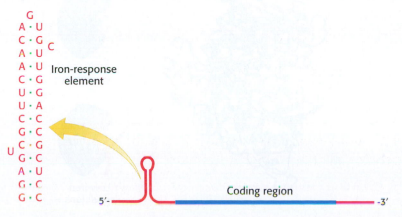

Iron-response element

Coding region

5′—   —3′

FIGURE 31.38 Iron-response element. Ferritin mRNA includes a stem-loop structure, termed an iron-response element (IRE), in its 5′ untranslated region. The IRE binds a specific protein that blocks the translation of this mRNA under low iron conditions.

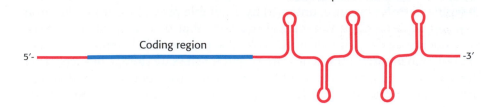

Coding region

5'-   -3'

substantially overlap. Thus, in the presence of iron, ferritin mRNA is released from the IRE-BP and translated to produce ferritin, which sequesters the excess iron.

An examination of the nucleotide sequence of transferrin-receptor mRNA reveals the presence of several IRE-like regions. However, these regions are located in the 3' untranslated region rather than in the 5' untranslated region (Figure 31.39). Under low-iron conditions, IRE-BP binds to these IREs. However, given the location of these binding sites, the transferrin-receptor mRNA can still be translated. What happens when the iron level increases and the IRE-BP no longer binds transferrin-receptor mRNA? Freed from the IRE-BP, transferrin-receptor mRNA is rapidly degraded. Thus, an increase in the cellular iron level leads to the destruction of transferrin-receptor mRNA and, hence, a reduction in the production of transferrin-receptor protein.

The purification of the IRE-BP and the cloning of its cDNA were sources of truly remarkable insight into evolution. The IRE-BP was found to be approximately 30% identical in amino acid sequence with the citric acid cycle enzyme aconitase from mitochondria. Further analysis revealed that the IRE-BP is, in fact, an active aconitase enzyme; it is a cytosolic aconitase that had been known for a long time, but its function was not well understood (Figure 31.40). The iron–sulfur center at the active site of the IRE-BP is rather unstable, and loss of the iron triggers significant changes in protein conformation. Thus, this protein can serve as an iron-sensing factor.

Other mRNAs, including those taking part in heme synthesis, have been found to contain IREs. Thus, genes encoding proteins required for iron metabolism acquired sequences that, when transcribed, provided binding sites for the iron-sensing protein. An environmental signal—the concentration of iron—controls the translation of proteins required for the metabolism of this metal. The IREs have evolved appropriately in the untranslated regions of mRNAs to lead to beneficial regulation by iron levels.

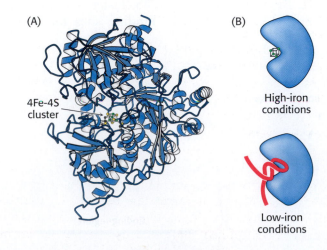

(A)

(B)

4Fe-4S
cluster

High-iron
conditions

Low-iron
conditions

SUMMARY

Prokaryotic DNA-Binding Proteins Bind Specifically to Regulatory Sites in Operons

In prokaryotes, many genes are clustered into operons, which are units of coordinated genetic expression. An operon consists of control sites (an operator and a promoter) and a set of structural genes. In addition, regulator genes encode proteins that interact with the operator and promoter sites to stimulate or inhibit transcription. The treatment of *E. coli* with lactose induces an increase in the production of β-galactosidase and two additional proteins that are encoded in the lactose operon. In the absence of lactose or a similar galactoside inducer, the *lac* repressor protein binds to an operator site on the DNA and blocks transcription. The binding of allolactose, a derivative of lactose, to the *lac* repressor induces a conformational change that leads to dissociation from DNA. RNA polymerase can then move through the operator to transcribe the *lac* operon.

Some proteins activate transcription by directly contacting RNA polymerase. For example, cyclic AMP, a hunger signal, stimulates the transcription of many catabolic operons by binding to the catabolite activator protein. The binding of the cAMP–CAP complex to a specific site in the promoter region of an inducible catabolic operon enhances the binding of RNA polymerse and the initiation of transcription. Many, but not all, prokaryotic DNA-binding proteins interact with DNA through helix-turn-helix motifs.

The Greater Complexity of Eukaryotic Genomes Requires Elaborate Mechanisms for Gene Regulation

Eukaryotic DNA is tightly bound to basic proteins called histones; the combination is called chromatin. DNA wraps around an octamer of core histones to form a nucleosome, blocking access to many potential DNA binding sites. Changes in chromatin structure play a major role in regulating gene expression. Enhancers can modulate gene expression from more than 1000 bp away from the start site of transcription by perturbing local chromatin structure. Enhancers are often specific for certain cell types, depending on which DNA-binding proteins are present.

Transcriptional Activation and Repression Are Mediated by Protein–Protein Interactions

Steroids such as estrogens bind to eukaryotic transcription factors called nuclear hormone receptors. These proteins are capable of binding DNA whether or not ligands are bound. The binding of ligands induces a conformational change that allows the recruitment of additional proteins called coactivators. Among the most important functions of coactivators is catalysis of the addition of acetyl groups to lysine residues in the tails of histone proteins. Histone acetylation decreases the affinity of the histones for DNA, making additional genes accessible for transcription. In addition, acetylated histones are targets for proteins containing specific binding units called bromodomains. Two classes of large complexes are eventually recruited: chromatin remodeling engines and RNA polymerase II and its associated factors. These complexes open up sites on chromatin and initiate transcription.

Gene Expression Can Be Controlled at Posttranscriptional Levels

Gene expression can also be regulated at the level of translation. In prokaryotes, many operons important in amino acid biosynthesis are regulated by attenuation, a process that depends on the formation of alternative structures in mRNA, one of which favors transcriptional termination. Attenuation is mediated by the translation of a leader region of mRNA. A ribosome stalled by the absence of an aminoacyl-tRNA needed to translate the leader mRNA alters the structure of mRNA so that RNA polymerase transcribes the operon beyond the attenuator site.

In eukaryotes, genes encoding proteins that transport and store iron are regulated at the translational level. Iron-response elements, structures that are present in certain mRNAs, are bound by an IRE-binding protein when this protein is not binding iron. Whether the expression of a gene is stimulated or inhibited in response to changes in the iron status of a cell depends on the location of the IRE within the mRNA.

KEY TERMS

cell type (p. 867)

β-galactosidase (p. 868)

operon model (p. 869)

repressor (p. 869)

lac operator (p. 870)

lac repressor (p. 870)

inducer (p. 871)

isopropylthioogalactoside (IPTG) (p. 871)

pur repressor (p. 872)

corepressor (p. 872)

catabolite activator protein (CAP) (p. 873)

catabolite repression (p. 873)

helix-turn-helix motif (p. 874)

histone (p. 875)

chromatin (p. 875)

nucleosome (p. 875)

nucleosome core particle (p. 875)

hypersensitive site (p. 877)

chromatin immunoprecipitation (ChIP) (p. 878)

enhancer (p. 878)

hypomethylation (p. 879)

CpG island (p. 879)

combinatorial control (p. 879)

nuclear hormone receptor (p. 880)

zinc-finger domain (p. 880)

estrogen response element (ERE) (p. 881)

coactivator (p. 881)

corepressor protein (p. 882)

agonist (p. 883)

anabolic steroid (p. 883)

antagonist (p. 883)

selective estrogen modulator (SERM) (p. 883)

histone acetyltransferase (HAT) (p. 884)

acetyllysine-binding domain (p. 884)

bromodomain (p. 884)

TATA-box-binding protein associated factor (TAF) (p. 884)

chromatin-remodeling engine (p. 885)

histone deacetylase (p. 885)

cyclic AMP-response element binding protein (CREB) (p. 886)

attenuation (p. 887)

transferrin (p. 889)

transferrin receptor (p. 889)

ferritin (p. 889)

iron-response element (IRE) (p. 889)

IRE-binding protein (IRE-BP) (p. 889)

SELECTED READINGS

Where to start

Pabo, C. O., and Sauer, R. T., 1984. Protein-DNA recognition *Annu. Rev. Biochem.* 53:293–321.

Struhl, K., 1989. Helix-turn-helix, zinc-finger, and leucine-zipper motifs for eukaryotic transcriptional regulatory proteins. *Trends Biochem. Sci.* 14:137–140.

Struhl, K., 1999. Fundamentally different logic of gene regulation in eukaryotes and prokaryotes. *Cell* 98:1–4.

Korzus, E., Torchia, J., Rose, D. W., Xu, L., Kurokawa, R., McInerney, E. M., Mullen, T. M., Glass, C. K., and Rosenfeld, M. G., 1998. Transcription factor-specific requirements for coactivators and their acetyltransferase functions. *Science.* 279:703–707.

Aalfs, J. D., and Kingston, R. E., 2000. What does "chromatin remodeling" mean? *Trends Biochem. Sci.* 25:548–555.

Books

Ptashne, M., 1992. *A Genetic Switch: Phage λ and Higher Organisms* (2nd ed.). Cell Press and Blackwell Scientific.

McKnight, S. L., and Yamamoto, K. R. (Eds.), 1992. *Transcriptional Regulation* (vols. 1 and 2). Cold Spring Harbor Laboratory Press.

Latchman, D. S., 1999. *Eukaryotic Transcription Factors* (3rd ed.). Academic Press.

Wolffe, A., 1992. *Chromatin Structure and Function.* Academic Press.

Lodish, H., Baltimore, D., Berk, A., Zipursky, S. L., Matsudaira, P., and Darnell, J., 1999. *Molecular Cell Biology* (4th ed.). Scientific American Books.

Prokaryotic gene regulation

Bell, C. E., and Lewis, M., 2001. The lac repressor: A second generation of structural and functional studies. *Curr. Opin. Struct. Biol.* 11:19–25.

Lewis, M., Chang, G., Horton, N. C., Kercher, M. A., Pace, H. C., Schumacher, M. A., Brennan, R. G., and Lu, P., 1996. Crystal structure of the lactose operon repressor and its complexes with DNA and inducer. *Science* 271:1247–1254.

Niu, W., Kim, Y., Tau, G., Heyduk, T., and Ebright, R. H., 1996. Transcription activation at class II CAP-dependent promoters: Two interactions between CAP and RNA polymerase. *Cell* 87:1123–1134.

Schultz, S. C., Shields, G. C., and Steitz, T. A., 1991. Crystal structure of a CAP-DNA complex: The DNA is bent by 90 degrees. *Science* 253:1001–1007.

Parkinson, G., Wilson, C., Gunasekera, A., Ebright, Y. W., Ebright, R. E., and Berman, H. M., 1996. Structure of the CAP-DNA complex at 2.5 angstroms resolution: A complete picture of the protein-DNA interface. *J. Mol. Biol.* 260:395–408.

Busby, S., and Ebright, R. H., 1999. Transcription activation by catabolite activator protein (CAP). *J. Mol. Biol.* 293:199–213.

Somers, W. S., and Phillips, S. E., 1992. Crystal structure of the met repressor-operator complex at 2.8 Å resolution reveals DNA recognition by beta-strands. *Nature* 359:387–393.

Nucleosomes and histones

Luger, K., Mader, A. W., Richmond, R. K., Sargent, D. F., and Richmond, T. J., 1997. Crystal structure of the nucleosome core particle at 2.8 Å resolution. *Nature* 389:251–260.

Arents, G., and Moudrianakis, E. N., 1995. The histone fold: A ubiquitous architectural motif utilized in DNA compaction and protein dimerization. *Proc. Natl. Acad. Sci. USA* 92:11170–11174.

Baxevanis, A. D., Arents, G., Moudrianakis, E. N., and Landsman, D., 1995. A variety of DNA-binding and multimeric proteins contain the histone fold motif. *Nucleic Acids Res.* 23:2685–2691.

Clements, A., Rojas, J. R., Trievel, R. C., Wang, L., Berger, S. L., and Marmorstein, R., 1999. Crystal structure of the histone acetyltransferase domain of the human PCAF transcriptional regulator bound to coenzyme A. *EMBO J.* 18:3521–3532.

Deckert, J., and Struhl, K., 2001. Histone acetylation at promoters is differentially affected by specific activators and repressors. *Mol. Cell. Biol.* 21:2726–2735.

Dutnall, R. N., Tafrov, S. T., Sternglanz, R., and Ramakrishnan, V., 1998. Structure of the histone acetyltransferase Hat1: A paradigm for the GCN5-related *N*-acetyltransferase superfamily. *Cell* 94: 427–438.

Finnin, M. S., Donigian, J. R., Cohen, A., Richon, V. M., Rifkind, R. A., Marks, P. A., Breslow, R., and Pavletich, N. P., 1999. Structures of a histone deacetylase homologue bound to the TSA and SAHA inhibitors. *Nature* 401:188–193.

Finnin, M. S., Donigian, J. R., and Pavletich, N. P., 2001. Structure of the histone deacetylase SIR2. *Nat. Struct. Biol.* 8:621–625.

Jacobson, R. H., Ladurner, A. G., King, D. S., and Tjian, R., 2000. Structure and function of a human TAFII250 double bromodomain module. *Science* 288:1422–1425.

Rojas, J. R., Trievel, R. C., Zhou, J., Mo, Y., Li, X., Berger, S. L., Allis, C. D., and Marmorstein, R., 1999. Structure of *Tetrahymena* GCN5 bound to coenzyme A and a histone H3 peptide. *Nature* 401:93–98.

Nuclear hormone receptors

Evans, R. M., 1988. The steroid and thyroid hormone receptor superfamily. *Science* 240:889–895.

Yamamoto, K. R., 1985. Steroid receptor regulated transcription of specific genes and gene networks. *Annu. Rev. Genet.* 19:209–252.

Tanenbaum, D. M., Wang, Y., Williams, S. P., and Sigler, P. B., 1998. Crystallographic comparison of the estrogen and progesterone receptor's ligand binding domains. *Proc. Natl. Acad. Sci. USA* 95: 5998–6003.

Schwabe, J. W., Chapman, L., Finch, J. T., and Rhodes, D., 1993. The crystal structure of the estrogen receptor DNA-binding domain bound to DNA: How receptors discriminate between their response elements. *Cell* 75:567–578.

Shiau, A. K., Barstad, D., Loria, P. M., Cheng, L., Kushner, P. J., Agard, D. A., and Greene, G. L., 1998. The structural basis of estrogen receptor/coactivator recognition and the antagonism of this interaction by tamoxifen. *Cell* 95:927–937.

Collingwood, T. N., Urnov, F. D., and Wolffe, A. P., 1999. Nuclear receptors: Coactivators, corepressors and chromatin remodeling in the control of transcription. *J. Mol. Endocrinol.* 23:255–275.

Chromatin and chromatin remodeling

Elgin, S. C., 1981. DNAase I-hypersensitive sites of chromatin. *Cell* 27:413–415.

Weintraub, H., Larsen, A., and Groudine, M., 1981. α-Globin-gene switching during the development of chicken embryos: Expression and chromosome structure. *Cell* 24:333–344.

Ren, B., Robert, F., Wyrick, J. J., Aparicio, O., Jennings, E. G., Simon, I., Zeitlinger, J., Schreiber, J., Hannett, N., Kanin, E., Volkert, T. L., Wilson, C. J., Bell, S. P., and Young, R. A., 2000. Genome-wide location and function of DNA-binding proteins. *Science* 290:2306–2309.

Goodrich, J. A., and Tjian, R., 1994. TBP-TAF complexes: Selectivity factors for eukaryotic transcription. *Curr. Opin. Cell. Biol.* 6:403–409.

Bird, A. P., and Wolffe, A. P., 1999. Methylation-induced repression: Belts, braces, and chromatin. *Cell* 99:451–454.

Cairns, B. R., 1998. Chromatin remodeling machines: Similar motors, ulterior motives. *Trends Biochem. Sci.* 23:20–25.

Albright, S. R., and Tjian, R., 2000. TAFs revisited: More data reveal new twists and confirm old ideas. *Gene* 242:1–13.

Urnov, F. D., and Wolffe, A. P., 2001. Chromatin remodeling and transcriptional activation: The cast (in order of appearance). *Oncogene* 20:2991–3006.

Posttranscriptional regulation

Kolter, R., and Yanofsky, C., 1982. Attenuation in amino acid biosynthetic operons. *Annu. Rev. Genet.* 16:113–134.

Yanofsky, C., 1981. Attenuation in the control of expression of bacterial operons. *Nature* 289:751–758.

Rouault, T. A., Stout, C. D., Kaptain, S., Harford, J. B., and Klausner, R. D., 1991. Structural relationship between an iron-regulated RNA-binding protein (IRE-BP) and aconitase: Functional implications. *Cell* 64:881–883.

Klausner, R. D., Rouault, T. A., and Harford, J. B., 1993. Regulating the fate of mRNA: The control of cellular iron metabolism. *Cell* 72:19–28.

Gruer, M. J., Artymiuk, P. J., and Guest, J. R., 1997. The aconitase family: Three structural variations on a common theme. *Trends Biochem. Sci.* 22:3–6.

Theil, E. C., 1994. Iron regulatory elements (IREs): A family of mRNA non-coding sequences. *Biochem. J.* 304:1–11.

Historical aspects

Jacob, F., and Monod, J., 1961. Genetic regulatory mechanisms in the synthesis of proteins. *J. Mol. Biol.* 3:318–356.

Ptashne, M., and Gilbert, W., 1970. Genetic repressors. *Sci. Am.* 222(6):36–44.

Lwoff, A., and Ullmann, A. (Eds.), 1979. *Origins of Molecular Biology: A Tribute to Jacques Monod.* Academic Press.

Judson, H., 1996. *The Eighth Day of Creation: Makers of the Revolution in Biology.* Cold Spring Harbor Laboratory Press.

——| **PROBLEMS** |——————————————————————

1. *Missing genes.* Predict the effects of deleting the following regions of DNA:

(a) The gene encoding *lac* repressor
(b) The *lac* operator
(c) The gene encoding CAP

2. *Minimal concentration.* Calculate the concentration of *lac* repressor, assuming that one molecule is present per cell. Assume that each *E. coli* cell has a volume of 10^{-12} cm^3. Would you expect the single molecule to be free or bound to DNA?

3. *Counting sites.* Calculate the expected number of times that a given 8-base-pair DNA site should be present in the *E. coli* genome. Assume that all four bases are equally probable. Repeat for a 10-base-pair site and a 12-base-pair site.

4. *Charge neutralization.* Given the histone amino acid sequences illustrated on the next page, estimate the charge of a histone octamer at pH 7. Assume that histidine residues are uncharged at this pH. How does this charge compare with the charge on 150 base pairs of DNA?

Histone H2A

MSGRGKQGGKARAKAKTRSSRAGLQFPVGRVHRLLRKGNYSERVGAGAPVYLAAVLEYLTAEILELAGNA

ARDNKKTRIIPRHLQLAIRNDEELNKLLGRVTIAQGGVLPNIQAVLLPKKTESHHKAKGK

Histone H2B

MPEPAKSAPAPKKGSKKAVTKAQKKDGKKRKRSRKESYSVYVYKVLKQVHPDTGISSKAMGIMNSFVNDI

FERIAGEASRLAHYNKRSTITSREIQTAVRLLLPGELAKHAVSEGTKAVTKYTSSK

Histone H3

MARTKQTARKSTGGKAPRKQLATKAARKSAPSTGGVKKPHRYRPGTVALREIRRYQKSTELLIRKLPFQR

LVREIAQDFKTDLRFQSAAIGALQEASEAYLVGLFEDTNLCAIHAKRVTIMPKDIQLARRIRGERA

Histone H4

MSGRGKGGKGLGKGGAKRHRKVLRDNIQGITKPAIRRLARRGGVKRISGLIYEETRGVLKVFLENVIRDA

VTYTEHAKRKTVTAMDVVYALKRQGRTLYGFGG

5. *Chromatin immunoprecipitation.* You have used the technique of chromatin immunoprecipitation to isolate DNA fragments containing a DNA-binding protein of interest. Suppose that you wish to know whether a particular known DNA fragment is present in the isolated mixture. How might you detect its presence? How many different fragments would you expect if you used antibodies to the *lac* repressor to perform a chromatin immunoprecipitation experiment in *E. coli*? If you used antibodies to the *pur* repressor?

6. *Nitrogen substitution.* Growth of mammalian cells in the presence of 5-azacytidine results in the activation of some normally inactive genes. Propose an explanation.

deoxyribose

5-Azacytosine

7. *A new domain.* A protein domain has been characterized that recognizes 5-methylcytosine in the context of double-stranded DNA. What role might proteins containing such a domain play in regulating gene expression? Where on a double-stranded DNA molecule would you expect such a domain to bind?

8. *The same but not the same.* The *lac* repressor and the *pur* operator are homologous proteins with very similar three-dimensional structures, yet they have different effects on gene expression. Describe two important ways in which the gene-regulatory properties of these proteins differ.

9. *The opposite direction.* Some compounds called anti-inducers bind to repressors such as the *lac* repressor and inhibit the action of inducers—that is, transcription is repressed and higher concentrations of inducer are required to induce transcription. Propose a mechanism of action for anti-inducers.

10. *Inverted repeats.* Suppose that a nearly perfect inverted repeat is observed in a DNA sequence over 20 base pairs. Provide two possible explanations.

Mechanism Problem

11. *Acetyltransferases.* Propose a mechanism for the transfer of an acetyl group from acetyl CoA to the amino group of lysine.

Chapter Integration Problem

12. *The biochemistry of memory.* Long-term memory requires the synthesis of new proteins. The neurotransmitter serotonin plays a key role in stimulating neurons that participate in the storage of long-term memory. Remarkably, microinjection of DNA fragments that contain binding sites for CREB, the cAMP-response element binding protein, blocks long-term memory. Propose an explanation for this effect. On the basis of your answer, propose a signaling pathway initiated by the arrival of serotonin at the surface of a neuron cell membrane and ending with the synthesis of new proteins having roles in memory storage.

Data Interpretation Problem

13. *Limited restriction.* The restriction enzyme HpaII is a powerful tool for analyzing DNA methylation. This enzyme cleaves sites of the form 5′-CCGG-3′ but will not cleave such sites if the DNA is methylated on any of the cytosine residues. Genomic DNA from different organisms is treated with HpaII and the results are analyzed by gel electrophoresis (see the adjoining patterns). Provide an explanation for the observed patterns.

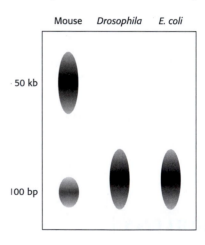

Appendices

Glossary of Compounds

Answers to Problems

Index

Appendix A   Physical Constants and Conversion of Units

Values of physical constants

Physical constant	Symbol	Value
Atomic mass unit (dalton)	amu	1.660×10^{-24} g
Avogadro's number	N	6.022×10^{23} mol^{-1}
Boltzmann's constant	k	1.381×10^{-23} J K^{-1}
		3.298×10^{-24} cal K^{-1}
Electron volt	eV	1.602×10^{-19} J
		3.828×10^{-20} cal
Faraday constant	F	9.649×10^{4} C mol^{-1}
		2.306×10^{4} cal volt^{-1} eq^{-1}
Curie	Ci	3.70×10^{10} disintegrations s^{-1}
Gas constant	R	8.315 J mol^{-1} K^{-1}
		1.987 cal mol^{-1} K^{-1}
Planck's constant	h	6.626×10^{-34} J s
		1.584×10^{-34} cal s
Speed of light in a vacuum	c	2.998×10^{10} cm s^{-1}

Abbreviations: C, coulomb; cal, calorie; cm, centimeter; K, Kelvin;
eq, equivalent; g, gram; J, joule; mol, mole; s, second.

Mathematical constants

$$\pi = 3.14159$$
$$e = 2.71828$$
$$\log_e x = 2.303 \log_{10} x$$

Conversion factors

Physical quantity	Equivalent
Length	1 cm $= 10^{-2}$ m $= 10$ mm $= 10^{4}$ μm $= 10^{7}$ nm
	1 cm $= 10^{8}$ Å $= 0.3937$ inch
Mass	1 g $= 10^{-3}$ kg $= 10^{-3}$ mg $= 10^{6}$ μg
	1 g $= 3.527 \times 10^{-2}$ ounce (avoirdupois)
Volume	1 cm^3 $= 10^{-6}$ m^3 $= 10^3$ mm^3
	1 ml $= 1$ cm^3 $= 10^{-3}$ liter $= 10^3$ μl
	1 cm^3 $= 6.1 \times 10^{-2}$ in^3 $= 3.53 \times 10^{-5}$ ft^3
Temperature	$K = {}^{\circ}C + 273.15$
	${}^{\circ}C = (5/9)({}^{\circ}F - 32)$
Energy	1 J $= 10^7$ erg $= 0.239$ cal $= 1$ watt s^{-}
Pressure	1 torr $= 1$ mm Hg $(0{}^{\circ}C)$
	$= 1.333 \times 10^2$ newtons m^{-2}
	$= 1.333 \times 10^2$ pascal
	$= 1.316 \times 10^{-3}$ atmospheres

Standard prefixes

Prefix	Symbol	Factor
kilo	k	10^{3}
hecto	h	10^{2}
deca	da	10^{1}
deci	d	10^{-1}
centi	c	10^{-2}
milli	m	10^{-3}
micro	μ	10^{-6}
nano	n	10^{-9}
pico	p	10^{-12}

Appendix B Acidity Constants

pK_a Values of Some Acids

Acid	pK' (at 25°C)	Acid	pK' (at 25°C)
Acetic acid	4.76	Lactic acid	3.86
Acetoacetic acid	3.58	Maleic acid, pK_1	1.83
Ammonium ion	9.25	pK_2	6.07
Ascorbic acid, pK_1	4.10	Malic acid, pK_1	3.40
pK_2	11.79	pK_2	5.11
Benzoic acid	4.20	Phenol	9.89
n-Butyric acid	4.81	Phosphoric acid, pK_1	2.12
Cacodylic acid	6.19	pK_2	7.21
Carbonic acid, pK_1	6.35	pK_3	12.67
pK_2	10.33	Pyridinium ion	5.25
Citric acid, pK_1	3.14	Pyrophosphoric acid, pK_1	0.85
pK_2	4.77	pK_2	1.49
pK_3	6.39	pK_3	5.77
Ethylammonium ion	10.81	pK_4	8.22
Formic acid	3.75	Succinic acid, pK_1	4.21
Glycine, pK_1	2.35	pK_2	5.64
pK_2	9.78	Trimethylammonium ion	9.79
Imidazolium ion	6.95	Tris (hydroxymethyl) aminomethane	8.08
		Water*	15.74

* $[H^+][OH^-] = 10^{-14}$; $[H_2O] = 55.5$ M.

Typical pK_a values of ionizable groups in proteins

Group	Acid	⇌	Base	Typical pKa	Group	Acid	⇌	Base	Typical pKa
Terminal α-carboxyl group		⇌		3.1	Cysteine		⇌		8.3
Aspartic acid Glutamic acid		⇌		4.1	Tyrosine		⇌		10.4
					Lysine		⇌		10.0
Histidine		⇌		6.0	Arginine		⇌		12.5
Terminal α-amino group		⇌		8.0					

Note: pK_a values depend on temperature, ionic strength, and the microenvironment of the ionizable group.

Appendix C Standard Bond Lengths

Bond	Structure	Length (Å)
C—H	R_2CH_2	1.07
	Aromatic	1.08
	RCH_3	1.10
C—C	Hydrocarbon	1.54
	Aromatic	1.40
C=C	Ethylene	1.33
C≡C	Acetylene	1.20
C—N	RNH_2	1.47
	O=C—N	1.34
C—O	Alcohol	1.43
	Ester	1.36
C=O	Aldehyde	1.22
	Amide	1.24
C—S	R_2S	1.82
N—H	Amide	0.99
O—H	Alcohol	0.97
O—O	O_2	1.21
P—O	Ester	1.56
S—H	Thiol	1.33
S—S	Disulfide	2.05

Glossary of Compounds

The following pages contain the structures of amino acids, common metabolic intermediates, nucleotide bases, and important cofactors. In many cases, two versions of the structure are shown: the Fisher structure (bottom) and a more stereochemically accurate version (top).

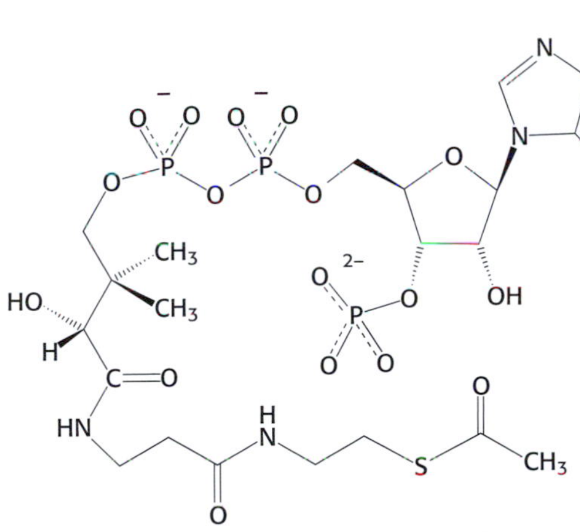

Acetyl coenzyme A (acetyl CoA)

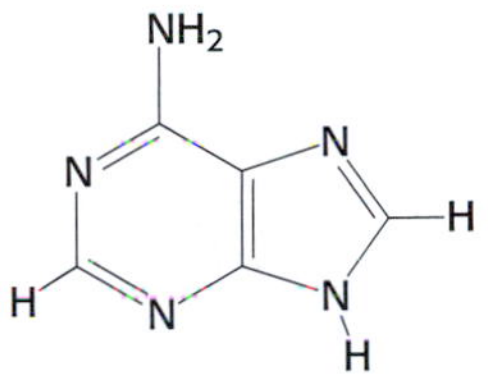

Adenine

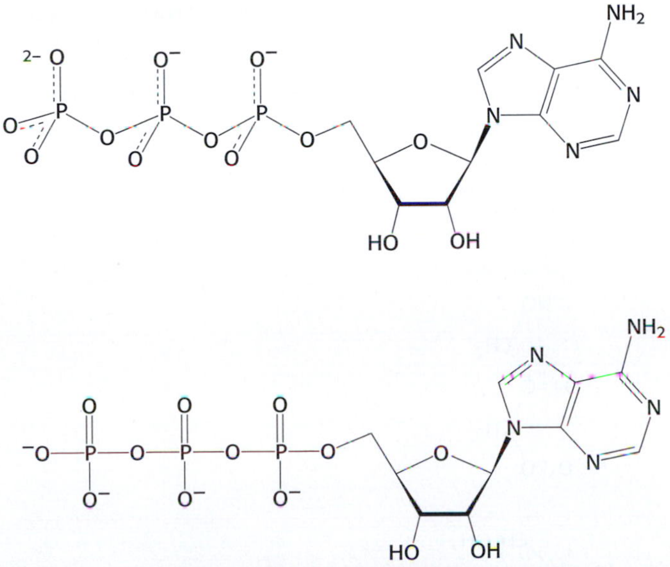

Adenosine triphosphate (ATP)

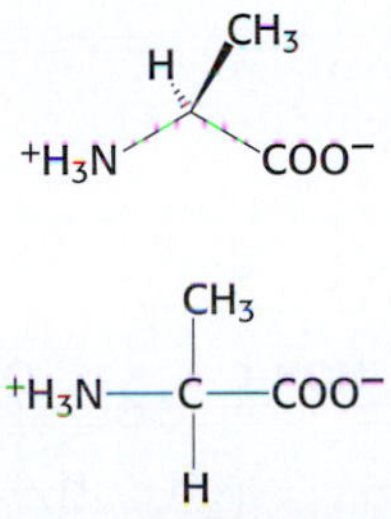

Alanine

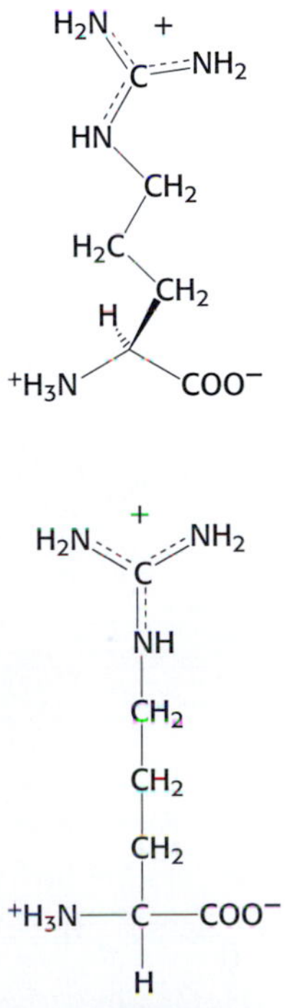

Arginine

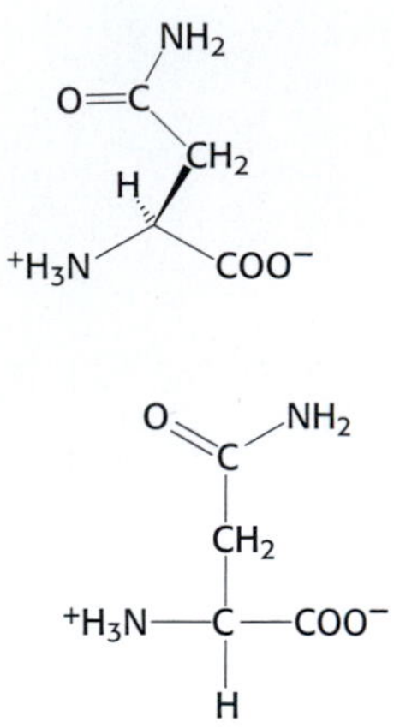

Asparagine

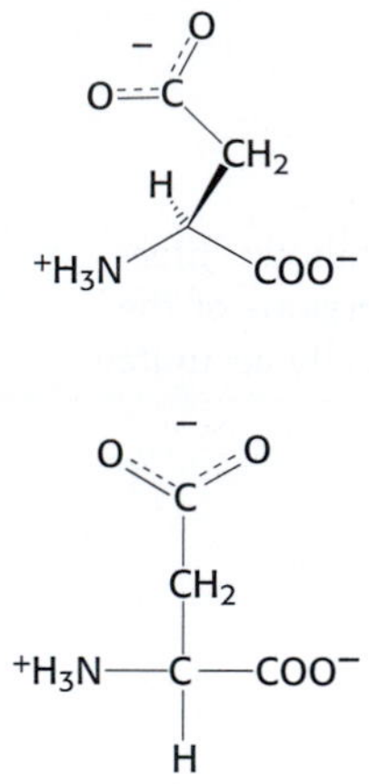

Aspartate

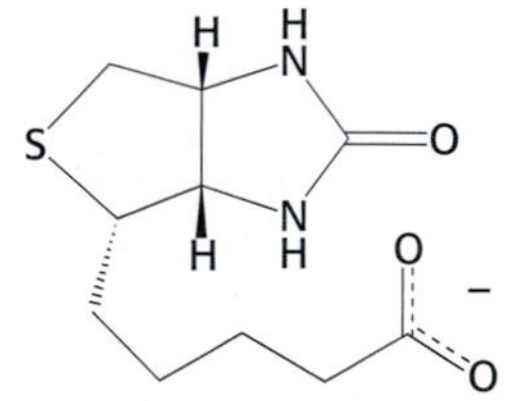

Biotin

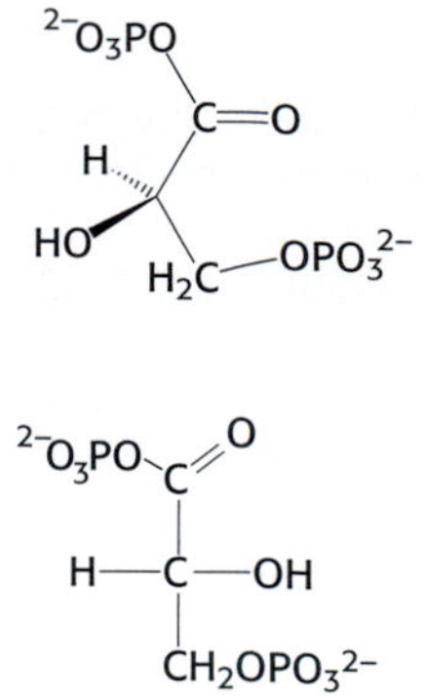

1,3-Bisphosphoglycerate

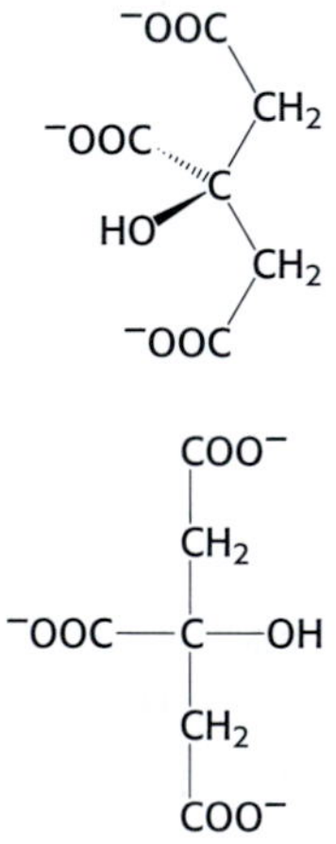

Citrate

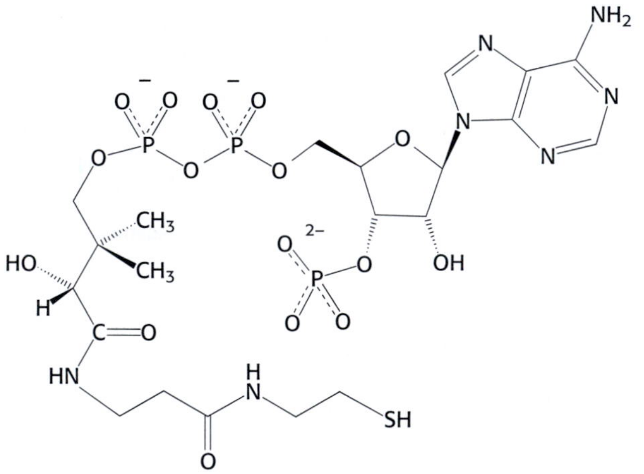

Coenzyme A

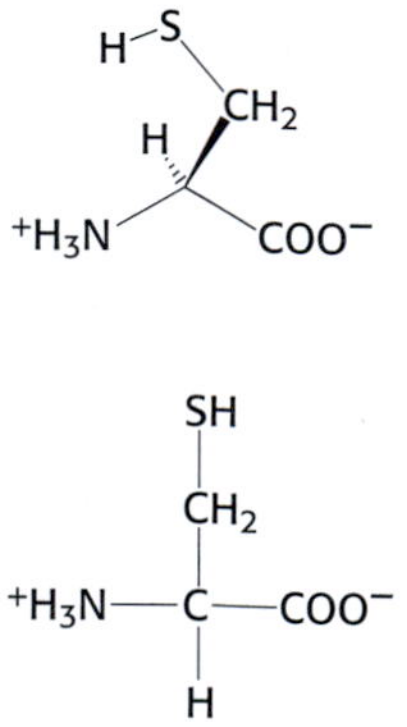

Cysteine

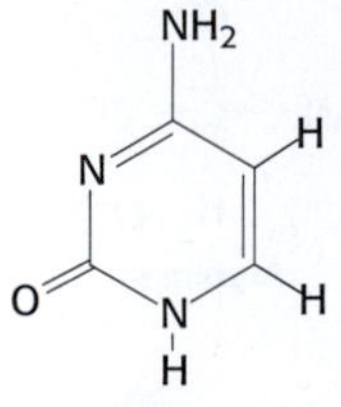

Cytosine

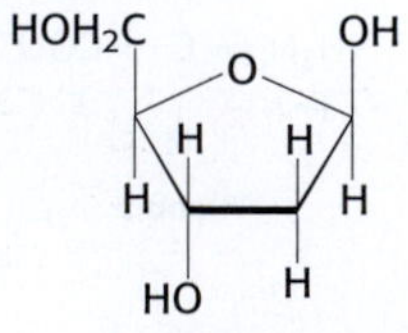

Deoxyribose

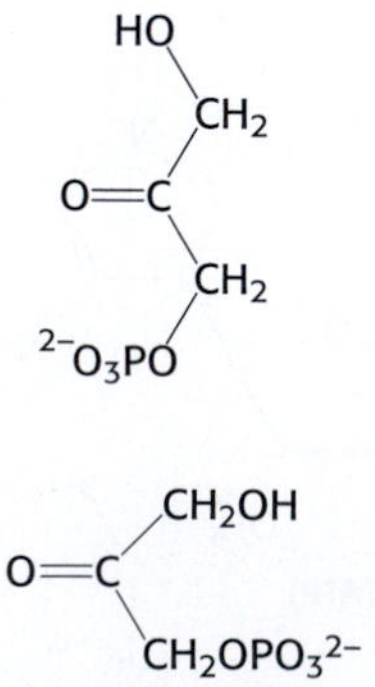

Dihydroxyacetone phosphate

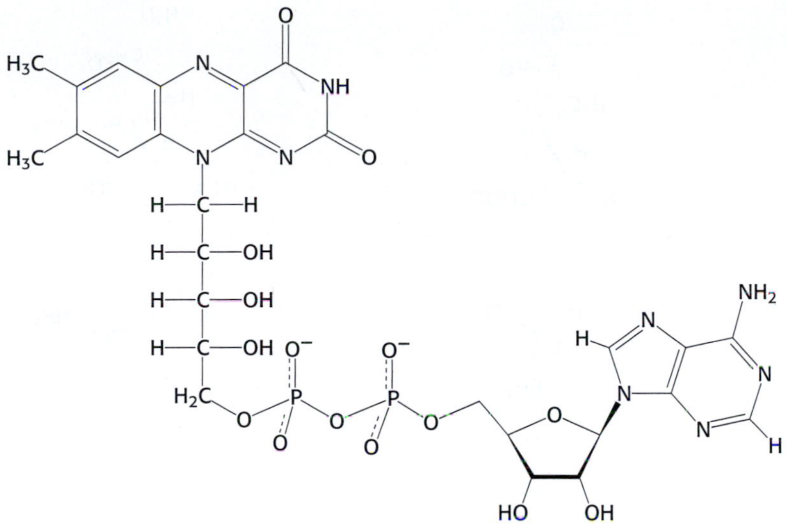

Flavin adenine dinucleotide (FAD)

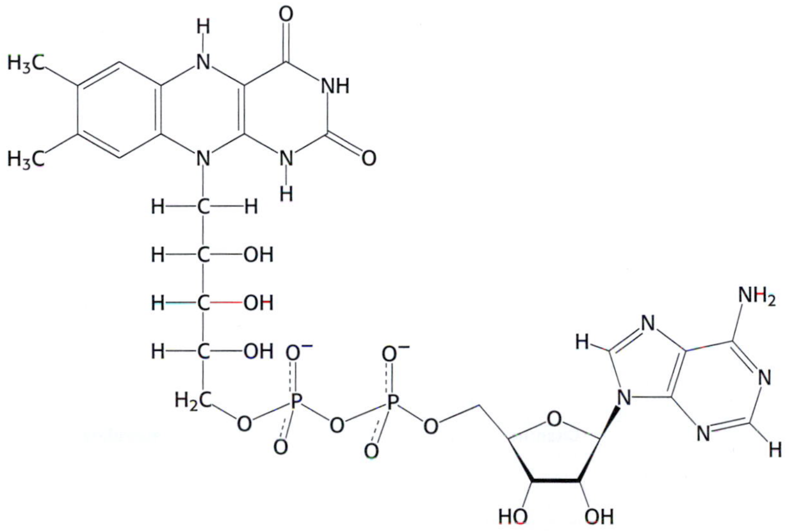

Flavin adenine dinucleotide (reduced) (FADH$_2$)

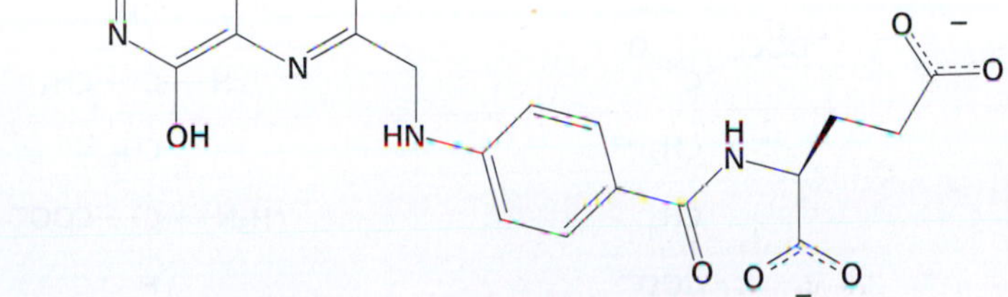

Folic acid

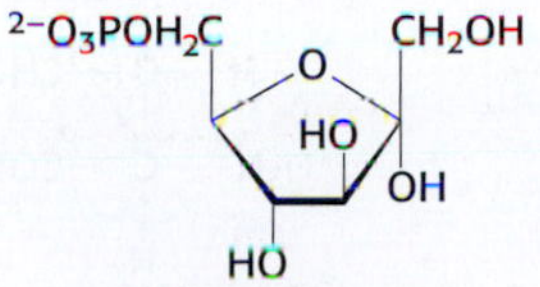

Fructose-6-phosphate

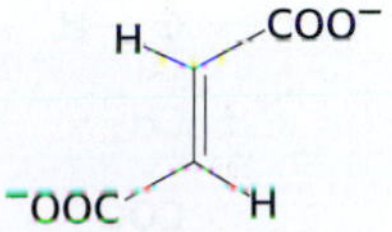

Fumarate

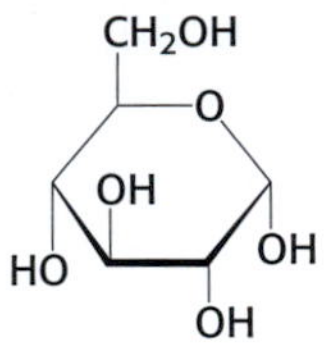

Glucose
(α-D-glucose)

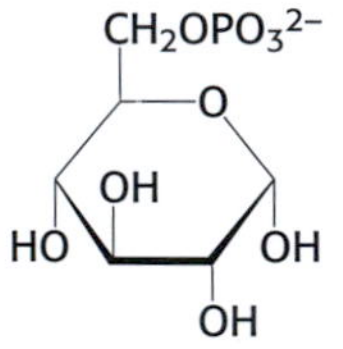

Glucose-6-phosphate

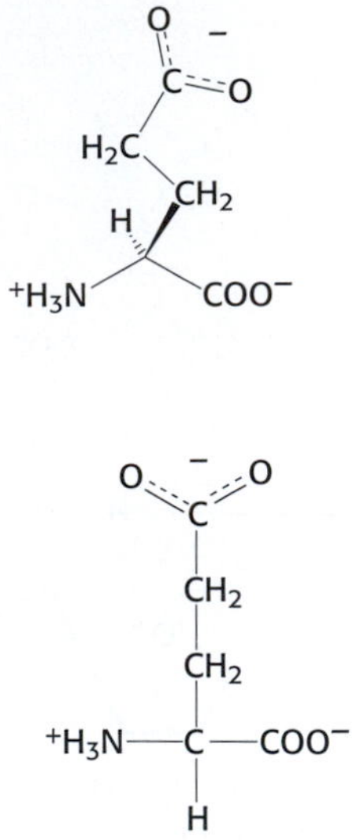

Glutamate

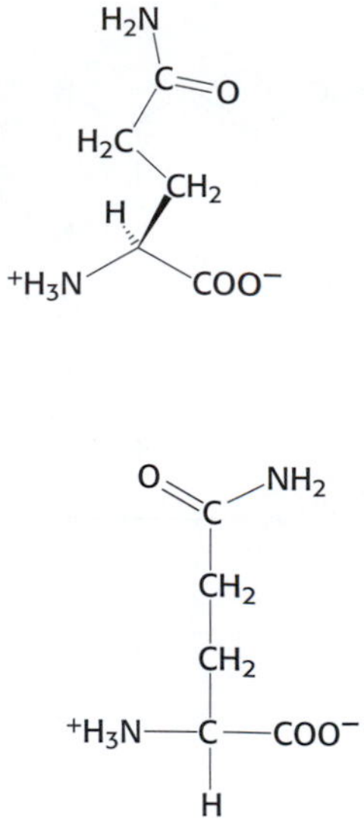

Glutamine

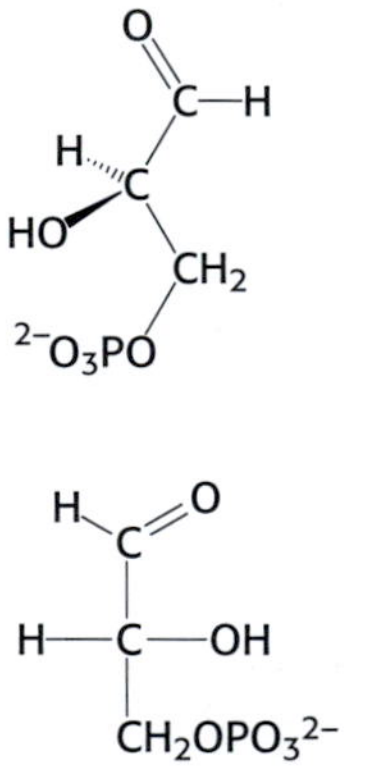

Glyceraldehyde 3-phosphate

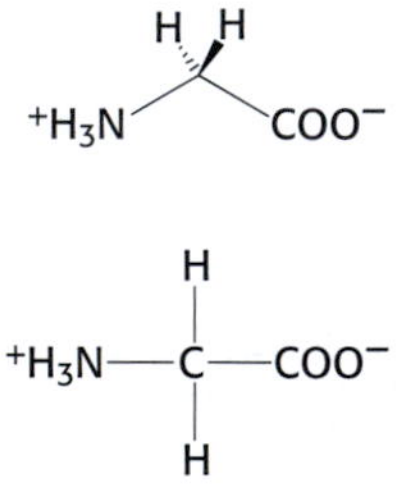

Glycine

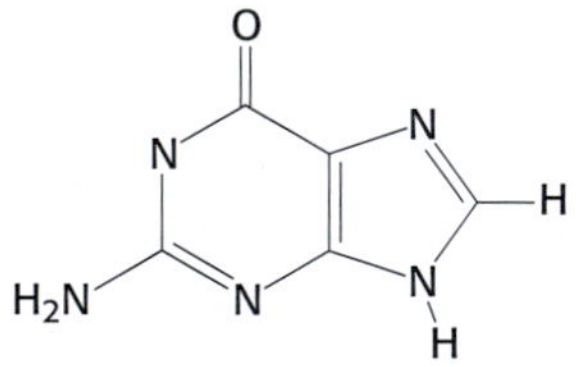

Guanine

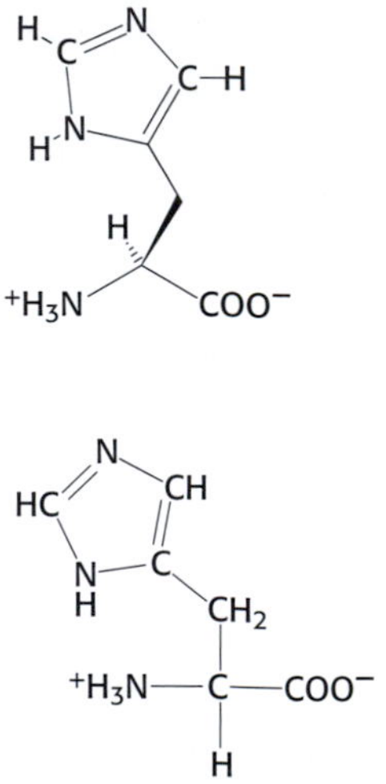

Histidine

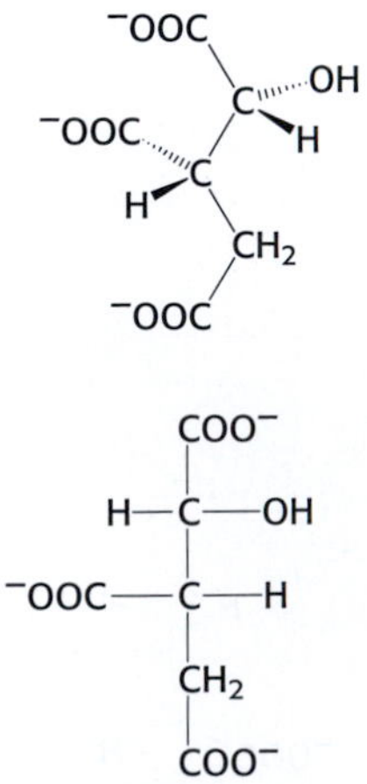

Isocitrate

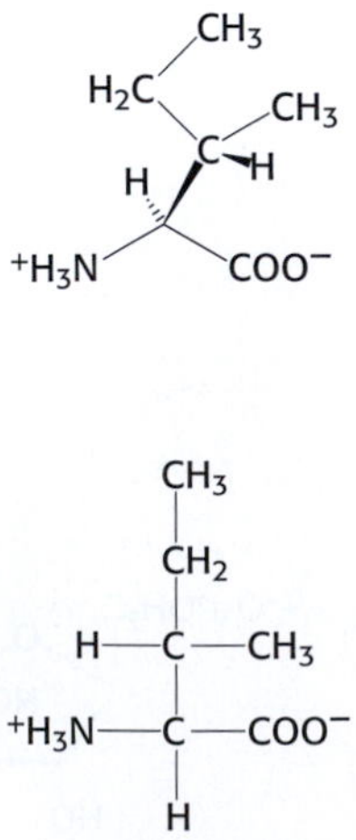

Isoleucine

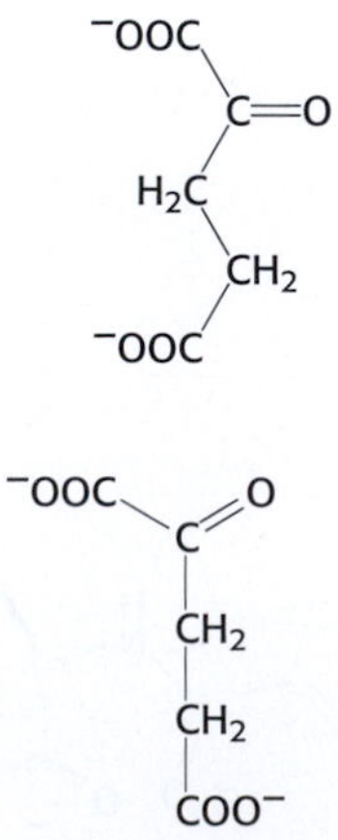

α-Ketoglutarate

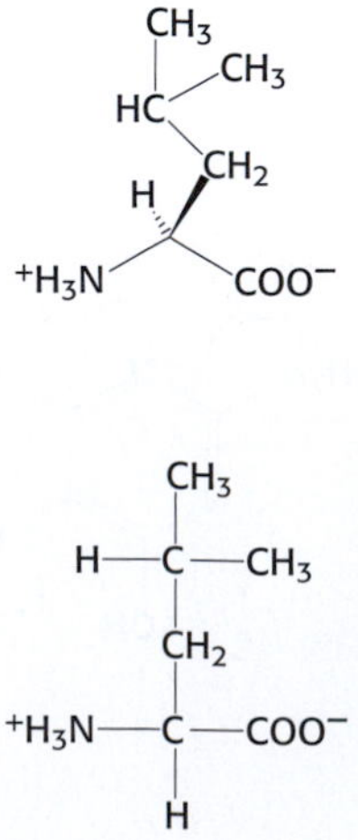

Leucine

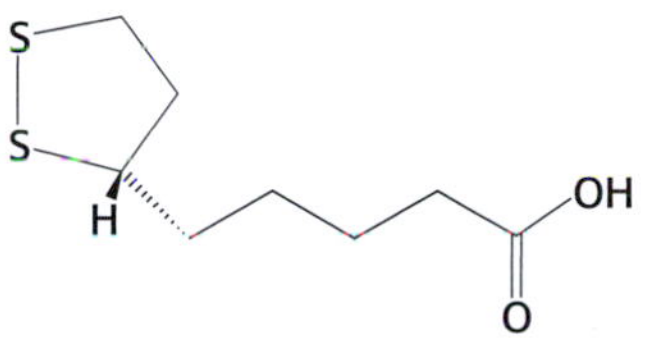

Lipoic acid

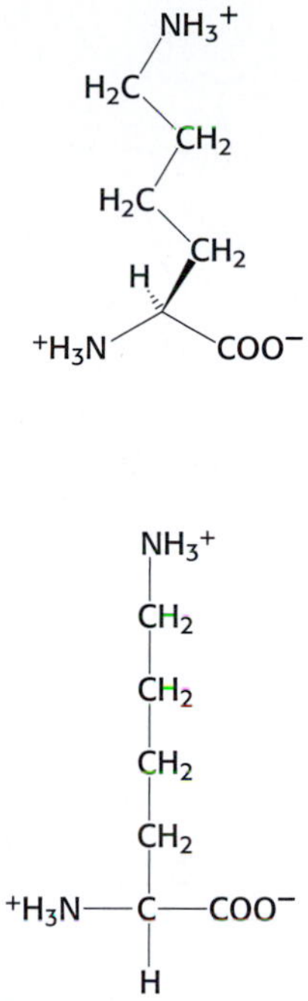

Lysine

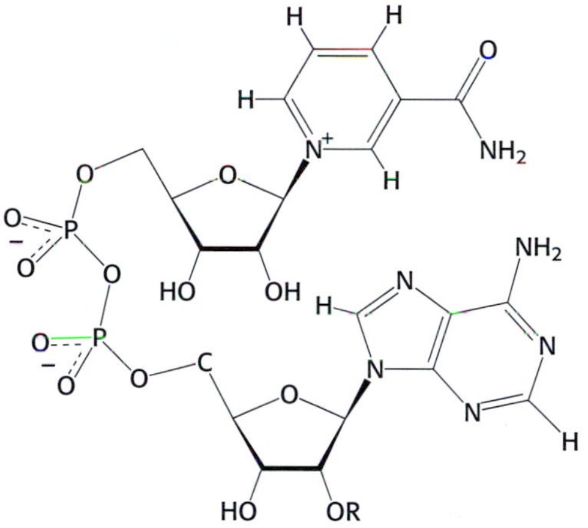

**Nicotinamide
adenine dinucleotide (R = H),
(NAD⁺)**

**Nicotinamide adenine
dinucleotide phosphate (R = PO₃²⁻)
(NADP⁺)**

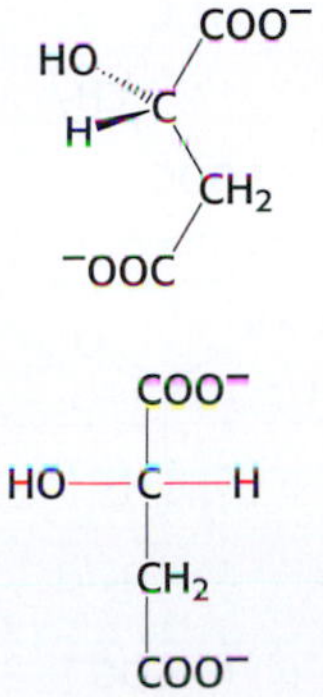

Malate

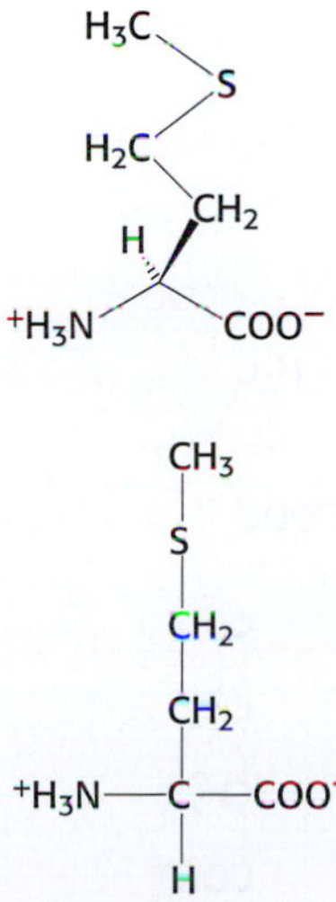

Methionine

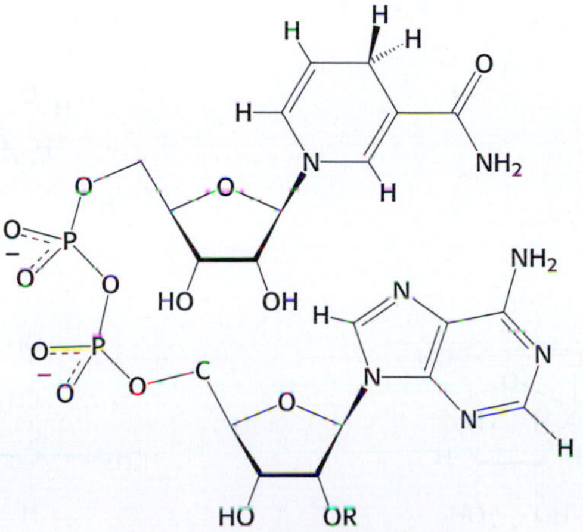

NADH (R = H), NADPH (R = PO₃²⁻)

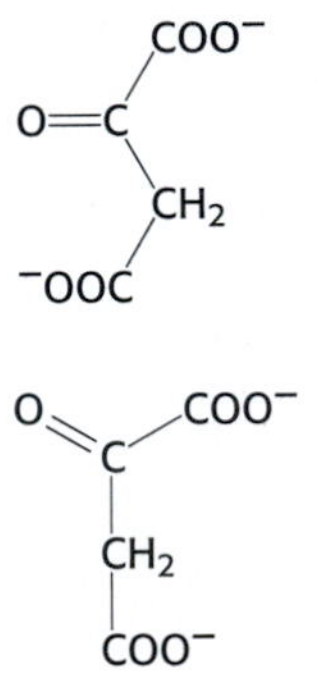

Oxaloacetate

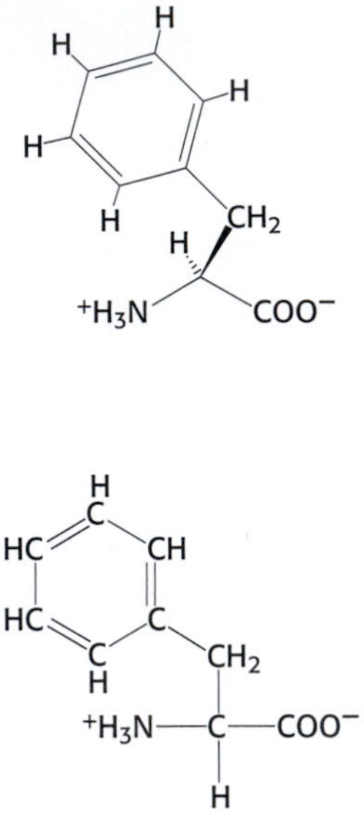

Phenylalanine

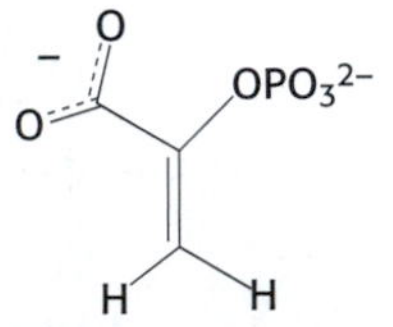

Phosphoenolpyruvate

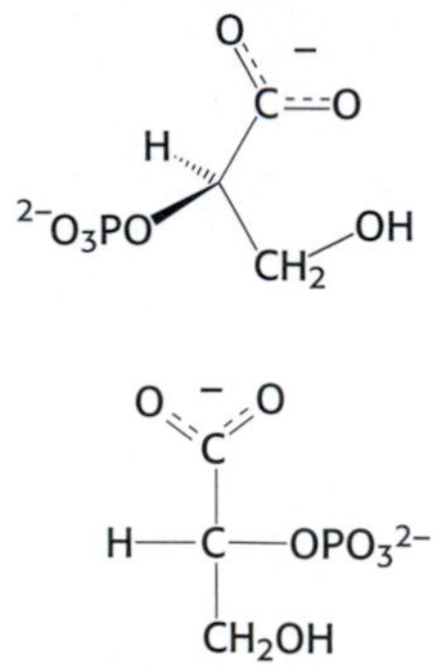

2-Phosphoglycerate

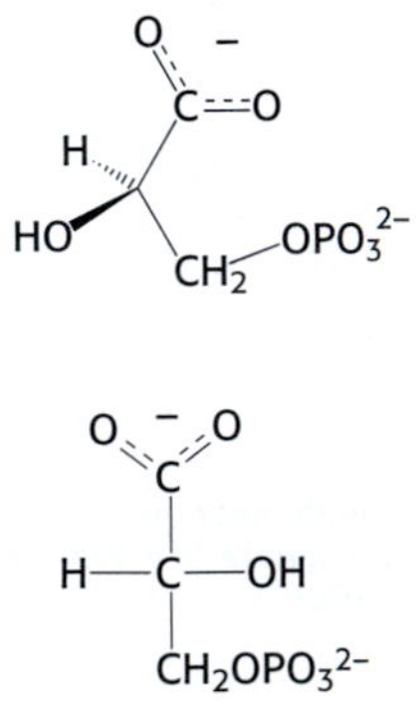

3-Phosphoglycerate

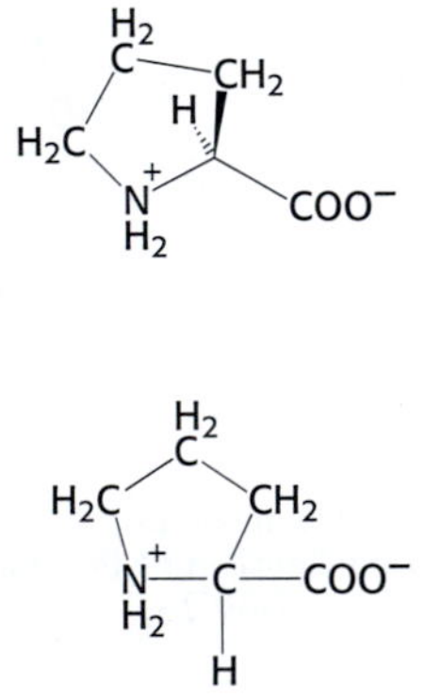

Proline

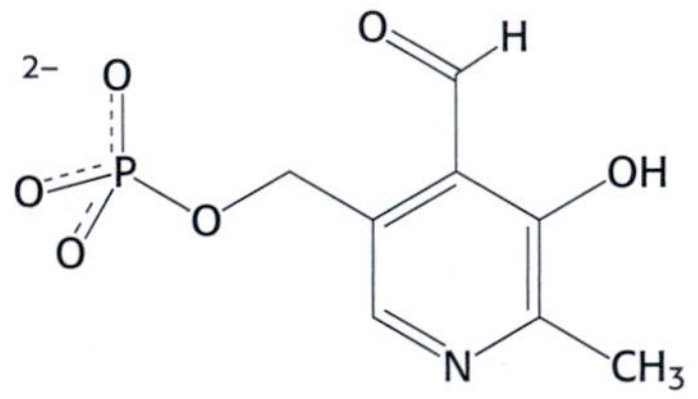

Pyridoxal phosphate

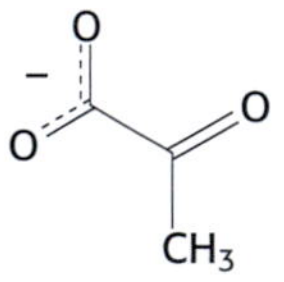

Pyruvate

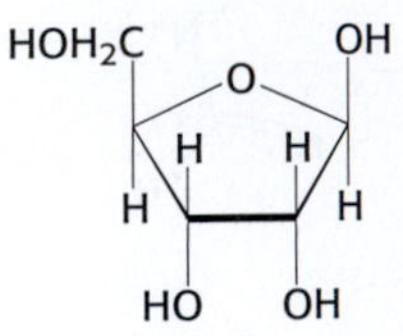

Ribose

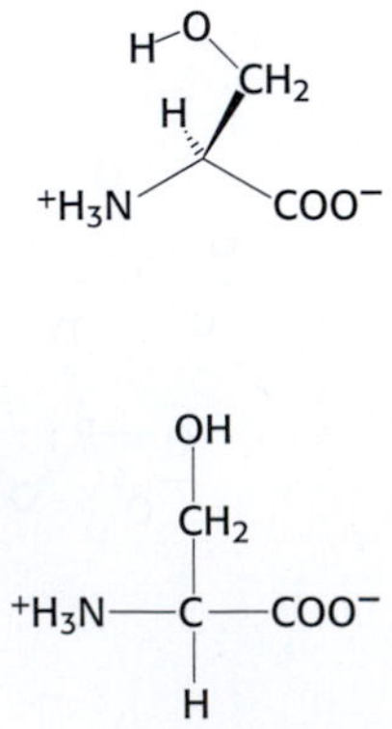

Serine

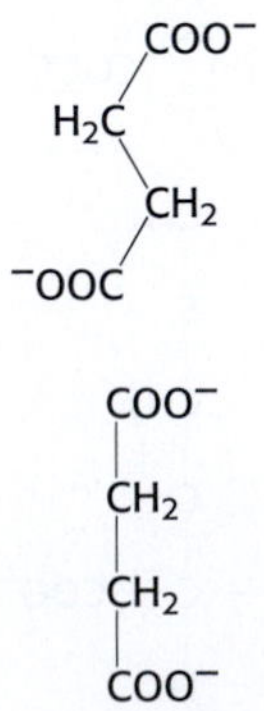

Succinate

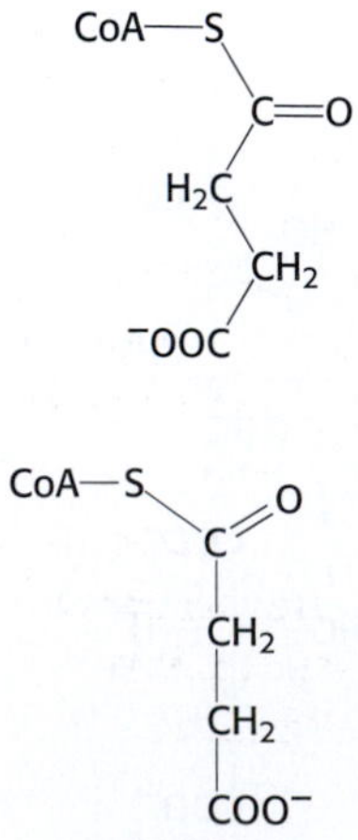

Succinyl CoA

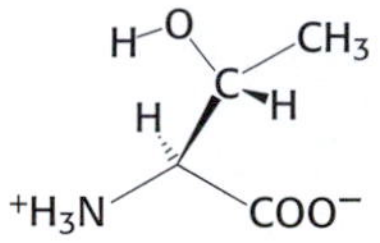

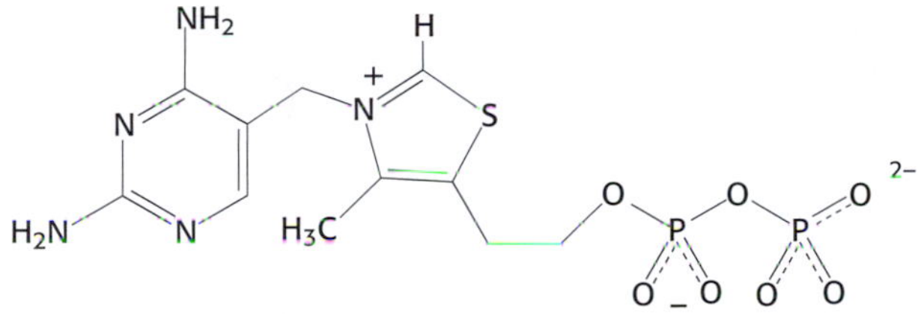

Thiamine pyrophosphate (TPP)

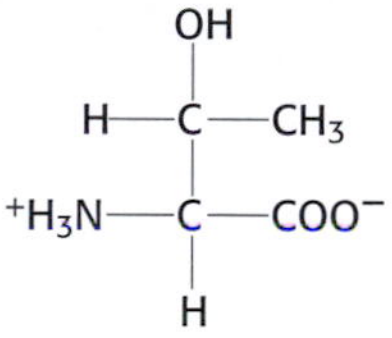

Threonine

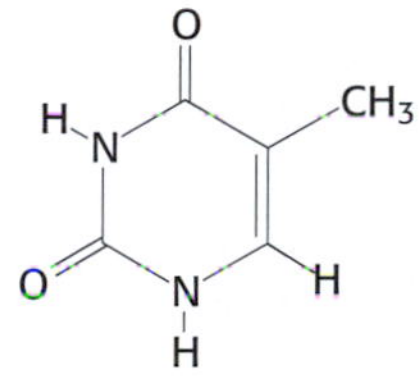

Thymine

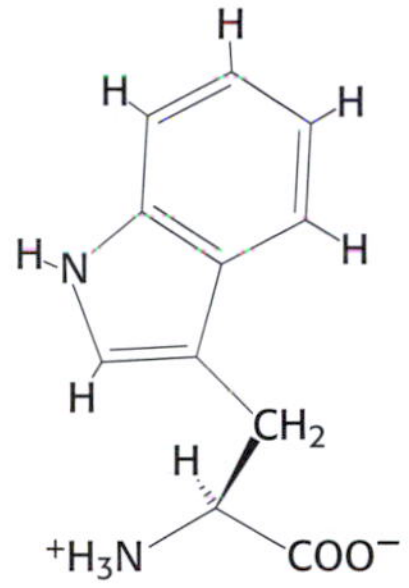

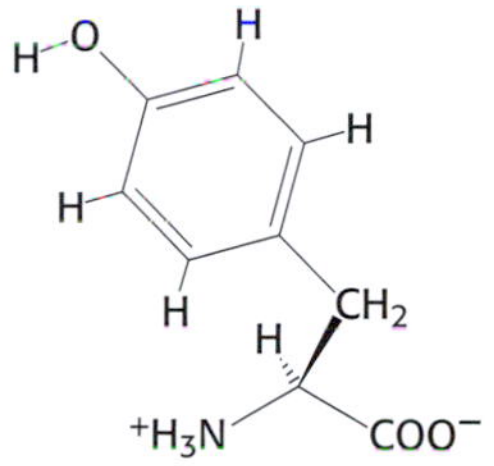

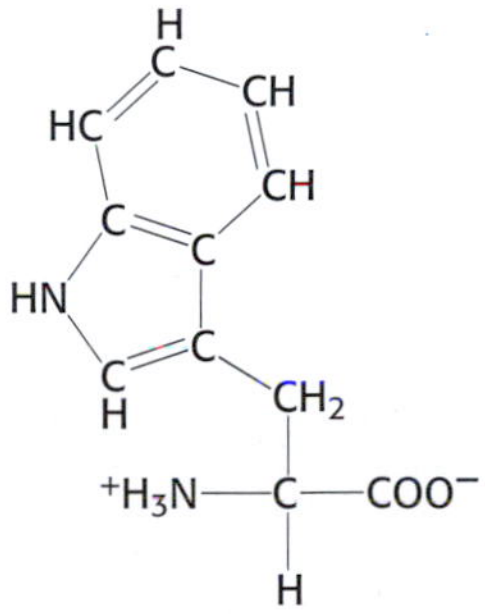

Tryptophan

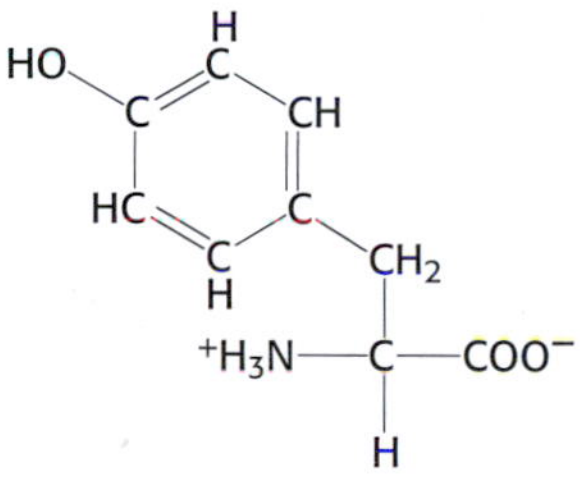

Tyrosine

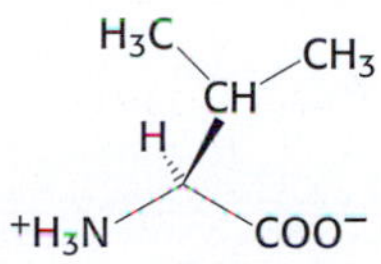

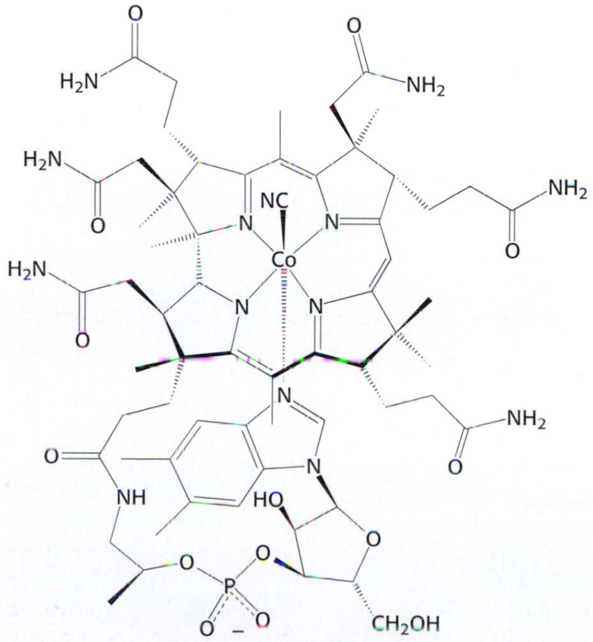

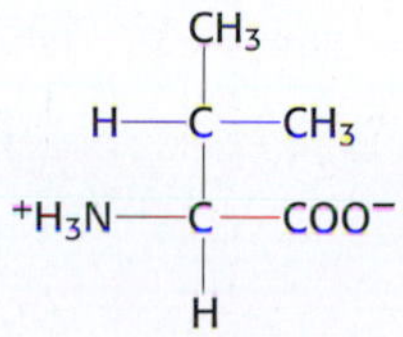

Valine

Uracil

Vitamin B₁₂ (cyanocobalamin)

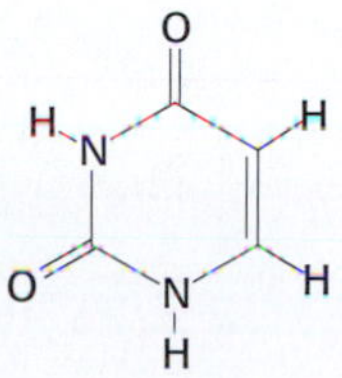

Answers to Problems

Need extra help? Purchase chapters of the *Student Companion* with complete solutions online at www.whfreeman. com/biochem5.

Chapter 2

1. The amino group comes from ammonia. All of the carbon atoms are derived from methane. The hydrogen atoms bonded to the carbon atoms remain with the methane during bond formation or they may come from hydrogen gas. The oxygen atoms of the carboxyl group are from water.

2. We start with 99 identical RNA molecules (which we will call L) that replicate in 15 minutes and 1 variant molecule (which we will call S) that replicates in 5 minutes. After 15 minutes, we will have $2 \times 99 = 198$ molecules of L and $2^3 \times 1 = 8$ molecules of S since it replicates 3 times in 15 minutes. Thus, the population now contains $8/(8 + 198) = 3.9\%$ S. After 10 generations, each molecule of L will have replicated 10 times while each molecule of S will have replicated 30 times. The population will contain $1 \times 2^{30}/(1 \times 2^{30} + 99 \times 2^{10}) = 99.991\%$ S. After 25 generations, the population will contain essentially all S and no L.

3. The mutation permits more efficient use of substrates and thus would be most beneficial when substrate is present in low concentrations.

4. The formation of an ion gradient requires a reduction in entropy, which requires an input of free energy.

5. The decrease in free energy that results when the protons run down the ion gradient could be used to pump ions out of the cell against a concentration gradient, an energy-requiring process.

6. Two protons per electron, or eight. The generation of four hydroxyl ions (OH^-) is equivalent to the generation of four protons (H^+) on the other side of the membrane from which the reaction is taking place. The oxidation of water produces four more protons.

7. Only the gene-control protein is necessary. The hydrophobic molecule can pass through the membrane on its own.

8. Approximately eight times.

Chapter 3

1. (a) Each strand is 35 kd and hence has about 318 residues (the mean residue mass is 110 daltons). Because the rise per residue in an α helix is 1.5 Å, the length is 477 Å. More precisely, for an α-helical coiled coil the rise per residue is 1.46 Å so that the length will be 464 Å.
(b) Eighteen residues in each strand (40 minus 4 divided by 2) are in a β-sheet conformation. Because the rise per residue is 3.5 Å, the length is 63 Å.

2. The methyl group attached to the β-carbon atom of isoleucine sterically interferes with α-helix formation. In leucine, this methyl group is attached to the γ-carbon atom, which is farther from the main chain and hence does not interfere.

3. The first mutation destroys activity because valine occupies more space than alanine does, and so the protein must take a different shape. The second mutation restores activity because

of a compensatory reduction of volume; glycine is smaller than isoleucine.

4. The native conformation of insulin is not the thermodynamically most stable form. Indeed, insulin is formed from proinsulin, a single-chain precursor containing 33 additional residues. In proinsulin, residue 30 of the future B chain of insulin is linked to residue 1 of the future A chain.

5. A segment of the main chain of the protease could hydrogen bond to the main chain of the substrate to form an extended parallel or antiparallel pair of β strands.

6. Glycine has the smallest side chain of any amino acid. Its size often is critical in allowing polypeptide chains to make tight turns or to approach one another closely.

7. Glutamate, aspartate, and the terminal carboxylate can form salt bridges with the guanidinium group of arginine. In addition, this group can be a hydrogen-bond donor to the side chains of glutamine, asparagine, serine, threonine, aspartate, and glutamate, and to the main-chain carbonyl group.

8. Disulfide bonds in hair are broken by adding a thiol and applying gentle heat. The hair is curled, and an oxidizing agent is added to re-form disulfide bonds to stabilize the desired shape.

9. The amino acids would be hydrophobic in nature. An α helix is especially suited to cross a membrane because all of the amide hydrogen atoms and carbonyl oxygen atoms of the peptide backbone take part in intrachain hydrogen bonds, thus stabilizing these polar atoms in a hydrophobic environment.

10. The energy barrier that must be crossed to go from the polymerized state to the hydrolyzed state is large even though the reaction is thermodynamically favorable.

11. Using the Henderson-Hasselbach equation, we find the ratio of alanine-COOH to alanine-COO$^-$ at pH 7 to be 10^{-4}. The ratio of NH_2-alanine to alanine-NH_3^+, determined in the same fashion, is 10^{-1}. Thus, the ratio of neutral alanine to zwitterionic species is $10^{-4} \times 10^{-1} = 10^{-5}$.

12. The assignment of absolute configuration requires the assignment of priorities to the four groups connected to a tetrahedral carbon. For all amino acids except cysteine, the priorities are: (1) amino group; (2) carbonyl group; (3) side chain; (4) hydrogen. For cysteine, because of the sulfur atom in its side chain, the side chain has a greater priority than does the carbonyl group, leading to the assignment of an R rather than S configuration.

13. SAVE ME I'M TRAPPED IN A GENE.

14. No, Pro–X would have the characteristics of any other peptide bond. The steric hindrance in X–Pro arises because the R group of Pro is bonded to the amino group. Hence, in X–Pro, the proline R group is near the R group of X. This would not be the case in Pro–X.

15. A, c; B, e; C, d; D, a; E, b.

16. With the use of Beer's law and the value of ε obtained from Section 3.1, the concentration of tyrosine is found to be ≈30 μM. Because there are three molecules of tyrosine per molecule of protein, the concentration of protein is ≈10 μM. There is 1 mg of protein per milliliter of solution.

Chapter 4

1. (a) Phenyl isothiocyanate; (b) dansyl chloride or dabsyl chloride; (c) urea; β-mercaptoethanol to reduce disulfides; (d) chymotrypsin; (e) CNBr; (f) trypsin

2. Each amino acid residue, except the carboxyl-terminal residue, gives rise to a hydrazide on reacting with hydrazine. The carboxyl-terminal residue can be identified because it yields a free amino acid.

3. The S-aminoethylcysteine side chain resembles that of lysine. The only difference is a sulfur atom in place of a methylene group.

4. A 1 mg/ml solution of myoglobin (17.8 kd) corresponds to 5.62×10^{-5} M. The absorbance of a 1-cm path length is 0.84, which corresponds to an I_0/I ratio of 6.96. Hence 14.4% of the incident light is transmitted.

5. Tropomyosin is rod shaped, whereas hemoglobin is approximately spherical.

6. The frictional coefficient f and the mass m determine S. Specifically, f is proportional to r (see equation 2 on p. 83). Hence, f is proportional to $m^{1/3}$, and so S is proportional to $m^{2/3}$ (see the equation on p. 88). An 80-kd spherical protein sediments 1.59 times as rapidly as a 40-kd spherical protein.

7. 50 kd.

8. The positions of disulfide bonds can be determined by diagonal electrophoresis (p. 96). The disulfide pairing is unaltered by the mutation if the off-diagonal peptides formed from the native and mutant proteins are the same.

9. A fluorescent-labeled derivative of a bacterial degradation product (e.g., a formylmethionyl peptide) would bind to cells containing the receptor of interest.

10. (a) Trypsin cleaves after arginine (R) and lysine (K), generating AVGWR, VK, and S. Because they differ in size, these products could be separated by molecular exclusion chromatography.

(b) Chymotrypsin, which cleaves after large aliphatic or aromatic R groups, generates two peptides of equal size (AVGW) and (RVKS). Separation based on size would not be effective. The peptide RVKS has two positive charges (R and K), whereas the other peptide is neutral. Therefore, the two products could be separated by ion-exchange chromatography.

11. An inhibitor of the enzyme being purified might have been present and subsequently removed by a purification step. This would lead to an apparent increase in the total amount of enzyme present.

13. Treatment with urea will disrupt noncovalent bonds. Thus the original 60-kd protein must be made of two 30-kd subunits. When these subunits are treated with urea and mercaptoethanol, a single 15-kd species results, suggesting that disulfide bonds link the 30-kd subunits.

14. (a) Electrostatic repulsion between positively charged ε-amino groups prevents α-helix formation at pH 7. At pH 10, the side chains become deprotonated, allowing α-helix formation.

(b) Poly-L-glutamate is a random coil at pH 7 and becomes α helical below pH 4.5 because the γ-carboxylate groups become protonated.

15. Light was used to direct the synthesis of these peptides. Each amino acid added to the solid support contained a photolabile protecting group instead of a t-Boc protecting group at its α-amino group. Illumination of selected regions of the solid support led to the release of the protecting group, which exposed the amino groups in these sites to make them reactive. The pattern of masks used in these illuminations and the sequence of reactants define the ultimate products and their locations.

16. AVRYSR

17. First amino acid: S
Last amino acid: L
Cyanogen bromide cleavage: M is 10th position,
C-terminal residues are: (2S,L,W)
Amino-terminal residues: (G,K,S,Y), tryptic peptide, ends in K
Amino-terminal sequence: SYGK
Chymotryptic peptide order: (S,Y), (G,K,L), (F,I,S), (M,T), (S,W), (S,L)
Sequence: SYGKLSIFTMSWSL

12.

Purification procedure	Total protein mg	Total activity (units)	Specific activity (units/mg)	Purification level	Yield (%)
Crude extract	20,000	4,000,000	200	1	100
(NH)$_4$SO$_4$ precipitation	5,000	3,000,000	600	3	75
DEAE–cellulose chromatography	1,500	1,000,000	667	3.3	25
Size-exclusion chromatography	500	750,000	1,500	7.5	19
Affinity chromatography	45	675,000	15,000	75	17

18.

Chapter 5

1. (a) TTGATC; (b) GTTCGA; (c) ACGCGT; (d) ATGGTA.
2. (a) [T] + [C] = 0.46. (b) [T] = 0.30, [C] = 0.24, and [A] + [G] = 0.46.
3. 5.7×10^3 base pairs.
4. In conservative replication, after 1.0 generation, half of the molecules would be ^{15}N-^{15}N, the other half ^{14}N-^{14}N. After 2.0 generations, one-quarter of the molecules would be ^{15}N-^{15}N, the other three-quarters ^{14}N-^{14}N. Hybrid ^{14}N-^{15}N molecules would not be observed in conservative replication.
5. (a) Tritiated thymine or tritiated thymidine. (b) dATP, dGTP, dCTP, and dTTP labeled with ^{32}P in the innermost (α) phosphorus atom.
6. Molecules in parts *a* and *b* would not lead to DNA synthesis because they lack a 3′-OH group (a primer). The molecule in part *d* has a free 3′-OH at one end of each strand but no template strand beyond. Only the molecule in part *d* would lead to DNA synthesis.
7. A deoxythymidylate oligonucleotide should be used as the primer. The poly(rA) template specifies the incorporation of dT; hence, radioactive dTTP should be used in the assay.
8. The ribonuclease serves to degrade the RNA strand, a necessary step in forming duplex DNA from the RNA–DNA hybrid.
9. Treat one aliquot of the sample with ribonuclease and another with deoxyribonuclease. Test these nuclease-treated samples for infectivity.

10. Deamination changes the original G · C base pair into a G · U pair. After one round of replication, one daughter duplex will contain a G · C pair, and the other duplex an A · U pair. After two rounds of replication, there would be two G · C pairs, one A · U pair, and one A · T pair.
11. (a) $4^8 = 65,536$. In computer terminology, there are 64K 8-mers of DNA.
(b) A bit specifies two bases (say, A and C) and a second bit specifies the other two (G and T). Hence, two bits are needed to specify a single nucleotide (base pair) in DNA. For example, 00, 01, 10, and 11 could encode A, C, G, and T. An 8-mer stores 16 bits ($2^{16} = 65,536$), the *E. coli* genome (4×10^6 bp) stores 8×10^6 bits, and the human genome (2.9×10^9 bases) stores 5.8×10^9 bits of genetic information.
(c) A floppy diskette stores about 1.5 megabytes, which is equal to 1.2×10^7 bits. A large number of 8-mer sequences could be stored on such a diskette. The DNA sequence of *E. coli*, once known, could be written on a single diskette. Nearly 500 diskettes would be needed to record the human DNA sequence.
12. (a) Deoxyribonucleoside triphosphates versus ribonucleoside triphosphates.
(b) 5′ → 3′ for both.
(c) Semiconserved for DNA polymerase I; conserved for RNA polymerase.
(d) DNA polymerase I needs a primer, whereas RNA polymerase does not.
13. (a) 5′-UAACGGUACGAU-3′.
(b) Leu-Pro-Ser-Asp-Trp-Met.
(c) Poly(Leu-Leu-Thr-Tyr).
14. The 2′-OH group in RNA acts as an intramolecular catalyst. In the alkaline hydrolysis of RNA, it forms a 2′-3′ cyclic intermediate.
15. Cordycepin terminates RNA synthesis. An RNA chain containing cordycepin lacks a 3′-OH group.
16. Only single-stranded RNA can serve as a template for protein synthesis.
17. Incubation with RNA polymerase and only UTP, ATP, and CTP led to the synthesis of only poly(UAC). Only poly(GUA) was formed when GTP was used in place of CTP.
18. These alternatives were distinguished by the results of studies of the sequence of amino acids in mutants. Suppose that the base C is mutated to C′. In a nonoverlapping code, only amino acid 1 will be changed. In a completely overlapping code, amino acids 1, 2, and 3 will all be altered by a mutation of C to C′. The results of amino acid sequence studies of tobacco mosaic virus mutants and abnormal hemoglobins showed that alterations usually affected only a single amino acid. Hence, it was concluded that the *genetic code is nonoverlapping*.
19. A peptide terminating with Lys (UGA is a stop codon), -Asn-Glu-, and -Met-Arg-.
20. Highly abundant amino acid residues have the most codons (e.g., Leu and Ser each have six), whereas the least-abundant amino acids have the fewest (Met and Trp each have only one). Degeneracy (a) allows variation in base composition and (b) decreases the likelihood that a substitution for a base will change the encoded amino acid. If the degeneracy were equally distributed, each of the 20 amino acids would have three codons. Both benefits (a and b) are maximized by the assignment of more codons to prevalent amino acids than to less frequently used ones.
21. Phe-Cys-His-Val-Ala-Ala.

22. Hydrogen cyanide. Adenine can be viewed as a pentamer of HCN.

23. (a) A codon for lysine cannot be changed to one for aspartate by the mutation of a single nucleotide.
(b) Arg, Asn, Gln, Glu, Ile, Met, or Thr.

24. The genetic code is degenerate. Of the 20 amino acids, 18 are specified by more than one codon. Hence, many nucleotide changes (especially in the third base of a codon) do not alter the nature of the encoded amino acid. Mutations leading to an altered amino acid are usually more deleterious than those that do not and hence are subject to more stringent selection.

Chapter 6

1. (a) 5′-GGCATAC-3′.
(b) The Sanger dideoxy method of sequencing would give the gel pattern shown here.

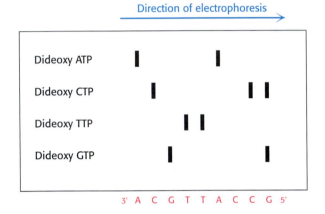

Direction of electrophoresis →

Dideoxy ATP

Dideoxy CTP

Dideoxy TTP

Dideoxy GTP

3′ A C G T T A C C G 5′

2. Ovalbumin cDNA should be used. *E. coli* lacks the machinery to splice the primary transcript arising from genomic DNA.

3. The presence of the *Alu*I sequence would, on average, be $(1/4)^4$, or 1/256, because the likelihood of any base being at any position is 1/4 and there are four positions. By the same reasoning, the presence of the *Not*I sequence would be $(1/4)^8$, or 1/65,536. Thus, the average product of digestion by *Alu*I would be ~250 base pairs (0.25 kb) in length, whereas that for *Not*I would be ~66,000 base pairs (66 kb) in length.

4. (a) No, because most human genes are much longer than 4 kb. One would obtain fragments containing only a small part of a complete gene.
(b) No, chromosome walking depends on having *overlapping* fragments. Exhaustive digestion with a restriction enzyme produces nonoverlapping, short fragments.

5. Southern blotting of an MstII digest would distinguish between the normal and the mutant genes. The loss of a restriction site would lead to the replacement of two fragments on the Southern blot by a single longer fragment. Such a finding would not prove that GTG replaced GAG; other sequence changes at the restriction site could yield the same result.

6. A simple strategy for generating many mutants is to synthesize a degenerate set of cassettes by using a mixture of activated nucleosides in particular rounds of oligonucleotide synthesis. Suppose that the 30-bp coding region begins with

GTT, which encodes valine. If a mixture of all four nucleotides is used in the first and second rounds of synthesis, the resulting oligonucleotides will begin with the sequence XYT (where X and Y denote A, C, G, or T). These 16 different versions of the cassette will encode proteins containing either Phe, Leu, Ile, Val, Ser, Pro, Thr, Ala, Tyr, His, Asn, Asp, Cys, Arg, or Gly at the first position. Likewise, one can make degenerate cassettes in which two or more codons are simultaneously varied.

7. Because PCR can amplify as little as one molecule of DNA, statements claiming the isolation of ancient DNA need to be greeted with some skepticism. The DNA would need to be sequenced. Is it similar to human, bacterial, or fungal DNA? If so, contamination is the likely source of the amplified DNA. Is it similar to that of birds or crocodiles? This sequence similarity would strengthen the case that it is dinosaur DNA, because these species are evolutionarily close to dinosaurs.

8. At high temperatures of hybridization, only very close matches between primer and target would be stable because all (or most) of the bases would need to find partners to stabilize the primer–target helix. As the temperature is lowered, more mismatches are tolerated; so the amplification is likely to yield genes with less sequence similarity. In regard to the yeast gene, you would need to synthesize primers corresponding to the ends of the gene, and then use these primers and human DNA as the target. If nothing is amplified at 54°C, the human gene differs from the yeast gene, but a counterpart may still be present. Repeat the experiment at a lower temperature of hybridization.

9. Digest genomic DNA with a restriction enzyme, and select the fragment that contains the known sequence. Circularize this fragment. Then carry out PCR with the use of a pair of primers that serve as templates for the synthesis of DNA away from the known sequence.

10. The encoded protein contains four repeats of a specific sequence.

11. Sequence-tagged sites (STSs) can serve as a common framework for establishing the relation between clones. An STS is a sequence from 200 to 500 bp long that is present only once in the entire genome. These sequences and those of a pair of PCR primers that can generate the STS are stored in a database. Laboratory 1 states that its YAC contains STS-34, STS-102, and STS-860. Laboratory 2 can then synthesize the corresponding PCR primers and learn whether its YAC contains any of these STSs.

12. T_m is the melting temperature of a double-stranded nucleic acid. If the melting temperatures of the primers are too different, the extent of hybridization with the target DNA will differ during the annealing phase. This would result in differential replications of the strands.

13. A mutation in person B has altered one of the genes for X, leaving the other intact. The fact that the mutated gene is smaller suggests that a deletion has occurred in one copy of the gene. The one functioning gene is transcribed and translated and apparently produces enough protein to render the person asymptomatic.

Person C has only the smaller version of the gene. This gene is neither transcribed (negative Northern blot) nor translated (negative Western blot).

Person D has a normal-size copy of the gene but no RNA or protein. There may be a mutation in the promoter region of the gene that prevents transcription.

Person E has a normal-size copy of the gene that is transcribed, but no protein is made, which suggests that a mutation prevents translation. There are a number of possible explanations, including a mutation that introduced an premature stop codon in the mRNA.

Person F has a normal amount of protein but still displays the metabolic problem. This finding suggests that the mutation affects the activity of the protein. For instance, a mutation that compromises the active site of enzyme Y.

14. Chongqing: residue 2, L → R, CTG → CGG
    Karachi: residue 5, A → P, GCC → CCC
    Swan River: residue 6, D → G, GAC → GGC

Chapter 7

1. There are 26 identities and two gaps for a score of 210. The two sequences are approximately 26% identical. For longer sequences, this percentage would suggest homology.

2. There is a very good probability that they are related, because three-dimensional structure is more conserved than is sequence identity.

3. (1) Identity score = −25; Blosum score = 7. (2) Identity score = 15; Blosum = −10.

4. U

U   **G**

5. There are 4^{40}, or 1.2×10^{24}, different molecules. Each molecule has a mass of 2.2×10^{-20}, because 1 mol of polymer has a mass of 330 g mol^{-1} × 40, and there are 6.02×10^{23} molecules mol^{-1}. Therefore, 26.4 kg of RNA would be required.

6. Because three-dimensional structure is much more closely associated with function than is sequence, tertiary structure is more evolutionarily conserved than is primary structure. In other words, protein function is the most important characteristic, and protein function is determined by structure. Thus, the structure must be conserved, but not necessarily a specific amino acid sequence.

7. Alignment score is 6 × 10 = 60. Many answers are possible, depending on the randomly reordered sequence. A possible result is

Shuffled sequence: TKADKAGEYL

Alignment: (1) ASNFLDRYGK

              TKADKAGEYL

Alignment score is 2 × 10 = 20.

8. (a) Almost certainly diverged from a common ancestor.
(b) Almost certainly diverged from a common ancestor.
(c) May have diverged from a common ancestor, but the sequence alignment does not provide supporting evidence.
(d) May have diverged from a common ancestor, but the sequence alignment does not provide supporting evidence.

9. Protein A is clearly homologous to protein B, given 65% sequence identity, and so A and B are expected to have quite

similar three-dimensional structures. Likewise, proteins B and C are clearly homologous, given 55% sequence identity, and so B and C are expected to have quite similar three-dimensional structures. Thus, one can conclude that proteins A and C are likely to have similar three-dimensional structures, even though they are only 15% identical in sequence alignment.

10. The likely secondary structure is

Chapter 8

1. (a) 31.1 µmol; (b) 0.05 µmol; (c) 622 s^{-1}.

2. (a) Yes. $K_M = 5.2 \times 10^{-6}$ M.
(b) $V_{max} = 6.8 \times 10^{-10}$ mol minute^{-1}.
(c) 337 s^{-1}.

3. Penicillinase, like glycopeptide transpeptidase, forms an acyl-enzyme intermediate with its substrate but transfers the intermediate to water rather than to the terminal glycine residue of the pentaglycine bridge.

4. (a) In the absence of inhibitor, V_{max} is 47.6 µmol minute^{-1}, and K_M is 1.1×10^{-5} M. In the presence of inhibitor, V_{max} is the same, and the apparent K_M is 3.1×10^{-5} M.
(b) Competitive.
(c) 1.1×10^{-3} M.
(d) f_{ES} is 0.243, and f_{EI} is 0.488.
(e) f_{ES} is 0.73 in the absence of inhibitor and 0.49 in the presence of 2×10^{-3} M inhibitor. The ratio of these values, 1.49, is the same as the ratio of the reaction velocities under these conditions.

5. (a) V_{max} is 9.5 µmol minute^{-1}. K_M is 1.1×10^{-5} M, the same as without inhibitor.
(b) Noncompetitive.
(c) 2.5×10^{-5} M.
(d) $f_{ES} = 0.73$, in the presence or absence of this noncompetitive inhibitor.

6. (a) $V = V_{max} - (V/[S]) K_M$.
(b) Slope = $-K_M$, y intercept = V_{max}, x intercept = V_{max}/K_M.
(c) An Eadie-Hofstee plot: here

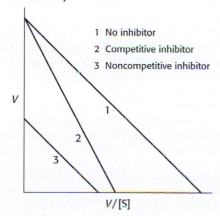

1 No inhibitor

2 Competitive inhibitor

3 Noncompetitive inhibitor

V

V/[S]

ANSWERS TO PROBLEMS

7. Potential hydrogen-bond donors at pH 7 are the side chains of the following residues: arginine, asparagine, glutamine, histidine, lysine, serine, threonine, tryptophan, and tyrosine.

8. The rates of utilization of substrates A and B are given by

$$V_A = \left(\frac{k_3}{K_M}\right)_A [E][A]$$

and

$$V_B = \left(\frac{k_3}{K_M}\right)_B [E][B]$$

Hence, the ratio of these rates is

$$V_A/V_B = \left(\frac{k_3}{K_M}\right)_A [A]/\left(\frac{k_3}{K_M}\right)_B [B]$$

Thus, an enzyme discriminates between competing substrates on the basis of their values of k_3/K_M rather than of K_M alone.

9. The mutation slows the reaction by a factor of 100 because the activation free energy is increased by $+2.72$ kcal mol^{-1} ($2.303RT$ log 100). Strong binding of the substrate relative to the transition state slows catalysis.

10. (a)

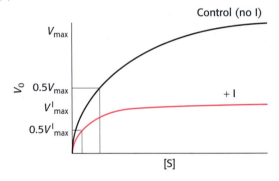

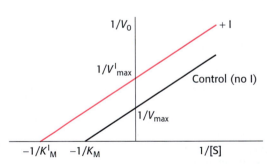

(b) The reaction shows that even the presence of very high [S] cannot drive all of the enzyme to the ES form; there will always be some ESI. Therefore, V_{max} will be lower in the presence of an uncompetitive inhibitor. K_M will decrease because the reaction with I decreases [ES], which causes the substrate-binding reaction to proceed farther to the right, increasing the apparent value of K_{+1}.

11. 11 μmol minute^{-1}.

12. If E_T is increased, V_{max} will increase, because $V_{max} = k_2[E_T]$. But $K_M = (k_{-1} + k_2)/k_1$; that is, it is independent of substrate concentration. The middle graph describes this situation.

13. (a)

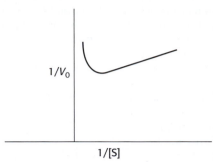

(b) This behavior is substrate inhibition—at high concentrations, the substrate forms unproductive complexes at the active site. The drawing below shows what might occur. Substrate normally binds in a defined orientation, shown in the drawing as red to red and blue to blue. At high concentrations, the substrate may bind at the active site such that the proper orientation is met for each end of the molecule, but two different substrate molecules are binding.

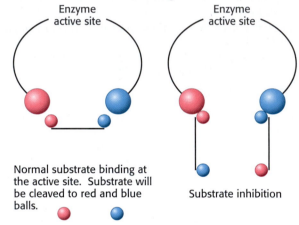

14. The first step will be the rate-limiting step. Enzymes E_A and E_B are operating at $1/2\ V_{max}$, whereas the K_M for enzyme E_A is greater than the substrate concentration. E_A would be operating at approximately $10^{-2}\ V_{max}$.

15. (a) When [S$^+$] is much greater than the value of K_M, pH will have a negligible effect on the enzyme because S$^+$ will interact with E$^-$ as soon as it becomes available.

(b)

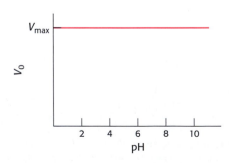

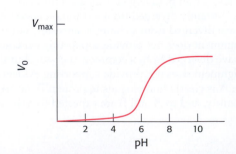

When $[S^+]$ is much less than the value of K_M, the plot of V_0 versus pH becomes essentially a titration curve for the ionizable groups, with enzyme activity being the titration marker. At low pH, the high concentration of H^+ will keep the enzyme in the EH form and inactive. As the pH rises, more and more of the enzyme will be in the E^- form and active. At high pH (low H^+), all of the enzyme is E^-.

(c) The midpoint on this curve will be the pK_a of the ionizable group, which is stated to be pH 6.

16. (a) Incubating the enzyme at 37°C leads to a denaturation of enzyme structure and a loss of activity. For this reason, most enzymes must be kept cool if they are not actively catalyzing their reaction. (b) The coenzyme apparently helps to stabilize the enzyme structure, because enzyme from pyridoxal phosphate-deficient cells denatures faster. Cofactors often help stabilize enzyme structure.

Chapter 9

1. The formation of the acyl enzyme intermediate is slower than the hydrolysis for this amide substrate, so no burst is observed. For ester substrates, the formation of the acyl-enzyme intermediate is faster.

2. The histidine residue in the substrate can substitute to some extent for the missing histidine residue of the catalytic triad of the mutant enzyme.

3. No. Once the catalytic activity has been reduced to almost zero by mutation of a required residue, the introduction of another deleterious mutation cannot further reduce activity.

4. The mutation would reduce the rate of catalysis. The nonpolar groups of chymotrypsin's normal substrate would be less likely to enter a substrate-binding pocket that has a charged residue in it.

5. Imidazole is apparently small enough to reach the active site of carbonic anhydrase. Buffers with large molecular components cannot do so, and the effects of the mutation are more evident.

6. No. The odds of such a sequence being present are approximately 1 in 1 million. Because a typical viral genome has only 50,000 bp, the target sequence would be unlikely to be present.

7. No, because the enzyme would destroy the host DNA before protective methylation could take place.

8. No. The bacteria receiving the enzyme would have its own DNA destroyed because it lacked the appropriate protective methylase.

9. (a) ATP and AMP. (b) Because the reaction has an equilibrium constant near 1, the concentrations of all components will be 0.05 mM.

10. EDTA will bind to Zn^{2+} and remove the ion, which is required for enzyme activity, from the enzyme.

11. Trypsin.

12. It reacts with the active site serine to form a hemiacetal.

13. (a) Cysteine protease: the same as Figure 9.8, except that cysteine replaces serine in the active site.

(b) Aspartyl protease:

(c) Metalloprotease:

Chapter 10

1. The protonated form of histidine probably stabilizes the negatively charged carbonyl oxygen atom of the scissile bond in the transition state. Deprotonation would lead to a loss of activity. Hence, the rate is expected to be half maximal at a pH of about 6.5 (the pK of an unperturbed histidine side chain in a protein) and to decrease as the pH is raised.

2. (a) 100. The change in the [R]/[T] ratio on binding one substrate molecule must be the same as the ratio of the substrate affinities of the two forms (see equations on p. 268). (b) 10. The binding of four substrate molecules changes the [R]/[T] by a factor of $100^4 = 10^8$. The ratio in the absence of substrate is 10^{-7}. Hence, the ratio in the fully liganded molecule is $10^8 \times 10^{-7} = 10$.

3. The fraction of molecules in the R form is 10^{-5}, 0.004, 0.615, 0.998, and 1 when 0, 1, 2, 3, and 4 ligands, respectively, are bound.

4. The sequential model can account for negative cooperativity, whereas the concerted model cannot.

5. The binding of PALA switches ATCase from the T to the R state because PALA acts as a substrate analog. An enzyme molecule containing bound PALA has fewer free catalytic sites than does an unoccupied enzyme molecule. However, the

PALA-containing enzyme will be in the R state and hence have higher affinity for the substrates. The dependence of the degree of activation on the concentration of PALA is a complex function of the allosteric constant L_0 and of the binding affinities of the R and T states for the analog and the substrates.

6. It would show simple Michaelis-Menten kinetics since it is essentially always in the R state.

7. (a) Increases, (b) decreases, (c) decreases, and (d) increases oxygen affinity.

8. (b) Inositol hexaphosphate.

9. The pK is (a) lowered, (b) raised, and (c) raised.

10. Activation is independent of zymogen concentration because the reaction is intramolecular.

11. Add blood from the second patient to a sample from the first. If the mixture clots, the second patient has a defect different from that of the first. This type of assay is called a complementation test.

12. Activated factor X remains bound to blood-platelet membranes, which accelerates its activation of prothrombin.

13. Antithrombin III is a very slowly hydrolyzed substrate of thrombin. Hence, its interaction with thrombin requires a fully formed active site on the enzyme.

14. Residues a and d are located in the interior of an α-helical coiled coil, near the axis of the superhelix. Hydrophobic interactions between these side chains contribute to the stability of the coiled coil.

15. Leucine would be a good choice. It is much more stable than methionine but has nearly the same volume and is very hydrophobic.

16. The simple sequential model predicts that the fraction of catalytic chains in the R state (f_R) is equal to the fraction containing bound substrate (Y). The concerted model, in contrast, predicts that f_R increases more rapidly than Y as the substrate concentration is increased. The change in f_R leads to the change in Y on addition of substrate, as predicted by the concerted model.

17. The binding of succinate to the functional catalytic sites of the native c_3 moiety changed the visible absorption spectrum of nitrotyrosine residues in the *other* c_3 moiety of the hybrid enzyme. Thus, the binding of substrate analog to the active sites of one trimer altered the structure of the other trimer.

18. According to the concerted model, an allosteric activator shifts the conformational equilibrium of all subunits toward the R state, whereas an allosteric inhibitor shifts it toward the T state. Thus, ATP (an allosteric activator) shifted the equilibrium to the R form, resulting in an absorption change similar to that obtained when substrate is bound. CTP had a different effect. Hence, this allosteric inhibitor shifted the equilibrium to the T form. Thus, the concerted model accounts for the ATP-induced and CTP-induced (heterotropic), as well as for the substrate-induced (homotropic), allosteric interactions of ATCase.

19. This mutation is the one that occurs in sickle-cell anemia. This alteration markedly reduces the solubility of the deoxygenated, but not the oxygenated, form of hemoglobin because the hemoglobin molecules interact with one another until a complex forms that is large enough that it cannot be solubilized. This complex precipitates and deforms the cell. Thus, sickling occurs when there is a high concentration of the deoxygenated form of hemoglobin S.

20.

21.

Groups expected in the active site include a base to remove the proton from the serine, groups such as Asp and Glu to bind the magnesium ion associated with the ATP, and other groups to stabilize the ADP leaving group.

Chapter 11

1. Carbohydrates were originally regarded as *hydrates of carbon* because the empirical formula of many of them is $(CH_2O)_n$.

2. Three amino acids can be linked by peptide bonds in only six different ways. However, three different monosaccharides can be linked in a plethora of ways. The monosaccharides can be linked in a linear or branched manner, with α or β linkages,

with bonds between C-1 and C-3, between C-1 and C-4, between C-1 and C-6, and so forth. Consequently, the number of possible trisaccharides greatly exceeds the number of tripeptides.

3. (a) Aldose-ketose; (b) epimers; (c) aldose-ketose; (d) anomers; (e) aldose-ketose; (f) epimers.

4. Aldoses are converted into aldonic acids; the aldehyde group of the sugar is oxidized to a carboxylate.

5. The proportion of the α-anomer is 0.36, and that of the β anomer is 0.64.

6. Glucose is reactive because of the presence of an aldehyde group in its open-chain form. The aldehyde group slowly condenses with amino groups to form Schiff-base adducts.

7. A pyranoside reacts with two molecules of periodate; formate is one of the products. A furanoside reacts with only one molecule of periodate; formate is not formed.

8. From methanol.

9. (a) β-D-Mannose; (b) β-D-galactose; (c) β-D-fructose; (d) β-D-glucosamine.

10. The trisaccharide itself should be a competitive inhibitor of cell adhesion if the trisaccharide unit of the glycoprotein is critical for the interaction.

11. Reducing ends would form 1,2,3,6-tetramethylglucose. The branch points would yield 2,3-dimethylglucose. The remainder of the molecule would yield 2,3,6-trimethylglucose.

12. (a) Not a reducing sugar. No open-chain forms are possible. (b) D-Galactose, D-glucose, D-fructose. (c) D-Galactose and sucrose.

13.

CH₂OH, O, OH, OH, OH, HO, H

β-D-Mannose

The hemiketal linkage of the α anomer is broken to form the open form. Rotation about the C-1 and C-2 bonds allows the formation of the β anomer, and a mixture of isomers results.

14. Heating converts the very sweet pyranose form to the more stable but less sweet furanose form. Consequently, it is difficult to accurately control the sweetness of the preparation, which also accounts for why honey loses sweetness with time. See Figure 11.6 for structures.

15. (a) Each glycogen molecule has one reducing end, whereas the number of nonreducing ends is determined by the number of branches, or α-1,6 linkages. (b) Because the number of nonreducing ends greatly exceeds the number of reducing ends in a collection of glycogen molecules, all of the degradation and synthesis of glycogen takes place at the nonreducing ends, thus maximizing the rate of degradation and synthesis.

16. 64. Each site either is or is not glycosylated so that there are $2^6 = 64$ possible proteins.

17. As discussed in Chapter 9, many enzymes display stereospecificity. Clearly, the enzymes of sucrose synthesis are able to distinguish between the isomers of the substrates and link only the correct pair.

Chapter 12

1. 2.86×10^6 molecules, because each leaflet of the bilayer contains 1.43×10^6 molecules.

2. 2×10^{-7} cm, 6×10^{-6} cm, and 2×10^{-4} cm.

3. The radius of this molecule is 3.1×10^{-7} cm, and its diffusion coefficient is 7.4×10^{-9} cm^2 s^{-1}. The average distances traversed are 1.7×10^{-7} cm in 1 μs, 5.4×10^{-6} in 1 ms, and 1.7×10^{-4} cm in 1 s.

4. The membrane underwent a phase transition from a highly fluid to a nearly frozen state when the temperature was lowered. A carrier can shuttle ions across a membrane only when the bilayer is highly fluid. A channel former, in contrast, allows ions to traverse its pore even when the bilayer is quite rigid.

5. The initial decrease in the amplitude of the paramagnetic resonance spectrum results from the reduction of spin-labeled phosphatidyl choline molecules in the outer leaflet of the bilayer. Ascorbate does not traverse the membrane under these experimental conditions; hence, it does not reduce the phospholipids in the inner leaflet. The slow decay of the residual spectrum is due to the reduction of phospholipids that have flipped over to the outer leaflet of the bilayer.

6. The addition of the carbohydrate introduces a significant energy barrier to the flip-flop because a hydrophilic carbohydrate moiety would need to be moved through a hydrophobic environment. This energetic barrier enhances membrane asymmetry.

7. The presence of a cis double bond introduces a kink that prevents packing of the fatty acid chains. Cis double bonds maintain fluidity. Trans fatty acids have no structural effect, relative to saturated fatty acids, and so they are rare.

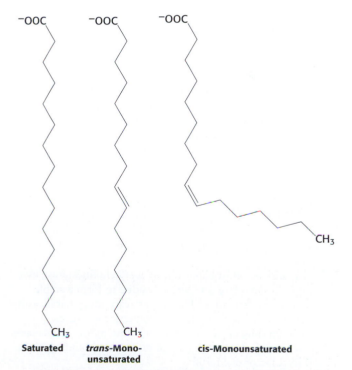

Saturated | *trans*-Mono-unsaturated | *cis*-Monounsaturated

8. In a hydrophobic environment, the formation of intrachain hydrogen bonds would stabilize the amide hydrogen atom and

carbonyl oxygen atoms of the polypeptide chain; so an α helix would form. In an aqueous environment, these groups would be stabilized by interaction with water; so there would be no energetic reason to form an α helix. Thus, the α helix would be most likely to form in an hydrophobic environment.

9. The shift to the lower temperature would decrease fluidity by enhancing packing of the hydrophobic chains by van der Waals interaction. To prevent this, new phospholipids would be synthesized having shorter chains and a greater number of cis double bonds. The shorter chains would reduce the amount of van der Waals interaction, and the cis double bonds, causing the kink in structure, would prevent packing of the fatty acid tails of the phospholipids.

10. (a) The graph shows that, as temperature increases, the phospholipid bilayer becomes more fluid. T_m is the temperature of the transition from the predominantly less fluid state to the predominantly more fluid state. Cholesterol broadens the transition from the less-fluid to the more-fluid state. In essence, cholesterol makes membrane fluidity less sensitive to temperature changes.
(b) This effect is important because the presence of cholesterol tends to stabilize membrane fluidity by preventing sharp transitions. Because protein function depends on the proper fluidity of the membrane, cholesterol maintains the proper environment for membrane-protein function.

11. Protein C is a transmembrane protein from *C. elegans*. It spans the membrane with four α helices that are prominently displayed as hydrophobic peaks in the hydropathy plot. Interestingly, protein A also is a membrane protein, a porin. This protein is made primarily of β strands, which lack the prominent hydrophobic window of membrane helices. This example shows that, although hydropathy plots are useful, they are not infallible.

12. To purify any protein, the protein must first be solubilized. For a membrane protein, solubilization usually requires a detergent—hydrophobic molecules that bind to the protein and thus replace the lipid environment of the membrane. If the detergent is removed, the protein aggregates and precipitates from solution. Often, performing purification steps, such as ion-exchange chromatography, in the presence of sufficient detergent to solubilize the protein is difficult. Crystals must be generated of appropriate protein–detergent complexes.

The two α chains are not in identical environments. Also, the presence of desensitized states in addition to the open and closed forms indicates that a more complex model is required.

4. (a) −22 mV; (b) +4.5; (c) 5.2 kcal mol^{-1} (22 kJ mol^{-1}).

5. An ion channel must transport ions in either direction at the same rate. The net flow of ions is determined only by the composition of the solutions on either side of the membrane.

6. The positively charge guanidinium group resembles sodium and binds to negatively charged carboxylate groups in the mouth of the channel.

7. The blockage of ion channels inhibits action potentials, leading to loss of nervous function. Like tetrodotoxin, these toxin molecules are useful for isolating and specifically inhibiting particular ion channels.

8. Because sodium ions are charged and because sodium channels carry only sodium ions (and not anions), the accumulation of excess positive charge on one side of the membrane dominates the chemical gradients.

9. Batrachotoxin blocks the transition from the open to the closed state.

10. (a) Chloride ions flow into the cell. (b) Chloride flux is inhibitory because it hyperpolarizes the membrane. (c) The channel consists of five subunits.

11. The free-energy cost is 7.6 kcal mol^{-1} (32 kJ mol^{-1}). The chemical work performed is 4.88 kcal mol^{-1} (20.4 kJ mol^{-1}) and the electrical work performed is 2.76 kcal mol^{-1} (11.5 kJ mol^{-1}).

12. (a) The conductance of the channel is 100 picosiemens (pS). (b) 3.1×10^4 ions flow through the channel during its mean open time of 1 ms. (c) The mean transit time for an ion is 32 ns.

13. Membrane vesicles containing a high concentration of lactose are formed. The binding of lactose to the inner face of the permease is followed by the binding of a proton. Both sites then evert. Because the lactose concentration on the outside is low, lactose and the proton will dissociate from the permease. The downhill flux of lactose will drive the uphill flux of protons in this in vitro system.

14. The catalytic prowess of acetylcholinesterase ensures that the duration of the nerve stimulus will be short.

Chapter 13

1. The ratio of closed to open forms of the channel is 10^5, 5000, 250, 12.5, and 0.625 when zero, one, two, three, and four ligands, respectively, are bound. Hence, the fraction of open channels is 10^{-5}, 2.0×10^{-4}, 4×10^{-4}, 7.4×10^{-2}, and 0.62.

2. These organic phosphates inhibit acetylcholinesterase by reacting with the active-site serine residue to form a stable phosphorylated derivative. They cause respiratory paralysis by blocking synaptic transmission at cholinergic synapses.

3. (a) The binding of the first acetylcholine molecule increases the open-to-closed ratio by a factor of 240, and binding of the second by a factor of 11,700.
(b) The free-energy contributions are 3.3 kcal mol^{-1} (14 kJ mol^{-1}) and 5.6 kcal mol^{-1} (23 kJ mol^{-1}) respectively.
(c) No. The MCW model predicts that the binding of each ligand will have the same effect on the open-to-closed ratio. The acetylcholine receptor channel is not perfectly symmetric.

15.

protein

Acetylcholine

$^+N(CH_3)_3$

protein

$^+N(CH_3)_3$

HO — $^+N(CH_3)_3$

H

HO

H_3C

16. (a) Only ASIC1a is inhibited by the toxin. (b) Yes, when the toxin was removed, the activity of the acid-sensing channel began to be restored. (c) 0.9 nM.

17. This mutation is one of a class of mutations that result in slow channel syndrome (SCS). The results suggest that there is a defect in channel closing; so the channel remains open for prolonged periods. Alternatively, the channel may have a higher affinity for acetylcholine than does the control channel.

18. The mutation reduces the affinity of acetylcholine for the receptor. The recordings would show the channel opening only infrequently.

19. Glucose displays a transport curve that suggests the participation of a carrier, because the initial rate is high but then levels off at higher concentrations, consistent with saturation of the carrier, which is reminiscent of Michaelis-Menten enzymes (Section 8.4.1). Indole shows no such saturation phenomenon, which implies that the molecule is lipophilic and simply diffuses across the membrane. Ouabain is a specific inhibitor the Na^+-K^+ pump. If ouabain were to inhibit glucose transport, then a Na^+-glucose cotransporter is assisting in transport.

Chapter 14

1. Reactions in parts *a* and *c*, to the left; reactions in parts *b* and *d*, to the right.

2. None whatsoever.

3. (a) $\Delta G^{\circ\prime} = +7.5$ kcal mol^{-1} ($+31.4$ kJ mol^{-1}) and $K'_{eq} = 3.2 \times 10^{-6}$. (b) 3.28×10^4.

4. $\Delta G^{\circ\prime} = 1.7$ kcal moll^{-1} (7.1 kJ mol^{-1}). The equilibrium ratio is 17.8.

5. (a) Acetate + CoA + H$^+$ goes to acetyl CoA + H$_2$O, $\Delta G^{\circ\prime} = +7.5$ kcal mol^{-1} (31.4 kJ mol^{-1}). ATP hydrolysis, $\Delta G^{\circ\prime} = -10.9$ kcal mol^{-1} (-45.6 kJ mol^{-1}). Overall reaction, $\Delta G^{\circ\prime} = -3.4$ kcal mol^{-1} (-14.2 kJ mol^{-1}).
(b) With pyrophosphate hydrolysis, $\Delta G^{\circ\prime} = -8.0$ kcal mol^{-1} (-33.4 kJ mol^{-1}).

6. (a) For an acid AH, $AH \rightleftharpoons A^- + H^+$, $K = \dfrac{[A^-][H^+]}{[AH]}$.
The pK is defined as $pK = -\log_{10} K$. $\Delta G^{\circ\prime}$ is the standard free energy change at pH 7. Thus, $\Delta G^{\circ\prime} = -RT \ln K = -2.303 \log_{10} K = -2.303 (pK - 7)$ kcal mol^{-1} since $[H^+] = 10^{-7}$ M.
(b) $\Delta G^{\circ\prime} = -2.303 (4.8 - 7) = -5.1$ kcal mol^{-1}.

7. Arginine phosphate in invertebrate muscle, like creatine phosphate in vertebrate muscle, serves as a reservoir of high-potential phosphoryl groups. Arginine phosphate maintains a high level of ATP in muscular exertion. NMR is a choice technique for monitoring the levels of arginine phosphate and ATP in contracting muscle.

8. An ADP unit.

9. (a) The rationale behind creatine supplementation is that it would be converted into creatine phosphate and thus serves as a rapid means of replenishing ATP after muscle contraction.
(b) If it is beneficial, it would affect activities that depend on short bursts of activity; any sustained activity would require ATP generation by fuel metabolism, which, as Figure 14.7 shows, requires more time.

10. Under standard conditions, $\Delta G^{\circ\prime} = -RT \ln$ [product]/[reactants]. Substituting $+5.7$ kcal mol^{-1} for $\Delta G^{\circ\prime}$ and solving for [products]/[reactants] yields 7×10^{-5}. In other

words, the forward reaction does not take place to a significant extent. Under intracellular conditions, ΔG is 0.3 kcal mol^{-1}. If one uses the equation $\Delta G = \Delta G°' + RT \ln$ [product]/[reactants] and solves for [products]/[reactants], the ratio is 3.7×10^{-5}. Thus, a reaction that is endergonic under standard conditions can be converted into an exergonic reaction by maintaining the [products]/[reactants] ratio below the equilibrium value. This conversion is usually attained by using the products in another coupled reaction as soon as they are formed.

11. Liver: -10.8 kcal mol^{-1} (-45.2 kJ mol^{-1}); muscle: -11.5 kcal mol^{-1} (-47.8 kJ mol^{-1}); brain: -11.6 kcal mol^{-1}(-48.4 kJ mol^{-1}). Brain

12. Recall that $\Delta G = \Delta G°' + RT \ln$ [products/reactants]. Altering the ratio of products to reactants will cause ΔG to vary. In glycolysis, the concentrations of the components of the pathway result in a value of ΔG greater than that of $\Delta G°'$.

13. The activated form of sulfate in most organisms is 3′-phosphoadenosine 5′-phosphosulfate.

14. (a) As the Mg^{2+} concentration falls, the ΔG of hydrolysis rises. Note that pMg is a logarithmic plot, and so each number on the x-axis represents a 10-fold change in $[Mg^{2+}]$.
(b) Mg^{2+} would bind to the phosphates of ATP and help to mitigate charge repulsion. As the $[Mg^{2+}]$ falls, charge stabilization of ATP would be less, leading to greater charge repulsion and an increase in ΔG on hydrolysis.

Chapter 15

1. Each activated β-adrenergic receptor activates many G proteins. Each G protein activates one molecule of adenylate cyclase, which generates many molecules of cAMP. Each activated human growth-factor receptor can activate more than one molecule of JAK2. Each JAK2 molecule can phosphorylate many STAT5 molecules. Each activated EGF receptor can activate Grb-2 and Sos. Each molecule of Sos can activate many molecules of Ras.

2. The negatively charged glutamate residues mimic the negatively charged phosphoserine or phosphothreonine residues and stabilize the active conformation of the enzyme.

3. No. Phosphoserine and phosphothreonine are considerably shorter than phosphotyrosine.

4. Removal of GDP from G proteins (heterotrimeric G proteins as well as the Ras family). Removal of the regulatory subunit from PKA with the binding of cAMP. Removal of the pseudosubstrate from PKC by diacylglycerol binding. Removal of the amphipathic helix from the active site of CaM kinase by calmodulin.

5. The mutant should always be active because the pseudosubstrate will not bind in the active site.

6. Growth-factor receptors can be activated by dimerization. If an antibody causes a receptor to dimerize, the signal-transduction pathway in a cell will be activated.

7. The α subunit should always be in the GTP form and, hence, should always be in the active form, stimulating its signaling pathway.

8. The mutated hormone would bind to the growth-hormone receptor but would not favor receptor dimerization. Thus, it would not stimulate the JAK-STAT signaling pathway. Such a mutated hormone might be useful as a competitive inhibitor for growth hormone. It would block the activity of native growth hormone.

9. Calcium ions diffuse slowly because they bind to many protein surfaces within a cell, impeding their free motion. cAMP does not bind as frequently, and so it diffuses more rapidly.

10. $G_{\alpha s}$ stimulates adenylate cyclase, leading to the generation of cAMP. This signal then leads to glucose mobilization (see Chapter 21). If cAMP phosphodiesterase were inhibited, then cAMP levels would remain high even after the termination of the epinephrine signal, and glucose mobilization would continue.

11. The formation of diacylglycerol implies the participation of phospholipase C. A simple pathway would entail receptor activation by cross-phosphorylation, followed by the binding of phospholipase C γ (through its SH2 domains). The participation of phospholipase C indicates that IP$_3$ should be formed and, hence, calcium concentrations should increase.

12. In the reaction catalyzed by adenylate cyclase, the 3′-OH group nucleophilically attacks the α-phosphorus atom attached to the 5′-OH group, leading to displacement of pyrophosphate. The reaction catalyzed by DNA polymerase is similar except that the 3′-OH group is on a different nucleotide.

13. (a) X $\approx 10^{-7}$ M; Y $\approx 5 \times 10^{-6}$ M; Z $\approx 10^{-3}$ M.
(b) Because much less X is required to fill half of the sites, X displays the highest affinity. (c) The binding affinity almost perfectly matches the ability to stimulate adenyl cyclase, suggesting that the hormone–receptor complex leads to the stimulation of adenyl cyclase. (d) Try performing the experiment in the presence of antibodies to $G_{\alpha s}$.

14. (a) The total binding does not distinguish between binding to a specific receptor, binding to different receptors, and nonspecific binding to the membrane.
(b) The rationale is that the receptor will have a high affinity for the ligand. Thus, in the presence of excess nonradioactive ligand, the receptor will bind to nonradioactive ligand. Therefore, any binding of the radioactive ligand must be nonspecific.
(c) The plateau suggests that the number of receptor-binding sites in the cell membrane is limited.

15. Number of receptors per cell =

$$\frac{10^4 \text{ cpm}}{\text{mg of membrane protein}} \times \frac{\text{mg of membrane protein}}{10^{10} \text{ cells}}$$

$$\times \frac{\text{mmol}}{10^{12} \text{ cpm}} \times \frac{6.023 \times 10^{20} \text{ molecules}}{\text{mmol}} = 600$$

Chapter 16

1. Glucose is reactive because its open-chain form contains an aldehyde group.

2. (a) The label is in the methyl carbon atom of pyruvate.
(b) 5 mCi/mM. The specific activity is halved because the number of moles of product (pyruvate) is twice that of the labeled substrate (glucose).

3. (a) Glucose + 2 P$_i$ + 2 ADP → 2 lactate + 2 ATP
(b) $\Delta G' = -27.2$ kcal mol^{-1} (114 kJ mol^{-1}).

4. 3.0×10^{-5}.

5. The equilibrium concentrations of fructose 1,6-bisphosphate, dihydroxyacetone phosphate, and glyceraldehyde 3-phosphate are 7.8×10^{-4} M, 2.2×10^{-4} M, and 2.2×10^{-4} M, respectively.

6. All three carbon atoms of 2,3-BPG are ^{14}C labeled. The phosphorus atom attached to the C-2 hydroxyl group is ^{32}P labeled.

7. Hexokinase has a low ATPase activity in the absence of a sugar because it is in a catalytically inactive conformation. The addition of xylose closes the cleft between the two lobes of the enzyme. However, xylose lacks a hydroxy-methyl group, and so it cannot be phosphorylated. Instead, a water molecule at the site normally occupied by the C-6 hydroxymethyl group acts as the acceptor of the phosphoryl group from ATP.

8. (a) The fructose 1-phosphate pathway forms glyceraldehyde 3-phosphate.
(b) Phosphofructokinase, a key control enzyme, is bypassed. Furthermore, fructose 1-phosphate stimulates pyruvate kinase.

9. (a)Increased; (b) increased; (c) increased; (d) decreased.

10. Fructose 2,6-bisphosphate, present at high concentration when glucose is abundant, normally inhibits gluconeogenesis by blocking fructose 1,6-bisphosphatase. In this genetic disorder, the phosphatase is active irrespective of the glucose level. Hence, substrate cycling is increased. The level of fructose 1,6-bisphosphate is consequently lower than normal. Less pyruvate is formed and thus less ATP is generated.

11. Reactions in parts *b* and *e* would be blocked.

12. There will be no labeled carbons. The CO_2 added to pyruvate (formed from the lactate) to form oxaloacetate is lost with the conversion of oxaloacetate into phosphoenolpyruvate.

13. The net reaction in the presence of arsenate is

Glyceraldehyde 3-phosphate + NAD^+ + H_2O →

3-phosphoglycerate + NADH + $2 H^+$

Glycolysis proceeds in the presence of arsenate, but the ATP normally formed in the conversion of 1,3-bisphosphoglycerate into 3-phosphoglycerate is lost. Thus, arsenate uncouples oxidation and phosphorylation by forming a highly labile acyl arsenate.

14. This example illustrates the difference between *stoichiometric* and *catalytic* utilization of a molecule. If cells used NAD^+ stoichiometrically, a new molecule of NAD^+ would be required each time a lactate is produced. As we will see, the synthesis of NAD^+ requires much ATP. On the other hand, if the NAD^+ that is converted into NADH could be recycled and reused, a small amount of the molecule could regenerate a vast amount of lactate. This is the case in the cell. NAD^+ is regenerated by the oxidation of NADH and reused. NAD^+ is thus used catalytically.

15. Consider the equilibrium equation of adenylate kinase.

$$K_{eq} = [ATP][AMP]/[ADP]^2 \qquad (1)$$

or

$$[AMP] = K_{eq}[ADP]^2/[ATP] \qquad (2)$$

Recall that [ATP] > [ADP] > [AMP] in the cell. As ATP is utilized, a small decrease in its concentration will result in a larger percentage increase in [ADP] because its concentration is greater than that of ADP. This larger percentage increase in [ADP] will result in an even greater percentage increase in [AMP] because its concentration is related to the square of [ADP]. In essence, equation 2 shows that monitoring the energy status with AMP magnifies small changes in [ATP], leading to tighter control.

16. The synthesis of glucose during intense exercise provides a good example of interorgan cooperation in higher organisms. When muscle is actively contracting, lactate is produced from glucose by glycolysis. The lactate is released into the blood and

absorbed by the liver, where it is converted by gluconeogenesis into glucose. The newly synthesized glucose is then released and taken up by the muscle for energy generation.

17. The input of four additional high-phosphoryl-transfer potential molecules in gluconeogenesis changes the equilibrium constant by a factor of 10^{32}, which makes the conversion of pyruvate into glucose thermodynamically feasible. Without this energetic input, gluconeogenesis would not take place.

18. The mechanism is analogous to that for triose phosphate isomerase (Figure 16.6). It proceeds through an enediol intermediate. The active site would be expected to have a general base (analogous to Glu 165 in TIM) and a general acid (analogous to His 95 in TIM).

19. Galactose is a component of glycoproteins. Possibly, the absence of galactose leads to the improper formation or function of glycoproteins required in the central nervous system. More generally, the fact that the symptoms arise in the absence of galactose suggests that galactose is required in some fashion.

20. (a) Curiously, the enzyme uses ADP as the phosphoryl donor rather than ATP.
(b) Both AMP and ATP behave as competitive inhibitors of ADP, the phosphoryl donor. Apparently, the *P. furiosus* enzyme is not allosterically inhibited by ATP.

Chapter 17

1. (a) After one round of the citric acid cycle, the label emerges in C-2 and C-3 of oxaloacetate. (b) The label emerges in CO_2 in the formation of acetyl CoA from pyruvate. (c) After one round of the citric acid cycle, the label emerges in C-1 and C-4 of oxaloacetate. (d and e) Same fate as that in part *a*.

2. (a) Isocitrate lyase and malate synthase are required in addition to the enzymes of the citric acid cycle.
(b) 2 Acetyl CoA + 2 NAD^+ + FAD + $3 H_2O$ → oxaloacetate + 2 CoA + 2 NADH + $FADH_2$ + $3 H^+$
(c) No. Hence, mammals cannot carry out the net synthesis of oxaloacetate from acetyl CoA.

3. -9.8 kcal mol^{-1} (-41 kJ mol^{-1}).

4. Enzymes or enzyme complexes are biological catalysts. Recall that a catalyst facilitates a chemical reaction without the catalyst itself being permanently altered. Oxaloacetate can be thought of as a catalyst because it binds to an acetyl group, leads to the oxidative decarboxylation of the two carbon atoms, and is regenerated at the completion of a cycle. In essence, oxaloacetate (and any cycle intermediate) acts as a catalyst.

5. The coenzyme stereospecificity of glyceraldehyde 3-phosphate dehydrogenase is the opposite of that of alcohol dehydrogenase

6. Thiamine thiazolone pyrophosphate is a transition-state analog. The sulfur-containing ring of this analog is uncharged, and so it closely resembles the transition state of the normal coenzyme in thiamine-catalyzed reactions (e.g., the uncharged resonance form of hydroxyethyl-TPP).

7. A decrease in the amount of O_2 will necessitate an increase in anaerobic glycolysis for energy production, leading to the generation of a large amount of lactic acid. Under conditions of shock, the kinase inhibitor is administered to ensure that pyruvate dehydrogenase is operating maximally.

8. (a) The steady-state concentrations of the products are low compared with those of the substrates. (b) The ratio of malate to oxaloacetate must be greater than 1.75×10^4 for oxaloacetate to be formed.

9.

$$\text{Pyruvate} + \text{CoA} + \text{NAD}^+ \xrightarrow{\text{Pyruvate dehydrogenase complex}} \text{acetyl CoA} + CO_2 + \text{NADH}$$

$$\text{Pyruvate} + CO_2 + \text{ATP} + H_2O \xrightarrow{\text{Pyruvate carboxylase}} \text{oxaloacetate} + \text{ADP} + P_i$$

$$\text{Oxaloacetate} + \text{acetyl CoA} + H_2O \xrightarrow{\text{Citrate synthase}} \text{citrate} + \text{CoA} + H^+$$

$$\text{Citrate} \xrightarrow{\text{Aconitase}} \text{isocitrate}$$

$$\text{Isocitrate} + \text{NAD}^+ \xrightarrow{\text{Isocitrate dehydrogenase}} \alpha\text{-ketoglutarate} + CO_2 + \text{NADH}$$

Net:

$$2\ \text{pyruvate} + 2\ \text{NAD}^+ + \text{ATP} + H_2O \longrightarrow$$
$$\alpha\text{-ketoglutarate} + CO_2 + \text{ADP} + P_i + 2\ \text{NADH} + 2\ H^+$$

10. We cannot get the net conversion of fats into glucose, because the only means to get the carbons from fats into oxaloacetate, the precursor to glucose, is through the citric acid cycle. However, although two carbon atoms enter the cycle as acetyl CoA, two carbon atoms are lost as CO_2 before oxaloacetate is formed. Thus, although some carbon atoms from fats may end up as carbon atoms in glucose, we cannot obtain a *net* synthesis of glucose from fats.

11. The enol intermediate of acetyl CoA attacks the carbonyl carbon atom of glyoxylate to form a C–C bond. This reaction is like the condensation of oxaloacetate with the enol intermediate of acetyl CoA in the reaction catalyzed by citrate synthase. Glyoxylate contains a hydrogen atom in place of the $-CH_2COO^-$ group of oxaloacetate; the reactions are otherwise nearly identical.

12. Citrate is a symmetric molecule. Consequently, it was assumed that the two $-CH_2COO^-$ groups in it would react identically. Thus, for every citrate molecule undergoing the reactions shown in path 1, it was thought that another citrate molecule would react as shown in path 2. If so, then only *half* the label should have emerged in the CO_2.

Path 2 (does not occur)

Path 1

13. Call one hydrogen atom A and the other B. Now suppose that an enzyme binds three groups of this substrate—X, Y, and H—at three complementary sites. The adjoining diagram shows X, Y, and H_A bound to three points on the enzyme. In contrast, X, Y, and H_B cannot be bound to this active site; two of these three groups can be bound, but not all three. Thus, H_A and H_B will have different fates.

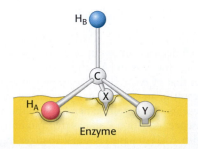

Sterically nonequivalent groups such as H_A and H_B will almost always be distinguished in enzymatic reactions. The essence of the differentiation of these groups is that the enzyme holds the substrate in a specific orientation. Attachment at three points, as depicted in the diagram, is a readily visualized way of achieving a particular orientation of the substrate, but it is not the only means of doing so.

14. (a) The complete oxidation of citrate requires 4.5 μmol of O_2 for every micromole of citrate.

$$C_6H_8O_7 + 4.5\,O_2 \rightarrow 6\,CO_2 + 4\,H_2O$$

Thus, 13.5 μmol of O_2 would be consumed by 3 μmol of citrate. (b) Citrate led to the consumption of far more O_2 than can be accounted for simply by the oxidation of citrate itself. Citrate thus facilitated O_2 consumption.

15. (a) In the absence of arsenite, the amount of citrate remained constant. In its presence, the concentration of citrate fell, suggesting that it was being metabolized.
(b) It is not altered. Citrate still disappears.
(c) Arsenite is preventing the regeneration of citrate. Recall (Section 17.3.2) that arsenite inhibits the pyruvate dehydrogenase complex.

16. (a) The initial infection is unaffected by the absence of isocitrate lyase, but the absence of this enzyme inhibits the latent phase of the infection.
(b) Yes.
(c) A critic could say that, in the process of deleting the isocitrate lyase gene, some other gene was damaged, and it is the absence of this other gene that prevents latent infection. Reinserting the isocitrate lyase gene into the bacteria from which it had been removed renders the criticism less valid.
(d) Isocitrate lyase enables the bacteria to synthesize carbohydrates that are necessary for survival, including carbohydrate components of the cell membrane.

Chapter 18

1. (a) 12.5; (b) 14; (c) 32; (d) 13.5; (e) 30; (f) 16.
2. Biochemists use E_0', the value at pH 7, whereas chemists use E_0, the value in 1 M H^+. The prime denotes that pH 7 is the standard state.
3. (a) Blocks electron transport and proton pumping at site 3. (b) Blocks electron transport and ATP synthesis by inhibiting the exchange of ATP and ADP across the inner mitochondrial membrane. (c) Blocks electron transport and proton pumping at site 1. (d) Blocks ATP synthesis without inhibiting electron transport by dissipating the proton gradient. (e) Blocks electron transport and proton pumping at site 3. (f) Blocks electron transport and proton pumping at site 2.
4. If the proton gradient is not dissipated by the influx of protons into a mitochondrion with the generation of ATP, eventually the outside of the mitochondrion develops such a large positive charge that the electron-transport chain can no longer pump protons against the gradient.
5. (a) No effect. Mitochondria cannot metabolize glucose.
(b) No effect. No fuel is present to power the synthesis of ATP.
(c) The $[O_2]$ falls because citrate is a fuel and ATP can be formed from ADP and P_i.
(d) Oxygen consumption stops because oligomycin inhibits ATP synthesis, which is coupled to the activity of the electron-transport chain.
(e) No effect for the reasons given in part d.
(f) $[O_2]$ falls rapidly because the system is uncoupled and does not require ATP synthesis to lower the proton-motive force.
(g) $[O_2]$ falls at a lower rate but still falls. Rotenone inhibits Complex I, but the presence of succinate will enable electrons to enter at Complex II.

(h) Oxygen consumption ceases because Complex IV is inhibited and the entire chain backs up.
6. (a) The P:O ratio is equal to the product of $(H^+/2\,e^-)$ and $(\sim P/H^+)$. Note that the P:O ratio is identical with the $(P:2\,e^-)$ ratio. (b) 2.5 and 1.5, respectively.
7. $\Delta G^{\circ\prime}$ is +16.1 kcal mol^{-1} (67 kJ mol^{-1}) for oxidation by NAD^+ and +1.4 kcal/mol^{-1} (5.9 kJ mol^{-1}) for oxidation by FAD. The oxidation of succinate by NAD^+ is not thermodynamically feasible.
8. Cyanide can be lethal because it binds to the ferric form of cytochrome oxidase and thereby inhibits oxidative phosphorylation. Nitrite converts ferrohemoglobin into ferrihemoglobin, which also binds cyanide. Thus, ferrihemoglobin competes with cytochrome oxidase for cyanide. This competition is therapeutically effective because the amount of ferrihemoglobin that can be formed without impairing oxygen transport is much greater than the amount of cytochrome oxidase.
9. The available free energy from the translocation of two, three, and four protons is −9.23, −13.8, and −18.5 kcal, respectively. The free energy consumed in synthesizing a mole of ATP under standard conditions is 7.3 kcal. Hence, the residual free energy of −1.93, −6.5, and −11.2 kcal can drive the synthesis of ATP until the [ATP]/[ADP][P_i] ratio is 26.2, 6.5×10^4, and 1.6×10^8, respectively. Suspensions of isolated mitochondria synthesize ATP until this ratio is greater than 10^4, which shows that the number of protons translocated per ATP synthesized is at least three.
10. Such a defect (called Luft syndrome) was found in a 38-year-old woman who was incapable of performing prolonged physical work. Her basal metabolic rate was more than twice normal, but her thyroid function was normal. A muscle biopsy showed that her mitochondria were highly variable and atypical in structure. Biochemical studies then revealed that oxidation and phosphorylation were not tightly coupled in these mitochondria. In this patient, much of the energy of fuel molecules was converted into heat rather than ATP. The development of mitochondrial medicine is lucidly reviewed in R. Luft, *Proc. Natl. Acad. Sci. U.S.A.* 91(1994):8731.
11. Dicylohexylcarbodiimide reacts readily with carboxyl groups, as discussed earlier in regard to its use in peptide synthesis (Section 4.4). Hence, the most likely targets are aspartate and glutamate side chains. In fact, aspartate 61 of subunit c of *E. coli* F_0 is specifically modified by this reagent. Conversion of this aspartate into asparagine by site-specific mutagenesis also eliminated proton conduction. See A. E. Senior, *Biochim. Biophys. Acta* 726(1983):81.
12. Triose phosphate isomerase converts dihydroxyacetone phosphate (a potential dead end) into glyceraldehyde 3-phosphate (a mainstream glycolytic intermediate).
13. This inhibitor (like antimycin A) blocks the reduction of cytochrome c_1 by QH_2, the crossover point.
14. If oxidative phosphorylation were uncoupled, no ATP could be produced. In a futile attempt to generate ATP, much fuel would be consumed. The danger lies in the dose. Too much uncoupling would lead to tissue damage in highly aerobic organs such as the brain and heart, which would have severe consequences for the organism as a whole. The energy that is normally transformed into ATP would be released as heat. To maintain body temperature, sweating might increase, although the very process of sweating itself depends on ATP.
15. Add the inhibitor with and without an uncoupler, and monitor the rate of O_2 consumption. If the O_2 consumption increases again in the presence of inhibitor and uncoupler, the

inhibitor must be inhibiting ATP synthase. If the uncoupler has no effect on the inhibition, the inhibitor is inhibiting the electron-transport chain.

16. The organic acids in the blood are indications that the mice are deriving a large part of their energy needs through anaerobic glycolysis. Lactate is the end product of anaerobic glycolysis. Alanine is an aminated transport form of lactate. Alanine formation plays a role in succinate formation, which is caused by the reduced state of the mitochondria.

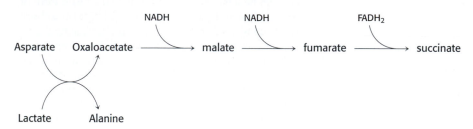

The electron-transport chain is slowed because the inner mitochondrial membrane is hyperpolarized. Without ADP to accept the energy of the proton-motive force, the membrane becomes polarized to such an extent that protons can no longer be pumped. The excess H_2O_2 is probably due to the fact the superoxide radical is present in higher concentration because the oxygen can no longer be effectively reduced.

$$O_2^{\cdot-} + O_2^{\cdot-} \rightarrow H_2O_2$$

Indeed, these mice provide evidence of such oxidative damage.

17. (a) Succinate is oxidized by Complex II, and the electrons are used to establish a proton-motive force that powers ATP synthesis.
(b) The ability to synthesize ATP is greatly reduced.
(c) Because the goal was to measure ATP hydrolysis. If succinate had been added in the presence of ATP, no reaction would have taken place, because of respiratory control.
(d) The mutation has little effect on the ability of the enzyme to catalyze the hydrolysis of ATP.
(e) They suggest two things: (1) the mutation did not affect the catalytic site on the enzyme, because the ATP synthase is still capable of catalyzing the reverse reaction; (2) the mutation did not affect the amount of enzyme present, given that the controls and patients had similar amounts of activity.

18. The absolute configuration of thiophosphate is opposite that of ATP in the reaction catalyzed by ATP synthase. This result is consistent with an in-line phosphoryl transfer reaction taking place in a single step. The retention of configuration in the Ca^{2+}-ATPase reaction points to two phosphoryl transfer reactions—inversion by the first and a return to the starting configuration by the second. The Ca^{2+}-ATPase reaction proceeds by a phosphorylated enzyme intermediate.

Chapter 19

1. $\Delta E_0' = +0.11$ V and $\Delta G^{\circ\prime} = -5.1$ kcal mol^{-1} (21.3 kJ mol^{-1}).
2. (a) All ecosystems require an energy source from outside the system, because the chemical-energy sources will ultimately be limited. The photosynthetic conversion of sunlight is one example of such a conversion.
(b) Not at all. Spock would point out that chemicals other than water can donate electrons and protons.

3. DCMU inhibits electron transfer in the link between photosystems II and I. O_2 can evolve in the presence of DCMU if an artificial electron acceptor such as ferricyanide can accept electrons from Q.
4. DCMU will have no effect, because it blocks photosystem II, and cyclic photophosphorylation uses photosystem I and the cytochrome bf complex.
5. (a) 28.7 kcal einstein^{-1} (120 kJ einstein^{-1}).
(b) 1.24 V.
(c) One 1000-nm photon has the free energy content of 2.4 molecules of ATP. A minimum of 0.42 photon is needed to drive the synthesis of a molecule of ATP.
6. At this distance, the expected rate is one electron per second.
7. Phycoerythrin, the most peripheral protein in the phycobilisome.
8. The distance doubles, and so the rate should decrease by a factor of 64 to 640 ps.
9. The electrons flow through photosystem II directly to ferricyanide. No other steps are required.
10. (a) Thioredoxin.
(b) The control enzyme is unaffected, but the mitochondrial enzyme with part of the chloroplast γ subunit increases activity as the concentration of DTT increases.
(c) The increase was even larger when thioredoxin was present. Thioredoxin is the natural reductant for the chloroplast enzyme, so presumably it operates more efficiently than would DTT, which probably functions to keep the thioredoxin reduced.
(d) It seems that they did.
(e) The enzyme is susceptible to control by the redox state. In plant cells, reduced thioredoxin is generated by photosystem I. Thus, the enzyme is active when photosynthesis is taking place.
(f) Cysteine.
(g) Group-specific modification or site-specific mutagenesis.

Chapter 20

1. Aldolase participates in the Calvin cycle, whereas transaldolase participates in the pentose phosphate pathway.
2. The concentration of 3-phosphoglycerate would increase, whereas that of ribulose 1,5-bisphosphate would decrease.
3. The concentration of 3-phosphoglycerate would decrease, whereas that of ribulose 1,5-bisphosphate would increase.

4. (a)

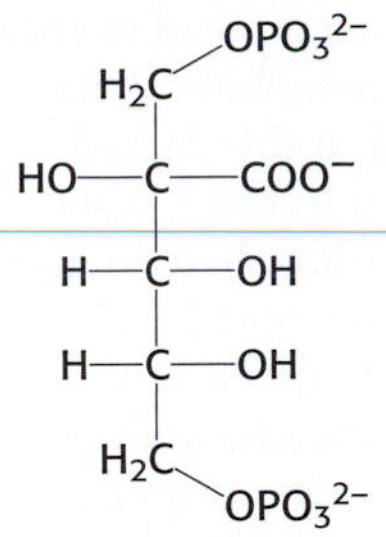

OPO$_3^{2-}$
H$_2$C
HO—C—COO$^-$
H—C—OH
H—C—OH
H$_2$C
OPO$_3^{2-}$

**2-Carboxyarabinitol
1,5-bisphosphate
(CABP)**

(b) CABP resembles the addition compound formed in the reaction of CO_2 and ribulose 1,5-bisphosphate.
(c) CABP is predicted to be a potent inhibitor of rubisco.
5. Aspartate + glyoxylate → oxaloacetate + glycine
6. ATP is converted into AMP. To convert this AMP back
into ATP, two molecules of ATP are required: one to form
ADP and another to form ATP from the ADP.
7. The oxygenase activity of rubisco increases with temperature. Crabgrass is a C_4 plant, whereas most grasses lack this
capability. Consequently, the crabgrass will thrive at the
hottest part of the summer because the C_4 pathway provides
an ample supply of CO_2.
8. As global warming progresses, C_4 plants will invade the
higher latitudes, whereas C_3 plants will retreat to cooler regions.
9. The label emerges at C-5 of ribulose 5-phosphate.
10. Oxidative decarboxylation of isocitrate to α-ketoglutarate.
A β-ketoacid intermediate is formed in both reactions.
11. C-1 and C-3 of fructose 6-phosphate are labeled, whereas
erythrose 4-phosphate is not labeled.
12. (a) 5 Glucose 6-phosphate + ATP → 6 ribose
5-phosphate + ADP + H$^+$
(b) Glucose 6-phosphate + 12 NADP$^+$ + 7 H$_2$O →
6 CO$_2$ + 12 NADPH + 12 H$^+$ + P$_i$
13. Form a Schiff base between a ketose substrate and transaldolase, reduce it with tritiated NaBH$_4$, and fingerprint the labeled enzyme.
14. ΔE_0 for the reduction of glutathione by NADPH is
$+0.09$ V. Hence, $\Delta G^{\circ}{}'$ is -4.2 kcal mol^{-1} (17.5 kJ mol^{-1}),
which corresponds to an equilibrium constant of 1126. The re-
quired [NADPH]/[NADP$^+$] ratio is 8.9×10^{-2}.

15.

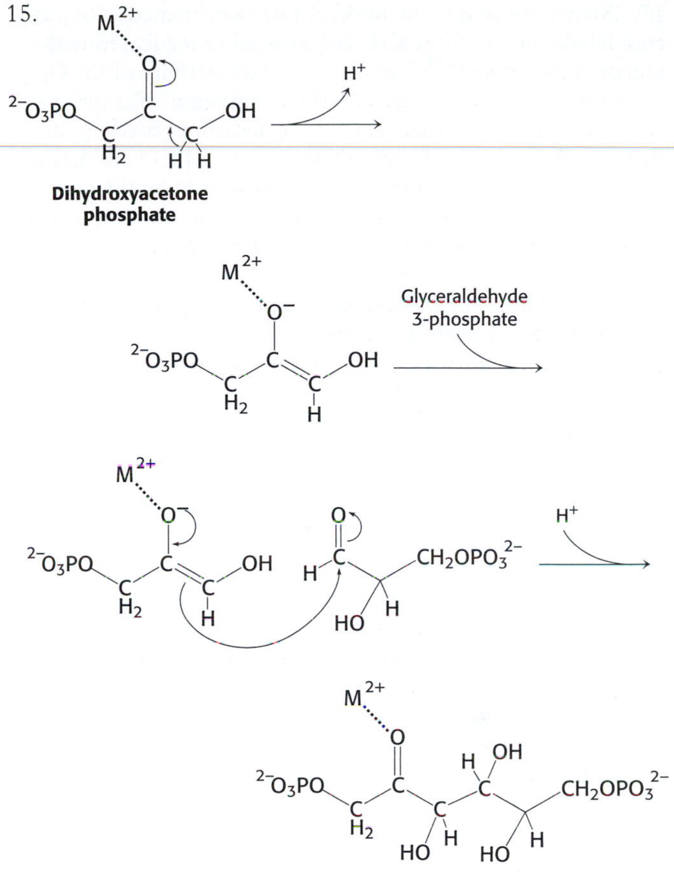

**Dihydroxyacetone
phosphate**

Glyceraldehyde
3-phosphate

Fructose 1,6-bisphosphate

16.

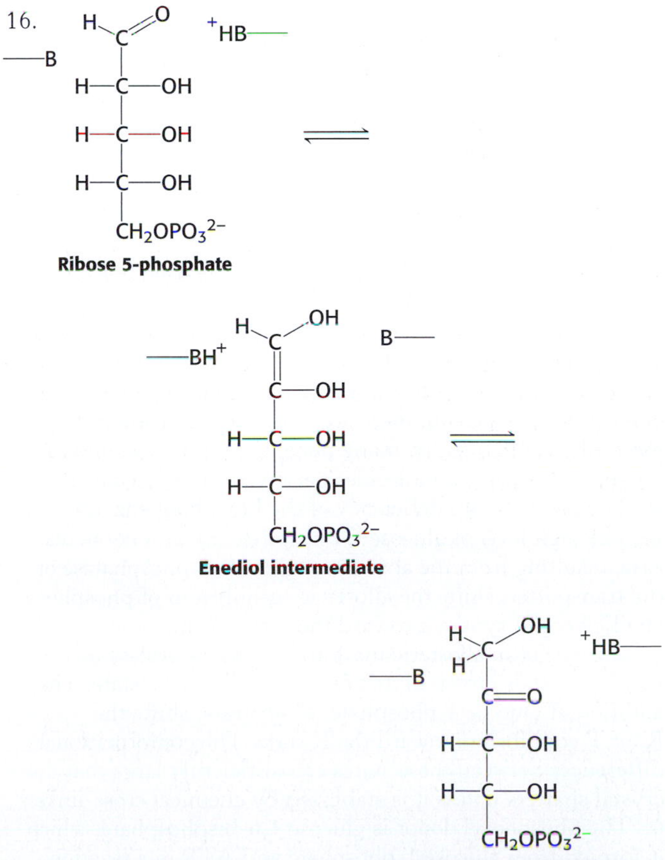

Ribose 5-phosphate

Enediol intermediate

Ribulose 5-phosphate

17. An aliquot of a tissue homogenate is incubated with glucose labeled with ^{14}C at C-1, and another is incubated with glucose labeled with ^{14}C at C-6. The radioactivity of the CO_2 produced by the two samples is then compared. The rationale of this experiment is that only C-1 is decarboxylated by the pentose phosphate pathway, whereas C-1 and C-6 are decarboxylated equally when glucose is metabolized by the glycolytic pathway, the pyruvate dehydrogenase complex, and the citric acid cycle. The reason for the equivalence of C-1 and C-6 in the latter set of reactions is that glyceraldehyde 3-phosphate and dihydroxyacetone phosphate are rapidly interconverted by triose phosphate isomerase.

18. The reduction of each CO_2 to the level of a hexose requires 2 moles of NADPH. The reduction of $NADP^+$ is a two-electron process. Hence, the formation of 2 moles of NADPH requires the pumping of four photons by photosystem I. The electrons given up by photosystem I are replenished by photosystem II, which needs to absorb an equal number of photons. Hence, eight photons are needed to generate the required NADPH. The energy input of 8 moles of photons is 381 kcal. Thus, the overall efficiency of photosynthesis under standard conditions is at least 114/381, or 30%.

19. (a) The curve on the right was generated by the C_4 plant. Recall that the oxygenase activity of rubisco increases with temperature more rapidly than does the carboxylase activity. Consequently, at higher temperatures, the C_3 plants would fix less carbon. Because C_4 plants can maintain a higher CO_2 concentration, the rise in temperature is less deleterious.
(b) The oxygenase activity will predominate. Additionally, when the temperature rise is very high, evaporation of water might become a problem. The higher temperatures can begin to damage protein structures as well.
(c) The C_4 pathway is a very effective active-transport system for concentrating CO_2, even when environmental concentrations are very low.
(d) With the assumption that the plants have approximately the same capability to fix CO_2, the C_4 pathway is apparently the rate-limiting step in C_4 plants.

Chapter 21

1. Galactose + ATP + UTP + H_2O + glycogen$_n$ →
glycogen$_{n+1}$ + ADP + UDP + 2 P_i + H^+
2. As an unbranched polymer, α-amylose has only one nonreducing end. Therefore, only one glycogen phosphorylase molecule could degrade each α-amylose molecule. Because glycogen is highly branched, there are many nonreducing ends per molecule. Consequently, many phosphorylase molecules can release many glucose molecules per glycogen molecule.
3. The patient has a deficiency of the branching enzyme.
4. The high level of glucose 6-phosphate in von Gierke disease, resulting from the absence of glucose 6-phosphatase or the transporter, shifts the allosteric equilibrium of phosphorylated glycogen synthase toward the active form.
5. Glucose is an allosteric inhibitor of phosphorylase a. Hence, crystals grown in its presence are in the T state. The addition of glucose 1-phosphate, a substrate, shifts the R ⇌ T equilibrium toward the R state. The conformational differences between these states are sufficiently large that the crystal shatters unless it is stabilized by chemical cross-links.
6. The phosphoryl donor is glucose 1,6-bisphosphate, which is formed from glucose 1-phosphate and ATP in a reaction catalyzed by phosphoglucokinase.

7. Water is excluded from the active site to prevent hydrolysis. The entry of water could lead to the formation of glucose rather than glucose 1-phosphate. A site-specific mutagenesis experiment is revealing in this regard. In phosphorylase, Tyr 573 is hydrogen bonded to the 2'-OH of a glucose residue. The ratio of glucose 1-phosphate to glucose product is 9000:1 for the wild-type enzyme, and 500:1 for the Phe 573 mutant. Model building suggests that a water molecule occupies the site normally filled by the phenolic OH of tyrosine and occasionally attacks the oxocarbonium ion intermediate to form glucose.
8. The amylase activity was necessary to remove all of the glycogen from the glycogenin. Recall that glycogenin synthesizes oligosaccharides of about eight glucose units, and then activity stops. Consequently, if the glucose residues are not removed by extensive amylase treatment, glycogenin would not function.
9. The substrate can be handed directly from the transferase site to the debranching site.
10. During exercise, [ATP] falls and [AMP] rises. Recall that AMP is an allosteric activator of glycogen phosphorylase b. Thus, even in the absence of covalent modification by phosphorylase kinase, glycogen is degraded.
11. (a) Muscle phosphorylase b will be inactive even when the AMP level is high. Hence, glycogen will not be degraded unless phosphorylase is converted into the a form by hormone-induced or Ca^{2+}-induced phosphorylation.
(b) Phosphorylase b cannot be converted into the much more active a form. Hence, the mobilization of liver glycogen will be markedly impaired.
(c) The elevated level of the kinase will lead to the phosphorylation and activation of glycogen phosphorylase. Because glycogen will be persistently degraded, little glycogen will be present in the liver.
(d) Protein phosphatase 1 will be continually active. Hence, the level of phosphorylase b will be higher than normal, and glycogen will be less readily degraded.
(e) Protein phosphatase 1 will be much less effective in dephosphorylating glycogen synthase and glycogen phosphorylase. Consequently, the synthase will stay in the less active b form, and the phosphorylase will stay in the more active a form. Both changes will lead to increased degradation of glycogen.
(f) The absence of glycogenin will block the initiation of glycogen synthesis. Very little glycogen will be synthesized in its absence.
12. (a) The α subunit is thus always on. cAMP is always produced. Glycogen degradation is always on, and glycogen synthesis is always off.
(b) Glycogen phosphorylase cannot be covalently activated. Glycogen degradation is off. Whether glycogen synthesis is on or off has no direct effect.
(c) Glycogen degradation is always on. Glycogen synthesis is turned off by the results of Glycogen degradation's constant activity.
(d) Glycogen degradation is always off; nothing can remain phosphorylated. Glycogen synthesis is always on; nothing can remain phosphorylated.
(e) Glycogen synthesis has nothing with which to build glycogen. Glycogen synthesis is off, which would eventually turn glycogen degradation off.
(f) Phosphodiesterase destroys cAMP. Therefore, glycogen degradation would always be on and glycogen synthesis would always be off.

13. The slow phosphorylation of the α subunits of phosphorylase kinase serves to prolong the degradation of glycogen. The kinase cannot be deactivated until its α subunits are phosphorylated. The slow phosphorylation of α subunits ensures that the kinase and, in turn, phosphorylase stay active for a defined interval.

14. Phosphorylation of the β subunit activates the kinase and leads to glycogen degradation. Subsequent phosphorylation of the α subunit make the β subunit and the α subunit substrates for protein phosphatase. Thus, if the α subunit were modified before the β subunit, the enzyme would be primed for shutdown before it was activated and little glycogen degradation would take place.

15.

Transferase reaction

α-1,6-glucosidase reaction

16. (a) The glycogen was too large to enter the gel and, because analysis was by Western blot with the use of an antibody specific to glycogenin, one would not expect to see background proteins.
(b) α-Amylase degrades glycogen, releasing the protein glycogenin so that it can be visualized by the Western blot.
(c) Glycogen phosphorylase, glycogen synthase, and protein phosphatase 1. These proteins might be visible if the gel were stained for protein, but a Western analysis reveals the presence of glycogenin only.

17. (a) The smear was due to molecules of glycogenin with increasingly large amounts of glycogen attached to them.
(b) In the absence of glucose in the medium, glycogen is metabolized, resulting in a loss of the high-molecular-weight material.
(c) Glycogen could be resynthesized and added to the glycogenin when the cells were fed glucose again.
(d) No difference between lanes 3 and 4 suggests that, by 1 hour, the glycogen molecules had attained maximum size in this cell line. Prolonged incubation does not apparently increase the amount of glycogen.
(e) α-Amylase removes essentially all of the glycogen, and so only the glycogenin remains.

Chapter 22

1. Glycerol + 2 NAD$^+$ + P$_i$ + ADP →
pyruvate + ATP + H$_2$O + 2 NADH + H$^+$
Glycerol kinase and glycerol phosphate dehydrogenase.

2. Stearate + ATP + $13\frac{1}{2}$ H$_2$O + 8 FAD + 8 NAD$^+$ →
$4\frac{1}{2}$ acetoacetate + $14\frac{1}{2}$ H$^+$ + 8 FADH$_2$ + 8 NADH +
AMP + 2 P$_i$
3. (a) Oxidation in mitochondria; synthesis in the cytosol.
(b) Acetyl CoA in oxidation; acyl carrier protein for synthesis.
(c) FAD and NAD$^+$ in oxidation; NADPH for synthesis.
(d) L isomer of 3-hydroxyacyl CoA in oxidation; D isomer in synthesis. (e) From carboxyl to methyl in oxidation; from methyl to carboxyl in synthesis. (f) The enzymes of fatty acid synthesis, but not those of oxidation, are organized in a multienzyme complex.
4. (a) Palmitoleate; (b) linoleate; (c) linoleate; (d) oleate;
(e) oleate; (f) linolenate.

5. C-1 is more radioactive.
6. Decarboxylation drives the condensation of malonyl ACP and acetyl ACP. In contrast, the condensation of two molecules of acetyl ACP is energetically unfavorable. In gluconeogenesis, decarboxylation drives the formation of phosphoenolpyruvate from oxaloacetate.
7. Adipose-cell lipase is activated by phosphorylation. Hence, overproduction of the cAMP-activated kinase will lead to an accelerated breakdown of triacylglycerols and a depletion of fat stores.
8. The mutant enzyme would be persistently active because it could not be inhibited by phosphorylation. Fatty acid synthesis would be abnormally active. Such a mutation might lead to obesity.
9. Carnitine translocase deficiency and glucose 6-phosphate transporter deficiency.
10. In the fifth round of β oxidation, cis-Δ^2-enoyl CoA is formed. Dehydration by the classic hydratase yields D-3-hydroxyacyl CoA, the wrong isomer for the next enzyme in β oxidation. This dead end is circumvented by a second hydratase that removes water to give trans-Δ^2-enoyl CoA. The addition of water by the classic hydratase then yields L-3-hydroxyacyl CoA, the appropriate isomer. Thus, hydratases of opposite stereospecificities serve to epimerize (invert the configuration of) the 3-hydroxyl group of the acyl CoA intermediate.
11. The probability of synthesizing an error-free polypeptide chain decreases as the length of the chain increases. A single mistake can make the entire polypeptide ineffective. In contrast, a defective subunit can be spurned in forming a non-

covalent multienzyme complex; the good subunits are not wasted.

12. The absence of ketone bodies is due to the fact that the liver, the source of blood-ketone bodies, cannot oxidize fatty acids to produce acetyl CoA. Moreover, because of the impaired fatty acid oxidation, the liver becomes more dependent on glucose as an energy source. This dependency results in a decrease in gluconeogenesis and a drop in blood-glucose levels, which is exacerbated by the lack of fatty acid oxidation in muscle and a subsequent increase in glucose uptake from the blood.

13. Peroxisomes enhance the degradation of fatty acids. Consequently, increasing the activity of peroxisomes could help to lower levels of blood triacylglycerides. In fact, clofibrate is rarely used because of serious side effects.

14. Citrate works by facilitating the formation of active filaments from inactive monomers. In essence, it increases the number of active sites available, or the concentration of enzyme. Consequently, its effect is visible as an increase in the value of V_{max}. Allosteric enzymes that alter their V_{max} values in response to regulators are sometimes called V-class enzymes. The more common type of allosteric enzyme, in which K_m is altered, comprises K-class enzymes. Palmitoyl CoA causes depolymerization and thus inactivation.

15. The thiolate anion of CoA attacks the 3-keto group to form a tetrahedral intermediate. This collapses to form acyl CoA and the enolate anion of acetyl CoA. Protonation of the enolate yields acetyl CoA.

16.

Malonyl-ACP

Acetyl-ACP

Acetoacetyl-ACP

17. (a) Fats burn in the flame of carbohydrates. Without carbohydrates, there would be no anapleurotic reactions to replenish the TCA-cycle components. With a diet of fats only, the acetyl CoA from fatty acid degradation would build up.
(b) Acetone from ketone bodies.
(c) Yes. Odd-chain fatty acids would lead to the production of propionyl CoA, which can be converted into succinyl CoA, a TCA-cycle component. It would serve to replenish the TCA cycle and mitigate the halitosis.

18. A labeled fat can enter the citric acid cycle as acetyl CoA and yield labeled oxaloacetate, but only after two carbon atoms have been lost as CO_2. Consequently, even though oxaloacetate may be labeled, there can be no net synthesis in the

amount of oxaloacetate and hence no net synthesis of glucose or glycogen.

19. (a) The V_{max} is decreased and the K_m is increased. V_{max} (wild type) = 13 nmol minute^{-1} mg^{-1}; K_m (wild type) = 45 μM; V_{max} (mutant) = 8.3 nmol minute^{-1} mg^{-1}; K_m (mutant) = 74 μM.
(b) Both the V_{max} and the K_m are decreased. V_{max} (wild type) = 41 nmol minute^{-1} mg^{-1}; K_m (wild type) = 104 μM; V_{max} (mutant) = 23 nmol minute^{-1} mg^{-1}; K_m (mutant) = 69 μM.
(c) The wild type is significantly more sensitive to malonyl CoA.
(d) With respect to carnitine, the mutant displays approximately 65% of the activity of the wild type; with respect to palmitoyl CoA, approximately 50% activity. On the other hand, 10 μM of malonyl CoA inhibits approximately 80% of the wild type but has essentially no effect on the mutant enzyme.
(e) The glutamate appears to play a more prominent role in regulation by malonyl CoA than in catalysis.

Chapter 23

1. (a) The ATPase activity of the 26S proteasome resides in the 19S subunit. The energy of ATP hydrolysis could be used to unfold the substrate, which is too large to enter the catalytic barrel. ATP may also be required for translocation of the substrate into the barrel.
(b) Substantiates the answer in part a. Because they are small, the peptides do not need to be unfolded. Moreover, small peptides could probably enter all at once and not require translocation.

2. (a) Pyruvate; (b) oxaloacetate; (c) α-ketoglutarate; (d) α-ketoisocaproate; (e) phenylpyruvate; (f) hydroxyphenyl-pyruvate.

3. (a) Aspartate + α-ketoglutarate + GTP + ATP + 2 H_2O + NADH + H$^+$ → $\frac{1}{2}$ glucose + glutamate + CO_2 + ADP + GDP + NAD$^+$ + 2 P$_i$
The required coenzymes are pyidoxal phosphate in the transamination reaction and NAD$^+$/NADH in the redox reactions.
(b) Aspartate + CO_2 + NH$_4$$^+$ + 3 ATP + NAD$^+$ + 4 H_2O → oxaloacetate + urea + 2 ADP + 4 P$_i$ + AMP + NADH + H$^+$

4. In the eukaryotic proteasome, the distinct β subunits have different substrate specificities, allowing proteins to be more thoroughly degraded.

5. The six subunits probably exist as a heterohexamer. Cross-linking experiments could test the model and help determine which subunits are adjacent to one another.

6. Thiamine pyrophosphate.

7. It acts as an electron sink.

8. CO_2 + NH$_4$$^+$ + 3 ATP + NAD$^+$ + aspartate + 3 H_2O → urea + 2 ADP + 2 P$_i$ + AMP + PP$_i$ + NADH + H$^+$ + oxaloacetate
Four ~P are spent.

9. Ornithine transcarbamoylase (analogous to PALA; see Chapter 10).

10. Ammonia could lead to the amination of α-ketoglutarate, producing a high concentration of glutamate in an unregulated fashion. α-Ketoglutarate for glutamate synthesis could be removed from the citric acid cycle, thereby diminishing the cell's respiration capacity.

11. The mass spectrometric analysis strongly suggests that three enzymes—pyruvate dehydrogenase, α-ketoglutarate dehydrogenase, and the branched-chain α-keto dehydrogenase—

are deficient. Most likely, the common E_3 component of these enzymes is missing or defective. This proposal could be tested by purifying these three enzymes and assaying their ability to catalyze the regeneration of lipoamide.

12. Benzoate, phenylacetate, and arginine would be given to supply a protein-restricted diet. Nitrogen would emerge in hippurate, phenylacetylglutamine, and citrulline.

13. Aspartame, a dipeptide ester (L-aspartyl-L-phenylalanine methyl ester), is hydrolyzed to L-aspartate and L-phenylalanine. High levels of phenylalanine are harmful in phenylketonurics.

14. *N*-Acetylglutamate is synthesized from acetyl CoA and glutamate. Once again, acetyl CoA serves as an activated acetyl donor. This reaction is catalyzed by *N*-acetylglutamate synthase.

15.

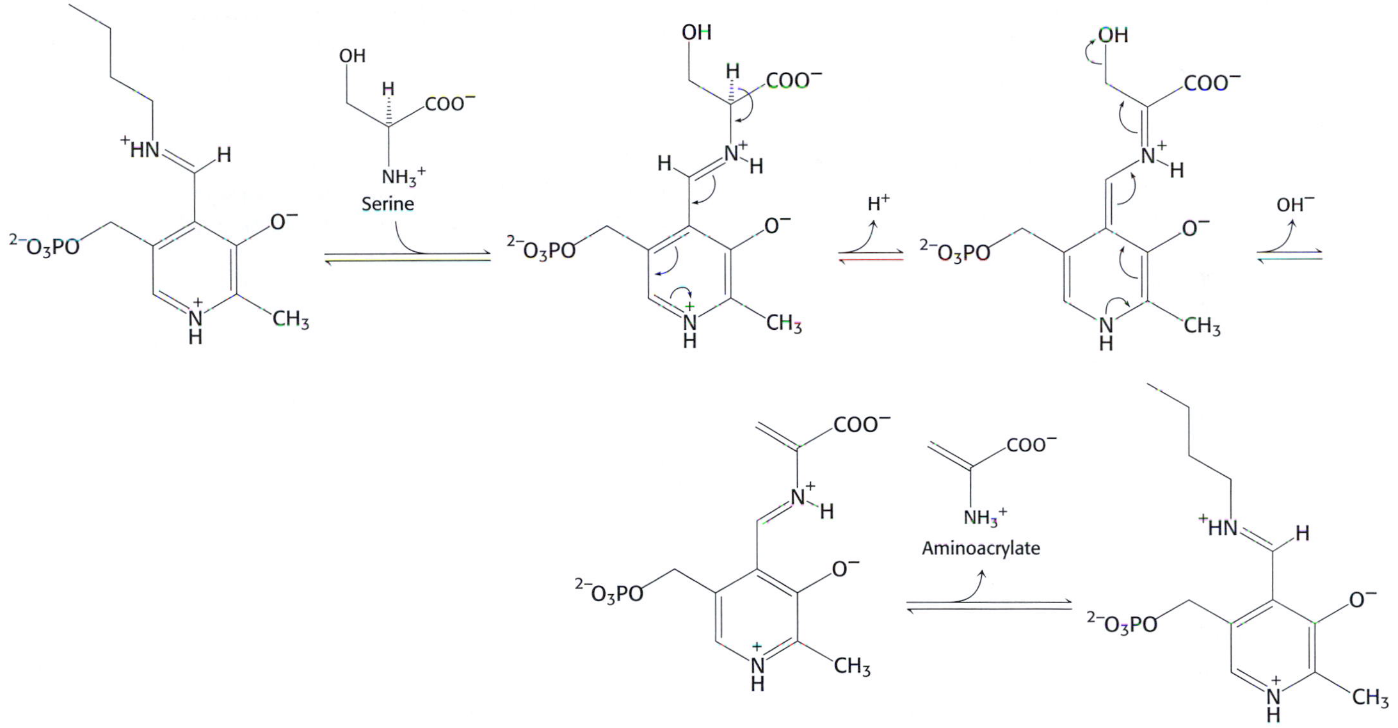

Serine

Aminoacrylate

16.

(Structure: pyridoxal phosphate Schiff base with L-Serine)

$^{2-}O_3PO$ — HN^+ ... H ... O^- ... N^+ / H / CH_3

OH H COO^- NH_3^+ L-Serine

H^+

OH H COO^- (PLP intermediate) $^{2-}O_3PO$ ... O^- ... N ... CH_3

H^+

OH COO^- (PLP intermediate) $^{2-}O_3PO$ ... O^- ... N / H / CH_3

OH H COO^- $^{2-}O_3PO$ ... O^- ... N / H / CH_3

OH H COO^- NH_3^+ D-Serine

^+HN H $^{2-}O_3PO$ ... O^- ... N^+ / H / CH_3

The equilibrium constant for the interconversion of L-serine and D-serine is exactly 1.

17. Exposure of such a domain would suggest that a component of a multiprotein complex has failed to form properly, or that one component has been synthesized in excess. This will lead to rapid degradation and restoration of appropriate stoichiometries.

18. (a) Depletion of glycogen stores. When they are gone, proteins must be degraded to meet the glucose needs of the brain. The resulting amino acids are deaminated, and the nitrogen atoms are excreted as urea.
(b) The brain has adapted to the use of ketone bodies, which are derived from fatty acid catabolism. In other words, the brain is being powered by fatty acid breakdown.
(c) When the glycogen and lipid stores are gone, the only available energy source is protein.

19. Deamination to α-keto-β-methylvalerate; oxidative decarboxylation to α-methylbutyryl CoA; oxidation to tiglyl CoA; hydration, oxidation and thiolysis yields acetyl CoA and propionyl CoA; propionyl CoA to succinyl CoA.

20. (a) Virtually no digestion in the absence of nucleotides. (b) Protein digestion is greatly stimulated by the presence of ATP. (c) AMP-PNP, a nonhydrolyzable analog of ATP, is no more effective than ADP. (d) The proteasome requires neither ATP nor PAN to digest small substrates. (e) PAN and ATP hydrolysis may be required to unfold the peptide and translocate it into the proteasome. (f) Although *Thermoplasma* PAN is not as effective with the other proteasomes, it nonetheless results in threefold to fourfold stimulation of digestion. (g) In light of the fact that the archaea and eukarya diverged several billion years ago, the fact that *Thermoplasma* PAN can stimulate rabbit muscle suggests homology not only between the proteasomes, but also between PAN and the 19S subunit (most likely the ATPases) of the mammalian 26S proteasome.

Chapter 24

1. Glucose + 2 ADP + 2 P_i + 2 NAD^+ + 2 glutamate → 2 alanine + 2 α-ketoglutarate + 2 ATP + 2 NADH + 2 H_2O + 2 H^+

2. N_2 → NH_4^+ → glutamate → serine → glycine → δ-aminolevulinate → porphobilinogen → heme

3. (a) N^5, N^{10}-Methylenetetrahydrofolate; (b) N^5-methyltetrahydrofolate.

4. γ-Glutamyl phosphate is a likely reaction intermediate.

5. The administration of glycine leads to the formation of isovalerylglycine. This water-soluble conjugate, in contrast with isovaleric acid, is excreted very rapidly by the kidneys.

6. They carry out nitrogen fixation. The absence of photosystem II provides an environment in which O_2 is not produced. Recall that the nitrogenase is very rapidly inactivated by O_2.

7. The cytosol is a reducing environment, whereas the extracellular milieu is an oxidizing environment.

8. Succinyl CoA is formed in the mitochondrial matrix.

9. Alanine from pyruvate; aspartate from oxaloacetate; glutamate from α-ketoglutarate.

10. Y could inhibit the C → D step, Z could inhibit the C → F step, and C could inhibit A → B. This scheme is an example of sequential feedback inhibition. Alternatively, Y could inhibit the C → D step, Z could inhibit the C → F step, and the A → B step would be inhibited only in the presence of both Y and Z. This scheme is called concerted feedback inhibition.

11. The rate of the A → B step in the presence of high levels of Y and Z would be 24 s^{-1} ($0.6 \times 0.4 \times 100\ s^{-1}$).

12. An external aldimine forms with SAM, which is deprotonated to form the quinonoid intermediate. The deprotonated carbon atom attacks the carbon atom adjacent to the sulfur atom to form the cyclopropane ring and release methyl thioadenosine, the other product.

13. An external aldimine forms with L-serine, which is deprotonated to form the quinonoid intermediate. This intermediate is reprotonated on its opposite face to form an aldimine with D-serine. This compound is cleaved to release D-serine. The equilibrium constant for a racemization reaction is 1 because the reactant and product are exact mirror images of each other.

14. Synthesis from oxaloacetate and α-ketoglutarate would deplete the citric acid cycle, which would decrease ATP production. Anaplerotic reactions would be required to replenish the citric acid cycle.

15. SAM is the donor for DNA methylation reactions that protect a host from digestion by its own restriction enzymes. A lack of SAM would render the bacterial DNA susceptible to digestion by the cell's own restriction enzymes.

16. (a) Aspargagine is much more common in the dark. More glutamine is present in the light. These amino acids show the most dramatic effects. Glycine also is more common in the light.
(b) Glutatmine is a more metabolically reactive amino acid, used in the synthesis of many other compounds. Consequently, when energy is available as light, glutamine will be preferentially synthesized. Asparagine, which carries more nitrogen per carbon atom and is thus a more efficient means of storing nitrogen when energy is short, is synthesized in the dark. Glycine is more prevalent in the light because of photorespiration.
(c) White asparagus has an especially high concentration of asparagine, which accounts for its intense taste. All asparagus has a large amount of asparagine. In fact, as suggested by its name, asparagine was first isolated from asparagus.

Chapter 25

1. Glucose + 2 ATP + 2 NADP$^+$ + H$_2$O → PRPP + CO$_2$ + ADP + AMP + 2 NADPH + 3 H$^+$
2. Glutamine + aspartate + CO$_2$ + 2 ATP + NAD$^+$ → orotate + 2 ADP + 2 P$_i$ + glutamate + NADH + H$^+$
3. (a, c, d, and e) PRPP; (b) carbamoyl phosphate.
4. PRPP and formylglycinamide ribonucleotide.
5. dUMP + serine + NADPH + H$^+$ → dTMP + NADP$^+$ + glycine
6. There is a deficiency of N^{10}-formyltetrahydrofolate. Sulfanilamide inhibits the synthesis of folate by acting as an analog of p-aminobenzoate, one of the precursors of folate.
7. PRPP is the activated intermediate in the synthesis of phosphoribosylamine in the de novo pathway of purine formation; of purine nucleotides from free bases by the salvage pathway; of orotidylate in the formation of pyrimidines; of nicotinate ribonucleotide; of phosphoribosyl ATP in the pathway leading to histidine; and of phosphoribosylanthranilate in the pathway leading to tryptophan.
8. (a) Cell A cannot grow in a HAT medium, because it cannot synthesize TMP either from thymidine or from dUMP. Cell B cannot grow in this medium, because it cannot synthesize purines by either the de novo pathway or the salvage pathway. Cell C can grow in a HAT medium because it contains active thymidine kinase from cell B (enabling it to phosphorylate thymidine to TMP) and hypoxanthine-guanine

phosphoribosyl transferase from cell A (enabling it to synthesize purines from hypoxanthine by the salvage pathway).
(b) Transform cell A with a plasmid containing foreign genes of interest and a functional thymidine kinase gene. The only cells that will grow in a HAT medium are those that have acquired a thymidylate kinase gene; nearly all of these transformed cells will also contain the other genes on the plasmid.
9. These patients have a high level of urate because of the breakdown of nucleic acids. Allopurinol prevents the formation of kidney stones and blocks other deleterious consequences of hyperuricemia by preventing the formation of urate (Section 25.6.1).
10. The free energies of binding are −13.8 (wild type), −11.9 (Asn 27), and −9.1 (Ser 27) kcal mol^{-1}. The loss in binding energy is 1.9 kcal mol^{-1} (7.9 kJ mol^{-1}) and 4.7 kcal mol^{-1} (19.7 kJ mol^{-1}).
11. Inosine or hypoxanthine could be administered.
12. N-1 in both cases, and the amine group linked to C-6 in ATP.
13. An oxygen atom is added to allopurinol to form alloxanthine.
14. The first reaction proceeds by phosphorylation of glycine to form an acyl phosphate followed by nucleophilic attack by the amine of phosphoribosylamine to displace orthophosphate. The second reaction consists of adenylation of the carbonyl group of xanthylate followed by nucleophilic attack by ammonia to displace AMP.
15. The –NH$_2$ group attacks the carbonyl carbon atom to form a tetrahedral intermediate. Removal of a proton leads to the elimination of water to form inosinate.
16. (a) cAMP; (b) ATP; (c) UDP-glucose; (d) acetyl CoA; (e) NAD$^+$, FAD; (f) dideoxynucleotides; (g) fluorouracil; (h) CTP inhibits ATCase.
17. In vitamin B$_{12}$ deficiency, methyltetrahydrofolate cannot donate its methyl group to homocysteine to regenerate methionine. Because the synthesis of methyltetrahydrofolate is irreversible, the cell's tetrahydrofolate will ultimately be converted into this form. No formyl or methylene tetrahydrofolate will be left for nucleotide synthesis. Pernicious anemia illustrates the intimate connection between amino acid and nucleotide metabolism.
18. The cytosolic level of ATP in the liver falls and that of AMP rises above normal in all three conditions. The excess AMP is degraded to urate.
19. Succinate → malate → oxaloacetate by the citric acid cycle. Oxaloacetate → aspartate by transamination, followed by pyrimidine synthesis. Carbons 4, 5, and 6.
20. (a) Some ATP can be salvaged from the ADP that is being generated. (b) There are equal numbers of high phosphoryl transfer potential groups on each side of the equation.
(c) Because the adenylate kinase reaction is at equilibrium, removing AMP would lead to the formation of more ATP. (d) Essentially, the cycle serves as an anaplerotic reaction for the generation of the citric acid cycle intermediate fumarate.

Chapter 26

1. Glycerol + 4 ATP + 3 fatty acids + 4 H$_2$O → triacylglycerol + ADP + 3 AMP + 7 P$_i$ + 4 H$^+$
2. Glycerol + 3 ATP + 2 fatty acids + 2 H$_2$O + CTP + serine → phosphatidyl serine + CMP + ADP + 2 AMP + 6 P$_i$ + 3 H$^+$

3. (a) CDP-diacylglycerol; (b) CDP-ethanolamine; (c) acyl CoA; (d) CDP-choline; (e) UDP-glucose or UDP-galactose; (f) UDP-galactose; (g) geranyl pyrophosphate.

4. (a and b) None, because the label is lost as CO_2.

5. The categories of mutations are: (1) no receptor is synthesized; (2) receptors are synthesized but do not reach the plasma membrane, because they lack signals for intracellular transport or do not fold properly; (3) receptors reach the cell surface, but they fail to bind LDL normally because of a defect in the LDL-binding domain; (4) receptors reach the cell surface and bind LDL, but they fail to cluster in coated pits because of a defect in their carboxyl-terminal regions.

6. Deamination of cytidine to uridine changes CAA (Gln) into UAA (stop).

7. Benign prostatic hypertrophy can be treated by inhibiting 5α-reductase. Finasteride, the 4-aza steroid analog of dihydrotestosterone, competitively inhibits the reductase but does not act on androgen receptors. Patients taking finasteride have a markedly lower plasma level of dihydrotestosterone and a nearly normal level of testosterone. The prostate gland becomes smaller, whereas testosterone-dependent processes such as fertility, libido, and muscle strength appear to be unaffected.

Finasteride

8. Patients who are most sensitive to debrisoquine have a deficiency of a liver P450 enzyme encoded by a member of the *CYP2* subfamily. This characteristic is inherited as an autosomal recessive trait. The capacity to degrade other drugs may be impaired in people who hydroxylate debrisoquine at a slow rate, because a single P450 enzyme usually handles a broad range of substrates.

9. Many hydrophobic odorants are deactivated by hydroxylation. Molecular oxygen is activated by a cytochrome P450 monooxygenase. NADPH serves as the reductant. One oxygen atom of O_2 goes into the odorant substrate, whereas the other is reduced to water.

10. Recall that dehydrotestosterone is crucial for the development of male characteristics in the embryo. If a pregnant woman were to be exposed to Propecia, the 5α-reductase of the male embryo would be inhibited, which could result in severe developmental abnormalities.

11. The oxygenation reactions catalyzed by the cytochrome P450 family permit greater flexibility in biosynthesis. Because plants are not mobile, they must rely on physical defenses, such as thorns, and chemical defenses, such as toxic alkaloids. The larger P450 array might permit greater biosynthetic versatility.

12. This knowledge would enable clinicians to characterize the likelihood of a patient's having an adverse drug reaction or being susceptible to chemical-induced illnesses. It would also permit a personalized and especially effective drug-treatment regime for diseases such as cancer.

13. The negatively charged phosphoserine residue interacts with the positively charged protonated histidine residue and decreases its ability to transfer a proton to the thiolate.

His

14. The methyl group is first hydroxylated. The hydroxymethylamine eliminated formaldehyde to form methylamine.

15. (a) There is no effect.
(b) Because actin is not controlled by cholesterol, the amount isolated should be the same in both experimental groups. A difference would suggest a problem in the RNA isolation.
(c) The presence of cholesterol in the diet dramatically reduces the amount of HMG-CoA protein.
(d) A common means of regulating the amount of a protein present is to regulate transcription, which is clearly not the case here.
(e) Translation of mRNA could be inhibited. The protein could be rapidly degraded.

Chapter 27

1. DNA polymerase I uses deoxyribonucleoside triphosphates; pyrophosphate is the leaving group. DNA ligase uses a DNA-adenylate (AMP joined to the 5′-phosphate) as a reaction partner; AMP is the leaving group. Topoisomerase I uses a DNA-tyrosyl intermediate (5′-phosphate linked to the phenolic OH group); the tyrosine residue of the enzyme is the leaving group.

2. FAD, CoA, and $NADP^+$ are plausible alternatives.

3. Positive supercoiling resists the unwinding of DNA. The melting temperature of DNA increases in proceeding from negatively supercoiled to relaxed to positively supercoiled DNA. Positive supercoiling is probably an adaptation to high temperature.

4. (a) There are long stretches of each because the transition is highly cooperative. (b) B–Z junctions are energetically highly unfavorable. (c) A–B transitions are less cooperative than B–Z transitions because the helix stays right-handed at an A–B junction but not at a B–Z junction.

5. (a) 96.2 revolutions per second (1000 nucleotides per second divided by 10.4 nucleotides per turn for B-DNA gives 96.2 rps). (b) 0.34 μm/s (1000 nucleotides per second corresponds to 3400 Å/s because the axial distance between nucleotides in B-DNA is 3.4 Å).

6. Eventually, the DNA would become so tightly wound that movement of the replication complex would be energetically impossible.

7. A hallmark of most cancer cells is prolific cell division, which requires DNA replication. If the telomerase were not activated, the chromosomes would shorten until they became nonfunctional, leading to cell death. Interestingly, telomerase is often, but not always, found to be activated in cancer cells.

8. Treat the DNA briefly with endonuclease to occasionally nick each strand. Add the polymerase with the radioactive dNTPs. At the broken bond, or nick, the polymerase will degrade the existing strand with its 5′ → 3′ exonuclease activity

and replace it with a radioactive complementary copy by using its polymerase activity. This reaction scheme is referred to as nick translation, because the nick is moved, or translated, along the DNA molecule without ever becoming sealed.

9. If replication were unidirectional, tracks with a low grain density at one end and a high grain density at the other end would be seen. On the other hand, if replication were bidirectional, the middle of a track would have a low density, as shown in the adjoining diagram. For *E. coli,* the grain tracks are denser on both ends than in the middle, indicating that replication is bidirectional.

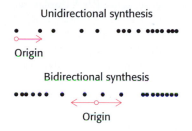

10. (a) Pro (CCC), Ser (UCC), Leu (CUC), and Phe (UUC). Alternatively, the last base of each of these codons could be U. (b) These C → U mutations were produced by nitrous acid.

11. Potentially deleterious side reactions are prevented. The enzyme itself might be damaged by light if it could be activated by light in the absence of bound DNA harboring a pyrimidine dimer.

12. DNA ligase relaxes supercoiled DNA by catalyzing the cleavage of a phosphodiester bond in a DNA strand. The attacking group is AMP, which becomes attached to the 5′-phosphoryl group at the site of scission. AMP is required because this reaction is the reverse of the final step in the joining of pieces of DNA (see Figure 27.28).

13. ATP hydrolysis is required to release DNA gyrase after the enzyme has acted on its DNA substrate. Negative supercoiling requires only the binding of ATP, not its hydrolysis.

14. (a) Size; the top is relaxed and the bottom is supercoiled DNA. (b) Topoisomers. (c) The DNA is becoming progressively more unwound, or relaxed, and thus slower moving.

15. (a) It was used to determine the number of spontaneous revertants—that is, the background mutation rate. (b) To firmly establish that the system was working. A known mutagen's failure to produce revertant would indicate that something was wrong with the experimental system. (c) The chemical itself has little mutagenic ability but is apparently activated into a mutagen by the liver homogenate. (d) Cytochrome P450 system.

Chapter 28

1. The sequence of the coding (+, sense) strand is

5′-ATGGGGAACAGCAAGAGTGGGGCCCTGTCCAAGGAG-3′

and the sequence of template (−, antisense) strand is

3′-TACCCCTTGTCGTTCTCACCCCGGGACAGGTTCCTC-5′

2. An error will only affect one molecule of mRNA of many synthesized from a gene. In addition, the errors do not become a permanent part of the genomic information.

3. At any given instant, only a fraction of the genome (total DNA) is being transcribed. Consequently, speed is not necessary.

4. Heparin, a glycosaminoglycan is highly anionic. Its negative charges, like the phosphodiester bridges of DNA templates, allow it to bind to lysine and arginine residues of RNA polymerase.

5. This mutant σ would competitively inhibit the binding of holoenzyme and prevent the specific initiation of RNA chains at promoter sites.

6. The core enzyme without σ binds more tightly to the DNA template than does the holoenzyme. The retention of σ after chain initiation would make the mutant RNA polymerase less processive. Hence, RNA synthesis would be much slower than normal.

7. A 100-kd protein contains about 910 residues, which are encoded by 2730 nucleotides. At a maximal transcription rate of 50 nucleotides per second, the protein would be synthesized in 54.6 s.

8. Initiation at strong promoters occurs every 2 s. In this interval, 100 nucleotides are transcribed. Hence, centers of transcription bubbles are 34 nm (340 Å) apart.

9. (a) The lowest band on the gel will be that of strand 3 alone (*i*), whereas the highest will be that of stands 1, 2, and 3 and core polymerase (*v*). Band *ii* will be at the same position as band *i* because the RNA is not complementary to the nontemplate strand, whereas band *iii* will be higher because a complex is formed between RNA and the template strand. Band *iv* will be higher than the others because strand 1 is complexed to 2, and strand 2 is complexed to 3. Band *v* is the highest because core polymerase associates with the three strands. (b) None, because rifampicin acts before the formation of the open complex. (c) RNA polymerase is processive. When the template is bound, heparin cannot enter the DNA-binding site. (d) When GTP is absent, synthesis stops when the first cytosine residue downstream of the bubble is encountered in the template strand. In contrast, with all four nucleoside triphosphates present, synthesis will continue to the end of the template.

10. The base-pairing energy of the di- and trinucleotide DNA–RNA hybrids formed at the very beginning of transcription is not sufficient to prevent strand separation and loss of product.

11. (a) Because cordycepin lacks a 3′-OH group, it cannot participate in 3′ → 5′ bond formation. (b) Because the poly(A) tail is a long stretch of adenosine nucleotides, the likelihood that a molecule of cordycepin would become incorporated is higher than with normal RNA. (c) Yes, it must be converted into cordycepin 5′-triphosphate.

12. Ser-Ile-Phe-His-Pro-Stop.

13. A mutation that disrupted the normal AAUAA recognition sequence for the endonuclease could account for this finding. In fact, a change from U to C in this sequence caused this defect in a thalassemic patient. Cleavage occurred at the AAUAAA 900 nucleotides downstream from this mutant AACAAA site.

14. One possibility is that the 3′ of the poly(U) donor strand cleaves the phosphodiester bond on the 5′ side of the insertion site. The newly formed 3′ terminus of the acceptor strand then cleaves the poly(U) strand on the 5′ side of the nucleotide that initiated the attack. In other words, a uridine residue could be added by two transesterification reactions. This postulated mechanism is similar to the one in RNA splicing.

15. Alternative splicing, RNA editing. Covalent modification of the proteins subsequent to synthesis.

16. Attach an oligo dT or oligo U sequence to an inert support to create an affinity column. When RNA is passed through the column, only poly(A) containing RNA will be retained.

17. (a) Different amounts of RNA are present for the various genes.
(b) Although all of the tissues have the same genes, the genes are expressed to different extents in different tissues.
(c) These genes are called housekeeping genes—genes that most tissues express. They might include genes for glycolytic or citric acid cycle enzymes.
(d) The point of the experiment is to determine which genes are initiated in vivo. The initiation inhibitor is added to prevent initiation at start sites that may have been activated during the isolation of the nuclei.

18. DNA is the single strand that forms the trunk of the tree. Strands of increasing length are RNA molecules; Beginning of transcription is where growing chains are the smallest; end is where chain growth stops. Direction is left to right. Many enzymes are actively transcribing each gene.

Chapter 29

1. (a) No; (b) no; (c) yes.

2. Four bands: light, heavy, a hybrid of light 30S and heavy 50S, and a hybrid of heavy 30S and light 50S.

3. Two hundred molecules of ATP are converted into AMP + 2 P_i to activate the 200 amino acids, which is equivalent to 400 molecules of ATP. One molecule of GTP is required for initiation, and 398 molecules of GTP are needed to form 199 peptide bonds.

4. (a, d, and e) Type 2; (b, c, and f) type 1.

5. A mutation caused by the insertion of an extra base can be suppressed by a tRNA that contains a fourth base in its anticodon. For example, UUUC rather than UUU is read as the codon for phenylalanine by a tRNA that contains 3′-AAAG-5′ as its anticodon.

6. One approach is to synthesize a tRNA that is acylated with a reactive amino acid analog. For example, bromoacetyl-phenylalanyl-tRNA is an affinity-labeling reagent for the P site of E. coli ribosomes.

7. The sequence GAGGU is complementary to a sequence of five bases at the 3′ end of 16S rRNA and is located several bases on the 5′ side of an AUG codon. Hence this region is a start signal for protein synthesis. The replacement of G by A would be expected to weaken the interaction of this mRNA with the 16S rRNA and thereby diminish its effectiveness as an initiation signal. In fact, this mutation results in a tenfold decrease in the rate of synthesis of the protein specified by this mRNA.

8. Proteins are synthesized from the amino to the carboxyl end on ribosomes, whereas they are synthesized in the reverse direction in the solid-phase method. The activated intermediate in ribosomal synthesis is an aminoacyl-tRNA; in the solid-phase method, it is the adduct of the amino acid and dicyclohexylcarbodiimide.

9. The error rates of DNA, RNA, and protein synthesis are of the order of 10^{-10}, 10^{-5}, and 10^{-4}, respectively, per nucleotide (or amino acid) incorporated. The fidelity of all three processes depends on the precision of base pairing to the DNA or mRNA template. No errors are corrected in RNA synthesis. In contrast, the fidelity of DNA synthesis is markedly increased by the 3′ → 5′ proofreading nuclease activity and by postreplicative repair. In protein synthesis, the

mischarging of some tRNAs is corrected by the hydrolytic action of aminoacyl-tRNA synthetase. Proofreading also takes place when aminoacyl-tRNA occupies the A site on the ribosome; the GTPase activity of EF-Tu sets the pace of this final stage of editing.

10. GTP is not hydrolyzed until aminoacyl-tRNA is delivered to the A site of the ribosome. An earlier hydrolysis of GTP would be wasteful because EF-Tu-GDP has little affinity for aminoacyl-tRNA.

11. The translation of an mRNA molecule can be blocked by antisense RNA, an RNA molecule with the complementary sequence. The antisense–sense RNA duplex cannot serve as a template for translation; single-stranded mRNA is required. Furthermore, the antisense–sense duplex is degraded by nucleases. Antisense RNA added to the external medium is spontaneously taken up by many cells. A precise quantity can be delivered by microinjection. Alternatively, a plasmid encoding the antisense RNA can be introduced into target cells.

12. (a) A_5. (b) $A_5 > A_4 > A_3 > A_2$, (c) Synthesis is from the amino terminus to the carboxyl terminus.

13. These enzymes convert nucleic acid information into protein information by interpreting the tRNA and linking it to the proper amino acid.

14. The rate would fall because the elongation step requires that the GTP be hydrolyzed before any further elongation can take place.

15. The nucleophile is the amino group of the aminoacyl-tRNA. This amino group attacks the carbonyl group of the ester of peptidyl-tRNA to form a tetrahedral intermediate, which eliminates the tRNA alcohol to form a new peptide bond.

16. The aminoacyl-tRNA can be initially synthesized. However, the side-chain amino group attacks the ester linkage to form a six-membered amide, releasing the tRNA.

17. EF-Ts catalyzes the exchange of GTP for GDP bound to EF-Tu. In G-protein cascades, an activated 7TM receptor catalyzes GTP–GDP exchange in a G protein.

18. The α subunits of G proteins are inhibited by a similar mechanism in cholera and whooping cough (Section 15.5.2).

19. (a) eIF4H had two effects: (1) the extent of unwinding was increased and (2) the rate of unwinding was increased, as indicated by the increased rise in activity at early reaction times.
(b) To firmly establish that the effect of eIFH4 was not due to any inherent helicase activity.
(c) Half-maximal activity was achieved at 0.11 μM of eIF4H. Therefore, maximal stimulation would be achieved at 0.4μM, or a ratio of 1:1.
(d) eIF4H enhances the rate of unwinding of all helices, but the effect is greater as the helices increase in stability.
(e) The results in graph C suggest that it increases the processivity.

Chapter 30

1. The liver and to a lesser extent the kidneys, contain glucose 6-phosphatase, whereas muscle and the brain do not. Hence, muscle and the brain, in contrast with the liver, do not release glucose. Another key enzymatic difference is that the liver has little of the transferase needed to activate acetoacetate to acetoacetyl CoA. Consequently, acetoacetate and 3-hydroxybutyrate are exported by the liver for use by heart muscle, skeletal muscle, and the brain.

2. (a) Adipose cells normally convert glucose into glycerol 3-phosphate for the formation of triacylglycerols. A deficiency

of hexokinase would interfere with the synthesis of triacyl-glycerols.

(b) A deficiency of glucose 6-phosphatase would block the export of glucose from the liver after glycogenolysis. This disorder (called von Gierke disease) is characterized by an abnormally high content of glycogen in the liver and a low blood-glucose level.

(c) A deficiency of carnitine acyltransferase I impairs the oxidation of long-chain fatty acids. Fasting and exercise precipitate muscle cramps.

(d) Glucokinase enables the liver to phosphorylate glucose even in the presence of a high level of glucose 6-phosphate. A deficiency of glucokinase would interfere with the synthesis of glycogen.

(e) Thiolase catalyzes the formation of two molecules of acetyl CoA from acetoacetyl CoA and CoA. A deficiency of thiolase would interfere with the utilization of acetoacetate as a fuel when the blood-sugar level is low.

(f) Phosphofructokinase will be less active than normal because of the lowered level of F-2,6-BP. Hence, glycolysis will be much slower than normal.

3. (a) A high proportion of fatty acids in the blood are bound to albumin. Cerebrospinal fluid has a low content of fatty acids because it has little albumin.

(b) Glucose is highly hydrophilic and soluble in aqueous media, in contrast with fatty acids, which must be carried by transport proteins such as albumin. Micelles of fatty acids would disrupt membrane structure.

(c) Fatty acids, not glucose, are the major fuel of resting muscle.

4. (a) A watt is equal to 1 joule (J) per second (0.239 calorie per second). Hence, 70 W is equivalent to 0.07 kJ s^{-1}, or 0.017 kcal s^{-1}.

(b) A watt is a current of 1 ampere (A) across a potential of 1 volt (V). For simplicity, let us assume that all the electron flow is from NADH to O_2 (a potential drop of 1.14 V). Hence, the current is 61.4 A, which corresponds to 3.86×10^{20} electrons per second (1 A = 1 coulomb s^{-1} = 6.28×10^{18} charge s^{-1}).

(c) Three molecules of ATP are formed per molecule of NADH oxidized (two electrons). Hence, one molecule of ATP is formed per 0.67 electron transferred. A flow of 3.86×10^{20} electrons per second therefore leads to the generation of 5.8×10^{20} molecules of ATP per second, or 0.96 mmol s^{-1}.

(d) The molecular weight of ATP is 507. The total body content of ATP of 50 g is equal to 0.099 mol. Hence, ATP turns over about once per 100 seconds when the body is at rest.

5. (a) The stoichiometry of the complete oxidation of glucose is

$$C_6H_{12}O_6 + 6\,O_2 \rightarrow 6\,CO_2 + 6\,H_2O$$

and that of tripalmitoylglycerol is

$$C_{51}H_{98}O_2 + 72.5\,O_2 \rightarrow 51\,CO_2 + 49\,H_2O$$

Hence, the RQ values are 1.0 and 0.703, respectively. (b) An RQ value reveals the relative usage of carbohydrate and fats as fuels. The RQ of a marathon runner typically decreases from 0.97 to 0.77 during a race. The lowering of the RQ indicates the shift in fuel from carbohydrate to fat.

6. One gram of glucose (molecular weight 180.2) is equal to 5.55 mmol, and one gram of tripalmitoylglycerol (molecular weight 807.3) is equal to 1.24 mmol. The reaction stoichiometries (see problem 5) indicate that 6 mol of H_2O are produced

per mole of glucose oxidized, and 49 mol of H_2O per mole of tripalmitoylglycerol oxidized. Hence, the H_2O yields per gram of fuel are 33.3 mmol (0.6 g) for glucose, and 60.8 mmol (1.09 g) for tripalmitoylglycerol. Thus, complete oxidation of this fat gives 1.82 times as much water as does glucose. Another advantage of triacylglycerols is that they can be stored in essentially anhydrous form, whereas glucose is stored as glycogen, a highly hydrated polymer. A hump consisting mainly of glycogen would be an intolerable burden—far more than the straw that broke the camel's back.

7. A typical macadamia nut has a mass of about 2 g. Because it consists mainly of fats (~9 kcal/g), a nut has a value of about 18 kcal. The ingestion of 10 nuts results in an intake of about 180 kcal. As stated in the answer to problem 4, a power consumption of 1 W corresponds to 0.239 cal s^{-1}, and so 400-W running requires 95.6 cal s^{-1}, or .0956 kcal s^{-1}. Hence, one would have to run 1882 s, or about 31 minutes, to spend the calories provided by 10 nuts.

8. A high blood-glucose level would trigger the secretion of insulin, which would stimulate the synthesis of glycogen and triacylglycerols. A high insulin level would impede the mobilization of fuel reserves during the marathon.

9. Lipid mobilization can occur so rapidly that it exceeds the ability of the liver to oxidize the lipids or convert them into ketone bodies. The excess is reesterified and released into the blood as VLDL.

10. A role of the liver is to provide glucose for other tissues. In the liver, glycolysis is used not for energy production but for biosynthetic purposes. Consequently, in the presence of glucagon, liver glycolysis stops so that the glucose can be released into the blood.

11. Urea cycle and gluconeogenesis.

12. (a) Insulin inhibits lipid utilization.

(b) Insulin stimulates protein synthesis, but there are no amino acids in the children's diet. Moreover, insulin inhibits protein breakdown. Consequently, muscle proteins cannot be broken down and used for the synthesis of essential proteins.

(c) Because proteins cannot be synthesized, blood osmolarity is too low. Consequently, fluid leaves the blood. An especially important protein for maintaining blood osmolarity is albumin.

13. The oxygen consumption at the end of exercise is used to replenish ATP and creatine phosphate and to oxidize any lactate produced.

14. Oxygen is used in oxidative phosphorylation to resynthesize ATP and creatine phosphate. The liver converts lactate released by the muscle into glucose. Blood must be circulated to return the body temperature to normal, and so the heart cannot return to its resting rate immediately. Hemoglobin must be reoxygenated to replace the oxygen used in exercise. The muscles that power breathing must continue working at the same time that the exercised muscles are returning to resting states. In essence, all the biochemical systems activated in intense exercise need increased oxygen to return to the resting state.

15. Ethanol may replace water that is hydrogen bonded to proteins and membrane surfaces. This alteration of the hydration state of the protein would alter its conformation and hence function. Ethanol may also alter phospholipid packing in membranes. The two effects suggest that integral membrane proteins would be most sensitive to ethanol, as indeed seems to be the case.

16. Cells from the type I fiber would be rich in mitochondria, whereas those of the type II fiber would have few mitochondria.

17. (a) The ATP expended during this race amounts to about 8380 kg, or 18,400 pounds.
(b) The cyclist would need about $1,260,000,000 to complete the race.

Chapter 31

1. (a) Cells will express β-galactosidase, lac permease, and thiogalactoside transacetylase even in the absence of lactose.
(b) Cells will express β-galactosidase, lac permease, and thio-galactoside transacetylase even in the absence of lactose.
(c) The levels of catabolic enzymes such as β-galactosidase and arabinose isomerase will remain low even at low levels of glucose.
2. The concentration is $1/(6 \times 10^{23})$ moles per 10^{-15} liter = 1.7×10^{-9} M. Because $K_d = 10^{-13}$ M, the single molecule should be bound to its specific binding site.
3. The number of possible 8-bp sites is $4^8 = 65,536$. In a genome of 4.6×10^6 base pairs, the average site should appear $4.6 \times 10^6/65,536 = 70$ times. Each 10-bp site should appear 4 times. Each 12-bp site should appear 0.27 times (many 12-bp sites will not appear at all).
4. The distribution of charged amino acids is H2A (13K, 13R, 2D, 7E, charge = +15), H2B (20K, 8R, 3D, 7E, charge = +18), H3 (13K, 18R, 4D, 7E, charge = +20), H4 (11K, 14R, 3D, 4E, charge = +18). The total charge of the histone oc-tamer is estimated to be $2 \times (15 + 18 + 20 + 18) = +142$. The total charge on 150 base pairs of DNA is −300. Thus, the histone octamer neutralizes approximately one-half of the charge.
5. The presence of a particular DNA fragment could be de-tected by hybridization or by PCR. For lac repressor, a single fragment should be isolated. For pur repressor, approximately 20 distinct fragments should be isolated.
6. 5-Azacytidine cannot be methylated. Some genes, normally repressed by methylation, will be active.

7. Proteins containing these domains will be targeted to methylated DNA in repressed promoter regions. They would likely bind in the major groove because that is where the methyl group is.
8. The lac repressor does not bind DNA when the repressor is bound to a small molecule (the inducer), whereas pur repressor binds DNA only when the repressor is bound to a small mole-cule (the corepressor). The E. coli genome contains only a sin-gle lac repressor-binding region, whereas it has many sites for the pur repressor.
9. Anti-inducers bind to the conformation of repressors, such as the lac repressor, that are capable of binding DNA. They occupy a site that overlaps that for the inducer and, therefore, compete for binding to the repressor.
10. The inverted repeat may be a binding site for a dimeric DNA-binding protein or it may correspond to a stem-loop structure in the encoded RNA.
11. The amino group of the lysine residue, formed from the protonated form by a base, attacks the carbonyl group of acetyl CoA to generate a tetrahedral intermediate. This inter-mediate collapses to form the amide bond and release CoA.
12. Long-term memory requires gene expression stimulated by CREB. The injection of many binding sites for this protein prevents CREB from binding to the necessary sites in chro-matin and, hence, blocks the activation of gene expression. A possible pathway begins with serotonin binding to a 7TM re-ceptor, which activates a G protein which in turn activates adenylate cyclase. This activation increases the concentration of cAMP-activiated protein kinase A, which phosphorylates CREB, thus activating gene expression necessary for memory storage.
13. In mouse DNA, most of the HpaII sites are methylated and therefore not cut by the enzyme, resulting in large frag-ments. Some small fragments are produced from CpG islands that are unmethylated. For Drosophila and E. coli DNA, there is no methylation and all sites are cut.

Index

References to structural formulas are given in **boldface** type.

Common Abbreviations in Biochemistry

A	adenine
ACP	acyl carrier protein
ADP	adenosine diphosphate
Ala	alanine
AMP	adenosine monophosphate
cAMP	cyclic AMP
Arg	arginine
Asn	asparagine
Asp	aspartate
ATP	adenosine triphosphate
ATPase	adenosine triphosphatase
C	cytosine
CDP	cytidine diphosphate
CMP	cytidine monophosphate
CoA	coenzyme A
CoQ	coenzyme Q (ubiquinone)
CTP	cytidine triphosphate
cAMP	adenosine 3′,5′-cyclic monophosphate
cGMP	guanosine 3′,5′-cyclic monophosphate
Cys	cysteine
Cyt	cytochrome
d	2′-deoxyribo-
DNA	deoxyribonucleic acid
cDNA	complementary DNA
DNase	deoxyribonuclease
*Eco*RI	*Eco*RI restriction endonuclease
EF	elongation factor
FAD	flavin adenine dinucleotide (oxidized form)
$FADH_2$	flavin adenine dinucleotide (reduced form)
fMet	formylmethionine
FMN	flavin mononucleotide (oxidized form)
$FMNH_2$	flavin mononucleotide (reduced form)
G	guanine
GDP	guanosine dephosphate
Gln	glutamine
Glu	glutamate
Gly	glycine
GMP	guanosine monophosphate
cGMP	cyclic GMP
GSH	redused glutathione
GSSG	oxidized glutathione
GTP	guanosine triphosphate
GTPase	guanosine triphosphatase
Hb	hemoglobin
HDL	high-density lipoprotein
HGPRT	hypoxanthine-guanine phosphoribosyl-transferase
His	histidine
Hyp	hydroxyproline
IgG	immunoglobulin G
Ile	isoleucine
IP_3	inositol 1,4,5,-trisphosphate
ITP	inosine triphosphate
LDL	low-density lipoprotein
Leu	leucine
Lys	lysine
Met	methionine
NAD^+	nicotinamide adenine dinucleotide (oxidized form)
NADH	nicotinamide adenine dinucleotide (reduced form)
$NADP^+$	nicotinamide adenine dinucleotide phosphate (oxidized form)
NADPH	nicotinamide adenine dinucleotide phosphate (reduced form)
PFK	phosphofructokinase
Phe	phenylalanine
P_i	inorganic orthophosphate
PLP	pyridoxal phosphate
PP_i	inorganic pyrophosphate
Pro	proline
PRPP	5-phosphoribosyl-1-pyrophosphate
Q	ubizuinone (or plastoquinone)
QH_2	ubiquinol (or plastoquinol)
RNA	ribonucleic acid
mRNA	messenger RNA
rRNA	ribosomal RNA
scRNA	small cytoplasmic RNA
snRNA	small nuclear RNA
tRNA	transfer RNA
RNase	ribonuclease
Ser	serine
T	thymine
Thr	threonine
TPP	thiamine pyrophosphate
Trp	tryptophan
TTP	thymidine triphosphate
Tyr	tyrosine
U	uracil
UDP	uridine diphosphate
UDP-galactose	uridine diphosphate galactose
UDP-glucose	uridine diphosphate glucose
UMP	uridine monophosphate
UTP	uridine triphosphate
Val	valine
VLDL	very low density lipoprotein